Digital Logic Circuits

with VHDL and Verilog Programs

for BE | BTech | BSc | MSc of all Indian Universities

Digital Logic Circuits

with VHDL and Verilog Programs

for BE | BTech | BSc | MSc of all Indian Universities

A Nagoor Kani
Founder
RBA Educational Group
Chennai

CBS Publishers & Distributors Pvt Ltd
New Delhi • Bengaluru • Chennai • Kochi • Kolkata • Lucknow • Mumbai
Hyderabad • Jharkhand • Nagpur • Patna • Pune • Uttarakhand

Disclaimer

Science and technology are constantly changing fields. New research and experience broaden the scope of information and knowledge. The author has tried his best in giving information available to him while preparing the material for this book. Although all efforts have been made to ensure optimum accuracy of the material, yet it is quite possible some errors might have been left uncorrected. The publisher, the printer and the author will not be held responsible for any inadvertent errors or inaccuracies.

Digital Logic Circuits

with VHDL and Verilog Programs

for BE | BTech | BSc | MSc of all Indian Universities

ISBN: 978-93-5466-960-6

First Edition: 2025

Published by **Satish Kumar Jain** and produced by **Varun Jain** for

CBS Publishers & Distributors Pvt Ltd

4819/XI Prahlad Street, 24 Ansari Road, Daryaganj, New Delhi 110 002, India.

Ph: 011-23289259, 23266868 Website: www.cbspd.com

e-mail: delhi@cbspd.com

Corporate Office: 204 FIE, Industrial Area, Patparganj, Delhi 110 092, India

Ph: 011-4934 4934 Fax: 011-4934 4935 e-mail: publishing@cbspd.com; publicity@cbspd.com

Branches

- **Bengaluru:** Seema House 2975, 17th Cross, K.R. Road, Banasankari 2nd Stage, Bengaluru 560 070, Karnataka, India
  Ph: +91-80-26771678/79 Fax: +91-80-26771680 e-mail: bangalore@cbspd.com
- **Chennai:** 7, Subbaraya Street, Shenoy Nagar, Chennai 600 030, Tamil Nadu, India
  Ph: +91-44-26680620, 26681266 Fax: +91-44-42032115 e-mail: chennai@cbspd.com
- **Kochi:** 42/1325, 36, Power House Road, Opposite KSEB, Ernakulam 682 018 Kochi, Kerala, India
  Ph: +91-484-4059061-65, 67 Fax: +91-484-4059065 e-mail: kochi@cbspd.com
- **Kolkata:** 147, Hind Ceramics Compound, 1st Floor, Nilgunj Road, Belghoria, Kolkata 700 056, West Bengal, India
  Ph: +91-33-25633055/56 e-mail: kolkata@cbspd.com
- **Lucknow:** Basement, Khushnuma Complex, 7-Meerabai Marg (Behind Jawahar Bhawan), Lucknow-226 001, Uttar Pradesh, India
  Ph: +910522-4000032 e-mail: tiwari.lucknow@cbspd.com
- **Mumbai:** PWD Shed, Gala No. 25/26, Ramchandra Bhatt Marg, Next to JJ Hospital Gate No. 2, Opp Union Bank of India, Noorbaug, Mumbai-400009, Maharashtra, India
  Ph: +91-22-66661880/89 e-mail: mumbai@cbspd.com

Representatives

• **Hyderabad**	0-9885175004	• **Jharkhand**	0-9811541605	• **Nagpur**	0-8692091830
• **Patna**	0-9334159340	• **Pune**	0-9664372571	• **Uttarakhand**	0-9716462459

Printed at: Mudrak, Noida, UP, India

to

Mr Satish K Jain CMD
CBS Publishers & Distributors

Preface

The main objective of this book is to explore the basic concepts of digital logic circuits and Systems in a simple and easy-to-understand manner. This text on digital logic circuits with VHDL and Verilog programs has been crafted for all students of engineering, arts and science colleges of various universities in India.

Considering the difficulty in logical analysis needed in this subject, more emphasis has been given on the problem-solving methodology. Considerable effort has been made to elucidate the logical explanations in a step-by-step manner. Exercise problems with various difficulty levels are given in the text to help students get an intuitive grasp on the subject.

This book with its lucid writing style and germane pedagogical features will prove to be a master text for engineering students.

Salient Features

- Proof of important concepts and theorems are clearly highlighted by shaded boxes
- Wherever required, problems are solved in multiple methods
- Additional explanations for solutions and proofs are provided in separate boxes
- Different types of fonts are used for text, proof, problems and VHDL/Verilog programs for better clarity
- Keywords are highlighted by bold, italic fonts
- Ample figures and examples to enhance students understanding
- Digital system or function implementation using standard ICs
- VHDL and Verilog programs which are useful for laboratory practice and other projects

This book is organized with 15 Chapters

- **Chapter 1** provides a simple introduction to various aspects of digital system. **Chapter 2** starts with a discussion of various number systems and basic arithmetic operations in these number systems. Representation of negative numbers by complement numbers and arithmetic in complement number systems are also presented in this chapter with examples.
- **Chapter 3** is devoted to various types of binary codes. BCD codes, alphanumeric codes and error detection and correction codes are clearly explained with appropriate examples. Addition and subtraction in excess-3 code is also explained with examples.
- **Chapter 4** includes Boolean algebra and theorems, Boolean functions and basic concepts of logic gates. The standard forms of Boolean functions are also presented in this chapter. Implementation of Boolean functions using universal gates and standard logic gate ICs, Boolean function minimization using theorems, K-maps and Quine-McCluskey methods are also discussed in this chapter. The formation of prime implicants using 1's and 0's are clearly explained with examples.
- **Chapter 5** is devoted to combinational logic circuit analysis and design. The analysis and design of code converters, decoders, encoders, multiplexers, demultiplexers, adders, subtractors, multipliers and magnitude comparator are presented with clear examples. Implementation of combinational logic circuits using multiplexers, decoders and standard ICs are also discussed in this chapter. Analysis and design of ALU (Arithmetic Logic Unit) using combinational circuits are included in this chapter. Some application examples of combinational logic systems like digital trans-receiver, Parity generator and checker and BCD to 7-segment decoder are included in this chapter.

- **Chapter 6** starts with detailed working of various types of flip-flops used to build synchronous circuits. The characteristics of flip-flops and timing analysis of inputs and outputs are presented with appropriate timing diagrams. The conversion of flip-flops from one type to another types are clearly explained in this chapter.
- **Chapter 7** is devoted to synchronous sequential circuit. Analysis and design of Mealy and Moore models of synchronous logic circuits are presented with step-by-step explanations. The basic concepts of state and state variables are explained clearly. Implementation of sequential logic circuits using flip-flops and standard ICs are also discussed in this chapter.
- The analysis and design of synchronous circuits like counters, sequence detectors and registers are explained with appropriate examples in **Chapter 8**. Design and implementation of various types of counters using flip-flops and standard ICs are presented in this chapter. Application examples of synchronous logic system like rolling display and real time clock are included in this chapter.
- **Chapter 9** is devoted to the analysis and design of asynchronous logic circuits. State reduction using implication table and merger diagram and race-free state assignment are clearly explained with appropriate examples. Design of fundamental mode asynchronous sequential circuits using latches and pulse mode sequential circuits using unclocked flip-flops are presented with step-by-step explanations.
- The detailed discussion on races and hazards are presented in **Chapter 10**. Also, various methods for eliminating races and hazards are discussed with examples. The characteristics of the devices of various logic families are presented in **Chapter 11**. The development of logic gates in various logic families are explained in detail in this chapter. Implementation of combinational logic circuit in CMOS is explained with examples in this chapter.
- A brief discussion on various types of semiconductor memory and programmable logic devices are presented in **Chapter 12**. The digital system design using programmable ICs like PROM, PLA and PAL are explained with appropriate examples. A short note on advanced programmable ICs CPLD and FPGA are also included in this chapter.
- Basic concepts of DAC and ADC are discussed in **Chapter 13**. Some popular types of DAC and ADC are explained clearly with appropriate diagrams. The standard and commonly used digital ICs are listed in Appendix for ready reference to implement digital systems using standard ICs
- **Chapters 14 and 15** are dedicated to Hardware Description Languages (HDLs). A brief introduction to VHDL language is presented in Chapter 14 and Verilog language in Chapter 15 for better understanding of the readers. VHDL and Verilog programs for combinational and sequential circuit design are presented in **chapters 14 and 15**. Test bench is also included for some of the VHDL and Verilog programs. A brief introduction about RTL design methodology is included with programming examples.

All VHDL/Verilog statement/program examples in this book are simulated and verified in license free Xilinx VHDL tool Vivado 2019.2 web edition which can be directly downloaded from the following link.

https://www.xilinx.com/support/download/index.html/content/xilinx/en/downloadNav/vivado-design-tools/archive.html

File name: Vivado HLx 2019.2: All OS installer Single-File Download (TAR/GZIP-26.55GB).

I have taken special care in presenting the concepts of digital logic circuits and systems in a simple manner supported by carefully chosen solved problems and VHDL/Verilog programs and hope that the teaching and student community will welcome the book. The readers can feel free to convey their criticism and suggestions to nagoorkani65@yahoo.com for further improvement of the book.

A Nagoor Kani

Acknowledgements

I express my heartfelt thanks to my wife Mrs C Gnanaparanjothi Nagoor Kani and my sons N Bharath Raj alias Chandrakani Allaudeen and N Vikram Raj, for the support, encouragement and cooperation they have extended to me throughout my career. I thank Ms T A Benazir, Manager, RBA Group, and all my office-staff for their cooperation in carrying out my day-to-day activities.

It's my pleasure to acknowledge the contributions of our technical editors, Ms S Saranya and Ms P Kanimozhi, Ms K G Sathyapriya and Mrs E Indhumathi for editing, proof-reading and type-setting of the manuscript and preparing the layout of the book.

My sincere thanks to all reviewers Dr R Swarnalatha, Associate Professor, BITS, Pilani (Dubai Campus), Mrs K Sivagami, research scholar, IIITDM, Kancheepuram, Mrs S Chitra Lakshmi and Mrs S G Vijayakumari, Assistant professor, SRMIST, for their valuable suggestions and comments which helps me to explore the subject to a greater depth.

I am also grateful to Mr Satish K Jain, CMD, CBS Publishers & Distributors, for his keen interest in publishing this work in CBS banner. My sincere thanks to all team members of CBS, for their concern and care in publishing this work.

Finally, a special note of appreciation is due to my sisters, brothers, relatives, friends, students and the entire teaching community for their overwhelming support and encouragement to my writing.

A Nagoor Kani

Contents

CHAPTER 1

INTRODUCTION TO DIGITAL SYSTEMS

1.1 Analog and Digital Electronics

The digital electronics refer to devices and circuits that process digital signals which are coded in binary. The real world signals are analog and hence the field of electronics was dominated by analog devices and circuits before the invention of digital devices. The real world analog signals are converted to digital signals and processed by digital systems and after processing the digital signals are used either directly or converted back to analog for real world applications. The ADCs (Analog-to-Digital Converters) are used for converting analog signals to digital signals and DACs (Digital-to-Analog Converters) are used to convert digital signals to analog signals.

The block diagram of a typical system with ADC and DAC is shown in Fig. 1.1. A system will have a number of control variables that can be modified or processed to achieve a desired task. The control variables are converted to electrical analog voltage signals by sensors and the analog voltage signal is converted to digital signal by ADC. The digital system process or modify the digital signal and output the processed digital signal which are converted back to analog by DAC. The processed analog signal is converted to physical variables by actuators which are used by the system to achieve the desired task.

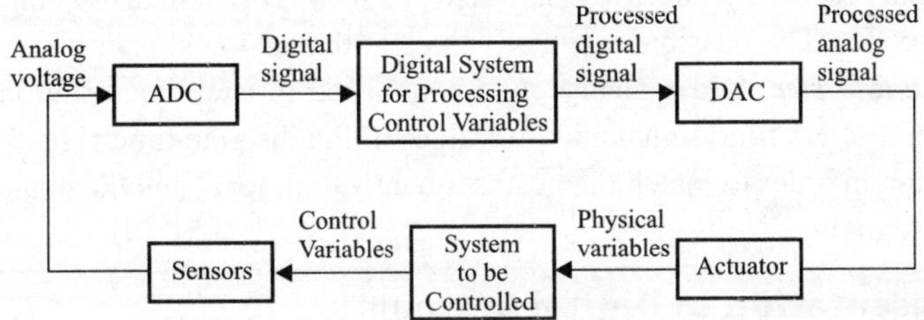

Fig. 1.1: A physical system interfaced to digital system using ADC and DAC.

The history of digital electronics dates back to 1900s during that period the digital circuits were made using mechanical relays and later replaced by vacuum tubes. The invention of semiconductor transistors in 1960s led to a revolution by replacing bulky vacuum tubes with transistors. The invention of integrated circuits (ICs) again led to minimize the size of digital devices. The invention of programmable ICs completely turns the world to digital era. In modern technology the processing and control of signals and systems are made digital.

Advantages of Digital Systems

- The digital hardware is compact, reliable, less expensive and programmable.
- Since the digital systems are programmable, the performance of the system can be easily upgraded/ modified.
- By employing high speed, sophisticated digital hardware higher precision can be achieved in processing of signals.
- The digital signals can be permanently stored in magnetic media so that they are transportable and can be processed in non-real-time or off-line.

1.2 Analog and Digital Signal

Any physical phenomenon that conveys or carries some information can be called a ***signal***. The music, speech, motion pictures, still photos, heart-beat, etc., are examples of signals that we normally encounter in day-to-day life.

When a signal is defined continuously for any value of an independent variable, it is called an ***analog*** or ***continuous signal***. Most of the signals encountered in science and engineering are analog in nature. The dependent variable of an analog signal is time and hence it is also called a ***continuous time signal***.

When a signal is defined for discrete intervals of an independent variable, it is called a ***discrete signal***. The dependent variable of a discrete signal is time and hence it is also called ***discrete time signal***. Most of the discrete signals are either sampled versions of analog signals for processing by digital systems or output of digital systems.

The quantized and coded version of the discrete time signals is called ***digital signal***. In digital signals the value of the signal for every discrete time instant of time is represented in binary codes. The process of conversion of a discrete time signal to digital signal involves quantization and coding.

Normally, for binary representation, a standard size of binary is chosen. In n-bit binary representation, we can have 2^n binary codes. The possible range of values of the analog signals are usually divided into 2^n steps called ***quantization step***, and a binary code is attached to each quantization step. The analog signal is sampled to get discrete time signal and the values of the discrete time signals are approximated by rounding or truncation in order to match the nearest quantization level and the quantized discrete time signal becomes digital signal.

1.3 Binary Representation in Digital Systems

In digital systems the signals are represented only in binary. The binary representation may be a binary number or a binary code. For example, the mathematical operations can be performed by representing the decimal numbers in binary.

The information other than numerals is represented by binary codes. For example, the English alphabets are represented by ASCII code, the digital information are coded for transmission, etc.

Moreover, any information is stored in a digital system only in binary. Human voice, music, images, videos, etc., are all stored and processed only in binary. Ultimately, in digital system everything is 1 and 0.

1.4 Digital Integrated Circuits (ICs)

Modern digital systems are built using digital integrated circuits (ICs). The digital ICs are ICs that can accept only digital signals in the form of binary and process the digital signal and output the digital signals in the form of binary.

Classification of Digital ICs Based on Number of Transistors

The fundamental electronic device in a digital IC is transistor. One way of classification of digital ICs are based on density of fundamental transistors used to fabricate an IC.

Based on number of transistors packed in a single IC, the digital ICs are classified as SSI, MSI, LSI, VLSI and UVLSI integrated circuits.

SSI (Small Scale Integration): SSI refers to integration of few (typically 4 to 10) transistors on a chip. It is the beginning of IC technology. This technology is used to fabricate logic gates AND, OR, NOT as ICs.

MSI (Medium Scale Integration): MSI refers to integration of 10 to 500 transistors on a single chip. This technology is used to fabricate small digital devices like decoders, encoders, multiplexers, adders, etc.

LSI (Large Scale Integration): LSI refers to integration of 500 to 20000 transistors on a single chip. This technology leads to fabrication of early version of programmable ICs and microprocessors.

VLSI (Very Large Scale Integration): VLSI refers to integration of 20000 to 1 Million transistors on a single chip. This technology is used to fabricate complex digital ICs and microprocessors and microcontrollers with advanced features

UVLSI (Ultra Very Large Scale Integration): Integration of more than 1 Million transistors on a single chip is called UVLSI or ULSI. This technology is currently used to fabricate modern processors with billions and trillions of transistors on a single chip.

Programmable and Nonprogrammable Digital ICs

In another way of classification, the digital ICs are classified into programmable and nonprogrammable ICs. The nonprogrammable ICs are designed to perform fixed functions. Examples of nonprogrammable digital ICs are decoders, encoders, multiplexers, adders, etc. The programmable digital ICs are programmable by the user to perform desired functions. The programmable ICs are again classified into one time programmable, multi-time programmable and field programmable or reconfigurable ICs.

The one time programmable ICs are used to implement small digital functions or systems by the user. The user can program these ICs one time to implement the digital system and once programmed the content or function of the IC cannot be altered. Examples of one time programmable digital ICs are PROM (Programmable Read Only Memory), PAL (Programmable Array Logic) and PLA (Programmable Logic Array).

The multi-time programmable ICs will have a set of instructions to write a program to implement the desired digital systems. These ICs will have various internals devices with programmable features whose functions can be altered by a program. The microprocessors and microcontrollers are examples of multi-time programmable ICs.

The CPLD (Complex Programmable Logic Device) and FPGA (Field Programmable Gate Arrays) are reconfigurable ICs. These devices will have programmable logic resources in the form of gate arrays which can be programmed to implement any digital systems. The logic resources are open resources without any internal connections. The programming of these devices will create internal connections and hence the process of program is called configuration. The internal connections made by a program can be altered by reprogramming the device and hence these ICs are referred to as reconfigurable ICs.

1.5 Digital Systems

Digital systems are designed using digital ICs. A digital system accepts digital inputs in the form of binary and after processing the inputs it generates digital outputs in the form of binary. The digital systems can be broadly classified into combinational systems and sequential systems.

Combinational Digital Systems

In the combinational systems, the outputs are generated based on present inputs. The outputs are not feedback to inputs and hence the present outputs do not depend on past output and therefore, the combinational systems do not require memory to store past outputs. The combinational systems are developed using logic gates, decoders, multiplexers, adders, etc.

Sequential Digital Systems

In the sequential systems, the outputs are generated based on present inputs and past outputs. The outputs are feedback to inputs and hence the present outputs depend on past output and therefore, the sequential systems require memory to store past outputs. The flip-flops are the fundamental memory device to store binary bits and hence flip-flops are employed in sequential systems to store the past outputs and fed back to input.

The sequential systems are again classified into synchronous and asynchronous sequential systems. The synchronous sequential systems are controlled by a clock whereas the asynchronous sequential systems do not have a clock.

In synchronous sequential systems the inputs and past outputs are recognized to generate outputs only when the clock signal is active. The time period (or frequency) of clock is selected such that the one time period of clock is sufficient for the sequential circuit to settle at a stable value for a change in input.

The asynchronous sequential circuits are not governed by clock and hence, in an asynchronous sequential circuit the change input at any time will generate a change in output. Therefore, in asynchronous sequential circuit any change in input should be made only after the circuit settle at a stable value for previous input.

1.6 Binary Storage Devices: Registers and Memory

In digital systems, the binary information is stored in the form of binary as 1's and 0's. The basic storage element to store one bit of information is called binary cell which is constructed using flip-flops by adding read and write control signals. Hence, the actual storage device inside a binary cell is a flip-flop. The storage devices can be broadly classified into registers and memory.

A Register is a group n binary cell to store n-bit binary word. Registers are temporary storage devices in a digital system. Registers are also used as input and output buffers in a digital system as interfacing devices to input and output device in order to match the current level of input and output signals.

Memory is a storage device with 1000s of registers on a single IC chip. The memory capacity is specified as $2^m \times n$ where m is identification number for a register in a memory and n is the size of binary word that can be stored in a register of memory.

The register in a memory is also referred to as memory location and the identification number for register is known as address of memory location. With m- bit binary it is possible to generate 2^m different binary values. Therefore, a memory of size $2^m \times n$ will have 2^m storage locations with an m-bit address for each location and a n-bit binary word can be stored in one location.

The basic human-machine communication in digital system is established using ASCII (**A**merican **S**tandard **C**ode for **I**nformation **I**nterchange) codes which is 7-bit code. By appending a 1-bit parity the ASCII code becomes an 8-bit code for numerals and English alphabets. Hence, the basic memory word size is retained as byte and standard memory capacity is specified in terms of kilo-bytes (KB).

A 10-bit binary will have $2^{10} = 1024$ different binary codes which can be used to address 1024 memory locations and hence one kilo in binary is 1024. The memory capacity will be multiple of 1024. For example 2 *kilo* is $2 \times 1024 = 2048$ and 4 *kilo* is $4 \times 1024 = 4096$ and so on.

Semiconductor Memories

The semiconductor memories are manufactured in the form of ICs and two major types of semiconductor ICs are RAM (Random Access Memory) and ROM (Read Only Memory).

The RAM is also called RWM (Read/Write Memory). The RAM memory can be read or write in the system it is fitted or connected. But ROM memory can be only read in the system it is fitted or connected. For writing ROM memory a separate system called ROM programmer is needed.

The RAM memory can be again classified into DRAM (Dynamic RAM) and SRAM (Static RAM). The DRAM has higher packing density than SRAM in a given size of chip area. Both DRAM and SRAM are volatile memories. In ***volatile memories***, the information will be retained as long as power supply is ON. When power supply is OFF, the stored information will be lost in RAM memory.

The ROM memory can be classified into OTP-ROM, EPROM and EEPROM/EAPROM. OTP-ROM is one-time programmable and once programmed the content cannot be altered and can only read by the system. For programming OTP-ROM separate programmer system is needed.

EPROM is Erasable Programmable ROM. For erasing the EPROM memory an UV (Ultra Violet) erasure is needed. It has to be exposed to UV light for 20 to 30 minutes. Then it can be reprogrammed using a separate programmer system.

Both ROM and EPROM can be read in the system it is fitted like a RAM memory, but differs only in write operation.

The EEPROM/EAPROM memory can be erased by passing an electrical current of specified value. They can be erased in the system they are fitted, and so no separate system is needed for erasing. The read and write operation is similar to RAM.

1.7 Binary Logic and Arithmetic

Binary logic consists of binary variables and a set of logical operations. The binary variables are denoted by letters of alphabet, A, B, C, ..., x, y, z, ... Each binary variable can take only two possible values 0 or 1. In positive logic, 0 is **low** and 1 is **high**. The basic logical operations are AND, OR and NOT.

The logical AND of two or more variables will be 1 if and only if the value of all the variables is 1. The logical OR of two or more variables will be 1 if the value of any one of the variables is 1. NOT operation is same as complement operation.

The arithmetic operations like addition, subtraction, multiplication and division can be implemented using logical operations AND, OR and NOT. The logical OR operation resembles binary addition but it does not generate carry. The logical AND operation resembles binary multiplication but it does not generates overflow. Therefore, additional logical operations are required apart from logical OR and logical AND operations to implement arithmetic operations.

1.8 Logic Families

Digital ICs (Integrated Circuits) are manufactured using different semiconductor technologies. The collection of different logical devices that perform different logic functions but manufactured using same semiconductor technology is called a logic family. The logic devices of a family will have similar input, output and internal circuit characteristics.

The digital ICs of the same logic family can be connected directly to implement a logic circuit. But, the digital ICs of different logic families cannot be connected directly to implement a logic circuit, because they may not have compatible characteristics. In such cases some interface circuit is required to match the characteristics of logic families.

Some of the popular digital logic families are,

- RTL (Resistor-transistor logic)
- DTL (Diode-transistor logic)
- TTL (Transistor-transistor logic)
- ECL (Emitter-coupled logic)
- CMOS (Complementary metal-oxide semiconductor)

The RTL, DTL, TTL and ECL digital logic family of ICs are constructed using bipolar NPN transistors as fundamental switching device. The CMOS digital logic family of devices are constructed using unipolar MOSFET (Metal Oxide Semiconductor Field Effect Transistor) as fundamental switching device.

1.9 Hardware Description Languages (HDLs)

Hardware **D**escription Languages (HDLs) are modelling language specially developed to describe digital hardware structures and to study the behavior of logic circuits. It can be used to implement logic circuits, truth tables, Boolean functions and state tables/diagrams as computer programs for simulation and verification.

Digital system design simulation, verification and hardware implementation are made easier by HDLs. VHDL and Verilog are the two popular IEEE standardized HDLs. VHDL is an acronym for **V**HSIC (Very High Speed Integrated Circuits) **H**ardware **D**escription **L**anguage. The name Verilog is the combination of two words "verification" and "logic".

CHAPTER 2

NUMBER SYSTEMS

2.1 Review of Number Systems Representation and Conversions

(JNTUH, Nov/Dec'20, 9 Marks)

There are four types of popular number systems.

1. Decimal
2. Binary
3. Octal
4. Hexadecimal

Every number system has a radix, r. The radix, r is also called ***base***. The ***radix*** or ***base*** is the number of unique symbols used to represent the number system. Hence the number systems are also called ***radix number systems*** or ***based number systems***.

Table 2.1: Radix/Base and Symbols of Number Systems

Number System	Radix/Base	Symbols
Decimal	10	0, 1, 2, 3, 4, 5, 6, 7, 8, 9
Binary	2	0, 1
Octal	8	0, 1, 2, 3, 4, 5, 6, 7
Hexadecimal	16	0, 1, 2, 3, 4, 5, 6, 7, 8, 9, A, B, C, D, E, F

Note: $A_{16} = 10_{10}$; $B_{16} = 11_{10}$; $C_{16} = 12_{10}$; $D_{16} = 13_{10}$; $E_{16} = 14_{10}$; $F_{16} = 15_{10}$

In order to differentiate the numbers of different number system the base or radix is marked as suffix as shown in following examples.

Examples of decimal number: 1024.765_{10} , 79645.23_{10}

Examples of binary number: 1011.101_2 , 11010.11_2

Examples of octal number: 7421.342_8 , 12467.57_8

Examples of hexadecimal number: $2A6B.7C4_{16}$, $1296B.CA_{16}$

2.1.1 Number Representation

A number in general is represented in the following format.

$$\underbrace{d_{n-1} \ldots d_4\, d_3\, d_2\, d_1\, d_0}_{\text{n digit integer part}} \cdot \underbrace{d_{-1}\, d_{-2}\, d_{-3} \ldots d_{-m}}_{\text{m digit fraction part}}$$

Here, d represents the digit in a number. The suffix in each digit represents the power in the weight of the digit. The dot is the seperation between integer part and fraction part.

Each symbol in the number is called ***digit***. Each digit has a ***weight*** which is given by the power of radix. In the above number representation, the digit d_{n-1} is Most Significant Digit (MSD) and the digit d_{-m} is Least Significant Digit (LSD). When the number has only integer part then d_{n-1} is MSD and d_0 is LSD. When the number has only fraction part then d_{-1} is MSD and d_{-m} is LSD.

Table 2.2: Weights of Digits in Number Systems

Number System	Weights of Digits .... d_4 d_3 d_2 d_1 d_0 . d_{-1} d_{-2} d_{-3} ....
Decimal	.... 10^4 10^3 10^2 10^1 10^0 . 10^{-1} 10^{-2} 10^{-3} ....
Binary	.... 2^4 2^3 2^2 2^1 2^0 . 2^{-1} 2^{-2} 2^{-3} ....
Octal	.... 8^4 8^3 8^2 8^1 8^0 . 8^{-1} 8^{-2} 8^{-3} ....
Hexadecimal	.... 16^4 16^3 16^2 16^1 16^0 . 16^{-1} 16^{-2} 16^{-3} ....

The following examples illustrate the relation between digits and weights of number system. These examples also shows the method to convert any number to decimal number.

$$786.35_{10} = 7 \times 10^2 + 8 \times 10^1 + 6 \times 10^0 + 3 \times 10^{-1} + 5 \times 10^{-2} = 786.35_{10}$$

$$572.24_8 = 5 \times 8^2 + 7 \times 8^1 + 2 \times 8^0 + 2 \times 8^{-1} + 4 \times 8^{-2} = 378.3125_{10}$$

$$110.11_2 = 1 \times 2^2 + 1 \times 2^1 + 0 \times 2^0 + 1 \times 2^{-1} + 1 \times 2^{-2} = 6.75_{10}$$

$$A79.B2_{16} = A \times 16^2 + 7 \times 16^1 + 9 \times 16^0 + B \times 16^{-1} + 2 \times 16^{-2}$$

$$= \underbrace{10_{10} \times 16^2 + 7 \times 16 + 9 \times 1}_{2681_{10}} + \underbrace{11_{10} \times 16^{-1} + 2 \times 16^{-2}}_{.6953125_{10}} = 2681.6953125_{10}$$

> ***Note:*** *Use calculator to compute the above calculations. In hexadecimal number the result has 11 digits and so in 10-digit calculator the calculation has to be performed separately for integer part and fraction part.*

2.1.2 Decimal Number System

Decimal number system is the common and widely used number system. The invention of digital systems leads to usage of binary number system. The octal and hexadecimal are useful for short-hand representation of binary numbers. Since decimal number system is the standard number system, the value of a number in any other number system is verified/cross-checked by converting to an equivalent decimal number.

The ***decimal number system*** is a radix number system in which the radix r is 10. Since the radix is 10, the number system uses 10 symbols which are chosen as 0, 1, 2, 3, 4, 5, 6, 7, 8 and 9.

A decimal number in general is represented in the following format.

$$\underbrace{d_{n-1} \,.....\, d_4\, d_3\, d_2\, d_1\, d_0}_{\text{n digit integer part}} \;\underset{\substack{\uparrow \\ \text{Decimal point}}}{.}\; \underbrace{d_{-1}\, d_{-2}\, d_{-3} \,.....\, d_{-m}}_{\text{m digit fraction part}}$$

In the above representation, d represents the digit in the decimal number. The suffix in each digit represents the power in the weight of the digit.

The dot in decimal number is called ***decimal point*** and it separates the integer part and fraction part.

Each digit of a decimal number has a weight which is given by the power of radix, 10. The digits and corresponding weights of a decimal number are shown in Table 2.3.

Table 2.3: Digits and Weights of Decimal Number

Digits	d_{n-1}	....	d_4	d_3	d_2	d_1	d_0	.	d_{-1}	d_{-2}	d_{-3}	....	d_{-m}
Weights	10^{n-1}	....	10^4	10^3	10^2	10^1	10^0	.	10^{-1}	10^{-2}	10^{-3}	....	10^{-m}

The value of a decimal number can be expressed as a sum of products of decimal digits and its corresponding weights as shown below:

$$d_{n-1} \times 10^{n-1} + ... + d_4 \times 10^4 + d_3 \times 10^3 + d_2 \times 10^2 + d_1 \times 10^1 + d_0 \times 10^0 + d_{-1} \times 10^{-1} + d_{-2} \times 10^{-2} + d_{-3} \times 10^{-3} + ... + d_{-m} \times 10^{-m}$$

Example: $1932.25_{10} = 1 \times 10^3 + 9 \times 10^2 + 3 \times 10^1 + 2 \times 10^0 + 2 \times 10^{-1} + 5 \times 10^{-2}$

2.1.3 Binary Number System

The ***binary number system*** is a radix number system in which the radix r is 2. Since the radix is 2, the number system uses only two symbols 0 and 1.

A binary number in general is represented in the following format.

$$\underbrace{d_{n-1} \,.....\, d_4\, d_3\, d_2\, d_1\, d_0}_{\text{n digit integer}} \;\underset{\substack{\uparrow \\ \text{Binary point}}}{.}\; \underbrace{d_{-1}\, d_{-2}\, d_{-3} \,.....\, d_{-m}}_{\text{m digit fraction}}$$

In the above representation, d represents the digit in the binary number. The suffix in each digit represents the power in the weight of the digit.

The dot in binary number is called ***binary point*** and it separates the integer part and fraction part.

Each digit of a binary number has a weight which is given by the power of radix, 2. The digits and corresponding weights of a binary number are shown in Table 2.4.

Table 2.4: Digits and Weights of Binary Number

Digits	d_{n-1}	....	d_4	d_3	d_2	d_1	d_0	.	d_{-1}	d_{-2}	d_{-3}	....	d_{-m}
Weights	2^{n-1}	....	2^4	2^3	2^2	2^1	2^0	.	2^{-1}	2^{-2}	2^{-3}	....	2^{-m}

The value of a binary number can be expressed as a sum of products of binary digits and its corresponding weights as shown ahead:

$$d_{n-1} \times 2^{n-1} + ... + d_4 \times 2^4 + d_3 \times 2^3 + d_2 \times 2^2 + d_1 \times 2^1 + d_0 \times 2^0 + d_{-1} \times 2^{-1} + d_{-2} \times 2^{-2} + d_{-3} \times 2^{-3} + ... + d_{-m} \times 2^{-m}$$

Example: $1011.01_2 = 1 \times 2^3 + 0 \times 2^2 + 1 \times 2^1 + 1 \times 2^0 + 0 \times 2^{-1} + 1 \times 2^{-2}$

Table 2.5: Decimal Numbers and Equivalent Binary Numbers

Decimal Number	Binary Number	Decimal Number	Binary Number	Decimal Number	Binary Number	Decimal Number	Binary Number
0	0	11	1011	21	10101	31	11111
1	1	12	1100	22	10110	32	100000
2	10	13	1101	23	10111	33	100001
3	11	14	1110	24	11000	34	100010
4	100	15	1111	25	11001	35	100011
5	101	16	10000	26	11010	36	100100
6	110	17	10001	27	11011	37	100101
7	111	18	10010	28	11100	38	100110
8	1000	19	10011	29	11101	39	100111
9	1001	20	10100	30	11110	40	101000
10	1010						

Decimal to Binary Conversion

(AKTU, Dec'22, 2 Marks)

Integer part: Divide the integer part of decimal number by 2 and list the quotient and remainder as shown below. Repeat division until quotient is less than 2. The remainders (d_{n-1}, d_{n-2}, ... d_1, d_0) are arranged in the order shown below to get the binary equivalent of decimal integer part.

Fraction Part: Multiply the fraction part of decimal number by 2 as shown below and record the overflow (d_{-1}, d_{-2}, ... $d_{-(m-1)}$, d_{-m}). Repeat multiplication of fraction part alone till the result is zero. The overflows are arranged in the order shown below to get the binary equivalent of decimal fraction.

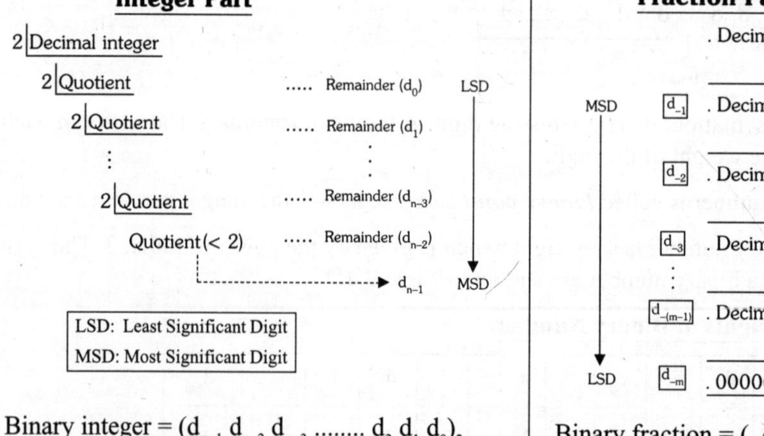

Binary integer = $(d_{n-1}\, d_{n-2}\, d_{n-3} \,.......\, d_2\, d_1\, d_0)_2$

Binary fraction = $(.\, d_{-1}\, d_{-2}\, d_{-3} \,.....\, d_{-(m-1)}\, d_{-m})_2$

Note: *In binary number system, the digit is also called bit. Hence, LSD is also called LSB (Least Significant Bit) and MSD is also called MSB (Most Significant Bit).*

Note: *In case of recurring fraction we will not get zero result in repeated multiplication and so stop multiplication when the required number of digits are obtained.*

Binary to Decimal Conversion

For binary to decimal conversion multiply each binary digit by its weight and sum up the product.

Consider the binary number,

$$.........d_3\, d_2\, d_1\, d_0 \,.\, d_{-1}\, d_{-2}\, d_{-3}\, ...$$

The decimal number can be obtained from the following calculation.

$$\text{Binary number} = ...d_3\, d_2\, d_1\, d_0 \,.\, d_{-1}\, d_{-2}\, d_{-3}......$$

$$= .....d_3 \times 2^3 + d_2 \times 2^2 + d_1 \times 2^1 + d_0 \times 2^0 + d_{-1} \times 2^{-1} + d_{-2} \times 2^{-2} + d_{-3} \times 2^{-3} + ...$$

$$= \text{Equivalent decimal number}$$

Example 2.1

(AKTU, Sep'20, 1 Mark) (RGPV, Jun'20, 3 Marks) (AKUB, Dec'20, 4 Marks)

Convert the following numbers to binary and check the result by reverse conversion.

a) 26.375_{10} b) 105_{10}

Solution

a) 26.375_{10}

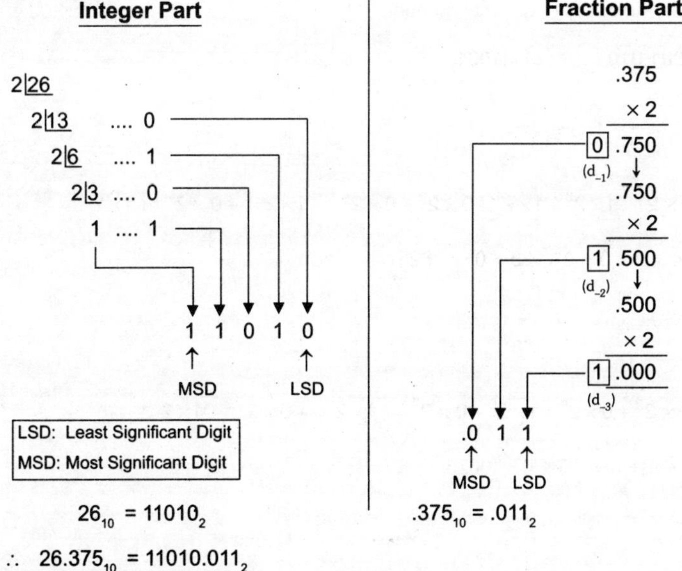

$26_{10} = 11010_2$ $\qquad .375_{10} = .011_2$

$\therefore\ 26.375_{10} = 11010.011_2$

Reverse Conversion

$$11010.011_2 = 1\times 2^4 + 1\times 2^3 + 0\times 2^2 + 1\times 2^1 + 0\times 2^0 + 0\times 2^{-1} + 1\times 2^{-2} + 1\times 2^{-3}$$

$$= 16 + 8 + 0 + 2 + 0 + 0 + .25 + .125$$

$$= 26.375_{10}$$

b) 105_{10}

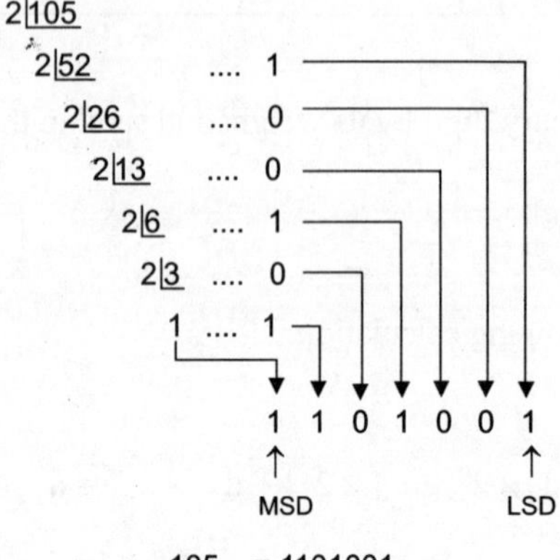

$\therefore \; 105_{10} = 1101001_2$

Reverse Conversion

$$1101001_2 = 1\times 2^6 + 1\times 2^5 + 0\times 2^4 + 1\times 2^3 + 0\times 2^2 + 0\times 2^1 + 1\times 2^0$$

$$= 64 + 32 + 0 + 8 + 0 + 0 + 1$$

$$= 105_{10}$$

Example 2.2

(RGPV, Jun'22, 3 Marks) *(JNTUH, Nov/Dec'20, 3 Marks)* *(GTU, Jun'19, 3 Marks)* *(RGPV, Jun'20, 3 Marks)*

Convert the following binary numbers to decimal number.

a) 10110.0101_2 **b)** 1001.0101_2 **c)** 11001_2

Solution

a) 10110.0101_2

$$10110.0101_2 = 1\times 2^4 + 0\times 2^3 + 1\times 2^2 + 1\times 2^1 + 0\times 2^0 + 0\times 2^{-1} + 1\times 2^{-2} + 0\times 2^{-3} + 1\times 2^{-4}$$

$$= 16 + 0 + 4 + 2 + 0 + 0 + .25 + 0 + .0625$$

$$= 22.3125_{10}$$

b) 1001.0101_2

$$1001.0101_2 = 1\times 2^3 + 0\times 2^2 + 0\times 2^1 + 1\times 2^0 + 0\times 2^{-1} + 1\times 2^{-2} + 0\times 2^{-3} + 1\times 2^{-4}$$

$$= 8 + 0 + 0 + 1 + 0 + .25 + 0 + .0625$$

$$= 9.3125_{10}$$

c) 11001_2

$$11001_2 = 1\times 2^4 + 1\times 2^3 + 0\times 2^2 + 0\times 2^1 + 1\times 2^0$$

$$= 16 + 8 + 0 + 0 + 1$$

$$= 25_{10}$$

2.1.4 Octal Number System

The ***octal number system*** is a radix number system in which the radix r is 8. Since the radix is 8, the number system uses 8 symbols which are chosen as 0, 1, 2, 3, 4, 5, 6 and 7.

An octal number in general is represented in the following format.

$$\underbrace{d_{n-1} \ldots d_4\, d_3\, d_2\, d_1\, d_0}_{\text{n digit integer part}} \;.\; \underbrace{d_{-1}\, d_{-2}\, d_{-3} \ldots d_{-m}}_{\text{m digit fraction part}}$$

In the above representation, d represents the digit in the octal number. The suffix in each digit represents the power in the weight of the digit.

Each digit of an octal number has a weight which is given by the power of radix, 8. The digits and corresponding weights of an octal number are shown in Table 2.6.

Table 2.6: Digits and Weights of Octal Number

Digits	d_{n-1}	....	d_4	d_3	d_2	d_1	d_0	.	d_{-1}	d_{-2}	d_{-3}	....	d_{-m}
Weights	8^{n-1}	....	8^4	8^3	8^2	8^1	8^0	.	8^{-1}	8^{-2}	8^{-3}	....	8^{-m}

The value of an octal number can be expressed as a sum of products of octal digits and its corresponding weights as shown below:

$$d_{n-1} \times 8^{n-1} + \ldots + d_4 \times 8^4 + d_3 \times 8^3 + d_2 \times 8^2 + d_1 \times 8^1 + d_0 \times 8^0 + d_{-1} \times 8^{-1} + d_{-2} \times 8^{-2} + d_{-3} \times 8^{-3} + \ldots + d_{-m} \times 8^{-m}$$

Example: $246.34_8 = 2 \times 8^2 + 4 \times 8^1 + 6 \times 8^0 + 3 \times 8^{-1} + 4 \times 8^{-2}$

Table 2.7: Decimal Numbers and Equivalent Octal Numbers

Decimal Number	Octal Number	Decimal Number	Octal Number	Decimal Number	Octal Number	Decimal Number	Octal Number
0	0	11	13	21	25	31	37
1	1	12	14	22	26	32	40
2	2	13	15	23	27	33	41
3	3	14	16	24	30	34	42
4	4	15	17	25	31	35	43
5	5	16	20	26	32	36	44
6	6	17	21	27	33	37	45
7	7	18	22	28	34	38	46
8	10	19	23	29	35	39	47
9	11	20	24	30	36	40	50
10	12						

Decimal to Octal Conversion

Integer part: Divide the integer part of decimal number by 8 and list the quotient and remainder as shown below. Repeat division until quotient is less than 8. The remainders (d_{n-1}, d_{n-2}, ... d_1, d_0) are arranged in the order shown below to get the octal equivalent of decimal integer part.

Fraction Part: Multiply the fraction part of decimal number by 8 as shown below and record the overflow (d_{-1}, d_{-2}, ... $d_{-(m-1)}$, d_{-m}). Repeat multiplication of fraction part alone till the result is zero. The overflows are arranged in the order shown below to get the octal equivalent of decimal fraction.

> ***Note:*** *In case of recurring fraction we will not get zero result in repeated multiplication and so stop multiplication when the required number of digits are obtained.*

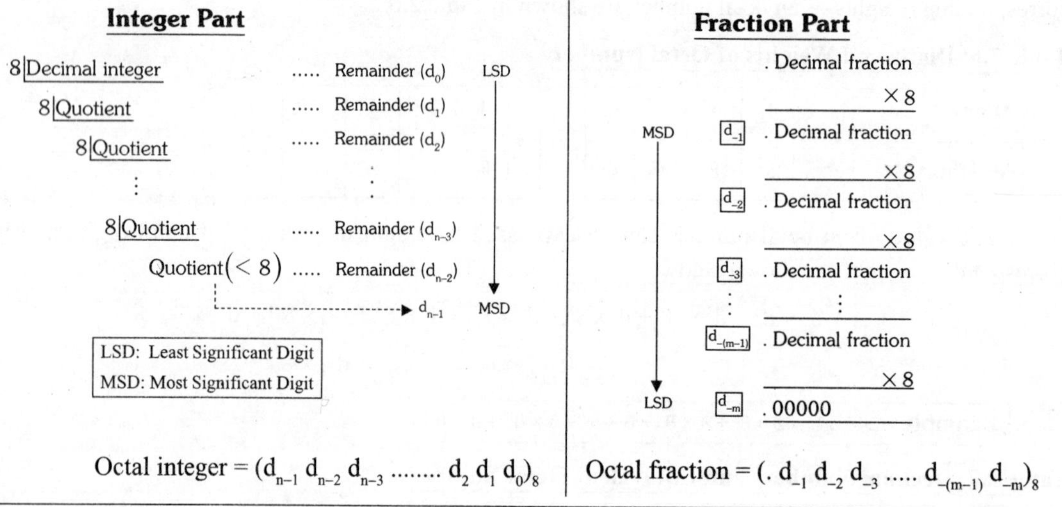

Octal to Decimal Conversion

For octal to decimal conversion multiply each octal digit by its weight and sum up the product.

Consider the octal number,

$$\ldots\ldots d_3\ d_2\ d_1\ d_0\ .\ d_{-1}\ d_{-2}\ d_{-3} \ldots\ldots$$

The decimal number can be obtained from following calculation.

$$\text{Octal number} = \ldots d_3\ d_2\ d_1\ d_0\ .\ d_{-1}\ d_{-2}\ d_{-3} \ldots$$

$$= \ldots d_3 \times 8^3 + d_2 \times 8^2 + d_1 \times 8^1 + d_0 \times 8^0 + d_{-1} \times 8^{-1} + d_{-2} \times 8^{-2} + d_{-3} \times 8^{-3} + \ldots$$

$$= \text{Equivalent decimal number}$$

Example 2.3

(RGPV, Jun'20, 3 Marks)

Convert the following numbers to octal and check the result by reverse conversion.

a) 1012.390625_{10} **b)** 204_{10}

Solution

a) 1012.390625_{10}

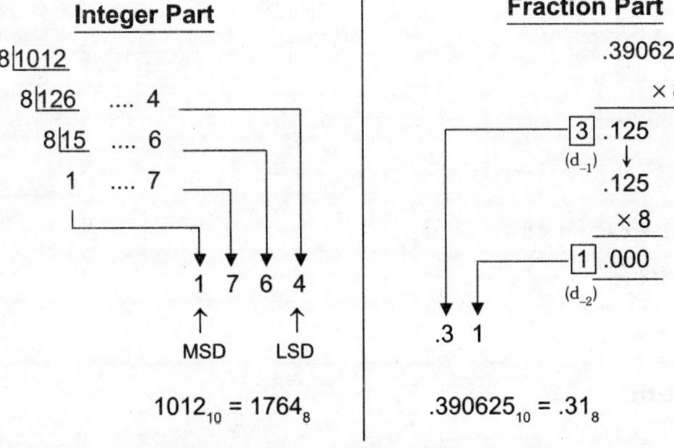

$1012_{10} = 1764_8$ $.390625_{10} = .31_8$

$\therefore \quad 1012\,.\,390625_{10} = 1764\,.\,31_8$

Reverse Conversion

$$1764.31_8 = 1\times 8^3 + 7\times 8^2 + 6\times 8^1 + 4\times 8^0 + 3\times 8^{-1} + 1\times 8^{-2}$$

Use calculator

$$= 512 + 448 + 48 + 4 + 0.375 + 0.015625$$

$$= 1012.390625_{10}$$

---

b) 204_{10}

8|204
8|25 4
3 1

3 1 4
MSD LSD

$\therefore \quad 204_{10} = 314_8$

Reverse Conversion

$$314_8 = 3\times 8^2 + 1\times 8^1 + 4\times 8^0$$

$$= 192 + 8 + 4$$

$$= 204_{10}$$

Example 2.4

(AKTU, Sep'20, 1 Mark) (RGPV, Jun'22, 3 Marks) (JNTUH, Nov/Dec'20, 3 Marks)

Convert the following numbers to decimal.

a) 26.24_8 b) 137.21_8

Solution

a) 26.24_8

$$26.24_8 = 2\times8^1 + 6\times8^0 + 2\times8^{-1} + 4\times8^{-2}$$
$$= 16 + 6 + 0.25 + 0.0625$$
$$= 22.3125_{10}$$

b) 137.21_8

$$137.21_8 = 1\times8^2 + 3\times8^1 + 7\times8^0 + 2\times8^{-1} + 1\times8^{-2}$$
$$= 64 + 24 + 7 + 0.25 + 0.015625$$
$$= 95.265625_{10}$$

2.1.5 Hexadecimal Number System

The ***hexadecimal number system*** is a radix number system in which the radix r is 16. Since the radix is 16, the number system uses 16 symbols which are chosen as 0, 1, 2, 3, 4, 5, 6, 7, 8, 9, A, B, C, D, E and F. The hexadecimal in short can be called ***hexa***.

A hexadecimal number in general is represented in the following format.

$$\underbrace{d_{n-1} \ldots\ldots d_4\, d_3\, d_2\, d_1\, d_0}_{\text{n digit integer part}} \;.\; \underbrace{d_{-1}\, d_{-2}\, d_{-3} \ldots\ldots d_{-m}}_{\text{m digit fraction part}}$$

In the above representation, d repesents the digit in the hexa number. The suffix in each digit represent power in the weight of the digit.

Each digit of a hexa number has a weight which is given by the power of radix,16. The digits and corresponding weights of a hexa number are shown in Table 2.8.

Table 2.8: Digits and Weights of Hexadecimal Number

Digits	d_{n-1}	....	d_4	d_3	d_2	d_1	d_0	.	d_{-1}	d_{-2}	d_{-3}	....	d_{-m}
Weights	16^{n-1}	....	16^4	16^3	16^2	16^1	16^0	.	16^{-1}	16^{-2}	16^{-3}	....	16^{-m}

The value of a hexadecimal number can be expressed as a sum of products of hexa digits and its corresponding weights as shown below:

$$d_{n-1} \times 16^{n-1} + \ldots + d_4 \times 16^4 + d_3 \times 16^3 + d_2 \times 16^2 + d_1 \times 16^1 + d_0 \times 16^0 + d_{-1} \times 16^{-1} + d_{-2} \times 16^{-2} + d_{-3} \times 16^{-3} + \ldots + d_{-m} \times 16^{-m}$$

Example: $AC2.7B = A \times 16^2 + C \times 16^1 + 2 \times 16^0 + 7 \times 16^{-1} + B \times 16^{-2}$

Table 2.9: Decimal Number and Equivalent Hexadecimal Number

Decimal Number	Hexadecimal Number	Decimal Number	Hexadecimal Number	Decimal Number	Hexadecimal Number	Decimal Number	Hexadecimal Number
0	0	11	B	21	15	31	1F
1	1	12	C	22	16	32	20
2	2	13	D	23	17	33	21
3	3	14	E	24	18	34	22
4	4	15	F	25	19	35	23
5	5	16	10	26	1A	36	24
6	6	17	11	27	1B	37	25
7	7	18	12	28	1C	38	26
8	8	19	13	29	1D	39	27
9	9	20	14	30	1E	40	28
10	A						

Decimal to Hexadecimal Conversion

Integer part: Divide the integer part of decimal number by 16 and list the quotient and remainder as shown below. Repeat division until quotient is less than 16. The remainders (d_{n-1}, d_{n-2}, ... d_1, d_0) are arranged in the order shown below to get the hexadecimal equivalent of decimal integer part.

Fraction Part: Multiply the fraction part of decimal number by 16 as shown below and record the overflow (d_{-1}, d_{-2}, ... $d_{-(m-1)}$, d_{-m}). Repeat multiplication of fraction part alone till the result is zero. The overflow are arranged in the order shown below to get the hexadecimal equivalent of decimal fraction.

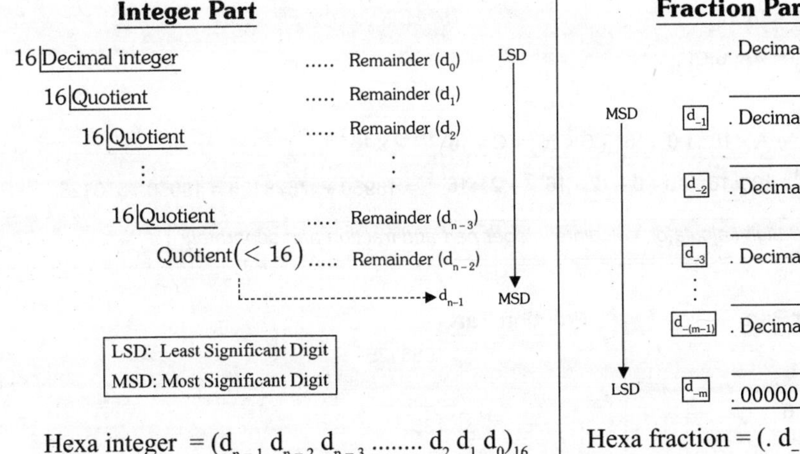

Hexa integer $= (d_{n-1}\, d_{n-2}\, d_{n-3} \,........\, d_2\, d_1\, d_0)_{16}$ Hexa fraction $= (.\, d_{-1}\, d_{-2}\, d_{-3} \,.....\, d_{-(m-1)}\, d_{-m})_{16}$

> ***Note:*** *In case of recurring fraction we will not get zero result in repeated multiplication and so stop multiplication when the required number of digits are obtained.*

Hexadecimal to Decimal Conversion

For hexadecimal to decimal conversion multiply each hexa digit by its weight and sum up the product.

Consider the hexadecimal number,

$$.......\, d_3\, d_2\, d_1\, d_0\, .\, d_{-1}\, d_{-2}\, d_{-3}\, ...$$

Now the decimal number can be obtained from the following calculation.

$$\text{Hexadecimal number} = ...d_3\,d_2\,d_1\,d_0\,.\,d_{-1}\,d_{-2}\,d_{-3}$$
$$= .....d_3 \times 16^3 + d_2 \times 16^2 + d_1 \times 16^1 + d_0 \times 16^0 + d_{-1} \times 16^{-1} + d_{-2} \times 16^{-2} + d_{-3} \times 16^{-3}$$
$$= \text{Equivalent decimal number}$$

Example 2.5

(AKTU, Sep'20, 3 Marks)

Convert the following numbers to hexadecimal number and cross-check result by reverse conversion.

a) 18950.7578125_{10} **b)** 22.1953125_{10}

Solution

a) 18950.7578125_{10}

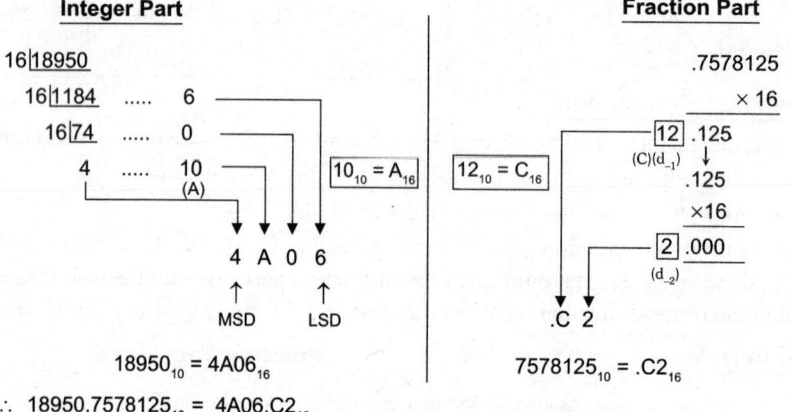

$18950_{10} = 4A06_{16}$ $\quad$ $.7578125_{10} = .C2_{16}$

$\therefore\ 18950.7578125_{10} = 4A06.C2_{16}$

Reverse Conversion

$$4A06.C2_{16} = 4 \times 16^3 + A \times 16^2 + 0 \times 16^1 + 6 \times 16^0 + C \times 16^{-1} + 2 \times 16^{-2}$$
$$= 4 \times 16^3 + 10 \times 16^2 + 0 + 6 + 12 \times 16^{-1} + 2 \times 16^{-2} = 18950 + .7578125 = 18950.7578125_{10}$$

> ***Note:*** *When using 10-digit calculator, calculate integer part and fraction part separately.*

b) 22.1953125_{10}

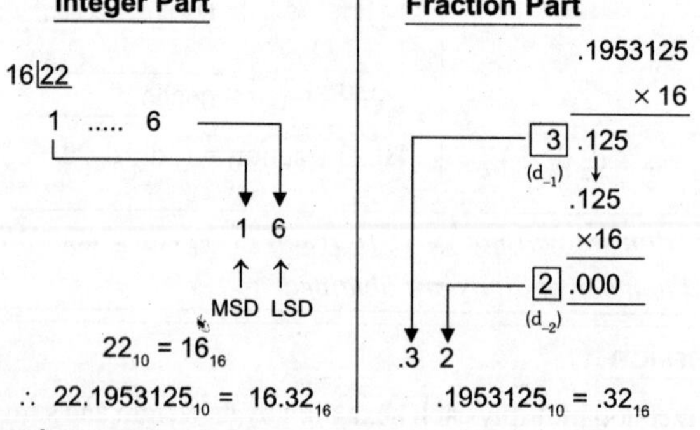

$22_{10} = 16_{16}$

$\therefore\ 22.1953125_{10} = 16.32_{16}$ $\quad$ $.1953125_{10} = .32_{16}$

Reverse Conversion

$$16.32_{16} = 1 \times 16^1 + 6 \times 16^0 + 3 \times 16^{-1} + 2 \times 16^{-2} = 22 + 0.1875 + 0.0078125 = 22.1953125_{16}$$

2.1.6 Relation between Binary, Octal and Hexadecimal Number Systems

The radix of octal is 8, the radix of hexa is 16.

Here, $8 = 2^3$ and $16 = 2^4$.

Therefore, an octal digit can be represented by 3 digit binary and a hexa digit by 4 digit binary. The octal and hexadecimal numbers are short hand form of binary.

An octal number can be directly converted to binary and vice-versa.

Similarly, a hexa number can be directly converted to binary and vice-versa.

Table 2.10: Binary Equivalent of Octal Digit

Octal	Binary
0	000
1	001
2	010
3	011
4	100
5	101
6	110
7	111

Table 2.11: Binary Equivalent of Hexa Digit

Hexa	Binary	Hexa	Binary
0	0000	8	1000
1	0001	9	1001
2	0010	A	1010
3	0011	B	1011
4	0100	C	1100
5	0101	D	1101
6	0110	E	1110
7	0111	F	1111

Example 2.6

(GTU, May'19, 2 Marks)

Convert 726.35_8 to binary and cross-check binary and octal by converting to decimal.

Solution

$726.35_8 = 111\ 010\ 110.011\ 101_2$

$= 111010110.011101_2$

Each octal digit is represented by three digit binary.

Cross-Check

$$726.35_8 = 7\times8^2+2\times8^1+6\times8^0+3\times8^{-1}+5\times8^{-2} = 470.453125_{10}$$

$$111010110.011101_2 = 1\times2^8+1\times2^7+1\times2^6+0\times2^5+1\times2^4+0\times2^3+1\times2^2+1\times2^1+0\times2^0$$

$$+0\times2^{-1}+1\times2^{-2}+1\times2^{-3}+1\times2^{-4}+0\times2^{-5}+1\times2^{-6}$$

$$= 2^8+2^7+2^6+2^4+2^2+2+2^{-2}+2^{-3}+2^{-4}+2^{-6}$$

$$= 256 + 128 + 64 + 16 + 4 + 2 + 0.25 + 0.125 + 0.0625 + 0.015625$$

$$= 470.453125_{10}$$

Example 2.7

(RGPV, Jun'20, 3 Marks) (GTU, May'19, 2 Marks)

Convert the following numbers to binary and cross-check binary and hexa by converting to decimal.

a) $3A7_{16}$ **b)** $A52.42_{16}$

Solution

a) $3A7_{16}$

$3A7_{16} = 0011\ 1010\ 0111_2$

$= 001110100111_2$

Each hexa digit is represented by four digit binary.

Cross-Check

$$3A7_{16} = 3\times16^2 + A\times16^1 + 7\times16^0 = 3\times16^2 + 10\times16^1 + 7\times16^0$$

$$= 768 + 160 + 7 = 935_{10}$$

$$001110100111_2 = 0\times2^{11} + 0\times2^{10} + 1\times2^9 + 1\times2^8 + 1\times2^7 + 0\times2^6 + 1\times2^5 + 0\times2^4$$
$$+ 0\times2^3 + 1\times2^2 + 1\times2^1 + 1\times2^0$$

$$= 2^9 + 2^8 + 2^7 + 2^5 + 2^2 + 2^1 + 2^0$$

$$= 512 + 256 + 128 + 32 + 4 + 2 + 1 = 935_{10}$$

---

b) $A52.42_{16}$

$$A52.42_{16} = 1010\ 0101\ 0010.0100\ 0010_2$$

$$= 101001010010.0100001_2$$

Each hexa digit is represented by four digit binary.

Cross-Check

$$A52.42_{16} = A\times16^2 + 5\times16^1 + 2\times16^0 + 4\times16^{-1} + 2\times16^{-2}$$

$$= 10\times16^2 + 5\times16 + 2 + 4\times16^{-1} + 2\times16^{-2}$$

$$= 2642 + 0.2578125 = 2642.2578125_{10}$$

$$101001010010.0100001_2 = 1\times2^{11} + 0\times2^{10} + 1\times2^9 + 0\times2^8 + 0\times2^7 + 1\times2^6 + 0\times2^5 + 1\times2^4$$
$$+ 0\times2^3 + 0\times2^2 + 1\times2^1 + 0\times2^0 + 0\times2^{-1} + 1\times2^{-2} + 0\times2^{-3}$$
$$+ 0\times2^{-4} + 0\times2^{-5} + 0\times2^{-6} + 1\times2^{-7}$$

$$= 2^{11} + 2^9 + 2^6 + 2^4 + 2 + 2^{-2} + 2^{-7} = 2048 + 512 + 64 + 16 + 2 + 0.25 + 0.0078125$$

$$= 2642 + 0.2578125 = 2642.2578125_{10}$$

Note: *When using 10-digit calculator, calculate integer part and fraction part separately.*

Example 2.8

Convert 10101001010.10101101_2 to octal and hexa number and cross-check binary, octal and hexa by converting to decimal.

Solution

Separate the binary into group of three digits to find octal number. Represent each three digit binary by an equivalent octal digit.

$$10101001010.10101101_2 = 10\ 101\ 001\ 010.101\ 011\ 01_2$$

$$= 010\ 101\ 001\ 010.101\ 011\ 010_2$$

$$= 2\quad 5\quad 1\quad 2\ .\ 5\quad 3\quad 2_8$$

$$= 2512.532_8$$

Append with leading and trailing zero.

Separate the given binary into group of four digit binary to find hexa number. Represent each four digit binary by an equivalent hexa digit.

$$10101001010.10101101_2 = 101\ 0100\ 1010.1010\ 1101_2$$

$$= 0101\ 0100\ 1010.1010\ 1101_2$$

$$= 5\quad 4\quad A\ .\ A\quad D_{16}$$

$$= 54A.AD_{16}$$

Append with leading zero.

Cross-Check

$$10101001010.10101101_2 = 1\times2^{10}+1\times2^8+1\times2^6+1\times2^3+1\times2^1+1\times2^{-1}+1\times2^{-3}+1\times2^{-5}+1\times2^{-6}+1\times2^{-8}$$

Weights of zeros are not shown.

$$= 2^{10} + 2^8 + 2^6 + 2^3 + 2^1 + 2^{-1} + 2^{-3} + 2^{-5} + 2^{-6} + 2^{-8}$$

$$= 1024 + 256 + 64 + 8 + 2 + 0.5 + 0.125 + 0.03125 + 0.015625 + 0.00390625$$

$$= 1354.67578125_{10}$$

$$2512.532_8 = 2\times8^3+5\times8^2+1\times8^1+2\times8^0+5\times8^{-1}+3\times8^{-2}+2\times8^{-3}$$

$$= 1024 + 320 + 8 + 2 + 0.625 + 0.046875 + 0.00390625 = 1354.67578125_{10}$$

$$54A.AD_{16} = 5\times16^2+4\times16^1+A\times16^0+A\times16^{-1}+D\times16^{-2}$$

$$= 5\times16^2+4\times16+10+10\times16^{-1}+13\times16^{-2}$$

$$= 1280 + 64 + 10 + 0.625 + 0.05078125 = 1354.67578125_{10}$$

Example 2.9

(AKUB, Dec'20, 4 Marks)

Convert 732.51_8 to binary and hexa and cross-check binary and hexa by converting to decimal.

Solution

$$732.51_8 = 111\ 011\ 010.101\ 001_2 = 111011010.101001_2$$

Represent each octal digit by three digit binary.

$$732.51_8 = 111\ 011\ 010.101\ 001_2 = 1\ 1101\ 1010.1010\ 01_2$$

Rearrange the binary representation into group of four digits.

$$= 0001\ 1101\ 1010.1010\ \ 0100_2$$

Append with leading and trailing zeros.

$$= 1 \quad D \quad A \ . \ A \quad 4_{16}$$

$$= 1DA\ .\ A4_{16}$$

Represent each group of four digit binary by an equivalent hexa digit.

Cross-Check

$$732.51_8 = 7\times8^2+3\times8^1+2\times8^0+5\times8^{-1}+1\times8^{-2}$$

$$= 448 + 24 + 2 + 0.625 + 0.015625 = 474.640625_{10}$$

$$111011010.101001_2 = 1\times2^8+1\times2^7+1\times2^6+1\times2^4+1\times2^3+1\times2^1+1\times2^{-1}+1\times2^{-3}+1\times2^{-6}$$

$$= 2^8 + 2^7 + 2^6 + 2^4 + 2^3 + 2^1 + 2^{-1} + 2^{-3} + 2^{-6}$$

$$= 256 + 128 + 64 + 16 + 8 + 2 + 0.5 + 0.125 + 0.015625 = 474.640625_{10}$$

$$1DA.A4_{16} = 1\times16^2+D\times16^1+A\times16^0+A\times16^{-1}+4\times16^{-2}$$

Weights of zeros are not shown.

$$= 1\times16^2+13\times16+10+10\times16^{-1}+4\times16^{-2}$$

$$= 256 + 208 + 10 + 0.625 + 0.015625 = 474.640625_{10}$$

2.2 Binary Arithmetic

Binary arithmetic is similar to decimal arithmetic of only two numbers 0 and 1. But in binary arithmetic, the result should be in binary. The rules for binary addition, subtraction and multiplication are listed in Table 2.12.

Table 2.12: Rules of Binary Arithmetic

Binary Addition	Binary Subtraction	Binary Multiplication
$0 + 0 = 0$	$0 - 0 = \ \ 0$	$0 \times 0 = 0$
$0 + 1 = 1$	$0 - 1 = 1 \ \ 1$	$0 \times 1 = 0$
$1 + 0 = 1$	$1 - 0 = \ \ 1$	$1 \times 0 = 0$
$1 + 1 = 1 \ \ 0$	$1 - 1 = \ \ 0$	$1 \times 1 = 1$
↑ Carry ↑ Sum	↑ Borrow ↑ Difference	

The binary division involves successive subtraction similar to decimal subtraction and therefore no separate rules for binary division.

Examples of binary arithmetic of unsigned binary are presented in this section. Practically, in digital systems the signed numbers are represented by 2's complement. The binary arithmetic using complement numbers are presented in Section 2.3.

2.2.1 Binary Addition

Addition of two binary numbers is performed by bit-by-bit addition starting from LSB (Least Significant Bit). The carry generated in one bit addition is considered in next bit addition. If carry is generated in the addition of MSB (Most Significant Bit) then it is considered as MSB of sum.

Example 2.10

Perform the following binary addition.

a) $10110_2 + 111_2$ b) $1010_2 + 11011_2$

Solution

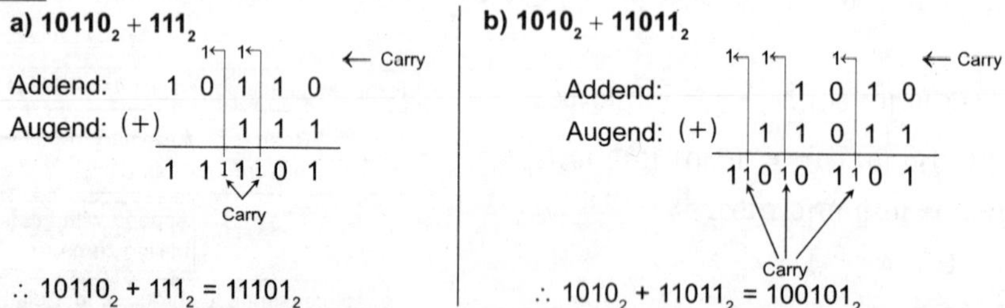

2.2.2 Binary Subtraction

Subtraction of two binary numbers is performed by bit-by-bit subtraction starting from LSB (Least Significant Bit). When a subtraction of two bits need borrow then a 1 is borrowed from next available higher order bit and the borrowed bit is made 0.

When a larger magnitude number has to be subtracted from smaller magnitude number then the subtraction is performed by subtracting smaller magnitude number from larger magnitude number and a minus sign is added in the result.

Example 2.11

Perform the following binary subtraction.

a) $10110_2 - 111_2$ b) $11000_2 - 1010_2$ c) $1011_2 - 11001_2$

Solution

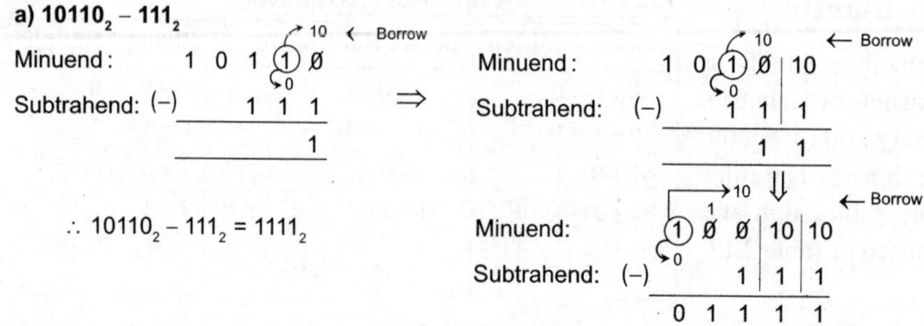

b) $11000_2 - 1010_2$

```
                 1→10   ← Borrow                        10        ← Borrow
Minuend:       1 ① Ø Ø 0        ⇒    Minuend:        ① Ø | 1 |10| 0
                 ↳0                                   ↳0
Subtrahend: (−)  1 0 1 0             Subtrahend: (−)   1 | 0 | 1 | 0
              ---------                              -------------
                 1 1 1 0                             0 1   1   1   0
              ---------                              -------------
```

$\therefore 11000_2 - 1010_2 = 1110_2$

---

c) $1011_2 - 11001_2$

Here, the subtrahend is greater than the minuend. Hence, the minuend and subtrahend are interchanged and then the subtraction is performed and the result is considered as negative.

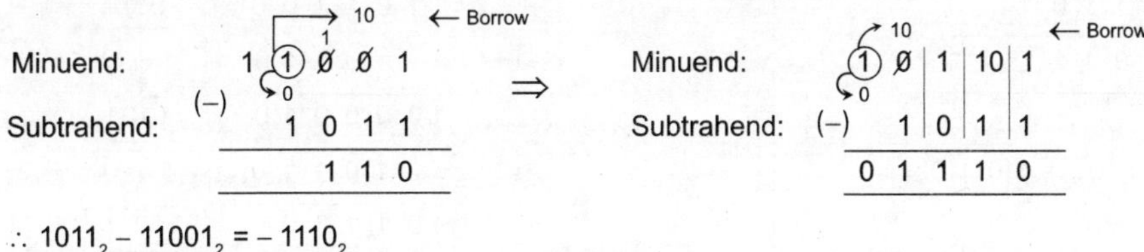

$\therefore 1011_2 - 11001_2 = -1110_2$

2.2.3 Binary Multiplication

Multiplication of two binary numbers is performed by multiplying each bit of multiplier (starting from LSB) to multiplicand to generate partial products. The partial products are arranged such that a partial product is left shifted by one bit position from previous product. Now, the sum of shifted products is the (final) product of the two binary numbers.

Alternatively, binary multiplication can be performed by successive addition. The multiplier is used as a count for number of additions. The initial product is considered as zero and the multiplicand is added repeatedly to initial product. After each addition the count is decremented by 1 and the addition is stopped when the count reaches 0.

Example 2.12

Multiply the binary numbers 10101_2 and 1101_2.

Solution

```
10101 × 1 = 10101   |            1 0 1 0 1
10101 × 0 = 00000   |    (×)       1 1 0 1
10101 × 1 = 10101   |        -------------
10101 × 1 = 10101   |            1 0 1 0 1
                    |    (+)   0 0 0 0 0
                    |    (+) 1 0 1 0 1
                    |    (+) 1 0 1 0 1
                    |        -----------------
                    |      1 0 0 0 1 0 0 0 1
                    |        -----------------
```

$\therefore 10101_2 \times 1101_2 = 100010001_2$

Example 2.13

Multiply the binary numbers 10110_2 and 101_2.

Solution

Method 1:

```
        1 0 1 1 0
(×)         1 0 1
    -------------
        1 0 1 1 0
(+)   0 0 0 0 0
(+) 1 0 1 1 0
    -------------
    1 1 0 1 1 1 0
    -------------
```

$\therefore 10110_2 \times 101_2 = 1101110_2$

Method 2:

	Product	Count
Initial Product:	0 0 0 0 0	101
(+)	1 0 1 1 0	−1
	1 0 1 1 0	100
(+)	1 0 1 1 0	−1
	1 0 1 1 0 0	011
(+)	1 0 1 1 0	−1
	1 0 0 0 0 1 0	010
(+)	1 0 1 1 0	−1
	1 0 1 1 0 0 0	001
(+)	1 0 1 1 0	−1
	1 1 0 1 1 1 0	000

$\therefore 10110_2 \times 101_2 = 1101110_2$

Shift and Add (or Computer) Method of Multiplication

In this method the multiplicand and multiplier are converted to same bit size by appending 0's to upper most bits. The standard bit sizes are 4-bits, 8-bits, 16-bits, etc. If the standard bit size is n-bits then the maximum size of product will be 2n-bits. In order to perform multiplication by this method three storage devices like registers are required. One of the register is a left shift-register of size 2n-bits to hold the partial sum/final product called P-register, another register of size n-bits to hold the multiplicand called M-register and a 1-bit register to hold the shifted bit called C-register.

The n-bit multiplier is converted to 2n-bits by appending 0's to lower most bits and stored in P-register. The n-bit multiplicand is stored in M-register. The content of P-register is shifted left n times such that most significant bit (MSB) is shifted to C-register. In each shift check for bit value in C-register, if the shifted bit is 1, then add M-register to P-register, otherwise add 0.

The content of P-register after n shift and add operations will be the final product.

Example 2.14

Perform the multiplication $1101_2 \times 1010_2$ by shift and add (or computer) method.

Solution

The given binary numbers are 4-bit numbers and hence the product will be of size 8-bits. First convert multiplier to 8-bits and save in P-register.

Multiplier = 1010_2 $\xrightarrow{\text{Convert to 8-bit}}$ $1010\ 0000_2$ $\xrightarrow{\text{Store in P-register}}$ P: [1010 0000]

Store the 4-bit multiplicand in M-register.

Multiplicand = 1101_2 $\xrightarrow{\text{Store in M-register}}$ M: [1101]

The content of P-register is shifted left 4 times such that MSB is shifted to C-register. In each shift check for value of bit in C-register, if value is 1 then add M-register to P-register otherwise add 0.

Shift-1

P [1010 0000] —Shift left→ C [1] P [0100 0000]

Since C = 1, add M register to P-register.

```
     0100 0000
(+)       1101
     ---------
     0100 1101  --Store in P-->  P [0100 1101]
```

Shift-2

P [0100 1101] —Shift left→ C [0] P [1001 1010]

Since C = 0, skip addition (or add 0 to P).

Shift-3

P [1001 1010] —Shift left→ C [1] P [0011 0100]

Since C = 1, add M-register to P-register.

```
       111 1    <- Carry
     0011 0100
(+)       1101
     ---------
     0100 0001  --Store in P-->  P [0100 0001]
```

Shift-4

P [0100 0001] —Shift left→ C [0] P [1000 0010]

Since C = 0, skip addition (or add 0 to P).

After 4 shift and add operations, the content of P-register is the final product.

$\therefore 1101_2 \times 1010_2 = 1000\ 0010_2$

Cross-check

```
          1 1 0 1
(×)       1 0 1 0
-----------------
          0 0 0 0
(+)     1 1 0 1
(+)   0 0 0 0
(+) 1 1 0 1
-----------------
  1 0 0 0 0 0 1 0
```

$\therefore 1101_2 \times 1010_2 = 1000\ 0010_2$

2.2.4 Binary Division

Binary division is performed similar to decimal division as shown in following examples.

Alternatively, binary division can be performed by successive subtraction. The divisor is subtracted from dividend repeatedly until the dividend becomes lesser than divisor. The initial quotient is considered as zero and incremented by 1 for each subtraction.

Example 2.15

Perform the following binary divisions.

a) $110101_2 \div 101_2$ **b)** $1101110_2 \div 1010_2$ **c)** $11010_2 \div 100_2$

Solution

a) $110101_2 \div 101_2$

```
            1 0 1 0   (Quotient)
     ---------------------
101 |       1 1 0 1 0 1
    | (–)   1 0 1
    |     ---------------
    |       0 0 1 1
    | (–)     0 0 0
    |     ---------------
    |       0 0 1 1 0
    | (–)       1 0 1
    |     ---------------
    |       0 0 0 0 1 1
    | (–)         0 0 0
    |     ---------------
    |       0 0 0 0 1 1   (Remainder)
```

$\therefore 110101_2 \div 101_2 = 1010_2$ (Remainder = 11_2)

b) $1101110_2 \div 1010_2$

```
             1 0 1 1   (Quotient)
      ----------------------
1010 |       1 1 0 1 1 1 0
     | (–)   1 0 1 0
     |     -----------------
     |       0 0 1 1 1
     | (–)     0 0 0 0
     |     -----------------
     |       0 0 1 1 1 1
     | (–)       1 0 1 0
     |     -----------------
     |       0 0 0 1 0 1 0
     | (–)         1 0 1 0
     |     -----------------
     |       0 0 0 0 0 0 0   (Remainder)
```

$\therefore 1101110_2 \div 1010_2 = 1011_2$ (Remainder = 0)

c) $11010_2 \div 100_2$

Method 1:

```
            1 1 0   (Quotient)
     -------------------
100 |       1 1 0 1 0
    | (–)   1 0 0
    |     -------------
    |       0 1 0 1
    | (–)     1 0 0
    |     -------------
    |       0 0 0 1 0
    | (–)       0 0 0
    |     -------------
    |       0 0 0 1 0   (Remainder)
```

$\therefore 11010_2 \div 100_2 = 110_2$

(Remainder = 10_2)

Method 2:

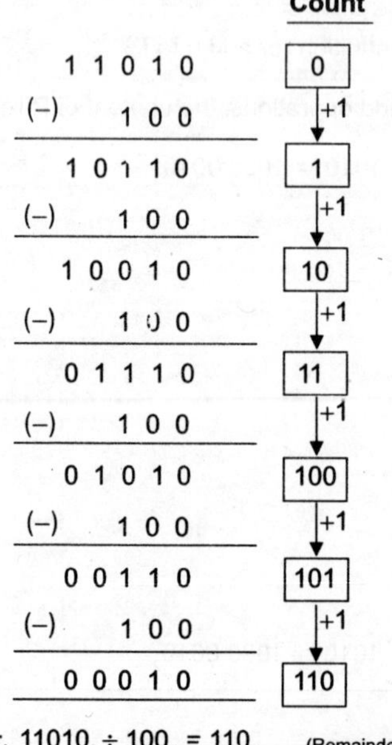

$\therefore 11010_2 \div 100_2 = 110_2$ (Remainder = 10_2)

Shift and Subtract (or Computer) Method of Division

This method supports division of 2n-bits binary by n-bit binary. Hence, the dividend is converted to 2n-bits and divisor is converted to n-bits by appending 0's to upper most bits. The standard bit sizes are 8-bit data divided by 4-bit data, 16-bit data divided by 8-bit data and so on. For maximum value of 2n-bits dividend and minimum value of n-bit divisor, the maximum value of quotient will be 2n-bits and remainder will be n-bits. For simplicity, consider 2n-bit by n-bit division with quotient not exceeding n-bits.

In order to perform 2n-bit by n-bit division when quotient not exceeding n-bits three storage devices like registers are required. One of the register is a left shift-register of size 2n-bits to hold the n-bit quotient and n-bit remainder called R-register, another register of size n-bits to hold the divisor called D-register and a 1-bit register to hold borrow in subtraction called C-register.

Initially, the 2n-bit dividend is stored in R-register. The n-bit divisor is stored in D-register. The content of D-register is subtracted from upper most n-bits of R-register and borrow is stored in C-register. If borrow is 1, then the subtraction is invalid and hence D-register is added back to R-register. If borrow is 0, then subtraction is valid and a 1 is added to lower most bit of R-register. Then the content of R-register is shifted left one time.

This process of subtract and shift are repeated n-times and after n such operations, the upper n-bits of R-register is the remainder and the lower n-bits of the R-register is the quotient.

Example 2.16

Perform the division $1000001_2 \div 110_2$ by shift and subtract (or computer) method.

Solution

The dividend is 7-bits and divisor is 3-bits.

Let us perform standard 8-bit by 4-bit division.

$\therefore$ Dividend = $100\ 0001_2$ —Convert to 8-bit→ $0100\ 0001_2$ —Store in R-register→ R: [0100 0001]

Divisor = 110_2 —Convert to 4-bit→ 0110_2 —Store in D-register→ D: [0110]

The subtract and shift operation is performed 4 times as shown below:

Subtract and Shift-1

Subtract D-register from upper most 4-bits of R-register.

```
   10         ← Borrow                 10              ← Borrow
 0 ① Ø 0 0001                  ⇒    ① Ø Ø 1 0 0 0001
   0                                   1
(-)0 1 1 0                         (-)  0 1 1 0
-------------                      --------------------
   1 0 0001                        C [1] 1 1 1 0 0001
```

Since C = 1, add divisor to R-register.

```
 1 1          ← Carry
 1110 0001
(+) 0110
-----------
[1] 0100 0001 —Shift left→ [0] 1000 0010 —Store in R→ R: [1000 0010]
 ↑                          ↑
Discard carry              Discard shifted bit
```

Subtract and Shift-2

Subtract D-register from upper most 4-bits of R-register.

```
  →10            ← Borrow
  1
 ①ØØ0 0010
 →0
(–) 0110
C [0] 0 010 0010
          + 1  (Add 1)
  0 010 0011  —Shift left→  [0] 0100 0110  —Store in R→  R [0100 0110]
                             ↑
                     Discard shifted bit
```

Subtract and Shift-3

Subtract D-register from upper most 4-bits of R-register.

```
  →10        ← Borrow            →10          ← Borrow
                                  1
 0①ØØ0 0110                     ①ØØ10 0 0110
  →0
(–)0 110             ⇒          (–)   0110
     10 0110                     C [1] 11 1 0 0110
```

Since C = 1, add divisor to R-register.

```
 111       ← Carry
 1110 0110
(+) 0110
[1] 0100 0110  —Shift left→  [0] 1000 1100  —Store in R→  R [1000 1100]
 ↑                            ↑
Discard carry         Discard shifted bit
```

Subtract and Shift-4

Subtract D-register from upper most 4-bits of R-register.

```
  →10            ← Borrow
  1
 ①ØØ0 1100
 →0
(–) 0110
C [0] 0 010 1100
          + 1  (Add 1)
  0 010 1101  —Shift left→  [0] 0101 1010  —Store in R→  [0101 1010]
                             ↑                         Remainder  Quotient
                     Discard shifted bit
```

$\therefore$ 100 0001 ÷ 110_2 = 1010_2 (Remainder = 101_2)

Cross-check

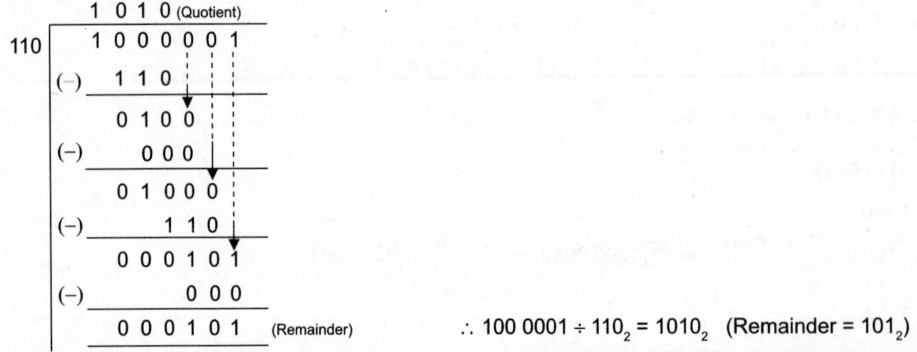

```
        1 0 1 0 (Quotient)
110 | 1 0 0 0 0 0 1
   (–)  1 1 0
        0 1 0 0
   (–)    0 0 0
        0 1 0 0 0
   (–)      1 1 0
        0 0 0 1 0 1
   (–)        0 0 0
        0 0 0 1 0 1  (Remainder)
```

$\therefore$ 100 0001 ÷ 110_2 = 1010_2 (Remainder = 101_2)

2.3 Unsigned and Signed Binary Number Systems

The binary number system can be unsigned binary or signed binary number system. The ***unsigned binary*** has only positive numbers. The ***signed binary*** has positive and negative numbers. There are three types of signed binary number system. They are,

1. Sign-magnitude Form
2. One's Complement Form
3. Two's Complement Form

2.3.1 Unsigned Binary

The general format of an unsigned binary number is given below:

$d_{n-1} \ldots\ldots d_4 d_3 d_2 d_1 d_0$

Each binary digit has a weight. The weight of digit, d_k is r^k. Here r = 2. Therefore weight is 2^k.

In a n-bit unsigned binary number the digit d_0 has the weight $2^0 = 1$, which is the lowest weight. Hence the digit d_0 is called ***LSD (Least Significant Digit)***.

In a n-bit unsigned binary number the digit d_{n-1} has the weight 2^{n-1}, which is the largest weight. Hence the digit d_{n-1} is called ***MSD (Most Significant Digit)***.

In binary number system, the digit is popularly known as bit. Therefore, LSD and MSD can also expressed as LSB and MSB respectively.

Range of Unsigned Binary Number

With n-bit binary it is possible to generate 2^n binary numbers.

When n = 1, $2^n = 2^1 = 2$, 1-bit binary numbers: 0, 1

When n = 2, $2^n = 2^2 = 4$, 2-bit binary numbers: 00, 01, 10, 11

When n = 3, $2^n = 2^3 = 8$, 3-bit binary numbers: 000, 001, 010, 011, 100, 101, 110, 111
and so on.

Therefore, the range of decimal number that can be represented by n-bit unsigned binary is 0 to $2^n - 1$.

The binary equivalent of decimal numbers 0 to 31 are listed in Table 2.13.

Sometime we have to write the binary number in a required size of 4-bit or 6-bit or 8-bit, etc.

In order to express the unsigned binary to required size, append required number of zeros before MSD. The 8-bit representation of the binary numbers are also shown in Table 2.13.

Table 2.13: Decimal Number and its Equivalent Unsigned Binary Number

Decimal Number	Unsigned Binary Number	8-bit Unsigned Binary Number	Decimal Number	Unsigned Binary Number	8-bit Unsigned Binary Number
0	0	0000 0000	16	10000	0001 0000
1	1	0000 0001	17	10001	0001 0001
2	10	0000 0010	18	10010	0001 0010
3	11	0000 0011	19	10011	0001 0011
4	100	0000 0100	20	10100	0001 0100
5	101	0000 0101	21	10101	0001 0101
6	110	0000 0110	22	10110	0001 0110
7	111	0000 0111	23	10111	0001 0111
8	1000	0000 1000	24	11000	0001 1000
9	1001	0000 1001	25	11001	0001 1001
10	1010	0000 1010	26	11010	0001 1010
11	1011	0000 1011	27	11011	0001 1011
12	1100	0000 1100	28	11100	0001 1100
13	1101	0000 1101	29	11101	0001 1101
14	1110	0000 1110	30	11110	0001 1110
15	1111	0000 1111	31	11111	0001 1111

2.3.2 Sign-Magnitude Form

In n-bit (or n-digit) binary representation of sign-magnitude form of binary, 1-bit is reserved for sign and n – 1 bits are used to represent the magnitude.

For positive number the sign bit is "0" and for negative number the sign bit is "1". The magnitude bits are same for a given number with opposite signs.

In order to represent the given decimal number in n-bit sign-magnitude form, first represent the magnitude as n – 1 bits unsigned number. Then put "**0**" in MSD for positive number and put "**1**" in MSD position for negative number. The drawback in sign-magnitude representation is that there will be two zeros + 0 and – 0. (Refer Table 2.14).

An example of representing $+5_{10}$ and -5_{10} in sign-magnitude form is shown below:

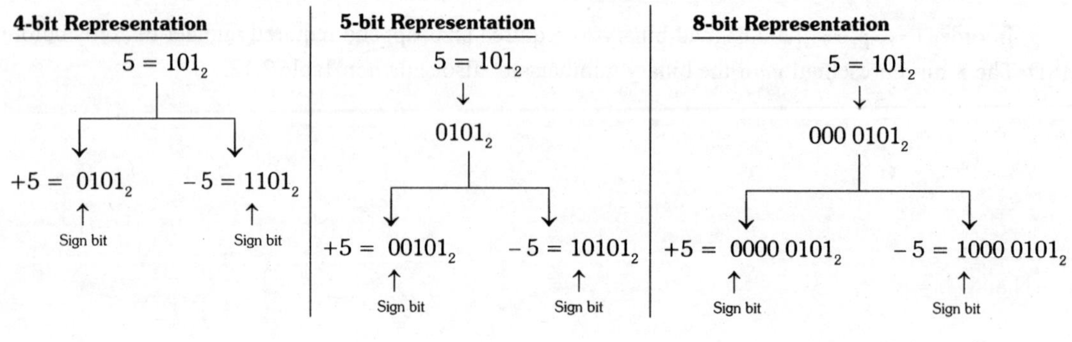

Range of Sign-Magnitude Binary Number

The range of decimal numbers that can be represented in n-bit sign-magnitude form of binary number is,

$$-(2^{n-1}-1) \quad \text{to} \quad +(2^{n-1}-1)$$

The range of decimal numbers that can be represented in sign-magnitude form of binary for n = 4 and 5 are shown below:

n = 4

Range = $-(2^{4-1}-1)$ to $+(2^{4-1}-1)$

= -7_{10} to $+7_{10}$

n = 5

Range = $-(2^{5-1}-1)$ to $+(2^{5-1}-1)$

= -15_{10} to $+15_{10}$

Table 2.14: Decimal and Sign-Magnitude form of Binary Number

Decimal Number	4-bit Sign-magnitude Binary Number	Standard 8-bit Sign-magnitude Binary Number
+0	0000	0000 0000
+1	0001	0000 0001
+2	0010	0000 0010
+3	0011	0000 0011
+4	0100	0000 0100
+5	0101	0000 0101
+6	0110	0000 0110
+7	0111	0000 0111
–0	1000	1000 0000
–1	1001	1000 0001
–2	1010	1000 0010
–3	1011	1000 0011
–4	1100	1000 0100
–5	1101	1000 0101
–6	1110	1000 0110
–7	1111	1000 0111

2.3.3 Complement of Number Systems (r's and (r – 1)'s Complement)

Complement of positive numbers are used to represent negative numbers in digital system.

In radix number system, two types of complement are used.

1. r's complement or radix complement.
2. (r – 1)'s complement or diminished radix complement.

Basically, ***complement*** is the number obtained by subtracting a number from it's base number. For an n-digit number the base number is r^n. Therefore, r's complement of an n-digit number N is given by $r^n - N$ and (r – 1)'s complement is given by $(r^n - 1) - N$.

Table 2.15: Summary of Complements in Number Systems

Number Systems	r's Complement	(r – 1)'s Complement
Decimal	10's complement	9's complement
Octal	8's complement	7's complement
Hexa	16's complement	15's complement
Binary	2's complement	1's complement

Importance of Complement

The basic mathematical operation is addition. Subtraction is addition of positive and negative number.

Example: 75 – 42 = 72 + (– 42)

Multiplication is successive addition.

Example: $6 \times 3 = 6 + 6 + 6 = 3 + 3 + 3 + 3 + 3 + 3$

Division is successive subtraction. Subtraction is again addition of positive and negative number.

Example: $10 \div 3 \Rightarrow 10 - 3 = 7 - 3 = 4 - 3 = 1$ (Remainder)
Quotient = 1 + 1 + 1 = 3 (Count of number of subtraction)

From the above examples we can infer that all mathematical operations are realized in terms of addition in a digital system.

Therefore, negative number is represented in complement form in digital system, so that every arithmetic operation is performed by addition.

Note: *The ALU (Arithmetic Logical Unit) of a microprocesser is basically an adder capable of performing arithmetic and logical operations in terms of addition.*

2.3.4 One's Complement

In one's complement form of binary number representation, the positive numbers will be same as that of unsigned binary. The negative of an n-bit (or n-digit) positive number N is given by $(2^n - 1) - N$. For any value of n, it can be observed that, $2^n - 1 = 1111....11$ and hence 1's complement can be obtained by subtracting every digit by 1. This is same as bit-by-bit complement of the positive number which is changing every 1 to 0 and 0 to 1 of the positive number.

In complement representation of binary numbers it can be observed that, for positive numbers the MSD is "0" and for negative numbers MSD is "1". (Refer Table 2.16).

In order to express the one's complement number in the required size, sign extension technique is employed. In this technique the upper most bit is extended to the required size.

The drawback in one's complement form is that there will be two zeros +0 and – 0. (Refer Table 2.16).

An example of representing $+5_{10}$ and -5_{10} in one's complement form is shown below:

$$5_{10} = 101_2$$

⇓ Append sign bit

$$0101_2 = +5$$

Sign bit ↑

⇓ Complement

$$1010_2 = -5 \text{ (One's complement form)}$$

Sign bit ↑

$\therefore +5_{10} = 0101_2$ (4-bit representation) $= 0000\ 0101_2$ (sign extended to 8-bit)

$-5_{10} = 1010_2$ (4-bit representation) $= 1111\ 1010_2$ (sign extended to 8-bit)

Range of One's Complement Binary Number

The range of decimal numbers that can be represented in n-bit one's complement form of binary number is,

$$-(2^{n-1}-1) \text{ to } +(2^{n-1}-1)$$

The range of decimal numbers that can be represented in one's complement fotm of binary for n = 4 and 5 are shown below:

n = 4

Range $= -(2^{4-1}-1)$ to $+(2^{4-1}-1)$

$= -7_{10}$ to $+7_{10}$

n = 5

Range $= -(2^{5-1}-1)$ to $+(2^{5-1}-1)$

$= -15_{10}$ to $+15_{10}$

Table 2.16: Decimal and One's Complement form of Binary Number

Decimal Number	4-bit One's Complement Binary Number	Sign-extended One's Complement 8-bit Binary Number
+0	0000	0000 0000
+1	0001	0000 0001
+2	0010	0000 0010
+3	0011	0000 0011
+4	0100	0000 0100
+5	0101	0000 0101
+6	0110	0000 0110
+7	0111	0000 0111
–7	1000	1111 1000
–6	1001	1111 1001
–5	1010	1111 1010
–4	1011	1111 1011
–3	1100	1111 1100
–2	1101	1111 1101
–1	1110	1111 1110
–0	1111	1111 1111

One's Complement of Binary Fraction Numbers

For one's complement of binary fraction numbers every bit is complemented like integer number.

Example: $1001.110_2 \xrightarrow{\text{1's complement}} 0110.001_2$

When a binary bit is complemented twice the value of bit will not change. Therefore, if one's complement is made twice we will get original number. Also, it can be shown that the sum of a binary number and its 1's complement is 1111.....11. These concepts are illustrated in the following examples.

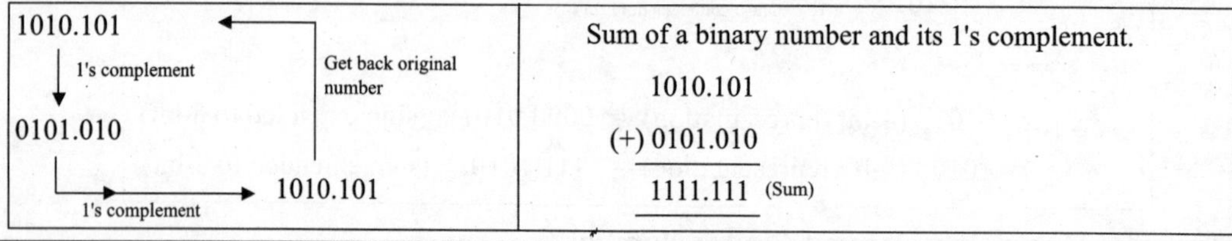

2.3.5 Two's Complement

In two's complement form of binary number representation, the positive numbers will be same as that of unsigned binary. The negative number is obtained by adding 1 to 1's complement of negative number. Therefore, negative of an n-bit (or n-digit) positive number N is given by $(2^n - 1) - N + 1 = 2^n - N$ except when N = 0. This is same as bit-by-bit complementing the positive number and then adding one. Similar to one's complement number, the MSD is "0" for positive number and "1" for negative number in two's complement number. The advantage in two's complement is that the 0 and its complement is same. (Refer Table 2.17). Due to unique representation for zero, two's complement binary number system is practically followed in all digital systems.

In two's complement representation the sign-extention technique is followed to represent the number in required size.

An example of representing $+5_{10}$ and -5_{10} in two's complement form is shown below:

$5_{10} = 101_2$

⇓ Append sign bit

$0101_2 = +5_{10}$

↑ Sign bit

⇓ Complement

1010

⇓ Add 1

$1011_2 = -5_{10}$ (Two's complement form)

↑ Sign bit

$\therefore +5_{10} = 0101_2$ (4-bit representation) $=$ $0000\ 0101_2$ (sign extended to 8-bit)

$-5_{10} = 1011_2$ (4-bit representation) $=$ $1111\ 1011_2$ (sign extended to 8-bit)

Range of Two's Complement Binary Number

The range of decimal numbers that can be represented in n-bit two's complement form of binary numbers is,

$$-(2^{n-1}) \quad \text{to} \quad +(2^{n-1}-1)$$

The range of decimal numbers that can be represented in two's complement form of binary for n = 4 and 5 are shown below:

n = 4	n = 5
Range $= -(2^{4-1})$ to $+(2^{4-1}-1)$	Range $= -(2^{5-1})$ to $+(2^{5-1}-1)$
$= -8_{10}$ to $+7_{10}$	$= -16_{10}$ to $+15_{10}$

Table 2.17: Decimal and Two's Complement form of Binary Number

Decimal Number	4-bit Two's Complement Binary Number	Sign-extended Two's Complement 8-bit Binary Number
+0	0000	0000 0000
+1	0001	0000 0001
+2	0010	0000 0010
+3	0011	0000 0011
+4	0100	0000 0100
+5	0101	0000 0101
+6	0110	0000 0110
+7	0111	0000 0111
–8	1000	1111 1000
–7	1001	1111 1001
–6	1010	1111 1010
–5	1011	1111 1011
–4	1100	1111 1100
–3	1101	1111 1101
–2	1110	1111 1110
–1	1111	1111 1111

Two's Complement of Binary Fraction Numbers

For two's complement of binary fraction number take one's complement of every bit and then add 1 to right most digit which is LSD of fraction part.

Example: 1011.110_2 $\xrightarrow{\text{1's complement}}$ 0100.001

$$\begin{array}{r} 0100.001 \\ +1 \\ \hline 0100.010 \end{array}$$ (2's complement)

When a binary bit is complemented twice the value of bit will not change. Therefore, if two's complement is made twice we will get original number. Also, it can be shown that the sum of a binary number and its 2's complement is 0. These concepts are illustrated in the following examples.

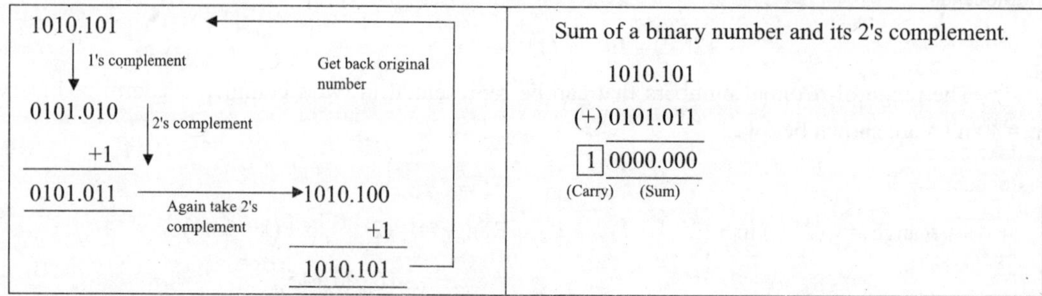

2.3.6 One's and Two's Complement Addition

Addition involves addition of two positive or negative numbers or addition of a positive and negative number. Usually, the binary numbers in complement form will have a fixed bit size which leads to a fixed range of positive and negative number representation for a specified bit size.

The addition of two positive numbers is same as unsigned binary addition except the handling of final carry. In one's complement addition the final carry is added to LSB (Least significant Bit) of sum but in two's complement addition the final carry is discarded. When negative numbers has to be added they are converted to complement form and then added. The final carry is handled similar to addition of positive numbers. But the sum can be positive or negative.

When addition involves negative numbers,

if MSB of sum is 0, then result is positive.

if MSB of sum is 1, then result is negative, take respective complement of sum and put "–"sign.

Example 2.17

Perform the following addition by one's and two's complement addition.

a) $01101_2 + (-01010_2)$ **b)** $00110_2 + (-01110_2)$ **c)** $(-01000_2) + (-01100_2)$ **d)** $00111_2 + 0010_2$

Solution

a) $01101_2 + (-01010_2)$

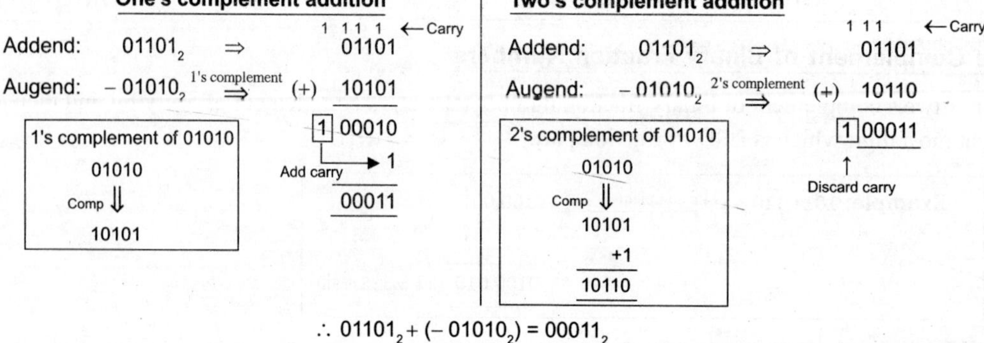

$\therefore 01101_2 + (-01010_2) = 00011_2$

b) $00110_2 + (-01110_2)$

One's complement addition

Addend: 00110_2 $\Rightarrow$ 00110

Augend: -01110_2 $\xRightarrow{\text{1's complement}}$ (+) 10001

10111 (Sum)

Since MSB of sum is 1, the result is negative. Hence take 1's complement of 10111 (sum) and put "–" sign for sum.

10111 $\xRightarrow{\text{complement}}$ 01000

1's complement of 01110
01110
Comp ⇓
10001

$\therefore$ Sum = -01000_2

Two's complement addition

1 1 ← Carry

Addend: 00110_2 $\Rightarrow$ 00110

Augend: -01110_2 $\xRightarrow{\text{2's complement}}$ (+) 10010

11000 (Sum)

Since MSB of sum is 1, the result is negative. Hence take 2's complement of 11000 (sum) and put "–" sign for sum.

2's complement of 01110
01110
Comp ⇓
10001
+1
10010

11000 $\xRightarrow{\text{complement}}$ 00111

+1

01000

$\therefore$ Sum = -01000_2

$\therefore 00110_2 + (-01110_2) = -01000_2$

c) $(-01000_2) + (-00100_2)$

One's complement addition

1 1 1 1 1 ← Carry

Addend: -01000_2 $\xRightarrow{\text{1's complement}}$ 10111

Augend: -00100_2 (+) 11011

[1] 10010

Add carry → 1

10011 (Sum)

1's complement of 01000
01000
Comp ⇓
10111

1's complement of 00100
00100
Comp ⇓
11011

Since MSB of sum is 1, the result is negative. Hence take 1's complement of 10011 (sum) and put "–" sign for sum.

10011 $\xRightarrow{\text{complement}}$ 01100

$\therefore$ Sum = -01100_2

Two's complement addition

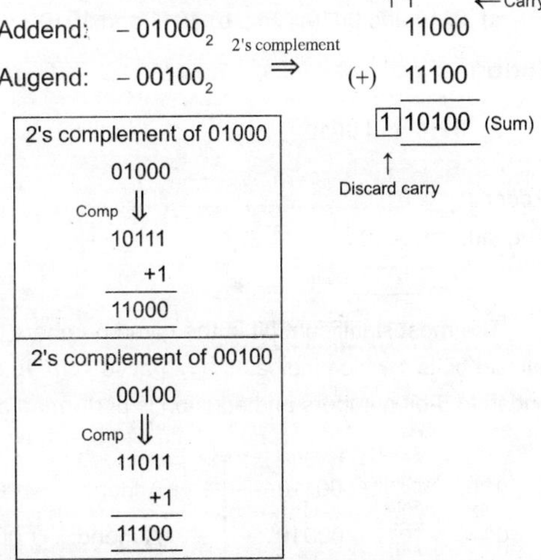

Since MSB of sum is 1, the result is negative. Hence take 2's complement of 10100 (sum) and put "–" sign for sum.

10100 $\xRightarrow{\text{complement}}$ 01011

+ 1

01100

$\therefore$ Sum = -01100_2

$\therefore -01000_2 + (-00100_2) = -01100_2$

d) $00111_2 + 00101_2$

One's and two's complement addition

```
                 1 1 1  ← Carry
Addend :         00111
Augend : (+)     00101
                 -----
                 01100
```

$\therefore 00111_2 + 00101_2 = 01100_2$

Errors in Complement Addition

In addition of two positive numbers the sum should not exceed maximum value of positive number representation for the bit size of the number and if it exceeds then the error is called overflow error. In addition of two negative numbers the sum should not go below the minimum value of negative number representation for the bit size of the number and if it goes below the minimum value then the error is called underflow error. In order to prevent overflow and underflow errors, the binary numbers should be sign extended to larger bit size before addition.

Example 2.18

Perform addition of two's complement numbers:

a) 0110 and 0010 **b)** 1011 and 1010

Solution

a) 0110 and 0010

```
              1 1   ← Carry
Addend:       0110
Augend: (+)   0010
              ----
              1000  (Sum)
```

The most significant bit in the given numbers is 0 and hence they are positive numbers. But in sum the most significant bit is 1 which indicates a negative sum. In order to prevent the overflow error, the given numbers are sign extended to 5-bit numbers and addition is performed as shown below:

```
0110  Sign extend   00110   |                   1 1   ← Carry
      to 5 bits             |   Addend:        00110
0010     ⟹          00010   |   Augend: (+)    00010
                            |                  -----
                            |                  01000  (Sum)
```

$\therefore 0110_2 + 0010_2 = 1000_2$

b) 1011 and 1010

```
                1 1    ← Carry
Addend:          1011
Augend: (+)      1010
                 ----
              [1] 0101  (Sum)
               ↑
          Discard carry
```

The most significant bit in the given numbers is 1 and hence they are negative numbers. But in sum the most significant bit is 0 which indicates a positive sum. In order to prevent the underflow error, the given numbers are sign extended to 5-bit numbers and addition is performed as shown ahead:

1011	Sign extend to 5 bits ⟹	11011	Addend: 11011 (Carry: 1 1 1)
1010		11010	Augend: (+) 11010
			[1] 10101 (Sum) ↑ Discard carry

$\therefore\ 1011_2 + 1010_2 = 10101_2$

2.3.7 One's and Two's Complement Subtraction

In one's and two's complement subtraction, the subtrahend is represented in complement form and added to minuend. In one's complement subtraction if there is a final carry then it is added to LSB (Least Significant Bit) and in two's complement subtraction the final carry is discarded.

The sum can be positive or negative. Therefore, in complement subtraction,

if MSB of sum is 0, then result is positive.

if MSB of sum is 1, then result is negative, take respective complement of sum and put "–" sign.

Example 2.19

(GTU, Jun'19, 2 Marks) ***(IKGPTU, May'20, 2 Marks)***

Perform the following subtraction by one's and two's complement subtraction.

a) $11101_2 - 10001_2$ **b)** $10001_2 - 11101_2$

Solution

a) $\mathbf{11101_2 - 10001_2}$

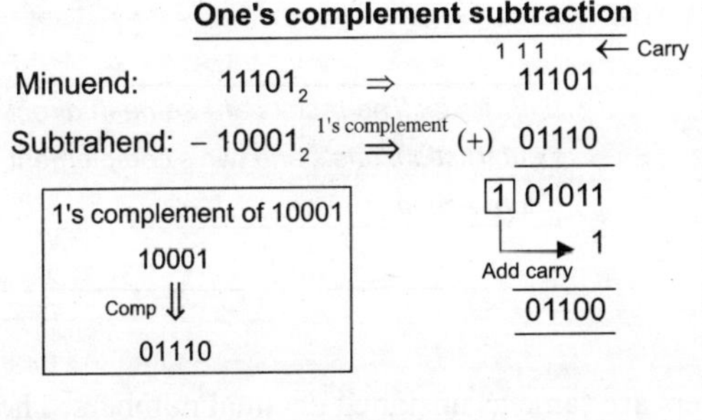

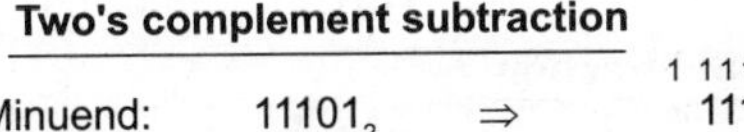

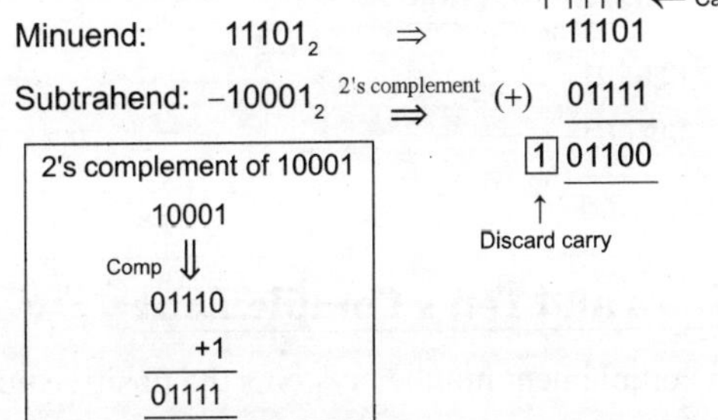

$\therefore\ 11101_2 - 10001_2 = 01100_2$

b) $\mathbf{10001_2 - 11101_2}$

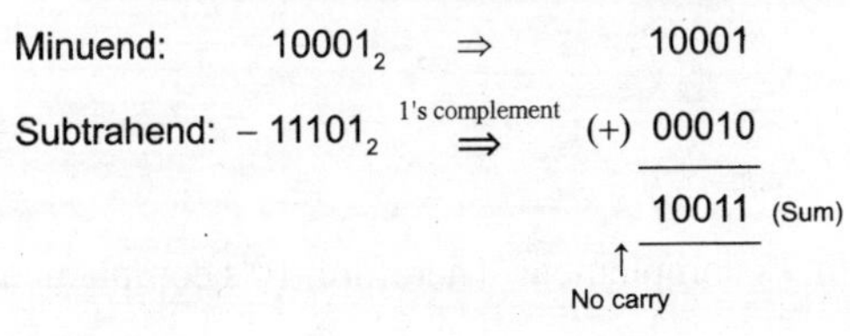

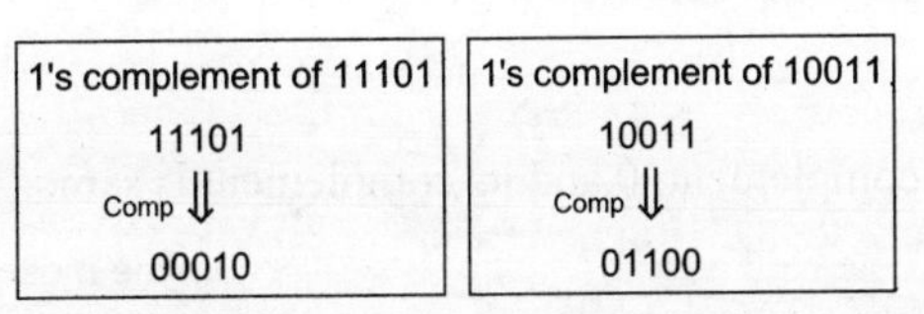

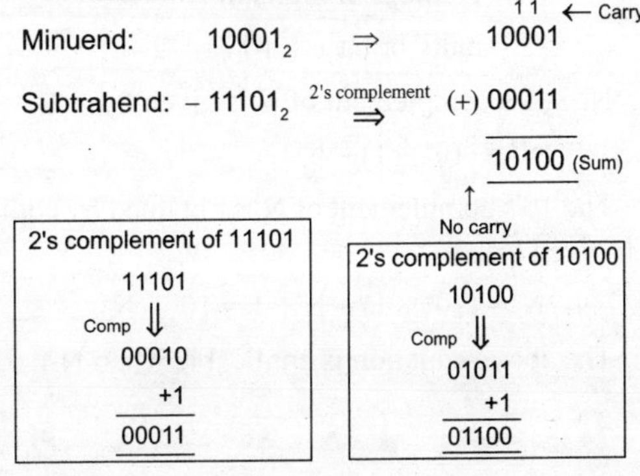

Since MSB of sum is 1, the result is negative. Hence take 1's complement of 10011 (sum) and put "–" sign for sum. $\therefore$ Sum = – 01100_2

Since MSB of sum is 1, the result is negative. Hence take 2's complement of 10100 (sum) and put "–" sign for sum. $\therefore$ Sum = – 01100_2

$$\therefore 10001_2 - 11101_2 = -01100_2$$

Example 2.20

Perform one's complement and two's complement subtraction of fractional numbers 1101.101_2 and 1000.001_2 and compare the results with actual subtraction.

Solution

Case i: One's complement subtraction

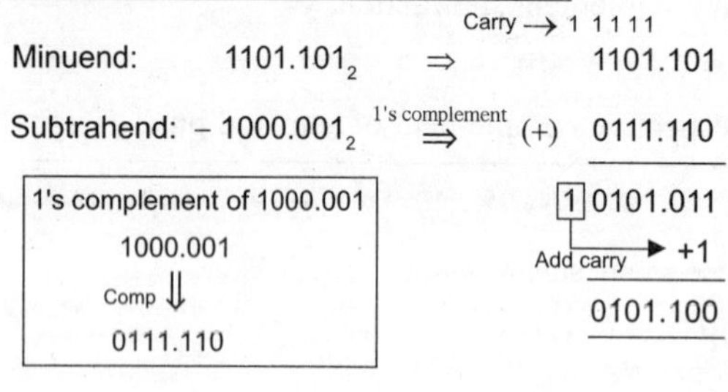

Case ii: Two's complement subtraction

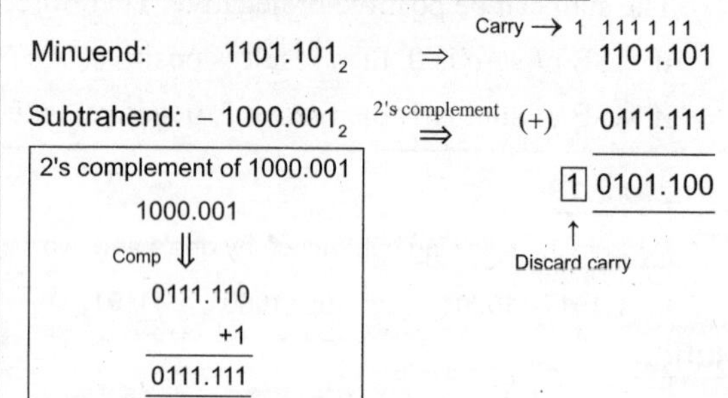

$$\therefore 1101.101_2 - 1000.001_2 = 0101.100_2$$

Case iii: Actual Subtraction

```
  1101.101
– 1000.001
  --------
  0101.100
```

> ***Note:*** *The results are same in direct subtraction, one's and two's complement subtraction.*

2.4 Nine's and Ten's Complement

In complement number systems the positive numbers are same as unsigned decimal numbers. The 9's and 10's complement are used to represent negative decimal numbers.

Let, N = Decimal number

N′ = Complement of N

n = Number of digits in N

r = radix or base = 10

Now, 9's complement of N is given by,

$$N' = (10^n - 1) - N$$

The 10's complement of N is obtained by adding 1 to 9's complement. Therefore, 10's complement of N is given by,

$$N' = (10^n - 1) - N + 1 = 10^n - N$$

The above equation is applicable when N ≠ 0. In 10's complement, 0 and its complement is same.

For any value of n it can be observed that $10^{n-1} = 999....99$. Hence 9's complement can be obtained by subtracting every digit from 9 and 10's complement is obtained by adding 1 to 9's complement.

In order to get 9's complement of decimal fraction subtract every digit from 9.

> **Example:** $127.68_{10} \xrightarrow{\text{9's complement}} 872.31_{10}$

It can be shown that sum of a decimal number and its 9's complement is 999...9.

Also, if 9's complement is made twice we will get original number. (Refer Example 2.22)

In order to get 10's complement of decimal fraction, first take 9's complement and then add 1 to right most digit which is LSD of fraction part.

> **Example:** $127.68_{10} \xrightarrow{\text{9's complement}} 872.31$
>
> $$\begin{array}{r} 872.31 \\ +1 \\ \hline 872.32 \text{ (10's complement)} \\ \hline \end{array}$$

It can be shown that sum of a number and its 10's complement is 0.

Also, if 10's complement is made twice we will get original number. (Refer Example 2.22)

Example 2.21

Find 9's and 10's complement of following decimal numbers.

a) 436_{10} **b)** 389.61_{10}

Solution

The 9's complement is obtained by subtracting every digit of decimal by 9. The 10's complement is obtained by adding 1 to 9's complement.

a) $\mathbf{436_{10}}$

9's Complement of 436	10's Complement of 436
$\begin{array}{r} 999 \\ -436 \\ \hline 563 \\ \hline \end{array}$	$\begin{array}{r} 999 \\ -436 \\ \hline 563 \\ +1 \\ \hline 564 \\ \hline \end{array}$
N' = 563	N' = 564

Note: *If we take complement twice we will get original number.*

9's Complement of 563	10's Complement of 564
$\begin{array}{r} 999 \\ -563 \\ \hline 436 \\ \hline \end{array}$	$\begin{array}{r} 999 \\ -564 \\ \hline 435 \\ +1 \\ \hline 436 \\ \hline \end{array}$

Note: *Sum of number and 10's complement is zero. Sum of number and 9's complement is 999....99.*

9's Complement	10's Complement
$N \Rightarrow 436$	$N \Rightarrow 436$
$N' \Rightarrow (+)\ 563$	$N' \Rightarrow (+)\ 564$
999 (Sum)	[1] 000 (Sum) (Carry)

b) 389.61$_{10}$

9's Complement of 389.61	10's Complement of 389.61
999.99	999.99
– 389.61	– 389.61
610.38	610.38
	+1
N′ = 610.38	610.39 N′ = 610.39

Example 2.22

Take 9's and 10's complement of following numbers twice and show that original number can be obtained.

a) 436.52$_{10}$ **b)** 389.61$_{10}$

Solution

a) 436.52$_{10}$

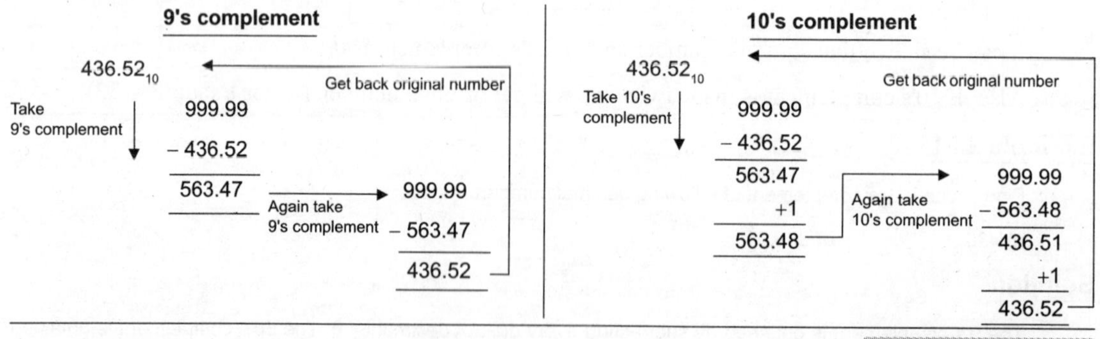

(JNTUK, Feb'22, 8 Marks)

b) 389.61$_{10}$

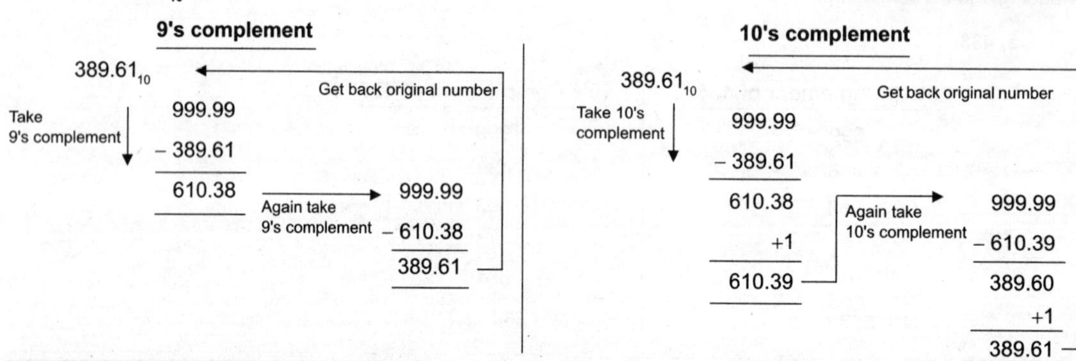

2.4.1 Nine's and Ten's Complement Addition

Addition involves addition of two positive or negative numbers or addition of a positive and negative number. The addition of two positive numbers is same as unsigned decimal addition except the handling of final carry. In 9's complement addition the final carry is added to LSD (Least Significant Digit) of sum but in 10's complement addition the final carry is discarded.

The 9's complement of 3 digit decimal numbers are listed in Table 2.18.

In 9's and 10's complement representation of decimal numbers it can be observed that in positive numbers the MSD is 0 to 4 and in negative numbers the MSD is 5 to 9. (Refer Table 2.18)

When negative numbers has to be added they are converted to complement form and then added. The final carry is handled similar to addition of positive numbers. But the sum can be positive or negative.

When addition involves negative numbers,

if MSD of sum is 0 to 4, then result is positive.

if MSD of sum is 5 to 9, then result is negative, take respective complement of sum and put "–"sign.

Table 2.18: The 9's and 10's Complement Form of Decimal Number

Decimal	9's Complement Decimal	10's Complement Decimal	Comments	Decimal	9's Complement Decimal	10's Complement Decimal	Comments
+ 000	000	000		– 500	---	500	
+ 001	001	001		– 499	500	501	
+ 002	002	002		– 498	501	502	
⋮	⋮	⋮		⋮	⋮	⋮	
+ 100	100	100		– 401	598	599	
+ 101	101	101		– 400	599	600	
⋮	⋮	⋮		⋮	⋮	⋮	
+ 200	200	200	Positive	– 301	698	699	Negative
+ 201	201	201	numbers	– 300	699	700	numbers
⋮	⋮	⋮		⋮	⋮	⋮	
+ 300	300	300		– 201	798	799	
+ 301	301	301		– 200	799	800	
⋮	⋮	⋮		⋮	⋮	⋮	
+ 400	400	400		– 101	898	899	
+ 401	401	401		– 100	899	900	
⋮	⋮	⋮		⋮	⋮	⋮	
				– 002	997	998	
+ 498	498	498		– 001	998	999	
+ 499	499	499		– 000	999	---	

Example 2.23

Perform the following additions by 9's and 10's complement addition.

a) $187_{10} + 249_{10}$ **b)** $467_{10} + (-376_{10})$

c) $238_{10} + (-469_{10})$ **d)** $(-176_{10}) + (-306_{10})$

Solution

a) $\mathbf{187_{10} + 249_{10}}$

9's and 10's complement addition

$$\begin{array}{lll} & & \ 11 \leftarrow \text{Carry} \\ \text{Addend:} & 187_{10} \Rightarrow & \ 187 \\ \text{Augend:} & +\ 249_{10} \Rightarrow & (+)\ \underline{249} \\ & & \ \underline{436} \end{array}$$

$\therefore 187_{10} + 249_{10} = 436_{10}$

b) $467_{10} + (-376_{10})$

9's complement addition

Addend: 467_{10} ⇒ 467 (Carry → 1 1)

Augend: -376_{10} $\xrightarrow{\text{9's complement}}$ (+) 623

= 1 090; Add carry +1 → 091

9's complement of 376: 999 − 376 = 623

10's complement addition

Addend: 467_{10} ⇒ 467 (1 1 ← Carry)

Augend: -376_{10} $\xrightarrow{\text{10's complement}}$ (+) 624

= 1 091 (↑ Discard carry)

10's complement of 376: 999 − 376 = 623; 623 + 1 = 624

$\therefore 467_{10} + (-376_{10}) = 091_{10}$

---

c) $238_{10} + (-469_{10})$

9's complement addition

Addend: 238_{10} ⇒ 238

Augend: -469_{10} $\xrightarrow{\text{9's complement}}$ (+) 530

768 (Sum)

9's complement of 469	9's complement of 768
999	999
− 469	− 768
530	231

Since most significant digit of sum is 5 to 9 the result is negative. Hence take 9's complement of 768 (sum) and put "–" sign for sum.

$\therefore$ Sum = -231_{10}

10's complement addition

Addend: 238_{10} ⇒ 238

Augend: -469_{10} $\xrightarrow{\text{10's complement}}$ (+) 531

769 (Sum)

10's complement of 469	10's complement of 769
999	999
− 469	− 769
530 + 1 = 531	230 + 1 = 231

Since most significant digit of sum is 5 to 9 the result is negative. Hence take 10's complement of 769 (sum) and put "–" sign for sum.

$\therefore$ Sum = -231_{10}

$\therefore 238_{10} + (-469_{10}) = -231_{10}$

---

d) $(-176_{10}) + (-306_{10})$

9's complement addition

Addend: -176_{10} ⇒ 823 (1 1 ← Carry)

Augend: -306_{10} $\xrightarrow{\text{9's complement}}$ (+) 693

= 1 516; Add carry +1 → 517

9's complement of 176: 999 − 176 = 823

9's complement of 306	9's complement of 517
999	999
− 306	− 517
693	482

Since most significant digit of sum is 5 to 9 the result is negative. Hence take 9's complement of 517 (sum) and put "–" sign for sum.

$\therefore$ Sum = -482_{10}

10's complement addition

Addend: -176_{10} ⇒ 824 (1 1 ← Carry)

Augend: -306_{10} $\xrightarrow{\text{10's complement}}$ (+) 694

= 1 518 (↑ Discard carry)

10's complement of 176: 999 − 176 = 823; 823 + 1 = 824

10's complement of 306	10's complement of 518
999	999
− 306	− 518
693 + 1 = 694	481 + 1 = 482

Since most significant digit of sum is 5 to 9 and so result is negative. Hence take 9's complement of 518 (sum) and put "–" sign for sum.

$\therefore$ Sum = -482_{10}

$\therefore (-176_{10}) + (-306_{10}) = -482_{10}$

2.4.2 Nine's and Ten's Complement Subtraction

In subtraction using complement, the negative number is represented in complement form and added to minuned.

Subtraction	**Complement Subtraction**
Minuend	Minuend
(–) Subtrahend	(+) Complement of Subtrahend
Difference	Carry Sum (Difference)

In 9's complement subtraction if there is a final carry then it is added to LSD (Least Significant Digit) of sum and in 10's complement subtraction, the final carry is discarded.

In complement subtraction,

if MSD of sum is 0 to 4 or final carry in addition then the result is positive.

if MSD of sum is 5 to 9 or no final carry in addition then the result is negative, take respective complement of sum and put " – " sign.

Example 2.24

Perform the decimal fraction subtraction 2928.54_{10} – 416.73_{10} by 9's and 10's complement subtraction and compare the result with actual subtraction.

Solution

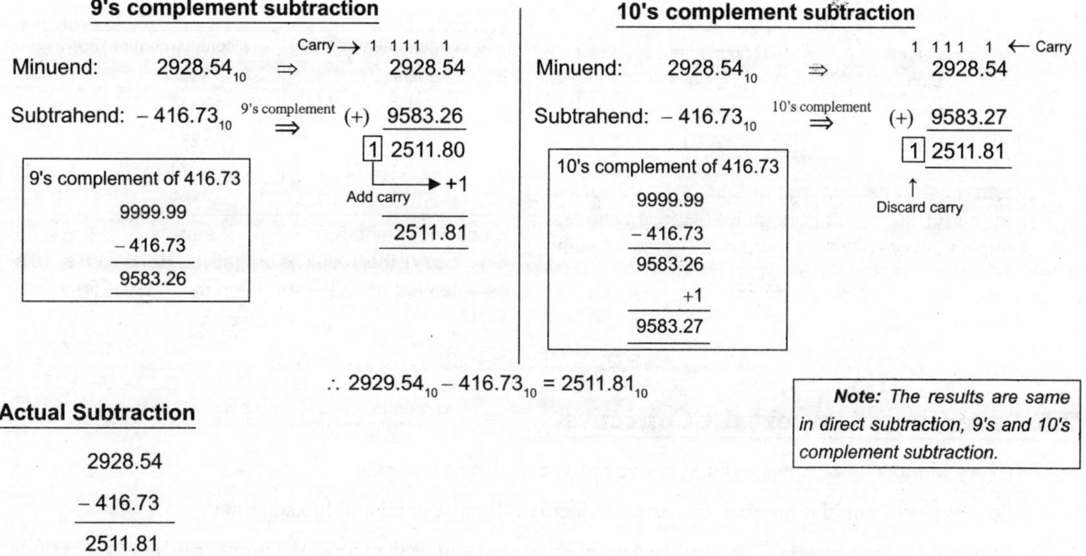

Example 2.25

Perform the following decimal subtraction by 9's and 10's complement method.

a) $7258_{10} - 4592_{10}$

b) $4592_{10} - 7258_{10}$

Solution

a) $7258_{10} - 4592_{10}$

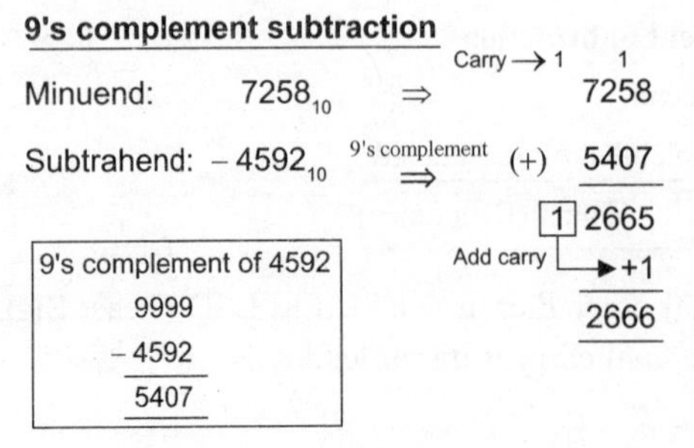

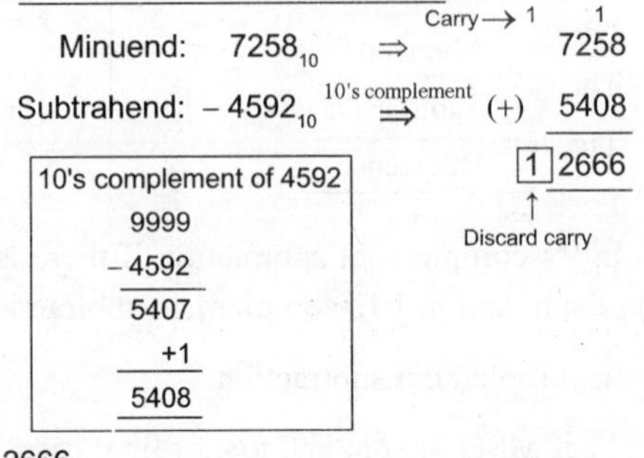

$\therefore 7258_{10} - 4592_{10} = 2666_{10}$

b) $4592_{10} - 7258_{10}$

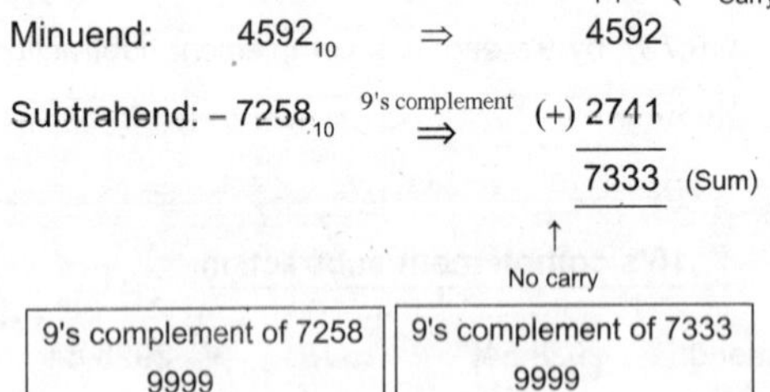

Since most significant digit of sum is 5 to 9 (or no final carry) the result is negative. Hence take 9's complement of 7333 (sum) and put "–" sign for sum.

$\therefore$ Sum = -2666_{10}

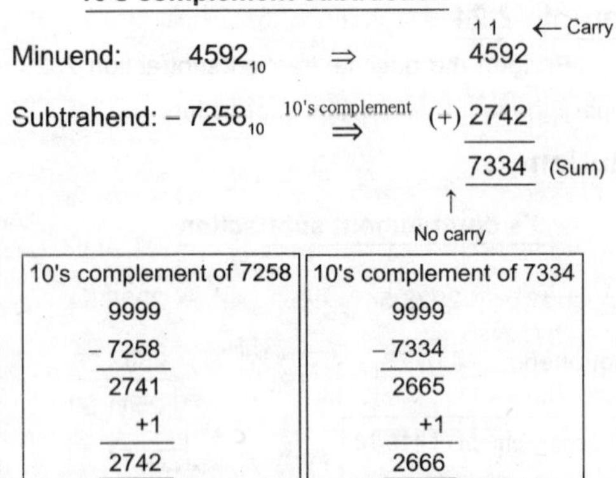

Since most significant digit of sum is 5 to 9 (or no final carry) the result is negative. Hence take 10's complement of 7334 (sum) and put "–" sign for sum.

$\therefore$ Sum = -2666_{10}

$\therefore 4592_{10} - 7258_{10} = -2666_{10}$

2.5 Summary of Important Concepts

1. Every number system has a radix, r. The radix, r is also called base.
2. Four types of popular number systems are decimal, binary, octal and hexadecimal.
3. Decimal number system is the standard number system and so the value of a number in any other number system is verified by converting to an equivalent decimal number.
4. The decimal number system is a radix number system in which the radix is 10.
5. The 10 symbols used in decimal number system are 0, 1, 2, 3, 4, 5, 6, 7, 8 and 9.
6. Each digit of a decimal number has a weight which is given by the power of radix, 10.
7. The dot in decimal number system is called decimal point and it separates the integer part and fraction part.
8. The binary number system is a radix number system in which the radix is 2.

9. The two symbols used in binary number system are 0 and 1.
10. The dot in binary number system is called binary point and it separates the integer part and fraction part.
11. Each digit of a binary number has a weight which is given by the power of radix, 2.
12. For binary to decimal conversion multiply each binary digit by its weight and sum up the product.
13. The octal number system is a radix number system in which the radix r is 8.
14. The eight symbols used in octal number system are 0, 1, 2, 3, 4, 5, 6 and 7.
15. Each digit of an octal number has a weight which is given by the power of radix, 8.
16. For octal to decimal conversion multiply each octal digit by its weight and sum up the product.
17. The hexadecimal number system is a radix number system in which the radix r is 16.
18. The 16 symbols used in hexadecimal number system are 0, 1, 2, 3, 4, 5, 6, 7, 8, 9, A, B, C, D, E and F.
19. Each digit of a hexadecimal number has a weight which is given by the power of radix, 16.
20. For hexadecimal to decimal conversion multiply each hexadecimal digit by its weight and sum up the product.
21. A hexadecimal digit can be represented by 4-digit binary and an octal digit by 3-digit binary.
22. The octal and hexa numbers can be directly converted to binary and vice-versa.
23. Binary arithmetic is similar to decimal arithmetic of only two numbers 0 and 1 with results in binary.
24. Binary multiplication can be performed by successive addition and binary division can be performed by succesive subtraction.
25. The three types of signed binary number system are Sign-magnitude form, One's complement form and Two's complement form.
26. The range of decimal number that can be represented by n-bit unsigned binary is 0 to $2^n - 1$.
27. The unsigned binary has only positive numbers. The signed binary has positive and negative numbers.
28. In an n-bit unsigned binary integer number the digit d_0 has the weight $2^0 = 1$, which is the lowest weight. Hence the digit d_0 is called LSD (Least Significant Digit).
29. In an n-bit unsigned binary integer number the digit d_{n-1} has the weight 2^{n-1}, which is the highest weight. Hence the digit d_{n-1} is called MSD (Most Significant Digit).
30. In n-bit representation of sign-magnitude form of binary, 1-bit is reserved for sign and n – 1 bits are used to represent the magnitude.
31. In sign-magnitude form of binary, for positive number the sign bit is "0" and for negative number the sign bit is "1". The magnitudes are same for a given number with opposite signs.
32. The range of decimal numbers that can be represented in n-bit sign-magnitude form of binary number is, $-(2^{n-1} - 1)$ to $+(2^{n-1} - 1)$.
33. The two types of complements in radix number systems are r's complements or radix complement and (r – 1)'s complement or diminished radix complement.
34. Complement of positive numbers are used to represent negative numbers in digital system.
35. In one's complement form of binary number representation, the positive numbers will be same as that of unsigned binary. The negative numbers are bit-by-bit complement of the positive numbers.
36. The range of decimal numbers that can be represented in n-bit one's complement form of binary number is, $-(2^{n-1} - 1)$ to $+(2^{n-1} - 1)$.

37. In two's complement form of binary number representation, the positive numbers will be same as that of unsigned binary. The negative number is obtained by bit-by-bit complementing the positive number and then adding one to LSB.

38. The range of decimal numbers that can be represented in n-bit two's complement form of binary number is, $-(2^{n-1})$ to $+(2^{n-1}-1)$.

39. In addition of complement numbers if the sum exceeds the maximum value of positive number representation for the specified bit size then the error is called overflow error.

40. In addition of complement numbers if the sum goes below the minimum value of negative number representation for the specified bit size then the error is called underflow error.

41. In order to prevent overflow and underflow errors in addition, the binary numbers should be sign extended to larger bit size before addition.

42. If one's or two's complement of a binary number is made twice then we will get back the original binary number.

43. The 9's complement of a decimal number is obtained by subtracting every digit of the number from 9.

44. The 10's complement of a decimal number is obtained by adding 1 to LSD of 9's complement of the number.

45. In 9's and 10's complement of decimal number representation, for positive numbers the MSD is 0 to 4 and for negative numbers the MSD is 5 to 9.

46. If 9's or 10's complement of a decimal number is made twice then we will get back the original decimal number.

2.6 Short-Answer Questions

Q2.1 ***List the various number systems?***

There are four types of popular number systems.

1. Decimal number system
2. Binary number system
3. Octal number system
4. Hexadecimal number system.

Q2.2 ***What is the largest binary number that can be expressed with 14 bits? Determine the equivalent decimal and hexadecimal numbers.***

The largest binary number that can be expressed with 14 bits = $11\ 1111\ 1111\ 1111_2$

$$11\ 1111\ 1111\ 1111_2 = 0011\ 1111\ 1111\ 1111_2 = 3FFF_{16}$$

$$3FFF_{16} = 3 \times 16^3 + F \times 16^2 + F \times 16^1 + F \times 16^0$$

$$= 3 \times 16^3 + 15 \times 16^2 + 15 \times 16^1 + 15 = 16383_{10}$$

Q2.3 ***Convert the following binary numbers to decimal number.*** **(AU, Apr/May'19, 2 Marks)**

a) 101.01_2 b) 11011.101_2

Solution:

$$101.01_2 = 1 \times 2^2 + 0 \times 2^1 + 1 \times 2^0 + 0 \times 2^{-1} + 1 \times 2^{-2} = 5.25_{10}$$

$$11011.101_2 = 1 \times 2^4 + 1 \times 2^3 + 0 \times 2^2 + 1 \times 2^1 + 1 \times 2^0 + 1 \times 2^{-1} + 0 \times 2^{-2} + 1 \times 2^{-3} = 27.625_{10}$$

Q2.4 ***Convert the following numbers to decimal number.*** ***(JNTUH, Nov/Dec'20, 2 Marks)***

a) 5432_8 ***b)*** 16.5_{16}

Solution:

$$5432_8 = 5 \times 8^3 + 4 \times 8^2 + 3 \times 8^1 + 2 \times 8^0 = 2842_{10}$$

$$16.5_{16} = 1 \times 16^1 + 6 \times 16^0 + 5 \times 16^{-1} = 22.3125_{10}$$

Q2.5 ***Perform 1's complement and 2's complement subtraction of binary numbers*** 1010_2 ***and*** 1000_2 ***and compare the results with actual subtraction.***

Solution:

Let, $1010_2 - 1000_2 = (+1010_2) + (-1000_2)$

One's complement subtraction

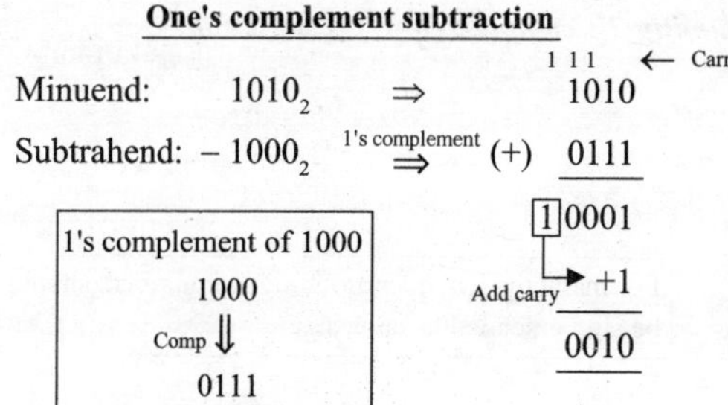

Two's complement subtraction

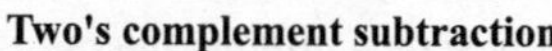

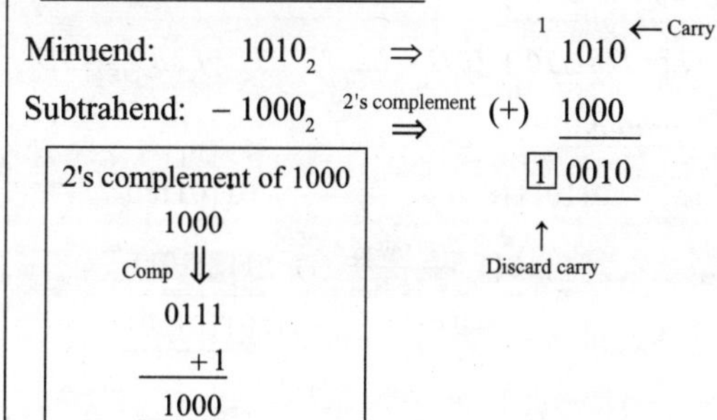

Actual Subtraction

```
    1010
(-) 1000
    0010
```

$\therefore 1010_2 - 1000_2 = 0010_2$

> ***Note:*** *The results are same in direct subtraction, one's complement and two's complement subtraction.*

Q2.6 ***Perform subtraction on the following unsigned binary numbers using the 2's complement of the subtrahend.***

a) $11011_2 - 11001_2$ ***b)*** $110100_2 - 10101_2$. ***(AU, Nov/Dec'18, 2 Marks)***

Solution:

a) $11011_2 - 11001_2 = (+11011_2) + (-11001_2)$

```
                                 1 1111  ← Carry
Minuend:      11011₂     ⇒        11011
                   2's complement
Subtrahend:  -11001₂     ⇒    (+) 00111
                                1|00010
                                ↑
                                Discard carry
```

2's complement of 11001:

```
     11001
Comp   ⇓
     00110
        +1
     00111
```

$\therefore 11011_2 - 11001_2 = 00010_2$

---

b) $110100_2 - 10101_2 = (+110100_2) + (-10101_2)$

```
                                 1        ← Carry
Minuend:     110100₂     ⇒        110100
                   2's complement
Subtrahend:  -10101₂     ⇒    (+) 101011
                                1|011111
                                ↑
                                Discard carry
```

2's complement of 10101:

```
10101 --Convert 6 bits--> 010101
                   Comp     ⇓
                          101010
                              +1
                          101011
```

$\therefore 110100_2 - 10101_2 = 011111_2$

Q2.7 Perform the arithmetic operation $11011100_2 - 10011011_2$ using 2's complement arithmetic

(JNTUA, Jun/July'22, 7 Marks)

Solution:

Let, $11011100_2 - 10011011_2 = (+11011100_2) + (-10011011_2)$

Two's complement subtraction

```
                                        1 11111      ← Carry
Minuend:     11011100₂       ⇒            11011100
Subtrahend: −10011011₂  2's complement ⇒ (+)  01100101
                                         ─────────
                                         [1] 01000001
                                          ↑
                                     Discard carry
```

2's complement of 10011011

10011011

Comp ⇓

01100100

+ 1

01100101

$\therefore 11011100_2 - 10011011_2 = 01000001_2$

Q2.8 Compute the following signed number operation assuming 2's complement representation.

a) $01101110 + 1011 = (.....)_2$ b) $3B7_{16} - 5D_{16} = (.....)_{16}$ c) $7_{10} - 11_{10} = (.....)_{10}$

Solution:

```
                                 1 111111   ← Carry
a)   01101110   ──────→           01101110
           Sign extend
     1011  ──────────→  (+)       11111011
            to 8-bits           ──────────
                              [1] 01101001
                               ↑
                          Discard carry
```

For mathematical operations signed numbers should be sign extended to same size.

$\therefore 01101110_2 + 1011_2 = 01101001_2$

---

b) $3B7_{16} - 5D_{16} = (.....)_{16} \Rightarrow 3B7_{16} + (-5D)_{16} = (.....)_{16}$

```
3B7 = 0011 1011 0111
                  1's                   +1
5D = 0101 1101 ──────────→ 1010 0010 ───────→ 1010 0011
               complement                        ↓ Sign extend to 12 bits
                                              1111 1010 0011

                        1 1111  1   111  ← Carry
      3B7  ──────→         0011 1011 0111
         2's complement
     −5D ──────────→ (+)   1111 1010 0011
            of 5D        ────────────────
                       [1] 0011 0101 1010  = 35A₁₆
                        ↑
                   Discard carry
```

−5D is obtained by taking 2's complement of +5D.

In 2's complement arithmetic carry should be discarded.

$\therefore 3B7_{16} - 5D_{16} = 35A_{16}$

---

c) $7_{10} - 11_{10} = (.....)_{10}$ **(IKGPTU, Dec'19, 2 Marks)**

Let us choose 5-bit binary representation.

```
7 = 00111
          1's             +1
11 = 01011 ──────────→ 10100 ──────→ 10101
          complement
                        111  ← Carry
      7  ──────→           00111
        2's complement
   − 11 ──────────→  (+)   10101
          of 11           ──────
                           11100

          1's             +1
11100 ──────────→ 00011 ──────→ 00100
      complement
```

$\therefore 7_{10} - 11_{10} = -00100_2 = -4_{10}$

Q2.9 ***Consider a decimal number – 132_{10}. Represent this number in the following data type.***

a) Signed binary ***b) 2's complement***

Solution:

First convert 132_{10} to binary

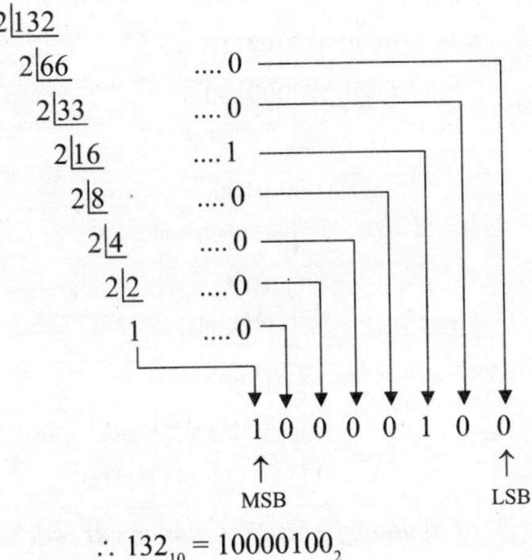

$\therefore 132_{10} = 10000100_2$

a) Signed binary

Append 1 before MSB to represent negative number.

$\therefore -132_{10} = \boxed{1}\ 10000100 = 110000100$

Append sign bit '1'

b) 2's complement

Append 0 before MSB to represent positive number and then convert to 2's complement.

$+132_{10} = \boxed{0}10000100 = 010000100$

Append sign bit '0'

↓ 1's complement

101111011

+ 1 (Add 1 to LSB)

101111100

A negative number is obtained by taking 2's complement of positive number.

$\therefore -132_{10} = 101111100_2$

Q2.10 ***What is the range of signed integer values in 2's complement representation, that can be represented using 16-bits?***

Solution:

The range of signed integer with 16 bits representation is,

$$-(2^{(16-1)}) \text{ to } (2^{(16-1)} - 1) \Rightarrow -2^{15} \text{ to } +(2^{15} - 1) \Rightarrow -32768_{10} \text{ to } +32767_{10}$$

Q2.11 ***Represent the signed binary number 10000100 as the following signed 16-bit binary number.***

a) Signed binary

b) 2's complement binary

Solution:

a) Signed binary

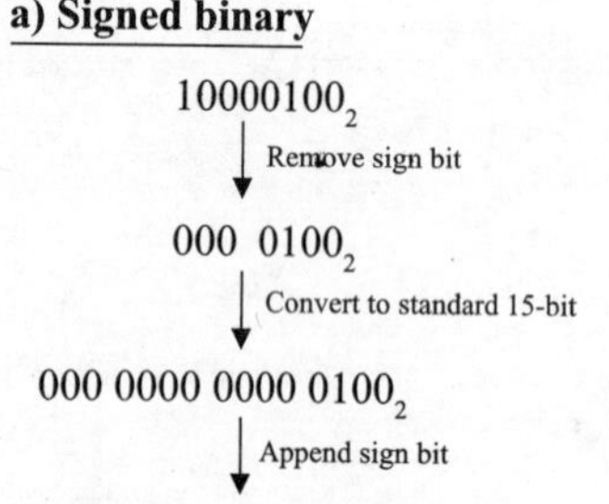

b) 2's complement binary

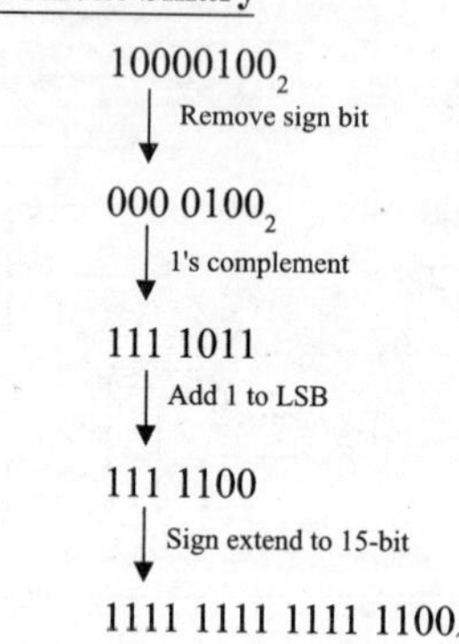

Q2.12 ***What is the largest positive number one can represent in an eight bit 2's complement code?***

Solution:

The largest positive number in 8-bit 2's complement representation is $2^{8-1}-1$.

$$\therefore\ 2^{8-1}-1 = 2^7 - 1 = 128 - 1 = +\ 127_{10}$$

Q2.13 ***Compute 0101 + 0110. Comment on result if the binary numbers are unsigned binary and 2's complement binary.***

Solution:

```
      1    ← Carry
      0101
  (+) 0110
      ----
      1011
```

$1011 = 11_{10}$ in unsigned binary

$1011 = -\ 5_{10}$ in 2's complement binary

Here, $0101 = 5_{10}$ and $0110 = 6_{10}$ and $5_{10} + 6_{10} = 11_{10}$

The range of unsigned 4-bit binary is 0 to 15_{10} and so 11_{10} is accommadated in the result and no error in addition.

The range of 4-bit 2's complement numbers are $-\ 8_{10}$ to $+7_{10}$ and so the result 11_{10} cannot be accommadated in 4-bit 2's complement binary and so the sum has an overflow error.

Q2.14 ***Which of the following binary numbers are even.***

a) 0101100010 ***b) 01000101***

c) 011111110 ***d) 11000001***

Solution:

The binary numbers ending with '0' are even and the binary numbers ending with '1' are odd. (i.e., if the least significant bit is zero then the number is even otherwise odd).

Therefore, **a)** 0101100010 and **c)** 011111110 are even numbers

and **b)** 01000101 and **d)** 11000001 are odd numbers

Q2.15 In 2's complement representation with 8-bits compute $-30_{10} - 16_{10}$. **(AKTU, Sep'20, 5 Marks)**

Solution:

$-30_{10} - 16_{10} = -30_{10} + (-16)_{10} = -46_{10}$

Division	Remainder		Division	Remainder
2 | 30			2 | 16	
2 | 15	0		2 | 8	0
2 | 7	1		2 | 4	0
2 | 3	1		2 | 2	0
1	1		1	0

1 1 1 1 0 (MSB ... LSB) $30 = 11110_2$ 1 0 0 0 0 (MSB ... LSB) $16 = 10000_2$

$30_{10} = 11110_2 \xrightarrow{\text{8-bit}} 0001\ 1110_2 \xrightarrow{\text{1's complement}} 1110\ 0001$

$-30_{10} \Leftarrow 1110\ 0010 \xleftarrow{+1} 1110\ 0001$

$16_{10} = 10000_2 \xrightarrow{\text{8-bit}} 0001\ 0000 \xrightarrow{\text{1's complement}} 1110\ 1111$

$-16_{10} \Leftarrow 1111\ 0000 \xleftarrow{+1} 1110\ 1111$

$$\begin{array}{llr} -30 & \xRightarrow{\text{2's complement}} & 1110\ 0010 \\ -16 & (+) & 1111\ 0000 \\ \hline & \text{Discard carry } \boxed{1} & 1101\ 0010 \text{ (Sum)} \end{array}$$

Since MSB of sum is 1, the sum is negative. Hence take two's complement of sum and put "–" sign.

$1101\ 0010 \xrightarrow{\text{1's complement}} 0010\ 1101 \xrightarrow{+1} 0010\ 1110_2 \xrightarrow[\text{decimal}]{\text{Convert to}} 46_{10}$

$\therefore 1101\ 0010 = -46_{10}$, $\therefore -30_{10} - 16_{10} = -46_{10}$

Q2.16 Solve a) $X_{16} + 1334_8 = 2000_8$

b) $X_8 \times X_8 = 10000\ 0000_8$

Solution:

a) $X_{16} + 1334_8 = 2000_8$

$1334_8 = 1 \times 8^3 + 3 \times 8^2 + 3 \times 8^1 + 4 \times 8^0 = 732_{10}$

$2000_8 = 2 \times 8^3 + 0 \times 8^2 + 0 \times 8^1 + 0 \times 8^0$

$= 2 \times 512 + 0 + 0 + 0 = 1024_{10}$

$X_{16} = 2000_8 - 1334_8 = 1024_{10} - 732_{10} = 292_{10}$

$X_{16} = 292_{10} = 124_{16}$

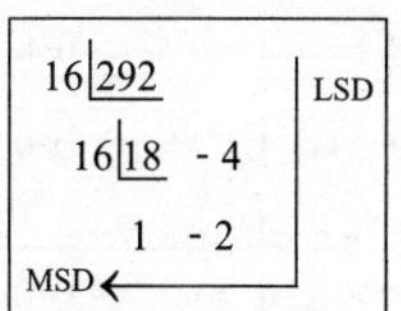

b) $X_8 \times X_8 = 10000\ 0000_8$

$$10000\ 0000_8 = (1 \times 8^8)_{10}$$

$$\therefore X_8^2 = (8^8)_{10}$$

$$X_8 = \sqrt{X_8^2} = \sqrt{8^8} = 8^{8\times\frac{1}{2}} = (8^4)_{10}$$

$$X_8 = 8^4 = 4096_{10} = 10000_8$$

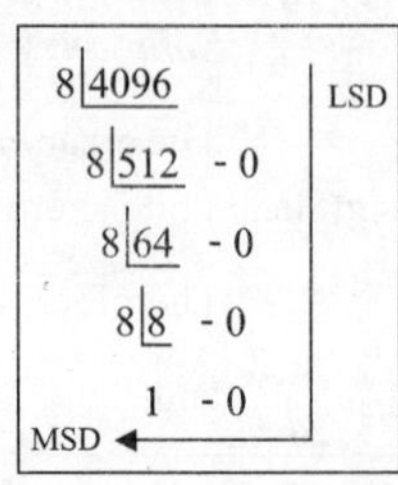

Q2.17 *Multiply the binary numbers 10101_2 and 100_2 in two methods.*

Solution:

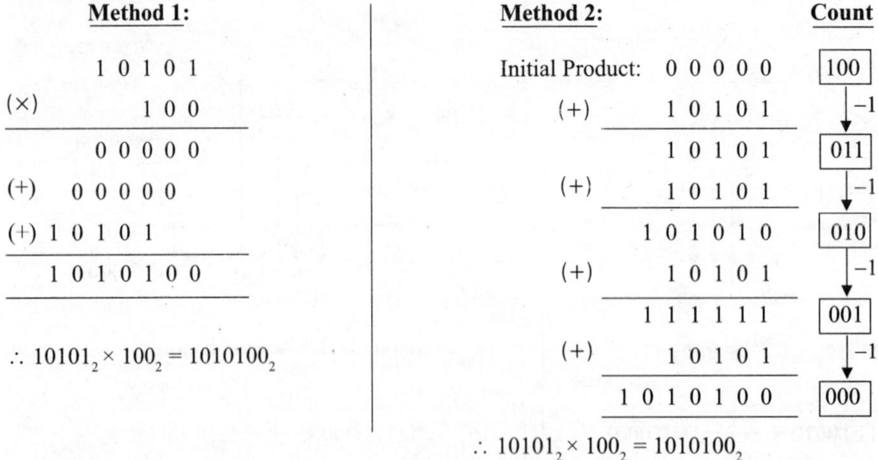

Q2.18 *Divide the binary numbers 1101_2 and 11 in two methods.*

Solution:

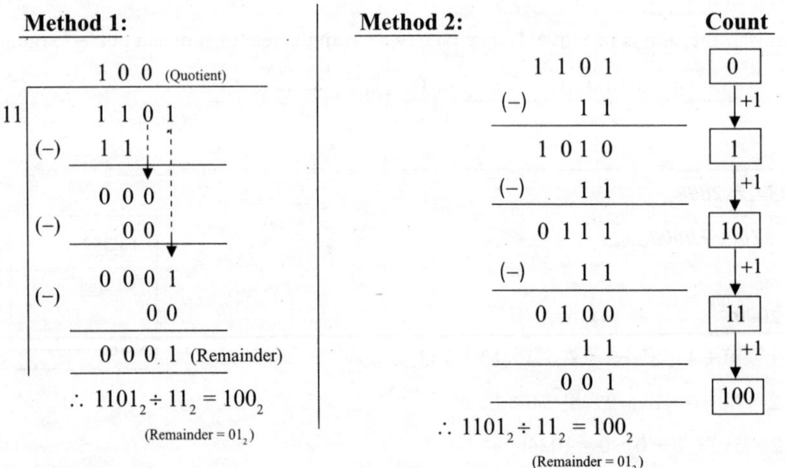

Q2.19 ***Find the 9's and 10's complement of the decimal number 146.21_{10}.***

Solution:

9's Complement of 146.21	10's Complement of 146.21
999.99	999.99
– 146.21	– 146.21
853.78	853.78
	+1
N′ = 853.78	853.79 N′ = 853.79

Q2.20 ***Find the 9's and 10's complement of the decimal number 23.35_{10} twice and show that original number can be obtained.***

Solution:

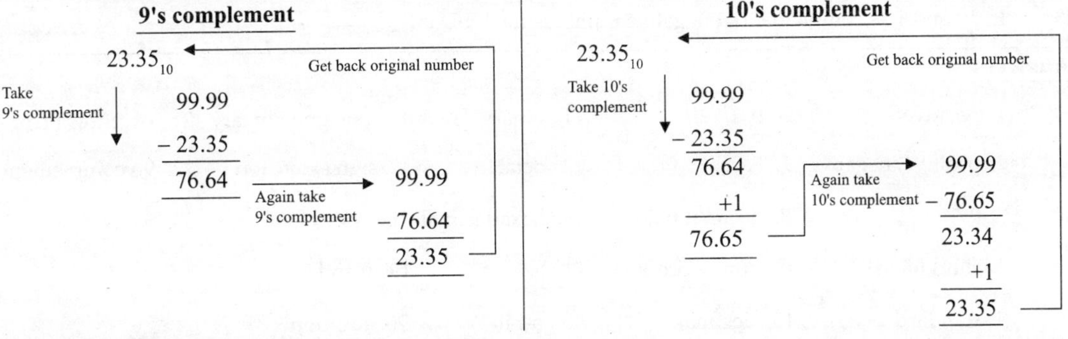

2.7 Exercises

I. Fill in the blanks

1. The unsigned binary has only ______ numbers.
2. The signed binary has ______ and ______ numbers.
3. For positive binary number, the sign bit is ______ and for negative number the sign bit is ______.
4. The octal and hexa decimal are useful for ______ representation of binary numbers.
5. Decimal number system is the ______ number system.
6. The binary number system uses only two symbols ______ and ______.
7. The radix, r is also called ______.
8. The dot in decimal number is called ______.
9. The value of a decimal number can be expressed as a ______ of decimal digit and its corresponding weight.
10. The ______ number system is a radix number system in which the radix r is 10.
11. The ______ number system is radix number system in which the radix r is 2.
12. The ______ number system is radix number system in which the radix r is 8.
13. The ______ number system is radix number system in which the radix r is 16.

14. Hexadecimal in short can be called ______.

15. In ______ binary representation of sign magnitude form of binary, 2^{n-1} positive numbers can be represented.

16. In one's complement the positive numbers will be same as that of ______.

17. In two's complement representation the ______ technique is followed to represent the number in required bit size.

18. If complement of a number is made ______ we will get original number.

19. Binary arithmetic is similar to decimal arithmetic of only two numbers ______ .

20. Binary multiplication is realized by successive ______.

21. In complement arithmetic of n-bit numbers, if sum exceeds maximum value in the n-bit representation, then the error is called______.

22. The______of decimal number is obtained by subtracting every digit from 9.

23. In 9's and 10's complement of negative numbers the MSD is ______.

Answers

1. positive	6. 0, 1	11. binary	16. unsigned binary	21. overflow error
2. positive, negative	7. base	12. octal	17. sign extension	22. 9's complement
3. 0, 1	8. decimal point	13. hexadecimal	18. twice	23. 5 to 9
4. short hand	9. sum of products	14. hexa	19. 0 and 1	
5. standard	10. decimal	15. n-bit	20. addition	

II. State whether the following statements are True or False

1. The hexadecimal number system is a radix number system in which the radix r is 16.
2. The octal number system is a radix number system in which the radix r is 2.
3. The unsigned binary has only negative numbers.
4. The complement of the numbers are used to represent the negative numbers in digital system.
5. Hexa digit can be represent by 4-digit binary and an octal digit by 3-digit binary.
6. An octal number cannot be directly converted to binary.
7. The drawback in one's complement form is that there will be two zero's + 0 and – 0.
8. The radix or base is the number of unique symbols used to represent the number system.
9. The octal and hexadecimal numbers are short hand form of decimal numbers.
10. The binary number system can be unsigned or signed binary number system.
11. The unsigned binary has positive and negative numbers.
12. In n-bit binary representation of sign-magnitude form of binary 1-bit is reserved for sign and n – 1 bits are used to represent the magnitude.
13. The (r – 1)'s complement is also called diminished radix complement.

14. In binary arithmetic the results can be in binary or decimal.
15. Binary division can be performed by successive subtraction.
16. In complement arithmetic of n-bit numbers, if sum goes below the minimum value in the n-bit representation, then the error is called overflow error.
17. The (10's complement) of decimal number is obtained by subtracting every digit from 9.
18. In 9's and 10's complement of positive numbers the MSD is 0 to 4.

Answers

1. True	5. True	9. False	13. True	17. False
2. False	6. False	10. True	14. False	18. True
3. False	7. True	11. False	15. True	
4. True	8. True	12. True	16. False	

III. Choose the right answer for the following questions

1. *The 4-bit 2's complement representation of a decimal number is 1000. The number is*

   a) +8 b) 0 c) – 7 d) – 8

2. *The range of signed decimal numbers that can be represented by 6 bit 1's complement number is*

   a) – 31 to + 31 b) – 63 to +64 c) – 64 to +63 d) – 32 to 31

3. *The 2's complement representation of – 7 is*

   a) 01110 b) 01111 c) 11110 d) 11001

4. *An equivalent 2's complement representation of the 2's complement number 1101 is*

   a) 110100 b) 001101 c) 110111 d) 111101

5. *The octal equivalent of the hexa number AB.CD is*

   a) 253.314 b) 253.632 c) 526.314 d) 526.632

6. *X = 0 1 1 1 0 and Y = 1 1 0 0 1 are two 5-bit binary numbers represented in two's complement format. The sum of X and Y represented in two's complement format using 6 bit is*

   a) 1 0 0 1 1 1 b) 0 0 1 0 0 0 c) 0 0 0 1 1 1 d) 1 0 1 0 0 1

7. The 7-bit sign-magnitude representation of a decimal number is 1000110. The number is

   a) +6 b) – 6 c) +134 d) – 134

8. The 1's complement representation of – 12 is

   a) 01100 b) 1100 c) 11100 d) 110011

9. The range of signed binary numbers that can be represented in 7-bit 2's complement number system is

   a) –128 to + 127 b) –127 to + 128 c) –128 to + 128 d) –127 to + 127

10. Which one of the following is not a 1's complement representation of – 9.

    a) 01001 b) 10110 c) 110110 d) 1110110

11. The 9's complement representation of -152_{10} is

a) 845 b) 846 c) 848 d) 847

12. The 10's complement representation of -9352_{10} is

a) 0647 b) 0648 c) 647 d) 648

Answers

1. d	5. b	9. a
2. a	6. c	10. a
3. d	7. b	11. d
4. d	8. d	12. b

IV. Answer the following questions

1. How many binary bits are needed to represent an octal and hexadecimal digit? Justify.
2. What is the range of decimal numbers that can be represent in n-bit binary representation of

   a) Sign-magnitude form of binary number **b)** Unsigned binary number **c)** 1's complement binary number **d)** 2's complement binary number?
3. How is carry handled in complement subtraction?
4. Define radix.
5. Compare signed and unsigned binary number system.
6. Explain sign magnitude form of binary number system.
7. Write a short note on binary number system.
8. Explain the conversion of decimal number with fraction part to binary number.
9. Write the rules for binary arithmetic.
10. How to determine the 1's and 2's complement of a binary number? Give an example.
11. How the sign of the result (sum) is determined in complement addition and subtraction?
12. How to determine the 9's and 10's complement of a decimal number? Give an example.

V. Solve the following problems

E2.1 Convert hexadecimal number $2C6B_{16}$ to binary and octal.

E2.2 Convert decimal number 27_{10} to binary in two ways.

a) Directly to binary.

b) Convert first to hexadecimal and then from hexa to binary.

E2.3 Add the following numbers without converting to decimal.

a) Binary numbers 1110_2 and 001_2.

b) Hexadecimal numbers $3E_{16}$ and 54_{16}.

E2.4 Multiply the binary numbers 101101_2 and 111_2 in two methods.

E2.5 Perform the multiplication $1011_2 \times 1001_2$ by shift and add (or computer) method.

E2.6 Divide the binary number 11101_2 by 110_2 in two methods.

E2.7 Perform the division $1001010_2 \div 101_2$ by shift and subtract (or computer) method.

E2.8 Obtain the 1's and 2's complement of the following binary numbers.

a) 11101111_2 b) 11111111_2 c) 01010101_2 d) 00100101_2

E2.9 Convert the following binary numbers to hexadecimal and decimal.

a) 1.11010 b) 100.101

E2.10 Convert decimal +49 and +39 to 2's complement binary representation and append enough digits to accommodate the negative numbers. Then perform the following additions in binary.

a) (+39) + (−49) b) (−39) + (+49) c) (−39) + (−49)

Convert the answers back to decimal and verify.

E2.11 Perform the following subtraction by 1's and 2's complement subtraction .

a) $100010_2 - 100110_2$ b) $10011_2 - 10010_2$

E2.12 Given: $X = 1101010_2$ and $Y = 0101011_2$

Find a) X − Y b) Y − X using 1's and 2's complement.

E2.13 Perform the following addition by 9's and 10's complement addition.

a) $287_{10} + 149_{10}$ b) $467_{10} + (-276_{10})$

E2.14 Perform the following subtraction by 9's and 10's complement subtraction.

a) $8258_{10} - 6592_{10}$ b) $6592_{10} - 8258_{10}$

E2.15 Convert the following decimal numbers to 9's and 10's complement form.

a) 4589_{10} b) 786.365_{10}

Answers

E2.1 0010 1100 0110 1011_2, 26153_8

E2.2 a) 11011_2 b) $1B_{16} = 00011011_2$

E2.3 a) 1111_2 b) 92_{16}

E2.4 100111011_2

E2.5 $1011_2 \times 1001_2 = 0110\ 0011_2$

E2.6 100_2 (Remainder = 101_2)

E2.7 $100\ 1010_2 \div 101_2 = 1110_2$ (Remainder = 100_2)

E2.8

1's Complement	2's Complement
a) 00010000	a) 00010001
b) 00000000	b) 00000001
c) 10101010	c) 10101011
d) 11011010	d) 11011011

E2.9	**a)** 1.8125_{10}, $1.D_{16}$ **b)** 4.625_{10}, $4.A_{16}$	
E2.10	**a)** $-0000\ 1010_2$, -10_{10} **b)** $0000\ 1010_2$, $+10_{10}$ **c)** $-0101\ 1000_2$, -88_{10}	
E2.11	**a)** -000100_2	**b)** 00001_2
E2.12	**a)** 0111111_2	**b)** -0111111_2
E2.13	**a)** 436_{10}	**b)** 291_{10}
E2.14	**a)** 1666_{10}	**b)** -1666_{10}
E2.15	**a)** 5410, 5411	**b)** 213.634, 213.635

CHAPTER 3

BINARY CODES

3.1 Introduction

Digital systems accept and process information only in binary and output the processed information in binary. Therefore, any information should be represented in binary to be processed by digital systems.

In the digital systems, each decimal number or piece of information is represented by a combination of binary digits called ***codes***. Representing the information in binary is called coding. Binary code is a set of n-bit binary strings used to represent different decimal numbers or different alphabets of English or different information in voice signal, image signal, video signal, etc.

The digital systems can perform any arithmetic or logical operations on data or process any information represented in binary. Codes are also used to represent the information in different forms of binary so that the secrecy of the information is maintained. Moreover, the binary codes are useful to store the information in memory devices and also for secure transmission of any information from one device/system to another device/system.

The size n, of binary code depends on number of codes needed to assign a unique code to each member/element of a set of information. An n-bit binary code will have a group of 2^n distinct binary combinations and hence used to code a set of 2^n information. Each binary code of the group is assigned to a member/element of a set of information that is being coded.

When $n = 1$, $2^1 = 2$, Two codes are possible

$n = 2$, $2^2 = 4$, Four codes are possible

$n = 3$, $2^3 = 8$, Eight codes are possible

$n = 4$, $2^4 = 16$, Sixteen codes are possible and so on.

The possible binary codes for $n = 0$ to 5 are presented in Table 3.1.

3.1.1 Classification of Binary Codes

Binary codes are used to code decimal numbers as well as alphabets and they can be broadly classified into following three categories.

- Codes for representing decimal numbers.
- Codes for representing alphanumeric characters.
- Codes for error correction and detection.

Table 3.1: Binary Codes for n = 1 to 5

Size of Binary Code (n)	Number of Codes 2^n	Binary Codes
1	$2^1 = 2$	0
		1
2	$2^2 = 4$	00
		01
		10
		11
3	$2^3 = 8$	000
		001
		010
		011
		100
		101
		110
		111
4	$2^4 = 16$	0000
		0001
		0010
		0011
		0100
		0101
		0110
		0111
		1000
		1001
		1010
		1011
		1100
		1101
		1110
		1111
5	$2^5 = 32$	00000
		00001
		00010
		00011
		00100
		00101
		00110
		00111
		01000
		01001
		01010
		01011
		01100
		01101
		01110
		01111
		10000
		10001
		10010
		10011
		10100
		10101
		10110
		10111
		11000
		11001
		11010
		11011
		11100
		11101
		11110
		11111

Codes for Representing Decimal Numbers

The codes that are used to represent decimal digits 0 to 9 of the decimal numbers are called BCD (**B**inary-Coded **D**ecimal) codes and the coded decimal numbers are called BCD. Some of the BCD codes are weighted codes. In weighted codes, each digit has a weight and on adding the weights we will get the decimal value represented by the code.

There are a large number BCD codes and some of the popular BCD codes are,

- BCD (or 8421 Code)
- 2421 Code
- 5421 Code
- 8, 4, −2, −1 Code
- Excess-3 Code
- Gray Code

Codes for Representing Alphanumeric Characters

The codes that are used to represent decimal digits 0 to 9 of the decimal numbers, English alphabets A to Z and special characters used in computer keyboards are called ***alphanumeric codes***.

The two popular alphanumeric codes are,

- ASCII (**A**merican **S**tandard **C**ode for **I**nformation **I**nterchange) Code
- EBCDIC (**E**xtended **BCD I**nterchange **C**ode) Code

Codes for Error Detection and Correction

The codes that are used to detect and correct errors in transmission and reception of digital information in the form of codes are grouped under error correction and detection codes.

Some of the popular error detection and correction codes are,

- Parity
- Hamming Code
- Checksum

In the codes listed above only Hamming code is error detection and correction code and parity and checksum can only detect error.

3.2 BCD Codes

(AKUB, Dec'20, 8 Marks)

The BCD Codes are used to represent decimal numbers. Usually, 4-bit binary is used to represent a decimal digit in BCD. The 4-bit code can generate $2^4 = 16$ codes and in these 10 codes are used to represent decimal digits 0 to 9 and the remaining 6 codes are unused. Different BCD representation uses different combination of 4-bit binary to represent the decimal digit. The popular BCD codes 8421 Code, 2421 Code, 5421 Code, 8, 4, −2, −1 Code, Excess-3 Code and Gray Code are discussed in this section.

The BCD codes can be classified into,

- Weighted and non-weighted BCD codes
- Reflective and non-reflective BCD codes
- Sequential and non-sequential BCD codes

Weighted and Non-weighted BCD Codes

(CSVTU, Nov/Dec'19, 2 Marks)

In weighted BCD codes, each digit of the binary code is assigned a specific weight according to its position. In these codes on adding the weights of 1's in the code, the value of the decimal number represented by the code is obtained. The 8421 Code, 2421 Code, 5421 Code, 8, 4, –2, –1 Code are weighted BCD codes.

Example: 5421 code: $8 \Rightarrow 1011 \Rightarrow 5+0+2+1=8$

$7 \Rightarrow 0111 \Rightarrow 0+4+2+1=7$

Example: 2421 code: $5 \Rightarrow 1011 \Rightarrow 2+0+2+1=5$

$6 \Rightarrow 1100 \Rightarrow 2+4+0+0=6$

Example: 8, 4, –2, –1: $4 \Rightarrow 0100 \Rightarrow 0+4+0+0=4$

$9 \Rightarrow 1111 \Rightarrow 8+4+(-2)+(-1)=9$

$7 \Rightarrow 1001 \Rightarrow 8+0+0+(-1)=7$

The non-weighted BCD codes are not positional weighted. In other words, each binary digit position within the binary code is not assigned a fixed value or weight. The excess-3 and Gray code are non-weighted codes.

Reflective and Non-reflective BCD Codes

(CSVTU, Nov/Dec'19, 3 Marks)

A BCD code is called reflective code when the code is self-complementing. When a BCD code is not self-complementing it is called non-reflective code. The 2421 and excess-3 are self-complementing BCD codes and other BCD codes are not self-complementing. *(CSVTU, Nov/Dec'19, 7 Marks)*

In ***self complementing code***, the complement of the code will give the code for 9's complement representation of decimal digit.

Some examples of self complementing code are given below:

$5 \xrightarrow[\text{BCD}]{2421} 1011 \xrightarrow{\text{Complement}} 0100 = 4$

(4 in 2421 BCD)

9's complement of 5
9
–5
4

$37 \xrightarrow[\text{BCD}]{2421} 0011\ 1101 \xrightarrow{\text{Complement}} 1100\ 0010 = 62$

(62 in 2421 code)

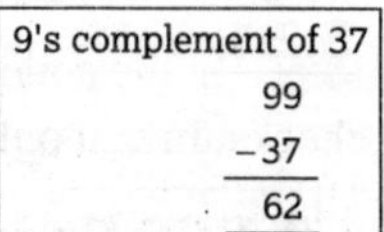

$6 \xrightarrow{\text{Excess-3}} 1001 \xrightarrow{\text{Complement}} 0110 = 3$

(3 in Excess-3)

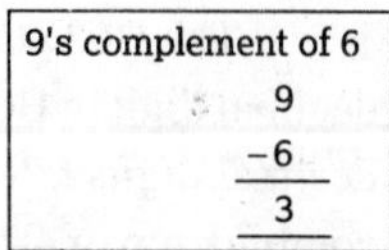

$452 \xrightarrow{\text{Excess-3}} 0111\ \ 1000\ \ 0101$

↓ Complement

$1000\ \ 0111\ \ 1010 = 547$

(547 in Excess-3)

9's complement of 452
999
–452
547

Sequential and Non-sequential BCD Codes

In sequential BCD codes, on incrementing the code for a decimal digit we will get the code for next decimal digit. In other words the binary codes for succeeding decimal digits differ by a value of 1. This property helps in arithmetic operations of data represented in BCD. The 8421 BCD and Excess-3 are sequential codes.

In non-sequential BCD codes, on incrementing the code for a decimal digit we will not get the code for next decimal digit. The 2421 Code, 5421 Code and 8, 4, – 2, –1 Code are non-sequential codes.

Table 3.2: Summary of Various BCD Codes for Decimal Digit 0 to 9

Decimal Digit	8421 BCD	2421 BCD	5421 BCD	8 4 –2 –1 BCD	Excess-3 BCD
0	0000	0000	0000	0000	0011
1	0001	0001	0001	0111	0100
2	0010	0010	0010	0110	0101
3	0011	0011	0011	0101	0110
4	0100	0100	0100	0100	0111
5	0101	1011	1000	1011	1000
6	0110	1100	1001	1010	1001
7	0111	1101	1010	1001	1010
8	1000	1110	1011	1000	1011
9	1001	1111	1100	1111	1100

Note: *In 4-bit binary code $2^4 = 16$ binary codes are possible. But in decimal representation using 4-bit codes, only 10 codes are used and 6 codes are not used. The unused codes are illegal codes in that respective codes.*

3.2.1 8421 BCD

The ***8421 BCD*** code is a binary coded decimal that uses 4-digit binary to represent each decimal digit. BCD is a weighted code with weight as 8, 4, 2 and 1 for the 4-digits of binary. In general, the 8421 BCD is simply called ***BCD***.

In BCD the 10 symbols of decimal number system is represented by a unique 4-bit code which is binary value of the decimal symbol itself.

In order to convert a decimal digit to BCD first convert the decimal digit to binary and then append with leading zeros to form 4-digit binary. The BCD for decimal digits 0 to 9 are listed in Table 3.2.

The weight of the four digits of BCD code are 8, 4, 2 and 1 and they are same as weight of binary number system.

$$8 \quad 4 \quad 2 \quad 1 \Rightarrow 2^3 \quad 2^2 \quad 2^1 \quad 2^0$$

The BCD code can be converted to decimal using binary to decimal conversion technique, in which add the weights of code wherever 1 is available as shown below:

0000 = 0 + 0 + 0 + 0 = 0	0101 = 0 + 4 + 0 + 1 = 5
0001 = 0 + 0 + 0 + 1 = 1	0110 = 0 + 4 + 2 + 0 = 6
0010 = 0 + 0 + 2 + 0 = 2	0111 = 0 + 4 + 2 + 1 = 7
0011 = 0 + 0 + 2 + 1 = 3	1000 = 8 + 0 + 0 + 0 = 8
0100 = 0 + 4 + 0 + 0 = 4	1001 = 8 + 0 + 0 + 1 = 9

In BCD representation of multi digit decimal number each digit of decimal number is represented by 4-digit BCD. Some examples of BCD representation of multi digit decimal numbers are shown in Table 3.3.

Table 3.3: Examples of BCD Representation of Multi Digit Decimal Number

Decimal Number	BCD
23	0010 0011
45	0100 0101
376	0011 0111 0110
498	0100 1001 1000
1746	0001 0111 0100 0110
8532	1000 0101 0011 0010

The advantage of BCD is that the digital system can directly perform mathematical operations on decimal number in BCD without converting them to binary number system.

3.2.2 BCD Addition

(AKTU, Nov/Dec'22, 2 Marks)

BCD addition is performed by adding the BCD representation of decimal numbers directly and then result is corrected to BCD.

In BCD addition, in order to correct the result, a correction 6_{10} (0110_2) is added to BCD sum of digits whenever sum of BCD digits exceeds 9 or if there is a carry from sum of BCD digits to next BCD digit addition.

Example 3.1

(IKGPTU, May'20, 2 Marks)

Perform BCD addition of following decimal numbers.

i) $6_{10} + 3_{10}$ **ii)** $7_{10} + 8_{10}$ **iii)** $42_{10} + 35_{10}$ **iv)** $72_{10} + 19_{10}$

v) $84_{10} + 63_{10}$ **vi)** $327_{10} + 915_{10}$ **vii)** $4831_{10} + 2974_{10}$

Solution

i) $6_{10} + 3_{10}$

BCD addition

```
                              1 1   ← Carry
      6                      0110
            Convert to BCD
  (+) 3     ==============> (+) 0011
  -----                     --------
      9                      1001  ---Convert to decimal--->  9₁₀
  -----                     --------
```

$\therefore 6_{10} + 3_{10} = 9_{10}$

ii) $7_{10} + 8_{10}$

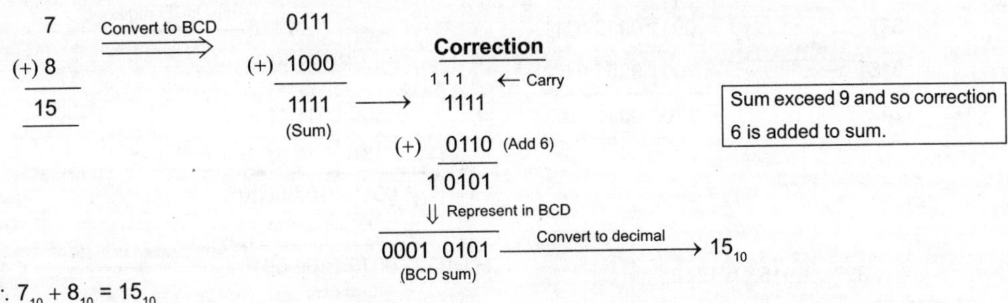

$\therefore 7_{10} + 8_{10} = 15_{10}$

iii) $42_{10} + 35_{10}$

BCD addition

```
   42    Convert to BCD        0100 0010
(+) 35   ==============>  (+)  0011 0101
   77                          0111 0111  --Convert to decimal-->  77₁₀
```

$\therefore 42_{10} + 35_{10} = 77_{10}$

iv) $72_{10} + 19_{10}$

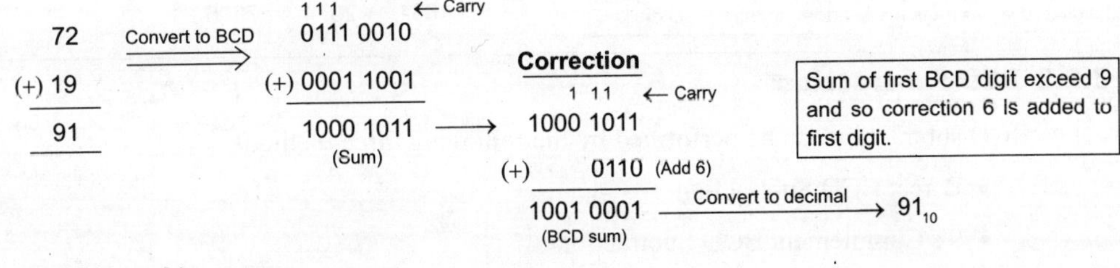

$\therefore 72_{10} + 19_{10} = 91_{10}$

v) $84_{10} + 63_{10}$

BCD addition

```
   84    Convert to BCD       1000 0100              Correction
(+) 63   ==============>  (+) 0110 0011              1 11      <- Carry
  147                         1110 0111   ------>    1110 0111
                              (Sum)              (+) 0110       (Add 6)
                                                   1 0100 0111
                                                     || Represent in BCD
                                                 0001 0100 0111  --Convert to decimal-->  147₁₀
                                                   (BCD sum)
```

Sum of second BCD digit exceed 9 and so correction 6 is added to second digit.

$\therefore 84_{10} + 63_{10} = 147_{10}$

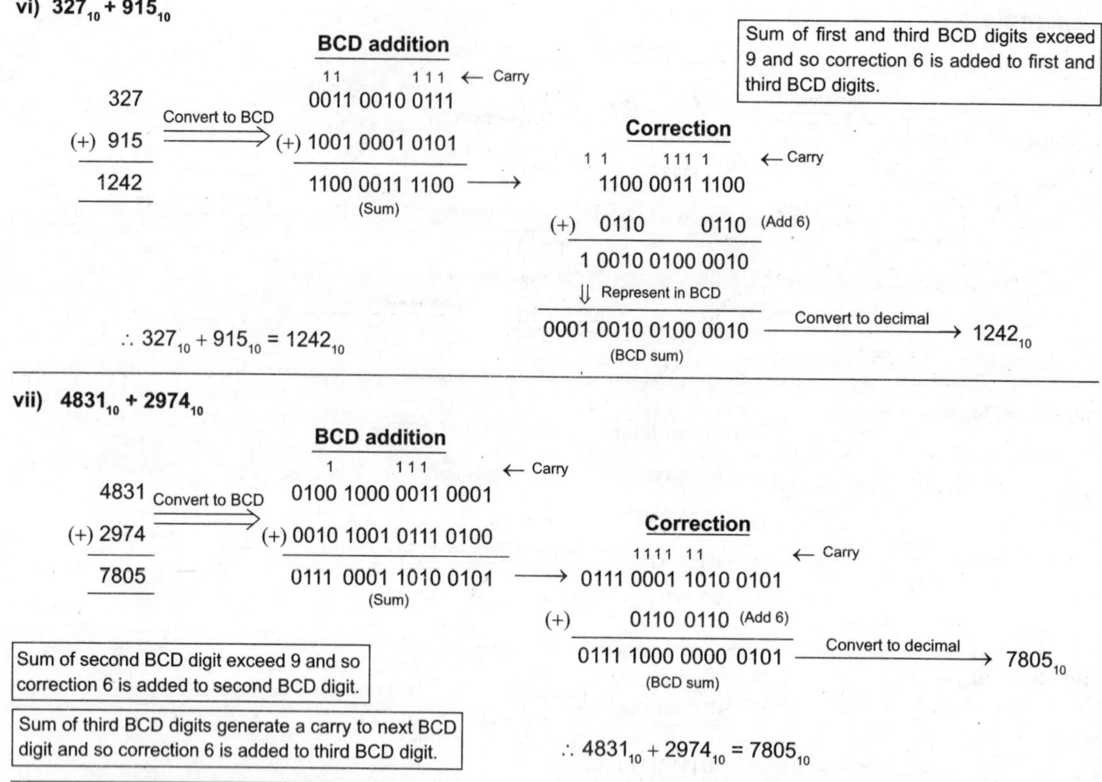

3.2.3 BCD Subtraction

BCD subtraction can be performed by the following three methods.

- Direct BCD Subtraction
- 9's Complement BCD Subtraction
- 10's Complement BCD Subtraction

Direct BCD Subtraction

In direct BCD subtraction the decimal numbers are expressed in BCD and the subtraction is performed similar to binary subtraction starting from lower most digit and the results are corrected to BCD.

In BCD subtraction, in order to correct the result, a correction 6_{10} (0110_2) is subtracted from the difference of BCD digits whenever the difference of BCD digits exceeds 9 or if there is borrow from previous BCD digit.

9's and 10's Complement BCD Subtraction

In 9's complement BCD subtraction the subtrahend is expressed in 9's complement of decimal number. The minuend and 9's complement of subtrahend are expressed in BCD and then added similar to binary addition. Now, the BCD subtraction is similar to BCD addition and hence the correction of sum is performed similar to BCD addition.

The 10's complement BCD subtraction is similar to 9's complement BCD subtraction except that the subtrahend is converted to 10's complement and added to minuend.

Example 3.2

(GTU, May'19, 4 Marks)

Perform the following subtraction by direct or actual BCD subtraction.

a) $65_{10} - 19_{10}$ **b)** $306.7_{10} - 247.8_{10}$ **c)** $75_{10} - 87_{10}$

Solution

a) $65_{10} - 19_{10}$

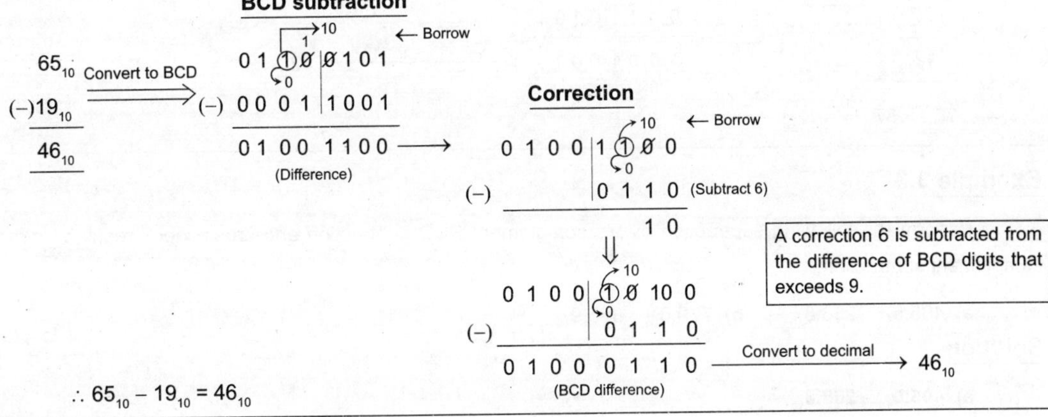

$\therefore 65_{10} - 19_{10} = 46_{10}$

b) $306.7_{10} - 247.8_{10}$

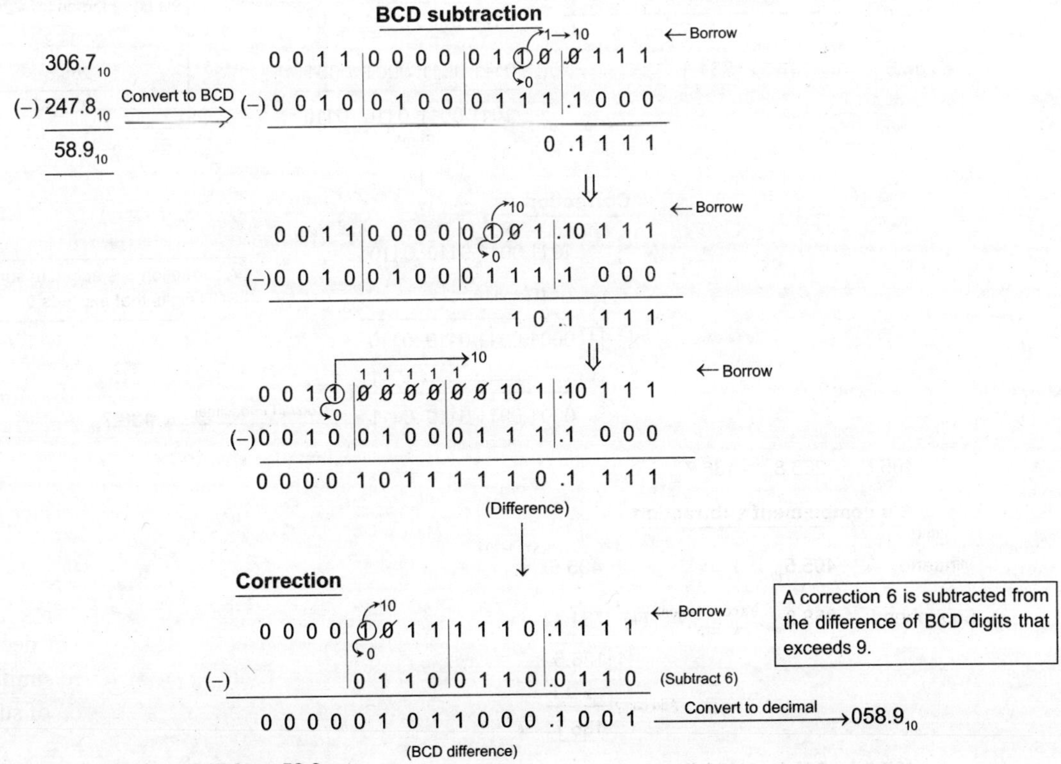

BCD subtraction

306.7_{10}

$(-)\ 247.8_{10}$ Convert to BCD

58.9_{10}

1 → 10 ← Borrow

0 0 1 1 | 0 0 0 0 | 0 1 ① Ø | .Ø 1 1 1

(−) 0 0 1 0 | 0 1 0 0 | 0 1 1 1 | .1 0 0 0

0 .1 1 1 1

⇓

10 ← Borrow

0 0 1 1 | 0 0 0 0 | 0 ① Ø 1 | .10 1 1 1

(−) 0 0 1 0 | 0 1 0 0 | 0 1 1 1 | .1 0 0 0

1 0 .1 1 1 1

⇓

1 1 1 1 1 → 10 ← Borrow

0 0 1 ① | Ø Ø Ø Ø | Ø Ø 10 1 | .10 1 1 1

(−) 0 0 1 0 | 0 1 0 0 | 0 1 1 1 | .1 0 0 0

0 0 0 0 1 0 1 1 1 1 1 0 .1 1 1 1

(Difference)

↓

Correction

10 ← Borrow

0 0 0 0 | ① Ø 1 1 | 1 1 1 0 | .1 1 1 1

(−) | 0 1 1 0 | 0 1 1 0 | .0 1 1 0 (Subtract 6)

0 0 0 0 0 1 0 1 1 0 0 0 .1 0 0 1 Convert to decimal → 058.9_{10}

(BCD difference)

A correction 6 is subtracted from the difference of BCD digits that exceeds 9.

$\therefore 306.7_{10} - 247.8_{10} = 58.9_{10}$

c) $75_{10} - 87_{10}$

Here, the subtrahend is greater than the minuend. Hence, the minuend and subtrahend are interchanged and then the subtraction is performed and the result is considered as negative.

BCD subtraction

Decimal		BCD	
		→10 ← Borrow	
87_{10}	Convert to BCD ⟹	1 0 0 0 | 0 1 1 1	
(−) 75_{10}		(−) 0 1 1 1 | 0 1 0 1	
12_{10}		0 0 0 1 0 0 1 0 (BCD difference)	Convert to decimal → 12_{10}

$\therefore 75_{10} - 87_{10} = -12_{10}$

Example 3.3

Perform the following subtraction by 9's complement BCD subtraction and cross-check result by direct 9's complement subtraction.

a) $405.5_{10} - 268.8_{10}$ **b)** $779.6_{10} - 885.9_{10}$

Solution

a) $405.5_{10} - 268.8_{10}$

9's complement BCD subtraction

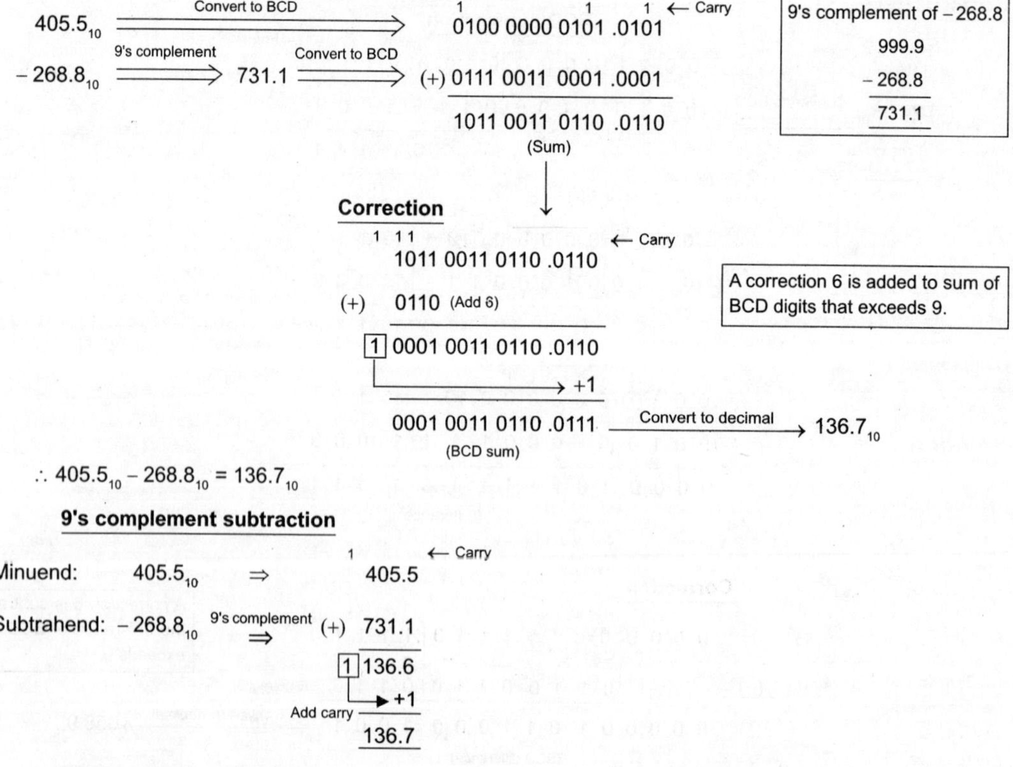

$\therefore 405.5_{10} - 268.8_{10} = 136.7_{10}$

b) $779.6_{10} - 885.9_{10}$

9's complement BCD subtraction

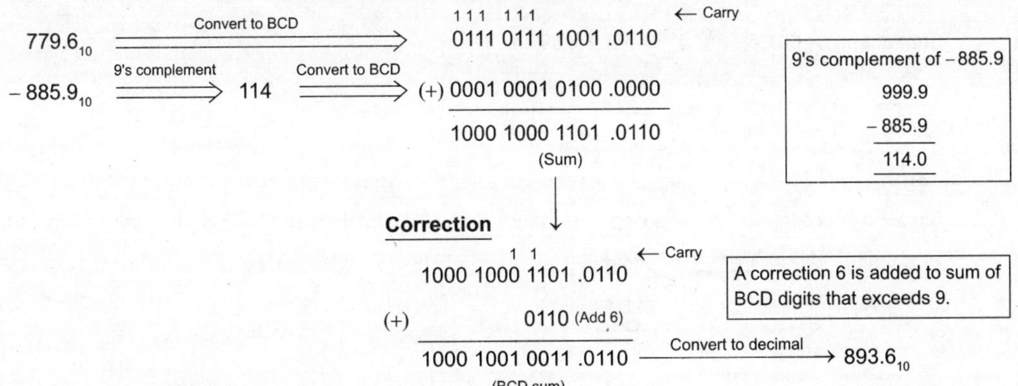

9's complement subtraction

Minuend: 779.6_{10} ⟹ 779.6

Subtrahend: -885.9_{10} $\xRightarrow{\text{9's complement}}$ (+) 114.0

893.6 (Sum)

↑ No carry

Conclusion

Since no final carry the result is negative. Hence take 9's complement of 893.6_{10} and put "–" sign.

893.6_{10} $\xrightarrow{\text{9's complement}}$ 999.9
− 893.6
106.3

$\therefore 779.6_{10} - 885.9_{10} = -106.3_{10}$

Example 3.4

Perform the following subtraction by 10's complement BCD subtraction and cross-check result by direct 10's complement subtraction.

a) $306.4_{10} - 607.6_{10}$ **b)** $442.7_{10} - 108.9_{10}$

Solution

a) $306.4_{10} - 607.6_{10}$

10's complement BCD subtraction

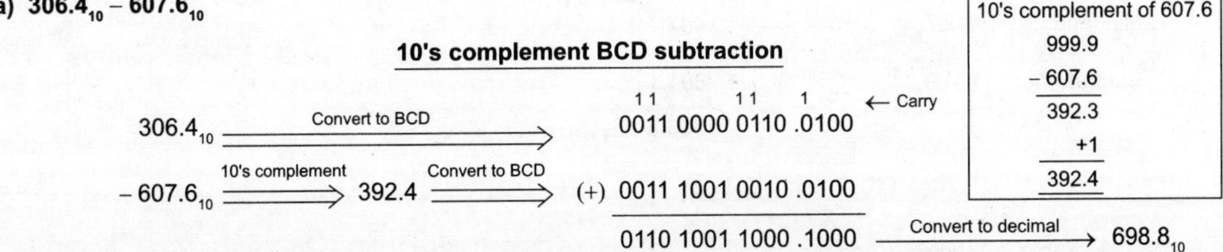

10's complement subtraction

Minuend: 306.4_{10} $\Rightarrow$ 306.4

Subtrahend: -607.6_{10} $\xRightarrow{\text{10's complement}}$ (+) 392.4

698.8 (Sum)

↑ No carry

Conclusion

Since no final carry the result is negative. Hence take 10's complement of 698.8_{10} and put "–" sign.

$$
698.8 \xrightarrow{\text{10's complement}} \begin{array}{r} 999.9 \\ -\ 698.8 \\ \hline 301.1 \\ +1 \\ \hline 301.2 \end{array}
$$

$\therefore 306.4_{10} - 607.6_{10} = -301.2_{10}$

---

b) $442.7_{10} - 108.9_{10}$

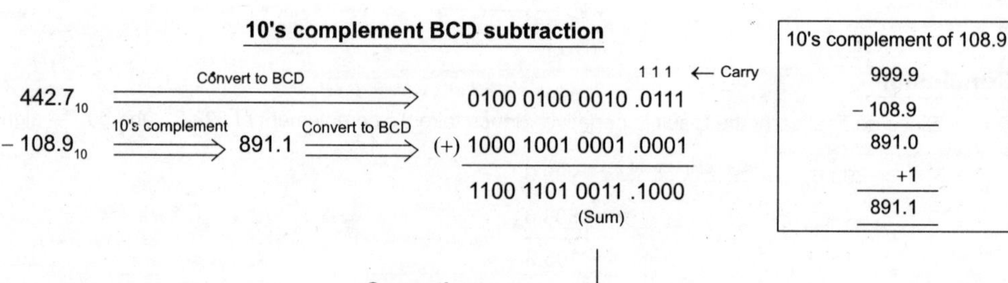

Correction

A correction 6 is added to sum of BCD digits that exceeds 9.

```
 1 1   1 1                 ← Carry
     1100 1101 0011 .1000
(+)  0110 0110  (Add 6)
  [1] 0011 0011 0011 .1000  ──Convert to decimal──→ 333.8₁₀
   ↑          (BCD sum)
Discard carry
```

$\therefore 442.7_{10} - 108.9_{10} = 333.8_{10}$

10's complement subtraction

```
                                1 1   ← Carry
Minuend:     442.7₁₀   ⇒          442.7
Subtrahend: – 108.9₁₀  ⇒ (10's complement)  (+) 891.1
                                [1] 333.8
                                 ↑
                            Discard carry
```

$\therefore 442.7_{10} - 108.9_{10} = 333.8_{10}$

3.2.4 2421 Code

The ***2421 code*** is a binary coded decimal that uses 4-digit binary to represent each decimal digit. It is a weighted code with weight as 2, 4, 2 and 1 for the 4-digits of binary.

The 2421 code for decimal digits 0 to 9 are listed in Table 3.2.

The decimal digit can be converted to 2421 code by realizing the decimal digit as sum of weights and then replacing the weight by 1 as shown below:

$$1_{10} \xrightarrow[\text{of 2421 to form 1}]{\text{Take weight}} 0+0+0+1 \xrightarrow[\text{in the place weight are considered}]{\text{Put 1}} 0001$$

$$2_{10} \xrightarrow[\text{of 2421 to form 2}]{\text{Take weight}} 0+0+2+0 \xrightarrow[\text{in the place weight are considered}]{\text{Put 1}} 0010$$

$$3_{10} \xrightarrow[\text{of 2421 to form 3}]{\text{Take weight}} 0+0+2+1 \xrightarrow[\text{in the place weight are considered}]{\text{Put 1}} 0011$$

$$4_{10} \xrightarrow[\text{of 2421 to form 4}]{\text{Take weight}} 0+4+0+0 \xrightarrow[\text{in the place weight are considered}]{\text{Put 1}} 0100$$

$$5_{10} \xrightarrow[\text{of 2421 to form 5}]{\text{Take weight}} 2+0+2+1 \xrightarrow[\text{in the place weight are considered}]{\text{Put 1}} 1011$$

$$6_{10} \xrightarrow[\text{of 2421 to form 6}]{\text{Take weight}} 2+4+0+0 \xrightarrow[\text{in the place weight are considered}]{\text{Put 1}} 1100$$

$$7_{10} \xrightarrow[\text{of 2421 to form 7}]{\text{Take weight}} 2+4+0+1 \xrightarrow[\text{in the place weight are considered}]{\text{Put 1}} 1101$$

$$8_{10} \xrightarrow[\text{of 2421 to form 8}]{\text{Take weight}} 2+4+2+0 \xrightarrow[\text{in the place weight are considered}]{\text{Put 1}} 1110$$

$$9_{10} \xrightarrow[\text{of 2421 to form 9}]{\text{Take weight}} 2+4+2+1 \xrightarrow[\text{in the place weight are considered}]{\text{Put 1}} 1111$$

The 2421 code can be converted to decimal by adding the weights wherever 1 is available in the code as shown below:

$0000 = 0 + 0 + 0 + 0 = 0$	$1011 = 2 + 0 + 2 + 1 = 5$
$0001 = 0 + 0 + 0 + 1 = 1$	$1100 = 2 + 4 + 0 + 0 = 6$
$0010 = 0 + 0 + 2 + 0 = 2$	$1101 = 2 + 4 + 0 + 1 = 7$
$0011 = 0 + 0 + 2 + 1 = 3$	$1110 = 2 + 4 + 2 + 0 = 8$
$0100 = 0 + 4 + 0 + 0 = 4$	$1111 = 2 + 4 + 2 + 1 = 9$

In 2421 code representation of multi digit decimal number, each digit of decimal number is represented by 4-digit 2421 code. Some examples of 2421 code representation of multi digit decimal numbers are shown in Table 3.4.

Table 3.4: Examples of 2421 and 5421 Code Representation of Multi Digit Decimal Number

Decimal Number	2421 Code				5421 Code			
23			0010	0011			0010	0011
45			0100	1011			0100	1000
376		0011	1101	1100		0011	1010	1001
498		0100	1111	1110		0100	1100	1011
1746	0001	1101	0100	1100	0001	1010	0100	1001
8532	1110	1011	0011	0010	1011	1000	0011	0010

Example 3.5

Convert 5_{10} to 2421 code and verify by back converting to decimal.

Solution

The 2421 code can be obtained expressing 5_{10} as sum of 2421 weight and then inserting 1 in the place of weight that are used to form the decimal value.

$$5_{10} \xrightarrow[\text{of 2421 to form 5}]{\text{Take weight}} 2+0+2+1 \xrightarrow[\text{in the place weight are considered}]{\text{Put 1}} 1011$$

$\therefore$ 2421 code for 5_{10} = 1011

To convert 2421 code to decimal add the weight wherever 1 is present in the code.

$$1011 \xrightarrow[\text{of weights where 1 is available}]{\text{Take sum}} 2+0+2+1 = 5_{10}$$

$\therefore$ Decimal value of 2421 code 1011 = 5_{10}

3.2.5 5421 Code

The ***5421 code*** is a binary coded decimal that uses 4-digit binary to represent each decimal digit. It is a weighted code with weight as 5, 4, 2 and 1 for the 4-digits of binary.

The 5421 code for decimal digits 0 to 9 are listed in Table 3.2.

The decimal digit can be converted to 5421 code by realizing the decimal digit as sum of weights and then replacing the weight by 1 as shown below:

$$1_{10} \xrightarrow[\text{of 5421 to form 1}]{\text{Take weight}} 0+0+0+1 \xrightarrow[\text{in the place weight are considered}]{\text{Put 1}} 0001$$

$$2_{10} \xrightarrow[\text{of 5421 to form 2}]{\text{Take weight}} 0+0+2+0 \xrightarrow[\text{in the place weight are considered}]{\text{Put 1}} 0010$$

$$3_{10} \xrightarrow[\text{of 5421 to form 3}]{\text{Take weight}} 0+0+2+1 \xrightarrow[\text{in the place weight are considered}]{\text{Put 1}} 0011$$

$$4_{10} \xrightarrow[\text{of 5421 to form 4}]{\text{Take weight}} 0+4+0+0 \xrightarrow[\text{in the place weight are considered}]{\text{Put 1}} 0100$$

$$5_{10} \xrightarrow[\text{of 5421 to form 5}]{\text{Take weight}} 5+0+0+0 \xrightarrow[\text{in the place weight are considered}]{\text{Put 1}} 1000$$

$$6_{10} \xrightarrow[\text{of 5421 to form 6}]{\text{Take weight}} 5+0+0+1 \xrightarrow[\text{in the place weight are considered}]{\text{Put 1}} 1001$$

$$7_{10} \xrightarrow[\text{of 5421 to form 7}]{\text{Take weight}} 5+0+2+0 \xrightarrow[\text{in the place weight are considered}]{\text{Put 1}} 1010$$

$$8_{10} \xrightarrow[\text{of 5421 to form 8}]{\text{Take weight}} 5+0+2+1 \xrightarrow[\text{in the place weight are considered}]{\text{Put 1}} 1011$$

$$9_{10} \xrightarrow[\text{of 5421 to form 9}]{\text{Take weight}} 5+4+0+0 \xrightarrow[\text{in the place weight are considered}]{\text{Put 1}} 1100$$

The 5421 code can be converted to decimal by adding the weights wherever 1 is available in the code as shown below:

$0000 = 0 + 0 + 0 + 0 = 0$

$0001 = 0 + 0 + 0 + 1 = 1$

$0010 = 0 + 0 + 2 + 0 = 2$

$0011 = 0 + 0 + 2 + 1 = 3$

$0100 = 0 + 4 + 0 + 0 = 4$

$1000 = 5 + 0 + 0 + 0 = 5$

$1001 = 5 + 0 + 0 + 1 = 6$

$1010 = 5 + 0 + 2 + 0 = 7$

$1011 = 5 + 0 + 2 + 1 = 8$

$1100 = 5 + 4 + 0 + 0 = 9$

In 5421 code representation of multi digit decimal number, each digit of decimal number is represented by 4-digit 5421 code. Some examples of 5421 code representation of multi digit decimal numbers are shown in Table 3.2.

Example 3.6

Convert 6_{10} to 5421 code and verify by back converting to decimal.

Solution

The 5421 code can be obtained by expressing 6_{10} as sum of 5421 weights and then inserting 1 in the place of weights that are used to form decimal value.

$$6_{10} \xrightarrow[\text{of 5421 to form 6}]{\text{Take weight}} 5+0+0+1 \xrightarrow[\text{in the place weight are considered}]{\text{Put 1}} 1001$$

$\therefore$ 5421 code for 6_{10} = 1001

To convert the 5421 code to decimal add the weights wherever 1 is present in the code.

$$1001 \xrightarrow[\text{of weights where 1is available}]{\text{Take sum}} 5+0+0+1 = 6_{10}$$

$\therefore$ Decimal value of 5421 code 1001 = 6_{10}

3.2.6 8, 4, −2, −1 Code

The ***8, 4, −2, −1 code*** is a binary coded decimal that uses 4-digit binary to represent each decimal digit. It is a weighted code with weight as 8, 4, −2 and −1 for the 4-digits of binary.

The 8, 4, −2, −1 code for decimal digits 0 to 9 are listed in Table 3.2.

The decimal digit can be converted to 8, 4, −2, −1 code by realizing the decimal digit as sum of weights and then replacing the weight by 1 as shown below:

$$1_{10} \xrightarrow[\text{of } 8, 4, -2, -1 \text{ to form } 1]{\text{Take weight}} 0+4-2-1 \xrightarrow[\substack{\text{in the place weight} \\ \text{are considered}}]{\text{Put 1}} 0111$$

$$2_{10} \xrightarrow[\text{of } 8, 4, -2, -1 \text{ to form } 2]{\text{Take weight}} 0+4-2+0 \xrightarrow[\substack{\text{in the place weight} \\ \text{are considered}}]{\text{Put 1}} 0110$$

$$3_{10} \xrightarrow[\text{of } 8, 4, -2, -1 \text{ to form } 3]{\text{Take weight}} 0+4+0-1 \xrightarrow[\substack{\text{in the place weight} \\ \text{are considered}}]{\text{Put 1}} 0101$$

$$4_{10} \xrightarrow[\text{of } 8, 4, -2, -1 \text{ to form } 4]{\text{Take weight}} 0+4+0+0 \xrightarrow[\substack{\text{in the place weight} \\ \text{are considered}}]{\text{Put 1}} 0100$$

$$5_{10} \xrightarrow[\text{of } 8, 4, -2, -1 \text{ to form } 5]{\text{Take weight}} 8+0-2-1 \xrightarrow[\substack{\text{in the place weight} \\ \text{are considered}}]{\text{Put 1}} 1011$$

$$6_{10} \xrightarrow[\text{of } 8, 4, -2, -1 \text{ to form } 6]{\text{Take weight}} 8+0-2+0 \xrightarrow[\substack{\text{in the place weight} \\ \text{are considered}}]{\text{Put 1}} 1010$$

$$7_{10} \xrightarrow[\text{of } 8, 4, -2, -1 \text{ to form } 7]{\text{Take weight}} 8+0+0-1 \xrightarrow[\substack{\text{in the place weight} \\ \text{are considered}}]{\text{Put 1}} 1001$$

$$8_{10} \xrightarrow[\text{of } 8, 4, -2, -1 \text{ to form } 8]{\text{Take weight}} 8+0+0+0 \xrightarrow[\substack{\text{in the place weight} \\ \text{are considered}}]{\text{Put 1}} 1000$$

$$9_{10} \xrightarrow[\text{of } 8, 4, -2, -1 \text{ to form } 9]{\text{Take weight}} 8+4-2-1 \xrightarrow[\substack{\text{in the place weight} \\ \text{are considered}}]{\text{Put 1}} 1111$$

The 8, 4, −2, −1 code can be converted to decimal by adding the weights wherever 1 is available in the code as shown below:

$$0000 = 0+0+0+0 = 0$$
$$0111 = 0+4-2-1 = 1$$
$$0110 = 0+4-2+0 = 2$$
$$0101 = 0+4+0-1 = 3$$
$$0100 = 0+4+0+0 = 4$$
$$1011 = 8+0-2-1 = 5$$
$$1010 = 8+0-2+0 = 6$$
$$1001 = 8+0+0-1 = 7$$
$$1000 = 8+0+0+0 = 8$$
$$1111 = 8+4-2-1 = 9$$

In 8, 4, −2, −1 code representation of multi digit decimal number, each digit of decimal number is represented by 4-digit 8, 4, −2, −1 code. Some examples of 8, 4, −2, −1 code representation of multi digit decimal numbers are shown in Table 3.5.

Table 3.5: Examples of 8, 4, –2, –1 and Excess-3 Code Representation of Multi Digit Decimal Number

Decimal Number	8, 4, –2, –1 Code	Excess-3 Code
23	0110 0101	0101 0110
45	0100 1011	0111 1000
376	0101 1001 1010	0110 1010 1001
498	0100 1111 1000	0111 1100 1011
1746	0111 1001 0100 1010	0100 1010 0111 1001
8532	1000 1011 0101 0110	1011 1000 0110 0101

Example 3.7

Convert 3_{10} to 8, 4, –2, –1 code and verify by back converting to decimal.

Solution

The 8, 4, –2, –1 code can be obtained by expressing 3_{10} as sum of 8, 4, –2, –1 weights and then inserting 1 in the place of weights that are used to form decimal value.

$$3_{10} \xrightarrow[\text{of 8, 4, -2, -1 to form 3}]{\text{Take weight}} 0+4+0-1 \xrightarrow[\text{in the place weight are considered}]{\text{Put 1}} 0101$$

$\therefore$ 8, 4, –2, –1 code for 3_{10} = 0101

To convert the 8, 4, –2, –1 code to decimal add the weights wherever 1 is present in the code.

$$0101 \xrightarrow[\text{of weights where 1 is available}]{\text{Take sum}} 0+4+0-1 = 3_{10}$$

$\therefore$ Decimal value of 8, 4, –2, –1 code 0101 = 3_{10}

3.2.7 Excess-3 Code

The ***excess-3 code*** is a binary coded decimal that uses 4-digit binary to represent each decimal digit. The excess-3 is non-weighted code. The excess-3 code for decimal digits 0 to 9 are listed in Table 3.2.

Excess-3 code for a decimal digit is obtained by adding 4-bit binary value of 3_{10} (0011_2) to 8421 BCD of the decimal digit as shown below:

$0 \Rightarrow 0000+0011 \Rightarrow 0011$	$5 \Rightarrow 0101+0011 \Rightarrow 1000$
$1 \Rightarrow 0001+0011 \Rightarrow 0100$	$6 \Rightarrow 0110+0011 \Rightarrow 1001$
$2 \Rightarrow 0010+0011 \Rightarrow 0101$	$7 \Rightarrow 0111+0011 \Rightarrow 1010$
$3 \Rightarrow 0011+0011 \Rightarrow 0110$	$8 \Rightarrow 1000+0011 \Rightarrow 1011$
$4 \Rightarrow 0100+0011 \Rightarrow 0111$	$9 \Rightarrow 1001+0011 \Rightarrow 1100$

The excess-3 code can be converted to decimal by subtracting 3_{10} (011_2) to get BCD and then converting BCD to decimal using binary to decimal convertion technique.

In excess-3 code representation of multi digit decimal number, each digit of decimal number is represented by 4-digit excess-3 code. Some examples of excess-3 code representation of multi digit decimal numbers are shown in Table 3.5.

Example 3.8

(GTU, May'19, 3 Marks)

Convert 8_{10} to excess-3 code and verify by back converting to decimal.

Solution

The excess-3 code can be obtained by adding 011_2 to binary value of the decimal digit.

$$8_{10} \xrightarrow{\text{Convert to binary}} 1000_2 \xrightarrow{\text{Add 011}} 1011$$

$\therefore$ Excess-3 code for 8_{10} = 1011

To convert excess-3 code to decimal, subtract 011_2 to get binary value and then convert binary to decimal.

$$1011 \xrightarrow{\text{Subtract } 011_2} 1000_2 \xrightarrow[\text{to decimal}]{\text{Convert}} 8_{10}$$

$\therefore$ Decimal value of excess-3 code 1011 = 8_{10}

Example 3.9

Represent the decimal number 2569 in **(a)** BCD **(b)** Excess-3 code **(c)** 2421 code **(d)** 5421 code **(e)** 8, 4, −2, −1 code

Solution

a) Convert to BCD

$$2_{10} \xrightarrow[\text{to 4-bit bianry}]{\text{Convert}} 0010_2 \qquad 6_{10} \xrightarrow[\text{to 4-bit bianry}]{\text{Convert}} 0110_2$$

$$5_{10} \xrightarrow[\text{to 4-bit bianry}]{\text{Convert}} 0101_2 \qquad 9_{10} \xrightarrow[\text{to 4-bit bianry}]{\text{Convert}} 1001_2$$

$\therefore$ BCD of 2569_{10} = 0010 0101 0110 1001

b) Convert to Excess-3

Add 011_2 to binary value of every decimal digit to get excess-3 code.

$2_{10} = 0010_2$	$5_{10} = 0101_2$	$6_{10} = 0110_2$	$9_{10} = 1001_2$
+011	+011	+011	+011
0101	1000	1001	1100

$\therefore$ Excess-3 code for 2569_{10} = 0101 1000 1001 1100

c) Convert to 2421 code

$$2_{10} \xrightarrow[\text{of 2421 to form 2}]{\text{Take weight}} 0+0+2+0 \xrightarrow[\text{in the place weight are considered}]{\text{Put 1}} 0010$$

$$5_{10} \xrightarrow[\text{of 2421 to form 5}]{\text{Take weight}} 2+0+2+1 \xrightarrow[\text{in the place weight are considered}]{\text{Put 1}} 1011$$

$$6_{10} \xrightarrow[\text{of 2421 to form 6}]{\text{Take weight}} 2+4+0+0 \xrightarrow[\text{in the place weight are considered}]{\text{Put 1}} 1100$$

$$9_{10} \xrightarrow[\text{of 2421 to form 9}]{\text{Take weight}} 2+4+2+1 \xrightarrow[\text{in the place weight are considered}]{\text{Put 1}} 1111$$

$\therefore$ 2421 code for 2569_{10} = 0010 1011 1100 1111

d) Convert to 5421 code

$$2_{10} \xrightarrow[\text{of 5421 to form 2}]{\text{Take weight}} 0+0+2+0 \xrightarrow[\substack{\text{in the place weight}\\\text{are considered}}]{\text{Put 1}} 0010$$

$$5_{10} \xrightarrow[\text{of 5421 to form 5}]{\text{Take weight}} 5+0+0+0 \xrightarrow[\substack{\text{in the place weight}\\\text{are considered}}]{\text{Put 1}} 1000$$

$$6_{10} \xrightarrow[\text{of 5421 to form 6}]{\text{Take weight}} 5+0+0+1 \xrightarrow[\substack{\text{in the place weight}\\\text{are considered}}]{\text{Put 1}} 1001$$

$$9_{10} \xrightarrow[\text{of 5421 to form 9}]{\text{Take weight}} 5+4+0+0 \xrightarrow[\substack{\text{in the place weight}\\\text{are considered}}]{\text{Put 1}} 1100$$

$\therefore$ 5421 code for 2569_{10} = 0010 1000 1001 1100

e) Convert to 8, 4, –2, –1 code

$$2_{10} \xrightarrow[\text{of 8, 4, – 2, – 1 to form 2}]{\text{Take weight}} 0+4-2-0 \xrightarrow[\substack{\text{in the place weight}\\\text{are considered}}]{\text{Put 1}} 0110$$

$$5_{10} \xrightarrow[\text{of 8, 4, – 2, – 1 to form 5}]{\text{Take weight}} 8+0-2-1 \xrightarrow[\substack{\text{in the place weight}\\\text{are considered}}]{\text{Put 1}} 1011$$

$$6_{10} \xrightarrow[\text{of 8, 4, – 2, – 1 to form 6}]{\text{Take weight}} 8+0-2+0 \xrightarrow[\substack{\text{in the place weight}\\\text{are considered}}]{\text{Put 1}} 1010$$

$$9_{10} \xrightarrow[\text{of 8, 4, – 2, – 1 to form 9}]{\text{Take weight}} 8+4-2-1 \xrightarrow[\substack{\text{in the place weight}\\\text{are considered}}]{\text{Put 1}} 1111$$

$\therefore$ 8, 4, –2, –1 code for 2569_{10} = 0110 1011 1010 1111

3.2.8 Excess-3 Addition

In Excess-3 addition the decimal numbers are expressed in Excess-3 code and the addition is performed similar to binary addition starting from lower most digit and the results are corrected to Excess-3 code.

In Excess-3 addition, in order to correct the result, a correction 3_{10} (0011_2) is added to sum of Excess-3 digits if there is a carry to next Excess-3 digit and a correction 3_{10} (0011_2) is subtracted from sum of Excess-3 digits when there is no carry to next Excess-3 digit.

Example 3.10

Perform Excess-3 addition of following decimal numbers.

a) $4_{10} + 5_{10}$

b) $47_{10} + 38_{10}$

c) $347.6_{10} + 459.4_{10}$

Solution

a) $4_{10} + 5_{10}$

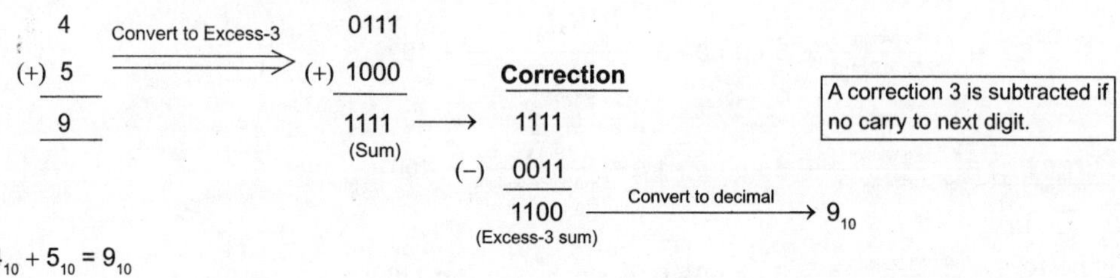

$\therefore 4_{10} + 5_{10} = 9_{10}$

b) $47_{10} + 38_{10}$

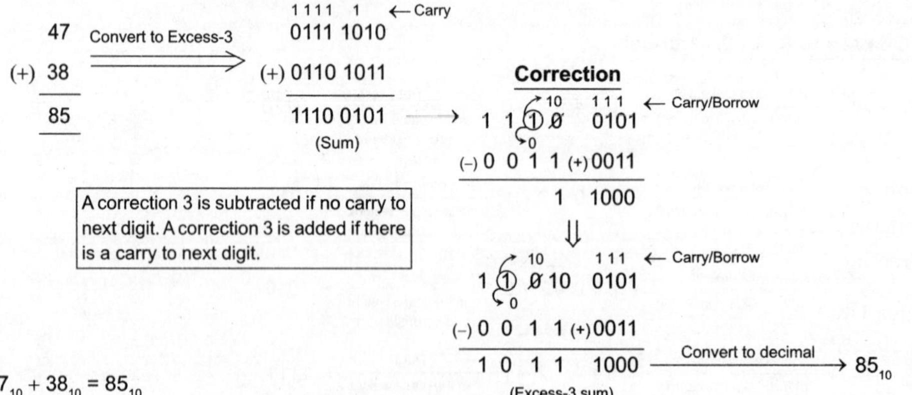

$\therefore 47_{10} + 38_{10} = 85_{10}$

c) $347.6_{10} + 459.4_{10}$

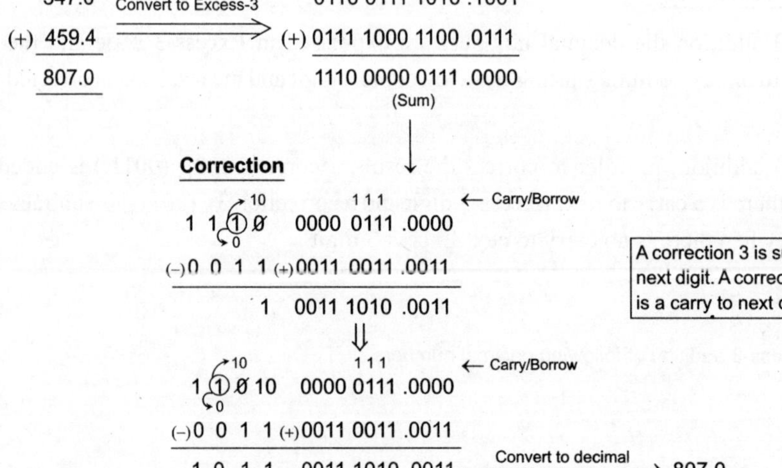

$\therefore 347.6_{10} + 459.4_{10} = 807.0_{10}$

3.2.9 Excess-3 Subtraction

Excess-3 subtraction can be performed by the following three methods.

- Direct Excess-3 Subtraction
- 9's Complement Excess-3 Subtraction
- 10's Complement Excess-3 Subtraction

Direct Excess-3 Subtraction

In direct Excess-3 subtraction the decimal numbers are expressed in Excess-3 code and the subtraction is performed similar to binary subtraction starting from lower most digit and the results are corrected to Excess-3 code.

In Excess-3 subtraction, in order to correct the result, a correction 3_{10} (0011_2) is added to the difference of Excess-3 digits if there is no borrow from previous digit and a correction 3_{10} (0011_2) is subtracted from the difference of Excess-3 digits if there is borrow from previous digit.

9's and 10's Complement Excess-3 Subtraction

In 9's complement Excess-3 subtraction the subtrahend is expressed in 9's complement of decimal number. The minuend and 9's complement of subtrahend are expressed in Excess-3 code and then added similar to binary addition. Now, the Excess-3 subtraction is similar to Excess-3 addition and hence the correction of sum is performed similar to Excess-3 addition.

The 10's complement Excess-3 subtraction is similar to 9's complement Excess-3 subtraction except that the subtrahend is converted to 10's complement and added to minuend.

Example 3.11

Perform the following subtraction by direct or actual Excess-3 subtraction.

a) $287_{10} - 195_{10}$ **b)** $75.4_{10} - 25.8_{10}$

Solution

a) $287_{10} - 195_{10}$

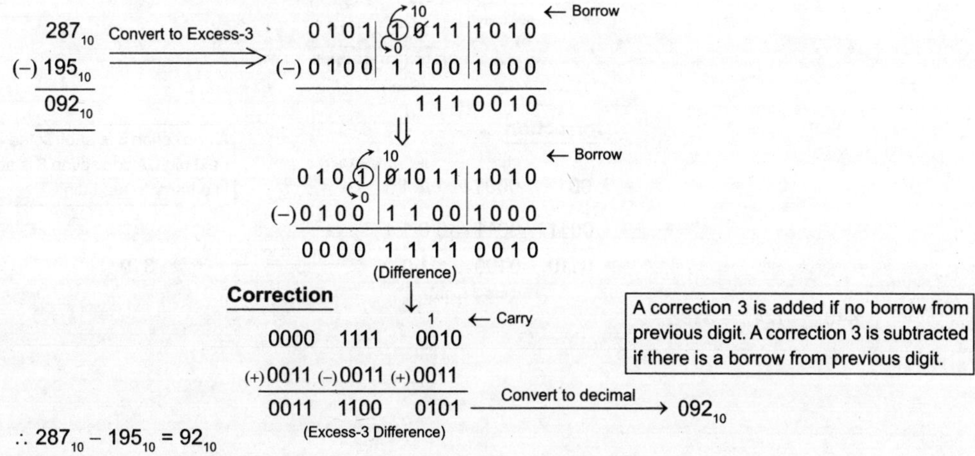

$\therefore 287_{10} - 195_{10} = 92_{10}$

b) $75.4_{10} - 25.8_{10}$

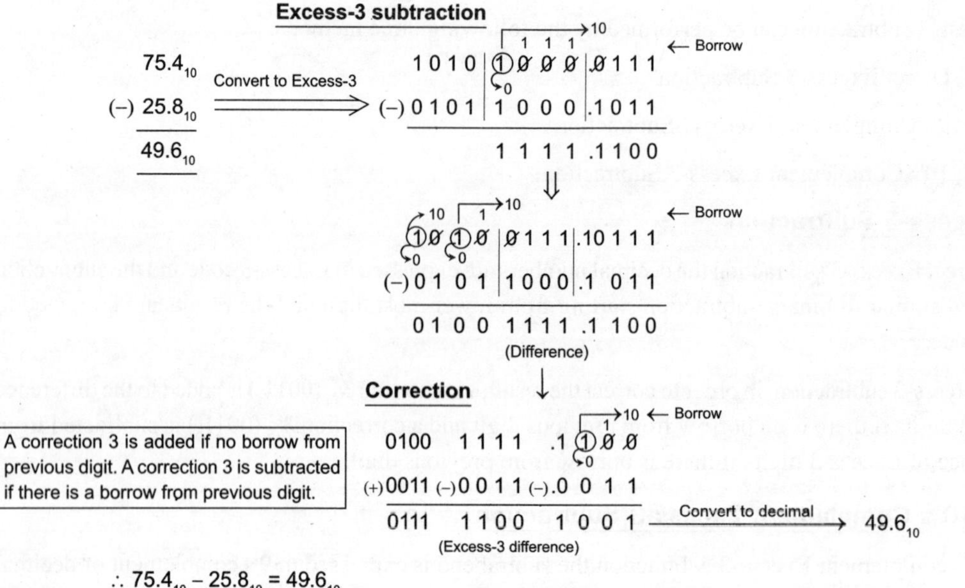

$\therefore 75.4_{10} - 25.8_{10} = 49.6_{10}$

Example 3.12

Perform the following subtraction by 9's complement Excess-3 subtraction and cross-check result by direct 9's complement subtraction.

a) $443_{10} - 124_{10}$ **b)** $123_{10} - 372_{10}$

Solution

a) $443_{10} - 124_{10}$

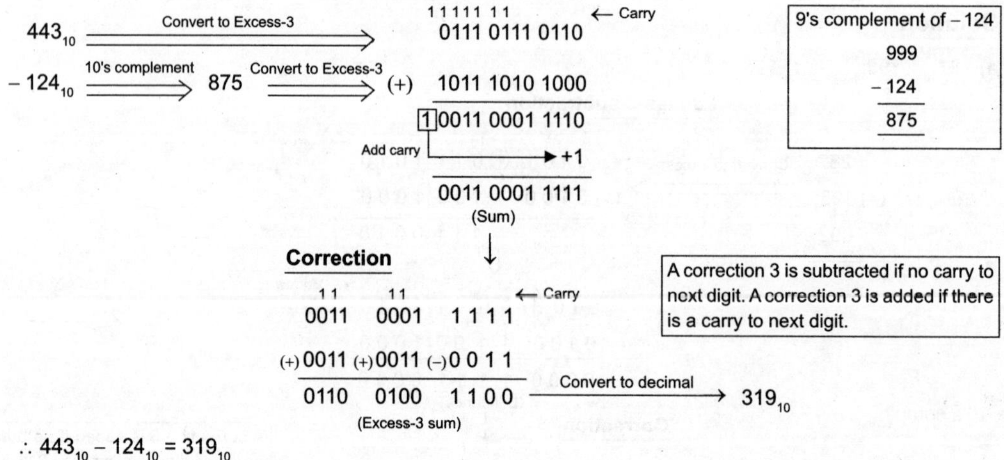

$\therefore 443_{10} - 124_{10} = 319_{10}$

9's complement subtraction

Minuend: 443_{10} ⇒ 443 (carry 1 1)

Subtrahend: -124_{10} ⇒ (9's complement) (+) 875

[1] 318

Add carry +1

319

$\therefore 443_{10} - 124_{10} = 319_{10}$

---

b) $123_{10} - 372_{10}$

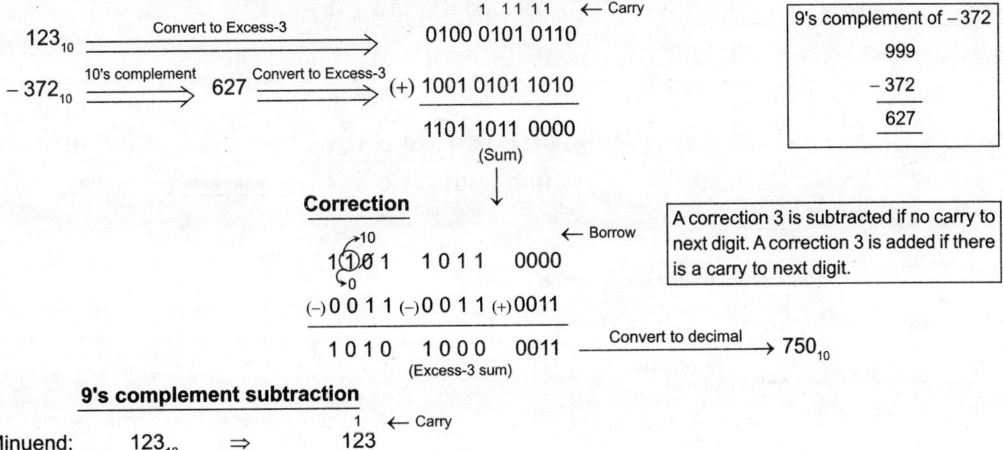

9's complement subtraction

Minuend: 123_{10} ⇒ 123 (carry 1)

Subtrahend: -372_{10} ⇒ (9's complement) (+) 627

750

No carry

Conclusion

Since no final carry the result is negative. Hence take 9's complement of 750_{10} and put "–" sign.

750 —10's complement→ 999

− 750

249

$\therefore 123_{10} - 372_{10} = -249_{10}$

Example 3.13

Perform the following subtraction by 10's complement Excess-3 subtraction and cross-check result by direct 10's complement subtraction.

a) $487_{10} - 259_{10}$

b) $132_{10} - 352_{10}$

Solution

a) $487_{10} - 259_{10}$

10's complement Excess-3 subtraction

			1 1111 111 ← Carry	
487_{10}	Convert to Excess-3 ⟹		0111 1011 1010	
-259_{10}	10's complement ⟹	741 Convert to Excesss-3 ⟹	(+) 1010 0111 0100	
			[1] 0010 0010 1110 (Sum)	

↑ Discard carry

10's complement of −259

999
−259
740
+1
741

Correction

A correction 3 is subtracted if no carry to next digit. A correction 3 is added if there is a carry to next digit.

		10 ← Borrow	
0010	0010	1 1 ① ∅	
(+)0011	(+)0011	(−)0 0 1 1	
		1	

⇓

1	1	10 ← Carry/Borrow	
0010	0010	1 ① ∅ 10	
(+) 0011	(+) 0011	(−) 0 0 1 1	
0101	0101	1 0 0 1	Convert to decimal ⟶ 228_{10}

(Excess-3 sum)

$\therefore 487_{10} - 259_{10} = 228_{10}$

10's complement subtraction

Minuend:	487_{10}	⇒	487
Subtrahend:	-259_{10}	10's complement ⇒	(+) 741
			[1] 228

↑ Discard carry

$\therefore 487_{10} - 259_{10} = 228_{10}$

---

b) $132_{10} - 325_{10}$

10's complement Excess-3 subtraction

			1111111 ← Carry
132_{10}	Convert to Excess-3 ⟹		0100 0110 0101
$+352_{10}$	10's complement ⟹	648 Convert to Excess-3 ⟹	(+) 1001 0111 1011
			1101 1110 0000 (Sum)

10's complement of −352

999
−352
647
+1
648

Correction

	10 ← Borrow		
1 1 0 1	1 1 ① ∅	0000	
(−)0 0 1 1	(−)0 0 1 1	(+)0011	
	1	0011	

A correction 3 is subtracted if no carry to next digit. A correction 3 is added if there is a carry to next digit.

⇓

10	10	← Borrow	
1 ① ∅ 1	1 ① ∅ 10	0000	
(−)0 0 1 1	(−)0 0 1 1	(+)0011	
1 0 1 0	1 0 1 1	0011	Convert to decimal ⟶ 780_{10}

(Excess-3 sum)

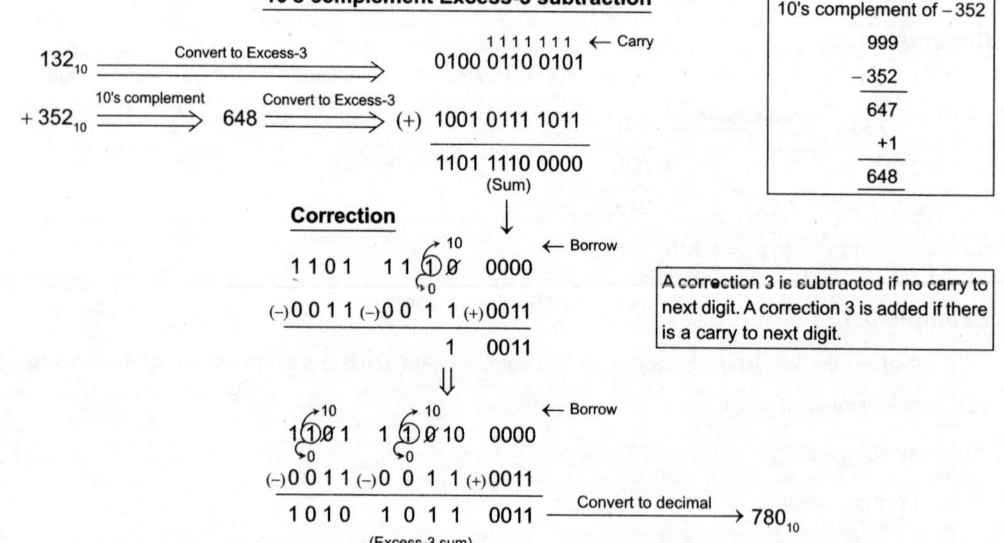

10's complement subtraction

```
                                     1 ← Carry
Minuend:      132₁₀        ⇒          132
                   10's complement
Subtrahend:  – 352₁₀        ⇒      (+) 648
                                     -----
                                      780
                                     -----
                                      ↑
                                   No carry
```

Conclusion

Since no final carry the result is negative. Hence take 10's complement of 780_{10} and put "–" sign.

$$780 \xrightarrow{\text{10's complement}} \begin{array}{r} 999 \\ -\ 780 \\ \hline 219 \\ +1 \\ \hline 220 \\ \hline \end{array}$$

$\therefore\ 132_{10} - 325_{10} = -\ 220_{10}$

3.3 Gray Code

(CSVTU, Nov/Dec'19, 4 Marks)

In ***Gray code*** only one bit changes from either "0" to "1" or "1" to "0" between two consecutive codes.

The 1-bit Gray code can be used to represent decimal numbers 0 and 1.

The 2-bit Gray code can be used to represent decimal numbers 0 to 3.

The 3-bit Gray code can be used to represent decimal numbers 0 to 7.

The 4-bit Gray code can be used to represent decimal numbers 0 to 15_{10}.

Table 3.6: Decimal and Equivalent Gray Code

Decimal Number	1-bit Gray Code	2-bit Gray Code	3-bit Gray Code	4-bit Gray Code
0	0	00	000	0000
1	1	01	001	0001
2		11	011	0011
3		10	010	0010
4			110	0110
5			111	0111
6			101	0101
7			100	0100
8				1100
9				1101
10				1111
11				1110
12				1010
13				1011
14				1001
15				1000

The Gray code is constructed as follows:

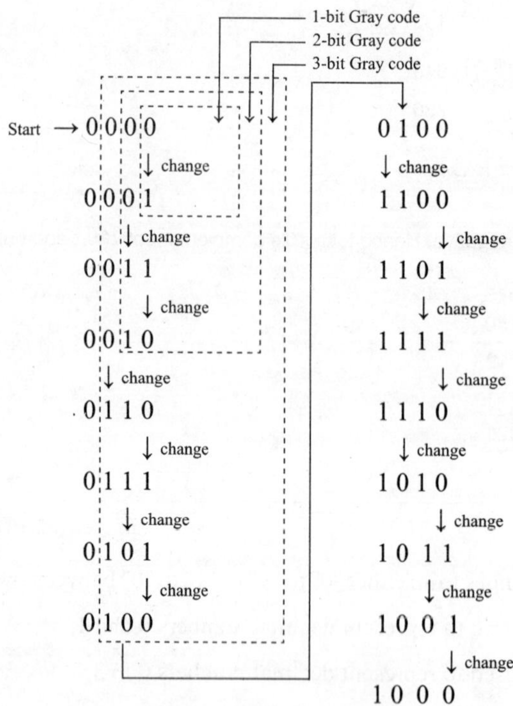

Gray code can be formed by another method as shown below:

Step-1: For 2^n Gray code first form 2^{n-1} Gray codes.

Step-2: Then write the Gray codes of step-1 in reverse order as next set of 2^{n-1} Gray codes.

Step-3: Append the Gray codes of step-1 with leading "0".

Step-4: Append the Gray codes of step-2 with leading "1".

We can start with 1-bit Gray code which is 0 and 1 then proceed to higher size as shown below:

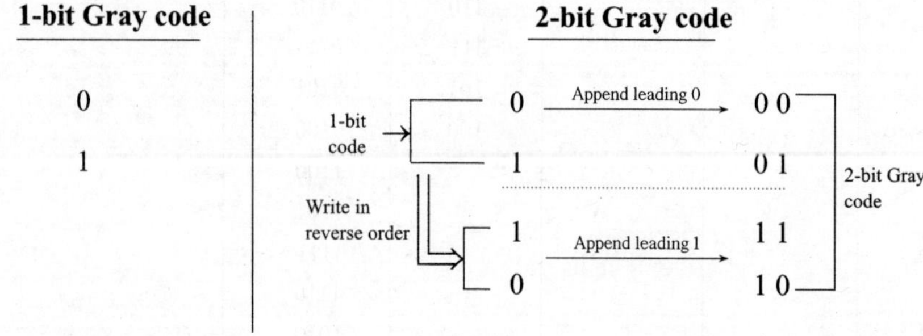

3-bit Gray code

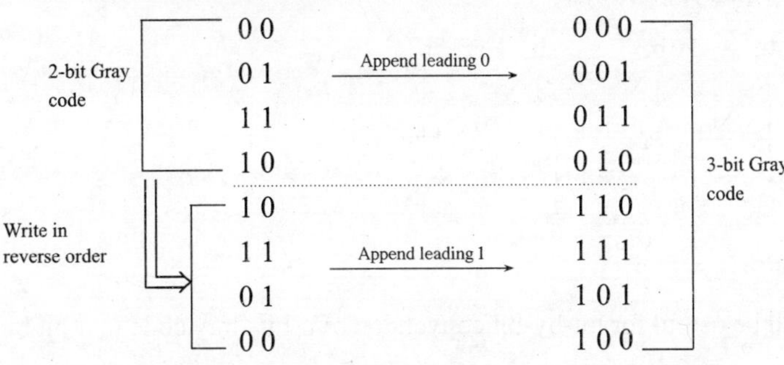

4-bit Gray code

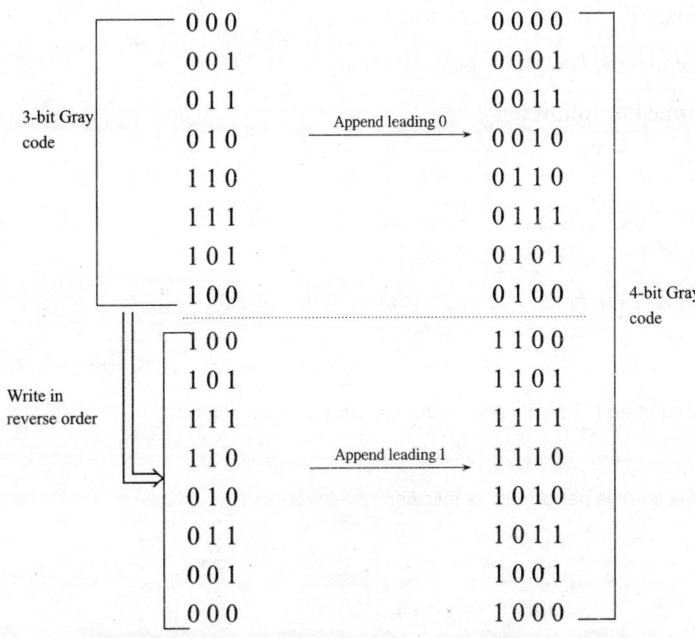

Similarly we can construct 5-bit, 6-bit, etc, Gray codes.

The two methods discussed above will be useful to find the full range of n-bit Gray code. But will not be useful to convert a given single binary code to Gray code. The following procedure can be followed to convert a given single binary code to Gray code and vice-versa.

Binary to Gray Code Conversion

The following procedure will be useful for bit-by-bit conversion of n-bit binary to n-bit Gray code.

Step-1: The most significant bit of binary and Gray code are same. Hence retain the most significant bit of binary as most significant bit of Gray code.

Step-2: Starting from most significant bit of binary, add two consecutive bits of binary and record the sum as bit of Gray code and discard the carry in every addition.

Let, $b_3\ b_2\ b_1\ b_0$ be 4-bit binary code and $g_3\ g_2\ g_1\ g_0$ be 4-bit Gray code

Now, the conversion is performed as follows:

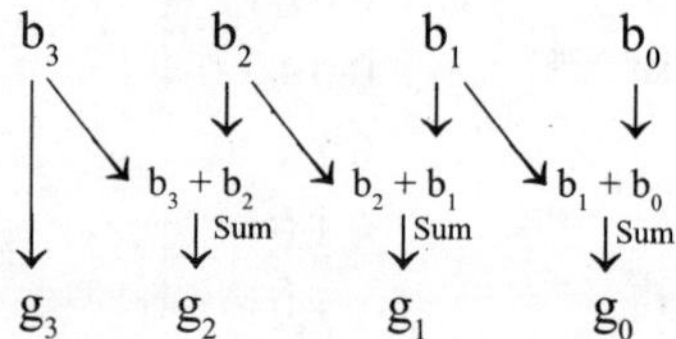

Gray Code to Binary Conversion

The following procedure will be useful for bit-by-bit conversion of n-bit Gray code to n-bit binary.

Step-1: The most significant bit of Gray code and binary are same. Hence retain the most significant bit of Gray code as most significant bit of binary.

Step-2: Starting from most significant bit of binary, add the generated binary bit to next bit of Gray code and record the sum as binary bit and discard the carry.

Let, $g_3\ g_2\ g_1\ g_0$ be 4-bit Gray code and $b_3\ b_2\ b_1\ b_0$ be 4-bit binary code

Now, the conversion is performed as follows:

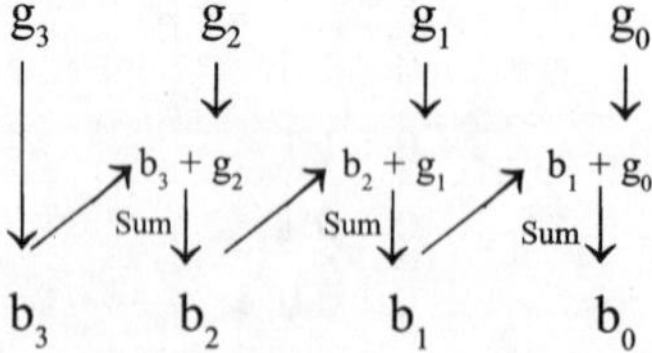

Example 3.14

(GTU, May'19, 3 Marks)

Convert the binary 1010 to Gray code and verify by converting the Gray code to binary.

Solution

Conversion of binary code to Gray code is performed as follows:

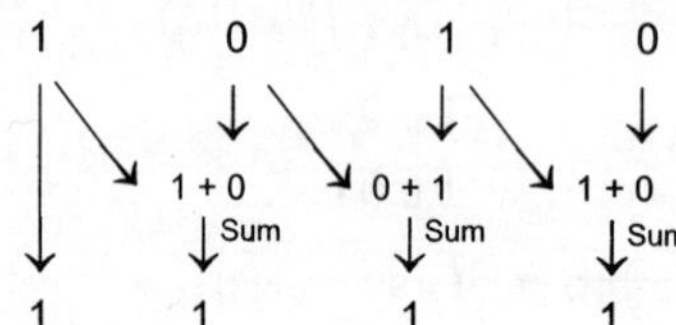

∴ Gray code for binary 1010 = 1111

Conversion of Gray code to binary is performed as follows:

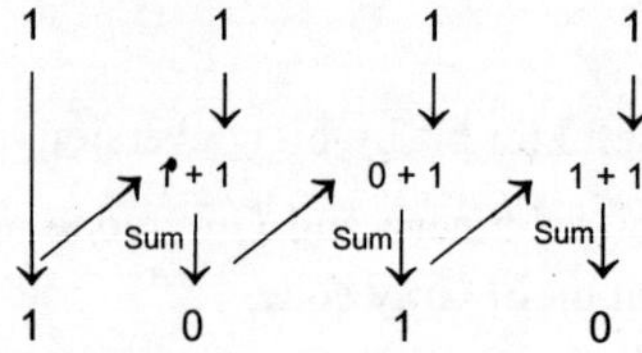

∴ Binary for Gray code 1111 = 1010

Application for Gray Code

In rotating machines the angular position of shaft and hence speed of rotating shaft are popularly measured using encoding disk shown in Fig. 3.1.

The disk is mounted at one end of shaft of rotating machine.

In this encoding disk 16 angular positions of rotating shaft are encoded in 4-bit Gray code.

Since Gray code changes only in one bit position when the rotating shaft move from one angular position to other, the errors if any in encoded binary can be easily detected.

The encoding disk is divided into 16-angular sectors and 4-circular regions to mark black and white pattern. The black and white pattern are created such that 4-bit Gray code is generated when sensed using four light sensors. Here white pattern generate "0" and black pattern generate "1", it can be vice-versa also.

For more precision the angular sectors can be divided in to 32 angular sectors so that 5-bit Gray code is generated and 64 angular sectors so that 6-bit Gray code generated and so on.

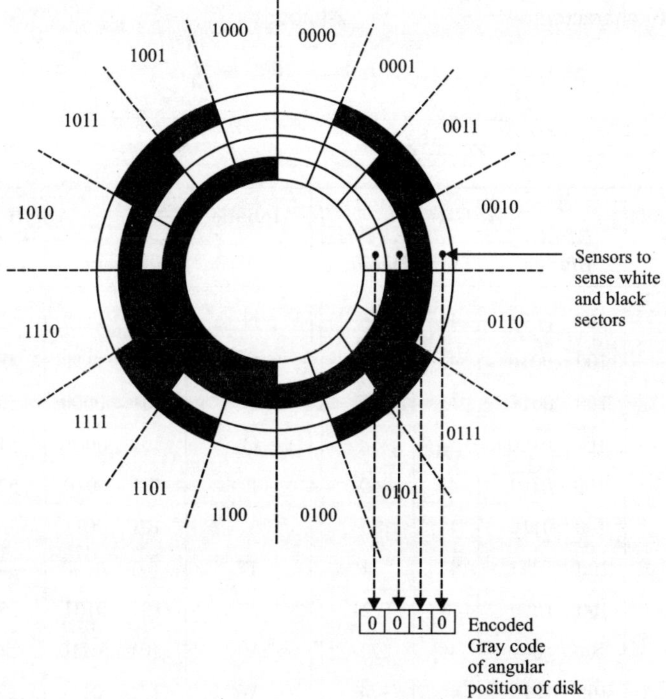

Fig. 3.1: Gray code encoding disk.

3.4 Alphanumeric Code

3.4.1 ASCII Code

ASCII: **A**merican **S**tandard **C**ode for **I**nformation **I**nterchange

ASCII was developed in the early stage of digital systems in order to standardize the coding system for human and machine interface.

ASCII is 7-bit binary code that includes most of the symbols in early type-writing machine.

ASCII includes codes for upper and lower case english alphabets and numerals (0 to 9).

With 7-bit binary code, 128 ($2^7 = 128$) ASCII codes are possible. They are assigned as follows:

Upper case alphabets,	A to Z	=	26 codes
Lower case alphabets,	a to z	=	26 codes
Numerals,	0 to 9	=	10 codes
Special printable characters		=	32 codes
Control characters		=	34 codes
			128 codes

Table 3.7: ASCII for Upper Case Alphabets

Alphabets	ASCII			Alphabets	ASCII		
	Binary	Hexa	Decimal		Binary	Hexa	Decimal
A	100 0001	41	65	N	100 1110	4E	78
B	100 0010	42	66	O	100 1111	4F	79
C	100 0011	43	67	P	101 0000	50	80
D	100 0100	44	68	Q	101 0001	51	81
E	100 0101	45	69	R	101 0010	52	82
F	100 0110	46	70	S	101 0011	53	83
G	100 0111	47	71	T	101 0100	54	84
H	100 1000	48	72	U	101 0101	55	85
I	100 1001	49	73	V	101 0110	56	86
J	100 1010	4A	74	W	101 0111	57	87
K	100 1011	4B	75	X	101 1000	58	88
L	100 1100	4C	76	Y	101 1001	59	89
M	100 1101	4D	77	Z	101 1010	5A	90

Table 3.8: ASCII for Lower Case Alphabets

Alphabet	ASCII			Alphabet	ASCII		
	Binary	Hexa	Decimal		Binary	Hexa	Decimal
a	110 0001	61	97	n	110 1110	6E	110
b	110 0010	62	98	o	110 1111	6F	111
c	110 0011	63	99	p	111 0000	70	112
d	110 0100	64	100	q	111 0001	71	113
e	110 0101	65	101	r	111 0010	72	114
f	110 0110	66	102	s	111 0011	73	115
g	110 0111	67	103	t	111 0100	74	116
h	110 1000	68	104	u	111 0101	75	117
i	110 1001	69	105	v	111 0110	76	118
j	110 1010	6A	106	w	111 0111	77	119
k	110 1011	6B	107	x	111 1000	78	120
l	110 1100	6C	108	y	111 1001	79	121
m	110 1101	6D	109	z	111 1010	7A	122

Table 3.9: ASCII for Special Printable Characters

Special Characters	ASCII			Special Characters	ASCII		
	Binary	Hexa	Decimal		Binary	Hexa	Decimal
!	010 0001	21	33	;	011 1011	3B	59
“	010 0010	22	34	<	011 1100	3C	60
#	010 0011	23	35	=	011 1101	3D	61
$	010 0100	24	36	>	011 1110	3E	62
%	010 0101	25	37	?	011 1111	3F	63
&	010 0110	26	38	@	100 0000	40	64
‘	010 0111	27	39	[	101 1011	5B	91
(	010 1000	28	40	\	101 1100	5C	92
)	010 1001	29	41	]	101 1101	5D	93
*	010 1010	2A	42	^	101 1110	5E	94
+	010 1011	2B	43	_	101 1111	5F	95
,	010 1100	2C	44	`	110 0000	60	96
–	010 1101	2D	45	{	111 1011	7B	123
.	010 1110	2E	46	|	111 1100	7C	124
/	010 1111	2F	47	}	111 1101	7D	125
:	011 1010	3A	58	~	111 1110	7E	126

Table 3.10: ASCII for Control Characters

Control Characters	ASCII			Control Characters	ASCII		
	Binary	Hexa	Decimal		Binary	Hexa	Decimal
NUL	000 0000	00	0	DC1	001 0001	11	17
SOH	000 0001	01	1	DC2	001 0010	12	18
STX	000 0010	02	2	DC3	001 0011	13	19
ETX	000 0011	03	3	DC4	001 0100	14	20
EOT	000 0100	04	4	NAK	001 0101	15	21
ENQ	000 0101	05	5	SYN	001 0110	16	22
ACK	000 0110	06	6	ETB	001 0111	17	23
BEL	000 0111	07	7	CAN	001 1000	18	24
BS	000 1000	08	8	EM	001 1001	19	25
HT	000 1001	09	9	SUB	001 1010	1A	26
LF	000 1010	0A	10	ESC	001 1011	1B	27
VT	000 1011	0B	11	FS	001 1100	1C	28
FF	000 1100	0C	12	GS	001 1101	1D	29
CR	000 1101	0D	13	RS	001 1110	1E	30
SO	000 1110	0E	14	US	001 1111	1F	31
SI	000 1111	0F	15	SP	010 0000	20	32
DLE	001 0000	10	16	DEL	111 1111	7F	127

Table 3.11: ASCII for Numerals

Numeral	ASCII			Numeral	ASCII		
	Binary	Hexa	Decimal		Binary	Hexa	Decimal
0	011 0000	30	48	5	011 0101	35	53
1	011 0001	31	49	6	011 0110	36	54
2	011 0010	32	50	7	011 0111	37	55
3	011 0110	33	51	8	011 1000	38	56
4	011 0100	34	52	9	011 1001	39	57

Table 3.12: Name/Meaning of ASCII Control Characters

Symbol	Name/Meaning	Symbol	Name/Meaning
NUL	Null	DC2	Device Control 2
SOH	Start of Heading	DC3	Device Control 3
STX	Start of Text	DC4	Device Control 4
ETX	End of Text	NAK	Negative Acknowledge
EOT	End of Transmission	SYN	Synchronous Idle
ENQ	Enquiry	ETB	End-of-Transmission Block
ACK	Acknowledge	CAN	Cancel
BEL	Bell	EM	End of Medium
BS	Backspace	SUB	Substitute
HT	Horizondal Tab	ESC	Escape
LF	Line Feed	FS	File Separator
VT	Vertical Tab	GS	Group Separator
FF	Form Feed	RS	Record Separator
CR	Carriage Return	US	Unit Separator
SO	Shift Out	SP	Space
SI	Shift In	DEL	Delete
DLE	Data-Link Escape		
DC1	Device Control 1		

3.4.2 EBCDIC Code

EBCDIC Code (Extended BCD Interchange Code): IBM computers introduce the EBCDIC (pronounced "ebb see dick") code to represent characters, letters, numerals, punctuation marks and other symbols in computerized text.

A character is represented in EBCDIC by eight bit or one byte. Each byte consists of two nibbles, each four bits wide. The first four bits define the class of character, while the second nibble defines the specific character inside that class. The EBCDIC code allows for 256 different characters in which only 139 codes are predefined and other codes can be assigned for specialized functions.

In EBCDIC, the lower nibble for all letters and decimal digits contains only hex values 0 to 9 which are same as BCD and hence the name extended BCD. (Refer Tables 3.14 and 3.17)

Table 3.13: EBCDIC for Upper Case Alphabets

Alphabets	EBCDIC		Alphabets	EBCDIC	
	Binary	Hexa		Binary	Hexa
A	1100 0001	C1	N	1101 0101	D5
B	1100 0010	C2	O	1101 0110	D6
C	1100 0011	C3	P	1101 0111	D7
D	1100 0100	C4	Q	1101 1000	D8
E	1100 0101	C5	R	1101 1001	D9
F	1100 0110	C6	S	1110 0010	E2
G	1100 0111	C7	T	1110 0011	E3
H	1100 1000	C8	U	1110 0100	E4
I	1100 1001	C9	V	1110 0101	E5
J	1101 0001	D1	W	1110 0110	E6
K	1101 0010	D2	X	1110 0111	E7
L	1101 0011	D3	Y	1110 1000	E8
M	1101 0100	D4	Z	1110 1001	E9

Table 3.14: EBCDIC for Lower Case Alphabets

Alphabets	EBCDIC		Alphabets	EBCDIC	
	Binary	Hexa		Binary	Hexa
a	1000 0001	81	n	1001 0101	95
b	1000 0010	82	o	1001 0110	96
c	1000 0011	83	p	1001 0111	97
d	1000 0100	84	q	1001 1000	98
e	1000 0101	85	r	1001 1001	99
f	1000 0110	86	s	1010 0010	A2
g	1000 0111	87	t	1010 0011	A3
h	1000 1000	88	u	1010 0100	A4
i	1000 1001	89	v	1010 0101	A5
j	1001 0001	91	w	1010 0110	A6
k	1001 0010	92	x	1010 0111	A7
l	1001 0011	93	y	1010 1000	A8
m	1001 0100	94	z	1010 1001	A9

Table 3.15: EBCDIC for Special Printable Characters

Special Characters	EBCDIC		Special Characters	EBCDIC	
	Binary	Hexa		Binary	Hexa
¢	0100 1010	4A	–	0110 0010	62
-	0100 1011	4B	%	0110 1100	6C
<	0100 1100	4C	>	0110 1110	6E
(	0100 1101	4D	?	0110 1111	6F
+	0100 1110	4E	▲	0111 1001	79
|	0100 1111	4F	:	0111 1010	7A
&	0101 0000	50	#	0111 1011	7B
!	0101 1010	5A	@	0111 1100	7C
$	0101 1011	5B	▲	0111 1101	7D
*	0101 1100	5C	–	0111 1110	7E
)	0101 1101	5D	„	0111 1111	7F
:	0101 1110	5E	~	1010 0001	A1
¬	0101 1111	5F	{	1100 0000	C0
–	0110 0000	60	}	1101 0000	D0
/	0110 0001	61	\	1110 0000	E0

Table 3.16: EBCDIC for Control Characters

Control Characters	EBCDIC		Control Characters	EBCDIC	
	Binary	Hexa		Binary	Hexa
NUL	0000 0000	00	SF	0001 1101	1D
SOH	0000 0001	01	FM	0001 1110	1E
STX	0000 0010	02	ITB	0001 1111	1F
ETX	0000 0011	03	ETB	0010 0110	26
PT	0000 0101	05	ESC	0010 0111	27
FF	0000 1100	0C	ENQ	0010 1101	2D
DLE	0001 0000	10	SYN	0011 0010	32
SBA	0001 0001	11	BOT	0011 0111	37
EUA	0001 0010	12	RA	0011 1100	3C
IC	0001 0011	13	NAK	0011 1101	3D
NL	0001 0101	15	SUB	0011 1111	3F
EM	0001 1001	19	SP	0100 0000	40
DUP	0001 1100	1C			

Table 3.17: EBCDIC for Numerals

Numeral	EBCDIC		Numeral	EBCDIC	
	Binary	Hexa		Binary	Hexa
0	1111 0000	F0	5	1111 0101	F5
1	1111 0001	F1	6	1111 0110	F6
2	1111 0010	F2	7	1111 0111	F7
3	1111 0011	F3	8	1111 1000	F8
4	1111 0100	F4	9	1111 1001	F9

Table 3.18: Name/Meaning of EBCDIC Control Characters

Symbol	Name/Meaning	Symbol	Name/Meaning
DLE	Data-link escape	ITB	End of intermediate transmission block
DUP	Duplicate	NUL	Null
EM	End of medium	PT	Program tab
ENQ	Enquiry	RA	Repeat to address
EOT	End of transmission	SBA	Set buffer address
ESC	Escape	SF	Start field
ETB	End of transmission block	SOH	Start of heading
ETX	End of text	SP	Space
EUA	Erase unprotected to address	STX	Start of text
FF	Form feed	SUB	Subtitute
FM	Field mark	SYN	Synchronous
IC	Insert cursor	NAK	Negative acknowledge

3.5 Error Detection and Correction

(CSVTU, Nov/Dec'19, 7 Marks) (GTU, May'19, 3 Marks)

In digital systems binary information is transmitted from one system to another system or one device to another device.While transmitting the binary information or codes, there may be errors. Different techniques have been developed to detect and correct errors in transmission of digital information or codes.

Parity and checksum are popular method of error detection and error correction is performed by sending a request to retransmit the data.

Hamming code is another popular error deduction and correction technique to deduct and correct 1-bit error in the received data.

3.5.1 Parity

(AKTU, Sep'20, 5 Marks)

Parity refer to number of 1's in an n-bit digital code. When the number of 1's in an n-bit code is even, then the code is said to have even parity. When the number of 1's in an n bit code is odd, then the code is said to have odd parity.

Parity of an n-bit code can be modified by appending an extra bit called ***parity bit***. The parity bit is appended at most significant bit position.

(CSVTU, Nov/Dec'19, 2 Marks)

In a digital transmission system, the information to be transmitted will be coded as n-bit binary. During transmisssion it is usual practice to transmit every n-bit code with same parity. When the transmission system adopt even parity then every n-bit code is converted to even parity by inserting a parity bit.

Similarly, when odd parity is adopted every n-bit code is converted to odd parity.

Converting the coded binary to even parity or odd parity is a popular and simple technique to detect errors in transmission of digital codes.

Example:

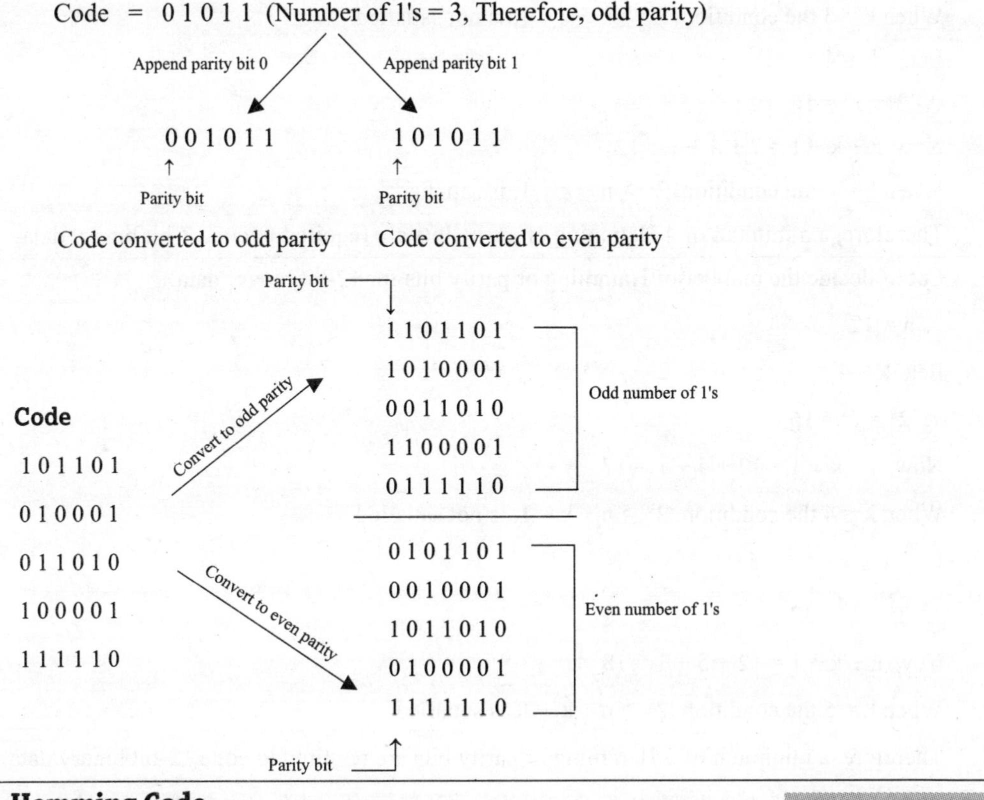

3.5.2 Hamming Code

(AU, Apr/May'19, 6 Marks)

Hamming code is developed by **Richard W Hamming** of Bell Telephone Laboratories. The Hamming code is error detecting and correcting code. It can accuratly detect 1-bit error.

The binary information to be coded is called data. In Hamming coding technique additional bits called Hamming or parity bits are inserted at various positions in between data bits. The combined data and Hamming bits is called Hamming code. In a particular transmission system, the position of data and Hamming bits are fixed and known to transmitter and receiver.

Number of Hamming or Parity Bits

In order to code n-bit binary data k number of Hamming or parity bits are needed. Here, n and k should satisfy the condition,

$$2^k \geq n + k + 1 \quad (\text{or } 2^k - k - 1 \geq n)$$

There will be multiple values of k that satisfy the above condition, but minimum value of k is chosen to restrict the size of coded data.

Let us decide the number of Hamming or parity bits for 7-bit binary data.

$\therefore n = 7$

Let, $k = 3$

$\therefore\ 2^k = 2^3 = 8$

Now, $n + k + 1 = 7 + 3 + 1 = 11$

When $k = 3$ the condition, $2^k \geq n + k + 1$, is not satisfied.

Let, $k = 4$

$\therefore\ 2^k = 2^4 = 16$

Now, $n + k + 1 = 7 + 4 + 1 = 12$

When $k = 4$ the condition, $2^k \geq n + k + 1$, is satisfied.

Therefore, a minimum of 4 Hamming or parity bits are required to code 7-bit binary data.

Let us decide the number of Hamming or parity bits for 12-bit binary data.

$\therefore\ n = 12$

Let, $k = 4$

$\therefore\ 2^k = 2^4 = 16$

Now, $n + k + 1 = 12 + 4 + 1 = 17$

When $k = 4$ the condition, $2^k \geq n + k + 1$, is not satisfied.

Let, $k = 5$

$\therefore\ 2^k = 2^5 = 32$

Now, $n + k + 1 = 12 + 5 + 1 = 18$

When $k = 5$ the condition, $2^k \geq n + k + 1$, is satisfied.

Therefore, a minimum of 5 Hamming or parity bits are required to code 12-bit binary data.

The range of data bits that satisfy the condition $2^k \geq n + k + 1$ for $k = 3, 4, 5, 6$ and 7 are listed in Table 3.19.

Table 3.19: Range of Data Bits for k Parity Bits

Number of Parity Bits k	Range of Data Bits n
3	2 to 4
4	5 to 11
5	12 to 26
6	27 to 57
7	58 to 120

Combining the Data and Hamming or Parity Bits

In order to code n-bit binary data, (d_n, d_{n-1}, ... d_3, d_2, d_1) determine k number of Hamming or parity bits (P_1, P_2, ... P_{k-1}, P_k) that satisfy the condition $2^k \geq n + k + 1$.

Let the bits of coded data is denoted as b_1, b_2, ... b_{n+k-1}, b_{n+k}. In the coded data the binary bit positions that corresponds to binary weights are reserved for parity bits. Therefore, b_1, b_2, b_4, b_8, b_{16}, ... are reserved for parity bits.

Consider a 5-bit data to be coded by 4 Hamming bits,

$\therefore\ n = 5,\ k = 4,\ 2^k = 2^4 = 16$

$\therefore\ n + k + 1 = 5 + 4 + 1 = 10$

Here the condition, $2^k \geq n + k + 1$, is satisfied.

Let, Binary data $= d_5\ d_4\ d_3\ d_2\ d_1$

Parity bits $= P_1\ P_2\ P_3\ P_4$

The size of coded data is $5 + 4 = 9$ bits

Let, coded data $= b_1\ b_2\ b_3\ b_4\ b_5\ b_6\ b_7\ b_8\ b_9$

Here, the binary bit positions that corresponds to binary weight are reserved for parity bits. Therefore, parity bits are inserted in b_1, b_2, b_4 and b_8 and data bits are inserted in other bit positions as shown below:

Coded data : $b_1\ b_2\ b_3\ b_4\ b_5\ b_6\ b_7\ b_8\ b_9$

$P_1\ P_2\ d_5\ P_3\ d_4\ d_3\ d_2\ P_4\ d_1$

The format of 6, 7, 8 and 9 bits coded data are shown below:

6-bit Data

$n = 6, k = 4, n + k = 10$

The size of coded data is 10-bits.

Data bits : $d_6\ d_5\ d_4\ d_3\ d_2\ d_1$

Parity bits : $P_1\ P_2\ P_3\ P_4$

Coded data bits : $b_1\ b_2\ b_3\ b_4\ b_5\ b_6\ b_7\ b_8\ b_9\ b_{10}$

Hamming code : $P_1\ P_2\ d_6\ P_3\ d_5\ d_4\ d_3\ P_4\ d_2\ d_1$

7-bit Data

$n = 7, k = 4, n + k = 11$

The size of coded data is 11-bits.

Data bits : $d_7\ d_6\ d_5\ d_4\ d_3\ d_2\ d_1$

Parity bits : $P_1\ P_2\ P_3\ P_4$

Coded data bits : $b_1\ b_2\ b_3\ b_4\ b_5\ b_6 b_7\ b_8\ b_9\ b_{10}\ b_{11}$

Hamming code : $P_1\ P_2\ d_7\ P_3\ d_6\ d_5\ d_4\ P_4\ d_3\ d_2\ d_1$

8-bit Data

$n = 8, k = 4, n + k = 12$

The size of coded data is 12-bits.

Data bits : $d_8\ d_7\ d_6\ d_5\ d_4\ d_3\ d_2\ d_1$

Parity bits : $P_1\ P_2\ P_3\ P_4$

Coded data bits : $b_1\ b_2\ b_3\ b_4\ b_5\ b_6\ b_7\ b_8\ b_9\ b_{10}\ b_{11}\ b_{12}$

Hamming code : $P_1\ P_2\ d_8\ P_3\ d_7\ d_6\ d_5\ P_4\ d_4\ d_3\ d_2\ d_1$

9-bit Data

$n = 9, k = 4, n + k = 13$

The size of coded data is 13-bits.

Data bits : $d_9\ d_8\ d_7\ d_6\ d_5\ d_4\ d_3\ d_2\ d_1$

Parity bits : $P_1\ P_2\ P_3\ P_4$

Coded data bits : $b_1\ b_2\ b_3\ b_4\ b_5\ b_6\ b_7\ b_8\ b_9\ b_{10}\ b_{11}\ b_{12}\ b_{13}$

Hamming code : $P_1\ P_2\ d_9\ P_3\ d_8\ d_7\ d_6\ P_4\ d_5\ d_4\ d_3\ d_2\ d_1$

Generation of Even Parity Bits

Consider the binary values of decimal numbers 1 to 13 shown in Table 3.20. The position of 1's in the column corresponds to binary weights w_1, w_2, w_3 and w_4 will determine the parity bits P_1, P_2, P_3 and P_4 respectively. The generation of parity bit for 9-bit code is shown below:

Table 3.20: Binary Value of Decimal Number

Decimal Number	4-bit Binary Values to be Considered for				
	9-bit Code	10-bit Code	11-bit Code	12-bit Code	13-bit Code
	$w_8w_4w_2w_1$	$w_8w_4w_2w_1$	$w_8w_4w_2w_1$	$w_8w_4w_2w_1$	$w_8w_4w_2w_1$
1	0 0 0 1	0 0 0 1	0 0 0 1	0 0 0 1	0 0 0 1
2	0 0 1 0	0 0 1 0	0 0 1 0	0 0 1 0	0 0 1 0
3	0 0 1 1	0 0 1 1	0 0 1 1	0 0 1 1	0 0 1 1
4	0 1 0 0	0 1 0 0	0 1 0 0	0 1 0 0	0 1 0 0
5	0 1 0 1	0 1 0 1	0 1 0 1	0 1 0 1	0 1 0 1
6	0 1 1 0	0 1 1 0	0 1 1 0	0 1 1 0	0 1 1 0
7	0 1 1 1	0 1 1 1	0 1 1 1	0 1 1 1	0 1 1 1
8	1 0 0 0	1 0 0 0	1 0 0 0	1 0 0 0	1 0 0 0
9	1 0 0 1	1 0 0 1	1 0 0 1	1 0 0 1	1 0 0 1
10		1 0 1 0	1 0 1 0	1 0 1 0	1 0 1 0
11			1 0 1 1	1 0 1 1	1 0 1 1
12				1 1 0 0	1 1 0 0
13					1 1 0 1

9-bit Code ($b_1\ b_2\ b_3\ b_4\ b_5\ b_6\ b_7\ b_8\ b_9$)

In Table 3.20, the following observations are made.

Excluding the number 1, the decimal numbers 3, 5, 7, 9 has 1 in binary weight position w_1. Hence, the parity bit P_1 along with b_3, b_5, b_7 and b_9 should be even parity.

$$\therefore P_1 = b_3 \oplus b_5 \oplus b_7 \oplus b_9$$

Excluding the numbers 1, 2, the decimal numbers 3, 6, 7 has 1 in binary weight position w_2. Hence, the parity bit P_2 along with b_3, b_6 and b_7 should be even parity.

$$\therefore P_2 = b_3 \oplus b_6 \oplus b_7$$

Excluding the numbers 1, 2, 3, 4, the decimal numbers 5, 6, 7 has 1 in binary weight position w_4. Hence, the parity bit P_3 along with b_5, b_6 and b_7 should be even parity.

$$\therefore P_3 = b_5 \oplus b_6 \oplus b_7$$

Excluding the numbers 1, 2, 3, 4, 5, 6, 7, 8, the decimal number 9 has 1 in binary weight position w_8. Hence, the parity bit P_4 along with b_9 should be even parity.

$$\therefore P_4 = b_9$$

By following the procedure for 9-bit code the parity bits of 9, 10, 11, 12 and 13 bits Hamming codes can be obtained using the following equations.

10-bit Code ($b_1\ b_2\ b_3\ b_4\ b_5\ b_6\ b_7\ b_8\ b_9\ b_{10}$)

$$P_1 = b_3 \oplus b_5 \oplus b_7 \oplus b_9$$
$$P_2 = b_3 \oplus b_6 \oplus b_7 \oplus b_{10}$$
$$P_3 = b_5 \oplus b_6 \oplus b_7$$
$$P_4 = b_9 \oplus b_{10}$$

11-bit Code ($b_1\ b_2\ b_3\ b_4\ b_5\ b_6\ b_7\ b_8\ b_9\ b_{10}\ b_{11}$)

$$P_1 = b_3 \oplus b_5 \oplus b_7 \oplus b_9 \oplus b_{11}$$
$$P_2 = b_3 \oplus b_6 \oplus b_7 \oplus b_{10} \oplus b_{11}$$
$$P_3 = b_5 \oplus b_6 \oplus b_7$$
$$P_4 = b_9 \oplus b_{10} \oplus b_{11}$$

12-bit Code ($b_1\ b_2\ b_3\ b_4\ b_5\ b_6\ b_7\ b_8\ b_9\ b_{10}\ b_{11}\ b_{12}$)

$$P_1 - b_3 \oplus b_5 \oplus b_7 \oplus b_9 \oplus b_{11}$$
$$P_2 = b_3 \oplus b_6 \oplus b_7 \oplus b_{10} \oplus b_{11}$$
$$P_3 = b_5 \oplus b_6 \oplus b_7 \oplus b_{12}$$
$$P_4 = b_9 \oplus b_{10} \oplus b_{11} \oplus b_{12}$$

13-bit Code ($b_1\ b_2\ b_3\ b_4\ b_5\ b_6\ b_7\ b_8\ b_9\ b_{10}\ b_{11}\ b_{12}\ b_{13}$)

$$P_1 = b_3 \oplus b_5 \oplus b_7 \oplus b_9 \oplus b_{11} \oplus b_{13}$$
$$P_2 = b_3 \oplus b_6 \oplus b_7 \oplus b_{10} \oplus b_{11}$$
$$P_3 = b_5 \oplus b_6 \oplus b_7 \oplus b_{12} \oplus b_{13}$$
$$P_4 = b_9 \oplus b_{10} \oplus b_{11} \oplus b_{12} \oplus b_{13}$$

Generation of Odd Parity Bits

The parity bits for odd parity Hamming code are complements of parity bits for even parity. Therefore, the parity bits for odd parity are determined using the complement of equation for even parity. Alternatively, the XOR operations in the equations for even parity can be replaced XNOR operations.

Error Detection and Correction

In order to detect error in transmitted data check bits are generated. The number of check bits will be same as number of parity bits. When there is no error in received data all check bits will be zero. When there is error the check bits will give the binary value of bit position that has error.

Let, check bits = $C_4\ C_3\ C_2\ C_1$

For example,

If $C_4\ C_3\ C_2\ C_1 = 0000$, then no error in received code.

If $C_4\ C_3\ C_2\ C_1 = 0011$, then error in b_3 bit in the received code.

If $C_4\ C_3\ C_2\ C_1 = 0111$, then error in b_7 bit in the received code.

The check for even parity Hamming code bits are generated by XOR of the parity bit position along with the bits that are used to generate parity bits.

Check Bits for 9-bit Code ($b_1\ b_2\ b_3\ b_4\ b_5\ b_6\ b_7\ b_8\ b_9$)

$$C_1 = b_1 \oplus b_3 \oplus b_5 \oplus b_7 \oplus b_9$$

$$C_2 = b_2 \oplus b_3 \oplus b_6 \oplus b_7$$

$$C_3 = b_4 \oplus b_5 \oplus b_6 \oplus b_7$$

$$C_4 = b_8 \oplus b_9$$

Check Bits for 10-bit Code ($b_1\ b_2\ b_3\ b_4\ b_5\ b_6\ b_7\ b_8\ b_9\ b_{10}$)

$$C_1 = b_1 \oplus b_3 \oplus b_5 \oplus b_7 \oplus b_9$$

$$C_2 = b_2 \oplus b_3 \oplus b_6 \oplus b_7 \oplus b_{10}$$

$$C_3 = b_4 \oplus b_5 \oplus b_6 \oplus b_7$$

$$C_4 = b_8 \oplus b_9 \oplus b_{10}$$

Check Bits for 11-bit Code ($b_1\ b_2\ b_3\ b_4\ b_5\ b_6\ b_7\ b_8\ b_9\ b_{10}\ b_{11}$)

$$C_1 = b_1 \oplus b_3 \oplus b_5 \oplus b_7 \oplus b_9 \oplus b_{11}$$

$$C_2 = b_2 \oplus b_3 \oplus b_6 \oplus b_7 \oplus b_{10} \oplus b_{11}$$

$$C_3 = b_4 \oplus b_5 \oplus b_6 \oplus b_7$$

$$C_4 = b_8 \oplus b_9 \oplus b_{10} \oplus b_{11}$$

Check Bits for 12-bit Code ($b_1\ b_2\ b_3\ b_4\ b_5\ b_6\ b_7\ b_8\ b_9\ b_{10}\ b_{11}\ b_{12}$)

$$C_1 = b_1 \oplus b_3 \oplus b_5 \oplus b_7 \oplus b_9 \oplus b_{11}$$

$$C_2 = b_2 \oplus b_3 \oplus b_6 \oplus b_7 \oplus b_{10} \oplus b_{11}$$

$$C_3 = b_4 \oplus b_5 \oplus b_6 \oplus b_7 \oplus b_{12}$$

$$C_4 = b_8 \oplus b_9 \oplus b_{10} \oplus b_{11} \oplus b_{12}$$

Check Bits for 13-bit Code ($b_1\ b_2\ b_3\ b_4\ b_5\ b_6\ b_7\ b_8\ b_9\ b_{10}\ b_{11}\ b_{12}\ b_{13}$)

$$C_1 = b_1 \oplus b_3 \oplus b_5 \oplus b_7 \oplus b_9 \oplus b_{11} \oplus b_{13}$$

$$C_2 = b_2 \oplus b_3 \oplus b_6 \oplus b_7 \oplus b_{10} \oplus b_{11}$$

$$C_3 = b_4 \oplus b_5 \oplus b_6 \oplus b_7 \oplus b_{12} \oplus b_{13}$$

$$C_4 = b_8 \oplus b_9 \oplus b_{10} \oplus b_{11} \oplus b_{12} \oplus b_{13}$$

Check Bits for Odd Parity Hamming Code

The check bits for odd parity Hamming code are complement of check bits for even parity Hamming code. Therefore, the check bits of odd parity code can be obtained from complement of the equations for even parity check bits. Alternatively, the XOR operations in the equations for even parity can be replaced by XNOR operations. Here, if the check bits are zero then no error in transmission. If there is an error then the check bits will give the value of the bit that has error.

Example 3.15

(AU, Apr/May'19, 2 Marks)

Generate the Hamming code for an 8-bit data string 10101101. Introduce error in a data bit and prove that Hamming code detect the error.

a) Choose even parity Hamming code **b)** Choose odd parity Hamming code

Solution

a) Even Parity Hamming Code

For 8 data bits, 4 parity bits are required.

$\therefore$ n = 8, k = 4

Here, $2^k = 2^4 = 16$

$n + k + 1 = 8 + 4 + 1 = 13$

The condition, $2^k \geq n + k + 1$ is satisfied.

The size of Hamming code = n + k = 8 + 4 = 12 bits.

Let, Hamming code = $b_1 b_2 b_3 b_4 b_5 b_6 b_7 b_8 b_9 b_{10} b_{11} b_{12}$

In Hamming code, the bit positions 1, 2, 4 and 8 are reserved for parity bits and remaining bits are used for data bits. The data and parity bits are combined to form Hamming code as shown below:

Hamming code : $b_1 b_2 b_3 b_4 b_5 b_6 b_7 b_8 b_9 b_{10} b_{11} b_{12}$

$P_1 P_2 d_8 P_3 d_7 d_6 d_5 P_4 d_4 d_3 d_2 d_1$

$\Downarrow$ Insert data bits

$P_1 P_2$ 1 P_3 0 1 0 P_4 1 1 0 1

Parity bits are calculated as follows.

$P_1 = b_3 \oplus b_5 \oplus b_7 \oplus b_9 \oplus b_{11} = 1 \oplus 0 \oplus 0 \oplus 1 \oplus 0 = 0$

$P_2 = b_3 \oplus b_6 \oplus b_7 \oplus b_{10} \oplus b_{11} = 1 \oplus 1 \oplus 0 \oplus 1 \oplus 0 = 1$

$P_3 = b_5 \oplus b_6 \oplus b_7 \oplus b_{12} = 0 \oplus 1 \oplus 0 \oplus 1 = 0$

$P_4 = b_9 \oplus b_{10} \oplus b_{11} \oplus b_{12} = 1 \oplus 1 \oplus 0 \oplus 1 = 1$

$\therefore P_1 P_2 P_3 P_4 = 0\,1\,0\,1$

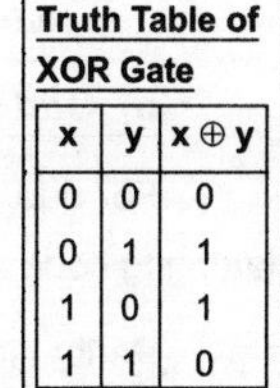

Truth Table of XOR Gate

x	y	$x \oplus y$
0	0	0
0	1	1
1	0	1
1	1	0

Substitute the parity bits in Hamming code.

Hamming code : $P_1 P_2$ 1 P_3 0 1 0 P_4 1 1 0 1

$\Downarrow$ Insert parity bits

0 1 1 0 0 1 0 1 1 1 0 1

Case i: Received Code has No Error

Received Hamming code : b_1 b_2 b_3 b_4 b_5 b_6 b_7 b_8 b_9 b_{10} b_{11} b_{12}

0 1 1 0 0 1 0 1 1 1 0 1

The check bits are generated as follows:

$$C_1 = b_1 \oplus b_3 \oplus b_5 \oplus b_7 \oplus b_9 \oplus b_{11} = 0 \oplus 1 \oplus 0 \oplus 0 \oplus 1 \oplus 0 = 0$$

$$C_2 = b_2 \oplus b_3 \oplus b_6 \oplus b_7 \oplus b_{10} \oplus b_{11} = 1 \oplus 1 \oplus 1 \oplus 0 \oplus 1 \oplus 0 = 0$$

$$C_3 = b_4 \oplus b_5 \oplus b_6 \oplus b_7 \oplus b_{12} = 0 \oplus 0 \oplus 1 \oplus 0 \oplus 1 = 0$$

$$C_4 = b_8 \oplus b_9 \oplus b_{10} \oplus b_{11} \oplus b_{12} = 1 \oplus 1 \oplus 1 \oplus 0 \oplus 1 = 0$$

$$\therefore C_4\, C_3\, C_2\, C_1 = 0\,0\,0\,0$$

Here all the check bits are zeros and hence no error in received Hamming code.

Case ii: Introduce Error in b_5 and Check for Error

Received Hamming code : b_1 b_2 b_3 b_4 b_5 b_6 b_7 b_8 b_9 b_{10} b_{11} b_{12}

0 1 1 0 1 1 0 1 1 1 0 1

The check bits are generated as follows:

$$C_1 = b_1 \oplus b_3 \oplus b_5 \oplus b_7 \oplus b_9 \oplus b_{11} = 0 \oplus 1 \oplus 1 \oplus 0 \oplus 1 \oplus 0 = 1$$

$$C_2 = b_2 \oplus b_3 \oplus b_6 \oplus b_7 \oplus b_{10} \oplus b_{11} = 1 \oplus 1 \oplus 1 \oplus 0 \oplus 1 \oplus 0 = 0$$

$$C_3 = b_4 \oplus b_5 \oplus b_6 \oplus b_7 \oplus b_{12} = 0 \oplus 1 \oplus 1 \oplus 0 \oplus 1 = 1$$

$$C_4 = b_8 \oplus b_9 \oplus b_{10} \oplus b_{11} \oplus b_{12} = 1 \oplus 1 \oplus 1 \oplus 0 \oplus 1 = 0$$

$$\therefore C_4\, C_3\, C_2\, C_1 = 0\,1\,0\,1$$

⇓

5 (In received Hamming code, error in bit b_5)

Now, complement the bit b_5 to get the correct code.

b_1 b_2 b_3 b_4 b_5 b_6 b_7 b_8 b_9 b_{10} b_{11} b_{12}

Received code = 0 1 1 0 1 1 0 1 1 1 0 1

↓ Complement b_5

Corrected code = 0 1 1 0 0 1 0 1 1 1 0 1

b) Odd Parity Hamming Code

For odd parity Hamming code, the parity bits are obtained by taking complement of parity bits of even parity Hamming code.

Parity bits of even parity hamming code = 0101 $\xrightarrow{\text{Complement}}$ 1010

$$\therefore P_1 P_2 P_3 P_4 = 1\,0\,1\,0$$

Substitute the above parity bits in Hamming code.

Hamming code : P_1 P_2 1 P_3 0 1 0 P_4 1 1 0 1 From (a)

⇓ Insert parity bits

1 0 1 1 0 1 0 0 1 1 0 1

Case i: Received Code has No Error

Received Hamming code : $b_1\ b_2\ b_3\ b_4\ b_5\ b_6\ b_7\ b_8\ b_9\ b_{10}\ b_{11}\ b_{12}$

$$1\ 0\ 1\ 1\ 0\ 1\ 0\ 0\ 1\ 1\ 0\ 1$$

The check bits for odd parity are complement of check bits of even parity.

The check bits are generated as follows:

$$C_1 = (b_1 \oplus b_3 \oplus b_5 \oplus b_7 \oplus b_9 \oplus b_{11})' = (1 \oplus 1 \oplus 0 \oplus 0 \oplus 1 \oplus 0)' = 0$$

$$C_2 = (b_2 \oplus b_3 \oplus b_6 \oplus b_7 \oplus b_{10} \oplus b_{11})' = (0 \oplus 1 \oplus 1 \oplus 0 \oplus 1 \oplus 0)' = 0$$

$$C_3 = (b_4 \oplus b_5 \oplus b_6 \oplus b_7 \oplus b_{12})' = (1 \oplus 0 \oplus 1 \oplus 0 \oplus 1)' = 0$$

$$C_4 = (b_8 \oplus b_9 \oplus b_{10} \oplus b_{11} \oplus b_{12})' = (0 \oplus 1 \oplus 1 \oplus 0 \oplus 1)' = 0$$

$$\therefore C_4\,C_3\,C_2\,C_1 = 0\,0\,0\,0$$

Here all the check bits are zeros and hence no error in received Hamming code.

Case ii: Introduce Error in b_5 and Check for Error

Received Hamming code : $b_1\ b_2\ b_3\ b_4\ b_5\ b_6\ b_7\ b_8\ b_9\ b_{10}\ b_{11}\ b_{12}$

$$1\ 0\ 1\ 1\ 1\ 1\ 0\ 0\ 1\ 1\ 0\ 1$$

The check bits are generated as follows:

$$C_1 = (b_1 \oplus b_3 \oplus b_5 \oplus b_7 \oplus b_9 \oplus b_{11})' = (1 \oplus 1 \oplus 1 \oplus 0 \oplus 1 \oplus 0)' = 1$$

$$C_2 = (b_2 \oplus b_3 \oplus b_6 \oplus b_7 \oplus b_{10} \oplus b_{11})' = (0 \oplus 1 \oplus 1 \oplus 0 \oplus 1 \oplus 0)' = 0$$

$$C_3 = (b_4 \oplus b_5 \oplus b_6 \oplus b_7 \oplus b_{12})' = (1 \oplus 1 \oplus 1 \oplus 0 \oplus 1)' = 1$$

$$C_4 = (b_8 \oplus b_9 \oplus b_{10} \oplus b_{11} \oplus b_{12})' = (0 \oplus 1 \oplus 1 \oplus 0 \oplus 1)' = 0$$

$$\therefore C_4\,C_3\,C_2\,C_1 = 0\,1\,0\,1$$

$$\Downarrow$$

5 (In received Hamming code, error in bit b_5)

Now, complement the bit b_5 to get the correct code.

$b_1\ b_2\ b_3\ b_4\ b_5\ b_6\ b_7\ b_8\ b_9\ b_{10}\ b_{11}\ b_{12}$

Received code = 1 0 1 1 1 1 0 0 1 1 0 1

$\downarrow$ Complement b_5

Corrected code = 1 0 1 1 0 1 0 0 1 1 0 1

Example 3.16

Generate the Hamming code for an 9-bit data string 100101001. Introduce error in a data bit and prove that Hamming code detect the error.

a) Choose even parity Hamming code **b)** Choose odd parity Hamming code

Solution

a) Even Parity Hamming Code

For 9 data bits, 4 parity bits are required.

$$\therefore n = 9,\ k = 4$$

Here, $2^k = 2^4 = 16$

$$n + k + 1 = 9 + 4 + 1 = 14$$

The condition, $2^k \geq n + k + 1$ is satisfied.

The size of Hamming code = n + k = 9 + 4 = 13 bits.

Let, Hamming code = $b_1 b_2 b_3 b_4 b_5 b_6 b_7 b_8 b_9 b_{10} b_{11} b_{12} b_{13}$

In Hamming code, the bit positions 1, 2, 4 and 8 are reserved for parity bits and remaining bits are used for data bits. The data and parity bits are combined to form Hamming code as shown below:

Hamming code : $b_1 b_2 b_3 b_4 b_5 b_6 b_7 b_8 b_9 b_{10} b_{11} b_{12} b_{13}$

$P_1 P_2 d_9 P_3 d_8 d_7 d_6 P_4 d_5 d_4 d_3\ d_2\ d_1$

⇓ Insert data bits

$P_1 P_2\ 1\ P_3 0\ 0\ 1\ P_4 0\ 1\ 0\ 0\ 1$

Parity bits are calculated as follows.

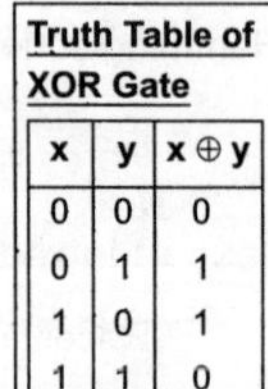

Truth Table of XOR Gate

x	y	x ⊕ y
0	0	0
0	1	1
1	0	1
1	1	0

$P_1 = b_3 \oplus b_5 \oplus b_7 \oplus b_9 \oplus b_{11} \oplus b_{13} = 1 \oplus 0 \oplus 1 \oplus 0 \oplus 0 \oplus 1 = 1$

$P_2 = b_3 \oplus b_6 \oplus b_7 \oplus b_{10} \oplus b_{11} = 1 \oplus 0 \oplus 1 \oplus 1 \oplus 0 = 1$

$P_3 = b_5 \oplus b_6 \oplus b_7 \oplus b_{12} \oplus b_{13} = 0 \oplus 0 \oplus 1 \oplus 0 \oplus 1 = 0$

$P_4 = b_9 \oplus b_{10} \oplus b_{11} \oplus b_{12} \oplus b_{13} = 0 \oplus 1 \oplus 0 \oplus 0 \oplus 1 = 0$

$\therefore P_1 P_2 P_3 P_4 = 1\ 1\ 0\ 0$

Substitute the parity bits in Hamming code.

Hamming code : $P_1 P_2\ 1\ P_3 0\ 0\ 1\ P_4 0\ 1\ 0\ 0\ 1$

⇓ Insert parity bits

1 1 1 0 0 0 1 0 0 1 0 0 1

Case i: Received Code has No Error

Received Hamming code : $b_1\ b_2\ b_3\ b_4\ b_5\ b_6\ b_7\ b_8\ b_9\ b_{10}\ b_{11}\ b_{12}\ b_{13}$

1 1 1 0 0 0 1 0 0 1 0 0 1

The check bits are generated as follows:

$C_1 = b_1 \oplus b_3 \oplus b_5 \oplus b_7 \oplus b_9 \oplus b_{11} \oplus b_{13} = 1 \oplus 1 \oplus 0 \oplus 1 \oplus 0 \oplus 0 \oplus 1 = 0$

$C_2 = b_2 \oplus b_3 \oplus b_6 \oplus b_7 \oplus b_{10} \oplus b_{11} = 1 \oplus 1 \oplus 0 \oplus 1 \oplus 1 \oplus 0 = 0$

$C_3 = b_4 \oplus b_5 \oplus b_6 \oplus b_7 \oplus b_{12} \oplus b_{13} = 0 \oplus 0 \oplus 0 \oplus 1 \oplus 0 \oplus 1 = 0$

$C_4 = b_8 \oplus b_9 \oplus b_{10} \oplus b_{11} \oplus b_{12} \oplus b_{13} = 0 \oplus 0 \oplus 1 \oplus 0 \oplus 0 \oplus 1 = 0$

$\therefore C_4\ C_3\ C_2\ C_1 = 0\ 0\ 0\ 0$

Here all the check bits are zeros and hence no error in received Hamming code.

Case ii: Introduce Error in b_6 and Check for Error

Received Hamming code : $b_1\ b_2\ b_3\ b_4\ b_5\ b_6\ b_7\ b_8\ b_9\ b_{10}\ b_{11}\ b_{12}\ b_{13}$

1 1 1 0 0 1 1 0 0 1 0 0 1

The check bits are generated as follows:

$C_1 = b_1 \oplus b_3 \oplus b_5 \oplus b_7 \oplus b_9 \oplus b_{11} \oplus b_{13} = 1 \oplus 1 \oplus 0 \oplus 1 \oplus 0 \oplus 0 \oplus 1 = 0$

$C_2 = b_2 \oplus b_3 \oplus b_6 \oplus b_7 \oplus b_{10} \oplus b_{11} = 1 \oplus 1 \oplus 1 \oplus 1 \oplus 1 \oplus 0 = 1$

$C_3 = b_4 \oplus b_5 \oplus b_6 \oplus b_7 \oplus b_{12} \oplus b_{13} = 0 \oplus 0 \oplus 1 \oplus 1 \oplus 0 \oplus 1 = 1$

$C_4 = b_8 \oplus b_9 \oplus b_{10} \oplus b_{11} \oplus b_{12} \oplus b_{13} = 0 \oplus 0 \oplus 1 \oplus 0 \oplus 0 \oplus 1 = 0$

$\therefore C_4\,C_3\,C_2\,C_1 = 0\,1\,1\,0$

$\Downarrow$

6 (In received Hamming code, error in bit b_6)

Now, complement the bit b_6 to get the correct code.

$b_1\ b_2\ b_3\ b_4\ b_5\ b_6\ b_7\ b_8\ b_9\ b_{10}\ b_{11}\ b_{12}\ b_{13}$

Received code = 1 1 1 0 0 1 1 0 0 1 0 0 1

$\downarrow$ Complement b_6

Corrected code = 1 1 1 0 0 0 1 0 0 1 0 0 1

b) Odd Parity Hamming Code

For odd parity Hamming code, the parity bits are obtained by taking complement of parity bits of even parity Hamming code.

Parity bits of even parity Hamming code = 1100 $\xrightarrow{\text{Complement}}$ 0011

$\therefore P_1\,P_2\,P_3\,P_4 = 0\,0\,1\,1$

Substitute the above parity bits in Hamming code.

Hamming code : $P_1\,P_2$ 1 P_3 0 0 1 P_4 0 1 0 0 1 From (a)

$\Downarrow$ Insert parity bits

0 0 1 1 0 0 1 1 0 1 0 0 1

Case i: Received Code has No Error

Received Hamming code : $b_1\ b_2\ b_3\ b_4\ b_5\ b_6\ b_7\ b_8\ b_9\ b_{10}\ b_{11}\ b_{12}\ b_{13}$

0 0 1 1 0 0 1 1 0 1 0 0 1

The check bits for odd parity are complement of check bits of even parity.

The check bits are generated as follows:

$C_1 = (b_1 \oplus b_3 \oplus b_5 \oplus b_7 \oplus b_9 \oplus b_{11} \oplus b_{13})' = (0 \oplus 1 \oplus 0 \oplus 1 \oplus 0 \oplus 0 \oplus 1)' = 0$

$C_2 = (b_2 \oplus b_3 \oplus b_6 \oplus b_7 \oplus b_{10} \oplus b_{11})' = (0 \oplus 1 \oplus 0 \oplus 1 \oplus 1 \oplus 0)' = 0$

$C_3 = (b_4 \oplus b_5 \oplus b_6 \oplus b_7 \oplus b_{12} \oplus b_{13})' = (1 \oplus 0 \oplus 0 \oplus 1 \oplus 0 \oplus 1)' = 0$

$C_4 = (b_8 \oplus b_9 \oplus b_{10} \oplus b_{11} \oplus b_{12} \oplus b_{13})' = (1 \oplus 0 \oplus 1 \oplus 0 \oplus 0 \oplus 1)' = 0$

$\therefore C_4\,C_3\,C_2\,C_1 = 0\,0\,0\,0$

Here all the check bits are zeros and hence no error in received Hamming code.

Case ii: Introduce Error in b_6 and Check for Error

Received Hamming code : $b_1\ b_2\ b_3\ b_4\ b_5\ b_6\ b_7\ b_8\ b_9\ b_{10}\ b_{11}\ b_{12}\ b_{13}$

0 0 1 1 0 1 1 1 0 1 0 0 1

The check bits are generated as follows:

$C_1 = (b_1 \oplus b_3 \oplus b_5 \oplus b_7 \oplus b_9 \oplus b_{11} \oplus b_{13})' = (0 \oplus 1 \oplus 0 \oplus 1 \oplus 0 \oplus 0 \oplus 1)' = 0$

$C_2 = (b_2 \oplus b_3 \oplus b_6 \oplus b_7 \oplus b_{10} \oplus b_{11})' = (0 \oplus 1 \oplus 0 \oplus 1 \oplus 1 \oplus 0)' = 1$

$C_3 = (b_4 \oplus b_5 \oplus b_6 \oplus b_7 \oplus b_{12} \oplus b_{13})' = (1 \oplus 0 \oplus 0 \oplus 1 \oplus 0 \oplus 1)' = 1$

$C_4 = (b_8 \oplus b_9 \oplus b_{10} \oplus b_{11} \oplus b_{12} \oplus b_{13})' = (1 \oplus 0 \oplus 1 \oplus 0 \oplus 0 \oplus 1)' = 0$

$\therefore C_4 C_3 C_2 C_1 = 0\,1\,1\,0$

$\Downarrow$

6 (In received Hamming code, error in bit b_6)

Now, complement the bit b_6 to get the correct code.

	b_1	b_2	b_3	b_4	b_5	b_6	b_7	b_8	b_9	b_{10}	b_{11}	b_{12}	b_{13}
Received code =	0	0	1	1	0	1	1	1	0	1	0	0	1
						↓ Complement b_6							
Corrected code =	0	0	1	1	0	0	1	1	0	1	0	0	1

Example 3.17

(CSVTU, Nov/Dec'19, 7 Marks)

Generate the Hamming code for an 7-bit data string 0010001. Introduce error in a data bit and prove that Hamming code detect the error.

a) Choose even parity Hamming code **b)** Choose odd parity Hamming code

Solution

a) Even Parity Hamming Code

For 7 data bits, 4 parity bits are required.

$\therefore n = 7, \; k = 4$

Here, $2^k = 2^4 = 16$

$n + k + 1 = 7 + 4 + 1 = 12$

The condition, $2^k \geq n + k + 1$ is satisfied.

The size of Hamming code = n + k = 7 + 4 = 11 bits.

Let, Hamming code = $b_1 b_2 b_3 b_4 b_5 b_6 b_7 b_8 b_9 b_{10} b_{11}$

In Hamming code, the bit positions 1, 2, 4 and 8 are reserved for parity bits and remaining bits are used for data bits. The data and parity bits are combined to form Hamming code as shown below:

Hamming code : $b_1 b_2 b_3 b_4 b_5 b_6 b_7 b_8 b_9 b_{10} b_{11}$

$P_1 P_2 d_7 P_3 d_6 d_5 d_4 P_4 \; d_3 d_2 d_1$

$\Downarrow$ Insert data bits

$P_1 P_2 \; 0 \; P_3 0 \; 1 \; 0 \; P_4 \; 0 \; 0 \; 1$

Parity bits are calculated as follows.

$P_1 = b_3 \oplus b_5 \oplus b_7 \oplus b_9 \oplus b_{11} = 0 \oplus 0 \oplus 0 \oplus 0 \oplus 1 = 1$

$P_2 = b_3 \oplus b_6 \oplus b_7 \oplus b_{10} \oplus b_{11} = 0 \oplus 1 \oplus 0 \oplus 0 \oplus 1 = 0$

$P_3 = b_5 \oplus b_6 \oplus b_7 = 0 \oplus 1 \oplus 0 = 1$

$P_4 = b_9 \oplus b_{10} \oplus b_{11} = 0 \oplus 0 \oplus 1 = 1$

$\therefore P_1 P_2 P_3 P_4 = 1\,0\,1\,1$

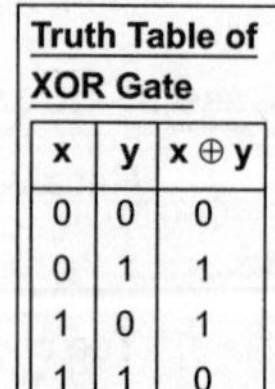

Truth Table of XOR Gate

x	y	$x \oplus y$
0	0	0
0	1	1
1	0	1
1	1	0

Substitute the parity bits in Hamming code.

Hamming code : $P_1 \; P_2 \; 0 \; P_3 \; 0 \; 1 \; 0 \; P_4 \; 0 \; 0 \; 1$

$\Downarrow$ Insert parity bits

1 0 0 1 0 1 0 1 0 0 1

Case i: Received Code has No Error

Received Hamming code : $b_1\ b_2\ b_3\ b_4\ b_5\ b_6\ b_7\ b_8\ b_9\ b_{10}\ b_{11}$

1 0 0 1 0 1 0 1 0 0 1

The check bits are generated as follows:

$C_1 = b_1 \oplus b_3 \oplus b_5 \oplus b_7 \oplus b_9 \oplus b_{11} = 1 \oplus 0 \oplus 0 \oplus 0 \oplus 0 \oplus 1 = 0$

$C_2 = b_2 \oplus b_3 \oplus b_6 \oplus b_7 \oplus b_{10} \oplus b_{11} = 0 \oplus 0 \oplus 1 \oplus 0 \oplus 0 \oplus 1 = 0$

$C_3 = b_4 \oplus b_5 \oplus b_6 \oplus b_7 = 1 \oplus 0 \oplus 1 \oplus 0 = 0$

$C_4 = b_8 \oplus b_9 \oplus b_{10} \oplus b_{11} = 1 \oplus 0 \oplus 0 \oplus 1 = 0$

$\therefore C_4\, C_3\, C_2\, C_1 = 0\,0\,0\,0$

Here all the check bits are zeros and hence no error in received Hamming code.

Case ii: Introduce Error in b_9 and Check for Error

Received Hamming code : $b_1\ b_2\ b_3\ b_4\ b_5\ b_6\ b_7\ b_8\ b_9\ b_{10}\ b_{11}$

1 0 0 1 0 1 0 1 1 0 1

The check bits are generated as follows:

$C_1 = b_1 \oplus b_3 \oplus b_5 \oplus b_7 \oplus b_9 \oplus b_{11} = 1 \oplus 0 \oplus 0 \oplus 0 \oplus 1 \oplus 1 = 1$

$C_2 = b_2 \oplus b_3 \oplus b_6 \oplus b_7 \oplus b_{10} \oplus b_{11} = 0 \oplus 0 \oplus 1 \oplus 0 \oplus 0 \oplus 1 = 0$

$C_3 = b_4 \oplus b_5 \oplus b_6 \oplus b_7 = 1 \oplus 0 \oplus 1 \oplus 0 = 0$

$C_4 = b_8 \oplus b_9 \oplus b_{10} \oplus b_{11} = 1 \oplus 1 \oplus 0 \oplus 1 = 1$

$\therefore C_4\, C_3\, C_2\, C_1 = 1\,0\,0\,1$

$\Downarrow$

9 (In received Hamming code, error in bit b_9)

Now, complement the bit b_9 to get the correct code.

$b_1\ b_2\ b_3\ b_4\ b_5\ b_6\ b_7\ b_8\ b_9\ b_{10}\ b_{11}$

Received code = 1 0 0 1 0 1 0 1 0 0 1

$\downarrow$ Complement b_9

Corrected code = 1 0 0 1 0 1 0 1 1 0 1

b) Odd parity Hamming code

For odd parity Hamming code, the parity bits are obtained by taking complement of parity bits of even parity Hamming code.

Parity bits of even parity Hamming code = 1011 $\xrightarrow{\text{Complement}}$ 0100

$\therefore P_1 P_2 P_3 P_4 = 0\,1\,0\,0$

Substitute the above parity bits in Hamming code.

Hamming code : $P_1\ P_2\ 0\ P_3\ 0\ 1\ 0\ P_4\ 0\ 0\ 1$ From (a)

$\Downarrow$ Insert parity bits

0 1 0 0 0 1 0 0 0 0 1

Case i: Received Code has No Error

Received Hamming code : b_1 b_2 b_3 b_4 b_5 b_6 b_7 b_8 b_9 b_{10} b_{11}

0 1 0 0 0 1 0 0 0 0 1

The check bits for odd parity are complement of check bits for even parity.

The check bits are generated as follows:

$$C_1 = (b_1 \oplus b_3 \oplus b_5 \oplus b_7 \oplus b_9 \oplus b_{11})' = (0 \oplus 0 \oplus 0 \oplus 0 \oplus 0 \oplus 1)' = 0$$

$$C_2 = (b_2 \oplus b_3 \oplus b_6 \oplus b_7 \oplus b_{10} \oplus b_{11})' = (1 \oplus 0 \oplus 1 \oplus 0 \oplus 0 \oplus 1)' = 0$$

$$C_3 = (b_4 \oplus b_5 \oplus b_6 \oplus b_7)' = (0 \oplus 0 \oplus 1 \oplus 0)' = 0$$

$$C_4 = (b_8 \oplus b_9 \oplus b_{10} \oplus b_{11})' = (0 \oplus 0 \oplus 0 \oplus 1)' = 0$$

$\therefore C_4\, C_3\, C_2\, C_1 = 0\,0\,0\,0$

Here all the check bits are zeros and hence no error in received Hamming code.

Case ii: Introduce Error in b_9 and Check for Error

Received Hamming code : b_1 b_2 b_3 b_4 b_5 b_6 b_7 b_8 b_9 b_{10} b_{11}

0 1 0 0 0 1 0 0 1 0 1

The check bits are generated as follows:

$$C_1 = (b_1 \oplus b_3 \oplus b_5 \oplus b_7 \oplus b_9 \oplus b_{11})' = (0 \oplus 0 \oplus 0 \oplus 0 \oplus 1 \oplus 1)' = 1$$

$$C_2 = (b_2 \oplus b_3 \oplus b_6 \oplus b_7 \oplus b_{10} \oplus b_{11})' = (1 \oplus 0 \oplus 1 \oplus 0 \oplus 0 \oplus 1)' = 0$$

$$C_3 = (b_4 \oplus b_5 \oplus b_6 \oplus b_7)' = (0 \oplus 0 \oplus 1 \oplus 0)' = 0$$

$$C_4 = (b_8 \oplus b_9 \oplus b_{10} \oplus b_{11})' = (0 \oplus 1 \oplus 0 \oplus 1)' = 1$$

$\therefore C_4\, C_3\, C_2\, C_1 = 1\,0\,0\,1$

⇓

9 (In received Hamming code, error in bit b_9)

Now, complement the bit b_9 to get the correct code.

b_1 b_2 b_3 b_4 b_5 b_6 b_7 b_8 b_9 b_{10} b_{11}

Received code = 0 1 0 0 0 1 0 0 1 0 1

↓ Complement b_9

Corrected code = 0 1 0 0 0 1 0 0 0 0 1

3.5.3 Checksum

Checksum is a method to check for error in the transmitted digital code or information. In this method, a checksum is generated at the transmitter and receiver side. The value of checksum at the receiver side is used to decide whether the information is received with or without error. This method is useful for large size digital code, usually 32-bits or higher size.

In the transmitter side, the digital code is divided into equal size frames of n-bit length by the checksum generator. This bit is generally of 8-bit length. These frames are then added together using one's complement method to generate an n-bit sum and the complement of this sum is called checksum. The checksum of each code is appended as last frame and transmitted to receiver.

In the receiver side, the received digital code with checksum is divide frames of n-bit length and added by one's complement method to generate a sum and the complement of this sum is checksum at receiver side. If the checksum is zero, then the code is received error-free. If the checksum at receiver side is non-zero then it means that the data contains an error and the receiver will reject the code and send a request to resend the code.

Example 3.18

Find the checksum for the following 32-bit binary code.

10110011 10101011 01011010 11010101

a) Transmit the code without error and check at receiver side.

b) Transmit the code with one bit error and check at receiver side.

Solution

The binary code to be transmitted has 32 bits and hence can be divided into four frames of 8 bits.

Given binary code = 10110011 10101011 01011010 11010101

Let, Frame 1 = 10110011

Frame 2 = 10101011

Frame 3 = 01011010

Frame 4 = 11010101

Addition on Transmitter Side

```
Frame 1:           10110011
Frame 2:       (+) 10101011
                 [1] 01011110
      Add carry          +1
Partial sum:       01011111
Frame 3:       (+) 01011010
Partial sum:       10111001
Frame 4:       (+) 11010101
                 [1] 10001110
      Add carry          +1
Total sum:         10001111
```

The total sum is complemented to get checksum at the transmitter side.

10001111 —Complement→ 01110000

Total sum   Checksum

The complemented total sum is the checksum at the transmitter side which is appended as frame 5 at the end of given 32-bit binary code and the 40-bit binary code is transmitted.

∴ Transmitted code = 10110011 10101011 01011010 11010101 01110000

	Frame 1	Frame 2	Frame 3	Frame 4	Frame 5
∴ Transmitted code =	10110011	10101011	01011010	11010101	01110000

a) Receiver Side has No Error

The transmitted code is received without error and hence the received code is same as transmitted code.

The received 40-bit binary code is divided into 5 frames of 8-bits each and added as shown below:

Addition on Receiver Side

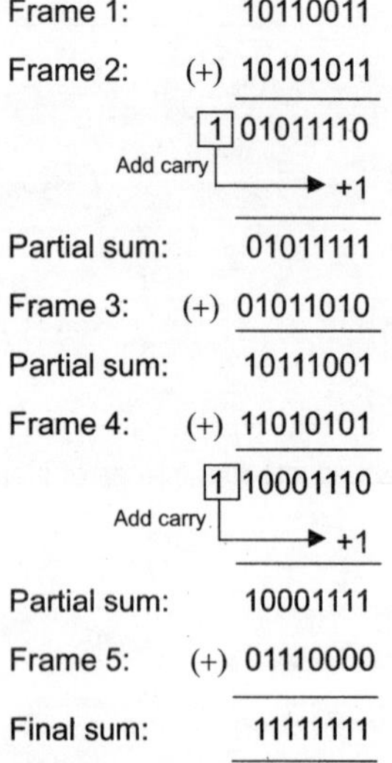

The final sum is complemented to get the checksum at receiver side.

11111111 —Complement→ 00000000
Final sum Checksum

Since the checksum at receiver end is zero, there is no error in the received code.

b) Receiver Side has One Bit Error

Let us introduce an error in bit b_4 of frame 1 in the received code. Therefore, the frame 1 of binary code with one bit error is 10100011.

∴ Received code = 10100011 (Frame 1) 10101011 (Frame 2) 01011010 (Frame 3) 11010101 (Frame 4) 01110000 (Frame 5)

The received 40-bit binary code is divided into 5 frames of 8-bits each and added as shown below:

Addition on Receiver Side

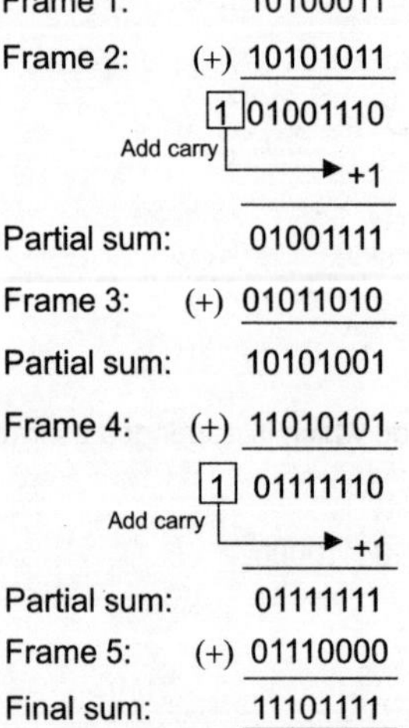

The final sum is complemented to get the checksum at receiver side.

11101111 —Complement→ 00010000
Final sum Checksum

The checksum at receiver side is non-zero and hence there is error in received code.

Example 3.19

Find the checksum for the following 16-bit binary code.

10101001 00111001

a) Transmit the code without error and check at receiver side.

b) Transmit the code with one bit error and check at receiver side.

Solution

The binary code to be transmitted has 16 bits and hence can be divided into two frames of 8 bits.

Given binary code = 10101001 00111001

Let, Frame 1 = 10101001

Frame 2 = 00111001

Addition on Transmitter Side

Frame 1: 10101001
Frame 2: (+) 00111001
Total sum: 11100010

The total sum is complemented to get checksum at receiver side.

11100010 —Complement→ 00011101
Total sum Checksum

The complemented total sum is checksum at the receiver side which is appended as frame 3 at the end of given 16-bit binary code and the 24-bit binary code is transmitted.

∴ Transmitted code = 10101001 00111001 00011101
Frame 1 Frame 2 Frame 3

a) Receiver Side has No Error

The transmitted code is received without error and hence the received code is same as transmitted code.

The received 24-bit binary code is divided into 3 frames of 8-bits each and added as shown below:

Addition on Receiver Side

Frame 1: 10101001
Frame 2: (+) 00111001
Partial sum: 11100010
Frame 3: (+) 00011101
Final sum: 11111111

The final sum is complemented to get the checksum at receiver side.

11111111 —Complement→ 00000000
Final sum Checksum

Since the checksum at receiver side is zero, there is no error in the received code.

b) Receiver Side has One Bit Error

Let us introduce an error in bit b_5 of frame 2 in the received code. Therefore, the frame 2 of binary code with one bit error is 00011001.

∴ Received code = 10100001 00011001 00011101
Frame 1 Frame 2 Frame 3

The received 24-bit binary code is divided into 3 frames of 8-bits each and added as shown below:

Addition on Receiver Side

Frame 1:	10101001
Frame 2:	(+) 00011001
Partial sum:	11000010
Frame 3:	(+) 00011101
Final sum:	11011111

The final sum is complemented to get the checksum at receiver side.

11011111 —Complement→ 00100000
Final sum Checksum

The checksum at receiver side is non-zero and hence there is error in received code.

3.6 Summary of Important Concepts

1. In the digital systems, each decimal number or piece of information is represented by a combination of binary digits called codes.
2. Binary code is a set of n-bit binary strings used to represent different decimal numbers or different alphabets of English or different information in voice signal, image signal, video signal, etc.
3. The digital systems can perform any arithmetic or logical operations on data or process any information represented in binary.
4. The binary codes are useful to store the information in memory devices and also for secure transmission of any information from one device/system to another device/system.
5. The n-bit binary code will have a group of 2^n distinct binary combinations and hence used to code a set of 2^n information.
6. The codes that are used to represent decimal digits 0 to 9 of the decimal numbers are called BCD (Binary-Coded Decimal) codes and the coded decimal numbers are called BCD.
7. In weighted codes, each digit has a weight and on adding the weights the decimal value represented by the code is obtained.
8. The codes that are used to represent decimal digits 0 to 9 of the decimal numbers, English alphabets A to Z and special characters used in computer keyboards are called alphanumeric codes.
9. The BCD Codes are used to represent decimal numbers.
10. Different types of BCD codes are, 8421 BCD (Generally called BCD), 2421 code, 5421 code, 8, 4, –2, – 1 code, Excess-3 code.
11. The 8421 Code, 2421 Code, 5421 Code, 8, 4, –2, –1 Code are weighted BCD codes.
12. In self complementing code, the complement of the code will give the code for 9's complement representation of decimal digit.

13. The 8421 BCD and Excess-3 are sequential codes.
14. The 2421 Code, 5421 Code and 8, 4, −2, −1 Code are non-sequential codes.
15. BCD is a weighted code with weight as 8, 4, 2 and 1 for the 4-digits of binary.
16. In BCD addition, in order to correct the result, a correction 6_{10} (0110_2) is added to BCD sum of digits whenever sum of BCD digits exceeds 9 or if there is a carry from sum of BCD digits to next BCD digit addition.
17. In BCD subtraction, in order to correct the result, a correction 6_{10} (0110_2) is subtracted from the difference of BCD digits whenever the difference of BCD digits exceeds 9 or if there is borrow from previous BCD digit.
18. Excess-3 code is obtained by adding 4-bit binary value of 3_{10} (0011_2) to 8421 BCD.
19. In Excess-3 subtraction, in order to correct the result, a correction 3_{10} (0011_2) is added to the difference of Excess-3 digits if there is no borrow from previous digit and a correction 3_{10} (0011_2) is subtracted from the difference of Excess-3 digits if there is borrow from previous digit.
20. In Gray code only one bit changes from either "0" to "1" or "1" to "0" between two consecutive codes.
21. ASCII is abbreviated form of American Standard Code for Information Interchange. ASCII was developed in the early stage of digital systems in order to standardize the coding system for human and machine interface.
22. ASCII is 7-bit binary code that includes most of the symbols in early type-writing machine.
23. With 7-bit binary code, 128 ($2^7 = 128$) ASCII codes are possible. They are assigned as 26 codes for upper case alphabets A to Z, 26 codes for lower case alphabets a to z, 10 codes for numerals 0 to 9, 32 codes for special printable characters and 34 codes for control characters.
24. EBCDIC Code is abbreviated form of Extended BCD Interchange Code.
25. Parity and checksum are popular method of error detection and error correction is performed by sending a request to retransmit the data.
26. Hamming code is another popular error deduction and correction technique to deduct and correct 1-bit error in the received data.
27. Parity refer to number of 1's in an n-bit code. If number of 1's is even then the code has even parity and number of 1's is odd then the code has odd parity.
28. Parity of an n-bit code can be modified by appending an extra bit called parity bit. The parity bit is appended at most significant bit position.
29. Hamming code is developed by Richard W Hamming of Bell Telephone Laboratories. The Hamming code is error detecting and correcting code.
30. The binary information to be coded is called data. In Hamming coding technique additional bits called Hamming or parity bits are inserted at various positions in between data bits.
31. Checksum is a method to check for error in the transmitted digital code or information.

3.7 Short-Answer Questions

Q3.1 ***Assume that there are 300 colours to be represented in binary digits. If every colour is to be assigned a unique bit pattern, what is the minimum number of bits required to do this? How many more colour's can be represented without requiring additional bits for these colour's unique bit pattern?***

With n-bit binary we can generate 2^n binary codes.

When $n = 8,\ 2^8 = 256_{10}$

When $n = 9,\ 2^9 = 512_{10}$

Therefore in order to represent 300 colours we require 9-bit binary bit pattern.

Here, 512 – 300 = 212, hence 212 more colours can be represented with 9-bit binary.

Q3.2 ***State the classification of binary codes.***

The binary codes are broadly classified into following four categories.

1. Binary coded decimal (BCD)
2. Excess-3 code
3. Gray code
4. Alphanumeric code (ASCII)

Q3.3 ***What do you mean by alphanumeric codes?***

The codes that are used to represent decimal digits 0 to 9 of the decimal numbers, English alphabets A to Z and special characters used in computer keyboards are called alphanumeric codes.

The two popular alphanumeric codes are,

- ASCII (**A**merican **S**tandard **C**ode for **I**nformation **I**nterchange) Code
- EBCDIC (Extended **BCD** **I**nterchange **C**ode) Code

Q3.4 ***What is meant by weighted and non-weighted codes?***

In weighted BCD codes, each digit of the binary code is assigned a specific weight according to its position. In these codes on adding the weights of 1's in the code we will get the value of the decimal number. The 8421 Code, 2421 Code, 5421 Code, 8, 4, –2, –1 Code are weighted BCD codes.

Example: 5421 code: $8 \Rightarrow 1011 \Rightarrow 5 + 0 + 2 + 1 = 8$

$7 \Rightarrow 0111 \Rightarrow 0 + 4 + 2 + 1 = 7$

Example: 2421 code: $5 \Rightarrow 1011 \Rightarrow 2 + 0 + 2 + 1 = 5$

$6 \Rightarrow 1100 \Rightarrow 2 + 4 + 0 + 0 = 6$

Example: 8, 4, –2, –1: $4 \Rightarrow 0100 \Rightarrow 0 + 4 + 0 + 0 = 4$

$9 \Rightarrow 1111 \Rightarrow 8 + 4 + (-2) + (-1) = 9$

$7 \Rightarrow 1001 \Rightarrow 8 + 0 + 0 + (-1) = 7$

Q3.5 ***What is meant by non-weighted codes?***

The non-weighted BCD codes are not positional weighted. In other words, each binary digit position within the binary code is not assigned a fixed value or weight. The excess-3 and Gray code are non-weighted codes.

Q3.6 ***What is meant by reflective and non-reflective BCD codes?***

A BCD code is called reflective code when the code is self-complementing. When a BCD code is not self-complementing it is called non-reflective code. The 2421 and excess-3 are self-complementing BCD codes and other BCD codes are not self-complementing.

In ***self complementing code***, the complement of the code will give the code for 9's complement representation of decimal digit.

Some examples of self complementing code are given below:

$$5 \xrightarrow[\text{BCD}]{\text{2421}} 1011 \xrightarrow{\text{Complement}} 0100 = 4$$

(4 in 2421 BCD)

9's complement of 5
9
−5
4

$$6 \longrightarrow 1001 \xrightarrow{\text{Complement}} 0110 = 3$$

(3 in Excess-3)

9's complement of 6
9
−6
3

Q3.7 ***Perform BCD addition of following decimal numbers.***

a) $3_{10} + 5_{10}$ ***b)*** $6_{10} + 7_{10}$

Solution:

a) $3_{10} + 5_{10}$

BCD Addition

```
                           1 1 1  ← Carry
     3                     0011
          Convert to BCD
(+)  5    ==============> (+) 0101
   ----                   ------
     8                     1000  --Convert to decimal-->  8_10
```

$\therefore 3_{10} + 5_{10} = 8_{10}$

---

b) $6_{10} + 7_{10}$

BCD Addition

```
                           1 1  ← Carry
     6                     0110
          Convert to BCD
(+)  7    ==============> (+) 0111          Correction
   ----                   ------           1 1  ← Carry
    13                     1101  ------>    1101
                           (Sum)
                                        (+) 0110 (Add 6)
                                        --------
                                          1 0011
                                            ⇓ Represent in BCD
                                        ---------
                                        0001 0011  --Convert to decimal-->  13_10
                                        (BCD sum)
```

Sum exceed 9 and so correction 6 is added to sum.

$\therefore 6_{10} + 7_{10} = 13_{10}$

Q3.8 ***Convert*** 7_{10} ***to 2421 code and verify by back converting to decimal.***

The 2421 code can be obtained expressing 7_{10} as sum of 2421 weight and then inserting 1 in the place of weight that are used to form the decimal value.

$$7_{10} \xrightarrow[\text{of 2421 to form 7}]{\text{Take weight}} 2 + 4 + 0 + 1 \xrightarrow[\text{in the place weight are considered}]{\text{Put 1}} 1101$$

$\therefore$ 2421 code for $7_{10} = 1101$

To convert 2421 code to decimal add the weight wherever 1 is present in the code.

$$1101 \xrightarrow[\text{of weights where 1 is available}]{\text{Take sum}} 2 + 4 + 0 + 1 = 7_{10}$$

$\therefore$ Decimal value of 2421 code $1101 = 7_{10}$

Q3.9 ***Convert 5_{10} to excess-3 code and verify by back converting to decimal.***

The excess-3 code can be obtained by adding 011_{10} to binary value of the decimal digit.

$$5_{10} \xrightarrow{\text{Convert to binary}} 1011_2 \xrightarrow{\text{Add 011}} 1110$$

$\therefore$ Excess-3 code for $5_{10} = 1110$

To convert excess-3 code to decimal, subtract 011_2 to get binary value and then convert binary to decimal.

$$1110 \xrightarrow{\text{Subtract 011}} 1011_2 \xrightarrow{\text{Convert to decimal}} 5_{10}$$

$\therefore$ Decimal value of Excess-3 code $1110 = 5_{10}$

Q3.10 ***Represent 3856_{10} in BCD and 2421 code?***

Solution:

$$3856_{10} \xrightarrow{\text{BCD}} 0011\ \ 1000\ \ 0101\ \ 0110$$

$$3856_{10} \xrightarrow{\text{2421 BCD}} 0011\ \ 1110\ \ 1011\ \ 1100$$

Refer Table 3.2

Q3.11 ***Convert decimal 8723_{10} to both BCD and ASCII with even parity.***

Solution:

$$8723 \xrightarrow{\text{BCD}} 1000\ \ 0111\ \ 0010\ \ 0011$$

ASCII is 7-bit binary code with reference to Table 3.11 the ASCII code for 8723_{10} are shown below. Here each decimal digit is represented by 7-bit ASCII code. The addition of even parity bit is also shown below:

$$8 \xrightarrow{\text{ASCII}} 38_H \xrightarrow{\text{Binary}} 011\ 1000 \xrightarrow{\text{Append parity bit}} 1011\ 1000$$

$$7 \xrightarrow{\text{ASCII}} 37_H \xrightarrow{\text{Binary}} 011\ 0111 \xrightarrow{\text{Append parity bit}} 1011\ 0111$$

$$2 \xrightarrow{\text{ASCII}} 32_H \xrightarrow{\text{Binary}} 011\ 0010 \xrightarrow{\text{Append parity bit}} 1011\ 0010$$

$$3 \xrightarrow{\text{ASCII}} 33_H \xrightarrow{\text{Binary}} 011\ 0011 \xrightarrow{\text{Append parity bit}} 0011\ 0011$$

$$8723_{10} \xrightarrow[\text{Append at left}]{\text{ASCII with even parity bit}} 1011\ 1000\ 1011\ 0111\ 1011\ 0010\ 0011\ 0011$$

Q3.12 Perform Excess-3 addition of decimal numbers. $2_{10} + 3_{10}$

Solution:

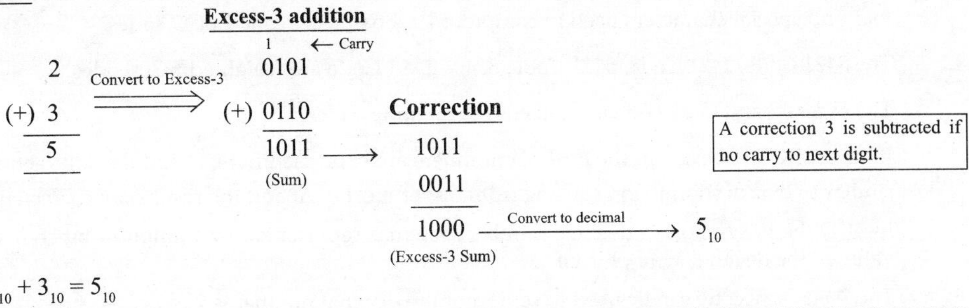

$\therefore 2_{10} + 3_{10} = 5_{10}$

3.8 Exercises

I. Fill in the blanks

1. In the digital systems, each decimal number or piece of information is represented by a combination of binary digits called ______.
2. The digital systems can perform any ______ or ______ operations on data or process any information represented in binary.
3. The codes that are used to represent decimal digits 0 to 9 of the decimal numbers are called ______ codes.
4. The codes that are used to represent decimal digits 0 to 9 of the decimal numbers, English alphabets A to Z and special characters used in computer keyboards are called ______ codes.
5. 8421, 2421, 5421 and 8, 4, –2, –1 are ______.
6. 2421 and excess-3 are ______.
7. ASCII was developed in the early stage of digital systems in order to standartize the coding system for ______ and ______ interface.
8. Gray code only one bit changes from either "0" to "1" or "1" to "0" between ______.
9. The black and white pattern are created such that 4-bit Gray code generated when sensed using ______ sensors.
10. Parity and checksum are popular method of error ______ and ______ is performed by sending a request to transmit the data.
11. Hamming code is another popular error deduction and correction technique to deduct and correct ______ error in the received data
12. Parity of an n-bit code can be modified by appending an extra bit is called ______.
13. The combined data and hamming bit is called ______.
14. ______ is a method to check for error in the transmitted digital code or information.
15. Checksum is useful for ______ size digital code, usually 32-bits or higher size.

Answers

1. codes	6. self complementing code	11. 1-bit
2. arithmetic and logical	7. human, machine	12. parity bit
3. BCD	8. two consecutive codes	13. hamming code
4. Alphanumeric	9. four light	14. checksum
5. weighted codes	10. detection and correction	15. large

II. State whether the following statements are True or False

1. The codes that are used to represent decimal digits 0 to 9 of the decimal numbers, English alphabets A to Z and special characters used in computer keyboards are called BCD codes.
2. The 8421 Code, 2421 Code, 5421 Code, 8, 4, –2, –1 Code are weighted BCD codes.
3. The excess-3 and Gray code are self complementing codes.
4. In direct BCD subtraction the decimal numbers are expressed in BCD and the subtraction is performed similar to binary subtraction starting from upper most digit and the results are corrected to BCD.
5. In BCD the 10 symbols of decimal number system is represented by a unique 4-bit code which is binary value of the decimal symbol itself.
6. The 2-bit Gray code can be used to represent the decimal number is 0 to 7.
7. In this Gray code, rotating machines the angular position of shaft and hence speed of rotating shaft of properly measured using encoding disk.
8. The hamming code is only error detecting code, it can accurately detect 1-bit error.
9. Parity and checksum are popular method of error detection and error correction is performed by sending a request to retransmit the data.
10. Checksum is useful for small size digit code, usually 32-bits or higher size.

Answers

1. False	6. False
2. True	7. True
3. False	8. False
4. False	9. True
5. True	10. False

III. Choose the right answer for the following questions

1. *The minimal function that can detect a 8421 BCD code (represented as $D_8\ D_4\ D_2\ D_1$) divisible by 3 is given by*

   a) $D_8D_1 + D_4D_2 + D_3D_2D_1$ b) $D_8D_1 + D_4D_2D_1' + D_4'D_2D_1$

   c) $D_4D_1 + D_4D_2 + D_3D_1'D_2'D_1$ d) $D_4 + D_2D_1' + D_4'D_2D_1 + D_8D_4D_2D_1$

2. *A new Binary Coded Pentary (BCP) number system is proposed in which every digit of a base-5 number is represented by its corresponding 3-bit binary code. For example, the base-5 number 24 will be represented by its BCP code 010100. In this numbering system, the BCP code 100010011001 corresponds to the following number in base-5 system.*

   a) 423   b) 1324

   c) 2201   d) 4231

3. *The following code is not a BCD code.*

   a) Excess-3 code   b) 2421 code

   c) 8421 code   d) Gray code

4. ***5421 code is***

   a) self complementing code
   b) weighted code
   c) non-weighted code
   d) alphanumeric code

5. ***ASCII and EBCDIC codes is***

   a) self complementing code
   b) BCD code
   c) error correcting codes
   d) alphanumeric code

6. ***ASCII code is***

   a) 5-bit code
   b) 7-bit code
   c) 8-bit code
   d) 10-bit code

7. ***EBCDIC code is***

   a) 5-bit code
   b) 7-bit code
   c) 8-bit code
   d) 10-bit code

8. ***The 7-bit Hamming code is used to transmit***

   a) 3 data bits
   b) 4 data bits
   c) 7 data bits
   d) 8 data bits

9. ***The word length of the Hamming code to represent decimal digits is***

   a) 4 b) 5 c) 6 d) 7

10. ***The number of parity bits in a 12-bit Hamming code is***

    a) 4 b) 5 c) 6 d) 7

Answers

1. b	6. b
2. d	7. c
3. d	8. b
4. b	9. d
5. d	10. a

IV. Answer the following questions

1. Define reflective and sequential codes.
2. Types of BCD codes.
3. Define BCD addition.
4. Convert 9_{10} to 5421 code and verify by back converting to decimal.
5. Convert 8_{10} to 8, 4, – 2, – 1 code and verify by back converting to decimal.
6. Define excess-3 code.
7. What is Gray code.
8. What is the advantage of Gray codes over the binary number sequence?

9. Applications for Gray code.
10. Define parity and parity bit and types.
11. Compare even and odd parity.
12. Explain the significance of parity?
13. What is Hamming code.
14. Write notes on error correcting codes.
15. What are error detecting codes?

V. Solve the following problems

E3.1 Represent the decimal number 5137 in a) BCD b) Excess-3 code c) 2421 code d) 8, 4, – 2, – 1 code.

E3.2 Generate the Hamming code for an 7-bit data string 1101010. Introduce error in a data bit and prove that Hamming code detect the error.

a) Choose even parity Hamming code

b) Choose odd parity Hamming code

E3.3 Generate the Hamming code for an 6-bit data string 101110. Introduce error in a data bit and prove that Hamming code detect the error.

a) Choose even parity Hamming code

b) Choose odd parity Hamming code

E3.4 Convert the binary to Gray code and verify by converting the Gray code to binary.

a) 111011 b) 10111011

E3.5 Perform BCD addition of following decimal numbers.

a) $33_{10} + 12_{10}$ b) $425_{10} + 634_{10}$

E3.6 Perform the following subtraction by direct or actual BCD subtraction.

a) $57_{10} - 29_{10}$ b) $503.7_{10} - 425.2_{10}$ c) $54_{10} - 95_{10}$

E3.7 Perform the following subtraction by 9's complement BCD subtraction and cross-check result by direct 9's complement subtraction.

a) $355.5_{10} - 188.8_{10}$ b) $513.3_{10} - 785.7_{10}$

E3.8 Perform the following subtraction by 10's complement BCD subtraction and cross-check result by direct 10's complement subtraction.

a) $324.5_{10} - 104.9_{10}$ b) $202.4_{10} - 309.3_{10}$

E3.9 Convert 8_{10} to 2421 code and verify by back converting to decimal.

E3.10 Convert 9_{10} to 5421 code and verify by back converting to decimal.

E3.11 Convert 5_{10} to 8, 4, – 2, – 1 code and verify by back converting to decimal.

E3.12 Convert 6_{10} to excess-3 code and verify by back converting to decimal.

E3.13 Perform Excess-3 addition of following decimal numbers.

a) $2_{10} + 4_{10}$ **b)** $43_{10} + 28_{10}$ **c)** $158.2_{10} + 379.6_{10}$

E3.14 Perform the following subtraction by direct or actual Excess-3 subtraction.

a) $220_{10} - 165_{10}$ **b)** $31.3_{10} - 12.1_{10}$

E3.15 Perform the following subtraction by 9's complement Excess-3 subtraction and cross-check result by direct 9's complement subtraction.

a) $458_{10} - 282_{10}$ **b)** $189_{10} - 309_{10}$

E3.16 Perform the following subtraction by 10's complement Excess-3 subtraction and cross-check result by direct 10's complement subtraction.

a) $355_{10} - 239_{10}$ **b)** $250_{10} - 536_{10}$

E3.17 Find the checksum for the following 32-bit binary code.

10011001 11100010 00100100 10000100

a) Transmit the code without error and check at receiver side.

b) Transmit the code with one bit error and check at receiver side.

E3.18 Find the checksum for the following 16-bit binary code.

01111100 00100100

a) Transmit the code without error and check at receiver side.

b) Transmit the code with one bit error and check at receiver side.

Answers

E3.1 a) 0101 0001 0011 0111 **c)** 1011 0001 0011 1101	**b)** 1000 0100 0110 1010 **d)** 1011 0111 0101 1001	
E3.2 a) 11101011010	**b)** 00111110010	
E3.3 a) 1110011110	**b)** 0010011010	
E3.4 a) 111011	**b)** 10111011	
E3.5 a) $33_{10} + 12_{10} = 45_{10}$	**b)** $425_{10} + 634_{10} = 1059_{10}$	
E3.6 a) $57_{10} - 29_{10} = 28_{10}$	**b)** $503.7_{10} - 425.2_{10} = 78.5_{10}$	**c)** $54_{10} - 95_{10} = -41_{10}$
E3.7 a) $355.5_{10} - 188.8_{10} = 166.7_{10}$	**b)** $513.3_{10} - 785.7_{10} = -272.4_{10}$	
E3.8 a) $324.5_{10} - 104.9_{10} = 219.6_{10}$	**b)** $202.4_{10} - 309.3_{10} = -106.9_{10}$	
E3.9 8_{10}	**E3.10** 9_{10}	
E3.11 5_{10}	**E3.12** 6_{10}	

E3.13	**a)** $2_{10} + 4_{10} = 6_{10}$	**b)** $43_{10} + 28_{10} = 71_{10}$	**c)** $158.2_{10} + 379.6_{10} = 537.8_{10}$
E3.14	**a)** $220_{10} - 165_{10} = 55_{10}$	**b)** $31.3_{10} - 12.1_{10} = 19.2_{10}$	
E3.15	**a)** $458_{10} - 282_{10} = 176_{10}$	**b)** $189_{10} - 309_{10} = 120_{10}$	
E3.16	**a)** $355_{10} - 239_{10} = 116_{10}$	**b)** $250_{10} - 534_{10} = 284_{10}$	
E3.17	Checksum = 11011010		
E3.18	Checksum = 01011111		

CHAPTER 4

BOOLEAN ALGEBRA AND LOGIC GATES

4.1 Boolean Algebra

(JNTUK, Feb'22, 7 Marks)

George Boole developed Boolean algebra in **1854**.

Boolean algebra is an algebraic structure that includes a set of elements consisting of

- Binary operators, " + " and " . "
- Binary variables, x, y, z .....
- Binary elements, 0 and 1
- Boolean postulates and theorems

Let x and y be two boolean variables that can take all possible combination of binary value. The rules for binary operators " + " and " . " are listed in Table 4.1 using the two binary variables x and y.

Table 4.1: Rules for Binary Operators " + " and " . "

x	y	x . y	x + y	x	x'
0	0	0	0	0	1
0	1	0	1	1	0
1	0	0	1		
1	1	1	1		

Operator Precedence while Evaluating Boolean Expression

A statement written with Boolean variables, constants and operators is called ***Boolean expression***.

Examples of Boolean expression: $x + y \,.\, z'$

$$(x \,.\, y) + z' + (x + y \,.\, z)'$$

In order to evaluate Boolean expression the operator precedence is as follows:

Step-1: Evaluate expression with in paranthesis, i.e, within ( ).

Step-2: Evaluate complement

Step-3: Evaluate " . "

Step-4: Evaluate " + "

> ***Note:*** *To simplify Boolean expression dot operation is represented without operator.*
>
> $x \,.\, y \Rightarrow xy$ $\qquad y \,.\, z \Rightarrow yz$ $\qquad x \,.\, y \,.\, z \Rightarrow xyz$

Example 4.1

Demonstrate the operator precedence in the evaluation of following Boolean expression.

a) x + y . z'   **b)** (x.y) + z' + (x + y . z)'

Solution

a) x + y . z'

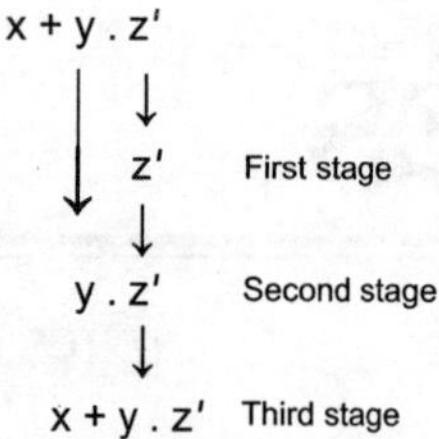

b) (x.y) + z' + (x + y . z)'

(x . y) + z' + (x + y . z)'

↓ ↓ ↓

(x . y)   z'   (y . z)   First stage

↓

x + (y . z)   Second stage

↓

(x + y . z)'   Third stage

↓ ↓ ↓

(x . y) + z' + (x + y . z)'   Fourth stage

4.1.1 Duality

The ***duality*** principle of Boolean algebra states that every algebraic expression deducible from the postulates of Boolean algebra remains valid if the operator and identity element are interchanged. Here, the operators are " + " and " . " and the identity elements are " 0 " and " 1 ".

Table 4.2: Basic Example of Duality

x	y	x + y
0	0	0
0	1	1
1	0	1
1	1	1

Change 0 to 1 and 1 to 0 →

1	1	1
1	0	0
0	1	0
0	0	0

Change "+" to "." →

x	y	x . y
0	0	0
0	1	0
1	0	0
1	1	1

4.1.2 Postulates of Boolean Algebra

Postulates of Boolean algebra are developed by **E. V. Huntington** in **1904.**

Boolean structure have to satisfy the following postulates.

1. a) The structure is closed with the respect to operator " + "

b) The structure is closed with respect to operator " . "

2. a) The element 0 is an identity element with respect to operator " + "

$\therefore$ $x + 0 = x$

$0 + x = x$

where x is a Boolean variable that can take a value either 0 or 1.

b) The element 1 is an identity element with respect to operator " . "

$\therefore$ $x . 1 = x$

$1 . x = x$

where x is Boolean variable that can take a value either 0 or 1.

3. a) The structure is commutative with respect to operator " + "

$\therefore$ $x + y = y + x$

where x and y are two Boolean variables.

b) The structure is commutative with respect to operator " ."

$\therefore$ $x . y = y . x$

4. a) The operator " . " is distributive over operator " + " Distributive Law

$\therefore$ $x . (y + z) = (x . y) + (x . z)$

b) The operator " + " is distributive over operator " . "

$\therefore$ $x + (y . z) = (x + y) . (x + z)$

5. a) For every variable x there exists an element x' called complement of x

$\therefore$ $x + x' = 1$ and $x . x' = 0$

6. There exists at least two variables x and y such that, $x \neq y$.

Verification of Postulates

1. From the Table 4.1 it is obvious that results of "+" and "." operations are either 0 or 1 and so structure is closed with respect to operators " + " and " . "

2. $x + 0 = x$, when $x = 0$, $x + 0 = 0 + 0 = 0 = x$

$x = 1$, $x + 0 = 1 + 0 = 1 = x$

$0 + x = x$, when $x = 0$, $0 + x = 0 + 0 = 0 = x$

$x = 1$, $0 + x = 0 + 1 = 1 = x$

3. $x . 1 = x$, when $x = 0$, $x . 1 = 0 . 1 = 0 = x$

$x = 1$, $x . 1 = 1 . 1 = 1 = x$

$1 . x = x$, when $x = 0$, $1 . x = 1 . 0 = 0 = x$

$x = 1$, $1 . x = 1 . 1 = 1 = x$

4. Let x, y, z be three Boolean variables that takes all possible combinations of binary value.

Construct a truth table for all possible combinations of x, y and z as shown in Table 4.3 to prove,

$x . (y + z) = (x . y) + (x . z)$

Table 4.3: Verification of (Distributive Law) x . (y + z) = (x . y) + (x . z) *(JNTUH, Nov/Dec'20, 7 Marks)*

x	y	z	y + z	x . (y + z)	x . y	x . z	(x . y) + (x . z)
0	0	0	0	0	0	0	0
0	0	1	1	0	0	0	0
0	1	0	1	0	0	0	0
0	1	1	1	0	0	0	0
1	0	0	0	0	0	0	0
1	0	1	1	1	0	1	1
1	1	0	1	1	1	0	1
1	1	1	1	1	1	1	1

Similarly, Table 4.4 is constructed to prove,

$x + (y \cdot z) = (x + y) \cdot (x + z)$

Table 4.4: Verification of (Distributive Law) x + (y . z) = (x + y) . (x + z)

x	y	z	y . z	x + (y . z)	(x + y)	(x + z)	(x + y) . (x + z)
0	0	0	0	0	0	0	0
0	0	1	0	0	0	1	0
0	1	0	0	0	1	0	0
0	1	1	1	1	1	1	1
1	0	0	0	1	1	1	1
1	0	1	0	1	1	1	1
1	1	0	0	1	1	1	1
1	1	1	1	1	1	1	1

5. $x + x' = 1$

when $x = 0,\ x' = 1, \quad \therefore\ x + x' = 0 + 1 = 1$

when $x = 1,\ x' = 0, \quad \therefore\ x + x' = 1 + 0 = 1$

$x \cdot x' = 0$

when $x = 0,\ x' = 1, \quad \therefore\ x \cdot x' = 0 \cdot 1 = 0$

when $x = 1,\ x' = 0, \quad \therefore\ x \cdot x' = 1 \cdot 0 = 0$

6. Boolean algebra has two elements 1 and 0, where $1 \neq 0$.

$\therefore$ If x and y are two variables, then there is a possibility that

$x = 0,\ y = 1, \quad \therefore\ x \neq y$

and $x = 1,\ y = 0, \quad \therefore\ x \neq y$

4.1.3 Boolean Theorems

(JNTUK, Feb'22, 7 Marks) (RGPV, Jun'20, 5 Marks) (GTU, Jun'19, 4 Marks)

Theorem 1: $x + x = x$

$x . x = x$

Theorem 2: $x + 1 = 1$

$x . 0 = 0$

Theorem 3: (Involution) $(x')' = x$

Theorem 4: (Associative)

$x + (y + z) = (x + y) + z$

$x . (y . z) = (x . y) . z$

Theorem 5: (DeMorgan's Law or DeMorgan's Theorem)

$(x + y)' = x' . y'$

$(x . y)' = x' + y'$

Theorem 6: (Absorption)

$x + (x . y) = x$

$x . (x + y) = x$

(y is absorbed)

Theorem 7: (Consensus Theorem)

$x . y + x' . z + y . z = x . y + x' . z$

$(x + y) . (x' + z) . (y + z) = (x + y) . (x' + z)$

Proof of Theorems

Theorem 1:

$x + x = x$	$x . x = x$
when, $x = 0$, $x + x = 0 + 0 = 0 = x$	when, $x = 0$, $x . x = 0 . 0 = 0 = x$
when, $x = 1$, $x + x = 1 + 1 = 1 = x$	when, $x = 1$, $x . x = 1 . 1 = 1 = x$

Theorem 2:

$x + 1 = 1$	$x . 0 = 0$
when, $x = 0$, $x + 1 = 0 + 1 = 1$	when, $x = 0$, $x . 0 = 0 . 0 = 0$
when, $x = 1$, $x + 1 = 1 + 1 = 1$	when, $x = 1$, $x . 0 = 1 . 0 = 0$

Theorem 3: Involution

$(x')' = x$

when, $x = 0$, $(x')' = (0')' = 1' = 0 = x$

when, $x = 1$, $(x')' = (1')' = 0' = 1 = x$

Theorem 4: Associative *(JNTUH, Nov/Dec'20, 7 Marks)*

$x + (y + z) = (x + y) + z$

Construct a truth table for all possible combinations of x, y and z as shown in Table 4.5 to prove,

$x + (y + z) = (x + y) + z$

$x . (y . z) = (x . y) . z$

Construct a truth table for all possible combinations of x, y and z as shown in Table 4.6 to prove,

$x . (y . z) = (x . y) . z$

Table 4.5: Verification of x + (y + z) = (x + y) + z

x	y	z	y + z	x + (y + z)	x + y	(x + y) + z
0	0	0	0	0	0	0
0	0	1	1	1	0	1
0	1	0	1	1	1	1
0	1	1	1	1	1	1
1	0	0	0	1	1	1
1	0	1	1	1	1	1
1	1	0	1	1	1	1
1	1	1	1	1	1	1

Table 4.6: Verification of x . (y . z) = (x . y) . z

x	y	z	y . z	x . (y . z)	x . y	(x . y) . z
0	0	0	0	0	0	0
0	0	1	0	0	0	0
0	1	0	0	0	0	0
0	1	1	1	0	0	0
1	0	0	0	0	0	0
1	0	1	0	0	0	0
1	1	0	0	0	1	0
1	1	1	1	1	1	1

Theorem 5: DeMorgan's Theorem *(RGPV, Jun'22, 5 Marks)* *(GTU, May'19, 7 Marks)*

$(x + y)' = x' . y'$

Construct a truth table for all possible combinations of x and y as shown in Table 4.7 to prove,

$(x + y)' = x' . y'$

$(x . y)' = x' + y'$

Construct a truth table for all possible combinations of x and y as shown in Table 4.8 to prove,

$(x . y)' = x' + y'$

Table 4.7: Verification of (x + y)′ = x′ . y′

x	y	x + y	(x + y)′	x′	y′	x′ . y′
0	0	0	1	1	1	1
0	1	1	0	1	0	0
1	0	1	0	0	1	0
1	1	1	0	0	0	0

Table 4.8: Verification of (x . y)′ = x′ + y′

x	y	x . y	(x . y)′	x′	y′	x′ + y′
0	0	0	1	1	1	1
0	1	0	1	1	0	1
1	0	0	1	0	1	1
1	1	1	0	0	0	0

Theorem 6: Absorption

$x + (x . y) = x$

Construct a truth table for all possible combinations of x and y as shown in Table 4.9 to prove,

$x + (x . y) = x$

$x . (x + y) = x$

Construct a truth table for all possible combinations of x and y as shown in Table 4.10 to prove,

$x . (x + y) = x$

Table 4.9: Verification of x + (x . y) = x

x	y	x . y	x + (x . y)
0	0	0	0
0	1	0	0
1	0	0	1
1	1	1	1

Table 4.10: Verification of x . (x + y) = x

x	y	x + y	x . (x + y)
0	0	0	0
0	1	1	0
1	0	1	1
1	1	1	1

Theorem 7: Consensus Theorem

$x . y + x' . z + y . z = x . y + x' . z$

Construct a truth table for all possible combinations of x, y and z as shown in Table 4.11 to prove,

$x . y + x' . z + y . z = x . y + x' . z$

$(x + y) . (x' + z) . (y + z) = (x + y) . (x' + z)$

Construct a truth table for all possible combinations of x, y and z as shown in Table 4.12 to prove,

$(x + y) . (x' + z) . (y + z) = (x + y) . (x' + z)$

Table 4.11: Verification of x . y + x′ . z + y . z = x . y + x′ . z

x	y	z	x′	x . y	x′ . z	y . z	x.y + x′.z + y.z	x.y + x′.z
0	0	0	1	0	0	0	0	0
0	0	1	1	0	1	0	1	1
0	1	0	1	0	0	0	0	0
0	1	1	1	0	1	1	1	1
1	0	0	0	0	0	0	0	0
1	0	1	0	0	0	0	0	0
1	1	0	0	1	0	0	1	1
1	1	1	0	1	0	1	1	1

Table 4.12: Verification of (x + y) . (x′ + z) . (y + z) = (x + y) . (x′ + z)

x	y	z	x′	x + y	x′ + z	y + z	(x + y) . (x′ + z) . (y + z)	(x + y) . (x′ + z)
0	0	0	1	0	1	0	0	0
0	0	1	1	0	1	1	0	0
0	1	0	1	1	1	1	1	1
0	1	1	1	1	1	1	1	1
1	0	0	0	1	0	0	0	0
1	0	1	0	1	1	1	1	1
1	1	0	0	1	0	1	0	0
1	1	1	0	1	1	1	1	1

4.2 Binary Logic

Binary logic consists of binary variables and a set of logical operations. The binary variables are denoted by letters of alphabet, A, B, C, ..., x, y, z, ...

Each binary variable can take only two possible values 0 or 1. In positive logic, 0 is **low** and 1 is **high**. The basic logical operations are AND, OR and NOT.

AND operation is same as " **.** " operation. In positive logic, the logical AND of two or more variables will be 1 if and only if the value of all the variables is 1.

OR operation is same as " + " operation. In positive logic, the logical OR of two or more variables will be 1 if the value of any one of the variables is 1.

NOT operation is same as complement operation.

The results or outputs of logical operations of two or more variables for all possible combinations of the variables can be listed in a table called ***truth table***. The truth tables of basic logical operations are shown in Table 4.13.

Table 4.13: Truth Tables of AND, OR and NOT Operation of Two Variables

AND operation

x	y	x . y
0	0	0
0	1	0
1	0	0
1	1	1

OR operation

x	y	x + y
0	0	0
0	1	1
1	0	1
1	1	1

NOT operation

x	x'
0	1
1	0

Positive and Negative Logic

Binary constants are 1 and 0. Physically in a digital circuit they represent two voltage levels.

The two voltage levels are called **high** and **low.**

The voltage level of **high** and **low** depends on technology used to fabricate the gates.

In TTL logic (Transistor Transistor Logic) **high** is + 5 *V* and **low** is 0 *V*.

In CMOS (Complementry Symmetry MOSFET) **high** is +3.3 *V* and **low** is 0 *V*.

If, 1 represent **high** voltage and 0 represent **low** voltage then logic system is called ***positive logic.***

If, 1 represent **low** voltage and 0 represent **high** voltage then the logic system is called ***negative logic***.

Table 4.14: Positive and Negative Logic Levels

Logic Type (or Logic System)	Technology	
	TTL	CMOS
Positive Logic	1 = High = +5 *V* 0 = Low = 0 *V*	1 = High = + 3.3 *V* 0 = Low = 0 *V*
Negative Logic	0 = High = + 5 *V* 1 = Low = 0 *V*	0 = High = + 3.3 *V* 1 = Low = 0 *V*

4.3 Logic Gates

(RGPV, Jun'20, 10 Marks)

Logic gates are electronic devices or circuits that perform logical operations on one or more input logical variables and produce a binary output. The basic logic gates are AND, OR and NOT gates.

The logic gates have standard symbols as shown in Figs. 4.1 to 4.5.

AND Gate

AND gate is an electronic device that performs logical AND operation of two or more variables.

Fig. 4.1: Two input AND gate.

Fig. 4.2: Three input AND gate.

OR Gate

OR gate is an electronic device that performs logical OR operation of two or more variables.

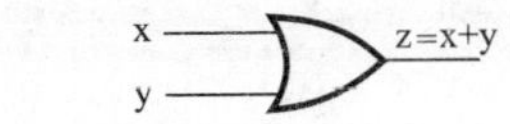

Fig. 4.3: Two input OR gate.

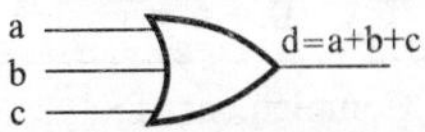

Fig. 4.4: Three input OR gate.

> ***Note:*** *Theoretically, AND and OR gates can have any number of inputs.*

NOT Gate (or inverter)

NOT gate is an electronic device that performs complement operation of a Boolean variable.

Fig. 4.5: NOT gate.

Other Logical Operations

With n variables it is possible to form 2^{2n} Boolean functions.

When, $n = 2$, $\quad 2^{2n} = 2^{2\times2} = 2^4 = 16$

Therefore, with two variables we can form 16 Boolean functions.

All the 16 possible function of two variables are listed in Table 4.15.

The output of 16 functions are 16 binary combinations of 4-bit binary.

Let, F_0, F_1, F_2, ..., F_{15} be 16 possible Boolean functions for two variables x, y.

Table 4.15: Two-Variable Boolean Functions

Functions	x	y	F	Functions	x	y	F
Null operation	0	0	0	NOR operation	0	0	1
$F = F_0 = 0$	0	1	0	$F = F_8 = (x + y)'$	0	1	0
	1	0	0		1	0	0
	1	1	0		1	1	0
AND operation	0	0	0	Exclusive NOR(XNOR)	0	0	1
$F = F_1 = x \cdot y$	0	1	0	(or equivalence)	0	1	0
	1	0	0	$F = F_9 = x \odot y = (x \oplus y)'$	1	0	0
	1	1	1	$= x \cdot y + x' \cdot y'$	1	1	1
Inhibition	0	0	0	Complement of y	0	0	1
$F = F_2 = x \cdot y'$	0	1	0	$F = F_{10} = y'$	0	1	0
(x true, but y false)	1	0	1		1	0	1
	1	1	0		1	1	0
Transfer	0	0	0	Implication	0	0	1
$F = F_3 = x$	0	1	0	$F = F_{11} = x + y'$	0	1	0
(Transfer x)	1	0	1	(If $y = 1, F = x$	1	0	1
	1	1	1	If $y = 0, F = y'$)	1	1	1
Inhibition	0	0	0	Complement of x	0	0	1
$F = F_4 = x' \cdot y$	0	1	1	$F = F_{12} = x'$	0	1	1
(y true, but x false)	1	0	0		1	0	0
	1	1	0		1	1	0
Transfer	0	0	0	Implication	0	0	1
$F = F_5 = y$	0	1	1	$F = F_{13} = x' + y$	0	1	1
(Transfer y)	1	0	0	(If $x = 1, F = y$	1	0	0
	1	1	1	If $x = 0, F = x'$)	1	1	1
XOR operation	0	0	0	NAND operation	0	0	1
$F = F_6 = x \oplus y$	0	1	1	$F = F_{14} = (x \cdot y)'$	0	1	1
$= x \cdot y' + x' \cdot y$	1	0	1		1	0	1
	1	1	0		1	1	0
OR operation	0	0	0	Identity	0	0	1
$F = F_7 = x + y$	0	1	1	$F = F_{15} = 1$	0	1	1
	1	0	1		1	0	1
	1	1	1		1	1	1

Summary of Logic Gates

The AND, OR and NOT are basic gates. Using these basic gates some more useful logical operations can be defined. They are NAND, NOR, Exclusive-OR (XOR) and Exclusive-NOR (XNOR).

Besides logic gates, buffers or drivers are also used in digital circuits to augment or increase current levels of signals where ever required.

The summary of popular logic gates used in digital electronics are listed in Table 4.16.

Table 4.16: Summary of Popular Logic Gates

Gate	Symbol	Truth Table
AND Gate	x y x.y	x y | x . y 0 0 | 0 0 1 | 0 1 0 | 0 1 1 | 1
OR Gate	x y x+y	x y | x + y 0 0 | 0 0 1 | 1 1 0 | 1 1 1 | 1
Inverter Gate (or NOT gate)	x x′	x | x′ 0 | 1 1 | 0
Buffer	x x	x 0 1
NAND Gate	x y (x.y)′	x | y | x.y | (x.y)′ 0 | 0 | 0 | 1 0 | 1 | 0 | 1 1 | 0 | 0 | 1 1 | 1 | 1 | 0

Table 4.16: Continued...

Gate	Symbol	Truth Table
NOR Gate	x, y → (x + y)′	x | y | x + y | (x + y)′ 0 | 0 | 0 | 1 0 | 1 | 1 | 0 1 | 0 | 1 | 0 1 | 1 | 1 | 0
Exclusive-OR Gate (XOR)	x, y → x.y′ + x′.y = x ⊕ y	x | y | x . y′ | x′ . y | x ⊕ y 0 | 0 | 0 | 0 | 0 0 | 1 | 0 | 1 | 1 1 | 0 | 1 | 0 | 1 1 | 1 | 0 | 0 | 0
Exclusive-NOR Gate (XNOR) (or equivalence)	x, y → x.y + x′.y′ = (x ⊕ y)′ = x ⊙ y	x | y | x⊕y | (x⊕y)′ = x ⊙ y 0 | 0 | 0 | 1 0 | 1 | 1 | 0 1 | 0 | 1 | 0 1 | 1 | 0 | 1

***Note:** Bubble at output of gate indicates NOT operation.*

Example 4.2

(JNTUK, Feb'22, 7 Marks)

Prove the following identity.

a) AB + A(B + C) + B (B + C) = B + AC

b) AB + A′B + A′B′ = A′ + B

c) AB′C + A′BC + ABC = AC + BC

Solution

a) AB + A(B + C) + B (B + C) = AB + ~~AB~~ + AC + BB + BC   [x + x = x]

= AB + AC + B + BC   [Repeated terms are considered once.] [xx = x]

= B(A + 1 + C) + AC = B.1 + AC   [1 + x = 1]

= B + AC

b) AB + A′B + A′B′ = AB + A′(B + B′)   [x + x′ = 1]

= AB + A′.1 = AB + A′

= (A′ + A) (A′ + B) = 1.(A′ + B)

= A′ + B

c) AB′C + A′BC + ABC = ABC + AB′C + A′BC + ABC

= AC (B + B′) + BC (A + A′)   [x + x′ = 1]

= AC + BC

Example 4.3

(IKGPTU, Dec'19, 5 Marks)

Prove the following identity using Boolean algebra:

$(A + B)(A + (AB)')C + A'(B + C') + A'B + ABC = C(A + B) + A'(B + C')$

Solution

$L.H.S = (A + B)(A + (AB)')C + A'(B + C') + A'B + ABC$

$R.H.S = C(A + B) + A'(B + C')$

$L.H.S = (A + B)(A + (AB)')C + A'(B + C') + A'B + ABC$

$= (A + B)(A + A' + B')C + A'B + A'C' + B(A' + AC)$ — Using DeMorgan's theorem

$= (A + B)(1 + B')C + A'B + A'C' + B((A' + A)(A' + C))$ — $x + x' = 1$

$= (A + B)1.C + A'B + A'C' + B(1.(A' + C))$

$= AC + BC + A'B + A'C' + B(A' + C)$

$= AC + BC + A'B + A'C' + \cancel{A'B} + \cancel{BC}$ — Repeated terms are considered once.

$= AC + BC + A'C' + A'B$

$= C(A + B) + A'(B + C')$

$= RHS$

Example 4.4

(AKUB, Dec'20, 6 Marks)

Prove the following:

a) $A \oplus B = A' \oplus B'$ **b)** $(A \oplus B)' = A \oplus B' = A' \oplus B$

Solution

a) $A \oplus B = A' \oplus B'$

$A' \oplus B' = A'(B')' + (A')'B'$ — $x \oplus y = xy' + x'y$

$= A'B + AB' = A \oplus B$

---

b) $(A \oplus B)' = A \oplus B' = A' \oplus B$

$A \oplus B' = A(B')' + A'B'$

$= AB + A'B' = (A \oplus B)'$

$A' \oplus B = A'B' + (A')'B$

$= A'B' + AB = (A \oplus B)'$

4.3.1 Universal Gates

(RGPV, Jun'22, 7 Marks) (GTU, May'19, 7 Marks)

NAND and NOR gates are called ***universal gates***, because any Boolean function can be realized only using NAND gates or only using NOR gates.

The NAND gate is a combination of AND followed by NOT gate. The NOR gate is a combination of OR followed by NOT gate.

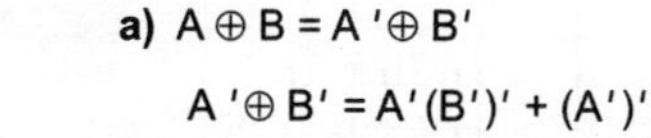

x, y → x.y → (x.y)' ⇒ x, y → (x.y)'

Fig. 4.6: NAND gate.

x, y → x + y → (x + y)' ⇒ x, y → (x + y)'

Fig. 4.7: NOR gate.

Table 4.17: Truth Table of NAND Gate

x	y	(x . y)′
0	0	1
0	1	1
1	0	1
1	1	0

Table 4.18: Truth Table of NOR Gate

x	y	(x + y)′
0	0	1
0	1	0
1	0	0
1	1	0

The realization of basic gates using only NAND gates are shown in Table 4.19.

The realization of basic gates using only NOR gates are shown in Table 4.20.

Table 4.19: Realization of Basic Logic Gates using NAND Gates

Gate	Logic Circuit using NAND	Truth Table
Inverter/NOT gate using NAND	 x → (x.x)′=x′	x | x′: 0 | 1; 1 | 0
AND using NAND	x, y → (x.y)′ → ((x.y)′)′ = x.y	x y | x . y: 0 0 | 0; 0 1 | 0; 1 0 | 0; 1 1 | 1
OR using NAND	x → x′; y → y′; (x′.y′)′ = x + y	x y | x + y: 0 0 | 0; 0 1 | 1; 1 0 | 1; 1 1 | 1
NOR using NAND	x → x′; y → y′; (x′.y′)′ → x′.y′=(x+y)′	x y | x + y | (x + y)′: 0 0 | 0 | 1; 0 1 | 1 | 0; 1 0 | 1 | 0; 1 1 | 1 | 0
XOR using NAND	x → x′; y → y′; (x′.y)′; (x.y′)′; ((x′.y)′.(x.y′)′)′ = x′.y + x.y′ = x ⊕ y	x y | x . y′ | x′ . y | x ⊕ y: 0 0 | 0 | 0 | 0; 0 1 | 0 | 1 | 1; 1 0 | 1 | 0 | 1; 1 1 | 0 | 0 | 0 $x \oplus y = x y' + x' y$

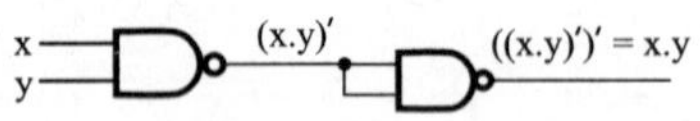

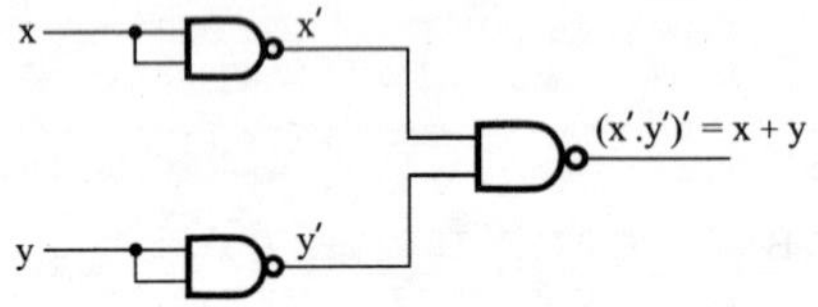

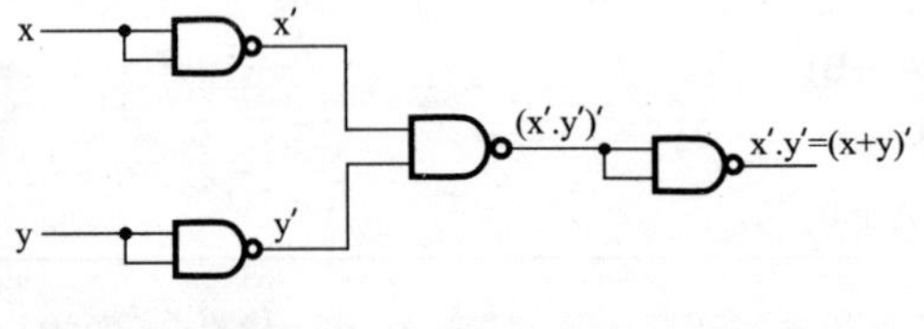

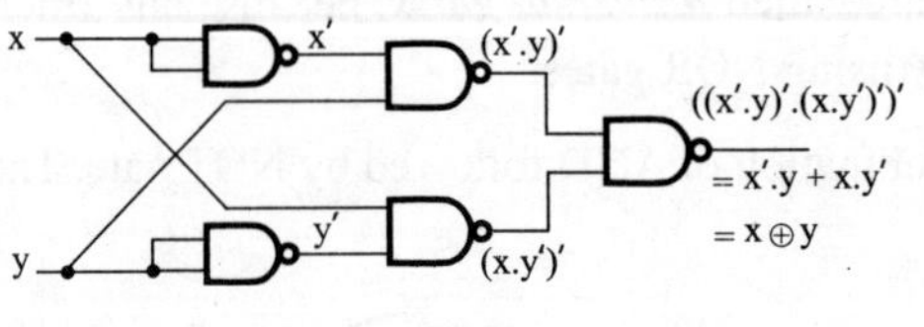

Table 4.20: Realization of Basic Logic Gates using NOR Gates

(RGPV, Jun'20, 10 Marks)

Gate	Logic Circuit using NOR	Truth Table
Inverter/NOT gate using NOR	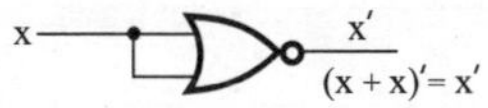 x → NOR → x′; $(x + x)' = x'$	x | x′: 0 | 1; 1 | 0
OR using NOR	x, y → NOR → $(x+y)'$ → NOR → $((x+y)')' = x+y$	x y | x + y: 0 0 | 0; 0 1 | 1; 1 0 | 1; 1 1 | 1
AND using NOR	x → NOR → x′; y → NOR → y′; x′, y′ → NOR → $(x'+y')' = x.y$	x y | x . y: 0 0 | 0; 0 1 | 0; 1 0 | 0; 1 1 | 1
NAND using NOR	x → NOR → x′; y → NOR → y′; x′, y′ → NOR → $(x'+y')'$ → NOR → $x'+y'=(x.y)'$	x y | x.y | (x.y)′: 0 0 | 0 | 1; 0 1 | 0 | 1; 1 0 | 0 | 1; 1 1 | 1 | 0
XOR using NOR	***(JNTUK, Feb'22, 8 Marks)*** x → NOR → x′; y → NOR → y′; $(x' + y)' = x.y'$; $(x + y')' = x'.y$; $(x.y' + x'.y)'$; $x.y' + x'.y = x\oplus y$	x y | x . y′ | x′. y | x ⊕ y: 0 0 | 0 | 0 | 0; 0 1 | 0 | 1 | 1; 1 0 | 1 | 0 | 1; 1 1 | 0 | 0 | 0. $x\oplus y = x.y' + x'.y$

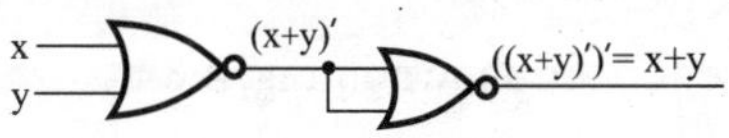

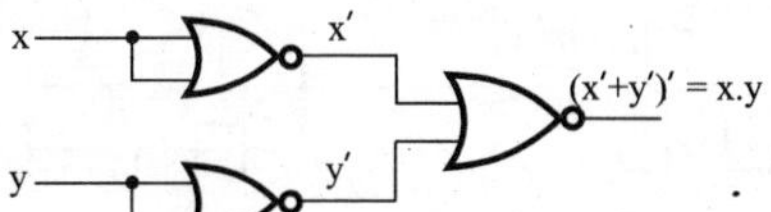

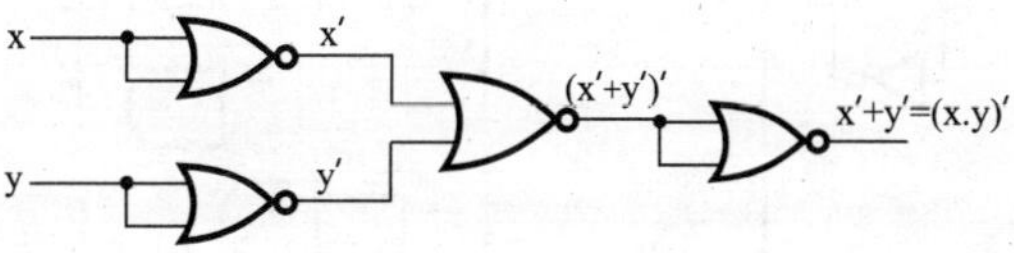

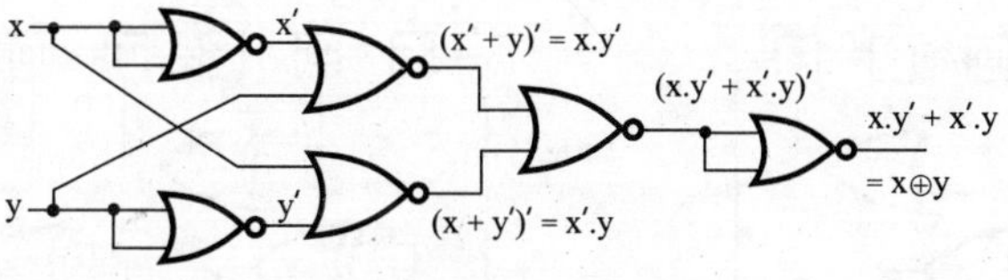

Commercially Available Logic Gates

Logic gate devices are available in the form of ICs manufactured using semiconductors. Usually 4 to 6 logic gates will be packed in a single IC package and will have a standard number. Some of the logic gate devices are,

7400 : 2-input NAND gate, pack of 4 gates

7402 : 2-input NOR gate, pack of 4 gates

7404 : Inverter gate, pack of 6 gates

7408 : 2-input AND gate, pack of 4 gates

7432 : 2-input OR gate, pack of 4 gates

7486 : 2-input XOR gate, pack of 4 gates

The pin diagram of the above ICs are shown in Fig. 4.8.

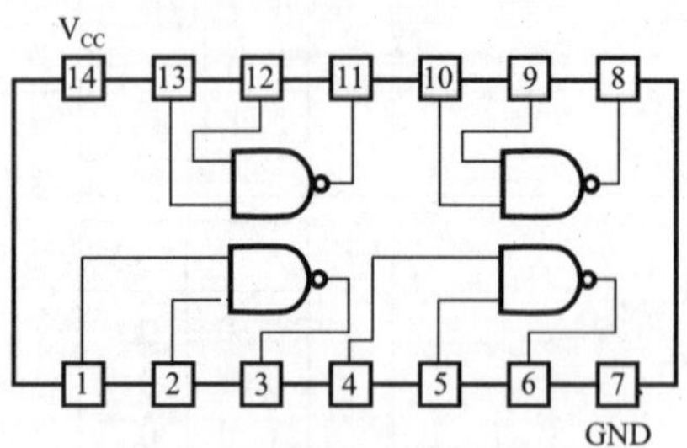

***Fig. a:** 2-input NAND gate 7400.*

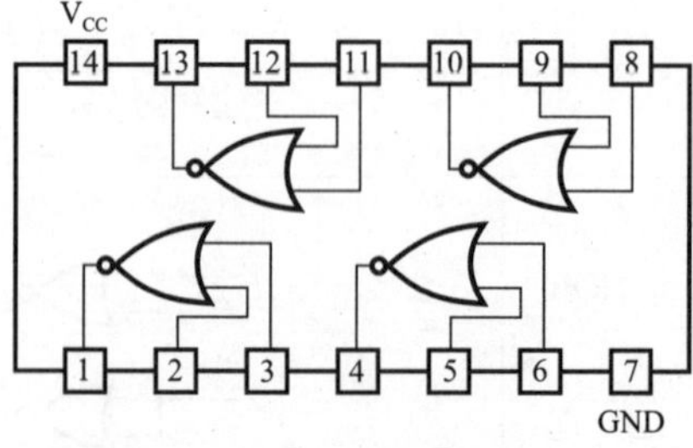

***Fig. b:** 2-input NOR gate 7402.*

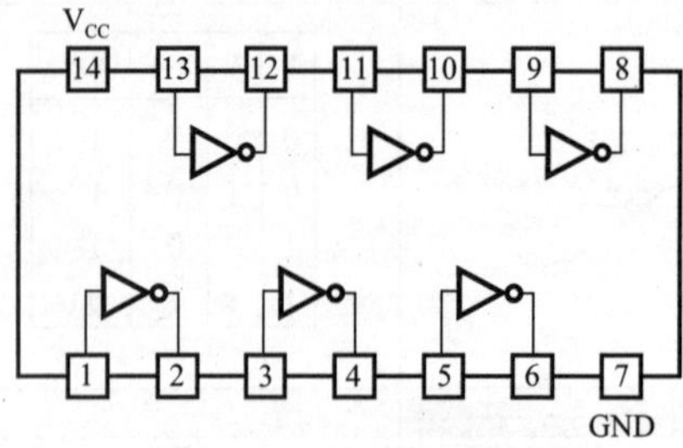

***Fig. c:** Inverters 7404.*

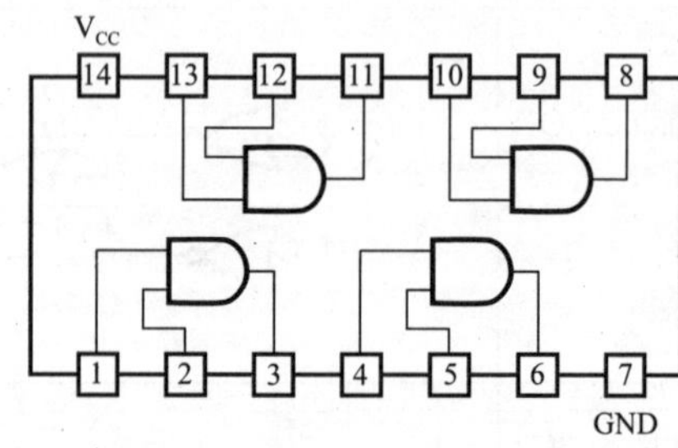

***Fig. d:** 2-input AND gate 7408.*

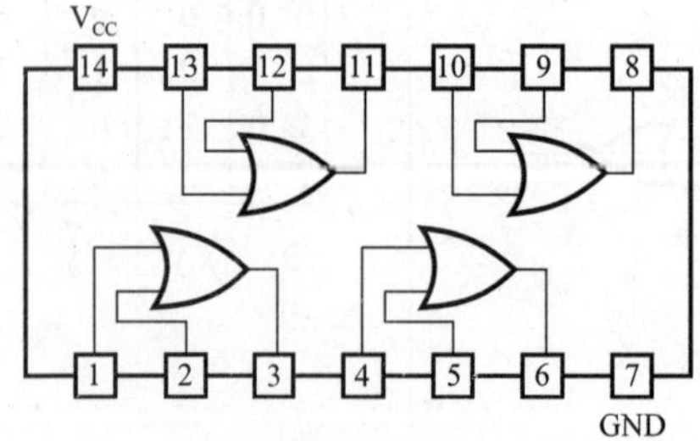

***Fig. e:** 2-input OR gate 7432.*

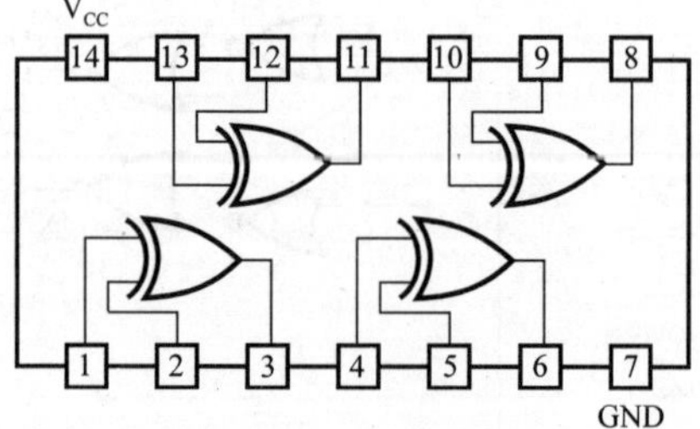

***Fig. f:** 2-input XOR gate 7486.*

***Fig. 4.8:** Digital gates in IC packages with identification numbers and pin assignments.*

Extension of Inputs of Logic Gates

The AND and OR logic gates theoretically, can have any number of inputs. But commercially available logic devices will have only few inputs. When larger number of inputs are required, the logic gate can be cascaded as shown in Figs. 4.9(a) and (b).

Since NAND and NOR logic do not satisfy associative property, they cannot be cascaded like AND and OR gates. But multi-input NAND gate can be realized by cascaded AND gate followed by NOT gate as shown in Fig. 4.9(c). Similarly, multi-input NOR gate can be realized by cascaded OR gate followed by NOT gate as shown in Fig. 4.9(d).

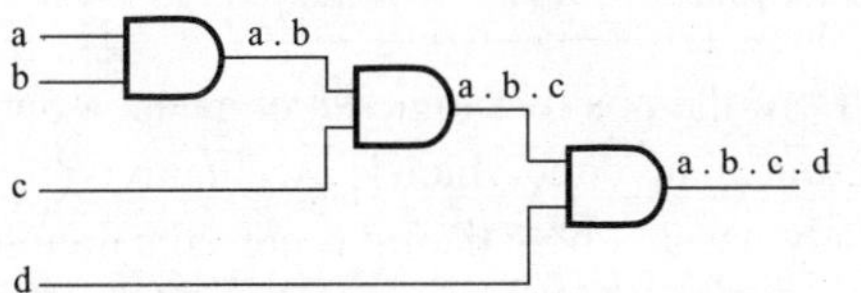

***Fig. a:** 4-input AND gate using 2-input AND gates.*

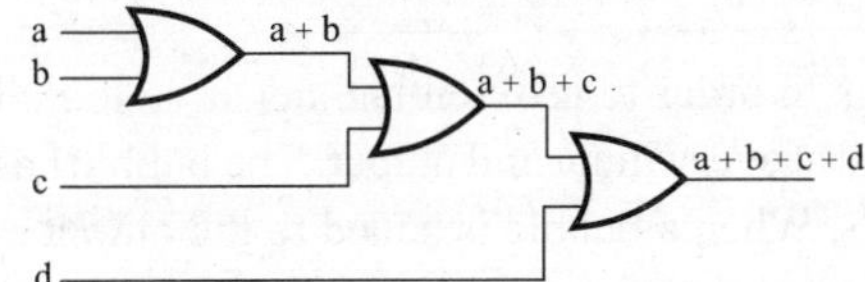

***Fig. b:** 4-input OR gate using 2-input OR gates.*

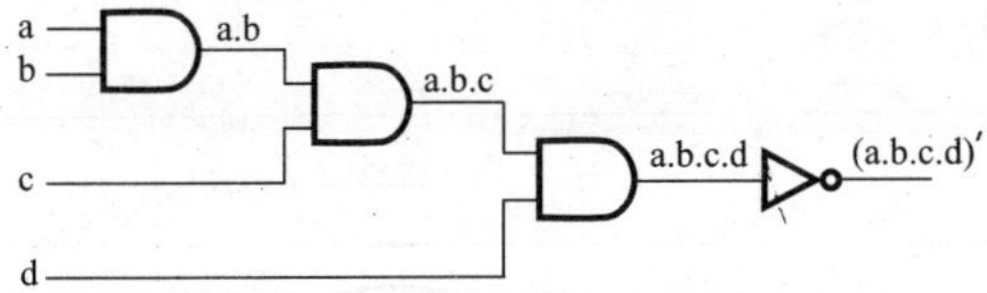

***Fig. c:** 4-input NAND gate using 2-input AND gates.*

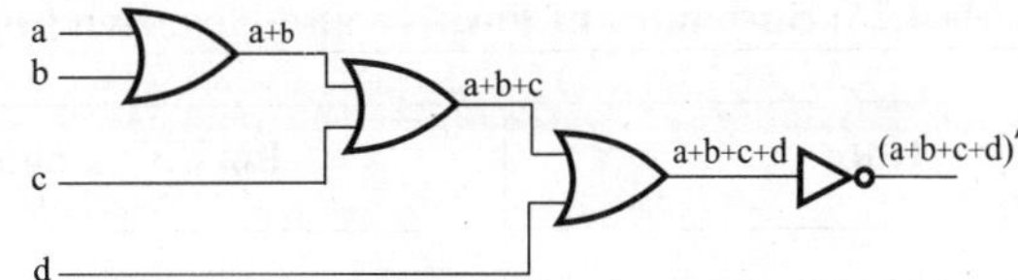

***Fig. d:** 4-input NOR gate using 2-input OR gates.*

***Fig. 4.9:** Multi-input logic gates.*

4.3.2 Positive and Negative Logic Gates

(AU, Nov/Dec'18, 13 Marks)

In positive logic, 1 is logic **high** and 0 is logic **low**. In negative logic, 0 is logic **high** and 1 is logic **low**. The logic gates discussed in Section 4.3 are positive logic gates which means that the logic levels of inputs and outputs are positive logic levels.

Logic gates can also be designed to work with negative logic levels. The AND and OR operation in positive and negative logic are given below:

In this book, only positive logic gates are used for analysis and design of digital logic circuit.

AND operation: AND operation of two or more variables is **high** only if all the variables are **high**.

Table 4.21: Truth Table of Positive Logic AND Gate

Inputs		Output
a	**b**	**a . b**
0 (low)	0 (low)	0 (low)
0 (low)	1 (high)	0 (low)
1 (high)	0 (low)	0 (low)
1 (high)	1 (high)	1 (high)

Table 4.22: Truth Table of Negative Logic AND Gate

Inputs		Output
a	**b**	**a . b**
0 (high)	0 (high)	0 (high)
0 (high)	1 (low)	1 (low)
1 (low)	0 (high)	1 (low)
1 (low)	1 (low)	1 (low)

OR operation: OR operation of two or more variables is **high** if one of the variable is **high**.

Table 4.23: Truth Table of Positive Logic OR Gate

Inputs		Output
a	b	a + b
0 (low)	0 (low)	0 (low)
0 (low)	1 (high)	1 (high)
1 (high)	0 (low)	1 (high)
1 (high)	1 (high)	1 (high)

Table 4.24: Truth Table of Negative Logic OR Gate

Inputs		Output
a	b	a + b
0 (high)	0 (high)	0 (high)
0 (high)	1 (low)	0 (high)
1 (low)	0 (high)	0 (high)
1 (low)	1 (low)	1 (low)

In order to differentiate negative logic level gates from the positive logic level gates, a bubble is added at every input and output. The bubble basically represent an inverter which is used to invert the logic levels. When a bubble is added to the output end that already has a bubble then it represents double time inversion and so that output end will not have a bubble in negative logic representation. The conversion of positive logic level gates to negative logic level gates are listed in Table 4.25.

Table 4.25: Summary of Positive and Negative Logic Gates

Gate	Positive Logic	Negative Logic
AND Gate	x, y; $z = x.y$	x, y; $z = (x'.y')'$
OR Gate	x, y; $z = x+y$	x, y; $z = (x'+y')'$
Inverter Gate	x; x'	x; x'
Buffer	x; x	x; x
NAND Gate	x, y; $z = (x.y)'$	x, y; $z = x'.y'$
NOR Gate	x, y; $z = (x+y)'$	x, y; $z = x'+y'$
XOR gate	x, y; $z = x \oplus y$	x, y; $z = x' \oplus y'$

Using the output equations of logic gates listed in Table 4.25, the truth table of positive and negative logic gates are obtained as shown in Tables 4.26 to 4.29.

Table 4.26: Truth Table of AND Gate

a) Positive Logic

Inputs		Output
x	y	z = x . y
0	0	0
0	1	0
1	0	0
1	1	1

b) Negative Logic

Inputs		Complement of Inputs		Output
x	y	x'	y'	z = (x' . y')'
0	0	1	1	0
0	1	1	0	1
1	0	0	1	1
1	1	0	0	1

Table 4.27: Truth Table of OR Gate

a) Positive Logic

Inputs		Output
x	y	z = x + y
0	0	0
0	1	1
1	0	1
1	1	1

b) Negative Logic

Inputs		Complement of Inputs		Output
x	y	x'	y'	z = (x' + y')'
0	0	1	1	0
0	1	1	0	0
1	0	0	1	0
1	1	0	0	1

In negative logic AND gate,

$z = (x' . y')' \Rightarrow z = x + y$ Using DeMorgan's theorem

Therefore, we can say that the negative logic AND gate is same as positive logic OR gate.

In negative logic OR gate,

$z = (x' + y')' \Rightarrow z = x . y$ Using DeMorgan's theorem

Therefore, we can say that the negative logic OR gate is same as positive logic AND gate.

Table 4.28: Truth Table of NAND Gate

a) Positive Logic

Inputs		Output
x	y	z = (x . y)'
0	0	1
0	1	1
1	0	1
1	1	0

b) Negative Logic

Inputs		Complement of Inputs		Output
x	y	x'	y'	z = x' . y'
0	0	1	1	1
0	1	1	0	0
1	0	0	1	0
1	1	0	0	0

Table 4.29: Truth Table of NOR Gate

a) Positive Logic

Inputs		Output
x	y	z = (x + y)′
0	0	1
0	1	0
1	0	0
1	1	0

b) Negative Logic

Inputs		Complement of Inputs		Output
x	y	x′	y′	z = x′ + y′
0	0	1	1	1
0	1	1	0	1
1	0	0	1	1
1	1	0	0	0

In negative logic NAND gate,

$z = x' . y' \Rightarrow z = (x + y)'$ Using DeMorgan's theorem

Therefore, we can say that the negative logic NAND gate is same as positive logic NOR gate.

In negative logic NOR gate,

$z = x' + y' \Rightarrow z = (x . y)'$ Using DeMorgan's theorem

Therefore, we can say that the negative logic NOR gate is same as positive logic NAND gate.

4.4 Boolean Functions

A ***Boolean function*** is described by a Boolean expression which consists of binary variables, binary constants 0 and 1 and logical operators, AND, OR and NOT. Binary variables are denoted by either lower case (a, b, c, ... x, y, z) or upper case (A, B, C, ... X, Y, Z) alphabets.

Some examples of Boolean functions are given below:

$F = (x . y) + z'$

$F_1 = (x . z) + y'$

$F_2 = (x . y . z) + (y . z')$

Alternatively,

$F(x, y, z) = (x . y) + z'$

$F_1(x, y, z) = (x . z) + y'$

$F_2(x, y, z) = (x . y . z) + (y . z')$

A Boolean function can be evaluated for all possible combinations of binary values of the variables of the function. The variables of a function are also known as ***input variables*** or ***inputs***.

Let, n = Number of variables in a Boolean function

2^n = Number of combinations of binary values of n variables.

Truth table for a Boolean function can be constructed with 2^n combinations of n input variables.

Note: *To simplify Boolean expression dot or AND operation is represented without operator.*

$x . y \Rightarrow xy$ $y . z \Rightarrow yz$ $x . y . z \Rightarrow xyz$

Example 4.5

Construct the truth table for the following functions.

a) $F_1 = xy + xy' + y'z$ **b)** $F_2 = bc + a'c'$

Solution

a) $F_1 = xy + xy' + y'z$

Table 1: Truth Table of Function, F_1

Input Variables and Complement				Product Terms			Function Output
x	y	z	y′	x y	x y′	y′z	F_1
0	0	0	1	0	0	0	0
0	0	1	1	0	0	1	1
0	1	0	0	0	0	0	0
0	1	1	0	0	0	0	0
1	0	0	1	0	1	0	1
1	0	1	1	0	1	1	1
1	1	0	0	1	0	0	1
1	1	1	0	1	0	0	1

b) $F_2 = bc + a'c'$

Table 1: Truth Table of Function, F_2

Input Variables and Complements					Product Terms		Function Output
a	b	c	a′	c′	bc	a′c′	F_2
0	0	0	1	1	0	1	1
0	0	1	1	0	0	0	0
0	1	0	1	1	0	1	1
0	1	1	1	0	1	0	1
1	0	0	0	1	0	0	0
1	0	1	0	0	0	0	0
1	1	0	0	1	0	0	0
1	1	1	0	0	1	0	1

Example 4.6

Simplify the Boolean expression $(A+B)'(A'+B')'$ to minimum number of literals and cross-check by constructing a truth table.

Solution

$$(A+B)'(A'+B')' = (A'B')((A')'(B')')$$ Using DeMorgan's theorem

$$= (A'B')(AB)$$

$$= (AA')(BB') = 0$$ $xx' = 0$

The given Boolean expression has 4 literals. On simplification it is observed that, there are no literals in the simplified Boolean expression.

The value of the Boolean expression is verified by constructing truth table shown in Table 1.

Table 1: Truth Table

Input Variables and Complement				Product Terms		Function Output
A	B	A′	B′	(A + B)′	(A′+ B′)′	(A + B)′(A′+ B′)′
0	0	1	1	1	0	0
0	1	1	0	0	0	0
1	0	0	1	0	0	0
1	1	0	0	0	1	0

Example 4.7

(IKGPTU, May'20, 2 Marks)

Draw the symbol and construct the truth table for three input XOR gate.

Solution

The symbol of three input XOR gate is shown in Fig. 1. The truth table is constructed as shown in Table 1. Here it is observed that if the input has odd number of 1's then the output is '1' otherwise '0'.

Table 1: Truth Table of XOR Gate

x	y	z	x ⊕ y	(x ⊕ y) ⊕ z
0	0	0	0	0
0	0	1	0	1
0	1	0	1	1
0	1	1	1	0
1	0	0	1	1
1	0	1	1	0
1	1	0	0	0
1	1	1	0	1

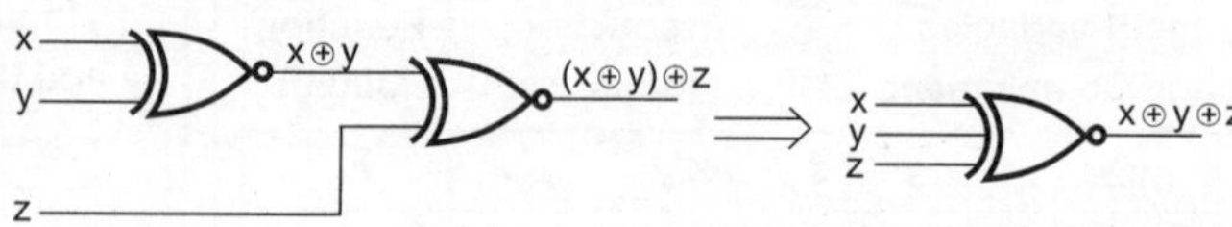

Fig. 1: 3-input XOR gate.

Example 4.8

(JNTUK, Feb'22, 4 Marks)

Realize 2-input XNOR gate using NOR gates.

Solution

The XNOR operation is given by the following Boolean equation.

$$x \odot y = xy + x'y'$$

The above equation can be implemented using NOR gate as shown in Fig. 1.

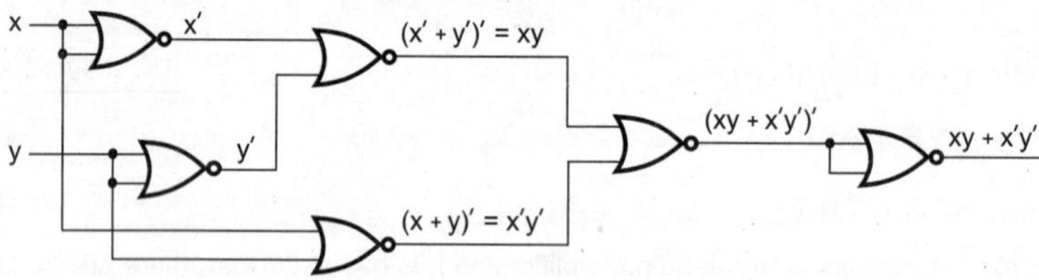

Fig. 1: XNOR operation using only NOR gates.

Example 4.9

(AKTU, Dec'22, 10 Marks)

Realize the 3-input gate using 2-input gates for the folowing gates:

a) AND **b)** OR **c)** NAND **d)** NOR

Solution

a) AND

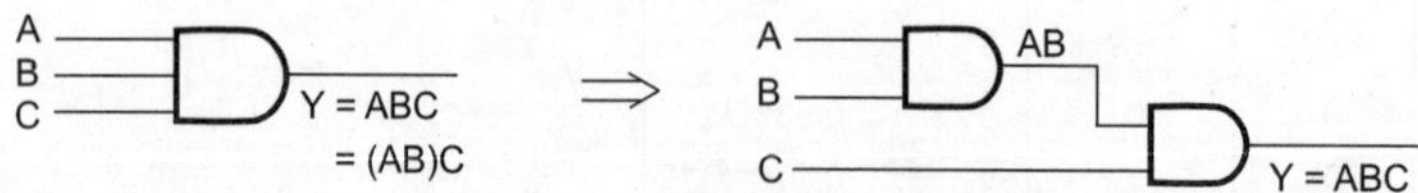

Fig. 1: 3-input AND gate using 2-input AND gate.

b) OR

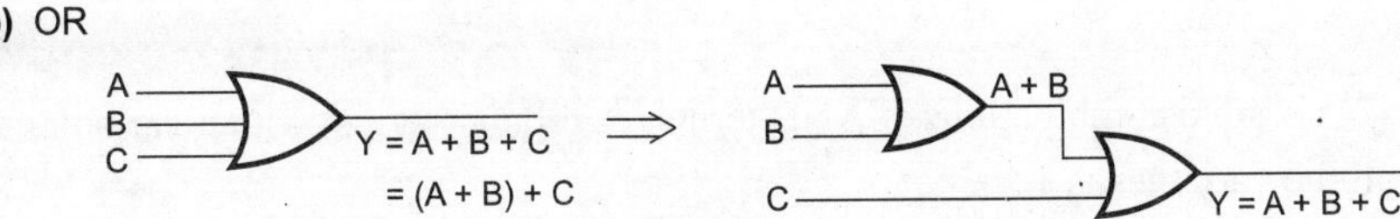

Fig. 2: *3-input OR gate using 2-input OR gate.*

c) NAND

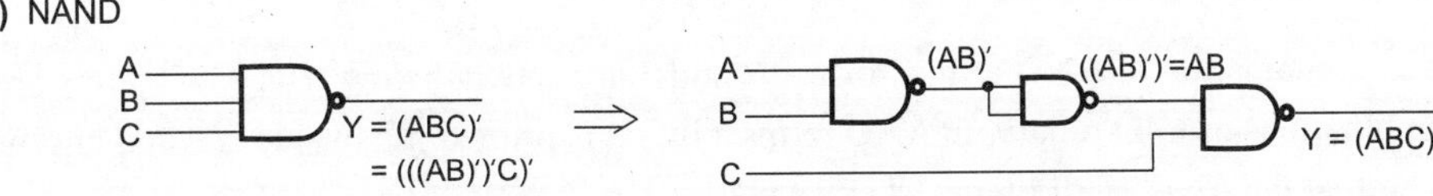

Fig. 3: *3-input NAND gate using 2-input NAND gate.*

d) NOR

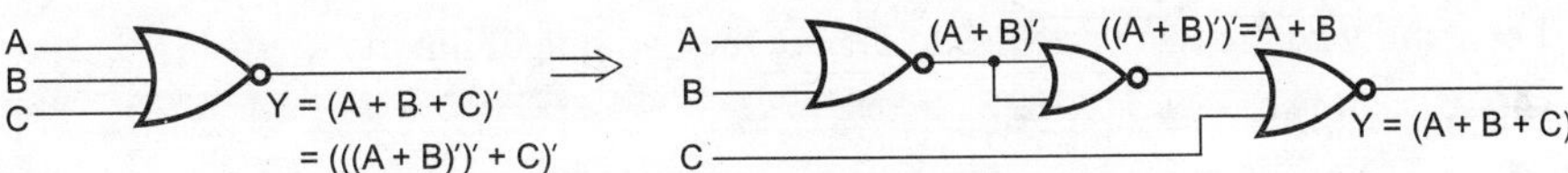

Fig. 4: *3-input NOR gate using 2-input NOR gate.*

4.4.1 Implementation of Boolean Functions by Logic Gates

A Boolean function basically involves product of variables, sum of variables and complement of variables. Therefore, the straight forward implementation of Boolean function can be made by using basic gates AND, OR and NOT. The complement of a variable is obtained using NOT gate. The product term is realized using AND gate. The sum term is realized using OR gate.

Example 4.10

Implement the following functions using basic logic gates.

a) $F_1 = xy + xy' + y'z$ **b)** $F_2 = bc + a'c'$

Solution

a) $F_1 = xy + xy' + y'z$

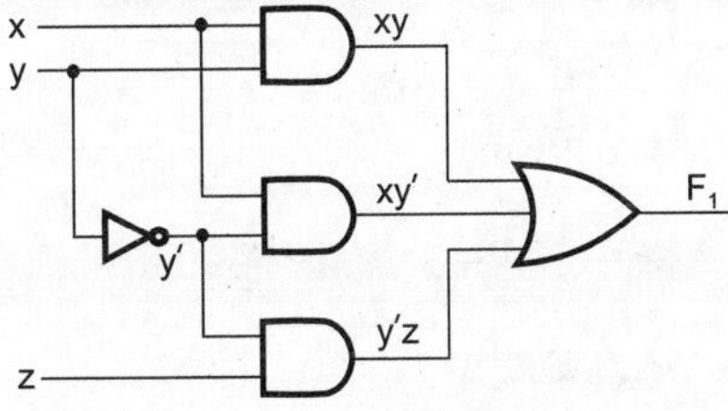

Fig. 1: *Logic circuit for function, F_1.*

b) $F_2 = bc + a'c'$

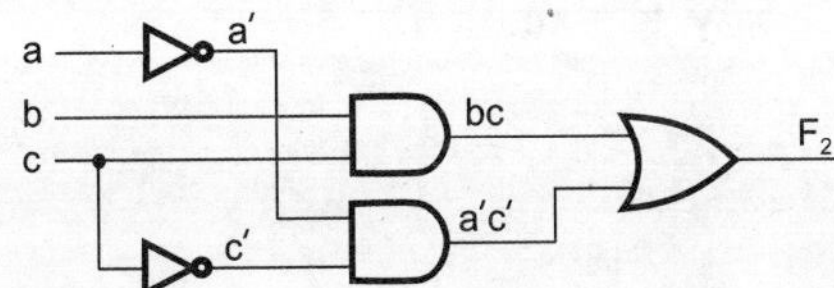

Fig. 2: *Logic circuit for function, F_2.*

4.4.2 Minterms

(RGPV, Jun'20, 5 Marks) (GTU, May'19, 3 Marks)

Minterms are 2^n possible combinations of AND terms (or product terms) with n variables such that the logical AND of all the variables is 1.

The minterms can be formed with two, three, four, ... variables. The 2^n minterms are denoted as m_0, m_1, m_2, m_3 ...., m_q where $q = 2^n - 1$.

The 2^n combination of AND terms, for n = 2 and 3 are shown below. The AND terms are formed from 2^n combinations of n-bit binary. In AND terms a literal is primed if its value is 0 and unprimed if its value is 1. So that the AND of all literals is always 1.

Let, n = 2

Let x and y be the two variables to denote the two bits of binary.

When n = 2, $2^n = 2^2 = 4$

Therefore, four combinations of AND terms with two variables are possible as shown below:

0 0		$x'.y'$	$(0'.0' = 1)$		$x'y' = m_0$
0 1	⇒	$x'.y$	$(0'.1 = 1)$	⇒	$x'y = m_1$
1 0		$x.y'$	$(1.0' = 1)$		$xy' = m_2$
1 1		$x.y$	$(1.1 = 1)$		$xy = m_3$

Let, n = 3

Let x, y and z be the three variables to denote the three bits of binary.

When n = 3, $2^n = 2^3 = 8$

Therefore, eight combinations of AND terms with three variables are possible as shown below:

0 0 0		$x'.y'.z'$	$(0'.0'.0' = 1)$		$x'y'z' = m_0$
0 0 1	⇒	$x'.y'.z$	$(0'.0'.1 = 1)$	⇒	$x'y'z = m_1$
0 1 0		$x'.y.z'$	$(0'.1.0' = 1)$		$x'yz' = m_2$
0 1 1		$x'.y.z$	$(0'.1.1 = 1)$		$x'yz = m_3$
1 0 0		$x.y'.z'$	$(1.0'.0' = 1)$		$xy'z' = m_4$
1 0 1	⇒	$x.y'.z$	$(1.0'.1 = 1)$	⇒	$xy'z = m_5$
1 1 0		$x.y.z'$	$(1.1.0' = 1)$		$xyz' = m_6$
1 1 1		$x.y.z$	$(1.1.1 = 1)$		$xyz = m_7$

Table 4.30: Two-Variable Minterms

x	y	Minterm	Notation
0	0	x'. y'	m_0
0	1	x'. y	m_1
1	0	x . y'	m_2
1	1	x . y	m_3

Table 4.31: Three-Variable Minterms

x	y	z	Minterm	Notation
0	0	0	x'. y'. z'	m_0
0	0	1	x'. y'. z	m_1
0	1	0	x'. y. z'	m_2
0	1	1	x'. y. z	m_3
1	0	0	x . y'. z'	m_4
1	0	1	x . y'. z	m_5
1	1	0	x . y . z'	m_6
1	1	1	x . y . z	m_7

Table 4.32: Four-Variable Minterms

a	b	c	d	Minterm	Notation
0	0	0	0	a'. b'. c'. d'	m_0
0	0	0	1	a'. b'. c'. d	m_1
0	0	1	0	a'. b'. c . d'	m_2
0	0	1	1	a'. b'. c . d	m_3
0	1	0	0	a'. b . c'. d'	m_4
0	1	0	1	a'. b . c'. d	m_5
0	1	1	0	a'. b . c . d'	m_6
0	1	1	1	a'. b . c . d	m_7
1	0	0	0	a . b'. c'. d'	m_8
1	0	0	1	a . b'. c'. d	m_9
1	0	1	0	a . b'. c . d'	m_{10}
1	0	1	1	a . b'. c . d	m_{11}
1	1	0	0	a . b . c'. d'	m_{12}
1	1	0	1	a . b . c'. d	m_{13}
1	1	1	0	a . b . c . d'	m_{14}
1	1	1	1	a . b . c . d	m_{15}

4.4.3 Maxterms

(RGPV, Jun'20, 5 Marks) (GTU, May'19, 3 Marks)

Maxterms are 2^n possible combinations of OR terms (or sum terms) with n variables such that the logical OR of all the variables is 0.

The maxterms can be formed with two, three, four, ... variables.

The 2^n maxterms are denoted as $M_0, M_1, M_2, M_3, \ldots\ldots, M_q$ where $q = 2^n - 1$.

The 2^n combinations of OR terms, for n = 2 and 3 are shown below. The OR terms are formed from 2^n combinations of n-bit binary. In OR terms a literal is primed if its value is 1, and unprimed if its value is 0 so that the OR of all literals is always 0.

Let, n = 2

Let x and y be the two variables to denote two bits of binary.

When n = 2, $2^n = 2^2 = 4$

Therefore, four combinations of OR terms with two variables are possible as shown below:

$$\begin{array}{llll} 0\ 0 & x+y & (0+0=0) & x+y = M_0 \\ 0\ 1 & x+y' & (0+1'=0) & x+y' = M_1 \\ 1\ 0 \Rightarrow & x'+y & (1'+0=0) \quad \Rightarrow & x'+y = M_2 \\ 1\ 1 & x'+y' & (1'+1'=0) & x'+y' = M_3 \end{array}$$

Let, n = 3

Let x, y and z be the three variables to denote three bits of binary.

When n = 3, $2^n = 2^3 = 8$

Therefore, eight combinations of OR terms with three variables are possible as shown below:

$$
\begin{array}{llll}
0\,0\,0 & x+y+z & (0+0+0=0) & x+y+z = M_0 \\
0\,0\,1 & x+y+z' & (0+0+1'=0) & x+y+z' = M_1 \\
0\,1\,0 & x+y'+z & (0+1'+0=0) & x+y'+z = M_2 \\
0\,1\,1 \Rightarrow & x+y'+z' & (0+1'+1'=0) \Rightarrow & x+y'+z' = M_3 \\
1\,0\,0 & x'+y+z & (1'+0+0=0) & x'+y+z = M_4 \\
1\,0\,1 & x'+y+z' & (1'+0+1'=0) & x'+y+z' = M_5 \\
1\,1\,0 & x'+y'+z & (1'+1'+0=0) & x'+y'+z = M_6 \\
1\,1\,1 & x'+y'+z' & (1'+1'+1'=0) & x'+y'+z' = M_7
\end{array}
$$

Table 4.33: Two-Variable Maxterms

x	y	Maxterm	Notation
0	0	x + y	M_0
0	1	x + y'	M_1
1	0	x'+ y	M_2
1	1	x'+ y'	M_3

Table 4.34: Three-Variable Maxterms

x	y	z	Maxterm	Notation
0	0	0	x + y + z	M_0
0	0	1	x + y + z'	M_1
0	1	0	x + y' + z	M_2
0	1	1	x + y' + z'	M_3
1	0	0	x' + y + z	M_4
1	0	1	x' + y + z'	M_5
1	1	0	x' + y' + z	M_6
1	1	1	x' + y' + z'	M_7

Table 4.35: Four-Variable Maxterms

a	b	c	d	Maxterm	Notation
0	0	0	0	a + b + c + d	M_0
0	0	0	1	a + b + c + d'	M_1
0	0	1	0	a + b + c' + d	M_2
0	0	1	1	a + b + c' + d'	M_3
0	1	0	0	a + b' + c + d	M_4
0	1	0	1	a + b' + c + d'	M_5
0	1	1	0	a + b' + c' + d	M_6
0	1	1	1	a + b' + c' + d'	M_7
1	0	0	0	a' + b + c + d	M_8
1	0	0	1	a' + b + c + d'	M_9
1	0	1	0	a' + b + c' + d	M_{10}
1	0	1	1	a' + b + c' + d'	M_{11}
1	1	0	0	a' + b' + c + d	M_{12}
1	1	0	1	a' + b' + c + d'	M_{13}
1	1	1	0	a' + b' + c' + d	M_{14}
1	1	1	1	a' + b' + c' + d'	M_{15}

4.4.4 Standard Forms

(GTU, Jun'19, 2 Marks)

There are three types of standard forms of expressing a Boolean function. They are,

1. Sum-of-products (SOP) form
2. Product-of-sums (POS) form
3. Canonical form

(JNTUK, Feb'22, 7 Marks)

The ***canonical form*** is also expressed in SOP or POS form. The SOP or POS form is said to be canonical only if all the variables are present in every term of the SOP or POS form.

In SOP or POS form each term of the Boolean expression need not have all the variables of the function. Hence SOP and POS forms are also simplified forms of Boolean functions.

4.4.5 Sum-of-Products and Product-of-Sums Simplification

(IKGPTU, Dec'19, 2 Marks)
(RGPV, Jun'19, 3 Marks)

Sum-of-products (SOP) can be realized by AND operation of literals followed by OR operation of output of AND. Here, AND operation of literals are the products and OR operation of output of AND is the sum of product terms. ***Product-of-sums (POS)*** can be realized by OR operation of literals followed by AND operation of output of OR. Here, OR operation of literals are the sums and AND operation of output of OR is the product of sum terms. The realizations in SOP and POS forms are called ***two-level realization of standard forms.***

The two-level realization of SOP form can be obtained by using AND gates to implement product terms followed by OR gate to get sum of product terms. The two-level realization of POS form can be obtained by using OR gates to implement sum terms followed by AND gate to get product of sum terms.

Since NAND and NOR gates are universal gates, the two-level realization of SOP form of a Boolean function can be obtained using only NAND gates. Similarly, two-level realization of POS form of a Boolean function can be obtained using only NOR gates.

Example 4.11

Draw the two-level realization of the following functions.

a) $F_1 = y' + xy + x'yz'$ **b)** $F_2 = x(y' + z)(x' + y + z')$

Solution

a) $\mathbf{F_1 = y' + xy + x'yz'}$ - Sum-of-products

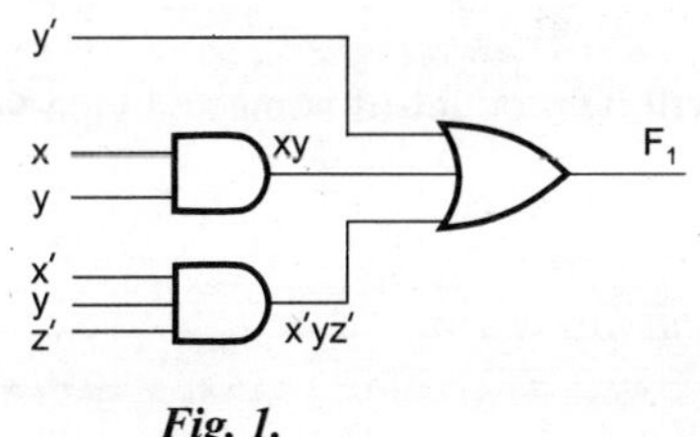

Fig. 1.

b) $\mathbf{F_2 = x(y' + z)(x' + y + z')}$ - Product-of-sums

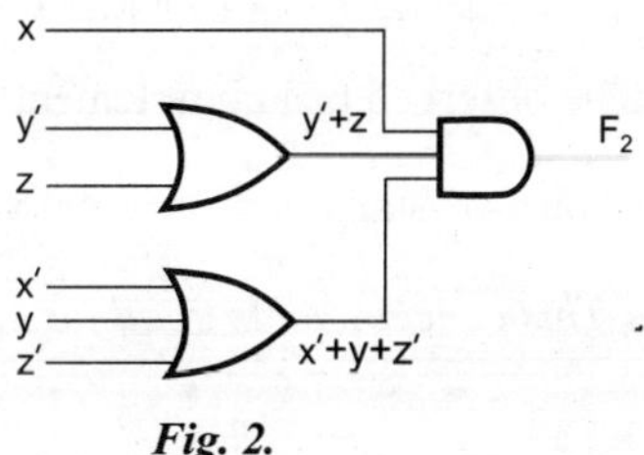

Fig. 2.

Example 4.12

Determine the complement of the following functions and show that SOP becomes POS and vice-versa.

a) $F_1 = y' + xy + x'yz$ **b)** $F_2 = x(y' + z)(x' + y + z')$

Solution

a) $\mathbf{F_1 = y' + xy + x'yz}$ - Sum-of-products

⇓ Complement

$F_1' = (y' + xy + x'yz)'$

$= (y')'(xy)'(x'yz)'$

$= y(x' + y')(x + y' + z')$ - Product-of-sums

Using DeMorgan's theorem
$(x + y)' = x'y'$
$(x \,.\, y)' = x' + y'$

b) $F_2 = x(y' + z)(x' + y + z')$ - Product-of-sums

$\Downarrow$ Complement

$$F_2' = (x(y'+z)(x'+y+z'))'$$
$$= x' + (y'+z)' + (x'+y+z')'$$
$$= x' + (yz') + (xy'z) \text{ - Sum-of-products}$$

Using DeMorgan's theorem
$(x + y)' = x'y'$
$(x \cdot y)' = x' + y'$

4.4.6 Canonical Form

(GTU, Jun'19, 2 Marks)

A Boolean function can be expressed algebraically from its truth table by forming a minterm for each combination of the variables that produces a 1 in the output and then taking the OR of all the minterms.

Similarly, a Boolean function can be expressed algebraically from its truth table by forming a maxterm for each combination of the variable that produces a 0 in the output and then taking the AND of all the maxterm. A Boolean function expressed as a sum of minterms or product of maxterms are said to be in canonical form.

The given function can be expressed in canonical form without using truth table. For sum of minterms insert sum of missing literal and its complement with "." operation and expand. Similarly for product of maxterms insert product of missing literal and its complement with "+" operation and expand.

4.4.7 Complement of a Function

If, F is a function then F' is complement of F. The complement of a function can be directly evaluated using Boolean theorems and postulates. Alternatively, the complement can also be obtained by using duality. In duality replace "+" by "." and "." by "+" and variables by its complement.

It can be observed that complement of sum-of-products will be product-of-sums and vice-versa

i.e, Product-of-sums $\xleftrightarrow{\text{Complement}}$ Sum-of-products

> ***Note:*** *This concept holds good for complement of canonical forms also.*

Example 4.13

Determine the complement of the following Boolean functions.

a) $F_1 = xy + x'z$ **b)** $F_2 = wx + yz$ **c)** $F_3 = xy' + x'y$

Solution

a) $F_1 = xy + x'z$

Case i: Direct Evaluation of Complement

$$F_1 = xy + x'z$$

$\Downarrow$ Complement

$$F_1' = (xy + x'z)'$$
$$= (xy)' \cdot (x'z)'$$
$$= (x' + y') \cdot ((x')' + z')$$
$$= (x' + y') \cdot (x + z')$$

Using DeMorgan's theorem
$(x + y)' = x'y'$
$(x \cdot y)' = x' + y'$

Case ii: Complement using Duality

$x \,.\, y + x' \,.\, z$

$\downarrow\ \downarrow\downarrow\ \downarrow\ \downarrow\ \downarrow\ \downarrow$ Replace by dual elements

$(x' + y') \,.\, (x + z')$

$\Downarrow$

$(x' + y') \,.\, (x + z') = F_1'$

b) $F_2 = w x + y z$

Case i: Direct Evaluation of Complement

$F_2 = w x + y z$

$\Downarrow$ Complement

$F_2' = (w x + y z)'$

$= (w x)' \,.\, (y z)'$

$= (w' + x') \,.\, (y' + z')$

Using DeMorgan's theorem
$(x + y)' = x'y'$
$(x \,.\, y)' = x' + y'$

Case ii: Complement using Duality

$w \,.\, x + y \,.\, z$

$\downarrow\ \downarrow\downarrow\ \downarrow\ \downarrow\ \downarrow\ \downarrow$ Replace by dual elements

$(w' + x') \,.\, (y' + z')$

$\Downarrow$

$(w' + x') \,.\, (y' + z') = F_2'$

c) $F_3 = x y' + x' y$

Case i: Direct Evaluation of Complement

$F_3 = x y' + x' y$

$\Downarrow$ Complement

$F_3' = (x y' + x' y)'$

$= (x y')' \,.\, (x' y)'$

$= (x' + y) \,.\, (x + y')$

Using DeMorgan's theorem
$(x + y)' = x'y'$
$(x \,.\, y)' = x' + y'$

Case ii: Complement using Duality

$x \,.\, y' + x' \,.\, y$

$\downarrow\ \downarrow\downarrow\ \downarrow\ \downarrow\ \downarrow\ \downarrow$ Replace by dual elements

$(x' + y) \,.\, (x + y')$

$\Downarrow$

$(x' + y) \,.\, (x + y') = F_3'$

4.4.8 Implementation of Boolean Functions using Universal Gates

The NAND and NOR gates are called ***universal gates***. When a Boolean function is expressed in SOP form, it can be realized using only NAND gates. When a Boolean function is expressed in POS form, it can be realized using only NOR gates.

Example 4.14

(JNTUA, Jun/July'22, 7 Marks)

Express the function F = x + y'z, in sum of minterms and product of maxterms and verify the result by simplification using Boolean theorems and postulates.

Solution

Table 1: Truth Table

Input Variables			Minterm		Maxterm		Complement of y	Product Term	Function Output
x	y	z					y'	y'z	F
0	0	0	m_0	x'y'z'	M_0	x + y + z	1	0	0
0	0	1	m_1	x'y'z	M_1	x + y + z'	1	1	1
0	1	0	m_2	x'y z'	M_2	x + y' + z	0	0	0
0	1	1	m_3	x'y z	M_3	x + y' + z'	0	0	0
1	0	0	m_4	x y'z'	M_4	x' + y + z	1	0	1
1	0	1	m_5	x y'z	M_5	x' + y + z'	1	1	1
1	1	0	m_6	x y z'	M_6	x' + y' + z	0	0	1
1	1	1	m_7	x y z	M_7	x' + y' + z'	0	0	1

A function can be expressed as a sum of minterms for which the function output is 1. Here, F is 1, when m_1, m_4, m_5, m_6 and m_7 are inputs.

$$\therefore\ F = m_1 + m_4 + m_5 + m_6 + m_7$$
$$= \Sigma m(1, 4, 5, 6, 7)$$
$$= x'y'z + xy'z' + xy'z + xyz' + xyz$$

A function can be expressed as a product of maxterms for which the function output is 0. Here, F is 0, when M_0, M_2 and M_3 are inputs

$$F = M_0 M_2 M_3$$
$$= \Pi M(0, 2, 3)$$
$$= (x + y + z)(x + y' + z)(x + y' + z')$$

Proof:

Case i:

$$F = m_1 + m_4 + m_5 + m_6 + m_7$$
$$= x'y'z + xy'z' + xy'z + xyz' + xyz$$
$$= x'y'z + xy'(z' + z) + xy(z' + z) \qquad \boxed{x + x' = 1}$$
$$= x'y'z + xy'.1 + xy.1$$
$$= x'y'z' + xy' + xy \qquad \boxed{x.1 = x}$$
$$= x'y'z + x(y' + y)$$
$$= x'y'z + x.1$$
$$= x'y'z + x$$
$$= (x' + x)(y'z + x)$$
$$= 1.y'z + x$$
$$= x + y'z$$

Case ii:

$$F = M_0\, M_2\, M_3$$
$$= (x + y + z)\,(x + y' + z)\,(x + y' + z')$$
$$= (x + z + y)(x + z + y')(x + y' + z')$$
$$= (x + z + (y\,y'))\,(x + y' + z') \qquad \boxed{x\,.\,x' = 0}$$
$$= (x + z + 0)\,(x + y' + z') \qquad \boxed{x + 0 = x}$$
$$= x + (z\,(y' + z'))$$
$$= x + zy' + zz'$$
$$= x + y'z + 0$$
$$= x + y'z$$

Example 4.15

(JNTUK, Feb'22, 7 Marks)

Express the Boolean function $F(p, q, r) = (p\,q + r)\,(q + p\,r)$ as sum of minterms and product of maxterms.

Solution:

i) Sum of Minterms

$$F(p, q, r) = (p\,q + r)(q + p\,r)$$
$$= p\,q\,q + p\,q\,p\,r + r\,q + r\,p\,r$$
$$= p\,q + p\,q\,r + r\,q + r\,p$$
$$= p\,q(r + r') + p\,q\,r + r\,q(p + p') + r\,p(q + q')$$
$$= p\,q\,r + p\,q\,r' + \cancel{p\,q\,r} + \cancel{p\,q\,r} + p'q\,r + \cancel{p\,q\,r} + p\,q'r$$

Repeated terms are considered once.

$$= p\,q\,r + p\,q\,r' + p'q\,r + p\,q'r$$
$$= p'q\,r + p\,q'r + p\,q\,r' + p\,q\,r$$
$$= m_3 + m_5 + m_6 + m_7 = \Sigma m\,(3, 5, 6, 7)$$

ii) Product of Maxterms

$$F(p, q, r) = (pq + r)\,(q + pr)$$
$$= (p + r)\,(q + r)\,(q + p)\,\cancel{(q + r)}$$
$$= (p + r)\,(q + r)\,(q + p)$$

Missing literal in first term is q, second term is p and third term is r.

$$= (p + r + (q\,q'))\,((p\,p') + q + r)\,(p + q + (r\,r'))$$
$$= (p + r + q)\,(p + r + q')\,\cancel{(p + q + r)}\,(p' + q + r)\,\cancel{(p + q + r)}\,(p + q + r')$$
$$= (p + q + r)\,(p + r + q')\,(p' + q + r)\,(p + q + r')$$
$$= (p + q + r)\,(p + q + r')\,(p + q' + r)\,(p' + q + r)$$
$$= M_0\, M_1\, M_2\, M_4 = \Pi M\,(0, 1, 2, 4)$$

Example 4.16

(VTU, Aug/Sep'20, 6 Marks) (VTU, Aug/Sep'20, 8 Marks)

Convert the following in the proper canonical form and write the decimal notation.

a) R = F(x, y, z) = (x + y) (x' + z) into maxterm canonical form.

b) Z = F(a, b, c) = a b + b'c + ac into minterm canonical form.

Solution

a) Maxterm canonical form (POS)

$$R = F(x, y, z) = (x + y)(x' + z)$$

$$= (x + y + (zz'))(x' + z + (yy'))$$

Missing literal in first term is z and second term is y.

$$= ((x + y + z)(x + y + z'))((x' + y + z)(x' + y' + z)$$

$$(M_0) \quad (M_1) \quad (M_4) \quad (M_6)$$

$$= M_0 M_1 M_4 M_6 = \prod M(0, 1, 4, 6)$$

b) Minterm canonical form (SOP)

$$Z = F(a, b, c) = (ab + b'c + ac)$$

Missing literal in first term is c, second term is a and third term is b.

$$= ab(c + c') + b'c(a + a') + ac(b + b')$$

$$= abc + abc' + ab'c + a'b'c + \cancel{abc} + \cancel{ab'c}$$

Repeated terms are considered once.

$$= abc + abc' + ab'c + a'b'c$$

$$(m_7) \quad (m_6) \quad (m_5) \quad (m_1)$$

$$= a'b'c + ab'c + abc' + abc$$

$$= m_1 + m_5 + m_6 + m_7 = \sum m(1, 5, 6, 7)$$

Example 4.17

(JNTUA, Jun/Jul'22, 7 Marks)

Express the following Boolean functions in canonical form and implement the Boolean functions using either only NAND or NOR gates.

a) $F_1 = xy + y'z$ **b)** $F_2 = (x + y)(y + z)$ **c)** $F_3 = x + y'z$

Solution

a) $F_1 = xy + y'z$

Case i: SOP

$$F_1 = xy + y'z$$

$$= xy(z + z') + (x + x')y'z$$

Missing literal in first term is z and second term is x.

$$= xyz + xyz' + xy'z + x'y'z$$

$$(m_7) \quad (m_6) \quad (m_5) \quad (m_1)$$

$x + x' = 1$

$x \cdot 1 = x$

$$= x'y'z + xy'z + xyz' + xyz$$

$$= m_1 + m_5 + m_6 + m_7$$

The SOP form of F_1 is realized using only NAND gates as shown in Fig. 1.

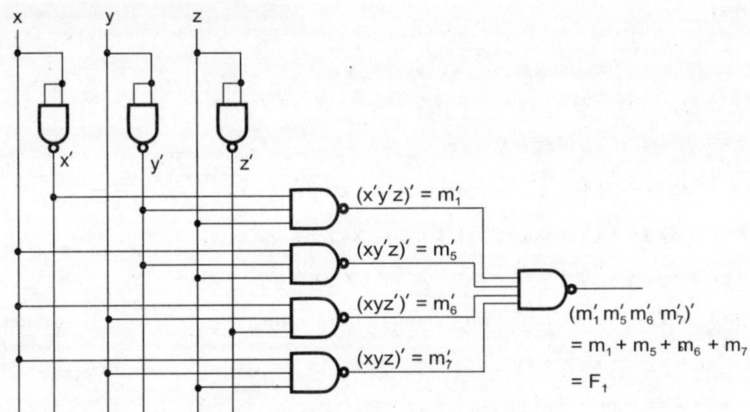

***Fig. 1:** Logic circuit of F_1 using only NAND gates.*

Case ii: POS

$F_1 = xy + y'z$

$= (x + y'z)\,(y + y'z)$

$= (x + y')\,(x + z)\,(y + y')\,(y + z)$

$= (x + y' + (zz'))\,(x + z + (yy'))\,((xx') + y + z)$

$= (x + y' + z)\,(x + y' + z')\,(x + z + y)\,(x + z + y')\,(x + y + z)\,(x' + y + z)$

$= (x + y' + z)\,(x + y' + z')\,(x + y + z)\,\cancel{(x + y' + z)}\,\cancel{(x + y + z)}\,(x' + y + z)$

$(M_2)\quad (M_3)\quad (M_0)\quad (M_2)\quad (M_0)\quad (M_4)$

$= (x + y + z)\,(x + y' + z)\,(x + y' + z')\,(x' + y + z)$

$= M_0 M_2 M_3 M_4$

$xx' = 0$
$x + 0 = x$

Missing literal in first term is z, second term is y and third term is x.

Repeated maxterms are considered only one time

The POS form of F_1 is realized using only NOR gates as shown in Fig. 2.

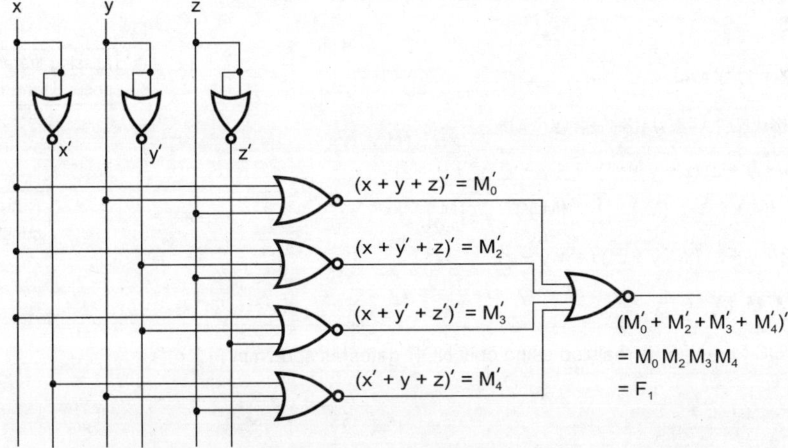

***Fig. 2:** Logic circuit of F_1 using only NOR gates.*

b) $F_2 = (x + y)(y + z)$

Case i: SOP

$F_2 = (x + y)(y + z) = x(y + z) + y(y + z)$

$= xy + xz + yy + yz = xy + xz + y + yz$ — yy = y

$= xy(z + z') + xz(y + y') + (x + x')y(z + z') + (x + x')yz$ — Insert missing literals; x + x' = 1; x . 1 = x

$= xyz + xyz' + xzy + xzy' + (xy + x'y)(z + z') + xyz + x'yz$

$= xyz + xyz' + \not{xyz} + xy'z + \not{xyz} + \not{xyz'} + x'yz + x'yz' + \not{xyz} + \not{x'yz}$

$(m_7)\ (m_6)\ (m_7)\ (m_5)\ (m_7)\ (m_6)\ (m_3)\ (m_2)\ (m_7)\ (m_3)$ — Repeated minterms are considered only one time

$= x'yz' + x'yz + xy'z + xyz' + xyz$

$= m_2 + m_3 + m_5 + m_6 + m_7$

The SOP form of F_2 is realized using only NAND gates as shown in Fig. 3.

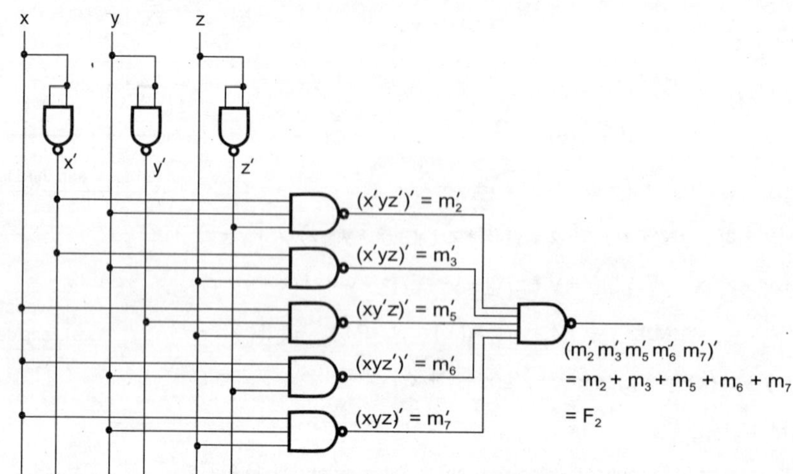

Fig. 3: Logic circuit of F_2 using only NAND gates.

Case ii: POS

$F_2 = (x + y)(y + z)$ — Missing literal in first term is z and second term is x.

$= ((x + y) + (zz'))((xx') + (y + z))$ — x x' = 0; x + 0 = x

$= (x + y + z)(x + y + z')\not{(x + y + z)}(x' + y + z)$

$(M_0)\ (M_1)\ (M_0)\ (M_4)$ — Repeated maxterms are considered only one time

$= (x + y + z)(x + y + z')(x' + y + z)$

$= M_0 M_1 M_4$

The POS form of F_2 is realized using only NOR gates as shown in Fig. 4.

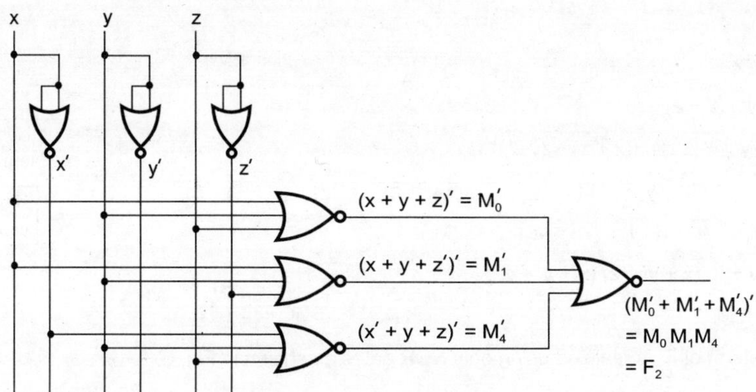

***Fig. 4:** Logic circuit of F_2 using only NOR gates.*

c) $F_3 = x + y'z$

Case i: SOP

$F_3 = x + y'z$

$= x(y + y') + (x + x')\, y'z$

$= xy + xy' + xy'z + x'y'z$

$= xy(z + z') + xy'(z + z') + xy'z + x'y'z$

$= xyz + xyz' + xy'z + xy'z' + xy'z + x'y'z$
$\quad (m_7) \quad (m_6) \quad (m_5) \quad (m_4) \quad (m_5) \quad (m_1)$

$= x'y'z + xy'z' + xy'z + xyz' + xyz$

$= m_1 + m_4 + m_5 + m_6 + m_7$

x . 1 = x
x + x′ = 1

Missing literal in first term is y and second term is x.

Missing literal in first term and second term is z.

Repeated minterms are considered only one time

The SOP form of F_3 is realized using only NAND gates as shown in Fig. 5.

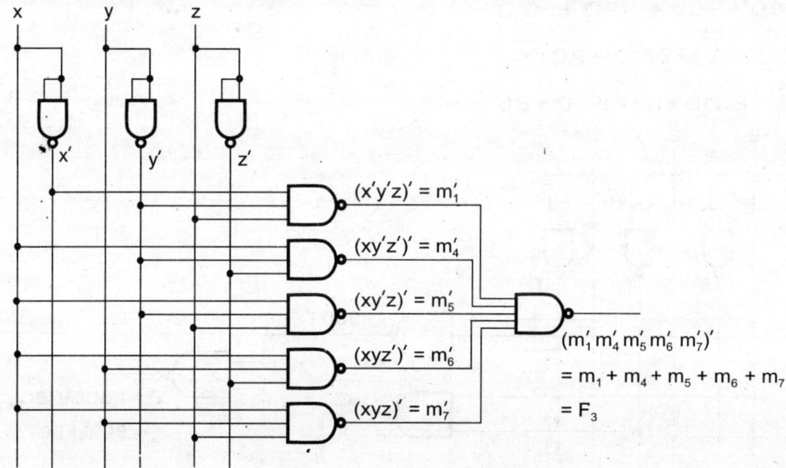

***Fig. 5:** Logic circuit of F_3 using only NAND gates.*

Case ii: POS

$F_3 = x + y'z$

$= (x + y')(x + z)$

$= ((x + y') + (zz'))((x + z) + (yy'))$

$= (x + y' + z)(x + y' + z')(x + y + z)\cancel{(x + y' + z)}$

$\quad (M_2) \quad (M_3) \quad (M_0) \quad (M_2)$

$= (x + y + z)(x + y' + z)(x + y' + z')$

$= M_0 M_2 M_3$

$x \cdot 1 = x$
$x + x' = 1$

Missing literal in first term is z and second term is y.

Repeated maxterms are considered only one time

The POS form of F_3 is realized using only NOR gates as shown in Fig. 6.

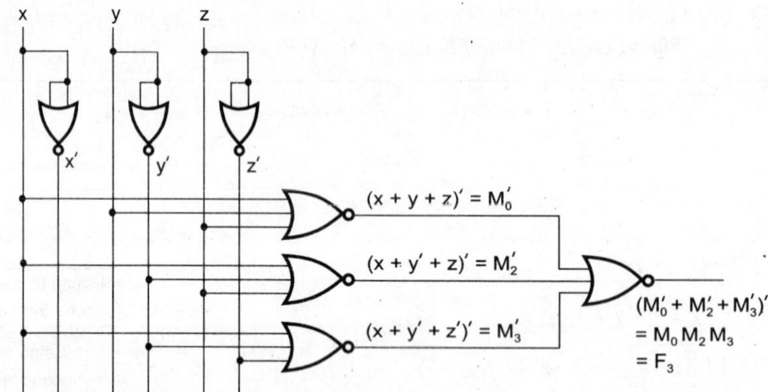

Fig. 6: Logic circuit of F_3 using only NOR gates.

Example 4.18

(GTU, May'19, 7 Marks)

Simplify $Y = A'BCD' + BCD' + BC'D' + BC'D$ and implement using NAND gate only.

Solution

$F = A'BCD' + BCD' + BC'D' + BC'D$

$= BCD'(1 + A') + BC'D' + BC'D$

$= BCD' + BC'(D + D') = BCD' + BC'$

$x + 1 = 1$

$x + x' = 1$

Fig. 1: Logic circuit of function, F using only NAND gates.

Using DeMorgan's theorem

Example 4.19

Express the following function in sum of minterms and product of maxterms

F(A, B, C, D) = A′B + BD + AC′

Solution

Table 1: Truth Table of F

Input Variables				Minterm		Maxterm		Complement of A and C		Product Terms			Function Output
A	B	C	D					A′	C′	A′B	BD	AC′	F
0	0	0	0	m_0	A′B′C′D′	M_0	A + B + C + D	1	1	0	0	0	0
0	0	0	1	m_1	A′B′C′D	M_1	A + B + C + D′	1	1	0	0	0	0
0	0	1	0	m_2	A′B′C D′	M_2	A + B + C′ + D	1	0	0	0	0	0
0	0	1	1	m_3	A′B′C D	M_3	A + B + C′ + D′	1	0	0	0	0	0
0	1	0	0	m_4	A′B C′D′	M_4	A + B′ + C + D	1	1	1	0	0	1
0	1	0	1	m_5	A′B C′D	M_5	A + B′ + C + D′	1	1	1	1	0	1
0	1	1	0	m_6	A′B C D′	M_6	A + B′ + C′ + D	1	0	1	0	0	1
0	1	1	1	m_7	A′B C D	M_7	A + B′ + C′ + D′	1	0	1	1	0	1
1	0	0	0	m_8	A B′C′D′	M_8	A′ + B + C + D	0	1	0	0	1	1
1	0	0	1	m_9	A B′C′D	M_9	A′ + B + C + D′	0	1	0	0	1	1
1	0	1	0	m_{10}	A B′C D′	M_{10}	A′ + B + C′ + D	0	0	0	0	0	0
1	0	1	1	m_{11}	A B′C D	M_{11}	A′ + B + C′ + D′	0	0	0	0	0	0
1	1	0	0	m_{12}	A B C′D′	M_{12}	A′ + B′ + C + D	0	1	0	0	1	1
1	1	0	1	m_{13}	A B C′D	M_{13}	A′ + B′ + C + D′	0	1	0	1	1	1
1	1	1	0	m_{14}	A B C D′	M_{14}	A′ + B′ + C′ + D	0	0	0	0	0	0
1	1	1	1	m_{15}	A B C D	M_{15}	A′ + B′ + C′ + D′	0	0	0	1	0	1

A function can be expressed as a sum of minterms for which the function output is 1. Here, F is 1, when m_4, m_5, m_6, m_7, m_8, m_9, m_{12}, m_{13} and m_{15} are inputs.

$$\therefore\ F = m_4 + m_5 + m_6 + m_7 + m_8 + m_9 + m_{12} + m_{13} + m_{15}$$

$$= A'BC'D' + A'BC'D + A'BCD' + A'BCD + AB'C'D' + AB'C'D + ABC'D' + ABC'D + ABCD$$

A function can be expressed as a product of maxterms for which the function output is 0. Here, F is 0, when M_0, M_1, M_2, M_3, M_{10}, M_{11} and M_{14} are inputs.

$$F = M_0 M_1 M_2 M_3 M_{10} M_{11} M_{14}$$

$$= (A + B + C + D)(A + B + C + D')(A + B + C' + D)(A + B + C' + D')(A' + B + C' + D)(A' + B + C' + D')$$
$$(A' + B' + C' + D)$$

Example 4.20

(AKTU, Sep'20, 5 Marks)

Express the Boolean function F = A + B′C in a sum of minterms.

Solution

Table 1: Truth Table of F = A + B′C

Input Variables				Minterm		Complement of B	Product Term	Function Output
A	B	C	D			B′	B′C	F
0	0	0	0	m_0	A′B′C′D′	1	0	0
0	0	0	1	m_1	A′B′C′D	1	0	0
0	0	1	0	m_2	A′B′C D′	1	1	1
0	0	1	1	m_3	A′B′C D	1	1	1
0	1	0	0	m_4	A′B C′D′	0	0	0
0	1	0	1	m_5	A′B C′D	0	0	0
0	1	1	0	m_6	A′B C D′	0	0	0
0	1	1	1	m_7	A′B C D	0	0	0
1	0	0	0	m_8	AB′C′D′	1	0	1
1	0	0	1	m_9	AB′C′D	1	0	1
1	0	1	0	m_{10}	AB′C D′	1	1	1
1	0	1	1	m_{11}	AB′C D	1	1	1
1	1	0	0	m_{12}	AB C′D′	0	0	1
1	1	0	1	m_{13}	AB C′D	0	0	1
1	1	1	0	m_{14}	AB C D′	0	0	1
1	1	1	1	m_{15}	AB C D	0	0	1

A function can be expressed as a sum of minterms for which the function output is 1. Here, F is 1, when m_2, m_3, m_8, m_9, m_{10}, m_{11}, m_{12}, m_{13}, m_{14} and m_{15} are inputs

$$\therefore\ F = m_2 + m_3 + m_8 + m_9 + m_{10} + m_{11} + m_{12} + m_{13} + m_{14} + m_{15}$$

$$= A'B'CD' + A'B'CD + AB'C'D' + AB'C'D + AB'CD' + AB'CD + ABC'D' + ABC'D + ABCD' + ABCD$$

4.4.9 Implementation of Logic Design using Standard Logic Gate IC's

A combinational logic design is represented by a set of Boolean equations or functions. A Boolean function basically involves product of variables, sum of variables and complement of variables. Therefore, the straight forward implementation of Boolean function can be made by using basic gates AND, OR and NOT. The complement of a variable is obtained using NOT gate. The product term is realized using AND gate. The sum term is realized using OR gate. The AND, OR and NOT gates that are available as standard ICs for implementing the Boolean functions are listed in Appendix 1.

Example 4.21

Implement the following functions using basic logic gates available as standard ICs.

a) $F_1 = A'B + C'$ **b)** $F_2 = x'y'z + x'yz + xy'z$ **c)** $F_3 = xy' + x'yz$

Solution

a) $F_1 = A'B + C'$

The given function can be implemented using NOT gate IC 7404, 2-input AND gate IC 7408 and 2-input OR gate IC 7432.

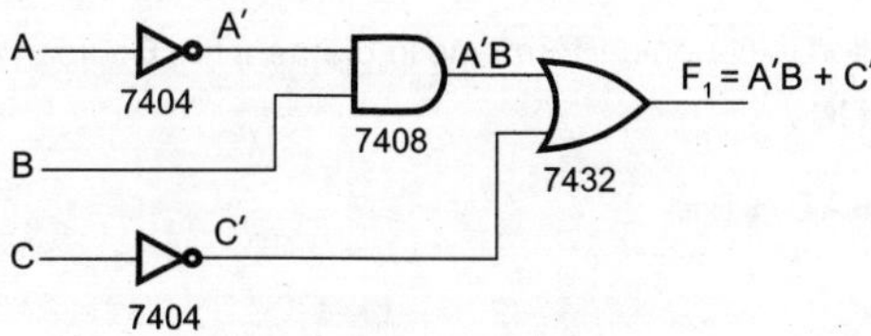

Fig. 1: Logic circuit for function, F_1.

b) $F_2 = x'y'z + x'yz + xy'z$

The given function can be implemented using NOT gate IC 7404, 3-input AND gate IC 7411 and 3-input OR gate IC 744075.

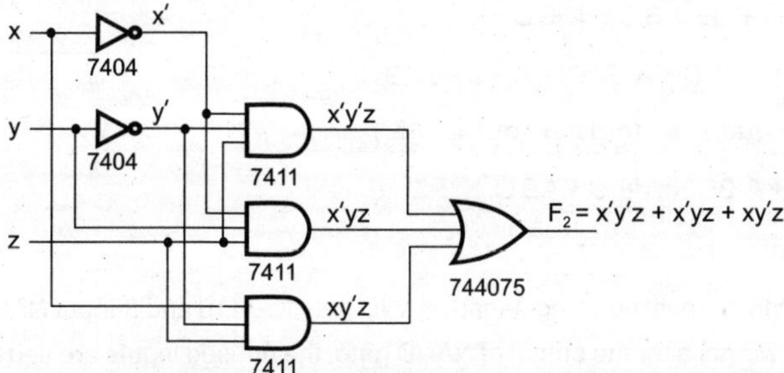

Fig. 2: Logic circuit for function, F_2.

c) $F_3 = xy' + x'yz$

The given function can be implemented using NOT gate IC 7404, 2-input AND gate IC 7408 and 2-input OR gate IC 7432.

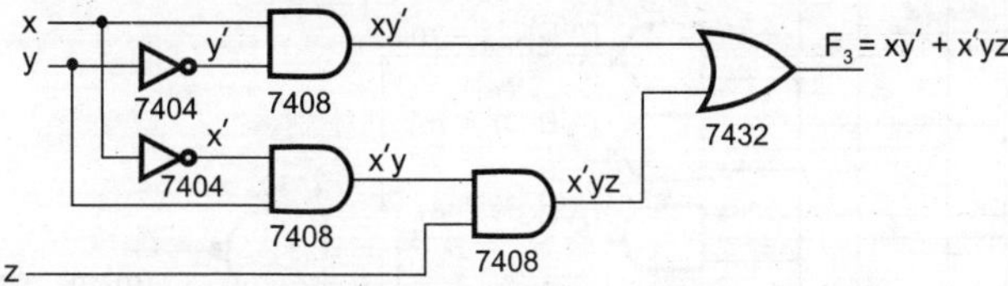

Fig. 3: Logic circuit for function, F_3.

4.4.10 Implementation of Logic Design using Standard Universal Gate IC's

The NAND and NOR gates are called ***universal gates***. Combinational logic design represented by SOP or POS form of Boolean functions can be implemented using universal gates. When a Boolean function is expressed in SOP form, it can be realized using only NAND gates. When a Boolean function is expressed in POS form, it can be realized using only NOR gates. The NAND and NOR gates that are available as standard ICs for implementation of SOP and POS form of Boolean functions are listed in Appendix 1.

Example 4.22

Express the following Boolean function in canonical form and implement the Boolean functions using either only NAND or NOR gates available as standard ICs.

a) $F_1 = A' + BC$ **b)** $F_2 = A'C + AB' + BC$

Solution

a) $F_1 = A' + BC$

Case i: SOP

$F_1 = A' + BC$

$= A'(B + B')(C + C') + (A + A')BC$

$= (A'B + A'B')(C + C') + ABC + A'BC$

$= A'BC + A'BC' + A'B'C + A'B'C' + ABC + A'BC$

$(m_3) \quad (m_2) \quad (m_1) \quad (m_0) \quad (m_7) \quad (m_3)$

$= A'B'C' + A'B'C + A'BC' + A'BC + ABC$

$= m_0 + m_1 + m_2 + m_3 + m_7$

$x + x' = 1$
$x \cdot 1 = x$

Missing literals in first term are A and B and in second term is A.

Repeated minterms are considered only one time.

The SOP form of F_1 can be realized using 3-input NAND gate IC 7410 and 8-input NAND gate IC 7430 as shown in Fig. 1. Since NAND with 1 will not alter the output of NAND gate, the unused inputs are tied to 1 in 8-input NAND gate.

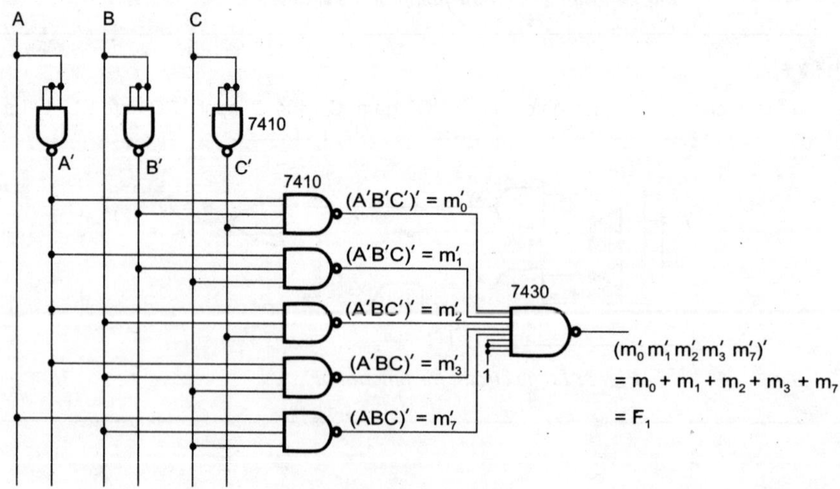

Fig. 1: Logic circuit of F_1 using only NAND gates.

Case ii: POS

$F_1 = A' + BC$

$= (A' + B)(A' + C) = (A' + B + (CC'))(A' + C + (BB'))$

$= (A' + B + C)(A' + B + C')(A' + C + B)(A' + C + B')$

$= (A' + B + C)(A' + B + C')(\cancel{A' + B + C})(A' + B' + C)$

$\quad (M_4) \quad (M_5) \quad (M_4) \quad (M_6)$

$= (A' + B + C)(A' + B + C')(A' + B' + C)$

$= M_4 M_5 M_6$

x . x' = 0
x + 0 = a

Missing literal in first term is C and in second term is B.

Repeated maxterms are considered only one time.

The POS form of F_1 can be realized using 3-input NOR gate IC 7427 as shown in Fig. 2.

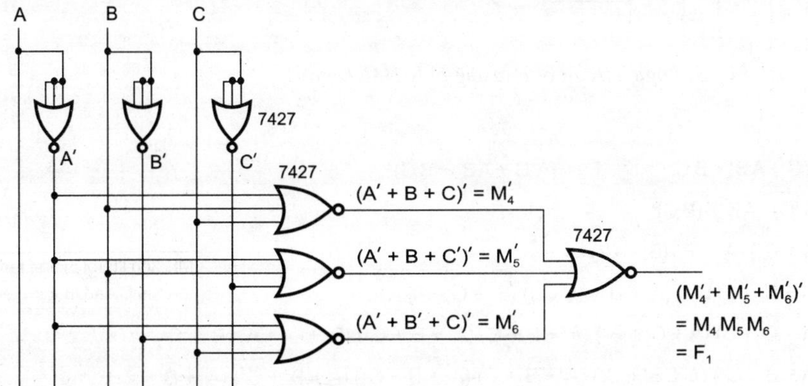

***Fig. 2:** Logic circuit of F_1 using only NOR gates.*

---

b) $F_2 = A'C + AB' + BC$

Case i: SOP

$F_2 = A'C + AB' + BC$

$= A'C(B + B') + AB'(C + C') + BC(A + A')$

$= A'CB + A'CB' + AB'C + AB'C' + BCA + BCA'$

$= A'BC + A'B'C + AB'C + AB'C' + ABC + \cancel{A'BC}$

$\quad (m_3) \quad (m_1) \quad (m_5) \quad (m_4) \quad (m_7) \quad (m_3)$

$= A'B'C + A'BC + AB'C' + AB'C + ABC$

$= m_1 + m_3 + m_4 + m_5 + m_7$

Missing literal in first term is B, second term is C and third term is A.

x + x' = 1
x . 1 = x

Repeated minterms are considered only one time.

The SOP form of F_2 can be realized using 3-input NAND gate IC 7410 and 8-input NAND gate IC 7430 as shown in Fig. 3. Since NAND with 1 will not alter the output of NAND gate, the unused inputs are tied to 1 in 8-input NAND gate.

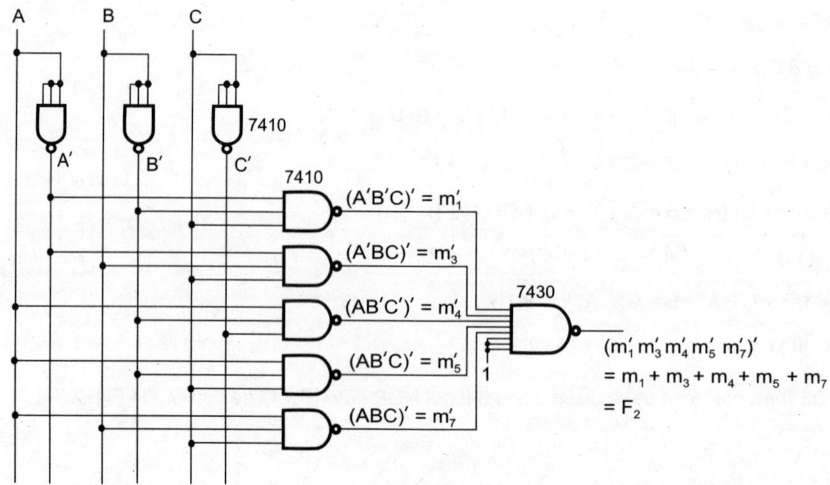

Fig. 3: Logic circuit of F_2 using only NAND gates.

Case ii: POS

$F_2 = A'C + AB' + BC \quad \Rightarrow \quad F_2 = (A'C + AB' + BC)'$

$\therefore\ F_2 = (A'C)'(AB')'(BC)'$

x . x′ = 0
x + 0 = x

$= (A + C')(A' + B)(B' + C')$

Missing literal in first term is B, second term is C and in third term is A.

$= (A + C' + (BB'))(A' + B + (CC'))(B' + C' + (AA'))$

$= (A + C' + B)(A + C' + B')(A' + B + C)(A' + B + C')(B' + C' + A)(B' + C' + A')$

$= (A + B + C')(A + B' + C')(A' + B + C)(A' + B + C')\ \cancel{(A + B' + C')}\ (A' + B' + C')$

$\quad (M_1) \quad (M_3) \quad (M_4) \quad (M_5) \quad (M_3) \quad (M_7)$

Repeated maxterms are considered only one time.

$= (A + B + C')(A + B' + C')(A' + B + C)(A' + B + C')(A' + B' + C')$

$= M_1\,M_3\,M_4\,M_5\,M_7$

$\therefore\ F_2 = (M_1\,M_3\,M_4\,M_5\,M_7)' = M_0\,M_2\,M_6$

$= (A + B + C)(A + B' + C)(A' + B' + C)$

The POS form of F_2 can be realized using 3-input NOR gate IC 7427 as shown in Fig. 4.

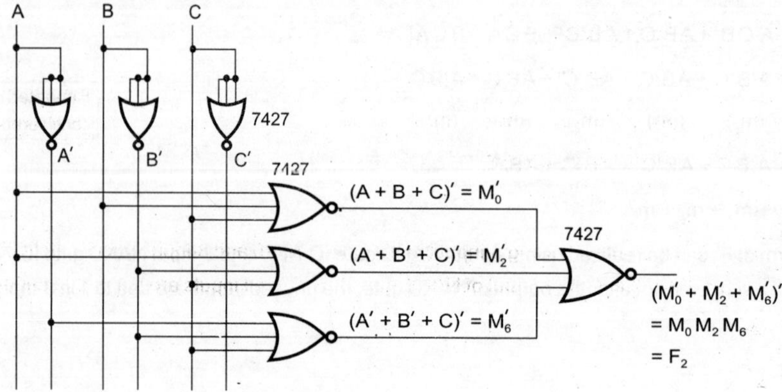

Fig. 4: Logic circuit of F_2 using only NOR gates.

4.5 Simplification of Boolean Expressions or Functions

Each variable with in a term of a Boolean expression is called a ***literal***. Example of 7 literals and 10 literals Boolean expression are shown below:

$$a'\,b + c\,d + a\,b\,d$$

↓ ↓ ↓ ↓ ↓ ↓ ↓

① ② ③ ④ ⑤ ⑥ ⑦ → 7 literals

$$a\ b\ c + b'c + a\ c' + a\ b'\ c$$

① ② ③ ④ ⑤ ⑥ ⑦ ⑧ ⑨ ⑩ → 10 literals

Reducing the number of literals in a Boolean expression or function will simplify the implementation of function by logic gates with minimum number of gates. The implementation of Boolean function using logic gates is also called MSI (Medium Scale Integration) Circuit.

The Boolean expression can be reduced or simplified using postulates and theorems of Boolean algebra but there is no standard procedure for simplification using theorems.

Alternatively, Boolean functions can be simplified using following two methods.

- Karnaugh map method
- Quine-McCluskey method or Table method

These two methods have standard procedure for simplification of Boolean functions.

Example 4.23

(JNTUH, Nov/Dec'20, 6 Marks)

Reduce the number of literals in following expressions.

a) $x\,y + x\,y'$ **b)** $x\,y + x\,(w\,z + w\,z')$ **c)** $(A + B)'(A' + B')'$

Solution

a) $xy + xy'$

$$x\,y + x\,y' = x\,(y + y')$$ [$y + y' = 1$]

$$= x.1$$ [$x.1 = x$]

$$= x$$

The given expression $x\,y + x\,y'$ is 4 literal expression which is reduced to single literal.

---

b) $xy + x\,(w\,z + w\,z')$

$$x\,y + x\,(w\,z + w\,z') = x\,y + x\,(w\,(z + z'))$$ [$z + z' = 1$]

$$= x\,y + x\,(w\,.\,1)$$ [$w.1 = w$]

$$= x\,y + x\,w$$

$$= x\,.\,(y + w)$$

The given expression $x\,y + x\,(w\,z + w\,z')$ is 7 literals expression which is reduced to three literals.

---

c) (A + B)′(A′+ B′)′

$$(A + B)'(A'+ B')' = (A + B)' . ((A')' . (B')')$$
$$= (A' . B') . (A . B)$$
$$= A' . B' . A . B$$
$$= A' . A . B . B' = 0 . 0 = 0$$

Using DeMorgan's law
$(x + y)' = x'.y'$

$x . x' = 0$

The given expression is a 4 literal expression which is reduced to null expression.

4.6 Karnaugh Map (or K-map) Minimization

K-map is a pictorial form of truth table and used to simplify Boolean functions. Simplification of Boolean function using K-map is simple and straight forward. A K-map is a diagram made up of squares with each square representing one minterm. Alternatively, maxterm can be used to construct K-map in which each square represent a maxterm.

When minterms are considered for simplification the resultant Boolean function will be in sum-of-products form. When maxterms are considered for simplification the resultant Boolean function will be in product-of-sums form.

Note: *The simplified Boolean function using K-map is not unique. Sometimes there may be multiple solutions.*

For n variable Boolean function the K-map will have 2^n squares. Each square represent a minterm or maxterm. The literals of minterm are split and arranged as rows and columns. While arranging the literal/minterms only one change is allowed if we move from one row to next row or from one column to next column.

4.6.1 Realization of Sum-of-Products Form using K-map

1. Construct the truth table of Boolean function.
2. Draw K-map with 2^n squares where n is number of variables in Boolean function.
3. Enter the value of function or output in the squares as either 1 or 0.
4. The squares with entry 1 are the minterms that represent the function.
5. Each square with 1 is an ***implicant***.
6. Combine adjacent 1's to form ***prime implicants***.
7. A prime implicant with single 1 represent n literal product term.
8. A prime implicant with two 1's represent n – 1 literal product term.
9. A prime implicant with four 1's represent n – 2 literal product term and so on.
10. The literals represented by a prime implicant are the literals that present in all the squares of the prime implicant. (or the literals common to all the squares of the prime implicant).
11. While forming prime implicants any number of over lapping is allowed in horizontal and vertical directions. But the diagonal elements cannot be combined to form prime implicants.
12. While forming prime implicants the K-map can be folded in any direction to get adjacent 1's.
13. The simplified Boolean function is sum of all the product terms of prime implicant.

14. If all minterms are 1's then all squares of K-map will be filled by 1 and function value is 1.

15. The logic or MSI (Medium Scale Integration) circuit of the simplified Boolean function can be drawn using logic gates.

Table 4.36: Relationship between Number of Adjacent Squares and Literals in the Product Term

Number of Adjacent Squares in Prime Implicants	Number of Literals in a Product Term in an n-variable Map			
	n = 2	n = 3	n = 4	n = 5
1	2	3	4	5
2	1	2	3	4
4	0	1	2	3
8		0	1	2
16			0	1
32				0

4.6.2 Two-Variable K-map

Let, n = Number of variables in Boolean function

x, y = Two variables of Boolean function

Therefore, n = 2, for two-variable K-map

Here, $2^n = 2^2 = 4$

Therefore, two-variable, K-map will have 4 squares. The two-variable K-map and arrangements of literals and minterms are shown in Fig. 4.10.

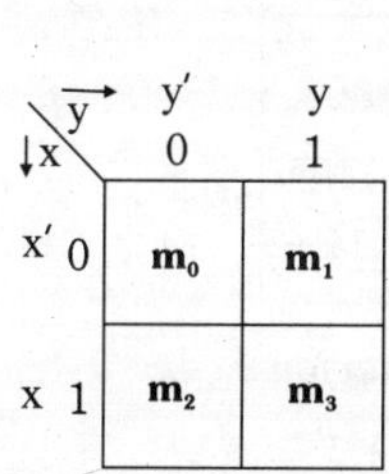

Fig. a: Arrangement of literals and minterms in 2-variable K-map.

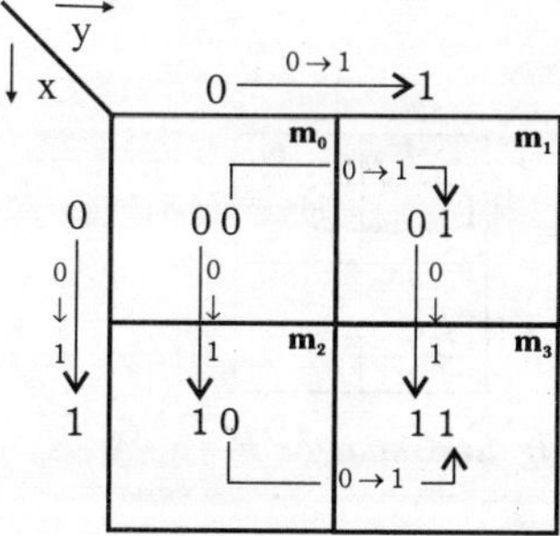

Fig. b: Change in bit values of literals and minterms in 2-variable K-map.

Fig. 4.10: *Two-variable K-map.*

Some examples of two-variable K-map with single prime implicant formed using single 1's is shown in Figs. 4.11 and 4.12. Since the prime implicants shown in Figs. 4.11 and 4.12 are formed using single 1's in two-variable K-map, each prime implicant represents a product term of two literals.

The K-maps in Fig. 4.11 have only one prime implicant and so the Boolean function, F is given by prime implicant of the K-map. The K-maps in Fig. 4.12 has two prime implicants and so the Boolean function, F is given by sum of the two prime implicants of the K-map.

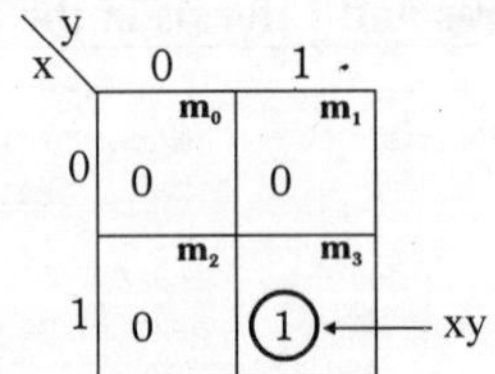

***Fig. a:** K-map for, F = xy.*

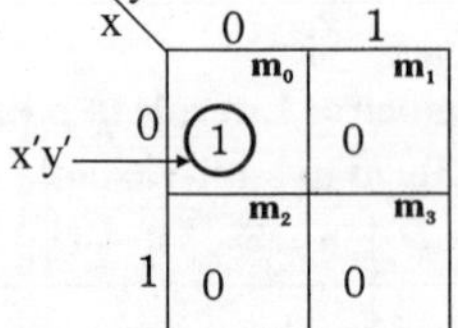

***Fig. b:** K-map for, F = x'y'.*

***Fig. 4.11:** Two-variable K-map with single prime implicant formed using single 1's.*

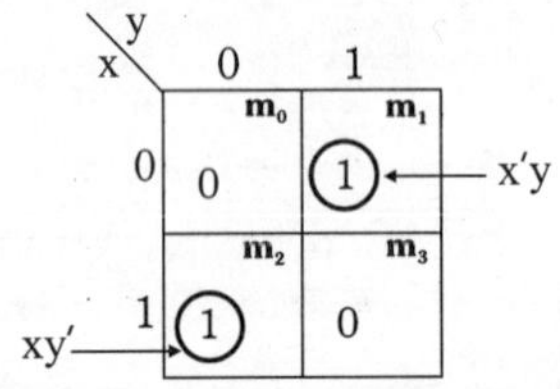

***Fig. a:** K-map for, F = x'y + xy'.*

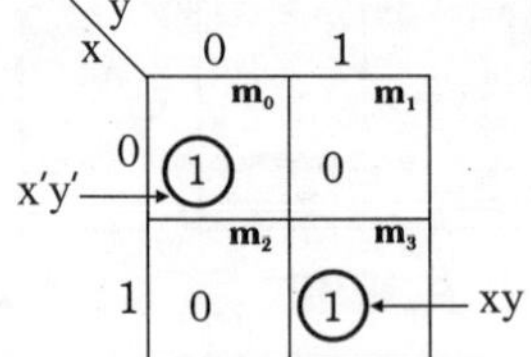

***Fig. b:** K-map for, F = x'y' + xy.*

***Fig. 4.12:** Two-variable K-map with two prime implicants formed using single 1's.*

Some examples of two-variable K-map with single prime implicant formed using two adjacent 1's is shown in Fig. 4.13. Since the prime implicants shown in Fig. 4.13 are formed using two adjacent 1's in two-variable K-map, each prime implicant represents a single literal. Since the K-map have only one prime implicant the Boolean function, F is given by prime implicant of the K-map.

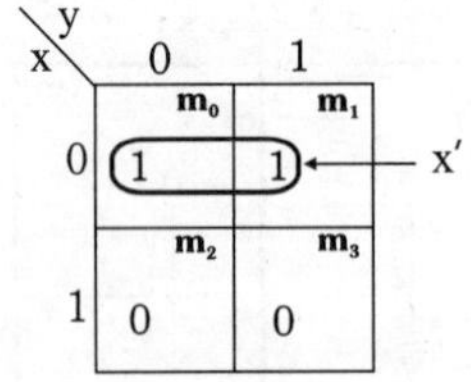

***Fig. a:** K-map for, F = x'.*

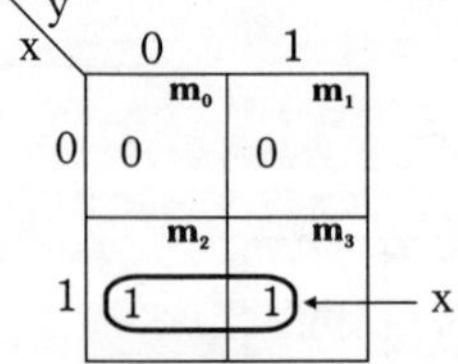

***Fig. b:** K-map for, F = x.*

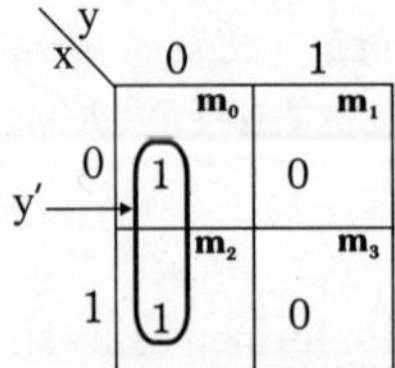

***Fig. c:** K-map for, F = y'.*

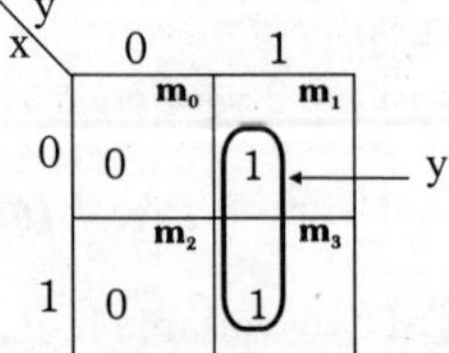

***Fig. d:** K-map for, F = y.*

***Fig. 4.13:** Two-variable K-map with prime implicant formed using two adjacent 1's.*

Some examples of two-variable K-map with two overlapping prime implicants formed using two adjacent 1's is shown in Fig. 4.14. Since the prime implicants shown in Fig. 4.14 are formed using two adjacent 1's in two-variable K-map, each prime implicant represents a single literal. Since the K-map has two prime implicant the Boolean function, F is given by sum of two prime implicants of the K-map.

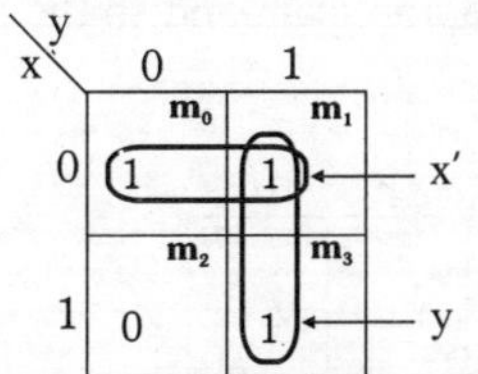

Fig. a: *K-map for, $F = x' + y$.*

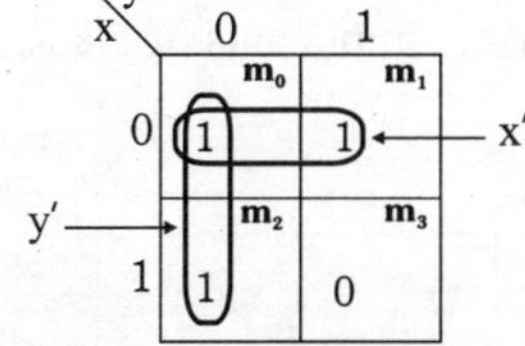

Fig. b: *K-map for, $F = x' + y'$.*

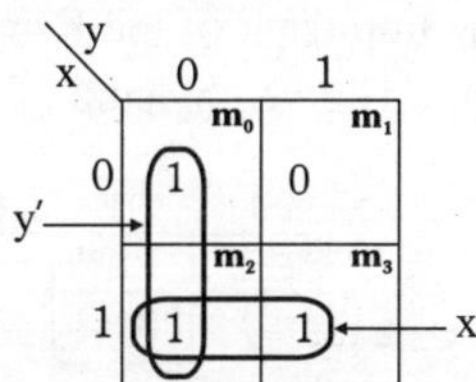

Fig. c: *K-map for, $F = x + y'$.*

Fig. 4.14: *Two-variable K-map with two overlapping prime implicants formed using two adjacent 1's.*

When all the entries of K-map (or all minterms) are 1's then the Boolean function value is 1 (F =1) as shown in Fig. 4.15.

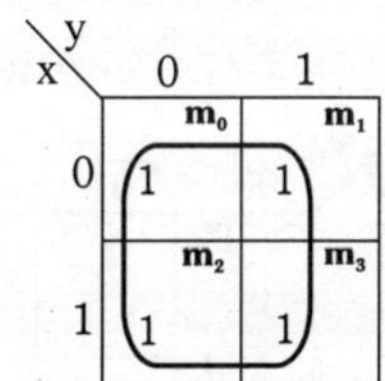

Fig. 4.15: *Two-variable K-map with all minterms equal to 1 and F=1.*

4.6.3 Three-Variable K-map

Let, n = Number of variables in Boolean function

x, y, z = Three variables of Boolean function

Therefore, n = 3, for three-variable K-map

Here, $2^n = 2^3 = 8$

Therefore, three-variable, K-map will have 8 squares. The three-variable K-map and arrangements of literals and minterms are shown in Fig. 4.16.

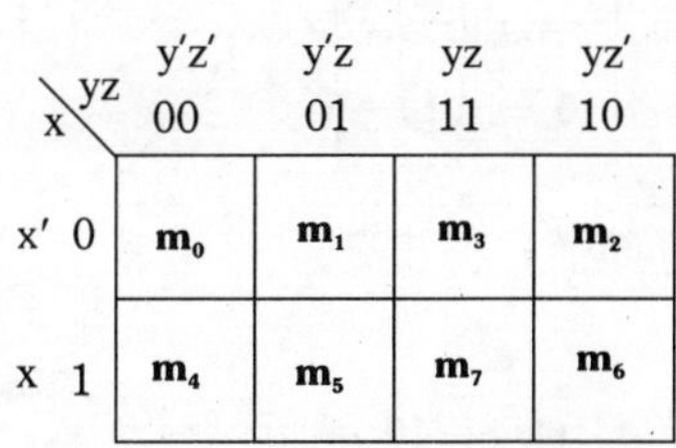

Fig. a: *Arrangement of literals and minterms in 3-variable K-map.*

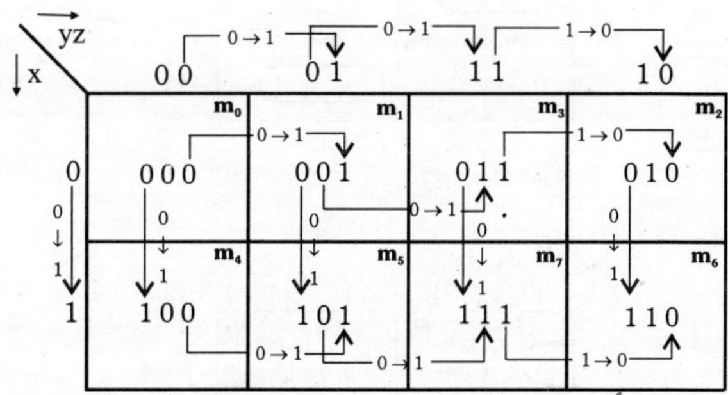

Fig. b: *Change in bit values of literals and minterms in 3-variable K-map.*

Fig. 4.16: *Three-variable K-map.*

Some examples of three-variable K-map with prime implicant formed using single 1's are shown in Figs. 4.17 and 4.18. Since the prime implicants shown in Figs. 4.17 and 4.18 are formed using single 1's in three-variable K-map, each prime implicant represents a product term of three literals.

The K-maps in Fig. 4.17 have only one prime implicant and so the Boolean function, F is given by prime implicant of the K-map. The K-maps in Fig. 4.18 has two prime implicant and so the Boolean function, F is given by sum of the two prime implicants of the K-map.

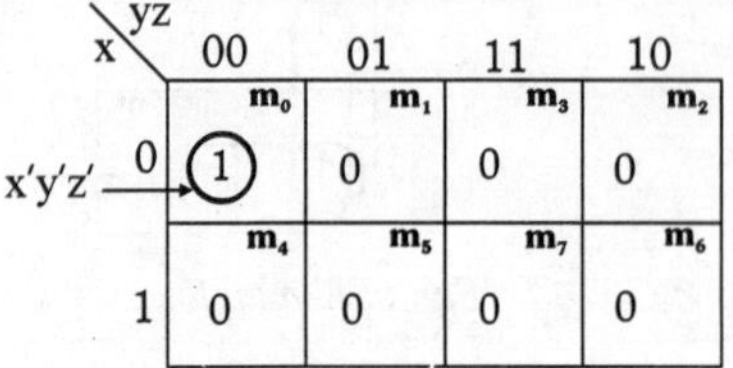

Fig. a: K-map for, F = x'y'z'.

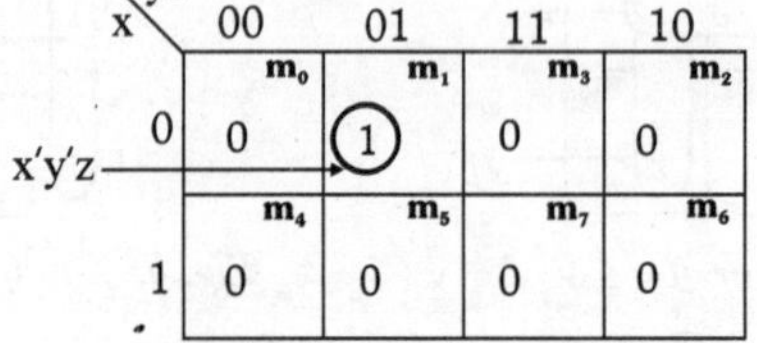

Fig. b: K-map for, F = x'y'z.

***Fig. 4.17:** Three-variable K-map with prime implicant formed using single 1's.*

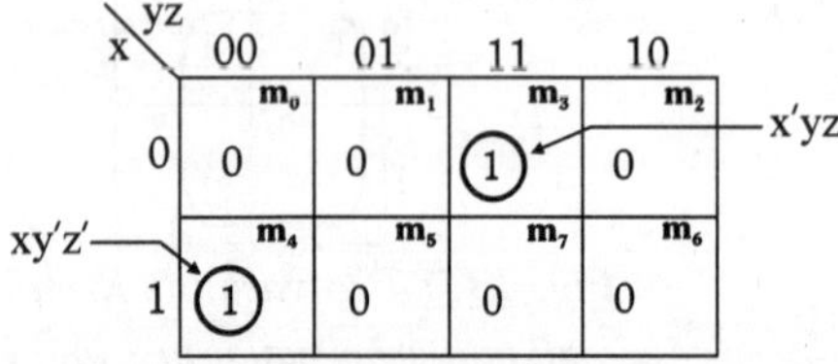

Fig. a: K-map for, F = x'yz + xy'z'.

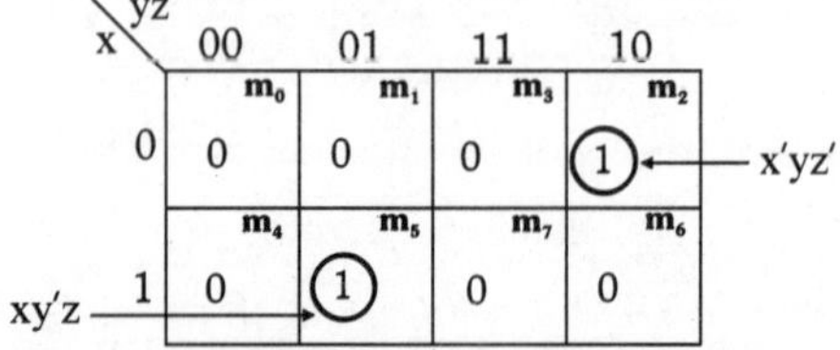

Fig. b: K-map for, F = x'yz' + xy'z.

***Fig. 4.18:** Three-variable K-map with two prime implicants formed using single 1's.*

Some examples of three-variable K-map with two prime implicants formed using two adjacent 1's are shown in Fig. 4.19. Since the prime implicants shown in Fig. 4.19 are formed using two adjacent 1's in three -variable K-map, each prime implicant represents a product term of two literals. Since the K-map has two prime implicants the Boolean function, F is given by sum of the two prime implicants of the K-map.

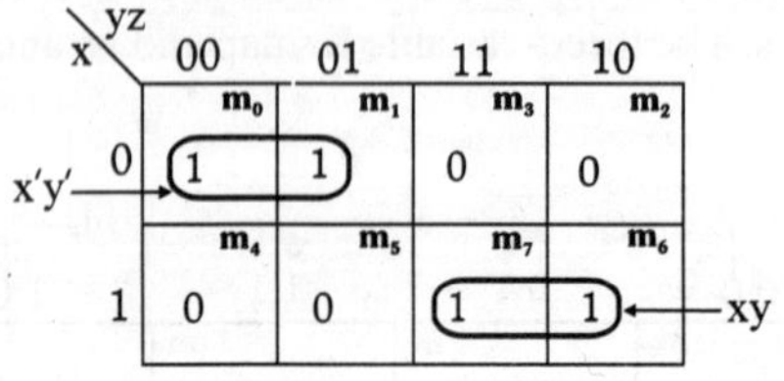

Fig. a: K-map for, F = x'y' + xy.

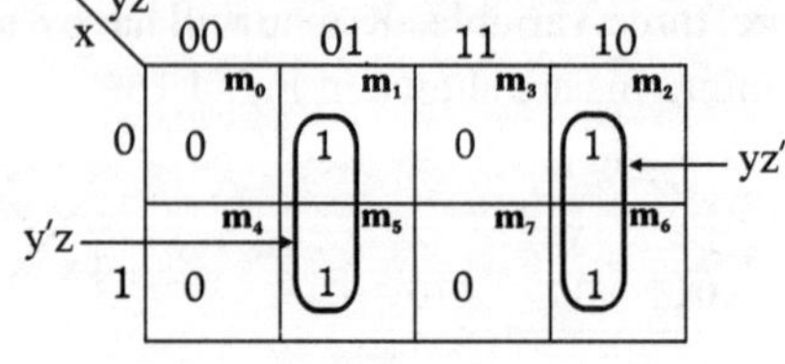

Fig. b: K-map for, F = y'z + yz'.

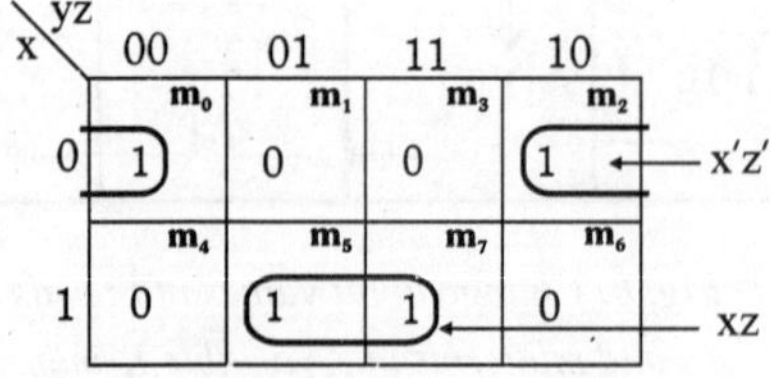

Fig. c: K-map for, F = x'z' + xz.

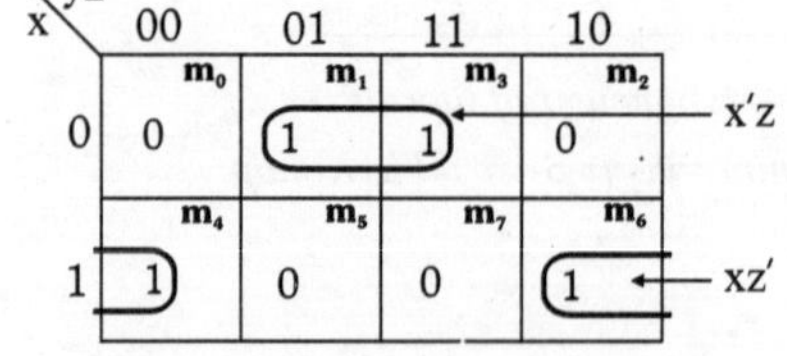

Fig. d: K-map for, F = x'z + xz'.

***Fig. 4.19:** Three-variable K-map with two prime implicants formed using two adjacent 1's.*

Some examples of three-variable K-map with two overlapping prime implicants formed using two adjacent 1's and a prime implicant with single 1 are shown in Fig. 4.20. The prime implicants formed using two adjacent 1's in three-variable K-map, represents a product term of two literals and the prime implicant with single 1 represents a product term of three literal. Since the K-map has three prime implicant the Boolean function, F is given by sum of three prime implicants of the K-map.

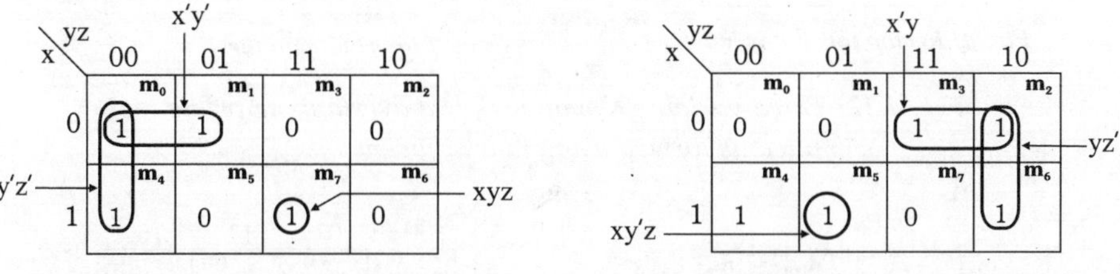

Fig. a: *K-map for, F = xyz + x'y' + y'z'.* ***Fig. b:*** *K-map for, F = xy'z + yz' + x'y.*

Fig. 4.20: *Three-variable K-map with prime implicants formed using single 1's and two overlapping two adjacent 1's.*

Some examples of three-variable K-map with prime implicant formed using four adjacent 1's are shown in Fig. 4.21. Since the prime implicants shown in Fig. 4.21 are formed using four adjacent 1's in three-variable K-map, each prime implicant represents a single literal. Since the K-map have only one prime implicant the Boolean function, F is given by prime implicant of the K-map.

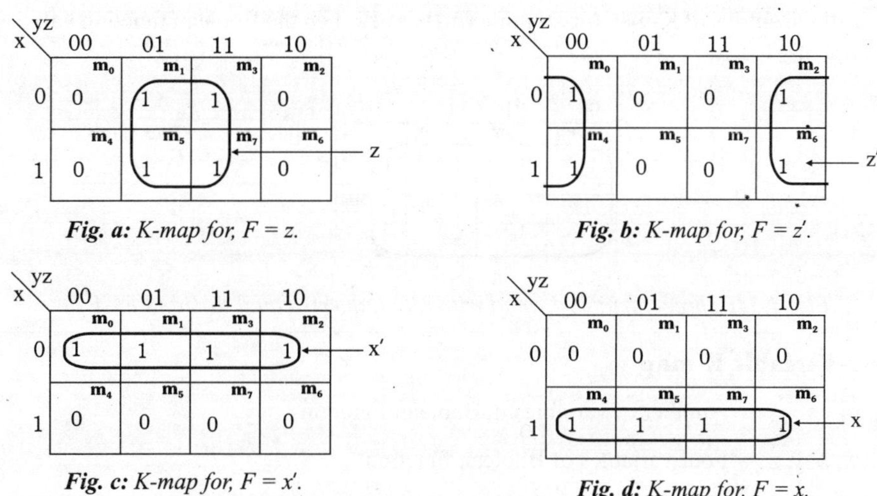

Fig. a: *K-map for, F = z.* ***Fig. b:*** *K-map for, F = z'.*

Fig. c: *K-map for, F = x'.* ***Fig. d:*** *K-map for, F = x.*

Fig. 4.21: *Three-variable K-map with prime implicant formed using four adjacent 1's.*

Some examples of three-variable K-map with two overlapping prime implicants formed using four adjacent 1's and two adjacent 1's are shown in Figs. 4.22 and 4.23. The prime implicants formed using two adjacent 1's in three-variable K-map, represents a product term of two literals and the prime implicant with four adjacent 1's represents a single literal. Since the K-map has two prime implicant the Boolean function, F is given by sum of two prime implicants of the K-map.

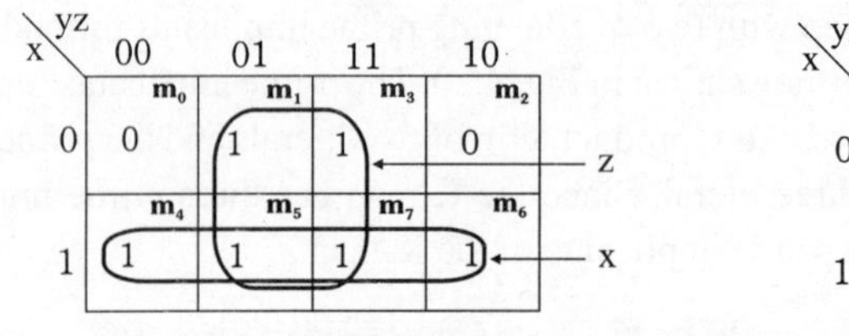

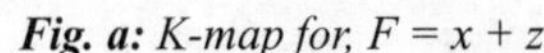

Fig. a: K-map for, F = x + z.

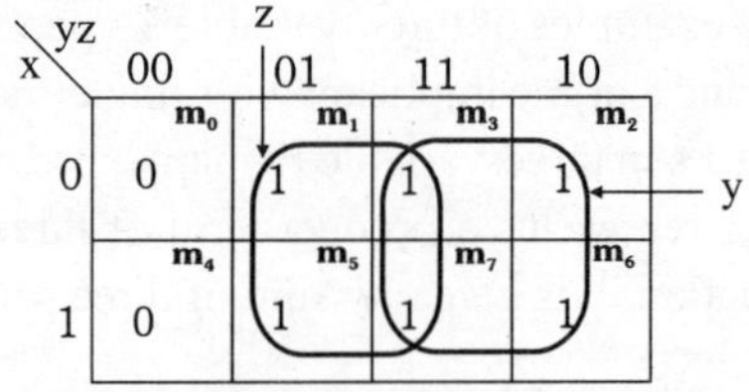

Fig. b: K-map for, F = y + z.

Fig. 4.22: Three-variable K-map with two overlapping prime implicants formed using four adjacent 1's.

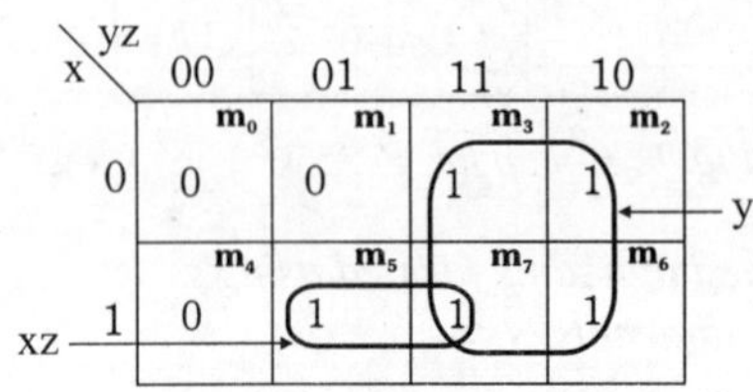

Fig. a: K-map for, F = y + xz.

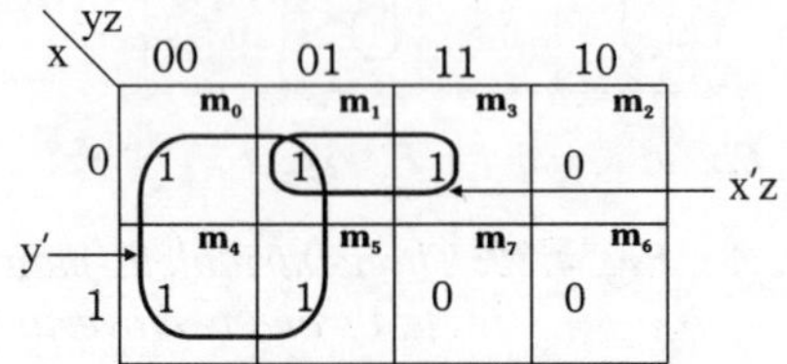

Fig. b: K-map for, F = y′ + x′z.

Fig. 4.23: Three-variable K-map with overlapping prime implicants formed using two adjacent 1's and four adjacent 1's.

When all the entries of K-map (or all minterms) are 1's then the Boolean function value is 1 (F = 1) as shown in Fig. 4.24.

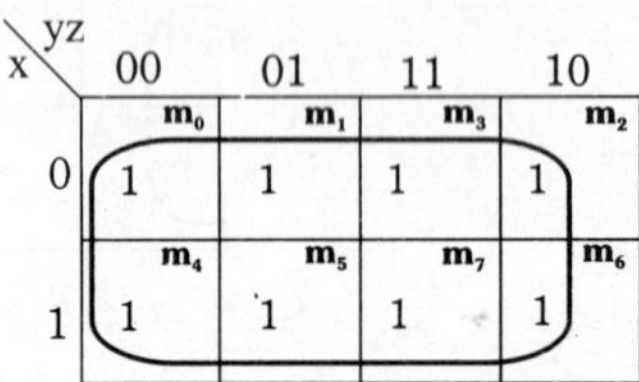

Fig. 4.24: Three-variable K-map with all minterms equal to 1 and F=1.

4.6.4 Four-Variable K-map

Let, n = Number of variables in Boolean function

w, x, y, z = Four variables of Boolean function

Therefore, n = 4, for four-variable K-map

Here, $2^n = 2^4 = 16$

Therefore, four-variable K-map will have 16 squares. The four-variable K-map and arrangements of literals and minterms are shown in Fig. 4.25.

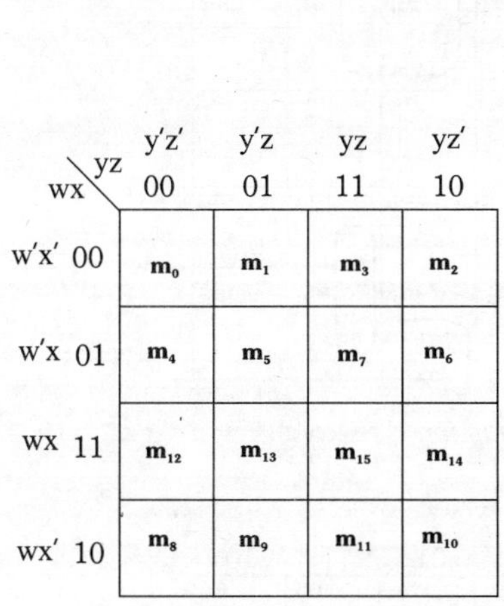

***Fig. a:** Arrangement of literals and minterms in four-variable K-map.*

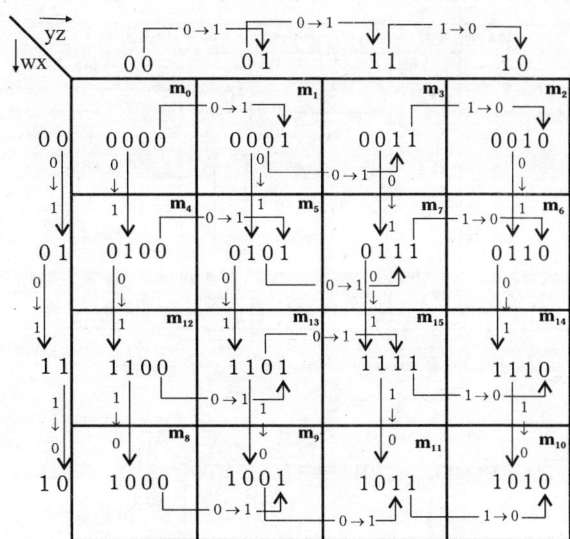

***Fig. b:** Change in bit values of literals and minterms in four-variable K-map.*

***Fig. 4.25:** Four-variable K-map.*

Some examples of four-variable K-map with prime implicants formed using four adjacent 1's, two adjacent 1's and single 1's are shown in Fig. 4.26. In four-variable K-map, the prime implicants formed using four adjacent 1's represents a product term of two literals, the prime implicants with two adjacent 1's represents a product term of three literals and the prime implicant with single 1 represents a product term of four literals. Boolean function, F is given by sum of prime implicants of the K-map.

wx \ yz	00	01	11	10
00	m_0: 0	m_1: (1) ← w'x'y'z	m_3: 0	m_2: 0
01	m_4: 0	m_5: 0	m_7: 0	m_6: 0
11	m_{12}: 0	m_{13}: 0	m_{15}: (1) ← wxyz	m_{14}: 0
10	m_8: (1) ← wx'y'z'	m_9: 0	m_{11}: 0	m_{10}: 0

***Fig. a:** K-map for, $F = w'x'y'z + wxyz + wx'y'z'$.*

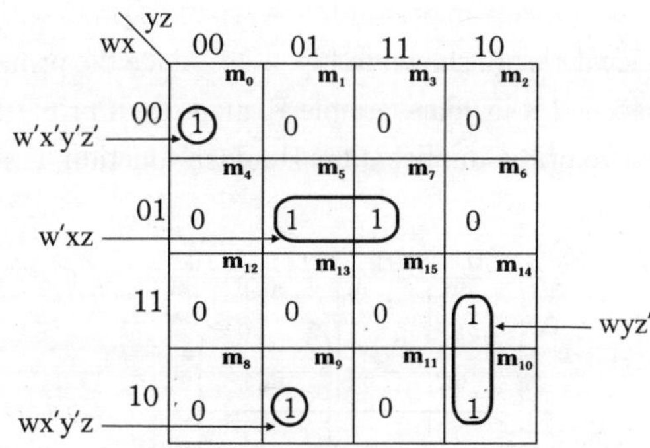

***Fig. b:** K-map for, $F = w'x'y'z' + wx'y'z + w'xz + wyz'$.*

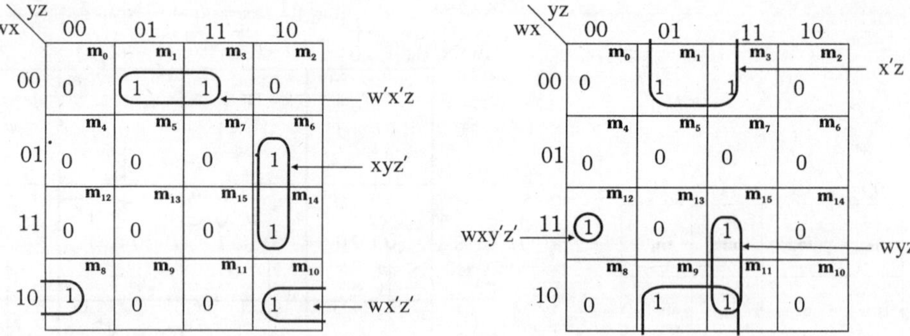

Fig. c: K-map for, F = w'x'z + xyz' + wx'z'.

Fig. d: K-map for, F = wxy'z' + wyz + x'z.

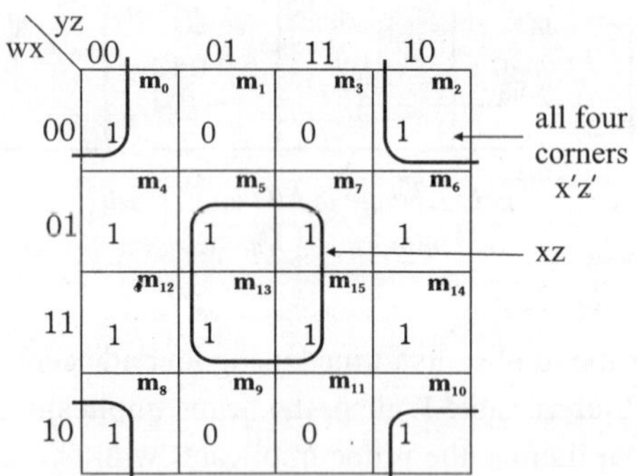

Fig. e: K-map for, F = xz + x'z'.

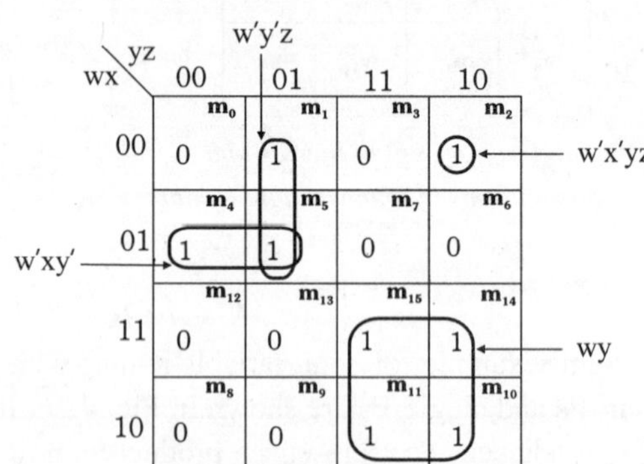

Fig. f: K-map for, F = w'x'yz' + w'y'z + w'xy' + wy.

***Fig. 4.26:** Four-variable K-maps.*

Some examples of four-variable K-map with two overlapping prime implicants formed using eight adjacent 1's are shown in Fig. 4.27. Since the prime implicants shown in Fig. 4.27 are formed using eight adjacent 1's in four-variable K-map, each prime implicant represents a single literal. Since the K-map has two prime implicant the Boolean function, F is given by sum of two prime implicants of the K-map.

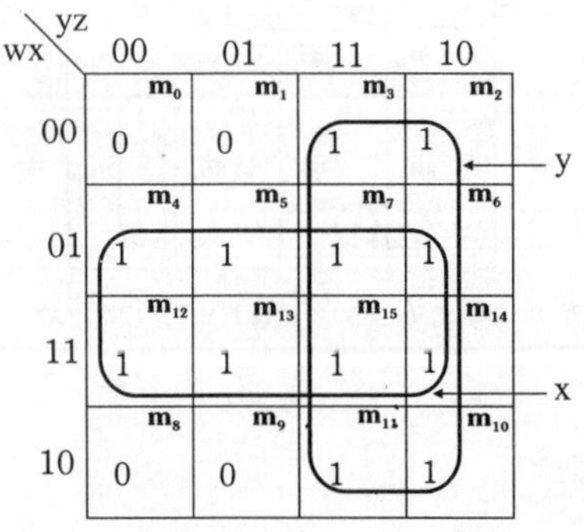

Fig. a: K-map for, F = x + y.

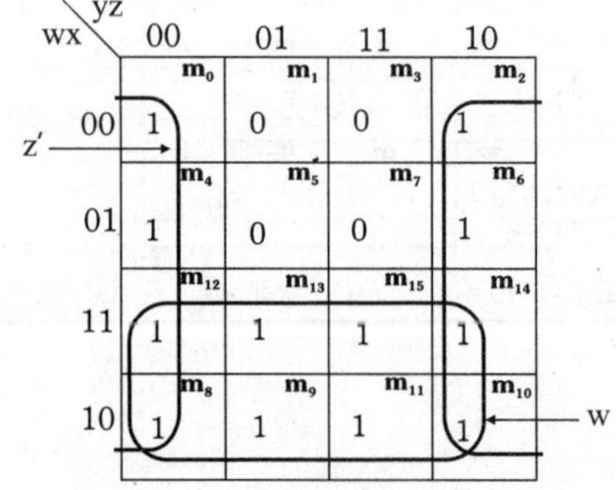

Fig. b: K-map for, F = w + z'.

***Fig. 4.27:** Four-variable K-map with overlapping prime implicants formed using eight adjacent 1's.*

When all the entries of K-map (or all minterms) are 1's then the Boolean function value is 1 (F = 1) as shown in Fig. 4.28.

wx \ yz	00	01	11	10
00	m_0 1	m_1 1	m_3 1	m_2 1
01	m_4 1	m_5 1	m_7 1	m_6 1
11	m_{12} 1	m_{13} 1	m_{15} 1	m_{14} 1
10	m_8 1	m_9 1	m_{11} 1	m_{10} 1

Fig. 4.28: Four-variable K-map with all minterms equal to 1 and F = 1.

4.6.5 Five-Variable K-map

(JNTUK, Feb'22, 8 Marks)

Let A, B, C, D and E be the five-variables.

Now, n = 5 and $2^5 = 32$

Therefore, 32 minterms are possible as shown in Table 4.37.

Table 4.37: Truth Table

A	B	C	D	E	Minterm	A	B	C	D	E	Minterm
0	0	0	0	0	m_0	1	0	0	0	0	m_{16}
0	0	0	0	1	m_1	1	0	0	0	1	m_{17}
0	0	0	1	0	m_2	1	0	0	1	0	m_{18}
0	0	0	1	1	m_3	1	0	0	1	1	m_{19}
0	0	1	0	0	m_4	1	0	1	0	0	m_{20}
0	0	1	0	1	m_5	1	0	1	0	1	m_{21}
0	0	1	1	0	m_6	1	0	1	1	0	m_{22}
0	0	1	1	1	m_7	1	0	1	1	1	m_{23}
0	1	0	0	0	m_8	1	1	0	0	0	m_{24}
0	1	0	0	1	m_9	1	1	0	0	1	m_{25}
0	1	0	1	0	m_{10}	1	1	0	1	0	m_{26}
0	1	0	1	1	m_{11}	1	1	0	1	1	m_{27}
0	1	1	0	0	m_{12}	1	1	1	0	0	m_{28}
0	1	1	0	1	m_{13}	1	1	1	0	1	m_{29}
0	1	1	1	0	m_{14}	1	1	1	1	0	m_{30}
0	1	1	1	1	m_{15}	1	1	1	1	1	m_{31}

The 32 minterms can be arranged in 2 numbers of 4 variable K-map, one for A = 0 and other for A = 1 as shown in Figs. 4.29 and 4.30. These two K-maps can be placed one over the other as shown in Fig. 4.31. While forming prime implicants any two square that lie one over the other are considered adjacent.

Apart from this, the guidelines for forming prime implicants of 4-variable K-map can be applied individually to Figs. 4.29 and 4.30 for the adjacent squares that do not lie one over the other in Fig. 4.31. Some examples of formation of prime implicants in 5-variable K-map are shown in Figs. 4.32 and 4.33.

A = 0

BC \ DE	00	01	11	10
00	m_0	m_1	m_3	m_2
01	m_4	m_5	m_7	m_6
11	m_{12}	m_{13}	m_{15}	m_{14}
10	m_8	m_9	m_{11}	m_{10}

Fig. 4.29: *4-variable K-map for A = 0.*

A = 1

BC \ DE	00	01	11	10
00	m_{16}	m_{17}	m_{19}	m_{18}
01	m_{20}	m_{21}	m_{23}	m_{22}
11	m_{28}	m_{29}	m_{31}	m_{30}
10	m_{24}	m_{25}	m_{27}	m_{26}

Fig. 4.30: *4-variable K-map for A = 1.*

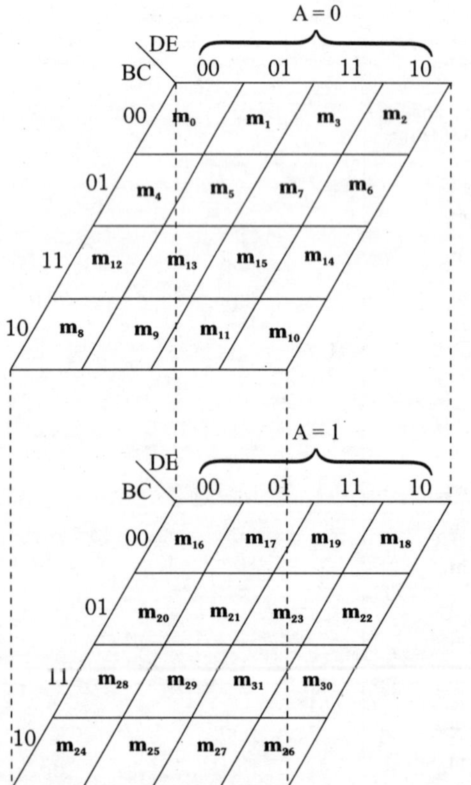

Fig. 4.31: *5-variable K-map.*

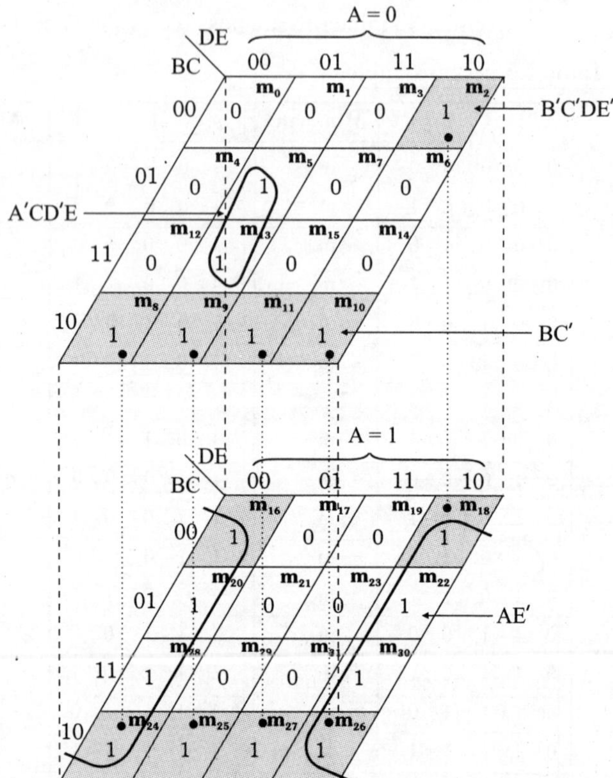

Fig. 4.32: *K-map for F = B′C′DE′ + A′CD′E + AE′ + BC′.*

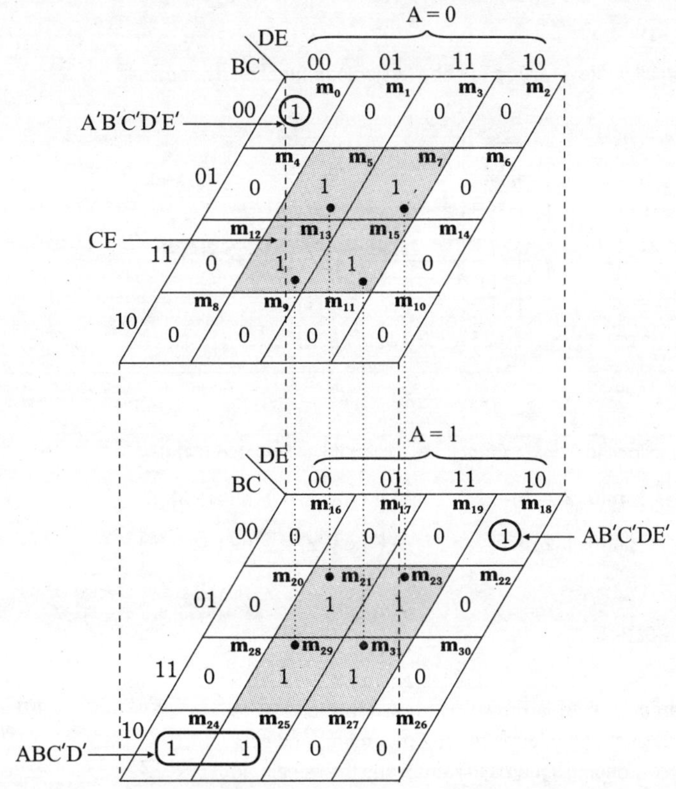

Fig. 4.33: K-map for F = A′B′C′D′E′ + CE + AB′C′DE′ + ABC′D′.

Example 4.24

(GTU, Jun'19, 7 Marks) *(AU, Apr/May'19, 7 Marks)*

Simplify and implement the logic function F(A, B, C) = Σ m(0, 1, 4, 5, 7) using logic gates.

Solution

The truth table of logic function is shown in Table 1 and the three variable K-map is constructed using Table 1 as shown in Fig. 1. The function output, F is 1 for minterms m_0, m_1, m_4, m_5 and m_7.

Table 1: Truth Table

Inputs			Minterm	Output
A	B	C		F
0	0	0	m_0	1
0	0	1	m_1	1
0	1	0	m_2	0
0	1	1	m_3	0
1	0	0	m_4	1
1	0	1	m_5	1
1	1	0	m_6	0
1	1	1	m_7	1

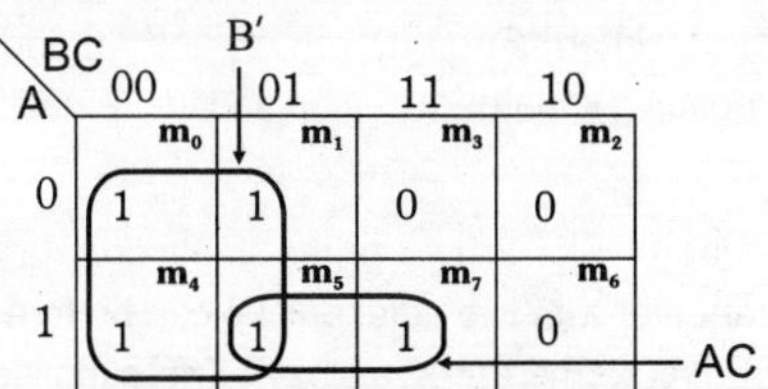

Fig. 1: K-map for F.

From the K-map we get the Boolean equation of the function as,

$$F = AC + B'$$

The logic circuit is drawn using the above Boolean function as shown in Fig. 2.

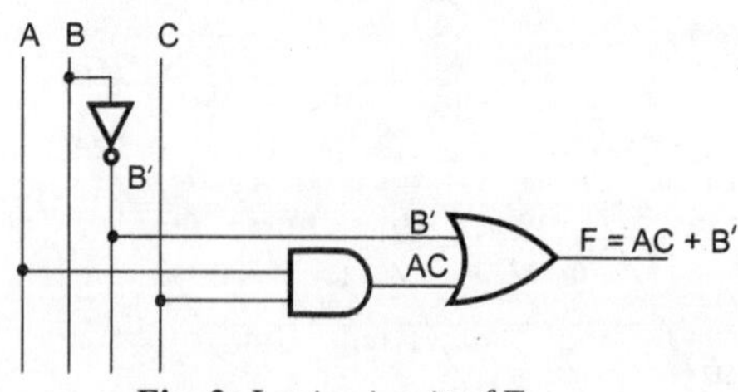

Fig. 2: Logic circuit of F.

Example 4.25

(JNTUA, Jun/July'22, 7 Marks)

Simplify the following Boolean functions, using three variable K-maps:

a) $F(x, y, z) = \sum m(0, 2, 4, 5)$ **b)** $F(x, y, z) = \sum m(0, 2, 4, 5, 6)$

c) $F(x, y, z) = \sum m(0, 1, 2, 3, 5)$ **d)** $F(x, y, z) = \sum m(1, 2, 3, 7)$

Solution

a) $F(x, y, z) = \sum m(0, 2, 4, 5)$

The given function has three variables and so a 3-variable K-map with 8 squares ($2^3 = 8$) is drawn as shown in Fig. 1. The function is defined as sum of minterms m_0, m_2, m_4 and m_5 and so a "1" is filled in the corresponding squares in K-map and the remaining squares are filled with "0".

x \ yz	00	01	11	10
0	m_0 1	m_1 0	m_3 0	m_2 1
1	m_4 1	m_5 1	m_7 0	m_6 0

(Groups: xy′ — m_4, m_5; x′z′ — m_0, m_2)

Fig. 1.

In the K-map of Fig. 1, two prime implicants each with two adjacent 1's can be formed and the corresponding product terms are xy′ and x′z′. The simplified Boolean function is given by sum of these two product terms.

$$\therefore F = xy' + x'z'$$

b) $F(x, y, z) = \sum m(0, 2, 4, 5, 6)$

The given function has three variables and so a 3-variable K-map with 8 squares ($2^3 = 8$) is drawn as shown in Fig. 2. The function is defined as sum of minterms m_0, m_2, m_4, m_5 and m_6 and so a "1" is filled in the corresponding squares in K-map and the remaining squares are filled with "0".

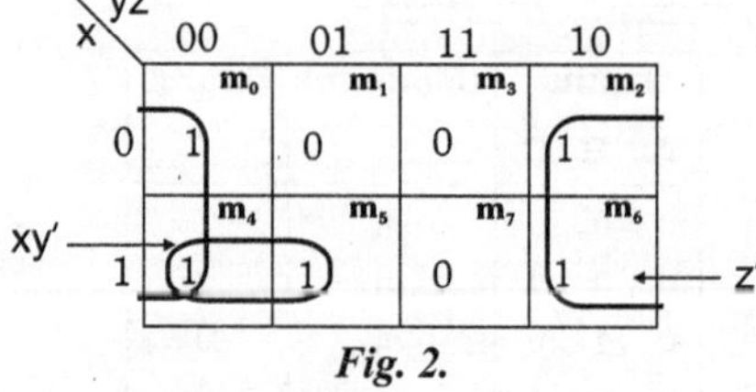

Fig. 2.

In the K-map of Fig. 2, two prime implicants one with four adjacent 1's and another with two adjacent 1's can be formed and the corresponding product terms are z′ and xy′. The simplified Boolean function is given by sum of these two product terms.

$$\therefore F = z' + xy'$$

c) F (x, y, z) = $\sum$ m(0, 1, 2, 3, 5)

The given function has three variables and so a 3-variable K-map with 8 squares ($2^3 = 8$) is drawn as shown in Fig. 3. The function is defined as sum of minterms m_0, m_1, m_2, m_3 and m_5 and so a "1" is filled in the corresponding squares in K-map and the remaining squares are filled with "0".

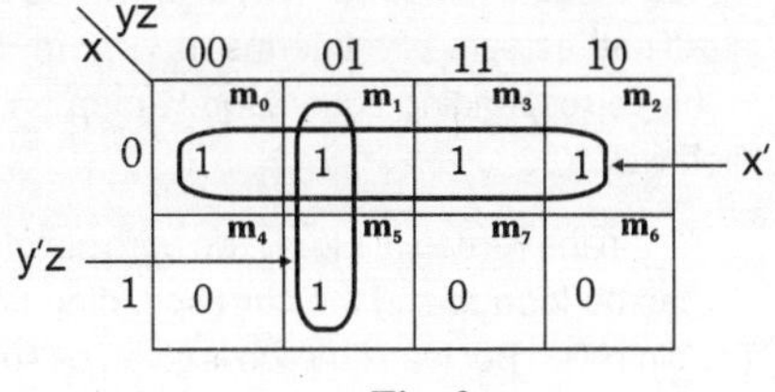

Fig. 3.

In the K-map of Fig. 3, two prime implicants one with four adjacent 1's and another with two adjacent 1's can be formed and the corresponding product terms are x' and y' z. The simplified Boolean function is given by sum of these two product terms.

$\therefore$ F = x' + y' z

d) F (x, y, z) = $\sum$ m(1, 2, 3, 7)

The given function has three variables and so a 3-variable K-map with 8 squares ($2^3 = 8$) is drawn as shown in Fig. 4. The function is defined as sum of minterms m_1, m_2, m_3 and m_7 and so a "1" is filled in the corresponding squares in K-map and the remaining squares are filled with "0".

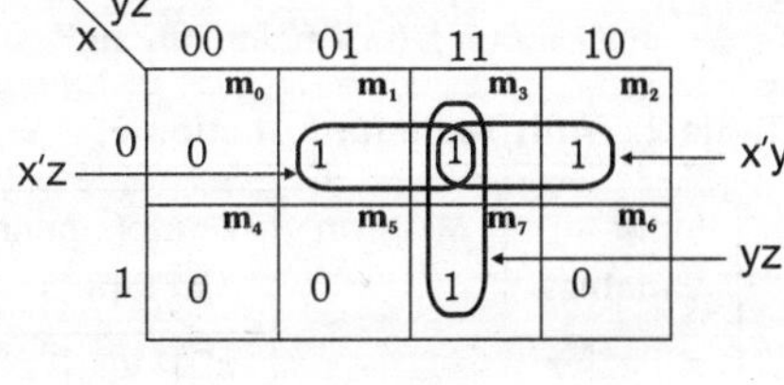

Fig. 4.

In the K-map of Fig. 4, three prime implicants each with two adjacent 1's can be formed and the corresponding product terms are x' z, x' y and yz. The simplified Boolean function is given by sum of these three product terms.

$\therefore$ F = x' z + x' y + y z

Example 4.26

(JNTUA, Jun/July'22, 7 Marks)

Simplify the following Boolean function, using three-variable K-maps:

a) F_1 = x y + x' y' z' + x' y z' b) F_2 (x, y, z) = x' y + y z' + y' z'

Solution

a) F_1 = x y + x' y' z' + x' y z'

The truth table of the given function is formed as shown in Table 1 and the minterms for which function output is "1" are determined and using these minterms the function F_1 can be expressed as,

$$F_1 = \Sigma (m_0, m_2, m_6, m_7)$$

Table 1: Truth Table for Functions F_1

Input Variables						Minterm	Product Terms			Function Output
x	y	z	x'	y'	z'		x y	x' y' z'	x' y z'	F_1
0	0	0	1	1	1	m_0	0	1	0	1
0	0	1	1	1	0	m_1	0	0	0	0
0	1	0	1	0	1	m_2	0	0	1	1
0	1	1	1	0	0	m_3	0	0	0	0
1	0	0	0	1	1	m_4	0	0	0	0
1	0	1	0	1	0	m_5	0	0	0	0
1	1	0	0	0	1	m_6	1	0	0	1
1	1	1	0	0	0	m_7	1	0	0	1

The given function has three variables and so a 3-variable K-map with 8 squares ($2^3 = 8$) is drawn as shown in Fig. 1. The function is defined as sum of minterms m_0, m_2, m_6 and m_7 and so a "1" is filled in the corresponding squares in K-map and the remaining squares are filled with "0".

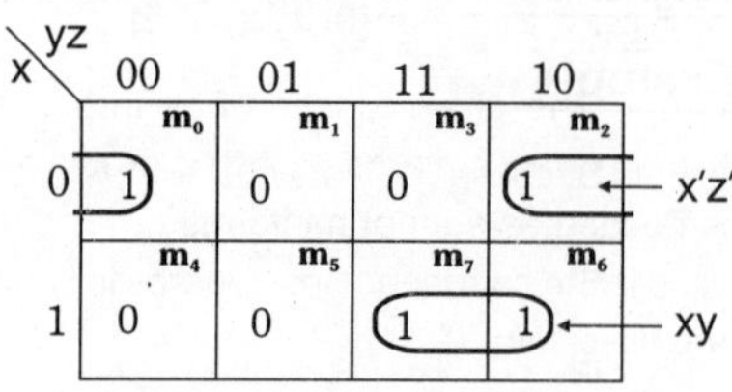

Fig. 1: K-map for function F_1.

In the K-map of Fig. 1, two prime implicants each with two adjacent 1's can be formed and the corresponding product terms are xy and x'z'. The simplified Boolean function is given by sum of these two product terms.

$\therefore F_1 = xy + x'z'$

---

b) $F_2(x, y, z) = x'y + yz' + y'z'$

The truth table of the given function is formed as shown in Table 2 and the minterms for which function output is "1" are determined and using these minterms the function F_2 can be expressed as,

$F_2 = \Sigma(m_0, m_2, m_3, m_4, m_6)$

Table 2: Truth Table for Function F_2

Input Variables			Minterm	Complement of Inputs			Product Terms			Function Output
x	y	z		x'	y'	z'	x'y	yz'	y'z'	F_2
0	0	0	m_0	1	1	1	0	0	1	1
0	0	1	m_1	1	1	0	0	0	0	0
0	1	0	m_2	1	0	1	1	1	0	1
0	1	1	m_3	1	0	0	1	0	0	1
1	0	0	m_4	0	1	1	0	0	1	1
1	0	1	m_5	0	1	0	0	0	0	0
1	1	0	m_6	0	0	1	0	1	0	1
1	1	1	m_7	0	0	0	0	0	0	0

The given function has three variables and so a 3-variable K-map with 8 squares ($2^3 = 8$) is drawn as shown in Fig. 2. The function is defined as sum of minterms m_0, m_2, m_3, m_4 and m_6 and so a "1" is filled in the corresponding squares in K-map and the remaining squares are filled with "0".

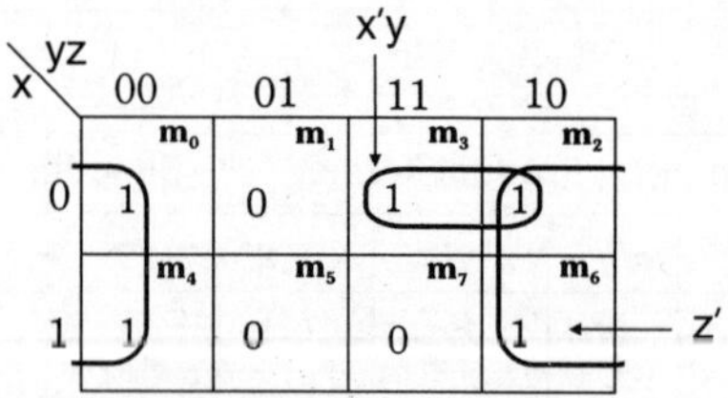

Fig. 2: K-map for function F_2.

In the K-map of Fig. 2, two prime implicants one with four adjacent 1's and another with two adjacent 1's can be formed and the corresponding product terms are z' and x'y. The simplified Boolean function is given by sum of these two product terms.

$\therefore F_2 = z' + x'y$

Example 4.27

(VTU, Aug/Sep'20, 6 Marks)

Simplify the following Boolean functions, using Karnaugh maps.

a) F_1 (A, B, C, D) = Σ m(3, 7, 11, 13, 14, 15) **b)** F_2(w, x, y, z) = Σ m(2, 3, 12, 13, 14, 15)

c) F_3 (A, B, C, D) = Σ m(0, 2, 4, 5, 8, 10, 14, 15)

Solution

a) F_1 (A, B, C, D) = Σ m(3, 7, 11, 13, 14, 15)

The given function has four variables and so a 4-variable K-map with 16 squares (2^4 = 16) is drawn as shown in Fig. 1. The function is defined as sum of minterms m_3, m_7, m_{11}, m_{13}, m_{14} and m_{15} and so a "1" is filled in the corresponding squares in K-map and the remaining squares are filled with "0".

In the K-map of Fig. 1, three prime implicants one with four adjacent 1's and the other two with two adjacent 1's can be formed and the corresponding product terms are CD, ABC and ABD. The simplified Boolean function is given by sum of these three product terms.

$\therefore$ F_1 = CD + ABC + ABD

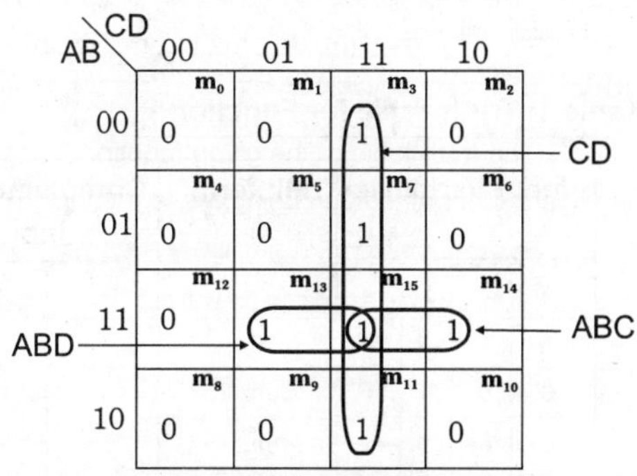

Fig. 1: K-map for F_1.

b) F_2 (w, x, y, z) = Σ m(2, 3, 12, 13, 14, 15)

The given function has four variables and so a 4-variable K-map with 16 squares (2^4 = 16) is drawn as shown in Fig. 2. The function is defined as sum of minterms m_2, m_3, m_{12}, m_{13}, m_{14} and m_{15} and so a "1" is filled in the corresponding squares in K-map and the remaining squares are filled with "0".

In the K-map of Fig. 2, two prime implicants one with four adjacent 1's and the other with two adjacent 1's can be formed and the corresponding product terms are wx and w'x'y. The simplified Boolean function is given by sum of these two product terms.

$\therefore$ F_2 = wx + w'x'y

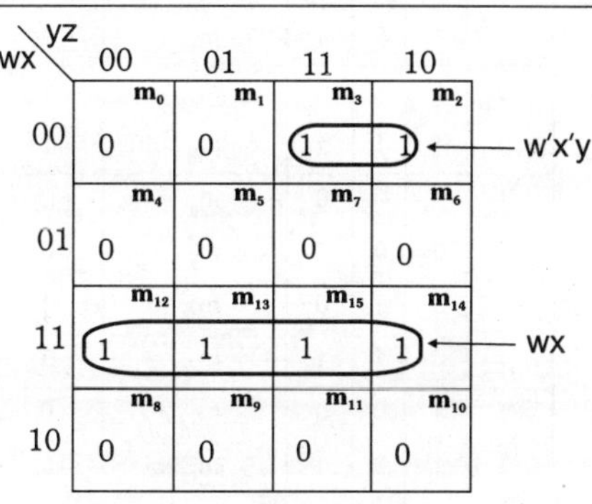

Fig. 2: K-map for F_2.

c) F_3 (A, B, C, D) = Σ m(0, 2, 4, 5, 8, 10, 14, 15)

The given function has four variables and so a 4-variable K-map with 16 squares (2^4 = 16) is drawn as shown in Fig. 3. The function is defined as sum of minterms m_0, m_2, m_4, m_5, m_8, m_{10}, m_{14} and m_{15} and so a "1" is filled in the corresponding squares in K-map and the remaining squares are filled with "0".

In the K-map of Fig. 3, three prime implicants one with four adjacent 1's and the other two with two adjacent 1's can be formed and the corresponding product terms are B'D', A'BC', and ABC. The simplified Boolean function is given by sum of these three product terms.

$\therefore$ F_3 = B'D' + A'BC' + ABC

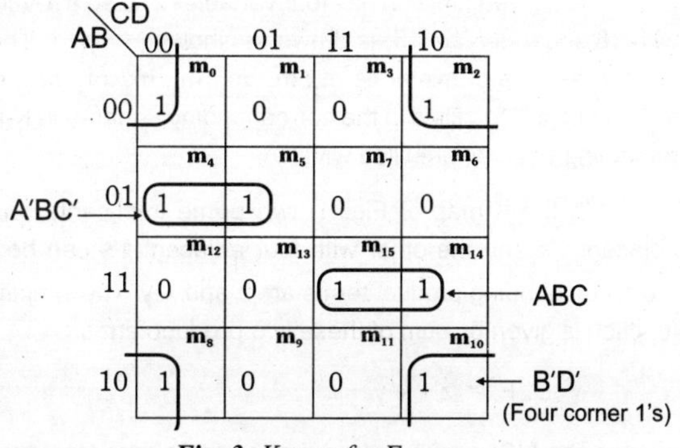

Fig. 3: K-map for F_3.

Example 4.28

(CSVTU, Nov/Dec'19, 7 Marks) (JNTUK, Feb'22, 8 Marks)

Simplify the following Boolean functions using K-maps and draw MSI circuits.

a) $F_1 = w'z + xz + x'y + wx'z$ b) $F_2 = AB'C + B'C'D' + BCD + ACD' + A'B'C + A'BC'D$

Solution

a) $F_1 = w'z + xz + x'y + wx'z$

The truth table of the given function is formed as shown in Table 1 and the minterms for which function output is "1" are determined and using these minterms the function F_1 can be expressed as,

$$F_1 = \Sigma\,(m_1, m_2, m_3, m_5, m_7, m_9, m_{10}, m_{11}, m_{13}, m_{15})$$

Table 1: Truth Table for Function F_1

Input Variables				Minterm	Complement of Inputs		Product Terms				Function Output
w	x	y	z		w'	x'	w'z	xz	x'y	wx'z	F_1
0	0	0	0	m_0	1	1	0	0	0	0	0
0	0	0	1	m_1	1	1	1	0	0	0	1
0	0	1	0	m_2	1	1	0	0	1	0	1
0	0	1	1	m_3	1	1	1	0	1	0	1
0	1	0	0	m_4	1	0	0	0	0	0	0
0	1	0	1	m_5	1	0	1	1	0	0	1
0	1	1	0	m_6	1	0	0	0	0	0	0
0	1	1	1	m_7	1	0	1	1	0	0	1
1	0	0	0	m_8	0	1	0	0	0	0	0
1	0	0	1	m_9	0	1	0	0	0	1	1
1	0	1	0	m_{10}	0	1	0	0	1	0	1
1	0	1	1	m_{11}	0	1	0	0	1	1	1
1	1	0	0	m_{12}	0	0	0	0	0	0	0
1	1	0	1	m_{13}	0	0	0	1	0	0	1
1	1	1	0	m_{14}	0	0	0	0	0	0	0
1	1	1	1	m_{15}	0	0	0	1	0	0	1

The given function has four variables and so a 4-variable K-map with 16 squares ($2^4 = 16$) is drawn as shown in Fig. 1. The function is defined as sum of minterms m_1, m_2, m_3, m_5, m_7, m_9, m_{10}, m_{11}, m_{13} and m_{15} and so a "1" is filled in the corresponding squares in K-map and the remaining squares are filled with "0".

In the K-map of Fig. 1, two prime implicants one with eight adjacent 1's and the other with four adjacent 1's can be formed and the corresponding product terms are z and x'y. The simplified Boolean function is given by sum of these two product terms.

$$\therefore F_1 = z + x'y$$

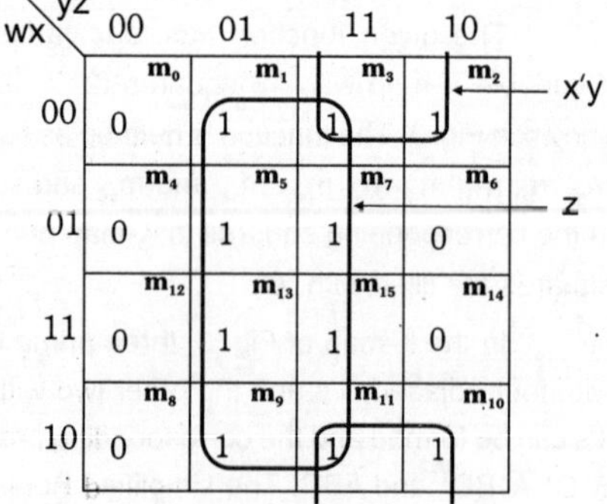

Fig. 1: K-map for function F_1.

The MSI circuit is drawn using the above Boolean function as shown in Fig. 2.

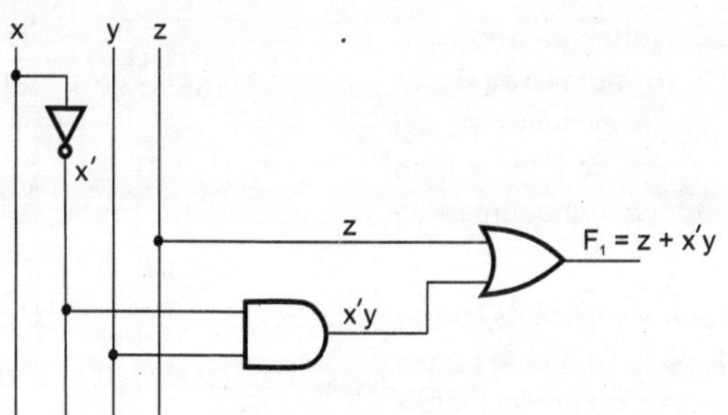

Fig. 2: MSI circuit of F_1.

b) $F_2 = A B'C + B'C'D' + B C D + A C D' + A'B'C + A'B C'D$

The truth table of the given function is formed as shown in Table 2 and the minterms for which function output is "1" are determined and using these minterms the function F_2 can be expressed as,

$$F_2 = \sum (m_0, m_2, m_3, m_5, m_7, m_8, m_{10}, m_{11}, m_{14}, m_{15})$$

Table 2: Truth Table for Function F_2

Input Variables	Minterm	Complement of Inputs				Product Terms						Function Output
A B C D		A'	B'	C'	D'	AB'C	B'C'D'	BCD	ACD'	A'B'C	A'BC'D	F_2
0 0 0 0	m_0	1	1	1	1	0	1	0	0	0	0	1
0 0 0 1	m_1	1	1	1	0	0	0	0	0	0	0	0
0 0 1 0	m_2	1	1	0	1	0	0	0	0	1	0	1
0 0 1 1	m_3	1	1	0	0	0	0	0	0	1	0	1
0 1 0 0	m_4	1	0	1	1	0	0	0	0	0	0	0
0 1 0 1	m_5	1	0	1	0	0	0	0	0	0	1	1
0 1 1 0	m_6	1	0	0	1	0	0	0	0	0	0	0
0 1 1 1	m_7	1	0	0	0	0	0	1	0	0	0	1
1 0 0 0	m_8	0	1	1	1	0	1	0	0	0	0	1
1 0 0 1	m_9	0	1	1	0	0	0	0	0	0	0	0
1 0 1 0	m_{10}	0	1	0	1	1	0	0	1	0	0	1
1 0 1 1	m_{11}	0	1	0	0	1	0	0	0	0	0	1
1 1 0 0	m_{12}	0	0	1	1	0	0	0	0	0	0	0
1 1 0 1	m_{13}	0	0	1	0	0	0	0	0	0	0	0
1 1 1 0	m_{14}	0	0	0	1	0	0	0	1	0	0	1
1 1 1 1	m_{15}	0	0	0	0	0	0	1	0	0	0	1

The given function has four variables and so a 4-variable K-map with 16 squares ($2^4 = 16$) is drawn as shown in Fig. 3. The function is defined as sum of minterms m_0, m_2, m_3, m_5, m_7, m_8, m_{10}, m_{11}, m_{14} and m_{15} and so a "1" is filled in the corresponding squares in K-map and the remaining squares are filled with "0".

In the K-map of Fig. 3, four prime implicants can be formed in which three with four adjacent 1's and the other with two adjacent 1's and the corresponding product terms are B'D', AC, B'C and A'BD. The simplified Boolean function is given by sum of these four product terms.

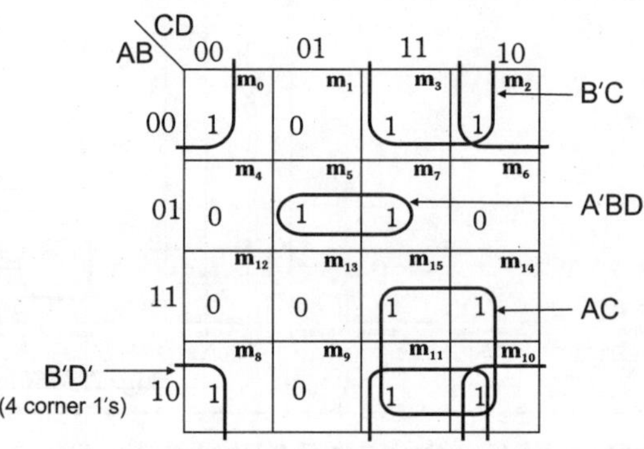

Fig. 3: K-map for function F_2.

$\therefore$ F_2 = B'D' + AC + B'C + A'BD

The MSI circuit is drawn using the above Boolean function as shown in Fig. 4.

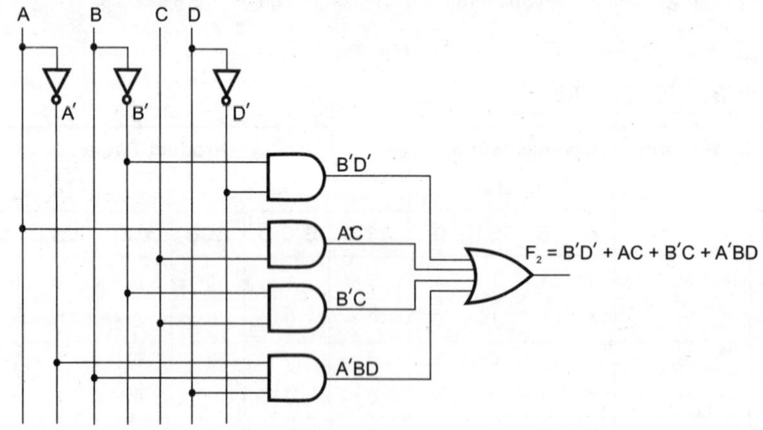

Fig. 4: MSI circuit of F_2.

Example 4.29

(GTU, May'19, 7 Marks)

Minimize F (A, B, C, D) = $\sum$ m(1, 3, 4, 6, 8, 11, 15) + d(0, 5, 7) using K-map and draw MSI circuit for the output.

Solution

The given function has four variables and so a 4-variable K-map with 16 squares ($2^4 = 16$) is drawn as shown in Fig. 1. The function is defined as sum of minterms m_1, m_3, m_4, m_6, m_8, m_{11} and m_{15} and so a "1" is filled in the corresponding squares in K-map the minterms m_0,m_5,m_7 and so a "×" is filled in corresponding squares in K-map and the remaining squares are filled with "0".

In the K-map of Fig. 1, three prime implicants one with four adjacent 1's and the other two with two adjacent 1's can be formed and the corresponding product terms are CD, A'B and B'C'D'. The simplified Boolean function is given by sum of these three product terms.

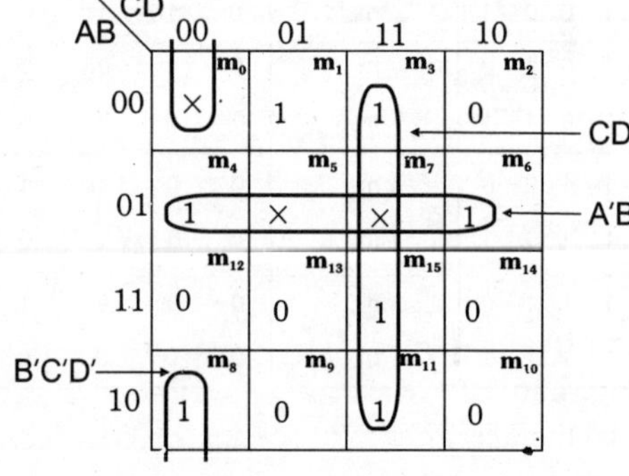

Fig. 1: K-map for F.

$\therefore$ F = CD + A'B + B'C'D'

The MSI circuit is drawn using the above Boolean function as shown in Fig. 2.

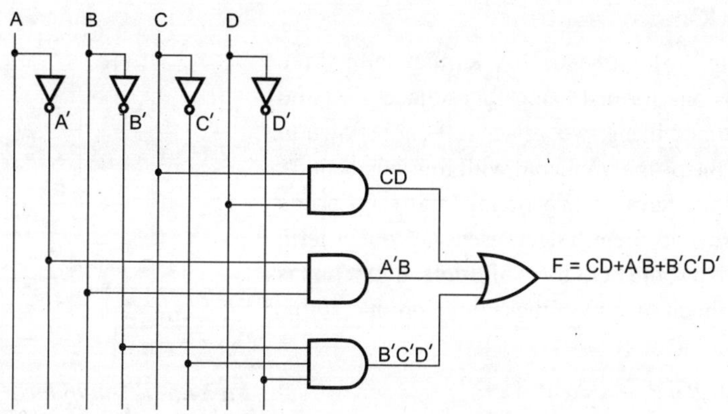

Fig. 2: MSI circuit of F.

4.6.6 Realization of Product-of-Sums Form using K-map

In truth table the function output value with "0" represent maxterms. Therefore in K-map if squares with 0's are considered for forming prime implicants and the complement of the resultant sum-of-products form will give product-of-sums form.

Two-Variable K-map

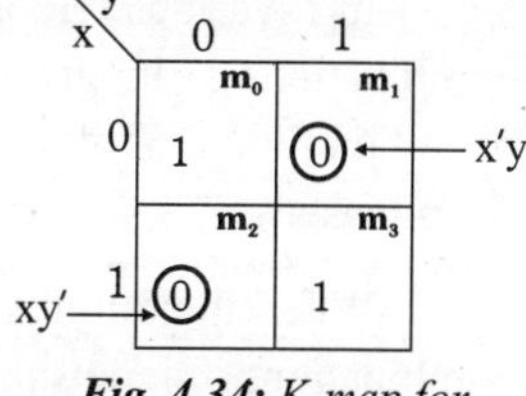

Fig. 4.34: K-map for $F = (x'y + xy')'$.

An example of two-variable K-map with two prime implicants formed using single 0's are shown in Fig. 4.34. Here, each prime implicant represents a product term of two literals. Now, the POS form of Boolean function is given by complement of sum of these two product terms as shown below:

$$F = (x'y + xy')' = (x'y)'(xy')'$$

$$= (x + y')(x' + y)$$ Using DeMorgan's Theorem

Three-Variable K-map

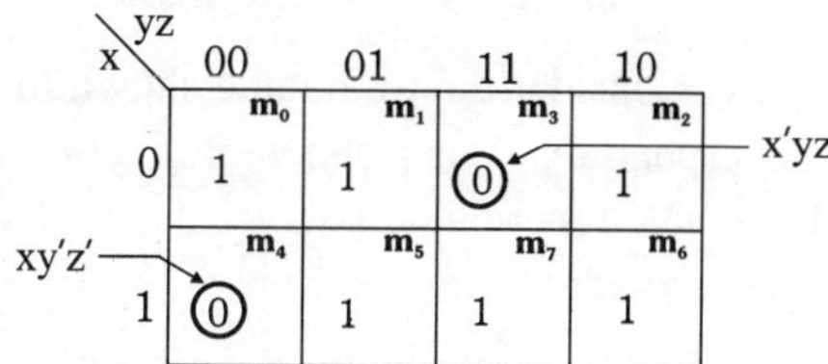

Fig. 4.35: K-map for $F = (x'yz + xy'z')'$.

An example of three-variable K-map with two prime implicants formed using single 0's are shown in Fig. 4.35. Here, each prime implicant represents a product term of three literals. Now, the POS form of Boolean function is given by complement of sum of these two product terms as shown below:

$$F = (x'yz + xy'z')'$$

$$= (x'yz)'(xy'z')'$$ Using DeMorgan's Theorem

$$= (x + y' + z')(x' + y + z)$$

An example of three-variable K-map with two prime implicants formed using two adjacent 0's are shown in Fig. 4.36. Here, each prime implicant represents a product term of two literals. Now, the POS form of Boolean function is given by complement of sum of these two product terms as shown below:

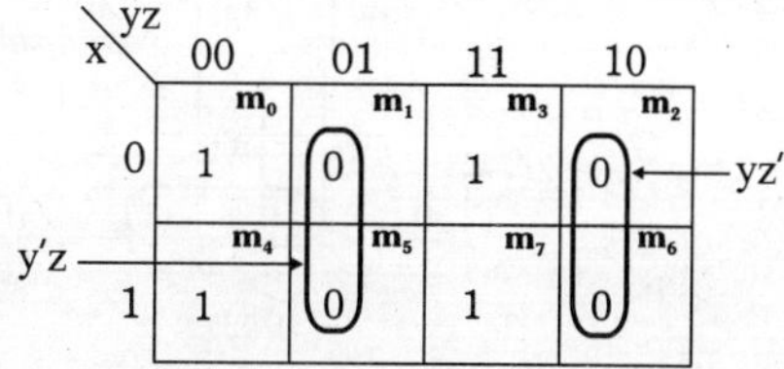

Fig. 4.36: K-map for $F = (yz' + y'z)'$.

$$F = (yz' + y'z)'$$ Using DeMorgan's Theorem

$$= (yz')'(y'z)' = (y' + z)(y + z')$$

Four-Variable K-map

An example of four-variable K-map with three prime implicants one formed using four adjacent 0's and the other two formed using two adjacent 0's are shown in Fig. 4.37. Here, the prime implicant with four adjacent 0's represent a product term of two literals and each prime implicant with two adjacent 0's represent a product term of three literals. Now, the POS form of Boolean function is given by complement of sum of these three product terms as shown below:

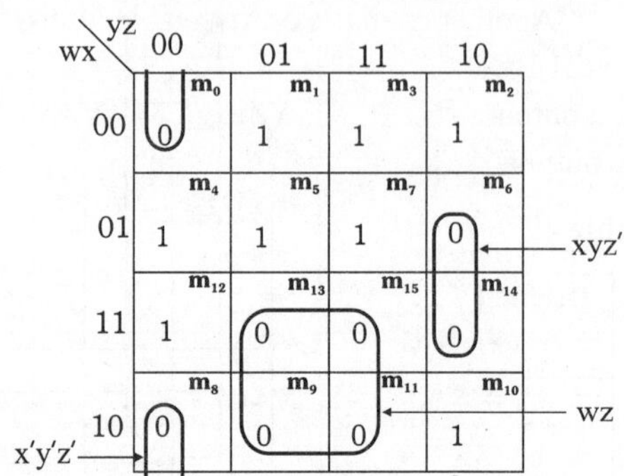

Fig. 4.37: K-map for F = (wz + x′y′z′+ xyz′)′.

$$F = (wz + x'y'z' + xyz')'$$

$$= (wz)'(x'y'z')'(xyz')'$$ Using DeMorgan's Theorem

$$= (w' + z')(x + y + z)(x' + y' + z)$$

4.6.7 Completely and Incompletely Specified Functions (K-map with Don't-Care Conditions)

In n-variable Boolean function there are 2^n possible combinations of inputs and so there will be 2^n outputs. But in certain functions all the 2^n possible outputs are not defined (or not used).

When the outputs are defined for all the possible 2^n combinations of inputs then the function is called ***completely specified function***. When the outputs are not defined for some of the possible combinations of inputs then the function is called ***incompletely specified function***.

Consider the truth table shown in Table 4.38 and the corresponding K-map shown in Fig. 4.38. In Table 4.38 the undefined function outputs are called ***don't-care conditions*** and denoted by ×.

Table 4.38: Truth Table with Don't-cares

Inputs x y z	Minterm	Output F
0 0 0	m_0	0
0 0 1	m_1	1
0 1 0	m_2	×
0 1 1	m_3	×
1 0 0	m_4	0
1 0 1	m_5	1
1 1 0	m_6	1
1 1 1	m_7	0

In simplification of Boolean functions using K-maps, don't-care conditions can be considered as either 0 or 1. It is not necessary that don't-care conditions to be included in prime implicants, but if inclusion of a don't-care condition leads to minimization of a literal then, don't-care can be included in the prime implicants as shown in Fig. 4.38.

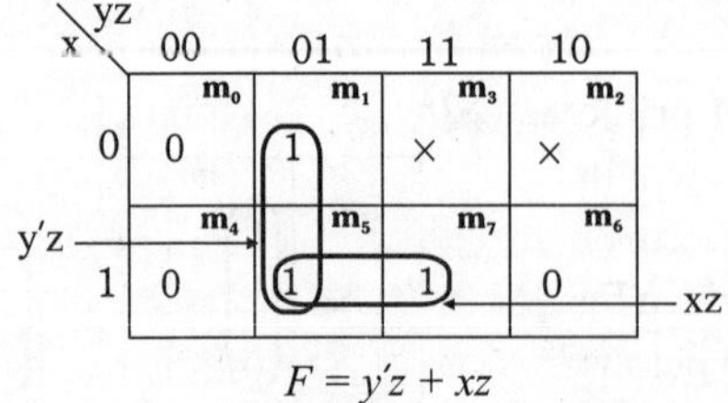

$F = y'z + xz$

***Fig. a:** Formation of prime implicants without including don't-care condition.*

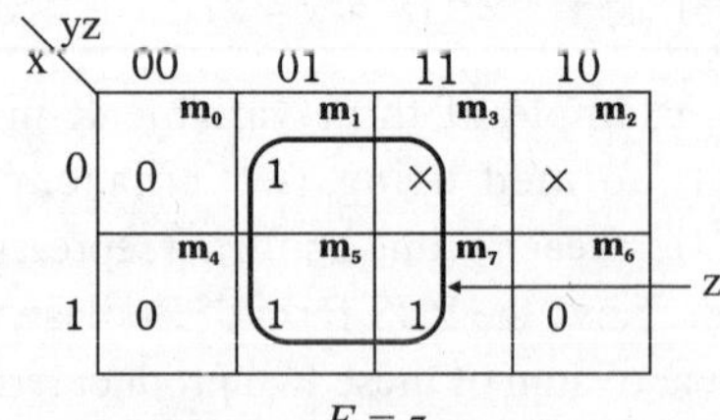

$F = z$

***Fig. b:** Formation of prime implicants including don't-care condition.*

***Fig. 4.38:** K-map with don't-care conditions.*

Another example of truth table with don't-cares is shown in Table 5.4 which is truth table of binary to 5421 BCD code conversion. K-maps for each output a_1, a_2, a_4 and a_5 are shown in Figs. 4.39 to 4.42. For each output two K-maps are drawn in order to show that the literals are reduced in Boolean expression of the outputs when don't-cares are included in formation of prime implicants.

Table 4.39: Truth Table of BCD

Decimal	Minterm	Input Binary w x y z	Output 5421 BCD a_5 a_4 a_2 a_1
0	m_0	0 0 0 0	0 0 0 0
1	m_1	0 0 0 1	0 0 0 1
2	m_2	0 0 1 0	0 0 1 0
3	m_3	0 0 1 1	0 0 1 1
4	m_4	0 1 0 0	0 1 0 0
5	m_5	0 1 0 1	1 0 0 0
6	m_6	0 1 1 0	1 0 0 1
7	m_7	0 1 1 1	1 0 1 0
8	m_8	1 0 0 0	1 0 1 1
9	m_9	1 0 0 1	1 1 0 0
-	m_{10}	1 0 1 0	× × × ×
-	m_{11}	1 0 1 1	× × × ×
-	m_{12}	1 1 0 0	× × × ×
-	m_{13}	1 1 0 1	× × × ×
-	m_{14}	1 1 1 0	× × × ×
-	m_{15}	1 1 1 1	× × × ×

(6 don't-care conditions)

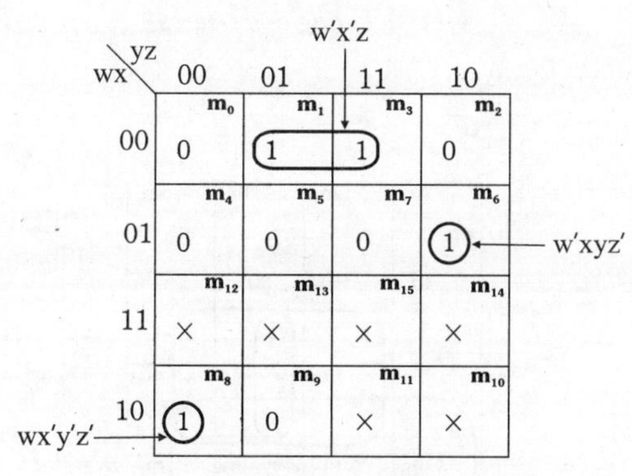

$a_1 = wx'y'z' + w'xyz' + w'x'z$

***Fig. a:** Prime implicants formed without including don't-cares.*

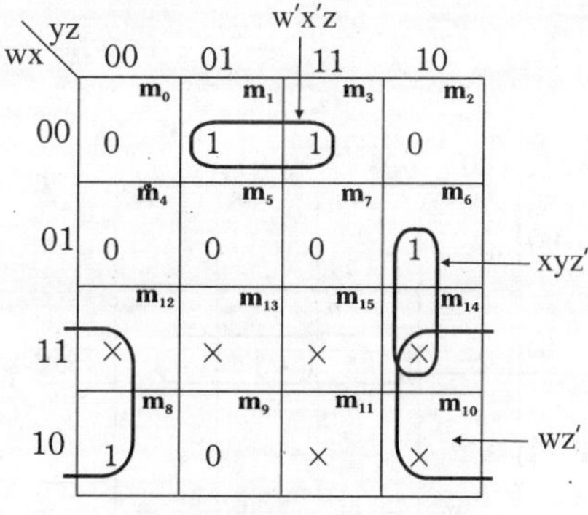

$a_1 = wz' + xyz' + w'x'z$

***Fig. b:** Prime implicants formed by including don't-cares.*

***Fig. 4.39:** K-map for a_1.*

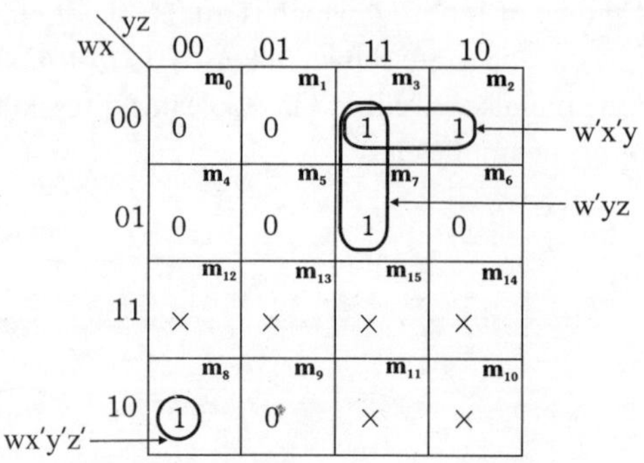

$a_2 = wx'y'z' + w'yz + w'x'y$

Fig. a: *Prime implicants formed without including don't-cares.*

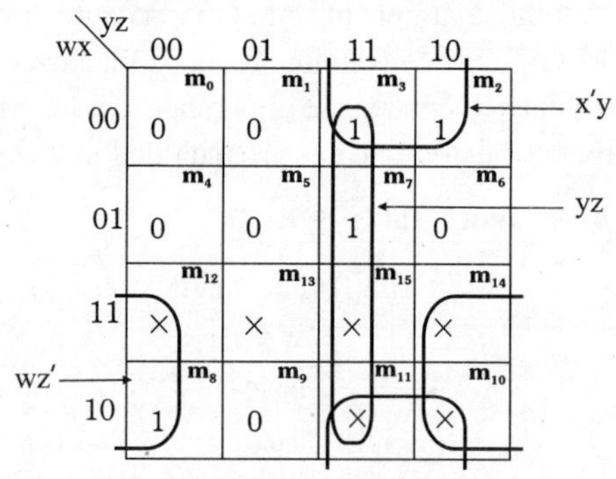

$a_2 = yz + x'y + wz'$

Fig. b: *Prime implicants formed by including don't-cares.*

Fig. 4.40: *K-map for* a_2.

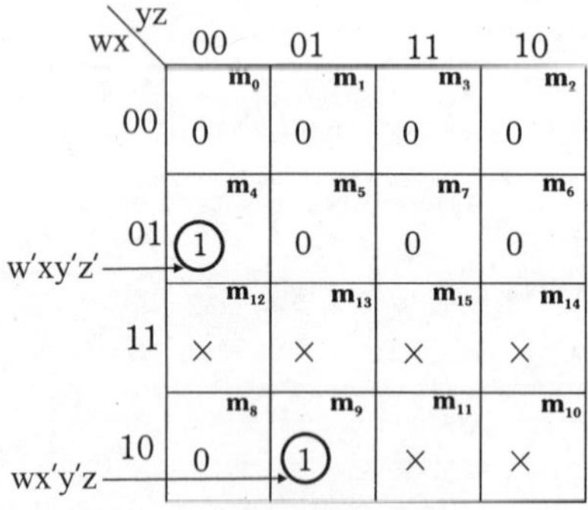

$a_4 = w'xy'z' + wx'y'z$

Fig. a: *Prime implicants formed without including don't-cares.*

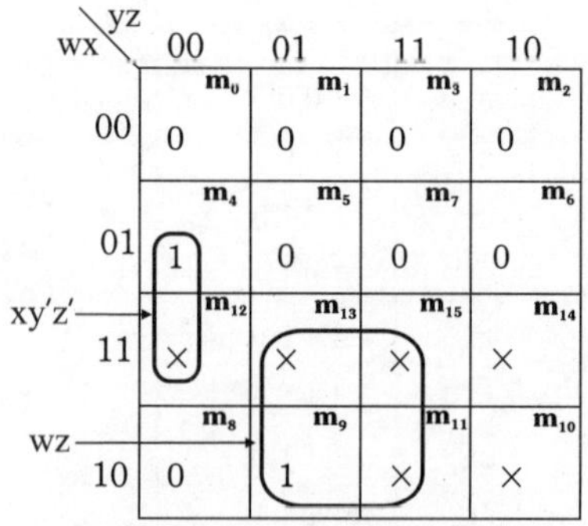

$a_4 = xy'z' + wz$

Fig. b: *Prime implicants formed by including don't-cares.*

Fig. 4.41: *K-map for* a_4.

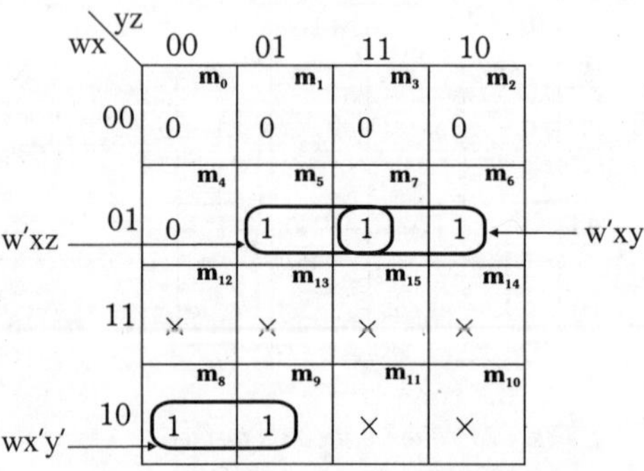

$a_5 = wx'y' + w'xz + w'xy$

Fig. a: *Prime implicants formed without including don't-cares.*

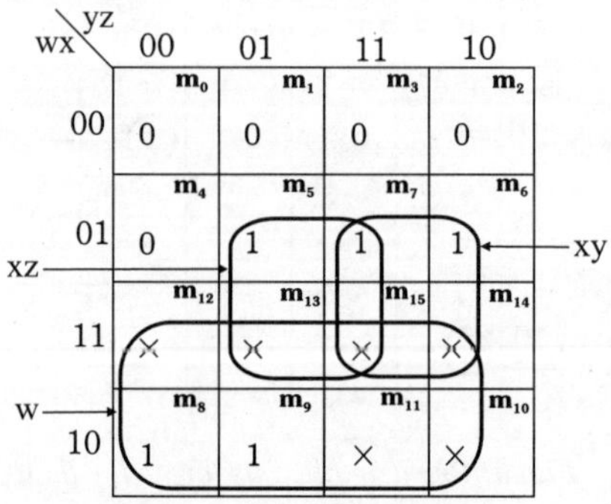

$a_5 = w + xz + xy$

Fig. b: *Prime implicants formed by including don't-cares.*

Fig. 4.42: *K-map for* a_5.

Boolean function for a_1, a_2, a_4 and a_5 when don't-cares are not considered for formation of prime implicants	Boolean function for a_1, a_2, a_4 and a_5 when don't-cares are included for formation of prime implicants
$a_1 = wx'y'z' + w'xyz' + w'x'z$	$a_1 = wz' + xyz' + w'x'z$
$a_2 = wx'y'z' + w'yz + w'x'y$	$a_2 = yz + x'y + wz'$
$a_4 = w'xy'z' + wx'y'z$	$a_4 = xy'z' + wz$
$a_5 = wx'y' + w'xz + w'xy$	$a_5 = w + xz + xy$

Example 4.30

(CSVTU, Nov/Dec'19, 7 Marks).

Simplify the following Boolean functions.

a) F_1 (A, B, C, D) = $\prod$ M(1, 3, 5, 7, 13, 15) **b)** F_2 (A, B, C, D) = $\prod$ M(1, 3, 6, 9, 11, 12, 14)

Solution

a) F_1 (A, B, C, D) = $\prod$ M(1, 3, 5, 7, 13, 15)

The given function has four variables and so a 4-variable K-map with 16 squares (2^4 = 16) is drawn as shown in Fig. 1. The function is defined as product of maxterms M_1, M_3, M_5, M_7, M_{13} and M_{15} and so a "0" is filled in the corresponding squares in K-map and the remaining squares are filled with "1".

In the K-map of Fig. 1, two prime implicants each with four adjacent 0's can be formed and the corresponding product terms are A'D and BD. The simplified Boolean function in POS form is given by complement of sum of these two product terms.

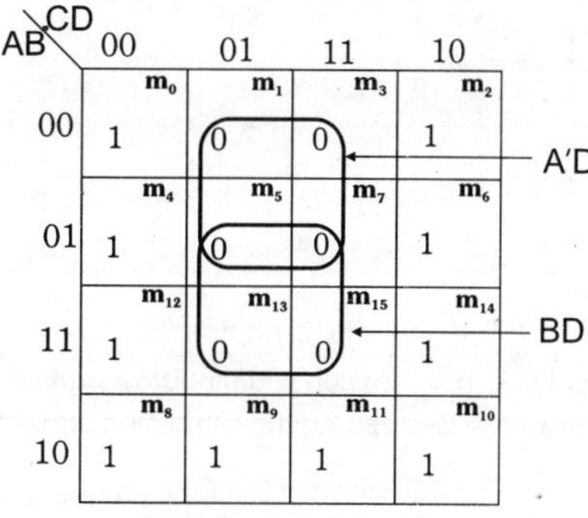

Fig. 1: K-map for F_1.

F_1 = (A'D + B D)'

= (A'D)'(B D)' [Using DeMorgan's Theorem]

= (A + D') (B' + D')

b) F_2 (A, B, C, D) = $\prod$ M(1, 3, 6, 9, 11, 12, 14)

The given function has four variables and so a 4-variable K-map with 16 squares (2^4 = 16) is drawn as shown in Fig. 2. The function is defined as product of maxterms M_1, M_3, M_6, M_9, M_{11}, M_{12} and M_{14} and so a "0" is filled in the corresponding squares in K-map and the remaining squares are filled with "1".

In the K-map of Fig. 2, three prime implicants one with four adjacent 0's and the other two each with two adjacent 0's can be formed and the corresponding product terms are B'D, BCD' and ABD'. The simplified Boolean function in POS form is given by complement of sum of these three product terms.

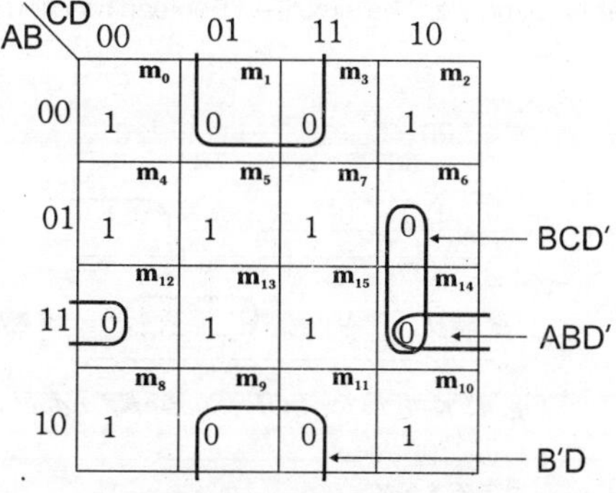

Fig. 2: K-map for F_2.

F_2 = (B'D + B C D' + A B D')'

= (B'D)' (B C D')' (A B D')' [Using DeMorgan's Theorem]

= (B + D') (B' + C' + D) (A' + B' + D)

Example 4.31

Simplify the following function to sum-of-products and product-of-sums.

F = x'z' + y'z' + y z' + x y

Solution

The truth table of the given function is formed as shown in Table 1 and the minterms for which function output is "1" are determined and using these minterms the function F can be expressed as,

$$F = \sum (m_0, m_2, m_4, m_6, m_7)$$

Table 1: Truth Table

Input Variables			Minterm	Complement of Inputs			Product Terms				Function Output
x	y	z		x'	y'	z'	x'z'	y'z'	y z'	x y	F
0	0	0	m_0	1	1	1	1	1	0	0	1
0	0	1	m_1	1	1	0	0	0	0	0	0
0	1	0	m_2	1	0	1	1	0	1	0	1
0	1	1	m_3	1	0	0	0	0	0	0	0
1	0	0	m_4	0	1	1	0	1	0	0	1
1	0	1	m_5	0	1	0	0	0	0	0	0
1	1	0	m_6	0	0	1	0	0	1	1	1
1	1	1	m_7	0	0	0	0	0	0	1	1

The given function has three variables and so a 3-variable K-map with 8 squares ($2^3 = 8$) is drawn as shown in Fig. 1. The function is defined as sum of minterms m_0, m_2, m_4, m_6 and m_7 and so a "1" is filled in the corresponding squares in K-map and the remaining squares are filled with "0".

For SOP form of Boolean function, the prime implicants are formed by considering the squares filled with 1's. In the K-map of Fig. 1, two prime implicants one with four adjacent 1's and the other with two adjacent 1's can be formed and the corresponding product terms are z' and xy. The simplified Boolean function in SOP form is given by sum of these two product terms.

For POS form of Boolean function, the prime implicants are formed by considering the squares filled with 0's. In the K-map of Fig. 2, two prime implicants each with two adjacent 0's can be formed and the corresponding product terms are x'z and y'z. The simplified Boolean function in POS form is given by complement of sum of these two product terms.

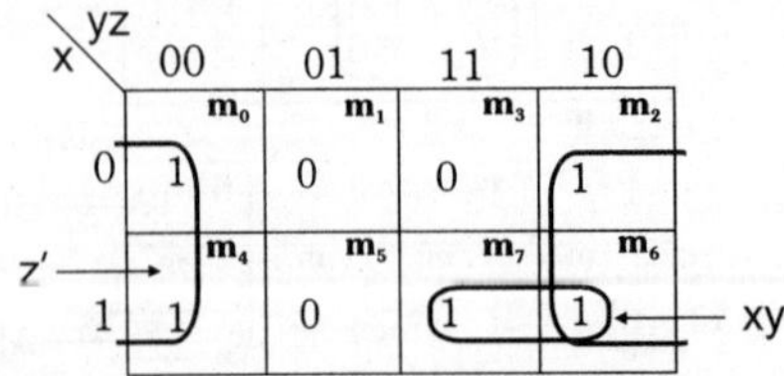

Fig. 1: K-map for SOP (Sum-of-Products).

F = z' + x y

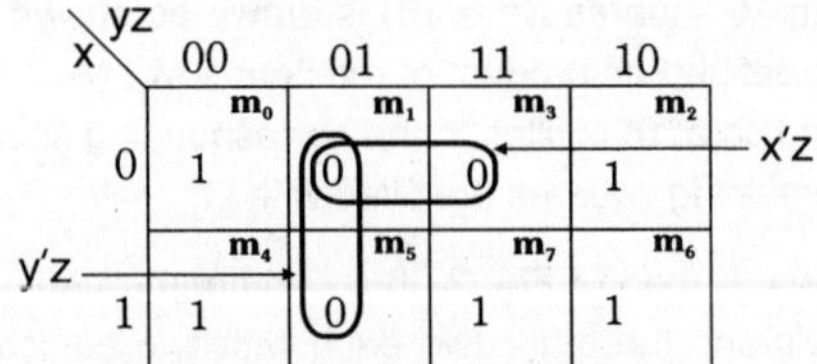

Fig. 2: K-map for POS (Product-of-Sums).

F = (x'z + y'z)'

= (x'z)' (y'z)' — Using DeMorgan's Theorem

= (x + z') (y + z')

Example 4.32

(AKTU, Sep'20, 5 Marks) (VTU, Aug/Sep'20, 5 Marks) (VTU, Aug/Sep'20, 7 Marks)

Simplify the following Boolean functions, together with the don't-care conditions and then express the simplified function in sum-of-products.

a) F_1 (A, B, C, D) = $\sum$ m(0, 6, 8, 13, 14) **b)** F_2 (A, B, C, D) = $\sum$ m(5, 6, 7, 12, 14, 15)

d_1 (A, B, C, D) = $\sum$ d(2, 4, 10) d_2 (A, B, C, D) = $\sum$ d(1, 3, 9, 11)

Solution

a) F_1 (A, B, C, D) = $\sum$ m(0, 6, 8, 13, 14)

d_2 (A, B, C, D) = $\sum$ d(2, 4, 10)

The given function has four variables and so a 4-variable K-map with 16 squares (2^4 = 16) is drawn as shown in Fig. 1. The function is defined as sum of minterms m_0, m_6, m_8, m_{13} and m_{14} and so a "1" is filled in the corresponding squares in K-map. The function also has don't-care outputs for the minterms m_2, m_4 and m_{10} and so a "x" is filled in the corresponding squares in K-map and the remaining squares are filled with "0".

In the K-map of Fig. 1, by considering the don't-care condition in m_2 and m_{10} as "1", three prime implicants can be formed in which two has four adjacent 1's and one have single 1 and the corresponding product terms are B'D', CD' and ABC'D. The simplified Boolean function is given by sum of these three product terms.

AB \ CD	00	01	11	10
00	m_0: 1	m_1: 0	m_3: 0	m_2: ×
01	m_4: ×	m_5: 0	m_7: 0	m_6: 1
11	m_{12}: 0	m_{13}: 1	m_{15}: 0	m_{14}: 1
10	m_8: 1	m_9: 0	m_{11}: 0	m_{10}: ×

ABC'D, CD', B'D' (All 4 corner)

***Fig. 1:** K-map for F_1.*

$\therefore$ F_1 = B'D' + CD' + ABC'D

> ***Note:** Don't-care in m_2 and m_{10} are considered as 1.*

---

b) F_2 (A, B, C, D) = $\sum$ m(5, 6, 7, 12, 14, 15)

d_2 (A, B, C, D) = $\sum$ d(1, 3, 9, 11)

The given function has four variables and so a 4-variable K-map with 16 squares (2^4 = 16) is drawn as shown in Fig. 2. The function is defined as sum of minterms m_5, m_6, m_7, m_{12}, m_{14} and m_{15} and so a "1" is filled in the corresponding squares in K-map. The function also has don't-care outputs for the minterms m_1, m_3, m_9 and m_{11} and so a "x" is filled in the corresponding squares in K-map and the remaining squares are filled with "0".

In the K-map of Fig. 2, by considering the don't-care condition in m_1 and m_3 as "1", three prime implicants can be formed in which each of the two prime implicants has four adjacent 1's and one prime implicant have two adjacent 1's and the corresponding product terms are A'D, BC and ABD'. The simplified Boolean function is given by sum of these three product terms.

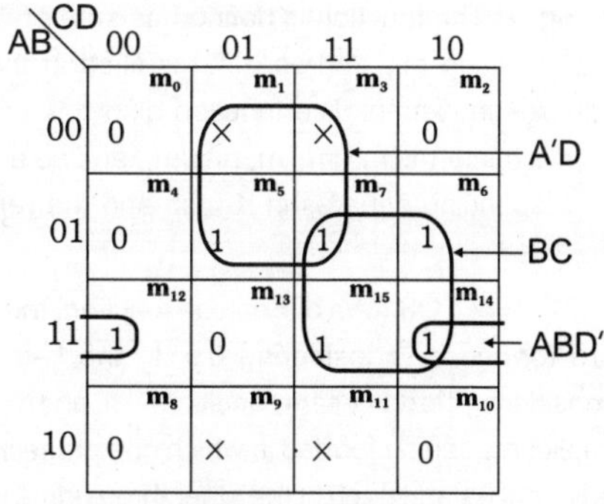

***Fig. 2:** K-map for F_2.*

$\therefore$ F_2 = A'D + BC + ABD'

> ***Note:** Don't-care in m_1 and m_3 are considered as 1.*

Example 4.33

(VTU, Aug/Sep'20, 8 Marks) (VTU, Aug/Sep'20, 6 Marks)

Simplify the following Boolean function with don't-care in POS form.

a) F_1 (A, B, C, D) = $\sum$ m(0, 2, 6, 12, 13, 14) **b)** F_2 (A, B, C, D) = $\sum$ m(5, 6, 12, 14, 15)

d_1 (A, B, C, D) = $\sum$ d(1, 4, 10) d_2 (A, B, C, D) = $\sum$ d(1, 3, 9, 11)

Solution

a) F_1 (A, B, C, D) = $\sum$ m(0, 2, 6, 12, 13, 14)

d_1 (A, B, C, D) = $\sum$ d(1, 4, 10)

The given function has four variables and so a 4-variable K-map with 16 squares (2^4 = 16) is drawn as shown in Fig. 1. The function is defined as sum of minterms m_0, m_2, m_6, m_{12}, m_{13} and m_{14} and so a "1" is filled in the corresponding squares in K-map. The function also has don't-care outputs for the minterms m_1, m_4 and m_{10} and so a "x" is filled in the corresponding squares in K-map and the remaining squares are filled with "0".

For POS form of Boolean function, the prime implicants are formed by considering 0's. In the K-map of Fig. 1, by considering the don't-care condition in m_1 and m_{10} as "0", three prime implicants can be formed in which all three has four adjacent 0's and the corresponding product terms are A'D, AB' and CD. The simplified Boolean function in POS form is given by complement of sum of these three product terms.

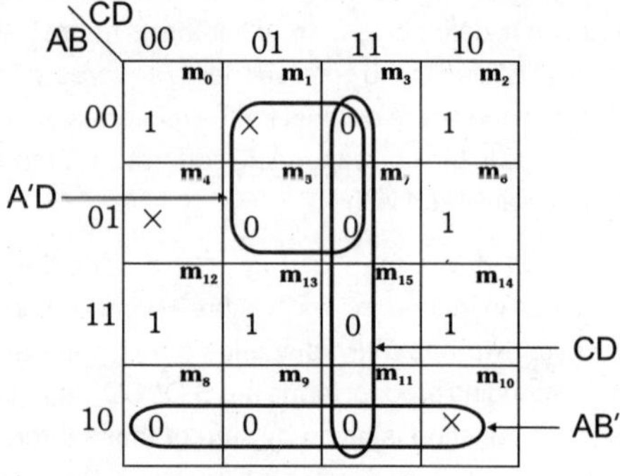

Fig. 1: K-map for F_1.

$$F_1 = (A'D + AB' + CD)'$$

$$= (A'D)' (AB')' (CD)'$$

$$= (A + D') (A' + B) (C' + D')$$ Using DeMorgan's Theorem

b) F_2 (A, B, C, D) = $\sum$ m(5, 6, 12, 14, 15)

d_2 (A, B, C, D) = $\sum$ d(1, 3, 9, 11)

The given function has four variables and so a 4-variable K-map with 16 squares (2^4 = 16) is drawn as shown in Fig. 2. The function is defined as sum of minterms m_5, m_6, m_{12}, m_{14} and m_{15} and so a "1" is filled in the corresponding squares in K-map. The function also has don't-care outputs for the minterms m_1, m_3, m_9 and m_{11} and so a "x" is filled in the corresponding squares in K-map and the remaining squares are filled with "0".

For POS form of Boolean function, the prime implicants are formed by considering 0's. In the K-map of Fig. 2, by considering the don't-care condition in m_3 and m_9 as "0", four prime implicants can be formed in which one prime implicant has four adjacent 0's and each of the other three prime implicants has two adjacent 0's and the corresponding product terms are B', A'C'D', AC'D and A'CD. The simplified Boolean function in POS form is given by complement of sum of these four product terms.

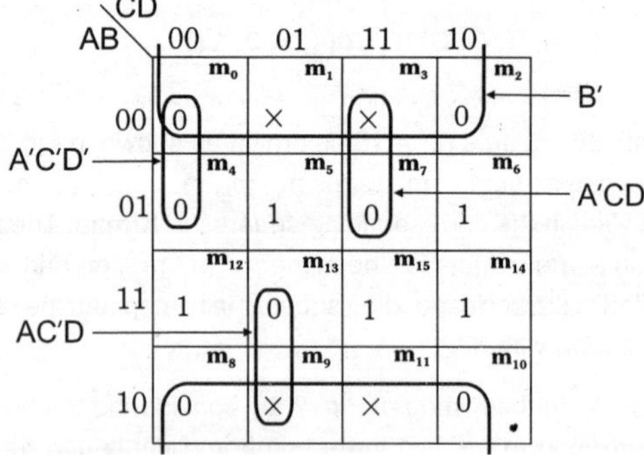

Fig. 2: K-map for F_2.

$$F_2 = (B' + A'C'D' + AC'D + A'CD)'$$

$$= B (A'C'D')' (AC'D)' (A'CD)'$$ Using DeMorgan's Theorem

$$= B (A + C + D) (A' + C + D') (A + C' + D')$$

Example 4.34

(IKGPTU, Dec'19, 10 Marks)

A combinational circuit has 3 inputs A, B, C and output F. F is true for the following input combinations.

A is false, B is true.

A is false, C is true.

A and B and C are false.

A and B and C are true.

a) Write the truth table for F. Use the convention True = 1 and false = 0.

b) Write the simplified expression for F in SOP form.

c) Write the simplified expression for F in POS form.

Solution

a) Truth Table for F

A is false, B is true ⟶ A'B

A is false, C is true ⟶ A'C

A and B and C are false ⟶ A'B'C'

A and B and C are true ⟶ ABC

The Boolean equation for F is given by sum of all the above product terms.

$$F = A'B + A'C + A'B'C' + ABC$$

$$= A'B(C + C') + A'C(B + B') + A'B'C' + ABC$$

$$= A'BC + A'BC' + \cancel{A'BC} + A'B'C + A'B'C' + ABC$$

$$= \underset{(m_3)}{A'BC} + \underset{(m_2)}{A'BC'} + \underset{(m_1)}{A'B'C} + \underset{(m_0)}{A'B'C'} + \underset{(m_7)}{ABC}$$

$$= \sum (m_0, m_1, m_2, m_3, m_7)$$

$x + x' = 1$

$x \cdot 1 = x$

Table 1: Truth Table of Function F

Input Variables			Minterm	Function Output
A	B	C		F
0	0	0	m_0	1
0	0	1	m_1	1
0	1	0	m_2	1
0	1	1	m_3	1
1	0	0	m_4	0
1	0	1	m_5	0
1	1	0	m_6	0
1	1	1	m_7	1

The truth table for the above function is shown in Table 1. The function output is 1 for the minterms m_0, m_1, m_2, m_3 and m_7 and 0 for minterms m_4, m_5 and m_6.

The given function has three variables and so a 3-variable K-map with 8 squares ($2^3 = 8$) is drawn as shown in Figs. 1 and 2. The function is defined as sum of minterms m_0, m_1, m_2, m_3 and m_7 and so a "1" is filled in the corresponding squares in K-map and the remaining squares are filled with "0".

b) SOP form

For SOP form of Boolean function, the prime implicants are formed by considering the squares filled with 1's. In the K-map of Fig. 1, one prime implicant with four adjacent 1's and another with two adjacent 1's can be formed and the corresponding product terms are A' and BC. The simplified Boolean function in SOP form is given by sum of these two product terms.

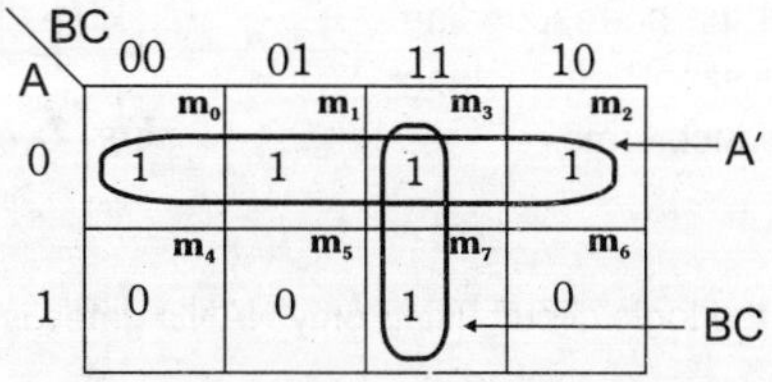

A \ BC	00	01	11	10
0	m_0: 1	m_1: 1	m_3: 1	m_2: 1
1	m_4: 0	m_5: 0	m_7: 1	m_6: 0

A' (row 0 group); BC (column 11 group)

Fig. 1: K-map for SOP (Sum-of-Products).

From the K-map the simplified Boolean equation for F in SOP form is given by,

F = A' + BC

C) POS form

For POS form of Boolean function, the prime implicants are formed by considering the squares filled with 0's. In the K-map of Fig. 2, two prime implicants each with two adjacent 0's can be formed and the corresponding product terms are AC' and AB'. The simplified Boolean function in POS form is given by complement of sum of these two product terms.

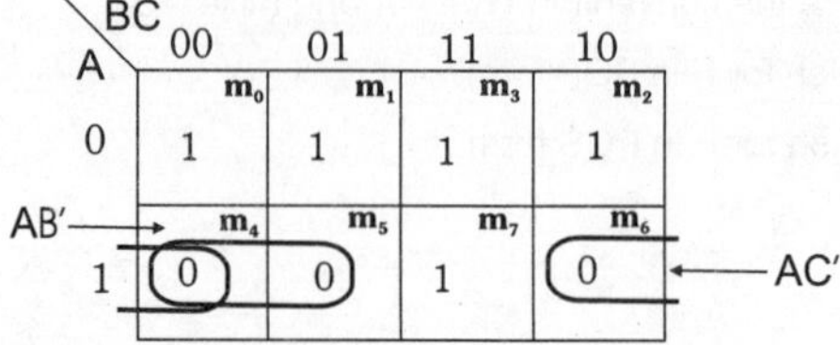

Fig. 2: K-map for POS (Product-of-Sum).

From the K-map the simplified Boolean equation for F in POS form is given by,

F = (AC' + AB')'

= (AC')'(AB')'

= (A' + C) (A' + B)

Example 4.35

(AU, Apr/May'19, 15 Marks) (AKTUB, Nov/Dec'22, 10 Marks) (CSVTU, Nov/Dec'19, 7 Marks)

Use Karnaugh Map method to simplify the following Boolean function

F(A, B, C, D) = Σm(2, 4, 6, 10, 12) + Σ d(0, 8, 9, 13)

a) Implement the Boolean function, F using only NAND gates.

b) Implement the Boolean function, F using only NOR gates.

Solution

a) To Implement the Boolean function, F using only NAND gates

The given function has four variables and so a 4-variable K-map with 16 squares (2^4 = 16) is drawn as shown in Fig. 1. The function is defined as sum of minterms m_2, m_4, m_6, m_{10} and m_{12} and so a "1" is filled in the corresponding squares in K-map. The function also has don't-care outputs for the minterms m_0, m_8, m_9 and m_{13} and so a "x" is filled in the corresponding squares in K-map and the remaining squares are filled with "0".

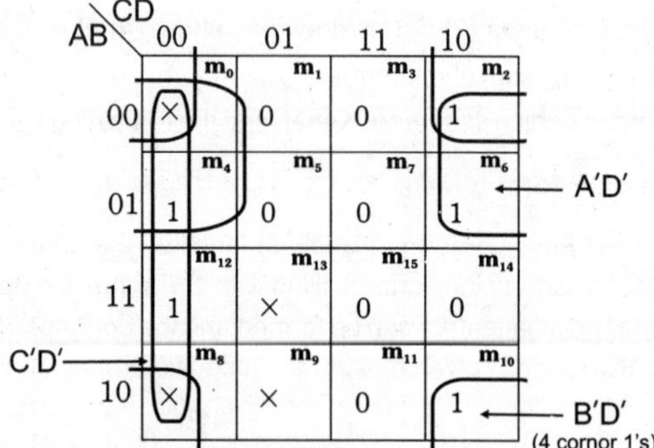

Fig. 1: K-map for F in SOP form.

For NAND implementation, the Boolean function should be formed in SOP form. In the K-map of Fig. 1, by considering the don't-care condition in m_0 and m_8 as "1", three prime implicants can be formed with four adjacent 1's and the corresponding product terms are A'D', C'D' and B'D'. The simplified Boolean function is given by sum of these three product terms.

∴ F = A'D' + C'D' + B'D'

Using the above Boolean function the logic circuit using only NAND gates is drawn as shown in Fig. 2.

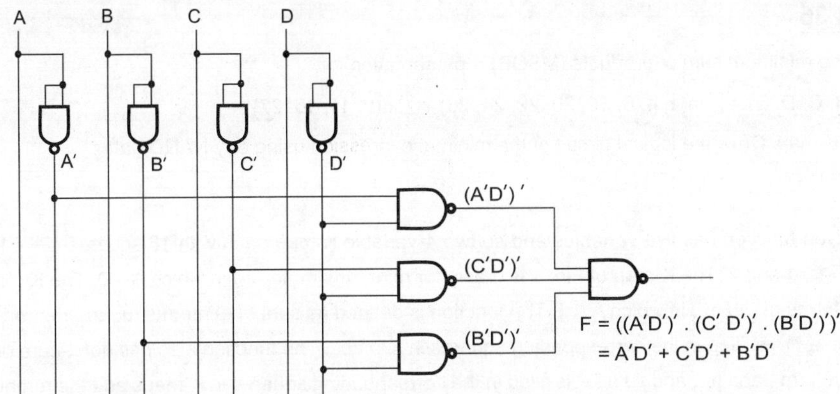

Fig. 2: Logic circuit of F using only NAND gates.

b) To Implement the Boolean function, F using only NOR gates

For NOR implementation, the Boolean function should be formed in POS form. For POS form of Boolean function, the prime implicants are formed by considering 0's. In the K-map of Fig. 3, by considering the don't-care condition in m_9 and m_{13} as "0" two prime implicants can be formed in which one prime implicant has eight adjacent 0's and one prime implicants has two adjacent 0's and the corresponding product terms are D and ABC. The simplified Boolean function in POS form is given by complement of sum of these two product terms.

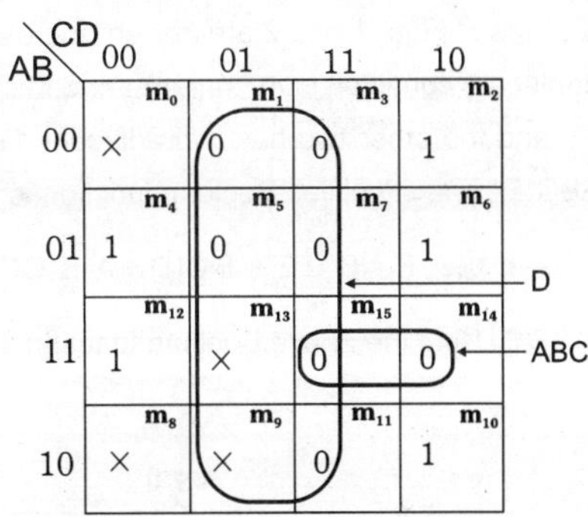

Fig. 3: K-map for F in POS form.

$$F = (D + ABC)'$$

$$= D'(ABC)'$$

$$= D'(A' + B' + C')$$

Using the above POS form of Boolean function, the logic circuit using only NOR gates is drawn as shown in Fig. 4.

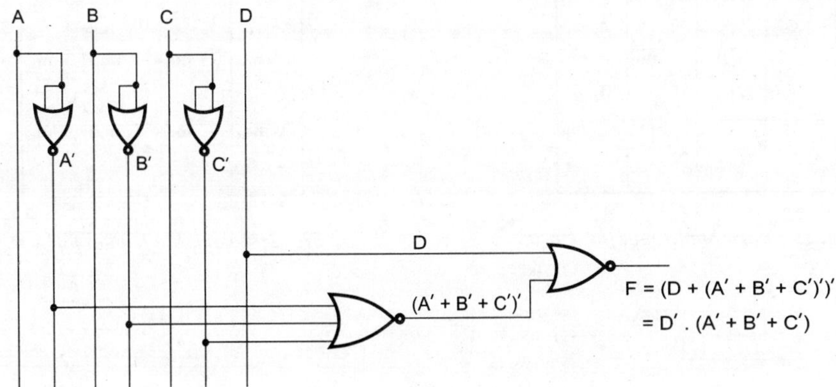

Fig. 4: Logic circuit of F using only NOR gates.

Example 4.36

(IKGPTU, May'20, 10 Marks)

Find the minimum sum of products (MSOP) representation for,

$F(A, B, C, D, E) = \sum m(1, 4, 6, 10, 20, 22, 24, 26) + \sum d(0, 11, 16, 27)$

using K-map. Draw the logical circuit of the minimal expression using ony NAND gates.

Solution

The given function has five variables and so two 4-variable K-map each with 16 squares ($2^4 = 16$) are drawn as shown in Figs. 1 and 2. The K-map of Fig. 1 is drawn for minterms m_0 to m_{15} in which A = 0. The K-map of Fig. 2 is drawn for minterms m_{16} to m_{31} in which A = 1. The function is defined as sum of minterms m_1, m_4, m_6, m_{10} m_{20}, m_{22}, m_{24} and m_{26} and so a "1" is filled in the corresponding squares in K-maps. The function also has don't-care outputs for the minterms m_0, m_{11}, m_{16} and m_{27} and so a "x" is filled in the corresponding squares in K-maps and the remaining squares are filled with "0".

For NAND realization the Boolean function should be formed in SOP form. In order to find adjacent squares the K-maps of Figs. 1 and 2 are placed one over the other as shown in Fig. 3. In the K-map of Fig. 3, by considering the don't-care condition in m_0, m_{11}, m_{16} and m_{27} as "1", four prime implicants can be formed in which two has four adjacent 1's and the other two has two adjacent 1's and the corresponding product terms are B′CE′, BC′D, A′B′C′D′ and ABC′E′. The simplified Boolean function is given by sum of these four product terms.

$\therefore \quad F = B'CE' + BC'D + A'B'C'D' + ABC'E'$

Using the above Boolean function the logic circuit using only NAND gate is drawn as shown in Fig. 4.

A = 0

BC \ DE	00	01	11	10
00	m_0 ×	m_1 1	m_3 0	m_2 0
01	m_4 1	m_5 0	m_7 0	m_6 1
11	m_{12} 0	m_{13} 0	m_{15} 0	m_{14} 0
10	m_8 0	m_9 0	m_{11} ×	m_{10} 1

Fig. 1: 4-variable K-map for A = 0.

A = 1

BC \ DE	00	01	11	10
00	m_{16} ×	m_{17} 0	m_{19} 0	m_{18} 0
01	m_{20} 1	m_{21} 0	m_{23} 0	m_{22} 1
11	m_{28} 0	m_{29} 0	m_{31} 0	m_{30} 0
0	m_{24} 1	m_{25} 0	m_{27} ×	m_{26} 1

Fig. 2: 4-variable K-map for A = 1.

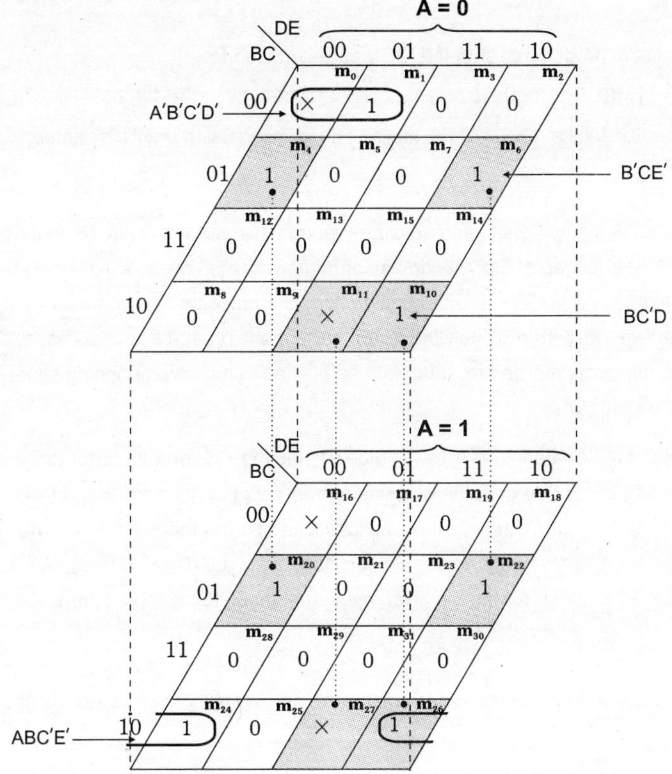

Fig. 3: 5-variable K-map.

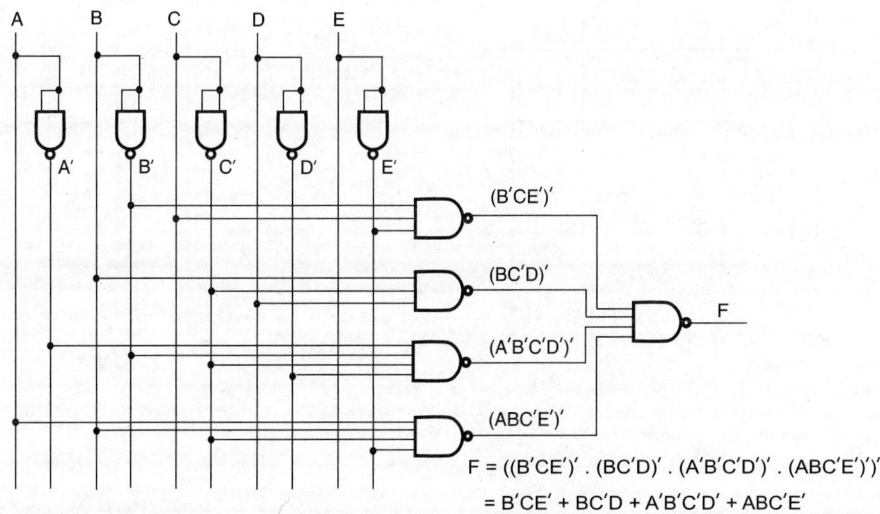

Fig. 4: Logic circuit of function, F using only NAND gates.

Example 4.37

Find the minimum sum products of sum (MPOS) representation for,

$F(A, B, C, D, E) = \prod M(0, 5, 7, 8, 9, 10, 11, 16, 20, 24, 25, 26, 27) + \sum d(23, 31)$

using K-map. Draw the logical circuit of the minimal expression using ony NOR gates.

Solution

The given function has five variables and so two 4-variable K-map each with 16 squares ($2^4 = 16$) are drawn as shown in Figs. 1 and 2. The K-map of Fig. 1 is drawn for minterms m_0 to m_{15} in which A = 0. The K-map of Fig. 2 is drawn for minterms m_{16} to m_{31} in which A = 1. The function is defined as sum of minterms m_0, m_5, m_7, m_8 m_9, m_{10}, m_{11} m_{16}, m_{20}, m_{24}, m_{25}, m_{26} and m_{27} and so a "0" is filled in the corresponding squares in K-maps. The function also has don't-care outputs for the minterms m_{23} and m_{31} and so a "x" is filled in the corresponding squares in K-maps and the remaining squares are filled with "0".

For NOR realization the Boolean function should be formed in POS form. In order to find adjacent squares the K-maps of Figs. 1 and 2 are placed one over the other as shown in Fig. 3. In the K-map of Fig. 3, by considering the don't-care condition in m_{23} and m_{31} as "1", four prime implicants can be formed in which two has four adjacent 1's and the other two has two adjacent 1's and the corresponding product terms are (B + C + D + E), (A + B + C'+ E'), (B' + C) and (A' + B + D + E). The simplified Boolean function is given by sum of these four product terms.

$\therefore$ F = (B + C + D + E) (A + B + C'+ E') (B'+ C) (A'+ B + D + E)

Using the above Boolean function the logic circuit using only NAND gate is drawn as shown in Fig. 4.

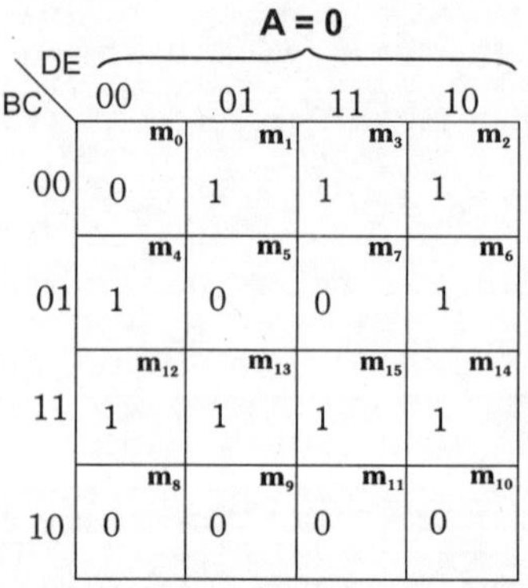

A = 0

BC \ DE	00	01	11	10
00	m_0: 0	m_1: 1	m_3: 1	m_2: 1
01	m_4: 1	m_5: 0	m_7: 0	m_6: 1
11	m_{12}: 1	m_{13}: 1	m_{15}: 1	m_{14}: 1
10	m_8: 0	m_9: 0	m_{11}: 0	m_{10}: 0

***Fig. 1:** 4-variable K-map for A = 0.*

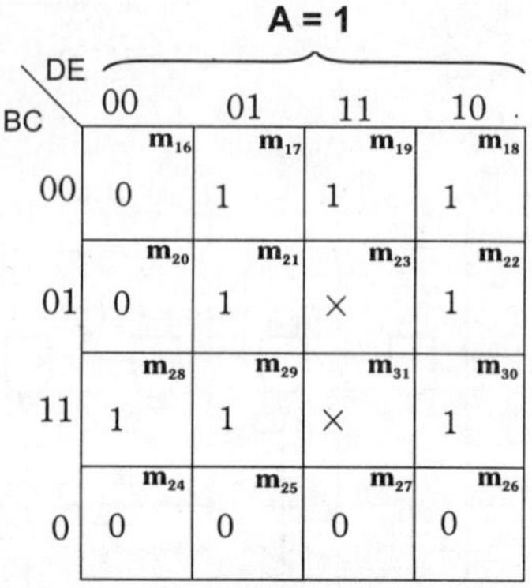

A = 1

BC \ DE	00	01	11	10
00	m_{16}: 0	m_{17}: 1	m_{19}: 1	m_{18}: 1
01	m_{20}: 0	m_{21}: 1	m_{23}: ×	m_{22}: 1
11	m_{28}: 1	m_{29}: 1	m_{31}: ×	m_{30}: 1
0	m_{24}: 0	m_{25}: 0	m_{27}: 0	m_{26}: 0

***Fig. 2:** 4-variable K-map for A = 1.*

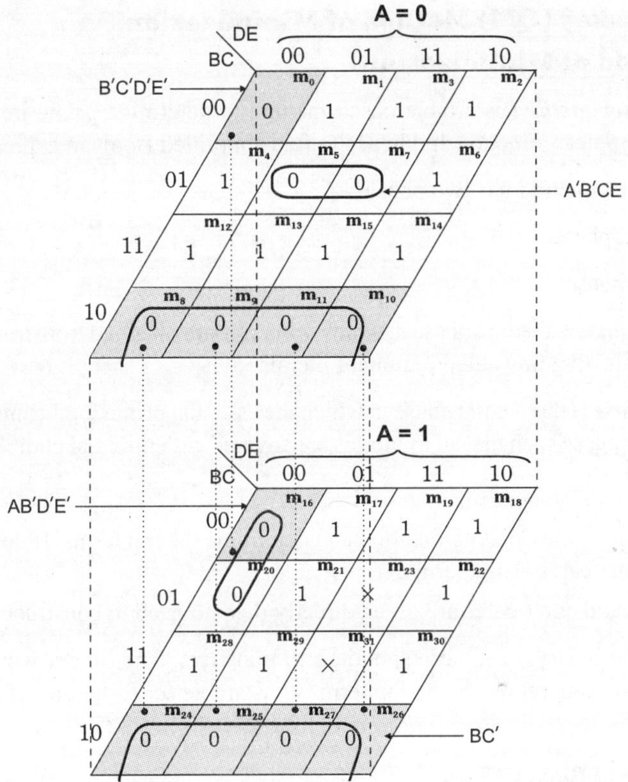

Fig. 3: 5-variable K-map.

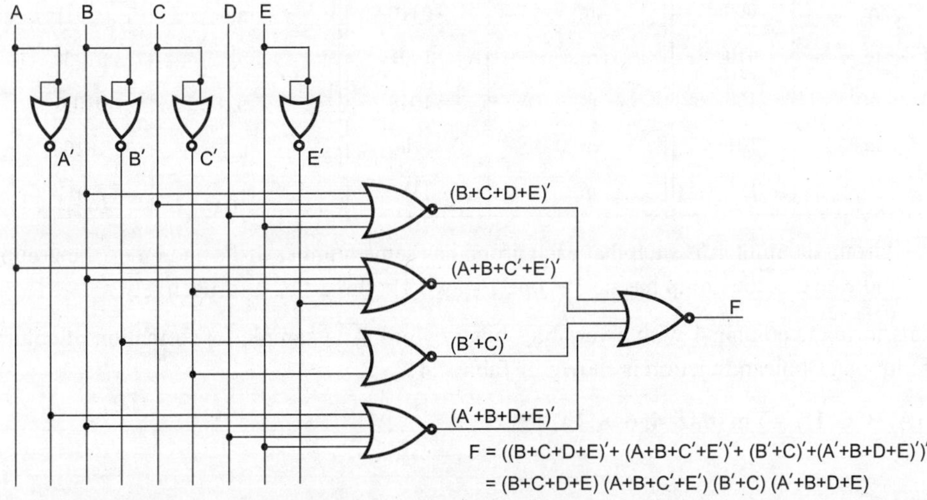

Fig. 4: Logic circuit of function, F using only NOR gates.

4.7 Quine-McCluskey (QM) Method of Minimization (Table Method of Minimization)

Quine-McCluskey method is a table based method to determine prime implicants and essential prime implicants of a Boolean function and hence to find simplified Boolean expression of a function.

Quine-McCluskey method has two phases:

1. Expansion phase
2. Covering phase.

The ***expansion phase*** is the first phase in which minterms are obtained from truth table and implicants are identified after eliminating unwanted/reduntant literals.

The ***covering phase*** is the second phase in which the essential prime implicants are selected from the prime implicants of first phase such that all minterms are covered with minimal number of prime implicants.

Procedure:

1. Construct truth table of given function and determine the minterms. If don't-cares are available then consider all the don't-cares as minterms.

(If minterms and don't-cares are given directly then no need to construct truth table).

2. Determine the binary word corresponding to minterms. The binary word is the binary value of the decimal number used to represent the minterm. An example for formation of binary word is shown in Table 4.40.

Table 4.40: Formation of Binary Word

Minterm (3 variables)	Binary word	Minterm (4 variables)	Binary word	Minterm (5 variables)	Binary word
m_0	000	m_6	0110	m_3	00011
m_2	010	m_9	1001	m_8	01000
m_4	100	m_{10}	1010	m_{11}	01011
m_5	101	m_{12}	1100	m_{20}	10100
m_7	111	m_{15}	1111	m_{27}	11011

3. Group the minterms such that each group has same number of 1's and index the group as 0, 1, 2, 3 ..... to indicate that the group has no 1's, one 1's, two 1's, three 1's,.... respectively.

This form the column-1 with n variables binary word. An example for formation of column-1 for the the following Boolean function is shown in Table 4.41.

$F(A, B, C, D) = \sum m\ (0, 2, 4, 6, 8, 14, 15)$

Table 4.41: Formation of Column-1 Grouping Minterms with Same Number of 1's

Minterm and binary word of function, F		Column-1		
		Minterm		**Binary word**
m_0	0000	Index 0	0	0000
m_2	0010			
m_4	0100	Index 1	2	0010
m_6	0110		4	0100
m_8	1000		8	1000
m_{14}	1110			
m_{15}	1111	Index 2	6	0110
		Index 3	14	1110
		Index 4	15	1111

4. Next, form column-2 with one variable/literal minimized. In order to reduce one variable, identify a change in any one digit position in the two binary words in consecutive group.

Starting from index 0, each binary word of an index is compared with a binary word of next index, if a change in any bit position from 0 to 1 or 1 to 0 is encountered, then that literal (or variable) can be removed and put an underscore ("_") in that position and write the reduced binary word in column-2.

The column-2 is formed after all possible comparisons. Whenever the comparison of a pair of minterms lead to reduction of a literal, put a √ mark (tick mark) in the corresponding binary words of column-1 to indicate that they are considered for reduction. The formation of column-2 using the column-1 in Table 4.41 is shown in Table 4.42.

Table 4.42: Formation of Column-2 after Comparison of Minterm Pairs in Column-1

Column-1			Column-2	
Minterm		**Binary word**	**Minterm**	**Binary word**
Index 0	0	0000 √	0, 2	00_0
			0, 4	0_00
Index 1	2	0010 √	0, 8	_000
	4	0100 √		
	8	1000 √	2, 6	0_10
			4, 6	01_0
Index 2	6	0110 √		
			6, 14	_110
Index 3	14	1110 √		
			14, 15	111_
Index 4	15	1111 √		

> ***Note:*** *When we compare minterms 8 and 6 there are three changes and so the minterm pairs 8 and 6 could not participate in reduction process shown in formation of column-2 in Table 4.42.*

Column-1 has 4 variables binary word which reduced to 3 variables binary word in column-2 and the binary word of column-2 is combination of two minterms.

5. Similarly form column-3 to get combination of 4 minterms and 2 variables binary word.

6. The reduction can be continued till 1 variable binary word or until no more reduction is possible.

7. After all possible reductions the unticked binary words are the prime implicants of the Boolean function. Mark the unticked binary words of all columns as P_0, P_1, P_2, P_3 ...... etc.

8. Then choose the ***essential prime implicants*** from P_0, P_1, P_2, P_3 ...... such that all minterms of function are covered with minimum number of prime implicants.

Example 4.38

(GTU, Jun'19, 7 Marks) (CSVTU, Nov/Dec'19, 7 Marks)

(VTU, Aug/Sep'20, 6 Marks)

Simplify the following Boolean function using table method.

$F(A, B, C, D) = \Sigma m(0, 2, 3, 6, 7, 8, 9, 10, 13)$

Solution

The minterms and corresponding binary words of the Boolean function are listed in Table 1. The given Boolean function is 4-variable function and so 4-bit binary words are used to represent minterms in Table 1.

Table 1: Minterms and Binary Words

Inputs A B C D	Minterm	Binary Word
0 0 0 0	m_0	0000
0 0 1 0	m_2	0010
0 0 1 1	m_3	0011
0 1 1 0	m_6	0110
0 1 1 1	m_7	0111
1 0 0 0	m_8	1000
1 0 0 1	m_9	1001
1 0 1 0	m_{10}	1010
1 1 0 1	m_{13}	1101

The minterms with no 1's are grouped as index 0, the minterms with one 1's are grouped as index 1, the minterms with two 1's are grouped as index 2 and the minterms with three 1's are grouped as index 3 and listed in column-1 of Table 2. The minterms of column-1 are 4-variable minterms.

The column-2 of Table 2 is formed by reducing one variable of binary words in column-1. In order to reduce a variable, starting from index 0, compare a binary word of an index with a binary word of next index and if a single change, either 1 to 0, or 0 to 1, is identified then remove that variable and form reduced binary word with an underscore in the place of removed variable. Also, a tick mark is made against the minterms of column-1 that are involved in comparison and reduction of a variable. The comparisons are performed with all possible pairs. The reduced binary words along with corresponding pair of minterms are listed in column-2. The minterms of column-2 are three variable minterms.

The column-3 of Table 2 is formed by reducing one variable of binary words in column-2. The reduced words are determined by following the procedure similar to that for reducing a literal of binary words of column-2. The minterms of column-3 are two variable minterms.

In this problem further reduction is not possible and so comparison is stopped with column-2. Now, the unticked binary words in columns-2 and 3 of Table 2 are the prime implicants. There are four unticked binary words in Table 2. Let us denote the four binary words as prime implicants P_0, P_1, P_2 and P_3. The reduced binary words along with minterms and prime implicants are listed in Table 3. Repeating words in column-3, can be considered once.

Table 2: Formation of Columns by Reducing a Variable of Binary Words

Column-1			Column-2 (First Reduction)		Column-3 (Second Reduction)	
	Minterm	Binary word	Minterm	Binary word	Minterm	Binary word
Index 0	0	0000 √	0, 2	00_0 √	0, 2, 8, 10	_0_0
			0, 8	_000 √	~~0, 8, 2, 10~~	~~_0_0~~
Index 1	2	0010 √				
	8	1000 √	2, 3	001_ √	2, 3, 6, 7	0_1_
			2, 6	0_10 √	~~2, 6, 3, 7~~	~~0_1_~~
	3	0011 √	2, 10	_010 √		
	6	0110 √	8, 9	100_		
Index 2	9	1001 √	8, 10	10_0 √		
	10	1010 √				
			3, 7	0_11 √		
	7	0111 √	6, 7	011_ √		
Index 3	13	1101 √	9, 13	1_01		

Unticked binary words

Table 3: Reduced Binary Words and Prime Implicants

Minterm	Reduced Binary Word	Prime Implicants
8, 9	100_	P_0 = AB′C′ (D is removed)
9, 13	1_01	P_1 = AC′D (B is removed)
0, 2, 8, 10	_0_0	P_2 = B′D′ (A and C are removed)
2, 3, 6, 7	0_1_	P_3 = A′C (B and D are removed)

The minterms of the function are 0, 2, 3, 6, 7, 8, 9, 10 and 13. Form Table 4 by taking prime implicants P_0, P_1, P_2 and P_3 on rows and minterms of function on columns.

Table 4: Selection of Essential Prime Implicants

Prime implicants	Minterm 0	2	3	6	7	8	9	10	13
P_0 8, 9						×	×		
P_1 9, 13							×		⊗
P_2 0, 2, 8, 10	⊗	×				×		⊗	
P_3 2, 3, 6, 7		×	⊗	⊗	⊗				

The following observations are made from Table 4.

1. The minterms 0 and 10 are covered only in P_2 and so P_2 is essential.
2. The minterms 3, 6 and 7 are covered only in P_3 and so P_3 is essential.
3. The minterms 13 is covered only in P_1 and so P_1 is essential.
4. The prime implicants P_0 has two minterms 8 and 9. And in this 8 is already covered in P_2 and 9 is already covered in P_1 and so P_0 is not essential.
5. Finally the reduced set of prime implicants are P_1, P_2 and P_3.

The function F is given by sum of the prime implicants P_1, P_2 and P_3.

$\therefore F = P_1 + P_2 + P_3$

$= AC'D + B'D' + A'C$

Example 4.39

(BATU, Jun'19, 6 Marks) (AKTU, Dec'22, 10 Marks) (VTU, Aug/Sep'20, 7 Marks)

Find all the prime implicants for the following Boolean function and determine the essential prime implicants and minimum SOP form of the function.

$F(w, x, y, z) = \Sigma\, m(0, 2, 4, 5, 6, 7, 8, 10, 13, 15)$

Solution

The minterms and corresponding binary words of the Boolean function are listed in Table 1. The given Boolean function is 4-variable function and so 4-bit binary words are used to represent minterms in Table 1.

Table 1: Minterms and Binary Words

Inputs	Minterm	Binary Word
w x y z		
0 0 0 0	m_0	0000
0 0 1 0	m_2	0010
0 1 0 0	m_4	0100
0 1 0 1	m_5	0101
0 1 1 0	m_6	0110
0 1 1 1	m_7	0111
1 0 0 0	m_8	1000
1 0 1 0	m_{10}	1010
1 1 0 1	m_{13}	1101
1 1 1 1	m_{15}	1111

The minterms with no 1's are grouped as index 0, the minterms with one 1's are grouped as index 1, the minterms with two 1's is grouped as index 2, the minterms with three 1's are grouped as index 3 and the minterms with four 1's are grouped as index 4 and listed in column-1 of Table 2. The minterms of column-1 are 4-variable minterms.

The column-2 of Table 2 is formed by reducing one variable of binary words in column-1. In order to reduce a variable, starting from index 0, compare a binary word of an index with a binary word of next index and if a single change, either 1 to 0, or 0 to 1, is identified then remove that variable and form reduced binary word with an underscore in the place of removed variable. Also, a tick mark is made against the minterms of column-1 that are involved in comparison and reduction of a variable. The comparisons are performed with all possible pairs. The reduced binary words along with corresponding pair of minterms are listed in column-2. The minterms of column-2 are three variable minterms.

The column-3 of Table 2 is formed by reducing one variable of binary words in column-2. The reduced words are determined by following the procedure similar to that for reducing a literal of binary words of column-2. The minterms of column-3 are two variable minterms.

In this problem further reduction is not possible and so comparison is stopped with column-2. Now, the unticked binary words in column-3 of Table 2 are the prime implicants. There are eight unticked binary words in Table 2 and in this each word is repeating twice and so four reduced binary words without repetition are selected as prime implicants. Let us denote the four binary words as prime implicants P_0, P_1, P_2 and P_3. The reduced binary words along with minterms and prime implicants are listed in Table 3.

Table 2: Formation of Columns by Reducing a Variable of Binary Words

Column-1			Column-2 (First Reduction)		Column-3 (Second Reduction)	
	Minterm	Binary word	Minterm	Binary word	Minterm	Binary word
Index 0	0	0000 √	0, 2	00_0 √	0, 2, 4, 6	0__0
Index 1	2	0010 √	0, 4	0_00 √	0, 2, 8, 10	_0_0
	4	0100 √	0, 8	_000 √	~~0, 4, 2, 6~~	~~0__0~~
	8	1000 √	2, 6	0_10 √	~~0, 8, 2, 10~~	~~_0_0~~
Index 2	5	0101 √	2, 10	_010 √	4, 5, 6, 7	01__
	6	0110 √	4, 5	010_ √	~~4, 6, 5, 7~~	~~01__~~
	10	1010 √	4, 6	01_0 √	5, 7, 13, 15	_1_1
Index 3	7	0111 √	8, 10	10_0 √	~~5, 13, 7, 15~~	~~_1_1~~
	13	1101 √	5, 7	01_1 √	Unticked binary words	
Index 4	15	1111 √	5, 13	_101 √		
			6, 7	011_ √		
			7, 15	_111 √		
			13, 15	11_1 √		

Table 3: Reduced Binary words and Prime Implicants

Minterm	Reduced Binary Word	Prime Implicants
0, 2, 4, 6	0__0	P_0 = w'z' (x and y are removed)
0, 2, 8, 10	_0_0	P_1 = x'z' (w and y are removed)
4, 5, 6, 7	01__	P_2 = w'x (y and z are removed)
5, 7, 13, 15	_1_1	P_3 = xz (w and y are removed)

Table 4: Selection of Essential Prime Implicants

Prime implicants	0	2	4	5	6	7	8	10	13	15
P_0 0, 2, 4, 6	×	×	×		×					
P_1 0, 2, 8, 10	×	×					⊗	⊗		
P_2 4, 5, 6, 7			×	×	×	×				
P_3 5, 7, 13, 15				×		×			⊗	⊗

The minterms of the function are 0, 2, 4, 5, 6, 7, 8, 10, 13 and 15. Form Table 4 by taking prime implicants P_0, P_1, P_2 and P_3 on rows and minterms of function on columns. The following observations are made from Table 4.

1. Minterms 8 and 10 are covered only in P_1 and so P_1 is an essential prime implicant.
2. Minterms 13 and 15 are covered only in P_3 and so P_3 is an essential prime implicant.
3. Apart from P_1 and P_3 if we select P_2 all minterms of the function are covered.

The minimum SOP form of the function, F is given by sum of prime implicants P_1, P_2 and P_3.

$$\therefore F = P_1 + P_2 + P_3 = x'z' + w'x + xz$$

Example 4.40

(CSVTU, Nov/Dec'19, 7 Marks) *(VTU, Aug/Sep'20, 10 Marks)*

Determine the minimum SOP form of the following Boolean function using Quine-McCulskey method

$$F(A, B, C, D) = \sum m(6, 7, 8, 9) + \sum d(10, 11, 12, 13, 14, 15)$$

Solution

The minterms and corresponding binary words of the Boolean function are listed in Table 1. Here the don't-cares are also considered as minterms. The given Boolean function is 4-variable function and so 4-bit binary words are used to represent minterms in Table 1.

Table 1: Minterms and Binary Words

Inputs A B C D	Minterm	Binary Word
0 1 1 0	m_6	0110
0 1 1 1	m_7	0111
1 0 0 0	m_8	1000
1 0 0 1	m_9	1001
1 0 1 0	m_{10}	1010
1 0 1 1	m_{11}	1011
1 1 0 0	m_{12}	1100
1 1 0 1	m_{13}	1101
1 1 1 0	m_{14}	1110
1 1 1 1	m_{15}	1111

The minterms with no 1's are grouped as index 0, the minterms with one 1's are grouped as index 1, the minterms with two 1's are grouped as index 2, the minterms with three 1's are grouped as index 3 and the minterms with four 1's are grouped as index 4 and listed in column-1 of Table 2. The minterms of column-1 are 4-variable minterms.

The column-2 of Table 2 is formed by reducing one variable of binary words in column-1. In order to reduce a variable, starting from index 0, compare a binary word of an index with a binary word of next index and if a single change, either 1 to 0, or 0 to 1, is identified then remove that variable and form reduced binary word with an underscore in the place of removed variable. Also, a tick mark is made against the minterms of column-1 that are involved in comparison and reduction of a variable. The comparisons are performed with all possible pairs. The reduced binary words along with corresponding pair of minterms are listed in column-2. The minterms of column-2 are three variable minterms.

The column-3 of Table 2 is formed by reducing one variable of binary words in column-2. The reduced words are determined by following the procedure similar to that for reducing a literal of binary words of column-2. The minterms of column-3 are two variable minterms.

The column-4 of Table 2 is formed by reducing one variable of binary words in column-3. The reduced words are determined by following the procedure similar to that for reducing a literal of binary words of column-2. The repeated reduced binary words of column-3 are ignored for comparison. The minterms of column-4 are one variable minterms.

In this problem further reduction is not possible and so comparison is stopped with column-4. Now, the unticked binary words in columns-3 and 4 of Table 2 are the prime implicants. There is one unticked binary word each in column-3 and 4 of Table 2. Let us denote the two binary words as prime implicants P_0 and P_1. The reduced binary words along with minterms and prime implicants are listed in Table 3.

Table 2: Formation of Columns by Reducing a Variable of Binary Words

Column-1		Column-2 (First Reduction)		Column-3 (Second Reduction)		Column-4 (Third Reduction)	
Minterm	**Binary word**	**Minterm**	**Binary word**	**Minterm**	**Binary word**		
Index 0 0	-						
Index 1 8	1000 √	8, 9	100_ √	8, 9, 10, 11	10__ √	8, 9, 10, 11	1___
		8, 10	10_0 √	8, 9, 12, 13	1_0_ √	12, 13, 14, 15	
Index 2 6	0110 √	8, 12	1_00 √	~~8, 10, 9, 11~~	~~10__~~	~~8, 9, 12, 13~~	~~1___~~
9	1001 √			8, 10, 12, 14	1__0 √	~~10, 11, 14, 15~~	
10	1010 √	6, 7	011_ √	~~8, 12, 9, 13~~	~~1_0_~~	~~8, 10, 12, 14~~	~~1___~~
12	1100 √	6, 14	_110 √	~~8, 12, 10, 14~~	~~1__0~~	~~9, 11, 13, 15~~	
		9, 11	10_1 √	6, 7, 14, 15	_11_		
Index 3 7	0111 √	9, 13	1_01 √	~~6, 14, 7, 15~~	~~_11_~~		
11	1011 √	10, 11	101_ √	9, 11, 13, 15	1__1 √	Unticked binary words	
13	1101 √	10, 14	1_10 √	~~9, 13, 11, 15~~	~~1__1~~		
14	1110 √	12, 13	110_ √	10, 11, 14, 15	1_1_ √		
		12, 14	11_0 √	~~10, 14, 11, 15~~	~~1_1_~~		
Index 4 15	1111 √			12, 13, 14, 15	11__ √		
		7, 15	_111 √	~~12, 14, 13, 15~~	~~11__~~		
		11, 15	1_11 √				
		13, 15	11_1 √				
		14, 15	111_ √				

Table 3: Reduced Binary words and Prime Implicants

Minterm	Reduced Binary Word	Prime Implicants
6, 7, 14, 15	_11_	P_0 = B C (A and D are removed)
8, 9, 10, 11 12, 13, 14, 15	1_ _ _	P_1 = A (B, C and D are removed)

From Table 3 it is observed that both the prime implicants P_0 and P_1 are essential so that all the minterms of the function are included.

The function F is given by sum of the prime implicants P_0 and P_1.

$\therefore$ F = B C + A

Example 4.41

(VTU, Aug/Sep'20, 5 Marks)

Represent the number of days in a month for a non-leap year by a truth table, indicating the output of invalid input if any by '0'.

Solution

In decimal number representation the twelve months of a year are represented by decimal numbers 1 to 12 and the binary representation requires 4-bit binary.

In a non-leap year the february month has 28 days and other months has either 30 or 31 days. Therefore, the binary representation of days require 5-bit binary.

The truth table with 4-bit representation for months and 5-bit representation for days is shown in Table 1. For valid input the output is 1 and for invalid input the output is 0.

Table 1: Truth Table

Inputs					Output
Month			Days		
Name	Decimal	Binary	Decimal	Binary	
-	0	0000	-	××××	0
January	1	0001	31	1 1111	1
February	2	0010	28	1 1100	1
March	3	0011	31	1 1111	1
April	4	0100	30	1 1110	1
May	5	0101	31	1 1111	1
June	6	0110	30	1 1110	1
July	7	0111	31	1 1111	1
August	8	1000	31	1 1111	1
September	9	1001	30	1 1110	1
October	10	1010	31	1 1111	1
November	11	1011	30	1 1110	1
December	12	1100	31	1 1111	1
-	13	1101	-	××××	0
-	14	1110	-	××××	0
-	15	1111	-	××××	0

4.8 Summary of Important Concepts

1. George Boole developed Boolean algebra in 1854.
2. Boolean algebra is an algebraic structure that includes a set of elements consisting of binary operators " + " and " . ", binary variables x, y, z, ... , binary elements 0 and 1 and Boolean postulates and theorems.
3. The duality principle of Boolean algebra states that every algebraic expression deducible from the postulates of Boolean algebra remains valid if the operator and identity element are interchanged. Here, the operators " + " and " . " and the identity elements are " 0 " and " 1 ".
4. A statement written with Boolean variables, constants and operators is called Boolean expression.
5. Postulates of Boolean algebra are developed by E. V. Huntington in 1904.
6. Binary logic consists of binary variables and a set of logical operations. The binary variables are denoted by letters of alphabet A, B, C, ..., x, y, z, ... Each binary variable can take only two possible values 0 or 1.
7. If, 1 represent **high** voltage and 0 represent **low** voltage then logic system is called positive logic.
8. If, 1 represent **low** voltage and 0 represent **high** voltage then the logic system is called negative logic.
9. Logic gates are electronic devices or circuits that perform logical operations on one or more input logical variables and produce a binary output.
10. The basic logical operations are AND, OR and NOT.
11. In positive logic, the logical AND of two or more variables will be 1 if and only if the value of all the variables are 1.
12. In positive logic, the logical OR of two or more variables will be 1 if the value of one of the variables is 1.
13. NOT operation is same as complement operation.
14. With n variables it is possible to form 2^{2n} Boolean functions. Therefore, with two variables we can form 16 ($2^{2 \times 2} = 16$) Boolean functions.
15. NAND and NOR gates are called universal gates, because any Boolean function can be realized only using NAND gates or only using NOR gates.
16. A Boolean function is described by a Boolean expression which consists of binary variables, binary constants 0 and 1 and logical operators AND, OR and NOT.
17. Minterms are 2^n possible combinations of AND terms with n variables such that the logical AND of all the variables is 1.
18. Maxterms are 2^n possible combinations of OR terms with n variables such that the logical OR of all the variables is 0.
19. The three types of standard forms of expressing a Boolean function are Sum-of-Products (SOP) form, Product-of-Sums (POS) form and Canonical form.
20. The SOP or POS form is said to be canonical only if all the variables are present in every term of the SOP or POS form.
21. Sum-of-products (SOP) can be realized by AND operation of literals followed by OR operation of output of AND.
22. Product-of-sums (POS) can be realized by OR operation of literals followed by AND operation of output. of OR.

23. The realizations in SOP and POS forms are called two level realization of standard forms.
24. It can be observed that complement of sum-of-products will be product-of-sums and vice-versa
25. A Boolean function expressed as a sum of minterms or product of maxterms are said to be in canonical form.
26. Each variable within a term of a Boolean expression is called a literal.
27. The complement of a function can be obtained by using duality by replacing "+" by "." and "." by "+" and variables by its complement
28. K-map is a pictorial form of truth table and used to simplify Boolean functions.
29. A K-map is a diagram made up of squares with each square representing a minterm.
30. Alternatively, maxterm can be used to construct K-map in which each square represent a maxterm.
31. When minterms are considered for simplification using K-map the resultant Boolean function will be in sum-of-products form.
32. When maxterms are considered for simplification using K-map the resultant Boolean function will be in product-of-sums form.
33. For n-variable Boolean function the K-map will have 2^n squares.
34. While arranging the literal/minterms in a K-map, only one change is allowed if we move from one row to next row or from one column to next column.
35. In a K-map, combine adjacent 1's to form prime implicants.
36. A prime implicant with single 1 represent n literal product term. A prime implicant with two 1's represent n – 1 literal product term. A prime implicant with four 1's represent n – 2 literal product term and so on.
37. The literals represented by a prime implicant are the literals that present in all the squares of the prime implicant (or the literals common to all the squares of prime implicants).
38. While forming prime implicants any number of over lapping is allowed in horizontal and vertical directions.
39. While forming prime implicants the K-map can be folded in any direction to get adjacent 1's.
40. When squares with 1's are considered for forming prime implicants, the simplified Boolean function is sum of all the product terms of prime implicant.
41. If all minterms are 1's then all squares of K-map will be filled by 1 and the function value is 1.
42. In K-map if squares with 0's are considered for forming prime implicants then the complement of the resultant sum-of-products form will give product-of-sums form.
43. The undefined function outputs are called don't-care conditions and denoted by ×.
44. In simplification of Boolean functions using K-maps, don't-care conditions can be considered as either 0 or 1.
45. If inclusion of a don't-care condition leads to minimization of a literal then don't-care can be included in the prime implicants.
46. Quine-McCluskey method is a table based method to determine prime implicants and essential prime implicants of a Boolean function and hence to find simplified Boolean expression of a function.
47. Quine-McCluskey method has two phases: Expansion phase and covering phase.

48. The expansion phase is the first phase in which minterms are obtained from truth table and prime implicants are identified after eliminating unwanted/redundant literals.

49. The covering phase is the second phase in which the essential prime implicants are selected from the prime implicants of first phase such that all minterms are covered with minimal number of prime implicants.

4.9 Short-Answer Questions

Q4.1 Prove that $A + A'B = A + B$ using Boolean algebra.

Solution:

$$A + A'B = A(1 + B) + A'B$$
$$= A + AB + A'B$$
$$= A + (A + A')B$$
$$= A + B$$

$x + x' = 1$

Q4.2 Simplify the three variable function $Y(A, B, C) = \sum m(1, 3, 5, 7)$ using Boolean algebra.

Solution:

$$Y(A, B, C) = \sum m(1, 3, 5, 7)$$
$$= A'B'C + A'BC + AB'C + ABC$$
$$= A'C(B' + B) + AC(B' + B)$$
$$= A'C + AC$$
$$= C(A' + A) = C$$

$x + x' = 1$

Q4.3 Find the complement of $F = wx + yz$ and then show that $FF' = 0$. ***(JNTUA, Jun/July'22, 7 Marks)***

$$F = wx + yz$$
$$F' = (wx + yz)' = (wx)'(yz)' = (w' + x')(y' + z')$$
$$FF' = (wx + yz)(w' + x')(y' + z')$$
$$= (wx(w' + x') + yz(w' + x'))(y' + z')$$
$$= (wxw' + wxx' + yzw' + yzx')(y' + z')$$
$$= (0 + 0 + yzw' + yzx')(y' + z')$$
$$= yzw'y' + yzw'z' + yzx'y' + yzx'z'$$
$$= 0 + 0 + 0 + 0 = 0$$

Using DeMorgan's theorem
$(x + y)' = x'y'$
$(x \cdot y)' = x' + y'$

$x \cdot x' = 0$

Q4.4 Reduce the following Boolean expressions.

a) $AB + A(B + C) + B'(B + D)$ b) $A + B + A'B'C$

Solution:

a) $AB + A(B + C) + B'(B + D)$

$$AB + A(B + C) + B'(B + D) = AB + \cancel{AB} + AC + B'B + B'D$$
$$= AB + AC + B'D$$
$$= A(B + C) + B'D$$

Repeated terms are considered once.

$x \cdot x' = 0$

b) A + B + A'B'C

$$A + B + A'B'C = A + (B + A')(B + B')(B + C) \qquad \boxed{x + x' = 1}$$

$$= A + (A' + B)(B + C)$$

$$= (A + A')(A + B)(B + C)$$

$$= (A + B)(B + C)$$

Q4.5 ***Demonstrate by means of truth table the validity of the DeMorgan's theorem for three variables:*** $(xyz)' = x' + y' + z'$

Solution:

Table 1: Truth Table

Input Variables			Complement of inputs			xyz	(xyz)'	x'+y'+z'
x	y	z	x'	y'	z'			
0	0	0	1	1	1	0	1	1
0	0	1	1	1	0	0	1	1
0	1	0	1	0	1	0	1	1
0	1	1	1	0	0	0	1	1
1	0	0	0	1	1	0	1	1
1	0	1	0	1	0	0	1	1
1	1	0	0	0	1	0	1	1
1	1	1	0	0	0	1	0	0

Q4.6 ***Write a truth table for, $Q = (X' + (X \cdot Y \cdot Z))'$.***

Solution:

Table 1: Truth Table of Q

Inputs			Outputs				
X	Y	Z	X'	XY	XYZ	X'+(XYZ)	Q
0	0	0	1	0	0	1	0
0	0	1	1	0	0	1	0
0	1	0	1	0	0	1	0
0	1	1	1	0	0	1	0
1	0	0	0	0	0	0	1
1	0	1	0	0	0	0	1
1	1	0	0	1	0	0	1
1	1	1	0	1	1	1	0

Q4.7 ***What are the basic digital logic gates?***

The three basic logic gates are AND, OR and NOT gate.

Q4.8 ***Write a truth table for the function, Z = (A + B)' (AB + C).***

Solution:

Table 1: Truth Table for Function, Z

Inputs			Outputs				
A	**B**	**C**	**A + B**	**(A + B)'**	**AB**	**(AB) + C**	**Z**
0	0	0	0	1	0	0	0
0	0	1	0	1	0	1	1
0	1	0	1	0	0	0	0
0	1	1	1	0	0	1	0
1	0	0	1	0	0	0	0
1	0	1	1	0	0	1	0
1	1	0	1	0	1	1	0
1	1	1	1	0	1	1	0

Q4.9 ***Realize the NAND operation using NOT, AND and OR gates.*** **(AU, Nov/Dec'18, 2 Marks)**

Solution:

Table 1: Truth Table of NAND Operation

x	y	Minterm	(x y)'
0	0	m_0	1
0	1	m_1	1
1	0	m_2	1
1	1	m_3	0

$F = m_0 + m_1 + m_2 = x'y' + x'y + xy'$

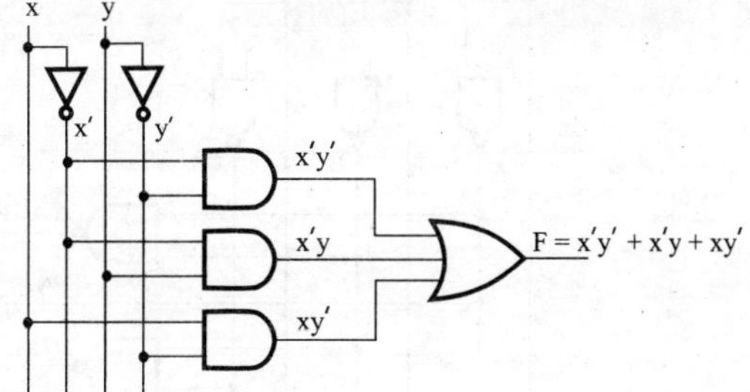

Fig. Q4.9: NAND operation using NOT, AND and OR gates.

Q4.10 ***Declare a module that describe a circuit that is specified with the following two Boolean expression:***

$$F = A + BC + B'D \quad ; \quad E = B'C + BC'D$$

Solution:

The given Boolean equations can be implemented using basic logic gate AND, OR and NOT as shown in Fig. Q4.10.

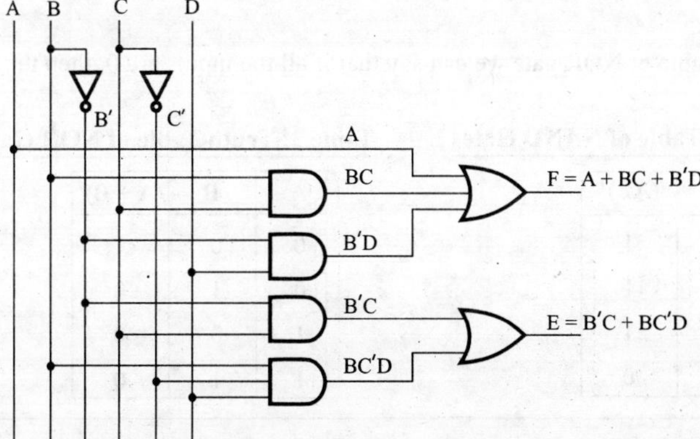

Fig. Q4.10: Logic circuit.

Q4.11 ***Implement the result of $A + A'D + AC'$.*** **(AU, Apr/May'19, 2 Marks)**

Solution:

Let, F be result of $A + A'D + AC'$ and it can be obtained from the logic circuit shown in Fig. Q4.11.

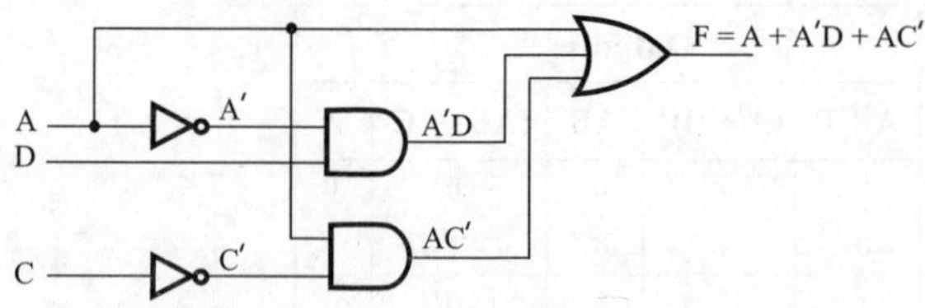

Fig. Q4.11: Logic circuit.

Q4.12 ***Implement $F = A'B' + A + (B + C)'$ using NAND gates only.*** **(AU, Nov/Dec'18, 5 Marks)** **(AKTU, Sep'20, 5 Marks)**

Solution:

$F = A'B' + A + (B + C)'$

$= A'B' + A + B'C'$ (Using DeMorgan's theorem)

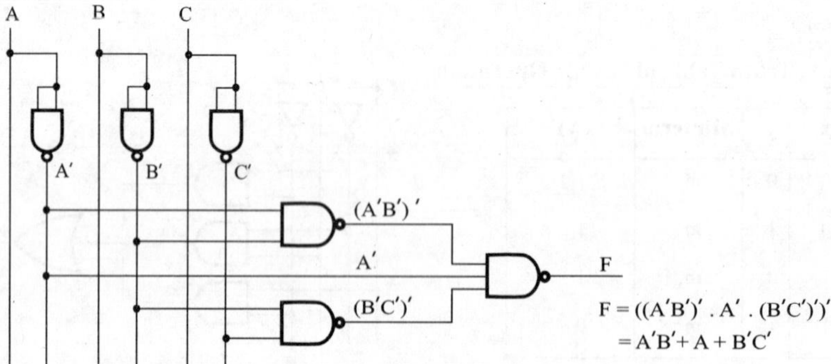

Fig. Q4.12: Logic circuit of F using only NAND gates.

Q4.13 ***i) When is the output of an NAND operation equal to 1?***

ii) When is the output of an NOR operation equal to 1?

From the truth of NAND gate we can say that if any one of the input is 0, then the output of NAND operation is equal to 1.

From the truth table of NOR gate we can say that if all the inputs are 0, then the output of NOR operation is equal to 1.

Table 1: Truth Table of NAND Gates

A	B	(AB)'
0	0	1
0	1	1
1	0	1
1	1	0

Table 2: Truth Table of NOR Gates

A	B	(A + B)'
0	0	1
0	1	0
1	0	0
1	1	0

Q4.14 ***Simplify the following Boolean function, $F = x'y' + xy + x'y$.***

Solution:

$$F = x'y' + xy + x'y = x'y' + x'y + xy$$
$$= x'(y + y') + xy = x'1 + xy$$
$$= x' + xy = (x' + x)(x' + y)$$
$$= 1(x' + y) = x' + y$$

$x + x' = 1$

Q4.15 ***Express $F = A + B'C$ in both canonical SOP and POS form.*** ***(AU, Nov/Dec'18, 2 Marks)***

Solution:

$F = A + B'C$

Canonical SOP: It is formed by taking sum of minterms for which F = 1 in truth table.

$$\therefore F = \Sigma (m_1, m_4, m_5, m_6, m_7)$$
$$= A'B'C + AB'C' + AB'C + ABC' + ABC$$

Canonical POS: It is formed by taking product of maxterms for which F = 0 in truth table.

$$\therefore F = \Pi (M_0, M_2, M_3)$$
$$= (A + B + C)(A + B' + C)(A + B' + C')$$

Table 1: Truth Table

A B C	Minterm	Maxterm	B'	B'C	F = A + B'C
0 0 0	m_0	M_0	1	0	0
0 0 1	m_1	M_1	1	1	1
0 1 0	m_2	M_2	0	0	0
0 1 1	m_3	M_3	0	0	0
1 0 0	m_4	M_4	1	0	1
1 0 1	m_5	M_5	1	1	1
1 1 0	m_6	M_6	0	0	1
1 1 1	m_7	M_7	0	0	1

Q4.16 ***Convert the following Boolean expression into standard SOP form:***
$AB'C + A'B' + ABC'D$

Solution:

$$F = AB'C + A'B' + ABC'D$$
$$= AB'C(D + D') + A'B'(C + C') + ABC'D$$
$$= AB'CD + AB'CD' + (A'B'C)(D + D') + (A'B'C')(D + D') + ABC'D$$
$$= AB'CD + AB'CD' + A'B'CD + A'B'CD' + A'B'C'D + A'B'C'D' + ABC'D$$
$$(m_{11}) \quad (m_{10}) \quad (m_3) \quad (m_2) \quad (m_1) \quad (m_0) \quad (m_{13})$$
$$= m_0 + m_1 + m_2 + m_3 + m_{10} + m_{11} + m_{13}$$
$$= \Sigma m(0, 1, 2, 3, 10, 11, 13)$$

Missing literals in first term is D and second term is C and D

Q4.17 ***Convert the function $Y = A + B'C$ in canonical POS form using Boolean algebra.***

Solution:

$$Y = A + B'C = (A + B')(A + C)$$
$$= ((A + B') + (CC'))((A + C) + (BB'))$$
$$= (A + B' + C)(A + B' + C')(A + B + C)\cancel{(A + B' + C)}$$
$$(M_2) \quad (M_3) \quad (M_0) \quad (M_2)$$
$$= (A + B + C)(A + B' + C)(A + B' + C')$$
$$= M_0 M_2 M_3$$

Missing literals in first term is C and second term is B.

Repeated maxterms are considered only one time.

Q4.18 *Compare SOP and POS.* *(GTU, May'19, 3 Marks) (IKGPTU, May'20, 2 Marks)*

SOP (Sum-of-Products)	POS (Product-of-Sums)
1. SOP form of Boolean function is formed using minterms.	1. POS form of Boolean function is formed using maxterms.
2. SOP form can be realized by AND operation of literals followed by OR operation of output of AND.	2. POS form can be realized by OR operation of literals followed by AND operation of output of OR.
3. SOP form of a Boolean function can be used to realize the function using only NAND gates.	3. POS form of a Boolean function can be used to realize the function using only NOR gates.

Q4.19 *Show that $AB'C + B + BD' + ABD' + A'C = B + C$.* *(GTU, Jun'19, 3 Marks)*

Solution:

$$AB'C + B + BD' + ABD' + A'C = B + BD' + ABD' + A'C + AB'C \quad \boxed{1 + x = 1}$$
$$= B(1 + D' + AD') + (A' + AB')C \quad \boxed{x.1 = x}$$
$$= B.1 + ((A' + A)(A' + B'))C \quad \boxed{x + x' = 1}$$
$$= B + (A' + B')C$$
$$= (B + A' + B')(B + C)$$
$$= (1 + A')(B + C)$$
$$= 1.(B + C) = B + C$$

Q4.20 *Using the Boolean algebra, simplify the expression:* *(IKGPTU, May'20, 5 Marks)*

$A'B'C + (A + B + C')' + A'B'C'D$

Solution:

$$A'B'C + (A + B + C')' + A'B'C'D = A'B'C + A'B'C + A'B'C'D \quad \boxed{\text{Using DeMorgan's theorem } (x + y)' = x'.y'}$$
$$= A'B'C + A'B'C'D \quad \boxed{\text{Repeated terms are considered once.}}$$
$$= A'B'C(1 + D) + A'B'C'D \quad \boxed{1 + x = 1}$$
$$= A'B'C + A'B'CD + A'B'C'D$$
$$= A'B'C + A'B'D(C + C') \quad \boxed{x + x' = 1}$$
$$= A'B'C + A'B'D$$

Q4.21 *Show that the dual of the XOR is equal to complement of XOR.* *(JNTUH, Nov/Dec'20, 6 Marks)*

Solution:

$$F = x \oplus y = xy' + x'y$$

Case i: Direct Complement

$$F' = (x \oplus y)' = (xy' + x'y)' \quad \boxed{\text{Using DeMorgan's theorem}}$$
$$= (xy')'(x'y)'$$
$$= (x' + y)(x + y')$$

Case ii: Duality

$x \oplus y = x \,.\, y' + x' \,.\, y$

↓ ↓ ↓ ↓ ↓ ↓ ↓ Replace by dual elements

$(x' + y) \,.\, (x + y')$

⇓

$(x' + y) \,.\, (x + y') = F'$

Q4.22 ***Find the complement of the function: F = x (y'z' + y z) by taking their duals and complementing each literal.***

Solution:

Case i: Direct Evaluation of Complement

$F = x(y'z' + yz)$

⇓ Complement

$F' = (x(y'z' + yz))' = x' + (y'z' + yz)'$

$= x' + ((y'z')'(yz)') = x' + ((y+z)(y'+z')) = (x'+y+z)(x'+y'+z')$

Using DeMorgan's theorem $(x+y)' = x'.y'$ $(x.y)' = x'+y'$

Case ii: Complement using Duality

$x.x' = 0$

$x \,.\, (y' \,.\, z' + y \,.\, z)$

↓ ↓ ↓ ↓ ↓ ↓ ↓ ↓ ↓ Replace by dual elements

$x' + ((y+z) \,.\, (y'+z'))$

⇓

$(x'+y+z').(x'+y'+z') = F'$

Q4.23 ***Draw the logic diagram of the following Boolean expression without simplification.***

B C' + A B + A C D ***(JNTUA, Jun/July'20, 4 Marks)***

Solution:

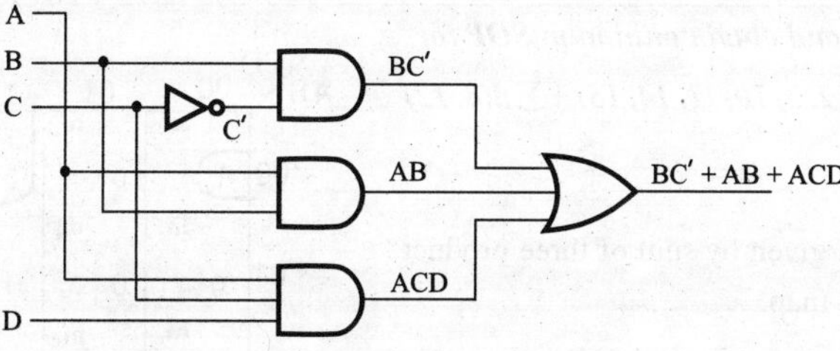

Fig. Q4.23: Logic circuit using basic gates.

Q4.24 ***Draw the multiple-level NAND circuit for the following expression:*** ***(JNTUA, Nov/Dec'20, 7 Marks)***

(AB' + CD')E + BC(A + B)

Solution:

$(AB' + CD')E + BC(A+B) = AB'E + CD'E + ABC + BBC$

$= AB'E + CD'E + ABC + BC$

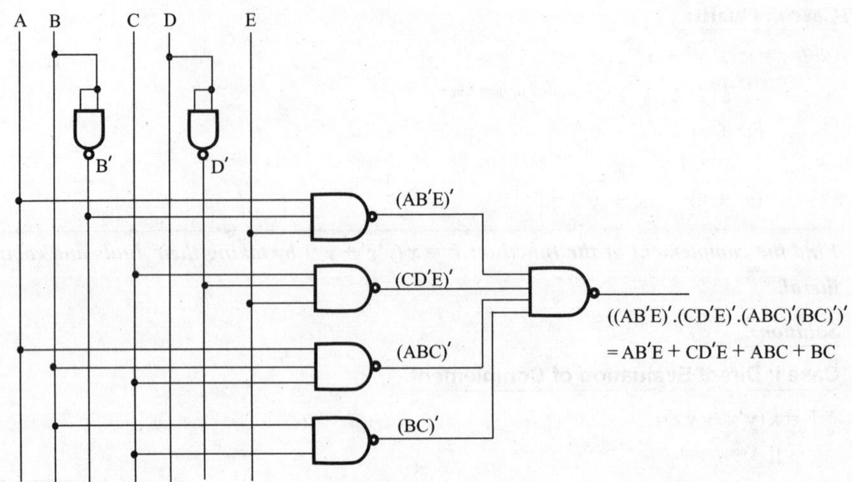

Fig. Q4.24: Logic circuit using multiple level NAND gates.

Q4.25 Draw the logic diagram of the following Boolean expression without simplification.
(A + B) (C + D) (A′ + B + D) ***(JNTUA, Jun/July'20, 4 Marks)***

Solution:

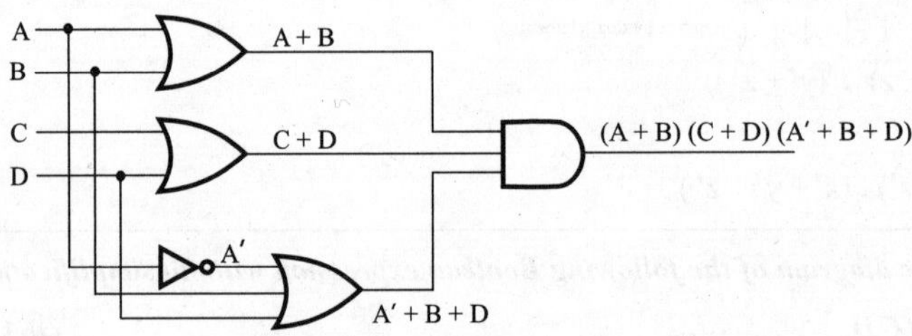

Fig. Q4.25: Logic circuit using basic gates.

Q4.26 Simplify the function and obtain minimum SOP for
F(A, B, C, D) = Σ m(0, 2, 3, 10, 11, 14, 15) + Σ d(8, 12)

Solution:

Minimum SOP form is given by sum of three product terms obtained from K-map.

∴ F = B′D′ + AC + B′C

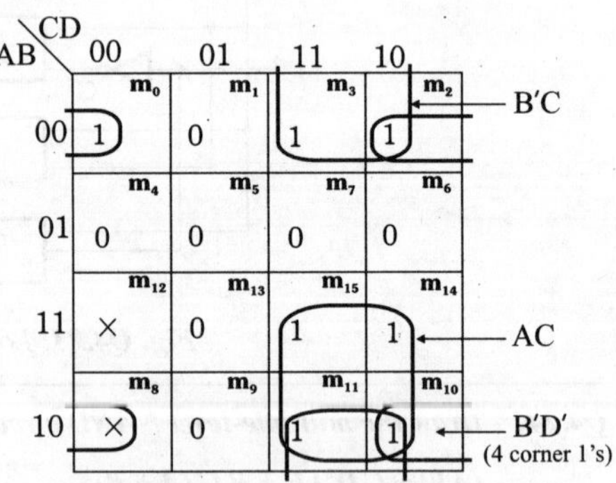

Fig. Q4.26: K-map for F.

Q4.27 Find the minimum POS expression for the function F(A, B, C) = ΠM (2, 6, 7)

Solution:

Minimum POS form is given by complement of sum of two product terms obtained from K-map.

$\therefore$ F = (BC'+AB)'

= (BC')' (AB)' Using DeMorgan's theorem

= (B'+C) (A'+B')

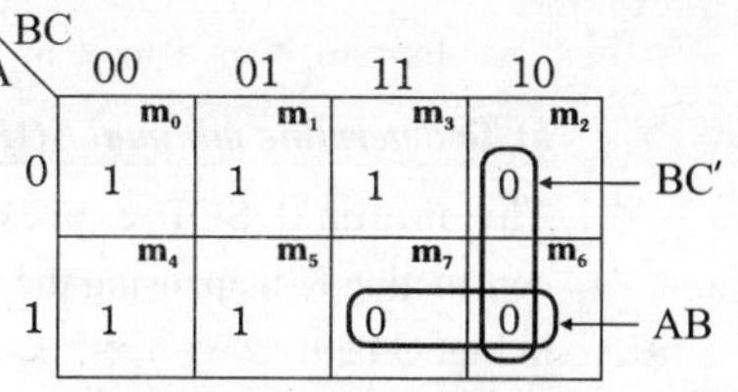

Fig. Q4.27: K-map for F.

Q4.28 Obtain the minimal POS for the function, F(A, B, C, D) = Σ m(2, 3, 8, 9, 12, 13)

Solution:

Minimum POS form is given by complement of sum of three product terms obtained from K-map.

F(A, B, C, D) = (A'B + A'C' + AC)'

= (A'B)'(A'C')'(AC)'

= (A + B') (A + C) (A' + C')

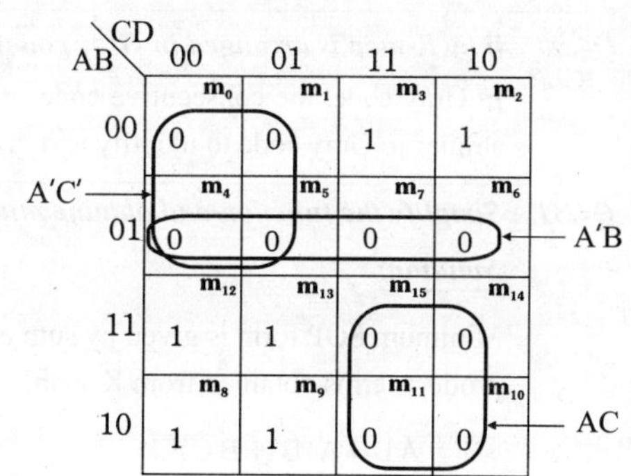

Fig. Q4.28: K-map for F.

Q4.29 Given the following boolean function F = A'C + A'B + AB'C + BC. **(AU, Nov/Dec'18, 13 Marks)**

a) Express it in sum of minterms.

b) Find the minimal sum of products expression.

Solution:

a) To express as sum of minterms

Table 1: Truth Table

Input Variables			Minterm		Complement of Inputs			Product Terms				Function Output
A	**B**	**C**			**A'**	**B'**	**C'**	**A'C**	**A'B**	**AB'C**	**BC**	**F**
0	0	0	m_0	A'B'C'	1	1	1	0	0	0	0	0
0	0	1	m_1	A'B'C	1	1	0	1	0	0	0	1
0	1	0	m_2	A'BC'	1	0	1	0	1	0	0	1
0	1	1	m_3	A'BC	1	0	0	1	1	0	1	1
1	0	0	m_4	AB'C'	0	1	1	0	0	0	0	0
1	0	1	m_5	AB'C	0	1	0	0	0	1	0	1
1	1	0	m_6	ABC'	0	0	1	0	0	0	0	0
1	1	1	m_7	ABC	0	0	0	0	0	0	1	1

A function can be expressed as a sum of minterms for which the function output is 1. Here, F is 1, when m_1, m_2, m_3, m_5 and m_7 are inputs.

$$\therefore\ F = m_1 + m_2 + m_3 + m_5 + m_7 = A'B'C + A'BC' + A'BC + AB'C + ABC$$

b) To determine minimal SOP

The minimal SOP expression can be obtained by constucting K-map, using the truth table (Table 1) as shown in Fig. Q4.29.

The minimal sum of products expression is given by,

$$F = A'B + C$$

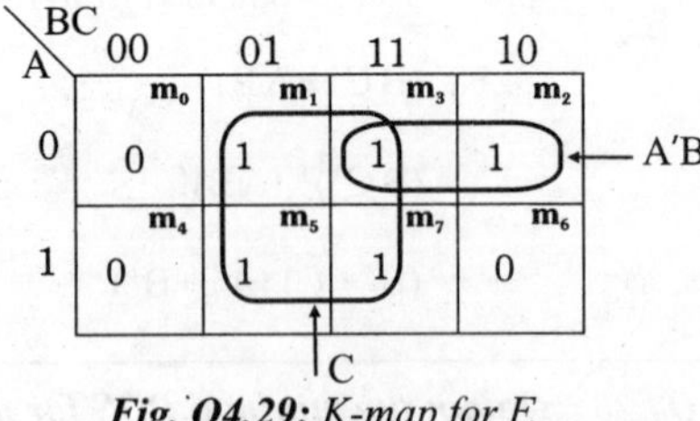

Fig. Q4.29: K-map for F.

Q4.30 *Why K-map is arranged in Gray code?* ***(CSVTU, Nov/Dec'19, 2 Marks)***

In Gray code, the consecutive codes differ only in value of one bit. In K-map, the minterms are arranged similar to Gray code to identify and group the minterms that differ only by one variable.

Q4.31 *Simplify the function and obtaine minimum SOP for F(A, B, C) = Σm(0, 1, 2, 6, 7)*

Solution:

Minimum SOP form is given by sum of three product terms obtained from K-map.

$$F = AB + A'B' + BC'$$

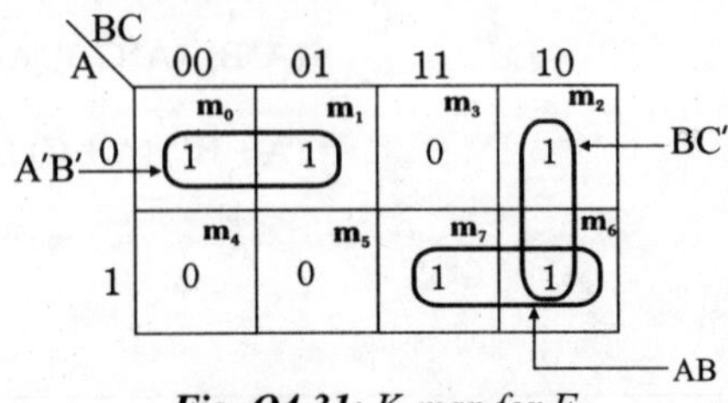

Fig. Q4.31: K-map for F.

Q4.32 *Minimize the following Boolean function F(A, B, C) = Σm(0, 1, 6, 7) + Σd(3, 5)*

Solution:

Minimized Boolean expression is,

$$F = AB + A'B'$$

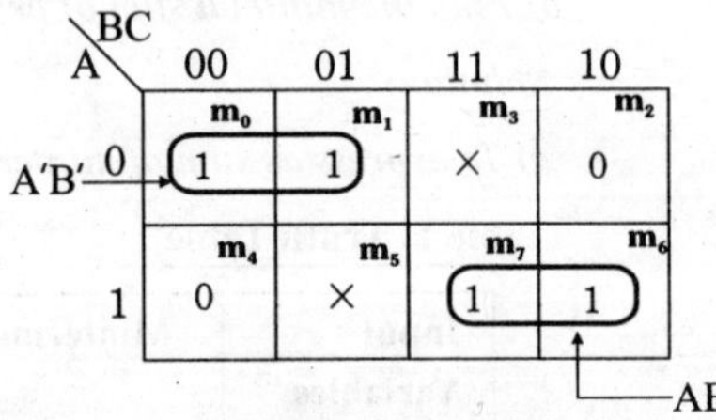

Fig. Q4.32: K-map for F.

Q4.33 *Obtain the minimum POS for the function F(A, B, C) = Σm(0, 2, 5, 7)*

Solution:

Minimum POS form is given by complement of sum of two product terms obtained from K-map.

$$F = (AC' + A'C)'$$

$$= (AC')'(A'C)'$$

$$= (A' + C)(A + C')$$ Using DeMorgan's theorem

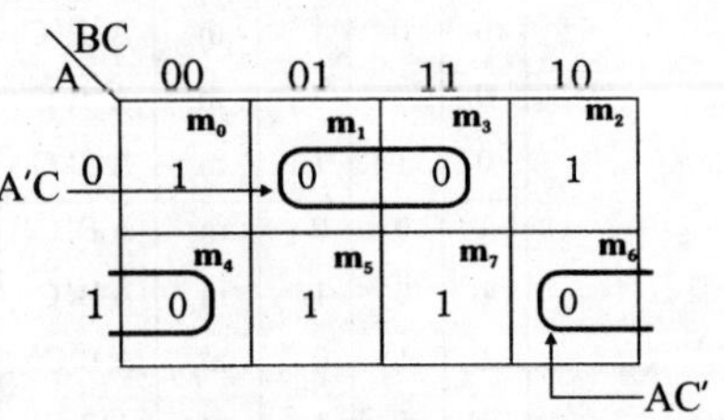

Fig. Q4.33: K-map for F.

Q4.34 *Simplify the following function using 2-variable K-map F(A, B, C) = Σm(0, 1, 3, 5, 7).*

a) Sum-of-products ***b) Product-of-sums***

Solution:

a) Sum-of-products

$$F = C + A'B'$$

b) Product-of-sums

$$F = (AC' + BC')'$$

$$= (AC')'(BC')' \quad \text{Using DeMorgan's theorem}$$

$$= (A' + C) + (B' + C)$$

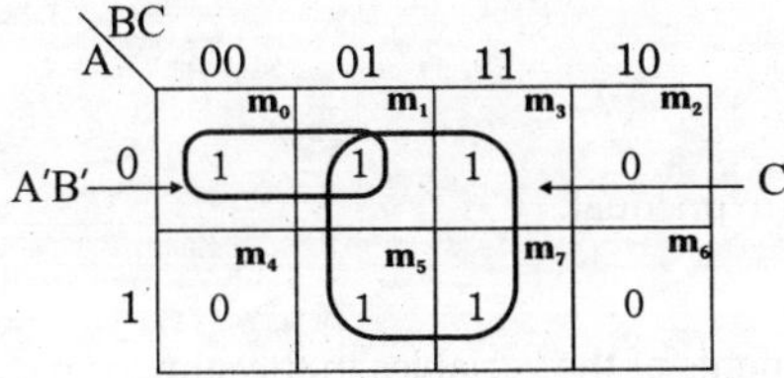

Fig. Q4.34a: K-map for F.

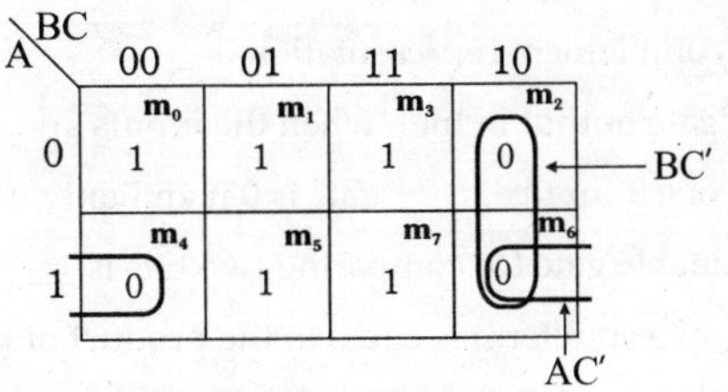

Fig. Q4.34b: K-map for F.

Q4.35 *Draw the two-variable K-map with single prime implicant formed using single 1's.*

Solution:

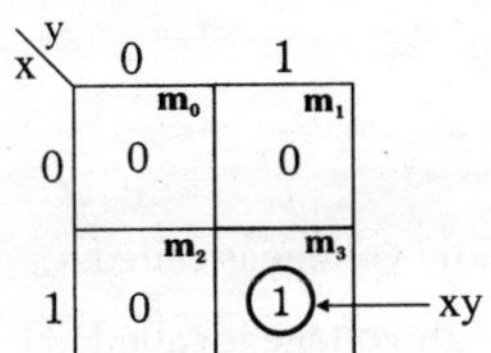

Fig. Q4.35a: K-map for F

$$F = xy$$

Fig. Q4.35b: K-map for F

$$F = x'y'$$

Q4.36 *Draw the two-variable K-map with two prime implicants formed using single 1's.*

Solution:

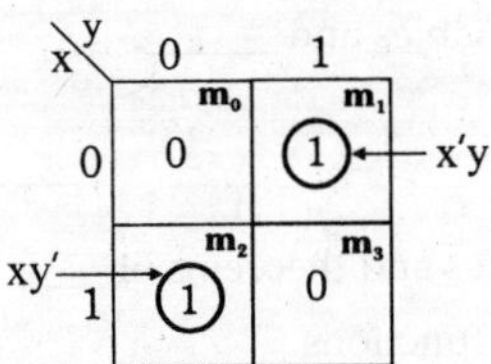

Fig. Q4.36a: K-map for F

$$F = x'y + xy'$$

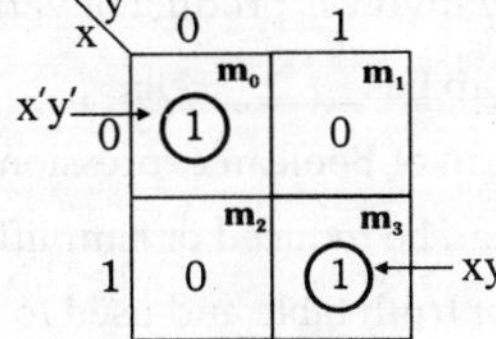

Fig. Q4.36b: K-map for F

$$F = xy + x'y'$$

4.10 Exercises

I. Fill in the blanks

1. The two voltage levels of logic circuits are called ______ and ______.
2. NAND and NOR gates are called ______ .
3. The realizations in SOP and POS forms are called ______ .
4. A gate in which any one input must be low to get a high output is called ______.
5. A gate in which any one input must be high to get a low output is called ______.
6. Binary ______ can be performed via addition by using the 1's or 2's complement method.
7. -19_{10} in 2's complement representation is ______.
8. The ______ gate output in high when the inputs are not the same.
9. The output of a 2-input ______ gate is 0 if and only if its inputs are unequal.
10. The most suitable gate for comparing two bits is ______ .
11. The ______ of a sum term is equal to the product of the complements of the variables in the sum term.
12. The operational symbol for exclusive-OR operation is ______.
13. AND-OR logic produces the output of an expression in ______ form.
14. To implement the expression AB'C + AC', it takes ______ OR gate(s), ______ AND gate(s) and ______ inverter(s).
15. The AND operation can be produced with ______ NAND gates.
16. Binary logic consist of binary variables and a set of ______.
17. NOT operation is same as ______.
18. The logic system in which '1' represent high voltage and '0' represent low voltage is called ______.
19. The logic system in which '1' represent low voltage and '0' represent high voltage is called ______.
20. Boolean function can be realized using only ______ or ______ gates.
21. Logic gate device are available in the form of IC's manufactured using ______.
22. The realizations in SOP and POS standard form are called ______.
23. The Two-level realization of SOP form of a Boolean function can be obtained using only ______.
24. Two-level realization of POS form of a Boolean function can be obtained using only ______.
25. Boolean function basically involves product of variables, sum of variables and ______.
26. A 3-variable Karnaugh map has ______ cells.
27. Each variable with in a term of Boolean expression is called ______.
28. The Boolean Expression can be reduced or simplified using postulates and theorems of ______.
29. ______ is a pictorial term of truth table and used to simplify Boolean functions.
30. The literal of minterm are split and arranged as ______ and ______.
31. Prime implicant with single 1 represent ______ literal product term.
32. Prime implicant with two 1's represent ______ literal product term.
33. Prime implicant with four 1's represent ______ literal product term.
34. ______ element cannot be combined to form prime implicants.
35. All minterms are 1's then all squares of K-map will be filled by 1 and function value is ______.

36. Two-variable, K-map will have ______ squares.
37. Three-variable, K-map will have ______ squares.
38. Four-variable, K-map will have ______ square.
39. Four-variable, K-map ______ minterms are possible.
40. n-variable Boolean function there are ______ possible combinations of inputs.

Answers

1. high, low	11. complement	21. semiconductors	31. n literal
2. universal gates	12. ⊕	22. two-level realization form	32. n−1 literal
3. standard form	13. SOP	23. NAND gates	33. n−2 literal
4. NAND gate	14. 1, 2, 2	24. NOR gates	34. diagonal
5. NOR gate	15. 2	25. complement of variables	35. 1
6. subtraction	16. logical operations	26. 8	36. 4
7. 11101101 (Standard 8-bit)	17. complement operation	27. literal	37. 8
8. exclusive-OR	18. positive logic	28. Boolean algebra	38. 16
9. XNOR	19. negative logic	29. K-map	39. 32
10. XOR	20. NAND, NOR	30. rows, colums	40. 2n

II. State whether the following statements are True or False

1. The output of NOR gate is high if and only if all its inputs are low.
2. AND gate can be used as an inverter.
3. $A + A'B = A + B$.
4. The complement of a product term is equal to the sum of the complement of the variables in the product term.
5. An example of a product-of-sums expression is $A(B + C) + AC'$.
6. The expression $A'BCD + ABCD' + AB'C'D$ cannot be simplified.
7. An inverter performs an operation known as complement.
8. Boolean algebra is an algebraic structure.
9. $(X + Y)' = X' + Y'$.
10. The basic logic gates are AND, OR and NOT gates.
11. The NAND gate is a combination of OR followed by NOT gate.
12. Logic gates can also be designed to work with negative logic levels.
13. Binary variables are denoted by either lower case or upper case alphabets.
14. Boolean function can't be evaluated for all possible combinations of binary values of the variables of the function.
15. Minterms are 2^n possible combinations of AND terms with n variables such that the logical AND of all the variables is 0.
16. On a Karnaugh map grouping the 0's produces AND-OR logic.
17. The Boolean expression can be reduced or simplified using postulates and theorems of Boolean algebra.
18. There are one methods have standard procedure for simplification of Boolean functions.

19. Maxterm can be used to construct K-map in which each squares represent a minterm.
20. Arranging the literal/minterms only one change is allowed if we move from one row to next row or from one column to next column.
21. While forming prime implicants any number of overlapping is allowed in horizontal and vertical directions.
22. The simplified Boolean function is sum of all the product terms of prime implicant
23. If all minterms are 1's then all squares of K-map will be filled by 1 and function value is 0.
24. Reducing the number of literals in a Boolean expression or function will simplify the implementation of function by logic gates with minimum number of gates.
25. The simplified Boolean function using K-map is not unique. Sometimes there may be multiple solutions.

Answers

1. True	6. True	11. False	16. False	21. True
2. False	7. True	12. True	17. True	22. True
3. True	8. True	13. True	18. False	23. False
4. True	9. False	14. False	19. False	24. True
5. False	10. True	15. False	20. True	25. True

III. Choose the right answer for the following questions

1. ***The Boolean expression A'B' + (A B)' + A B is equivalent to***

   a) A' + B'A' b) A'B

   c) A B + (AB)' d) A B

2. ***The Boolean expression A ⊕ B ⊕ B is equivalent to***

   a) A B + A'B' b) A'B

   c) A d) B

3. ***In the logic circuit shown in Fig. 3, the output F is***

   a) xy' + yz' + zx'

   b) x + z

   c) xy + yz + zx

   d) (xy)' + (yz)' + (zx)'

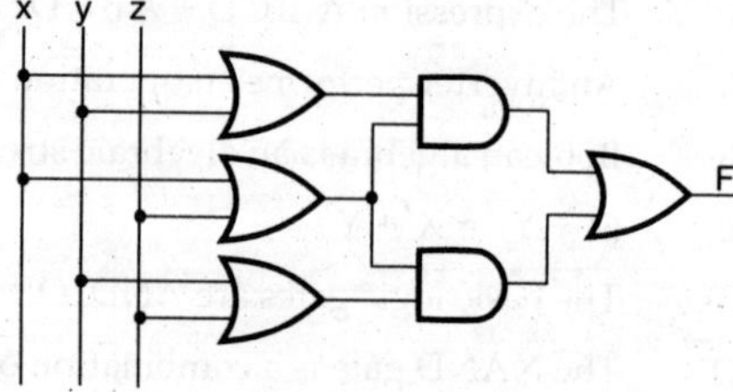

Fig. 3.

4. ***An XOR gate with 6-variables is given by A⊕B⊕C⊕D⊕E⊕F. The number of minterms in the Boolean expression***

   a) 6 b) 12 c) 64 d) 32

5. ***Which of the following equation is equivalent to NOT***

   a) X NAND X b) X NOR X' c) X NAND 1 d) X NOR 1

6. ***The Boolean function Y = AB + CD is to be realized using only 2-input NAND gates. The minimum number of gates required is***

   a) 2 b) 3 c) 4 d) 5

7. ***Y = AB'C' + A'BC' is equivalent to***

a) Y = A b) Y = AB c) Y = AC d) Y = C'(A ⊕ B)

8. ***The circuit shown in Fig. 8 realizes the function***

a) [(A' + B') + C] . (D' + E')

b) (A + B) C + D + E

c) AB + C + DE

d) AB + C(D + E)

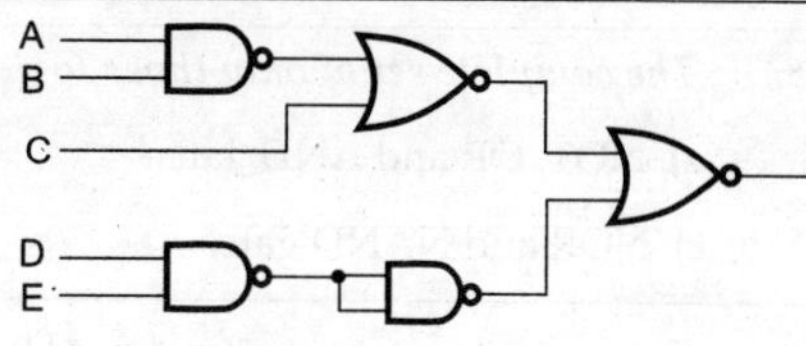

Fig. 8.

9. ***The output Y of the given function in Fig. 9 represents***

a) OR b) AND

c) NAND d) NOR

Fig. 9.

10. ***Which of the following is equivalent of AND-OR realization***

a) NAND-NOR realization b) NOR-NOR realization

c) NOR-NAND realization d) NAND-NAND realization

11. ***The number of minterms in the function f(x, y, z) = xy' + z' is***

a) 4 b) 5 c) 3 d) 6

12. ***If X = 1 in the logic equation, (X + Z (Y' + (Z' + XY'))) (X' + Z'(X + Y)) = 1, then***

a) Y = Z b) Y = Z' c) Z = 1 d) Z = 0

13. ***The circuit shown in Fig. 13 realises the function***

a) (A' + B').C + (DE)'

b) (A + B) (C + D + E)

c) AB + C + DE

d) AB + C(D + E)

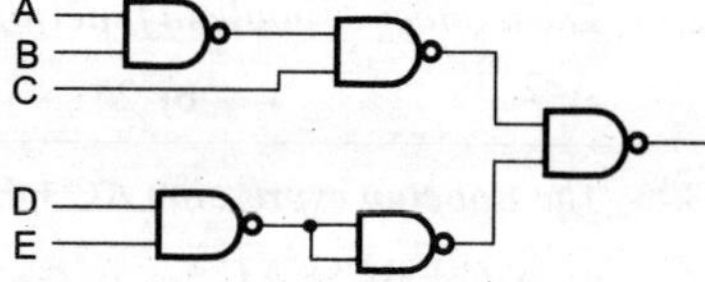

Fig. 13.

14. ***In the Fig. 14 X_0, X_1, X_2 will be 1's complement of ABC if***

a) Y = 1

b) Y = 0

c) Y = A' = B' = C'

d) Y = A = B = C

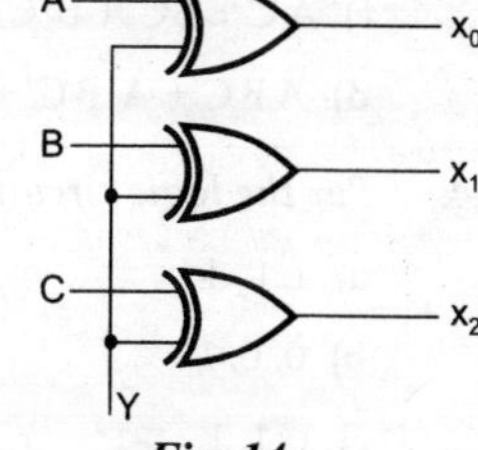

Fig. 14.

15. ***The circuit in Fig. 15 works as***

a) XOR

b) XNOR

c) NOR

d) NAND

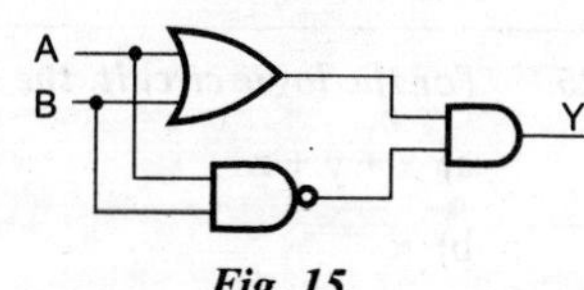

Fig. 15.

16. ***The Boolean function A + BC is a reduced form of***

a) AB + BC b) (A + B) (A + C) c) A'B + AB'C d) (A' + C) B

17. *The number of distinct Boolean expression of 4 variables is*

a) 16 b) 256 c) 1024 d) 65536

18. *The complete set of only those logic gates designated as universal gates is*

a) NOT, OR and AND gates b) XNOR, NOR and NAND gates

c) NOR and NAND gates d) XOR, NOR and NAND gates

19. *The logical expression Y = A + A'B is equivalent to*

a) Y = AB + A b) Y = A'B + A'

c) Y = AB' + B d) Y = AB + A'

20. *The minimized form of logical expression Y = A'B'C' + A'B C' + A'B C + ABC' is*

a) A'C' + BC' + A'B b) AC' + B'C + A'B

c) A'C + B'C + A'B d) AC' + B'C + AB'

21. *For the identity AB + AC + BC = AB + AC the dual form is*

a) (A + B) (A' + C) (B + C) = (A + B) (A' + C)

b) (A' + B') (A' + C') (B' + C') = (A' + B') (A' + C')

c) (A + B) (A' + C) (B + C) = (A' + B') (A + C')

d) A'B' + AC' + B'C' = A'B' + AC'

22. *The number of boolean functions that can be generated by n variable is equal to*

a) 2^{2^n} b) 2^{2n} c) 2^{n-1} d) 2^n

23. *The Booelan expression AC + BC' is equivalent to*

a) A'C + BC' + AC

b) B'C + AC + BC' + A'CB'

c) AC + BC' + B'C + ABC

d) ABC + A'BC' + ABC' + AB'C

24. *For the logic circuit in Fig. 24 the required input condition (x, y, z) to make the output F = 1 is*

a) 1, 0, 1

b) 0, 0, 1

c) 1, 1, 1

d) 0, 1, 1

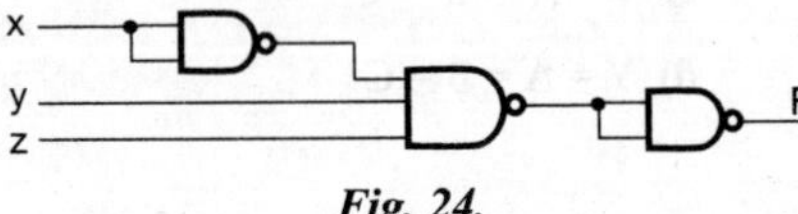

Fig. 24.

25. *For the logic circuit, the simplified Boolean expression for the output F is*

a) x + y + z

b) x

c) y

d) z

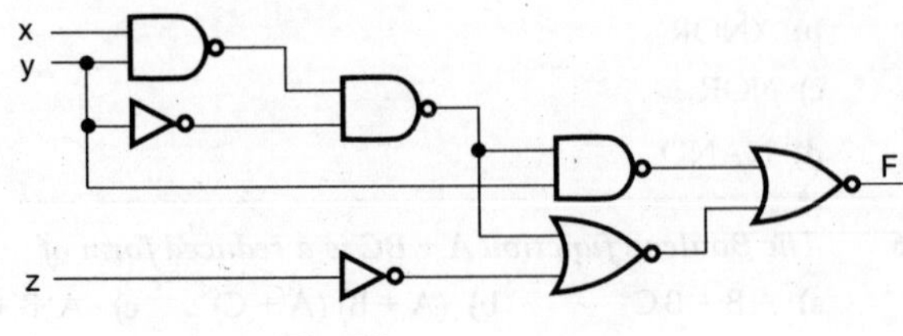

Fig. 25.

26. ***The minimum number of 2 input NAND gates required to implement the Boolean function AB'C assuming that A, B and C are available is***

a) 2 b) 3 c) 5 d) 6

27. ***For the logic circuit shown in Fig. 27 the output F is***

a) (xyz)'

b) x'+y'+z'

c) (xy)'+(yz)'+x'+z'

d) x'+z'

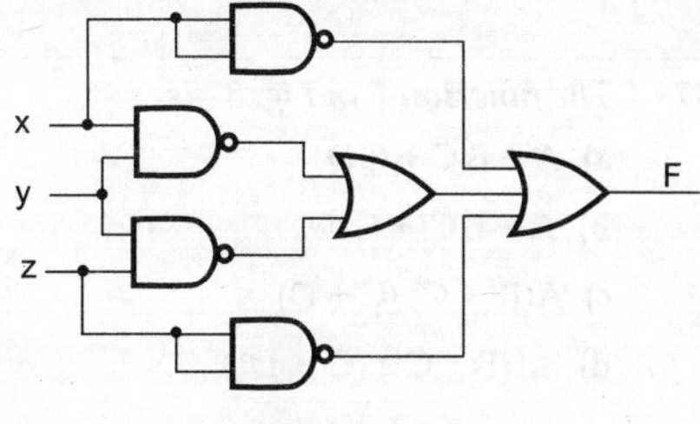

Fig. 27.

28. ***Boolean expression for output of XNOR equivalent logic gate in dual form with inputs A and B is***

a) AB'+A'B b) (AB)'+AB

c) (A'+B)(A+B) d) (A'+B')(A+B)

29. ***The output of a logic gate is 1 when all its inputs are at logic 0. Then gate is either***

a) a NAND or an XOR gate b) a NOR or an XNOR gate

c) an OR or an XNOR gate d) an AND or an XOR gate

30. ***The minimum number of NAND gates required to implement the boolean function A + AB' + AB'C is equal to***

a) 0 b) 1 c) 4 d) 7

31. ***The output of the logic gate in Fig. 31 is***

a) 0 b) 1 c) A d) A'

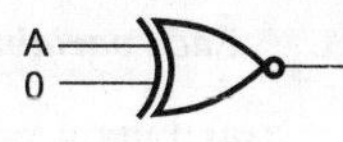

Fig. 31.

32. ***What is output Z of an XOR gate whose all inputs are set at A***

a) Z = A b) Z = 1 c) Z = 0 d) Z = A

33. ***In the circuit shown the value of input for Y = 0 is***

a) 00 b) 01 c) 10 d) 11

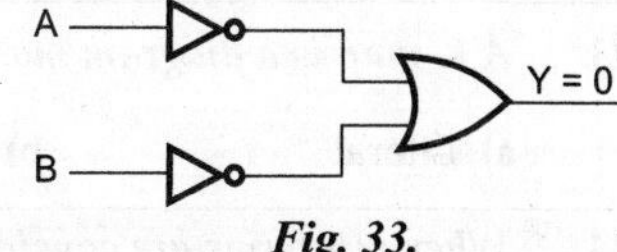

Fig. 33.

34. ***The simplified form of the boolean expression Y = (A'BC + D) (A'D + B'C') can be written as***

a) A'D + B'C'D b) AD + BC'D

c) (A'+D)(B'C+D') d) AD'+BCD'

35. ***The boolean expression x'yz' + (xy)'z + xyz' + xy'z + xyz can be simplified to***

a) x'+z b) y+z

c) y'+z' d) y'+z

36. ***The output F in Fig. 36 is***

a) (A' + B') b) AB

c) (A'. B') d) A + B

Fig. 36.

37. ***The function F in Fig. 37 is***

a) A + BC + CD

b) A + BC' + C'D'

c) A(B' + C) (C + D)

d) A'(B + C') (C' + D')

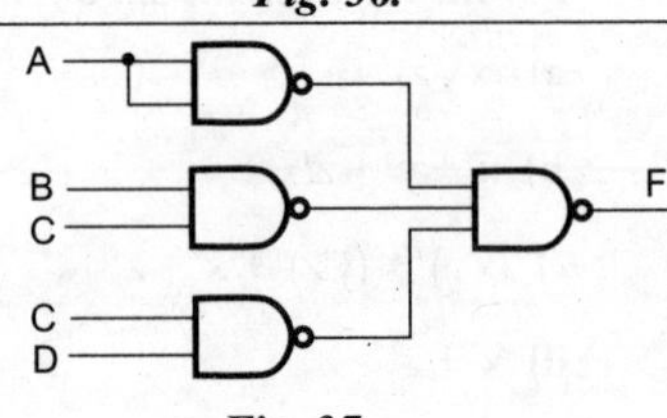

Fig. 37.

38. ***The dual of the function is A(B'C + BC + BC')***

a) A' + (B + C') (B' + C') (B' + C) b) A'(BC' + B'C' + B'C)

c) C (B'A + CA + CA') d) AB'C + ABC + ABC'

39. ***The expression (A + A'B) (B' + BA) can be simplified to***

a) AB' + A'B b) AB' + A'B' c) 1 d) A

40. ***The circuit in Fig. 40 is works as***

a) XOR b) XNOR

c) NOR d) NAND

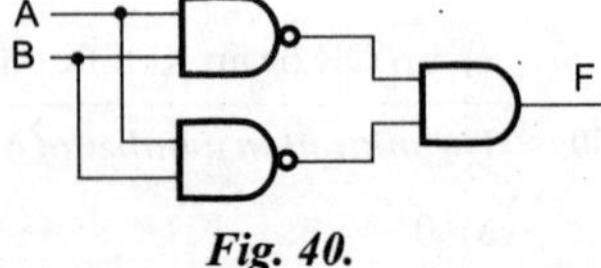

Fig. 40.

41. ***Each variable within a term of a Boolean expression is called***

a) Literal b) Minterm c) Maxterm d) Prime implicant

42. ***A prime implicant with single 1 represent***

a) n literal product term b) n−1 literal product term

c) n−2 literal product term d) n−3 literal product term

43. ***A K-map is a diagram made up of squares with each square representing***

a) Literal b) Minterm c) Maxterm d) Prime implicant

44. ***When minterms are considered for simplification using K-map the resultant Boolean function will be in***

a) Sum-of-product form b) Product-of-sum form

c) MSOP form d) Prime implicants

45. ***For n-variable Boolean function the K-map will be***

a) 2n square b) 2^n square c) n square d) 2^{2n} square

46. ***A prime implicant with two 1's represent***

a) n literal product term b) n − 1 literal product term

c) n − 2 literal product term d) n − 3 literal product term

47. ***While forming prime implicants any number of overlapping is allowed in***

a) Horizontal directions **b)** Vertical directions

c) Both (a) and (b) **d)** None of the above

---

48 ***Undefined function outputs are called***

a) Literal **b)** Minterm **c)** Maxterm **d)** Don't-care

---

49. ***The expansion phase is the first phase in which minterms are obtained from truth table are prime implicants are identified after eliminating***

a) Literal **b)** $n-1$ literal **c)** $n-2$ literal **d)** Redundant literals

---

50. ***The literals of minterm are split and arranged***

a) Only rows **b)** Only columns

c) Both (a) and (b) **d)** None of the above

---

51. ***The K-map for a Boolean function is given. The number of essential prime implicants for this function is***

a) 4

b) 5

c) 6

d) 8

AB \ CD	00	01	11	10
00	m_0 1	m_1 1	m_3 0	m_2 1
01	m_4 0	m_5 0	m_7 0	m_6 1
11	m_{12} 1	m_{13} 0	m_{15} 0	m_{14} 0
10	m_8 1	m_9 0	m_{11} 0	m_{10} 1

Fig. 51.

---

52. ***The minimum POS expression for K-map given in the Fig. 52 is***

a) (B + C) (C' + D)

b) (B + D) (A + C)

c) (A' + C') (C + D')

d) (B + D) (A + D')

AB \ CD	00	01	11	10
00	m_0 0	m_1 0	m_3 1	m_2 0
01	m_4 1	m_5 1	m_7 1	m_6 0
11	m_{12} 1	m_{13} 1	m_{15} 1	m_{14} 0
10	m_8 0	m_9 0	m_{11} 1	m_{10} 0

Fig. 52.

---

53. ***While obtaining minimal SOP expression***

a) All don't-cares are ignored

b) All don't-cares are treated as logic ones

c) All don't-cares are treated as logic zeros

d) Only such don't-cares that aid minimization are treated as logic ones

Answers

1. c	11. b	21. b	31. d	41. a	51. a
2. c	12. d	22. b	32. c	42. a	52. a
3. b	13. a	23. d	33. d	43. b	53. d
4. a	14. a	24. d	34. a	44. a	
5. a	15. a	25. c	35. b	45. b	
6. b	16. b	26. c	36. b	46. b	
7. d	17. b	27. b	37. a	47. c	
8. a	18. c	28. d	38. a	48. d	
9. a	19. c	29. b	39. d	49. d	
10. d	20. a	30 a	40. d	50. c	

IV. Answer the following questions

1. Explain absorption theorem with truth table.
2. Which are universal gates, why are they called so?
3. What are minterms and maxterms?
4. What is the difference between canonical and SOP/POS form?
5. Explain the operator precedence while evaluating the Boolean expression with an example.
6. What is truth table. Give an example.
7. What is duality?
8. What are postulates of Boolean algebra. Give two examples.
9. What is DeMorgan's theorem.
10. What is positive and negative logic?
11. How will be the resultant function of K-map when minterm and maxterm are used?
12. How do we handle don't-care conditions?
13. What are the methods available to simplify a Boolean expression?
14. Explain K-map.
15. Define expansion phase and covering phase.

V. Solve the following problems

E4.1 Find the complement of the following expression:

a) $x'y + x'y'$ **b)** $xyz + x'y + xyz'$

E4.2 Given two Boolean functions F_1 and F_2, show that (a) the Boolean function $E = F_1 + F_2$ contains the sum of the minterms of F_1 and F_2. (b) The Boolean function $G = F_1 F_2$ contains only the minterms that are common to F_1 and F_2.

E4.3 List the truth table of the function:

a) $F_1 = x'y' + x'y + yz'$ **b)** $F_2 = b'c' + ac$

E4.4 Demonstrate the validity of the following identities by means of truth tables:

DeMorgan's theorem for 3 variables: a) $(x + y + z)' = x'y'z'$ b) $(xyz)' = x' + y' + z'$

The associate law: c) $x + (y + z) = (x + y) + z$ d) $x \cdot (y \cdot z) = (x \cdot y) \cdot z$

E4.5 Express the complement of the following functions in sum of minterms form,

a) $F(A, B, C, D) = \sum m(3, 5, 9, 11, 15)$ b) $F(x, y, z) = \prod M(2, 4, 5, 7)$

E4.6 Convert each of the following expression into sum of products and product of sums.

a) $(AB + C)(B + C'D)$ b) $x + x'(x' + y')(y' + z')$

E4.7 Show that the complement of XOR is equal to XNOR.

E4.8 Determine the complement of Boolean expression $(b + d)(a' + b' + c)$ and show that POS becomes SOP.

E4.9 Determine the complement of Boolean expression $a'b + a'c' + abc$ and show that SOP becomes POS.

E4.10 Implement the Boolean function, $F = xy + x'y + y'z'$

a) With AND, OR and NOT gates. b) With OR and NOT gates.

c) With AND and NOT gates. d) With NAND and NOT gate.

e) With NOR and NOT gates.

E4.11 Obtain the truth table of the following functions and express each function in sum of minterms and product of maxterms form:

a) $(xy + z)(y + xz)$ b) $(x + y')(y' + z)$ c) $x'z + wx'y + wyz' + w'y'$ d) $(xy + yz' + x'z)(x + z)$

E4.12 For the Boolean function, $F = x'yz + xyz' + wx'y' + w'xy + wxy$

a) Obtain the truth table of F.

b) Draw the logic diagram, using the original Boolean expressions.

c) Use Boolean algebra to simplify the function to a minimum number of literals.

d) Obtain the truth table of the function from the simplified expression and show that it is the same as the one in part (a).

e) Draw the logic diagram from the simplified expression and compare the total number of gates with the diagram in part (b).

E4.13 Simplify the following Boolean functions using 3-variable K-maps:

a) $F_1(x, y, z) = \sum m(0, 1, 5, 7)$ b) $F_2(x, y, z) = \sum m(2, 3, 4, 5)$

E4.14 Simplify the following Boolean functions using K-maps:

a) $F_1(x, y, z) = \sum m(2, 3, 6, 7)$ b) $F_2(A, B, C, D) = \sum m(4, 6, 7, 15)$

E4.15 Simplify the following Boolean functions using 4-variable K-map:

a) $F_1(w, x, y, z) = \sum m(1, 4, 5, 6, 12, 14, 15)$

b) $F_2(A, B, C, D) = \sum m(2, 3, 6, 7, 12, 13, 14)$

E4.16 Simplify the following Boolean functions using 3-variable K-maps:

a) $F_1 = x'y' + yz' + x'yz'$ b) $F_2 = x'yz + xy'z' + xy'z$

E4.17 Simplify the Boolean function using 4-variable K-maps: $F = A'B'C'D' + AC'D' + B'CD' + A'BCD + BC'D$

E4.18 Find the minterms of the following Boolean expression by first plotting each function in a K-map:

a) $F_1 = xy + yz + xy'z$ **b)** $F_2 = C'D + ABC' + ABD' + A'B'D$

E4.19 Using K-map find the minimum sum of products (MSOP) representation for,

F(A, B, C, D, E) = Σ m(2, 5, 7, 11, 21, 23, 25, 27) + Σ d(1, 12, 17, 28)

Draw the logic circuit of the minimal expression using only NAND gates.

E4.20 Using K-map find the minimum product of sum (MPOS) representation for,

F(A, B, C, D, E) = Π(0, 1, 2, 3, 4, 5, 16, 17, 18, 19, 24, 25) + Σ d(26, 27)

Draw the logic circuit of the minimal expression using only NOR gates.

E4.21 Find the prime implicants for the following Boolean function and determine the essential prime implicants.

F(A, B, C, D) = Σ m(0, 2, 3, 5, 7, 8, 10, 11, 14, 15)

E4.22 Implement the Boolean function with NAND gate and draw the logic diagram.

a) F(x, y, z) = (1, 2, 3, 4, 5, 7) **b) F(x, y, z) = (0, 1, 3, 5, 6, 7)**

Answers

E4.1 a) $(x + y')(x + y)$ **b)** $(x' + y' + z')(x + y')(x' + y' + z)$

E4.3 a)

Input Variables			Complement of Inputs			Product Terms			Function Output
x	y	z	x'	y'	z'	x'y'	x'y	yz'	F_1
0	0	0	1	1	1	1	0	0	1
0	0	1	1	1	0	1	0	0	1
0	1	0	1	0	1	0	1	1	1
0	1	1	1	0	0	0	1	0	1
1	0	0	0	1	1	0	0	0	0
1	0	1	0	1	0	0	0	0	0
1	1	0	0	0	1	0	0	1	1
1	1	1	0	0	0	0	0	0	0

b)

Input Variables			Complement of Inputs			Product Terms		Function Output
a	b	c	a'	b'	c'	b'c'	ac	F_2
0	0	0	1	1	1	1	0	1
0	0	1	1	1	0	0	0	0
0	1	0	1	0	1	0	0	0
0	1	1	1	0	0	0	0	0
1	0	0	0	1	1	1	0	1
1	0	1	0	1	0	0	1	1
1	1	0	0	0	1	0	0	0
1	1	1	0	0	0	0	1	1

E4.4a)

x	y	z	x + y + z	(x + y + z)'	x'	y'	z'	x'y'z'
0	0	0	0	1	1	1	1	1
0	0	1	1	0	1	1	0	0
0	1	0	1	0	1	0	1	0
0	1	1	1	0	1	0	0	0
1	0	0	1	0	0	1	1	0
1	0	1	1	0	0	1	0	0
1	1	0	1	0	0	0	1	0
1	1	1	1	0	0	0	0	0

b)

x	y	z	x y z	(x y z)'	x'	y'	z'	x' + y' + z'
0	0	0	0	1	1	1	1	1
0	0	1	0	1	1	1	0	1
0	1	0	0	1	1	0	1	1
0	1	1	0	1	1	0	0	1
1	0	0	0	1	0	1	1	1
1	0	1	0	1	0	1	0	1
1	1	0	0	1	0	0	1	1
1	1	1	1	0	0	0	0	0

c)

x	y	z	y + z	x + (y + z)	x + y	(x + y) + z
0	0	0	0	0	0	0
0	0	1	1	1	0	1
0	1	0	1	1	1	1
0	1	1	1	1	1	1
1	0	0	0	1	1	1
1	0	1	1	1	1	1
1	1	0	1	1	1	1
1	1	1	1	1	1	1

d)

x	y	z	y . z	x . (y . z)	x . y	(x . y) . z
0	0	0	0	0	0	0
0	0	1	0	0	0	0
0	1	0	0	0	0	0
0	1	1	1	0	0	0
1	0	0	0	0	0	0
1	0	1	0	0	0	0
1	1	0	0	0	1	0
1	1	1	1	1	1	1

E4.5 **a)** $\sum m(0, 1, 2, 4, 6, 7, 8, 10, 12, 13, 14)$ **b)** $\sum m(2, 4, 5, 7)$

E4.6 **a)** $AB + BC$ (SOP), $B(A + C)$ (POS) **b)** $x + x'y' + x'z'$ (SOP), $x + y' + z'$ (POS)

E4.8 $b'd' + abc'$ (Sum of products)

E4.9 $(a + b')(a + c)(a' + b' + c')$ (Product of sums)

E4.10 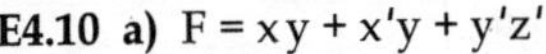
a) $F = xy + x'y + y'z'$

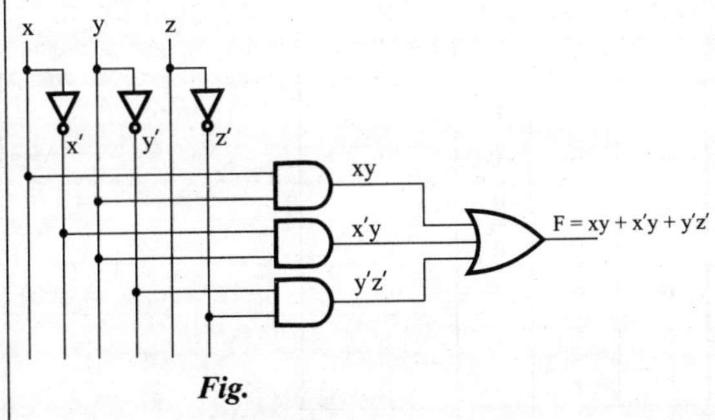

Fig.

b) $F = (x' + y')' + (x + y')' + (y + z)'$

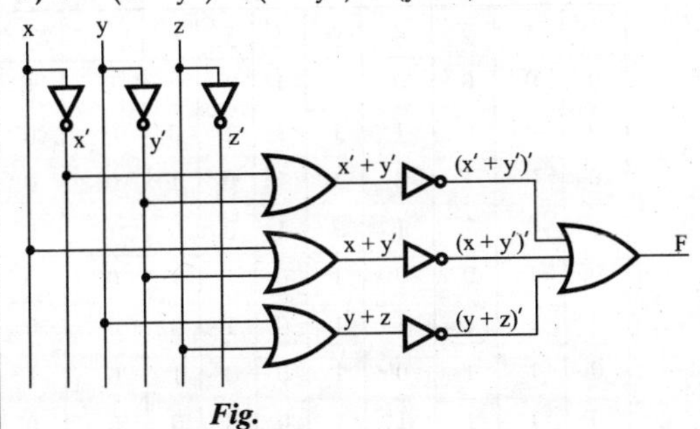

Fig.

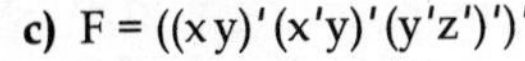
c) $F = ((xy)'(x'y)'(y'z')')'$

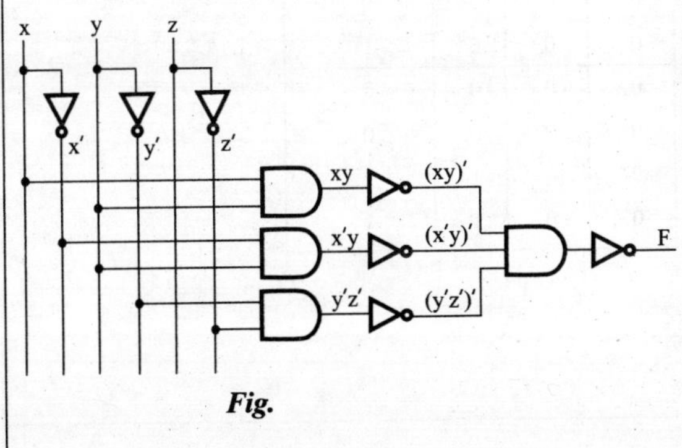

Fig.

d) $F = ((xy)'(x'y)'(y'z')')'$

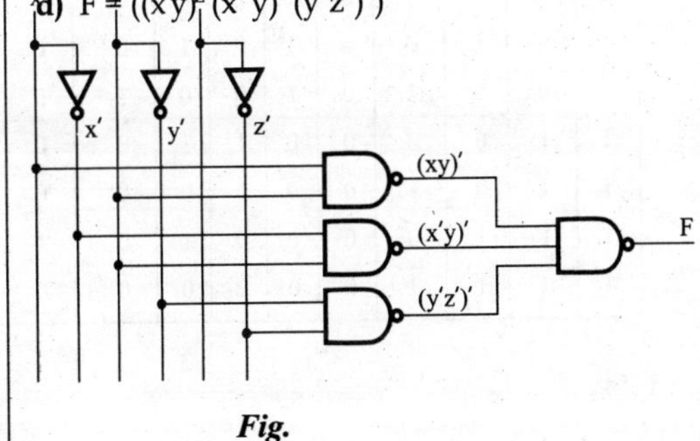

Fig.

e) $F = (x' + y')' + (x + y')' + (y + z)'$

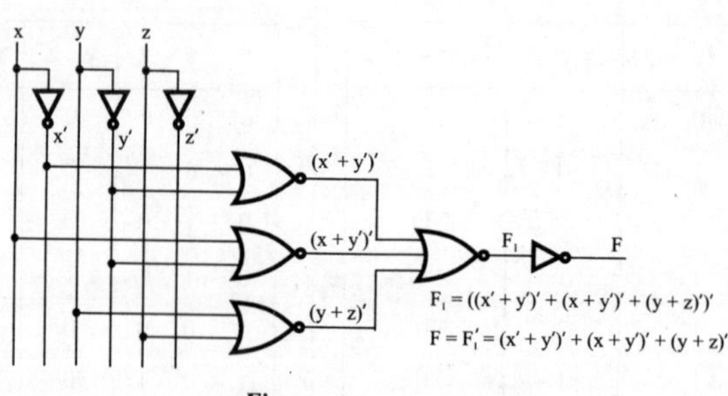

Fig.

E4.11 a) $\sum m(3, 5, 6, 7) = \prod M(0, 1, 2, 4)$

b) $\sum m(0, 1, 4, 5, 7) = \prod M(2, 3, 6)$

c) $\sum m(0, 1, 3, 4, 5, 9, 10, 11, 14) = \prod M(2, 6, 7, 8, 12, 13, 15)$

d) $\sum m(1, 3, 6, 7) = \prod M(0, 2, 4, 5)$

E4.12 a)

Input Variables				Complement of Inputs				Product Terms					Function Output
w	x	y	z	w'	x'	y'	z'	x'yz	xyz'	wx'y'	w'xy	wxy	F
0	0	0	0	1	1	1	1	0	0	0	0	0	0
0	0	0	1	1	1	1	0	0	0	0	0	0	0
0	0	1	0	1	1	0	1	0	0	0	0	0	0
0	0	1	1	1	1	0	0	1	0	0	0	0	1
0	1	0	0	1	0	1	1	0	0	0	0	0	0
0	1	0	1	1	0	1	0	0	0	0	0	0	0
0	1	1	0	1	0	0	1	0	1	0	1	0	1
0	1	1	1	1	0	0	0	0	0	0	1	0	1
1	0	0	0	0	1	1	1	0	0	1	0	0	1
1	0	0	1	0	1	1	0	0	0	1	0	0	1
1	0	1	0	0	1	0	1	0	0	0	0	0	0
1	0	1	1	0	1	0	0	1	0	0	0	0	1
1	1	0	0	0	0	1	1	0	0	0	0	0	0
1	1	0	1	0	0	1	0	0	0	0	0	0	0
1	1	1	0	0	0	0	1	0	1	0	0	1	1
1	1	1	1	0	0	0	0	0	0	0	0	1	1

b)

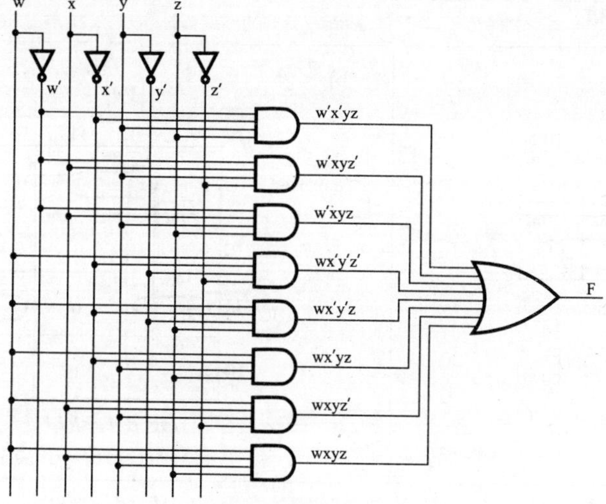

***Fig:** Logic circuit of F.*

c) $F_{simplified} = xy + yz + wx'y'$

d)

Input Variables				Product Terms			Function Output
w	x	y	z	xy	yz	wx'y'	$F_{simplified}$
0	0	0	0	0	0	0	0
0	0	0	1	0	0	0	0
0	0	1	0	0	0	0	0
0	0	1	1	0	1	0	1
0	1	0	0	0	0	0	0
0	1	0	1	0	0	0	0
0	1	1	0	1	0	0	1
0	1	1	1	1	1	0	1
1	0	0	0	0	0	1	1
1	0	0	1	0	0	1	1
1	0	1	0	0	0	0	0
1	0	1	1	0	1	0	1
1	1	0	0	0	0	0	0
1	1	0	1	0	0	0	0
1	1	1	0	1	0	0	1
1	1	1	1	1	1	0	1

e)

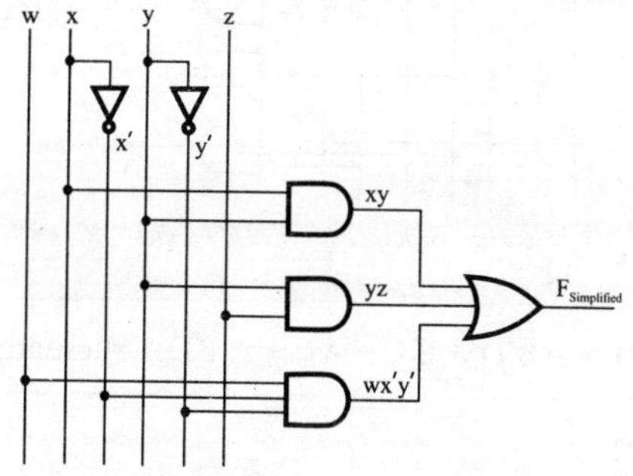

***Fig:** Logic circuit of $F_{Simplified}$.*

E4.13 a) $F_1 = x'y' + xz$ **b)** $F_2 = x'y + xy'$

E4.14 a) $F_1 = y$ **b)** $F_2 = A'BD' + BCD$

E4.15 a) $F_1 = w'y'z + wxy + xz'$ **b)** $F_2 = ABC' + A'C + BCD'$

E4.16 a) $F_1 = x'y' + yz'$ **b)** $F_2 = xy' + x'yz$

E4.17 F = A B C′ + A′B D + B′D′

E4.18 a)

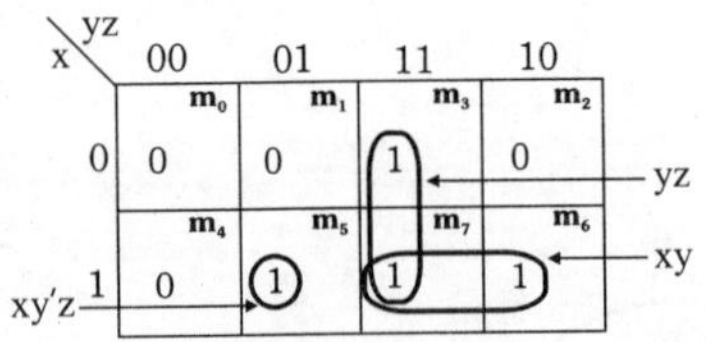

Fig: K-map for F_1.

$F_1 = \Sigma m(3, 5, 6, 7)$

b)

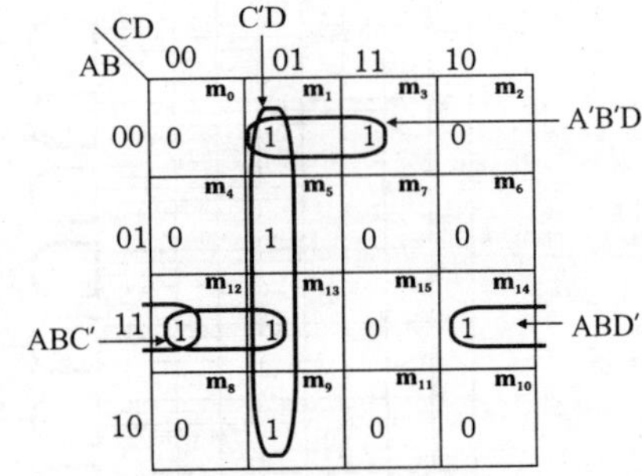

Fig: K-map for F_2.

$F_2 = \Sigma m(1, 3, 5, 9, 12, 13, 14)$

E4.19 F = A′B′C′DE′ + BC′DE + B′CE + ABC′E

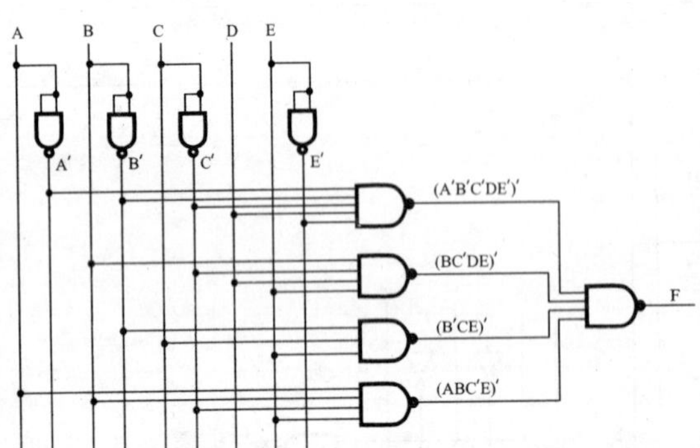

Fig: Logic circuit of function, F using only NAND gates.

E4.20 F = (B + C) (A + B + D) (A′ + C)

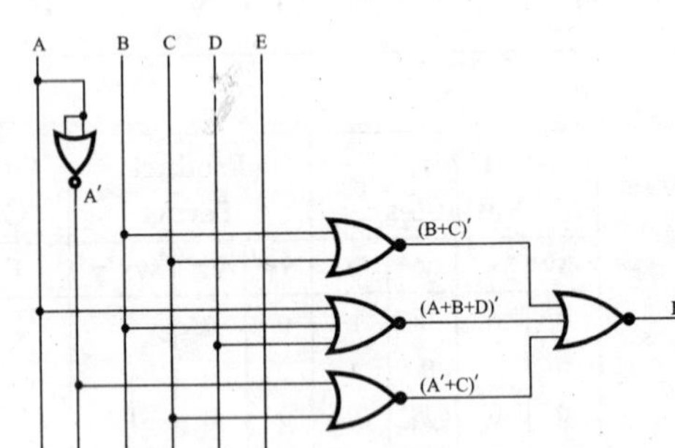

Fig: Logic circuit of function, F using only NOR gates.

E4.21 F = B′D′ + B′C + A C + A′B D (Alternatively, F = B′D′ + C D + A C + A′B D)

E4.22 a)

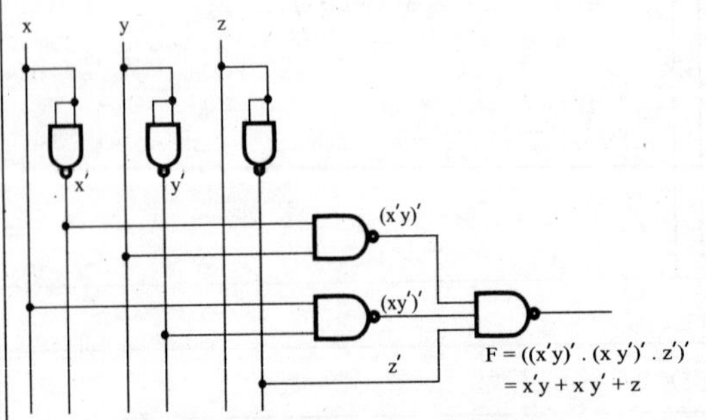

Fig: Logic circuit of F using only NAND gates.

b)

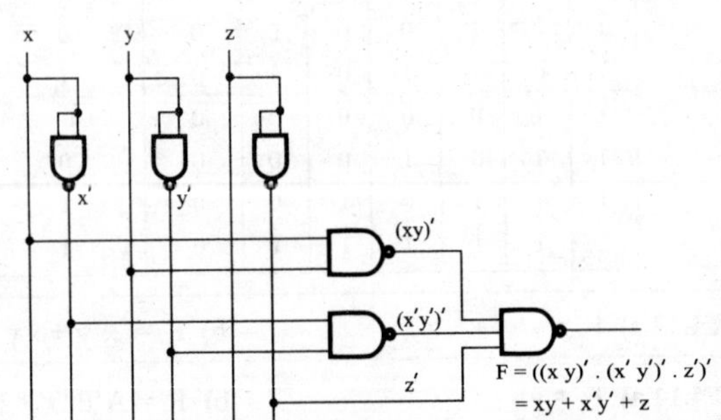

Fig: Logic circuit of F using only NAND gates.

CHAPTER 5

COMBINATIONAL LOGIC CIRCUITS

5.1 Introduction

Electronic circuits constructed using digital logic gates and devices and designed to operate on digital inputs and outputs are called ***digital logic circuits***. The digital logic circuits can be broadly classified into combinational circuits and sequential circuits.

The combinational circuits are digital logic circuits without feedbacks from output to input. Therefore, the outputs of combinational circuit will depend only on present inputs.

***Fig. 5.1:** Block diagram representation of a combinational circuit.*

The sequential circuits are digital logic circuits with feedback from output to input, the sequential circuits are discussed in Chapter-7.

(VTU, Aug/Sep'20, 4 Marks)

5.2 Combinational Circuits

The combinational logic does not depend on feedback signals or previous output. The logical operations are performed using present inputs. The circuits that perform combinational logic operations are called ***combinational logic circuits*** and they are constructed using logic gates. In combinational circuits the output at any time depends on input at that time. In combinational circuits there is no storage element and there is no feedback from output to input.

Therefore, combinational circuits are designed for applications which do not require a record of previous outputs and for applications in which the present outputs do not depend on previous outputs. Some examples of combinational circuits are shown in Fig. 5.2.

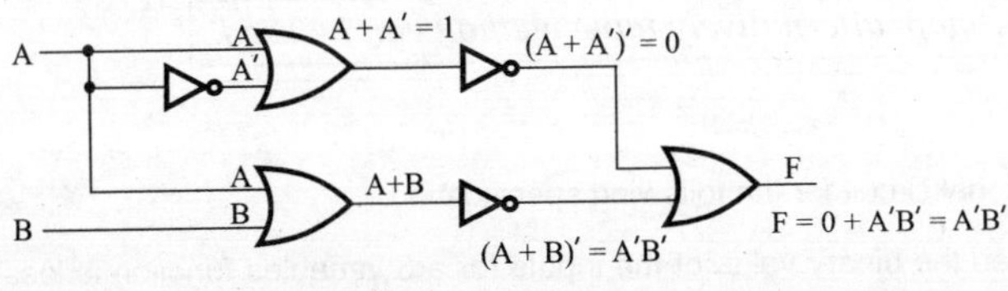

Fig. a.

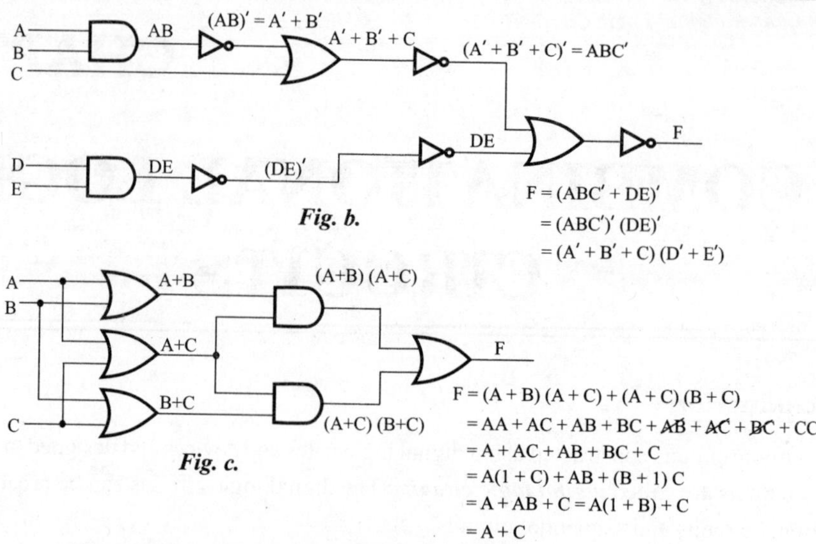

Fig. 5.2: Examples of combinational circuit.

5.2.1 K-map Representation and Minimization using K-maps

K-map representation is a pictorial form of truth table of combinational logic and used to simplify functions. Simplification of functions using K-map is simple and straight forward. A K-map is a diagram made up of squares with each square representing one minterm. Alternatively, maxterm can be used to construct K-map in which each square represent a maxterm.

When minterms are considered for simplification the resultant combinational logic function will be in sum-of-products form. When maxterms are considered for simplification the resultant combinational logic function will be in product-of-sums form.

For detailed discussion on K-map, refer Chapter-4, Section 4.6.

5.2.2 Problem Formulation and Design of Combinational Circuits

The combinational circuits are designed to provide hardware based solution for logical problems. The design starts with problem specifications. The procedure to design a combinational circuit is given below:

1. Determine the required inputs and outputs from the problem specifications.
2. Assign a symbol to each input and output.
3. Derive the truth table.
4. Draw K-map for every output and form the prime implicants.
5. Determine the simplified Boolean function for every output from the prime implicants.
6. Implement the Boolean functions of all the outputs as a digital circuit using logic gates.

Note: *Instead of K-map, alternatively table method can be used.*

Example 5.1

Design a combinational circuit for the following specification.

The output is 1 when the binary value of the inputs for a 3 variables function is less than 3. The output is 0 otherwise.

Solution

The output is 1, when the input is 0, 1 or 2. The output will be 0 if the input is greater than or equal to 3. Let us consider 3-bit binary inputs which can take values from 0 to 7. Let x, y and z be input variables and F be function output. The truth table is shown in Table 1 and the three-variable K-map is shown in Fig. 1.

Table 1: Truth Table

Inputs			Minterm	Output
x	y	z		F
0	0	0	m_0	1
0	0	1	m_1	1
0	1	0	m_2	1
0	1	1	m_3	0
1	0	0	m_4	0
1	0	1	m_5	0
1	1	0	m_6	0
1	1	1	m_7	0

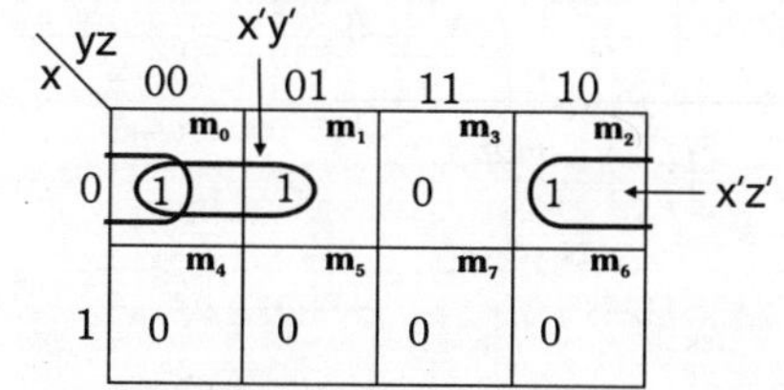

Fig. 1: K-map for F.

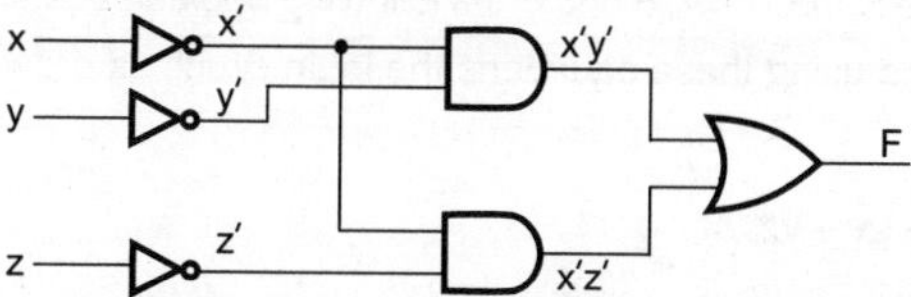

Fig. 2: Logic circuit of F.

In the K-map of Fig. 1, two overlapping prime implicants are formed using two adjacent 1's and the corresponding product terms are x' y' and x' z'. Therefore, the Boolean function, F is sum of these two prime implicants as shown below. The combinational logic circuit is drawn using logic gates as shown in Fig. 2.

$$F = x'y' + x'z'$$

Example 5.2

(AU, Nov/Dec'18, 15 Marks)

Design a combinational circuit with three inputs x, y and z and the 3 outputs, A, B and C. When the binary input is 0, 1, 2 or 3 the binary output is one greater than the input. When the binary input is 4, 5, 6 or 7, the binary output is one less than the input.

Solution

The truth table of the given problem is developed as shown in Table 1. Using the truth table the K-maps are constructed as shown in Figs. 1 to 3.

Table 1: Truth Table

Inputs				Outputs		
x	y	z	Minterm	A	B	C
0	0	0	m_0	0	0	1
0	0	1	m_1	0	1	0
0	1	0	m_2	0	1	1
0	1	1	m_3	1	0	0
1	0	0	m_4	0	1	1
1	0	1	m_5	1	0	0
1	1	0	m_6	1	0	1
1	1	1	m_7	1	1	0

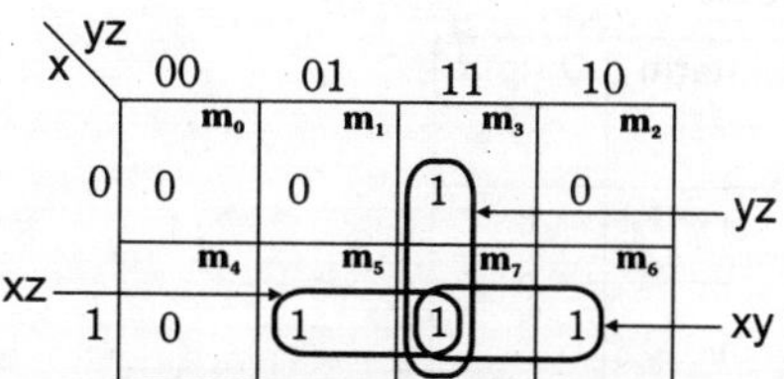

Fig. 1: K-map for A.

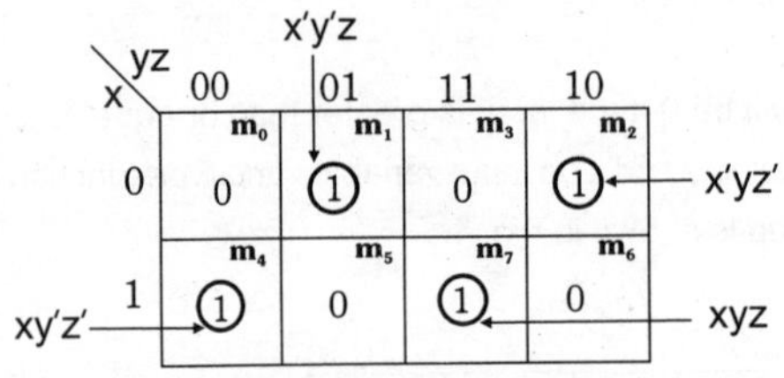

Fig. 2: K-map for B.

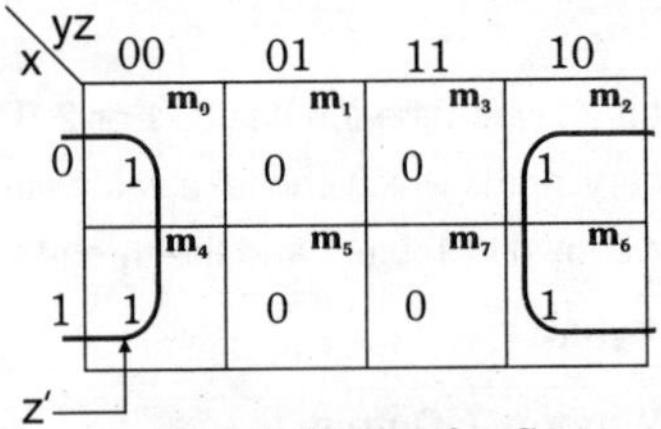

Fig. 3: K-map for C.

From the K-map for A, B and C the following Boolean equations are obtained and using these equations the logic circuit is drawn in Fig. 4.

A = xz + xy + yz

B = xy'z' + xyz + x'y'z + x'yz'

C = z'

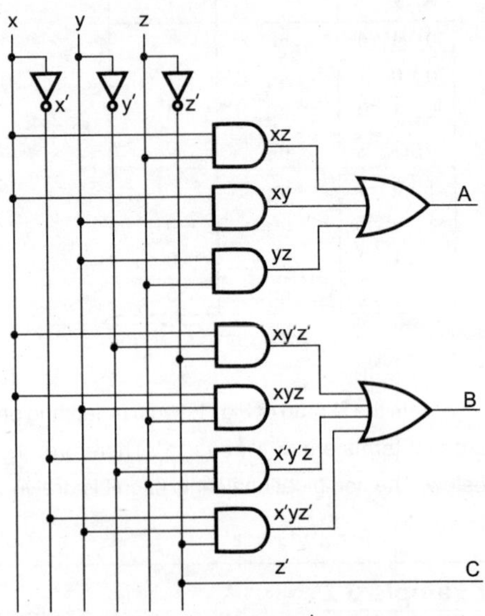

Fig. 4: Logic circuit of A, B and C.

Example 5.3

(VTU, Aug/Sep'20, 6 Marks)

Design a 3-input majority circuit, in which the output is equal to 1 if the input variables have more 1's than 0's, otherwise, output is 0.

Solution

The truth table of majority circuit described in the problem is shown in Table 1 and the three variable K-map constructed using Table 1 is shown in Fig. 1.

Table 1: Truth Table

Inputs			Minterm	Output
x	y	z		F
0	0	0	m_0	0
0	0	1	m_1	0
0	1	0	m_2	0
0	1	1	m_3	1
1	0	0	m_4	0
1	0	1	m_5	1
1	1	0	m_6	1
1	1	1	m_7	1

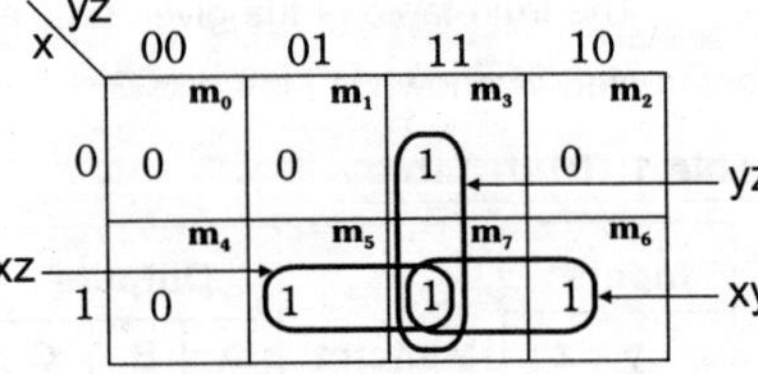

Fig. 1: K-map for F.

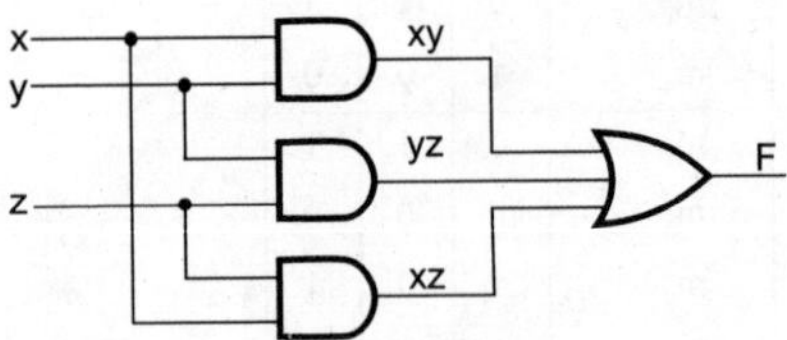

Fig. 2: Logic circuit of F.

In the K-map of Fig. 1, three overlapping prime implicants are formed using two adjacent 1's and the corresponding product terms are xy, yz and xz. Therefore, the Boolean function, F is sum of these three prime implicants as shown below. The combinational logic circuit is drawn using logic gates as shown in Fig. 2.

$$F = xy + yz + xz$$

Example 5.4

(AU, Nov/Dec'18, 13 Marks)

Assume a 3-input AND gate with output F and a 3-input OR gate with output G. Show the signals of the outputs F and G as functions of the three inputs A, B and C. Use all 8 possible combinations of inputs ABC.

Solution

It's given that,

$$F = ABC$$

$$G = A + B + C$$

The truth table for F and G for all possible combinations of A, B and C is developed as shown in Table 1.

Using the truth table the K-map for F and G are constructed as shown in Figs. 1 and 2.

Table 1: Truth Table for 3-input AND and OR gate

Inputs			Minterm	Outputs	
A	B	C		F	G
0	0	0	m_0	0	0
0	0	1	m_1	0	1
0	1	0	m_2	0	1
0	1	1	m_3	0	1
1	0	0	m_4	0	1
1	0	1	m_5	0	1
1	1	0	m_6	0	1
1	1	1	m_7	1	1

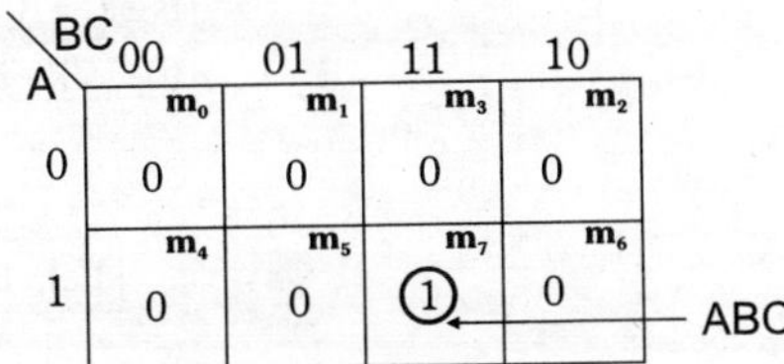

Fig. 1: K-map for F.

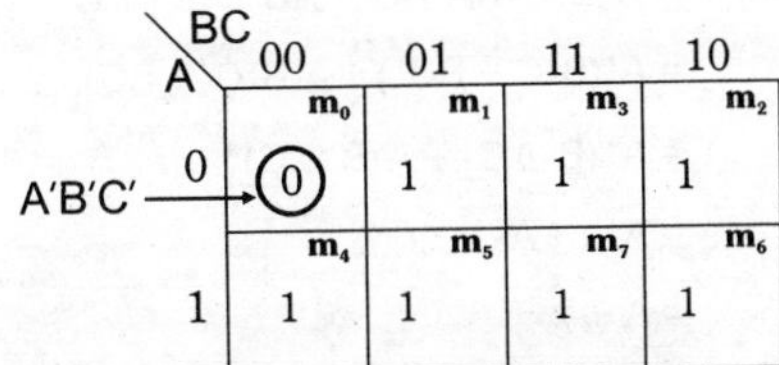

Fig. 2: K-map for G.

From the K-map we get the following Boolean equations.

$$F = ABC$$

$$G = (A'B'C')' = A + B + C$$

Using DeMorgan's law

Using the above Boolean equations, the logic circuit for F and G is drawn as shown in Fig. 3.

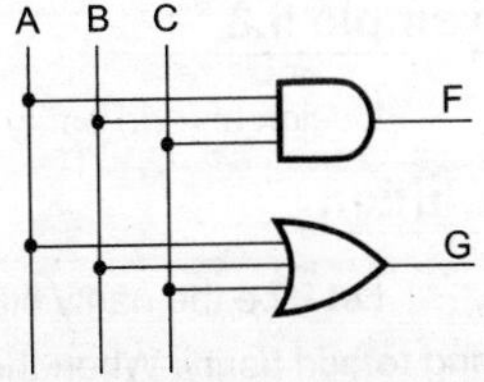

Fig. 3: Logic circuit of F and G.

Example 5.5

Design an even parity generator, that generates an even parity bit for every input string of 3-bits.

Solution

Let P be the parity bit. When the input has odd number of 1's parity bit P is assigned 1 in order to convert input string to even parity. When the input string has even number of 1's parity bit is assigned 0.

The truth table is shown in Table 1 and using truth table the K-map for P is drawn as shown in Fig. 1.

Table 1: Truth Table

Inputs			Minterm	Output
A	B	C		P
0	0	0	m_0	0
0	0	1	m_1	1
0	1	0	m_2	1
0	1	1	m_3	0
1	0	0	m_4	1
1	0	1	m_5	0
1	1	0	m_6	0
1	1	1	m_7	1

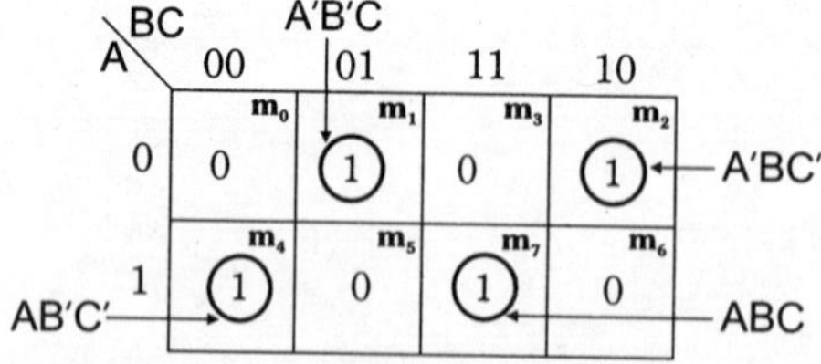

Fig. 1: K-map for parity bit P.

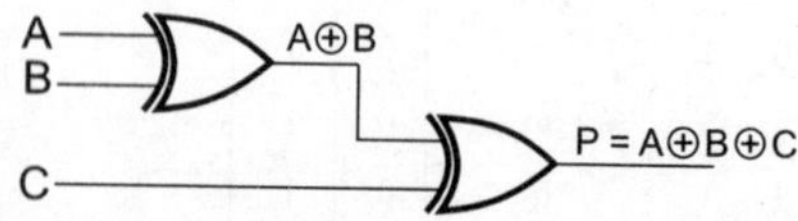

Fig. 2: Logic circuit of even parity bit generator.

From the K-map we get the following Boolean function for P, which is again simplified and using the simplified Boolean function the logic circuit of even parity generator is drawn as shown in Fig. 2.

$$P = A'B'C + A'BC' + AB'C' + ABC$$
$$= A'(B'C + BC') + A(B'C' + BC)$$
$$= A'(B \oplus C) + A(B \oplus C)'$$
$$= A'x + Ax'$$
$$= A \oplus x$$
$$= A \oplus (B \oplus C)$$
$$= A \oplus B \oplus C$$

$B \oplus C = B'C + BC'$
$(B \oplus C)' = BC + B'C'$

Let, $B \oplus C = x$

Example 5.6

Design an odd parity generator that generates an odd parity bit for every input string of 3 bits.

Solution

Let P be the parity bit. When the input has odd number of 1's parity bit P is assigned 0 in order to convert input string to odd parity. When the input string has even number of 1's parity bit is assigned 1. The truth table is shown in Table 1 and using truth table the K-map for P is drawn as shown in Fig. 1.

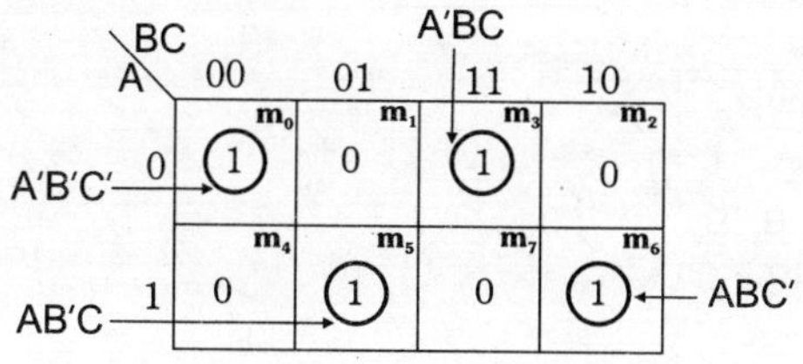

Fig. 1: K-map for parity bit P.

Table 1: Truth Table

Inputs			Minterm	Output
A	B	C		P
0	0	0	m_0	1
0	0	1	m_1	0
0	1	0	m_2	0
0	1	1	m_3	1
1	0	0	m_4	0
1	0	1	m_5	1
1	1	0	m_6	1
1	1	1	m_7	0

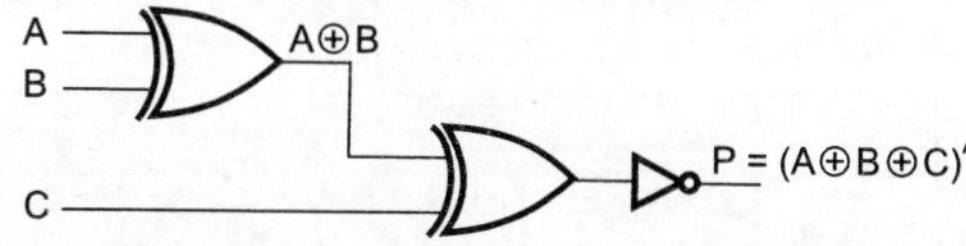

Fig. 2: Logic circuit of odd parity bit generator.

From the K-map we get the following Boolean function for P, which is again simplified and using the simplified Boolean function the logic circuit of odd parity generator is drawn as shown in Fig. 2.

$$P = A'B'C' + AB'C + A'BC + ABC'$$

$$= (AB + A'B')C' + (AB' + A'B)C$$

$$= (A \oplus B)'C' + (A \oplus B)C$$

$$= x'C' + xC$$

$$= (x \oplus C)'$$

$$= (A \oplus B \oplus C)'$$

$A \oplus B = A'B + AB'$

$(A \oplus B)' = AB + A'B'$

Let, $A \oplus B = x$

5.3 Code Converters

The various types of codes are discussed in Chapter-3, Section 3.1.1 The ***code converters*** are combinational circuits that convert one type of code to another type of code.

The design of code converter starts with formation of truth table by taking one type of code as inputs and another type of code as outputs. K-maps are constructed for output codes and from the K-map simplified Boolean expressions are obtained and they can be implemented using logic gates. Two examples of code conversion are presented in Examples 5.7, 5.8 and 5.9.

Example 5.7

(AKUB, Dec'20, 8 Marks)

Design excess-3 to BCD code converter using minimum number of NAND gates.

Solution

The BCD code is 4-bit binary code. In 4-bit binary code, $2^4 = 16$ codes are possible. But the BCD has only 10 valid codes. The excess-3 code is obtained by adding 0011_2 (3_{10}) to BCD code. The excess-3 code will also have only 10 codes and the remaining six codes are don't-care outputs. The truth table of excess-3 to BCD code converter is shown in Table 1.

Using the truth table, K-map for 4 bits of BCD code B_0, B_1, B_2 and B_3 are drawn as shown in Figs. 1 to 4.

Table 1: Truth Table

Decimal Digit	Inputs: Excess-3 Code E_3 E_2 E_1 E_0	Minterm	Outputs: BCD B_3 B_2 B_1 B_0
-	0 0 0 0	m_0	× × × ×
-	0 0 0 1	m_1	× × × ×
-	0 0 1 0	m_2	× × × ×
0	0 0 1 1	m_3	0 0 0 0
1	0 1 0 0	m_4	0 0 0 1
2	0 1 0 1	m_5	0 0 1 0
3	0 1 1 0	m_6	0 0 1 1
4	0 1 1 1	m_7	0 1 0 0
5	1 0 0 0	m_8	0 1 0 1
6	1 0 0 1	m_9	0 1 1 0
7	1 0 1 0	m_{10}	0 1 1 1
8	1 0 1 1	m_{11}	1 0 0 0
9	1 1 0 0	m_{12}	1 0 0 1
-	1 1 0 1	m_{13}	× × × ×
-	1 1 1 0	m_{14}	× × × ×
-	1 1 1 1	m_{15}	× × × ×

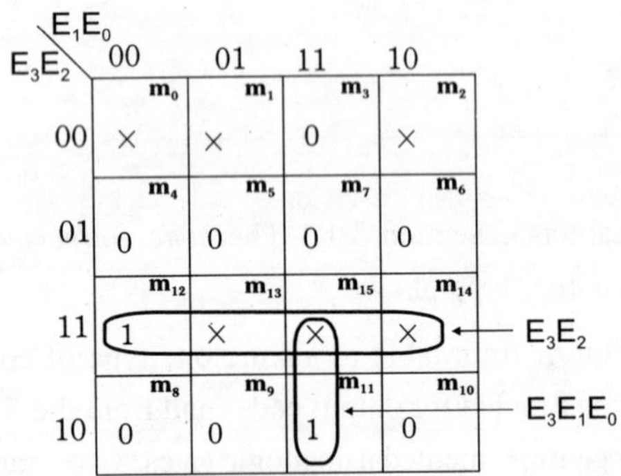

Fig. 1: K-map for B_3.

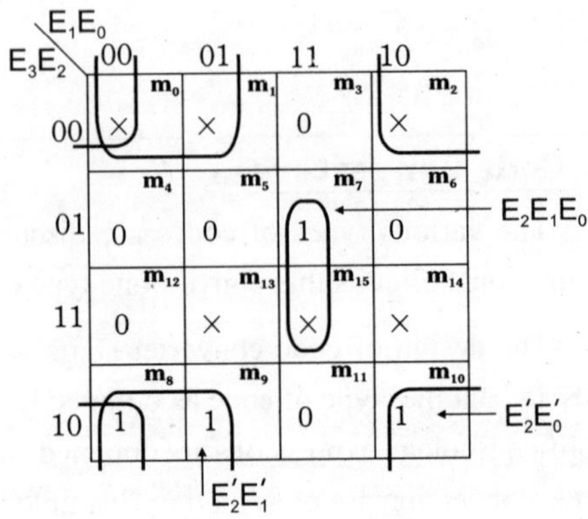

Fig. 2: K-map for B_2.

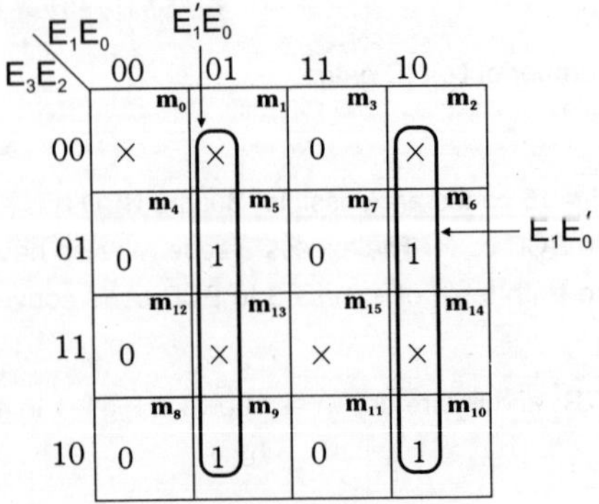

Fig. 3: K-map for B_1.

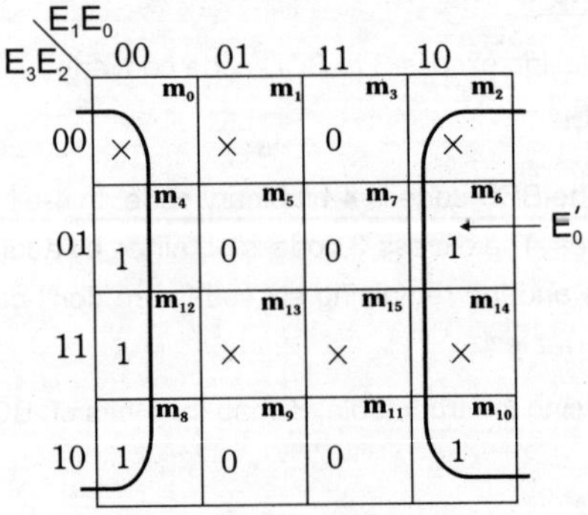

Fig. 4: K-map for B_0.

From the K-map we get the following Boolean equations and for implementation using NAND gates the Boolean equations are modified as shown below:

$$B_3 = E_3E_2 + E_3E_1E_0 = ((E_3E_2 + E_3E_1E_0)')' = ((E_3E_2)' . (E_3E_1E_0)')'$$

$$B_2 = E_2'E_1' + E_2'E_0' + E_2E_1E_0 = ((E_2'E_1' + E_2'E_0' + E_2E_1E_0)')' = ((E_2'E_1')' . (E_2'E_0')' . (E_2E_1E_0)')'$$

$$B_1 = E_1'E_0 + E_1E_0' = ((E_1'E_0 + E_1E_0')')' = ((E_1'E_0)' . (E_1E_0')')'$$

$$B_0 = E_0'$$

Using the above Boolean equations the logic circuit for excess-3 to BCD code converter using NAND gates is drawn as shown in Fig. 5.

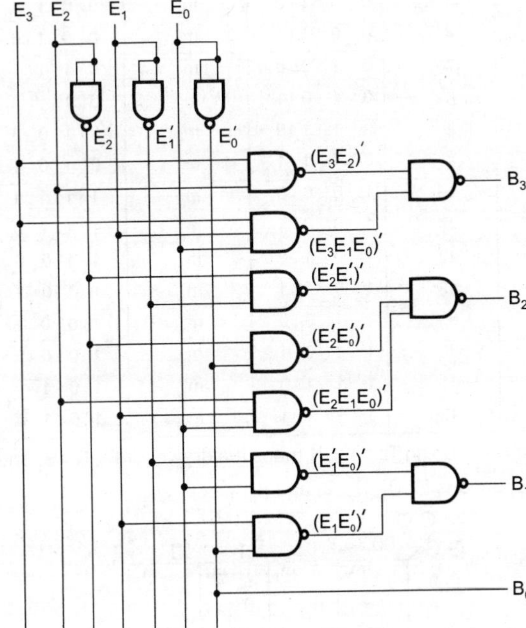

Fig. 5: Logic circuit of excess-3 to BCD code converter using NAND gates.

Example 5.8

(GTU, Jun'19, 4 Marks)

Design 4-bit Gray code to binary code converter.

Solution

The Gray and binary code are 4-digit binary codes. Hence the code converter will have 4 inputs and 4 outputs.

Table 1: 4-Bit Gray Code

Decimal Number	Gray Code	Decimal Number	Gray Code
0	0000	8	1100
1	0001	9	1101
2	0011	10	1111
3	0010	11	1110
4	0110	12	1010
5	0111	13	1011
6	0101	14	1001
7	0100	15	1000

Table 2: Truth Table of Gray to Binary Code Converter

Decimal Number	Inputs: 4-bit Gray $G_3\ G_2\ G_1\ G_0$	Minterm	Outputs: 4-bit Binary $B_3\ B_2\ B_1\ B_0$
0	0 0 0 0	m_0	0 0 0 0
1	0 0 0 1	m_1	0 0 0 1
3	0 0 1 0	m_2	0 0 1 1
2	0 0 1 1	m_3	0 0 1 0
7	0 1 0 0	m_4	0 1 1 1
6	0 1 0 1	m_5	0 1 1 0
4	0 1 1 0	m_6	0 1 0 0
5	0 1 1 1	m_7	0 1 0 1
15	1 0 0 0	m_8	1 1 1 1
14	1 0 0 1	m_9	1 1 1 0
12	1 0 1 0	m_{10}	1 1 0 0
13	1 0 1 1	m_{11}	1 1 0 1
8	1 1 0 0	m_{12}	1 0 0 0
9	1 1 0 1	m_{13}	1 0 0 1
11	1 1 1 0	m_{14}	1 0 1 1
10	1 1 1 1	m_{15}	1 0 1 0

Using Table 2, K-maps are drawn for B_3, B_2, B_1 and B_0 as shown in Figs. 1 to 4 in order to find simplified Boolean equations for B_3, B_2, B_1 and B_0.

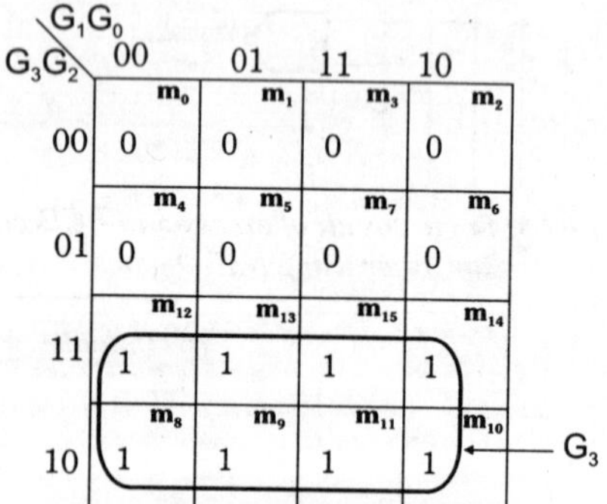

Fig. 1: K-map for B_3.

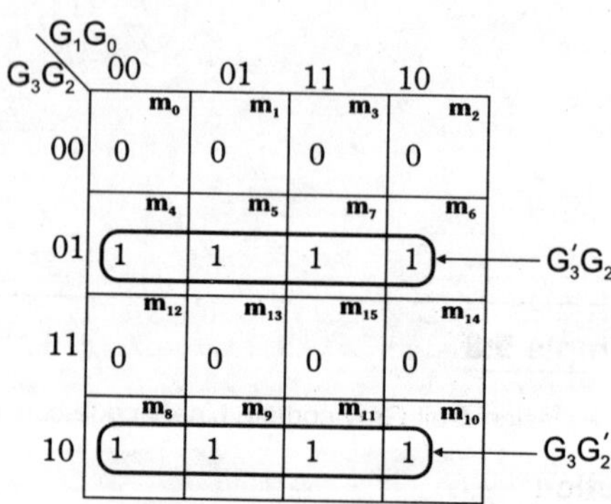

Fig. 2: K-map for B_2.

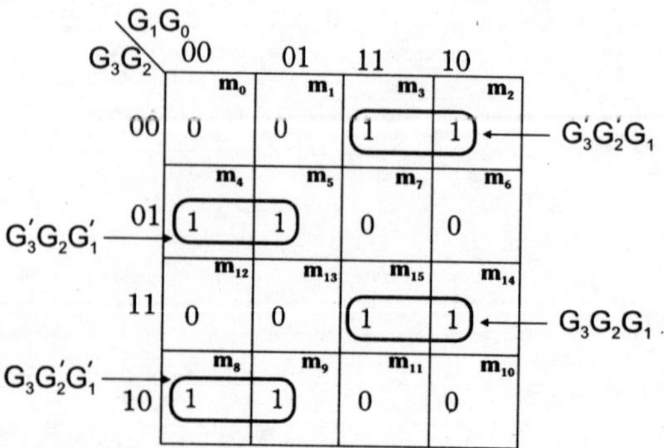

Fig. 3: K-map for B_1.

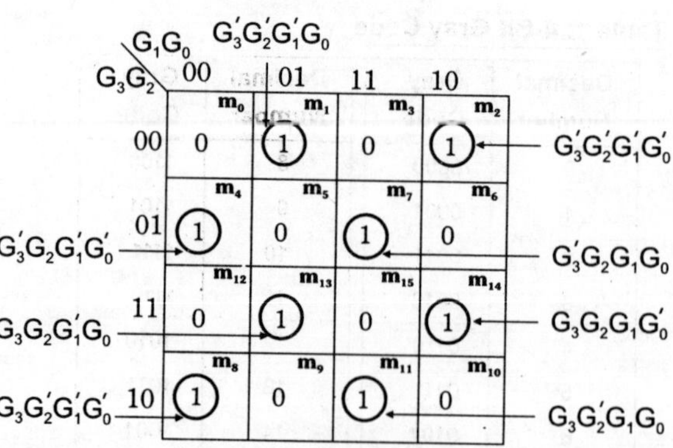

Fig. 4: K-map for B_0.

The following Boolean equations are obtained from K-map of Figs. 1 to 4.

$$B_3 = G_3$$

$$B_2 = G_3G_2' + G_3'G_2$$

$$B_1 = G_3'G_2'G_1 + G_3'G_2G_1' + G_3G_2'G_1' + G_3G_2G_1$$

$$B_0 = G_3'G_2'G_1'G_0 + G_3'G_2'G_1'G_0 + G_3'G_2G_1'G_0' + G_3'G_2G_1G_0 + G_3G_2G_1'G_0 + G_3G_2G_1G_0' + G_3G_2'G_1'G_0' + G_3G_2'G_1G_0$$

The logic circuit for 4-bit Gray to binary code converter is drawn using the above Boolean equations as shown in Fig. 5.

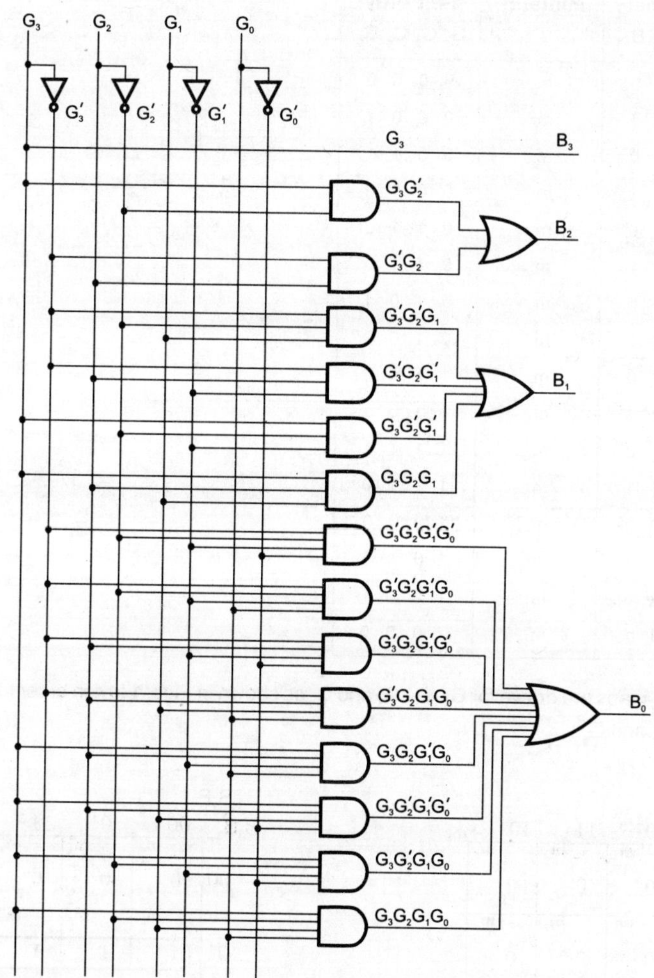

Fig. 5: Logic circuit of Gray to binary code converter.

Example 5.9

(BATU, Jun'19, 6 Marks)

Design 4-bit binary to Gray code converter.

Solution

The binary and Gray code are 4-digit binary codes. Hence the code converter will have 4 inputs and 4 outputs.

Table 1: Truth Table of Gray to Binary Code Converter

Decimal Number	Inputs: 4-bit Binary $B_3\ B_2\ B_1\ B_0$	Minterm	Outputs: 4-bit Gray $G_3\ G_2\ G_1\ G_0$
0	0 0 0 0	m_0	0 0 0 0
1	0 0 0 1	m_1	0 0 0 1
2	0 0 1 0	m_2	0 0 1 1
3	0 0 1 1	m_3	0 0 1 0
4	0 1 0 0	m_4	0 1 1 0
5	0 1 0 1	m_5	0 1 1 0
6	0 1 1 0	m_6	0 1 0 1
7	0 1 1 1	m_7	0 1 0 0
8	1 0 0 0	m_8	1 1 0 0
9	1 0 0 1	m_9	1 1 0 1
10	1 0 1 0	m_{10}	1 1 1 1
11	1 0 1 1	m_{11}	1 1 1 0
12	1 1 0 0	m_{12}	1 0 1 0
13	1 1 0 1	m_{13}	1 0 1 1
14	1 1 1 0	m_{14}	1 0 0 1
15	1 1 1 1	m_{15}	1 0 0 0

Using Table 1, K-maps are drawn for G_3, G_2, G_1 and G_0 as shown in Figs. 1 to 4 in order to find simplified Boolean equations for G_3, G_2, G_1 and G_0.

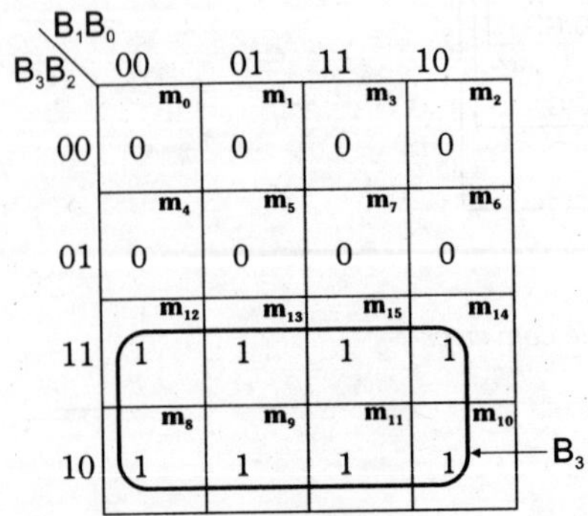

Fig. 1: K-map for G_3.

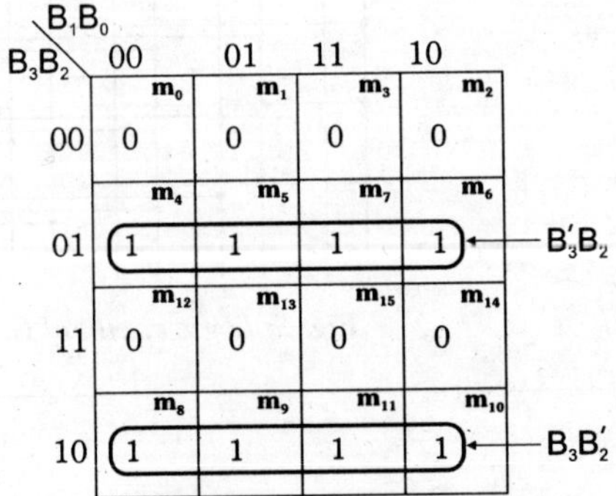

Fig. 2: K-map for G_2.

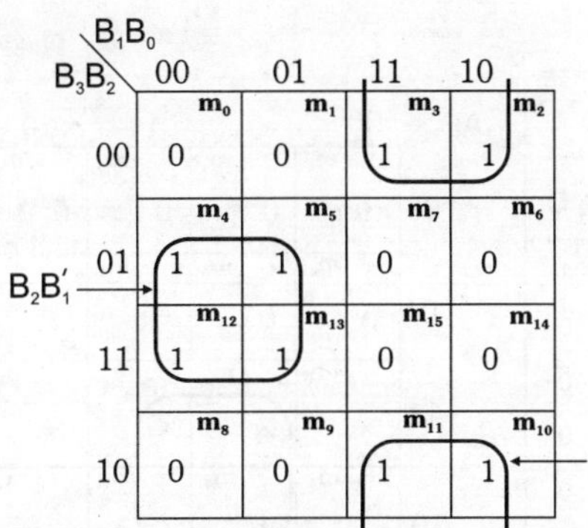

Fig. 3: K-map for G_1.

Fig. 4: K-map for G_0.

The following Boolean equations are obtained from K-map of Figs. 1 to 4.

$$G_3 = B_3$$

$$G_2 = B_3'B_2 + B_3B_2'$$

$$G_1 = B_2B_1' + B_2'B_1$$

$$G_0 = B_1'B_0 + B_1B_0'$$

The logic circuit for 4-bit binary to Gray code converter is drawn using the above Boolean equations as shown in Fig. 5.

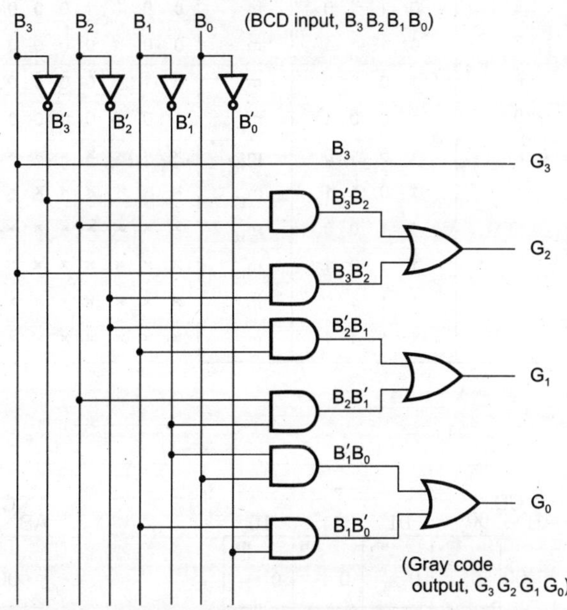

Fig. 5: Logic circuit of binary to Gray code conveter.

Example 5.10

(AKTU, Dec'22, 10 Marks)

Design the logic circuit for a BCD to decimal decoder.

Solution

The BCD code is 4-bit binary code used to represent decimal digits 0 to 9. In 4-bit binary code, 2^4 = 16 codes are possible. But the BCD has only 10 valid codes. The BCD to decimal decoder will generate an output similar to a decoder. The decoder will have 10 outputs corresponding decimal digits 0 to 9. Let the 10 outputs be denoted as D_0, D_1, D_2, D_3, D_4, D_5, D_6, D_7, D_8 and D_9. The input BCD is decoded and the corresponding output is asserted high. The truth table of BCD to decimal decoder is shown in Table 1.

Using the truth table, K-map for 10 bits of decoder D_0, D_1, D_2, D_3, D_4, D_5, D_6, D_7, D_8 and D_9 are drawn as shown in Figs. 1 to 10.

Table 1: Truth Table

Decimal Number	Inputs BCD $B_3\ B_2\ B_1\ B_0$	Minterm	Outputs Decoder $D_9\ D_8\ D_7\ D_6\ D_5\ D_4\ D_3\ D_2\ D_1\ D_0$
0	0 0 0 0	m_0	0 0 0 0 0 0 0 0 0 1
1	0 0 0 1	m_1	0 0 0 0 0 0 0 0 1 0
2	0 0 1 0	m_2	0 0 0 0 0 0 0 1 0 0
3	0 0 1 1	m_3	0 0 0 0 0 0 1 0 0 0
4	0 1 0 0	m_4	0 0 0 0 0 1 0 0 0 0
5	0 1 0 1	m_5	0 0 0 0 1 0 0 0 0 0
6	0 1 1 0	m_6	0 0 0 1 0 0 0 0 0 0
7	0 1 1 1	m_7	0 0 1 0 0 0 0 0 0 0
8	1 0 0 0	m_8	0 1 0 0 0 0 0 0 0 0
9	1 0 0 1	m_9	1 0 0 0 0 0 0 0 0 0
-	1 0 1 0	m_{10}	× × × × × × × × × ×
-	1 0 1 1	m_{11}	× × × × × × × × × ×
-	1 1 0 0	m_{12}	× × × × × × × × × ×
-	1 1 0 1	m_{13}	× × × × × × × × × ×
-	1 1 1 0	m_{14}	× × × × × × × × × ×
-	1 1 1 1	m_{15}	× × × × × × × × × ×

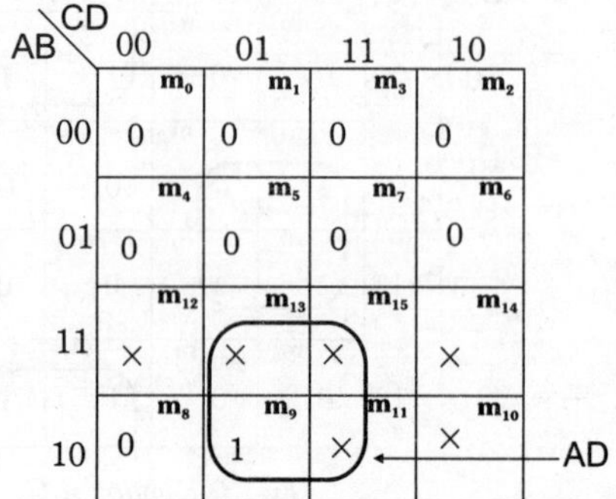

Fig. 1: K-map for D_9.

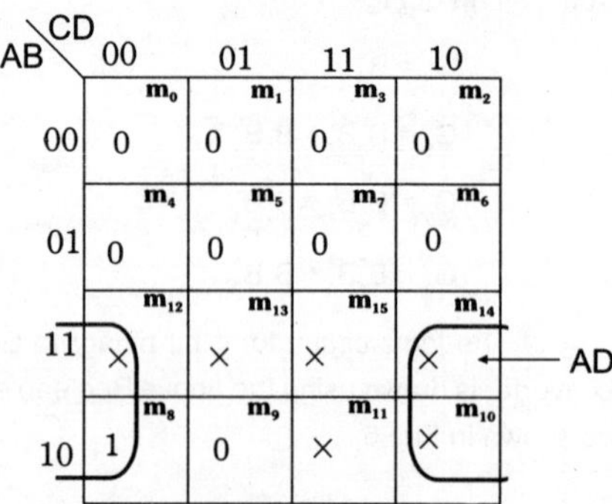

Fig. 2: K-map for D_8.

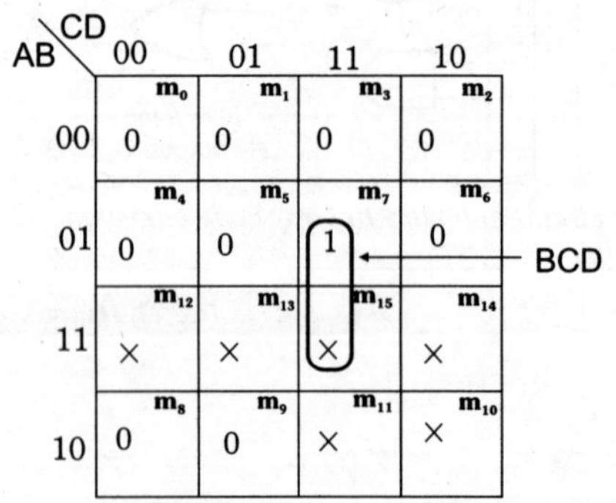

Fig. 3: K-map for D_7.

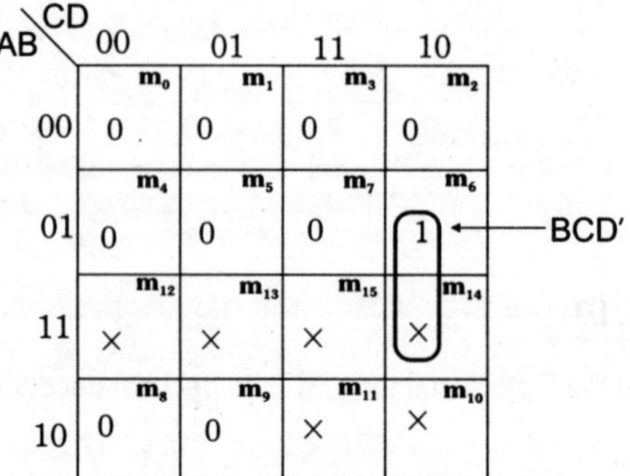

Fig. 4: K-map for D_6.

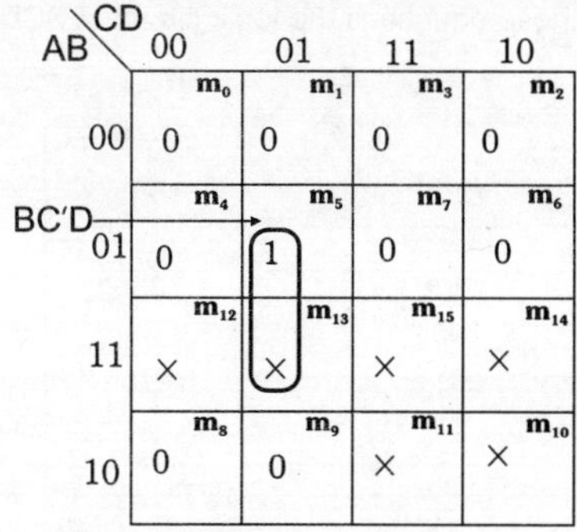

Fig. 5: *K-map for D_5.*

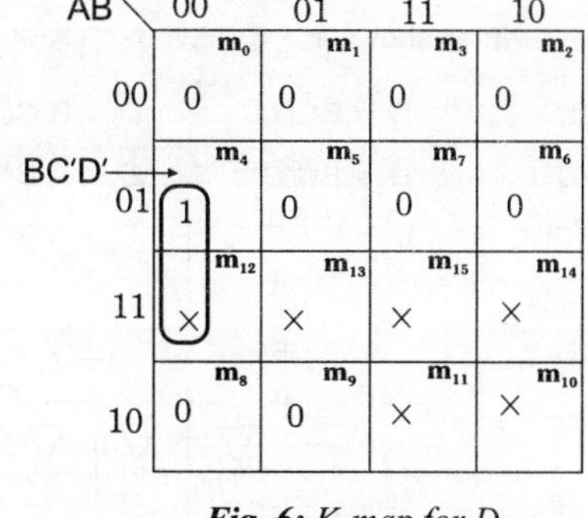

Fig. 6: *K-map for D_4.*

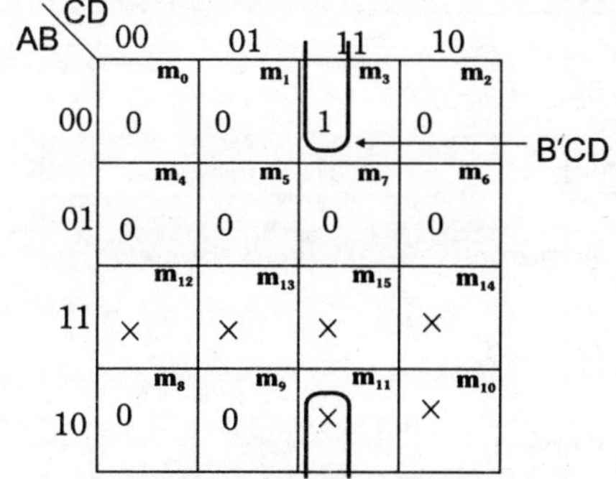

Fig. 7: *K-map for D_3.*

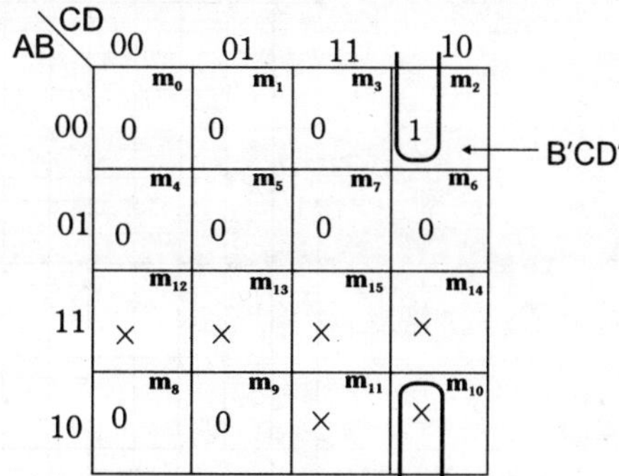

Fig. 8: *K-map for D_2.*

A'B'C'D

AB \ CD	00	01	11	10
00	m_0: 0	m_1: (1)	m_3: 0	m_2: 0
01	m_4: 0	m_5: 0	m_7: 0	m_6: 0
11	m_{12}: ×	m_{13}: ×	m_{15}: ×	m_{14}: ×
10	m_8: 0	m_9: 0	m_{11}: ×	m_{10}: ×

Fig. 9: *K-map for D_1.*

A'B'C'D'

AB \ CD	00	01	11	10
00	m_0: (1)	m_1: 0	m_3: 0	m_2: 0
01	m_4: 0	m_5: 0	m_7: 0	m_6: 0
11	m_{12}: ×	m_{13}: ×	m_{15}: ×	m_{14}: ×
10	m_8: 0	m_9: 0	m_{11}: ×	m_{10}: ×

Fig. 10: *K-map for D_0.*

From the K-map we get the following Boolean equations and using these equations the logic circuit of BCD to decimal decoder is drawn as shown in Fig. 11.

$D_9 = AD$	$D_5 = BC'D$	$D_2 = B'CD'$
$D_8 = AD'$	$D_4 = BC'D'$	$D_1 = A'B'C'D$
$D_7 = BCD$	$D_3 = B'CD$	$D_0 = A'B'C'D'$
$D_6 = BCD'$		

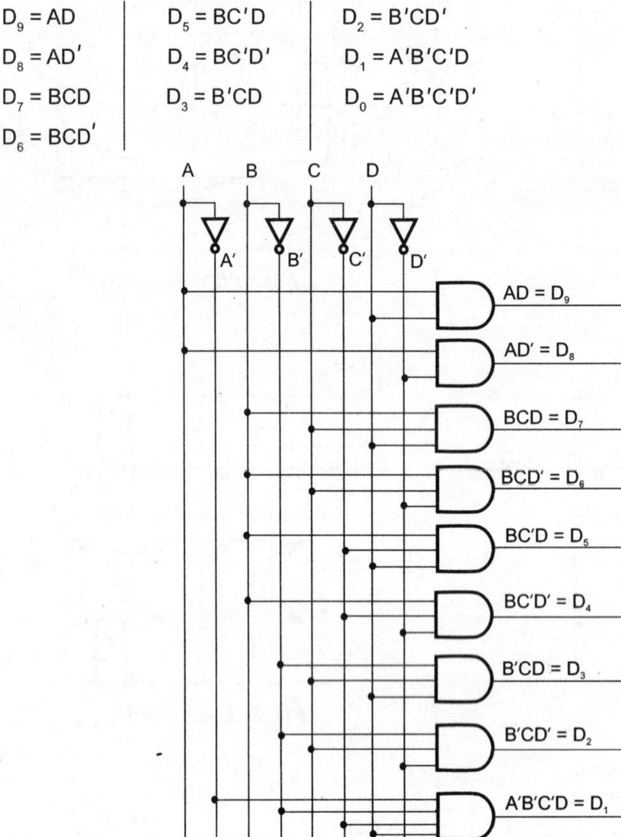

Fig. 11: Logic circuit of BCD to Decimal Decoder.

5.4 Decoders

(IKGPTU, May'20, 2 Marks)

A ***decoder*** is a combinational logic device that decodes one of the 2^n binary information depending on n-bit binary input. The n-bit binary information is decoded into 2^n binary information.

In other words, for an n-bit binary input one of the 2^n output is activated. The activated output may be logic **high** or logic **low**. In active **high** or logic **high** decoder, one of the output is **high** for an n-bit input and all other outputs are **low**. In active **low** or logic **low** decoder, one of the output is **low** for an n-bit input and all other outputs are **high**.

In general, a decoder is referred to as n-to-2^n decoder and the decoder outputs may be logic **high** or **low**.

For example, consider a 3-to-2^3 (3-to-8) decoder. Let the 8 outputs of decoder be Y_0, Y_1, Y_2, Y_3, Y_4, Y_5, Y_6 and Y_7. In a 3-bit binary input, the possible minterms are m_0, m_1, m_2, m_3, m_4, m_5, m_6 and m_7.

In logic **high** 3-to-8 decoder,

When input is m_0 then output Y_0 is active **high**,

When input is m_1 then output Y_1 is active **high**,

and so on.

Therefore, we can say that output decodes the input binary by asserting the output as **high**.

Similarly, in a 3-bit binay input, the possible maxterms are M_0, M_1, M_2, M_3, M_4, M_5, M_6 and M_7.

In logic **low** 3-to-8 decoder,

When input is M_0 then output Y_0 is active **low**,

When input is M_1 then output Y_1 is active **low**,

and so on.

Therefore, we can say that output decodes the input binary by asserting the output as **low**.

Most of the decoders are provided with an enable signal. The input of decoder is recognized only when enable signal is active. The enable can be active **high** or **low**. In decoder with active **high** enable, the input is recognized only when enable is tied to **high**. In decoder with active **low** enable, the input is recognized only when enable is tied to **low**.

Therefore, six types of n-to-2^n decoders are available as shown in Fig. 5.3. A bubble in the input/output line indicates that the corresponding input/output is active **low**.

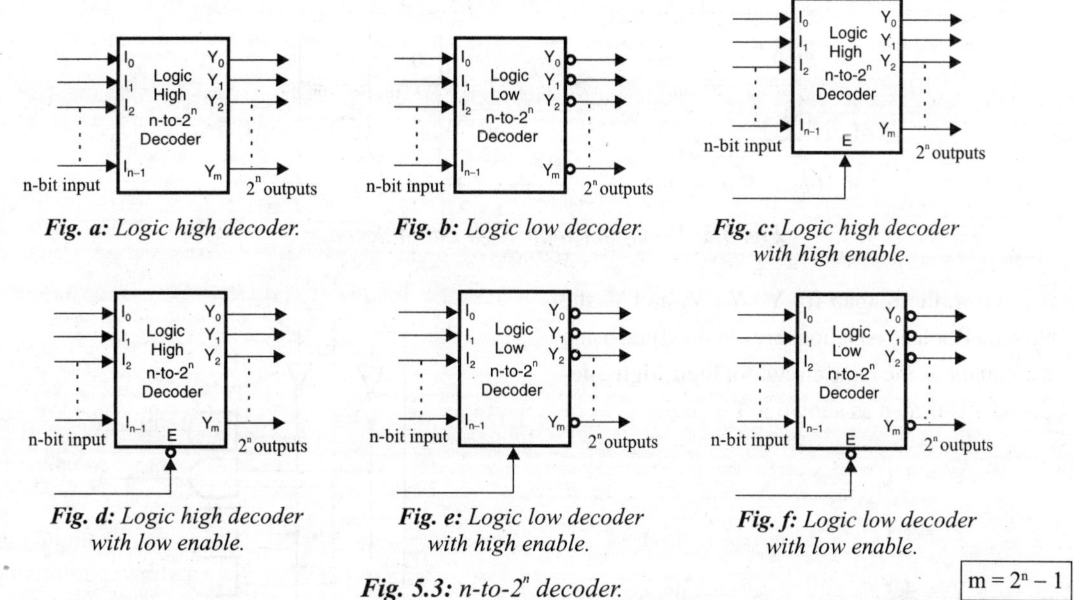

Fig. a: Logic high decoder. *Fig. b: Logic low decoder.* *Fig. c: Logic high decoder with high enable.*

Fig. d: Logic high decoder with low enable. *Fig. e: Logic low decoder with high enable.* *Fig. f: Logic low decoder with low enable.*

Fig. 5.3: n-to-2^n decoder.

5.4.1 Logic High 2-to-4 Decoder

(AU, Nov/Dec'18, 13 Marks) (GTU, Jun'19, 3 Marks)

A 2-to-4 (2-to-2^2) decoder can generate 4 decoded outputs from the two bit input.

The ***logic high 2-to-4 decoder*** will generate 4 unique outputs in which only one will be **high** at any one time and all other output will be zero. The block diagram representation of logic **high** 2-to-4 decoder is shown in Fig. 5.4.

Table 5.1: Truth Table of Logic High 2-to-4 Decoder

Inputs		Minterm	Outputs			
I_1	I_0		Y_0	Y_1	Y_2	Y_3
0	0	m_0	1	0	0	0
0	1	m_1	0	1	0	0
1	0	m_2	0	0	1	0
1	1	m_3	0	0	0	1

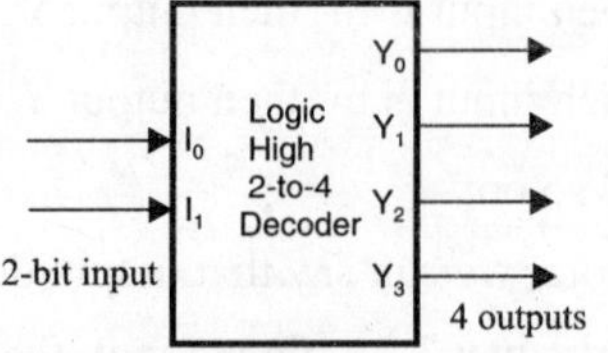

Fig. 5.4: Logic high 2-to-4 decoder.

The truth table of logic **high** 2-to-4 decoder is shown in Table 5.1 and using this truth table, the K-maps for the decoder design are drawn as shown in Fig. 5.5.

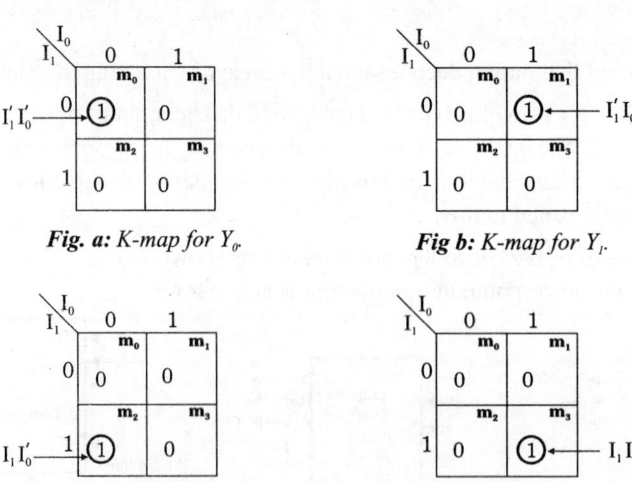

Fig. 5.5: K-map for logic high 2-to-4 decoder.

From the K-map for Y_0, Y_1, Y_2 and Y_3 the following Boolean equations are obtained and using these equations the logic circuit for logic **high** 2-to-4 decoder is drawn as shown in Fig. 5.5.

$$Y_0 = I_1' I_0'$$

$$Y_1 = I_1' I_0$$

$$Y_2 = I_1 I_0'$$

$$Y_3 = I_1 I_0$$

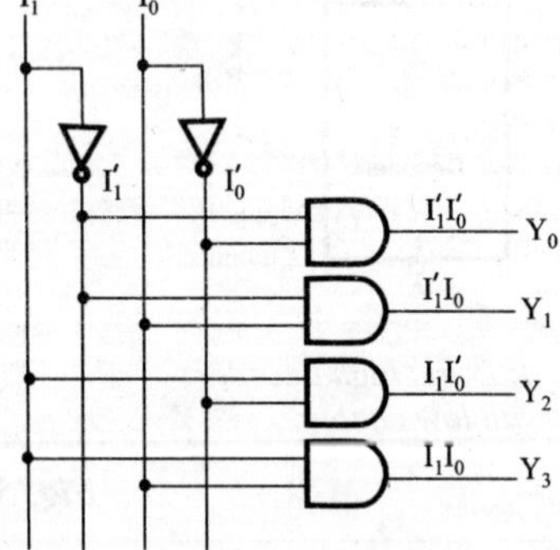

Fig. 5.6: Logic circuit for logic high 2-to-4 decoder.

> ***Note:** Alternatively, the output equations of logic **high** 2-to-4 decoder can be directly written from the truth table which are minterms for which output is **high**.*

5.4.2 Logic Low 2-to-4 Decoder

The ***logic low 2-to-4 decoder*** will generate 4 unique outputs in which only one of the output will be **low** at any one time. The output of logic **low** decoder will be complement to that of logic **high** decoder.

The block diagram representation of logic **low** 2-to-4 decoder is shown in Fig. 5.7. The bubble at the outputs indicate that the outputs are active **low**. The dual 2-to-4 decoder is available as a standard IC with number 74139. The pin configuration of 74139 is shown in Fig. 5.8.

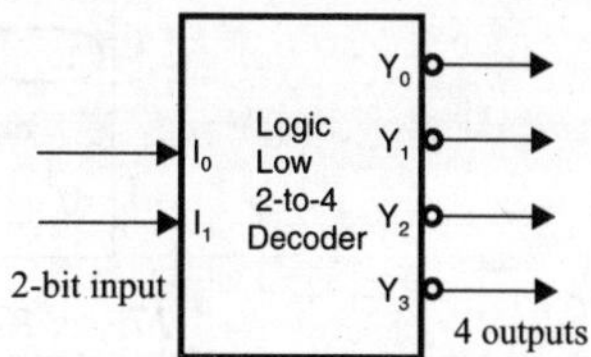

***Fig. 5.7:** Logic low 2-to-4 decoder.*

The truth table of logic **low** 2-to-4 decoder is shown in Table 5.2 and using this truth table, the K-maps for the design of decoder are drawn as shown in Fig. 5.9.

Table 5.2: Truth Table of Logic Low 2-to-4 Decoder

Inputs		Minterm	Outputs			
I_1	I_0		Y_0	Y_1	Y_2	Y_3
0	0	m_0	0	1	1	1
0	1	m_1	1	0	1	1
1	0	m_2	1	1	0	1
1	1	m_3	1	1	1	0

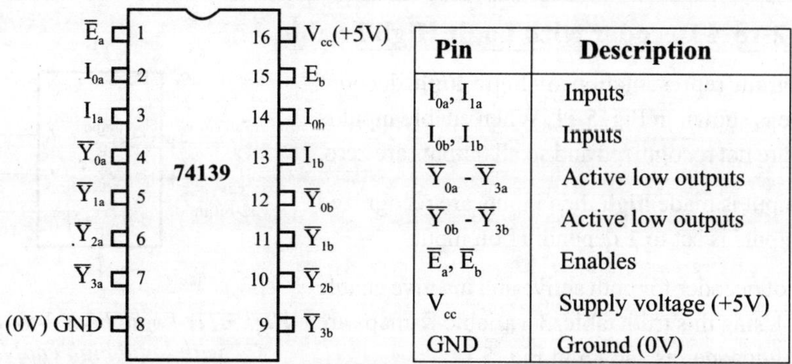

Pin	Description
I_{0a}, I_{1a}	Inputs
I_{0b}, I_{1b}	Inputs
$\overline{Y}_{0a}$ - $\overline{Y}_{3a}$	Active low outputs
$\overline{Y}_{0b}$ - $\overline{Y}_{3b}$	Active low outputs
$\overline{E}_a$, $\overline{E}_b$	Enables
V_{cc}	Supply voltage (+5V)
GND	Ground (0V)

***Fig. 5.8:** Pin configuration of dual 2-to-4 decoder IC 74139.*

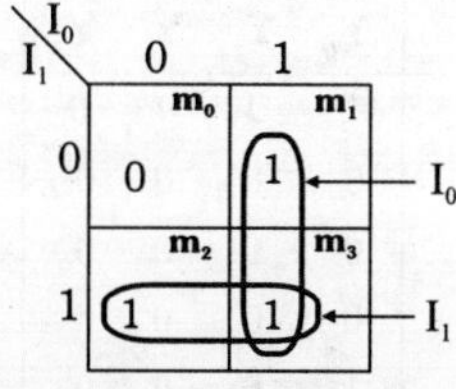

***Fig. a:** K-map for Y_0.*

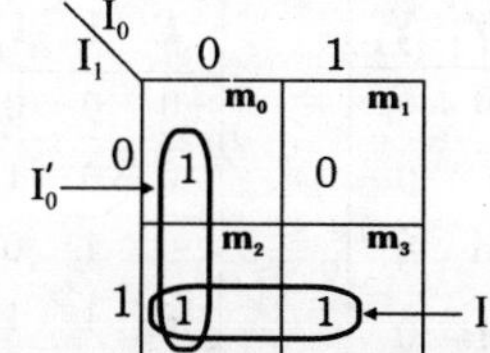

***Fig. b:** K-map for Y_1.*

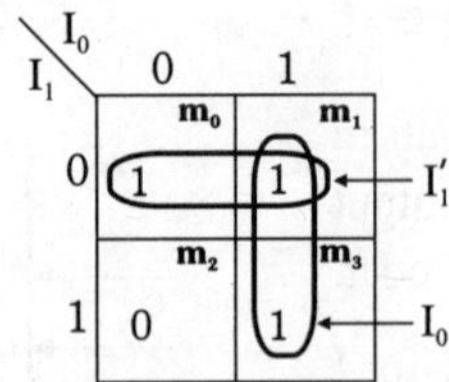

Fig. c: K-map for Y_2.

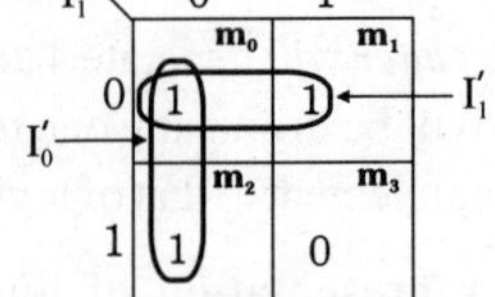

Fig. d: K-map for Y_3.

Fig. 5.9: K-map for logic low 2-to-4 decoder.

From the K-map for Y_0, Y_1, Y_2 and Y_3 the following Boolean equations are obtained and using these equations the logic circuit for logic **low** 2-to-4 decoder is drawn in Fig. 5.10.

$$Y_0 = I_1 + I_0$$

$$Y_1 = I_1 + I_0'$$

$$Y_2 = I_1' + I_0$$

$$Y_3 = I_1' + I_0'$$

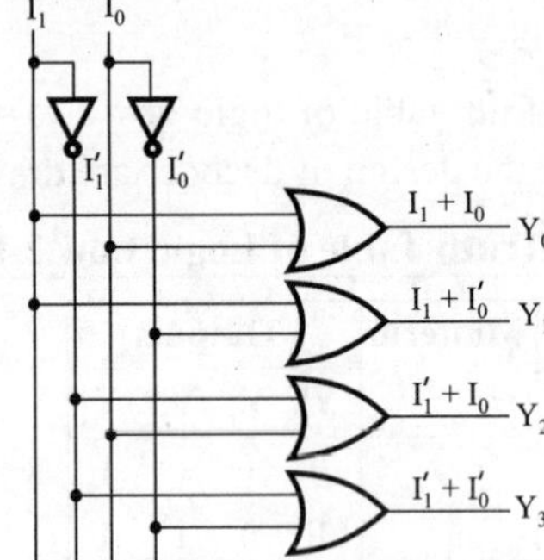

Fig. 5.10: Logic circuit for logic low 2-to-4 decoder.

> ***Note:*** *Alternatively, the output equations of logic* ***low*** *2-to-4 decoder can be directly written from the truth table which are maxterms for which output is* ***low****.*

5.4.3 Logic High 2-to-4 Decoder with Logic High Enable

The block diagram representation of logic **high** decoder with logic **high** enable is shown in Fig. 5.11. When enable input is made **low** then inputs are not recognized and so all outputs are zero.

When enable input is made **high** then inputs are recognized and any one of the outputs is set to 1 depending on input.

The truth table of decoder for both active and inactive enable is shown in Table 5.3. Using this truth table, 3-variable K-maps are drawn for 4-outputs of decoder as shown in Fig. 5.12.

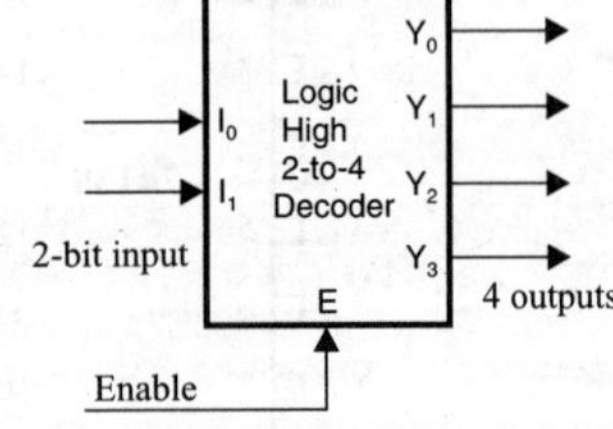

Fig. 5.11: Logic high 2-to-4 decoder with logic high enable.

Table 5.3: Truth Table of Logic High 2-to-4 Decoder with Logic High Enable

Inputs			Outputs			
E	I_1	I_0	Y_0	Y_1	Y_2	Y_3
0	×	×	0	0	0	0
1	0	0	1	0	0	0
1	0	1	0	1	0	0
1	1	0	0	0	1	0
1	1	1	0	0	0	1

Inputs			Minterm	Outputs				Comment
E	I_1	I_0		Y_0	Y_1	Y_2	Y_3	
0	0	0	m_0	0	0	0	0	When enable is not active
0	0	1	m_1	0	0	0	0	
0	1	0	m_2	0	0	0	0	
0	1	1	m_3	0	0	0	0	
1	0	0	m_4	1	0	0	0	When enable is active
1	0	1	m_5	0	1	0	0	
1	1	0	m_6	0	0	1	0	
1	1	1	m_7	0	0	0	1	

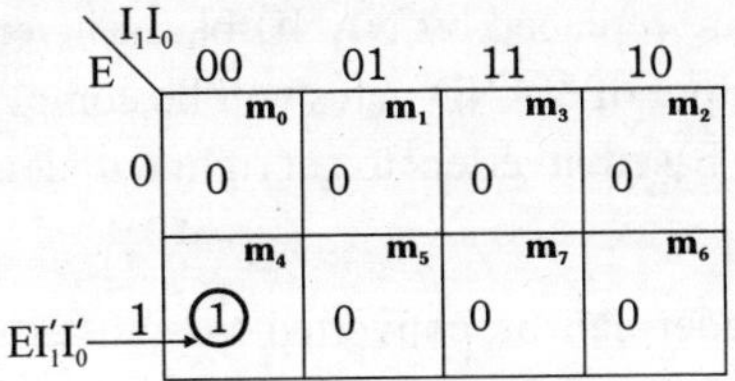

Fig. a: *K-map for Y_0.*

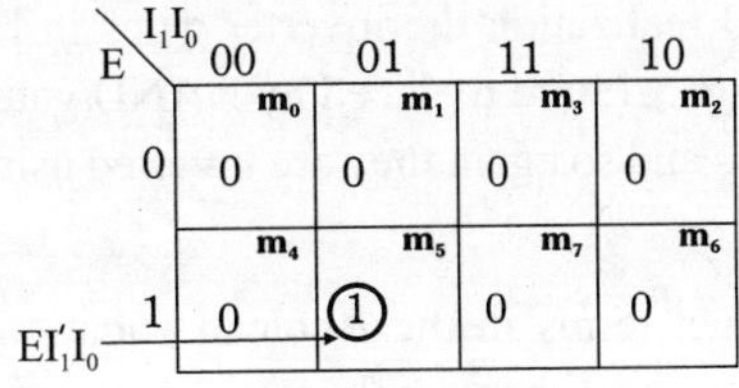

Fig. b: *K-map for Y_1.*

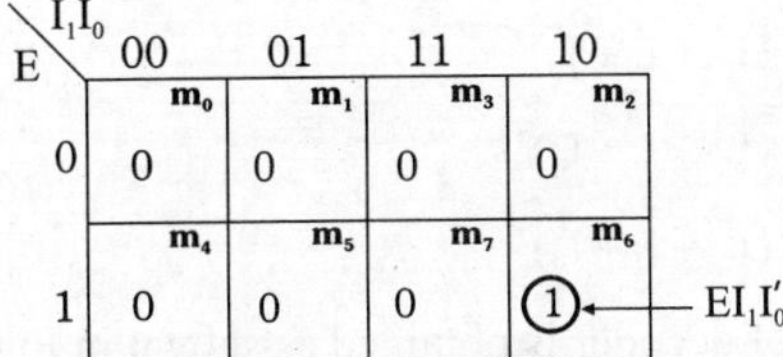

Fig. c: *K-map for Y_2.*

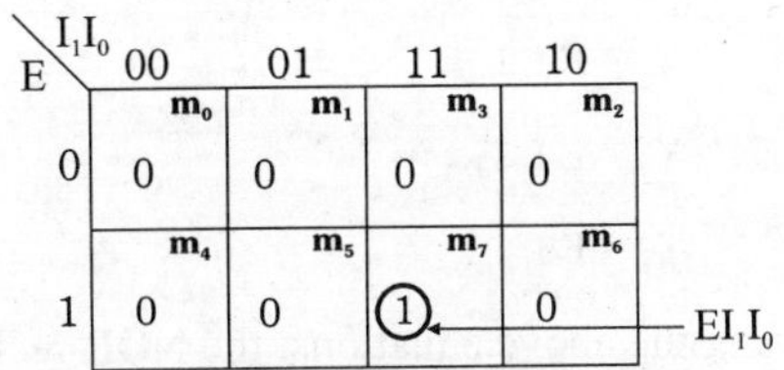

Fig. d: *K-map for Y_3.*

Fig. 5.12: *K-map for logic high 2-to-4 decoder with logic high enable.*

From the K-map for Y_0, Y_1, Y_2 and Y_3 the following Boolean equations are obtained and using these equations the logic circuit of decoder is drawn in Fig. 5.13.

$$Y_0 = EI_1'I_0'$$

$$Y_1 = EI_1'I_0$$

$$Y_2 = EI_1I_0'$$

$$Y_3 = EI_1I_0$$

> ***Note:*** *Alternatively, the output equations of logic **high** 2-to-4 decoder can be directly written from the truth table which are minterms for which output is **high**.*

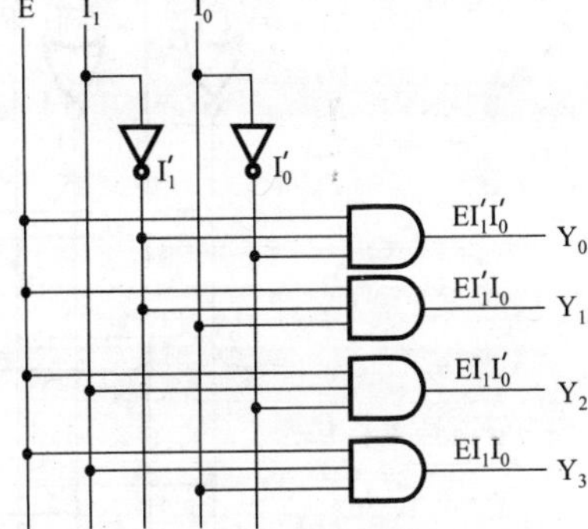

Fig. 5.13: *Logic circuit for 2-to-4 logic high decoder with logic high enable.*

Using the above equations it is possible to realize the decoder using only NAND gates as shown in Fig. 5.14.

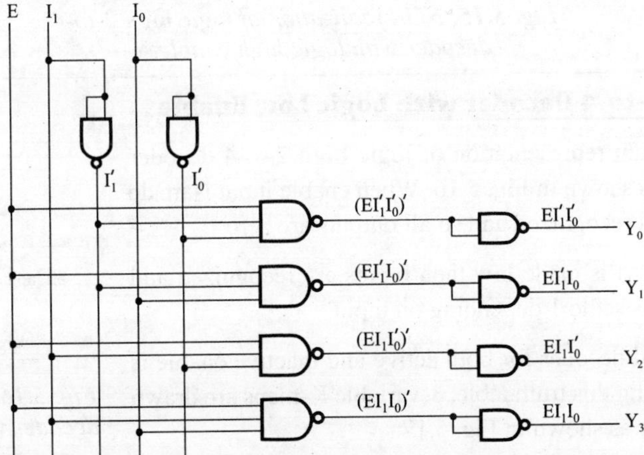

Fig. 5.14: *NAND realization of logic high 2-to-4 decoder with logic high enable.*

In NAND realization the inverter gates in Fig. 5.13 are replaced by NAND based inverters and AND gates in Fig. 5.13 are replaced by NAND gates. The output of NAND gates will be complement of the product terms and so again they are inverted using NAND based inverters to get right output as shown in Fig. 5.14.

The product terms in the Boolean equations of decoder can be converted to sum terms using DeMorgan's theorem as shown below:

$$Y_0 = E\, I_1'\, I_0' \qquad \xrightarrow{\text{Using DeMorgan's Theorem}} \qquad Y_0 = (E' + I_1 + I_0)'$$

$$Y_1 = E\, I_1'\, I_0 \qquad Y_1 = (E' + I_1 + I_0')'$$

$$Y_2 = E\, I_1\, I_0' \qquad Y_2 = (E' + I_1' + I_0)'$$

$$Y_3 = E\, I_1\, I_0 \qquad Y_3 = (E' + I_1' + I_0')'$$

Using the above equations, the NOR realization of decoder is obtained as shown in Fig. 5.15.

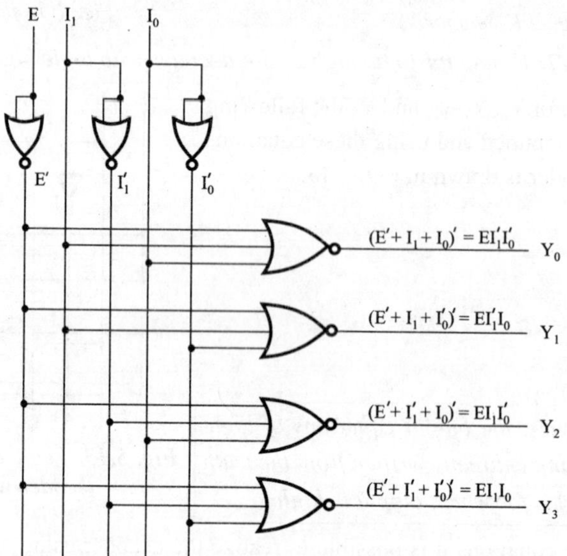

Fig. 5.15: NOR realization of logic high 2-to-4 decoder with logic high enable.

5.4.4 Logic High 2-to-4 Decoder with Logic Low Enable

The block diagram representation of logic **high** 2-to-4 decoder with logic **low** enable is shown in Fig. 5.16. When enable input is made **high** then inputs are not recognized and so all outputs are zero.

When enable input is made **low** then inputs are recognized and any one of the outputs is set to 1 depending on input.

The truth table of decoder for both active and inactive enable is shown in Table 5.4. Using this truth table, 3-variable K-maps are drawn for 4 outputs of decoder as shown in Fig. 5.17.

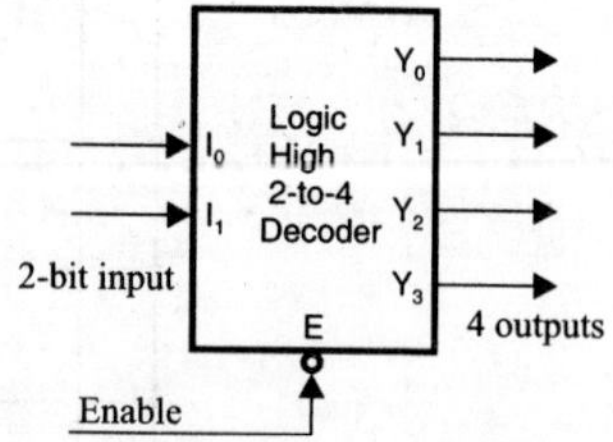

Fig. 5.16: Logic high 2-to-4 decoder with logic low enable.

Table 5.4: Truth Table of Logic High 2-to-4 Decoder with Logic Low Enable

Inputs			Outputs			
E	I_1	I_0	Y_0	Y_1	Y_2	Y_3
1	×	×	0	0	0	0
0	0	0	1	0	0	0
0	0	1	0	1	0	0
0	1	0	0	0	1	0
0	1	1	0	0	0	1

Inputs			Minterm	Outputs				Comment
E	I_1	I_0		Y_0	Y_1	Y_2	Y_3	
0	0	0	m_0	1	0	0	0	Output when enable is active
0	0	1	m_1	0	1	0	0	
0	1	0	m_2	0	0	1	0	
0	1	1	m_3	0	0	0	1	
1	0	0	m_4	0	0	0	0	Output when enable is inactive
1	0	1	m_5	0	0	0	0	
1	1	0	m_6	0	0	0	0	
1	1	1	m_7	0	0	0	0	

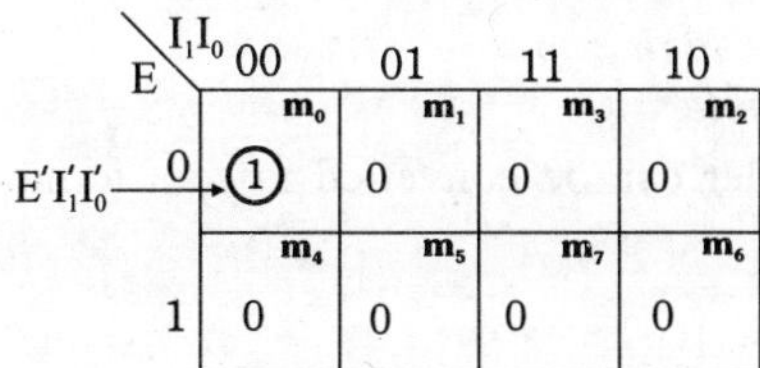

Fig. a: K-map for Y_0.

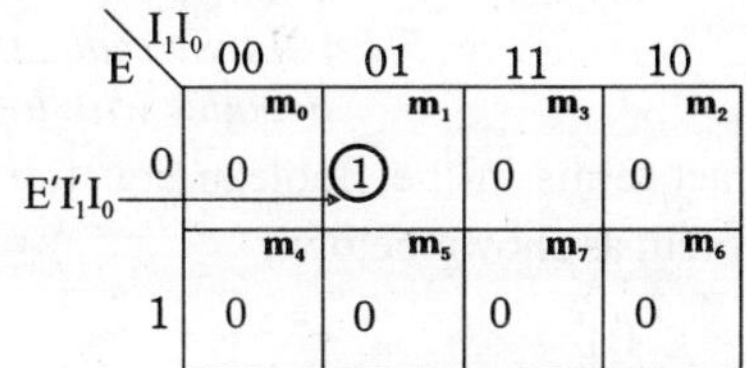

Fig. b: K-map for Y_1.

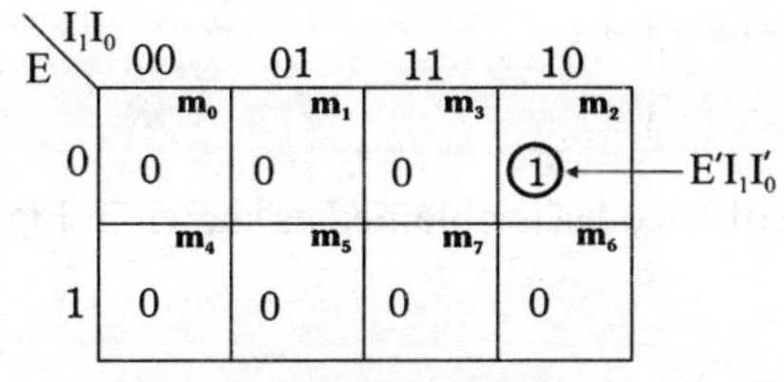

Fig. c: K-map for Y_2.

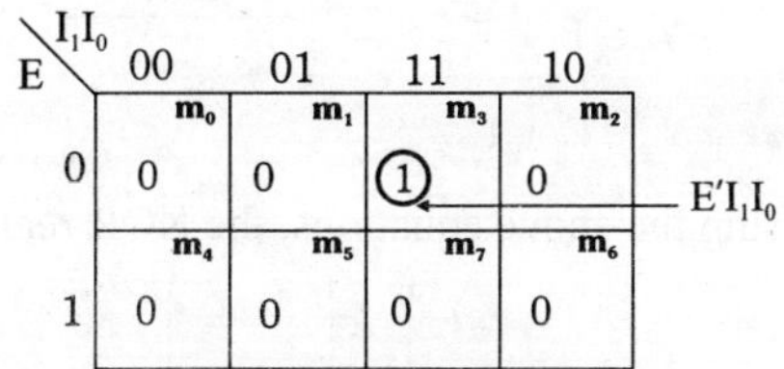

Fig. d: K-map for Y_3.

***Fig. 5.17:** K-map for logic high 2-to-4 decoder with logic low enable.*

From the K-map for Y_0, Y_1, Y_2 and Y_3 the following Boolean equations are obtained and using these equations the logic circuit of decoder is drawn in Fig. 5.18.

$$Y_0 = E'\,I_1'I_0'$$

$$Y_1 = E'\,I_1'I_0$$

$$Y_2 = E'\,I_1I_0'$$

$$Y_3 = E'\,I_1I_0$$

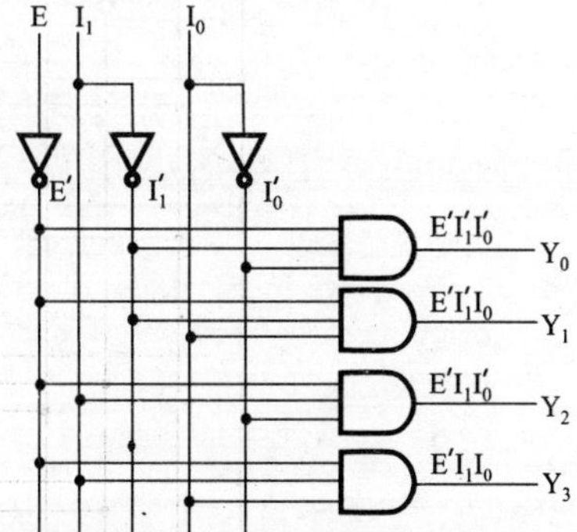

***Fig. 5.18:** Logic circuit for 2-to-4 logic high decoder with logic low enable.*

Using the above equations it is possible to realize the decoder using only NAND gates as shown in Fig. 5.19.

> ***Note:** Alternatively, the output equations of logic **high** 2-to-4 decoder can be directly written from the truth table which are minterms for which output is **high**.*

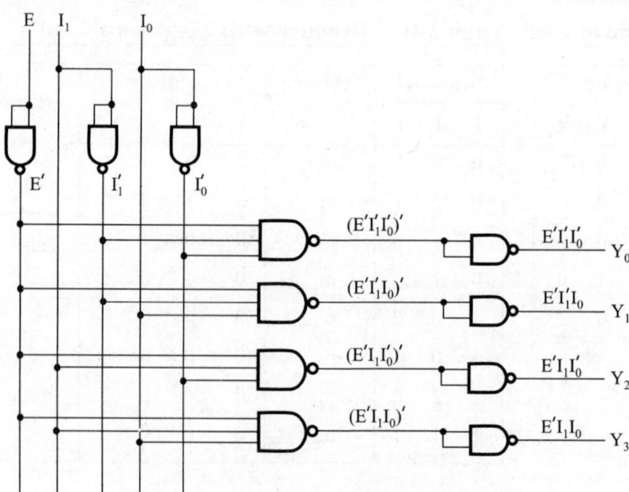

Fig. 5.19: *NAND realization of logic high 2-to-4 decoder with logic low enable.*

The product terms in the Boolean equations of decoder can be converted to sum terms using DeMorgan's theorem as shown below:

$$Y_0 = E'\,I_1'\,I_0' \qquad\qquad Y_0 = (E + I_1 + I_0)'$$

$$Y_1 = E'\,I_1'\,I_0 \qquad\qquad Y_1 = (E + I_1 + I_0')'$$

$$Y_2 = E'\,I_1\,I_0' \quad \xrightarrow{\text{Using DeMorgan's Theorem}} \quad Y_2 = (E + I_1' + I_0)'$$

$$Y_3 = E'\,I_1\,I_0 \qquad\qquad Y_3 = (E + I_1' + I_0')'$$

Using the above equations, the NOR realization of decoder is obtained as shown in Fig. 5.20.

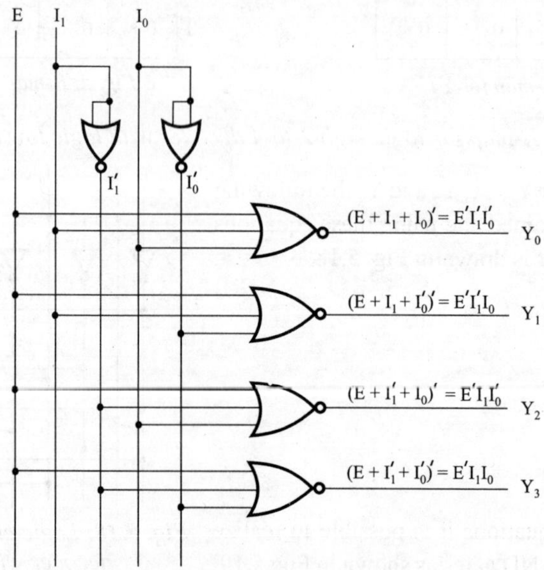

Fig. 5.20: *NOR realization of logic high 2-to-4 decoder with logic low enable.*

5.4.5 Logic Low 2-to-4 Decoder with Logic High Enable

The block diagram representation of logic **low** 2-to-4 decoder with logic **high** enable is shown in Fig. 5.21. When enable input is made **low** then inputs are not recognized and so all outputs are set to 1.

When enable input is made **high** then inputs are recognized and any one of the outputs is made **low** depending on input.

The truth table of decoder for both active and inactive enable is shown in Table 5.5. Using this truth table, 3-variable K-maps are drawn for 4 outputs of decoder as shown in Fig. 5.22.

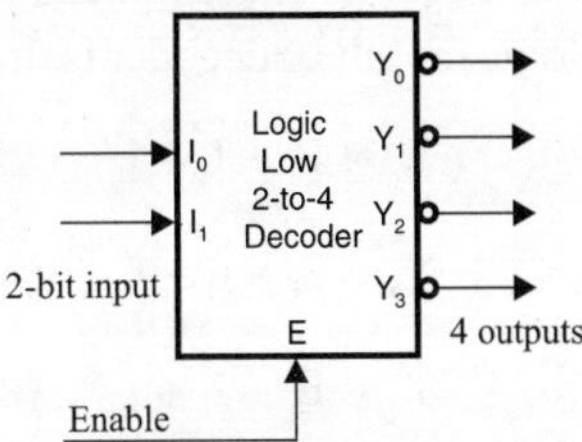

Fig. 5.21: Logic low 2-to-4 decoder with logic high enable.

Table 5.5: Truth Table of Logic low 2-to-4 Decoder with Logic High Enable

Inputs			Minterm	Outputs			
E	I_1	I_0		Y_0	Y_1	Y_2	Y_3
0	×	×	m_0	1	1	1	1
1	0	0	m_4	0	1	1	1
1	0	1	m_5	1	0	1	1
1	1	0	m_6	1	1	0	1
1	1	1	m_7	1	1	1	0

Inputs			Minterm	Outputs				Comment
E	I_1	I_0		Y_0	Y_1	Y_2	Y_3	
0	0	0	m_0	1	1	1	1	Output when enable is inactive
0	0	1	m_1	1	1	1	1	
0	1	0	m_2	1	1	1	1	
0	1	1	m_3	1	1	1	1	
1	0	0	m_4	0	1	1	1	Output when enable is active
1	0	1	m_5	1	0	1	1	
1	1	0	m_6	1	1	0	1	
1	1	1	m_7	1	1	1	0	

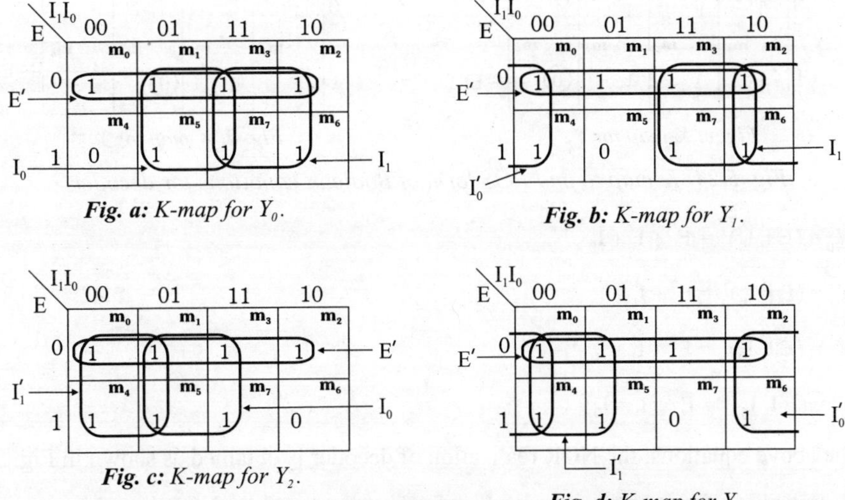

Fig. a: K-map for Y_0.

Fig. b: K-map for Y_1.

Fig. c: K-map for Y_2.

Fig. d: K-map for Y_3.

***Fig. 5.22:** K-map for logic low 2-to-4 decoder with logic high enable.*

From the K-map for Y_0, Y_1, Y_2 and Y_3 the following Boolean equations are obtained and using these equations the logic circuit of decoder is drawn as shown in Fig. 5.23.

$$Y_0 = E' + I_1 + I_0$$

$$Y_1 = E' + I_1 + I_0'$$

$$Y_2 = E' + I_1' + I_0$$

$$Y_3 = E' + I_1' + I_0'$$

> ***Note:** Alternatively, the output equations of logic **low** 2-to-4 decoder can be directly written from the truth table which are maxterms for which output is **low**.*

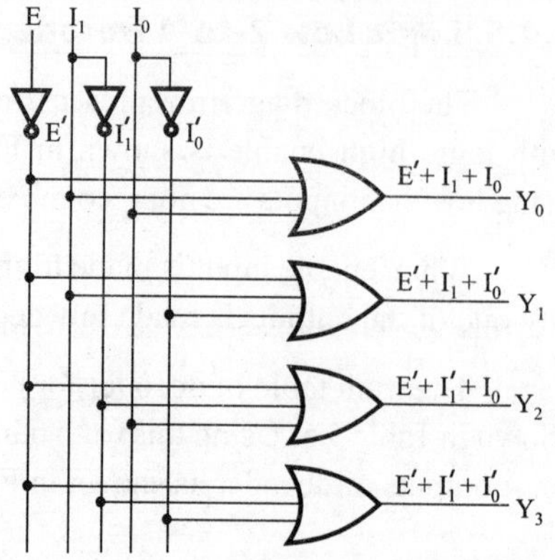

Fig. 5.23: Logic circuit for logic low 2-to-4 decoder with logic high enable.

Alternatively, the following POS form of Boolean equations can be obtained by forming prime implicants using 0's as shown in Fig. 5.24.

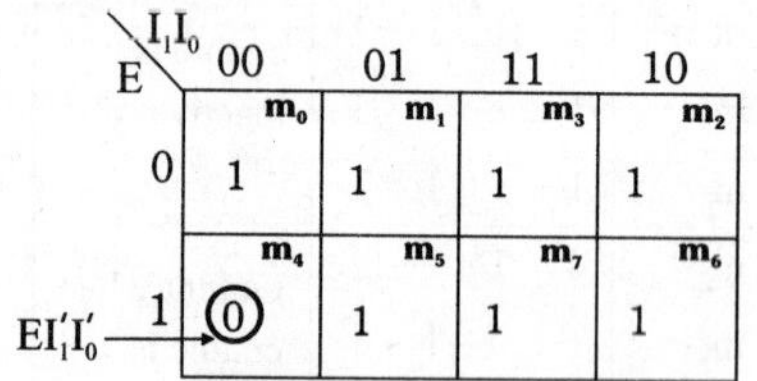

E \ I_1I_0	00	01	11	10
0	1 (m_0)	1 (m_1)	1 (m_3)	1 (m_2)
1	0 (m_4)	1 (m_5)	1 (m_7)	1 (m_6)

$EI_1'I_0'$

Fig. a: K-map for Y_0.

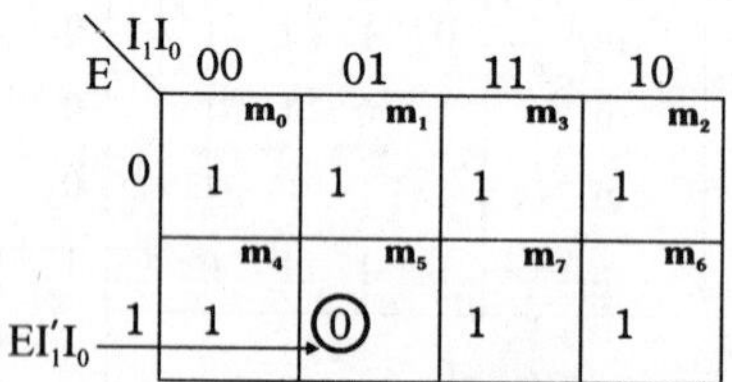

E \ I_1I_0	00	01	11	10
0	1 (m_0)	1 (m_1)	1 (m_3)	1 (m_2)
1	1 (m_4)	0 (m_5)	1 (m_7)	1 (m_6)

$EI_1'I_0$

Fig. b: K-map for Y_1.

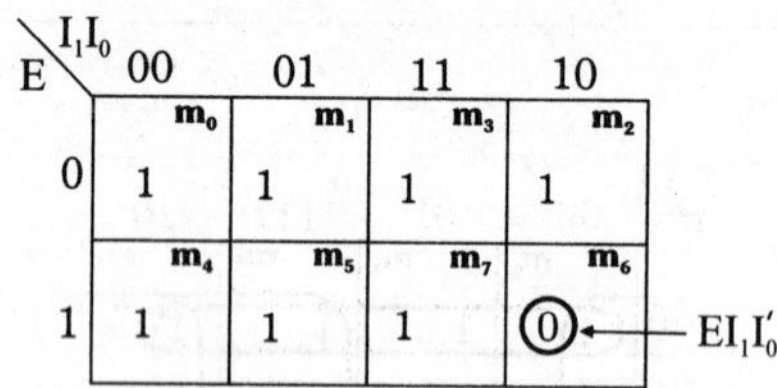

E \ I_1I_0	00	01	11	10
0	1 (m_0)	1 (m_1)	1 (m_3)	1 (m_2)
1	1 (m_4)	1 (m_5)	1 (m_7)	0 (m_6)

EI_1I_0'

Fig. c: K-map for Y_2.

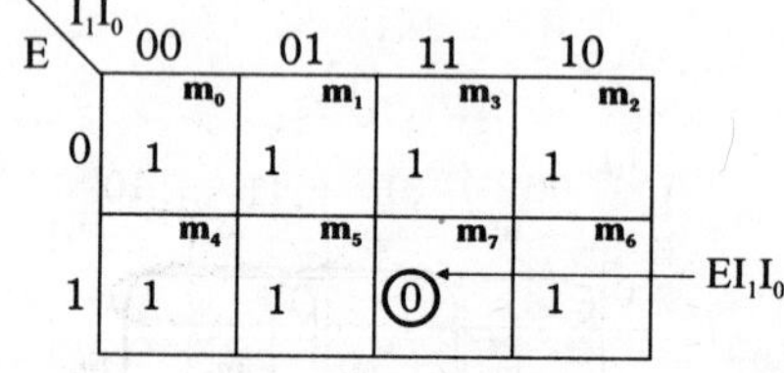

E \ I_1I_0	00	01	11	10
0	1 (m_0)	1 (m_1)	1 (m_3)	1 (m_2)
1	1 (m_4)	1 (m_5)	0 (m_7)	1 (m_6)

EI_1I_0

Fig. d: K-map for Y_3.

***Fig. 5.24:** K-map to find POS form of Boolean equations for decoder.*

$$Y_0 = (E\,I_1'\,I_0')' = E' + I_1 + I_0$$

$$Y_1 = (E\,I_1'\,I_0)' = E' + I_1 + I_0'$$

$$Y_2 = (E\,I_1\,I_0')' = E' + I_1' + I_0$$

$$Y_3 = (E\,I_1\,I_0)' = E' + I_1' + I_0'$$

Using the above equations, the NOR realization of decoder is obtained as shown in Fig. 5.25.

In NOR realization the inverter gates in Fig. 5.23 are replaced by NOR based inverters and OR gates in Fig. 5.23 are replaced by NOR gates. The output of NOR gates will be complement of the sum terms and so again they are inverted using NOR based inverters to get right output as shown in Fig. 5.25.

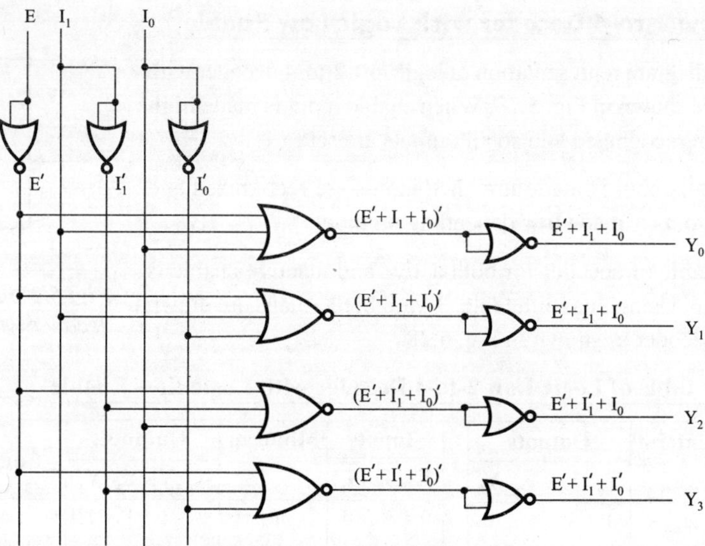

***Fig. 5.25:** NOR realization of logic low 2-to-4 decoder with logic high enable.*

The sum terms in the Boolean equations of decoder can be converted to product terms using DeMorgan's theorem as shown below:

$$Y_0 = E' + I_1 + I_0 \qquad Y_0 = (E\, I_1'\, I_0')'$$

$$Y_1 = E' + I_1 + I_0' \quad \xrightarrow{\text{Using DeMorgan's Theorem}} \quad Y_1 = (E\, I_1'\, I_0)'$$

$$Y_2 = E' + I_1' + I_0 \qquad Y_2 = (E\, I_1\, I_0')'$$

$$Y_3 = E' + I_1' + I_0' \qquad Y_3 = (E\, I_1\, I_0)'$$

Using the above equations, the NAND realization of decoder is obtained as shown in Fig. 5.26.

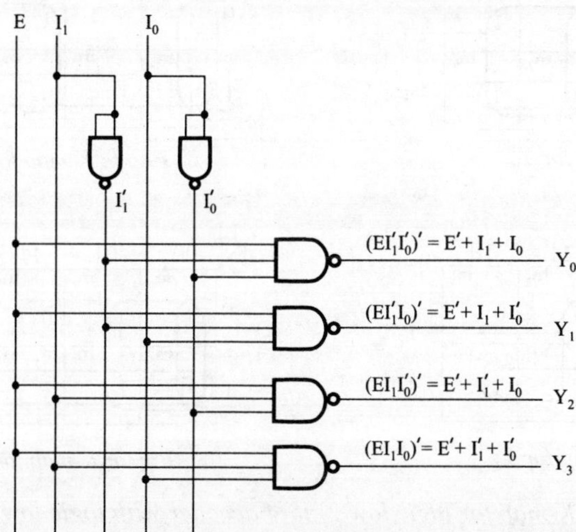

***Fig. 5.26:** NAND realization of logic low 2-to-4 decoder with logic high enable.*

5.4.6 Logic Low 2-to-4 Decoder with Logic Low Enable

The block diagram representation of logic **low** 2-to-4 decoder with logic **low** enable is shown in Fig. 5.27. When enable input is made **high** then inputs are not recognized and so all outputs are set to 1.

When enable input is made **low** then inputs are recognized and any one of the outputs is made **low** depending on input.

The truth table of decoder for both active and inactive enable is shown in Table 5.6. Using this truth table, 3-variable K-maps are drawn for 4 outputs of decoder as shown in Fig. 5.28.

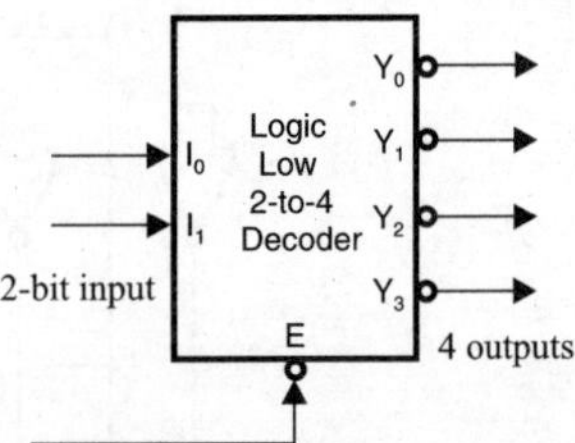

Fig. 5.27: Logic low 2-to-4 decoder with logic low enable.

Table 5.6: Truth Table of Logic Low 2-to-4 Decoder with Logic Low Enable

Inputs			Minterm	Outputs			
E	I_1	I_0		Y_0	Y_1	Y_2	Y_3
1	×	×	m_4	1	1	1	1
0	0	0	m_0	0	1	1	1
0	0	1	m_1	1	0	1	1
0	1	0	m_2	1	1	0	1
0	1	1	m_3	1	1	1	0

Inputs			Minterm	Outputs				Comment
E	I_1	I_0		Y_0	Y_1	Y_2	Y_3	
0	0	0	m_0	0	1	1	1	Output when enable is active
0	0	1	m_1	1	0	1	1	
0	1	0	m_2	1	1	0	1	
0	1	1	m_3	1	1	1	0	
1	0	0	m_4	1	1	1	1	Output when enable is inactive
1	0	1	m_5	1	1	1	1	
1	1	0	m_6	1	1	1	1	
1	1	1	m_7	1	1	1	1	

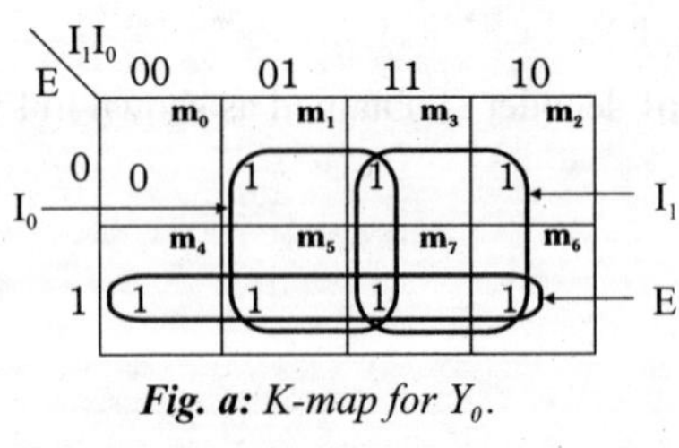

Fig. a: K-map for Y_0.

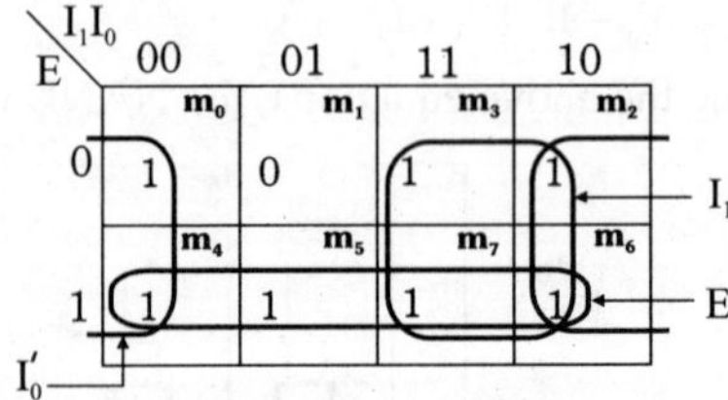

Fig. b: K-map for Y_1.

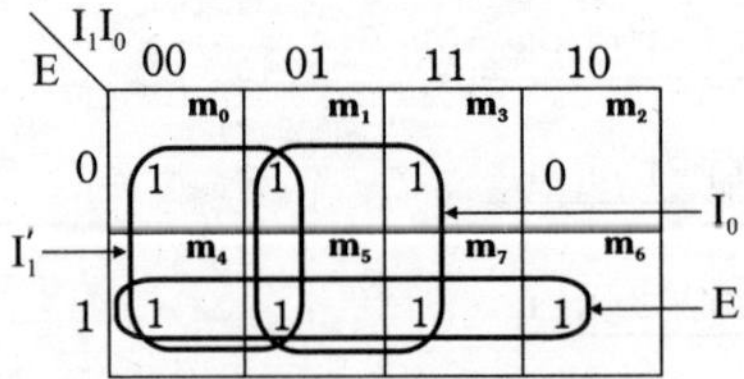

Fig. c: K-map for Y_2.

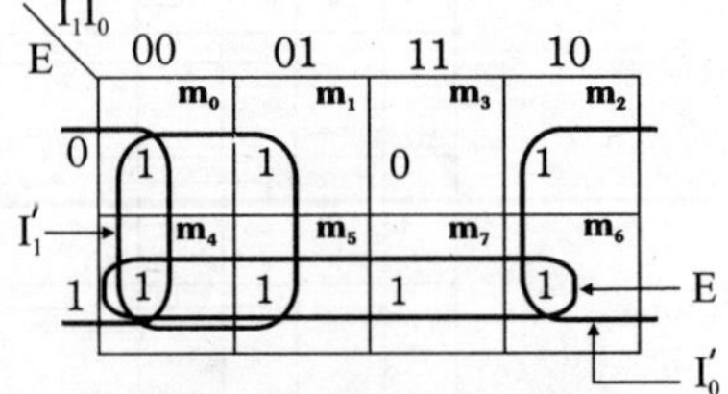

Fig. d: K-map for Y_3.

Fig. 5.28: K-map for logic low 2- to-4 decoder with logic low enable.

From the K-map of Fig. 5.28 for Y_0, Y_1, Y_2 and Y_3 the following Boolean equations are obtained and using these equations the logic circuit of decoder is drawn as shown in Fig. 5.29.

$$Y_0 = E + I_1 + I_0$$

$$Y_1 = E + I_1 + I_0'$$

$$Y_2 = E + I_1' + I_0$$

$$Y_3 = E + I_1' + I_0'$$

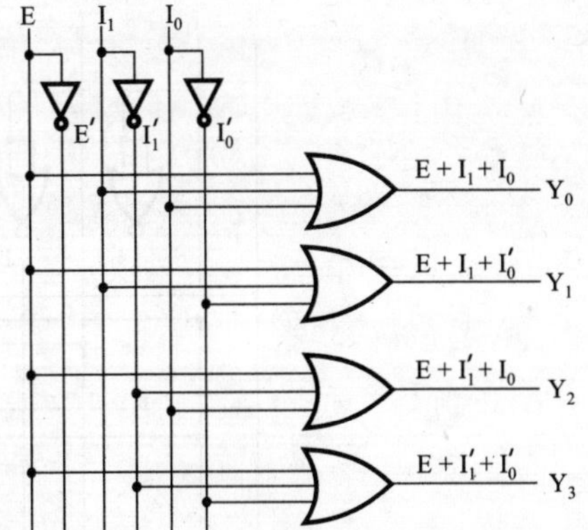

***Fig. 5.29:** Logic circuit for logic low 2-to-4 decoder with logic low enable.*

> ***Note:** Alternatively, the output equations of logic **low** 2-to-4 decoder can be directly written from the truth table which are maxterms for which output is **low**.*

Alternatively, the following POS form of Boolean equations can be obtained by forming prime implicants using 0's as shown in Fig. 5.30.

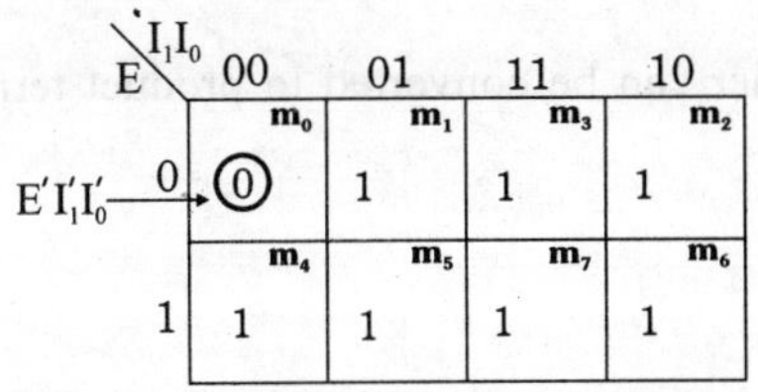

***Fig. a:** K-map for Y_0.*

E \ I_1I_0	00	01	11	10
0	m_0: 1	m_1: (0) ← $E'I_1'I_0$	m_3: 1	m_2: 1
1	m_4: 1	m_5: 1	m_7: 1	m_6: 1

***Fig. b:** K-map for Y_1.*

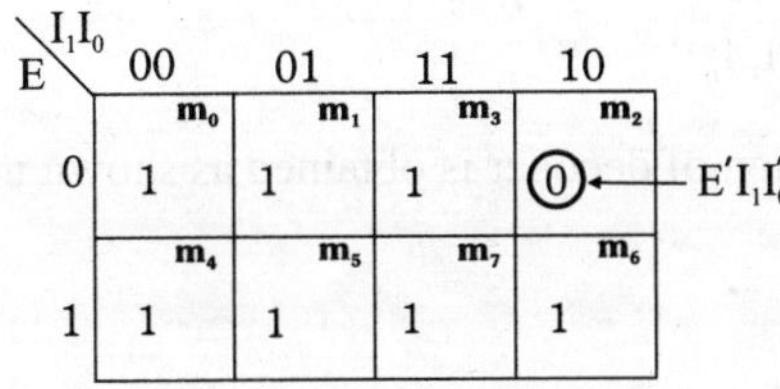

***Fig. c:** K-map for Y_2.*

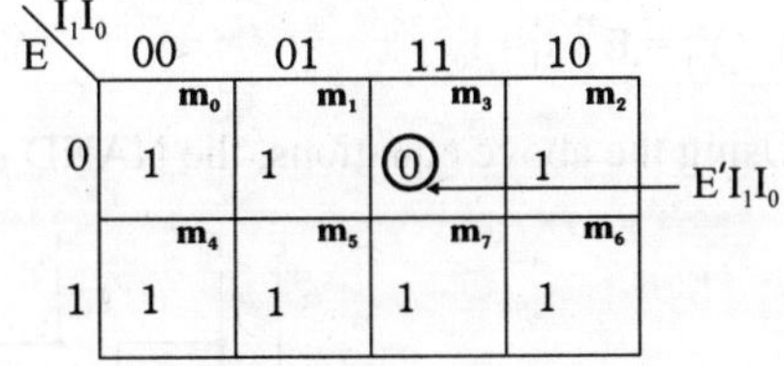

***Fig. d:** K-map for Y_3.*

***Fig. 5.30:** K-map to find POS form of Boolean equations for decoder.*

$$Y_0 = (E' I_1' I_0')' = E + I_1 + I_0$$

$$Y_1 = (E' I_1' I_0)' = E + I_1 + I_0'$$

$$Y_2 = (E' I_1 I_0')' = E + I_1' + I_0$$

$$Y_3 = (E' I_1 I_0)' = E + I_1' + I_0'$$

Using the above equations it is possible to realize the decoder using only NOR gates as shown in Fig. 5.31.

In NOR realization the inverter gates in Fig. 5.29 are replaced by NOR based inverters and OR gates in Fig. 5.29 are replaced by NOR gates. The output of NOR gates will be complement of the sum terms and so again they are complemented using NOR based inverters to get right output as shown in Fig. 5.31.

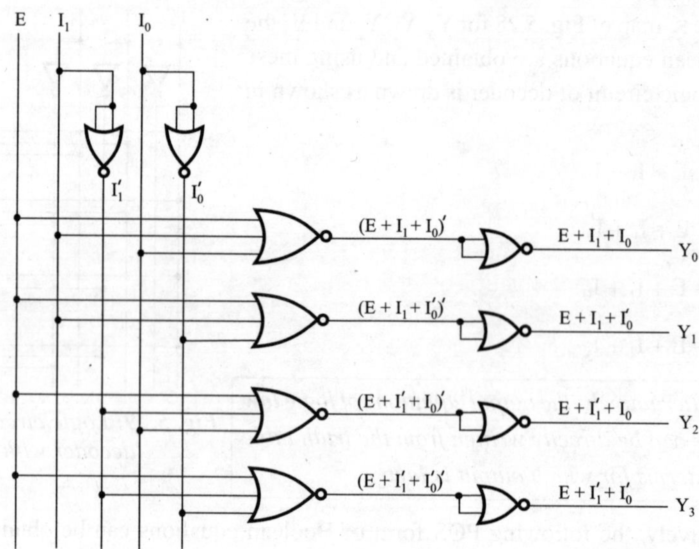

Fig. 5.31: NOR realization of logic low 2-to-4 decoder with logic low enable.

The sum terms in the Boolean equations of decoder can be converted to product terms using DeMorgan's theorem as shown below:

$$Y_0 = E + I_1 + I_0 \qquad Y_0 = (E'\, I_1'\, I_0')'$$

$$Y_1 = E + I_1 + I_0' \quad \xrightarrow{\text{Using DeMorgan's Theorem}} \quad Y_1 = (E'\, I_1'\, I_0)'$$

$$Y_2 = E + I_1' + I_0 \qquad Y_2 = (E'\, I_1\, I_0')'$$

$$Y_3 = E + I_1' + I_0' \qquad Y_3 = (E'\, I_1\, I_0)'$$

Using the above equations, the NAND realization of decoder is obtained as shown in Fig. 5.32.

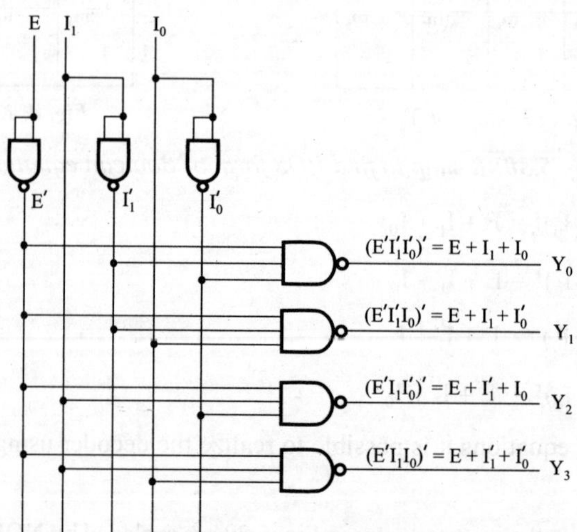

Fig. 5.32: NAND realization of logic low 2-to-4 decoder with logic low enable.

5.4.7 Logic High 3-to-8 Decoder *(IKGPTU, May'20, 7 Marks) (GTU, May'19, 7 Marks) (GTU, Jun'19, 3 Marks)*

A ***3-to-8 (3-to-2^3) decoder*** can generate 8 decoded outputs from the three bit inputs.

The ***logic high 3-to-8 decoder*** will generate 8 unique outputs in which only one will be **high** at any one time and all other outputs will be **low**.

The block diagram representation of logic **high** 3-to-8 decoder is shown in Fig. 5.33. The truth table of logic **high** 3-to-8 decoder is shown in Table 5.7.

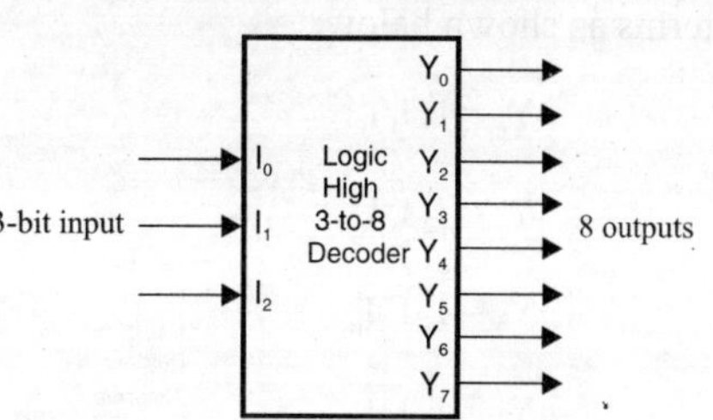

***Fig. 5.33:** Logic high 3-to-8 decoder.*

From the truth table we can directly write Boolean equations for outputs of logic **high** 3-to-8 decoder which are the minterms for which output is **high** as shown below:

$$Y_0 = I_2' I_1' I_0'$$

$$Y_1 = I_2' I_1' I_0$$

$$Y_2 = I_2' I_1 I_0'$$

$$Y_3 = I_2' I_1 I_0$$

$$Y_4 = I_2 I_1' I_0'$$

$$Y_5 = I_2 I_1' I_0$$

$$Y_6 = I_2 I_1 I_0'$$

$$Y_7 = I_2 I_1 I_0$$

Table 5.7: Truth Table of Logic High 3-to-8 Decoder

Inputs			Minterm	Outputs							
I_2	I_1	I_0		Y_0	Y_1	Y_2	Y_3	Y_4	Y_5	Y_6	Y_7
0	0	0	m_0	1	0	0	0	0	0	0	0
0	0	1	m_1	0	1	0	0	0	0	0	0
0	1	0	m_2	0	0	1	0	0	0	0	0
0	1	1	m_3	0	0	0	1	0	0	0	0
1	0	0	m_4	0	0	0	0	1	0	0	0
1	0	1	m_5	0	0	0	0	0	1	0	0
1	1	0	m_6	0	0	0	0	0	0	1	0
1	1	1	m_7	0	0	0	0	0	0	0	1

The logic **high** 3-to-8 decoder is available as a standard IC with number 74237. The pin configuration of 74237 is shown in Fig. 5.34.

74237

Pin	Name	Pin	Name
1	I_0	16	V_{cc}(+5V)
2	I_1	15	Y_0
3	I_2	14	Y_1
4	$\overline{E}_0$	13	Y_2
5	E_1	12	Y_3
6	E_2	11	Y_4
7	Y_7	10	Y_5
8	(0V) GND	9	Y_6

Pin	Description
I_0 - I_2	Inputs
$\overline{E}_0$	Logic low enable
E_1, E_2	Logic high enables
Y_0 - Y_7	Active high outputs
V_{cc}	Supply voltage (+5V)
GND	Ground (0V)

***Fig. 5.34:** Pin configuration of logic high 3-to-8 decoder IC 74237.*

The product terms in the Boolean equations of logic **high** 3-to-8 decoder can be converted to sum terms as shown below:

(Used for AND Realization)		(Used for NOR Realization)
$Y_0 = I_2' I_1' I_0'$		$Y_0 = (I_2 + I_1 + I_0)'$
$Y_1 = I_2' I_1' I_0$		$Y_1 = (I_2 + I_1 + I_0')'$
$Y_2 = I_2' I_1 I_0'$	Using DeMorgan's Theorem	$Y_2 = (I_2 + I_1' + I_0)'$
$Y_3 = I_2' I_1 I_0$	$\longrightarrow$	$Y_3 = (I_2 + I_1' + I_0')'$
$Y_4 = I_2 I_1' I_0'$		$Y_4 = (I_2' + I_1 + I_0)'$
$Y_5 = I_2 I_1' I_0$		$Y_5 = (I_2' + I_1 + I_0')'$
$Y_6 = I_2 I_1 I_0'$		$Y_6 = (I_2' + I_1' + I_0)'$
$Y_7 = I_2 I_1 I_0$		$Y_7 = (I_2' + I_1' + I_0')'$

Using the above Boolean equations the logic circuit of logic **high** 3-to-8 decoder using AND gates is drawn as shown in Fig. 5.35. It is possible to realize the logic **high** 3-to-8 decoder using only NOR gates as shown in Fig. 5.36.

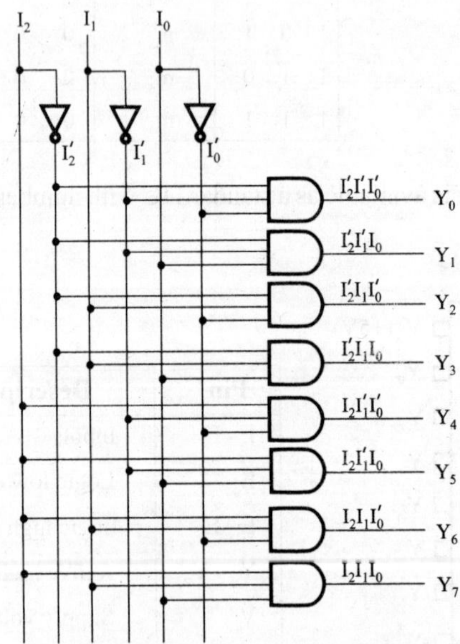

***Fig. 5.35:** AND realization of logic high of 3-to-8 decoder.*

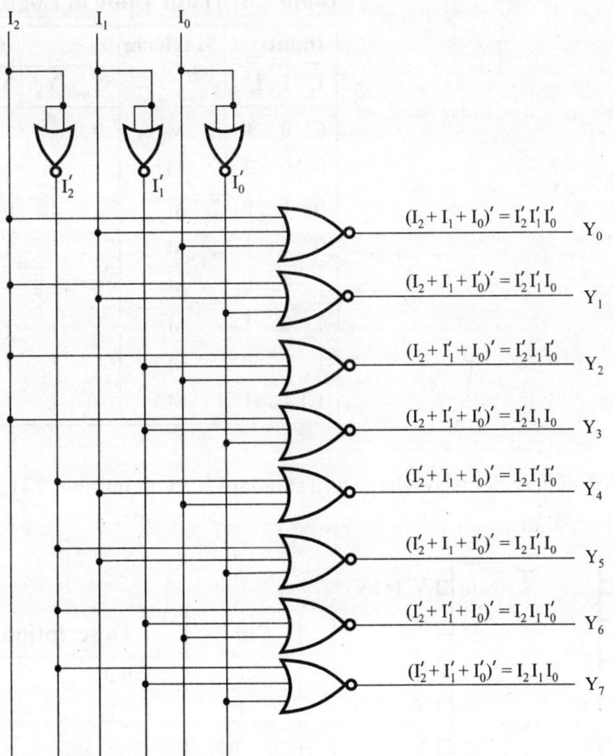

Fig. 5.36: *NOR realization of logic high 3-to-8 decoder.*

5.4.8 Logic Low 3-to-8 Decoder

The ***logic low 3-to-8 decoder*** will generate 8 unique outputs in which only one will be **low** at any one time and all other outputs will be **high**.

The block diagram representation of logic **low** 3-to-8 decoder is shown in Fig. 5.37. The truth table of logic **low** 3-to-8 decoder is shown in Table 5.8.

From the truth table we can directly write Boolean equations for outputs of logic **low** 3-to-8 decoder which are the maxterms for which output is **low** as shown ahead.

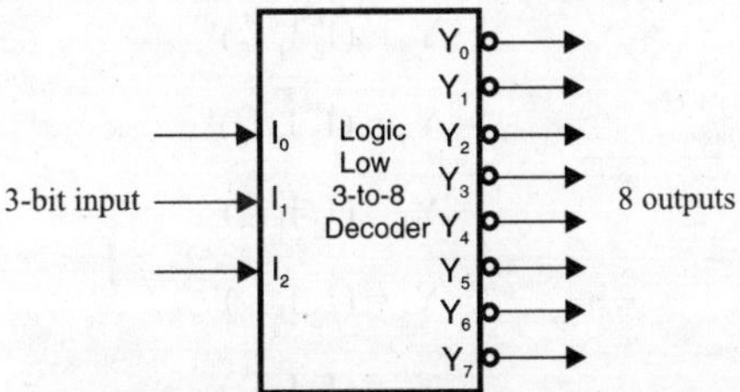

Fig. 5.37: *Logic low 3-8 decoder.*

$Y_0 = I_2 + I_1 + I_0$

$Y_1 = I_2 + I_1 + I_0'$

$Y_2 = I_2 + I_1' + I_0$

$Y_3 = I_2 + I_1' + I_0'$

$Y_4 = I_2' + I_1 + I_0$

$Y_5 = I_2' + I_1 + I_0'$

$Y_6 = I_2' + I_1' + I_0$

$Y_7 = I_2' + I_1' + I_0'$

Table 5.8: Truth Table of Logic Low 3-to-8 Decoder

Inputs			Maxterm	Outputs							
I_2	I_1	I_0		Y_0	Y_1	Y_2	Y_3	Y_4	Y_5	Y_6	Y_7
0	0	0	M_0	0	1	1	1	1	1	1	1
0	0	1	M_1	1	0	1	1	1	1	1	1
0	1	0	M_2	1	1	0	1	1	1	1	1
0	1	1	M_3	1	1	1	0	1	1	1	1
1	0	0	M_4	1	1	1	1	0	1	1	1
1	0	1	M_5	1	1	1	1	1	0	1	1
1	1	0	M_6	1	1	1	1	1	1	0	1
1	1	1	M_7	1	1	1	1	1	1	1	0

The logic **low** 3-to-8 decoder is available as a standard IC with number 74138. The pin configuration of 74138 is shown in Fig. 5.38.

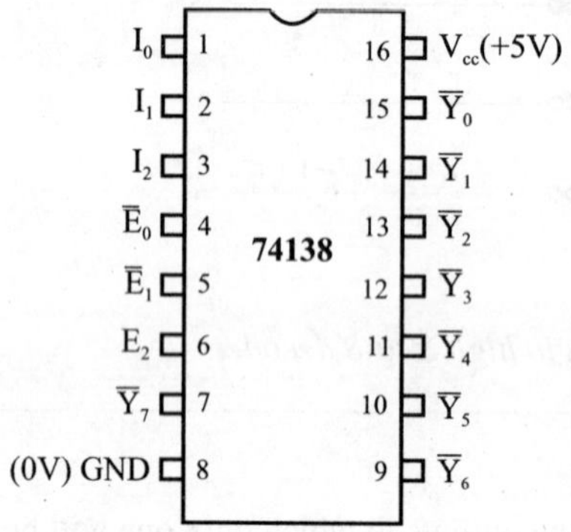

74138

Pin	Signal	Pin	Signal
1	I_0	16	V_{cc}(+5V)
2	I_1	15	$\overline{Y}_0$
3	I_2	14	$\overline{Y}_1$
4	$\overline{E}_0$	13	$\overline{Y}_2$
5	$\overline{E}_1$	12	$\overline{Y}_3$
6	E_2	11	$\overline{Y}_4$
7	$\overline{Y}_7$	10	$\overline{Y}_5$
8	(0V) GND	9	$\overline{Y}_6$

Pin	Description
I_0 - I_2	Inputs
$\overline{E}_0$, $\overline{E}_1$	Logic low enables
E_2	Logic high enable
$\overline{Y}_0$ - $\overline{Y}_7$	Active low outputs
V_{cc}	Supply voltage (+5V)
GND	Ground (0V)

Fig. 5.38: Pin configuration of logic low 3-to-8 decoder IC 74138.

The sum terms in the Boolean equations of 3-to-8 logic **low** decoder can be converted to product terms as shown below:

Using DeMorgan's Theorem →

(Used for OR Realization)	(Used for NAND Realization)
$Y_0 = I_2 + I_1 + I_0$	$Y_0 = (I_2' I_1' I_0')'$
$Y_1 = I_2 + I_1 + I_0'$	$Y_1 = (I_2' I_1' I_0)'$
$Y_2 = I_2 + I_1' + I_0$	$Y_2 = (I_2' I_1 I_0')'$
$Y_3 = I_2 + I_1' + I_0'$	$Y_3 = (I_2' I_1 I_0)'$
$Y_4 = I_2' + I_1 + I_0$	$Y_4 = (I_2 I_1' I_0')'$
$Y_5 = I_2' + I_1 + I_0'$	$Y_5 = (I_2 I_1' I_0)'$
$Y_6 = I_2' + I_1' + I_0$	$Y_6 = (I_2 I_1 I_0')'$
$Y_7 = I_2' + I_1' + I_0'$	$Y_7 = (I_2 I_1 I_0)'$

Using the above Boolean equations the logic circuit of logic **low** 3-to-8 decoder using OR gates is drawn as shown in Fig. 5.39. It is possible to realize the logic **low** 3-to-8 decoder using only NAND gates as shown in Fig. 5.40.

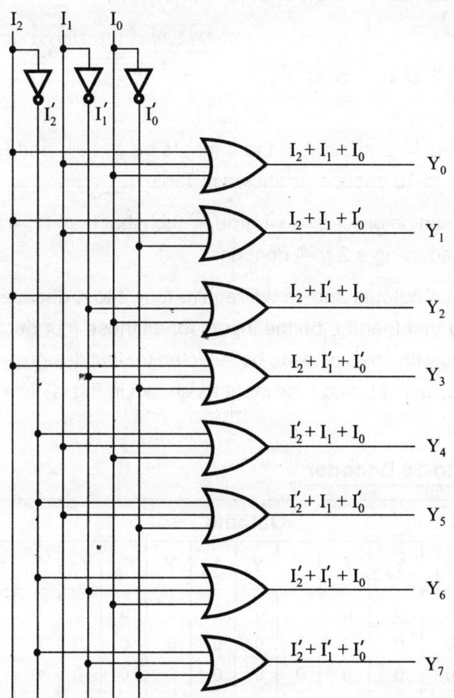

***Fig. 5.39:** OR realization of logic low 3-to-8 decoder.*

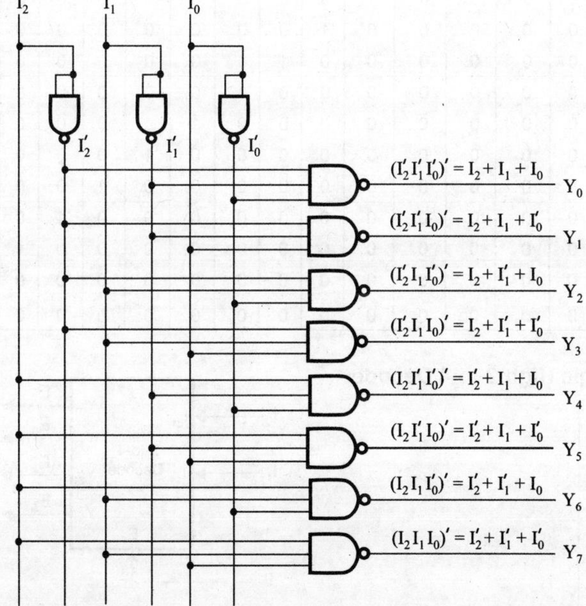

***Fig. 5.40:** NAND realization of logic low 3-to-8 decoder.*

5.4.9 Expanding the Decoder

Decoder of smaller size can be connected in cascade to expand the size of decoder. Three examples of decoder expansion are presented in Examples 5.11, 5.12 and 5.13.

Example 5.11

(BATU, Jun'19, 6 Marks) (JNTUA, Jun/July'22, 7 Marks)

Construct a 4-to-16 decoder using 2-to-4 decoder.

Solution

A 4-to-16 decoder has 4 inputs and 16 outputs. Let the 4 inputs be I_0, I_1, I_2 and I_3 and the 16 outputs be Y_0, Y_1, ..., Y_{15}. The truth table of logic **high** 4-to-16 decoder is shown in Table 1.

The truth table of 4-to-16 decoder can be divided into 4 groups such that I_2 and I_3 are same in each group. Now, the outputs in each group can be generated using a 2-to-4 decoder.

Each 2-to-4 decoder can generate 4 outputs and so we require four 2-to-4 decoder to generate 16 outputs as shown in Fig. 2. From Table 1 we can say that I_0 and I_1 be the inputs for all these four decoders but only one decoder should be active at a time. Hence decoders with enable has to be selected for this design and another 2-to-4 decoder is required to generate four enable signals by using I_2 and I_3 as input as shown in Fig. 1. The working of 4-to-16 decoder using 2-to-4 decoder is verified in Table 3.

Table 1: Truth Table of Logic High 4-to-16 Decoder

Inputs				Outputs															
I_3	I_2	I_1	I_0	Y_0	Y_1	Y_2	Y_3	Y_4	Y_5	Y_6	Y_7	Y_8	Y_9	Y_{10}	Y_{11}	Y_{12}	Y_{13}	Y_{14}	Y_{15}
0	0	0	0	1	0	0	0	0	0	0	0	0	0	0	0	0	0	0	0
0	0	0	1	0	1	0	0	0	0	0	0	0	0	0	0	0	0	0	0
0	0	1	0	0	0	1	0	0	0	0	0	0	0	0	0	0	0	0	0
0	0	1	1	0	0	0	1	0	0	0	0	0	0	0	0	0	0	0	0
0	1	0	0	0	0	0	0	1	0	0	0	0	0	0	0	0	0	0	0
0	1	0	1	0	0	0	0	0	1	0	0	0	0	0	0	0	0	0	0
0	1	1	0	0	0	0	0	0	0	1	0	0	0	0	0	0	0	0	0
0	1	1	1	0	0	0	0	0	0	0	1	0	0	0	0	0	0	0	0
1	0	0	0	0	0	0	0	0	0	0	0	1	0	0	0	0	0	0	0
1	0	0	1	0	0	0	0	0	0	0	0	0	1	0	0	0	0	0	0
1	0	1	0	0	0	0	0	0	0	0	0	0	0	1	0	0	0	0	0
1	0	1	1	0	0	0	0	0	0	0	0	0	0	0	1	0	0	0	0
1	1	0	0	0	0	0	0	0	0	0	0	0	0	0	0	1	0	0	0
1	1	0	1	0	0	0	0	0	0	0	0	0	0	0	0	0	1	0	0
1	1	1	0	0	0	0	0	0	0	0	0	0	0	0	0	0	0	1	0
1	1	1	1	0	0	0	0	0	0	0	0	0	0	0	0	0	0	0	1

Table 2: Truth Table of Logic High 2-to-4 Decoder

Inputs		Outputs			
I_1	I_0	Y_0	Y_1	Y_2	Y_3
0	0	1	0	0	0
0	1	0	1	0	0
1	0	0	0	1	0
1	1	0	0	0	1

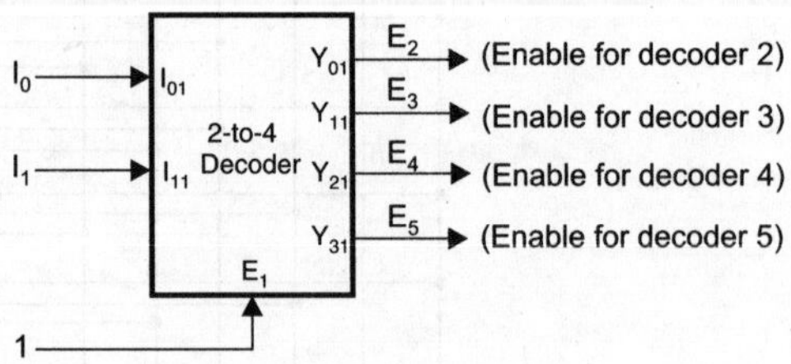

Fig. 1: Generation of enable signal.

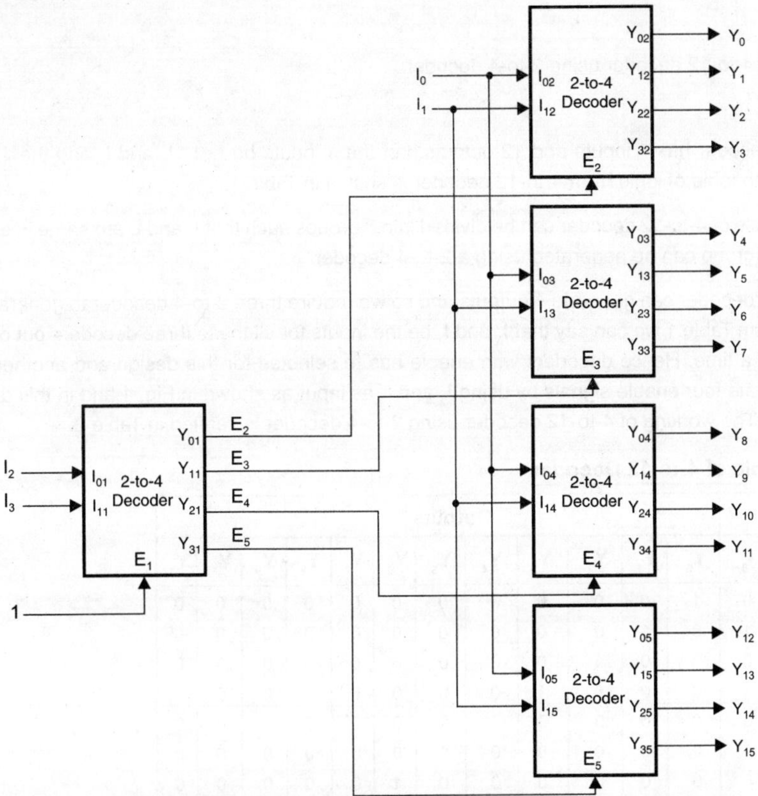

Fig. 2: 4-to-16 decoder using 2-to-4 decoder.

Table 3: Verification of 4-to-16 Decoder Shown in Fig. 2

Inputs				Enables				Outputs															
I_3	I_2	I_1	I_0	$E_2 = Y_{01}$	$E_3 = Y_{11}$	$E_4 = Y_{21}$	$E_5 = Y_{31}$	$Y_0 = Y_{02}$	$Y_1 = Y_{12}$	$Y_2 = Y_{22}$	$Y_3 = Y_{32}$	$Y_4 = Y_{03}$	$Y_5 = Y_{13}$	$Y_6 = Y_{23}$	$Y_7 = Y_{33}$	$Y_8 = Y_{04}$	$Y_9 = Y_{14}$	$Y_{10} = Y_{24}$	$Y_{11} = Y_{34}$	$Y_{12} = Y_{05}$	$Y_{13} = Y_{15}$	$Y_{14} = Y_{25}$	$Y_{15} = Y_{35}$
0	0	0	0	1	0	0	0	1	0	0	0	0	0	0	0	0	0	0	0	0	0	0	0
0	0	0	1	1	0	0	0	0	1	0	0	0	0	0	0	0	0	0	0	0	0	0	0
0	0	1	0	1	0	0	0	0	0	1	0	0	0	0	0	0	0	0	0	0	0	0	0
0	0	1	1	1	0	0	0	0	0	0	1	0	0	0	0	0	0	0	0	0	0	0	0
0	1	0	0	0	1	0	0	0	0	0	0	1	0	0	0	0	0	0	0	0	0	0	0
0	1	0	1	0	1	0	0	0	0	0	0	0	1	0	0	0	0	0	0	0	0	0	0
0	1	1	0	0	1	0	0	0	0	0	0	0	0	1	0	0	0	0	0	0	0	0	0
0	1	1	1	0	1	0	0	0	0	0	0	0	0	0	1	0	0	0	0	0	0	0	0
1	0	0	0	0	0	1	0	0	0	0	0	0	0	0	0	1	0	0	0	0	0	0	0
1	0	0	1	0	0	1	0	0	0	0	0	0	0	0	0	0	1	0	0	0	0	0	0
1	0	1	0	0	0	1	0	0	0	0	0	0	0	0	0	0	0	1	0	0	0	0	0
1	0	1	1	0	0	1	0	0	0	0	0	0	0	0	0	0	0	0	1	0	0	0	0
1	1	0	0	0	0	0	1	0	0	0	0	0	0	0	0	0	0	0	0	1	0	0	0
1	1	0	1	0	0	0	1	0	0	0	0	0	0	0	0	0	0	0	0	0	1	0	0
1	1	1	0	0	0	0	1	0	0	0	0	0	0	0	0	0	0	0	0	0	0	1	0
1	1	1	1	0	0	0	1	0	0	0	0	0	0	0	0	0	0	0	0	0	0	0	1

Example 5.12

Construct a 4-to-12 decoder using 2-to-4 decoder.

Solution

A 4-to-12 decoder has 4 inputs and 12 outputs. Let the 4 inputs be I_0, I_1, I_2 and I_3 and the 12 outputs be Y_0, Y_1, ..., Y_{11}. The truth table of logic **high** 4-to-12 decoder is shown in Table 1.

The truth table of 4-to-12 decoder can be divided into 3 groups such that I_2 and I_3 are same in each group. Now, the outputs in each group can be generated using a 2-to-4 decoder.

Each 2-to-4 decoder can generate 4 outputs and so we require three 2-to-4 decoder to generate 12 outputs as shown in Fig. 2. From Table 1 we can say that I_0 and I_1 be the inputs for all these three decoders but only one decoder should be active at a time. Hence decoders with enable has to selected for this design and another 2-to-4 decoder is required to generate four enable signals by using I_2 and I_3 as input as shown in Fig. 1 and in this design one of the enable is not used. The working of 4-to-12 decoder using 2-to-4 decoder is verified in Table 2.

Table 1: Truth Table of 4-to-12 Decoder

Inputs				Outputs											
I_3	I_2	I_1	I_0	Y_0	Y_1	Y_2	Y_3	Y_4	Y_5	Y_6	Y_7	Y_8	Y_9	Y_{10}	Y_{11}
0	0	0	0	1	0	0	0	0	0	0	0	0	0	0	0
0	0	0	1	0	1	0	0	0	0	0	0	0	0	0	0
0	0	1	0	0	0	1	0	0	0	0	0	0	0	0	0
0	0	1	1	0	0	0	1	0	0	0	0	0	0	0	0
0	1	0	0	0	0	0	0	1	0	0	0	0	0	0	0
0	1	0	1	0	0	0	0	0	1	0	0	0	0	0	0
0	1	1	0	0	0	0	0	0	0	1	0	0	0	0	0
0	1	1	1	0	0	0	0	0	0	0	1	0	0	0	0
1	0	0	0	0	0	0	0	0	0	0	0	1	0	0	0
1	0	0	1	0	0	0	0	0	0	0	0	0	1	0	0
1	0	1	0	0	0	0	0	0	0	0	0	0	0	1	0
1	0	1	1	0	0	0	0	0	0	0	0	0	0	0	1

Table 2: Verification of 4-to-12 Decoder Shown in Fig. 1

Inputs				Enables			Outputs											
I_3	I_2	I_1	I_0	$E_2 = Y_{01}$	$E_3 = Y_{11}$	$E_4 = Y_{21}$	$Y_0 = Y_{02}$	$Y_1 = Y_{12}$	$Y_2 = Y_{22}$	$Y_3 = Y_{32}$	$Y_4 = Y_{03}$	$Y_5 = Y_{13}$	$Y_6 = Y_{23}$	$Y_7 = Y_{33}$	$Y_8 = Y_{04}$	$Y_9 = Y_{14}$	$Y_{10} = Y_{24}$	$Y_{11} = Y_{34}$
0	0	0	0	1	0	0	1	0	0	0	0	0	0	0	0	0	0	0
0	0	0	1	1	0	0	0	1	0	0	0	0	0	0	0	0	0	0
0	0	1	0	1	0	0	0	0	1	0	0	0	0	0	0	0	0	0
0	0	1	1	1	0	0	0	0	0	1	0	0	0	0	0	0	0	0
0	1	0	0	0	1	0	0	0	0	0	1	0	0	0	0	0	0	0
0	1	0	1	0	1	0	0	0	0	0	0	1	0	0	0	0	0	0
0	1	1	0	0	1	0	0	0	0	0	0	0	1	0	0	0	0	0
0	1	1	1	0	1	0	0	0	0	0	0	0	0	1	0	0	0	0
1	0	0	0	0	0	1	0	0	0	0	0	0	0	0	1	0	0	0
1	0	0	1	0	0	1	0	0	0	0	0	0	0	0	0	1	0	0
1	0	1	0	0	0	1	0	0	0	0	0	0	0	0	0	0	1	0
1	0	1	1	0	0	1	0	0	0	0	0	0	0	0	0	0	0	1

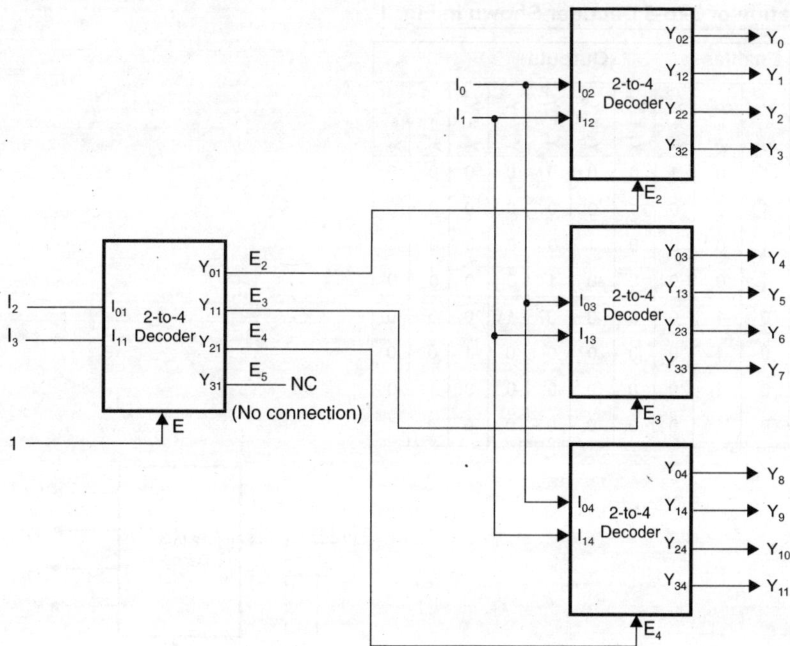

Fig. 1: 4-to-12 decoder using 2-to-4 decoder.

Example 5.13

(AU, Apr/May'19, 7 Marks)

Design 3-to-8 decoder using 2-to-4 decoders. Draw the truth table.

Solution

A 3-to-8 decoder has 3 inputs and 8 outputs. Let the 3 inputs be I_0, I_1 and I_2 and the 8 outputs be Y_0, Y_1, ..., Y_7. The truth table of logic **high** 3-to-8 decoder is shown in Table 1. The truth table of 3-to-8 decoder can be divided into 2 groups such that I_0 and I_1 are same in each group. Now, the outputs in each group can be generated using a 2-to-4 decoder.

Each 2-to-4 decoder can generate 4 outputs and so we require two 2-to-4 decoder to generate 8 outputs as shown in Fig. 1. From Table 1 we can say that I_0 and I_1 be the inputs for all these two decoders but only one decoder should be active at a time. Hence decoders with enable has to be selected for this design and another 2-to-4 decoder is required to generate four enable signals by using I_2 and 0 as input as shown in Fig. 1 and in this two enables are not used. The working of 3-to-8 decoder using 2-to-4 decoder is verified in Table 2.

Table 1: Truth Table of 3-to-8 Decoder

Inputs			Outputs							
I_2	I_1	I_0	Y_0	Y_1	Y_2	Y_3	Y_4	Y_5	Y_6	Y_7
0	0	0	1	0	0	0	0	0	0	0
0	0	1	0	1	0	0	0	0	0	0
0	1	0	0	0	1	0	0	0	0	0
0	1	1	0	0	0	1	0	0	0	0
1	0	0	0	0	0	0	1	0	0	0
1	0	1	0	0	0	0	0	1	0	0
1	1	0	0	0	0	0	0	0	1	0
1	1	1	0	0	0	0	0	0	0	1

Table 2: Verification of 3-to-8 Decoder Shown in Fig. 1

Inputs			Enables		Outputs							
I_2	I_1	I_0	$E_2 = Y_{01}$	$E_3 = Y_{21}$	$Y_0 = Y_{02}$	$Y_1 = Y_{12}$	$Y_2 = Y_{22}$	$Y_3 = Y_{32}$	$Y_4 = Y_{03}$	$Y_5 = Y_{13}$	$Y_6 = Y_{23}$	$Y_7 = Y_{33}$
0	0	0	1	0	1	0	0	0	0	0	0	0
0	0	1	1	0	0	1	0	0	0	0	0	0
0	1	0	1	0	0	0	1	0	0	0	0	0
0	1	1	1	0	0	0	0	1	0	0	0	0
1	0	0	0	1	0	0	0	0	1	0	0	0
1	0	1	0	1	0	0	0	0	0	1	0	0
1	1	0	0	1	0	0	0	0	0	0	1	0
1	1	1	0	1	0	0	0	0	0	0	0	1

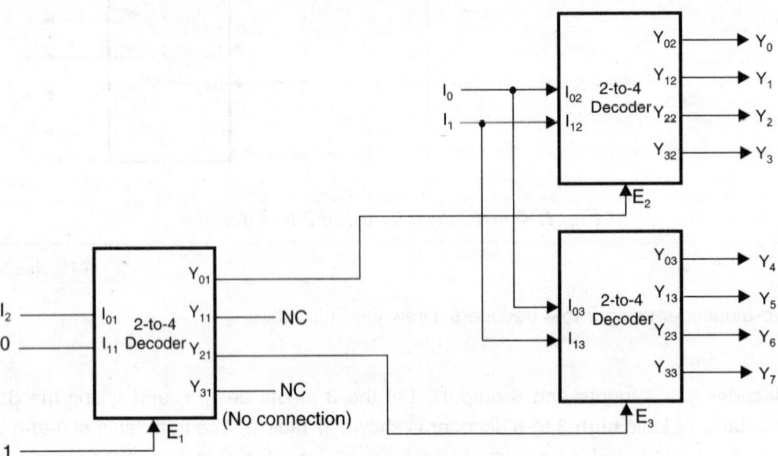

Fig. 1: 3-to-8 decoder using 2-to-4 decoder.

5.4.10 Combinational Circuit Design using Decoder

The logic **high** decoder generate output 1 for only one particular minterm as input. Therefore, a Boolean function expressed as a sum of minterms can be obtained by logical OR of the outputs of decoder for which the minterms of function generates logic **high** or 1.

The complement of logic **low** decoder output will be same as output of logic **high** decoder. Hence, the complement of output of logic **low** decoder can be used to realize SOP form of Boolean function.

Similarly, the logic **low** decoder generates output 0 for only one particular maxterm as input. Therefore, a Boolean function expressed as a product of maxterms can be obtained by logical AND of the outputs of decoder for which the maxterms of function generates logic **low** or 0.

The complement of logic **high** decoder output will be same as output of logic **low** decoder. Hence the complement of output of logic **high** decoder can be used to realize POS form of Boolean function.

Example 5.14

Realize the following functions,

$F_1 = \Sigma m(0, 1, 3, 6)$

$F_2 = \Pi M(0, 2, 4, 7)$

(a) Using logic **high** 3-to-8 decoder and external OR gates only

(b) Using logic **high** 3-to-8 decoder and external NOR gates only.

Solution

Given that, $F_1 = \Sigma m(0, 1, 3, 6)$

$F_2 = \Pi M(0, 2, 4, 7) = \Sigma m(1, 3, 5, 6)$

Since the given functions has minterms m_0 to m_7, the given functions are 3-variable functions and so can be realized using a 3-to-8 decoder. The 3-to-8 decoder has three inputs and the minterms are formed from all possible combination of inputs as shown in Table 1. Here, x, y and z are input variables of the functions.

The truth table of F_1 and F_2 is shown in Table 2.

Table 1: Truth Table of Logic High 3-to-8 Decoder

Inputs				Outputs							
x	y	z	Minterm	Y_0	Y_1	Y_2	Y_3	Y_4	Y_5	Y_6	Y_7
0	0	0	m_0	1	0	0	0	0	0	0	0
0	0	1	m_1	0	1	0	0	0	0	0	0
0	1	0	m_2	0	0	1	0	0	0	0	0
0	1	1	m_3	0	0	0	1	0	0	0	0
1	0	0	m_4	0	0	0	0	1	0	0	0
1	0	1	m_5	0	0	0	0	0	1	0	0
1	1	0	m_6	0	0	0	0	0	0	1	0
1	1	1	m_7	0	0	0	0	0	0	0	1

Table 2: Truth Table of F_1 and F_2

Minterm	F_1	F_2
m_0 = 0 0 0	1	0
m_1 = 0 0 1	1	1
m_2 = 0 1 0	0	0
m_3 = 0 1 1	1	1
m_4 = 1 0 0	0	0
m_5 = 1 0 1	0	1
m_6 = 1 1 0	1	1
m_7 = 1 1 1	0	0

(a) Using Logic High 3-to-8 Decoder and External OR Gates Only

In function F_1, the output is "1" for minterms 0, 1, 3 and 6. In logic **high** decoder the output Y_n is "1" when the input corresponds to minterm m_n and so the outputs Y_0, Y_1, Y_3 and Y_6 are logically ORed for OR implementation.

Similarly, for function, F_2 the decoder outputs Y_1, Y_3, Y_5 and Y_6 are logically ORed for OR implementation.

The Boolean equations of functions F_1 and F_2 in terms of decoder output are given below:

$$F_1 = Y_0 + Y_1 + Y_3 + Y_6 \quad .....(1)$$

$$F_2 = Y_1 + Y_3 + Y_5 + Y_6 \quad .....(2)$$

The equations (1) and (2) are used to implement functions F_1 and F_2 using logic **high** 3-to-8 decoder and external OR gates as shown in Fig. 1.

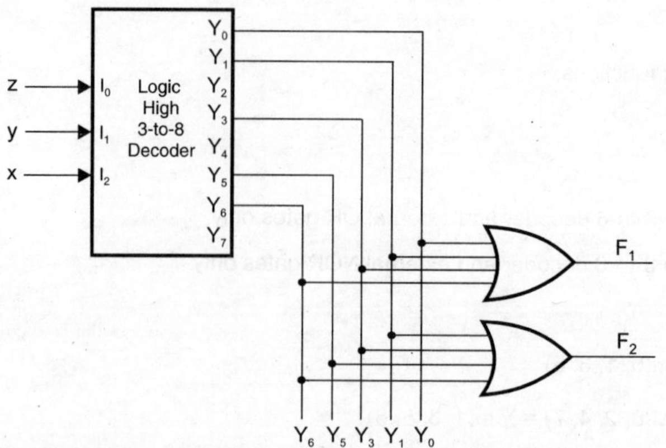

Fig. 1: OR implementation of F_1 and F_2.

(b) Using Logic High 3-to-8 Decoder and External NOR Gates Only

The equations (1) and (2) can be modified for NOR implementation as shown below:

$$\therefore F_1 = Y_0 + Y_1 + Y_3 + Y_6$$

$$= \left((Y_0 + Y_1 + Y_3 + Y_6)'\right)' = (F_1')' \qquad .....(3)$$

$$F_2 = Y_1 + Y_3 + Y_5 + Y_6$$

$$= \left((Y_1 + Y_3 + Y_5 + Y_6)'\right)' = (F_2')' \qquad .....(4)$$

The equations (3) and (4) are used to implement functions F_1 and F_2 using logic **high** 3-to-8 decoder and external NOR gates as shown in Fig. 2.

In NOR implementation the inverters are constructed using NOR gates and the OR gate is constructed using NOR gate followed by an inverter as shown in Fig. 2.

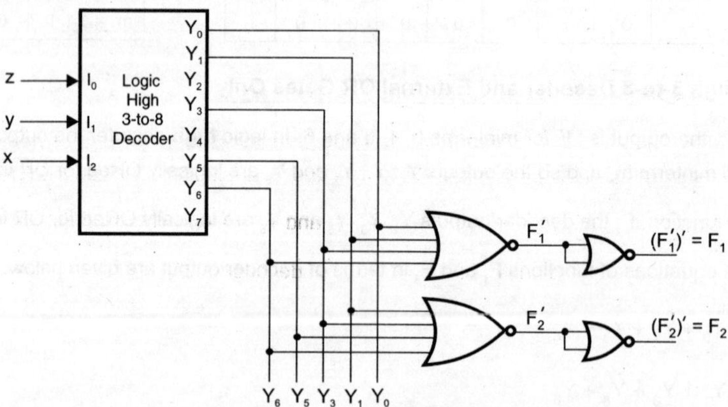

Fig. 2: NOR implementation of F_1 and F_2.

Example 5.15

Using decoder and external gates, design the combination circuit defined by the following Boolean functions:

i) $F_1 = (y + x')z$

ii) $F_2 = yz + x'y + y'z$

iii) $F_3 = (x + y')z$

Solution

The truth table of Boolean functions F_1, F_2 and F_3 are shown in Table 1. Here, x, y and z are input variables of the functions.

Table 1: Truth Table for F_1, F_2 and F_3

x y z	Minterm	Maxterm	x' y' z'	y + x'	F_1	yz	x'y	y'z	F_2	x + y'	F_3
0 0 0	m_0	M_0	1 1 1	1	0	0	0	0	0	1	0
0 0 1	m_1	M_1	1 1 0	1	1	0	0	1	1	1	1
0 1 0	m_2	M_2	1 0 1	1	0	0	1	0	1	0	0
0 1 1	m_3	M_3	1 0 0	1	1	1	1	0	1	0	0
1 0 0	m_4	M_4	0 1 1	0	0	0	0	0	0	1	0
1 0 1	m_5	M_5	0 1 0	0	0	0	0	1	1	1	1
1 1 0	m_6	M_6	0 0 1	1	0	0	0	0	0	1	0
1 1 1	m_7	M_7	0 0 0	1	1	1	0	0	1	1	1

From the truth table the following Boolean equations in SOP form are obtained. The function is given by sum of minterms for which output is 1.

$$F_1 = m_1 + m_3 + m_7$$

$$F_2 = m_1 + m_2 + m_3 + m_5 + m_7$$

$$F_3 = m_1 + m_5 + m_7$$

On taking into account the minterms for which function outputs are 1 the following Boolean equations are obtained using the logic **high** decoder output.

$F_1 = Y_1 + Y_3 + Y_7$ $\qquad$ $F_2 = Y_1 + Y_2 + Y_3 + Y_5 + Y_7$ $\qquad$ $F_3 = Y_1 + Y_5 + Y_7$

Using the above Boolean equations the logic circuit for the given functions using logic **high** 3-to-8 decoder and OR gate is drawn as shown in Fig. 1.

From the truth table the following Boolean equations in POS form are obtained. The function is given by product of maxterms for which function output is 0.

$F_1 = M_0 M_2 M_4 M_5 M_6$; $F_2 = M_0 M_4 M_6$; $F_3 = M_0 M_2 M_3 M_4 M_6$

On taking into account the maxterms for which the functions outputs are 0, the following Boolean equations are obtained using logic **low** decoder output.

$F_1 = Y_0 Y_2 Y_4 Y_5 Y_6$ $\qquad$ $F_2 = Y_0 Y_4 Y_6$ $\qquad$ $F_3 = Y_0 Y_2 Y_3 Y_4 Y_6$

Using the above Boolean equations the logic circuit for the given functions using logic **low** 3-to-8 decoder and AND gate is drawn as shown in Fig. 2.

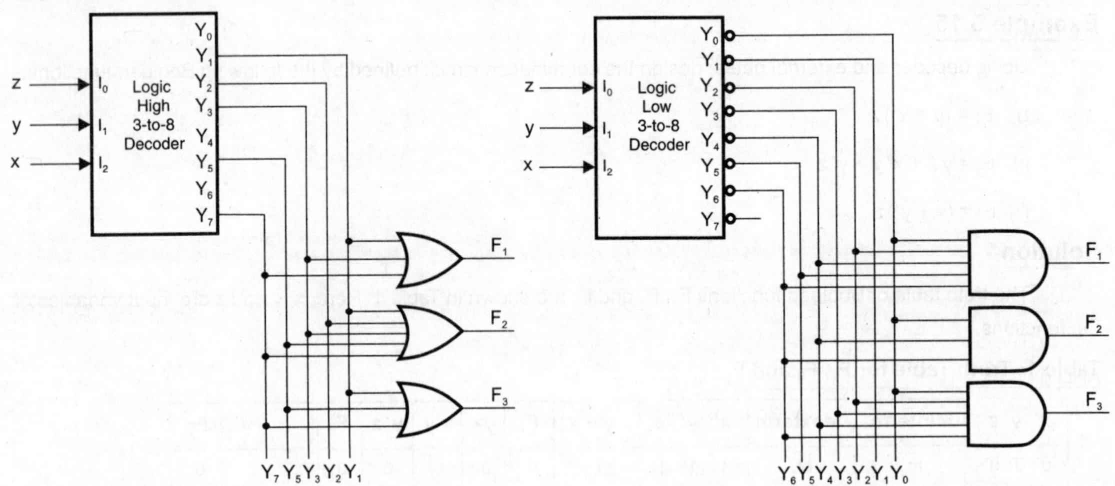

Fig. 1: *Logic circuit of F_1, F_2, F_3 using logic high decoder and OR gate.*

Fig. 2: *Logic circuit of F_1, F_2, F_3 using logic low decoder and AND gate.*

Example 5.16

Realize the following functions,

$F_1 = \sum m(0, 1, 3, 6)$

$F_2 = \prod M(0, 2, 4, 5)$

(a) Using logic **low** 3-to-8 decoder and external AND gates only

(b) Using logic **low** 3-to-8 decoder and external NAND gates only.

Solution

Given that, $F_1 = \sum m(0, 1, 3, 6) = \prod M(2, 4, 5, 7)$

$F_2 = \prod M(0, 2, 4, 5)$

Since the given function has maxterms M_0 to M_7 the given functions are 3-variable functions and so can be realized using a 3-to-8 decoder. The 3-to-8 decoder has three inputs and the maxterms are formed from all possible combination of inputs as shown in Table 1. Here, x, y and z are input variables of the functions.

The truth table of F_1 and F_2 is shown in Table 2.

Table 1: Truth Table of Logic Low 3-to-8 Decoder

Inputs			Maxterm	Outputs							
x	y	z		Y_0	Y_1	Y_2	Y_3	Y_4	Y_5	Y_6	Y_7
0	0	0	M_0	0	1	1	1	1	1	1	1
0	0	1	M_1	1	0	1	1	1	1	1	1
0	1	0	M_2	1	1	0	1	1	1	1	1
0	1	1	M_3	1	1	1	0	1	1	1	1
1	0	0	M_4	1	1	1	1	0	1	1	1
1	0	1	M_5	1	1	1	1	1	0	1	1
1	1	0	M_6	1	1	1	1	1	1	0	1
1	1	1	M_7	1	1	1	1	1	1	1	0

Table 2: Truth Table of F_1 and F_2

Maxterm	F_1	F_2
$M_0 = 0 + 0 + 0$	1	0
$M_1 = 0 + 0 + 1'$	1	1
$M_2 = 0 + 1' + 0$	0	0
$M_3 = 0 + 1' + 1'$	1	1
$M_4 = 1' + 0 + 0$	0	0
$M_5 = 1' + 0 + 1'$	0	0
$M_6 = 1' + 1' + 0$	1	1
$M_7 = 1' + 1' + 1'$	0	1

(a) Using Logic Low 3-to-8 Decoder and External AND Gates Only

In function, F_1 the output is "0" for maxterms 2, 4, 5 and 7. In logic low decoder the output Y_n is "0" when the input corresponds to maxterm M_n and so the outputs Y_2, Y_4, Y_5 and Y_7 are logically ANDed in order to realize F_1.

Similarly, for function, F_2 the decoder outputs Y_0, Y_2, Y_4 and Y_5 are logically ANDed in order to realize F_2.

The Boolean equations of functions F_1 and F_2 in terms of logic **low** decoder output are given below:

$$F_1 = \prod M(2, 4, 5, 7) = Y_2\, Y_4\, Y_5\, Y_7 \quad .....(1)$$

$$F_2 = \prod M(0, 2, 4, 5) = Y_0\, Y_2\, Y_4\, Y_5 \quad .....(2)$$

The equations (1) and (2) are used to implement functions F_1 and F_2 using logic **low** 3-to-8 decoder and external AND gate as shown in Fig. 1.

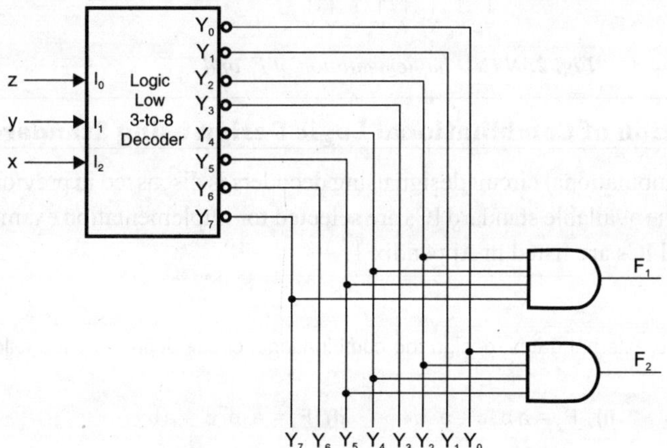

Fig. 1: AND implementation of F_1 and F_2.

(b) Using Logic Low 3-to-8 Decoder and External NAND Gates Only

The equations (1) and (2) can be modified for NAND implementation as shown below:

$$F_1 = \prod M(2, 4, 5, 7) = Y_2\, Y_4\, Y_5\, Y_7 = ((Y_2\, Y_4\, Y_5\, Y_7)')' = (F_1')' \quad .....(3)$$

$$F_2 = \prod M(0, 2, 4, 5) = Y_0\, Y_2\, Y_4\, Y_5 = ((Y_0\, Y_2\, Y_4\, Y_5)')' = (F_2')' \quad .....(4)$$

The equations (3) and (4) are used to implement functions F_1 and F_2 using logic **low** 3-to-8 decoder and external NAND gate as shown in Fig. 2.

In NAND implementation the inverters are constructed using NAND gates and AND gate is constructed using NAND gate followed by an inverter as shown in Fig. 2.

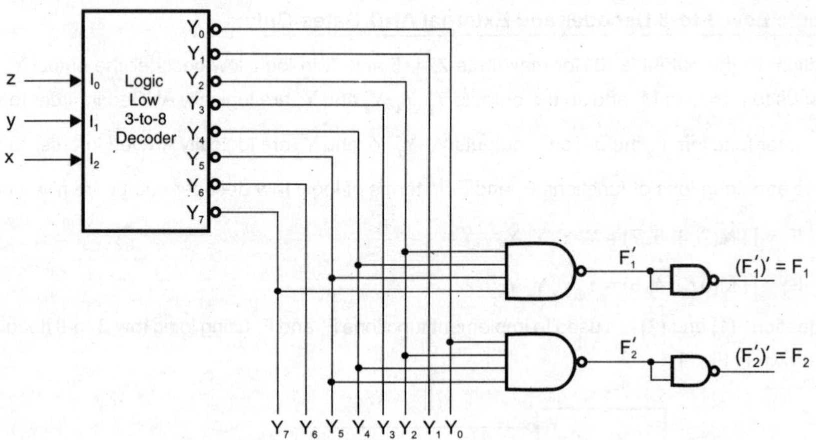

Fig. 2: NAND implementation of F_1 and F_2.

5.4.11 Implementation of Combinational Logic Design using Standard Decoder IC's

The concept of combinational circuit design using decoders is discussed in previous section (Section 5.4.10.). In this section, the available standard ICs are selected for implementation examples. The decoders that available as standard ICs are listed in Appendix 1.

Example 5.17

Using a decoder and external gates, design the combinational circuit defined by the following three Boolean functions:

i) $F_1 = a'\,b\,c' + a\,c$ **ii)** $F_2 = a\,b'\,c + a'\,b$ **iii)** $F_3 = a'\,b'\,c' + a\,b$

Solution

The truth table of Boolean functions F_1, F_2 and F_3 are shown in Table 1. Here, a, b and c are input variables of the functions.

Table 1: Truth Table for F_1, F_2 and F_3

a	b	c	Minterm	Maxterm	a'	b'	c'	a'bc'	ac	F_1	ab'c	a'b	F_2	a'b'c'	ab	F_3
0	0	0	m_0	M_0	1	1	1	0	0	0	0	0	0	1	0	1
0	0	1	m_1	M_1	1	1	0	0	0	0	0	0	0	0	0	0
0	1	0	m_2	M_2	1	0	1	1	0	1	0	1	1	0	0	0
0	1	1	m_3	M_3	1	0	0	0	0	0	0	1	1	0	0	0
1	0	0	m_4	M_4	0	1	1	0	0	0	0	0	0	0	0	0
1	0	1	m_5	M_5	0	1	0	0	1	1	1	0	1	0	0	0
1	1	0	m_6	M_6	0	0	1	0	0	0	0	0	0	0	1	1
1	1	1	m_7	M_7	0	0	0	0	1	1	0	0	0	0	1	1

From the truth table the following Boolean equations in SOP form are obtained.

$F_1 = m_2 + m_5 + m_7$; $F_2 = m_2 + m_3 + m_5$; $F_3 = m_0 + m_6 + m_7$

On taking into account the minterms for which function outputs are 1 the following Boolean equations are obtained using the logic **high** decoder output.

$$F_1 = Y_2 + Y_5 + Y_7$$

$$F_2 = Y_2 + Y_3 + Y_5$$

$$F_3 = Y_0 + Y_6 + Y_7$$

Using the above Boolean equations the logic circuit for the given functions using logic **high** 3-to-8 decoder IC 74237 and 3-input OR gate IC 744075 is implemented as shown in Fig. 1.

From the truth table the following Boolean equations in POS form are obtained.

$$F_1 = M_0 \, M_1 \, M_3 M_4 M_6$$

$$F_2 = M_0 \, M_1 \, M_4 M_6 M_7$$

$$F_3 = M_1 \, M_2 \, M_3 M_4 M_5$$

On taking into account the maxterms for which the functions outputs are 0, the following Boolean equations are obtained using logic **low** decoder output.

$$F_1 = Y_0 Y_1 Y_3 Y_4 Y_6$$

$$F_2 = Y_0 Y_1 Y_4 Y_6 Y_7$$

$$F_3 = Y_1 Y_2 Y_3 Y_4 Y_5$$

Using the above Boolean equations the logic circuit for the given functions using logic **low** 3-to-8 decoder IC 74138 and 3-input AND gate IC 7411 is implemented as shown in Fig. 2.

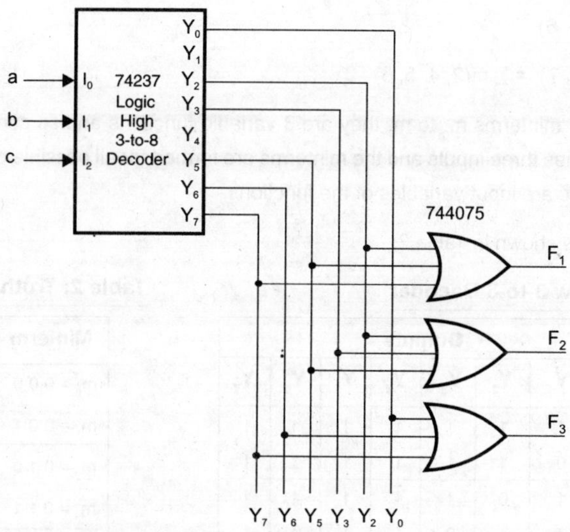

***Fig. 1:** Logic circuit of F_1, F_2, F_3 using logic high decoder and OR gate.*

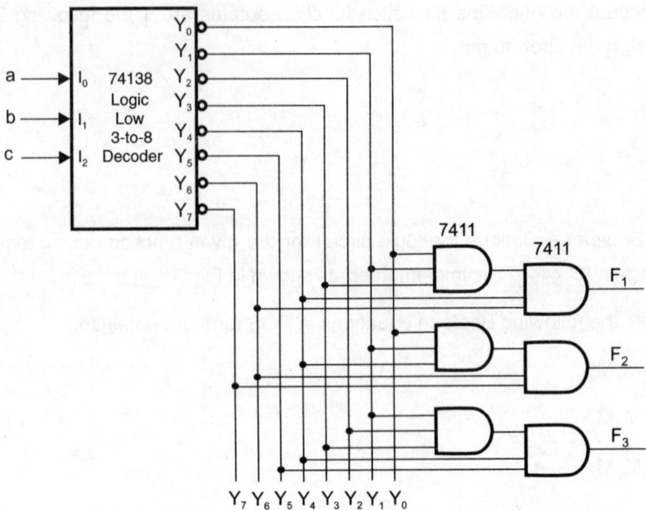

Fig. 2: Logic circuit of F_1, F_2, F_3 using logic low decoder and AND gate.

Example 5.18

Realize the following functions,

$F_1 = \Sigma m(0, 2, 5, 6) \qquad F_2 = \Pi M(0, 1, 3, 7)$

(a) Using logic **low** 3-to-8 decoder and external OR gates only

(b) Using logic **low** 3-to-8 decoder and external NOR gates only.

Solution

Given that, $F_1 = \Sigma m(0, 2, 5, 6)$

$F_2 = \Pi M(0, 1, 3, 7) = \Sigma m(2, 4, 5, 6)$

Since the given function has minterms m_0 to m_7 they are 3-variable functions and so can be realized using a 3-to-8 decoder. The 3-to-8 decoder has three inputs and the minterms are formed for all possible combination of inputs as shown in Table 1. Here, a, b and c are input variables of the functions.

The truth table of F_1 and F_2 is shown in Table 2.

Table 1: Truth Table of Logic Low 3-to-8 Decoder

Inputs			Minterm	Outputs							
a	b	c		Y_0	Y_1	Y_2	Y_3	Y_4	Y_5	Y_6	Y_7
0	0	0	m_0	0	1	1	1	1	1	1	1
0	0	1	m_1	1	0	1	1	1	1	1	1
0	1	0	m_2	1	1	0	1	1	1	1	1
0	1	1	m_3	1	1	1	0	1	1	1	1
1	0	0	m_4	1	1	1	1	0	1	1	1
1	0	1	m_5	1	1	1	1	1	0	1	1
1	1	0	m_6	1	1	1	1	1	1	0	1
1	1	1	m_7	1	1	1	1	1	1	1	0

Table 2: Truth Table of F_1 and F_2

Minterm	F_1	F_2
m_0 = 0 0 0	1	0
m_1 = 0 0 1	0	0
m_2 = 0 1 0	1	1
m_3 = 0 1 1	0	0
m_4 = 1 0 0	0	1
m_5 = 1 0 1	1	1
m_6 = 1 1 0	1	1
m_7 = 1 1 1	0	0

(a) Using Logic Low 3-to-8 Decoder and External OR Gates Only

In function, F_1 the output is "1" for minterms 0, 2, 5 and 6. In logic **low** decoder the output Y_n is "0" when the input corresponds to minterm m_n and so the outputs Y_0, Y_2, Y_5 and Y_6 are inverted and then logically ORed for OR implementation as shown in Fig. 1.

Similarly, for function, F_2 the decoder outputs Y_2, Y_4, Y_5 and Y_6 are inverted and then logically ORed for OR implementation as shown in Fig. 1.

The Boolean equations of functions F_1 and F_2 in terms of decoder output are given below:

$$F_1 = Y_0' + Y_2' + Y_5' + Y_6' \quad .....(1)$$

$$F_2 = Y_2' + Y_4' + Y_5' + Y_6' \quad .....(2)$$

The equations (1) and (2) are used to implement functions F_1 and F_2 using logic **low** 3-to-8 decoder IC 74138 and external 4-input OR gate IC 744072 as shown in Fig. 1. The standard NOT gate IC 7404 is used to invert the decoder outputs.

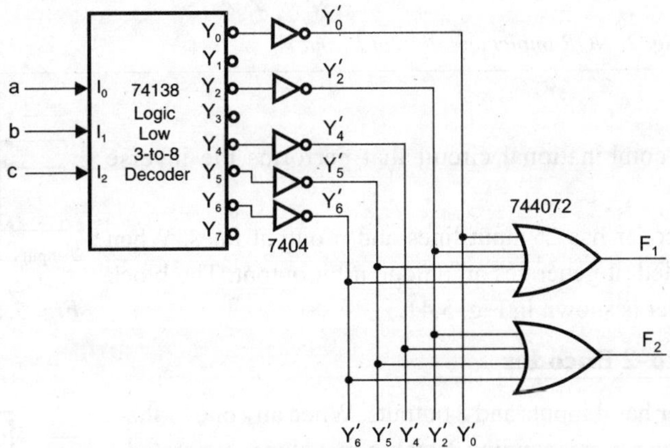

Fig. 1: OR implementation of F_1 and F_2.

(b) Using Logic Low 3-to-8 Decoder and External NOR Gates Only

In NOR implementation the inverters are constructed using NOR gates and OR gate is constructed using NOR gate followed by an inverter as shown in Fig. 2.

The equations (1) and (2) can be modified for NOR implementation as shown below:

$$\therefore F_1 = Y_0' + Y_2' + Y_5' + Y_6'$$

$$= \left(\left(Y_0' + Y_2' + Y_5' + Y_6'\right)'\right)' \quad .....(3)$$

$$F_2 = Y_2' + Y_4' + Y_5' + Y_6'$$

$$= \left(\left(Y_2' + Y_4' + Y_5' + Y_6'\right)'\right)' \quad .....(4)$$

The equations (3) and (4) are used to implement functions F_1 and F_2 using logic **low** 3-to-8 decoder IC 74138 and external 4-input NOR gate IC 7429 as shown in Fig. 2. The inverters are implemented using 2-input NOR gate IC 7402.

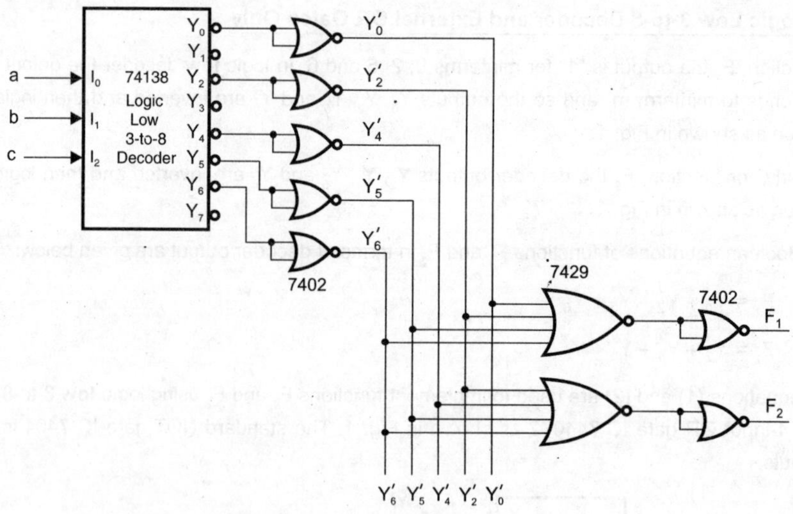

Fig. 2: NOR implementation of F_1 and F_2.

5.5 Encoders

An ***encoder*** is a combinational circuit that performs the inverse operation of a decoder.

In general, an encoder has 2^n input lines and n output lines. When one of 2^n inputs is activated, it generates an unique n-bit output. The block diagram of 2^n-to-n encoder is shown in Fig. 5.41.

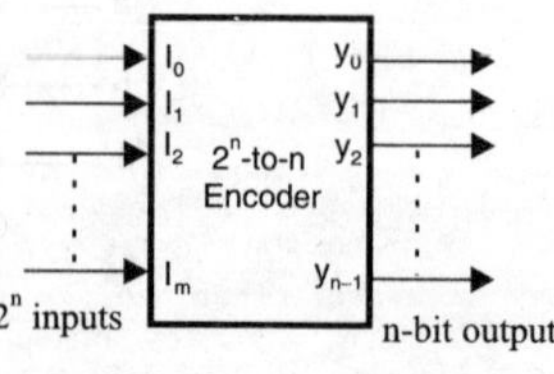

***Fig. 5.41:** 2^n-to-n encoder.*

m = $2^n - 1$

5.5.1 Logic High 4-to-2 Encoder

The ***4-to-2 encoder*** has 4 inputs and 2 outputs. When any one of the inputs is asserted **high** then a corresponding 2-bit binary value is generated in the output.

Let, I_0, I_1, I_2, I_3 = Inputs

Y_0, Y_1 = Outputs

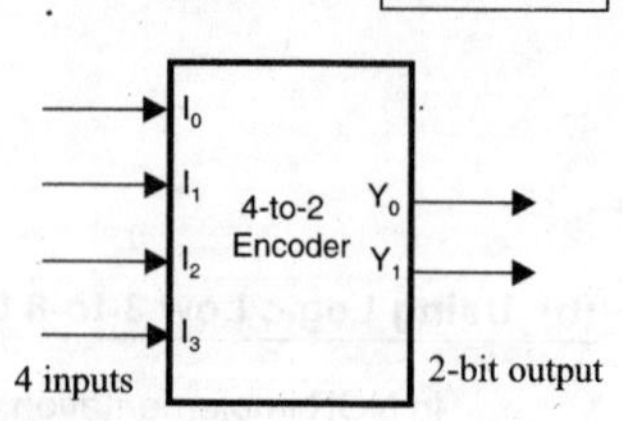

***Fig. 5.42:** 4-to-2 encoder.*

The block diagram of 4-to-2 encoder is shown in Fig. 5.42 and the truth table is shown in Table 5.9. Three different types of encoder design is presented here.

Table 5.9: Truth Table of 4-to-2 Logic High Encoder

Inputs				Minterm	Outputs	
I_0	I_1	I_2	I_3		Y_1	Y_0
1	0	0	0	m_8	0	0
0	1	0	0	m_4	0	1
0	0	1	0	m_2	1	0
0	0	0	1	m_1	1	1

Design 1: 4-to-2 Encoder Design using K-map

Since the 4-to-2 encoder has 4 inputs the possible minterms are m_0, m_1, m_2, ..., m_{15}. But only 4 minterms are required for valid outputs as shown in truth table (Table 5.9). The outputs for remaining minterms are don't-care outputs.

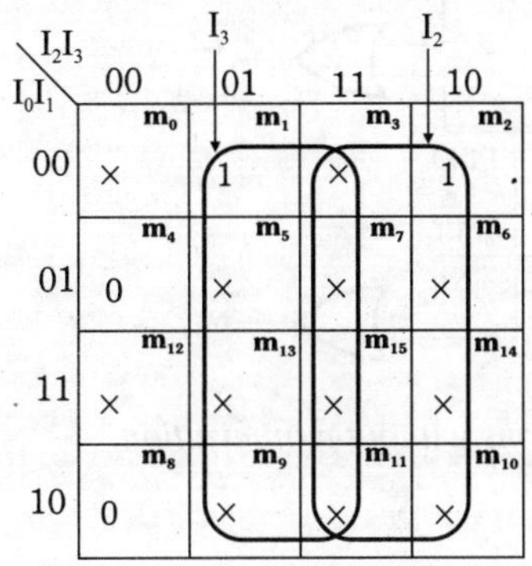

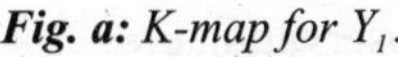

Fig. a: K-map for Y_1.

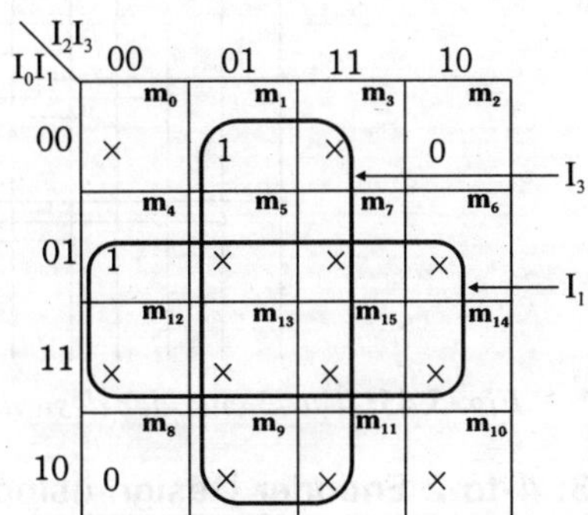

Fig. b: K-map for Y_0.

Fig. 5.43: 4-to-2 encoder design using K-map.

Using Table 5.9, the K-maps for Y_0 and Y_1 are constructed as shown in Fig. 5.43. From the K-map we get the following Boolean equations for encoder outputs.

$$Y_1 = I_2 + I_3$$

$$Y_0 = I_1 + I_3$$

Using the above Boolean equations the logic circuit of 4-to-2 encoder is drawn as shown in Fig. 5.44.

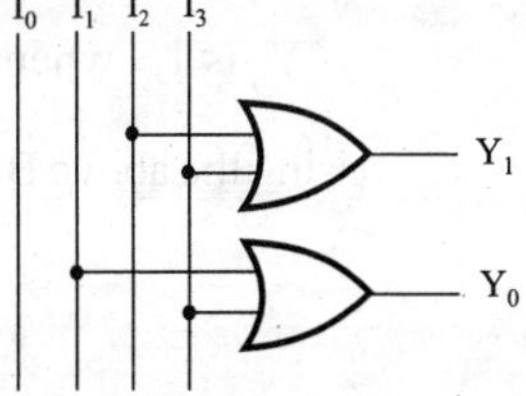

Fig. 5.44: 4-to-2 encoder.

***Note:** Since we use minimum literals I_0 is not used in generating output.*

Design 2: 4-to-2 Encoder Design using Minterms without Minimization

In this design a SOP form of Boolean equation can be directly formed for each output by considering the minterms for which outputs are 1.

The truth table of 4-to-2 encoder is shown in Table 5.10. In this table the minterms for which outputs are 1 are identified and using these minterms the following SOP form of Boolean equations are obtained for outputs.

Table 5.10: Truth Table of 4-to-2 Encoder

I_0	I_1	I_2	I_3	Minterm	Y_1	Y_0
1	0	0	0	m_8	0	0
0	1	0	0	m_4	0	1
0	0	1	0	m_2	1	0
0	0	0	1	m_1	1	1

$$Y_1 = m_2 + m_1 = I_0' I_1' I_2 I_3' + I_0' I_1' I_2' I_3$$

$$Y_0 = m_4 + m_1 = I_0' I_1 I_2' I_3' + I_0' I_1' I_2' I_3$$

Using the above Boolean equations the logic circuit of 4-to-2 encoder is drawn as shown in Fig. 5.45.

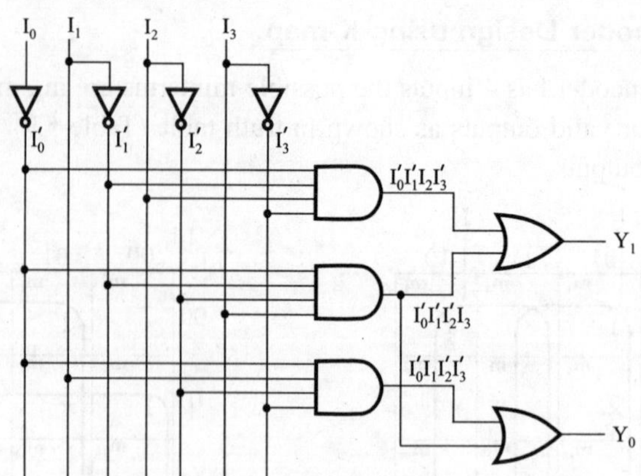

Fig. 5.45: 4-to-2 encoder design using minterms without minimization.

Design 3: 4-to-2 Encoder Design using OR Gate

From the truth table (Table 5.9) we can make following observations.

Y_1 is 1, when $I_2 = 1$ or $I_3 = 1$; $\therefore$ $Y_0 = I_2 + I_3$

Y_0 is 1, when $I_1 = 1$ or $I_3 = 1$; $\therefore$ $Y_1 = I_1 + I_3$

Using the above Boolean equations the logic circuit of 4-to-2 encoder is drawn as shown in Fig. 5.46.

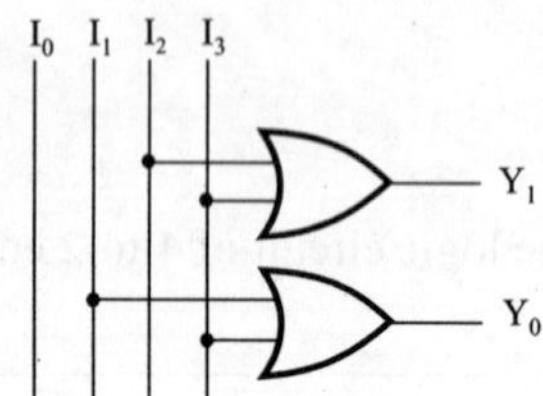

Fig. 5.46: 4-to-2 encoder design using OR gate.

5.5.2 Logic High 8-to-3 Encoder

The ***8-to-3 encoder*** has eight inputs and 3 outputs. When any one of the inputs is asserted **high** then a corresponding 3-bit binary value is generated in the output.

Let, $I_0, I_1, ..., I_7$ = Inputs

$Y_0, Y_1\ Y_2$ = Outputs

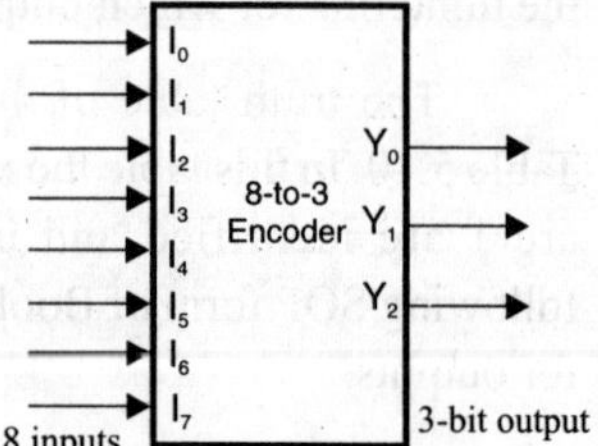

Fig. 5.47: 8-to-3 encoder.

The block diagram of 8-to-3 encoder is shown in Fig. 5.47 and the truth table is shown in Table 5.11. Two different types of 8-to-3 encoder design is presented here.

Since the 8-to-3 encoder has 8 inputs the possible minterms are m_0, m_1, m_2, ..., m_{255}. But only 8 minterms are required for valid outputs as shown in Table 5.11. The outputs for remaining minterms are don't-care outputs.

Table 5.11: Truth Table of 8-to-3 Logic High Encoder

Inputs								Minterms	Outputs		
I_0	I_1	I_2	I_3	I_4	I_5	I_6	I_7		Y_2	Y_1	Y_0
1	0	0	0	0	0	0	0	m_{128}	0	0	0
0	1	0	0	0	0	0	0	m_{64}	0	0	1
0	0	1	0	0	0	0	0	m_{32}	0	1	0
0	0	0	1	0	0	0	0	m_{16}	0	1	1
0	0	0	0	1	0	0	0	m_8	1	0	0
0	0	0	0	0	1	0	0	m_4	1	0	1
0	0	0	0	0	0	1	0	m_2	1	1	0
0	0	0	0	0	0	0	1	m_1	1	1	1

Design 1: 8-to-3 Encoder Design using OR gates

From the truth table (Table 5.11) we can make following observations.

Y_0 is 1 when $I_1 = 1, I_3 = 1, I_5 = 1, I_7 = 1$; $\therefore\ Y_0 = I_1 + I_3 + I_5 + I_7$

Y_1 is 1 when $I_2 = 1, I_3 = 1\ I_6 = 1, I_7 = 1$; $\therefore\ Y_1 = I_2 + I_3 + I_6 + I_7$

Y_2 is 1 when $I_4 = 1, I_5 = 1, I_6 = 1, I_7 = 1$; $\therefore\ Y_2 = I_4 + I_5 + I_6 + I_7$

Using the above Boolean equations the logic circuit of 8-to-3 encoder is drawn using OR gates as shown in Fig. 5.48.

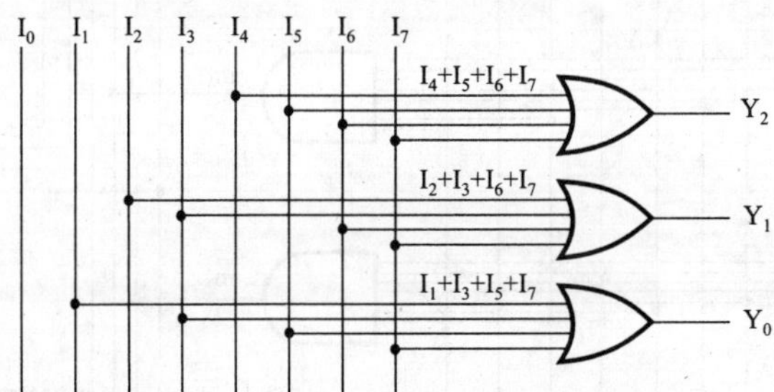

***Fig. 5.48:** 8-to-3 encoder using OR gates.*

Design 2: 8-to-3 Encoder Design using Minterms and without Minimization

In this design a SOP form of Boolean equation can be directly formed for each output by considering the minterms for which output is 1.

In the truth table (Table 5.11) the minterms for which outputs are 1 are identified and using these minterms the following SOP form of Boolean equations are obtained for encoder outputs.

$$Y_0 = \Sigma(m_1, m_4, m_{15}, m_{64}) = I_0' I_1' I_2' I_3' I_4' I_5' I_6' I_7 + I_0' I_1' I_2' I_3' I_4' I_5 I_6' I_7' + I_0' I_1' I_2' I_3 I_4' I_5' I_6' I_7' + I_0' I_1 I_2' I_3' I_4' I_5' I_6' I_7'$$

$$Y_1 = \Sigma(m_1, m_2, m_{15}, m_{32}) = I'_0 I'_1 I'_2 I'_3 I'_4 I'_5 I'_6 I_7 + I'_0 I'_1 I'_2 I'_3 I'_4 I'_5 I_6 I'_7 + I'_0 I'_1 I'_2 I_3 I'_4 I'_5 I'_6 I'_7$$
$$+ I'_0 I'_1 I_2 I'_3 I'_4 I'_5 I'_6 I'_7$$

$$Y_2 = \Sigma(m_1, m_2, m_4, m_8) = I'_0 I'_1 I'_2 I'_3 I'_4 I'_5 I'_6 I_7 + I'_0 I'_1 I'_2 I'_3 I'_4 I'_5 I_6 I'_7 + I'_0 I'_1 I'_2 I'_3 I'_4 I_5 I'_6 I'_7$$
$$+ I'_0 I'_1 I'_2 I'_3 I_4 I'_5 I'_6 I'_7$$

Using the above Boolean equations the logic circuit of 8-to-3 encoder is drawn as shown in Fig. 5.49.

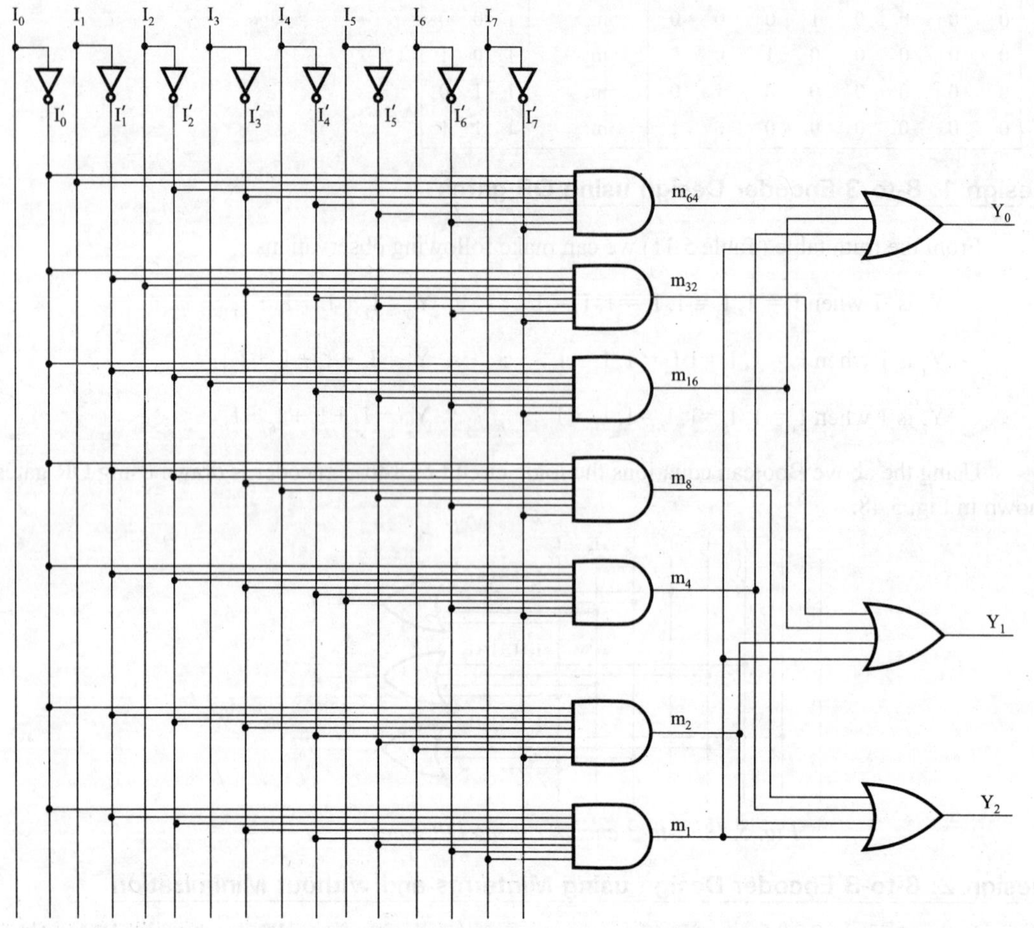

Fig. 5.49: 8-to-3 encoder.

5.6 Priority Encoders

A ***priority encoder*** is an encoder circuit that includes a priority function in order to recognize only one input when multiple inputs are asserted **high**.

The operation of the priority encoder is such that if two or more inputs are asserted **high** (or equal to 1) then the output corresponds to the input having highest priority.

5.6.1 Logic High 4-to-2 Priority Encoder

(AU, Apr/May'19, 6 Marks)

The 4-to-2 encoder has 4 inputs and 2 outputs.

Let, I_0, I_1, I_2, I_3 = Inputs

Y_0, Y_1 = Outputs

Let the order of priority from highest to lowest be,

I_3 - Highest

I_2

I_1

I_0 - Lowest

Now, the above priority works as follows:

If I_3 and I_2 are set **high** then output will correspond to only I_3.

If I_2 and I_1 are set **high** then output will correspond to only I_2.

If I_1 and I_0 are set **high** then output will correspond to only I_1.

To implement the above priority we have to write a priority function.

The priority functions for the above example are,

$H_3 = I_3$; I_3 is recognized

$H_2 = I_2 I_3'$; I_2 is recognized only if $I_3 = 0$

$H_1 = I_1 I_2' I_3'$; I_1 is recognized only if $I_2 = I_3 = 0$

$H_0 = I_0 I_1' I_2' I_3'$; I_0 is recognized only if $I_1 = I_2 = I_3 = 0$

The truth table of priority encoder with above priority functions is shown in Table 5.12.

Table 5.12: Truth Table of 4-to-2 Priority Encoder

Inputs				Outputs	
I_0	I_1	I_2	I_3	Y_1	Y_0
1	0	0	0	0	0
×	1	0	0	0	1
×	×	1	0	1	0
×	×	×	1	1	1

From the truth table (Table 5.12) we can note that Y_1 is 1 when I_2 and I_3 are 1.

$$\therefore\ Y_1 = H_2 + H_3 = I_2 I_3' + I_3$$

From the truth table (Table 5.12) we can note that Y_0 is 1 when I_1 and I_3 are 1.

$$\therefore\ Y_0 = H_1 + H_3 = I_1 I_2' I_3' + I_3$$

Therefore, the Boolean equations of 4-to-2 priority encoder with priority order from I_3 to I_0 are,

$$Y_1 = I_2\, I_3' + I_3$$

$$Y_0 = I_1\, I_2' I_3' + I_3$$

The above equations are verified in Table 5.13 for same possible combinations of inputs.

Table 5.13: Verification of 4-to-2 Priority Encoder

I_0	I_1	I_2	I_3	I_2'	I_3'	$I_2 I_3'$	$I_1 I_2' I_3'$	Y_1	Y_0	Comment
1	0	0	0	1	1	0	0	0	0	Only one input is asserted high and so the output corresponds to the input asserted high.
0	1	0	0	1	1	0	1	0	1	
0	0	1	0	0	1	1	0	1	0	
0	0	0	1	1	0	0	0	1	1	
0	0	1	1	0	0	0	0	1	1	2 and 3 asserted high ; output corresponds to 3
0	1	1	0	0	1	1	0	1	0	1 and 2 asserted high ; output corresponds to 2
1	1	0	0	1	1	0	1	0	1	0 and 1 asserted high ; output corresponds to 1

Usually a priority encoder will have a valid bit v, to indicate that one or more input is set **high**. The valid bit is zero if all inputs are zero. The truth table of 4-to-2 priority encoder with valid bit v is shown in Table 5.14. Since, value of valid bit is 1 if any one input is 1, the equation of valid bit is given by sum of all inputs as shown below:

$$\therefore \; v = I_0 + I_1 + I_2 + I_3$$

Table 5.14: Truth Table of 4-to-2 Priority Encoder with Valid Bit v

Inputs				Outputs		
I_0	I_1	I_2	I_3	Y_1	Y_0	v
0	0	0	0	×	×	0
1	0	0	0	0	0	1
×	1	0	0	0	1	1
×	×	1	0	1	0	1
×	×	×	1	1	1	1

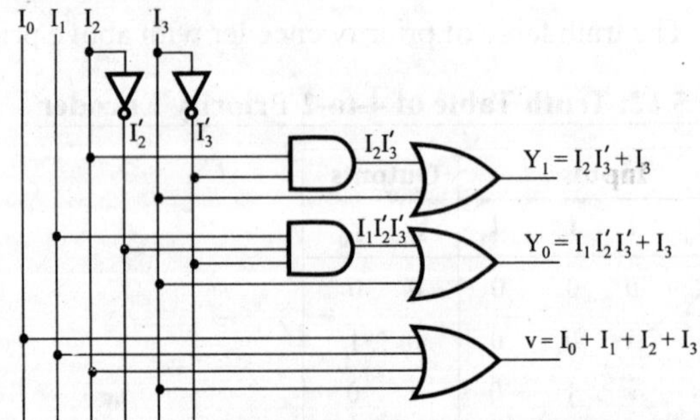

***Fig. 5.50:** 4-to-2 priority encoder with I_3 having highest priority.*

The logic circuit of 4-to-2 priority encoder with priority order I_3 to I_0 is drawn as shown in Fig. 5.50 using the Boolean equations of outputs and valid bit.

Alternatively, the priority encoder can be designed using K-maps.

5.6.2 Logic High 4-to-2 Priority Encoder Design using K-maps

The don't-care inputs are expanded with all possible combination of inputs and listed in Table 5.15.

Table 5.15: Truth Table of Priority Encoder with all Possible Combination of Inputs

Inputs					Outputs		
I_0	I_1	I_2	I_3	Minterm	Y_1	Y_0	v
0	0	0	0	m_0	×	×	0
1	0	0	0	m_8	0	0	1
×	1	0	0	m_4, m_{12}	0	1	1
×	×	1	0	m_2, m_6, m_{10}, m_{14}	1	0	1
×	×	×	1	m_1, m_3, m_5, m_7, m_9, m_{11}, m_{13}, m_{15}	1	1	1

Inputs					Outputs		
I_0	I_1	I_2	I_3	Minterm	Y_1	Y_0	v
0	0	0	0	m_0	×	×	0
1	0	0	0	m_8	0	0	1
0	1	0	0	m_4	0	1	1
1	1	0	0	m_{12}	0	1	1
0	0	1	0	m_2	1	0	1
0	1	1	0	m_6	1	0	1
1	0	1	0	m_{10}	1	0	1
1	1	1	0	m_{14}	1	0	1
0	0	0	1	m_1	1	1	1
0	0	1	1	m_3	1	1	1
0	1	0	1	m_5	1	1	1
0	1	1	1	m_7	1	1	1
1	0	0	1	m_9	1	1	1
1	0	1	1	m_{11}	1	1	1
1	1	0	1	m_{13}	1	1	1
1	1	1	1	m_{15}	1	1	1

The K-map for Y_0, Y_1 and v are constructed using the truth table (Table 5.15) as shown in Fig. 5.51.

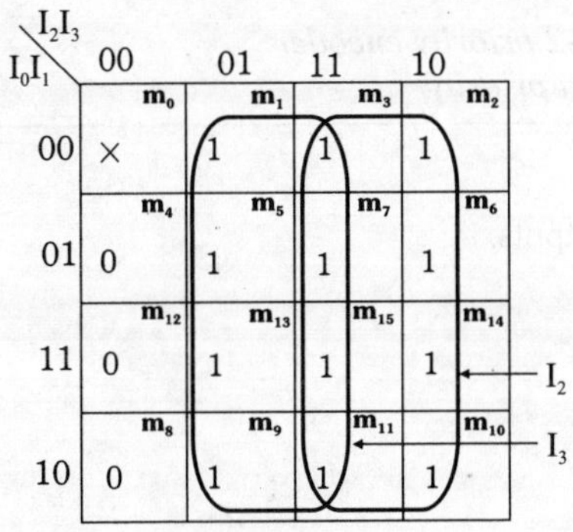

Fig. a: K-map for Y_1.

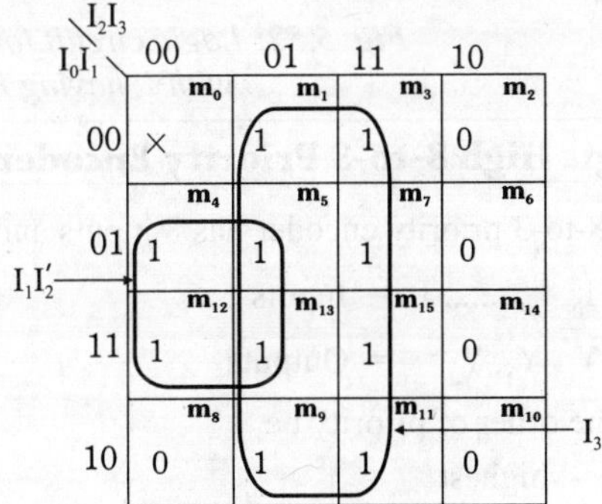

Fig. b: K-map for Y_0.

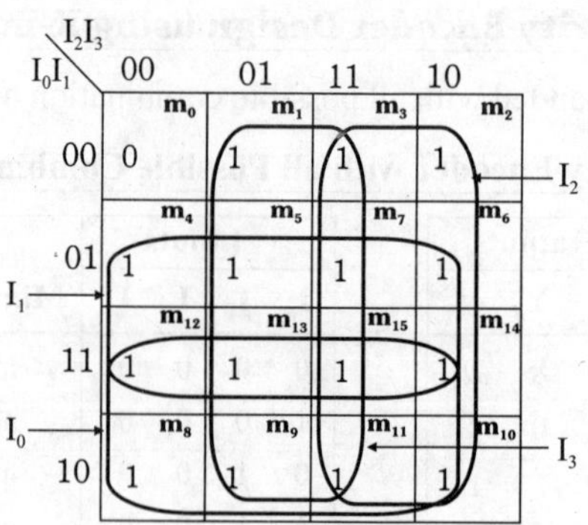

Fig. c: K-map for v.

***Fig. 5.51:** K-map for design of 4-to-2 encoder.*

From the K-map we get the following Boolean equations.

$$Y_1 = I_2 + I_3 \quad ; \quad Y_0 = I_1 I_2' + I_3 \quad ; \quad v = I_0 + I_1 + I_2 + I_3$$

Using the above Boolean equations the logic circuit for 4-to-2 priority encoder is drawn as shown in Fig. 5.52.

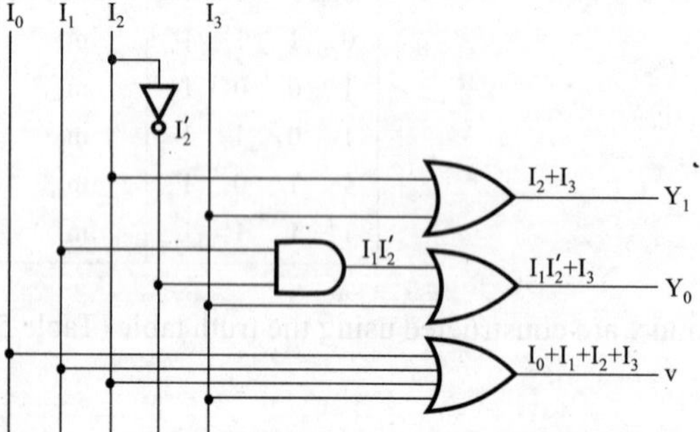

***Fig. 5.52:** Logic circuit for 4-to-2 priority encoder with I_3 having highest priority.*

5.6.3 Logic High 8-to-3 Priority Encoder

The 8-to-3 priority encoder has 8 inputs and 3 outputs.

Let, $I_0, I_1, ..., I_7$ = Inputs

Y_0, Y_1, Y_2 = Outputs

Let the order of priority be,

I_7 - highest
I_6 - .
I_5 - .
I_4 - .
I_3 - .
I_2 - .
I_1 - .
I_0 - lowest

The priority functions for the above priority order are,

$H_7 = I_7$; I_7 is recognized

$H_6 = I_6\ I_7'$; I_5 is recognized only if I_7 is 0

$H_5 = I_5\ I_6'\ I_7'$; I_5 is recognized only if I_6 and I_7 are 0

$H_4 = I_4\ I_5'\ I_6'\ I_7'$; I_4 is recognized only if I_5, I_6 and I_7 are 0

$H_3 = I_3\ I_4'\ I_5'\ I_6'\ I_7'$; I_3 is recognized only if I_4, I_5, I_6 and I_7 are 0

$H_2 = I_2\ I_3'\ I_4'\ I_5'\ I_6'\ I_7'$; I_2 is recognized only if I_3, I_4, I_5, I_6 and I_7 are 0

$H_1 = I_1\ I_2'\ I_3'\ I_4'\ I_5'\ I_6'\ I_7'$; I_1 is recognized only if I_2, I_3, I_4, I_5, I_6 and I_7 are 0

$H_0 = I_0\ I_1'\ I_2'\ I_3'\ I_4'\ I_5'\ I_6'\ I_7'$; I_0 is recognized only if I_1, I_2, I_3, I_4, I_5, I_6 and I_7 are 0

The truth table of priority encoder with above priority functions is shown in Table 5.16.

From the truth table we can note that Y_2 is 1 when, I_4, I_5, I_6 and I_7 are 1.

$$\therefore Y_2 = H_4 + H_5 + H_6 + H_7$$
$$= I_4\ I_5'\ I_6'\ I_7' + I_5\ I_6'\ I_7' + I_6\ I_7' + I_7$$

From the truth table we can note that Y_1 is 1 when I_2, I_3, I_6 and I_7 are 1.

$$\therefore Y_1 = H_2 + H_3 + H_6 + H_7$$
$$= I_2\ I_3'\ I_4'\ I_5'\ I_6'\ I_7' + I_3\ I_4'\ I_5'\ I_6'\ I_7' + I_6\ I_7' + I_7$$

From the truth table we can note that Y_0 is 1 when I_1, I_3, I_5 and I_7 are 1

$$\therefore Y_0 = H_1 + H_3 + H_5 + H_7$$
$$= I_1\ I_2'\ I_3'\ I_4'\ I_5'\ I_6'\ I_7' + I_3\ I_4'\ I_5'\ I_6'\ I_7' + I_5\ I_6'\ I_7' + I_7$$

The valid bit v equation is,

$$v = I_0 + I_1 + I_2 + I_3 + I_4 + I_5 + I_6 + I_7$$

Using the above Boolean equations the logic circuit for 8-to-3 priority encoder is drawn as shown in Fig. 5.53.

Table 5.16: Truth Table of 8-to-3 Priority Encoder

Inputs								**Outputs**			
I_0	I_1	I_2	I_3	I_4	I_5	I_6	I_7	Y_2	Y_1	Y_0	v
0	0	0	0	0	0	0	0	×	×	×	0
1	0	0	0	0	0	0	0	0	0	0	1
×	1	0	0	0	0	0	0	0	0	1	1
×	×	1	0	0	0	0	0	0	1	0	1
×	×	×	1	0	0	0	0	0	1	1	1
×	×	×	×	1	0	0	0	1	0	0	1
×	×	×	×	×	1	0	0	1	0	1	1
×	×	×	×	×	×	1	0	1	1	0	1
×	×	×	×	×	×	×	1	1	1	1	1

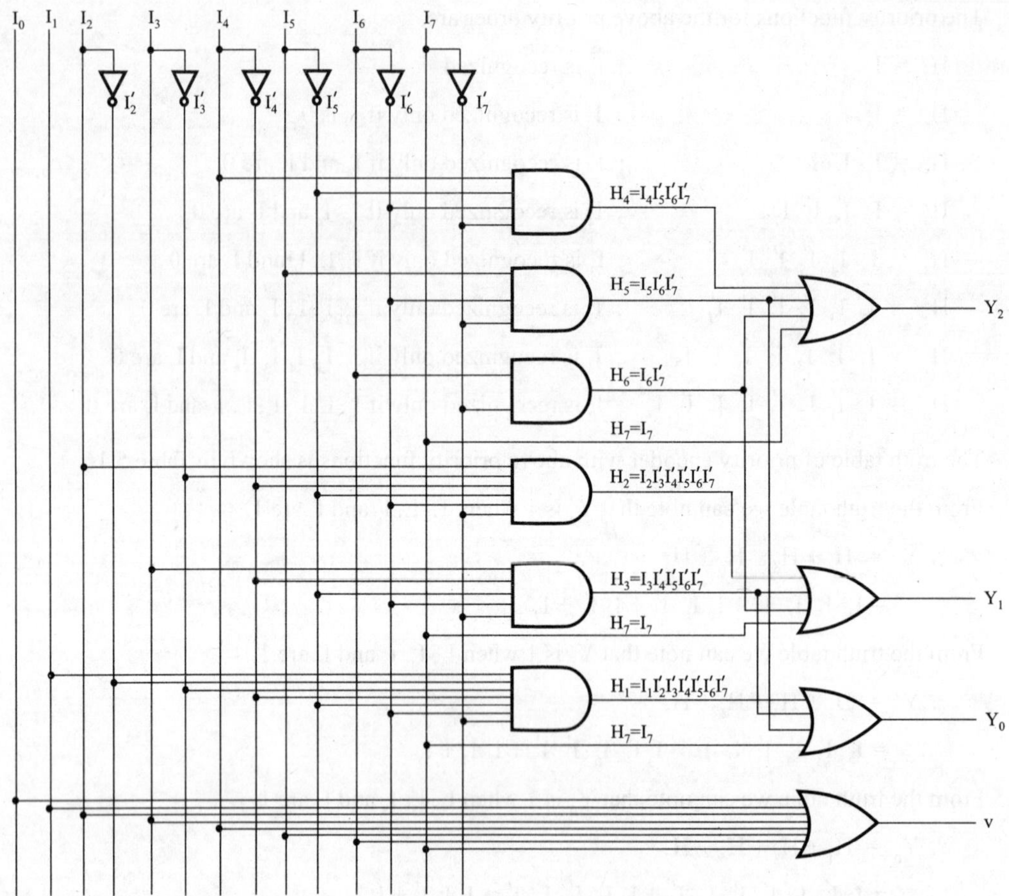

Fig. 5.53: *8-to-3 priority encoder with I_7 having highest priority.*

The 8-to-3 priority encoder is available as a standard IC with number 74148. The pin configuration of 74148 is shown in Fig. 5.54.

74148

Pin name	Pin No.	Pin No.	Pin name
I_4	1	16	V_{cc}(+5V)
I_5	2	15	EO
I_6	3	14	v
I_7	4	13	I_3
EI	5	12	I_2
Y_2	6	11	I_1
Y_1	7	10	I_0
(0V) GND	8	9	Y_0

Pin	Description
I_0 - I_7	Inputs
Y_0 - Y_2	Outputs
v	Output
V_{cc}	Supply voltage (+5V)
GND	Ground (0V)
EI	Input Enable
EO	Output Enable

Fig. 5.54: *Pin configuration of 8-to-3 priority encoder IC 74148.*

Example 5.19

Design a 4-to-2 priority encoder with input I_0 having the highest priority and input I_3 the lowest priority.

Solution

Let, I_0, I_1, I_2, I_3 = Inputs

Y_0, Y_1, v = Outputs

The truth table of priority encoder is shown in Table 1.

Table 1: Truth Table of 4-to-2 Priority Encoder with I_0 as Highest Priority

Inputs				Outputs		
I_0	I_1	I_2	I_3	Y_1	Y_0	v
0	0	0	0	×	×	0
1	×	×	×	0	0	1
0	1	×	×	0	1	1
0	0	1	×	1	0	1
0	0	0	1	1	1	1

Given that the order of priority highest to lowest are,

I_0 - highest

I_1 - .

I_2 - .

I_3 - lowest

The priority functions are,

$H_0 = I_0$; I_0 is recognized

$H_1 = I_1 I_0'$; I_1 is recognized only if I_0 is 0

$H_2 = I_2 I_1' I_0'$; I_2 is recognized only if I_1 and I_0 are 0

$H_3 = I_3 I_2' I_1' I_0'$; I_3 is recognized only if I_2, I_1 and I_0 are 0

From the truth table we can note that Y_1 is 1 when, I_2 and I_3 are 1.

$$\therefore Y_1 = H_2 + H_3 = I_2 I_1' I_0' + I_3 I_2' I_1' I_0'$$

From the truth table we can note that Y_0 is 1 when I_1 and I_3 are 1.

$$\therefore Y_0 = H_1 + H_3$$

$$= I_1 I_0' + I_3 I_2' I_1' I_0'$$

The equation for valid bit, v is,

$$v = I_0 + I_1 + I_2 + I_3$$

The logic circuit of 4-to-2 priority encoder with priority order I_0 and I_3 is drawn as shown in Fig. 1, using Boolean equations of output and valid bit, v.

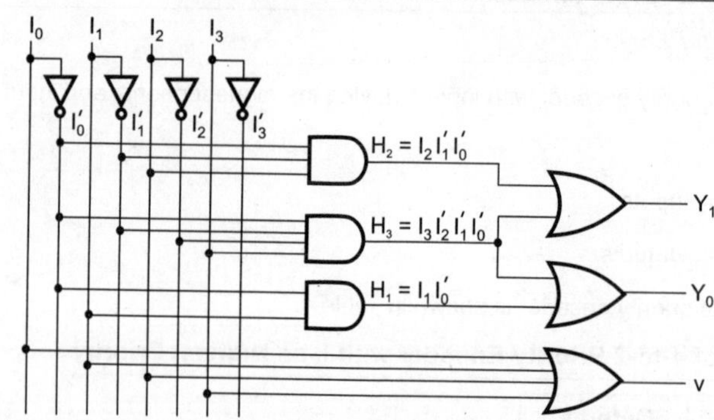

Fig. 1: Logic circuit of 4-to-2 priority encoder with I_0 having highest priority.

The output Boolean equation are verified for some possible combinations of inputs in Table 2.

Table 2: Verification of Output Equations

Inputs				Product Terms						Outputs		Comments
I_0	I_1	I_2	I_3	I_0'	I_1'	I_2'	$I_2 I_1' I_0'$	$I_3 I_2' I_1' I_0'$	$I_1 I_0'$	Y_1	Y_0	
1	0	0	0	0	1	1	0	0	0	0	0	Only one input is asserted high and so the output corresponds to the input asserted high
0	1	0	0	1	0	1	0	0	1	0	1	
0	0	1	0	1	1	0	1	0	0	1	0	
0	0	0	1	1	1	1	0	1	0	1	1	
1	1	0	0	0	0	1	0	0	0	0	0	0 and 1 are asserted high ; output corresponds to 0
0	1	1	0	1	0	0	0	0	1	0	1	1 and 2 are asserted high ; output corresponds to 1
0	0	1	1	1	1	0	1	0	0	1	0	2 and 3 are asserted high ; output corresponds to 2

5.7 Multiplexer

(RGPV, Jun'20, 5 Marks) (GTU, May'19, 3 Marks)

A ***multiplexer*** is a combinational circuit that selects one binary information from many inputs. For this reason a multiplexer is otherwise called a ***data selector***. In general, a multiplexer will have 2^n inputs and one output and so it allows only one input to output at any one time. In order to select one of the 2^n inputs the multiplexer will have n selection lines (or address). In short multiplexer is denoted as MUX.

The block diagram and symbolic representation of $2^n : 1$ multiplexer with n selection lines to select one of the 2^n inputs are shown in Fig. 5.55.

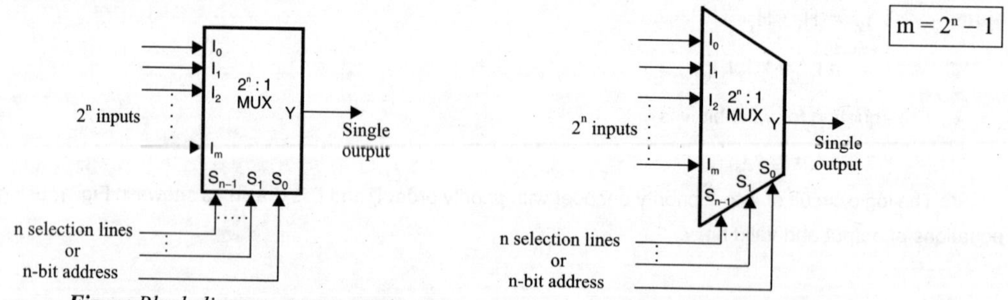

Fig. a: Block diagram representation.

Fig. b: Standard symbol.

Fig. 5.55: Multiplexer.

2:1 Multiplexer

The ***2:1 multiplexer*** is used to select one of the two inputs using 1-bit address or selection input. Since $2 = 2^1$, for 2:1 multiplexer one selection line is required to select one of the two inputs.

Let, I_0, I_1 = Inputs

S_0 = Selection input/Address

Y = Output

In 2:1 multiplexer,

When, $S_0 = 0$, I_0 is passed to output, $\therefore$ $Y = I_0$

When, $S_0 = 1$, I_1 is passed to output, $\therefore$ $Y = I_1$

The above selection logic are listed as truth table in Table 5.17.

Table 5.17: Truth Table of 2:1 Multiplexer

Selection Input S_0	Output Y
0	I_0
1	I_1

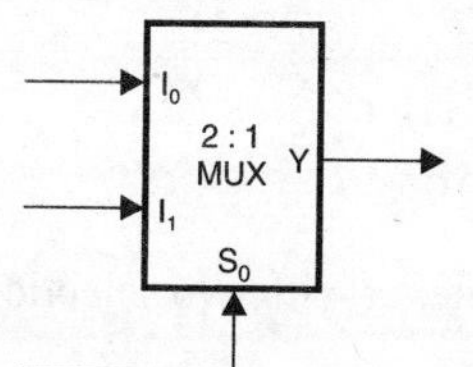

Fig. a: Block diagram representation.

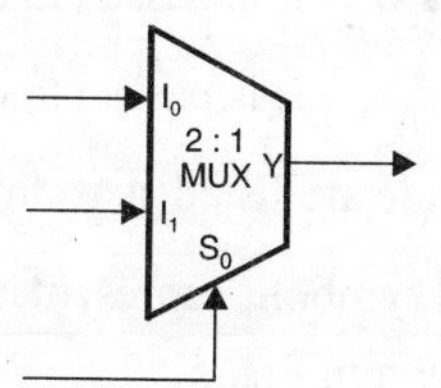

Fig. b: Standard symbol.

***Fig. 5.56:** 2:1 multiplexer.*

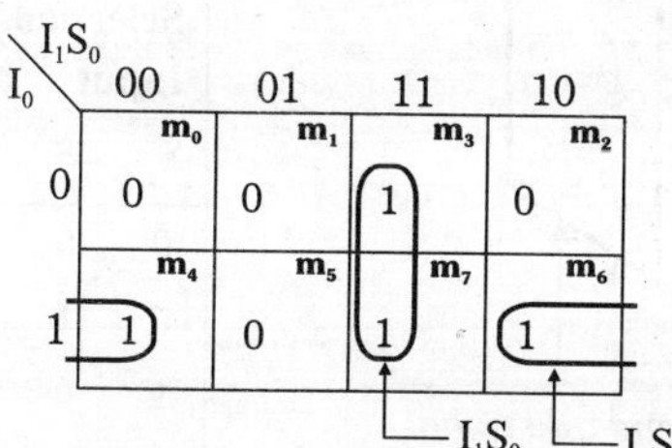

***Fig. 5.57:** K-map for design of 2:1 multiplexer.*

Table 5.18: Truth Table of 2:1 Multiplexer for all Combination of Inputs

Inputs I_0	Inputs I_1	Selection Input S_0	Minterm	Output Y
0	0	0	m_0	$Y = I_0 = 0$
0	0	1	m_1	$Y = I_1 = 0$
0	1	0	m_2	$Y = I_0 = 0$
0	1	1	m_3	$Y = I_1 = 1$
1	0	0	m_4	$Y = I_0 = 1$
1	0	1	m_5	$Y = I_1 = 0$
1	1	0	m_6	$Y = I_0 = 1$
1	1	1	m_7	$Y = I_1 = 1$

The truth table of 2:1 multiplexer for all combination of inputs is shown in Table 5.18. The K-map for the design of 2:1 multiplexer is shown in Fig. 5.57 and from the K-map we get the following Boolean equation.

$$Y = I_0S_0' + I_1S_0$$

The logic circuit for 2:1 multiplexer is drawn using the above equation as shown in Fig. 5.58.

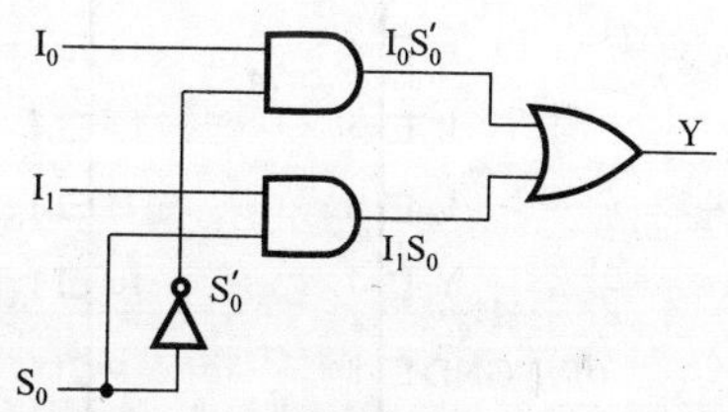

***Fig. 5.58:** Logic circuit of 2:1 multiplexer.*

4:1 Multiplexer

The ***4:1 multiplexer*** is used to select one of the four inputs using 2-bit address or selection input.

Since $4 = 2^2$, for 4:1 multiplexer two selection lines are required to select one of the four inputs.

Let, I_0, I_1, I_2, I_3 = Inputs

S_0, S_1 = Selection inputs/Address

Y = Output

In 4:1 multiplexer,

When, $S_1 = 0$, $S_0 = 0$, I_0 is passed to output, $\therefore Y = I_0$

When, $S_1 = 0$, $S_0 = 1$, I_1 is passed to output, $\therefore Y = I_1$

When, $S_1 = 1$, $S_0 = 0$, I_2 is passed to output, $\therefore Y = I_2$

When, $S_1 = 1$, $S_0 = 1$, I_3 is passed to output, $\therefore Y = I_3$

The above selection logic are listed as truth table in Table 5.19.

The block diagram and symbolic representation of 4:1 multiplexer with two selection lines to select one of the four inputs are shown in Fig. 5.59.

Table 5.19: Truth Table of 4:1 Multiplexer

Selection Input		Output
S_1	S_0	Y
0	0	I_0
0	1	I_1
1	0	I_2
1	1	I_3

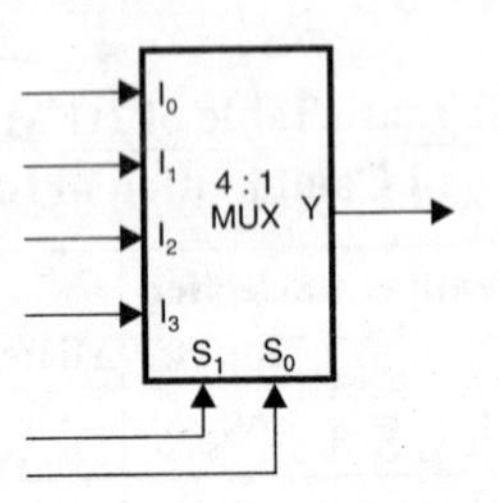

Fig. a: *Block diagram representation.*

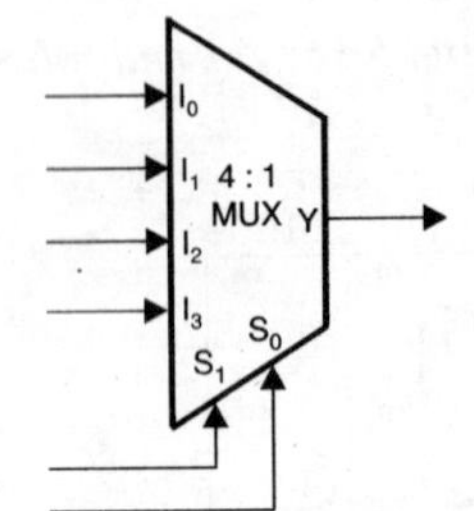

Fig. b: *Standard symbol.*

Fig. 5.59: *4:1 multiplexer.*

The ***dual 4:1 multiplexer*** is available as a standard IC with number 7453. The pin configuration of 7453 is shown in Fig. 5.60.

7453

Pin	Signal	Pin	Signal
1	$\overline{E}_a$	16	V_{cc}(+5V)
2	S_1	15	$\overline{E}_b$
3	I_{3a}	14	S_0
4	I_{2a}	13	I_{3b}
5	I_{1a}	12	I_{2b}
6	I_{0a}	11	I_{1b}
7	Y_a	10	I_{0b}
8	(0V) GND	9	Y_b

Pin	Description
S_0 , S_1	Selection inputs
$\overline{E}_a$, $\overline{E}_b$	Enable (Active low) inputs
I_{0a} - I_{3a}	Multiplexer inputs
I_{0b} - I_{3b}	Multiplexer inputs
Y_a, Y_b	Multiplexer output
V_{cc}	Supply voltage (+5V)
GND	Ground (0V)

Fig. 5.60: *Pin configuration of dual 4:1 multiplexer IC 7453.*

The logic circuit of 4:1 multiplexer can be realized using AND and OR gates as shown in Fig. 5.61. Table 5.20 can be used to check the logic circuit of 4:1 multiplexer.

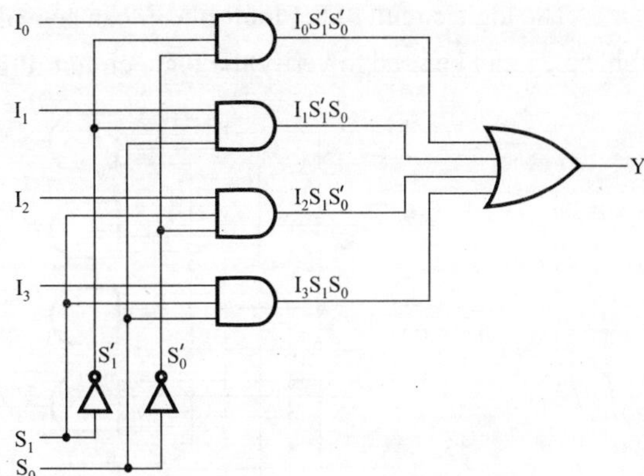

Fig. 5.61: *Logic circuit of 4:1 multiplexer.*

Table 5.20: Verification of Output of 4:1 MUX shown in Fig. 5.51

S_1	S_0	S_1'	S_0'	$I_0S_1'S_0'$	$I_1S_1'S_0$	$I_2S_1S_0'$	$I_3S_1S_0$	Y
0	0	1	1	I_0	0	0	0	I_0
0	1	1	0	0	I_1	0	0	I_1
1	0	0	1	0	0	I_2	0	I_2
1	1	0	0	0	0	0	I_3	I_3

8:1 Multiplexer

The ***8:1 multiplexer*** is used to select one of the eight inputs using 3-bit address or selection input.

Since $8 = 2^3$, for 8:1 multiplexer 3 selection lines are required to select one of the eight inputs. The input selection logic are listed in truth table shown in Table 5.21. The block diagram and symbolic representation of 8:1 multiplexer with three selection lines to select one of the eight inputs are shown in Fig. 5.62.

Let, $I_0, I_1, I_2, \ldots\ldots I_7$ = Inputs

S_0, S_1, S_2 = Selection inputs/Address

Y = Output

Table 5.21: Truth Table of 8:1 Multiplexer

Selection Inputs			Output Y
S_2	S_1	S_0	
0	0	0	I_0
0	0	1	I_1
0	1	0	I_2
0	1	1	I_3
1	0	0	I_4
1	0	1	I_5
1	1	0	I_6
1	1	1	I_7

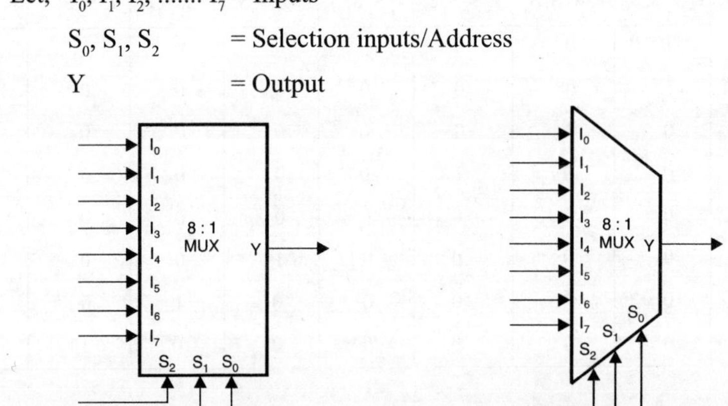

Fig. a: *Block diagram representation.* ***Fig. b:*** *Standard symbol.*

Fig. 5.62: *8:1 multiplexer.*

The logic circuit of 8:1 multiplexer can be realized using AND and OR gates as shown in Fig. 5.63. Table 5.22 can be used to verify the logic circuit of 8:1 multiplexer.

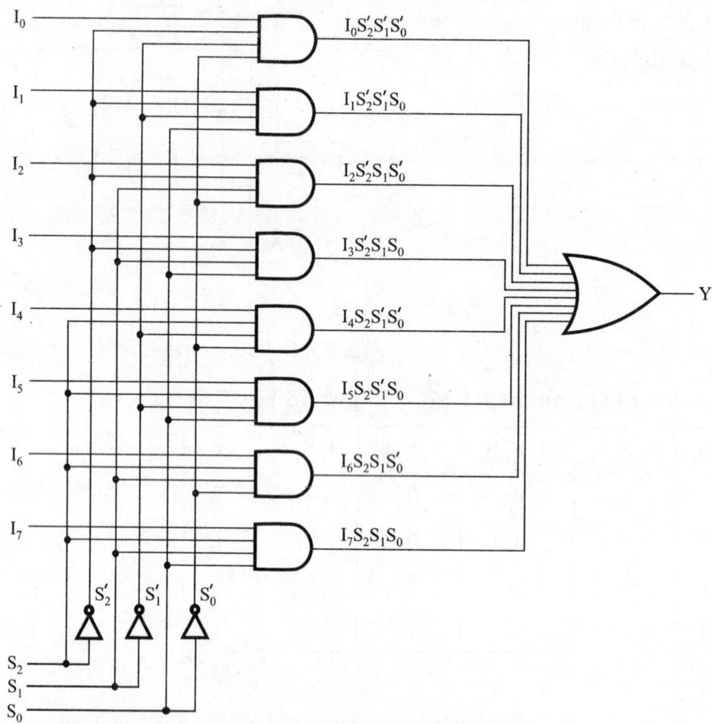

Fig. 5.63: Logic circuit of 8:1 multiplexer.

Table 5.22: Verification of Output of 8:1 Multiplexer of Fig. 5.63

Selection Inputs	Complement of Selection Inputs	Product Terms								Output
S_2 S_1 S_0	S'_2 S'_1 S'_0	$I_0S'_2S'_1S'_0$	$I_1S'_2S'_1S_0$	$I_2S'_2S_1S'_0$	$I_3S'_2S_1S_0$	$I_4S_2S'_1S'_0$	$I_5S_2S'_1S_0$	$I_6S_2S_1S'_0$	$I_7S_2S_1S_0$	Y
0 0 0	1 1 1	I_0	0	0	0	0	0	0	0	I_0
0 0 1	1 1 0	0	I_1	0	0	0	0	0	0	I_1
0 1 0	1 0 1	0	0	I_2	0	0	0	0	0	I_2
0 1 1	1 0 0	0	0	0	I_3	0	0	0	0	I_3
1 0 0	0 1 1	0	0	0	0	I_4	0	0	0	I_4
1 0 1	0 1 0	0	0	0	0	0	I_5	0	0	I_5
1 1 0	0 0 1	0	0	0	0	0	0	I_6	0	I_6
1 1 1	0 0 0	0	0	0	0	0	0	0	I_7	I_7

The 8:1 multiplexer is available as a standard IC with number 74151. The pin configuration of 74151 is shown in Fig. 5.64.

Left pin	No.	74151	No.	Right pin
I_3	1		16	V_{cc}(+5V)
I_2	2		15	I_4
I_1	3		14	I_5
I_0	4		13	I_6
Y	5		12	I_7
$\overline{Y}$	6		11	S_0
$\overline{E}$	7		10	S_1
(0V) GND	8		9	S_2

Pin	Description
S_0 - S_2	Selection inputs
$\overline{E}$	Enable (Active low) input
I_0 - I_7	Multiplexer inputs
Y	Multiplexer output
$\overline{Y}$	Complement of output
V_{cc}	Supply voltage (+5V)
GND	Ground (0V)

Fig.5.64: Pin configuration of 8:1 multiplexer IC 74151.

5.7.1 Expanding the Multiplexer

Multiplexer of smaller size can be connected in cascade to expand the size of multiplexer. For example 2:1 multiplexers can be connected in cascade to form 4:1 multiplexer or multiplexers of higher order.

Similarly, 4:1 multiplexers can be connected in cascade to form 8:1 multiplexer or multiplexers of higher order.

4:1 MUX using 2:1 MUX

(AKTU, Nov/Dec'22, 12 Marks)

An example of constructing 4:1 multiplexer using 2:1 multiplexer is shown in Fig. 5.65.

A ***4:1 multiplexer*** will have 4 inputs I_0, I_1, I_2 and I_3 and two selection inputs S_0 and S_1. Let, Y be the output of 4:1 MUX.

The 4:1 MUX using 2:1 is constructed in two stages with two 2:1 MUX in first stage and one 2:1 MUX in second stage.

The inputs I_0 and I_1 are connected to one of the 2:1 MUX of first stage and another two inputs I_2 and I_3 are connected to another 2:1 MUX of first stage. The outputs of first stage MUXs are connected to inputs of second stage 2:1 MUX.

The selection input S_0 is used to select first stage MUXs and the selection input S_1 is used to select second stage MUX. The output of second stage 2:1 MUX will be same as output of 4:1 MUX which is verified in Table 5.23.

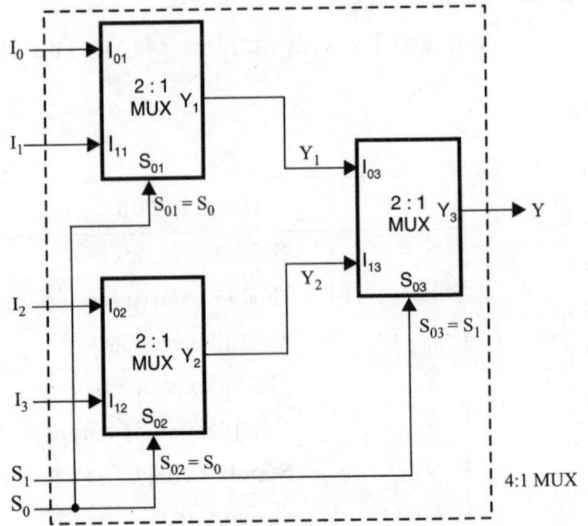

Fig. 5.65: 4:1 MUX using 2:1 MUX.

Truth Table of 2:1 Multiplexer

Selection Input S_0	Output Y
0	I_0
1	I_1

Truth Table of 4:1 Multiplexer

Selection Input S_1	S_0	Output Y
0	0	I_0
0	1	I_1
1	0	I_2
1	1	I_3

Table 5.23: Verification of Output of 4:1 Multiplexer of Fig. 5.65

Selection Inputs		2:1 MUX Outputs		4:1 MUX Output
S_1	S_0	Y_1	Y_2	Y
0	0	$Y_1 = I_{01} = I_0$	$Y_2 = I_{02} = I_2$	$Y = I_{03} = Y_1 = I_0$
0	1	$Y_1 = I_{11} = I_1$	$Y_2 = I_{12} = I_3$	$Y = I_{03} = Y_1 = I_1$
1	0	$Y_1 = I_{01} = I_0$	$Y_2 = I_{02} = I_2$	$Y = I_{13} = Y_2 = I_2$
1	1	$Y_1 = I_{11} = I_1$	$Y_2 = I_{12} = I_3$	$Y = I_{13} = Y_2 = I_3$

8:1 MUX using 4:1 and 2:1

An example of constructing 8:1 MUX using 4:1 and 2:1 MUX is shown in Fig. 5.66.

A ***8:1 MUX*** will have eight inputs I_0, I_1, ... I_7 and three selection inputs S_0, S_1 and S_2. Let, Y be the output of 8:1 MUX.

The 8:1 MUX is constructed in two stages with two 4:1 MUX in first stage and one 2:1 MUX in second stage. The inputs I_0, I_1, I_2 and I_3 are connected to inputs of one of the 4:1 MUX of first stage and other 4 inputs I_4, I_5, I_5 and I_7 are connected to input of another 4:1 MUX of first stage. The outputs of first stage MUXs are connected to input of second stage 2:1 MUX.

The selection inputs S_1 and S_0 are used to select first stage MUXs and the selction input S_2 is used to select second stage MUX. The output of second stage 2:1 MUX will be same as output of 8:1 MUX which is verified in Table 5.24.

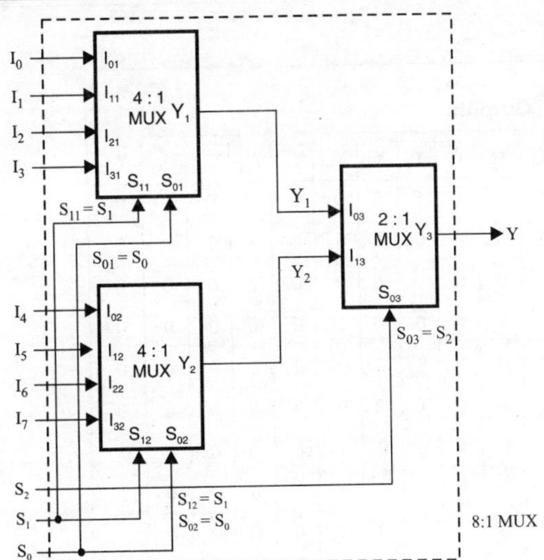

Fig. 5.66: 8:1 MUX using 4:1 and 2:1 MUX.

Truth Table of 4:1 Multiplexer

Selection Input		Output
S_1	S_0	Y
0	0	I_0
0	1	I_1
1	0	I_2
1	1	I_3

Truth Table of 8:1 Multiplexer

Selection Inputs			Output Y
S_2	S_1	S_0	
0	0	0	I_0
0	0	1	I_1
0	1	0	I_2
0	1	1	I_3
1	0	0	I_4
1	0	1	I_5
1	1	0	I_6
1	1	1	I_7

Table 5.24: Verification of Output of 8:1 MUX of Fig. 5.66

Selection Inputs			4:1 MUX Outputs		8:1 MUX Output
S_2	S_1	S_0	Y_1	Y_2	Y
0	0	0	$Y_1 = I_{01} = I_0$	$Y_2 = I_{02} = I_4$	$Y = I_{03} = Y_1 = I_0$
0	0	1	$Y_1 = I_{11} = I_1$.	$Y_2 = I_{12} = I_5$	$Y = I_{03} = Y_1 = I_1$
0	1	0	$Y_1 = I_{21} = I_2$	$Y_2 = I_{22} = I_6$	$Y = I_{03} = Y_1 = I_2$
0	1	1	$Y_1 = I_{31} = I_3$	$Y_2 = I_{32} = I_7$	$Y = I_{03} = Y_1 = I_3$
1	0	0	$Y_1 = I_{01} = I_0$	$Y_2 = I_{02} = I_4$	$Y = I_{13} = Y_2 = I_4$
1	0	1	$Y_1 = I_{11} = I_1$	$Y_2 = I_{12} = I_5$	$Y = I_{13} = Y_2 = I_5$
1	1	0	$Y_1 = I_{21} = I_2$	$Y_2 = I_{22} = I_6$	$Y = I_{13} = Y_2 = I_6$
1	1	1	$Y_1 = I_{31} = I_3$	$Y_2 = I_{32} = I_7$	$Y = I_{13} = Y_2 = I_7$

Example 5.20

Design 32:1 multiplexer using four 8:1 multiplexer and 2-to-4 decoder.

Solution

A 32:1 multiplexer has 32 inputs and 1 output. Let the 32 inputs be I_0, I_1, I_2, ..., I_{31} and the output be Y. The 32:1 multiplexer require 5-bit ($2^5 = 32$) selection or address input to select any one of the 32 inputs. Let the selection inputs be S_0, S_1, S_2, S_3 and S_4. The truth table of 32:1 multiplexer is shown in Table 1.

A 8:1 multiplexer has 8 inputs, 3 selection inputs and 1 output. To implement 32:1 multiplexer using 8:1 multiplexer we require 4 numbers of 8:1 multiplexer, so that we get $4 \times 8 = 32$ inputs. But at any one time only one 8:1 multiplexer is allowed to operate and so 4 enable signals are required and they are generated using a 2-to-4 decoder with S_3 and S_4 of 32:1 multiplexer as inputs. The outputs of decoder are used as enables for 8:1 multiplexers.

The selection inputs S_0, S_1 and S_2 of 32:1 multiplexer are used as selection inputs of all the 8:1 multiplexer.

The outputs of 8:1 multiplexers are again connected as inputs of 4:1 MUX so that one of the output of 8:1 MUX is selected as input for 4:1 MUX. The selection input of decoder and 4:1 MUX are same.

Table 1: Truth Table of 32:1 Multiplexer

Selection Inputs					Outputs															
S_4	S_3	S_2	S_1	S_0	Y_0	Y_1	Y_2	Y_3	Y_4	Y_5	Y_6	Y_7	Y_8	Y_9	Y_{10}	Y_{11}	Y_{12}	Y_{13}	Y_{14}	Y_{15}
0	0	0	0	0	I_0	0	0	0	0	0	0	0	0	0	0	0	0	0	0	0
0	0	0	0	1	0	I_1	0	0	0	0	0	0	0	0	0	0	0	0	0	0
0	0	0	1	0	0	0	I_2	0	0	0	0	0	0	0	0	0	0	0	0	0
0	0	0	1	1	0	0	0	I_3	0	0	0	0	0	0	0	0	0	0	0	0
0	0	1	0	0	0	0	0	0	I_4	0	0	0	0	0	0	0	0	0	0	0
0	0	1	0	1	0	0	0	0	0	I_5	0	0	0	0	0	0	0	0	0	0
0	0	1	1	0	0	0	0	0	0	0	I_6	0	0	0	0	0	0	0	0	0
0	0	1	1	1	0	0	0	0	0	0	0	I_7	0	0	0	0	0	0	0	0
0	1	0	0	0	0	0	0	0	0	0	0	0	I_8	0	0	0	0	0	0	0
0	1	0	0	1	0	0	0	0	0	0	0	0	0	I_9	0	0	0	0	0	0
0	1	0	1	0	0	0	0	0	0	0	0	0	0	0	I_{10}	0	0	0	0	0
0	1	0	1	1	0	0	0	0	0	0	0	0	0	0	0	I_{11}	0	0	0	0
0	1	1	0	0	0	0	0	0	0	0	0	0	0	0	0	0	I_{12}	0	0	0
0	1	1	0	1	0	0	0	0	0	0	0	0	0	0	0	0	0	I_{13}	0	0
0	1	1	1	0	0	0	0	0	0	0	0	0	0	0	0	0	0	0	I_{14}	0
0	1	1	1	1	0	0	0	0	0	0	0	0	0	0	0	0	0	0	0	I_{15}

Table 1: Continued....

Selection Inputs					Outputs															
S_4	S_3	S_2	S_1	S_0	Y_{16}	Y_{17}	Y_{18}	Y_{19}	Y_{20}	Y_{21}	Y_{22}	Y_{23}	Y_{24}	Y_{25}	Y_{26}	Y_{27}	Y_{28}	Y_{29}	Y_{30}	Y_{31}
1	0	0	0	0	I_{16}	0	0	0	0	0	0	0	0	0	0	0	0	0	0	0
1	0	0	0	1	0	I_{17}	0	0	0	0	0	0	0	0	0	0	0	0	0	0
1	0	0	1	0	0	0	I_{18}	0	0	0	0	0	0	0	0	0	0	0	0	0
1	0	0	1	1	0	0	0	I_{19}	0	0	0	0	0	0	0	0	0	0	0	0
1	0	1	0	0	0	0	0	0	I_{20}	0	0	0	0	0	0	0	0	0	0	0
1	0	1	0	1	0	0	0	0	0	I_{21}	0	0	0	0	0	0	0	0	0	0
1	0	1	1	0	0	0	0	0	0	0	I_{22}	0	0	0	0	0	0	0	0	0
1	0	1	1	1	0	0	0	0	0	0	0	I_{23}	0	0	0	0	0	0	0	0
1	1	0	0	0	0	0	0	0	0	0	0	0	I_{24}	0	0	0	0	0	0	0
1	1	0	0	1	0	0	0	0	0	0	0	0	0	I_{25}	0	0	0	0	0	0
1	1	0	1	0	0	0	0	0	0	0	0	0	0	0	I_{26}	0	0	0	0	0
1	1	0	1	1	0	0	0	0	0	0	0	0	0	0	0	I_{27}	0	0	0	0
1	1	1	0	0	0	0	0	0	0	0	0	0	0	0	0	0	I_{28}	0	0	0
1	1	1	0	1	0	0	0	0	0	0	0	0	0	0	0	0	0	I_{29}	0	0
1	1	1	1	0	0	0	0	0	0	0	0	0	0	0	0	0	0	0	I_{30}	0
1	1	1	1	1	0	0	0	0	0	0	0	0	0	0	0	0	0	0	0	I_{31}

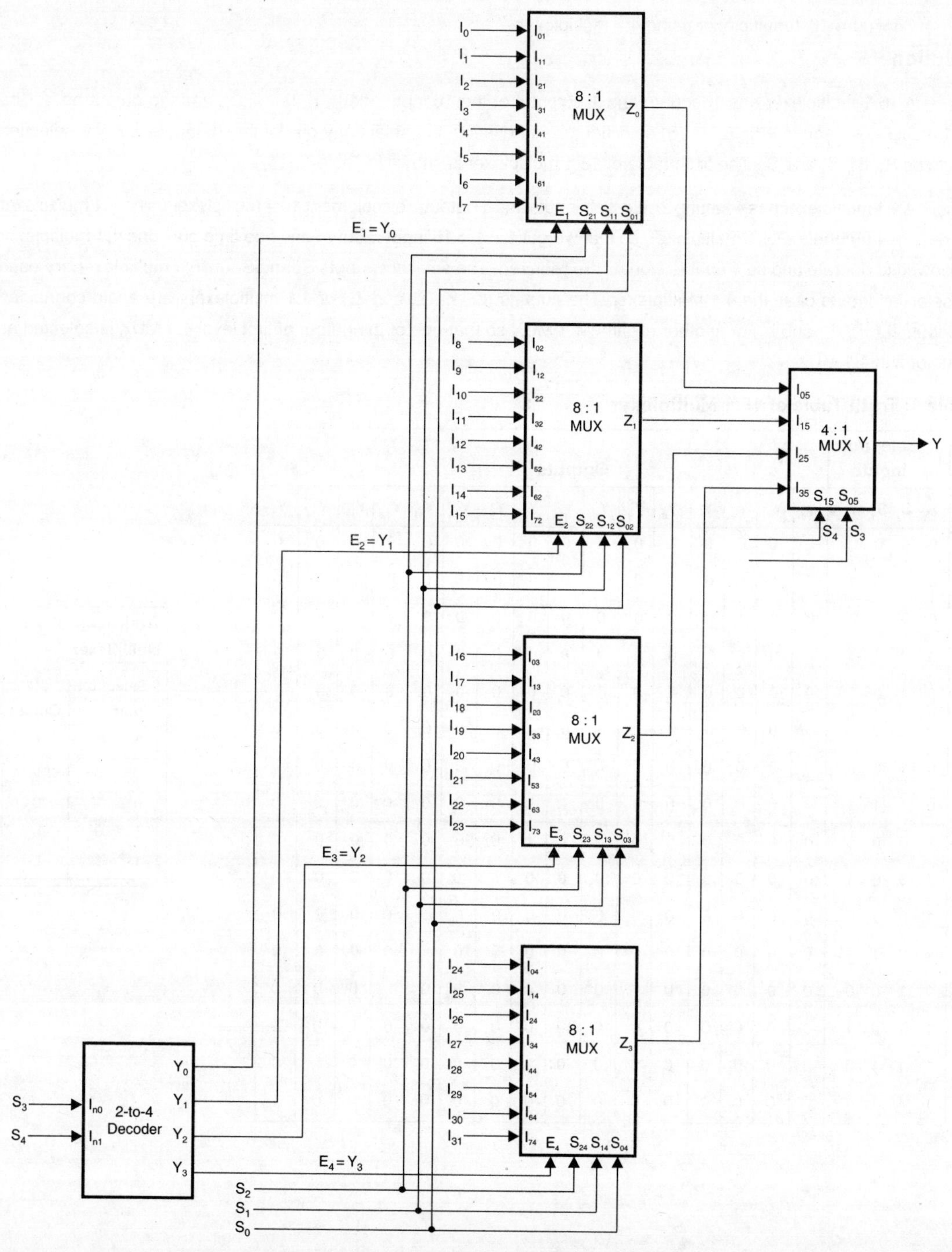

Fig. 1: *32:1 MUX with four 8:1 and the 2-to-4 decoder.*

Example 5.21

(VTU, Aug/Sep'20, 4 Marks)

Configure 16:1 multiplexer using 4:1 multiplexer.

Solution

A 16:1 multiplexer has 16 inputs and 1 output. Let the 16 inputs be I_0, I_1, I_2, ..., I_{15} and the output be Y. The 16:1 multiplexer require 4-bit (2^4 = 16) selection or address input to select any one of the 16 inputs. Let the selection inputs be S_0, S_1, S_2 and S_3. The truth table of 16:1 multiplexer is shown in Table 1.

A 4:1 multiplexer has 4 inputs, 2 selection inputs and 1 output. To implement 16:1 multiplexer using 4:1 multiplexer we require 4 numbers of 4:1 multiplexer, so that we get 4 × 4 = 16 inputs. But at any one time only one 4:1 multiplexer is allowed to operate and so 4 enable signals are required. The selection inputs S_0 and S_1 of 16:1 multiplexer are used as selection inputs of all the 4:1 multiplexer. The outputs (Z_0, Z_1, Z_2 and Z_3) of 4:1 multiplexers are again connected as inputs (I_{05}, I_{15}, I_{25} and I_{35}) of another (or fifth) 4:1 MUX so that one of the output of first level 4:1 MUX is selected as input for fifth 4:1 MUX.

Table 1: Truth Table of 16:1 Multiplexer

Selection Inputs				Outputs															
S_3	S_2	S_1	S_0	Y_0	Y_1	Y_2	Y_3	Y_4	Y_5	Y_6	Y_7	Y_8	Y_9	Y_{10}	Y_{11}	Y_{12}	Y_{13}	Y_{14}	Y_{15}
0	0	0	0	I_0	0	0	0	0	0	0	0	0	0	0	0	0	0	0	0
0	0	0	1	0	I_1	0	0	0	0	0	0	0	0	0	0	0	0	0	0
0	0	1	0	0	0	I_2	0	0	0	0	0	0	0	0	0	0	0	0	0
0	0	1	1	0	0	0	I_3	0	0	0	0	0	0	0	0	0	0	0	0
0	1	0	0	0	0	0	0	I_4	0	0	0	0	0	0	0	0	0	0	0
0	1	0	1	0	0	0	0	0	I_5	0	0	0	0	0	0	0	0	0	0
0	1	1	0	0	0	0	0	0	0	I_6	0	0	0	0	0	0	0	0	0
0	1	1	1	0	0	0	0	0	0	0	I_7	0	0	0	0	0	0	0	0
1	0	0	0	0	0	0	0	0	0	0	0	I_8	0	0	0	0	0	0	0
1	0	0	1	0	0	0	0	0	0	0	0	0	I_9	0	0	0	0	0	0
1	0	1	0	0	0	0	0	0	0	0	0	0	0	I_{10}	0	0	0	0	0
1	0	1	1	0	0	0	0	0	0	0	0	0	0	0	I_{11}	0	0	0	0
1	1	0	0	0	0	0	0	0	0	0	0	0	0	0	0	I_{12}	0	0	0
1	1	0	1	0	0	0	0	0	0	0	0	0	0	0	0	0	I_{13}	0	0
1	1	1	0	0	0	0	0	0	0	0	0	0	0	0	0	0	0	I_{14}	0
1	1	1	1	0	0	0	0	0	0	0	0	0	0	0	0	0	0	0	I_{15}

Truth Table of 4:1 Multiplexer

Selection Input		Output
S_1	S_0	Y
0	0	I_0
0	1	I_1
1	0	I_2
1	1	I_3

Table 2: Verification of 16:1 MUX using 4:1 MUX

Selection Inputs I_3 I_2 I_1 I_0	Enables E_1 E_2 E_3 E_4	4:1 MUX Outputs	16:1 MUX Outputs	Selection Inputs I_3 I_2 I_1 I_0	Enables E_1 E_2 E_3 E_4	4:1 MUX Outputs	16:1 MUX Outputs
0 0 0 0	1 0 0 0	$Z_0 = I_{01} = I_0$	$Y = I_{05} = Z_0 = I_{01} = I_0$	1 0 0 0	0 0 1 0	$Z_2 = I_{03} = I_8$	$Y = I_{05} = Z_2 = I_{03} = I_8$
0 0 0 1	1 0 0 0	$Z_0 = I_{11} = I_1$	$Y = I_{15} = Z_0 = I_{11} = I_1$	1 0 0 1	0 0 1 0	$Z_2 = I_{13} = I_9$	$Y = I_{15} = Z_2 = I_{13} = I_9$
0 0 1 0	1 0 0 0	$Z_0 = I_{21} = I_2$	$Y = I_{25} = Z_0 = I_{21} = I_2$	1 0 1 0	0 0 1 0	$Z_2 = I_{23} = I_{10}$	$Y = I_{25} = Z_2 = I_{23} = I_{10}$
0 0 1 1	1 0 0 0	$Z_0 = I_{31} = I_3$	$Y = I_{35} = Z_0 = I_{31} = I_3$	1 0 1 1	0 0 1 0	$Z_2 = I_{33} = I_{11}$	$Y = I_{35} = Z_2 = I_{33} = I_{11}$
0 1 0 0	0 1 0 0	$Z_1 = I_{02} = I_4$	$Y = I_{05} = Z_1 = I_{02} = I_4$	1 1 0 0	0 0 0 1	$Z_3 = I_{04} = I_{12}$	$Y = I_{05} = Z_3 = I_{04} = I_{12}$
0 1 0 1	0 1 0 0	$Z_1 = I_{12} = I_5$	$Y = I_{15} = Z_1 = I_{12} = I_5$	1 1 0 1	0 0 0 1	$Z_3 = I_{14} = I_{13}$	$Y = I_{15} = Z_3 = I_{14} = I_{13}$
0 1 1 0	0 1 0 0	$Z_1 = I_{22} = I_6$	$Y = I_{25} = Z_1 = I_{22} = I_6$	1 1 1 0	0 0 0 1	$Z_3 = I_{24} = I_{14}$	$Y = I_{25} = Z_3 = I_{24} = I_{14}$
0 1 1 1	0 1 0 0	$Z_1 = I_{32} = I_7$	$Y = I_{35} = Z_1 = I_{32} = I_7$	1 1 1 1	0 0 0 1	$Z_4 = I_{34} = I_{15}$	$Y = I_{35} = Z_3 = I_{34} = I_{15}$

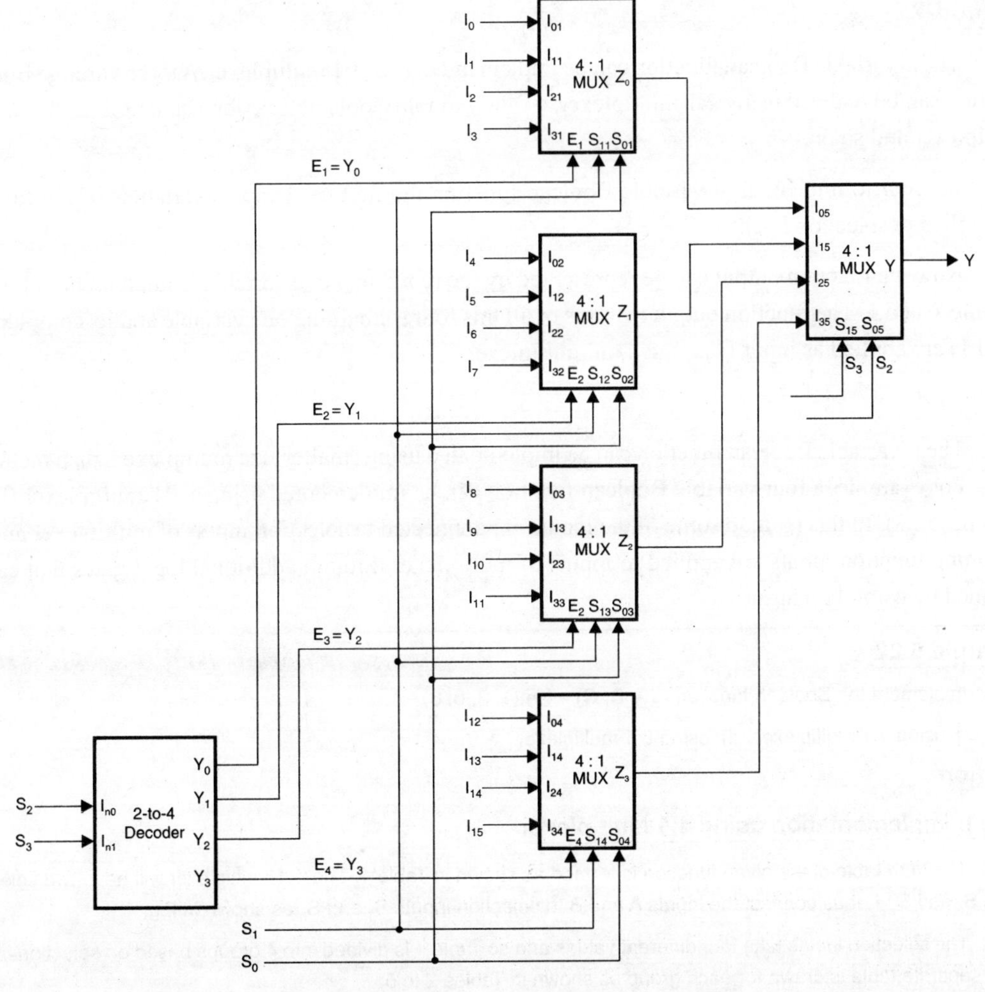

***Fig. 1:** 16:1 MUX using 4:1 MUX.*

5.7.2 Combinational Circuit Design using Multiplexer

In multiplexer, the possible binary values of selection inputs are same as minterms. Therefore, the logic circuit of Boolean functions in SOP form can be realized using multiplexer. Since a multiplexer has a single output, a separate multiplexer is required for every output of Boolean function.

There are different methods of implementing combinational circuits using multiplexer.

Method-1

The n-variable Boolean function can be implemented directly by 2^n:1 multiplexer. The n inputs of Boolean function are connected to n selection inputs. The inputs I_0, I_1, I_2, ..., I_{2n-1} are tied to either 0 or 1 from the knowledge of output in the truth table of Boolean function. If a minterm m_p generates 1 in output of Boolean function then input I_p of multiplexer is tied to 1. If a minterm m_q generates 0 in output of Boolean function then input I_q of multiplexer is tied to 0.

Method-2

The n-variable Boolean function can be implemented by 2^{n-1}:1 multiplexer. A three variable Boolean function can be realized using 2:1 multiplexer, a four variable Boolean function can be realized using 8:1 multiplexer and so on.

In order to implement n-variable Boolean function the first n – 1 inputs/variables of the function are applied to selection lines.

Now the function output can be represented by the remaining one variable, complement of this one variable, 0 and 1. The function output in terms of all this four (remaining one variable and its complement, 0 and 1) are applied as input I_0 .... I_{2n-1} of multiplexer.

Method-3

The n-variable Boolean function can be implemented using smaller size multiplexers than method-1 and 2. For example, a four variable Boolean function can be implemented using a 4:1 multiplexer (Refer Example 5.23). In this method some of the inputs are connected to selection inputs of multiplexer and the remaining function inputs are applied to inputs of multiplexer through additional logic gates that can be designed by using K-maps.

Example 5.22

(AKTU, Sep'20, 10 Marks) (AKTU, Nov/Dec'22, 10 Marks)

Implement the Boolean function F(A, B, C) = Σm(1, 3, 5, 6).

i) using 4:1 multiplexer ii) using 8:1 multiplexer.

Solution

Case i: Implementation using 4:1 Multiplexer

The truth table of the given function is formed as shown in Table 1. The 4:1 multiplexer will have two selection inputs S_0 and S_1. Let us connect the inputs A and B to selection inputs S_1 and S_0 as shown in Fig. 1.

The selection inputs take four different values and so Table 1 is divided into 4 groups based on selection inputs and a separate table is drawn for each group as shown in Tables 2 to 5.

Table 1: Truth Table of F

Inputs A	B	C	Minterm	Output F	Comments
0	0	0	m_0	0	When, A = B = C = 0, F = 0
0	0	1	m_1	1	When, A = B = 0, C = 1, F = 1
0	1	0	m_2	0	When, A = C = 0, B = 1, F = 0
0	1	1	m_3	1	When, A = 0, B = C = 1, F = 1
1	0	0	m_4	0	When, A = 1, B = C = 0, F = 0
1	0	1	m_5	1	When, A = C = 1, B = 0, F = 1
1	1	0	m_6	1	When, A = B = 1, C = 0, F = 1
1	1	1	m_7	0	When, A = B = C = 1, F = 0

Truth Table of 4:1 Multiplexer

Selection Input S_1	S_0	Output Y
0	0	I_0
0	1	I_1
1	0	I_2
1	1	I_3

Table 2:

Inputs A B	C	Selection Inputs S_1	S_0	F = C
0 0	0	0	0	0
0 0	1	0	0	1

Table 3:

Inputs A B	C	Selection Inputs S_1	S_0	F = C
0 1	0	0	1	0
0 1	1	0	1	1

Table 4:

Inputs A B	C	Selection Inputs S_1	S_0	F = C
1 0	0	1	0	0
1 0	1	1	0	1

Table 5:

Inputs A B	C	Selection Inputs S_1	S_0	F = C′
1 1	0	1	1	1
1 1	1	1	1	0

The input I_0, I_1, I_2 and I_3 of multiplexer are obtained from Tables 2 to 5.

From Tables 2, 3 and 4 it can be observed that function output is C when selection inputs are 00, 01 and 10 and so I_0, I_1 and I_2 are tied to C as shown in Fig. 1. From Table 5 it can be observed that function output is complement of C when selection input is 11 and so C is inverted to get C′ and connected to I_3 as shown in Fig. 1. The working of 4:1 MUX is verified in Table 6.

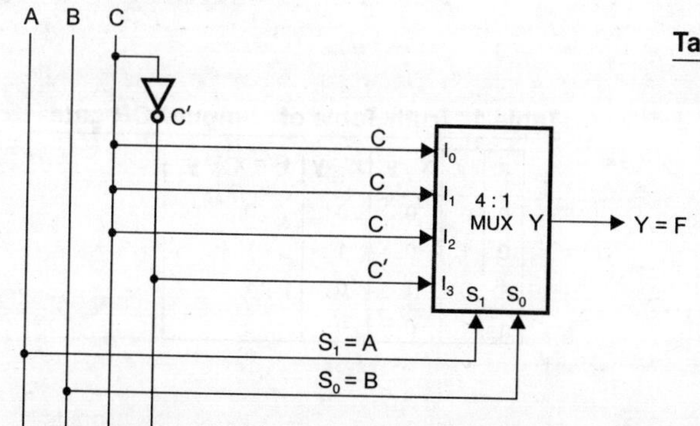

Fig. 1: Implementation of Boolean function using 4:1 MUX.

Table 6: Verification of Logic Circuit of F in Fig. 1

A	B	C	S_1	S_0	F	Y = F
0	0	0	0	0	0	$Y = I_0 = C = 0$
0	0	1	0	0	1	$Y = I_0 = C = 1$
0	1	0	0	1	0	$Y = I_1 = C = 0$
0	1	1	0	1	1	$Y = I_1 = C = 1$
1	0	0	1	0	0	$Y = I_2 = C = 0$
1	0	1	1	0	1	$Y = I_2 = C = 1$
1	1	0	1	1	1	$Y = I_3 = C' = 1$
1	1	1	1	1	0	$Y = I_3 = C' = 0$

Case ii: Implementation using 8:1 Multiplexer

The function F is 3-variable function and so a 2^3:1 multiplexer (8:1 multiplexer) can be used to implement the function as shown in Fig. 2. The function inputs A, B and C are connected to selection inputs S_2, S_1 and S_0. The function output F is 1 for minterms m_1, m_3, m_5 and m_6 and so the MUX inputs I_1, I_3, I_5 and I_6 are permanently tied to 1. The function output F is 0 for minterms m_0, m_2, m_4 and m_7 and so the MUX inputs I_0, I_2, I_4 and I_7 are permanently tied to 0. The working of 8:1 MUX is verified in Table 8.

Table 7: Truth Table of F

Inputs			Minterm	Output	Comments
A	B	C		F	
0	0	0	m_0	0	When, A = B = C = 0, F = 0
0	0	1	m_1	1	When, A = B = 0, C = 1, F = 1
0	1	0	m_2	0	When, A = C = 0, B = 1, F = 0
0	1	1	m_3	1	When, A = 0, B = C = 1, F = 1
1	0	0	m_4	0	When, A = 1, B = C = 0, F = 0
1	0	1	m_5	1	When, A = C = 1, B = 0, F = 0
1	1	0	m_6	1	When, A = B = 1, C = 0, F = 1
1	1	1	m_7	0	When, A = B = C = 1, F = 0

Truth Table of 8:1 Multiplexer

Selection Inputs			Output Y
S_2	S_1	S_0	
0	0	0	I_0
0	0	1	I_1
0	1	0	I_2
0	1	1	I_3
1	0	0	I_4
1	0	1	I_5
1	1	0	I_6
1	1	1	I_7

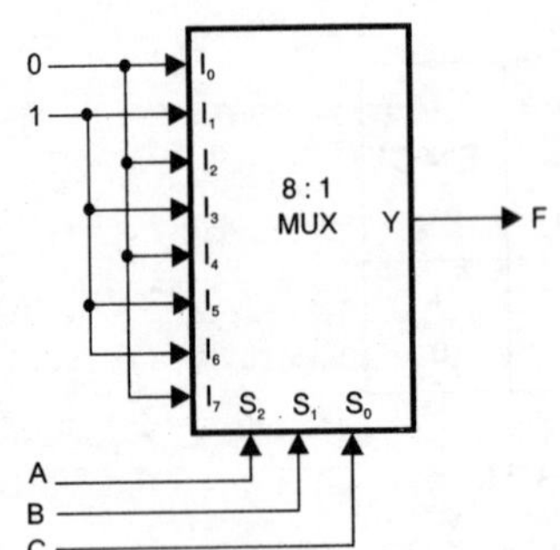

Fig. 2: Realization of F using 8:1 MUX.

Table 8: Verification of Logic Circuit of F in Fig. 2

A	B	C	S_2	S_1	S_0	F	Y = F
0	0	0	0	0	0	0	$Y = I_0 = 0$
0	0	1	0	0	1	1	$Y = I_1 = 1$
0	1	0	0	1	0	0	$Y = I_2 = 0$
0	1	1	0	1	1	1	$Y = I_3 = 1$
1	0	0	1	0	0	0	$Y = I_4 = 0$
1	0	1	1	0	1	1	$Y = I_5 = 1$
1	1	0	1	1	0	1	$Y = I_6 = 1$
1	1	1	1	1	1	0	$Y = I_7 = 0$

Example 5.23

(GTU, Jun'19, 4 Marks)

Design two input XOR gate using 4:1 MUX.

Solution

x y $z = x \oplus y$

Fig. 1: Symbol of 2-input XOR gate.

$\therefore x \oplus y = x y' + x' y$

Table 1: Truth Table of 2-input XOR gate

x	y	x . y'	x' . y	$F = x \oplus y$
0	0	0	0	0
0	1	0	1	1
1	0	1	0	1
1	1	0	0	0

Implementation using 4:1 Multiplexer

The truth table of the given function is formed as shown in Table 2. The 4:1 multiplexer will have two selection inputs S_0 and S_1. Let us connect the inputs x and y to selection inputs S_1 and S_0 as shown in Fig. 2.

The selection inputs take two different values and so Table 2 is divided into 2 groups based on selection inputs and a separate table is drawn for each group as shown in Tables 3 and 4.

Truth Table of 4:1 Multiplexer

Selection Input		Output
S_1	S_0	Y
0	0	I_0
0	1	I_1
1	0	I_2
1	1	I_3

Table 2: Truth Table of F

Inputs		Minterm	Output	Comments
x	y		F	
0	0	m_0	0	When, x = y = 0, F = 0
0	1	m_1	1	When, x = 0, y = 1, F = 1
1	0	m_2	1	When, x = 1, y = 0, F = 1
1	1	m_3	0	When, x = y = 1, F = 0

Table 3:

Inputs	Selection Inputs		F = y
x y	S_1	S_0	
0 0	0	0	0
0 1	0	1	1

Table 4:

Inputs	Selection Inputs		F = y'
x y	S_1	S_0	
1 0	1	0	1
1 1	1	1	0

The input I_0, I_1, I_2 and I_3 of multiplexer are obtained from Tables 3 and 4.

From Table 3, it can be observed that function output is y when selection inputs are 00 and 01 and so I_0 and I_1 are tied to y as shown in Fig. 2. From Table 4, it can be observed that function output is complement of y when selection input is 10 and 11 and so y is inverted to get y' and connected to I_2 and I_3 as shown in Fig. 2. The working of 4:1 MUX is verified in Table 5.

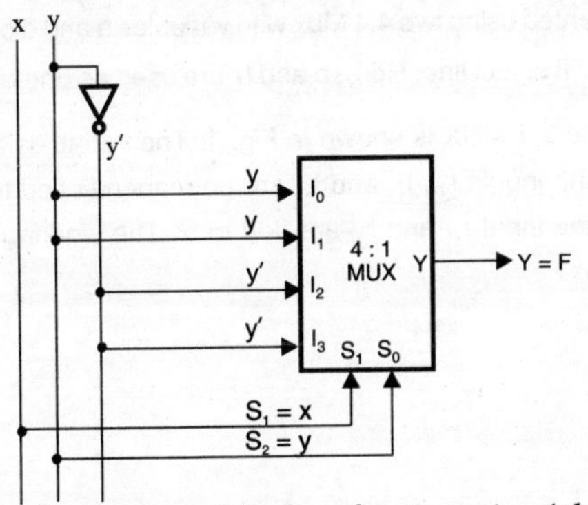

Fig. 2: Implementation of Boolean function using 4:1 MUX.

Table 5: Verification of Logic Circuit of F in Fig. 2

x	y	S_1	S_0	F	Y = F
0	0	0	0	0	$Y = I_0 = y = 0$
0	1	0	1	1	$Y = I_1 = y = 1$
1	0	1	0	1	$Y = I_2 = y' = 1$
1	1	1	1	0	$Y = I_3 = y' = 0$

Example 5.24

(VTU, Aug/Sep'20, 10 Marks)

Implement the Boolean function F(a, b, c, d) = Σm(0, 2, 4, 5, 9, 10, 14) using two 4:1 MUX with variable a, d connected to their select lines in the first level and one 2:1 MUX with variable c connected to its select line in the second level.

Solution

Table 1: Truth Table of Function F

Inputs Variables				Minterm	Output F
a	b	c	d		
0	0	0	0	m_0	1
0	0	0	1	m_1	0
0	0	1	0	m_2	1
0	0	1	1	m_3	0
0	1	0	0	m_4	1
0	1	0	1	m_5	1
0	1	1	0	m_6	0
0	1	1	1	m_7	0
1	0	0	0	m_8	0
1	0	0	1	m_9	1
1	0	1	0	m_{10}	1
1	0	1	1	m_{11}	0
1	1	0	0	m_{12}	0
1	1	0	1	m_{13}	0
1	1	1	0	m_{14}	1
1	1	1	1	m_{15}	0

Table 2 is truth table of given Boolean function implemented using two 4:1 Mux with variables a and d connected to their select lines and one 2:1 Mux with variable c connected to it select line. Here, b and b′are used as one of the inputs.

The implementation of function, F using 4:1 MUX and 2:1 MUX is shown in Fig. 1. The variables a and d are used as selection inputs of 4:1 MUX. In 4:1 MUX of Fig. 1, the inputs I_{21}, I_{12} and I_{32} are permanently tied to 0 and the inputs I_{01} and I_{22} are tied to 1. The input I_{11} is tied to b and the input I_{31} and I_{02} are tied to b′. The working of function implemented using multiplexers is verified in Table 2.

Table 2: Verification of Function F Implemented using Multiplexers

a	b	c	d	Minterm	Output F	Output in terms of b, b′, 0, 1	Comments
0	0	0	0	m_0	1	F = 1	When, a = c = d = 0
0	1	0	0	m_4	1	F = 1	Output, F= I_{01} = 1
0	0	0	1	m_1	0	F = b	When, a = c = 0, d = 1
0	1	0	1	m_5	1	F = b	Output, F = I_{11} = b
1	0	0	0	m_8	0	F = 0	When, c = d = 0, a = 1
1	1	0	0	m_{12}	0	F = 0	Output, F = I_{21} = 0
1	0	0	1	m_9	1	F = b′	When, a = d = 1, c = 0
1	1	0	1	m_{13}	0	F = b′	Output, F = I_{31} = b′
0	0	1	0	m_2	1	F = b′	When c = 1, a = d = 0
0	1	1	0	m_6	0	F = b′	Output, F = I_{02} = b′
0	0	1	1	m_3	0	F = 0	When, a = 0, c = d = 1
0	1	1	1	m_7	0	F = 0	Output, F = I_{12} = 0
1	0	1	0	m_{10}	1	F = 1	When, a = c = 1, d = 0
1	1	1	0	m_{14}	1	F = 1	Output, F = I_{22} = 1
1	0	1	1	m_{11}	0	F = 0	When, a = c = d = 1
1	1	1	1	m_{15}	0	F = 0	Output, F = I_{32} = 0

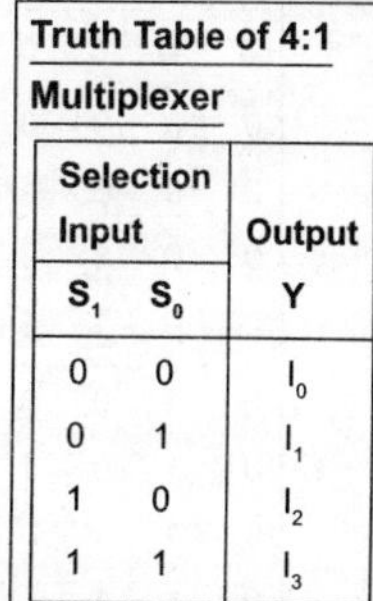

Truth Table of 2:1 Multiplexer

Selection Input S_0	Output Y
0	I_0
1	I_1

Truth Table of 4:1 Multiplexer

Selection Input S_1	S_0	Output Y
0	0	I_0
0	1	I_1
1	0	I_2
1	1	I_3

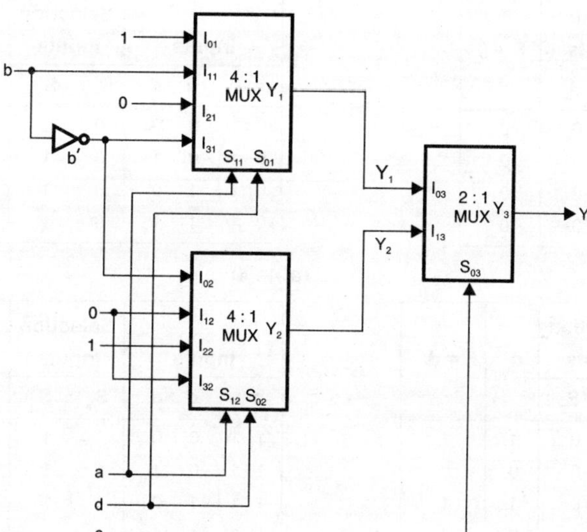

Fig. 1: Logic circuit for Boolean function using two 4:1 and one 2:1 MUX.

Example 5.25

(VTU, Aug/Sep'20, 10 Marks)

Implement Boolean function F(a, b, c, d) = ∑m(4, 5, 7, 8, 10, 12, 15) using 4:1 MUX and external gates:

a) a, b are connected to select lines S_1 and S_0 respectively.

b) c, d are connected to select lines S_1 and S_0 respectively.

Solution

a) a, b are connected to select lines S_1 and S_0

Table 1: Truth Table

a	b	c	d	Minterm	Output F	a	b	c	d	Minterm	Output F
0	0	0	0	m_0	0	1	0	0	0	m_8	1
0	0	0	1	m_1	0	1	0	0	1	m_9	0
0	0	1	0	m_2	0	1	0	1	0	m_{10}	1
0	0	1	1	m_3	0	1	0	1	1	m_{11}	0
0	1	0	0	m_4	1	1	1	0	0	m_{12}	1
0	1	0	1	m_5	1	1	1	0	1	m_{13}	0
0	1	1	0	m_6	0	1	1	1	0	m_{14}	0
0	1	1	1	m_7	1	1	1	1	1	m_{15}	1

***Note:** The Boolean equations for F in Tables 3 and 5 are obtained by drawing 2 variable K-map by taking c and d as inputs.*

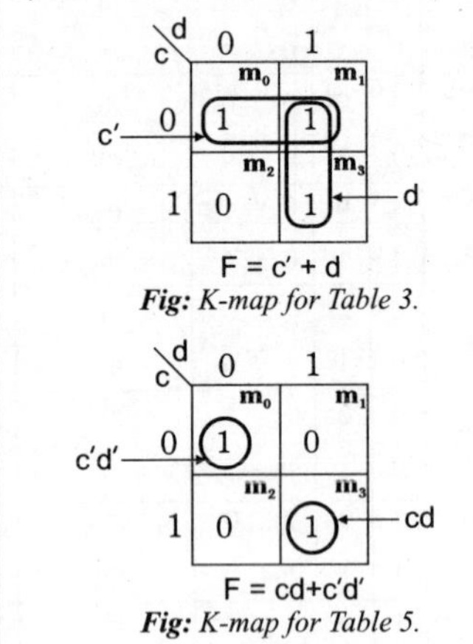

Fig: K-map for Table 3.

Fig: K-map for Table 5.

Table 2:

Inputs a b	c	d	Selection Inputs S_1	S_0	F = 0
0 0	0	0	0	0	0
0 0	0	1	0	0	0
0 0	1	0	0	0	0
0 0	1	1	0	0	0

Table 3:

Inputs a b	c	d	Selection Inputs S_1	S_0	c′	c′ + d	F = c′ + d
0 1	0	0	0	1	1	1	1
0 1	0	1	0	1	1	1	1
0 1	1	0	0	1	0	0	0
0 1	1	1	0	1	0	1	1

Table 4:

Inputs a b	c	d	Selection Inputs S_1	S_0	d′	F = d′
1 0	0	0	1	0	1	1
1 0	0	1	1	0	0	0
1 0	1	0	1	0	1	1
1 0	1	1	1	0	0	0

Table 5:

Inputs a b	c	d	Selection Inputs S_1	S_0	cd	c′d′	F = cd + c′d′
1 1	0	0	1	1	0	1	1
1 1	0	1	1	1	0	0	0
1 1	1	0	1	1	0	0	0
1 1	1	1	1	1	1	0	1

Table 6: Verification

a	b	c	d	S_1	S_0	F	Y = F
0	0	0	0	0	0	0	$I_0 = 0$
0	0	0	1	0	0	0	$I_0 = 0$
0	0	1	0	0	0	0	$I_0 = 0$
0	0	1	1	0	0	0	$I_0 = 0$
0	1	0	0	0	1	1	$I_1 = c' + d = 1$
0	1	0	1	0	1	1	$I_1 = c' + d = 1$
0	1	1	0	0	1	0	$I_1 = c' + d = 0$
0	1	1	1	0	1	1	$I_1 = c' + d = 1$
1	0	0	0	1	0	1	$I_2 = d' = 1$
1	0	0	1	1	0	0	$I_2 = d' = 0$
1	0	1	0	1	0	1	$I_2 = d' = 1$
1	0	1	1	1	0	0	$I_2 = d' = 0$
1	1	0	0	1	1	1	$I_3 = cd + c'd' = 1$
1	1	0	1	1	1	0	$I_3 = cd + c'd' = 0$
1	1	1	0	1	1	0	$I_3 = cd + c'd' = 0$
1	1	1	1	1	1	1	$I_3 = cd + c'd' = 1$

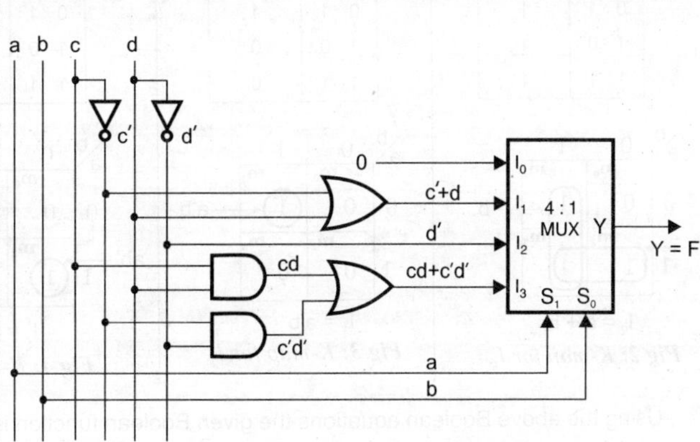

Fig. 1: Implementation of Boolean function using 4:1 MUX by taking a and b as selection inputs..

b) c, d are connected to select lines S_1 and S_0

Table 7:

Inputs				Selection Inputs		F
a	b	c	d	S_1	S_0	
0	0	0	0	0	0	$I_0 = 0$
0	0	0	1	0	1	$I_1 = 0$
0	0	1	0	1	0	$I_2 = 0$
0	0	1	1	1	1	$I_3 = 0$

Table 8:

Inputs				Selection Inputs		F
a	b	c	d	S_1	S_0	
0	1	0	0	0	0	$I_0 = 1$
0	1	0	1	0	1	$I_1 = 1$
0	1	1	0	1	0	$I_2 = 0$
0	1	1	1	1	1	$I_3 = 1$

Table 9:

Inputs				Selection Inputs		F
a	b	c	d	S_1	S_0	
1	0	0	0	0	0	$I_0 = 1$
1	0	0	1	0	1	$I_1 = 0$
1	0	1	0	1	0	$I_2 = 1$
1	0	1	1	1	1	$I_3 = 0$

Table 10:

Inputs				Selection Inputs		F
a	b	c	d	S_1	S_0	
1	1	0	0	0	0	$I_0 = 1$
1	1	0	1	0	1	$I_1 = 0$
1	1	1	0	1	0	$I_2 = 0$
1	1	1	1	1	1	$I_3 = 1$

In order to find Boolean equations for 4:1 MUX inputs I_0, I_1, I_2 and I_3 the Tables 11 to 14 are formed as shown ahead and K-maps are constructed to determine the Boolean equations for I_0, I_1, I_2 and I_3.

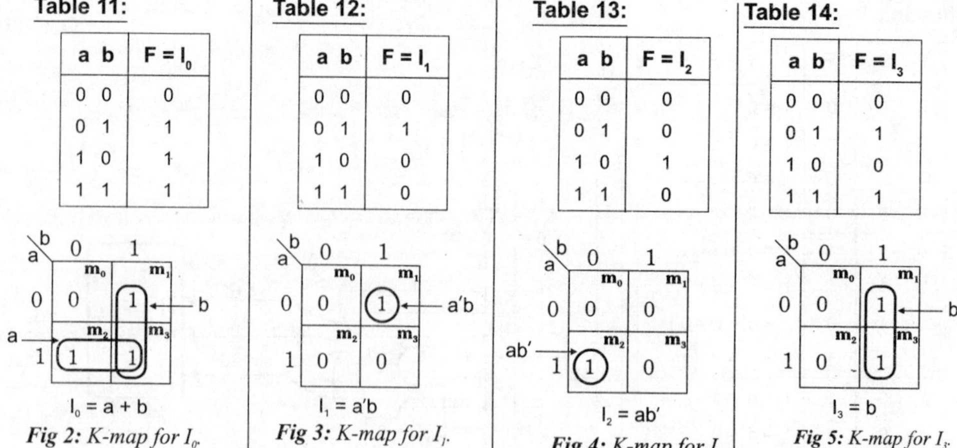

Table 11:

a b	F = I_0
0 0	0
0 1	1
1 0	1
1 1	1

Table 12:

a b	F = I_1
0 0	0
0 1	1
1 0	0
1 1	0

Table 13:

a b	F = I_2
0 0	0
0 1	0
1 0	1
1 1	0

Table 14:

a b	F = I_3
0 0	0
0 1	1
1 0	0
1 1	1

Fig 2: K-map for I_0. *Fig 3: K-map for I_1.* *Fig 4: K-map for I_2.* *Fig 5: K-map for I_3.*

Using the above Boolean equations the given Boolean function is implemented using 4:1 MUX by taking c and d as selection inputs as shown in Fig. 6.

Table 11: Verification

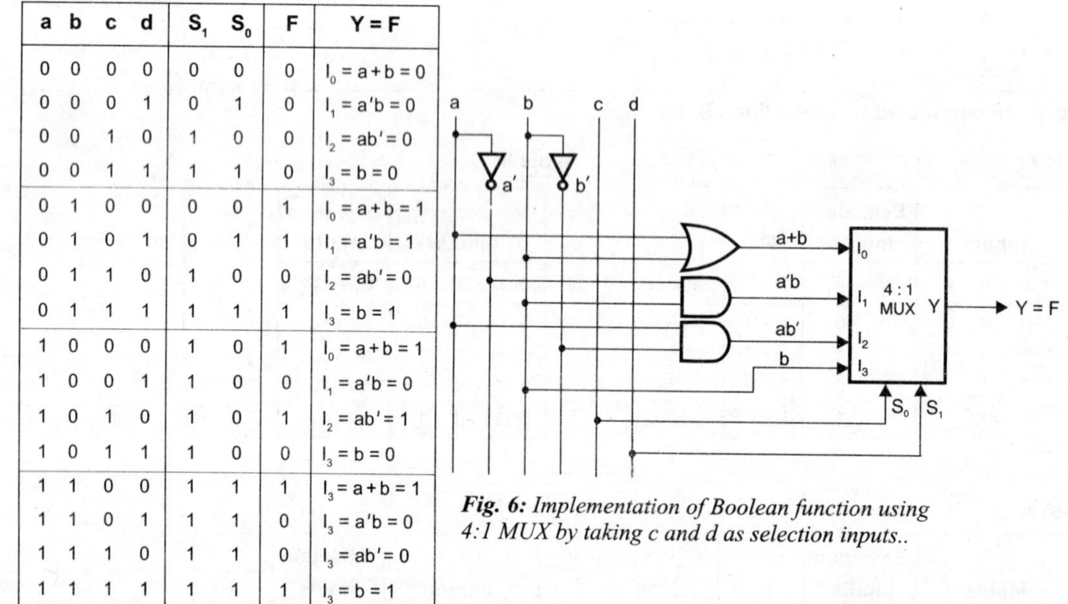

a	b	c	d	S_1	S_0	F	Y = F
0	0	0	0	0	0	0	I_0 = a + b = 0
0	0	0	1	0	1	0	I_1 = a'b = 0
0	0	1	0	1	0	0	I_2 = ab' = 0
0	0	1	1	1	1	0	I_3 = b = 0
0	1	0	0	0	0	1	I_0 = a + b = 1
0	1	0	1	0	1	1	I_1 = a'b = 1
0	1	1	0	1	0	0	I_2 = ab' = 0
0	1	1	1	1	1	1	I_3 = b = 1
1	0	0	0	1	0	1	I_0 = a + b = 1
1	0	0	1	1	0	0	I_1 = a'b = 0
1	0	1	0	1	0	1	I_2 = ab' = 1
1	0	1	1	1	0	0	I_3 = b = 0
1	1	0	0	1	1	1	I_3 = a + b = 1
1	1	0	1	1	1	0	I_3 = a'b = 0
1	1	1	0	1	1	0	I_3 = ab' = 0
1	1	1	1	1	1	1	I_3 = b = 1

Fig. 6: Implementation of Boolean function using 4:1 MUX by taking c and d as selection inputs..

Example 5.26

(VTU, Aug/Sep'20, 6 Marks) (IKGPTU, Dec'19, 10 Marks) (CSVTU, Nov/Dec'19, 7 Marks) (JNTUK, Feb'22, 8 Marks)

Realize F(A, B, C, D) = $\sum m(0, 1, 3, 4, 8, 9, 15)$ using 8:1 MUX by selecting A as an input.

Solution

It is given that function output F is 1 when the inputs are equal to minterms m_0, m_1, m_3, m_4, m_8, m_9 and m_{15}. For all other inputs the function output F is 0.

The given function is analyzed using Table 1 and observed that the function output can be expressed in terms of A, A′, 0 and 1. Therefore, the given 4-variable function can be realized using 2^{4-1}:1 multiplexer (8:1 multiplexer) by taking B, C and D as selection inputs as shown in Fig. 1.

Table 1: Truth Table of F

Inputs				Minterm	Output	Output in terms of	Comments
A	B	C	D		F	A, A′, 0, 1	
0	0	0	0	m_0	1	F = 1	When, B = C = D = 0
1	0	0	0	m_8	1	F = 1	Output, F = 1
0	0	0	1	m_1	1	F = 1	When, B = C = 0, D = 1
1	0	0	1	m_9	1	F = 1	Output, F = 1
0	0	1	0	m_2	0	F = 0	When, B = D = 0, C = 1
1	0	1	0	m_{10}	0	F = 0	Output, F = 0
0	0	1	1	m_3	1	F = A′	When, B = 0, C = D = 1
1	0	1	1	m_{11}	0	F = A′	Output, F = A′
0	1	0	0	m_4	1	F = A′	When B = 1, C = D = 0
1	1	0	0	m_{12}	0	F = A′	Output, F = A′
0	1	0	1	m_5	0	F = 0	When, B = D =1, C = 0
1	1	0	1	m_{13}	0	F = 0	Output, F = 0
0	1	1	0	m_6	0	F = 0	When, B = C = 1, D = 0
1	1	1	0	m_{14}	0	F = 0	Output, F = 0
0	1	1	1	m_7	0	F = A	When, B = C = D = 1
1	1	1	1	m_{15}	1	F = A	Output, F = A

The implementation of function, F using 8:1 MUX is shown in Fig. 1. The variables B, C and D are used as selection inputs S_2, S_1 and S_0 of 8:1 MUX. In 8:1 multiplexer of Fig. 1, the inputs I_2, I_5 and I_6 are permanentely tied to 0 and the inputs I_0 and I_1 are tied to 1. The input I_7 is tied to A and the inputs I_3 and I_4 are tied to A′. The working of 8:1 multiplexer is verified in Table 2.

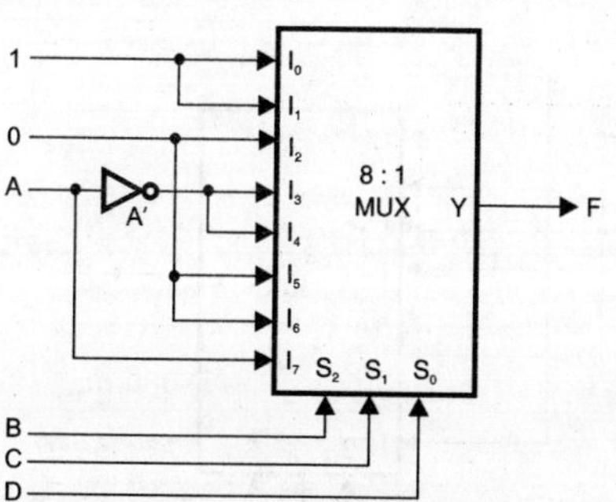

Fig. 1: Realization of F using 8:1 MUX.

Table 2: Verification of Logic Circuit of F in Fig. 1

Inputs				Selection Inputs			Function Output	MUX Output
A	B	C	D	S_2	S_1	S_0	F	Y = F
0	0	0	0	0	0	0	1	$Y = I_0 = 1$
1	0	0	0	0	0	0	1	$Y = I_0 = 1$
0	0	0	1	0	0	1	1	$Y = I_1 = 1$
1	0	0	1	0	0	1	1	$Y = I_1 = 1$
0	0	1	0	0	1	0	0	$Y = I_2 = 0$
1	0	1	0	0	1	0	0	$Y = I_2 = 0$
0	0	1	1	0	1	1	1	$Y = I_3 = A' = 1$
1	0	1	1	0	1	1	0	$Y = I_3 = A' = 0$
0	1	0	0	1	0	0	1	$Y = I_4 = A' = 1$
1	1	0	0	1	0	0	0	$Y = I_4 = A' = 0$
0	1	0	1	1	0	1	0	$Y = I_5 = 0$
1	1	0	1	1	0	1	0	$Y = I_5 = 0$
0	1	1	0	1	1	0	0	$Y = I_6 = 0$
1	1	1	0	1	1	0	0	$Y = I_6 = 0$
0	1	1	1	1	1	1	0	$Y = I_7 = A = 0$
1	1	1	1	1	1	1	1	$Y = I_7 = A = 1$

Example 5.27

(CSVTU, Nov/Dec'19, 7 Marks) (JNTUK, Feb'22, 8 Marks) (IKGPTU, May'20, 5 Marks)

Implement the following Boolean function

$F(w, x, y, z) = \Sigma m(2, 3, 5, 6, 11, 14, 15)$

i) using mulitplexer, ii) using decoder, iii) using multiplexer and decoder

Solution

Case i: Implementation using Multiplexer

It is given that function output F is 1 when the inputs are equal to minterms m_2, m_3, m_5, m_6, m_{11}, m_{14} and m_{15}. For all other inputs the function output F is 0.

The given function is analyzed using Table 1 and observed that the function output can be expressed in terms of z, z′, 0 and 1. Therefore, the given 4-variable function can be realized using 2^{4-1}:1 multiplexer (8:1 multiplexer).

Table 1: Truth Table of F

Inputs w	x	y	z	Minterm	Output F	Output in terms of z, z′, 0, 1	Comments
0	0	0	0	m_0	0	F = 0	When, w = x = y = 0
0	0	0	1	m_1	0	F = 0	Output, F = 0
0	0	1	0	m_2	1	F = 1	When, w = x = 0, y = 1
0	0	1	1	m_3	1	F = 1	Output, F = 1
0	1	0	0	m_4	0	F = z	When, w = y = 0, x = 1
0	1	0	1	m_5	1	F = z	Output, F = z
0	1	1	0	m_6	1	F = z′	When, w = 0, x = y = 1
0	1	1	1	m_7	0	F = z′	Output, F = z′
1	0	0	0	m_8	0	F = 0	When w = 1, x = y = 0
1	0	0	1	m_9	0	F = 0	Output, F = 0
1	0	1	0	m_{10}	0	F = z	When, w = y =1, x = 0
1	0	1	1	m_{11}	1	F = z	Output, F = z
1	1	0	0	m_{12}	0	F = 0	When, w = x = 1, y = 0
1	1	0	1	m_{13}	0	F = 0	Output, F = 0
1	1	1	0	m_{14}	1	F = 1	When, w = x = y = 1
1	1	1	1	m_{15}	1	F = 1	Output, F = 1

Table 2: Verification of Logic Circuit of F in Fig. 1

w	x	y	z	S_2	S_1	S_0	F	Y = F
0	0	0	0	0	0	0	0	$Y = I_0 = 0$
0	0	0	1	0	0	0	0	$Y = I_0 = 0$
0	0	1	0	0	0	1	1	$Y = I_1 = 1$
0	0	1	1	0	0	1	1	$Y = I_1 = 1$
0	1	0	0	0	1	0	0	$Y = I_2 = z = 0$
0	1	0	1	0	1	0	1	$Y = I_2 = z = 1$
0	1	1	0	0	1	1	1	$Y = I_3 = z' = 1$
0	1	1	1	0	1	1	0	$Y = I_3 = z' = 0$
1	0	0	0	1	0	0	0	$Y = I_4 = 0$
1	0	0	1	1	0	0	0	$Y = I_4 = 0$
1	0	1	0	1	0	1	0	$Y = I_5 = z = 0$
1	0	1	1	1	0	1	1	$Y = I_5 = z = 1$
1	1	0	0	1	1	0	0	$Y = I_6 = 0$
1	1	0	1	1	1	0	0	$Y = I_6 = 0$
1	1	1	0	1	1	1	1	$Y = I_7 = 1$
1	1	1	1	1	1	1	1	$Y = I_7 = 1$

The implementation of function, F using 8:1 MUX is shown in Fig. 1. The variables w, x and y are used as selection inputs S_2, S_1 and S_0 of 8:1 MUX. In 8:1 MUX of Fig. 1, the inputs I_0, I_4 and I_5 are permanently tied to 0 and the inputs I_1 and I_7 are tied to 1. The inputs I_2 and I_5 are tied to z and the input I_3 is tied to z′. The working of 8:1 MUX is verified in Table 2.

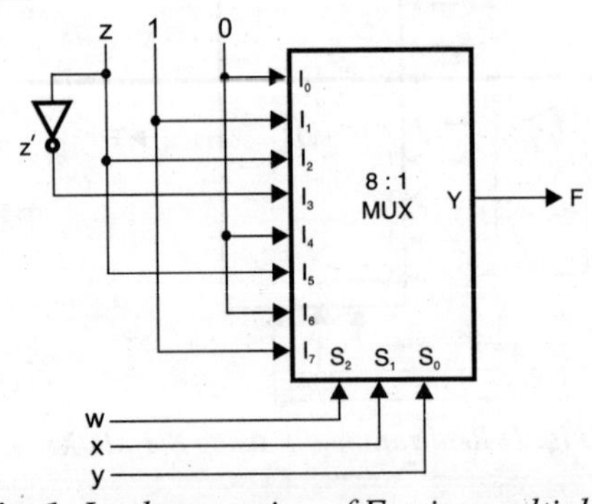

Fig. 1: Implementation of F using multiplexer.

Case ii: Implementation using Decoder

From Table 1 it can be observed that when w = 0, the inputs x, y and z can take 8 combinaitons of binary. Similarly, when w = 1, the inputs x, y and z can take 8 combinations of binary. Hence, two 3-to-8 decoders can be used to implement the given function. One decoder is enabled by w' and other by w.

The inputs of both the decoders are connected to x, y and z. The output of decoder enabled by w' will be same as function output when inputs are minterms m_0 to m_7. The output of decoder enabled by w will be same as function output when inputs are minterms m_8 to m_{15}. The required outputs are logically ORed to get function output as shown in Fig. 2.

On taking into account the minterms for which function outputs are 1, the following Boolean equations are obtained using the logic **high** decoder output.

$F_1 = Y_{21} + Y_{31} + Y_{51} + Y_{61}$; $F_2 = Y_{32} + Y_{62} + Y_{72}$

$F = F_1 + F_2$

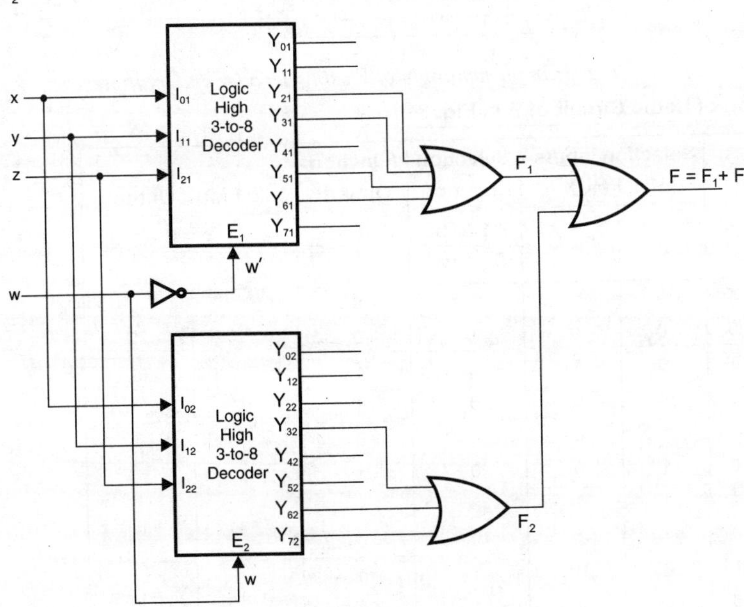

Fig. 2: Implementation of F using decoder.

Case iii: Implementation using Multiplexer and Decoder

In order to implement the given 4-variable Boolean function, one 1-to-2 decoder, two 4:1 MUX and one 2:1 MUX are used as shown in Fig. 3. A 4:1 MUX has 4 inputs, 2 selection inputs,1 enable input and 1 output. The enable inputs are generated using 1-to-2 decoder. The outputs of decoder are used as enable inputs for 4:1 MUX.

The inputs x and y are used as selection inputs S_1 and S_0 of 4:1 MUX.

The output of 4:1 MUX are again connected as inputs of 2:1 MUX. The input w is used as selection input S_{02} of 2:1 MUX.

In the first 4:1 MUX of Fig. 3, the input I_{01} is tied to 0 and I_{11} is tied to 1. The input I_{21} is tied to z and I_{31} is tied to z'. In the second 4:1 MUX of Fig. 3, the inputs I_{02} and I_{22} are tied to 0, the input I_{32} is tied to 1 and the input I_{12} is tied to 1. The working of the circuit of Fig. 3, is verified in Table 3.

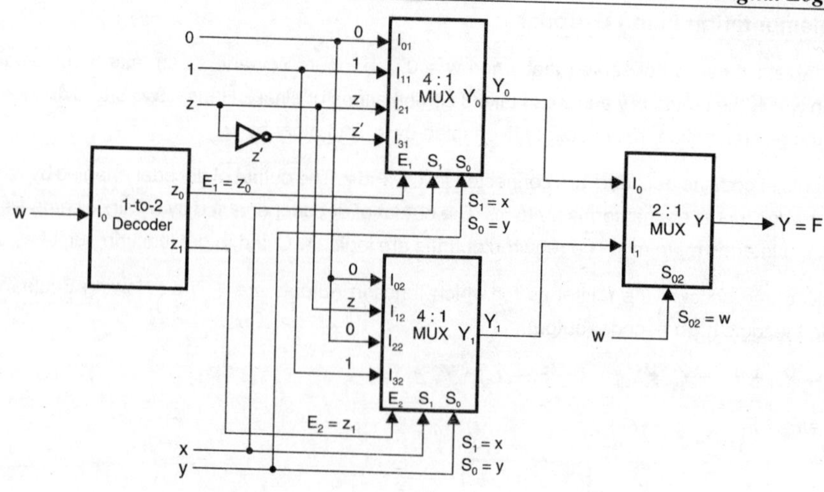

Fig. 3: Implementation of F using decoder and multiplexer.

Table 3: Verification of Logic Circuit of F in Fig. 3

Inputs w x y z	z'	Selection Inputs of 4:1 MUX $S_1 = x$	$S_0 = y$	Selection Input of 2:1 MUX $S_{02} = w$	Function Output F	2:1 MUX Output Y = F	
0 0 0 0	1	0	0	0	0	$Y = I_0 = Y_0$	$Y_0 = I_{01} = 0$
0 0 0 1	0	0	0	0	0		
0 0 1 0	1	0	1	0	1		$Y_0 = I_{11} = 1$
0 0 1 1	0	0	1	0	1		
0 1 0 0	1	1	0	0	0		$Y_0 = I_{21} = z$
0 1 0 1	0	1	0	0	1		
0 1 1 0	1	1	1	0	1		$Y_0 = I_{31} = z'$
0 1 1 1	0	1	1	0	0		
1 0 0 0	1	0	0	1	0	$Y = I_1 = Y_1$	$Y_1 = I_{02} = 0$
1 0 0 1	0	0	0	1	0		
1 0 1 0	1	0	1	1	0		$Y_1 = I_{12} = z$
1 0 1 1	0	0	1	1	1		
1 1 0 0	1	1	0	1	0		$Y_1 = I_{22} = 0$
1 1 0 1	0	1	0	1	0		
1 1 1 0	1	1	1	1	1		$Y_1 = I_{23} = 1$
1 1 1 1	0	1	1	1	1		

5.7.3 Implementation of Combinational Logic Design using Standard Multiplexer IC's

The concept of combinational circuit design using multiplexer is discussed in previous section (Section 5.7.2.). In this section, the available standard ICs are selected for implementation examples. The multiplexers that are available as standard ICs are listed in Appendix 1.

Example 5.28

Implement the following Boolean function using a 4:1 multiplexer available as standard ICs.

$F = \sum m(0, 2, 5, 8, 10, 14)$

Solution

The truth table of the given function is formed as shown in Table 1. Let us select 4:1 multiplexer IC 7453 to implement the Boolean function. The 4:1 multiplexer will have two selection inputs S_0 and S_1. Let us connect the inputs A and B to selection inputs S_1 and S_0 as shown in Fig. 1.

The selection inputs take four different values and so Table 1 is divided into 4 groups based on selection inputs and a separate table is drawn for each group as shown in Tables 2 to 5.

***Note:** The Boolean equations for F in Tables 3 to 5 can be obtained by drawing 2-variable K-maps using C and D as inputs.*

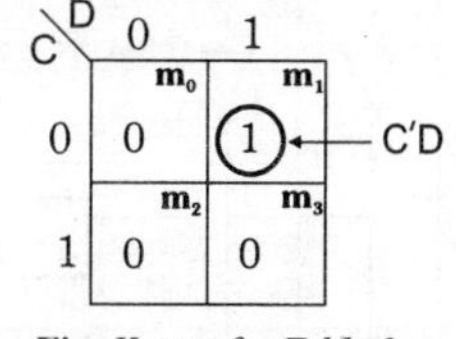

Fig: K-map for Table 3.

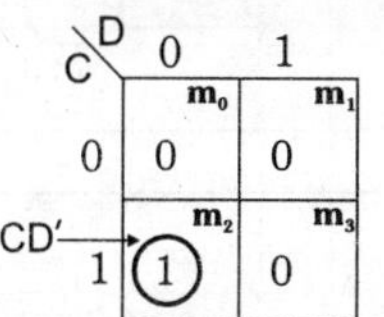

Fig: K-map for Table 5.

Table 1: Truth Table

A	B	C	D	Minterm	Output F	A	B	C	D	Minterm	Output F
0	0	0	0	m_0	1	1	0	0	0	m_8	1
0	0	0	1	m_1	0	1	0	0	1	m_9	0
0	0	1	0	m_2	1	1	0	1	0	m_{10}	1
0	0	1	1	m_3	0	1	0	1	1	m_{11}	0
0	1	0	0	m_4	0	1	1	0	0	m_{12}	0
0	1	0	1	m_5	1	1	1	0	1	m_{13}	0
0	1	1	0	m_6	0	1	1	1	0	m_{14}	1
0	1	1	1	m_7	0	1	1	1	1	m_{15}	0

Table 2: Truth Table When Selection Input 00

Inputs				Selection Inputs		D'	F = D'
A	B	C	D	S_1	S_0		
0	0	0	0	0	0	1	1
0	0	0	1	0	0	0	0
0	0	1	0	0	0	1	1
0	0	1	1	0	0	0	0

Table 3: Truth Table When Selection Input 01

Inputs				Selection Inputs		C'	C'D	F = C'D
A	B	C	D	S_1	S_0			
0	1	0	0	0	1	1	0	0
0	1	0	1	0	1	1	1	1
0	1	1	0	0	1	0	0	0
0	1	1	1	0	1	0	0	0

Table 4: Truth Table When Selection Input 10

Inputs				Selection Inputs		D'	F = D'
A	B	C	D	S_1	S_0		
1	0	0	0	1	0	1	1
1	0	0	1	1	0	0	0
1	0	1	0	1	0	1	1
1	0	1	1	1	0	0	0

Table 5: Truth Table When Selection Input 11

Inputs				Selection Inputs		D'	CD'	F = CD'
A	B	C	D	S_1	S_0			
1	1	0	0	1	1	1	0	0
1	1	0	1	1	1	0	0	0
1	1	1	0	1	1	1	1	1
1	1	1	1	1	1	0	0	0

The input I_0, I_1, I_2 and I_3 of multiplexer are obtained from Tables 2 to 5. The NOT and AND gates are needed to generate inputs for I_0 and I_1 as shown in Fig. 1. The NOT gate IC 7404 and 2-input AND gate IC 7408 are used in this design.

From Tables 2 and 4 it can be observed that function output is D′ when selection inputs are 00 and 10 and so I_0 and I_2 are tied to D′ in Fig. 1. From Table 3 it can be observed that function output is logical AND of C′ and D when selection inputs are 01 and so I_1 is tied to C′D in Fig. 1. From Table 5 it can be observed that function output is logical AND of C and D′ when selection inputs are 11 and so I_3 is tied to CD′ in Fig. 1. The working of 4:1 MUX is verified in Table 6.

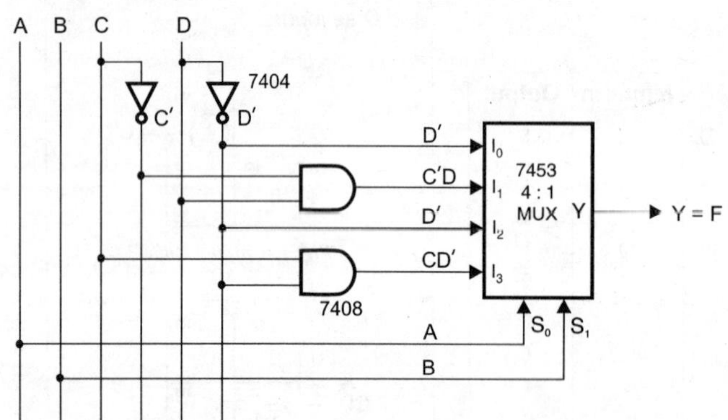

Fig. 1: Implementation of Boolean function using 4:1 MUX.

Table 6: Verification

A B C D	S_1	S_0	F	Y = F
0 0 0 0	0	0	1	$Y = I_0 = D' = 1$
0 0 0 1	0	0	0	$Y = I_0 = D' = 0$
0 0 1 0	0	0	1	$Y = I_0 = D' = 1$
0 0 1 1	0	0	0	$Y = I_0 = D' = 0$
0 1 0 0	0	1	0	$Y = I_1 = C'D = 0$
0 1 0 1	0	1	1	$Y = I_1 = C'D = 1$
0 1 1 0	0	1	0	$Y = I_1 = C'D = 0$
0 1 1 1	0	1	0	$Y = I_1 = C'D = 0$
1 0 0 0	1	0	1	$Y = I_2 = D' = 1$
1 0 0 1	1	0	0	$Y = I_2 = D' = 0$
1 0 1 0	1	0	1	$Y = I_2 = D' = 1$
1 0 1 1	1	0	0	$Y = I_2 = D' = 0$
1 1 0 0	1	1	0	$Y = I_3 = CD' = 0$
1 1 0 1	1	1	0	$Y = I_3 = CD' = 0$
1 1 1 0	1	1	1	$Y = I_3 = CD' = 1$
1 1 1 1	1	1	0	$Y = I_3 = CD' = 0$

Example 5.29

Implement the following Boolean function using multiplexers available as standard ICs.

i) $F_1(x, y) = \Sigma m(0, 2)$ **ii)** $F_2(A, B, C) = \Sigma m(1, 3, 5, 7)$ **iii)** $F_3(A, B, C, D) = \Sigma m(1, 5, 8, 9, 11, 12)$

Solution

i) $F_1(x, y) = \Sigma m(0, 2)$

The function F_1 is 2-variable function and so a 2^2:1 multiplexer (4:1 multiplexer IC 7453) can be used to implement the function. The function inputs x and y are connected to selection inputs S_1 and S_0 as shown in Fig. 1. The function output F_1 is 1 for minterms m_0 and m_2 and so the inputs I_0 and I_2 of MUX are permanently tied to 1 as shown in Fig. 1. The function output F_1 is 0 for minterms m_1 and m_3 and so the inputs I_1 and I_3 are permanently tied to 0 as shown in Fig. 1. The working of 4:1 MUX is verified in Table 1.

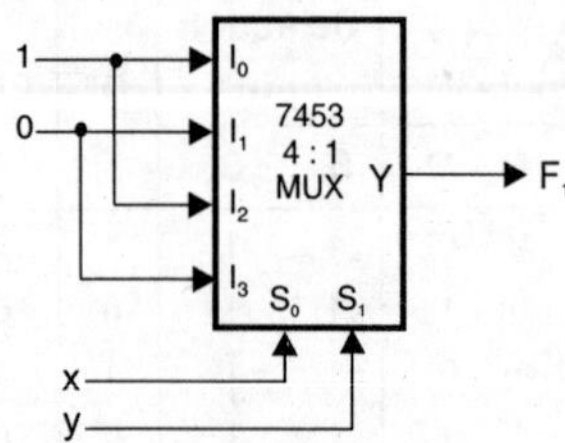

Fig. 1: Realization of F_1 using 4:1 MUX.

Table 1: Verification

x y	Minterm	S_1	S_0	F_1	$Y = F_1$
0 0	m_0	0	0	1	$Y = I_0 = 1$
0 1	m_1	0	1	0	$Y = I_1 = 0$
1 0	m_2	1	0	1	$Y = I_2 = 1$
1 1	m_3	1	1	0	$Y = I_3 = 0$

ii) $F_2(A, B, C) = \sum m(1, 3, 5, 7)$

The function F_2 is 3-variable function and so a 2^3:1 multiplexer (8:1 multiplexer IC 74151) can be used to implement the function. The function inputs A, B and C are connected to selection inputs S_2, S_1 and S_0 as shown in Fig. 2. The function output F_2 is 1 for minterms m_1, m_3, m_5 and m_7 and so the MUX inputs I_1, I_3, I_5 and I_7 are permanently tied to 1 as shown in Fig. 2. The function output F_2 is 0 for minterms m_0, m_2, m_4 and m_5 and so the MUX inputs I_0, I_2, I_4 and I_6 are permanently tied to 0 as shown in Fig. 2. The working of 8:1 MUX is verified in Table 2.

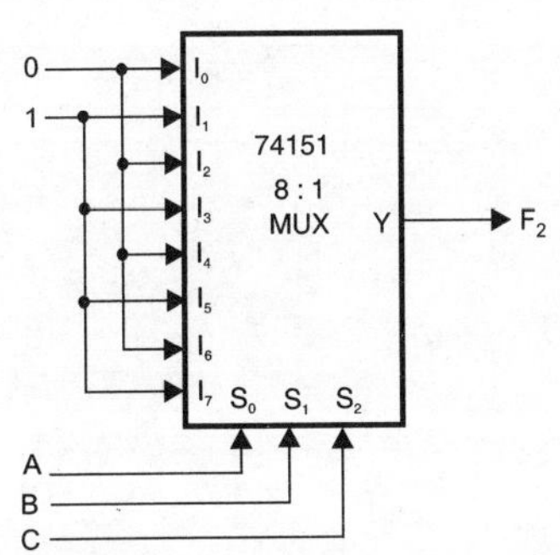

Fig. 2: Realization of F_2 using 8:1 MUX.

Table 2: Verification

A	B	C	Minterm	S_2	S_1	S_0	F_2	$Y = F_2$
0	0	0	m_0	0	0	0	0	$Y = I_0 = 0$
0	0	1	m_1	0	0	1	1	$Y = I_1 = 1$
0	1	0	m_2	0	1	0	0	$Y = I_2 = 0$
0	1	1	m_3	0	1	1	1	$Y = I_3 = 1$
1	0	0	m_4	1	0	0	0	$Y = I_4 = 0$
1	0	1	m_5	1	0	1	1	$Y = I_5 = 1$
1	1	0	m_6	1	1	0	0	$Y = I_6 = 0$
1	1	1	m_7	1	1	1	1	$Y = I_7 = 1$

iii) $F_3(A, B, C, D) = \sum m(1, 5, 8, 9, 11, 12)$

The function F_3 is 4-variable function and so a 2^4:1 multiplexer (16:1 multiplexer IC 74150) can be used to implement the function. The function inputs A, B, C and D are connected to selection inputs S_3, S_2, S_1 and S_0 of MUX as shown in Fig. 3.

The function output F_3 is 1 for minterms m_1, m_5, m_8, m_9, m_{11} and m_{12} and so the MUX inputs I_1, I_5, I_8, I_9, I_{11} and I_{12} are permanently tied to 1 as shown in Fig. 3. The function output F_3 is 0 for minterms m_0, m_2, m_3, m_4, m_6, m_7, m_{10}, m_{13}, m_{14} and m_{15} and so the MUX inputs I_0, I_2, I_3, I_4, I_6, I_7, I_{10}, I_{13}, I_{14} and I_{15} are permanently tied to 0 as shown in Fig. 3. The working of 16:1 MUX is verified in Table 3.

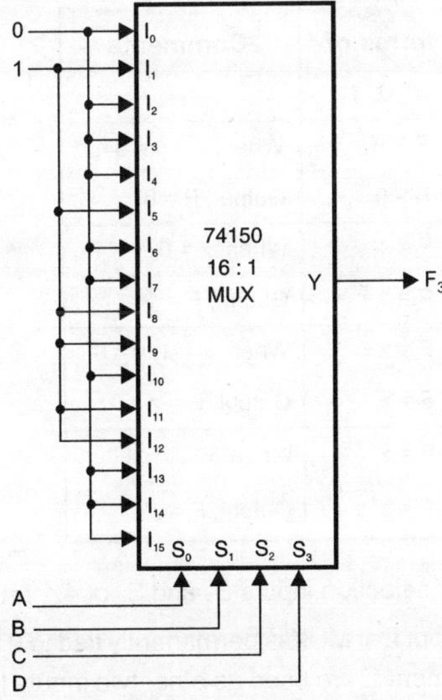

Fig. 3: Realization of F_3 using 16:1 MUX.

Table 3: Verification

A	B	C	D	Minterm	S_3	S_2	S_1	S_0	F_3	$Y = F_3$
0	0	0	0	m_0	0	0	0	0	0	$Y = I_0 = 0$
0	0	0	1	m_1	0	0	0	1	1	$Y = I_1 = 1$
0	0	1	0	m_2	0	0	1	0	0	$Y = I_2 = 0$
0	0	1	1	m_3	0	0	1	1	0	$Y = I_3 = 0$
0	1	0	0	m_4	0	1	0	0	0	$Y = I_4 = 0$
0	1	0	1	m_5	0	1	0	1	1	$Y = I_5 = 1$
0	1	1	0	m_6	0	1	1	0	0	$Y = I_6 = 0$
0	1	1	1	m_7	0	1	1	1	0	$Y = I_7 = 0$
1	0	0	0	m_8	1	0	0	0	1	$Y = I_8 = 1$
1	0	0	1	m_9	1	0	0	1	1	$Y = I_9 = 1$
1	0	1	0	m_{10}	1	0	1	0	0	$Y = I_{10} = 0$
1	0	1	1	m_{11}	1	0	1	1	1	$Y = I_{11} = 1$
1	1	0	0	m_{12}	1	1	0	0	1	$Y = I_{12} = 1$
1	1	0	1	m_{13}	1	1	0	1	0	$Y = I_{13} = 0$
1	1	1	0	m_{14}	1	1	1	0	0	$Y = I_{14} = 0$
1	1	1	1	m_{15}	1	1	1	1	0	$Y = I_{15} = 0$

Example 5.30

Implement the Boolean function, $F(x, y, z) = \Sigma(m_2, m_3, m_5, m_6)$ using multiplexer available as standard ICs.

Solution

It is given that function output F is 1, when the inputs are equal to minterms m_2, m_3, m_5 and m_6. Function output F is 0 for all other inputs. The given function is analyzed using Table 1 and observed that the output F can be expressed in terms of z, z′, 0 and 1. Therefore, the given 3-variable function can be realized using 2^{3-1}:1 multiplexer (4:1 multiplexer IC 7453).

Table 1: Truth Table

Inputs			Minterm	Output	Output	
x	y	z		F	in terms of z, z′, 0, 1	Comments
0	0	0	m_0	0	F = 0	When, x = y = 0
0	0	1	m_1	0	F = 0	Output, F = 0
0	1	0	m_2	1	F = 1	When, x = 0, y = 1
0	1	1	m_3	1	F = 1	Output, F = 1
1	0	0	m_4	0	F = z	When, x = 1, y = 0
1	0	1	m_5	1	F = z	Output, F = z
1	1	0	m_6	1	F = z′	When, x = y = 1
1	1	1	m_7	0	F = z′	Output, F = z′

Let the variables x and y be used as selection inputs S_1 and S_0 of 4:1 MUX as shown in Fig. 1.

In 4:1 multiplexer of Fig. 1, the first input I_0 of MUX is permanently tied to 0 and the second input I_1 is permanently tied to 1. The third variable z and its complement z′ are used as other two inputs I_2 and I_3 respectively. The complement of z is obtained by using a standard NOT gate IC 7404. The working of 4:1 multiplexer is verified in Table 2.

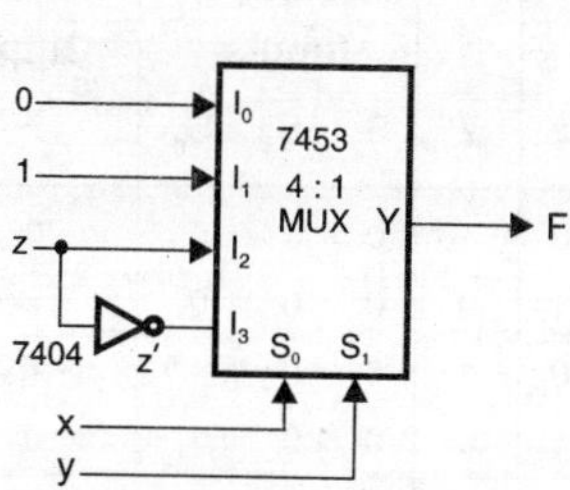

Fig. 1: Implementation of Boolean function using 4:1 MUX.

Table 2: Verification

Inputs				Selection Inputs		Function Output	Output
x	y	z	z'	S_1	S_0	F	Y = F
0	0	0	1	0	0	0	$Y = I_0 = 0$
0	0	1	0	0	0	0	$Y = I_0 = 0$
0	1	0	1	0	1	1	$Y = I_1 = 1$
0	1	1	0	0	1	1	$Y = I_1 = 1$
1	0	0	1	1	0	0	$Y = I_2 = z = 0$
1	0	1	0	1	0	1	$Y = I_2 = z = 1$
1	1	0	1	1	1	1	$Y = I_3 = z' = 1$
1	1	1	0	1	1	0	$Y = I_3 = z' = 0$

Example 5.31

Implement the Boolean function, $F(w, x, y, z) = \sum(m_1, m_2, m_4, m_6, m_7, m_{10}, m_{13})$.

Solution

It is given that function output F is 1 when the inputs are equal to minterms m_1, m_2, m_4, m_6, m_7, m_{10} and m_{13}. For all other inputs the function output F is 0.

The given function is analyzed using Table 1 and observed that the function output can be expressed in terms of z, z', 0 and 1. Therefore, the given 4-variable function can be realized using 2^{4-1}:1 multiplexer (8:1 multiplexer IC 74151).

Table 1: Truth Table

Inputs				Minterm	Output F	Output in terms of z, z', 0, 1	Comments
w	x	y	z				
0	0	0	0	m_0	0	F = z	When, w = x = y = 0
0	0	0	1	m_1	1	F = z	Output, F = z
0	0	1	0	m_2	1	F = z'	When, w = x = 0, y = 1
0	0	1	1	m_3	0	F = z'	Output, F = z'
0	1	0	0	m_4	1	F = z'	When, w = y = 0, x = 1
0	1	0	1	m_5	0	F = z'	Output, F = z'
0	1	1	0	m_6	1	F = 1	When, w = 0, x = y = 1
0	1	1	1	m_7	1	F = 1	Output, F = 1
1	0	0	0	m_8	0	F = 0	When w = 1, x = y = 0
1	0	0	1	m_9	0	F = 0	Output, F = 0
1	0	1	0	m_{10}	1	F = z'	When, w = y =1, x = 0
1	0	1	1	m_{11}	0	F = z'	Output, F = z'
1	1	0	0	m_{12}	0	F = z	When, w = x = 1, y = 0
1	1	0	1	m_{13}	1	F = z	Output, F = z
1	1	1	0	m_{14}	0	F = 0	When, w = x = y = 1
1	1	1	1	m_{15}	0	F = 0	Output, F = 0

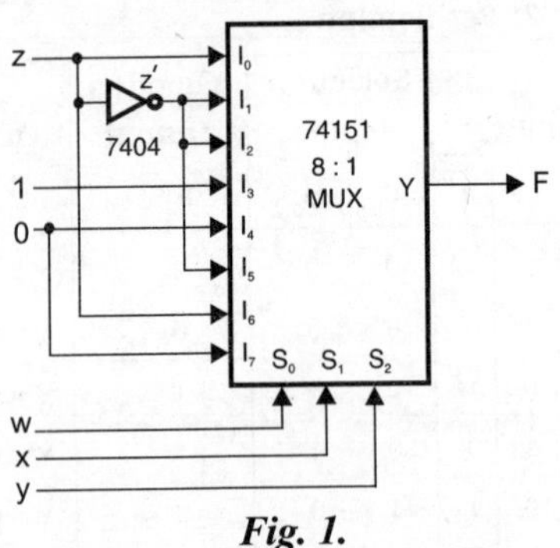

Fig. 1.

Let the variables w, x and y be used as selection inputs S_2, S_1 and S_0 of 8:1 MUX as shown in Fig. 1. In 8:1 multiplexer of Fig. 1, the inputs I_4 and I_7 are permanentely tied to 0 and the input I_3 is tied to 1. The input I_0 and I_5 are tied to z and the inputs I_1, I_2 and I_5 are tied to z'. The complement of z is obtained by using a standard NOT gate IC 7404. The working of 8:1 multiplexer is verified in Table 2.

Table 2: Verification

Inputs					Selection Inputs			Function Output	MUX Output
w	x	y	z	z'	S_2	S_1	S_0	F	Y = F
0	0	0	0	1	0	0	0	0	$Y = I_0 = z = 0$
0	0	0	1	0	0	0	0	1	$Y = I_0 = z = 1$
0	0	1	0	1	0	0	1	1	$Y = I_1 = z' = 1$
0	0	1	1	0	0	0	1	0	$Y = I_1 = z' = 0$
0	1	0	0	1	0	1	0	1	$Y = I_2 = z' = 1$
0	1	0	1	0	0	1	0	0	$Y = I_2 = z' = 0$
0	1	1	0	1	0	1	1	1	$Y = I_3 = 1$
0	1	1	1	0	0	1	1	1	$Y = I_3 = 1$
1	0	0	0	1	1	0	0	0	$Y = I_4 = 0$
1	0	0	1	0	1	0	0	0	$Y = I_4 = 0$
1	0	1	0	1	1	0	1	1	$Y = I_5 = z' = 1$
1	0	1	1	0	1	0	1	0	$Y = I_5 = z' = 0$
1	1	0	0	1	1	1	0	0	$Y = I_6 = z = 0$
1	1	0	1	0	1	1	1	1	$Y = I_6 = z = 1$
1	1	1	0	1	1	1	1	0	$Y = I_7 = 0$
1	1	1	1	0	1	1	1	0	$Y = I_7 = 0$

Example 5.32

Implement the following Boolean function using 2 : 1 multiplexer available as standard ICs.

$F = \sum m(0, 2, 3, 5)$

Solution

The truth table of given function can be formed as shown in Table 1. Let us select a standard 2:1 multiplexer IC 74157 to implement the Boolean function. The 2:1 multiplexer will have one selection input S_0. Let us connect input x to selection input S_0 as shown in Fig. 1. The selection input take two different values and so Table 1 is divided into two groups based on selection input and a separate table is drawn for each group as shown in Tables 2 and 3.

Table 1: Truth Table of Function, F

Inputs				Output
x	y	z	Minterm	F
0	0	0	m_0	1
0	0	1	m_1	0
0	1	0	m_2	1
0	1	1	m_3	1
1	0	0	m_4	0
1	0	1	m_5	1
1	1	0	m_6	0
1	1	1	m_7	0

Table 2: Truth Table When Selection Input 0

Inputs			Selection Input	y + z'	F = y + z'
x	y	z	S_0		
0	0	0	0	1	1
0	0	1	0	0	0
0	1	0	0	1	1
0	1	1	0	1	1

Table 3: Truth Table When Selection Input 1

Inputs			Selection Input	y'z	F = y'z
x	y	z	S_0		
1	0	0	1	0	0
1	0	1	1	1	1
1	1	0	1	0	0
1	1	1	1	0	0

Table 4: Verification

x	y	z	S_0	F	Y = F
0	0	0	0	1	$Y = I_0 = y + z' = 1$
0	0	1	0	0	$Y = I_0 = y + z' = 0$
0	1	0	0	1	$Y = I_0 = y + z' = 1$
0	1	1	0	1	$Y = I_0 = y + z' = 1$
1	0	0	1	0	$Y = I_1 = y'z = 0$
1	0	1	1	1	$Y = I_1 = y'z = 1$
1	1	0	1	0	$Y = I_1 = y'z = 0$
1	1	1	1	0	$Y = I_1 = y'z = 0$

***Note:** The Boolean equations for F in Tables 2 and 3 can be obtained by drawing 2-variable K-maps using y and z as inputs.*

The input I_0 and I_1 for 2:1 MUX are obatined from Tables 2 and 3. The AND, OR and NOT gates are required to generate inputs for I_0 as are shown in Fig. 1. The AND gate IC 7408, OR gate IC 7432 and NOT gate IC 7404 are used in this design.

From Table 2 we get, when $S_0 = 0$, $F = y + z'$ and so I_0 is tied to $y + z'$ in Fig. 1.

From Table 3 we get, when $S_0 = 1$, $F = y'z$ and so I_1 is tied to $y'z$ in Fig. 1.

The working of 2:1 MUX is verified in Table 4.

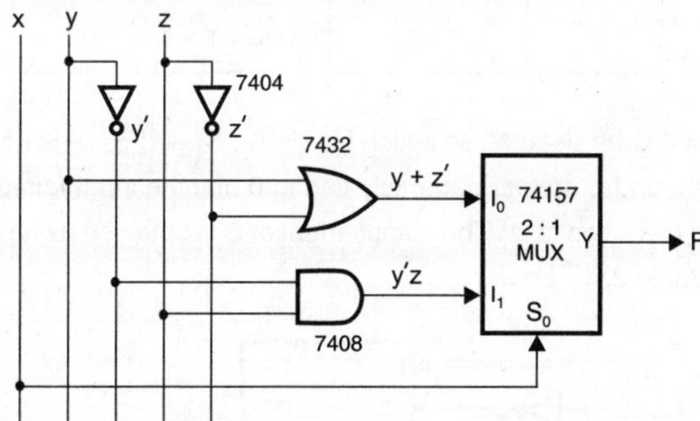

Fig. 1: Implementation of Boolean function using 2:1 MUX.

Example 5.33

Implement the following boolean function with an 8-to-1 line multiplexer and an inverter.

$F(A, B, C, D) = \Sigma m(2, 4, 6, 9, 10, 11, 15)$

Solution

It is given that function output F is 1 when the inputs are equal to minterms m_2, m_4, m_6, m_9, m_{10}, m_{11} and m_{15}. For all other inputs the function output F is 0.

The given function is analyzed using Table 1 and observed that the function output can be expressed in terms of D, D', 0 and 1. Therefore, the given 4-variable function can be realized using 2^{4-1}:1 multiplexer (8:1 multiplexer IC 74151).

Table 1: Truth Table

Inputs A	B	C	D	Minterm	Output F	Output in terms of D, D′, 0, 1	Comments
0	0	0	0	m_0	0	F = 0	When, A = B = C = 0
0	0	0	1	m_1	0	F = 0	Output, F = 0
0	0	1	0	m_2	1	F = D′	When, A = B = 0, C = 1
0	0	1	1	m_3	0	F = D′	Output, F = D′
0	1	0	0	m_4	1	F = D′	When, A = C = 0, B = 1
0	1	0	1	m_5	0	F = D′	Output, F = D′
0	1	1	0	m_6	1	F = D′	When, A = 0, B = C = 1
0	1	1	1	m_7	0	F = D′	Output, F = D′
1	0	0	0	m_8	0	F = D	When A = 1, B = C = 0
1	0	0	1	m_9	1	F = D	Output, F = D
1	0	1	0	m_{10}	1	F = 1	When, A = C =1, B = 0
1	0	1	1	m_{11}	1	F = 1	Output, F = 1
1	1	0	0	m_{12}	0	F = 0	When, A = B = 1, C = 0
1	1	0	1	m_{13}	0	F = 0	Output, F = 0
1	1	1	0	m_{14}	0	F = D	When, A = B = C = 1
1	1	1	1	m_{15}	1	F = D	Output, F = D

Let the variables A, B and C be used as selection inputs S_2, S_1 and S_0 of 8:1 MUX as shown in Fig. 1. In 8:1 multiplexer of Fig. 1, the inputs I_0 and I_5 are permanentely tied to 0 and the input I_5 is tied to 1. The input I_4 and I_7 are tied to D and the inputs I_1, I_2 and I_3 are tied to D′. The complement of D is obtanied using standard IC 7404. The working of 8:1 multiplexer is verified in Table 2.

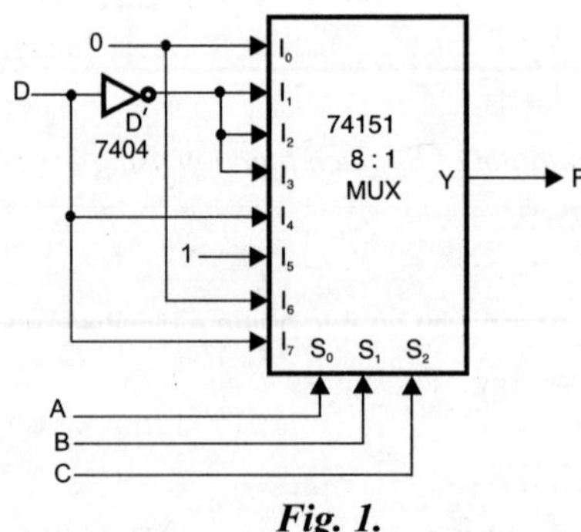

Fig. 1.

Table 2: Verification

Inputs					Selection Inputs			Function Output	MUX Output
A	B	C	D	D'	S_2	S_1	S_0	F	Y = F
0	0	0	0	1	0	0	0	0	$Y = I_0 = 0$
0	0	0	1	0	0	0	0	0	$Y = I_0 = 0$
0	0	1	0	1	0	0	1	1	$Y = I_1 = D' = 1$
0	0	1	1	0	0	0	1	0	$Y = I_1 = D' = 0$
0	1	0	0	1	0	1	0	1	$Y = I_2 = D' = 1$
0	1	0	1	0	0	1	0	0	$Y = I_2 = D' = 0$
0	1	1	0	1	0	1	1	1	$Y = I_3 = D' = 1$
0	1	1	1	0	0	1	1	0	$Y = I_3 = D' = 0$
1	0	0	0	1	1	0	0	0	$Y = I_4 = D = 0$
1	0	0	1	0	1	0	0	1	$Y = I_4 = D = 1$
1	0	1	0	1	1	0	1	1	$Y = I_5 = 1$
1	0	1	1	0	1	0	1	1	$Y = I_5 = 1$
1	1	0	0	1	1	1	0	0	$Y = I_6 = 0$
1	1	0	1	0	1	1	1	0	$Y = I_6 = 0$
1	1	1	0	1	1	1	1	0	$Y = I_7 = D = 0$
1	1	1	1	0	1	1	1	1	$Y = I_7 = D = 1$

5.8 Demultiplexer

(GTU, Jun'19, 7 Marks)

A demultiplexer is a combinational circuit that send one binary information to many outputs. For this reason a demultiplexer is otherwise called a ***data distributor***.

A demultiplexer will perform the reverse operation of multiplexer. Therefore, a demultiplexer is a combinational circuit which can transmit a binary input to any one of 2^n output lines using n selection lines or n-bit address. The demultiplexer is shortly called ***demux***.

The block diagram and symbolic representation of $1{:}2^n$ demultiplexer with n selection inputs is shown in Fig. 5.67.

$m = 2^n - 1$

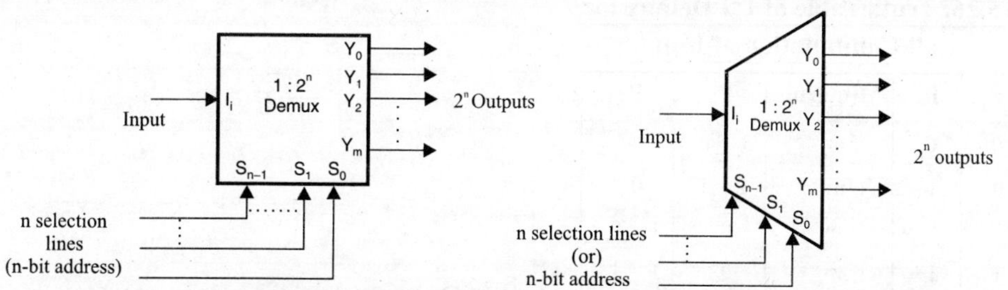

Fig. a: *Block diagram representation.* ***Fig. b:*** *Standard symbol.*

Fig. 5.67: *$1{:}2^n$ demultiplexer.*

1:2 Demux

The *1:2 demux* will transmit one binary input to any one of the two outputs depending on the value of selection input.

Let, I_i = Input

S_0 = Selection input/Address

Y_0, Y_1 = Outputs

In 1:2 demultiplexer,

When, $S_0 = 0$, I_i is passed to Y_0, $\therefore$ $Y_0 = I_i$

When, $S_0 = 1$, I_i is passed to Y_1, $\therefore$ $Y_1 = I_i$

The above selection logic is listed as truth table in Table 5.25. The block diagram and symbolic representation of 1:2 demux with selection input S_0 are shown in Fig. 5.68.

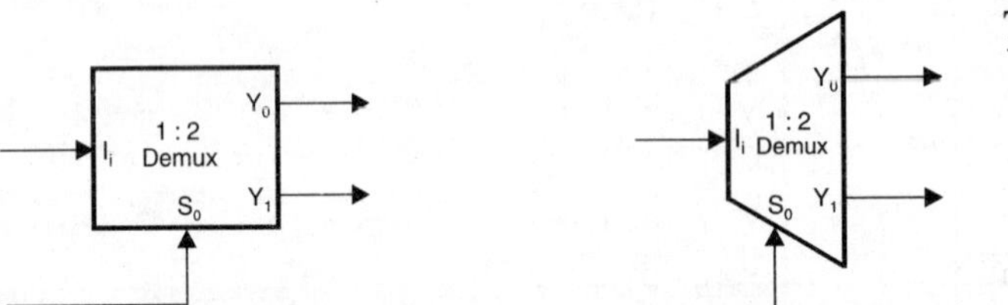

Fig. a: Block diagram representation. *Fig. b: Standard symbol.*

Fig. 5.68: 1:2 demultiplexer.

Table 5.25: Truth Table of 1:2 Demultiplexer

S_0	Y_0	Y_1
0	I_i	0
1	0	I_i

The truth table of 1:2 demux for all possible combination of inputs is shown in Table 5.26. The K-maps for design of 1:2 demux are shown in Fig. 5.69 and from the K-map we get the following Boolean equations,

$$Y_0 = I_i\, S_0'$$

$$Y_1 = I_i\, S_0$$

The logic circuit of 1:2 demux is drawn using the above Boolean equations as shown in Fig. 5.70.

Table 5.26: Truth Table of 1:2 Demux for all Combination of Inputs

S_0	I_i	Minterm	Y_0	Y_1
0	0	m_0	$I_i = 0$	0
0	1	m_1	$I_i = 1$	0
1	0	m_2	0	$I_i = 0$
1	1	m_3	0	$I_i = 1$

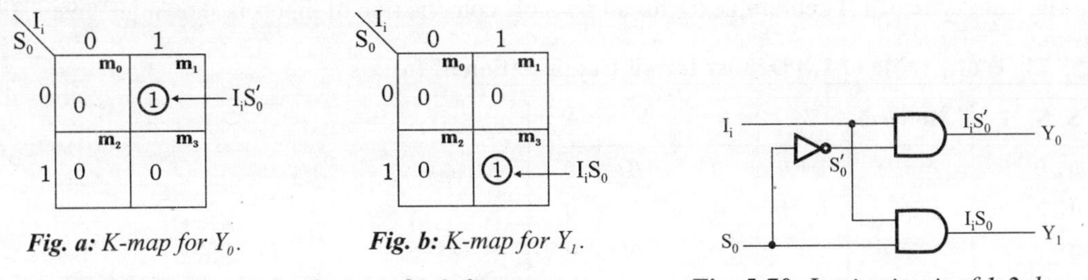

Fig. a: *K-map for Y_0.* ***Fig. b:*** *K-map for Y_1.*

Fig. 5.69: *K-map for design of 1:2 demux.* ***Fig. 5.70:*** *Logic circuit of 1:2 demux.*

1:4 Demux

The ***1:4 demultiplexer*** will transmit one binary input to any one of the four outputs depending on the value of selection input.

Let, I_i = Input

S_0, S_1 = Selection inputs

Y_0, Y_1, Y_2, Y_3 = Outputs

In 1:4 demultiplexer,

When, $S_1 = 0$ and $S_0 = 0$, I_i is passed to Y_0, $\therefore \; Y_0 = I_i$

When, $S_1 = 0$ and $S_0 = 1$, I_i is passed to Y_1, $\therefore \; Y_1 = I_i$

When, $S_1 = 1$ and $S_0 = 0$, I_i is passed to Y_2, $\therefore \; Y_2 = I_i$

When, $S_1 = 1$ and $S_0 = 1$, I_i is passed to Y_3, $\therefore \; Y_3 = I_i$

The above selection logic is listed as truth table in Table 5.27. The block diagram and symbolic representation of 1:4 demux with selection inputs S_0 and S_1 are shown in Fig. 5.71.

Table 5.27: Truth Table of 1:4 Demux

S_1	S_0	Y_0	Y_1	Y_2	Y_3
0	0	I_i	0	0	0
0	1	0	I_i	0	0
1	0	0	0	I_i	0
1	1	0	0	0	I_i

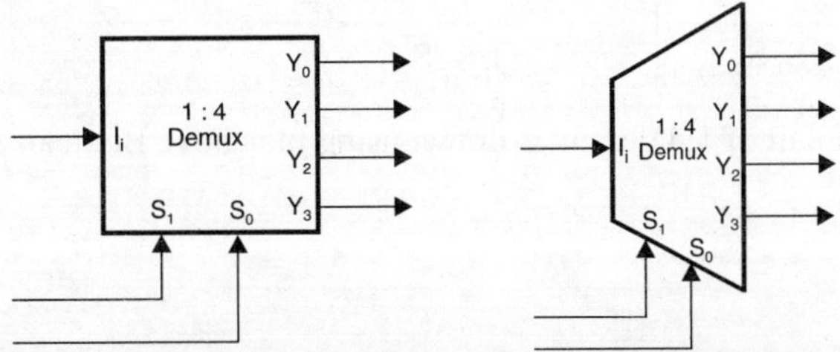

Fig. a: *Block diagram representation.* ***Fig. b:*** *Standard symbol.*

Fig. 5.71: *1:4 demultiplexer.*

The truth table of 1:4 demultiplexer for all possible combination of inputs is shown in Table 5.28.

Table 5.28: Truth Table of 1:8 Demux for all Combination of Inputs

S_1 S_0 I_i	Minterm	Y_0	Y_1	Y_2	Y_3
0 0 0	m_0	$I_i = 0$	0	0	0
0 0 1	m_1	$I_i = 1$	0	0	0
0 1 0	m_2	0	$I_i = 0$	0	0
0 1 1	m_3	0	$I_i = 1$	0	0
1 0 0	m_4	0	0	$I_i = 0$	0
1 0 1	m_5	0	0	$I_i = 1$	0
1 1 0	m_6	0	0	0	$I_i = 0$
1 1 1	m_7	0	0	0	$I_i = 1$

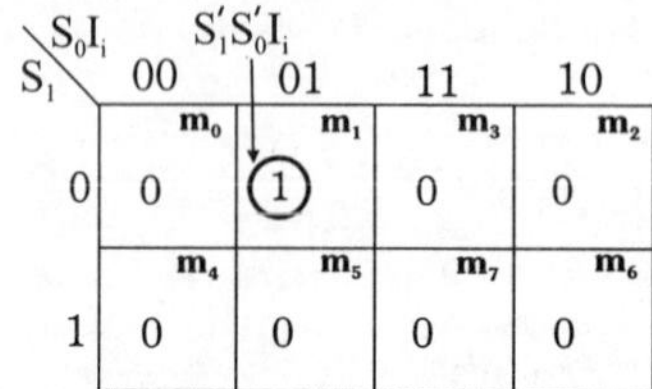

Fig. a: K-map for Y_0.

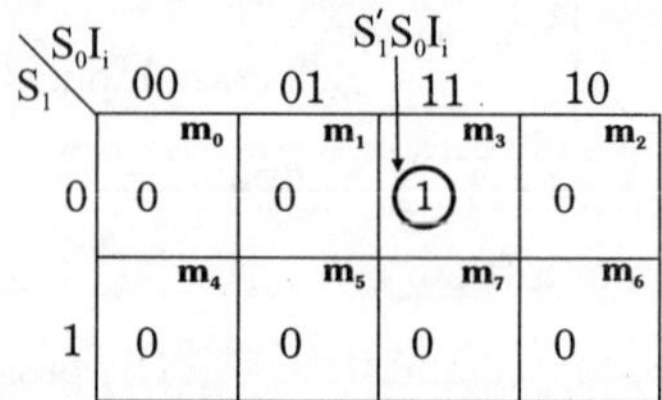

Fig. b: K-map for Y_1.

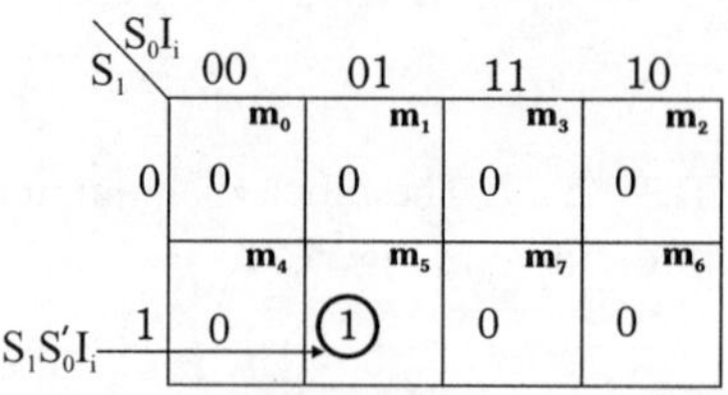

Fig. c: K-map for Y_2.

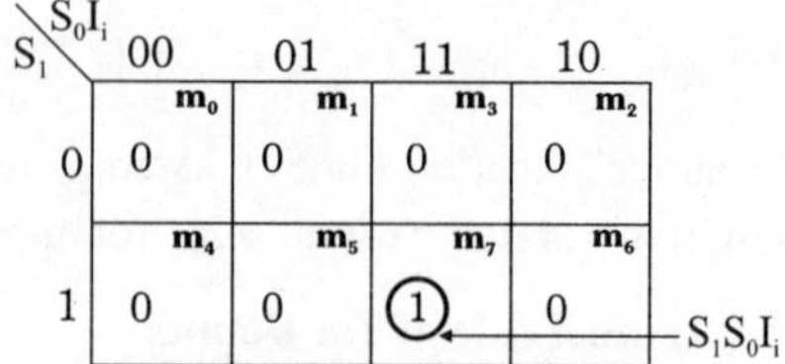

Fig. d: K-map for Y_3.

Fig. 5.72: K-map for design of 1:4 demux.

The K-maps for design of 1:4 demultiplexer are drawn in Fig. 5.72 and from the K-map we get the following Boolean equations,

$$Y_0 = I_i\, S_1'\, S_0' \qquad Y_2 = I_i\, S_1\, S_0'$$

$$Y_1 = I_i\, S_1'\, S_0 \qquad Y_3 = I_i\, S_1\, S_0$$

The logic circuit of 1:4 demux is drawn using the above Boolean equations as shown in Fig. 5.73.

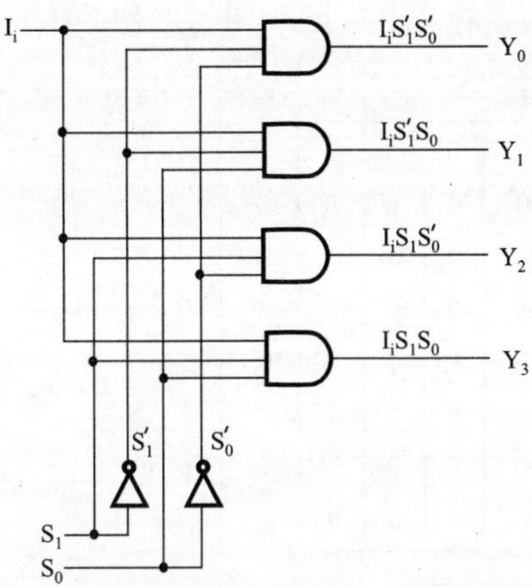

Fig. 5.73: *Logic circuit of 1:4 demux.*

The dual 1:4 demux is available as a standard IC with number 74155. The pin configuration of 74155 is shown in Fig. 5.74.

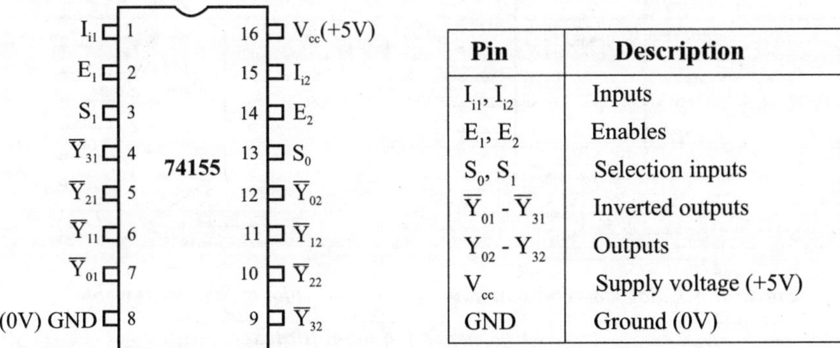

Pin	Description
I_{i1}, I_{i2}	Inputs
E_1, E_2	Enables
S_0, S_1	Selection inputs
$\overline{Y}_{01}$ - $\overline{Y}_{31}$	Inverted outputs
Y_{02} - Y_{32}	Outputs
V_{cc}	Supply voltage (+5V)
GND	Ground (0V)

Fig. 5.74: *Pin configuration of dual 1:4 demultiplexer IC 74155.*

1:8 Demux

The ***1:8 demultiplexer*** will transmit one binary input to any one of the eight outputs depending on the value of selection inputs.

Let, I_i = Input

S_0, S_1, S_2 = Selection inputs

$Y_0, Y_1, Y_2, \ldots, Y_7$ = Outputs

The logic to pass the input to output in 1:8 demux is listed in Table 5.29.

Table 5.29: Truth Table of 1:8 Demux

S_2	S_1	S_0	Y_0	Y_1	Y_2	Y_3	Y_4	Y_5	Y_6	Y_7
0	0	0	I_i	0	0	0	0	0	0	0
0	0	1	0	I_i	0	0	0	0	0	0
0	1	0	0	0	I_i	0	0	0	0	0
0	1	1	0	0	0	I_i	0	0	0	0
1	0	0	0	0	0	0	I_i	0	0	0
1	0	1	0	0	0	0	0	I_i	0	0
1	1	0	0	0	0	0	0	0	I_i	0
1	1	1	0	0	0	0	0	0	0	I_i

The block diagram and symbolic representation of 1:8 demux is shown in Fig. 5.75.

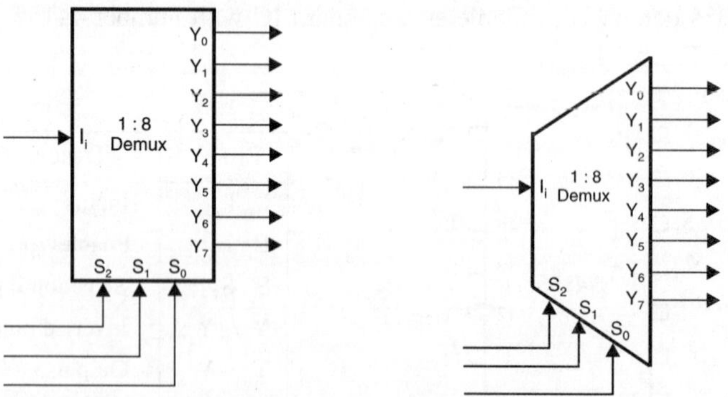

Fig. a: Block diagram representation. *Fig. b: Standard symbol.*

Fig. 5.75: 1:8 demultiplexer.

Table 5.30: Verification of Output of 1:8 Demux of Fig. 5.75

Selection Inputs S_2 S_1 S_0	Complement of Selection Inputs S_2' S_1' S_0'	$Y_0 = I_i S_2' S_1' S_0'$	$Y_1 = I_i S_2' S_1' S_0$	$Y_2 = I_i S_2' S_1 S_0'$	$Y_3 = I_i S_2' S_1 S_0$	$Y_4 = I_i S_2 S_1' S_0'$	$Y_5 = I_i S_2 S_1' S_0$	$Y_6 = I_i S_2 S_1 S_0'$	$Y_7 = I_i S_2 S_1 S_0$
0 0 0	1 1 1	I_i	0	0	0	0	0	0	0
0 0 1	1 1 0	0	I_i	0	0	0	0	0	0
0 1 0	1 0 1	0	0	I_i	0	0	0	0	0
0 1 1	1 0 0	0	0	0	I_i	0	0	0	0
1 0 0	0 1 1	0	0	0	0	I_i	0	0	0
1 0 1	0 1 0	0	0	0	0	0	I_i	0	0
1 1 0	0 0 1	0	0	0	0	0	0	I_i	0
1 1 1	0 0 0	0	0	0	0	0	0	0	I_i

The logic circuit of 1:8 demux can be realized using AND gate as shown in Fig. 5.76 and Table 5.30 can be used to verify the logic circuit of Fig. 5.76.

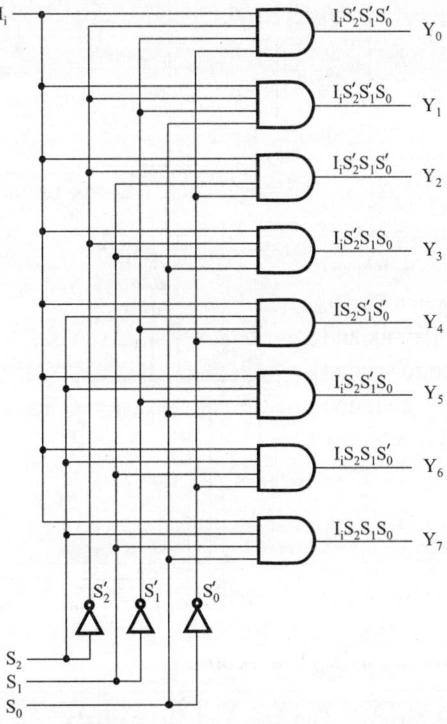

Fig. 5.76: Logic circuit of 1:8 demux.

5.8.1 Expanding the Demultiplexer

Demultiplexers of smaller size can be connected in parallel to expand the size of demultiplexer. For example two 1:2 demultiplexer can be connected in parallel to form 1:4 demultiplexer and two 1:4 demultiplexers can be connected in parallel to form 1:8 demultiplexer and so on.

An example of two 1:2 demux connected in parallel to form 1:4 demux is shown in Fig. 5.77. In this circuit the selection input S_1 is used as enable for the input of demux. When $S_1 = 0$, the input I_i is allowed to first 1:2 demux and when $S_1 = 1$, the input I_i is allowed to second 1:2 demux. The selection input S_0 is common for both 1:2 demux. The working of this demux is verified in Table 5.31.

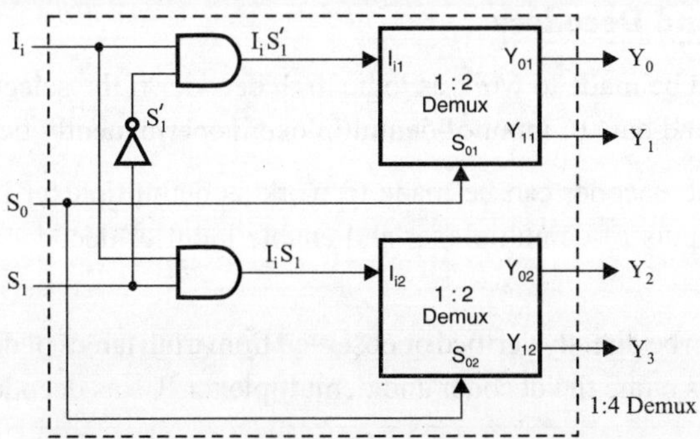

Fig. 5.77: 1:4 demux using 1:2 demux.

Table 5.31: Verification of 1:4 Demux using 1:2 Demux

S_1	S_0	I_iS_1'	I_iS_1	$Y_0 = Y_{01}$	$Y_1 = Y_{11}$	$Y_2 = Y_{02}$	$Y_3 = Y_{12}$
0	0	I_i	0	$I_{i1} = I_i$	0	0	0
0	1	I_i	0	0	$I_{i1} = I_i$	0	0
1	0	0	I_i	0	0	$I_{i2} = I_i$	0
1	1	0	I_i	0	0	0	$I_{i2} = I_i$

An example of two 1:4 demux connected in parallel to form 1:8 demux is shown in Fig. 5.78. In this circuit the selection input S_2 is used as enable for the input of demux. When $S_2 = 0$, the input I_i is allowed to first 1:4 demux and when $S_2 = 1$, the input I_i is allowed to second 1:4 demux. The selection input S_1 is common for both 1:4 demux. The working of this demux is verified in Table 5.32.

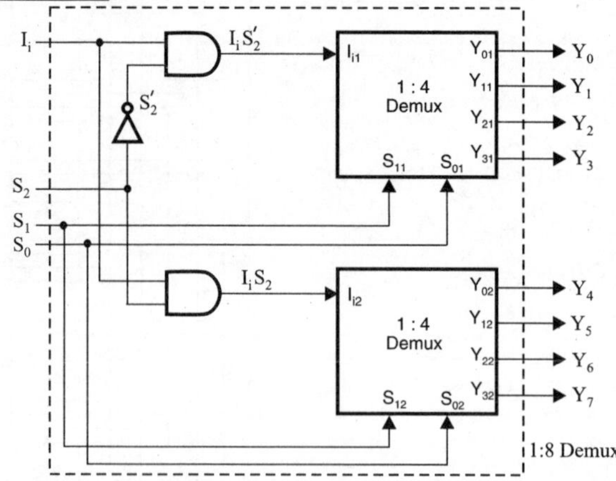

***Fig. 5.78:** 1:8 demux using 1:4 demux.*

Table 5.32: Verification of 1:8 Demux using 1:4 Demux

S_2	S_1	S_0	I_iS_2'	I_iS_2	$Y_0 = Y_{01}$	$Y_1 = Y_{11}$	$Y_2 = Y_{21}$	$Y_3 = Y_{31}$	$Y_4 = Y_{02}$	$Y_5 = Y_{12}$	$Y_6 = Y_{22}$	$Y_7 = Y_{32}$
0	0	0	I_i	0	$I_{i1} = I_i$	0	0	0	0	0	0	0
0	0	1	I_i	0	0	$I_{i1} = I_i$	0	0	0	0	0	0
0	1	0	I_i	0	0	0	$I_{i1} = I_i$	0	0	0	0	0
0	1	1	I_i	0	0	0	0	$I_{i1} = I_i$	0	0	0	0
1	0	0	0	I_i	0	0	0	0	$I_{i2} = I_i$	0	0	0
1	0	1	0	I_i	0	0	0	0	0	$I_{i2} = I_i$	0	0
1	1	0	0	I_i	0	0	0	0	0	0	$I_{i2} = I_i$	0
1	1	1	0	I_i	0	0	0	0	0	0	0	$I_{i2} = I_i$

5.8.2 Demultiplexers and Decoders

(AKTU, Dec'22, 2 Marks)

The demultiplexer can be made to work as logic high decoder if the select inputs of demultiplexer are made inputs of decoder and data of input of demultiplexer is permanently tied to 1.

On the other hand, the decoder can be made to work as demultiplexer if the input of logic high decoder is made as select inputs of demultiplexer and enable input of decoder is made as data input of demultiplexer.

The above concepts can be directly verified or observed from truth tables of decoder and demultiplexer. Many of the IC manufactures name the decoder and demultiplexer ICs as decoder/demultiplexer.

5.9 Adders

Adders have been developed to perform arithmetic addition operation on binary numbers. Half adder can perform addition of two 1-bit binary number. The possible additions of two 1-bit binary numbers are shown below:

0 + 0 =	0		0 0
0 + 1 =	1	2-bit standard representation →	0 1
1 + 0 =	1		0 1
1 + 1 =	1 0		1 0

(In 1 0: the right bit → Sum, the left bit → Carry)

In the above additions, it can be observed that result is either 1-bit or 2-bit. The 1-bit result is called ***sum***. In the 2-bit result, the first bit is called ***sum*** and the second bit is called ***carry***. Hence, half adder is designed to generate 2-bit standard output.

In addition of n-bit binary numbers, the addition is performed bit-by-bit. In this the carry generated in an addition should be considered in next addition to get correct result and so full adder is developed which can add three binary input bits in which one of the bits is carry generated in previous addition. Therefore, n-bit addition can be performed by using n full adders in parallel.

5.9.1 Half Adder

(RGPV, Jun'22, 7 Marks)

Half adder (HA) is a combinational circuit that performs arithmetic sum of two binary bits. The outputs are sum and carry. The truth table of half adder is shown in Table 5.33. The block diagram and symbolic representation of half adder are shown in Figs. 5.79 and 5.82 respectively.

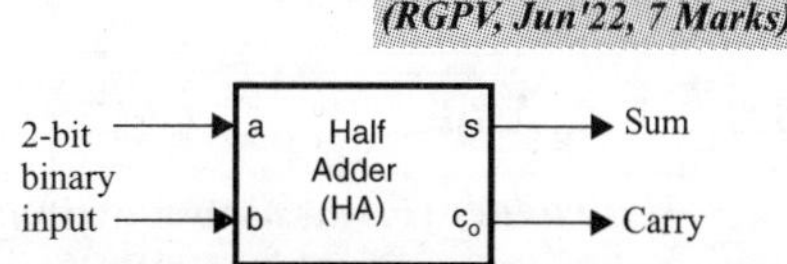

Fig. 5.79: *Block digaram of half adder.*

Table 5.33: Truth Table of Half Adder

Inputs		Minterm	Outputs	
a	**b**		**s**	$\mathbf{c_o}$
0	0	m_0	0	0
0	1	m_1	1	0
1	0	m_2	1	0
1	1	m_3	0	1

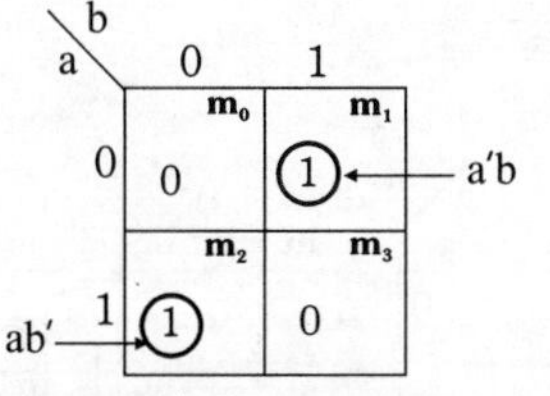

Fig. a: *K-map for s.*

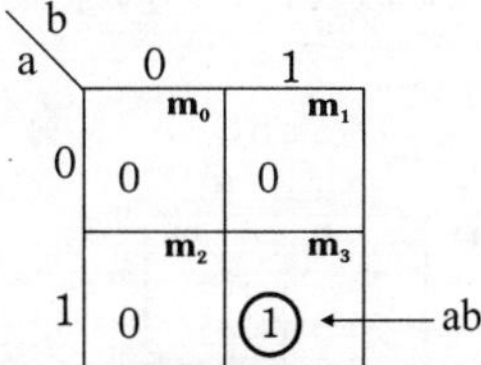

Fig. b: *K-map for c_o.*

Fig. 5.80: *K-map for design of half adder.*

Using truth table (Table 5.33) the K-maps for designing half adder are drawn as shown in Fig. 5.80. From the K-map we get the following Boolean equations and using these equations the logic circuit of half adder is drawn as shown in Fig. 5.81.

$$s = ab' + a'b \quad ; \quad c_o = ab$$

$$\boxed{s = ab' + a'b = a \oplus b}$$

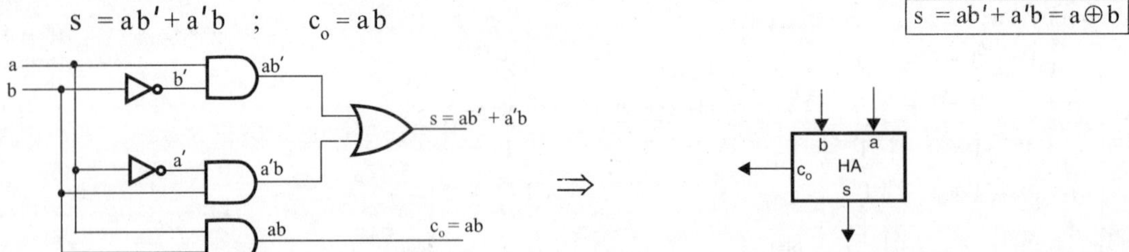

Fig. 5.81: Logic circuit of half adder.

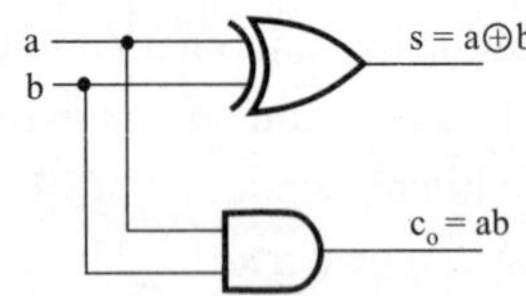

Fig. 5.82: Symbolic representation of half adder.

Alternatively, the sum can be realized from XOR gate. If we compare the sum and output of XOR gate we observe that both are same. The logic circuit of half adder using XOR gate is shown in Fig. 5.83.

Table 5.34: Truth Table of XOR Gate

a	b	$a \oplus b$
0	0	0
0	1	1
1	0	1
1	1	0

Fig. 5.83: Half adder using XOR gate.

5.9.2 Full Adder

Full adder (FA) is a combinational circuit that perform arithmetic sum of three binary bits in which one of the bit is carry generated in previous addition. Hence, full adder will be useful in n-bit addition to add carry of an addition in the next bit addition.

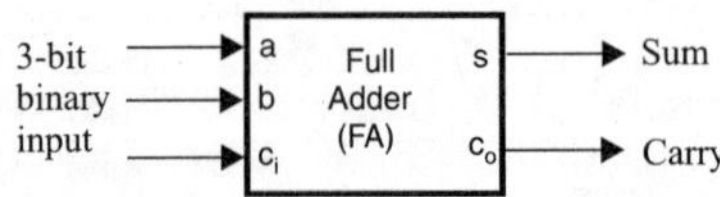

Fig. 5.84: Block diagram of full adder.

The outputs are sum and carry. The truth table of full adder is shown in Table 5.35. The block diagram and symbolic representation of full adder are shown in Figs. 5.84 and 5.89 respectively.

Table 5.35: Truth Table of Full Adder

Inputs			Minterm	Outputs	
a	b	c_i		s	c_o
0	0	0	m_0	0	0
0	0	1	m_1	1	0
0	1	0	m_2	1	0
0	1	1	m_3	0	1
1	0	0	m_4	1	0
1	0	1	m_5	0	1
1	1	0	m_6	0	1
1	1	1	m_7	1	1

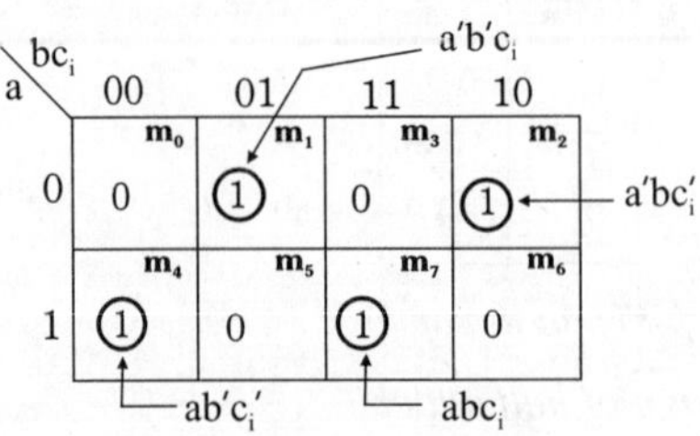

Fig. 5.85: K-map for s.

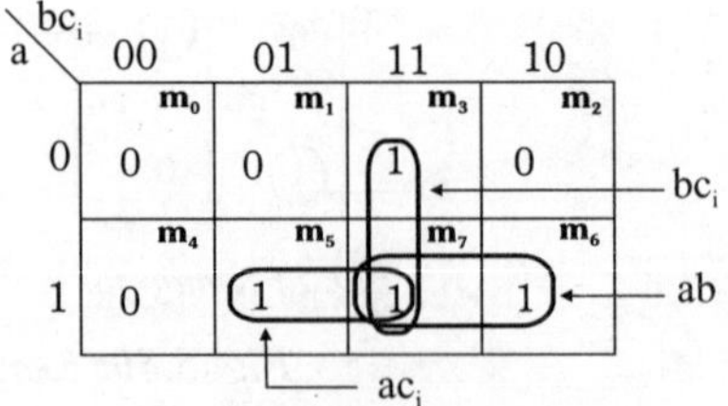

Fig. 5.86: K-map for c_o.

Using truth table (Table 5.35) the K-maps for designing full adder are drawn as shown in Figs. 5.85 and 5.86. From the K-map we get the following Boolean equations and using these equations the logic circuit of full adder is drawn as shown in Fig. 5.87.

> ***Note:*** *With reference to Example 5.5, sum, s can be further simplified to XOR of a, b and c_i.*
>
> $\therefore s = a \oplus b \oplus c_i$
>
> a, b, c_i → $s = a \oplus b \oplus c_i$

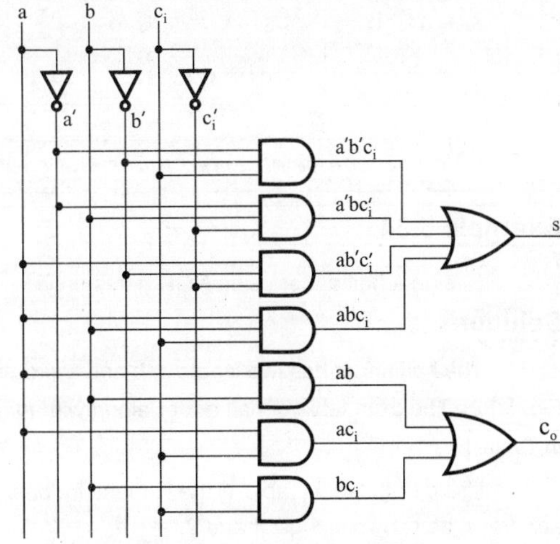

Fig. 5.87: *Logic circuit of full adder.*

Alternatively, full adder can be realized using two half adders as shown in Fig. 5.88. The first half adder is used to add two binary input bits and the second half adder is used to add sum of first half adder and third binary input.

The carry is obtained by logical OR of the carry of both the half adders.

The combinational circuit of Fig. 5.88 can be verified using the truth table (Table 5.36).

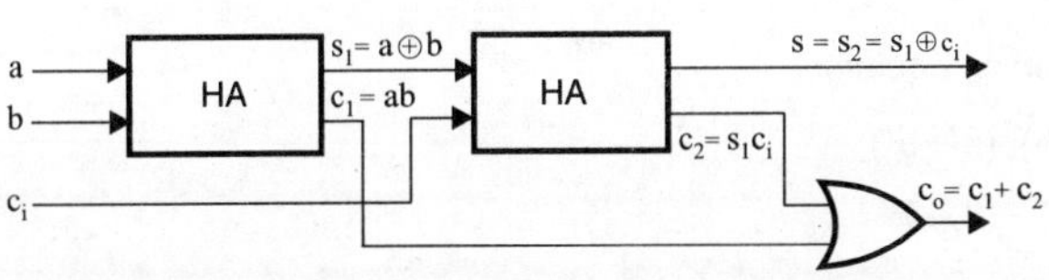

Fig. 5.88: *Full adder using two half adders.*

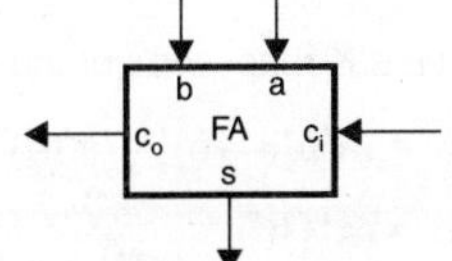

Fig. 5.89: *Symbolic representation of full adder.*

Table 5.36: Truth Table to Verify Logic Circuit of Fig. 5.88

Inputs			s_1	$s = s_2$	c_1	c_2	c_o
a	b	c_i	$a \oplus b$	$s_1 \oplus c_i$	ab	s_1c_i	$c_1 + c_2$
0	0	0	0	0	0	0	0
0	0	1	0	1	0	0	0
0	1	0	1	1	0	0	0
0	1	1	1	0	0	1	1
1	0	0	1	1	0	0	0
1	0	1	1	0	0	1	1
1	1	0	0	0	1	0	1
1	1	1	0	1	1	0	1

The following equations are obtained from design of half adder in Section 5.9.1, for the sum and carry of the logic circuit of Fig. 5.88.

$s_1 = a \oplus b$	$s = s_2 = s_1 \oplus c_i$
$c_1 = ab$	$c_2 = s_1\, c_i$

Note: *It can be observed that s and c_o of Table 5.36 is same as s and c_o of full adder.*

Example 5.34

(CSVTU, Nov/Dec'19, 2 Marks)

Design a half adder using NOR gates only.

Solution

The half adder has two inputs a, b and two outputs s, c_o (Refer Fig. 5.82). The truth table of half adder along with minterms is shown in Table 1.

Table 1: Truth Table of Half Adder

Inputs		Minterm	Outputs	
a	b		s	c_o
0	0	m_0	0	0
0	1	m_1	1	0
1	0	m_2	1	0
1	1	m_3	0	1

Using truth table (Table 1) the K-maps for designing half adder are drawn as shown in Figs. 1 and 2.

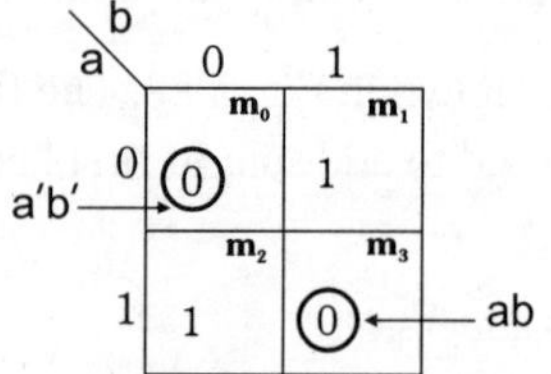

Fig. 1: K-map for s.

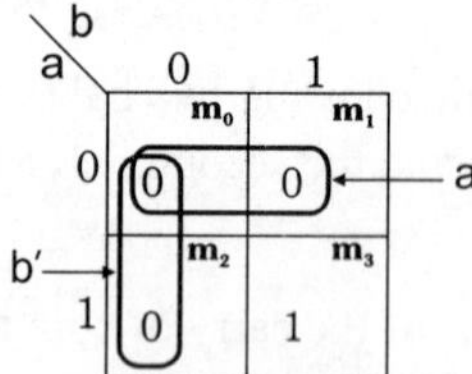

Fig. 2: K-map for c_o.

From the K-maps we get the following Boolean equations,

$$s = (a'b' + ab)' = (a'b').(ab)' = (a+b).(a'+b')$$

$$c_o = (a'+b')'$$

Using the above Boolean equations the logic circuit of half adder is drawn using only NOR gates as shown in Fig. 3.

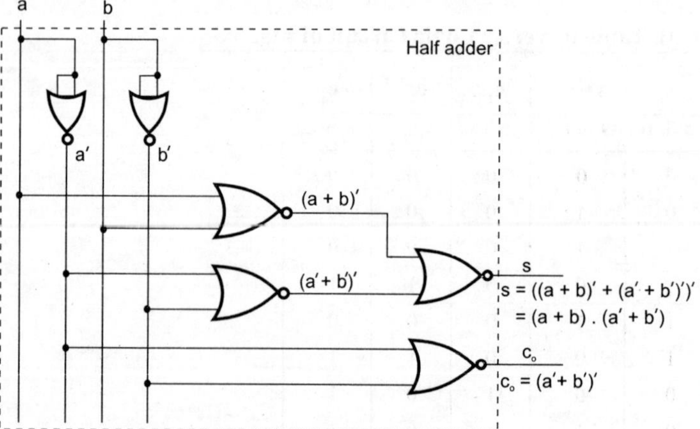

Fig. 3: Logic circuit of half adder using NOR gates.

Example 5.35

Design a half adder using logic **high** decoder.

Solution

The half adder has two inputs a, b and two outputs s, c_o (Refer Fig. 5.82). The truth table of half adder along with minterms is shown in Table 1.

Table 1: Truth Table of Half Adder

Inputs		Minterm	Outputs	
a	b		s	c_0
0	0	m_0	0	0
0	1	m_1	1	0
1	0	m_2	1	0
1	1	m_3	0	1

From the truth table of half adder we can write the following SOP form of Boolean equations for sum and carry of half adder.

Sum, $s = m_1 + m_2$; Carry, $c_o = m_3$

Since the half adder has two inputs, a 2-to-4 decoder can be selected to implement the half adder. The truth table of logic **high** 2-to-4 decoder is shown in Table 2.

On taking into account the minterms for which sum and carry are 1, the following Boolean equations can be obtained for sum and carry using the logic **high** decoder output.

$s = Y_1 + Y_2$; $c_o = Y_3$

Using the above Boolean equations the logic circuit of half adder using logic **high** 2-to-4 decoder and 2-input OR gate is drawn as shown in Fig. 1.

Table 2: Truth Table of Logic High 2-to-4 Decoder

Inputs		Minterm	Decoder Outputs			
a	b		Y_0	Y_1	Y_2	Y_3
0	0	m_0	1	0	0	0
0	1	m_1	0	1	0	0
1	0	m_2	0	0	1	0
1	1	m_3	0	0	0	1

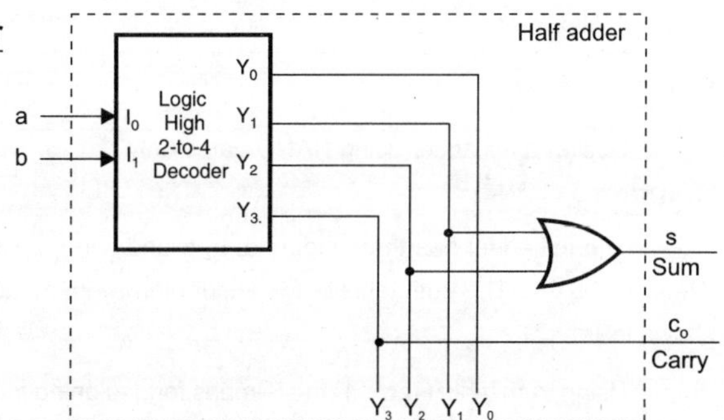

Fig. 1: Half adder using logic high 2-to-4 decoder.

Example 5.36

Design a half adder using a logic **low** decoder and NOR gates.

Solution

The half adder has two inputs a, b and two outputs s, c_o (Refer Fig. 5.82). The truth table of half adder along with maxterms is shown in Table 1.

Table 1: Truth Table of Half Adder

Inputs		Maxterm	Outputs	
a	b		s	c_0
0	0	M_0	0	0
0	1	M_1	1	0
1	0	M_2	1	0
1	1	M_3	0	1

From the truth table of half adder we can write the following POS form of Boolean equations for sum and carry of half adder.

Sum, $s = M_0 M_3$

Carry, $c_o = M_0 M_1 M_2$

Since the half adder has two inputs, a 2-to-4 decoder can be selected to implement the half adder. The truth table of logic **low** 2-to-4 decoder is shown in Table 2.

On taking into account the maxterm for which sum and carry are 0, the following Boolean equations can be obtained for sum and carry using the logic **low** decoder output.

$$s = Y_0Y_3 = (Y_0 + Y_3)'$$

$$c_o = Y_0Y_1Y_2 = (Y_0 + Y_1 + Y_2)'$$

Using the above Boolean equations the logic circuit of half adder using logic **low** 2-to-4 decoder and NOR gates is drawn as shown in Fig. 1.

Table 2: Truth Table of Logic Low 2-to-4 Decoder

Inputs		Maxterm	Decoder Outputs			
a	b		Y_0	Y_1	Y_2	Y_3
0	0	M_0	0	1	1	1
0	1	M_1	1	0	1	1
1	0	M_2	1	1	0	1
1	1	M_3	1	1	1	0

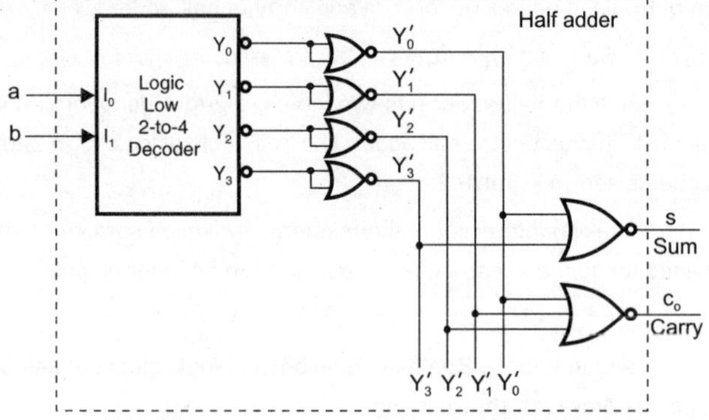

Fig. 1: Half adder using logic low 2-to-4 decoder.

Example 5.37

Design a full adder using NAND gates only.

Solution

The full adder has three inputs a, b, c_i and two outputs s, c_o (Refer Fig. 5.89). The truth table of full adder along with minterms is shown in Table 1.

Using truth table (Table 1) the K-maps for designing full adder are drawn as shown in Figs. 1 and 2.

Table 1: Truth Table of Full Adder

Inputs			Minterm	Outputs	
a	b	c_i		Sum s	Carry c_o
0	0	0	m_0	0	0
0	0	1	m_1	1	0
0	1	0	m_2	1	0
0	1	1	m_3	0	1
1	0	0	m_4	1	0
1	0	1	m_5	0	1
1	1	0	m_6	0	1
1	1	1	m_7	1	1

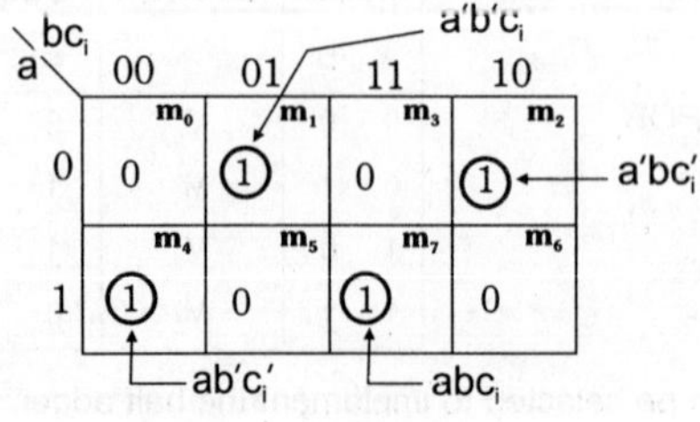

Fig. 1: K-map for s.

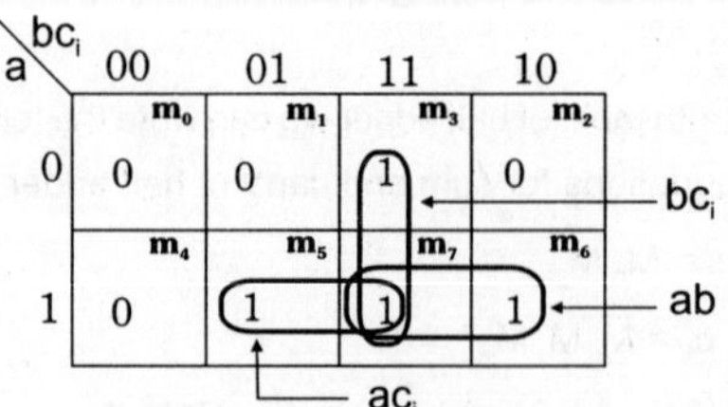

Fig. 2: K-map for c_o.

From the K-maps we get the following Boolean equations,

$s = a'b'c_i + a'bc_i' + ab'c_i' + abc_i$

$c_o = ab + ac_i + bc_i$

Using the above Boolean equations the logic circuit of full adder using only NAND gates is drawn as shown in Fig. 3.

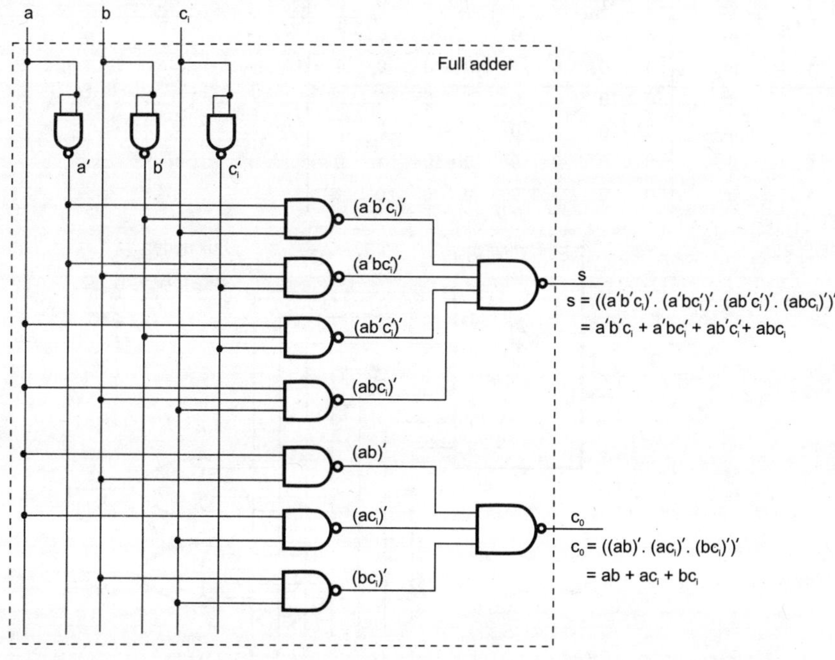

Fig. 3: Logic circuit of full adder using NAND gates.

Example 5.38

(CSVTU, Nov/Dec'19, 7 Marks)

Design a full adder using logic **high** decoder.

Solution

The full adder has three inputs a, b, c_i and two outputs s, c_o (Refer Fig. 5.89). The truth table of full adder along with minterms is shown in Table 1.

Table 1: Truth Table of Full Adder

Inputs			Minterm	Outputs	
a	b	c_i		Sum s	Carry c_o
0	0	0	m_0	0	0
0	0	1	m_1	1	0
0	1	0	m_2	1	0
0	1	1	m_3	0	1
1	0	0	m_4	1	0
1	0	1	m_5	0	1
1	1	0	m_6	0	1
1	1	1	m_7	1	1

From the truth table of full adder we can write the following SOP form of Boolean equations for sum and carry of full adder.

Sum, $s = m_1 + m_2 + m_4 + m_7$; Carry, $c_o = m_3 + m_5 + m_6 + m_7$

Since the full adder has three inputs, a 3-to-8 decoder can be selected to implement the full adder. The truth table of logic **high** 3-to-8 decoder is shown in Table 2. On taking into account the minterms for which sum and carry are 1, the following Boolean equations can be obtained for sum and carry using the logic **high** decoder output.

$s = Y_1 + Y_2 + Y_4 + Y_7$; $c_o = Y_3 + Y_5 + Y_6 + Y_7$

Using the above Boolean equations the logic circuit of full adder using logic **high** 3-to-8 decoder and 4-input OR gate is drawn as shown in Fig. 1.

Table 2: Truth Table of Logic High 3-to-8 Decoder

Inputs			Minterm	Decoder Outputs							
a	b	c_i		Y_0	Y_1	Y_2	Y_3	Y_4	Y_5	Y_6	Y_7
0	0	0	m_0	1	0	0	0	0	0	0	0
0	0	1	m_1	0	1	0	0	0	0	0	0
0	1	0	m_2	0	0	1	0	0	0	0	0
0	1	1	m_3	0	0	0	1	0	0	0	0
1	0	0	m_4	0	0	0	0	1	0	0	0
1	0	1	m_5	0	0	0	0	0	1	0	0
1	1	0	m_6	0	0	0	0	0	0	1	0
1	1	1	m_7	0	0	0	0	0	0	0	1

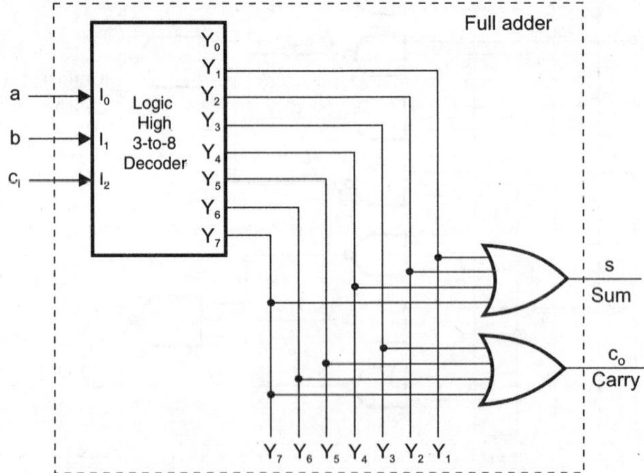

Fig. 1: Full adder using logic high 3-to-8 decoder.

Example 5.39

(AKUB, Dec'20, 6 Marks)

Design a full adder using multiplexer.

a) Use 4:1 MUX **b)** Use 8:1 MUX

Solution

a) Full Adder using 4:1 MUX

The full adder has three inputs a, b, c_i and two outputs s, c_o (Refer Fig. 5.89). The truth table of full adder along with minterms is shown in Table 1.

The full adder has two outputs and hence two separate multiplexers are needed for outputs. Since the full adder has 3-bit input, let us select 2^2:1 (4:1) multiplexers. The full adder using 4:1 multiplexer is shown in Fig. 1. The inputs a and b of full adder are connected to selection inputs of both the multiplexers. The input c_i of full adder and its complement are used as some of the inputs of multiplexer. The other inputs of multiplexers are permanently tied to 0 or 1. The working of full adder using 4:1 multiplexer is verified in Table 2.

Table 1: Truth Table of Full Adder

Inputs			Minterm	Outputs	
a	b	c_i		Sum s	Carry c_o
0	0	0	m_0	0	0
0	0	1	m_1	1	0
0	1	0	m_2	1	0
0	1	1	m_3	0	1
1	0	0	m_4	1	0
1	0	1	m_5	0	1
1	1	0	m_6	0	1
1	1	1	m_7	1	1

Table 2: Verification of Logic Circuit of s, c_o in Fig. 1

Inputs			Selection Inputs		Full adder Outputs		MUX Outputs	
a	b	c_i	S_1	S_0	s	c_o	$Y_1 = s$	$Y_2 = c_o$
0	0	0	0	0	0	0	$Y_1 = I_{01} = c_i$	$Y_2 = I_{02} = 0$
0	0	1	0	0	1	0		
0	1	0	0	1	1	0	$Y_1 = I_{11} = c_i'$	$Y_2 = I_{12} = c_i$
0	1	1	0	1	0	1		
1	0	0	1	0	1	0	$Y_1 = I_{21} = c_i'$	$Y_2 = I_{22} = c_i$
1	0	1	1	0	0	1		
1	1	0	1	1	0	1	$Y_1 = I_{31} = c_i$	$Y_3 = I_{32} = 1$
1	1	1	1	1	1	1		

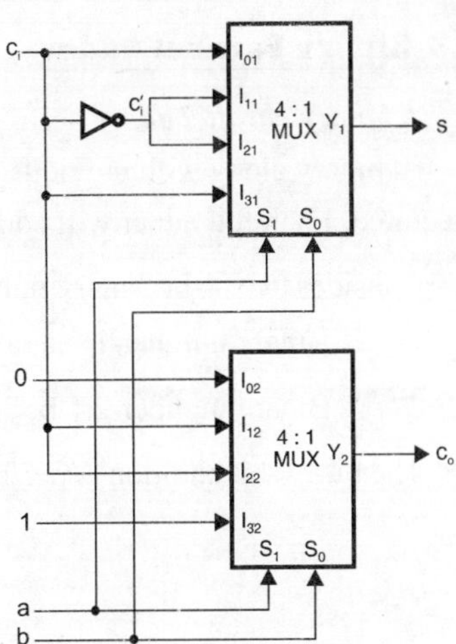

Fig. 1: *Full adder using two 4:1 MUX.*

b) Full Adder using 8:1 MUX

In 8:1 MUX if selection input S_2 is tied to 0 permanently then it will ignore the last four inputs and behave as 4:1 MUX. Now adder design using 8:1 MUX is same as that of design using 4:1 MUX. The logic circuit of full adder using two 8:1 MUX is shown in Fig. 2.

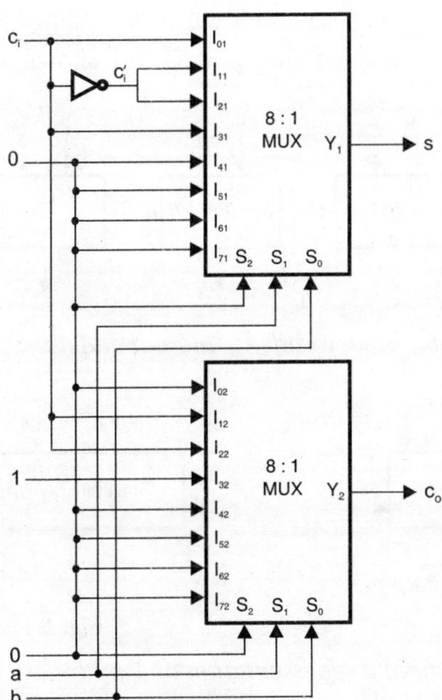

Fig. 2: *Full adder using two 8:1 MUX.*

5.9.3 Binary Parallel Adder

A ***binary parallel adder*** performs arithmetic sum of two n-bit binary numbers. A full adder can add two 1-bit binary along with previous carry. Hence, in order to add two n-bit binary numbers, n full adders are required. Each full adder will add one bit of binary numbers.

Consider two, 4-bit binary numbers A and B shown below:

Binary number-1: $a_3\ a_2\ a_1\ a_0$ (A)

Binary number-2: $b_3\ b_2\ b_1\ b_0$ (B)

The bit-by-bit addition is performed as follows.

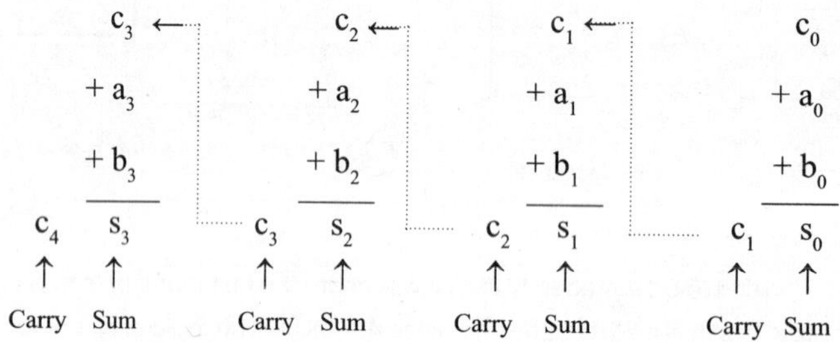

In the above addition the initial carry $c_o = 0$. The carry generated in an addition is considered in the next addition. A 4-bit binary adder to perform above addition is constructed using 4 full adders as shown in Fig. 5.90. The symbolic representations of 4-bit binary adders are shown in Fig. 5.91.

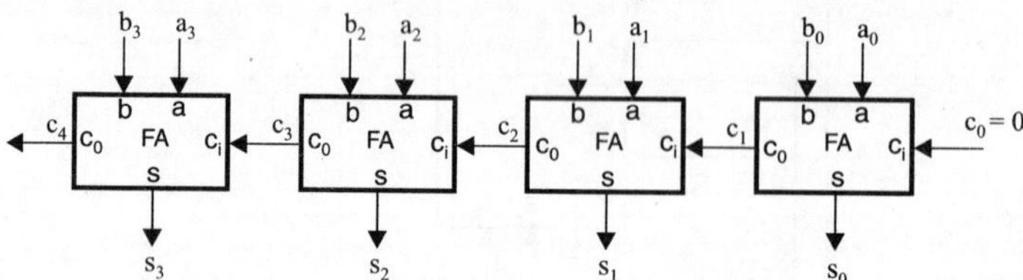

Fig. 5.90: *4-bit binary adder using full adders.*

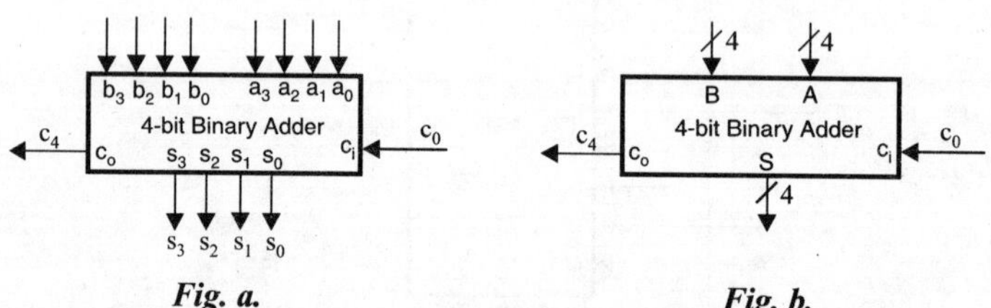

Fig. a. ***Fig. b.***

Fig. 5.91: *Symbolic representation of 4-bit binary adder.*

The 4-bit binary adder is available as a standard IC with number 7483. The pin configuration of 7483 is shown in Fig. 5.92.

Left pin	No.		No.	Right pin
s_2	1	**7483**	16	V_{cc}(+5V)
b_2	2		15	b_3
a_2	3		14	a_3
s_1	4		13	s_3
a_1	5		12	a_4
b_1	6		11	b_4
c_o	7		10	s_4
(0V) GND	8		9	c_4

Pin	Description
a_1 - a_4	Operand a inputs
b_1 - b_4	Operand b inputs
c_o	Carry input
s_1 - s_4	Sum outputs
c_4	Carry output
V_{cc}	Supply voltage (+5V)
GND	Ground (0V)

***Fig. 5.92:** Pin configuration of 4-bit binary adder IC 7483.*

The 4-bit binary adders can be connected in parallel to perform addition of higher bit size binary. Two 4-bit adders can be connected as shown in Fig. 5.93 to add 8-bit binary numbers. Three 4-bit adders can be connected as shown in Fig. 5.94 to add 12-bit binary numbers and so on.

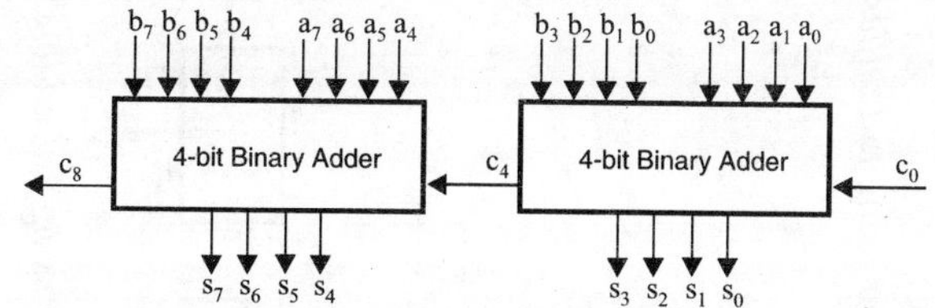

***Fig. 5.93:** 8-bit binary adder using 4-bit binary adders.*

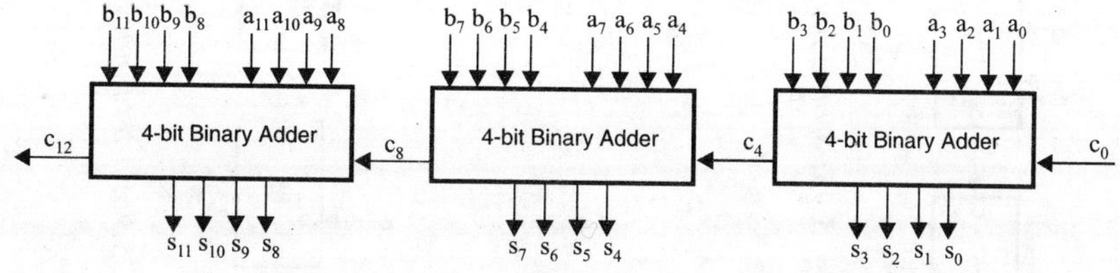

***Fig. 5.94:** 12-bit binary adder using 4-bit binary adders.*

Alternatively, 8-bit binary adder can be constructed by connecting 8 full adders in parallel as shown in Fig. 5.95. In general, n-bit binary adder can be constructed by connecting n full adders in parallel as shown in Fig. 5.96.

> ***Note:** n-bit binary adder is called parallel adder because it add all bits of data simultaneously.*

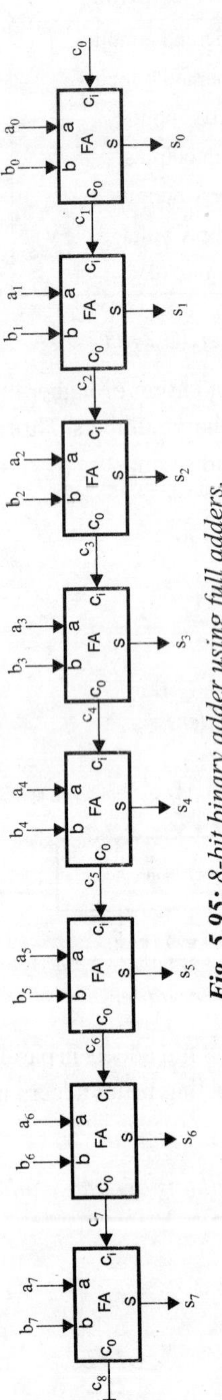

***Fig. 5.95:** 8-bit binary adder using full adders.*

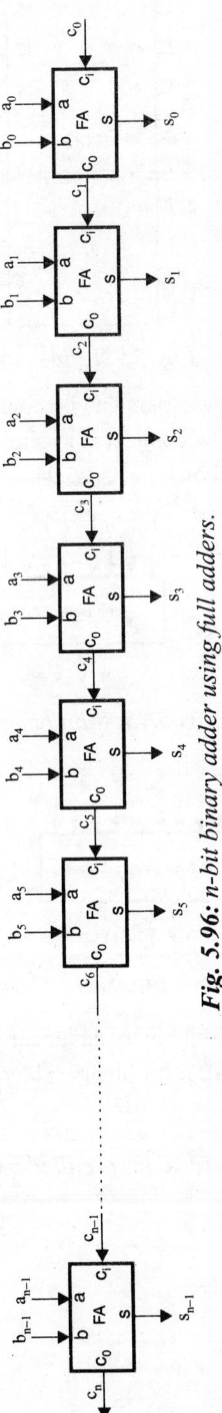

***Fig. 5.96:** n-bit binary adder using full adders.*

5.9.4 Carry Look Ahead Adder

(JNTUK, Feb'22, 7 Marks) (GTU, May'19, 7 Marks) (CSVTU, Nov/Dec'19, 7 Marks) (AKTU, Sep'20, 10 Marks) (RGPV, Jun'20, 6 Marks)

In n-bit binary adders the carry of ith adder propagates to $(i + 1)^{th}$ adder. This creates a delay in getting a stable result. The sum will not be stable until the carry propagates to n^{th} adder. This delay can be avoided if an additional circuit called carry look ahead circuit is designed to predict the carry before addition and the adder that includes carry look ahead circuit is called ***carry look ahead adder***.

With reference to half adder circuit shown in Fig. 5.83 and full adder circuit using half adders shown in Fig. 5.88 we can construct a full adder circuit as shown in Fig. 5.97. From Fig. 5.97 we get the following equations for i^{th} sum and $(i + 1)^{th}$ carry.

$$s_i = p_i \oplus c_i$$

$$c_{i+1} = g_i + p_i\, c_i$$

$$p_i = a_i \oplus b_i$$

$$g_i = a_i\, b_i$$

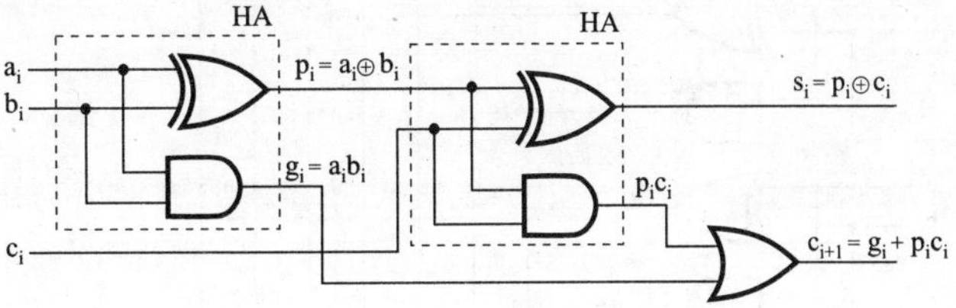

Where, p_i = Carry propagate from i^{th} stage

g_i = Carry generate of ith stage

HA HA
a_i b_i c_i
$p_i = a_i \oplus b_i$
$g_i = a_i b_i$
$p_i c_i$
$s_i = p_i \oplus c_i$
$c_{i+1} = g_i + p_i c_i$

***Fig. 5.97:** Full adder circuit using two XOR gate based half adders.*

The carry propagate decides whether a carry propagates from i^{th} to $(i + 1)^{th}$ stage. The g_i generates a carry when both a_i and b_i are 1.

In a 4-bit binary adder, the i takes values from 0 to 3. The above equations for values of i = 0, 1, 2, 3 are evaluated as shown below:

When i = 0;

$$p_0 = a_0 \oplus b_0$$

$$g_0 = a_0\, b_0$$

$$s_0 = p_0 \oplus c_0$$

$$c_1 = g_0 + p_0\, c_0$$

When i = 1;

$$p_1 = a_1 \oplus b_1$$

$$g_1 = a_1\, b_1$$

$$s_1 = p_1 \oplus c_1$$

$$c_2 = g_1 + p_1\, c_1$$

$$= g_1 + p_1\,(g_0 + p_0\, c_0)$$

$$= g_1 + p_1\, g_0 + p_1\, p_0\, c_0$$

When i = 2;

$$p_2 = a_2 \oplus b_2$$

$$g_2 = a_2\, b_2$$

$$s_2 = p_2 \oplus c_2$$

$$c_3 = g_2 + p_2\, c_2$$

$$= g_2 + p_2\,(g_1 + p_1\, g_0 + p_1\, p_0\, c_0)$$

$$= g_2 + p_2\, g_1 + p_2\, p_1\, g_0 + p_2\, p_1\, p_0\, c_0$$

When i = 3;

$$p_3 = a_3 \oplus b_3$$

$$g_3 = a_3\, b_3$$

$$s_3 = p_3 \oplus c_3$$

$$c_4 = g_3 + p_3\, c_3$$

$$= g_3 + p_3\,(g_2 + p_2\, g_1 + p_2\, p_1\, g_0 + p_2\, p_1\, p_0\, c_0)$$

$$= g_3 + p_3\, g_2 + p_3\, p_2\, g_1 + p_3\, p_2\, p_1\, g_0 + p_3\, p_2\, p_1\, p_0\, c_0$$

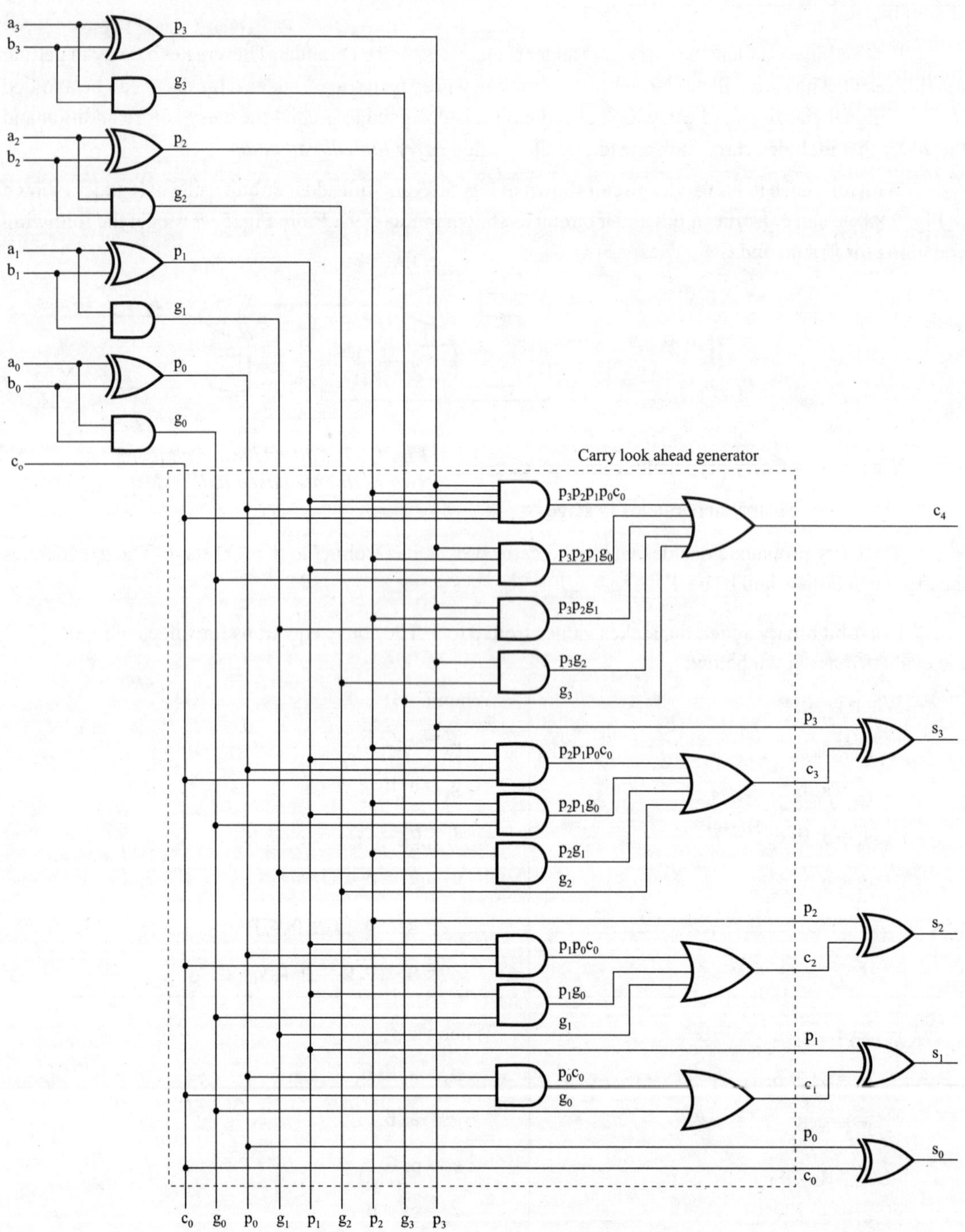

***Fig. 5.98:** Carry look ahead 4-bit binary adder.*

The ***carry look ahead generator*** is a combinational circuit that generate carry bits before generation of sum bits and such a circuit is developed using the equations for c_1, c_2, c_3 and c_4. The 4-bit binary adder with carry look ahead generator is drawn as shown in Fig. 5.98. In this adder the propagation delay is minimized at the expense of additional hardware. Also it can be observed that the computation of each bit of sum has equal number of stages and so propagation delay time will be equal for all outputs.

The carry look ahead generator is available as a standard IC with number 74182. The pin configuration of 74182 is shown in Fig. 5.99. The carry look ahead 4-bit binary adder using 74182 is shown in Fig. 5.100.

Pin	74182		Pin
g_1	1	16	V_{cc}(+5V)
p_1	2	15	p_2
g_0	3	14	g_2
p_0	4	13	c_0
g_3	5	12	c_1
p_3	6	11	c_2
p	7	10	c_4
(0V) GND	8	9	c_3

Pin	Description
g_0 - g_3	Carry generate inputs
p_0 - p_3	Carry propagate inputs
c_o	Carry input
c_1, c_2, c_3	Carry output
c_4	Carry generate output
p	Carry propagate output
V_{cc}	Supply voltage (+5V)
GND	Ground (0V)

Fig. 5.99: *Pin configuration of carry look ahead adder IC 74182.*

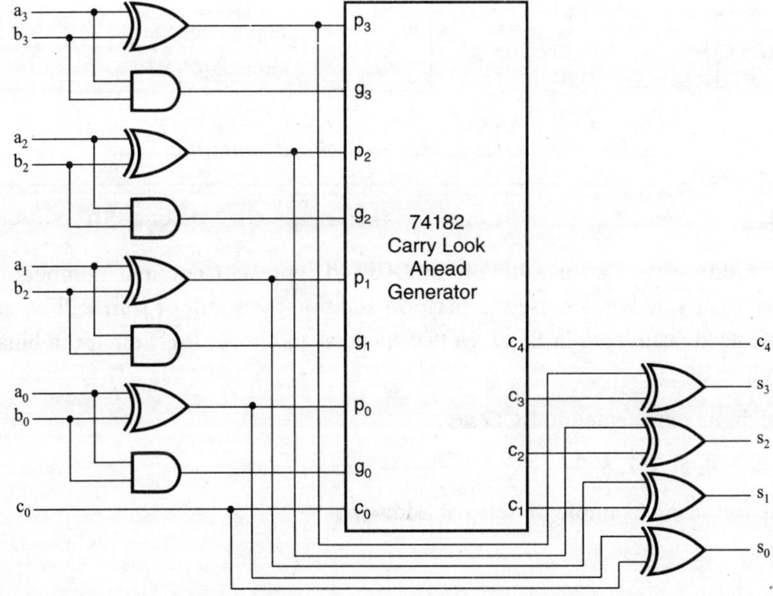

Fig. 5.100: *Carry look ahead 4-bit binary adder using 74182 carry look ahead generator.*

The 4-bit binary full adder with internal carry look ahead generator is available as standard ICs 7483 and 74283. The two ICs are functionally same and differ only in pin configurations. The pin description of IC 7483 and 74283 are shown in Fig. 5.101 and 5.102.

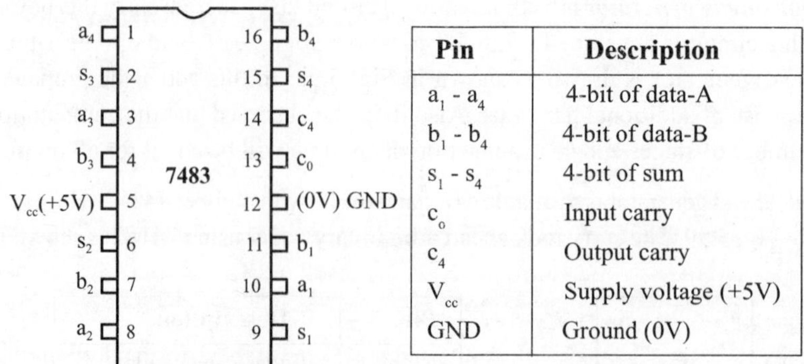

Pin	Description
a_1 - a_4	4-bit of data-A
b_1 - b_4	4-bit of data-B
s_1 - s_4	4-bit of sum
c_o	Input carry
c_4	Output carry
V_{cc}	Supply voltage (+5V)
GND	Ground (0V)

Fig. 5.101: *Pin description of 7483 4-bit binary adder with internal carry look ahead generator.*

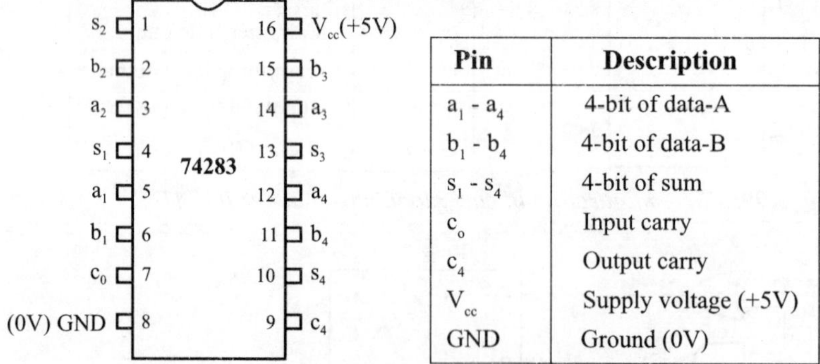

Pin	Description
a_1 - a_4	4-bit of data-A
b_1 - b_4	4-bit of data-B
s_1 - s_4	4-bit of sum
c_o	Input carry
c_4	Output carry
V_{cc}	Supply voltage (+5V)
GND	Ground (0V)

Fig. 5.102: *Pin description of 74283 4-bit binary adder with internal carry look ahead generator.*

5.9.5 BCD Adder

(CSVTU, Nov/Dec'19, 7 Marks) (AKTU, Nov/Dec'22, 10 Marks)

In decimal addition we can use binary or BCD to represent decimal numbers. In both the representation when binary adders are used to perform addition the result or sum will be in binary. But in BCD addition we need result/sum in BCD. Hence we need additional logic circuit in binary adders to convert binary result to BCD.

The decimal digits represented in BCD are,

0, 1, 2, 3, 4, 5, 6, 7, 8, 9

Consider the following example of decimal addition.

```
      1 1      ← Carry
    4 2 9 8
  + 5 3 9 7
  ---------
    9 6 9 5
  ---------
        ↳ Maximum possible sum = 1 + 9 + 9 = 19
```

From the above example it is evident that maximum possible sum of two BCD digits is,

$$1 + 9 + 9 = 19$$

↑

Previous carry

Therefore in BCD addition the sum of any two digits will be in the range 0 to 19_{10}. When sum is in the range 0 to 9_{10} no correction is needed and it can be represented by 4 digit BCD. But for sum in the range 10_{10} to 19_{10} a correction 6_{10} is added to sum to get the result in BCD. (Refer Chapter-2, Section 2.2.1 for examples of BCD addition.)

The ***BCD adder*** is a binary adder with additional logic circuit to perform addition of correction 6_{10} when the sum of a BCD digit exceeds 9_{10}.

A truth table (Table 5.37) is formed to show the range of binary sum and correction needed to convert binary sum to BCD sum. In Table 5.37, the binary sum is denoted as $z_8\ z_4\ z_2\ z_1$ and BCD sum as $s_8\ s_4\ s_2\ s_1$. Here, k is carry in binary sum and c is carry in BCD sum.

Table 5.37: Truth Table for Conversion of Binary to BCD Sum

Decimal Sum	Binary Addition		Correction	BCD Addition	
	Carry k	Binary Sum $z_8\ z_4\ z_2\ z_1$	Binary Sum $+ 6_{10}$ (0110)	Carry c	BCD Sum $s_8\ s_4\ s_2\ s_1$
0	0	0 0 0 0	↑	0	0 0 0 0
1	0	0 0 0 1		0	0 0 0 1
2	0	0 0 1 0	No	0	0 0 1 0
3	0	0 0 1 1	Correction	0	0 0 1 1
4	0	0 1 0 0		0	0 1 0 0
5	0	0 1 0 1		0	0 1 0 1
6	0	0 1 1 0		0	0 1 1 0
7	0	0 1 1 1		0	0 1 1 1
8	0	1 0 0 0		0	1 0 0 0
9	0	1 0 0 1	↓	0	1 0 0 1
10	0	1 0 1 0	0 1 0 1 0 + 0 1 1 0	1	0 0 0 0
11	0	1 0 1 1	0 1 0 1 1 + 0 1 1 0	1	0 0 0 1
12	0	1 1 0 0	0 1 1 0 0 + 0 1 1 0	1	0 0 1 0
13	0	1 1 0 1	0 1 1 0 1 + 0 1 1 0	1	0 0 1 1
14	0	1 1 1 0	0 1 1 1 0 + 0 1 1 0	1	0 1 0 0
15	0	1 1 1 1	0 1 1 1 1 + 0 1 1 0	1	0 1 0 1
16	1	0 0 0 0	1 0 0 0 0 + 0 1 1 0	1	0 1 1 0
17	1	0 0 0 1	1 0 0 0 1 + 0 1 1 0	1	0 1 1 1
18	1	0 0 1 0	1 0 0 1 0 + 0 1 1 0	1	1 0 0 0
19	1	0 0 1 1	1 0 0 1 1 + 0 1 1 0	1	1 0 0 1

From the truth table (Table 5.37) we can observe that the correction by adding + 6_{10} (0110_2) can be achieved using following three conditions.

1. When binary sum is 10_{10} and 11_{10} correction 6 can be added if both z_2 and z_8 are 1.
2. When binary sum is in the range 12_{10} to 15_{10} correction 6 can be added if both z_4 and z_8 are 1.
3. When binary sum is in the range 16_{10} to 19_{10} correction 6 can be added if carry k in binary sum is 1.

Therefore, during BCD addition, first binary addition is performed using 4-bit binary adder then additional logic circuit is used to generate correction 6 for the conditions mentioned above and another 4-bit binary adder is used to add correction 6 to binary sum so that the binary sum is converted to BCD sum.

The logic circuit to perform 1-digit BCD addition using two 4-bit binary adders is shown in Fig. 5.103. The symbolic representation of BCD adder is shown in Fig. 5.104.

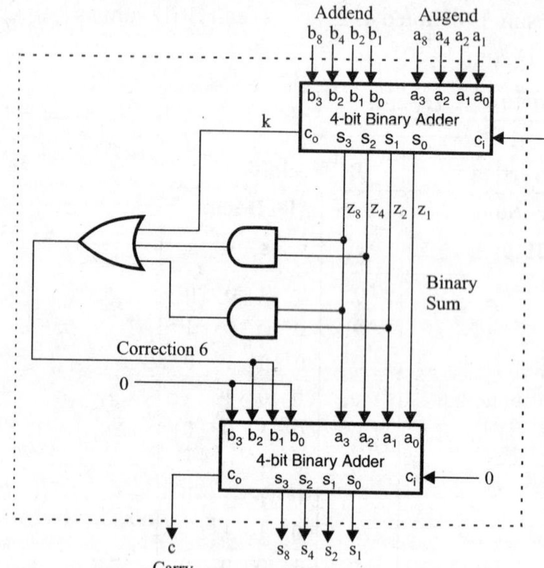

Fig. 5.103: BCD adder using binary adders.

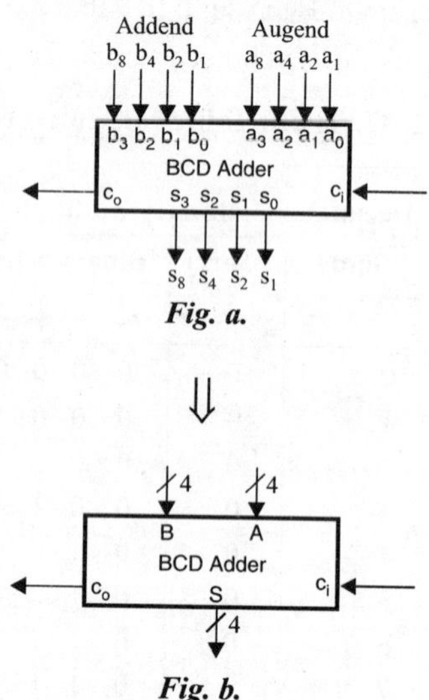

Fig. 5.104: Symbolic representation of BCD adder.

4-digit BCD Adder

The BCD adder shown in Fig. 5.104 is used to perform BCD addition of 1-digit decimal numbers. In order to perform 4-digit BCD addition we need 4 BCD adders shown in Fig. 5.105.

Consider two, 4-digit decimal numbers represented in BCD

Decimal number 1: $A_3\,A_2\,A_1\,A_0$

Decimal number 2: $B_3\,B_2\,B_1\,B_0$

Let, BCD Sum = $c\ S_3\ S_2\ S_1\ S_0$

$\uparrow$ Final carry (c)

Let the format of two number and sum be as follows.

A_3	A_2	A_1	A_0
$a_8\ a_4\ a_2\ a_1$	$a_8\ a_4\ a_2\ a_1$	$a_8\ a_4\ a_2\ a_1$	$a_8\ a_4\ a_2\ a_1$
B_3	B_2	B_1	B_0
$b_8\ b_4\ b_2\ b_1$	$b_8\ b_4\ b_2\ b_1$	$b_8\ b_4\ b_2\ b_1$	$b_8\ b_4\ b_2\ b_1$
S_3	S_2	S_1	S_0
$s_8\ s_4\ s_2\ s_1$	$s_8\ s_4\ s_2\ s_1$	$s_8\ s_4\ s_2\ s_1$	$s_8\ s_4\ s_2\ s_1$

Fig. 5.105 shows, 4 BCD adders connected in parallel to perform addition of 4-digit BCD.

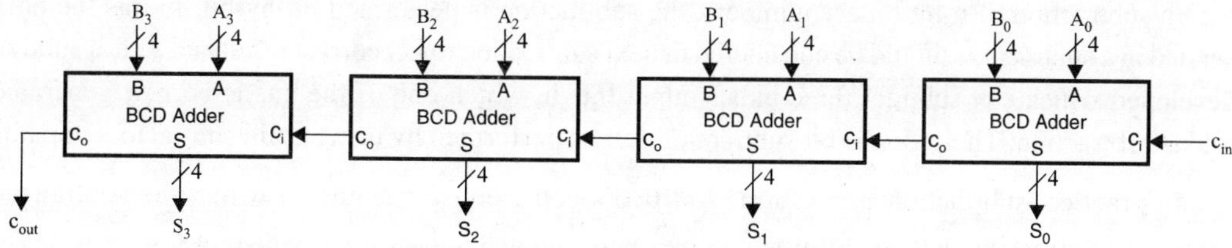

***Fig. 5.105:** 4-digit BCD adder.*

In general, n BCD adders can be connected in parallel to perform addition of n-digit BCD.

The 4-bit BCD adder is available as a standard IC with number 74583. The pin configuration of 74583 is shown in Fig. 5.106.

Pin (74583)		Pin	
b_1	1	16	V_{cc}(+5V)
b_2	2	15	a_2
b_3	3	14	a_1
a_3	4	13	a_0
c_i	5	12	b_0
c_{out}	6	11	s_0
s_2	7	10	s_1
(0V) GND	8	9	s_3

Pin	Description
a_0 - a_3	Operand a inputs
b_0 - b_3	Operand b inputs
c_i	Carry input
s_0 - s_3	Sum outputs
c_{out}	Carry output
V_{cc}	Supply voltage (+5V)
GND	Ground (0V)

***Fig. 5.106:** Pin configuration of 4-bit BCD adder IC 74583.*

5.10 Subtractors

Subtractors have been developed to perform arithmetic subtraction operation on binary numbers. Half subtractor can perform subtraction of two 1-bit binary number. The possible subtractions of two 1-bit binary numbers are shown below:

0 – 0 =	0		0 0
0 – 1 =	1 1	2-bit standard representation →	1 1
1 – 0 =	1		0 1
1 – 1 =	0		0 0

(→ Difference; → Borrow)

In the above subtractions, it can be observed that result is either 1-bit or 2-bit. The 1-bit result is called ***difference***. In the 2-bit result, the first bit is called ***difference*** and the second bit is called ***borrow***. Hence, half adder is designed to generate 2-bit standard output.

In subtraction of n-bit binary numbers, the subtraction is performed bit-by-bit. In this the borrow generated in a subtraction should be considered in next subtraction to get correct result and so full subtractor is developed which can subtract three binary input bits in which one of the bits is borrow generated in previous subtraction. Therefore, n-bit subtraction can be performed by using n full subtractors in parallel.

In practice, subtraction is performed in two's complement method in which the subtrahend is converted to two's complement form and added with minuend. Therefore, adders are used to perform subtraction as an addition of positive and negative number. The addition of positive number and two's complement of negative number will give the result of subtraction.

5.10.1 Half Subtractor

Half subtractor is a combination circuit that performs arithmetic subtraction of two binary bits.

The outputs are difference and borrow. The truth table of half subtractor is shown in Table 5.38. The block diagram of half subtractor is shown in Fig. 5.107.

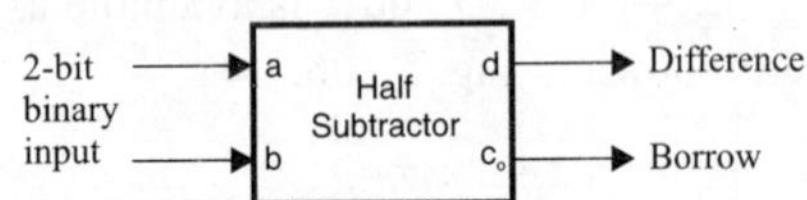

Fig. 5.107: *Block diagram of half subtractor.*

Table 5.38: Truth Table of Half Subtractor

Inputs		Minterm	Outputs	
a	**b**		**d**	c_o
0	0	m_0	0	0
0	1	m_1	1	1
1	0	m_2	1	0
1	1	m_3	0	0

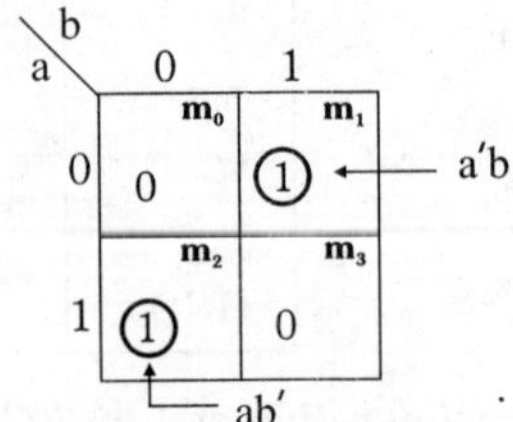

Fig. a: *K-map for d.*

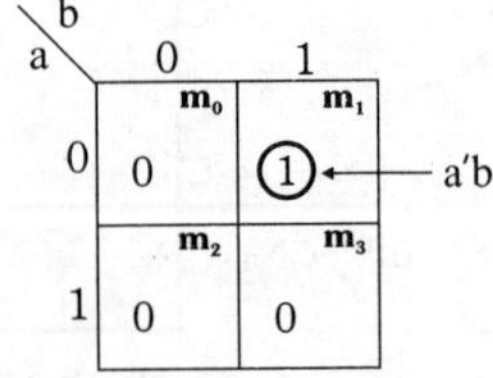

Fig. b: *K-map for c_o.*

Fig. 5.108: *K-map for design of half subtractor.*

Using truth table (Table 5.38) the K-maps for designing half subtractor are drawn as shown in Fig. 5.108.

From the K-map we get the following Boolean equations,

$$d = ab' + a'b$$

$$\boxed{d = ab' + a'b = a \oplus b}$$

$$c_o = a'b$$

Using the above Boolean equations the logic circuit of half subtractor is drawn as shown in Fig. 5.109.

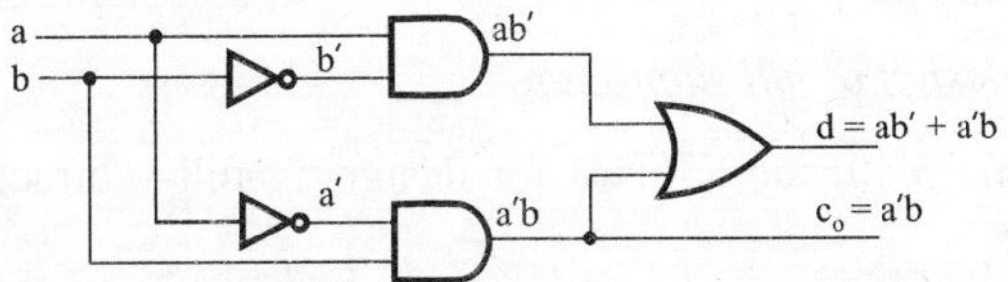

Fig. 5.109: *Logic circuit of half subtractor.*

Alternatively, the difference can be realized from XOR gate. If we compare the difference and output of XOR gate, we observe that both are same. The logic circuit of half subtractor using XOR gate is shown in Fig. 5.110.

Table 5.39: Truth Table of XOR Gate

a	b	$a \oplus b$
0	0	0
0	1	1
1	0	1
1	1	0

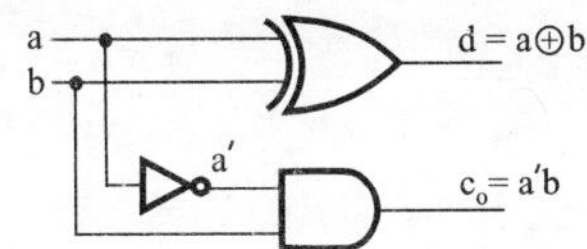

Fig. 5.110: *Half subtractor using XOR gate.*

5.10.2 Full Subtractor

(AKTU, Sep'20, 5 Marks) (GTU, Jun'19, 7 Marks)

Full subtractor is a combinational circuit that perform arithmetic subtraction of three binary bits in which one of the bit is borrow generated in previous subtraction.

Hence, full subtractor will be useful in n-bit subtraction to subtract borrow of a subtraction in next bit subtraction. The outputs are difference and borrow. The truth table of full subtractor is shown in Table 5.40. The block diagram of full subtractor is shown in Fig. 5.111.

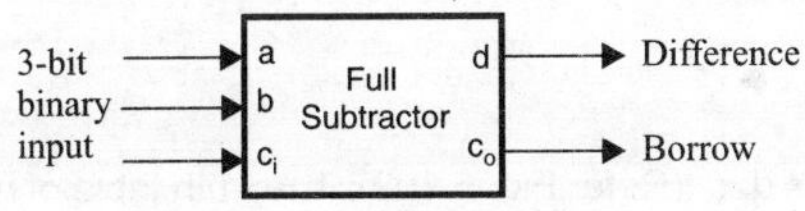

Fig. 5.111: *Block diagram of full subtractor.*

Table 5.40: Truth Table of Full Subtractor

Inputs			Minterm	Outputs	
a	b	c_i		d	c_o
0	0	0	m_0	0	0
0	0	1	m_1	1	1
0	1	0	m_2	1	1
0	1	1	m_3	0	1
1	0	0	m_4	1	0
1	0	1	m_5	0	0
1	1	0	m_6	0	0
1	1	1	m_7	1	1

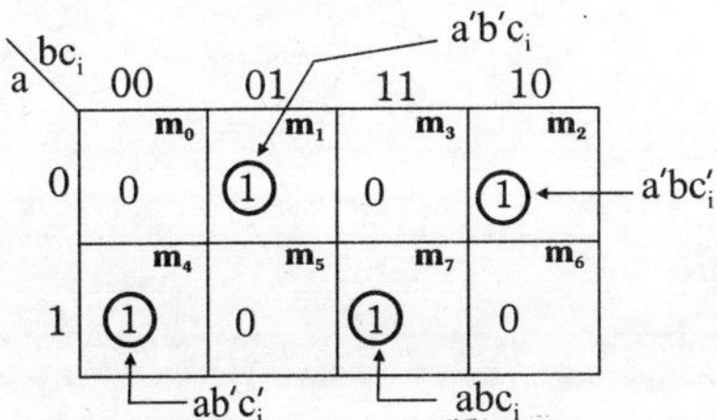

Fig. a: *K-map for d.*

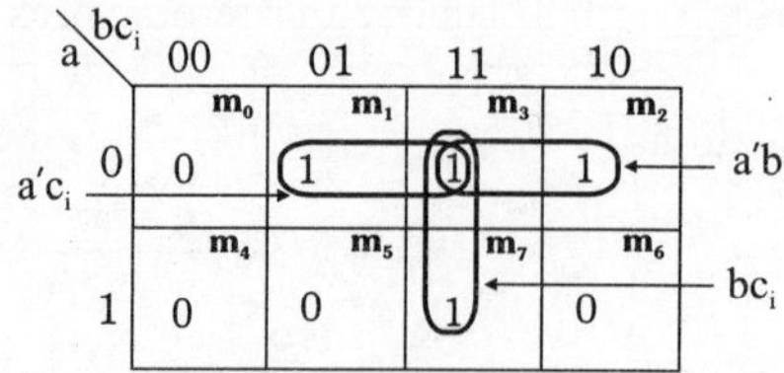

Fig. b: K-map for c_o.

> ***Note:** With reference to Example 5.5, difference, d can be further simplified to XOR of a, b and c_i.*
>
> $$\therefore d = a \oplus b \oplus c_i$$
>
> a, b, c_i → $d = a \oplus b \oplus c_i$

***Fig. 5.112:** K-map for design of full subtractor.*

Using truth table (Table 5.40) the K-maps for designing full subtractor are drawn as shown in Fig. 5.112.

From the K-maps we get the following Boolean equations,

$$d = a'b'c_i + a'bc_i' + ab'c_i' + abc_i$$

$$c_o = a'b + a'c_i + bc_i$$

Using the above Boolean equations the logic circuit of full subtractor is drawn as shown in Fig. 5.113.

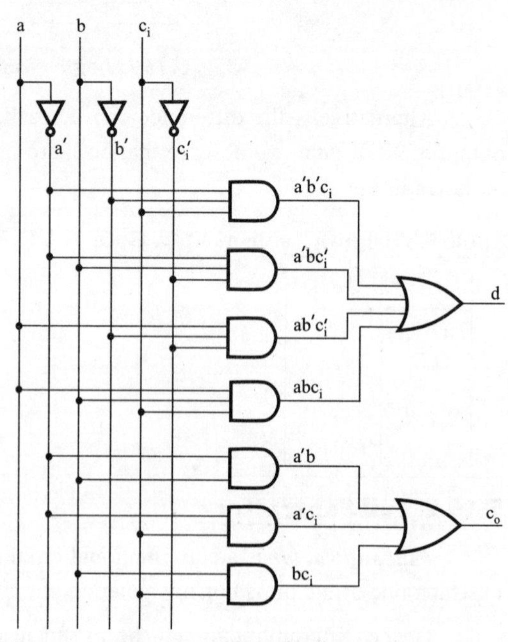

***Fig. 5.113:** Logic circuit of full subtractor.*

Example 5.40

Design a half subtractor using NAND gates only.

Solution

The half subtractor has two inputs a, b and two outputs d, c_o (Refer Fig. 5.109). The truth table of half subtractor adder along with minterms is shown in Table 1.

Table 1: Truth Table of Half Subtractor

Inputs		Minterm	Outputs	
a	**b**		**d**	**c_o**
0	0	m_0	0	0
0	1	m_1	1	1
1	0	m_2	1	0
1	1	m_3	0	0

Using truth table (Table 1) the K-maps for designing half subtractor are drawn as shown in Figs. 1 and 2.

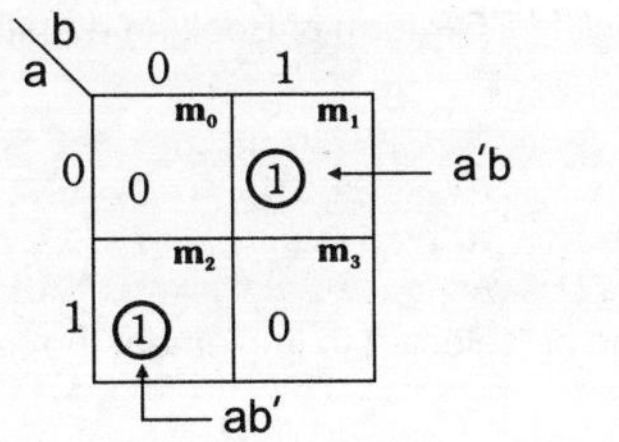

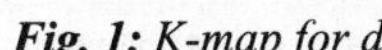

Fig. 1: K-map for d.

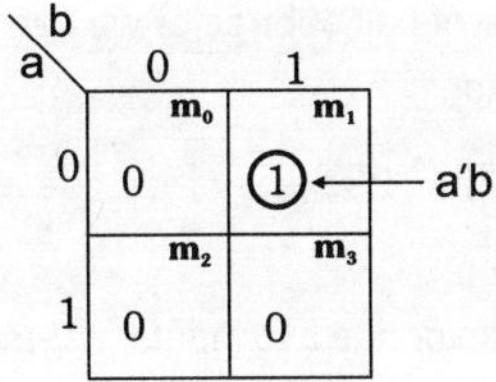

Fig. 2: K-map for c_o.

From the K-maps we get the following Boolean equations,

$d = ab' + a'b$; $c_o = a'b$

Using the above Boolean equations the logic circuit of half subtractor using only NAND gates is drawn as shown in Fig. 3.

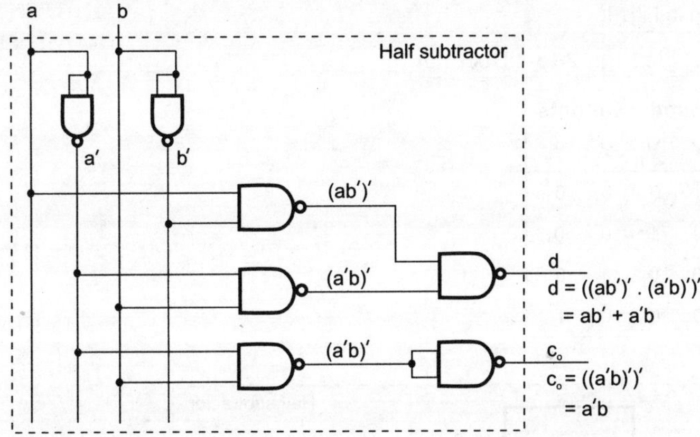

Fig. 3: Logic circuit of half subtractor using NAND gates.

Example 5.41

Design a half subtractor using logic **high** decoder.

Solution

The half subtractor has two inputs a, b and two outputs s, c_o (Refer Fig. 5.109). The truth table of half subtractor adder along with minterms is shown in Table 1.

Table 1: Truth Table of Half Subtractor

Inputs		Minterm	Outputs	
a	**b**		**d**	**c_o**
0	0	m_0	0	0
0	1	m_1	1	1
1	0	m_2	1	0
1	1	m_3	0	0

From the truth table of half subtractor we can write the following SOP form of Boolean equations for difference and borrow of half subtractor.

Difference, $d = m_1 + m_2$

Borrow, $c_o = m_1$

Since the full subtractor has two inputs, a 2-to-4 decoder can be selected to implement the half subtractor. The truth table of logic **high** 2-to-4 decoder is shown in Table 2.

On taking into account the minterms for which difference and borrow are 1, the following Boolean equations can be obtained for difference and borrow using the logic **high** decoder output.

$d = Y_1 + Y_2$

$c_o = Y_1$

Using the above Boolean equations the logic circuit of half subtractor using logic **high** 2-to-4 decoder and 2-input OR gate is drawn as shown in Fig. 1.

Table 2: Truth Table of Logic High 2-to-4 Decoder

Inputs		Minterm	Decoder Outputs			
a	b		Y_0	Y_1	Y_2	Y_3
0	0	m_0	1	0	0	0
0	1	m_1	0	1	0	0
1	0	m_2	0	0	1	0
1	1	m_3	0	0	0	1

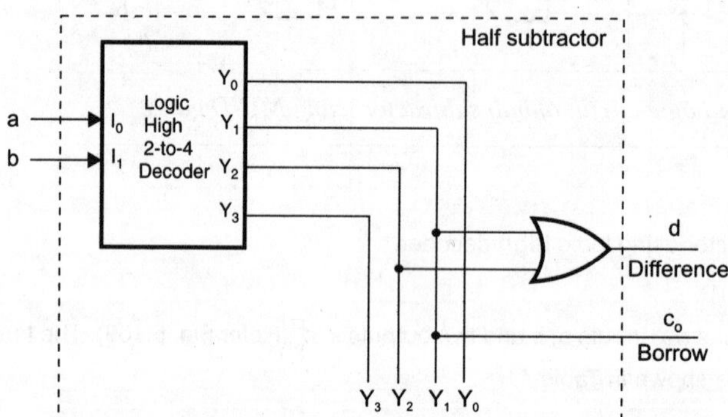

Fig. 1: Half subtractor using logic high 2-to-4 decoder.

Example 5.42

Design a half subtractor using a logic **low** decoder and NOR gates.

Solution

The half subtractor has two inputs a, b and two outputs d, c_o (Refer Fig. 5.109). The truth table of half subtractor along with maxterms is shown in Table 1.

Table 1: Truth Table of Half Subtractor

Inputs		Maxterm	Outputs	
a	b		d	c_o
0	0	M_0	0	0
0	1	M_1	1	1
1	0	M_2	1	0
1	1	M_3	0	0

From the truth table of half subtractor we can write the following POS form of Boolean equations for difference and borrow of half subtractor.

Difference, $d = M_0 M_3$; Borrow, $c_o = M_0 M_2 M_3$

Since the half subtractor has two inputs, a 2-to-4 decoder can be selected to implement the half subtractor. The truth table of logic **low** 2-to-4 decoder is shown in Table 2.

On taking into account the maxterm for which difference and borrow are 0, the following Boolean equations can be obtained for difference and borrow using the logic **low** decoder output.

$s = Y_0 Y_3 = (Y_0' + Y_3')'$; $c_o = Y_0 Y_2 Y_3 = (Y_0' + Y_2' + Y_3')'$

Using the above Boolean equations the logic circuit of half subtractor using logic **low** 2-to-4 decoder and NOR gates is drawn as shown in Fig. 1.

Table 2: Truth Table of Logic Low 2-to-4 Decoder

Inputs		Maxterm	Decoder Outputs			
a	b		Y_0	Y_1	Y_2	Y_3
0	0	M_0	0	1	1	1
0	1	M_1	1	0	1	1
1	0	M_2	1	1	0	1
1	1	M_3	1	1	1	0

Fig. 1: Half subtractor using logic low 2-to-4 decoder.

Example 5.43

(AKTU, Dec'22, 10 Marks)

Design a full subtractor using NAND gates.

Solution

The full subtractor has three inputs a, b, c_i and two outputs d, c_o (Refer Fig. 5.113). The truth table of full subtractor along with minterms is shown in Table 1.

Table 1: Truth Table of Full Subtractor

Inputs			Minterm	Outputs	
a	b	c		d	c_o
0	0	0	m_0	0	0
0	0	1	m_1	1	1
0	1	0	m_2	1	1
0	1	1	m_3	0	1
1	0	0	m_4	1	0
1	0	1	m_5	0	0
1	1	0	m_6	0	0
1	1	1	m_7	1	1

Using truth table (Table 1) the K-maps for designing full subtractor are drawn as shown in Figs. 1 and 2.

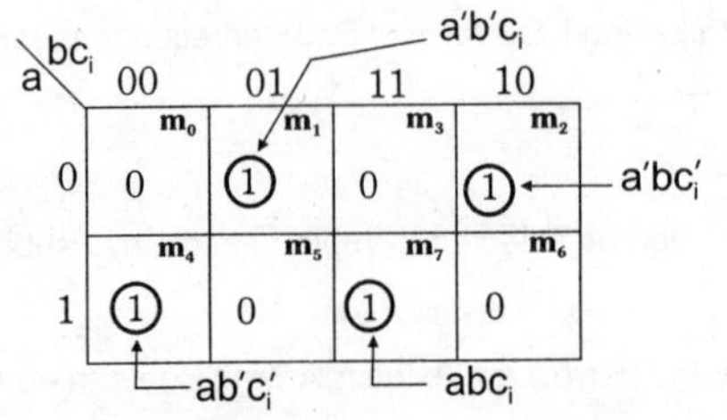

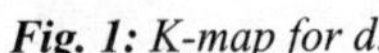

Fig. 1: K-map for d.

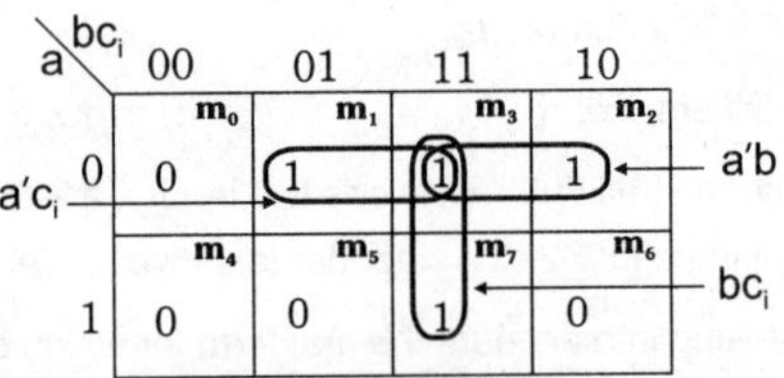

Fig. 2: K-map for c_o.

From the K-maps we get the following Boolean equations,

$$d = a'b'c_i + a'bc_i' + ab'c_i' + abc_i \quad ; \quad c_o = a'b + a'c_i + bc_i$$

Using the above Boolean equations the logic circuit of full subtractor using only NAND gates is drawn as shown in Fig. 3.

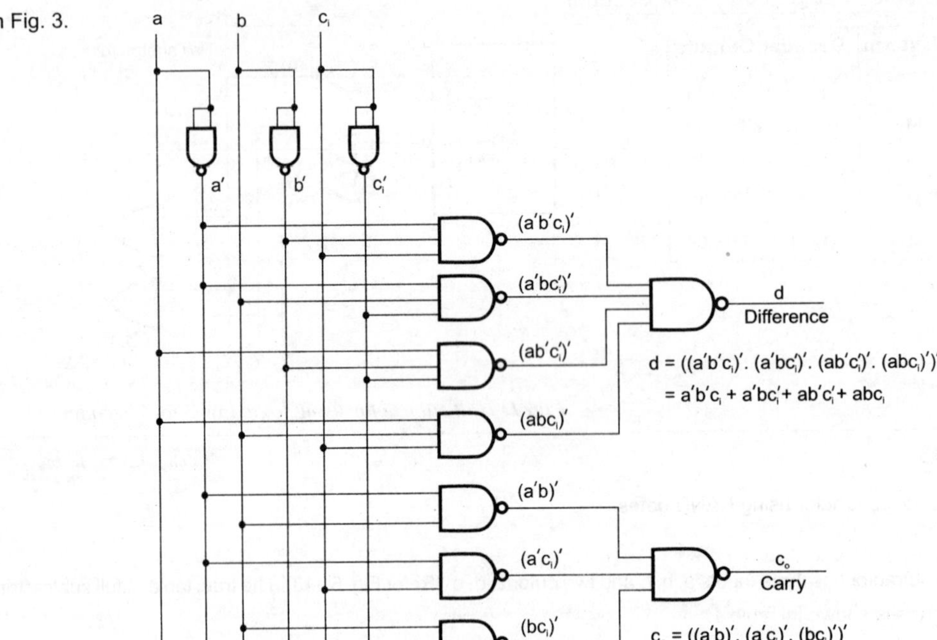

Fig. 3: Logic circuit of full subtractor using NAND gates.

5.11 Parallel Adder/Subtractor

5.11.1 Binary Parallel Adder/Subtractor

(JNTUA, Jun/July'22, 7 Marks)

Practically subtraction is performed only in 2's complement form in which the subtraction is considered as addition of positive and negative numbers and the negative number is represented in 2's complement form. Therefore, subtractor circuit is not of much use and subtraction can be performed using adder itself.

Therefore, subtraction of two BCD digit A and B is considered as,

$$A - B = A + (-B)$$

↑ 2's complement of B

The complement of binary can be easily obtained using XOR gate. Consider the truth table of XOR gate (Table 5.41).

Table 5.41: Truth Table of XOR Gate

x	y	$x \oplus y$
0	0	0
0	1	1
1	0	1
1	1	0

x	y	$x \oplus y = y$
0	0	0
0	1	1

x	y	$x \oplus y = y'$
1	0	1
1	1	0

With reference to Table 5.41 we can say that,

If $x = 0$, then $x \oplus y = y$

If $x = 1$, then $x \oplus y = y'$

Therefore in XOR gate, input x can be used as control or selection and input y can be data.

If $x = 0$, the data will be output as such.

If $x = 1$, the output will be complement of data.

The n-bit binary adder using full adder can be modified to perform both addition and subtraction as shown in Fig. 5.114.

One input can be given directly to full adders, another input can be given through XOR gate to full adders, so that second data can be sent to full adders with or without complement.

The control input M is used to perform complement operation. When M = 1, the output of XOR gate is complement of the input.

Consider addition and subtraction of two, 4-bit binary numbers.

Binary number-1: $a_3\ a_2\ a_1\ a_0$ (data-A)

Binary number-2: $b_3\ b_2\ b_1\ b_0$ (data-B)

Let us perform, $A + B$ and $A - B = A + (-B)$

For addition, B is input to full adder without complement as shown in Fig. 5.115.

$\therefore$ For addition, $M = c_o = 0$

For subtraction, B is input to full adder after complement and adding 1 for 2's complement can be performed by making $c_0 = 1$ for subtraction as shown in Fig. 5.116.

$\therefore$ For subtraction, $M = c_o = 1$

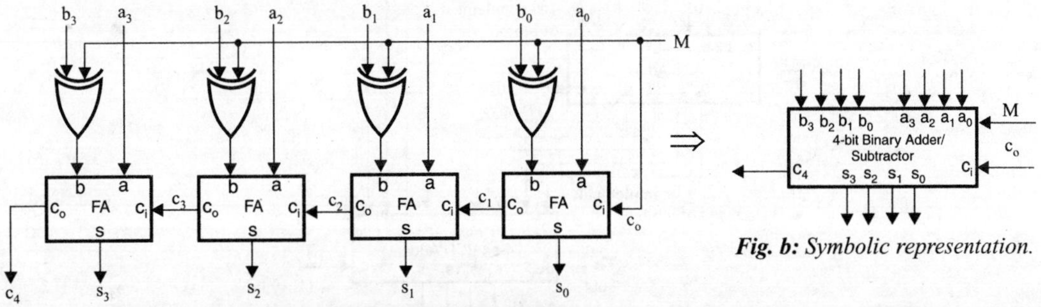

Fig. a: Binary adder/subtractor using full adders.

Fig. b: Symbolic representation.

***Fig. 5.114:** 4-bit binary adder/subtractor.*

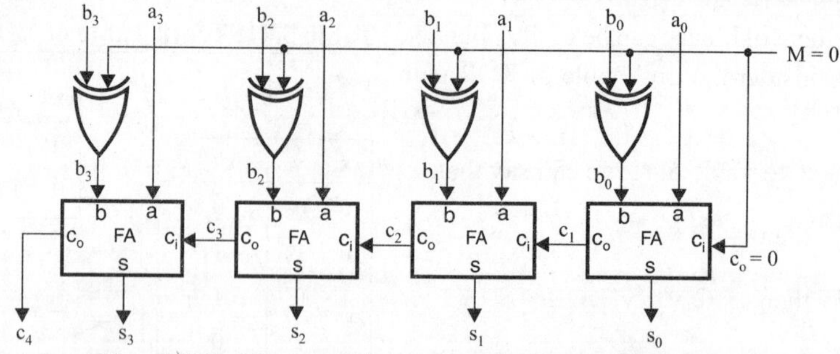

Fig. 5.115: Binary addition using adder/subtractor.

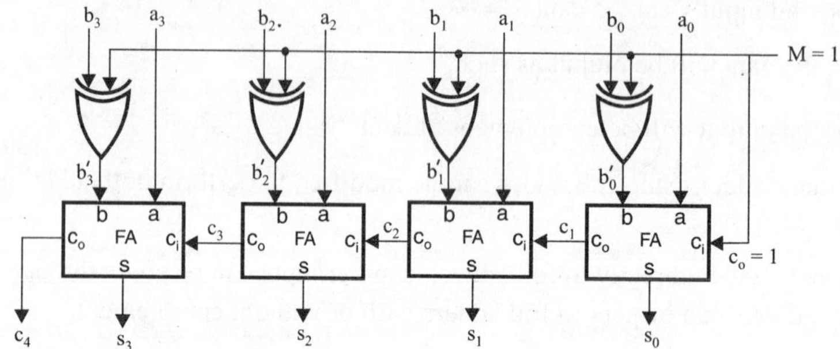

Fig. 5.116: Binary subtraction using adder/subtractor.

***Note:** Sum is difference and carry is borrow.*

5.11.2 BCD Subtractor

(JNTUK, Feb'22, 8 Marks)

BCD subtraction is performed by 9's complement subtraction. (Refer Chapter-2, Section 2.4.2 for 9's complement BCD subtraction). The digital logic circuit to perform 9's complement BCD subtraction is shown in Fig. 5.117. In this circuit a 4-bit binary adder is used to obtain 9's complement of subtrahend. The 9's complement of subtrahend is obtained by taking 1's complement of subtrahend by logical XOR with 1 and adding to 1010_2. A 4-bit BCD adder is used to add the minuend and 9's complement of subtrahend.

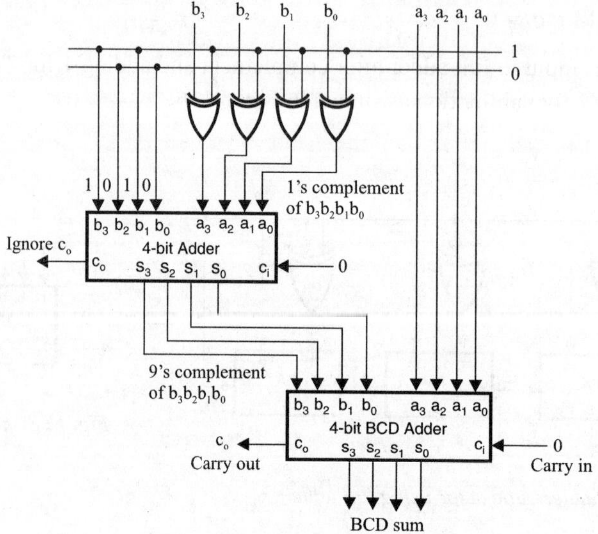

Fig. 5.117: BCD subtractor.

5.11.3 BCD Adder/Subtractor

(JNTUK, Nov/Dec'20, 8 Marks)

A mode control signal, M is introduced in the BCD subtractor as shown in Fig. 5.118, so that the BCD subtractor circuit can be made to perform both BCD addition and subtraction. For addition operation, M is tied to 0, so that 0 is added to subtrahend by the 4-bit binary adder and hence the subtrahend is passed to 4-bit BCD adder without 9's complement.

For subtraction, M is tied to 1, so that the 4-bit binary adder will add 1010_2 to 1's complement of subtrahend in order to generate 9's complement of subtrahend and pass to BCD adder. The BCD adder will add the 9's complement of subtrahend to minuend to perform 9's complement BCD subtraction.

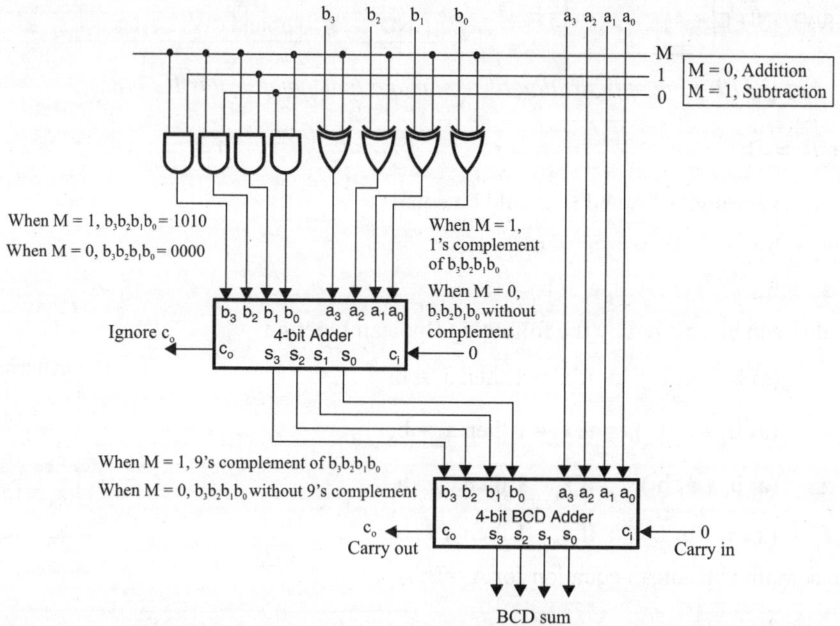

Fig. 5.118: BCD adder/subtractor.

5.12 Magnitude Comparator

(CSVTU, Nov/Dec'19, 7 Marks) (RGPV, Jun'20, 2 Marks)

Magnitude comparator is a combinational circuit used to compare two binary numbers and determine whether they are equal or unequal and if unequal then it can make a decision on larger or smaller magnitude.

5.12.1 4-bit Magnitude Comparator

(GTU, May'19, 7 Marks) (AKTU, Dec'22, 10 Marks)

(GTU, Jun'19, 4 Marks)

Consider two 4-bit binary numbers A and B.

Let, $A = a_3\ a_2\ a_1\ a_0$ and $B = b_3\ b_2\ b_1\ b_0$

On comparing the two numbers using magnitude comparator it is possible to determine any one of the following condition.

1. $A = B$
2. $A < B$
3. $A > B$

The 4-bit magnitude comparator is available as a standard IC with number 7485. The pin configuration of 7485 is shown in Fig. 5.119.

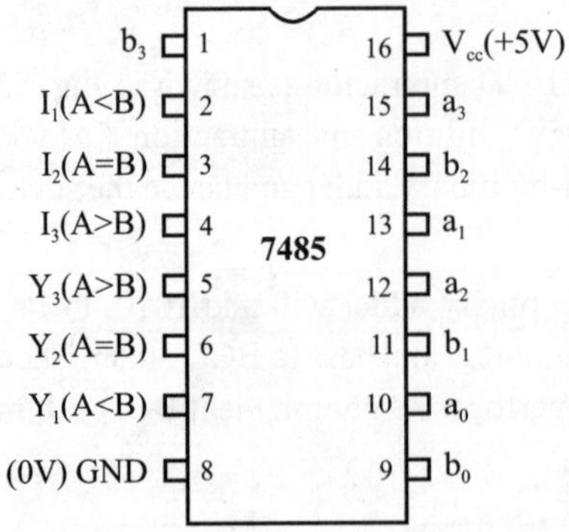

Left pins	Pin	7485	Pin	Right pins
b_3	1		16	V_{cc}(+5V)
I_1(A<B)	2		15	a_3
I_2(A=B)	3		14	b_2
I_3(A>B)	4		13	a_1
Y_3(A>B)	5		12	a_2
Y_2(A=B)	6		11	b_1
Y_1(A<B)	7		10	a_0
(0V) GND	8		9	b_0

Pin	Description
a_0 - a_3	4-bit data-A
b_0 - b_3	4-bit data-B
Y_1, Y_2, Y_3	Outputs
I_1, I_2, I_3	Cascading inputs
V_{cc}	Supply voltage (+5V)
GND	Ground (0V)

***Fig. 5.119:** Pin configuration of 4-bit magnitude comparator IC 7485.*

Checking for A = B

For, A = B, every digit of A and B should be equal.

i.e., $a_3 = b_3$ | $a_1 = b_1$

$a_2 = b_2$ | $a_0 = b_0$

The equality can be ensured by the following Boolean function.

Let, $x_3 = (a_3' b_3 + a_3 b_3')'$; if $x_3 = 1$ then $a_3 = b_3$

$x_2 = (a_2' b_2 + a_2 b_2')'$; if $x_2 = 1$ then $a_2 = b_2$

$x_1 = (a_1' b_1 + a_1 b_1')'$; if $x_1 = 1$ then $a_1 = b_1$

$x_0 = (a_0' b_0 + a_0 b_0')'$; if $x_0 = 1$ then $a_0 = b_0$

Alternatively

$x_3 = (a_3 \oplus b_3)'$

$x_2 = (a_2 \oplus b_2)'$

$x_1 = (a_1 \oplus b_1)'$

$x_0 = (a_0 \oplus b_0)'$

Now the combined Boolean equation for A = B, is,

$$x_3 x_2 x_1 x_0 = 1$$

Let, $F_1 = x_3 x_2 x_1 x_0$

Now, if A = B, then $F_1 = 1$

and, if $A \neq B$, then $F_1 = 0$

Every digit of the binary numbers can take two possible values 0 and 1. Therefore, the value of $x_3 = 1$, when the digits a_3 and b_3 take either, both 0 or both 1. This can be verified in Table 5.42. Similarly the equality of other bits when they take either, both 0 or both 1 are verified in Tables 5.42 to 5.45.

Table 5.42: Verification of x_3 for $a_3 = b_3$

a_3 b_3	a_3' b_3'	a_3' b_3	a_3 b_3'	$a_3' b_3 + a_3 b_3'$	$x_3 = (a_3' b_3 + a_3 b_3')'$
0 0	1 1	0	0	0	1
0 1	1 0	1	0	1	0
1 0	0 1	0	1	1	0
1 1	0 0	0	0	0	1

Table 5.43: Verification of x_2 for $a_2 = b_2$

a_2 b_2	a_2' b_2'	a_2' b_2	a_2 b_2'	a_2' $b_2 + a_2$ b_2'	$x_2 = (a_2'$ $b_2 + a_2$ $b_2')'$
0 0	1 1	0	0	0	1
0 1	1 0	1	0	1	0
1 0	0 1	0	1	1	0
1 1	0 0	0	0	0	1

Table 5.44: Verification of x_1 for $a_1 = b_1$

a_1 b_1	a_1' b_1'	a_1' b_1	a_1 b_1'	a_1' $b_1 + a_1$ b_1'	$x_1 = (a_1'$ $b_1 + a_1$ $b_1')'$
0 0	1 1	0	0	0	1
0 1	1 0	1	0	1	0
1 0	0 1	0	1	1	0
1 1	0 0	0	0	0	1

Table 5.45: Verification of x_0 for $a_0 = b_0$

a_0 b_0	a_0' b_0'	a_0' b_0	a_0 b_0'	a_0' $b_0 + a_0$ b_0'	$x_0 = (a_0'$ $b_0 + a_0$ $b_0')'$
0 0	1 1	0	0	0	1
0 1	1 0	1	0	1	0
1 0	0 1	0	1	1	0
1 1	0 0	0	0	0	1

The logic circuit to check only equality of two 4-bit binary number is shown in Fig. 5.120.

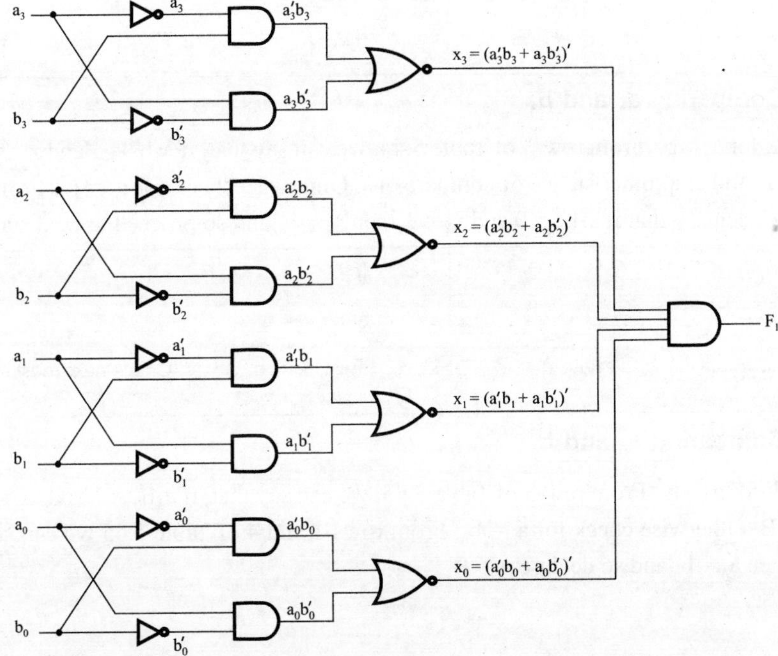

Fig. 5.120: *Logic circuit to check equality of two 4-bit binary data.*

Checking for A > B

In order to determine A > B, check bit-by-bit starting from most significant bit. If most significant bits of A and B are not equal and if most significant bit of A is greater than most significant bit of B then A > B.

If most significant bits are equal then proceed to next most significant bit and so on. For comparison of 4-digit binary numbers, 4 stages of comparison is required.

Stage-1: Comparing a_3 and b_3

First check for $a_3 > b_3$. From row-3 of Table 5.42, we can say that, if $a_3 b_3' = 1$ and $x_3 = 0$ then $a_3 > b_3$ and declare A > B and skip other stages of comparisons. Otherwise check for $a_3 = b_3$. From rows-1 and 4 of Table 5.42 we can say that if $a_3 b_3' = 0$ and $x_3 = 1$ then $a_3 = b_3$ and so proceed to next stage.

If $a_3 b_3' = 1$	If $a_3 b_3' = 0$
and $x_3 = 0$	and $x_3 = 1$
then $a_3 > b_3$; Declare A > B	then $a_3 = b_3$; Check next most significant bits

Stage-2: Comparing a_2 and b_2

Check for $a_2 > b_2$. From row-3 of Table 5.43, we can say that, if $a_2 b_2' = 1$ and $x_2 = 0$ then $a_2 > b_2$ and declare A > B and skip other stages of comparisons. Otherwise check for $a_2 = b_2$. From rows-1 and 4 of Table 5.43 we can say that if $a_2 b_2' = 0$ and $x_2 = 1$ then $a_2 = b_2$ and so proceed to next stage.

If $a_2 b_2' = 1$	If $a_2 b_2' = 0$
and $x_2 = 0$	and $x_2 = 1$
then $a_2 > b_2$; Declare A > B	then $a_2 = b_2$; Check next most significant bits

Stage-3: Comparing a_1 and b_1

Check for $a_1 > b_1$. From row-3 of Table 5.44, we can say that, if $a_1 b_1' = 1$ and $x_1 = 0$ then $a_1 > b_1$ and declare A > B and skip other stages of comparisons. Otherwise check for $a_1 = b_1$. From rows-1 and 4 of Table 5.44 we can say that if $a_1 b_1' = 0$ and $x_1 = 1$ then $a_1 = b_1$ and so proceed to next stage.

If $a_1 b_1' = 1$	If $a_1 b_1' = 0$
and $x_1 = 0$	and $x_1 = 1$
then $a_1 > b_1$; Declare A > B	then $a_1 = b_1$; Check next most significant bits

Stage-4: Comparing a_0 and b_0

Check for $a_0 > b_0$. From row-3 of Table 5.45, we can say that, if $a_0 b_0' = 1$ and $x_0 = 0$ then $a_0 > b_0$ and declare A > B. Otherwise check for $a_0 = b_0$. From rows-1 and 4 of Table 5.45 we can say that if $a_0 b_0' = 0$ and $x_0 = 1$ then $a_0 = b_0$ and so declare A = B.

If $a_0 b_0' = 1$	If $a_0 b_0' = 0$
and $x_0 = 0$	and $x_0 = 1$
then $a_0 > b_0$; Declare A > B	then $a_0 = b_0$; Declare A = B

Conclusion

Now the combined Boolean equation for A > B is,

$$F_2 = a_3\, b_3' + x_3\, a_2 b_2' + x_3 x_2 a_1 b_1' + x_3 x_2 x_1\, a_0\, b_0'$$

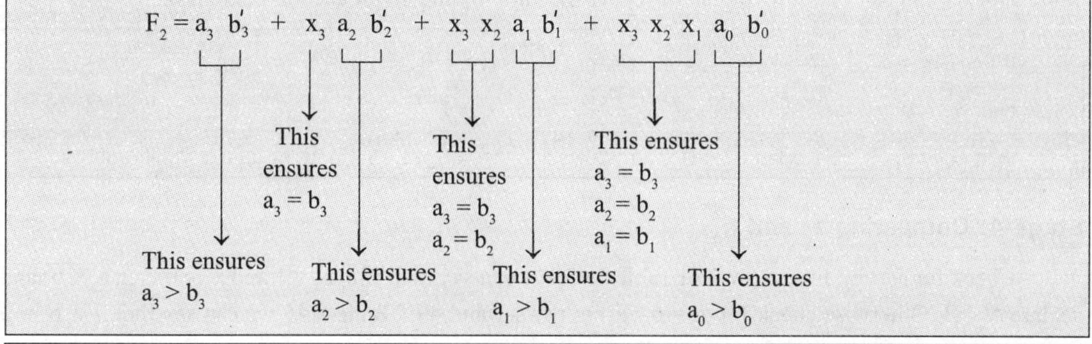

> ***Note:*** *In any stage of comparison if greater than or equality conditions are not satisfied then obviously A < B.*

Check for A < B

In order to determine A < B, check bit-by-bit starting from most significant bit. If most significant bits of A and B are not equal and if most significant bit of A is less than most significant bit of B then A < B.

If most significant bits are equal then proceed to next most significant bit and so on. For comparison of 4-digit binary numbers, 4 stages of comparison is required.

Stage-1: Comparing a_3 and b_3

First check for $a_3 < b_3$. From row-2 of Table 5.42, we can say that, if $a_3' b_3 = 1$ and $x_3 = 0$ then $a_3 < b_3$ and declare A < B and skip other stages of comparisons. Otherwise check for $a_3 = b_3$. From rows-1 and 4 of Table 5.42 we can say that if $a_3' b_3 = 0$ and $x_3 = 1$ then $a_3 = b_3$ and so proceed to next stage.

If $a_3' b_3 = 1$	If $a_3' b_3 = 0$
and $x_3 = 0$	and $x_3 = 1$
then $a_3 < b_3$; Declare A < B	then $a_3 = b_3$; Check next most significant bits

Stage-2: Comparing a_2 and b_2

Check for $a_2 < b_2$. From row-2 of Table 5.43, we can say that, if $a_2' b_2 = 1$ and $x_2 = 0$ then $a_2 < b_2$ and declare A < B and skip other stages of comparisons. Otherwise check for $a_2 = b_2$. From rows-1 and 4 of Table 5.43 we can say that if $a_2' b_2 = 0$ and $x_2 = 1$ then $a_2 = b_2$ and so proceed to next stage.

If $a_2' b_2 = 1$	If $a_2' b_2 = 0$
and $x_2 = 0$	and $x_2 = 1$
then $a_2 < b_2$; Declare $A < B$	then $a_2 = b_2$; Check next most significant bits

Stage-3: Comparing a_1 and b_1

Check for $a_1 < b_1$. From row-2 of Table 5.44, we can say that, if $a_1' b_1 = 1$ and $x_1 = 0$ then $a_1 < b_1$ and declare $A < B$ and skip other stages of comparisons. Otherwise check for $a_1 = b_1$. From rows-1 and 4 of Table 5.44 we can say that if $a_1' b_1 = 0$ and $x_1 = 1$ then $a_1 = b_1$ and so proceed to next stage.

If $a_1' b_1 = 1$	If $a_1' b_1 = 0$
and $x_1 = 0$	and $x_1 = 1$
then $a_1 < b_1$; Declare $A < B$	then $a_1 = b_1$; Check next most significant bits

Stage-4: Comparing a_0 and b_0

Check for $a_0 < b_0$. From row-2 of Table 5.45, we can say that, if $a_0' b_0 = 1$ and $x_0 = 0$ then $a_0 < b_0$ and declare $A < B$. Otherwise check for $a_0 = b_0$. From rows-1 and 4 of Table 5.45 we can say that if $a_0' b_0 = 0$ and $x_0 = 1$ then $a_0 = b_0$ and so declare $A = B$.

If $a_0' b_0 = 1$	If $a_0' b_0 = 0$
and $x_0 = 0$	and $x_0 = 1$
then $a_0 < b_0$; Declare $A < B$	then $a_0 = b_0$; Declare $A = B$

Conclusion

Now the combined Boolean equation for $A < B$ is,

$$F_3 = a_3' b_3 + x_3 a_2' b_2 + x_3 x_2 a_1' b_1 + x_3 x_2 x_1 a_0' b_0$$

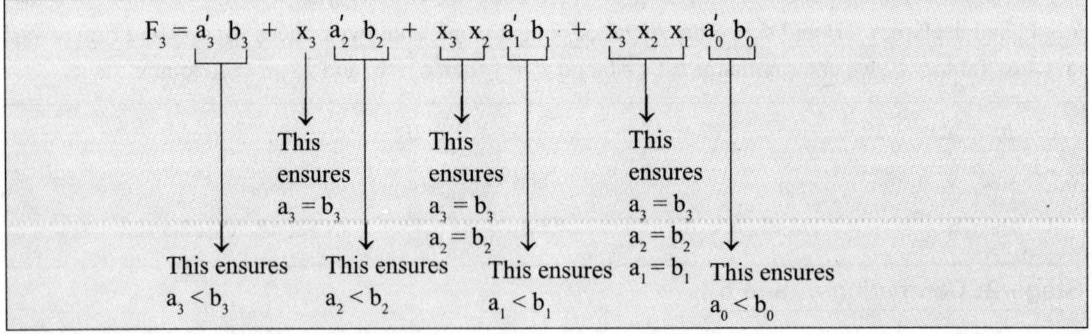

***Note:** In any stage of comparison if less than or equality conditions are not satisfied then obviously $A > B$.*

Logic Circuit to Check A = B or A > B or A < B

The following three Boolean equations are developed for the comparison of two 4-bit binary numbers.

$$F_1 = x_3 x_2 x_1 x_0 \quad ; \quad F_2 = a_3 b_3' + x_3 a_2 b_2' + x_3 x_2 a_1 b_1' + x_3 x_2 x_1 a_0 b_0'$$

$$F_3 = a_3' b_3 + x_3 a_2' b_2 + x_3 x_2 a_1' b_1 + x_3 x_2 x_1 a_0' b_0$$

where, $x_3 = (a_3 \oplus b_3)' = (a_3' b_3 + a_3 b_3')'$; $x_2 = (a_2 \oplus b_2)' = (a_2' b_2 + a_2 b_2')'$

$x_1 = (a_1 \oplus b_1)' = (a_1' b_1 + a_1 b_1')'$; $x_0 = (a_0 \oplus b_0)' = (a_0' b_0 + a_0 b_0')'$

Using the above equations the logic circuit to compare two 4-bit binary numbers is drawn as shown in Fig. 5.121. The circuit has three outputs, which can be interpretted as follows:

If $F_1 = 1$, then $A = B$; If $F_2 = 1$, then $A > B$; If $F_3 = 1$, then $A < B$

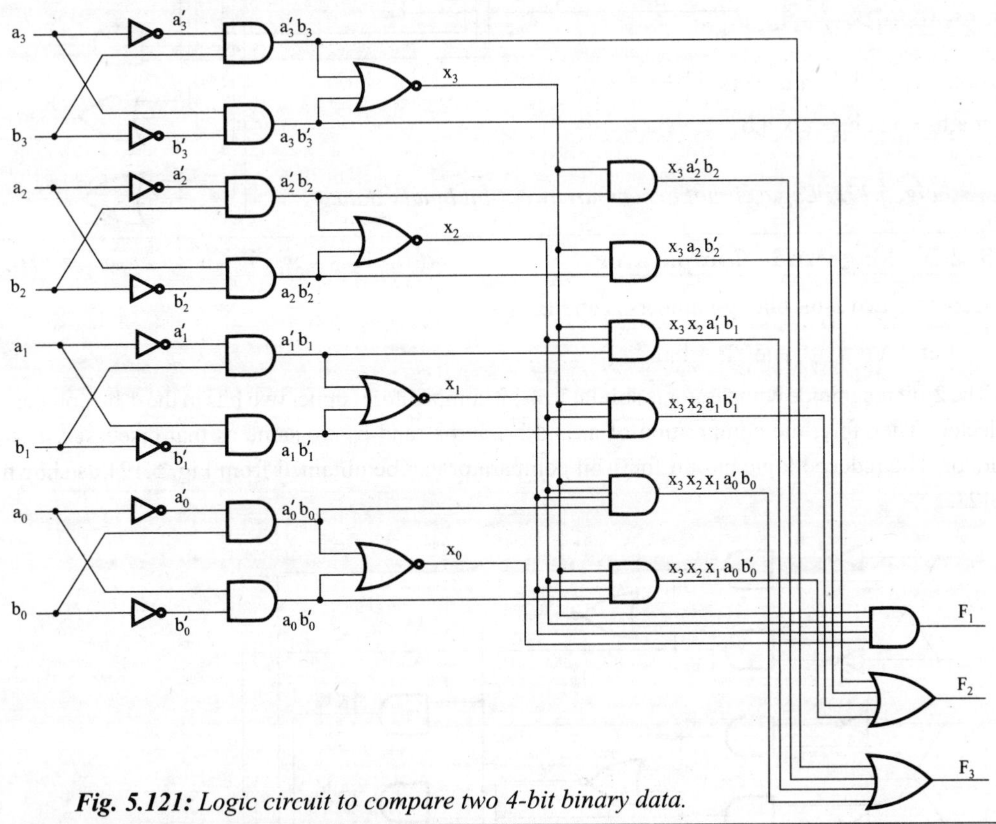

Fig. 5.121: Logic circuit to compare two 4-bit binary data.

5.12.2 3-bit Magnitude Comparator

Consider two 3-bit binary numbers A and B.

Let, $A = a_2 a_1 a_0$ and $B = b_2 b_1 b_0$

The 3-bit magnitude comparator is similar to 4-bit comparator if the MSB (Most Significant Bit) in the 4-bit comparator is neglected. Therefore, the comparition of a_2 and b_2, a_1 and b_1, and a_0 and b_0 are same as that discussed in 4-bit comparator. The reduced logic circuit for 3-bit comparator can be obtained from Fig. 5.121 as shown in Fig. 5.122.

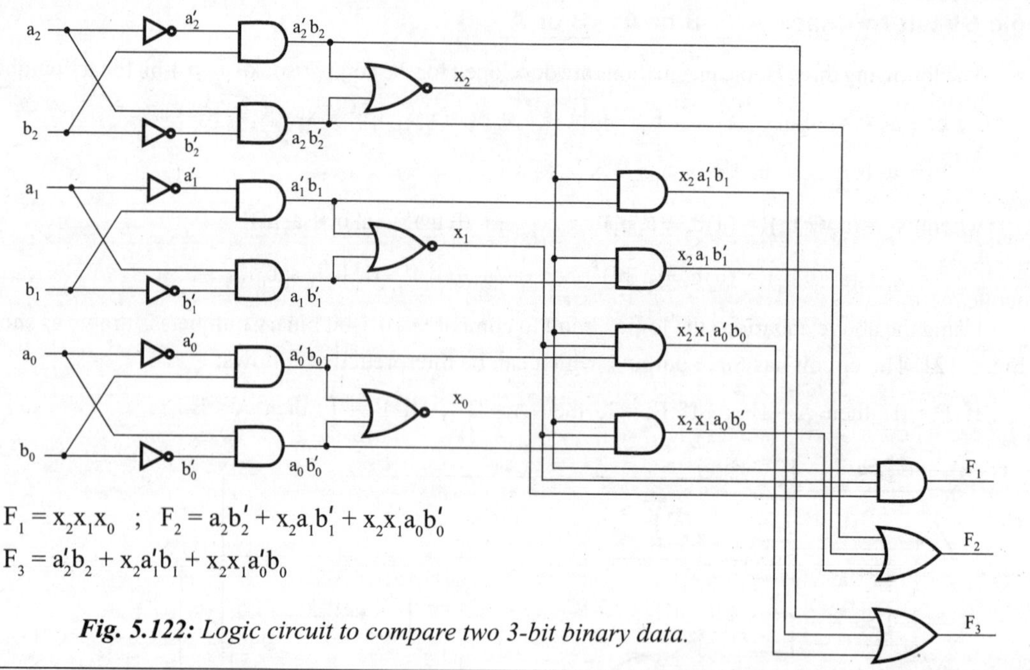

$F_1 = x_2x_1x_0$; $F_2 = a_2b_2' + x_2a_1b_1' + x_2x_1a_0b_0'$

$F_3 = a_2'b_2 + x_2a_1'b_1 + x_2x_1a_0'b_0$

Fig. 5.122: Logic circuit to compare two 3-bit binary data.

5.12.3 2-bit Magnitude Comparator

(VTU, Aug/Sep'20, 8 Marks) (RGPV, Jun'22, 5 Marks)

Consider two 2-bit binary numbers A and B.

Let, $A = a_1 a_0$ and $B = b_1 b_0$

The 2-bit magnitude comparator is similar to 4-bit comparator if upper two bits in the 4-bit comparator is neglected. Therefore, the comparition of a_1 and b_1, and a_0 and b_0 are same as that discussed in 4-bit comparator. The reduced logic circuit for 2-bit comparator can be obtained from Fig. 5.121 as shown in Fig. 5.123.

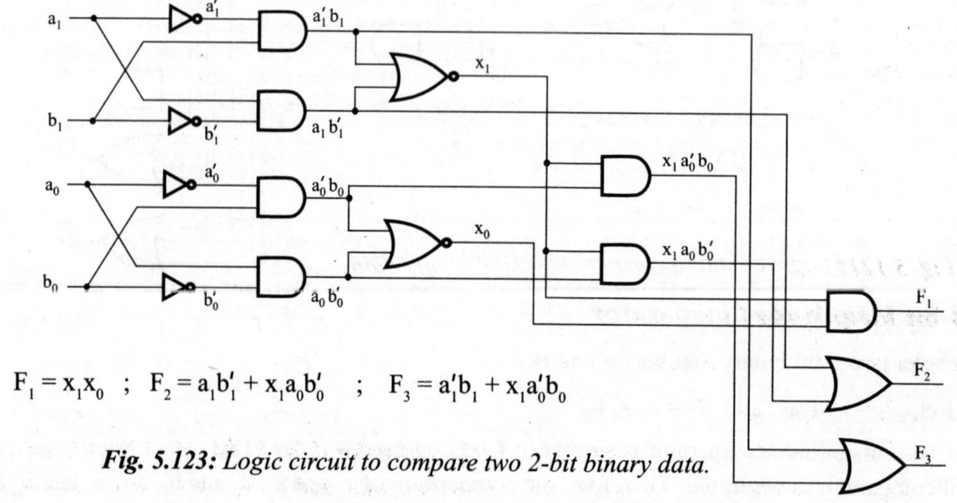

$F_1 = x_1x_0$; $F_2 = a_1b_1' + x_1a_0b_0'$; $F_3 = a_1'b_1 + x_1a_0'b_0$

Fig. 5.123: Logic circuit to compare two 2-bit binary data.

5.13 Combinational Multiplier

Combinational multiplier is used to multiply two unsigned binary numbers similar to decimal multiplication.

Multiplication of two binary numbers is performed by multiplying each bit of multiplier (starting from LSB) to multiplicand to generate partial products. The partial products are arranged such that a partial product is left shifted by one bit position from previous product. Now, the sum of shifted products is the (final) product of the two binary numbers. The multiplication of n-bit by n-bit binary numbers will generate a product of size 2n bits.

5.13.1 Combinational 2-bit Binary Multiplier

(JNTUA, Jun/Jul'22, 7 Marks)

Consider two 2-bit binary numbers. The multiplication of two 2-bit binary numbers will generate a 4-bit product.

Let, Multiplier $= x_1 x_0$

Multiplicand $= y_1 y_0$

Product $= p_3 p_2 p_1 p_0$

The process of multiplication of multiplicand by each bit of multiplier and addition of partial products is shown below:

Multiplication:

$$\begin{array}{rrrr} & & y_1 & y_0 \\ (\times) & & x_1 & x_0 \\ \hline & & x_0y_1 & x_0y_0 \\ & x_1y_1 & x_1y_0 & \\ \hline p_3 & p_2 & p_1 & p_0 \end{array}$$

Addition:

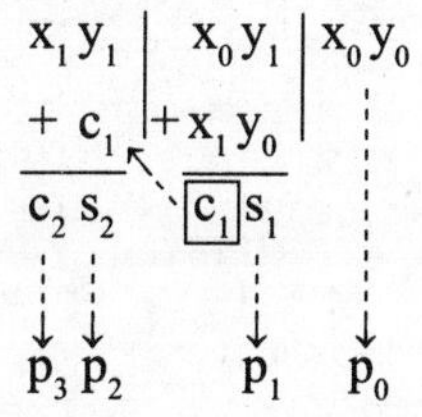

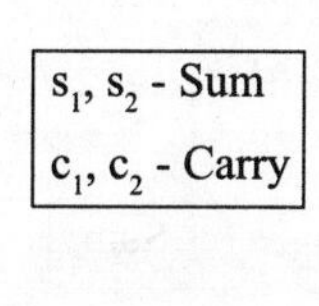

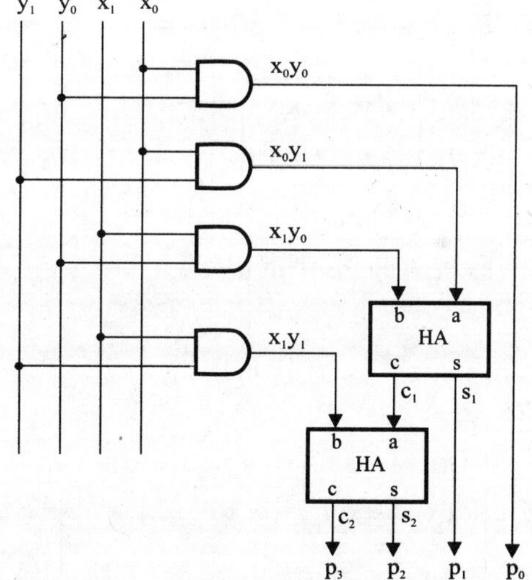

Fig. 5.124: 2-bit × 2-bit combinational multiplier.

A combinational logic circuit to perform the multiplication two 2-bit binary numbers is shown in Fig. 5.124. In this combinational multiplier, each multiplication of two binary bits is realized by a two input AND gate and the addition of partial products are realized by half adders.

5.13.2 Combinational 4-bit Binary Multiplier

(VTU, Aug/Sep'20, 10 Marks)

Consider two 4-bit binary numbers. The multiplication of two 4-bit binary numbers will generate a 8-bit product.

Let, Multiplier $= x_3 x_2 x_1 x_0$

Multiplicand $= y_3 y_2 y_1 y_0$

Product $= p_7 p_6 p_5 p_4 p_3 p_2 p_1 p_0$

The process of multiplication of multiplicand by each bit of multiplier and addition of partial products is shown below:

Multiplication:

$$
\begin{array}{cccccccc}
 & & & & y_3 & y_2 & y_1 & y_0 \\
 & & & & x_3 & x_2 & x_1 & x_0 \\
\hline
 & & & & x_0y_3 & x_0y_2 & x_0y_1 & x_0y_0 \\
 & & & x_1y_3 & x_1y_2 & x_1y_1 & x_1y_0 & \\
 & & x_2y_3 & x_2y_2 & x_2y_1 & x_2y_0 & & \\
 & x_3y_3 & x_3y_2 & x_3y_1 & x_3y_0 & & & \\
\hline
p_7 & p_6 & p_5 & p_4 & p_3 & p_2 & p_1 & p_0 \\
\hline
\end{array}
$$

Addition-1:

$$
\begin{array}{c|c|c|c|c}
 & x_0y_3 & x_0y_2 & x_0y_1 & x_0y_0 \\
x_1y_3 & +\,x_1y_2 & +\,x_1y_1 & +\,x_1y_0 & \\
+\,c_{13} & c_{12} & c_{11} & & \\
\hline
c_{14}\,s_{14} & \boxed{c_{13}}\,s_{13} & \boxed{c_{12}}\,s_{12} & \boxed{c_{11}}\,s_{11} & x_0y_0 \\
 & & & \downarrow & \downarrow \\
 & & & p_1 & p_0
\end{array}
$$

$s_{11}, s_{12}, s_{13}, s_{14}$ - Sum

$c_{11}, c_{12}, c_{13}, c_{14}$ - Carry

Addition-2:

$$
\begin{array}{c|c|c|c}
c_{14} & s_{13} & s_{12} & s_{11} \\
+\,x_2y_3 & +\,x_2y_2 & +\,x_2y_1 & +\,x_2y_0 \\
c_{23} & c_{22} & c_{21} & \\
\hline
c_{24}\,s_{24} & \boxed{c_{23}}\,s_{23} & \boxed{c_{22}}\,s_{22} & \boxed{c_{21}}\,s_{21} \\
 & & & \downarrow \\
 & & & p_2
\end{array}
$$

$s_{21}, s_{22}, s_{23}, s_{24}$ - Sum

$c_{21}, c_{22}, c_{23}, c_{24}$ - Carry

A combinational logic circuit to perform the multiplication two 4-bit binary numbers is shown in Fig. 5.125. In this combinational multiplier, each multiplication of two binary bits is realized by a two input AND gate and the addition of partial products are realized by 4-bit binary adders.

Addition-3:

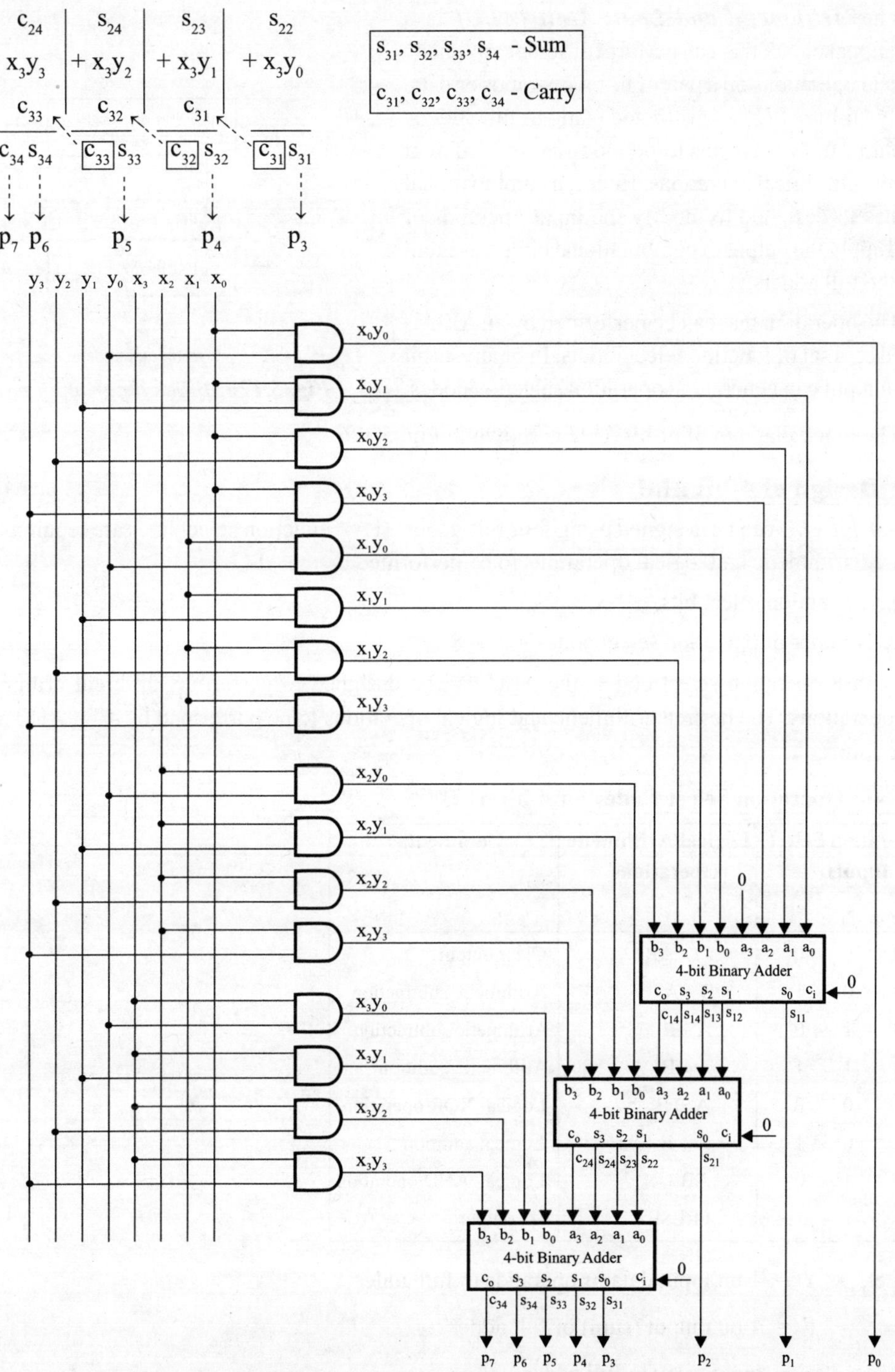

***Fig. 5.125:** 4-bit × 4-bit combinational multiplier.*

5.14 Arithmetic and Logic Unit

The ***Arithmetic and Logic Unit (ALU)*** is a combinational circuit that can perform different arithmetic and logical operations on a pair of m-bit operands or data. Basically, m-bit ALU is constructed using m full adders. The input to full adders has to be modified for different arithmetic and logical operations. Hence, a combinational circuit has to designed to modify the input operands or data and apply the output of combinational circuit as input data to the full adders.

The operation that can be performed by an ALU is specified by a set of function select inputs. In binary, n-bit selection input can generate 2^n operation selection codes.

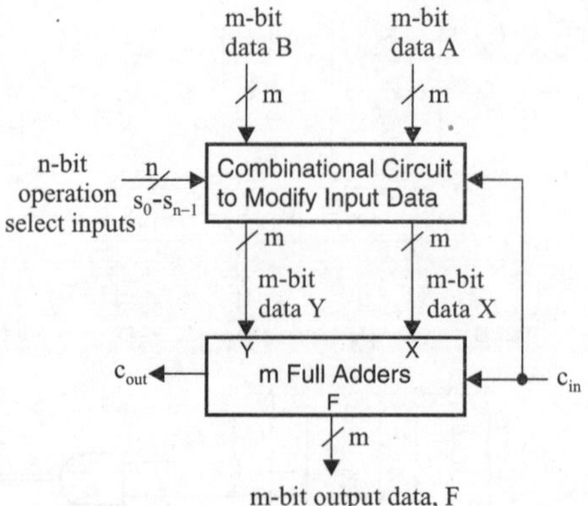

Fig. 5.126: Block diagram of m-bit ALU.

The block diagram of m-bit ALU is shown in Fig. 5.126.

5.14.1 Design of 4-bit ALU

(GTU, Jun'19, 4 Marks)

A ***4-bit ALU*** can be designed using four full-adders (FA). Function select lines are required to select the desired arithmetic and logical operations to be performed by the ALU.

Let, function select bits, n = 3

$\therefore$ Number of operation select codes = $2^3 = 8$

With 8 operation select codes, the ALU can be designed to perform 8 different arithmetic and logical operations. The desired arithmetic and logical operations to be performed by 4-bit ALU are listed in Table 5.46.

Table 5.46: Operation Select Codes for 4-bit ALU

Operation Select Inputs			Logical/Arithmetic Operations	Comments
s_2	s_1	s_0		
0	0	0	CLEAR	Clear output
0	0	1	B – A	Arithmetic subtraction
0	1	0	A – B	Arithmetic subtraction
0	1	1	A + B	Arithmetic addition
1	0	0	$A \oplus B$	Logical XOR operation
1	0	1	A + B	Logical addition
1	1	0	AB	Logical AND operation
1	1	1	PRESET	Set output

Let, X, Y = 4-bit input data or operands to full adder

F = 4-bit output (sum) of full adder

c_{in} = Input carry to full adder

Now, the output (sum) of full adder is given by XOR of the input operands. (Refer section 5.9.2)

$\therefore F = X \oplus Y \oplus c_{in}$

If x_j, y_j and f_j are j^{th} bit of input and output, the above equation can be modified as shown below:

$f_j = x_j \oplus y_j \oplus c_{i\_j}$

where, j = 0, 1, 2, 3

c_{i_j} = Input carry of jth full adder

Let, A, B = 4-bit input data or operands to ALU

a_0, a_1, a_2, a_3 = 4-bits of data-A

b_0, b_1, b_2, b_3 = 4-bits of data-B

An excitation table is developed for input to full adders for various logical and arithmetic operations and listed in Table 5.47.

The selection input bit s_2 is used to decide the arithmetic or logical operations. When $s_2 = 0$, an arithmetic operation is performed based on the value of s_1 and s_0. Similarly, when $s_2 = 1$, a logical operation is performed.

Table 5.47: Excitation Table of Full Adders

Operation Select Inputs			Minterm	Full Adder Inputs			Output of Full Adder	Operation Performed by ALU
				Carry	Data			
s_2	s_1	s_0		c_{i_j}	x_j	y_j	$f_j = x_j \oplus y_j \oplus c_{i\_j}$	
0	0	0	m_0	c_{i_j}	a_j	a_j	$f_j = a_j \oplus a_j \oplus c_{i\_j}$	CLEAR output, F = 0000
0	0	1	m_1	c_{i_j}	b_j	a'_j	$f_j = b_j \oplus a'_j \oplus c_{i\_j}$	Arithmetic subtraction, F = B – A
0	1	0	m_2	c_{i_j}	a_j	b'_j	$f_j = a_j \oplus b'_j \oplus c_{i\_j}$	Arithmetic subtraction, F = A – B
0	1	1	m_3	c_{i_j}	a_j	b_j	$f_j = a_j \oplus b_j \oplus c_{i\_j}$	Arithmetic addition, F = A + B
1	0	0	m_4	0	a_j	b_j	$f_j = a_j \oplus b_j$	Logical XOR operation, $F = A \oplus B$
1	0	1	m_5	0	$a_j + b_j$	0	$f_j = (a_j + b_j) \oplus 0$	Logical OR operation, F = A + B
1	1	0	m_6	0	$a_j + b'_j$	b'_j	$f_j = (a_j + b'_j) \oplus b'_j$	Logical AND operation, F = A . B
1	1	1	m_7	0	a_j	a_j	$f_j = a_j \oplus a'_j$	SET output, F = 1111

Explanation for Full Adder Inputs in Table 5.47

(GTU, Jun'20, 4 Marks)

Case i: $s_2 s_1 s_0 = 000$, CLEAR output, F = 0000

The XOR operation of a data with the data itself will generate output 0. Hence for clearing the output one of the data is XORed with that data itself.

$s_2 s_1 s_0$	Minterm	$s_2 s_1 s_0$	Minterm
0 0 0	m_0	1 0 0	m_4
0 0 1	m_1	1 0 1	m_5
0 1 0	m_2	1 1 0	m_6
0 1 1	m_3	1 1 1	m_7

Case ii: $s_2 s_1 s_0 = 001$, Arithmetic subtraction, B – A

The 2's complement subtraction is performed by adding minuend to 2's complement of subtrahend. Hence b_j is applied as x input and a_j is complemented and applied as y input.

Case iii: $s_2 s_1 s_0 = 010$, Arithmetic subtraction, A – B

The 2's complement subtraction is performed by adding minuend to 2's complement of subtrahend. Hence a_j is applied as x input and b_j is complemented and applied as y input.

Note: For 2's complement of subtrahend, a 1 is added to least significant bit of y_j (y_0).

Case iv: $s_2s_1s_0 = 011$, Arithmetic addition, A + B

The XOR of the two data will generate the output sum of adder. Hence a_j and b_j are applied directly as two inputs of full adder.

Case v: $s_2s_1s_0 = 100$, Logical XOR operation, $A \oplus B$

The full adder sum is given by XOR operation and so the input data are directly applied to input of full adders.

Case vi: $s_2s_1s_0 = 101$, Logical OR operation, A + B

The logical OR operation is performed externally and applied as one of the input to full adder and the other input to full adder is 0.

Case vii: $s_2s_1s_0 = 110$, Logical AND operation, A . B

It can be proved that XOR operation of $(a_j + b_j')$ and b_j' will be same as AND operation of a_j and b_j. Hence logical OR operation of a_j and b_j' is performed externally and applied as one of the input to full adder and the other input is b_j'.

Proof:

$$\therefore f_j = (a_j + b_j') \oplus b_j'$$

$$= (a_j + b_j')(b_j')' + (a_j + b_j')' b_j' \qquad \boxed{x \oplus y = xy' + x'y}$$

$$= (a_j + b_j') b_j + a_j' b_j b_j' \qquad \boxed{(x + y)' = x'y'}$$

$$= a_j b_j + b_j b_j' + 0 \qquad \boxed{b_j . b_j' = 0}$$

$$= a_j b_j + 0 + 0$$

$$= a_j b_j$$

Case viii: $s_2s_1s_0 = 111$, SET output, F = 1111

The XOR operation of a data and its complement will generate output 1. Hence to set the output, one of the data and its complement are applied as two inputs to the full adder.

K-maps are constructed for full adder inputs x_j and y_j as shown in Figs. 5.127 and 5.128. Here, the minterms are formed using selection inputs s_2, s_1 and s_0.

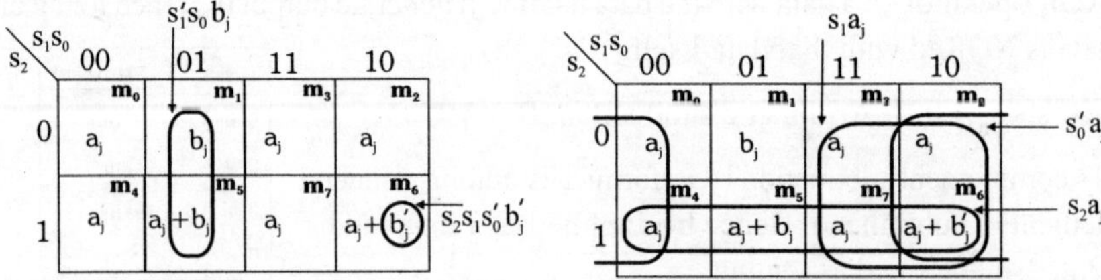

Fig. a: K-map to form prime implicants using b_j.

Fig. b: K-map to form prime implicants using a_j.

***Fig. 5.127:** K-map for x_j.*

From the K-map we get the following boolean equations for x_j and y_j

$$x_j = s_0'a_j + s_1a_j + s_2a_j + s_1's_0b_j + s_2s_1s_0'b_j'$$

$$= a_j(s_0' + s_1 + s_2) + s_1's_0b_j + s_2s_1s_0'b_j'$$

$$y_j = s_1s_0'b_j' + s_2's_1's_0'a_j + s_2's_1's_0a_j' + s_2s_1s_0a_j' + s_2's_1s_0b_j + s_2s_1's_0'b_j$$

$$= s_1s_0'b_j' + s_2's_1's_0'a_j + a_j'(s_2's_1's_0 + s_2s_1s_0) + b_j(s_2's_1s_0 + s_2s_1's_0')$$

For i = 0, 1, 2, 3, the x_j is given by,

$$x_0 = a_0(s_0' + s_1 + s_2) + s_1's_0b_0 + s_2s_1s_0'b_0'$$

$$x_1 = a_1(s_0' + s_1 + s_2) + s_1's_0b_1 + s_2s_1s_0'b_1'$$

$$x_2 = a_2(s_0' + s_1 + s_2) + s_1's_0b_2 + s_2s_1s_0'b_2'$$

$$x_3 = a_3(s_0' + s_1 + s_2) + s_1's_0b_3 + s_2s_1s_0'b_3'$$

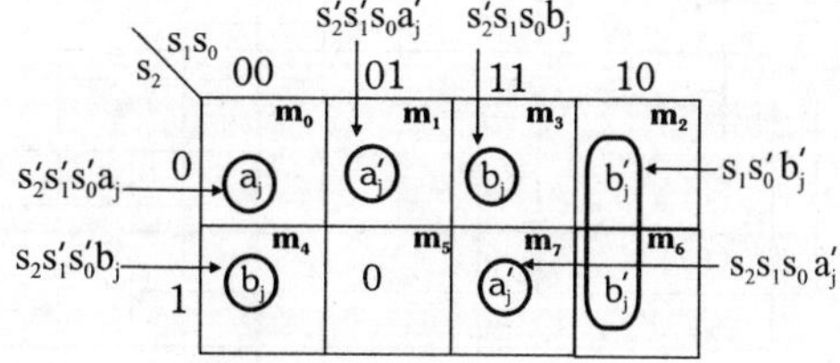

Fig. 5.128: K-map for y_j.

For i = 0, 1, 2, 3, the y_j is given by,

$$y_0 = s_1s_0'b_0' + s_2's_1's_0'a_0 + a_0'(s_2's_1's_0 + s_2s_1s_0) + b_0(s_2's_1s_0 + s_2s_1's_0') + s_2's_1's_0 + s_2's_1s_0'$$

$$y_1 = s_1s_0'b_1' + s_2's_1's_0'a_1 + a_1'(s_2's_1's_0 + s_2s_1s_0) + b_1(s_2's_1s_0 + s_2s_1's_0')$$

$$y_2 = s_1s_0'b_2' + s_2's_1's_0'a_2 + a_2'(s_2's_1's_0 + s_2s_1s_0) + b_2(s_2's_1s_0 + s_2s_1's_0')$$

$$y_3 = s_1s_0'b_3' + s_2's_1's_0'a_3 + a_3'(s_2's_1's_0 + s_2s_1s_0) + b_3(s_2's_1s_0 + s_2s_1's_0')$$

> For 2's complement of subtrahend a 1 is added to y_0 when $s_2's_1's_0 = 1$ for B – A and $s_2's_1s_0' = 1$ for A – B.

For mathematical operations, the carry propagates from one full adder to another but for logical operations the carry propagation is disabled.

Hence the complement of function select input s_2 (s_2') is logically ANDed with carry output of one full adder and connected to carry input of next full adder.

Let, $c_{i\_0}, c_{i\_1}, c_{i\_2}, c_{i\_3}$ = Input carry of full adders.

Let, $c_{o\_0}, c_{o\_1}, c_{o\_2}, c_{o\_3}$ = Output carry of full adders.

Now, the equation for input carry of full adders are,

$c_{i\_0} = s_2'c_{in}$	$c_{i\_2} = s_2'c_{o\_1}$	$c_{out} = s_2'c_{o\_3}$
$c_{i\_1} = s_2'c_{o\_0}$	$c_{i\_3} = s_2'c_{o\_2}$	

The logic diagram of 4-bit ALU is drawn as shown in Fig. 5.129, using the equations for x_j, y_j and c_{i_j}

> ***Note:*** *The equations for 4-bit ALU can be extended to 8-bit ALU by taking values of j from 0 to 7. Using these equations the logic circuit for 8-bit ALU can be drawn by taking 8 full adders. The logic circuit will be cascading of two 4-bit ALUs shown in Fig. 5.129.*

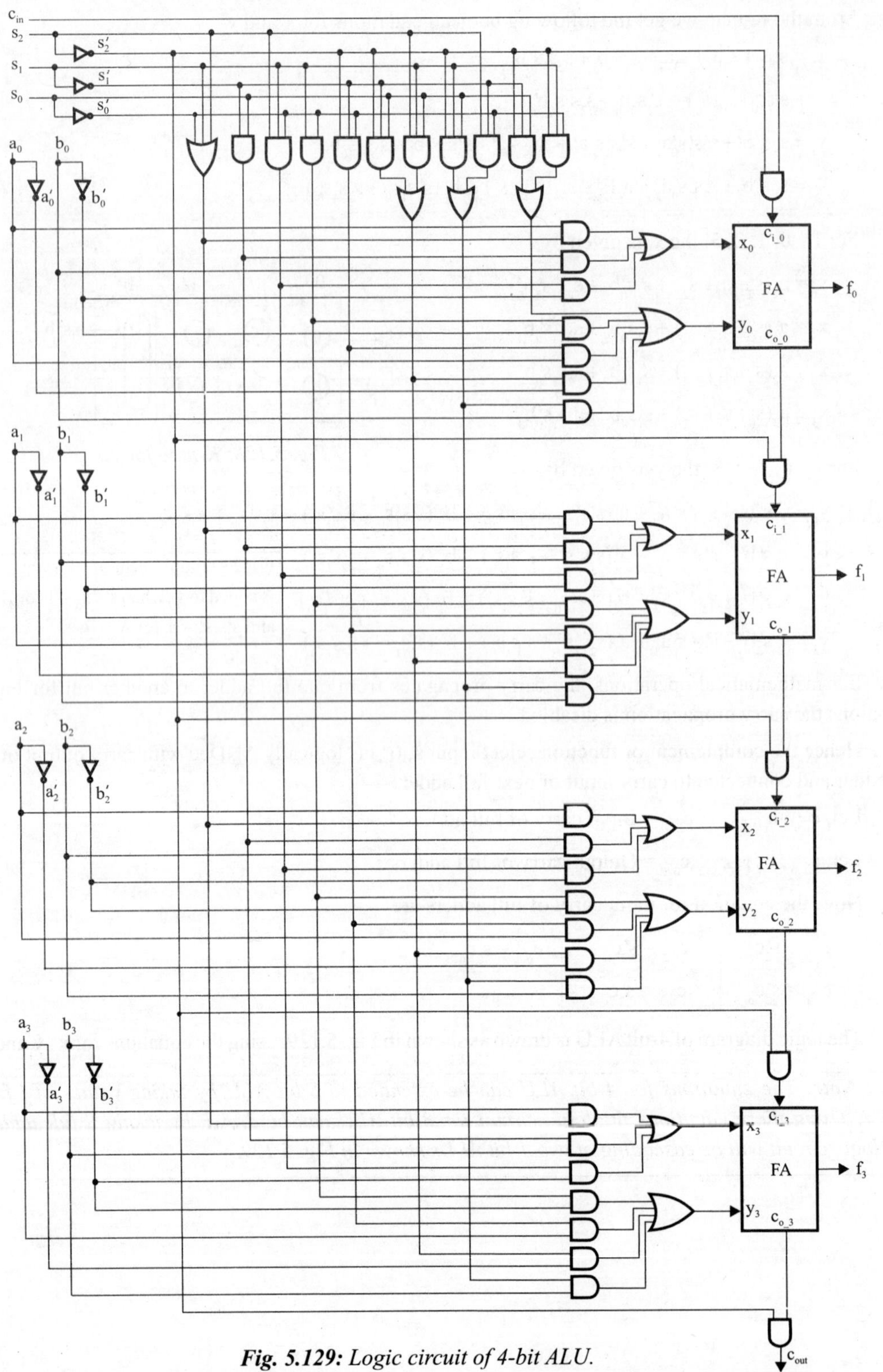

Fig. 5.129: Logic circuit of 4-bit ALU.

5.14.2 ALU using Standard IC's

A 4-bit ALU that can perform 8 different arithmetic and logical operations are available as standard IC 74381/74382. A 4-bit ALU that can perform 16 different arithmetic and logical operations are available as standard IC 74181. These 4-bit ALU's can be cascaded to form 8-bit and 16-bit ALU's.

5.14.3 8-bit and 16-bit ALU using 74381/382

The pin configuration of 4-bit ALU IC 74381 and 74382 are shown in Figs. 7.130 and 7.131 and the function table is shown in Table 5.48. The function table is same for ICs 74381 and 74382.

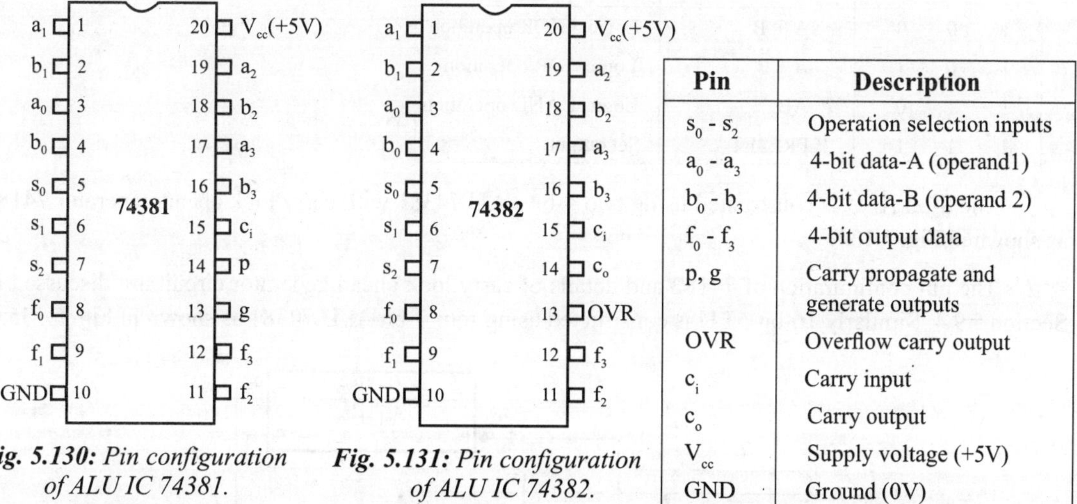

Fig. 5.130: *Pin configuration of ALU IC 74381.*

Fig. 5.131: *Pin configuration of ALU IC 74382.*

Pin	Description
s_0 - s_2	Operation selection inputs
a_0 - a_3	4-bit data-A (operand1)
b_0 - b_3	4-bit data-B (operand 2)
f_0 - f_3	4-bit output data
p, g	Carry propagate and generate outputs
OVR	Overflow carry output
c_i	Carry input
c_o	Carry output
V_{cc}	Supply voltage (+5V)
GND	Ground (0V)

Note: *1. Carry propagate and generate are not available in 74382.*
2. OVR and c_o are not available in 74381.

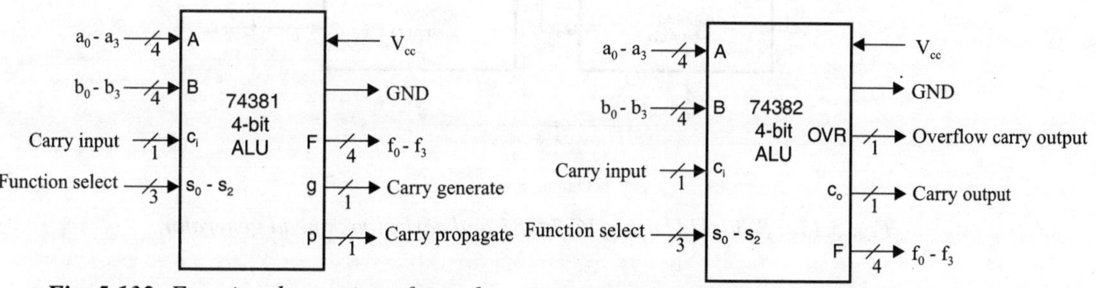

Fig. 5.132: *Functional grouping of signals of 4-bit ALU IC 74381.*

Fig. 5.133: *Functional grouping of signals of 4-bit ALU IC 74382.*

Table 5.48: Function Table of IC 74381/IC 74382

Operation Select Inputs			Logic/Arithmetic Operations	Comments
s_2	s_1	s_0		
0	0	0	CLEAR	Clear output
0	0	1	B – A	Arithmetic subtraction
0	1	0	A – B	Arithmetic subtraction
0	1	1	A + B	Arithmetic addition
1	0	0	A ⊕ B	Logical XOR operation
1	0	1	A + B	Logical OR operation
1	1	0	AB	Logical AND operation
1	1	1	PRESET	Set output

The 8-bit ALU is constructed using two 4-bit ALU 74381 with carry look ahead generator 74182 as shown in Fig. 5.134.

The pin configuration of 74182 and details of carry look ahead generator circuit are discussed in Section 5.9.4. Similarly 16-bit ALU is constructed using four 4-bit ALU 74381 as shown in Fig. 5.135.

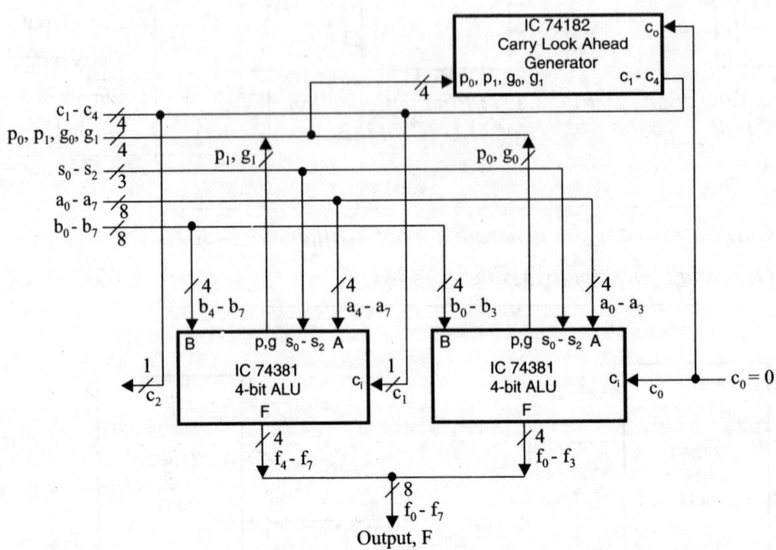

Fig. 5.134: 8-bit ALU using IC 74381 and carry look ahead generator.

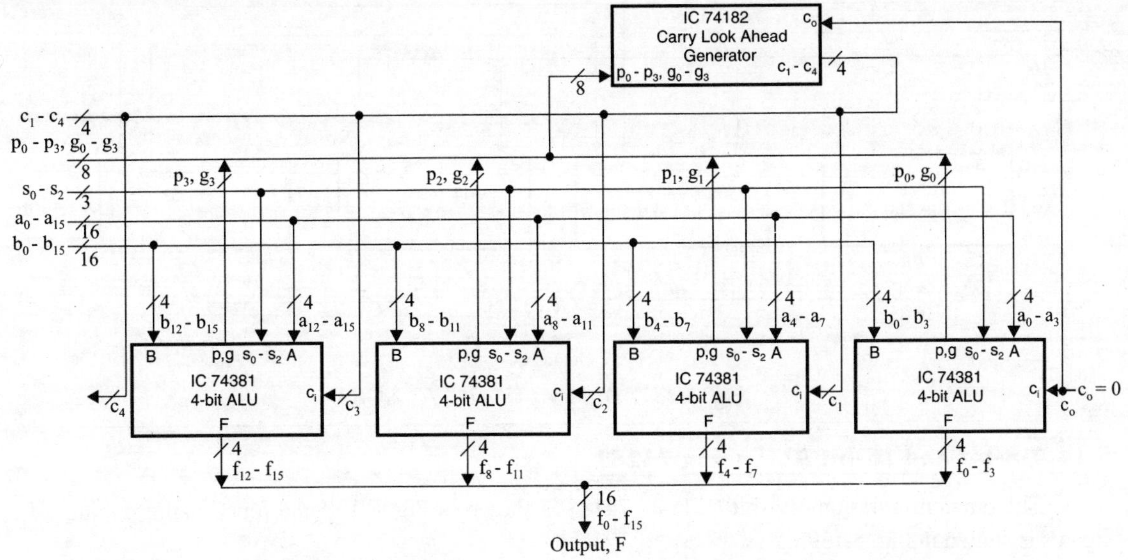

***Fig. 5.135:** 16-bit ALU using IC 74381 and carry look ahead generator.*

The ripple carry 8 and 16-bit ALUs without using carry look ahead generator can be constructed using 4-bit ALU 74382 as shown in Figs. 5.136 and 5.137. The ripple carry ALUs will take longer delay to perform computation when compared to ALUs with carry look ahead generator.

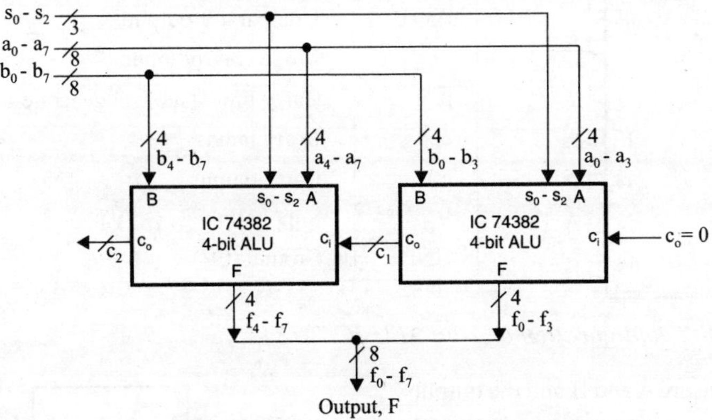

***Fig. 5.136:** 8-bit ALU using IC 74382.*

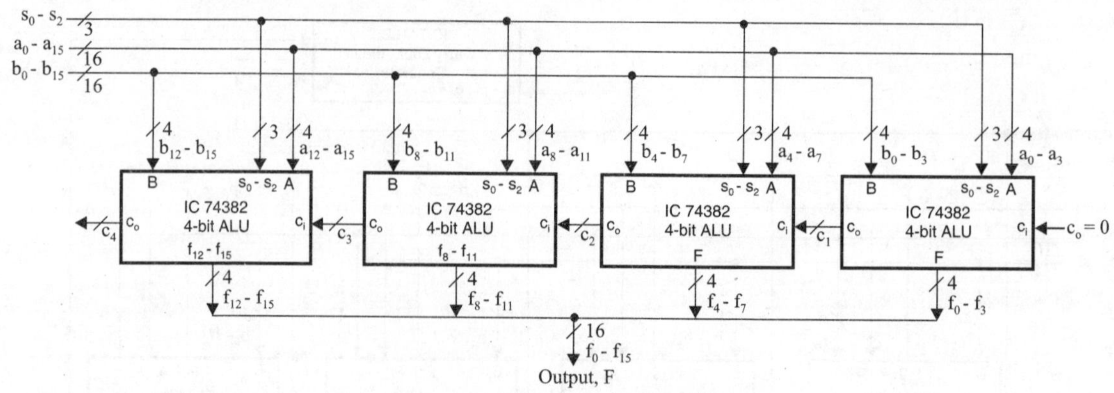

***Fig. 5.137:** 16-bit ALU using IC 74382.*

5.14.4 8-bit and 16-bit ALU using 74181

The pin configuration of 4-bit ALU IC 74181 is shown in Fig. 5.138 and functional grouping of signals is shown in Fig. 5.139.

Signal	Pin	74181	Pin	Signal
b_0	1		24	V_{cc}(+5V)
a_0	2		23	a_1
s_3	3		22	b_1
s_2	4		21	a_2
s_1	5		20	b_2
s_0	6		19	a_3
c_i	7		18	b_3
m	8		17	g
f_0	9		16	c_o
f_1	10		15	p
f_2	11		14	A = B
(0V) GND	12		13	f_3

Pin	Description
s_0 - s_3	Operation Selection Inputs
a_0 - a_3	4-bit data-A (operand 1)
b_0 - b_3	4-bit data-B (operand 2)
f_0 - f_3	4-bit output data
A = B	Comparator output
m	Mode control input
p, g	Carry Propagate and generate outputs
c_i	Carry input
c_o	Carry output
V_{cc}	Supply voltage (+5V)
GND	Ground (0V)

***Fig. 5.138:** Pin configuration of 4-bit ALU IC 74181.*

The two input data are A and B and the output data is F. The four bits of A is denoted as a_0, a_1, a_2 and a_3 and the four bits of B is denoted as b_0, b_1, b_2, b_3. The 4 bits of output data is denoted as f_0, f_1, f_2 and f_3.

The operation performed by 74181 is selected by m and s_0 - s_3 inputs. The mode control input m is used to select arithmetic or logical operations.

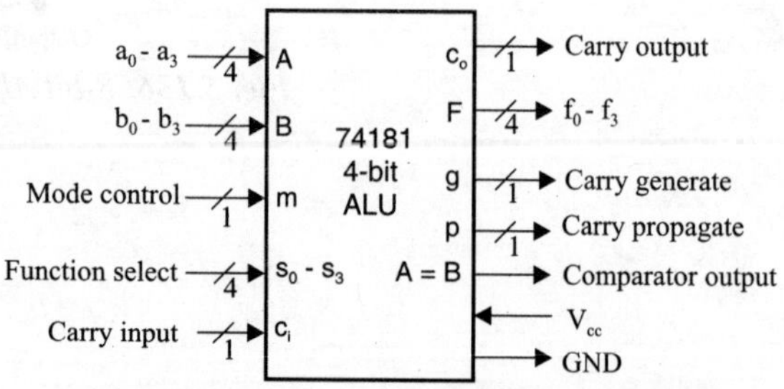

***Fig. 5.139:** Functional grouping of signals of 4-bit ALU IC74181.*

Case i: When m = 1, logical operations are selected and no carries propogate between stages and c_i is ignored. The logical operations selected by s_0 - s_3 bits are listed in Table 5.49.

Case ii: When m = 0, arithmetic operations are selected, carries propagate between the stages, and c_i is used as a carry input to the least significant stage. The arithmetic operations selected by s_0 - s_3 bits are listed in Table 5.49.

Table 5.49: Function Table of IC 74181

Operation Select Inputs				Logic Operations		Arithmetic Operations	
				Mode Input	**Comments**	**Mode Input**	**Comments**
s_3	s_2	s_1	s_0	**m = 1**		**m = 0**	
0	0	0	0	F = A′	Complement of A	F = A **+** c_i	Add c_i to A
0	0	0	1	F = (A + B)′	NOR operation	F = (A + B) **+** c_i	Add c_i to logical sum of A and B
0	0	1	0	F = A′B	F = 1, if A alone false	F = (A + B′) **+** c_i	F = 0 + c_i, if B alone true
0	0	1	1	F = 0000	Clear output	F = 0000	Clear output
0	1	0	0	F = (A B)′	NAND operation	F = A **+** AB′ **+** c_i	F = 1 + A + c_i, if A alone true
0	1	0	1	F = B′	Complement of B	F = (A + B) **+** AB′ **+** c_i	F = Logical sum of A and B + 1 + c_i, if A alone true
0	1	1	0	F = A ⊕ B	XOR operation	F = A – B – c_i	Arithmetic subtraction
0	1	1	1	F = A B′	F = 1, if A alone true	F = AB′ – c_i	F = 1 – c_i, if A alone true
1	0	0	0	F = A′ + B	F = 0, if B alone false	F = A **+** (A + B) **+** c_i	Add A and c_i to logical sum of A and B
1	0	0	1	F =(A ⊕ B)′	XNOR operation	F = A **+** B **+** c_i	Arithmetic sum
1	0	1	0	F = B	Output B	F = (A + B′)**+** AB	F = 0 + Logical product of A and B, if B alone true
1	0	1	1	F = A B	AND operation	F = AB – c_i	Subtract c_i from logical product of A and B
1	1	0	0	F = 1111	Set output	F = A **+** A **+** c_i	Arithmetic shift left
1	1	0	1	F = A + B′	F = 0, if B alone true	F = A B **+** A **+** c_i	Add A and c_i to logical product of A and B
1	1	1	0	F = A + B	OR operation	F = (A + B′) **+** A **+** c_i	F = 0 + A + c_i, if B alone true
1	1	1	1	F = A	Output A	F = A – c_i	Subtract c_i from A

Note: *Bold + (**+**) indicates arithmetic sum.*

The 8-bit ALU is constructed using two 4-bit ALU 74181 with carry look ahead adder 74182 as shown in Fig. 5.140.

The pin configuration of 74182 and details of carry look ahead adder circuit are discussed in Section 5.9.4. Similarly 16-bit ALU is constructed using four 4-bit ALU 74181 as shown in Fig. 5.141.

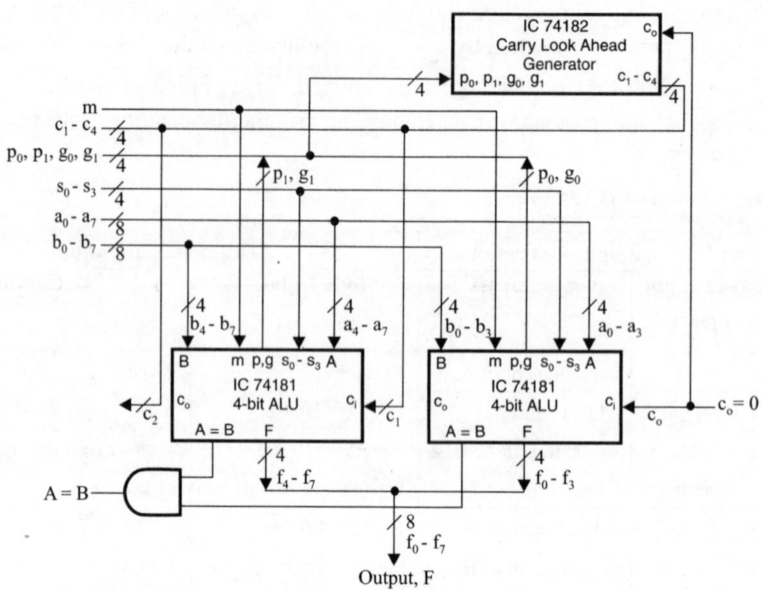

Fig. 5.140: *8-bit ALU using IC 74181 and carry look ahead generator.*

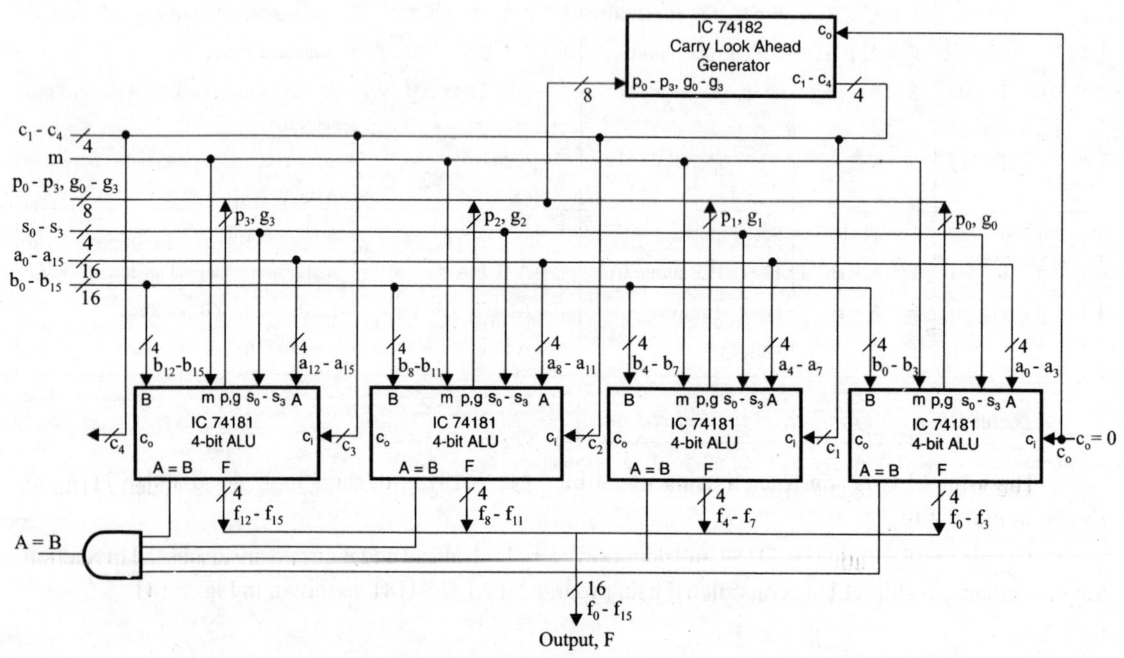

Fig. 5.141: *16-bit ALU using IC 74181 and carry look ahead generator.*

The ripple carry 8 and 16-bit ALUs without carry look ahead generator can also be constructed using 74181. The circuits will be similar to the ALUs shown in Figs. 5.134 and 5.135.

5.15 Combinational Systems Case Study

5.15.1 Digital Trans-Receiver

In digital systems the informations are coded in binary and stored or transmitted as a group of m-bit words or data. Each m-bit word or data can be transmitted either in parallel or serial. In parallel transmission m conducting lines are needed for data transmission but in serial transmission, the data can be transmitted bit-by-bit using single conducting line. In general, the data transmitters and receivers used in digital systems for data transmission and reception are called ***digital trans-receiver***.

5.15.2 Parallel Data Trans-Receiver

The digital data can be transmitted using data transeivers between two data buses in either direction. Consider two data buses A and B. A m-bit transeiver can transmit data from bus-A to bus-B and vice-versa as shown in Fig. 5.142.

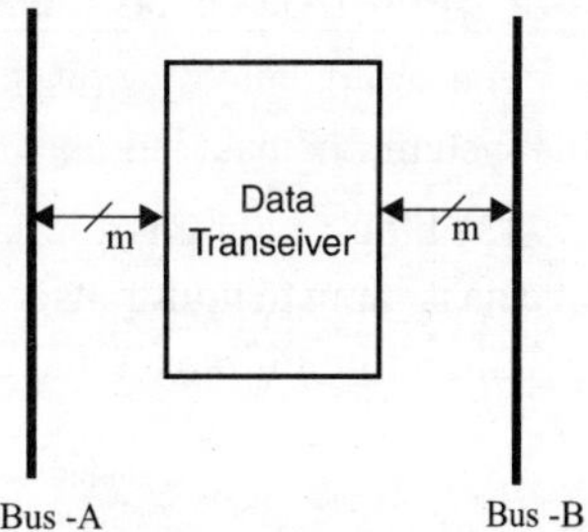

Fig. 5.142: *Parallel data transeiver.*

Data transeivers are available as standard ICs. IC 74243 is a 4-bit data transeiver and IC 74245 is an 8-bit data transeiver.The pin configuration of IC 74243 and IC 74245 are shown in Figs. 5.143 and 5.144 respectively.

Both these ICs can transmit parallel data either from A to B or B to A asynchronously. By proper choice of control inputs listed in function tables (Tables 5.50 and 5.51), either the data transmission from A to B or B to A can be acheived at a time.

74243

Pin	Name	Pin	Name
1	$\bar{E}_{AB}$	14	V_{cc}(+5V)
2	NC	13	E_{BA}
3	A_1	12	NC
4	A_2	11	B_1
5	A_3	10	B_2
6	A_4	9	B_3
7	(0V) GND	8	B_4

Pin	Description
A_1 - A_4	Data lines for bus-A
B_1 - B_4	Data lines for bus-B
E_{BA}, $\bar{E}_{AB}$	Enable input
V_{cc}	Supply voltage (+5V)
GND	Ground (0V)

Fig. 5.143: *Pin configuration of IC 74243.*

74245

Pin	Name	Pin	Name
1	DIR	20	V_{cc}(+5V)
2	A_1	19	$\bar{E}$
3	A_2	18	B_1
4	A_3	17	B_2
5	A_4	16	B_3
6	A_5	15	B_4
7	A_6	14	B_5
8	A_7	13	B_6
9	A_8	12	B_7
10	(0V) GND	11	B_8

Pin	Description
A_1 - A_8	Data lines for bus-A
B_1 - B_8	Data lines for bus-B
$\bar{E}$	Enable input
DIR	Direction control input
V_{cc}	Supply voltage (+5V)
GND	Ground (0V)

Fig. 5.144: *Pin configuration of IC 74245.*

Table 5.50: Function Table of IC 74243

$\overline{E}_{AB}$	E_{BA}	Operation
0	0	A to B transmission enabled
1	1	B to A transmission enabled
1	0	Isolation
0	1	Latch A and B, A = B

Table 5.51: Function Table of IC 74245

$\overline{E}$	DIR	Operation
0	0	B to A transmission enabled
0	1	A to B transmission enabled
1	×	Isolation

5.15.3 Serial Data Trans-Receiver

The serial data transmission within a system can be achieved using multiplexer and demultiplexer. In this system, the data is transmitted bit-by-bit serially.

The block diagram of serial data transmission using MUX and Demux is shown in Fig. 5.145. Here, m-bit data is input to multiplexer and transmitted bit-by-bit to demultiplexer. The selection or address input are common to multiplexer and demultiplexer.

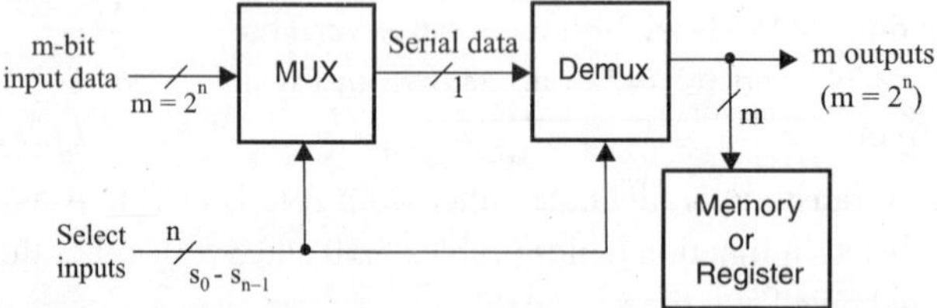

Fig. 5.145: Serial data trans-receiver using MUX and Demux.

The advantage in this transmission system is that the data can be transmitted using only one data line. But we require n select lines to transmit and receive data in the right order. This transmission system can be used to store data in a distant memory or register located in the same system.

Parity error checking can also be introduced in this system as shown in Fig. 5.146. The ***parity generator*** will generate a parity bit that can be appended with data as most significant bit. The ***parity checker*** will check the parity of received data and generate a error signal if there is a parity error in the transmitted data. The parity generator and receiver should be selected for same parity either both even parity or both odd parity.

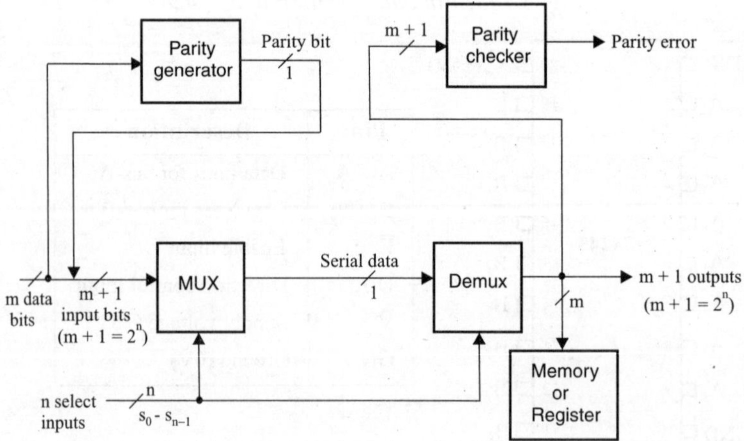

Fig. 5.146: Serial digital trans-receiver using MUX and Demux with parity bit.

Serial data transmission between two systems can be acheieved by a serial data transmitter and receiver connected via a cable or by any wireless communication method.

For bidirectional serial data transmission between two systems, each system should have a serial data transmitter and receiver as shown in Fig. 5.147. The serial data transmitter and receiver are constructed using shift registers. Please refer Chapter-8, Section 8.6, for shift registers.

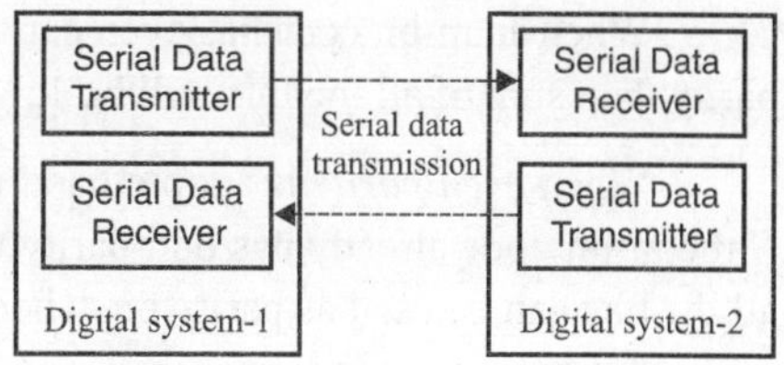

Fig. 5.147: Block diagram of serial data transmission between two systems.

The digital data trans-receiver using shift register is shown in Fig. 5.148. The transmitter consists of a transmit buffer, parity generator and parallel to serial shift register. The receiver consists of serial to parallel shift register, receive buffer and parity checker.

In this system, the digital information is coded as group of m-bit data or word. A parity bit is appended to each m-bit data and the m+1 bit data is loaded in transmit buffer register which inturn load the data in parallel-to-serial shift register. The m+1 data will be shifted out bit-by-bit by the parallel to serial shift register for transmission to another system.

On the receiver side the m+1 bit data is received bit-by-bit and converted to parallel data using a serial to parallel shift register and the parallel data is loaded in receive buffer register. Also the parallel data is loaded in parity checker to check the parity of received data and generate an error signal if there is a parity error in transmission.

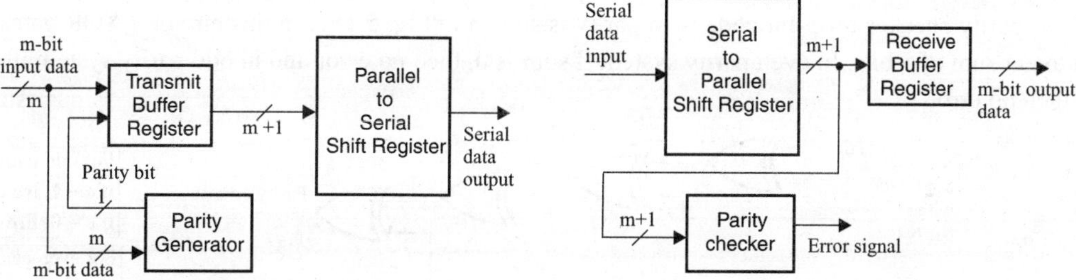

Fig. a: Serial data transmitter. *Fig. b: Serial data receiver.*

Fig. 5.148: Digital data trans-receiver using shift register.

5.15.4 Parity Generator and Checker

(RGPV, Jun'22, 7 Marks)

The concept of parity is discussed in Chapter-3, Section 3.5.1. The parity is used to check error in transmission of digital information.

During transmission, a parity bit is appended to m-bit code and the code along with parity bit is trasmitted from transmitter to the receiver. At the receiving end, the number of 1's in the code received are counted and the parity of received code is verified.

A ***parity generator/checker*** is a combinational logic circuit that generates the parity bit to be appended during transmission and at the receiving end it is used to verify the parity of received information.

For a m-bit code, the parity generation and checking is performed by adding all the m-bits of the code. In addition, the sum bit (or simply sum) is the least significant bit of the result and the other bits are called ***carry***.

When an m-bit code has even parity then sum of all the bits will be 0. When the m-bit code has odd parity then sum of all the bits will be 1.

When ***even parity*** is adopted and if the m-bit code already has even parity then parity bit is 0 and if the m-bit code already has odd parity then parity bit is 1. Therefore, in even parity system the sum of all the bits can be used as parity bit. The sum of 4-bit code is generated using three XOR gates as shown in Fig. 5.149.

When ***odd parity*** is adopted the parity bit of circuit in Fig. 5.149 is inverted and used as parity bit for odd parity system as shown in Fig. 5.150.

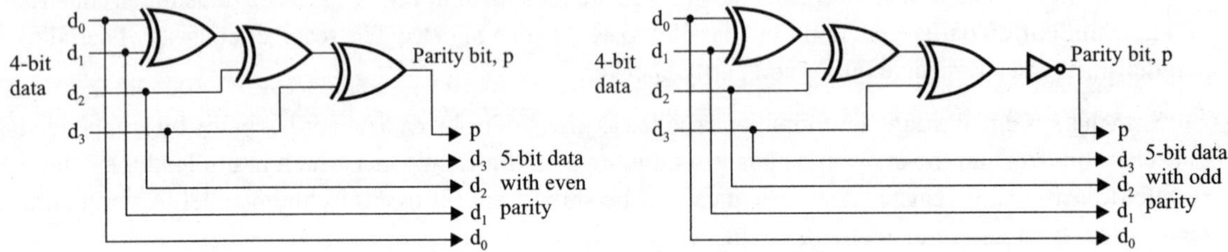

Fig. 5.149: *Even parity generator.*

Fig. 5.150: *Odd parity generator.*

On the receiver side the error can be checked by same technique as that of parity generation by adding all the bits of codes and verify the sum. When even parity is adopted the sum of received bits will be 0 and if odd parity is adopted then sum of received bits will be 1.

A parity checker for 5-bit code with parity is shown in Fig. 5.151. In this circuit 4 XOR gates are used to get sum of 5 bits. In even parity system if sum is 0, then no error and in odd parity system if sum is 1, then no error.

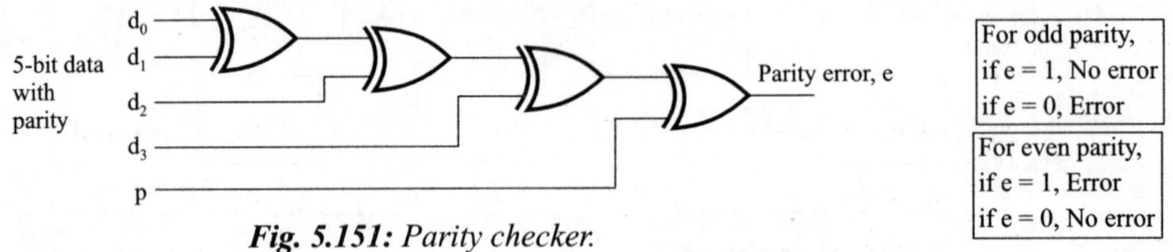

Fig. 5.151: *Parity checker.*

A standard IC 74180 is available to generate parity bit or verify parity of the system when the information is transmitted as 9-bit binary code consisting of 8 data bits and 1 parity bit. The function table of 74180 is shown in Table 5.52.

Pin			Pin
d_6	1	14	V_{cc}(+5V)
d_7	2	13	d_5
EI	3	12	d_4
OI	4	11	d_3
ΣE	5	10	d_2
ΣO	6	9	d_1
(0V) GND	7	8	d_0

74180 Parity Bit Generator/ Checker

Pin	Description
d_0 - d_7	8-bit data input
ΣE	Even sum output
ΣO	Odd sum output
EI	Even parity select input
OI	Odd parity select input
V_{cc}	Supply voltage (+5V)
GND	Ground (0V)

Fig. 5.152: *Pin configuration of IC 74180.*

Table 5.52: Function Table of 74180

Number of 1's in Data Bits $d_0 - d_7$	Inputs		Outputs		Comment
	EI	OI	ΣE	ΣO	
EVEN	1	0	1	0	Even parity system
ODD	1	0	0	1	
EVEN	0	1	0	1	Odd parity system
ODD	0	1	1	0	
×	1	1	0	0	Not allowed
×	0	0	1	1	

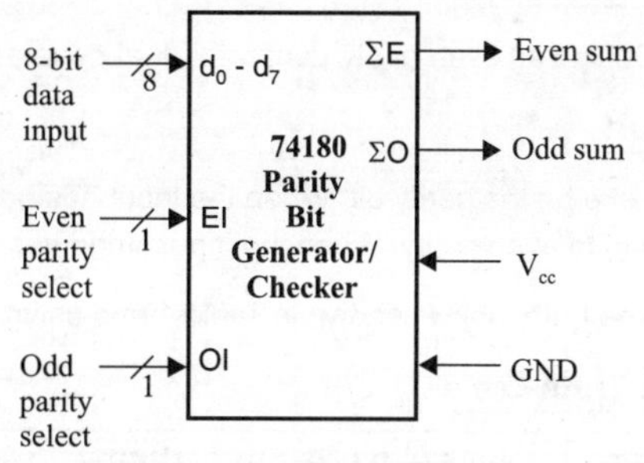

Fig. 5.153: Functional grouping of IC 74180.

The parity bit generation using IC 74180 for even and odd parity system are shown in Fig. 5.154. In even parity bit generation system, EI is tied to 1 and OI is tied to 0. Here, odd sum output is 0 if the data already has even parity and odd sum ouput is 1 if the data already has odd parity and so the parity bit is obtained from odd sum output, ΣO. In odd parity bit generation system, EI is tied to 0 and OI is tied 1. Here, even sum output is 1 if the data already has odd parity and even sum output is 0 if the data already has even parity and so the parity bit is obtained from even sum output, ΣE.

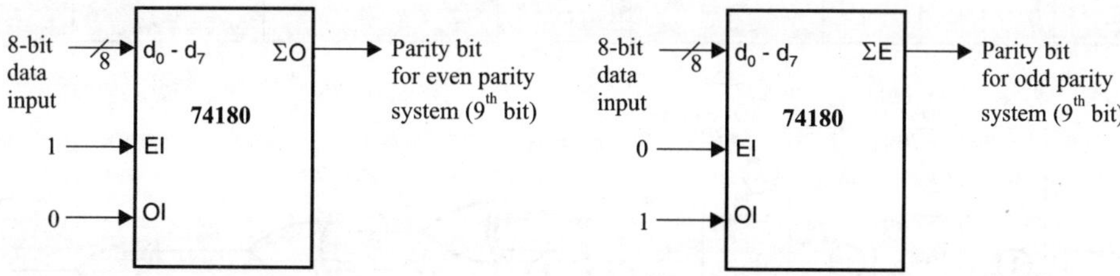

Fig. a: Even parity system. *Fig. b: Odd parity system.*

Fig. 5.154: Parity bit generation using 74180.

The parity error detection using 74180 for even and odd parity system are shown in Fig. 5.155. In order to detect error in even parity system, the parity bit in the received data is connected to EI and OI is tied to 0. In even parity system, when error occurs, ΣE = 1 and when no error, ΣE = 0.

In order to detect error in odd parity system, the parity bit in the received data is connected to OI and EI is tied to 0. In odd parity system, when error occurs, ΣO = 0 and when no error, ΣO = 1.

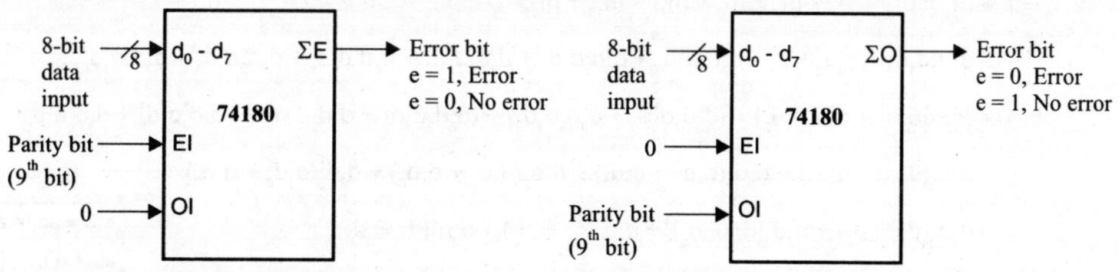

Fig. a: Even parity system. *Fig. b: Odd parity system.*

Fig. 5.155: Parity error detection using 74180.

Example 5.44

Design an even parity generator, that generates an even parity bit for every input string of 4-bits.

Solution

Let p be the parity bit. When the input has odd number of 1's then parity bit p is assigned 1 in order to convert input string to even parity. When the input string has even number of 1's then parity bit is assigned 0.

The truth table is shown in Table 1 and using truth table the K-map for p is drawn as shown in Fig. 1.

Table 1: Truth Table

Minterm	Inputs (4-bit Data)				Output	Minterm	Inputs (4-bit Data)				Output
	d_3	d_2	d_1	d_0	Parity Bit, p		d_3	d_2	d_1	d_0	Parity Bit, p
m_0	0	0	0	0	0	m_8	1	0	0	0	1
m_1	0	0	0	1	1	m_9	1	0	0	1	0
m_2	0	0	1	0	1	m_{10}	1	0	1	0	0
m_3	0	0	1	1	0	m_{11}	1	0	1	1	1
m_4	0	1	0	0	1	m_{12}	1	1	0	0	0
m_5	0	1	0	1	0	m_{13}	1	1	0	1	1
m_6	0	1	1	0	0	m_{14}	1	1	1	0	1
m_7	0	1	1	1	1	m_{15}	1	1	1	1	0

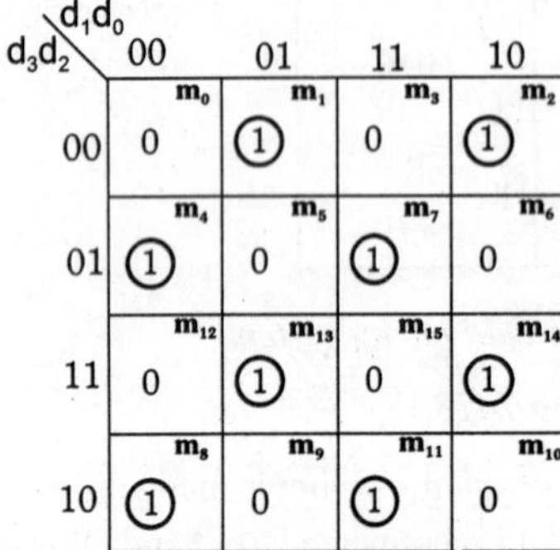

Fig. 1: K-map for p.

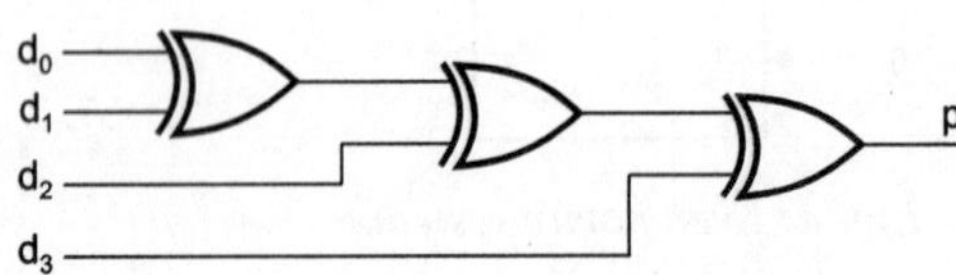

Fig. 2: Logic circuit of even parity bit generator.

All the prime implicants in the K-map are single 1's and so simplification using K-map is not possible. Hence the Boolean equation for p is given by sum of all the minterms for which output is 1.

$$\therefore p = m_1 + m_2 + m_4 + m_7 + m_8 + m_{11} + m_{13} + m_{14}$$

$$= d_3'd_2'd_1'd_0 + d_3'd_2'd_1d_0' + d_3'd_2d_1'd_0' + d_3'd_2d_1d_0 + d_3d_2'd_1'd_0' + d_3d_2'd_1d_0 + d_3d_2d_1'd_0 + d_3d_2d_1d_0'$$

$$= (d_3'd_2'd_1'd_0 + d_3'd_2'd_1d_0') + (d_3'd_2d_1'd_0' + d_3'd_2d_1d_0) + (d_3d_2d_1'd_0 + d_3d_2d_1d_0') + (d_3d_2'd_1'd_0' + d_3d_2'd_1d_0)$$

$$= d_3'd_2'(d_1'd_0 + d_1d_0') + d_3'd_2(d_1'd_0' + d_1d_0) + d_3d_2(d_1'd_0 + d_1d_0') + d_3d_2'(d_1'd_0' + d_1d_0)$$

$$= d_3'd_2'(d_1 \oplus d_0) + d_3'd_2(d_1 \oplus d_0)' + d_3d_2(d_1 \oplus d_0) + d_3d_2'(d_1 \oplus d_0)'$$

$a'b + ab' = a \oplus b$
$ab + a'b' = (a \oplus b)'$

$$= (d_1 \oplus d_0)(d_3'd_2' + d_3d_2) + (d_1 \oplus d_0)'(d_3'd_2 + d_3d_2')$$

$$= (d_1 \oplus d_0)(d_3 \oplus d_2)' + (d_1 \oplus d_0)'(d_3 \oplus d_2)$$

$$= d_3 \oplus d_2 \oplus d_1 \oplus d_0$$

Using the above simplified Boolean function the logic circuit of even parity generator is drawn as shown in Fig. 2.

Example 5.45

Design an odd parity generator that generates an odd parity bit for every input string of 4 bits.

Solution

Let p be the parity bit. When the input has odd number of 1's then parity bit p is assigned 0 in order to convert input string to odd parity. When the input string has even number of 1's then parity bit is assigned 1. The truth table is shown in Table 1 and using truth table the K-map for p is drawn as shown in Fig. 1.

Table 1: Truth Table

Minterm	Inputs (4-bit Data)				Output	Minterm	Inputs (4-bit Data)				Output
	d_3	d_2	d_1	d_0	Parity Bit, p		d_3	d_2	d_1	d_0	Parity Bit, p
m_0	0	0	0	0	1	m_8	1	0	0	0	0
m_1	0	0	0	1	0	m_9	1	0	0	1	1
m_2	0	0	1	0	0	m_{10}	1	0	1	0	1
m_3	0	0	1	1	1	m_{11}	1	0	1	1	0
m_4	0	1	0	0	0	m_{12}	1	1	0	0	1
m_5	0	1	0	1	1	m_{13}	1	1	0	1	0
m_6	0	1	1	0	1	m_{14}	1	1	1	0	0
m_7	0	1	1	1	0	m_{15}	1	1	1	1	1

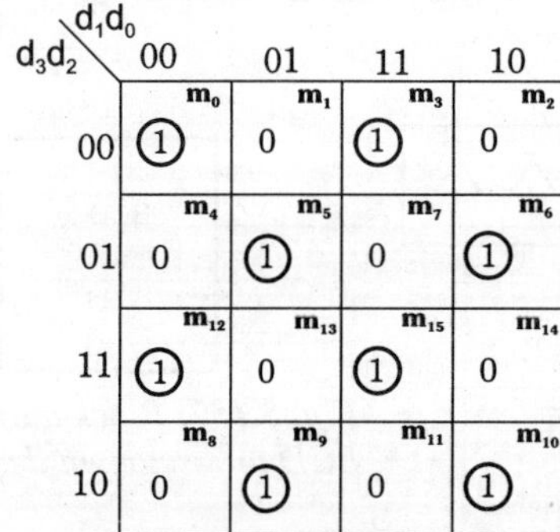

Fig. 1: K-map for F.

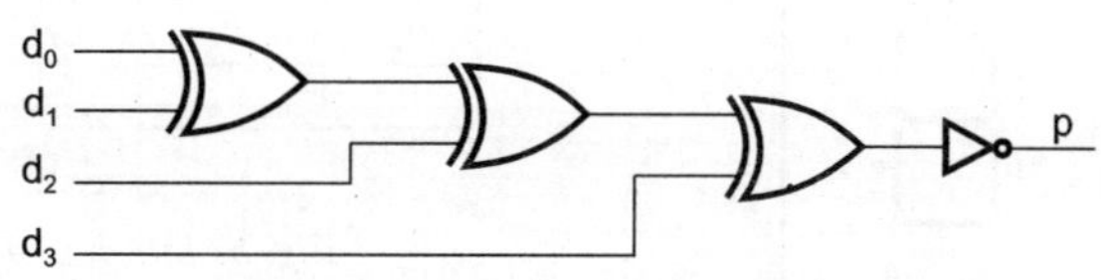

Fig. 2: Logic circuit of odd parity bit generator.

All the prime implicants in the K-map are single 1's and so simplication using K-map is not possible. Hence the Boolean equation for p is given by sum of all the minterms for which output is 1.

$$\therefore\ p = m_0 + m_3 + m_5 + m_6 + m_9 + m_{10} + m_{12} + m_{15}$$

$$= d_3'd_2'd_1'd_0' + d_3'd_2'd_1d_0 + d_3'd_2d_1'd_0 + d_3'd_2d_1d_0' + d_3d_2'd_1'd_0 + d_3d_2'd_1d_0' + d_3d_2d_1'd_0' + d_3d_2d_1d_0$$

$$= (d_3'd_2'd_1'd_0' + d_3'd_2'd_1d_0) + (d_3d_2d_1'd_0' + d_3d_2d_1d_0) + (d_3'd_2d_1'd_0 + d_3'd_2d_1d_0') + (d_3d_2'd_1'd_0 + d_3d_2'd_1d_0')$$

$$= d_3'd_2'(d_1'd_0' + d_1d_0) + d_3d_2(d_1'd_0' + d_1d_0) + d_3'd_2(d_1'd_0 + d_1d_0') + d_3d_2'(d_1'd_0 + d_1d_0')$$

$$= d_3'd_2'(d_1 \oplus d_0)' + d_3d_2(d_1 \oplus d_0)' + d_3'd_2(d_1 \oplus d_0) + d_3d_2'(d_1 \oplus d_0)$$

$a'b + ab' = a \oplus b$
$ab + a'b' = (a \oplus b)'$

$$= (d_1 \oplus d_0)'(d_3'd_2' + d_3d_2) + (d_1 \oplus d_0)(d_3'd_2 + d_3d_2')$$

$$= (d_1 \oplus d_0)'(d_3 \oplus d_2)' + (d_1 \oplus d_0)(d_3 \oplus d_2)$$

$$= (d_3 \oplus d_2 \oplus d_1 \oplus d_0)'$$

Using the above simplified Boolean function the logic circuit of odd parity generator is drawn as shown in Fig. 2.

5.15.5 BCD to 7-Segment Decoder

(AKUB, Nov/Dec'20, 8 Marks)

The 7-segment LEDs are the most popular display devices used in digital systems. Each ***7-segment LED*** will have seven **L**ight **E**mitting **D**iodes (LEDs) arranged in the form of small rectangular segments and another LED as a dot point in a single package. The 7-segment LEDs can be either common anode type or common cathode type.

In common anode type, all the anode terminals of LEDs are internally shorted and one/two pins are provided for external connection. The cathode of LEDs are terminated on separate pins for external connection. The pin configuration and the internal connection of the common anode 7-segment LED LT542 are shown in Fig. 5.156. In a 7-segment LED, a segment will glow or emit light when it is forward biased. Therefore, a segment can be made to glow by applying a logic **high** (+5 V) to the anode and a logic **low** (0 V) to cathode. The alphabetic/numeric characters can be displayed on the 7-segment LED by forward biasing the appropriate segments.

In a common anode 7-segment LED, the common point is tied to logic **high**. To display a character, logic **low** is applied to the cathode of segments which has to emit light and logic **high** is applied to the cathode of segments which should not emit light. The binary and hex codes for displaying the decimal digits 0 to 9 in the common anode 7-segment LED are listed in Table 5.53. The binary code to display the alpha-numeric characters in 7-segment display is also known as 7-segment code.

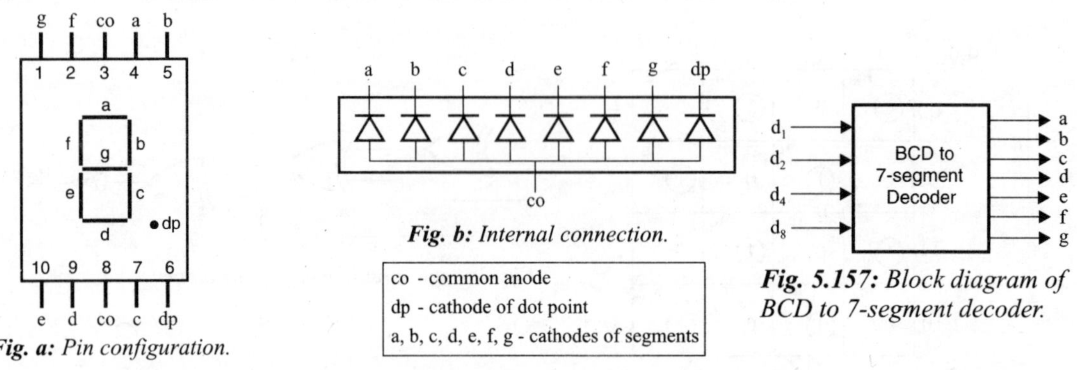

***Fig. a:** Pin configuration.*

***Fig. b:** Internal connection.*

***Fig. 5.157:** Block diagram of BCD to 7-segment decoder.*

***Fig. 5.156:** Common anode 7-segment LED LT 542.*

In order to dispaly the decimal digits on the 7-segment LED, a code converter (or BCD to 7-segment decoder) which converts the BCD to 7-segment code has to be designed. The block diagram of BCD to 7-segment decoder is shown in Fig. 5.157.

The BCD to 7-segment decoder is designed by taking BCD as minterms and construting K-maps for segment outputs a to g as shown in Figs. 5.158 to 5.164.

Table 5.53: Truth Table of BCD 7-Segment Common Anode Decoder

Minterm	Input BCD	BCD Digit	Output of 7-Segment Decoder						
	d_8 d_4 d_2 d_1		a	b	c	d	e	f	g
m_0	0 0 0 0		0	0	0	0	0	0	1
m_1	0 0 0 1		1	0	0	1	1	1	1
m_2	0 0 1 0		0	0	1	0	0	1	0
m_3	0 0 1 1		0	0	0	0	1	1	0
m_4	0 1 0 0		1	0	0	1	1	0	0
m_5	0 1 0 1		0	1	0	0	1	0	0
m_6	0 1 1 0	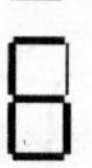	0	1	0	0	0	0	0
m_7	0 1 1 1		0	0	0	1	1	1	1
m_8	1 0 0 0		0	0	0	0	0	0	0
m_9	1 0 0 1		0	0	0	0	1	0	0
m_{10}	1 0 1 0	-	×	×	×	×	×	×	×
m_{11}	1 0 1 1	-	×	×	×	×	×	×	×
m_{12}	1 1 0 0	-	×	×	×	×	×	×	×
m_{13}	1 1 0 1	-	×	×	×	×	×	×	×
m_{14}	1 1 1 0	-	×	×	×	×	×	×	×
m_{15}	1 1 1 1	-	×	×	×	×	×	×	×

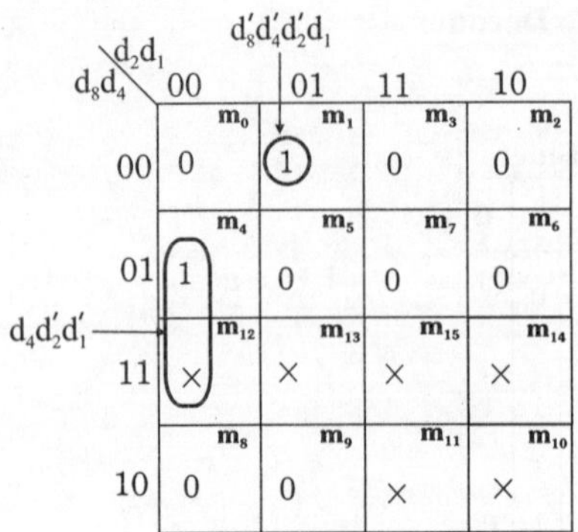

***Fig. 5.158:** K-map for a.*

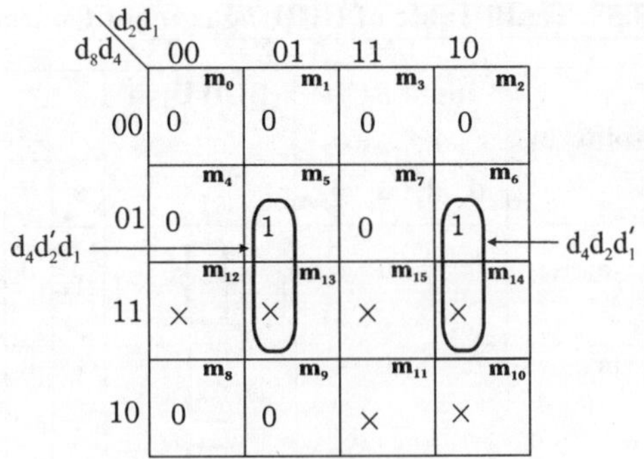

***Fig. 5.159:** K-map for b.*

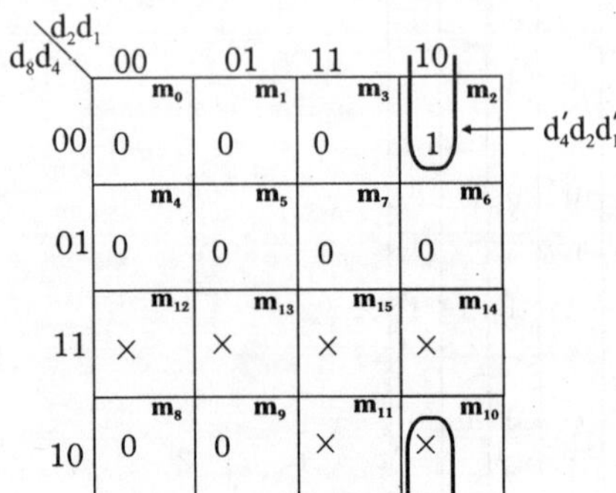

***Fig. 5.160:** K-map for c.*

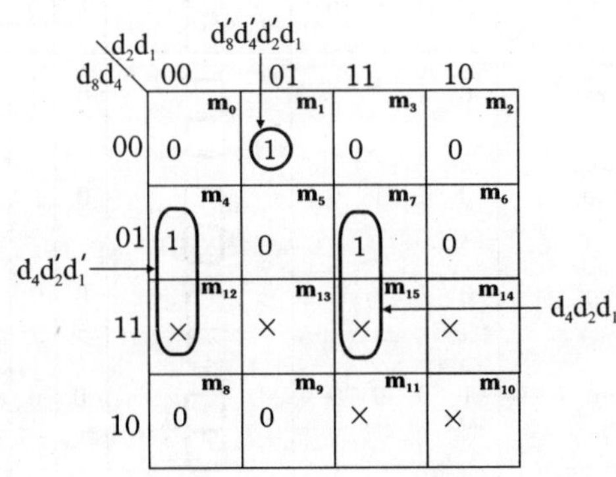

***Fig. 5.161:** K-map for d.*

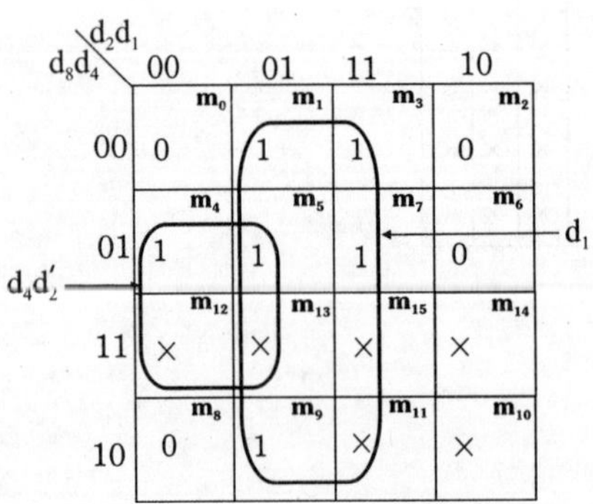

***Fig. 5.162:** K-map for e.*

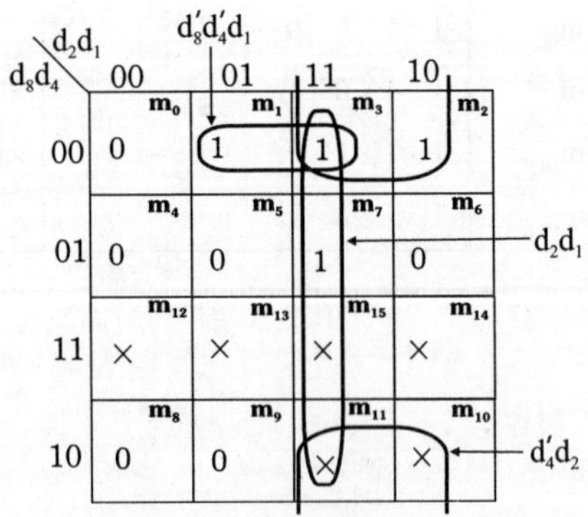

***Fig. 5.163:** K-map for f.*

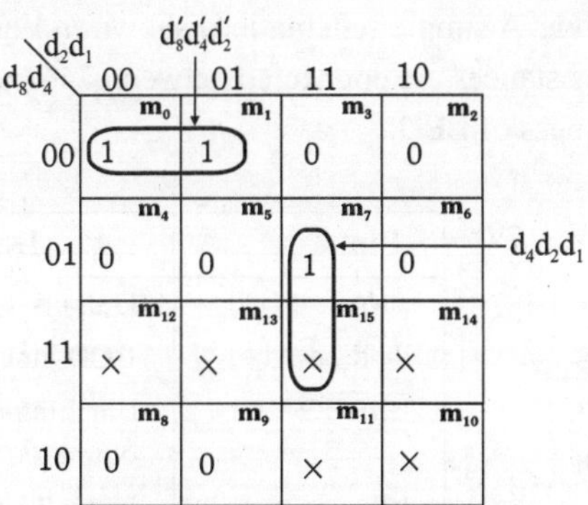

Fig. 5.164: *K-map for g.*

From the K-map for a, b, c, d, e, f and g the following Boolean equations are obtained and using these equations the logic circuit of seven segment decoder is drawn as shown in Fig. 5.165.

$$a = d_8' d_4' d_2' d_1 + d_4 d_2' d_1'$$

$$b = d_4 d_2' d_1 + d_4 d_2 d_1'$$

$$c = d_4' d_2 d_1'$$

$$d = d_8' d_4' d_2' d_1 + d_4 d_2' d_1' + d_4 d_2 d_1$$

$$e = d_1 + d_4 d_2'$$

$$f = d_2 d_1 + d_4' d_2 + d_8' d_4' d_1$$

$$g = d_8' d_4' d_2' + d_4 d_2 d_1$$

The BCD to 7-segment decoder for common anode 7-segment LED is available as a standard IC 7447. The 7447 accept 4-bit BCD code and generate 7-bit code to turn on 7-segments of common anode LED. The dot point dp is turned on directly by applying logic low.

The 7447 has blanking input and display test input. The ***blanking input***, BI is activated to clear the previous diaplay value before sending next display value to prevent ghosting of information. The ***lamp test input***, $\overline{LT}$ is made low to check the healthiness of display.

The display codes for LEDs can be generated by using the BCD to 7-segment decoder, IC 7447. When a BCD code is sent to the input of the 7447, it outputs **low** on the segments required to display the

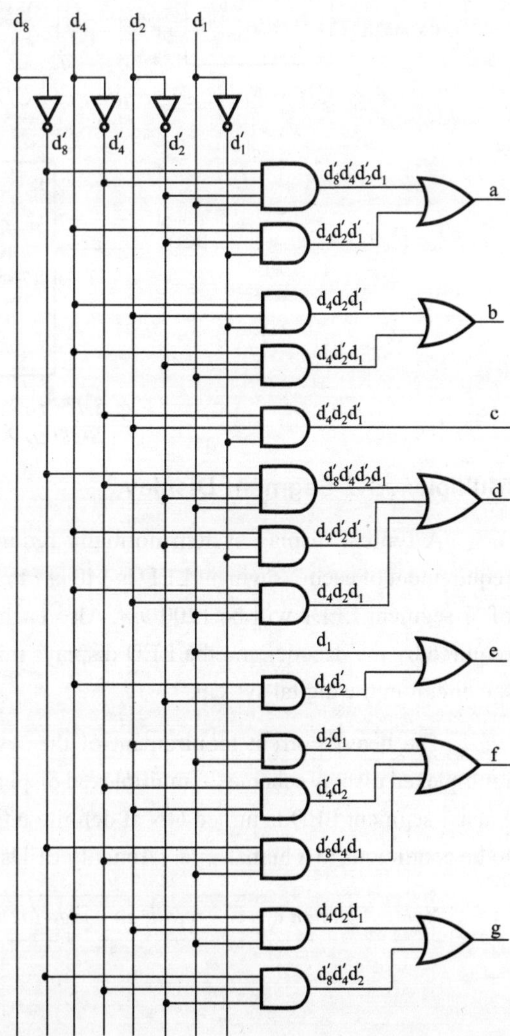

Fig. 5.165: *Logic circuit of seven-segment decoder.*

number represented by the BCD code. A simple schematic is shown in Fig. 5.167, to interface a common anode 7-segment LED to 7447. Resistances are connected between 7447 and 7-segment LED to limit the current and hence adjust the brightness of LED.

Pin name	Pin	7447	Pin	Pin name
d_2	1		16	V_{cc}(+5V)
d_4	2		15	f
$\overline{LT}$	3		14	g
$\overline{BI}/\overline{RBO}$	4		13	a
$\overline{RBI}$	5		12	b
d_8	6		11	c
d_1	7		10	d
(0V) GND	8		9	e

Pin	Description
a - g	Outputs
d_8, d_4, d_2, d_1	BCD input
$\overline{BI}/\overline{RBO}$	Blanking input/Ripple Blanking output
$\overline{LT}$	Lamp Test
$\overline{RBI}$	Ripple Blanking input
V_{cc}	Supply voltage (+5V)
GND	Ground (0V)

Fig. 5.166: Pin configuration of BCD to 7-segment decoder IC 7447.

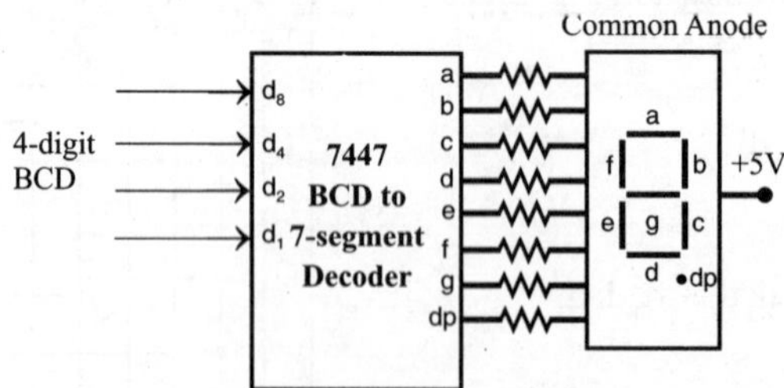

Fig. 5.167: 7-Segment LED display using BCD to 7-segment decoder 7447.

Multiplexed 7-Segment Display

A typical display system normally requires 5 to 8 numbers of 7-segment LEDs. The current requirement of each 7-segment LED is 140 *mA* to 200 *mA*. Hence, the total current requirement for 5 units of 7-segment LEDs will be 1200 *mA*. Also each 7-segment LED requires a 7447 decoder. The current required by the decoder and the LED displays might be several times the current required by the rest of the circuit in the digital system.

The heavy current requirement of the 7-segment displays can be reduced drastically by using multiplexed display scheme. In multiplexed display, only one 7-segment display is made to glow at a time. Each 7-segment LED is turned ON at definite intervals. Due to persistence of vision the display appears to be continuous to a human eye. (Actually LEDs are turned ON and OFF.)

> ***Note:*** *A human eye can retain an image for 125 milliseconds.*

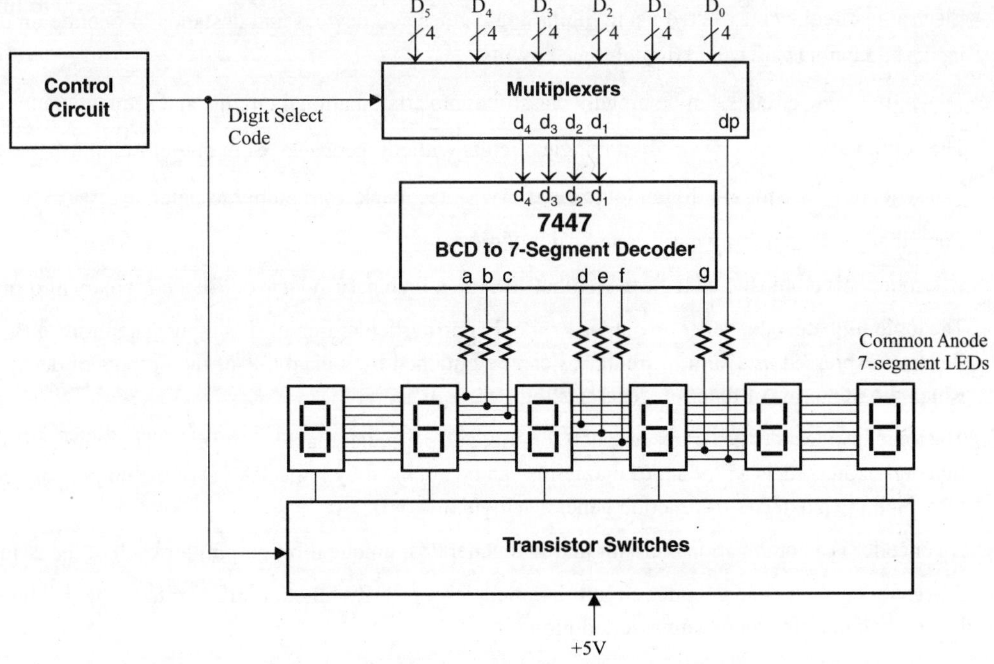

Fig. 5.168: *Multiplexed 7-segment display.*

The advantages of a multiplexed display are as follows:

1. Only one 7447 is needed for all the 7-segment LEDs.
2. In a current requirement of one-7-segment LED, 5 to 8 LEDs can be interfaced.

Figure 5.168 shows a multiplexed display of 5 units of 7-segment common anode LEDs. The segment pins (cathodes) of 7-segment LEDs are connected to a common bus. The output of the decoder (7447) is connected to this common bus. The BCD code for the character to be displayed is sent to 7447 through multiplexers. The common anode of each 7-segment LED has a driver transistor (PNP type). A driver transistor can be turned ON by sending **low** to the base of the transistor.

The trick of ***multiplexed display*** is that the segment information is sent out to all of the digits on the common bus, but only one display digit is turned on at a time. The PNP transistors in series with the common-anode of each digit acts as an ON and OFF switch for that digit.

A control circuit should be designed to generate digit selection code to turn on 7-segment display one-by-one and this code is common for multiplexers and transistor switches. At any one time, the digit select code will enable the multiplxers to allow the BCD code of one digit to 7447 and at the same time will turn ON the corresponding transistor switch. In this technique, each display will be made to turn-on one-by-one for 5 to 15 *ms* so that in a scan time of 30 to 90 *ms*, each display will be turned ON one time.

5.16 Summary of Important Concepts

1. Electronic circuits constructed using digital logic gates and devices and designed to operate on digital inputs and outputs are called digital logic circuits.
2. The digital logic circuits can be broadly classified into combinational circuits and sequential circuits.
3. The combinational circuits are digital logic circuits without feedbacks from output to input.
4. The sequential circuits are digital logic circuits with feedback from output to input.
5. Combinational circuits are constructed using logic gates.
6. A decoder is a combinational logic device that decodes n-bit binary input to one of the 2^n binary information.
7. The logic **high** decoder generate output 1 for only one particular minterm as input. Therefore, a Boolean function expressed as a sum of minterms can be obtained by logical OR of the outputs of decoder for which the minterms of function generates logic **high** or 1.
8. The logic **low** decoder generates output 0 for only one particular maxterm as input. Therefore, a Boolean function expressed as a product of maxterms can be obtained by logical AND of the outputs of decoder for which the maxterms of function generates logic **low** or 0.
9. An encoder is a combinational circuit that will generate a unique n-bit output, for each of the 2^n inputs.
10. A priority encoder is an encoder circuit that includes a priority function in order to recognize only one input when multiple inputs are asserted **high**.
11. A multiplexer is a combinational circuit which can select one of the 2^n inputs as output using n selection lines.
12. In multiplexer the possible binary values of selection inputs are same as minterms. Therefore, the logic circuit of Boolean functions in SOP form can be realized using multiplexer.
13. Multiplexer of smaller size can be connected in cascade to expand the size of multiplexer.
14. A demultiplexer will perform the reverse operation of multiplexer.
15. A demultiplexer is a combinational circuit which can transmit a binary input to any one of 2^n output lines using n selection lines or n-bit address.
16. Demultiplexers of smaller size can be connected in parallel to expand the size of demultiplexer.
17. Adders have been developed to perform arithmetic addition operation on binary numbers.
18. Half adder can perform addition of two 1-bit binary number.
19. Full adder is a combinational circuit that perform the arithmetic sum of three binary bits.
20. Full adder can also be realized using two half adders.
21. The outputs of half and full adders are sum and carry.
22. A binary parallel adder performs arithmetic sum of two n-bit binary numbers.
23. In carry look ahead adder, it can be observed that the computation of each bit of sum has equal number of stages and so propagation delay time will be equal for all outputs.
24. In decimal addition we can use binary or BCD to represent decimal numbers.
25. In BCD addition the sum of any two digits will be in the range 0 to 19_{10}.

26. In BCD addition, when sum is in the range 0 to 9_{10} no correction is needed and it can be represented by 4 digit BCD.
27. In BCD addition, when sum in the range 10_{10} to 19_{10} a correction 6_{10} is added to binary sum to get the result in BCD.
28. The BCD adder is a binary adder with additional logic circuit to perform addition of correction 6_{10} when the sum of a BCD digit exceeds 9_{10}.
29. Subtractors have been developed to perform arithmetic subtraction operation on binary numbers.
30. Half subtractor can perform subtraction of two 1-bit binary number.
31. Full subtractor is a combinational circuit that perform arithmetic subtraction of three binary bits.
32. The outputs of half and full subtractors are difference and borrow.
33. Magnitude comparator is a combinational circuit used to compare two binary numbers.
34. Combinational multiplier is used to multiply two unsigned binary numbers similar to decimal multiplication.
35. Multiplication of two binary numbers is performed by multiplying each bit of multiplier (starting from LSB) to multiplicand to generate partial products.
36. In combinational 2-bit binary multiplier, each multiplication of two binary bits is realized by a two input AND gate and the addition of partial products are realized by half adders.
37. In combinational 4-bit binary multiplier, each multiplication of two binary bits is realized by a two input AND gate and the addition of partial products are realized by 4-bit binary adders.
38. The Arithmetic and Logic Unit (ALU) is a combinational circuit that can perform different arithmetic and logical operations on a pair of m-bit operands or data.
39. In binary, n-bit selection input can generate 2^n operation selection codes.
40. A 4-bit ALU can be designed using four full-adders (FA).
41. Function select lines are required to select the desired arithmetic and logical operations to be performed by the ALU.

5.17 Short-Answer Questions

Q5.1 Mention the dependency of output in combinational circuits. ***(AU, Nov/Dec'18, 2 Marks)***

The output depends only on the preset or current value of input. Since there is no feedback of output to input in combinational circuits, the present output will not depend on previous output.

Q5.2 Define digtal logic circuits.

Electronic circuits constructed using digital logic gates and devices and designed to operate on digital inputs and outputs are called digital logic circuits.

Q5.3 What are the types of digtal logic circuits.

The digital logic circuits can be broadly classified into,

1. Combinational circuits
2. Sequential circuits.

Q5.4 Explain the design procedure for combinational circuits.

1. Determine the required inputs and outputs from the problem specifications.
2. Assign a symbol to each input and output.
3. Derive the truth table.
4. Draw K-map for every output and form the prime implicants.
5. Determine the simplified Boolean function for every output from the prime implicants.
6. Implement the Boolean functions of all the outputs as a digital circuit using logic gates.

Q5.5 Define Decoder.

A decoder is a combinational circuit that converts binary information from n input lines to a maximum of 2^n unique output lines. In a decoder, for an n-bit binary input one of the 2^n output is activated.

Q5.6 How multiplexer differ from decoder?

A decoder accept n-bit binary input to activate one of the 2^n outputs whereas the multiplexer accept n-bit address to allow one of the 2^n input on the output line.

Q5.7 Write the truth table of 2-to-4 line decoder and draw its logic diagram. **(IKGPTU, May'20, 2 Marks)**

Solution:

Table 1: Truth Table

Inputs		Outputs			
I_1	I_0	Y_0	Y_1	Y_2	Y_3
0	0	1	0	0	0
0	1	0	1	0	0
1	0	0	0	1	0
1	1	0	0	0	1

Boolean equations:

$$Y_0 = I_1' I_0' \quad ; \quad Y_1 = I_1' I_0 \quad ; \quad Y_2 = I_1 I_0' \quad ; \quad Y_3 = I_1 I_0$$

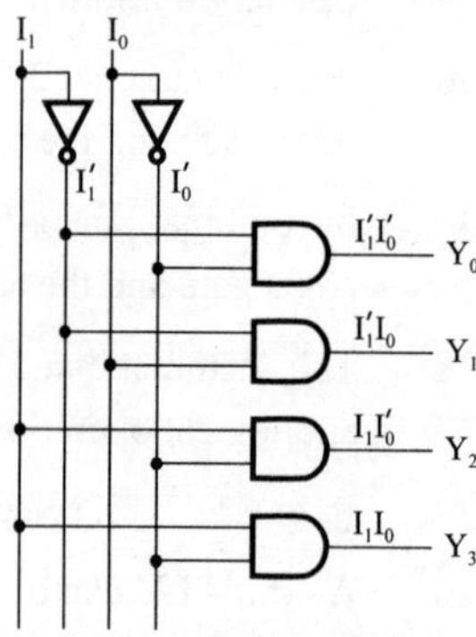

Fig. Q5.7: Logic diagram.

Q5.8 Define Encoder. **(RGPV, Jun'22, 2 Marks)**

An encoder is a combinational circuit that perform the inverse operation of a decoder. An encoder has 2^n input lines and n output lines. When one of 2^n inputs is activated, it generates an unique n-bit output.

Q5.9 What is priority Encoder. **(RGPV, Jun'22, 2 Marks)**

A priority encoder is an encoder circuit that includes the priority function. In priority encoder, if two or more inputs are asserted high (or equal to 1) then the output corresponds to the input having highest priority.

Q5.10 What is Demultiplexer.

Demultiplexer is a combinational circuit which can transmit a binary input to any one of 2^n output lines using n selection lines or n-bit address.

Q5.11 An odd parity operation results in 1 if an odd number of bits in a given sequence of bits are 1. Write a truth table for 4-input odd parity operation.

Solution:

Inputs				Output	Inputs				Output
a	b	c	d		a	b	c	d	
0	0	0	0	0	1	0	0	0	1
0	0	0	1	1	1	0	0	1	0
0	0	1	0	1	1	0	1	0	0
0	0	1	1	0	1	0	1	1	1
0	1	0	0	1	1	1	0	0	0
0	1	0	1	0	1	1	0	1	1
0	1	1	0	0	1	1	1	0	1
0	1	1	1	1	1	1	1	1	0

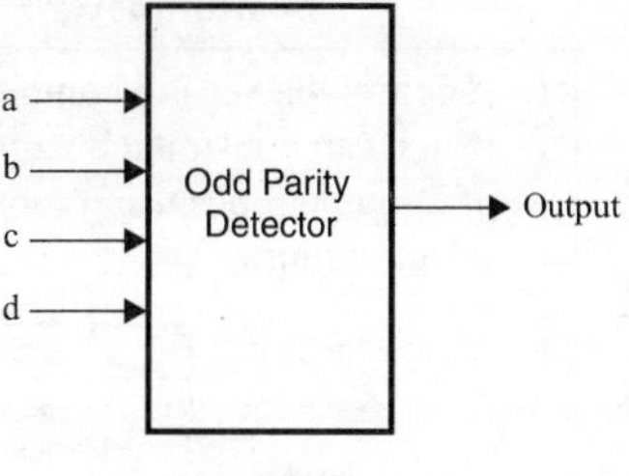

Fig. Q5.11.

Output = 1, When number of 1's odd
Output = 0, When number of 1's even

Q5.12 Convert a 2-to-4 decoder with enable input to 1:4 demux.

Solution:

The truth table of 2-to-4 decoder working as 1:4 MUX is shown in Table 1.

When the 2 inputs I_0, I_1 of decoder are used as selection inputs S_0 and S_1 of demux and enable of decoder used as demux input, a 2-to-4 decoder will work as a 1:4 demux.

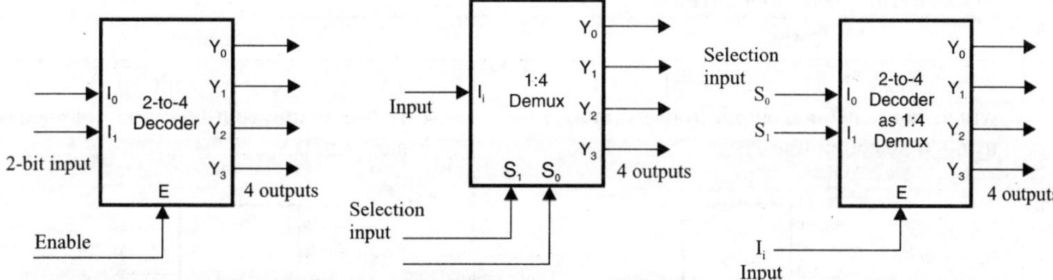

Fig. Q5.12.1: 2-to-4 decoder. *Fig. Q5.12.2: 1:4 demux.* *Fig. Q5.12.3: 2-to-4 decoder as 1:4 demux.*

Table 1: Truth Table of 2-to-4 Decoder with 1:4 Demux

Inputs	Selection Inputs		Outputs			
$I_i = E$	S_1	S_0	Y_0	Y_1	Y_2	Y_3
$I_i = 0$	0	0	0	0	0	0
$I_i = 0$	0	1	0	0	0	0
$I_i = 0$	1	0	0	0	0	0
$I_i = 0$	1	1	0	0	0	0
$I_i = 1$	0	0	I_i	0	0	0
$I_i = 1$	0	1	0	I_i	0	0
$I_i = 1$	1	0	0	0	I_i	0
$I_i = 1$	1	1	0	0	0	I_i

Q5.13 Compare demux and decoder.

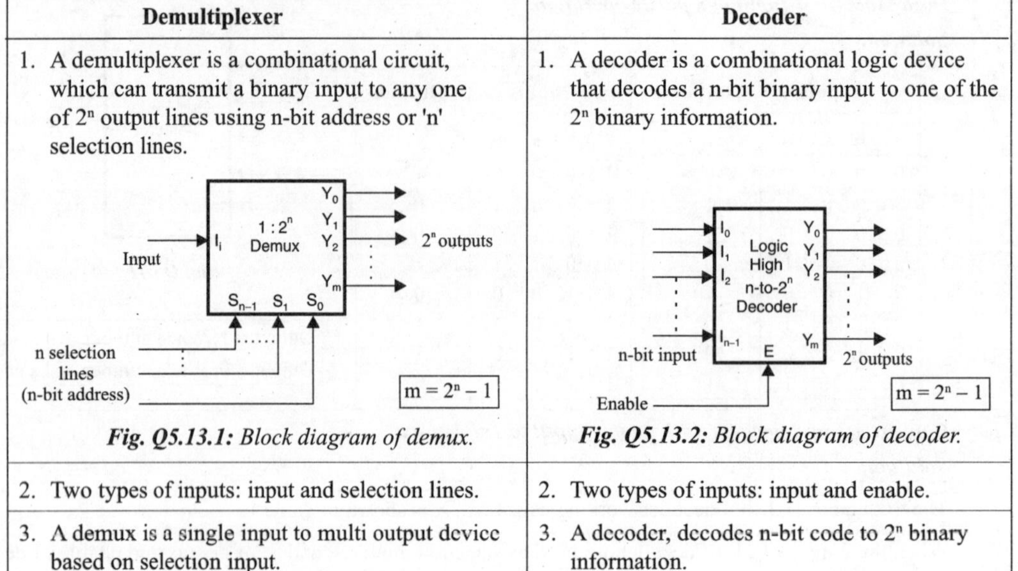

Demultiplexer	Decoder
1. A demultiplexer is a combinational circuit, which can transmit a binary input to any one of 2^n output lines using n-bit address or 'n' selection lines. *Fig. Q5.13.1: Block diagram of demux.*	1. A decoder is a combinational logic device that decodes a n-bit binary input to one of the 2^n binary information. *Fig. Q5.13.2: Block diagram of decoder.*
2. Two types of inputs: input and selection lines.	2. Two types of inputs: input and enable.
3. A demux is a single input to multi output device based on selection input.	3. A decoder, decodes n-bit code to 2^n binary information.

Q5.14 Compare multiplexer and encoder.

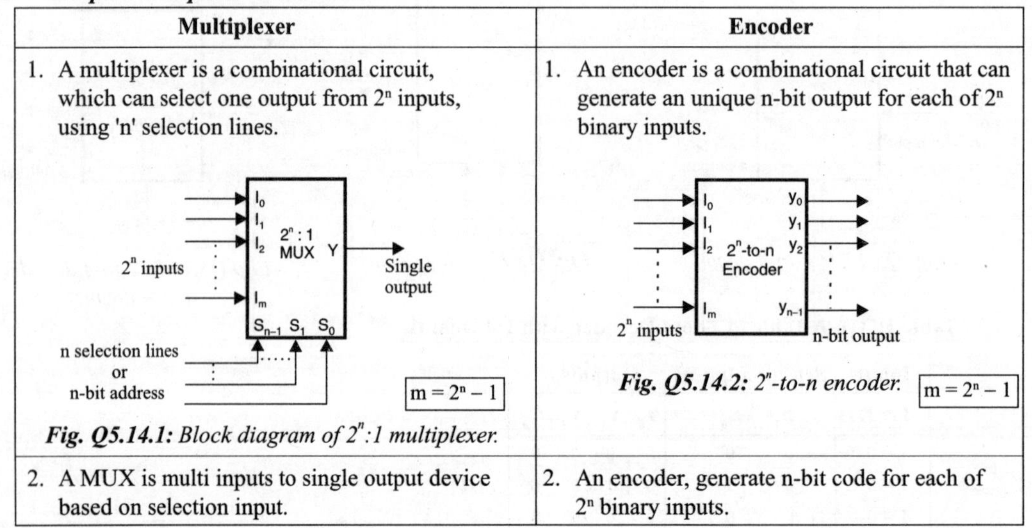

Multiplexer	Encoder
1. A multiplexer is a combinational circuit, which can select one output from 2^n inputs, using 'n' selection lines. *Fig. Q5.14.1: Block diagram of 2^n:1 multiplexer.*	1. An encoder is a combinational circuit that can generate an unique n-bit output for each of 2^n binary inputs. *Fig. Q5.14.2: 2^n-to-n encoder.*
2. A MUX is multi inputs to single output device based on selection input.	2. An encoder, generate n-bit code for each of 2^n binary inputs.

Q5.15 Differentiate between multiplexer and decoder.

Multiplexer	Decoder
1. MUX accepts several inputs and allow only one data output.	1. It takes n input binary code and convert it into a corresponding outputs.
2. Select line are used to select data inputs and allow only one of them.	2. Enable inputs are used to control the operation of the decoder.
3. Multiplexer converts the unary code into binary code.	3. Decoder converts binary code into unary.

Q5.16 *Differentiate between multiplexer and demultiplexer.* ***(JNTUA, Jun/Jul'22, 7 Marks)***

Multiplexer	Demultiplexer
1. A multiplexer is a combinational circuit that select one binary information from many inputs.	1. A demultiplxer will perform the reverse operation of multiplexer.
2. In general, a MUX will have 2^n inputs and one output. In order to select one of the 2^n inputs the MUX will have n selection lines or address.	2. A demux is a combinational circuit which can transmit a binary input of any one of 2^n output lines using n selection lines or n-bit address.

Q5.17 *What do you mean by parity checker?*

A parity bit is an extra bit added to the binary code to make the number of 1's either odd or even. It is mainly used for detecting the errors in transmission of binary information. Parity checker is the logic circuit that checks the parity of binary code received by the receiver. In even parity system, the parity checker will check for even parity in the received codes.

In odd parity system, the parity checker will check for odd parity in the received codes. If the parity of a codes is not correct then an error signal is generated to indicate error in transmitted code.

Q5.18 *Define Half adder and Full adder.*

Half adder is a combinational circuit that performs arithmetic sum of 2-bit binary.

Full adder is a combinational circuit that performs arithmetic sum of 3-bit binary.

Q5.19 *Design a half adder using basic logic gates.*

Solution:

$s = ab' + a'b$; $c_o = ab$

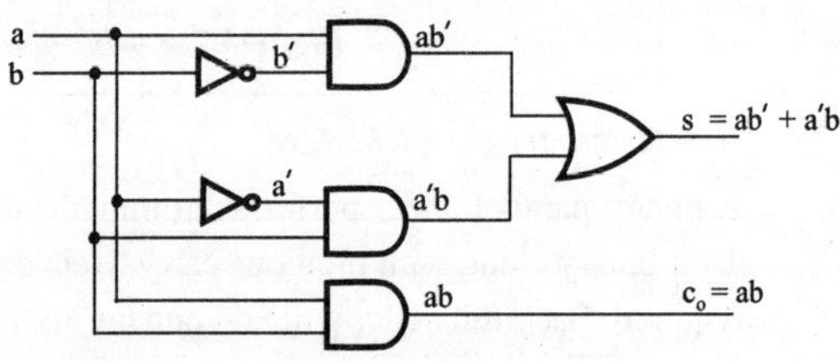

***Fig. Q5.19:** Logic circuit of half adder.*

Q5.20 *Draw a full adder using two half adders.*

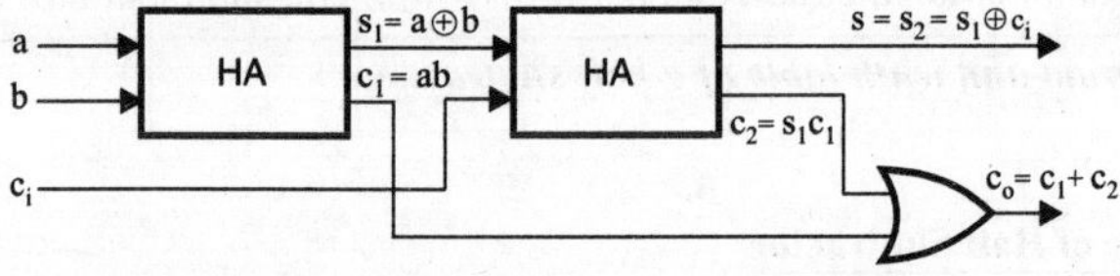

***Fig. Q5.20:** Full adder using two half adders.*

Q5.21 *Give the logical expression for half adder and a full adder.*

Half adder: Sum, $s = ab' + a'b$

Carry, $c_o = ab$

Full adder: Sum, $s = a'b'c_i + a'bc_i' + ab'c_i' + abc_i$

Carry, $c_o = ab + ac_i + bc_i$

Q5.22 ***Draw a full adder circuit using two XOR gate based half adders.***

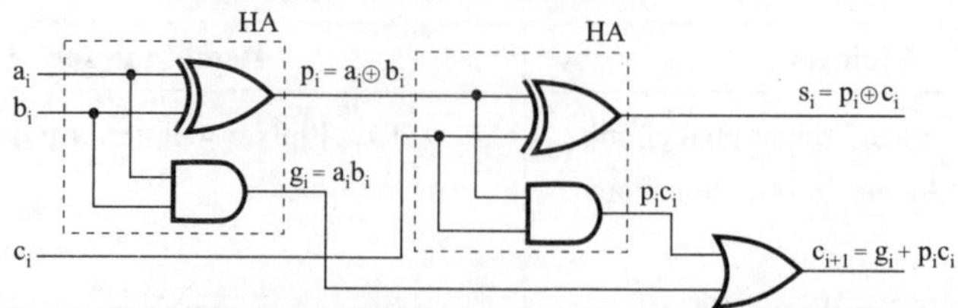

Fig. Q5.22: Full adder circuit using two XOR gate based half adders.

Q5.23 ***Determine the exact number of half adders and full adders required for performing the addition of two binary numbers of 5 bits length each.*** ***(AU, Apr/May'19, 2 Marks)***

When addition is performed using half adders, we require two half adders to add one bit binary number. Therefore, for addition of 5-bit binary numbers 10 half adders are required.

When addition is performed using full adders we require one full adder to add one bit binary number. Therefore, for addition of 5-bit binary numbers we require 5 full adders.

Q5.24 ***Design a 4-bit parallel adder using full adders.***

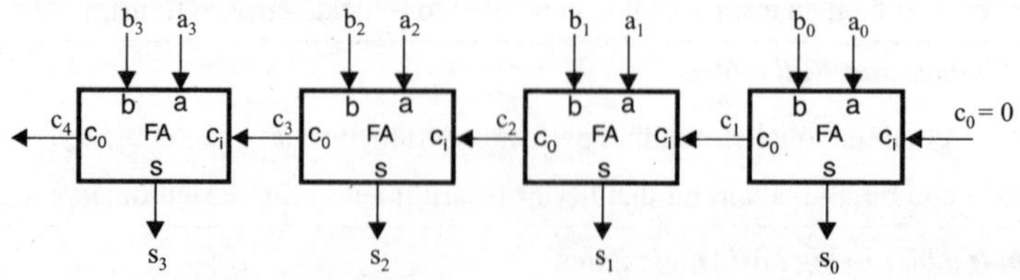

Fig. Q5.24: 4-bit binary adder using full adders.

Q5.25 ***What is binary parallel adder.***

A binary parallel adder performs arithmetic sum of two n-bit binary numbers. A full adder can add two 1-bit binary along with previous carry. Hence, in order to add two n-bit binary numbers, n full adders are required. Each full adder will add one bit of binary numbers.

Q5.26 ***Define Half subtractor and Full subtractor.***

Half subtractor is a combinational circuit that performs arithmetic subtraction of two binary bits.

Full subtractor is a combinational circuit that performs arithmetic subtraction of three binary bits.

Q5.27 ***Draw the logic diagram and truth table of a half subtractor.***

Solution:

Table 1: Truth Table of Half Subtractor

Inputs		Minterm	Outputs	
a	b		d	c_o
0	0	m_0	0	0
0	1	m_1	1	1
1	0	m_2	1	0
1	1	m_3	0	0

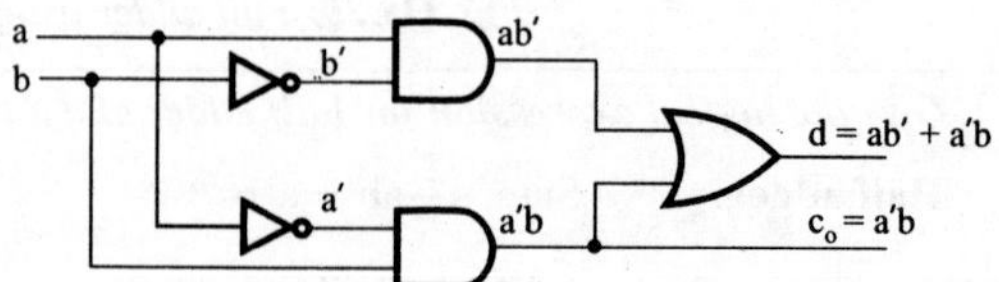

Fig. Q5.27: Logic circuit of half subtractor.

Q5.28 Give the logical expression for half subtractor and a full subtractor.

Half subtractor: Difference, $d = ab' + a'b$; Borrow, $c_o = a'b$

Full subtractor: Difference, $d = a'b'c_i + a'bc_i' + ab'c_i' + abc_i$; Borrow, $c_o = a'b + a'c_i + bc_i$

Q5.29 Design a binary addition using adder/subtractor.

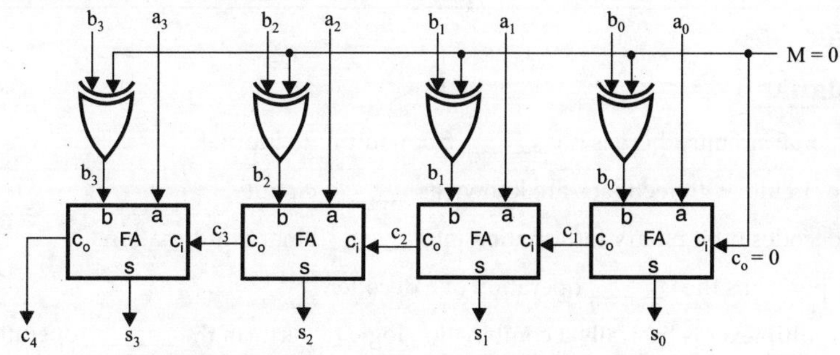

***Fig. Q5.29:** Binary addition using adder/subtractor.*

Q5.30 Design a binary subtraction using adder/subtractor.

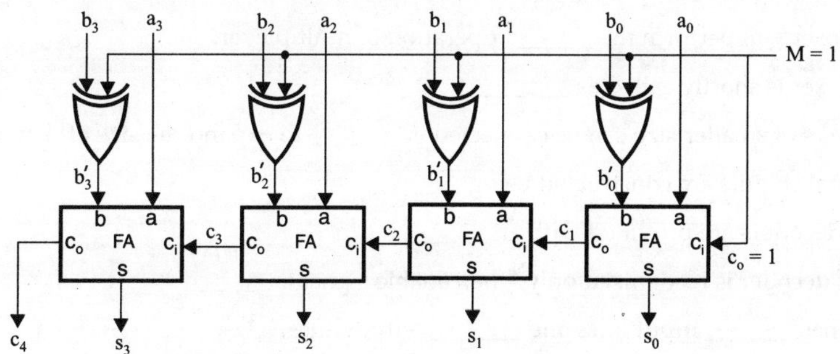

***Fig. Q5.30:** Binary subtraction using adder/subtractor.*

Q5.31 What is a magnitude comparator.

Magnitude comparator is a combinational circuit used to compare two binary numbers and determine whether they are equal or unequal and if unequal then it can make a decision on larger or smaller magnitude.

Q5.32 Draw the block diagram of ALU.

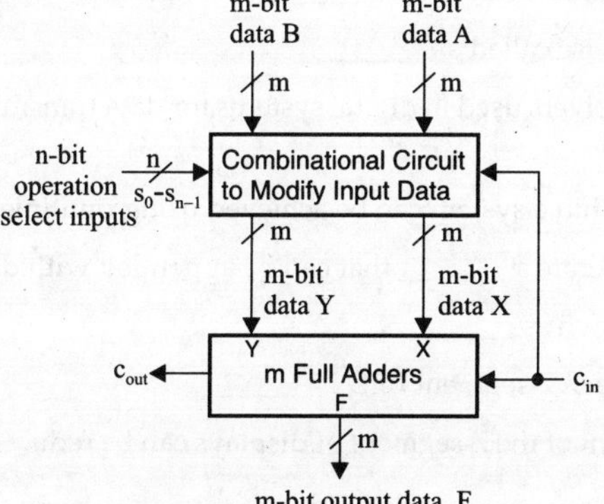

***Fig. Q5.32:** Block diagram of m-bit ALU.*

Q5.33 What is ALU and its operation?

The Arithmetic and Logic Unit (ALU) is a combinational circuit that can perform different arithmetic and logical operations on a pair of m-bit operands or data. The operation that can be performed by an ALU is specified by a set of function select inputs. In binary, n-bit selection input can generate 2^n operation selection codes.

5.18 Exercises

I. Fill in the blanks

1. In combinational circuits, there is no ______ from output to input.
2. Digital logic circuits with feedback are known as ______ circuits.
3. A decoder decodes n-bit binary information into ______ binary information.
4. An encoder performs the ______ operation of a decoder.
5. The digital multiplexer is basically a combination logic to perform the ______ operation.
6. The multiplexer is otherwise called ______ .
7. A multiplexer will have ______ inputs and ______ output.
8. A demultiplexer will perform the ______ operation of multiplexer.
9. A demultiplexer is shortly called as ______ .
10. Demultiplexers of smaller size can be connected in ______ to expand the size of demux.
11. Combinational circuits are constructed using ______.
12. Most of the decoders are provided with an ______.
13. The input of decoder is recognized only when enable signal is ______.
14. An encoder has ______ input lines and ______ output lines.
15. The operation the priority encoder is such that if two or more inputs are asserted ______.
16. A multiplexer is shortly called as ______.
17. The 4:1 multiplexer is used to select one of the four inputs using ______ selection input.
18. The 8:1 multiplexer is used to select one of the eight inputs using ______ selection input.
19. In multiplexer, the possible binary values of selection inputs are same as ______.
20. Multiplexer of smaller size can be connected in ______ to expand the size of multiplexer.
21. The data distributor is otherwise called as ______.
22. The data transmitters and receivers used in digital systems for data transmission and reception are called as digital ______.
23. The ______ transmission within a system can be achieved using multiplexer and demultiplexer.
24. The parity generator will generate a ______ that can be appended with data as most significant bit.
25. 7-segment display is also known as ______.
26. The current requirement of each 7-segment LED is ______.
27. The heavy current requirement of the 7-segment of displays can be reduced drastically by using ______ display.
28. ______ is a combinational circuit that perform arithmetic sum of three 1-bit binary.

29. _______ is a combinational circuit that performs arithmetic sum of two 1-bit binary.
30. The two outputs of a half adder are _______ and _______.
31. In a half adder, the sum can be realized using _______ gate.
32. In a half adder, the carry can be realized using _______ gate.
33. Full adder is a combinational circuit that perform arithmetic sum of _______ binary bits.
34. One of the input bit in full adder is _______ generated in previous addition.
35. A full adder has _______ outputs.
36. Practically _______ is considered as addition of positive and negative numbers.
37. _______ is a combinational circuit used to compare two binary numbers.
38. Adders have been developed to perform _______ operation on binary numbers.
39. _______ is designed to generate 2-bit standard output.
40. Half subtractor can perform subtraction of _______ 1-bit binary number.
41. A _______ performs arithmetic sum of two-n-bit binary numbers.
42. In decimal addition we can use _______ to represent decimal numbers.
43. The logic circuit of half subtractor using _______ gate.
44. A half subtractor has _______ outputs.
45. A full subtractor has _______ inputs.
46. The Arithmetic and logic unit (ALU) is a combinational circuit that can perform different arithmetic and logical operations on a pair of _______ operands or data.
47. M-bit ALU is constructed using _______.

Answers

1. feedback	11. logic gates	21. demux	31. XOR	41. binary parallel adder
2. sequential	12. enable signal	22. trans-receiver	32. AND	42. binary or BCD
3. 2^n	13. active	23. serial data	33. three	43. XOR
4. inverse	14. 2^n and n	24. parity bit	34. carry	44. two
5. AND-OR	15. high	25. 7-segmnt code	35. two	45. three
6. data selector	16. mux	26. 140 to 200 mA	36. subtraction	46. m-bit
7. 2^n, one	17. 2-bit	27. multiplexed	37. magnitude comparator	47. m full adder
8. reverse	18. 3-bit	28. full adder	38. arithmetic addition	
9. demux	19. minterms	29. half adder	39. half adder	
10. parallel	20. cascade	30. sum, carry	40. two	

II. State whether the following statements are True or False

1. A logic gate is a data storage device.
2. In general, a multiplexer has several data inputs, several data outputs and selection inputs.
3. A comparator compares for the equality of two input numbers.
4. A demultiplexer can transmit a binary input to any one of 2^n output lines using n-bit address.
5. Demultiplexer of smaller size can be connected in series to expand the size of demultiplexer.

6. In a decoder, n-bit binary information is decoded into 2^n binary information.
7. A n-to-2^n decoder outputs may be logic **high** or **low**.
8. An encoder is a combinational circuit that performs the reverse of decoder function.
9. A priority encoder is a decoder.
10. A demultiplexer is a data distributor.
11. The code 10011000 exhibits even parity.
12. In combinational circuits the output at any time depends on input at that time.
13. In combinational circuits there is a storage element and there is a feedback from output to input.
14. The code converters are combinational circuits that convert one type of code to another type of codes.
15. The design of code converter starts with formation of truth table by taking one type of code as inputs and another type of code as outputs.
16. A decoder is a combinational logic device that decodes one of the 2n binary information depending on n-bit binary input.
17. The n-bit binary information is decoded into n binary information.
18. The complement of logic high decoder output will be same as output of logic low decoder.
19. The complement of output of logic high decoder can be used to realize POS form of Boolean function.
20. A multiplexer is otherwise called data distributor.
21. 2:1 Multiplexers can be connected in series to form 4:1 multiplexer or multiplexers of higher order.
22. The logic circuit of Boolean functions in SOP form can be realised using multiplexer.
23. A demultiplexer is a combinational circuit that send one binary information to many outputs.
24. The parity checker will check the parity of transmitted data and generate a error signal if there is a parity error in the transmitted data.
25. The parity generator and receiver should be selected for same parity either both even or both odd parity.
26. A full adder is characterized by two inputs and two outputs.
27. A 4-bit parallel adder can add two 4-bit binary numbers.
28. A comparator compares for the equality of two input numbers.
29. Half adder has two inputs and two outputs.
30. In addition of n-bit binary numbers, the addition is performed bit-by-bit.
31. n-bit binary addition can be performed by using n full adder in parallel.
32. Subtractor have been developed to perform arithmetic subtraction operation on binary numbers.
33. In the subtractor 2-bit result, first bit is called borrow and the second bit is called difference.
34. The addition of negative number and 2's complement of positive number will give the result of subtraction.
35. The BCD adder is a binary adder with additional logic circuit to perform addition of correction 6_{10} when the sum of a BCD digit exceeds 9_{10}.
36. Full subtractor is a combinational circuit that perform arithmetic subtraction of two binary bits in which one of the bit is borrow generated in previous subtraction.
37. The input to full adders has to be modified for different arithmetic and logical operations.

38. Half subtractor has two inputs and two outputs.

39. Full subtractor has two inputs and two outputs.

40. A 4-bit ALU can be designed using four full-subtractors.

Answers

1. False	5. True	11. False	15. True	21. True	26. False	31. True	36. False
2. False	7. True	12. True	17. True	22. True	27. True	32. True	37. True
3. True	8. True	13. False	18. False	23. True	28. True	33. False	38. True
4. True	9. False	14. False	19. True	24. True	29. True	34. False	39. False
5. False	10. True	15. True	20. True	25. True	30. True	35. True	40. False

III. Choose the right answer for the following questions

1. ***A combinational circuit is one in which the output depends on the***

a) Input combination at that time

b) Input combination and the previous output

c) Input combination at that time and previous input combination

d) Present output and previous output

2. ***The following switching functions are to be implemented using a decoder***

$f_1 = \Sigma m(1, 2, 4, 8, 10, 14)$; $f_2 = \Sigma m(2, 5, 9, 11)$; $f_3 = \Sigma m(2, 4, 5, 5, 7)$

The minimum configuration of the decoder should be

a) 2 to 4 line **b)** 3 to 8 line

c) 4 to 15 line **d)** 5 to 32 line

3. ***The circuit of Fig. 3 is equivalent to,***

a) AND gate **b)** OR gate

c) XOR gate **d)** XNOR gate

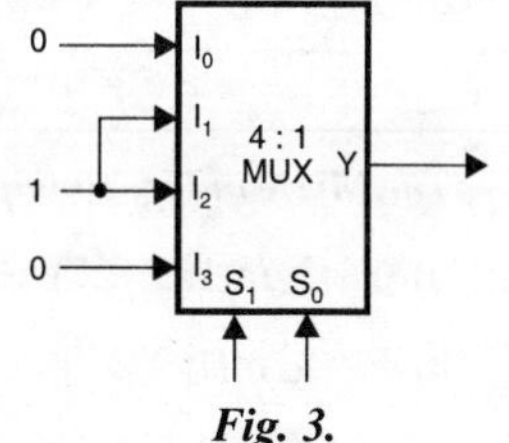

Fig. 3.

4. ***The minimum number of 2×1 MUX needed (with no added gates) to form a 3×1 MUX is***

a) 2 **b)** 3 **c)** 4 **d)** 1

5. ***The circuit in Fig. 5 works as***

a) binary to Gray converter

b) Gray to binary converter

c) odd parity generator

d) even parity generator

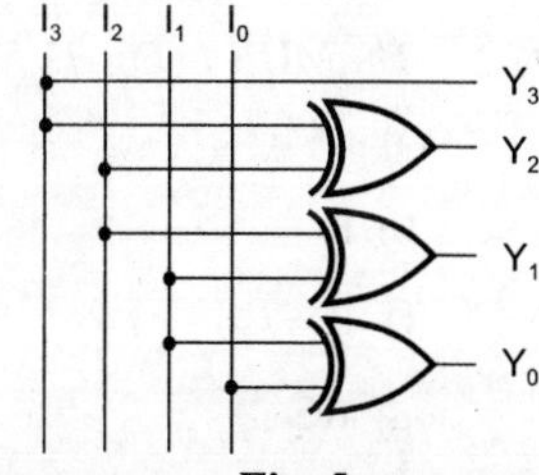

Fig. 5.

6. *The function "F" implemented by the multiplexer shown in Fig. 6 is*

a) A

b) B

c) A′B

d) AB′+ A′B

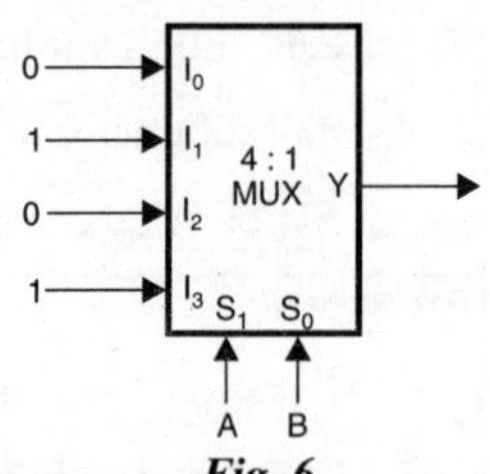

Fig. 6.

7. *The MUX in Fig. 7 implements the function*

a) F = BC′

b) F = BC

c) F = B′C′

d) F = B′C

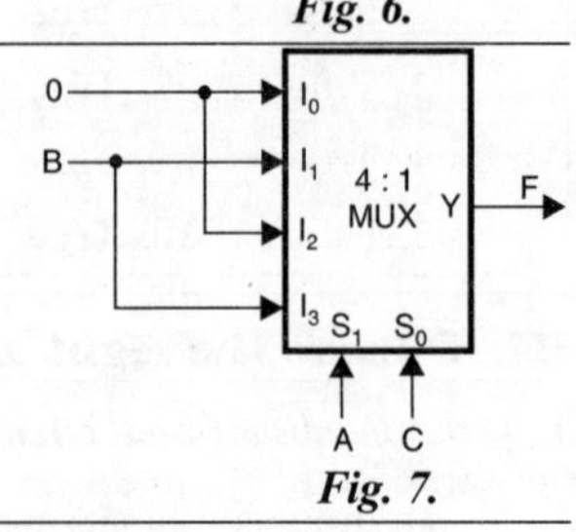

Fig. 7.

8. *The circuit shown in Fig. 8 is a*

a) Odd parity generator b) Even parity generator

c) Full adder d) Comparator

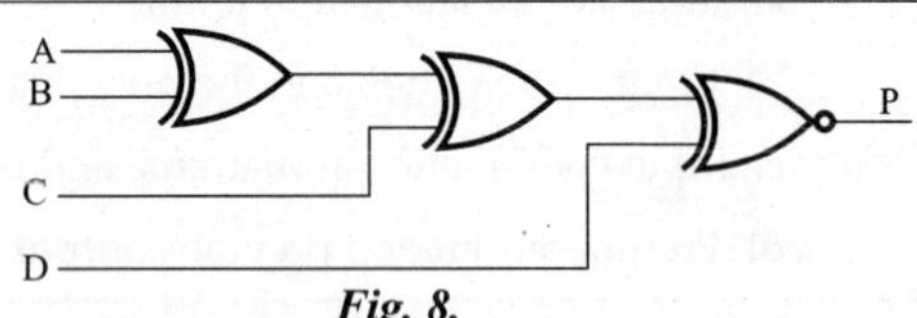

Fig. 8.

9. *The multiplexer in Fig. 9 implements the function*

a) Y = AB

b) Y = A′B + B′C

c) Y = A + BC

d) Y = BC

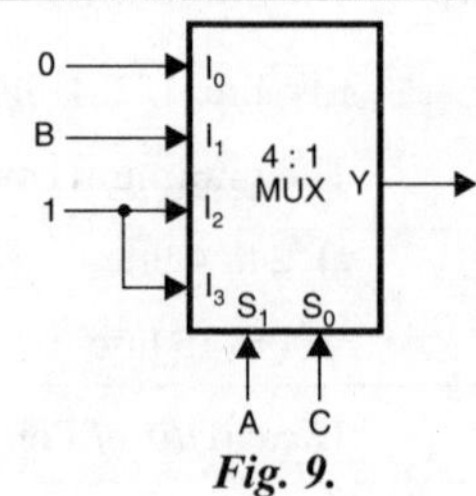

Fig. 9.

10. *The MUX in Fig. 10 implements the function*

a) $Y = \sum m(1, 3, 5, 7)$

b) $Y = \sum m(1, 2, 3, 5)$

c) $Y = \sum m(2, 5, 5, 7)$

d) $Y = \sum m(1, 4, 5, 5)$

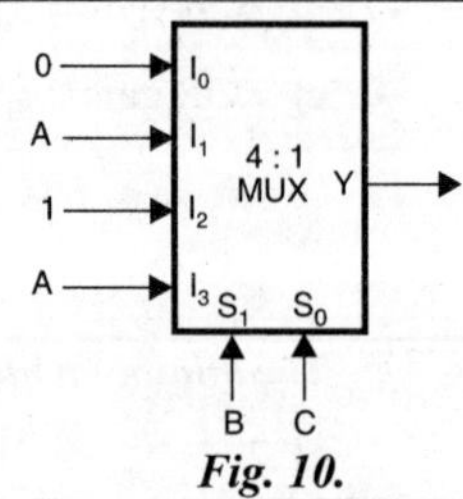

Fig. 10.

11. *The MUX in Fig. 11 implements the function*

a) B′

b) B

c) $B \oplus C$

d) $B \odot C$

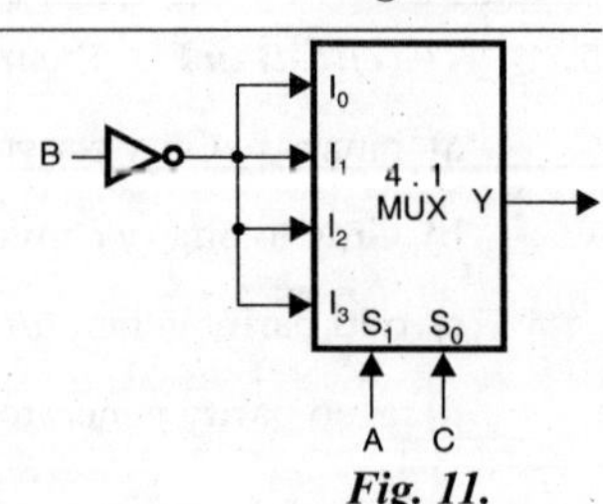

Fig. 11.

12. ***The output of the 4:1 MUX shown in Fig. 12 is***

    a) $x'y' + x$

    b) $x + y$

    c) $x' + y'$

    d) $xy + x'$

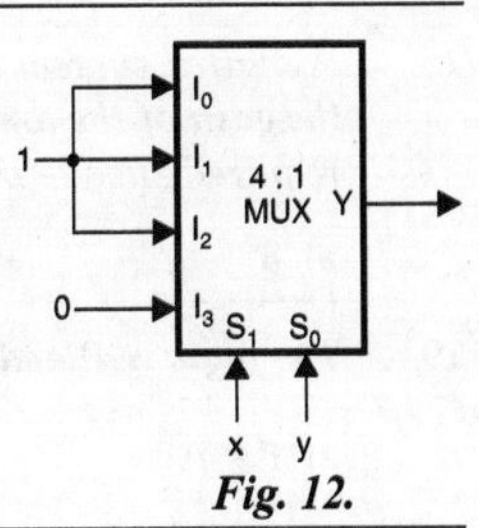

Fig. 12.

13. ***The boolean function realized by the given multiplexer is given by,***

    a) Σ m(0, 3, 5, 7, 8, 11, 12, 14)

    b) Σ m(0, 3, 5, 7, 12, 14)

    c) Σ m(2, 4, 5, 9, 10, 13, 15)

    d) Σ m(0, 1, 5, 7, 9, 12)

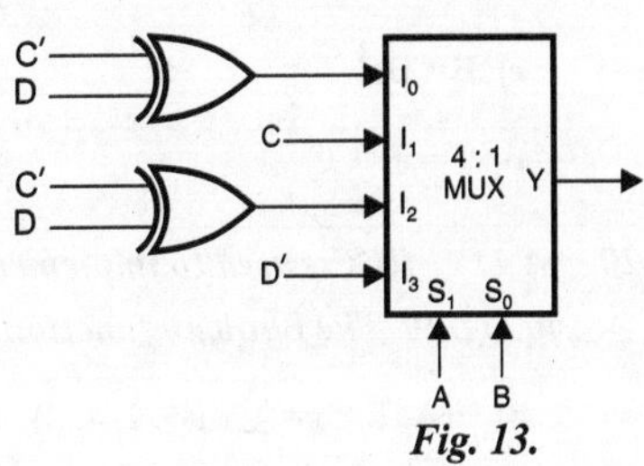

Fig. 13.

14. ***The boolean function implemented by the MUX is given as,***

    a) Σ m(0, 3, 4, 7)

    b) ΠM(3, 4, 7)

    c) ΠM(1, 2, 5, 5)

    d) None of the above

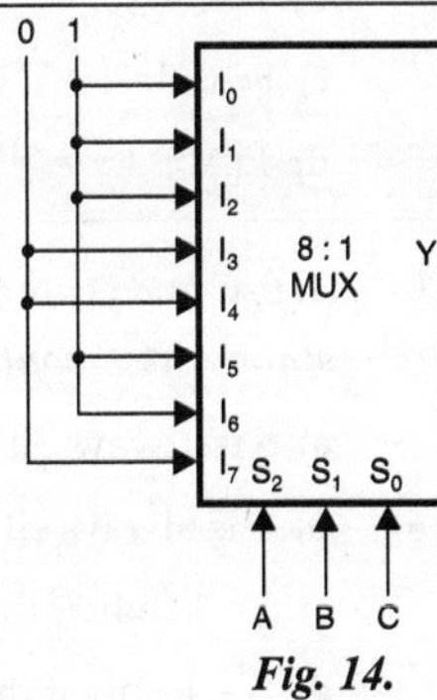

Fig. 14.

15. ***The circuit perform***

    a) OR function

    b) AND function

    c) NOR function

    d) Inverter

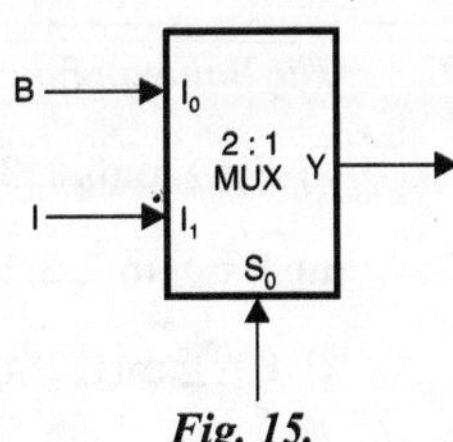

Fig. 15.

16. ***Without any additional circuitry, an 8 : 1 MUX can be used to obtain***

    a) Some but not all boolean function of 3 variables

    b) All functions of 3 variables but none of 4 variables

    c) All functions of 3 variables and some but not all of 4 variables

    d) All functions of 4 variables

17. ***The minimum number of 2:1 MUX required to realize a 4:1 MUX is***

    a) 1 b) 2 c) 3 d) 4

18. *A digital system is required to amplify a binary encoded audio signal, the user should be able to control the gain of the amplifier from a minimum to a maximum in 248 increments. The minimum number of bits required encode in straight binary is*

a) 8 **b)** 5 **c)** 5 **d)** 7

19. *The logic realized by the circuit of Fig. 19 is*

a) F = AC

b) F = A ⊕ C

c) F = BC

d) F = B + C

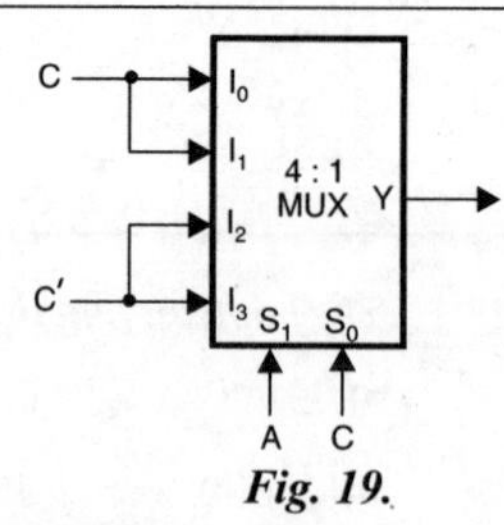

Fig. 19.

20. *A 4 : 1 MUX is used to implement a 3-input Boolean function as shown in Fig. 20. The Boolean function F(A, B, C) implemented is*

a) F(A, B,C) = Σ m(1, 2, 4, 5)

b) F(A, B,C) = Σ m(1, 2, 5)

c) F(A, B,C) = Σ m(2, 4, 5, 5)

d) F(A, B,C) = Σ m(1, 5, 5)

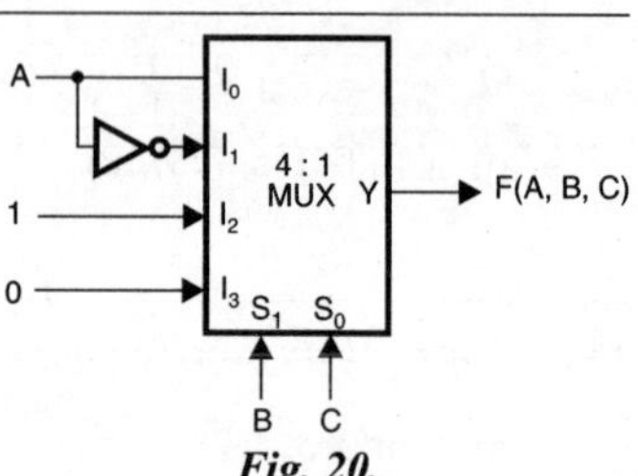

Fig. 20.

21. *A, B, C and D are input bits and Y is the output bit in the XOR gate circuit of Fig. 21. Which of the following statements about the sum S of A, B, C, D is correct?*

a) S is always either zero or odd

b) S is always either zero or even

c) S = 1 only if the sum of A, B, C and D is even

d) S = 1 only if the sum of A, B, C and D is odd

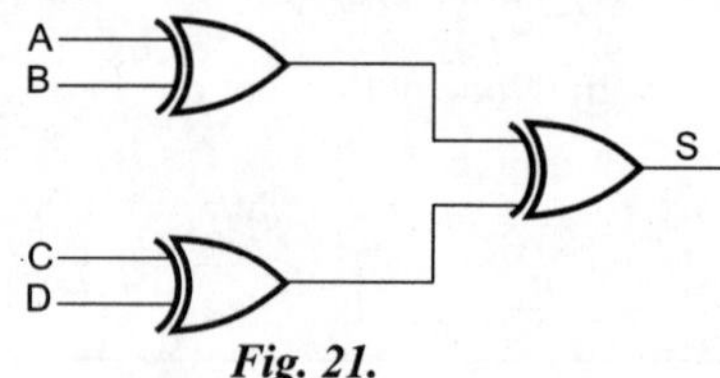

Fig. 21.

22. *The Boolean function realized by the logic circuit shown in Fig. 22 is*

a) F = Σ m(0, 1, 3, 5, 9, 10, 14)

b) F = Σ m(2, 3, 5, 7, 8, 12, 13)

c) F = Σ m(1, 2, 4, 5, 11, 14, 15)

d) F = Σ m(2, 3, 5, 7, 8, 9, 12)

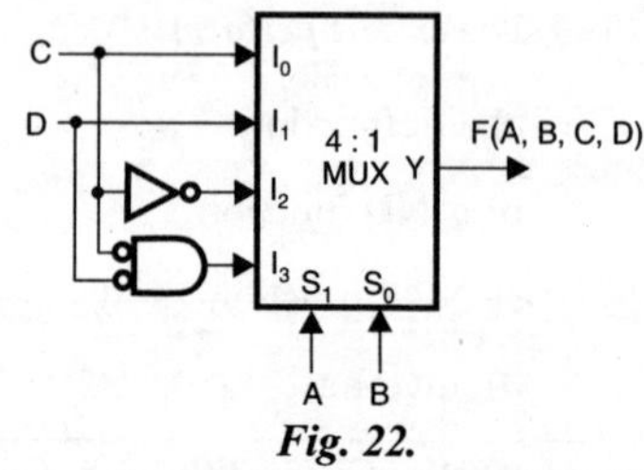

Fig. 22.

23. *In the circuit of Fig. 23, X is given by*

a) X = AB'C + A'BC' + A'B'C + ABC

b) X = A'BC + AB'C + ABC' + A'B'C'

c) X = AB + BC + AC

d) X = A'B' + B'C' + A'C'

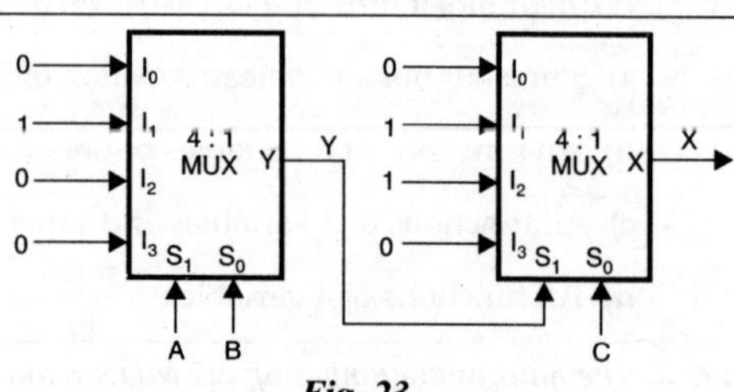

Fig. 23.

24. ***The circuit shown in Fig. 24 works as***

a) binary to Gray converter

b) Gray to binay converter

c) binary to its 2's complement converter

d) a parity generator

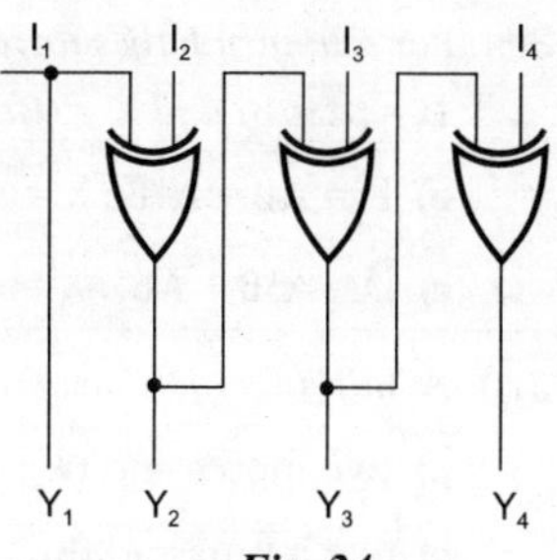

Fig. 24.

25. ***The output of MUX shown in Fig. 25 is***

a) $A \oplus B$

b) $(A \oplus B)'$

c) $(A . B) C$

d) $C' (A \oplus B)' + C (A \oplus B)$

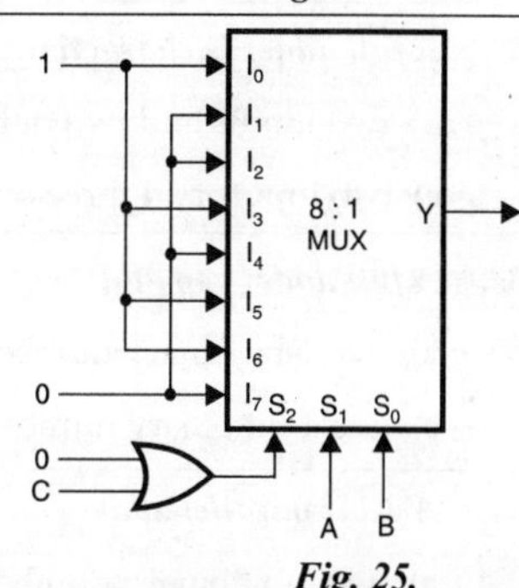

Fig. 25.

26. ***The given logic diagram in Fig. 26 represents a***

a) Multiplexer

b) Full adder

c) Half adder

d) None of the above

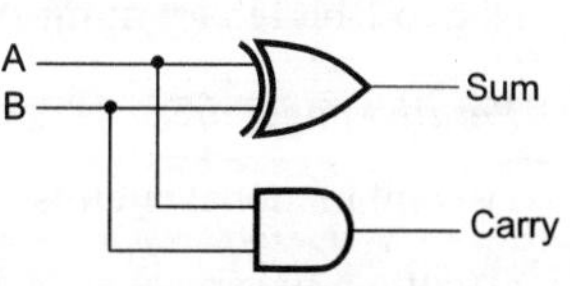

Fig. 26.

27 ***An instruction used to set the carry flag in a computer can be classified as***

a) data transfer **b)** arithmetic

c) logical **d)** program control

28 ***Two 2's complement number having sign bit x and y are added and the sign bit of the result is 2. If z is input carry, then the occurance of overflow is indicated by the Boolean function,***

a) xyz **b)** x'y'z'

c) x'y'z + xyz' **d)** xy + yz + zx

29. ***The circuit in Fig. 29 works as***

a) a full subractor

b) a full adder

c) a binary to gray converter

d) a gray to binary converter

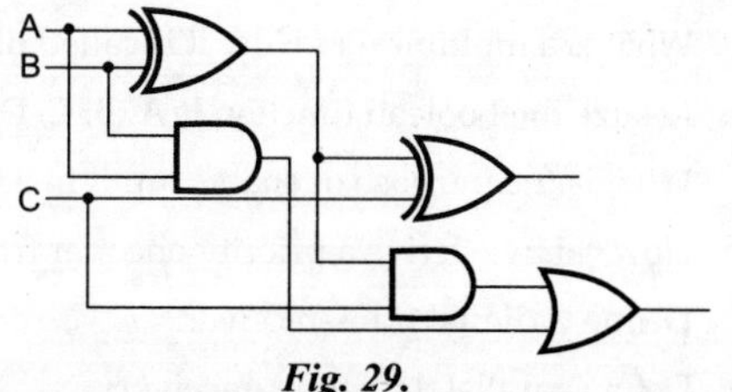

Fig. 29.

30. ***For a binary half subtractor having 2 input A and B, the correct set of logical expression for the outputs D = (A – B) and X = (Borrow) are,***

a) D = AB + A'B ; X = A'B b) D = A'B + AB'+ AB'; X = AB'

c) D = A'B + AB'; X = A'B d) D = AB + A'B ; X = AB'

31. ***A half adder is characterized by***

a) two inputs and two outputs b) three inputs and two outputs

c) two inputs and three outputs d) two inputs and one output

32. ***A full adder is characterized by***

a) two inputs and two outputs b) three inputs and two outputs

c) two inputs and three outputs d) two inputs and one output

33. ***A full adder can add***

a) two 1-bit binary number b) two 2-bit binary number

c) two 4-bit binary number d) four bits at a time

34. ***A 4-bit parallel adder can add***

a) two 2-bit binary number b) two 4-bit binary number

c) two 1-bit binary number d) four bits at a time

35. ***Parallel adder are***

a) combinational circuits b) sequential circuits

c) both of the above d) None of the above

Answers

1. a	5. b	11. a	15. d	21. d	26. c	31. a
2. c	7. b	12. c	17. c	22. d	27. b	32. b
3. c	8. a	13. a	18. a	23. a	28. d	33. a
4. a	9. c	14. b	19. b	24. b	29. b	34. b
5. a	10. c	15. a	20. a	25. d	30. c	35. a

IV. Answer the following questions

1. Give examples for combinational circuit.
2. What is a multiplexer? How it is called otherwise? Why?
3. Realize the Boolean function F(A, B, C, D) = $\Sigma(0, 1, 5, 5)$ using appropriate multiplexer.
4. What is the purpose of enable input in a decoder?
5. How can we derive a priority encoder from an encoder?
6. Define is digital trans-receiver.
7. Define parallel data trans-receiver.
8. Define serial data trans-receiver.
9. What is even parity generator and draw the diagram.

10. What is odd parity generator and draw the diagram.
11. Write short notes on BCD to 7-segment decoder.
12. Explain is multiplexed 7-segment display.
13. Write advantages of multiplexed display.
14. Realize half adder using K-map.
15. Draw a block diagram of half adder. Write a truth table.
16. Draw logic circuit and truth table of full adder.
17. Explain the operation of half subtractor.
18. Realize full subtractor using K-map.
19. Design full subtractor circuit. Write a truth table.
20. Explain the working of 4-bit parallel adder.
21. Realize a carry look ahead adder.
22. Design BCD adder using binary adders.
23. Design binary adder/subtractor using full adders.
24. Draw a logic circuit of 2-bit magnitude comparator.
25. What is combinational multiplier.
26. Draw a diagram of 8-bit ALU using IC 74382.

V. Solve the following problems

E5.1 Obtain the simplified Boolean expression for output F interms of the input variables in the circuit of Fig. E5.1.

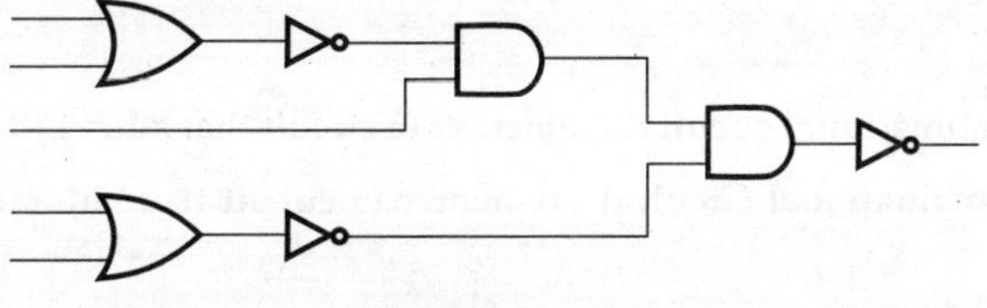

Fig. E5.1.

E5.2 Design a Boolean function with 3 inputs and 1 output.

a) The output is 0 when the binary value of the input is less than 3. The output is 1 otherwise.

b) The output is 1 when the binary value of the input is an odd number.

E5.3 Design a combinational circuit that compares two 3-bit numbers to check if they are equal. The circuit output is 0 if the two numbers are equal and 1 otherwise.

E5.4 Design a 3-bit combinational circuit that generates the 2's complement of the input binary number. Draw the simplified logic circuit.

E5.5 Design a majority circuit whose output is equal to 1 if the input variables have more 0's than 1's. The output is 0 otherwise.

E5.6 Design a code converter that converts a BCD to Gray code.

E5.7 Design a code converter that converts a decimal digit from the 8, 4, – 2, – 1 code to BCD.

E5.8 Design a combinational circuits that converts a decimal digit from the 2, 4, 2, 1 code to 8, 4, – 2, – 1 code.

E5.9 A combinational circuit is specified by the following Boolean functions.

$F_1(A, B, C) = \sum m(3, 5, 6)$; $F_2(A, B, C) = \sum m(1, 4)$; $F_3(A, B, C) = \sum m(2, 3, 5, 6, 7)$

Implement the circuit with a logic low decoder and NAND or AND gates connected to the decoder outputs. Use a block diagram of the decoder. Minimize the number of inputs in the external gates.

E5.10 Using decoder and external gates, design the combination circuit defined by the following Boolean functions.

$F_1 = x'y'z' + xz + yz$

$F_2 = y'z' + xy' + yz'$

$F_3 = x'y'z + xy$

E5.11 Construct a 16:1 multiplexer with two 8:1 and a 2:1 MUX.

E5.12 Implement the Boolean function with a MUX

a) $F(A, B, C, D) = \sum m(0, 2, 5, 7, 11, 14)$

b) $F(A, B, C, D) = \prod M(3, 8, 12, 15)$

E5.13 Implement a following Boolean function with 4:1 MUX and gates.

$F_1\ (A, B, C, D) = \sum m(2, 4, 5, 12, 13, 14, 15)$

$F_2\ (A, B, C, D) = \sum m(2, 3, 6, 8, 9, 14, 15)$

Connect inputs A and B to the selection lines. The input requirements for the four data lines will be a function of the variables C and D. These values are obtained by expressing F as a function of C and D for each of the four cases when AB = 00, 01, 10, 11. The functions may have to be implemented with external gates and with connections to power (1) and ground (0).

E5.14 Using adders,

a) Design a 3-bit combinational circuit incrementer (a circuit that adds 1 to a 3-bit binary number).

b) Design a 3-bit combinational circuit decrementer (a circuit that subtracts 1 from a 3-bit binary number).

E5.15 Design a half adder using NAND gates.

E5.16 Design a half adder using logic low decoder.

E5.17 Design a half adder using a logic high decoder and NAND gates.

E5.18 Design a half adder using multiplexer.

E5.19 Design a full adder using NOR gates.

E5.20 Design a full adder using logic low decoder.

E5.21 Design a half subtractor using NOR gates.

E5.22 Design a half subtractor using logic low decoder.

E5.23 Design a half subtractor using a logic high decoder and NAND gates.

E5.24 Design a half subtractor using multiplexer.

E5.25 Design a full subtractor using logic low decoder.

E5.26 Design a full subtractor using multiplexer.

Answers

E5.1 $F = A + B + C' + D + E$ **E5.2** **a)** $x + yz$ **b)** z

E5.3

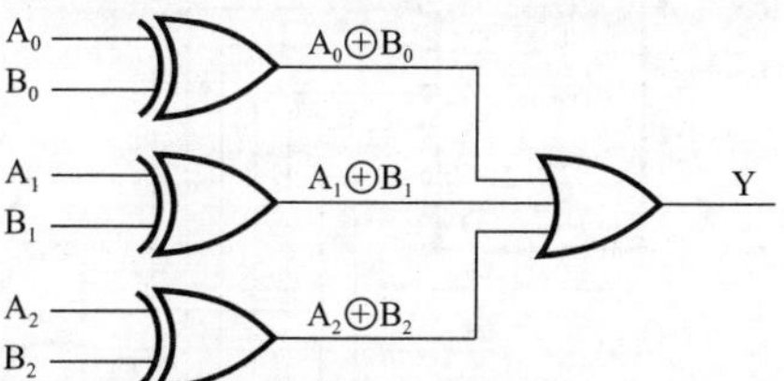

Fig: *Combinational circuit to compare two 3-bit numbers for equality.*

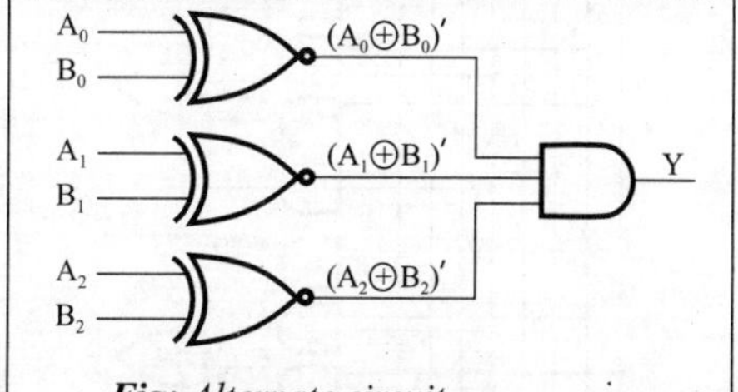

Fig: *Alternate circuit.*

E5.4

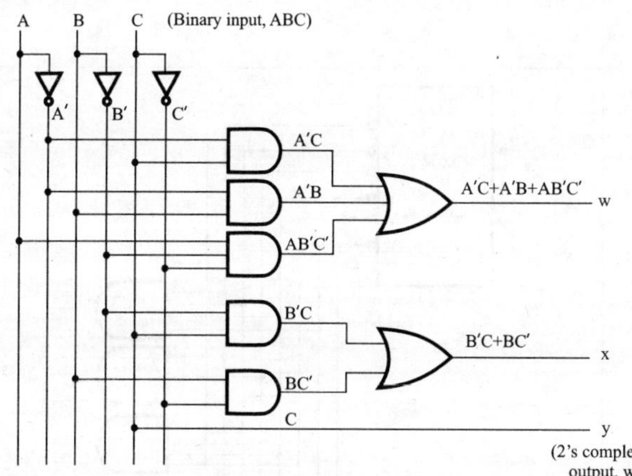

Fig: *Logic circuit.*

E5.5 $F = x'y' + x'z' + y'z'$

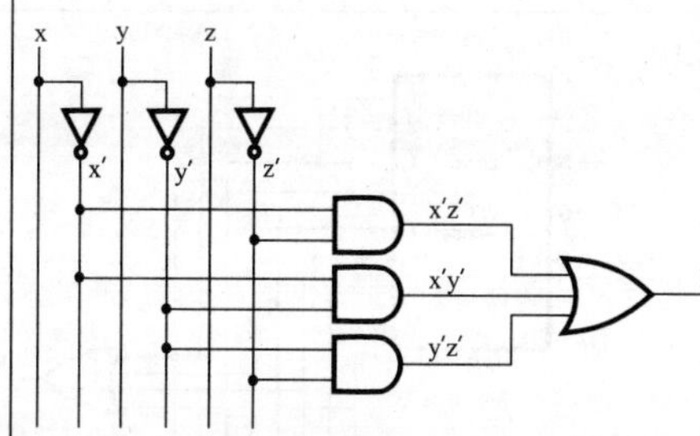

Fig: *Logic circuit of F.*

E5.6

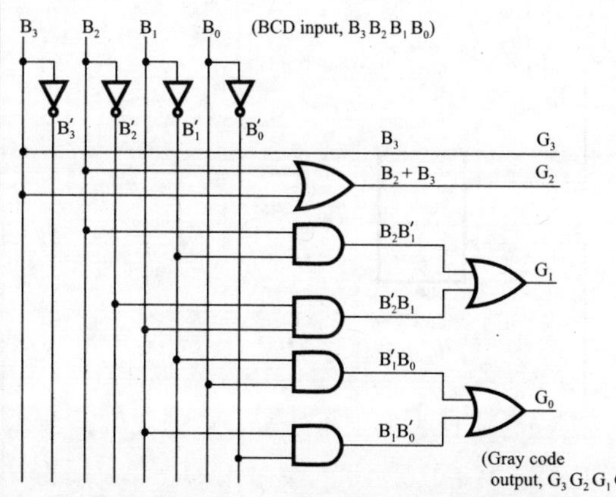

Fig: *Logic circuit of BCD to gray code conveter.*

E5.7

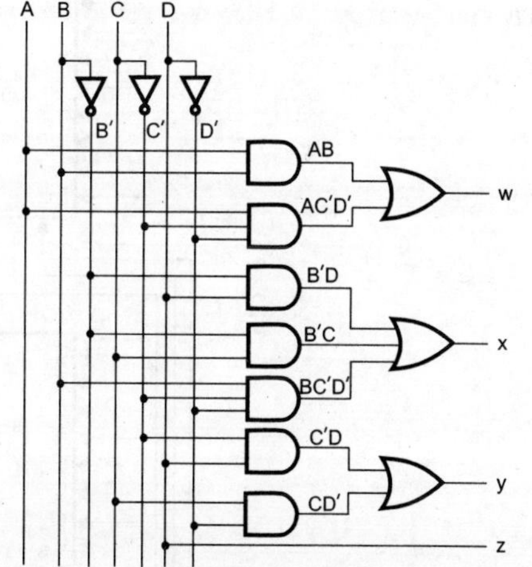

Fig: *Logic circuit of 8, 4, −2, −1 code to BCD code converter.*

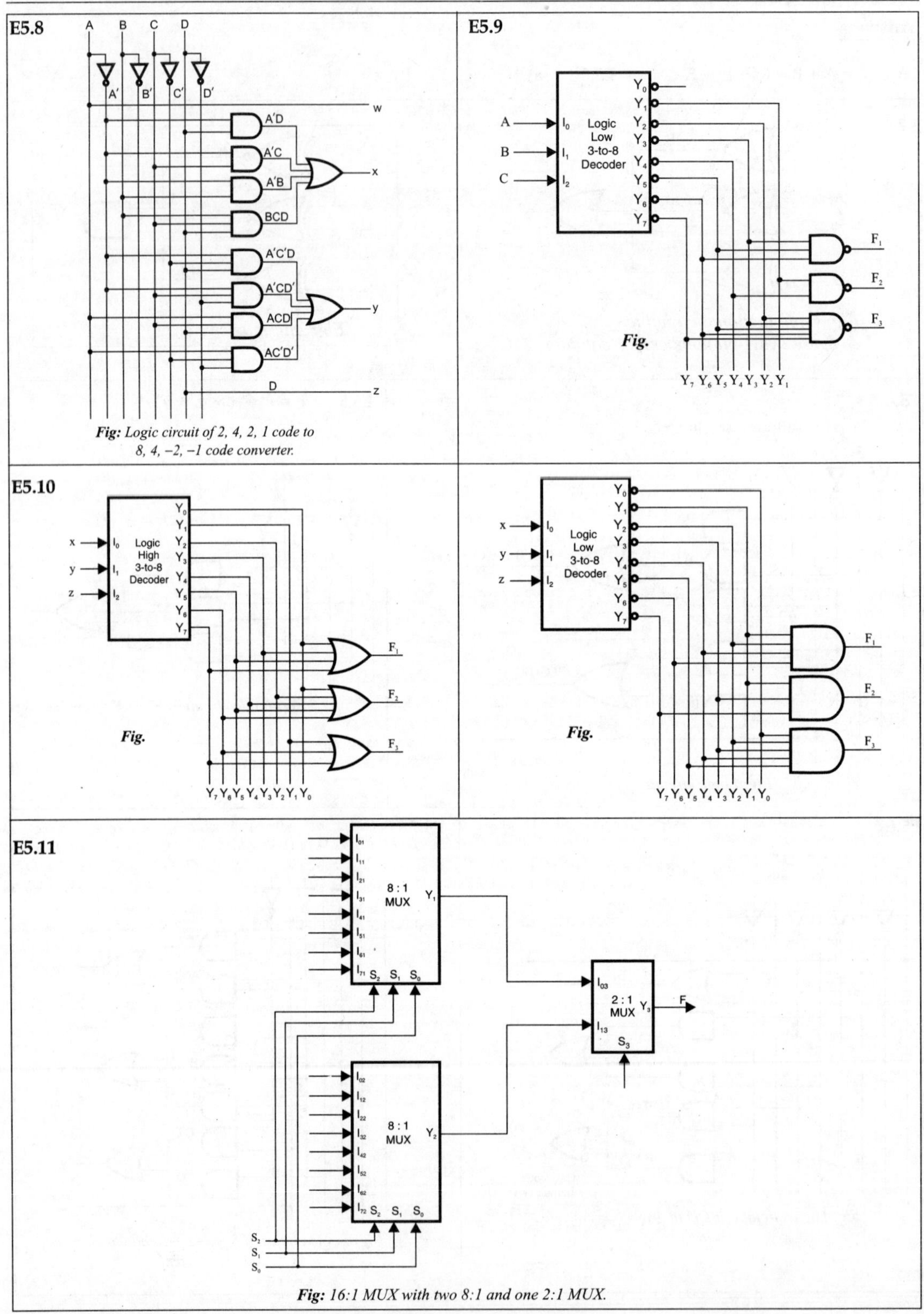

Fig: *Logic circuit of 2, 4, 2, 1 code to 8, 4, −2, −1 code converter.*

Fig.

Fig.

Fig.

Fig: *16:1 MUX with two 8:1 and one 2:1 MUX.*

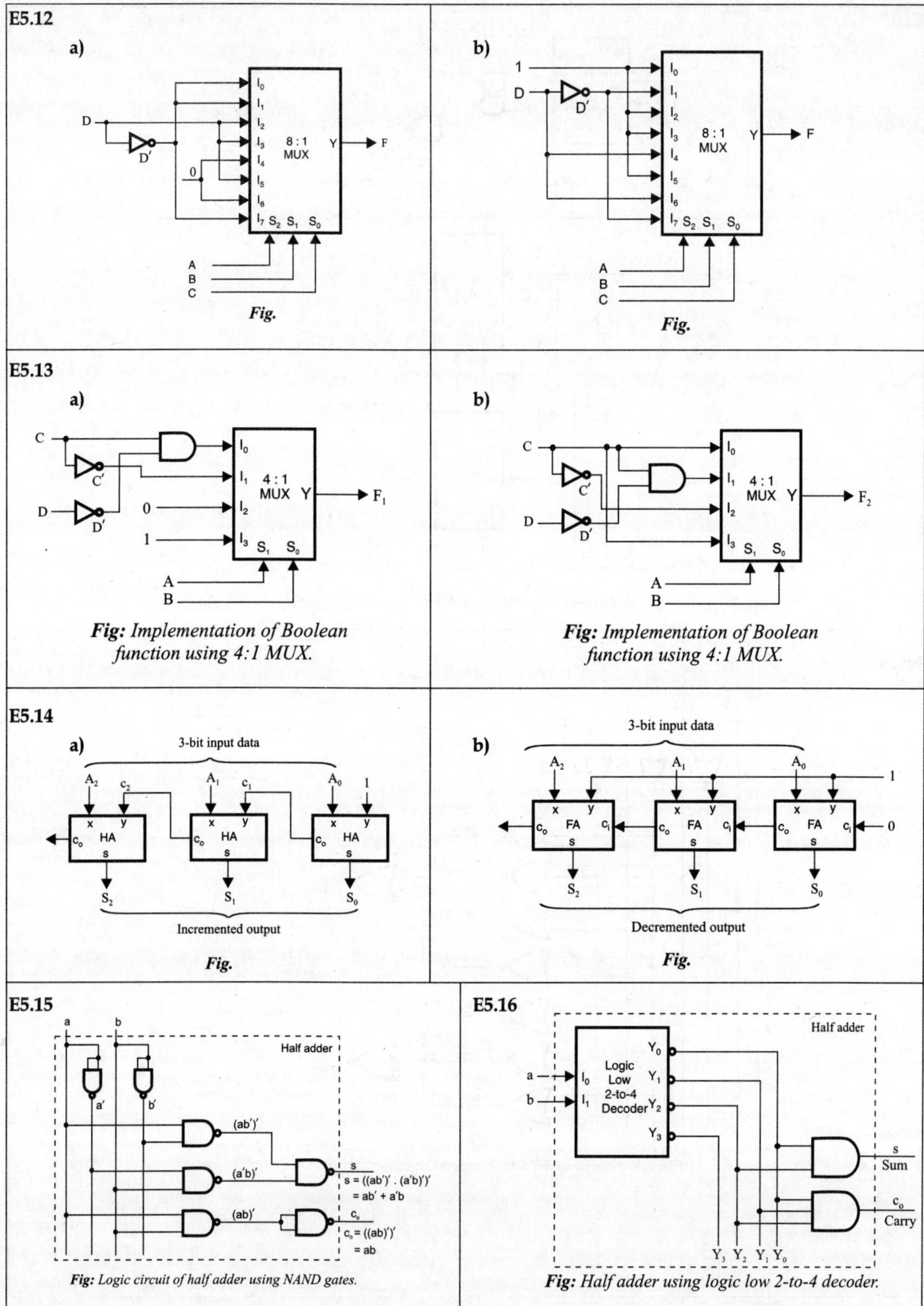

Fig.

Fig.

Fig: Implementation of Boolean function using 4:1 MUX.

Fig: Implementation of Boolean function using 4:1 MUX.

Fig.

Fig.

Fig: Logic circuit of half adder using NAND gates.

Fig: Half adder using logic low 2-to-4 decoder.

E5.17

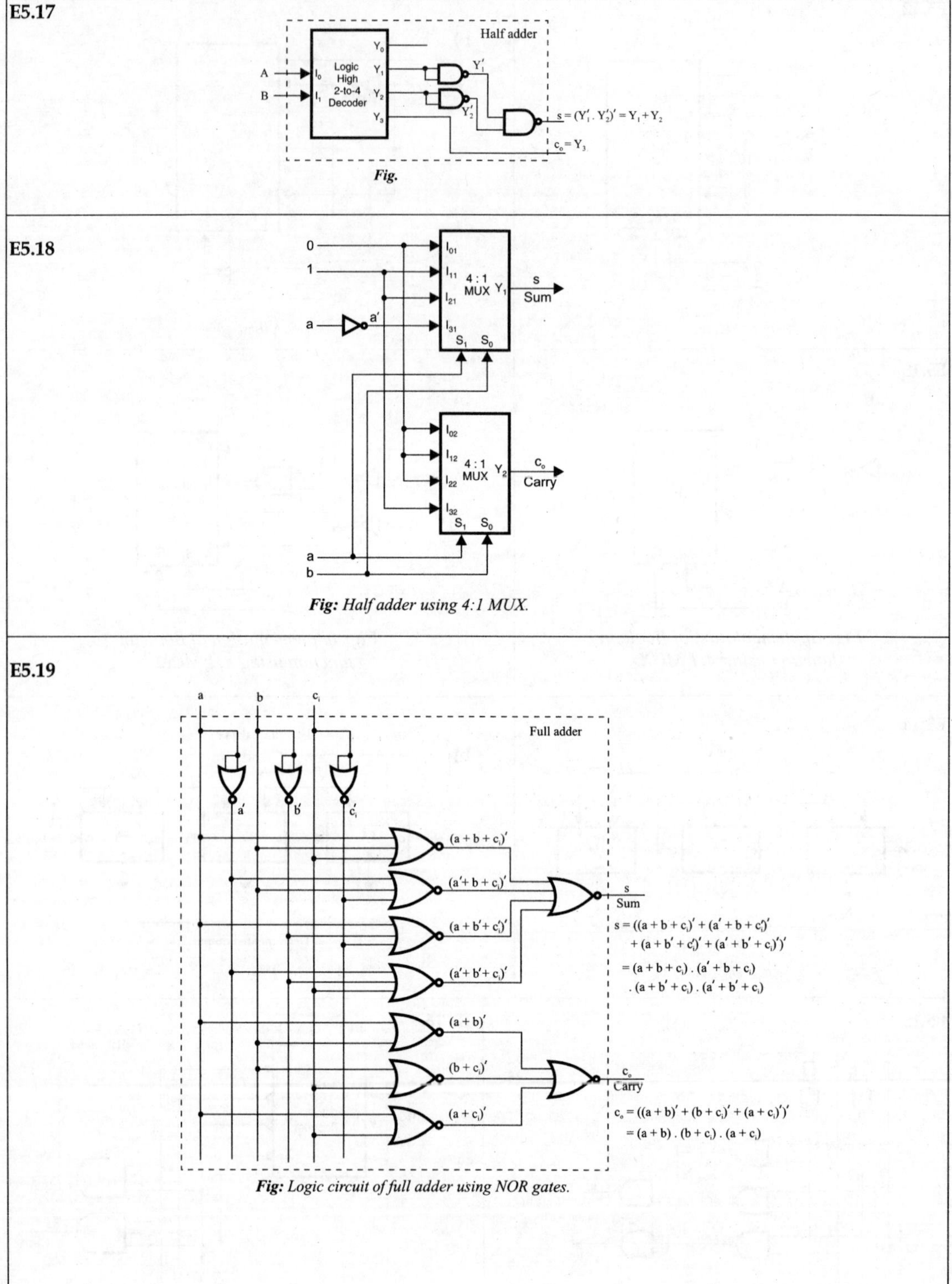

Fig.

E5.18

***Fig:** Half adder using 4:1 MUX.*

E5.19

***Fig:** Logic circuit of full adder using NOR gates.*

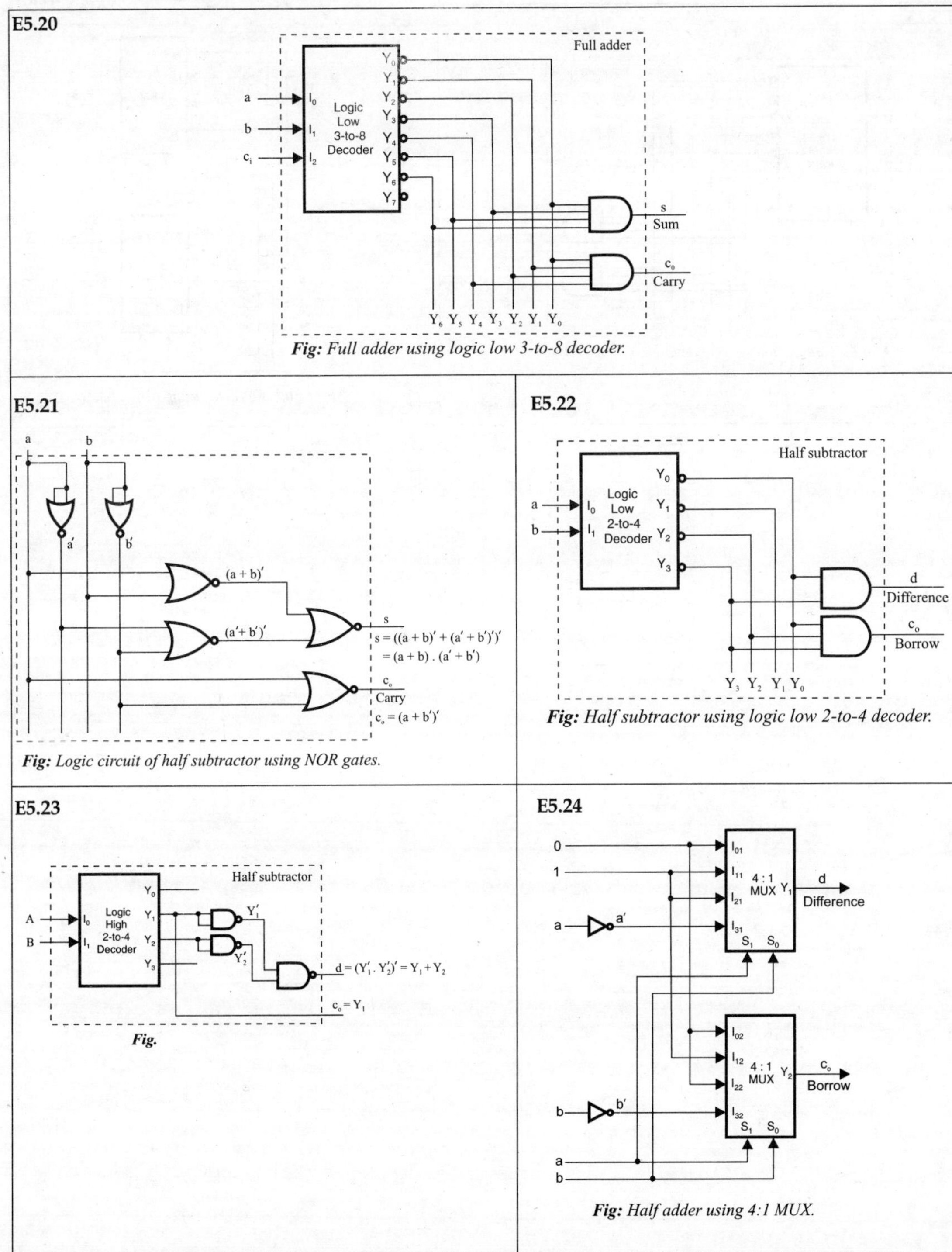

Fig: Full adder using logic low 3-to-8 decoder.

Fig: Logic circuit of half subtractor using NOR gates.

Fig: Half subtractor using logic low 2-to-4 decoder.

Fig.

Fig: Half adder using 4:1 MUX.

E5.25

E5.26

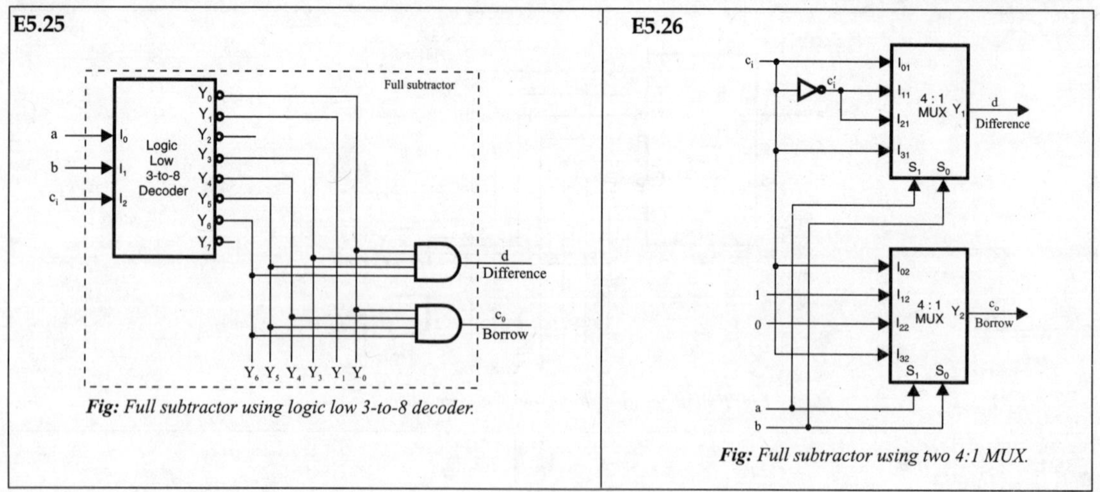

Fig: Full subtractor using logic low 3-to-8 decoder.

Fig: Full subtractor using two 4:1 MUX.

CHAPTER 6

LATCHES AND FLIP-FLOPS

6.1 Bistable Element and Latches

The simplest ***bistable element*** consists of two inverters connected in a series loop as shown in Fig. 6.1. It has no inputs but has two outputs Q and Q′. Since the bistable element has no inputs, the value of outputs Q and Q′ cannot be changed. However, Q will take either 1 or 0 during power ON.

During power ON, if Q = 0 then Q′ = 1 and if Q = 1 then Q′ = 0.

The above two conditions of the bistable element can be defined as two states and hence the circuit is called bistable element.

State-1: Q = 0 then Q′ = 1

State-2: Q = 1 then Q′ = 0

The bistable element will take any one of the two states during power ON and remain stable in that state. Therefore, the bistable element has memory in the sense that it can remember the content (or state) of the circuit indefinitely. But the drawback is that the state it can take up during power ON is not assured and we cannot change the state. This draw back can be eliminated by forcing some inputs and such a circuit is called latch. Therefore, the bistable element forms a basic circuit of a latch.

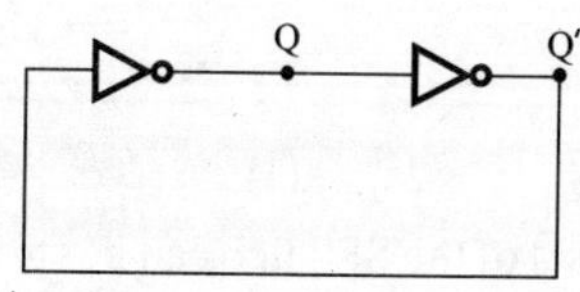

***Fig. a:** Two inverters in series loop.*

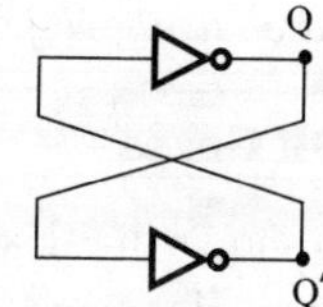

***Fig. b:** Modified structure of Fig. a.*

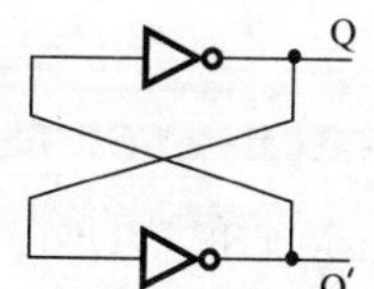

***Fig. c:** Bistable element with two outputs.*

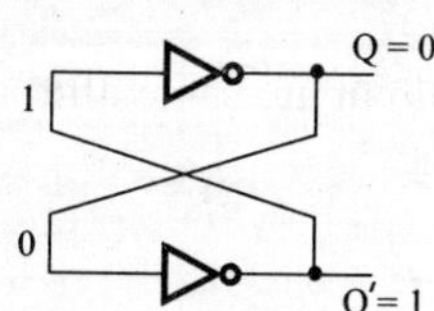

***Fig. d:** Stable state-1.*

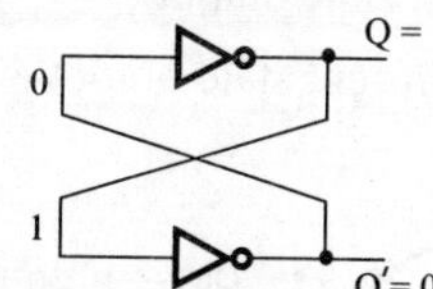

***Fig. e:** Stable state-2.*

***Fig. 6.1:** Bistable element.*

(RGPV, Jun'22, 2 Marks)

A ***latch*** is a basic storage device to store a binary bit. A latch is designed with two stable states in order to store two possible values of a binary bit 1 and 0. Latches are designed with complementary inputs and outputs so that positive or negative logic can be implemented easily. The latches are sequential circuits without a clock. The two types of basic latches are SR latch and D latch.

6.1.1 SR Latch

The ***SR latch*** is a digital circuit designed to store 1 or 0 applied as two inputs called set and reset and it is provided with an output and complement of the output.

Set is an input used to set the output to 1 and it is denoted by S.

Reset is an input used to clear the output to 0 and it is denoted by R.

The two outputs are denoted as Q and Q′. Here, Q and Q′ are complements. The standard symbol of SR latch is shown in Fig. 6.2.

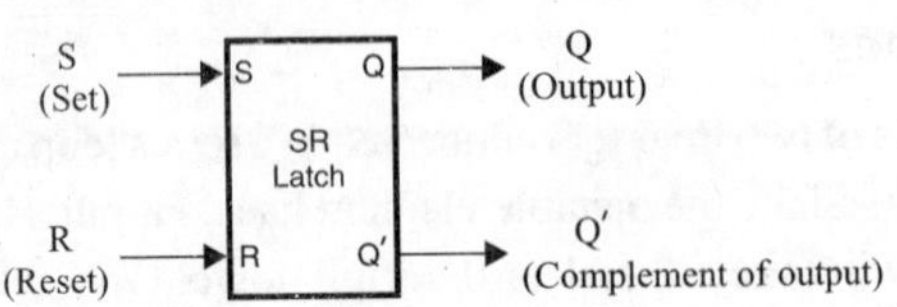

Fig. 6.2: Symbol of SR latch.

Table 6.1: Stable States of SR Latch

Set State	Reset State
Q = 1	Q = 0
Q′ = 0	Q′ = 1

The SR latch is designed to have two stable states as shown in Table 6.1. A transition from one state to another can be initiated by applying appropriate input. The SR latch can be constructed using two cross-coupled NOR gates or using two cross-coupled NAND gates as shown in Fig. 6.3.

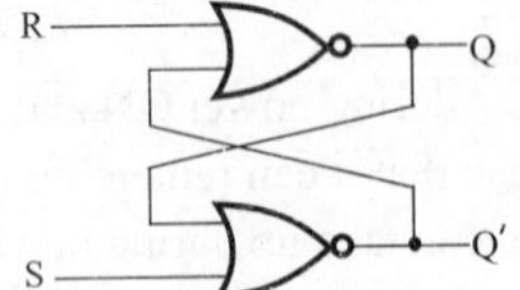

Fig. a: SR latch using NOR gates.

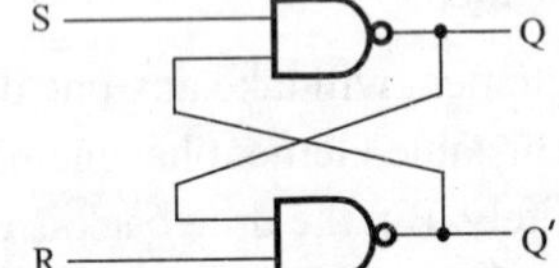

Fig. b: SR latch using NAND gates.

Fig. 6.3: SR latch using NOR and NAND gates.

6.1.2 Analysis of SR Latch using NOR Gates

(AU, Nov/Dec'18, 13 Marks)

Consider SR latch using NOR gates as shown in Fig. 6.4. The analysis of the SR latch can be carried for driving the output from set to reset state and vice-versa.

Let, Q(t) and Q′(t) be present state output.

Let, Q(t + 1) and Q′(t + 1) be next state output.

A ***stable state*** is the state when present state output and next state output are same after a change in input.

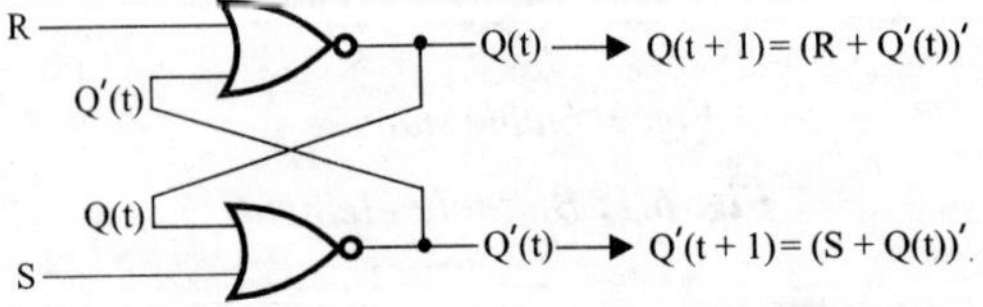

Fig. 6.4: SR latch using NOR gates.

Case i: Drive the output from reset to set state

Initially both inputs are 0 and the output is in reset state which is one of the stable state. In NOR based SR latch, when both inputs are 0 there is no change in output.

When the output is in reset state, a high pulse (or momentary 1) applied at set input (S = 1) drives the output to set state (Q = 1, Q′ = 0).

From Fig. 6.4 we get the following equations for next state output.

$$Q(t+1) = (R + Q'(t))'$$

$$Q'(t+1) = (S + Q(t))'$$

The various transitions that occur in the NOR based SR latch when it is driven from reset to set state are listed in Table 6.2.

When the output settle at stable set state after a momentary 1 applied at S, the inputs can remain at both 0 and this will maintain the previous state (set state) and will not make any change in output.

Table 6.2: Transition from Reset to Set State in NOR Based SR Latch

Inputs		Outputs				Comment
S	R	Present State		Next State		
				Q(t + 1)	Q′(t + 1)	
		Q(t)	Q′(t)	(R + Q′(t))′	(S + Q(t))′	
0	0	0	1	-	-	Reset state
1	0	0	1	0	0	S → 1
1	0	0	0	1	0	Transition
1	0	1	0	1	0	Stable set state
0	0	1	0	1	0	S → 0 No change Remain in set state

S should remain at 1 until a stable set state is achieved.

Note: *Q(t) = Previous Q(t + 1) ; Q′(t) = Previous Q′(t + 1)*

Case ii: Drive the output from set to reset state

Initially both inputs are 0 and the output is in set state which is one of the stable state. In NOR based SR latch, when both inputs are 0 there is no change in output.

When the output is in set state, a high pulse (or momentary 1) applied at reset input (R = 1) drives the output to reset state (Q = 0, Q′ = 1).

The equations for next state output are same as that of case i.

The various transitions that occur in the NOR based SR latch when it is driven from set to reset state are listed in Table 6.3.

When the output settle at stable reset state after a momentary 1 applied at R, the inputs can remain at both 0 and this will maintain the previous state (reset state) and will not make any change in output.

Table 6.3: Transition from Set to Reset State in NOR Based SR Latch

Inputs		Outputs				Comment
S	R	Present State		Next State		
		Q(t)	Q′(t)	Q(t + 1) (R + Q′(t))′	Q′(t + 1) (S + Q(t))′	
0	0	1	0	-	-	Set state
0	1	1	0	0	0	R →1
0	1	0	0	0	1	Transition
0	1	0	1	0	1	Stable reset state
0	0	0	1	0	1	R → 0 No change Remain in reset state

R should remain at 1 until a stable reset state is achieved.

Note: *Q(t) = Previous Q(t + 1) ; Q′(t) = Previous Q′(t + 1)*

Based on the two cases of analysis of NOR based SR latch, the truth table of SR latch using NOR gates is obtained as shown in Table 6.4.

Note: *Applying both R = 1 and S = 1 is illegal and lead to indeterminate or unpredictable output.*

Table 6.4: Truth Table of SR latch using NOR Gates

Inputs		Outputs	
S	R	Q	Q′
0	0	No change	
0	1	0	1
1	0	1	0
1	1	Indeterminate	

Timing Diagram of NOR based SR Latch

The timing diagram or waveform of SR latch is drawn by taking various possible combinations of binary inputs and predicting the outputs from the knowledge of truth table of SR latch.

The timing diagram of NOR based SR latch shown in Fig. 6.5 is drawn by taking SR inputs as 00, 01, 00, 10, 00 and 11. The outputs are predicted from the truth table shown in Table 6.4. In the timing diagram, the following changes occur at various timings. Here, any change in inputs will immediately initiate a change in output.

Time t_0: S and R inputs are both 0. Initial outputs are assumed as Q = 1 and Q′ = 0.

Time t_1: R input is made 1 and so output Q is reset to 0 and Q′ becomes 1.

Time t_2: R input is made 0. Since S and R inputs are both 0, no change in outputs.

Time t_3: S input is made 1 and so output Q is set to 1 and Q′ becomes 0.

Time t_4: S input is made 0. Since S and R inputs are both 0, no change in outputs.

Time t_5: S and R inputs are both 1 and hence the outputs are indeterminate.

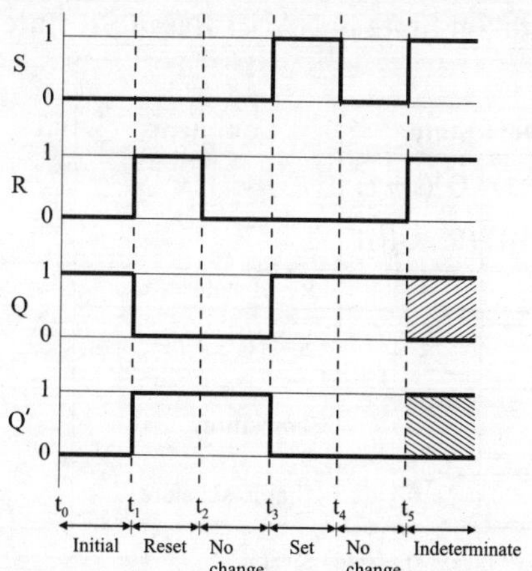

Fig. 6.5: Timing diagram (or waveform) of NOR based SR latch.

6.1.3 Analysis of SR Latch using NAND Gates *(AU, Nov/Dec'18, 13 Marks) (VTU, Aug/Sep'20, 8 Marks)*

Consider SR latch using NAND gates as shown in Fig. 6.6. The analysis of the SR latch can be carried for driving the output from set to reset state and vice-versa.

Let, Q(t) and Q′(t) be present state output.

Let, Q(t + 1) and Q′(t + 1) be next state output.

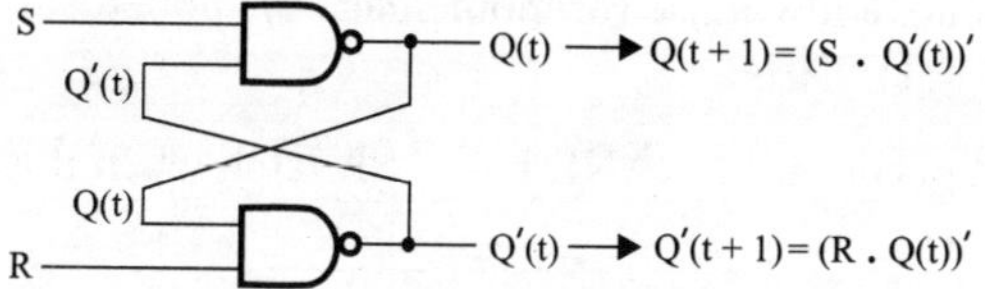

Fig. 6.6: SR latch using NAND gates.

Case i: Drive the output from reset to set state

Initially both inputs are 1 and the output is in reset state which is one of the stable state. In NAND based SR latch, when both inputs are 1 there is no change in output.

When the output is in reset state, a low pulse (or momentary 0) applied at set input (S = 0) drives the output to set state (Q = 1, Q′ = 0).

From Fig. 6.6 we get the following equations for next state output.

$$Q(t+1) = (S\,.\,Q'(t))'$$

$$Q'(t+1) = (R\,.\,Q(t))'$$

The various transitions that occur in the NAND based SR latch when it is driven from reset to set state are listed in Table 6.5.

When the output settle at stable set state after a momentary 0 applied at S, the inputs can remain at both 1, and this will maintain the previous state (set state) and will not make any change in output.

Table 6.5: Transition from Reset to Set State in NAND Based SR Latch

Inputs		Outputs				Comment
S	R	Present State		Next State		
		Q(t)	Q'(t)	Q(t + 1) (S . Q'(t))'	Q'(t + 1) (R . Q(t))'	
1	1	0	1	-	-	Reset state
0	1	0	1	1	1	S → 0
0	1	1	1	1	0	Transition
0	1	1	0	1	0	Stable set state
1	1	1	0	1	0	S → 1 No change Remain in set state

S should remain at 0 until a stable set state is achieved.

Note: Q(t) = Previous Q(t + 1) ; Q'(t) = Previous Q'(t + 1)

Case ii: Drive the output from set to reset state

Initially both inputs are 1 and the output is in set state which is one of the stable state. In NAND based SR latch, when both inputs are 1 there is no change in output.

When the output is in set state, a low pulse (or momentary 0) applied at reset input (R = 0) drives the output to reset state (Q = 0, Q' = 1).

The various transitions that occur in the NAND based SR latch when it is driven from set to reset state are listed in Table 6.6.

When the output settle at stable reset state after a momentary 0 applied at R, the inputs can remain at both 1, and this will maintain the previous state (reset state) and will not make any change in output.

Table 6.6: Transition from Set to Reset State in NAND Based SR Latch

Inputs		Outputs				Comment
S	R	Present State		Next State		
		Q(t)	Q'(t)	Q(t + 1) (S . Q'(t))'	Q'(t + 1) (R . Q(t))'	
1	1	1	0	-	-	Set state
1	0	1	0	1	1	R → 0
1	0	1	1	0	1	Transition
1	0	0	1	0	1	Stable reset state
1	1	0	1	0	1	R → 1 No change Remain in reset state

R should remain at 0 until a stable reset state is achieved.

Note: Q(t) = Previous Q(t + 1) ; Q'(t) = Previous Q'(t + 1)

Based on the two cases of analysis of NAND based SR latch, the truth table of SR latch using NAND gates is obtained as shown in Table 6.7. The NAND based SR latch is available as a standard IC 74279. The pin configuration of 74279 is shown in Fig. 6.7.

***Note:** Applying both R = 0 and S = 0 is illegal and lead to indeterminate or unpredictable output.*

Table 6.7: Truth Table of SR Latch using NAND Gates

Inputs		Outputs	
S	R	Q	Q′
0	0	Indeterminate	
0	1	1	0
1	0	0	1
1	1	No change	

74279

Pin	Signal	Pin	Signal
1	$1\overline{R}$	16	V_{cc}(+5V)
2	$1\overline{S}_1$	15	$4\overline{S}$
3	$1\overline{S}_2$	14	$4\overline{R}$
4	1Q	13	4Q
5	$2\overline{R}$	12	$3\overline{S}_2$
6	$2\overline{S}$	11	$3\overline{S}_1$
7	2Q	10	$3\overline{R}$
8	GND (0V)	9	3Q

Pin	Description
$1\overline{S}_1$, $1\overline{S}_2$, $3\overline{S}_1$, $3\overline{S}_2$	Dual set inputs
$2\overline{S}$, $4\overline{S}$	Set inputs
$1\overline{R}$, $2\overline{R}$, $3\overline{R}$, $4\overline{R}$	Reset inputs
1Q, 2Q, 3Q, 4Q	Outputs
V_{cc}	Supply voltage (+5V)
GND	Ground (0V)

***Note:** Latches 1 and 3 has two set inputs $\overline{S}_1$, $\overline{S}_2$.*

Fig. 6.7: Pin configuration of quad SR latch IC 74279.

Timing Diagram of NAND based SR Latch

The timing diagram of NAND based SR latch shown in Fig. 6.8 is drawn by taking SR inputs as 11, 10, 11, 01, 11 and 00. The outputs are predicted from the truth table shown in Table 6.7. In the timing diagram, the following changes occur at various timings. Here, any change in inputs will immediately initiate a change in output.

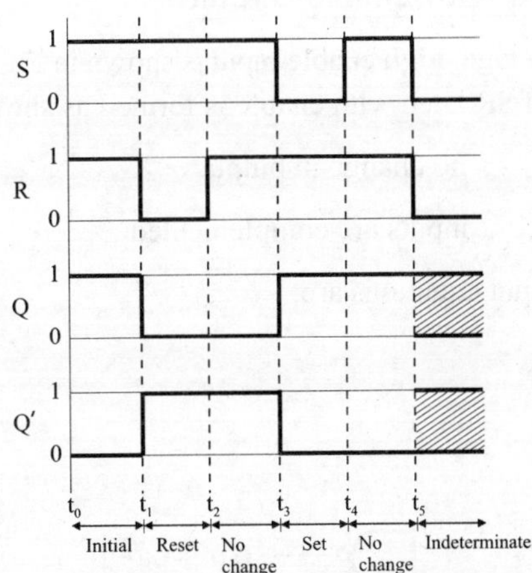

Fig. 6.8: Timing diagram (or waveform) of NAND based SR latch.

Time t_0: S and R inputs are both 1. Initial outputs are assumed as Q = 1 and Q′ = 0.

Time t_1: R input is made 0 and so output Q is reset to 0 and Q′ becomes 1.

Time t_2: R input is made 1. Since S and R inputs are both 1, no change in outputs.

Time t_3: S input is made 0 and so output Q is set to 1 and Q′ becomes 0.

Time t_4: S input is made 1. Since S and R inputs are both 1, no change in outputs.

Time t_5: S and R inputs are both 0 and hence the outputs are indeterminate.

Enable Circuit for SR Latch using NAND Gate

An enable circuit can be added to NAND based SR latch such that the inputs are allowed only when enable is active. The NAND operation of each input with a logic **high** enable signal, E as shown in Fig. 6.9 can be used as enable circuit. The truth table of enable circuit is shown in Table 6.8.

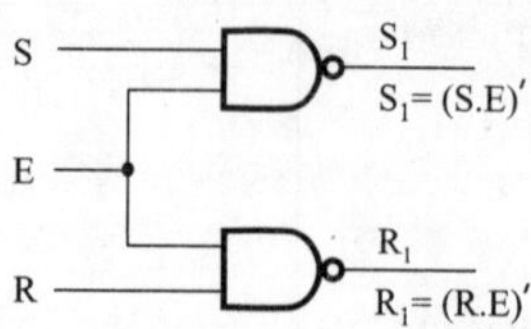

Fig. 6.9: Enable circuit for SR latch.

Table 6.8: Truth Table of Enable Circuit

E	S	R	S_1	R_1
0	0	0	1	1
0	0	1	1	1
0	1	0	1	1
0	1	1	1	1
1	0	0	1	1
1	0	1	1	0
1	1	0	0	1
1	1	1	0	0

From Table 6.8 it can be observed that,

As long as E = 0, both S_1 and R_1 are both 1.

As long as E = 1, both S_1 and R_1 are complement of S and R.

On comparing Table 6.7 and 6.8 we can conclude that on connecting enable circuit to NAND based SR latch as shown in Fig. 6.10, there is no change in output as long as E = 0 and output is allowed to change only when E = 1.

SR Latch using NAND Gates and with Enable Circuit

The NAND based SR latch with logic **high** enable input is shown in Fig. 6.10. From the knowledge of Tables 6.7 and 6.8, the truth table of SR latch with enable is formed as shown in Table 6.9.

When $E = 0$, $S_1 = R_1 = 1$; no change in output

When $E = 1$, $S_1 = S', R_1 = R'$; inputs are complemented

When $E = 1$, the next state output equations are,

$$Q(t+1) = (S_1 \,.\, Q'(t))'$$

$$Q'(t+1) = (R_1 \,.\, Q(t))'$$

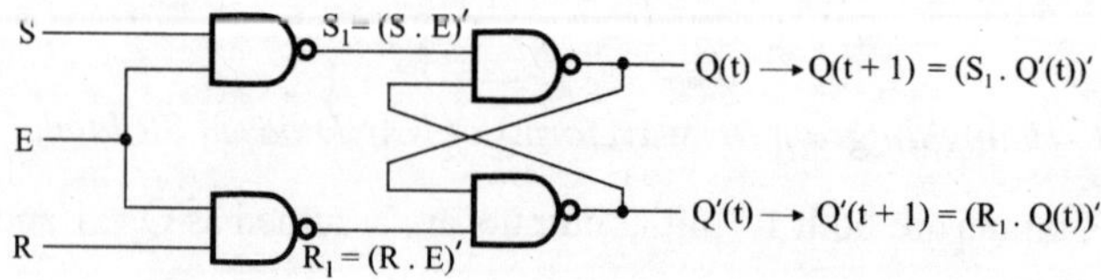

Fig. 6.10: NAND based SR latch with enable.

Table 6.9: Truth Table of NAND based SR Latch with Enable

Inputs			Intermediate Signal		Outputs
E	S	R	S_1	R_1	Q Q′
0	×	×	1	1	No change
1	0	0	1	1	No change
1	0	1	1	0	0 1 (Reset state)
1	1	0	0	1	1 0 (Set state)
1	1	1	0	0	Indeterminate

On comparing Tables 6.4 and 6.9, it can be observed that the truth table of NAND based SR latch with enable is same as NOR based SR latch without enable. Therefore, working of SR latch of Fig. 6.10 is same as working of NOR based SR latch of Fig. 6.4.

Timing Diagram of NAND based SR Latch with Enable

The timing diagram of NAND based SR latch with enable, E is shown in Fig. 6.11. Here, enable is made 0 from time t_0 to t_4 and hence changes in inputs are not recognized during this time. Then at t_4, enable is made 1 and SR inputs as made as 00, 01, 00, 10, 00 and 11 and the outputs are predicted from the truth table shown in Table 6.9. In the timing diagram, the following changes occur at various timings. Here, when enable E = 1, any change in inputs will immediately initiate a change in output.

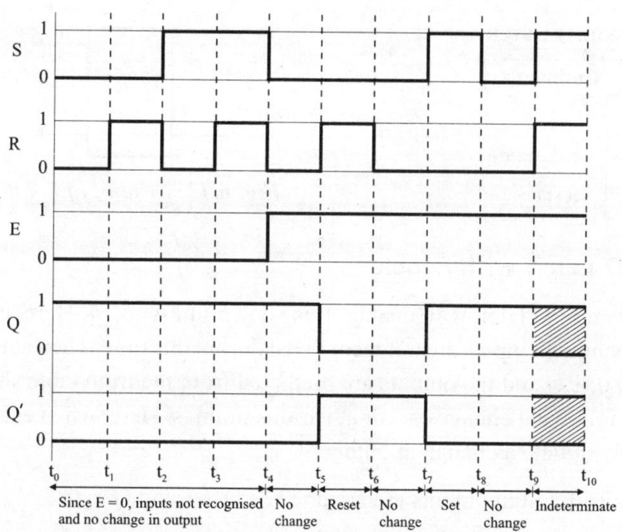

Fig. 6.11: Timing diagram (or waveform) of SR latch using NAND gates and with enable circuit.

Time t_0: S and R inputs are both 0. Initial outputs are assumed as Q = 1 and Q′ = 0.

Time t_0 to t_4: Enable, E = 0, S and R inputs are ignored, output remains same as initial value.

Time t_5: R input is made 1 and so output Q is reset to 0 and Q′ becomes 1.

Time t_6: R input is made 0. Since S and R inputs are both 0, no change in outputs.

Time t_7: S input is made 1 and so output Q is set to 1 and Q′ becomes 0.

Time t_8: S input is made 0. Since S and R inputs are both 0, no change in outputs.

Time t_9: S and R inputs are both 1 and hence the outputs are indeterminate.

6.1.4 D Latch

A single input latch can be developed from SR latch, if the inputs to SR latch are always complementary inputs. The complementary input conditions can be achieved by adding an inverter to one of the inputs of SR latch. The input D is applied to S and complement of D is applied to R as shown in Fig. 6.12. Now SR latch is single input latch called D latch.

Table 6.10: Stable States of D Latch

Inputs	Outputs	State
D = S = 1, R = 0	Q = 1, Q′ = 0	Set state
D = S = 0, R = 1	Q = 0, Q′ = 1	Reset state

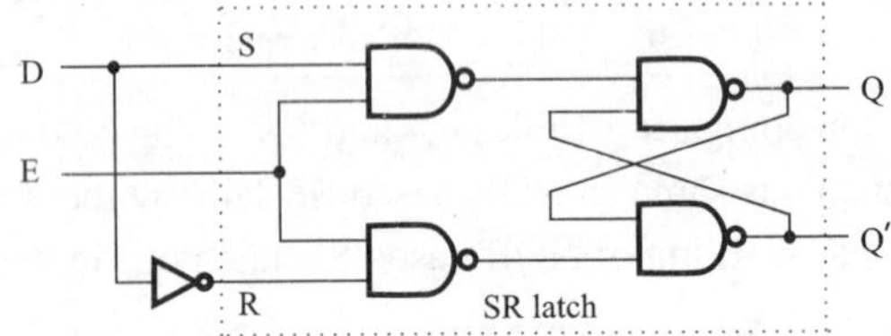

Fig. 6.12: D latch.

Therefore, D latch is a single input latch with two stable states which are used to store two different values 0 and 1 of a binary data. Hence, the D latch is used as a basic storage device in digital systems.

The stable states of D latch are shown in Table 6.10. The truth table of D latch is shown in Table 6.11 and the standard symbol of D latch is shown in Fig. 6.13.

Table 6.11: Truth Table of D Latch

E	D	Q	Q′	Comment
0	×	No change		-
1	0	0	1	Reset state
1	1	1	0	Set state

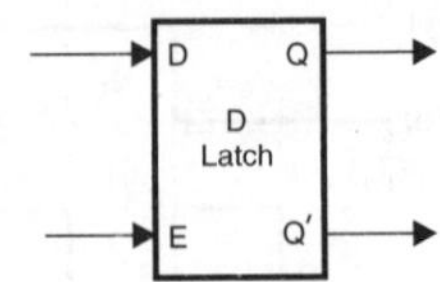

Fig. 6.13: Symbol of D latch.

Timing Diagram of D Latch with Enable

The timing diagram of D latch with enable, E is shown in Fig. 6.14. Here, enable is made 0 from time t_0 to t_2 and hence changes in inputs are not recognized during this time. Then at t_2 enable is made 1 and D input is made as 0, 1, 0, 1… and the outputs are predicted from the truth table shown in Table 6.11. In the timing diagram, the following changes occur at various timings. Here, when enable E = 1, any change in input will immediately initiate a change in output.

Time t_0: D input is 0. Initial outputs are assumed as Q = 1 and Q′ = 0.

Time t_0 to t_2: Enable, E = 0, D input is ignored, output remains same as initial value.

Time t_2: D input is made 0, and so output Q is reset to 0 and Q′ becomes 1.

Time t_3: D input is made 1, and so output Q is set to 1 and Q′ becomes 0.

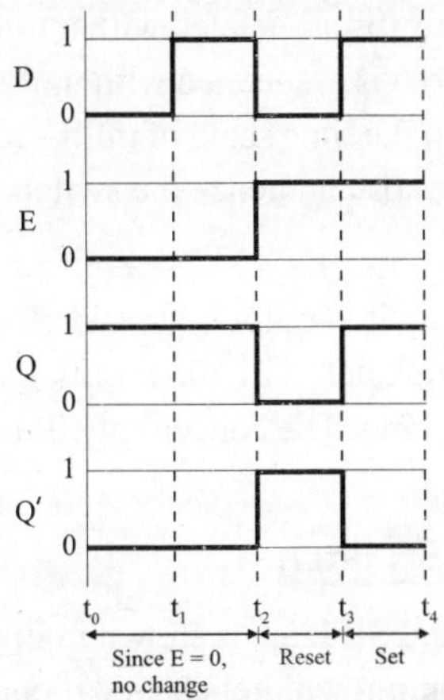

Fig. 6.14: Timing diagram (or waveform) of D latch with enable.

6.1.5 Application of Latches-Switch Debouncing

(VTU, Aug/Sep'20, 7 Marks)

In digital circuits mechanical switches are used to apply logical inputs 1 and 0 which are connection to +5 Volts and 0 Volt (Ground) power supply. A mechanical switch will have two fixed contacts for connection to +5 (logic 1) and ground (logic 0) and one movable contact connected to input of digital circuit. At any one time the movable contact will connect to one of the fixed contact and hence, the mechanical switch will connect the digital circuit input to either 0 or 1 at any one time.

The two fixed contacts are also called position-1 and 2. The change in input is made by moving the contact from one fixed position to another fixed position. This movement is associated with bouncing action due to its nature of construction which creates multiple contacts before settling at final position. The bouncing action of switches when changed from position-1 to 2 and position-2 to 1are shown in Fig. 6.15

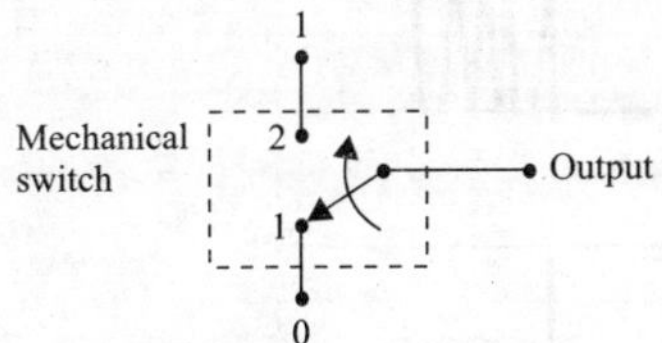

Fig. a: Mechanical switch in position-1.

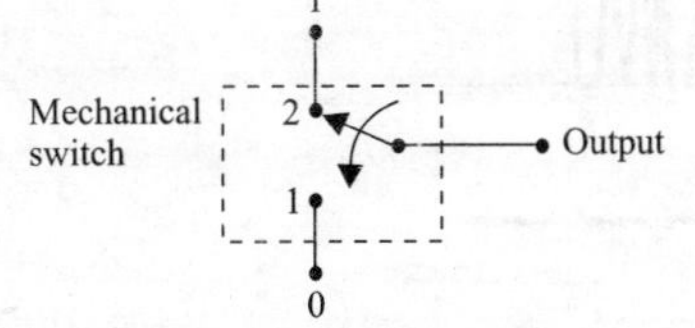

Fig. b: Mechanical switch in position-2.

1 = +5 Volt
= Logic high
0 = 0 Volt
= Logic low

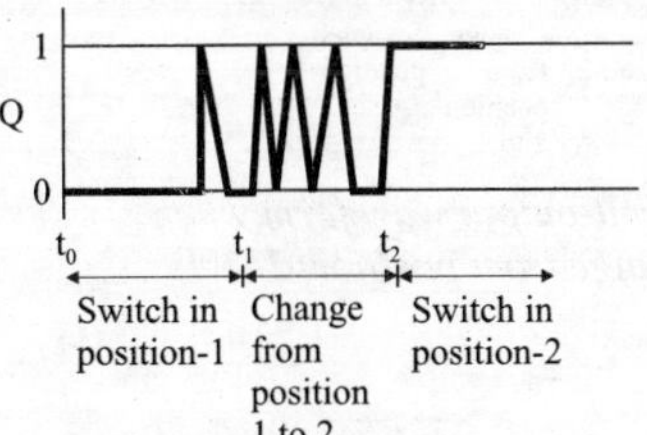

Fig. c: Output when switch change from position-1 to 2.

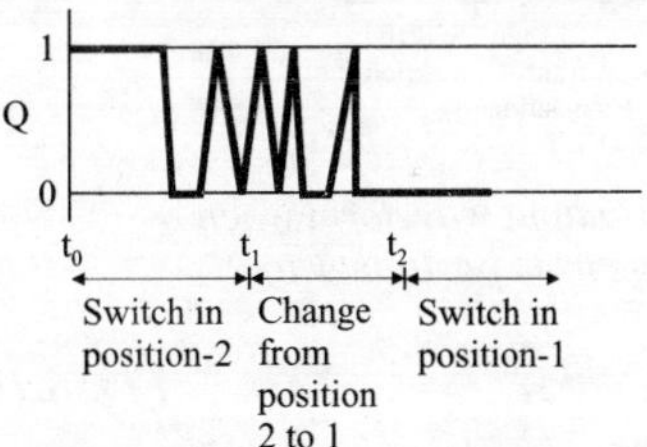

Fig. c: Output when switch change from position-2 to 1.

Fig. 6.15.

Due to bouncing the change from 0 to 1 is associated with multiple changes between 0 and 1 before settling at 1. Similarly, the change from 1 to 0 is associated with multiple changes between 1 and 0 before settling at 0. These oscillations will occur in the time scale of milli-seconds but the digital circuit response time is in the order of micro and Nano-seconds and hence the switch will create multiple inputs to digital circuits.

In order to avoid the oscillations in switch output, the switch output contact (or movable contact) is connected to digital input via a latch. The latch will allow only the very first change in switch output and ignore subsequent changes due to bouncing. The concept of eliminating multiple inputs due to switch bouncing is called debouncing.

Switch-debouncing by NOR based SR latch

(VTU, Aug/Sep'20, 8 Marks)

The switch debouncing by NOR based SR latch is shown in Fig. 6.16. From the truth table of NOR based SR latch it can be observed that the output will not change if both inputs are 0. Hence normally both inputs are tied to 0. Applying 1 to S input will set the output and applying 1 to R input will reset the output.

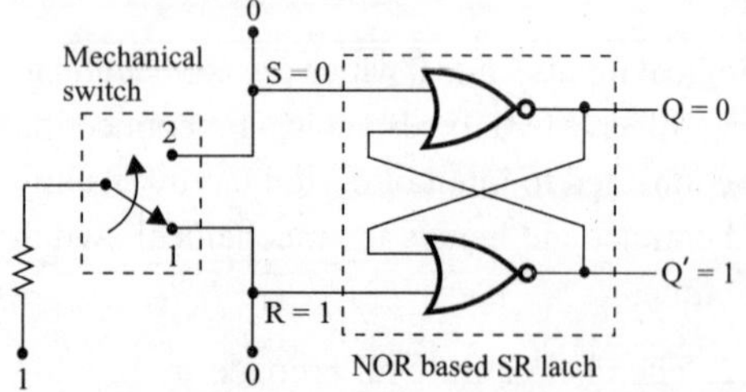

Fig. a: Mechanical switch in position-1.

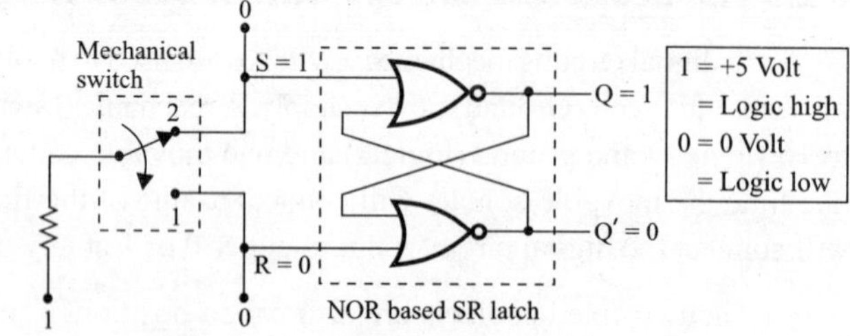

Fig. b: Machanical switch in position-2.

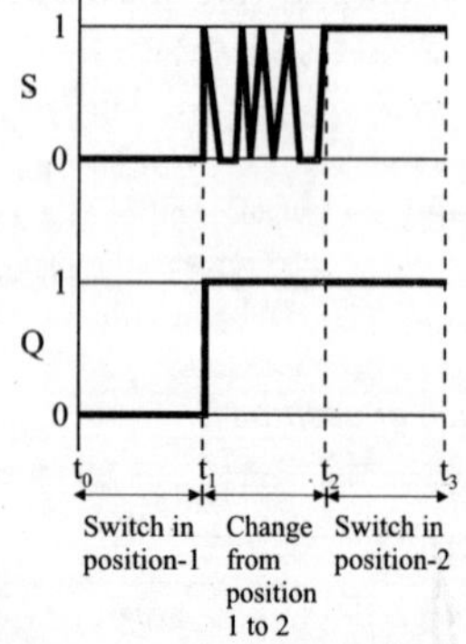

Fig. c: Input-output waveform when switch change from position-1 to 2.

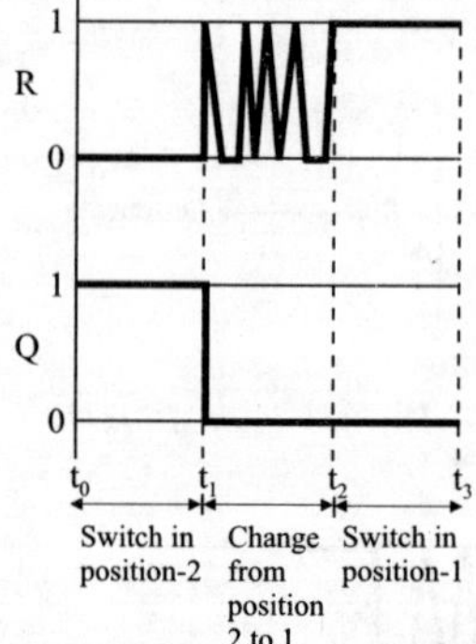

Fig. d: Input-output waveform when switch change from position-2 to 1.

Fig. 6.16.

Therefore, when S input is changed from 0 to 1, by keeping the R input as 0, the output will set. Since, R input is 0, after applying 1 to S, if the S input oscillates between 1 and 0, there will not be any change in latch output.

Similarly, when R input is changed from 0 to 1, by keeping the S input as 0, the output will reset. Since, S input is 0, after applying 1 to R, if the R input oscillates between 1 and 0, there will not be any change in latch output.

Truth Table of SR latch using NOR Gates

Inputs S	Inputs R	Outputs Q	Outputs Q'
0	0	No change	
0	1	0	1
1	0	1	0
1	1	Indeterminate	

Switch-debouncing by NAND based SR latch

The switch debouncing by NAND based SR latch is shown in Fig. 6.17. From the truth table of NAND based SR latch it can be observed that the output will not change if both inputs are 1. Hence normally both inputs are tied to 1. Applying 0 to S input will set the output and applying 0 to R input will reset the output.

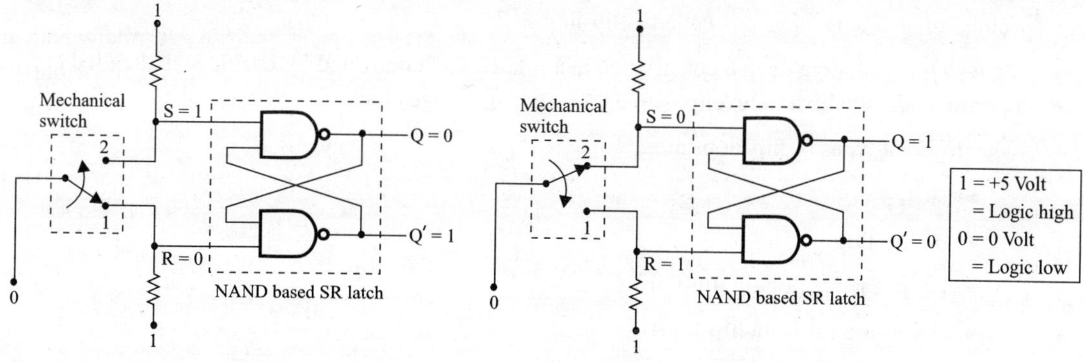

Fig. a: Mechanical switch in position-1.

Fig. b: Mechanical switch in position-2.

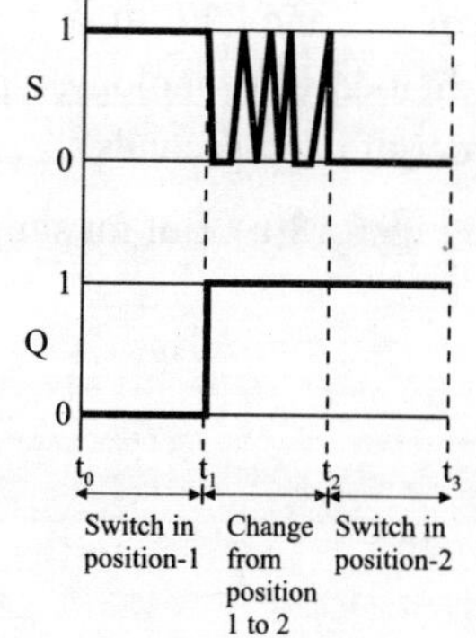

Fig. c: Input-output waveform when switch change from position-1 to 2.

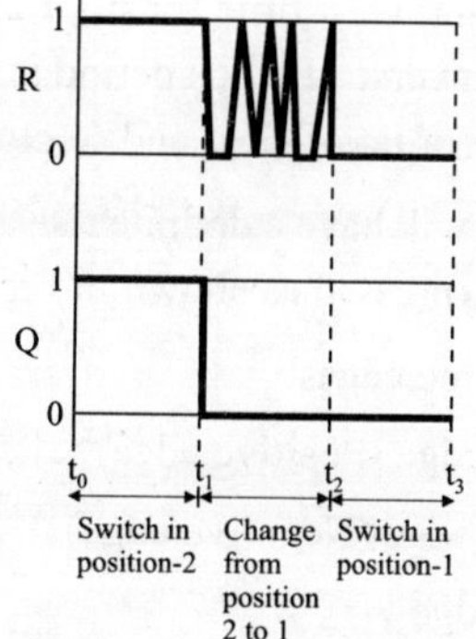

Fig. d: Input-output waveform when switch change from position-2 to 1.

Fig. 6.17.

Therefore, when S input is changed from 1 to 0, by keeping the R input as 1, the output will set. Since, R input is 1, after applying 0 to S, if the S input oscillates between 0 and 1, there will not be any change in latch output.

Similarly, when R input is changed from 1 to 0, by keeping the S input as 1, the output will reset. Since, S input is 1, after applying 0 to R, if the R input oscillates between 0 and 1, there will not be any change in latch output.

Truth Table of SR latch using NAND Gates

Inputs		Outputs	
S	R	Q	Q'
0	0	Indeterminate	
0	1	1	0
1	0	0	1
1	1	No change	

6.2 Flip-Flops

(RGPV, Jun'22, 7 Marks)

In latches with enable, it is observed that inputs are recognized only when enable is active. Therefore, it is possible to replace enable by a momentary pulse called ***clock***, so that inputs can be recognized only at a specified time. Such a device is called a flip-flop.

Flip-flop is a storage device constructed using latch and controlled by enable signal called clock. Flip-flops are made sensitive to level/positive edge/negative edge clock.

The different types of flip-flops are,

- SR flip-flop
- JK flip-flop
- T flip-flop (or Toggle flip-flop)
- D flip-flop (or Data flip-flop)

6.2.1 Clock and Triggering of Flip-Flops

The clock (or clock signal) is basically a periodic square pulse signal. Usually the clock pulses will have equal ON and OFF time (or **high** and **low** time). The sum of ON and OFF time is called ***one period of clock***. The inverse of clock period is called ***clock frequency***. In today's technology, typical clock frequencies are in the order of GHz and so clock periods are in the order of nano-seconds.

A clock signal will have a definite rise and fall time as shown in Fig. 6.18(b). But for simplicity, the clock signal can be expressed as shown in Fig. 6.18(a).

A clock has three edges.

1. Rising edge (Positive edge)
2. Falling edge (Negative edge)
3. Level edge

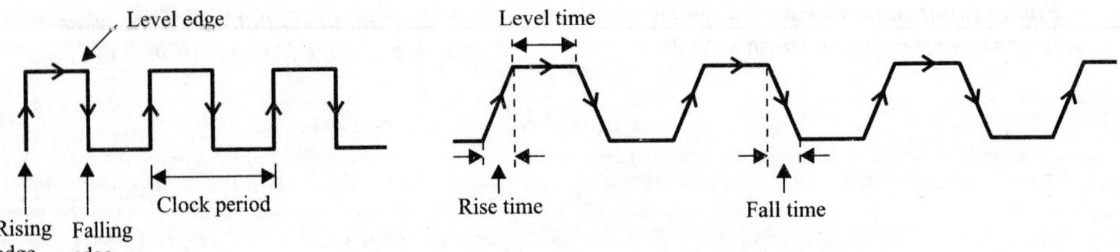

Fig. a: *Various edges of a clock signal.*

Fig. b: *Practical clock signal.*

Fig. 6.18: *Clock signal.*

Clock is used in sequential logic circuits in order to have precise timing control over changes in logic circuits. Usually, the clocked sequential circuits or synchronous sequential circuits are designed as edge sensitive device. In edge sensitive device or circuits the timing of signal (or input) recognization is more precise. Inputs are recognized only at the specified edge of the clock.

Therefore, the synchronous sequential circuits can be classified as,

- Level sensitive devices/circuits
- Rising/Positive edge sensitive devices/circuits
- Falling/Negative edge sensitive devices/circuits

> ***Note: 1. Rising edge** is also called **positive edge.***
> ***2. Falling edge** is also called **negative edge.***
> ***3. Level sensitive devices** are also known as **level triggered devices.***
> ***4. Edge sensitive devices** are also known as **edge triggered devices.***

Specific clock symbol is used to denote positive/negative/level edge sensitivity in sequential devices as shown in Fig. 6.19.

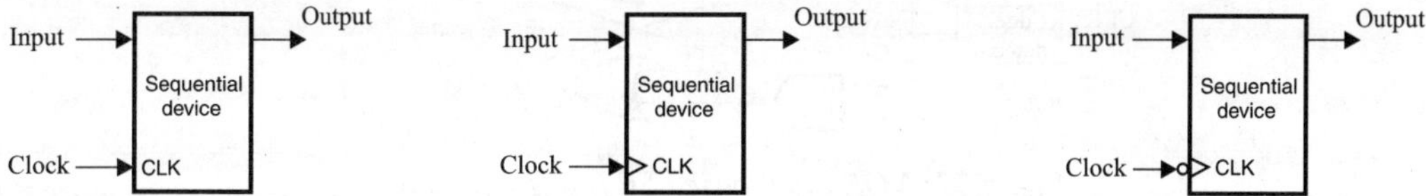

***Fig. a:** Level sensitive device.* ***Fig. b:** Positive edge sensitive device.* ***Fig. c:** Negative edge sensitive device.*

***Fig. 6.19:** Clock symbol for clock sensitivity in sequential device.*

Triggering of Flip-Flops

(IKGPTU, May'20, 2 Marks) (IKGPTU, Dec'19, 2 Marks)

In flip-flops the process of applying a control signal to change from one stable state to another stable state is called ***triggering***. Usually, the control signal is a clock signal shown in Fig. 6.18.

Mostly the triggering is performed by edge sensitive clock. A typical clock edge detector circuit is shown in Fig. 6.20 which generates a narrow pulse during a positive edge or negative edge. In this circuit a small propagation delay for clock pulse to travel through inverter will create a time difference between CLK and $\overline{\text{CLK}}$ which is used to generate a very narrow clock pulse.

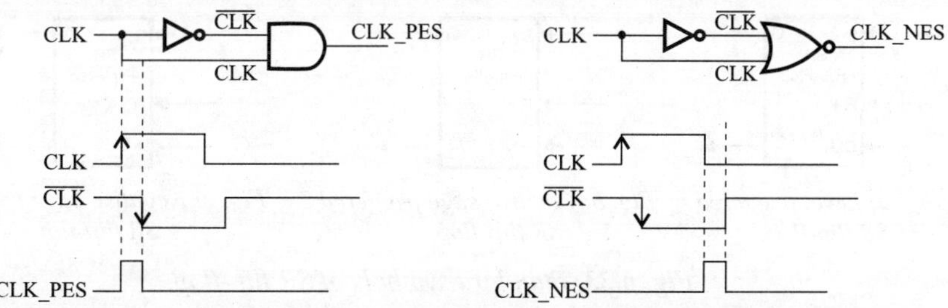

***Fig. a:** Positive edge detection.* ***Fig. b:** Negative edge detection.*

***Fig. 6.20:** Clock edge detector circuits.*

6.2.2 SR Flip-Flop

(RGPV, Jun'20, 7 Marks)

The ***SR flip-flop*** is a digital device with two stable states: set and reset states. Hence it can be used as a switch or as a storage device. The SR latch with enable shown in Fig. 6.10 becomes a SR flip-flop, if the enable is replaced by a clock as shown in Fig. 6.21.

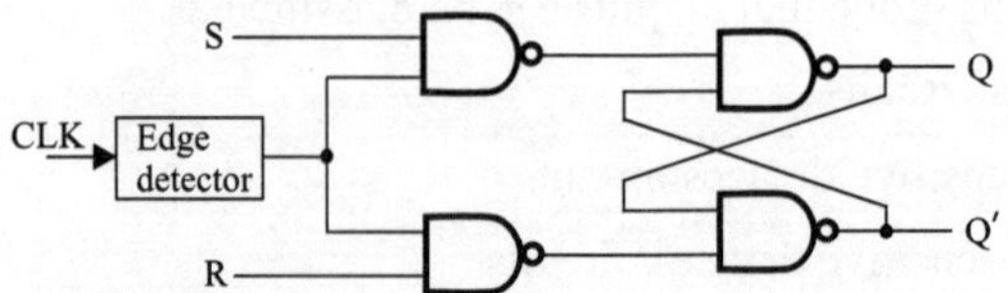

***Fig. 6.21:** NAND latch based SR flip-flop.*

Similarly, if an AND gate based clock circuit is added to NOR based SR latch of Fig. 6.4 it will work as SR flip-flop. The NOR latch based SR flip-flop is shown in Fig. 6.22. The truth table of SR flip-flop is shown in Table 6.12 and standard symbols are shown in Fig. 6.23. The drawback in SR flip-flop is that it has an indeterminate output state.

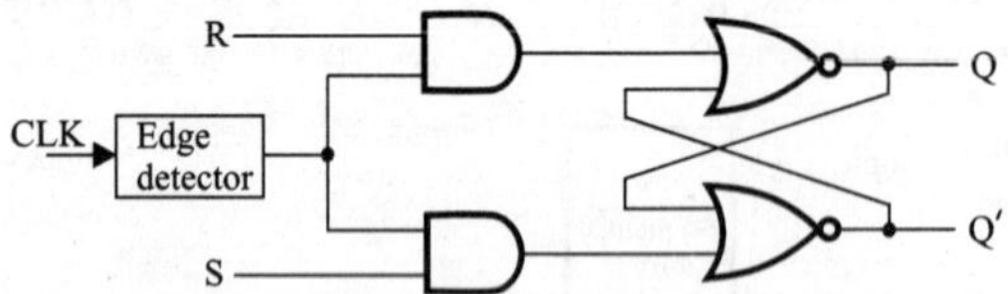

***Fig. 6.22:** NOR latch based SR flip-flop.*

Table 6.12: Truth Table of SR Flip-Flop

Level Triggered

Inputs		Clock	Outputs	
S	R		Q	Q'
0	0	1	No change	
0	1	1	0	1
1	0	1	1	0
1	1	1	Indeterminate	

Positive Edge Triggered

Inputs		Clock	Outputs	
S	R		Q	Q'
0	0	↑	No change	
0	1	↑	0	1
1	0	↑	1	0
1	1	↑	Indeterminate	

Negative Edge Triggered

Inputs		Clock	Outputs	
S	R		Q	Q'
0	0	↓	No change	
0	1	↓	0	1
1	0	↓	1	0
1	1	↓	Indeterminate	

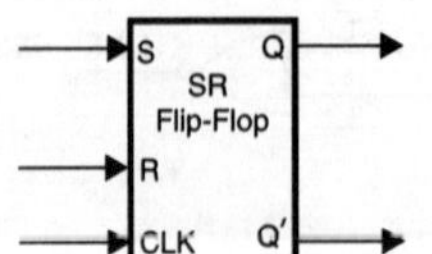

***Fig. a:** Level triggered SR flip-flop.*

***Fig. b:** Positive edge triggered SR flip-flop.*

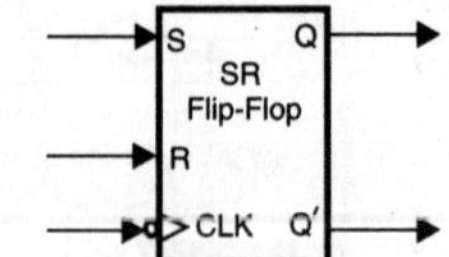

***Fig. c:** Negative edge triggered SR flip-flop.*

***Fig. 6.23:** Standard symbols of SR flip-flop.*

Verification of SR Flip-Flop Truth Table

The input-output relations of SR flip-flop shown in truth table are verified by taking following six cases of inputs and initial outputs.

Case i: S = R = 0, Q = 0, Q′ = 1

Case ii: S = R = 0, Q = 1, Q′ = 0

Case iii: S = 0, R = 1, Q = 1, Q′ = 0

Case iv: S = 1, R = 0, Q = 0, Q′ = 1

Case v: S = R = 1, Q = 0, Q′ = 1

Case vi: S = R = 1, Q = 1, Q′ = 0

The various state transitions in the above six cases of inputs and outputs are diagrammatically shown in Figs. 6.24 to 6.29. In each case, the condition of flip-flop when a clock pulse is applied is shown in these figures. The condition of flip-flop before and after clock are also shown in these figures.

The clock is a narrow pulse with duration of **high** level, equal to a transition time of a NAND gate, so that the clock is valid only for a short time to allow the flip-flop inputs to reach the input of latch. When clock is **low** the flip-flop inputs are not recognized.

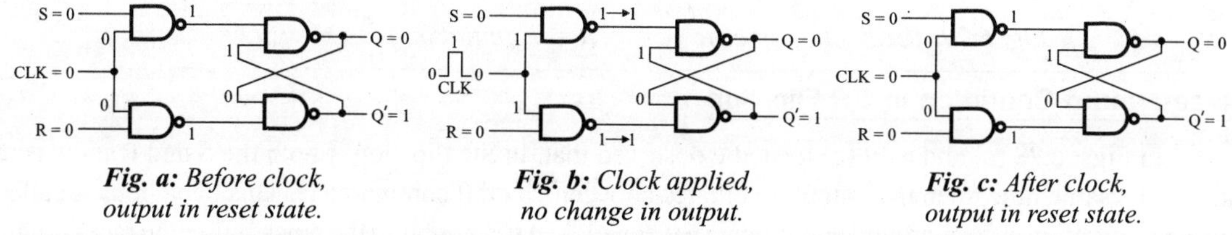

Fig. a: Before clock, output in reset state. | *Fig. b: Clock applied, no change in output.* | *Fig. c: After clock, output in reset state.*

***Fig. 6.24:** Verification of case i: S = R = 0, initial output Q = 0, Q′ = 1.*

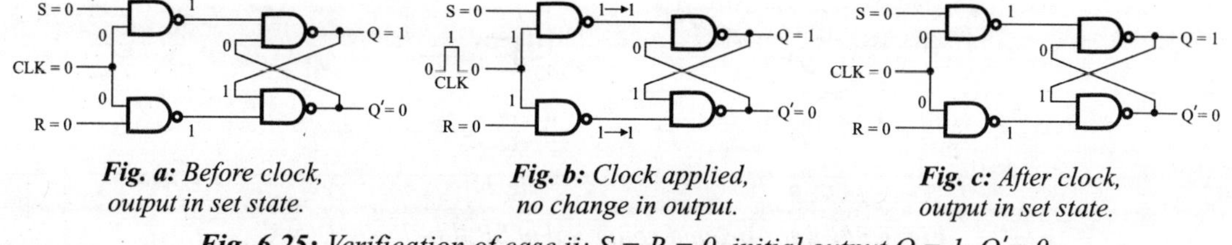

Fig. a: Before clock, output in set state. | *Fig. b: Clock applied, no change in output.* | *Fig. c: After clock, output in set state.*

***Fig. 6.25:** Verification of case ii: S = R = 0, initial output Q = 1, Q′ = 0.*

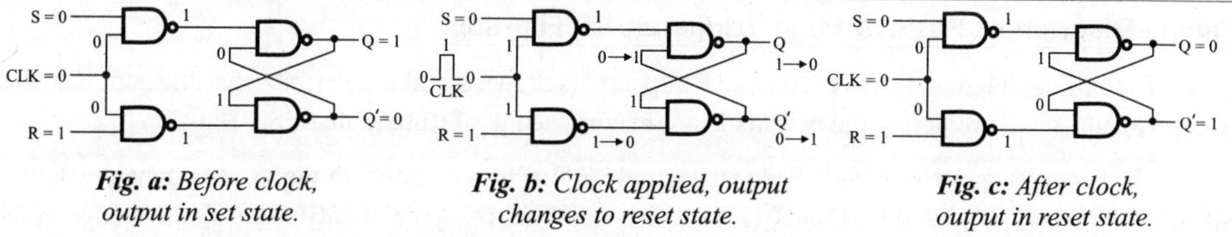

Fig. a: Before clock, output in set state. | *Fig. b: Clock applied, output changes to reset state.* | *Fig. c: After clock, output in reset state.*

***Fig. 6.26:** Verification of case iii: S = 0, R = 1, initial output Q = 1, Q′ = 0.*

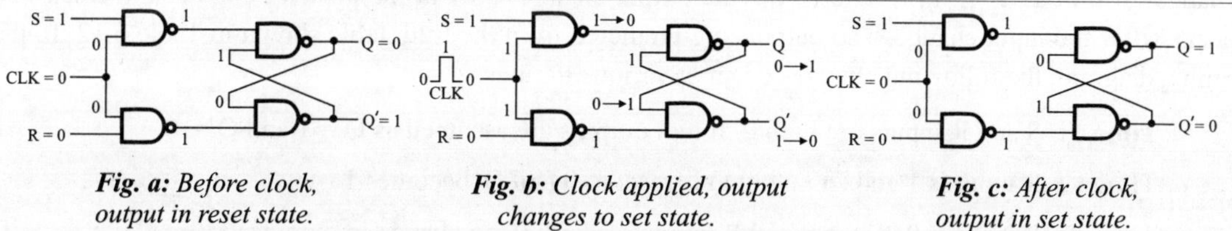

Fig. a: Before clock, output in reset state. | *Fig. b: Clock applied, output changes to set state.* | *Fig. c: After clock, output in set state.*

***Fig. 6.27:** Verification of case iv: S = 1, R = 0, initial output Q = 0, Q′ = 1.*

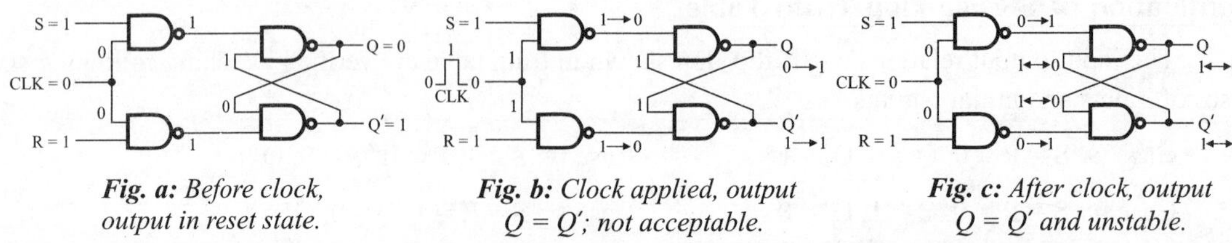

Fig. a: Before clock, output in reset state.

Fig. b: Clock applied, output Q = Q′; not acceptable.

Fig. c: After clock, output Q = Q′ and unstable.

***Fig. 6.28:** Verification of case v: S = R = 1, initial output Q = 0, Q′= 1.*

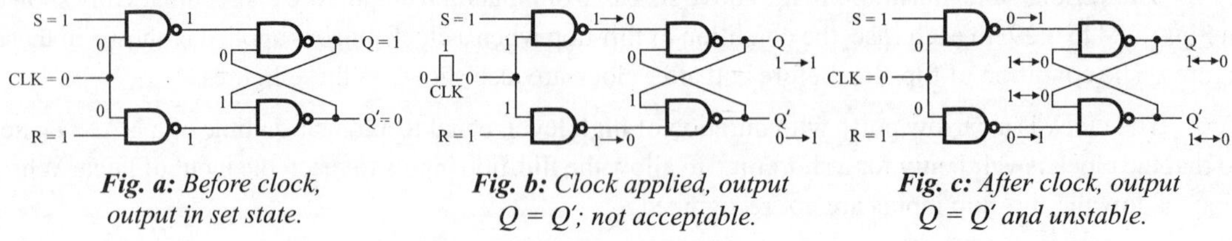

Fig. a: Before clock, output in set state.

Fig. b: Clock applied, output Q = Q′; not acceptable.

Fig. c: After clock, output Q = Q′ and unstable.

***Fig. 6.29:** Verification of case vi: S = R = 1, initial output Q = 1, Q′= 0.*

Race-around Condition in SR Flip-flop

In Figs. 6.28 (c) and 6.29 (c) it can be observed that, in SR flip-flop if both the S and R inputs are high at the same time, then the outputs oscillates between 1 and 0 continuously. This behaviour is called unstable condition or race-around condition in SR flip-flop. After applying the clock pulse, the race-around condition is established in the NAND latch as shown in Figs. 6.30 (a) and (b).

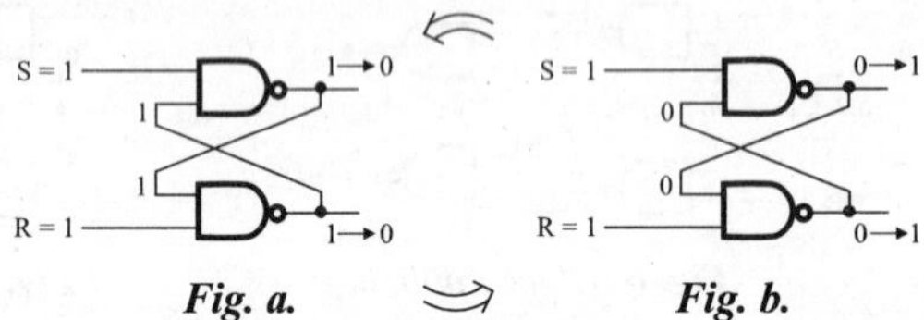

Fig. a. ***Fig. b.***

***Fig. 6.30:** Race-around condition in SR flip-flop.*

Timing Diagram of Positive Edge Triggered SR Flip-flop

The timing diagram or waveform of SR flip-flop is drawn by taking various possible combinations of binary inputs and predicting the outputs from the knowledge of truth table of SR flip-flop.

The timing diagram of positive edge triggered SR flip-flop is shown in Fig. 6.31 is drawn by taking SR inputs as 00, 01, 00, 10, 00, 11 and 01. Here, the inputs are recognized only at the positive edge of the clock and the output is decided by the value of the inputs at the positive edge of the clock. Here, input changes at time t_2, t_4, t_6, t_8, t_{10} and t_{12} but the output changes only at the positive edge of the clock that occurs after the input change. The outputs are predicted from the truth table shown in Table 6.12. In the timing diagram, the following changes occur at various timings.

Time t_0: S and R inputs are both 0. Initial outputs are assumed as Q = 1 and Q′= 0.

Time t_3: R input is 1 and so output Q is reset to 0 and Q′ becomes 1.

Time t_5: R input is 0. Since S and R inputs are both 0, no change in outputs.

Time t_7: S input is 1 and so output Q is set to 1 and Q′ becomes 0.

Time t_9: S input is 0. Since S and R inputs are both 0, no change in outputs.

Time t_{11}: S and R inputs are both 1 and hence the outputs are indeterminate.

Time t_{13}: R input is 1 and so output Q is reset to 0 and Q′ becomes 1.

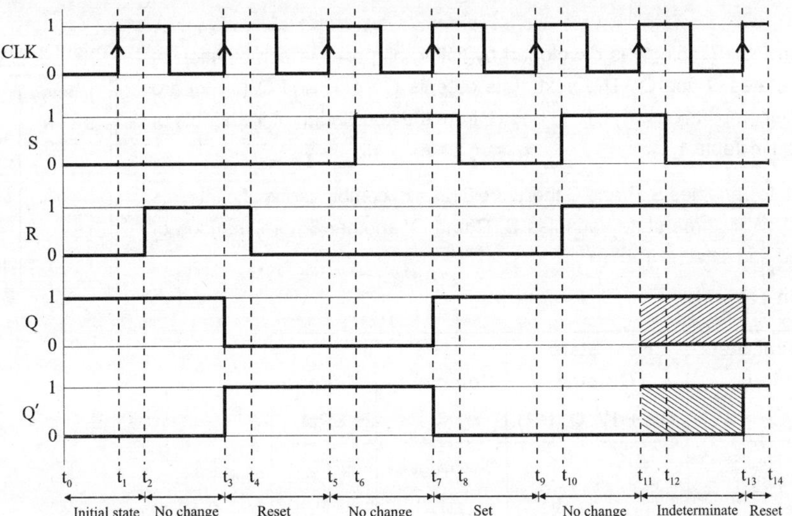

***Fig. 6.31:** Timing diagram (or waveform) of positive edge triggered SR flip-flop.*

Timing Diagram of Negative Edge Triggered SR Flip-flop

The timing diagram of negative edge triggered SR flip-flop is shown in Fig. 6.32. Here, the inputs and outputs are same as that positive edge triggered flip-flop except that the output changes at the negative edge of the clock.

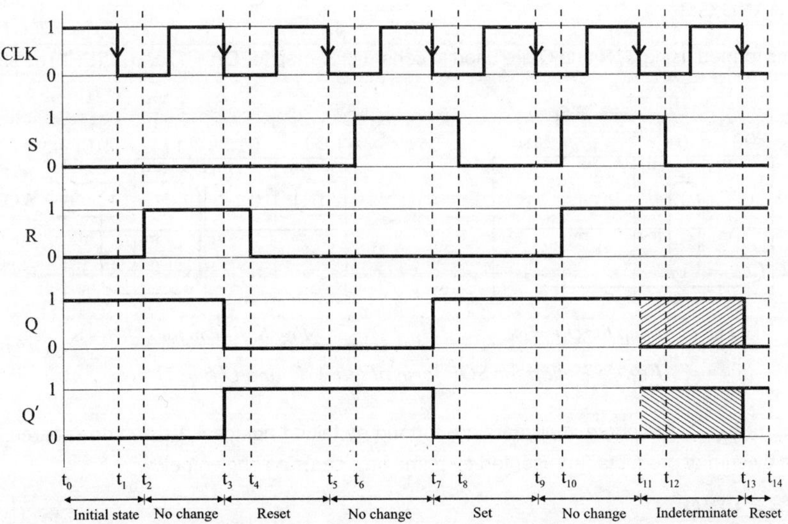

***Fig. 6.32:** Timing diagram (or waveform) of negative edge triggered SR flip-flop.*

Example 6.1

Design and implement SR flip-flop.

a) NAND gate **b)** Using NOR gate

Solution

An excitation table (Table 1) is developed by listing all possible combination of inputs and present states Q and Q'. The next state outputs Q(t + 1) and Q'(t + 1) are determined for all the combinations of S, R, Q and Q' from the knowledge of truth table of SR flip-flop and listed in Table 1.

Truth Table of SR Flip-Flop

Inputs		Outputs	
S	R	Q	Q'
0	0	No change	
0	1	0	1
1	0	1	0
1	1	Indeterminate	

With three state variables S, R and Q there are 8 ($2^3 = 8$) combinations of minterms as shown in Table 1. With three state variables S, R and Q' another 8 combinations of minterms are formed and listed in Table 1.

Table 1: Excitation Table

Inputs		Present State Output		Next State Output		Comment	Minterm
S	R	Q	Q'	Q(t + 1)	Q'(t+1)		SRQ
0	0	0	1	0	1	No change	m_0
0	0	1	0	1	0		m_1
0	1	0	1	0	1	Reset state	m_2
0	1	1	0	0	1		m_3
1	0	0	1	1	0	Set state	m_4
1	0	1	0	1	0		m_5
1	1	0	1	×	×	Indeterminate	m_6
1	1	1	0	×	×		m_7

a) Design using NAND Gate

The minterms formed using S, R and Q are used to construct K-map for Q(t + 1) and Q'(t + 1) as shown in Fig. 1.

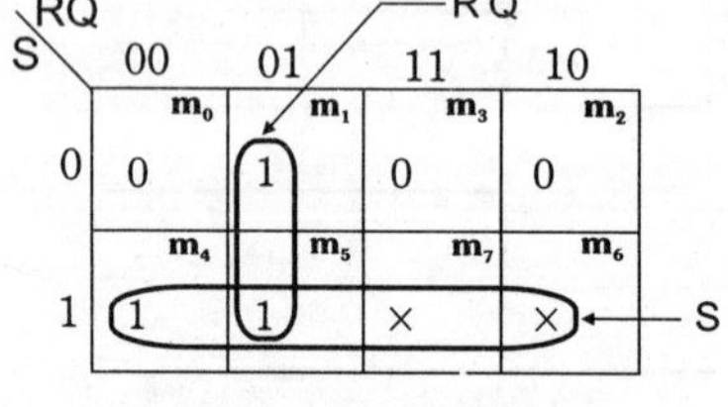

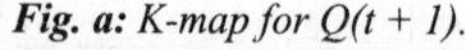

Fig. a: K-map for Q(t + 1).

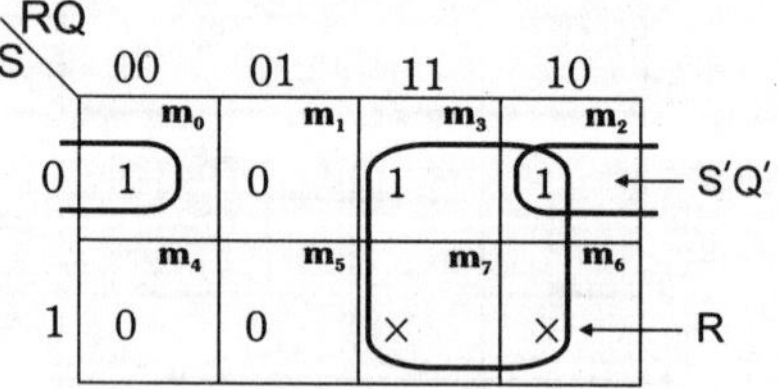

Fig. b: K-map for Q'(t + 1).

Fig. 1: K-map for SOP form of Q(t + 1) and Q'(t + 1).

In the K-map of Fig. 1, the prime implicants are formed by taking adjacent 1's and don't-cares. The Boolean equations are given by sum of products represented by prime implicants as shown below:

Q(t + 1) = S + R'Q

Q'(t + 1) = R + S'Q'

The SR flip-flop that satisfy the above equations is shown in Fig. 2. The SR flip-flop output is verified in Table 2.

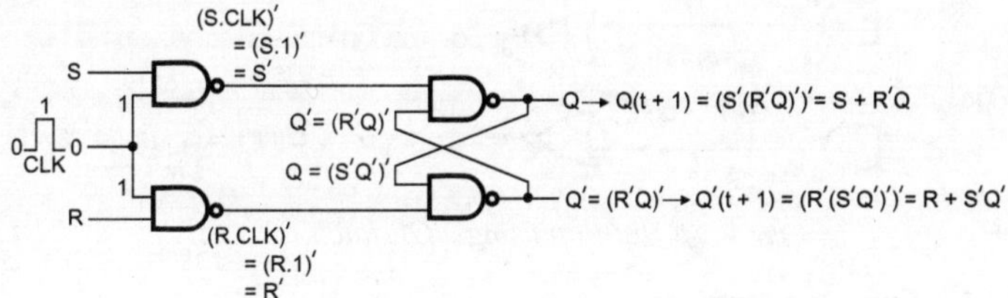

Fig. 2: SR flip-flop using NAND latch.

Table 2: Verification of SR Flip-Flop Outputs

Inputs		Present State Output						Next State Output		Comment
S	R	Q	Q′	S′	R′	R′Q	S′Q′	Q(t + 1) = S + R′Q	Q′(t + 1) = R + S′Q′	
0	0	0	1	1	1	0	1	0	1	No change
0	0	1	0	1	1	1	0	1	0	
0	1	0	1	1	0	0	1	0	1	Reset state
0	1	1	0	1	0	0	0	0	1	
1	0	0	1	0	1	0	0	1	0	Set state
1	0	1	0	0	1	1	0	1	0	
1	1	0	1	0	0	0	0	1	1	Not acceptable
1	1	1	0	0	0	0	0	1	1	

b) Design using NOR Gate

The excitation table (Table 1) formed for NAND based SR latch design is also used for NOR based SR latch. The minterms formed using S, R and Q are used to construct K-map for Q(t + 1) and Q′(t + 1) as shown in Fig. 3.

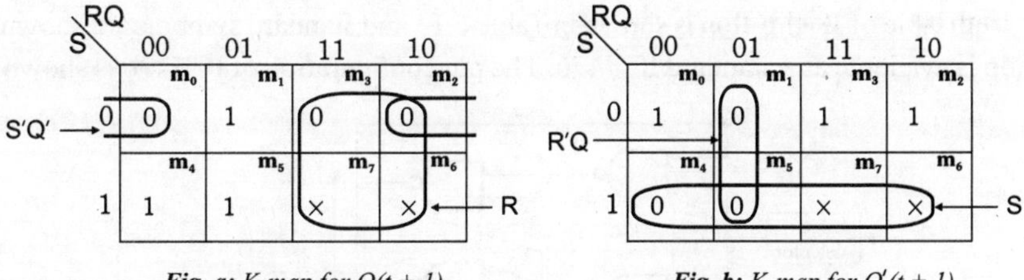

Fig. a: K-map for Q(t + 1). *Fig. b: K-map for Q′(t + 1).*

Fig. 3: K-map for POS form of Q(t + 1) and Q′(t + 1).

In the K-map of Fig. 3, the prime implicants are formed by taking adjacent 0's and don't-cares. The Boolean equations are given by complement of sum of products represented by prime implicants as shown below:

Q(t + 1) = (R + S′Q′)′	Q′(t + 1) = (S + R′Q)′
= R′(S′Q′)′	= S′(R′Q)′
= R′(S + Q)	= S′(R + Q′)

The SR flip-flop that satisfy the above equations is shown in Fig. 4. The SR flip-flop output is verified in Table 3.

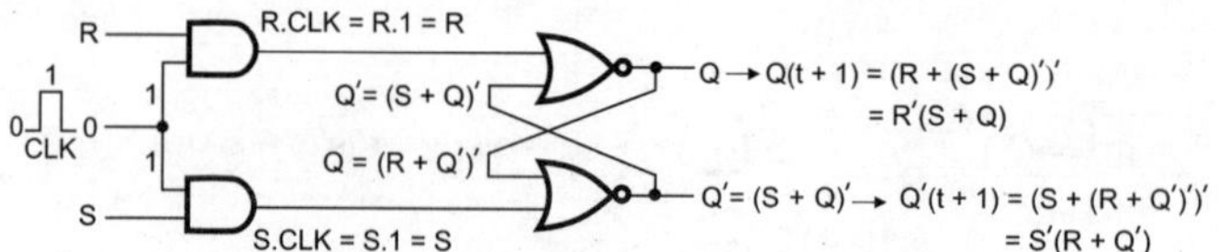

Fig. 4: SR flip-flop using NOR latch.

Table 3: Verification of SR Flip-Flop Outputs

Inputs		Present State Output						Next State Output		Comment
S	R	Q	Q′	R′	S + Q	S′	R + Q′	Q(t + 1) = R′(S + Q)	Q′(t + 1) = S′(R + Q′)	
0	0	0	1	1	0	1	1	0	1	No change
0	0	1	0	1	1	1	0	1	0	
0	1	0	1	0	0	1	1	0	1	Reset state
0	1	1	0	0	1	1	1	0	1	
1	0	0	1	1	1	0	1	1	0	Set state
1	0	1	0	1	1	0	0	1	0	
1	1	0	1	0	1	0	1	0	0	Not acceptable
1	1	1	0	0	1	0	1	0	0	

6.2.3 JK Flip-Flop

(AKTU, Sep'20, 5 Marks)

The JK flip-flop is a digital device with two stable states: set and reset states, similar to SR flip-flop but without indeterminate output.

In JK flip-flop, the indeterminate output of SR flip-flop is converted to another useful output called toggle output. In ***toggle output***, the present output is complement of previous output. This can be achieved by providing an additional feedback from output to input clock circuit in SR flip-flop as shown in Fig. 6.33.The truth table of JK flip-flop is shown in Table 6.13 and standard symbols are shown in Fig. 6.34. The JK flip-flop is available as a standard IC 7476. The pin configuration of IC 7476 is shown in Fig. 6.35.

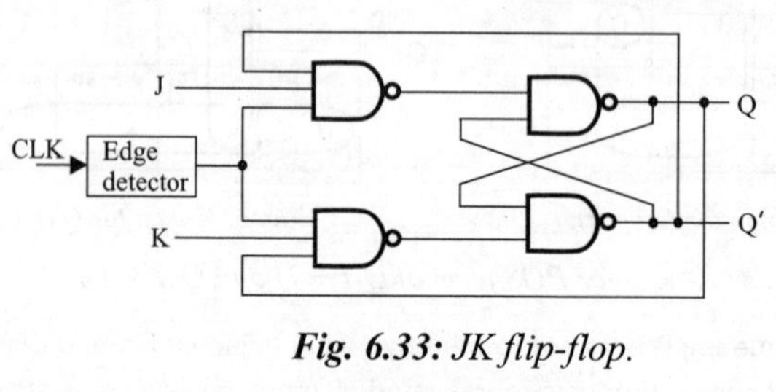

Fig. 6.33: JK flip-flop.

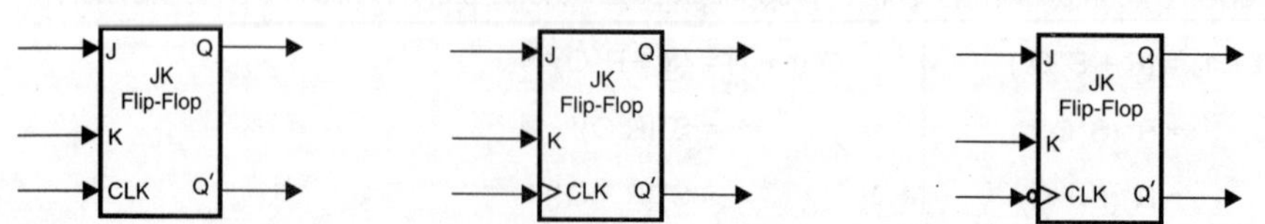

Fig. a: *Level triggered.* ***Fig. b:*** *Positive edge triggered.* ***Fig. c:*** *Negative edge triggered.*

Fig. 6.34: *Standard symbols of JK flip-flop.*

Table 6.13: Truth Table of JK Flip-Flop

Level Triggered

Inputs J	K	Clock	Outputs Q	Q′	Comment
0	0	1	No change		-
0	1	1	0	1	Reset state
1	0	1	1	0	Set state
1	1	1	Toggle		Toggle previous output

Positive Edge Triggered

Inputs J	K	Clock	Outputs Q	Q′	Comment
0	0	↑	No change		-
0	1	↑	0	1	Reset state
1	0	↑	1	0	Set state
1	1	↑	Toggle		Toggle previous output

Negative Edge Triggered

Inputs J	K	Clock	Outputs Q	Q′	Comment
0	0	↓	No change		-
0	1	↓	0	1	Reset state
1	0	↓	1	0	Set state
1	1	↓	Toggle		Toggle previous output

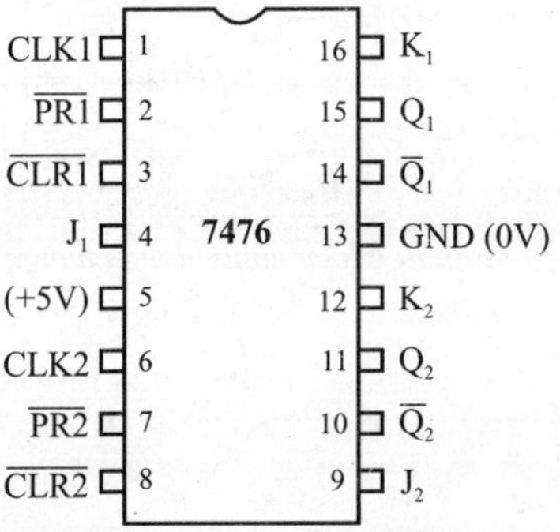

Pin	Description
J, K	Inputs
$\overline{CLR}_1$, $\overline{CLR}_2$	Clear
CLK_1, CLK_2	Clock
$\overline{PR}_1$, $\overline{PR}_2$	Set output
Q_1, Q_2	Outputs
$\overline{Q}_1$, $\overline{Q}_2$	Complement of outputs
V_{cc}	Supply voltage (+5V)
GND	Ground (0V)

***Fig. 6.35:** Pin configuration of dual JK flip-flop IC 7476.*

Verification of JK Flip-Flop Truth Table

The input-output relations of JK flip-flop shown in truth table are verified by taking following six cases of inputs and initial outputs.

Case i: J = K = 0, Q = 0, Q′ = 1

Case ii: J = K = 0, Q = 1, Q′ = 0

Case iii: J = 0, K = 1, Q = 1, Q′ = 0

Case iv: J = 1, K = 0, Q = 0, Q′ = 1

Case v: J = K = 1, Q = 0, Q′ = 1

Case vi: J = K = 1, Q = 1, Q′ = 0

The various state transitions in the above six cases of inputs and outputs are diagrammatically shown in Figs. 6.36 to 6.41. In each case, the condition of flip-flop when a clock pulse is applied is shown in these figures. The condition of flip-flop before and after clock are also shown in these figures. The clock is a narrow pulse with duration of **high** level, equal to a transition time of a NAND gate, so that the clock is valid only for a short time to allow the flip-flop inputs to reach the input of latch. When clock is **low** the flip-flop inputs are not recognized.

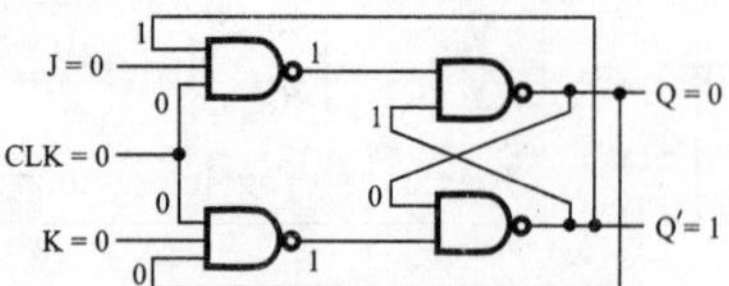

Fig. a: Before clock, output in reset state.

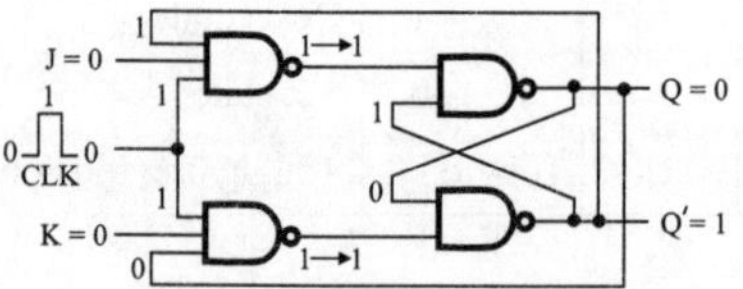

Fig. b: Clock applied, no change in output.

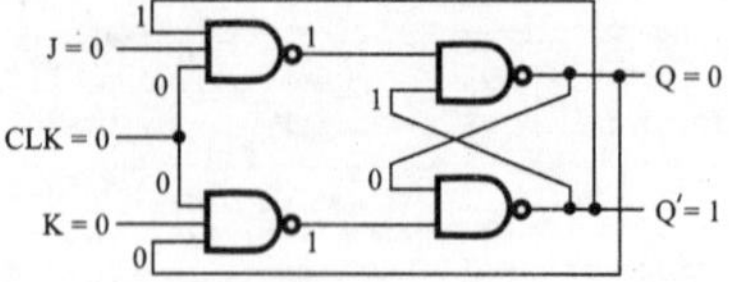

Fig. c: After clock, output in reset state.

***Fig. 6.36:** Verification of case i: $J = K = 0$, initial output $Q = 0$, $Q' = 1$.*

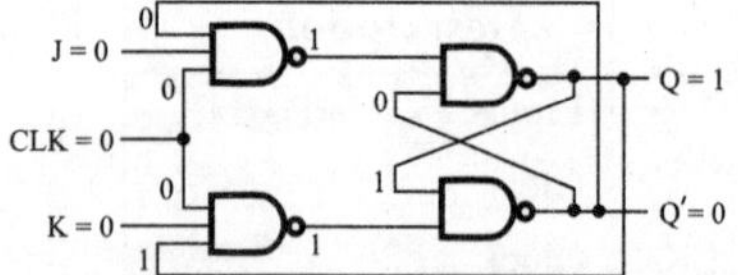

Fig. a: Before clock, output in set state.

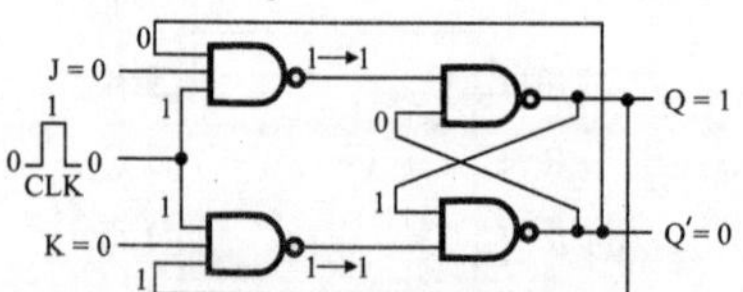

Fig. b: Clock applied, no change in output.

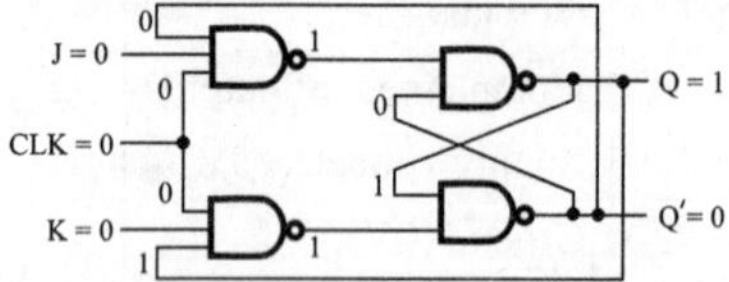

Fig. c: After clock, output in set state.

***Fig. 6.37:** Verification of case ii: $J = K = 0$, initial output $Q = 1$, $Q' = 0$.*

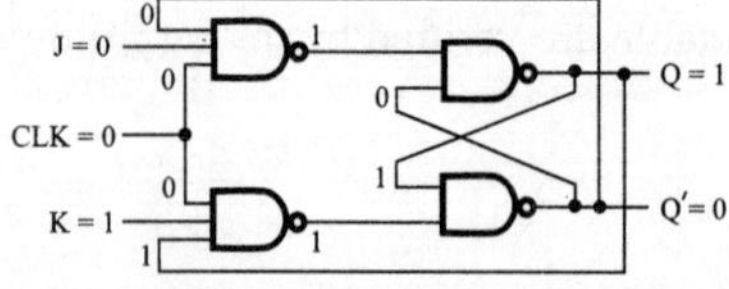

Fig. a: Before clock, output in set state.

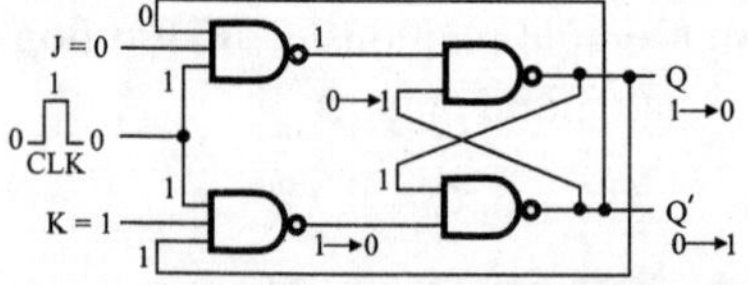

Fig. b: Clock applied, output changes to reset state.

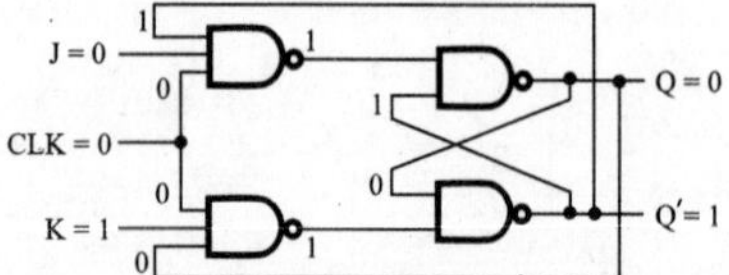

Fig. c: After clock, output in reset state.

***Fig. 6.38:** Verification of case iii: $J = 0$, $K = 1$, initial output $Q = 1$, $Q' = 0$.*

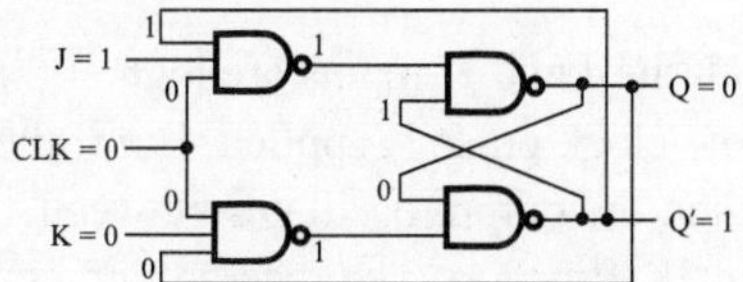

Fig. a: *Before clock, output in reset state.*

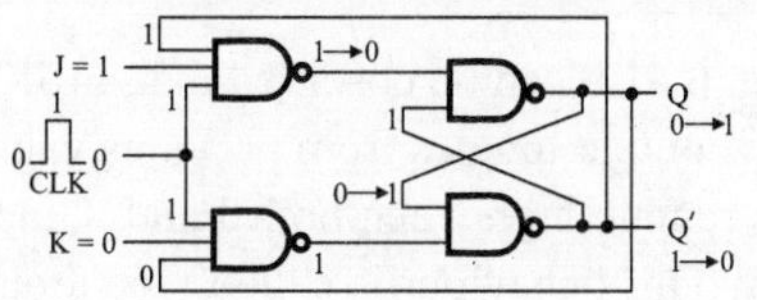

Fig. b: *Clock applied, output changes to set state.*

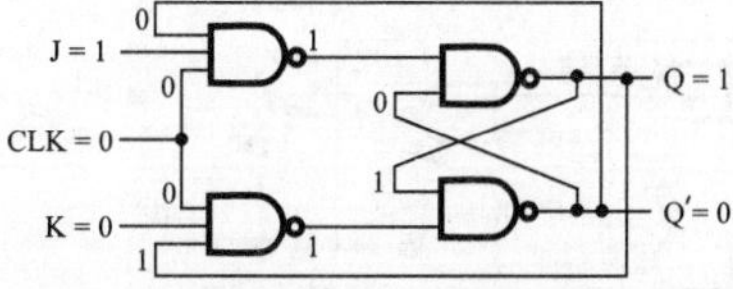

Fig. c: *After clock, output in set state.*

Fig. 6.39: *Verification of case iv: J = 1, K = 0, initial output Q = 0, Q′ = 1.*

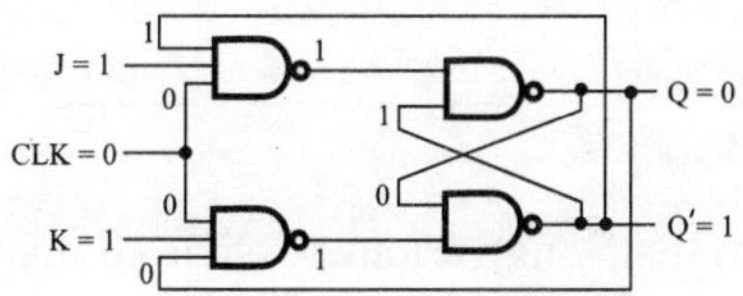

Fig. a: *Before clock, output in reset state.*

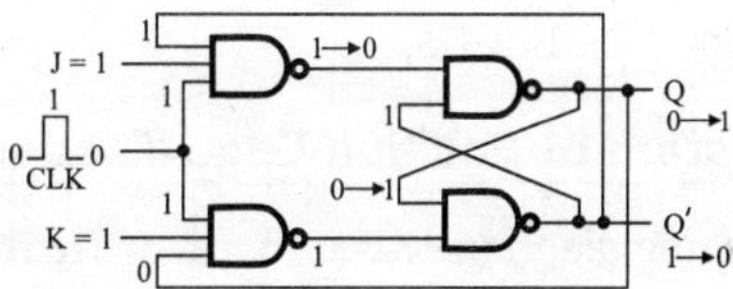

Fig. b: *Clock applied, output toggles.*

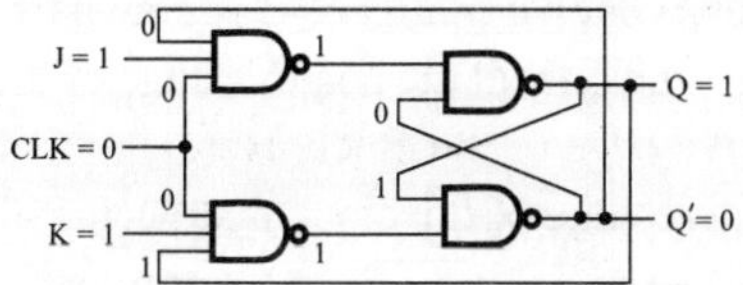

Fig. c: *After clock, output in set state.*

Fig. 6.40: *Verification of case v: J = K = 1, initial output Q = 0, Q′ = 1.*

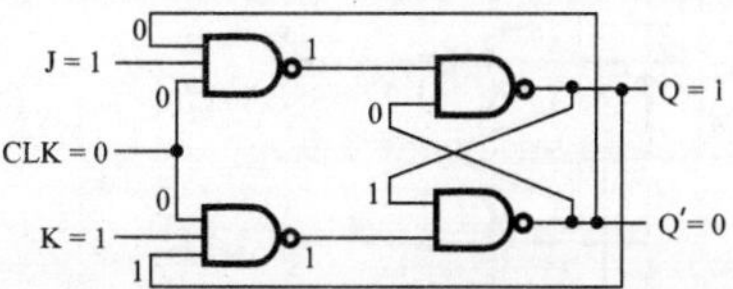

Fig. a: *Before clock, output in set state.*

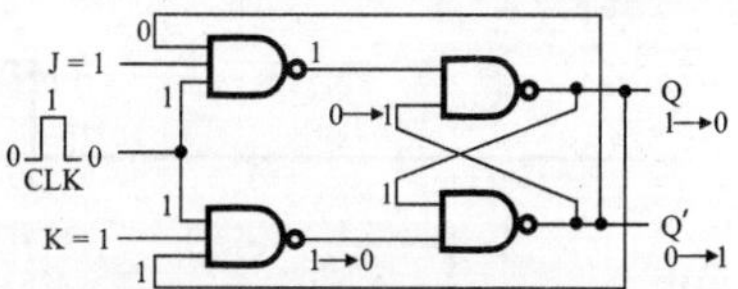

Fig. b: *Clock applied, output toggles.*

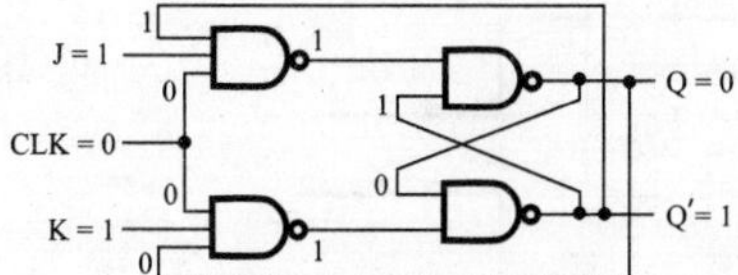

Fig. c: *After clock, output in reset state.*

Fig. 6.41: *Verification of case vi: J = K = 1, initial output Q = 1, Q′ = 0.*

Race-around Condition in JK Flip-flop

In Fig. 6.41 it can be observed that, in JK flip-flop if both the J and K inputs are high at the same time, then the output toggles from previous value when a narrow clock pulse is applied. But if the clock continues to remain high when both J and K inputs are high, then the output oscillates between 1 and 0 continuously. This behaviour is called race-around condition in JK flip-flop. The race-around condition of JK flip-flop is shown in Figs. 6.42 (a) and (b).

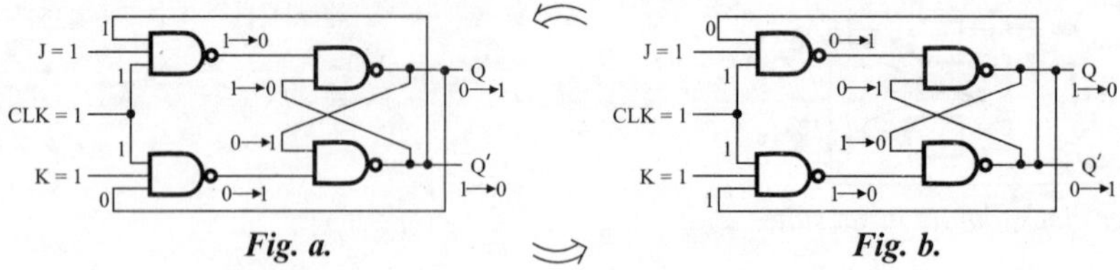

Fig. 6.42: Race-around condition in JK flip-flop.

The race-around condition in JK flip-flop can be avoided by using either master-slave JK flip-flop or edge-triggered JK flip-flop.

Timing Diagram of Positive Edge Triggered JK Flip-flop

The timing diagram or waveform of JK flip-flop is drawn by taking various possible combinations of binary inputs and predicting the outputs from the knowledge of truth table of JK flip-flop.

The timing diagram of positive edge triggered JK flip-flop is shown in Fig. 6.43 is drawn by taking JK inputs as 00, 01, 00, 10, 00 and 11. Here, the inputs are recognized only at the positive edge of the clock and the output is decided by the value of the inputs at the positive edge of the clock. Here, input changes at time t_2, t_4, t_6, t_8 and t_{10} but the output changes only at the positive edge of the clock that occurs after the input change. The outputs are predicted from the truth table shown in Table 6.13. In the timing diagram, the following changes occur at various timings.

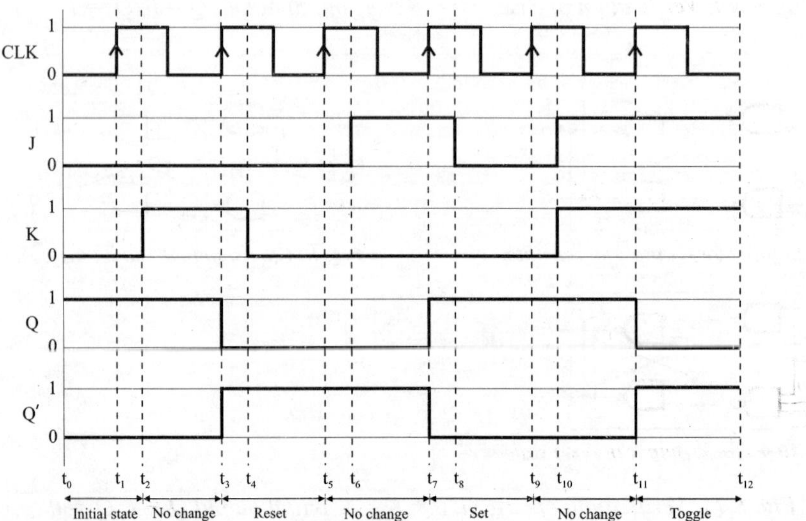

Fig. 6.43: Timing diagram (or waveform) of positive edge triggered JK flip-flop.

Time t_0: J and K inputs are both 0. Initial outputs are assumed as Q = 1 and Q′= 0.

Time t_3: K input is 1 and so output Q is reset to 0 and Q′ becomes 1.

Time t_5: K input is 0. Since J and K inputs are both 0, no change in outputs.

Time t_7: J input is 1 and so output Q is set to 1 and Q′ becomes 0.

Time t_9: J input is 0. Since J and K inputs are both 0, no change in outputs.

Time t_{11}: J and K inputs are both 1 and hence the outputs toggles from previous value.

Timing Diagram of Negative Edge Triggered JK Flip-flop

The timing diagram of negative edge triggered JK flip-flop is shown in Fig. 6.44. Here, the inputs and outputs are same as that positive edge triggered flip-flop except that the output changes at the negative edge of the clock.

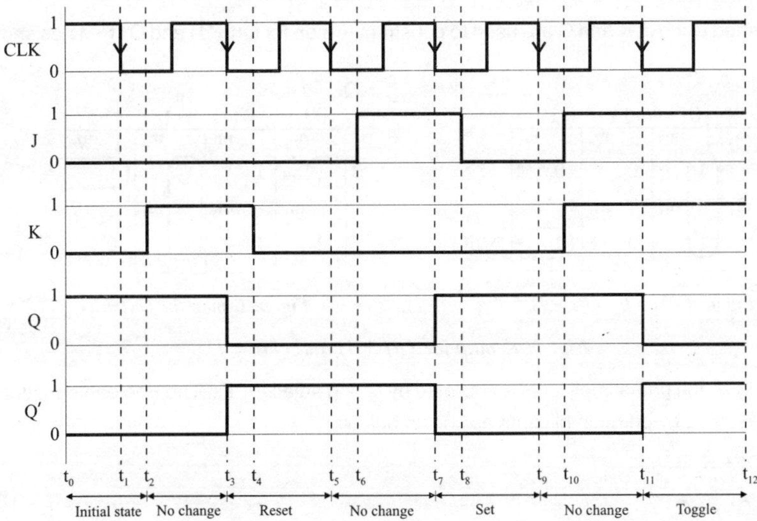

Fig. 6.44: Timing diagram (or waveform) of negative edge triggered JK flip-flop.

Example 6.2

Design and implement JK flip-flop.

Solution

An excitation table (Table 1) is developed by listing all possible combination of inputs and present states Q and Q′. The next state outputs Q(t + 1) and Q′(t + 1) are determined for all the combinations of J, K, Q and Q′ from the knowledge of truth table of JK flip-flop and listed in Table 1.

With three state variables J, K and Q there are 8 ($2^3 = 8$) combinations of minterms as shown in Table 1. With three state variables J, K and Q′ another 8 combinations of minterms are formed and listed in Table 1.

Truth Table of JK Flip-Flop

Inputs		Outputs	
J	K	Q	Q′
0	0	No change	
0	1	0	1
1	0	1	0
1	1	Toggle	

Table 1: Excitation Table

Inputs		Present State Output		Next State Output		Comment	Minterm
J	K	Q	Q′	Q(t + 1)	Q′(t+1)		JKQ
0	0	0	1	0	1	No change	m_0
0	0	1	0	1	0		m_1
0	1	0	1	0	1	Reset state	m_2
0	1	1	0	0	1		m_3
1	0	0	1	1	0	Set state	m_4
1	0	1	0	1	0		m_5
1	1	0	1	1	0	Toggle	m_6
1	1	1	0	0	1		m_7

The minterms formed using J, K and Q are used to construct K-map for Q(t + 1) and Q′(t + 1) as shown in Fig. 1.

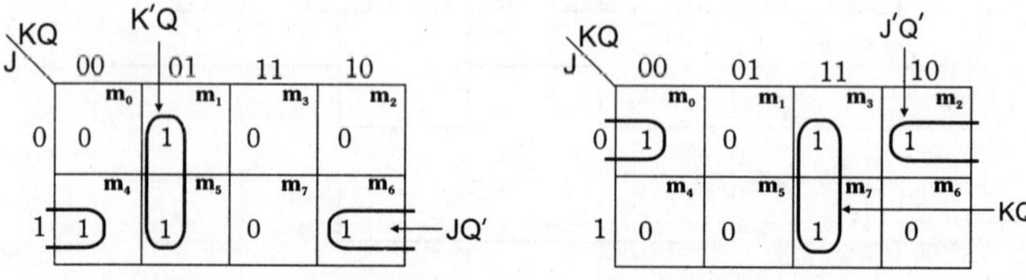

Fig. a: K-map for Q(t + 1).

Fig. b: K-map for Q′(t + 1).

Fig. 1: K-map for Q(t + 1) and Q′(t + 1).

In the K-map of Fig. 1, the prime implicants are formed by taking adjacent 1's. The Boolean equations are given by sum of products represented by prime implicants as shown below:

$$Q(t+1) = J\,Q' + K'\,Q$$

$$Q'(t+1) = J'\,Q' + KQ$$

The JK flip-flop that satisfy the above equations is shown in Fig. 2. The JK flip-flop output is verified in Table 2.

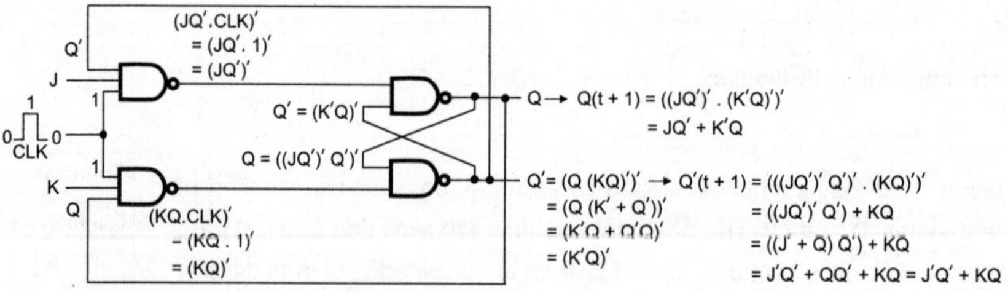

Fig. 2: JK flip-flop using NAND latch.

Table 2: Verification of JK Flip-Flop Outputs

Inputs		Present State Output								Next State Output		
J	K	Q	Q′	K′	JQ′	K′Q	J′	J′Q′	KQ	Q(t + 1) = JQ′ + K′Q	Q′(t + 1) = J′Q′ + KQ	Comment
0	0	0	1	1	0	0	1	1	0	0	1	No change
0	0	1	0	1	0	1	1	0	0	1	0	
0	1	0	1	0	0	0	1	1	0	0	1	Reset state
0	1	1	0	0	0	0	1	0	1	0	1	
1	0	0	1	1	1	0	0	0	0	1	0	Set state
1	0	1	0	1	0	1	0	0	0	1	0	
1	1	0	1	0	1	0	0	0	0	1	0	Toggle
1	1	1	0	0	0	0	0	0	1	0	1	

6.2.4 T Flip-Flop (Toggle Flip-Flop)

(CSVTU, Nov/Dec'19, 7 Marks) (JNTUA, Jun/July'22, 7 Marks)

In JK flip-flop when J = K = 1 the output toggles for every clock input and when J = K = 0 no change in output. This property can be used to construct T flip-flop as shown in Fig. 6.45. Therefore, T flip-flop is a single input flip-flop in which the present output is complement of previous output when clock is applied, provided input is 1. The T flip-flops are used to design binary counters. The truth table of T flip-flop is shown in Table 6.14 and standard symbols of T flip-flop are shown in Fig. 6.46.

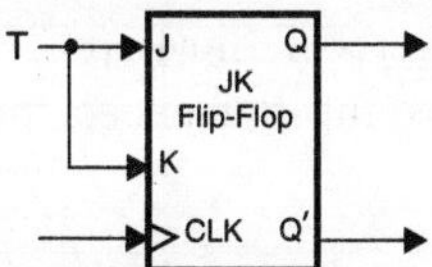

***Fig. 6.45:** Converting JK flip-flop to T flip-flop.*

Table 6.14: Truth Table of T Flip-Flop

Level Triggered

T	Clock	Q Q′
0	1	No change
1	1	Toggles previous output

Positive Edge Triggered

T	Clock	Q Q′
0	↑	No change
1	↑	Toggles previous output

Negative Edge Triggered

T	Clock	Q Q′
0	↓	No change
1	↓	Toggles previous output

***Fig. a:** Level triggered T flip-flop.*

***Fig. b:** Positive edge triggered T flip-flop.*

***Fig. c:** Negative edge triggered T flip-flop.*

***Fig. 6.46:** Standard symbols of T flip-flop.*

Timing Diagram of Positive Edge Triggered T Flip-flop

The timing diagram of positive edge triggered T flip-flop is shown in Fig. 6.47. In T flip-flop, if the input T = 0, then there is no change in output. But if the input T = 1, then the output toggles at every positive edge of the clock. Here, from t_0 to t_3, input T = 0 and hence no change in output but after t_3, input T = 1 and hence the output Q is alternatively set and reset at every positive edge of the clock.

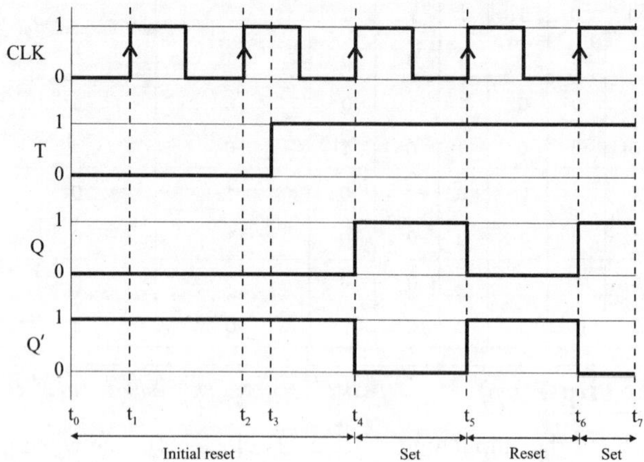

Fig. 6.47: Timing diagram (or waveform) of positive edge triggered T flip-flop.

Timing Diagram of Negative Edge Triggered T Flip-flop

The timing diagram of negative edge triggered T flip-flop is shown in Fig. 6.48. Here, the inputs and outputs are same as that positive edge triggered flip-flop except that the output changes at the negative edge of the clock.

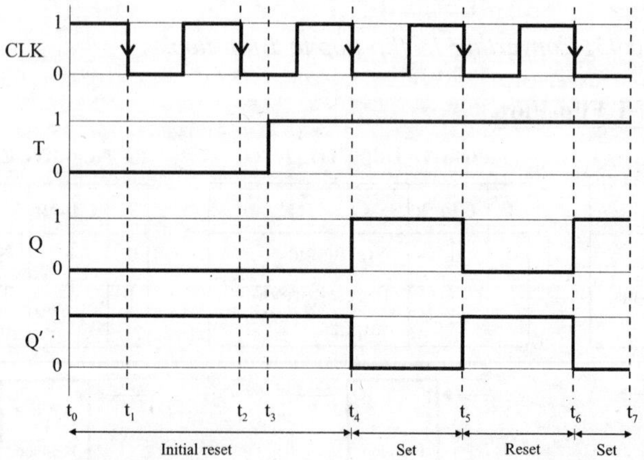

Fig. 6.48: Timing diagram (or waveform) of negative edge triggered T flip-flop.

6.2.5 D Flip-Flop (Data Flip-Flop) *(JNTUA, Jun/July'22, 7 Marks) (GTU, Jun'19, 3 Marks)*

A single input flip-flop can be developed from SR flip-flop, if the input to SR flip-flops are always complementary inputs, so that if input is 1 then output is also 1 and if input is 0 then output is also 0. Therefore, D flip-flop is a single input flip-flop in which the input is stored in the output when clock is applied.

The D latch with enable shown in Fig. 6.49 becomes a D flip-flop if the enable is replaced by a clock as shown in Fig. 6.49. The truth table of D flip-flop is shown in Table 6.15 and standard symbols are shown in Fig. 6.50. The D flip-flop is available as a standard IC 7474. The pin configuration of IC 7474 is shown in Fig. 6.51.

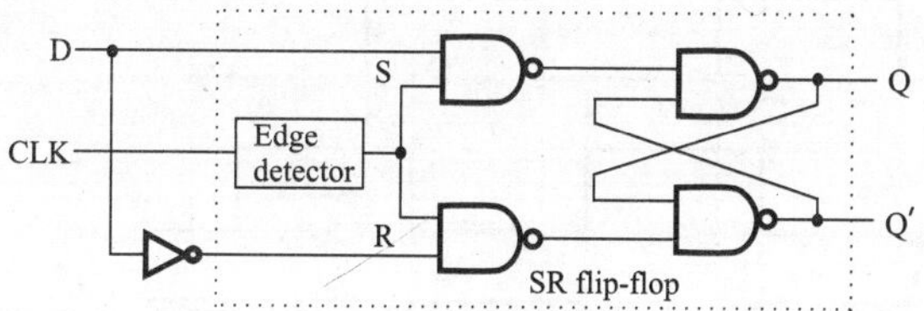

Fig. 6.49: D flip-flop.

Table 6.15: Truth Table of D Flip-Flop

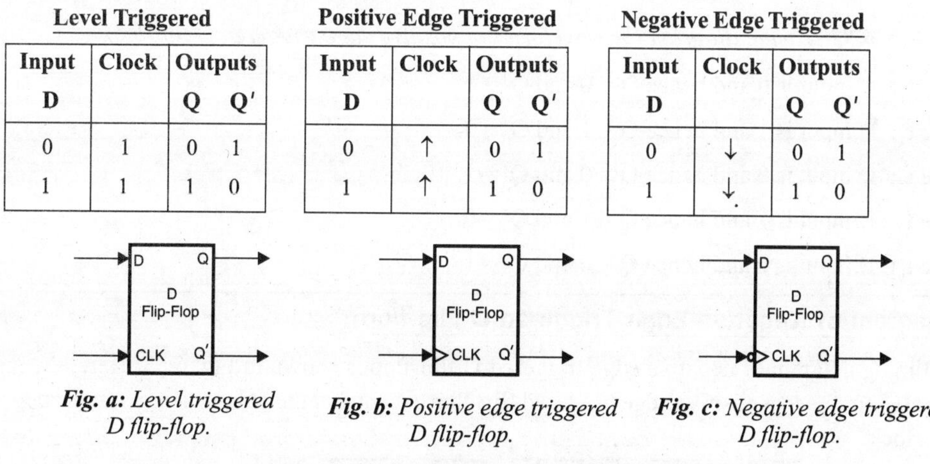

Level Triggered

Input D	Clock	Outputs Q	Q′
0	1	0	1
1	1	1	0

Positive Edge Triggered

Input D	Clock	Outputs Q	Q′
0	↑	0	1
1	↑	1	0

Negative Edge Triggered

Input D	Clock	Outputs Q	Q′
0	↓	0	1
1	↓	1	0

Fig. a: Level triggered D flip-flop. *Fig. b: Positive edge triggered D flip-flop.* *Fig. c: Negative edge triggered D flip-flop.*

Fig. 6.50: Standard symbol of D flip-flop.

Pin	Name	Pin	Name
1	$\overline{CLR1}$	14	V_{cc}(+5V)
2	D_1	13	$\overline{CLR2}$
3	CLK1	12	D_2
4	$\overline{PR1}$	11	CLK2
5	Q_1	10	$\overline{PR2}$
6	$\overline{Q}_1$	9	Q_2'
7	(0V) GND	8	$\overline{Q}_2$

7474

Pin	Description
D_1, D_2	Inputs
$\overline{PR}_1$, $\overline{PR}_2$	Preset
$\overline{CLR}_1$, $\overline{CLR}_2$	Clear
CLK_1, CLK_2	Clock
Q_1, Q_2, $\overline{Q}_1$, $\overline{Q}_2$	Outputs
V_{cc}	Supply voltage (+5V)
GND	Ground (0V)

Fig. 6.51: Pin configuration of dual D flip-flop IC 7474.

Timing Diagram of Positive Edge Triggered D Flip-flop

(VTU, Aug/Sep'20, 8 Marks)

The timing diagram of positive edge triggered D flip-flop is shown in Fig. 6.52. Here, the inputs are recognized only at the positive edge of the clock and the output is decided by the value of the inputs at the positive edge of the clock. Here, input changes at time t_2, t_4, t_6, t_8 and t_{10} but the output changes only at the positive edge of the clock that occurs after the input change. The outputs are predicted from the truth table shown in Table 6.15. In the timing diagram, the following changes occur at various timings.

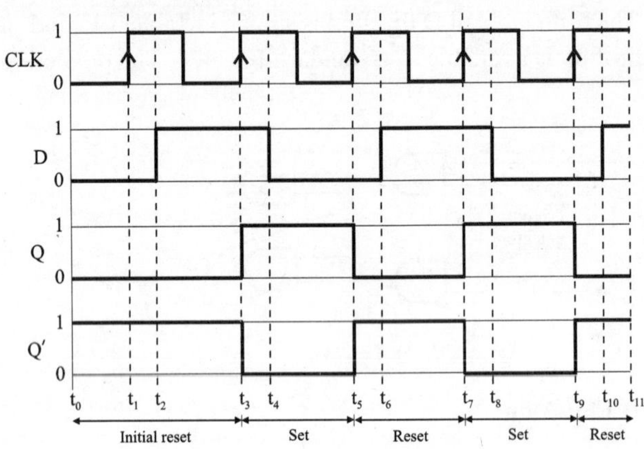

Fig. 6.52: Timing diagram (or waveform) of positive edge triggered D flip-flop.

Time t_1: D input is 0 and hence Q = 0 and Q′ = 1.

Time t_3: D input is 1 and hence Q = 1 and Q′ = 0.

Time t_5: D input is 0 and hence Q = 0 and Q′ = 1.

Time t_7: D input is 1 and hence Q = 1 and Q′ = 0.

Time t_9: D input is 0 and hence Q = 0 and Q′ = 1.

Timing Diagram of Negative Edge Triggered D Flip-flop

The timing diagram of negative edge triggered D flip-flop is shown in Fig. 6.53. Here, the inputs and outputs are same as that positive edge triggered flip-flop except that the output changes at the negative edge of the clock.

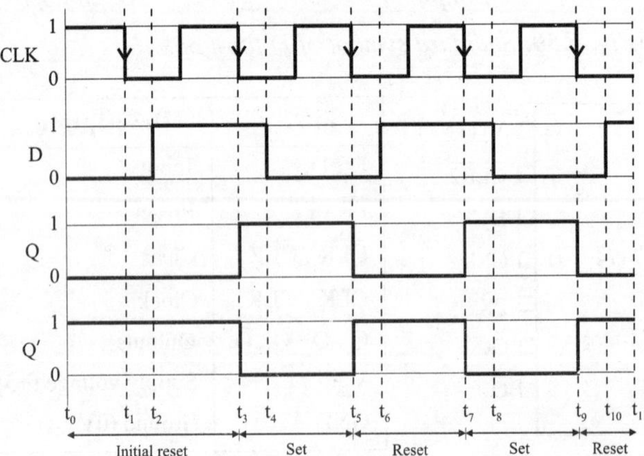

Fig. 6.53: Timing diagram (or waveform) of negative edge triggered D flip-flop.

6.2.6 Flip-Flop with Preset and Reset Facility *(CSVTU, Nov/Dec'19, 2 Marks) (AKUB, Nov/Dec'20, 3 Marks)*

When power is switched ON flip-flop outputs will be set or reset at random.

In order to ensure a definite set state, preset facility is provided in flip-flops.

In order to ensure a definite reset state, reset facility is provided in flip-flops.

The preset and reset are logic **low** input signals denoted by P and R respectively. As long as preset is 0, the output is set to 1 (set state). Only when P is made 1, the input and clock are allowed to change the output.

As long as reset is 0 the output is reset to 0 (reset state). Only when R is made 1, the input and clock are allowed to change the output. Therefore, for output to change depending on input and clock, preset and reset should be tied to 1.

Note: *Both P = 0 and R = 0 is illegal.*

D Flip-Flop with Preset and Reset

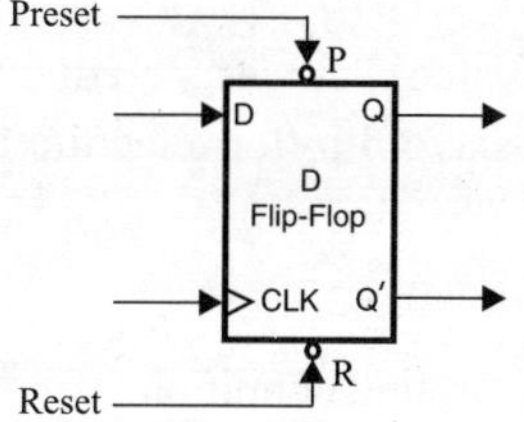

Fig. 6.54: *D flip-flop with preset and reset.*

Table 6.16: Truth Table D Flip-Flop with Preset and Reset

P	R	CLK	D	Q	Q'
0	1	×	×	1	0
1	0	×	×	0	1
1	1	↑	0	0	1
1	1	↑	1	1	0

SR Flip-Flop with Preset and Reset

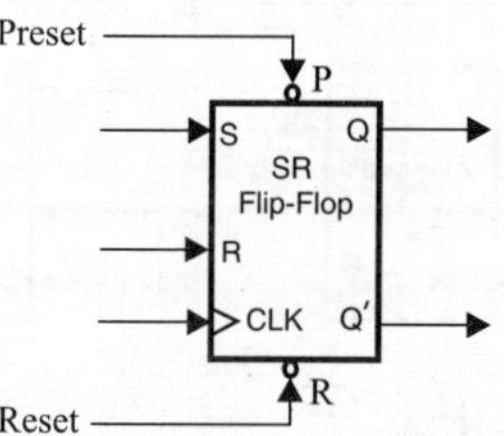

Fig. 6.55: *SR flip-flop with preset and reset.*

Table 6.17: Truth Table of SR Flip-Flop with Preset and Reset

P	R	CLK	S	R	Q	Q'
0	1	×	×	×	1	0
1	0	×	×	×	0	1
1	1	↑	0	0	No change	
1	1	↑	0	1	0	1
1	1	↑	1	0	1	0
1	1	↑	1	1	Indeterminate	

JK Flip-Flop with Preset and Reset

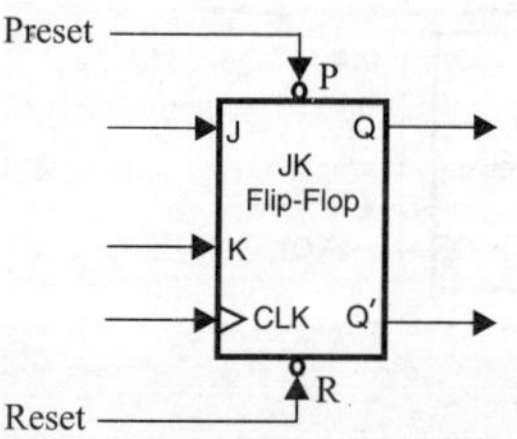

Fig. 6.56: *JK flip-flop with preset and reset.*

Table 6.18: Truth Table of JK Flip-Flop with Preset and Reset

P	R	CLK	J	K	Q	Q'
0	1	×	×	×	1	0
1	0	×	×	×	0	1
1	1	↑	0	0	No change	
1	1	↑	0	1	0	1
1	1	↑	1	0	1	0
1	1	↑	1	1	Toggle previous output	

T Flip-Flop with Preset and Reset

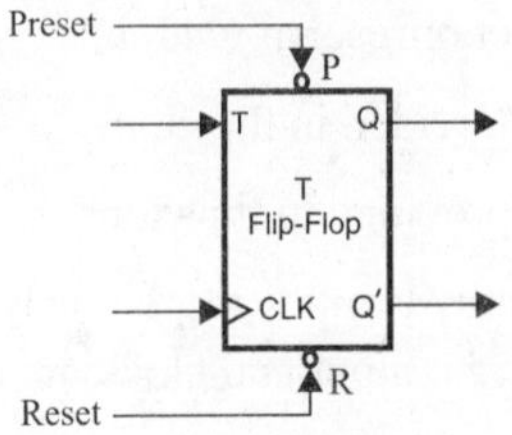

Fig. 6.57: T flip-flop with preset and reset.

Table 6.19: Truth Table of T Flip-Flop with Preset and Reset

P	R	CLK	T	Q	Q'
0	1	×	×	1	0
1	0	×	×	0	1
1	1	↑	0	1	0
1	1	↑	1	0	1

6.3 Master-Slave Flip-Flop *(JNTUK, Feb'22, 7 Marks) (AU, Nov/Dec'18, 13 Marks) (VTU, Aug/Sep'20, 10 Marks)*

A ***master-slave flip-flop*** is cascade connection of two level triggered flip-flops, as shown in Fig. 6.58. The first flip-flop is called master and the second flip-flop is called slave.

When clock is high, the input is passed to output of master flip-flop. The master output is passed to slave output on the negative edge of clock.

Therefore, the value at the output of the flip-flop is the value that was stored in the master stage immediately before the negative edge of clock. Hence master-slave flip-flop function as negative edge triggered flip-flop. *(VTU, Aug/Sep'20, 7 Marks)*

The behaviour of the master-slave flip-flop ensure the following,

1. The output may change only once and there is no race around condition.
2. A change in the output is triggered by the negative edge of the clock.

The master-slave flip-flops are shown in Fig. 6.58.

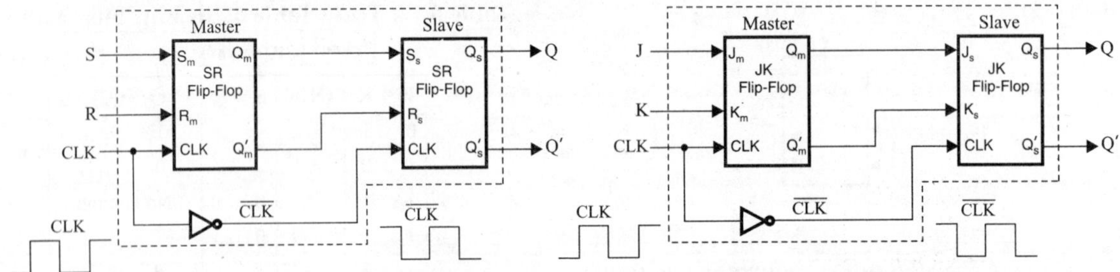

Fig. a: Master slave SR flip-flop *Fig. b: Master slave JK flip-flop*

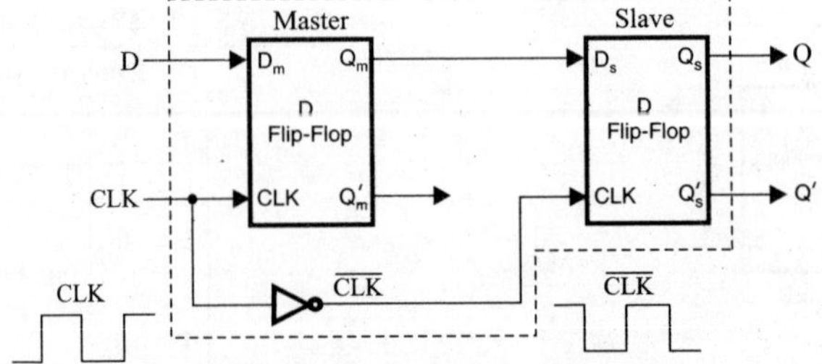

Fig. b: Master slave D flip-flop

Fig. 6.58: Master-slave flip-flops.

Timing Diagram of Master-slave SR/JK Flip-flop

The timing diagram or waveform of master slave SR/JK flip-flop is drawn by taking alternate set and reset conditions of binary inputs and predicting the outputs from the knowledge of truth table of SR/JK flip-flop.

The timing diagram of master-slave SR/JK flip-flop is shown in Fig. 6.59 is drawn by taking SR/JK inputs as 01, 10 and 01. Here, the flip-flop inputs are recognized by the master flip-flop during high value of clock and outputs of master are decided by flip-flop inputs as long as clock is high. The outputs of master are applied as inputs to slave and outputs of slave will change immediately on the negative edge of the clock.

The outputs are predicted from the truth table shown in Table 6.12 and 6.13. In the timing diagram, the following changes occur at various timings.

Time t_0: Input corresponds to no change condition and output is assumed to be in set state.

Time t_1: Input changes to reset condition. Since CLK is low master does not accept inputs.

Time t_2: CLK goes high and master input corresponds to reset and so master output is reset.

Time t_3: CLK goes low, the master output is passed to slave output and slave is reset.

Time t_4: Input changes to set condition. Since CLK is low master does not accept inputs.

Time t_5: CLK goes high and master input corresponds to set and so master output is set.

Time t_6: CLK goes low, the master output is passed to slave output and slave is set.

Time t_7: Input changes to reset condition. Since CLK is low master does not accept inputs.

Time t_8: CLK goes high and master input corresponds to reset and so master output is reset.

Time t_9: CLK goes low, the master output is passed to slave output and slave is reset.

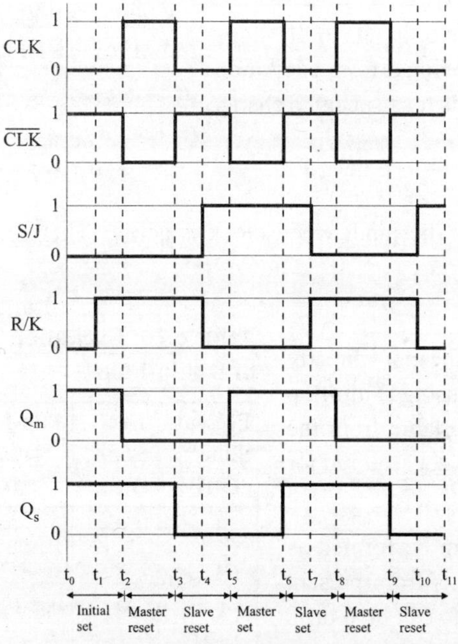

Fig. 6.59: Timing diagram (or waveform) of master-slave SR/JK flip-flop.

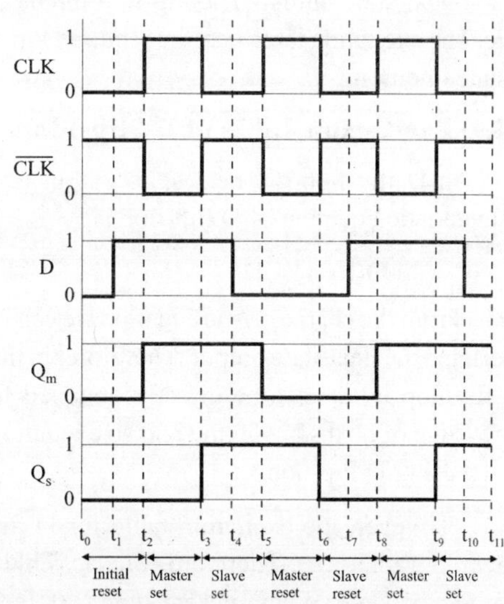

Fig. 6.60: Timing diagram (or waveform) of master-slave D flip-flop.

Timing Diagram of Master-slave D Flip-flop

The timing diagram or waveform of master slave D flip-flop is drawn by taking alternate set and reset conditions of binary input and predicting the output from the knowledge of truth table of D flip-flop.

The timing diagram of master-slave D flip-flop is shown in Fig. 6.60 is drawn by taking D input as 1, 0 and 1. Here, the flip-flop input is recognized by the master flip-flop during high value of clock and output of master is decided by flip-flop input as long as clock is high. The output of master is applied as input to slave and output of slave will change immediately on the negative edge of the clock.

The output is predicted from the truth table shown in Table 6.15. In the timing diagram, the following changes occur at various timings.

Time t_0: Input corresponds to no change condition and output is assumed to be in reset state.

Time t_1: Input changes to set condition. Since CLK is low master does not accept input.

Time t_2: CLK goes high and master input corresponds to set and so master output is set.

Time t_3: CLK goes low, the master output is passed to slave output and slave is set.

Time t_4: Input changes to reset condition. Since CLK is low master does not accept input.

Time t_5: CLK goes high and master input corresponds to reset and so master output is reset.

Time t_6: CLK goes low, the master output is passed to slave output and slave is reset.

Time t_7: Input changes to set condition. Since CLK is low master does not accept input.

Time t_8: CLK goes high and master input corresponds to set and so master output is set.

Time t_9: CLK goes low, the master output is passed to slave output and slave is set.

6.4 Excitation Table and Characteristic Equation of Flip-Flops

(RGPV, Jun'20, 5 Marks)
(BATU, Jun'19, 6 Marks)
(VTU, Aug/Sep'20, 10 Marks)

The design using flip-flops require the knowledge of inputs to flip-flops for the next state output. The flip-flop inputs can be obtained from excitation table. The excitation table lists the required flip-flop inputs for a specified change of state in output.

6.4.1 Excitation Table of D Flip-Flop

In D flip-flop the next state output is given by the input when clock is active. Therefore, the characteristic equation of D flip-flop is,

$$Q(t+1) = D$$

From the characteristic equation we can infer that the input directly determine the next state output. Therefore, in the design using D flip-flop the flip-flop inputs can be directly obtained from state table from the knowledge of next state output and so excitation table is not required for design using D flip-flop.

However, the excitation table for D flip-flop can be formed as shown in Table 6.20. There are four possible transitions from present state to next state. When present state is 0, the next state can be 0 or 1. When present state is 1, the next state can be 0 or 1.

Table 6.20: Excitation Table of D Flip-Flop

Present State Q(t)	Next State Q(t+1)	Flip-Flop Input D
0	0	0
0	1	1
1	0	0
1	1	1

The explanation for formation of each row of excitation table is given below:

Row 1 of Excitation Table: Present and Next States are 0

Now, $Q(t) = 0$; $Q(t+1) = 0$

Output in reset state. No change from present state to next state.

From truth table of D flip-flop we can say that,

if $D = 0$, then there is no change in output

Therefore, input to flip-flop is,

$D = 0$

Truth Table of D Flip-Flop

Input	Outputs	
D	Q	Q′
0	0	1
1	1	0

Row 2 of Excitation Table: Present State is 0 and Next State is 1

Now, $Q(t) = 0$; $Q(t+1) = 1$

Output changes from reset to set state.

From the truth table of D flip-flop we can say that,

if $D = 1$, then the output will be driven to set state.

Therefore, input to flip-flop is,

$D = 1$

Row 3 of Excitation Table: Present State is 1 and Next State is 0

Now, $Q(t) = 1$; $Q(t+1) = 0$

Output changes from set to reset state.

From the truth table of D flip-flop we can say that,

if $D = 0$, then the output will be driven to reset state.

Therefore, input to flip-flop is,

$D = 0$

Row 4 of Excitation Table: Present State and Next States are 1

Now, $Q(t) = 1$; $Q(t+1) = 1$

Output in set state. No change from present state to next state.

From the truth table of D flip-flop we can say that,

if $D = 1$, there is no change in output.

Therefore, input to flip-flop is,

$D = 1$

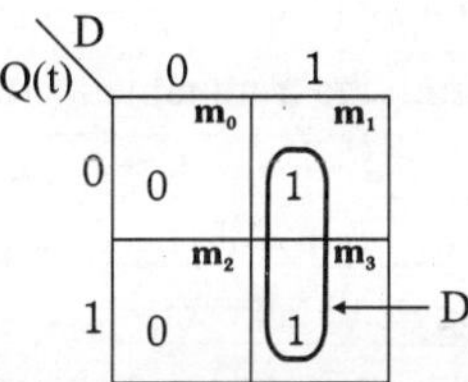

Fig. 6.61: K-map for Q(t + 1) of D flip-flop.

The excitation table (Table 6.20) can be used to develop characteristic equation of D flip-flop by forming minterms using present state, Q(t) and input, D as shown below and constructing K-map for next state Q(t + 1) as shown in Fig. 6.61.

Q(t) D	Minterm	Q(t + 1)
0 0	m_0	0
0 1	m_1	1
1 0	m_2	0
1 1	m_3	1

From the K-map we get the following characteristic equation for D flip-flop.

$$Q(t+1) = D$$

Using the characteristic equation the characteristic table of D flip-flop is formed as shown in Table 6.21.

Table 6.21: Characteristic Table of D Flip-Flop

Input D	Present State Q	Next State Q(t+1)
0	0	0
0	1	0
1	0	1
1	1	1

6.4.2 Excitation Table of JK Flip-Flop

In the design using JK flip-flops an excitation table is formed to determine the inputs of flip-flops for next state output.

The excitation table of JK flip-flop is shown in Table 6.22. There are four possible transitions from present state to next state. When present state is 0 the next state can be 0 or 1. When present state is 1 the next state can be 0 or 1.

The explanation for formation of each row of excitation table is given below:

Table 6.22: Excitation Table of JK Flip-Flop

Present State Q(t)	Next State Q(t+1)	Flip-Flop Inputs J	K
0	0	0	×
0	1	1	×
1	0	×	1
1	1	×	0

Row 1 of Excitation Table: Present and Next States are 0

Now, $Q(t) = 0$; $Q(t+1) = 0$

Output in reset state. No change from present state to next state.

From truth table of JK flip-flop we can say that,

if J = 0 and K = 0, then there is no change in output

Also, if J = 0 and K = 1 then output remain in reset state.

Therefore, inputs to flip-flop are,

J = 0 ⇒ J = 0

K = 0 or 1 (don't-care) K = ×

Row 2 of Excitation Table: Present State is 0 and Next State is 1

Truth Table of JK Flip-Flop

Inputs J	K	Outputs Q	Q'
0	0	No change	
0	1	0	1 (Reset)
1	0	1	0 (Set)
1	1	Toggle	

Now, $Q(t) = 0$; $Q(t+1) = 1$

Output changes from reset to set state.

From the truth table of JK flip-flop we can say that,

if J = 1 and K = 0, then the output will be driven to set state.

Also, if J = 1 and K = 1 then the output toggles. Here present state 0 will toggle to next state 1.

Therefore, inputs to flip-flop are,

J = 1 ⇒ J = 1

K = 0 or 1 (don't-care) K = ×

Row 3 of Excitation Table: Present State is 1 and Next State is 0

Now, $Q(t) = 1$; $Q(t+1) = 0$

Output changes from set to reset state.

From the truth table of JK flip-flop we can say that,

if J = 0 and K = 1, then the output will be driven to reset state.

Also, if J = 1 and K = 1 then the output toggles. Here present state 1 will toggle to next state 0.

Therefore, inputs to flip-flop are,

J = 0 or 1 (don't-care) ⇒ J = ×

K = 1 K = 1

Row 4 of Excitation Table: Present State and Next States are 1

Now, Q(t) = 1 ; Q(t + 1) = 1

Output in set state. No change from present state to next state.

From the truth table of JK flip-flop we can say that,

if J = 0 and K = 0, there is no change in output.

Also, if J = 1 and K = 0 then output remain in set state.

Therefore, inputs to flip-flop are,

J = 0 or 1 (don't-care) ⇒ J = ×

K = 0 K = 0

The excitation table (Table 6.22) can be used to develop characteristic equation of JK flip-flop by forming minterms using Q(t), J and K as shown below and constructing K-map for next state Q(t + 1) as shown in Fig. 6.62.

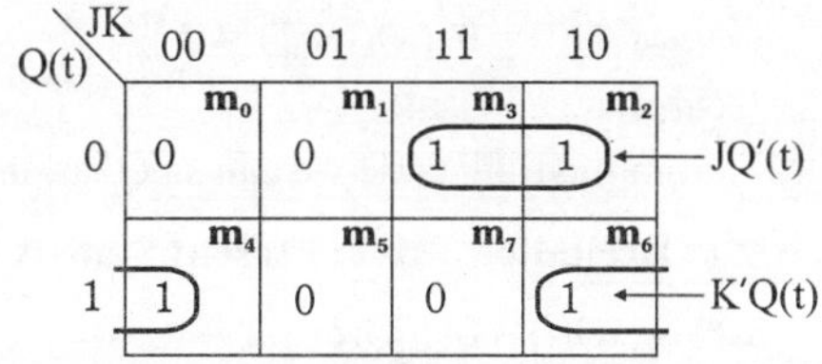

Fig. 6.62: K-map for Q(t + 1) of JK flip-flop.

Q(t)	J	K	Minterm	Q(t + 1)	Q(t)	J	K	Minterm	Q(t + 1)
0	0	0	m_0	0	1	0	1	m_5	0
0	0	1	m_1	0	1	1	1	m_7	0
0	1	0	m_2	1	1	0	0	m_4	1
0	1	1	m_3	1	1	1	0	m_6	1

***Note:** Two possible values 0 and 1 are taken for don't-care, × of J and K in excitation table.*

From the K-map we get the following characteristic equation for JK flip-flop.

$$Q(t+1) = JQ'(t) + K'Q(t)$$

Using the characteristic equation the characteristic table of JK flip-flop is formed as shown in Table 6.23.

Table 6.23: Characteristic Table of JK Flip-Flop ***(AU, Apr/May'19, 2 Marks)***

Inputs J K	Present State Q	Next State Q(t+1)
0 0	0	0
0 0	1	1
0 1	0	0
0 1	1	0
1 0	0	1
1 0	1	1
1 1	0	1
1 1	1	0

6.4.3 Excitation Table of T Flip-Flop

In design using T flip-flops an excitation table is formed to determine the inputs to flip-flops for the next state output.

The excitation table of T flip-flop is shown in Table 6.24. There are four possible transitions from present to next state. When present state is 0 the next state can be 0 or 1. When present state is 1 the next state can be 0 or 1. The explanation for formation of each row of excitation table is given below:

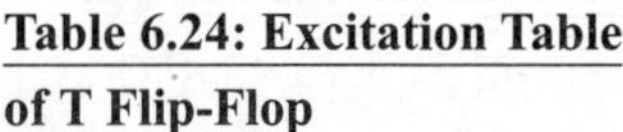

Table 6.24: Excitation Table of T Flip-Flop

Present State Q(t)	Next State Q(t + 1)	Flip-Flop Input T
0	0	0
0	1	1
1	0	1
1	1	0

Row 1 of Excitation Table: Present and Next States are 0

Now, Q(t) = 0 ; Q(t + 1) = 0

No change in state.

From the truth table we can say that, input T = 0 for this output condition.

Truth Table of T Flip-Flop

Input T	Outputs Q Q′
0	No change
1	Toggle

Row 2 of Excitation Table: Present State is 0 and Next State are 1

Now, Q(t) = 0 ; Q(t + 1) = 1

Output toggles.

From the truth table we can say that, input T = 1 for this output condition.

Row 3 of Excitation Table: Present State is 1 and Next State are 0

Now, Q(t) = 1 ; Q(t + 1) = 0

Output toggles.

From the truth table we can say that, input T = 1 for this output condition.

Row 4 of Excitation Table: Present and Next States are 1

Now, Q(t) = 1 ; Q(t + 1) = 1

No change in state.

From the truth table we can say that, input T = 0 for this output condition.

The excitation table (Table 6.24) can be used to develop characteristic equation of T flip-flop by forming minterms using present state, Q(t) and input, T as shown below and constructing K-map for next state Q(t + 1) as shown in Fig. 6.63.

Q(t) T	Minterm	Q(t + 1)
0 0	m_0	0
0 1	m_1	1
1 0	m_2	1
1 1	m_3	0

Q(t) \ T	0	1
0	m_0: 0	m_1: (1) ← TQ′(t)
1	T′Q(t) → m_2: (1)	m_3: 0

Fig. 6.63: K-map for Q(t + 1) of T flip-flop.

From the K-map we get the following characteristic equation for T flip-flop.

$$Q(t+1) = TQ'(t) + T'Q(t)$$

Using the characteristic equation the characteristic table of T flip-flop is formed as shown in Table 6.25.

Table 6.25: Characteristic Table of T Flip-Flop

Input T	Present State Q	Next State Q(t+1)
0	0	0
0	1	1
1	0	1
1	1	0

6.4.4 Excitation Table of SR Flip-Flop

In the design using SR flip-flops an excitation table is formed to determine the inputs of flip-flops for next state output.

The excitation table of SR flip-flop is shown in Table 6.26. There are four possible transitions from present state to next state. When present state is 0 the next state can be 0 or 1. When present state is 1 the next state can be 0 or 1.

Table 6.26: Excitation Table of SR Flip-Flop

Present State Q(t)	Next State Q(t+1)	Flip-Flop Inputs S	R
0	0	0	×
0	1	1	0
1	0	0	1
1	1	×	0
0	×	1	1
1	×	1	1

Truth Table of SR Flip-Flop

Inputs S	R	Outputs Q	Q'
0	0	No change	
0	1	0	1 (Reset)
1	0	1	0 (Set)
1	1	Indeterminate	

The explanation for formation of each row of excitation table is given below:

Row 1 of Excitation Table: Present and Next States are 0

Now, $Q(t) = 0$; $Q(t+1) = 0$

Output in reset state. No change from present state to next state.

From truth table of SR flip-flop we can say that,

if S = 0 and R = 0, then there is no change in output

Also, if S = 0 and R = 1 then output remain in reset state.

Therefore, inputs to flip-flop are,

$S = 0 \quad \Rightarrow \quad S = 0$

$R = 0$ or 1 (don't-care) $\quad R = \times$

Row 2 of Excitation Table: Present State is 0 and Next State is 1

Now, $Q(t) = 0$; $Q(t+1) = 1$

Output changes from reset to set state.

From the truth table of SR flip-flop we can say that,

if S = 1 and R = 0, then the output will be driven to set state.

Therefore, inputs to flip-flop are,

$S = 1, R = 0$

Row 3 of Excitation Table: Present State is 1 and Next State is 0

Now, $Q(t) = 1$; $Q(t+1) = 0$

Output changes from set to reset state.

From the truth table of SR flip-flop we can say that,

if S = 0 and R = 1, then the output will be driven to reset state.

Therefore, inputs to flip-flop are,

$S = 0, R = 1$

Row 4 of Excitation Table: Present State and Next States are 1

Now, $Q(t) = 1$; $Q(t+1) = 1$

Output in set state. No change from present state to next state.

From the truth table of SR flip-flop we can say that,

if S = 0 and R = 0, there is no change in output.

Also, if S = 1 and R = 0 then output remain in set state.

Therefore, inputs to flip-flop are,

S = 0 or 1 (don't-care) ⇒ S = ×

R = 0 R = 0

Rows 5 and 6 of Excitation Table: Indeterminate Next State

Now, Q(t) = 0 or 1 ; Q(t + 1) = ×

Output in indeterminate state.

From the truth table of SR flip-flop we can say that,

if S = 1 and R = 1, then the output will be indeterminate.

Therefore, inputs to flip-flop are,

S = 1, R = 1

The excitation table (Table 6.26) can be used to develop characteristic equation of SR flip-flop by forming minterms using Q(t), S and R as shown below and constructing K-map for next state Q(t + 1) as shown in Fig. 6.64.

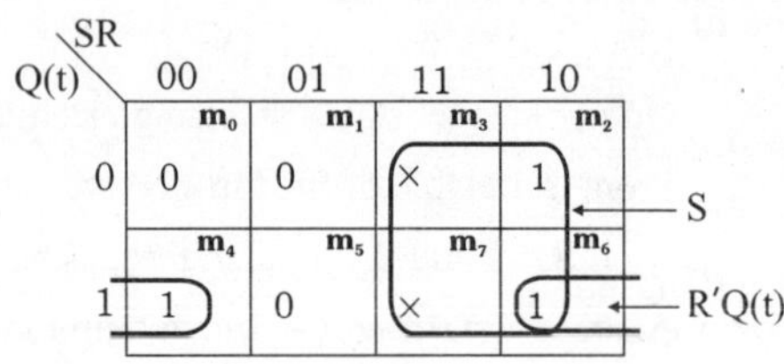

Fig. 6.64: K-map for Q(t + 1) of SR flip-flop.

Q(t) S R	Minterm	Q(t + 1)	Q(t) S R	Minterm	Q(t + 1)
0 0 0	m_0	0	1 0 0	m_4	1
0 0 1	m_1	0	1 1 0	m_6	1
0 1 0	m_2	1	0 1 1	m_3	×
1 0 1	m_5	0	1 1 1	m_7	×

> ***Note:*** *Two possible values 0 and 1 are taken for don't-care, × in S and R in excitation table.*

From the K-map we get the following characteristic equation for SR flip-flop.

Q(t + 1) = S + R′Q(t)

Using the characteristic equation the characteristic table of SR flip-flop is formed as shown in Table 6.27.

Table 6.27: Characteristic Table of SR Flip-Flop

Inputs S R	Present State Q	Next State Q(t+1)
0 0	0	0
0 0	1	1
0 1	0	0
0 1	1	0
1 0	0	1
1 0	1	1
1 1	0	×
1 1	1	×

Table 6.28: Summary of Characteristic Equation of Flip-Flops

Flip-Flop	Characteristic Equation
D	Q(t + 1) = D
JK	Q(t + 1) = JQ′(t) + K′Q(t)
T	Q(t + 1) = TQ′(t) + T′Q(t)
SR	Q(t + 1) = S + R′Q(t)

6.5 Conversion of Flip-Flops

There are four different types of flip-flops and they are,

- SR flip-flop
- JK flip-flop
- D flip-flop (or Data flip-flop)
- T flip-flop (or Toggle flip-flop)

It is possible to convert one type of flip-flop to any other type of flip-flop. For example, SR flip-flop can be converted to JK flip-flop or D flip-flop or T flip-flop.

The flip-flop outputs are commonly denoted as Q and Q′. The different types of inputs will drive the flip-flop to get different values of Q and Q′. Therefore, the conversion of flip-flop requires design of a combinational circuit to supply the required inputs to a flip-flop in order to get the output of another desired flip-flop.

The design of combinational circuit requires a table that lists possible combinations of inputs, present state and next state of the desired flip-flop. These informations are obtained from truth table of desired flip-flop. Using the excitation table of flip-flop to be converted, the inputs of desired flip-flop to drive the output to next state are determined and listed in same table.

The minterms are formed for all possible combinations of input and present state of desired flip-flop. Using these minterms the K-maps are constructed for input of flip-flop to be converted. From the K-map the simplified equations are obtained for flip-flop inputs.

6.5.1 Conversion of SR to JK Flip-Flop

(RGPV, Jun'22, 7 Marks) (JNTUK, Feb'22, 7 Marks)

Using the truth table of JK flip-flop, a conversion table is formed by listing the possible combinations of inputs, present and next states of JK flip-flop as shown in Table 6.29.

Table 6.29: Conversion Table to Convert SR to JK Flip-Flop

JK Flip-Flop				Comment	SR Flip-Flop Inputs		Minterm
Inputs		Present State	Next State				
J	K	Q	Q(t+1)		S	R	JKQ
0	0	0	0	No change	0	×	m_0
0	0	1	1		×	0	m_1
0	1	0	0	Reset state	0	×	m_2
0	1	1	0		0	1	m_3
1	0	0	1	Set state	1	0	m_4
1	0	1	1		×	0	m_5
1	1	0	1	Toggle	1	0	m_6
1	1	1	0		0	1	m_7

Truth Table of JK Flip-Flop

Inputs		Outputs	
J	K	Q	Q′
0	0	No change	
0	1	0	1 (Reset)
1	0	1	0 (Set)
1	1	Toggles	

In order to drive the present state to next state listed in Table 6.29, the SR flip-flop inputs are obtained using the excitation table of SR flip-flop and listed in Table 6.29. Minterms are formed using J, K and Q and the K-maps are constructed for S and R as shown in Fig. 6.65.

Excitation Table of SR Flip-Flop

Present State	Next State	Flip-Flop Inputs	
Q(t)	Q(t + 1)	S	R
0	0	0	×
0	1	1	0
1	0	0	1
1	1	×	0
0	×	1	1
1	×	1	1

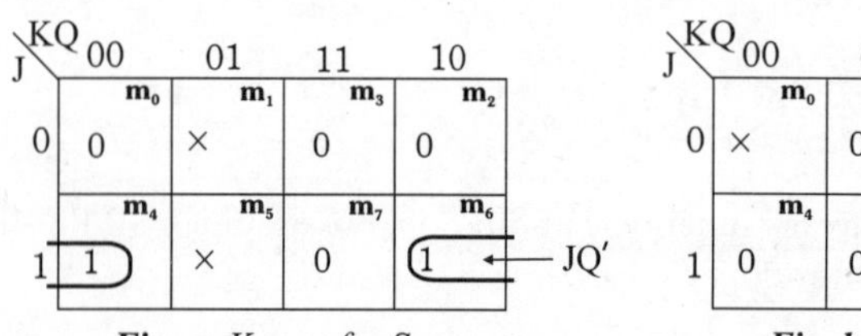

Fig. a: K-map for S. *Fig. b: K-map for R.*

***Fig. 6.65:** K-map for SR flip-flop inputs.*

From the K-map we get the following Boolean equations for SR flip-flop inputs and using these equations, the JK flip-flop logic circuit is drawn as shown in Fig. 6.66.

$$S = JQ' \quad ; \quad R = KQ$$

The working of JK flip-flop shown in Fig. 6.66 is verified for all possible combinations of inputs and present state as shown in Table 6.30. The next state output of JK flip-flop is obtained from SR flip-flop characteristic table.

***Fig. 6.66:** Logic circuit of JK flip-flop using SR flip-flop.*

Table 6.30: Verification of JK Flip-Flop of Fig. 6.66

JK Flip-Flop Inputs		Present State		SR Flip-Flop Inputs		JK Flip-Flop Next State		Comment
J	K	Q	Q′	S = JQ′	R = KQ	Q(t+1)	Q′(t+1)	
0	0	0	1	0	0	0	1	No change
0	0	1	0	0	0	1	0	
0	1	0	1	0	0	0	1	Reset state
0	1	1	0	0	1	0	1	
1	0	0	1	1	0	1	0	Set state
1	0	1	0	0	0	1	0	
1	1	0	1	1	0	1	0	Toggle
1	1	1	0	0	1	0	1	

Characteristic Table of SR Flip-Flop

Inputs S R	Present State Q	Next State Q(t+1)
0 0	0	0
0 0	1	1
0 1	0	0
0 1	1	0
1 0	0	1
1 0	1	1
1 1	0	×
1 1	1	×

6.5.2 Conversion of SR to D Flip-Flop

From the truth table of SR flip-flop we can observe that when the inputs to SR flip-flop are always complementary then the output will be same as D flip-flop. Therefore, when the D input is directly applied to S and complement of D is applied to R as shown in Fig. 6.67, then the SR flip-flop will behave as D flip-flop.

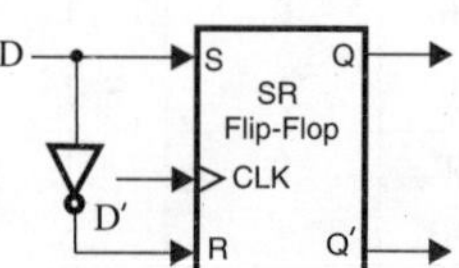

***Fig. 6.67:** Logic circuit of D flip-flop using SR flip-flop.*

6.5.3 Conversion of SR to T Flip-Flop

(GTU, Jun'19, 4 Marks)

Using the truth table of T flip-flop, a conversion table is formed by listing the possible combinations of input, present and next states of T flip-flop as shown in Table 6.31.

Table 6.31: Conversion Table to Convert SR to T Flip-Flop

T Flip-Flop				SR Flip-Flop Inputs		Minterm
Input	Present State	Next State	Comment			
T	Q	Q(t+1)		S	R	TQ
0	0	0	No change	0	×	m_0
0	1	1		×	0	m_1
1	0	1	Toggle	1	0	m_2
1	1	0		0	1	m_3

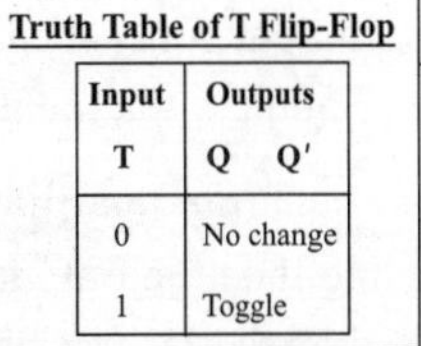

Truth Table of T Flip-Flop

Input	Outputs
T	Q Q′
0	No change
1	Toggle

Excitation Table of SR Flip-Flop

Present State	Next State	Flip-Flop Inputs	
Q(t)	Q(t + 1)	S	R
0	0	0	×
0	1	1	0
1	0	0	1
1	1	×	0
0	×	1	1
1	×	1	1

In order to drive the present state to next state listed in Table 6.31, the SR flip-flop inputs are obtained using the excitation table of SR flip-flop and listed in Table 6.31. Minterms are formed using T and Q and the K-maps for S and R are constructed as shown in Fig. 6.68.

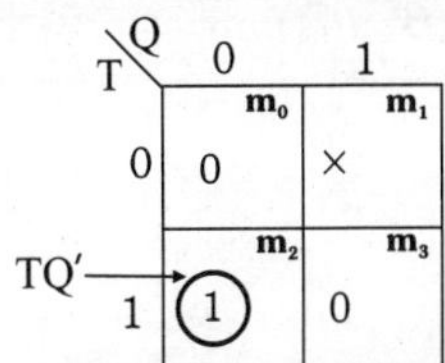

Fig. a: K-map for S.

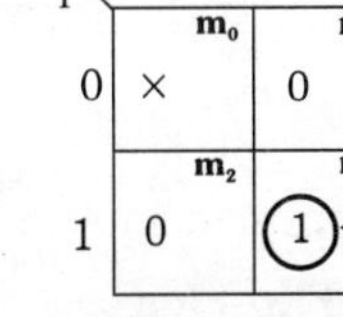

Fig. b: K-map for R.

Fig. 6.68: *K-map for SR flip-flop inputs.*

From the K-map we get the following Boolean equations for SR flip-flop inputs and using these equations, the T flip-flop logic circuit is drawn as shown in Fig. 6.69.

$$S = TQ' \quad ; \quad R = TQ$$

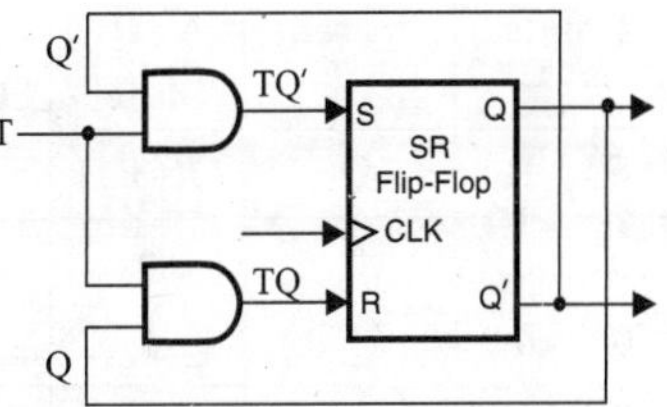

Fig. 6.69: *Logic circuit of T flip-flop using SR flip-flop.*

The working of T flip-flop shown in Fig. 6.69 is verified for all possible combinations of input and present state as shown in Table 6.32. The next state output of T flip-flop is obtained from SR flip-flop characteristic table.

Table 6.32: Verification of T Flip-Flop of Fig. 6.69

T Flip-Flop					T Flip-Flop		
Input	Present State		SR Flip-Flop Inputs		Next State		Comment
T	Q	Q′	S = TQ′	R = TQ	Q(t+1)	Q′(t+1)	
0	0	1	0	0	0	1	No change
0	1	0	0	0	1	0	
1	0	1	1	0	1	0	Toggle
1	1	0	0	1	0	1	

Characteristic Table of SR Flip-Flop

Inputs	Present State	Next State
S R	Q	Q(t+1)
0 0	0	0
0 0	1	1
0 1	0	0
0 1	1	0
1 0	0	1
1 0	1	1
1 1	0	×
1 1	1	×

6.5.4 Conversion of JK Flip-Flop to SR, D and T Flip-Flops

(VTU, Aug/Sep'20, 4 Marks)

The difference between JK and SR flip-flop is the type of output when the inputs are same. In JK flip-flop the output toggles when the inputs are same, whereas in SR flip-flop the output is indeterminate when the inputs are same. Practically, the indeterminate state is not used in SR flip-flop and so if we ignore the toggle state of JK flip-flop, then the behaviour of JK flip-flop will be same as SR flip-flop.

From the truth table of JK flip-flop we can observe that when the inputs to JK flip-flop are always complementary then the output will be same as D flip-flop. Therefore, when the D input is directly applied to J and complement of D is applied to K then the JK flip-flop will behave as D flip-flop.

From the truth table of JK flip-flop we can observe that when the inputs of JK flip-flop are always same then the output will be same as T flip-flop. Therefore, when the T input is directly applied to J and K then the JK flip-flop will behave as T flip-flop.

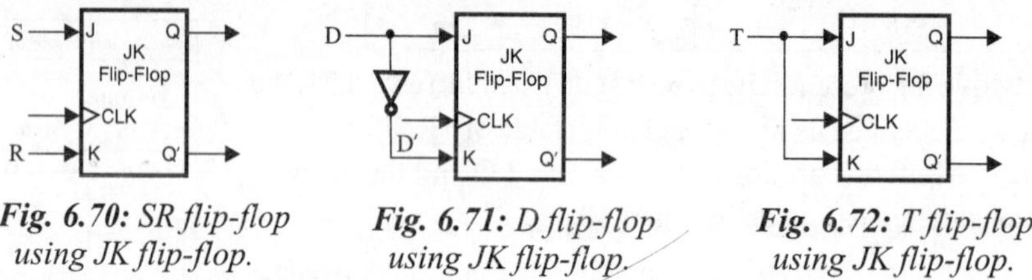

Fig. 6.70: SR flip-flop using JK flip-flop.

Fig. 6.71: D flip-flop using JK flip-flop.

Fig. 6.72: T flip-flop using JK flip-flop.

6.5.5 Conversion of D to SR Flip-Flop

Using the truth table of SR flip-flop, a conversion table is formed by listing the possible combinations of inputs, present and next states of SR flip-flop as shown in Table 6.33.

Table 6.33: Conversion Table to Convert D to SR Flip-Flop

SR Flip-Flop					D Flip-Flop Input	Minterm
Inputs		Present State	Next State	Comment		
S	R	Q	Q(t+1)			SRQ
0	0	0	0	No change	0	m_0
0	0	1	1		1	m_1
0	1	0	0	Reset state	0	m_2
0	1	1	0		0	m_3
1	0	0	1	Set state	1	m_4
1	0	1	1		1	m_5
1	1	0	×	Indeterminate	×	m_6
1	1	1	×		×	m_7

Truth Table of SR Flip-Flop

Inputs		Outputs	
S	R	Q	Q′
0	0	No change	
0	1	0	1 (Reset)
1	0	1	0 (Set)
1	1	Indeterminate	

In order to drive the present state to next state listed in Table 6.33, the D flip-flop input is obtained using the excitation table of D flip-flop and listed in Table 6.33. Minterms are formed using S, R and Q and the K-map is constructed for D as shown in Fig. 6.73.

S \ RQ	00	01	11	10
0	m_0: 0	m_1: 1	m_3: 0	m_2: 0
1	m_4: 1	m_5: 1	m_7: ×	m_6: ×

(Groupings: R'Q — column 01; S — row 1)

Fig. 6.73: K-map for D flip-flop input.

Excitation Table of D Flip-Flop

Present State Q(t)	Next State Q(t + 1)	Flip-Flop Input D
0	0	0
0	1	1
1	0	0
1	1	1

From the K-map we get the following Boolean equation for D flip-flop input and using this equation, the SR flip-flop logic circuit is drawn as shown in Fig. 6.74.

$$D = S + R'Q$$

The working of SR flip-flop shown in Fig. 6.74 is verified for all possible combinations of inputs and present state as shown in Table 6.34. The next state output of SR flip-flop is obtained from D flip-flop characteristic table.

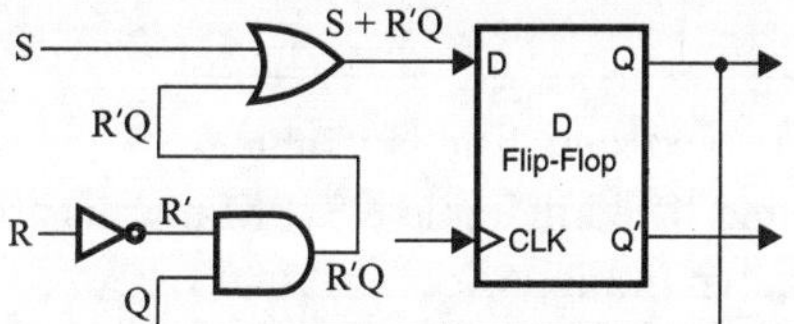

Fig. 6.74: SR flip-flop using D flip-flop.

Table 6.34: Verification of SR Flip-Flop of Fig. 6.74

SR Flip-Flop Inputs S	R	SR Flip-Flop Present State Q	Q'	R'	R'Q	D Flip-Flop Input D = S + R'Q	SR Flip-Flop Next State Q(t+1)	Q'(t+1)	Comment
0	0	0	1	1	0	0	0	1	No change
0	0	1	0	1	1	1	1	0	
0	1	0	1	0	0	0	0	1	Reset state
0	1	1	0	0	0	0	0	1	
1	0	0	1	1	0	1	1	0	Set state
1	0	1	0	1	1	1	1	0	
1	1	0	1	0	0	1	1	0	Not acceptable
1	1	1	0	0	0	1	1	0	

Characteristic Table of D Flip-Flop

Input D	Present State Q	Next State Q(t+1)
0	0	0
0	1	0
1	0	1
1	1	1

6.5.6 Conversion of D to JK Flip-Flop

Using the truth table of JK flip-flop, a conversion table is formed by listing the possible combinations of inputs, present and next states of JK flip-flop as shown in Table 6.35.

Table 6.35: Conversion Table to Convert D to JK Flip-Flop

JK Flip-Flop					D Flip-Flop Input	Minterm
Inputs		Present State	Next State	Comment		
J	K	Q	Q(t+1)			JKQ
0	0	0	0	No change	0	m_0
0	0	1	1		1	m_1
0	1	0	0	Reset state	0	m_2
0	1	1	0		0	m_3
1	0	0	1	Set state	1	m_4
1	0	1	1		1	m_5
1	1	0	1	Toggle	1	m_6
1	1	1	0		0	m_7

Truth Table of JK Flip-Flop

Inputs		Outputs	
J	K	Q	Q′
0	0	No change	
0	1	0	1 (Reset)
1	0	1	0 (Set)
1	1	Toggle	

In order to drive the present state to next state listed in Table 6.35, the D flip-flop input is obtained using the excitation table of D flip-flop and listed in Table 6.35. Minterms are formed using J, K and Q and the K-map is constructed for D as shown in Fig. 6.75.

Excitation Table of D Flip-Flop

Present State	Next State	Flip-Flop Input
Q(t)	Q(t + 1)	D
0	0	0
0	1	1
1	0	0
1	1	1

J \ KQ	00	01	11	10
0	0 (m_0)	1 (m_1)	0 (m_3)	0 (m_2)
1	1 (m_4)	1 (m_5)	0 (m_7)	1 (m_6)

K′Q (group m_1, m_5); JQ′ (group m_4, m_6)

Fig. 6.75: K-map for D flip-flop input.

From the K-map we get the following Boolean equation for D flip-flop input and using this equation, the JK flip-flop logic circuit is drawn as shown in Fig. 6.76.

$$D = JQ' + K'Q$$

The working of JK flip-flop shown in Fig. 6.76 is verified for all possible combinations of inputs and present state as shown in Table 6.36. The next state output of JK flip-flop is obtained from D flip-flop characteristic table.

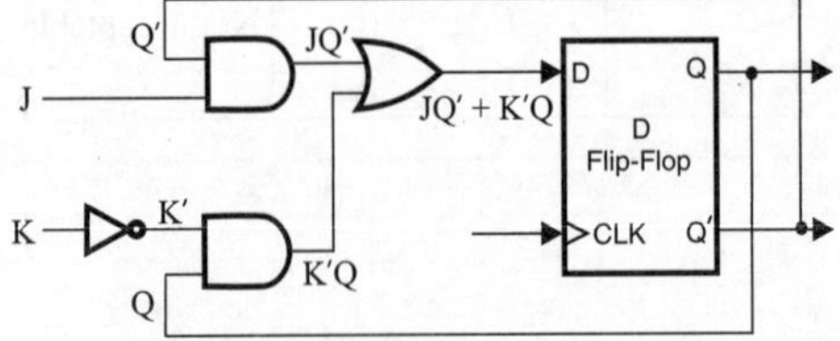

Fig. 6.76: JK flip-flop using D flip-flop.

Table 6.36: Verification of JK Flip-Flop of Fig. 6.76

JK Flip-Flop								JK Flip-Flop		
Inputs		Present State					D Flip-Flop Input	Next State		Comment
J	K	Q	Q′	K′	JQ′	K′Q	D = JQ′ + K′Q	Q(t+1)	Q′(t+1)	
0	0	0	1	1	0	0	0	0	1	No change
0	0	1	0	1	0	1	1	1	0	
0	1	0	1	0	0	0	0	0	1	Reset state
0	1	1	0	0	0	0	0	0	1	
1	0	0	1	1	1	0	1	1	0	Set state
1	0	1	0	1	0	1	1	1	0	
1	1	0	1	0	1	0	1	1	0	Toggle
1	1	1	0	0	0	0	0	0	1	

Characteristic Table of D Flip-Flop

Input D	Present State Q	Next State Q(t+1)
0	0	0
0	1	0
1	0	1
1	1	1

6.5.7 Conversion of D to T Flip-Flop

Using the truth table of T flip-flop, a conversion table is formed by listing the possible combinations of input, present and next states of T flip-flop as shown in Table 6.37.

Table 6.37: Conversion Table to Convert D to T Flip-Flop

T Flip-Flop					
Input	Present State	Next State	Comment	D Flip-Flop Input	Minterm
T	Q	Q(t+1)			TQ
0	0	0	No change	0	m_0
0	1	1		1	m_1
1	0	1	Toggle	1	m_2
1	1	0		0	m_3

Truth Table of T Flip-Flop

Input T	Outputs Q Q′
0	No change
1	Toggle

In order to drive the present state to next state listed in Table 6.37, the D flip-flop input is obtained using the excitation table of D flip-flop and listed in Table 6.37. Minterms are formed using T and Q and the K-map is constructed for D as shown in Fig. 6.77.

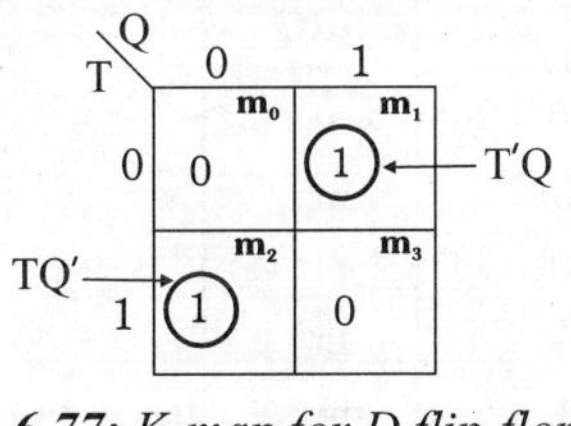

T \ Q	0	1
0	m_0: 0	m_1: (1) ← T′Q
1	TQ′ → m_2: (1)	m_3: 0

Fig. 6.77: K-map for D flip-flop input.

Excitation Table of D Flip-Flop

Present State Q(t)	Next State Q(t + 1)	Flip-Flop Input D
0	0	0
0	1	1
1	0	0
1	1	1

From the K-map we get the following Boolean equation for D flip-flop input and using this equation, the T flip-flop logic circuit is drawn as shown in Fig. 6.78.

$$D = TQ' + T'Q$$

The working of T flip-flop shown in Fig. 6.78 is verified for all possible combinations of input and present state as shown in Table 6.38. The next state output of T flip-flop is obtained from D flip-flop characteristic table.

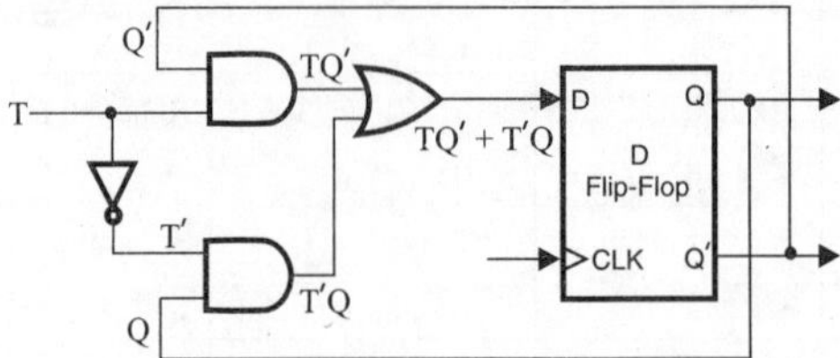

Fig. 6.78: T flip-flop using D flip-flop.

Table 6.38: Verification of T Flip-Flop of Fig. 6.78

T Flip-Flop							T Flip-Flop		
Input	Present State					D Flip-Flop Input	Next State		Comment
T	Q	Q'	T'	TQ'	T'Q	D = TQ' + T'Q	Q(t+1)	Q'(t+1)	
0	0	1	1	0	0	0	0	1	No change
0	1	0	1	0	1	1	1	0	
1	0	1	0	1	0	1	1	0	Toggle
1	1	0	0	0	0	0	0	1	

Characteristic Table of D Flip-Flop

Input D	Present State Q	Next State Q(t+1)
0	0	0
0	1	0
1	0	1
1	1	1

6.5.8 Conversion of T to SR Flip-Flop

Using the truth table of SR flip-flop, a conversion table is formed by listing the possible combinations of inputs, present and next states of SR flip-flop as shown in Table 6.39.

Table 6.39: Conversion Table to Convert T to SR Flip-Flop

SR Flip-Flop					T Flip-Flop Input	Minterm
Inputs		Present State	Next State	Comment		
S	R	Q	Q(t+1)			SRQ
0	0	0	0	No change	0	m_0
0	0	1	1		0	m_1
0	1	0	0	Reset state	0	m_2
0	1	1	0		1	m_3
1	0	0	1	Set state	1	m_4
1	0	1	1		0	m_5
1	1	0	×	Indeterminate	×	m_6
1	1	1	×		×	m_7

Truth Table of SR Flip-Flop

Inputs		Outputs	
S	R	Q	Q'
0	0	No change	
0	1	0	1 (Reset)
1	0	1	0 (Set)
1	1	Indeterminate	

In order to drive the present state to next state listed in Table 6.39, the T flip-flop input is obtained using the excitation table of T flip-flop and listed in Table 6.39. Minterms are formed using S, R and Q and the K-map is constructed for T as shown in Fig. 6.79.

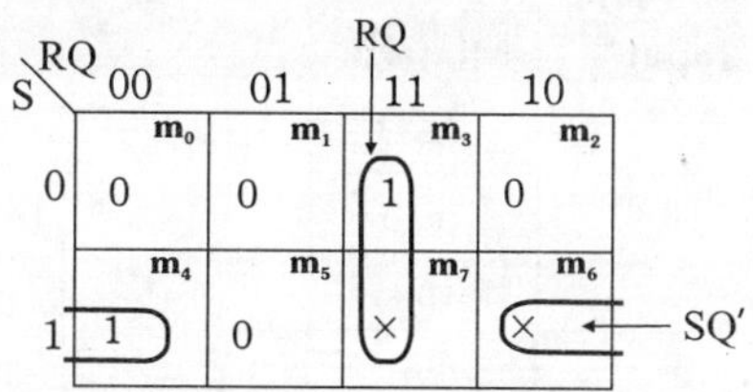

Fig. 6.79: K-map for T flip-flop input.

Excitation Table of T Flip-Flop

Present State	Next State	Flip-Flop Input
Q(t)	Q(t + 1)	T
0	0	0
0	1	1
1	0	1
1	1	0

From the K-map we get the following Boolean equation for T flip-flop input and using this equation, the SR flip-flop logic circuit is drawn as shown in Fig. 6.80.

$$T = SQ' + RQ$$

Fig. 6.80: SR flip-flop using T flip-flop.

The working of SR flip-flop shown in Fig. 6.80 is verified for all possible combinations of inputs and present state as shown in Table 6.40. The next state output of SR flip-flop is obtained from T flip-flop characteristic table.

Table 6.40: Verification of SR Flip-Flop of Fig. 6.80

SR Flip-Flop							SR Flip-Flop		
Inputs		Present State				T Flip-Flop Input	Next State		Comment
S	R	Q	Q'	SQ'	RQ	T = SQ' + RQ	Q(t+1)	Q'(t+1)	
0	0	0	1	0	0	0	0	1	No change
0	0	1	0	0	0	0	1	0	
0	1	0	1	0	0	0	0	1	Reset state
0	1	1	0	0	1	1	0	1	
1	0	0	1	1	0	1	1	0	Set state
1	0	1	0	0	0	0	1	0	
1	1	0	1	1	0	1	1	0	Toggle
1	1	1	0	0	1	1	0	1	

Characteristic Table of T Flip-Flop

Input	Present State	Next State
T	Q	Q(t+1)
0	0	0
0	1	1
1	0	1
1	1	0

6.5.9 Conversion of T to JK Flip-Flop

Using the truth table of JK flip-flop, a conversion table is formed by listing the possible combinations of inputs, present and next states of JK flip-flop as shown in Table 6.41.

Table 6.41: Conversion Table to Convert T to JK Flip-Flop

JK Flip-Flop						
Inputs		Present State	Next State	Comment	T Flip-Flop Input	Minterm
J	K	Q	Q(t+1)			JKQ
0	0	0	0	No change	0	m_0
0	0	1	1		0	m_1
0	1	0	0	Reset state	0	m_2
0	1	1	0		1	m_3
1	0	0	1	Set state	1	m_4
1	0	1	1		0	m_5
1	1	0	1	Toggle	1	m_6
1	1	1	0		1	m_7

Truth Table of JK Flip-Flop

Inputs		Outputs	
J	K	Q	Q′
0	0	No change	
0	1	0	1 (Reset)
1	0	1	0 (Set)
1	1	Toggle	

In order to drive the present state to next state listed in Table 6.41, the T flip-flop input is obtained using the excitation table of T flip-flop and listed in Table 6.41. Minterms are formed using J, K and Q and the K-map is constructed for T as shown in Fig. 6.81.

Excitation Table of T Flip-Flop

Present State	Next State	Flip-Flop Input
Q(t)	Q(t + 1)	T
0	0	0
0	1	1
1	0	1
1	1	0

J \ KQ	00	01	11	10
0	m_0: 0	m_1: 0	m_3: 1	m_2: 0
1	m_4: 1	m_5: 0	m_7: 1	m_6: 1

KQ; JQ′

Fig. 6.81: K-map for T flip-flop input.

From the K-map we get the following Boolean equation for T flip-flop input and using this equation, the JK flip-flop logic circuit is drawn as shown in Fig. 6.82.

$$T = JQ' + KQ$$

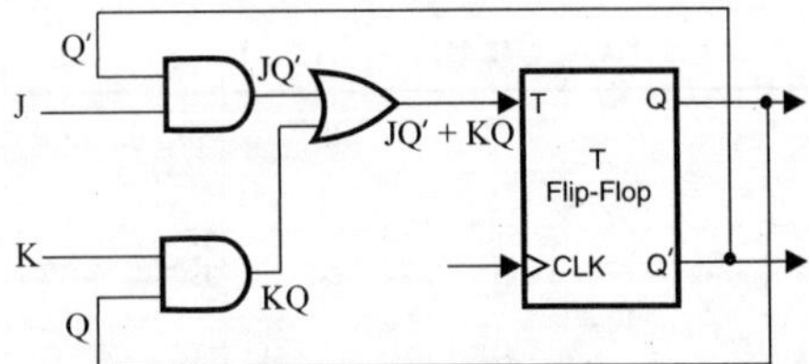

Fig. 6.82: JK flip-flop using T flip-flop.

The working of JK flip-flop shown in Fig. 6.82 is verified for all possible combinations of inputs and present state as shown in Table 6.42. The next state output of JK flip-flop is obtained from T flip-flop characteristic table.

Table 6.42: Verification of JK Flip-Flop of Fig. 6.82

JK Flip-Flop							SR Flip-Flop		
Inputs		Present State				T Flip-Flop Input	Next State		Comment
J	K	Q	Q'	JQ'	KQ	T = JQ' + KQ	Q(t+1)	Q'(t+1)	
0	0	0	1	0	0	0	0	1	No change
0	0	1	0	0	0	0	1	0	
0	1	0	1	0	0	0	0	1	Reset state
0	1	1	0	0	1	1	0	1	
1	0	0	1	1	0	1	1	0	Set state
1	0	1	0	0	0	0	1	0	
1	1	0	1	1	0	1	1	0	Toggle
1	1	1	0	0	1	1	0	1	

Characteristic Table of T Flip-Flop

Input T	Present State Q	Next State Q(t+1)
0	0	0
0	1	1
1	0	1
1	1	0

6.5.10 Conversion of T to D Flip-Flop

Using the truth table of D flip-flop, a conversion table is formed by listing the possible combinations of input, present and next states of D flip-flop as shown in Table 6.43.

Table 6.43: Conversion Table to Convert T to D Flip-Flop

D Flip-Flop					
Input	Present State	Next State	Comment	T Flip-Flop Input	Minterm
D	Q	Q(t+1)			TQ
0	0	0	Next state output is same as input	0	m_0
0	1	0		1	m_1
1	0	1		1	m_2
1	1	1		0	m_3

Truth Table of D Flip-Flop

Input D	Outputs Q	Q'
0	0	1
1	1	0

In order to drive the present state to next state listed in Table 6.43, the T flip-flop input is obtained using the excitation table of T flip-flop and listed in Table 6.43. Minterms are formed using D and Q and the K-map is constructed for T as shown in Fig. 6.83.

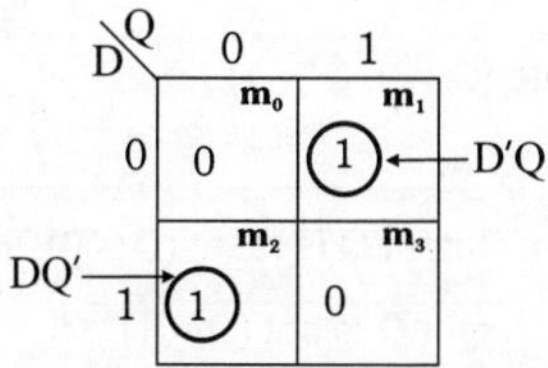

Fig. 6.83: K-map for T flip-flop input.

Excitation Table of T Flip-Flop

Present State Q(t)	Next State Q(t + 1)	Flip-Flop Input T
0	0	0
0	1	1
1	0	1
1	1	0

From the K-map we get the following Boolean equation for T flip-flop input and using this equation, the D flip-flop logic circuit is drawn as shown in Fig. 6.84.

$$T = DQ' + D'Q$$

The working of D flip-flop shown in Fig. 6.84 is verified for all possible combinations of input and present state as shown in Table 6.44. The next state output of D flip-flop is obtained from T flip-flop characteristic table.

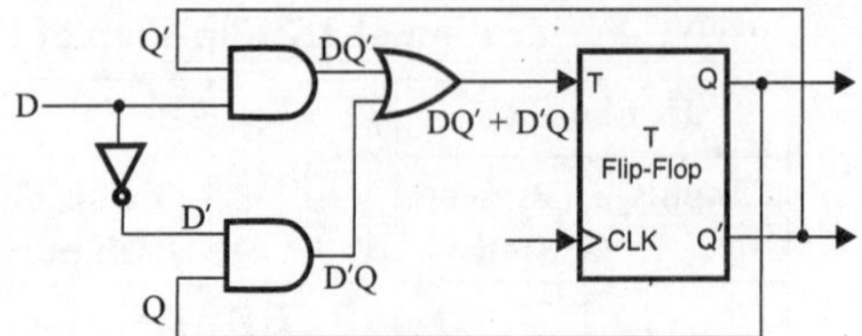

Fig. 6.84: D flip-flop using T flip-flop.

Table 6.44: Verification of D Flip-Flop of Fig. 6.84

D Flip-Flop							D Flip-Flop		
Input		Present State				T Flip-Flop Input	Next State		Comment
D	D'	Q	Q'	DQ'	D'Q	T = DQ' + D'Q	Q(t+1)	Q'(t+1)	
0	1	0	1	0	0	0	0	1	Next state output is same as input
0	1	1	0	0	1	1	0	1	
1	0	0	1	1	0	1	1	0	
1	0	1	0	0	0	0	1	0	

Characteristic Table of T Flip-Flop

Input T	Present State Q	Next State Q(t+1)
0	0	0
0	1	1
1	0	1
1	1	0

Example 6.3

The input waveforms of NOR based SR latch is shown in Fig. 1. Draw the output waveform. Assume that initial output is reset.

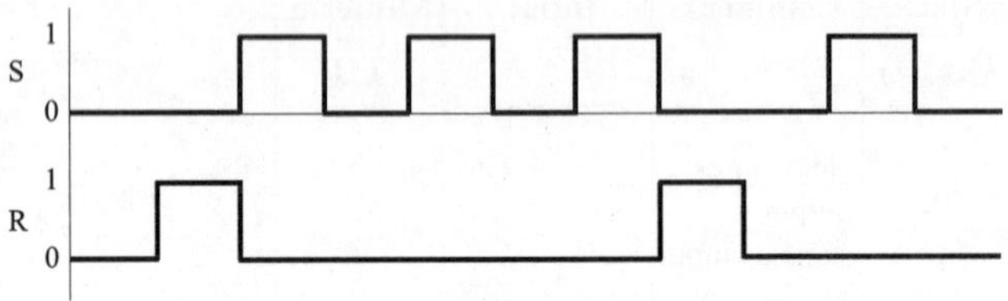

Fig. 1.

Solution

Let Q be the output of SR latch and Q' is complement output. The outputs corresponding to inputs at various timings are determined using the truth table of NOR based SR latch. The input and output timing diagrams are drawn as shown in Fig. 2 and from the timing diagram the following outputs are observed at various timings.

Truth Table of SR latch using NOR Gates

Inputs		Outputs	
S	R	Q	Q'
0	0	No change	
0	1	0	1
1	0	1	0
1	1	Indeterminate	

Time t_0: Initial outputs are assumed to be in reset state (Q = 0, Q' = 1).

Time t_1: Both inputs are 0, and hence no change in output.

Time t_2: S = 0 and R = 1. Inputs are in reset condition. Output Q is reset (Q = 0).

Time t_3: S = 1 and R = 0, Inputs are in set condition. Output Q is set (Q = 1).

In the time interval from t_3 to t_8 the inputs are in either set or no change condition and hence the output remains in set state.

Time t_8: S = 0 and R = 1. Inputs are in reset condition. Output Q is reset (Q = 0).

Time t_9: Both inputs are 0, and hence no change in output.

Time t_{10}: S = 1 and R = 0, Inputs are in set condition. Output Q is set (Q = 1).

Time t_{11}: Both inputs are 0, and hence no change in output.

Note: *Q' is complement of Q.*

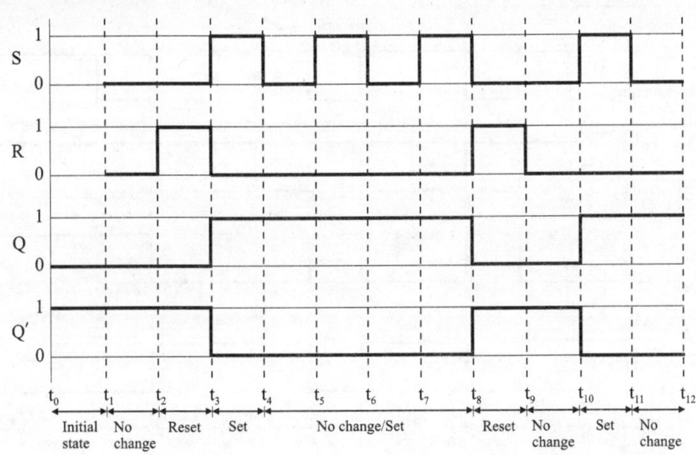

***Fig. 2:** Timing diagram (or waveform) of NOR based SR latch.*

Example 6.4

The waveform are applied to NAND based SR latch without enable. Sketch the resulting output waveform in relation to the inputs. Assume that initial output is reset.

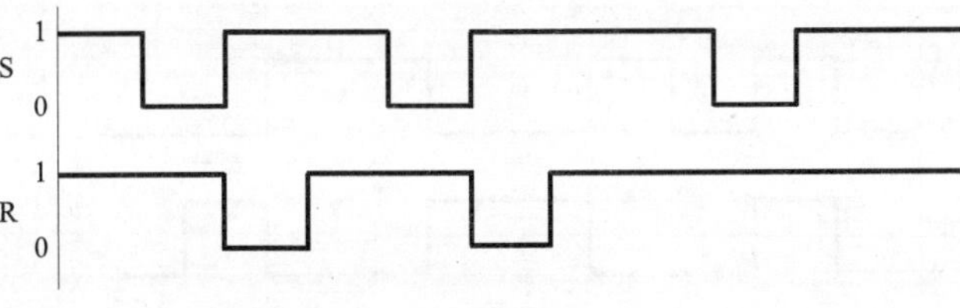

Fig. 1.

Solution

Let Q be the output of SR latch and Q' is complement output. The outputs corresponding to inputs at various timings are determined using the truth table of NAND based SR latch. The input and output timing diagrams are drawn as shown in Fig. 2 and from the timing diagram the following outputs are observed at various timings.

Truth Table of SR latch using NAND Gates

Inputs		Outputs	
S	R	Q	Q'
0	0	Indeterminate	
0	1	1	0
1	0	0	1
1	1	No change	

Time t_0: Initial outputs are assumed to be in reset state (Q = 0, Q' = 1).

Time t_1: Both inputs are 1, and hence no change in output.

Time t_2: S = 0 and R = 1. Inputs are in set condition. Output Q is set (Q = 1).

Time t_3: S = 1 and R = 0, Inputs are in reset condition. Output Q is reset (Q = 0).

Time t_4: Both inputs are 1, and hence no change in output.

Time t_5: S = 0 and R = 1. Inputs are in set condition. Output Q is set (Q = 1).

Time t_6: S = 1 and R = 0, Inputs are in reset condition. Output Q is reset (Q = 0).

Time t_7: Both inputs are 1, and hence no change in output.

Time t_8: S = 0 and R = 1. Inputs are in set condition. Output Q is set (Q = 1).

Time t_9: Both inputs are 1, and hence no change in output.

> ***Note:*** *Q' is complement of Q.*

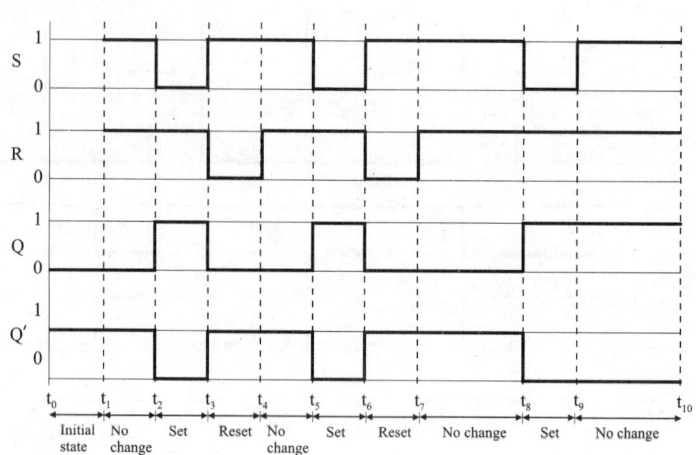

Fig. 2: Timing diagram (or waveform) of NAND based SR latch.

Example 6.5

For the NAND based SR latch with enable determine the outputs for the inputs shown in Fig. 1. Assume that initial output is reset.

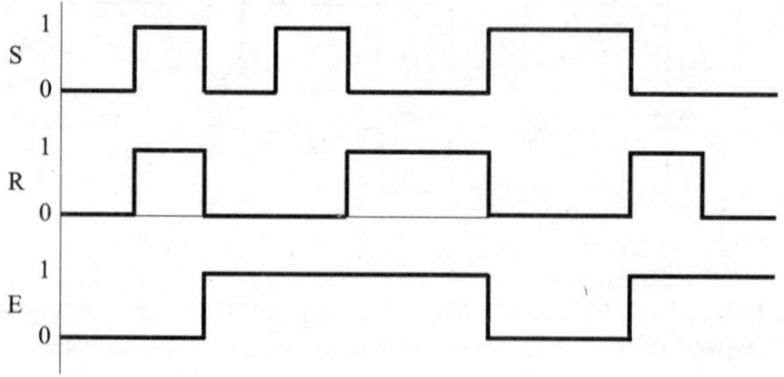

Fig. 1.

Solution

Let Q be the output of SR latch and Q' is complement output. The outputs corresponding to inputs at various timings are determined using the truth table of NAND based SR latch. The input and output timing diagrams are drawn as shown in Fig. 2 and from the timing diagram the following outputs are observed at various timings.

Truth Table of NAND based SR latch with Enable

Inputs			Intermediate Signal		Outputs	
E	S	R	S_1	R_1	Q	Q'
0	×	×	1	1	No change	
1	0	0	1	1	No change	
1	0	1	1	0	0	1 (Reset state)
1	1	0	0	1	1	0 (Set state)
1	1	1	0	0	Indeterminate	

Time t_0: Initial outputs are assumed to be in reset state (Q = 0, Q' = 1).

In the time interval from t_1 to t_3 the enable, E = 0, and hence the inputs are not recognized and output continue to remain in previous state.

In the time interval from t_3 to t_4 the enable, E = 1, but both inputs are 0 and hence no change in output.

Time t_4: E = 1, S = 1 and R = 0. Inputs are in set condition. Output Q is set (Q = 1).

Time t_5: E = 1, S = 0 and R = 1. Inputs are in reset condition. Output Q is reset (Q = 0).

Time t_6: E = 0. Since enable is 0 no change in output.

Time t_7: E = 1, S = 0 and R = 1. Inputs are in reset condition. Output Q is reset (Q = 0).

Time t_8: E = 1, but both inputs are 0, and hence no change in output.

Note: *Q' is complement of Q.*

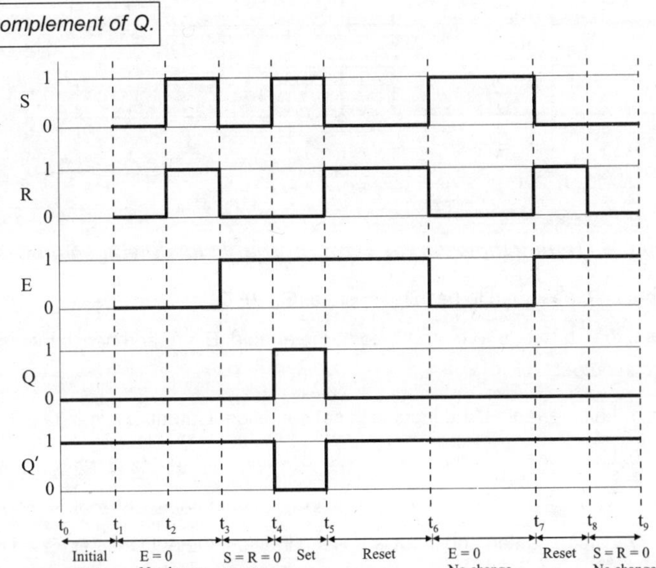

***Fig. 2:** Timing diagram (or waveform) of SR latch using NAND gates and with enable circuit.*

Example 6.6

For D latch the waveform shown in Fig. 1 are applied on its inputs. Sketch the waveform expected on the output if initial output is reset.

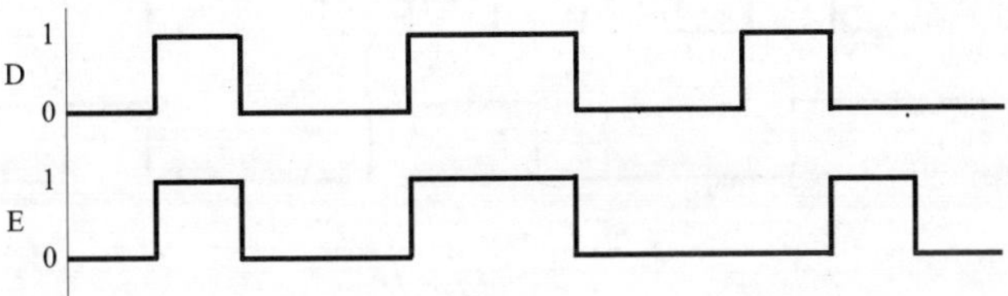

Fig. 1.

Solution

Let Q be the output of D latch and Q' is complement output. The output corresponding to input at various timings are determined using the truth table of D latch. The input and output timing diagrams are drawn as shown in Fig. 2 and from the timing diagram the following outputs are observed at various timings.

Truth Table of D Latch

E	D	Q	Q'	Comment
0	×	No change		-
1	0	0	1	Reset state
1	1	1	0	Set state

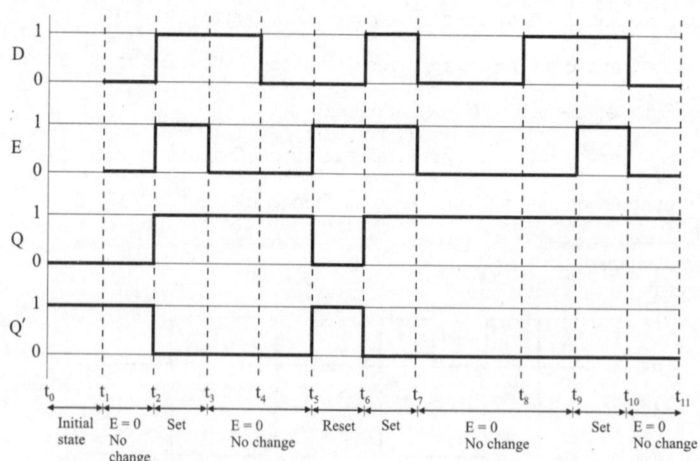

Fig. 2: Timing diagram (or waveform) of D latch with enable.

Time t_0: Initial outputs are assumed to be in reset state (Q = 0, Q' = 1).

In the time intervals t_1 to t_2, t_3 to t_5, t_7 to t_9, and t_{10} to t_{11} the enable, E = 0, and hence the input is not recognized and there will be no change in output.

Time t_2: E = 1, D = 1. Latch enabled and input is in set condition. Output Q is set (Q = 1).

Time t_5: E = 1, D = 0. Latch enabled and input in reset condition. Output Q is reset (Q = 0).

Time t_6: E = 1, D = 1. Latch enabled and input is in set condition. Output Q is set (Q = 1).

Time t_9: E = 1, D = 1. Latch enabled and input is in set condition. Output Q is set (Q = 1).

Note: *Q' is complement of Q.*

Example 6.7

(AKUB, Dec'20, 6 Marks)

On the following graph, inputs CLK and D are shown:

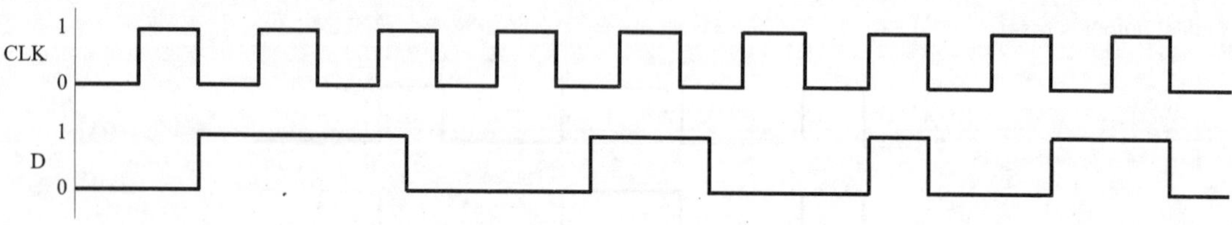

Fig. 1.

They are inputs to a D latch and a positive edge triggered D flip-flop. Assuming initial output 0, draw the output waveform for flip-flop and latch. Do the two outputs differ? If so, why?

Solution

a) D Latch with Clock as Enable Signal

Let Q be the output of D latch and Q' is complement output. The output corresponding to input at various timings are determined using the truth table of D latch with enable. The input and output timing diagrams are drawn as shown in Fig. 2 and from the timing diagram the following outputs are observed at various timings.

Truth Table of D Latch

E	D	Q	Q'	Comment
0	×	No change		-
1	0	0	1	Reset state
1	1	1	0	Set state

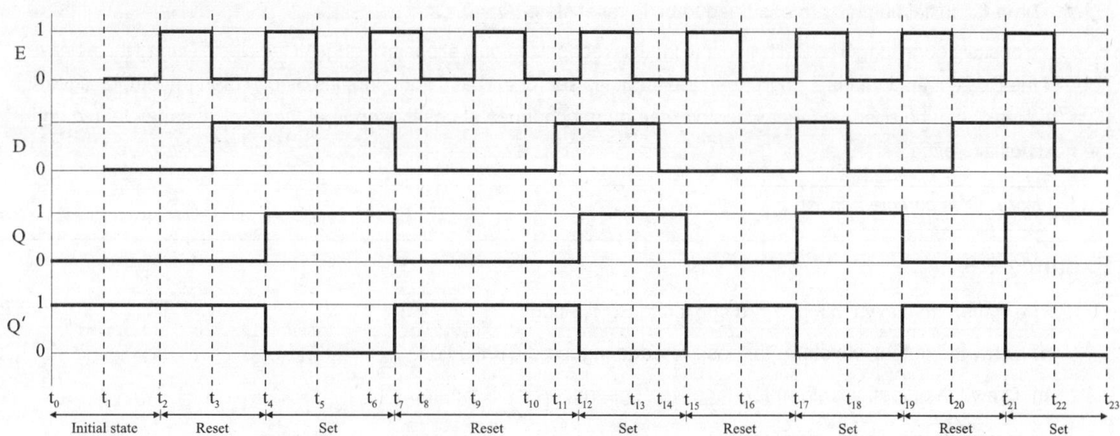

Fig. 2: Timing diagram (or waveform) of D latch with enable.

Time t_0: Initial outputs are assumed to be in reset state (Q = 0, Q' = 1).

In the timing diagram, the clock is considered as enable and hence the output follows the input when enable is 1 (high). When the enable is 0 (low) the input is not recognized and hence previous output is maintained.

***Note:** Q' is complement of Q.*

b) D Flip-flop with Positive Edge Triggered Clock

Let Q be the output of D flip-flop and Q' is complement output. The output corresponding to input at various timings are determined using the truth table of D flip-flop with positive edge clock. The input and output timing diagrams are drawn as shown in Fig. 3 and from the timing diagram the following outputs are observed at various timings.

Truth Table of Positive Edge Triggered D Flip-flop

Input	Clock	Outputs	
D		Q	Q'
0	↑	0	1
1	↑	1	0

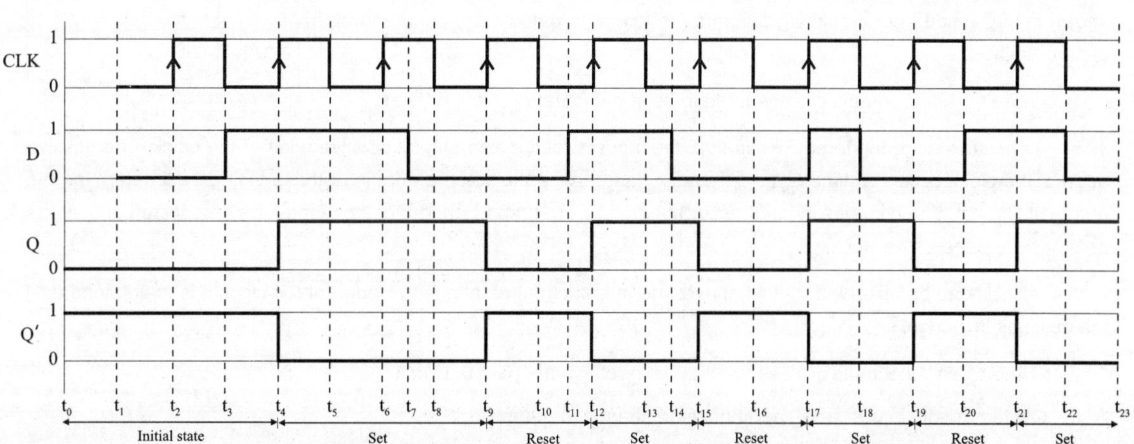

Fig. 3: Timing diagram (or waveform) of positive edge triggered D flip-flop.

Time t_0: Initial outputs are assumed to be in reset state (Q = 0, Q' = 1).

In positive edge triggered D flip-flop, the input is recognized only at positive edge of the clock. During the positive edge of the clock, if the input is 1 (high) then the output is set (Q = 1) and if the input is 0 (low) then the output is reset (Q = 0). Also, it can be observed that whenever the output changes at positive edge of the clock it is maintained until the next positive edge.

Note: *Q' is complement of Q.*

Example 6.8

The waveform shown in Fig. 1 are applied to SR flip-flop.

a) Draw the output waveform for positive edge triggered SR flip-flop.

b) Draw the output waveform for negative edge triggered SR flip-flop.

Assume that initial output is reset.

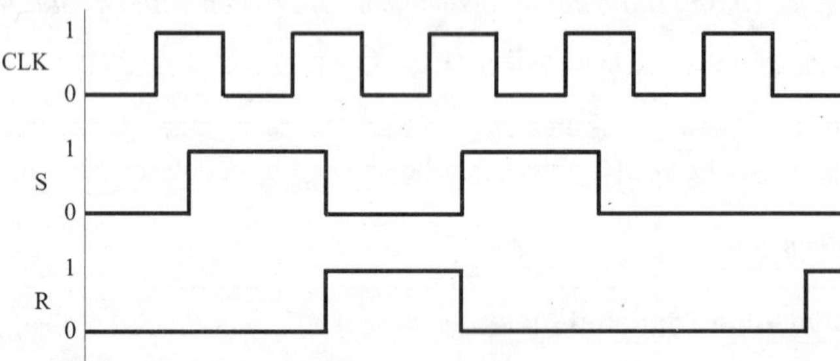

Fig. 1.

Solution

a) SR Flip-flop with Positive Edge Triggered Clock

Let Q be the output of SR flip-flop and Q' is complement output. The output corresponding to input at various timings are determined using the truth table of SR flip-flop with positive edge clock. The input and output timing diagrams are drawn as shown in Fig. 2 and from the timing diagram the following outputs are observed at various timings.

Truth Table of Positive Edge Triggered SR Flip-flop

Inputs S	Inputs R	Clock	Outputs Q	Outputs Q'
0	0	↑	No change	
0	1	↑	0	1
1	0	↑	1	0
1	1	↑	Indeterminate	

Time t_0: Initial outputs are assumed to be in reset state (Q = 0, Q' = 1).

In positive edge triggered SR flip-flop, the input is recognized only at positive edge of the clock. The positive edge of the clock occurs at time t_2, t_4, t_6, t_8 and t_{10} and hence the value of inputs at these timings will determine the output of flip-flop. The changes in input occur at time t_3, t_5, t_7, t_9 and t_{11} and these changes are recognized only at the next positive edge of the clock.

Also, it can be observed that whenever the output changes at positive edge of the clock it is maintained until the next positive edge.

Time t_0: Initial outputs are assumed to be in reset state (Q = 0, Q' = 1).

Time t_2: Both inputs are 0, and hence no change in output.

Time t_4: S = 1 and R = 0. Inputs are in set condition. Output Q is set (Q = 1).

Time t_6: S = 0 and R = 1. Inputs are in reset condition. Output Q is reset (Q = 0).

Time t_8: S = 1 and R = 0. Inputs are in set condition. Output Q is set (Q = 1).

Time t_{10}: Both inputs are 0, and hence no change in output.

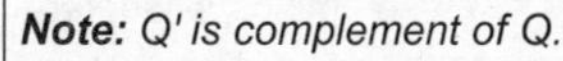

Note: *Q' is complement of Q.*

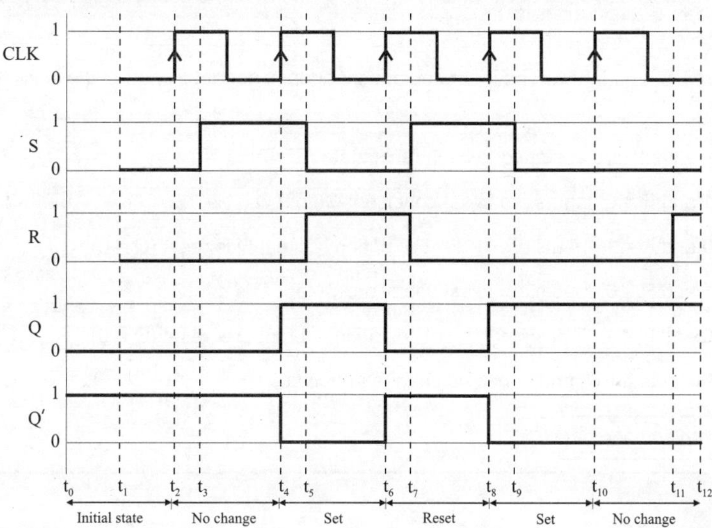

Fig. 2: *Timing diagram (or waveform) of positive edge triggered SR flip-flop.*

b) SR Flip-flop with Negative Edge Triggered Clock

Let Q be the output of SR flip-flop and Q′ is complement output. The output corresponding to input at various timings are determined using the truth table of SR flip-flop with negative edge clock. The input and output timing diagrams are drawn as shown in Fig. 3 and from the timing diagram the following outputs are observed at various timings.

Truth Table of Negative Edge Triggered SR Flip-flop

Inputs		Clock	Outputs	
S	R		Q	Q′
0	0	↓	No change	
0	1	↓	0	1
1	0	↓	1	0
1	1	↓	Indeterminate	

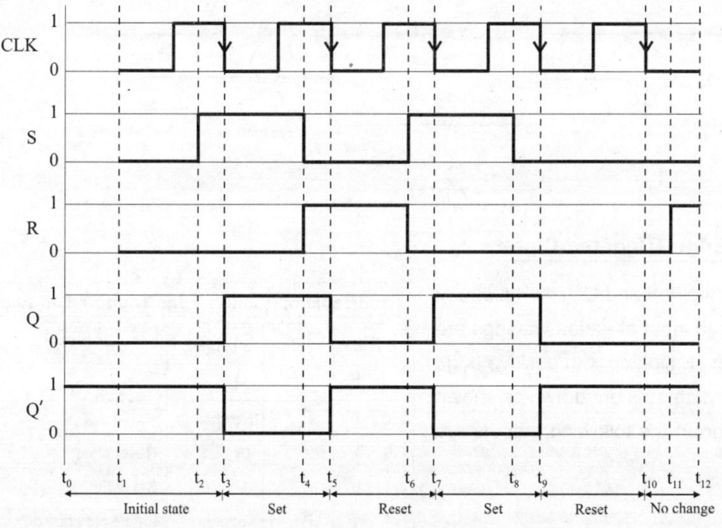

Fig. 3: *Timing diagram (or waveform) of negative edge triggered SR flip-flop.*

Time t_0: Initial outputs are assumed to be in reset state (Q = 0, Q' = 1).

In negative edge triggered SR flip-flop, the input is recognized only at negative edge of the clock. The negative edge of the clock occurs at time t_3, t_5, t_7, t_9 and t_{10} and hence the value of inputs at these timings will determine the output of flip-flop. The changes in input occur at time t_2, t_4, t_6, t_8 and t_{11} and these changes are recognized only at the next negative edge of the clock.

Also, it can be observed that whenever the output changes at negative edge of the clock it is maintained until the next negative edge.

Time t_0: Initial outputs are assumed to be in reset state (Q = 0, Q' = 1).

Time t_3: S = 1 and R = 0. Inputs are in set condition. Output Q is set (Q = 1).

Time t_5: S = 0 and R = 1. Inputs are in reset condition. Output Q is reset (Q = 0).

Time t_7: S = 1 and R = 0. Inputs are in set condition. Output Q is set (Q = 1).

Time t_9: S = 0 and R = 1. Inputs are in reset condition. Output Q is reset (Q = 0).

Time t_{10}: Both inputs are 0, and hence no change in output.

***Note:** Q' is complement of Q.*

Example 6.9

The waveform shown in Fig. 1 are applied to JK flip-flop.

a) Draw the output waveform for positive edge triggered JK flip-flop.

b) Draw the output waveform for negative edge triggered JK flip-flop.

Assume that initial output is reset.

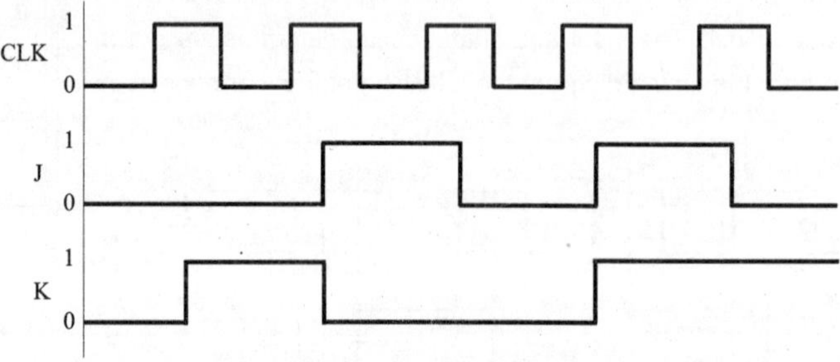

Fig. 1.

Solution

a) JK Flip-flop with Positive Edge Triggered Clock

Let Q be the output of JK flip-flop and Q' is complement output. The output corresponding to input at various timings are determined using the truth table of JK flip-flop with positive edge clock. The input and output timing diagrams are drawn as shown in Fig. 2 and from the timing diagram the following outputs are observed at various timings.

Truth Table of Positive Edge Triggered JK Flip-flop

Inputs J	Inputs K	Clock	Outputs Q	Outputs Q'	Comment
0	0	↑	No change		-
0	1	↑	0	1	Reset state
1	0	↑	1	0	Set state
1	1	↑	Toggle		Toggle previous output

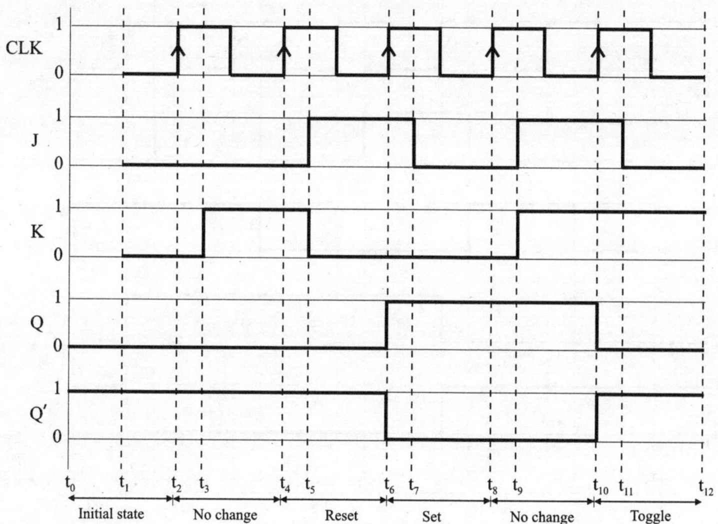

Fig. 2: Timing diagram (or waveform) of positive edge triggered JK flip-flop.

Time t_0: Initial outputs are assumed to be in reset state (Q = 0, Q' = 1).

In positive edge triggered JK flip-flop, the input is recognized only at positive edge of the clock. The positive edge of the clock occurs at time t_2, t_4, t_6, t_8 and t_{10} and hence the value of inputs at these timings will determine the output of flip-flop. The changes in input occur at time t_3, t_5, t_7, t_9 and t_{11} and these changes are recognized only at the next positive edge of the clock.

Also, it can be observed that whenever the output changes at positive edge of the clock it is maintained until the next positive edge.

Time t_0: Initial outputs are assumed to be in reset state (Q = 0, Q' = 1).

Time t_2: Both inputs are 0, and hence no change in output.

Time t_4: J = 0 and K = 1. Inputs are in reset condition. Output Q is reset (Q = 0).

Time t_6: J = 1 and K = 0. Inputs are in set condition. Output Q is set (Q = 1).

Time t_8: Both inputs are 0, and hence no change in output.

Time t_{10}: Both inputs are 1, and hence output toggles.

Note: *Q' is complement of Q.*

b) JK Flip-flop with Negative Edge Triggered Clock

Let Q be the output of JK flip-flop and Q' is complement output. The output corresponding to input at various timings are determined using the truth table of JK flip-flop with negative edge clock. The input and output timing diagrams are drawn as shown in Fig. 3 and from the timing diagram the following outputs are observed at various timings.

Truth Table of Negative Edge Triggered JK Flip-flop

Inputs		Clock	Outputs		Comment
J	K		Q	Q'	
0	0	↓	No change		-
0	1	↓	0	1	Reset state
1	0	↓	1	0	Set state
1	1	↓	Toggle		Toggle previous output

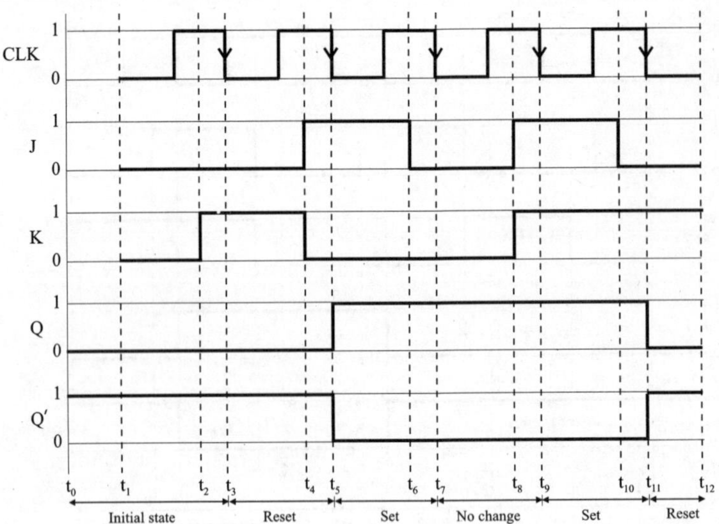

Fig. 3: Timing diagram (or waveform) of negative edge triggered JK flip-flop.

Time t_0: Initial outputs are assumed to be in reset state (Q = 0, Q' = 1).

In negative edge triggered JK flip-flop, the input is recognized only at negative edge of the clock. The negative edge of the clock occurs at time t_3, t_5, t_7, t_9 and t_{11} and hence the value of inputs at these timings will determine the output of flip-flop. The changes in input occur at time t_2, t_4, t_6, t_8 and t_{10} and these changes are recognized only at the next negative edge of the clock.

Also, it can be observed that whenever the output changes at negative edge of the clock it is maintained until the next negative edge.

Time t_0: Initial outputs are assumed to be in reset state (Q = 0, Q' = 1).

Time t_3: J = 0 and K = 1. Inputs are in reset condition. Output Q is reset (Q = 0).

Time t_5: J = 1 and K = 0. Inputs are in set condition. Output Q is set (Q = 1).

Time t_7: Both inputs are 0, and hence no change in output.

Time t_9: J = 1 and K = 0. Inputs are in set condition. Output Q is set (Q = 1).

Time t_{11}: J = 0 and K = 1. Inputs are in reset condition. Output Q is reset (Q = 0).

***Note:** Q' is complement of Q.*

Example 6.10

The waveform shown in Fig. 1 are applied to T flip-flop.

a) Draw the output waveform for positive edge triggered T flip-flop.

b) Draw the output waveform for negative edge triggered T flip-flop.

Assume that initial output is reset.

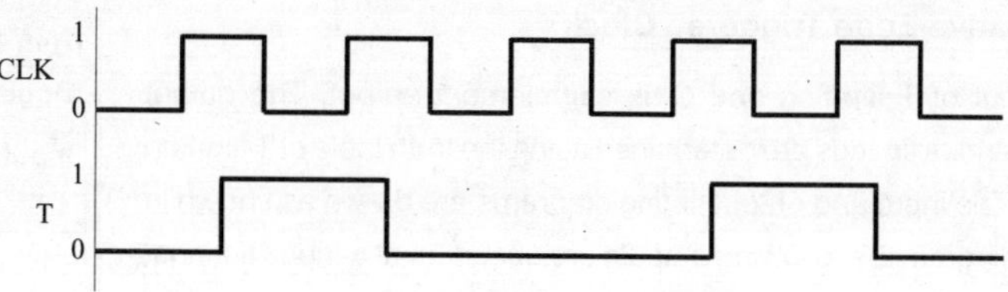

Fig. 1.

Solution

a) T Flip-flop with Positive Edge Triggered Clock

Let Q be the output of T flip-flop and Q′ is complement output. The output corresponding to input at various timings are determined using the truth table of T flip-flop with positive edge clock. The input and output timing diagrams are drawn as shown in Fig. 2 and from the timing diagram the following outputs are observed at various timings.

Truth Table of Positive Edge Triggered T Flip-flop

Input T	Clock	Outputs Q Q′
0	↑	No change
1	↑	Toggle previous output

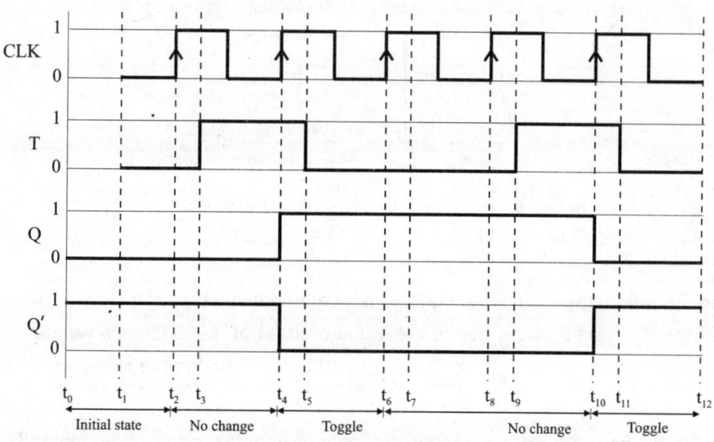

Fig. 2: Timing diagram (or waveform) of positive edge triggered T flip-flop.

Time t_0: Initial outputs are assumed to be in reset state (Q = 0, Q′ = 1).

In positive edge triggered T flip-flop, the input is recognized only at positive edge of the clock. The positive edge of the clock occurs at time t_2, t_4, t_6, t_8 and t_{10} and hence the value of input at these timings will determine the output of flip-flop. The changes in input occur at time t_3, t_5, t_7, t_9 and t_{11} and these changes are recognized only at the next positive edge of the clock.

In positive edge triggered T flip-flop, the input is recognized only at positive edge of the clock. During the positive edge of the clock, if the input is 1 (high) then the output toggles from previous value and if the input is 0 (low) then there is no change in output. Also, it can be observed that whenever the output changes at positive edge of the clock it is maintained until the next positive edge.

***Note:** Q′ is complement of Q.*

b) T Flip-flop with Negative Edge Triggered Clock

Let Q be the output of T flip-flop and Q' is complement output. The output corresponding to input at various timings are determined using the truth table of T flip-flop with negative edge clock. The input and output timing diagrams are drawn as shown in Fig. 3 and from the timing diagram the following outputs are observed at various timings.

Truth Table of Negative Edge Triggered T Flip-flop

Input T	Clock	Outputs Q Q'
0	↓	No change
1	↓	Toggle previous output

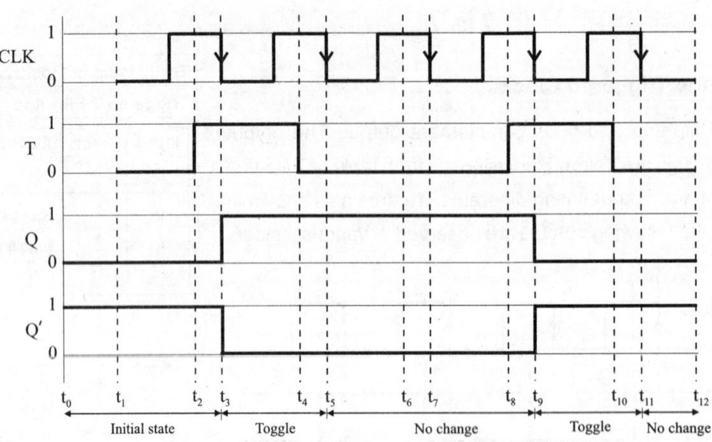

Fig. 3: Timing diagram (or waveform) of negative edge triggered T flip-flop.

Time t_0: Initial outputs are assumed to be in reset state (Q = 0, Q' = 1).

In negative edge triggered T flip-flop, the input is recognized only at negative edge of the clock. The negative edge of the clock occurs at time t_3, t_5, t_7, t_9 and t_{11} and hence the value of input at these timings will determine the output of flip-flop. The changes in input occur at time t_2, t_4, t_6, t_8 and t_{10} and these changes are recognized only at the next negative edge of the clock.

In negative edge triggered T flip-flop, the input is recognized only at negative edge of the clock. During the negative edge of the clock, if the input is 1 (high) then the output toggles from previous value and if the input is 0 (low) then there is no change in output. Also, it can be observed that whenever the output changes at positive edge of the clock it is maintained until the next negative edge.

Note: *Q' is complement of Q.*

Example 6.11

The waveform shown in Fig. 1 are applied to D flip-flop.

a) Draw the output waveform for positive edge triggered D flip-flop.

b) Draw the output waveform for negative edge triggered D flip-flop.

Assume that initial output is reset.

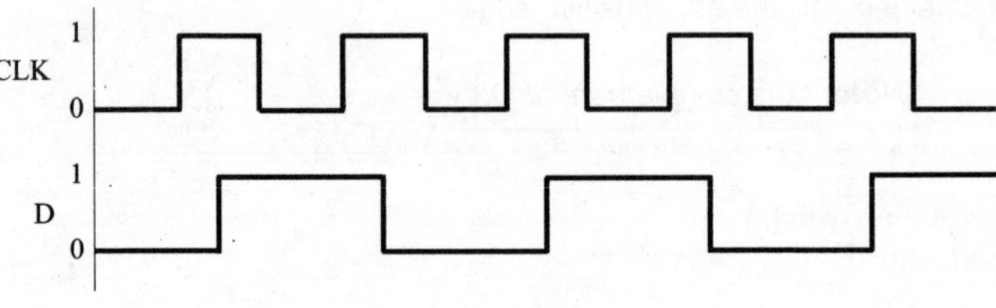

Fig. 1.

Solution

a) D Flip-flop with Positive Edge Triggered Clock

Truth Table of Positive Edge Triggered D Flip-flop			
Input	Clock	Outputs	
D		Q	Q'
0	↑	0	1
1	↑	1	0

Let Q be the output of D flip-flop and Q' is complement output. The output corresponding to input at various timings are determined using the truth table of D flip-flop with positive edge clock. The input and output timing diagrams are drawn as shown in Fig. 2 and from the timing diagram the following outputs are observed at various timings.

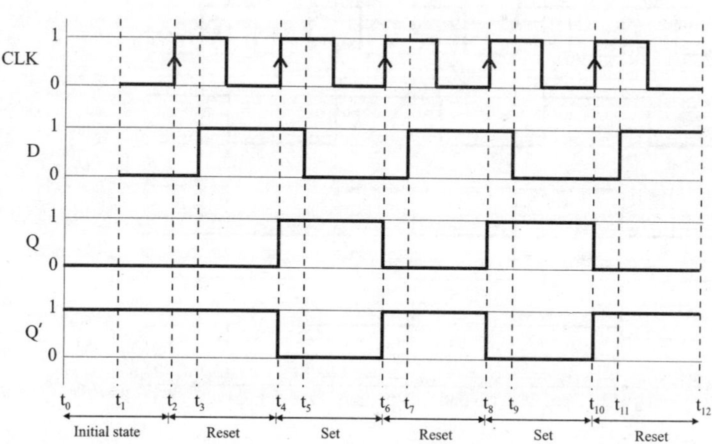

Fig. 2: Timing diagram (or waveform) of positive edge triggered D flip-flop.

Time t_0: Initial outputs are assumed to be in reset state (Q = 0, Q' = 1).

In positive edge triggered D flip-flop, the input is recognized only at positive edge of the clock. The positive edge of the clock occurs at time t_2, t_4, t_6, t_8 and t_{10} and hence the value of input at these timings will determine the output of flip-flop. The changes in input occur at time t_3, t_5, t_7, t_9 and t_{11} and these changes are recognized only at the next positive edge of the clock.

In positive edge triggered D flip-flop, the input is recognized only at positive edge of the clock. During the positive edge of the clock, if the input is 1 (high) then the output is set (Q = 1) and if the input is 0 (low) then the output is reset (Q = 0). Also, it can be observed that whenever the output changes at positive edge of the clock it is maintained until the next positive edge.

Note: *Q' is complement of Q.*

b) D Flip-flop with Negative Edge Triggered Clock

Truth Table of Negative Edge Triggered D Flip-flop			
Input	Clock	Outputs	
D		Q	Q'
0	↓	0	1
1	↓	1	0

Let Q be the output of D flip-flop and Q' is complement output. The output corresponding to input at various timings are determined using the truth table of D flip-flop with negative edge clock. The input and output timing diagrams are drawn as shown in Fig. 3 and from the timing diagram the following outputs are observed at various timings.

Time t_0: Initial outputs are assumed to be in reset state (Q = 0, Q' = 1).

In negative edge triggered D flip-flop, the input is recognized only at negative edge of the clock. The negative edge of the clock occurs at time t_3, t_5, t_7, t_9 and t_{11} and hence the value of input at these timings will determine the output of flip-flop. The changes in input occur at time t_2, t_4, t_6, t_8 and t_{10} and these changes are recognized only at the next negative edge of the clock.

In negative edge triggered D flip-flop, the input is recognized only at negative edge of the clock. During the negative edge of the clock, if the input is 1 (high) then the output is set (Q = 1) and if the input is 0 (low) then the output is reset (Q = 0). Also, it can be observed that whenever the output changes at negative edge of the clock it is maintained until the next negative edge.

Note: *Q' is complement of Q.*

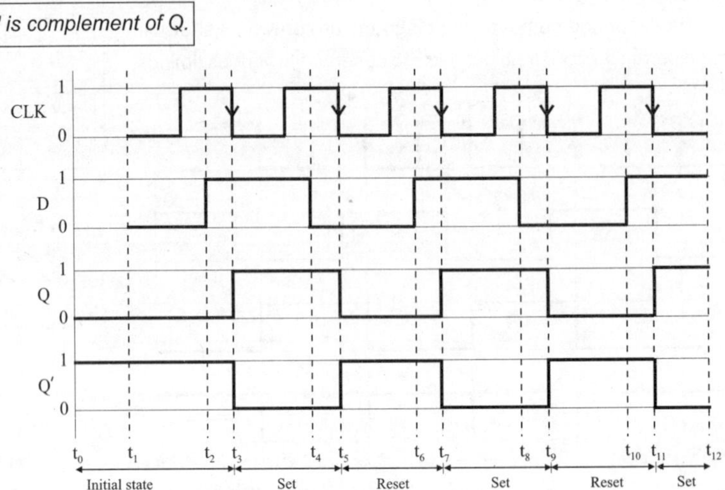

Fig. 3: *Timing diagram (or waveform) of negative edge triggered D flip-flop.*

6.6 Summary of Important Concepts

1. A latch is the basic storage device to store a binary bit.
2. A latch is designed with two stable states in order to store two possible values of a binary bit 1 and 0.
3. The SR latch is a digital circuit designed to store 1 or 0 applied as two inputs called set and reset and it is provided with an output and complement of the output.
4. A stable state is the state when present state output and next state output are same after a change in input.
5. D latch is a single input latch with two stable states which are used to store two different values 0 and 1 of a binary data.
6. Flip-flop is a storage device constructed using latch and controlled by enable signal called clock.
7. The four types of flip-flops are SR, JK, D and T flip-flops.
8. The SR flip-flop is a digital device with two stable states: set and reset states. The draw back in SR flip flop is that it has an indeterminate output when S = R = 1.
9. The JK flip-flop is a digital device with two stable states: set and reset states, similar to SR flip-flop but the indeterminate output of SR flip-flop is converted to toggle output in which the present output will be complement of previous output.
10. The D flip-flop is a single input flip-flop in which the input is stored in the output when clock is applied.
11. The T flip-flop is a single input flip-flop in which the present output is complement of previous output when clock is applied provided input is 1.
12. A master/slave flip-flop is cascade connection of two flip-flops, in which the first flip-flop is called master and the second flip-flop is called slave.

13. In flip-flops the process of applying a control signal to change from one stable state to another stable state is called triggering. Usually, the control signal is a clock pulse.
14. In flip-flops preset is provided to set the output and reset is provided to reset the output.
15. In flip-flops, when preset is 0, the output is set to 1 and when reset is 0, the output is reset to 0.
16. In flip-flops, the preset and reset should be tied 1, for input and clock to make changes in output.
17. In flip-flops, the excitation table lists the required inputs for a specified change of state in output.
18. To convert a flip-flop from one type to another type, a combinational circuit is designed to supply the inputs of the flip-flop in order to get the output of another desired flip-flop.
19. In SR and JK flip-flops if the inputs are always complementary then they behave as D flip-flop.
20. In SR and JK flip-flops if the inputs are always same then they behave as T flip-flop.

6.7 Short-Answer Questions

Q6.1 Define clock and clock frequency.

The clock is basically a periodic square pulse signal. The sum of ON and OFF time (or **high** and **low** time) is called one period of clock. The inverse of clock period is called clock frequency. The flip-flop is possible to replace enable by a momentary pulse called clock.

Q6.2 Define latch and flip-flop. ***(VTU, Aug/Sep'20, 2 Marks)***

Latch: A latch is the basic storage device to store a binary bit. A latch is designed with two stable states in order to store two possible values of a binary bit 1 and 0. Latches are designed with complementary inputs and outputs so that positive or negative logic can be implemented easily. The latches are sequential circuits without a clock.

Flip-flop: Flip-flop is a storage device constructed using latch and controlled by enable signal called clock. Flip-flops are made sensitive to level/positive edge/negative edge clock.

Q6.3 State the difference between latches and flip-flops.

The main difference between latches and flip-flops is that the flip-flops are clocked storage devices but the latches are unclocked storage devices.

Q6.4 Write the function of set and reset inputs of a flip-flop.

Set is an input used to set the output to 1 and it is denoted by S.

Reset is an input used to clear the output to 0 and it is denoted by R.

Q6.5 What is meant by stable state.

A ***stable state*** is the state when present state output and next state output are same after a change in input.

Q6.6 Define data flip-flop and toggle flip-flop.

D flip-flop is a single input flip-flop in which the input is stored in the output when clock is applied. It is also called as data flip-flop.

T flip-flop is a single input flip-flop in which the present output is complement of previous output when clock is applied, provided input is 1. It is also called as toggle flip-flop.

Q6.7 Define master-slave flip-flop.

A master/slave flip-flop is cascade connection of two flip-flops. The first flip-flop is called master and the second flip-flop is called slave. The input is passed to output of master during positive edge of clock. The master output is passed to slave output on the negative edge of clock.

The behaviour of the master-slave flip-flop ensure the following,

1. The output may change only once and there is no race around condition.
2. A change in the output is triggered by the negative edge of the clock.

Q6.8 ***What is mean by triggering of flip-flops?***

In flip-flops the process of applying a control signal to change from one stable state to another stable state is called triggering. Mostly the triggering is performed by edge sensitive clock.

Q6.9 ***Give the difference between positive and negative edge triggering.*** ***(AKTU, Dec'22, 2 Marks)***

In positive edge triggered flip-flop the inputs are recognized at positive edge (i.e, when clock take a transition from low to high) whereas in negative edge triggered flip-flop the inputs are recognized at negative edge (i.e, when clock take a transition from high to low).

6.8 Exercises

I. Fill in the blanks

1. The SR latch is a digital circuit designed to store 1 or 0 applied as two inputs called _______ and _______.
2. A JK flip-flop is in toggle condition when ______.
3. A gated D latch does not have ______.
4. A latch is the basic storage device to store a _______.
5. Latches are designed ______ inputs and outputs so that positive or negative logic can be implemented easily.
6. The SR latches can be constructed using two cross-coupled _______ gates or using two cross-coupled _______ gate.
7. The clock is basically a periodic square _______.
8. The clock pulses will have equal _______ and _______ time.
9. The sum of ON and OFF time is called _______.
10. The inverse of clock period is called _______.
11. Rising edge is also called _______.
12. Falling edge is also called as _______.
13. In flip-flops the process of applying a control signal to change from one stable state to another stable state is called _______.
14. The triggering is performed by _______ sensitive clock.
15. Level sensitive devices are also called as _______ triggered devices.
16. Edge sensitive devices are also called as _______ triggered devices.
17. The SR flip-flop is a digital device with _______.
18. In JK flip-flop, the indeterminate output of SR flip-flop is converted to another useful output called _______ output.
19. The_______ are used to design binary counters.
20. _______ is cascade connection of two flip-flops.

Answers

1. set, reset	5. complementary	9. one period of clock	13. triggering	17. two stable states
2. J = 1, K = 1	6. NOR, NAND	10. clock frequency	14. edge	18. toggle
3. a clock input	7. pulse signal	11. positive edge	15. level	19. T flip-flop
4. binary bit	8. ON, OFF	12. negative edge	16. edge	20. master slave flip-flop

II. State whether the following statements are True or False

1. An SR latch is a one bit memory element.
2. A stable state is the state when present state output and next state output are same after a change in input.
3. In flip-flops, the preset and reset should be tied 0, for input and clock to make changes in output.
4. In SR and JK flip-flops if the inputs are always complementary then they behave as T flip-flop.
5. In flip-flops, the excitation table lists the required inputs for a specified change of state in output.
6. A latch is the basic storage device to store a binary bit.
7. A latch is designed with three stable states in order to store three possible values of a binary bit 1 and 0.
8. A latches is designed with two stable states in order to store two possible values of a binary bit 1 and 0.
9. The latches are sequential circuits without a clock.
10. D latch is a single input latch with two stable states which are used to store two different values 0 and 1 of a binary data.
11. In latches with enable, it is observed that inputs are recognized only when enable is in active.
12. Flip-flop is a storage device constructed using latch and controlled by enable signal called clock.
13. Typical clock frequencies are in order of GHz and so clock periods are in the order of pico-seconds.
14. The clocked sequential circuits or synchronous sequential circuits are designed as edge sensitive device.
15. JK flip-flop is similar to SR flip-flop but with indeterminate output.
16. The difference between JK and SR flip-flops is the type of output when the inputs are same.

Answers

1. True	5. True	9. True	13. False
2. True	6. True	10. True	14. True
3. False	7. False	11. False	15. False
4. False	8. True	12. True	16. True

III. Choose the right answer for the following questions

1. ***A master slave flip-flop has the characteristics that***

   a) Change in the input is immediately reflected in the output

   b) Change in the output occurs when the state of the master is affected

   c) Change in the output occurs when the state of the slave is affected

   d) both the master and the slave states are affected at the same time.

2. *The characteristics equation of a JK flip-flop is*

a) JQ' + K'Q   b) JQ + K'Q   c) J'Q + K'Q   d) JQ + K'Q

3. *A master slave flip-flop is made up of*

a) two flip-flop connected in series
b) two flip-flop connected in parallel
c) debouncer circuit
d) D-latch

4. *The output of a toggle flip-flop is _____ the frequency of its input.*

a) twice   b) half   c) thrice   d) one-third

5. *Assume that initially Q = 1 in Fig. 5, with clock pulses being given, the subsequence states of Q will be*

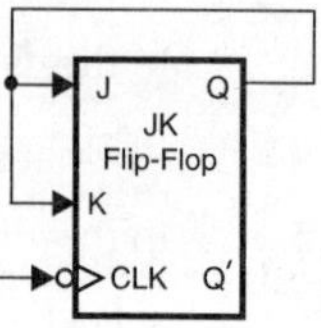

a) 10101
b) 0000
c) 1111
d) 01010

Fig. 5.

6. *The Q output of a JK flip-flop is '1'. The output doesn't change when a clock pulse is applied. The inputs J and K will be respectively (where X is don't care input)*

a) 0 and X   b) X and 0   c) 1 and 0   d) 0 and 1

7. *The circuit shown in Fig. 7 behave as*

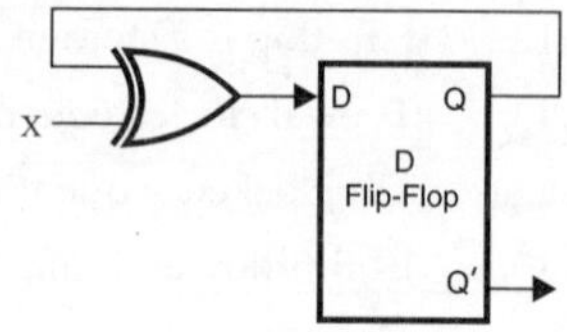

a) T flip-flop
b) D flip-flop
c) JK flip-flop with K connected to ground and X connected to J
d) none of the above

Fig. 7.

8. *In a JK flip-flop we have J = Q' and K = 1. (Refer Fig. 8) Assuming the flip-flop was initially cleared and then clocked for 6 pulses. The sequence at the Q output will be.*

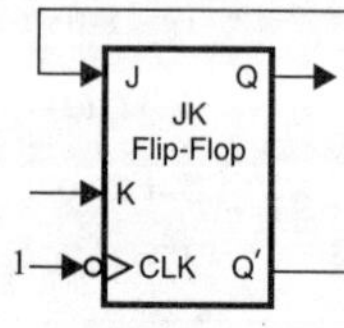

a) 010000
b) 011001
c) 010010
d) 010101

Fig. 8.

9. *The present output Q(t) of an edge triggered JK flip-flop is logic 0. If J = 1, then Q(t+1)*

a) Cannot be determined
b) will be logic 0
c) will be logic 1
d) will race around

10. *In Fig. 10 the input clock frequency is 12 kHz. The frequency at the output is*

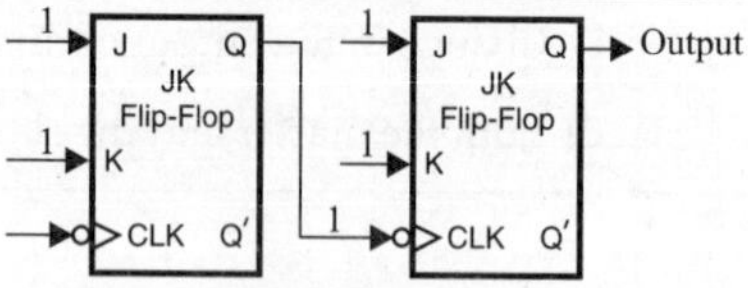

a) 24 kHz
b) 12 kHz
c) 3 kHz
d) 6 kHz

Fig. 10.

11. ***If the clock frequency is 60 kHz, the frequency at 'B' of Fig. 11 is***

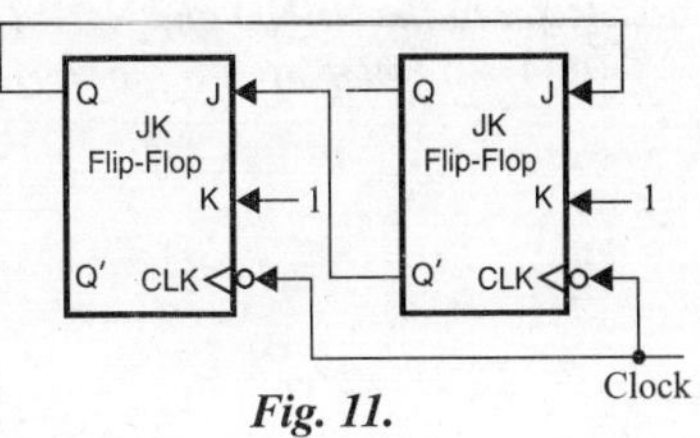

Fig. 11.

a) 12.5 kHz

b) 15 kHz

c) 30 kHz

d) 16.67 kHz

---

12. ***A sequential circuit using D flip-flop and logic gates is shown in Fig. 12 where X and Y are the inputs and Z is the output. The circuit is***

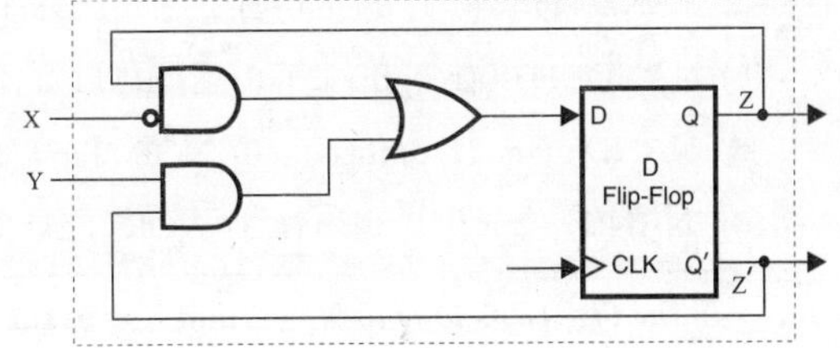

Fig. 12.

a) SR flip-flop with inputs X = R and Y = S

b) SR flip-flop with inputs X = S and Y = R

c) JK flip-flop with inputs X = J and Y = K

d) JK flip-flop with inputs X = K and Y = J

---

13. ***The following binary values were applied to the X and Y inputs of the NAND latch shown in the sequence indicated below.***

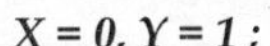

X = 0, Y = 1 ; X = 0, Y = 0 ; X = 1, Y = 1

The corresponding stable P, Q outputs will be

X
P
Y
Q

Fig. 13.

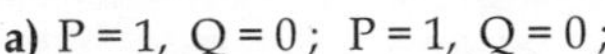

a) P = 1, Q = 0 ; P = 1, Q = 0 ; P = 1, Q = 0 (or) P = 0, Q = 1

b) P = 1, Q = 0 ; P = 1, Q = 1 (or) P = 0, Q = 1 ; P = 0, Q = 1

c) P = 1, Q = 0 ; P = 1, Q = 1 ; P = 1, Q = 0 (or) P = 0, Q = 1

d) P = 1, Q = 0 ; P = 1, Q = 1 ; P = 1, Q = 1

---

14. ***Assume that all flip-flops are in reset condition initially. The count sequence observed at Q_A in the circuit shown is***

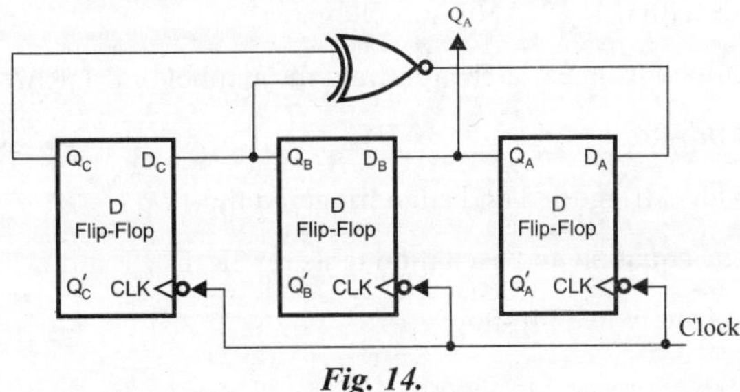

Fig. 14.

a) 0010111

b) 0001011

c) 0101111

d) 0110100

15. ***Refer to the NAND and NOR latches shown in Fig. 15. The inputs (P_1, P_2) for both the lathces are first made (0, 1) and then after a few seconds made (1, 1). The corresponding stable outputs (Q_1, Q_2) are***

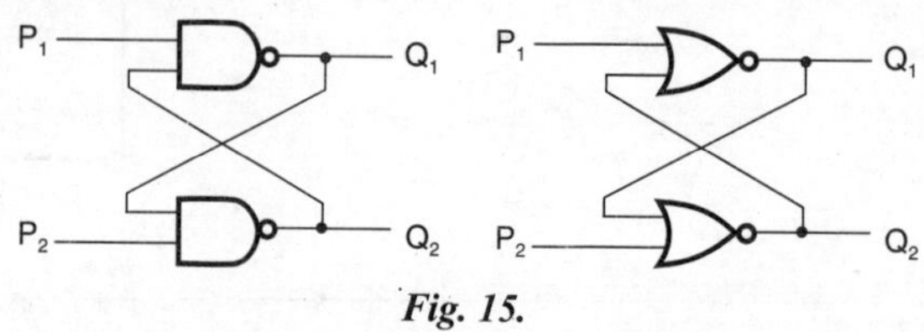

Fig. 15.

a) NAND : First (0, 1) then (0, 1) NOR : First (1, 0) then (0, 0)

b) NAND : First (1, 0) then (1, 0) NOR : First (1, 0) then (1, 0)

c) NAND : First (1, 0) then (1, 0) NOR : First (1, 0) then (0, 0)

d) NAND : First (1, 0) then (1, 1) NOR : First (0, 1) then (0, 1)

16. ***In the Fig. 16 as long as $X_1 = 1$ and $X_2 = 1$, the output Q remains***

a) at 1 **b)** at 0

c) at its initial value **d)** unstable

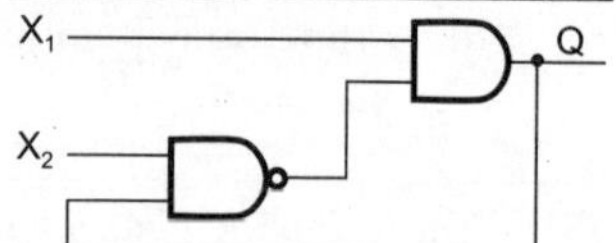

Fig. 16.

Answers

1. c	6. d	11. c	16. c
2. a	7. a	12. d	
3. a	8. d	13. d	
4. b	9. c	14. d	
5. b	10. c	15. a	

IV. Answer the following questions

1. Draw the symbol of SR latch with using NAND and OR gates?
2. How does the D latch developed from SR latch and draw the symbol of D latch?
3. Write a few applications of flip-flops.
4. List the difference between level triggeried and edge triggered flip-flop.
5. Write down the characteristic equation and excitation table for SR, JK , T and D flip-flop.
6. Explain the preset and reset facility in a flip-flop.
7. Define SR latches.
8. Define D latches.
9. Types of flip-flop?
10. Explain one period of clock and clock frequency.
11. Define SR flip-flop and JK flip-flop with truth table.
12. Explain the behaviour of master-slave flip-flop.

V. Solve the following problems

E6.1 The input waveforms of NOR based SR latch is shown in Fig. E6.1. Draw the output waveform. Assume that initial output is reset.

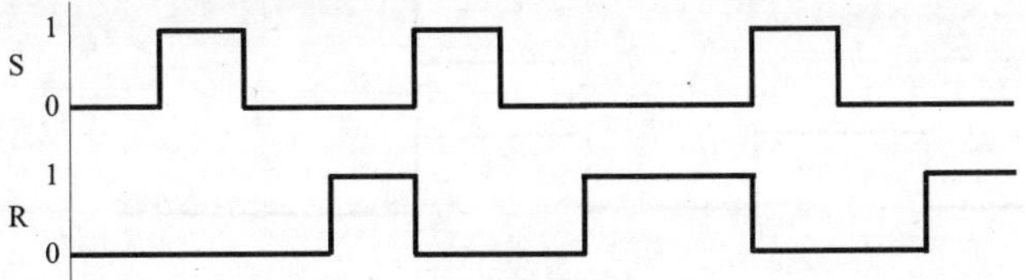

Fig. E6.1.

E6.2 The waveform are applied to NAND based SR latch without enable. Sketch the resulting output waveform in relation to the inputs. Assume that initial output is reset.

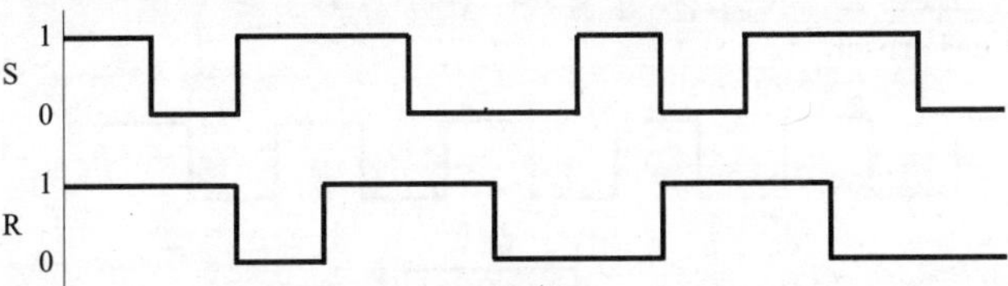

Fig. E6.2.

E6.3 For the NAND based SR latch with enable determine the outputs for the inputs shown in Fig. E6.3. Assume that initial output is reset.

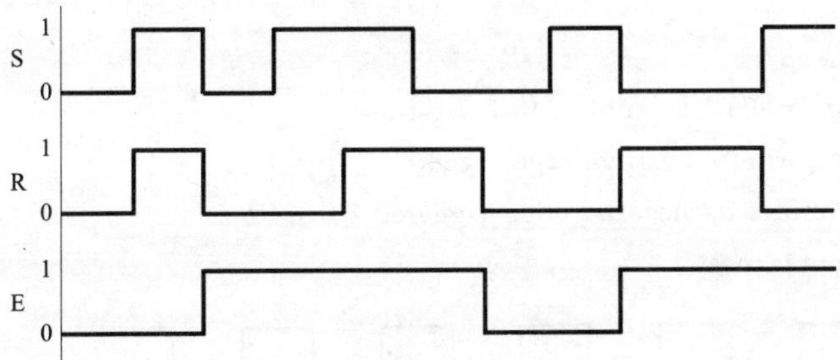

Fig. E6.3.

E6.4 For D latch the waveform shown in Fig. E6.4 are applied on its inputs. Sketch the waveform expected on the output if initial output is reset.

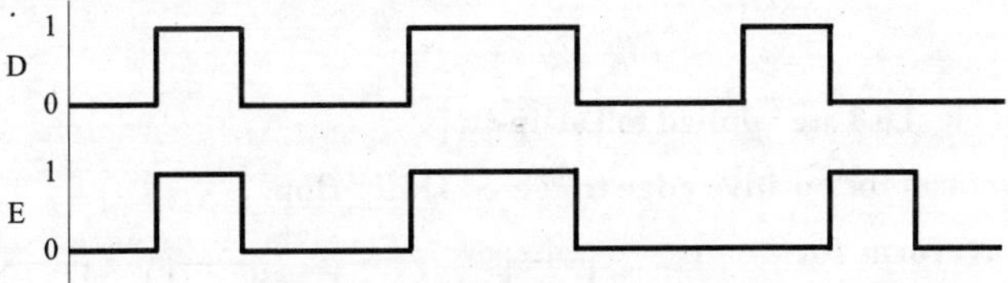

Fig. E6.4.

E6.5 The waveform shown in Fig. E6.5 are applied to SR flip-flop.

a) Draw the output waveform for positive edge triggered SR flip-flop.

b) Draw the output waveform for negative edge triggered SR flip-flop.

Assume that initial output is reset.

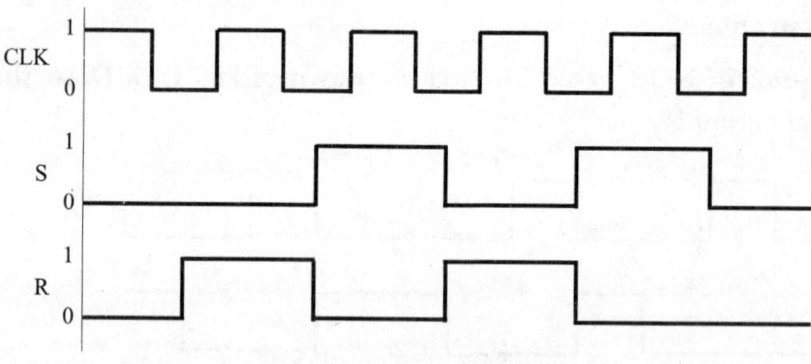

Fig. E6.5.

E6.6 The waveform shown in Fig. E6.6 are applied to JK flip-flop.

a) Draw the output waveform for positive edge triggered JK flip-flop.

b) Draw the output waveform for negative edge triggered JK flip-flop.

Assume that initial output is reset.

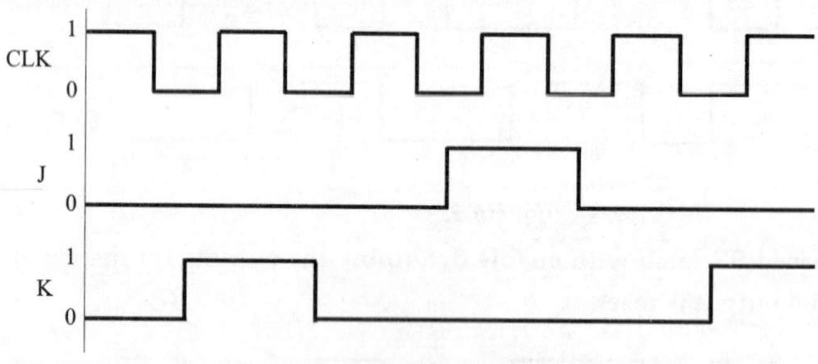

Fig. E6.6.

E6.7 The waveform shown in Fig. E6.7 are applied to T flip-flop.

a) Draw the output waveform for positive edge triggered T flip-flop.

b) Draw the output waveform for negative edge triggered T flip-flop.

Assume that initial output is reset.

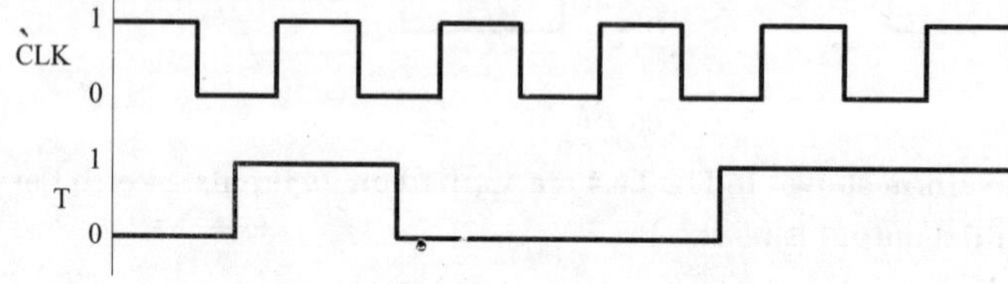

Fig. E6.7.

E6.8 The waveform shown in Fig. E6.8 are applied to D flip-flop.

a) Draw the output waveform for positive edge triggered D flip-flop.

b) Draw the output waveform for negative edge triggered D flip-flop.

Assume that initial output is reset.

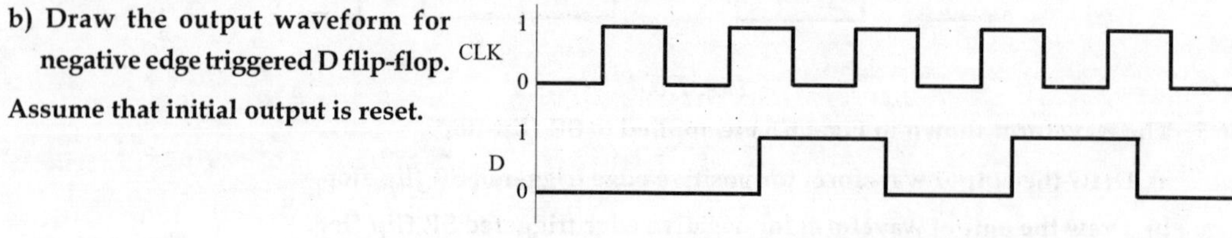

Fig. E6.8.

Answers

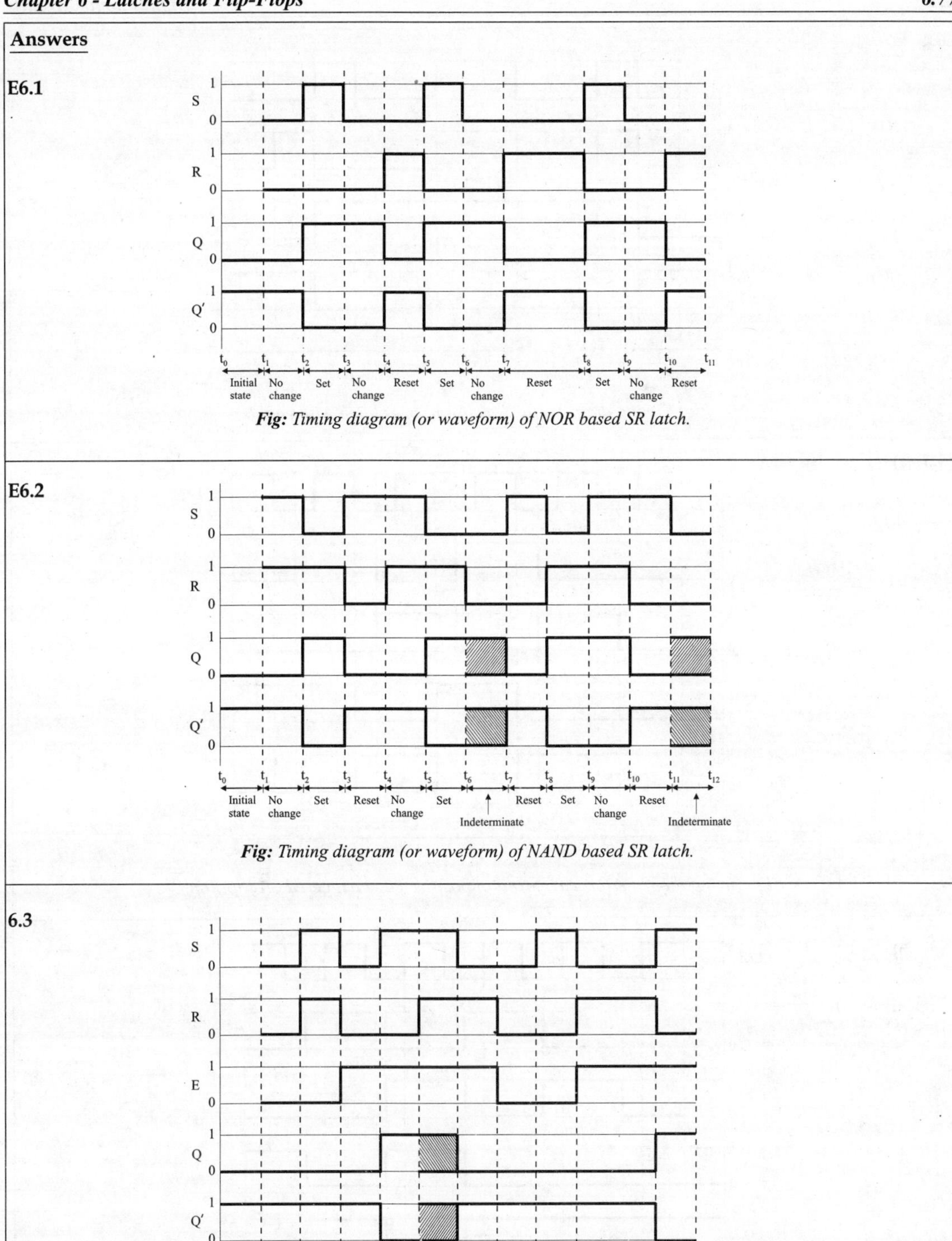

Fig: *Timing diagram (or waveform) of NOR based SR latch.*

Fig: *Timing diagram (or waveform) of NAND based SR latch.*

Fig: *Timing diagram (or waveform) of SR latch using NAND gates and with enable circuit.*

E6.4

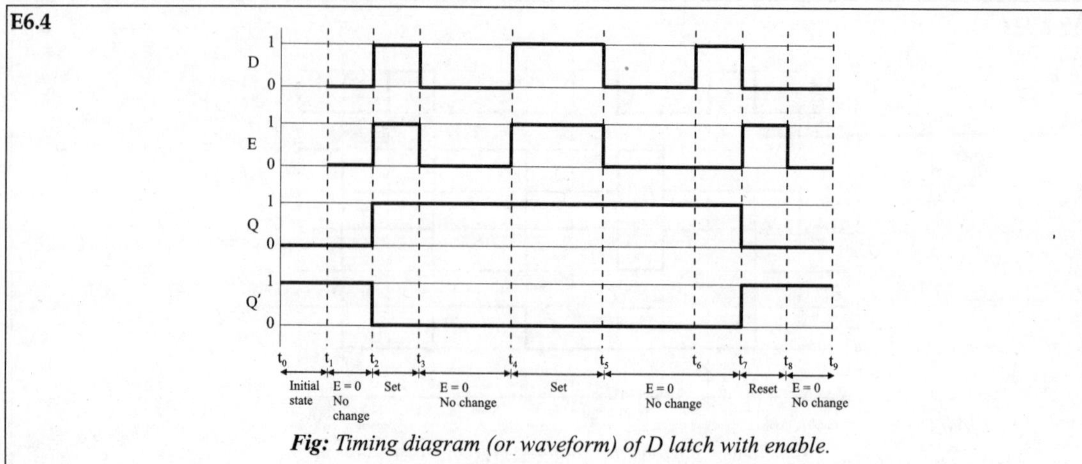

Fig: Timing diagram (or waveform) of D latch with enable.

E6.5 a)

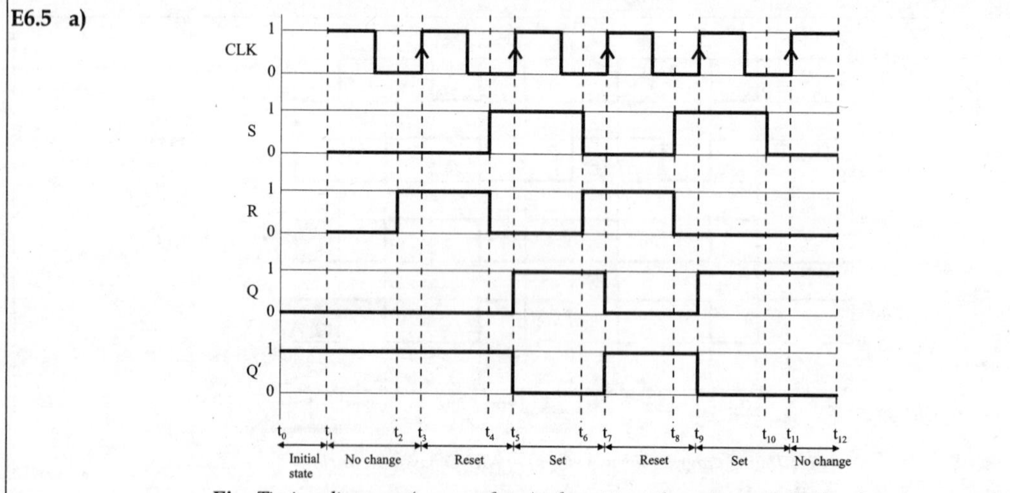

Fig: Timing diagram (or waveform) of positive edge triggered SR flip-flop.

b)

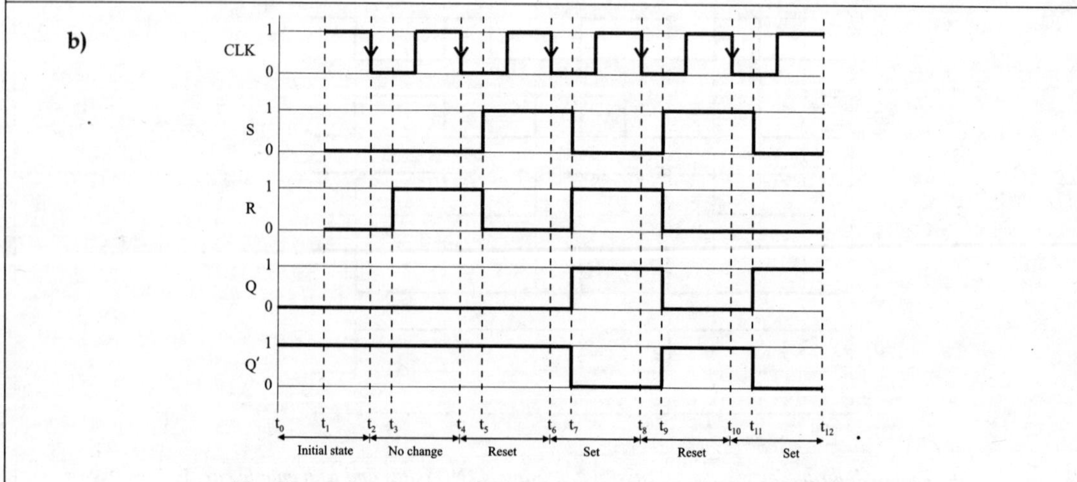

Fig: Timing diagram (or waveform) of negative edge triggered SR flip-flop.

E6.6 a)

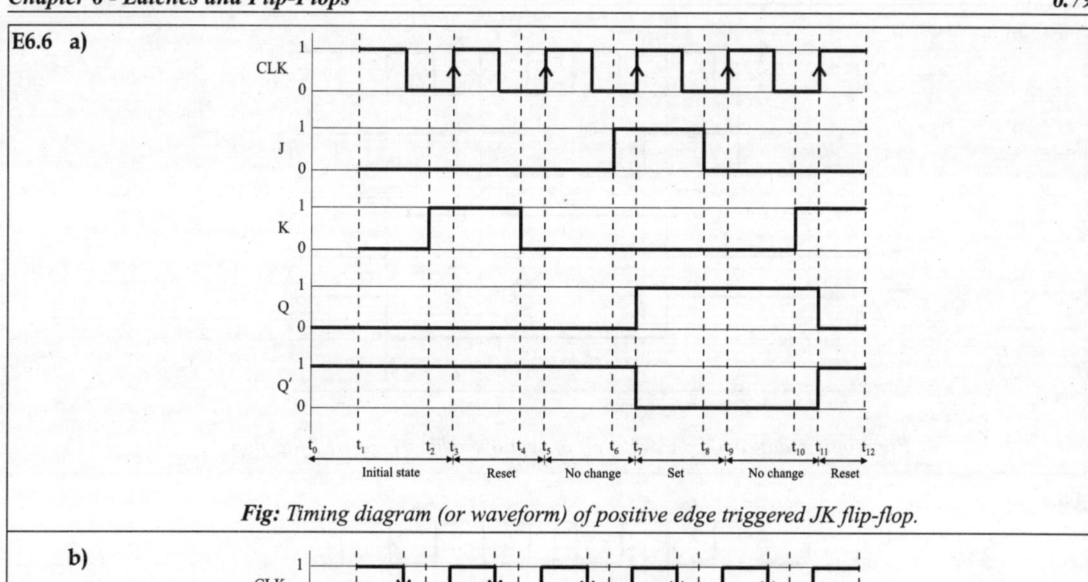

***Fig:** Timing diagram (or waveform) of positive edge triggered JK flip-flop.*

b)

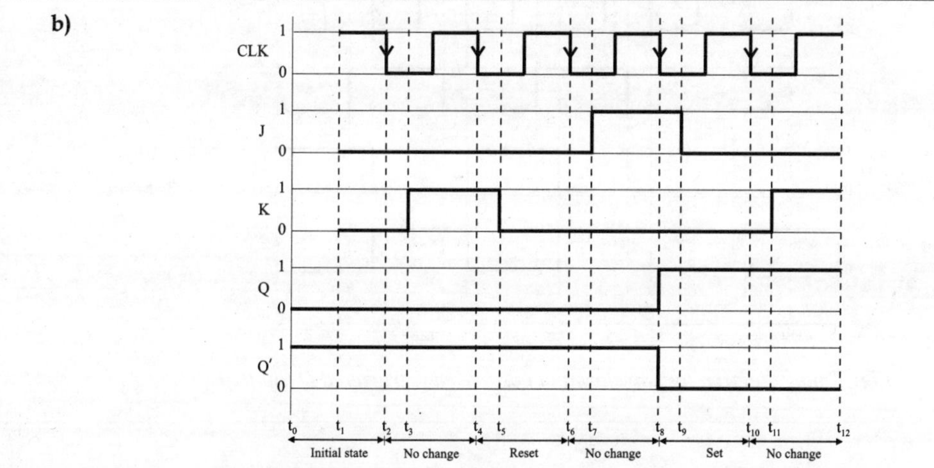

***Fig:** Timing diagram (or waveform) of negative edge triggered JK flip-flop.*

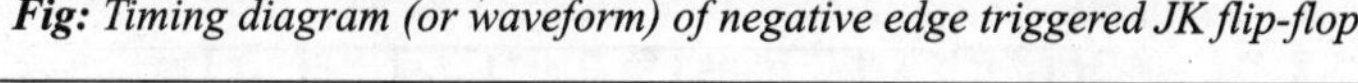

E6.7 a)

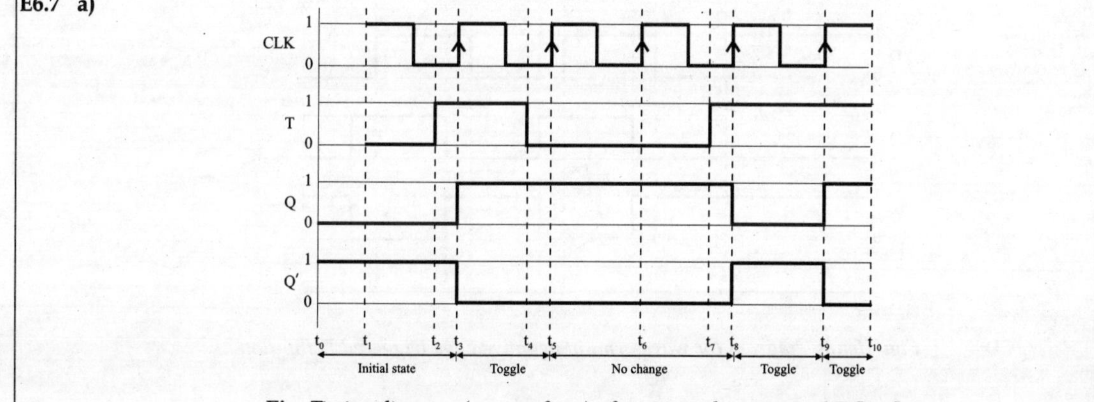

***Fig:** Timing diagram (or waveform) of positive edge triggered T flip-flop.*

b)

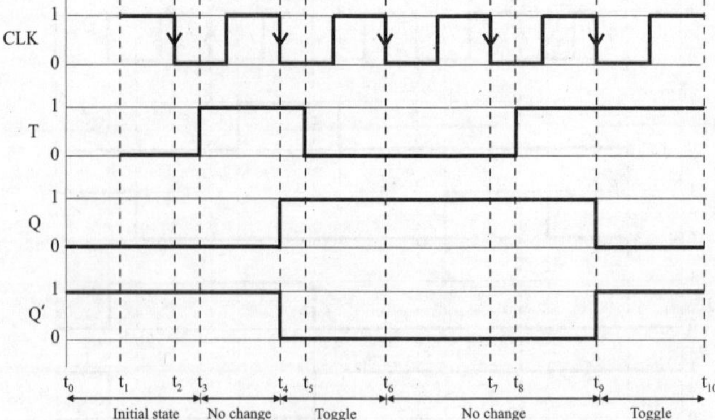

Fig: Timing diagram (or waveform) of negative edge triggered T flip-flop.

E6.8 a)

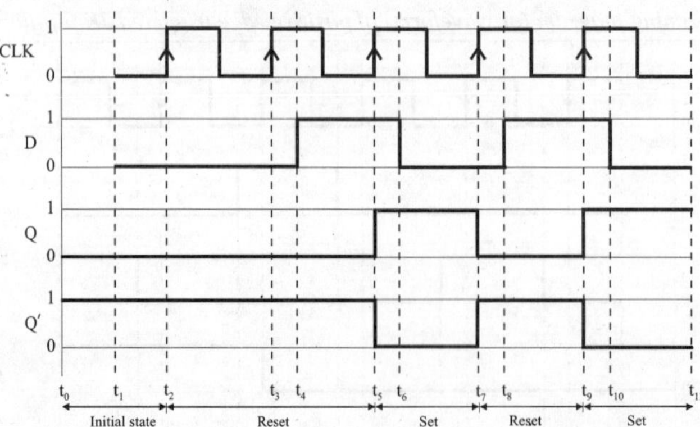

Fig: Timing diagram (or waveform) of positive edge triggered D flip-flop.

b)

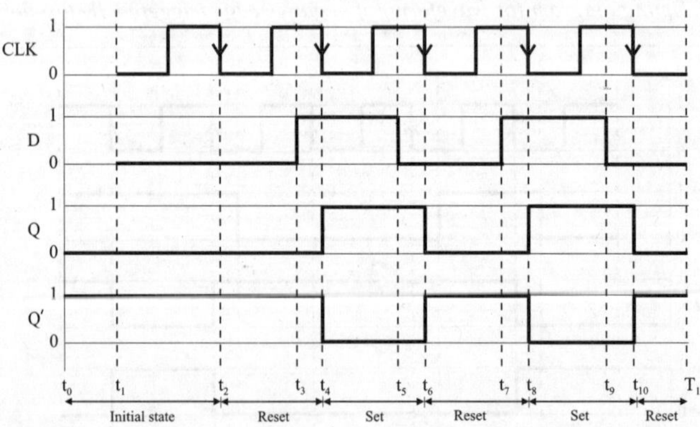

Fig: Timing diagram (or waveform) of negative edge triggered D flip-flop.

CHAPTER 7

SYNCHRONOUS SEQUENTIAL CIRCUITS

7.1 Introduction

The sequential logic circuits are digital circuits whose output depends on present inputs and past outputs. The sequential circuits will have storage elements to store the past outputs, so that they can be fed back to input. Therefore, a sequential logic circuit can be expressed as a combinational circuit with storage elements and feedback paths as shown in Fig. 7.1.

Fig. 7.1: Sequential logic circuit.

There are two types of sequential circuits.

1. Asynchronous sequential circuits
2. Synchronous sequential circuits

The sequential circuits without a clock are called ***asynchronous sequential circuits***.

The sequential circuits with a clock are called ***synchronous sequential circuits***. In clocked sequential circuits the inputs are recognized only when clock signal is applied otherwise ignored. Therefore, output also changes only when clock is applied.

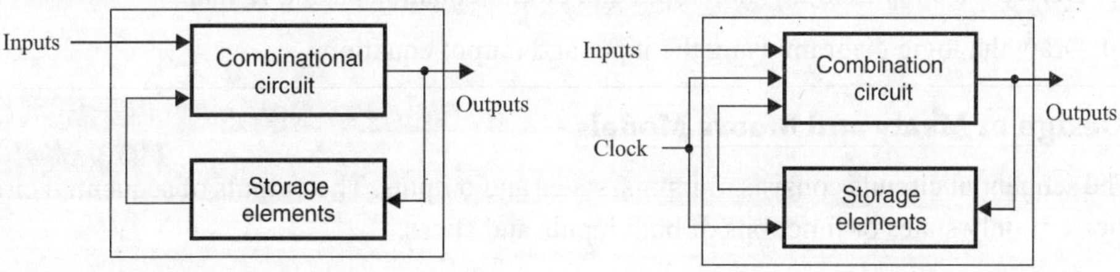

Fig. a: *Asynchronous sequential circuit.* ***Fig. b:*** *Synchronous sequential circuit.*

Fig. 7.2: *Asynchronous and synchronous sequential circuits.*

The basic storage elements in sequential circuits are latches and flip-flops. Latches are sequential circuits without clock. Flip-flops are sequential circuits with clock.

7.2 Analysis and Design of Synchronous or (Clocked) Sequential Circuits

The sequential circuits can be designed using flip-flops with combinational circuits at the input and output of flip-flops. The sequential circuit involves feedback from output to input. The circuit can be designed using either D or JK or T flip-flop. Usually the circuit is designed with only any one of the type of flip-flops.

The T flip-flops do not have set state and reset state and so they cannot be used to design general storage devices or sequential circuits like D flip-flop and JK flip-flop. But T flip-flops are popularly used to design counters which is also a type of sequential circuits. The combinational circuit at the input is designed by constructing K-map for flip-flop inputs. The combinational circuit at the output is designed by constructing K-map for outputs.

The design starts with specifications which includes a word description, inputs and outputs of the circuit.

(JNTUA, Jun/July'22, 7 Marks)

The steps in design of sequential circuits are,

1. Draw the state diagram from specifications.
2. Determine the state table from state diagram.
3. If necessary, reduce the number of states and obtain the reduced state table.
4. Assign binary values or codes to states. If there are m state variables and n inputs then 2^{m+n} binary codes are needed to code the states.
5. Form the binary coded state table.
6. Choose the type of flip-flop to be used in design (either D or JK or T). The number of flip-flops is equal to number of state variables.
7. The flip-flops inputs are same as next state outputs, when D flip-flops are used for design. In order to design using JK or T flip-flops determine the flip-flop inputs using excitation table.
8. Form the expanded state table with flip-flop inputs.
9. Derive the simplified flip-flop input and output equations using K-map.
10. Draw the logic diagram using the input and output equations.

7.2.1 Design of Mealy and Moore Models

(JNTUK, Feb'22, 7 Marks) (VTU, Aug/Sep'20, 6, 10 Marks) (AU, Apr/May'19, 7 Marks)

The sequential circuits consists of inputs, states and outputs. The outputs of sequential circuits may be functions of only states or functions of both inputs and states.

When the outputs of a sequential circuit are functions of inputs and states (or present states) then the sequential circuit is called ***Mealy type*** or ***Mealy model*** of sequential circuits. The sequential circuits in Examples 7.3, 7.4, 7.6, 7.7 and 7.8 are Mealy models.

When the outputs of a sequential circuit are functions of states (or present states) alone, then the sequential circuit is called ***Moore type*** or ***Moore model*** of sequential circuits. The sequential circuits in Examples 7.1, 7.2, 7.9, 7.12 and 7.13 are Moore models.

Since the output depends only on states in Moore model of sequential circuits, the state diagram of Moore type models will have circles with state and associated output. The lables on the directed lines will have only inputs that initiate transitions from one state to another. (Refer examples 7.1, 7.2 and 7.9 for state diagram of Moore model sequential circuits)

The functional behaviour of sequential circuits can be represented using a finite number of states. Hence the sequential circuits are also called ***finite state machines (FSM)***. Therefore, the Mealy model is called ***Mealy FSM*** or ***Mealy Machine***. The moore model is called ***Moore FSM*** or ***Moore Machine***.

The general block diagrams of Mealy and Moore Machine are shown in Figs. 7.3 and 7.4 respectively. In the block diagram the state registers are flip-flops that are used to store the value of states.

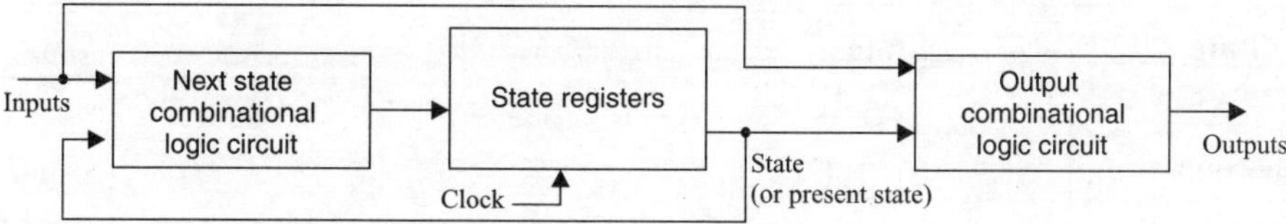

***Fig.** 7.3: Block diagram of Mealy Machine (Mealy FSM).*

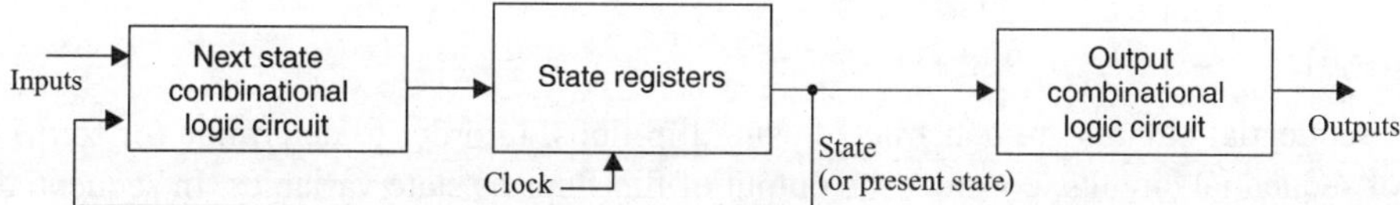

***Fig.** 7.4: Block diagram of Moore Machine (Moore FSM).*

The logic circuits on the input side of flip-flops are called next state combination logic circuit, because they are needed to develop the inputs for flip-flops before arrival of clock to change the state from present state to next state. The logic circuits on the output side of flip-flops are used to construct outputs, and so they are called output combinational logic circuit.

In a Moore model, the outputs are synchronized with clock. But in a Mealy model the inputs may change any time and so outputs are not synchronized with clock. The synchronization of inputs with clock in Mealy models is necessary to ensure right outputs and to overcome the difference in delays in input changes and state changes.

7.3 State and State Equations

A ***state*** is the condition of a sequential circuit described by a set of variables called ***state variables***. Flip-flops are used to store state variables and so the output of flip-flops constitute the states of a sequential circuit. With m state variables we can form 2^m different states.

The ***state equations*** (or transition equations) are a set of equations that describe the next state as a function of the present states and inputs.

Let, x, y, z, ... are used to denote inputs.

X, Y, Z, ... are used to denote outputs of a sequential circuit.

a, b, c, ... are used to denote states

A, B, C, ... are used to denote the state variables or flip-flop outputs.

Since the inputs, outputs and state variables are all functions of time we can differentiate the present and next state variables by two different notations as function of time.

Present State Variables: A(t), B(t), C(t) ....

Next State Variables: A(t + 1), B(t + 1), C(t + 1) ....

> ***Note:*** *For simplicity the " t " and " t+1" can be dropped and variables in present and next state can be represented in different columns in state table, refer examples or solved problems.*

Example: Let A, B and C be three state variables. Now, with 3 state variables, number of different states are $2^3 = 8$ states. Let the 8 states be denoted as a, b, c, d, e, f, g and h. The states and state variables are related by assigning 3-bit code for state variables as shown below:

State ABC	State ABC
a = 000	e = 100
b = 001	f = 101
c = 010	g = 110
d = 011	h = 111

The sequential circuits are constructed using flip-flops. In order to determine (or form) the state equations of sequential circuits, consider the output of flip-flops as state variables. In sequential circuits one flip-flop is used to store one state variable and so number of state variable will be equal to number of flip-flops.

The present state is the output of flip-flops before clock pulse and next state is the output of flip-flop after clock pulse. The next state can be obtained from the truth table of flip-flop or characteristic equations of flip-flops. Thus state equations are obtained from characteristic equations.

Refer examples 7.3 and 7.4 for the formation of state equations of sequential circuits.

7.3.1 State Table and State Diagram *(RGPV, Jun'20, 7 Marks) (JNTUK, Feb'22, 5 Marks) (GTU, Jun'19, 3 Marks)*

The ***state table*** is listing of outputs and next states for all possible combinations of inputs and present states. In general, if there are m state variables and n inputs then possible combinations of inputs and states are 2^{m+n}. The state table list the outputs and next states for 2^{m+n} combinations of inputs and states.

The ***state diagram*** is graphical representation of all transitions from present state to next state for all possible combination of inputs and outputs. Therefore, the state diagram is the graphical representation of information available in a state stable. Hence, the state digram is also called state graph.

The state diagram consists of circles and directed lines. The states are represented by circles and directed lines indicate the transition from one state to another. The associated inputs and outputs of every state transition is marked as label in the directed lines. The format of label is input/output.

Guidelines for Construction of State Diagram or State Graph *(VTU, Aug/Sep'20, 10 Marks)*

1. First form the state table by listing outputs and next states for all possible combinations of present states and inputs.
2. Represent each state by a circle and mark the assigned binary code inside the circle.
3. For each state, represent all state transitions from that state to all other states by directed lines and mark the inputs and outputs on each directed line.

4. Finally combine all the state transitions in to a single state diagram.

> ***Note:*** *Refer examples 7.3 and 7.4 for the formation of state table and state diagram of sequential circuits.*

7.3.2 State Reduction

(JNTUA, Jun/July'22, 7 Marks)

The synthesis or design of sequential circuit start from a state diagram. There is a possibility of realizing the same input-output relationship of the given state diagram with an equivalent or alternate state diagram with reduced number of states. An example of reduced state diagram is shown in Fig. 7.5.

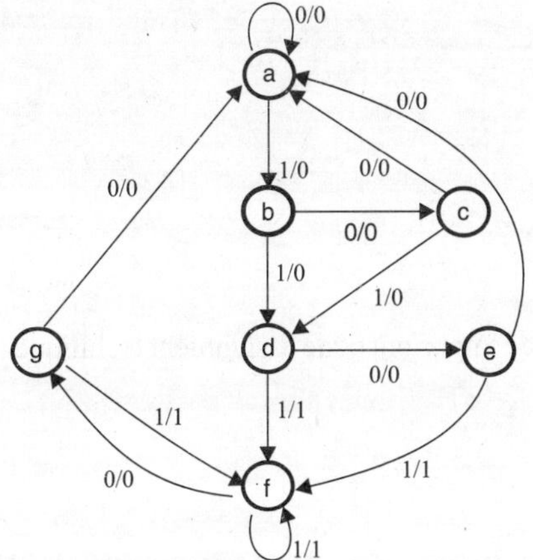

Fig. a: *State diagram with seven states.*

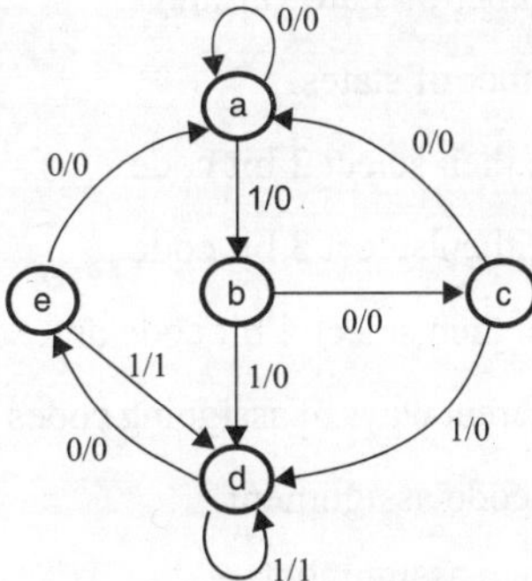

Fig. b: *Reduced equivalent state diagram with 5 states.*

The reduction is explained in the Example 9.5.

Fig. 7.5: *Example of state diagram and reduced state diagram.*

In order to check the possibility of reducing the number of states in a given state diagram, first form the state table. Then check for equivalent states in state table. Two states are said to be equivalent if the input-output relationship are same with identical present state and changes in input, which drive the present state to identical next state. When two states are equivalent, one of them can be removed without altering the input-output relationship.

In the state reduction problem first check for a pair of equivalent states in the state table and remove or eliminate one of the identical state and form a reduced state table. Then again check for a pair of equivalent states in the reduced state table and remove or eliminate one of the identical state and again form another reduced state table.

The above procedure is repeated until all the equivalent pair of states are identified and one of the state is removed in each pair. The original (or specified or given) state diagram and the reduced state diagram are verified using an input-output sequence. If both the original and reduced state diagram satisfy the input-output sequence then the reduced state diagram is equivalent to specified state diagram.

The advantage in reducing the states is that the sequential circuit can be realized with less number of flip-flops, so that the hardware cost will be reduced.

7.3.3 State Assignment

(AKTU, Nov/Dec'22, 2 Marks) (JNTUK, Feb'22, 2 Marks)

In the design specifications when the states are specified by alphabets it is necessary to assign binary codes for practical implementations. Therefore, an n-bit binary code is assigned to each state. Assigning binary codes to states represented by alphabets is called ***state assignment*** (or ***code assignment***).

With n-bits we can form 2^n binary codes.

When $n = 2$, $2^n = 2^2 = 4$

When $n = 3$, $2^n = 2^3 = 8$

When $n = 4$, $2^n = 2^4 = 16$ and so on.

Let, m = Number of state variables.

2^m = Number of states

If, $2^m \leq 4$, then select 2 bit code

If, $2^m \leq 8$, then select 3 bit code

If, $2^m \leq 16$, then select 4 bit code and so on.

There are different ways of assigning codes to states. Three important code assignment techniques are,

1. Binary code assignment
2. Gray code assignment
3. One hot code assignment

Binary Code Assignment

In binary code assignment, the codes are assigned in binary order.

Consider assigning three bit binary code to 5 states a, b, c, d and e. With 3-bit code we get 8 ($2^3 = 8$) binary codes. The first 5 codes are assigned to states a, b, c, d and e as shown below and the remaining 3 codes are unused or don't-care codes.

State		**Three bit binary code**			
a	⟶	0 0 0	Assigned codes	1 0 1	Don't-care codes
b	⟶	0 0 1		1 1 0	
c	⟶	0 1 0		1 1 1	
d	⟶	0 1 1			
e	⟶	1 0 0			

Gray Code Assignment

Consider assigning three bit gray code to 5 states a, b, c, d and e. With 3-bit gray code we get 8 ($2^3 = 8$) gray codes. The first 5 codes are assigned to states a, b, c, d and e as shown below and the remaining 3 codes are unused or don't-care codes.

State	Three bit gray code	
a ⟶	0 0 0	Assigned codes
b ⟶	0 0 1	
c ⟶	0 1 1	
d ⟶	0 1 0	
e ⟶	1 1 0	
	1 1 1	Don't-care codes
	1 0 1	
	1 0 0	

One Hot Code Assignment

The one-hot assignment will not satisfy 2^n logic. Here for n state assignments, an n-bit code is chosen. In n-bit code only one digit is 1 and remaining are zeros. The 1 is assigned starting from lower order bit.

The assignment of 5 bit one-hot code to 5 states a, b, c, d and e is shown below:

State	One-hot code
a ⟶	0 0 0 0 1
b ⟶	0 0 0 1 0
c ⟶	0 0 1 0 0
d ⟶	0 1 0 0 0
e ⟶	1 0 0 0 0

7.3.4 Lock-out Condition Circuit Implementation

In sequential circuits with some don't-care or unused states, the sequential circuit may find itself in some unused state during starting or switch ON. If next state of the unused state is again some unused one and if the sequential circuit cycle through only unused states and never arrives at in a used state then this condition is called ***lock-out***. An example of lock-out state diagram is shown in Fig. 7.6. In the state diagram of Fig. 7.6, the sequential circuit has to detect the sequence 1, 2, 4 and 7. Here, unused states are 0, 3, 5 and 6.

During switch ON if the sequential circuit enter an unused state and it cycles through only unused states as shown in Fig. 7.6(b) then the circuit will be in lock-out condition.

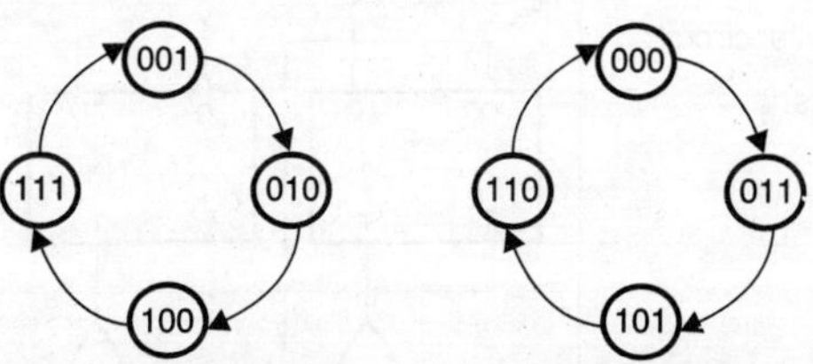

Fig. a: *Used states.* ***Fig. b:*** *Unused states.*

Fig. 7.6: *State diagram with lock-out condition.*

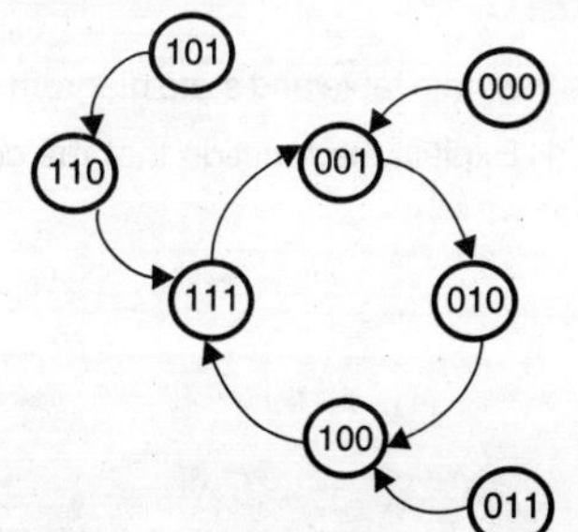

Fig. 7.7: *State diagram to prevent lock-out condition.*

In order to ensure that lock-out does not occur, the counter should be designed such that if the initial state is an unused state, then the state transition occurs from unused state to a used state. Therefore, every unused state is introduced in front of a used state. The example of Fig. 7.6, is modified as shown in Fig. 7.7, to prevent lock-out condition. The solution of this state diagram is discussed in Example 8.2 in Chapter-8. Alternatively, special circuit can be designed to reset the counter whenever it got struck into an unused state.

Example 7.1

(AU, Apr/May'19, 6 Marks)

Draw the state stable for the following state diagram.

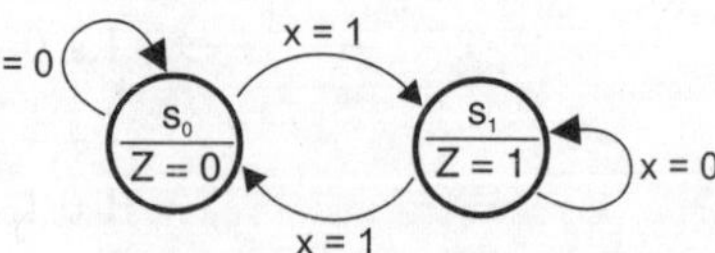

Fig. 1: State diagram.

Solution

From the knowledge of state transitions shown in Fig. 1, the state table is constructed as shown in Table 1.

Let us choose two state variables A and B and assign two bit binary code for state variables.

∴ Possible states, AB = 0 0

0 1

1 0

1 1

Let, S_0 = 00 and S_1 = 01

Here, 10 and 11 are unused states.

Table 1: State Table

Present States	Next States		Outputs	
A B	x = 0	x = 1	x = 0	x = 1
	A(t+1)	B(t+1)	A(t+1)	B(t+1)
S_0	S_0	S_1	0	0
S_1	S_1	S_0	1	1

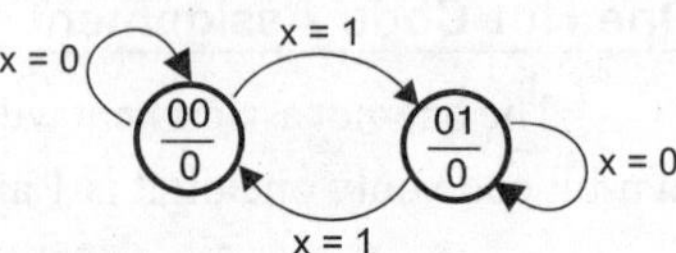

Fig. 2: State diagram with binary code assignment.

Table 2: State Table with Code Assignment

Present States		Input	Next States		Output
A	B	x	A(t+1)	B(t + 1)	Z
0	0	0	0	0	0
0	0	1	0	1	0
0	1	0	0	1	1
0	1	1	0	0	1
1	0	0	×	×	×
1	0	1	×	×	×
1	1	0	×	×	×
1	1	1	×	×	×

The state diagram with binary code assignment is shown in Fig. 2 and using this state diagram the state table is constructed as shown in Table 2.

Here, the output depends only on states and so it is a Moore model sequential circuit.

Example 7.2

Derive the state table and state diagram of the sequential circuit shown in Fig. 1. Explain the function that the circuit performs.

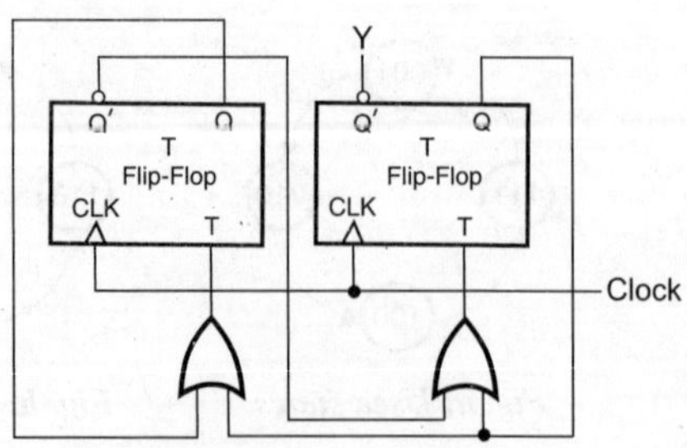

Fig. 1: Sequential circuit.

Solution

The sequential circuit is redrawn as shown in Fig. 2.

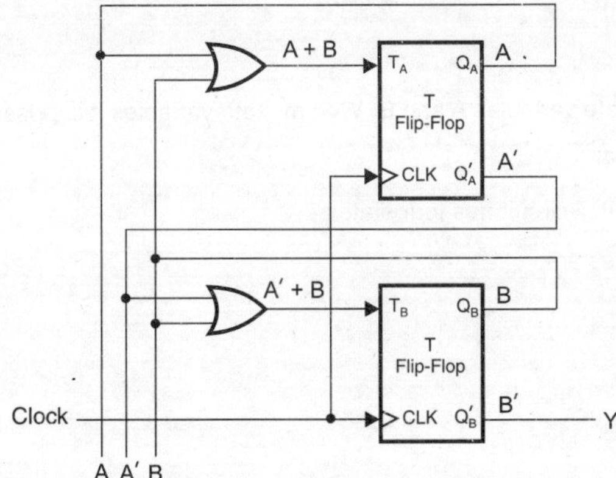

Fig. 2: Sequential circuit.

In the given sequential circuit, the output depends on only present states and so it is a Moore model.

The sequential circuit is constructed using two T flip-flops.

Since there are two flip-flops let us take two state variables A and B and equate the state variables to flip-flop output.

$\therefore\ A = Q_A \ \Rightarrow\ A(t+1) = Q_A(t+1)$

$B = Q_B \ \Rightarrow\ B(t+1) = Q_B(t+1)$

The characteristic equations of the two T flip-flops are,

$Q_A(t+1) = T_A A' + T_A' A$

$Q_B(t+1) = T_B B' + T_B' B$

Characteristic equation of T flip-flop, $Q(t+1) = TQ' + T'Q$

Now, the characteristic equations, in terms of state variables are,

$A(t+1) = T_A A' + T_A' A$

$B(t+1) = T_B B' + T_B' B$

From Fig. 2 we get the following Boolean equations for flip-flop inputs.

$T_A = A + B$

$T_B = A' + B$

On substituting the above Boolean equations in the characteristic equations we get the following state equations.

State equations: $A(t+1) = T_A A' + T_A' A$

$= (A+B)A' + (A+B)'A = AA' + A'B + A'B'A$ Using DeMorgan's law

$= 0 + A'B + 0 = A'B$

$B(t+1) = T_B B' + T_B' B$

$= (A'+B)B' + (A'+B)'B = A'B' + BB' + AB'B$

$= A'B' + 0 + 0 = A'B'$

Here, Y = Output

From Fig. 2 we get the output equation as,

Output equation : Y = B′

State Table

The circuit has two state variables A and B. With m state variables it is possible to have 2^m different states.

Here, m = 2, ∴ $2^2 = 4$

Therefore, the sequential circuit has four states.

The four possible states are,

AB = 0 0

= 0 1

= 1 0

= 1 1

The state table is listing of output and next states for 4 possible present states.

Table 1: State Table

Present States		Complement of Present States		Next States		Output
A	B	A′	B′	A(t + 1)	B(t + 1)	Y
0	0	1	1	0	1	1
0	1	1	0	1	0	0
1	0	0	1	0	0	1
1	1	0	0	0	0	0

A(t + 1) = A′B
B(t + 1) = A′B′
Y = B′

State Diagram

In the state diagram the states are represented by circles. The transition from one state to another is represented by directed lines (i.e, lines with arrow). The arrow indicates the transition from one state to another.

In this example, there are 4 possible states and so the state diagram has four circles with values of state variables inside the circle as shown below:

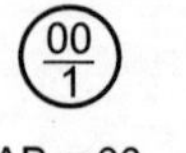

AB = 00 | 01/0 AB = 01 | 10/1 AB = 10 | 11/0 AB = 11

Fig. 3: State diagram.

The state diagram is drawn using state table as shown in Fig. 3.

Working of Circuit

From the state diagram we can say that the output is 1 when both states are reset to 0 or A is set when B is reset and in all other cases the output is 0. Therefore, the circuit detects a condition whether both states are reset or A alone is set after a reset.

Example 7.3

(AKTU, Dec'22, 10 Marks)

A sequential circuit has one flip-flop Q, two inputs x and y, and one output s. It consist of a full adder circuit connected to a D flip-flop, as shown in Fig. 1. Derive the state table and state diagram of the sequential circuit. What does this circuit represent?

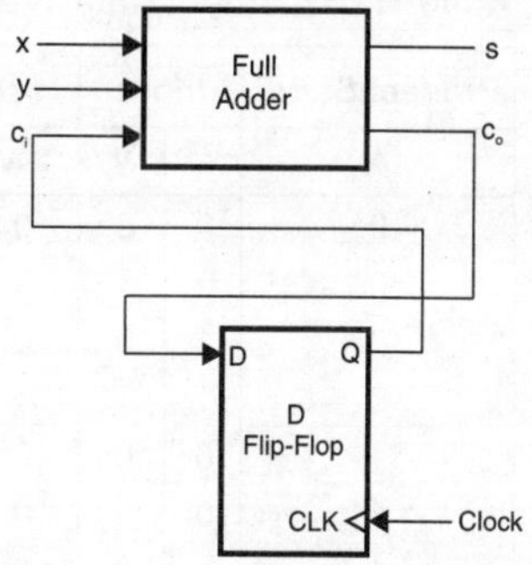

Fig. 1.

Solution

State Table

With n-bit input it is possible to have 2^n different inputs.

Here, n = 2, $\therefore 2^2 = 4$, Four possible inputs.

Four possible inputs: xy = 0 0

= 0 1

= 1 0

= 1 1

The circuit has one state variable A. With m state variables it is possible to have 2^m different states.

Here, m = 1, $\therefore 2^1 = 2$

Therefore, the sequential circuit has two states.

The two possible states are,

A = 0

= 1

When there are m state variables and n inputs then possible combinations of inputs and states are 2^{m+n}.

Here, m = 1, n = 2

$\therefore 2^{m+n} = 2^{1+2} = 2^3 = 8$

The 8 combinations of inputs and states are given below:

when, xy = 0 0, A = 0	when xy = 0 1, A = 0	when xy = 1 0, A = 0	when xy = 1 1, A = 0
= 1	= 1	= 1	= 1

The state table is listing of outputs and next states for 8 combinations of inputs and present states.

Here, present state variable : A

next state variable : A(t + 1)

The characteristic equation is, $Q(t + 1) = D \Rightarrow A(t + 1) = D$

Since, $D = c_o$, $A(t + 1) = c_o$

From the Boolean equation of full adder we can write,

$$A(t + 1) = c_o = xy + xc_i + yc_i$$

$c_i = A$

Boolean equations of full adder, $s = x \oplus y \oplus c_i$

$c_o = xy + xc_i + yc_i$

$$\therefore A(t + 1) = xy + xA + yA$$

The next states are computed using above equation and state table is constructed as shown in Table 1. Here, the flip-flop input is same as next state output.

Table 1: Expanded State Table

Present State	Input		Product Terms			Next State	Flip-Flop Input	Output
A	x	y	x′y	x′A	yA	A(t+1)	D	s
0	0	0	0	0	0	0	0	0
0	0	1	0	0	0	0	0	1
0	1	0	0	0	0	0	0	1
0	1	1	1	0	0	1	1	0
1	0	0	0	0	0	0	0	1
1	0	1	0	0	1	1	1	0
1	1	0	0	1	0	1	1	0
1	1	1	1	1	1	1	1	1

$A(t+1) = x'y + x'A + yA$

$D = A(t+1)$

$s = x \oplus y \oplus A$

In the state table, the output depends on input and states and so it is a Melay model.

State Diagram

In this example, there is one state variable and so the state diagram has two states represented by circles with values of state variables inside the circle as shown below:

(0) A = 0 | (1) A = 1

The possible state transitions are shown in Figs. 2 and 3.

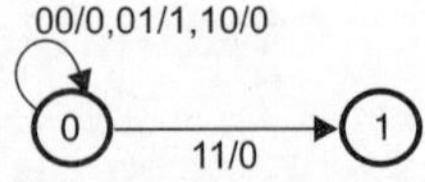

Fig. 2: Transition from state 0.

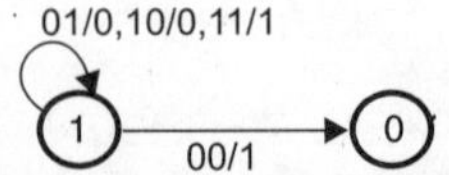

Fig. 3: Transition from state 1.

On combining the above state transitions into one single diagram, we get the state diagram as shown in Fig. 4.

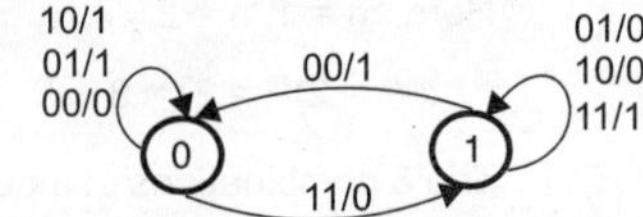

Fig. 4: State diagram.

Conclusion: The circuit of Fig. 1 work as serial adder. Refer Section 8.6.6 in Chapter 8.

Example 7.4

Construct the state table and state diagram for the sequential circuit using D flip-flops shown in Fig. 1.

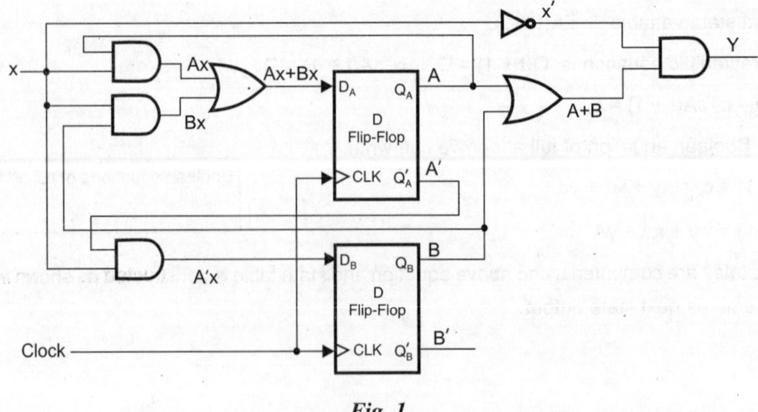

Fig. 1.

Solution

In the given sequential circuit, the output depends on input and states and so it is a Mealy model.

The sequential circuit is constructed using two D flip-flops.

Since there are two flip-flops let us take two state variables A and B and equate the state variables to flip-flop output.

$\therefore\ A = Q_A \ \Rightarrow\ A(t+1) = Q_A(t+1)$

$B = Q_B \ \Rightarrow\ B(t+1) = Q_B(t+1)$

The characteristic equations of the two D flip-flops are,

Characteristic equation of D flip-flop, $Q(t+1) = D$

$Q_A(t+1) = D_A$

$Q_B(t+1) = D_B$

Now, the characteristic equations, in terms of state variables are,

$A(t+1) = D_A$

$B(t+1) = D_B$

From Fig. 1 we get the following Boolean equations for flip-flop inputs.

$D_A = Ax + Bx$

$D_B = A'x$

On substituting the above Boolean equations in the characteristic equations we get the following state equations.

State equations: $A(t+1) = Ax + Bx$

$B(t+1) = A'x$

Here, Y = Output

From Fig. 1 we get the output equation as,

Output equation : $Y = (A + B)\,x'$

State Table

With n-bit input it is possible to have 2^n different inputs.

Here, n = 1, $\therefore\ 2^1 = 2$, Two possible inputs.

Two possible inputs: x = 0

x = 1

The circuit has two state variables A and B. With m state variables it is possible to have 2^m different states.

Here, m = 2, $\therefore\ 2^2 = 4$

Therefore, the sequential circuit has four states.

The four possible states are,

AB = 0 0

= 0 1

= 1 0

= 1 1

Here, when x = 0, AB can take four possible states.

Similarly, when x = 1, AB can take four possible states.

i.e, when, x = 0, AB = 0 0	when x = 1, AB = 0 0
= 0 1	= 0 1
= 1 0	= 1 0
= 1 1	= 1 1

When there are m state variables and n inputs then possible combinations of inputs and states are 2^{m+n}.

Here, m = 2, n = 1

$\therefore 2^{m+n} = 2^{2+1} = 2^3 = 8$

The evaluation of next state and output are shown in Table 1.

The state table is listing of outputs and next states for 8 combinations of inputs and present states as shown in Table 2 which can be represented in the simplified form as shown in Table 3.

Table 1: Evaluation of Next State and Output

Present States			Input		Product and Sum Terms				Next States		Output
A	B	A′	x	x′	Ax	Bx	A′x	A + B	A(t + 1)	B(t + 1)	Y
0	0	1	0	1	0	0	0	0	0	0	0
0	0	1	1	0	0	0	1	0	0	1	0
0	1	1	0	1	0	0	0	1	0	0	1
0	1	1	1	0	0	1	1	1	1	1	0
1	0	0	0	1	0	0	0	1	0	0	1
1	0	0	1	0	1	0	0	1	1	0	0
1	1	0	0	1	0	0	0	1	0	0	1
1	1	0	1	0	1	1	0	1	1	0	0

A(t + 1) = A x + B x
B(t + 1) = A′x
Y = (A + B) x′

Table 2: State Table

Present States		Input	Next States		Output
A	B	x	A(t+1)	B (t+1)	Y
0	0	0	0	0	0
0	0	1	0	1	0
0	1	0	0	0	1
0	1	1	1	1	0
1	0	0	0	0	1
1	0	1	1	0	0
1	1	0	0	0	1
1	1	1	1	0	0

Table 3: Simplified form of State Table

Present States		Next States x = 0		Next States x = 1		Output x = 0	Output x = 1
A	B	A(t+1)	B(t+1)	A(t+1)	B(t+1)	Y	Y
0	0	0	0	0	1	0	0
0	1	0	0	1	1	1	0
1	0	0	0	1	0	1	0
1	1	0	0	1	0	1	0

State Diagram

In the state diagram the states are represented by circles. The transition from one state to another is represented by directed lines (i.e, lines with arrow). The arrow indicates the transition from one state to another.

Every transition has associated inputs and outputs, those inputs and outputs are marked as labels on the directed lines. The format of label is "input/output".

In this example, there are 4 possible states and so the state diagram has four circles with values of state variables inside the circle as shown below:

(00)	(01)	(10)	(11)
AB = 00	AB = 01	AB = 10	AB = 11

In this example, input = x

output = Y

Therefore, format of label on the directed lines is x/Y.

In this example there is a possibility of two transitions from one state to another for x = 0 and x = 1.

The possible transitions are shown in Fig. 2.

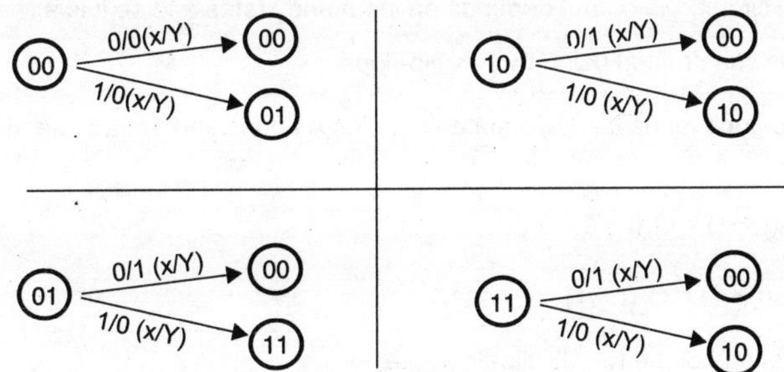

***Fig. 2:** Transitions from one state to another state.*

The possible transitions shown in Fig. 2 has four states and 8 directed lines. On combining all the state transitions into a single diagram, we get the state diagram as shown in Fig. 3.

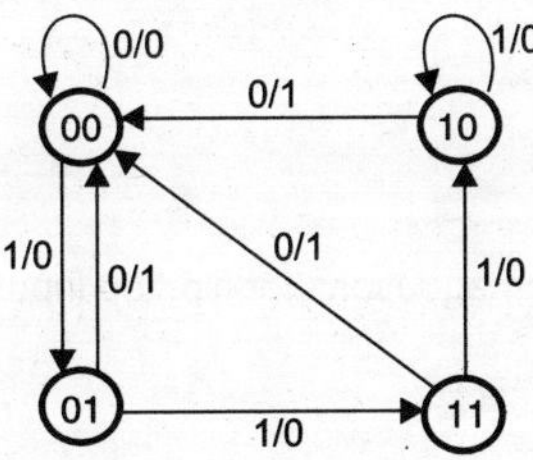

***Fig. 3:** State diagram.*

Example 7.5

Develop the state table and diagram for the sequential circuit using JK flip-flop shown in Fig. 1.

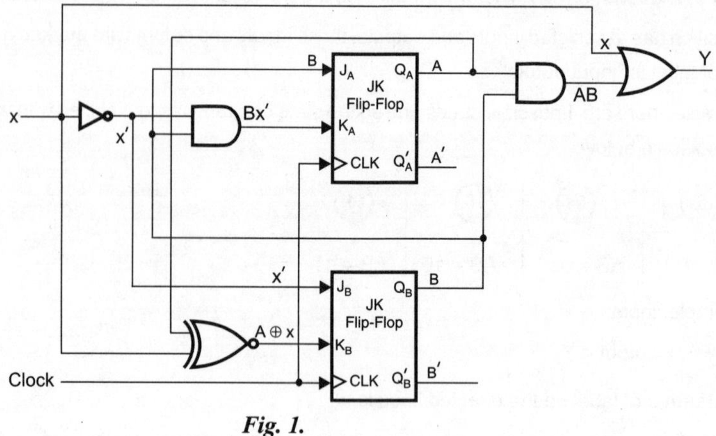

Fig. 1.

Solution

In the given sequential circuit, the output depends on input and states and so it is a Mealy model.

The sequential circuit is constructed using two JK flip-flops.

Since there are two flip-flops let us take two state variables A and B and equate the state variables to flip-flop output.

$\therefore\ A = Q_A \quad \Rightarrow \quad A(t+1) = Q_A(t+1)$

$B = Q_B \quad \Rightarrow \quad B(t+1) = Q_B(t+1)$

The characteristic equations of the two JK flip-flops are,

$$Q_A(t+1) = J_A\,Q_A' + K_A'\,Q_A$$

$$Q_B(t+1) = J_B\,Q_B' + K_B'\,Q_B$$

Characteristic equation of JK flip-flop, $Q(t+1) = J\,Q' + K'Q$

Now, the characteristic equations of JK flip-flops, in terms of state variables are,

$$A(t+1) = J_A\,A' + K_A'\,A$$

$$B(t+1) = J_B\,B' + K_B'\,B$$

From Fig. 1, we get the following Boolean equations for flip-flop inputs.

$J_A = B$

$K_A = B\,x'$

$J_B = x'$

$K_B = A \oplus x$

On substituting the above Boolean equations in characteristic equations we get the following state equations.

$A(t + 1) = J_A A' + K_A' A$

$= B A' + (B x')' A$ — Using DeMorgan's law

$= B A' + (B' + x) A$

$= A'B + AB' + Ax$

$B(t + 1) = J_B B' + K_B' B$

$= x'B' + (A \oplus x)'B$ — $(a \oplus b)' = a \odot b = ab + a'b'$

$= x'B' + (Ax + A'x')B$

$= B'x' + ABx + A'Bx'$

$\therefore$ State equations: $A(t + 1) = A'B + AB' + Ax$

$B(t + 1) = B'x' + ABx + A'Bx'$

From Fig. 1, we get the output equation as,

Output equation: $Y = AB + x$

State Table

Here we have 1 input and 2 state variables.

$\therefore \quad n = 1, m = 2$

Number of inputs $= 2^n = 2^1 = 2$ (x = 0 and x = 1)

Number of states $= 2^m = 2^2 = 4$

Combinations of inputs and states $= 2^{n+m} = 2^{1+2} = 2^3 = 8$

The 8 combinations of inputs and states are,

x = 0, AB = 0 0

= 0 1

= 1 0

= 1 1

x = 1, AB = 0 0

= 0 1

= 1 0

= 1 1

The next states are decided by the state equations and the output is determined by the output equation. The next states and output are evaluated for all possible combinations of inputs and states and listed in Table 1 and the state table is shown in Table 2.

Table 1: Evaluation of Next State and Output

Present States				Input		Product Terms							Next States		Output
A	B	A′	B′	x	x′	A′B	AB′	Ax	B′x′	AB x	A′B x′	A B	A(t+1)	B(t+1)	Y
0	0	1	1	0	1	0	0	0	1	0	0	0	0	1	0
0	0	1	1	1	0	0	0	0	0	0	0	0	0	0	1
0	1	1	0	0	1	1	0	0	0	0	1	0	1	1	0
0	1	1	0	1	0	1	0	0	0	0	0	0	1	0	1
1	0	0	1	0	1	0	1	0	1	0	0	0	1	1	0
1	0	0	1	1	0	0	1	1	0	0	0	0	1	0	1
1	1	0	0	0	1	0	0	0	0	0	0	1	0	0	1
1	1	0	0	1	0	0	0	1	0	1	0	1	1	1	1

$A(t+1) = A'B + AB' + Ax$; $B(t+1) = B'x' + ABx + A'Bx'$; $Y = AB + x$

Table 2: State Table

Present States		Input	Next States		Output
A	B	x	A(t+1)	B(t+1)	Y
0	0	0	0	1	0
0	0	1	0	0	1
0	1	0	1	1	0
0	1	1	1	0	1
1	0	0	1	1	0
1	0	1	1	0	1
1	1	0	0	0	1
1	1	1	1	1	1

State Diagram

This example has 4 states. Therefore, the state diagram has four circles with values of state variables inside the circle as shown below:

(00)	(01)	(10)	(11)
AB = 00	AB = 01	AB = 10	AB = 11

In this example, input = x

output = Y

Therefore, format of label on the directed lines is x/Y.

There is a possibility of two transitions from one state to another for x = 0 and x = 1. The possible transitions are shown in Fig. 2.

The possible transitions shown in Fig. 2 has four states and 8 directed lines. On combining all the transitions into one single diagram we get the state diagram as shown in Fig. 3.

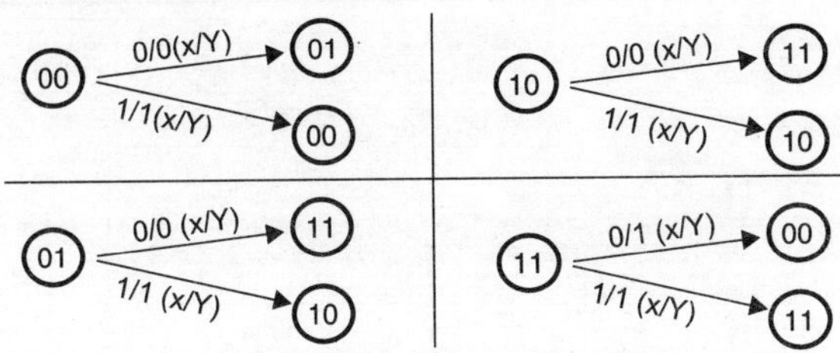

***Fig. 2:** Transitions from one state to another state.*

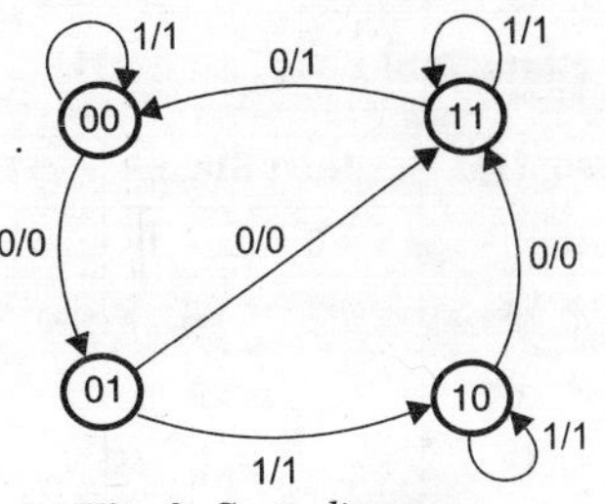

***Fig. 3:** State diagram.*

Alternate Method for State Equations

K-maps are constructed for A(t+1), B(t+1) and Y using the values listed in Table 1. From the K-map we get the following state equations.

$$A(t+1) = A'B + AB' + Ax$$

$$B(t+1) = B'x' + ABx + A'Bx'$$

$$Y = AB + x$$

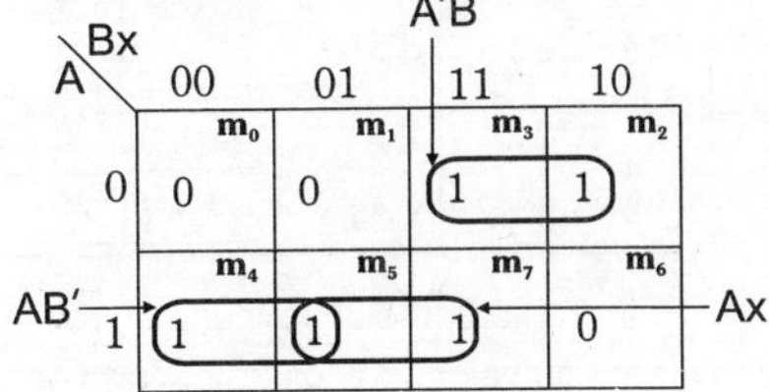

***Fig. 4:** K-map for A(t+1).*

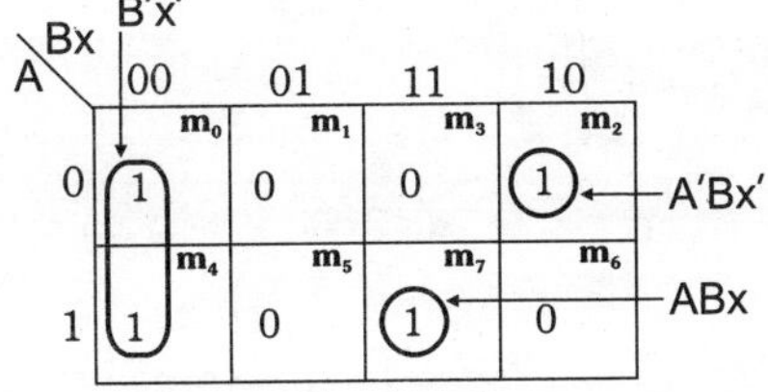

***Fig. 5:** K-map for B(t+1).*

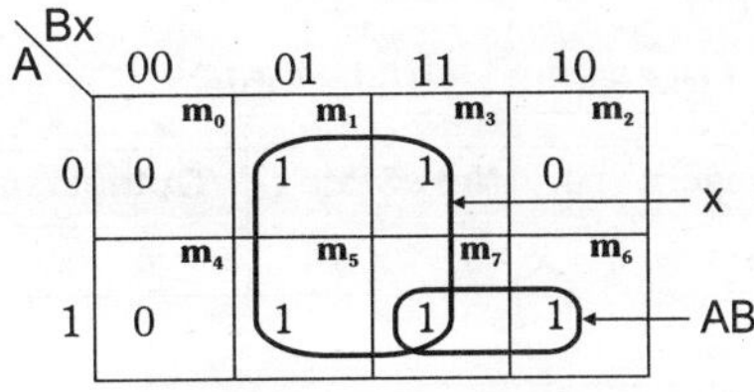

***Fig. 6:** K-map for Y.*

Example 7.6

Reduce the states in the state diagram shown in Fig. 1 and draw the reduced state diagram. Verify the state diagram with the following input sequence. Also assign binary code to states in reduced state diagram.

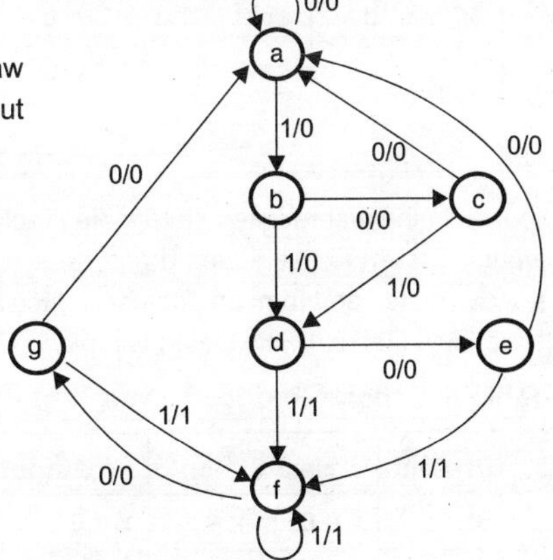

***Fig 1:** State diagram.*

Table 1: Input Sequence for Verification

	Column Number											
	1	2	3	4	5	6	7	8	9	10	11	
State	a	a	b	c	d	e	f	f	g	f	g	a
Input	0	1	0	1	0	1	1	0	1	0	0	
Output	0	0	0	0	0	1	1	0	1	0	0	

Solution

In the given state diagram the following state transitions are identified.

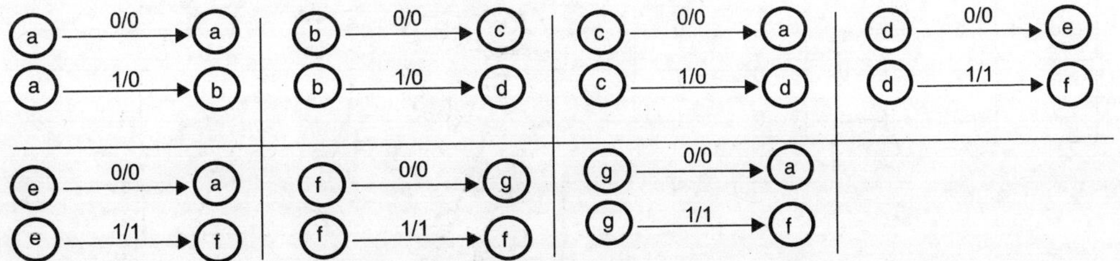

Using the above transitions, the state table is constructed as shown in Table 2.

Table 2: State Table

Present State	Next State		Output	
	x = 0	x = 1	x = 0	x = 1
a	a	b	0	0
b	c	d	0	0
c	a	d	0	0
d	e	f	0	1
5th row, e	a	f	0	1
f	g	f	0	1
7th row, g	a	f	0	1

Equivalent pair

In Table 2, the 5th row and 7th row has identical next states and outputs for same input, Therefore, the states e and g are equivalent pair of states, and so one of the state can be removed. Let us remove state g and form a reduced state table with last row eliminated. Also in the reduced state table, the state g is replaced by e in all entries in the retained rows.

Table 3: First Reduced State Table

Present State	Next State		Output	
	x = 0	x = 1	x = 0	x = 1
a	a	b	0	0
b	c	d	0	0
c	a	d	0	0
4th row, d	e	f	0	1
e	a	f	0	1
6th row, f	g → e	f	0	1

Equivalent pair

In the first reduced state table (Table 3) the 4th row and 6th row has identical next states and outputs for same inputs. Therefore, the states d and f are equivalent pair of states, and so one of the state can be removed. Let us remove state f and form an another reduced state table (Table 4), with last row eliminated. Also in the second reduced state table (Table 4), the state f is replaced by d in all entries in the retained rows.

Table 4: Second Reduced State Table

Present State	Next State		Output	
	x = 0	x = 1	x = 0	x = 1
a	a	b	0	0
b	c	d	0	0
c	a	d	0	0
d	e	f → d	0	1
e	a	f → d	0	1

The reduced state diagram is drawn using the second reduced state Table (Table 4) as shown below:

In the second reduced state diagram the following state transitions are identified.

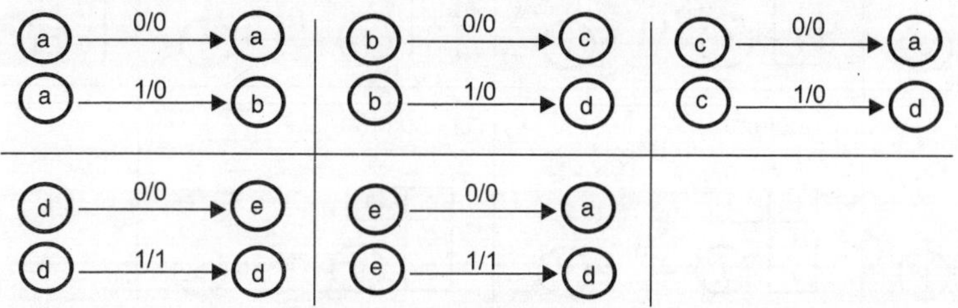

On combining the above state transitions the reduced state diagram is drawn as shown in Fig. 2.

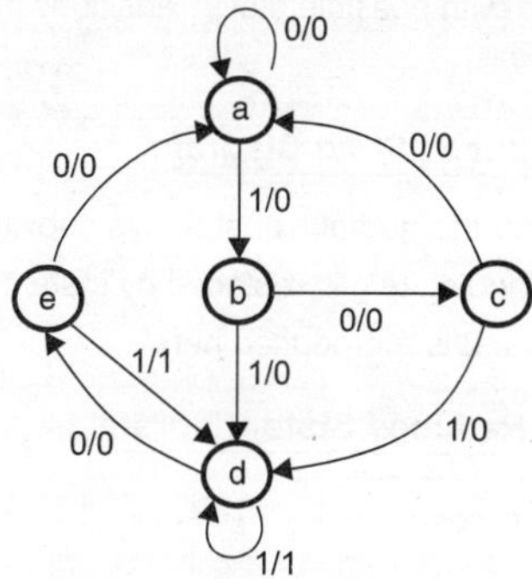

Fig. 2: Reduced state diagram.

Verification of Given Input Sequence in Original State Diagram

For convenience, the Table 1 and Fig. 1 are reproduced here. In Table 1, each column represent a state transition due to change in input.

Table 1: Input Sequence for Verification

	Column Number											
	1	2	3	4	5	6	7	8	7	10	11	
State	a	a	b	c	d	e	f	f	g	f	g	a
Input	0	1	0	1	0	1	1	0	1	0	0	
Output	0	0	0	0	0	1	1	0	1	0	0	

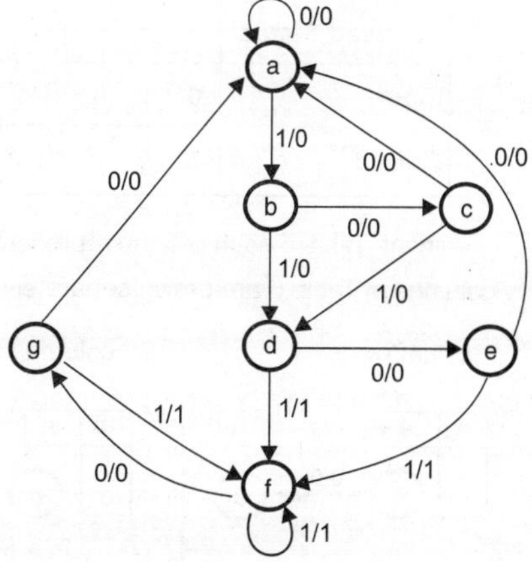

Fig 1: State diagram.

The state transitions represented by columns of Table 1 are drawn separately as shown below:

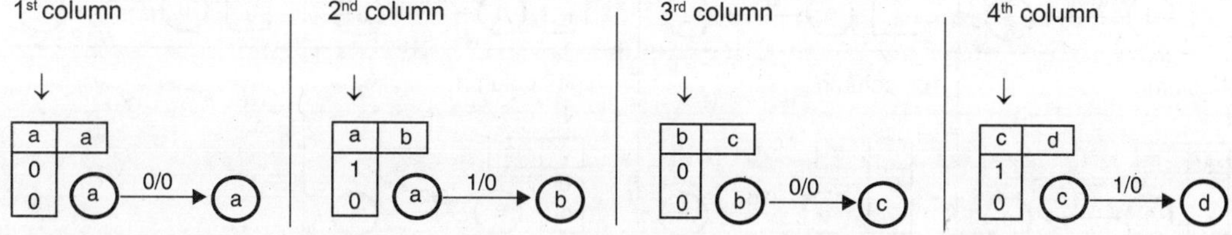

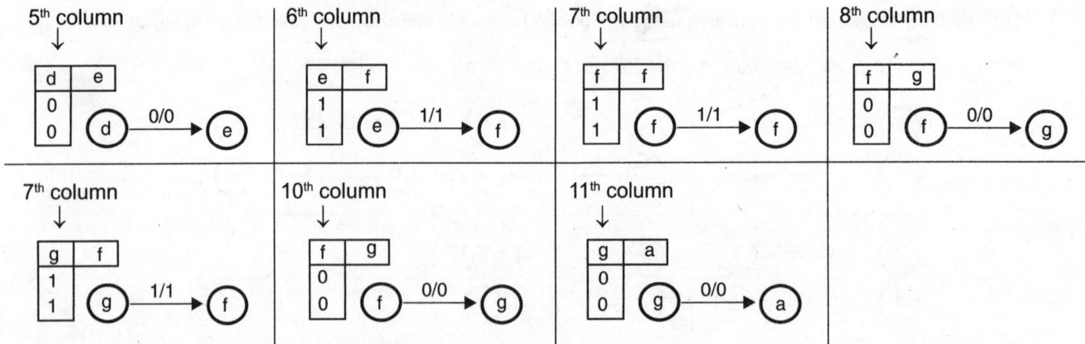

The above state transitions are compared with possible state transitions in Fig. 1 and verified that all the state transitions of Table 1 are valid transitions.

Verification of Given Input Sequence in Reduced State Diagram

The sequence to be verified after reducing the number of states is shown in Table 5. Here the state f is replaced by state d and state g is replaced by state e. For convenience the reduced state diagram of Fig. 2 is reproduced here.

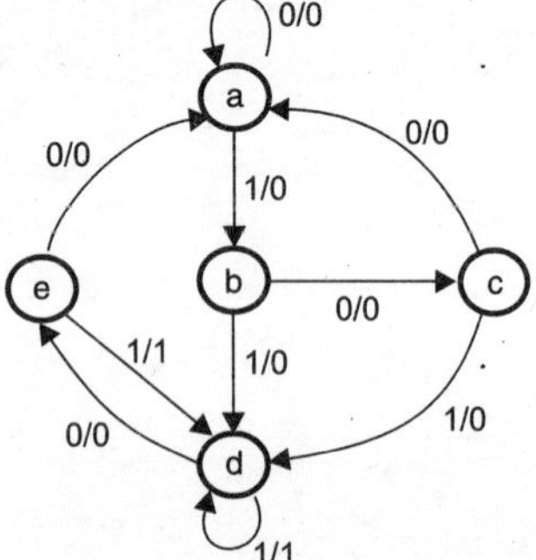

Fig. 2: Reduced state diagram.

Table 5: Input Sequence for Verification in Reduced State Diagram

States	a	a	b	c	d	e	f	f	g	f	g	a
							↓	↓	↓	↓	↓	
Reduced States	a	a	b	c	d	e	d	d	e	d	e	a
Input	0	1	0	1	0	1	1	0	1	0	0	
Output	0	0	0	0	0	1	1	0	1	0	0	

In the Table 5 each column represent a state transition due to change in input. The state transition represented by columns of Table 5 are drawn separately as shown below:

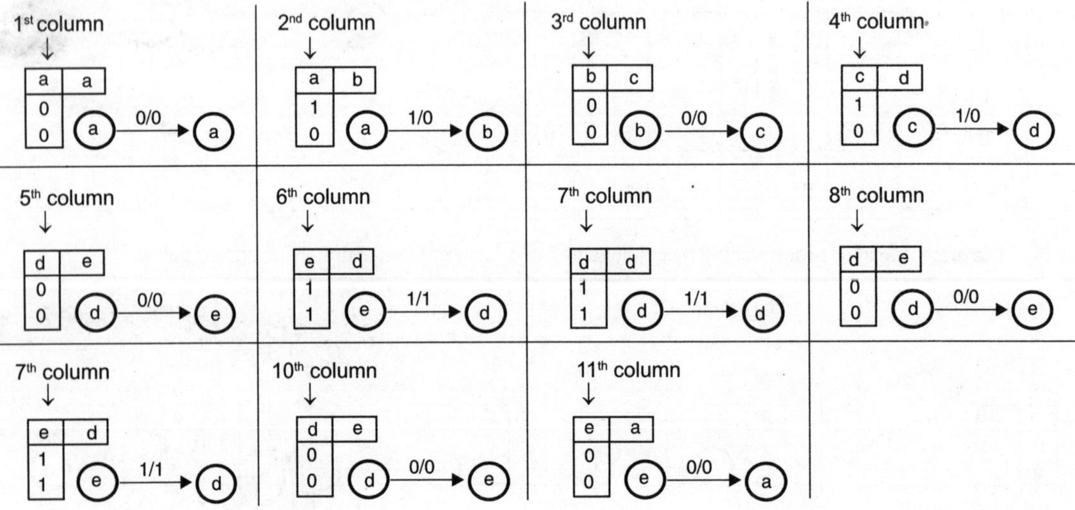

The above state transitions are compared with possible state transitions in Fig. 2 and verified that all the state transitions of Table 5 are valid transitions.

State Assignments

In this example the reduced state diagram has 5 states and so let us assign 3 bit code for states.

Binary State Assignments

The 3-bit binary codes are 000, 001, 010, 011, 100, 101, 110, 111. Let us assign first five codes to a, b, c, d and e. The remaining (or last three) are don't-cares.

State		Three bit binary code	
a	⟶	0 0 0	Assigned codes
b	⟶	0 0 1	
c	⟶	0 1 0	
d	⟶	0 1 1	
e	⟶	1 0 0	
		1 0 1	Don't-care codes
		1 1 0	
		1 1 1	

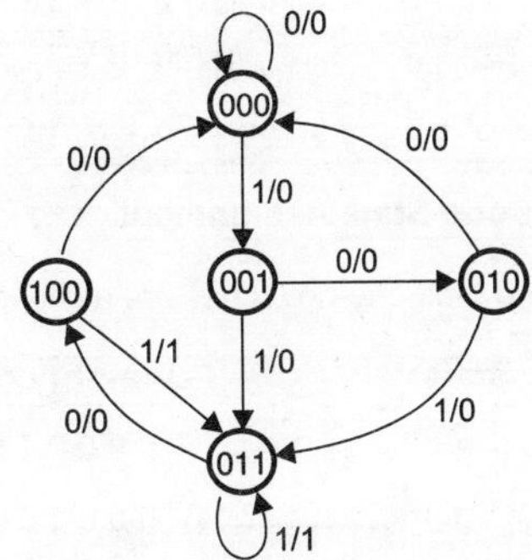

Fig. 3: State diagram with binary code assignment.

Table 6: State Table with Binary Code Assignment

Present State	Next State x = 0	Next State x = 1	Output x = 0	Output x = 1
0 0 0	0 0 0	0 0 1	0	0
0 0 1	0 1 0	0 1 1	0	0
0 1 0	0 0 0	0 1 1	0	0
0 1 1	1 0 0	0 1 1	0	1
1 0 0	0 0 0	0 1 1	0	1

Gray Code State Assignment

Here 3-bit gray codes are assigned to states. There are eight possible combinations of three bit gray code and in this first five codes are assigned to states a, b, c, d and e. The remaining are don't-cares.

State		Three bit gray code	
a	⟶	0 0 0	Assigned codes
b	⟶	0 0 1	
c	⟶	0 1 1	
d	⟶	0 1 0	
e	⟶	1 1 0	
		1 1 1	Don't-care codes
		1 0 1	
		1 0 0	

Table 7: State Table with Gray Code Assignment

Present	Next State		Output	
State	x = 0	x = 1	x = 0	x = 1
000	000	001	0	0
001	011	010	0	0
011	000	010	0	0
010	110	010	0	1
110	000	010	0	1

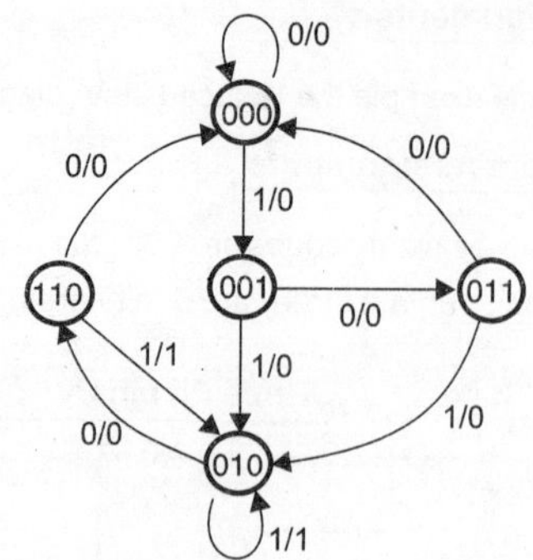

Fig. 4: State diagram with gray code assignement.

One-hot Code State Assignment

Here 5-bit one-hot codes are assigned to states.

State		**One-hot code**
a	⟶	00001
b	⟶	00010
c	⟶	00100
d	⟶	01000
e	⟶	10000

Table 8: StateTable with One-hot Code Assignement

Present	Next State		Output	
State	x = 0	x = 1	x = 0	x = 1
00001	00001	00010	0	0
00010	00100	01000	0	0
00100	00001	01000	0	0
01000	10000	01000	0	1
10000	00001	01000	0	1

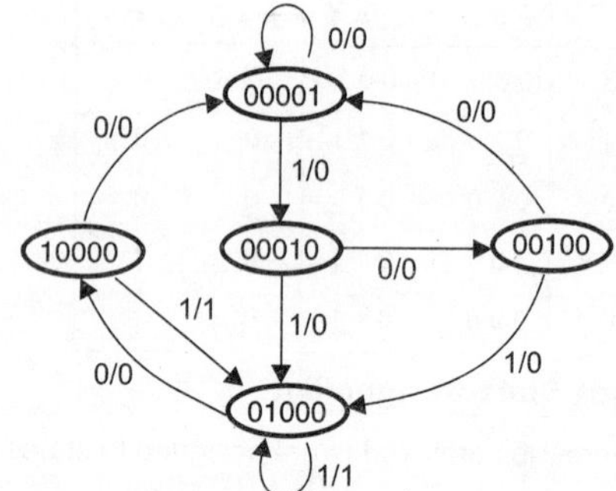

Fig. 5: State diagram with one-hot code assignment.

Example 7.7

The state diagram of a digital lock with password and two keys 0 and 1 is shown in Fig. 1.

a) Determine the password of the digital lock.

b) Modify the design by adding an additional security feature in digital lock.

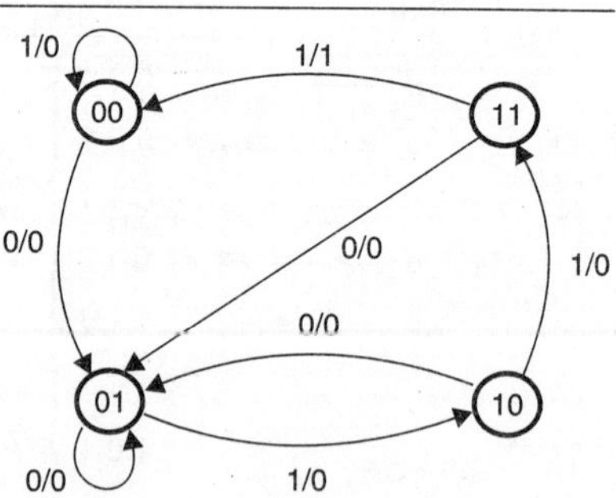

Fig. 1.

Solution

a) To determine the password

The given state diagram has 4 states, single input and output. The two keys in the lock is provided to input two possible values 0 and 1 of single input.

In the state diagram there are 8 transitions and the output is 1 only in one transition, when state changes from 11 to 00. Therefore, we have to start from 00 and come back to 00 with output 1.

This is possible only if we go through following state transitions,

$$00 \xrightarrow[0/0]{} 01 \xrightarrow[1/0]{} 10 \xrightarrow[1/0]{} 11 \xrightarrow[1/1]{} 00$$

The state transitions with input alone are shown below:

$$00 \xrightarrow[\text{Input 0}]{} 01 \xrightarrow[\text{Input 1}]{} 10 \xrightarrow[\text{Input 1}]{} 11 \xrightarrow[\text{Input 1}]{} 00$$

The sequence of inputs in the state transitions is the password.

$\therefore$ Password = 0111

b) Add New Security Feature

New security feature can be added by introducing an additional output. Additional or second output is defined such that output 1 in all wrong path and output 0 in all right path.

The state transitions with two outputs in the right path and wrong path are shown separately in Figs. 2 and 3 respectively. The state diagram with two output is shown in Fig. 4.

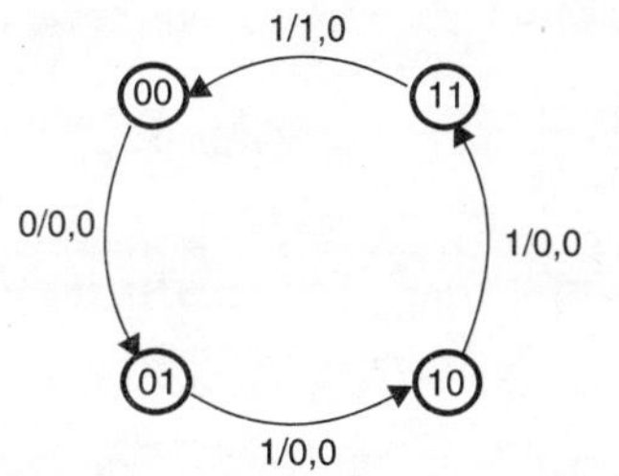

***Fig. 2:** State transitions in right path.*

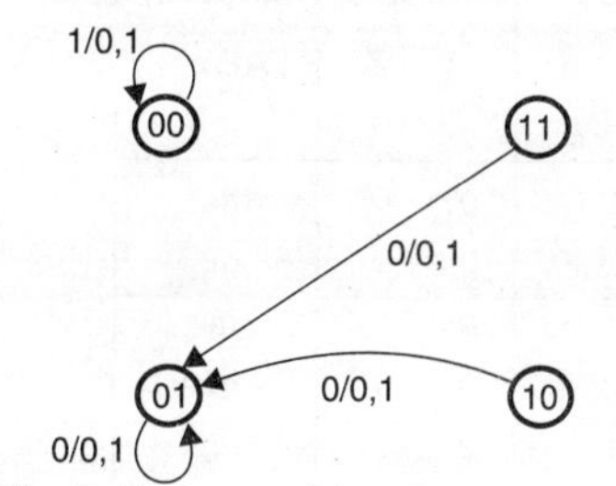

***Fig. 3:** State transitions in wrong path.*

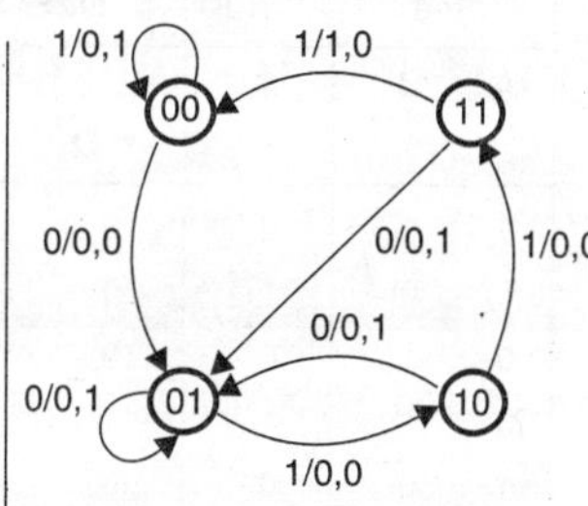

***Fig. 4:** State diagram with additional output.*

The four ($2^2 = 4$) states can be obtained from two state variables. Let A and B be the two state variables.

In sequential circuit, each state variable is stored in the output of a flip-flop. Hence, choose two D flip-flops to store two state variables A and B as shown in Fig. 6.

$\therefore$ Present state variables : A(t), B(t)

$\Downarrow$

A, B (For simplicity)

Next state variables : A(t + 1), B(t + 1)

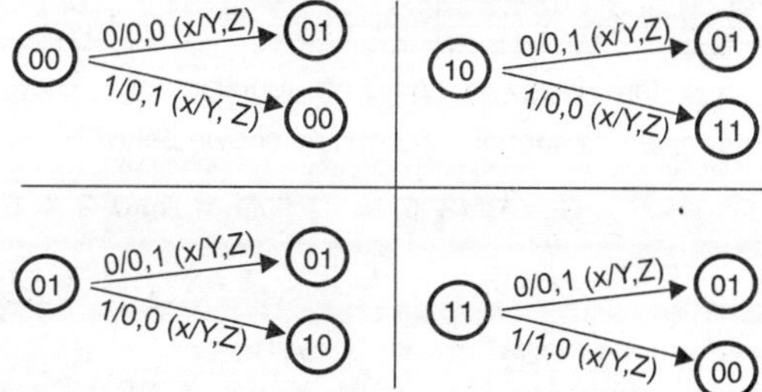

***Fig. 5:** Transitions from one state to another state.*

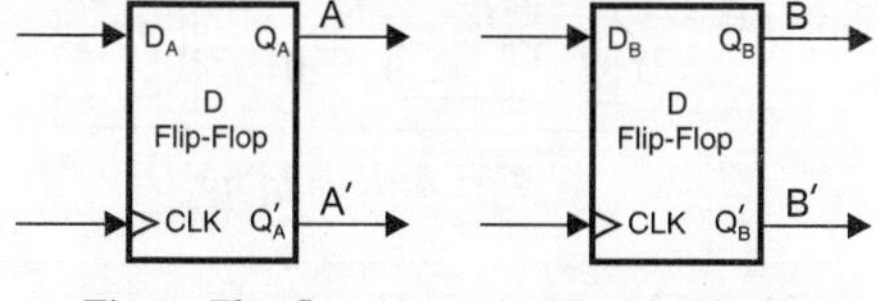

***Fig. a:** Flip-flop A.* ***Fig. b:** Flip-flop B.*

***Fig. 6:** D flip-flop to store state variables.*

The characteristic equations of two D flip-flops are,

$$Q_A(t + 1) = D_A$$

$$Q_B(t + 1) = D_B$$

Characteristic equation of D flip-flop.
Q(t + 1) = D

The two state variables can be equated to flip-flop outputs,

$\therefore\ A = Q_A \quad \Rightarrow \quad A(t+1) = Q_A(t+1)$

$B = Q_B \quad \Rightarrow \quad B(t+1) = Q_B(t+1)$

Now the characteristic equations of D flip-flop in terms of state variables are,

$A(t+1) = D_A \quad ; \quad B(t+1) = D_B$

From the above equations we can say that the D flip-flop input is same as next state output.

Let, x = Input

Y, Z = Outputs

Using the state transitions shown in Fig. 5 and the characteristic equations, the state table is constructed as shown in Table 1.

Table 1: State Table

Present States		Input	Next States		Outputs		Minterm
A	B	x	A(t + 1) = D_A	B(t + 1) = D_B	Y	Z	ABx
0	0	0	0	1	0	0	m_0
0	0	1	0	0	0	1	m_1
0	1	0	0	1	0	1	m_2
0	1	1	1	0	0	0	m_3
1	0	0	0	1	0	1	m_4
1	0	1	1	1	0	0	m_5
1	1	0	0	1	0	1	m_6
1	1	1	0	0	1	0	m_7

Here, D_A, D_B, Y and Z are functions of A, B, x. Using Table 1, the Boolean equations for D_A, D_B, Y and Z are determined as sum of minterms as shown below:

$\therefore\ D_A = \Sigma m(3, 5)$ | $D_B = \Sigma m(0, 2, 4, 5, 6)$

$Y = m_7$ | $Z = \Sigma m(1, 2, 4, 6)$

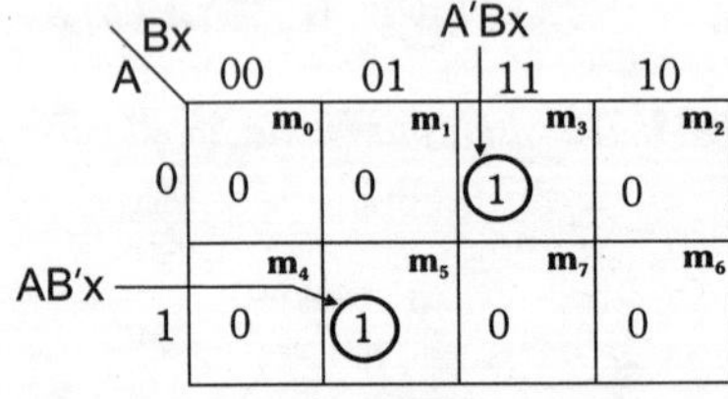

Fig. 7: K-map for D_A.

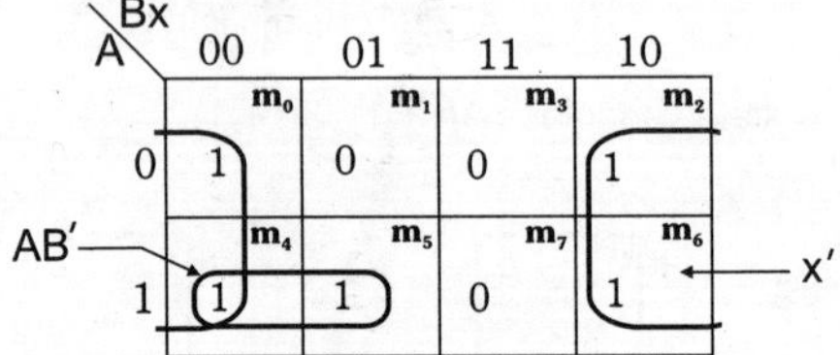

Fig. 8: K-map for D_B.

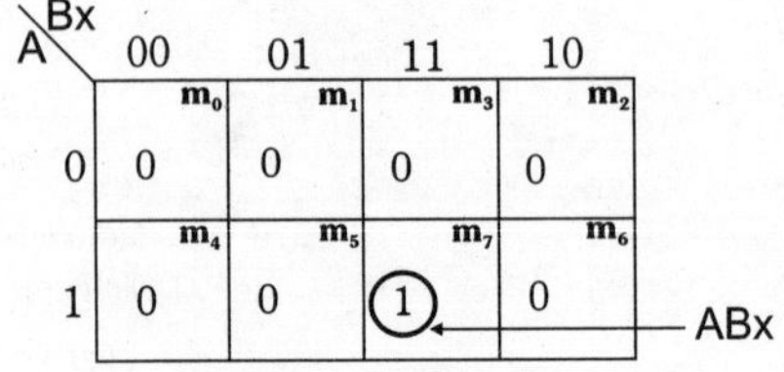

Fig. 9: K-map for Y.

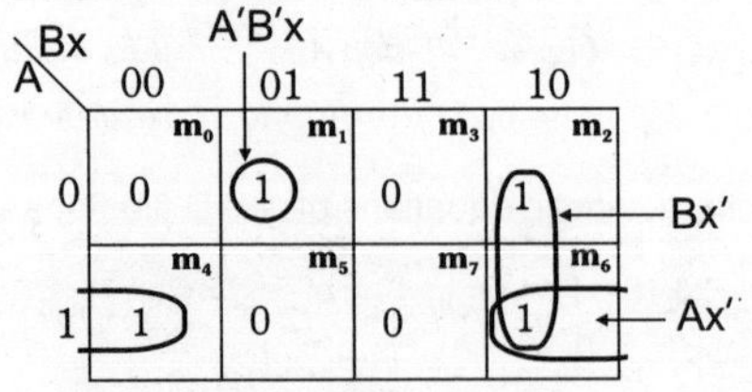

Fig. 10: K-map for Z.

Using the state table, the K-maps for D_A, D_B, Y and Z are constructed as shown in Figs. 7 to 10. From the K-map we get the following Boolean equations for D_A, D_B, Y and Z.

$D_A = A'Bx + AB'x$

$D_B = AB' + x'$

$Y = ABx$

$Z = A'B'x + Ax' + Bx'$

Using the Boolean equations for D_A, D_B, Y and Z the sequential logic circuit of digital lock with two outputs is drawn as shown in Fig. 11. The output depends on input and states and so this model is Mealy model.

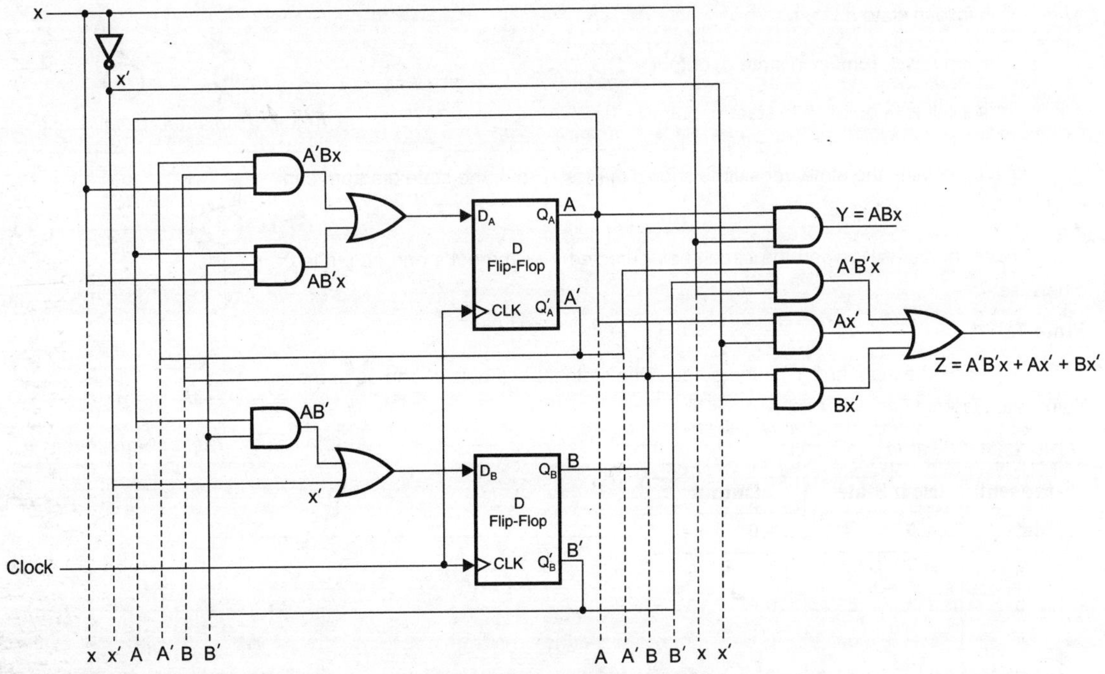

Fig. 11: Sequential circuit to implement digital lock with two outputs.

Example 7.8

Design a sequential circuit to detect a sequence of four or more consecutive 1's in a string of bits coming through an input line.

Case 1: Design using D flip-flop

Case 2: Design using JK flip-flop

Solution

The sequential circuit is designed to detect four or more consecutive 1's. Hence we can choose four states a, b, c and d. The possible state transitions from one state to other states are shown in Figs. 1 to 4.

1. While in state a

   if input = 1, go to next state b, output = 0

   if input = 0, remain in state a, output = 0

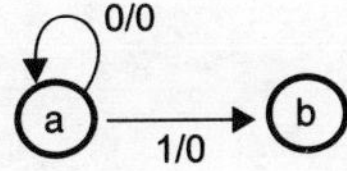

Fig. 1: Transitions from state a.

2. While in state b

   if input = 1, go to next state c output = 0

   if input = 0, go back to state a, output = 0

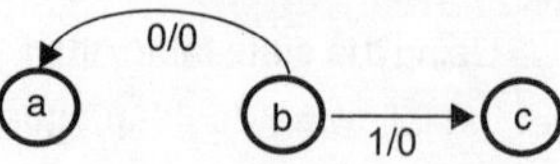

Fig. 2: Transitions from state b.

3. While in state c

   if input = 1, go to next state d, output = 0

   if input = 0, go back to state a, output = 0

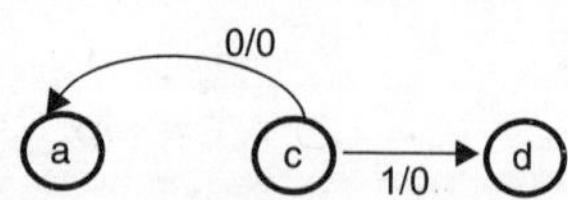

Fig. 3: Transitions from state c.

4. While in state d

   if input is 1, remain in state d, output = 1

   if input is 0, go back to state a, output = 0

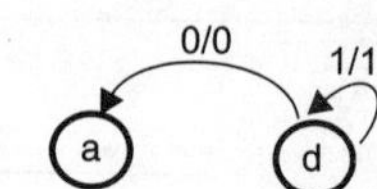

Fig. 4: Transitions from state d.

On combining the state transitions shown in Figs. 1 to 4, the state diagram can be drawn as shown in Fig. 5.

Here, output is 1, when the circuit has detected 4 or more 1's and output is 0 otherwise.

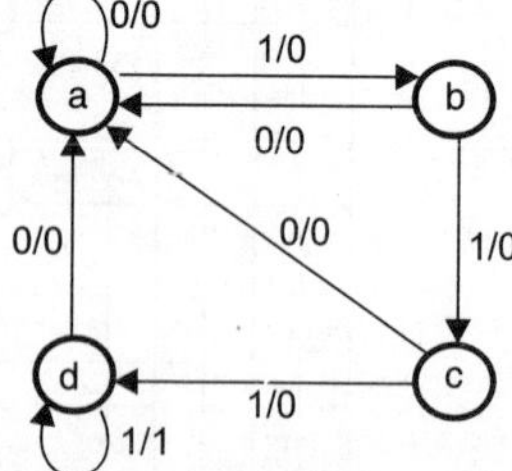

Fig. 5: State diagram.

State Table

Let, x be the input and Y be the output. The state table (Table 1) can be formed from state diagram.

Table 1: State Table

Present State	Next State		Output	
	x = 0	x = 1	x = 0	x = 1
a	a	b	0	0
b	a	c	0	0
c	a	d	0	0
d	a	d	0	1

There is no identical or equivalent pair of states in the state table with same output and same next state, and so there is no need for state reduction. The four ($2^2 = 4$) states can be formed with two state variables. Let A and B be the two state variables.

Since there are four states, let us choose 2-bit code for states. The state diagram is redrawn as shown in Fig. 6 and state table is modified as shown in Table 2 with following state assignment.

a ⟶ 0 0
b ⟶ 0 1
c ⟶ 1 0
d ⟶ 1 1

} Assigned codes

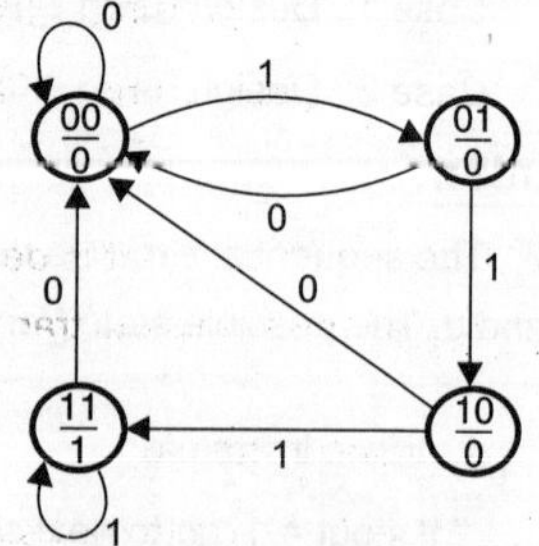

Fig. 6: State diagram with state assignment.

Table 2: State Table with Binary Code Assignment for States

Present States		Next States				Output	
		x = 0		x = 1		x = 0	x = 1
A	B	A(t+1)	B(t+1)	A(t+1)	B(t+1)	Y	Y
0	0	0	0	0	1	0	0
0	1	0	0	1	0	0	0
1	0	0	0	1	1	0	0
1	1	0	0	1	1	0	1

Case i: Design using D Flip-Flop

In the sequential circuit, each state variable is stored in the output of a flip-flop. Hence, choose two D flip-flops to store the two state variables A and B as shown in Fig. 7.

∴ Present state variables : A(t), B(t) ⇒ A, B (For simplicity)

Next state variables : A(t + 1), B(t + 1)

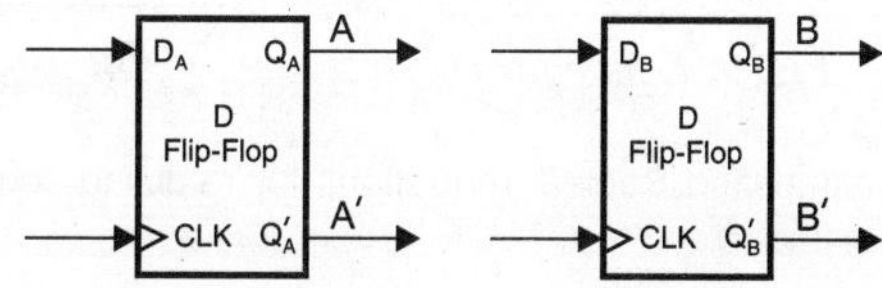

Fig. a: *Flip-flop A.* ***Fig. b:*** *Flip-flop B.*

Fig. 7: *D flip-flop to store state variables.*

The characteristic equation of two D flip-flops are,

> Characteristic equation of D flip-flop.
> $Q(t + 1) = D$

$Q_A(t + 1) = D_A$

$Q_B(t + 1) = D_B$

The two state variables can be equated to flip-flop outputs.

$\therefore A = Q_A \Rightarrow A(t + 1) = Q_A(t + 1)$

$B = Q_A \Rightarrow B(t + 1) = Q_B(t + 1)$

Now the characteristic equations in terms of state variables are,

$A(t + 1) = D_A$

$B(t + 1) = D_B$

Using the above equations the flip-flop inputs are determined from the next state and listed in Table 3. The D flip-flop the input is same as next state output.

Table 3: Expanded State Table

Present States		Input	Next States		Flip-Flop Inputs		Output	Minterm
A	B	x	A(t+1)	B(t+1)	D_A	D_B	Y	ABx
0	0	0	0	0	0	0	0	m_0
0	0	1	0	1	0	1	0	m_1
0	1	0	0	0	0	0	0	m_2
0	1	1	1	0	1	0	0	m_3
1	0	0	0	0	0	0	0	m_4
1	0	1	1	1	1	1	0	m_5
1	1	0	0	0	0	0	0	m_6
1	1	1	1	1	1	1	1	m_7

Here, D_A, D_B, Y are functions of A, B, x. Using Table 3, the Boolean equations for D_A, D_B and Y are determined as sum of minterms as shown below:

$$\therefore\ D_A = \Sigma m(3, 5, 7)$$

$$D_B = \Sigma m(1, 5, 7)$$

$$Y = m_7$$

Let us draw K-map for D_A, D_B and Y as shown in Figs. 8 to 10 to develop simplified Boolean equations for D_A, D_B and Y.

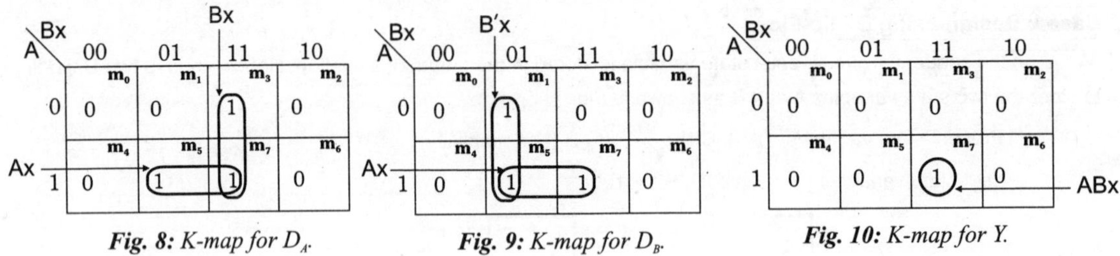

Fig. 8: K-map for D_A. *Fig. 9: K-map for D_B.* *Fig. 10: K-map for Y.*

From the K-map we get the following simplified Boolean equations for realizing sequence detector using D flip-flops.

$$D_A = Ax + B\,x$$

$$D_B = Ax + B'x$$

$$Y = ABx$$

The logic diagram of sequence detector using the above equations is drawn as shown in Fig. 11. Here, the output depends on input and states and so this model is Mealy model.

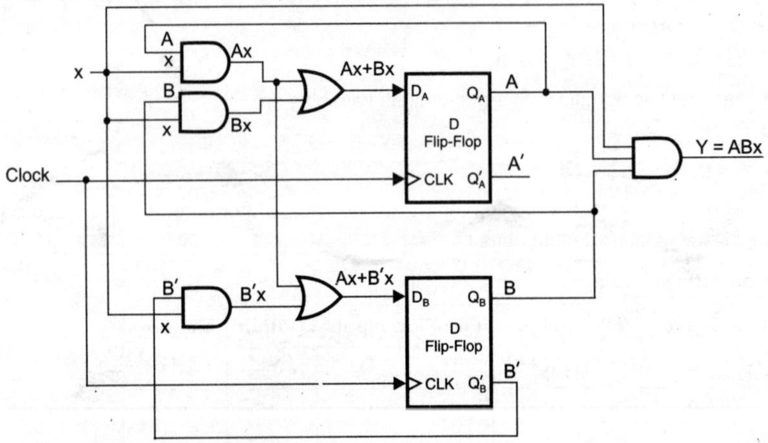

Fig. 11: Logic diagram of sequence detector using D flip-flop.

Case ii: Design using JK Flip-Flop

The state table and state assignment in the design using D flip-flop can be used for design using JK flip-flops. But the expanded state table require two columns for flip-flop inputs. Let us choose two JK flip-flops to store two state variables A and B as shown in Fig. 12. The state table is shown in Table 4.

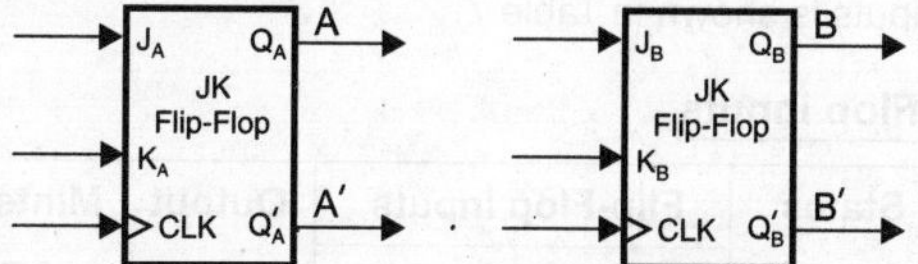

Fig. a: Flip-flop A. *Fig. a: Flip-flop B.*

Fig. 12: JK flip-flops to store state variables.

Table 4: State Table

Present States		Input	Next States	
A	B	x	A(t+1)	B(t + 1)
0	0	0	0	0
0	0	1	0	1
0	1	0	0	0
0	1	1	1	0
1	0	0	0	0
1	0	1	1	1
1	1	0	0	0
1	1	1	1	1

Excitation Table of JK Flip-Flop

Q(t)	Q(t + 1)	J	K
0	0	0	×
0	1	1	×
1	0	×	1
1	1	×	0

The inputs J_A, J_B, K_A, K_B of flip-flops can be determined using excitation table as shown below:

State A

For state A the Q(t), Q(t + 1), J and K of excitation table are considered as A, A(t + 1), J_A and K_A.

$$\therefore Q(t) \rightarrow A \quad ; \quad Q(t+1) \rightarrow A(t+1)$$

$$J \rightarrow J_A \quad ; \quad K \rightarrow K_A$$

The flip-flop inputs J_A and K_A that drives the present state A to next states A(t + 1) are obtained from excitation table and listed in Table 5.

Table 5:

A	A(t+1)	J_A	K_A
0	0	0	×
0	0	0	×
0	0	0	×
0	1	1	×
1	0	×	1
1	1	×	0
1	0	×	1
1	1	×	0

State B

For state B the Q(t), Q(t + 1), J and K of excitation table are considered as B, B(t + 1), J_B and K_B.

$$\therefore Q(t) \rightarrow B \quad ; \quad Q(t+1) \rightarrow B(t+1)$$

$$J \rightarrow J_B \quad ; \quad K \rightarrow K_B$$

The flip-flop inputs J_B and K_B that drives the present state B to next states B(t + 1) are obtained from excitation table and listed in Table 6.

Table 6:

B	B(t+1)	J_B	K_B
0	0	0	×
0	1	1	×
1	0	×	1
1	0	×	1
0	0	0	×
0	1	1	×
1	0	×	1
1	1	×	0

The state table with flip-flop inputs is shown in Table 7.

Table 7: State Table with JK Flip-Flop Inputs

Present States		Input	Next States		Flip-Flop Inputs				Output	Minterm
A	B	x	A(t+1)	B(t+1)	J_A	K_A	J_B	K_B	Y	ABx
0	0	0	0	0	0	×	0	×	0	m_0
0	0	1	0	1	0	×	1	×	0	m_1
0	1	0	0	0	0	×	×	1	0	m_2
0	1	1	1	0	1	×	×	1	0	m_3
1	0	0	0	0	×	1	0	×	0	m_4
1	0	1	1	1	×	0	1	×	0	m_5
1	1	0	0	0	×	1	×	1	0	m_6
1	1	1	1	1	×	0	×	0	1	m_7

Note: Output is same as in D flip-flop design.

The inputs J_A, K_A, J_B, K_B are functions of present state and input x. Therefore, minterms are same as in design using D flip-flops.

J_A, K_A, J_B, K_B are functions of A, B, x and using Table 7, J_A, K_A, J_B, K_B are expressed as sum of minterms and don't-cares as shown below:

$$J_A = m_3 + \Sigma d(4, 5, 6, 7)$$

$$K_A = \Sigma m(4, 6) + \Sigma d(0, 1, 2, 3)$$

$$J_B = \Sigma m(1, 5) + \Sigma d(2, 3, 6, 7)$$

$$K_B = \Sigma m(2, 3, 6) + \Sigma d(0, 1, 4, 5)$$

Let us draw K-map for J_A, K_A, J_B, K_B as shown in Figs. 13 to 16 to develop simplified equations for J_A, K_A, J_B, K_B. The equation for Y and K-map for Y are same as in design using D flip-flop.

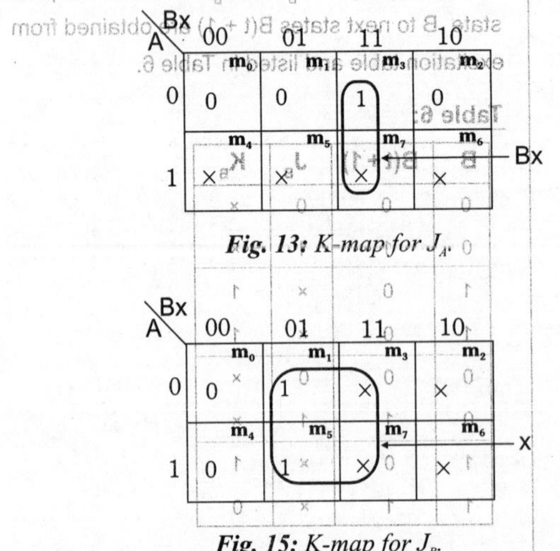

Fig. 13: K-map for J_A.

Fig. 15: K-map for J_B.

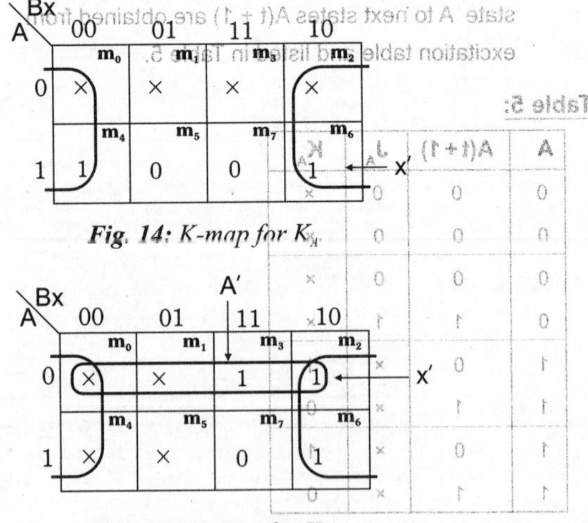

Fig. 14: K-map for K_A.

Fig. 16: K-map for K_B.

From the K-map we get the following simplified Boolean equations for realizing sequence detector using JK flip-flops.

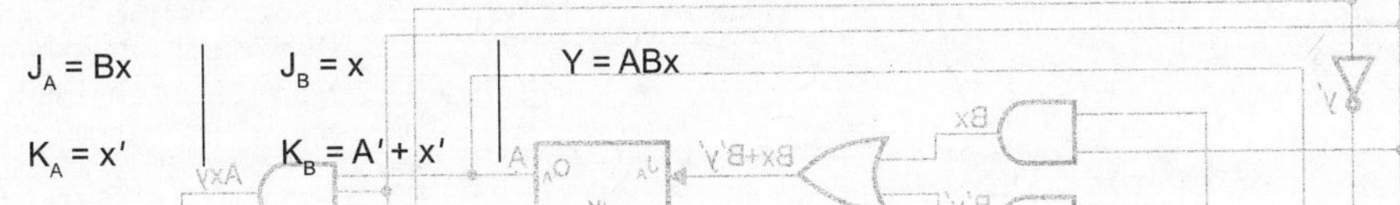

$J_A = Bx$	$J_B = x$	$Y = ABx$
$K_A = x'$	$K_B = A' + x'$	

The logic diagram of sequence detector using above equations is drawn as shown in Fig. 17. Here, the output depends on input and states and so this model is Mealy model.

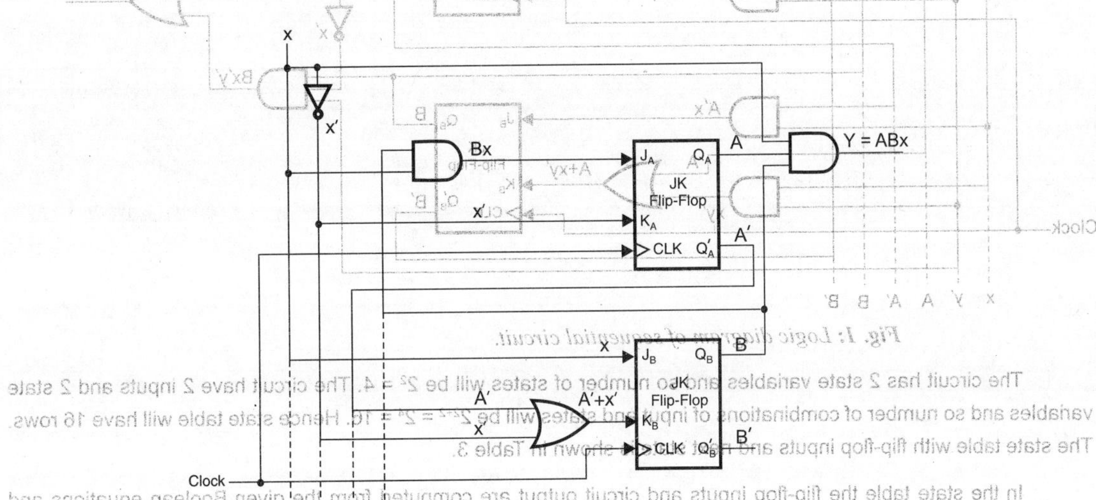

Fig. 17: Logic diagram of sequence detector using JK flip-flop.

Example 7.9

A sequential circuit has two JK flip-flops A and B, two inputs x and y and one output Z. The flip-flop input equations and circuit output equation are,

$$J_A = Bx + B'y'$$

$$K_A = B'xy'$$

$$J_B = A'x$$

$$K_B = A + xy'$$

$$Z = Axy + Bx'y'$$

Draw the logic diagram and state table of the circuit. Also derive the state equations for A and B.

Solution

The input and output equations are directly implemented using AND, OR and NOT gates as shown in Fig. 1. The sequential circuit has two flip-flops and so the circuit has two state variables A and B which are output of flip-flops. From the logic diagram of sequential circuit in Fig. 1, the output depends on inputs and states and so this model is Mealy model.

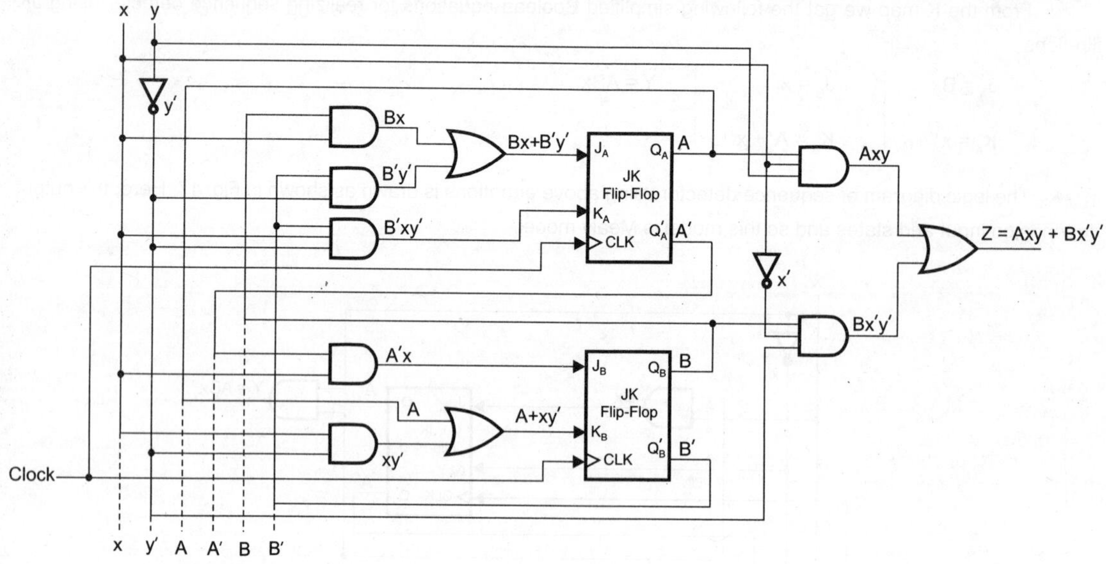

Fig. 1: Logic diagram of sequential circuit.

The circuit has 2 state variables and so number of states will be $2^2 = 4$. The circuit have 2 inputs and 2 state variables and so number of combinations of input and states will be $2^{2+2} = 2^4 = 16$. Hence state table will have 16 rows. The state table with flip-flop inputs and next state is shown in Table 3.

In the state table the flip-flop inputs and circuit output are computed from the given Boolean equations and listed in Table 1.

$J_A = Bx + B'y'$ | $K_A = B'xy'$

$J_B = A'x$ | $K_B = A + xy'$

$Z = Axy + Bx'y'$

Table 1: Flip-Flop Inputs and Circuit Output

Present States		Complement of Present States		Inputs		Complement of Inputs		Product Terms							Flip-Flop Inputs				Output
A	B	A'	B'	x	y	x'	y'	Bx	B'y'	B'xy'	A'x	xy'	Axy	Bx'y'	J_A	K_A	J_B	K_B	Z
0	0	1	1	0	0	1	1	0	1	0	0	0	0	0	1	0	0	0	0
0	0	1	1	0	1	1	0	0	0	0	0	0	0	0	0	0	0	0	0
0	0	1	1	1	0	0	1	0	1	1	1	1	0	0	1	1	1	1	0
0	0	1	1	1	1	0	0	0	0	0	1	0	0	0	0	0	1	0	0
0	1	1	0	0	0	1	1	0	0	0	0	0	0	1	0	0	0	0	1
0	1	1	0	0	1	1	0	0	0	0	0	0	0	0	0	0	0	0	0
0	1	1	0	1	0	0	1	1	0	0	1	1	0	0	1	0	1	1	0
0	1	1	0	1	1	0	0	1	0	0	1	0	0	0	1	0	1	0	0
1	0	0	1	0	0	1	1	0	1	0	0	0	0	0	1	0	0	1	0
1	0	0	1	0	1	1	0	0	0	0	0	0	0	0	0	0	0	1	0
1	0	0	1	1	0	0	1	0	1	1	0	1	0	0	1	1	0	1	0
1	0	0	1	1	1	0	0	0	0	0	0	0	1	0	0	0	0	1	1
1	1	0	0	0	0	1	1	0	0	0	0	0	0	1	0	0	0	1	1
1	1	0	0	0	1	1	0	0	0	0	0	0	0	0	0	0	0	1	0
1	1	0	0	1	0	0	1	1	0	0	0	1	0	0	1	0	0	1	0
1	1	0	0	1	1	0	0	1	0	0	0	0	1	0	1	0	0	1	1

The next states A(t + 1) and B(t + 1) are determined using the following JK flip-flop characteristics equations and listed in Table 2.

$$A(t+1) = J_A A' + K'_A A$$

$$B(t+1) = J_B B' + K'_B B$$

Characteristic equation of JK flip-flop.

$$Q(t+1) = JQ' + K'Q$$

Table 2: Next State Flip-Flop Outputs

Present States		Complement of Present States		Flip-Flop Inputs				Complement of K_A, K_B		Product Terms				Next States	
A	B	A′	B′	J_A	K_A	J_B	K_B	K'_A	K'_B	$J_A A'$	$K'_A A$	$J_B B'$	$K'_B B$	A(t + 1)	B(t + 1)
0	0	1	1	1	0	0	0	1	1	1	0	0	0	1	0
0	0	1	1	0	0	0	0	1	1	0	0	0	0	0	0
0	0	1	1	1	1	1	1	0	0	1	0	1	0	1	1
0	0	1	1	0	0	1	0	1	1	0	0	1	0	0	1
0	1	1	0	0	0	0	0	1	1	0	0	0	1	0	1
0	1	1	0	0	0	0	0	1	1	0	0	0	1	0	1
0	1	1	0	1	0	1	1	1	0	1	0	0	0	1	0
0	1	1	0	1	0	1	0	1	1	1	0	0	1	1	1
1	0	0	1	1	0	0	1	1	0	0	1	0	0	1	0
1	0	0	1	0	0	0	1	1	0	0	1	0	0	1	0
1	0	0	1	1	1	0	1	0	0	0	0	0	0	0	0
1	0	0	1	0	0	0	1	1	0	0	1	0	0	1	0
1	1	0	0	0	0	0	1	1	0	0	1	0	0	1	0
1	1	0	0	0	0	0	1	1	0	0	1	0	0	1	0
1	1	0	0	1	0	0	1	1	0	0	1	0	0	1	0
1	1	0	0	1	0	0	1	1	0	0	1	0	0	1	0

Table 3: State Table with JK Flip-Flop Inputs

Present States		Inputs		Next States		Flip-Flop Inputs				Output
A	B	x	y	A(t + 1)	B(t + 1)	J_A	K_A	J_B	K_B	Z
0	0	0	0	1	0	1	0	0	0	0
0	0	0	1	0	0	0	0	0	0	0
0	0	1	0	1	1	1	1	1	1	0
0	0	1	1	0	1	0	0	1	0	0
0	1	0	0	0	1	0	0	0	0	1
0	1	0	1	0	1	0	0	0	0	0
0	1	1	0	1	0	1	0	1	1	0
0	1	1	1	1	1	1	0	1	0	0
1	0	0	0	1	0	1	0	0	1	0
1	0	0	1	1	0	0	0	0	1	0
1	0	1	0	0	0	1	1	0	1	0
1	0	1	1	1	0	0	0	0	1	1
1	1	0	0	1	0	0	0	0	1	1
1	1	0	1	1	0	0	0	0	1	0
1	1	1	0	1	0	1	0	0	1	0
1	1	1	1	1	0	1	0	0	1	1

From the knowledge of inputs, outputs and present and next states obtained from state table, the state diagram is drawn as shown in Fig. 2

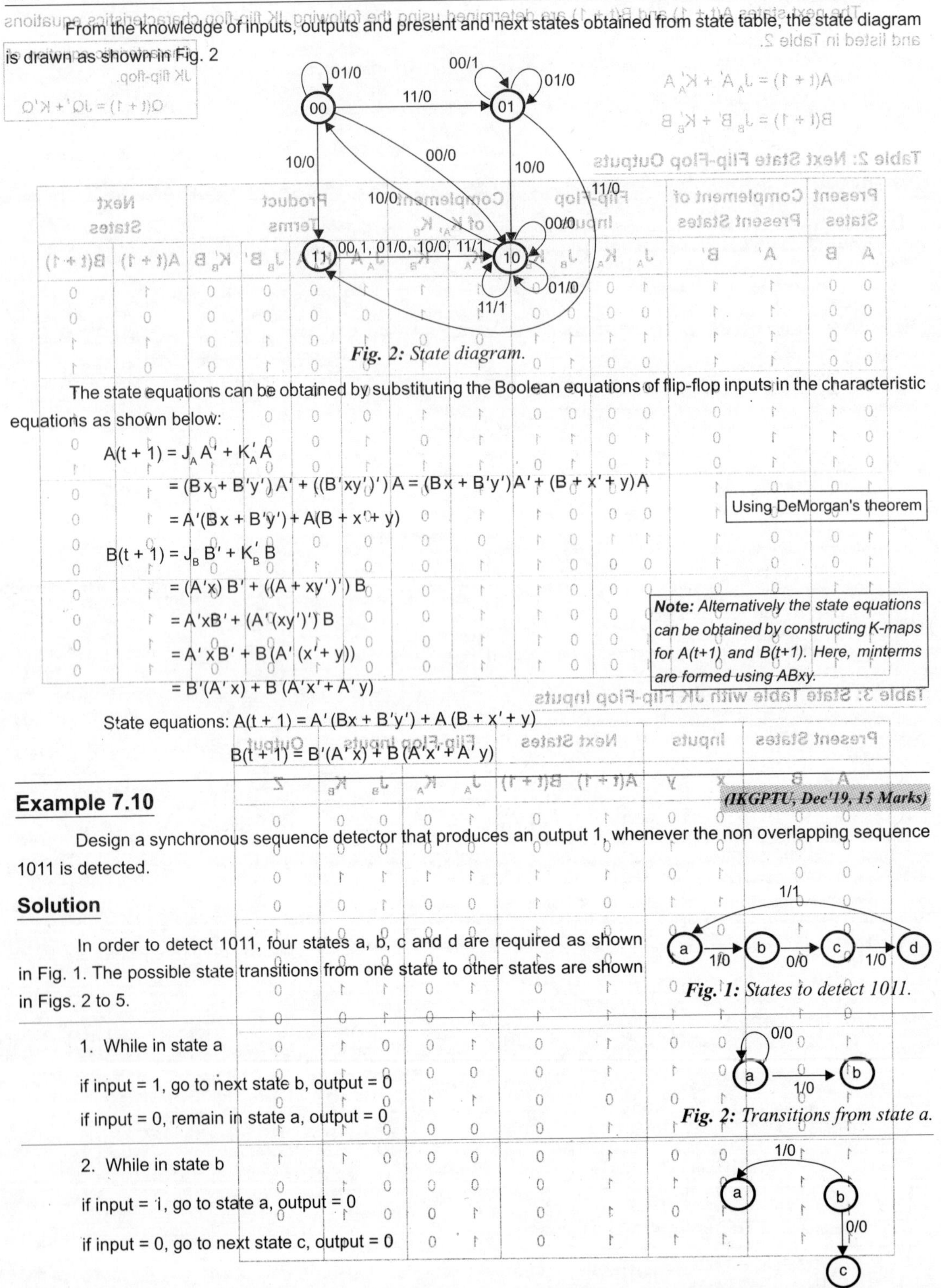

Fig. 2: State diagram.

The state equations can be obtained by substituting the Boolean equations of flip-flop inputs in the characteristic equations as shown below:

$$A(t+1) = J_A A' + K_A' A$$

$$= (Bx + B'y')A' + ((B'xy')')A = (Bx + B'y')A' + (B + x' + y)A \quad \text{Using DeMorgan's theorem}$$

$$= A'(Bx + B'y') + A(B + x' + y)$$

$$B(t+1) = J_B B' + K_B' B$$

$$= (A'x)B' + ((A + xy')')B$$

$$= A'xB' + (A'(xy')')B$$

$$= A'xB' + B(A'(x' + y))$$

$$= B'(A'x) + B(A'x' + A'y)$$

***Note:** Alternatively the state equations can be obtained by constructing K-maps for A(t+1) and B(t+1). Here, minterms are formed using ABxy.*

State equations: $A(t+1) = A'(Bx + B'y') + A(B + x' + y)$

$$B(t+1) = B'(A'x) + B(A'x' + A'y)$$

Example 7.10

(IKGPTU, Dec'19, 15 Marks)

Design a synchronous sequence detector that produces an output 1, whenever the non overlapping sequence 1011 is detected.

Solution

In order to detect 1011, four states a, b, c and d are required as shown in Fig. 1. The possible state transitions from one state to other states are shown in Figs. 2 to 5.

a → b (1/0), b → c (0/0), c → d (1/0), d → a (1/1)

Fig. 1: States to detect 1011.

1. While in state a

if input = 1, go to next state b, output = 0

if input = 0, remain in state a, output = 0

a (0/0), a → b (1/0)

Fig. 2: Transitions from state a.

2. While in state b

if input = 1, go to state a, output = 0

if input = 0, go to next state c, output = 0

b → a (1/0), b → c (0/0)

Fig. 3: Transitions from state b.

3. While in state c

   if input = 1, go to next state d, output = 0

   if input = 0, go to state a, output = 0

Fig. 4: Transitions from state c.

4. While in state d

   if input = 1, go to state a, output = 1

   if input = 0, go to state a, output = 0

Fig. 5: Transitions from state d.

On combining the state transitions shown in Figs. 1 to 5, the state diagram can be drawn as shown in Fig. 6. Here, the output is 1, when the circuit detected the sequence 1011, otherwise output is 0.

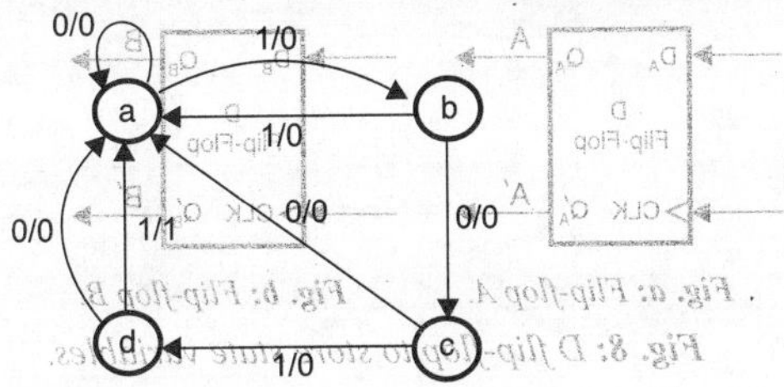

Fig. 6: State diagram.

State Table

Let, x be the input and Y be the output. The state table (Table 1) can be formed from state diagram. The next state and output are determined from state diagram.

Table 1: State Table

Present State	Next State x = 0	Next State x = 1	Output x = 0	Output x = 1
a	a	b	0	0
b	c	a	0	0
c	a	d	0	0
d	a	a	0	1

There is no identical or equivalent pair of states in the state table with same output and same next state and so, there is no need for state reduction. The four ($2^2 = 4$) states can be formed with two state variables. Let A and B be the two state variables. Since there are four states, let us choose 2-bit code for states as shown below:

	AB	
a ⟶	0 0	Assigned codes
b ⟶	0 1	
c ⟶	1 0	
d ⟶	1 1	

The state diagram is redrawn as shown in Fig. 7 with the above code assignments. Also the state table with binary codes for states is shown in Table 2.

Table 2: State Table with Binary Code Assignment for States

Present States		Next States				Output	
		x = 0		x = 1		x = 0	x = 1
A	B	A(t+1)	B(t+1)	A(t+1)	B(t+1)	Y	Y
0	0	0	0	0	1	0	0
0	1	1	0	0	0	0	0
1	0	0	0	1	1	0	0
1	1	0	0	0	0	0	1

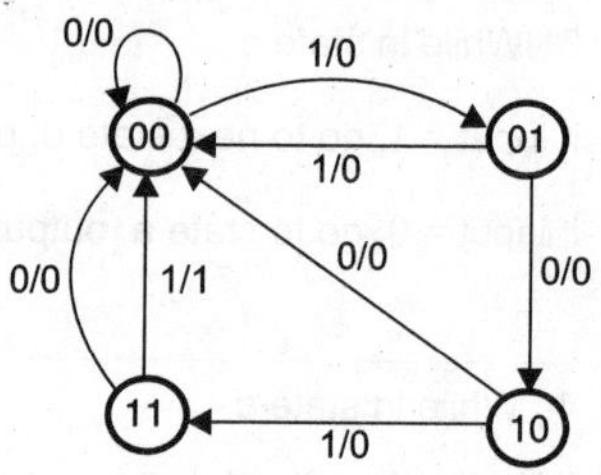

Fig. 7: State diagram with state assignment.

In sequential circuit, each state variable is stored in the output of a flip-flop. Hence, choose two D flip-flops to store two state variables A and B as shown in Fig. 8.

$\therefore$ Present state variables : A(t), B(t) $\Rightarrow$ A, B (for simplicity)

Next state variables : A(t + 1), B(t + 1)

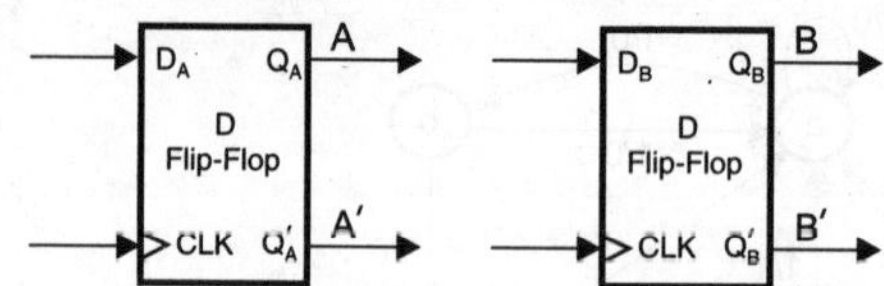

Fig. a: Flip-flop A. *Fig. b: Flip-flop B.*

Fig. 8: D flip-flop to store state variables.

The characteristic equation of two D flip-flops are,

$$Q_A(t + 1) = D_A$$

$$Q_B(t + 1) = D_B$$

The two state variables can be equated to flip-flop outputs.

$$\therefore A = Q_A \Rightarrow A(t + 1) = Q_A(t + 1)$$

$$B = Q_B \Rightarrow B(t + 1) = Q_B(t + 1)$$

Now the characteristic equations in terms of state variables are,

$$A(t + 1) = D_A$$

$$B(t + 1) = D_B$$

> Characteristic equation of D flip-flop.
>
> Q(t + 1) = D

Using the above equations the flip-flop inputs are determined for the next state and listed in Table 3. In D flip-flop, the input is same as next state output.

Table 3: Expanded State Table

Present States		Input	Next States		Flip-Flop Inputs		Output	Minterm
A	B	x	A(t+1)	B(t+1)	D_A	D_B	Y	ABx
0	0	0	0	0	0	0	0	m_0
0	0	1	0	1	0	1	0	m_1
0	1	0	1	0	1	0	0	m_2
0	1	1	0	0	0	0	0	m_3
1	0	0	0	0	0	0	0	m_4
1	0	1	1	1	1	1	0	m_5
1	1	0	0	0	0	0	0	m_6
1	1	1	0	0	0	0	1	m_7

Using Table 3, the Boolean equations for D_A, D_B and Y are determined as sum of minterms as shown below:

$D_A = \Sigma m(2, 5)$; $D_B = \Sigma m(1, 5)$; $Y = m_7$

Let us draw K-map for D_A, D_B and Y to develop simplified Boolean equations for D_A, D_B and Y.

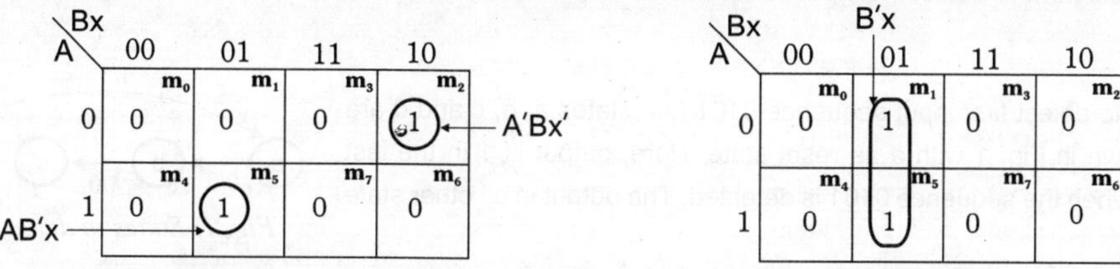

Fig. 9: K-map for D_A.

Fig. 10: K-map for D_B.

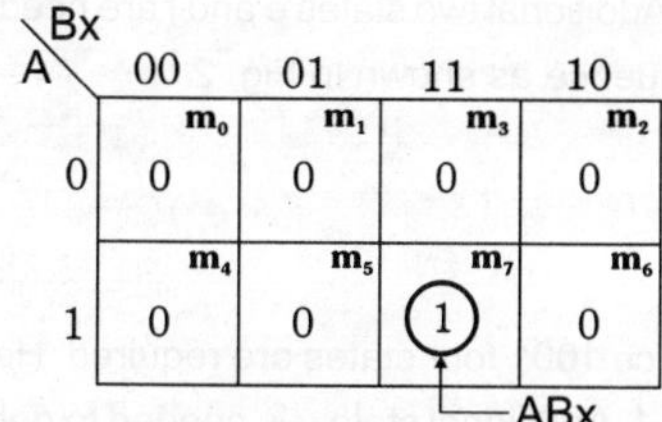

Fig. 11: K-map for Y.

From the K-map we get the following simplified Boolean equations for realizing sequence detector using D flip-flops,

$D_A = A'Bx' + AB'x$

$D_B = B'x$

$Y = ABx$

The logic diagram of sequence detector using the above equations is drawn as shown in Fig. 12. Here, the output depends on input and states and so this model is Mealy model.

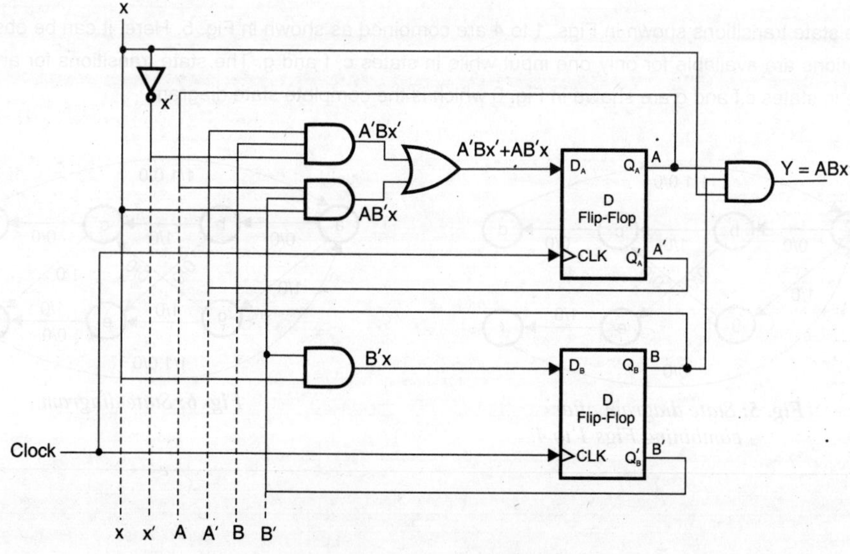

Fig. 12: Logic diagram of sequence detector.

Example 7.11

(VTU, Aug/Sep'20, 10 Marks)

A sequential circuit has one input (x) and output (z) the circuit examines gropus of four consecutive inputs and produces an output z = 1 if the input sequence 0101 and 1001 occurs. The circuit resets after every four inputs. Draw the mealy state diagram to detect the input sequences 0101,0010,1001 and 0100.

Solution

In order to detect first input sequence 0101 four states a, b, c and d are required as shown in Fig. 1 with a as reset state. Here, output is 1 in the last state transition when the sequence 0101 is detected. The output in all other state transitions is 0.

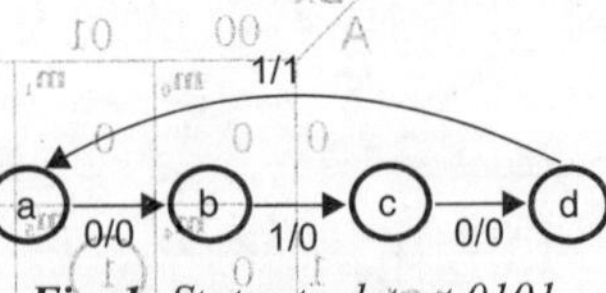

Fig. 1: States to detect 0101.

In order to detect second input sequence 0010 four states are required. Here, first two states can be same as Fig. 1. Additional two states e and f are needed to detect the last three bits of the input sequence as shown in Fig. 2.

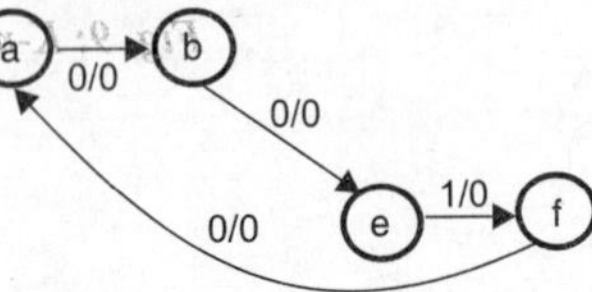

Fig. 2: States to detect 0010.

In order to detect third input sequence 1001 four states are required. Here, first and last two states can be same as Fig. 1. Additional state g is needed to detect the first two input bits as shown in Fig. 3.

Fig. 3: States to detect 1001.

The fourth input sequence 0100 can be detected using the same four states a, b, c and d as shown in Fig. 1. Here, in the last state transition also the output is 0.

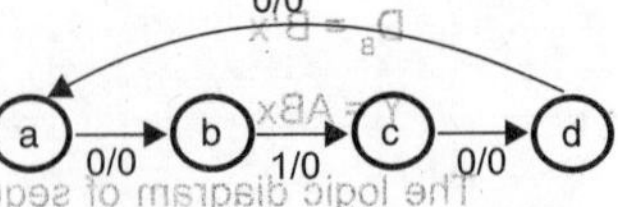

Fig. 4: States to detect 0100.

The state transitions shown in Figs. 1 to 4 are combined as shown in Fig. 5. Here, it can be observed that the state transitions are available for only one input while in states c, f and g. The state transitions for another possible input, while in states c,f and g are shown in Fig. 6 which is the complete state diagram.

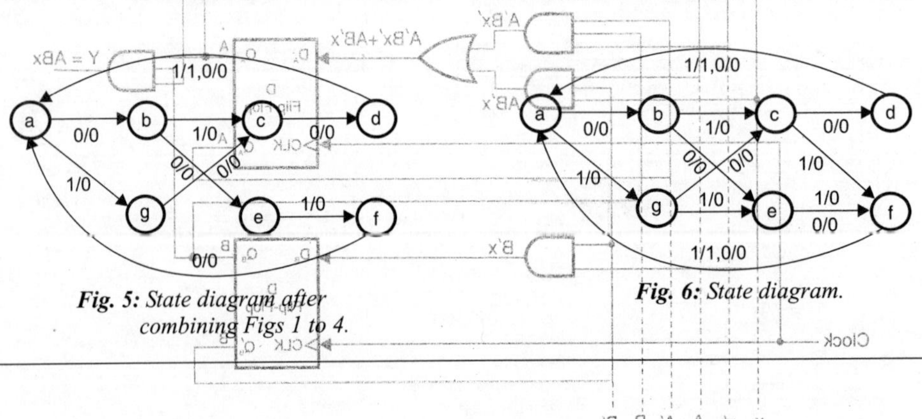

Fig. 5: State diagram after combining Figs 1 to 4.

Fig. 6: State diagram.

Example 7.12

(JNTUA, Jun/July'20, 7 Marks)

Design a 5 state sequential machine to detect the input sequence 01101 whose sequential states are : 000, 001, 010, 110, 111. Assume initial state is 000.

Solution

The sequential machine has five states 000, 001, 010, 110, 111. The state transitions to detect the sequence 01101 are shown in Fig. 1. All possible state transitions from one state to other states are shown in Figs. 2 to 6.

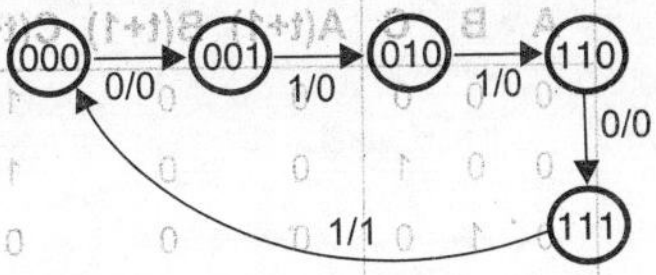

Fig. 1: States to detect 01101.

1. While in state 000

   if input = 0, go to next state 001, output = 0

   if input = 1, remain in state 000, output = 0

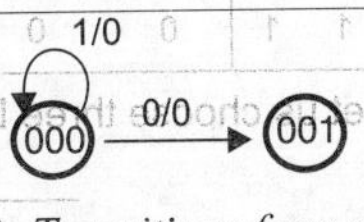

Fig. 2: Transitions from state 000.

2. While in state 001

   if input = 1, go to next state 010 output = 0

   if input = 0, remain in state 001, output = 0

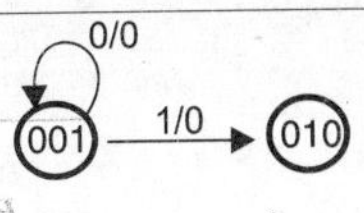

Fig. 3: Transitions from state 001.

3. While in state 010

   if input = 1, go to next state 110, output = 0

   if input = 0, go back to state 000, output = 0

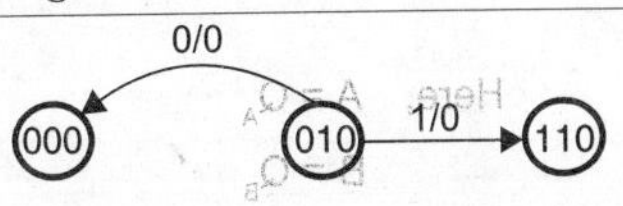

Fig. 4: Transitions from state 010.

4. While in state 110

   if input is 0, go to next state 111, output = 0

   if input is 1, go back to state 000, output = 0

Fig. 5: Transitions from state 110.

5. While in state 111

   if input is 0, go to state 001, output = 0

   if input is 1, go back to state 000, output = 1

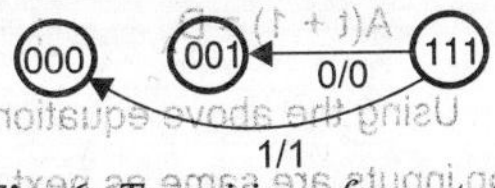

Fig. 6: Transitions from state 111.

On combining the state transitions shown in Figs. 1 to 6, the state diagram can be drawn as shown in Fig. 7.

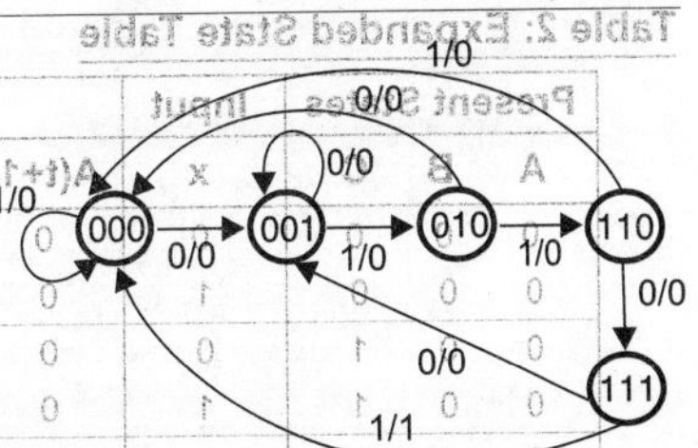

State Table

Let, x be the input and Y be the output. The state table (Table 1) can be formed from state diagram. Here, each bit of binary code is a state variable. Hence the sequential circuit has 3 state variables. Let the state variables be A, B and C.

$\therefore$ Present state variables : A(t), B(t), C(t) $\Rightarrow$ A, B, C (For simplicity)

Next state variables : A(t + 1), B(t + 1), C(t + 1)

Fig. 7: State diagram.

The state table (Table 1) is constructed from the knowledge of state transitions shown in Figs. 2 to 6.

Table 1: State Table

Present States			Next States						Output	
			x = 0			x = 1			x = 0	x = 1
A	B	C	A(t+1)	B(t+1)	C(t+1)	A(t+1)	B(t+1)	C(t+1)	Y	Y
0	0	0	0	0	1	0	0	0	0	0
0	0	1	0	0	1	0	1	0	0	0
0	1	0	0	0	0	1	1	0	0	0
1	1	0	1	1	1	0	0	0	0	0
1	1	1	0	0	1	0	0	0	0	1

Let us choose three D flip-flops as shown in Fig. 8. Here, A, B and C be output of flip-flops.

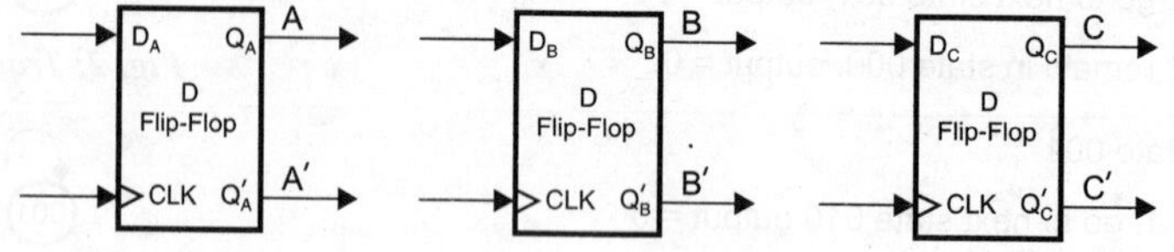

Fig. a: Flip-flop A. *Fig. b: Flip-flop B.* *Fig. c: Flip-flop C.*

Fig. 8: D flip-flop to store state variables.

Here, $A = Q_A$ $\qquad A(t + 1) = Q_A(t + 1)$

$B = Q_B$ $\qquad B(t + 1) = Q_B(t + 1)$

$C = Q_C$ $\qquad C(t + 1) = Q_C(t + 1)$

The characteristic equation of three D flip-flops are,

$$Q_A(t + 1) = D_A \quad ; \quad Q_B(t + 1) = D_B \quad ; \quad Q_C(t + 1) = D_C$$

Characteristic equation of D flip-flop.

$Q(t + 1) = D$

Using the above equations we can write,

$$A(t + 1) = D_A \quad ; \quad B(t + 1) = D_B \quad ; \quad C(t + 1) = D_C$$

Using the above equations the flip-flop inputs are determined for the next state and listed in Table 2. Here, the flip-flop inputs are same as next state outputs.

Table 2: Expanded State Table

Present States			Input	Next States			Flip-Flop Inputs			Output	Minterm
A	B	C	x	A(t+1)	B(t+1)	C(t+1)	D_A	D_B	D_C	Y	ABCx
0	0	0	0	0	0	1	0	0	1	0	m_0
0	0	0	1	0	0	0	0	0	0	0	m_1
0	0	1	0	0	0	1	0	0	1	0	m_2
0	0	1	1	0	1	0	0	1	0	0	m_3
0	1	0	0	0	0	0	0	0	0	0	m_4
0	1	0	1	1	1	0	1	1	0	0	m_5
0	1	1	0	×	×	×	×	×	×	×	m_6
0	1	1	1	×	×	×	×	×	×	×	m_7
1	0	0	0	×	×	×	×	×	×	×	m_8
1	0	0	1	×	×	×	×	×	×	×	m_9
1	0	1	0	×	×	×	×	×	×	×	m_{10}
1	0	1	1	×	×	×	×	×	×	×	m_{11}
1	1	0	0	1	1	1	1	1	1	0	m_{12}
1	1	0	1	0	0	0	0	0	0	0	m_{13}
1	1	1	0	0	0	1	0	0	1	0	m_{14}
1	1	1	1	0	0	0	0	0	0	1	m_{15}

> ***Note:*** *Binary codes 011, 100 and 101 are unused codes and so next state and outputs are considered as don't-care for this codes.*

Here, D_A, D_B, D_C and Y are functions of A, B, C, x. Using Table 2, the Boolean equations for D_A, D_B, D_C and Y are determined as sum of minterms as shown below:

$\therefore$ $D_A = \Sigma m(5, 12) + \Sigma d(6, 7, 8, 9, 10, 11)$ | $D_B = \Sigma m(3, 5, 12) + \Sigma d(6, 7, 8, 9, 10, 11)$

$D_C = \Sigma m(0, 2, 12, 14) + \Sigma d(6, 7, 8, 9, 10, 11)$ | $Y = m_{15} + \Sigma d(6, 7, 8, 9, 10, 11)$

Let us draw K-map for D_A, D_B, D_C and Y as shown in Figs. 9 to 12 to develop simplified Boolean equations for D_A, D_B, D_C and Y.

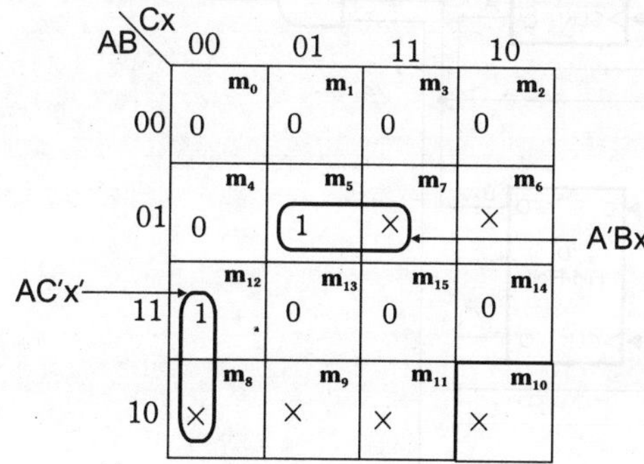

Fig. 9: K-map for D_A.

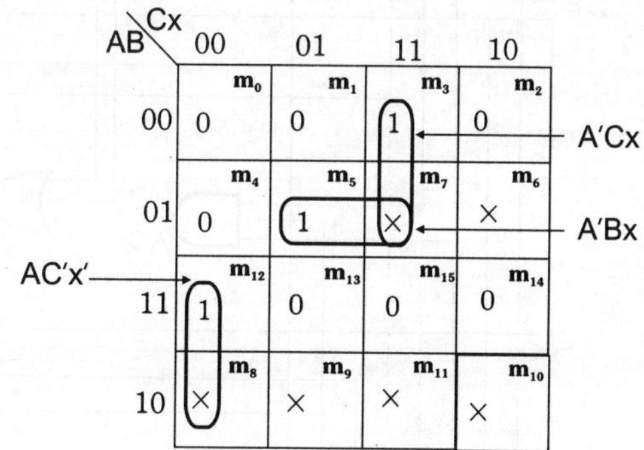

Fig. 10: K-map for D_B.

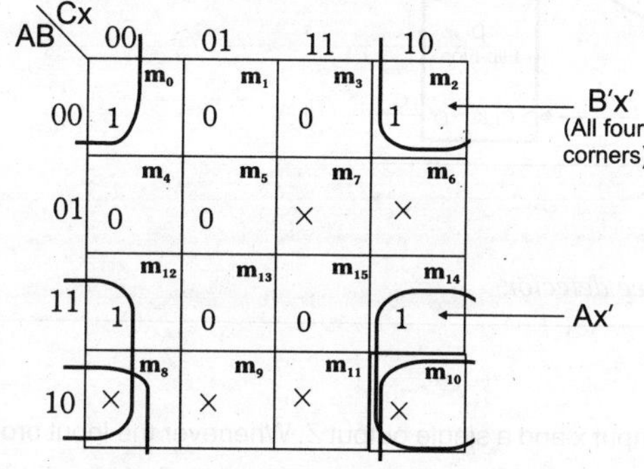

Fig. 11: K-map for D_C.

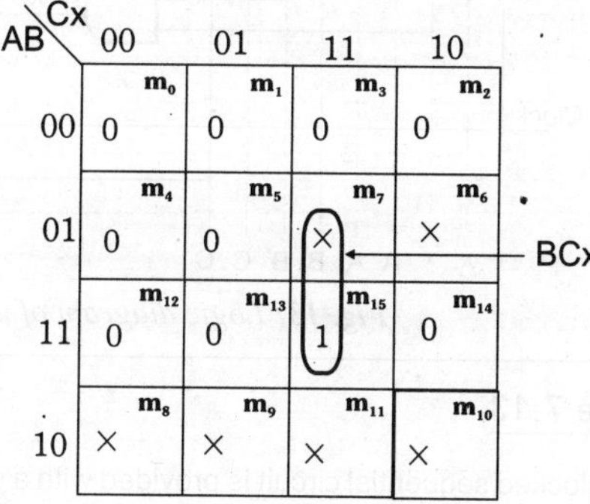

Fig. 12: K-map for Y.

From the K-map we get the following simplified Boolean equations for realizing sequence detector using D flip-flops.

$D_A = AC'x' + A'Bx$ | $D_B = AC'x' + A'Bx + A'Cx$

$D_C = Ax' + B'x'$ | $Y = BCx$

The logic diagram of sequence detector using the above equations is drawn as shown in Fig. 13. Here, the output depends on input and states and so this model is Mealy model.

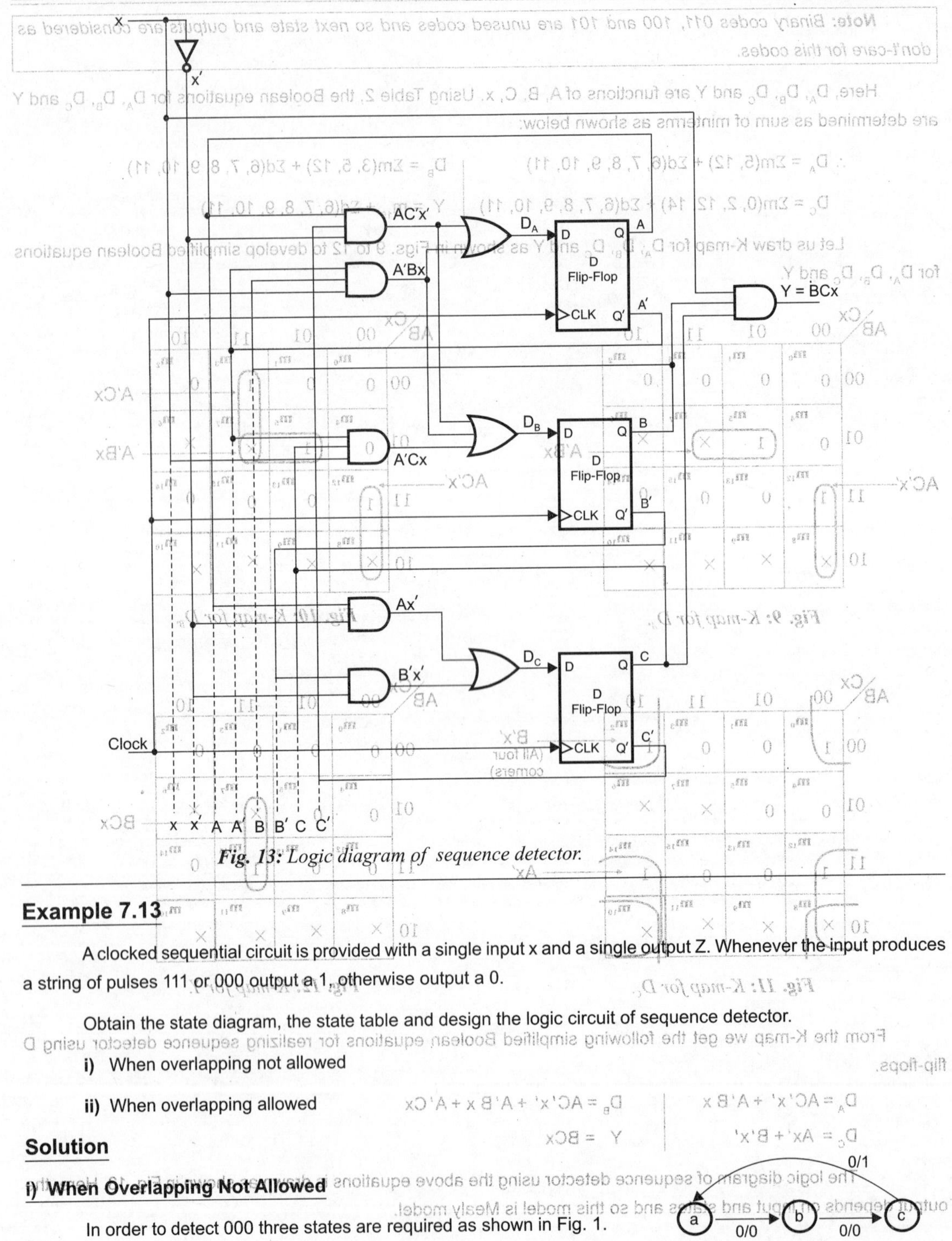

Fig. 13: Logic diagram of sequence detector.

Example 7.13

A clocked sequential circuit is provided with a single input x and a single output Z. Whenever the input produces a string of pulses 111 or 000 output a 1, otherwise output a 0.

Obtain the state diagram, the state table and design the logic circuit of sequence detector.

i) When overlapping not allowed

ii) When overlapping allowed

Solution

i) When Overlapping Not Allowed

In order to detect 000 three states are required as shown in Fig. 1.

Fig. 1: States to detect 000.

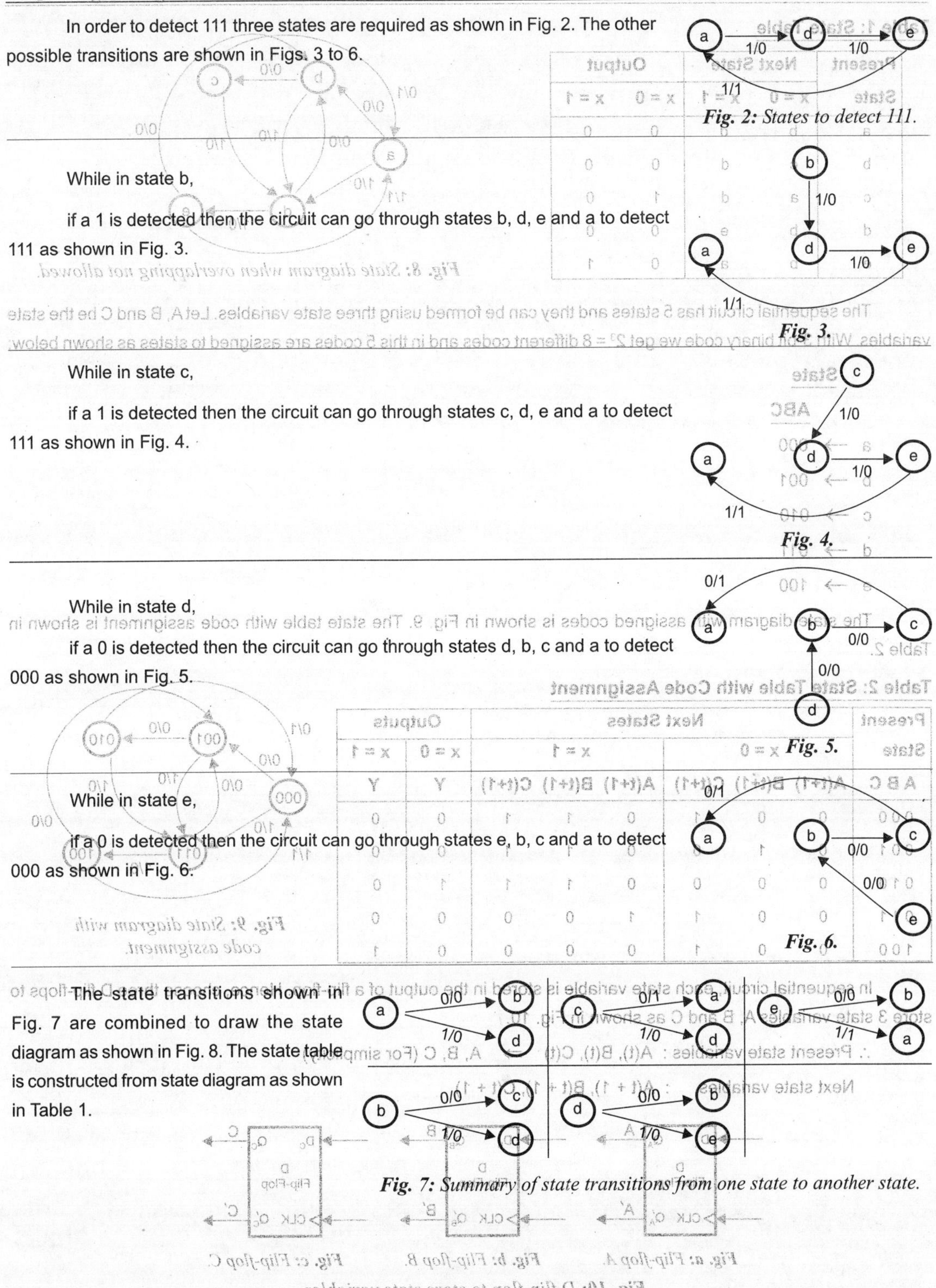

In order to detect 111 three states are required as shown in Fig. 2. The other possible transitions are shown in Figs. 3 to 6.

Fig. 2: States to detect 111.

While in state b,

if a 1 is detected then the circuit can go through states b, d, e and a to detect 111 as shown in Fig. 3.

Fig. 3.

While in state c,

if a 1 is detected then the circuit can go through states c, d, e and a to detect 111 as shown in Fig. 4.

Fig. 4.

While in state d,

if a 0 is detected then the circuit can go through states d, b, c and a to detect 000 as shown in Fig. 5.

Fig. 5.

While in state e,

if a 0 is detected then the circuit can go through states e, b, c and a to detect 000 as shown in Fig. 6.

Fig. 6.

The state transitions shown in Fig. 7 are combined to draw the state diagram as shown in Fig. 8. The state table is constructed from state diagram as shown in Table 1.

Fig. 7: Summary of state transitions from one state to another state.

Table 1: State Table

Present State	Next State		Output	
	x = 0	x = 1	x = 0	x = 1
a	b	d	0	0
b	c	d	0	0
c	a	d	1	0
d	b	e	0	0
e	b	a	0	1

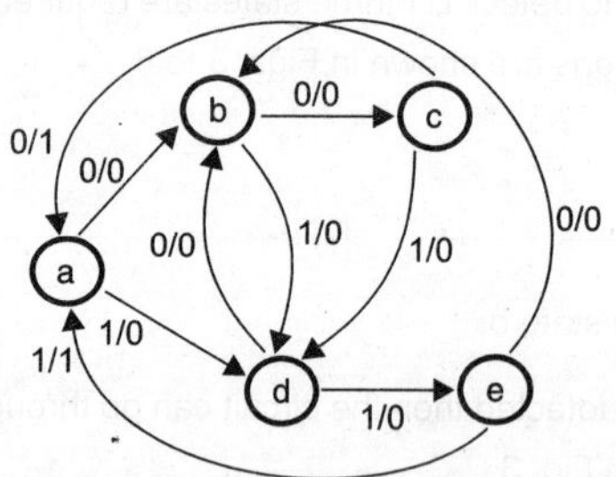

Fig. 8: State diagram when overlapping not allowed.

The sequential circuit has 5 states and they can be formed using three state variables. Let A, B and C be the state variables. With 3-bit binary code we get $2^3 = 8$ different codes and in this 5 codes are assigned to states as shown below:

State

ABC

a → 000

b → 001

c → 010

d → 011

e → 100

The state diagram with assigned codes is shown in Fig. 9. The state table with code assignment is shown in Table 2.

Table 2: State Table with Code Assignment

Present State	Next States						Outputs	
	x = 0			x = 1			x = 0	x = 1
A B C	A(t+1)	B(t+1)	C(t+1)	A(t+1)	B(t+1)	C(t+1)	Y	Y
0 0 0	0	0	1	0	1	1	0	0
0 0 1	0	1	0	0	1	1	0	0
0 1 0	0	0	0	0	1	1	1	0
0 1 1	0	0	1	1	0	0	0	0
1 0 0	0	0	1	0	0	0	0	1

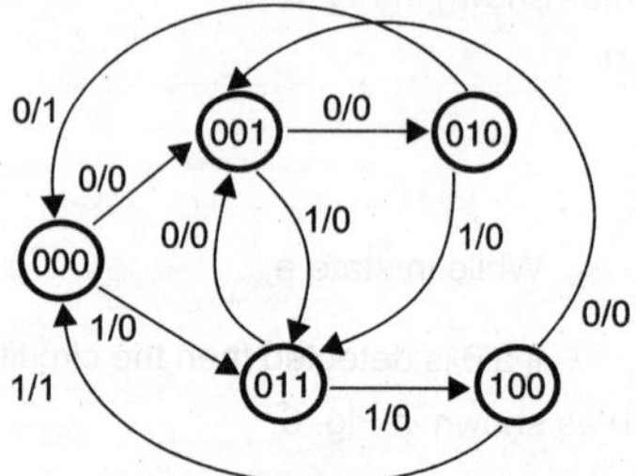

Fig. 9: State diagram with code assignment.

In sequential circuit, each state variable is stored in the output of a flip-flop. Hence, choose three D flip-flops to store 3 state variables A, B and C as shown in Fig. 10.

∴ Present state variables : A(t), B(t), C(t) ⇒ A, B, C (For simplicity)

Next state variables : A(t + 1), B(t + 1), C(t + 1)

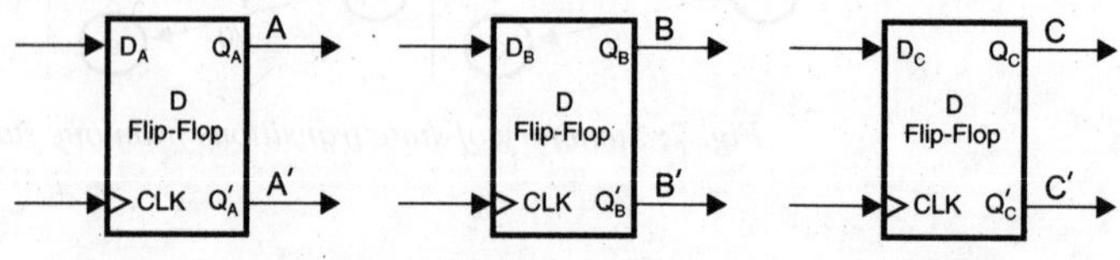

Fig. a: Flip-flop A. *Fig. b: Flip-flop B.* *Fig. c: Flip-flop C.*

Fig. 10: D flip-flop to store state variables.

The characteristic equation of two D flip-flops are,

$Q_A(t+1) = D_A$

$Q_B(t+1) = D_B$

$Q_C(t+1) = D_C$

Characteristic equation of D flip-flop.
$Q(t+1) = D$

The three state variables can be equated to flip-flop outputs.

$\therefore A = Q_A \Rightarrow A(t+1) = Q_A(t+1)$

$B = Q_B \Rightarrow B(t+1) = Q_B(t+1)$

$C = Q_C \Rightarrow C(t+1) = Q_C(t+1)$

Now the characteristic equations in terms of state variables are,

$A(t+1) = D_A$

$B(t+1) = D_B$

$C(t+1) = D_C$

Using the above equations the flip-flop inputs are determined for the next state and listed in Table 3. In D flip-flop the input is same as next state output.

Table 3: Expanded State Table

Present States			Input	Next States			Flip-Flop Inputs			Output	Minterm
A	B	C	x	A(t+1)	B(t+1)	C(t+1)	D_A	D_B	D_C	Y	ABCx
0	0	0	0	0	0	1	0	0	1	0	m_0
0	0	0	1	0	1	1	0	1	1	0	m_1
0	0	1	0	0	1	0	0	1	0	0	m_2
0	0	1	1	0	1	1	0	1	1	0	m_3
0	1	0	0	0	0	0	0	0	0	1	m_4
0	1	0	1	0	1	1	0	1	1	0	m_5
0	1	1	0	0	0	1	0	0	1	0	m_6
0	1	1	1	1	0	0	1	0	0	0	m_7
1	0	0	0	0	0	1	0	0	1	0	m_8
1	0	0	1	0	0	0	0	0	0	1	m_9
1	0	1	0	×	×	×	×	×	×	×	m_{10}
1	0	1	1	×	×	×	×	×	×	×	m_{11}
1	1	0	0	×	×	×	×	×	×	×	m_{12}
1	1	0	1	×	×	×	×	×	×	×	m_{13}
1	1	1	0	×	×	×	×	×	×	×	m_{14}
1	1	1	1	×	×	×	×	×	×	×	m_{15}

> ***Note:*** *Binary codes 101, 110 and 111 are unused codes and so next states and output are considered as don't-care for these codes.*

Here, D_A, D_B, D_C and Y are functions of A, B, C and x. Using Table 3, the Boolean equations for D_A, D_B, D_C and Y are determined as sum of minterms and don't-cares as shown below:

$$D_A = m_7 + \Sigma d(10, 11, 12, 13, 14, 15)$$

$$D_B = \Sigma m(1, 2, 3, 5) + \Sigma d(10, 11, 12, 13, 14, 15)$$

$$D_C = \Sigma m(0, 1, 3, 5, 6, 8) + \Sigma d(10, 11, 12, 13, 14, 15)$$

$$Y = \Sigma m(4, 9) + \Sigma d(10, 11, 12, 13, 14, 15)$$

Using the above equations K-maps are drawn for D_A, D_B, D_C and Y in order to obtain simplified Boolean equations.

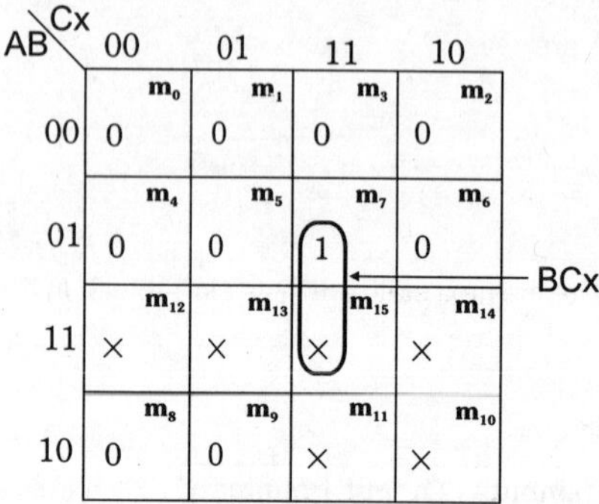

Fig. 11: K-map for D_A.

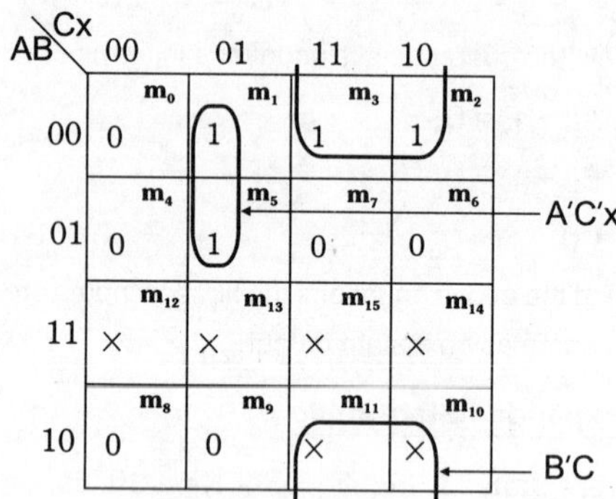

Fig. 12: K-map for D_B.

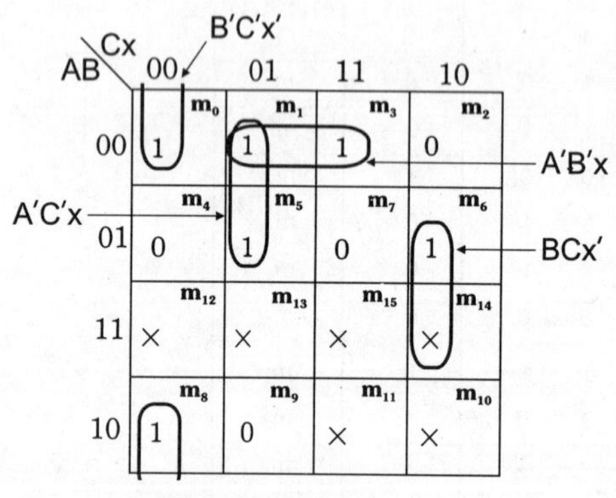

Fig. 13: K-map for D_C.

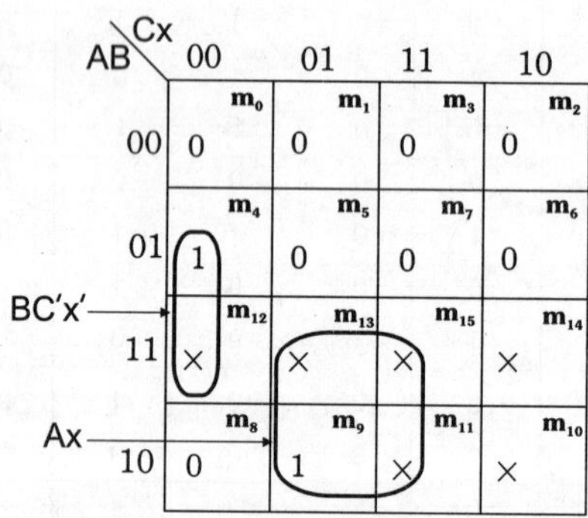

Fig. 14: K-map for Y.

From the K-map we get the following simplified Boolean equations for detecting the sequence 111 or 000 when overlapping not allowed using D flip-flops,

$$D_A = BCx$$

$$D_B = B'C + A'C'x$$

$$D_C = B'C'x' + A'C'x + A'B'x + BCx'$$

$$Y = BC'x' + Ax$$

Using the above equations the logic diagram of sequence detector is drawn as shown in Fig. 15. Here, the output depends on input and states and so this model is Mealy model.

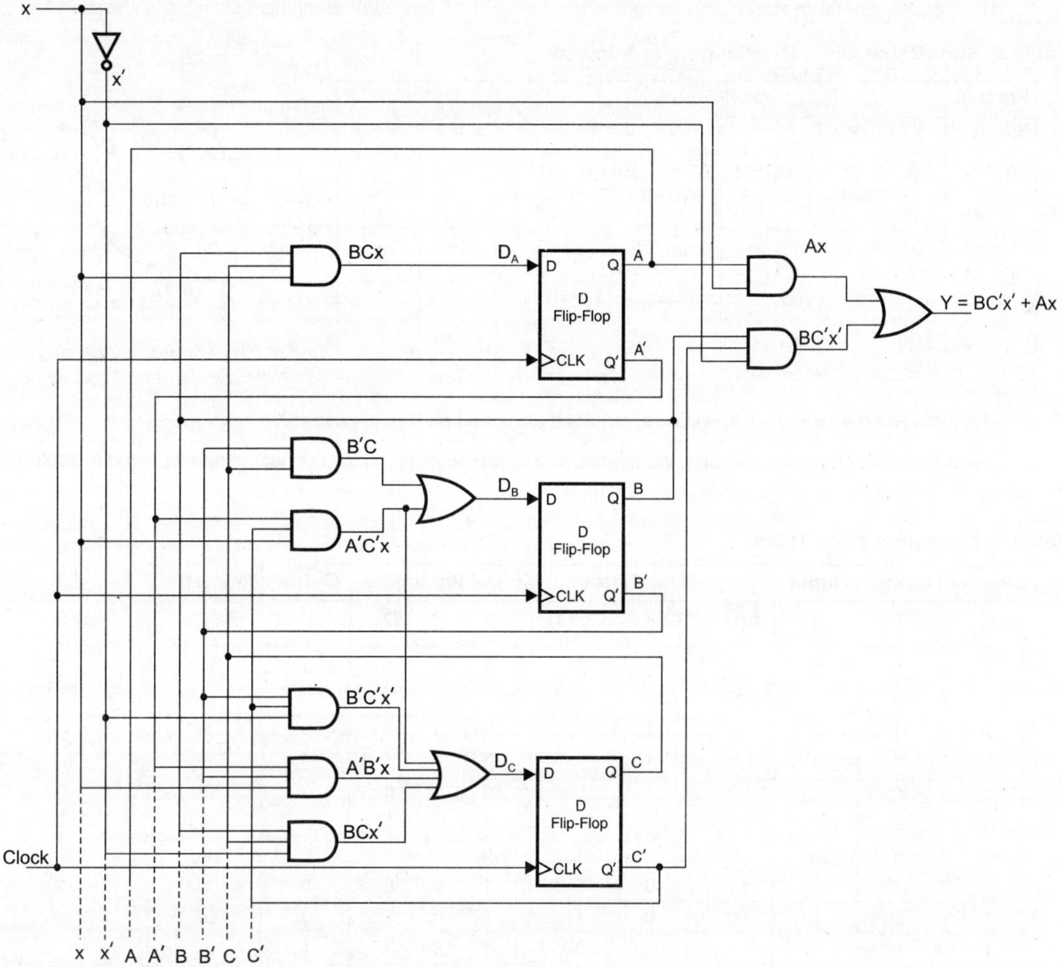

***Fig. 15:** Logic diagram to detect 111 or 000 when overlapping not allowed.*

ii) When Overlapping Allowed

When overlapping is allowed the following two changes are to be made in state diagram of Fig. 7 and redrawn as in Fig. 18.

1. With reference to Fig. 1, while in state c (010), if input is 0 then output 1 and remain in state c itself instead of going to state a (000) as shown in Fig. 16.

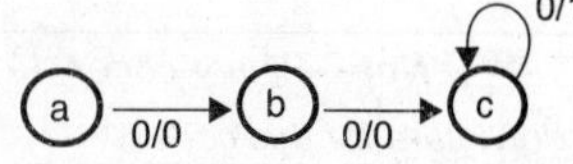

***Fig. 16:** States to detect 000 in overlapping sequence.*

2. With reference to Fig. 2, while in state e (100), if input is 1 then output 1 and remain in state e itself instead of going to state a (000) as shown in Fig. 17.

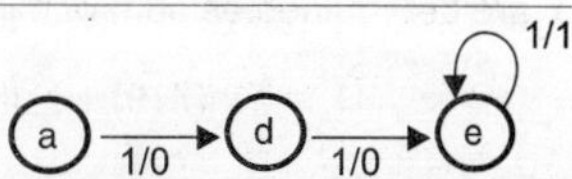

***Fig. 17:** States to detect 111 in overlapping sequence.*

The state diagram with above two changes is shown in Fig. 18 and corresponding state table is shown in Table 4.

Table 4: State Table when Overlapping is Allowed

Present States	Next States						Outputs	
	x = 0			x = 1			x = 0	x = 1
A B C	A(t+1)	B(t+1)	C(t+1)	A(t+1)	B(t+1)	C(t+1)	Y	Y
0 0 0	0	0	1	0	1	1	0	0
0 0 1	0	1	0	0	1	1	0	0
0 1 0	0	1	0	0	1	1	1	0
0 1 1	0	0	1	1	0	0	0	0
1 0 0	0	0	1	1	0	0	0	1

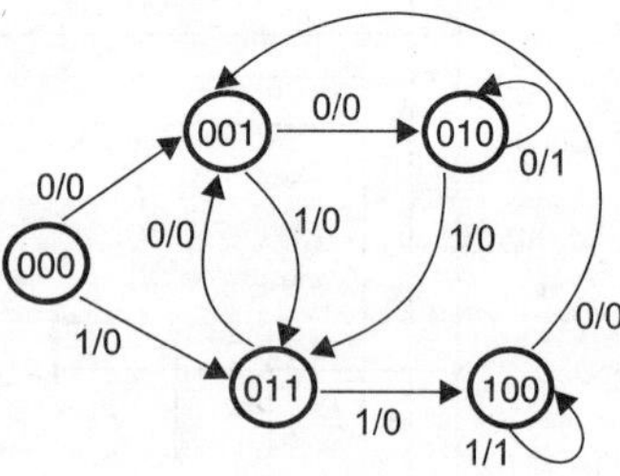

Fig. 18: State diagram when overlapping is allowed.

The state assignment and the choice of flip-flops are same as that of case (i).

Using the characteristic equations, the flip-flop inputs are determined for the next state and listed in Table 5. In D flip-flop the input is same as that of next state output.

Table 5: Expanded State Table

Present States			Input	Next States			Flip-Flop Inputs			Output	Minterm
A	B	C	x	A(t+1)	B(t+1)	C(t+1)	D_A	D_B	D_C	Y	ABCx
0	0	0	0	0	0	1	0	0	1	0	m_0
0	0	0	1	0	1	1	0	1	1	0	m_1
0	0	1	0	0	1	0	0	1	0	0	m_2
0	0	1	1	0	1	1	0	1	1	0	m_3
0	1	0	0	0	1	0	0	1	0	1	m_4
0	1	0	1	0	1	1	0	1	1	0	m_5
0	1	1	0	0	0	1	0	0	1	0	m_6
0	1	1	1	1	0	0	1	0	0	0	m_7
1	0	0	0	0	0	1	0	0	1	0	m_8
1	0	0	1	1	0	0	1	0	0	1	m_9
1	0	1	0	×	×	×	×	×	×	×	m_{10}
1	0	1	1	×	×	×	×	×	×	×	m_{11}
1	1	0	0	×	×	×	×	×	×	×	m_{12}
1	1	0	1	×	×	×	×	×	×	×	m_{13}
1	1	1	0	×	×	×	×	×	×	×	m_{14}
1	1	1	1	×	×	×	×	×	×	×	m_{15}

> ***Note:*** *Binary codes 101, 110 and 111 are unused codes and so next states and output are considered as don't-care for this codes.*

Here, D_A, D_B, D_C and Y are functions of A, B, C and x. Using Table 5, the Boolean equations for D_A, D_B, D_C and Y are determined as sum of minterms and don't-cares as shown below:

$$D_A = \Sigma m(7, 9) + \Sigma d(10, 11, 12, 13, 14, 15)$$

$$D_B = \Sigma m(1, 2, 3, 4, 5) + \Sigma d(10, 11, 12, 13, 14, 15)$$

$$D_C = \Sigma m(0, 1, 3, 5, 6, 8) + \Sigma d(10, 11, 12, 13, 14, 15)$$

$$Y = \Sigma m(4, 9) + \Sigma d(10, 11, 12, 13, 14, 15)$$

The equations for D_C and Y are same as in case (i), hence K-maps are constructed for D_A and D_B as shown in Figs. 19 and 20 to get simplified equations for D_A and D_B.

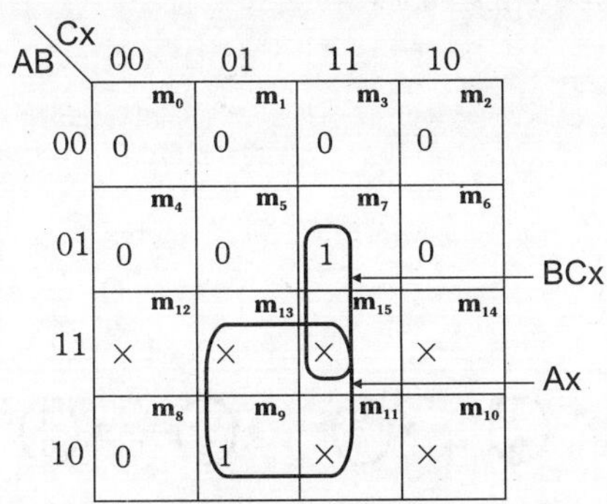

***Fig. 19:** K-map for D_A.*

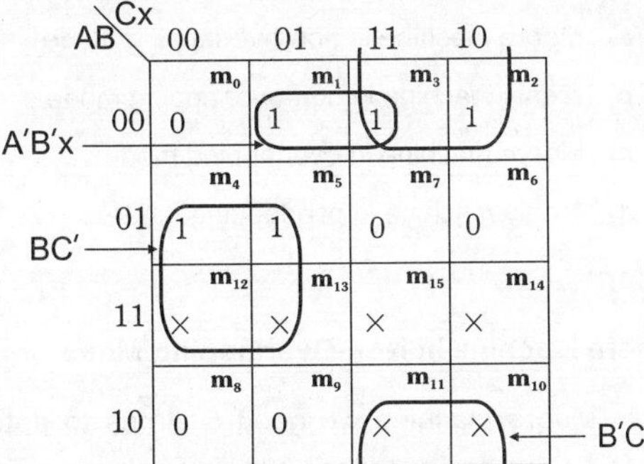

***Fig. 20:** K-map for D_B.*

From the K-map we get the following simplified Boolean equations for D_A, D_B, D_C and Y.

D_A = BCx + Ax

D_B = A'B'x + B'C + BC'

D_C = B'C'x' + A'C'x + A'B'x + BCx'

Y = BC'x' + Ax

Using the above equations the logic circuit using D flip-flops to detect 111 or 000 when overlapping is allowed is drawn as shown in Fig. 21. Here, the output depends on input and states and so this model is Mealy model.

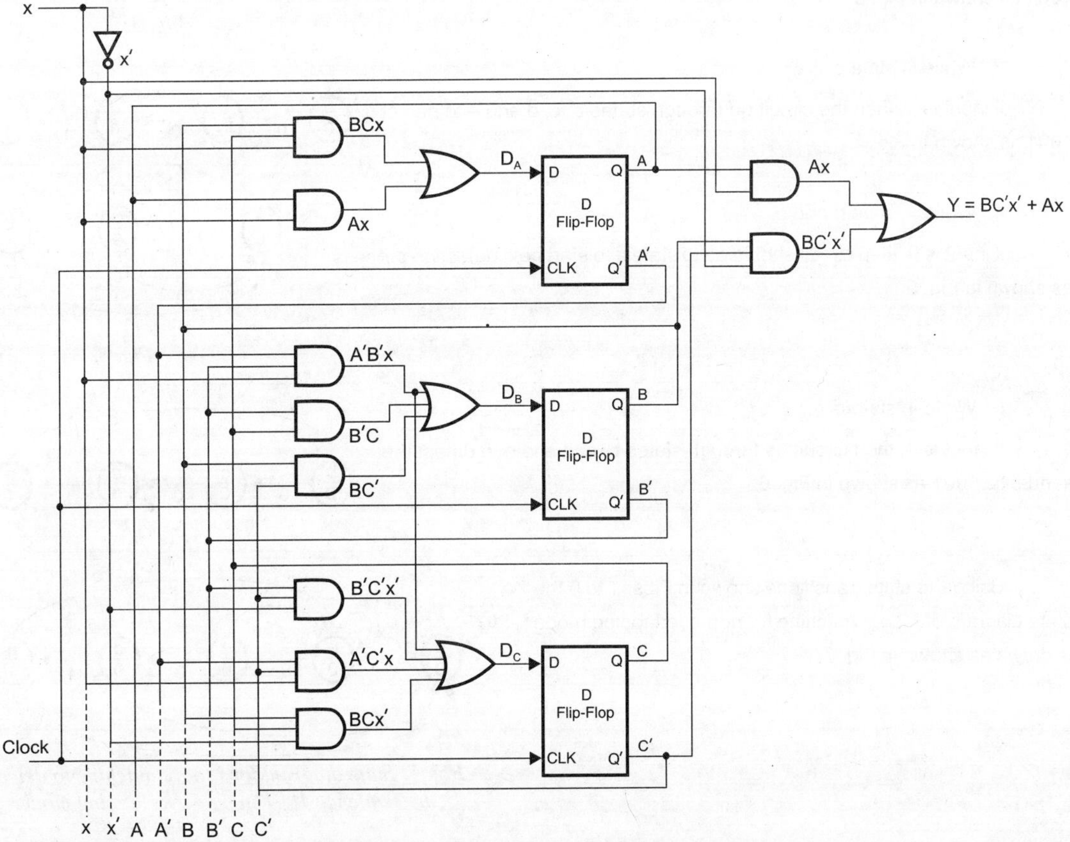

***Fig. 21:** Logic diagram to deduct 111 or 000 when overlapping allowed.*

Example 7.14

Draw the state diagram for sequence detector 1001 for the following cases.

a) Moore machine in non-overlapping mode

b) Mealy machine in non-overlapping mode

c) Moore machine in overlapping mode

d) Mealy machine in overlapping mode

Solution

a) Moore Machine in Non-Overlapping Mode

In Moore machine we need 5 states to detect the sequence 1001 in non-overlapping mode as shown in Fig. 1.

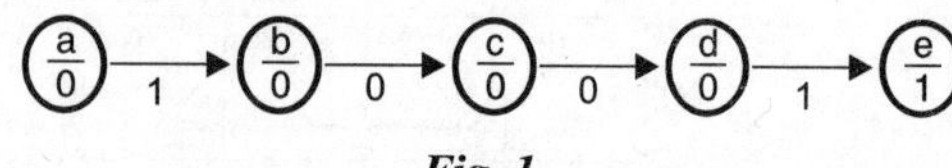

Fig. 1.

1. While in state a,

   if input is 0, then circuit should remain in state a itself as shown in Fig. 2.

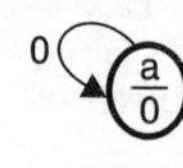

Fig. 2.

2. While in state b,

   if input is 1, then the circuit go through states b, c, d and e to detect 1001 as shown in Fig. 3.

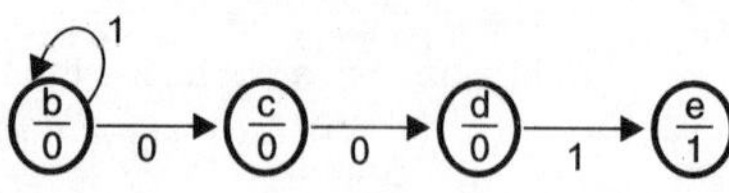

Fig. 3.

3. While in state c,

   if input is 1, then the circuit go through states b, c, d and e to detect 1001 as shown in Fig. 4.

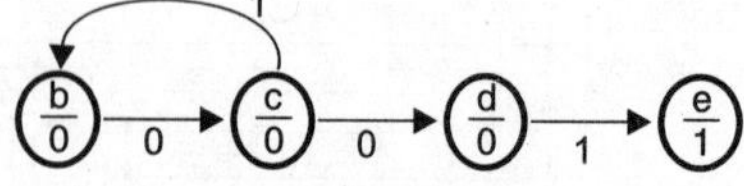

Fig. 4.

4. While in state d and e,

   if input is 0, then circuit should go to state a to start next detection process as shown in Fig. 5.

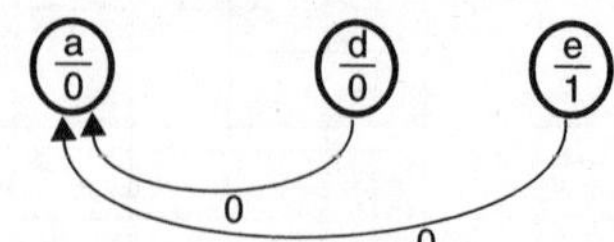

Fig. 5.

5. While in state e,

   if input is 1, then circuit go through states b, c, d and e to detect the sequence 1001 as shown in Fig. 6.

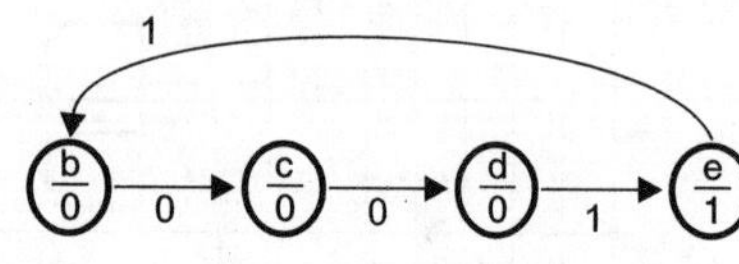

Fig. 6.

Using the state transitions shown in Figs. 1 to 6 the state diagram of Moore machine for non-overlapping mode is drawn as shown in Fig. 7.

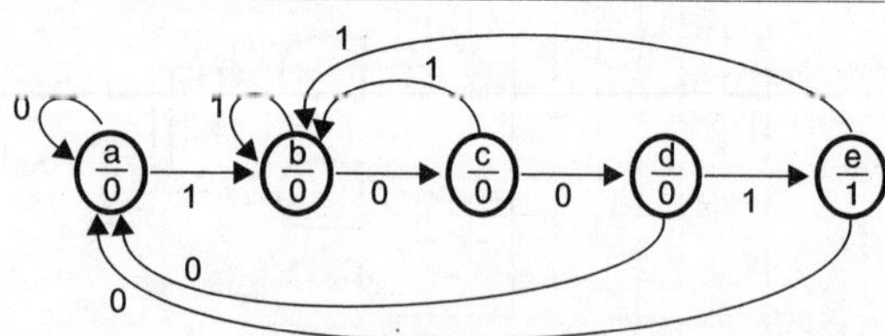

Fig. 7: State diagram for Moore machine to detect the sequence 1001 in non-overlapping mode.

b) Mealy Machine in Non-Overlapping Mode

In Mealy machine we need 4 states to detect the sequence 1001 in non-overlapping mode as shown in Fig. 8.

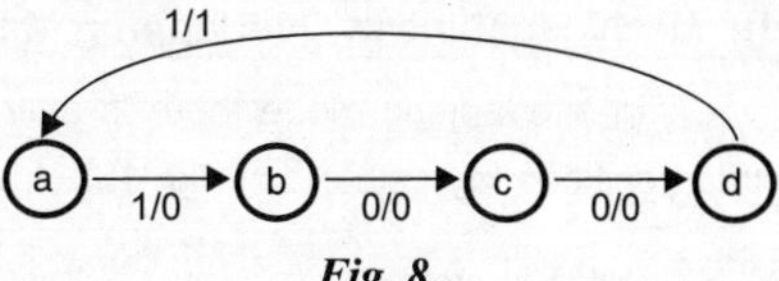

Fig. 8.

1. While in state a,

   if input is 0, then circuit should remain in state a itself as shown in Fig. 9.

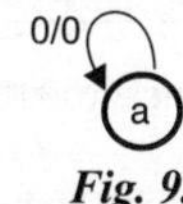

Fig. 9.

2. While in state b,

   if input is 1, then the circuit go through states b, c, d and a to detect 1001 as shown in Fig. 10.

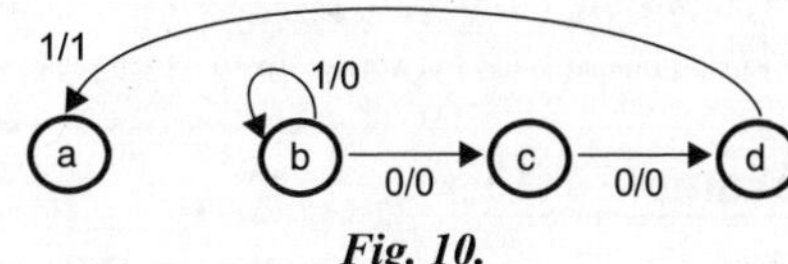

Fig. 10.

3. While in state c,

   if input is 1, then the circuit go through states b, c, d and a to detect 1001 as shown in Fig. 11.

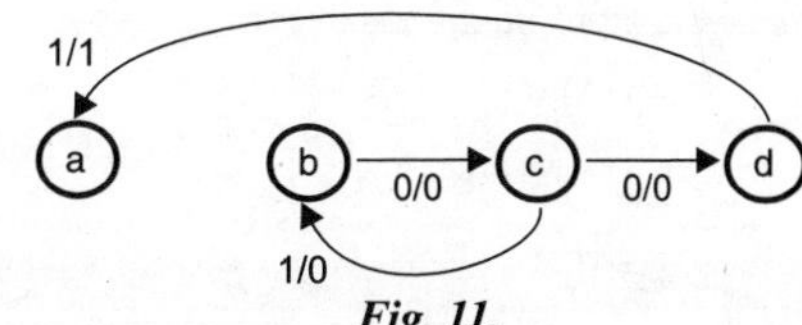

Fig. 11.

4. While in state d,

   if input is 0, then circuit should go to state a to start next detection process as shown in Fig. 12.

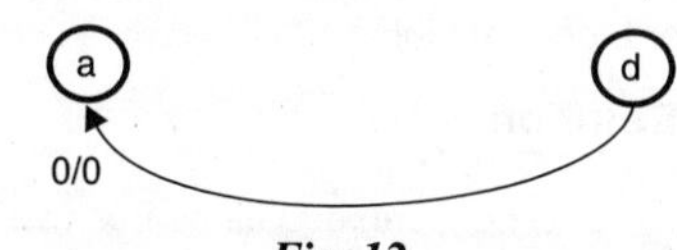

Fig. 12.

Using the state transitions shown in Figs. 8 to 12 the state diagram of Mealy machine for non-overlapping mode is drawn as shown in Fig. 13.

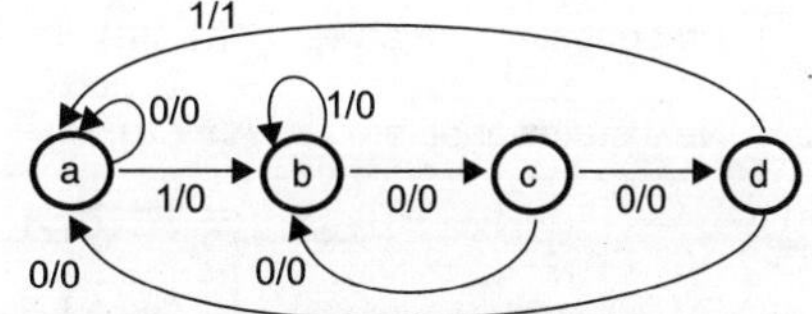

Fig. 13: State diagram for Mealy machine to detect the sequence 1001 in non-overlapping mode.

c) Moore Machine in Overlapping Mode

In overlapping mode there is only one change in state transition from state e of Fig. 7.

While in state e,

if input is 0, then circuit go to state c to detect overlapped sequence 1001001.

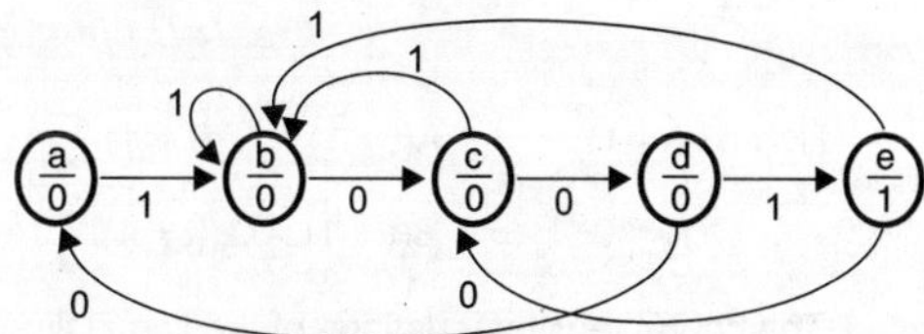

Fig. 14: State diagram for Moore machine to detect the sequence 1001 in overlapping mode.

The overlapped sequence is detected when passing through states a, b, c, d, e, c, d and e. The state diagram for Moore machine in overlapping mode is shown in Fig. 14.

d) Mealy Machine in Overlapping Mode

In overlapping mode there is only one change in state transition from state d of Fig. 13.

While in state d,

if input is 1, then circuit go to state b to detect overlapped sequence 1001001.

The overlapped sequence is detected when passing through states a, b, c, d, b, c, d and b. The state diagram for Mealy machine in overlapping mode is shown in Fig. 15.

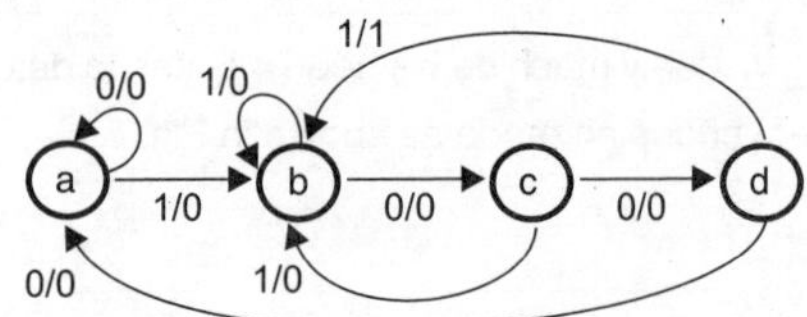

Fig. 15: State diagram for Mealy machine to detect the sequence 1001 in overlapping mode.

Example 7.15

A sequential circuit with two D flip-flops A and B and two inputs x and y, and one output Z is specified by the following input equations :

$A(t+1) = x'y + xA$

$B(t+1) = x'B + xA$

$Z = B$

Draw the logic diagram of the circuit. Derive the state table and state diagram and state whether it is a Mealy or Moore machine.

Solution

The given equations has two state variables A and B. Hence the sequential circuit can be designed using two D flip-flops shown in Fig. 1 to store two state variables.

∴ Present state variables : A(t), B(t) ⇒ A, B (For simplicity)

Next state variables : A(t + 1), B(t + 1)

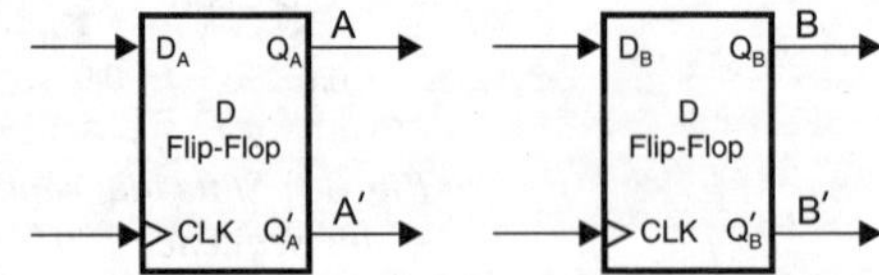

Fig. a: Flip-flop A. *Fig. b: Flip-flop B.*

Fig. 1: D flip-flop to store state variables.

Here, $A = Q_A \Rightarrow A(t + 1) = Q_A(t + 1)$

$B = Q_B \Rightarrow B(t + 1) = Q_B(t + 1)$

The characteristic equations of the two D flip-flops are,

$Q_A(t + 1) = D_A$

$Q_B(t + 1) = D_B$

It is given that,

$A(t + 1) = x'y + xA$

$B(t + 1) = x'B + xA$

Characteristic equation of D flip-flop,

$Q(t + 1) = D$

From the knowledge of above equations we get the following equations for flip-flop input.

$A(t + 1) = D_A \Rightarrow D_A = x'y + xA$

$B(t + 1) = D_B \Rightarrow D_B = x'B + xA$

Using the flip-flop input equations and given output equation the logic diagram of sequential circuit is drawn as shown in Fig. 2.

Here, it is given that output, Z = B

Therefore, the output depends only on states and so it is a Moore machine.

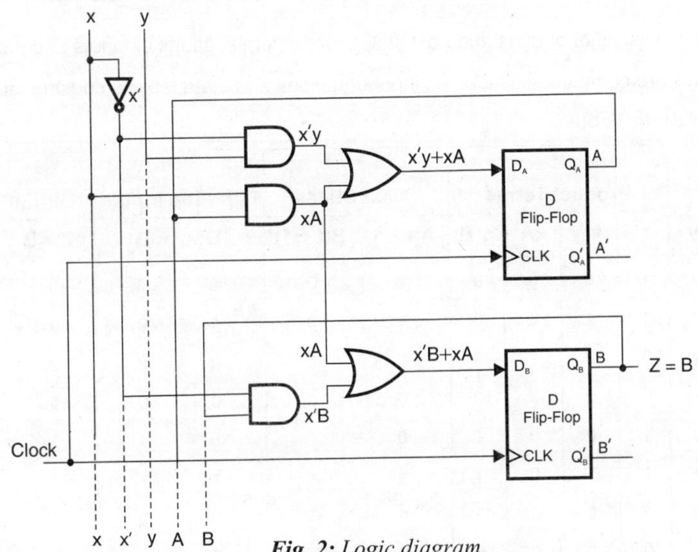

Fig. 2: Logic diagram.

State Table

With n-bit input it is possible to have 2^n different inputs.

Here, n = 2, $\therefore$ $2^2 = 4$, Four possible inputs.

Four possible inputs: xy = 0 0

= 0 1

= 1 0

= 1 1

The circuit has two state variables A and B. With m state variables it is possible to have 2^m different states.

Here, m = 2, $\therefore$ $2^2 = 4$

Therefore, the sequential circuit has four states.

The four possible states are, A B = 0 0

= 0 1

= 1 0

= 1 1

When there are m state variables and n inputs then possible combinations of inputs and states are 2^{m+n}.

Here, m = 2, n = 2

$\therefore$ $2^{m+n} = 2^{2+2} = 2^4 = 16$

The 16 combinations of inputs and states are given below:

when, xy = 0 0, AB = 0 0	when xy = 0 1, AB = 0 0	when xy = 1 0, AB = 0 0	when xy = 1 1, AB = 0 0
= 0 1	= 0 1	= 0 1	= 0 1
= 1 0	= 1 0	= 1 0	= 1 0
= 1 1	= 1 1	= 1 1	= 1 1

The state table is listing of outputs and next states for 16 combinations of inputs and present states.

The next states are computed using given input equations and state table is constructed as shown in Table 1.

Table 1: Expanded State Table

Present States		Inputs		Product Terms				Next States		Flip-Flop Inputs		Output
A	B	x	y	x′	x′y	xA	x′B	A(t + 1)	B(t + 1)	D_A	D_B	Z = B
0	0	0	0	1	0	0	0	0	0	0	0	0
0	0	0	1	1	1	0	0	1	0	1	0	0
0	0	1	0	0	0	0	0	0	0	0	0	0
0	0	1	1	0	0	0	0	0	0	0	0	0
0	1	0	0	1	0	0	1	0	1	0	1	1
0	1	0	1	1	1	0	1	1	1	1	1	1
0	1	1	0	0	0	0	0	0	0	0	0	1
0	1	1	1	0	0	0	0	0	0	0	0	1
1	0	0	0	1	0	0	0	0	0	0	0	0
1	0	0	1	1	1	0	0	1	0	1	0	0
1	0	1	0	0	0	1	0	1	1	1	1	0
1	0	1	1	0	0	1	0	1	1	1	1	0
1	1	0	0	1	0	0	1	0	1	0	1	1
1	1	0	1	1	1	0	1	1	1	1	1	1
1	1	1	0	0	0	1	0	1	1	1	1	1
1	1	1	1	0	0	1	0	1	1	1	1	1

$A(t + 1) = x'y + xA$

$B(t + 1) = x'B + xA$

$D_A = A(t + 1)$

$D_B = B(t + 1)$

***Note:** In D flip-flop the inputs are same as next state flip-flop outputs.*

State Diagram

The possible state transitions from one state to other states are obtained from Table 1 and state transitions are diagrammatically shown in Figs. 3 to 6.

1. While in state 00,

   if input = 00, 10,11, remain in state 00.

   if input = 01, go to state 10.

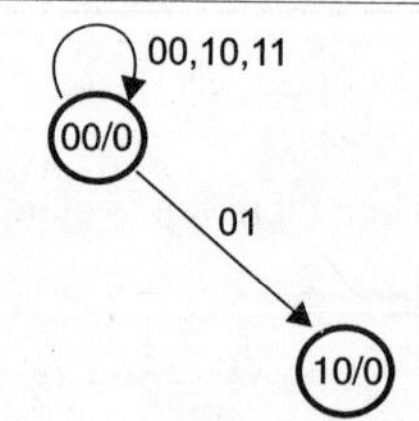

***Fig. 3:** Transitions from state 00.*

2. While in state 01,

if input = 00, remain in state state 01.

if input = 10,11, go to state 00.

if input = 01, go to state 11.

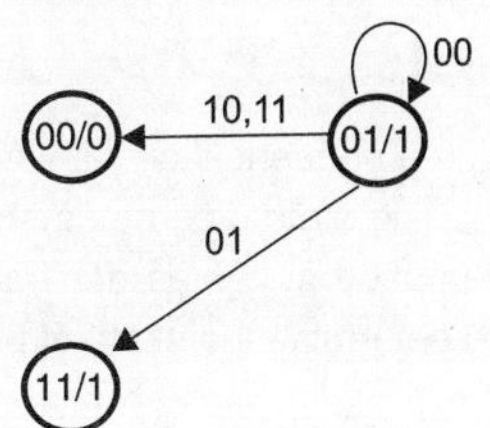

Fig. 4: Transitions from state 01.

3. While in state 10,

if input = 00, go to state 00

if input = 01, go to state 10

if input = 10, 11, go to state 11

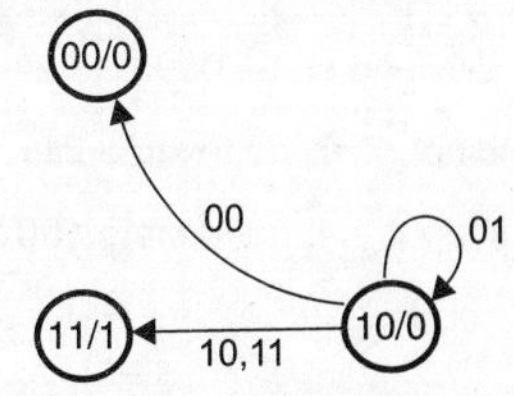

Fig. 5: Transitions from state 10.

4. While in state 11,

if input = 00, go to state 01

if input = 01, 10,11, remain in state 11

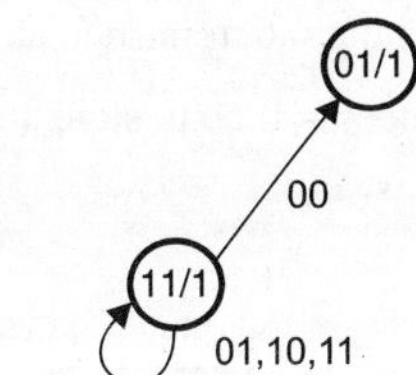

Fig. 6: Transitions from state 11.

On combining the state transitions shown in Figs. 3 to 6, the state diagram can be drawn as shown in Fig. 7. The standard form of state stable is shown in Table 2.

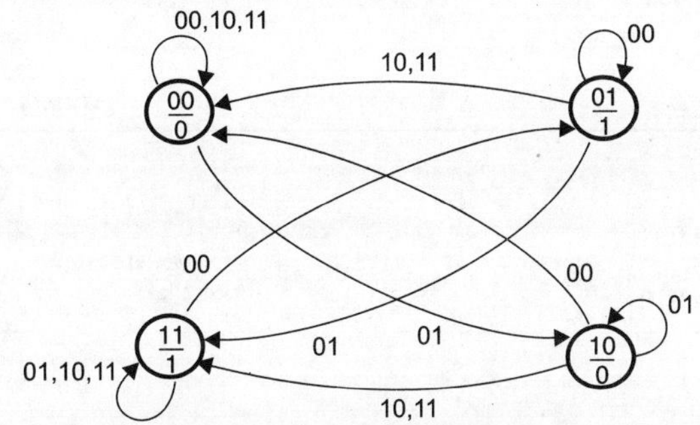

Fig. 7: State diagram.

Table 2: State Table

Present States		Next States								Output			
		xy = 00		xy = 01		xy = 10		xy = 11		xy = 00	xy = 01	xy = 10	xy = 11
A	B	A(t + 1)	B(t + 1)	A(t + 1)	B(t + 1)	A(t + 1)	B(t + 1)	A(t + 1)	B(t + 1)	Z	Z	Z	Z
0	0	0	0	1	0	0	0	0	0	0	0	0	0
0	1	0	1	1	1	0	0	0	0	1	1	1	1
1	0	0	0	1	0	1	1	1	1	0	0	0	0
1	1	0	1	1	1	1	1	1	1	1	1	1	1

Example 7.16

Design a sequential circuit which has three flip-flops A, B, C, one input x and one output Y. The state diagram is shown in Fig. 1. The circuit is to be designed by treating the unused states as don't-care conditions. Analyze the circuit obtain from the design to determine the effect of the unused states. Use D flip-flops in the design.

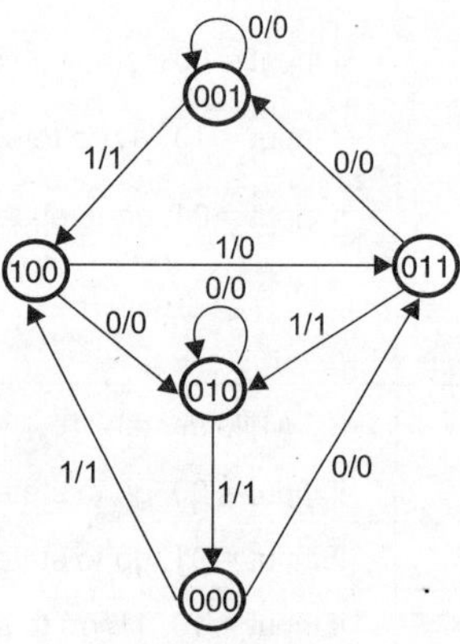

Fig. 1: State diagram.

Solution

From the given state diagram the following state transitions are observed.

1. While in state, 000
   if input = 0, go to state 011
   if input = 1, go to state 100

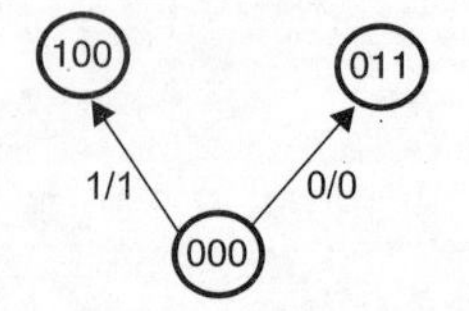

Fig. 2: Transitions from state 000.

2. While in state, 001
   if input = 0, remain in state 001
   if input = 1, go to state 100

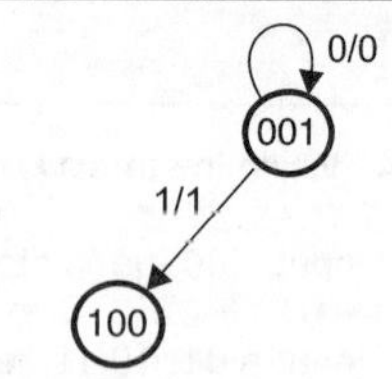

Fig. 3: Transitions from state 001.

3. While in state, 010
   if input = 0, remain in state 010
   if input = 1, go to state 000

Fig. 4: Transitions from state 010.

4. While in state, 011
   if input = 0, go to state 001
   if input = 1, go to state 010

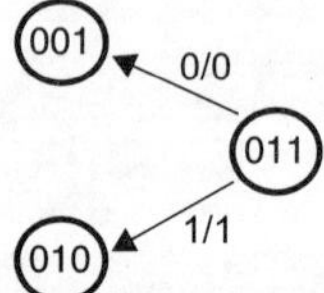

Fig. 5: Transitions from state 011.

5. While in state, 100
   if input = 0, go to state 010
   if input = 1, go to state 011

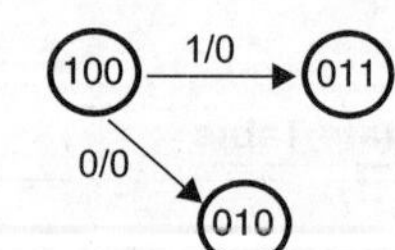

Fig. 6: Transitions from state 100.

Here, each bit of binary code is a state variable. Hence the sequential circuit has 3 state variables. Let the state variables be A, B and C.

∴ Present state variables : A(t), B(t), C(t) $\Rightarrow$ A, B, C (For simplicity)

Next state variables : A(t + 1), B(t + 1), C(t + 1)

The state table (Table 1) is constructed from the knowledge of state transitions shown in Figs. 2 to 6.

Table 1: State Table

Present States			Input	Next States			Output
A	B	C	x	A(t + 1)	B(t + 1)	C(t + 1)	Y
0	0	0	0	0	1	1	0
0	0	0	1	1	0	0	1
0	0	1	0	0	0	1	0
0	0	1	1	1	0	0	1
0	1	0	0	0	1	0	0
0	1	0	1	0	0	0	1
0	1	1	0	0	0	1	0
0	1	1	1	0	1	0	1
1	0	0	0	0	1	0	0
1	0	0	1	0	1	1	0
1	0	1	0	×	×	×	×
1	0	1	1	×	×	×	×
1	1	0	0	×	×	×	×
1	1	0	1	×	×	×	×
1	1	1	0	×	×	×	×
1	1	1	1	×	×	×	×

Let us choose three D flip-flops as shown in Fig. 7. Here, A, B and C be output of flip-flops.

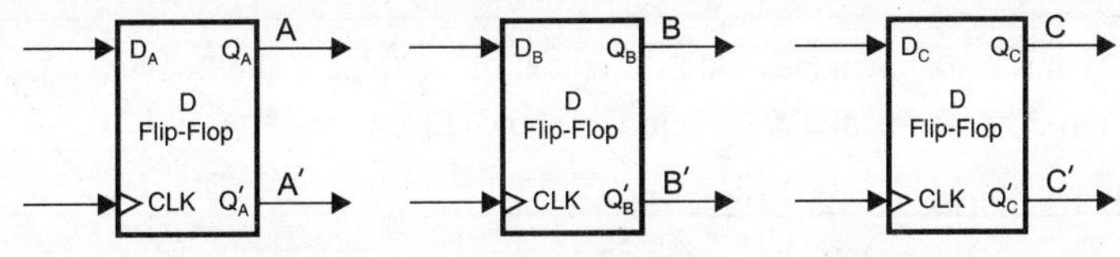

Fig. a: Flip-flop A. *Fig. b: Flip-flop B.* *Fig. c: Flip-flop C.*

Fig. 7: D flip-flop to store state variables.

Here, $A = Q_A$ $\qquad A(t+1) = Q_A(t+1)$

$B = Q_B \quad \Rightarrow \quad B(t+1) = Q_B(t+1)$

$C = Q_C \qquad C(t+1) = Q_C(t+1)$

The characteristic equation of three D flip-flops are,

$Q_A(t+1) = D_A$

$Q_B(t+1) = D_B$

$Q_C(t+1) = D_C$

Characteristic equation of D flip-flop,
$Q(t+1) = D$

Using the above equations we can write,

$A(t+1) = D_A$

$B(t+1) = D_B$

$C(t+1) = D_C$

Using the above equations the flip-flop inputs are determined for the next state and listed in Table 2. Here, the flip-flop inputs are same as next state outputs.

Table 2: Expanded State Table

Present States			Input	Next States			Flip-Flop Inputs			Output	Minterm
A	B	C	x	A(t+1)	B(t+1)	C(t+1)	D_A	D_B	D_C	Y	ABCx
0	0	0	0	0	1	1	0	1	1	0	m_0
0	0	0	1	1	0	0	1	0	0	1	m_1
0	0	1	0	0	0	1	0	0	1	0	m_2
0	0	1	1	1	0	0	1	0	0	1	m_3
0	1	0	0	0	1	0	0	1	0	0	m_4
0	1	0	1	0	0	0	0	0	0	1	m_5
0	1	1	0	0	0	1	0	0	1	0	m_6
0	1	1	1	0	1	0	0	1	0	1	m_7
1	0	0	0	0	1	0	0	1	0	0	m_8
1	0	0	1	0	1	1	0	1	1	0	m_9
1	0	1	0	×	×	×	×	×	×	×	m_{10}
1	0	1	1	×	×	×	×	×	×	×	m_{11}
1	1	0	0	×	×	×	×	×	×	×	m_{12}
1	1	0	1	×	×	×	×	×	×	×	m_{13}
1	1	1	0	×	×	×	×	×	×	×	m_{14}
1	1	1	1	×	×	×	×	×	×	×	m_{15}

Note: *Binary codes 101, 110 and 111 are unused codes and so next state and outputs are considered as don't-care for this codes.*

Here, D_A, D_B, D_C and Y are functions of A, B, C and x. Using Table 2, the Boolean equations for D_A, D_B, D_C and Y are determined as sum of minterms and don't-cares as shown below:

$$D_A = \Sigma m(1, 3) + \Sigma d(10, 11, 12, 13, 14, 15)$$

$$D_B = \Sigma m(0, 4, 7, 8, 9) + \Sigma d(10, 11, 12, 13, 14, 15)$$

$$D_C = \Sigma m(0, 2, 6, 9) + \Sigma d(10, 11, 12, 13, 14, 15)$$

$$Y = \Sigma m(1, 3, 5, 7) + \Sigma d(10, 11, 12, 13, 14, 15)$$

Using the above equations K-maps are drawn for D_A, D_B, D_C and Y in order to obtain simplified Boolean equations.

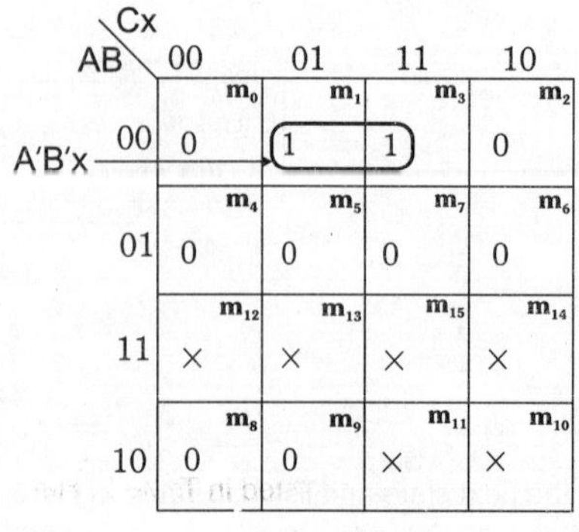

Fig. 8: K-map for D_A.

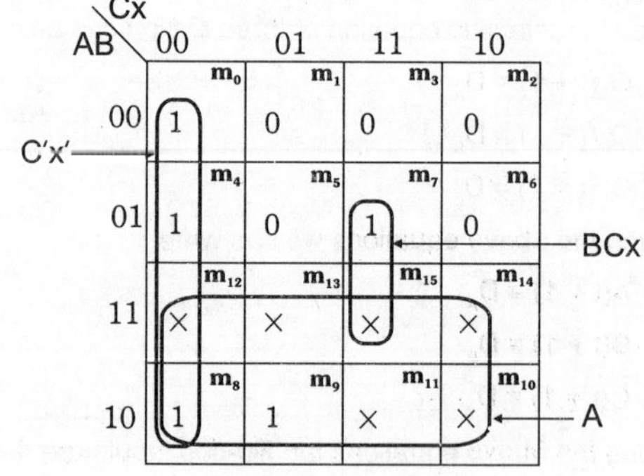

Fig. 9: K-map for D_B.

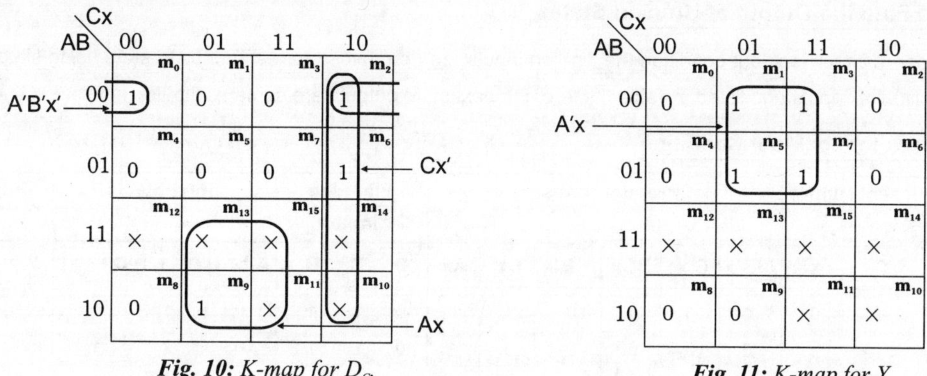

Fig. 10: K-map for D_C. *Fig. 11: K-map for Y.*

From the K-map we get the following simplified Boolean equations to realize sequential circuit using D flip-flops.

$D_A = A'B'x$

$D_B = A + C'x' + BCx$

$D_C = A'B'x' + Cx' + Ax$

$Y = A'x$

Using the above equations the logic diagram of sequential circuit is drawn as shown in Fig. 12. Here, the output depends on input and states and so this model is Mealy model.

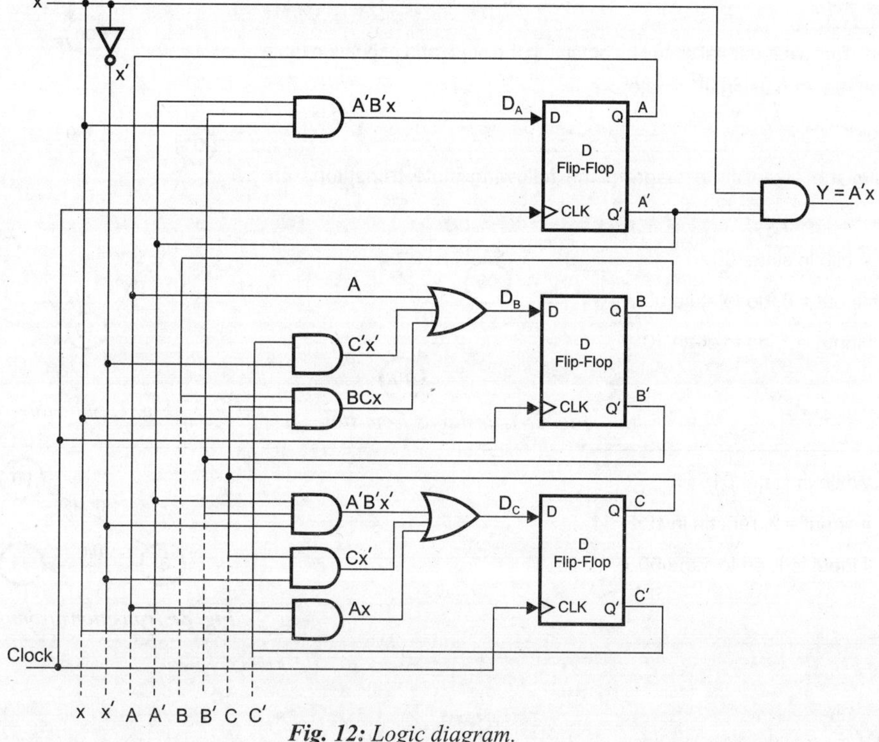

Fig. 12: Logic diagram.

Analysis to Find the Output of Unused States

In order to check lock-out condition the flip-flop inputs are determined for each unused state using simplified equations of flip-flop inputs and listed in Table 3. Here, the next state outputs are same as flip-flop inputs.

Table 3: Check for Lock-out Condition

Unused Present States			Input	Product Terms						Flip-Flop Inputs			Next States			Output
A	B	C	x	A'B'x	C'x'	BCx	A'B'x'	Cx'	Ax	D_A	D_B	D_C	A(t+1)	B(t + 1)	C(t + 1)	Y = A'x
1	0	1	0	0	0	0	0	1	0	0	1	1	0	1	1	0
1	0	1	1	0	0	0	0	0	1	0	1	1	0	1	1	0
1	1	0	0	0	1	0	0	0	0	0	1	0	0	1	0	0
1	1	0	1	0	0	0	0	0	1	0	1	1	0	1	1	0
1	1	1	0	0	0	0	0	1	0	0	1	1	0	1	1	0
1	1	1	1	0	0	1	0	0	1	0	1	1	0	1	1	0

$D_A = A'B'x$; $D_B = A + C'x' + BCx$; $D_C = A'B'x' + Cx' + Ax$

From the Table 3 we can observe the following state transitions from unused state to next state.

$101 \xrightarrow{0/0} 011,\ 101 \xrightarrow{1/0} 011,\ 110 \xrightarrow{0/0} 010,\ 110 \xrightarrow{1/0} 011,\ 111 \xrightarrow{0/0} 011,\ 111 \xrightarrow{1/0} 011$

The unused states becomes used state in the next transition as shown above and so there is no lock-out condition in this design.

Example 7.17

(VTU, Aug/Sep'20, 10 Marks)

Construct a sequential circuit by obtaining the state and excitation table for the given diagram using JK flip-flop.

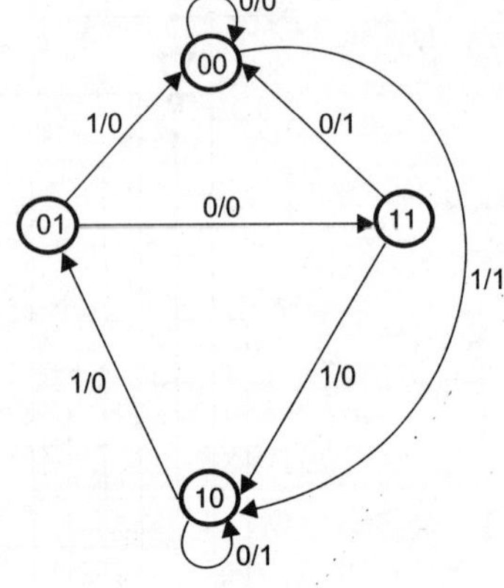

Fig. 1: *State diagram.*

Solution

From the given state diagram the following state transitions are observed.

1. While in state, 00

   if input = 0, go to state 00

   if input = 1, go to state 10

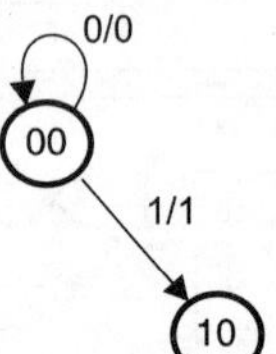

Fig. 2: *Transitions from state 00.*

2. While in state, 01

   if input = 0, remain in state 11

   if input = 1, go to state 00

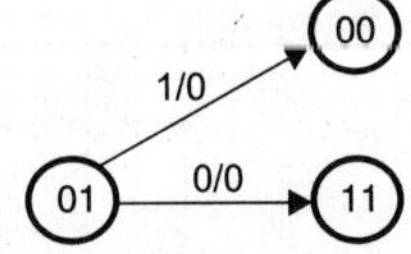

Fig. 3: *Transitions from state 01.*

3. While in state, 10

   if input = 0, remain in state 10

   if input = 1, go to state 01

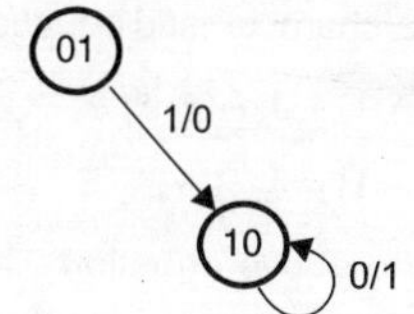

Fig. 4: *Transitions from state 10.*

4. While in state, 11

   if input = 0, go to state 00

   if input = 1, go to state 10

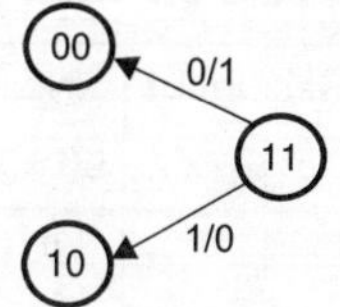

Fig. 5: *Transitions from state 11.*

State Table

Let, x be the input and Y be the output. The state table (Table 1) can be formed from state diagram.

Table 1: State Table

Present States		Next States				Output	
		x = 0		x = 1		x = 0	x = 1
A	B	A(t+1)	B(t+1)	A(t+1)	B(t+1)	Y	Y
0	0	0	0	1	0	0	1
0	1	1	1	0	0	0	0
1	0	1	0	0	1	1	0
1	1	0	0	1	0	1	0

Design using JK Flip-Flop

Here, there are two state variables. Let A and B be the two state variables.

∴ Present state variables : A(t), B(t) ⇒ A, B (For simplicity)

∴ Next state variables : A(t + 1), B(t + 1)

Let us choose two flip-flops to store two state variables A and B as shown in Fig. 6. Here, A and B are output of flip-flops.

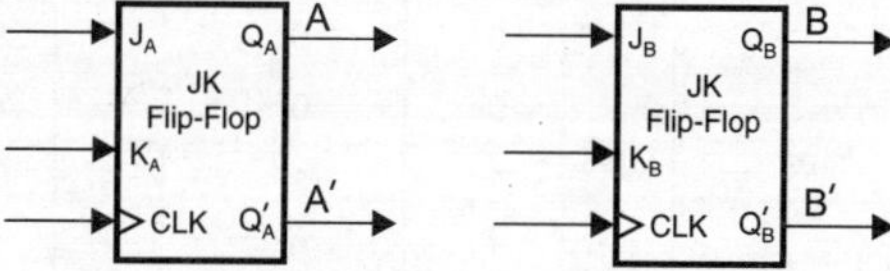

Fig. a: *Flip-flop A.* ***Fig. a:*** *Flip-flop B.*

Fig. 6: *JK flip-flops to store state variables.*

Excitation Table of JK Flip-Flop

Present State Q(t)	Next State Q(t + 1)	Flip-Flop Inputs J	K
0	0	0	×
0	1	1	×
1	0	×	1
1	1	×	0

The characteristic equation of two JK flip-flops are,

$$Q_A(t + 1) = J_A Q_A + K_A Q_A$$

$$Q_B(t + 1) = J_B Q_B + K_B Q_B$$

The two state variables can be equated to flip-flop outputs.

$$\therefore A = Q_A \Rightarrow A(t + 1) = Q_A(t + 1)$$

$$B = Q_B \Rightarrow B(t + 1) = Q_B(t + 1)$$

Now the characteristic equations in terms of state variables are,

$A(t + 1) = J_A A' + K_A A$

$B(t + 1) = J_B B' + K_B B$

Using the above equations the flip-flop inputs are determined from the next state and listed in Table 2. The D flip-flop the input is same as next state output.

Table 2: Expanded State Table

Present States		Input	Next States		Output	Minterm
A	B	x	A(t+1)	B(t+1)	Y	ABx
0	0	0	0	0	0	m_0
0	0	1	1	0	1	m_1
0	1	0	1	1	0	m_2
0	1	1	0	0	0	m_3
1	0	0	1	0	1	m_4
1	0	1	0	1	0	m_5
1	1	0	0	0	1	m_6
1	1	1	1	0	0	m_7

The inputs J_A, J_B, K_A, K_B of flip-flops can be determined using excitation table as shown below:

State A

For state A the Q(t), Q(t + 1), J and K of excitation table are considered as A, A(t + 1), J_A and K_A.

$\therefore Q(t) \to A \quad ; \quad Q(t+1) \to A(t+1)$

$J \to J_A \quad ; \quad K \to K_A$

The flip-flop inputs J_A and K_A that drives the present state A to next states A(t + 1) are obtained from excitation table and listed in Table 3.

Table 3:

A	A(t + 1)	J_A	K_A
0	0	0	×
0	1	1	×
0	1	1	×
0	0	0	×
1	1	×	0
1	0	×	1
1	0	×	1
1	1	×	0

State B

For state B the Q(t), Q(t + 1), J and K of excitation table are considered as B, B(t + 1), J_B and K_B.

$\therefore Q(t) \to B \quad ; \quad Q(t+1) \to B(t+1)$

$J \to J_B \quad ; \quad K \to K_B$

The flip-flop inputs J_B and K_B that drives the present state B to next states B(t + 1) are obtained from excitation table and listed in Table 4.

Table 4:

B	B(t + 1)	J_B	K_B
0	0	0	×
0	0	0	×
1	1	×	0
1	0	×	1
0	0	0	×
0	1	1	×
1	0	×	1
1	0	×	1

The state table with flip-flop inputs is shown in Table 5.

Table 5: State Table with JK Flip-Flop Inputs

Present States		Input	Next States		Flip-Flop Inputs				Output	Minterm
A	B	x	A(t+1)	B(t+1)	J_A	K_A	J_B	K_B	Y	ABx
0	0	0	0	0	0	×	0	×	0	m_0
0	0	1	1	0	1	×	0	×	1	m_1
0	1	0	1	1	1	×	×	0	0	m_2
0	1	1	0	0	0	×	×	1	0	m_3
1	0	0	1	0	×	0	0	×	1	m_4
1	0	1	0	1	×	1	1	×	0	m_5
1	1	0	0	0	×	1	×	1	1	m_6
1	1	1	1	0	×	0	×	1	0	m_7

The inputs J_A, K_A, J_B, K_B are functions of present state and input x. Therefore, minterms are same as in design using D flip-flops. J_A, K_A, J_B, K_B are functions of A, B, x and using Table 6, J_A, K_A, J_B, K_B are expressed as sum of minterms and don't-cares as shown below:

$J_A = \Sigma m(1, 2) + \Sigma d(4, 5, 6, 7)$

$K_A = \Sigma m(5, 6) + \Sigma d(0, 1, 2, 3)$

$Y = \Sigma m(1, 4, 6)$

$J_B = \Sigma m(5) + \Sigma d(2, 3, 6, 7)$

$K_B = \Sigma m(3, 6, 7) + \Sigma d(0, 1, 4, 5)$

Using the above equations, let us draw K-map for J_A, K_A, J_B, K_B and Y to develop simplified Boolean equations for J_A, K_A, J_B, K_B and Y.

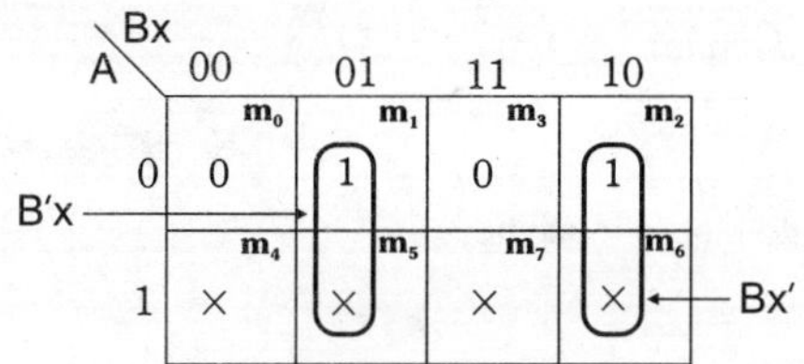

Fig. 7: K-map for J_A.

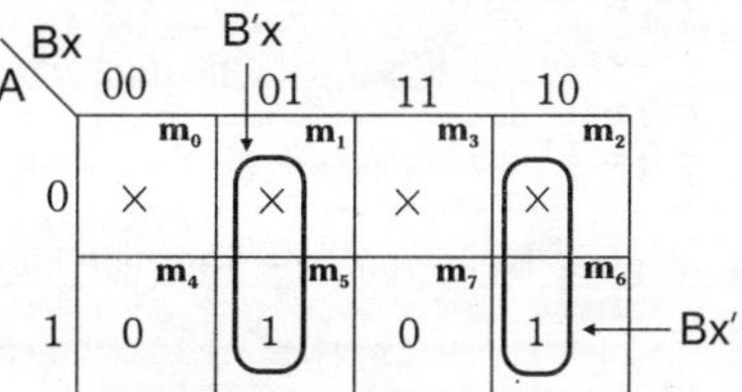

Fig. 8: K-map for K_A.

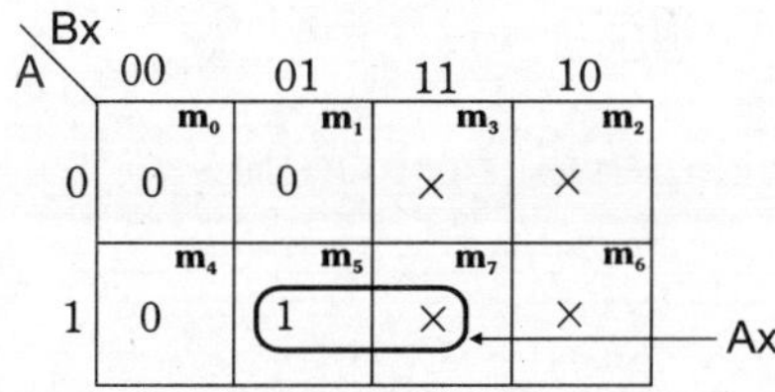

Fig. 9: K-map for J_B.

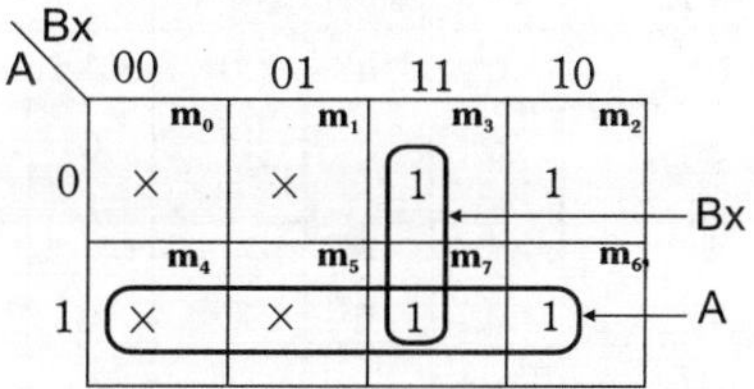

Fig. 10: K-map for K_B.

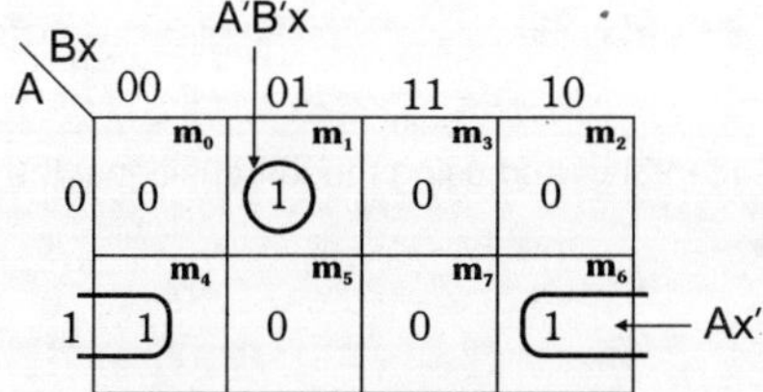

Fig. 11: K-map for Y.

From the K-map we get the following simplified Boolean equations for realizing sequence detector using JK flip-flops.

$J_A = K_A = Bx' + B'x$ | $K_B = A + Bx$

$J_B = Ax$ | $Y = Ax' + A'B'x$

The logic diagram of sequence detector using above equations is drawn as shown in Fig. 12.

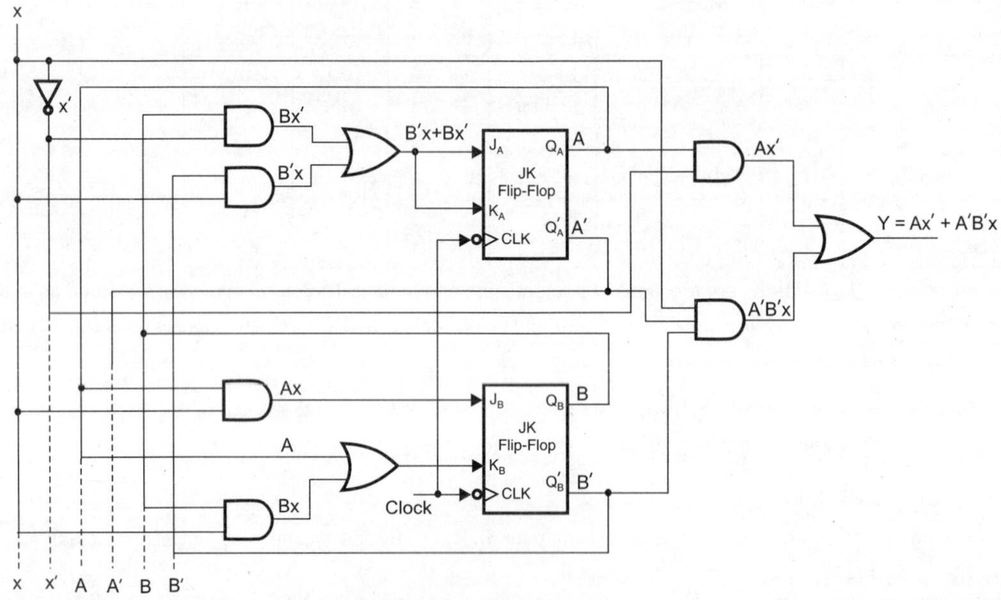

Fig. 12: Logic diagram of sequential circuit.

Example 7.18

(JNTUK, Feb'22, 8 Marks)

For the given Mealy machine transition table convert into Moore machine.

Table 1:

Present State	Next State		Output	
Q(t)	**x = 0**	**x = 1**	**x = 0**	**x = 1**
q^0	q^3	q^{11}	0	1
q^{10}	q^0	q^3	1	0
q^{11}	q^0	q^3	1	0
q^{20}	q^{21}	q^{20}	1	0
q^{21}	q^{21}	q^{20}	1	0
q^3	q^{10}	q^0	0	1

Solution

From the state table the following state transitions are identified and using these transitions, the state diagram Mealy machine is drawn as shown in Fig. 1.

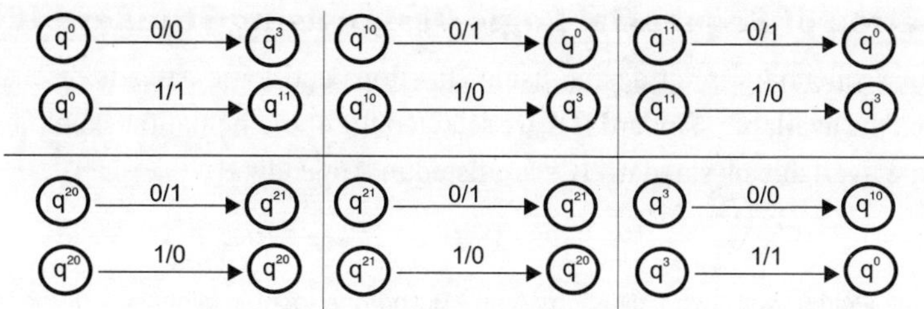

Fig. 1: Transitions from one state to another state.

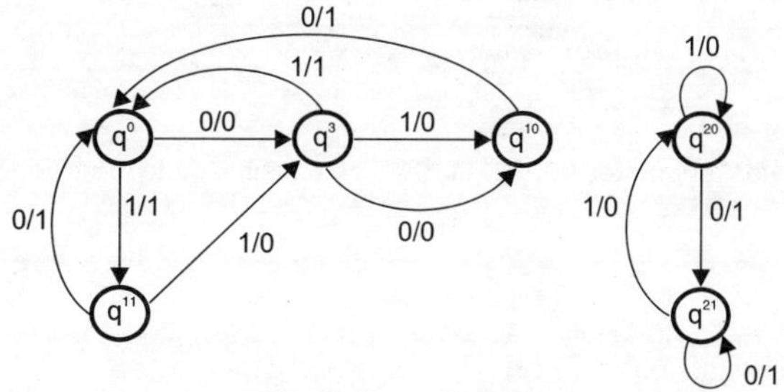

Fig. 2: State diagram.

In the state diagram of Fig. 2, the following observations are made and based on the observations the outputs are assigned to states in order to draw Moore machine state diagram.

1. When the next state is q^0, the outputs are 1. Hence the state q^0 can be replaced by $q^0/1$.
2. When the next state is q^3, the outputs are 0. Hence the state q^3 can be replaced by $q^3/0$.
3. When the next state is q^{10}, the outputs are 0. Hence the state q^{10} can be replaced by $q^{10}/0$.
4. When the next state is q^{11}, the output is 1. Hence the state q^{11} can be replaced by $q^{11}/1$.
5. When the next state is q^{20}, the outputs are 0. Hence the state q^{20} can be replaced by $q^{20}/0$.
6. When the next state is q^{21}, the outputs are 1. Hence the state q^{21} can be replaced by $q^{21}/1$.

The Moore machine state diagram with output assigned to states is drawn as shown in Fig. 3.

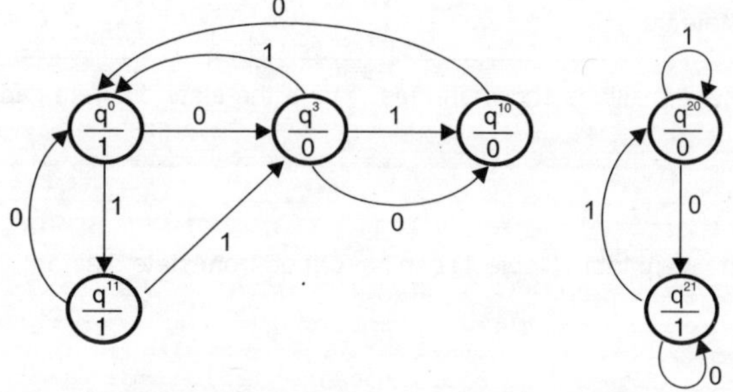

Fig. 3: State diagram for Moore machine.

7.4 Implementation of Sequential Logic Design using Standard IC's

The concept of sequential circuit design using flip-flops and logic gates is discussed in previous section. In this section, the available standard ICs are selected for implementation examples. The flip-flops and logic gates that are available as standard ICs are listed in Appendix-1.

Example 7.19

Design a sequential circuit with two D filp-flops A and B and one input x. When x = 0, the state of the circuit remains the same. When x = 1, the circuit goes through the state transistions from 00 to 10 to 11 to 01, back to 00 and repeats. Implement the logic circuit using available standard ICs.

Solution

State Diagram

The sequential circuit has four states 00, 10, 11, 01. The possible state transitions from one state to other states are shown in Figs. 1 to 4.

1. While in state, 00

   if input = 0, remain in state 00

   if input = 1, go to state 10

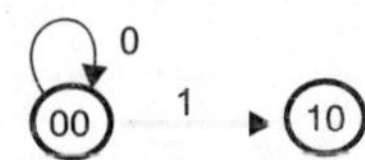

Fig. 1: Transitions from state 00.

2. While in state, 10

   if input = 0, remain in state 10

   if input = 1, go to state 11

Fig. 2: Transitions from state 10.

3. While in state, 11

   if input = 0, remain in state 11

   if input = 1, go to state 01

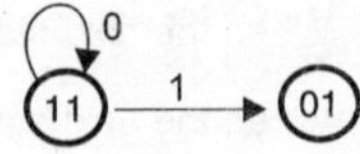

Fig. 3: Transitions from state 11.

4. While in state, 01

   if input = 0, remain in state 01

   if input = 1, go to state 00

Fig. 4: Transitions from state 01.

On combining the state transitions shown in Figs. 1 to 4, the state diagram can be drawn as shown in Fig. 5.

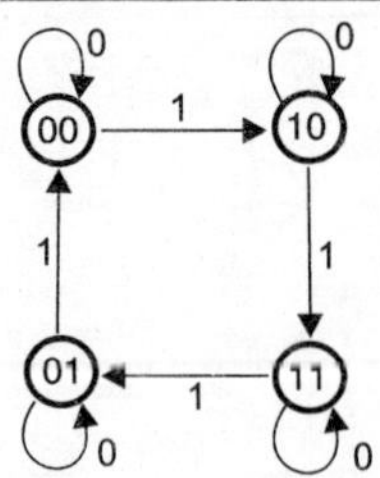

Fig. 5: State diagram.

State Table

Let, x be the input. The state table (Table 1) can be formed from state diagram.

Table 1: State Table

Present States	Next State x = 0	Next State x = 1
0 0	0 0	1 0
0 1	0 1	0 0
1 0	1 0	1 1
1 1	1 1	0 1

Design using D Flip-Flop

Here, there are two state variables. Let A and B be the two state variables.

$\therefore$ Present state variables : A(t), B(t) $\Rightarrow$ A, B (For simplicity)

$\therefore$ Next state variables : A(t + 1), B(t + 1)

Let us choose a standard dual D flip-flop IC 7474 to store two state variables A and B as shown in Fig. 6. Here, A and B are output of flip-flops.

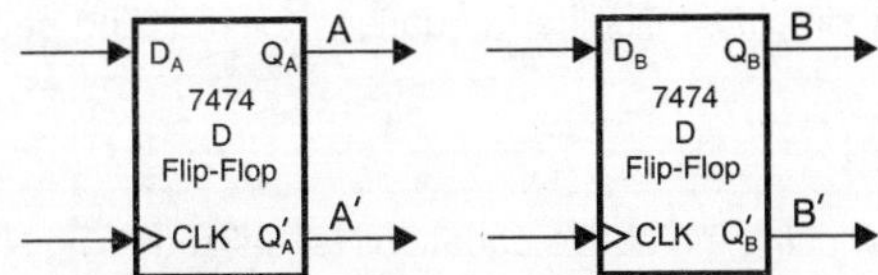

Fig. a: Flip-flop A. *Fig. b: Flip-flop B.*

Fig. 6: D flip-flop to store state variables.

Here, $A = Q_A \Rightarrow A(t + 1) = Q_A(t + 1)$

$B = Q_B \Rightarrow B(t + 1) = Q_B(t + 1)$

The characteristic equation of two D flip-flops are,

$Q_A(t + 1) = D_A$

$Q_B(t + 1) = D_B$

Characteristic equation of D flip-flop,
$Q(t + 1) = D$

Using the above equation we can write,

$A(t + 1) = D_A$

$B(t + 1) = D_B$

Using the above equations the flip-flop inputs are determined for the next state and listed in Table 3. Here, the flip-flop inputs are same as next state outputs.

Table 3: Expanded State Table

Present States		Input	Next States		Flip-Flop Inputs		Minterm
A	**B**	**x**	**A(t+1)**	**B(t+1)**	D_A	D_B	**ABx**
0	0	0	0	0	0	0	m_0
0	0	1	1	0	1	0	m_1
0	1	0	0	1	0	1	m_2
0	1	1	0	0	0	0	m_3
1	0	0	1	0	1	0	m_4
1	0	1	1	1	1	1	m_5
1	1	0	1	1	1	1	m_6
1	1	1	0	1	0	1	m_7

Here, D_A and D_B are functions of A, B, x. Using Table 3, the Boolean equations for D_A and D_B are determined as sum of minterms as shown below:

$\therefore D_A = \Sigma m(1, 4, 5, 6)$; $D_B = \Sigma m(2, 5, 6, 7)$

Using the above equations, let us draw K-map for D_A and D_B to develop simplified Boolean equations for D_A and D_B.

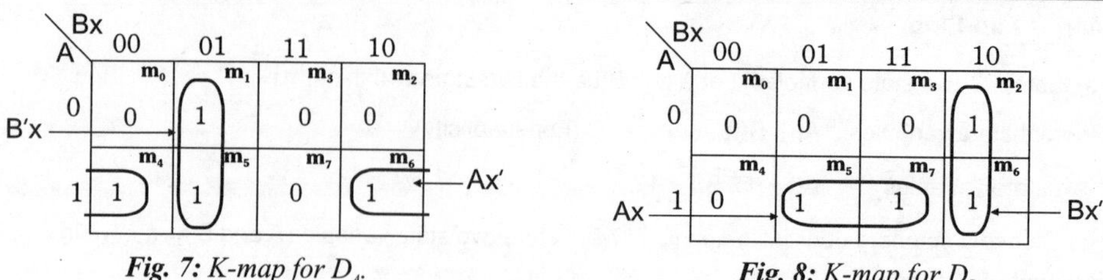

Fig. 7: K-map for D_A. *Fig. 8: K-map for D_B.*

From the K-map we get the following simplified Boolean equations for realizing sequence detector using D flip-flops,

$$D_A = Ax' + B'x \quad ; \quad D_B = Ax + Bx'$$

The logic diagram of sequence detector using the above equations is realized using NOT gate IC 7404, 2-input AND gate IC 7408 and OR gate IC 7432 and dual D flip-flop IC 7474 as shown in Fig. 9. Here the sequence detector output is the present state output of flip-flops as shown in Fig. 9.

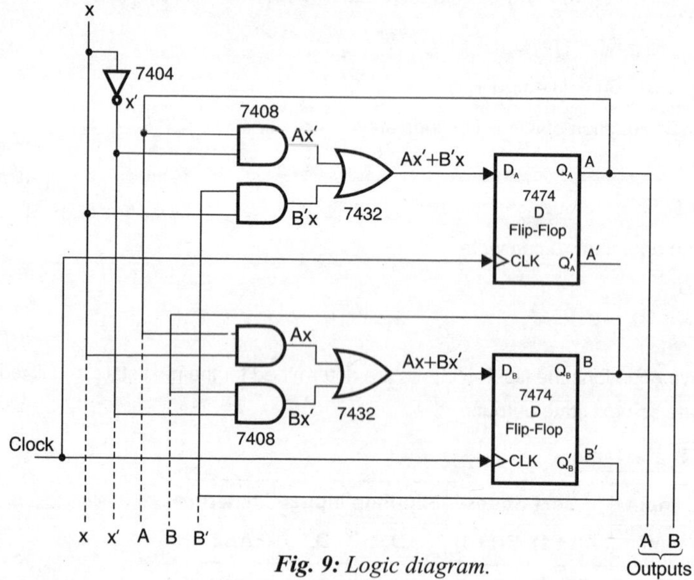

Fig. 9: Logic diagram.

Example 7.20

Draw the state diagram for the sequential circuit whose state table is shown in Table 1. Also, design a logic circuit using JK flip-flop. Implement the logic circuit using available standard ICs.

Table 1: State Table of a Sequential Circuit

Present States		Input	Next States		Output
A	B	x	A(t + 1)	B(t + 1)	Y
0	0	0	0	0	0
0	0	1	0	1	0
0	1	0	1	0	0
0	1	1	0	1	0
1	0	0	1	0	0
1	0	1	1	1	0
1	1	0	1	1	1
1	1	1	0	0	1

Solution

In the given state stable it can be observed that the output does not depends on input and so output can be associated with states. Hence the given model is Moore model.

The various state transitions listed in Table 1 are diagrammatically shown in Fig. 1.

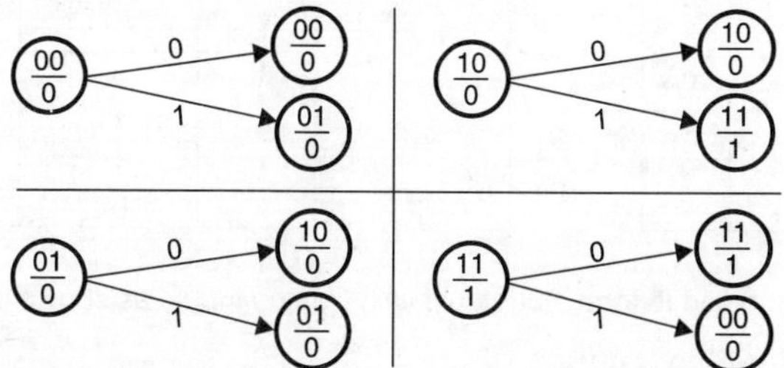

Fig. 1: Transitions from one state to another state.

On combining the state transitions of Fig. 1 into single diagram we get the Moore model state diagram shown in Fig. 2.

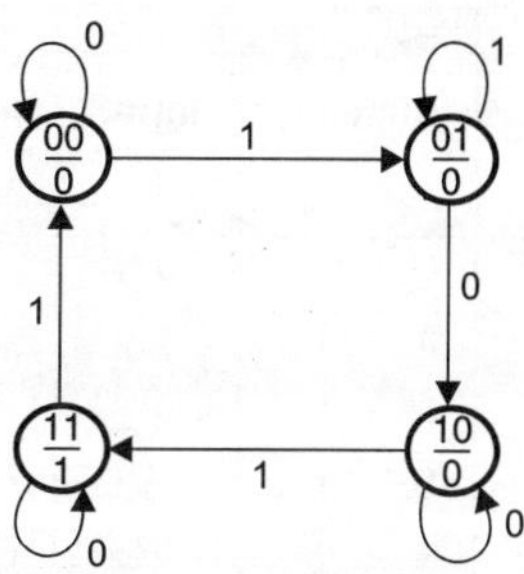

Fig. 2: State diagram.

The given state stable has two state variables A_0 and A_1. In sequential circuits, each state variable is stored in a flip-flop. Hence, choose two JK flip-flops IC 74112 to store two state variables A_0 and A_1 as shown in Fig. 3.

$\therefore$ Present state variables : $A_0(t), A_1(t) \Rightarrow A_0, A_1$ (For simplicity)

Next state variables : $A_0(t + 1), A_1(t + 1)$

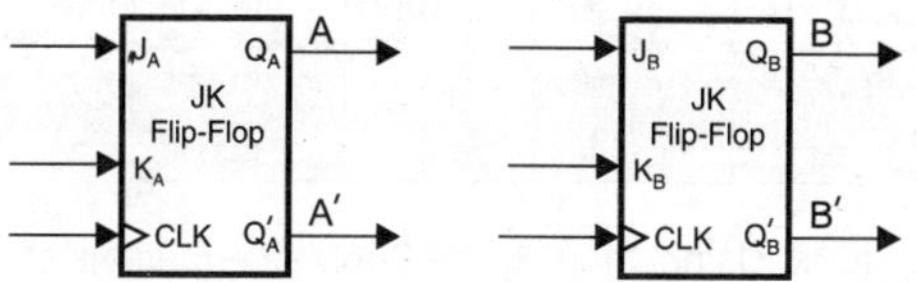

Fig. a: Flip-flop A. *Fig. b: Flip-flop B.*

Fig. 3: JK flip-flops to store state variables.

Excitation Table of JK Flip-Flop

Present State Q(t)	Next State Q(t + 1)	Flip-Flop Inputs J	K
0	0	0	×
0	1	1	×
1	0	×	1
1	1	×	0

The flip-flops inputs can be obtained from excitation table as shown below:

State A_1

For state A_1 the Q(t), Q(t + 1), J and K of excitation table are considered as A_1, $A_1(t + 1)$, J_1 and K_1.

$$\therefore Q(t) \to A_1 \quad ; \quad Q(t+1) \to A_1(t+1)$$

$$J \to J_1 \quad ; \quad K \to K_1$$

The flip-flop inputs J_1 and K_1 that drives the present state A_1 to next state $A_1(t+1)$ are obtained from excitation table and listed in Table 2.

State A_0

For state A_0 the Q(t), Q(t + 1), J and K of excitation table are considered as A_0, $A_0(t + 1)$, J_0 and K_0.

$$\therefore Q(t) \to A_0 \quad ; \quad Q(t+1) \to A_0(t+1)$$

$$J \to J_0 \quad ; \quad K \to K_0$$

The flip-flop inputs J_B and K_B that drives the present state A_0 to next state $A_0(t + 1)$ are obtained from excitation table and listed in Table 3.

Table 2:

A_1	$A_1(t+1)$	J_1	K_1
0	0	0	×
0	0	0	×
0	1	1	×
0	·0	0	×
1	1	×	0
1	1	×	0
1	1	×	0
1	0	×	1

Table 3:

A_0	$A_0(t+1)$	J_0	K_0
0	0	0	×
0	1	1	×
1	0	×	1
1	1	×	0
0	0	0	×
0	1	1	×
1	1	×	0
1	0	×	1

The Tables 2 and 3 are combined to form state table with flip-flop inputs as shown in Table 4.

Table 4: State Table with JK Flip-Flop Inputs

Present States		Input	Next States		Flip-Flop Inputs				Output	Minterm
A_1	A_0	x	$A_1(t+1)$	$A_0(t+1)$	J_1	K_1	J_0	K_0	Y	ABx
0	0	0	0	0	0	×	0	×	0	m_0
0	0	1	0	1	0	×	1	×	0	m_1
0	1	0	1	0	1	×	×	1	0	m_2
0	1	1	0	1	0	×	×	0	0	m_3
1	0	0	1	0	×	0	0	×	0	m_4
1	0	1	1	1	×	0	1	×	0	m_5
1	1	0	1	1	×	0	×	0	1	m_6
1	1	1	0	0	×	1	×	1	1	m_7

Now, J_1, K_1, J_0, K_0 and Y are functions of A_1, A_0, x. Using Table 4, the Boolean equations of J_1, K_1, J_0, K_0 and Y are obtained as sum of minterms and don't-cares as shown below:

$J_1 = m_2 + \Sigma d(4, 5, 6, 7)$

$K_1 = m_7 + \Sigma d(0, 1, 2, 3)$

$Y = \Sigma m(6, 7)$

$J_0 = \Sigma m(1, 5) + \Sigma d(2, 3, 6, 7)$

$K_0 = \Sigma m(2, 7) + \Sigma d(0, 1, 4, 5)$

Let us draw K-map for J_1, K_1, J_0, K_0 and Y as shown in Figs. 4 to 8 and develop simplified Boolean equations to implement logic diagram.

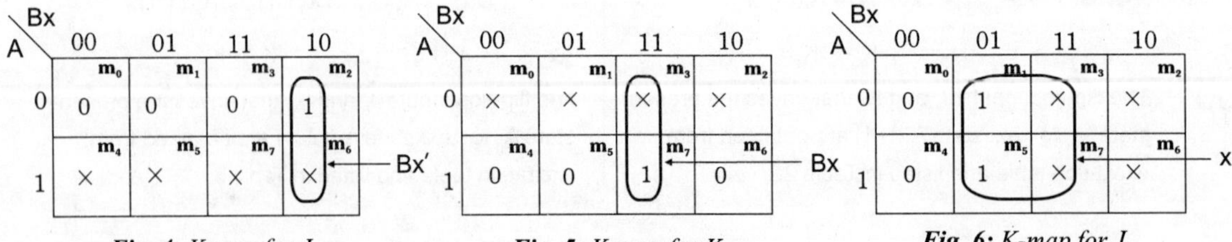

Fig. 4: K-map for J_1. *Fig. 5: K-map for K_1.* *Fig. 6: K-map for J_0.*

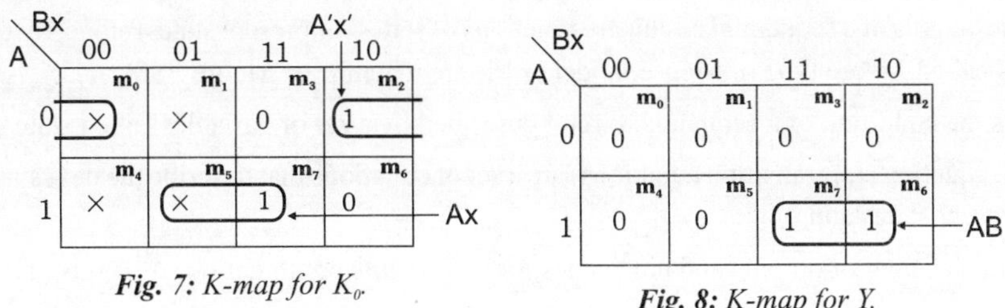

Fig. 7: K-map for K_0.

Fig. 8: K-map for Y.

From the K-maps we get the following simplified Boolean equations for realizing the given state diagrams using JK flip-flops,

$J_1 = Bx'$	$K_1 = Bx$
$J_0 = x$	$K_0 = A'x' + Ax$
$Y = AB$	

The logic diagram using above equations is drawn as shown in Fig. 9.

Using the simplified Boolean equations obtained from K-map the logic diagram of sequential circuit is drawn as shown in Fig. 9. The logic circuit is implemented using 2-input AND gate IC 7408 and OR gate IC 7432 and dual JK flip-flop IC 74112.

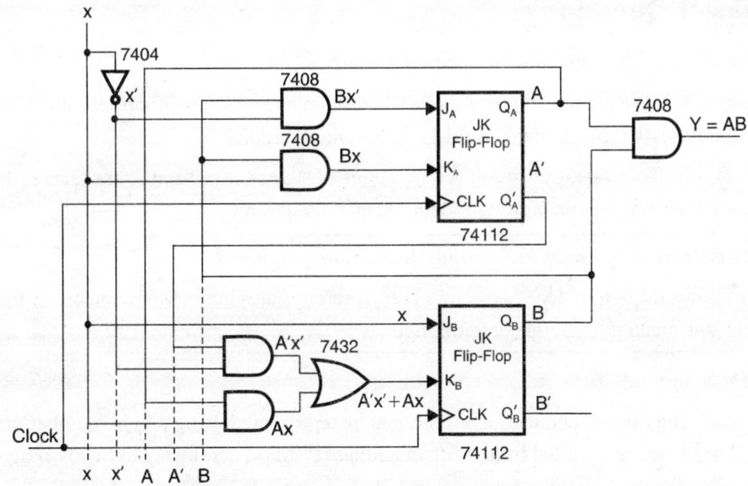

Fig. 9: Logic diagram of sequential circuit.

7.5 Summary of Important Concepts

1. Sequential logic circuits are digital circuits whose output depends on present inputs and past outputs.
2. A sequential logic circuit is a combinational circuit with storage elements and feedback paths.
3. The sequential circuits without a clock are called asynchronous sequential circuits.
4. The sequential circuits with a clock are called synchronous sequential circuits.
5. Since, the functional behaviour of sequential circuits are represented using a finite number of states they are also called finite state machines (FSM).
6. When the outputs of a sequential circuit are functions of inputs and states (or present states) then the sequential circuit is called Mealy type or Mealy model or Mealy machine or Mealy FSM.

7. When the outputs of a sequential circuit are functions of states (or present states) alone, then the sequential circuit is called Moore type or Moore model or Moore machine or Moore FSM.
8. A state is the condition of a sequential circuit described by a set of variables called state variables.
9. The state equations (or transition equations) are a set of equations that describe the next state as a function of the present states and inputs.
10. State table is listing of outputs and next states for all possible combinations of inputs and present states.
11. If there are m state variables and n inputs then the state table list the outputs and next states for 2^{m+n} combinations of inputs and states.
12. State diagram is graphical representation of all transitions from present state to next state for all possible combination of inputs and outputs. Therefore, the state diagram is the graphical representation of information available in a state stable.
13. Two states are said to be equivalent if the input-output relationship are same with identical present state and changes in input, which drive the present state to identical next state.
14. Assigning binary codes to states represented by alphabets is called state assignment (or code assignment).
15. In sequential circuits with unused states, when clock pulses are applied, if the sequential circuit cycle through only unused states and never arrives at in a used state then this condition is called lock-out.

7.6 Short-Answer Questions

Q7.1 ***Define Moore and Melay model of sequential circuits.***

When the outputs of a sequential circuit are functions of inputs and states (or present states) then the sequential circuit is called Mealy type or Mealy model of sequential circuits.

When the outputs of a sequential circuit are functions of states (or present states) alone, then the sequential circuit is called Moore type or Moore model of sequential circuits.

Q7.2 ***What is the difference between Melay and Moore state machines?***

In Mealy machine, the output depends on inputs and present states but in Moore machine, the outputs depends only present states and does not depend on input.

Q7.3 ***Define FSM.*** *(AU, Apr/May'19, 2 Marks)*

The functional behaviour of sequential circuits can be represented using a finite number of states. Hence the sequential circuits are also called finite state machines (FSM). Therefore, the Mealy model is called Mealy FSM or Mealy Machine. The moore model is called Moore FSM or Moore Machine.

Q7.4 ***Define state and state equations.***

A state is the condition of a sequential circuit described by a set of variables called state variables. Flip-flops are used to store state variables and so the output of flip-flops constitute the states of a sequential circuit. With m state variables we can form 2^m different states.

The state equations (or transition equations) are a set of equations that describe the next state as a function of the present states and inputs.

Q7.5 ***What is state diagram?***

The state diagram is graphical representation of all transitions from present state to next state for all possible combination of inputs and outputs. Therefore, the state diagram is the graphical representation of information available in a state stable.

Q7.6 What is state table?

The state table is listing of outputs and next states for all possible combinations of inputs and present states. In general, if there are m state variables and n inputs then possible combinations of inputs and states are 2^{m+n}. The state table list the outputs and next states for 2^{m+n} combinations of inputs and states.

Q7.7 Define state assignment.

In the design specifications when the states are specified by alphabets it is necessary to assign binary codes for practical implementations. Therefore, an n-bit binary code is assigned to each state. Assigning binary codes to states represented by alphabets is called state assignment (or code assignment).

Q7.8 Define lock-out condition?

In sequential circuits with some don't-care or unused states, the sequential circuit may find itself in some unused state during starting or switch ON. If next state of the unused state is again some unused one and if the sequential circuit cycle through only unused states and never arrives at in a used state then this condition is called lock-out.

Q7.9 Difference between combinational logic circuit and sequential logic circuit. ***(GTU, May'19, 4 Marks)***

Combinational Logic Circuit	Sequential Logic Circuit
1. The combinational logic circuits are digital logic circuits without feedback from output to input.	1. The combinational logic circuits are digital logic circuits with feedback from output to input.
2. The outputs of combinational circuit depend only on present inputs.	2. The outputs of sequential circuit depend on present inputs and past outputs.
3. Since output is not fed back to input no storage or memory elements are required to store the feedback signals.	3. Since output is fed back to input storage or memory elements are required to store the feedback signals.
4. Combinational logic circuits are constructed using all logic devices like logic gates, decoders, multiplexers, adders, etc., except flip-flops or latches.	4. Sequential logic circuits are constructed using combinational logic circuits and flip-flops or latches.

7.7 Exercises

I. Fill in the blanks

1. A state is the condition of a sequential circuit described by a set of variables called ______ .
2. The graphical representation of information available in state stable is called ______.
3. The sequential circuits will have storage elements to store the ______.
4. The sequential circuits without a clock are called ______ sequential circuits.
5. The sequential circuits with a clock are called ______ sequential circuits.
6. The basic elements in sequential circuits are ______ and ______.
7. The state diagram consists of ______ and ______.
8. In binary code assignment, the codes are assigned in ______.
9. The one-hot assignment will not satisfy ______ logic.

10. In clocked sequential circuits the inputs are recognized only when clock signal is applied otherwise ______.

11. The functional behaviour of sequential circuits can be represented using a finite number of states. Hence the sequential circuits are called ______.

12. The ______ is listing of outputs and next states for all possible combinations of inputs and present states

Answers

1. state variables	6. latches and flip-flops	11. finite state machine (FSM)
2. state diagram	7. circles and directed lines	12. state table
3. past outputs	8. binary order	
4. asynchronous	9. 2^n	
5. synchronous	10. ignored	

II. State whether the following statements are True or False

1. State table is listing of outputs and next states for all possible combinations of inputs and present states.
2. Latches are sequential circuits without clock. Flip-flops are sequential circuits with clock
3. The sequential circuits can be designed using flip-flops with combinational circuits at the input and output of flip-flops
4. The combinational circuit at the input is designed by constructing K-map for flip-flop outputs
5. When the outputs of a sequential circuit are functions of inputs and sates then the sequential circuit is called Mealy type or Mealy model
6. When the outputs of a sequential circuit are functions of inputs and sates alone, then the sequential circuit is called Moore type or Moore model
7. The lableson the directed lines will have only outputs that initiate transitions from one state to another
8. In Moore model, the outputs are synchronized with clock
9. In Mealy model the inputs may change any time and so outputs are synchronized with clock
10. The synchronization of inputs with clock in Moore models is necessary to ensure right outputs and to overcome the difference in delays in input changes and state changes
11. The state equations are a set of equations that describe the next state as a function of the present states and outputs

Answers

1. True	6. True	11. False
2. True	7. False	
3. True	8. True	
4. False	9. False	
5. True	10. False	

III. Choose the right answer for the following questions

1. *The state diagram is also called*

a) State variable **b)** State graph

c) State table **d)** State reduction

2. ***The state diagram consist of***

a) Only circle b) Only direct line

c) Circle and direct line d) None of the above

3. ***Assigning binary codes to states represented by alphabets is called***

a) State variable b) State graph

c) State table d) State assignment

4. ***Which codes are assigned in binary order***

a) Binary code assignment b) Gray code assignment

c) One hot code assignment d) None of the above

5. ***Which codes are assignment will not satisfy 2^n logic***

a) Binary code assignment b) Gray code assignment

c) One hot code assignment d) None of the above

6. ***Sequential logic circuits are digital circuits whose outputs depends on***

a) Present inputs and past inputs b) Present outputs and past inputs

c) Present inputs and past outputs d) Present outputs and past outputs

7. ***A state is the condition of a sequential circuit described by a set of variable called***

a) State variable b) State graph

c) State table d) State reduction

8. ***In Mealy machine, the outputs depends***

a) Only present state b) Only inputs

c) Inputs and present state d) None of the above

9. ***In Moore machine, the outputs depends***

a) Only present state b) Only inputs

c) Inputs and present state d) None of the above

10. ***When the outputs of a sequential circuit are functions of inputs and states then the sequential circuit is***

a) Mealy model b) Moore model

c) Mealy and Moore model d) None of the above

11. ***When the outputs of a sequential circuit are functions of states alone, then the sequential circuit is***

a) Mealy model b) Moore model

c) Mealy and Moore model d) None of the above

Answers

1. b	5. c	9. a
2. c	6. c	10. a
3. d	7. a	11. b
4. a	8. c	

IV. Answer the following questions

1. What are the types of sequential circuit?
2. Draw the block diagram of asynchronous and synchronous sequential circuits.
3. Draw the block diagram of Mealy machine (Mealy FSM).
4. Draw the block diagram of Moore machine (Moore FSM).
5. List the guidelines for construction of state graphs.

V. Solve the following problems

E7.1 A sequential circuit with two D flip-flops A and B, two inputs x and y and one output Z is specified, by the following next state and output equations.

$$A(t+1) = xy' + xB$$

$$B(t+1) = xA + xB'$$

$$Z = A$$

a) Draw the logic diagram of the circuit.

b) List the state table for the sequential circuit.

c) Draw the corresponding state diagram.

E7.2 A sequential circuit has one D flip-flop, two inputs x and y, and one output d. The full-subtractor output carry c_o is connected to the D flip-flop, as shown in Fig. E7.2. Derive the state table and state diagram of the sequential circuit.

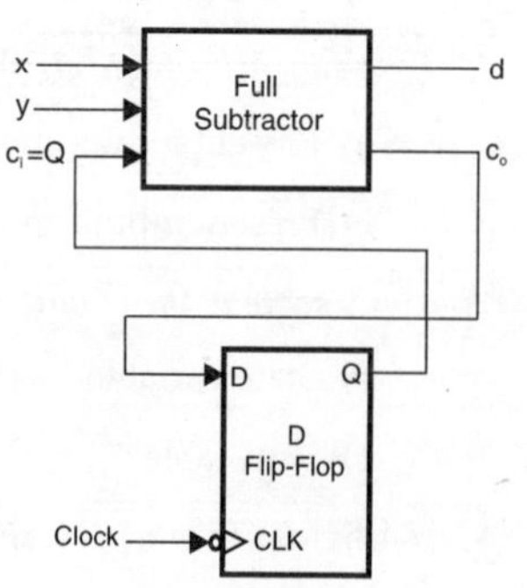

Fig. E7.2.

E7.3 For the state table shown in Table E7.3,

a) Draw the corresponding state diagram.

b) Tabulate the reduced state table.

c) Draw the state diagram corresponding to the reduced state table.

Table E7.3: State Table

Present State	Next State		Output	
Q(t)	x = 0	x = 1	x = 0	x = 1
a	f	b	0	0
b	d	c	0	0
c	f	e	0	0
d	g	a	1	0
e	d	c	0	0
f	f	b	1	1
g	g	h	0	1
h	g	a	1	0

E7.4 A sequential circuit has two JK flip-flops A and B and one input x. The circuit is described by the following flip-flop input equations.

$$J_A = x \quad ; \quad K_A = B' \quad ; \quad J_B = x \quad ; \quad K_B = A$$

a) Derive the state equations A(t + 1) and B(t + 1) by substituting the input equations for the J and K variables.

b) Draw the state diagram of the circuit.

E7.5 A sequential circuit has three flip-flops A, B, and C one input x and one output Y. The state diagram is shown in Fig. E7.5. The circuit is to be designed by treating the unused states as don't-care conditions. Analyze the circuit obtained from the design to determine the effect of the unused states.

a) Use JK flip-flops in the design.

b) Use T flip-flops in the design.

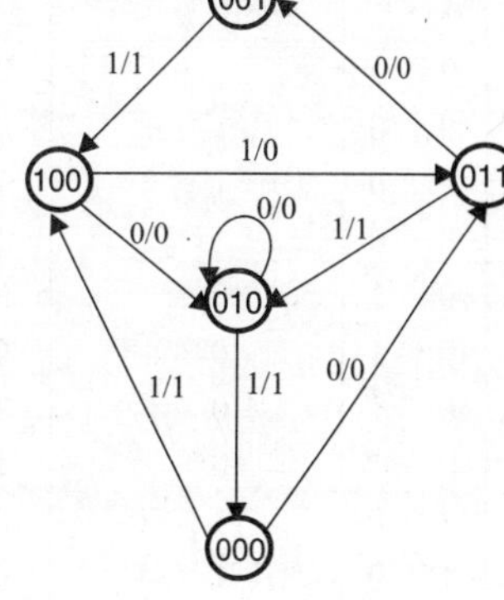

Fig. E7.5.

E7.6 Design a sequential circuit with two D flip-flops A and B and one input x.

a) When x = 0, the state of the circuit remains the same. When x = 1, the circuit goes through the state transistions.

00 → 01 → 10 → 11 → 00

b) When x = 0, the state of the circuit remains the same. When x = 1, the circuit goes through the state transitions.

00 → 11 → 01 → 10 → 00

E7.7 Develop the state diagram for a Mealy state machine that detects a sequence of three or more consecutive 1's in a string of bits coming through an input line using D flip-flops.

E7.8 Design a synchronous sequence detector that produces an output 1, whenever the non overlapping sequence 1010 is detected.

E7.9 Draw the state diagram for the sequential circuit whose state table is shown in Table 1. Also design a logic circuit using D flip-flop. Implement the logic circuit using available standard ICs.

E7.10 Design a sequential circuit to detect a sequence of four or more consecutive 1's in a string of bits coming through an input line using T flip-flop.

Answers

E7.1 a)

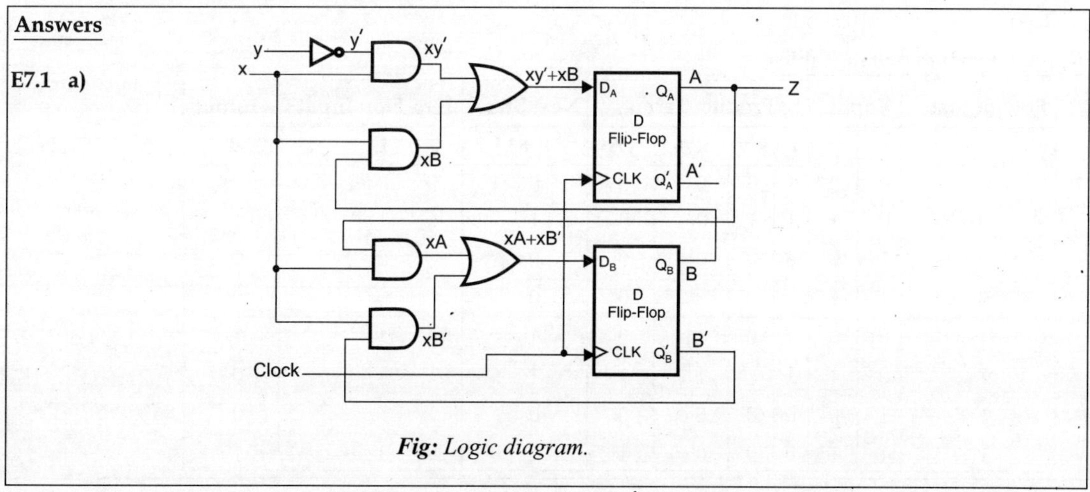

Fig: Logic diagram.

b)

Table 1: Expanded State Table

Present States		Complement of Present State B	Inputs		Product terms					Next States		Flip-Flop Inputs		Output
A	B	B'	x	y	y'	xy'	xB	xA	xB'	A(t+1)	B(t+1)	D_A	D_B	Z = A
0	0	1	0	0	1	0	0	0	0	0	0	0	0	0
0	0	1	0	1	0	0	0	0	0	0	0	0	0	0
0	0	1	1	0	1	1	0	0	1	1	1	1	1	0
0	0	1	1	1	0	0	0	0	1	0	1	0	1	0
0	1	0	0	0	1	0	0	0	0	0	0	0	0	0
0	1	0	0	1	0	0	0	0	0	0	0	0	0	0
0	1	0	1	0	1	1	1	0	0	1	0	1	0	0
0	1	0	1	1	0	0	1	0	0	1	0	1	0	0
1	0	1	0	0	1	0	0	0	0	0	0	0	0	1
1	0	1	0	1	0	0	0	0	0	0	0	0	0	1
1	0	1	1	0	1	1	0	1	1	1	1	1	1	1
1	0	1	1	1	0	0	0	1	1	0	1	0	1	1
1	1	0	0	0	1	0	0	0	0	0	0	0	0	1
1	1	0	0	1	0	0	0	0	0	0	0	0	0	1
1	1	0	1	0	1	1	1	1	0	1	1	1	1	1
1	1	0	1	1	0	0	1	1	0	1	1	1	1	1

c)

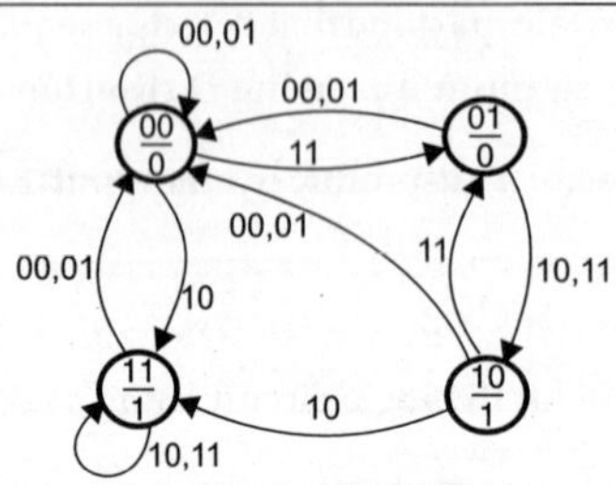

Fig: State diagram.

E7.2 a)

Table 1: Expanded State Table

Present State	Input		Product Terms				Next State	Flip-Flop Input	Output
A	x	y	x'	x'y	x'A	yA	A(t+1)	D	d
0	0	0	1	0	0	0	0	0	0
0	0	1	1	1	0	0	1	1	1
0	1	0	0	0	0	0	0	0	1
0	1	1	0	0	0	0	0	0	0
1	0	0	1	0	1	0	1	1	1
1	0	1	1	1	1	1	1	1	0
1	1	0	0	0	0	0	0	0	0
1	1	1	0	0	0	1	1	1	1

b)

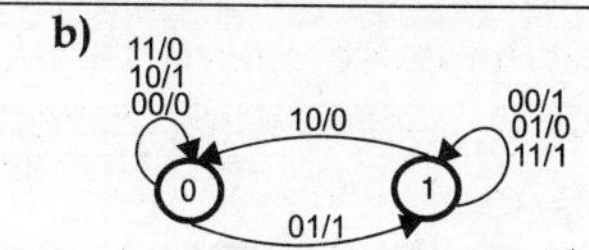

Fig: State diagram.

E7.3 a)

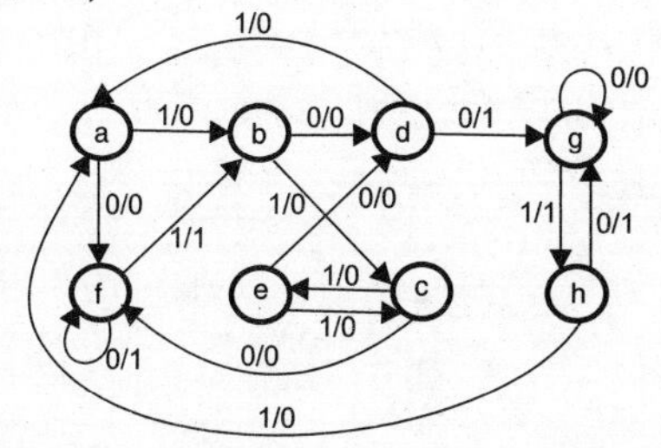

Fig: State diagram.

b) Table: Third Reduced State Table

Present State	Next State		Output	
Q(t)	x = 0	x = 1	x = 0	x = 1
a	f	b	0	0
b	d	c → a	0	0
d	g	a	1	0
f	f	b	1	1
g	g	d	0	1

c)

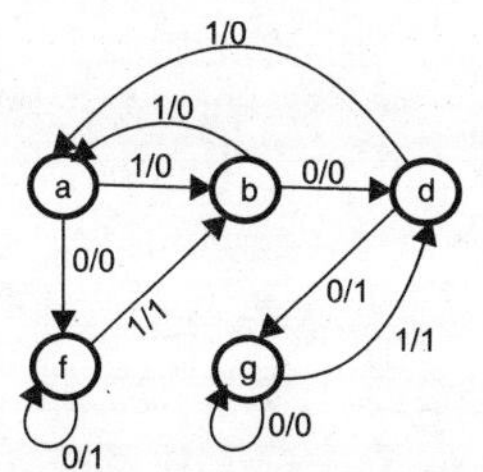

Fig: State diagram with reduced states.

E7.4 a) State equations: $A(t+1) = A'x + AB$

$B(t+1) = B'x + A'B$

b)

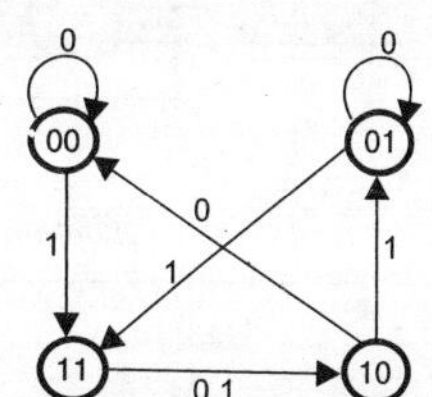

Fig: State diagram.

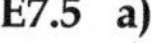

E7.5 a)

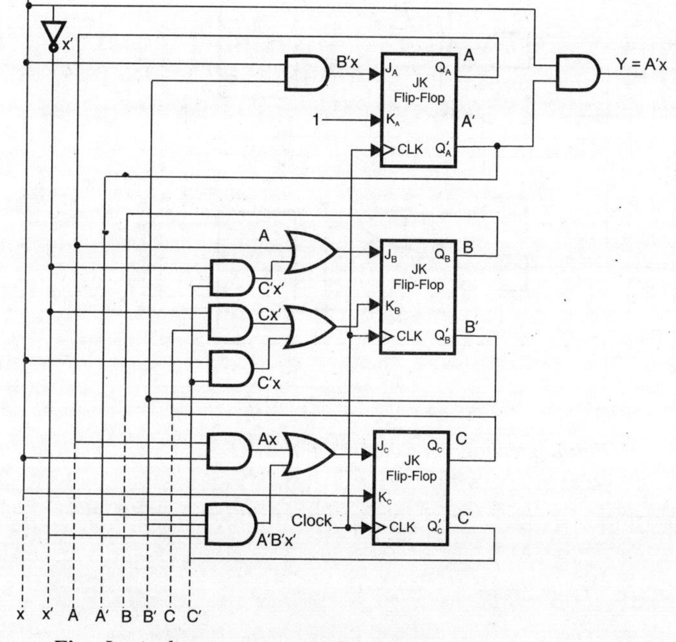

Fig: Logic diagram using JK flip-flops.

b)

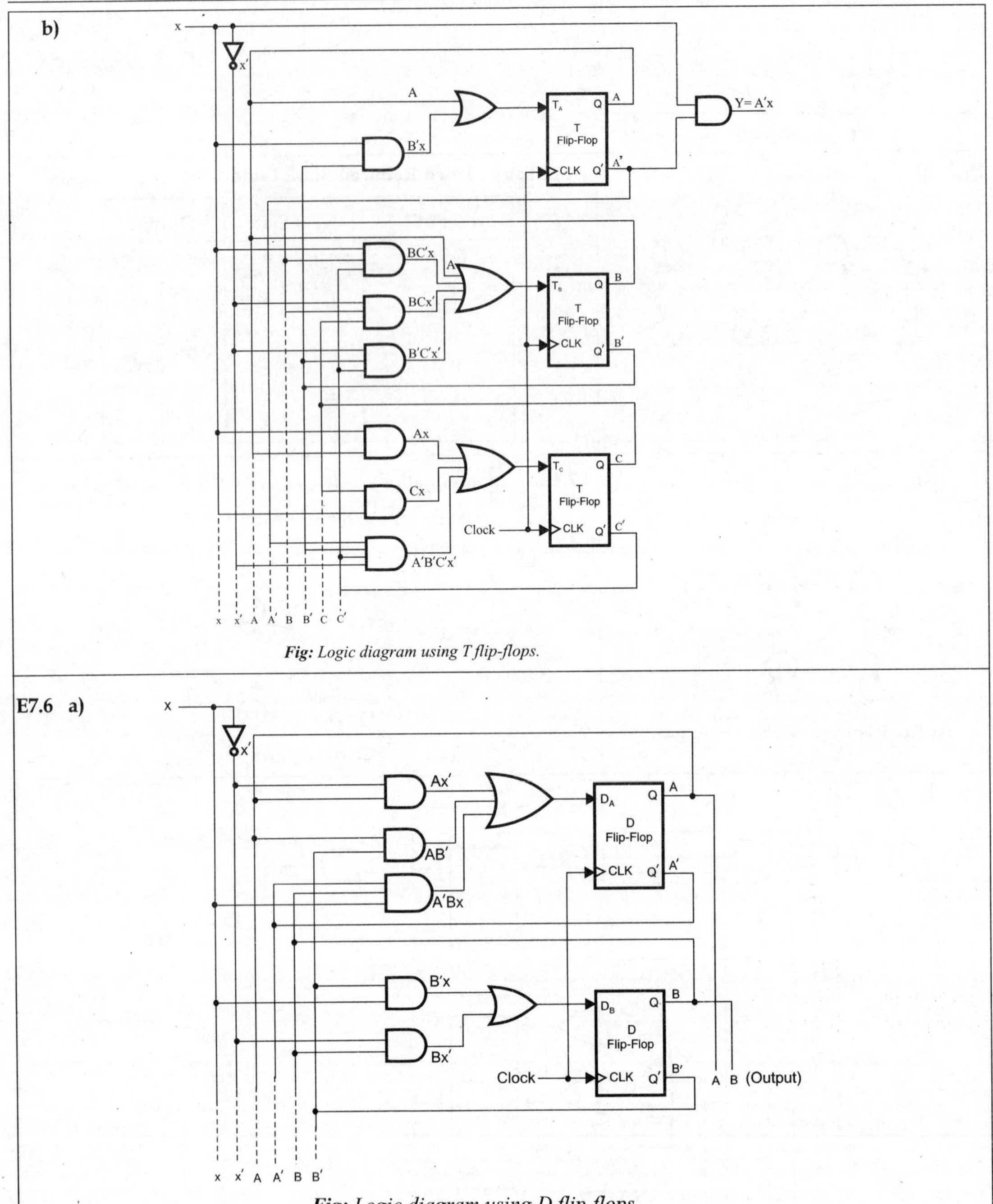

***Fig:** Logic diagram using T flip-flops.*

***Fig:** Logic diagram using D flip-flops.*

b)

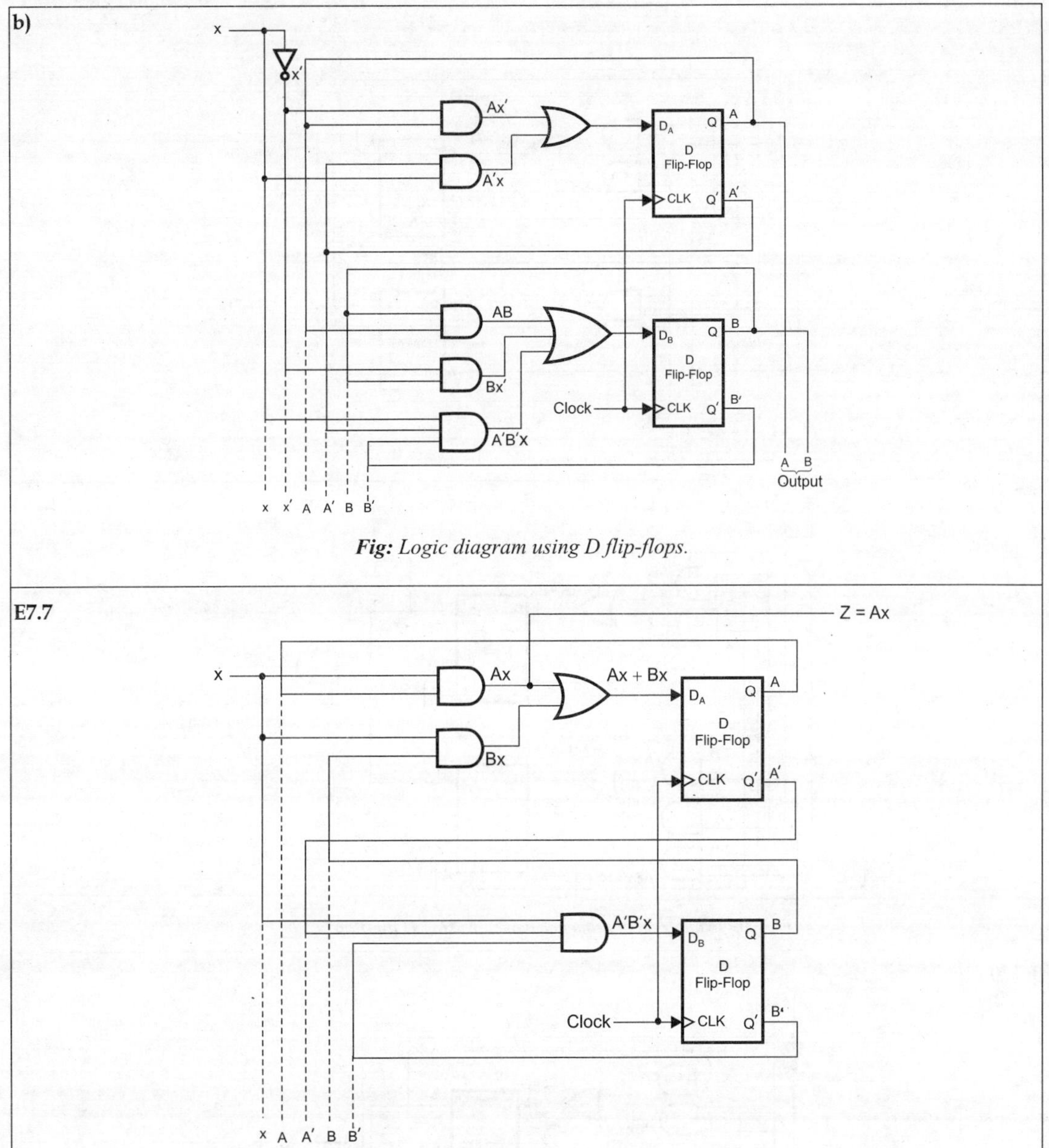

Fig: *Logic diagram using D flip-flops.*

E7.7

Fig: *Logic diagram using D flip-flops.*

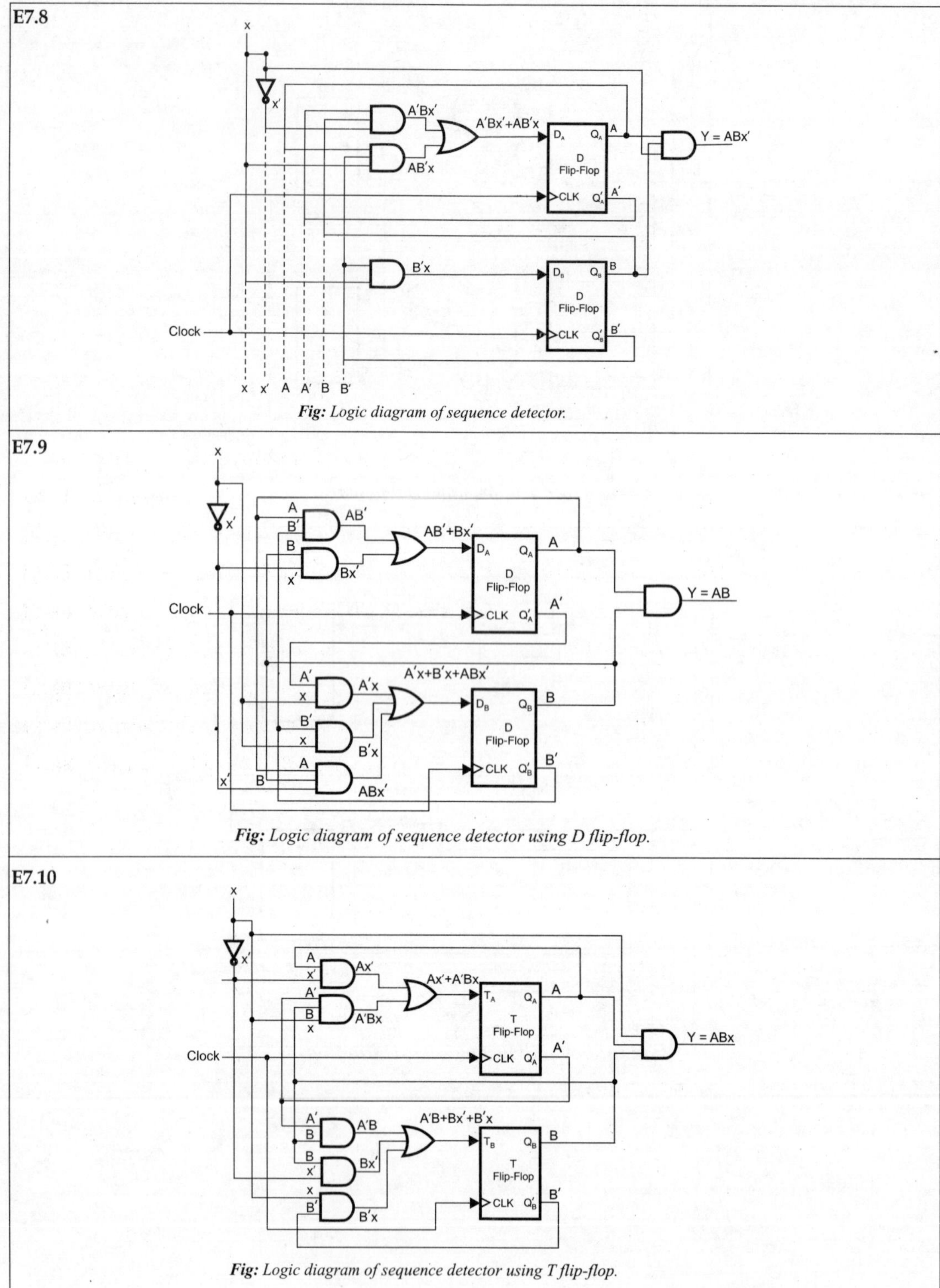

Fig: *Logic diagram of sequence detector.*

Fig: *Logic diagram of sequence detector using D flip-flop.*

Fig: *Logic diagram of sequence detector using T flip-flop.*

CHAPTER 8

SEQUENTIAL LOGIC-COUNTERS AND REGISTERS

8.1 Design of Counters

A register that goes through prescribed (or predefined) sequence of states upon the application of clock pulses is called a ***counter***. The counters are used to generate or go through a sequence of binary code known as ***count sequence***.

A counter that follows a binary count sequence is called ***binary counter***. An n-bit binary counter consists of n flip-flops and can count binary from 0 to $2^n - 1$. In binary up counter, the output will increment for every clock pulse and in binary down counter the output will decrement for every clock pulse. Mostly, counters will have a reset facility to clear the output at any desired time and also will have a preload facility to load a known value during starting.

The modulus of the counter refers to number states a counter will pass through before coming back to initial state. An n-bit counter will have 2^n possible states and if the n-bit counter passes through all possible states before coming to initial state then it is called ***MOD-2^n counter***. A 3-bit binary counter has 8 possible states and so it is called MOD-8 counter.

Also, the counter divides the input clock frequency. An n-bit binary counter divides the clock frequency by 2^n and so n-bit counter is also called divide-by-2^n counter. An n-bit counter is constructed using n flip-flops and the output of each flip-flop is the counter output which is also a clock.

If f is the frequency of the clock then frequency of outputs of flip-flop is f/2, f/4, f/8, ……, $f/2^n$. A 3-bit counter will have three flip-flops and the frequency of output clock of flip-flops is f/2, f/4 and f/8.

An n-bit counter can be designed to have modulus less than 2^n. Such counters use less number of states than full modulus counter. Let, MOD-N counter be a reduced modulus n-bit counter, where $N<2^n$.

Now n-bit MOD-N counter uses only N states in the 2^n possible states of n-bit counter and the remaining states are unused states. Also, the MOD-N counter divides the clock frequency by N and so it is also called ***divide-by-N counter***.

(IKGPTU, Dec'19, 2 Marks) (RGPV, Jun'22, 2 Marks)

The counters can be broadly classified into two categories: Asynchronous counters and synchronous counters. The asynchronous counter are aslso called ripple counter.

In a ripple counter, a flip-flop output will act as clock for another flip-flop.

In synchronous counters all the flip-flops are triggered simultaneously by a common clock.

8.2 Asynchronous or Ripple Counter

(JNTUK, Feb'22, 8 Marks) (VTU, Aug/Sep'20, 6 Marks) (JNTUH, Nov/Dec'20, 15 Marks)

A binary ripple counter is constructed using a series of complementing flip-flops. The binary ripple counter can be constructed using any flip-flop, that are connected to generate complement of present state as next state output. The pulses to be counted or the clock signal is applied to clock of first flip-flop. The output of first flip-flop is connected to clock of second flip-flop. The output of second flip-flp is connected to clock of third and so on. In general, the output of low order flip-flop is connected to clock of next higher order flip-flop.

8.2.1 Up Counter

(AKTU, Sep'20, 5 Marks) (BATU, Jun'19, 6 Marks)

In up counter, the output will increment for every clock pulse and in n-bit up counter, the count sequence will be from 0 to $2^n - 1$. In generally, counter refers to up counter.

Binary Ripple Counter using T Flip-Flop

In T-flip-flop, if the input is tied to 1 then the output will complement on every clock pulse and so in counter using T-flip-flop the input of all flip-flops are tied to 1. The count sequence or clock signal is applied as clock to first flip-flop.

The output of low order flip-flop is connected to clock of next higher order flip-flop. Also, the flip-flop outputs are counter outputs. A 4-bit binary ripple counter using T flip-flop is shown in Fig. 8.1.

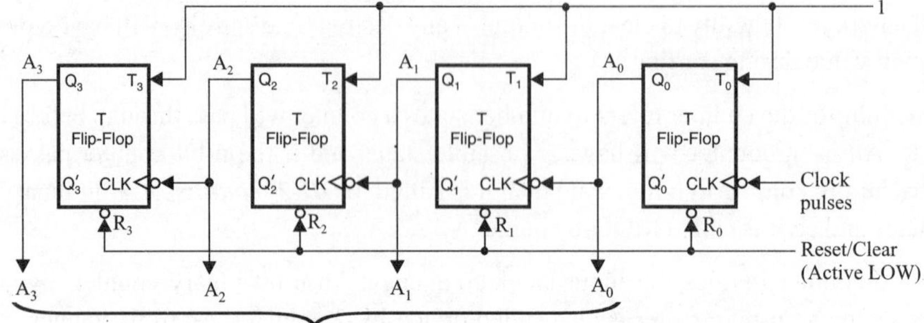

***Fig. 8.1:** 4-bit binary ripple counter using T flip-flop.*

Binary Ripple Counter using JK Flip-Flop

When the inputs J and K are tied to 1 then the JK flip-flop will behave as T-flip-flop. Therefore, in ripple counter using JK flip-flops, the J and K inputs of all flip-flops are tied to 1 and all other are same as T flip-flop based ripple counter. A 4-bit binary ripple counter using JK flip-flop is shown in Fig. 8.2.

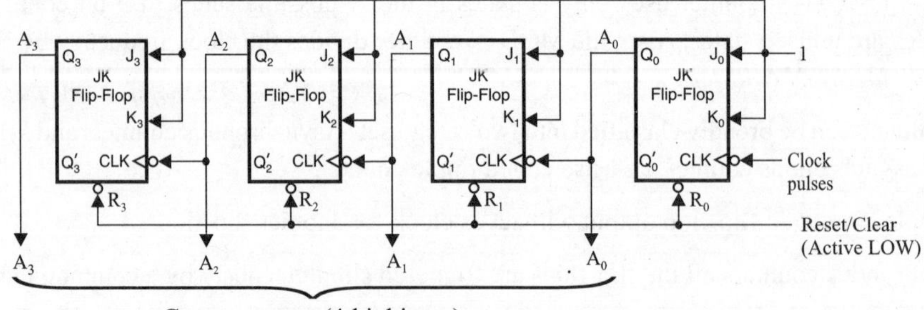

***Fig. 8.2:** 4-bit binary ripple counter using JK flip-flop.*

Binary Ripple Counter using D Flip-Flop

In D flip-flop, after a reset, Q = 0, and Q′ = 1, and so if Q′ is connected to D, then for every clock pulse output toggles. This concept is used in construction of binary ripple counter using D flip-flop. A 4-bit binary ripple counter using D flip-flop is shown in Fig. 8.3.

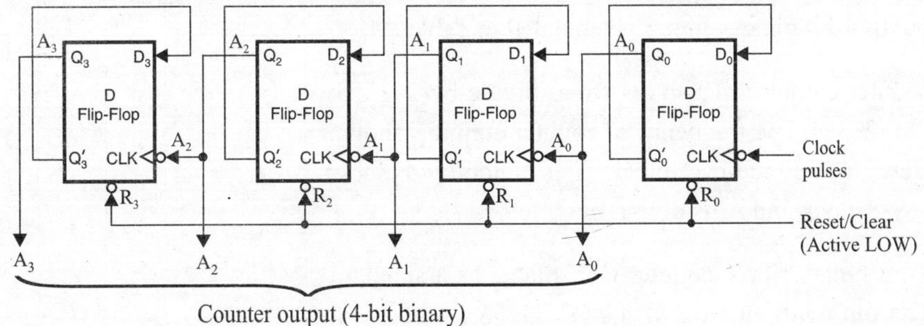

***Fig. 8.3:** 4-bit binary counter using D flip-flop.*

Working of 4-bit Binary Ripple Counter

(JNTUA, Jun/Jul'22, 7 Marks)

In binary ripple counter, the input to flip-flops are tied to 1. The clock is negative edge triggered and so on every negative edge of clock, the output of flip-flop toggles.

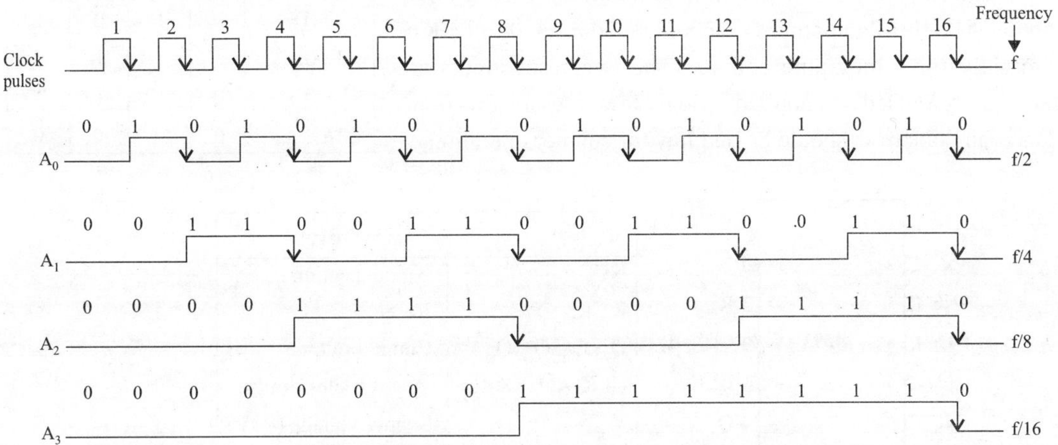

***Fig. 8.4:** Clock and output signals of 4-bit binary ripple counter.*

Initially all the outputs are made zero using reset/clear signal. Then pulses to be counted or clock pulse is applied to clock of first flip-flop. So first flip-flop output A_0 complements or toggle on negative edge of every clock pulse.

Since A_0 is connected as clock of second flip-flop, the output A_1 of second flip-flop complements on every negative edge of A_0.

Since A_1 is connected as clock of third flip-flop, the output A_2 of third flip-flop complements on every negative edge of A_1.

Since A_2 is connected as clock of fourth flip-flop, the output A_3 of fourth flip-flop, complements on every negative edge of A_2. The clock and output signals of 4-bit binary ripple counter are shown in Fig. 8.4.

In Fig. 8.4, it can be observed that after 15th count pulses, all the outputs become zero. Therefore the count sequence again start after 16th clock pulse.

Hence, the 4-bit binary ripple counter can count from 0 to 15. The output of 4-bit binay counter is tabulated in Table 8.1.

The counter outputs are also known as divide-by-2^n clock signal for n = 1, 2, 3, 4. The frequency of counter output is shown in Fig. 8.4. Here, A_0 is divide-by-2 clock, A_1 is divide-by-4 clock, A_2 is divide-by-8 clock and A_3 is divide-by-16 clock.

The 4-bit binary ripple counter is available as a standard IC 74293. The pin configuration of the IC 74293 is shown in Fig. 8.5. Here, Q_A, Q_B, Q_C and Q_D are output of internal flip-flops A, B, C and D respectively. Separate clock signals can be applied for flip-flops A and B. The flip-flop B, C and D are internally connected to function as 3-bit ripple counter.

The flip-flop A can function as divide-by-2 or MOD-2 counter. The flip-flop output Q_A should be connected to CLK-B externally for 4-bit ripple counter. The two reset inputs are internally NANDed and applied as logic **low** reset for flip-flops. Hence, any one reset should be tied **low** for counter operation.

Table 8.1: Output of 4-bit Binary Ripple Counter

Count or Clock Pulse	Counter Outputs			
	A_3	A_2	A_1	A_0
-	0	0	0	0
1	0	0	0	1
2	0	0	1	0
3	0	0	1	1
4	0	1	0	0
5	0	1	0	1
6	0	1	1	0
7	0	1	1	1
8	1	0	0	0
9	1	0	0	1
10 (A)	1	0	1	0
11 (B)	1	0	1	1
12 (C)	1	1	0	0
13 (D)	1	1	0	1
14 (E)	1	1	1	0
15 (F)	1	1	1	1
16	0	0	0	0

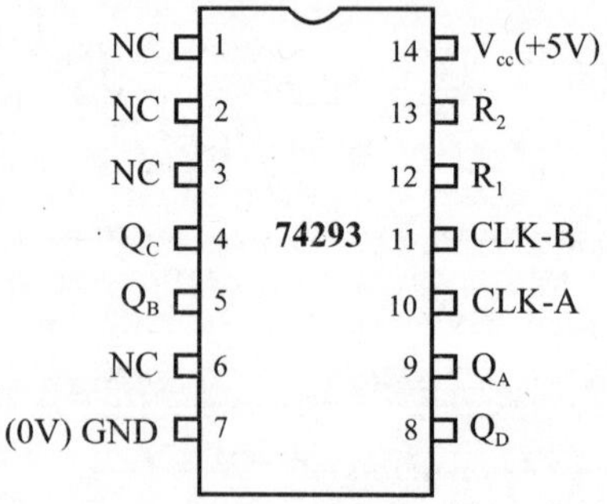

Pin (left)		74293		Pin (right)
NC	1		14	V_{cc}(+5V)
NC	2		13	R_2
NC	3		12	R_1
Q_C	4		11	CLK-B
Q_B	5		10	CLK-A
NC	6		9	Q_A
(0V) GND	7		8	Q_D

Pin	Description
NC	No connection
R_1, R_2	Reset
Q_A, Q_B, Q_C, Q_D	Counter outputs
CLK-A, CLK-B	A and B clock pulse
V_{cc}	Supply voltage (+5V)
GND	Ground (0V)

Fig. 8.5: Pin configuration of 4-bit binary counter IC 74293.

8.2.2 Down Counter

(GTU, May'19, 7 Marks)

In down counter, the counter output decrement for every clock pulse and in an n-bit down counter the count sequence will be from 2^{n-1} to 0. The complement of up counter output will be the down count. A 4-bit down counter using T flip-flop is shown in Fig. 8.6. The up counter and complement of up counter output signals are shown in Fig. 8.7. The up and down counters outputs are tabultated in Table 8.2.

The clock of down counter is similar to up counter, but output is taken from complement of output Q′ of each flip-flop.

The down counter using JK and D flip-flops are same as up counter except that output is taken from Q′ of every flip-flop.

Table 8.2: Up and Down Counter Outputs

Count or Clock Pulse	Up Counter Outputs				Down Counter Outputs			
	A_3	A_2	A_1	A_0	A'_3	A'_2	A'_1	A'_0
					D_3	D_2	D_1	D_0
-	0	0	0	0	1	1	1	1
1	0	0	0	1	1	1	1	0
2	0	0	1	0	1	1	0	1
3	0	0	1	1	1	1	0	0
4	0	1	0	0	1	0	1	1
5	0	1	0	1	1	0	1	0
6	0	1	1	0	1	0	0	1
7	0	1	1	1	1	0	0	0
8	1	0	0	0	0	1	1	1
9	1	0	0	1	0	1	1	0
10 (A)	1	0	1	0	0	1	0	1
11 (B)	1	0	1	1	0	1	0	0
12 (C)	1	1	0	0	0	0	1	1
13 (D)	1	1	0	1	0	0	1	0
14 (E)	1	1	1	0	0	0	0	1
15 (F)	1	1	1	1	0	0	0	0
16	0	0	0	0	1	1	1	1

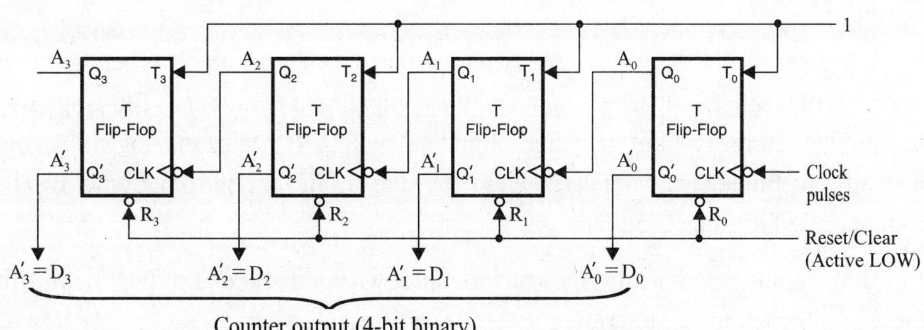

***Fig. 8.6:** 4-bit down counter using T flip-flop.*

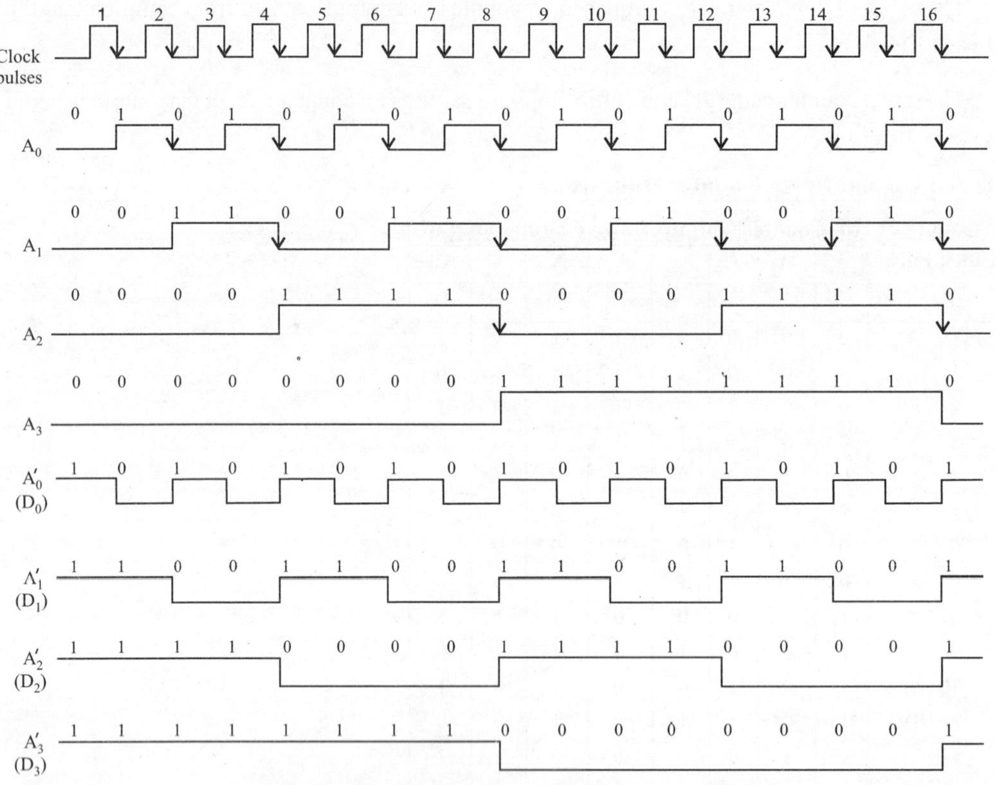

Fig. 8.7: Output signals of up and down counter.

8.2.3 Up/Down Counter using T Flip-Flop

A binary counter can be used to perform both up/down counting by adding a selection logic in the output as shown in Fig. 8.8.

When up/down = 1, only A_0, A_1, A_2 and A_3 are allowed to reach output. When up/down = 0 only A_0', A_1', A_2' and A_3' will reach output. The output and the complement of output of flip-flops are separately ANDed with up/down control signal and then output of both AND gates are logically ORed. In this arrangement, at any one time only one of the input to OR gate is valid and the other input will be 0, so that the active input is passed to counter output.

The flip-flop outputs are logically ANDed with up/down control signal so that the flip-flop outputs A_0, A_1, A_2 and A_3 will reach the counter outputs D_0, D_1, D_2 and D_3 when up/down = 1. Now the counter behave as up counter.

The complement of flip-flop outputs are logically ANDed with complement of up/down control signal so that the complement of flip-flop outputs A_0', A_1', A_2' and A_3' will reach counter outputs D_0, D_1, D_2 and D_3 when up/down = 0. Now the counter behave as down counter.

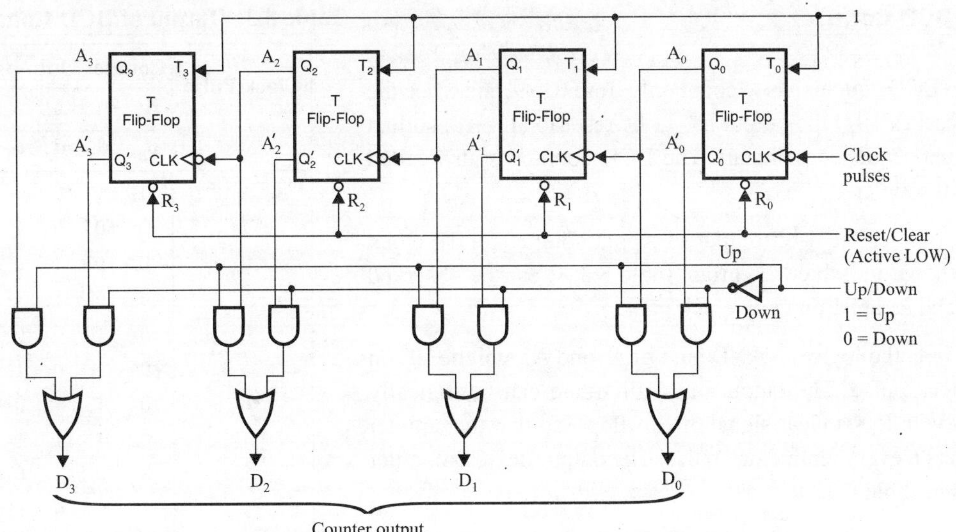

Fig. 8.8: 4-bit up/down counter using T flip-flop.

8.2.4 Down Counter with Positive Clock

In the up counter discussed in Section 8.2.1, if the clock of flip-flops are positive edge triggered then the up counter will behave as down counter.

This can be observed from the clock and output signals shown in Fig. 8.9.

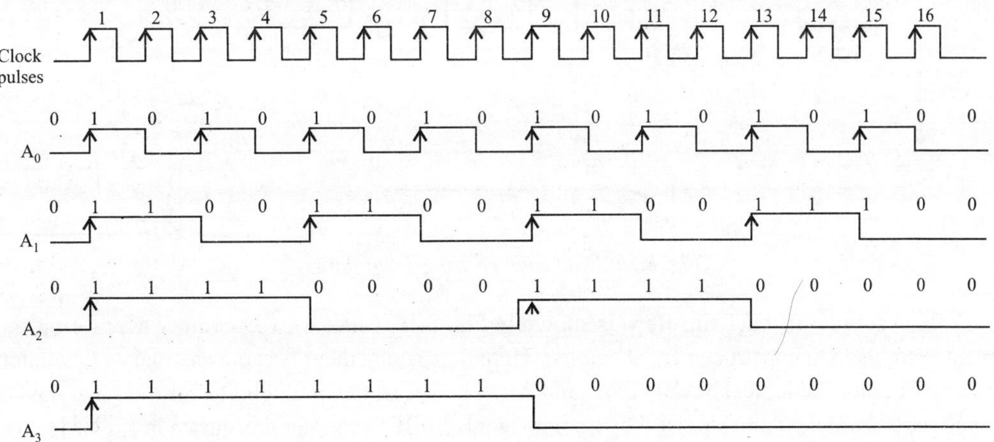

Fig. 8.9: Output signals of down counter with positive edge triggered clock.

Also it can be observed that the complements of A_0, A_1, A_2 and A_3 will be up count.

Therefore the outputs A_0', A_1', A_2' and A_3' will be output of up counter, when the clock in positive edge triggered.

This concept is applicable for binary ripple counter using, T or JK or D flip-flops.

8.2.5 BCD Counter

(RGPV, Jun'20,22, 7 Marks)
(JNTUA, Jun/Jul'22, 7 Marks)

A BCD counter has to count only from 0 to 9, and when the count reaches 10_{10}, the output has to be reset to all zero, so that next count sequence will start. The BCD counter is also called MOD-10 counter.

Consider the 4-bit output of ripple counter for 10 clock pulses shown in Table 8.3. From Table 8.3, it can be observed that A_1 and A_3 are 1 on tenth clock pulse.

Here, the logical NAND gate of A_1 and A_3, will be "0" on tenth clock pulse. The output of NAND gate can be logically ANDed with Reset/clear signal of flip-flops, so that the counter will reset on every tenth clock pulse. The output of BCD counter is listed in Table 8.3.

Table 8.3: Output of BCD Counter

Clock Pulse	Counter Outputs			
	A_3	A_2	A_1	A_0
	d_8	d_4	d_2	d_1
-	0	0	0	0
1	0	0	0	1
2	0	0	1	0
3	0	0	1	1
4	0	1	0	0
5	0	1	0	1
6	0	1	1	0
7	0	1	1	1
8	1	0	0	0
9	1	0	0	1
10	1	0	1	0
	↓ Reset			
	0	0	0	0

BCD Ripple Counter using T Flip-Flop

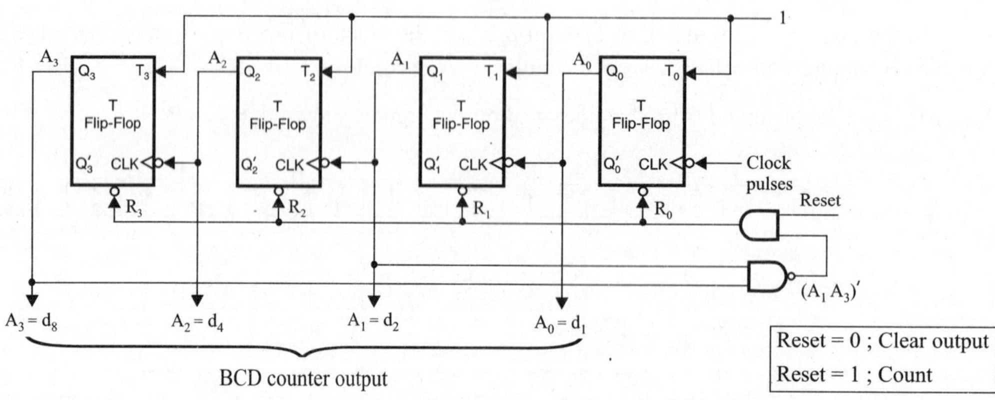

Fig. 8.10: BCD counter using T flip-flops.

A BCD counter using T flip-flops is shown in Fig. 8.10. Since a BCD counter needs 4-bit output, 4 flip-flops are used to construct a BCD counter. Here the arrangement is same as binary up counter, but on tenth clock pulse the logical NAND of A_1 and A_3 will reset all the output bit to zero. So that next count sequence starts from next clock pulse.The output signals of BCD counter are shown in Fig. 8.11.

In binary ripple counter, the input to flip-flops are tied to 1. The clock is negative edge triggered and so on every negative edge of clock, the output of flip-flop toggles.

The BCD counter shown in Fig. 8.10 can count one digit of BCD in the range 0 to 9. In order to count m digit BCD we need m BCD counters shown in Fig. 8.10. The BCD counters should be connected in cascade or series.

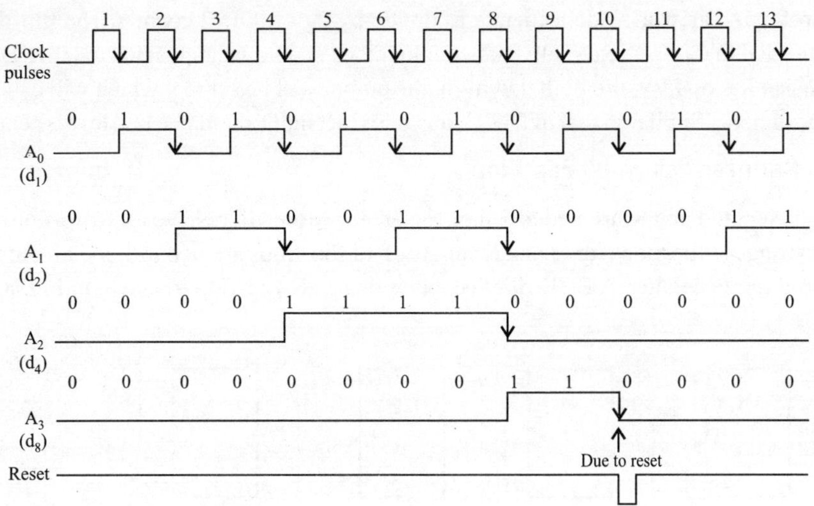

Fig. 8.11: *Output signals of BCD counter.*

The cascade connection of 4 BCD counters to count 4-digit BCD is shown in Fig. 8.12. The upper most bit in each BCD counter is used as clock for next digit BCD counter.

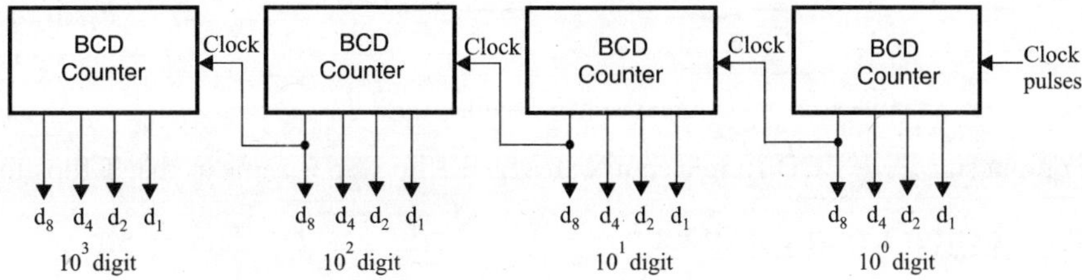

Fig. 8.12: *4-digit BCD counter.*

BCD counter is available as a standard IC 74290. Pin configuration of 74290 is shown in Fig. 8.13.

Pin (left)	No.	74290	No.	Pin (right)
$R9_1$	1		14	V_{cc}(+5V)
NC	2		13	R_1
$R9_2$	3		12	R_2
Q_C	4		11	CLK-A
Q_B	5		10	CLK-B
NC	6		9	Q_A
(0V) GND	7		8	Q_B

Pin	Description
NC	No connection
R_1, R_2	Reset
Q_A, Q_B, Q_C, Q_D	Counter outputs
CLK-A, CLK-B	A and B clock pulse
$R9_1$, $R9_2$	Enable to set output to 9
V_{cc}	Supply voltage (+5V)
GND	Ground (0V)

Fig. 8.13: *Pin configuration of BCD counters IC 74290.*

Here, Q_A, Q_B, Q_C and Q_D are output of internal flip-flops A, B, C and D respectively. Separate clock signals can be applied for flip-flops A and B. The flip-flops B, C and D are internally connected to function as 3-bit BCD ripple counter.

The flip-flop A can work independently as divide-by-2 or MOD-2 counter. The flip-flop output Q_D should be connected to CLK-A externally and counter clock should be connected to CLK-B for fout digit BCD count. When $R9_1$ and $R9_2$ are both tied **high**, the output will be set to 9 which will can be used in 9's complement arithmetic. Similar to that of 74293, any one reset input should be tied **low** for count operation.

BCD Ripple Counter using JK Flip-Flop

(VTU, Aug/Sep'20, 6 Marks)

When the inputs J and K are tied to 1 then the JK flip-flop will behave as T-flip-flop. Therefore, in ripple counter using JK flip-flops, the J and K inputs of all flip-flops are tied to 1 and all other are same as T flip-flop based ripple counter. A BCD ripple counter using JK flip-flop is shown in Fig. 8.14.

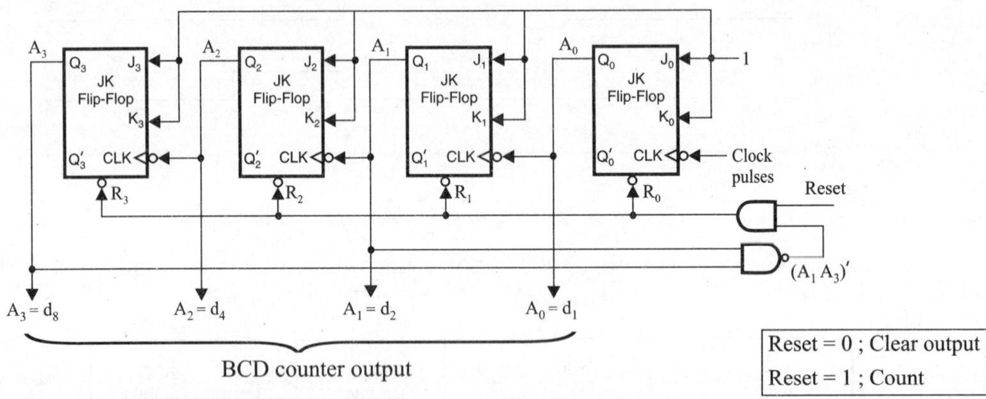

Fig. 8.14: BCD Ripple counter using JK flip-flop.

The output signals of BCD Ripple counter using JK flip-flop is same as that T flip-flop.

BCD Ripple Counter using D Flip-Flop

In D flip-flop, after a reset, Q = 0, and Q′ = 1, and so if Q′ is connected to D, then for every clock pulse output toggles. This concept is used in construction of binary ripple counter using D flip-flop. A BCD ripple counter using D flip-flop is shown in Fig. 8.15.

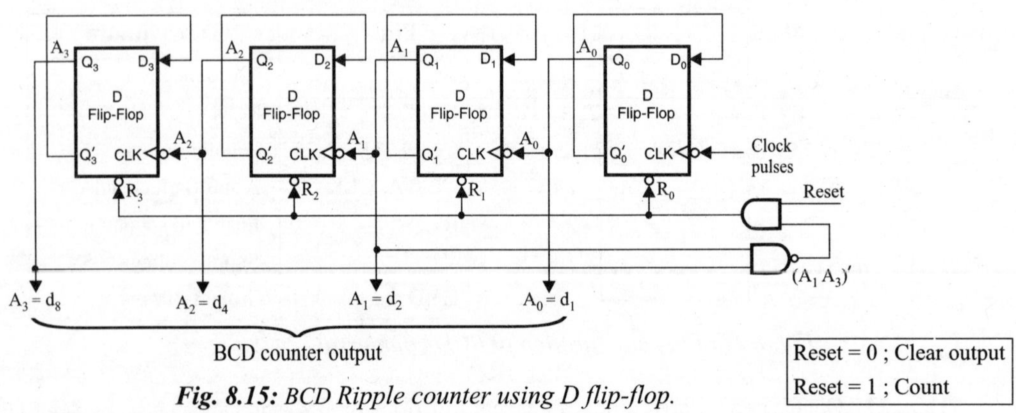

Fig. 8.15: BCD Ripple counter using D flip-flop.

The output signals of BCD Ripple counter using D flip-flop is same as that T flip-flop.

8.2.6 MOD Counter

(IKGPTU, Dec'19, 2 Marks)

An n-bit binary counter will have 2^n states and implemented using n flip-flops. MOD-N counters can be designed using n-bit binary counter by using only N states of n-bit counter provided $N < 2^n$. For example, MOD-7 counter uses 7 states of 3-bit counter which has $2^3 = 8$ states and can be implemented using 3 flip-flops.

Consider 3-bit binary counter which has $2^3 = 8$ states. Here, counter output and states are same. The output of 3-bit binary counter is shown in Table 8.4. The 3-bit binary counter is shown in Fig. 8.16. The 3-bit counter can be modified to work as MOD-2, MOD-3, MOD-4, MOD-5 and MOD-6 counters as shown in Figs. 8.17 to 8.26.

Table 8.4: Output of 3-bit Binary Counter

Clock Pulse	Counter Outputs or States		
	A_2	A_1	A_0
-	0	0	0
1	0	0	1
2	0	1	0
3	0	1	1
4	1	0	0
5	1	0	1
6	1	1	0
7	1	1	1

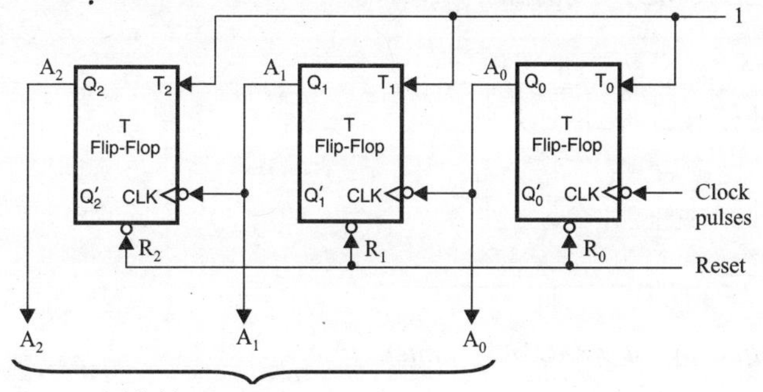

Fig. 8.16: 3-bit binary counter.

MOD-2 Counter

The 3-bit binary counter will work as MOD-2 counter if A_1 is inverted and logically ANDed with reset as shown in Fig. 8.17. Now the counter will reset in every second clock pulse.

The clock and output signal are shown in Fig. 8.18 and the counter outputs are listed in Table 8.5.

Table 8.5: Output of MOD-2 Counter

Clock Pulse	Counter Outputs		
	A_2	A_1	A_0
-	0	0	0
1	0	0	1
2	0	0	0
3	0	0	1
4	0	0	0
5	0	0	1
6	0	0	0

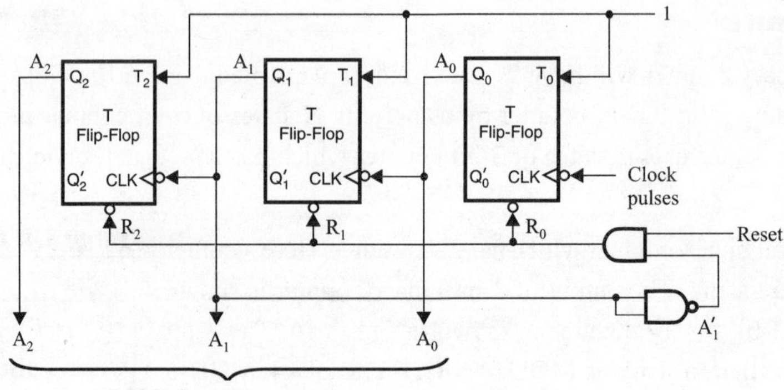

Fig. 8.17: MOD-2 counter.

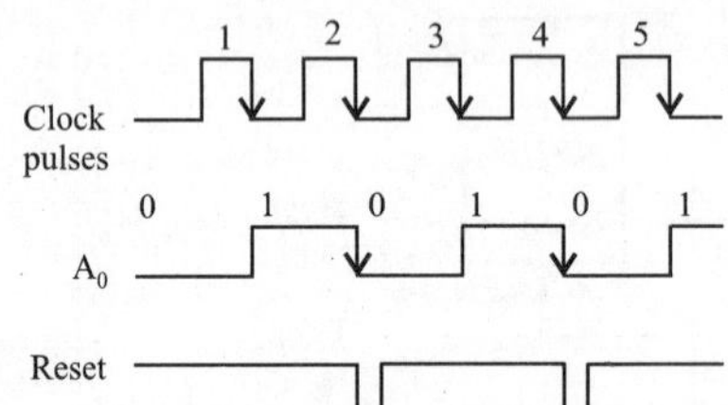

Fig. 8.18: Output signal of MOD-2 counter.

MOD-3 Counter

The 3-bit binary counter will work as MOD-3 counter if A_1 and A_0 are logically NANDed and output of NAND gate is logically ANDed with reset as shown in Fig. 8.19. Now the counter will reset in every third clock pulse.

The clock and output signals are shown in Fig. 8.20 and the counter outputs are listed in Table 8.6.

Table 8.6: Output of MOD-3 Counter

Clock Pulse	Counter Outputs		
	A_2	A_1	A_0
-	0	0	0
1	0	0	1
2	0	1	0
3	0	0	0
4	0	0	1
5	0	1	0
6	0	0	0
7	0	0	1
8	0	1	0
9	0	0	0

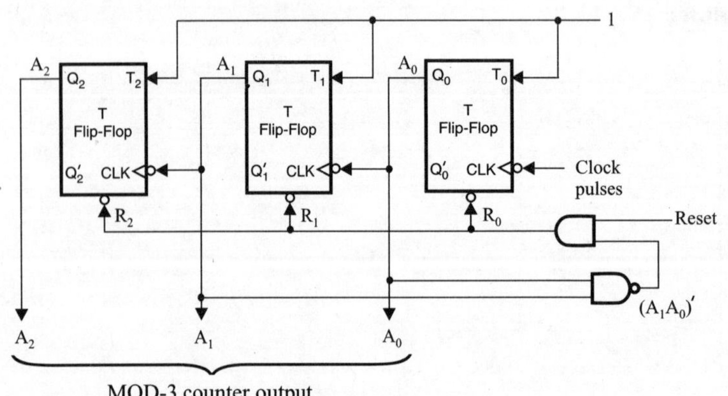

Fig. 8.19: MOD-3 counter.

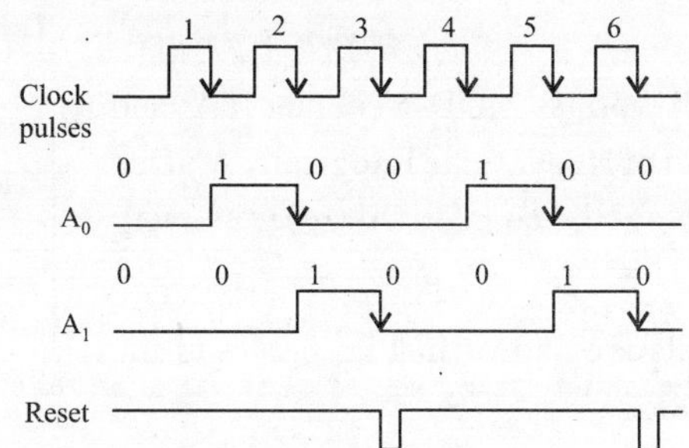

***Fig. 8.20:** Output signals of MOD-3 counter.*

MOD-4 Counter

The 3-bit binary counter will work as MOD-4 counter if A_2 is inverted and logically ANDed with reset as shown in Fig. 8.21. Now the counter will reset in every fourth clock pulse.

The clock and output signals are shown in Fig. 8.22 and the counter outputs are listed in Table 8.7.

Table 8.7: Output of MOD-4 Counter

Clock Pulse	Counter Outputs		
	A_2	A_1	A_0
-	0	0	0
1	0	0	1
2	0	1	0
3	0	1	1
4	0	0	0
5	0	0	1
6	0	1	0
7	0	1	1
8	0	0	0

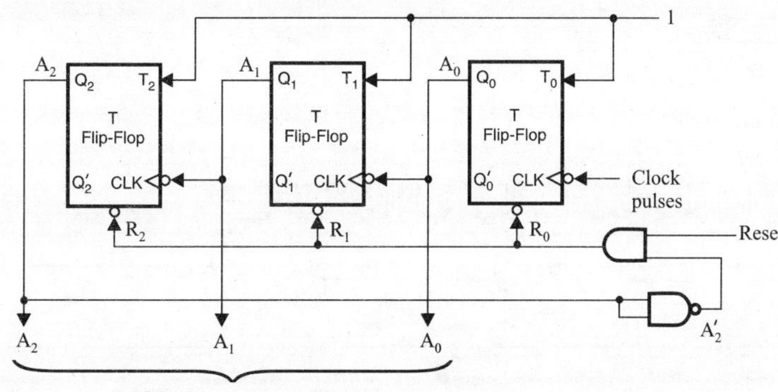

***Fig. 8.21:** MOD-4 counter.*

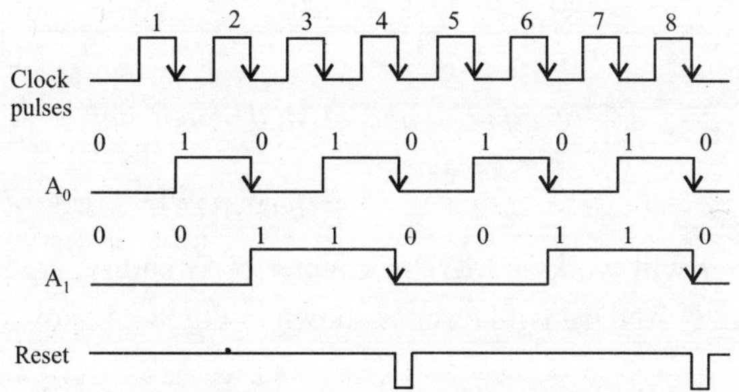

***Fig. 8.22:** Outputs signal of MOD-4 counter.*

MOD-5 Counter

(VTU, Aug/Sep'20, 6 Marks)

The 3-bit binary counter will work as MOD-5 counter if A_0 and A_2 are logically NANDed and output of NAND gate is logically ANDed with reset as shown in Fig. 8.23. Now the counter will reset in every fifth clock pulse.

The clock and output signals are shown in Fig. 8.24 and the counter outputs are listed in Table 8.8.

Table 8.8: Output of MOD-5 Counter

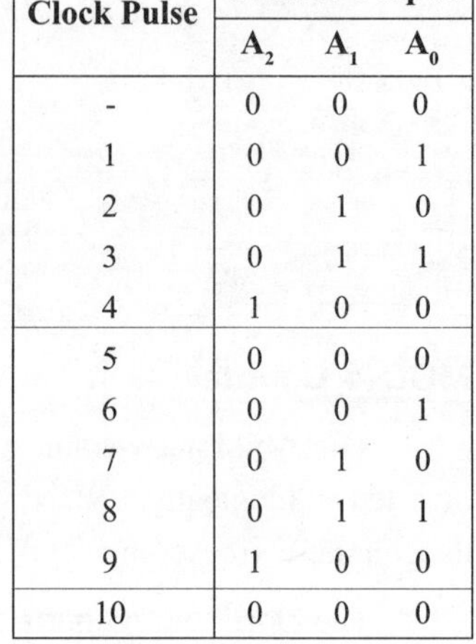

Clock Pulse	Counter Outputs		
	A_2	A_1	A_0
-	0	0	0
1	0	0	1
2	0	1	0
3	0	1	1
4	1	0	0
5	0	0	0
6	0	0	1
7	0	1	0
8	0	1	1
9	1	0	0
10	0	0	0

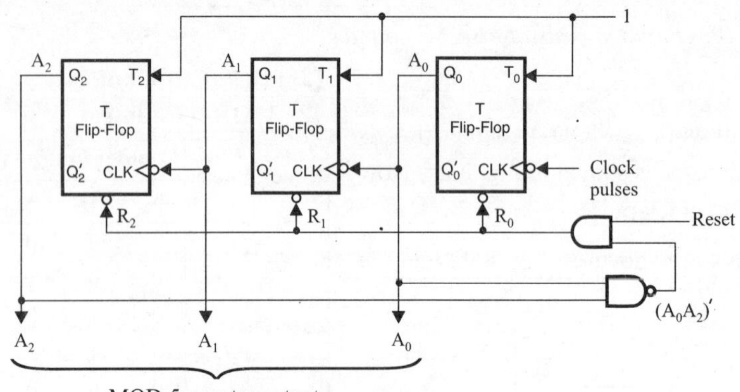

Fig. 8.23: MOD-5 counter.

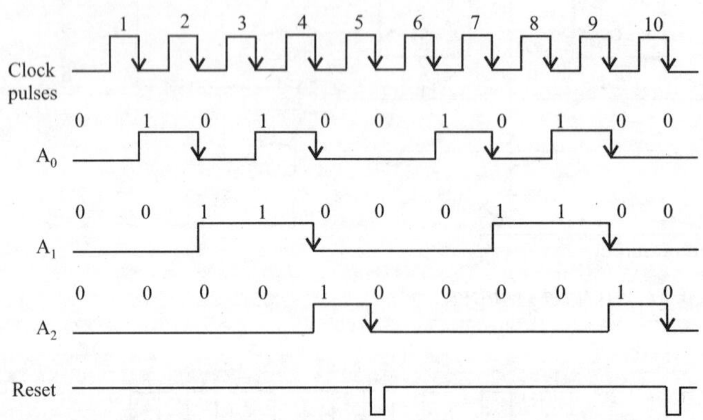

Fig. 8.24: Output signals of MOD-5 counter.

MOD-6 Counter

(AU, Nov/Dec'19, 7 Marks) (VTU, Aug/Sep'20, 8 Marks)

The 3-bit binary counter will work as MOD-6 counter if A_1 and A_2 are logically NANDed and output of NAND gate is logically ANDed with reset as shown in Fig. 8.25. Now the counter will reset in every sixth clock pulse.

The clock and output signals are shown in Fig. 8.26 and the counter outputs are listed in Table 8.9.

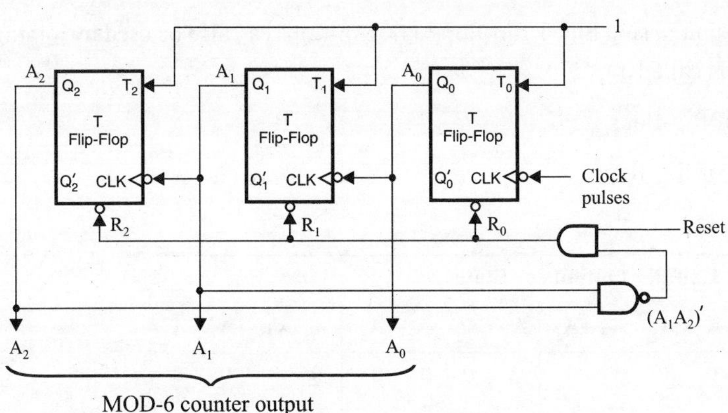

Fig. 8.25: MOD-6 counter.

Table 8.9: Output of MOD-6 Counter

Clock Pulse	Counter Outputs		
	A_2	A_1	A_0
-	0	0	0
1	0	0	1
2	0	1	0
3	0	1	1
4	1	0	0
5	1	0	1
6	0	0	0
7	0	0	1
8	0	1	0
9	0	1	1
10	1	0	0
11	1	0	1
12	0	0	0

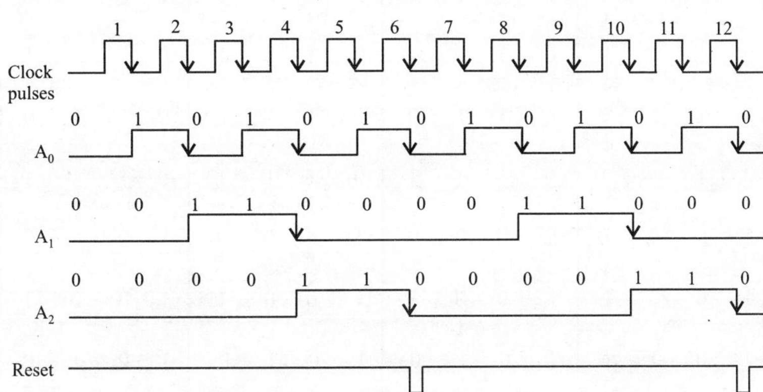

Fig. 8.26: Output signals of MOD-6 counter.

8.3 Synchronous Counter

The synchronous counters are designed using complementing flip-flops similar to ripple counters, but in synchronous counters, the clock is applied simultaneously to all the flip-flops and the inputs to the flip-flops are applied through some combinational logic circuits.

The combinational logic circuits will consider the output of one or more flip-flops to decide the input to flip-flops for next state outputs. Since the counters are also sequential circuits, the design of counters are same as that of design of sequential circuits. In design of counters, the counter outputs are considered as states.

8.3.1 Synchronous Binary Up Counter

Let us consider 4-bit binary counter using four T flip-flops. (JK flip-flops can also be used by joining J and K to form single input, which is called T).

Let, T_0, T_1, T_2, T_3 = Input of flip-flops

Q_0, Q_1, Q_2, Q_3 = Output of flip-flops

Table 8.8: Output of Up Counter

Clock Pulse	Counter Outputs or States															
	A_3	A_2	A_1	A_0	A_3	A_2	A_1	A_0	A_3	A_2	A_1	A_0	A_3	A_2	A_1	A_0
-	0	0	0	0	0	0	0	0	0	0	0	0	0	0	0	0
				↓												
1	0	0	0	1	0	0	0	1	0	0	0	1	0	0	0	1
				↓			↓									
2	0	0	1	0	0	0	1	0	0	0	1	0	0	0	1	0
				↓												
3	0	0	1	1	0	0	1	1	0	0	1	1	0	0	1	1
				↓			↓			↓ Complement						
4	0	1	0	0	0	1	0	0	0	1	0	0	0	1	0	0
				↓												
5	0	1	0	1	0	1	0	1	0	1	0	1	0	1	0	1
				↓			↓									
6	0	1	1	0	0	1	1	0	0	1	1	0	0	1	1	0
				↓												
7	0	1	1	1	0	1	1	1	0	1	1	1	0	1	1	1
				↓			↓			↓ Complement			↓ Complement			
8	1	0	0	0	1	0	0	0	1	0	0	0	1	0	0	0
				↓												
9	1	0	0	1	1	0	0	1	1	0	0	1	1	0	0	1
				↓			↓									
10	1	0	1	0	1	0	1	0	1	0	1	0	1	0	1	0
				↓												
11	1	0	1	1	1	0	1	1	1	0	1	1	1	0	1	1
				↓			↓			↓ Complement						
12	1	1	0	0	1	1	0	0	1	1	0	0	1	1	0	0
				↓												
13	1	1	0	1	1	1	0	1	1	1	0	1	1	1	0	1
				↓			↓									
14	1	1	1	0	1	1	1	0	1	1	1	0	1	1	1	0
				↓												
15	1	1	1	1	1	1	1	1	1	1	1	1	1	1	1	1
				↓			↓			↓ Complement			↓ Complement			
16	0	0	0	0	0	0	0	0	0	0	0	0	0	0	0	0

Note: *1. The down arrow (↓) indicates complement.*

2. Here the counter output is listed 4 times to show the complement operation of A_0, A_1, A_2 and A_3 separately.

Here, A_3, A_2, A_1 and A_0 be the output of binary counter which are same as flip-flop outputs. The output of binary counter for 16 clock pulses are listed in Table 8.8. Here, counter outputs and states are same. The count sequence of 4-bit up counter is from 0 to $2^4 - 1$ (i.e, 0 to 15_8).

In Table 8.8 the following observations are made,

The bit A_0 is complemented for every clock pulse.

The bit A_1 is complemented when A_0 is 1 in previous state.

The bit A_2 is complemented when both A_0 and A_1 are 1 in previous state.

The bit A_3 is complemented when A_0, A_1 and A_2 are 1 in previous state.

A 4-bit synchronous binary up counter using T flip-flop is shown in Fig. 8.27.

Since the counter output is 4-bit, 4 flip-flops are selected. A control signal called count enable is provided to start and stop the count operation. When count enable is 0, the flip-flop inputs are 0 and so no change in flip-flop outputs. The counter will start counting only when count enable is tied to 1.

The clock and output signals of binary counter are shown in Fig. 8.28. The synchronous binary up counter is same as binary ripple up counter except the following two differences.

1. The clock is directly applied to all flip-flops.
2. The inputs to flip-flops are applied through a combinational circuit.

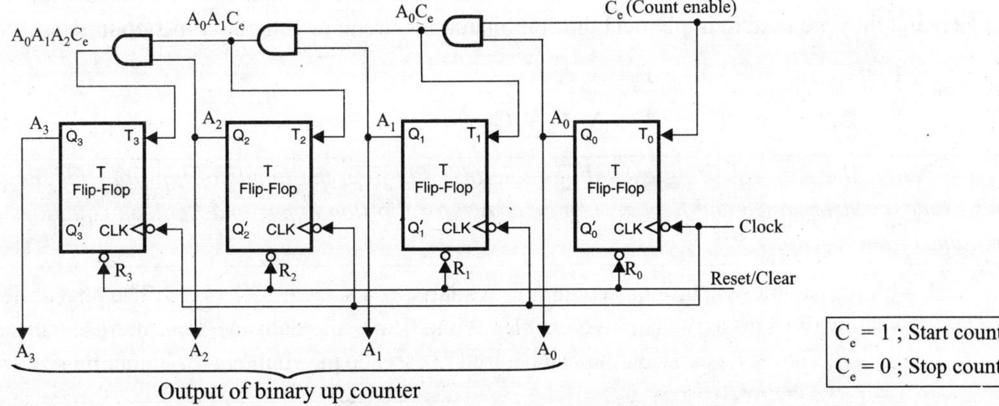

***Fig. 8.27:** 4-bit synchronous binary up counter.*

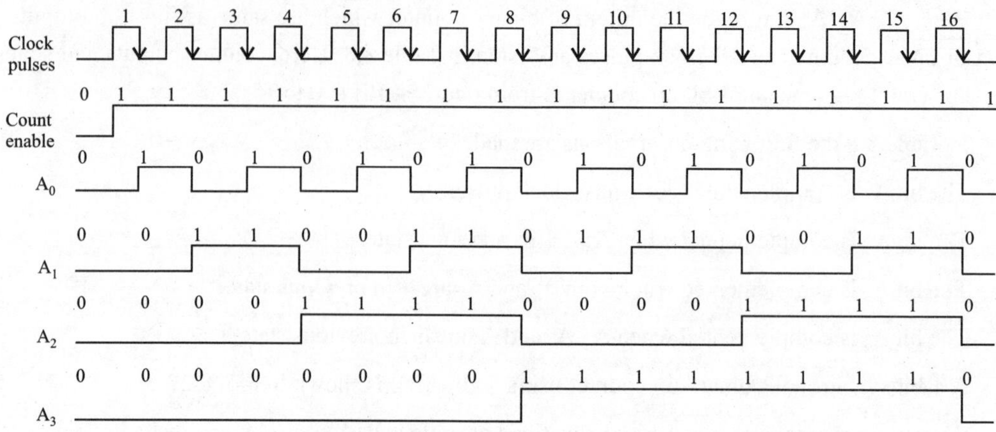

Fig. 8.28: Signals of synchronous binary counter.

The logic circuit of counter is designed as follows,

Count enable signal is applied directly to first flip-flop.

A_0 and count enable are logically ANDed and applied as input to second flip-flop.

A_0, A_1 and count enable are logically ANDed and applied as input to third flip-flop.

A_0, A_1, A_2 and count enable are logically ANDed and applied as input to fourth flip-flop.

Clock signal is applied simultaneously to all the flip-flops.

Based on the above discussion, we get the following Boolean equations for flip-flop inputs of up counter and they are used to implement the combinational circuit to generate flip-flop inputs.

$$T_0 = C_e \qquad T_2 = A_0 A_1 C_e$$

$$T_1 = A_0 C_e \qquad T_3 = A_0 A_1 A_2 C_e$$

> ***Note:*** *Alternatively, using excitation table of T flip-flop, the inputs of flip-flops for the desired next state are determined and K-maps can be drawn for flip-flop inputs and Boolean equations can be obtained from K-map.*

A 4-bit presettable synchronous counter is available as a standard IC 74161. The pin configuration of 74161 is shown in Fig. 8.29. The two enables P and T are internally ANDed and used as start/stop control operation. The ripple carry output can be used to cascade the counters for higher bit size counters.

74161

Signal	Pin	Pin	Signal
$\overline{R}$	1	16	V_{cc}(+5V)
$\overline{CLK}$	2	15	RCO
d_0	3	14	Q_0
d_1	4	13	Q_1
d_2	5	12	Q_2
d_3	6	11	Q_3
P	7	10	T
(0V) GND	8	9	$\overline{Load}$

Pin	Description
Q_0, Q_1, Q_2, Q_3	Counter outputs
d_0, d_1, d_2, d_3	Preload count inputs
$\overline{R}$	Reset
$\overline{CLK}$	Clock
P, T	Enable
RCO	Ripple carry output
V_{cc}	Supply voltage (+5V)
GND	Ground (0V)

Fig. 8.29: *Pin configuration of synchronous 4-bit counters IC 74161.*

8.3.2 Synchronous Binary Down Counter

(AU, Apr/May'19, 7 Marks) (AKTU, Sep'20, 5 Marks)

Let us consider 4-bit binary counter using four T flip-flops. (JK flip-flops can also be used by joining J and K to form single input, which is called T).

Let, T_0, T_1, T_2, T_3 = Input of flip-flops

Q_0, Q_1, Q_2, Q_3 = Output of flip-flops

Here, A_3, A_2, A_1 and A_0 be the output of binary counter which are same as flip-flop outputs. The output of binary counter for 16 clock pulses are listed in Table 8.11. Here, counter outputs and states are same. The count sequence of 4-bit up counter is from $2^4 - 1$ to 0 (i.e, 15_{10} to 0).

The output of down counter for 16 clock pulses are listed in Table 8.11, and in this table A_0, A_1, A_2 and A_3 are output of down counter and A_0', A_1', A_2' and A_3' are complement of output which are used to design the combinational circuit to generate flip-flop inputs.

Table 8.11: Output of Down Counter

Clock Pulse	Down Counter Outputs				Complement of Outputs			
	A_3	A_2	A_1	A_0	A_3'	A_2'	A_1'	A_0'
-	0	0	0	0	1	1	1	1
1	1	1	1	1	0	0	0	0
2	1	1	1	0	0	0	0	1
3	1	1	0	1	0	0	1	0
4	1	1	0	0	0	0	1	1
5	1	0	1	1	0	1	0	0
6	1	0	1	0	0	1	0	1
7	1	0	0	1	0	1	1	0
8	1	0	0	0	0	1	1	1
9	0	1	1	1	1	0	0	0
10	0	1	1	0	1	0	0	1
11	0	1	0	1	1	0	1	0
12	0	1	0	0	1	0	1	1
13	0	0	1	1	1	1	0	0
14	0	0	1	0	1	1	0	1
15	0	0	0	1	1	1	1	0
16	0	0	0	0	1	1	1	1

In Table 8.11, the following observations are made.

A_0' complemented for every clock pulse.

A_1' is complemented when A_0' is 1 in previous state.

A_2' is complemented when A_0' and A_1' are 1 in previous state.

A_3' is complemented when A_0', A_1' and A_2' are 1 in previous state.

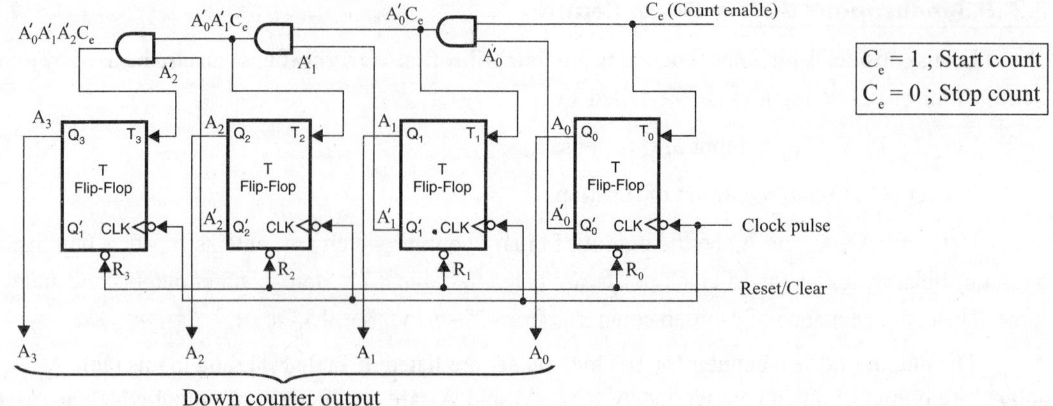

***Fig. 8.30:** 4-bit synchronous binary down counter.*

A 4-bit synchronous binary down counter is shown in Fig. 8.30. Since the counter output is 4-bit, 4 flip-flops are selected. A count enable is provided to similar to up counter. The output signals of synchronous down counter will be same as D_0, D_1, D_2 and D_3 shown in Fig. 8.7.

The synchronous binary down counter is same as binary ripple down counter except the following three differences.

1. Clock is directly applied to all flip-flops.
2. The inputs to flip-flops are applied through a combinational circuit.
3. The counter output is obtained from flip-flop outputs similar to up counter.

The logic circuit of down counter is designed as follows,

Count enable is applied directly to first flip-flop.

A'_0 and count enable are logically ANDed and applied as input to second flip-flop.

A'_0, A'_1 and count enable are logically ANDed and applied as input to third flip-flop.

A'_0, A'_1, A'_2 and count enable are logically ANDed and applied as input to fourth flip-flop.

Based on the above discussion, we get the following Boolean equations for flip-flop inputs of down counter and they are used to implement the combinational circuit to generate flip-flop inputs in Fig. 8.30.

$$T_0 = C_e$$

$$T_1 = A'_0\, C_e$$

$$T_2 = A'_0\, A'_1\, C_e$$

$$T_3 = A'_0\, A'_1\, A'_2\, C_e$$

***Note:** Alternatively, using excitation table of T flip-flop, the inputs of flip-flops for the desired next state are determined and K-maps can be drawn for flip-flop inputs and Boolean equations can be obtained from K-map.*

8.3.3 Synchronous Up/Down Counter

(RGPV, Jun'20, 8 Marks)

The up and down count operation can achieved in a single counter by logical OR operation of input to flip-flops in up and down counter and with two control inputs as shown in Fig. 8.31, so that only one input is allowed to flip-flops at any one time.

Let, C_u = Up count enable

C_d = Down count enable

Now, the control logic for up/down count is,

$C_u = 1$ and $C_d = 0$ for up count

$C_u = 0$ and $C_d = 1$ for down count

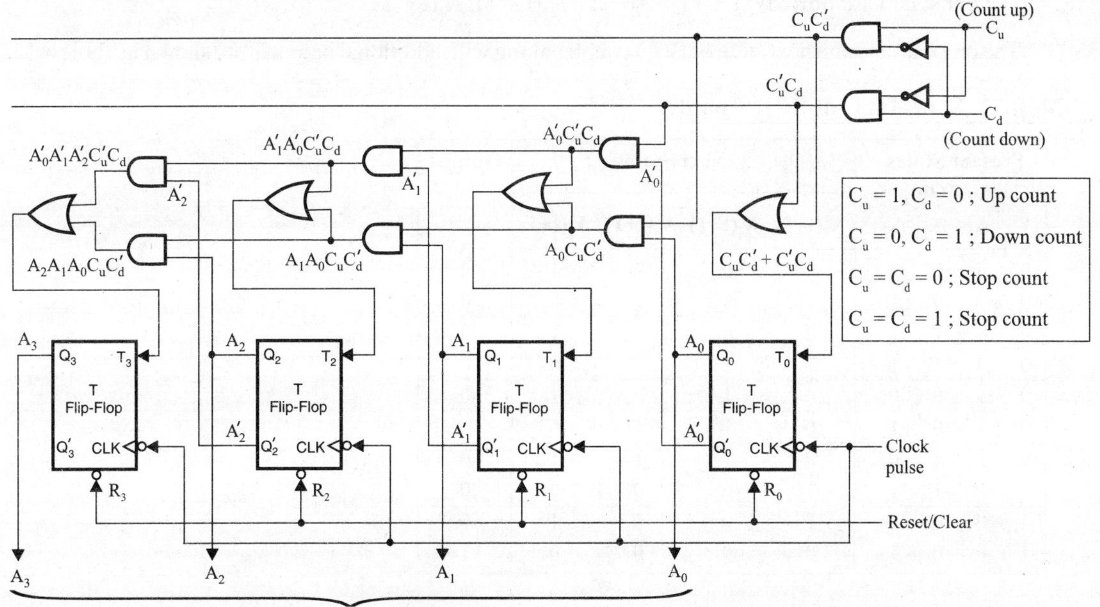

***Fig. 8.31:** Synchronous 4-bit up/down counter.*

The Boolean equations of up/down counter are given by sum of Boolean equations of up and down counters after replacing C_e of up counter by $C_uC'_d$ and C_e of down counter by C'_uC_d.

The following are Boolean equations for input of flip-flops of up/down counter and they are used to implement the combinational circuit to generate flip-flop inputs.

$$T_0 = C_u\,C'_d + C'_u\,C_d$$

$$T_1 = A_0\,C_u\,C'_d + A'_0\,C'_u\,C_d$$

$$T_2 = A_0\,A_1\,C_u\,C'_d + A'_0\,A'_1\,C'_u\,C_d$$

$$T_3 = A_0\,A_1\,A_2\,C_u\,C'_d + A'_0\,A'_1\,A'_2\,C'_u\,C_d$$

8.3.4 Synchronous BCD Counter

BCD counter will count only from 0 to 9 for 9 clock pulses and upon reaching tenth pulse output is reset to 0, so that next count sequence will start. Also, an additional output 1 is needed to use as carry to next stage.

The design of synchronous BCD counter using T flip-flop is presented in this section.

The flip-flops inputs to achieve the desired next state are determined from the excitation table of flip-flop. The combinational circuit to generate flip-flops are designed using K-maps.

Let, A_8, A_4, A_2 and A_1 be 4 state variables of BCD counter.

$\therefore$ Present state variables : $A_8(t), A_4(t), A_2(t), A_1(t) \Rightarrow A_8, A_4, A_2, A_1$ (For simplicity)

Next state variables : $A_8(t+1), A_4(t+1), A_2(t+1), A_1(t+1)$

The present state and next state of BCD counter along with additional output Y are shown in Table 8.12.

Table 8.12: State Table of BCD Counter

Present States (Counter Outputs)				Next States				Output
A_8	A_4	A_2	A_1	$A_8(t+1)$	$A_4(t+1)$	$A_2(t+1)$	$A_1(t+1)$	Y
0	0	0	0	0	0	0	1	0
0	0	0	1	0	0	1	0	0
0	0	1	0	0	0	1	1	0
0	0	1	1	0	1	0	0	0
0	1	0	0	0	1	0	1	0
0	1	0	1	0	1	1	0	0
0	1	1	0	0	1	1	1	0
0	1	1	1	1	0	0	0	0
1	0	0	0	1	0	0	1	0
1	0	0	1	0	0	0	0	1

Let us choose four T flip-flops to store four state variables A_1, A_2, A_4 and A_8 as shown in Fig. 8.32. Here, A_1, A_2, A_4 and A_8 are output of flip-flops.

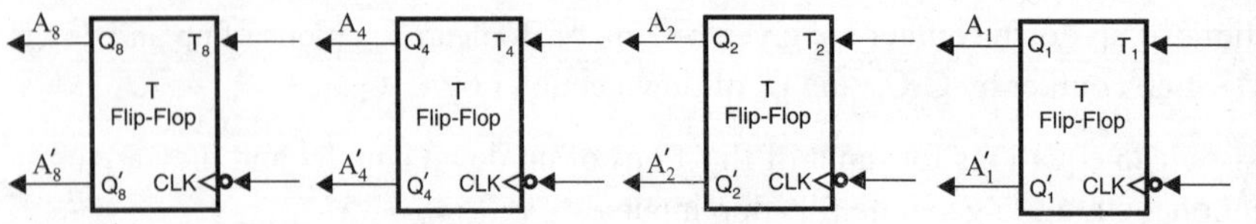

Fig. a: Flip-flop A_8. *Fig. b: Flip-flop A_4.* *Fig. c: Flip-flop A_2.* *Fig. d: Flip-flop A_1.*

***Fig. 8.32:** T flip-flops to store state variables.*

Excitation Table of T Flip-Flop

Present State	Next State	Flip-Flop Input
Q(t)	Q(t + 1)	T
0	0	0
0	1	1
1	0	1
1	1	0

The flip-flop inputs can be obtained from excitation table as shown below:

State A_8

For state A_8, the Q(t), Q(t + 1) and T of excitation table are considered as A_8, $A_8(t + 1)$ and T_8.

$\therefore Q(t) \rightarrow A_8 \;;\; Q(t+1) \rightarrow A_8(t+1)$

$T \rightarrow T_8$

The flip-flop input T_8 that drives the present state A_8 to next state is obtained from excitation table and listed in Table 8.13.

State A_4

For state A_4, the Q(t), Q(t + 1) and T of excitation table are considered as A_4, $A_4(t + 1)$ and T_4.

$\therefore Q(t) \rightarrow A_4 \;;\; Q(t+1) \rightarrow A_4(t+1)$

$T \rightarrow T_4$

The flip-flop input T_4 that drives the present state A_4 to next state is obtained from excitation table and listed in Table 8.14.

State A_2

For state A_2, the Q(t), Q(t + 1) and T of excitation table are considered as A_2, $A_2(t + 1)$ and T_2.

$\therefore Q(t) \rightarrow A_2 \;;\; Q(t+1) \rightarrow A_2(t+1)$

$T \rightarrow T_2$

The flip-flop input T_2 that drives the present state A_2 to next state is obtained from excitation table and listed in Table 8.15.

State A_1

For state A_1, the Q(t), Q(t + 1) and T of excitation table are considered as A_1, $A_1(t + 1)$ and T_1.

$\therefore Q(t) \rightarrow A_1 \;;\; Q(t+1) \rightarrow A_1(t+1)$

$T \rightarrow T_1$

The flip-flop input T_1 that drives the present state A_1 to next state is obtained from excitation table and listed in Table 8.16.

Table 8.13:

A_8	$A_8(t+1)$	T_8
0	0	0
0	0	0
0	0	0
0	0	0
0	0	0
0	0	0
0	0	0
0	1	1
1	1	0
1	0	1

Table 8.14:

A_4	$A_4(t+1)$	T_4
0	0	0
0	0	0
0	0	0
0	1	1
1	1	0
1	1	0
1	1	0
1	0	1
0	0	0
0	0	0

Table 8.15:

A_2	$A_2(t+1)$	T_2
0	0	0
0	1	1
1	1	0
1	0	1
0	0	0
0	1	1
1	1	0
1	0	1
0	0	0
0	0	0

Table 8.16:

A_1	$A_1(t+1)$	T_1
0	1	1
1	0	1
0	1	1
1	0	1
0	1	1
1	0	1
0	1	1
1	0	1
0	1	1
1	0	1

The tables 8.12 to 8.16 are combined to form state table with flip-flop inputs as shown in Table 8.17.

Table 8.17: State Table with Flip-Flop Inputs

Present States (Counter Outputs)				Next States				Output	Flip-Flop Inputs				Minterm
A_8	A_4	A_2	A_1	$A_8(t+1)$	$A_4(t+1)$	$A_2(t+1)$	$A_1(t+1)$	Y	T_8	T_4	T_2	T_1	$A_8A_4A_2A_1$
0	0	0	0	0	0	0	1	0	0	0	0	1	m_0
0	0	0	1	0	0	1	0	0	0	0	1	1	m_1
0	0	1	0	0	0	1	1	0	0	0	0	1	m_2
0	0	1	1	0	1	0	0	0	0	1	1	1	m_3
0	1	0	0	0	1	0	1	0	0	0	0	1	m_4
0	1	0	1	0	1	1	0	0	0	0	1	1	m_5
0	1	1	0	0	1	1	1	0	0	0	0	1	m_6
0	1	1	1	1	0	0	0	0	1	1	1	1	m_7
1	0	0	0	1	0	0	1	0	0	0	0	1	m_8
1	0	0	1	0	0	0	0	1	1	0	0	1	m_9
1	0	1	0	×	×	×	×	×	×	×	×	×	m_{10}
1	0	1	1	×	×	×	×	×	×	×	×	×	m_{11}
1	1	0	0	×	×	×	×	×	×	×	×	×	m_{12}
1	1	0	1	×	×	×	×	×	×	×	×	×	m_{13}
1	1	1	0	×	×	×	×	×	×	×	×	×	m_{14}
1	1	1	1	×	×	×	×	×	×	×	×	×	m_{15}

Note: *1. The states 1010 to 1111 are unused states and so next states, output and flip-flop inputs are considered as don't-care for these states.*

2. Minterms are formed using state variables A_8, A_4, A_2 and A_1.

Here, T_8, T_4, T_2, T_1 and Y are functions of A_8, A_4, A_2 and A_1. Therefore, the Boolean equations for T_8, T_4, T_2, T_1 and Y are obtained from Table 8.17 as sum of minterms and don't-cares as shown below:

$$T_8 = \Sigma m(7, 9) + \Sigma d(10, 11, 12, 13, 14, 15)$$

$$T_4 = \Sigma m(3, 7) + \Sigma d(10, 11, 12, 13, 14, 15)$$

$$T_2 = \Sigma m(1, 3, 5, 7) + \Sigma d(10, 11, 12, 13, 14, 15)$$

$$T_1 = \Sigma m(0, 1, 2, 3, 4, 5, 6, 7, 8, 9) + \Sigma d(10, 11, 12, 13, 14, 15)$$

$$Y = m_9 + \Sigma d(10, 11, 12, 13, 14, 15)$$

Let us draw K-maps for T_8, T_4, T_2, T_1 and Y in order to determine the simplified Boolean equations for flip-flop inputs and output Y as shown in Figs. 8.33 to 8.37.

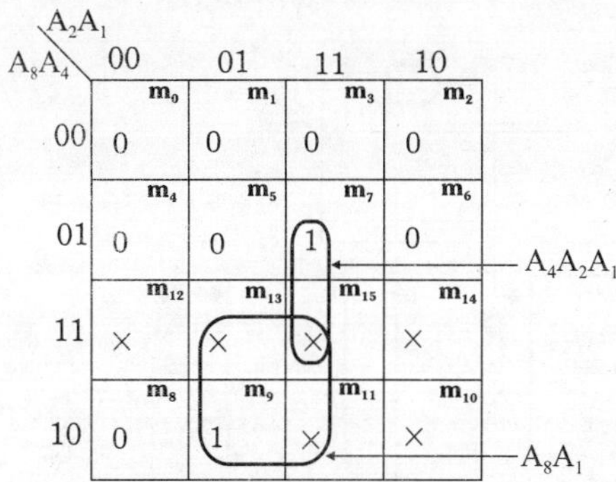

Fig. 8.33: *K-map for T_8.*

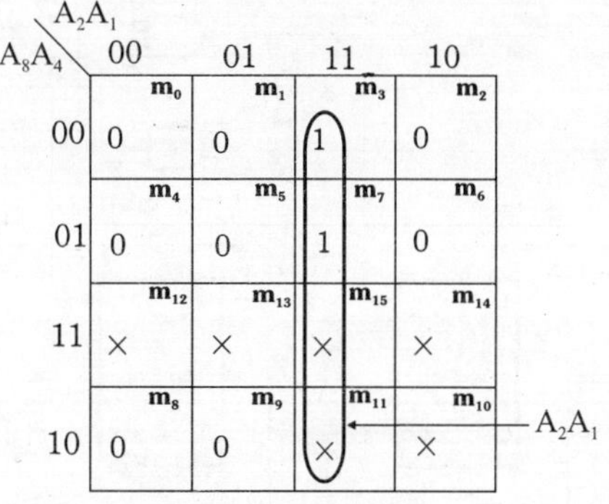

Fig. 8.34: *K-map for T_4.*

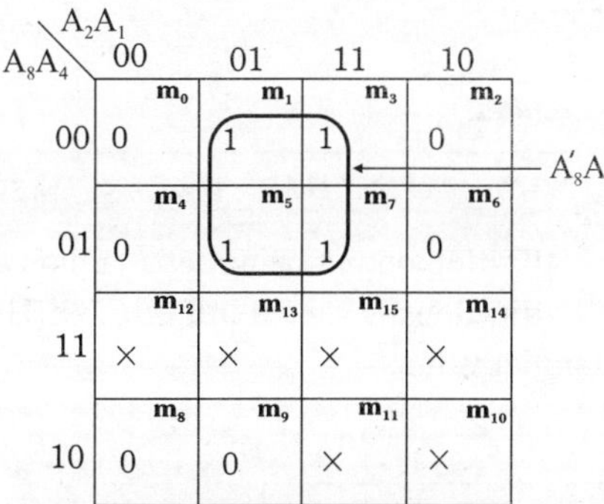

Fig. 8.35: *K-map for T_2.*

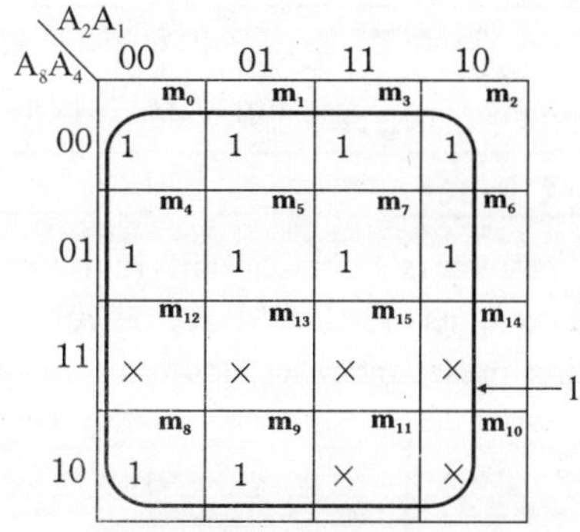

Fig. 8.36: *K-map for T_1.*

From the K-map we get the following simplified Boolean equations for T_8, T_4, T_2, T_1 and Y. Using these equations the logic circuit of synchronous BCD counter is drawn as shown in Fig. 8.38.

$$T_8 = A_8A_1 + A_4A_2A_1$$

$$T_4 = A_2\,A_1$$

$$T_2 = A'_8\,A_1$$

$$T_1 = 1$$

$$Y = A_8\,A_1$$

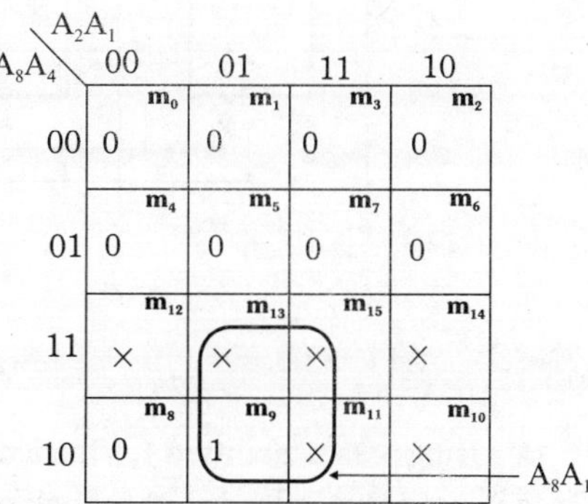

Fig. 8.37: *K-map for Y.*

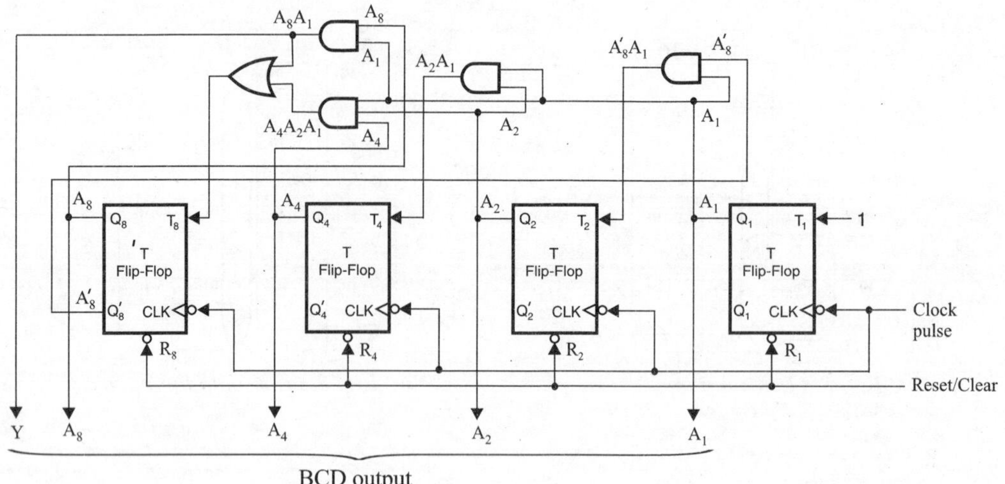

Fig. 8.38: Synchronous BCD counter.

8.3.5 Ring Counter

(RGPV, Jun'22, 2 Marks) (AKTU, Sep'20, 5 Marks)

A ring counter is a circular shift register with only one flip-flop output being set at a time and all other flip-flop outputs are cleared. A ring counter with four D flip-flops is shown in Fig. 8.39. The ring counter requires preset and reset facility in flip-flops to load initial value.

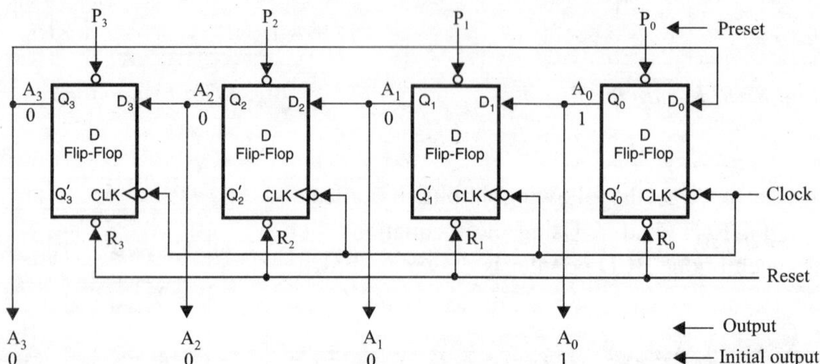

Fig. 8.39: Ring counter.

Using preset facility, a 1 is loaded in one flip-flop output and all other flip-flop outputs are loaded 0. Once initial value is loaded, the clock pulses can be applied to start the counting operation. Here, all the flip-flops are clocked simultaneosly. For every clock pulse, the flip-flop input is passed to output. Here, only one flip-flop input is 1 and all other flip-flop inputs are 0 and so 1 will shift by one position.

This produces a sequence of timing signals that are shifted by one clock time period. The time delay between start of any two consecutive output will be one clock time period. The ring counter output for 12 clock pulses are listed in Table 8.18. The output signals of ring counter are shown in Fig. 8.40.

Table 8.18: Ring Counter Output

Clock Pulse	A_3	A_2	A_1	A_0
-	0	0	0	1
1	0	0	1	0
2	0	1	0	0
3	1	0	0	0
4	0	0	0	1
5	0	0	1	0
6	0	1	0	0
7	1	0	0	0
8	0	0	0	1
9	0	0	1	0
10	0	1	0	0
11	1	0	0	0
12	0	0	0	1

Fig. 8.40: Clock and output signals of ring counter.

Advantages of Ring Counter

(IKGPTU, Dec'19, 2 Marks)

The n-bit Ring counter will generate n distinct states and hence it can be used in sequential control systems to initiate n events one by one.

The output of ring counter can also be used as delay lines, by allowing one clock pulse at a time, so that the output will be a delayed input with a delay of one clock time period.

The output of n-bit ring counter can be used as n-phase clock, in which each output will be high for one clock period in a total time of n clock periods.

8.3.6 Johnson Counter

Johnson counter is basically a ring counter with complement of last flip-flop connected to input of first flip-flop. This arrangement is called ***switch tail ring counter.***

Johnson counter using D flip-flops is shown in Fig. 8.41. Johnson counter operation starts with clear operation. When the flip-flops are cleared or reset all the outputs will be zero, and complement of outputs will be one. So that input to first flip-flop is 1 due to the connection of last flip-flop complemented output to input of first flip-flop.

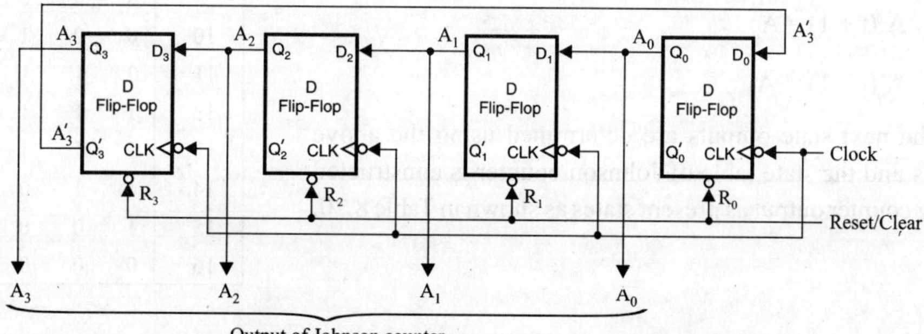

Fig. 8.41: Johnson counter using D flip-flop.

When clock pulses are applied, the counter generates 8 distinguishable outputs and this output repeats for every 8 clock pulses. The output signals of Johnson counter are shown in Fig. 8.42. The Johnson counter output for 16 clock pulses are listed in Table 8.19.

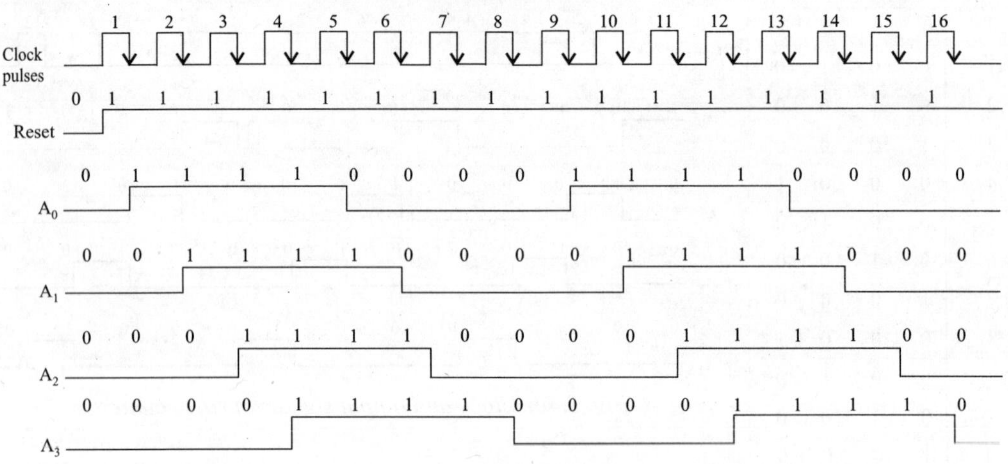

Fig. 8.42: *Clock and output signals of Johnson counter.*

The counter output are state variables. Hence the counter has 4 state variables A_3, A_2, A_1 and A_0.

$\therefore$ Present state variables : $A_3(t)$, $A_2(t)$, $A_1(t)$, $A_0(t)$

$$\Downarrow$$

$$A_3, A_2, A_1, A_0 \text{ (For simplicity)}$$

Next state variables : $A_3(t+1)$, $A_2(t+1)$, $A_1(t+1)$, $A_0(t+1)$

In D flip-flops the next state output is given by input of the flip-flop at the time of applying clock pulse. Hence with reference to Fig. 8.41, the following equations can be obtained for next state outputs.

$$A_3(t+1) = A_2$$

$$A_2(t+1) = A_1$$

$$A_1(t+1) = A_0$$

$$A_0(t+1) = A_3'$$

The next state outputs are determined using the above equations and the state table of Johnson counter is constructed by taking counter output as present states as shown in Table 8.20.

Table 8.19: Johnson Counter Output

Clock Pulse	Counter Outputs A_3	A_2	A_1	A_0
-	0	0	0	0
1	0	0	0	1
2	0	0	1	1
3	0	1	1	1
4	1	1	1	1
5	1	1	1	0
6	1	1	0	0
7	1	0	0	0
8	0	0	0	0
9	0	0	0	1
10	0	0	1	1
11	0	1	1	1
12	1	1	1	1
13	1	1	1	0
14	1	1	0	0
15	1	0	0	0
16	0	0	0	0

Table 8.20: State Table of Johnson Counter

Present States					Next States			
A_3	A_2	A_1	A_0		$A_3(t+1) = A_2$	$A_2(t+1) = A_1$	$A_1(t+1) = A_0$	$A_0(t+1) = A_3'$
0	0	0	0	8 distinguishable states	0	0	0	1
0	0	0	1		0	0	1	1
0	0	1	1		0	1	1	1
0	1	1	1		1	1	1	1
1	1	1	1		1	1	1	0
1	1	1	0		1	1	0	0
1	1	0	0		1	0	0	0
1	0	0	0		0	0	0	0

(IKGPTU, Dec'19, 2 Marks)

The 4-bit Johnson counter generates 8 distinguishable states. In general, k bit Johnson counter will generate 2k distinguishable states.

The 8 distinguishable states of 4-bit Johnson counter can be decoded to generate 8 unique output signals as shown in Table 8.21. The required decoded signals with respect to clock signal is shown in Fig. 8.43.

Table 8.21: Decoded Output Signals of Johnson Counter

Present States				Decoded Outputs								Minterm
A_3	A_2	A_1	A_0	Y_7	Y_6	Y_5	Y_4	Y_3	Y_2	Y_1	Y_0	$A_3A_2A_1A_0$
0	0	0	0	0	0	0	0	0	0	0	1	m_0
0	0	0	1	0	0	0	0	0	0	1	0	m_1
0	0	1	1	0	0	0	0	0	1	0	0	m_3
0	1	1	1	0	0	0	0	1	0	0	0	m_7
1	1	1	1	0	0	0	1	0	0	0	0	m_{15}
1	1	1	0	0	0	1	0	0	0	0	0	m_{14}
1	1	0	0	0	1	0	0	0	0	0	0	m_{12}
1	0	0	0	1	0	0	0	0	0	0	0	m_8

Note: *The states corresponding to minterms m_2, m_4, m_5, m_6, m_9, m_{10}, m_{11} and m_{13} are unused states and so corresponding decoded outputs are don't-cares.*

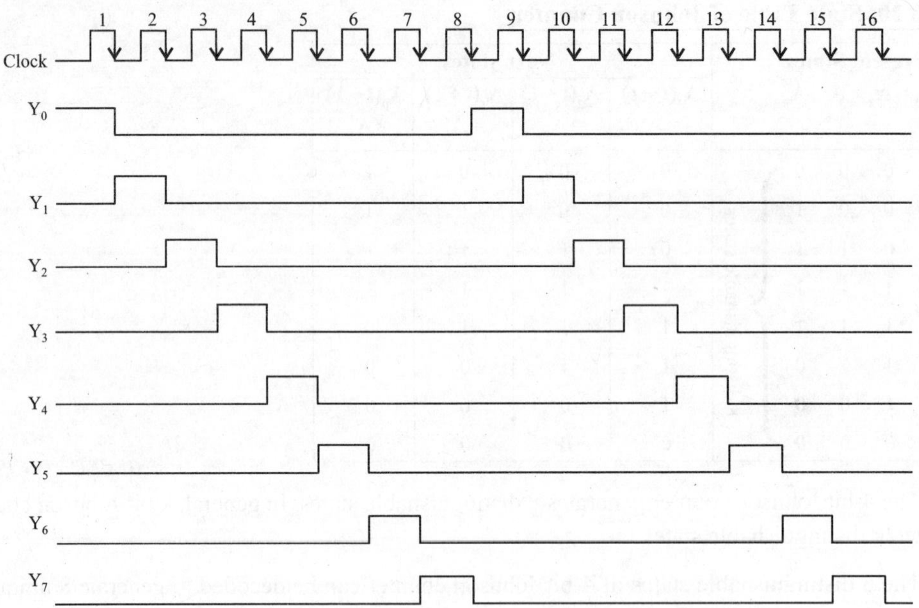

Fig. 8.43: Decoded signals of 8 states of Johnson counter.

In Table 8.21, the minterms are formed using present state variables A_3, A_2, A_1 and A_0. From Table 8.21 the following Boolean equations are obtained for the decoded output of Johnson counter as sum of minterms and don't-cares. Using these equations, K-maps are constructed as shown in Figs. 8.44 to 8.51 to get simplified Boolean equations.

$Y_0 = m_0 + \Sigma d(2, 4, 5, 6, 9, 10, 11, 13)$

$Y_1 = m_1 + \Sigma d(2, 4, 5, 6, 9, 10, 11, 13)$

$Y_2 = m_3 + \Sigma d(2, 4, 5, 6, 9, 10, 11, 13)$

$Y_3 = m_7 + \Sigma d(2, 4, 5, 6, 9, 10, 11, 13)$

$Y_4 = m_{15} + \Sigma d(2, 4, 5, 6, 9, 10, 11, 13)$

$Y_5 = m_{14} + \Sigma d(2, 4, 5, 6, 9, 10, 11, 13)$

$Y_6 = m_{12} + \Sigma d(2, 4, 5, 6, 9, 10, 11, 13)$

$Y_7 = m_8 + \Sigma d(2, 4, 5, 6, 9, 10, 11, 13)$

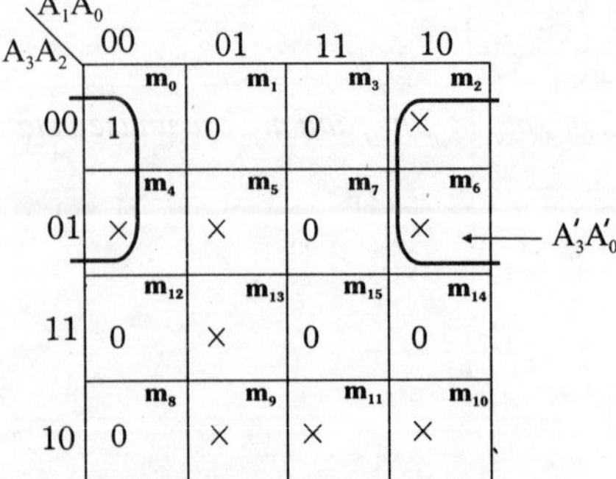

Fig. 8.44: K-map for Y_0.

A_3A_2 \ A_1A_0	00	01	11	10
00	m_0 0	m_1 1	m_3 0	m_2 ×
01	m_4 ×	m_5 ×	m_7 0	m_6 ×
11	m_{12} 0	m_{13} ×	m_{15} 0	m_{14} 0
10	m_8 0	m_9 ×	m_{11} ×	m_{10} ×

$A_1'A_0$

Fig. 8.45: K-map for Y_1.

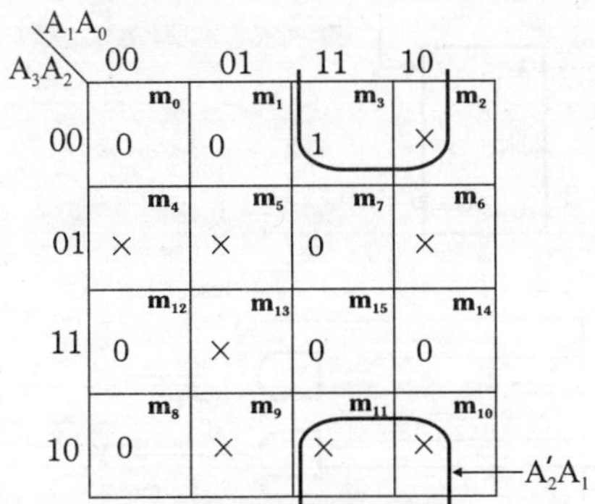
Fig. 8.46: K-map for Y_2.

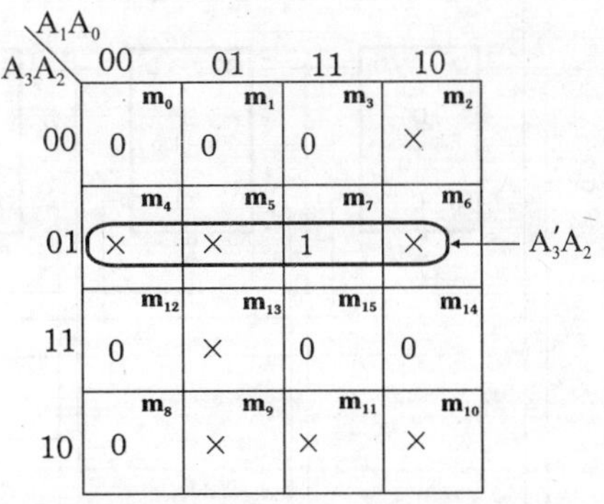
Fig. 8.47: K-map for Y_3.

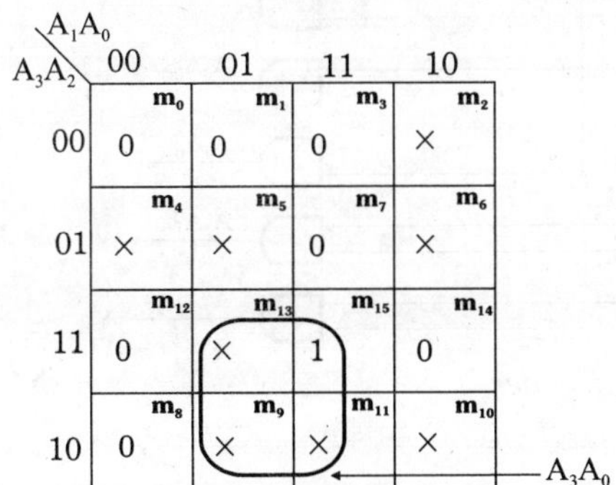
Fig. 8.48: K-map for Y_4.

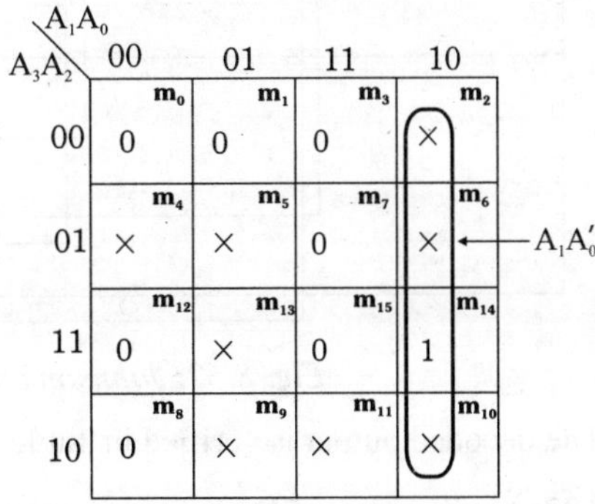
Fig. 8.49: K-map for Y_5.

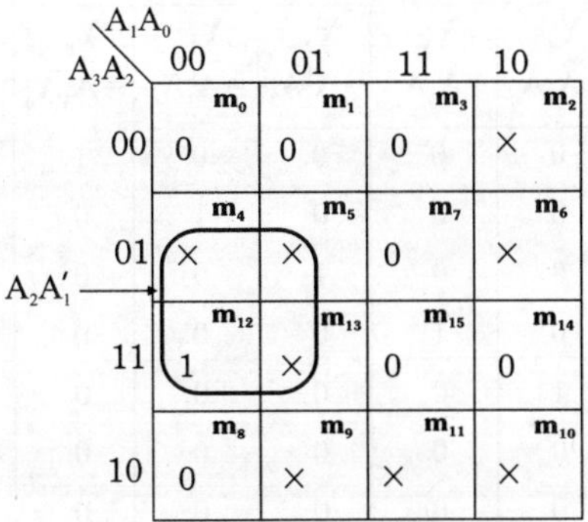
Fig. 8.50: K-map for Y_6.

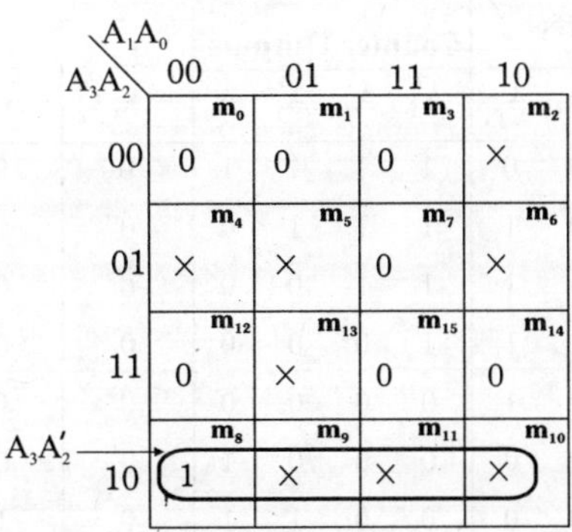
Fig. 8.51: K-map for Y_7.

From the K-map the following simplified Boolean equations are obtained for decoded output Y_0 to Y_7. Using these equations the logic circuits of Johnson counter as decoder is drawn as shown in Fig. 8.52.

$$Y_0 = A_3'A_0' \qquad Y_4 = A_3A_0$$
$$Y_1 = A_1'A_0 \qquad Y_5 = A_1A_0'$$
$$Y_2 = A_2'A_1 \qquad Y_6 = A_2A_1'$$
$$Y_3 = A_3'A_2 \qquad Y_7 = A_3A_2'$$

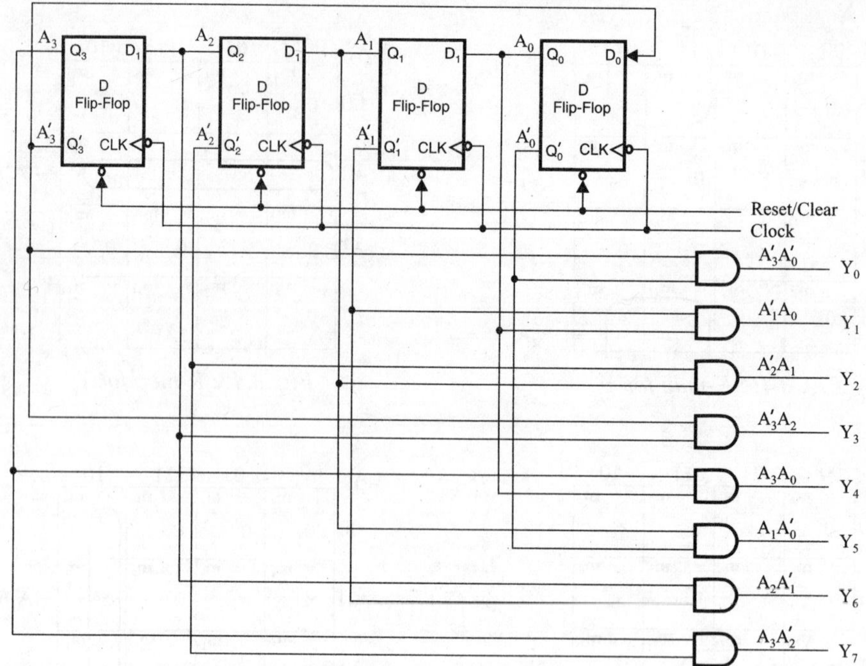

Fig. 8.52: Johnson counter as decoder.

The decoded output is verified in Table 8.22.

Table 8.22: Verification of Decoding of Johnson Counter Output

Counter Outputs				Complement of Counter Outputs				Decoded Outputs							
								Y_7	Y_6	Y_5	Y_4	Y_3	Y_2	Y_1	Y_0
A_3	A_2	A_1	A_0	A_3'	A_2'	A_1'	A_0'	$=A_3A_2'$	$=A_2A_1'$	$=A_1A_0'$	$=A_3A_0$	$=A_3'A_2$	$=A_2'A_1$	$=A_1'A_0$	$=A_3'A_0'$
0	0	0	0	1	1	1	1	0	0	0	0	0	0	0	1
0	0	0	1	1	1	1	0	0	0	0	0	0	0	1	0
0	0	1	1	1	1	0	0	0	0	0	0	0	1	0	0
0	1	1	1	1	0	0	0	0	0	0	0	1	0	0	0
1	1	1	1	0	0	0	0	0	0	0	1	0	0	0	0
1	1	1	0	0	0	0	1	0	0	1	0	0	0	0	0
1	1	0	0	0	0	1	1	0	1	0	0	0	0	0	0
1	0	0	0	0	1	1	1	1	0	0	0	0	0	0	0

Example 8.1

(GTU, Jun'19, 7 Marks)

Design a 3-bit binary counter using T flip-flops

Solution

The 3-bit binary counter can count decimal values in the range 0 to 7. Let us assign binary value of decimal number as code for states. With 3-bit code 8 states ($2^3 = 8$) are possible and in this all the 8 states are used states.

Binary Code Assignment

0	000
1	001
2	010
3	011
4	100
5	101
6	110
7	111

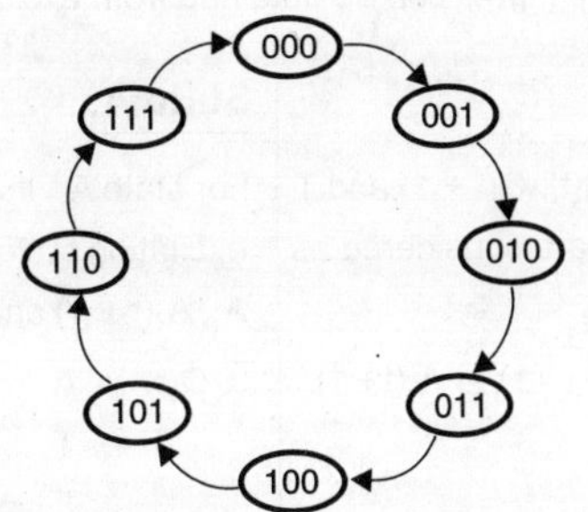

Fig. 1: State diagram of 3-bit binary counter.

The state diagram using all the 8 states is shown in Fig. 1.

Here, each bit of binary code is a state variable. Hence the counter has 3 state variables. Let the state variables be A_0, A_1 and A_2.

$\therefore$ Present state variables : $A_0(t), A_1(t), A_2(t) \Rightarrow A_0, A_1, A_2$ (For simplicity)

Next state variables : $A_0(t + 1), A_1(t + 1), A_2(t + 1)$

Using the state diagram of Fig. 1, the state table is constructed as shown in Table 1.

Table 1: State Table

Present States			Next States		
A_2	A_1	A_0	$A_2(t+1)$	$A_1(t+1)$	$A_0(t+1)$
0	0	0	0	0	1
0	0	1	0	1	0
0	1	0	0	1	1
0	1	1	1	0	0
1	0	0	1	0	1
1	0	1	1	1	0
1	1	0	1	1	1
1	1	1	0	0	0

Let us choose three T flip-flops to store three state variables A_0, A_1 and A_2 as shown in Fig. 2. Here, A_0, A_1 and A_2 are output of flip-flops.

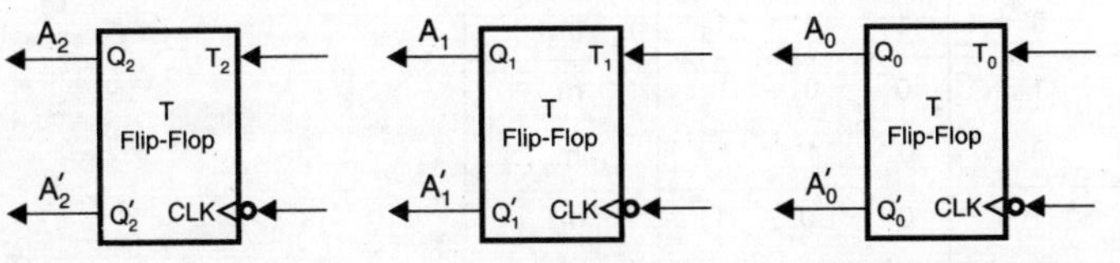

Fig. a: Flip-flop A_2. *Fig. b: Flip-flop A_1.* *Fig. c: Flip-flop A_0.*

Fig. 2: T flip-flops to store state variables.

Excitation Table of T Flip-Flop

Present State Q(t)	Next State Q(t + 1)	Flip-Flop Input T
0	0	0
0	1	1
1	0	1
1	1	0

The flip-flop inputs can be obtained from excitation table as shown below:

State A_2

For state A_2, the Q(t), Q(t + 1) and T of excitation table are considered as A_2, $A_2(t + 1)$ and T_2.

$\therefore Q(t) \rightarrow A_2$; $Q(t+1) \rightarrow A_2(t+1)$

$T \rightarrow T_2$

The flip-flop input T_2 that drives the present state A_2 to next state is obtained from excitation table and listed in Table 2.

Table 2:

A_2	$A_2(t+1)$	T_2
0	0	0
0	0	0
0	0	0
0	1	1
1	1	0
1	1	0
1	1	0
1	0	1

State A_1

For state A_1, the Q(t), Q(t + 1) and T excitation table are considered as A_1, $A_1(t + 1)$ and T_1.

$\therefore Q(t) \rightarrow A_1$; $Q(t+1) \rightarrow A_1(t+1)$

$T \rightarrow T_1$

The flip-flop input T_1 that drives the present state A_1 to next state is obtained from excitation table and listed in Table 3.

Table 3:

A_1	$A_1(t+1)$	T_1
0	0	0
0	1	1
1	1	0
1	0	1
0	0	0
0	1	1
1	1	0
1	0	1

State A_0

For state A_0, the Q(t), Q(t + 1) and T of excitation table are considered as A_0, $A_0(t + 1)$ and T_0.

$\therefore Q(t) \rightarrow A_0$; $Q(t+1) \rightarrow A_0(t+1)$

$T \rightarrow T_0$

The flip-flop input T_0 that drives the present state A_0 to next state is obtained from excitation table and listed in Table 4.

Table 4:

A_0	$A_0(t+1)$	T_0
0	1	1
1	0	1
0	1	1
1	0	1
0	1	1
1	0	1
0	1	1
1	0	1

The Tables 1 to 4 are combined to form state table with flip-flop inputs as shown in Table 5.

Table 5: Expanded State Table

Present States			Next States			Flip-Flop Inputs			Minterm
A_2	A_1	A_0	$A_2(t+1)$	$A_1(t+1)$	$A_0(t+1)$	T_2	T_1	T_0	$A_2A_1A_0$
0	0	0	0	0	1	0	0	1	m_0
0	0	1	0	1	0	0	1	1	m_1
0	1	0	0	1	1	0	0	1	m_2
0	1	1	1	0	0	1	1	1	m_3
1	0	0	1	0	1	0	0	1	m_4
1	0	1	1	1	0	0	1	1	m_5
1	1	0	1	1	1	0	0	1	m_6
1	1	1	0	0	0	1	1	1	m_7

***Note:** Here, present state $A_2A_1A_0$ is the counter output.*

The Boolean equations for T_2, T_1 and T_0 are obtained from Table 5 by taking present state as minterms.

$T_2 = \Sigma m(3, 7)$

$T_1 = \Sigma m(1, 3, 5, 7)$

$T_0 = \Sigma m(0, 1, 2, 3, 4, 5, 6, 7)$

The simplified Boolean equations are obtained by constructing K-map using the above equation for T_2, T_1 and T_0 as shown in Figs. 3 to 5.

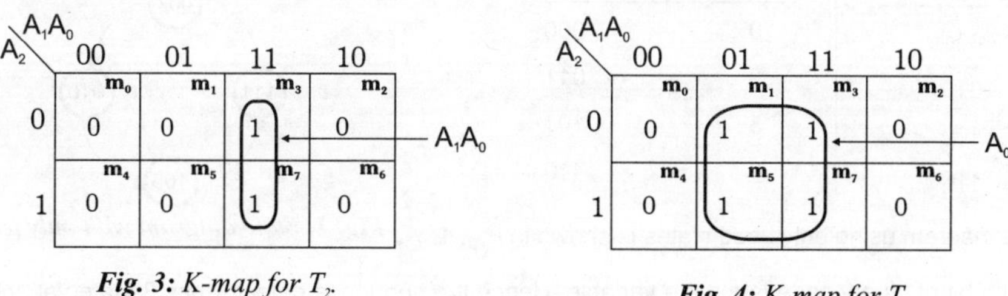

Fig. 3: K-map for T_2. *Fig. 4: K-map for T_1.*

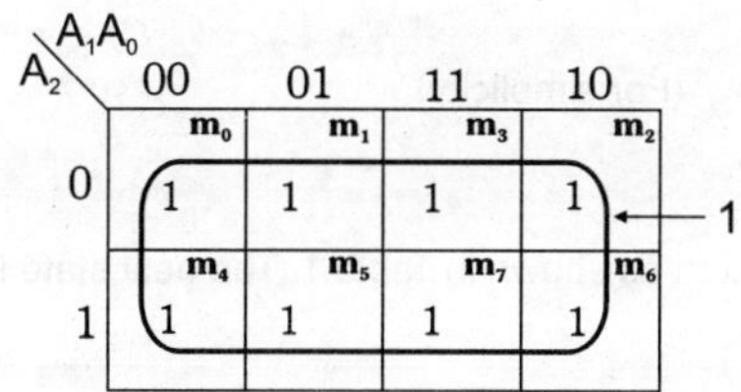

Fig. 5: K-map for T_0.

From the K-map we get the following simplified Boolean equations to realize 3-bit binary counter using T flip-flops and they are used to construct the logic diagram of 3-bit counter as shown in Fig. 6.

$T_2 = A_1A_0$

$T_1 = A_0$

$T_0 = 1$

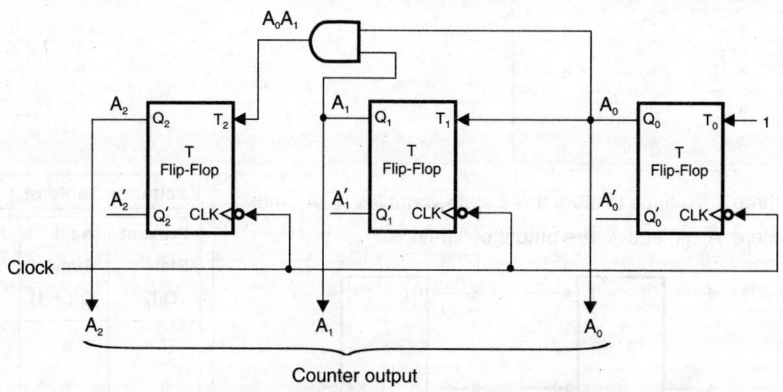

Fig. 6: Logic diagram of 3-bit binary counter.

Example 8.2

Design a counter using T flip-flop that goes through states 1, 2, 4, 7, 1. Is the counter is self starting?

Solution

The sequential circuit has to detect decimal values in the range 0 to 7. Let us assign binary value of decimal number as code for states. Hence we need 3-bit binary code to represent the decimal sequence. With 3-bit code 8 states ($2^3 = 8$) are possible and in this 4 states are used states and 4 states are unused states.

Binary Code Assignment		**Unused Binary Codes**	
1	001	0	000
2	010	3	011
4	100	5	101
7	111	6	110

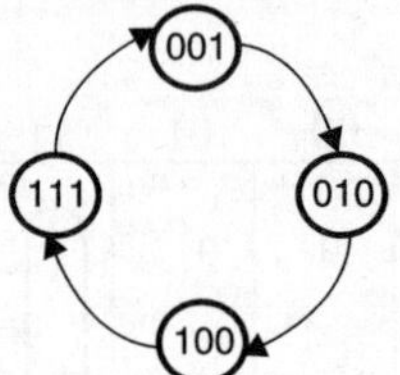

Fig. 1: State diagram with only used states.

The state diagram using only used states is shown in Fig. 1.

Here, each bit of binary code is a state variable. Hence the sequence detector has 3 state variables. Let the state variables be A_0, A_1 and A_2.

∴ Present state variablels : $A_0(t), A_1(t), A_2(t) \Rightarrow A_0, A_1, A_2$ (For simplicity)

Next state variablels : $A_0(t + 1), A_1(t + 1), A_2(t + 1)$

Using the state diagram of Fig. 1, the state table is constructed as shown in Table 1. The next state for unused states are considered as don't-cares.

Table 1: State Table

Present States			Next States		
A_2	A_1	A_0	$A_2(t+1)$	$A_1(t+1)$	$A_0(t+1)$
0	0	0	×	×	×
0	0	1	0	1	0
0	1	0	1	0	0
0	1	1	×	×	×
1	0	0	1	1	1
1	0	1	×	×	×
1	1	0	×	×	×
1	1	1	0	0	1

Let us choose three T flip-flops to store three state variables A_0, A_1 and A_2 as shown in Fig. 2. Here, A_0, A_1 and A_2 are output of flip-flops.

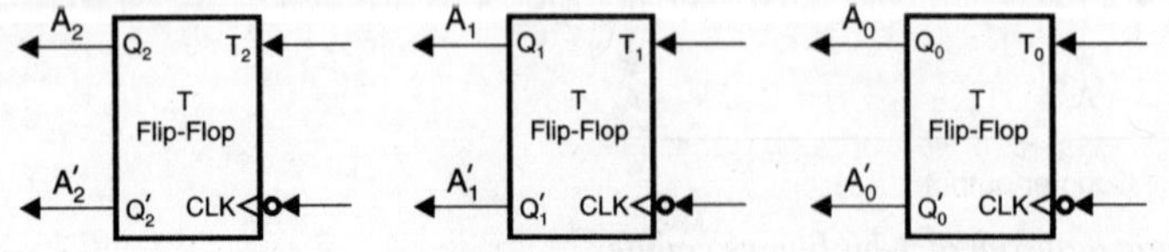

Fig. a: Flip-flop A_2. *Fig. b: Flip-flop A_1.* *Fig. c: Flip-flop A_0.*

Fig. 2: T flip-flops to store state variables.

Excitation Table of T Flip-Flop

Present State Q(t)	Next State Q(t + 1)	Flip-Flop Input T
0	0	0
0	1	1
1	0	1
1	1	0

The flip-flop inputs can be obtained from excitation table as shown below:

State A_2

For state A_2, the Q(t), Q(t + 1) and T of excitation table are considered as A_2, $A_2(t + 1)$ and T_2.

$\therefore Q(t) \rightarrow A_2 \;;\; Q(t+1) \rightarrow A_2(t+1)$

$T \rightarrow T_2$

The flip-flop input T_2 that drives the present state A_2 to next state is obtained from excitation table and listed in Table 2.

State A_1

For state A_1, the Q(t), Q(t + 1) and T of excitation table are considered as A_1, $A_1(t + 1)$ and T_1.

$\therefore Q(t) \rightarrow A_1 \;;\; Q(t+1) \rightarrow A_1(t+1)$

$T \rightarrow T_1$

The flip-flop input T_1 that drives the present state A_1 to next state is obtained from excitation table and listed in Table 3.

State A_0

For state A_0, the Q(t), Q(t + 1) and T of excitation table are considered as A_0, $A_0(t + 1)$ and T_0.

$\therefore Q(t) \rightarrow A_0 \;;\; Q(t+1) \rightarrow A_0(t+1)$

$T \rightarrow T_0$

The flip-flop input T_0 that drives the present state A_0 to next state is obtained from excitation table and listed in Table 4.

Table 2:

A_2	$A_2(t+1)$	T_2
0	×	×
0	0	0
0	1	1
0	×	×
1	1	0
1	×	×
1	×	×
1	0	1

Table 3:

A_1	$A_1(t+1)$	T_1
0	×	×
0	1	1
1	0	1
1	×	×
0	1	1
0	×	×
1	×	×
1	0	1

Table 4:

A_0	$A_0(t+1)$	T_0
0	×	×
1	0	1
0	0	0
1	×	×
0	1	1
1	×	×
0	×	×
1	1	0

The Tables 1 to 4 are combined to form state table with flip-flop inputs as shown in Table 5.

Table 5: Expanded State Table

Present States			Next States			Flip-Flop Inputs			Minterm
A_2	A_1	A_0	$A_2(t+1)$	$A_1(t+1)$	$A_0(t+1)$	T_2	T_1	T_0	$A_2 A_1 A_0$
0	0	0	×	×	×	×	×	×	m_0
0	0	1	0	1	0	0	1	1	m_1
0	1	0	1	0	0	1	1	0	m_2
0	1	1	×	×	×	×	×	×	m_3
1	0	0	1	1	1	0	1	1	m_4
1	0	1	×	×	×	×	×	×	m_5
1	1	0	×	×	×	×	×	×	m_6
1	1	1	0	0	1	1	1	0	m_7

***Note:** Here, present state $A_2 A_1 A_0$ is the counter output.*

The Boolean equations for T_2, T_1 and T_0 are obtained from Table 5 by taking present state as minterms.

$$T_2 = \Sigma m(2, 7) + \Sigma d(0, 3, 5, 6) \quad ; \quad T_1 = \Sigma m(1, 2, 4, 7) + \Sigma d(0, 3, 5, 6) \quad ; \quad T_0 = \Sigma m(1, 4) + \Sigma d(0, 3, 5, 6)$$

The simplified Boolean equations are obtained by constructing K-map using the above equations for T_2, T_1 and T_0 as shown in Figs. 3 to 6.

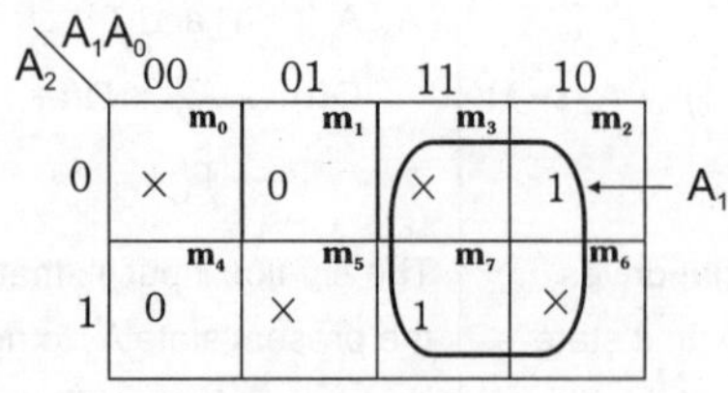

Fig. 3: K-map for T_2.

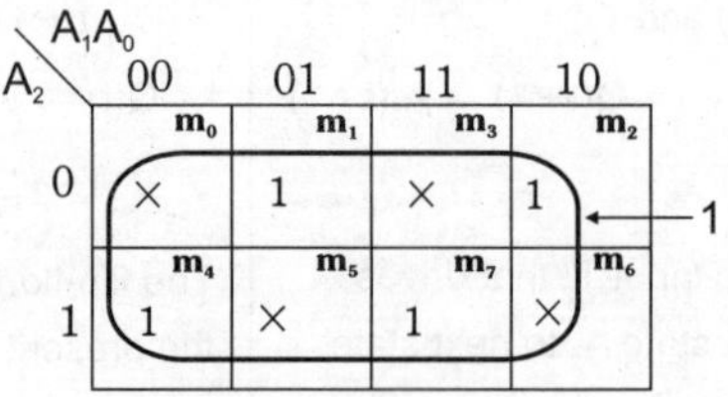

Fig. 4: K-map for T_1.

From the K-map we get the following simplified Boolean equations to realize 3-bit binary counter using T flip-flops and they are used to construct the logic diagram of 3-bit counter as shown in Fig. 6.

$$T_2 = A_1 \quad ; \quad T_1 = 1 \quad ; \quad T_0 = A_1'$$

A_2 \ A_1A_0	00	01	11	10
0	m_0 ×	m_1 1	m_3 ×	m_2 0
1	m_4 1	m_5 ×	m_7 0	m_6 ×

(Group: A_1')

Fig. 5: K-map for T_0.

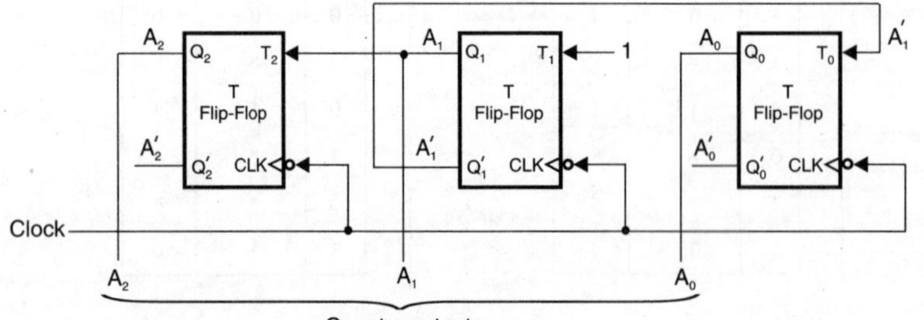

Fig. 6: Logic diagram.

Check for Self Starting

In order to check self starting, the flip-flop inputs are determined for each unused state using simplified equations of flip-flop inputs and listed in Table 6. The next state is obtained using characteristic table of T flip-flop.

Characteristic Table of T Flip-Flop

Input T	Present State Q	Next State Q(t+1)
0	0	0
0	1	1
1	0	1
1	1	0

Table 6: Check for Self Starting

Unused Present States			Flip-Flop Inputs			Next States		
			T_2	T_1	T_0	$A_2(t+1)$	$A_1(t+1)$	$A_0(t+1)$
A_2	A_1	A_0	$= A_1$	$= 1$	$= A_1'$			
0	0	0	0	1	1	0	1	1
0	1	1	1	1	0	1	0	1
1	0	1	0	1	1	1	1	0
1	1	0	1	1	0	0	0	0

000 → 011 → 101 → 110 → 000

Fig. 7: State diagram of unused states.

Using Table 6 we can construct the state diagram of unused states as shown in Fig. 7 and from this state diagram we can observe that when the counter enter an unused state then it enter a lock-out condition so that the counter will cycle through only unused states and will never enter a used state. Hence the counter is not self starting.

Counter Design for Self Starting

In order to make the counter self starting the unused states can be connected to enter used states as shown in Fig. 8. Once the counter enter an used state it will circle through used states and counter becomes self starting.

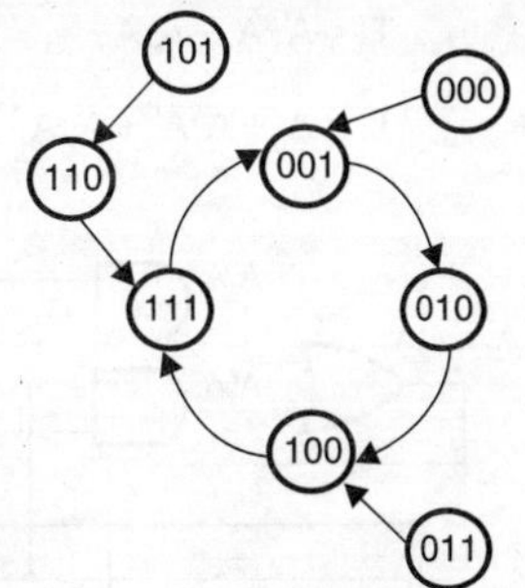

Fig. 8: State diagram to prevent lock-out condition.

From the state diagram of Fig. 8, the state table without don't-cares is formed as shown in Table 7. Here, the next state is obtained from state diagram and flip-flop inputs are obtained from excitation table.

Table 7: Expanded State Table without don't-cares

Present States			Next States			Flip-Flop Inputs			Minterm
A_2	A_1	A_0	$A_2(t+1)$	$A_1(t+1)$	$A_0(t+1)$	T_2	T_1	T_0	$A_2\,A_1\,A_0$
0	0	0	0	0	1	0	0	1	m_0
0	0	1	0	1	0	0	1	1	m_1
0	1	0	1	0	0	1	1	0	m_2
0	1	1	1	·0	0	1	1	1	m_3
1	0	0	1	1	1	0	1	1	m_4
1	0	1	1	1	0	0	1	1	m_5
1	1	0	1	1	1	0	0	1	m_6
1	1	1	0	0	1	1	1	0	m_7

***Note:** Here, present state $A_2\,A_1\,A_0$ is the counter output.*

In Table 7, the minterms are formed using present states. The Boolean equations for T_2, T_1 and T_0 are obtained as sum of minterms as shown below:

$$T_2 = \Sigma m(2, 3, 7)$$

$$T_1 = \Sigma m(1, 2, 3, 4, 5, 7)$$

$$T_0 = \Sigma m(0, 1, 3, 4, 5, 6)$$

The simplified Boolean equations are obtained by constructing K-map using the above equations for T_2, T_1 and T_0 as shown in Figs. 9 to 11.

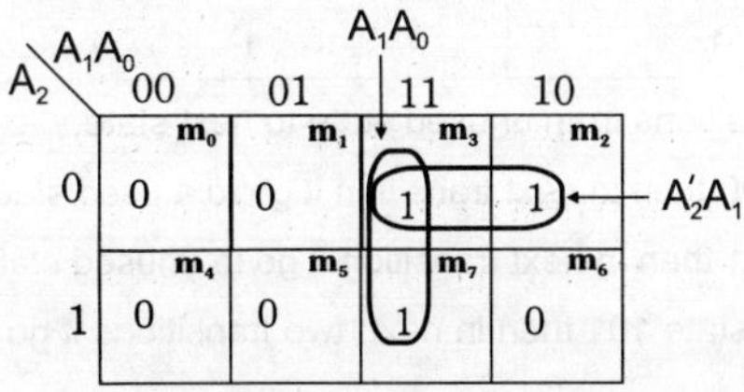

Fig. 9: K-map for T_2.

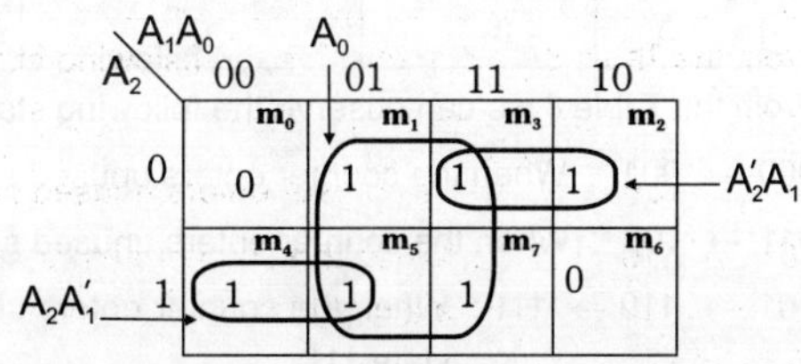

Fig. 10: K-map for T_1.

From the K-map we get the following simplified Boolean equations to realize 3-bit binary counter using T flip-flops and they are used to realize the logic diagram of 3-bit counter as shown in Fig. 12.

$T_2 = A_2' A_1 + A_1 A_0$

$T_1 = A_0 + A_2 A_1' + A_2' A_1$

$T_0 = A_1' + A_2' A_0 + A_2 A_0'$

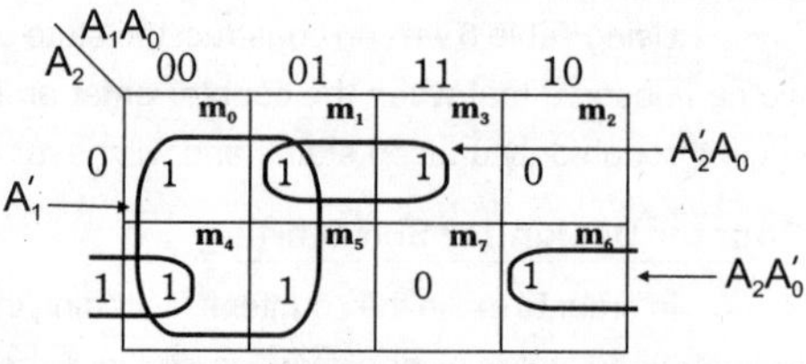

Fig. 11: K-map for T_0.

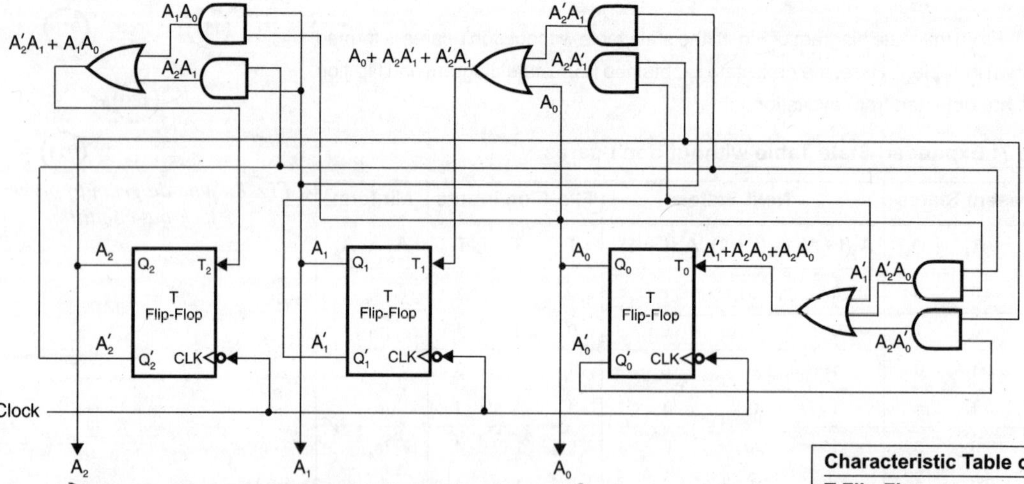

Counter output

Fig. 12: Logic diagram of (1, 2, 4, 7, 1) sequence counter.

Characteristic Table of T Flip-Flop

Input T	Present State Q	Next State Q(t+1)
0	0	0
0	1	1
1	0	1
1	1	0

Check for Self Starting

In order to check for self starting, the flip-flop inputs are determined for each unused state using simplified equations of flip-flop inputs and listed in Table 8. The next states are obtained from characteristic table of T flip-flop.

Table 8: Check for Self Starting

Unused Present States			Flip-Flop Inputs			Next States		
			T_2	T_1	T_0	$A_2(t+1)$	$A_1(t+1)$	$A_0(t+1)$
A_2	A_1	A_0	$= A_2' A_1 + A_1 A_0$	$= A_0 + A_2 A_1' + A_2' A_1$	$= A_1' + A_2' A_0 + A_2 A_0'$			
0	0	0	0	0	1	0	0	1
0	1	1	1	1	1	1	0	0
1	0	1	0	1	1	1	1	0
1	1	0	0	0	1	1	1	1

From the Table 8 we can observe the following state transitions from unused state to next state.

000 → 001 ; When the counter enters unused state 000 then in next transition it go to a used state 001.

011 → 100 ; When the counter enters unused state 011 then in next transition it go to unused state 100

101 → 110 → 111 ; When the counter enters unused state 101 then in next two transitions it go to a used state 111.

110 → 111 ; When the counter enters unused state 110 then in next transition it go to a used state 111.

The unused states enters used state in the next transition as shown above and so the counter is self starting.

Example 8.3

(VTU, Aug/Sep'20, 8 Marks)

Design a synchronous counter that goes through states $5 \to 7 \to 4 \to 3 \to 1 \to 5$ by avoiding lock-out condition. Use JK flip-flops.

Solution

The sequential circuit has to detect decimal values in the range 0 to 7. Let us assign binary value of decimal number as code for states. Hence we need 3-bit binary code to represent the decimal sequence. With 3-bit code 8 states ($2^3 = 8$) are possible and in this 5 states are used states 3 states are unused states.

Binary Code Assignment		**Unused Binary Codes**	
1	001	0	000
3	011	2	010
4	100	6	110
5	101		
7	111		

The state diagram using only used states is shown in Fig. 1. In order to prevent lock-out condition the unused states are included in state diagram such that the next state is a nearest used state as shown in Fig. 2.

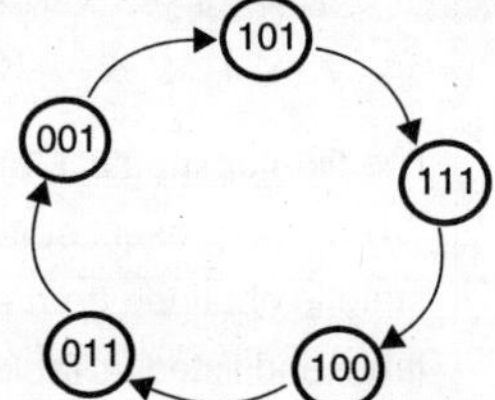

Fig. 1: State diagram with only used states.

Fig. 2: State diagram to prevent lock-out condition.

Here, each bit of binary code is a state variable. Hence the sequence detector has 3 state variables. Let the state variables be A_0, A_1, A_2.

$\therefore$ Present state variables : $A_0(t), A_1(t), A_2(t) \Rightarrow A_0, A_1, A_2$ (For simplicity)

Next state variables : $A_0(t+1), A_1(t+1), A_2(t+1)$

Using the state diagram of Fig. 2, the state table is constructed as shown in Table 1.

Table 1: State Table

Present States			**Next States**		
A_2	A_1	A_0	$A_2(t+1)$	$A_1(t+1)$	$A_0(t+1)$
0	0	0	0	0	1
0	0	1	1	0	1
0	1	0	0	1	1
0	1	1	0	0	1
1	0	0	0	1	1
1	0	1	1	1	1
1	1	0	1	1	1
1	1	1	1	0	0

Let us choose three JK flip-flops to store three state variables A_0, A_1 and A_2 as shown in Fig. 3. Here, A_0, A_1 and A_2 are output of flip-flops.

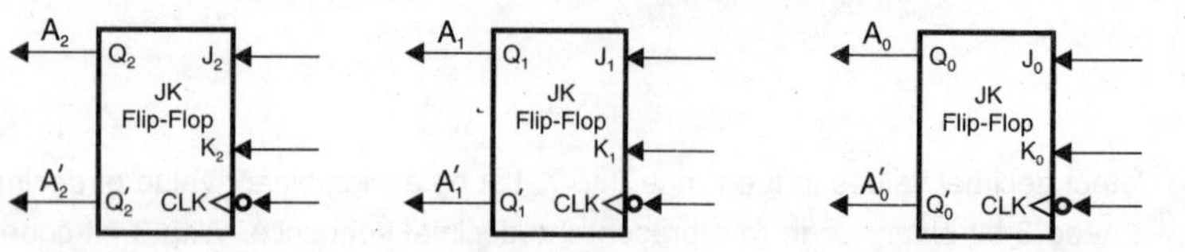

Fig. a: Flip-flop A_2. *Fig. b: Flip-flop A_1.* *Fig. b: Flip-flop A_0.*

Fig. 3: JK flip-flops to store state variables.

Excitation Table of JK Flip-Flop

Present State Q(t)	Next State Q(t + 1)	Flip-Flop Inputs J	K
0	0	0	×
0	1	1	×
1	0	×	1
1	1	×	0

The flip-flop inputs can be obtained from excitation table as shown below:

State A_2

For state A_2, the Q(t), Q(t + 1), J and K of excitation table are considered as A_2, $A_2(t + 1)$, J_2 and K_2.

$$\therefore Q(t) \rightarrow A_2 \;;\; Q(t+1) \rightarrow A_2(t+1)$$

$$J \rightarrow J_2 \;;\; K \rightarrow K_2$$

The flip-flop inputs J_2 and K_2 that drives the present state A_2 to next state is obtained from excitation table and listed in Table 2.

Table 2:

A_2	$A_2(t + 1)$	J_2	K_2
0	0	0	×
0	1	1	×
0	0	0	×
0	0	0	×
1	0	×	1
1	1	×	0
1	1	×	0
1	1	×	0

State A_1

For state A_1, the Q(t), Q(t + 1), J and K of excitation table are considered as A_1, $A_1(t + 1)$, J_1 and K_1.

$$\therefore Q(t) \rightarrow A_1 \;;\; Q(t+1) \rightarrow A_1(t+1)$$

$$J \rightarrow J_1 \;;\; K \rightarrow K_1$$

The flip-flop inputs J_1 and K_1 that drives the present state A_1 to next state is obtained from excitation table and listed in Table 3.

Table 3:

A_1	$A_1(t + 1)$	J_1	K_1
0	0	0	×
0	0	0	×
1	1	×	0
1	0	×	1
0	1	1	×
0	1	1	×
1	1	×	0
1	0	×	1

State A_0

For state A_0, the Q(t), Q(t + 1), J and K of excitation table are considered as A_0, $A_0(t + 1)$, J_0 and K_0.

$$\therefore Q(t) \rightarrow A_0 \;;\; Q(t+1) \rightarrow A_0(t+1)$$

$$J \rightarrow J_0 \;;\; K \rightarrow K_0$$

The flip-flop inputs J_0 and K_0 that drives the present state A_0 to next state is obtained from excitation table and listed in Table 4.

Table 4:

A_0	$A_0(t + 1)$	J_0	K_0
0	1	1	×
1	1	×	0
0	1	1	×
1	1	×	0
0	1	1	×
1	1	×	0
0	1	1	×
1	0	×	1

The Tables 1 to 4 are combined to form state table with flip-flop inputs as shown in Table 5.

Table 5: Expanded State Table

Present States			Next States			Flip-Flop Inputs						Minterm
A_2	A_1	A_0	$A_2(t+1)$	$A_1(t+1)$	$A_0(t+1)$	J_2	K_2	J_1	K_1	J_0	K_0	$A_2A_1A_0$
0	0	0	0	0	1	0	×	0	×	1	×	m_0
0	0	1	1	0	1	1	×	0	×	×	0	m_1
0	1	0	0	1	1	0	×	×	0	1	×	m_2
0	1	1	0	0	1	0	×	×	1	×	0	m_3
1	0	0	0	1	1	×	1	1	×	1	×	m_4
1	0	1	1	1	1	×	0	1	×	×	0	m_5
1	1	0	1	1	1	×	0	×	0	1	×	m_6
1	1	1	1	0	0	×	0	×	1	×	1	m_7

Note: *Here, present state $A_2A_1A_0$ is the counter output.*

The flip-flop inputs are functions of present state and so minterms are formed using present state and the Boolean equations for J_2, K_2, J_1, K_1, J_0 and K_0 are obtained from Table 5 as sum of minterms and don't-cares as shown below:

$J_2 = m_1 + \Sigma d(4, 5, 6, 7)$

$K_2 = m_4 + \Sigma d(0, 1, 2, 3)$

$J_1 = \Sigma m(4, 5) + \Sigma d(2, 3, 6, 7)$

$K_1 = \Sigma m(3, 7) + \Sigma d(0, 1, 4, 5)$

$J_0 = \Sigma m(0, 2, 4, 6) + \Sigma d(1, 3, 5, 7)$

$K_0 = m_7 + \Sigma d(0, 2, 4, 6)$

The simplified Boolean equations are obtained by constructing K-map using the above equations for J_2, K_2, J_1, K_1, J_0 and K_0 as shown in Figs. 4 to 9.

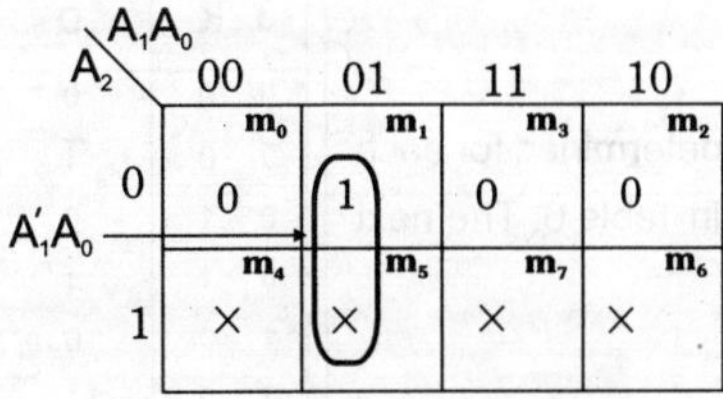

Fig. 4: K-map for J_2.

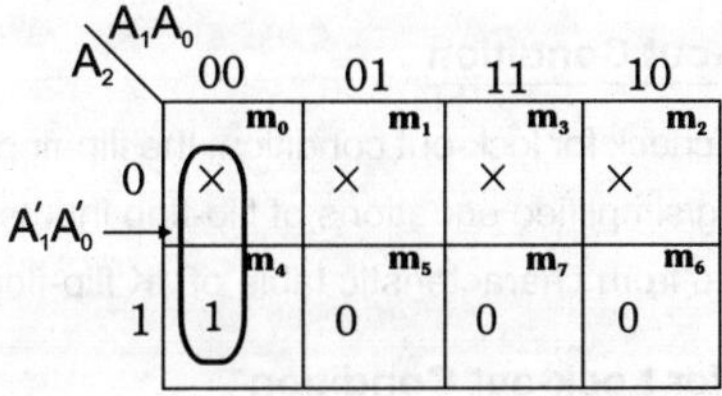

Fig. 5: K-map for K_2.

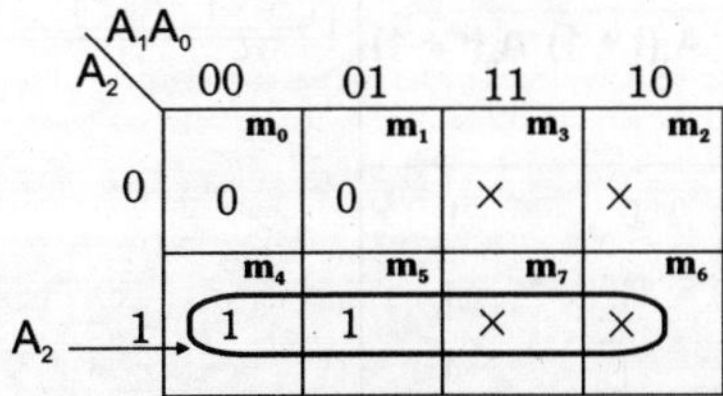

Fig. 6: K-map for J_1.

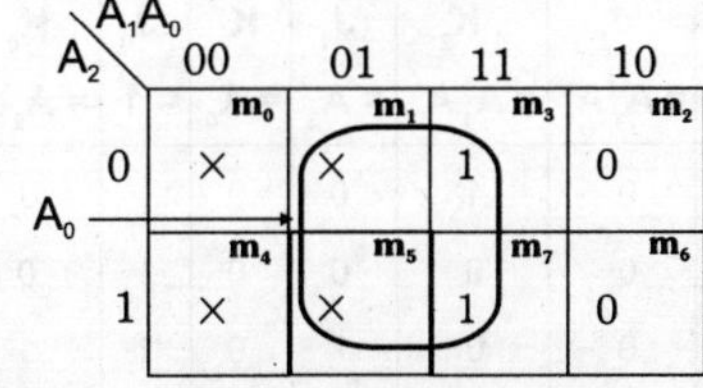

Fig. 7: K-map for K_1.

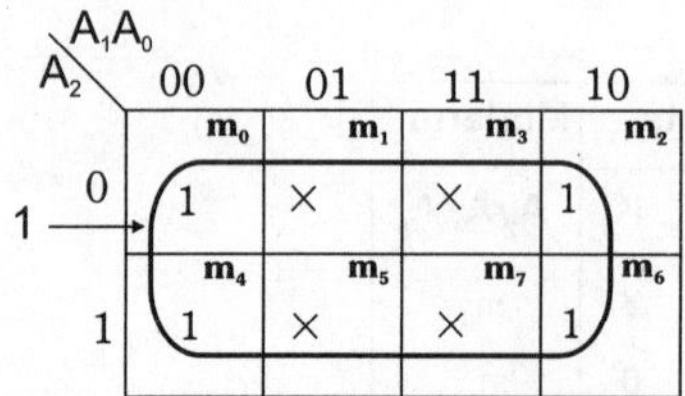

Fig. 8: K-map for J_0.

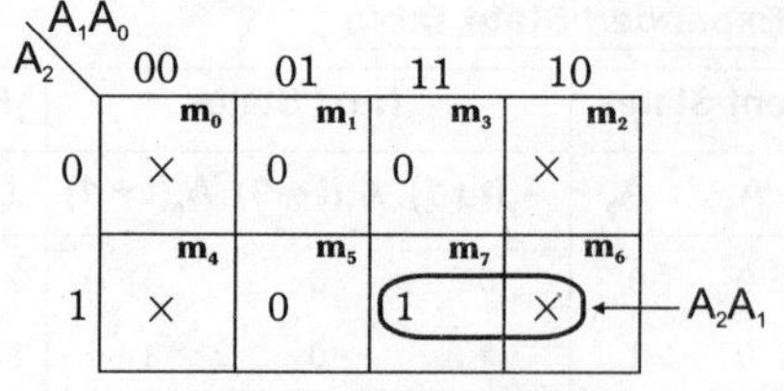

Fig. 9: K-map for K_0.

From the K-map we get the following simplified Boolean equations to realize 3-bit binary counter using JK flip-flops.

$J_2 = A_1' A_0$	$K_2 = A_1' A_0'$
$J_1 = A_2$	$K_1 = A_0$
$J_0 = 1$	$K_0 = A_2 A_1$

Using the simplified Boolean equations obtained from K-map the logic diagram of 3-bit counter is drawn as shown in Fig. 10.

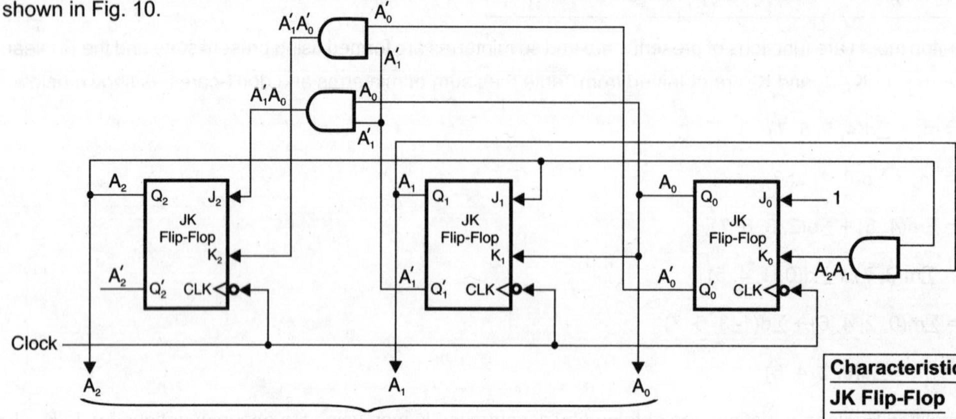

Fig. 10: Logic diagram of (5, 7, 4, 3, 1, 5) sequence counter.

Characteristic Table of JK Flip-Flop

Input J	Input K	Present State Q	Next State Q(t+1)
0	0	0	0
0	0	1	1
0	1	0	0
0	1	1	0
1	0	0	1
1	0	1	1
1	1	0	1
1	1	1	0

Check for Lock-out Condition

In order to check for lock-out condition, the flip-flop inputs are determined for each unused state using simplified equations of flip-flop inputs and listed in Table 6. The next states are obtained from characteristic table of JK flip-flop.

Table 6: Check for Lock-out Condition

Unused Present States			Flip-Flop Inputs						Next States		
			J_2	K_2	J_1	K_1	J_0	K_0	$A_2(t+1)$	$A_1(t+1)$	$A_0(t+1)$
A_2	A_1	A_0	$= A_1' A_0$	$= A_1' A_0'$	$= A_2$	$= A_0$	$= 1$	$= A_2 A_1$			
0	0	0	0	1	0	0	1	0	0	0	1
0	1	0	0	0	0	0	1	0	0	1	1
1	1	0	0	0	1	0	1	1	1	1	1

From the Table 6 we can observe the following state transitions from unused state to next state.

000 → 001, 010 → 011, 110 → 111

The unused states becomes used state in the next transition as shown above and hence the lock-out condition is avoided.

Example 8.4

(AKTU, Sep'20, 10 Marks) *(IKGPTU, Dec'19, 8 Marks)*

Design a 3-bit Gray code counter using T flip-flop and JK flip-flop.

Solution

The 3-bit gray code is used to represent decimal numbers in the range 0 to 7. Let us assign gray code of decimal number as binary code for states. With 3-bit gray code 8 states ($2^3 = 8$) are possible and in this all the 8 states are used states.

Gray Code Assignment

0	000
1	001
2	011
3	010
4	110
5	111
6	101
7	100

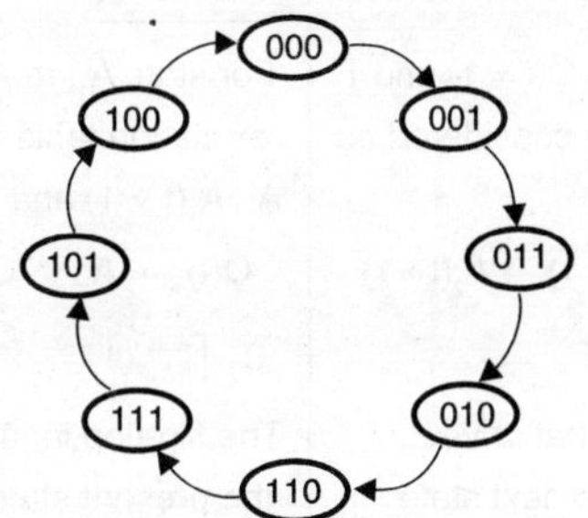

Fig. 1: State diagram of 3-bit gray code counter.

The state diagram using all the 8 states is shown in Fig. 1.

Here, each bit of gray code is a state variable. Hence the gray code counter has 3 state variables. Let the state variables be A_0, A_1 and A_2.

∴ Present state variables : $A_0(t), A_1(t), A_2(t) \Rightarrow A_0, A_1, A_2$ (For simplicity)

Next state variables : $A_0(t+1), A_1(t+1), A_2(t+1)$

Using the state diagram of Fig. 1, the state table is constructed as shown in Table 1.

Table 1: State Table

Present States			Next States		
A_2	A_1	A_0	$A_2(t+1)$	$A_1(t+1)$	$A_0(t+1)$
0	0	0	0	0	1
0	0	1	0	1	1
0	1	1	0	1	0
0	1	0	1	1	0
1	1	0	1	1	1
1	1	1	1	0	1
1	0	1	1	0	0
1	0	0	0	0	0

a) Design using T Flip-Flop

Let us choose three T flip-flops to store three state variables A_0, A_1 and A_2 as shown in Fig. 2. Here, A_0, A_1 and A_2 are output of flip-flops.

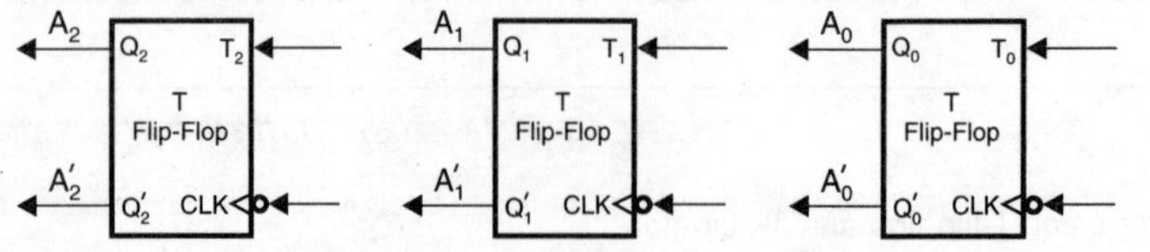

Fig. a: Flip-flop A_2. *Fig. b: Flip-flop A_1.* *Fig. c: Flip-flop A_0.*

Fig. 2: T flip-flops to store state variables.

Excitation Table of T Flip-Flop

Present State Q(t)	Next State Q(t + 1)	Flip-Flop Input T
0	0	0
0	1	1
1	0	1
1	1	0

The flip-flop inputs can be obtained from excitation table as shown below:

State A_2

For state A_2, the Q(t), Q(t + 1) and T of excitation table are considered as A_2, $A_2(t + 1)$ and T_2.

$\therefore Q(t) \rightarrow A_2$; $Q(t+1) \rightarrow A_2(t+1)$

$T \rightarrow T_2$

The flip-flop input T_2 that drives the present state A_2 to next state is obtained from excitation table and listed in Table 2.

Table 2:

A_2	$A_2(t+1)$	T_2
0	0	0
0	0	0
0	0	0
0	1	1
1	1	0
1	1	0
1	1	0
1	0	1

State A_1

For state A_1, the Q(t), Q(t + 1) and T excitation table are considered as A_1, $A_1(t + 1)$ and T_1.

$\therefore Q(t) \rightarrow A_1$; $Q(t+1) \rightarrow A_1(t+1)$

$T \rightarrow T_1$

The flip-flop input T_1 that drives the present state A_1 to next state is obtained from excitation table and listed in Table 3.

Table 3:

A_1	$A_1(t+1)$	T_1
0	0	0
0	1	1
1	1	0
1	1	0
1	1	0
1	0	1
0	0	0
0	0	0

State A_0

For state A_0, the Q(t), Q(t + 1) and T of excitation table are considered as A_0, $A_0(t + 1)$ and T_0.

$\therefore Q(t) \rightarrow A_0$; $Q(t+1) \rightarrow A_0(t+1)$

$T \rightarrow T_0$

The flip-flop input T_0 that drives the present state A_0 to next state is obtained from excitation table and listed in Table 4.

Table 4:

A_0	$A_0(t+1)$	T_0
0	1	1
1	1	0
1	0	1
0	0	0
0	1	1
1	1	0
1	0	1
0	0	0

The Tables 1 to 4 are combined to form state table with flip-flop inputs as shown in Table 5.

Table 5: Expanded State Table

Present States			Next States			Flip-Flop Inputs			Minterm
A_2	A_1	A_0	$A_2(t+1)$	$A_1(t+1)$	$A_0(t+1)$	T_2	T_1	T_0	$A_2A_1A_0$
0	0	0	0	0	1	0	0	1	m_0
0	0	1	0	1	1	0	1	0	m_1
0	1	1	0	1	0	0	0	1	m_2
0	1	0	1	1	0	1	0	0	m_3
1	1	0	1	1	1	0	0	1	m_4
1	1	1	1	0	1	0	1	0	m_5
1	0	1	1	0	0	0	0	1	m_6
1	0	0	0	0	0	1	0	0	m_7

Note: *Here, present state $A_2A_1A_0$ is the counter output.*

The Boolean equations for T_2, T_1 and T_0 are obtained from Table 5 by taking present state as minterms.

$$T_2 = \Sigma m(3, 7) \quad ; \quad T_1 = \Sigma m(1, 5) \quad ; \quad T_0 = \Sigma m(0, 2, 4, 6)$$

The simplified Boolean equations are obtained by constructing K-map using the above equation for T_2, T_1 and T_0 as shown in Figs. 3 to 5.

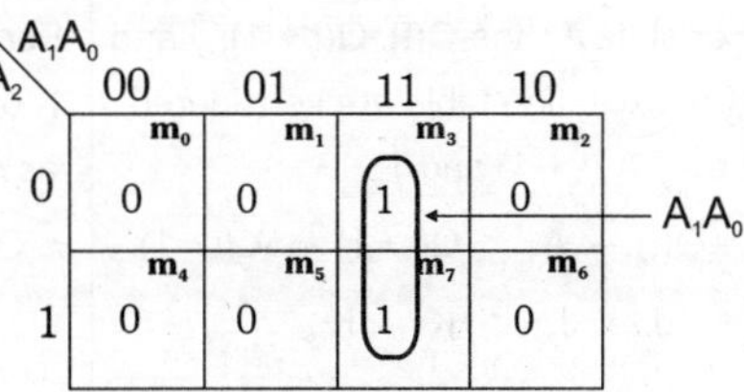

Fig. 3: *K-map for T_2.*

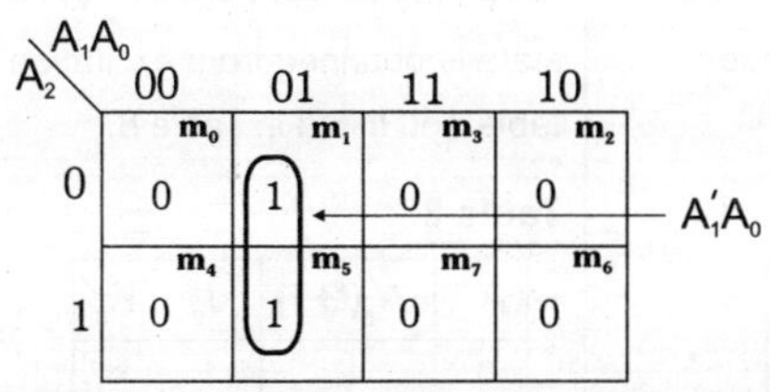

Fig. 4: *K-map for T_1.*

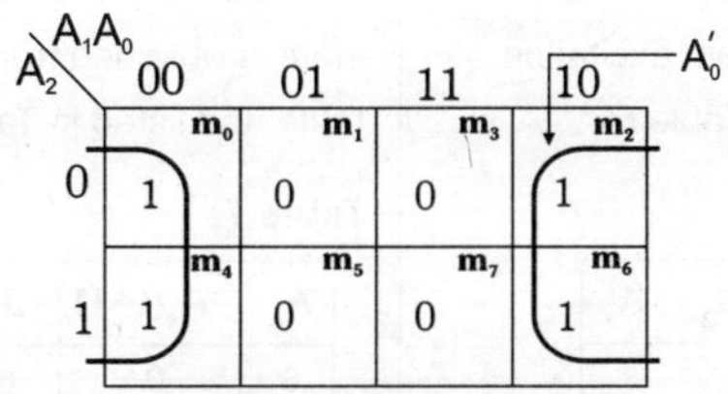

Fig. 5: *K-map for T_0.*

From the K-map we get the following simplified Boolean equations to realize 3-bit Gray code counter using T flip-flops and they are used to construct the logic diagram of 3-bit counter as shown in Fig. 6.

$$T_2 = A_1A_0$$

$$T_1 = A_1'A_0$$

$$T_0 = A_0'$$

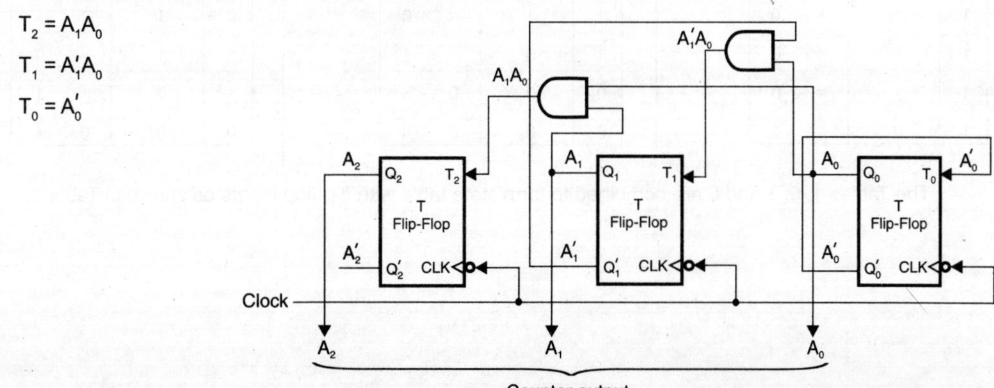

Fig. 6: *Logic diagram of 3-bit gray code counter using T flip-flop.*

b) Design using JK Flip-Flop

Let us choose three JK flip-flops to store three state variables A_0, A_1 and A_2 as shown in Fig. 7. Here, A_0, A_1 and A_2 are output of flip-flops.

Excitation Table of JK Flip-Flop

Present State Q(t)	Next State Q(t + 1)	Flip-Flop Inputs J	K
0	0	0	×
0	1	1	×
1	0	×	1
1	1	×	0

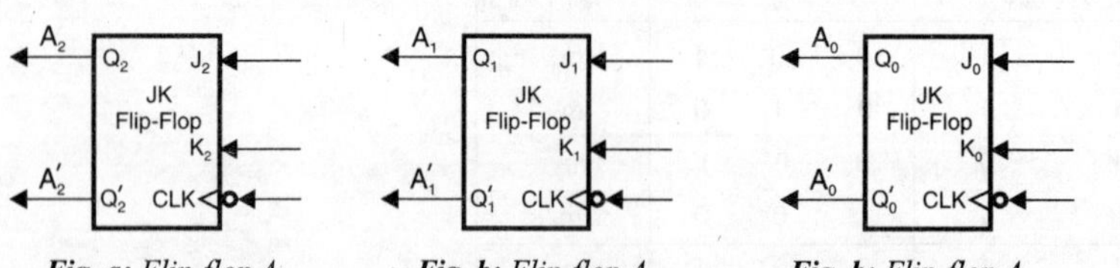

Fig. a: Flip-flop A_2. *Fig. b: Flip-flop A_1.* *Fig. b: Flip-flop A_0.*

Fig. 7: JK flip-flops to store state variables.

The flip-flop inputs can be obtained from excitation table as shown below:

State A_2

For state A_2, the Q(t), Q(t + 1), J and K of excitation table are considered as A_2, $A_2(t + 1)$ and T_2.

$\therefore Q(t) \rightarrow A_2 \;;\; Q(t+1) \rightarrow A_2(t+1)$

$J \rightarrow J_2 \;;\; K \rightarrow K_2$

The flip-flop input J_2 and K_2 that drives the present state A_2 to next state is obtained from excitation table and listed in Table 6.

Table 6:

A_2	$A_2(t+1)$	J_2	K_2
0	0	0	×
0	0	0	×
0	0	0	×
0	1	1	×
1	1	×	0
1	1	×	0
1	1	×	0
1	0	×	1

State A_1

For state A_1, the Q(t), Q(t + 1), J and K of excitation table are considered as A_1, $A_1(t + 1)$ and T_1.

$\therefore Q(t) \rightarrow A_1 \;;\; Q(t+1) \rightarrow A_1(t+1)$

$J \rightarrow J_1 \;;\; K \rightarrow K_1$

The flip-flop input J_1 and K_1 that drives the present state A_1 to next state is obtained from excitation table and listed in Table 7.

Table 7:

A_1	$A_1(t+1)$	J_1	K_1
0	0	0	×
0	1	1	×
1	1	×	0
1	1	×	0
1	1	×	0
1	0	×	1
0	0	0	×
0	0	0	×

State A_0

For state A_0, the Q(t), Q(t + 1), J and K of excitation table are considered as A_0, $A_0(t + 1)$ and T_0.

$\therefore Q(t) \rightarrow A_0 \;;\; Q(t+1) \rightarrow A_0(t+1)$

$J \rightarrow J_0 \;;\; K \rightarrow K_0$

The flip-flop input J_0 and K_0 that drives the present state A_0 to next state is obtained from excitation table and listed in Table 8.

Table 8:

A_0	$A_0(t+1)$	J_0	K_0
0	1	1	×
1	1	×	0
1	0	×	1
0	0	0	×
0	1	1	×
1	1	×	0
1	0	×	1
0	0	0	×

The Tables 1, 6, 7 and 8 are combined to form state table with flip-flop inputs as shown in Table 9.

Table 9: Expanded State Table

Present States			Next States			Flip-Flop Inputs						Minterm
A_2	A_1	A_0	$A_2(t+1)$	$A_1(t+1)$	$A_0(t+1)$	J_2	K_2	J_1	K_1	J_0	K_0	$A_2A_1A_0$
0	0	0	0	0	1	0	×	0	×	1	×	m_0
0	0	1	0	1	1	0	×	1	×	×	0	m_1
0	1	1	0	1	0	0	×	0	×	×	1	m_2
0	1	0	1	1	0	1	×	0	×	0	×	m_3
1	1	0	1	1	1	×	0	0	×	1	×	m_4
1	1	1	1	0	1	×	0	×	1	×	0	m_5
1	0	1	1	0	0	×	0	0	×	×	1	m_6
1	0	0	0	0	0	×	1	0	×	0	×	m_7

***Note:** Here, present state $A_2A_1A_0$ is the counter output.*

The flip-flop inputs are functions of present state and so minterms are formed using present state and the Boolean equations for J_2, K_2, J_1, K_1, J_0 and K_0 are obtained from Table 5 as sum of minterms and don't-cares as shown below:

$J_2 = m_3 + \Sigma d(4, 5, 6, 7)$ | $K_2 = m_7 + \Sigma d(0, 1, 2, 3)$

$J_1 = m_1 + \Sigma d(5)$ | $K_1 = m_5 + \Sigma d(0, 1, 2, 3, 4, 6, 7)$

$J_0 = \Sigma m(0, 4) + \Sigma d(1, 2, 5, 6)$ | $K_0 = \Sigma m(2, 6) + \Sigma d(0, 3, 4, 7)$

The simplified Boolean equations are obtained by constructing K-map using the above equations for J_2, K_2, J_1, K_1, J_0 and K_0 as shown in Figs. 8 to 13.

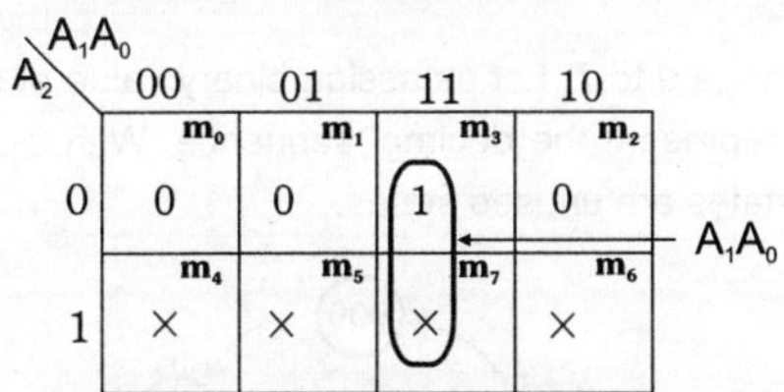

Fig. 8: K-map for J_2.

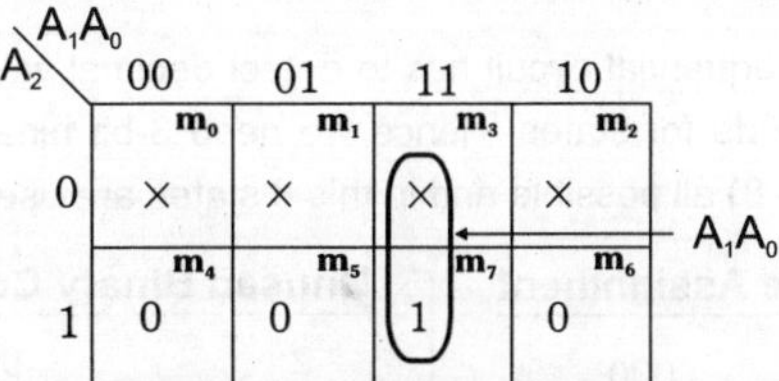

Fig. 9: K-map for K_2.

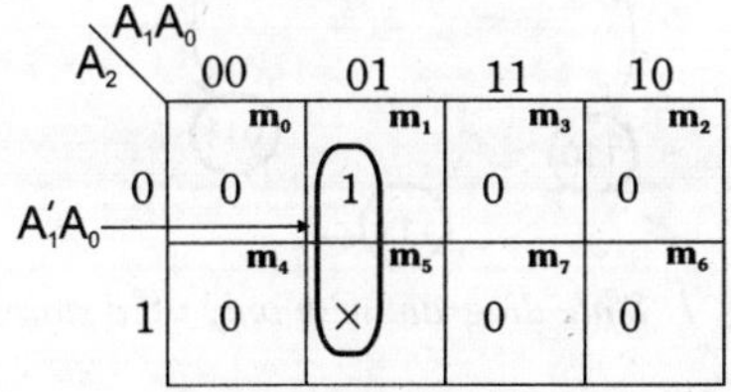

Fig. 10: K-map for J_1.

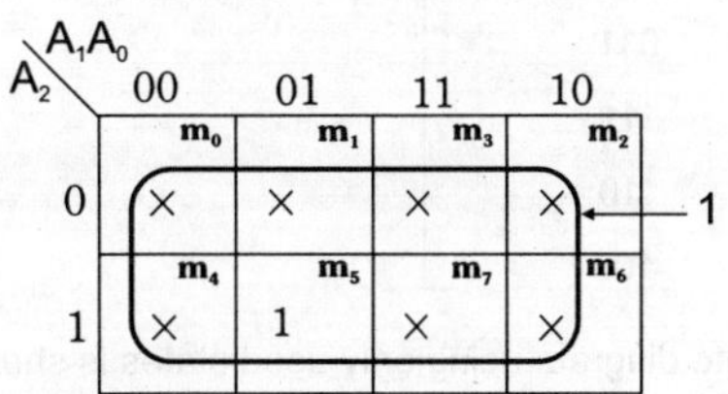

Fig. 11: K-map for K_1.

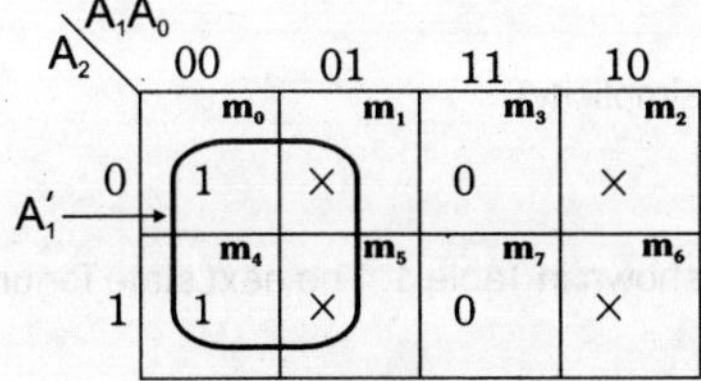

Fig. 12: K-map for J_0.

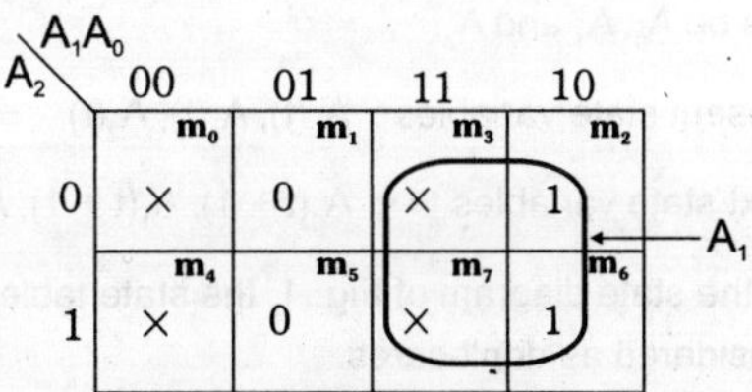

Fig. 13: K-map for K_0.

From the K-map we get the following simplified Boolean equations to realize 3-bit Gray code counter using JK flip-flops and they are used to construct the logic diagram of 3-bit counter as shown in Fig. 14.

$J_2 = A_1A_0$	$K_2 = A_1A_0$
$J_1 = A_1'A_0$	$K_2 = 1$
$T_0 = A_1'$	$K_2 = A_1$

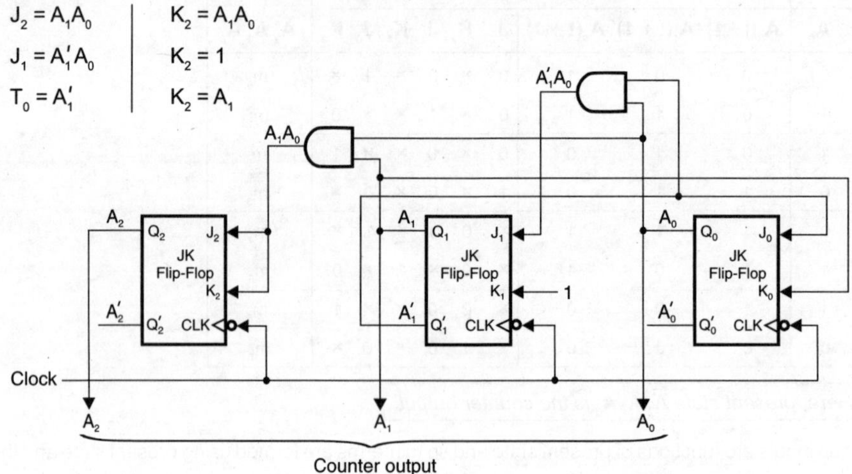

Fig. 14: Logic diagram of 3-bit Gray code counter using JK flip-flop.

Example 8.5

Design a counter with T flip-flops that goes through the binary sequence : 0, 1, 3, 7, 6, 4. Show that when binary states 010 and 101 are considered as don't-care conditions, the counter may not operate properly. Find a way to correct the design.

Solution

The sequential circuit has to detect decimal values in the range 0 to 7. Let us assign binary value of decimal number as code for states. Hence we need 3-bit binary code to represent the decimal sequence. With 3-bit code 8 states ($2^3 = 8$) all possible and in this 6 states are used states 2 states are unused states.

Binary Code Assignment

0	000
1	001
3	011
7	111
6	110
4	100

Unused Binary Codes

2	011
5	101

Fig. 1: State diagram with only used states.

The state diagram using only used states is shown in Fig. 1.

Here, each bit of binary code is a state variable. Hence the sequence detector has 3 state variables. Let the state variables be A_0, A_1 and A_2.

$\therefore$ Present state variables : $A_0(t), A_1(t), A_2(t) \Rightarrow A_0, A_1, A_2$ (For simplicity)

Next state variables : $A_0(t+1), A_1(t+1), A_2(t+1)$

Using the state diagram of Fig. 1, the state table is constructed as shown in Table 1. The next state for unused states are considered as don't-cares.

Table 1: State Table

Present States			Next States		
A_2	A_1	A_0	$A_2(t+1)$	$A_1(t+1)$	$A_0(t+1)$
0	0	0	0	0	1
0	0	1	0	1	1
0	1	0	×	×	×
0	1	1	1	1	1
1	0	0	0	0	0
1	0	1	×	×	×
1	1	0	1	0	0
1	1	1	1	1	0

Let us choose three T flip-flops to store three state variables A_0, A_1 and A_2 as shown in Fig. 2. Here, A_0, A_1 and A_2 are output of flip-flops.

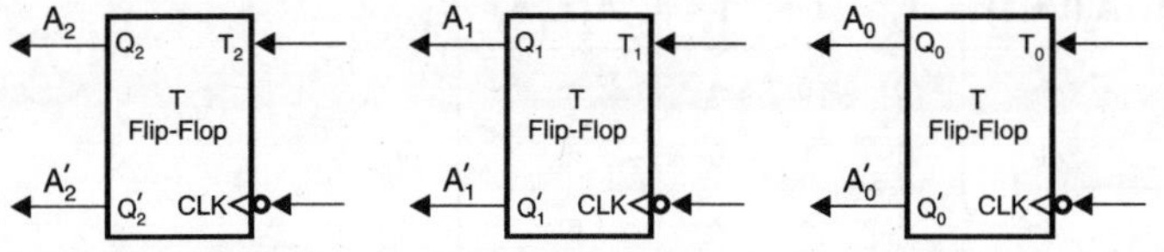

Fig. a: *Flip-flop A_2.* ***Fig. b:*** *Flip-flop A_1.* ***Fig. c:*** *Flip-flop A_0.*

Fig. 2: *T flip-flops to store state variables.*

Excitation Table of T Flip-Flop

Present State Q(t)	Next State Q(t + 1)	Flip-Flop Input T
0	0	0
0	1	1
1	0	1
1	1	0

The flip-flop inputs can be obtained from excitation table as shown below:

State A_2

For state A_2, the Q(t), Q(t + 1) and T of excitation table are considered as A_2, $A_2(t + 1)$ and T_2.

$\therefore Q(t) \to A_2$; $Q(t+1) \to A_2(t+1)$

$T \to T_2$

The flip-flop input T_2 that drives the present state A_2 to next state is obtained from excitation table and listed in Table 2.

State A_1

For state A_1, the Q(t), Q(t + 1) and T of excitation table are considered as A_1, $A_1(t + 1)$ and T_1.

$\therefore Q(t) \to A_1$; $Q(t+1) \to A_1(t+1)$

$T \to T_1$

The flip-flop input T_1 that drives the present state A_1 to next state is obtained from excitation table and listed in Table 3.

State A_0

For state A_0, the Q(t), Q(t + 1) and T of excitation table are considered as A_0, $A_0(t + 1)$ and T_0.

$\therefore Q(t) \to A_0$; $Q(t+1) \to A_0(t+1)$

$T \to T_0$

The flip-flop input T_0 that drives the present state A_0 to next state is obtained from excitation table and listed in Table 4.

Table 2:

A_2	$A_2(t+1)$	T_2
0	0	0
0	0	0
0	×	×
0	1	1
1	0	1
1	×	×
1	1	0
1	1	0

Table 3:

A_1	$A_1(t+1)$	T_1
0	0	0
0	1	1
1	×	×
1	1	0
0	0	0
0	×	×
1	0	1
1	1	0

Table 4:

A_0	$A_0(t+1)$	T_0
0	1	1
1	1	0
0	×	×
1	1	0
0	0	0
1	×	×
0	1	0
1	1	1

The Tables 1 to 4 are combined to form state table with flip-flop inputs as shown in Table 5.

Table 5: Expanded State Table

Present States			Next States			Flip-Flop Inputs			Minterm
A_2	A_1	A_0	$A_2(t+1)$	$A_1(t+1)$	$A_0(t+1)$	T_2	T_1	T_0	$A_2A_1A_0$
0	0	0	0	0	1	0	0	1	m_0
0	0	1	0	1	1	0	1	0	m_1
0	1	0	×	×	×	×	×	×	m_2
0	1	1	1	1	1	1	0	0	m_3
1	0	0	0	0	0	1	0	0	m_4
1	0	1	×	×	×	×	×	×	m_5
1	1	0	1	0	0	0	1	0	m_6
1	1	1	1	1	0	0	0	1	m_7

***Note:** Here, present state $A_2\,A_1\,A_0$ is the counter output.*

The Boolean equations for T_2, T_1 and T_0 are obtained from Table 5 by taking present state as minterms.

$$T_2 = \Sigma m(3, 4) + \Sigma d(2, 5)$$

$$T_1 = \Sigma m(1, 6) + \Sigma d(2, 5)$$

$$T_0 = \Sigma m(0, 7) + \Sigma d(2, 5)$$

The simplified Boolean equations are obtained by constructing K-map using the above equation for T_2, T_1 and T_0 as shown in Figs. 3 to 5.

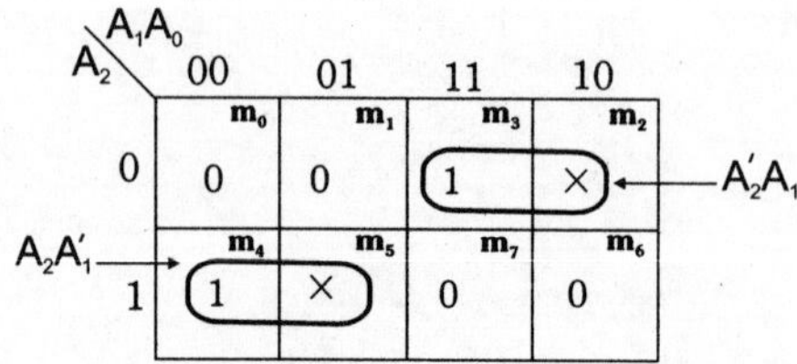

Fig. 3: K-map for T_2.

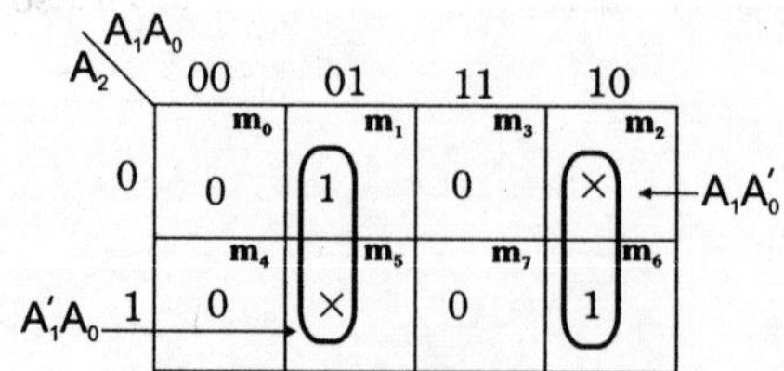

Fig. 4: K-map for T_1.

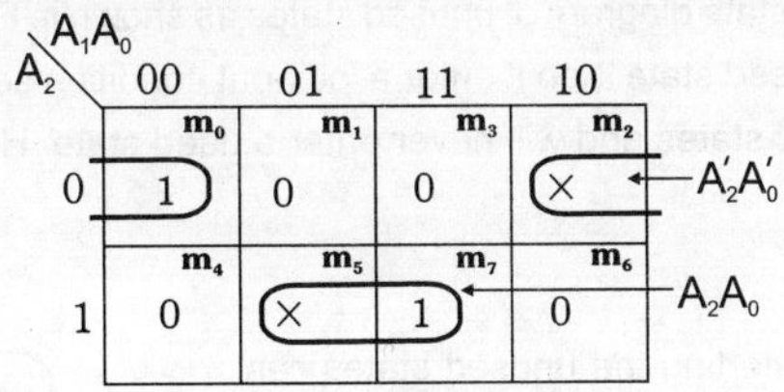

Fig. 5: *K-map for T_0.*

From the K-map we get the following simplified Boolean equations to realize 3-bit binary counter using T flip-flops and they are used to construct the logic diagram of 3-bit counter as shown in Fig. 6.

$$T_2 = A_2'A_1 + A_2A_1'$$

$$T_1 = A_1'A_0 + A_1A_0'$$

$$T_0 = A_2'A_0' + A_2A_0$$

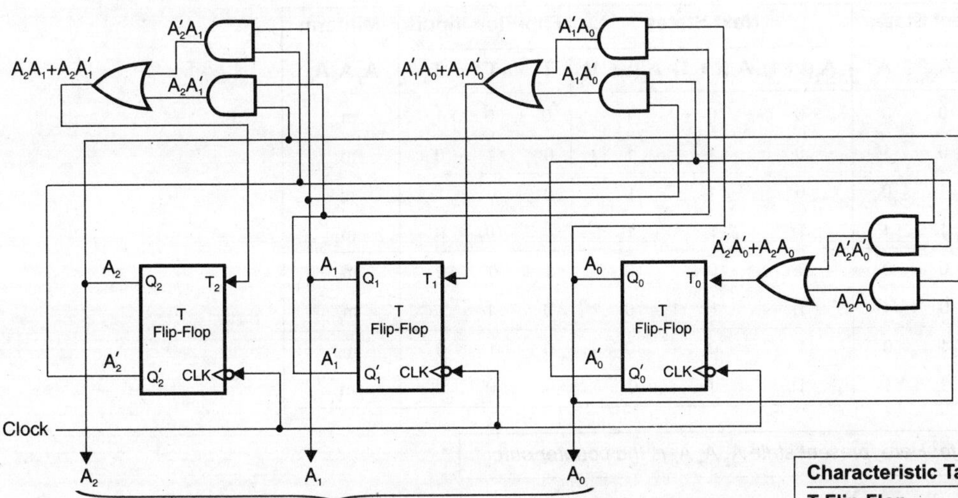

Fig. 6: *Logic diagram of (0, 1, 3, 7, 6, 4) sequence counter.*

Characteristic Table of T Flip-Flop

Input T	Present State Q	Next State Q(t+1)
0	0	0
0	1	1
1	0	1
1	1	0

Check for Self Starting

In order to check for self starting, the flip-flop inputs are determined for each unused state using simplified equations of flip-flop inputs and listed in Table 6. The next states are obtained from characteristic table of T flip-flop.

Table 6: Check for Self Starting

Unused Present States			Flip-Flop Inputs			Next States		
			T_2	T_1	T_0	$A_2(t+1)$	$A_1(t+1)$	$A_0(t+1)$
A_2	A_1	A_0	$= A_2'A_1+A_2A_1'$	$= A_1'A_0+A_1A_0'$	$= A_2'A_0'+A_2A_0$			
0	1	0	1	1	1	1	0	1
1	0	1	1	1	1	0	1	0

From Table 6 we can construct the state diagram of unused states as shown in Fig. 7. Therefore, when the counter enter an unused state then it enter a lock-out condition so that the counter will cycle through only unused states and will never enter a used state. Hence the counter is not self starting.

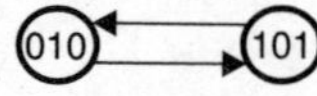

Fig. 7: State diagram of unused states.

Self Starting Counter

In order to make the counter self starting the unused states can be connected to enter used states as shown in Fig. 8. Once the counter enter an used state it will circle through used states and counter becomes self starting.

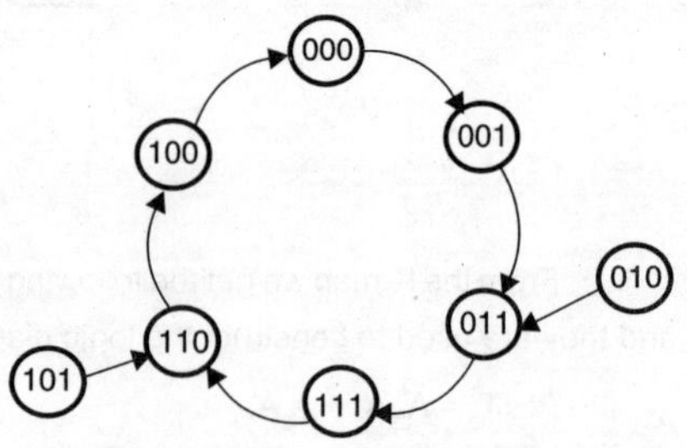

Fig. 8: State diagram to prevent lock-out condition.

From the state diagram of Fig. 8, the state table without don't-cares is formed as shown in Table 7. Here, the next state is obtained from state diagram and flip-flop inputs are obtained from excitation table.

Table 7: Expanded State Table

Present States			Next States			Flip-Flop Inputs			Minterm
A_2	A_1	A_0	$A_2(t+1)$	$A_1(t+1)$	$A_0(t+1)$	T_2	T_1	T_0	$A_2A_1A_0$
0	0	0	0	0	1	0	0	1	m_0
0	0	1	0	1	1	0	1	0	m_1
0	1	0	0	1	1	0	0	1	m_2
0	1	1	1	1	1	1	0	0	m_3
1	0	0	0	0	0	1	0	0	m_4
1	0	1	1	1	0	0	1	1	m_5
1	1	0	1	0	0	0	1	0	m_6
1	1	1	1	1	0	0	0	1	m_7

***Note:** Here, present state $A_2\, A_1\, A_0$ is the counter output.*

The Boolean equations for T_2, T_1 and T_0 are obtained from Table 7 by taking present state as minterms.

$$T_2 = \Sigma m(3, 4)$$

$$T_1 = \Sigma m(1, 5, 6)$$

$$T_0 = \Sigma m(0, 2, 5, 7)$$

The simplified Boolean equations are obtained by constructing K-map using the above equation for T_2, T_1 and T_0 as shown in Figs. 9 to 11.

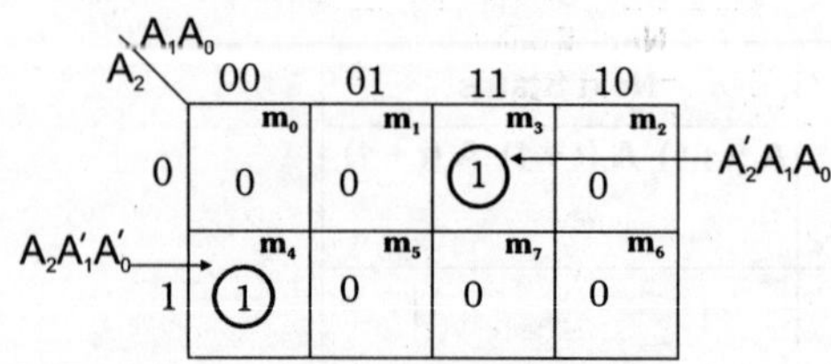

Fig. 9: K-map for T_2.

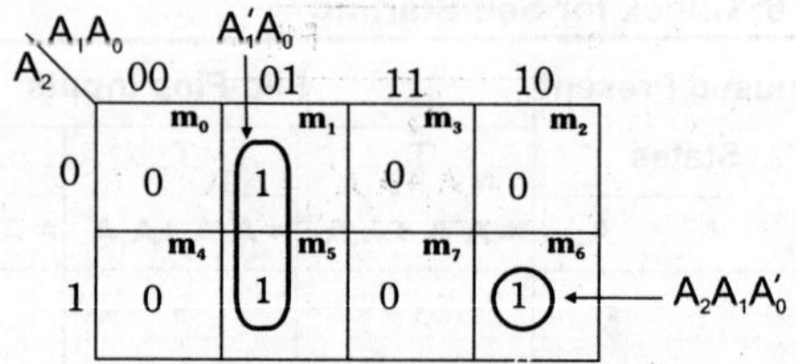

Fig. 10: K-map for T_1.

From the K-map we get the following simplified Boolean equations to realize 3-bit binary counter using T flip-flops and they are used to construct the logic diagram of 3-bit counter as shown in Fig. 12.

$$T_2 = A_2'A_1A_0 + A_2A_1'A_0'$$

$$T_1 = A_1'A_0 + A_2A_1A_0'$$

$$T_0 = A_2'A_0' + A_2A_0$$

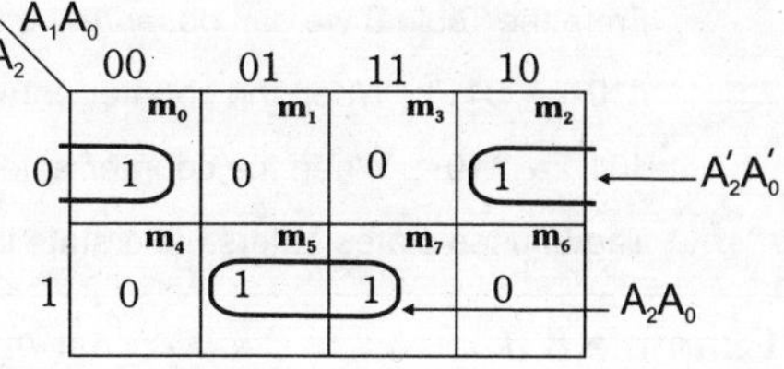

***Fig. 11:** K-map for T_0.*

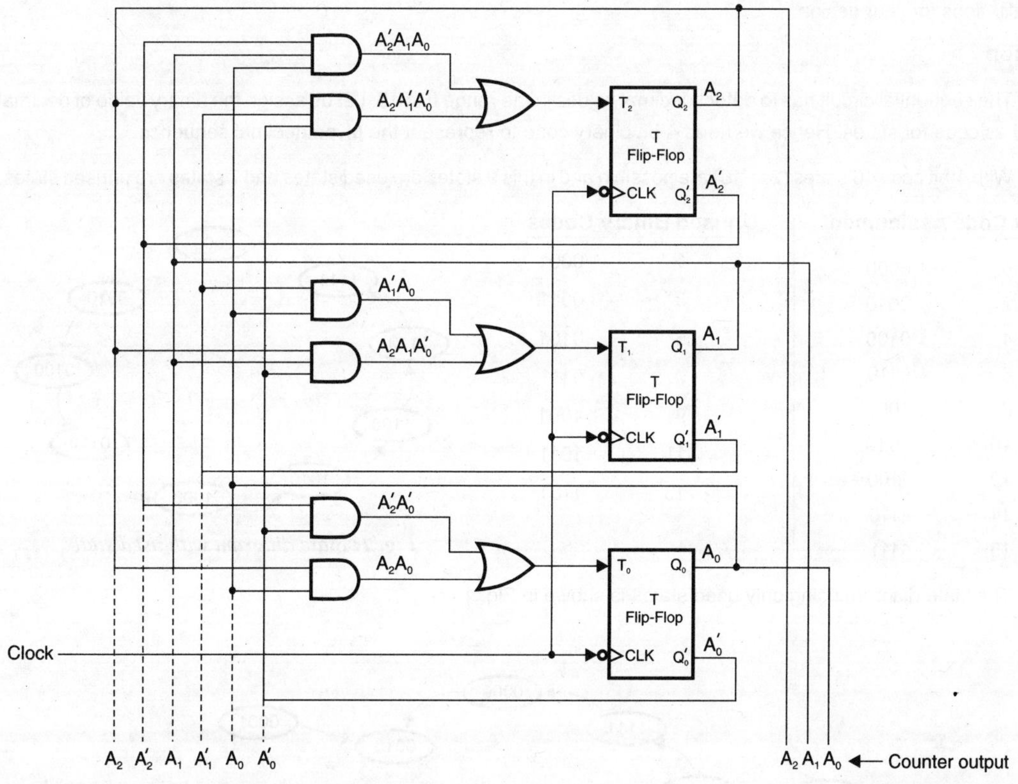

***Fig. 12:** Logic diagram of self starting (0, 1, 3, 7, 6, 4) sequence counter.*

Characteristic Table of T Flip-Flop

Input T	Present State Q	Next State Q(t+1)
0	0	0
0	1	1
1	0	1
1	1	0

Check for Self Starting

In order to check for self starting, the flip-flop inputs are determined for each unused state using simplified equations of flip-flop inputs and listed in Table 8. The next states are obtained from characteristic table of T flip-flop.

Table 8: Check for Self Starting

Unused Present States			Flip-Flop Inputs			Next States		
			T_2	T_1	T_0	$A_2(t+1)$	$A_1(t+1)$	$A_0(t+1)$
A_2	A_1	A_0	$= A_2'A_1A_0 + A_2A_1'A_0'$	$= A_1'A_0 + A_2A_1A_0'$	$= A_2'A_0' + A_2A_0$			
0	1	0	0	0	1	0	1	1
1	0	1	0	1	1	1	1	0

From the Table 8 we can observe the following state transitions from unused state to next state.

010 → 011 ; When the counter enters unused state 010 then in next transition it go to a used state 011.

101 → 110 ; When the counter enters unused state 101 then in next transition it go to unused state 110.

The unused states enters used state in the next transition as shown above and so the counter is self starting.

Example 8.6

(AKTU, Nov/Dec'22, 8 Marks) (JNTUH, Nov/Dec'20, 15 Marks) (IKGPTU, May'20, 7 Marks)

Design a synchronous sequential circuit that goes through the sequence 0, 2, 4, 6, 8, 10, 12, 14, 15 repeatedly. Use D flip-flops for your design.

Solution

The sequential circuit has to detect decimal values in the range 0 to 15. Let us assign the binary value of decimal number as code for states. Hence we need 4-bit binary code to represent the given decimal sequence.

With 4-bit code 16 states ($2^4 = 16$) are possible and in this 9 states are used states and 7 states are unused states.

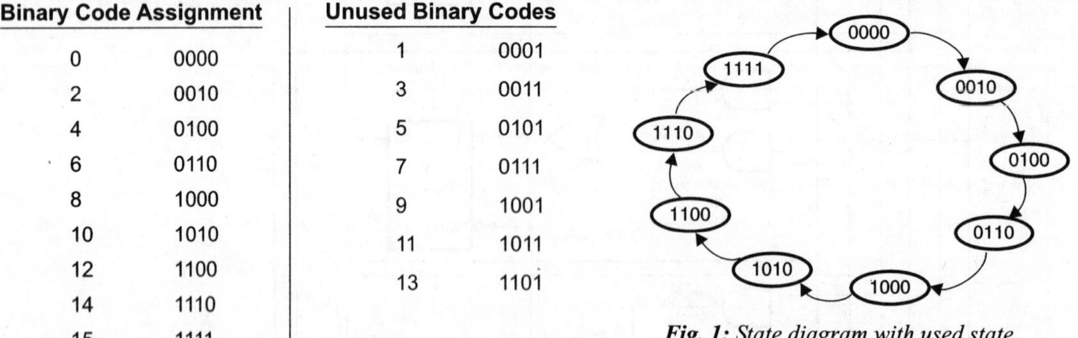

Binary Code Assignment

0	0000
2	0010
4	0100
6	0110
8	1000
10	1010
12	1100
14	1110
15	1111

Unused Binary Codes

1	0001
3	0011
5	0101
7	0111
9	1001
11	1011
13	1101

Fig. 1: State diagram with used state.

The state diagram using only used states is shown in Fig. 1.

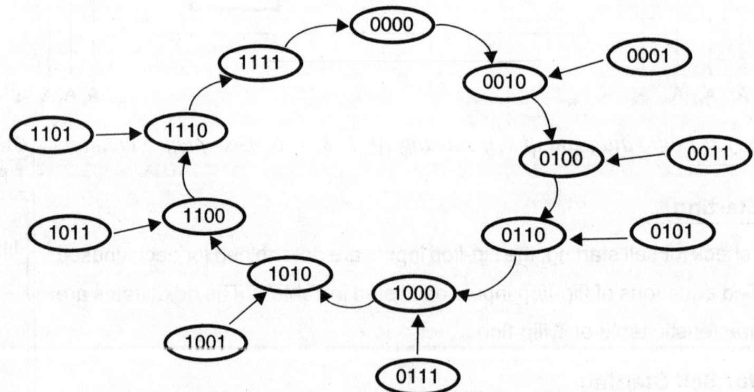

Fig. 2: State diagram to prevent lock-out condition.

Here, each bit of binary code is a state variable. Hence the sequence counter has 4 state variables. Let the state variables be A_0, A_1, A_2 and A_3.

$\therefore$ Present state variables : $A_0(t), A_1(t), A_2(t), A_3(t) \Rightarrow A_0, A_1, A_2, A_3$ (For simplicity)

Next state variables : $A_0(t+1), A_1(t+1), A_2(t+1), A_3(t+1)$

Using the state diagram of Fig. 2, the state table is constructed as shown in Table 1.

Table 1: State Table

Present States				Next States			
A_3	A_2	A_1	A_0	$A_3(t+1)$	$A_2(t+1)$	$A_1(t+1)$	$A_0(t+1)$
0	0	0	0	0	0	1	0
0	0	0	1	0	0	1	0
0	0	1	0	0	1	0	0
0	0	1	1	0	1	0	0
0	1	0	0	0	1	1	0
0	1	0	1	0	1	1	0
0	1	1	0	1	0	0	0
0	1	1	1	1	0	0	0
1	0	0	0	1	0	1	0
1	0	0	1	1	0	1	0
1	0	1	0	1	1	0	0
1	0	1	1	1	1	0	0
1	1	0	0	1	1	1	0
1	1	0	1	1	1	1	0
1	1	1	0	1	1	1	1
1	1	1	1	0	0	0	0

Let us choose four D flip-flops to store four state variables A_0, A_1, A_2 and A_3 as shown in Fig. 3. Here, A_0, A_1, A_2 and A_3 are output of flip-flops.

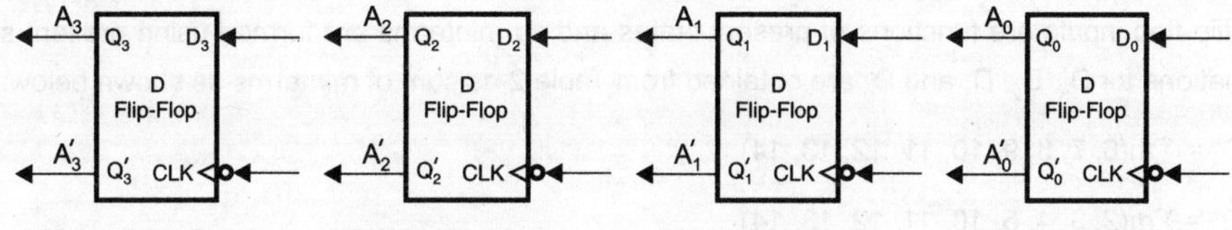

Fig. a: *Flip-flop A_3.* ***Fig. b:*** *Flip-flop A_2.* ***Fig. c:*** *Flip-flop A_1.* ***Fig. d:*** *Flip-flop A_0.*

Fig. 3: *D flip-flop to store state variables.*

Here, $A_0 = Q_0$, $A_1 = Q_1$, $A_2 = Q_2$, $A_3 = Q_3$ $\Rightarrow$ $A_0(t+1) = Q_0(t+1)$, $A_1(t+1) = Q_1(t+1)$, $A_2(t+1) = Q_2(t+1)$, $A_3(t+1) = Q_3(t+1)$

The characteristic equation of four D flip-flop are,

$$Q_0(t+1) = D_0$$

$$Q_1(t+1) = D_1$$

$$Q_2(t+1) = D_2$$

$$Q_3(t+1) = D_3$$

Characteristic equation of D flip-flop,

$$Q(t+1) = D$$

Using the above equations we can write,

$A_3(t + 1) = D_3$ | $A_1(t + 1) = D_1$

$A_2(t + 1) = D_2$ | $A_0(t + 1) = D_0$

Using the above equations the flip-flop inputs are determined for the next state and listed in Table 2. Here, the flip-flop inputs are same as next state outputs.

Table 2: Expanded State Table

Present States				Next States				Flip-Flop Inputs				Minterm
A_3	A_2	A_1	A_0	$A_3(t+1)$	$A_2(t+1)$	$A_1(t+1)$	$A_0(t+1)$	D_3	D_2	D_1	D_0	$A_3A_2A_1A_0$
0	0	0	0	0	0	1	0	0	0	1	0	m_0
0	0	0	1	0	0	1	0	0	0	1	0	m_1
0	0	1	0	0	1	0	0	0	1	0	0	m_2
0	0	1	1	0	1	0	0	0	1	0	0	m_3
0	1	0	0	0	1	1	0	0	1	1	0	m_4
0	1	0	1	0	1	1	0	0	1	1	0	m_5
0	1	1	0	1	0	0	0	1	0	0	0	m_6
0	1	1	1	1	0	0	0	1	0	0	0	m_7
1	0	0	0	1	0	1	0	1	0	1	0	m_8
1	0	0	1	1	0	1	0	1	0	1	0	m_9
1	0	1	0	1	1	0	0	1	1	0	0	m_{10}
1	0	1	1	1	1	0	0	1	1	0	0	m_{11}
1	1	0	0	1	1	1	0	1	1	1	0	m_{12}
1	1	0	1	1	1	1	0	1	1	1	0	m_{13}
1	1	1	0	1	1	1	1	1	1	1	1	m_{14}
1	1	1	1	0	0	0	0	0	0	0	0	m_{15}

***Note:** Here, present state $A_3\,A_2\,A_1\,A_0$ is the sequence detector output.*

The flip-flop inputs are functions of present states and so minterms are formed using present states and the Boolean equations for D_3, D_2, D_1 and D_0 are obtained from Table 2 as sum of minterms as shown below:

$$D_3 = \Sigma m(6, 7, 8, 9, 10, 11, 12, 13, 14)$$

$$D_2 = \Sigma m(2, 3, 4, 5, 10, 11, 12, 13, 14)$$

$$D_1 = \Sigma m(0, 1, 4, 5, 8, 9, 12, 13, 14)$$

$$D_0 = m_{14}$$

The simplified Boolean equations are obtained by constructing K-map using the above equations for D_3, D_2, D_1 and D_0 as shown in Figs. 4 to 7.

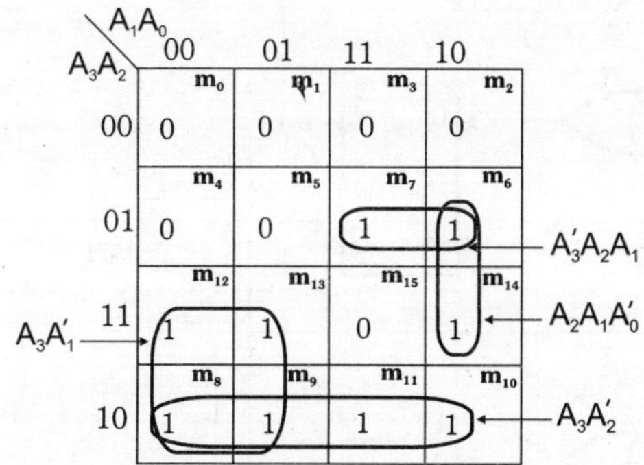

Fig. 4: K-map for D_3.

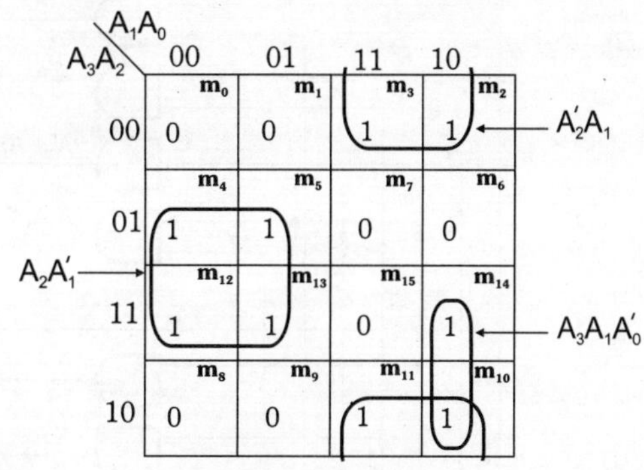

Fig. 5: K-map for D_2.

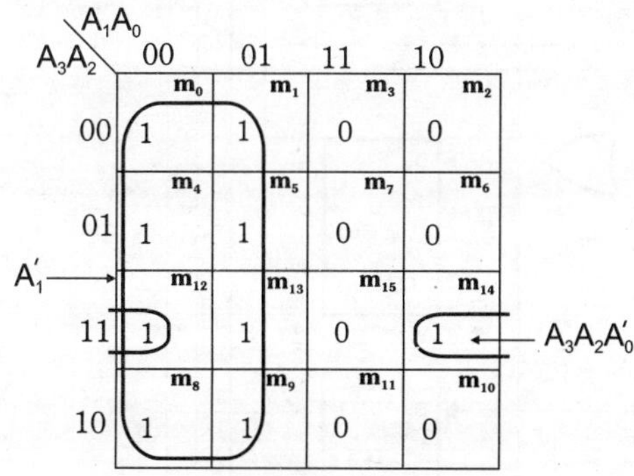

Fig. 6: K-map for D_1.

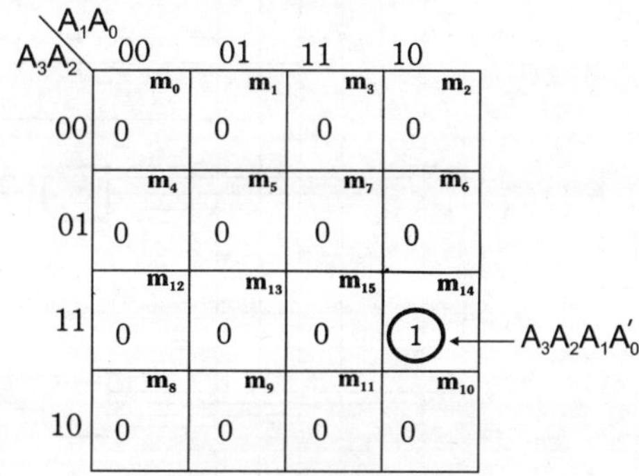

Fig. 7: K-map for D_0.

From the K-map we get the following simplified Boolean equations to realize 4-bit binary sequence counter using D flip-flops and they are used to realize the sequence counter as shown in Fig. 8.

$$D_3 = A_3A_1' + A_3A_2' + A_3'A_2A_1 + A_2A_1A_0'$$

$$D_2 = A_2A_1' + A_2'A_1 + A_3A_1A_0'$$

$$D_1 = A_1' + A_3A_2A_0'$$

$$D_0 = A_3A_2A_1A_0'$$

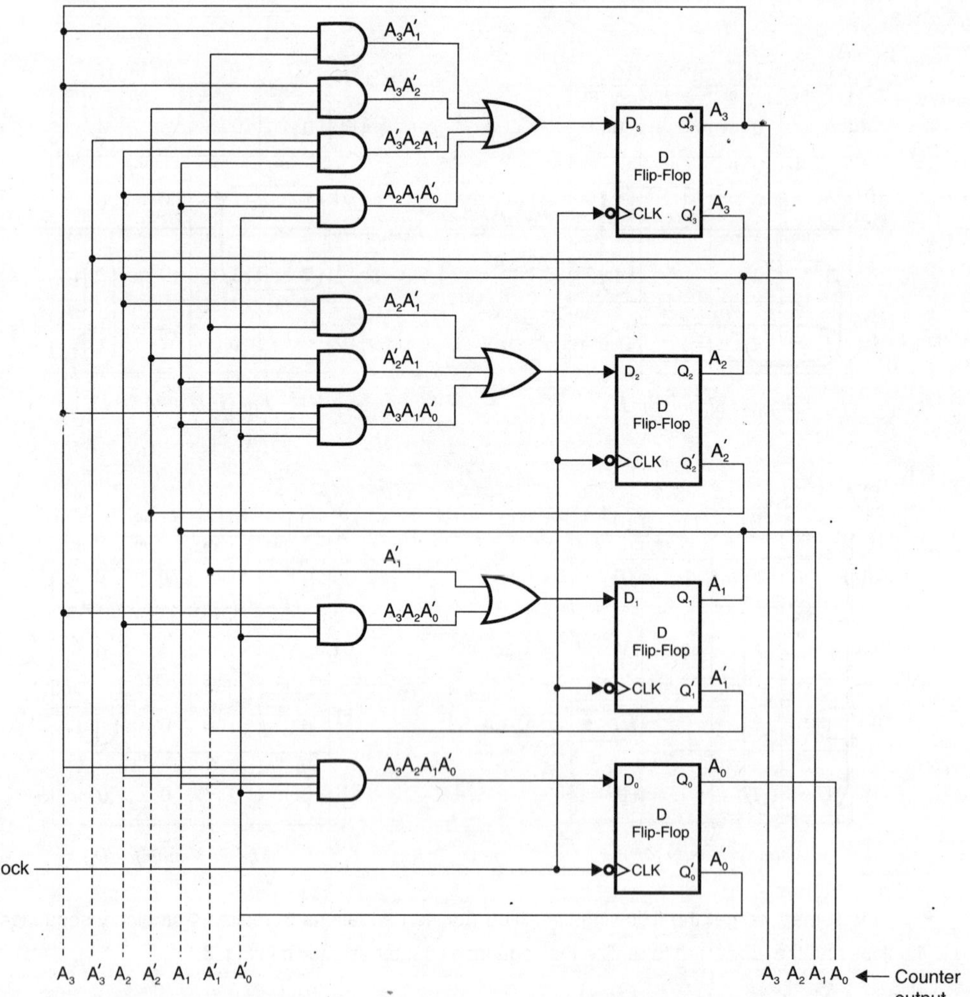

***Fig. 8:** Logic diagram of (0, 2, 4, 6, 8, 10, 12, 14, 15) sequence counter.*

Example 8.7

(CSVTU, Non/Dec'19, 7 Marks)

Design a synchronous BCD counter using JK flip-flop.

Solution

BCD counter will count only from 0 tc 9 for 9 clock pulses and upon reaching tenth pulse output is reset to 0, so that next count sequence will start. Also, an additional output 1 is needed to use as carry to next stage.

Let, A_8, A_4, A_2 and A_1 be 4 state variables of BCD counter.

$\therefore$ Present state variables : $A_8(t)$, $A_4(t)$, $A_2(t)$, $A_1(t)$ $\Rightarrow$ A_8, A_4, A_2, A_1 (For simplicity)

Next state variables : $A_8(t + 1)$, $A_4(t + 1)$, $A_2(t + 1)$, $A_1(t + 1)$

The present state and next state of BCD counter along with additional output Y are shown in Table 1.

Table 1: State Table of BCD Counter

Present States (Counter Outputs)				Next States				Output
A_8	A_4	A_2	A_1	$A_8(t+1)$	$A_4(t+1)$	$A_2(t+1)$	$A_1(t+1)$	Y
0	0	0	0	0	0	0	1	0
0	0	0	1	0	0	1	0	0
0	0	1	0	0	0	1	1	0
0	0	1	1	0	1	0	0	0
0	1	0	0	0	1	0	1	0
0	1	0	1	0	1	1	0	0
0	1	1	0	0	1	1	1	0
0	1	1	1	1	0	0	0	0
1	0	0	0	1	0	0	1	0
1	0	0	1	0	0	0	0	1

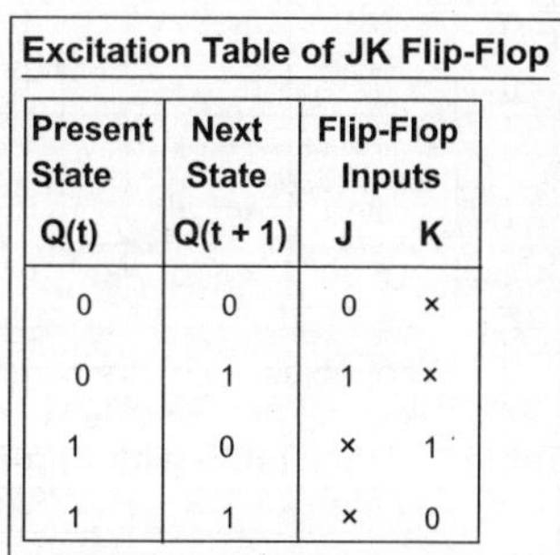

Excitation Table of JK Flip-Flop

Present State Q(t)	Next State Q(t + 1)	Flip-Flop Inputs J	K
0	0	0	×
0	1	1	×
1	0	×	1
1	1	×	0

Let us choose four JK flip-flops to store four state variables A_1, A_2, A_4 and A_8 as shown in Fig. 1. Here, A_1, A_2, A_4 and A_8 are output of flip-flops.

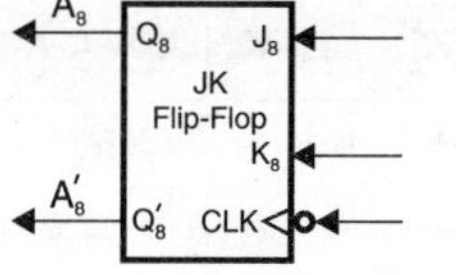

Fig. a: Flip-flop A_8.

Fig. b: Flip-flop A_4.

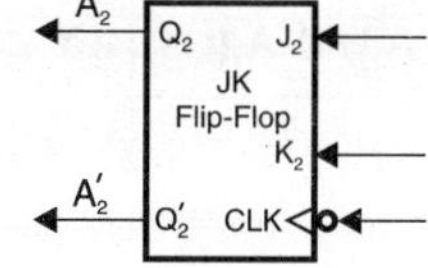

Fig. c: Flip-flop A_2.

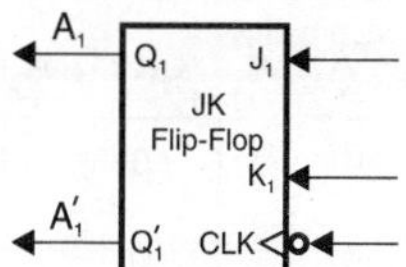

Fig. d: Flip-flop A_1.

Fig. 1: JK flip-flops to store state variables.

The flip-flop inputs can be obtained from excitation table as shown below:

State A_8

For state A_8, the Q(t), Q(t + 1), J and K of excitation table are considered as A_8, $A_8(t + 1)$, J_8 and K_8.

$\therefore Q(t) \rightarrow A_8 \;\; ; \;\; Q(t+1) \rightarrow A_8(t+1)$

$J \rightarrow J_8 \;\; ; \;\; K \rightarrow K_8$

The flip-flop input J_8 and K_8 that drives the present state A_8 to next state is obtained from excitation table and listed in Table 2.

State A_4

For state A_4, the Q(t), Q(t + 1), J and K of excitation table are considered as A_4, $A_4(t + 1)$, J_4 and K_4.

$\therefore Q(t) \rightarrow A_4 \;\; ; \;\; Q(t+1) \rightarrow A_4(t+1)$

$J \rightarrow J_4 \;\; ; \;\; K \rightarrow K_4$

The flip-flop input J_4 and K_4 that drives the present state A_4 to next state is obtained from excitation table and listed in Table 3.

State A_2

For state A_2, the Q(t), Q(t + 1), J and K of excitation table are considered as A_2, $A_2(t + 1)$, J_2 and K_2.

$\therefore Q(t) \rightarrow A_2 \;\; ; \;\; Q(t+1) \rightarrow A_2(t+1)$

$J \rightarrow J_2 \;\; ; \;\; K \rightarrow K_2$

The flip-flop input J_2 and K_2 that drives the present state A_2 to next state is obtained from excitation table and listed in Table 4 .

State A_1

For state A_1, the Q(t), Q(t + 1), J and K of excitation table are considered as A_1, $A_1(t + 1)$, J_1 and K_1.

$\therefore Q(t) \rightarrow A_1 \;\; ; \;\; Q(t+1) \rightarrow A_1(t+1)$

$J \rightarrow J_1 \;\; ; \;\; K \rightarrow K_1$

The flip-flop input J_1 and K_1 that drives the present state A_1 to next state is obtained from excitation table and listed in Table 5.

Table 2:

A_8	$A_8(t+1)$	J_8	K_8
0	0	0	×
0	0	0	×
0	0	0	×
0	0	0	×
0	0	0	×
0	0	0	×
0	0	0	×
0	1	1	×
1	1	×	0
1	0	×	1

Table 3:

A_4	$A_4(t+1)$	J_4	K_4
0	0	0	×
0	0	0	×
0	0	0	×
0	1	1	×
1	1	×	0
1	1	×	0
1	1	×	0
1	0	×	1
0	0	0	×
0	0	0	×

Table 4:

A_2	$A_2(t+1)$	J_2	K_2
0	0	0	×
0	1	1	×
1	1	×	0
1	0	×	1
0	0	0	×
0	1	1	×
1	1	×	0
1	0	×	1
0	0	0	×
0	0	0	×

Table 5:

A_1	$A_1(t+1)$	J_1	K_1
0	1	1	×
1	0	×	1
0	1	1	×
1	0	×	1
0	1	1	×
1	0	×	1
0	1	1	×
1	0	×	1
0	1	1	×
1	0	×	1

The tables 1 to 5 are combined to form state table with flip-flop inputs as shown in Table 6

Table 6: State Table with Flip-Flop Inputs

Present state				Next state				Output	JK Flip-Flop inputs								Minterms
A_8	A_4	A_2	A_1	$A_8(t+1)$	$A_4(t+1)$	A_2t+1)	$A_1(t+1)$	Y	J_8	K_8	J_4	K_4	J_2	K_2	J_1	K_1	$A_8A_4A_2A_1$
0	0	0	0	0	0	0	1	0	0	×	0	×	0	×	1	×	m_0
0	0	0	1	0	0	1	0	0	0	×	0	×	1	×	×	1	m_1
0	0	1	0	0	0	1	1	0	0	×	0	×	×	0	1	×	m_2
0	0	1	1	0	1	0	0	0	0	×	1	×	×	1	×	1	m_3
0	1	0	0	0	1	0	1	0	0	×	×	0	0	×	1	×	m_4
0	1	0	1	0	1	1	0	0	0	×	×	0	1	×	×	1	m_5
0	0	1	0	0	1	1	1	0	0	×	×	0	×	0	1	×	m_6
0	1	1	1	1	0	0	0	0	1	×	×	1	×	1	×	1	m_7
1	0	0	0	1	0	0	1	0	×	0	0	×	0	×	1	×	m_8
1	0	0	1	0	0	0	0	1	×	1	0	×	0	×	×	1	m_9

Note: *1. The states 1010 to 1111 are unused states and so next states, output and flip-flop inputs are considered as don't-care for these states.*

2. Minterms are formed using state variables A_8, A_4, A_2 and A_1.

The flip-flop inputs are functions of present state and so minterms are formed using present state and the Boolean equations for J_8, K_8, J_4, K_4, J_2, K_2, J_1, K_1 and Y are obtained from Table 6 as sum of minterms and don't-cares as shown below:

$$J_8 = m_7 + \Sigma d(8, 9, 10, 11, 12, 13, 14, 15)$$

$$K_8 = m_9 + \Sigma d(0, 1, 2, 3, 4, 5, 6, 7, 10, 11, 12, 13, 14, 15)$$

$J_4 = m_3 + \sum d(4, 5, 6, 7, 10, 11, 12, 13, 14, 15)$

$K_4 = m_7 + \sum d(0, 1, 2, 3, 8, 9, 10, 11, 12, 13, 14, 15)$

$J_2 = \sum m(1, 5) + \sum d(2, 3, 6, 7, 10, 11, 12, 13, 14, 15)$

$K_2 = \sum m(3, 7) + \sum d(0, 1, 4, 5, 8, 9, 10, 11, 12, 13, 14, 15)$

$J_1 = \sum m(0, 2, 4, 6, 8) + \sum d(1, 3, 5, 7, 9, 10, 11, 12, 13, 14, 15)$

$K_1 = \sum m(1, 3, 5, 7, 9) + \sum d(0, 2, 4, 6, 8, 10, 11, 12, 13, 14, 15)$

$Y = m_9 + \sum d(10, 11, 12, 13, 14, 15)$

The simplified Boolean equations are obtained by constructing K-map using the above equations for J_8, K_8, J_4, K_4, J_2, K_2, J_1, K_1 and Y as shown in Figs. 2 to 10.

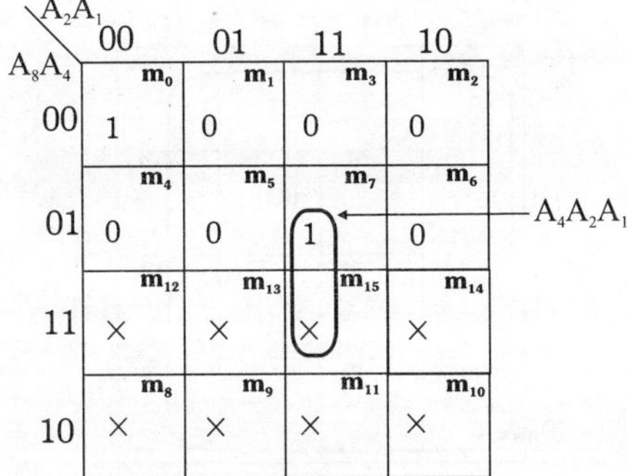

Fig. 2: K-map for J_8.

A_8A_4 \ A_2A_1	00	01	11	10
00	× (m_0)	× (m_1)	× (m_3)	× (m_2)
01	× (m_4)	× (m_5)	× (m_7)	× (m_6)
11	× (m_{12})	× (m_{13})	× (m_{15})	× (m_{14})
10	0 (m_8)	1 (m_9)	× (m_{11})	× (m_{10})

$\leftarrow A_1$

Fig. 3: K-map for K_8.

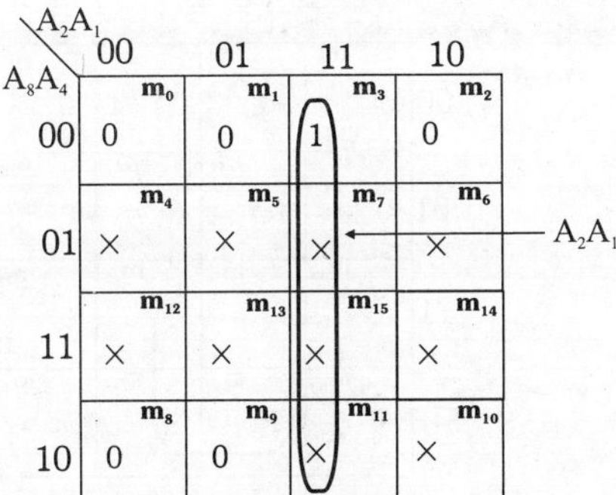

Fig. 4: K-map for J_4.

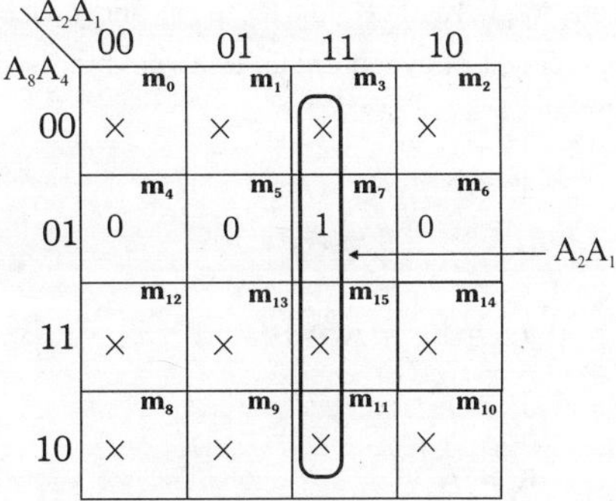

Fig. 5: K-map for K_4.

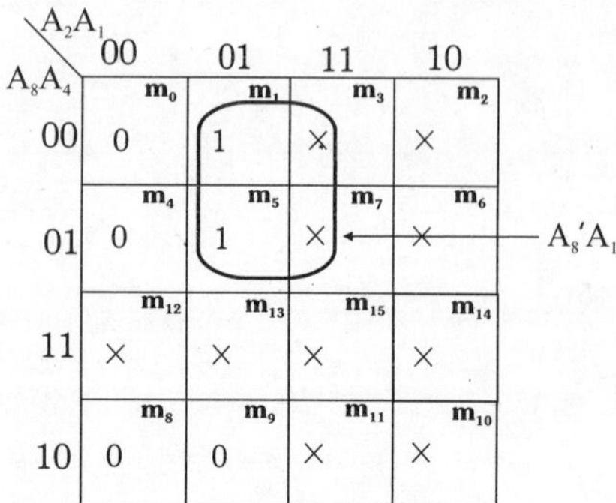

Fig. 6: K-map for J_2.

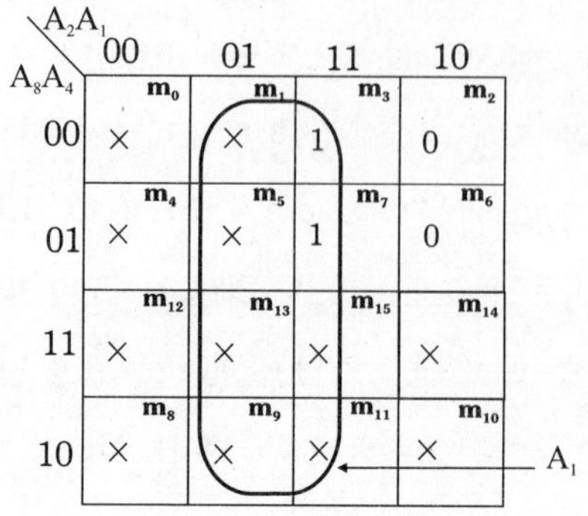

Fig. 7: K-map for K_2.

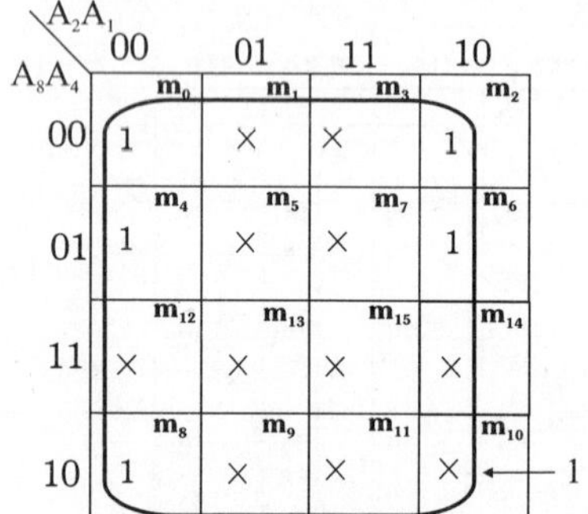

Fig. 8: K-map for J_1.

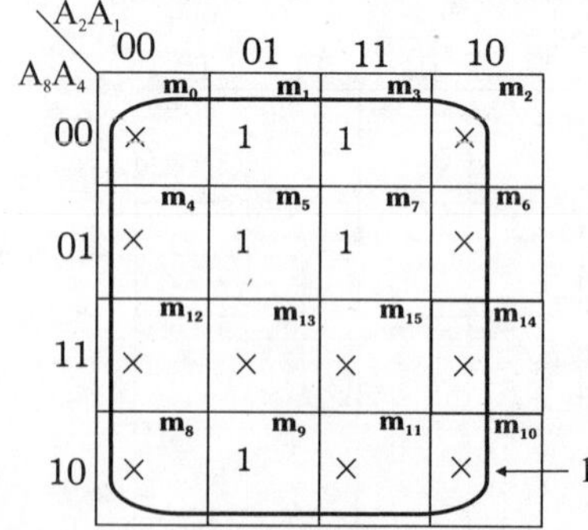

Fig. 9: K-map for K_1.

From the K-maps we get the following simplified Boolean equations for J_8, K_8, J_4, K_4, J_2, K_2, J_1, K_1 and Y. Using these equations the logic circuit of synchronous BCD counter is drawn as shown in Fig. 11.

$J_8 = A_4 A_2 A_1$

$K_8 = A_1$

$J_4 = K_4 = A_2 A_1$

$J_2 = A_8' A_1$

$K_2 = A_1$

$J_1 = K_1 = 1$

$Y = A_8 A_1$

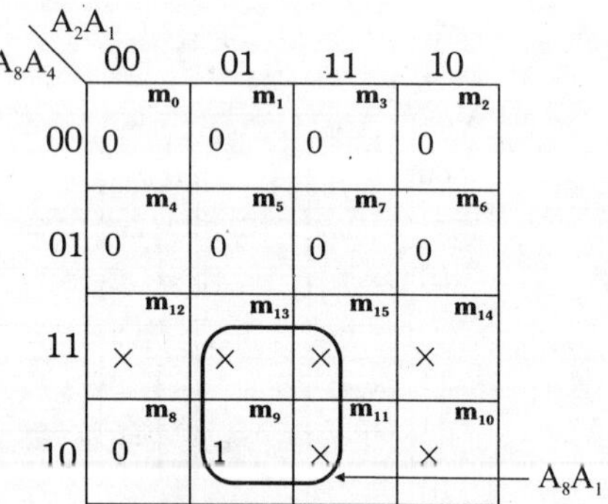

Fig. 10: K-map for Y.

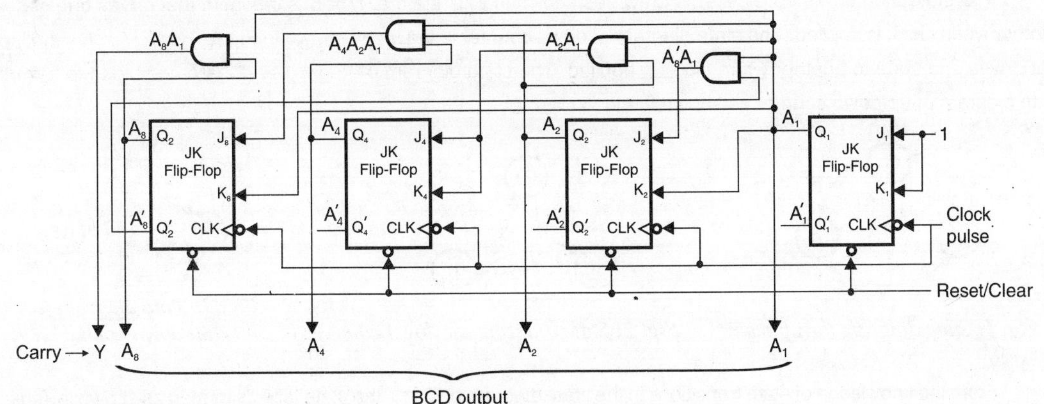

Fig. 11: Synchronous BCD counter using JK Flip-Flop.

Example 8.8

(VTU, Aug/Sep'20, 8 Marks)

Design a synchronous circuit using positive edge triggered JK flip-flop with minimal combinational gating to generate the sequence: 0 - 1 - 2 - 0 if input x = 0 ; 0 - 2 - 1 - 0 if input x = 1

Solution

Binary output size = 2 bits.

The 2 bits are represented by two state variables A_1 and A_0.

$\therefore$ Present state variables : $A_1(t), A_0(t) \Rightarrow A_1, A_0$ (For simplicity)

Next state variables : $A_1(t+1), A_0(t+1)$

With 2 state variables it is possible to form 4 states ($2^2 = 4$).

The 4 states are 00, 01, 10 and 11.

The truth table of up and down count are shown in Tables 1 and 2 respectively. Here, the present states are the counter outputs.

Table 1: Truth Table of Up Count

Clock	Counter Outputs	
	A_1	A_0
-	0	0
↑	0	1
↑	1	0
↑	1	1
↑	0	0

Table 2: Truth Table of Down Count

Clock	Counter Outputs	
	A_1	A_0
-	1	1
↑	1	0
↑	0	1
↑	0	0
↑	1	1

Let, M be a mode control bit to select up and down count.

$\therefore$ When, M = 0, select up count

When, M = 1, select down count

The state diagram of up counter is drawn as shown in Fig. 1. Here, M = 0 is the input that drives one state to another when clock is applied. The state diagram of down counter is drawn as shown in Fig. 2. Here, M = 1 is the input that drives one state to another when clock is applied. The state diagram of Figs. 1 and 2 are combined to form the state diagram of up/down counter as shown in Fig. 3.

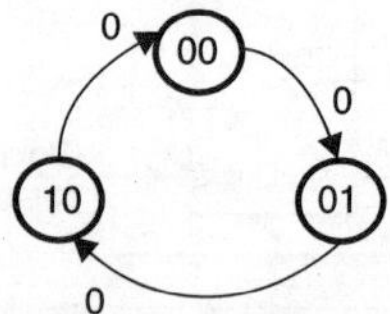

Fig. 1: State diagram for up count.

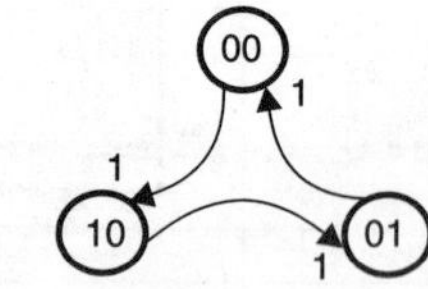

Fig. 2: State diagram for down count.

Fig. 3: State diagram for up/down count.

From the knowledge of state transitions in the state diagram of Fig. 3, the state table is formed as shown in Table 3.

Table 3: State Table

Present States		Mode Control	Next States	
A_1	A_0	M	$A_1(t+1)$	$A_0(t+1)$
0	0	0	0	1
0	0	1	1	0
0	1	0	1	0
0	1	1	0	0
1	0	0	0	0
1	0	1	0	1
1	1	0	×	×
1	1	1	×	×

Let us choose two JK flip-flops to store two state variables A_0 and A_1 as shown in Fig. 4. Here, A_0 and A_1 are output of flip-flops.

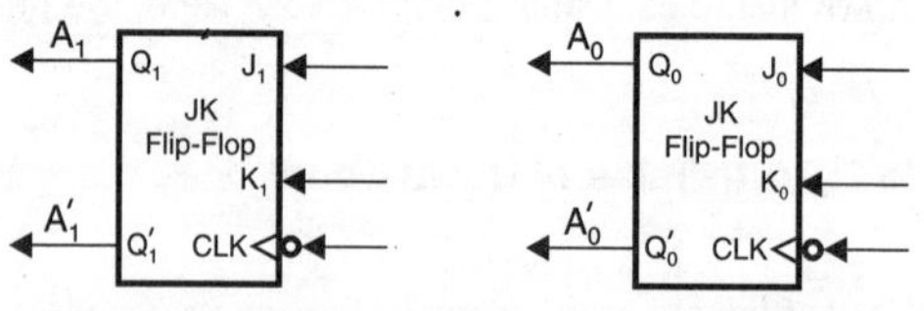

Fig. a: Flip-flop A_1. *Fig. b: Flip-flop A_0.*

Fig. 4: JK flip-flops to store state variables.

Excitation Table of JK Flip-Flop

Present State Q(t)	Next State Q(t + 1)	Flip-Flop Inputs J	K
0	0	0	×
0	1	1	×
1	0	×	1
1	1	×	0

The flip-flop inputs can be obtained from excitation table as shown below:

State A_1

For state A_1, the Q(t), Q(t + 1), J and K of excitation table are considered as A_1, $A_1(t+1)$, J_1 and K_1.

$$\therefore Q(t) \rightarrow A_1 \quad ; \quad Q(t+1) \rightarrow A_1(t+1)$$

$$J \rightarrow J_1 \quad ; \quad K \rightarrow K_1$$

The flip-flop inputs J_1 and K_1 that drives the present state A_1 to next state is obtained from excitation table and listed in Table 4.

State A_0

For state A_0, the Q(t), Q(t + 1), J and K of excitation table are considered as A_0, $A_0(t+1)$, $J_0$and K_0

$$\therefore Q(t) \rightarrow A_0 \quad ; \quad Q(t+1) \rightarrow A_0(t+1)$$

$$J \rightarrow J_0 \quad ; \quad K \rightarrow K_0$$

The flip-flop inputs J_0 and K_0 thatdrives the present state A_0 to next state is obtained from excitation table and listed in Table 5.

Table 4:

A_1	$A_1(t + 1)$	J_1	K_1
0	0	0	×
0	1	1	×
0	1	1	×
0	0	0	×
1	0	×	1
1	0	×	1
1	×	×	×
1	×	×	×

Table 5:

A_0	$A_0(t + 1)$	J_0	K_0
0	1	1	×
0	0	0	×
1	0	×	1
1	0	×	1
0	0	0	×
0	1	1	×
1	×	×	×
1	×	×	×

The Tables 3 to 5 are combined to form state table with flip-flop inputs as shown in Table 6.

Table 6: Expanded State Table

Present States		Mode Control	Next States		Flip-Flop Inputs				Minterm
A_1	A_0	M	$A_1(t+1)$	$A_0(t+1)$	J_1	K_1	J_0	K_0	A_1A_0M
0	0	0	0	1	0	×	1	×	m_0
0	0	1	1	0	1	×	0	×	m_1
0	1	0	1	0	1	×	×	1	m_2
0	1	1	0	0	0	×	×	1	m_3
1	0	0	0	0	×	1	0	×	m_4
1	0	1	0	1	×	1	1	×	m_5
1	1	0	×	×	×	×	×	×	m_6
1	1	1	×	×	×	×	×	×	m_7

> ***Note**: Here, present state A_1 A_0 is the counter output.*

The flip-flop inputs are functions of present states and mode control bit, M. Therefore, minterms are formed using present states and M and the Boolean equations for flip-flop inputs are obtained as sum of minterms and don't cares as shown below:

$J_1 = \Sigma m(1, 2) + \Sigma d(4, 5, 6, 7)$ | $K_1 = \Sigma m(4, 5) + \Sigma d(0, 1, 2, 3, 6, 7)$

$J_0 = \Sigma m(0, 5) + \Sigma d(2, 3, 6, 7)$ | $K_0 = \Sigma m(2, 3) + \Sigma d(0, 1, 4, 5, 6,7)$

The simplified Boolean equations for flip-flop inputs are obtained by constructing K-map using the above equations as shown in Figs. 5 to 8.

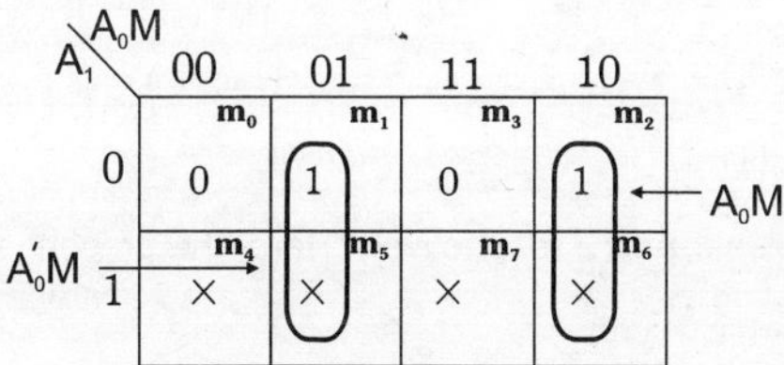

Fig. 5: K-map for J_1.

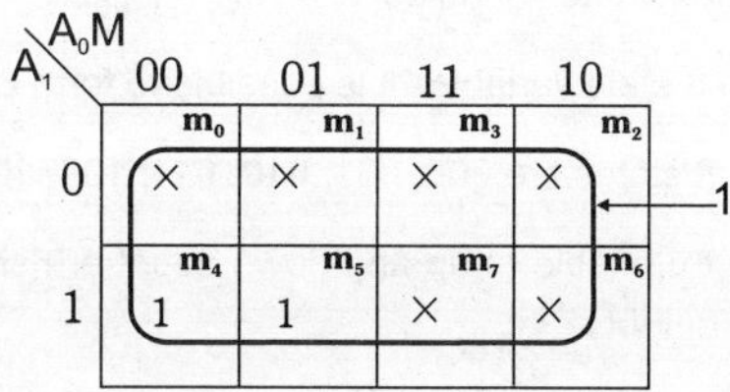

Fig. 6: K-map for K_1.

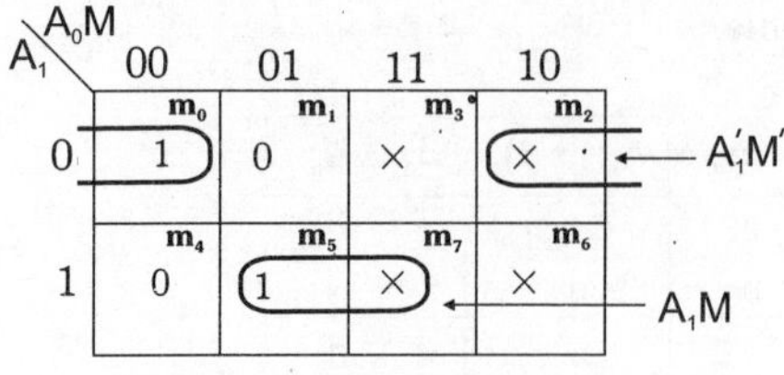

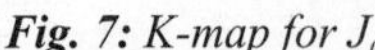

Fig. 7: K-map for J_0.

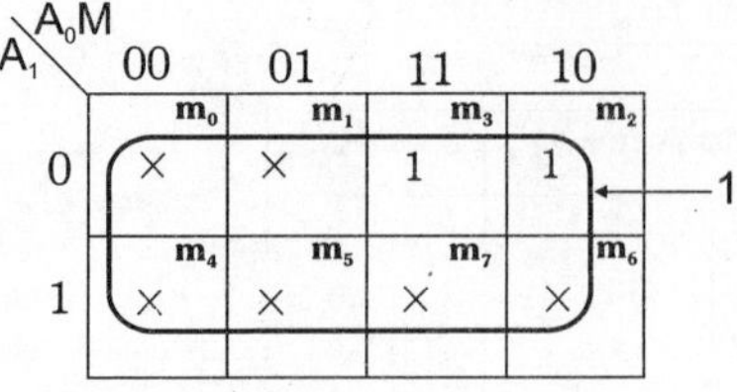

Fig. 8: K-map for K_0.

From the K-map we get the following simplified Boolean equations for J_1, K_1, J_0 and K_0. Using these equations the logic circuit of 2-bit counter is drawn as

$J_1 = A_0M' + A_0'M$	$K_1 = K_0 = 1$
$J_0 = A_1M + A_1'M'$	

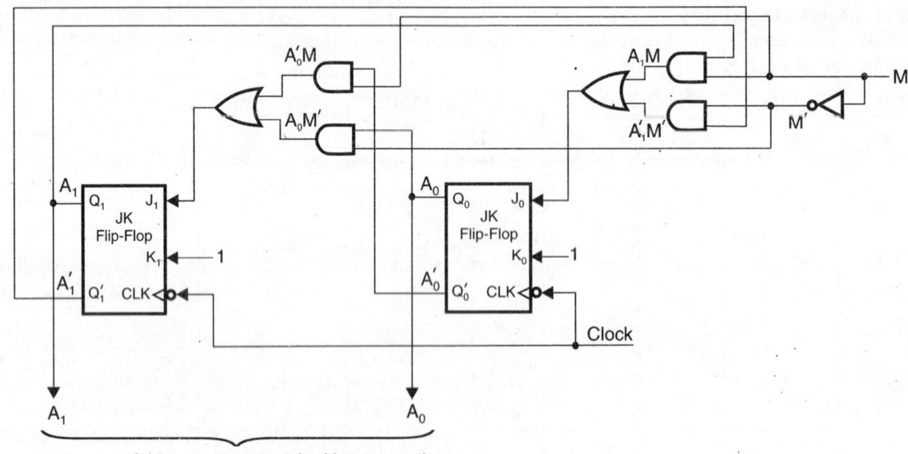

Fig. 9: Logic diagram of 2-bit synchronous up/down counter.

Example 8.9

(AKTU, Dec'22, 8 Marks) (JNTUH, Nov/Dec'20, 9 Marks) (VTU, Aug/Sep'20, 8 Marks)

Design of 3-bit synchronous up/down counter using JK flip-flops.

Solution

Binary output size = 3 bits.

The 3 bits are represented by three state variables A_2, A_1 and A_0.

$\therefore$ Present state variables : $A_2(t), A_1(t), A_0(t) \Rightarrow A_2, A_1, A_0$ (For simplicity)

Next state variables : $A_2(t+1), A_1(t+1), A_0(t+1)$

With 3 state variables it is possible to form 8 states ($2^3 = 8$).

The 8 states are 000, 001, 010, 011, 100, 101, 110 and 111.

The truth table of up and down count are shown in Tables 1 and 2 respectively. Here, the present states are the counter outputs.

Table 1: Truth Table of Up Count

Clock	Counter Outputs		
	A_2	A_1	A_0
-	0	0	0
↓	0	0	1
↓	0	1	0
↓	0	1	1
↓	1	0	0
↓	1	0	1
↓	1	1	0
↓	1	1	1
↓	0	0	0

Table 2: Truth Table of Down Count

Clock	Counter Outputs		
	A_2	A_1	A_0
-	1	1	1
↓	1	1	0
↓	1	0	1
↓	1	0	0
↓	0	1	1
↓	0	1	0
↓	0	0	1
↓	0	0	0
↓	1	1	1

Let, M be a mode control bit to select up and down count.

∴ When, M = 1, select up count

When, M = 0, select down count

The state diagram of up counter is drawn as shown in Fig. 1. Here, M = 1 is the input that drives one state to another when clock is applied.

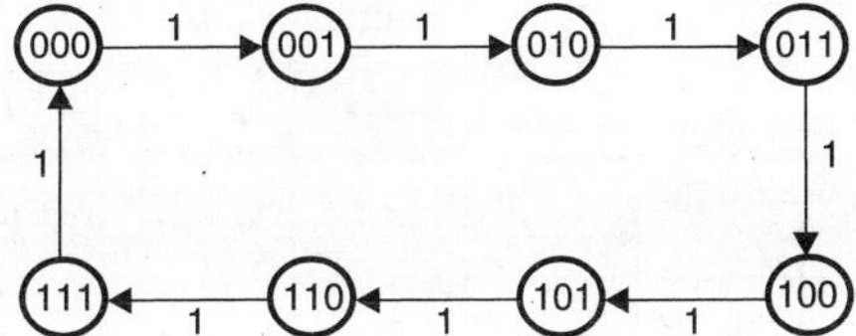

Fig. 1: State diagram for up count.

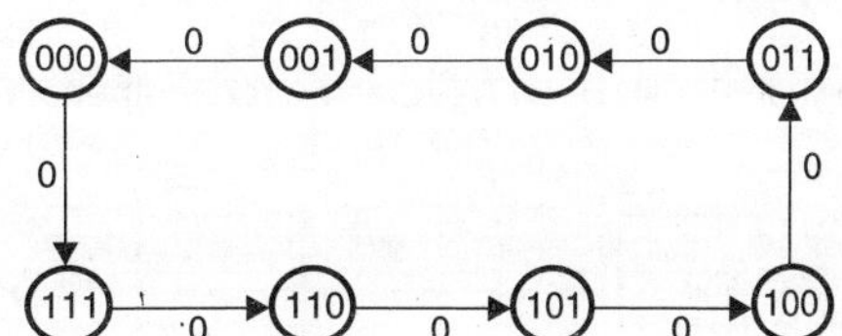

Fig. 2: State diagram for down count.

The state diagram of down counter is drawn as shown in Fig. 2. Here, M = 0 is the input that drives one state to another when clock is applied.

The state diagram of Figs. 1 and 2 are combined to form the state diagram of up/down counter as shown in Fig. 3.

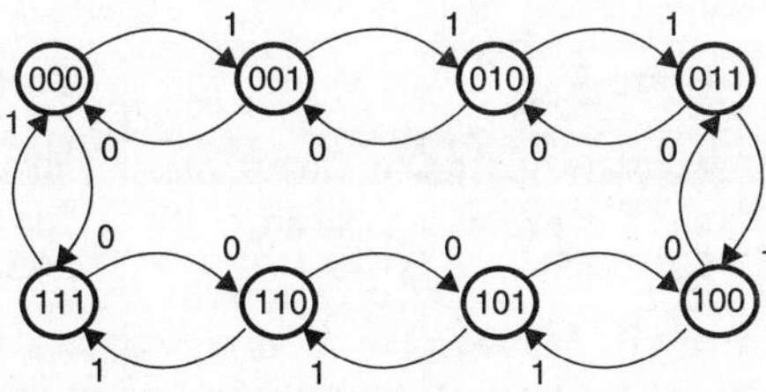

Fig. 3: State diagram for up/down count.

> ***Note:*** *Mode control can be changed at any state (Need not be changed only at initial state). If mode is changed from 1 to 0, when circuit is in state 101, then it will start down count starting from 101.*

From the knowledge of state transitions in the state diagram of Fig. 3, the state table is formed as shown in Table 3.

Table 3: State Table

Present States			Mode Control	Next States		
A_2	A_1	A_0	M	$A_2(t+1)$	$A_1(t+1)$	$A_0(t+1)$
0	0	0	0	1	1	1
0	0	0	1	0	0	1
0	0	1	0	0	0	0
0	0	1	1	0	1	0
0	1	0	0	0	0	1
0	1	0	1	0	1	1
0	1	1	0	0	1	0
0	1	1	1	1	0	0
1	0	0	0	0	1	1
1	0	0	1	1	0	1
1	0	1	0	1	0	0
1	0	1	1	1	1	0
1	1	0	0	1	0	1
1	1	0	1	1	1	1
1	1	1	0	1	1	0
1	1	1	1	0	0	0

Let us choose three JK flip-flops to store three state variables A_0, A_1 and A_2 as shown in Fig. 4. Here, A_0, A_1 and A_2 are output of flip-flops.

Excitation Table of JK Flip-Flop

Present State Q(t)	Next State Q(t + 1)	Flip-Flop Inputs J	K
0	0	0	×
0	1	1	×
1	0	×	1
1	1	×	0

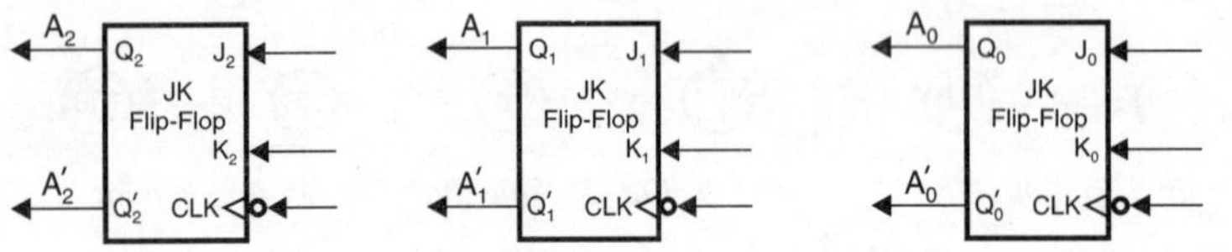

Fig. a: Flip-flop A_2. *Fig. b: Flip-flop A_1.* *Fig. c: Flip-flop A_0.*

Fig. 4: JK flip-flops to store state variables.

The flip-flop inputs can be obtained from excitation table as shown below:

State A_2

For state A_2, the Q(t), Q(t + 1), J and K of excitation table are considered as A_2, $A_2(t + 1)$, J_2 and K_2.

$$\therefore Q(t) \rightarrow A_2 \;;\; Q(t+1) \rightarrow A_2(t+1)$$

$$J \rightarrow J_2 \;;\; K \rightarrow K_2$$

The flip-flop inputs J_2 and K_2 that drives the present state A_2 to next state is obtained from excitation table and listed in Table 4.

State A_1

For state A_1, the Q(t), Q(t + 1), J and K of excitation table are considered as A_1, $A_1(t + 1)$, J_1 and K_1.

$$\therefore Q(t) \rightarrow A_1 \;;\; Q(t+1) \rightarrow A_1(t+1)$$

$$J \rightarrow J_1 \;;\; K \rightarrow K_1$$

The flip-flop inputs J_1 and K_1 that drives the present state A_1 to next state is obtained from excitation table and listed in Table 5.

State A_0

For state A_0, the Q(t), Q(t + 1), J and K of excitation table are considered as A_0, $A_0(t + 1)$, J_0 and K_0.

$$\therefore Q(t) \rightarrow A_0 \;;\; Q(t+1) \rightarrow A_0(t+1)$$

$$J \rightarrow J_0 \;;\; K \rightarrow K_0$$

The flip-flop inputs J_0 and K_0 that drives the present state A_0 to next state is obtained from excitation table and listed in Table 6.

Table 4:

A_2	$A_2(t+1)$	J_2	K_2
0	1	1	×
0	0	0	×
0	0	0	×
0	0	0	×
0	0	0	×
0	0	0	×
0	0	0	×
0	1	1	×
1	0	×	1
1	1	×	0
1	1	×	0
1	1	×	0
1	1	×	0
1	1	×	0
1	1	×	0
1	0	×	1

Table 5:

A_1	$A_1(t+1)$	J_1	K_1
0	1	1	×
0	0	0	×
0	0	0	×
0	1	1	×
1	0	×	1
1	1	×	0
1	1	×	0
1	0	×	1
0	1	1	×
0	0	0	×
0	0	0	×
0	1	1	×
1	0	×	1
1	1	×	0
1	1	×	0
1	0	×	1

Table 6:

A_0	$A_0(t+1)$	J_0	K_0
0	1	1	×
0	1	1	×
1	0	×	1
1	0	×	1
0	1	1	×
0	1	1	×
1	0	×	1
1	0	×	1
0	1	1	×
0	1	1	×
1	0	×	1
1	0	×	1
0	1	1	×
0	1	1	×
1	0	×	1
1	0	×	1

The Tables 3 to 6 are combined to form state table with flip-flop inputs as shown in Table 7.

Table 7: Expanded State Table

Present States			Mode Control	Next States			Flip-Flop Inputs						Minterm
A_2	A_1	A_0	M	$A_2(t+1)$	$A_1(t+1)$	$A_0(t+1)$	J_2	K_2	J_1	K_1	J_0	K_0	$A_2A_1A_0M$
0	0	0	0	1	1	1	1	×	1	×	1	×	m_0
0	0	0	1	0	0	1	0	×	0	×	1	×	m_1
0	0	1	0	0	0	0	0	×	0	×	×	1	m_2
0	0	1	1	0	1	0	0	×	1	×	×	1	m_3
0	1	0	0	0	0	1	0	×	×	1	1	×	m_4
0	1	0	1	0	1	1	0	×	×	0	1	×	m_5
0	1	1	0	0	1	0	0	×	×	0	×	1	m_6
0	1	1	1	1	0	0	1	×	×	1	×	1	m_7
1	0	0	0	0	1	1	×	1	1	×	1	×	m_8
1	0	0	1	1	0	1	×	0	0	×	1	×	m_9
1	0	1	0	1	0	0	×	0	0	×	×	1	m_{10}
1	0	1	1	1	1	0	×	0	1	×	×	1	m_{11}
1	1	0	0	1	0	1	×	0	×	1	1	×	m_{12}
1	1	0	1	1	1	1	×	0	×	0	1	×	m_{13}
1	1	1	0	1	1	0	×	0	×	0	×	1	m_{14}
1	1	1	1	0	0	0	×	1	×	1	×	1	m_{15}

Note: *Here, present state $A_2 A_1 A_0$ is the counter output.*

The flip-flop inputs are functions of present states and mode control bit, M. Therefore, minterms are formed using present states and M and the Boolean equations for flip-flop inputs are obtained as sum of minterms and don't cares as shown below:

$J_2 = \Sigma m(0, 7) + \Sigma d(8, 9, 10, 11, 12, 13, 14, 15)$

$J_1 = \Sigma m(0, 3, 8, 11) + \Sigma d(4, 5, 6, 7, 12, 13, 14, 15)$

$J_0 = \Sigma m(0, 1, 4, 5, 8, 9, 12, 13) + \Sigma d(2, 3, 6, 7, 10, 11, 14, 15)$

$K_2 = \Sigma m(8, 15) + \Sigma d(0, 1, 2, 3, 4, 5, 6, 7)$

$K_1 = \Sigma m(4, 7, 12, 15) + \Sigma d(0, 1, 2, 3, 8, 9, 10, 11)$

$K_0 = \Sigma m(2, 3, 6, 7, 10, 11, 14, 15) + \Sigma d(0, 1, 4, 5, 8, 9, 12, 13)$

The simplified Boolean equations for flip-flop inputs are obtained by constructing K-map using the above equations as shown in Figs. 5 to 8.

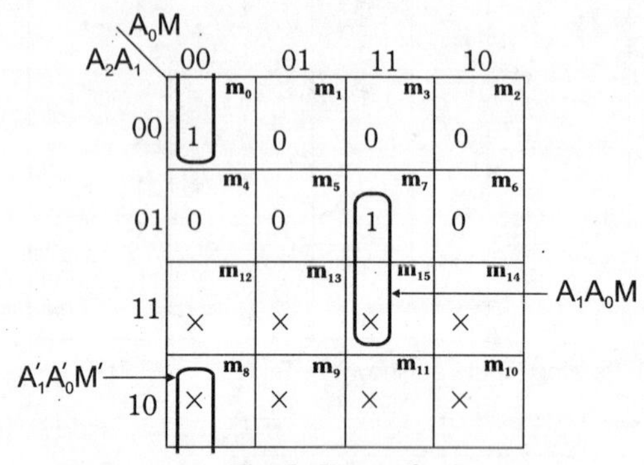

Fig. 5: K-map for J_2.

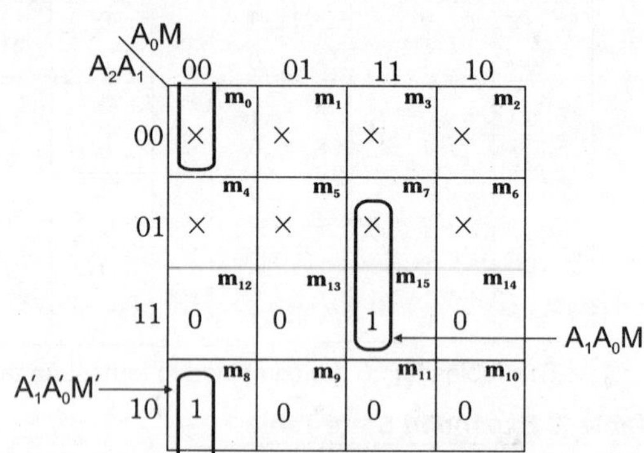

Fig. 6: K-map for K_2.

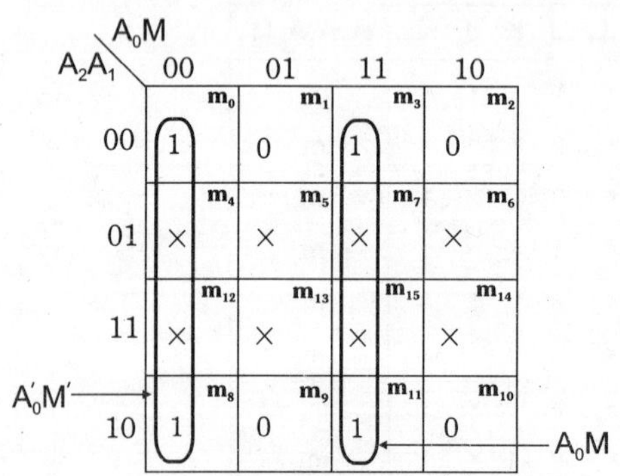

Fig. 7: K-map for J_1.

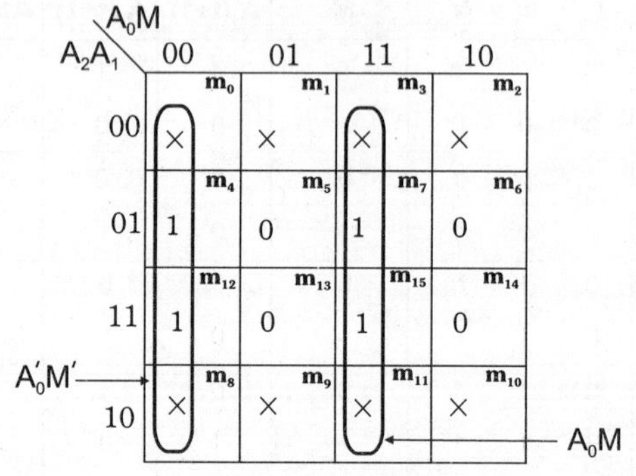

Fig. 8: K-map for K_1.

Here, J_0 and K_0 are always either 1 or don't-care and so no need of K-map and they can be permanently tied to 1.

$\therefore J_0 = K_0 = 1$

From the K-map we get the following Boolean equations for flip-flop inputs.

$J_1 = K_1 = A_0 M + A_0' M'$

$J_2 = K_2 = A_1 A_0 M + A_1' A_0' M'$

Using the above Boolean equations the logic diagram of 3-bit counter is drawn as shown in Fig. 9.

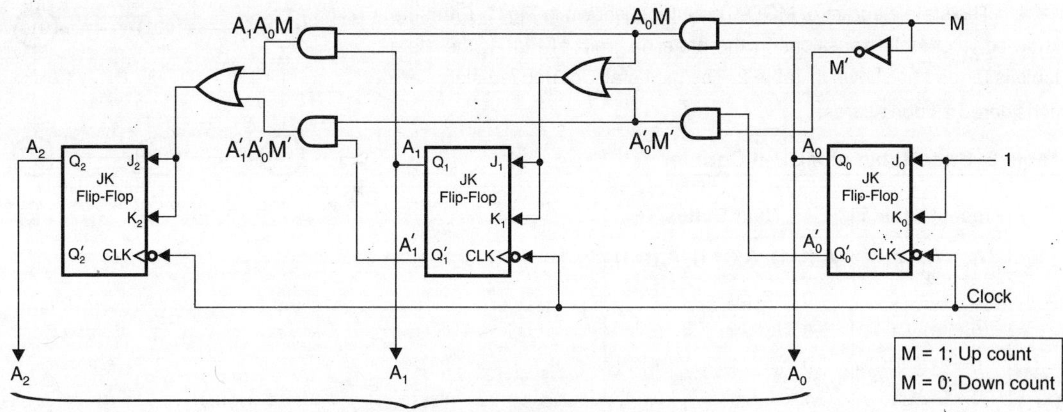

Fig. 9: Logic diagram of 3-bit synchronous up/down counter.

Example 8.10

(CSVTU, Nov/Dec'19, 7 Marks) (JNTUK, Feb'22, 8 Marks)

Design a synchronous MOD-6 counter using JK flip-flops.

Solution

MOD-6 is a binary counter that count from 0 to 5 and on reaching 6 the output should be reset to zero, in order to start next count sequence.

In order to count from 0 to 5 we need 3-bit binary output.

The three bits are represented by three state variables A_2, A_1 and A_0.

$\therefore$ Present state variables : $A_2(t), A_1(t), A_0(t) \Rightarrow A_2, A_1, A_0$ (For simplicity)

Next state variables : $A_2(t+1), A_1(t+1), A_0(t+1)$

With 3 state variables it is possible to form 8 states ($2^3 = 8$).

The 8 states are 000, 001, 010, 011, 100, 101, 110 and 111.

In this, only first 6 states are used by MOD-6 counter and the last two are unused states. The truth table of MOD-6 counter is shown in Table 1. Here, the present states are the counter outputs.

Table 1: Truth Table of MOD-6 Counter

Clock	Counter Outputs		
	A_2	A_1	A_0
-	0	0	0
↓	0	0	1
↓	0	1	0
↓	0	1	1
↓	1	0	0
↓	1	0	1
↓	0	0	0

The state diagram of MOD-6 counter is shown in Fig. 1. From the knowledge of state transitions in the state diagram of Fig. 1, the state table is formed as shown in Table 2. The next state for unused states are considered as don't-cares.

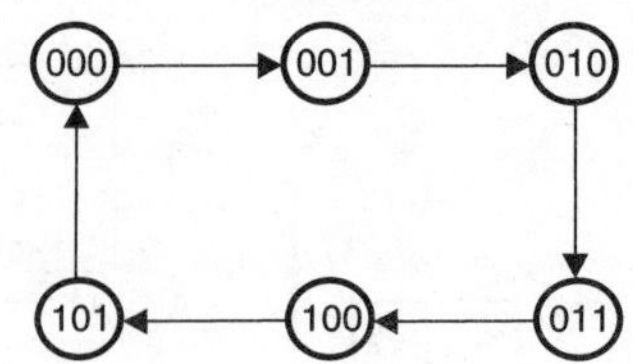

Fig. 1: State diagram of MOD-6 counter.

Table 2: State Table of MOD-6 Counter

Present States			Next States		
A_2	A_1	A_0	$A_2(t+1)$	$A_1(t+1)$	$A_0(t+1)$
0	0	0	0	0	1
0	0	1	0	1	0
0	1	0	0	1	1
0	1	1	1	0	0
1	0	0	1	0	1
1	0	1	0	0	0
1	1	0	×	×	×
1	1	1	×	×	×

Let us choose three JK flip-flops to store three state variables A_0, A_1 and A_2 as shown in Fig. 2. Here, A_0, A_1 and A_2 be output of flip-flops.

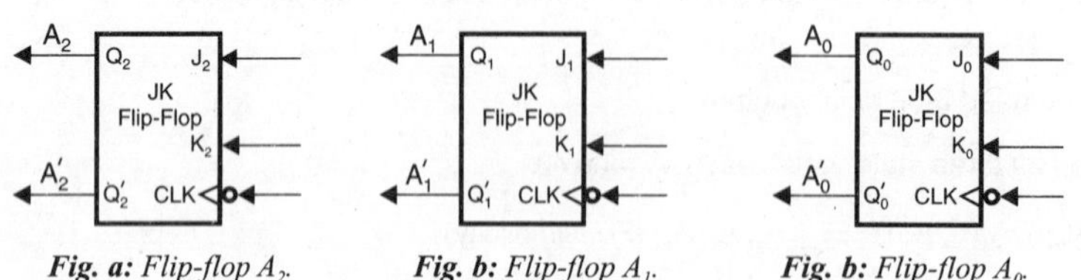

Fig. a: Flip-flop A_2. *Fig. b: Flip-flop A_1.* *Fig. b: Flip-flop A_0.*

Fig. 2: JK flip-flops to store state variables.

Excitation Table of JK Flip-Flop

Present State	Next State	Flip-Flop Inputs	
Q(t)	Q(t + 1)	J	K
0	0	0	×
0	1	1	×
1	0	×	1
1	1	×	0

The flip-flop inputs can be obtained from excitation table as shown below:

State A_2

For state A_2, the Q(t), Q(t + 1), J and K of excitation table are K of considered as A_2, $A_2(t + 1)$, J_2 and K_2.

$$\therefore Q(t) \rightarrow A_2 \;\; ; \;\; Q(t+1) \rightarrow A_2(t+1)$$

$$J \rightarrow J_2 \;\; ; \;\; K \rightarrow K_2$$

The flip-flop inputs J_2 and K_2 that drives the present state A_2 to next state is obtained from excitation table and listed in Table 3.

State A_1

For state A_1, the Q(t), Q(t + 1), J and excitation table are considered as A_1, $A_1(t + 1)$, J_1 and K_1.

$$\therefore Q(t) \rightarrow A_1 \;\; ; \;\; Q(t+1) \rightarrow A_1(t+1)$$

$$J \rightarrow J_1 \;\; ; \;\; K \rightarrow K_1$$

The flip-flop inputs J_1 and K_1 that drives the present state A_1 to next state is obtained from excitation table and listed in Table 4.

State A_0

For state A_0, the Q(t), Q(t + 1), J and K of excitation table are considered as A_0, $A_0(t + 1)$, J_0 and K_0.

$$\therefore Q(t) \rightarrow A_0 \;\; ; \;\; Q(t+1) \rightarrow A_0(t+1)$$

$$J \rightarrow J_0 \;\; ; \;\; K \rightarrow K_0$$

The flip-flop inputs J_0 and K_0 that drives the present state A_0 to next state is obtained from excitation table and listed in Table 5.

Table 3:

A_2	$A_2(t+1)$	J_2	K_2
0	0	0	×
0	0	0	×
0	0	0	×
0	1	1	×
1	1	×	0
1	0	×	1

Table 4:

A_1	$A_1(t+1)$	J_1	K_1
0	0	0	×
0	1	1	×
1	1	×	0
1	0	×	1
0	0	0	×
0	0	0	×

Table 5:

A_0	$A_0(t+1)$	J_0	K_0
0	1	1	×
1	0	×	1
0	1	1	×
1	0	×	1
0	1	1	×
1	0	×	1

The tables 2 to 5 are combined to form state table with flip-flop inputs as shown in Table 6.

Table 6: Expanded State Table

Present States			Next States			Flip-Flop Inputs						Minterm
A_2	A_1	A_0	$A_2(t+1)$	$A_1(t+1)$	$A_0(t+1)$	J_2	K_2	J_1	K_1	J_0	K_0	$A_2A_1A_0$
0	0	0	0	0	1	0	×	0	×	1	×	m_0
0	0	1	0	1	0	0	×	1	×	×	1	m_1
0	1	0	0	1	1	0	×	×	0	1	×	m_2
0	1	1	1	0	0	1	×	×	1	×	1	m_3
1	0	0	1	0	1	×	0	0	×	1	×	m_4
1	0	1	0	0	0	×	1	0	×	×	1	m_5
1	1	0	×	×	×	×	×	×	×	×	×	m_6
1	1	1	×	×	×	×	×	×	×	×	×	m_7

The flip-flop inputs are functions of present states and so minterms are formed using present states and the Boolean equations for flip-flop inputs are obtained as sum of minterms and don't-cares as shown below:

$$J_2 = m_3 + \Sigma d(4, 5, 6, 7)$$

$$K_2 = m_5 + \Sigma d(0, 1, 2, 3, 6, 7)$$

$$J_1 = m_1 + \Sigma d(2, 3, 6, 7)$$

$$K_1 = m_3 + \Sigma d(0, 1, 4, 5, 6, 7)$$

$$J_0 = \Sigma m(0, 2, 4) + \Sigma d(1, 3, 5, 6, 7)$$

$$K_0 = \Sigma m(1, 3, 5) + \Sigma d(0, 2, 4, 6, 7)$$

The simplified Boolean equations for flip-flop inputs are determined by constructing K-maps using the above equations as shown in Figs. 3 to 6.

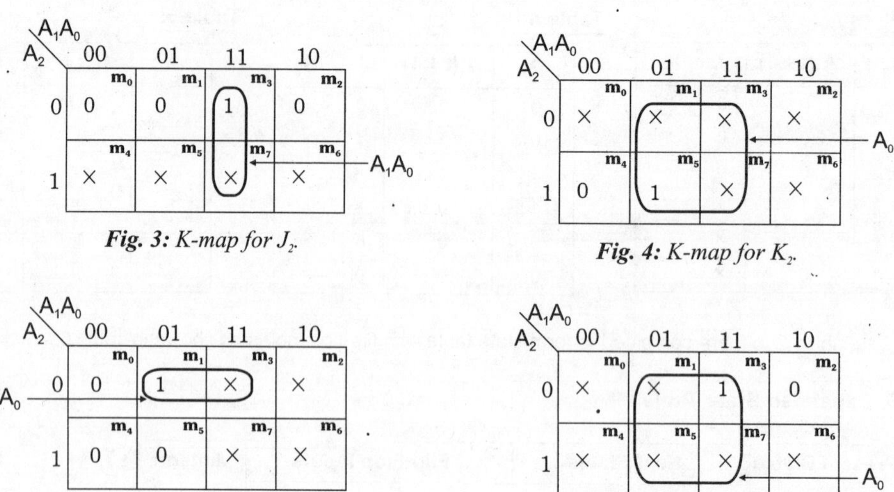

Fig. 3: K-map for J_2.

Fig. 4: K-map for K_2.

Fig. 5: K-map for J_1.

Fig. 6: K-map for K_1.

Here, J_0 and K_0 are always either 1 or don't-cares and so no need of K-map and they can be permanently tied to 1.

$\therefore\ J_0 = K_0 = 1$

From the K-map we get the following Boolean equations for flip-flop inputs and using these equation the logic circuit of MOD-6 counter is drawn as shown in Fig. 7.

$J_1 = A_2'A_0$	$K_1 = A_0$
$J_2 = A_1A_0$	$K_2 = A_0$

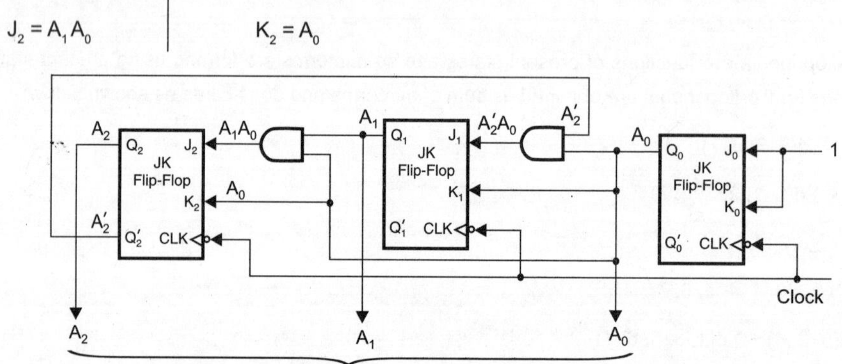

Fig. 7: Logic diagram of MOD-6 synchronous counter.

Check for Self Starting

In order to check for self starting, the flip-flop inputs are determined for each unused state using simplified equations of flip-flop inputs and listed in Table 7. The next state is obtained using characteristic table of JK flip-flop.

Table 7: Check for Self Starting

Unused Present States			Flip-Flop Inputs						Next States		
			J_2	K_2	J_1	K_1	J_0	K_0	$A_2(t+1)$	$A_1(t+1)$	$A_0(t+1)$
A_2	A_1	A_0	$=A_2'A_0$	$=A_0$	$=A_1A_0$	$=A_0$	$=1$	$=1$			
1	1	0	0	0	0	0	1	1	1	1	1
1	1	1	0	1	1	1	1	1	0	0	0

Characteristic Table of JK Flip-Flop

Input J K	Present State Q	Next State Q(t+1)
0 0	0	0
0 0	1	1
0 1	0	0
0 1	1	0
1 0	0	1
1 0	1	1
1 1	0	1
1 1	1	0

From the Table 7 we can observe the following state transitions from unused state to next state.

$$110 \rightarrow 111 \rightarrow 000 \;;\; 111 \rightarrow 000$$

The unused states enters used state in the next transition as shown above. Hence the counter is self-starting.

Example 8.11

(VTU, Aug/Sep'20, 8 Marks)

Design a synchronous counter using D flip-flop to count the sequence 0-2-3-6-5-1.

Solution

The sequential circuit has to detect decimal values in the range 0 to 7. Let us assign binary value of decimal number as code for states. Hence we need 3-bit binary code to represent the decimal sequence. With 3-bit code 8 states ($2^3 = 8$) all possible and in this 6 states are used states 2 states are unused states.

Binary code assignment

0	000
1	001
2	010
3	011
5	101
6	110

Unused binary code

4	100
7	111

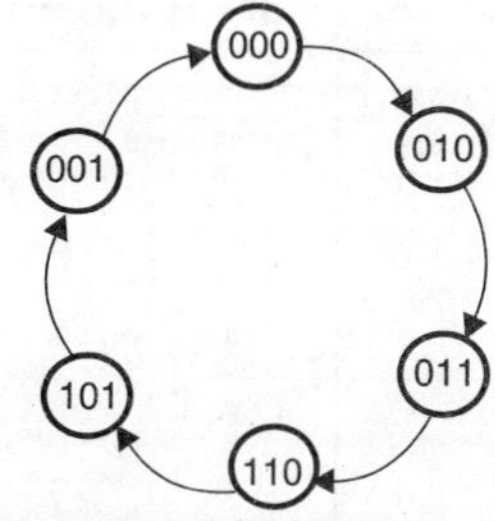

Fig. 1: State diagram with only used states.

In order to count the given sequence 3-bit binary counter is needed and let us represent the three bits by three state variables A_2, A_1 and A_0.

∴ Present state variables : $A_2(t), A_1(t), A_0(t)$

⇓

A_2, A_1, A_0 (For simplicity)

Next state variables : $A_2(t+1), A_1(t+1), A_0(t+1)$

With 3 state variables it is possible to form 8 ($2^3 = 8$) states.

The 8 states are 000, 001, 010, 011, 100, 101, 110 and 111.

In this 6 states are used by the counter and the 2 states are unused.

The state diagram of the sequence counter is shown in Fig. 1. From the knowledge of state transitions in the state diagram of Fig. 1, the state table is formed as shown in Table 1. The next state for unused states are considered as don't-cares.

Table 1: State Table

Present States			Next States		
A_2	A_1	A_0	$A_2(t+1)$	$A_1(t+1)$	$A_0(t+1)$
0	0	0	0	1	0
0	0	1	0	0	0
0	1	0	0	1	1
0	1	1	1	1	0
1	0	0	×	×	×
1	0	1	0	0	1
1	1	0	1	0	1
1	1	1	×	×	×

Let us choose three D flip-flops to store three state variables A_0, A_1 and A_2 as shown in Fig. 2. Here, A_0, A_1 and A_2 are output of flip-flops.

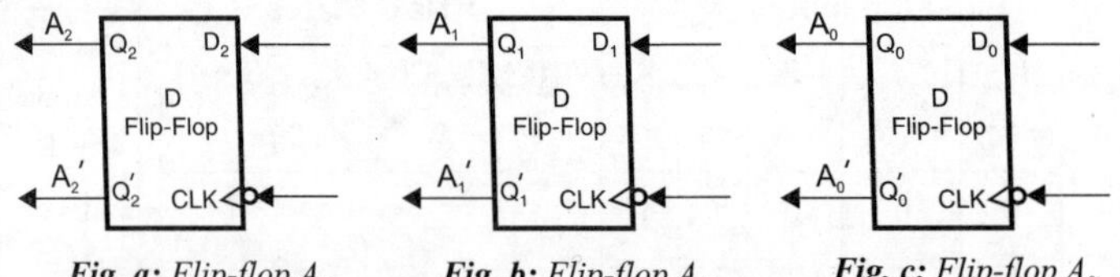

Fig. a: Flip-flop A_2. *Fig. b: Flip-flop A_1.* *Fig. c: Flip-flop A_0.*

Fig. 2: D flip-flop to store state variables.

Here, $A_0 = Q_0$ $\quad A_0(t + 1) = Q_0(t + 1)$

$A_1 = Q_1$ $\quad A_1(t + 1) = Q_1(t + 1)$

$A_2 = Q_2$ $\quad A_2(t + 1) = Q_2(t + 1)$

The characteristic equation of four D flip-flop are,

$$Q_0(t + 1) = D_0$$
$$Q_1(t + 1) = D_1$$
$$Q_2(t + 1) = D_2$$

Characteristic equation of D flip-flop,
$Q(t + 1) = D$

Using the above equations we can write,

$$A_2(t + 1) = D_2 \ ; \ A_1(t + 1) = D_1 \ ; \ A_0(t + 1) = D_0$$

Using the above equations the flip-flop inputs are determined for the next state and listed in Table 2. Here, the flip-flop inputs are same as next state outputs.

Table 2: Expanded State Table

Present States			Next States			Flip-Flop Inputs			Minterm
A_2	A_1	A_0	$A_2(t+1)$	$A_1(t+1)$	$A_0(t+1)$	D_2	D_1	D_0	$A_2A_1A_0$
0	0	0	0	1	0	0	1	0	m_0
0	0	1	0	0	0	0	0	0	m_1
0	1	0	0	1	1	0	1	1	m_2
0	1	1	1	1	0	1	1	0	m_3
1	0	0	×	×	×	×	×	×	m_4
1	0	1	0	0	1	0	0	1	m_5
1	1	0	1	0	1	1	0	1	m_6
1	1	1	×	×	×	×	×	×	m_7

***Note**: Here, present state $A_2 A_1 A_0$ is the sequence detector output.*

The Boolean equations for D_2, D_1 and D_0 are obtained from Table 2 by taking present state as minterms.

$$D_2 = \Sigma m(3, 6) + \Sigma d(4, 7)$$

$$D_1 = \Sigma m(0, 2, 3) + \Sigma d(4, 7)$$

$$D_0 = \Sigma m(2, 5, 6) + \Sigma d(4, 7)$$

The simplified Boolean equations are obtained by constructing K-map using the above equation for D_2, D_1 and D_0 as shown in Figs. 3 to 5.

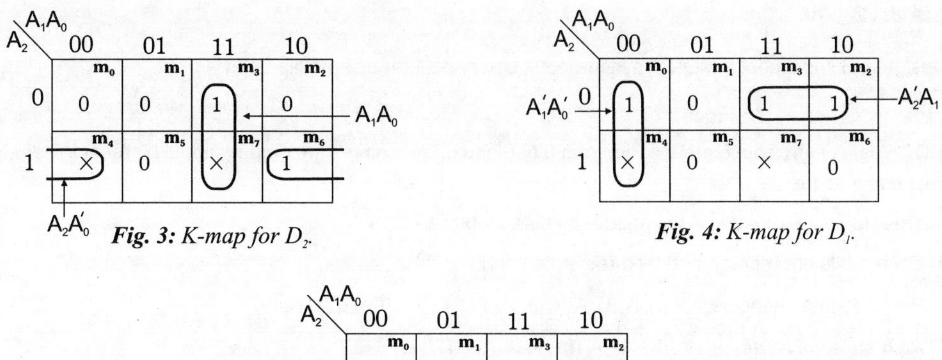

***Fig. 3:** K-map for D_2.*

***Fig. 4:** K-map for D_1.*

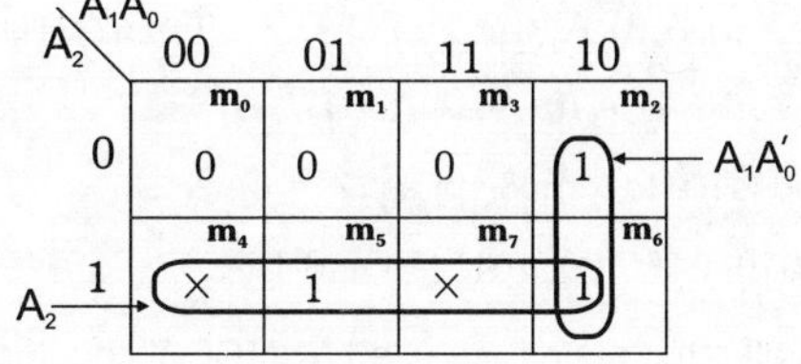

***Fig. 5:** K-map for D_0.*

From the K-map we get the following simplified Boolean equations to realize the synchronous sequence counter using D flip-flops and they are used to realize the sequence counter as shown in Fig. 6.

$$D_2 = A_1A_0 + A_2A_0' \quad ; \quad D_1 = A_1'A_0' + A_2'A_1 \quad ; \quad D_0 = A_2 + A_1A_0'$$

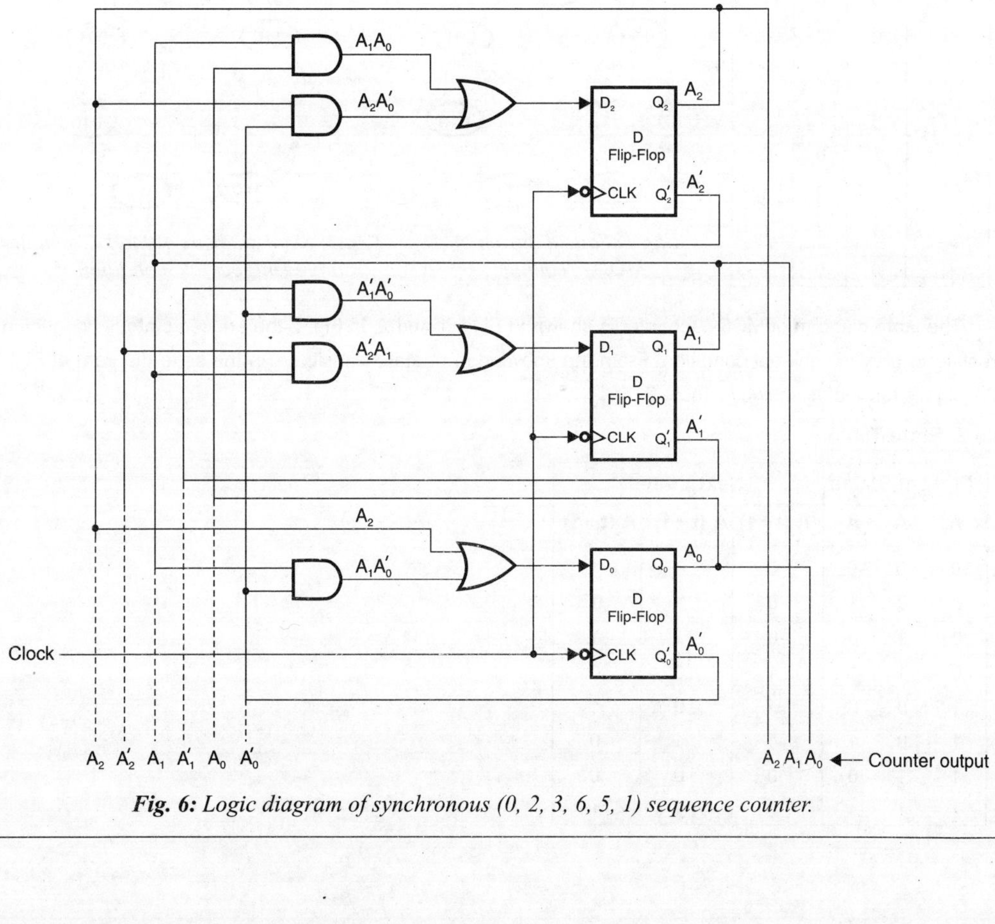

***Fig. 6:** Logic diagram of synchronous (0, 2, 3, 6, 5, 1) sequence counter.*

Example 8.12

Design and implement a MOD-7 synchronous binary counter using T flip-flops.

Solution

MOD-7 is a binary counter that count from 0 to 6 and on reaching 7 the output should be reset to zero, in order to start next count sequence.

In order to count from 0 to 6 we need 3-bit binary output.

The three bits are represented by three state variables A_2, A_1 and A_0.

$\therefore$ Present state variables : $A_2(t), A_1(t), A_0(t) \Rightarrow A_2, A_1, A_0$ (For simplicity)

Next state variables : $A_2(t+1), A_1(t+1), A_0(t+1)$

With 3 state variables it is possible to form 8 states ($2^3 = 8$).

The 8 states are 000, 001, 010, 011, 100, 101, 110 and 111.

In this except the state 111 all other states are used by MOD-7 counter. The truth table of MOD-7 counter is shown in Table 1. Here, the present states are the counter outputs.

Table 1: Truth Table of MOD-7 Counter

Clock	Counter Outputs		
	A_2	A_1	A_0
-	0	0	0
↓	0	0	1
↓	0	1	0
↓	0	1	1
↓	1	0	0
↓	1	0	1
↓	1	1	0
↓	0	0	0

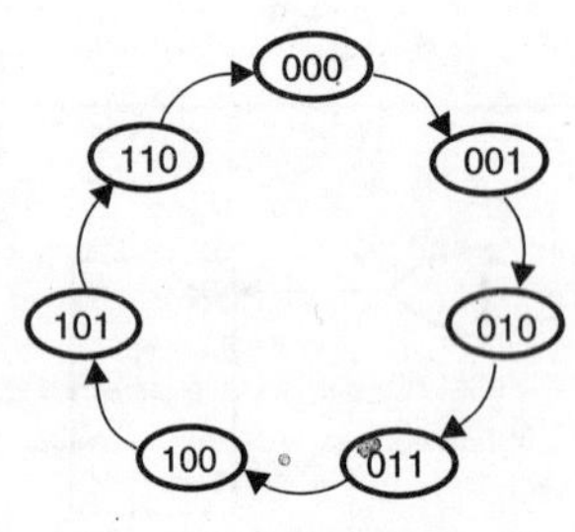

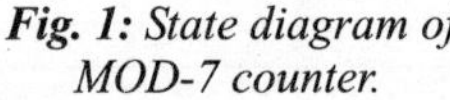

Fig. 1: State diagram of MOD-7 counter.

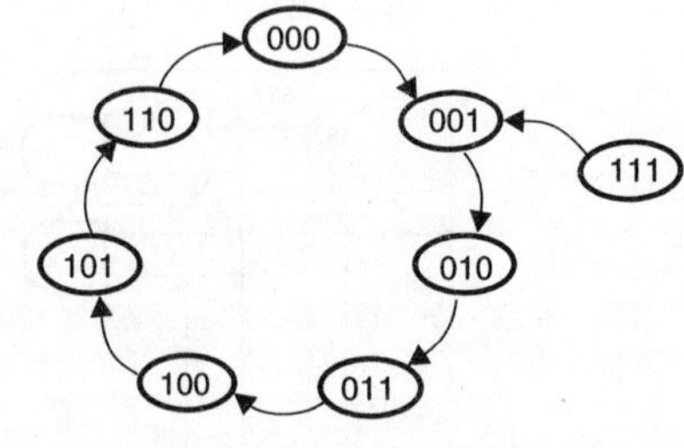

Fig. 2: State diagram of MOD-7 counter to prevent lock-out condition.

The state diagram of MOD-7 counter is shown in Figs. 1 and 2. In Fig. 2 the unused state is connected to an used state to prevent lock-out condition. From the knowledge of state transitions in the state diagram of Fig. 2, the stable table is formed as shown in Table 2.

Table 2: State Table

Present States			Next States		
A_2	A_1	A_0	$A_2(t+1)$	$A_1(t+1)$	$A_0(t+1)$
0	0	0	0	0	1
0	0	1	0	1	0
0	1	0	0	1	1
0	1	1	1	0	0
1	0	0	1	0	1
1	0	1	1	1	0
1	1	0	0	0	0
1	1	1	0	0	1

Let us choose three T flip-flops to store three state variables A_0, A_1 and A_2 as shown in Fig. 3. Here, A_0, A_1 and A_2 are output of T flip-flops.

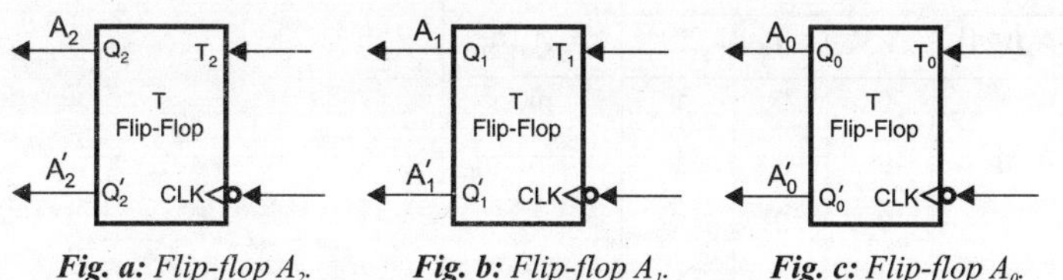

Fig. a: Flip-flop A_2. *Fig. b: Flip-flop A_1.* *Fig. c: Flip-flop A_0.*

***Fig. 3:** T flip-flop to store state variables.*

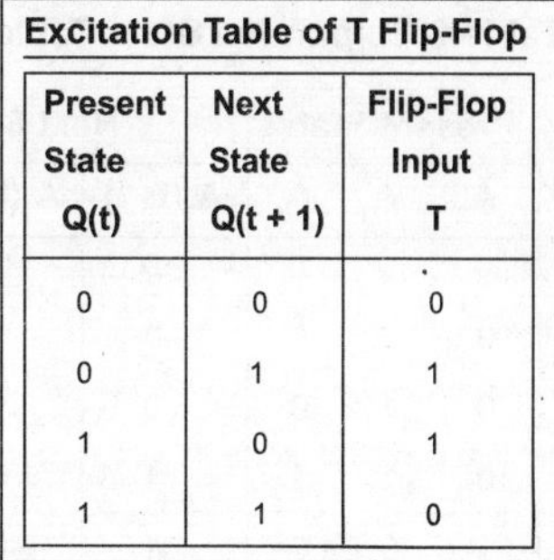

Excitation Table of T Flip-Flop

Present State Q(t)	Next State Q(t + 1)	Flip-Flop Input T
0	0	0
0	1	1
1	0	1
1	1	0

The flip-flop inputs can be determined from excitation table of T flip-flop as shown below:

State A_2

For state A_2, the Q(t), Q(t + 1) and T of excitation table are considered as A_2, $A_2(t + 1)$ and T_2.

$\therefore\ Q(t) \rightarrow A_2\ ;\ Q(t+1) \rightarrow A_2(t+1)$

$T \rightarrow T_2$

The flip-flop input T_2 that drives the present state A_2 to next states is obtained from excitation table and listed in Table 3.

Table 3:

A_2	$A_2(t+1)$	T_2
0	0	0
0	0	0
0	0	0
0	1	1
1	1	0
1	1	0
1	0	1
1	0	1

State A_1

For state A_1, the Q, Q(t + 1) and T of excitation table are considered as A_1, $A_1(t + 1)$ and T_1.

$\therefore\ Q(t) \rightarrow A_1\ ;\ Q(t+1) \rightarrow A_1(t+1)$

$T \rightarrow T_1$

The flip-flop input T_1 that drives the present state A_1 to next states is obtained from excitation table and listed in Table 4.

Table 4:

A_1	$A_1(t+1)$	T_1
0	0	0
0	1	1
1	1	0
1	0	1
0	0	0
0	1	1
1	0	1
1	0	1

State A_0

For state A_0, the Q(t), Q(t + 1) and T of excitation table are considered as A_0, $A_0(t + 1)$ and T_0.

$\therefore\ Q(t) \rightarrow A_0\ ;\ Q(t+1) \rightarrow A_0(t+1)$

$T \rightarrow T_0$

The flip-flop input T_0 that drives the present state A_0 to next states is obtained from excitation table and listed in Table 5.

Table 5:

A_0	$A_0(t+1)$	T_0
0	1	1
1	0	1
0	1	1
1	0	1
0	1	1
1	0	1
0	0	0
1	1	0

The tables 2 to 5 are combined to form expanded state table with flip-flop inputs as shown in Table 6.

Table 6: Expanded State Table

Present States			Next States			Flip-Flop Inputs			Minterm
A_2	A_1	A_0	$A_2(t+1)$	$A_1(t+1)$	$A_0(t+1)$	T_2	T_1	T_0	$A_2A_1A_0$
0	0	0	0	0	1	0	0	1	m_0
0	0	1	0	1	0	0	1	1	m_1
0	1	0	0	1	1	0	0	1	m_2
0	1	1	1	0	0	1	1	1	m_3
1	0	0	1	0	1	0	0	1	m_4
1	0	1	1	1	0	0	1	1	m_5
1	1	0	0	0	0	1	1	0	m_6
1	1	1	0	0	1	1	1	0	m_7

***Note:** Here, present state $A_2\,A_1\,A_0$ is the counter output.*

The flip-flop inputs are functions of present states. Hence, minterms are formed using present states and the Boolean equations for T_2, T_1 and T_0 are obtained from Table 6 as sum as minterms as shown below:

$$T_2 = \Sigma m(3, 6, 7)$$

$$T_1 = \Sigma m(1, 3, 5, 6, 7)$$

$$T_0 = \Sigma m(0, 1, 2, 3, 4, 5)$$

The simplified Boolean equations are obtained by constructing K-map for T_2, T_1 and T_0 as shown in Figs. 4 to 6.

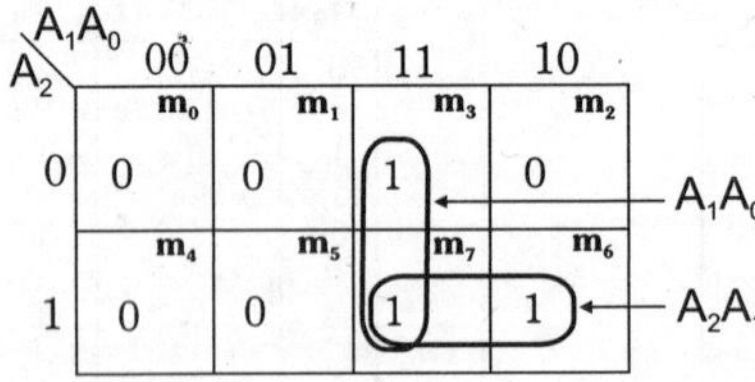

Fig. 4: K-map for T_2.

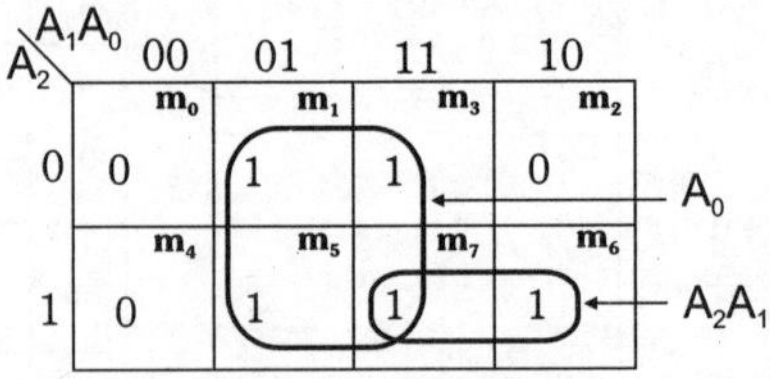

Fig. 5: K-map for T_1.

From the K-map we get the following simplified Boolean equations to realize 4-bit binary counter using T flip-flops and they are used to construct the logic circuit of synchronous MOD-7 counter as shown in Fig. 7.

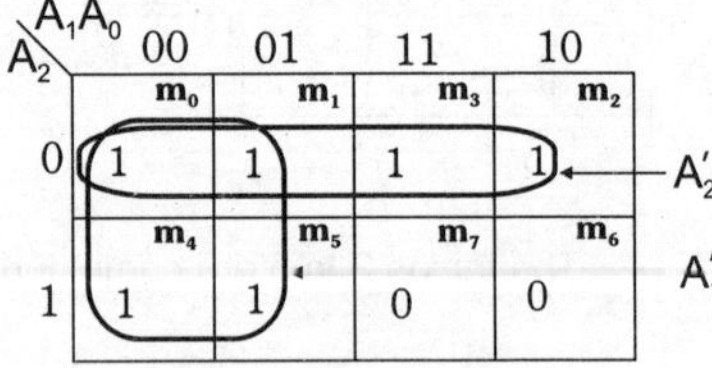

Fig. 6: K-map for T_0.

$$T_2 = A_2A_1 + A_1A_0$$

$$T_1 = A_2A_1 + A_0$$

$$T_0 = A_2' + A_1'$$

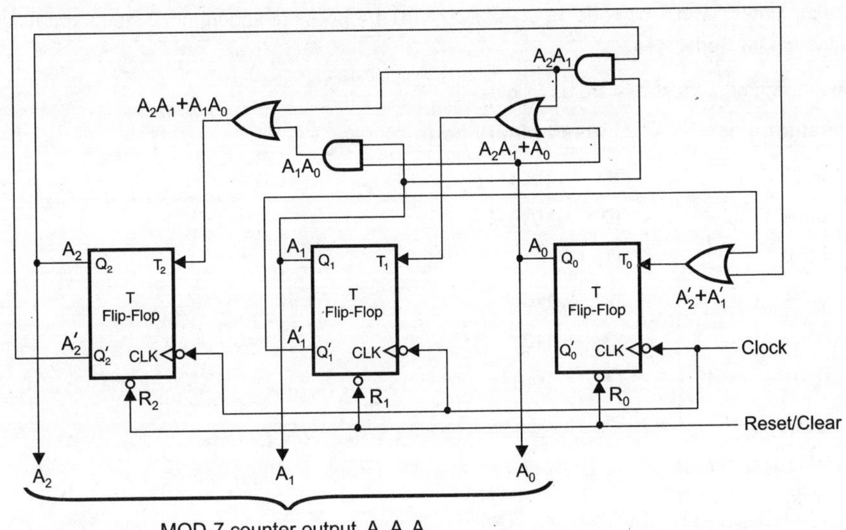

Fig. 7: Logic diagram of MOD-7 synchronous counter using T flip-flop.

Check for Lock-out Condition

In order to check lock-out condition the flip-flop inputs are determined for each unused state using simplified equation of flip-flop inputs and listed in Table 7. The next state are obtained from characteristic table of T flip-flop.

Table 7: Check for Lock-out Condition

Unused Present States			Flip-Flop Inputs			Next States		
			T_2	T_1	T_0	$A_2(t+1)$	$A_1(t+1)$	$A_0(t+1)$
A_2	A_1	A_0	$= A_2A_1 + A_1A_0$	$= A_2A_1 + A_0$	$= A_2' + A_1'$			
1	1	1	1	1	0	0	0	1

Characteristic Table of T Flip-Flop

Input T	Present State Q	Next State Q(t+1)
0	0	0
0	1	1
1	0	1
1	1	0

From the Table 7 we can observe the following state transition from unused state to next state.

$111 \rightarrow 001$

The unused state enter an used state in the next transition and so the counter is self starting.

Example 8.13

(AKUB, Dec'20, 8 Marks)

Design a MOD-9 counter using T flip-flops.

Solution

MOD-9 is a binary counter that count from 0 to 8_{10} and on reaching 9_{10} the output should be reset to zero in order to start next count sequence.

In order to count from 0 to 8_{10} we need 4-bit binary output.

Hence we need 4 bit binary code to represent the decimal sequence. Let us assign binary value of decimal number as code for states. With 4-bit code 16 states ($2^4 = 16$) are possible and in this 9 states are used states and remaining 7 states are unused states.

The code assignment for states are as follows,

Binary code assignment		Unused binary code	
0	0000	9	1001
1	0001	10	1010
2	0010	11	1011
3	0011	12	1100
4	0100	13	1101
5	0101	14	1110
6	0110	15	1111
7	0111		
8	1000		

The four bits of binary codes are represented by four state variables A_3, A_2, A_1 and A_0.

$\therefore$ Present state variables : $A_3(t), A_2(t), A_1(t), A_0(t) \Rightarrow A_3, A_2, A_1, A_0$ (For simplicity)

Next state variables : $A_3(t + 1), A_2(t + 1), A_1(t + 1), A_0(t + 1)$

The truth table of MOD-9 counter is shown in Table 1. Here, the present states are counter outputs. The state diagram using only used states is shown in Fig. 1.

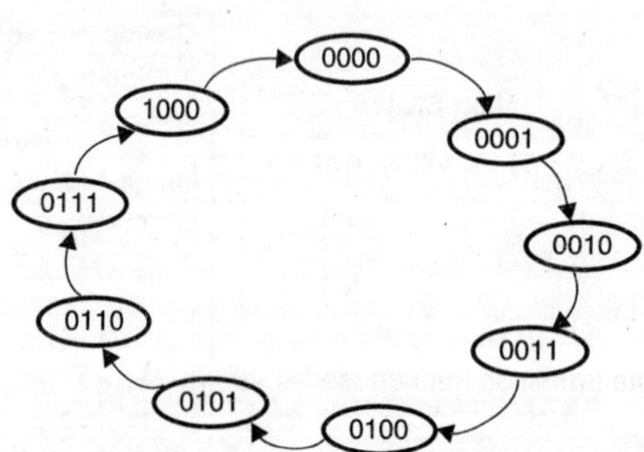

Fig. 1: State diagram with used state.

Table 1: Truth Table of MOD-9 Counter

Clock	Counter Outputs			
	A_3	A_2	A_1	A_0
-	0	0	0	0
↓	0	0	0	1
↓	0	0	1	0
↓	0	0	1	1
↓	0	1	0	0
↓	0	1	0	1
↓	0	1	1	0
↓	0	1	1	1
↓	1	0	0	0
↓	0	0	0	0

From the knowledge of state transitions in the state diagram of Fig. 1, the state table is constructed as shown in Table 2. Here, the next state for unused states are considered as don't-cares.

Table 2: State Table

Present States				Next States			
A_3	A_2	A_1	A_0	$A_3(t+1)$	$A_2(t+1)$	$A_1(t+1)$	$A_0(t+1)$
0	0	0	0	0	0	0	1
0	0	0	1	0	0	1	0
0	0	1	0	0	0	1	1
0	0	1	1	0	1	0	0
0	1	0	0	0	1	0	1
0	1	0	1	0	1	1	0
0	1	1	0	0	1	1	1
0	1	1	1	1	0	0	0
1	0	0	0	0	0	0	0
1	0	0	1	×	×	×	×
1	0	1	0	×	×	×	×
1	0	1	1	×	×	×	×
1	1	0	0	×	×	×	×
1	1	0	1	×	×	×	×
1	1	1	0	×	×	×	×
1	1	1	1	×	×	×	×

Let us choose four T flip-flops to store four state variables A_0, A_1, A_2 and A_3 as shown in Fig. 2. Here, A_0, A_1, A_2 and A_3 are output of T flip-flops.

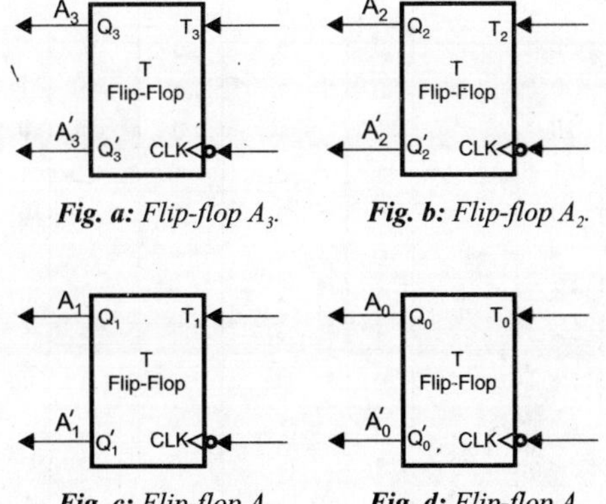

Fig. a: Flip-flop A_3. *Fig. b: Flip-flop A_2.*

Fig. c: Flip-flop A_1. *Fig. d: Flip-flop A_0.*

***Fig. 2:** T flip-flop to store state variables.*

Excitation Table of T Flip-Flop

Present State Q(t)	Next State Q(t + 1)	Flip-Flop Input T
0	0	0
0	1	1
1	0	1
1	1	0

The flip-flop inputs are determined using excitation table of T flip-flop as shown below:

State A_3

For state A_3, the Q(t), Q(t + 1) and T of excitation table are considered as A_3, $A_3(t + 1)$ and T_3.

$\therefore Q(t) \rightarrow A_3 \;;\; Q(t+1) \rightarrow A_3(t+1) \;;\; T \rightarrow T_3$

The flip-flop input T_3 that drives the present state A_3 to next states is obtained from excitation table and listed in Table 3.

State A_2

For state A_2, the Q(t), Q(t + 1) and T of excitation table are considered as A_2, $A_2(t + 1)$ and T_2.

$\therefore Q(t) \rightarrow A_2 \;;\; Q(t+1) \rightarrow A_2(t+1) \;;\; T \rightarrow T_2$

The flip-flop input T_2 that drives the present state A_2 to next states is obtained from excitation table and listed in Table 4.

State A_1

For state A_1, the Q(t), Q(t + 1) and T of excitation table are considered as A_1, $A_1(t + 1)$ and T_1.

$\therefore Q(t) \rightarrow A_1 \;;\; Q(t+1) \rightarrow A_1(t+1) \;;\; T \rightarrow T_1$

The flip-flop input T_1 that drives the present state A_1 to next states is obtained from excitation table and listed in Table 5.

State A_0

For state A_0, the Q(t), Q(t + 1) and T of excitation table are considered as A_0, $A_0(t + 1)$ and T_0.

$\therefore Q(t) \rightarrow A_0 \;;\; Q(t+1) \rightarrow A_0(t+1) \;;\; T \rightarrow T_0$

The flip-flop input T_0 that drives the present state A_0 to next states is obtained from excitation table and listed in Table 6.

Table 3:

A_3	$A_3(t+1)$	T_3
0	0	0
0	0	0
0	0	0
0	0	0
0	0	0
0	0	0
0	0	0
0	1	1
1	0	1
1	×	×
1	×	×
1	×	×
1	×	×
1	×	×
1	×	×
1	×	×

Table 4:

A_2	$A_2(t+1)$	T_2
0	0	0
0	0	0
0	0	0
0	1	1
1	1	0
1	1	0
1	1	0
1	0	1
0	0	0
0	×	×
0	×	×
0	×	×
1	×	×
1	×	×
1	×	×
1	×	×

Table 5:

A_1	$A_1(t+1)$	T_1
0	0	0
0	1	1
1	1	0
1	0	1
0	0	0
0	1	1
1	1	0
1	0	1
0	0	0
0	×	×
1	×	×
1	×	×
0	×	×
0	×	×
1	×	×
1	×	×

Table 6:

A_0	$A_0(t+1)$	T_0
0	1	1
1	0	1
0	1	1
1	0	1
0	1	1
1	0	1
0	1	1
1	0	1
0	0	0
1	×	×
0	×	×
1	×	×
0	×	×
1	×	×
0	×	×
1	×	×

The Tables 2 to 6 are combined to form expanded state table with flip-flop inputs as shown in Table 7.

Table 7: Expanded State Table

Present States				Next States				Flip-Flop Inputs				Minterm
A_3	A_2	A_1	A_0	$A_3(t+1)$	$A_2(t+1)$	$A_1(t+1)$	$A_0(t+1)$	T_3	T_2	T_1	T_0	$A_3A_2A_1A_0$
0	0	0	0	0	0	0	1	0	0	0	1	m_0
0	0	0	1	0	0	1	0	0	0	1	1	m_1
0	0	1	0	0	0	1	1	0	0	0	1	m_2
0	0	1	1	0	1	0	0	0	1	1	1	m_3
0	1	0	0	0	1	0	1	0	0	0	1	m_4
0	1	0	1	0	1	1	0	0	0	1	1	m_5
0	1	1	0	0	1	1	1	0	0	0	1	m_6
0	1	1	1	1	0	0	0	1	1	1	1	m_7
1	0	0	0	0	0	0	0	1	0	0	0	m_8
1	0	0	1	×	×	×	×	×	×	×	×	m_9
1	0	1	0	×	×	×	×	×	×	×	×	m_{10}
1	0	1	1	×	×	×	×	×	×	×	×	m_{11}
1	1	0	0	×	×	×	×	×	×	×	×	m_{12}
1	1	0	1	×	×	×	×	×	×	×	×	m_{13}
1	1	1	0	×	×	×	×	×	×	×	×	m_{14}
1	1	1	1	×	×	×	×	×	×	×	×	m_{15}

Note: *Here, present state $A_3A_2A_1A_0$ is the counter output.*

The flip-flop inputs are functions of present states and so minterms are formed using present state and the Boolean equations for T_3, T_2, T_1 and T_0 are obtained from Table 7 as sum of minterms and don't-cares as shown below:

$T_3 = \Sigma m(7, 8) + \Sigma d(9, 10, 11, 12, 13, 14, 15)$

$T_2 = \Sigma m(3, 7) + \Sigma d(9, 10, 11, 12, 13, 14, 15)$

$T_1 = \Sigma m(1, 3, 5, 7) + \Sigma d(9, 10, 11, 12, 13, 14, 15)$

$T_0 = \Sigma m(0, 1, 2, 3, 4, 5, 6, 7) + \Sigma d(9, 10, 11, 12, 13, 14, 15)$

The simplified Boolean equations are obtained by constructing K-map for T_3, T_2, T_1 and T_0 as shown in Figs. 3 to 6.

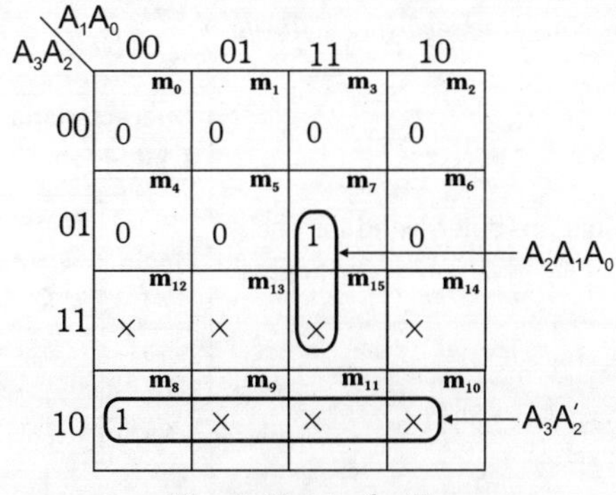

Fig. 3: K-map for T_3.

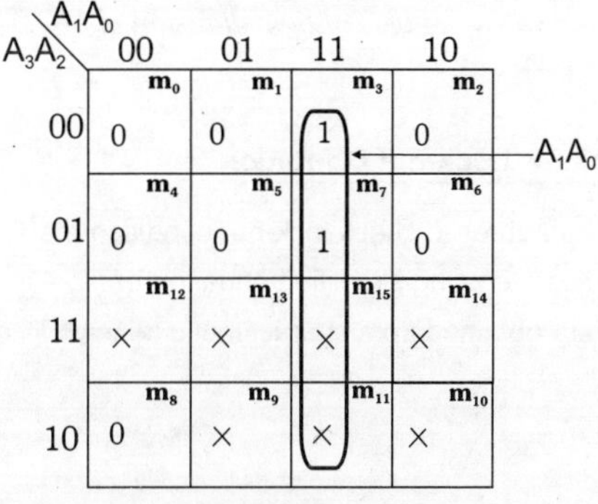

Fig. 4: K-map for T_2.

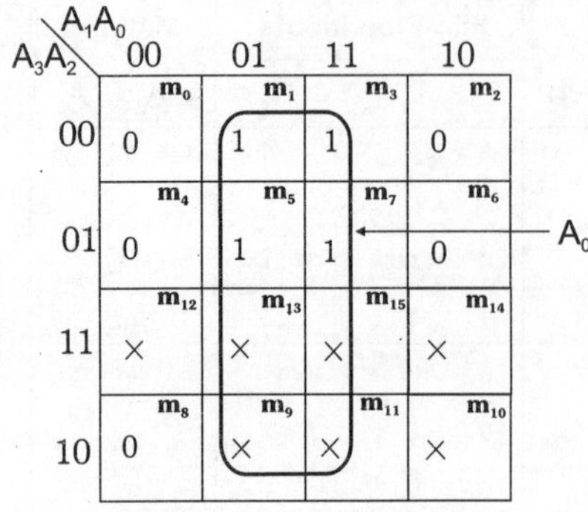

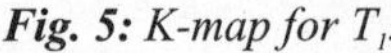
Fig. 5: K-map for T_1.

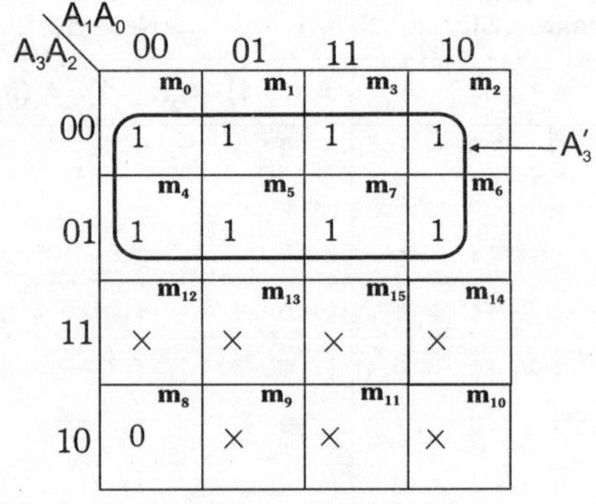

Fig. 6: K-map for T_0.

From the K-map we get the following simplified Boolean equations to realize MOD-9 counter using T flip-flops. Using the following equations, the MOD-9 synchronous counter is implemented as shown in Fig. 7.

$$T_3 = A_2A_1A_0 + A_3A_2' \qquad T_1 = A_0$$

$$T_2 = A_1A_0 \qquad T_0 = A_3'$$

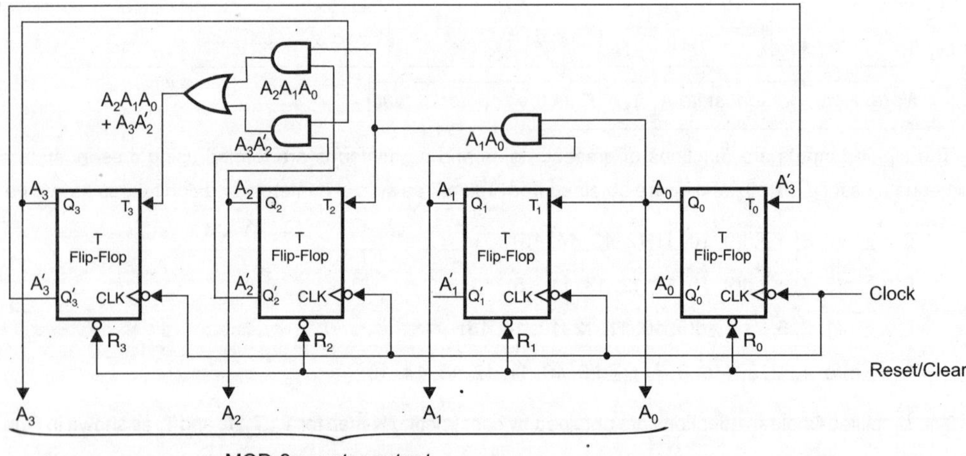

Fig. 7: Logic diagram of MOD-9 synchronous counter using T flip-flop.

Check for Lock-out Condition

In order to check lock-out condition the flip-flop inputs are determined for each unused state using simplified equation of flip-flop inputs and listed in Table 8. The next state are obtained from characteristic table of T flip-flop.

Characteristic Table of T Flip-Flop

Input T	Present State Q	Next State Q(t+1)
0	0	0
0	1	1
1	0	1
1	1	0

Table 8: Check for Lock-out Condition

Unused Present States A_3 A_2 A_1 A_0	Flip-Flop Inputs				Next States			
	$T_3 = A_2A_1A_0 + A_3A_2'$	$T_2 = A_1A_0$	$T_1 = A_0$	$T_0 = A_3'$	$A_3(t+1)$	$A_2(t+1)$	$A_1(t+1)$	$A_0(t+1)$
1 0 0 1	1	0	1	0	0	0	1	1
1 0 1 0	1	0	0	0	0	0	1	0
1 0 1 1	1	1	1	0	0	1	0	1
1 1 0 0	0	0	0	0	1	1	0	0
1 1 0 1	0	0	1	0	1	1	1	1
1 1 1 0	0	0	0	0	1	1	1	0
1 1 1 1	1	1	1	0	0	0	0	1

From the Table 8 we can observe the following state transitions from unused state to next state.

1001 → 0011, 1010 → 0010, 1011 → 0101, 1100 ⇔ 1100, 1101 → 1111 → 0001, 1110 ⇔ 1110

Here, it is observed that, whenever counter enters the unused states 1100 and 1110 then it go to a lock-out condition and so the counter is not self starting.

Example 8.14

(JNTUA, Jun/July'22, 7 Marks)

Design a MOD-12 synchronous binary up counter using T flip-flops.

Solution

MOD-12 is a binary counter that count from 0 to 11_{10} and on reaching 12_{10} the output should be reset to zero in order to start next count sequence.

In order to count from 0 to 11_{10} we need 4-bit binary output.

Hence we need 4 bit binary code to represent the decimal sequence. Let us assign binary value of decimal number as code for states. With 4-bit code 16 states ($2^4 = 16$) are possible and in this 12 states are used states and remaining 4 states are unused states.

The code assignment for states are as follows,

Binary code assignment

0	0000
1	0001
2	0010
3	0011
4	0100
5	0101
6	0110
7	0111
8	1000
9	1001
10	1010
11	1011

Unused binary code

12	1100
13	1101
14	1110
15	1111

The four bits of binary codes are represented by four state variables A_3, A_2, A_1 and A_0.

$\therefore$ Present state variables : $A_3(t), A_2(t), A_1(t), A_0(t) \Rightarrow A_3, A_2, A_1, A_0$ (For simplicity)

Next state variables : $A_3(t+1), A_2(t+1), A_1(t+1), A_0(t+1)$

The truth table of MOD-12 counter is shown in Table 1. Here, the present states are counter outputs. The state diagram using only used states is shown in Fig. 1.

Table 1: Truth Table of MOD-12 Counter

Clock	Counter Outputs			
	A_3	A_2	A_1	A_0
-	0	0	0	0
↓	0	0	0	1
↓	0	0	1	0
↓	0	0	1	1
↓	0	1	0	0
↓	0	1	0	1
↓	0	1	1	0
↓	0	1	1	1
↓	1	0	0	0
↓	1	0	0	1
↓	1	0	1	0
↓	1	0	1	1
↓	0	0	0	0

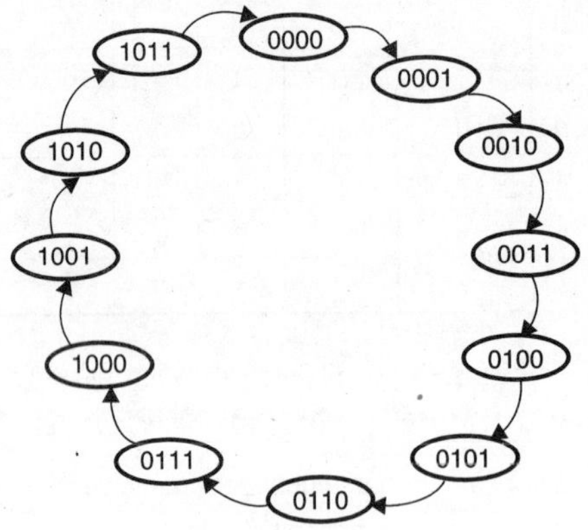

Fig. 1: Used states.

From the knowledge of state transitions in the state diagram of Fig. 1, the state table is constructed as shown in Table 2. Here, the next state for unused states are considered as don't-cares.

Table 2: State Table

Present States				Next States			
A_3	A_2	A_1	A_0	$A_3(t+1)$	$A_2(t+1)$	$A_1(t+1)$	$A_0(t+1)$
0	0	0	0	0	0	0	1
0	0	0	1	0	0	1	0
0	0	1	0	0	0	1	1
0	0	1	1	0	1	0	0
0	1	0	0	0	1	0	1
0	1	0	1	0	1	1	0
0	1	1	0	0	1	1	1
0	1	1	1	1	0	0	0
1	0	0	0	1	0	0	1
1	0	0	1	1	0	1	0
1	0	1	0	1	0	1	1
1	0	1	1	0	0	0	0
1	1	0	0	×	×	×	×
1	1	0	1	×	×	×	×
1	1	1	0	×	×	×	×
1	1	1	1	×	×	×	×

Excitation Table of T Flip-Flop

Present State Q(t)	Next State Q(t + 1)	Flip-Flop Input T
0	0	0
0	1	1
1	0	1
1	1	0

Let us choose four T flip-flops to store four state variables A_0, A_1, A_2 and A_3 as shown in Fig. 2. Here, A_0, A_1, A_2 and A_3 are output of T flip-flops.

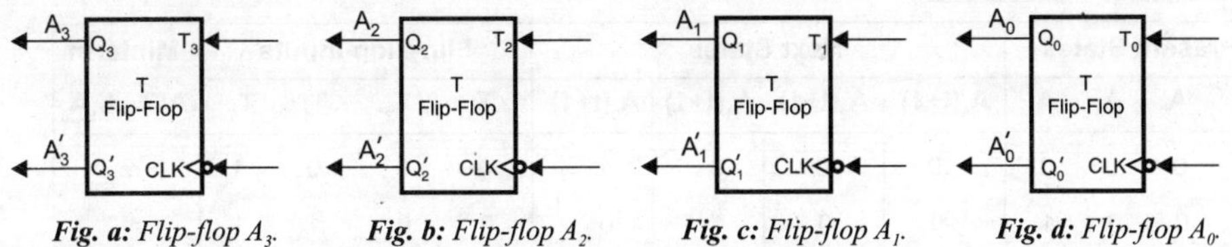

Fig. a: *Flip-flop A_3.* ***Fig. b:*** *Flip-flop A_2.* ***Fig. c:*** *Flip-flop A_1.* ***Fig. d:*** *Flip-flop A_0.*

Fig. 2: *T flip-flop to store state variables.*

The flip-flop inputs are determined using excitation table of T flip-flop as shown below:

State A_3

For state A_3, the Q(t), Q(t + 1) and T of excitation table are considered as A_3, $A_3(t + 1)$ and T_3.

$\therefore Q(t) \rightarrow A_3 \;;\; Q(t+1) \rightarrow A_3(t+1) \;;\; T \rightarrow T_3$

The flip-flop input T_3 that drives the present state A_3 to next states is obtained from excitation table and listed in Table 3.

State A_2

For state A_2, the Q(t), Q(t + 1) and T of excitation table are considered as A_2, $A_2(t + 1)$ and T_2.

$\therefore Q(t) \rightarrow A_2 \;;\; Q(t+1) \rightarrow A_2(t+1) \;;\; T \rightarrow T_2$

The flip-flop input T_2 that drives the present state A_2 to next states is obtained from excitation table and listed in Table 4.

State A_1

For state A_1, the Q(t), Q(t + 1) and T of excitation table are considered as A_1, $A_1(t + 1)$ and T_1.

$\therefore Q(t) \rightarrow A_1 \;;\; Q(t+1) \rightarrow A_1(t+1) \;;\; T \rightarrow T_1$

The flip-flop input T_1 that drives the present state A_1 to next states is obtained from excitation table and listed in Table 5.

State A_0

For state A_0, the Q(t), Q(t + 1) and T of excitation table are considered as A_0, $A_0(t + 1)$ and T_0.

$\therefore Q(t) \rightarrow A_0 \;;\; Q(t+1) \rightarrow A_0(t+1) \;;\; T \rightarrow T_0$

The flip-flop input T_0 that drives the present state A_0 to next states is obtained from excitation table and listed in Table 6.

Table 3:

A_3	$A_3(t+1)$	T_3
0	0	0
0	0	0
0	0	0
0	0	0
0	0	0
0	0	0
0	0	0
0	1	1
1	1	0
1	1	0
1	1	0
1	0	1
1	×	×
1	×	×
1	×	×
1	×	×

Table 4:

A_2	$A_2(t+1)$	T_2
0	0	0
0	0	0
0	0	0
0	1	1
1	1	0
1	1	0
1	1	0
1	0	1
0	0	0
0	0	0
0	0	0
0	0	0
1	×	×
1	×	×
1	×	×
1	×	×

Table 5:

A_1	$A_1(t+1)$	T_1
0	0	0
0	1	1
1	1	0
1	0	1
0	0	0
0	1	1
1	1	0
1	0	1
0	0	0
0	1	1
1	1	0
1	0	1
0	×	×
0	×	×
1	×	×
1	×	×

Table 6:

A_0	$A_0(t+1)$	T_0
0	1	1
1	0	1
0	1	1
1	0	1
0	1	1
1	0	1
0	1	1
1	0	1
0	1	1
1	0	1
0	1	1
1	0	1
0	×	×
1	×	×
0	×	×
1	×	×

The Tables 2 to 6 are combined to form expanded state table with flip-flop inputs as shown in Table 7.

Table 7: Expanded State Table

Present States				Next States				Flip-Flop Inputs				Minterm
A_3	A_2	A_1	A_0	$A_3(t+1)$	$A_2(t+1)$	$A_1(t+1)$	$A_0(t+1)$	T_3	T_2	T_1	T_0	$A_3A_2A_1A_0$
0	0	0	0	0	0	0	1	0	0	0	1	m_0
0	0	0	1	0	0	1	0	0	0	1	1	m_1
0	0	1	0	0	0	1	1	0	0	0	1	m_2
0	0	1	1	0	1	0	0	0	1	1	1	m_3
0	1	0	0	0	1	0	1	0	0	0	1	m_4
0	1	0	1	0	1	1	0	0	0	1	1	m_5
0	1	1	0	0	1	1	1	0	0	0	1	m_6
0	1	1	1	1	0	0	0	1	1	1	1	m_7
1	0	0	0	1	0	0	1	0	0	0	1	m_8
1	0	0	1	1	0	1	0	0	0	1	1	m_9
1	0	1	0	1	0	1	1	0	0	0	1	m_{10}
1	0	1	1	0	0	0	0	1	0	1	1	m_{11}
1	1	0	0	×	×	×	×	×	×	×	×	m_{12}
1	1	0	1	×	×	×	×	×	×	×	×	m_{13}
1	1	1	0	×	×	×	×	×	×	×	×	m_{14}
1	1	1	1	×	×	×	×	×	×	×	×	m_{15}

***Note:** Here, present state $A_3\ A_2\ A_1\ A_0$ is the counter output.*

The flip-flop inputs are functions of present states and so minterms are formed using present state and the Boolean equations for T_3, T_2, T_1 and T_0 are obtained from Table 7 as sum of minterms and don't-cares as shown below:

$$T_3 = \Sigma m(7, 11) + \Sigma d(12, 13, 14, 15)$$

$$T_2 = \Sigma m(3, 7) + \Sigma d(12, 13, 14, 15)$$

$$T_1 = \Sigma m(1, 3, 5, 7, 9, 11) + \Sigma d(12, 13, 14, 15)$$

$$T_0 = \Sigma m(0, 1, 2, 3, 4, 5, 6, 7, 8, 9, 10, 11) + \Sigma d(12, 13, 14, 15)$$

The simplified Boolean equations are obtained by constructing K-map for T_3, T_2, T_1 and T_0 as shown in Figs. 3 to 6.

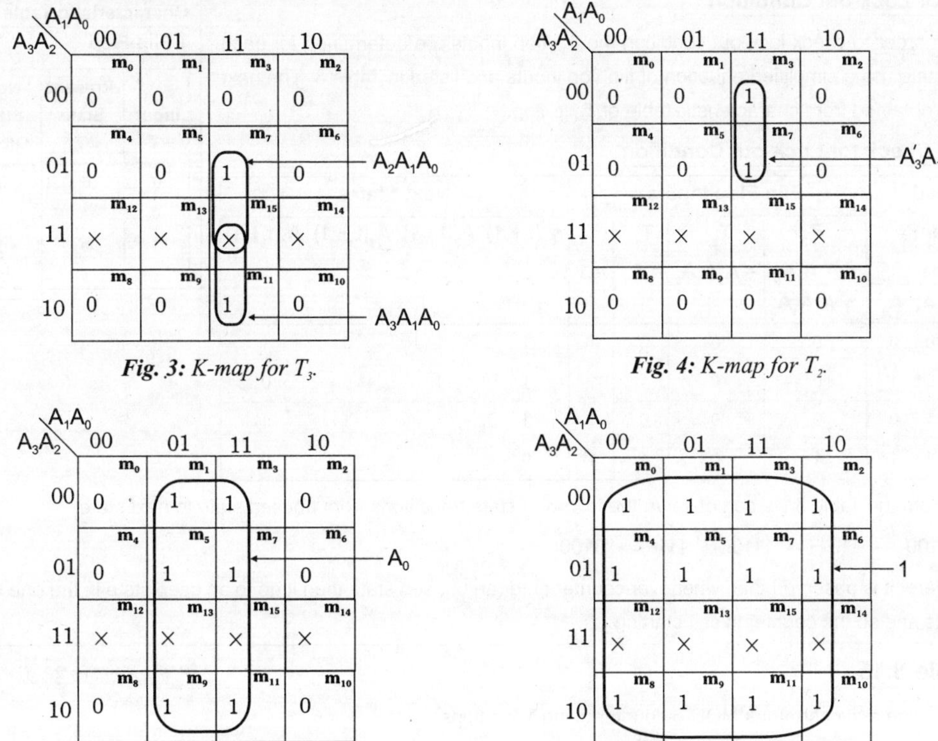

Fig. 3: K-map for T_3. *Fig. 4: K-map for T_2.*

Fig. 5: K-map for T_1. *Fig. 6: K-map for T_0.*

From the K-map we get the following simplified Boolean equations to realize MOD-12 counter using T flip-flops.

Using the following equations, the MOD-12 synchronous counter is implemented as shown in Fig. 7.

$$T_3 = A_2A_1A_0 + A_3A_1A_0 \qquad T_1 = A_0$$

$$T_2 = A_3'A_1A_0 \qquad T_0 = 1$$

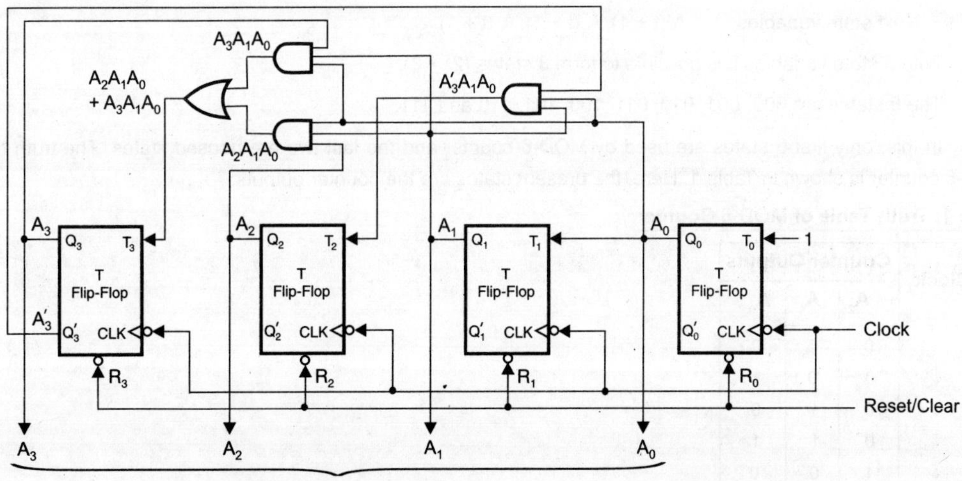

Fig. 7: Logic diagram of MOD-12 synchronous counter using T flip-flop.

Check for Lock-out Condition

In order to check lock-out condition the flip-flop inputs are determined for each unused state using simplified equation of flip-flop inputs and listed in Table 8. The next state are obtained from characteristic table of T flip-flop.

Characteristic Table of T Flip-Flop

Input T	Present State Q	Next State Q(t+1)
0	0	0
0	1	1
1	0	1
1	1	0

Table 8: Check for Lock-out Condition

Unused Present States $A_3\ A_2\ A_1\ A_0$	$T_3 = A_2A_1A_0 + A_3A_1A_0$	$T_2 = A_3'A_1A_0$	$T_1 = A_0$	$T_0 = 1$	$A_3(t+1)$	$A_2(t+1)$	$A_1(t+1)$	$A_0(t+1)$
1 1 0 0	0	0	0	1	1	1	0	1
1 1 0 1	0	0	1	1	1	1	1	0
1 1 1 0	0	0	0	1	1	1	1	1
1 1 1 1	1	0	1	1	0	1	0	0

From the Table 8 we can observe the following state transitions from unused state to next state.

1100 → 1101 → 1110 → 1111 → 0100

Here, it is observed that, whenever counter enter an unused state then it go to an used state in the one to four transitions and so the counter is self starting.

Example 8.15

(VTU, Aug/Sep'20, 8 Marks)

Design a synchronous MOD-6 counter using T flip-flops.

Solution

MOD-6 is a binary counter that count from 0 to 5 and on reaching 6 the output should be reset to zero, in order to start next count sequence.

In order to count from 0 to 5 we need 3-bit binary output.

The three bits are represented by three state variables A_2, A_1 and A_0.

∴ Present state variables : $A_2(t), A_1(t), A_0(t) \Rightarrow A_2, A_1, A_0$ (For simplicity)

Next state variables : $A_2(t+1), A_1(t+1), A_0(t+1)$

With 3 state variables it is possible to form 8 states ($2^3 = 8$).

The 8 states are 000, 001, 010, 011, 100, 101, 110 and 111.

In this, only first 6 states are used by MOD-6 counter and the last two are unused states. The truth table of MOD-6 counter is shown in Table 1. Here, the present states are the counter outputs.

Table 1: Truth Table of MOD-6 Counter

Clock	Counter Outputs A_2	A_1	A_0
-	0	0	0
↓	0	0	1
↓	0	1	0
↓	0	1	1
↓	1	0	0
↓	1	0	1
↓	0	0	0

The state diagram of MOD-6 counter is shown in Fig.1. From the knowledge of state transitions in the state diagram of Fig. 1, the stable table is formed as shown in Table 2. The next state for unused states are considered as don't-cares.

Fig. 1: State diagram of MOD-6 counter.

Table 2: State Table of MOD-6 Counter

Present States			Next States		
A_2	A_1	A_0	$A_2(t+1)$	$A_1(t+1)$	$A_0(t+1)$
0	0	0	0	0	1
0	0	1	0	1	0
0	1	0	0	1	1
0	1	1	1	0	0
1	0	0	1	0	1
1	0	1	0	0	0
1	1	0	×	×	×
1	1	1	×	×	×

Let us choose three T flip-flops to store three state variables A_0, A_1 and A_2 as shown in Fig. 2. Here, A_0, A_1 and A_2 be output of flip-flops.

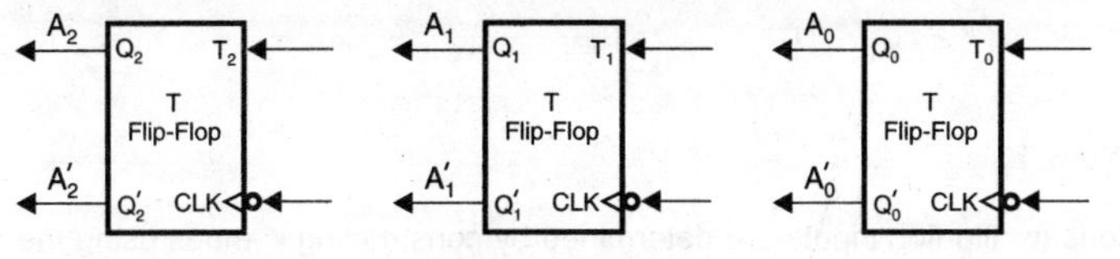

Fig. a: Flip-flop A_2. *Fig. b: Flip-flop A_1.* *Fig. c: Flip-flop A_0.*

Fig. 2: T flip-flops to store state variables.

Excitation Table of T Flip-Flop

Present State Q(t)	Next State Q(t + 1)	Flip-Flop Input T
0	0	0
0	1	1
1	0	1
1	1	0

The flip-flop inputs can be determined from excitation table of T flip-flop as shown below:

State A_2

For state A_2, the Q(t), Q(t + 1) and T of excitation table are considered as A_2, $A_2(t + 1)$ and T_2.

$\therefore Q(t) \rightarrow A_2$; $Q(t+1) \rightarrow A_2(t+1)$

$T \rightarrow T_2$

The flip-flop input T_2 that drives the present state A_2 to next states is obtained from excitation table and listed in Table 3.

Table 3:

A_2	$A_2(t+1)$	T_2
0	0	0
0	0	0
0	0	0
0	1	1
1	1	0
1	0	1

State A_1

For state A_1, the Q, Q(t + 1) and T of excitation table are considered as A_1, $A_1(t + 1)$ and T_1.

$\therefore Q(t) \rightarrow A_1$; $Q(t+1) \rightarrow A_1(t+1)$

$T \rightarrow T_1$

The flip-flop input T_1 that drives the present state A_1 to next states is obtained from excitation table and listed in Table 4.

Table 4:

A_1	$A_1(t+1)$	T_1
0	0	0
0	1	1
1	1	0
1	0	1
0	0	0
0	0	0

State A_0

For state A_0, the Q(t), Q(t + 1) and T of excitation table are considered as A_0, $A_0(t + 1)$ and T_0.

$\therefore Q(t) \rightarrow A_0$; $Q(t+1) \rightarrow A_0(t+1)$

$T \rightarrow T_0$

The flip-flop input T_0 that drives the present state A_0 to next states is obtained from excitation table and listed in Table 5.

Table 5:

A_0	$A_0(t+1)$	T_0
0	1	1
1	0	1
0	1	1
1	0	1
0	1	1
1	0	1

The tables 2 to 5 are combined to form state table with flip-flop inputs as shown in Table 6.

Table 6: Expanded State Table

Present States			Next States			Flip-Flop Inputs			Minterm
A_2	A_1	A_0	$A_2(t+1)$	$A_1(t+1)$	$A_0(t+1)$	T_2	T_1	T_0	$A_2A_1A_0$
0	0	0	0	0	1	0	0	1	m_0
0	0	1	0	1	0	0	1	1	m_1
0	1	0	0	1	1	0	0	1	m_2
0	1	1	1	0	0	1	1	1	m_3
1	0	0	1	0	1	0	0	1	m_4
1	0	1	0	0	0	1	0	1	m_5
1	1	0	×	×	×	×	×	×	m_6
1	1	1	×	×	×	×	×	×	m_7

The flip-flop inputs are functions of present states and so minterms are formed using present states and the Boolean equations for flip-flop inputs are obtained as sum of minterms and don't-cares as shown below:

$$T_2 = \sum m(3, 5) + \sum d(6, 7)$$

$$T_1 = \sum m(1, 3) + \sum d(6, 7)$$

$$T_0 = \sum m(0, 1, 2, 3, 4, 5) + \sum d(6, 7)$$

The simplified Boolean equations for flip-flop inputs are determined by constructing K-maps using the above equations as shown in Figs. 3 to 4.

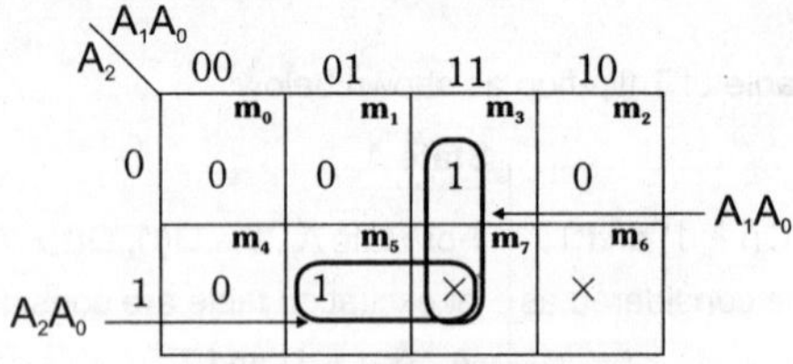

Fig. 3: K-map for T_2.

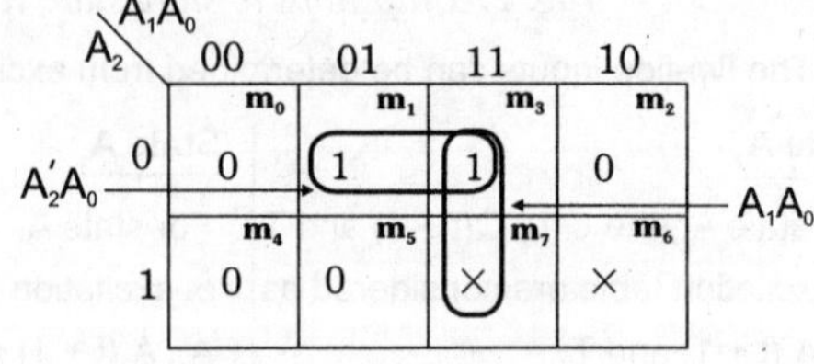

Fig. 4: K-map for T_1.

Here, T_0 is always 1 or × and so no need for K-map and they can be permanently tied to 1.

$$\therefore T_0 = 1$$

From the K-map we get the following Boolean equations for flip-flop inputs and using these equation the logic circuit of MOD-6 counter is drawn as shown in Fig. 5.

$$T_2 = A_1A_0 + A_2A_0$$

$$T_1 = A_1A_0 + A_2'A_0$$

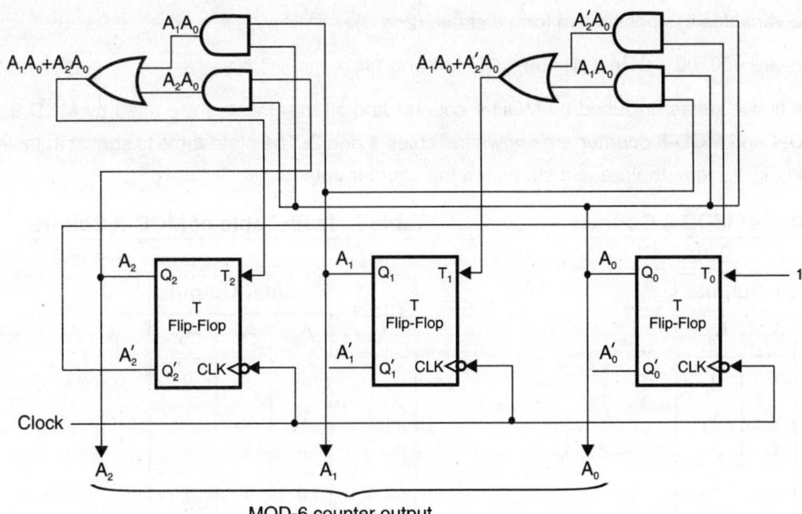

Fig. 5: Logic diagram of MOD-6 synchronous counter..

Check for Self Starting

In order to check for self starting, the flip-flop inputs are determined for each unused state using simplified equations of flip-flop inputs and listed in Table 7. The next state is obtained using characteristic table of T flip-flop.

Characteristic Table of T Flip-Flop

Input T	Present State Q	Next State Q(t+1)
0	0	0
0	1	1
1	0	1
1	1	0

Table 7: Check for Self Starting

Unused Present States			Flip-Flop Inputs			Next States		
			T_2	T_1	T_0	$A_2(t+1)$	$A_1(t+1)$	$A_0(t+1)$
A_2	A_1	A_0	$= A_1A_0 + A_2A_0$	$= A_1A_0 + A_2'A_0$	$= 1$			
1	1	0	0	0	1	1	1	1
1	1	1	1	1	1	0	0	0

From the Table 7 we can observe the following state transitions from unused state to next state.

$110 \rightarrow 111 \rightarrow 000 \quad ; \quad 111 \rightarrow 000$

The unused states enters used state in the next transition as shown above. Hence the counter is self-starting.

Example 8.16

(JNTUA, Jun/Jul'22, 7 Marks)

Design a synchronous counter is controlled by two input signals A and B. The counter does not operate, if A = 0 and B = 0. When A = 0 and B = 1, the counter operates as a mod 4 counter. if A = 1 and B = 0, the counter operates as a mod 8 counter.

Solution

In order to count from 0 to 7 we need 3-bit binary output.

The three bits are represented by three state variables A_2, A_1 and A_0.

$\therefore$ Present state variables : $A_2(t), A_1(t), A_0(t) \Rightarrow A_2, A_1, A_0$ (For simplicity)

Next state variables : $A_2(t+1), A_1(t+1), A_0(t+1)$

With 3 state variables it is possible to form 8 states ($2^3 = 8$).

The 8 states are 000, 001, 010, 011, 100, 101, 110 and 111.

In this, only first 4 states are used by MOD-4 counter and all the 8 states are used by MOD-8 counter. The truth tables of MOD-4 and MOD-8 counter are shown in Tables 1 and 2. The state table is shown in table 3 and state diagram is shown in Fig 1. Here, the present states are the counter outputs.

Table 1: Truth Table of MOD-4 Counter

Clock	Counter Outputs		
	A_2	A_1	A_0
-	0	0	0
↓	0	0	1
↓	0	1	0
↓	0	1	1
↓	0	0	0

Table 2: Truth Table of MOD-8 Counter

Clock	Counter Outputs		
	A_2	A_1	A_0
-	0	0	0
↓	0	0	1
↓	0	1	0
↓	0	1	1
↓	1	0	0
↓	1	0	1
↓	1	1	0
↓	1	1	1
↓	0	0	0

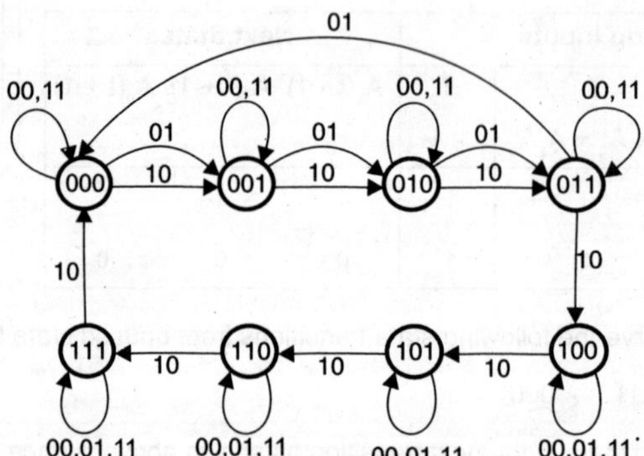

Fig. 1: State diagram .

Table 3: State Table

Present States			Input		Next States		
A_2	A_1	A_0	x	y	$A_2(t+1)$	$A_1(t+1)$	$A_0(t+1)$
0	0	0	0	0	0	0	0
0	0	0	0	1	0	0	1
0	0	0	1	0	0	0	1
0	0	0	1	1	0	0	0
0	0	1	0	0	0	0	1
0	0	1	0	1	0	1	0
0	0	1	1	0	0	1	0
0	0	1	1	1	0	0	1
0	1	0	0	0	0	1	0
0	1	0	0	1	0	1	1
0	1	0	1	0	0	1	1
0	1	0	1	1	0	1	0
0	1	1	0	0	0	1	1
0	1	1	0	1	0	0	0
0	1	1	1	0	1	0	0
0	1	1	1	1	0	1	1
1	0	0	0	0	1	0	0
1	0	0	0	1	1	0	0
1	0	0	1	0	1	0	1
1	0	0	1	1	1	0	0
1	0	1	0	0	1	0	1
1	0	1	0	1	1	0	1
1	0	1	1	0	1	1	0
1	0	1	1	1	1	0	1
1	1	0	0	0	1	1	0
1	1	0	0	1	1	1	0
1	1	0	1	0	1	1	1
1	1	0	1	1	1	1	1
1	1	1	0	0	1	1	1
1	1	1	0	1	1	1	1
1	1	1	1	0	0	0	0
1	1	1	1	1	1	1	1

Let us choose three T flip-flops to store three state variables A_0, A_1 and A_2 as shown in Fig. 2. Here, A_0, A_1 and A_2 are output of flip-flops.

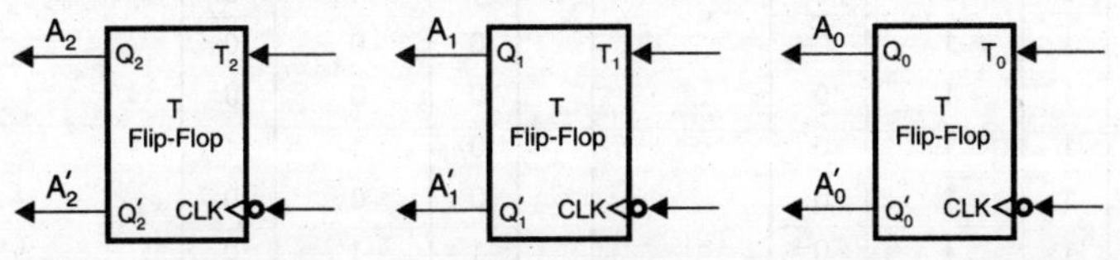

Fig. a: Flip-flop A_2. *Fig. b: Flip-flop A_1.* *Fig. c: Flip-flop A_0.*

Fig. 2: T flip-flops to store state variables.

Excitation Table of T Flip-Flop

Present State Q(t)	Next State Q(t + 1)	Flip-Flop Input T
0	0	0
0	1	1
1	0	1
1	1	0

The flip-flop inputs can be obtained from excitation table of T flip-flops as shown below:

State A_2

For state A_2, the Q(t), Q(t + 1) and T of excitation table are considered as A_2, $A_2(t + 1)$ and T_2.

$$\therefore Q(t) \rightarrow A_2 \quad ; \quad Q(t+1) \rightarrow A_2(t+1)$$
$$T \rightarrow T_2$$

The flip-flop input T_2 that drives the present state A_2 to next state is obtained from excitation table and listed in Table 4.

Table 4:

A_2	$A_2(t+1)$	T_2
0	0	0
0	0	0
0	0	0
0	0	0
0	0	0
0	0	0
0	0	0
0	0	0
0	0	0
0	0	0
0	0	0
0	0	0
0	0	0
0	0	0
0	1	1
0	0	0
1	1	0
1	1	0
1	1	0
1	1	0
1	1	0
1	1	0
1	1	0
1	1	0
1	1	0
1	1	0
1	1	0
1	1	0
1	1	0
1	1	0
1	0	1
1	1	0

State A_1

For state A_1, the Q(t), Q(t + 1) and T excitation table are considered as A_1, $A_1(t + 1)$ and T_1.

$$\therefore Q(t) \rightarrow A_1 \quad ; \quad Q(t+1) \rightarrow A_1(t+1)$$
$$T \rightarrow T_1$$

The flip-flop input T_1 that drives the present state A_1 to next state is obtained from excitation table and listed in Table 5.

Table 5:

A_1	$A_1(t+1)$	T_1
0	0	0
0	0	0
0	0	0
0	0	0
0	0	0
0	1	1
0	1	1
0	0	0
1	1	0
1	1	0
1	1	0
1	1	0
1	1	0
1	0	1
1	0	1
1	1	0
0	0	0
0	0	0
0	0	0
0	0	0
0	0	0
0	0	0
0	1	1
0	0	0
1	1	0
1	1	0
1	1	0
1	1	0
1	1	0
1	1	0
1	0	1
1	1	0

State A_0

For state A_0, the Q(t), Q(t + 1) and T of excitation table are considered as A_0, $A_0(t + 1)$ and T_0.

$$\therefore Q(t) \rightarrow A_0 \quad ; \quad Q(t+1) \rightarrow A_0(t+1)$$
$$T \rightarrow T_0$$

The flip-flop input T_0 that drives the present state A_0 to next state is obtained from excitation table and listed in Table 6.

Table 6:

A_0	$A_0(t+1)$	T_0
0	0	0
0	1	1
0	1	1
0	0	0
1	1	0
1	0	1
1	0	1
1	1	0
0	0	0
0	1	1
0	1	1
0	0	0
1	1	0
1	0	1
1	0	1
1	1	0
0	0	0
0	0	0
0	1	1
0	0	0
1	1	0
1	1	0
1	0	1
1	1	0
0	0	0
0	0	0
0	1	1
0	0	0
1	1	0
1	1	0
1	0	1
1	1	0

The Tables 3 to 6 are combined to form expanded state table with flip-flop inputs as shown in Table 7.

Table 5: Expanded State Table

Present States			Input		Next States			Flip-Flop Inputs			Minterm
A_2	A_1	A_0	X	Y	$A_2(t+1)$	$A_1(t+1)$	$A_0(t+1)$	T_2	T_1	T_0	$A_2A_1A_0XY$
0	0	0	0	0	0	0	0	0	0	0	m_0
0	0	0	0	1	0	0	1	0	0	1	m_1
0	0	0	1	0	0	0	1	0	0	1	m_2
0	0	0	1	1	0	0	0	0	0	0	m_3
0	0	1	0	0	0	0	1	0	0	0	m_4
0	0	1	0	1	0	1	0	0	1	1	m_5
0	0	1	1	0	0	1	0	0	1	1	m_6
0	0	1	1	1	0	0	1	0	0	0	m_7
0	1	0	0	0	0	1	0	0	0	0	m_8
0	1	0	0	1	0	1	1	0	0	1	m_9
0	1	0	1	0	0	1	1	0	0	1	m_{10}
0	1	0	1	1	0	1	0	0	0	0	m_{11}
0	1	1	0	0	0	1	1	0	0	0	m_{12}
0	1	1	0	1	0	0	0	0	1	1	m_{13}
0	1	1	1	0	1	0	0	1	1	1	m_{14}
0	1	1	1	1	0	1	1	0	0	0	m_{15}
1	0	0	0	0	1	0	0	0	0	0	m_{16}
1	0	0	0	1	1	0	0	0	0	0	m_{17}
1	0	0	1	0	1	0	1	0	0	1	m_{18}
1	0	0	1	1	1	0	0	0	0	0	m_{19}
1	0	1	0	0	1	0	1	0	0	0	m_{20}
1	0	1	0	1	1	0	1	0	0	0	m_{21}
1	0	1	1	0	1	1	0	0	1	1	m_{22}
1	0	1	1	1	1	0	1	0	0	0	m_{23}
1	1	0	0	0	1	1	0	0	0	0	m_{24}
1	1	0	0	1	1	1	0	0	0	0	m_{25}
1	1	0	1	0	1	1	1	0	0	1	m_{26}
1	1	0	1	1	1	1	0	0	0	0	m_{27}
1	1	1	0	0	1	1	1	0	0	0	m_{28}
1	1	1	0	1	1	1	1	0	0	0	m_{29}
1	1	1	1	0	0	0	0	1	1	1	m_{30}
1	1	1	1	1	1	1	1	0	0	0	m_{31}

***Note:** Here, present state $A_2 A_1 A_0$ is the counter output.*

The Boolean equations for T_2, T_1 and T_0 are obtained from Table 7 by taking present state and inputs as minterms.

$T_2 = \Sigma m(14, 30)$

$T_1 = \Sigma m(5, 6, 13, 14, 22, 30)$

$T_0 = \Sigma m(1, 2, 5, 6, 9, 10, 13, 14, 18, 22, 26, 30)$

The simplified Boolean equations are obtained by constructing K-map using the above equation for T_2, T_1 and T_0 as shown in Figs. 3 to 5.

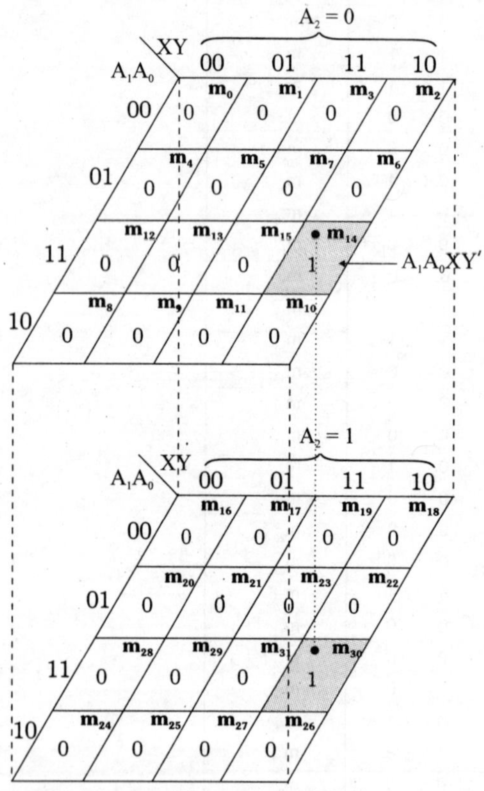

Fig. 3: K-map for T_2.

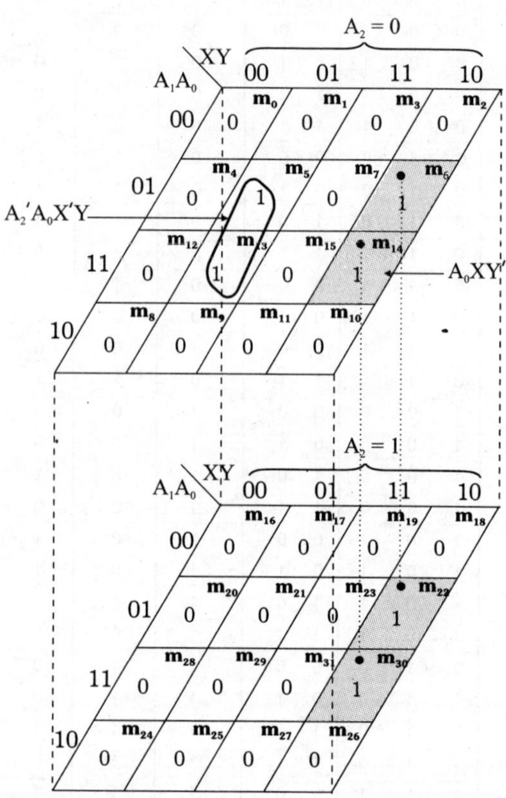

Fig. 4: K-map for T_1.

From the K-map we get the following simplified Boolean equations to realize 3-bit MOD-4 and MOD-8 counter with two bit control inputs using T flip-flops and they are used to construct the logic diagram of 3-bit counter as shown in Fig. 6.

$$T_2 = A_1 A_0 X Y'$$

$$T_1 = A_2' A_0 X' Y + A_0 X Y'$$

$$T_0 = A_2 X' Y + X Y'$$

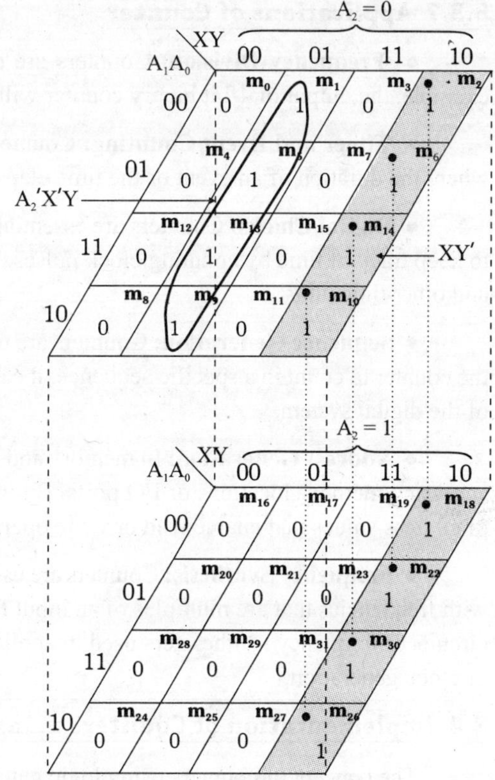

Fig. 5: K-map for T_0.

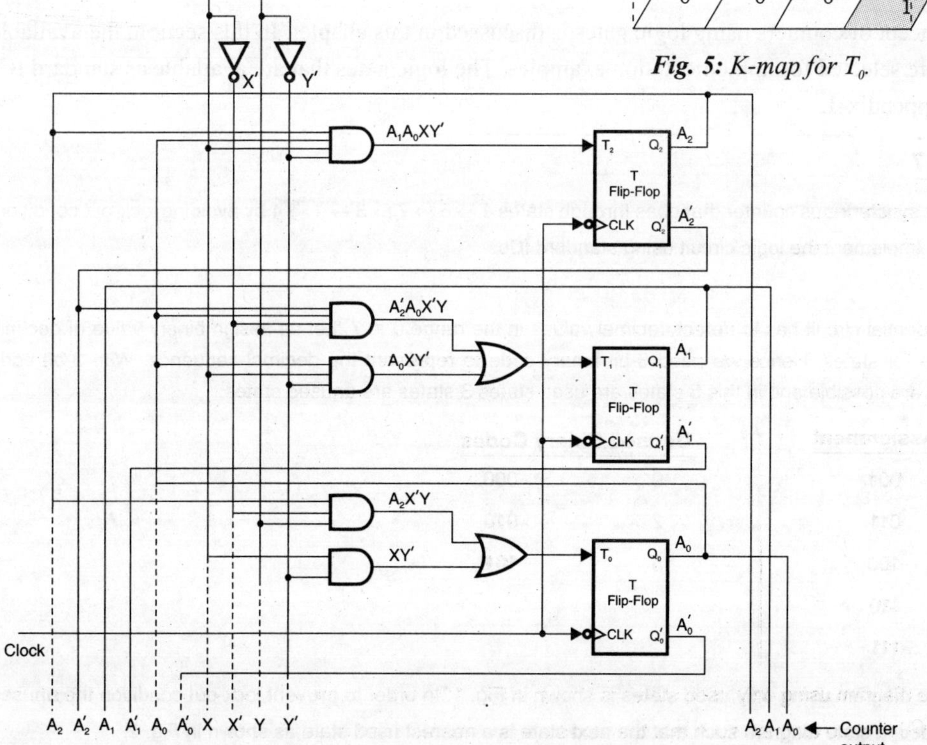

Fig. 6: Logic diagram of 3-bit MOD-4 and MOD-8 counter.

8.3.7 Applications of Counter

(RGPV, Jun'22, 5 Marks)

- **Frequency Division:** Counters are often used to divide the frequency of clock signals. For example, the output of 4-bit binary counter will be f/2, f/4, f/8 f/16 of input clock frequency,f .

- **Timer and Event Counting:** Counters can be used as timers or event counters in applications where the duration of an event or the time elapsed between events needs to be measured.

- **Digital Clocks:** Counters are essential components in digital clock circuits, where they are used to keep track of time by counting clock pulses. They can be configured to count seconds, minutes, hours, and other time units.

- **Sequence Generation:** Counters are used to generate sequences of digital states. By configuring the counter to count in a specific sequence, it can be used to control the sequential operation of other parts of the digital system.

- **Address Generation:** In memory and I/O interfacing, counters are used to generate addresses for accessing memory locations or I/O ports. By incrementing the counter sequentially, it generates a series of address values that enable read or write operations to specific memory locations or peripheral devices.

- **Frequency Synthesis:** Counters are used in frequency synthesis circuits to generate output signals with frequencies that are multiples of an input frequency. They are often part of phase-locked loop (PLL) circuits or frequency synthesizers used in applications such as telecommunications, RF communications, or clock generation.

8.4 Implementation of Counters using Standard IC's

The concept of counters using logic gates is discussed in this chapter. In this section, the available standard ICs are selected for implementation examples. The logic gates that are available as standard ICs are listed in Appendix-1.

Example 8.17

Design a synchronous counter that goes through states $4 \rightarrow 6 \rightarrow 7 \rightarrow 3 \rightarrow 1 \rightarrow 4$ by avoiding lock-out condition. Use JK flip-flop. Implement the logic circuit using standard ICs.

Solution

The sequential circuit has to detect decimal values in the range 0 to 7. Let us assign binary value of decimal number as code for states. Hence we need 3-bit binary code to represent the decimal sequence. With 3-bit code 8 states ($2^3 = 8$) are possible and in this 5 states are used states 3 states are unused states.

Binary Code Assignment

1	001
3	011
4	100
6	110
7	111

Unused Binary Codes

0	000
2	010
5	101

The state diagram using only used states is shown in Fig. 1. In order to prevent lock-out condition the unused states are included in state diagram such that the next state is a nearest used state as shown in Fig. 2.

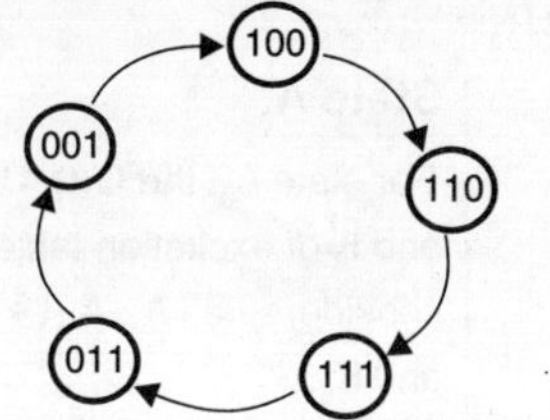

Fig. 1: State diagram with only used states.

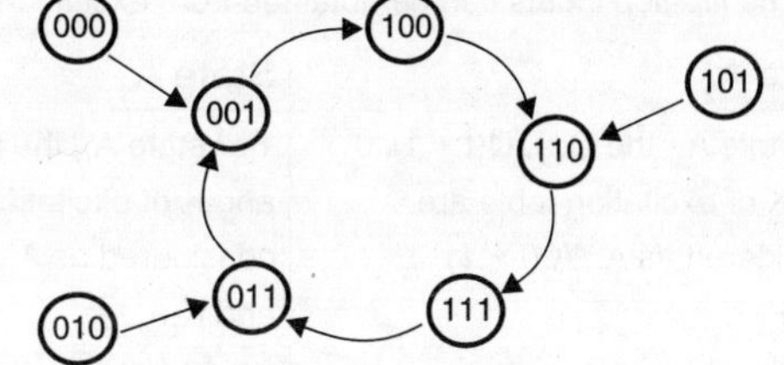

Fig. 2: State diagram to prevent lock-out condition.

Here, each bit of binary code is a state variable. Hence the sequence counter has 3 state variables. Let the state variables be A_0, A_1 and A_2.

$\therefore$ Present state variables : $A_0(t)$, $A_1(t)$, $A_2(t)$ $\Rightarrow$ A_0, A_1, A_2 (For simplicity)

Next state variables : $A_0(t+1)$, $A_1(t+1)$, $A_2(t+1)$

Using the state diagram of Fig. 2, the state table is constructed as shown in Table 1.

Table 1: State Table

Present States			Next States		
A_2	A_1	A_0	$A_2(t+1)$	$A_1(t+1)$	$A_0(t+1)$
0	0	0	0	0	1
0	0	1	1	0	0
0	1	0	0	1	1
0	1	1	0	0	1
1	0	0	1	1	0
1	0	1	1	1	0
1	1	0	1	1	1
1	1	1	0	1	1

Let us choose two numbers of standard dual JK flip-flop IC 74112 to store three state variables A_0, A_1 and A_2 as shown in Fig. 3. Here, A_0, A_1 and A_2 are output of flip-flops and one of the flip-flop in one 74112 is not used.

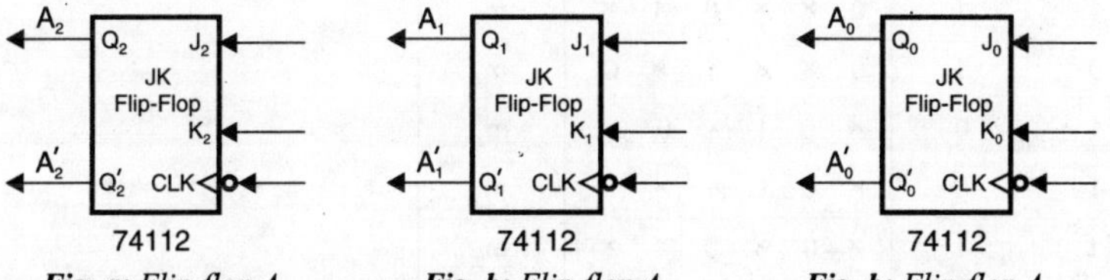

Fig. a: Flip-flop A_2. *Fig. b: Flip-flop A_1.* *Fig. b: Flip-flop A_0.*

Fig. 3: JK flip-flops to store state variables.

Excitation Table of JK Flip-Flop

Present State Q(t)	Next State Q(t + 1)	Flip-Flop Inputs J	K
0	0	0	×
0	1	1	×
1	0	×	1
1	1	×	0

The flip-flop inputs can be obtained from excitation table as shown below:

State A_2

For state A_2, the Q(t), Q(t + 1), J and K of excitation table are considered as A_2, $A_2(t + 1)$, J_2 and K_2.

$$\therefore Q(t) \rightarrow A_2 \ ; \ Q(t+1) \rightarrow A_2(t+1)$$
$$J \rightarrow J_2 \ ; \ K \rightarrow K_2$$

The flip-flop inputs J_2 and K_2 that drives the present state A_2 to next state is obtained from excitation table and listed in Table 2.

Table 2:

A_2	$A_2(t + 1)$	J_2	K_2
0	0	0	×
0	1	1	×
0	0	0	×
0	0	0	×
1	1	×	0
1	1	×	0
1	1	×	0
1	0	×	1

State A_1

For state A_1, the Q(t), Q(t + 1), J and K of excitation table are considered as A_1, $A_1(t + 1)$, J_1 and K_1.

$$\therefore Q(t) \rightarrow A_1 \ ; \ Q(t+1) \rightarrow A_1(t+1)$$
$$J \rightarrow J_1 \ ; \ K \rightarrow K_1$$

The flip-flop inputs J_1 and K_1 that drives the present state A_1 to next state is obtained from excitation table and listed in Table 3.

Table 3:

A_1	$A_1(t + 1)$	J_1	K_1
0	0	0	×
0	0	0	×
1	1	×	0
1	0	×	1
0	1	1	×
0	1	1	×
1	1	×	0
1	1	×	0

State A_0

For state A_0, the Q(t), Q(t + 1), J and K of excitation table are considered as A_0, $A_0(t + 1)$, J_0 and K_0.

$$\therefore Q(t) \rightarrow A_0 \ ; \ Q(t+1) \rightarrow A_0(t+1)$$
$$J \rightarrow J_0 \ ; \ K \rightarrow K_0$$

The flip-flop inputs J_0 and K_0 that drives the present state A_0 to next state is obtained from excitation table and listed in Table 4.

Table 4:

A_0	$A_0(t + 1)$	J_0	K_0
0	1	1	×
1	0	×	1
0	1	1	×
1	1	×	0
0	0	0	×
1	0	×	1
0	1	1	×
1	1	×	0

The tables 1 to 4 are combined to form state table with flip-flop inputs as shown in Table 5.

Table 5: Expanded State Table

Present States			Next States			Flip-Flop Inputs						Minterm
A_2	A_1	A_0	$A_2(t+1)$	$A_1(t+1)$	$A_0(t+1)$	J_2	K_2	J_1	K_1	J_0	K_0	$A_2A_1A_0$
0	0	0	0	0	1	0	×	0	×	1	×	m_0
0	0	1	1	0	0	1	×	0	×	×	1	m_1
0	1	0	0	1	1	0	×	×	0	1	×	m_2
0	1	1	0	0	1	0	×	×	1	×	0	m_3
1	0	0	1	1	0	×	0	1	×	0	×	m_4
1	0	1	1	1	0	×	0	1	×	×	1	m_5
1	1	0	1	1	1	×	0	×	0	1	×	m_6
1	1	1	0	1	1	×	1	×	0	×	0	m_7

***Note:** Here, present state $A_2A_1A_0$ is the counter output.*

The flip-flop inputs are functions of present state and so minterms are formed using present states and the Boolean equations for J_2, K_2, J_1, K_1, J_0 and K_0 are obtained from Table 5 as sum of minterms and don't-cares as shown below:

$$J_2 = m_1 + \Sigma d(4, 5, 6, 7)$$

$$K_2 = m_7 + \Sigma d(0, 1, 2, 3)$$

$J_1 = m(4, 5) + \Sigma d(2, 3, 6, 7)$

$K_1 = m_3 + \Sigma d(0, 1, 4, 5)$

$J_0 = \Sigma m(0, 2, 6) + \Sigma d(1, 3, 5, 7)$

$K_0 = \Sigma m(1, 5) + \Sigma d(0, 2, 4, 6)$

The simplified Boolean equations are obtained by constructing K-map using the above equations for J_2, K_2, J_1, K_1, J_0 and K_0 as shown in Figs. 4 to 9.

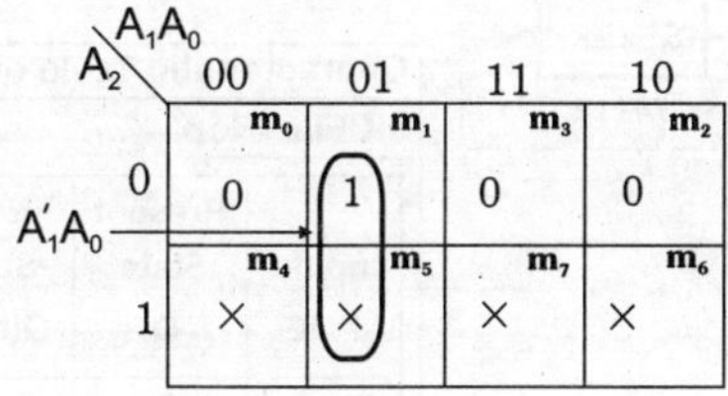

***Fig. 4:** K-map for J_2.*

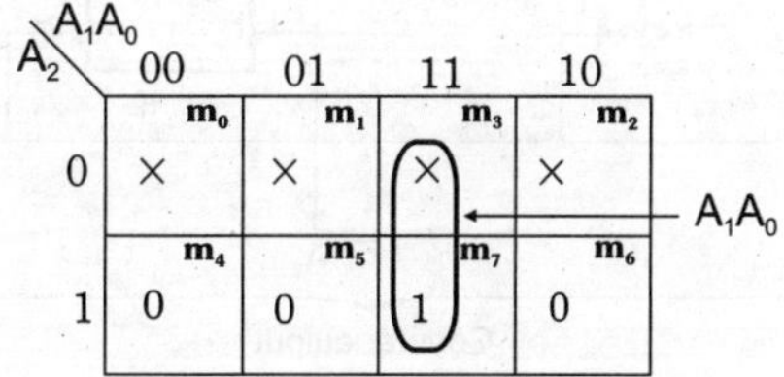

***Fig. 5:** K-map for K_2.*

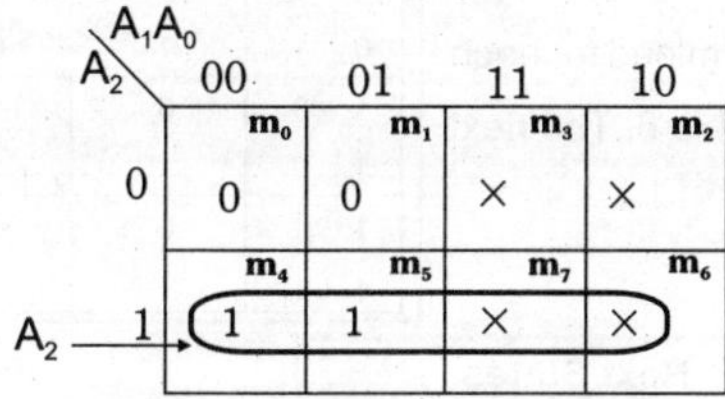

***Fig. 6:** K-map for J_1.*

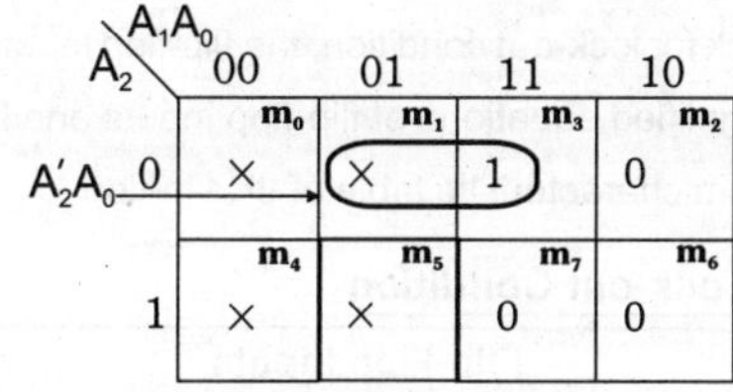

***Fig. 7:** K-map for K_1.*

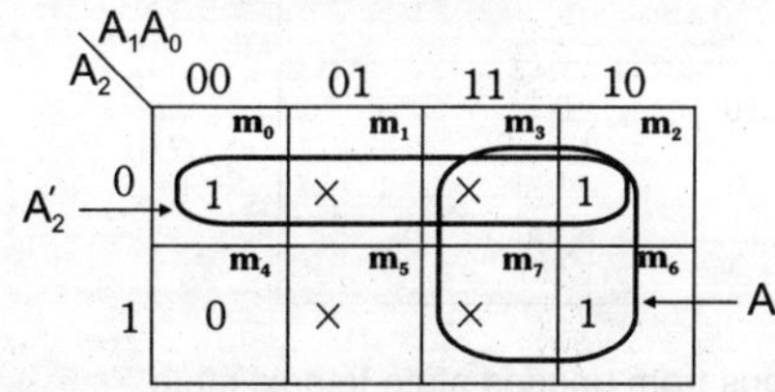

***Fig. 8:** K-map for J_0.*

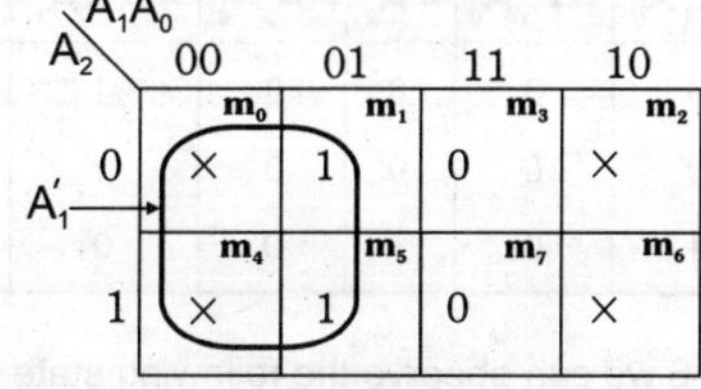

***Fig. 9:** K-map for K_0.*

From the K-map we get the following simplified Boolean equations to realize 3-bit binary counter using JK flip-flops.

$J_2 = A_1' A_0$ | $K_2 = A_1 A_0$

$J_1 = A_2$ | $K_1 = A_2' A_0$

$J_0 = A_2' + A_1$ | $K_0 = A_1'$

Using the simplified Boolean equations obtained from K-map the logic diagram of 3-bit counter is drawn as shown in Fig. 10. The logic circuit is implemented using 2-input AND gate IC 7408 and OR gate IC 7432 and dual JK flip-flop IC 74112.

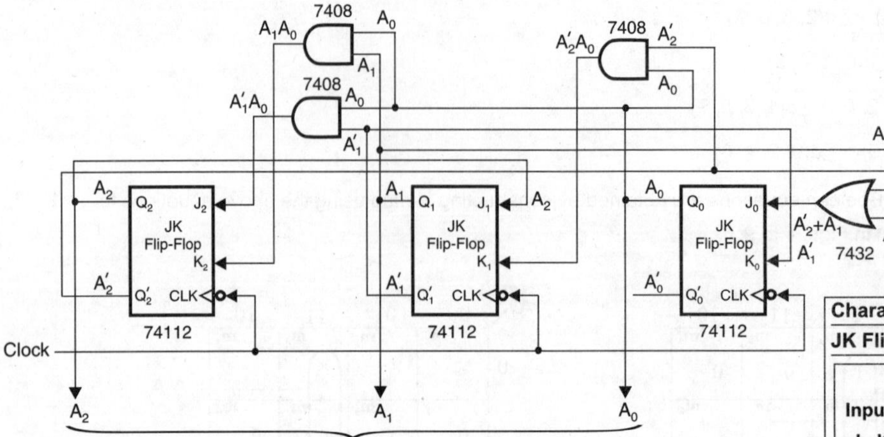

Fig. 10: Logic diagram.

Characteristic Table of JK Flip-Flop

Inputs J	Inputs K	Present State Q	Next State Q(t+1)
0	0	0	0
0	0	1	1
0	1	0	0
0	1	1	0
1	0	0	1
1	0	1	1
1	1	0	1
1	1	1	0

Check for Lock-out Condition

In order to check for lock-out condition, the flip-flop inputs are determined for each unused state using simplified equations of flip-flop inputs and listed in Table 6. The next states are obtained from characteristic table of JK flip-flop

Table 6: Check for Lock-out Condition

Unused Present States			Flip-Flop Inputs						Next States		
A_2	A_1	A_0	$J_2 = A'_1A_0$	$K_2 = A_1A_0$	$J_1 = A_2$	$K_1 = A'_2A_0$	$J_0 = A'_2+A_1$	$K_0 = A'_1$	$A_2(t+1)$	$A_1(t+1)$	$A_0(t+1)$
0	0	0	0	0	0	0	1	1	0	0	1
0	1	0	0	0	0	0	1	0	0	1	1
1	0	1	1	0	1	0	0	1	1	1	0

From the Table 6 we can observe the following state transitions from unused state to next state.

000 → 001, 010 → 011, 101 → 110

The unused states enter used state in the next transition as shown above and hence the lock-out condition is avoided.

Example 8.18

Design a sequential circuit for a 3-bit synchronous binary counter that counts clock pulses and generate output in the range 000_2 to 111_2. Implement the logic circuit using standard ICs.

Solution

In counters the states are directly taken as outputs.

Since 3-bit output is needed, there are eight possible combination of states formed using 3 state variables. Let the three state variables be A_0, A_1 and A_2.

∴ Present state variables : $A_0(t), A_1(t), A_2(t) \Rightarrow A_0, A_1, A_2$ (For simplicity)

Next state variables : $A_0(t+1), A_1(t+1), A_2(t+1)$

Since T flip-flops are not available as standard IC, the JK flip-flop is converted to T flip-flop by connecting the J and K inputs to function as T input. The standard dual JK flip-flop IC 74112 is selected for storing state variables A_0, A_1 and A_2 as shown in Fig. 1. Here A_0, A_1 and A_2 are output of flip-flops and J and K inputs are connected to function as T input.

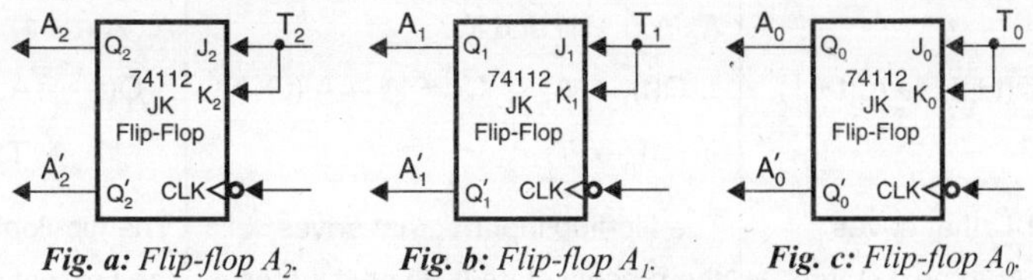

Fig. a: Flip-flop A_2. *Fig. b: Flip-flop A_1.* *Fig. c: Flip-flop A_0.*

Fig. 1: T flip-flops to store state variables.

The binary counter has to increment by 1 for every clock pulse. The initial output is 000, then for every clock pulse it has to increment by 1. So that the output sequence is,

000, 001, 010, 011, 100, 101, 110, 111

Upon reaching the upper most count 111 the count should start from 000 for next clock pulse. Using this logic, the state diagram is drawn as shown in Fig. 2.

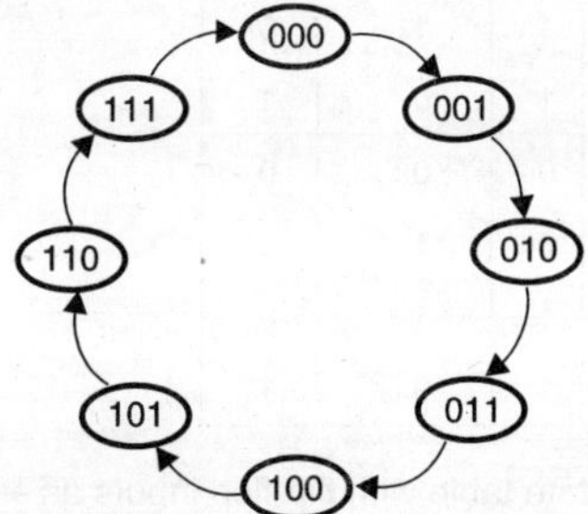

Fig. 2: State diagram of 3-bit binary counter.

The state table is formed using state diagram as shown in Table 1. The flip-flop inputs can be obtained from excitation table.

Table 1: State Table

Present States			Next States		
A_2	A_1	A_0	$A_2(t+1)$	$A_1(t+1)$	$A_0(t+1)$
0	0	0	0	0	1
0	0	1	0	1	0
0	1	0	0	1	1
0	1	1	1	0	0
1	0	0	1	0	1
1	0	1	1	1	0
1	1	0	1	1	1
1	1	1	0	0	0

Excitation Table of T Flip-Flop

Present State Q(t)	Next State Q(t + 1)	Flip-Flop Input T
0	0	0
0	1	1
1	0	1
1	1	0

State A_2

For state A_2, the Q(t), Q(t + 1) and T of excitation table are considered as A_2, $A_2(t + 1)$ and T_2.

$\therefore Q(t) \rightarrow A_2$; $Q(t+1) \rightarrow A_2(t+1)$

$T \rightarrow T_2$

The flip-flop input T_2 that drives the present state A_2 to next states is obtained from excitation table and listed in Table 2.

Table 2:

A_2	$A_2(t+1)$	T_2
0	0	0
0	0	0
0	0	0
0	1	1
1	1	0
1	1	0
1	1	0
1	0	1

State A_1

For state A_1, the Q(t), Q(t + 1) and T of excitation table are considered as A_1, $A_1(t + 1)$ and T_1.

$\therefore Q(t) \rightarrow A_1$; $Q(t+1) \rightarrow A_1(t+1)$

$T \rightarrow T_1$

The flip-flop input T_1 that drives the present state A_1 to next states is obtained from excitation table and listed in Table 3.

Table 3:

A_1	$A_1(t+1)$	T_1
0	0	0
0	1	1
1	1	0
1	0	1
0	0	0
0	1	1
1	1	0
1	0	1

State A_0

For state A_0, the Q(t), Q(t + 1) and T of excitation table are considered as A_0, $A_0(t + 1)$ and T_0.

$\therefore Q(t) \rightarrow A_0$; $Q(t+1) \rightarrow A_0(t+1)$

$T \rightarrow T_0$

The flip-flop input T_0 that drives the present state A_0 to next states is obtained from excitation table and listed in Table 4.

Table 4:

A_0	$A_0(t+1)$	T_0
0	1	1
1	0	1
0	1	1
1	0	1
0	1	1
1	0	1
0	1	1
1	0	1

The tables 1 to 4 are combined to form state table with flip-flop inputs as shown in Table 5.

Table 5: Expanded State Table

Present States			Next States			Flip-Flop Inputs			Minterm
A_2	A_1	A_0	$A_2(t+1)$	$A_1(t+1)$	$A_0(t+1)$	T_2	T_1	T_0	$A_2A_1A_0$
0	0	0	0	0	1	0	0	1	m_0
0	0	1	0	1	0	0	1	1	m_1
0	1	0	0	1	1	0	0	1	m_2
0	1	1	1	0	0	1	1	1	m_3
1	0	0	1	0	1	0	0	1	m_4
1	0	1	1	1	0	0	1	1	m_5
1	1	0	1	1	1	0	0	1	m_6
1	1	1	0	0	0	1	1	1	m_7

***Note:** Here, present state $A_2A_1A_0$ is the counter output.*

The flip-flop inputs are functions of present states and so minterms are formed using present states and Boolean equations for flip-flop inputs are obtained as sum of minterms as shown below:

$T_2 = \Sigma m(3, 7)$; $T_1 = \Sigma m(1, 3, 5, 7)$; $T_0 = \Sigma m(0, 1, 2, 3, 4, 5, 6, 7)$

The simplified Boolean equations are obtained by constructing K-map for T_2, T_1 and T_0 as shown in Figs. 3 to 5.

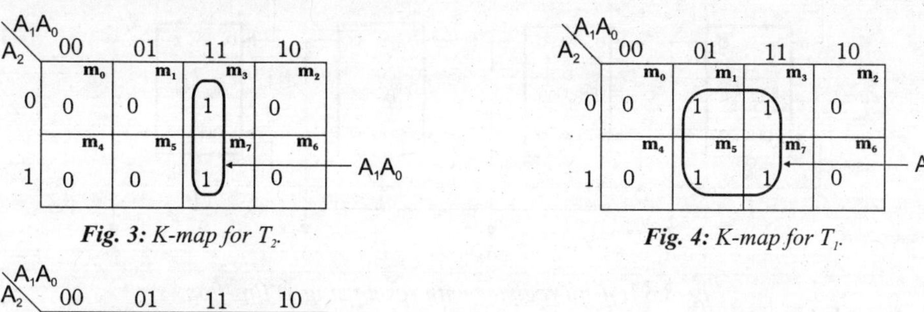

Fig. 3: K-map for T_2.

Fig. 4: K-map for T_1.

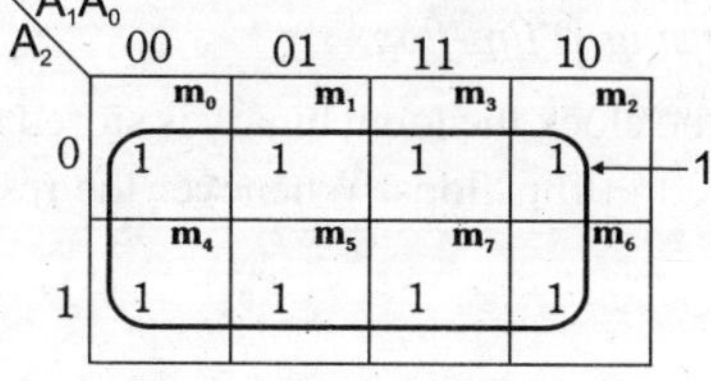

Fig. 5: K-map for T_0.

From the K-map we get the following simplified Boolean equations to realize 3-bit binary counter using T flip-flops.

$T_2 = A_1 A_0$

$T_1 = A_0$

$T_0 = 1$

Using the above equations the logic diagram of 3-bit counter is drawn as shown in Fig. 6. Here, 2-input AND gate IC 7408 and dual JK flip-flop ICs 74112 with J and K inputs connected to work as T input are used to implement the logic circuit.

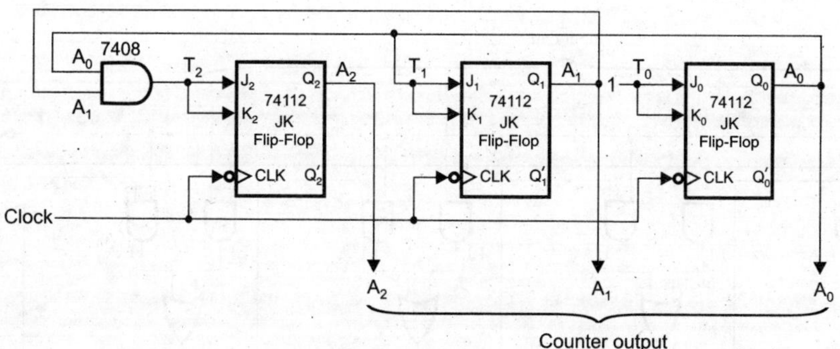

Fig. 6: Logic diagram of 3-bit synchronous binary counter.

8.5 Register

A ***register*** is a group of flip-flops that can store binary information. Each flip-flop in a register can store one bit binary and so, in order to store n-bit binary the register should have n flip-flops.

All the flip-flops in a register are controlled or synchronized to a common clock. A simple 4-bit register using D flip-flops is shown in Fig. 8.53.

Let, I_3, I_2, I_1 and I_0 are 4-bit input and A_3, A_2, A_1 and A_0 are 4-bit output.

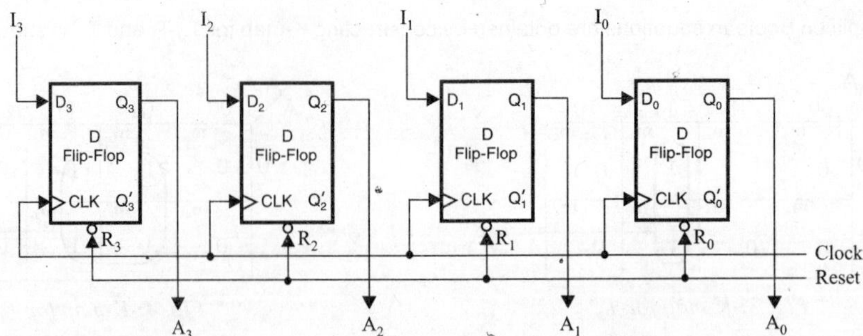

Fig. 8.53: 4-bit register with reset using D flip-flop.

When clock is applied then, on every positive edge of the clock the input binary is stored in flip-flop output. In order to clear the output, a reset facility is provided in flip-flops. Whenever the reset is made low, all the output bits are cleared or reset to zero.

Register with Parallel Load Control

A 4-bit register with logic **high** load control is shown in Fig. 8.54. Generally in a digital system the clock will be used by or all the devices in the system. Therefore, an individual control is needed for every register in order to load data or store data. A practical register will have a load control (or write control) so that the input data will be stored in the register only when load control is activated.

Also, the D flip-flop require a facility to rewrite the output data to input so that the output data is preserved on every clock pulse when the load control is inactive.

Let, L = Load control input

L′= Complement of load control input

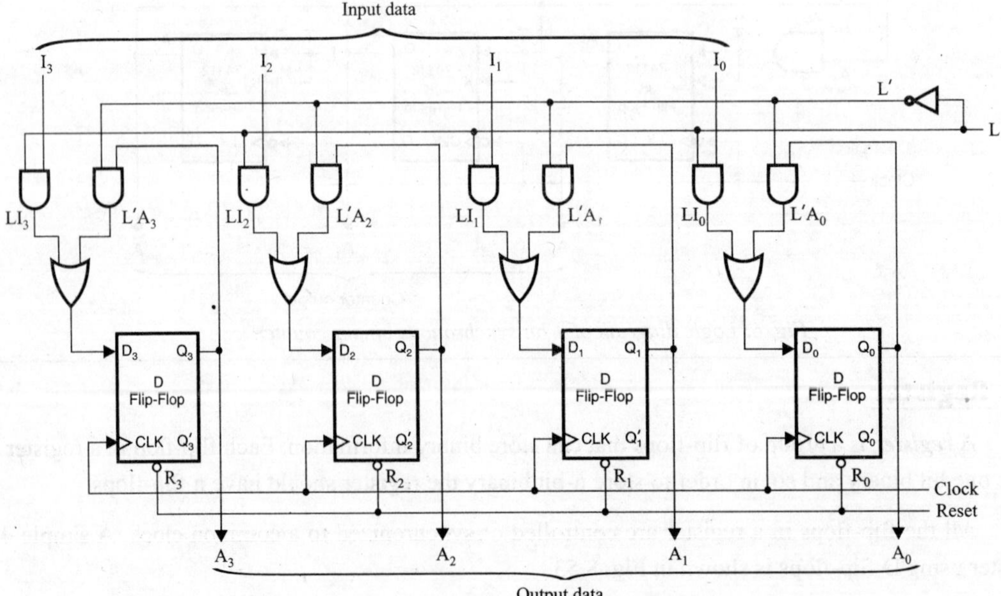

Fig. 8.54: 4-bit register with high parallel load control.

The load control L is logically ANDed with input data and complement of load control L′ is logically ANDed with flip-flop output separately. The output of both AND gates are logically ORed and applied as input to D flip-flop.

When L = 1 and L′ = 0 then input data is passed to flip-flop input and this logic is illustrated in Fig. 8.55.

When L = 0 and L′ = 1 then flip-flop output is rewritten as input and this logic is illustrated in Fig. 8.56.

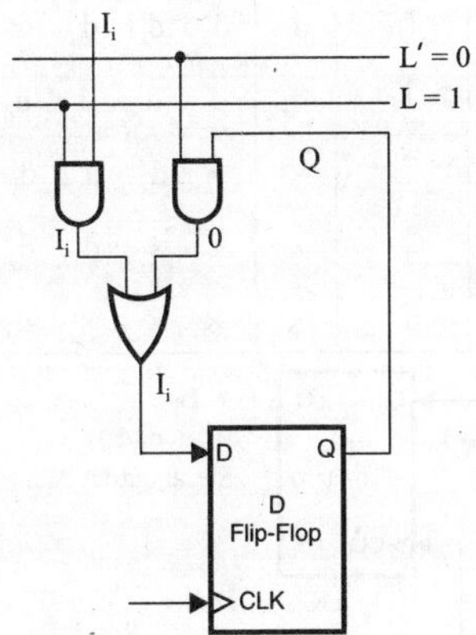

***Fig. 8.55:** Load input data.*

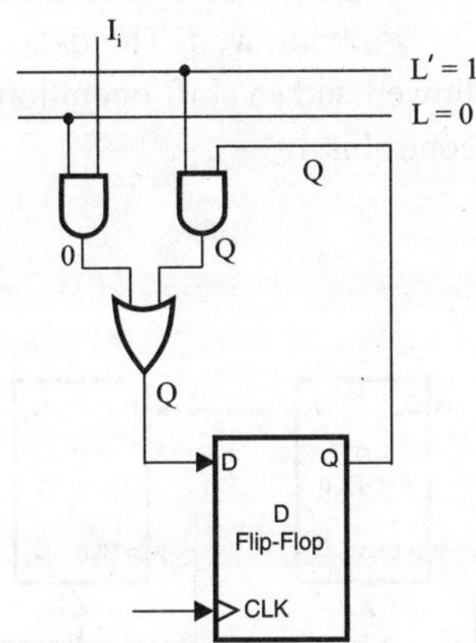

***Fig. 8.56:** Rewriting output to input.*

Buffer and Controlled Buffer Register

In a digital system, during data transmission, when an n-bit register is used in between two digital devices in order to provide signal current level matching then the register is called buffer register. When control signals are provided to load and clear the content of buffer register then the buffer register is called controlled buffer register.

8.6 Shift Register

(AU, Nov/Dec'18, 13 Marks) (RGPV, Jun'20, 7 Marks) (VTU, Aug/Sep'20, 8 Marks)

A register capable of shifting the binary information from one flip-flop to another (or one cell to another) is called ***shift register***.

There are four types of shift registers.

1. Serial-in Serial-out shift register
2. Serial-in Parallel-out shift register
3. Parallel-in Serial-out shift register
4. Parallel-in Parallel-out shift register.

8.6.1 Serial-In Serial-Out Shift Register

(AKTU, Sep'20, 5 Marks) (AKUB, Nov/Dec'20, 4 Marks)

Right Shift Register

A 4-bit serial-in serial-out shift register for right shift operation is shown in Fig. 8.57. The input data is given to left most flip-flop and output data is obtained from right most flip-flop. The output of flip-flops are cascaded, i.e, output of one becomes input for next and so on. For every clock pulse the data bit is shifted by one position from left to right as shown in Table 8.23

Reset facility and shift control are also provided in the shift register.

When reset is 0 all the flip-flop outputs will be cleared. Usually, the output of flip-flops are cleared in the beginning by making reset **low** and then reset is tied to **high**.

The shift control is provided to allow or disallow the clock. When shift control is 1, clock is allowed. When shift control is 0, clock is not allowed. The data will shift only when clock is allowed and so shift operation is performed as long as shift control is 1.

Table 8.23: Data Transfer in Right Shift Serial-in Serial-out Shift Register

Clock	D_{in}	Q_3	Q_2	Q_1	Q_0	D_{out}
-	d_0	×	×	×	×	×
↑	d_0	d_0	×	×	×	×
↑	d_1	d_1	d_0	×	×	×
↑	d_2	d_2	d_1	d_0	×	×
↑	d_3	d_3	d_2	d_1	d_0	d_0
↑	×	×	d_3	d_2	d_1	d_1
↑	×	×	×	d_3	d_2	d_2
↑	×	×	×	×	d_3	d_3

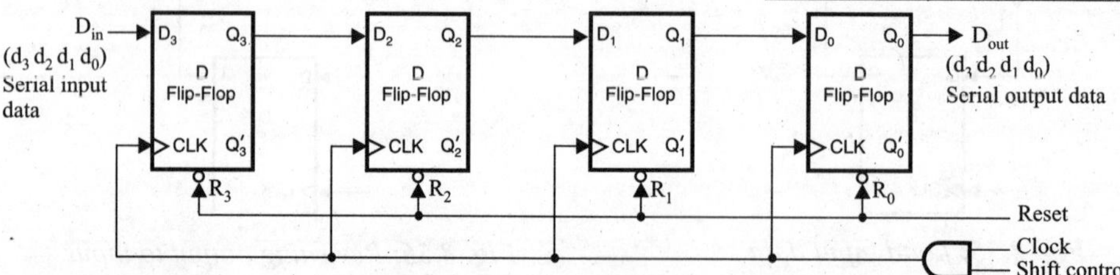

Fig. 8.57: Serial-in serial-out shift register for right shift (shift left to right).

Left Shift Register

For left shift operation the shifting is performed from right to left and the connections in Fig. 8.57 should be modified as shown in Fig. 8.58.

The input data is given to right most flip-flop and the output data is obtained from left most flip-flop. The shift operation from right to left is shown in Table 8.24. The reset and shift control are similar to left to right shift register.

Table 8.24: Data Transfer in Left Shift Serial-in Serial-out Shift Register

Clock	D_{out}	Q_3	Q_2	Q_1	Q_0	D_{in}
-	×	×	×	×	×	d_0
↑	×	×	×	×	d_0	d_0
↑	×	×	×	d_0	d_1	d_1
↑	×	×	d_0	d_1	d_2	d_2
↑	d_0	d_0	d_1	d_2	d_3	d_3
↑	d_1	d_1	d_2	d_3	×	×
↑	d_2	d_2	d_3	×	×	×
↑	d_3	d_3	×	×	×	×

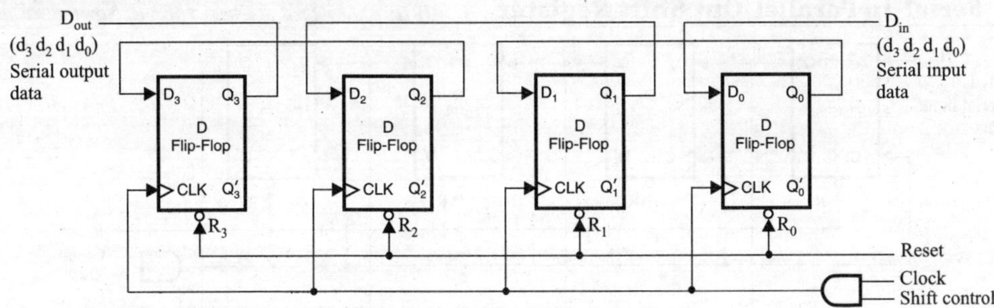

Fig. 8.58: *Serial-in serial-out shift register for left shift (shift right to left).*

Shift Register for Right and Left Shift Operation

The left to right and right to left shift operation can be performed in a single serial-in serial-out shift register with additional logic circuits and two shift control signals as shown in Fig. 8.59.

The input of flip-flops are applied through AND-OR combinational circuit. The flip-flop input for left to right shift is logically ANDed with RS control signal. The flip-flop input for right to left shift is logically ANDed with LS control signal. The output of both AND gates are logically ORed and applied as flip-flop inputs. The arrangement will allow only one shift data to flip-flop inputs.

When RS = 1 and LS = 0, the left to right shift data is allowed to flip-flop inputs to perform right shift operation. When RS = 0 and LS = 1, the right to left shift data is allowed to flip-flop inputs to perform left shift operation.

The RS and LS control signals are also used to control the input and output data and also clock signal. When both RS and LS are either 0 or 1 then clock is not allowed and no shift operation is performed.

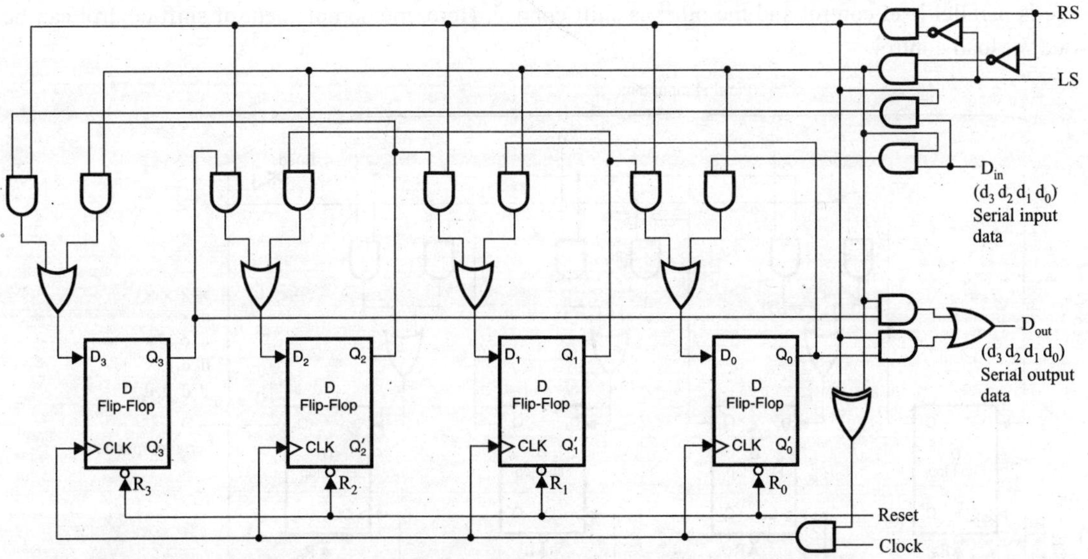

Fig. 8.59: *Serial-in serial-out shift register for both right and left shift.*

8.6.2 Serial-In Parallel-Out Shift Register

(RGPV, Jun'20, 8 Marks) (AKUB, Nov/Dec'20, 4 Marks)

Fig. 8.60: Serial-in parallel-out shift register.

A 4-bit serial-in parallel-out shift register is shown in Fig. 8.60. The input data is applied to left most flip-flop. The reset and shift control are similar to serial-in serial-out shift register. Here, 4 clock pulses are needed to load the data in the shift register by right shift operation.

Here input data bit is shifted one position from left to right for every clock pulse provided shift control is **high**. But the all the 4-bit output are available for read operation at any one time. The left shift of input data is possible if input connections are modified as shown in left shift register of Fig. 8.58.

8.6.3 Parallel-In Serial-Out Shift Register

(CSVTU, Nov/Dec'19, 7 Marks)

A 4-bit parallel-in serial-out shift register is shown in Fig. 8.61. Here, the flip-flops are connected to output data by right shift operation. The shift register require two control inputs for proper operation. One is parallel load control and the other is shift control. Here, the complement of shift control can be used for load control.

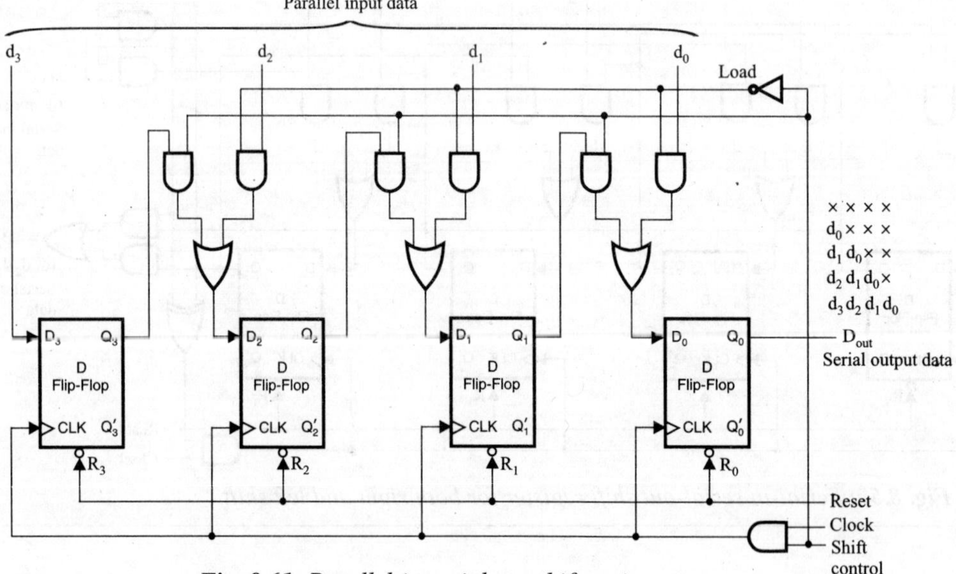

Fig. 8.61: Parallel-in serial-out shift register.

When shift control is 0, the data can be loaded in parallel and clock also not allowed. When shift control is 1, the data is shifted left to right for every positive edge of clock. The data output by left shift is possible if connections are modified for left shift of output as shown in Fig. 8.58.

8.6.4 Parallel-In Parallel-Out Shift Register

A 4-bit parallel in parallel out shift register is shown in Fig. 8.62. Here, a load control signal is logically ANDed with clock to allow or disallow the clock. When load control is 1, the clock is allowed and input data is loaded as the output. When load control is 0, the clock is not allowed and input data is not loaded in output.

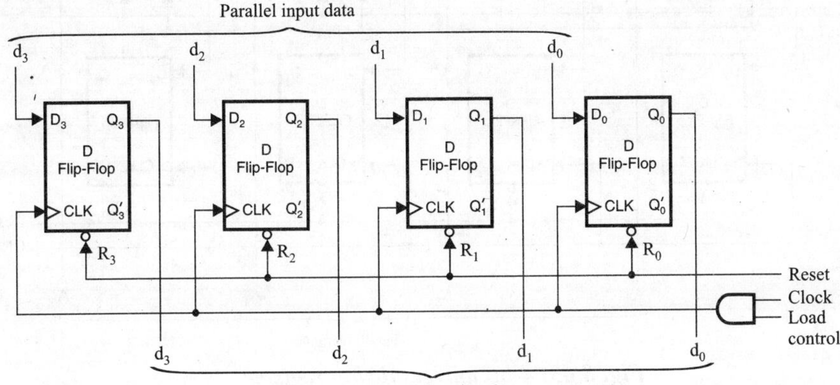

Fig. 8.62: Parallel-in parallel-out shift register.

8.6.5 Universal Shift Register

(AU, Apr/May'19, 6 Marks) (RGPV, Jun'22, 7 Marks) (VTU, Aug/Sep'20, 8 Marks) (JNTUH, Nov/Dec'20, 6 Marks) (AKTU, Dec'22, 8 Marks)

A shift register that can accept serial or parallel input data and output data in parallel along with right or left shift operation is called ***universal shift register***.

Therefore, an universal shift register can be used to implement any type of shift register with right/ left shift selection.

The universal shift register will have the following features.

- A reset/clear control in order to clear the output to 0.
- A clock input to synchronize the operations.
- A shift right control.
- A shift left control.
- A parallel load control.
- Parallel output lines.
- No-change state.

A 4-bit universal shift register using D flip-flops is shown in Fig. 8.63. The input to D flip-flops are applied through 4:1 MUX which has two selection lines which are used as operation mode control.

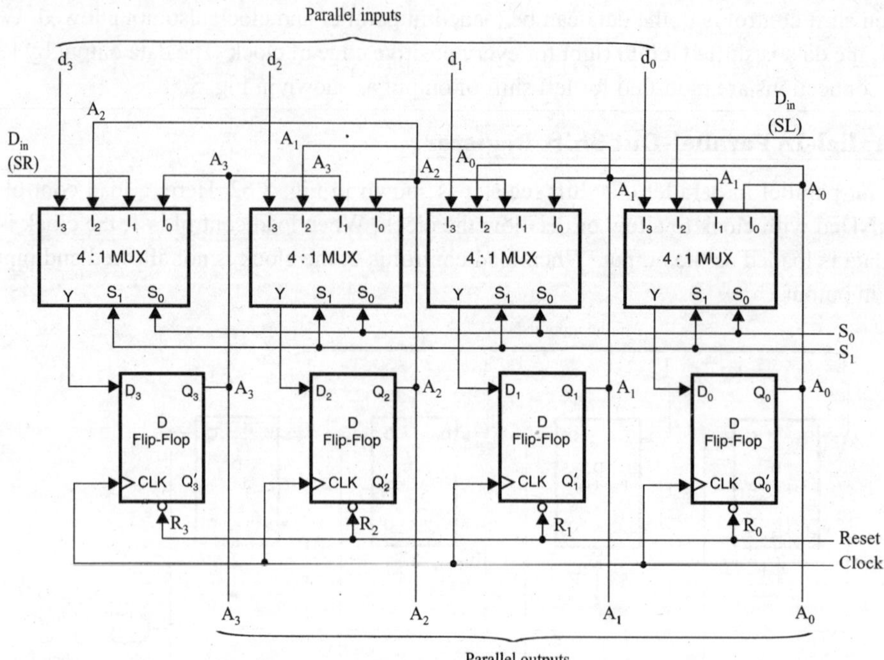

Fig. 8.63: 4-bit universal shift register.

Table 8.25: Operation Modes of Universal Shift Register

Operation Mode	Mode Selection S_1	S_0	Operation
Mode-0	0	0	No change
Mode-1	0	1	Shift right
Mode-2	1	0	Shift left
Mode-3	1	1	Parallel load

Mode-0 : $S_1 = S_0 = 0$, No Change in Output

In the universal shift register the, output of each flip-flop is connected to input 4:1 MUX that is used to input data to D flip-flop as shown in Fig. 8.64.

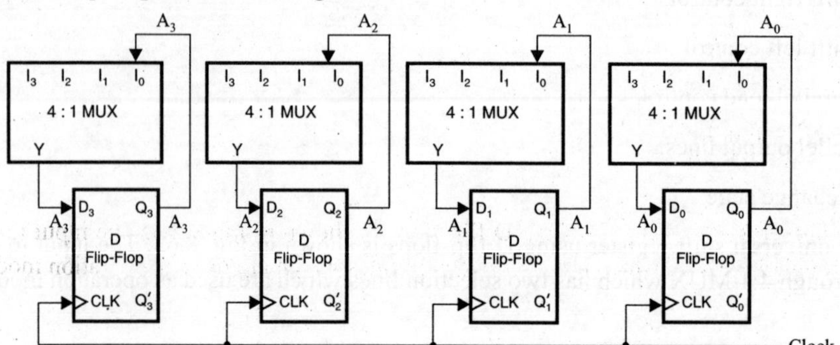

Fig. 8.64: Mode-0 (no change) operation of universal shift register.

Therefore, when $S_1 = S_0 = 0$, the MUX input I_0 is passed to output Y of MUX and so the input of D flip-flops are same as outputs. Hence, for every clock the output is rewritten as input and so there is no change in output.

Mode-1 : $S_1 = 0$, $S_0 = 1$, Shift Right

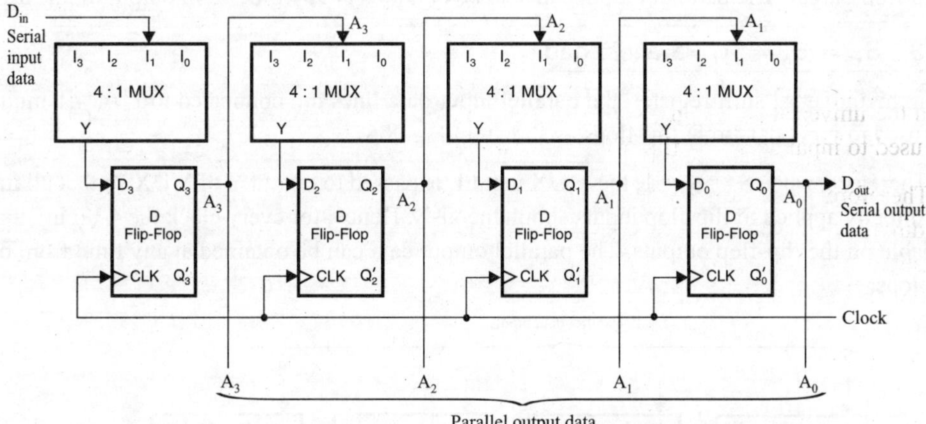

Fig. 8.65: *Mode-1 (shift right) operation of universal shift register.*

In the universal shift register, the output of a higher order flip-flop is connected to input I_1 of 4:1 MUX that is used to input data to D flip-flop as shown in Fig. 8.65.

Therefore, when $S_1 = 0$ and $S_0 = 1$, the MUX input I_1 is passed to output Y of MUX so that the output of higher order flip-flop is applied as input to lower order flip-flop. Hence, for every clock the output is shifted right by one position.

The serial input is applied to left most flip-flop input and serial output can be obtained from right most flip-flop output. The parallel output can also be obtained at any time from output of all flip-flops.

Mode-2 : $S_1 = 1$, $S_0 = 0$, Shift Left

In the universal shift register, the output of a lower order flip-flop is connected to input I_2 of 4:1 MUX that is used to input data to D flip-flop as shown in Fig. 8.66.

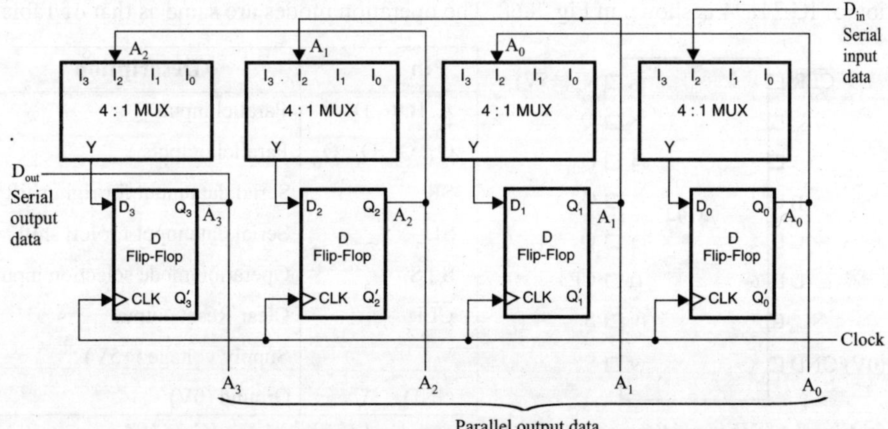

Fig. 8.66: *Mode-2 (shift left) operation of universal shift register.*

Therefore, when $S_1 = 1$ and $S_0 = 0$, the MUX input I_2 is passed to output Y of MUX so that the output of a lower order flip-flop is applied as input to higher order flip-flop. Hence, for every clock the output is shifted left by one position.

The serial input is applied to right most flip-flop input and serial output can be obtained from left most flip-flop output. The parallel output can also be obtained at any time from output of all flip-flops.

Mode-3 : $S_1 = S_0 = 1$, Parallel Load

In the universal shift register, the parallel input data lines are connected to I_3 of 4:1 multiplexers that are used to input data to D flip-flops as shown in Fig. 8.67.

Therefore, when $S_1 = S_0 = 1$, the MUX input I_3 is passed to output Y of MUX so that all input data bits are directly applied to flip-flop inputs simultaneously. Hence, for every clock the 4-bit input data will be available on the flip-flop outputs. The parallel output data can be obtained at any time from output of all flip-flops.

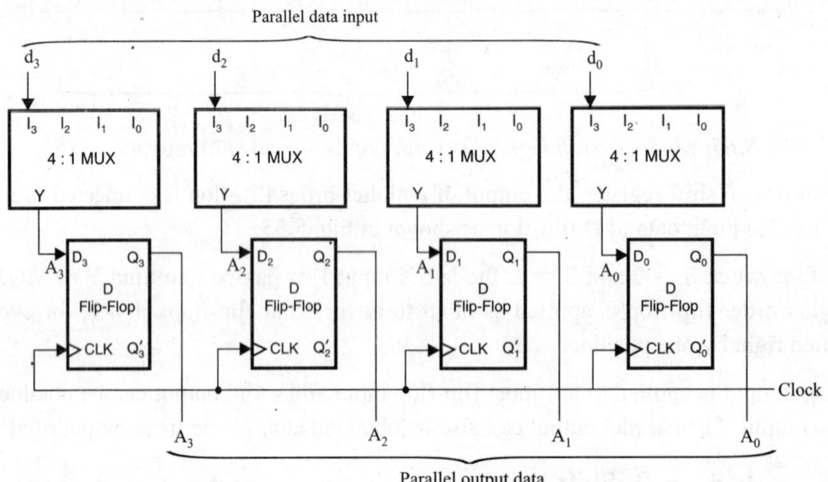

Fig. 8.67: Mode-3 (parallel load) operation of universal shift register.

The universal shift register discussed in this section is available as a standard IC 74194. The pin configuration of IC 74194 is shown in Fig. 8.68. The operation modes are same as that of Table 8.25.

Pin	Name	Pin	Name
1	$\overline{CLR}$	16	V_{cc}(+5V)
2	SR_{in}	15	Q_A
3	A	14	Q_B
4	B	13	Q_C
5	C	12	Q_D
6	D	11	CP
7	SL_{in}	10	S_1
8	(0V) GND	9	S_0

74194

Pin	Description
A, B, C, D	Parallel inputs
Q_A, Q_B, Q_C, Q_D	Parallel outputs
SR_{in}	Serial data input for right shift
SL_{in}	Serial data input for left shift
S_0, S_1	Operation mode selection inputs
$\overline{CLR}$	Clear/Reset output
V_{cc}	Supply voltage (+5V)
GND	Ground (0V)

Fig. 8.68: Pin configuration of 4-bit high speed shift register IC 74194.

8.6.6 Applications of Shift Register

(VTU, Aug/Sep'20, 8 Marks)

- **Serial-to-Parallel Conversion:** Shift registers can convert serial data into parallel data. This is useful for interfacing serial devices with parallel devices in microcontrollers, display devices or memory.

- **Parallel-to-Serial Conversion:** Shift registers can convert parallel data into serial data. This is useful for interfacing parallel devices with serial devices in microcontrollers, display devices or memory.

- **Data Storage and Memory Expansion:** Shift registers can be used as temporary storage elements to store data temporarily.

- **Data Serializing and de-serializing:** In communication systems, shift registers are often used to serialize parallel data for transmission over a serial link and de-serialize serial data back into parallel form at the receiving end.

- **Shift and Rotate Operations:** Shift registers can perform shift and rotate operations on data. This is useful in applications such as cryptography and error detection/correction algorithms and to implement arithmetic and logical operations in digital systems.

- **Frequency Division and Counting:** Shift registers can be configured as frequency dividers or counters.

- **LED and Display Drivers:** Shift registers can be used to implement multiplexed display drivers in LED based display devices to reduce power requirements.

- **Pattern Generation and Pattern Detection:** Shift registers can generate complex patterns by loading specific bit sequences into the register. They can also be used for pattern detection in applications such as data compression, image processing, or sequence recognition.

Serial Adder

A serial adder is a digital circuit that performs addition operation on binary numbers serially. The serial adder adds one bit at a time, starting from the least significant bit (LSB) to the most significant bit (MSB). Serial adders are commonly used in applications where resources are limited or when sequential processing is required. A 4-bit serial adder is shown in Fig. 8.69. The two data to be added are stored in shift registers and shifted to a full adder.

The binary digits (bits) are added one pair at a time using a full-adder circuit. The carry generated from the full adder is transferred to a D-flip flop. Thus, the output of this D-flip flop is then used as the carry input for the next pair of addition. The sum bit S is transferred to a third shift register. The input data are loaded parallely in shift registers and addition is performed serially and after all bits addition the sum can be read parallely.

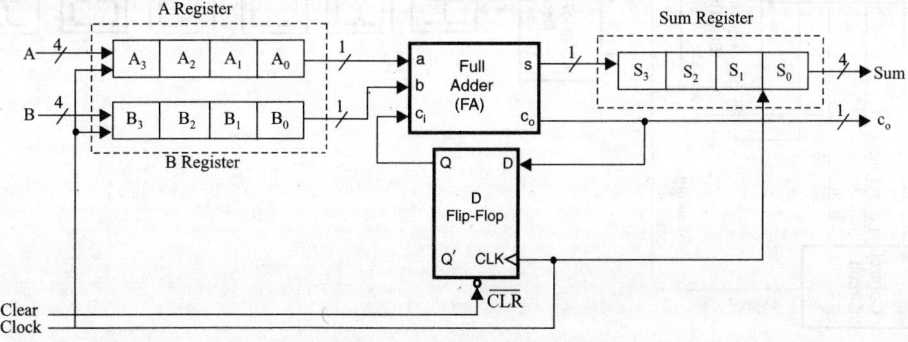

***Fig 8.69:** 4-bit serial adder.*

8.7 Synchronous Systems Model Development

Design and development of models of digital systems for specific applications are presented in this section. The digital systems are designed using logic gates, combinational and sequential devices.

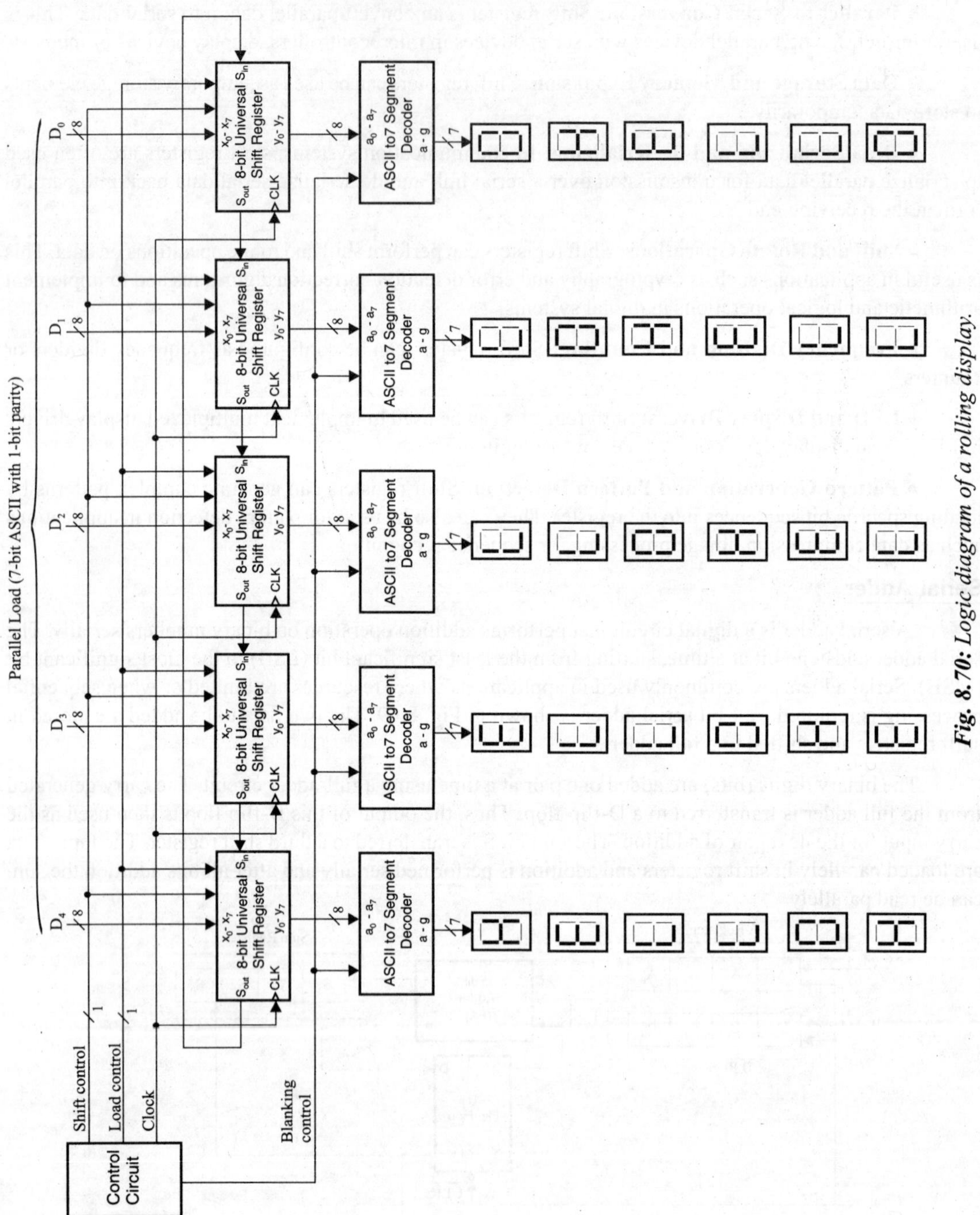

Fig. 8.70: Logic diagram of a rolling display.

8.7.1 Designing of Rolling Display

The rolling display is rolling a message with alpha-numeric symbols from left to right or vice-versa. In digital system, the alpha-numeric symbols are expressed as ASCII codes. In rolling display the message to be displayed is input as ASCII codes. Each character of message consists of 7-bit ASCII code and 1 parity bit. The 7-segment LED displays are selected to display the message.

Rolling display system to roll a 5 character word or message is developed using universal shift registers and ASCII-to-7-segment decoder as shown in Fig. 8.70. Here, the serial output of a lower order shift register is connected to serial input of next higher order shift register in order to left shift the characters serially from one register to another register.

The universal shift register has the facility to input and output the data both in parallel and serial. The message to be displayed is initially loaded in parallel to shift registers and then output in parallel to ASCII-to-7-segment decoder for display. After a delay called shift time the 8-bit characters loaded in shift registers are serially shifted bit-by-bit left 8 times so that the character data in one shift register move to next register.

During this shift time the 7-segment display has to be blanked using a blanking control signal to prevent ghosting of information. After shifting the ASCII characters the data is output in parallel for display. This process is repeated again and again, so that the message will roll from left to right as shown in Fig. 8.70. A control circuit has to be designed to take care of parallel loading, shifting and blanking control.

8.7.2 Designing of Real Time Clock

The real time clock is a digital clock that maintains the time in hours, minutes and seconds. The 12 hours real time clock is developed using ripple carry MOD counters and 7-segment LED displays as shown in Fig. 8.71. The clock is designed to show seconds using two 7-segment displays in order to display 0 to 59 seconds. Similarly, two 7-segment displays are used to display minutes from 0 to 59 and two 7-segment displays are used to display hours from 0 to 12. MOD-10 and MOD-6 counters are used to count seconds and minutes. MOD-10 and MOD-2 counters are used to count hours. The basic timing signal is selected as 1 *Hz* clock signal. The inverse of frequency is time period and so the time period or duration of every clock pulse is one second.

The 1 *Hz* clock is directly applied to clock of MOD-10 seconds counter, so that it is increment by 1 in every one second. The MOD-10 seconds counter will generate a reset signal in every 10th clock pulse to reset the counter and this reset is also divide-by-10 clock of 1 *Hz* clock signal. The frequency of this clock is 1/10 = 0.1 *Hz* and time period is 1/0.1 = 10 seconds. The 0.1 *Hz* clock out of MOD-10 seconds counter is used as input clock of the MOD-6 seconds counter, so that it increment by 1 in every 10 seconds.

The MOD-6 seconds counter will generate a reset signal in every 6th clock pulse to reset the counter and this reset is also divide-by-6 clock of 0.1 *Hz* clock signal. The frequency of this clock is 0.1/6 *Hz* and the time period is 6/0.1 = 60 seconds or 1 minute. The 0.1/6 *Hz* clock out of MOD-6 seconds counter is used as input clock of the MOD-10 minutes counter, so that it increment by 1 in every 60 seconds.

The divide-by-10 clock out of MOD-10 minutes counter will have a frequency of (0.1/6)/10 = 0.1/60 *Hz* and the time period is 60/0.1 = 600 seconds or 600/60 = 10 minutes and this clock is used as input clock of the MOD-6 minutes counter, so that it increment by 1 in every 10 minutes.

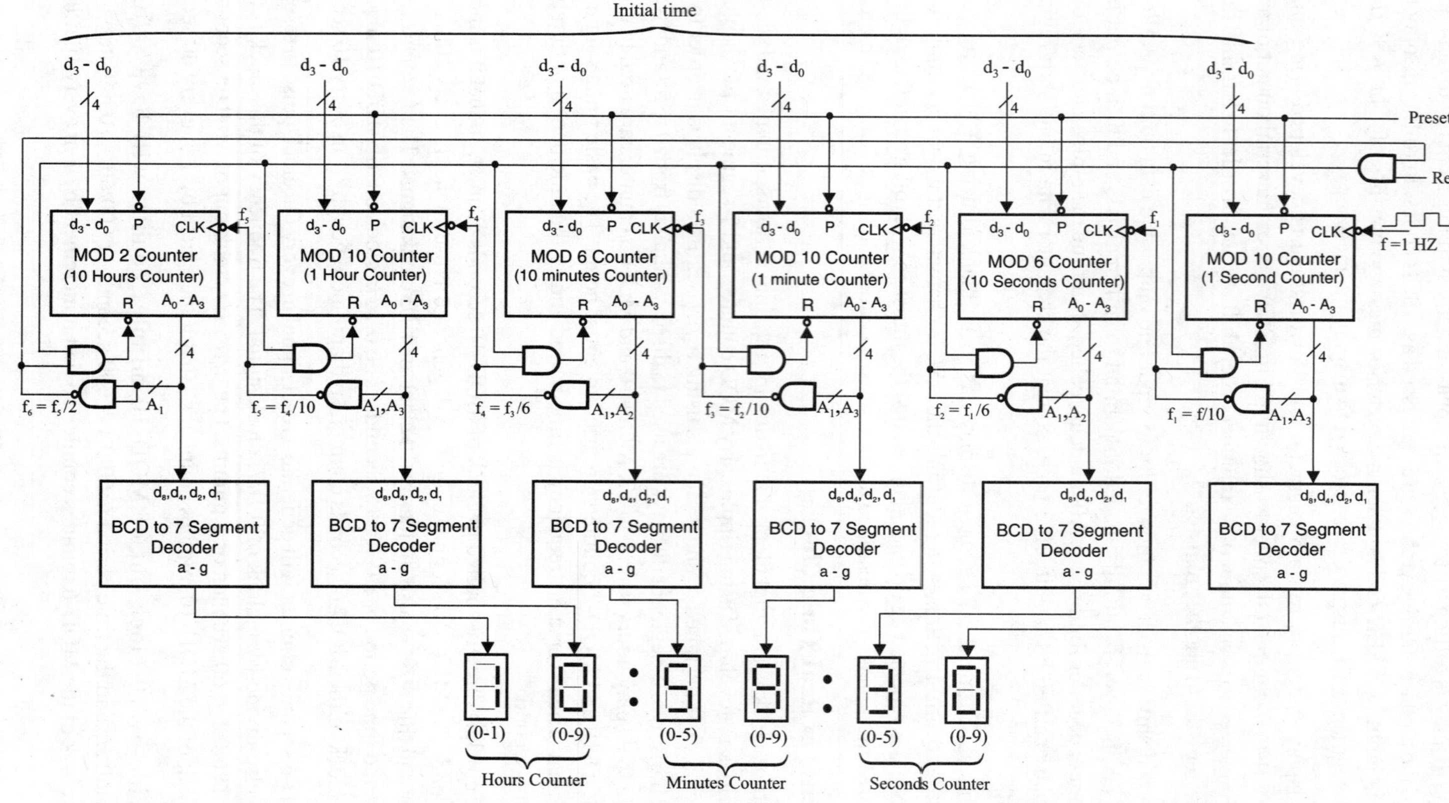

***Fig. 8.71:** Logic diagram of a 12 hours real time digital clock which displays seconds, minutes and hours.*

The divide-by-6 clock out of MOD-6 minutes counter will have a frequency of (0.1/60)/6 = 0.1/360 *Hz* and the time period is 360/0.1 = 3600 seconds or 3600/60 = 60 minutes = 1 hour and this clock is used as input clock of the MOD-8 hours counter, so that it increment by 1 in every 60 minutes.

The divide-by-10 clock out of MOD-10 hour's counter will have a frequency of (0.1/360)/10 = 0.1/3600 *Hz* and the time period is 3600/0.1 = 36000 seconds or 36000/60 = 600 minutes or 600/60 = 10 hours and this clock is used as input clock of the MOD-2 hour's counter, so that it increment by 1 in every 10 hours. The MOD-2 counter generates a divide-by-2 output clock signal with a frequency of (0.1/3600)/2 = 0.1/7200 *Hz* and with a time period 7200/0.1 = 72000 seconds or 72000/6000 = 12 hours, which is used as master reset to reset all counters when the time reaches 12 hours.

The real time clock needs a facility to load or preset initial timing in the beginning or during switch ON, so that the clock will maintain the time from this initial value.

8.8 Summary of Important Concepts

1. In sequential circuits with unused states, when clock pulses are applied, if the sequential circuit cycle through only unused states and never arrives at in a used state then this condition is called lock-out.
2. A register that goes through prescribed (or predefined) sequence of states upon the application of clock pulses is called a counter.
3. A counter that follows a binary count sequence is called binary counter.
4. An n-bit binary counter consists of n flip-flops and can count binary from 0 to $2^n - 1$.
5. The modulus of the counter refers to number states a counter will pass through before coming back to initial state.
6. An n-bit counter will have 2^n possible states and if the n-bit counter passes through all possible states before coming to initial state then it is called MOD-2^n counter.
7. An n-bit binary counter divides the clock frequency by 2^n and so n-bit counter is also called divide-by-2^n counter
8. The n-bit MOD-N counter uses only N states in the 2^n possible states of n-bit counter and the remaining states are unused states.
9. MOD-N counters can be designed using n-bit binary counter by using only N states of n-bit counter provided $N < 2^n$
10. The MOD-N counter divides the clock frequency by N and so it is also called divide-by-N counter.
11. The counters can be broadly classified into two categories: Ripple counters and synchronous counters.
12. In a ripple counter, a flip-flop output will act as clock for another flip-flop.
13. In synchronous counters all the flip-flops are triggered simultaneously by a common clock.
14. A binary ripple counter is constructed using a series of complementing flip-flops.
15. In up counter, the output will increment for every clock pulse and the count sequence will be from 0 to $2^n–1$.
16. In down counter, the output will decrement for every clock pulse and counter the count sequence will be from $2^n - 1$ to 0.
17. A BCD counter has to count only from 0 to 9, and when the count reaches 1010, the output has to be reset to all zero, so that next count sequence will start.
18. The BCD counter is also called MOD-10 counter.

19. The synchronous counters are designed using complementing flip-flops similar to ripple counters, but the clock is applied simultaneously to all the flip-flops.
20. In synchronous counters, the flip-flop inputs are applied through some combinational logic circuits.
21. A ring counter is a circular shift register with only one flip-flop output being set at a time and all other flip-flop outputs are cleared.
22. A ring counter produces a sequence of timing signals that are shifted by one clock time period.
23. Johnson counter is basically a ring counter with complement of last flip-flop connected to input of first flip-flop. This arrangement is called switch tail ring counter.
24. In general, k bit Johnson counter will generate 2k distinguishable states and they decoded to generate 2k unique output signals that resembles the output of a decoder.
25. A register is a group of flip-flops that can store binary information.
26. Each flip-flop in a register can store one bit binary and so, in order to store n-bit binary the register should have n flip-flops.
27. A register capable of shifting the binary information from one flip-flop to another (or one cell to another) is called shift register.
28. The four types of shift registers are serial-in serial-out, serial-in parallel-out, parallel-in serial out and parallel-in parallel out shift register.
29. A shift register that can accept serial or parallel input data and output data in parallel along with right or left shift is called universal shift register.

8.9 Short-Answer Questions

Q8.1 Define lock-out condition?

In sequential circuits with some don't-care or unused states, the sequential circuit may find itself in some unused state during starting or switch ON. If next state of the unused state is again some unused one and if the sequential circuit cycle through only unused states and never arrives at in a used state then this condition is called lock-out.

Q8.2 Define counter and its classifications.

A register that goes through prescribed (or predefined) sequence of states upon the application of clock pulses is called a counter. The counters are used to generate or go through a sequence binary code known as count sequence.

The counters can be broadly classified into two categories: Ripple counters and synchronous counters.

In a ripple counter, a flip-flop output will act as clock for another flip-flop.

In synchronous counters all the flip-flops are triggered simultaneously by a common clock.

Q8.3 Define binary counter.

A counter that follows a binary count sequence is called binary counter. An n-bit binary counter consists of n flip-flops and can count binary from 0 to $2^n - 1$.

Q8.4 What is switch tail ring counter?

Johnson counter is basically a ring counter with complement of last flip-flop connected to input of first flip-flop. This arrangement is called switch tail ring counter.

Q8.5 Explain ring counter.

A ring counter is a circular shift register with only one flip-flop output being set at a time and all other flip-flop outputs are cleared. Using preset facility, a 1 is loaded in one flip-flop output and all other outputs are loaded 0. When clock pulses are applied the 1 will shift by one position. This produces a sequence of timing signals that are shifted by one clock time period.

Q8.6 How to make the counter self starting?

In order to make the counter self starting the unused states can be connected to enter used states.Once the counter enter an used state it will circle through used states and counter becomes self starting.

Q8.7 How many fiip-flops will be complemented in a 10-bit ripple counter to reach the next count after the count value of 1001100111?

Add 1 to the binary value 1001100111 in order to get next count value.

$$\begin{array}{r} 1001100111 \\ +\ 1 \\ \hline 1001101000 \\ \hline \end{array}$$

In the result of above addition we observe that 4 bits are complemented. In a counter, each bit is stored in a flip-flop. Therefore, 4 flip-flops will be complemented in a 10-bit ripple counter to reach the next count after the count of 1001100111.

Q8.8 Define BCD or MOD-10 counter.

A BCD counter has to count only from 0 to 9, and when the count reaches 10_{10}, the output has to be reset to all zero, so that next count sequence will start. The BCD counter is also called MOD-10 counter.

Q8.9 Comment about a presettable counter and ripple counter. *(AU, Nov/Dec'18, 2 Marks)*

A presettable counter will have a preload facility to load a count value before the start of the count and count sequence starts from this preloaded value. But in a ripple counter without a preload facility the count sequence will be all possible combinations of binary.

Q8.10 Differentiate synchronous and asynchronous counter *(VTU, Aug/Sep'20, 4 Marks)*

Asynchronous Counter	Synchronous Counter
1. In asynchronous counter output of a flip-flop is used as clock for the next flip-flop.	1. In synchronous counter output of a flip-flop is not used as clock of any other flip-flop.
2. All the flip-flops are not clocked simultaneously.	2. All the flip-flops are clocked simultaneously.
3. Logic circuit is very simple even if there are large number of states.	3 Logic circuit becomes complex with increase in number of states.
4. Since the clock propagates through number of flip-flops, the counters are very slow in reaching final output.	4. Since the clock is simultaneously applied to the all flip-flops, there is no long delay in reaching final output.

Q8.11 Define register.

A register is a group of flip-flops that can store binary information. Each flip-flop in a register can store one bit binary and so, in order to store n-bit binary the register should have n flip-flops.

Q8.12 Define shift register. ***(AU, Apr/May'18, 2 Marks)***

A register capable of shifting the binary information from one flip-flop to another (or one cell to another) is called shift register.

Q8.13 Write a types of shift register.

There are four types of shift registers.

1. Serial-in Serial-out shift register
2. Serial-in Parallel-out shift register
3. Parallel-in Serial-out shift register
4. Parallel-in Parallel-out shift register.

Q8.14 A flip-flop has 5 ns delay from the time the clock edge occur to the time the output is complemented. What is the maximum delay in a 10-bit binary ripple counter that uses these flip-flops? What is maximum frequency the counter can operate reliably? ***(AKTU, Dec'22, 2 Marks)***

For a 10-bit binarry ripple counter there are 10 number of flip-flops.

Maximum delay of one flip-flop = 5 *ns*

∴ Maximum the delay of 10 flip-flop = 10 × 5 *ns* = 50 *ns*

∴ The maximum frequency of 10-bit binary ripple counter $= \frac{1}{\text{Maximum delay}} = \frac{1}{50 \times 10^{-9}}\ Hz$

$$= 0.2 \times 10^{8}\ Hz$$

$$= 20 \times 10^{6}\ Hz = 20\ MHz$$

8.10 Exercises

I. Fill in the blanks

1. A register capable of shifting the binary information from one flip-flop to another is called ______.
2. A counter that follows a binary counter sequence is called ______.
3. A gated D latch does not have ______.
4. A register is a group of ______.
5. The number of flip-flops needed to construct a MOD-17 synchronous counter is ______.
6. The BCD counter is also called ______.
7. A ______ is a group of flip-flops that can store binary information.
8. A ______ counter produces a sequence of timing signals that are shifted by one clock time period.
9. MOD-N counters can be designed using n-bit binary counter by using only N states of n-bit counter provided ______.
10. A ______ is constructed using a series of complementing flip-flops.
11. A n-bit binary counter consists of n flip-flops and can count binary from ______.
12. A register that goes through prescribed sequence of states upon the applications of clock pulses is called a ______.
13. The counters are used to generate or go through a sequence of binary code known as ______.
14. A 3-bit binary counter has 8 possible states and so it is called ______.

15. A binary ripple counter is constructed using a series of complementing _______.
16. Johnson counter is basically a ring counter with complement of last flip-flop connected to input of first flip-flop this arrangement called _______.
17. All the flip-flops in a _______ are controlled or synchronized to a common clock.
18. The rolling display is rolling a message with _______ symbols from left to right.
19. The alpha-numeric symbols are expressed as _______ codes.
20. A shift register that can accept serial or parallel input data and output data in parallel along with right or left shift is called _______ .

Answers

1. shift register	6. MOD-10 counter	11. 0 to $2^n - 1$	16. swing tail ring counter
2. binary counter	7. register	12. counter	17. register
3. a clock input	8. ring	13. counter sequence	18. alpha-numeric
4. flip-flops	9. $N < 2^n$	14. MOD-8 counter	19. ASCII
5. 5	10. binary ripple counter	15. flip-flops	20. universal shift register

II. State whether the following statements are True or False

1. The serial-in parallel-out shift registers transfers data from one parallel data bus to another parallel data bus by transferring one bit at a time across a single line.
2. A ripple counter is an asynchronous counter.
3. Counters are common components in digital clocks.
4. A pulse train can be delayed by a finite number of clock periods using a serial in serial out shift register.
5. A ring counter consisting of seven flip-flops will have 5 states.
6. A BCD counter has to count only from 0 to 8, when the count reaches 88, the output has to be reset to all zero, so that next count sequence will start.
7. In up counter, the output will increment for every clock pulse and the count sequence will be from 0 to $2^n - 1$.
8. The modulus of the counter refers to number states a counter will pass through before coming back to initial state.
9. An a n-bit counter will have 2n possible states and if the n-bit counter passes through all possible states before coming to initial state then it is called MOD-2n counter.
10. An n-bit binary counter divides the clock frequency by 2n and so n-bit counter is also called divide-by-2n counter.
11. In a synchronous counter, a flip-flop output will act as clock for another flip-flop.
12. In ripple counters all the flip-flops are triggered simultaneously by common clock.
13. In down counter, the counter output decrement for every clock pulse and in an n-bit down counter the
14. The BCD counter is also called MOD-6 counter
15. The combinational logic circuits will consider the output of one or more flip-flops to decide the input to flip-flops for next state outputs.
16. A ring counter is a circular shift register with only one flip-flop output being set at a time and all other flip-flop outputs are cleared.
17. Johnson counter is basically a ring counter with complement of last flip-flop connected to input of first flip-flop.

18. The digital systems are designed using only logic gates.

19. During this shift time the 7-segment display has to be blanked using a blanking control signal to prevent ghosting of information.

Answers

1. False	5. False	9. True	13. True	17. True
2. True	6. False	10. True	14. False	18. False
3. True	7. True	11. False	15. True	19. True
4. False	8. True	12. False	16. True	

III. Choose the right answer for the following questions

1. *A 0 to 6 counter consists of 3 flip-flops and a combination circuit of 2 input gate. The combination circuits consists of*

   a) one AND gate b) one OR gate
   c) one AND gate and one OR gate d) two AND gate

2. *A 4 bit ripple counter and a 4 bit synchronous counter are made using flip-flops having a propagation delay of 10 ns each. If the worst case delay in the ripple counter and synchronous counter be R and S respectively then.*

   a) R = 10 ns, S = 40 ns
   b) R = 40 ns, S = 10 ns
   c) R = 10 ns, S = 30 ns
   d) R = 30 ns, S = 10 ns

3. *A ring counter consisting of seven flip-flops will have*

   a) 5 states b) 7 states c) 32 states d) Infinie states

4. *The initial state of a MOD-16 down counter is 0110. After 37 clock pulses, the state of the counter will be*

   a) 1011 b) 0110 c) 0101 d) 0001

5. *A 4-bit binary ripple counter uses flip-flops with a propagation delay time of 20 ns each. The maximum possible time required for change of state will be,*

   a) 25 ns b) 80 ns c) 75 ns d) 100 ns

6. *The minimum number of flip-flops needed to make MOD-2 counter is*

   a) 1 b) 2 c) 3 d) 4

7. *A decade counter has a maximum of*

   a) 8 states b) 10 states c) 9 states d) 7 states

8. *The input clock frequency of the MOD-5 counter is 25 KHz. Then the output has the frequency of*

   a) 5 KHz b) 2 KHz c) 0.2 KHz d) 1 KHz

9. *The number of flip-flops needed to construct a MOD-15 synchronous counter is*

   a) 9 b) 4 c) 3 d) 2

10. ***The modulus of the counter shown in Fig. 10***

a) 4

b) 3

c) 2

d) 1

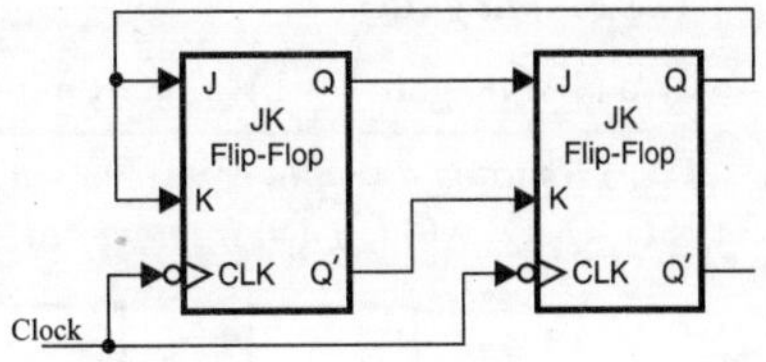

Fig. 10.

11. ***The number of flip-flops needed to construct a MOD-17 synchronous counter is***

a) 9 **b)** 5 **c)** 3 **d)** 4

12. ***A switch-tail ring counter is made by using a single D flip-flop. The resulting circuit is a***

a) SR flip-flop **b)** JK flip-flop **c)** D flip-flop **d)** T flip-flop

13. ***The counter shown in Fig. 13 is***

a) MOD-3 ripple counter

b) MOD-7 synchronous counter

c) MOD-5 ripple counter

d) MOD-7 ripple counter

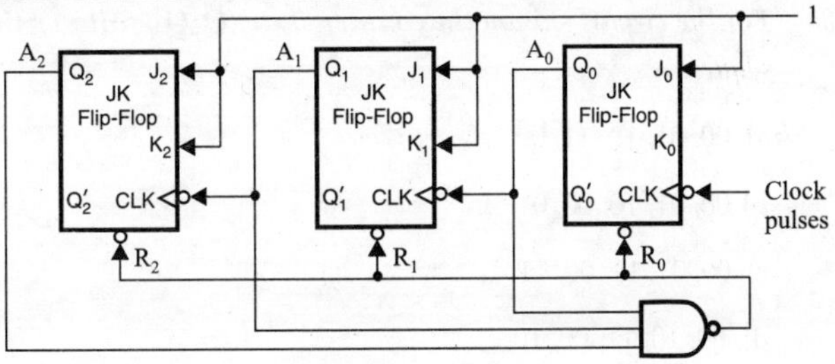

Fig. 13.

14. ***The modulus of the counter in Fig. 14 is***

a) 2

b) 3

c) 4

d) 5

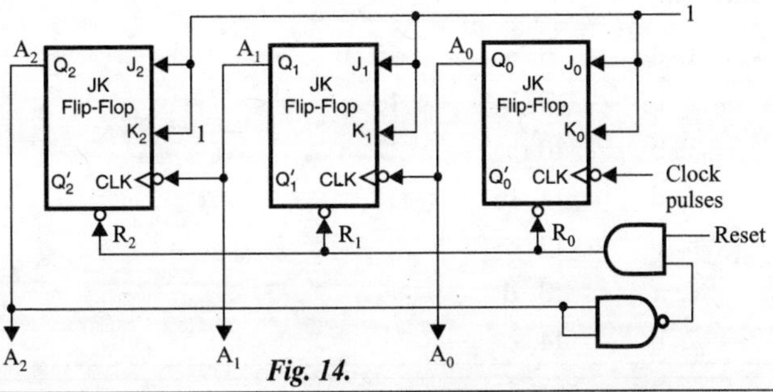

Fig. 14.

15. ***Fig. 15 shows a MOD K counter, here K is equal to***

a) 1

b) 2

c) 3

d) 4

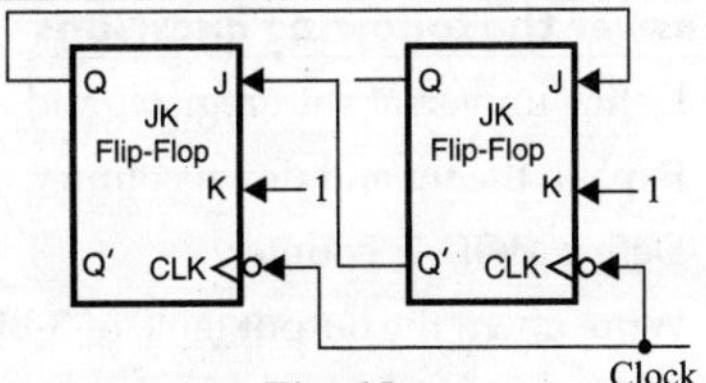

Fig. 15.

16. ***In the module-6 ripple counter, the output of the 2 input gate is used to clear the JK flip-flops.***

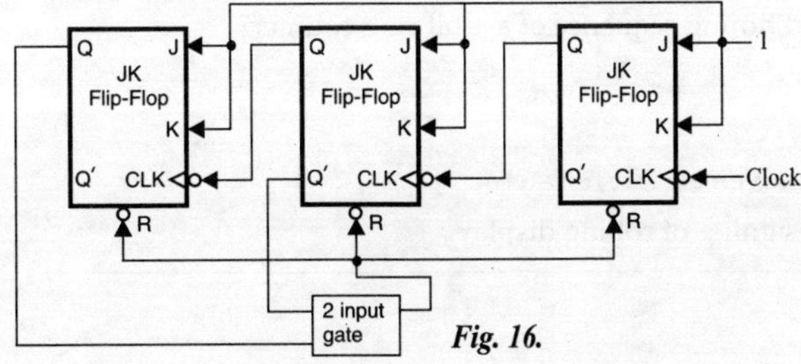

Fig. 16.

The 2 input gate is

a) a NAND gate **b)** a NOR gate **c)** an OR gate **d)** an AND gate

17. *Fig. 17 shows a ripple counter using positive edge triggered flip-flops. If the present state of the counter is $Q_2\ Q_1\ Q_0 = 011$. Then its next state ($Q_2\ Q_1\ Q_0$) will be*

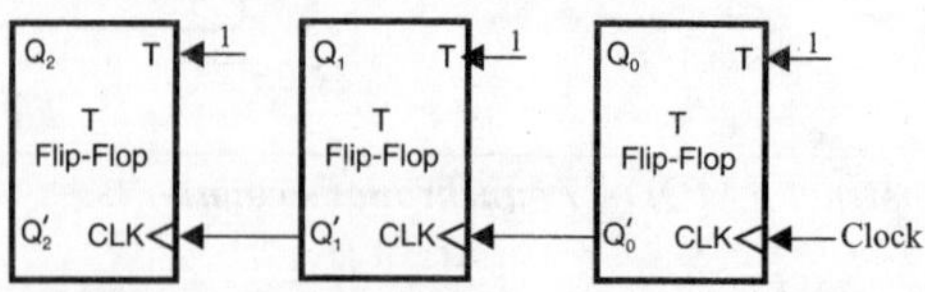

Fig. 17.

a) 010 **b)** 100 **c)** 111 **d)** 101

18. *For the circuit shown the counter state ($Q_1\ Q_2$) follows the sequence*

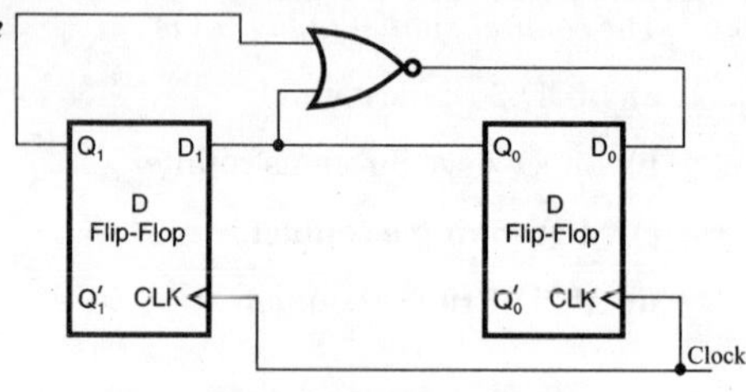

Fig. 18.

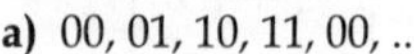

a) 00, 01, 10, 11, 00, ...

b) 00, 01, 10, 00, 01, ...

c) 00, 01, 11, 00, 11, ...

d) 00, 10, 11, 00, 10, ...

Answers

1. d	8. a	15. b
2. b	9. b	16. a
3. b	10. a	17. a
4. d	11. b	18. b
5. b	12. c	
6. a	13. d	
7. b	14. c	

V. Answer the following questions

1. Define universal shift register and its operation mode.
2. Explain the up and down counter.
3. Define MOD-2^n counter.
4. Write down the output table of 3-bit binary counter.
5. Write down the output table of 4-bit binary counter.
6. Explain synchronous counter.
7. Draw the logic diagram and timing sequence of a 4-bit ring counter?
8. Define divide-by-N counter.
9. Explain ripple counter.
10. List out of the features for universal shift register.
11. Write short notes on the designing of rolling display.

VI. Solve the following problems

E8.1 Design a counter to generate the binary sequence : 0, 1, 2, 4, 6 using D flip-flops.

E8.2 Design a counter using T flip-flops that goes through states 0, 2, 4, 6, 7, 0. Is the counter is self starting?

E8.3 Using JK flip-flops, design a counter to generate the binary sequence 0, 1, 2, 4, 5, 6. Draw the logic circuit of the counter.

E8.4 Design a 4-bit binary synchronous counter using D flip-flops.

E8.5 The flip-flop input equations for a BCD counter using T flip-flops are given in below. Obtain the input equations for a BCD counter that uses a) JK flip-flops and b) D flip-flops. Compare the three designs to determine which is most efficient.

$T_{A1} = 1$; $T_{A2} = A_8'A_1$; $T_{A4} = A_2A_1$; $T_{A8} = A_8A_1 + A_4A_2A_1$

E8.6 Design a MOD-5 counter using JK flip-flops.

Answers

E8.1

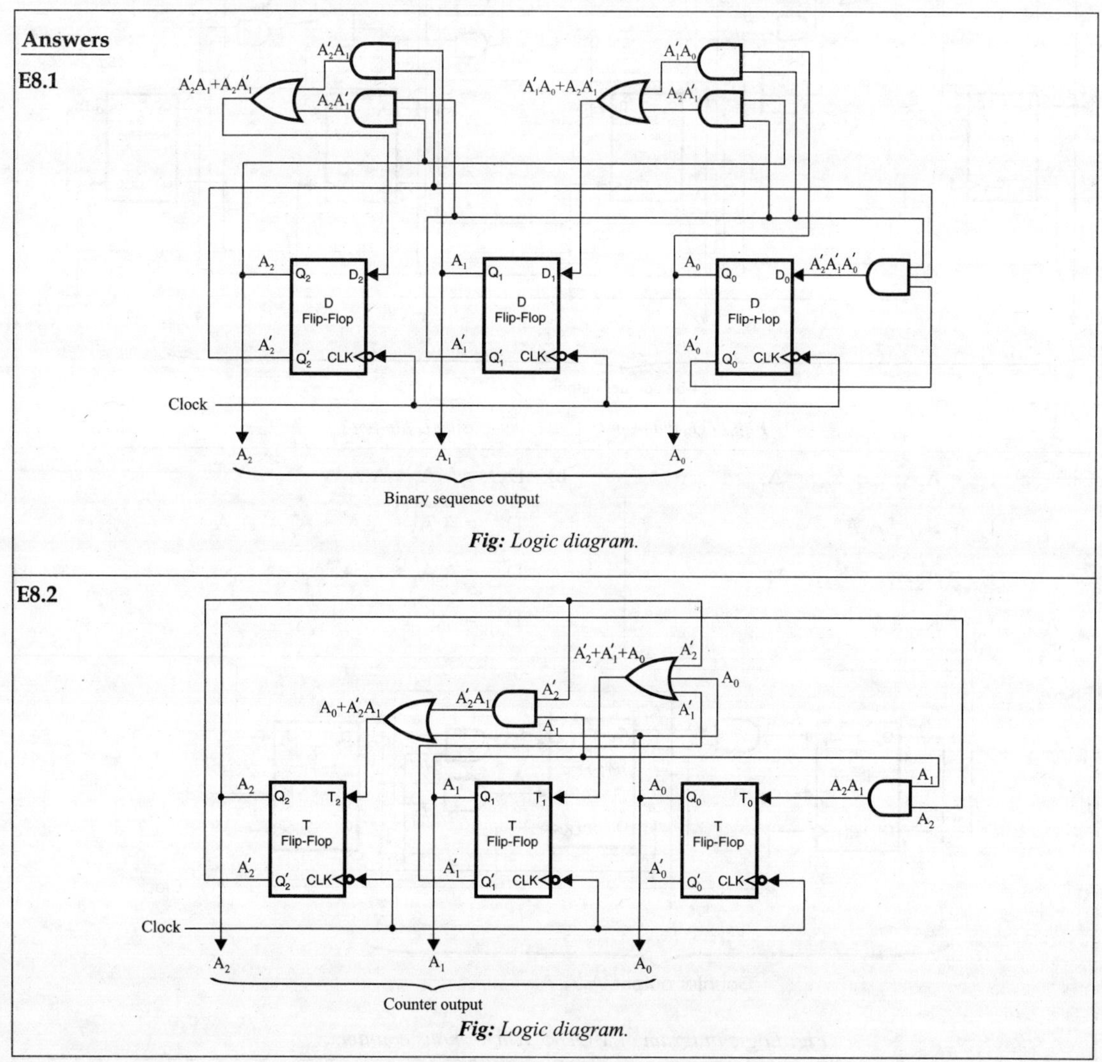

***Fig:** Logic diagram.*

E8.2

***Fig:** Logic diagram.*

E8.3

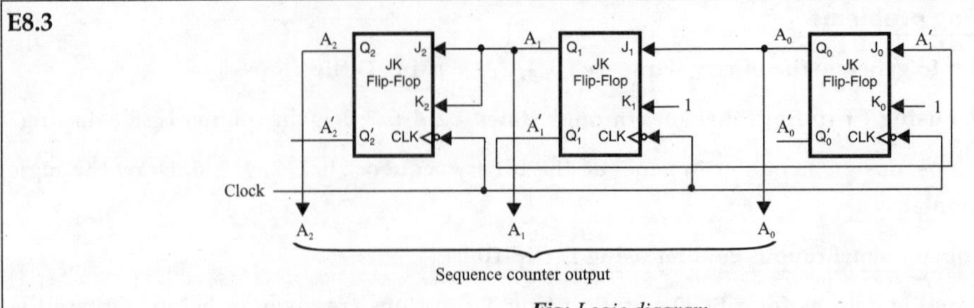

***Fig:** Logic diagram.*

E8.4

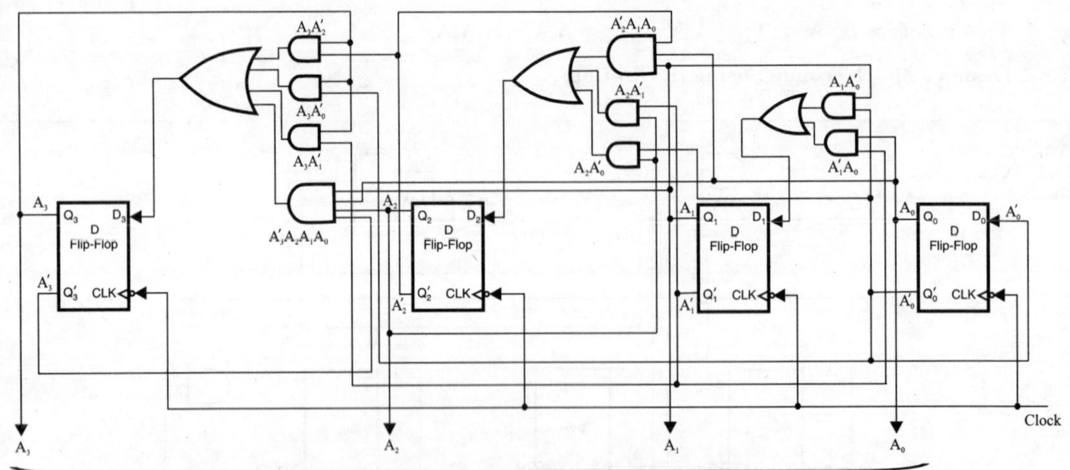

***Fig:** Logic diagram of 4-bit counter using D flip-flops.*

E8.5 a) $J_{A8} = A_4 A_2 A_1$; $K_{A8} = A_1$

$J_{A4} = K_{A4} = A_2 A_1$

$J_{A2} = A_8' A_1$; $K_{A2} = A_1$

$J_{A1} = K_{A1} = 1$

b) $D_{A8} = A_8 A_1' + A_4 A_2 A_1$

$D_{A4} = A_4 A_2' + A_4 A_1' + A_8' A_4' A_2 A_1$

$D_{A2} = A_2 A_1' + A_8' A_2' A_1$

$D_{A1} = A_1'$

E8.6

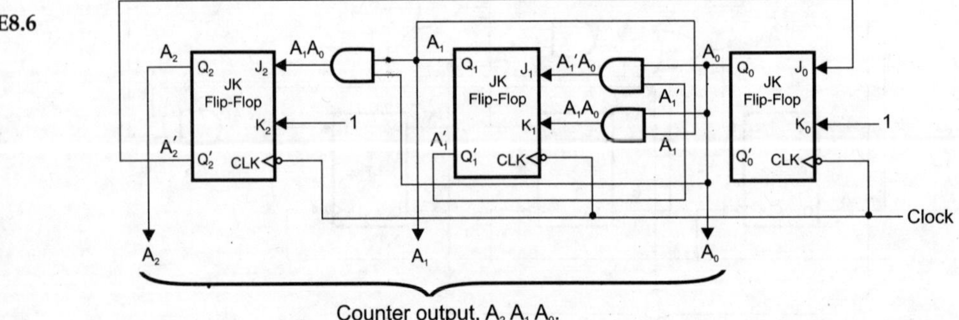

***Fig:** Logic diagram of MOD-5 synchronous counter..*

CHAPTER 9

ASYNCHRONOUS SEQUENTIAL CIRCUITS

9.1 Introduction to Asynchronous Sequential Logic Circuits

The combinational logic circuits with feedback and without clock are called ***asynchronous sequential circuits***. Since there is no clock, the changes in internal states occur immediately when there is a change in input and so the response in an asynchronous sequential circuit will be fast. In the design of asynchronous sequential circuit sufficient care must be taken to ensure that every change in state leads to a stable condition.

Table 9.1: Differences between Synchronous and Asynchronous Sequential Circuits

Synchronous Sequential Circuit	Asynchronous Sequential Circuit
1. Provided with clock	1. Do not have a clock
2. Changes in internal states occur in response to clock.	2. Changes in internal states occur when there is a change in input variables.
3. Memory elements are clocked flip-flops.	3. Memory elements are either latches (unclocked flip-flops) or time-delay elements.

Basically, an asynchronous sequential circuit can be expressed as a combinational circuit with feedback paths having delay elements. The asynchronous sequential circuits will have input, output and state variables. The present and next state variables of asynchronous sequential circuit are called secondary and excitation variables respectively. The block diagram of asynchronous sequential circuit is shown in Fig. 9.1.

Here, $x_1, x_2, \ldots, x_n$ = n input variables

$Z_1, Z_2, \ldots, Z_m$ = m output variables

$Y_1, Y_2, \ldots, Y_k$ = k excitation variables (next state variables)

$y_1, y_2, \ldots, y_k$ = k secondary variables (present state variables)

In asynchronous circuits the state transitions occur when there is a change in input. The circuit will attain a stable condition when excitation (or next state) variables are same as secondary (or present) state variables. Therefore, a change in input is allowed only after the circuit attains a stable state due to previous change in input. Also, when there are multiple inputs, only one input is allowed to change at any one time in order to ensure proper operation of circuit and this is called fundamental mode of operation. Therefore, the ***fundamental mode*** of operation is defined as the operating mode in which only one input is allowed to change when the circuit is in stable state.

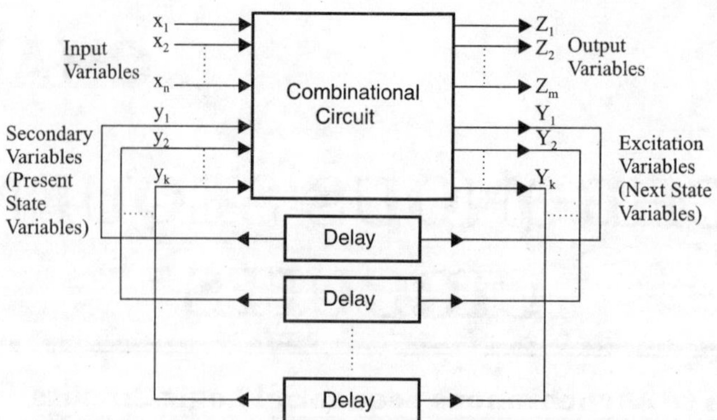

Fig. 9.1: Block diagram representation of an asynchronous sequential circuit.

9.1.1 Fundamental and Pulse Mode Sequential Circuits

In fundamental mode asynchronous sequential circuits the inputs are level triggered inputs, whereas in pulse mode asynchronous sequential circuits the inputs are pulses of sufficient duration for the output to change for a given input.

Design of Fundamental Mode Asynchronous Sequential Circuits

The design of fundamental asynchronous sequential logic circuit is designing a race-free and hazard free logic circuit from the given system specifications. The following procedural steps can be followed for the design.

- Develop a primitive flow table from the word description of asynchronous circuit
- Minimize the number of states in the flow table.
- Assign race-free binary codes to states.
- Construct the transition table.
- Construct excitation maps and find Boolean equations for a hazard free realization.
- Check for essential hazards.
- Draw the logic circuit by implementing methods to eliminate essential hazards.

The examples presented in sections 9.1 to 9.6 in this chapter are related to design of fundamental mode asynchronous circuits.

9.1.2 Transition Table

In asynchronous sequential circuit, the state table is called ***transition table***. Practically, the state table itself is listing of all possible transitions between the states of a circuit. In asynchronous circuit state table, there should be at least one next state that is equal to present state in each row.

The procedure to form transition table is as follows,

1. Identify all feedback paths in the circuit. Let the circuit has k number of feedback paths.
2. Assign secondary (or present state) and excitation (or next state) variables to each feedback path. Let, y_1, y_2, ..., y_k be secondary variables and Y_1, Y_2, ..., Y_k be excitation variables.

3. Now, the excitation variables are functions of secondary variables and circuit external inputs. Determine the Boolean equations for excitation variables in terms of secondary variables and inputs. Note that, the secondary variables are inputs to a combinational circuit and excitation variables are output from the combinational circuit. The excitation variables are given as inputs to feedback paths.
4. Draw maps similar to K-maps for every excitation variable, Y_1, Y_2, ..., Y_k. In these maps, the rows are formed by secondary variables and columns by inputs.
5. Now, there will be k number of maps. Combine all the k number of maps into a single map, by combining the corresponding entries in each square of individual map. Now each square will have k-bit binary representing next state.
6. In each row mark the next states ($Y_1 Y_2 \ldots Y_k$) that are same as present state ($y_1 y_2 \ldots y_k$) by circles. The circled entries represent stable states and uncircled entries represent unstable states.

The ***transition table*** lists the possible state transitions for all possible changes in inputs and so it can be used to analyze the circuit behavior.

Consider the asynchronous sequential circuit shown in Fig. 9.2. This circuit has one input and output and two feedback paths. Let us assume that the delay in feedback is due to propagation delay in the conducting wires that connect the output and input.

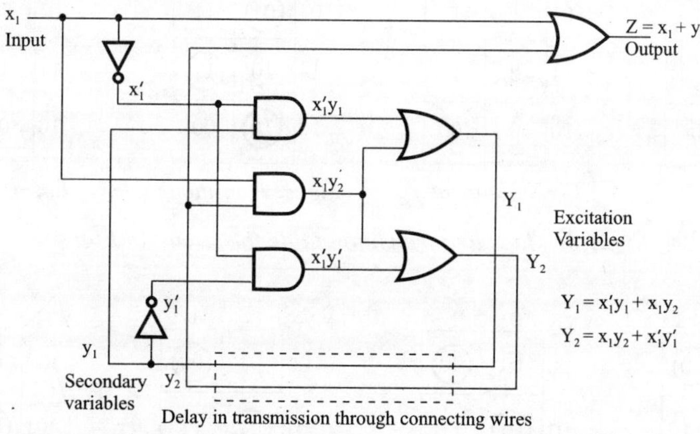

***Fig. 9.2:** Example of an asynchronous sequential circuit.*

Let, x_1 = Input

y_1, y_2 = Secondary variables

Y_1, Y_2 = Excitation variables

Z = Output

Now, with reference to Fig. 9.2, we can write the following Boolean equations for excitation variables, Y_1 and Y_2 and output Z.

$$Y_1 = x_1' y_1 + x_1 y_2$$

$$Y_2 = x_1 y_2 + x_1' y_1'$$

$$Z = x_1 + y_2$$

Using the above Boolean equations the excitation variables, Y_1 and Y_2 are evaluated for all possible combinations input x_1 and secondary variables y_1 and y_2 and listed in Table 9.2.

Table 9.2: Evaluation of Excitation Variables and Output

y_1	y_2	x_1	y_1'	y_2'	x_1'	$x_1'y_1$	x_1y_2	$x_1'y_1'$	Y_1	Y_2	$Z = x_1 + y_2$
0	0	0	1	1	1	0	0	1	0	1	0
0	0	1	1	1	0	0	0	0	0	0	1
0	1	0	1	0	1	0	0	1	0	1	1
0	1	1	1	0	0	0	1	0	1	1	1
1	0	0	0	1	1	1	0	0	1	0	0
1	0	1	0	1	0	0	0	0	0	0	1
1	1	0	0	0	1	1	0	0	1	0	1
1	1	1	0	0	0	0	1	0	1	1	1

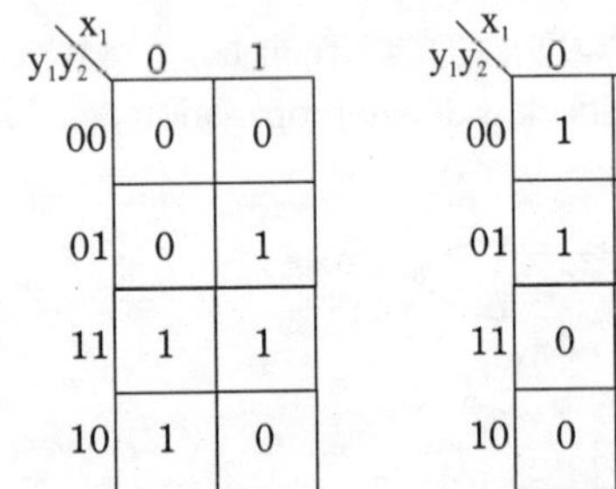

Fig. a: Map for Y_1. *Fig. b: Map for Y_2.*

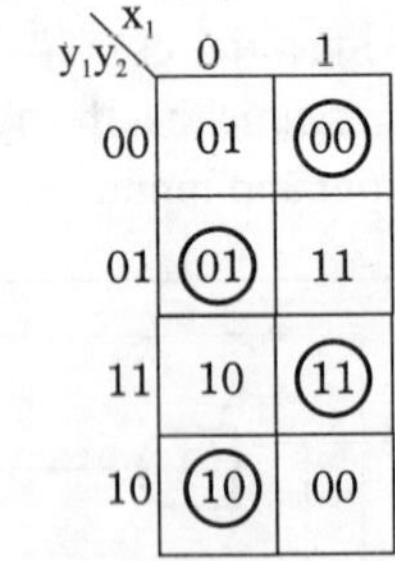

Fig. c: Transition table.

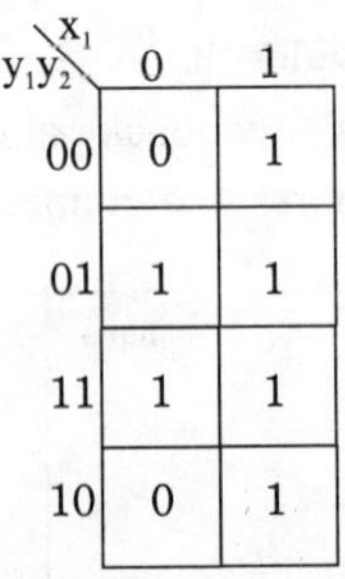

Fig. d: Output map.

***Fig. 9.3:** Maps and transition table for circuit of Fig. 9.2.*

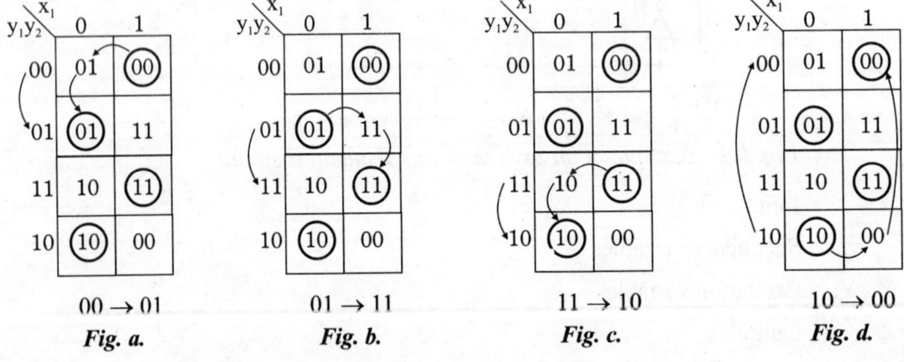

Fig. a. ***Fig. b.*** ***Fig. c.*** ***Fig. d.***

***Fig. 9.4:** State transitions due to change in input.*

Maps are constructed for Y_1 and Y_2 by taking possible values of secondary variables y_1 and y_2 on rows and input x_1 on columns as shown in Figs. 9.3(a) and (b) and the squares in the maps are filled with values of Y_1 and Y_2 listed in Table 9.2. The transition table is formed by combining the maps of Figs. 9.3(a) and (b) into a single map as shown in Fig. 9.3(c). The output map is shown in Fig. 9.3(d).

9.1.3 Stable and Unstable States

A state is a circuit condition represented by a set of variables called ***state variables***.

The states are represented either by letter symbols a, b, c, d, ... or by binary codes. The circuit of Fig. 9.2 has two state variables and so states are represented by 2-bit binary codes 00, 01, 11 and 10.

The circuit will change from one state to another state when there is a change in input. During a state transition when the next state is same as present state then the circuit attains a stable condition and the corresponding state is called ***stable state***. On the other hand in a state transition, if the next state is not equal to present state then the corresponding state is called ***unstable state***.

In the transition table shown in Fig. 9.3(c), the entries in each square represents next state. Therefore, in each row, the squares that have identical values with present state are marked with circles. The circled entries represent stable states and uncircled entries represent unstable states.

Normally, an asynchronous sequential circuit will remain in a stable state for a given input. The transition from one stable state to another stable state occurs due to changes in input.

Consider the circuit shown in Fig. 9.2 and its transition table shown in Fig. 9.3. In this circuit, the following four state transitions due to changes in input are possible as shown in Fig. 9.4.

When in state 00, if the input changes from 1 to 0, then state transitions occur as shown in Fig. 9.4(a).

When in state 01, if the input changes from 0 to 1, then state transitions occur as shown in Fig. 9.4(b).

When in state 11, if the input changes from 1 to 0, then state transitions occur as shown in Fig. 9.4(c).

When in state10, if the input changes from 0 to 1, then state transitions occur as shown in Fig. 9.4(d).

In an asynchronous circuit, the state changes occurs immediately when there is a change in input, and so the input value can be combined with state to declare a condition called total state of the circuit. All possible combinations of state variables and inputs are called ***total states***. The total state in the circuit of Fig. 9.2 is given by the possible binary values of $y_1y_2x_1$. With three variables 8 different total states can be formed, in which the four states that are circled in transition table are stable states and the other four states are unstable

Stable total states: 001, 010, 111, 100

Unstable total states: 000, 011, 110, 101

Using the transition table, the state table for asynchronous circuit similar to that of a state table of synchronous circuit can be constructed as shown in Table 9.3. In state table of asynchronous circuit, at least one entry of each row should have same present state and next state.

Table 9.3: State Table of the Circuit of Fig. 9.2

Present State		Next State				Output	
		$x_1 = 0$		$x_1 = 1$		$x_1 = 0$	$x_1 = 1$
y_1	y_2	Y_1	Y_2	Y_1	Y_2	Z	Z
0	0	0	1	0	0	0	1
0	1	0	1	1	1	1	1
1	0	1	0	0	0	0	1
1	1	1	0	1	1	1	1

Unstable Condition in Asynchronous Circuit

Consider the transition table of an asynchronous circuit shown in Fig. 9.5(a). In this transition table, there, is no stable state when input is 11. Hence, when input changes from 01 to 11 or 10 to 11 the circuit will enter an unstable condition with next state keeps on changing from 1 to 0 and 0 to 1 continuously as shown in Figs. 9.5(b) and (c).

While designing asynchronous sequential circuits such unstable conditions should be avoided and this can be achieved by ensuring one stable state for every input.

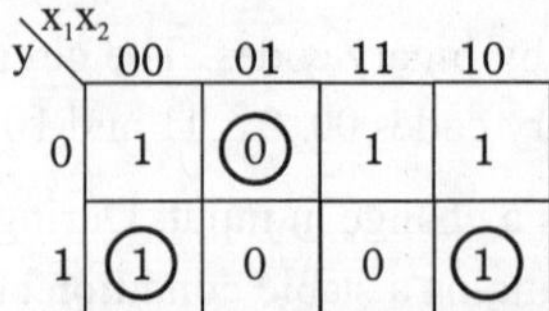

Fig. a: Transition table.

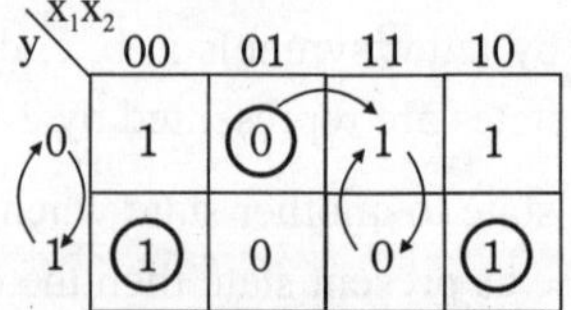

Fig. b: Unstable condition when input change from 01 to 11.

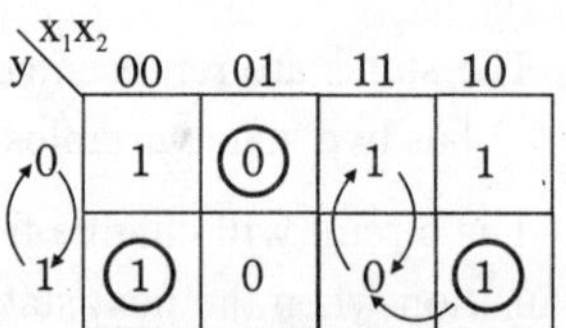

Fig. c: Unstable condition when input change from 10 to 11.

***Fig. 9.5:** Example of transition table of unstable circuit.*

The continuous change between two unstable states is also called ***metastability*** which refers to an equilibrium condition in an unstable circuit.

9.1.4 Flow Table

The ***flow table*** lists all state transitions from present state to next state along with output for all possible changes in input. It is similar to transition table, but has two differences. One is the states are represented by letter symbols and not by binary values. Another difference is that the output is also listed in each square along with next state.

A flow table may have one or more stable states in a row. When there is only one stable state in each row of a flow table then it is called a ***primitive flow table***. The flow table for the circuit of Fig. 9.2 is shown in Fig. 9.6. Here, the following letter symbols are assigned to binary values of states.

$y_1y_2 \rightarrow y$

	$00 \rightarrow a$
	$01 \rightarrow b$
	$11 \rightarrow c$
	$10 \rightarrow d$

y \ x_1	0	1
a	b, 0	(a),1
b	(b),1	c, 1
c	d, 1	(c),1
d	(d),0	a, 1

***Fig. 9.6:** Flow table of circuit shown in Fig. 9.2.*

9.1.5 Cycles and Races

In an asynchronous sequential circuit, when there are changes in two or more state variables due to change in input, a ***race condition*** may exist, in which changes in variables may occur at different timings due to difference in delay in the feedback paths. This leads to unpredictable transition paths during a change in state.

For example, consider the state change from 011 to 100. Some of the possible transition paths are shown below:

$011 \rightarrow 111 \rightarrow 101 \rightarrow 100$

$011 \rightarrow 010 \rightarrow 000 \rightarrow 100$

$011 \rightarrow 001 \rightarrow 101 \rightarrow 100$

The race condition may be classified into critical and noncritical race. In a state transition, if the final stable state depends on the order of change in state variables, then the race is called ***critical race***. On the other hand, in a state transition, if the final stable state does not depends on the order of change in state variables, then the race is called ***noncritical race***.

Examples of noncritical race are shown in Fig. 9.7 and examples of critical race are shown in Fig. 9.8.

Races are identified by tracing possible state transitions through unstable states with one bit change while going from one stable state to another stable state for a change in input. The actual race is through internal states or secondary variables.

In the transition table shown in Fig. 9.7(a), while in stable state 00, if the input changes from 0 to 1, there are two possible transition paths and in both the transition paths the circuit go to same stable next state 11 as shown in Fig. 9.7(b). Hence they are noncritical race. Similarly, two more state transitions from stable state 01 to 11 and 10 to 11 for change in input from 0 to 1 are shown in Figs. 9.7(c) and (d). In these transitions also races are noncritical.

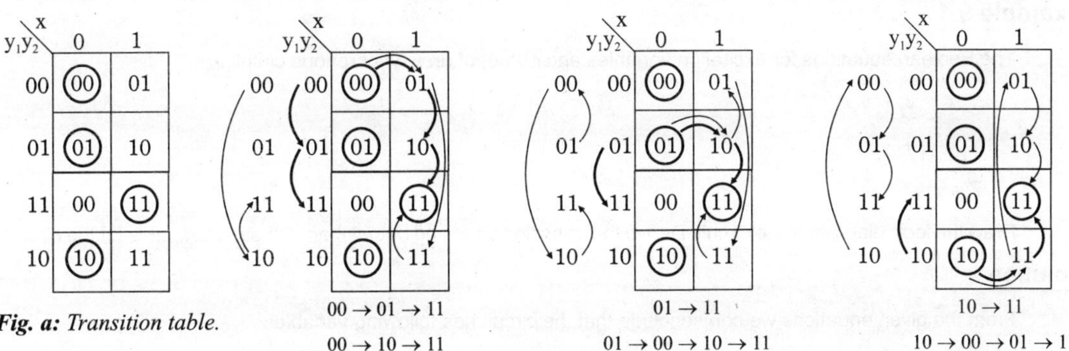

***Fig. a:** Transition table.*

***Fig. b:** Noncritical race in state transition from 00 to 11.*

***Fig. c:** Noncritical race in state transition from 01 to 11.*

***Fig. d:** Noncritical race in state transition from 10 to 11.*

***Fig. 9.7:** Examples of noncritical race when input change from 0 to 1.*

In the transition table shown in Fig. 9.8(a), while in stable state 00, if the input changes from 0 to 1, there are three possible state transition paths and in each transition path the circuit go to different stable next state as shown in Fig. 9.8(b). Hence they are critical race.

Also, it can be observed that when a column has only one stable state the critical race do not occur.

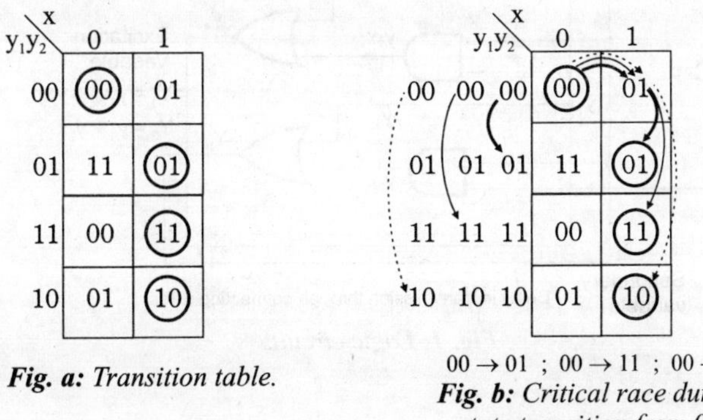

***Fig. a:** Transition table.*

***Fig. b:** Critical race during state transition from 00.*

***Fig. 9.8:** Examples of critical race when input change from 0 to 1.*

Races can be avoided by proper choice of binary values to states. While assigning binary values to states, if the values are assigned such that only one variable change at a time during a state transition in a flow table then the races can be avoided. The race free state assignment is discussed in this chapter in Section 9.1.6. **(IKGTU, Dec'19, 2 Marks)**

When the state transitions in a circuit go through a unique sequence of unstable states then that condition is called a ***cycle***. Races can also be avoided by using cycles, which terminate on a stable state. When a cycle does not terminate on a stable state then the entire circuit will be unstable.

The following state transitions shown in Fig. 9.7 is also examples of cycles because while going from one stable state to another stable state, the transitions go through unstable states.

$00 \rightarrow 01 \rightarrow 11$	$01 \rightarrow 00 \rightarrow 10 \rightarrow 11$	$10 \rightarrow 00 \rightarrow 01 \rightarrow 11$
$00 \rightarrow 10 \rightarrow 11$	$01 \rightarrow 10 \rightarrow 11$	$10 \rightarrow 01 \rightarrow 11$
$00 \rightarrow 01 \rightarrow 10 \rightarrow 11$		

Example 9.1

The Boolean equations for excitation variables and output of an asynchronous circuit are,

$Y_1 = y_1 y_2 + y_2 x'$

$Y_2 = y_1 + y_2 x$

$Z = y_1 y_2$

Draw the logic diagram of the circuit. Derive the transition table and output map. Obtain a flow table for the circuit.

Solution

From the given equations we can conclude that the circuit has following variables.

x = Input	y_1, y_2 = Secondary variables (Present state variables)
Z = Output	Y_1, Y_2 = Excitation variables (Next state variables)

Using the given Boolean equations the logic circuit is drawn as shown in Fig. 1 and the excitation variables Y_1 and Y_2 is evaluated for possible values of input, x and secondary variables, y_1 and y_2 and listed in Table 1.

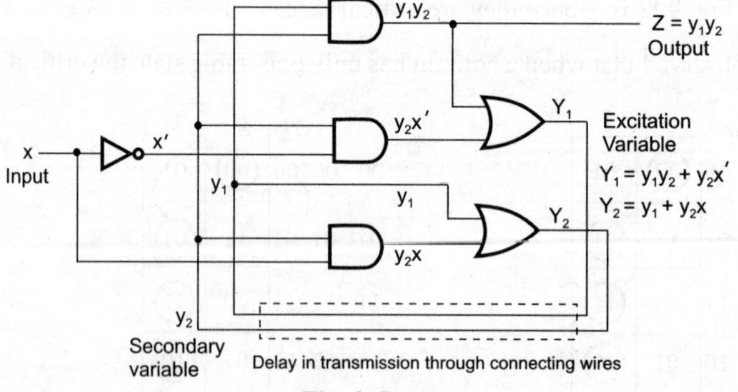

Fig. 1: Logic circuit.

Table 1: Evaluation of Excitation Variables and Output

y_1 y_2 x	y_1' y_2' x'	y_1y_2	y_2x'	y_2x	Y_1	Y_2	$Z = y_1y_2$
0 0 0	1 1 1	0	0	0	0	0	0
0 0 1	1 1 0	0	0	0	0	0	0
0 1 0	1 0 1	0	1	0	1	0	0
0 1 1	1 0 0	0	0	1	0	1	0
1 0 0	0 1 1	0	0	0	0	1	0
1 0 1	0 1 0	0	0	0	0	1	0
1 1 0	0 0 1	1	1	0	1	1	1
1 1 1	0 0 0	1	0	1	1	1	1

y_1y_2 \ x	0	1
00	0	0
01	1	0
11	1	1
10	0	0

Fig. a: *Map for Y_1.*

y_1y_2 \ x	0	1
00	0	0
01	0	1
11	1	1
10	1	1

Fig. b: *Map for Y_2.*

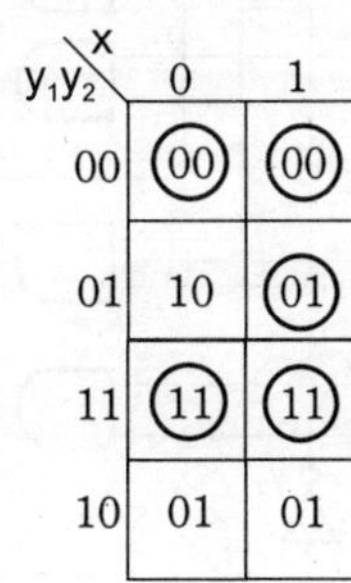

y_1y_2 \ x	0	1
00	(00)	(00)
01	10	(01)
11	(11)	(11)
10	01	01

Fig. c: *Transition table.*

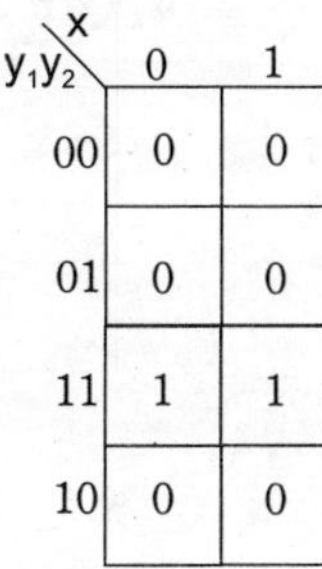

y_1y_2 \ x	0	1
00	0	0
01	0	0
11	1	1
10	0	0

Fig. d: *Output map.*

Fig. 2: *Maps and transition table.*

Maps are constructed for Y_1 and Y_2 by taking possible values of secondary variable y_1 and y_2 on rows and input x on columns as shown in Figs. 2 (a) and (b) and the squares in the maps are filled with values of Y_1 and Y_2 listed in Table 1. The transition table is formed by combining the maps of Figs. 2(a) and (b) into a single map as shown in Fig. 2(c). The output map is constructed as shown in Fig. 2(d) and the squares are filled with values of Z.

In each row of the transition table, the squares that have identical values with secondary values are marked with circles. The circled entries represent stable states and uncircled entries represent unstable states.

In order to draw flow table let us assign letter symbols to binary value of states as shown below:

$y_1y_2 \rightarrow y$
- $00 \rightarrow a$
- $01 \rightarrow b$
- $11 \rightarrow c$
- $10 \rightarrow d$

Using the above symbol assignment for states the flow table is obtained by combining the transition table and output map as shown in Fig. 3.

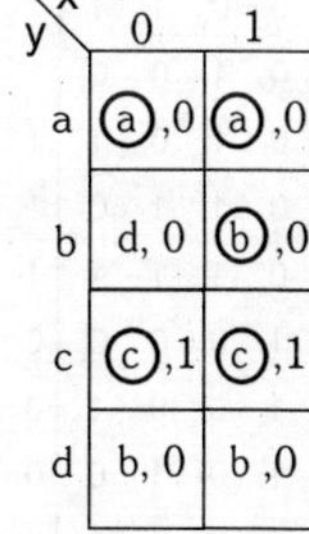

y \ x	0	1
a	(a),0	(a),0
b	d, 0	(b),0
c	(c),1	(c),1
d	b, 0	b ,0

Fig. 3: *Flow table.*

Example 9.2

The Boolean equations for excitation variables and output of an asynchronous circuit are,

$$Y_1 = x_1x_2 + x_1y_2' + x_2'y_1 \quad ; \quad Y_2 = x_2' + x_1y_1'y_2 + x_1'y_1 \qquad Z = x_2 + y_1$$

Draw the logic diagram of the circuit. Derive the transition table and output map. Obtain a flow table for the circuit.

Solution

From the given equations we can conclude that the circuit has following variables.

x = Input	y_1 , y_2 = Secondary variables (Present state variables)
Z = Output	Y_1, Y_2 = Excitation variables (Next state variables)

Using the given Boolean equations the logic circuit is drawn as shown in Fig. 1 and the excitation variables Y_1 and Y_2 are evaluated for all possible combinations for inputs (x_1 and x_2) and secondary variables (y_1 and y_2) and listed in Table 1.

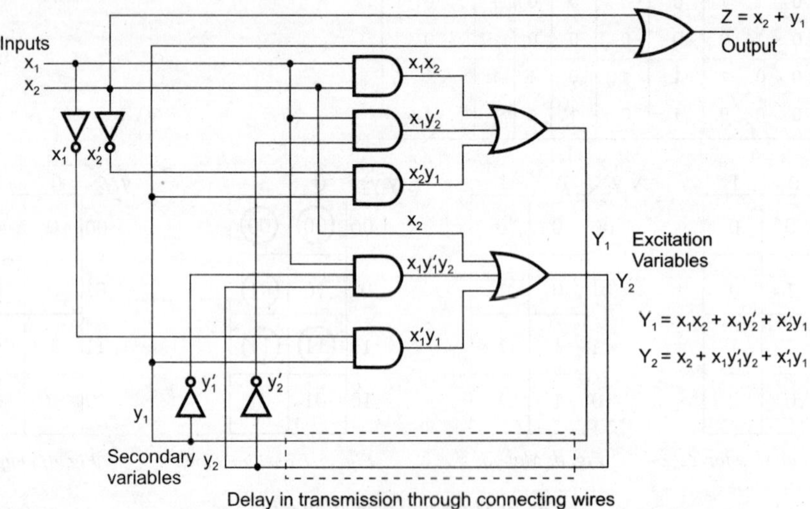

Fig. 1: Logic circuit.

Table 1: Evaluation of Excitation Variables and Output

y_1	y_2	x_1	x_2	y_1'	y_2'	x_1'	x_2'	x_1x_2	x_1y_2'	$x_2'y_1$	$x_1y_1'y_2$	$x_1'y_1$	Y_1	Y_2	$Z = x_2 + y_1$
0	0	0	0	1	1	1	1	0	0	0	0	0	0	0	0
0	0	0	1	1	1	1	0	0	0	0	0	0	0	1	1
0	0	1	0	1	1	0	1	0	1	0	0	0	1	0	0
0	0	1	1	1	1	0	0	1	1	0	0	0	1	1	1
0	1	0	0	1	0	1	1	0	0	0	0	0	0	0	0
0	1	0	1	1	0	1	0	0	0	0	0	0	0	1	1
0	1	1	0	1	0	0	1	0	0	0	1	0	0	1	0
0	1	1	1	1	0	0	0	1	0	0	1	0	1	1	1
1	0	0	0	0	1	1	1	0	0	1	0	1	1	1	1
1	0	0	1	0	1	1	0	0	0	0	0	1	0	1	1
1	0	1	0	0	1	0	1	0	1	1	0	0	1	0	1
1	0	1	1	0	1	0	0	1	1	0	0	0	1	1	1
1	1	0	0	0	0	1	1	0	0	1	0	1	1	1	1
1	1	0	1	0	0	1	0	0	0	0	0	1	0	1	1
1	1	1	0	0	0	0	1	0	0	1	0	0	1	0	1
1	1	1	1	0	0	0	0	1	0	0	0	0	1	1	1

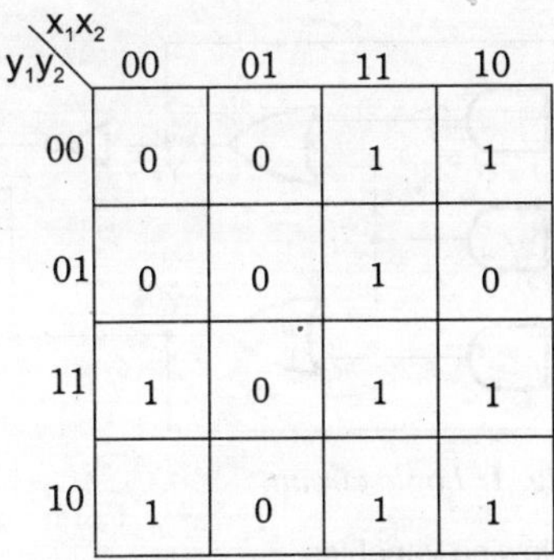

y_1y_2 \ x_1x_2	00	01	11	10
00	0	0	1	1
01	0	0	1	0
11	1	0	1	1
10	1	0	1	1

Fig. 2: Map for Y_1.

y_1y_2 \ x_1x_2	00	01	11	10
00	0	1	1	0
01	0	1	1	1
11	1	1	1	0
10	1	1	1	0

Fig. 3: Map for Y_2.

y_1y_2 \ x_1x_2	00	01	11	10
00	(00)	01	11	10
01	00	(01)	11	(01)
11	(11)	01	(11)	10
10	11	01	11	(10)

Fig. 4: Transition table.

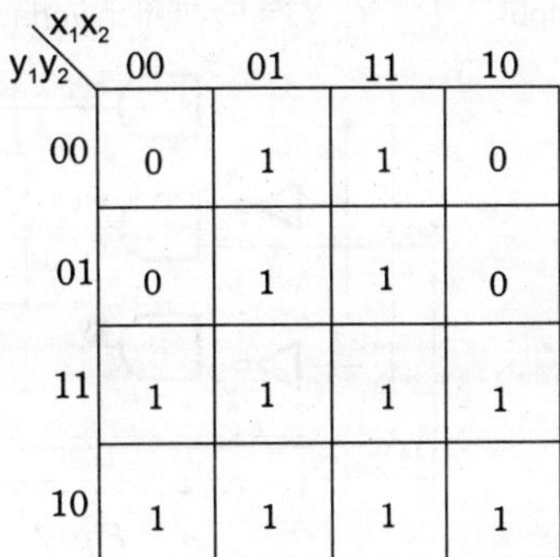

y_1y_2 \ x_1x_2	00	01	11	10
00	0	1	1	0
01	0	1	1	0
11	1	1	1	1
10	1	1	1	1

Fig. 5: Output map.

Maps are constructed for Y_1 and Y_2 by taking possible values of secondary variables y_1 and y_2 on rows and input x_1 and x_2 on columns as shown in Figs. 2 and 3 and the squares in the maps are filled with values of Y_1 and Y_2 listed in Table 1. The transition table is formed by combining the maps of Figs. 2 and 3 into a single map as shown in Fig. 4. The output map is constructed as shown in Fig. 5 and the squares are filled with values of Z.

In each row of the transition table, the squares that have identical values with secondary values are marked with circles. The circled entries represent stable states and uncircled entries represent unstable states.

In order to draw flow table let us assign letter symbols to binary value of states as shown below:

$y_1y_2 \rightarrow y$	
	$00 \rightarrow a$
	$01 \rightarrow b$
	$11 \rightarrow c$
	$10 \rightarrow d$

Using the above symbol assignment for states the flow table is obtained by combining the transition table and output map as shown in Fig. 6.

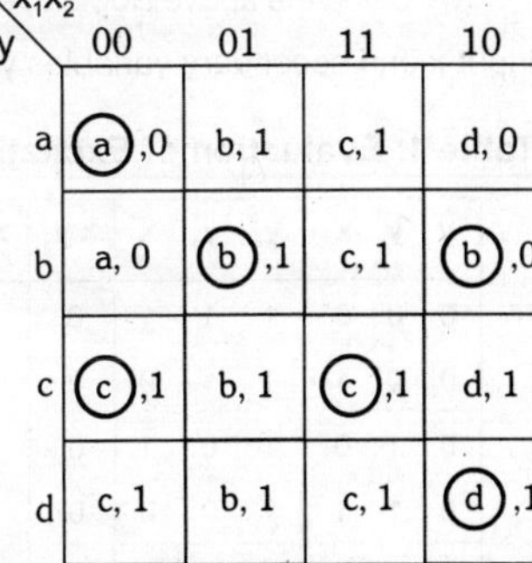

y \ x_1x_2	00	01	11	10
a	(a),0	b, 1	c, 1	d, 0
b	a, 0	(b),1	c, 1	(b),0
c	(c),1	b, 1	(c),1	d, 1
d	c, 1	b, 1	c, 1	(d),1

Fig. 6: Flow table.

Example 9.3

Write logical equation and construct transition table for the circuit output in terms of the circuit input and secondary variables. Show the state transitions for changes in input and hence find the races and cycles. Also, construct the state and flow table.

Fig. 1: Logic circuit.

Solution

From the given circuit we can conclude that the circuit has following variables.

x = Input	y_1, y_2 = Secondary variables (Present state variables)
Z = Output	Y_1, Y_2 = Excitation variables (Next state variables)

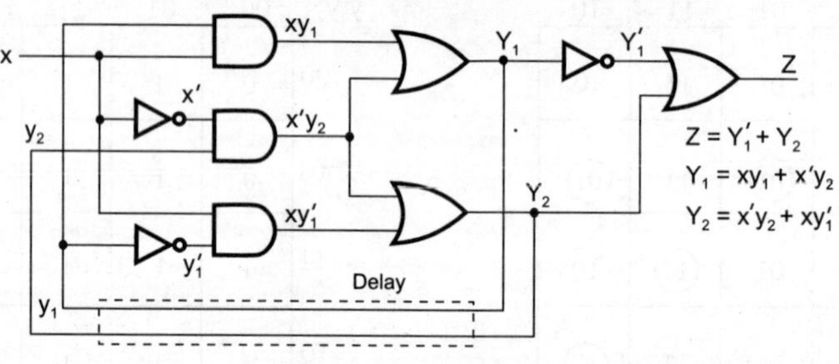

Fig. 2.

From the circuit shown in Fig. 2, we get the following Boolean equations for excitation variables and output of the asynchronous circuit.

$$Y_1 = xy_1 + x'y_2$$

$$Y_2 = x'y_2 + xy_1'$$

$$Z = Y_1' + Y_2$$

Using the above Boolean equations the excitation variables, Y_1 and Y_2 are evaluated for all possible combinations input x and secondary variables y_1 and y_2 and listed in Table 1.

Table 1: Evaluation of Excitation Variables and Output

y_1	y_2	x	y_1'	y_2'	x'	xy_1	$x'y_2$	xy_1'	Y_1	Y_2	Y_1'	$Z = Y_1' + Y_2$
0	0	0	1	1	1	0	0	0	0	0	1	1
0	0	1	1	1	0	0	0	1	0	1	1	1
0	1	0	1	0	1	0	1	0	1	1	0	1
0	1	1	1	0	0	0	0	1	0	1	1	1
1	0	0	0	1	1	0	0	0	0	0	1	1
1	0	1	0	1	0	1	0	0	1	0	0	0
1	1	0	0	0	1	0	1	0	1	1	0	1
1	1	1	0	0	0	1	0	0	1	0	0	0

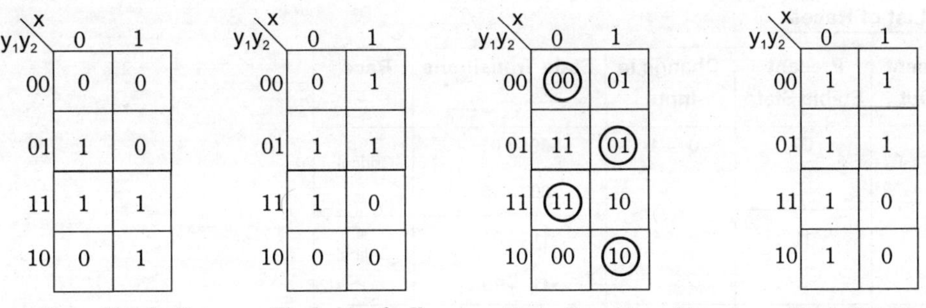

Fig. a: Map for Y_1. *Fig. b:* Map for Y_2. *Fig. c:* Transition table. *Fig. d:* Output map.

Fig. 3: *Maps and transition table.*

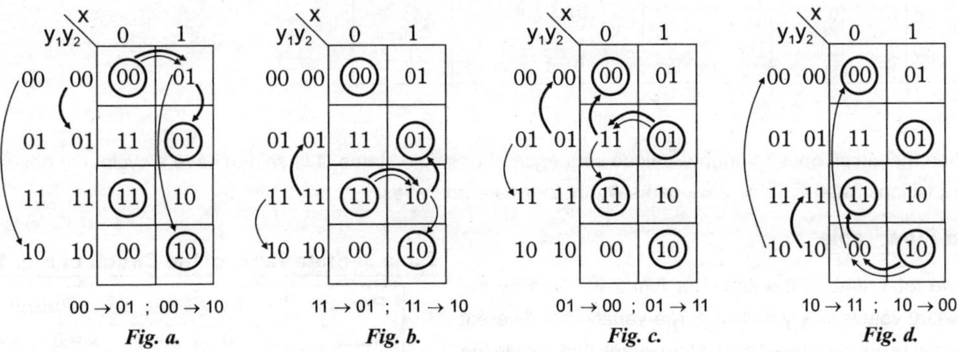

Fig. a. ***Fig. b.*** ***Fig. c.*** ***Fig. d.***

Fig. 4: *State transitions due to change in input.*

Maps are constructed for Y_1 and Y_2 by taking possible values of secondary variables y_1 and y_2 on rows and input x on columns as shown in Figs. 3(a) and (b) and the squares in the maps are filled with values of Y_1 and Y_2 listed in Table 1. The transition table is formed by combining the maps of Figs. 3(a) and (b) into a single map as shown in Fig. 3(c). The output map is constructed as shown in Fig. 3(d) and the squares are filled with values of Z.

In each row of the transition table, the squares that have identical values with secondary values are marked with circles. The circled entries represent stable states and uncircled entries represent unstable states.

In the transition table it can be observed that each column has two stable states for every input. But, for a given input the circuit will remain in one of the stable state.

The transition from one stable state to another stable state occurs due to changes in input.

Since there are two stable states for each input the circuit can go to any one of the stable state for a change in input as shown in Fig. 4 and hence every change in input leads to a critical race.

When in state 00, if the input changes from 0 to 1, then state transitions occur as shown in Fig. 4(a).

When in state 11, if the input changes from 0 to 1, then state transitions occur as shown in Fig. 4(b).

When in state 01, if the input changes from 1 to 0, then state transitions occur as shown in Fig. 4(c).

When in state 10, if the input changes from 1 to 0, then state transitions occur as shown in Fig. 4(d).

The state transitions in Fig. 4 are also listed in Table 2.

Table 2: List of Races

Present Input	Present Stable State	Change in Input	State Transitions	Race
0	00	0 → 1	00 → 01 00 → 10	Critical
	11		11 → 01 11 → 10	Critical
1	01	1 → 0	01 → 00 01 → 11	Critical
	10		10 → 00 10 → 11	Critical

Cycles

When the circuit goes through a unique sequence of unstable states, it is said to have a cycle. On observing the state transitions listed in Table 2 we can say that there are no cycles.

State and Flow Table

The total state in the circuit of Fig. 2 is given by the possible binary values of y_1y_2x. With three variables 8 different total states can be formed, in which the four states that are circled in transition table are stable states and the other four states are unstable.

Stable states: 000, 011, 110, 101

Unstable states: 001, 010, 111, 100

Using the transition table, the state table for asynchronous circuit similar to that of a state table of synchronous circuit can be constructed as shown in Table 3.

Table 3: State Table of the Circuit of Fig. 1

Present State		Next State				Output	
		x = 0		x = 1		x = 0	x = 1
y_1	y_2	Y_1	Y_2	Y_1	Y_2	Z	Z
0	0	0	0	0	1	1	1
0	1	1	1	0	1	1	1
1	0	0	0	1	0	1	0
1	1	1	1	1	0	1	0

In order to draw flow table let us assign letter symbols to binary value of states as shown below:

$y_1y_2 \rightarrow y$

00 → a
01 → b
11 → c
10 → d

Using the above symbol assignment for states the flow table is obtained by combining the transition table and output map as shown in Fig. 5.

y \ x	0	1
a	(a),1	b, 1
b	c, 1	(b),1
c	(c),1	d, 0
d	a, 1	(d),0

Fig. 5: Flow table.

Example 9.4

An asynchronous sequential circuit is described by the excitation functions, $Y_1 = y_1 y_2' + (y_1 + y_2') x$ and $Y_2 = y_2' x + y_1 x$ and the output function, $Z = Y_2$. Draw the logic diagram of the circuit. Derive the transition table and output map. Also discuss about the behaviour of the circuit.

Solution

From the given equations we can conclude that the circuit has following variables.

x = Inputs	y_1, y_2 = Secondary variables (Present state variables)
Z = Output	Y_1, Y_2 = Excitation variables (Next state variables)

Given that, $Y_1 = y_1 y_2' + (y_1 + y_2') x$

$Y_2 = y_2' x + y_1 x$

$Z = Y_2$

Using the given Boolean equations the logic circuit is drawn as shown in Fig. 1 and the excitation variables, Y_1 and Y_2 are evaluated for all possible combinations of input x and secondary variables y_1 and y_2 and listed in Table 1.

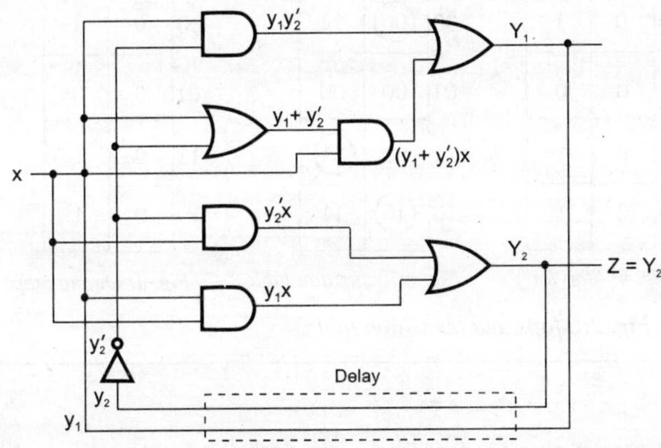

Fig. 1: Logic circuit.

Table 1: Evaluation of Excitation Variables and Output

y_1	y_2	x	y_1'	y_2'	x'	y_1y_2'	y_1+y_2'	$(y_1+y_2')x$	$y_2'x$	y_1x	Y_1	Y_2	$Z=Y_2$
0	0	0	1	1	1	0	1	0	0	0	0	0	0
0	0	1	1	1	0	0	1	1	1	0	1	1	1
0	1	0	1	0	1	0	0	0	0	0	0	0	0
0	1	1	1	0	0	0	0	0	0	0	0	0	0
1	0	0	0	1	1	1	1	0	0	0	1	0	0
1	0	1	0	1	0	1	1	1	1	1	1	1	1
1	1	0	0	0	1	0	1	0	0	0	0	0	0
1	1	1	0	0	0	0	1	1	0	1	1	1	1

The transition table is formed by taking possible values of secondary variable on rows and input x on columns as shown in Fig. 2(c) and the squares in the transition table are filled with values of Y_1 and Y_2 listed in Table 1. The output map is constructed as shown in Fig. 2(d) and the squares are filled with values of Z.

In each row of the transition table, the squares that have identical values with secondary values are maked with circles. The circled entires represent stable states and uncircled entries represent unstable states.

In order to draw flow table let us assign letter symbols to binary value of states as shown below:

$y_1 y_2 \rightarrow y$ | $00 \rightarrow a$
 | $01 \rightarrow b$
 | $11 \rightarrow c$
 | $10 \rightarrow d$

Using the above symbol assignment for states the flow table is obtained by combining the transition table and output map as shown in Fig. 3.

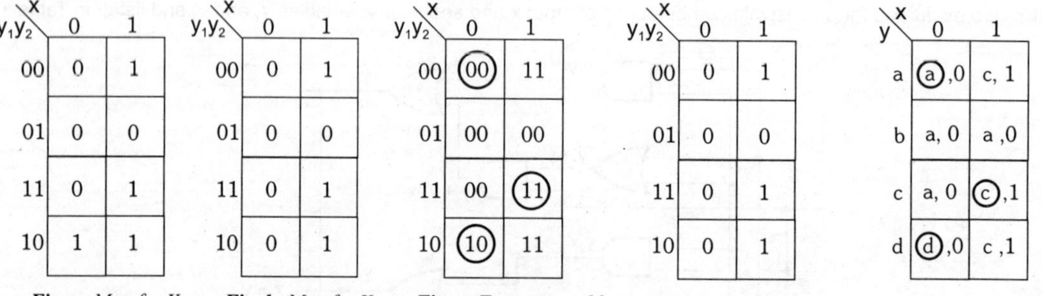

y_1y_2 \ x	0	1
00	0	1
01	0	0
11	0	1
10	1	1

Fig. a: Map for Y_1.

y_1y_2 \ x	0	1
00	0	1
01	0	0
11	0	1
10	0	1

Fig. b: Map for Y_2.

y_1y_2 \ x	0	1
00	(00)	11
01	00	00
11	00	(11)
10	(10)	11

Fig. c: Transition table.

y_1y_2 \ x	0	1
00	0	1
01	0	0
11	0	1
10	0	1

Fig. d: Output map.

y \ x	0	1
a	(a),0	c, 1
b	a, 0	a ,0
c	a, 0	(c),1
d	(d),0	c ,1

Fig. 3: Flow table.

Fig. 2: Maps and transition table.

Example 9.5

An asynchronous sequential circuit is described by the excitation function, $Y = x_1 x_2' + (x_1 + x_2')y$ and the output function, Z = Y. Draw the logic diagram of the circuit. Derive the transition table and output map. Also discuss about the behaviour of the circuit.

Solution

From the given equations we can conclude that the circuit has following variables.

x_1, x_2 = Inputs | y = Secondary variable (Present state variable)
Z = Output | Y = Excitation variable (Next state variable)

Given that, $Y = x_1 x_2' + (x_1 + x_2')y$

$Z = Y$

Using the given Boolean equations the logic circuit is drawn as shown in Fig. 1 and the excitation variable, Y is evaluated for all possible combinations of input x and secondary variables y_1 and y_2 and listed in Table 1.

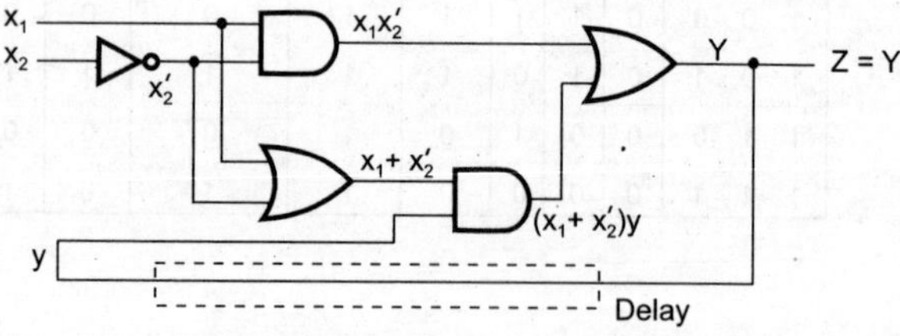

Fig. 1: Logic circuit.

Table 1: Evaluation of Excitation Variables and Output

y	x_1	x_2	y'	x_1'	x_2'	x_1x_2'	x_1+x_2'	$(x_1+x_2')y$	Y	Z = Y
0	0	0	1	1	1	0	1	0	0	0
0	0	1	1	1	0	0	0	0	0	0
0	1	0	1	0	1	1	1	0	1	1
0	1	1	1	0	0	0	1	0	0	0
1	0	0	0	1	1	0	1	1	1	1
1	0	1	0	1	0	0	0	0	0	0
1	1	0	0	0	1	1	1	1	1	1
1	1	1	0	0	0	0	1	1	1	1

The transition table is formed by taking possible values of secondary variable on rows and inputs x_1 and x_2 on columns as shown in Fig. 2 and the squares in the transition table are filled with values of Y listed in Table 1. The output map is constructed as shown in Fig. 3 and the squares are filled with values of Z.

In each row of the transition table, the squares that have identical values with secondary values are maked with circles. The circled entires represent stable states and uncircled entries represent unstable states.

The flow table is formed by combining transition table and output map as shown in Fig. 4.

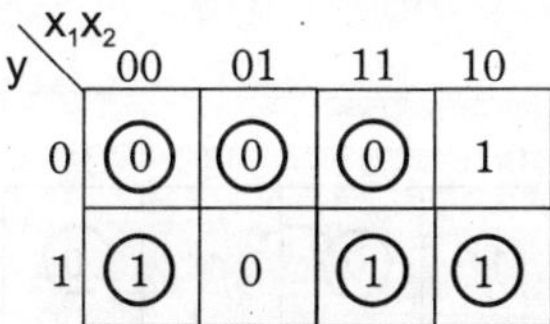

Fig. 2: Transition table.

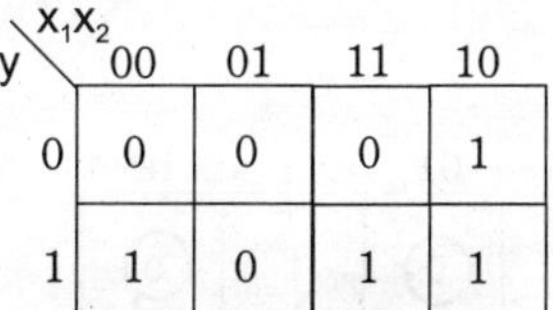

Fig. 3: Output map.

y \ x_1x_2	00	01	11	10
0	(0),0	(0),0	(0),0	1, 1
1	(1),1	0, 0	(1),1	(1),1

Fig. 4: Flow table.

y \ x_1x_2	00	01	11	10
0	(0),0	(0),0	(0),0	1, 1
1	(1),1	0, 0	(1),1	(1),1

Fig. 5: State transitions when the input changes to either 00 or 11.

Comment : From the flow table we can conclude that when the input is 01, the output is 0. When the input is 10 and the output is 1. Whenever the input assumes one of the other two combinations (either 00 or 11), the output retains its previous value as shown in Fig. 5.

Example 9.6

Investigate the transition table of Fig. 1, and determine all race conditions and whether they are critical or noncritical. Determine also whether there are any cycles.

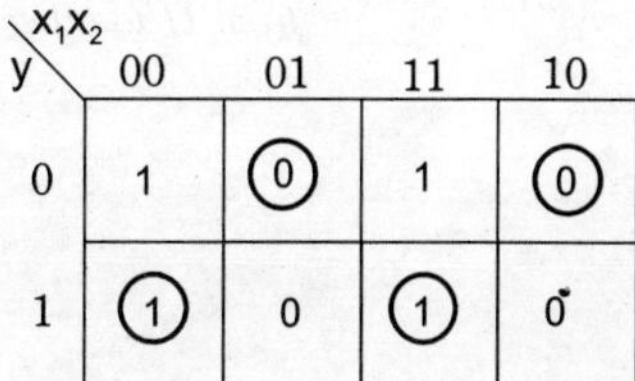

Fig. 1: Transition table.

Solution

State Transitions Due to Change in Input

Here, each input has one stable state. State transitions occur only when there is a change in input. Normally, when the input is stable or when no change in input the circuit remain in the stable state corresponding to the present input.

Here, there are four possible combinations of inputs. When the circuit remain in a stable state at any one input then state transitions will occur when the input changes to other possible inputs with one bit change.

When input is 00, the circuit remain in state 1. The input can change from 00 to 01 or 00 to 10 and the corresponding state transitions will occur as shown in Fig. 2.

When input is 01, the circuit remain in state 0. The input can change from 01 to 11 or 01 to 00 and the corresponding state transitions will occur as shown in Fig. 3.

When input is 11, the circuit remain in state 1. The input can change from 11 to 10 or 11 to 01 and the corresponding state transitions will occur as shown in Fig. 4.

When input is 10, the circuit remain in state 0. The input can change from 10 to 00 or 10 to 11 and the corresponding state transitions will occur as shown in Fig. 5.

The state transitions shown in Figs. 2 to 5 are listed in Table 1. It can be observed that there are no races and cycles in the state transitions.

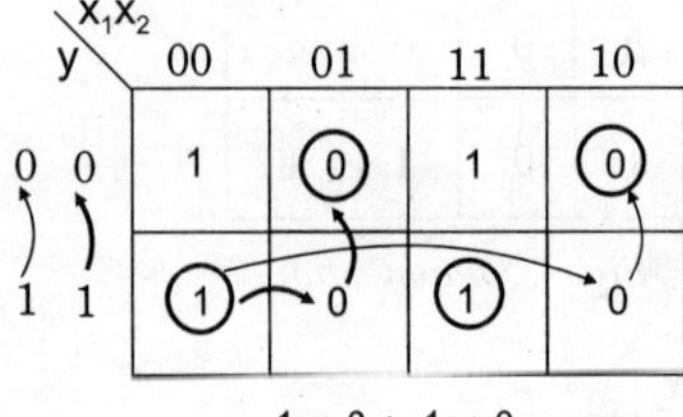

Fig. 2: *Transitions due to change in input from 00 to other possible inputs.*

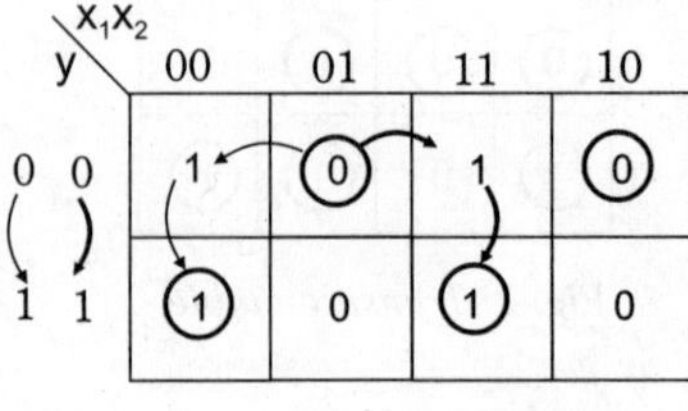

Fig. 3: *Transitions due to change in input from 01 to other possible inputs.*

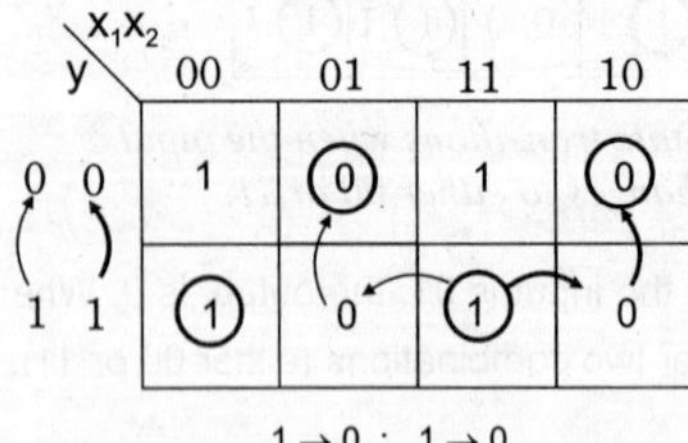

Fig. 4: *Transitions due to change in input from 11 to other possible inputs.*

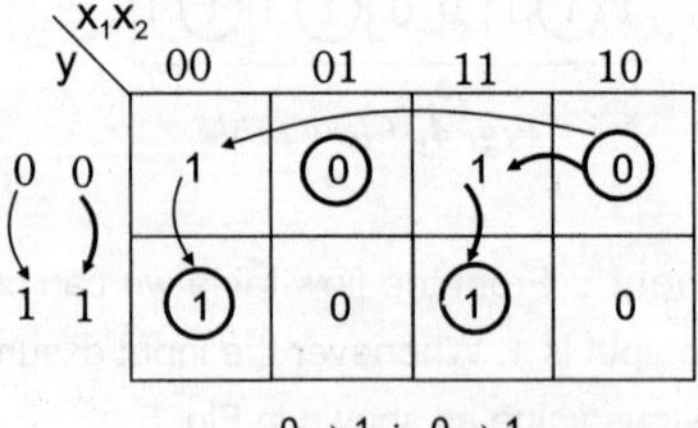

Fig. 5: *Transitions due to change in input from 10 to other possible inputs.*

Table 1: List of Races and Cycles

Present Input	Present Stable State	Change in Input	State Transitions
00	1	00 → 01	1 → 0
		00 → 10	1 → 0
01	0	01 → 00	0 → 1
		01 → 11	0 → 1
11	1	11 → 01	1 → 0
		11 → 10	1 → 0
10	0	10 → 11	0 → 1
		10 → 00	0 → 1

Example 9.7

Investigate the transition table of Fig. 1, and determine all race conditions and whether they are critical or noncritical. Determine also whether there are any cycles.

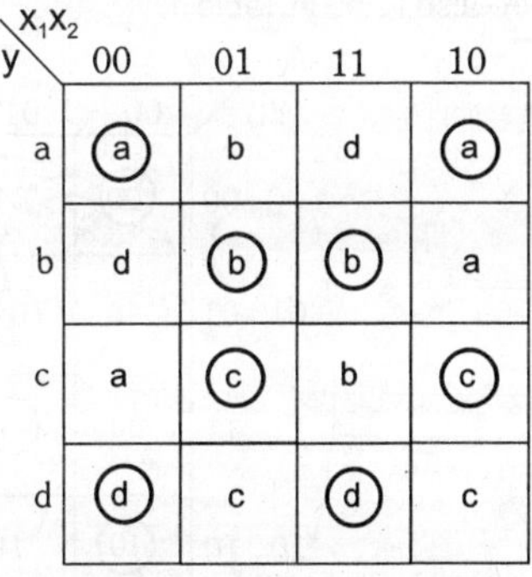

$y \backslash x_1x_2$	00	01	11	10
a	(a)	b	d	(a)
b	d	(b)	(b)	a
c	a	(c)	b	(c)
d	(d)	c	(d)	c

Fig. 1: Transition table.

Solution

Binary State Assignment

Let us assign binary codes to states as shown below. Here there are 4 states and hence 2-bit binary codes are selected, so that we get, $2^2 = 4$ codes.

Let the two bits of the binary codes be represented by the two state variables y_1 and y_2.

a → 00	c → 11	$y = y_1 y_2$
b → 01	d → 10	

The transition table with binary code assignment is shown in Fig. 2.

$y_1y_2 \backslash x_1x_2$	00	01	11	10
00	(00)	01	10	(00)
01	10	(01)	(01)	00
11	00	(11)	01	(11)
10	(10)	11	(10)	11

Fig. 2: Transition table with binary code assignment.

State Transitions Due to Change in Input

State transitions occur only when there is a change in input. Normally, when the input is stable or when no change in input the circuit remain in one of the stable state corresponding to the present input.

Here, there are four possible combinations of inputs. When the circuit remain in a stable state at any one input then state transitions will occur when the input changes to other possible inputs with one bit change.

Each input has two stable states. Hence at any one input the circuit can remain in one of the stable state.

When input is 00, the circuit remain in either state 00 or 10. The input can change from 00 to 01 or 00 to 10 and the corresponding state transitions will occur as shown in Fig. 3.

When input is 01, the circuit remain in either state 01 or 11. The input can change from 01 to 11 or 01 to 00 and the corresponding state transitions will occur as shown in Fig. 4.

When input is 9, the circuit remain in either state 00 or 10. The input can change from 11 to 10 or 11 to 01 and the corresponding state transitions will occur as shown in Fig. 5.

When input is 10, the circuit remain in either state 00 or 11. The input can change from 10 to 00 or 10 to 11 and the corresponding state transitions will occur as shown in Fig. 6.

The state transitions shown in Figs. 3 to 6 are listed in Table 1. The races and cycles involved in state transitions are also listed in Table 1.

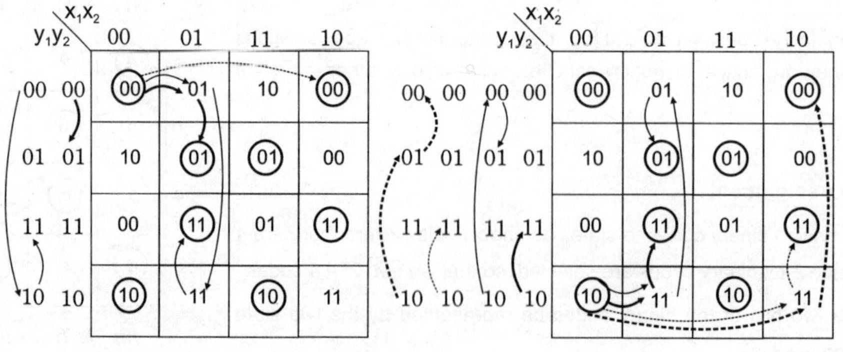

Fig. a: *Transitions from stable state 00.*

Fig. b: *Transitions from stable state 10.*

Fig. 3: *Transitions due to change in input from 00 to other possible inputs.*

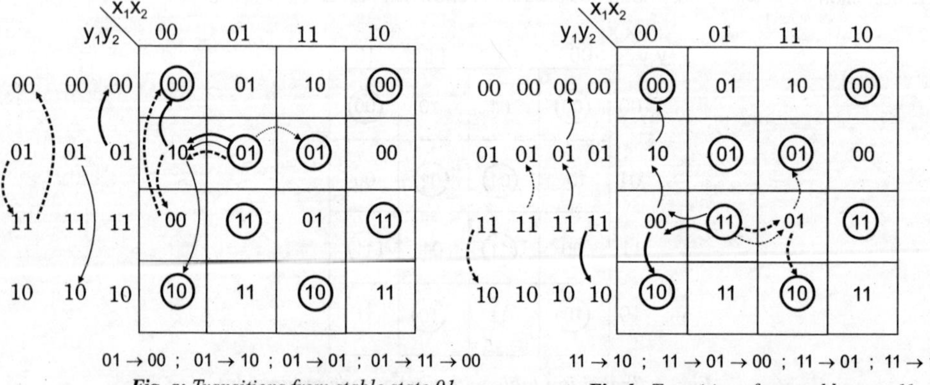

Fig. a: *Transitions from stable state 01.*

Fig. b: *Transitions from stable state 11.*

Fig. 4: *Transitions due to change in input from 01 to other possible inputs.*

Table 1: List of Races and Cycles

Present Input	Present Stable State	Change in Input	State Transitions	Race	Cycles
00	00	00 → 01	00 → 01 00 → 10 → 11	Critical	- 00 → 10 → 11
		00 → 10	00 → 00	-	-
	10	00 → 01	10 → 11 10 → 00 → 01	Critical	- 10 → 00 → 01
		00 → 10	10 → 11 10 → 01 → 00	Critical	- 10 → 01 → 00
01	01	01 → 00	01 → 00 01 → 11 → 10	Critical	- 01 → 11 → 10
		01 → 11	01 → 01	-	-
	11	01 → 00	11 → 10 11 → 01 → 00	Critical	- 11 → 01 → 00
		01 → 11	11 → 01 11 → 10	Critical	- -
11	01	11 → 01	01 → 01	-	-
		11 → 10	01 → 00 01 → 11	Critical	- -
	10	11 → 01	10 → 11 10 → 00 → 01	Critical	- 10 → 00 → 01
		11 → 10	10 → 11 10 → 00	Critical	- -
10	00	10 → 00	00 → 00	-	-
		10 → 11	00 → 01 00 → 10	Critical	- -
	11	10 → 00	11 → 01 → 00 11 → 10	Critical	11 → 01 → 00
		10 → 11	11 → 10 11 → 01	Critical	- -

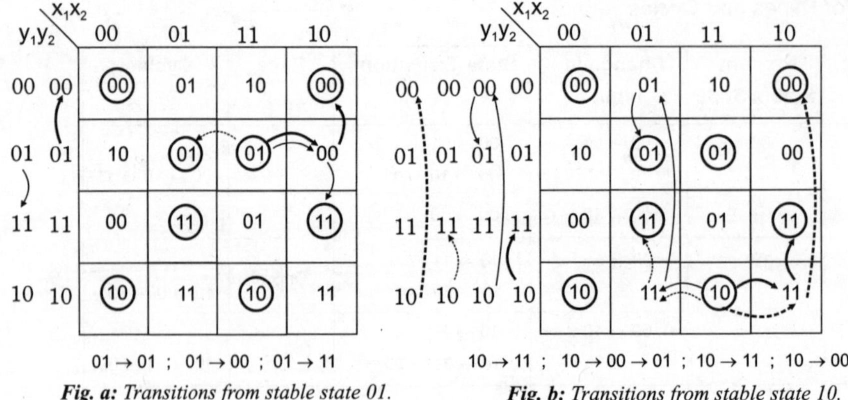

***Fig. a:** Transitions from stable state 01.*

***Fig. b:** Transitions from stable state 10.*

***Fig. 5:** Transitions due to change in input from 11 to other possible inputs.*

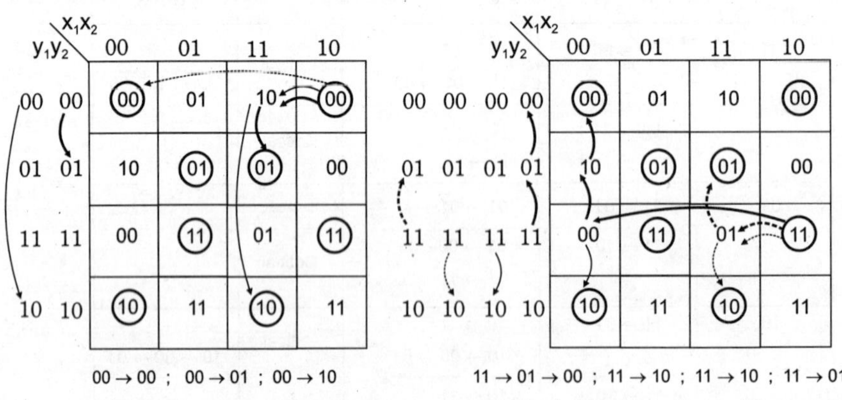

***Fig. a:** Transitions from stable state 00.*

***Fig. b:** Transitions from stable state 11.*

***Fig. 6:** Transitions due to change in input from 10 to other possible inputs.*

Cycles

When the circuit goes through a unique sequence of unstable states, it is said to have a cycle and they are listed in Table 1.

The following cycles are observed during state transition for changes in inputs,

Cycles during change in input from 00.	Cycles during change in input from 01.	Cycle during change in input from 11.	Cycle during change in input from 10.
00 → 10 → 11	01 → 11 → 10	10 → 00 → 01	11 → 01 → 00
10 → 00 → 01	11 → 01 → 00		

Example 9.8

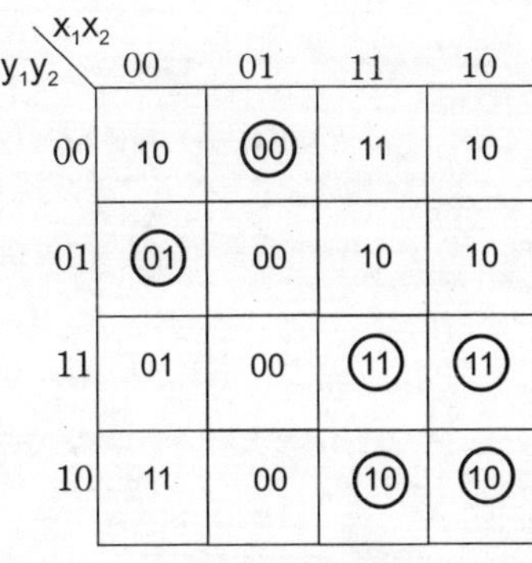

y_1y_2 \ x_1x_2	00	01	11	10
00	10	(00)	11	10
01	(01)	00	10	10
11	01	00	(11)	(11)
10	11	00	(10)	(10)

Fig. 1: Transition table.

Investigate the transition table of Fig. 1, and determine all race conditions and whether they are critical or noncritical. Determine also whether there are any cycles.

Solution

State Transitions Due to Change in Input

State transitions occur only when there is a change in input. Normally, when the input is stable or when no change in input the circuit remain in one of the stable state corresponding to the input.

Here, there are four possible combinations of inputs. When the circuit remain in a stable state at any one input then state transitions will occur when the input changes to other possible inputs with one bit change.

Here, two inputs has one stable state each and the other two inputs has two stable states each. Hence at any one input the circuit can remain in one of the stable state.

When input is 00, the circuit remain in state 01. The input can change from 00 to 01 or 00 to 10 and the corresponding state transitions will occur as shown in Fig. 2.

When input is 01, the circuit remain in state 00. The input can change from 01 to 11 or 01 to 00 and the corresponding state transitions will occur as shown in Fig. 3.

When input is 11, the circuit remain in either state 11 or 10. The input can change from 11 to 10 or 11 to 01 and the corresponding state transitions will occur as shown in Fig. 4.

When input is 10, the circuit remain in either state 11 or 10. The input can change from 10 to 00 or 10 to 11 and the corresponding state transitions will occur as shown in Fig. 5.

The state transitions shown in Figs. 2 to 5 are listed in Table 1. The races and cycles that are involved in state transitions are also listed in Table 1.

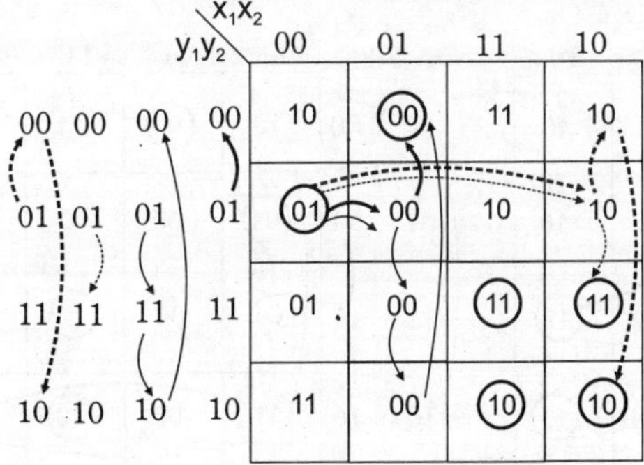

01 → 00 ; 01 → 11 → 10 → 00 ; 01 → 11 ; 01 → 00 → 10

Fig. 2: Transitions due to change in input from 00 to other possible inputs.

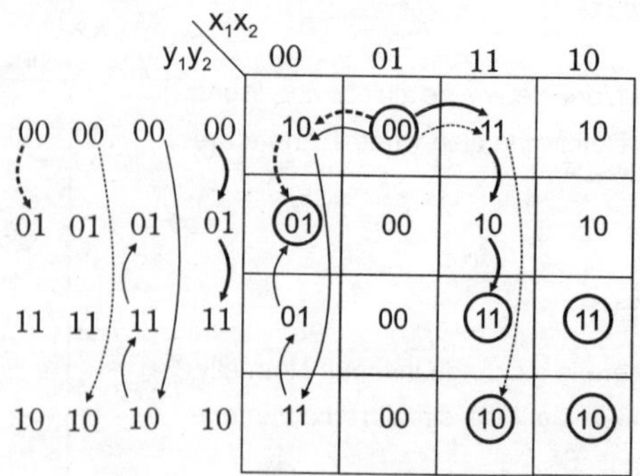

00 → 01 ; 00 → 10 → 11 → 01 ; 00 → 10 ; 00 → 01 → 11

Fig. 3: Transitions due to change in input from 01 to other possible inputs.

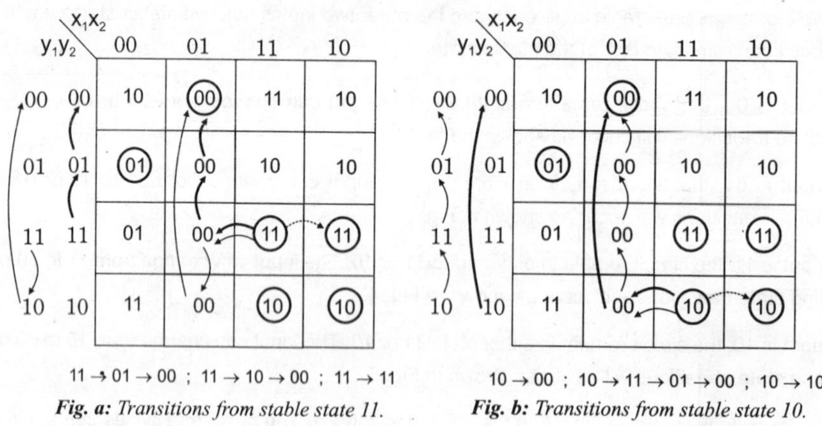

11 → 01 → 00 ; 11 → 10 → 00 ; 11 → 11

Fig. a: Transitions from stable state 11.

10 → 00 ; 10 → 11 → 01 → 00 ; 10 → 10

Fig. b: Transitions from stable state 10.

Fig. 4: Transitions due to change in input from 11 to other possible inputs.

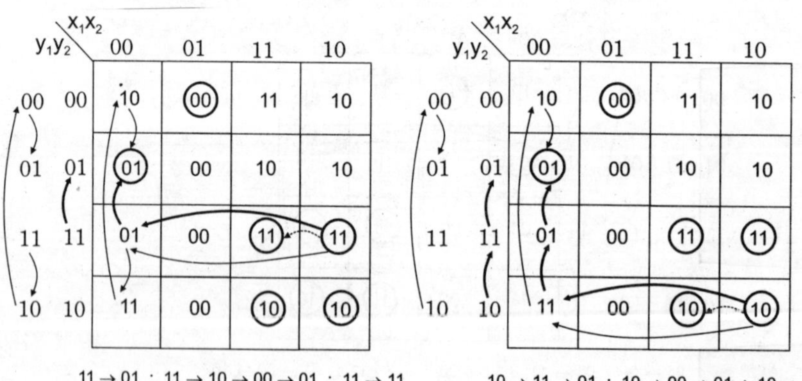

11 → 01 ; 11 → 10 → 00 → 01 ; 11 → 11

Fig. a: Transitions from stable state 11.

10 → 11 → 01 ; 10 → 00 → 01 ; 10 → 10

Fig. b: Transitions from stable state 10.

Fig. 5: Transitions due to change in input from 10 to other possible inputs.

Table 1: List of Races and Cycles

Present Input	Present Stable State	Change in Input	State Transitions	Race	Cycles
00	01	00 → 01	01 → 00 01 → 11 → 10 → 00	Noncritical	- 01 → 11 → 10 → 00
		00 → 10	01 → 11 01 → 00 → 10	Critical	- 01 → 00 → 10
01	00	01 → 00	00 → 01 00 → 10 → 11 → 01	Noncritical	- 00 → 10 → 11 → 01
		01 → 11	00 → 10 00 → 01 → 11	Critical	- 00 → 01 → 11
11	11	11 → 01	11 → 01 → 00 11 → 10 → 00	Noncritical	11 → 01 → 00 11 → 10 → 00
		11 → 10	11 → 11	-	-
	10	11 → 01	10 → 00 10 → 11 → 01 → 00	Noncritical	- 10 → 11 → 01 → 00
		11 → 10	10 → 10	-	-
10	11	10 → 00	11 → 01 11 → 10 → 00 → 01	Noncritical	- 11 → 10 → 00 → 01
		10 → 11	11 → 11	-	-
	10	10 → 00	10 → 11 → 01 10 → 00 → 01	Noncritical	10 → 11 → 01 10 → 00 → 01
		10 → 11	10 → 10	-	-

Cycles

When the circuit goes through a unique sequence of unstable states, it is said to have a cycle and they are listed in Table 1.

The following cycles are observed during state transitions for changes in inputs,

Cycles during change in input from 00.	Cycles during change in input from 01.	Cycles during change in input from 11.	Cycles during change in input from 10.
01 → 11 → 10 → 00	00 → 10 → 11 → 01	11 → 01 → 00	11 → 10 → 00 → 01
01 → 00 → 10	00 → 01 → 11	11 → 10 → 00	10 → 11 → 01
		10 → 11 → 01 → 00	10 → 00 → 01

9.1.6 Race-Free Assignment

In order to avoid critical races, the state assignment or binary code assignment should be such that only one variable change at a time when a state transition occurs in a flow table. Two binary codes are said to be adjacent if they differ by only one variable. For example the 3-bit code 011 and 001 are adjacent, because they differ by only one variable. In order to avoid critical races, the two states between which a transition occurs should be assigned adjacent codes. Assigning adjacent codes to states to avoid critical races is called ***race-free assignment***. Sometimes, race free assignment needs addition of extra rows in flow table which can be achieved by the following two methods.

Method 1: Additional Rows with Extra States

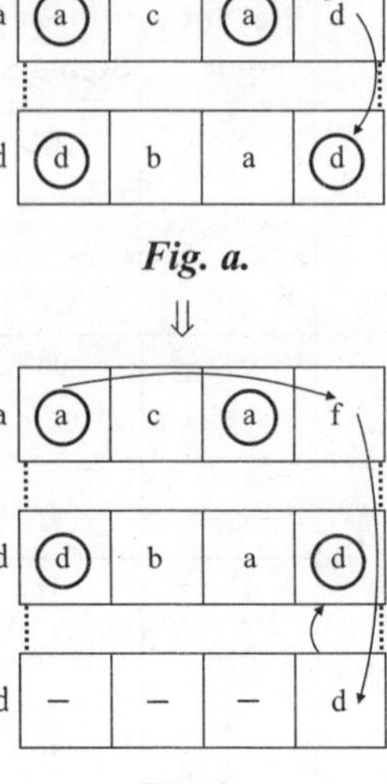

Fig. b.

Fig. 9.9.

Using the flow table, a transition diagram can be constructed that represents all possible state transitions. Binary code assignment for states can be made in the transition diagram such that states that are involved in every transition are assigned adjacent codes.

When adjacent code assignment is not possible for a pair of states involved in a state transition insert an additional state such that the state transition between the pair of states go via the additional states. The concept of inserting additional state can be extended to all state pair involved in state transitions that do not have adjacent state assignment.

The insertion of additional states leads to extra rows in flow table. For every extra state one row is added in flow table.

Consider a state transition between two rows a and d as shown in Fig. 9.9(a).

Let a and d are not adjacent. Now an additional state f is added such that the transition from a to d go via state f as shown below:

$$a \rightarrow d \quad \Rightarrow \quad a \rightarrow f \rightarrow d$$

In order to assure this transition an additional state named state f is selected which leads to an additional row for state f in flow table table as shown in Fig. 9.9(b). Now, the state d in row a is replaced by f and the corresponding column in row f is filled by d and all other columns in row f are don't-cares.

Method 2: Multiple Row with Complement States

In multiple row method each row in a flow table is represented by two or more equivalent rows. In two rows representation, each row is represented by two rows and they are known as original row and equivalent row.

The rows corresponding to original flow table are assigned binary codes such that they are adjacent. The complement of binary codes are assigned to equivalent rows. The stable states are filled with subscript 1 in original rows and subscript 2 in the equivalent rows. The unstable states are filled such that they are adjacent to stable states. (Refer Example 9.9(b) for multiple row method).

9.1.7 Transition Diagram

The transition table lists all possible state transitions from present to next state. The state transitions between unstable states in each row of transition table can be converted to transition diagram which will be more useful for race free state assignment. The diagrammetic way of representing all possible state transitions between unstable states in every row of transition table is called ***transition diagram***.

An example of drawing the transition diagram from a transition table is shown in Fig. 9.10.

From the transition table shown in Fig. 9.10(a) we can observe the following state transition.

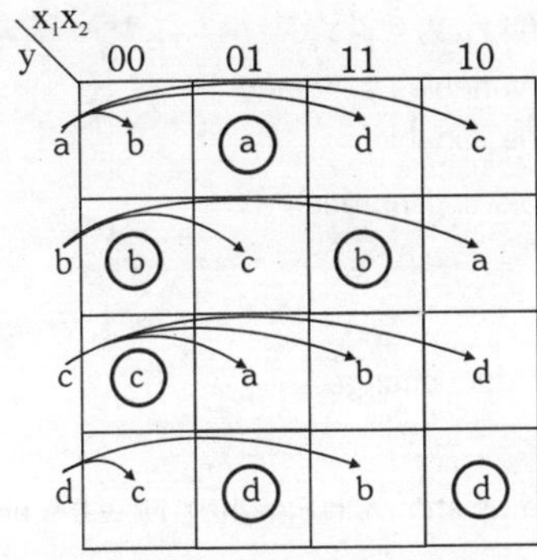

Fig. a: Transition table.

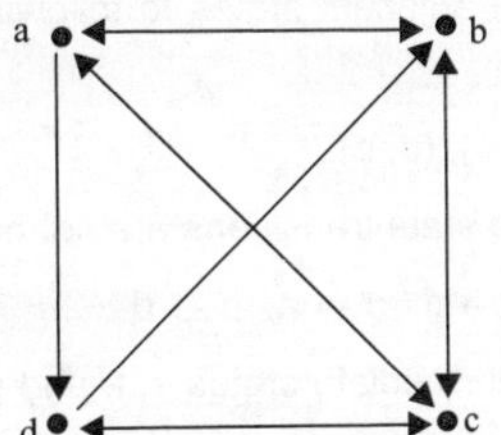

Fig. b: Transition diagram.

***Fig. 9.10:** Example of drawing transition diagram from a transition table.*

From state a	From state b	From state c	From state d
a → b	b → c	c → a	d → c
a → d	b → a	c → b	d → b
a → c		c → d	

The above state transitions can be combined as a single state transition diagram shown in Fig. 9.10(b).

Example 9.9

Consider the reduced flow table shown in Fig. 1.

a) Obtain the transition diagram, and show that three state variables are needed for a race-free binary state assignment.

b) Obtain the expanded flow table, using the multiple row method of assignment.

y \ x_1x_2	00	01	11	10
a	(a)	c	(a)	d
b	a	(b)	c	(b)
c	(c)	(c)	(c)	d
d	(d)	b	a	(d)

***Fig. 1:** Reduced flow table.*

Solution

a) Transition Diagram and Race Free State Assignment

The following transitions from present state to next state are observed in transition table in Fig. 2.

From a	From b	From c	From d
a → c	b → a	c → d	d → b
a → d	b → c		d → a

The above transitions are combined as a single diagram in the transition diagram shown in Fig. 3.

For a critical race-free state assignment the state transitions in Fig. 2 should have adjacent state assignments. Here, there are 4 states and so a minimum of 4 binary codes are required for state assignment. But for race free state assignment additional rows are required and so let us choose 8 binary codes which is possible with 3-bit code (2^3 = 8 possible codes). The possible 3-bit codes are,

000, 001, 010, 011, 100, 101, 110, 111

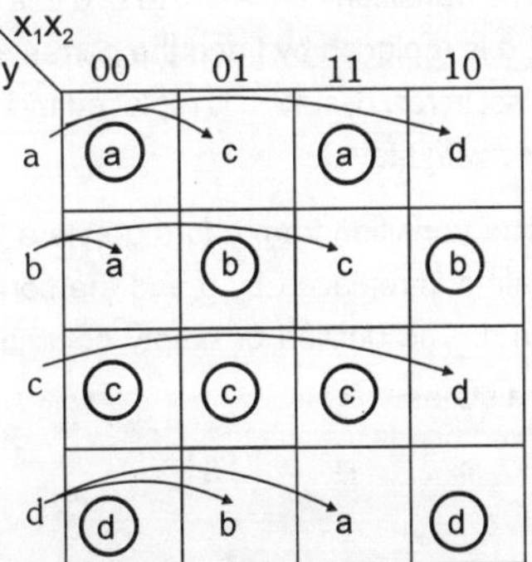

***Fig. 2:** Transitions from present state to next state.*

Let us represent the three binary bits by three state variables y_1, y_2 and y_3.

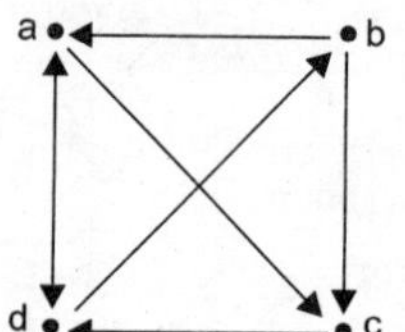

Fig. 3: Transition diagram.

Let us construct a binary assignment table with three state variables y_3, y_2 and y_1. Two binary values are said to be adajcent if they differ in only one variable.

Let us assign adjacent states to following state pair to avoid critical races as shown in Fig. 4(a).

(b, a), (b, c), (d, b)

The following state transitions are not possible with one bit state change,

$a \rightarrow c, \; a \rightarrow d, \; d \rightarrow a, \; c \rightarrow d$

Therefore, intermediate states e, f, and g can be introduced as shown in Fig. 4(b). Now the above transitions will be as follows,

$a \rightarrow c \Rightarrow a \rightarrow e \rightarrow c$	$a \rightarrow d \Rightarrow a \rightarrow f \rightarrow d$
$c \rightarrow d \Rightarrow c \rightarrow g \rightarrow d$	$d \rightarrow a \Rightarrow d \rightarrow f \rightarrow a$

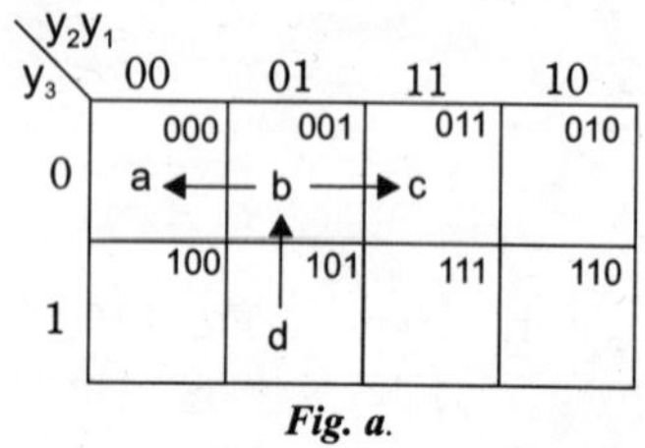

Fig. a.

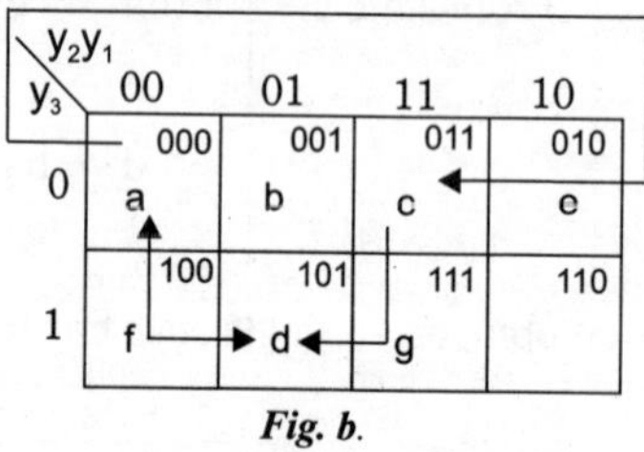

Fig. b.

Fig. 4: Binary state assignment.

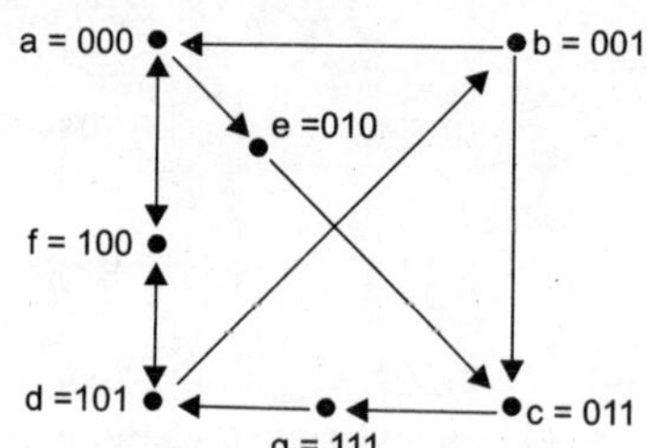

Fig. 5: Transition diagram with binary assignment.

The transition diagram with critical race free binary assignment is shown in Fig. 5. The introduction of three additional states e, f and g leads to three extra rows in transition table so that 4 rows table is modified to 7 rows table as shown in Fig. 6. The rows are arranged such that their state assignments are adjacent. Here, one the of code is not used and so there is a row with all don't-care entries in columns. The state transition from state b does not require additional state and so the correspond row in modified table is same as original table.

The transition from a to c go via a new state e. Therefore, in row a, c is replaced by e and in row e the corresponding column is filled with c.

The transition from a to d and d to a go via a new state f. Therefore, in row a, d is replaced by f and the corresponding column in row f is filled with d. Also, in row d, state a is replaced by f and the corresponding column in f is filled with state a.

The transition from c to d go via a new state g. Therefore, in row c, the state d is replaced by g and the corresponding column in row g is filled with d. The unfilled or empty columns in Fig. 6 are considered as don't-care states.

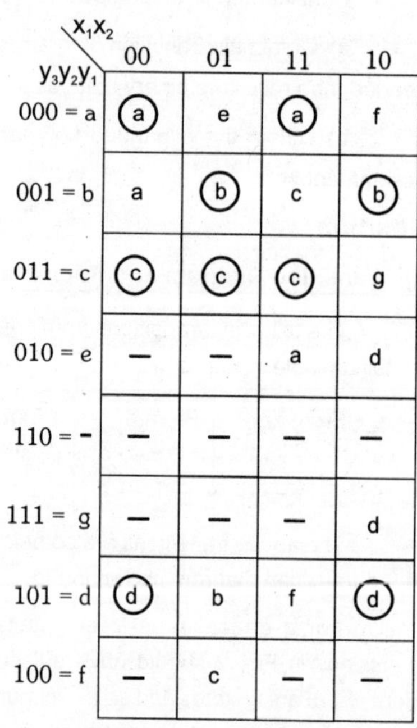

$y_3y_2y_1$ \ x_1x_2	00	01	11	10
000 = a	(a)	e	(a)	f
001 = b	a	(b)	c	(b)
011 = c	(c)	(c)	(c)	g
010 = e	—	—	a	d
110 = -	—	—	—	—
111 = g	—	—	—	d
101 = d	(d)	b	f	(d)
100 = f	—	c	—	—

Fig. 6: Modified flow table with race free state assignment.

b) Multiple Row Method

In multpile row method each row in original flow table is represented by two or more equivalent rows. Let us represent each row by two equivalent rows so that,

- a is represented by a_1 and a_2
- b is represented by b_1 and b_2
- c is represented by c_1 and c_2
- d is represented by d_1 and d_2

Now, there are 8 states and so a 3-bit code is required for 8 binary codes ($2^3 = 8$). The 8 possible codes are,

000, 001, 010, 011, 100,101, 110, 111

Let us represent three bits of binary code by three state variables y_1, y_2 and y_3.

Let us assign 4 adjacent codes to a_1, b_1, c_1 and d_1 as shown in Fig. 7(a). In multiple row method the equivalent states are assigned with complement codes. Hence complement of codes for a_1, b_1, c_1 and d_1 are assigned to states a_2, b_2, c_2 and d_2 as shown in Fig. 7(b).

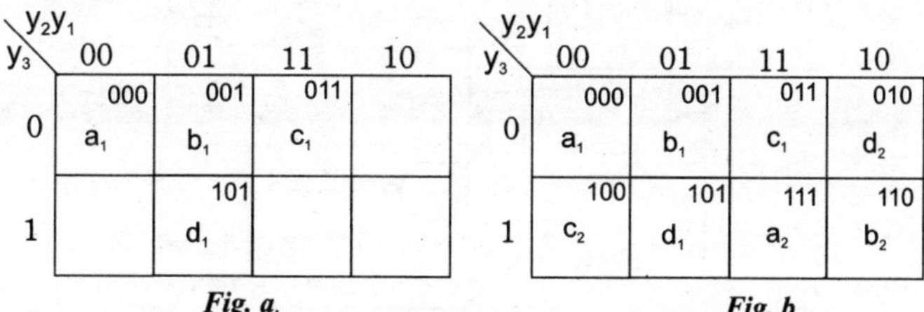

Fig. 7: Binary state assignment in multiple row method.

$y_3y_2y_1$ \ x_1x_2	00	01	11	10
000 = a_1	(a_1)		(a_1)	
111 = a_2	(a_2)		(a_2)	
001 = b_1		(b_1)		(b_1)
110 = b_2		(b_2)		(b_2)
011 = c_1	(c_1)	(c_1)	(c_1)	
100 = c_2	(c_2)	(c_2)	(c_2)	
101 = d_1	(d_1)			(d_1)
010 = d_2	(d_2)			(d_2)

Fig. a.

$y_3y_2y_1$ \ x_1x_2	00	01	11	10
000 = a_1	(a_1)	c_2	(a_1)	d_2
111 = a_2	(a_2)	c_1	(a_2)	d_1
001 = b_1	a_1	(b_1)	c_1	(b_1)
110 = b_2	a_2	(b_2)	c_2	(b_2)
011 = c_1	(c_1)	(c_1)	(c_1)	d_2
100 = c_2	(c_2)	(c_2)	(c_2)	d_1
101 = d_1	(d_1)	b_1	a_2	(d_1)
010 = d_2	(d_2)	b_2	a_1	(d_2)

Fig. b.

Fig. 8: Expanded flow table.

The expanded flow table is formed by representing each row of original flow table by their two equivalent rows as shown in Fig. 8.

First fill the table shown in Fig. 8 with equivalent stable states as shown in Fig. 8(a). The equivalent stable states are same as original stable states but with subscript 1 and 2. Next, the unstable states are marked with reference to state assignment table such that they are adjacent.

In rows a_1 and a_2 the unstable states c_1, c_2, d_1 and d_2 are filled such that they are adjacent to a_1 and a_2.

In rows b_1 and b_2 the unstable states a_1, a_2, c_1 and c_2 are filled such that they are adjacent to b_1 and b_2.

In rows c_1 and c_2 the unstable states d_1 and d_2 are filled such that they are adjacent to c_1 and c_2.

In rows d_1 and d_2 the unstable states a_1, a_2, b_1 and b_2 are filled such that they are adjacent to d_1 and d_2.

Example 9.10

(AKTU, Dec'22, 10 Marks)

Derive the transition table for the asynchronous sequential circuit shown in Fig. 1. Determine the sequence of the internal states $Y_1 Y_2$ for the following sequence of inputs x_1x_2 : 00, 10, 11, 01, 11, 00.

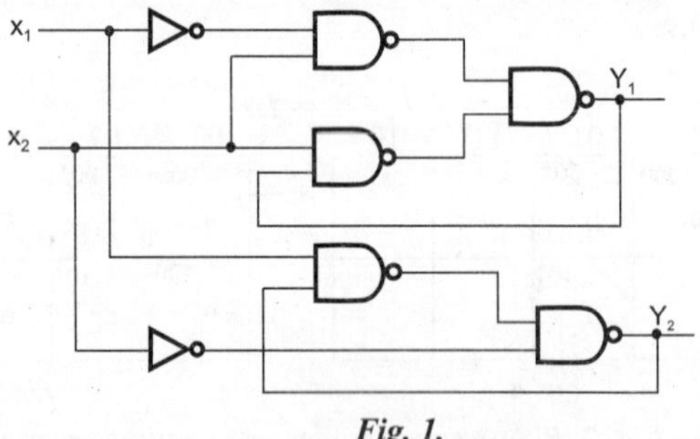

Fig. 1.

Solution

From the given circuit we can conclude that the circuit has following variables.

x_1, x_2 = Input

y_1, y_2 = Secondary variables (Present state variables)

Y_1, Y_2 = Excitation variables (Next state variables)

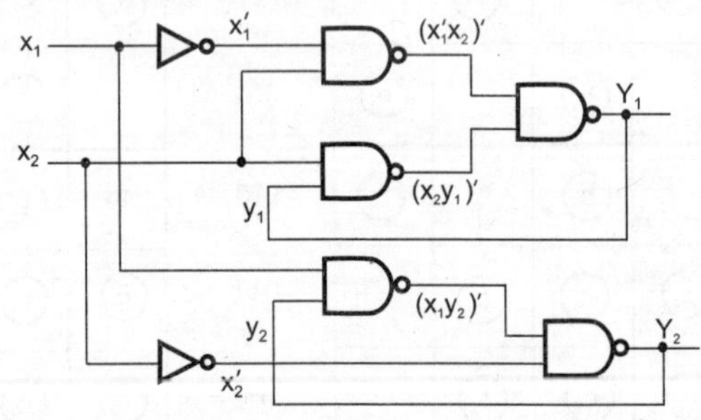

Fig. 2.

From the asynchronous sequential circuit shown in Fig. 2, we get the following equations,

$$Y_1 = ((x_1' x_2)' . (x_2y_1)')' = x_1'x_2 + x_2y_1$$

$$Y_2 = ((x_1y_2)' . x_2')' = x_1y_2 + x_2$$

Using the above equations the excitation variables Y_1 and Y_2 are evaluated for all possible combinations for inputs (x_1 and x_2) and secondary variables (y_1 and y_2) and listed in Table 1.

Table 1: Evaluation of Excitation Variables

Present states				Product Terms			Next States	
y_1	y_2	x_1	x_2	$x_1'x_2$	x_2y_1	x_1y_2	Y_1	Y_2
0	0	0	0	0	0	0	0	0
0	0	0	1	1	0	0	1	1
0	0	1	0	0	0	0	0	0
0	0	1	1	0	0	0	0	1
0	1	0	0	0	0	0	0	0
0	1	0	1	1	0	0	1	1
0	1	1	0	0	0	1	0	1
0	1	1	1	0	0	1	0	1
1	0	0	0	0	0	0	0	0
1	0	0	1	1	1	0	1	1
1	0	1	0	0	0	0	0	0
1	0	1	1	0	1	0	1	1
1	1	0	0	0	0	0	0	0
1	1	0	1	1	1	0	1	1
1	1	1	0	0	0	1	0	1
1	1	1	1	0	1	1	1	1

Maps are constructed for Y_1 and Y_2 by taking possible values of secondary variables y_1 and y_2 on rows and input x_1 and x_2 on columns as shown in Figs. 3(a) and (b) and the squares in the maps are filled with values of Y_1 and Y_2 listed in Table 1. The transition table is formed by combining the maps of Figs. 3(a) and (b) into a single map as shown in Fig. 4.

In the transition table of each row, the squares that have identical values with secondary values are marked with circles. The circled entries represent stable states and uncircled entries represent unstable states.

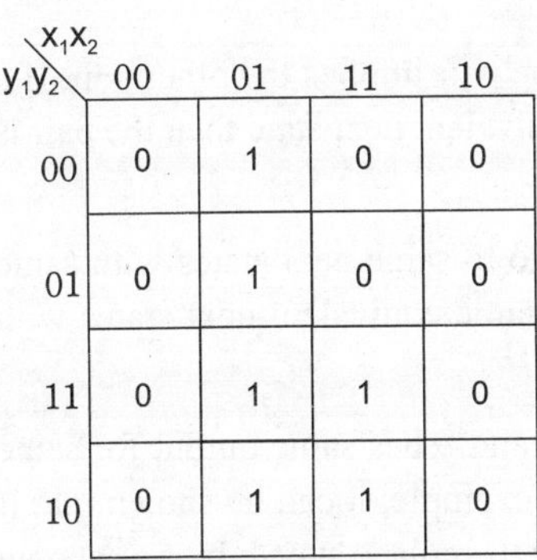

Fig. a: K-map for Y_1.

y_1y_2 \ x_1x_2	00	01	11	10
00	0	1	1	0
01	0	1	1	1
11	0	1	1	1
10	0	1	1	0

Fig. b: K-map for Y_2.

Fig. 3: Map for two state variables Y_1, Y_2.

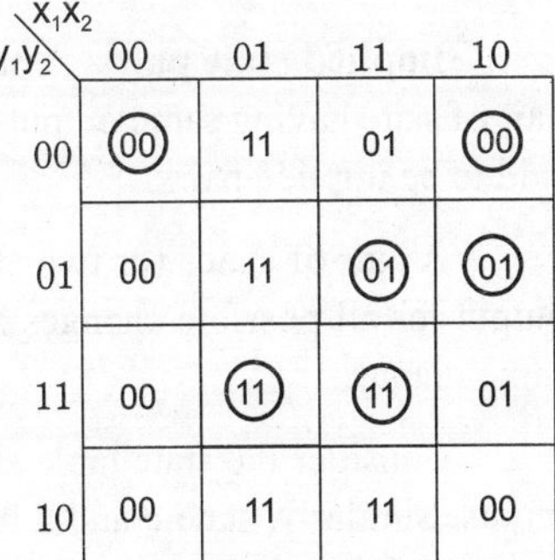

Fig. 4: Transition table.

It is given that the sequence of input changes is,

$00 \rightarrow 10 \rightarrow 11 \rightarrow 01 \rightarrow 11 \rightarrow 00$

The state transitions for the above input changes are shown in Fig. 5. The present and next state for the sequence of changes in input are also listed in Table 2.

Table 2:

Change in Input x_1 x_2	Present State y_1 y_2	State Transitions	Next State Y_1 Y_2
0 0	0 0	–	0 0
00 → 10	0 0	00 → 00	0 0
10 → 11	0 0	00 → 01 → 01	0 1
11 → 01	0 1	01 → 11 → 11	1 1
01 → 11	1 1	11 → 11	1 1
11 → 00	1 1	11 → 00 → 00	0 0

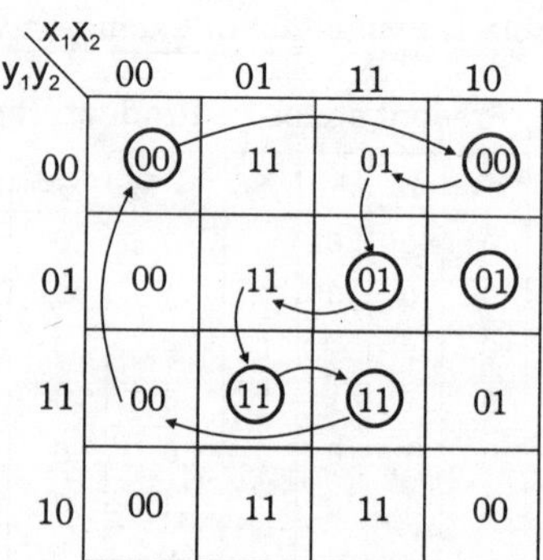

Fig. 5: Transitions through the sequence of inputs x_1 x_2.

Conclusion: For the input sequence of 00, 10, 11, 01, 11 and 00, the sequence of Y_1Y_2 is,

Y_1Y_2 : 00, 00, 01, 11, 11, 00.

9.2 State Reduction

(JNTUA, Jun/Jul'22, 7 Marks)

In asynchronous sequential circuits, the state reduction is performed via an implication table. In the original state table, the equivalent states are identified and each pair of equivalent states is merged as a single state.

The rows of state tables can be classified in to following three groups.

Equivalent States: State pairs that have same next state and output for the same input are equivalent states.

Not equivalent states: State pairs having different next state and output for same input are not equivalent states.

Implied state pairs: A pair of states can be equivalent only if one pair implies the other pair. If a pair of state having same outputs for the same inputs and go to same equivalent next state then the pair is said to be implied pair.

A pair of states (or two states) are said to be equivalent if they go to same next states with same output for all possible changes in inputs. Alternatively, they can go to same equivalent next states with same output.

Consider the state table shown in Table 9.4. In this table, state a and b has same output for same inputs. Similarly, state c and d has same output for same inputs. In this example, it can be shown that if state c and d are equivalent then state a and b will also be equivalent. Let us replace state b by a as shown in Table 9.5. Now, in Table 9.5, if state d is equal to c then state a and b will be equivalent as well as state c and d are also equivalent. Hence we can say that (a, b) implies (c, d).

The state table with state d replaced by state c is shown in Table 9.6 and from this table we can say that state a and b are equivalent and also, state c and d are equivalent. Hence, one of the equivalent states can be removed and reduced state table can be formed as shown in Table 9.7.

Table 9.4: State Table

Present State	Next State x = 0	Next State x = 1	Output x = 0	Output x = 1
a	d	a	1	0
b	c	b	1	0
c	b	d	0	1
d	a	d	0	1

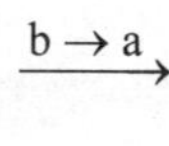

Table 9.5: State Table with State b Replaced by State a

Present State	Next State x = 0	Next State x = 1	Output x = 0	Output x = 1
a	d	a	1	0
a (b → a)	c	a (b → a)	1	0
c	a (b → a)	d	0	1
d	a	d	0	1

d → c

Table 9.6: State Table with State d Replaced by State c

Present State	Next State x = 0	Next State x = 1	Output x = 0	Output x = 1
a	c (d → c)	a	1	0
a	c	a	1	0
c	a	c (d → c)	0	1
c (d → c)	a	c (d → c)	0	1

Reduced state table ←

Table 9.7: Reduced State Table

Present State	Next State x = 0	Next State x = 1	Output x = 0	Output x = 1
a	c	a	1	0
c	a	c	0	1

9.2.1 Implication Table

(JNTUA, Jun/Jul'22, 7 Marks)

When the state table has large number of rows it is difficult to find equivalent and implied states and in this situation implication table is used to find equivalent and implied states and then form the reduced state table.

The implication table is constructed by listing the states on rows and columns in order to form squares at the intersection of every row and column. Each square represent a pair of state. While forming rows, the first state is not considered and while forming column the last state is not considered so that only state pairs are represented by squares. The implication table is also known as ***merger table***.

Consider the state table of asynchronous sequential circuit with 7 states a, b, c, d, e, f and g shown in Table 9.8. Now, the implication table is formed by listing states b, c, d, e, f and g on rows and states a, b, c, d, e, and f on columns in order to form squares at the intersection of each row and column as shown in Fig. 9.11(a). The state pair represented by each square is shown in Fig. 9.11(b).

Table 9.8: State Table

Present State	Next State		Output	
	x = 0	x = 1	x = 0	x = 1
a	e	a	0	0
b	d	b	0	0
c	g	f	1	0
d	a	d	0	1
e	a	d	0	1
f	b	c	0	0
g	a	e	0	1

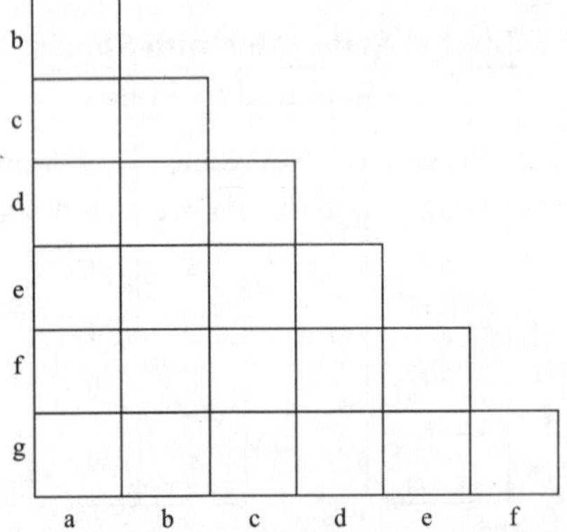

Fig. a: Listing of states in rows and columns.

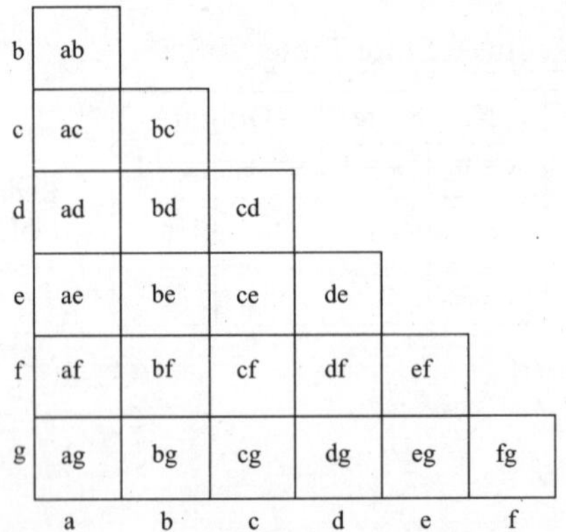

Fig. b: Pair of states represented by squares.

Fig. 9.11: Structure of implication table.

First mark × in the squares in which the pair of states do not have same output for same inputs as shown in Fig. 9.12(a). The state pairs marked by × are not equivalent pairs. Next mark a tick (√) in the squares in which the state pairs are equivalent as shown in Fig. 9.12(b). Here, the state pair (d, e) is equivalent because they go to same next state and output for same input.

The next step is to identify the implied pairs among the state pairs that are not marked by × and tick (√) in Figs. 9.12(a) and (b). Mark the next state pairs corresponding to the state pair represented by the squares that are blank in Figs. 9.12(a) and (b), as shown in Fig. 9.12(c).

When next state pair and present state pair are same then those state pairs can be ignored in Fig. 9.12(c). Similarly, when a present state pair go to same next state then those next states can also be ignored in Fig. 9.12(c).

The state pair entries in Fig. 9.12(c), are marked × if the pair is not equivalent and marked with tick (√) if the state pair is equivalent. The three implication tables shown in Figs. 9.12(a), (b) and (c) are combined to form the implication table shown in Fig. 9.12(d). Now, the equivalent state pairs are determined from implication table of Fig. 9.12(d).

When a square representing a state pair has an equivalent state pair entry then that state pair is also equivalent. In Fig. 9.12(d), the squares representing (a, b), (d, g) and (e, g) has equivalent state pair entires and so (a, b), (d, g) and (e, g) are equivalent state pairs.

When a square representing a state pair has not equivalent state pair entry then that state pair is also not equivalent. In Fig. 9.12(d), the squares representing (a, f) and (b, f) has not equivalent state pairs and so (a, f) and (b, f) are not equivalent state pairs.

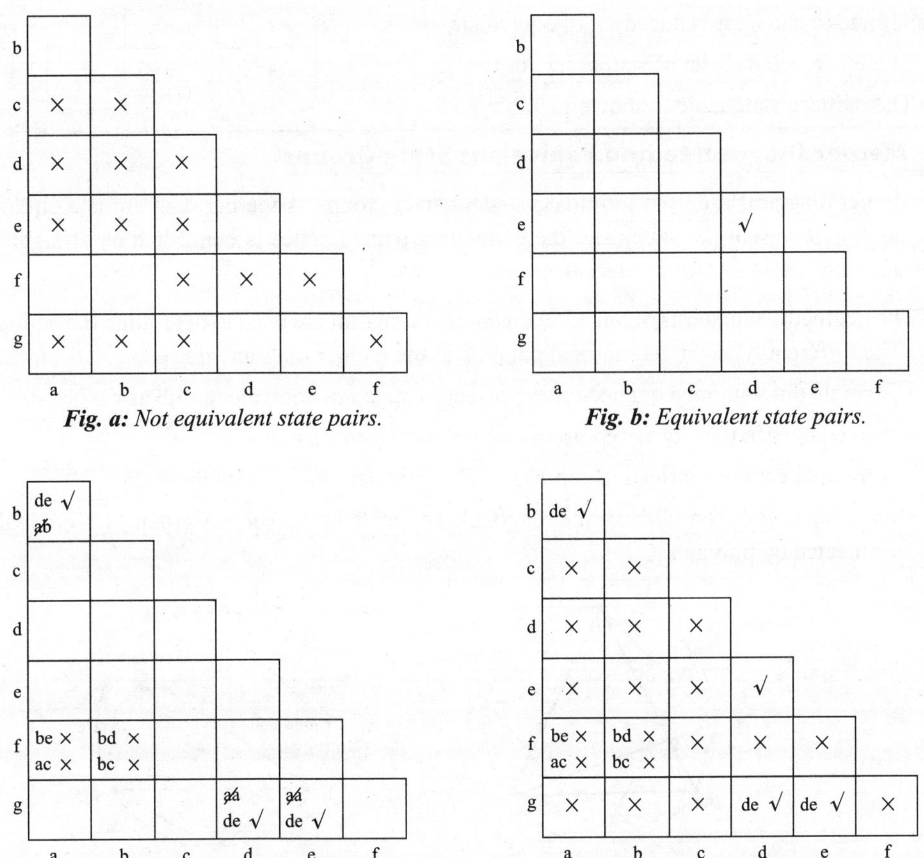

Fig. a: *Not equivalent state pairs.*

Fig. b: *Equivalent state pairs.*

Fig. c: *Implied state pairs.*

Fig. d: *Implication table.*

Fig. 9.12: *Formation of implication table.*

Now, the equivalent state pairs are,

(a, b), (d, e), (d, g), (e, g)

The state pairs (d, e), (d, g) and (e, g) can be merged to larger group (d, e, g). Now, the equivalent state groups are,

(a, b), (d, e, g)

In the equivalent state groups shown above only one state is selected in each group. Let us select state a in group (a, b) and state d in group (d, e, g).

Now the reduced states are given by one of the state in each equivalent group and the states not listed in the equivalent pairs.

Here, the states c and f are not included in equivalent pairs and states a and d are reduced set of equivalent states.

Therefore, reduced set of states: a, c, d, f.

The reduced state table is shown in Table 9.9.

Table 9.9: Reduced State Table

Present State	Next State		Output	
	x = 0	x = 1	x = 0	x = 1
a	d (e → d)	a	0	0
c	d (g → d)	f	1	0
d	a	d	0	1
f	a (b → a)	c	0	0

9.2.2 Merger Diagram to find Equivalent State Groups

Merger diagrams are used to find equivalent state groups. A merger diagram is a circle with dots where the dots represent the states and the equivalent pair of states is connected by straight lines. The merger diagram is also known as ***merger graph***.

The geometrical patterns obtained in the merger diagram are used to determine the equivalent state groups. The different types of geometrical patterns in the merger diagram are,

- A single dot without a connection represents a state not equivalent with any other state.
- A line between two states represents a equivalent pair of states.
- A triangle between three dots represents an equivalent group with three states.
- An n-sided polygon with all the diagonals connected represents a group of n-equivalent states connected by polygon.

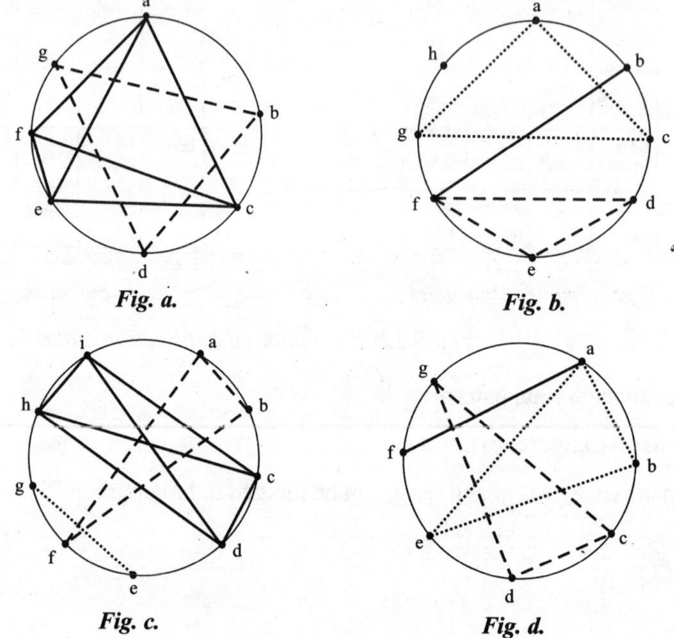

Fig. 9.13: Examples of merger diagrams.

Some examples of merger diagrams with different groups of equivalent states are shown in Fig. 9.13.

Equivalent state groups in Fig. 9.13(a) : (b, d, g), (a, c, e, f)

Equivalent state groups in Fig. 9.13(b) : (b, f), (a, c, g), (d, e, f)

Equivalent state groups in Fig. 9.13(c) : (e, g), (a, b, f), (c, d, h, i)

Equivalent state groups in Fig. 9.13(d) : (a, f), (a, b, e), (c, d, g)

Example 9.11

Reduce the state table shown in Table 1 to a minimum number of states using the implication table method.

Table 1: State Table

Present State	Next State		Output	
	x = 0	x = 1	x = 0	x = 1
a	a	e	0	0
b	c	b	0	1
c	a	f	0	0
d	c	b	0	1
e	f	e	0	0
f	a	f	0	0

Solution

The given state table (Table 1) of asynchronous sequential circuit has 6 states a, b, c, d, e and f. Now, the implication table is formed by listing states b, c, d, e and f on rows and states a, b, c, d, and e on columns and forming squares at the intersection of each row and column as shown in Fig. 1(a). The state pair represented by each square is shown in Fig. 1(b).

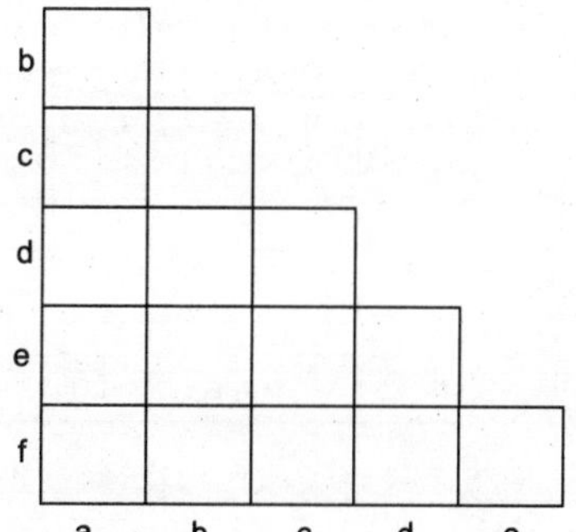

Fig. a: Listing of states in rows and columns.

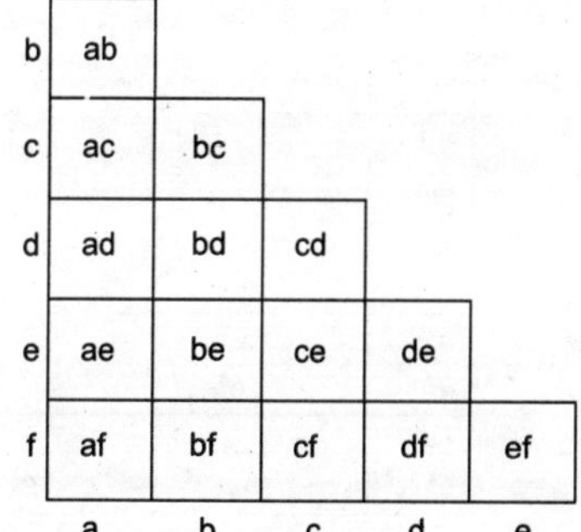

Fig. b: Pair of states represented by squares.

Fig. 1: Structure of implication table.

Here, three implication tables (Figs. 2(a), (b) and (c)) are drawn to separately show the not equivalent, equivalent and implied state pairs.

First mark × in the squares in Fig. 2(a) in which the pair of states do not have same output for same inputs. The state pairs marked by × are not equivalent pairs. The not equivalent state pairs are listed in Table 2.

Next mark a tick (√) in the squares in Fig. 2(b) in which the state pairs are equivalent. Here, the state pair (b, d) and (c, f) is equivalent because they go to same next state and output for same input. The equivalent state pairs are listed in Table 3.

The next step is to identify the implied pairs among the state pairs that are not marked by × and tick (√) in Figs. 2(a) and (b). Mark the next state pairs corresponding to the state pair represented by the squares that are blank in Figs. 2(a) and (b), as shown in Fig. 2(c). When a present state pair go to same next state then those next states can be ignored in Fig. 2(c).

Table 2: Not Equivalent State Pairs

Present State	Next State		Output	State Pair
a	a	e	0 0	(a, b)
b	c	b	0 1	
a	a	e	0 0	(a, d)
d	c	b	0 1	
b	c	b	0 1	(b, c)
c	a	f	0 0	
b	c	b	0 1	(b, e)
e	f	e	0 0	
b	c	b	0 1	(b, f)
f	a	f	0 0	
c	a	f	0 0	(c, d)
d	c	b	0 1	
d	c	b	0 1	(d, e)
e	f	e	0 0	
d	c	b	0 1	(d, f)
f	a	f	0 0	

Table 3: Equivalent State Pairs

Present State	Next State		Output	State Pair
b	c	b	0 1	(b, d)
d	c	b	0 1	
c	a	f	0 0	(c, f)
f	a	f	0 0	

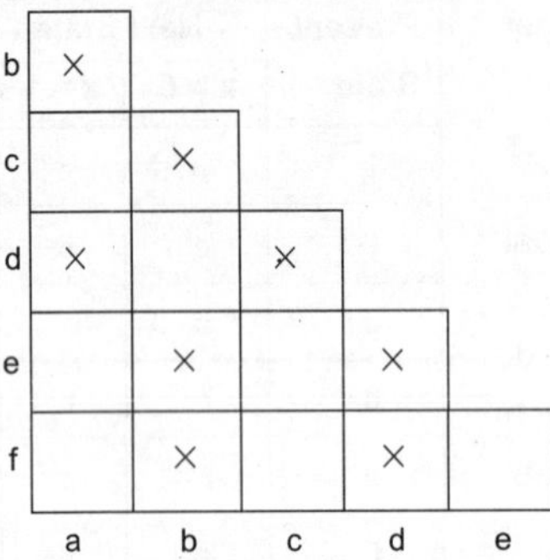

Fig. a: Not equivalent state pairs.

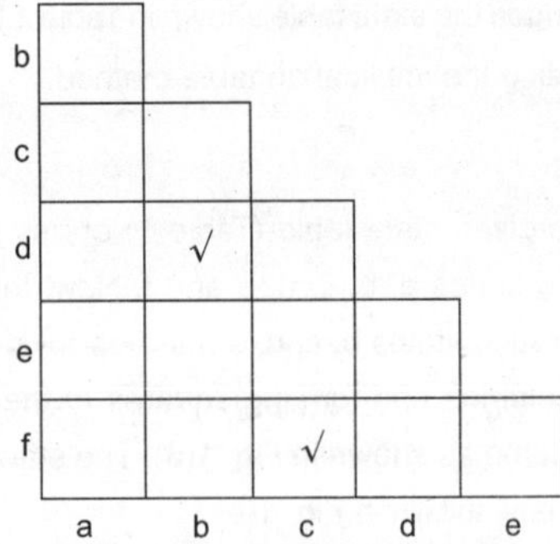

Fig. b: Equivalent state pairs.

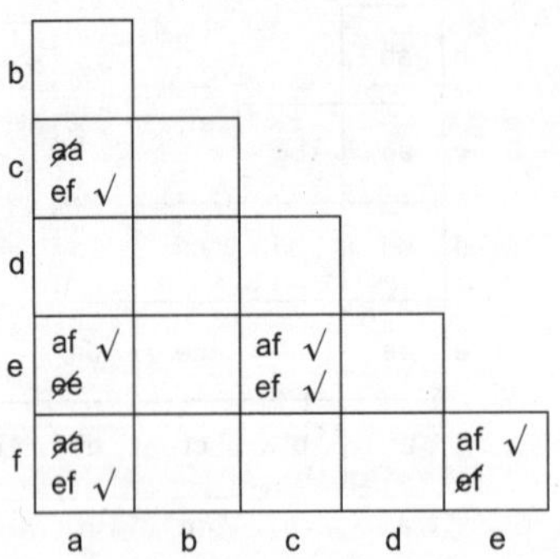

Fig. c: Implied state pairs.

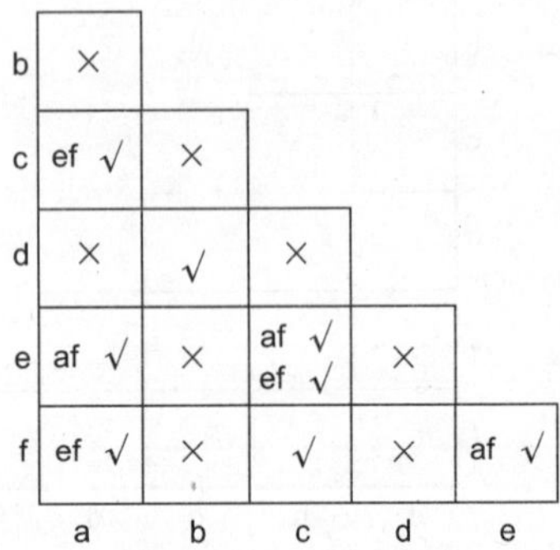

Fig. d: Implication table.

Fig. 2: Formation of implication table.

Here, all entries in Fig. 2(c) are either state pair (a, f) or (e, f). The state pair (a, f) go to same next state for same input. Similarly, state pair (e, f) go to same next state for same input. Hence, state pairs (a, f) and (e, f) are implied pairs.

With reference to Tables 4 and 5 when f = a then state pair (a, f) implies (e, f). Hence state pairs (a, f) and (e, f) are equivalent pairs and they are marked with tick (√).

Table 4: State Table for State Pairs (a, f) and (e, f)

Present State	Next State x = 0	Next State x = 1	Output x = 0	Output x = 1
a	a	e	0	0
f	a	f	0	0
e	f	e	0	0
f	a	f	0	0

$\xrightarrow{f \rightarrow a}$

Table 5: Table 4 with State f Replaced by State a

Present State	Next State x = 0	Next State x = 1	Output x = 0	Output x = 1
a	a	e	0	0
a (f → a)	a	a (f → a)	0	0
e	a (f → a)	e	0	0
a (f → a)	a	a (f → a)	0	0

The three implication tables in Figs. 2(a), (b) and (c) are combined as a single implication table in Fig. 2(d) and the equivalent state pairs are determined from Fig. 2(d).

When a square representing a state pair has an equivalent state pair entry then that state pair is also equivalent. In Fig. 2(d), the squares representing (a, c), (a, e), (a, f), (c, e) and (e, f) has equivalent state pair entires and so (a, c), (a, e), (a, f), (c, e) and (e, f) are equivalent state pairs.

With reference to Fig. 2(d), the equivalent state pairs are,

(a, c), (a, e), (a, f), (c, e) (e, f), (c, f), (b, d)

The state pairs (a, c), (a, e), (a, f), (c, e) (e, f) and (c, f) can be merged to larger group (a, c, e, f). Now, the equivalent state groups are,

(a, c, e, f), (b, d)

Alternatively, the above equivalent groups can be obtained from the merger diagram shown in Fig. 3. In this diagram a 4 sided polygon with diagonals connected form an equivalent group (a, c, e, f) and a line betweem two states form an equivalent pair (b, d)

In the equivalent state groups shown above only one state is selected in each group. Let us select state a in group (a, c, e, f) and state b in group (b, d).

Now the reduced states are given by one of the state in each equivalent group and the states not listed in the equivalent pairs. Here, all states are included in equivalent state groups and so the states a and b are reduced set of equivalent states.

Therefore, reduced set of states: a, b.

The reduced state table is shown in Table 6.

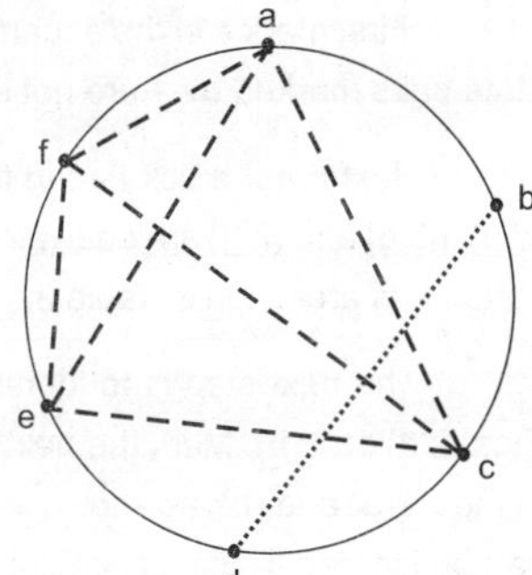

Fig. 3: Merger diagram.

Table 6: Reduced State Table

Present State	Next State		Output	
	x = 0	x = 1	x = 0	x = 1
a	a	a (e → a)	0	0
b	a (c → a)	b	0	1

Example 9.12

Reduce the state table shown in Table 1 to a minimum number of states using the implication table method.

Table 1: State Table

Present State	Next State		Output	
	x = 0	x = 1	x = 0	x = 1
a	d	b	0	0
b	f	a	0	0
c	g	f	0	1
d	a	d	1	0
e	c	b	0	0
f	a	d	1	0
g	a	d	1	0

Solution

The given state table (Table 1) of asynchronous sequential circuit has 7 states a, b, c, d, e, f and g. Now, the implication table is formed by listing states b, c, d, e, f and g on rows and states a, b, c, d, e and f on columns and forming squares at the intersection of each row and column as shown in Fig. 1(a). The state pair represented by each square is shown in Fig. 1(b).

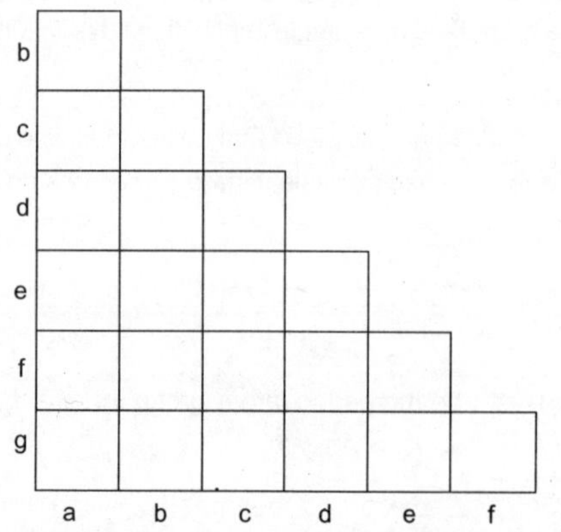

Fig. a: Listing of states in rows and columns.

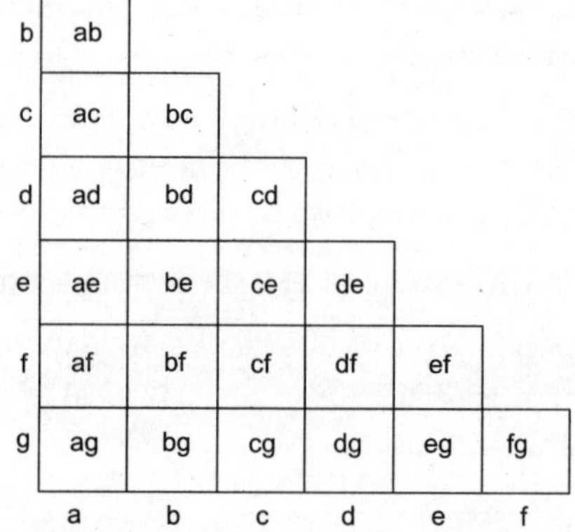

Fig. b: Pair of states represented by squares.

Fig. 1: Structure of implication table.

Here, three implication tables (Figs. 2(a), (b) and (c)) are drawn to separately show the not equivalent, equivalent and implied state pairs.

First mark × in the squares in Fig. 2(a) in which the pair of states do not have same output for same inputs. The state pairs marked by × are not equivalent pairs. The not equivalent state pairs are listed in Table 2.

Next mark a tick (√) in the squares in Fig. 2(b) in which the state pairs are equivalent. Here, the state pairs (d, f), (d, g) and (f, g) are equivalent pairs because they go to same next state and output for same input. The equivalent state pairs are listed in Table 3.

The next step is to identify the implied pairs among the state pairs that are not marked by × and tick (√) in Figs. 2(a) and (b). Mark the next state pairs corresponding to the state pair represented by the squares that are blank in Figs. 2(a) and (b), as shown in Fig. 2(c). When next state pair and present state pair are same then those state pairs can be ignored in Fig. 2(c). Similarly, when a present state pair go to same next state then those next states can also be ignored in Fig. 2(c).

The state pair entries in Fig. 2(c), are marked × if the pair is not equivalent and marked with tick (√) if the state pair is equivalent.

The three implication tables in Figs. 2(a), (b) and (c) are combined as a single implication table in Fig. 2(d) and the equivalent state pairs are determined from Fig. 2(d).

When a square representing a state pair has an equivalent state pair entry then that state pair is also equivalent. In Fig. 2(d), the square representing (a, b) has equivalent state pair entry and so (a, b) is equivalent state pair. When a square representing a state pair has not equivalent state pair entry then that state pair is also not equivalent. In Fig. 2(d), the squares representing (a, e) and (b, e) have not equivalent state pairs and so (a, e) and (b, e) are not equivalent state pairs.

Table 2: Not Equivalent State Pairs

State	Next states	Outputs	Pair	State	Next states	Outputs	Pair
a	d b	0 0	(a, c)	c	g f	0 1	(c, d)
c	g f	0 1		d	a d	1 0	
a	d b	0 0	(a, d)	c	g f	0 1	(c, e)
d	a d	1 0		e	c b	0 0	
a	d b	0 0	(a, f)	c	g f	0 1	(c, f)
f	a d	1 0		f	a d	1 0	
a	d b	0 0	(a, g)	c	g f	0 1	(c, g)
g	a d	1 0		g	a d	1 0	
b	f a	0 0	(b, c)	d	a d	1 0	(d, e)
c	g f	0 1		e	c b	0 0	
b	f a	0 0	(b, d)	e	c b	0 0	(e, f)
d	a d	1 0		f	a d	1 0	
b	f a	0 0	(b, f)	e	c b	0 0	(e, g)
f	a d	1 0		g	a d	1 0	
b	f a	0 0	(b, g)				
g	a d	1 0					

Table 3: Equivalent State Pairs

State	Next states	Outputs	Pair
d	a d	1 0	(d, f)
f	a d	1 0	
d	a d	1 0	(d, g)
g	a d	1 0	
f	a d	1 0	(f, g)
g	a d	1 0	

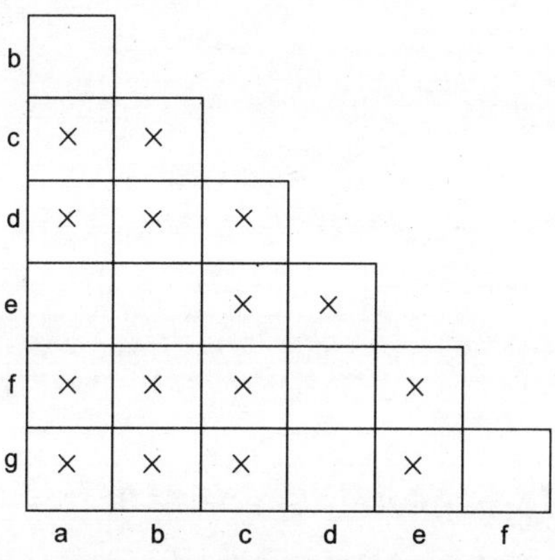

Fig. a: *Not equivalent state pairs.*

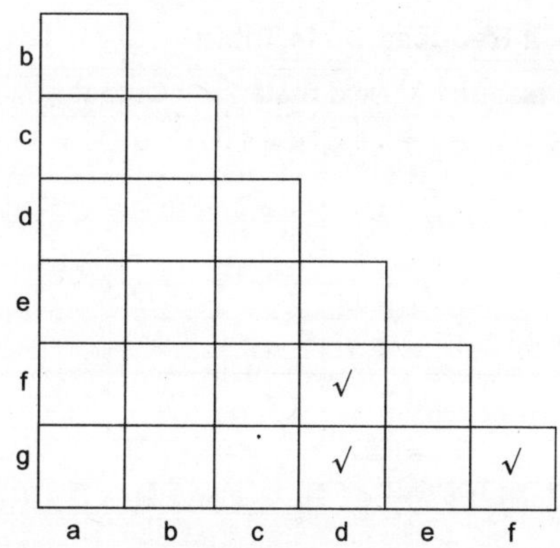

Fig. b: *Equivalent state pairs.*

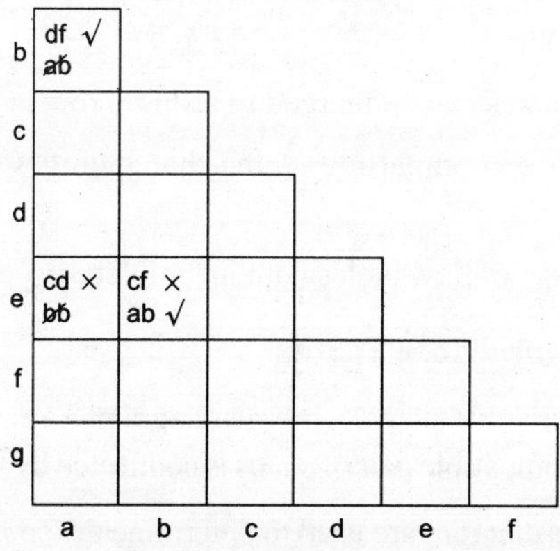

Fig. c: *Implied state pairs.*

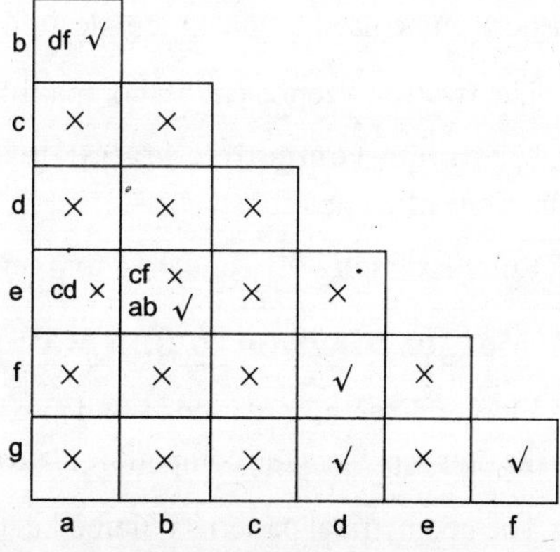

Fig. d: *Implication table.*

Fig. 2: Formation of implication table.

With reference to Fig. 2(d), the equivalent state pairs are,

(a, b), (d, f), (d, g), (f, g)

The state pairs (d, f), (d, g) and (f, g) can be merged to larger group (d, f, g). Now, the equivalent state groups are,

(a, b), (d, f, g)

Alternatively, the above equivalent states can be obtained from merger diagram shown in Fig. 3. In this diagram, a triangle form an equivalent group (d, f, g) and a line between two states form an equivalent pair (a, b). In the equivalent state groups shown above only one state is selected in each group. Let us select state a in group (a, b) and state d in group (d, f, g).

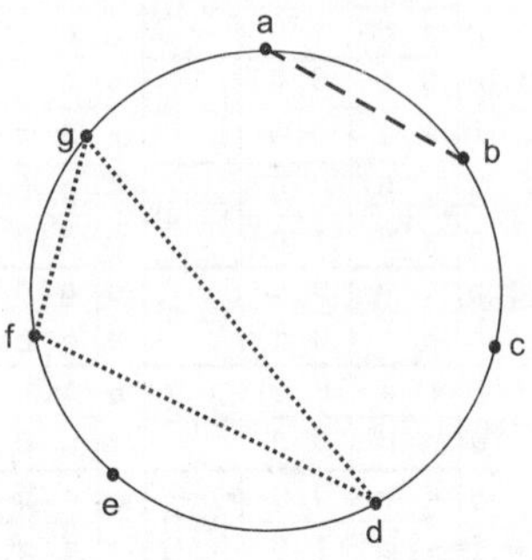

Fig. 3: Merger diagram.

Now the reduced states are given by one of the state in each equivalent group and the states not listed in the equivalent pairs. Here, the states selected from equivalent groups are a and d. The states not included in state groups are c and e.

Therefore, reduced set of states: a, c, d, e.

The reduced state table with 4 states is shown in Table 4.

Table 4: Reduced State Table

Present State	Next State		Output	
	x = 0	x = 1	x = 0	x = 1
a	d	a (b → a)	0	0
c	d (g → d)	d (f → d)	0	1
d	a	d	1	0
e	c	a (b → a)	0	0

9.2.3 Merging of Rows of Flow Tables with Incompletely Specified States

In asynchronous circuits with incompletely specified states, the states can be reduced by identifying compatible pairs. Two states are said to be a compatible pair if the specified outputs and next states are same. Here, dashes are used to represent unspecified states or outputs. While identifying compatible pairs the dashes can be used to match any desired state or output.

The two rows representing compatible pair of states can be merged to a single row in a flow table.

Maximum compatible states: It is a group of compatible states that includes all possible combinations of compatible states.

Refer Examples 9.13 and 9.14 for merging of rows of flow tables with incompletely specified states.

9.2.4 Merger Diagram to find Maximum Compatible States

Merger diagrams are used to find maximum compatible states. A merger diagram is a circle with dots where the dots represent the compatible states and the compatible pair of states is connected by straight lines.

The geometrical patterns obtained in the merger diagram are used to determine the compatible state groups. The different types of geometrical patterns in the merger diagram are,

- A single dot without a connection represents a state not compatible with any other state.

- A line between two states represents a compatible pair of states.
- A triangle between three dots represents a state compatibility with the three connected states.
- An n-sided polygon with all the diagonals connected represents a group of n-compatible states connected by polygon.

Examples of merger diagram shown in Fig. 9.13 are applicable for determining compatible groups. Here, the equivalent groups are called ***compatible groups***.

9.2.5 Maximum Compatible

The maximum compatible is a group of compatible states that includes all the possible combinations of compatible states. The maximum compatible states can be determined from merger diagram. The states forming various geometrical patterns in the merger diagram are the maximum compatibles.

The reduced flow table can be formed by taking one member of each maximum compatible set as a row in reduced flow table.

In many cases, the maximum compatibles will not be a minimal set of compatibles. A minimum set of compatibles can be determined from closed-covering condition.

9.2.6 Closed-Covering Condition

In implication table, the compatible pairs may or may not have implied pairs. Hence the closed condition can be defined for two different types of implication table.

1. In implication table, if the compatible pairs does not have implied pairs then the closed-condition refers to a set of compatibles that does not include implied states.

2. In implication table, when the compatible pairs has implied states then the implied pairs should be included in selected compatible pairs.

The covering-condition refers to inclusion of all states of original state table in the compatible sets. The closed-covering condition refers to set of compatibles that covers all the states and satisfy covering condition. The set of compatibles that satisfy the closed-covering condition form the minimal set of compatibles. The reduced flow table can be formed by taking one element of each of the minimal compatible set to form the rows of reduced flow table.

9.2.7 Assigning Outputs in Flow Table with Incompletely Specified States

In flow tables with incompletely specified states, every stable state will be associated with an output. Normally, output will not specified for unstable states. But in order to find compatible pairs and reduce the states, output is necessary for unstable states too. Hence, output has to be assigned to all unstable states such that no transient outputs are generated during a transition from one stable state to another stable state. Therefore, the following rules are followed to assign output to unstable states in a flow table.

Rule 1 : Assign 0 to an output of an unstable state which is transient state during a transition between two stable states that have a 0 as output.

Rule 2 : Assign 1 to an output of an unstable state which is transient state during a transition between two stable states that have a 1 as output.

Rule 3 : Assign a don't-care condition to an output of an unstable state which is transient state during a transition between two stable states that have a different values (0 and 1 or 1 and 0) as output.

Examples of output assignment using the above rules are shown in Fig. 9.14.

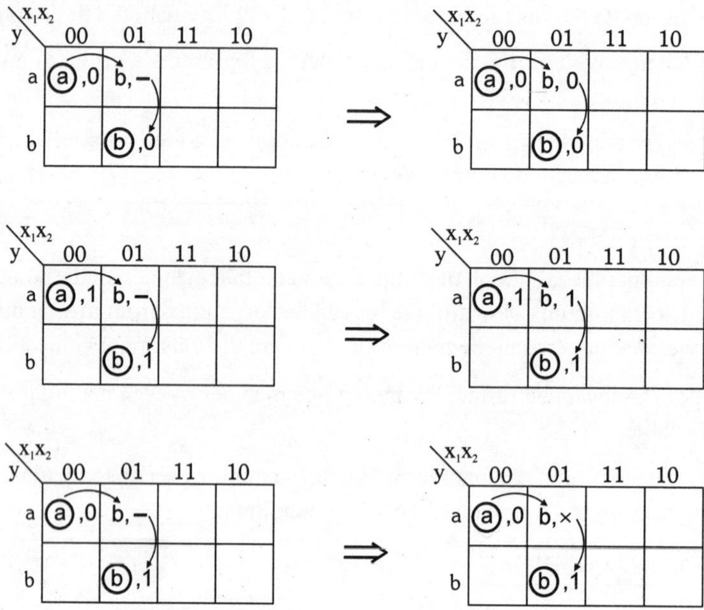

Fig. 9.14: Examples of output assignment.

Refer Examples 9.16, 9.18 in Chapter 9 and 10.5 in Chapter 10 for output assignment.

Example 9.13

Merge the primitive flow table shown in Fig. 1. Find all compatible pairs by means of an implication table. Find the maximal compatibles by means of a merger diagram. Find a minimal set of compatibles that covers all the states and is closed.

Solution

The given flow table (Table 1) has 8 states a, b, c, d, e, f, g and h. The states b, c, d, e, f, g and h are listed to form rows and the states a, b, c, d, e, f and g are listed to form columns in implication table.

Here, three implication tables (Figs. 2(a), (b) and (c)) are drawn to separately show the not compatible, compatible and implied pairs.

The given flow table has 4 combinations of inputs. Hence compatibility of states for all the four inputs should be marked in each square of implication table.

The dashes in the given flow table represents unspecified states or outputs. While identifying compatible pairs the dashes can be used to match any desired state or output.

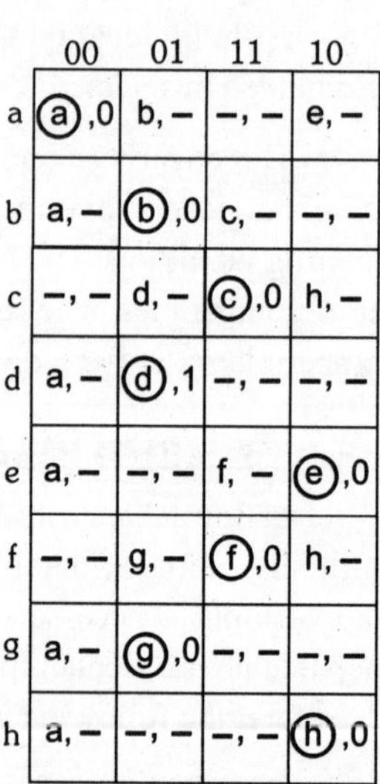

	00	01	11	10
a	(a),0	b, –	–, –	e, –
b	a, –	(b),0	c, –	–, –
c	–, –	d, –	(c),0	h, –
d	a, –	(d),1	–, –	–, –
e	a, –	–, –	f, –	(e),0
f	–, –	g, –	(f),0	h, –
g	a, –	(g),0	–, –	–, –
h	a, –	–, –	–, –	(h),0

Fig. 1: Primitive flow table.

First mark × in squares in Fig. 2(a) in which the state pairs do not have same outputs for same inputs.

- When input is 01, present state pair (b, d), go to next state (b, d) with different output.
- When input is 01, present state pair (d, g), go to next state (d, g) with different output.

Hence the squares representing (b, d) and (d, g) pairs are marked × in Fig. 2(a). The not compatible state pairs that has transitions to stable states with different output are listed in Table 1.

Next mark tick (√) in squares in Fig. 2(b) in which state pairs are compatible. Here, dashes are matched with states and outputs to form compatible pairs. The state pair (a, b) goes to same next state with same output when dashes are considered as same state and output and so (a, b) is compatible pair.

Similarly, (a, e), (b, h), (c, d), (c, h), (d, e), (d, h), (e, g), (f, g), (f, h) and (g, h) are compatible pairs. The compatible state pairs are listed in Table 2.

The next step is to identify the implied pairs among the state pairs that are not marked by × and tick (√) in Figs. 2(a) and (b). Mark the next state pairs corresponding to the state pair represented by the squares that are blank in Figs. 2(a) and (b), as shown in Fig. 2(c). When next state pair and present state pair are same then those state pairs can be ignored in Fig. 2(c). Similarly, when a present state pair goes to same next state then those next states can also be ignored in Fig. 2(c).

The state pair entries in Fig. 2(c), are marked × if the pair is not compatible and marked with tick (√) if the state pair is compatible.

Table 1: Not Compatible State Pairs

	00	01	11	10	
b	a, –	ⓑ,0	c, –	–, –	(b, d)
d	a, –	ⓓ,1	–, –	–, –	
d	a, –	ⓓ,1	–, –	–, –	(d, g)
g	a, –	ⓖ,0	–, –	–, –	

Table 2: Compatible State Pairs

	00	01	11	10	
a	ⓐ,0	b, 0	c, –	e, –	(a, b)
b	a, 0	ⓑ,0	c, –	e, –	
a	ⓐ,0	b, –	f, –	e, 0	(a, e)
e	a, 0	b, –	f, –	ⓔ,0	
b	a, –	ⓑ,0	c, –	h, 0	(b, h)
h	a, –	b, 0	c, –	ⓗ,0	
c	a, –	d, 1	ⓒ,0	h, –	(c, d)
d	a, –	ⓓ,1	c, 0	h, –	
c	a, –	d, –	ⓒ,0	h, 0	(c, h)
h	a, –	d, –	c, 0	ⓗ,0	
d	a, –	ⓓ,1	f, –	e, 0	(d, e)
e	a, –	d, 1	f, –	ⓔ,0	

	00	01	11	10	
d	a, –	ⓓ,1	–, –	h, 0	(d, h)
h	a, –	d, 1	–, –	ⓗ,0	
e	a, –	g, 0	f, –	ⓔ,0	(e, g)
g	a, –	ⓖ,0	f, –	e, 0	
f	a, –	g, 0	ⓕ,0	h, –	(f, g)
g	a, –	ⓖ,0	f, 0	h, –	
f	a, –	g, –	ⓕ,0	h, 0	(f, h)
h	a, –	g, –	f, 0	ⓗ,0	
g	a, –	ⓖ,0	–, –	h, 0	(g, h)
h	a, –	g, 0	–, –	ⓗ,0	

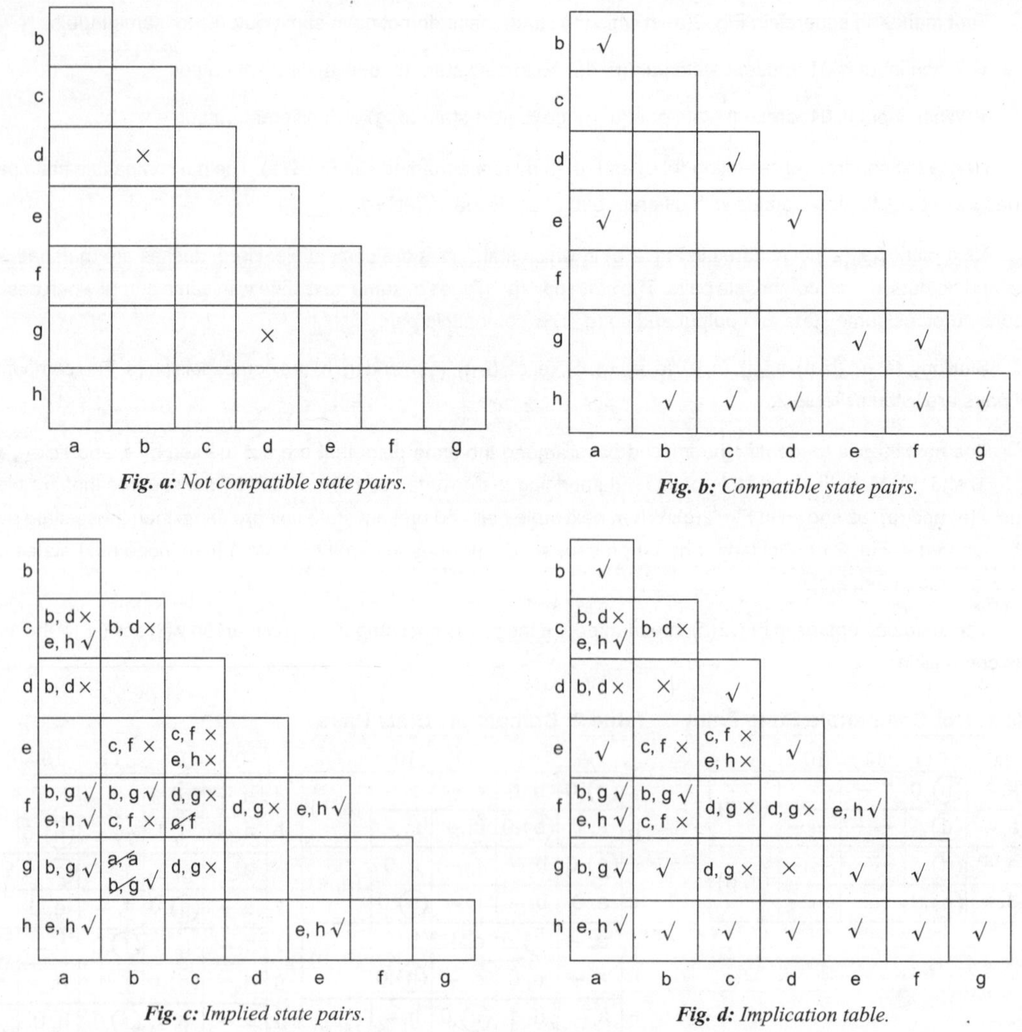

Fig. a: Not compatible state pairs.

Fig. b: Compatible state pairs.

Fig. c: Implied state pairs.

Fig. d: Implication table.

***Fig. 2:** Formation of implication table.*

The three implication tables in Figs. 2(a), (b) and (c) are combined as a single implication table in Fig. 2(d) and the compatible state pairs are determined from Fig. 2(d).

When a square representing a state pair has a compatible state pair entry then that state pair is also compatible. In Fig. 2(d), the square representing (a, f), (a, g), (a, h), (e, f) and (e, h) has compatible state pair entry and so they are compatible state pairs. Here, (b, g) implies (b, g) itself and so (b, g) is also a compatible state pair.

When a square representing a state pair has not compatible state pair entry then that state pair is also not compatible. In Fig. 2(d), the squares representing (a, c), (a, d), (b, c), (b, e), (b, f), (c, e), (c, f), (c, g) and (d, f) have not compatible state pairs and so they are not compatible state pairs.

The compatible pairs obtained from implication table of Fig. 2(d) are,

(a, b), (a, e), (a, f), (a, g), (a, h), (b, g), (b, h), (c, d), (c, h), (d, e), (d, h), (e, f), (e, g), (e, h), (f, g), (f, h), (g, h)

The above compatible pairs can be merged to form maximum compatibles using a merger diagram shown in Fig. 3. The following compatible groups are obtained from merger diagram.

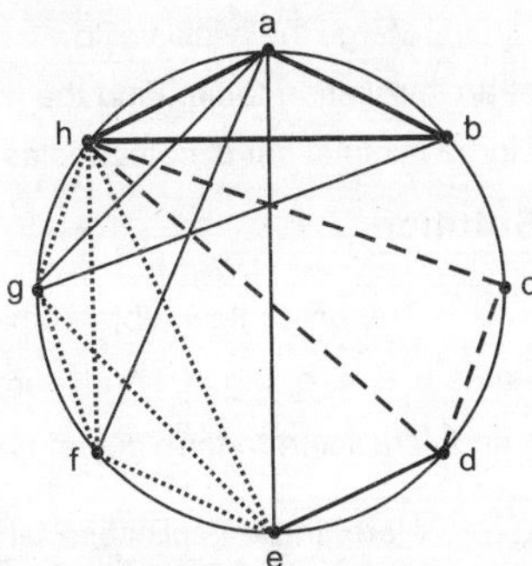

Fig. 3: Merger diagram.

4-sided polygon : (e, f, g, h)

Triangles : (a, b, h), (c, d, h)

Straight lines : (a, g), (a, f), (a, e), (b, g), (e, d)

The compatible groups obtained from merger diagram are,

(e, f, g, h), (a, b, h), (c, d, h), (a, g), (a, f), (a, e), (b, g), (e, d)

For covering condition all states should be included in compatible groups. Let us select the following compatible groups to cover all the 8 states a, b, c, d, e, f, g and h.

(e, f, g, h), (a, b, h), (c, d, h)

Since, state h is covered in (e, f, g, h), the state h can be eliminated in the groups (a, b, h) and (c, d, h) to form the compatible pairs (a, b) and (c, d).

Now, the modified compatible groups that satisfy covering condition are,

(a, b) (c, d) (e, f, g, h)

The possible compatible pairs in the above group are (a, b), (c, d), (e, f), (e, g), (e, h), (f, g), (f, h) and (g, h).

With reference to Fig. 2(d) we can say that the compatible pairs has no implied pair except (e, f). The pair (e, f) has an implied pair (e, h) which is included in (e, f, g, h).

Therefore, the closure condition is also satisfied by the selected compatible pairs.

Conclusion: The compatible pairs that satisfy closed-covering condition are,

(a, b), (c, d), (e, f, g, h)

Now the reduced flow table can be obtained by merging the rows in the compatible group.

- Compatible group (a, b) : Rows b is merged to row a.
- Compatible group (c, d) : Row d is merged to row c.
- Compatible group (e, f, g, h) : Row f, g and h are merged to row e.

The reduced flow table with only three rows is shown in Fig. 4.

	00	01	11	10
a, b	(a),0	(b),0	c, –	e, –
c, d	a, –	(d),1	(c),0	h, –
e, f, g, h	a, –	(g),0	(f),0	(h),0

Fig. a.

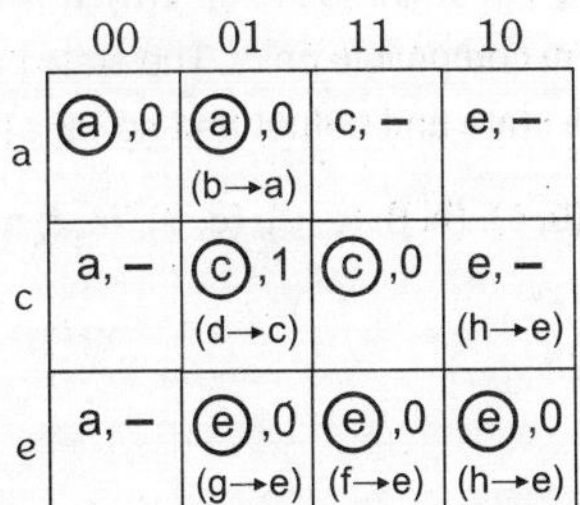

	00	01	11	10
a	(a),0	(a),0 (b→a)	c, –	e, –
c	a, –	(c),1 (d→c)	(c),0	e, – (h→e)
e	a, –	(e),0 (g→e)	(e),0 (f→e)	(e),0 (h→e)

Fig. b.

Fig. 4: Reduced flow table.

Example 9.14

Merge the primitive flow table shown in Fig. 1. Find all compatible pairs by means of an implication table. Find the maximal compatibles by means of a merger diagram. Find a minimal set of compatibles that covers all the states and is closed.

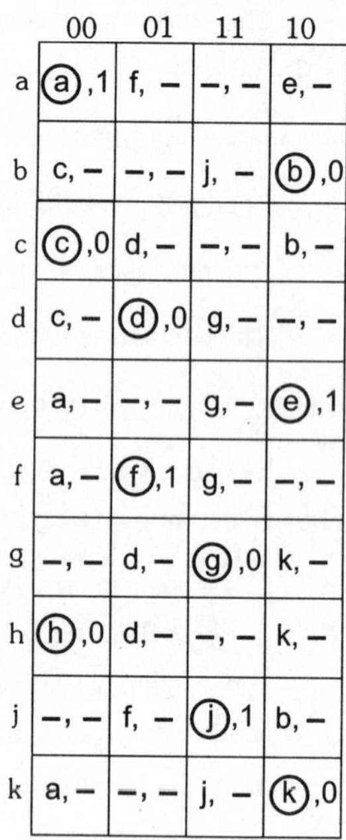

	00	01	11	10
a	ⓐ,1	f, –	–, –	e, –
b	c, –	–, –	j, –	ⓑ,0
c	ⓒ,0	d, –	–, –	b, –
d	c, –	ⓓ,0	g, –	–, –
e	a, –	–, –	g, –	ⓔ,1
f	a, –	ⓕ,1	g, –	–, –
g	–, –	d, –	ⓖ,0	k, –
h	ⓗ,0	d, –	–, –	k, –
j	–, –	f, –	ⓙ,1	b, –
k	a, –	–, –	j, –	ⓚ,0

Fig. 1: Primitive flow table.

Solution

The given flow table (Table 1) has 10 states a, b, c, d, e, f, g, h, j and k. The states b, c, d, e, f, g, h, j and k are listed to form rows and the states a, b, c, d, e, f, g, h and j are listed to form columns in implication table.

Here, three implication tables (Figs. 2(a), (b) and (c)) are drawn to separately show the not compatible, compatible and implied pairs.

The given flow table has 4 combinations of inputs. Hence compatibility of states for all the four inputs should be marked in each square of implication table.

The dashes in the given flow table represents unspecified states or outputs. While identifying compatible pairs the dashes can be used to match any desired state or output.

First mark × in squares in Fig. 2(a) in which the state pairs do not have same outputs for same inputs.

- When input is 00, present state pair (a, c), go to next state (a, c) with different output.
- When input is 00, present state pair (a, h), go to next state (a, h) with different output.
- When input is 01, present state pair (d, f), go to next state (d, f) with different output.
- When input is 11, present state pair (g, j), go to next state (g, j) with different output.
- When input is 10, present state pair (b, e), go to next state (b, e) with different output.
- When input is 10, present state pair (e, k), go to next state (e, k) with different output.

Hence squares representing (a, c), (a, h), (b, e), (d, f), (e, k) and (g, j) are marked × in Fig. 2(a). The not compatible state pairs that has transitions to stable states with different output are listed in Table 1.

Next mark tick (√) in squares in Fig. 2(b) in which state pairs are compatible. Here, dashes are matched with states and outputs to form compatible pairs. The state pair (a, e) goes to same next state with same output when dashes are considered as same state and output and so (a, e) is compatible pair.

Similarly, (a, f), (b, c), (b, j), (c, d), (d, g), (e, f) and (g, h) are compatible state pairs. The compatible state pairs are listed in Table 2.

Table 1: Not Compatible State Pairs

	00	01	11	10	
a	ⓐ,1	f, –	–, –	e, –	(a, c)
c	ⓒ,0	d, –	–, –	b, –	
a	ⓐ,1	f, –	–, –	e, –	(a, h)
h	ⓗ,0	d, –	–, –	k, –	
b	c, –	–, –	j, –	ⓑ,0	(b, e)
e	a, –	–, –	g, –	ⓔ,1	
d	c, –	ⓓ,0	g, –	–, –	(d, f)
f	a, –	ⓕ,1	g, –	–, –	
e	a, –	–, –	g, –	ⓔ,1	(e, k)
k	a, –	–, –	j, –	ⓚ,0	
g	–, –	d, –	ⓖ,0	k, –	(g, j)
j	–, –	f, –	ⓙ,1	b, –	

Table 2: Compatible State Pairs

	00	01	11	10	
a	ⓐ,1	f, –	g, –	e, 1	(a, e)
e	a, 1	f, –	g, –	ⓔ,1	
a	ⓐ,1	f, 1	g, –	e, –	(a, f)
f	a, 1	ⓕ,1	g, –	e, –	
b	c, 0	d, –	j, –	ⓑ,0	(b, c)
c	ⓒ,0	d, –	j, –	b, 0	
b	c, –	f, –	j, 1	ⓑ,0	(b, j)
j	c, –	f, –	ⓙ,1	b, 0	
c	ⓒ,0	d, 0	g, –	b, –	(c, d)
d	c, 0	ⓓ,0	g, –	b, –	
d	c, –	ⓓ,0	g, 0	k, –	(d, g)
g	c, –	d, 0	ⓖ,0	k, –	
e	a, –	f, 1	g, –	ⓔ,1	(e, f)
f	a, –	ⓕ,1	g, –	e, 1	
g	h, 0	d, –	ⓖ,0	k, –	(g, h)
h	ⓗ,0	d, –	g, 0	k, –	

The next step is to identify the implied pairs among the state pairs that are not marked by × and tick (√) in Figs. 2(a) and (b). Mark the next state pairs corresponding to the state pair represented by the squares that are blank in Figs. 2(a) and (b), as shown in Fig. 2(c). When next state pair and present state pair are same then those state pairs can be ignored in Fig. 2(c). Similarly, when a present state pair goes to same next state then those next states can also be ignored in Fig. 2(c). The state pair entries in Fig. 2(c), are marked × if the pair is not compatible and marked with tick (√) if the state pair is compatible.

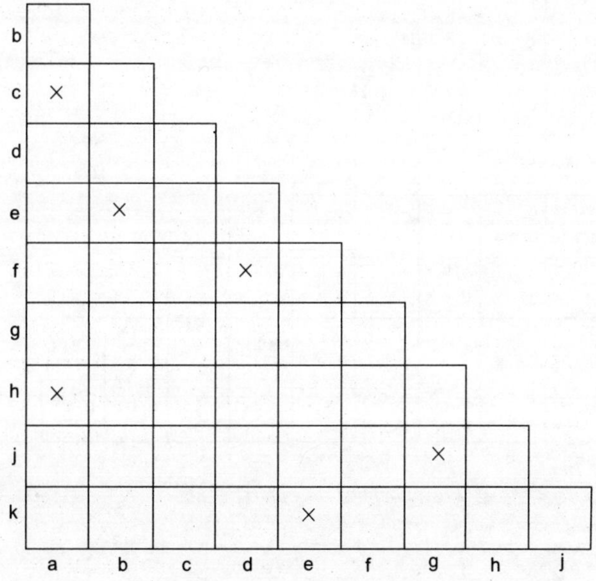

Fig. a: Not compatible state pairs.

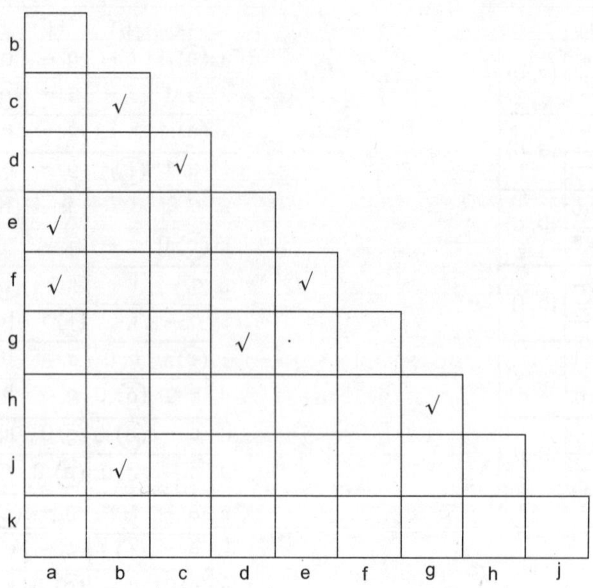

Fig. b: *Compatible state pairs.*

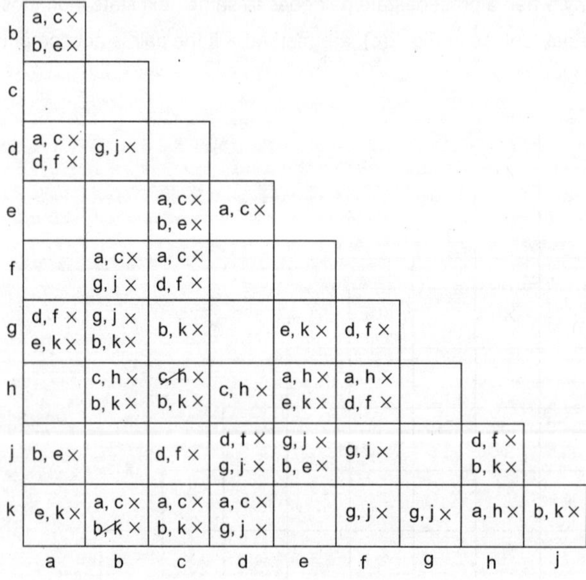

Fig. c: *Implied state pairs.*

b	a, c × b, e ×								
c	×	√							
d	a, c × d, f ×	g, j ×	√						
e	√	×	a, c × b, e ×	a, c ×					
f	√	a, c × g, j ×	a, c × d, f ×	×	√				
g	d, f × e, k ×	g, j × b, k ×	b, k ×	√	e, k ×	d, f ×			
h	×	c, h × b, k ×	b, k ×	c, h ×	a, h × e, k ×	a, h × d, f ×	√		
j	b, e ×	√	d, f ×	d, f × g, j ×	g, j × b, e ×	g, j ×	×	d, f × b, k ×	
k	e, k ×	a, c ×	a, c × b, k ×	a, c × g, j ×	×	g, j ×	g, j ×	a, h ×	b, k ×
	a	b	c	d	e	f	g	h	j

Fig. d: Implication table.

Fig. 2: Formation of implication table.

The three implication tables in Figs. 2(a), (b) and (c) are combined as a single implication table in Fig. 2(d) and the compatible state pairs are determined from Fig. 2(d).

When a square representing a state pair has not compatible state pair entry then that state pair is also not compatible. In Fig. 2(d), the squares representing (a, b), (a, d), (a, g), (a, j), (a, k), (b, d), (b, f), (b, g), (b, h), (b, k), (c, e), (c, f), (c, g), (c, h), (c, j), (c, k), (d, e), (d, h), (d, j), (d, k), (e, g), (e, h), (e, j), (f, g), (f, h), (f, j), (f, k), (g, k), (h, j), (h, k) and (j, k) have not compatible state pairs and so they are not compatible state pairs.

The compatible pairs obtained from implication table of Fig. 2(d) are,

(a, e), (a, f), (b, c), (b, j), (c, d), (d, g), (e, f), (g, h)

The above compatible pairs can be merged to form maximum compatibles using a merger diagram shown in Fig. 3. The following compatible groups are obtained from merger diagram.

Triangle : (a, e, f)

Straight lines : (b, j), (b, c), (c, d), (d, g), (g, h)

The compatible groups obtained from merger diagram are,

(a, e, f), (b, j), (b, c), (c, d) (d, g), (g, h), (k)

For covering condition all states should be included in compatible groups. Let us select the following compatible groups to cover all the 10 states a, b, c, d, e, f, g, h, j and k.

(a, e, f), (b, j), (c, d), (g, h), (k)

The possible compatible pairs in the above group are (a, e), (a, f), (e, f), (b, j), (c, d) and (g, h). With reference to Fig. 2(d) we can say that the compatible pairs has no implied pair. Therefore, the closure condition is also satisfied by the selected compatible pairs.

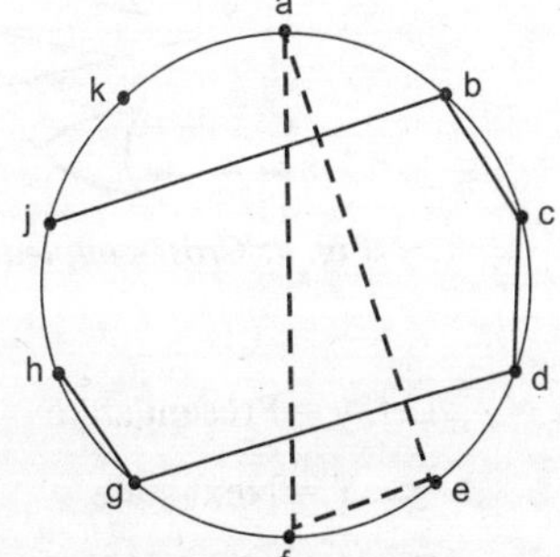

Fig. 3: Merger diagram.

Conclusion: The compatible pairs that satisfy closed-covering condition are,

(a, e, f), (b, j), (c, d), (g, h), (k)

Now the reduced flow table can be obtained by merging the rows of each compatible group. Here, the following row mergers are made to obtain reduced flow table.

- Compatible group (a, e, f) : Rows e and f are merged to row a.
- Compatible group (b, j) : Row j is merged to row b.
- Compatible group (c, d) : Row d is merged to row c.
- Compatible group (g, h) : Row h is merged to row g.

The reduced flow table with five rows is shown in Fig. 4.

	00	01	11	10
a, e, f	ⓐ,1	ⓕ,1	g, –	ⓔ,1
b, j	c, –	f, –	ⓙ,1	ⓑ,0
c, d	ⓒ,0	ⓓ,0	g, –	b, –
g, h	ⓗ,0	d, –	ⓖ,0	k, –
k	a, –	–, –	j, –	ⓚ,0

Fig. a.

	00	01	11	10
a	ⓐ,1	ⓐ,1 (f→a)	g, –	ⓐ,1 (e→a)
b	c, –	a, – (f→a)	ⓑ,1 (j→b)	ⓑ,0
c	ⓒ,0	ⓒ,0 (d→c)	g, –	b, –
g	ⓖ,0 (h→g)	c, – (d→c)	ⓖ,0	k, –
k	a, –	–, –	b, – (j→b)	ⓚ,0

Fig. b.

Fig. 4: Reduced flow table.

9.3 Asynchronous Circuits with SR Latches

The latches are used as memory elements in asynchronous circuits. Therefore, in an n-variable asynchronous circuit, n number of SR latches are employed to store the n variables. There are two types of SR latches: one type is constructed using two cross coupled NOR gate and another type is constructed using two cross coupled NAND gates. Both these two latches has two inputs S and R. The working of SR latch is discussed in detail in Chapter-6, Section 6.1.

9.3.1 NOR Based SR Latch

The NOR based SR latch is constructed using two cross coupled NOR gates as shown in Fig. 9.15(a) and the feedback path circuit is shown in Fig. 9.15(b).

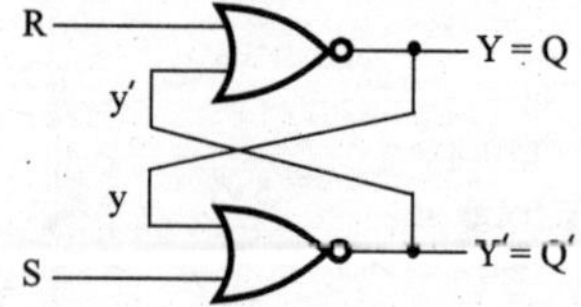

Fig. a: Cross-coupled NOR gates.

Fig. b: Feedback circuit.

Fig. 9.15: SR latch using NOR gates.

Let, y = Present state

Y = Next state

From the feedback circuit of Fig. 9.15(b), we get the following equation for next state or output of NOR based SR latch.

$$Y = (R + (S + y)')' = R'(S + y) = R'S + R'y$$

$$\therefore Y = SR' + R'y$$

Truth Table of NOR Based SR Latch

Inputs S	Inputs R	Outputs Q	Outputs Q'
0	0	No change	
0	1	0	1
1	0	1	0
1	1	Indeterminate	

From the truth table of NOR based SR latch we can say that both S and R should not be equal to 1 simultaneously and this can be transferred to a condition that logical AND of S and R should not be 1 or should be always 0. Therefore, in NOR based SR latch, $S . R = 0$.

Consider the Boolean expression, $SR' + SR$.

Now, $SR' + SR = S(R' + R)$

Since $R' + R = 1$, we can write, $SR' + SR = S$

From the above equation we can say that if $SR = 0$, then $SR' = S$.

On substituting, $SR' = S$ in the next state equation of SR latch we get the following simplified equation for NOR based SR latch.

$$Y = SR' + R'y \quad \Rightarrow \quad Y = S + R'y$$

(Simplified equation of SR latch)

The simplified equation of SR latch can be used to analyze and design the asynchronous circuit using NOR based SR latch when $SR = 0$.

9.3.2 NAND Based SR Latch

The NAND based SR latch is constructed using two cross coupled NAND gates as shown in Fig. 9.16(a) and the feedback path circuit is shown in Fig. 9.16(b).

Let, y = Present state

Y = Next state

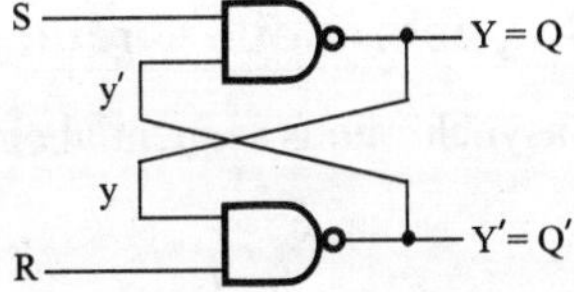

Fig. a: Cross-coupled NAND gates.

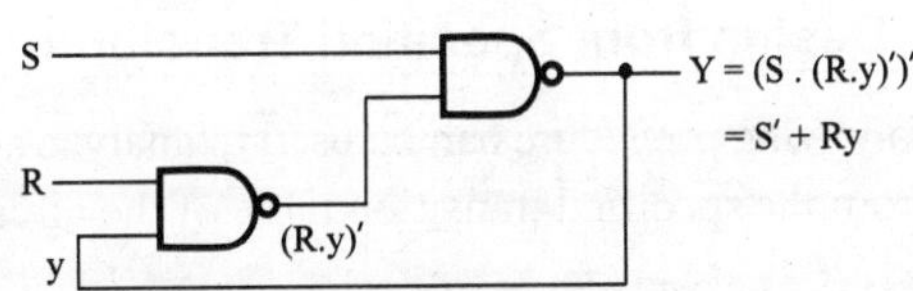

Fig. b: Feedback circuit.

Fig. 9.16: SR latch using NAND gates.

Truth Table of NAND Based SR Latch

Inputs S	Inputs R	Outputs Q	Outputs Q'
0	0	Indeterminate	
0	1	1	0
1	0	0	1
1	1	No change	

From the truth table of NAND based SR latch we can say that both S and R should not be equal to 0 simultaneously. This can be transferred to a condition that logical AND of S' and R' should not be 1 or should be always 0. Therefore, in NAND based SR latch, $S' . R' = 0$. From the feedback circuit of Fig. 9.16(b), we get the following equation for next state or output of NAND based SR latch.

$$Y = S' + Ry$$

On comparing the above equation with NOR based SR latch we can observe that S is replaced by S' and R' is replaced by R and so the above equation will be valid only when $S'R' = 0$. Therefore, the above equation can be used to analyze and design the asynchronous circuit using NAND based SR latch when $S'R' = 0$.

9.3.3 Design using SR Latches

The design of asynchronous circuits can start with either Boolean equations of asynchronous circuit or with transition table.

Method-1: Design from Specified Boolean Equations of Asynchronous Circuit

The following procedure can be used to analyze and design asynchronous sequential circuits using SR latches from the specified Boolean equations of the circuit.

Let, $Y_1, Y_2, ... Y_n$ = n variables of the asynchronous circuit.

Now, $y_1, y_2, ... y_n$ are present state variables and $Y_1, Y_2, ... Y_n$ are next state variables.

In order to store n variables, choose n number of NOR or NAND based SR latches.

Determine the Boolean functions for the R and S input of each latch by comparing the given equations with following next state equation of SR latch.

For NOR based SR latch: $Y_j = S_j + R_j' y_j$; for j = 1, 2, ... n.

For NAND based SR latch: $Y_j = S_j' + R_j y_j$; for j = 1, 2, ... n.

When NOR based SR latch is selected, check for $S_j R_j = 0$ for j = 1, 2, ... n.

When NAND based SR latch is selected, check for $S_j' R_j' = 0$ for for j = 1, 2, ... n.

When the above condition is not satisfied, the circuit designed using given equation may not work properly and so design is not possible. When the above condition is satisfied, draw the logic circuit using equations of Y_j for j = 1, 2, ... n.

The asynchronous circuit can be analyzed for cycles and races by constructing transition table using the equations of Y_j for j = 1, 2, ... n.

Method-2: Design from Specified Transition Table of Asynchronous Circuit

The following procedure can be used to analyze and design asynchronous sequential circuits using SR latches from the specified transition table of the circuit.

Let, $Y_1, Y_2, ... Y_n$ = n variables of the asynchronous circuit.

Now, $y_1, y_2, ... y_n$ are present state variables and $Y_1, Y_2, ... Y_n$ are next state variables.

In order to store n variables, choose n number of NOR or NAND based SR latches.

When the states are denoted by letter symbols, assign race-free binary codes for states.

The design starts from the transition table with binary codes for states. The entries in the squares of transition tables are binary values for next states $Y_1, Y_2, ..., Y_n$. From the knowledge of present state and next state values, the excitation table of SR latch is used to determine the binary values of S and R inputs of all SR latches and list in a table.

Then construct K-map for every S and R inputs and determine the simplified Boolean equations for latch inputs.

It can be observed that the S and R inputs in the excitation table of NOR and NAND latches are complements. The equations obtained from K-map for NOR implemented can be complemented and used for NAND implementation.

Table 9.10: Excitation Table of NOR based SR Latch

Present State y	Next State Y	Latch Inputs S R
0	0	0 ×
0	1	1 0
1	0	0 1
1	1	× 0

Table 9.11: Excitation Table of NAND based SR Latch

Present State y	Next State Y	Latch Inputs S R
0	0	1 ×
0	1	0 1
1	0	1 0
1	1	× 1

When NOR based SR latch is selected, check for $S_jR_j = 0$ for $j = 1, 2, \ldots n$.

When NAND based SR latch is selected, check for $S_j'R_j' = 0$ for for $j = 1, 2, \ldots n$.

When the above condition is not satisfied, the circuit designed using given equation may not work properly and so design is not possible. When the above condition is satisfied, draw the combinational logic circuit to generate latch inputs, using equations of S_j and R_j for $j = 1, 2, \ldots n$. The complete logic circuit is given by this combinational logic circuit connected to SR latches.

Short-cut Method for Deriving SR Latch Input Equations

The following theorem is stated without proof.

In realization of a flow table with race-free state assignment using SR latches there exists equations for inputs S_i and R_i of latch-i independent of y_i.

The above theorem can be used to find equations for latch inputs from the maps of next state variables. For example, consider a flow table with two state varaibles Y_1 and Y_2. The maps of Y_1 and Y_2 are shown in Figs. 9.17 and 9.18.

From the map of Y_1 we can get equations for S_1 and R_1 independent of y_1. Hence divide the map of Y_1 into two maps: One map when $y_1 = 0$ and another map when $y_1 = 1$.

In the map of Fig. 9.17(b), the prime implicants are formed by considering adjacent 1's. The sum of prime implicants gives the equation for S_1. In the map of Fig. 9.17(c), the prime implicants are formed by considering adjacent 0's. The complement of sum of prime implicants gives the equation for R_1'.

Similarly, from the map of Y_2 we get the equations for S_2 and R_2 independent of y_2. Here, the map of Y_2 is divided into two: One map when $y_2 = 0$ and another map when $y_2 = 1$. The formation of equations for S_2 and R_2' is similar to that of formation of equations for S_1 amd R_1'.

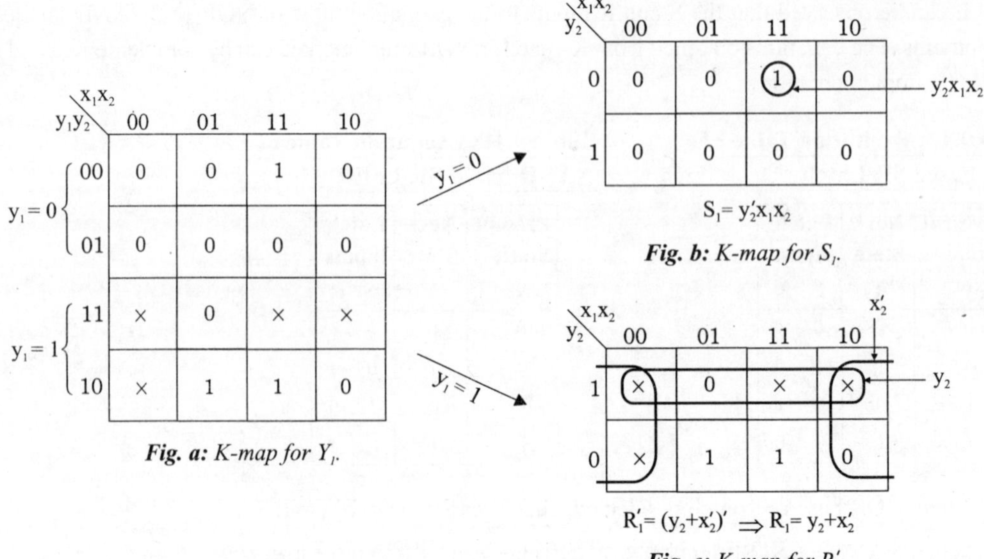

Fig. a: K-map for Y_1.

Fig. b: K-map for S_1.

Fig. c: K-map for R_1'.

***Fig. 9.17:** K-maps to find equations for S_1 and R_1.*

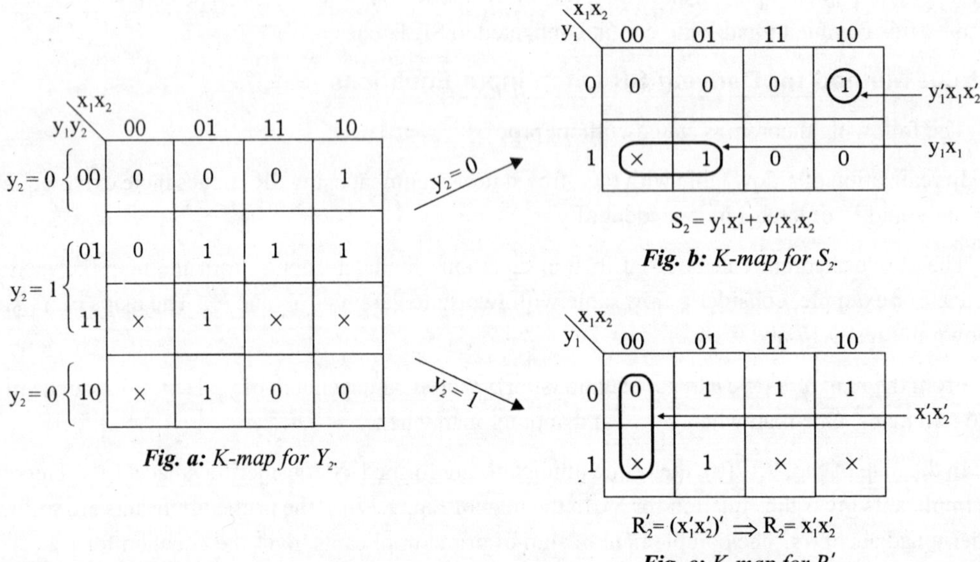

Fig. a: K-map for Y_2.

Fig. b: K-map for S_2.

Fig. c: K-map for R_2'.

***Fig. 9.18:** K-maps to find equations for S_2 and R_2.*

Example 9.15

An asynchronous sequential circuit is described by the following excitation and output functions

$Y = x_1x_2' + (x_1 + x_2')\,y \quad ; \quad Z = Y$

Implement the above equations in a logic circuit

a) using NOR latch

b) using NAND latch

Solution

a) Implementation using NOR Latch

The excitation function of the asynchronous sequential circuit is,

$$Y = x_1x_2' + (x_1 + x_2')\,y \qquad .....(1)$$

The excitation function of SR latch using NOR gate is,

$$Y = S + R'y \qquad .....(2)$$

On comparing equations (1) and (2) we get,

$S = x_1x_2'$

$R' = x_1 + x_2' \quad \Rightarrow \quad R = (x_1 + x_2')' = x_1'x_2$ [a.a′ = 0]

Let us check for SR = 0.

$SR = x_1x_2' \cdot x_1'x_2 = (x_1x_1') \cdot (x_2x_2')$

$= 0.0 = 0$

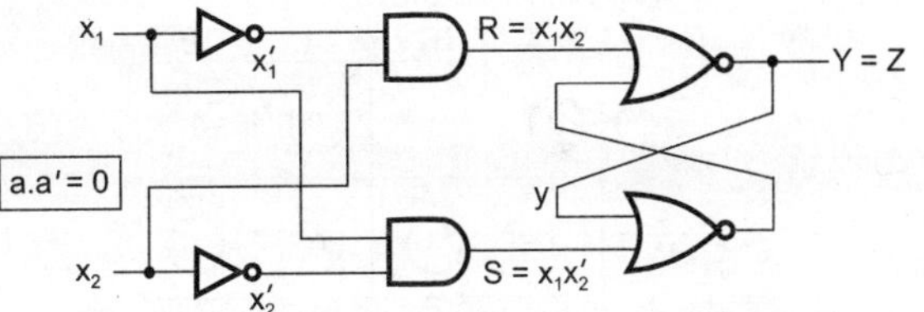

Fig. 1: Logic implementation using NOR based SR latch.

Since, SR = 0, the given equation can be implemented using NOR based SR latch as shown in Fig. 1.

b) Implementation using NAND Latch

The excitation function of the asynchronous sequential circuit is,

$$Y = x_1x_2' + (x_1 + x_2')\,y \qquad .....(3)$$

The excitation function of SR latch using NAND gate is,

$$Y = S' + Ry \qquad .....(4)$$

On comparing equatibns (3) and (4) we get,

$S' = x_1x_2' \quad \Rightarrow \quad S = (x_1x_2')' = x_1' + x_2$

$R = x_1 + x_2'$

Let us check for S′R′ = 0.

$S'R' = x_1x_2'(x_1 + x_2')' = x_1x_2' \cdot x_1'x_2 = (x_1x_1') \cdot (x_2x_2') = 0.0 = 0$ [a.a′ = 0]

Since, S′R′ = 0, the given equation can be implemented using NAND based SR latch as shown in Fig. 2.

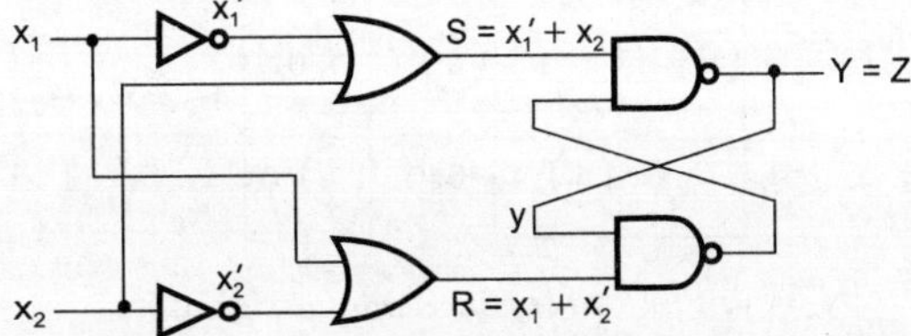

Fig. 2: Logic implementation using NAND based SR latch.

Example 9.16

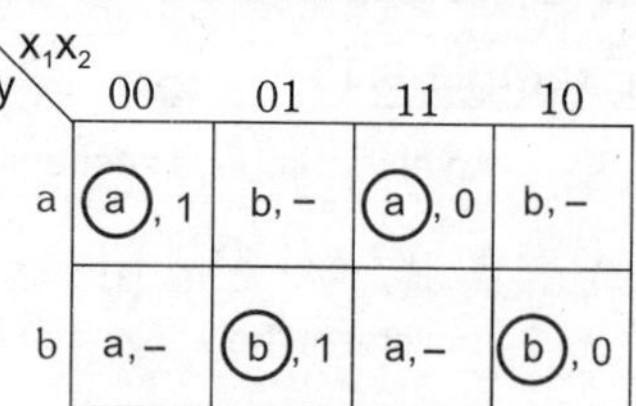

Fig. 1: Reduced flow table.

a) Obtain a binary state assignment for the reduced flow table shown in Fig. 1 to avoid critical race condition.

b) Assign output to unstable states and obtain the output equation.

c) Implement the logic circuit using NOR based SR latch.

d) Implement the logic circuit using NAND based SR latch.

Solution

a) Race-Free State Assignment

The given flow table has only two states a and b. Therefore, state transitions can occur either from a to b or b to a. For two states, 1-bit state assignment is sufficient and in 1-bit state assignment there will not be any race condition.

Let us assign binary code 0 to state a and binary code 1 to state b. The transition and flow table with binary code assignment is shown in Fig. 2.

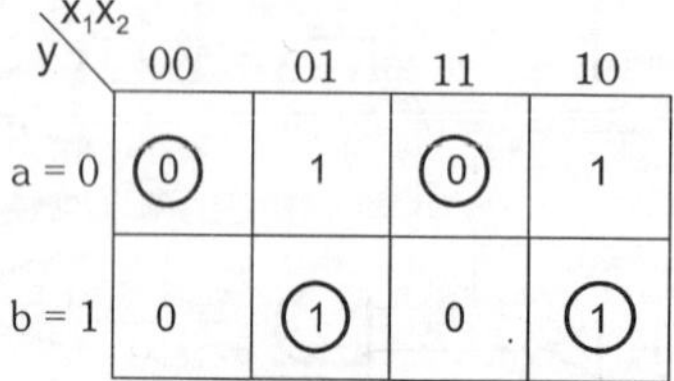

Fig. a: Transition table.

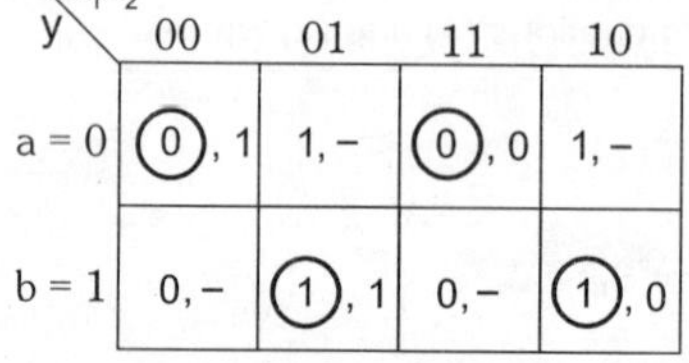

Fig. b: Flow table.

Fig. 2: Transition and flow table with code assignment.

b) To Assign Output and Determine the Output Equation

In flow table, the dashes in the output are assigned values according to the following rules to prevent transient output during a transition.

Rule 1 : Assign 0 to an output of an unstable state which is transient state during a transition between two stable states that have a 0 as output.

Rule 2 : Assign 1 to an output of an unstable state which is transient state during a transition between two stable states that have a 1 as output.

Rule 3 : Assign a don't-care condition to an output of an unstable state which is transient state during a transition between two stable states that have a different values (0 and 1 or 1 and 0) as output.

The state transitions that are used to assign output using the above rules are shown in Fig. 3 and the flow table with output assigned to unstable states is shown in Fig. 4. The output map of Z is shown in Fig. 5.

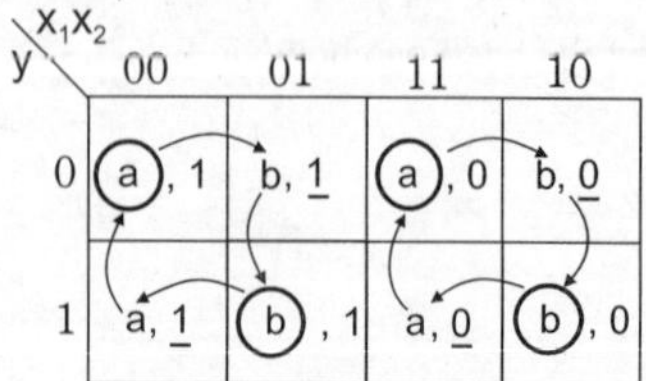

Fig. 3: State transitions used to assign output.

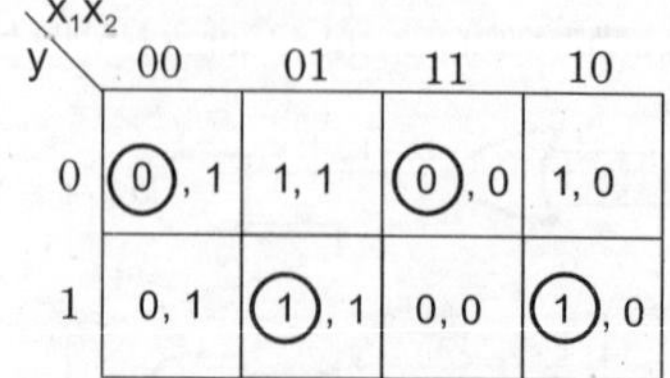

Fig. 4: Flow table with output assigned to unstable states.

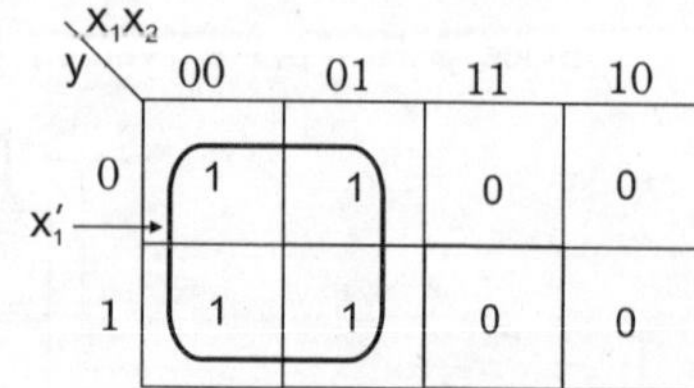

Fig. 5: Output map.

Let, Z = output

The output map is constructed as shown in Fig. 8 and from output map we get the following simplified Boolean equation for output.

Output, $Z = x_1'$

c) Logic Implementation using NOR based SR Latch

The binary values inside each square of transition table is the value of next state Y. Hence, using the transition table the map of Y can be drawn as shown in Fig. 6.

In order to store the one state variable Y, one SR latch is required. Let us choose one NOR based SR latch.

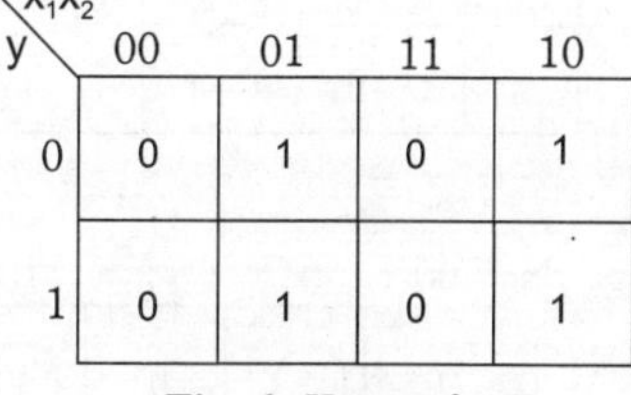

Fig. 6: K-map for Y.

Let, S, R = Inputs for SR latch

To find Equations for Latch Inputs using Map of Y

In this method the equations for S and R that are independent of y can be obtained from map of Y. The map for y = 0 is used to find equation for S by forming prime implicants using adjacent 1's. The map for y = 1 is used to find equation for R′ by forming prime implicants using adjacent 0's. Hence the map of Y is divided into two: One for y = 0 and another for y = 1 as shown in Fig. 7, to form map of S and R′.

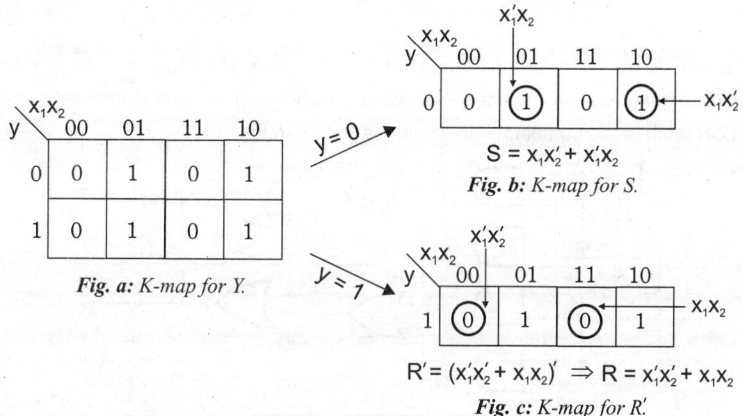

Fig. 7: K-maps to find equations for S and R.

From the maps of S and R′ we get the following equations for SR latch inputs.

$S = x_1x_2' + x_1'x_2$

$R = x_1'x_2' + x_1x_2$

To find Equations for Latch Inputs using Excitation Table of SR Latch

Using the excitation table of NOR based SR latch, the inputs for SR latch can be obtained from the circuit inputs, present and next state in the map of Y and listed in Table 1.

Using the binary values of SR latch inputs listed in Table 1, the K-map for latch inputs are constructed as shown in Figs. 7 and 8, in order to determine the simplified equations for latch inputs.

Table 1: Inputs of SR Latch

	$x_1x_2 = 00$				$x_1x_2 = 01$				$x_1x_2 = 11$				$x_1x_2 = 10$			
y	y	Y	S	R	y	Y	S	R	y	Y	S	R	y	Y	S	R
0	0	0	0	×	0	1	1	0	0	0	0	×	0	1	1	0
1	1	0	0	1	1	1	×	0	1	0	0	1	1	1	×	0

Excitation Table of NOR based SR Latch

Present State y	Next State Y	Latch Inputs S	R
0	0	0	×
0	1	1	0
1	0	0	1
1	1	×	0

y \ x_1x_2	00	01	11	10
0	0	1	0	1
1	0	×	0	×

Groups: $x_1'x_2$ (column 01), x_1x_2' (column 10)

Fig. 8: K-map for S.

y \ x_1x_2	00	01	11	10
0	×	0	×	0
1	1	0	1	0

Groups: $x_1'x_2'$ (column 00), x_1x_2 (column 11)

Fig. 9: K-map for R.

From the K-maps shown in Figs. 8 and 9, we get the following simplified equations for SR latch inputs. The output equation is obtained from output map.

$$S = x_1x_2' + x_1'x_2 \quad ; \quad R = x_1'x_2' + x_1x_2 \quad ; \quad Z = x_1'$$

Logic Diagram

The Boolean equations obtained for latch inputs in both methods are same and using these equations the combinational logic circuit to generate latch inputs are drawn as shown in Fig. 10. The complete logic circuit is obtained by adding NOR based SR latch to the combinational circuit as shown in Fig. 10.

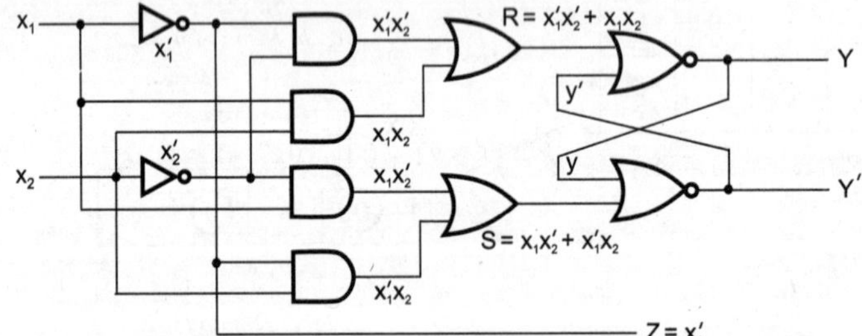

Fig. 10: Logic circuit using NOR based SR latch.

d) Logic Implementation using NAND based SR Latch

The Boolean equations to generate inputs for NAND based SR latches are given by complemented of the equations for NOR based SR latch.

Therefore, the boolean equations to generate inputs for NAND based SR latches are,

$$S = (x_1x_2' + x_1'x_2)' = (x_1x_2')' \cdot (x_1'x_2)' = (x_1' + x_2) \cdot (x_1 + x_2')$$

$$R = (x_1'x_2' + x_1x_2)' = (x_1'x_2')' \cdot (x_1x_2)' = (x_1 + x_2) \cdot (x_1' + x_2')$$

$$Z = x_1'$$

The output equation is obtained from output map. Using the above equations the combinational logic circuit to generate latch inputs are drawn as shown in Fig. 11. The complete logic is obtained by adding NAND based SR latch to the combinational circuit as shown in Fig. 11.

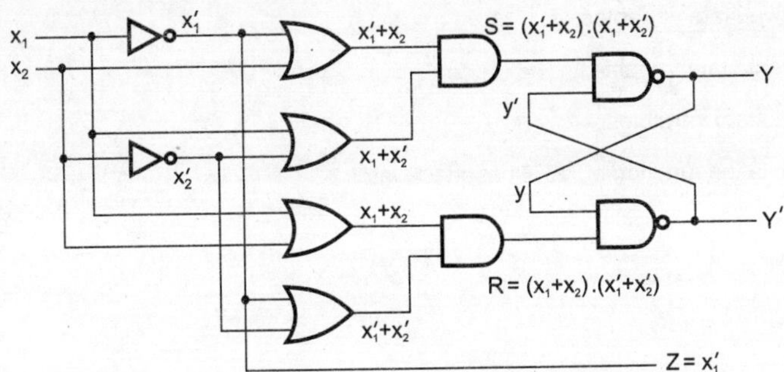

Fig. 11: Logic circuit using NAND based SR latch.

Example 9.17

The Boolean functions for the S and R inputs of SR latch based asynchronous logic circuit are,

$S_1 = x_1y_2$	$S_2 = x_1x_2$
$R_1 = x_1'\,x_2'$	$R_2 = x_2'\,y_1$

Analyze the circuit and find the stability and race condition.

Solution

The given Boolean equations are,

$S_1 = x_1y_2$	$S_2 = x_1x_2$
$R_1 = x_1'\,x_2'$	$R_2 = x_2'\,y_1$

Let us check for $S_1R_1 = 0$ and $S_2R_2 = 0$

$$S_1R_1 = x_1y_2 \,.\, x_1'x_2' = (x_1x_1') \,.\, y_2x_2' = 0 \,.\, y_2x_2' = 0$$ a.a' = 0

$$S_2R_2 = x_1x_2 \,.\, x_2'y_1 = x_1 \,.\, (x_2x_2') \,.\, y_1 = x_1.0.y = 0$$

Since the given equations satisfy the condition SR = 0, the equations can be implemented by a logic circuit using NOR based SR latch as shown in Fig. 1.

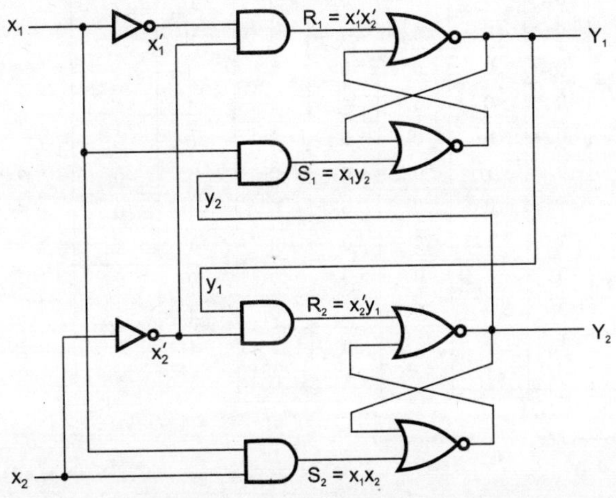

Fig. 1: Circuit with SR latch.

The circuit has two state variables.

Let, y_1 y_2 = Present state variables

Y_1Y_2 = Next state varaibles

The reduced excitation function of NOR based SR latch is,

$$Y = S + R'y$$

$$\therefore Y_1 = S_1 + R_1' y_1$$

$$= x_1y_2 + (x_1'x_2')' y_1$$

$$= x_1y_2 + (x_1 + x_2)\, y_1$$

$$= x_1y_2 + x_1y_1 + x_2y_1$$

$$Y_2 = S_2 + R_2'y_2$$

$$= x_1x_2 + (x_2'y_1)' y_2$$

$$= x_1x_2 + (x_2 + y_1')\, y_2$$

$$= x_1x_2 + x_2y_2 + y_1'y_2$$

Using the above two equations the next state variables are evaluated for all possible choice of present state and inputs and listed in Table 1. Using Table1 the maps for Y_1 and Y_2 are drawn as shown in Figs. 2 and 3, and they are combined to form transition table as shown in Fig. 4.

Table 1: Evaluation of Excitation Variables and Outputs

Present States				Product Terms						Next States	
y_1	y_2	x_1	x_2	x_1y_2	x_1y_1	x_2y_1	x_1x_2	x_2y_2	$y_1'y_2$	Y_1	Y_2
0	0	0	0	0	0	0	0	0	0	0	0
0	0	0	1	0	0	0	0	0	0	0	0
0	0	1	0	0	0	0	0	0	0	0	0
0	0	1	1	0	0	0	1	0	0	0	1
0	1	0	0	0	0	0	0	0	1	0	1
0	1	0	1	0	0	0	0	1	1	0	1
0	1	1	0	1	0	0	0	0	1	1	1
0	1	1	1	1	0	0	1	1	1	1	1
1	0	0	0	0	0	0	0	0	0	0	0
1	0	0	1	0	0	1	0	0	0	1	0
1	0	1	0	0	1	0	0	0	0	1	0
1	0	1	1	0	1	1	1	0	0	1	1
1	1	0	0	0	0	0	0	0	0	0	0
1	1	0	1	0	0	1	0	1	0	1	1
1	1	1	0	1	1	0	0	0	0	1	0
1	1	1	1	1	1	1	1	1	0	1	1

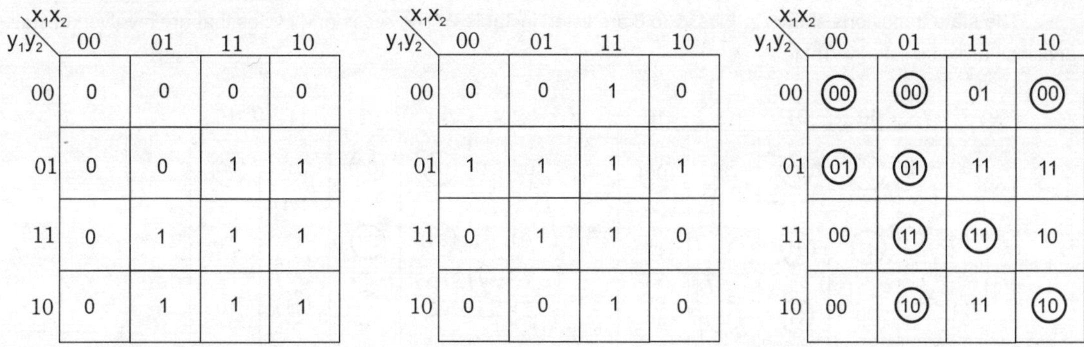

$y_1y_2 \backslash x_1x_2$	00	01	11	10
00	0	0	0	0
01	0	0	1	1
11	0	1	1	1
10	0	1	1	1

***Fig. 2:** K-map for Y_1.*

$y_1y_2 \backslash x_1x_2$	00	01	11	10
00	0	0	1	0
01	1	1	1	1
11	0	1	1	0
10	0	0	1	0

***Fig. 3:** K-map for Y_2.*

$y_1y_2 \backslash x_1x_2$	00	01	11	10
00	(00)	(00)	01	(00)
01	(01)	(01)	11	11
11	00	(11)	(11)	10
10	00	(10)	11	(10)

***Fig. 4:** Transition table.*

Stability Condition

In the transition table of Fig. 4, it can be observed that every column has at least one stable state, so that the circuit remain in a stable state for every possible input and so the circuit is stable.

State Transitions due to Change in Input

State transitions occurs only when there is a change in input. Normally, when the input is stable or when no change in input the circuit remain in one of the stable state corresponding to the input.

Here, there are four possible combinations of inputs. When the circuit remain in a stable state at any one input then state transitions will occur when the input changes to other possible inputs with one bit change.

When input is 00, the circuit remain in either state 00 or 01. The input can change from 00 to 01 or 00 to 10 and the corresponding state transitions will occur as shown in Fig. 5.

When input is 01, the circuit remain in either state 00 or 01. The input can change from 01 to 11 or 01 to 00 and the corresponding state transitions will occur as shown in Fig. 6.

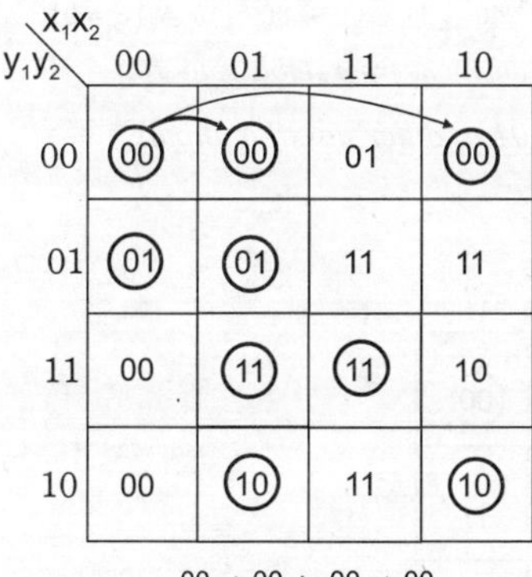

$y_1y_2 \backslash x_1x_2$	00	01	11	10
00	(00)	(00)	01	(00)
01	(01)	(01)	11	11
11	00	(11)	(11)	10
10	00	(10)	11	(10)

***Fig.a:** Transitions from stable state 00.*

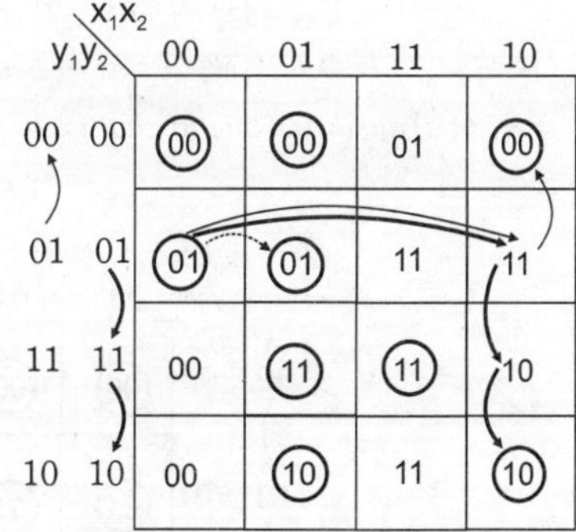

y_1y_2	$y_1y_2 \backslash x_1x_2$	00	01	11	10
00	00	(00)	(00)	01	(00)
01	01	(01)	(01)	11	11
11	11	00	(11)	(11)	10
10	10	00	(10)	11	(10)

***Fig.b:** Transitions from stable state 01.*

***Fig. 5:** Transitions due to change in input from 00 to other possible inputs.*

When input is 11, the circuit remain in state 11. The input can change from 11 to 10 or 11 to 01 and the corresponding state transitions will occur as shown in Fig. 7.

When input is 10, the circuit remain in state 00. The input can change from 10 to 00 or 10 to 11 and the corresponding state transitions will occur as shown in Fig. 8.

The state transitions shown in Figs. 5 to 8 are listed in Table 2. The races and cycles that are involved in state transitions are also listed in Table 2.

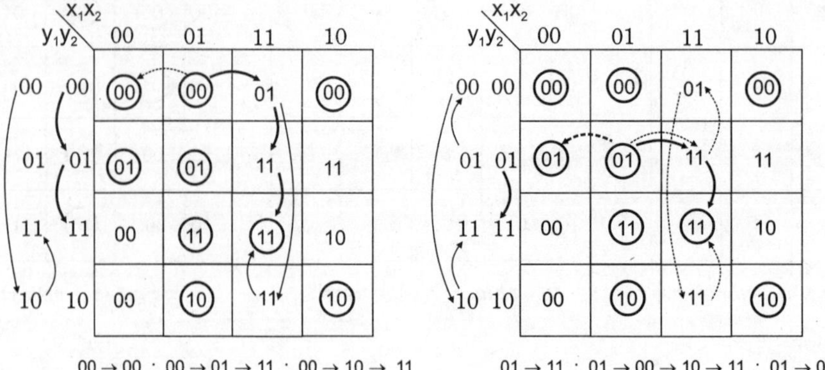

00 → 00 ; 00 → 01 → 11 ; 00 → 10 → 11

Fig. a: Transitions from stable state 00.

01 → 11 ; 01 → 00 → 10 → 11 ; 01 → 01

Fig. b: Transitions from stable state 01.

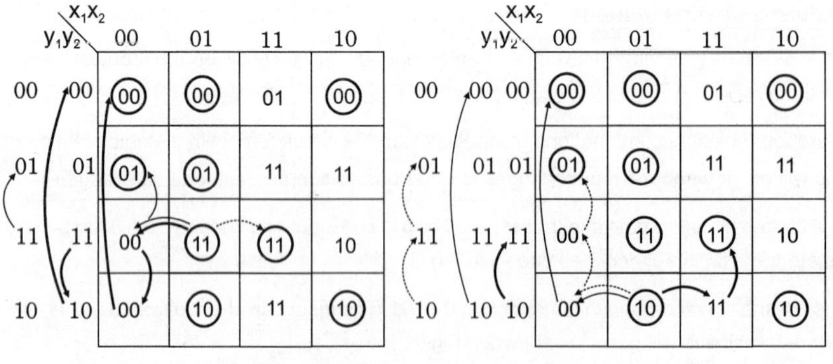

11 → 11 ; 11 → 10 → 00 ; 11 → 01

Fig. c: Transitions from stable state 11.

10 → 11 ; 10 → 00 ; 10 → 11 → 01

Fig. d: Transitions from stable state 10.

Fig. 6: *Transitions due to change in input from 01 to other possible inputs.*

y_1y_2 \ x_1x_2	00	01	11	10
00	(00)	(00)	01	(00)
01	(01)	(01)	11	11
11	00	(11)	(11)	10
10	00	(10)	11	(10)

11 → 11 ; 11 → 10 ; 11 → 01 → 00

Fig. 7: *Transitions due to change in input from 11 to other possible inputs.*

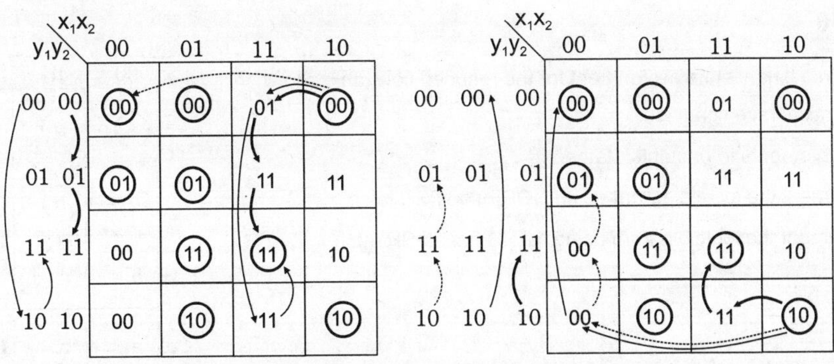

00 → 00 ; 00 → 01 → 11 ; 00 → 10 → 11

Fig. a: Transitions from stable state 00.

10 → 11 ; 10 → 00 ; 10 → 11 → 01

Fig. b: Transitions from stable state 10.

Fig. 8: Transitions due to change in input from 10 to other possible inputs.

Table 2: List of Races and Cycles

Present Input	Present Stable State	Change in Input	State Transitions	Race	Cycles
00	00	00 → 01	00 → 00	-	-
		00 → 10	00 → 00	-	-
	01	00 → 01	01 → 01	-	-
		00 → 10	01 → 11 → 10	Critical	00 → 11 → 10
			01 → 00		-
01	00	01 → 11	00 → 01 → 11 00 → 10 → 11	Noncritical	00 → 01 → 11 00 → 10 → 11
		01 → 00	00 → 00	-	-
	01	01 → 11	01 → 11	Noncritical	-
			01 → 00 → 10 → 11		01 → 00 → 10 → 11
		01 → 00	01 → 01	-	-
	11	01 → 11	11 → 11	-	-
		01 → 00	11 → 10 → 00	Critical	11 → 10 → 00
			11 → 01		-
	10	01 → 11	10 → 11	-	-
		01 → 00	10 → 00	Critical	-
			10 → 11 → 01		10 → 11 → 01
11	01	11 → 01	11 → 11	-	-
		11 → 10	11 → 10	Critical	-
			11 → 01 → 00		11 → 01 → 00
10	00	10 → 11	00 → 01 → 11 00 → 10 → 11	Noncritical	00 → 01 → 11 00 → 10 → 11
		10 → 00	00 → 00	-	-
	10	10 → 11	10 → 11	-	-
		10 → 00	10 → 00	Critical	-
			10 → 11 → 01		10 → 11 → 01

Example 9.18

a) Obtain a binary state assignment for the reduced flow table shown in Fig. 1 to avoid critical race condition.

b) Assign outputs to unstable states.

c) Implement the logic diagram using NOR based SR latch.

d) Implement the logic diagram using NAND based SR latch.

y \ x_1x_2	00	01	11	10
a	(a),0	(a), 1	b, –	d, –
b	a, –	(b), 0	(b), 0	c, –
c	b, –	–, –	d, –	(c), 0
d	a, –	a, –	(d), 1	(d), 1

Fig. 1: Reduced flow table.

Solution

a) Transition Table

The following transitions from present state to next state are observed in transition table in Fig. 2.

From a	From b	From c	From d
a → b	b → a	c → b	d → a
a → d	b → c	c → d	

The above transitions are combined as a single diagram in the transition diagram shown in Fig. 3.

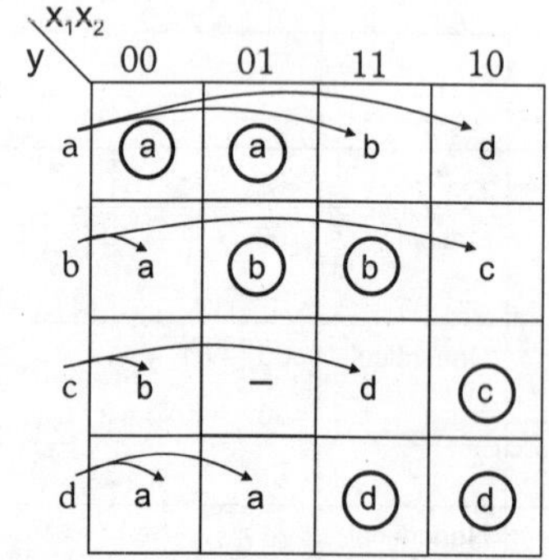

y \ x_1x_2	00	01	11	10
a	(a)	(a)	b	d
b	a	(b)	(b)	c
c	b	–	d	(c)
d	a	a	(d)	(d)

Fig. 2: Transitions from present to next state.

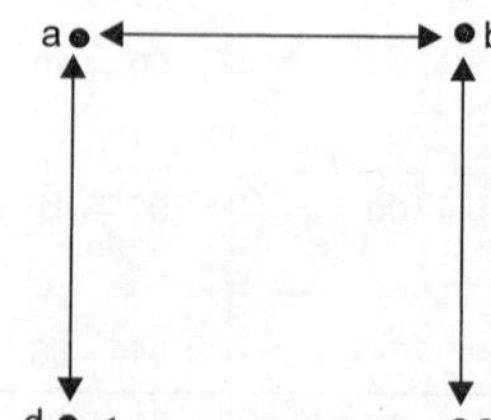

Fig. 3: Transition diagram.

For a critical race-free state assignment the state transitions in Fig. 2 should have adjacent state assignments. Here, it is possible to achieve adjacent state assignments with the following 2-bit binary codes as shown in Fig. 4.

a → 00	c → 11	$y = y_1 y_2$
b → 01	d → 10	

Let the two bits of binary code be represented by the state variables y_1 and y_2. The transition table with binary code assignment is shown in Fig. 5.

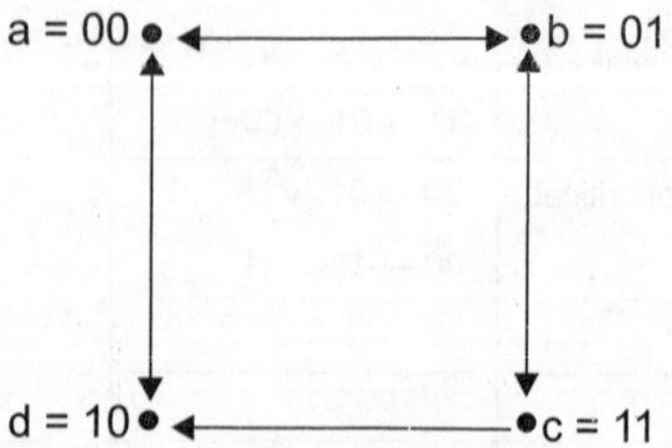

Fig. 4: Transition diagram with binary with assignment.

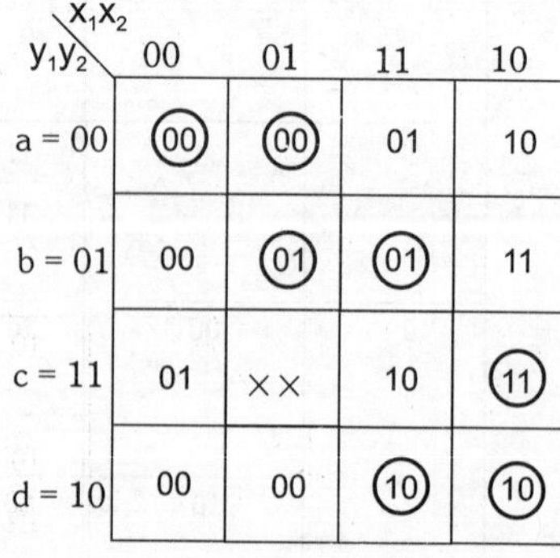

y_1y_2 \ x_1x_2	00	01	11	10
a = 00	(00)	(00)	01	10
b = 01	00	(01)	(01)	11
c = 11	01	××	10	(11)
d = 10	00	00	(10)	(10)

Fig. 5: Transition table.

b) To Assign Output for Unstable States

The dashes in the output are assigned values according to the following rules.

Rule 1 : Assign 0 to an output of an unstable state which is transient state during a transition between two stable states that have a 0 as output.

Rule 2 : Assign 1 to an output of an unstable state which is transient state during a transition between two stable states that have a 1 as output.

Rule 3 : Assign a don't-care condition to an output of an unstable state which is transient state during a transition between two stable states that have a different values (0 and 1 or 1 and 0) as output.

The state transitions that are used to assign output using the above rules are shown in Fig. 6 and the flow table with output assigned to unstable states is shown in Fig. 7.

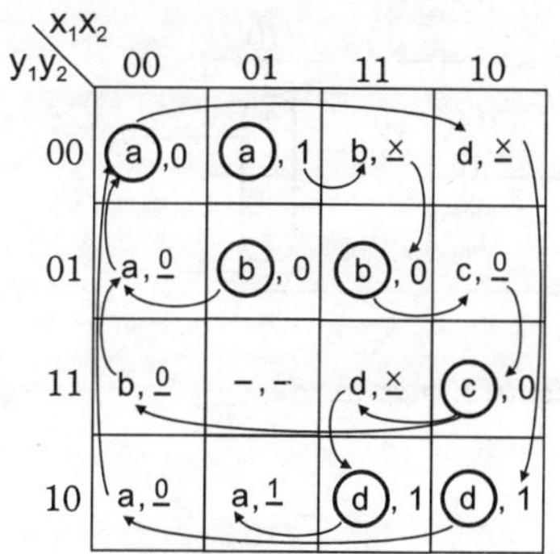

y_1y_2 \ x_1x_2	00	01	11	10
00	a, 0	a, 1	b, ×	d, ×
01	a, 0	b, 0	b, 0	c, 0
11	b, 0	–, –	d, ×	c, 0
10	a, 0	a, 1	d, 1	d, 1

***Fig. 6:** State transitions used to assign output.*

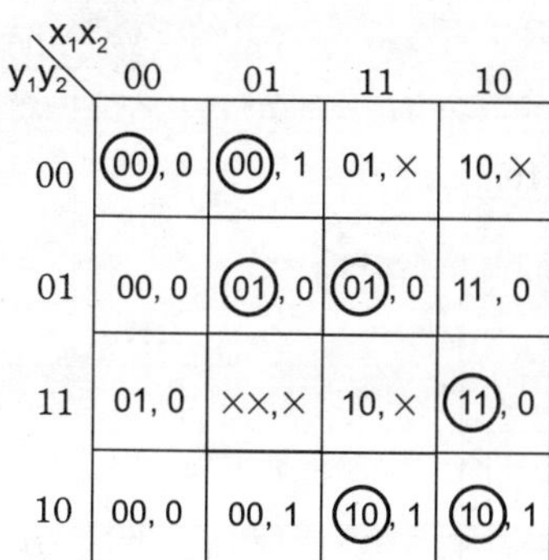

y_1y_2 \ x_1x_2	00	01	11	10
00	00, 0	00, 1	01, ×	10, ×
01	00, 0	01, 0	01, 0	11, 0
11	01, 0	××, ×	10, ×	11, 0
10	00, 0	00, 1	10, 1	10, 1

***Fig. 7:** Flow table with output assigned to unstable states.*

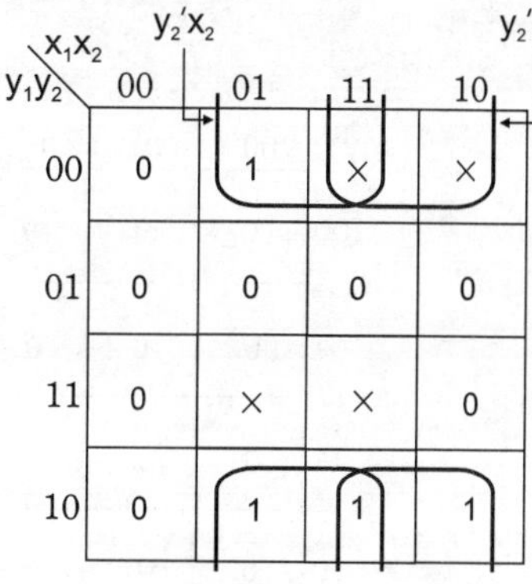

y_1y_2 \ x_1x_2	00	01	11	10
00	0	1	×	×
01	0	0	0	0
11	0	×	×	0
10	0	1	1	1

***Fig. 8:** Output map.*

Let, Z = Output

The output map is constructed as shown in Fig. 8 and from output map we get the following simplified Boolean equation for output.

Output, $Z = y_2'x_2 + y_2'x_1$

c) Logic Implementation using NOR based SR Latch

The binary values inside each square of transition table is the value of next state Y_1Y_2. Hence, using the transition table the maps of Y_1 and Y_2 are drawn as shown in Figs. 9 and 10.

y_1y_2 \ x_1x_2	00	01	11	10
00	0	0	0	1
01	0	0	0	1
11	0	×	1	1
10	0	0	1	1

***Fig. 9:** K-map for Y_1.*

y_1y_2 \ x_1x_2	00	01	11	10
00	0	0	1	0
01	0	1	1	1
11	1	×	0	1
10	0	0	0	0

***Fig. 10:** K-map for Y_2.*

In order to store the two state variables Y_1 and Y_2, two SR latches are required. Let us choose two NOR based SR latches.

Let, S_1, R_1 = Inputs for SR latch-1

S_2, R_2 = Inputs for SR latch-2

To Find Equations for Latch Inputs using Map of Y_1 and Y_2

In latch-1, the input equations for S_1 and R_1 will be independent y_1. Therefore, the map of Y_1 can be split into two maps: One map when y_1 = 0 and other map when y_1 = 1 as shown in Fig. 11.

The map y_1 = 0 is used to find equation for S_1 by forming prime implicants using 1's.

The map y_1 = 1 is used to find equation for R_1' by forming prime implicants using 0's.

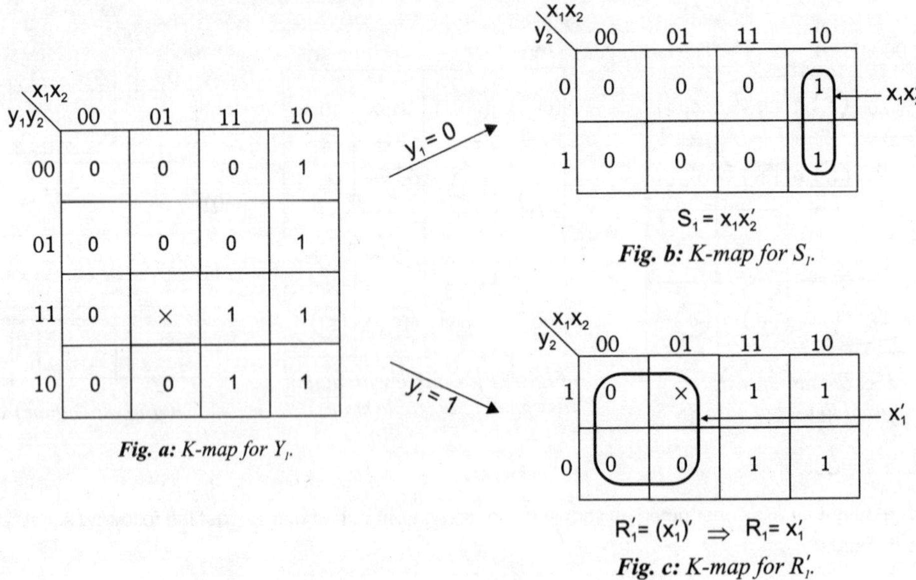

Fig. a: K-map for Y_1.

Fig. b: K-map for S_1.

Fig. c: K-map for R_1'.

***Fig. 11:** K-maps to find equations for S_1 and R_1.*

In latch-2, the input equations for S_2 and R_2 will be independent y_2. Therefore, the map of Y_2 can be split into two maps: One map when y_2 = 0 and other map when y_2 = 1 as shown in Fig. 12.

The map y_2 = 0 is used to find equation for S_2 by forming prime implicants using 1's.

The map y_2 = 1 is used to find equation for R_2' by forming prime implicants using 0's.

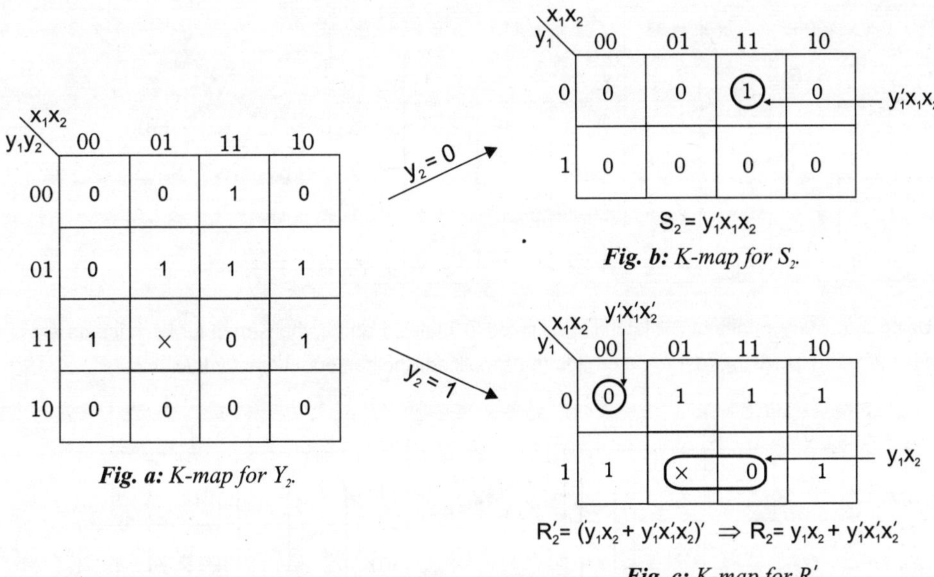

Fig. a: K-map for Y_2.

Fig. b: K-map for S_2.

Fig. c: K-map for R_2'.

Fig. 12: K-maps to find equations for S_2 and R_2.

Form the maps of S_1, R_1', S_2 and R_2' we get the following equations for SR latch inputs. The input equations is obtained from output map.

$S_1 = x_1x_2'$ | $R_1 = x_1'$

$S_2 = y_1'x_1x_2$ | $R_2 = y_1x_2 + y_1'x_1'x_2'$

To find Equations for Latch Inputs using Excitation Table of SR Latch

Using the excitation table of NOR based SR latch, the inputs for SR latch can be obtained for the circuit inputs, present and next states in the maps of Y_1 and Y_2 and listed in Tables 1 and 2.

Table 1: Inputs of SR Latch-1

	$x_1x_2 = 00$		$x_1x_2 = 01$		$x_1x_2 = 11$		$x_1x_2 = 10$	
y_1y_2	y_1 Y_1	S_1 R_1	y_1 Y_1	S_1 R_1	y_1 Y_1	S_1 R_1	y_1 Y_1	S_1 R_1
00	0 0	0 ×	0 0	0 ×	0 0	0 ×	0 1	1 0
01	0 0	0 ×	0 0	0 ×	0 0	0 ×	0 1	1 0
11	1 0	0 1	1 ×	× ×	1 1	× 0	1 1	× 0
10	1 0	0 1	1 0	0 1	1 1	× 0	1 1	× 0

Excitation Table of NOR based SR Latch

Present State y	Next State Y	Latch Inputs S	R
0	0	0	×
0	1	1	0
1	0	0	1
1	1	×	0

Table 2: Inputs of SR Latch-2

	$x_1x_2 = 00$				$x_1x_2 = 01$				$x_1x_2 = 11$				$x_1x_2 = 10$			
y_1y_2	y_2	Y_2	S_2	R_2	y_2	Y_2	S_2	R_2	y_2	Y_2	S_2	R_2	y_2	Y_2	S_2	R_2
00	0	0	0	×	0	0	0	×	0	1	1	0	0	0	0	×
01	1	0	0	1	1	1	×	0	1	1	×	0	1	1	×	0
11	1	1	×	0	1	×	×	×	1	0	0	1	1	1	×	0
10	0	0	0	×	0	0	0	×	0	0	0	×	0	0	0	×

Using the binary values of SR latch inputs listed in Tables 1 and 2, the K-map for latch inputs are constructed as shown in Figs. 13 to 16, in order to determine the simplified Boolean equations for latch inputs.

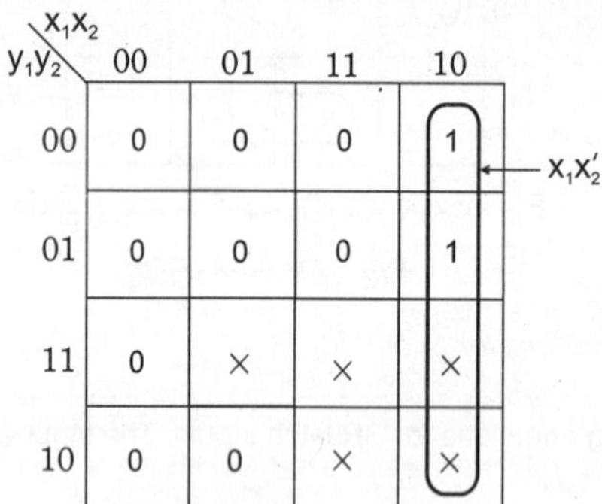

Fig. 13: K-map for S_1.

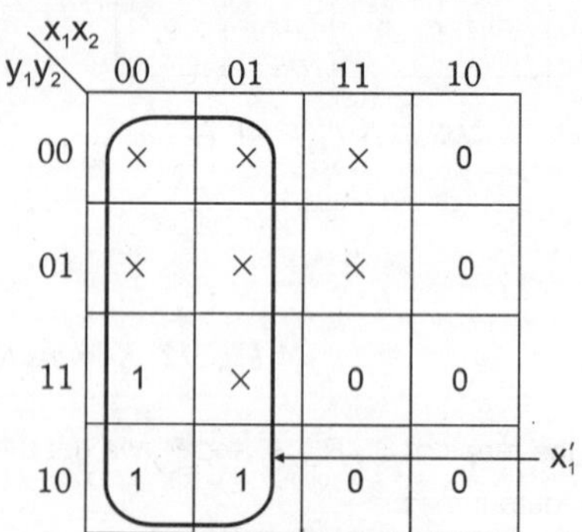

Fig. 14: K-map for R_1.

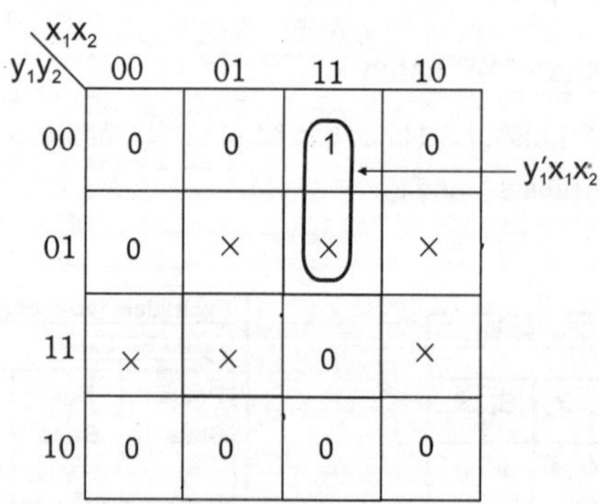

Fig. 15: K-map for S_2.

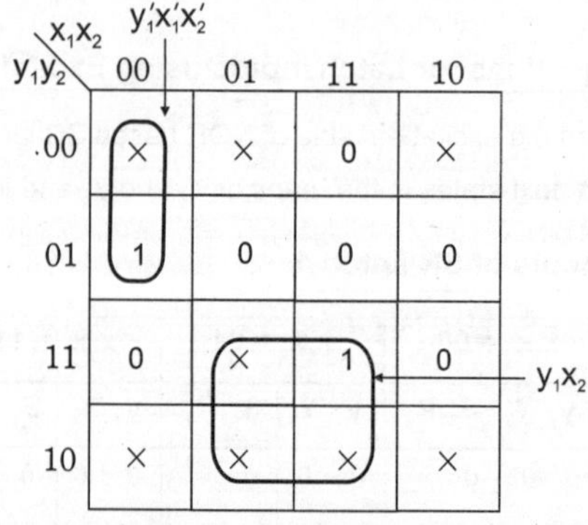

Fig. 16: K-map for R_2.

From the K-maps shown in Figs. 13 to 16, we get the following simplified equations for SR latch inputs. The output equation is obtained from output map.

$S_1 = x_1x_2'$	$S_2 = y_1'x_1x_2$	$Z = y_2'x_2 + y_2'x_1$
$R_1 = x_1'$	$R_2 = y_1'x_1'x_2' + y_1x_2$	

Logic Diagram

The Boolean equations obtained for latch inputs in both methods are same and using these equations the combinational logic circuit to generate the latch inputs are drawn as shown in Fig. 17. The complete logic circuit is obtained by adding NOR based SR latch to the combinational circuit as shown in Fig. 17.

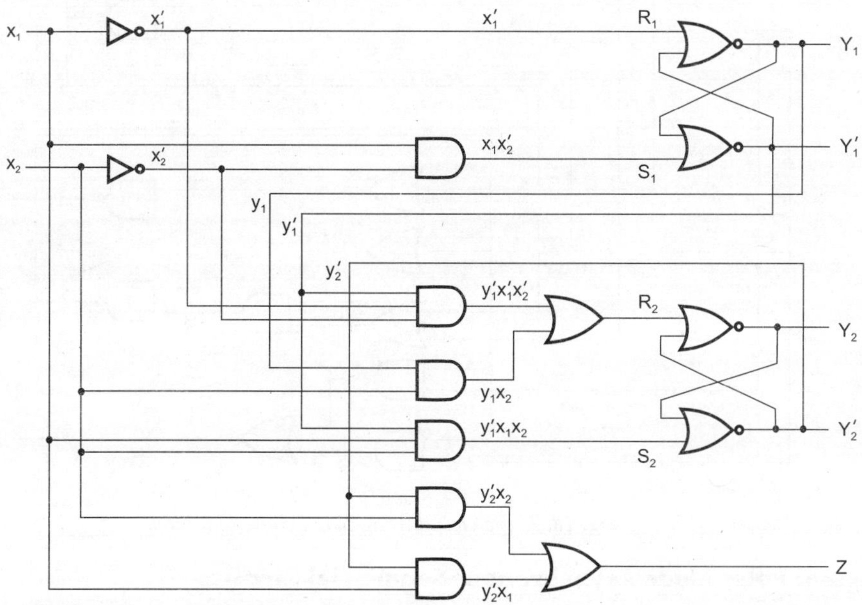

Fig. 17: Logic circuit using NOR based SR latch.

d) Logic Implementation using NAND based SR Latch

The Boolean equations to generate inputs for NAND based SR latches are given by complement of the equations for NOR based SR latch.

Therefore, the Boolean equations to generate inputs for NAND based SR latches are,

$S_1 = (x_1 x_2')'$	$S_2 = (y_1' x_1 x_2)'$	$R_1 = (x_1')'$	$R_2 = (y_1' x_1' x_2' + y_1 x_2)'$	$Z = y_2' x_2 + y_2' x_1$
$= x_1' + x_2$	$= y_1 + x_1' + x_2'$	$= x_1$	$= (y_1' x_1' x_2')' \cdot (y_1 x_2)'$	
			$= (y_1 + x_1 + x_2) \cdot (y_1' + x_2')$	

The output equation is obtained from output map. Using the above equations the combinational logic circuit to generate latch inputs are drawn as shown in Fig. 18. The complete logic circuit is obtained by adding NAND based SR latches to the combinational circuit as shown in Fig. 18.

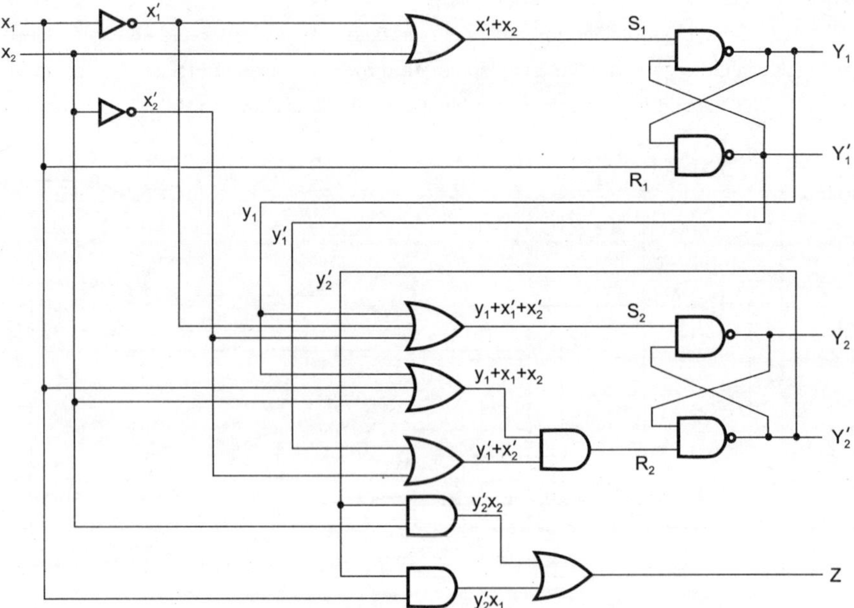

Fig. 18: Logic circuit using NAND based SR latch.

9.4 Design of Pulse Mode Asynchronous Sequential Circuits

In pulse mode asynchronous sequential circuits the inputs are pulses of sufficient duration for the output to change for a given input pulse. Here, also only one input is allowed to change at a time, but the input pulse will be available only for a short duration sufficient for transition from one stable state to another stable state.

The following procedural steps can be followed for the design of pulse mode asynchronous sequential circuits from system specifications.

- Develop a state diagram and hence state table from the system specifications.
- If necessary minimize the number of states in the state table.
- Assign race-free n-bit binary codes to states.
- The n-bit code will have n state variables and so choose n flip-flops to the state variables.
- Determine the flip-flop inputs for next state output using excitation table of flip-flop.
- Draw maps for flip-flop inputs.
- Form prime implicants by considering adjacent 1's in each column.
- Determine the SOP form of Boolean equations for flip-flop inputs.
- Draw the logic circuit using the Boolean equations.

Example 9.19

Design a pulse mode asynchronous sequential circuit with two inputs x_1 and x_2 and an output Z such that output pulse is produced only when an input sequence x_1-x_2-x_1 is detected and output is 0 for any other input sequence.

Implement the logic circuit i) using D filp-flop, ii) T flip-flop, iii) JK flip-flop, iv) SR flip-flop.

Solution

The circuit requires three states S_0, S_1 and S_2 to detect the sequence x_1-x_2-x_1 as shown in Fig. 1.

Fig. 1.

While in S_0, if an input pulse x_2 is received the circuit remains in S_0 itself. But if an input pulse x_1 is received the circuit go to next state S_1. In both the cases the output is 0 as shown in Fig. 2.

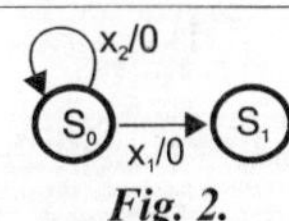

Fig. 2.

While in S_1 if an input pulse x_1 is detected the circuit remains in state S_1 itself. But if an input pulse x_2 is detected the circuit go to state S_2. In both the cases output is 0 as shown in Fig. 3.

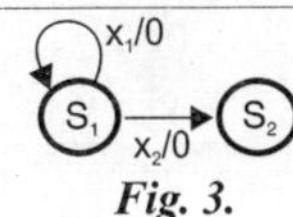

Fig. 3.

While in S_2 if an input pulse x_1 is received then the circuit go to state S_0 and output 1 to indicate the detection of sequence x_1-x_2-x_1. But if an input pulse x_2 is detected then the circuit go to state S_0 with output 0 as shown in Fig. 4.

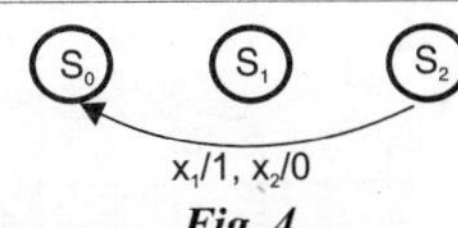

Fig. 4.

The transitions in Figs. 1 to 4 are combined to get the state diagram of the circuit as shown in Fig. 5.

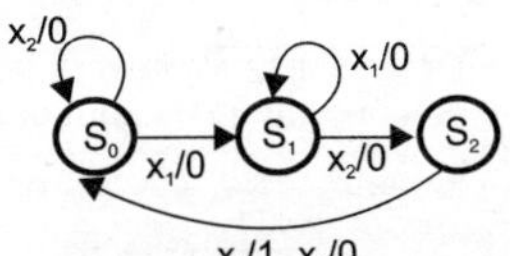

Fig. 5: State diagram.

State Table

Using the state diagram shown in Fig. 5, the state table is formed as shown in Table 1.

Table 1: State Table

Present State	Input	Next State	Output Z
S_0	x_1	S_1	0
	x_2	S_0	0
S_1	x_1	S_1	0
	x_2	S_2	0
S_2	x_1	S_0	1
	x_2	S_0	0

There is no identical or equivalent pair of states in the state table with same output and same next state, and so there is no need for state reduction. The three states can be formed with two state variables. Let y_1 and y_2 be the two state variables.

$\therefore$ Present state variables : y_1, y_2

Next state variables : Y_1, Y_2

Since there are three states, let us choose 2-bit code for states. The state diagram is redrawn as shown in Fig. 6 and state table is modified as shown in Table 2 with following state assignment.

$S_0 \longrightarrow 0\,0$
$S_1 \longrightarrow 0\,1$ Assigned codes
$S_2 \longrightarrow 1\,1$

Here, the code 10 is not used and so it is considered as don't-care.

$x_2/0$, $x_1/0$, 00, 01, 11, $x_1/0$, $x_2/0$, $x_1/1$, $x_2/0$

Fig. 6: State diagram with state assignment.

Table 2: State Table with Binary Code Assignment for States

Present State y_1 y_2	Input	Next State Y_1 Y_2	Output Z
0 0	x_1	0 1	0
0 0	x_2	0 0	0
0 1	x_1	0 1	0
0 1	x_2	1 1	0
1 1	x_1	0 0	1
1 1	x_2	0 0	0

Case i: Design using D Flip-Flop

In the sequential circuit, each state variable is stored in the output of a flip-flop. Hence, choose two D flip-flops to store the two state variables Y_1 and Y_2 as shown in Fig. 7.

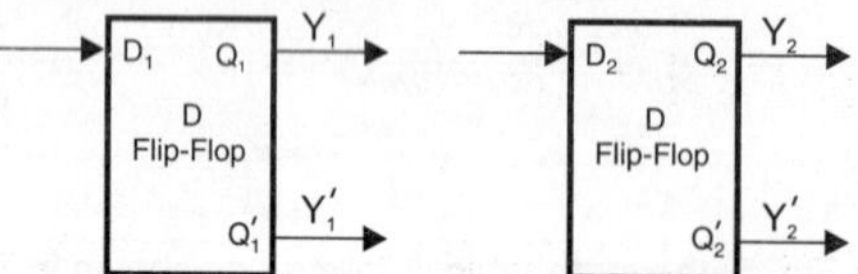

Fig. a: Flip-flop Y_1. *Fig. b: Flip-flop Y_2.*

Fig. 7: D flip-flop to store state variables.

The two state variables can be equated to flip-flop outputs.

$$\therefore\ y_1 = Q_1 \Rightarrow Y_1 = Q_1(t+1)$$

$$y_2 = Q_2 \Rightarrow Y_2 = Q_2(t+1)$$

The characteristic equation of two D flip-flops are,

$$Q_1(t+1) = D_1$$

$$Q_2(t+1) = D_2$$

Now the characteristic equations in terms of state variables are,

$$Y_1 = D_1$$

$$Y_2 = D_2$$

Using the above equations the flip-flop inputs are determined for the next state and listed in Table 3. Here, in the D flip-flop the input is same as next state output.

Table 3: Expanded State Table

Present State		Input	Next State		Flip-Flop Inputs		Output
y_1	y_2		Y_1	Y_2	D_1	D_2	Z
0	0	x_1	0	1	0	1	0
0	0	x_2	0	0	0	0	0
0	1	x_1	0	1	0	1	0
0	1	x_2	1	1	1	1	0
1	1	x_1	0	0	0	0	1
1	1	x_2	0	0	0	0	0
×	×	x_1	×	×	×	×	×
×	×	x_2	×	×	×	×	×

Let us draw K-map for D_1, D_2 and Z as shown in Figs. 8 to 10 to develop simplified Boolean equations for D_1, D_2 and Z.

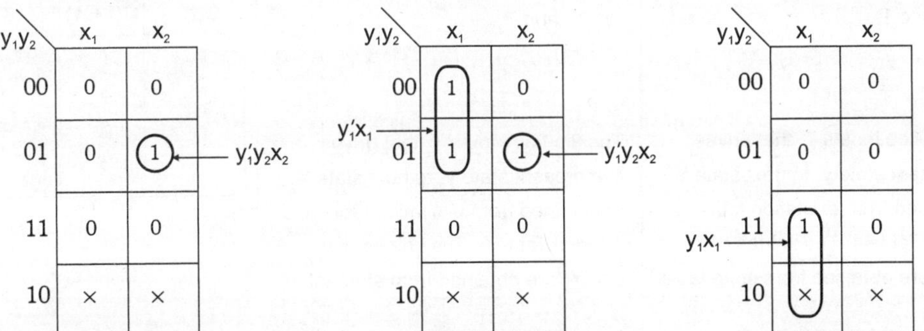

Fig. 8: K-map for D_1. *Fig. 9: K-map for D_2.* *Fig. 10: K-map for Z.*

From the K-map we get the following simplified Boolean equations for realizing sequence detector using D flip-flops.

$D_1 = y_1' y_2 x_2$; $D_2 = y_1' x_1 + y_1' y_2 x_2$; $Z = y_1 x_1$

The logic diagram of sequence detector using the above equations is drawn as shown in Fig. 11.

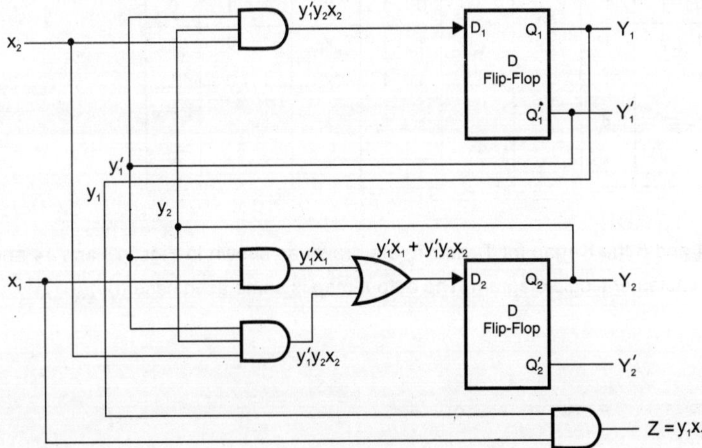

Fig. 11: Logic diagram of sequence detector using D flip-flop.

Case ii: Design using T Flip-Flop

Let us choose two T flip-flops to store two state variables Y_1 and Y_2 as shown in Fig. 12. Here, Y_1 and Y_2 be output of flip-flops.

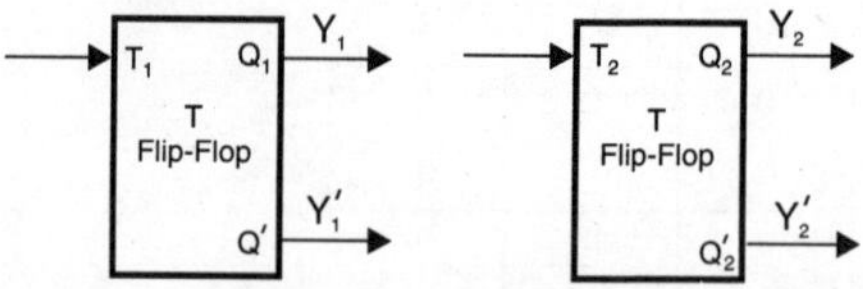

Fig. a: Flip-flop Y_1. *Fig. b: Flip-flop Y_2.*

Fig. 12: T flip-flops to store state variables.

The flip-flop inputs can be obtained from excitation table as shown below:

Excitation Table of T Flip-Flop

Present State Q(t)	Next State Q(t + 1)	Flip-Flop Input T
0	0	0
0	1	1
1	0	1
1	1	0

State Y_1

For state Y_1, the Q(t), Q(t + 1), T of excitation table are considered as y_1, Y_1 and T_1.

$\therefore Q(t) \rightarrow y_1$; $Q(t+1) \rightarrow Y_1$

$T \rightarrow T_1$

The flip-flop inputs T_1 that drives the present state y_1 to next state Y_1 is obtained from excitation table and listed in Table 4. The values of y_1 and Y_1 are obtained from state table.

Table 4:

y_1 y_2	Input	y_1 Y_1	T_1
0 0	x_1	0 0	0
0 0	x_2	0 0	0
0 1	x_1	0 0	0
0 1	x_2	0 1	1
1 1	x_1	1 0	1
1 1	x_2	1 0	1
× ×	x_1	× ×	×
× ×	x_2	× ×	×

State Y_2

For state Y_2, the Q(t), Q(t + 1), T of excitation table are considered as y_2, Y_2 and T_2.

$\therefore Q(t) \rightarrow y_2$; $Q(t+1) \rightarrow Y_2$

$T \rightarrow T_2$

The flip-flop inputs T_2 that drives the present state y_2 to next state Y_2 is obtained from excitation table and listed in Table 5. The values of y_2 and Y_2 are obtained from state table.

Table 5:

y_1 y_2	Input	y_2 Y_2	T_2
0 0	x_1	0 1	1
0 0	x_2	0 0	0
0 1	x_1	1 1	0
0 1	x_2	1 1	0
1 1	x_1	1 0	1
1 1	x_2	1 0	1
× ×	x_1	× ×	×
× ×	x_2	× ×	×

Using tables 4 and 5 the K-map for T_1 and T_2 are drawn as shown in Figs. 13 and 14 and develop simplified Boolean equations to implement logic diagram. The output map is same as in case (i).

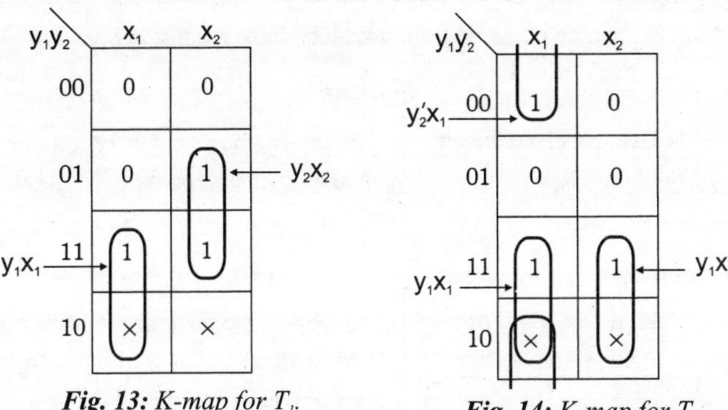

Fig. 13: K-map for T_1.

Fig. 14: K-map for T_2.

From the K-map we get the following simplified Boolean equations for realizing the logic circuit using T flip-flops.

$T_1 = y_1x_1 + y_2x_2$; $T_2 = y_2'x_1 + y_1x_1 + y_1x_2$; $Z = y_1x_1$ (Equation for Z is same as case (i))

The logic diagram using above equations is drawn as shown in Fig. 15.

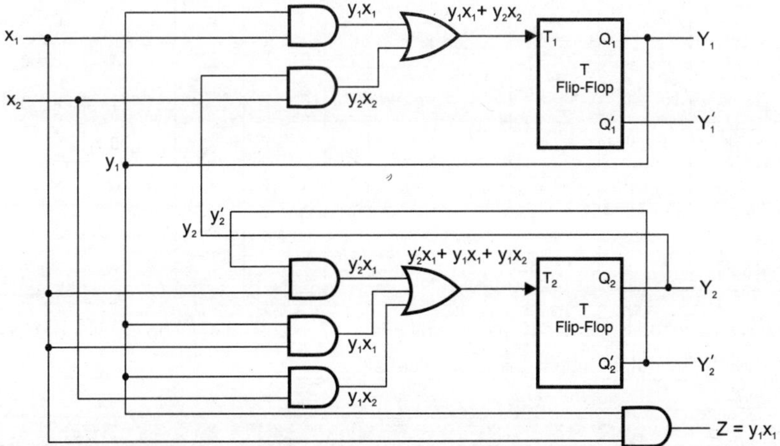

Fig. 15: Logic diagram of sequence detector using T flip-flop.

Case iii: Design using JK Flip-Flop

The given state stable has two state variables Y_1 and Y_2. In sequential circuits, each state variable is stored in a flip-flop. Hence, choose two JK flip-flops to store two state variables Y_1 and Y_2 as shown in Fig. 16.

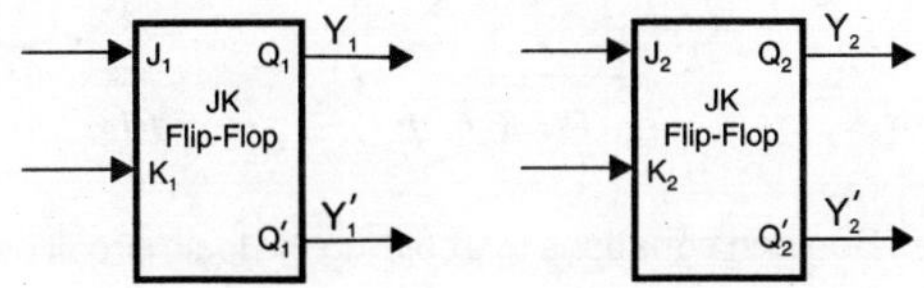

Fig. a: Flip-flop Y_1. *Fig. a: Flip-flop Y_2.*

Fig. 16: JK flip-flops to store state variables.

Excitation Table of JK Flip-Flop

Present State Q(t)	Next State Q(t + 1)	Flip-Flop Inputs J	K
0	0	0	×
0	1	1	×
1	0	×	1
1	1	×	0

The flip-flops inputs J_1, K_1, J_2, K_2 can be obtained from excitation table as shown below:

State Y_1

For state Y_1 the Q(t), Q(t + 1), J and K of excitation table are considered as y_1, Y_1, J_1 and K_1.

$$\therefore Q(t) \rightarrow y_1 \quad ; \quad Q(t+1) \rightarrow Y_1$$

$$J \rightarrow J_1 \quad ; \quad K \rightarrow K_1$$

The flip-flop inputs J_1 and K_1 that drives the present state y_1 to next state Y_1 are obtained from excitation table and listed in Table 6.

Table 6:

y_1 y_2	Input	y_1 Y_1	J_1 K_1
0 0	x_1	0 0	0 ×
0 0	x_2	0 0	0 ×
0 1	x_1	0 0	0 ×
0 1	x_2	0 1	1 ×
1 1	x_1	1 0	× 1
1 1	x_2	1 0	× 1
× ×	x_1	× ×	× ×
× ×	x_2	× ×	× ×

State Y_2

For state Y_2 the Q(t), Q(t + 1), J and K of excitation table are considered as y_2, Y_2, J_2 and K_2.

$$\therefore Q(t) \rightarrow y_2 \quad ; \quad Q(t+1) \rightarrow Y_2$$

$$J \rightarrow J_2 \quad ; \quad K \rightarrow K_2$$

The flip-flop inputs J_2 and K_2 that drives the present state y_2 to next state Y_2 are obtained from excitation table and listed in Table 7.

Table 7:

y_1 y_2	Input	y_2 Y_2	J_2 K_2
0 0	x_1	0 1	1 ×
0 0	x_2	0 0	0 ×
0 1	x_1	1 1	× 0
0 1	x_2	1 1	× 0
1 1	x_1	1 0	× 1
1 1	x_2	1 0	× 1
× ×	x_1	× ×	× ×
× ×	x_2	× ×	× ×

Let us draw K-map for J_1, K_1, J_2 and K_2 as shown in Figs. 17 to 20 and develop simplified Boolean equations to implement logic diagram. The output map is same as in case (i).

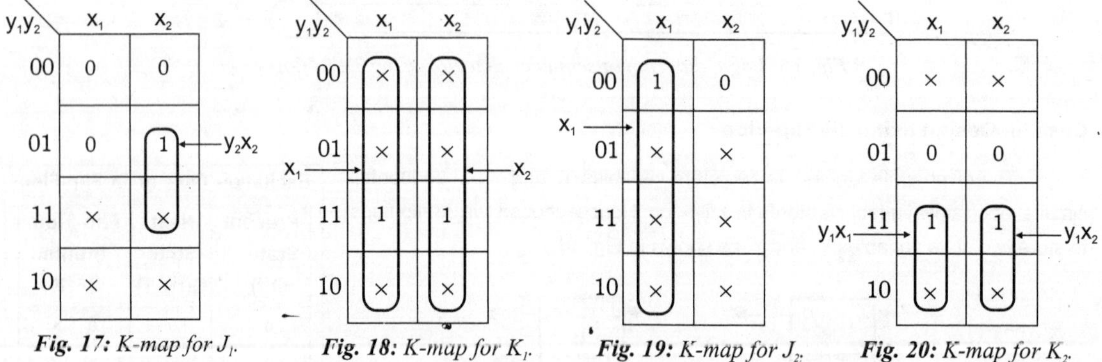

Fig. 17: K-map for J_1. *Fig. 18: K-map for K_1.* *Fig. 19: K-map for J_2.* *Fig. 20: K-map for K_2.*

From the K-maps we get the following simplified Boolean equations for realizing the logic circuit using JK flip-flops,

$J_1 = y_2x_2$ | $J_2 = x_1$ | $Z = y_1x_1$ (Equation for Z is same as case (i))

$K_1 = x_1 + x_2$ | $K_2 = y_1x_1 + y_1x_2$

The logic diagram using above equations is drawn as shown in Fig. 21.

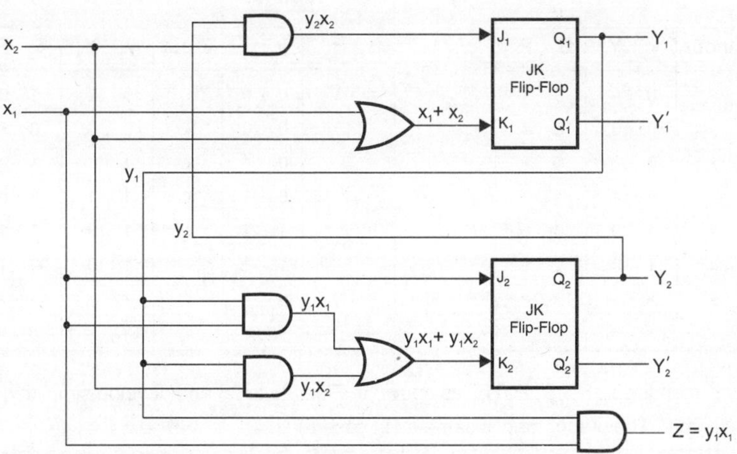

***Fig. 21:** Logic diagram of sequence detector using SR flip-flop.*

Case iv: Design using SR Flip-Flop

The given state stable has two state variables Y_1 and Y_2. In sequential circuits, each state variable is stored in a flip-flop. Hence, choose two SR flip-flops to store two state variables Y_1 and Y_2 as shown in Fig. 22.

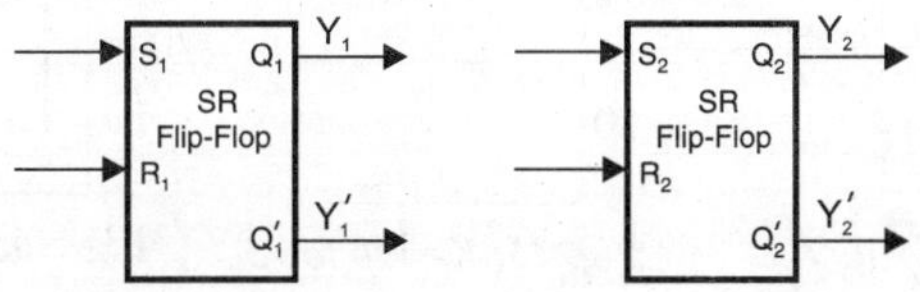

***Fig. a:** Flip-flop Y_1.* ***Fig. a:** Flip-flop Y_2.*

***Fig. 22:** SR flip-flops to store state variables.*

Excitation Table of SR Flip-Flop

Present State	Next State	Flip-Flop Inputs	
Q(t)	Q(t + 1)	S	R
0	0	0	×
0	1	1	0
1	0	0	1
1	1	×	0
0	×	1	1
1	×	1	1

The flip-flops inputs S_1, R_1, S_2, R_2 can be obtained from excitation table as shown below:

State Y_1

For state Y_1 the Q(t), Q(t + 1), S and R of excitation table are considered as y_1, Y_1, S_1 and R_1.

$$\therefore Q(t) \rightarrow y_1 \quad ; \quad Q(t+1) \rightarrow Y_1$$

$$S \rightarrow S_1 \quad ; \quad R \rightarrow R_1$$

The flip-flop inputs S_1 and R_1 that drives the present state y_1 to next state Y_1 are obtained from excitation table and listed in Table 8.

State Y_2

For state Y_2 the Q(t), Q(t + 1), S and R of excitation table are considered as y_2, Y_2, S_2 and R_2.

$$\therefore Q(t) \rightarrow y_2 \quad ; \quad Q(t+1) \rightarrow Y_2$$

$$S \rightarrow S_2 \quad ; \quad R \rightarrow R_2$$

The flip-flop inputs S_2 and R_2 that drives the present state y_2 to next state Y_2 are obtained from excitation table and listed in Table 9.

Table 8:

y_1 y_2	Input	y_1 Y_1	S_1 R_1
0 0	x_1	0 0	0 ×
0 0	x_2	0 0	0 ×
0 1	x_1	0 0	0 ×
0 1	x_2	0 1	1 0
1 1	x_1	1 0	0 1
1 1	x_2	1 0	0 1
× ×	x_1	× ×	× ×
× ×	x_2	× ×	× ×

Table 9:

y_1 y_2	Input	y_2 Y_2	S_2 R_2
0 0	x_1	0 1	1 0
0 0	x_2	0 0	0 ×
0 1	x_1	1 1	× 0
0 1	x_2	1 1	× 0
1 1	x_1	1 0	0 1
1 1	x_2	1 0	0 1
× ×	x_1	× ×	× ×
× ×	x_2	× ×	× ×

Let us draw K-map for S_1, R_1, S_2 and R_2 as shown in Figs. 23 to 26 and develop simplified Boolean equations to implement logic diagram. The output map is same as in case (i).

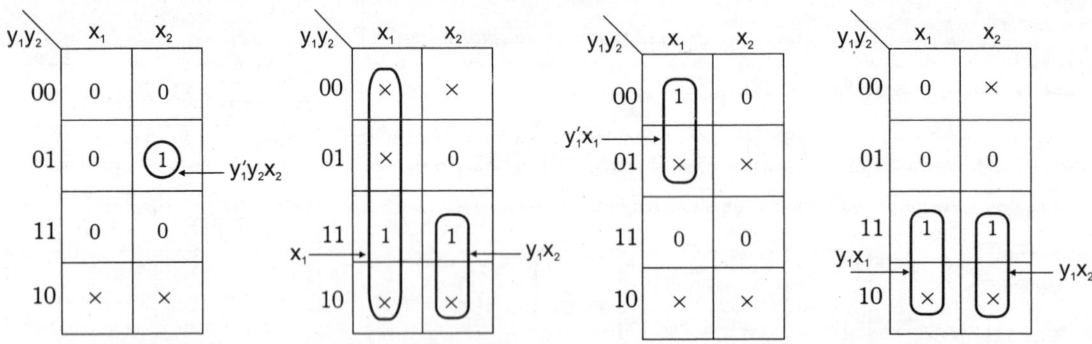

Fig. 23: K-map for S_1. *Fig. 24: K-map for R_1.* *Fig. 25: K-map for S_2.* *Fig. 26: K-map for R_2.*

From the K-maps we get the following simplified Boolean equations for realizing the logic circuit using JK flip-flops,

$S_1 = y_1' y_2 x_2$ | $S_2 = y_1'x_1$ | $Z = y_1x_1$ (Equation for Z is same as case (i))

$R_1 = x_1 + y_1x_2$ | $R_2 = y_1x_1 + y_1x_2$

The logic diagram using above equations is drawn as shown in Fig. 27.

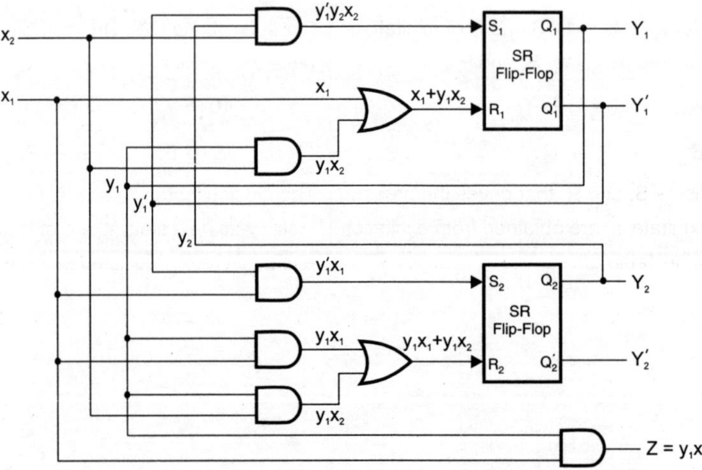

Fig. 27: Logic diagram of sequence detector using SR flip-flop.

Example 9.20

Design a pulse mode asynchronous sequential circuit with three inputs x_1, x_2 and x_3 and an output Z such that output pulse is produced only when an input sequence x_1-x_2-x_3 is detected and output is 0 for any other input sequence.

Implement the logic circuit,

i) using D filp-flop

ii) T flip-flop

iii) JK flip-flop

iv) SR flip-flop

Solution

The circuit requires four states S_0, S_1, S_2 and S_3. Let initial state be S_0,

While in S_0, if an input pulse x_2 and x_3 is received the circuit remains in S_0 itself. But if an input pulse x_1 is received the circuit go to next state S_1 to detect next input pulse of the sequence. In all the cases the output is 0 as shown in Fig. 1.

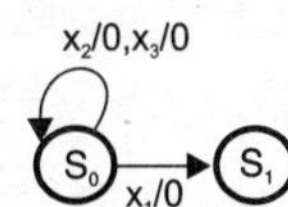

Fig. 1.

While in S_1 if an input pulse x_2 is received then the circuit go to next state S_2 to detect next input pulse of the sequence and if an input pulse x_3 is received then the circuit go to state S_0. In all the cases the output is 0. But if an input pulse x_1 is received the circuit remains in S_1 itself to start detecting next sequence and the output is 0 as shown in Fig. 2.

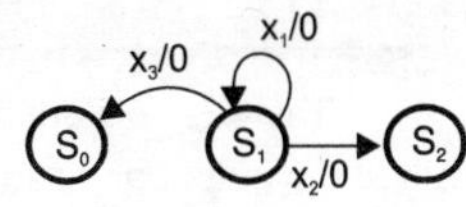

Fig. 2.

While in S_2 if an input pulse x_3 is received then the circuit go to next state S_3 and output 1 to indicate the detection of sequence x_1-x_2-x_3. But if an input pulse x_1 is detected then the circuit go to state S_1 with output 0 and if an input pulse x_2 is detected then the circuit go to state S_0 with output 0 to start detecting next sequence as shown in Fig. 3.

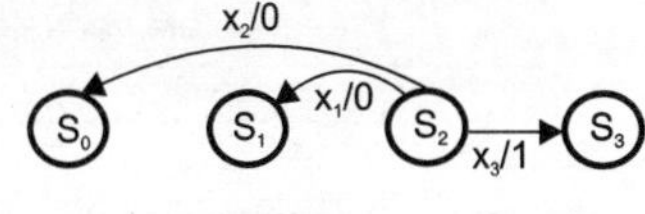

Fig. 3.

While in S_3, if an input pulse x_2 and x_3 is received the circuit remains in S_3 itself. But if an input pulse x_1 is received the circuit go to next state S_1 to start detecting next sequence. In all the cases the output is 0 as shown in Fig. 4.

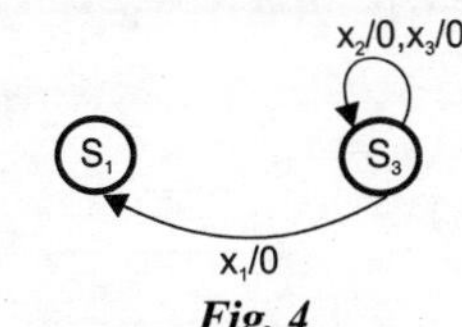

Fig. 4.

The transitions in Figs.1 to 4 are combined to get the state diagram of the circuit.

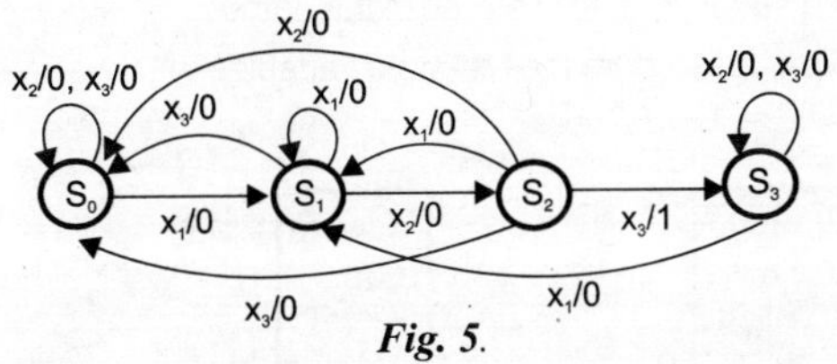

Fig. 5.

State Table

Using the state diagram shown in Fig. 5, the state table is formed as shown in Table 1.

Table 1: State Table

Present State	Input	Next State	Output Z
S_0	x_1	S_1	0
	x_2	S_0	0
	x_3	S_0	0
S_1	x_1	S_1	0
	x_2	S_2	0
	x_3	S_0	0
S_2	x_1	S_1	0
	x_2	S_0	0
	x_3	S_3	1
S_3	x_1	S_1	0
	x_2	S_3	0
	x_3	S_3	0

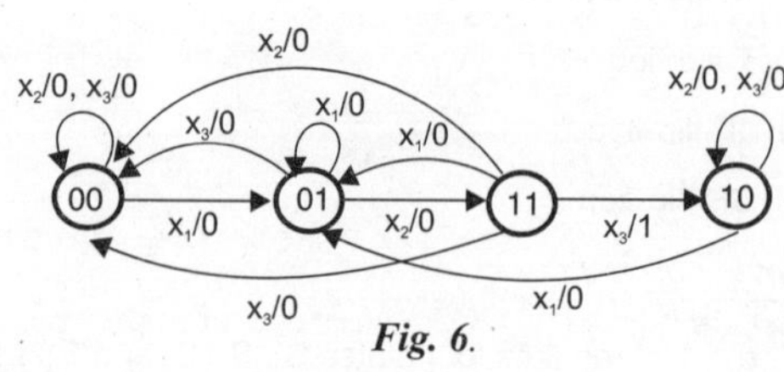

Fig. 6.

There is no identical or equivalent pair of states in the state table with same output and same next state, and so there is no need for state reduction. The four states can be formed with two state variables. Let Y_1 and Y_2 be the two state variables.

$\therefore$ Present state variables : y_1, y_2

Next state variables : Y_1, Y_2

Since there are four states, let us choose 2-bit code for states. The state diagram is redrawn as shown in Fig. 6 and state table is modified as shown in Table 2 with following state assignment.

$S_0 \longrightarrow 0\,0$
$S_1 \longrightarrow 0\,1$
$S_2 \longrightarrow 1\,1$
$S_3 \longrightarrow 1\,0$
} Assigned codes

Table 2: State Table with Binary Code Assignment for States

Present State y_1	y_2	Input	Next State Y_1	Y_2	Output Z
0	0	x_1	0	1	0
0	0	x_2	0	0	0
0	0	x_3	0	0	0
0	1	x_1	0	1	0
0	1	x_2	1	1	0
0	1	x_3	0	0	0
1	1	x_1	0	1	0
1	1	x_2	0	0	0
1	1	x_3	1	0	1
1	0	x_1	0	1	0
1	0	x_2	1	0	0
1	0	x_3	1	0	0

Case i: Design using D Flip-Flop

In the sequential circuit, each state variable is stored in the output of a flip-flop. Hence, choose two D flip-flops to store the two state variables Y_1 and Y_2 as shown in Fig. 7.

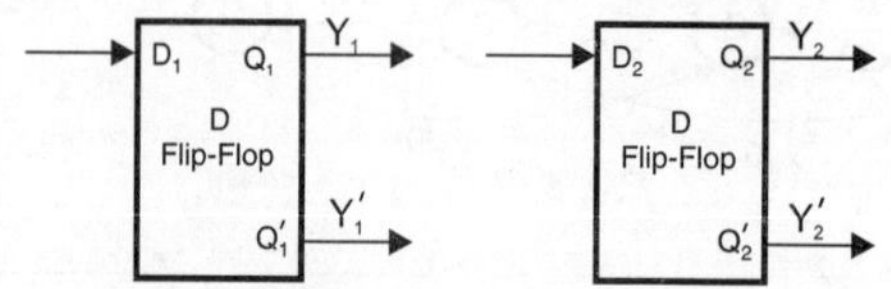

Fig. a: Flip-flop Y_1. *Fig. b: Flip-flop Y_2.*

Fig. 7: D flip-flop to store state variables.

The two state variables can be equated to flip-flop outputs.

$\therefore y_1 = Q_1 \;\Rightarrow Y_1 = Q_1(t+1)$

$y_2 = Q_2 \;\Rightarrow Y_2 = Q_2(t+1)$

The characteristic equation of two D flip-flops are,

$Q_1(t+1) = D_1$

$Q_2(t+1) = D_2$

The two state variables can be equated to flip-flop outputs.

$\therefore y_1 = Q_1 \;\Rightarrow Y_1 = Q_1(t+1)$

$y_2 = Q_2 \;\Rightarrow Y_2 = Q_2(t+1)$

Now the characteristic equations in terms of state variables are,

$Y_1 = D_1$

$Y_2 = D_2$

Using the above equations the flip-flop inputs are determined from the next state and listed in Table 3. Here,the D flip-flop the input is same as next state output.

Table 3: Expanded State Table

Present State		Input	Next State		Flip-Flop Inputs		Output
y_1	y_2		Y_1	Y_2	D_1	D_2	Z
0	0	x_1	0	1	0	1	0
0	0	x_2	0	0	0	0	0
0	0	x_3	0	0	0	0	0
0	1	x_1	0	1	0	1	0
0	1	x_2	1	1	1	1	0
0	1	x_3	0	0	0	0	0
1	1	x_1	0	1	0	1	0
1	1	x_2	0	0	0	0	0
1	1	x_3	1	0	1	0	1
1	0	x_1	0	1	0	1	0
1	0	x_2	1	0	1	0	0
1	0	x_3	1	0	1	0	0

Let us draw K-map for D_1, D_2 and Z as shown in Figs. 8 to 10 to develop simplified Boolean equations for D_1, D_2 and Z.

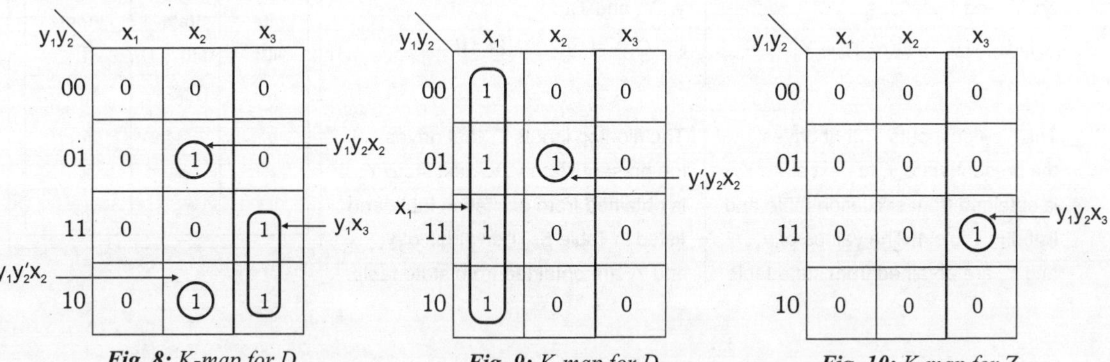

Fig. 8: K-map for D_1. *Fig. 9: K-map for D_2.* *Fig. 10: K-map for Z.*

From the K-map we get the following simplified Boolean equations for realizing sequence detector using D flip-flops.

$$D_1 = y_1' y_2 x_2 + y_1 y_2' x_2 + y_1 x_3 \quad ; \quad D_2 = y_1' y_2 x_2 + x_1 \quad ; \quad Z = y_1 y_2 x_3$$

The logic diagram of sequence detector using the above equations is drawn as shown in Fig. 11.

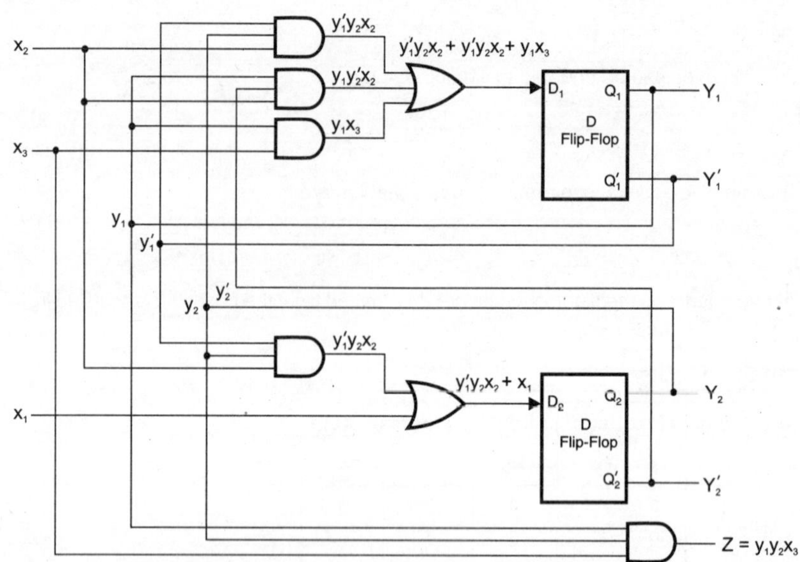

Fig. 11: Logic diagram of sequence detector using D flip-flop.

Case ii: Design using T Flip-Flop

Let us choose two T flip-flops to store two state variables Y_1 and Y_2 as shown in Fig. 12. Here, Y_1 and Y_2 be output of flip-flops.

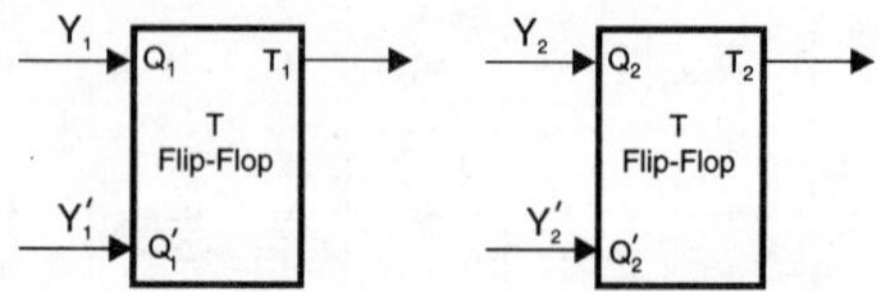

Fig. a: Flip-flop Y_1. *Fig. b: Flip-flop Y_2.*

Fig. 12: T flip-flops to store state variables.

The flip-flop inputs can be obtained from excitation table as shown below:

State Y_1

For state Y_1, the Q(t), Q(t + 1), T of excitation table are considered as y_1, Y_1 and T_1.

$$\therefore Q(t) \rightarrow y_1 \; ; \; Q(t+1) \rightarrow Y_1$$

$$T \rightarrow T_1$$

The flip-flop inputs T_1 that drives the present state y_1 to next state Y_1 is obtained from excitation table and listed in Table 4. The values of y_1 and Y_1 are obtained from state table.

State Y_2

For state Y_2, the Q(t), Q(t + 1), T of excitation table are considered as y_2, Y_2 and T_2.

$$\therefore Q(t) \rightarrow y_2 \; ; \; Q(t+1) \rightarrow Y_2$$

$$T \rightarrow T_2$$

The flip-flop inputs T_2 that drives the present state y_2 to next state Y_2 is obtained from excitation table and listed in Table 5. The values of y_2 and Y_2 are obtained from state table.

Excitation Table of T Flip-Flop

Present State Q(t)	Next State Q(t + 1)	Flip-Flop Input T
0	0	0
0	1	1
1	0	1
1	1	0

Table 4:

y_1 y_2	Input	y_1	Y_1	T_1
0 0	x_1	0	0	0
0 0	x_2	0	0	0
0 0	x_3	0	0	0
0 1	x_1	0	0	0
0 1	x_2	0	1	1
0 1	x_3	0	0	0
1 1	x_1	1	0	1
1 1	x_2	1	0	1
1 1	x_3	1	1	0
1 0	x_1	1	0	1
1 0	x_2	1	1	0
1 0	x_3	1	1	0

Table 5:

y_1 y_2	Input	y_2	Y_2	T_2
0 0	x_1	0	1	1
0 0	x_2	0	0	0
0 0	x_3	0	0	0
0 1	x_1	1	1	0
0 1	x_2	1	1	0
0 1	x_3	1	0	1
1 1	x_1	1	1	0
1 1	x_2	1	0	1
1 1	x_3	1	0	1
1 0	x_1	0	1	1
1 0	x_2	0	0	0
1 0	x_3	0	0	0

Using tables 4 and 5 the K-map for T_1 and T_2 are drawn as shown in Figs. 13 and 14 and develop simplified Boolean equations to implement logic diagram.The output map is same as in case (i).

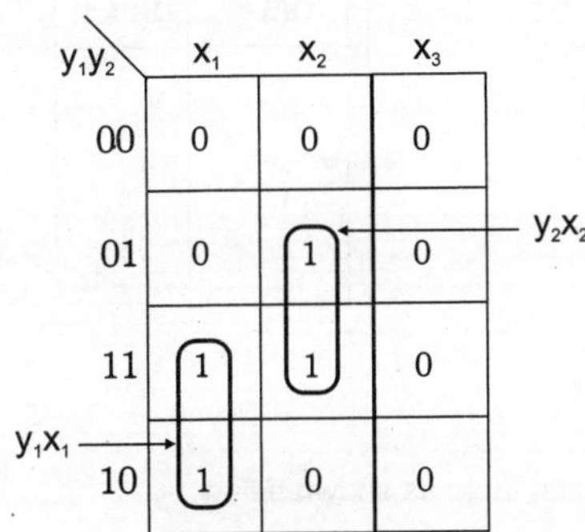

Fig. 13: K-map for T_1.

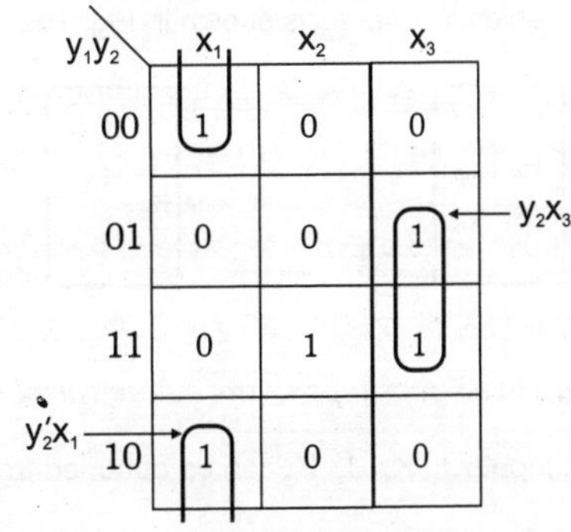

Fig. 14: K-map for T_2.

From the K-map we get the following simplified Boolean equations for realizing the sequence detector using T flip-flops.

$$T_1 = y_2 x_2 + y_1 x_1$$

$$T_2 = y_2' x_1 + y_2 x_3$$

$Z = y_1\, y_2\, x_3$ (Equation for Z is same as case (i))

The logic diagram using above equations is drawn as shown in Fig. 15.

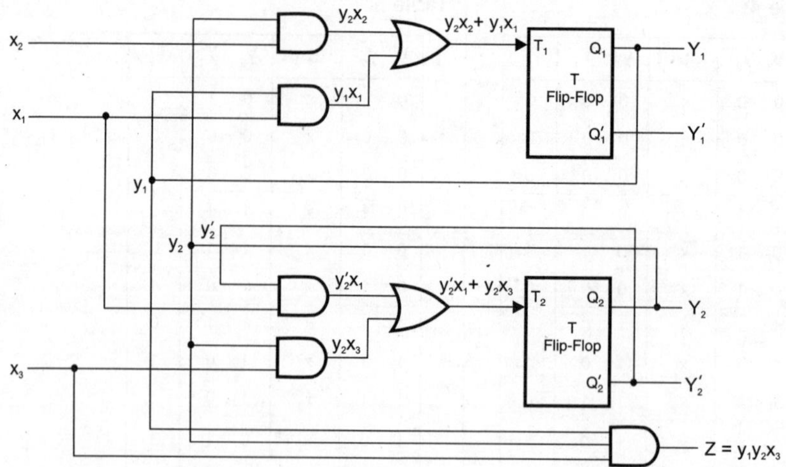

Fig. 15: Logic diagram of sequence detector using T flip-flop.

Case iii: Design using JK Flip-Flop

The given state stable has two state variables Y_1 and Y_2. In sequential circuits, each state variable is stored in a flip-flop. Hence, choose two JK flip-flops to store two state variables Y_1 and Y_2 as shown in Fig. 16.

Excitation Table of JK Flip-Flop

Present State Q(t)	Next State Q(t + 1)	Flip-Flop Inputs J K
0	0	0 ×
0	1	1 ×
1	0	× 1
1	1	× 0

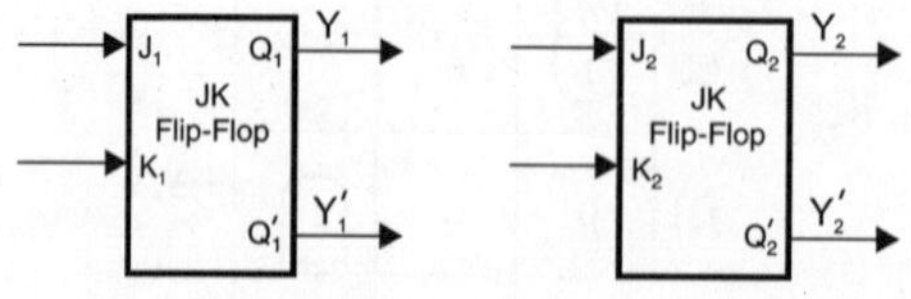

Fig. a: Flip-flop Y_1. *Fig. a: Flip-flop Y_2.*

Fig. 16: JK flip-flops to store state variables.

The flip-flops inputs J_1, K_1, J_2, K_2 can be obtained from excitation table as shown below:

State Y_1

For state Y_1 the Q(t), Q(t + 1), J and K of excitation table are considered as y_1, Y_1, J_1 and K_1.

$$\therefore Q(t) \rightarrow y_1 \quad ; \quad Q(t+1) \rightarrow Y_1$$

$$J \rightarrow J_1 \quad ; \quad K \rightarrow K_1$$

The flip-flop inputs J_1 and K_1 that drives the present state y_1 to next state Y_1 are obtained from excitation table and listed in Table 6.

State Y_2

For state Y_2 the Q(t), Q(t + 1), J and K of excitation table are considered as y_2, Y_2, J_2 and K_2.

$$\therefore Q(t) \rightarrow y_2 \quad ; \quad Q(t+1) \rightarrow Y_2$$

$$J \rightarrow J_2 \quad ; \quad K \rightarrow K_2$$

The flip-flop inputs J_2 and K_2 that drives the present state y_2 to next state Y_2 are obtained from excitation table and listed in Table 7.

Table 6:

y_1 y_2	Input	y_1 Y_1	J_1 K_1
0 0	x_1	0 0	0 ×
0 0	x_2	0 0	0 ×
0 0	x_3	0 0	0 ×
0 1	x_1	0 0	0 ×
0 1	x_2	0 1	1 ×
0 1	x_3	0 0	0 ×
1 1	x_1	1 0	× 1
1 1	x_2	1 0	× 1
1 1	x_3	1 1	× 0
1 0	x_1	1 0	× 1
1 0	x_2	1 1	× 0
1 0	x_3	1 1	× 0

Table 7:

y_1 y_2	Input	y_2 Y_2	J_2 K_2
0 0	x_1	0 1	1 ×
0 0	x_2	0 0	0 ×
0 0	x_3	0 0	0 ×
0 1	x_1	1 1	× 0
0 1	x_2	1 1	× 0
0 1	x_3	1 0	× 1
1 1	x_1	1 1	× 0
1 1	x_2	1 0	× 1
1 1	x_3	1 0	× 1
1 0	x_1	0 1	1 ×
1 0	x_2	0 0	0 ×
1 0	x_3	0 0	0 ×

Let us draw K-map for J_1, K_1, J_2 and K_2 as shown in Figs. 17 to 20 and develop simplified Boolean equations to implement logic diagram. The output map is same as in case (i).

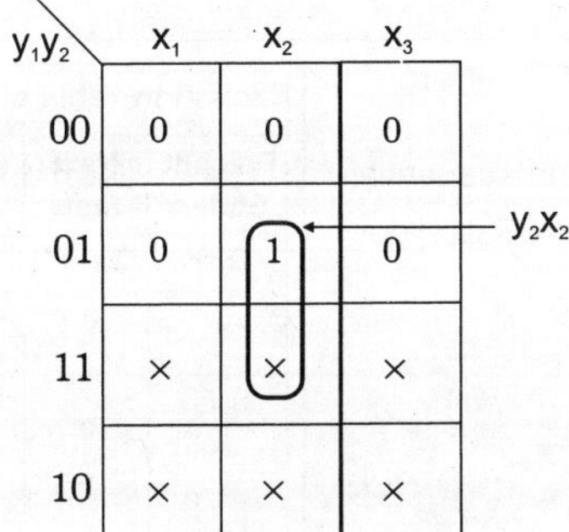

***Fig. 17:** K-map for J_1.*

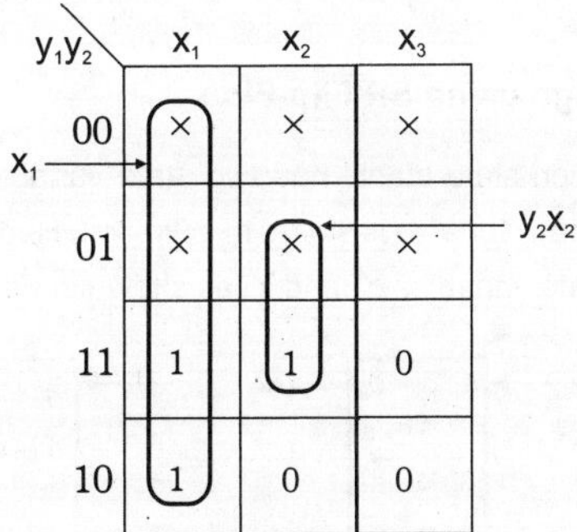

***Fig. 18:** K-map for K_1.*

y_1y_2	x_1	x_2	x_3
00	1	0	0
01	×	×	×
11	×	×	×
10	1	0	0

(x_1 grouped)

***Fig. 19:** K-map for J_2.*

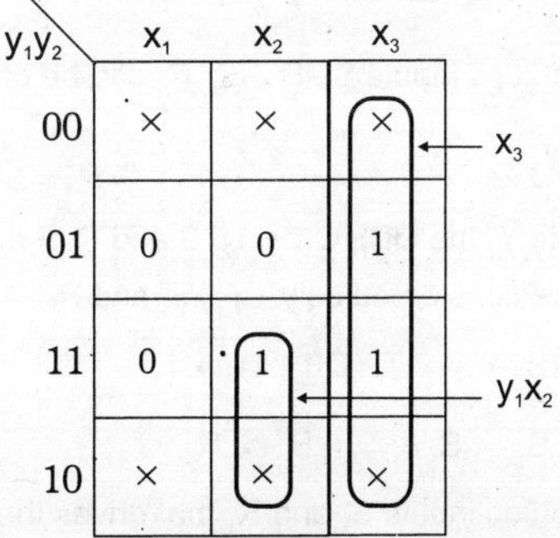

***Fig. 20:** K-map for K_2.*

From the K-maps we get the following simplified Boolean equations for realizing the sequence detector using JK flip-flops,

$J_1 = y_2 x_2$	$J_2 = x_1$	$Z = y_1\, y_2\, x_3$ (Equation for Z is same as case (ii))
$K_1 = y_2 x_2 + x_1$	$K_2 = y_1 x_2 + x_3$	

The logic diagram using above equations is drawn as shown in Fig. 21.

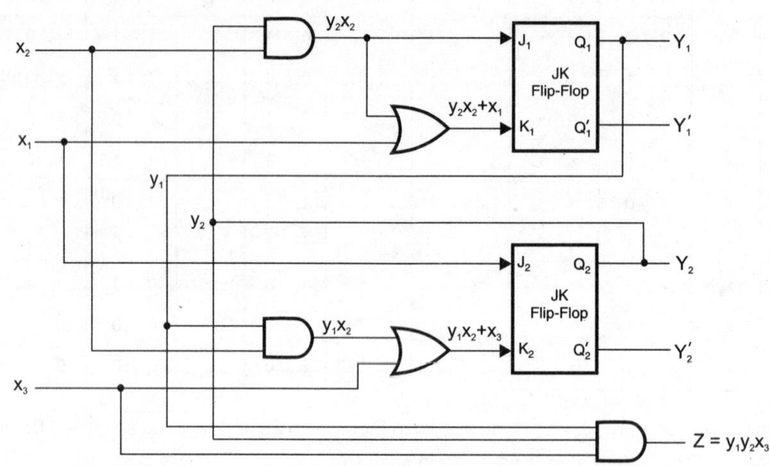

Fig. 21: Logic diagram of sequence detector using JK flip-flop.

Case iv: Design using SR Flip-Flop

The given state stable has two state variables Y_1 and Y_2. In sequential circuits, each state variable is stored in a flip-flop. Hence, choose two SR flip-flops to store two state variables Y_1 and Y_2 as shown in Fig. 22.

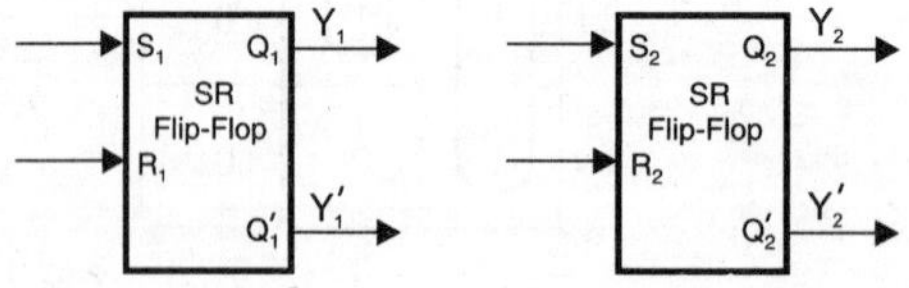

Fig. a: Flip-flop Y_1. *Fig. a: Flip-flop Y_2.*

Fig. 22: SR flip-flops to store state variables.

Excitation Table of SR Flip-Flop

Present State	Next State	Flip-Flop Inputs	
Q(t)	Q(t + 1)	S	R
0	0	0	×
0	1	1	0
1	0	0	1
1	1	×	0
0	×	1	1
1	×	1	1

The flip-flops inputs S_1, R_1, S_2, R_2 can be obtained from excitation table as shown below:

State Y_1

For state Y_1 the Q(t), Q(t + 1), S and R of excitation table are considered as y_1, Y_1, S_1 and R_1.

$$\therefore Q(t) \rightarrow y_1 \quad ; \quad Q(t+1) \rightarrow Y_1$$

$$S \rightarrow S_1 \quad ; \quad R \rightarrow R_1$$

The flip-flop inputs S_1 and R_1 that drives the present state y_1 to next state Y_1 are obtained from excitation table and listed in Table 8.

State Y_2

For state Y_2 the Q(t), Q(t + 1), S and R of excitation table are considered as y_2, Y_2, S_2 and R_2.

$$\therefore Q(t) \rightarrow y_2 \quad ; \quad Q(t+1) \rightarrow Y_2$$

$$S \rightarrow S_2 \quad ; \quad R \rightarrow R_2$$

The flip-flop inputs S_2 and R_2 that drives the present state y_2 to next state Y_2 are obtained from excitation table and listed in Table 9.

Table 8:

y_1 y_2	Input	y_1 Y_1	S_1 R_1
0 0	x_1	0 0	0 ×
0 0	x_2	0 0	0 ×
0 0	x_3	0 0	0 ×
0 1	x_1	0 0	0 ×
0 1	x_2	0 1	1 0
0 1	x_3	0 0	0 ×
1 1	x_1	1 0	0 1
1 1	x_2	1 0	0 1
1 1	x_3	1 1	× 0
1 0	x_1	1 0	0 1
1 0	x_2	1 1	× 0
1 0	x_3	1 1	× 0

Table 9:

y_1 y_2	Input	y_2 Y_2	S_2 R_2
0 0	x_1	0 1	1 0
0 0	x_2	0 0	0 ×
0 0	x_3	0 0	0 ×
0 1	x_1	1 1	× 0
0 1	x_2	1 1	× 0
0 1	x_3	1 0	0 1
1 1	x_1	1 1	× 0
1 1	x_2	1 0	0 1
1 1	x_3	1 0	0 1
1 0	x_1	0 1	1 0
1 0	x_2	0 0	0 ×
1 0	x_3	0 0	0 ×

Let us draw K-map for S_1, R_1, S_2 and R_2 as shown in Figs. 23 to 26 and develop simplified Boolean equations to implement logic diagram. The output map is same as in case (i).

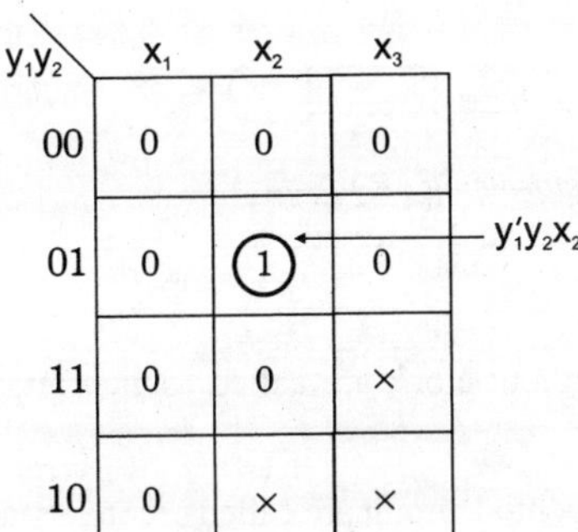

Fig. 23: K-map for S_1.

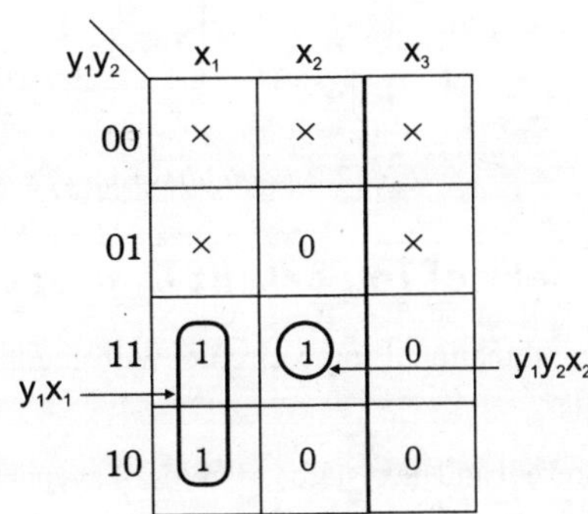

Fig. 24: K-map for R_1.

y_1y_2	X_1	X_2	X_3
00	1	0	0
01	×	×	0
11	×	0	0
10	1	0	0

x_1 →

Fig. 25: K-map for S_2.

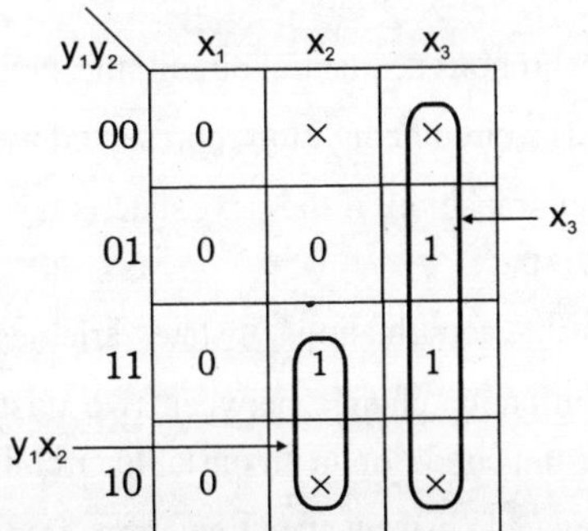

Fig. 26: K-map for R_2.

From the K-maps we get the following simplified Boolean equations for realizing the sequence detector using JK flip-flops,

$S_1 = y_1' y_2 x_2$	$S_2 = x_1$	$Z = y_1 y_2 x_3$ (Equation for Z is same as case (ii))
$R_1 = y_1 y_2 x_2 + y_1 x_1$	$R_2 = y_1 x_2 + x_3$	

The logic diagram using above equations is drawn as shown in Fig. 27.

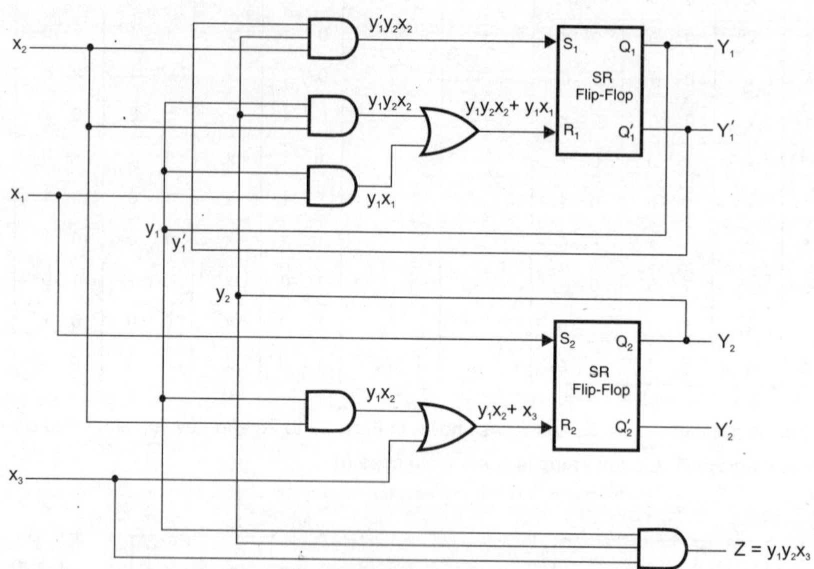

Fig. 27: Logic diagram of sequence detector using SR flip-flop.

9.5 Summary of Important Concepts

1. The combinational logic circuits with feedback and without clock are called asynchronous sequential circuits.
2. The present and next state variables of asynchronous sequential circuit are called secondary and excitation variables.
3. The circuit will attain a stable condition when excitation (or next state) variables are same as secondary (or present state) variables.
4. In asynchronous sequential circuit, the state table is called transition table.
5. A state is a circuit condition represented by a set of variables called state variables.
6. In a state transition, if the next state is not equal to present state then the corresponding state is called unstable state.
7. All possible combinations of state variables and inputs are called total states of the circuit.
8. The continuous change between two unstable states is also called metastability which refers to an equilibrium condition in an unstable circuit.
9. The flow table lists all state transitions from present state to next state along with output for all possible changes in input.

10. When there is only one stable state in each row of a flow table then it is called a primitive flow table.
11. The diagrammatic way of representing all possible state transitions between unstable states in every row of transition table is called transition diagram.
12. The race condition is classified into critical and noncritical race condition.
13. In a state transition, if the final stable state depends on the order of change in state variables, then the race is called critical race.
14. In a state transition, if the final stable state does not depends on the order of change in state variables, then the race is called noncritical race.
15. When the state transitions in a circuit go through a unique sequence of unstable states then that condition is called a cycle.
16. Races can also be avoided by using cycles, which terminate on a stable state.
17. Two binary codes are said to be adjacent if they differ by only one variable.
18. Assigning adjacent codes to states to avoid critical races is called race-free assignment.
19. In fundamental mode asynchronous sequential circuits the inputs are level triggered inputs.
20. In asynchronous sequential circuits, the state reduction is performed via an implication table or merger table.
21. State pairs that have same next state and output for the same input are equivalent states.
22. State pairs having different next state and output for same input are not equivalent states.
23. If a pair of state having same outputs for the same inputs and go to same equivalent next state then the pair is said to be implied pair.
24. A merger diagram or merger graph is a circle with dots representing the states and every equivalent pair of states is connected by a straight line.
25. Asynchronous circuits with incompletely specified states, the states can be reduced by identifying compatible pairs.
26. Maximum compatible states is a group of compatible states that includes all possible combinations of compatible states.
27. The closed-covering condition refers to set of compatibles that covers all the states and satisfy covering condition.
28. The set of compatibles that satisfy the closed-covering condition form the minimal set of compatibles.
29. Rows of reduced flow table can be formed by taking one element of each of the minimal compatible set.
30. The latches are used as memory elements in asynchronous circuits.
31. SR latch is constructed using two cross coupled NOR gates or two cross coupled NAND gates.
32. In fundamental mode asynchronous sequential circuits the inputs are level triggered inputs.
33. In pulse mode asynchronous sequential circuits the inputs are pulses of sufficient duration for the output to change for a given input pulse.

9.6 Short-Answer Questions

Q9.1 Define asynchronous sequential circuits.

The combinational logic circuits with feedback and without clock are called asynchronous sequential circuits.

Q9.2 What are the characteristic of asynchronous sequential circuits?

The changes in internal states occur immediately when there is a change in input and hence the response in an asynchronous sequential circuit will be fast. In the asynchronous sequential circuit the change in input may lead to races and unstable condition and therefore, sufficient care must be taken to avoid races and instablity.

Q9.3 List the difference between asynchronous and synchronous sequential circuits.

Synchronous Sequential Circuit	Asynchronous Sequential Circuit
1. Provided with clock	1. Do not have a clock
2. Changes in internal states occur in response to clock.	2. Changes in internal states occur when there is a change in input variables.
3. Memory elements are clocked flip-flops.	3. Memory elements are either latches (unclocked flip-flops) or time-delay elements.

Q9.4 What are secondary and excitation variables.

The present and next state variables of asynchronous sequential circuits are called secondary and excitation variables.

Q9.5 Draw the general model of asynchronous sequential machine (ASM). *(AU, Nov/Dec'18, 2 Marks)*

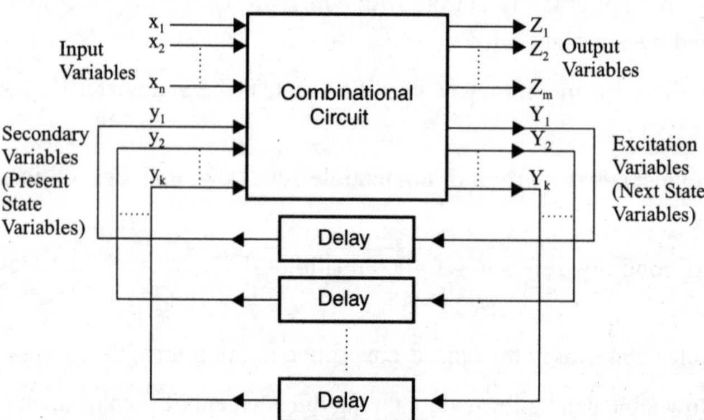

Fig. Q9.5: Block diagram representation of an asynchronous sequential circuit.

Q9.6 Explain stable and unstable states.

The asynchronous sequential circuit will change from one state to another state when there is a change in input. During a state transition when the next state is same as present state then the circuit attains a stable condition and the corresponding state is called a stable stable. On the other hand in a state transition, if the next state is not equal to present state then the corresponding state is called an unstable state.

Q9.7 Define total states.

All possible combinations of state variables and inputs are called total states of the circuit.

Q9.8 Define metastable state. **(AU, Apr/May'19, 2 Marks)**

The continuous change between two unstable states is also called metastability (or metastable state) which refers to an equilibrium condition in an unstable circuit.

Q9.9 What is difference between an internal state and a total state?

When the excitation variables are fedback to input they are called internal states or secondary variables. The excitation variables are called external states.

The internal states along with inputs are called total states.

Q9.10 Define flow table in asynchronous sequential circuit.

The listing of all state transitions from present state to next state along with output for all possible changes in input is called flow table.

Q9.11 What is primitive flow table.

A flow table may have one or more stable states in a row. When there is only one stable state in each row of a flow table then it is called a primitive flow table.

Q9.12 What is cycle?

When the state transitions in a circuit go through a unique sequence of unstable states then that condition is called a cycle.

Q9.13 What is meant by race? (or What is race around condition?) **(IKGTU, Dec'19, 2 Marks)** **(CSVTU, Nov/Dec'19, 2 Marks)**

In an asynchronous sequential circuit, when there are changes in two or more state variables due to change in input, a race condition may exist, in which changes in variables may occur at different timings due to difference in delay in the feedback paths. This leads to unpredictable transition paths during a change in state which in turn settle at different stable state.

Q9.14 Define critical race and non critical race. **(AKTU, Dec'22, 2 Marks)**

The race condition may be classified into critical and noncritical race.

Critical race : In a state transition, if the final stable state depends on the order of change in state variables, then the race is called ***critical race***.

Noncritical race : In a state transition, if the final stable state does not depends on the order of change in state variables, then the race is called ***noncritical race***.

Q9.15 Distinguish between noncritical and critical race.

In noncritical race the final stable state will be same irrespective of the order of changes in state variables and hence final stable state will be same for a change in input.

But in critical race the final stable state depends on the order of changes in state variables and hence the final stable state will be different for a change in input.

Q9.16 What is meant by race-free condition in sequential circuit?

During state transitions if the order of changes in internal states does not affect the final stable state then the condition is called race-free condition.

Q9.17 What is race-free state assignment?

Assigning adjacent codes to states to avoid critical races is called race-free assignment. Two binary codes are said to be adjacent if they differ only in one variable.

Q9.18 An asynchronous sequential circuit is described by the excitation function and output function,

$$Y_1 = x_1x_2 + x_1y_2 + x_2'y_1 \quad ; \quad Y_2 = x_2' + x_1y_1y_2 + x_1y_1 \quad ; \quad Z = x_2' + y_2$$

Draw the logic diagram of the circuit.

Solution:

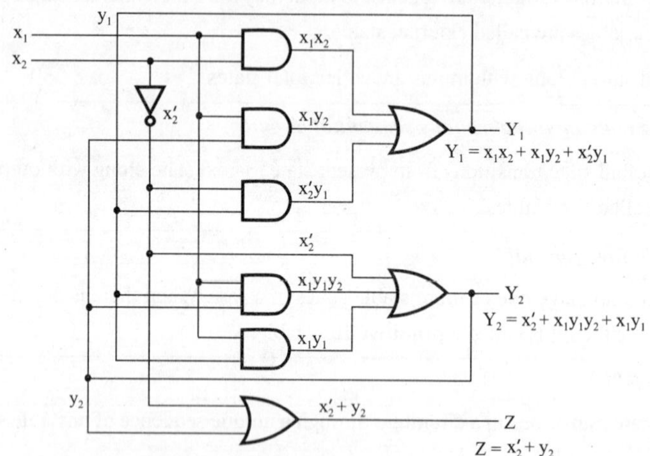

Fig. Q9.18: Logic diagram.

Q9.19 Write the Boolean equation for the given circuit.

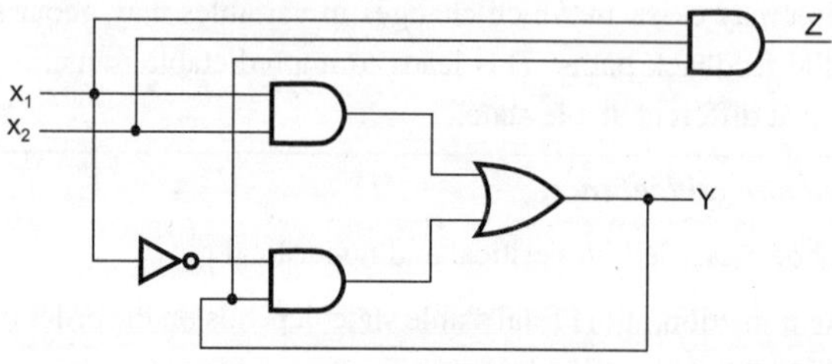

Fig. Q9.19.1: Logic circuit.

Solution:

From the circuit shown in Fig. Q9.19.2 we get the following Boolean equations for excitation variable and output of the asynchronous circuit.

$$Y = x_1x_2 + x_1'y$$

$$Z = x_1x_2y$$

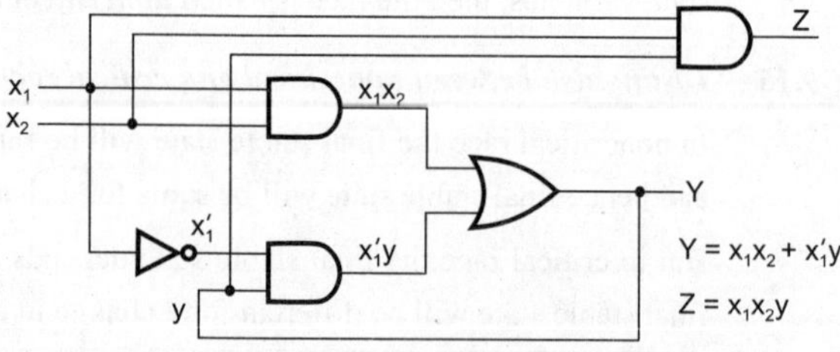

Fig. Q9.19.2: Logic circuit with intermediate signals.

Q9.20 Give an example for critical race.

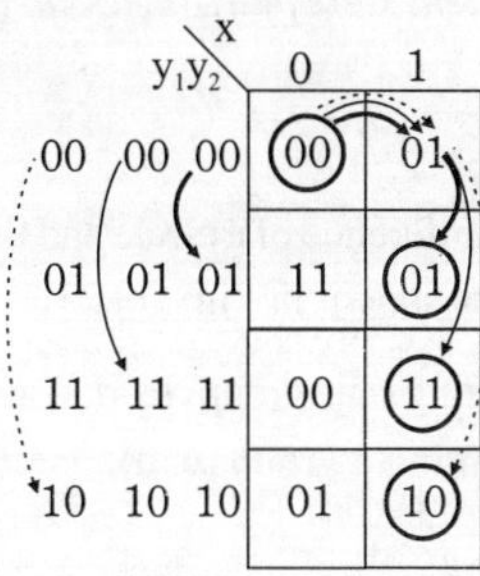

$00 \to 01$; $00 \to 11$; $00 \to 10$

Fig. Q9.20: *Critical race during state transition from 00.*

Q9.21 Define merger graph and merger table.

Merger graph : Merger graph is a tool used to reduce number of states in the incompletely specified machine. Merger graph is also known as merger diagram.

Merger table : The table in which each cell represents a compatible pair and their implications is called merger table. Merger table is also known as implication table.

Q9.22 Draw the transition diagram for the transition table shown in Fig. Q9.22.1.

y \ x_1x_2	00	01	11	10
a	(a)	b	c	(a)
b	a	(b)	(b)	c
c	a	(c)	(c)	(c)
d	(d)	a	c	(d)

Fig. Q9.22.1: *Transition table.*

Solution:

The following state transitions from present state to next state are observed in transition table of Fig. Q9.22.2.

From a	From b	From c	From d
$a \to b$	$b \to a$	$c \to a$	$d \to a$
$a \to c$	$b \to c$		$d \to c$

The above transitions are combined as a single diagram as shown in Fig. Q9.22.3 to get transition diagram.

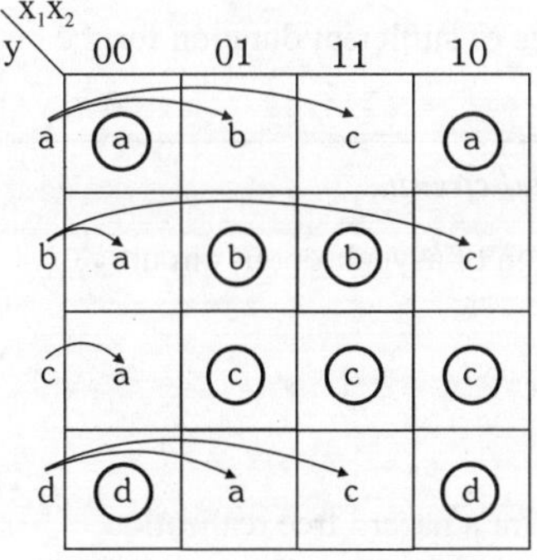

Fig. Q9.22.2: *Transition table.*

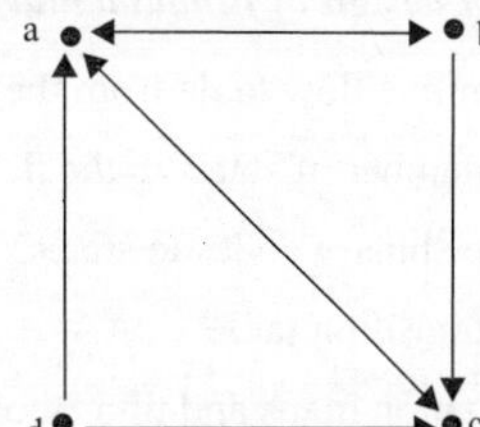

Fig. Q9.22.3: *Transition diagram.*

Q9.23 Draw the the merger diagram for the equivalent state pairs (a, d), (a, f), (b, d), (b, e), (c, d), (d, f), (d, e) and hence find the equivalent state groups.

Solution:

The states are represented by dots on the circumference of a circle and the equivalent pairs are joined by straight lines as shown in Fig. Q9.23.

The merger diagram has two triangles representing state groups (a, d, f) and (b, d, e). Also, there is a straight line representing a group (c, d).

∴ Equivalent state groups : (a, d, f), (b, d, e), (c, d)

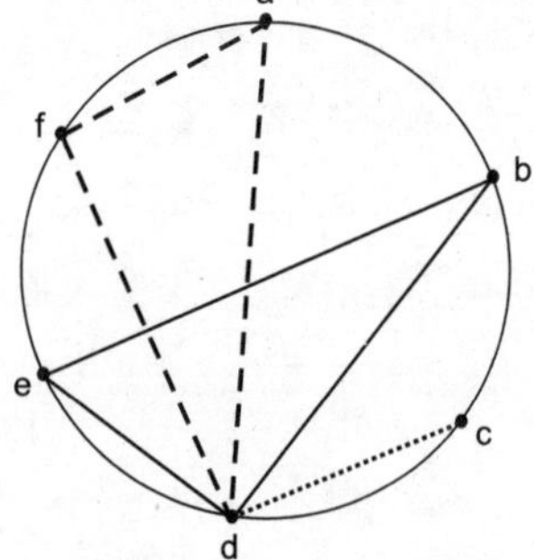

Fig. Q9.23: *Merger diagram.*

Q9.24 What are the types of asynchronous circuits?

There are two types of asynchronous sequential circuits.

1. Fundamental mode circuits. 2. Pulse mode circuits.

Q9.25 What is fundamental mode asynchronous sequential circuit?

In fundamental mode circuits the inputs are level triggered inputs and only one input is allowed to change at any one time when the circuit is in a stable state.

Q9.26 What is pulse mode sequential circuit?

In pulse mode asynchronous sequential circuits the inputs are pulses of suffieient duration for the output to change for a given input pulse. Here also, only one input is allowed to change at a time, but the input pulse will be available only for a short duration sufficient for transition from one stable state to another stable state.

Q9.27 Distinguish between fundamental mode asynchronous sequential circuit and pulse mode asynchronous sequential circuits.

In fundamental mode asynchronous sequential circuits the inputs are level triggered inputs, whereas in pulse mode asynchronous sequential circuits the inputs are pulse of sufficient duration for the output to change for a given input.

Q9.28 What are steps for design of fundamental mode sequential circuit.

1. Develop a primitive flow table from the word description of asynchronous circuit
2. Minimize the number of states in the flow table.
3. Assign race-free binary codes to states.
4. Construct the transition table.
5. Construct excitation maps and find Boolean equations for a hazard free realization.
6. Check for essential hazards.
7. Draw the logic circuit by implementing methods to eliminate essential hazards.

Q9.29 What are steps for design of pulse mode sequential circuit.

1. Develop a state diagram and hence state table from the system specifications.
2. If necessary minimize the number of states in the state table.
3. Assign race-free n-bit binary codes to states.
4. The n-bit code will have n state variables and so choose n flip-flops to the state variables.
5. Determine the flip-flop inputs for next state output using excitation table of flip-flop.
6. Draw maps for flip-flop inputs.
7. Form prime implicants by considering adjacent 1's in each column.
8. Determine the SOP form of Boolean equations for flip-flop inputs.
9. Draw the logic circuit using the Boolean equations.

9.7 Exercises

I. Fill in the blanks

1. The combinational logic circuits with feedback and without clock are called ______ circuits.
2. The present and next state variables of asynchronous sequential circuit are called ______ and ______ variables.
3. In asynchronous sequential circuit, the state table is called ______ .
4. A state is a circuit condition represented by a set of variables called ______ .
5. The continuous change between two unstable states is also called ______ .
6. When there is only one stable state in each row of a flow table then it is called a ______ .
7. The diagrammetic way of representing all possible state transitions between unstable states in every row of transition table is called ______ .
8. The race condition may be classified into ______ and ______ race.
9. In a state transition, if the final stable state depends on the order of change in state variables, then the race is called ______ .
10. In a state transition, if the final stable state does not depends on the order of change in state variables, then the race is called ______ .
11. When the state transitions in a circuit go through a unique sequence of unstable states then that condition is called a ______ .
12. In asynchronous sequential circuits, the state reduction is performed via an ______ .
13. State pairs that have same next state and output for the same input are ______ states.
14. State pairs having different next state and output for same input are ______ states.
15. The ______ is a group of compatible states that includes all the possible combinations of compatible states.
16. A minimum set of compatibles can be determined from ______ condition.
17. The ______ are used as memory element in asynchronous circuits.
18. Assigning adjacent code to states to avoid critical races is called ______.

Answers

1. asynchronous sequential	7. transition diagram	13. equivalent
2. secondary, excitation	8. critical, noncritical	14. not equivalent
3. transition table	9. critical race	15. maximum compatible
4. state variables	10. noncritical race	16. closed-covering
5. metastable	11. cycle	17. latches
6. primitive flow table	12. implication table	18. race-free assignment

II. State whether the following statements are True or False

1. The combinational logic circuits without feedback and with clock are called asynchronous sequential circuits.
2. The combination of output values and state variables are called total states of the circuit.
3. In state table of asynchronous circuit, at least one entry of each row should have same present state and next state.
4. When the state transitions in a circuit go through a unique sequence of stable states then that condition is called a cycle.
5. Races can also be avoided by using cycles, which terminate on a stable state.
6. Two binary codes are said to be adjacent codes if they differ only in one variable.
7. State pairs that have different next state and output for the same input are equivalent states.
8. State pairs having same next state and output for same input are not equivalent states.
9. In implication table each square represent a pair of state.
10. In asynchronous circuits with incompletely specified states, the states can be reduced by identifying compatible pairs.
11. The reduced flow table can be formed by taking one member of each maximum compatible set as a row in reduced flow table.
12. The latches are used as memory element in asynchronous circuits.
13. Races can also be avoided by using cycles, which terminate on a unstable state.
14. The closed-covering condition refers to set of compatibles that covers all the states and satisfy covering condition.

Answers

1. False	5. True	9. True	13. False
2. False	6. True	10. True	14. True
3. True	7. False	11. True	
4. False	8. False	12. True	

III. Choose the right answer for the following questions

1. *An asynchronous sequential circuit,*

   a) does not use clock pulses **b)** only one input can change at a time

   c) consists of combinational circuit and latches **d)** all the above

2. ***In an asynchronous sequential circuit, _______ number of inputs can change at a time.***

a) 1 **b)** 2 **c)** 3 **d)** n

3. ***The total states are formed with***

a) Inputs and outputs **b)** Outputs and state variables

c) Inputs and state variables **d)** Internal and external states

4. ***After a change in input the asynchronous circuit attains a stable condition when***

a) Internal state and output are equal **b)** Internal and external states are equal

c) Internal state and input are equal **d)** Input and output are equal

5. ***State pairs are said to be equivalent if the states have,***

a) Same next state and output for same input

b) Same next data and output for different inputs

c) Same present and next states for same input

d) Same present and next states for different inputs

6. ***We can design an asynchronous sequential circuit using***

a) Truth table b) Excitation table **c)** Primitive state table **d)** State table

7. ***Asynchronous sequential circuit can be designed by using***

a) T flip-flop **b)** JK flip-flop **c)** D flip-flop **d)** all the above

8. ***Races occur due to***

a) Clock input **b)** Pulse input

c) Delay in one feedback path **d)** Delay in more than two feedback path

9. ***In the fundamental mode asynchronous sequential circuit, the inputs are,***

a) Level triggered **b)** Positive edge triggered

c) Negative edge triggered **d)** Positive or negative edge triggered

10. ***In the pulse mode asynchronous sequential circuit, the inputs are***

a) Level triggered **b)** Positive edge triggered

c) Negative edge triggered **d)** Positive or negative edge triggered

Answers

1. d	6. c
2. a	7. d
3. c	8. d
4. b	9. a
5. a	10. d

V. Answer the following questions

1. What is the difference between flow table and transition table?
2. Define transition diagram.
3. What is state diagram? Give an example.
4. What are the significance of state assignment?
5. What is state reduction?
6. How races are avoided in digital circuits?
7. Define closed-covering condition
8. What is maximum compatible

VI. Solve the following problems

E9.1 The Boolean equations for excitation variable and output of an asynchronous circuit are,

$$Y = x_1x_2 + x_2'y$$

$$Z = x_2 + y'$$

Draw the logic diagram of the circuit. Derive the transition table and output map. Obtain a flow table for the circuit.

E9.2 The Boolean equations for excitation variables and output of an asynchronous circuit are,

$$Y_1 = x'y_1 + xy_2' \quad ; \quad Y_2 = xy_1' + x'y_2 \quad ; \quad Z = Y_2$$

Draw the logic diagram of the circuit. Derive the transition table and output map. Obtain a flow table for the circuit.

E9.3 Derive the flow table for the circuit shown in Fig. E9.3.

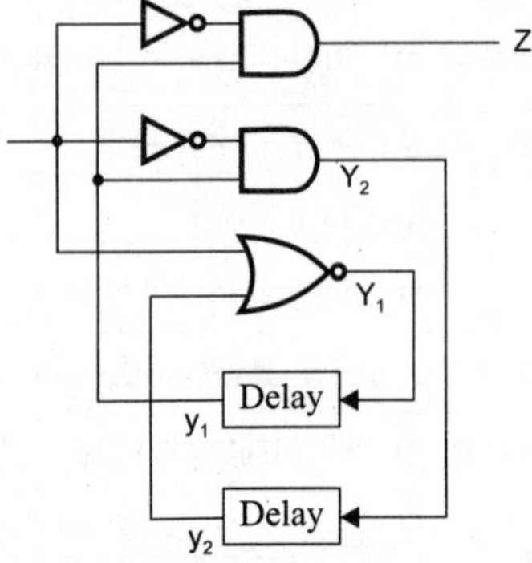

Fig. E9.3: Logic circuit.

E9.4 Investigate the transition table of Fig. E9.4, and determine all race conditions and whether they are critical or noncritical. Determine also whether there are any cycles.

E9.5 Investigate the transition table of Fig. E9.5, and determine all race conditions and whether they are critical or noncritical. Determine also whether there are any cycles.

y \ x_1x_2	00	01	11	10
a	(a)	b	b	(a)
b	a	(b)	(b)	a

Fig. E9.4: Transition table.

y \ x_1x_2	00	01	11	10
a	(a)	c	b	(a)
b	c	(b)	(b)	d
c	(c)	(c)	b	a
d	c	b	(d)	(d)

Fig. E9.5: Reduced flow table.

E9.6 **Reduce the state table shown in Table E9.6 to a minimum number of states using the implication table method.**

Table E9.6: State Table

Present State	Next State		Output	
	x = 0	x = 1	x = 0	x = 1
a	a	a	1	0
b	a	f	1	1
c	b	e	0	0
d	a	g	0	0
e	d	c	1	1
f	b	e	0	0
g	d	c	1	1

Table E9.7: State Table

Present State	Next State		Output	
	x = 0	x = 1	x = 0	x = 1
a	a	e	0	1
b	a	e	1	1
c	b	f	1	1
d	b	f	1	1
e	c	g	0	0
f	c	g	0	0
g	d	h	0	0
h	d	h	0	0

E9.7 **Reduce the state table shown in Table E9.7 to a minimum number of states using the implication table method.**

E9.8 **Merge the primitive flow table shown in Fig. E9.8. Find all compatible pairs by means of an implication table. Find the maximal compatibles by means of a merger diagram. Find a minimal set of compatibles that covers all the states and is closed.**

	00	01	11	10
a	–,–	f, –	(a),0	b, –
b	g, –	–,–	c, –	(b),1
c	–,–	h, –	(c),1	d, –
d	e, –	–,–	a, –	(d),0
e	(e),0	f, –	–,–	d, –
f	e, –	(f),0	a, –	–,–
g	(g),0	h, –	–,–	b, –
h	g, –	(h),1	c, –	–,–

Fig. E9.8: *Primitive flow table.*

	00	01	11	10
a	c, –	(a),0	b, –	–, –
b	–, –	a, –	(b),1	e, –
c	(c),0	a, –	–, –	d, –
d	c, –	–, –	b, –	(d),1
e	f, –	–, –	b, –	(e),1
f	(f),1	a, –	–, –	e, –

Fig. E9.9: *Primitive flow table.*

E9.9 **Merge the primitive flow table shown in Fig. E9.9. Find all compatible pairs by means of an implication table. Find the maximal compatibles by means of a merger diagram. Find a minimal set of compatibles that covers all the states and is closed.**

E9.10 **Find a critical race-free state assignment for the given transition table shown in Fig. E9.10.**

E9.11 **a) Obtain a binary state assignment for the reduced flow table shown in Fig. E9.11 to avoid critical race condition.**

b) Assign output to unstable states and obtain the output equation.

c) Implement the logic circuit using NOR based SR latch.

d) Implement the logic circuit using NAND based SR latch.

E9.12 **Consider the reduced flow table shown in Fig. E9.12.**

a) Obtain the transition diagram, and show that three state variables are needed for a race-free binary state assignment.

b) Obtain the expanded flow table, using the multiple row method of assignment.

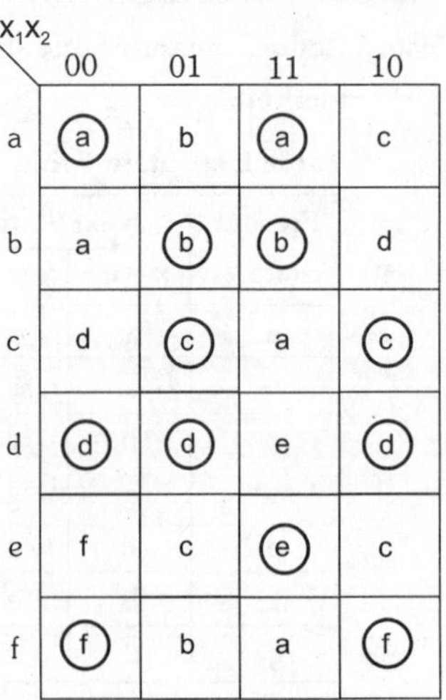

y \ x_1x_2	00	01	11	10
a	(a)	b	(a)	c
b	a	(b)	(b)	d
c	d	(c)	a	(c)
d	(d)	(d)	e	(d)
e	f	c	(e)	c
f	(f)	b	a	(f)

Fig. E9.10: Transition table.

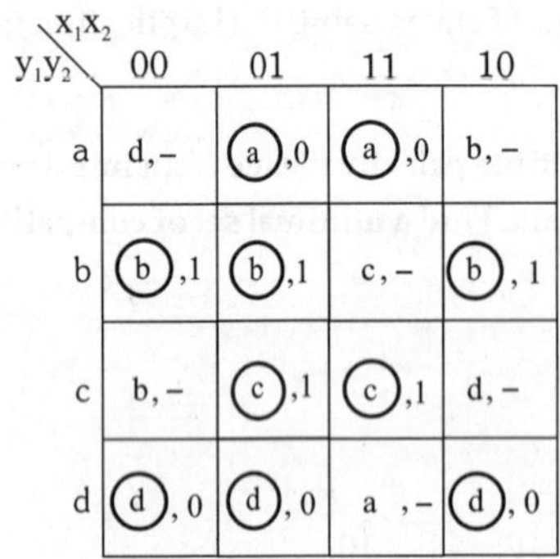

y_1y_2 \ x_1x_2	00	01	11	10
a	d, –	(a), 0	(a), 0	b, –
b	(b), 1	(b), 1	c, –	(b), 1
c	b, –	(c), 1	(c), 1	d, –
d	(d), 0	(d), 0	a, –	(d), 0

Fig. E9.11: Reduced flow table.

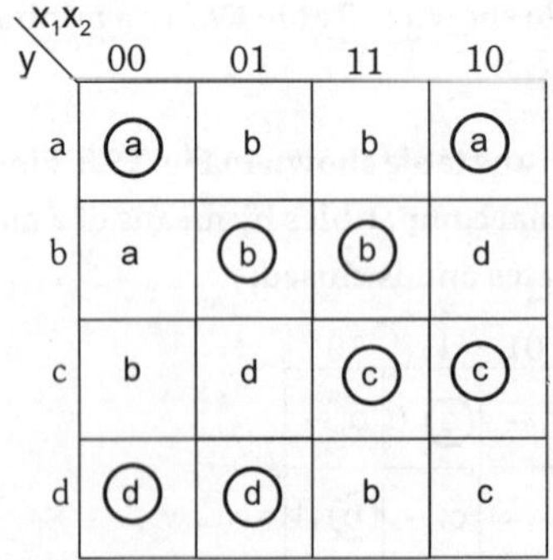

y \ x_1x_2	00	01	11	10
a	(a)	b	b	(a)
b	a	(b)	(b)	d
c	b	d	(c)	(c)
d	(d)	(d)	b	c

Fig. E9.12: Reduced flow table.

E9.13 **a) Obtain a binary state assignment for the reduced flow table shown in Fig. E9.13 to avoid critical race condition.**

b) Assign output to unstable states and obtain the output equation.

c) Implement the logic diagram using NOR based SR latch.

d) Implement the logic diagram using NAND based SR latch.

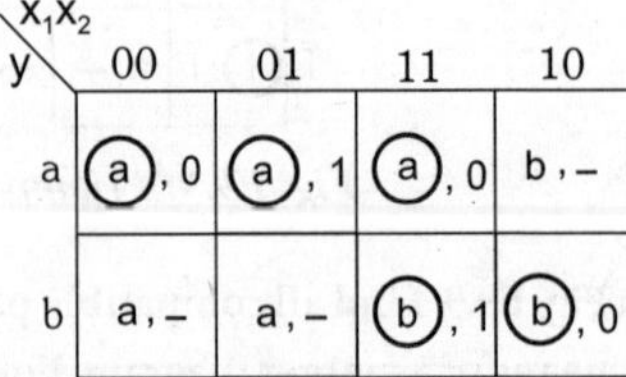

y \ x_1x_2	00	01	11	10
a	(a), 0	(a), 1	(a), 0	b, –
b	a, –	a, –	(b), 1	(b), 0

Fig. E9.13: Reduced flow table.

E9.14 **The Boolean functions for the inputs of an SR latch are,**

$$S = x_1' x_2' x_3 + x_1x_2x_3 \quad ; \quad R = x_1x_2' + x_2x_3'$$

Obtain the circuit diagram using only NAND gates.

E9.15 Implement the circuit defined in Fig. E9.15 with NOR SR latch. Repeat with NAND SR latch.

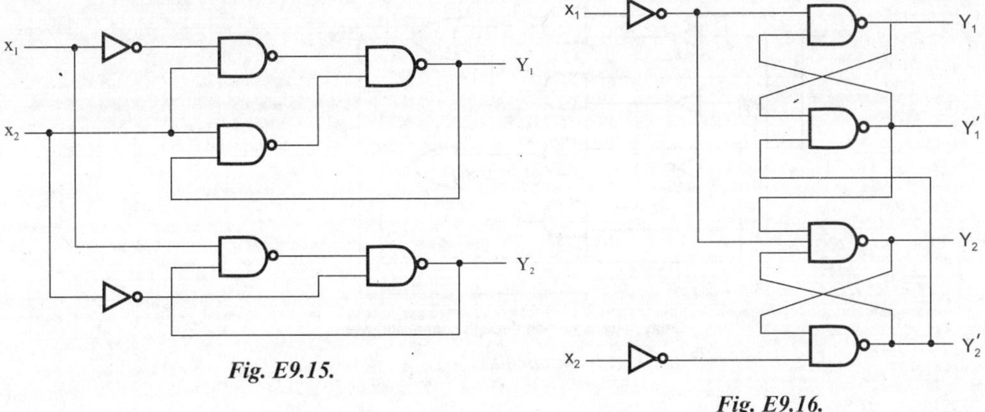

Fig. E9.15.

Fig. E9.16.

E9.16 For the asynchronous sequential circuit shown in Fig. E9.16. Derive the Boolean functions for the outputs of the two SR latches Y_1 and Y_2.

E9.17 Design a pulse mode asynchronous sequential circuit with two inputs x_1 and x_2 and an output Z such that output pulse is produced only when an input sequence x_1-x_1-x_2 is detected and output is 0 for any other input sequence. Implement the logic circuit using D flip-flop.

E9.18 Design a pulse mode asynchronous sequential circuit with three inputs x_1, x_2 and x_3 and an output Z such that output pulse is produced only when an input sequence x_1-x_3-x_2 is detected and output is 0 for any other input sequence. Implement the logic circuit using D flip-flop.

Answers

E9.1

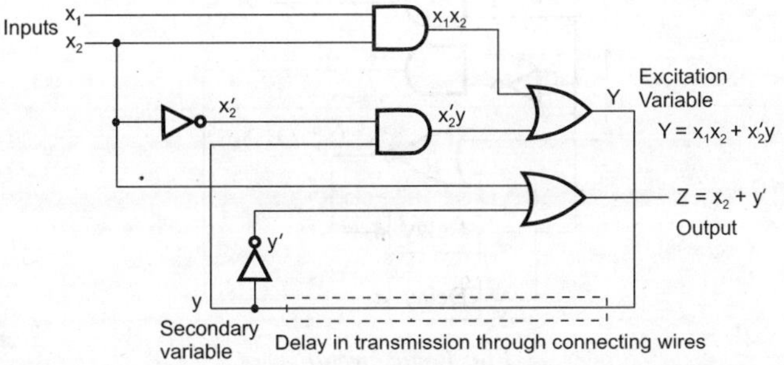

Inputs x_1, x_2; x_1x_2; x_2'; $x_2'y$; Excitation Variable Y; $Y = x_1x_2 + x_2'y$; $Z = x_2 + y'$ Output; y'; y Secondary variable; Delay in transmission through connecting wires

Fig: Logic circuit.

y \ x_1x_2	00	01	11	10
0	(0)	(0)	1	(0)
1	(1)	0	(1)	(1)

Fig: Transition table.

y \ x_1x_2	00	01	11	10
0	1	1	1	1
1	0	1	1	0

Fig: Output map.

y \ x_1x_2	00	01	11	10
a	(a),1	(a),1	b,1	(a),1
b	(b),0	a, 1	(b),1	(b),0

Fig: Flow table.

E9.2

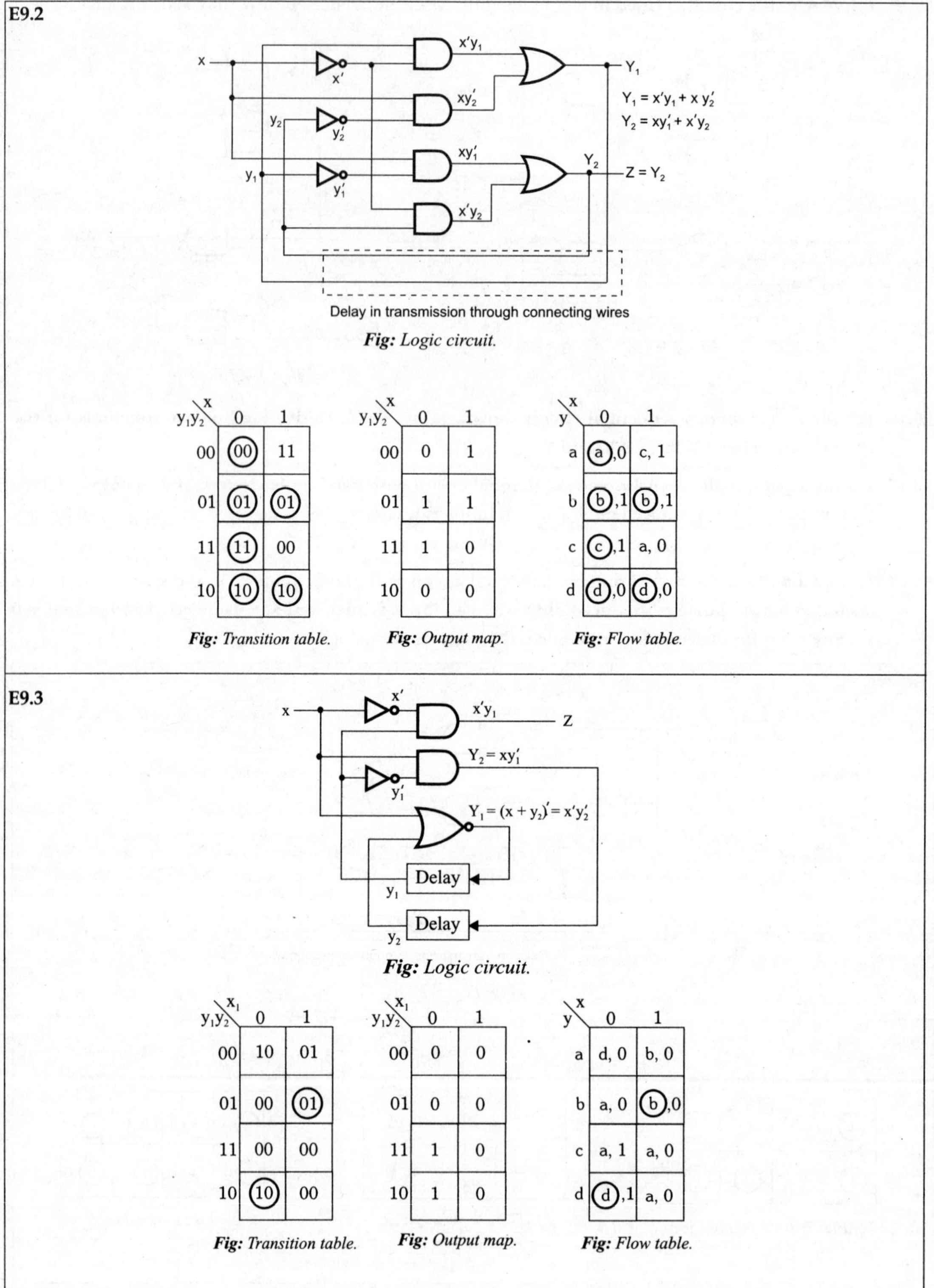

***Fig:** Logic circuit.*

y_1y_2 \ x	0	1
00	00	11
01	01	01
11	11	00
10	10	10

***Fig:** Transition table.*

y_1y_2 \ x	0	1
00	0	1
01	1	1
11	1	0
10	0	0

***Fig:** Output map.*

y \ x	0	1
a	a,0	c, 1
b	b,1	b,1
c	c,1	a, 0
d	d,0	d,0

***Fig:** Flow table.*

E9.3

***Fig:** Logic circuit.*

y_1y_2 \ x_1	0	1
00	10	01
01	00	01
11	00	00
10	10	00

***Fig:** Transition table.*

y_1y_2 \ x_1	0	1
00	0	0
01	0	0
11	1	0
10	1	0

***Fig:** Output map.*

y \ x	0	1
a	d, 0	b, 0
b	a, 0	b,0
c	a, 1	a, 0
d	d,1	a, 0

***Fig:** Flow table.*

E9.4 No races and cycles.

Table 1: List of State Transitions

Present Input	Present Stable State	Change in Input	State Transitions
00	0	00 → 01	0 → 1
		00 → 10	0 → 0
01	1	01 → 00	1 → 0
		01 → 11	1 → 1
11	1	11 → 01	1 → 1
		11 → 10	1 → 0
10	0	10 → 11	0 → 1
		10 → 00	0 → 0

E9.5 **Table 1: List of Races and Cycles**

Present Input	Present Stable State	Change in Input	State Transitions	Races	Cycles
00	00	00 → 01	00 → 01 00 → 10 → 11	Critical	- 00 → 10 → 11
		00 → 10	00 → 00	-	-
00	11	00 → 01	11 → 10 11 → 01 → 00	Critical	- 11 → 01 → 00
		00 → 10	11 → 11	-	-
01	01	01 → 00	01 → 00 01 → 11	Critical	- -
		01 → 11	01 → 01	-	-
01	11	00 → 00	11 → 11	-	-
		00 → 11	11 → 10 11 → 01	Critical	- -
11	01	11 → 01	01 → 00 01 → 11 → 10	Critical	- 01 → 11 → 10
		11 → 10	01 → 01	-	-
11	10	11 → 01	10 → 11 10 → 00 → 01	Critical	- 10 → 00 → 01
		11 → 10	10 → 10	-	-
10	00	10 → 00	00 → 00	-	-
		10 → 11	00 → 01 00 → 10	Critical	- -
10	10	10 → 00	10 → 11 10 → 00	Critical	- -
		10 → 11	10 → 10	-	-

E9.6

Table: Reduced State Table

Present State	Next State		Output	
	x = 0	x = 1	x = 0	x = 1
a	a	a	1	0
b	a	c (f → c)	1	1
c	b	e	0	0
d	a	e (g → e)	0	0
e	d	c	1	1

E9.7

Table: Reduced State Table

Present State	Next State		Output	
	x = 0	x = 1	x = 0	x = 1
a	a	e	0	1
b	a	e	1	1
c	b	f (f → e)	1	1
e	c	e (g → e)	0	0

E9.8 The compatible pairs that satisfy closed-covering condition are, (a, f), (b, g), (c, h), (d, e)

	00	01	11	10
a, f	e, –	(f),0	(a),0	b, –
b, g	(g),0	h, –	c, –	(b),1
c, h	g, –	(h),1	(c),1	d, –
d, e	(e),0	f, –	a, –	(d),0

Fig. a.

	00	01	11	10
a	d, – (e→d)	(a),0 (f→a)	(a),0	b, –
b	(b),0 (g→b)	c, – (h→c)	c, –	(b),1
c	b, – (g→b)	(c),1 (h→c)	(c),1	d, –
d	(d),0 (e→d)	a, – (f→a)	a, –	(d),0

Fig. b.

Fig: Reduced flow table.

E9.9 The compatible pairs that satisfy closed-covering condition are, (a, c, d), (b, e, f)

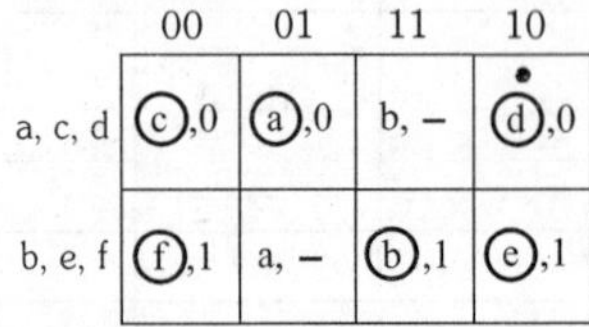

	00	01	11	10
a, c, d	(c),0	(a),0	b, –	(d),0
b, e, f	(f),1	a, –	(b),1	(e),1

Fig. a.

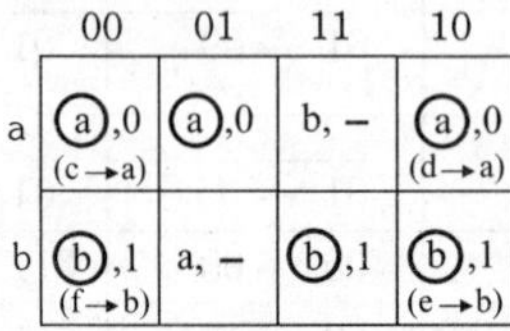

	00	01	11	10
a	(a),0 (c→a)	(a),0	b, –	(a),0 (d→a)
b	(b),1 (f→b)	a, –	(b),1	(b),1 (e→b)

Fig. b.

Fig: Reduced flow table.

E9.10

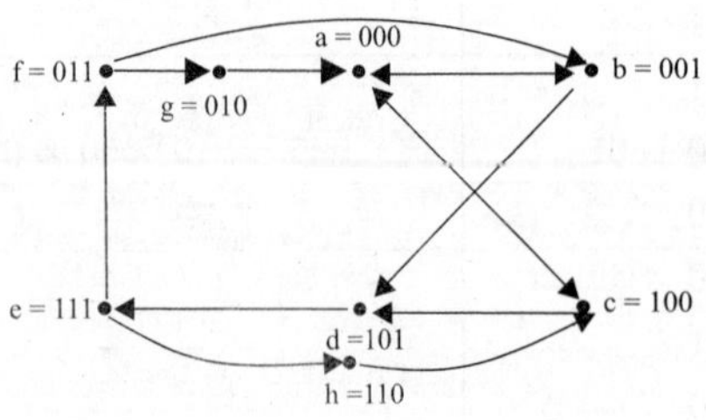

Fig: Transition diagram with binary assignment.

$y_3y_2y_1$ \ x_1x_2	00	01	11	10
000 = a	(a)	b	(a)	c
001 = b	a	(b)	(b)	d
011 = f	(f)	b	g	(f)
010 = g	–	–	a	–
110 = h	–	c	–	c
111 = e	f	h	(e)	h
101 = d	(d)	(d)	e	(d)
100 = c	d	(c)	a	(c)

Fig: Modified flow table with race free state assignment.

E9.11

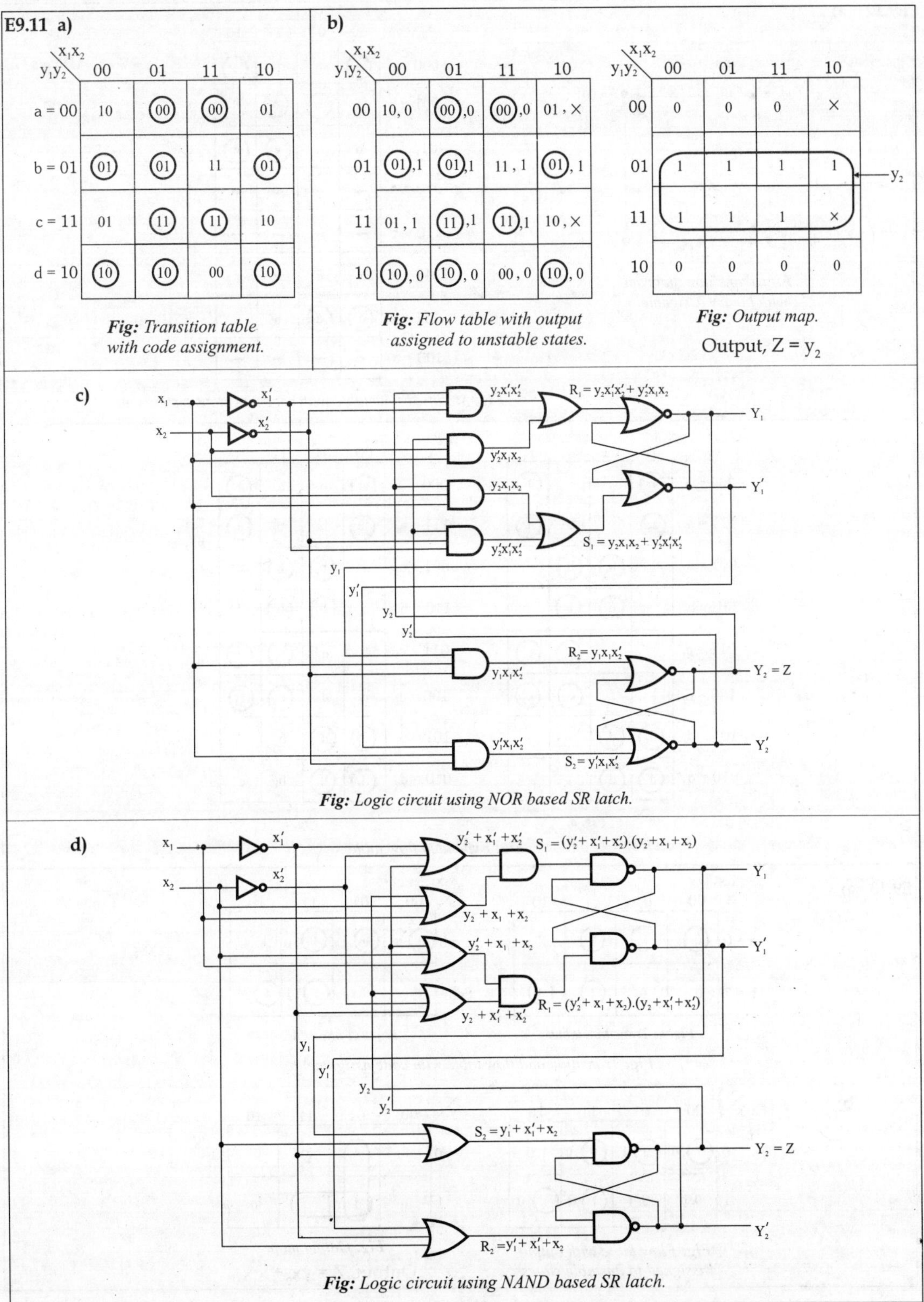

Fig: *Transition table with code assignment.*

Fig: *Flow table with output assigned to unstable states.*

Fig: *Output map.*

Fig: *Logic circuit using NOR based SR latch.*

Fig: *Logic circuit using NAND based SR latch.*

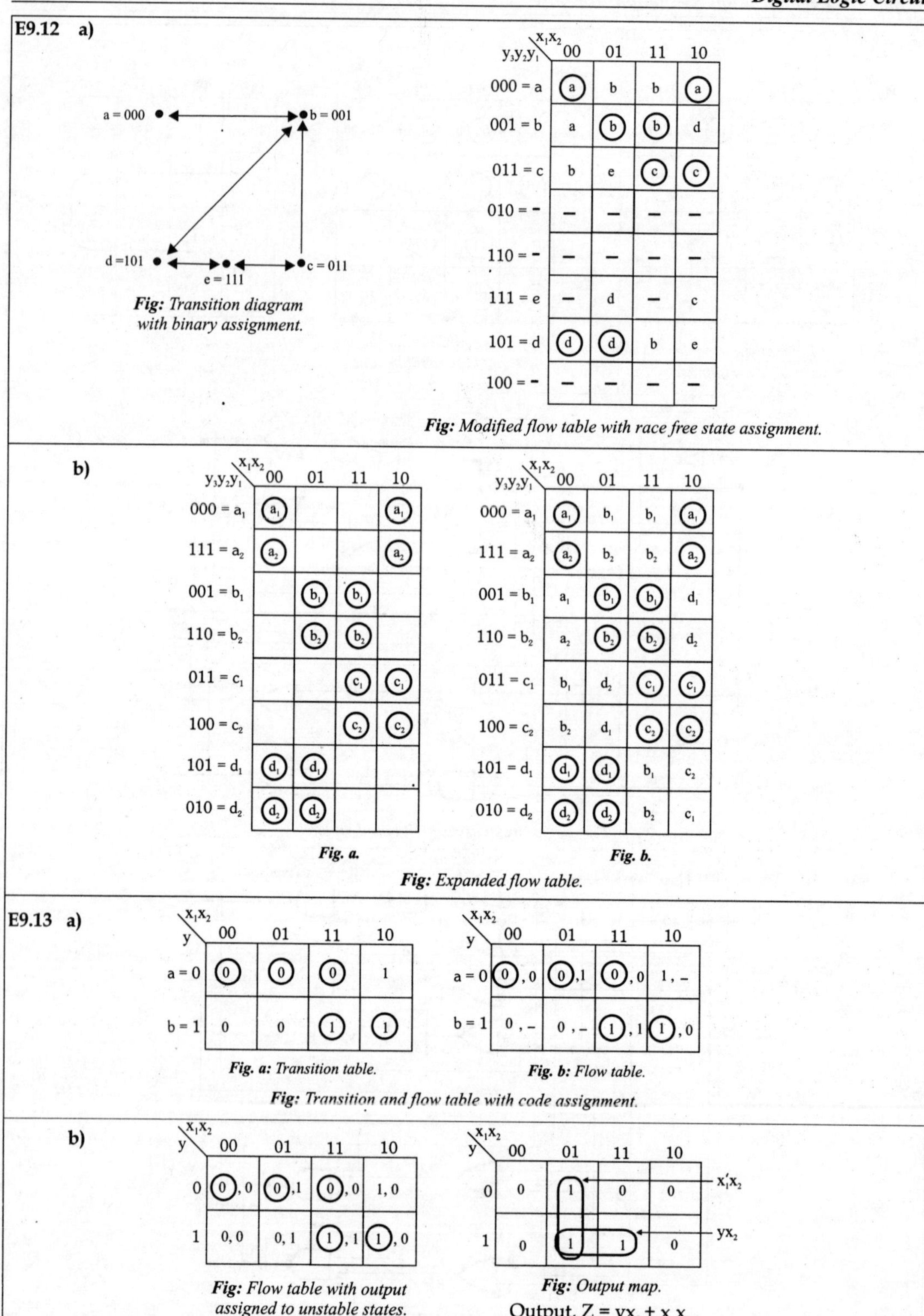

E9.12 a)

Fig: Transition diagram with binary assignment.

$y_3y_2y_1$ \ x_1x_2	00	01	11	10
000 = a	(a)	b	b	(a)
001 = b	a	(b)	(b)	d
011 = c	b	e	(c)	(c)
010 = –	–	–	–	–
110 = –	–	–	–	–
111 = e	–	d	–	c
101 = d	(d)	(d)	b	e
100 = –	–	–	–	–

Fig: Modified flow table with race free state assignment.

b)

$y_3y_2y_1$ \ x_1x_2	00	01	11	10
000 = a_1	(a_1)			(a_1)
111 = a_2	(a_2)			(a_2)
001 = b_1		(b_1)	(b_1)	
110 = b_2		(b_2)	(b_2)	
011 = c_1			(c_1)	(c_1)
100 = c_2			(c_2)	(c_2)
101 = d_1	(d_1)	(d_1)		
010 = d_2	(d_2)	(d_2)		

Fig. a.

$y_3y_2y_1$ \ x_1x_2	00	01	11	10
000 = a_1	(a_1)	b_1	b_1	(a_1)
111 = a_2	(a_2)	b_2	b_2	(a_2)
001 = b_1	a_1	(b_1)	(b_1)	d_1
110 = b_2	a_2	(b_2)	(b_2)	d_2
011 = c_1	b_1	d_2	(c_1)	(c_1)
100 = c_2	b_2	d_1	(c_2)	(c_2)
101 = d_1	(d_1)	(d_1)	b_1	c_2
010 = d_2	(d_2)	(d_2)	b_2	c_1

Fig. b.

Fig: Expanded flow table.

E9.13 a)

y \ x_1x_2	00	01	11	10
a = 0	(0)	(0)	(0)	1
b = 1	0	0	(1)	(1)

***Fig. a:** Transition table.*

y \ x_1x_2	00	01	11	10
a = 0	(0), 0	(0), 1	(0), 0	1, –
b = 1	0, –	0, –	(1), 1	(1), 0

***Fig. b:** Flow table.*

Fig: Transition and flow table with code assignment.

b)

y \ x_1x_2	00	01	11	10
0	(0), 0	(0), 1	(0), 0	1, 0
1	0, 0	0, 1	(1), 1	(1), 0

Fig: Flow table with output assigned to unstable states.

y \ x_1x_2	00	01	11	10
0	0	1	0	0
1	0	1	1	0

Fig: Output map.

Output, $Z = yx_2 + x_1x_2$

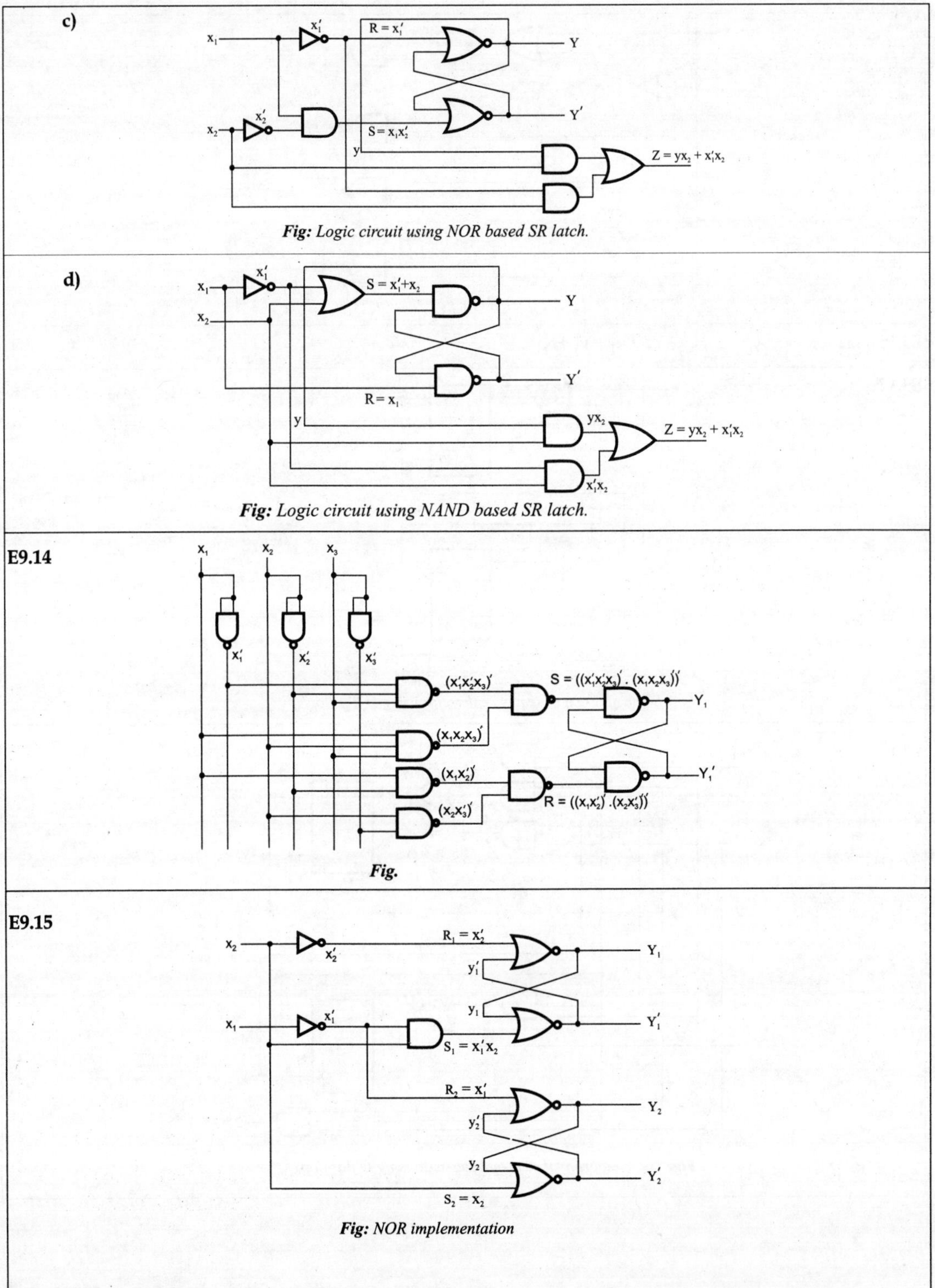

Fig: *Logic circuit using NOR based SR latch.*

Fig: *Logic circuit using NAND based SR latch.*

Fig.

Fig: *NOR implementation*

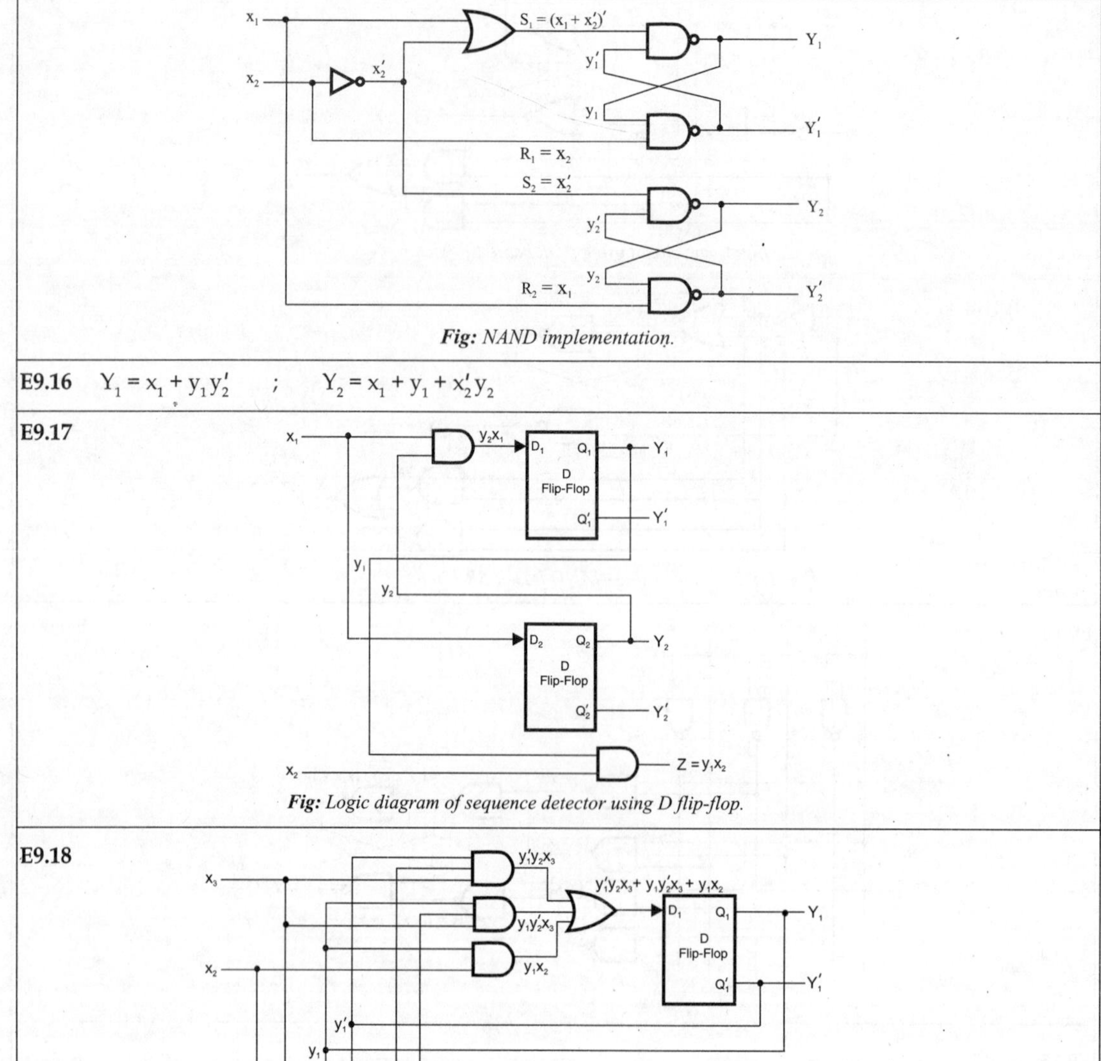

Fig: *NAND implementation.*

E9.16 $Y_1 = x_1 + y_1 y_2'$; $Y_2 = x_1 + y_1 + x_2' y_2$

E9.17

Fig: *Logic diagram of sequence detector using D flip-flop.*

E9.18

Fig: *Logic diagram of sequence detector using D flip-flop.*

CHAPTER 10

HAZARDS

10.1 Hazards

Hazards are unwanted transient outputs in the digital circuits that may arise due to gates and connecting wires with different propagation delay.

The different types of hazards are static-1 hazard, static-0 hazard and dynamic hazard.

Static-1 hazard: When the expected output is 1 for a change in input, if the circuit generates a temporary 0 as shown in Fig. 10.1(a) then the hazard is called static-1 hazard.

Static-0 hazard: When the expected output is 0 for a change in input, if the circuit generates a temporary 1 as shown in Fig. 10.1(b) then the hazard is called static-0 hazard.

Dynamic hazard: When the expected output is either 1 or 0 for a change in input, if the circuit output temporarily changes from 0 to 1 and 1 to 0 three or more times as shown in Fig. 10.1(c), then the hazard is called dynamic hazard.

1 0 1 0 1 0 1 0

Fig. a: *Static-1 hazard.* ***Fig. b:*** *Static-0 hazard.* ***Fig. c:*** *Dynamic hazard.*

Fig. 10.1: *Types of hazards.*

Hazards may occur in combinational circuits as well as asynchronous sequential circuits. The synchronous sequential circuits will not have hazards because in those circuits the changes are synchronized with a clock.

10.2 Hazards in Combinational Circuits

In combinational circuits the output may have one of the following possible values for a change in input.

- The output may change from 1 to 0
- The output may change from 0 to 1
- The output may remain in previous value, either 1 or 0.

When the combinational circuit has a hazard some unwanted temporary output are generated than one of the expected output listed above.

10.2.1 Static-0 and Static-1 Hazards

Some examples of combinational circuits with static-0 and static-1 hazards are shown in Figs. 10.2, 10.4, 10.6 and 10.8. In all these examples the hazards are due to longer propagation delay in inverter or NOT gate. In each example, circuit with and without hazard are drawn and change in signals are shown in a timing diagram. Also, the signal values at various timings are listed in a table.

In the example shown in Fig. 10.2, initially all the inputs are 1 and at time, t_2 the input x_2 changes from 1 to 0. When the inverter or NOT gate has negligible delay the signal propagation through gates 2 and 3 will be same as in the path through gates 1 and 3 and so the output will not have hazard. On the other hand, if the inverter or NOT gate has longer delay then the signal propagation through the two different paths will take different timings which creates a temporary 0 output as shown in Fig. 10.2(b). The signal timings are shown in Fig. 10.3 and values of signals at various timings are listed in Table 10.1.

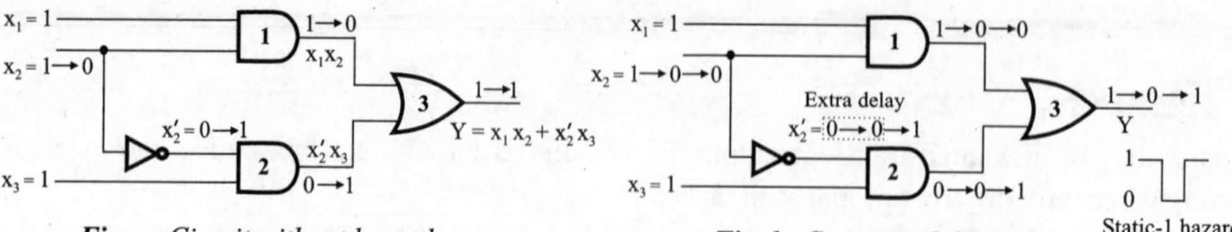

Fig. a: Circuit without hazard. *Fig. b: Circuit with hazard.*

***Fig. 10.2:** Example of combinational circuit with static-1 hazard.*

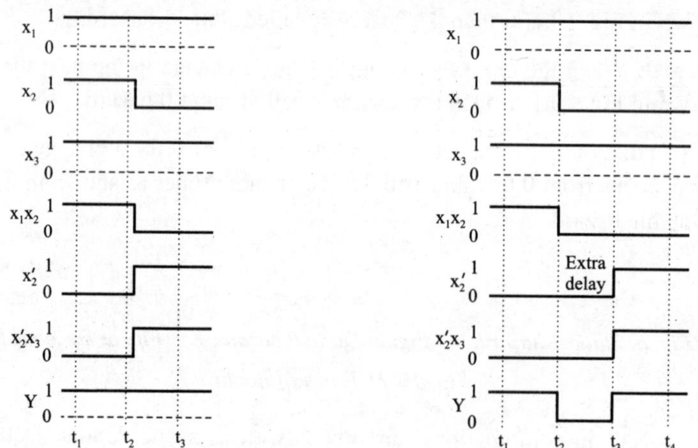

Fig. a: When inverter has negligible delay. *Fig. b: When inverter has longer delay.*

***Fig. 10.3:** Signal timings in the circuit of Fig. 10.2.*

Table 10.1: Signal Values at Various Timings in Circuit of Fig. 10.2

	Signal Values at Various Timings						
	When Inverter has Negligible Delay			When Inverter has Longer Delay			
	t_1	t_2	t_3	t_1	t_2	t_3	t_4
x_1	1	1	1	1	1	1	1
x_2	1	0	0	1	0	0	0
x_3	1	1	1	1	1	1	1
$x_1 x_2$	1	0	0	1	0	0	0
x_2'	0	1	1	0	0	1	1
$x_2' x_3$	0	1	1	0	0	1	1
Y	1	1	1	1	0	1	1

In the example shown in Fig. 10.4, initially all the inputs are 0 and at time, t_2 the input x_2 changes from 0 to 1. When the inverter or NOT gate has negligible delay the signal propagation through gates 1 and 3 will be same as in the path through gates 2 and 3 and so the output will not have hazard. On the other hand, if the inverter or NOT gate has longer delay then the signal propagation through the two different paths will take different timings which creates a temporary 1 output as shown in Fig. 10.4(b). The signal timings are shown in Fig. 10.5 and values of signals at various timings are listed in Table 10.2.

(AKTU, Dec'22, 3 Marks)

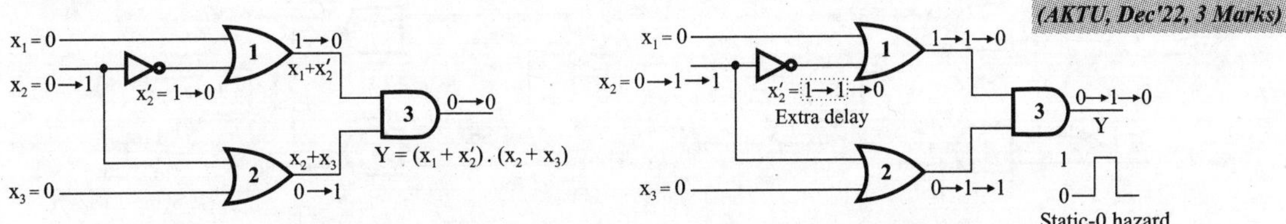

Fig. a: Circuit without hazard. *Fig. b: Circuit with hazard.*

***Fig. 10.4:** Example of combinational circuit with static-0 hazard.*

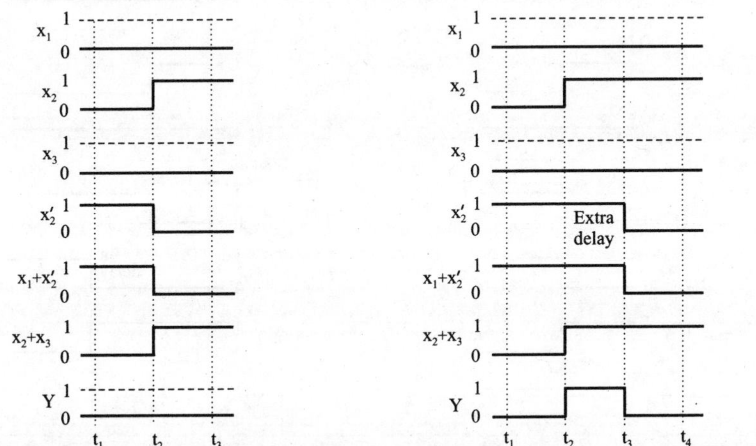

Fig. a: When inverter has negligible delay. *Fig. b: When inverter has longer delay.*

***Fig. 10.5:** Signal timings in the circuit of Fig. 10.4.*

Table 10.2: Signal Values at Various Timings in Circuit of Fig. 10.4

	Signal Values at Various Timings						
	When Inverter has Negligible Delay			When Inverter has Longer Delay			
	t_1	t_2	t_3	t_1	t_2	t_3	t_4
x_1	0	0	0	0	0	0	0
x_2	0	1	1	0	1	1	1
x_3	0	0	0	0	0	0	0
x_2'	1	0	0	1	1	0	0
$x_1 + x_2'$	1	0	0	1	1	0	0
$x_2 + x_3$	0	1	1	0	1	1	1
Y	0	0	0	0	1	0	0

In the example shown in Fig. 10.6, initially all the inputs are 1 and at time, t_2 the input x_2 changes from 1 to 0. When the inverter or NOT gate has negligible delay the signal propagation through gates 2 and 3 will be same as in the path through gates 1 and 3 and so the output will not have hazard. On the other hand, if the inverter or NOT gate has longer delay then the signal propagation through the two different paths will take different timings which creates a temporary 0 output as shown in Fig. 10.6(b). The signal timings are shown in Fig. 10.7 and values of signals at various timings are listed in Table 10.3.

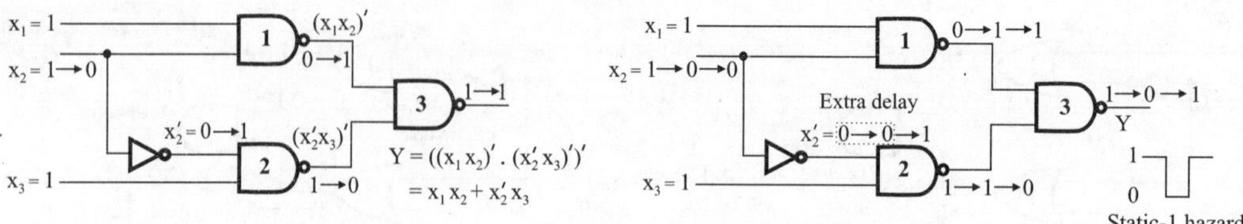

Fig. a: Circuit without hazard. *Fig. b: Circuit with hazard.*

Fig. 10.6: Example of combinational circuit with static-1 hazard.

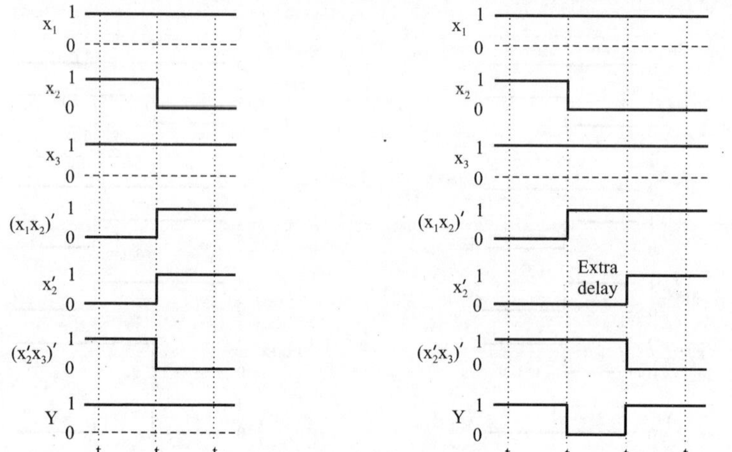

Fig. a: When inverter has negligible delay. *Fig. b: When inverter has longer delay.*

Fig. 10.7: Signal timings in the circuit of Fig. 10.6.

Table 10.3: Signal Values at Various Timings in Circuit of Fig. 10.6

	Signal Values at Various Timings						
	When Inverter has Negligible Delay			When Inverter has Longer Delay			
	t_1	t_2	t_3	t_1	t_2	t_3	t_4
x_1	1	1	1	1	1	1	1
x_2	1	0	0	1	0	0	0
x_3	1	1	1	1	1	1	1
$(x_1 x_2)'$	0	1	1	0	1	1	1
x_2'	0	1	1	0	0	1	1
$(x_2' x_3)'$	1	0	0	1	1	0	0
Y	1	1	1	1	0	1	1

In the example shown in Fig. 10.8, initially all the inputs are 0 and at time, t_2 the input x_2 changes from 0 to 1. When the inverter or NOT gate has negligible delay the signal propagation through gates 1 and 3 will be same as in the path through gates 2 and 3 and so the output will not have hazard. On the other hand, if the inverter or NOT gate has longer delay then the signal propagation through the two different paths will take different timings which creates a temporary 0 output as shown in Fig. 10.8(b). The signal timings are shown in Fig. 10.9 and values of signals at various timings are listed in Table 10.4.

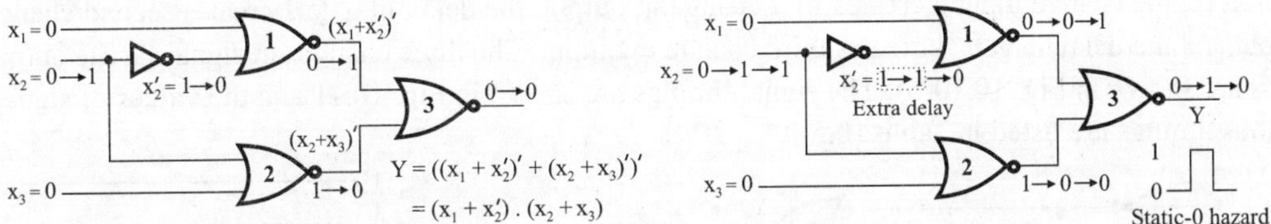

Fig. a: *Circuit without hazard.* ***Fig. b:*** *Circuit with hazard.*

Fig. 10.8: *Example of combinational circuit with static-0 hazard.*

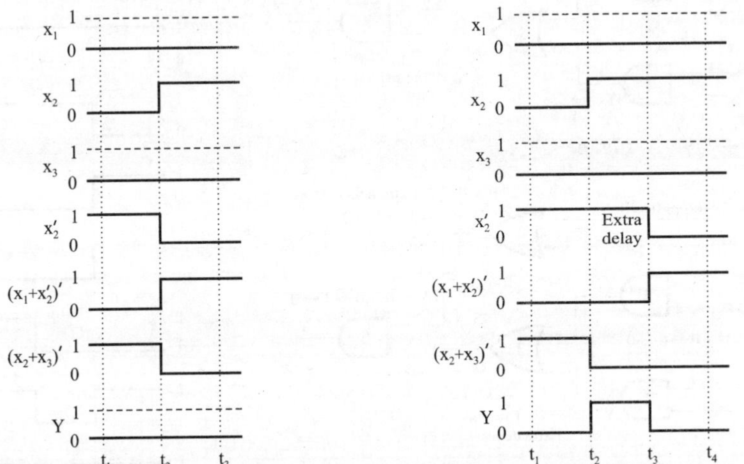

Fig. a: *When inverter has negligible delay.* ***Fig. b:*** *When inverter has longer delay.*

Fig. 10.9: *Signal timings in the circuit of Fig. 10.8.*

Table 10.4: Signal Values at Various Timings in Circuit of Fig. 10.8

	Signal Values at Various Timings						
	When Inverter has Negligible Delay			**When Inverter has Longer Delay**			
	t_1	t_2	t_3	t_1	t_2	t_3	t_4
x_1	0	0	0	0	0	0	0
x_2	0	1	1	0	1	1	1
x_3	0	0	0	0	0	0	0
x_2'	1	0	0	1	1	0	0
$(x_1 + x_2')'$	0	1	1	0	0	1	1
$(x_2 + x_3)'$	1	0	0	1	0	0	0
Y	0	0	0	0	1	0	0

10.2.2 Dynamic Hazard

Dynamic hazard occurs only if there are three or more paths between input and output. An example of dynamic hazard with three paths between input and output is shown in Fig. 10.10.

In the example shown in Fig. 10.10, initially a and b are 1 and c is 0. At time, t_2 the input c changes from 0 to 1. Here, the paths through gates 2 and 4 has delay and the delay in gate 4 is longer than gate 2. Therefore, the change input c creates first change in output, the delay in gate 2 creates second change in output and the delay in gate 4 creates third change in output. The three changes in signals at the output of gates are shown in Fig. 10.10(b). The signal timings are shown in Fig. 10.11 and the values of signals at various timings are listed in Table 10.5.

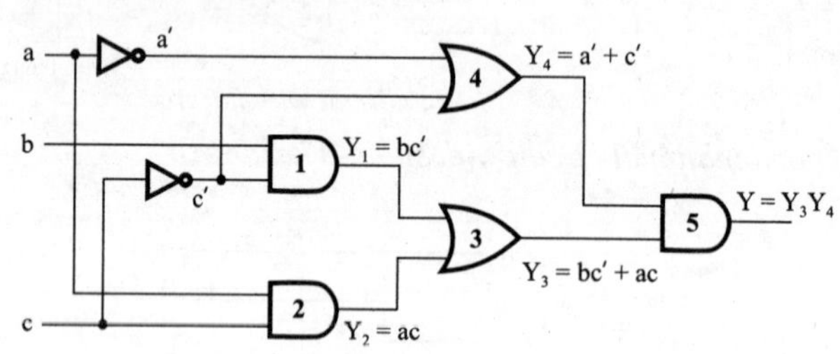

Fig. a.

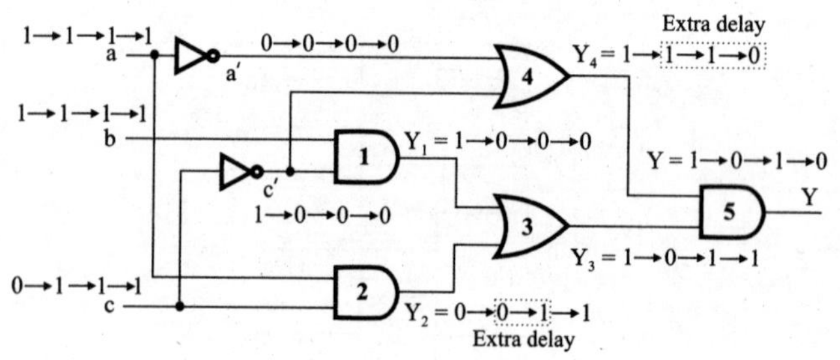

Fig. b.

Fig. 10.10: Example of dynamic hazard.

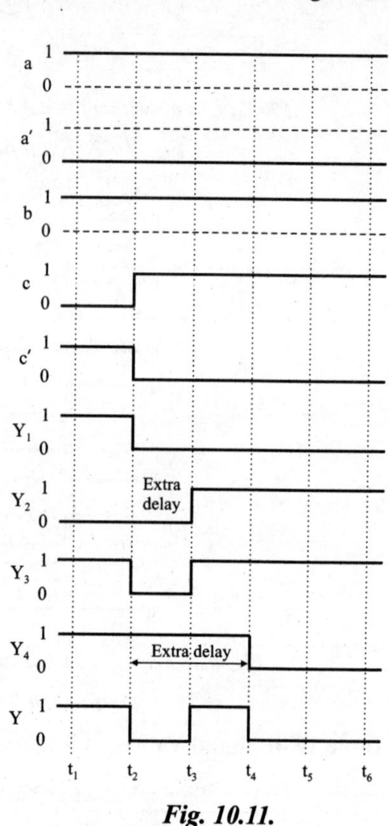

Fig. 10.11.

Table 10.5:

	t_1	t_2	t_3	t_4	t_5	t_6
a	1	1	1	1	1	1
a'	0	0	0	0	0	0
b	1	1	1	1	1	1
c	0	1	1	1	1	1
c'	1	0	0	0	0	0
Y_1	1	0	0	0	0	0
Y_2	0	0	1	1	1	1
Y_3	1	0	1	1	1	1
Y_4	1	1	1	0	0	0
Y	1	0	1	0	0	0

10.3 Hazards in Sequential Circuits

In synchronous sequential circuits the changes in circuit conditions are synchronized by a clock and the clock time period is selected long enough to overcome internal delays in signal transitions. Hence hazards will not occur in synchronous sequential circuits.

In asynchronous sequential circuits the internal changes will occur immediately after an input change and so if there are different delays in various signal paths then hazards will occur in asynchronous sequential circuits. The static and dynamic hazards in asynchronous sequential circuits can be eliminated similar to combinational circuit by adding redundant gates or by using SR latch based circuit design.

There is another type of hazard in asynchronous sequential circuit known as essential hazard. The essential hazard is due to different delays in different paths that start from a same input.

10.4 Essential Hazards

An essential hazard is due to unequal delay in two or more paths that start from same input and also due to delay in input path longer than delay in feedback path. Therefore, due to essential hazard, the feedback signals will reach the input of excitation circuit much before the changes in input reach the input of excitation circuit. Hence feedback inputs will change two or more times before reaching a stable value.

Alternatively, we can say that an essential hazard exist if the total stable state reached after three changes in an input is different from stable state reached after a single change in that input.

The essential hazard cannot be eliminated by redundant gate. In order to eliminate essential hazard each path has to be studied individually and additional delay is added at appropriate path.

Identifying Essential Hazards

The presence of essential hazard can be identified from the flow table. The essential hazard exists if there are multiple paths for state transitions due to change in input. An example of flow table with possibilities of essential hazard is shown Fig. 10.12.

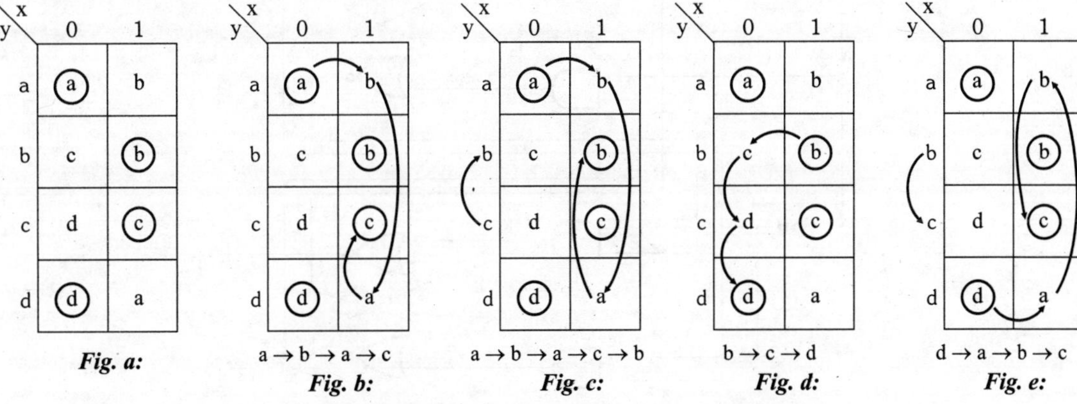

Fig. a: *Fig. b:* *Fig. c:* *Fig. d:* *Fig. e:*

Fig. 10.12: Flow table with essential hazards.

Example of Essential Hazard

Consider the following excitation functions of an asynchronous circuit.

$$Y_1 = xy_2' + (x' + y_2')y_1$$

$$Y_2 = x'y_1 + (x + y_1)y_2$$

Here, y_1, y_2 = Present state variables

Y_1, Y_2 = Next state varaibles

x = Input

The circuit has two state variables Y_1 and Y_2 and so let us choose two NOR based SR latches to store Y_1 and Y_2.

Consider the simplified excitation function of NOR based SR latch, $Y = S + R'y$

Now the excitation function of two NOR based SR latches are,

$$Y_1 = S_1 + R_1' y_1$$

$$Y_2 = S_2 + R_2' y_2$$

On comparing the excitation functions of the circuit and SR latches, we get the following equations for SR latch inputs.

$S_1 = xy_2'$	$S_2 = x'y_1$
$R_1 = (x' + y_2')' = xy_2$	$R_2 = (x + y_1)' = x'y_1'$

Using the above equations, the combinational circuit for latch inputs are constructed as shown in Fig. 10.13. The complete logic circuit is obtained by adding SR latches as shown in Fig. 10.13.

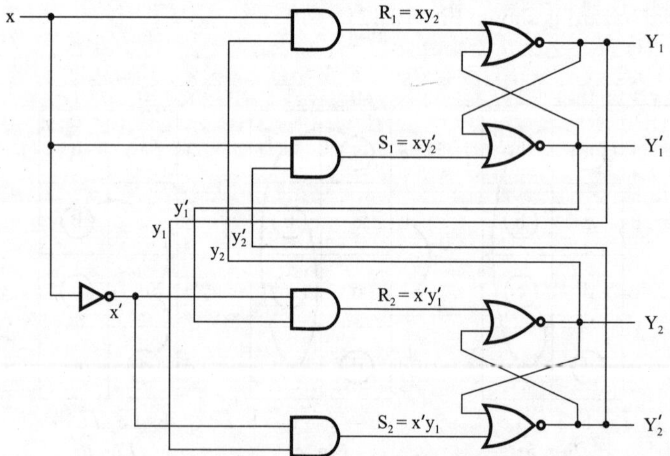

Fig. 10.13: Example of asynchronous sequential circuit with essential hazard.

Table 10.6:

y_1	y_2	x	y_1'	y_2'	x'	xy_2'	$x'y_1$	$(x'+y_2')$	$(x+y_1)$	$(x'+y_2')y_1$	$(x+y_1)y_2$	Y_1	Y_2
0	0	0	1	1	1	0	0	1	0	0	0	0	0
0	1	0	1	0	1	0	0	1	0	0	0	0	0
1	0	0	0	1	1	0	1	1	1	1	0	1	1
1	1	0	0	0	1	0	1	1	1	1	1	1	1
0	0	1	1	1	0	1	0	1	1	0	0	1	0
0	1	1	1	0	0	0	0	0	1	0	1	0	1
1	0	1	0	1	0	1	0	1	1	1	0	1	0
1	1	1	0	0	0	0	0	0	1	0	1	0	1

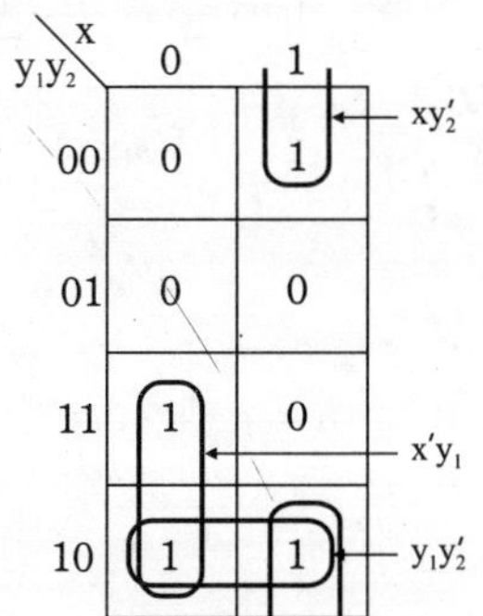

***Fig. a:** K-map for Y_1.*

y_1y_2 \ x	0	1
00	0	0
01	0	1
11	1	1
10	1	0

xy_2, y_1y_2, $x'y_1$

***Fig. b:** K-map for Y_2.*

y_1y_2 \ x	0	1
00	(00)	10
01	00	(01)
11	(11)	01
10	11	(10)

***Fig. c:** Transition table.*

Fig. 10.14.

The values of next state Y_1 and Y_2 are evaluated and listed in Table 10.6 and using these values the maps of Y_1 and Y_2 are drawn as shown in Figs. 10.14(a) and (b). The maps of Y_1 and Y_2 are combined to form transition table as shown in Fig. 10.14(c).

Let us mark the product terms of Y_1 and Y_2 in the maps of Y_1 and Y_2. Now, we can observed that all the product terms are overlapped and so there is no possibility of static hazards.

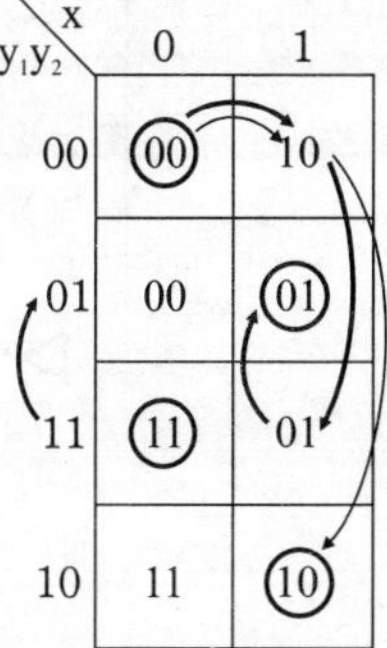

***Fig. 10.15:** State transitions from stable state 00 due to change in input from 0 to 1.*

Let us examine the essential hazard when the input change from 0 to 1 with circuit in stable state 00. If there is no essential hazard the state transition will occur from 00 to 10

Due to essential hazard there is a possibility of state transitions from 00 → 10 → 11 → 01 and the circuit will settle in state 01 instead of 10. The signal changes during these transitions are shown in Fig. 10.16.

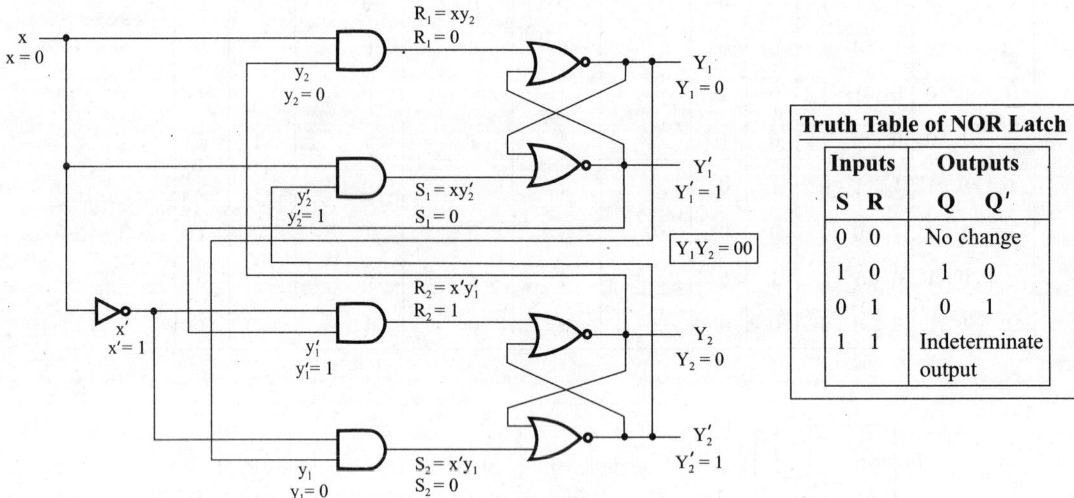

Truth Table of NOR Latch

Inputs		Outputs	
S	R	Q	Q'
0	0	No change	
1	0	1	0
0	1	0	1
1	1	Indeterminate output	

***Fig. a:** Initial value of signals when circuit in stable state 00.*

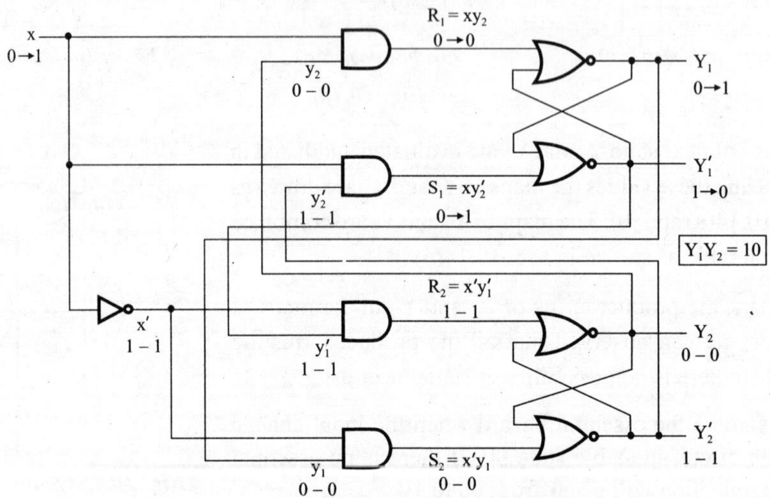

***Fig. b:** Signal changes during state transition from 00 to 10.*

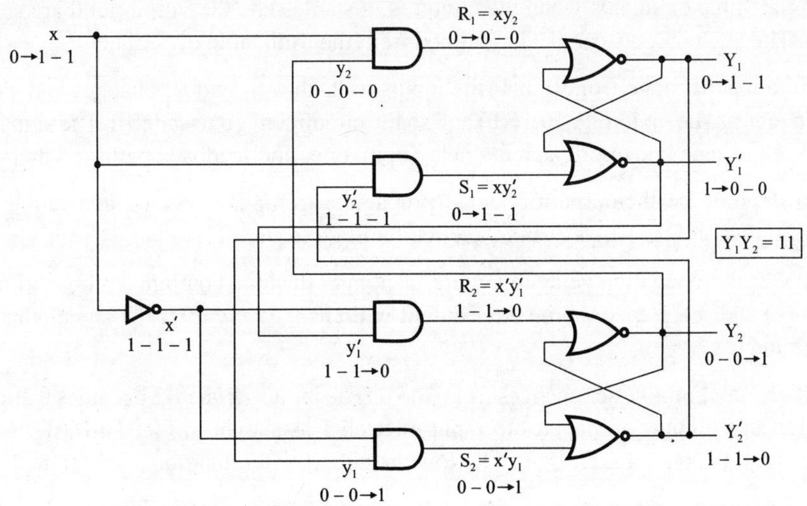

***Fig. c:** Signal changes during state transition from 10 to 11.*

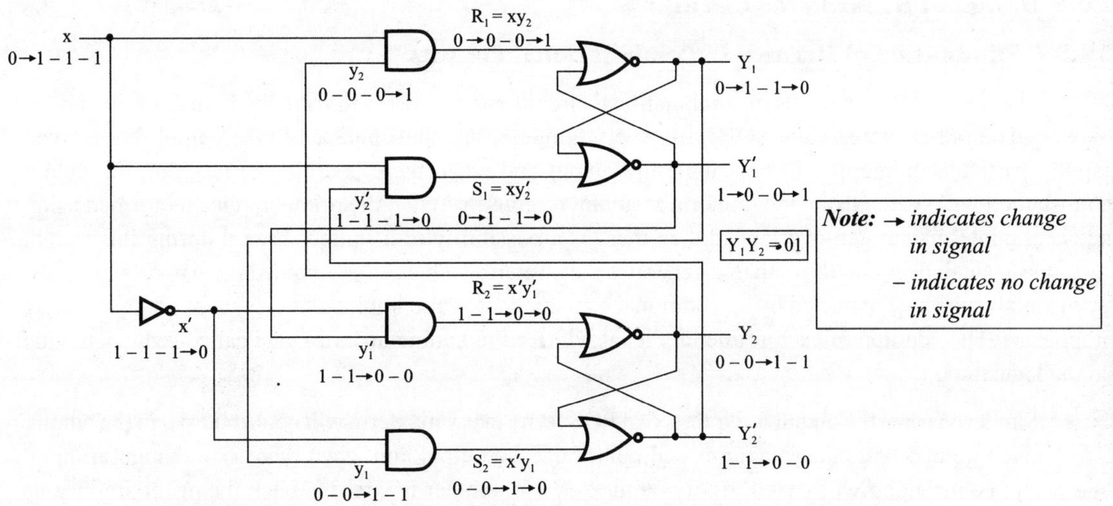

***Fig. d:** Signal changes during state transition from 11 to 01.*

***Fig. 10.16:** Signal changes during state transitions when the input is changed from 0 to 1.*

Table 10.7: Signals during State Transition 00 → 10 → 11 → 01

x	x'	y_1	y_2	y'_1	y'_2	R_1 xy_2	S_1 xy'_2	R_2 $x'y'_1$	S_2 $x'y_1$	Y_1	Y_2	Y_1	Y_2
0	1	0	0	1	1	0	0	1	0	0	0	1 1	1 1
1	1	0	0	1	1	0	1	1	0	1	0	0 0	0 0
1	1	1	0	0	1	0	1	0	1	1·	1	Hazard	No hazard
1	0	1	1	0	0	1	0	0	0	0	1	in Y_1	in Y_2

No change due to delay

The initial value of signals when the circuit is in stable state 00 with input 0 are shown shown in Fig. 10.16(a). The latch outputs can be determined from the truth table of SR latch.

When the input change from 0 to 1, the inputs to latch-1 S_1 and R_1 changes and so the output of latch-1 will change as shown in Fig. 10.16(b) and so the circuit will go to state 10. The signals at other part of circuit will not change due to propagation delay in inverter and feedback paths in latch-1.

After a small delay, the output of latch-1 propagates to input of latch-2 and so the output latch-2 will change as shown in Fig. 10.16(c) and the circuit will go to state 11.

Let the signal propagation delay in inverter be longer than the combined propagation delay shown in Figs. 10.16(b) and (c). Hence the inverter output will change only after the signal changes shown in Figs. 10.16(b) and (c).

When the output of inverter changes from 1 to 0. The inputs to latch-2 becomes 0 and so no change in its output but this outputs propagates to input of latch-1 as shown in Fig. 10.16(d) and so output of latch-1 changes and the circuit go to state 01. Now the circuit will remain in stable state 01.

The signals during state transition $00 \rightarrow 10 \rightarrow 11 \rightarrow 01$ are listed in Table 10.7.

10.5 Design of Hazard Free Circuits

(AKTU, Dec'22, 5 Marks)

10.5.1 Elimination of Hazards in Combinational Circuits

The possibility of hazards in combinational circuits can be observed from the K-map of the inputs. When the output is expressed in SOP form, every minterm will contribute a 1 in the output. For a given input a particular minterm will contribute the output and when there is a change in input, the output contribution will switch from one minterm to another minterm. When the prime implicants formed using adjacent minterms are non-overlapped then there is a possibility of a static-1 hazard during this switch over due to small delays in the path that generates the minterms. This can be avoided if the non-overlapped prime implicants are connected by an additional or redundant prime implicant to create overlapped prime implicants. The addition of each prime implicant will lead to adding an extra gate called redundant gate in the logic circuit.

Similarly, when the output is expressed in POS form, every maxterm will contribute a 0 in the output. For a given input a particular maxterm will contribute the output and when there is a change in input, the output contribution will switch from one maxterm to another maxterm. When the prime implicants formed using adjacent maxterms are non-overlapped then there is a possibility of a static-0 hazard during this switch over due to small delays in the path that generates the maxterms. This can be avoided if the non-overlapped prime implicants are connected by an additional or redundant prime implicant to create overlapped prime implicants. The addition of each prime implicant will lead to adding an extra gate called redundant gate in the logic circuit.

Consider the combinational circuit shown in Fig. 10.2. The output equation is,

$$Y = x_1 x_2 + x_2' x_3$$

The K-map for the above equation is shown in Fig. 10.17. Here, the two prime implicants are non-overlapped prime implicants and so a hazard may occur when the output contribution switch from minterm of a prime implicant to minterm of another prime implicant. The hazard is due to delay in the path that generates minterm. This can be avoided by introducing an additional prime implicant $x_1 x_3$ that makes overlapped prime implicants as shown in Fig. 10.17.

Now, the output equation is,

$$Y = x_1x_2 + x_2'x_3 + x_1x_3$$

The additional product term x_1x_3 can be implemented by an additional AND gate as shown in Fig. 10.18.

x_1 \ x_2x_3	00	01	11	10
0	m_0 0	m_1 1	m_3 0	m_2 0
1	m_4 0	m_5 1	m_7 1	m_6 1

Groups: $x_2'x_3$ (m_1, m_5); x_1x_2 (m_7, m_6); x_1x_3 (m_5, m_7).

Fig. 10.17.

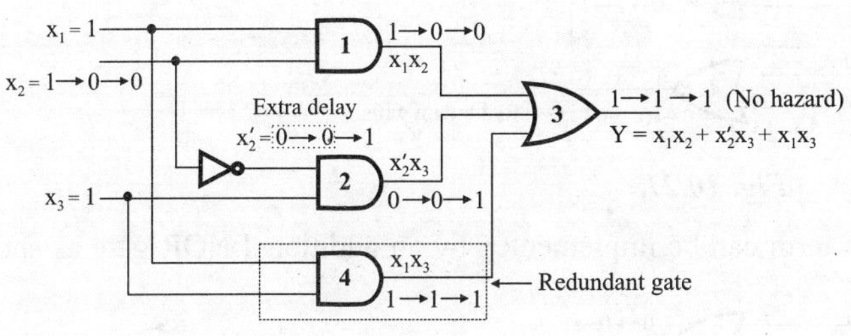

Fig. 10.18.

Alternatively, the additional product term can be implemented by an additional NAND gate as shown in Fig. 10.19.

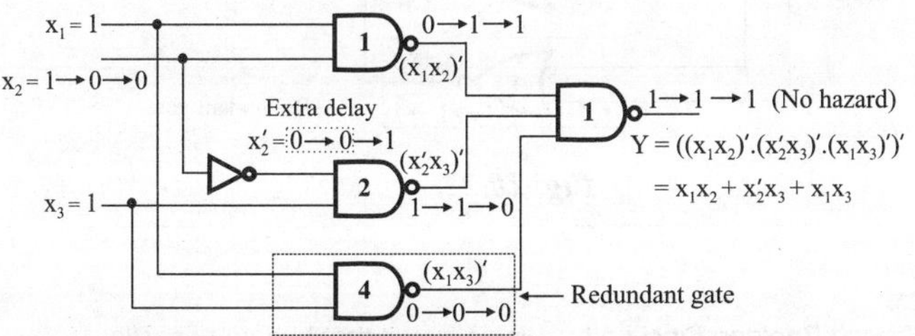

Fig. 10.19.

Consider the combinational circuit shown in Fig. 10.4. The output equation is,

(AKTU, Dec'22, 10 Marks)

$$Y = (x_1 + x_2') \cdot (x_2 + x_3)$$
$$= (((x_1 + x_2') \cdot (x_2 + x_3))')'$$
$$= ((x_1 + x_2')' + (x_2 + x_3)')'$$
$$= (x_1'x_2 + x_2'x_3')'$$

The K-map for the above equation is shown in Fig. 10.20. Here, the two prime implicants are non-overlapped prime implicants and so a hazard may occur when the output contribution switch from maxterm of a prime implicant to maxterm of another prime implicant. The hazard is due to delay in the path that generates maxterm. This can be avoided by introducing an additional prime implicant $(x_1'x_3')' = (x_1 + x_3)$ that makes overlapped prime implicants as shown in Fig. 10.20.

Now, the output equation is,

$$Y = (x_1'x_2 + x_2'x_3' + x_1'x_3')'$$
$$= (x_1'x_2)' \cdot (x_2'x_3')' \cdot (x_1'x_3')'$$
$$= (x_1 + x_2') \cdot (x_2 + x_3) \cdot (x_1 + x_3)$$

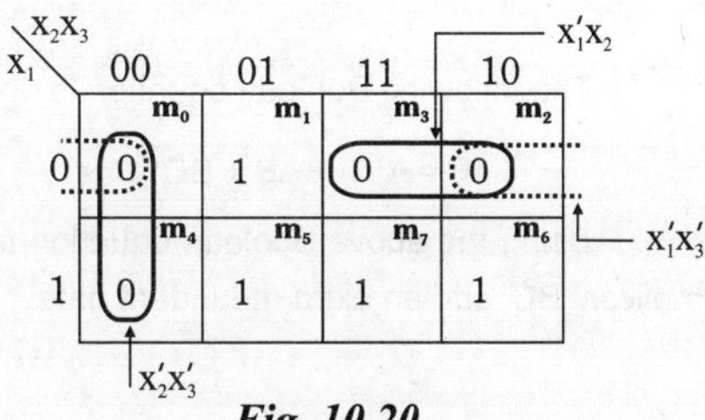

Fig. 10.20.

The additional sum term can be implemented by an additional OR gate as shown in Fig. 10.21.

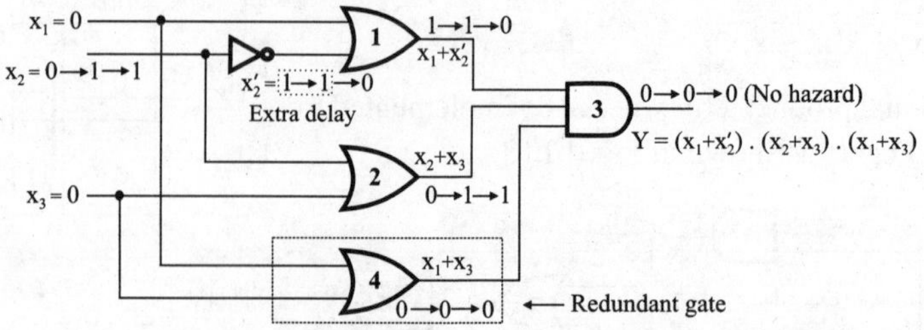

Fig. 10.21.

Alternatively, the additional sum term can be implemented by an additional NOR gate as shown in Fig. 10.22.

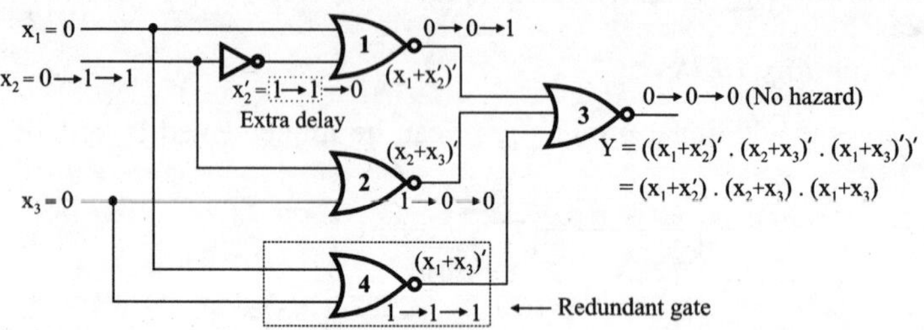

Fig. 10.22.

Example 10.1

Implement the following Boolean function by a logic circuit that has no hazards.

F(A, B, C) = Σ(0, 2, 6, 7)

Solution

The K-map of the given function is shown in Fig. 1. Here, the prime implicants A′C′ and AB are non-overlapped prime implicants that can create a static-1 hazard. Hence, let us add an additional prime implicant BC′ that creates overlapped prime implicants as shown in Fig. 2.

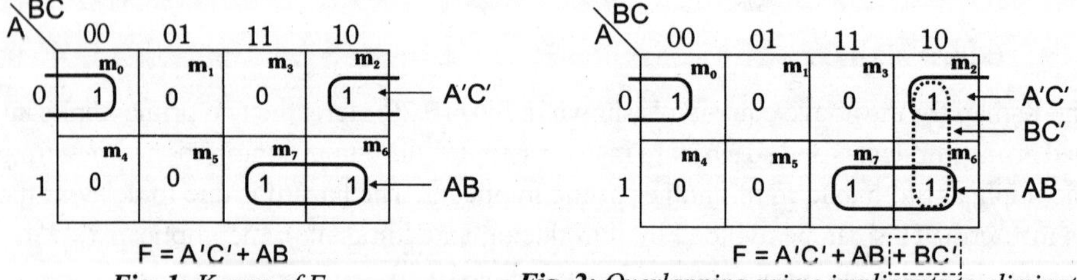

Fig. 1: K-map of F. *Fig. 2: Overlapping prime implicants to eliminate hazard.*

The following Boolean equation is obtained from K-map of Fig. 2.

F = A′C′ + AB + BC′

Using the above Boolean equation the hazard free logic circuit is drawn as shown in Fig. 3. Here, the prime implicant BC′ add an extra redundant gate.

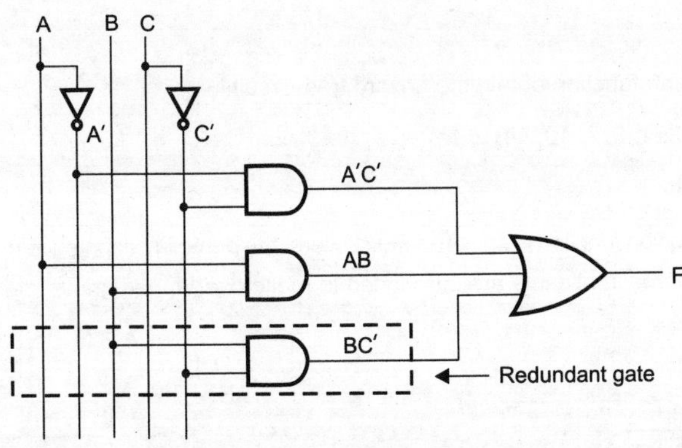

Fig. 3: Hazard free logic circuit.

Example 10.2

Implement the following Boolean function by a logic circuit that has no hazards.

$$F(A, B, C, D) = \Sigma(0, 2, 6, 7, 8, 10, 12)$$

Solution

The K-map of the given function is shown in Fig. 1. Here, the prime implicants B′D′, AC′D′ and A′BC are non-overlapped prime implicants that can create a static-1 hazard. Hence, let us add an additional prime implicant A′CD′ that creates overlapped prime implicant as shown in Fig. 2.

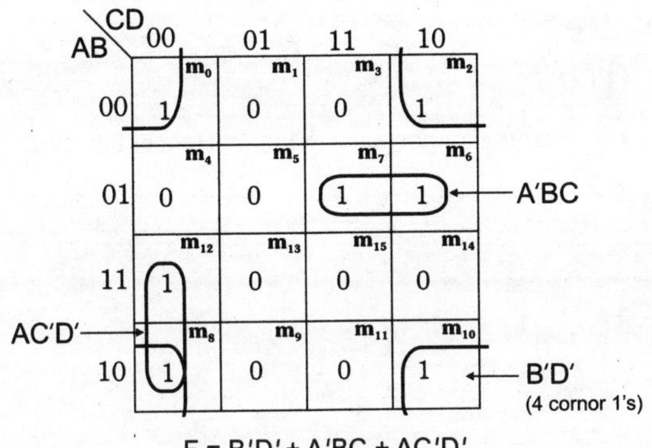

Fig. 1: K-map of F.

Fig. 2: Overlapping prime implicants to eliminate hazard.

The following Boolean equation is obtained from K-map of Fig. 2.

$$F = B'D' + A'BC + AC'D' + A'CD'$$

Using the above Boolean equation the hazard free logic circuit is drawn as shown in Fig. 3. Here, the prime implicant A′CD′ add an extra redundant gate.

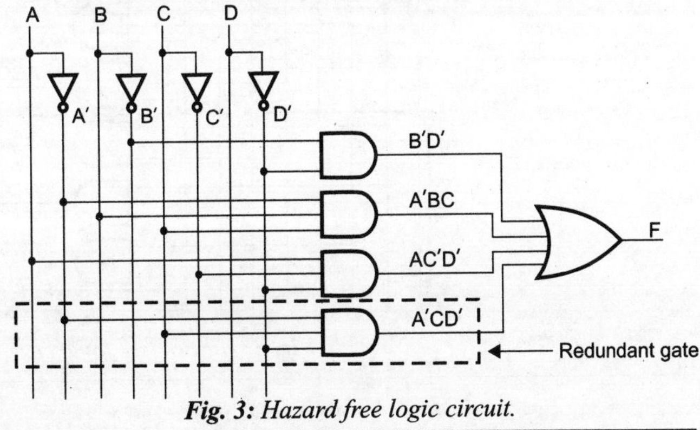

Fig. 3: Hazard free logic circuit.

Example 10.3

For a given Boolean function obtain the hazard free circuit.

$$F = \Sigma m(0, 1, 3, 4, 6, 7, 10, 14)$$

Solution

The K-map of the given function is shown in Fig. 1. Here, the prime implicants are non-overlapped prime implicants. Hence, three additional prime implicants are introduced to make overlapped prime implicants as shown in Fig. 2.

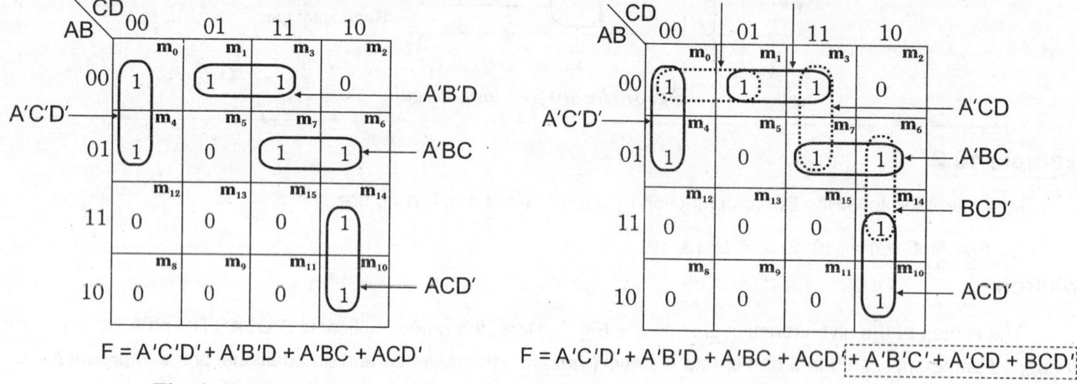

F = A'C'D' + A'B'D + A'BC + ACD'

Fig. 1: K-map of F.

F = A'C'D' + A'B'D + A'BC + ACD' + A'B'C' + A'CD + BCD'

Fig. 2: Overlapping prime implicants to eliminate hazard.

The following Boolean equation is obtained from K-map of Fig. 2.

F = A'C'D' + A'B'D + A'BC + ACD' + A'B'C' + A'CD + BCD'

Using the above Boolean equation the hazard free logic circuit is drawn as shown in Fig. 3. Hence, the thee extra prime implicants adds three redundant gates in the logic circuit.

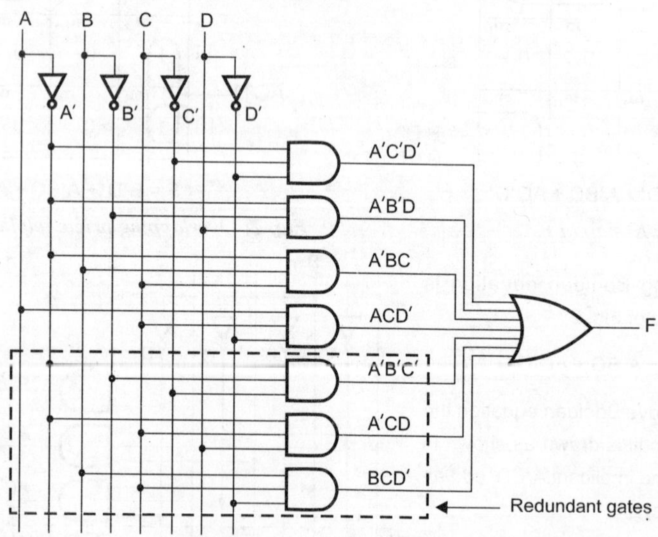

Fig. 3: Hazard free logic circuit.

Example 10.4

Implement the following Boolean function by a logic circuit that has no hazards.

$$Y = \prod M(0, 1, 5, 7, 14, 15)$$

Solution

The K-map of the given function is shown in Fig. 1. Here, the prime implicants B′D′ and A′BC are non-overlapped prime implicants that can create a static-0 hazard. Hence, let us add an additional prime implicant A′CD′ that creates overlapped prime implicants as shown in Fig. 2.

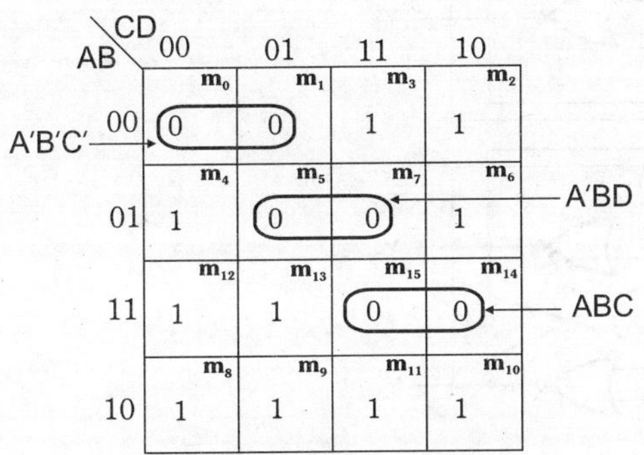

Fig. 1: K-map of F.

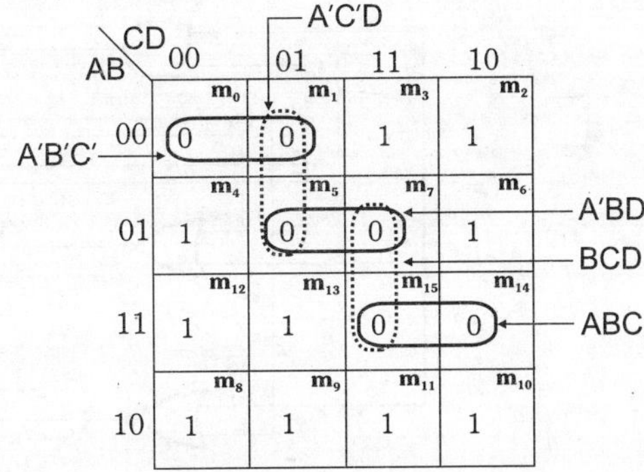

Fig. 2: Overlapping prime implicants to eliminate hazard.

The following Boolean equation is obtained from K-map of Fig. 2.

$$F = (ABC + A'B'C' + A'BD + A'C'D + BCD)'$$

$$= (ABC)' . (A'B'C')' . (A'BD)' . (A'C'D)' . (BCD)'$$

$$= (A' + B' + C') . (A + B + C) . (A + B' + D') . (A + C + D') . (B' + C' + D')$$

Using the above Boolean equation the hazard free logic circuit is drawn as shown in Fig. 3. Here, the prime implicants (A + C + D′) and (B′ + C′ + D′) adds an extra redundant gate each.

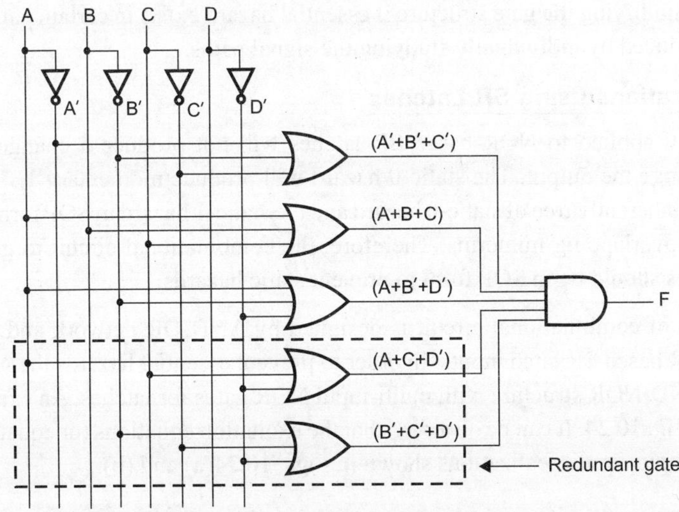

Fig. 3: Hazard free logic circuit.

10.5.2 Elimination of Hazards in Sequential Circuits

(AU, Nov/Dec'18, 13 Marks)

The essential hazard can be eliminated by rearranging the gate structure of the combinational circuit that generates inputs to latches such that the effect of input changes propagates to all latch inputs before propagation of feedback signals (or next state variables) to latch inputs.

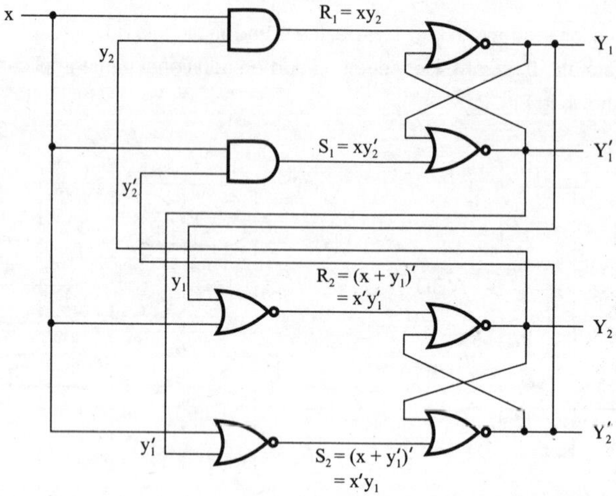

Fig. 10.23: *Essential hazard free realization of the circuit shown in Fig. 10.16(a).*

The essential hazard in the circuit of Fig. 10.16 can be eliminated by replacing the inverter and AND gates in the circuit of latch-2 by NOR gates as shown in Fig. 10.23.

Here, it is assumed that the gate delay is concentrated at the output. The replacement of NOT-AND combination by NOR gate will ensure that any change input, x will propagate to inputs of latch-2 before latch-1 change state. This ensure that the change in circuit inputs and feedback from latch-1 output arrive at same time to latch-2 inputs and so essential hazard is eliminated.

In general, the essential hazard is minimized or eliminated by reducing the complexity of gate structure. Even after modifying the gate structure if essential hazard exists in certain path then appropriate delays has to be introduced by individually studying the signal paths.

Hazard Free Realizations using SR Latches

A momentary 0 applied to NOR based SR latches will not produce a change in output. But a momentary 1 can change the output. The static-0 hazard will produce momentary 1. The combinational circuit in SOP form is inherently free of static-0 hazard and the static-1 hazard in SOP form can be prevented by adding additional overlapping minterms. Therefore, the combinational circuit to generate inputs for NOR based SR latches should be in SOP form to prevent static hazards.

The SOP form of combinational circuit is designed by AND-OR network and the output of OR gate is applied to NOR based SR latch inputs. In order to prevent essential hazards the AND-OR structure can be replaced by AND-NOR structure with multi-input NOR gates for latches. An example of modified structure is shown in Fig. 10.24. It can be observed that the excitation equations (or equations for Y and Y′) are same in both the logic circuit realizations shown in Figs. 10.24(a) and (b).

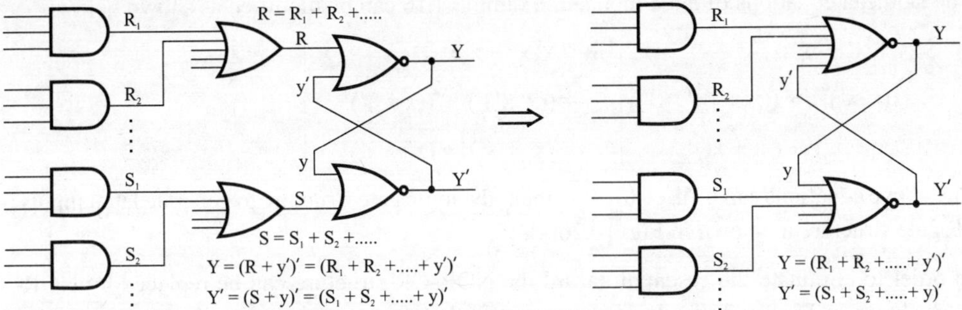

***Fig. a:** Combinational circuit in SOP form.* ***Fig. b:** Equivalent circuit to eliminate essential hazard.*

***Fig. 10.24:** Elimination of essential hazard with modified gate structure.*

Consider the logic circuit implementation using NOR based SR latch shown in Example 9.16.

The Boolean equations for latch inputs are,

$$S = x_1 x_2' + x_1' x_2 \quad ; \quad R = x_1' x_2' + x_1 x_2 \quad ; \quad Z = x_1'$$

The above equations are implemented using AND-OR gate structure as shown in Fig. 10.25(a). In this implementation the OR gate can be eliminated by using multi-input NOR gates for latch as shown in Fig. 10.25(b). It can be observed that the excitation equations (or equations for Y and Y′) are same in both the logic circuits shown in Figs. 10.25(a) and (b).

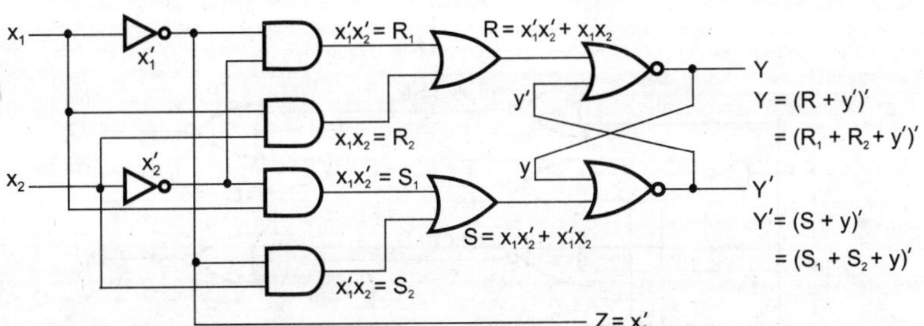

***Fig. a:** Combinational circuit with AND-OR gate structure.*

⇓

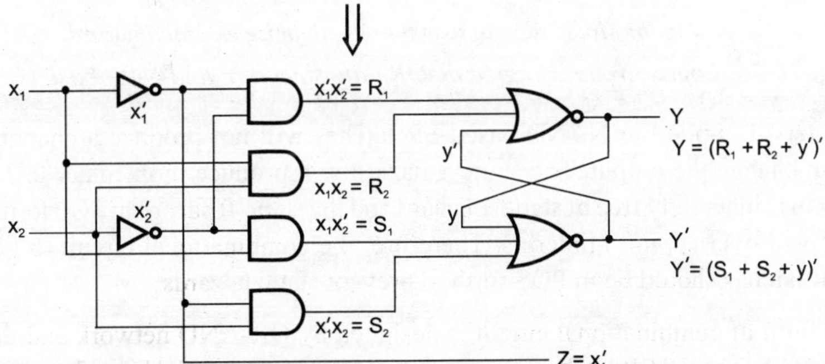

***Fig. b:** Modified gate structure to minimize essential hazard.*

***Fig. 10.25:** Logic circuit using AND-OR gate structure and NOR based SR latch.*

The Boolean equations of latch inputs in Example 4.16 can be modified as shown below:

$$S = x_1 x_2' + x_1' x_2 = ((x_1 x_2')')' + ((x_1' x_2)')' = (x_1' + x_2)' + (x_1 + x_2')'$$

$$R = x_1' x_2' + x_1 x_2 = ((x_1' x_2')')' + ((x_1 x_2)')' = (x_1 + x_2)' + (x_1' + x_2')'$$

$$Z = x_1'$$

The above equations can be used to implement the logic gate structure to generate latch inputs using NOR-OR gate structure as shown in Fig. 10.26(a).

In order to eliminate the essential hazard the NOR-OR structure can be replaced by NOR-NOR structure as shown in Fig. 10.26(b). It can be observed that the excitation equations (or equations for Y and Y') are same in both the logic circuits shown in Figs. 10.26(a) and (b).

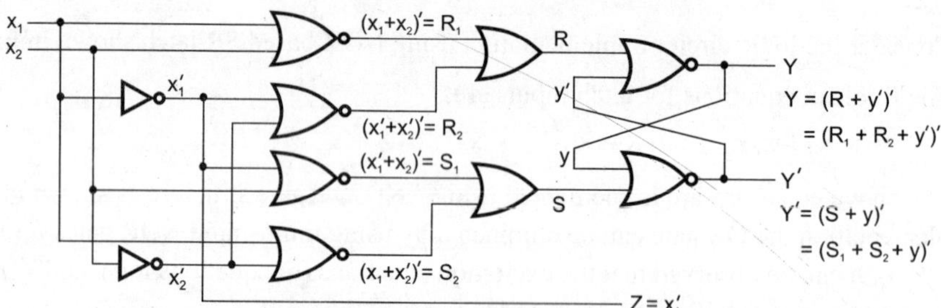

***Fig. a:** Combinational circuit with NOR-OR gate structure.*

⇓

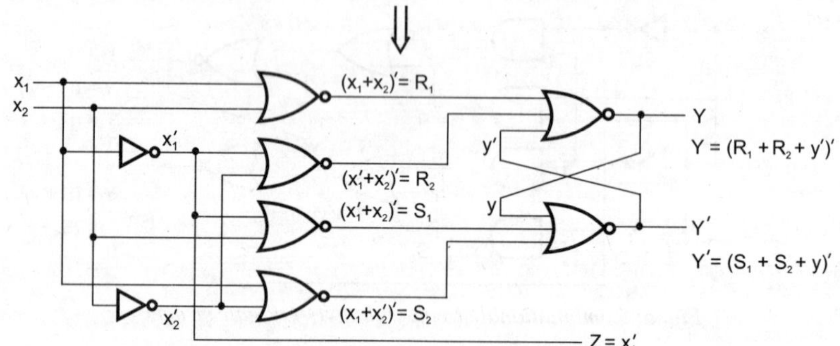

***Fig. b:** Modified gate structure to minimize essential hazard.*

***Fig. 10.26:** Logic circuit using NOR-OR gate structure and NOR based SR latch.*

A momentary 1 applied to NAND based SR latches will not produce a change in output. But a momentary 0 can change the output. The static-1 hazard will produce momentary 0. The combinational circuit in POS form is inherently free of static-1 hazard and the static-0 hazard in SOP form can be prevented by adding additional overlapping minterms. Therefore, the combinational circuit to generate inputs for NAND based SR latches should be in POS form to prevent static hazards.

The POS form of combinational circuit is designed by OR-AND network and the output of AND gate is applied to NAND based SR latch inputs. In order to prevent essential hazards the OR-AND structure can be replaced by OR-NAND structure with multi-input NAND gates for latches. An example of modified structure is shown in Fig. 10.27. It can be observed that the excitation equations (or equations for Y and Y') are same in both the logic circuit realizations shown in Figs. 10.27(a) and (b).

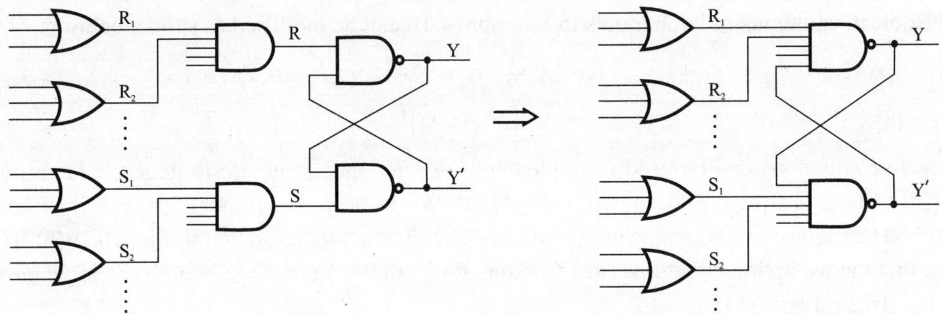

Fig. a: Combinational circuit in POS form. *Fig. b: Equivalent circuit to eliminate essential hazard.*

Fig. 10.27: *Elimination of essential hazard with modified gate structure.*

Consider the logic circuit implementation using NAND based SR latch shown in Example 9.16. The Boolean equations for latch inputs in Example 9.16 are modified as shown below:

$$S = (x_1' + x_2) \cdot (x_1 + x_2') \quad ; \quad R = (x_1 + x_2)(x_1' + x_2') \quad ; \quad Z = x_1'$$

The above equations are implemented using AND-OR gate structure as shown in Fig. 10.28(a). In this implementation the AND gate can be eliminated by using multi-input NAND gates for latch as shown in Fig. 10.28(b). It can be observed that the excitation equations (or equations for Y and Y′) are same in both the logic circuits shown in Figs. 10.28(a) and (b).

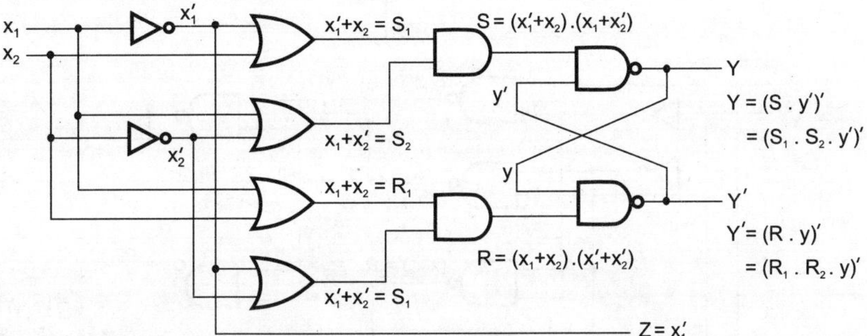

Fig. a: *Combinational circuit with OR-AND gate structure.*

⇓

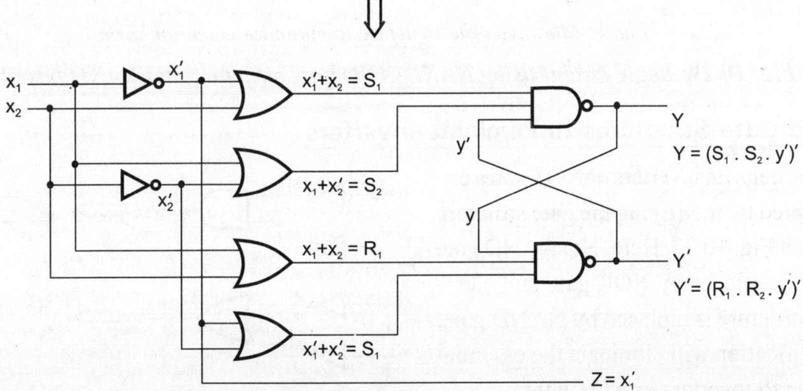

Fig. b: *Modified gate structure to minimize essential hazard.*

Fig. 10.28: *Logic circuit using OR-AND gate structure and NAND based SR latch.*

The Boolean equations of latch inputs in Example 4.16 can be modified as shown below:

$$S = (x_1 x_2' + x_1' x_2)' = (x_1 x_2')' . (x_1' x_2)'$$

$$R = (x_1' x_2' + x_1 x_2)' = (x_1' x_2')' . (x_1 x_2)'$$

$$Z = x_1'$$

The above equations can be used to implement the logic gate structure to generate latch inputs using NAND-AND gate structure as shown in Fig. 10.29(a). In order to eliminate the essential hazard the NAND-AND structure can be replaced by NAND-NAND structure as shown in Fig. 10.29(b). It can be observed that the excitation equations (or equations for Y and Y′) are same in both the logic circuits shown in Figs. 10.29(a) and (b).

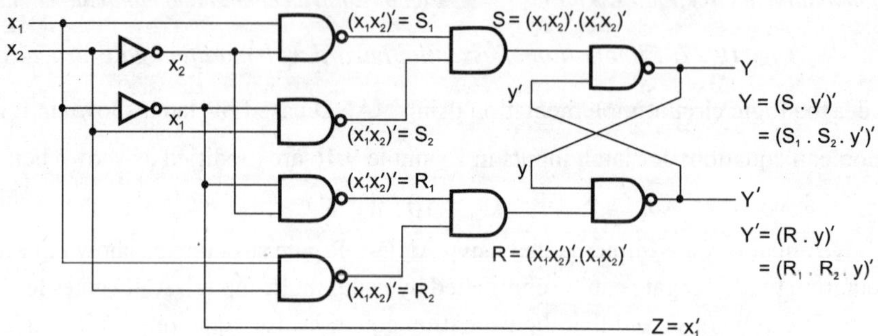

Fig. a: Combinational circuit with NAND-AND gate structure.

⇓

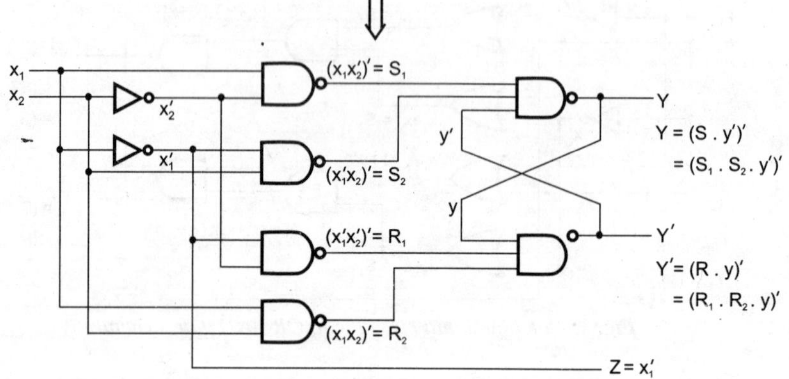

Fig. b: Modified gate structure to minimize essential hazard.

***Fig. 10.29:** Logic circuit using NAND-AND gate structure and NAND based SR latch.*

Modified Gate Structures to Eliminate Inverters

The delay in inverters or NOT gates can be eliminated by modifying the gate structure as shown in Fig. 10.30. Here, NOT-AND gate structure is replaced by NOR gate and NOT-OR gate structure is replaced by NAND gate. This modification will eliminate the essential hazard due to inverters or NOT gate.

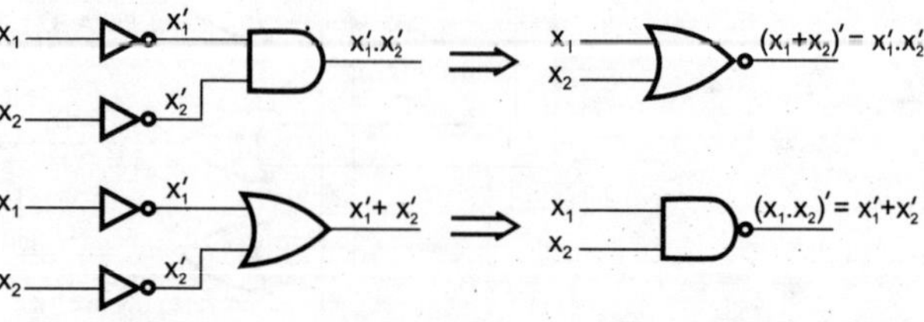

***Fig. 10.30:** Gate structure to replace inverters.*

Example 10.5

Assign output values to the don't-care states in the flow tables of Fig. 1 in such a way to avoid tranisent output pulses or hazard.

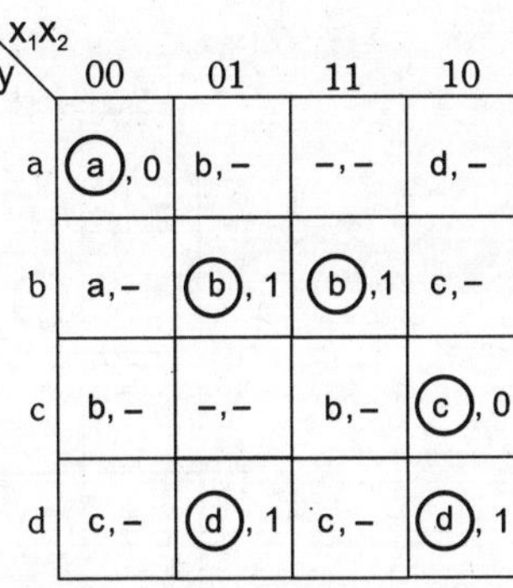

y \ x_1x_2	00	01	11	10
a	(a), 0	b, –	–, –	d, –
b	a, –	(b), 1	(b), 1	c, –
c	b, –	–, –	b, –	(c), 0
d	c, –	(d), 1	c, –	(d), 1

***Fig. 1:** Reduced flow table.*

Solution

In the reduced flow table of Fig. 1, the dashes in the output are assigned values according to the following rules.

Rule 1: Assign 0 to an output variable of an unstable state which is transient state between two stable states that have a 0 in the corresponding output variable.

Rule 2: Assign 1 to an output variable of an unstable state which is transient state between two stable states that have a 1 in the corresponding output variable.

Rule 3: Assign a don't-care condition to an output of an unstable state which is transient state during a transition between two stable states that have a different values (0 and 1 or 1 and 0) as output.

The state transitions that are used to assign output using the above rules are shown in Figs. 2 to 5. Here, there are four combinations of inputs. The input is allowed to change only one bit at any one time. Here, the outputs assigned in Fig. 2(b), are retained during next state transitions in Fig. 3(b). Similarly, the outputs assigned in previous transitions are retained during next state transitions in Fig. 3(d) and so on. The state transitions and the assigned outputs are listed in Table 1 and from this table it can be observed that the assigned output does not produce any hazard.

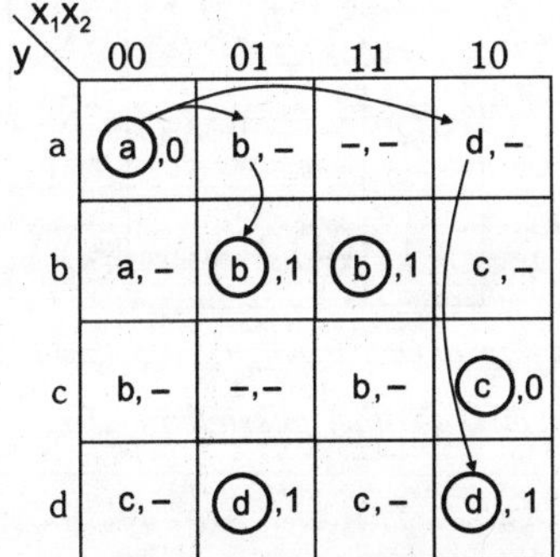

y \ x_1x_2	00	01	11	10
a	(a),0	b, –	–, –	d, –
b	a, –	(b),1	(b),1	c, –
c	b, –	–,–	b, –	(c),0
d	c, –	(d),1	c, –	(d), 1

***Fig. a:** Transitions from stable state a.*

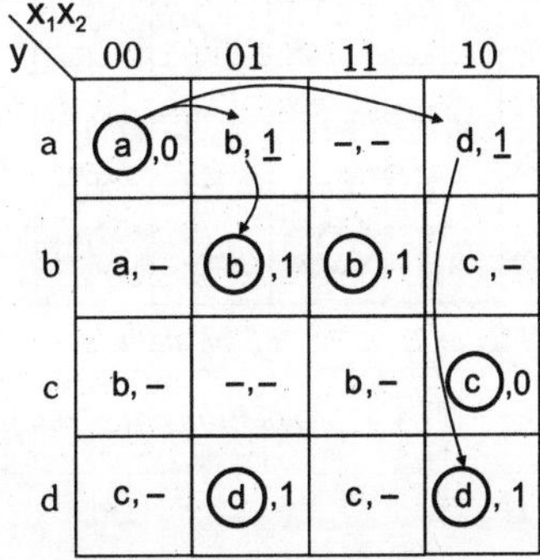

y \ x_1x_2	00	01	11	10
a	(a),0	b, 1	–, –	d, 1
b	a, –	(b),1	(b),1	c, –
c	b, –	–,–	b, –	(c),0
d	c, –	(d),1	c, –	(d), 1

***Fig. b:** Output assigned to avoid hazard.*

***Fig. 2:** Transitions due to change in input from 00 to other possible inputs.*

y \ x_1x_2	00	01	11	10
a	(a), 0	b, 1	–, –	d, 1
b	a, –	(b),1	(b),1	c, –
c	b, –	–, –	b, –	(c), 0
d	c, –	(d), 1	c, –	(d), 1

***Fig. a:** Transitions from stable state b.*

y \ x_1x_2	00	01	11	10
a	(a), 0	b, 1	–, –	d, 1
b	a, ×	(b),1	(b),1	c, –
c	b, –	–, –	b, –	(c), 0
d	c, –	(d), 1	c, –	(d), 1

***Fig. b:** Output assigned to aviod hazard.*

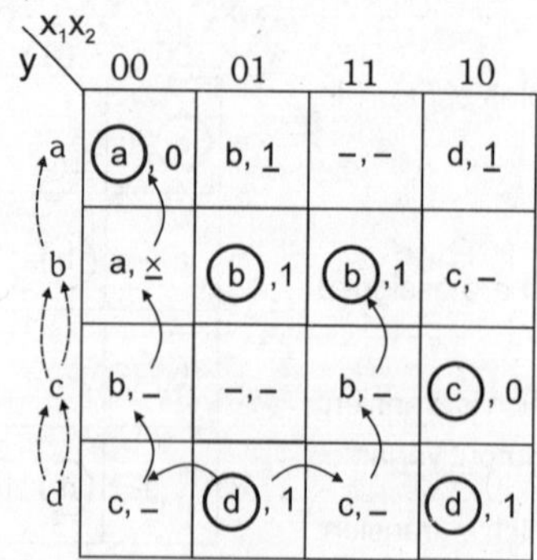

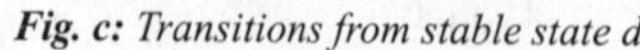
Fig. c: Transitions from stable state d.

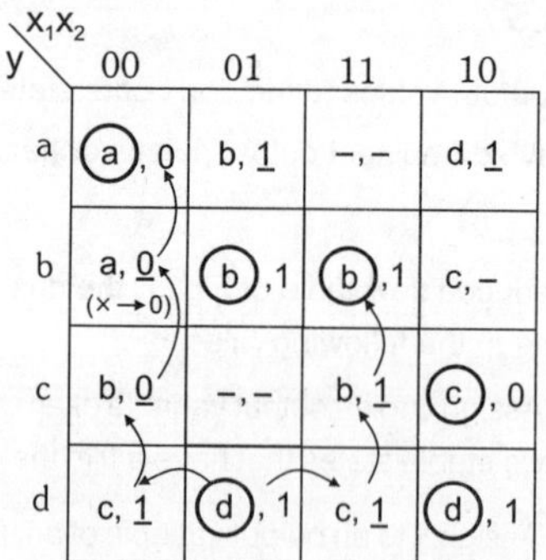

Fig. d: Output assigned to avoid hazard.

***Fig. 3:** Transitions due to change in input from 01 to other possible inputs.*

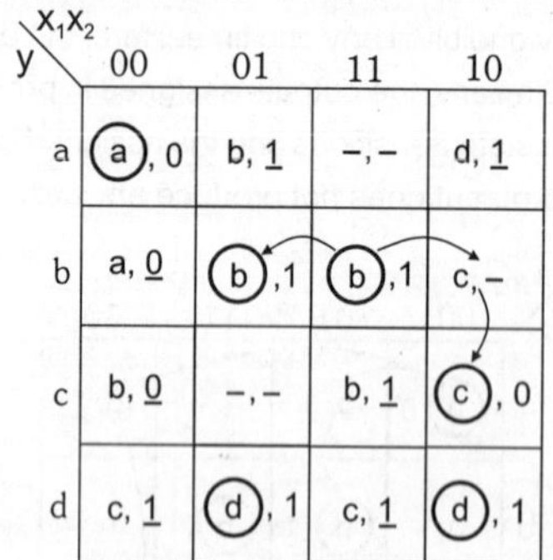

Fig. a: Transitions from stable state b.

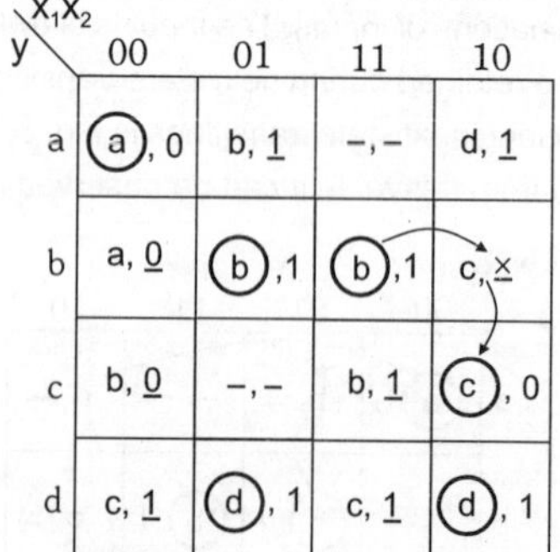

Fig. b: Output assigned to aviod hazard.

***Fig. 4:** Transitions due to change in input from 11 to other possible inputs.*

y \ x_1x_2	00	01	11	10
a	(a), 0	b, 1	–,–	d, 1
b	a, 0	(b),1	(b),1	c, ×
c	b, 0	–,–	b, 1	(c),0
d	c, 1	(d), 1	c, 1	(d), 1

Fig. a: Transitions from stable state c.

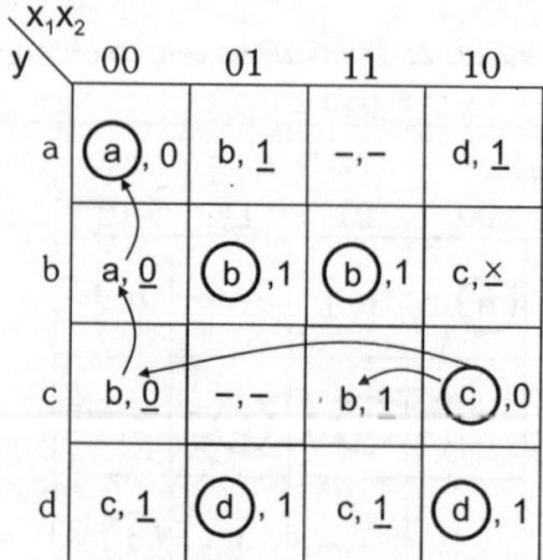

Fig. b: Output assigned to avoid hazard.

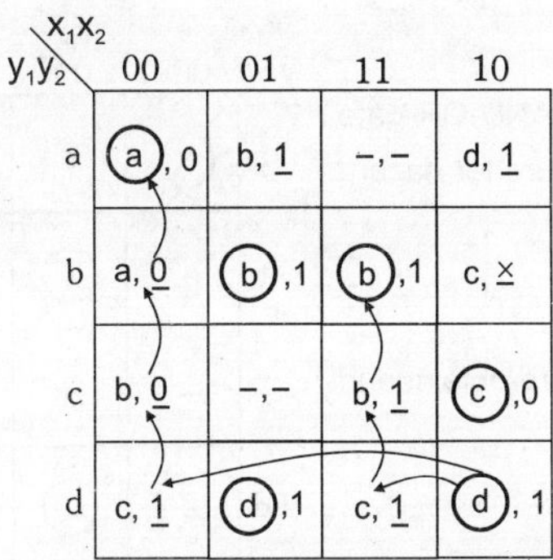

y_1y_2 \ x_1x_2	00	01	11	10
a	(a), 0	b, 1	–, –	d, 1
b	a, 0	(b), 1	(b), 1	c, ×
c	b, 0	–, –	b, 1	(c), 0
d	c, 1	(d), 1	c, 1	(d), 1

Fig. c: Transitions from stable state d.

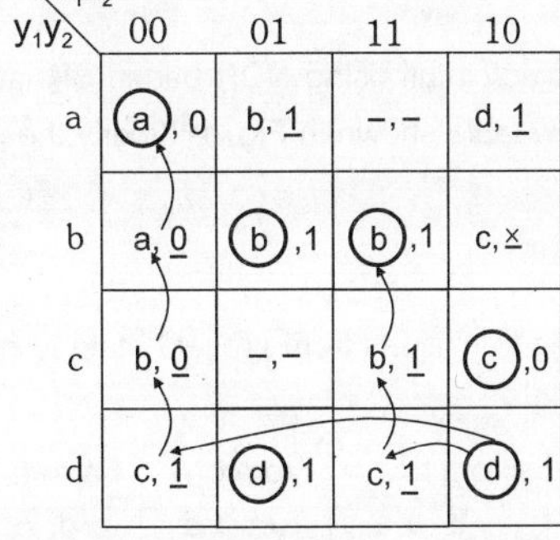

y_1y_2 \ x_1x_2	00	01	11	10
a	(a), 0	b, 1	–, –	d, 1
b	a, 0	(b), 1	(b), 1	c, ×
c	b, 0	–, –	b, 1	(c), 0
d	c, 1	(d), 1	c, 1	(d), 1

Fig. d: Output assigned to avoid hazard.

Fig. 5: Transitions due to change in input from 10 to other possible inputs.

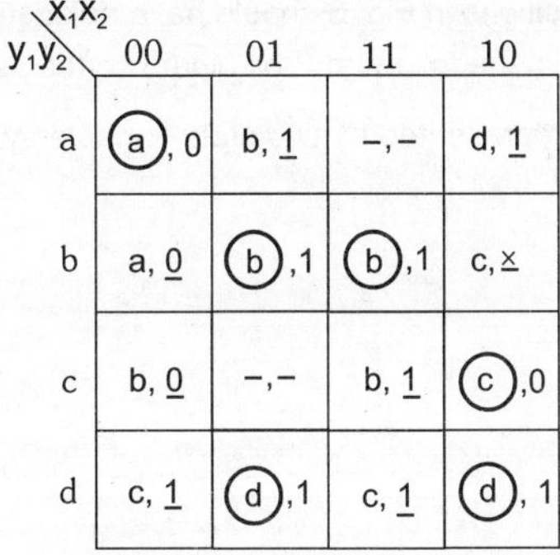

y_1y_2 \ x_1x_2	00	01	11	10
a	(a), 0	b, 1	–, –	d, 1
b	a, 0	(b), 1	(b), 1	c, ×
c	b, 0	–, –	b, 1	(c), 0
d	c, 1	(d), 1	c, 1	(d), 1

Fig. 6: Flow table with output assigned to avoid hazards.

Table 1:

Present Input	Present Stable State	Change in Input	State Transitions	Output Changes	Hazard
00	a	00 → 01	a → b → b	0 → 1 → 1	No hazard
		00 → 10	a → d → d	0 → 1 → 1	No hazard
01	b	01 → 00	b → a → a	1 → 0 → 0	No hazard
01	d	01 → 00	d → c → b → a → a	1 → 1 → 0 → 0 → 0	No hazard
		01 → 11	d → c → b → b	1 → 1 → 1 → 1	No hazard
11	b	11 → 10	b → c → c	1 → 0 → 0 (× → 0)	No hazard
10	c	10 → 00	c → b → a → a	0 → 0 → 0 → 0	No hazard
		10 → 11	c → b → b	0 → 1 → 1	No hazard
10	d	10 → 00	d → c → b → a → a	1 → 1 → 0 → 0 → 0	No hazard
		10 → 11	d → c → b → b	1 → 1 → 1 → 1	No hazard

Example 10.6

Realize a logic circuit using NOR based SR latches and AND-OR gate structure for the flow table shown in Fig. 1. Modify the gate structure for hazard free realization.

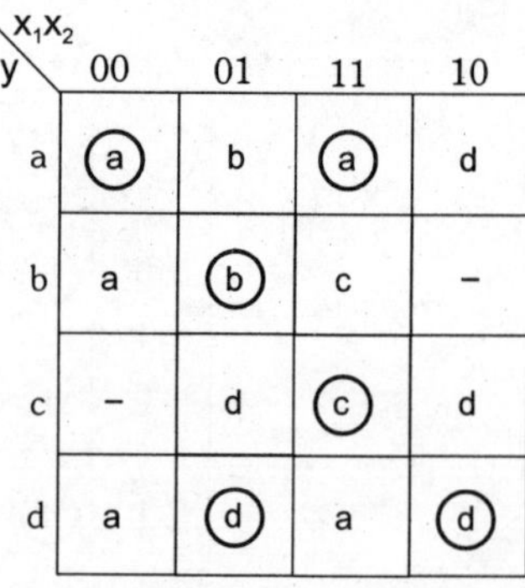

y \ x_1x_2	00	01	11	10
a	(a)	b	(a)	d
b	a	(b)	c	–
c	–	d	(c)	d
d	a	(d)	a	(d)

Fig. 1: Reduced flow table.

Solution

The following transitions from present state to next state are obtained in transition table in Fig. 2.

From a	From b	From c	From d
a → b	b → a	c → d	d → a
a → d	b → c		

The above transitions are combined as a single diagram in the transition diagram shown in Fig. 3. For a critical race free state assignment the state transitions in Fig. 3 should have adjacent state assignments. Here, it is possible to achieve adjacent state assignments with the following 2-bit binary codes as shown in Fig. 4.

The two bits of binary codes are represented by two state variables y_1 and y_2. The transition table with binary code assignment is shown in Fig. 5.

a → 00	c → 11	$y = y_1 y_2$
b → 01	d → 10	

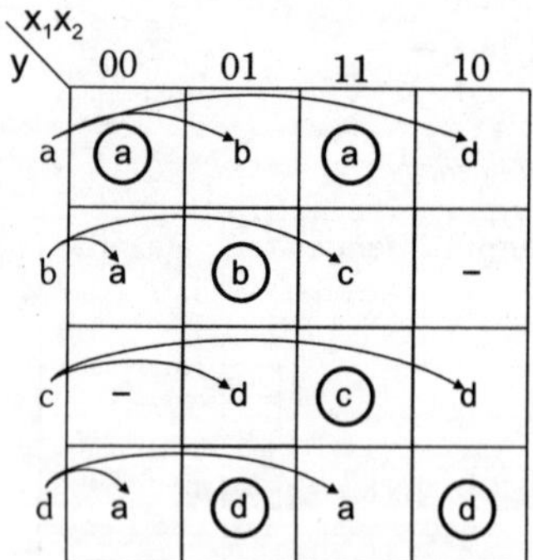

y \ x_1x_2	00	01	11	10
a	(a)	b	(a)	d
b	a	(b)	c	–
c	–	d	(c)	d
d	a	(d)	a	(d)

Fig. 2: Transitions from present state to next state.

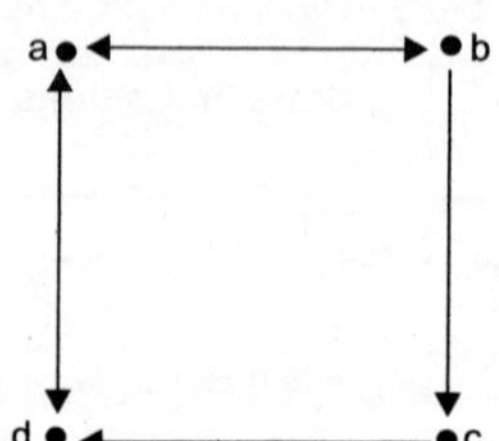

Fig. 3: Transition diagram.

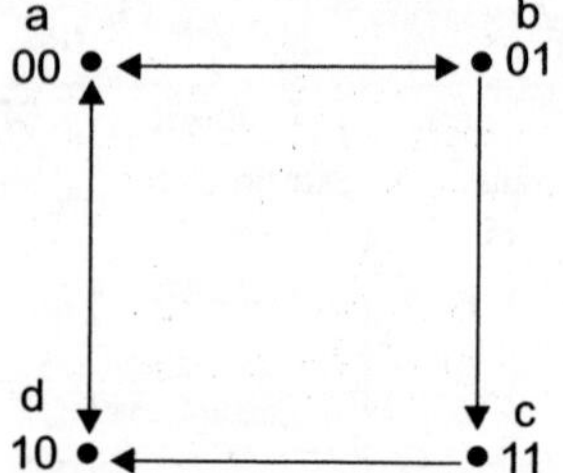

Fig. 4: Transition diagram with binary code assignment.

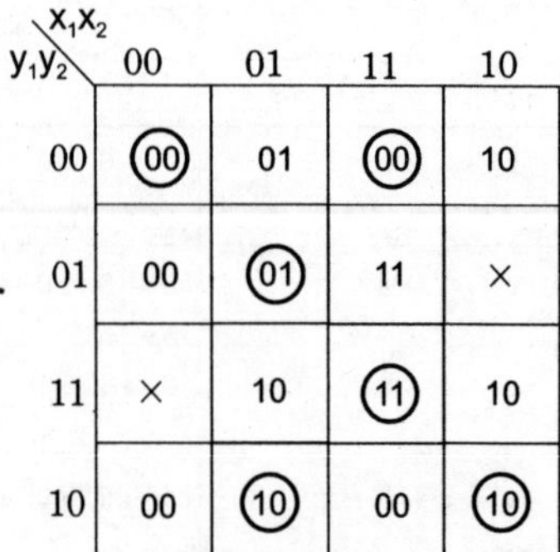

y_1y_2 \ x_1x_2	00	01	11	10
00	(00)	01	(00)	10
01	00	(01)	11	×
11	×	10	(11)	10
10	00	(10)	00	(10)

Fig. 5: Transition table with binary code assignment.

Logic Implementation using NOR based SR Latch

The binary values inside each square of transition table is the value of next state $Y_1 Y_2$. Hence, using the transition table the maps of Y_1 and Y_2 are drawn as shown in Figs. 6 and 7.

$y_1y_2 \backslash x_1x_2$	00	01	11	10
00	0	0	0	1
01	0	0	1	×
11	×	1	1	1
10	0	1	0	1

***Fig. 6:** K-map for Y_1.*

$y_1y_2 \backslash x_1x_2$	00	01	11	10
00	0	1	0	0
01	0	1	1	×
11	×	0	1	0
10	0	0	0	0

***Fig. 7:** K-map for Y_2.*

In order to store the two state variables Y_1 and Y_1, two SR latches are required. Let us choose two NOR based SR latches.

Let, S_1, R_1 = Inputs for SR latch-1

S_2, R_2 = Inputs for SR latch-2

To Find Equations for Latch Inputs using Maps of Y_1 and Y_2

In latch-1, the input equations for S_1 and R_1 will be independent of y_1. Therefore, the map of Y_1 can be split into two maps: One map when y_1 = 0 and other map when y_1 = 1 as shown in Fig. 8.

The map y_1 = 0 is used to find equation for S_1 by forming prime implicants using 1's.

The map y_1 = 1 is used to find equation for R_1' by forming prime implicants using 0's.

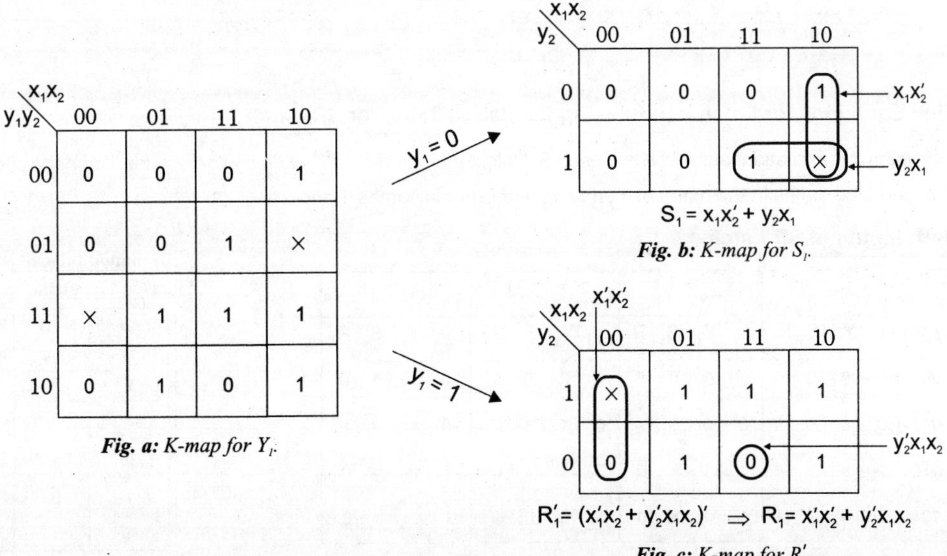

***Fig. a:** K-map for Y_1.*

***Fig. b:** K-map for S_1.*

***Fig. c:** K-map for R_1'.*

***Fig. 8:** K-maps to find equations for S_1 and R_1.*

In latch-2, the input equations for S_2 and R_2 will be independent of y_2. Therefore, the map of Y_2 can be split into two maps: One map when $y_2 = 0$ and other map when $y_2 = 1$ as shown in Fig. 9.

The map $y_2 = 0$ is used to find equation for S_2 by forming prime implicants using 1's.

The map $y_2 = 1$ is used to find equation for R_2' by forming prime implicants using 0's.

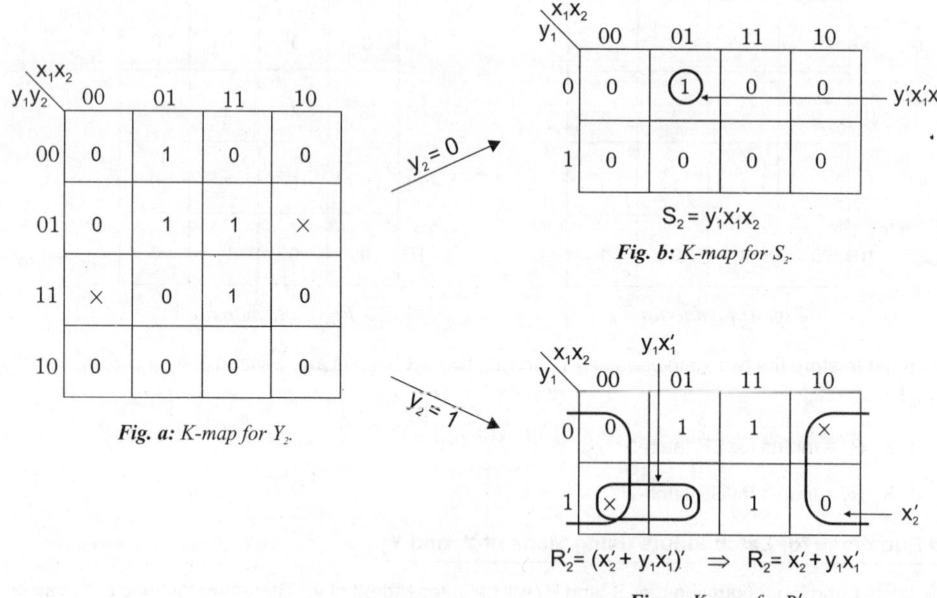

Fig. a: K-map for Y_2.

Fig. b: K-map for S_2.

Fig. c: K-map for R_2'.

Fig. 9: K-maps to find equations for S_2 and R_2.

Form the maps of S_1, R_1', S_2 and R_2' we get the following equations for SR latch inputs.

$S_1 = x_1x_2' + y_2x_1$	$R_1 = x_1'x_2' + y_2'x_1x_2$
$S_2 = y_1'x_1'x_2$	$R_2 = x_2' + y_1x_1'$

To Find Equations for Latch Inputs using Excitation Table for SR Latch

Using the excitation table of NOR based SR latch, the inputs for SR latch can be obtained from the circuit inputs, present and next states in the maps of Y_1 and Y_2 and listed in Tables 1 and 2.

Table 1: Inputs of SR Latch-1

	$x_1x_2 = 00$				$x_1x_2 = 01$				$x_1x_2 = 11$				$x_1x_2 = 10$			
y_1y_2	y_1	Y_1	S_1	R_1	y_1	Y_1	S_1	R_1	y_1	Y_1	S_1	R_1	y_1	Y_1	S_1	R_1
00	0	0	0	×	0	0	0	×	0	0	0	×	0	1	1	0
01	0	0	0	×	0	0	0	×	0	1	1	0	0	×	×	×
11	1	×	×	×	1	1	×	0	1	1	×	0	1	1	×	0
10	1	0	0	1	1	1	×	0	1	0	0	1	1	1	×	0

Excitation Table of NOR based SR Latch

Present State y	Next State Y	Inputs for Next state S	R
0	0	0	×
0	1	1	0
1	0	0	1
1	1	×	0

Table 2: Inputs of SR Latch-2

	$x_1x_2 = 00$				$x_1x_2 = 01$				$x_1x_2 = 11$				$x_1x_2 = 10$			
y_1y_2	y_2	Y_2	S_2	R_2	y_2	Y_2	S_2	R_2	y_2	Y_2	S_2	R_2	y_2	Y_2	S_2	R_2
00	0	0	0	×	0	1	1	0	0	0	0	×	0	0	0	×
01	1	0	0	1	1	1	×	0	1	1	×	0	1	×	×	×
11	1	×	×	×	1	0	0	1	1	1	×	0	1	0	0	1
10	0	0	0	×	0	0	0	×	0	0	0	×	0	0	0	×

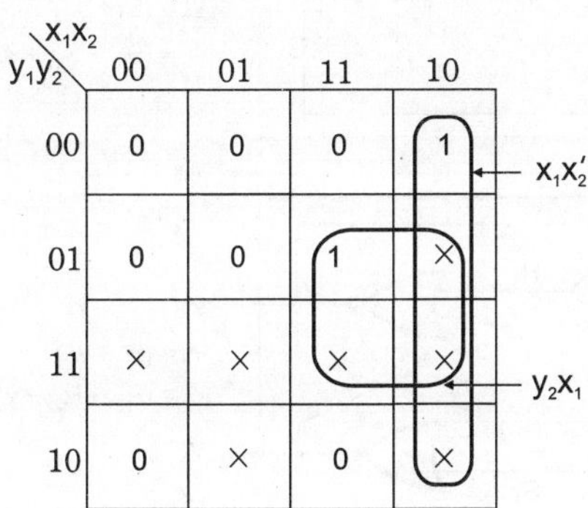

Fig. 10: K-map for S_1.

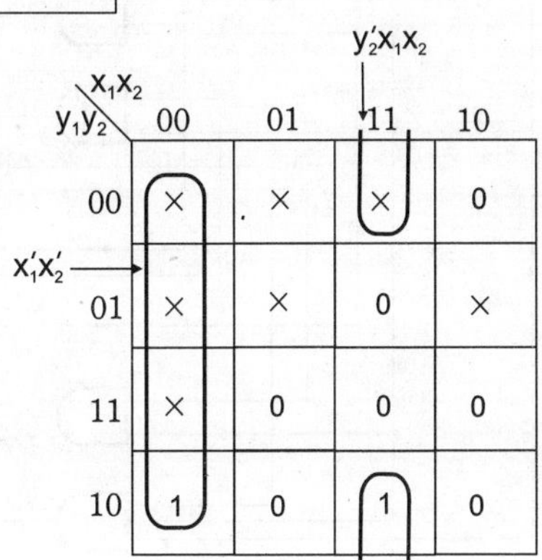

Fig. 11: K-map for R_1.

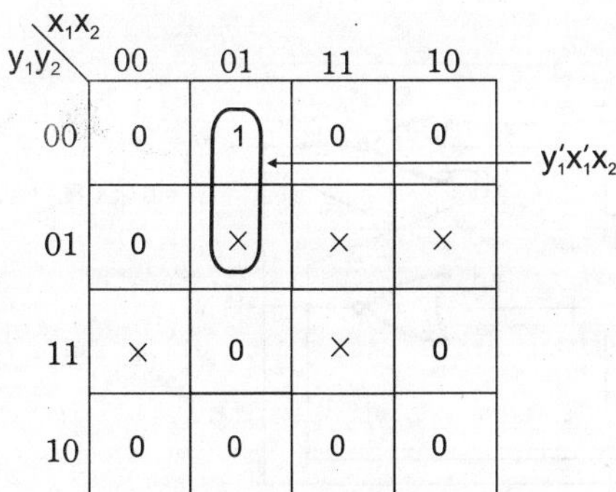

Fig. 12: K-map for S_2.

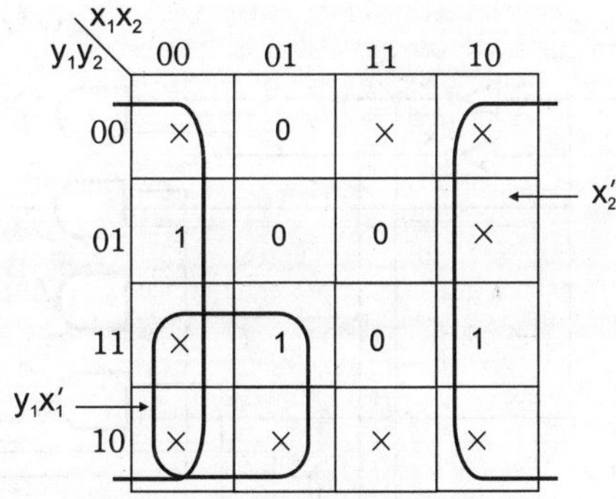

Fig. 13: K-map for R_2.

From Figs. 10 to 13, the maps for S_1, R_1, S_2, R_2 expressions are as shown below:

$S_1 = x_1x_2' + y_2x_1$ | $R_1 = x_1'x_2' + y_2'x_1x_2$

$S_2 = y_1'x_1'x_2$ | $R_2 = x_2' + y_1x_1'$

Logic Diagram

The Boolean equations obtained for latch inputs in both methods are same and using these equations the combinational logic circuit to generate latch inputs are drawn as shown in Fig. 14(a). The complete logic circuit is obtained by adding NOR based SR latches to the combinational circuit as shown in Fig. 14(a).

In Fig. 14(a), the combinational circuit is implemented using AND-OR gate structure. In order to prevent essential hazards the AND-OR structure can be replaced by AND-NOR structure with multi-input NOR gates for latches as shown in Fig. 14(b).

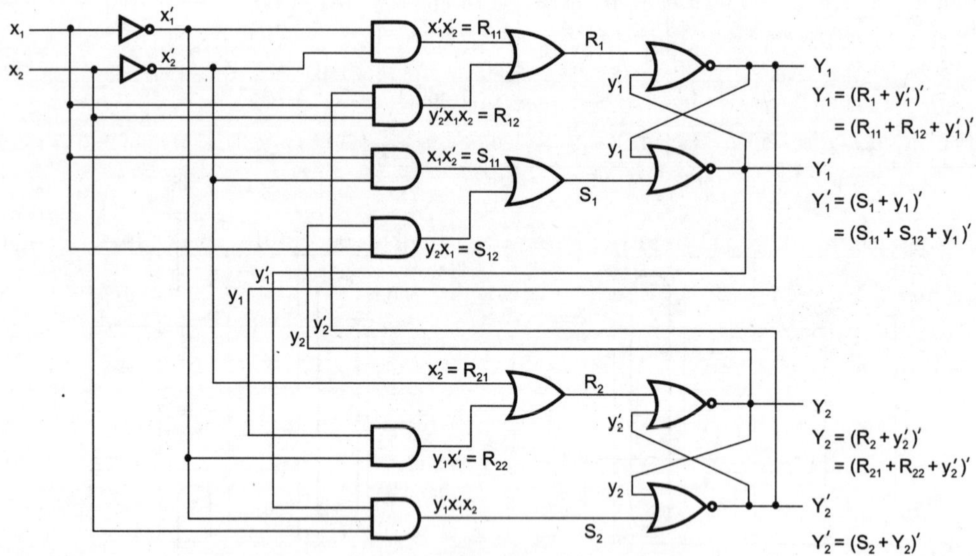

Fig. a: Combinational circuit with AND-OR gate structure.

⇓

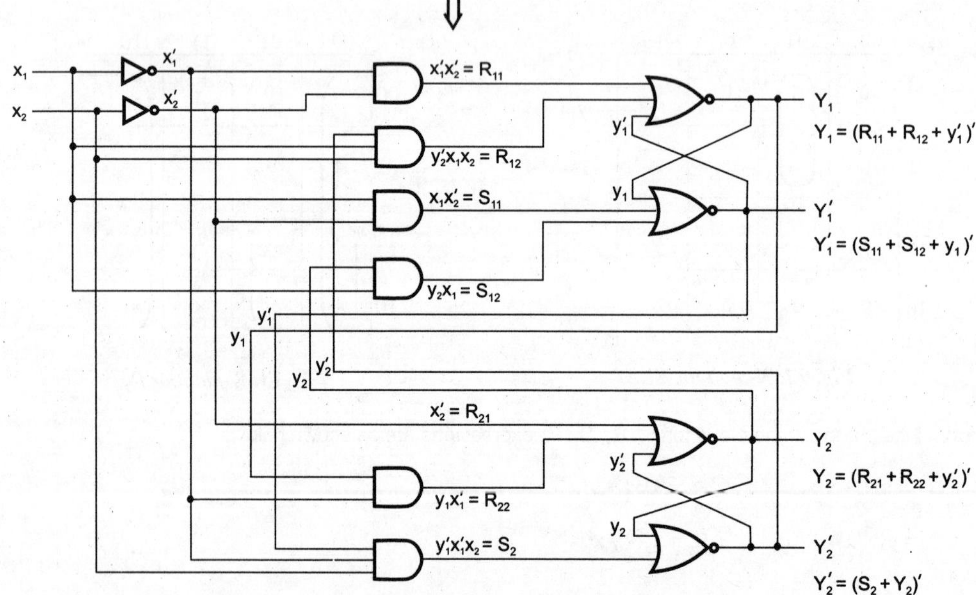

Fig. b: Modified gate structure to minimize essential hazard.

Fig. 14: Logic circuit using AND gate structure and NOR based SR latch.

Modification of Gate Structure to Eliminate Essential Hazard Due to Inverters

In the combinational logic circuit to generate latch inputs in Fig. 14(b), the inverters can be replaced by modified gate structure shown in Fig. 15. This will eliminate the essential hazards due to inveters. The modified complete logic circuit is shown in Fig. 16.

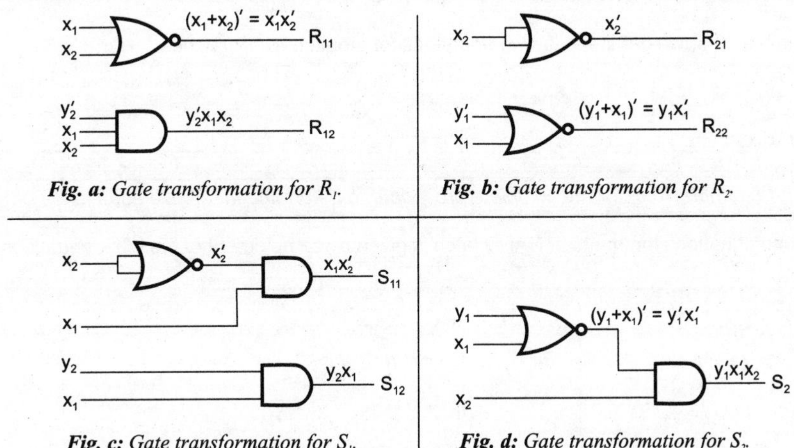

Fig. a: *Gate transformation for R_1.*

Fig. b: *Gate transformation for R_2.*

Fig. c: *Gate transformation for S_1.*

Fig. d: *Gate transformation for S_2.*

Fig. 15: *Gate transformation to eliminate essential hazards.*

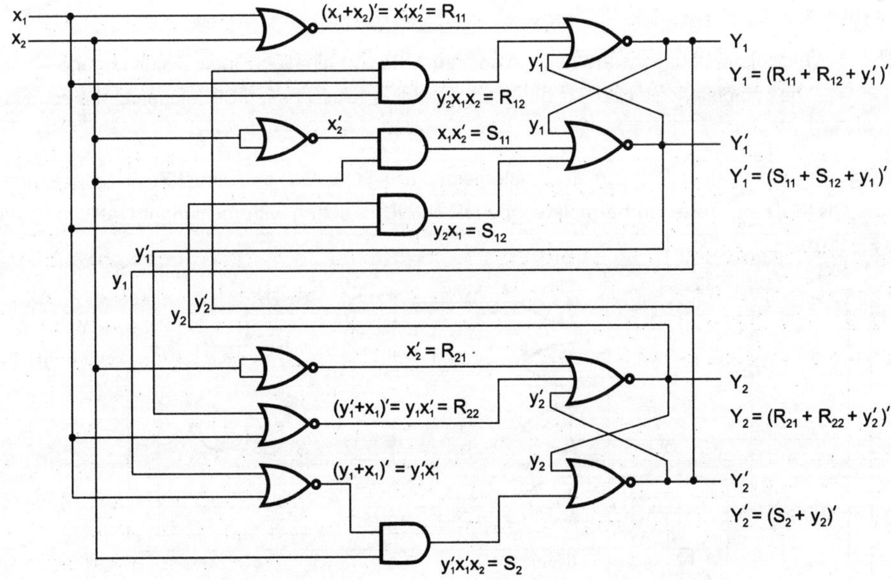

Fig. 16: *Logic circuit using NOR based SR latch with modified gate structure to eliminate essential hazard due to inverters.*

Example 10.7

Realize a logic circuit using NAND based SR latches and OR-AND gate structure for the flow table of Example 10.6. Modify the gate structure for hazard free realization.

Solution

From Example 10.6, we get the following equations for inputs of NOR latches.

$$S_1 = x_1x_2' + y_2x_1 \qquad R_1 = x_1'x_2' + y_2'x_1x_2$$

$$S_2 = y_1'x_1'x_2 \qquad R_2 = x_2' + y_1x_1'$$

***Note:** Refer Example 10.6, for the complete procedure to determine the above equations from the flow table.*

The Boolean equations for inputs of NAND latch is given by complement of the input equations of NOR based SR latch.

$$\therefore S_1 = (x_1x_2' + y_2x_1)' = (x_1x_2')' \cdot (y_2x_1)' = (x_1' + x_2) \cdot (y_2' + x_1')$$

$$S_2 = (y_1'x_1'x_2)' = y_1 + x_1 + x_2'$$

$$R_1 = (x_1'x_2' + y_2'x_1x_2)' = (x_1'x_2')' \cdot (y_2'x_1x_2)' = (x_1 + x_2) \cdot (y_2 + x_1' + x_2')$$

$$R_2 = (x_2' + y_1x_1')' = (x_2')' \cdot (y_1x_1')' = x_2 \cdot (y_1' + x_1)$$

Logic Diagram

The above Boolean equations are used to construct the combinational logic circuit to generate latch inputs as shown in Fig. 1. The complete logic circuit is obtained by adding NAND based SR latches to the combinational circuit as shown in Fig. 1.

In Fig. 1(a), the combinational circuit is implemented using OR-AND gate structure. In order to prevent essential hazards the OR-AND structure can be replaced by OR-NAND structure with multi-input NAND gates for latches as shown in Fig. 1(b).

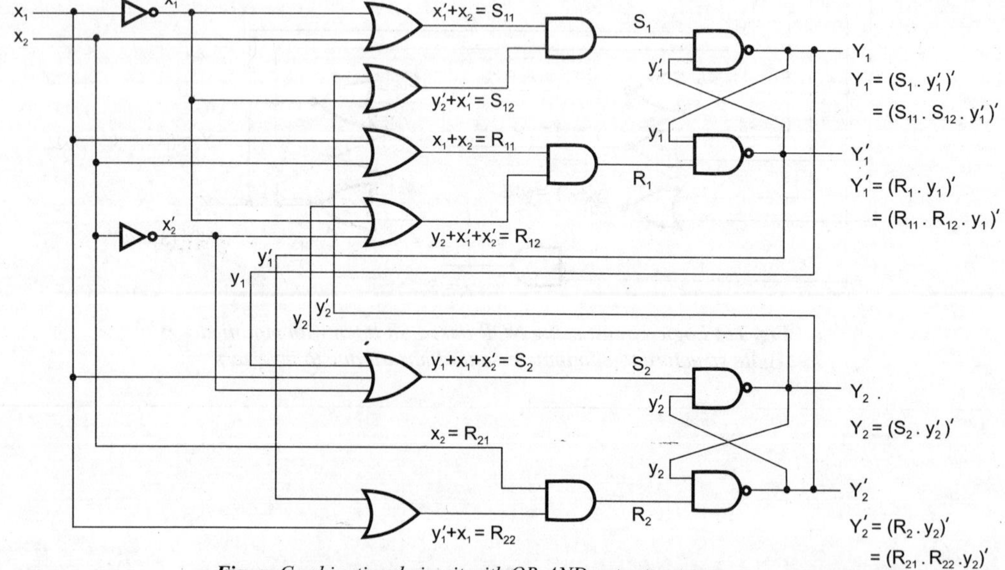

Fig. a: Combinational circuit with OR-AND gate structure.

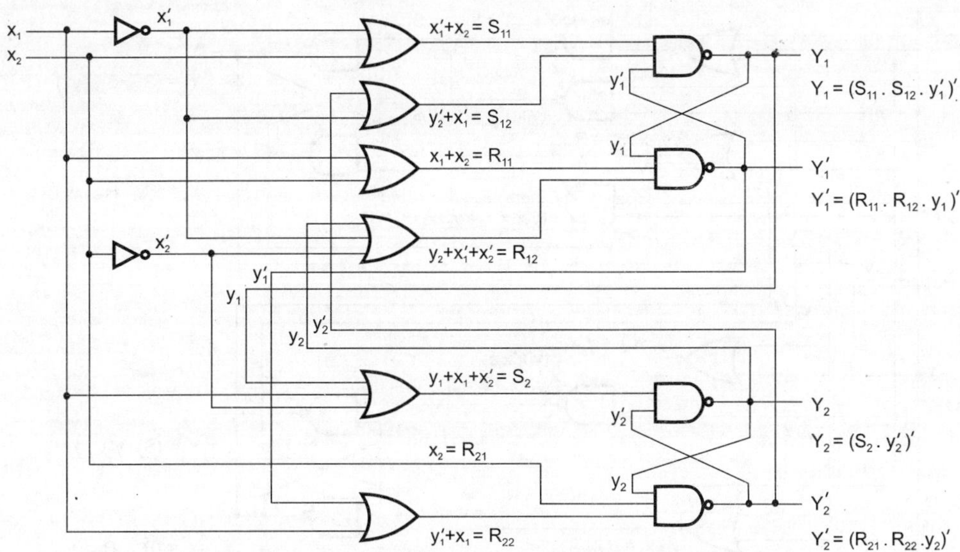

Fig. b: Modified gate structure to minimize essential hazard.

Fig. 1: Logic circuit using OR gate structure and NAND based SR latch.

Modification of Gate Structure to Eliminate Essential Hazard

In the combinational circuit to generate latch input in the circuit of Fig. 1(b), the inverters can be replaced by modified gate structure shown in Fig. 2. This will eliminate the essential hazards due to inveters. The modified complete logic circuit is shown in Fig. 3.

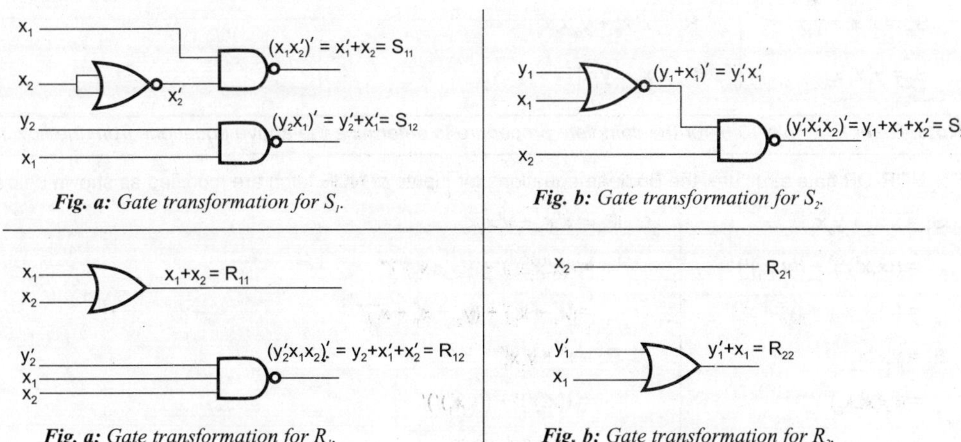

Fig. a: Gate transformation for S_1. *Fig. b: Gate transformation for S_2.*

Fig. a: Gate transformation for R_1. *Fig. b: Gate transformation for R_2.*

Fig. 2: Gate transformation to eliminate essential hazards.

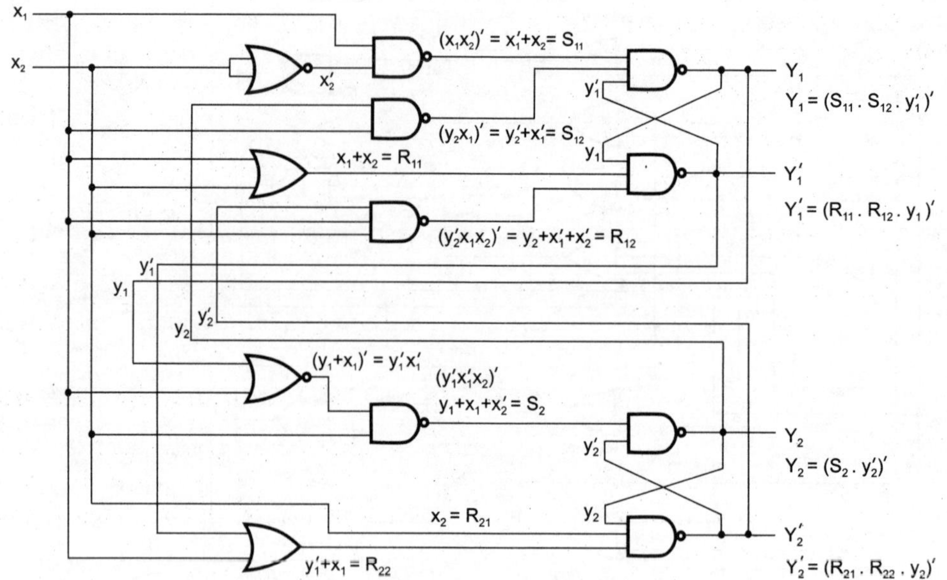

***Fig. 3:** Logic circuit using NAND based SR latch with modified gate structure to eliminate essential hazard due to inverters.*

Example 10.8

Realize a logic circuit using NOR based SR latches and NOR-OR gate structure for the flow table of Example 10.6. Modify the gate structure for hazard free realization.

Solution

From Example 10.6, we get the following equations for inputs of NOR latches.

$S_1 = x_1x_2' + y_2x_1$	$R_1 = x_1'x_2' + y_2'x_1x_2$
$S_2 = y_1'x_1'x_2$	$R_2 = x_2' + y_1x_1'$

Note: *Refer Example 10.6, for the complete procedure to determine the above equations from the flow table.*

For NOR-OR gate structure, the Boolean equations for inputs of NOR latch are modified as shown below:

$$\therefore S_1 = x_1x_2' + y_2x_1 = ((x_1x_2')')' + ((y_2x_1)')' = (x_1' + x_2)' + (y_2' + x_1')'$$

$$R_1 = x_1'x_2' + y_2'x_1x_2 = ((x_1'x_2')')' + ((y_2'x_1x_2)')' = (x_1 + x_2)' + (y_2 + x_1' + x_2')'$$

$$S_2 = y_1'x_1'x_2 = ((y_1'x_1'x_2)')' = (y_1 + x_1 + x_2')'$$

$$R_2 = x_2' + y_1x_1' = ((x_2')')' + ((y_1x_1')')' = x_2' + (y_1' + x_1)'$$

Logic Diagram

The above Boolean equations are used to construct the combinational logic circuit to generate latch inputs as shown in Fig. 1. The complete logic circuit is obtained by adding NOR based SR latches to the combinational circuit as shown in Fig. 1.

In Fig. 1(a), the combinational logic circuit is implemented using NOR-OR gate structure. In order to prevent essential hazards the NOR-OR structure can be replaced by NOR-NOR structure with multi-input NOR gates for latches as shown in Fig. 1(b).

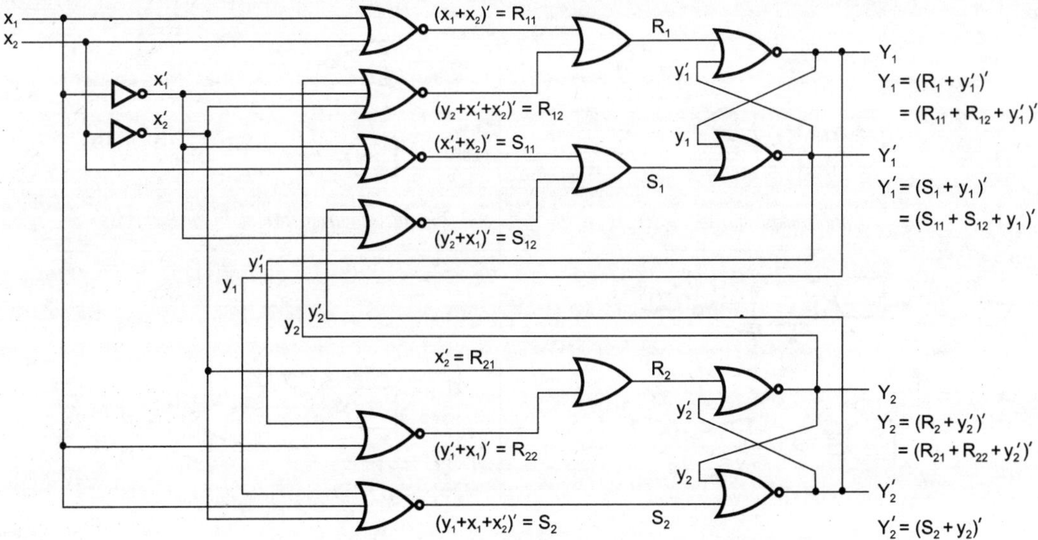

***Fig. a:** Combinational circuit with NOR-OR gate structure.*

⇓

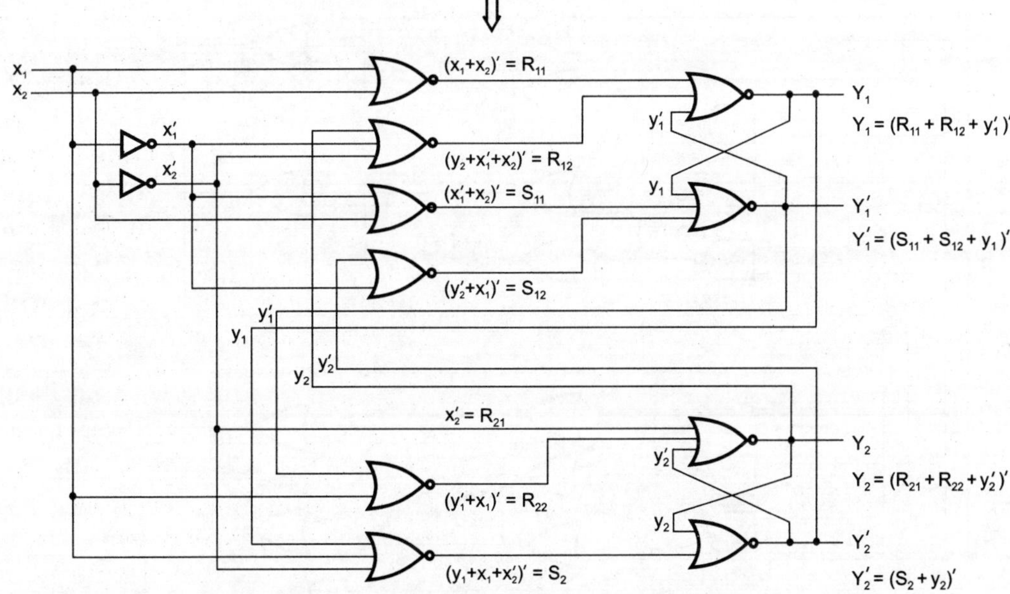

***Fig. b:** Modified gate structure to minimize essential hazard.*

***Fig. 1:** Logic circuit using NOR gate structure and NOR based SR latch.*

Modification of Gate Structure to Eliminate Essential Hazard Due to Inverters

In the combinational logic circuit to generate latch inputs in Fig. 1(b), the inverters can be replaced by modified gate structure shown in Fig. 2. This will eliminate the essential hazards due to inveters. The modified complete logic circuit is shown in Fig. 3.

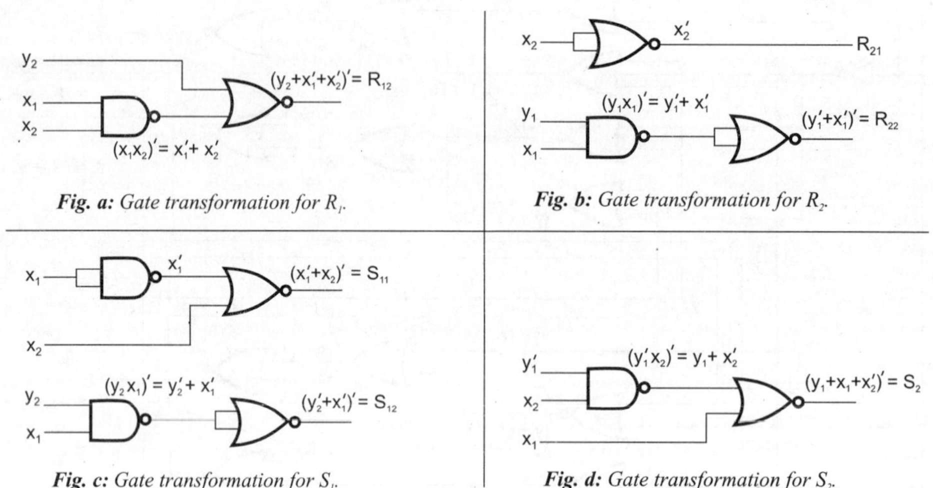

Fig. a: Gate transformation for R_1.

Fig. b: Gate transformation for R_2.

Fig. c: Gate transformation for S_1.

Fig. d: Gate transformation for S_2.

Fig. 2: Gate transformation to eliminate essential hazards.

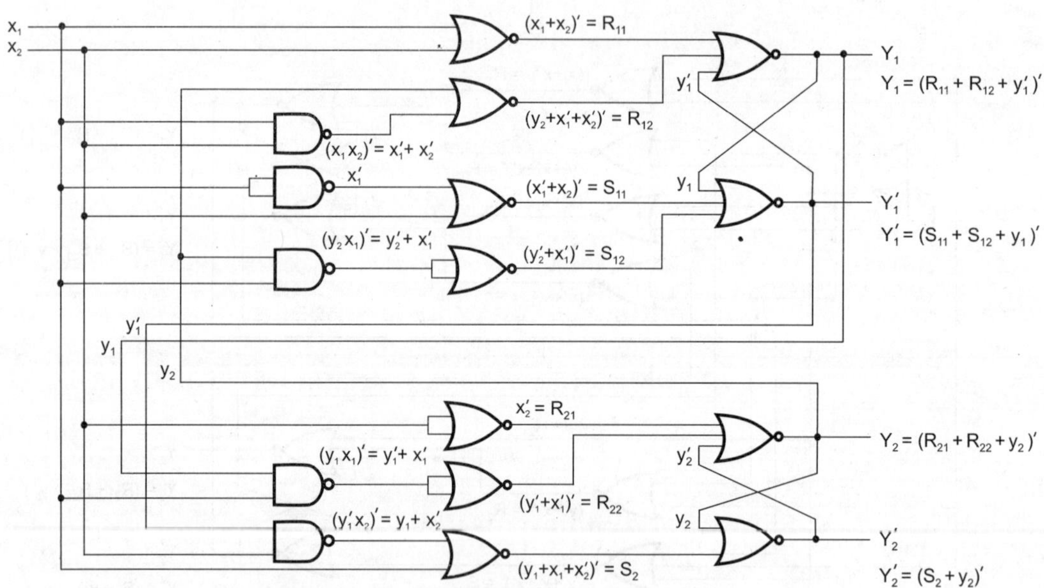

Fig. 3: Logic circuit using NOR based SR latch with modified gate structure to eliminate essential hazard.

Example 10.9

Realize a logic circuit using NAND based SR latches and NAND-AND gate structure for the flow table of Example 10.6. Modify the gate structure for hazard free realization.

Solution

From Example 10.6, we get the following equations for inputs of NOR latches.

$S_1 = x_1x_2' + y_2x_1$ | $R_1 = x_1'x_2' + y_2'x_1x_2$

$S_2 = y_1'x_1'x_2$ | $R_2 = x_2' + y_1x_1'$

> ***Note:*** *Refer Example 10.6, for the complete procedure to determine the above equations from the flow table.*

The Boolean equations for inputs of NAND latch is given by complement of the input equations of NOR based SR latch.

$\therefore S_1 = (x_1x_2' + y_2x_1)'$

$= (x_1x_2')' . (y_2x_1)'$

$S_2 = y_1'x_1'x_2$

$= (y_1'x_1'x_2)'$

$R_1 = (x_1'x_2' + y_2'x_1x_2)'$

$= (x_1'x_2')' . (y_2'x_1x_2)'$

$R_2 = x_2' + y_1x_1'$

$= (x_2' + y_1x_1')'$

$= (x_2')' . (y_1x_1')' = x_2 . (y_1x_1')'$

Logic Diagram

The above Boolean equations are used to construct the combinational logic circuit to generate latch inputs as shown in Fig. 1. The complete logic circuit is obtained by adding NAND based SR latches to the combinational circuit as shown in Fig. 1.

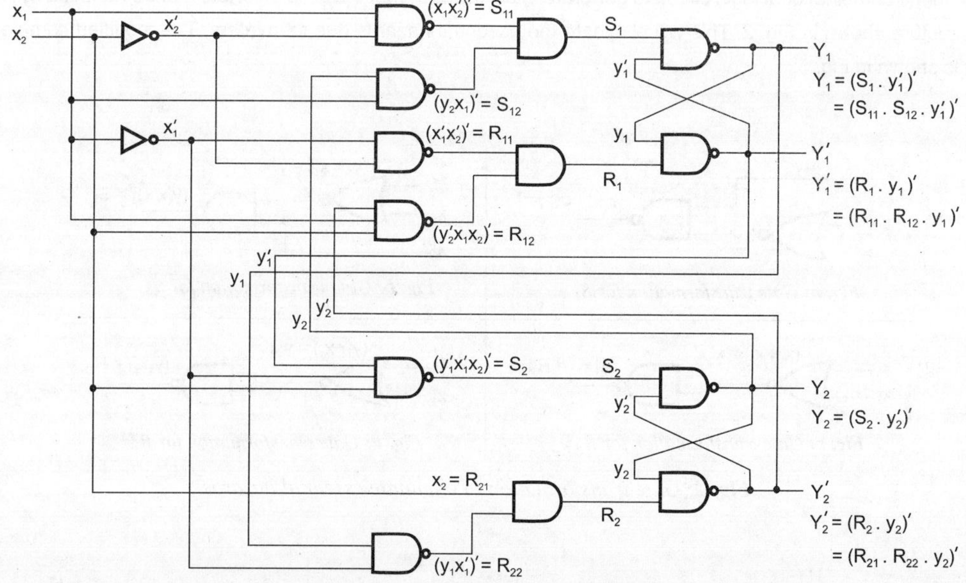

Fig. a: Combinational circuit with NAND-AND gate structure.

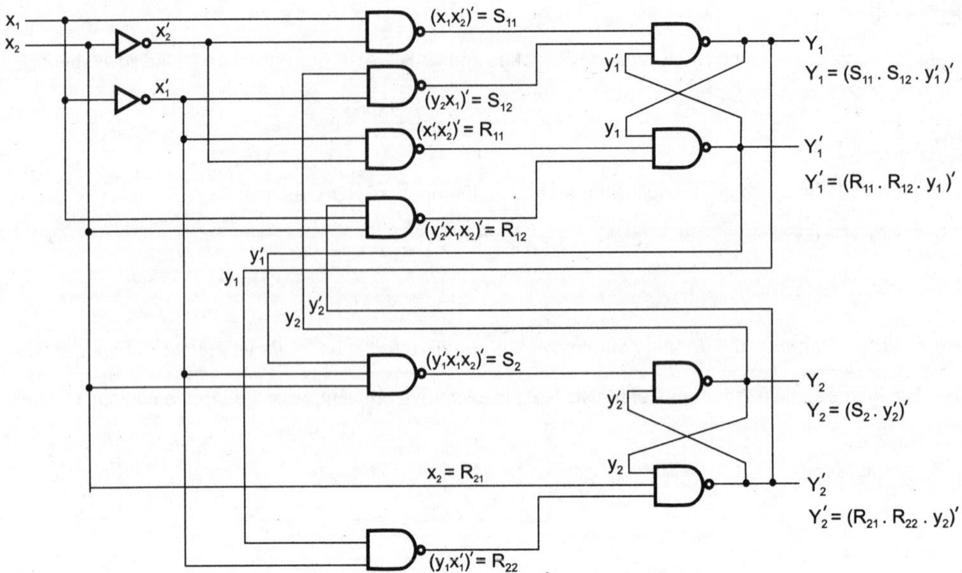

Fig. b: Modified gate structure to minimize essential hazard.

Fig. 1: Logic circuit using NAND gate structure and NAND based SR latch.

In Fig. 1(a), the combinational logic circuit is implemented using NAND-AND gate structure. In order to prevent essential hazards the NAND-AND structure can be replaced by NAND-NAND structure with multi-input NAND gates for latches as shown in Fig. 1(b).

Modification of Gate Structure to Eliminate Essential Hazard Due to Inverters

In the combinational logic circuit to generate latch inputs in Fig. 1(b), the inverters can be replaced by modified gate structure shown in Fig. 2. This will eliminate the essential hazards due to inveters. The modified complete logic circuit is shown in Fig. 3.

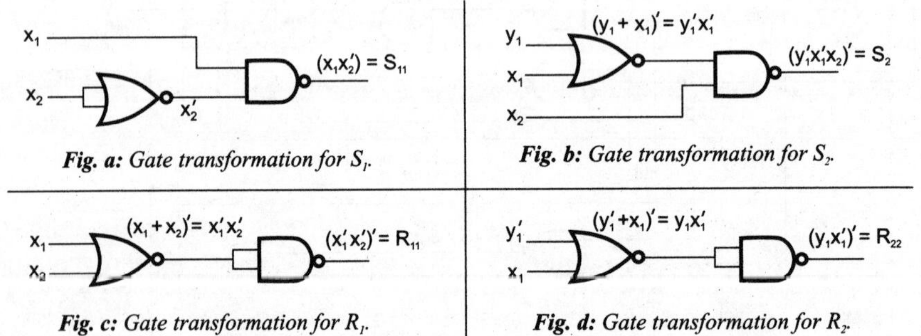

Fig. 2: Gate transformation to eliminate essential hazards.

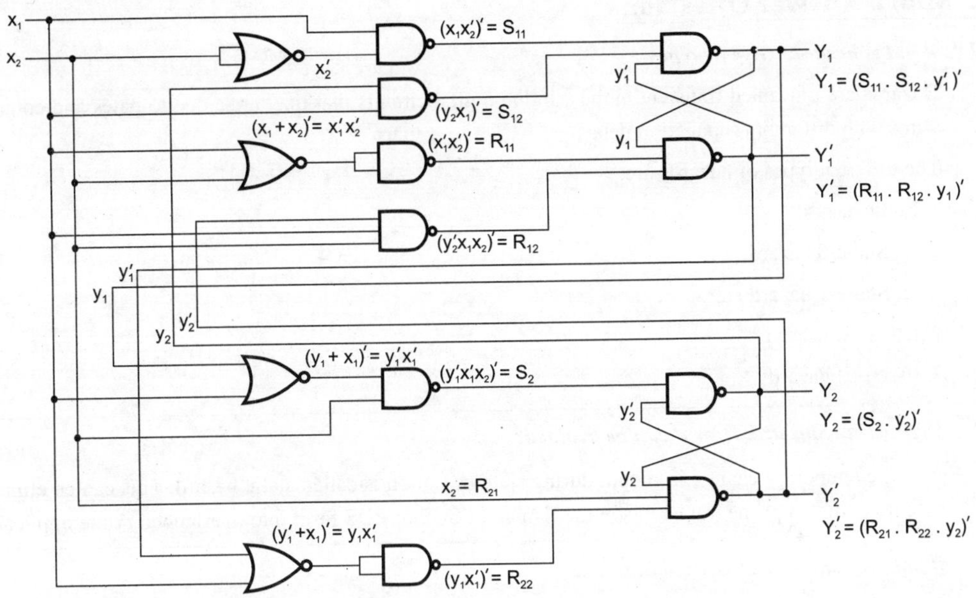

Fig. 3: Logic circuit using NAND based SR latch with modified gate structure to eliminate essential hazard.

10.6 Summary of Important Concepts

1. Hazards are unwanted transient outputs in the digital circuits that may arise due to gates and connecting wires with different propagation delay.
2. When the expected output is 1 for a change in input, if the circuit generates a temporary 0 then the hazard is called static-1 hazard.
3. When the expected output is 0 for a change in input, if the circuit generates a temporary 1 then the hazard is called static-0 hazard.
4. When the expected output is either 1 or 0 for a change in input, if the circuit output temporarily changes from 0 to 1 and 1 to 0 three or more times, then the hazard is called dynamic hazard.
5. The synchronous sequential circuits will not have hazards because in those circuits the changes are synchronized with a clock.
6. In synchronous sequential circuits the changes in circuit conditions are synchronized by a clock with time period long enough to overcome internal delays in signal transitions.
7. Hazards will occur in asynchronous sequential circuits if there are different delays in various signal paths.
8. The static and dynamic hazards in asynchronous sequential circuits can be eliminated by adding redundant gates or by using SR latch based circuit design.
10. The essential hazard is due to different delays in different paths that start from a same input.
11. An essential hazard is due to unequal delay in two or more paths that start from same input and also due to delay in input path longer than delay in feedback path.
12. The essential hazard can be eliminated by rearranging the gate structure of the combinational circuit that generates inputs to latches.

10.7 Short-Answer Questions

Q10.1 *What is hazard? Give its types.*

Hazards are unwanted transient outputs in the digital circuits that may arise due to gates and connecting wires with different propagation delay.

The different types of hazards are,

1. Static hazard
   1. Static-0 hazard
   2. Static-1 hazard
2. Dynamic hazard.
3. Essential hazard

Q10.2 *Define static hazard. How it can be avoided?*

A single temporary change in output during state transition is called static hazard. This can be eliminated by converting all prime implicants in the K-map of SOP or POS form into overlapped prime implicants.

Q10.3 *What are static-1 and static-0 hazards?*

Static-1 hazard: When the expected output is 1 for a change in input, if the circuit generates a temporary 0 as shown in Fig. Q10.3(a) then the hazard is called static-1 hazard.

Static-0 hazard: When the expected output is 0 for a change in input, if the circuit generates a temporary 1 as shown in Fig. Q10.3(b) then the hazard is called static-0 hazard.

Fig. a: *Static-1 hazard.*

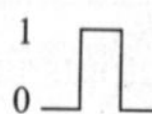

Fig. b: *Static-0 hazard.*

Fig. Q10.3.

Q10.4 *Explain dynamic hazard.*

When the expected output is either 1 or 0 for a change in input, if the circuit output temporarily changes from 0 to 1 and 1 to 0 three or more times as shown in Fig. Q10.4, then the hazard is called dynamic hazard.

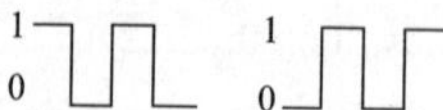

Fig. Q10.4: *Dynamic hazard.*

Q10.5 *What is an essential hazard?*

An essential hazard is due to unequal delay in two or more paths that start from same input.

Alternatively, we can say that an essential hazard exist if the total stable state reached after three changes in an input is different from stable state reached after a single change in that input.

Q10.6 *What is the cause of essential hazard?*

The essential hazard is due to delay in changes in input reaching the input of excitation circuit before the feedback output reaching the input of excitation circuit. Hence feedback inputs will change two or more times before reaching a stable value.

Q10.7 How to eliminate essential hazard?

In general, the essential hazard is minimized or eliminated by reducing the complexity of gate structure. Even after modifying the gate structure if essential hazard exists in certain path then appropriate delays has to be introduced by individually studying the signal paths.

Q10.8 Draw the hazard free circuit, $F = \sum m(1, 5, 6, 7)$.

Solution:

The K-map of the given function is shown in Fig. Q10.8.1. Here, the prime implicants are non-overlapped prime implicants. Hence, one additional prime implicant is added to create overlapped prime implicants and to prevent hazard as shown in Fig. Q10.8.2.

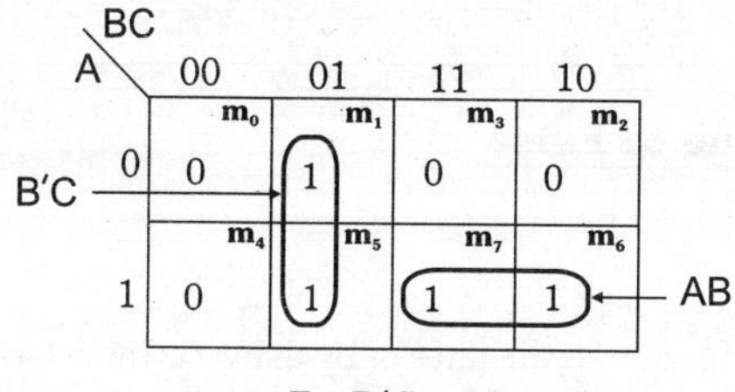

***Fig. Q10.8.1:** K-map of F.*

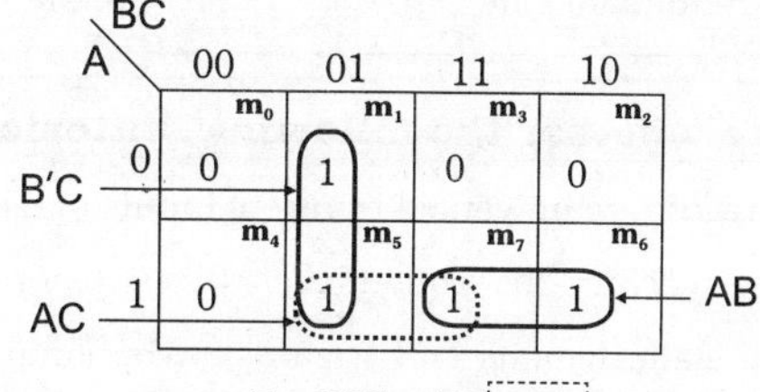

***Fig. Q10.8.2:** Overlapping minterms to eliminate hazard.*

The following Boolean equations are obtained from K-map of Fig. Q10.8.2.

$$F = B'C + AB + AC$$

Using the above Boolean equation the hazard free logic circuit is drawn as shown in Fig. Q10.8.3.

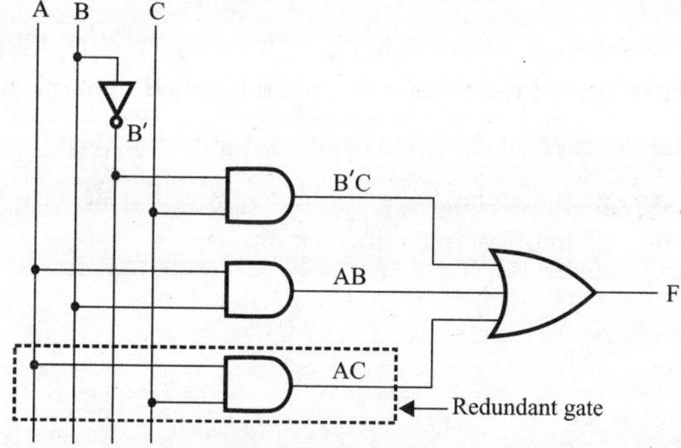

***Fig. Q10.8.3:** Logic circuit.*

10.8 Exercises

I. Fill in the blanks

1. ______ occurs only if there are three or more paths between input and output.
2. Hazards will not occur in ______ circuits.
3. Another type of hazard in asynchronous sequential circuit is known as ______.
4. The essential hazard cannot be eliminated by ______.
5. The presence of essential hazard can be identified from the ______.

6. Hazards are transient outputs in the digital circuits that may arise due to gates and connecting wire with different ______.
7. Hazard may occur in ______ as well as asynchronous sequential circuits.
8. The ______ and ______ hazards in asynchronous sequential circuits can be eliminated similar to combinational circuit by adding redundant gates or by using SR latch based circuit design.

Answers

1. dynamic hazard
2. synchronous sequential
3. essential hazard
4. redundant gate
5. flow table
6. propagation delay
7. combinational circuits
8. static, dynamic

II. State whether the following statements are True or False

1. Hazards are unwanted transient inputs in the digital circuits.
2. The synchronous sequential circuits will not have hazards.
3. The static dynamic and hazards can be eliminated by adding redundant gates or by using SR latch based circuit design.
4. The essential hazard is due to different delays in different paths that start from a same input.
5. When the expected output is 1 for a change in input, If the circuit generates a temporary 0 then the hazard is called static-1 hazard.
6. When the expected output is 0 for a change in input, If the circuit generates a temporary 1then the hazard is called static-0 hazard.
7. When the expected output is either 1 or 0 for a change in input, if the circuit output temporarily changes from 0 to 1 and 1 to 0 three or more times, then the hazard is called dynamic hazard.
8. The dynamic hazard is due to different delays in different paths that start from a sample input.
9. An essential hazard is due to unequal delay in two or more paths that start from same input and also due to delay in output path longer than delay in feedback path.

Answers

1. False
2. True
3. True
4. True
5. True
6. True
7. True
8. False
9. True

III. Choose the right answer for the following questions

1. *Static-0 hazard occurs when*

   a) The output becomes '1' for sometime when it should be '0'

   b) The output becomes '0' for sometime when it should be '1'

   c) Output is always 1

   d) Output is always 0

2. ***Static-1 hazard occurs when***

   a) The output becomes '1' for sometime when it should be '0'

   b) The output becomes '0' for sometime when it should be '1'

   c) Output is always 1

   d) Output is always 0

3. ***Static hazard can be avoided by***

   a) Nonoverlapped prime implicants

   b) Overlapped prime implicants

   c) Adding don't-cares to some inputs

   d) Taking POS equation

4. ***The hazard in which the output changes from 0 to 1, three or more times is,***

   a) Static b) Dynamic

   c) Essential d) Functional

5. ***_______ hazard occurs only in asynchronous sequential circuits with 2 or more feedbacks.***

   a) Static b) Dynamic

   c) Essential d) Functional

6. ***Static hazard can be eliminated by,***

   a) Converting POS to SOP equation

   b) Using redundant gate

   c) Using NAND gates only

   d) Using NOR gates only

7. ***Essential hazard can be eliminated by,***

   a) Converting POS to SOP equation

   b) Converting SOP to POS equation

   c) Rearranging gate structure

   d) Using NOR or NAND gates only

8. ***Essential hazard cannot be eliminated by,***

   a) Rearranging gate structure

   b) Using NOR and NAND gates only

   c) Converting, SOP to POS equation

   d) Redundant gate

9. ***The presence of essential hazard can be identified from the***

a) Flow table b) Propagation delay

c) Prime implicants d) Redundant gate

10. ***The addition of each prime implicant will lead to adding an extra gate called***

a) Rearranging gate b) Redundant gate

c) NOR gate d) NAND gate

Answers

1. a	6. b
2. b	7. c
3. b	8. d
4. b	9. a
5. c	10. b

V. Answer the following questions

1. What are hazard free digital circuits?
2. Write a short note on hazards in asynchronous sequnetial circuits.
3. Write detailed notes on hazards in combinational circuits and sequential circuits.
4. Draw hazard free circuit for $F = \sum m(0, 1, 5, 7)$
5. For the given Boolean function, obtain a hazard free circuit. $F(A, B, C, D) = \sum m(1, 3, 6, 7, 13, 15)$
6. Find a circuit that has no static hazards and implements the Boolean function.

   $F(A, B, C, D) = \sum m(0, 2, 8, 10, 10)$
7. When do hazards occur?
8. How to avoid dynamic hazard?

VI. Solve the following problems

E10.1 Implement the switching function, $F = \sum m(1, 3, 5, 7, 10, 11, 12, 13)$ by a static hazard free two level AND-OR gate network.

E10.2 For a given Boolean function obtain the hazard free circuit.

$$F = \sum m(0, 1, 3, 4, 6, 7)$$

E10.3 Implement the following Boolean function by a logic circuit that has no hazards.

$$Y = \prod M(2, 3, 4, 6)$$

E10.4 Assign output values to the unstable states in the flow tables of Fig. E10.4 in such a way to avoid tranisent outpût pulses or hazard.

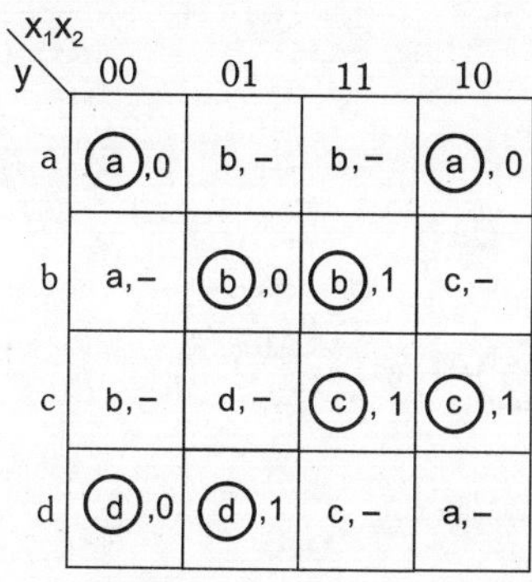

$y \backslash x_1x_2$	00	01	11	10
a	(a), 0	b, –	b, –	(a), 0
b	a, –	(b), 0	(b), 1	c, –
c	b, –	d, –	(c), 1	(c), 1
d	(d), 0	(d), 1	c, –	a, –

***Fig. E10.4:** Reduced flow table.*

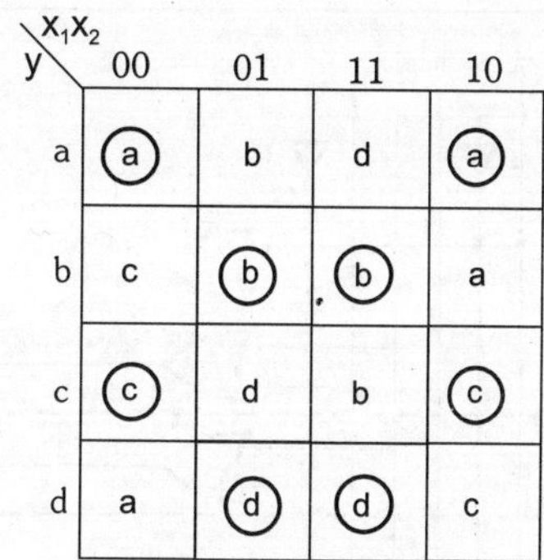

$y \backslash x_1x_2$	00	01	11	10
a	(a)	b	d	(a)
b	c	(b)	(b)	a
c	(c)	d	b	(c)
d	a	(d)	(d)	c

***Fig. E10.5:** Reduced flow table.*

E10.5 Realize a logic circuit using NOR based SR latches and AND-OR gate structure for the flow table shown in Fig. E10.5. Modify the gate structure for hazard free realization.

E10.6 Realize a logic circuit using NOR based SR latches and NOR-OR gate structure for the flow table of exercise problem E10.5. Modify the gate structure for hazard free realization.

E10.7 Realize a logic circuit using NAND based SR latches and OR-AND gate structure for the flow table of exercise problem Fig. E10.5. Modify the gate structure for hazard free realization.

E10.8 Realize a logic circuit using NAND based SR latches and NAND-AND gate structure for the flow table of exercise problem Fig. E10.5. Modify the gate structure for hazard free realization.

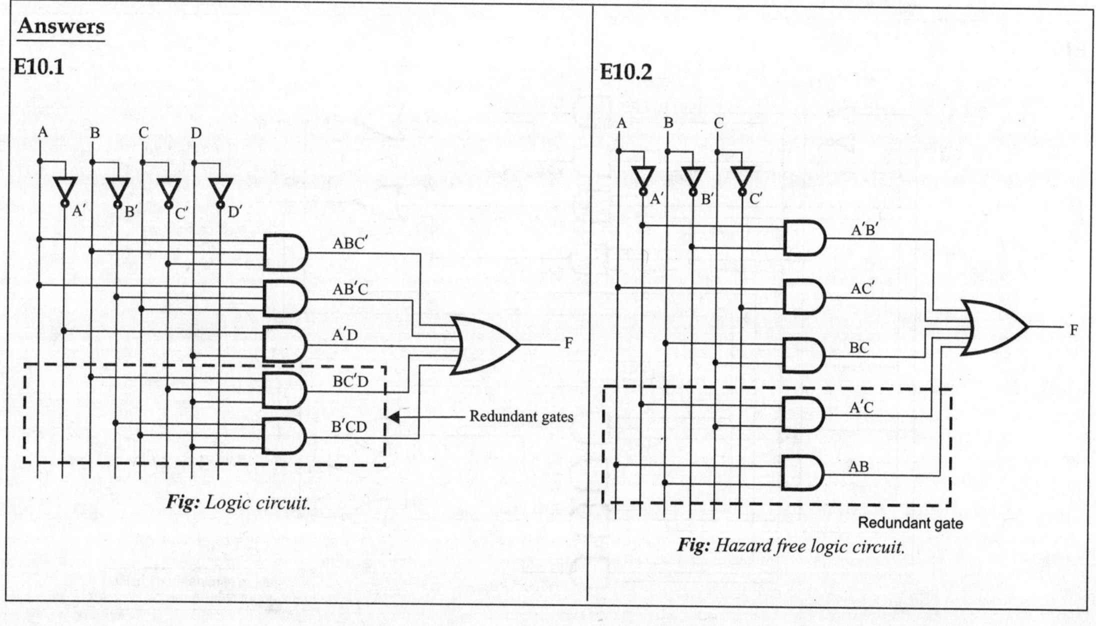

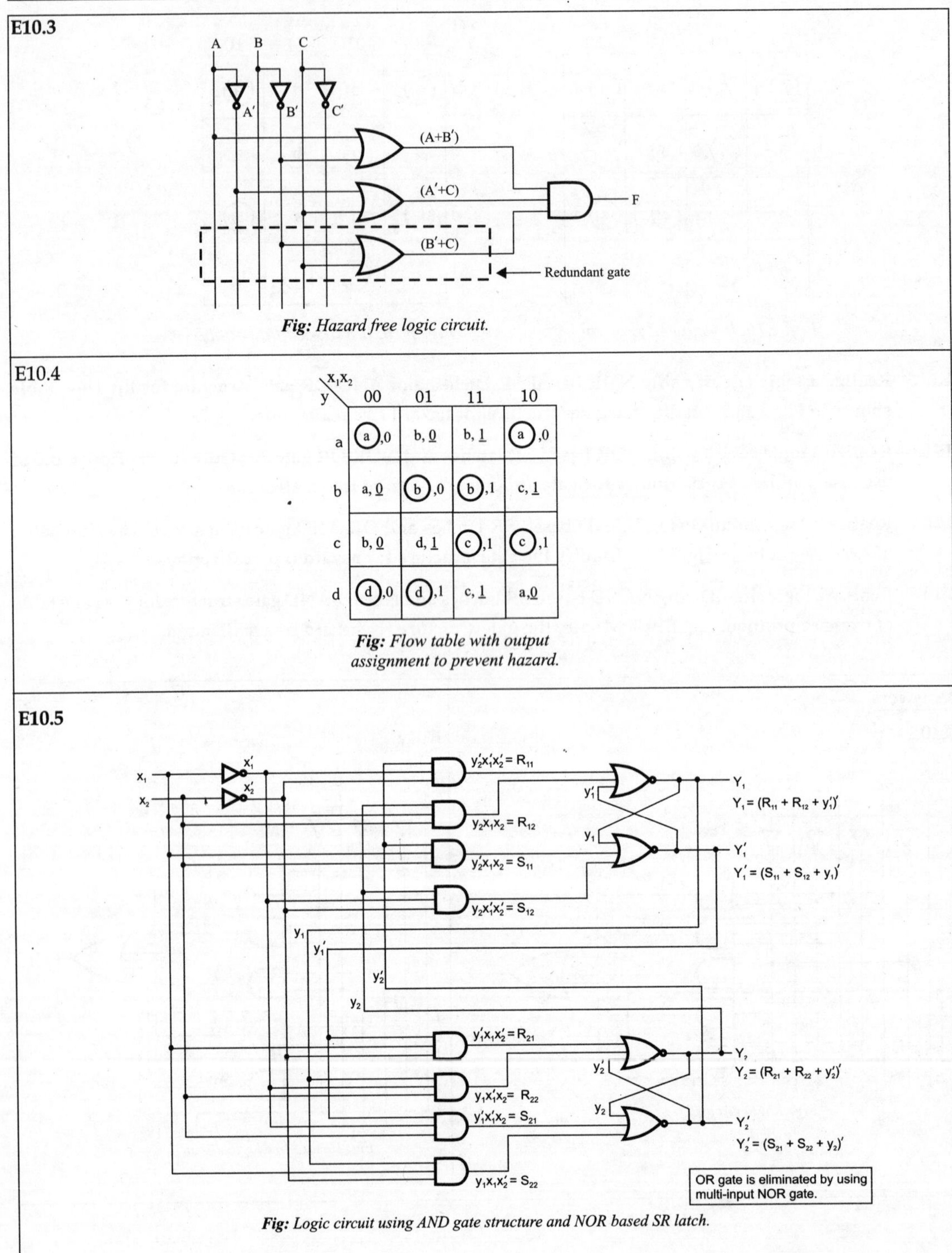

Fig: Hazard free logic circuit.

y \ x_1x_2	00	01	11	10
a	(a),0	b, 0	b, 1	(a),0
b	a, 0	(b),0	(b),1	c, 1
c	b, 0	d, 1	(c),1	(c),1
d	(d),0	(d),1	c, 1	a,0

Fig: Flow table with output assignment to prevent hazard.

Fig: Logic circuit using AND gate structure and NOR based SR latch.

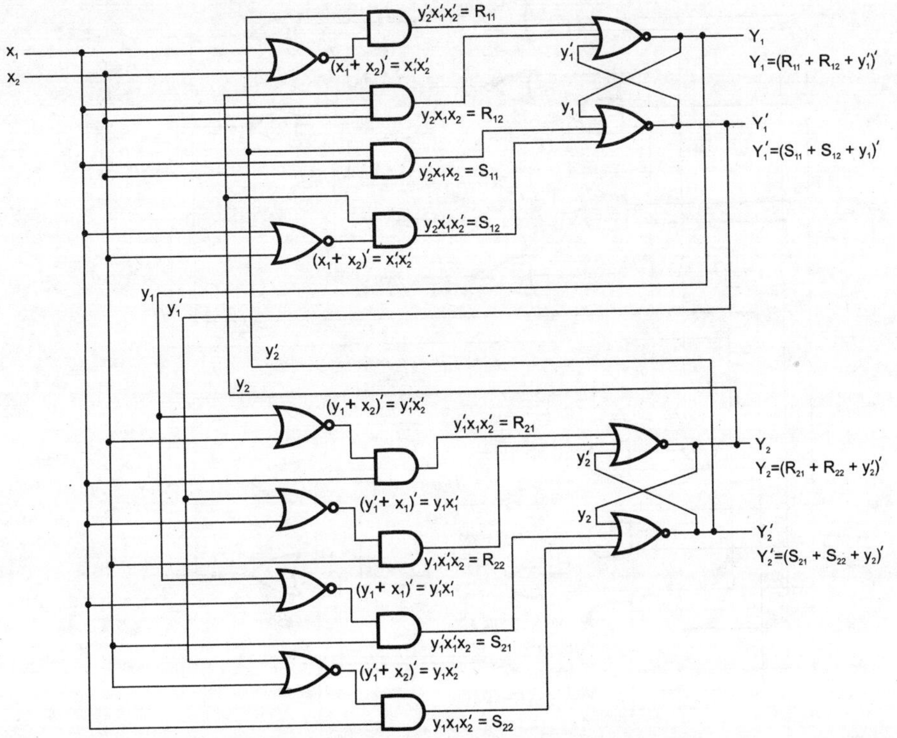

x_1
x_2
$y_2'x_1'x_2' = R_{11}$
$(x_1 + x_2)' = x_1'x_2'$
$y_2x_1x_2 = R_{12}$
$y_2'x_1x_2 = S_{11}$
$y_2x_1'x_2' = S_{12}$
$(x_1 + x_2)' = x_1'x_2'$
y_1'
y_1
Y_1
$Y_1 = (R_{11} + R_{12} + y_1')'$
Y_1'
$Y_1' = (S_{11} + S_{12} + y_1)'$
y_1
y_1'
y_2'
y_2
$(y_1 + x_2)' = y_1'x_2'$
$y_1'x_1x_2' = R_{21}$
$(y_1' + x_1)' = y_1x_1'$
$y_1x_1'x_2 = R_{22}$
$(y_1 + x_1)' = y_1'x_1'$
$y_1'x_1'x_2 = S_{21}$
$(y_1' + x_2)' = y_1x_2'$
$y_1x_1x_2' = S_{22}$
y_2'
y_2
Y_2
$Y_2 = (R_{21} + R_{22} + y_2')'$
Y_2'
$Y_2' = (S_{21} + S_{22} + y_2)'$

***Fig:** Logic circuit using NOR based SR latch with modified gate structure to eliminate essential hazard due to inverters.*

E10.6

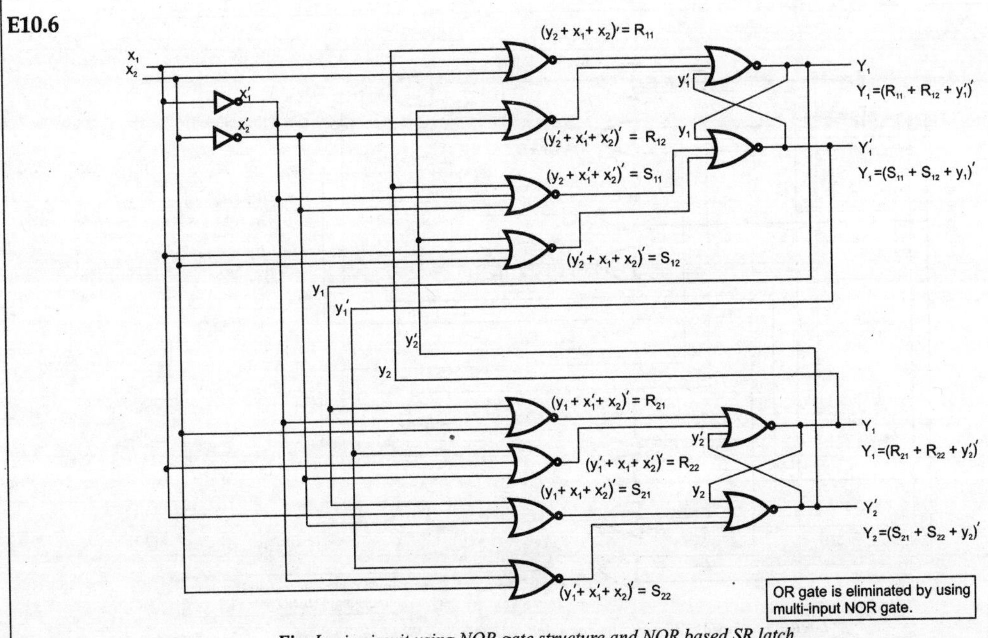

***Fig:** Logic circuit using NOR gate structure and NOR based SR latch.*

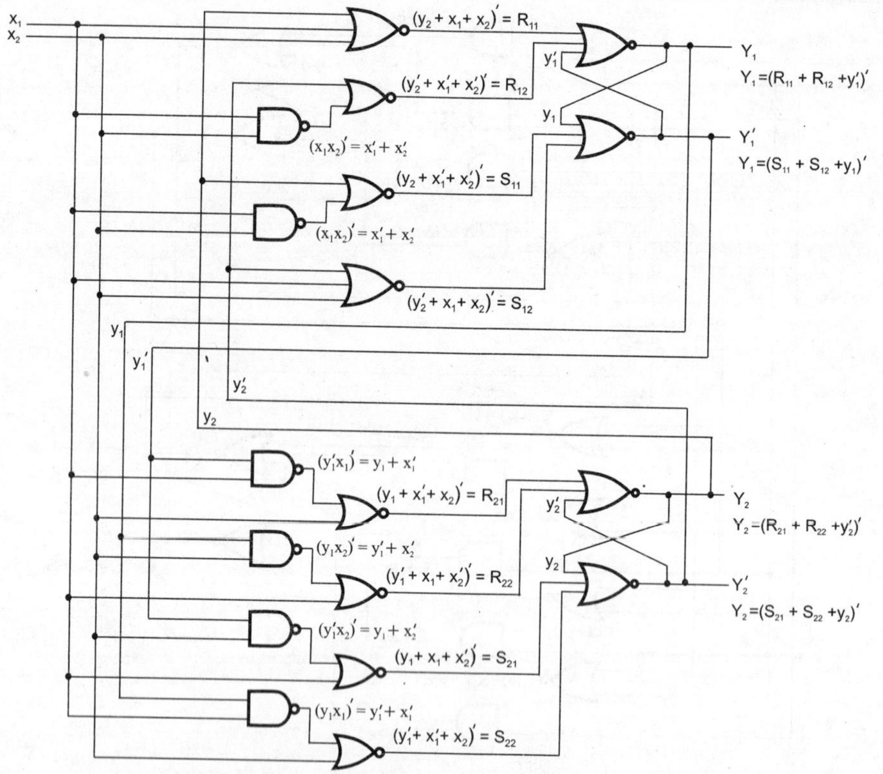

x₁
x₂
$(y_2 + x_1 + x_2)' = R_{11}$
$(y_2' + x_1' + x_2')' = R_{12}$
$(x_1x_2)' = x_1' + x_2'$
$(y_2 + x_1' + x_2')' = S_{11}$
$(x_1x_2)' = x_1' + x_2'$
$(y_2' + x_1 + x_2)' = S_{12}$
y_1'
y_1
Y_1
$Y_1 = (R_{11} + R_{12} + y_1')'$
Y_1'
$Y_1 = (S_{11} + S_{12} + y_1)'$
y_1
y_1'
y_2'
y_2
$(y_1'x_1)' = y_1 + x_1'$
$(y_1 + x_1' + x_2)' = R_{21}$
$(y_1x_2)' = y_1' + x_2'$
$(y_1' + x_1 + x_2')' = R_{22}$
$(y_1'x_2)' = y_1 + x_2'$
$(y_1 + x_1 + x_2')' = S_{21}$
$(y_1x_1)' = y_1' + x_1'$
$(y_1' + x_1' + x_2)' = S_{22}$
y_2'
y_2
Y_2
$Y_2 = (R_{21} + R_{22} + y_2')'$
Y_2'
$Y_2 = (S_{21} + S_{22} + y_2)'$

***Fig:** Logic circuit using NOR based SR latch with modified gate structure to eliminate essential hazard due to inverters.*

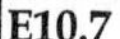

E10.7

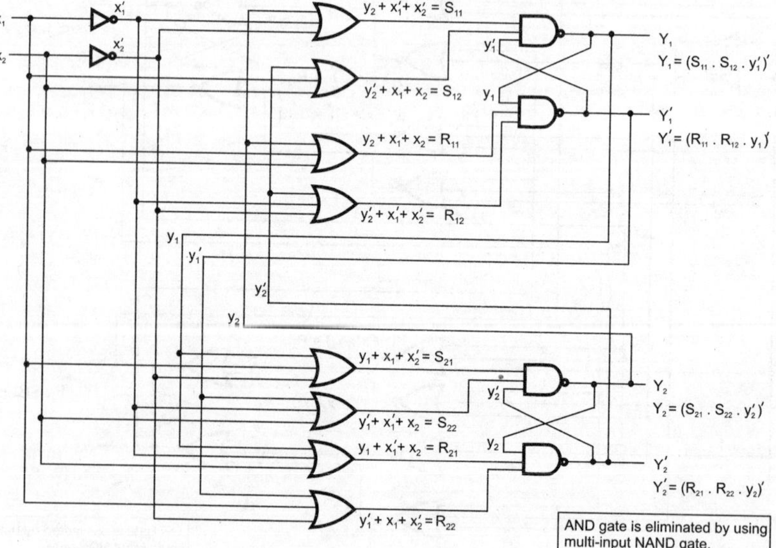

x₁
x₂
x_1'
x_2'
$y_2 + x_1' + x_2' = S_{11}$
$y_2' + x_1 + x_2 = S_{12}$
$y_2 + x_1 + x_2 = R_{11}$
$y_2' + x_1' + x_2' = R_{12}$
y_1'
y_1
Y_1
$Y_1 = (S_{11} . S_{12} . y_1')'$
Y_1'
$Y_1' = (R_{11} . R_{12} . y_1)'$
y_1
y_1'
y_2'
y_2
$y_1 + x_1 + x_2' = S_{21}$
$y_1' + x_1' + x_2 = S_{22}$
$y_1 + x_1' + x_2 = R_{21}$
$y_1' + x_1 + x_2' = R_{22}$
y_2'
y_2
Y_2
$Y_2 = (S_{21} . S_{22} . y_2')'$
Y_2'
$Y_2' = (R_{21} . R_{22} . y_2)'$
AND gate is eliminated by using multi-input NAND gate.

***Fig:** Logic circuit using OR gate structure and NAND based SR latch.*

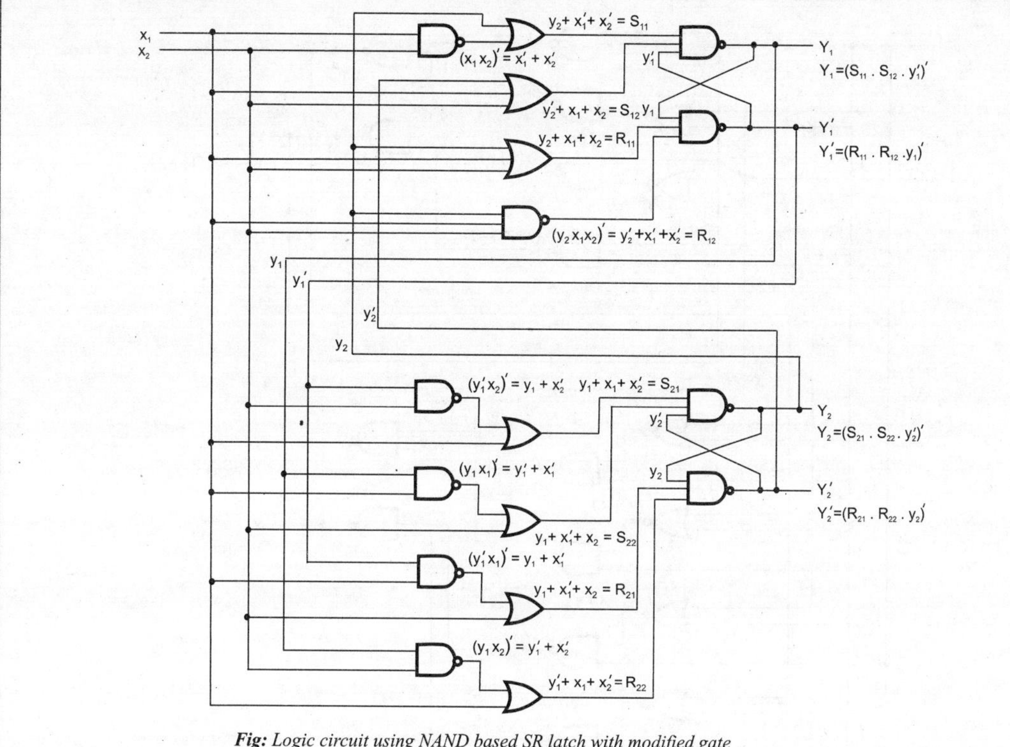

Fig: *Logic circuit using NAND based SR latch with modified gate structure to eliminate essential hazard due to inverters.*

E10.8

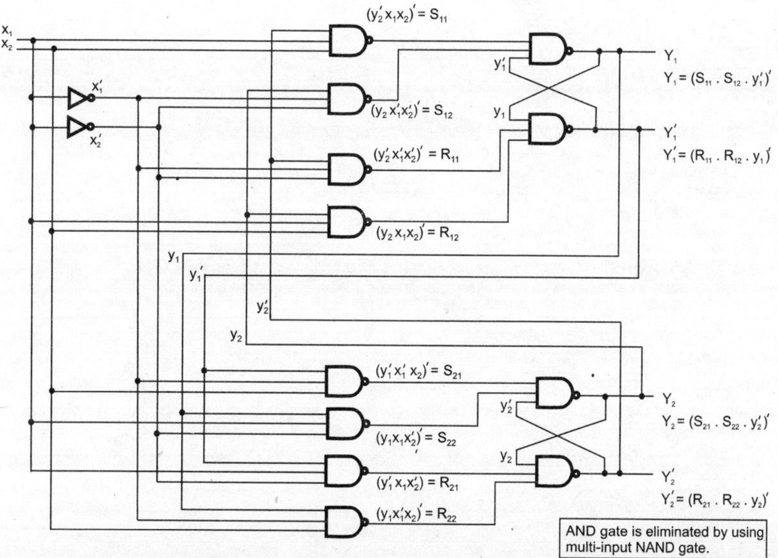

Fig: *Logic circuit using NAND gate structure and NAND based SR latch.*

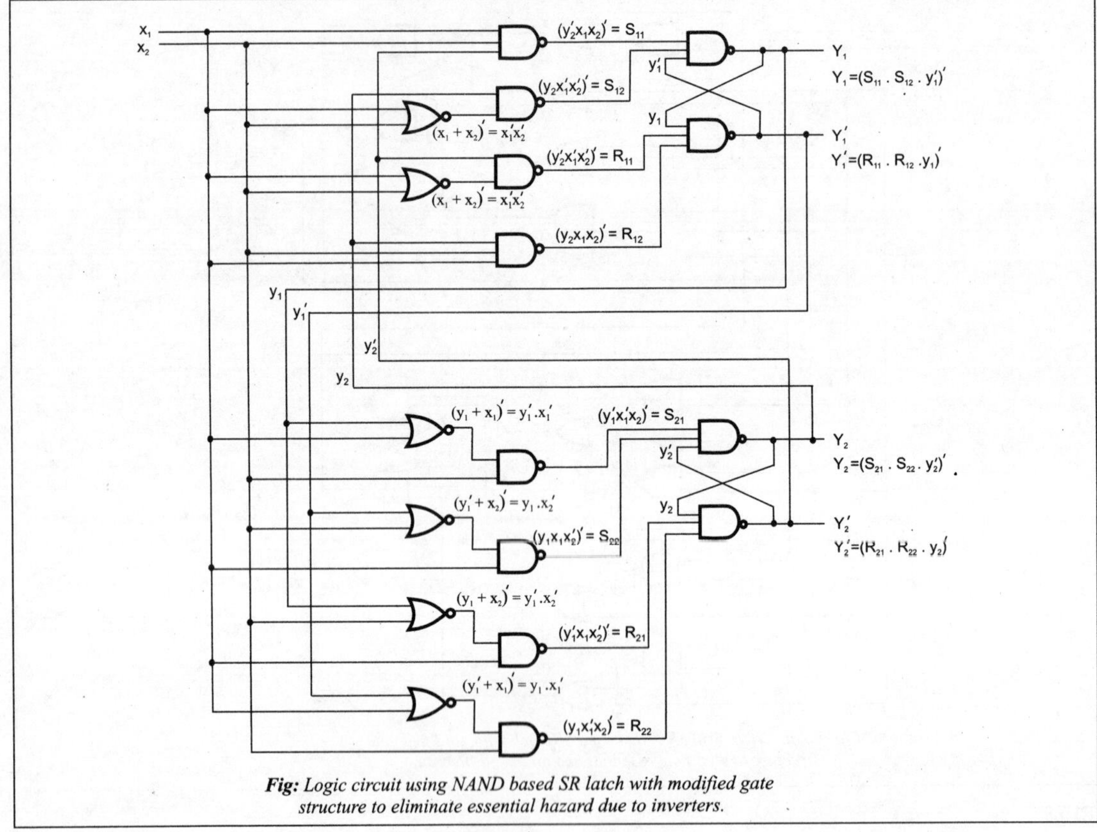

***Fig:** Logic circuit using NAND based SR latch with modified gate structure to eliminate essential hazard due to inverters.*

CHAPTER 11

LOGIC FAMILIES

11.1 Logic Families

(CSVTU, Nov/Dec'19, 7 Marks)

Digital ICs (Integrated Circuits) are manufactured using different semiconductor technologies. The collection of different logical devices that perform different logic functions but manufactured using same semiconductor technology is called a ***logic family***. The logic devices of a family will have similar input, output and internal circuit characteristics.

The digital ICs of the same logic family can be connected to implement a logic circuit. Whereas, the digital ICs of different logic families cannot be connected to implement a logic circuit, because they may not have compatible characteristics.

Some of the popular digital logic families are,

- RTL (Resistor-transistor logic)
- DTL (Diode-transistor logic)
- TTL (Transistor-transistor logic)
- ECL (Emitter-coupled logic)
- CMOS (Complementary metal-oxide semiconductor)

The RTL, DTL, TTL and ECL digital logic family of ICs are constructed using bipolar NPN transistors as fundamental switching device. The CMOS digital logic family of devices are constructed using unipolar MOSFET (Metal Oxide Semiconductor Field Effect Transistor) as fundamental switching device.

The important characteristics of digital logic family of integrated circuits (ICs) that are used to compare the performance of various family of devices are,

- Logic levels
- Propagation delay
- Power dissipation
- Fan-out and Fan-in
- Noise margin

11.1.1 Evolution of Logic Family

The first logic family developed was Resistor-transistor logic (RTL). The RTL (Resistor-Transistor Logic) family of devices uses resistors to receive inputs and use transistor as switching device and so the name Resistor-Transistor Logic. The primary advantage of RTL technology was that it involved a minimum number of transistors. The disadvantage of RTL is its high power dissipation during conduction state.

Diode-Transistor Logic (DTL) is a class of digital circuits built from diodes, resistors and bipolar junction transistors (BJT). In DTL (Diode-Transistor Logic) family of devices the resistances of RTL family are replaced by diodes. Therefore, DTL family of devices uses diodes to receive inputs and use transistor as switching device and so the name diode-transistor logic.

The TTL logic family has been developed to overcome the disadvantages in RTL and DTL family. In TTL (Transistor-Transistor Logic) family of devices the diodes of DTL family are replaced by transistors.

The p-n junction of a diode has considerable capacitance, so changing the logic level of an input connected to a diode, as in DTL, requires considerable time and energy. In TTL, the base-emitter junction is used to receive inputs instead of diodes in DTL. Therefore, TTL family of devices uses transistors to receive inputs as well as switching device and so the name transistor-transistor logic.

In RTL, DTL and TTL families the transistor switches are driven from saturation to cut-off and vice-versa, during switching operation. This switching action takes larger propagation time and power dissipation.

The need for fast switching leads to development of ECL (Emitter-Coupled Logic) family which works in active region with a small difference in two logic voltage levels which is decided by the currents. Hence, ECL is non-saturated logic family of devices and has low power dissipation and very low propagation delay. The, other bipolar logic families are based on voltage mode logic, ECL utilizes current mode logic. The invention of MOSFET leads to development very low power CMOS switches. The modern trend is using various versions of CMOS family of devices with low power and high-speed switching.

11.2 Operation and Characteristics of Digital Logic Families *(AKTU, Nov/Dec'22, 2 Marks)*

The basic devices in the logic families are BJT and CMOS transistor and they are operated in saturation and cut-off region. In saturation the transistor is considered as ON and in cut-off the transistor is considered as OFF. The ON-OFF characteristics of BJT and CMOS transistors are discussed in Sections 11.2.6 and 11.2.7.

The performance of logic family of devices are measured in terms of logic levels, propagation delay, power dissipation, noise margin, fan-in and fan-out and they are also called performance characteristics.

11.2.1 Logic Levels *(IKGPTU, Dec'19, 5 Marks)*

The digital ICs operates on binary constants 1 and 0, which are physically two different voltages in a digital circuit or device. The two different voltage levels are called ***logic levels***. In positive logic, the binary constant 1 is **high** and 0 is **low**. Different logic families will have different values of voltages for logic levels which is decided by the fundamental device used in the logic families.

In general, the logic levels will have a range of voltages and also associated with currents. The **high** and **low** logic levels will have a minimum and maximum value of voltage and current. The definition of minimum and maximum voltage and current levels of digital logic devices are given ahead:

V_{IH_min}**:** It is the minimum value for a **high** logic level input voltage at an input of a logical device to recognize a logical 1 or **high**.

V_{IL_max}**:** It is the maximum value for a **low** logic level input voltage at an input of a logical device to recognize a logical 0 or **low**.

V_{OH_min}**:** It is the minimum value for a **high** logic level output voltage at an output of a logical device to recognize a logical 1 or **high** at a standard load condition.

V_{OH_max}: It is the maximum value for a **high** logic level output voltage at an output of a logical device to recognize a logical 0 or **low** at a standard load condition.

I_{IH}: It is the current that flows into an input when a **high** logic level voltage is applied to that input.

I_{IL}: It is the current that flows into an input when a **low** logic level voltage is applied to that input.

I_{OH}: It is the current that flows from an output when a **high** logic level voltage exists at output at a standard load condition.

I_{OL}: It is the current that flows from the output when a **low** logic level voltage exists at output at a standard load condition.

The currents and voltages during a **high** state and a **low** state are shown in Fig. 11.1.

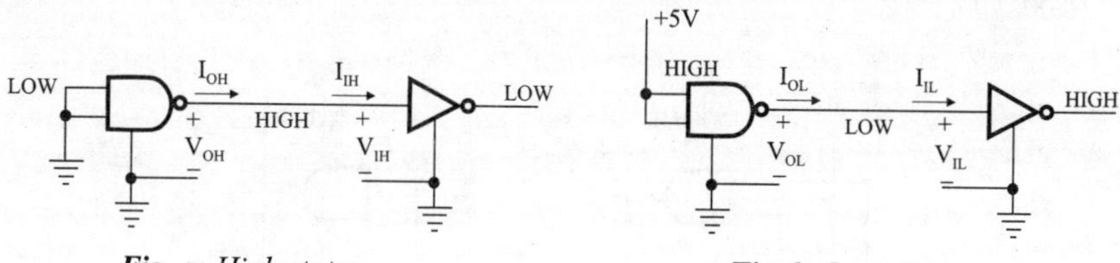

***Fig. a:** High state.* ***Fig. b:** Low state.*

***Fig. 11.1:** Currents and voltages in the high and low states.*

11.2.2 Propagation Delay

(CSVTU, Nov/Dec'19, 2 Marks)

In digital circuits or devices, the logic levels change from **high** to **low** or vice-versa. The logic voltage level changes always experience a delay for a change in input to be available as a change in the output. The propagation delay is defined as the time required for a change in input to be available in the output. Two delay times are defined in order to estimate the propagation delay with reference to an inverter or NOT gate.

t_{PLH}: Delay time for a **low** at input to realize as a **high** in output.

t_{PHL}: Delay time for a **high** at input to realize as a **low** in output.

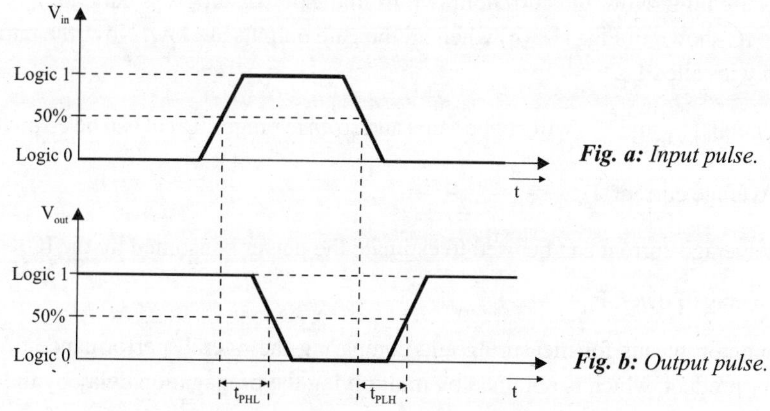

***Fig. a:** Input pulse.*

***Fig. b:** Output pulse.*

***Fig. 11.2:** Propagation delay for a change in logic levels in an inverter.*

Figure 11.2 shows the two delay times of an inverter. The delay times are measured between the 50 percent points on the input and output transitions. In general, t_{PHL} and t_{PLH} will not be same and so the propagation delay is defined as the average of the delay times, t_{PHL} and t_{PLH}.

$$\text{Propagation delay, } t_{pd} = \frac{t_{PHL} + t_{PLH}}{2}$$

The propagation delay time will be in the order of nano seconds.

11.2.3 Power Requirement

Every Integrated circuit (IC) require an electrical power supply connection at the prescribed voltage level. Hence, every IC will consume a definite power during its working time.

The amount of power that an IC consumes is given by the product of current (I_{CC}) drawn by the IC, and the voltage level (V_{cc}) of power supply connected to that IC.

$$\text{Power} = I_{CC}\, V_{CC}$$

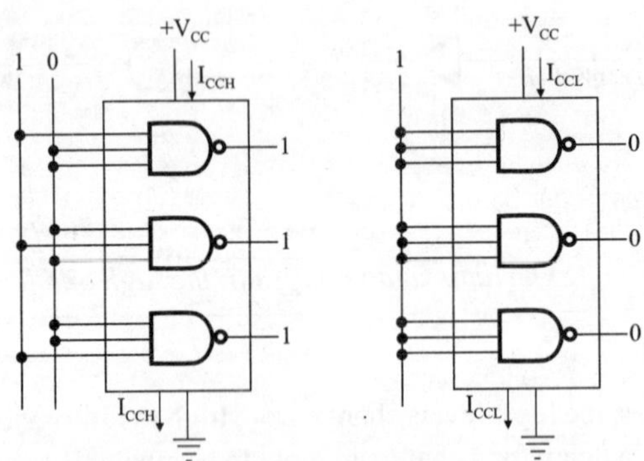

Fig. a: *High state.* ***Fig. b:*** *Low state.*

Fig. 11.3: *High and low state currents in a NAND gate.*

In digital ICs the current drawn from the power supply will vary and depends on the logic states of the devices in the chip. For example, consider a NAND gate IC shown in Fig. 11.3(a) when all the gate outputs are **high**. Now, the current drawn from the power supply is called I_{CCH}. Likewise, consider a NAND gate IC shown in Fig. 11.3(b) when all the gate outputs are **low**. Now, the current drawn from the power supply is called I_{CCL}.

In general, I_{CCH} and I_{CCL} will not be same and so an average current can be estimated as shown below:

$$\text{Average current, } I_{CC\_av} = \frac{I_{CCH} + I_{CCL}}{2}$$

This average current can be used to estimate the power consumed by the IC.

$$\text{Average Power, } P_{av} = V_{CC}\, I_{CC\_av}$$

A common means for measuring and comparing the overall performance of an IC family is the speed-power product, which is obtained by multiplying the propagation delay by the power dissipation.

For example, suppose an IC family has an average propagation delay of 2 *ns* and an average power dissipation of 4 *milli-Watt*.

Now, Speed-power product $= 2\ ns \times 5\ mW = 2 \times 10^{-9} \times 5 \times 10^{-3}$ *Watt-second*

$= 10 \times 10^{-12}$ *Watt-second* or *Joules* $= 10$ *pico-Joules (pJ)*

11.2.4 Fan-out and Fan-in

(CSVTU, Nov/Dec'19, 2 Marks)

In general, a logic-circuit output is connected to several other logic inputs. The fan-out is defined as the maximum number of standard load that an output can drive reliably. A standard load is usually defined as the amount of current needed by an input of another gate in the same logic family. Therefore, fan-out is the maximum number of inputs of other logic devices that can be connected to the output of a gate, and is expressed by a number.

When the output of a gate is in **high** state it will source a current, I_{OH} to the connected inputs as shown in Fig. 11.4(a). When the output of gate 1 is in the **high** state, it supplies a current I_{IH} to the input of gate 2. Thus, the output of gate 1 is acting as a source of current for the gate 2 input.

When the output of a gate is in **low** state it will sink a current, I_{OL} from connected inputs as shown in Fig. 11.4(b). When the gate 1 output goes to its **low** state, current will flow in the direction shown from the input circuit of gate 2 back through the output resistance of gate 1. In other words, in the **low** state the circuit output that drives the input of gate 2 must be able to sink a current, I_{IL}, coming from that input.

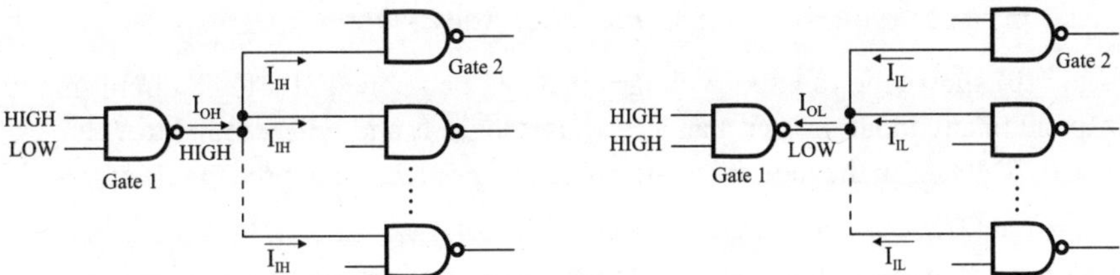

Fig. a: Current sourcing in high state. *Fig. b: Current sinking in low state.*

Fig. 11.4: Current sourcing and sinking.

The fan-out of a gate is calculated from the ratio I_{OH}/I_{IH} or I_{OL}/I_{IL}. The lowest among the two ratio is chosen as fan-out.

$$\text{Fan-out (Low)} = \frac{I_{OL}}{I_{IL}}$$

$$\text{Fan-out (High)} = \frac{I_{OH}}{I_{IH}}$$

Fan-in is the number of inputs provided (or available) in a logic gate and it is decided during manufacturing itself which in turn depends on technology used for manufacturing the IC. Larger number of inputs will increase the propagation delay and so number of inputs are normally limited to 2 or 3.

11.2.5 Noise Margin

(CSVTU, Nov/Dec'19, 2 Marks)

Stray electric and magnetic fields can induce voltages on the connecting wires between logic circuits. These unwanted, spurious signals are called noise and can sometimes cause the voltage at the input of a logic circuit to drop below V_{IH_min} or rise above V_{IL_max}, which could produce unpredictable operation. The noise immunity of a logic circuit refers to the circuit's ability to tolerate noise voltages without causing unwanted changes in the output voltage. A quantitative measure of noise immunity in terms of voltage is called ***noise margin***.

Figure 11.5 shows the range of voltages that can occur at a logic-circuit output and in this range, any voltages greater than V_{OH_min} are considered as logic 1, and any voltages lower than V_{OL_max} are considered as logic 0. Voltages in the disallowed range should not appear at a logic circuit output under normal conditions.

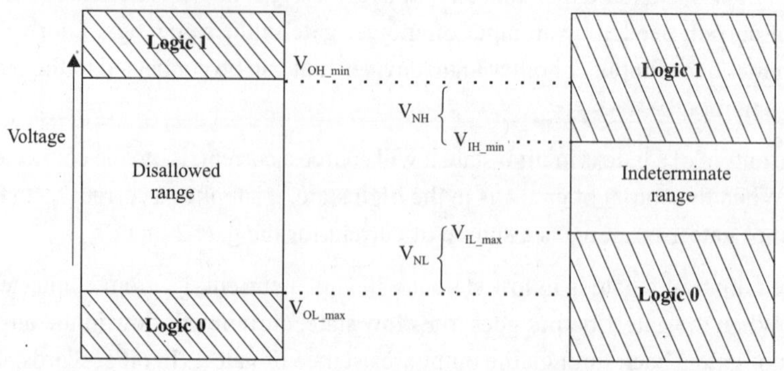

***Fig. a:** Output voltage ranges.* ***Fig. b:** Input voltage ranges.*

***Fig. 11.5:** Noise margin.*

Figure 11.5(b) shows the range of voltages that can be applied at a logic circuit input. The logic circuit will respond to any input greater than V_{IH_min} as a logic 1, and will respond to voltages lower than V_{IL_max} as a logic 0. Voltages in the indeterminate range will produce an unpredictable response and should be avoided.

Based on the minimum voltage range of **high** input and output voltage levels, a high-state noise margin can be defined as shown below:

High-state noise margin, $V_{NH} = V_{OH\_min} - V_{IH\_min}$

Based on the maximum voltage range of **low** input and output voltage levels, a low-state margin can be defined as shown below:

Low-state noise margin, $V_{NL} = V_{IL\_max} - V_{OL\_max}$

Table 11.1: Typical Values of Performance Characteristics of Logic Families

Logic Family	Propagation Delay *(ns)*	Power Dissipation per Gate *(mW)*	Noise Margin *(V)*	Fan-in	Fan-out
RTL	12	30	0.2	5	5
DTL	30	12	0.4	5	8
TTL	9	10	0.4	8	10
ECL	1	50	0.25	5	10
CMOS	20 - 50	0.01	1	10	50

11.2.6 ON-OFF Characteristics of Bipolar Junction Transistors

In logic devices the transistors are used as switches in which they are operated either in saturation region/state or cut-off region/state. In saturation state, collector and emitter terminals will behave as short circuit and maximum current flows through collector and emitter terminals of the transistor and so it is called ON state. In cut-off state, collector and emitter terminals will behave as open circuit and no current flows through collector and emitter terminals of the transistor and so it is called OFF state.

In logic families, NPN transistors are widely used and the input is applied at base and output is taken either from collector or emitter of the transistors. The NPN transistor is turned ON by a logic **high** voltage and turned OFF by a logic **low** voltage, applied at the base of the transistor with respect to emitter.

The collector is biased through a resistance either in collector or in emitter. The NPN transistor voltages and currents during ON and OFF state with a biasing resistance in collector are shown in Fig. 11.6. The NPN transistor voltages and currents during ON and OFF state with a biasing resistance in emitter are shown in Fig. 11.7.

During ON state the transistor is driven to saturation and collector-emitter voltage, $V_{CE} = 0.2\ V$, which is approximately equal to zero and so the collector and emitter terminals will behave as short circuit (S.C.). During OFF state the transistor is driven to cut-off and collector current, I_C is approximately zero and so the collector and emitter terminals will behave as open circuit (O.C.).

When output is obtained from collector the input logic is available as inverted logic in the output. But when the output is obtained from emitter the output logic is same as input logic.

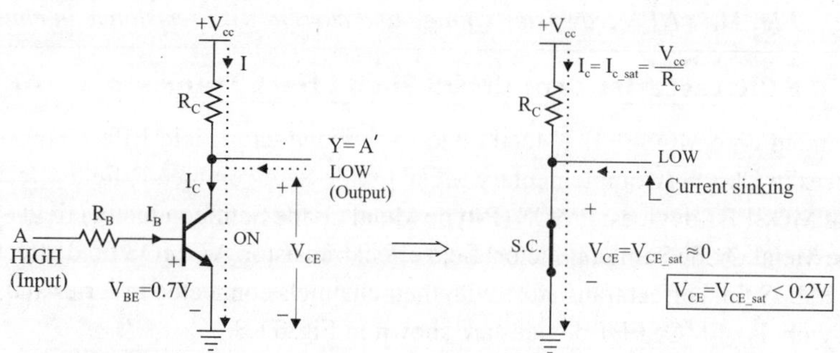

Fig. a: ON state.

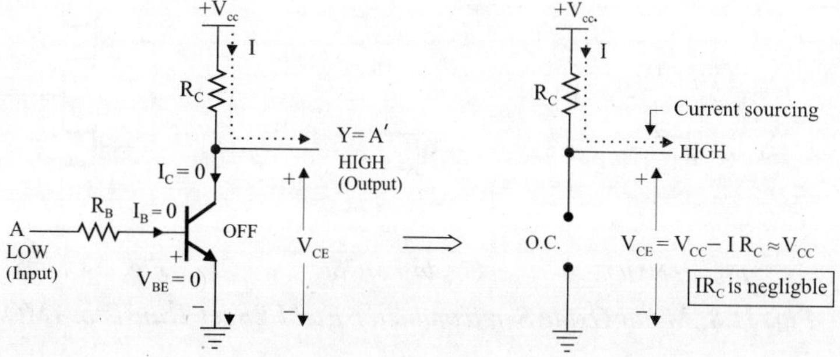

Fig. b: OFF state.

Fig. 11.6: *NPN transistor voltage and current with resistance in collector.*

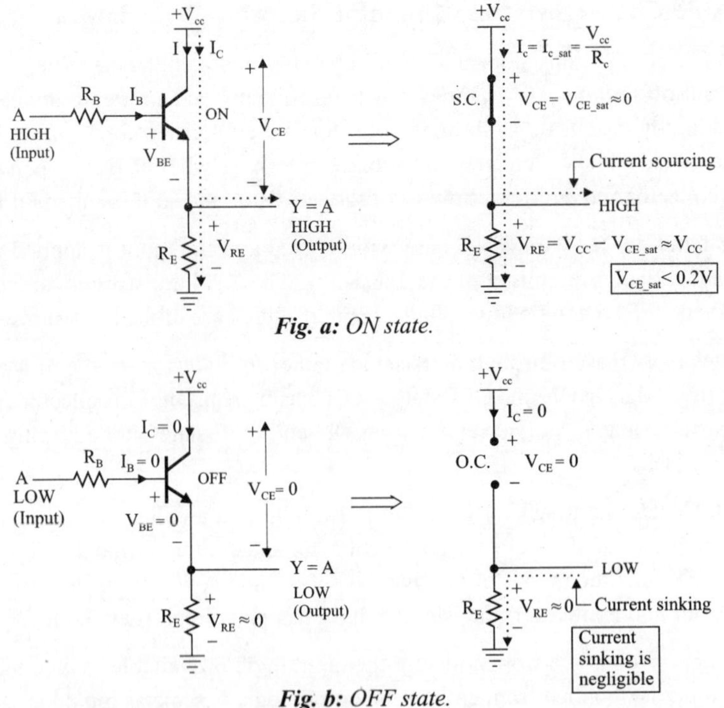

Fig. a: ON state.

Fig. b: OFF state.

***Fig. 11.7:** NPN transistor voltage and current with resistance in emitter.*

11.2.7 ON-OFF Characteristics of CMOS Field Effect Transistors *(AU, Nov/Dec'18, 2 Marks)*

Enhancement type MOSFETs (Metal Oxide Semiconductor Field Effect Transistor) are the basic switching devices in CMOS (Complementary Metal Oxide Semiconductor) field effect transistors. There are two types of MOSFET devices: PMOS (P-type Metal Oxide Semiconductor) field effect transistor and NMOS (N-type Metal Oxide Semiconductor) field effect transistor. A CMOS field effect transistor is a pair of NMOS and PMOS field effect transistors with their channels connected in series and gate terminals tied together. The symbols of MOSFET devices are shown in Fig. 11.8.

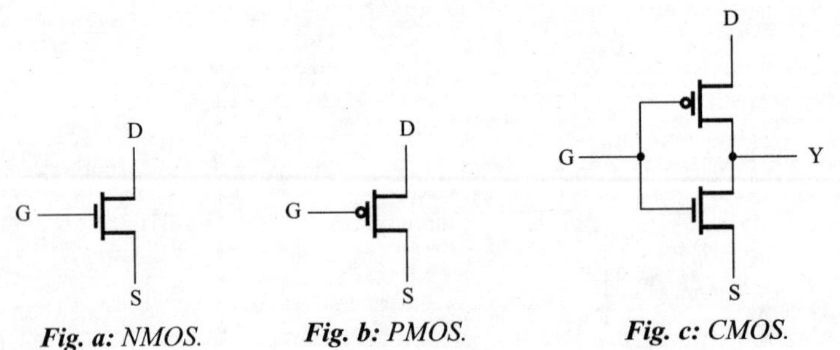

Fig. a: NMOS. *Fig. b: PMOS.* *Fig. c: CMOS.*

***Fig. 11.8:** Metal Oxide Semiconductor Field Effect Transistor (MOSFET).*

In logic devices the MOSFETs are used as switches in which they are operated either in saturation region/state or cut-off region/state. In saturation state, drain and source terminals will behave as short circuit and maximum current flows through the channel and so it is called ON state. In cut-off state, drain and source terminal will behave as open circuit and no current flows through the channel and so it is called OFF state.

In COMS logic family of devices, the input is applied at gate terminal of MOSFET and output is taken from the drain terminal of NMOS field effect transistor.

The PMOS field effect transistor is turned ON by a logic **low** voltage and turned OFF by a logic **high** voltage, applied at the gate of the transistor with respect to source. A bubble in the gate of PMOS indicates that it is turned ON by a logic **low** voltage.

The NMOS field effect transistor is turned ON by a logic **high** voltage and turned OFF by a logic **low** voltage, applied at the gate of the transistor with respect to source.

The CMOS is biased by directly connecting a DC supply voltage between drain terminal of the PMOS and the source terminal of NMOS. The MOSFET is tuned ON or OFF by the logic voltage input applied at the gate terminal. The CMOS field effect transistor voltages and currents during ON and OFF state are shown in Fig. 11.9.

During ON state the MOSFET is driven to saturation and drain-source voltage is approximately equal to zero and so the drain and source terminals will behave as short circuit (S.C.). During OFF state the MOSFET is driven to cut-off and the drain current is approximately zero and so the drain and source terminals will behave as open circuit (O.C.). The output of CMOS transistor is inverted logic of the input. Hence, CMOS switch basically an inverter.

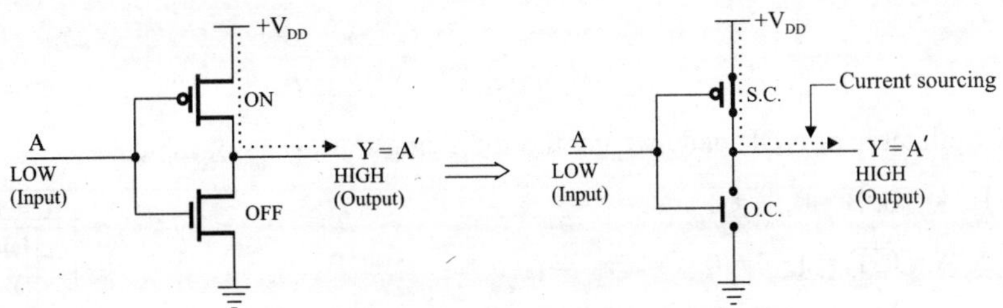

Fig. a: PMOS ON and NMOS OFF.

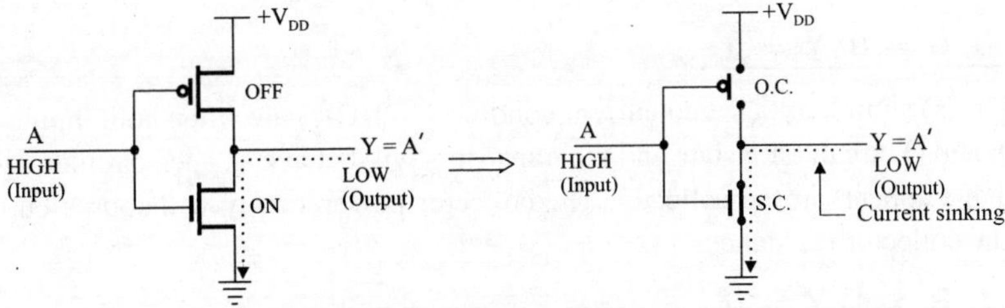

Fig. b: PMOS OFF and NMOS ON.

***Fig. 11.9:** CMOS voltage and current during ON and OFF state.*

11.3 RTL (Resistor-Transistor Logic)

In RTL (Resistor-Transistor Logic) family, input network is made of resistors and switching devices are transistors and so the name RTL. The logic functions are realized by parallel or series connection of switching transistors.

In RTL family, the inputs are applied to base of switching transistors through resistors and the output is obtained from the collector of switching transistor.

The switching transistors operate in cut-off (OFF state) or saturation (ON state). In RTL family the logic **low** is 0 to 0.2 *V* and the logic **high** is 1 *V* to 3.6 *V*.

11.3.1 RTL NOR Gate

A two input NOR gate is shown in Fig. 11.10. Here, two transistors Q_1 and Q_2 are connected in parallel such that collectors are tied together to form the output and emitters are grounded. Inputs are applied to base of the transistors through resistance.

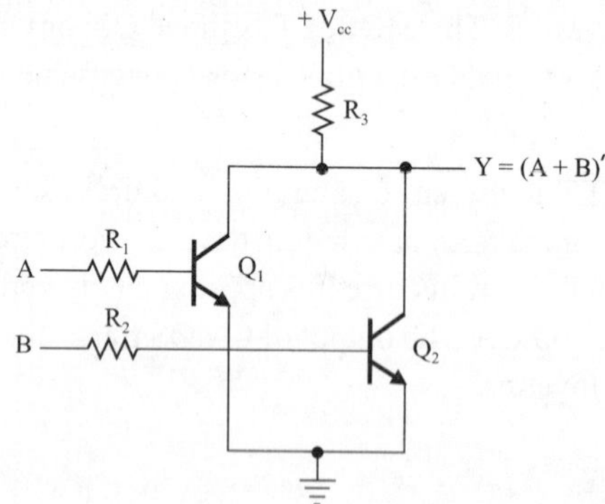

Fig. 11.10: Two input RTL NOR gate.

From the truth table of NOR gate, we get the following four cases of inputs.

Case 1: A = 0, B = 0, Y = 1

Case 2: A = 0, B = 1, Y = 0

Case 3: A = 1, B = 0, Y = 0

Case 4: A = 1, B = 1, Y = 0

Truth Table of NOR Gate

Inputs		Output
A	B	Y
0	0	1
0	1	0
1	0	0
1	1	0

Case 1: A = 0, B = 0, Y = 1

Figure 11.11(a) shows voltage and current condition of NOR gate when both inputs are **low**. Now, the transistors Q_1 and Q_2 are in OFF state and the transistors will behave as open circuit to ground. Hence, a logic **high** voltage appears in the collector. The collector power supply will source current to load at output through the collector resistance.

Case 4: A = 1, B = 1, Y = 0

Figure 11.11(d) shows voltage and current condition of NOR gate when both inputs are **high**. Now, the transistors Q_1 and Q_2 are in ON state (or saturation state) and so collector-emitter terminals of transistors will behave as short circuit. Hence a logic **low** voltage appears in the collector and output load is shorted to ground through the switching transistors and so current will sink to output.

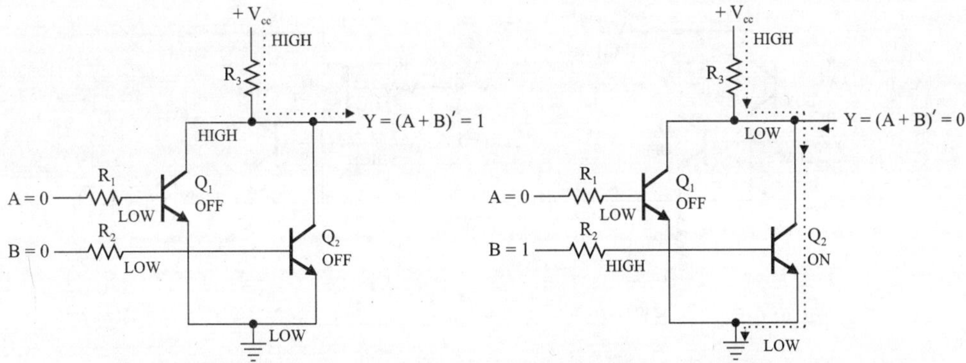

Fig. a: Case 1, A = B = 0, Y = 1.   *Fig. b: Case 2, A = 0, B = 1, Y = 0.*

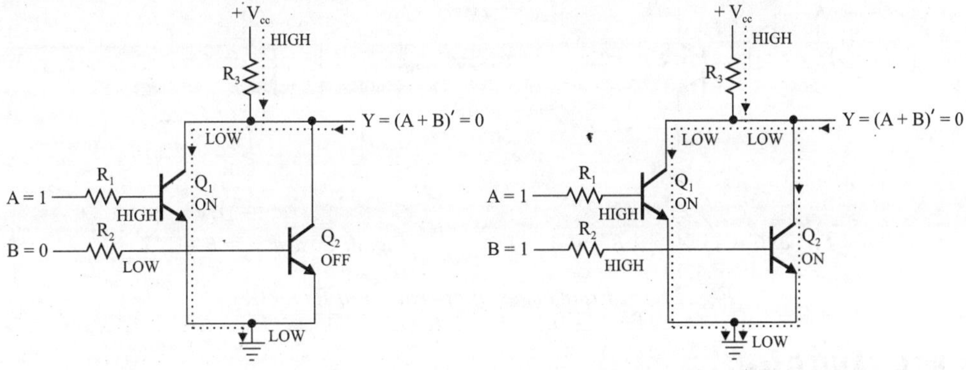

Fig. c: Case 3, A = 1, B = 0, Y = 0.   *Fig. d: Case 4, A = B = 1, Y = 0.*

***Fig. 11.11:** Four cases of operation of RTL NOR gate.*

Case 2 and 3: A = 0, B = 1, Y = 0 and A = 1, B = 0, Y = 0

In these two cases one of the base of transistors is **high** and the other is **low**. Hence, one of the transistors will be ON and the other is OFF. The ON transistor will pull the output to **low**. Therefore, the working of NOR gate in these two cases will be similar to case 4. Figures 11.11(b) and 11.11(c) shows voltage and current condition of NOR gate when any one of the inputs is **high**.

11.3.2 RTL OR Gate

The OR logic can be obtained by inverting the NOR logic. The inverted output of NOR gate can be obtained by connecting the biasing resistance on emitter side and taking output from emitter terminal as shown in Fig. 11.12. The voltage and current conditions for various inputs are shown in Figs. 11.13(a) to 11.13(d).

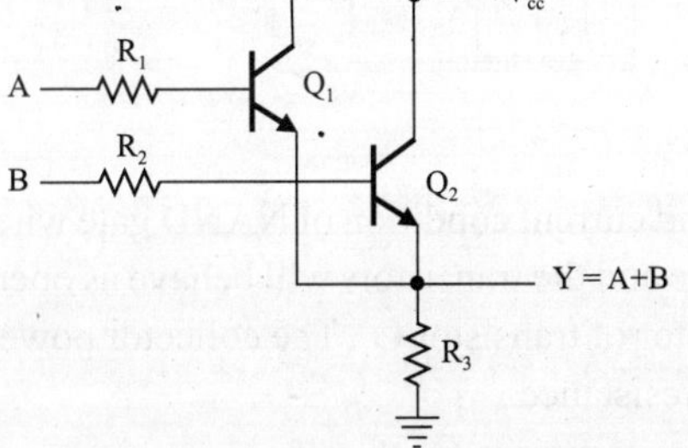

Truth Table of OR Gate

Inputs		Output
A	B	Y
0	0	0
0	1	1
1	0	1
1	1	1

***Fig. 11.12:** Two input RTL OR gate.*

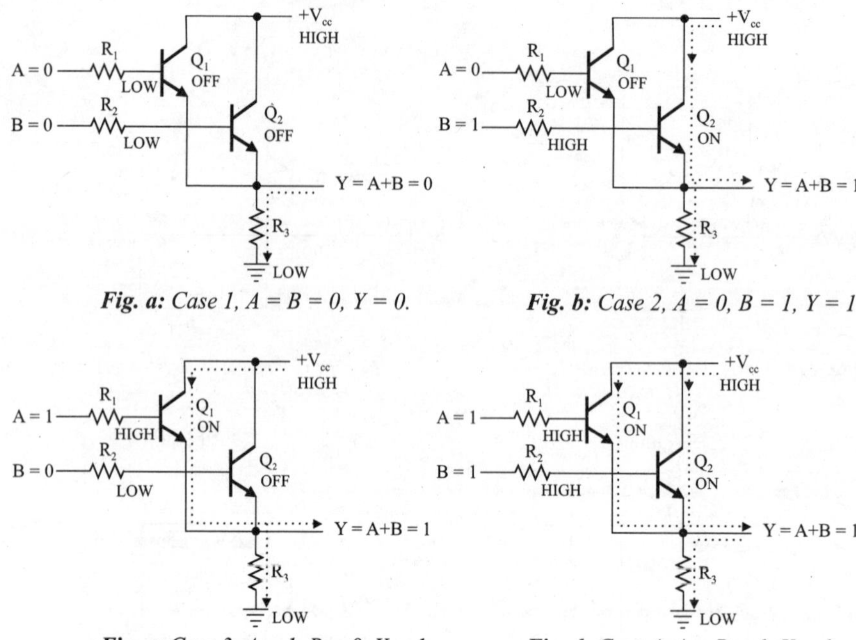

Fig. a: Case 1, A = B = 0, Y = 0.

Fig. b: Case 2, A = 0, B = 1, Y = 1.

Fig. c: Case 3, A = 1, B = 0, Y = 1.

Fig. d: Case 4, A = B = 1, Y = 1.

***Fig. 11.13:** Four cases of operation of RTL OR gate.*

11.3.3 RTL NAND Gate

A two input NAND gate is shown in Fig. 11.14. Here, two transistors Q_1 and Q_2 are connected in series such that emitter of Q_1 is connected to collector of Q_2 and emitter of Q_2 is grounded. Inputs are applied to base of the transistors through resistance. Output is taken from collector of Q_1.

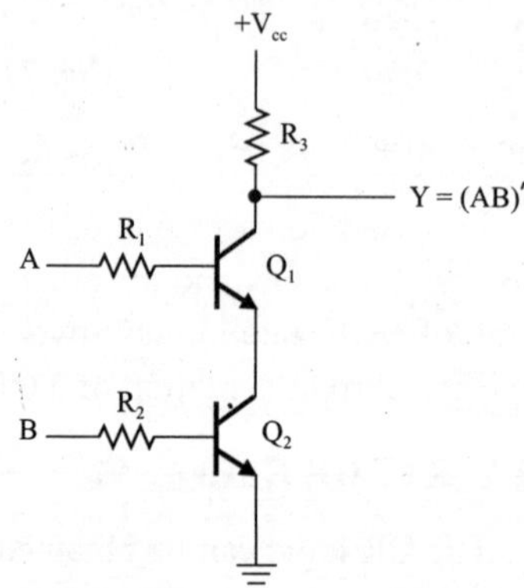

***Fig. 11.14:** Two input RTL NAND gate.*

From the truth table of NAND gate, we get the following four cases of inputs.

Case 1: A = 0, B = 0, Y = 1

Case 2: A = 0, B = 1, Y = 1

Case 3: A = 1, B = 0, Y = 1

Case 4: A = 1, B = 1, Y = 0

Truth Table of NAND Gate

Inputs		Output
A	B	Y
0	0	1
0	1	1
1	0	1
1	1	0

Case 1: A = 0, B = 0, Y = 1

Figure 11.15(a) shows voltage and current condition of NAND gate when both inputs are **low**. Now, the transistors Q_1 and Q_2 are in OFF state and the transistors will behave as open circuit to ground. Hence, a logic **high** voltage appears in the collector of transistor Q_1. The collector power supply will source current to load at output through the collector resistance.

Case 4: A = 1, B = 1, Y = 0

Figure 11.15(d) shows voltage and current condition of NAND gate when both inputs are **high**. Now, the transistors Q_1 and Q_2 are in ON state (or saturation state) and so collector-emitter terminals of transistors will behave as short circuit. Hence a logic **low** voltage appears in the collector and output load is shorted to ground through the switching transistors and so current will sink to output.

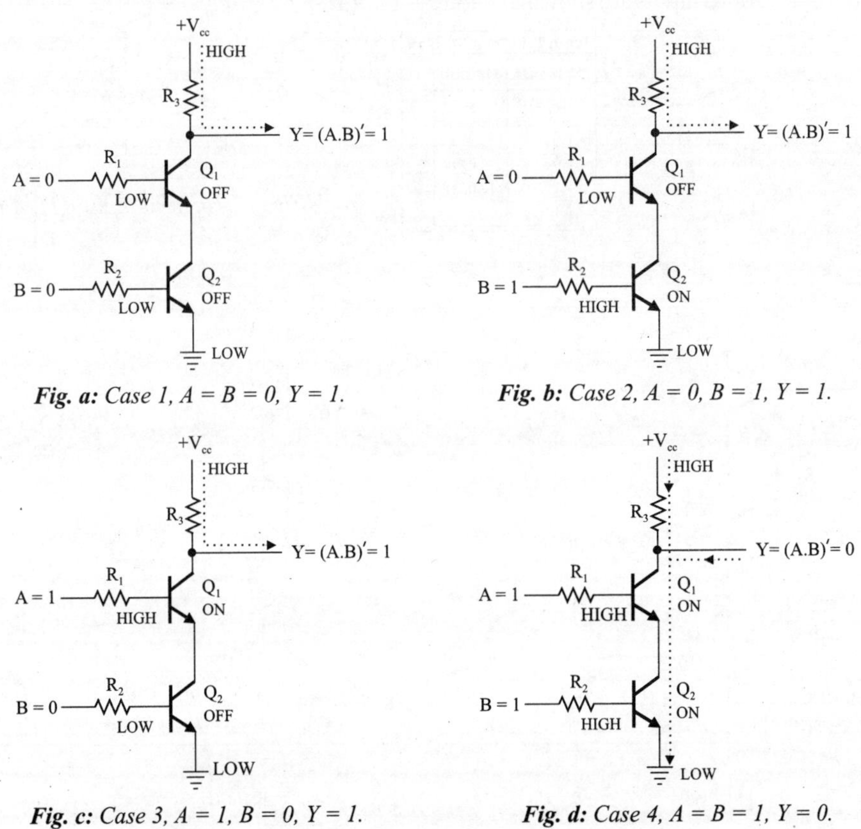

***Fig. 11.15:** Four cases of operation of RTL NAND gate.*

Case 2 and 3: A = 0, B = 1, Y = 1 and A = 1, B = 0, Y = 1

In these two cases one of the base of transistors is **high** and the other is **low**. Hence, one of the transistors will be ON and the other is OFF. The OFF transistor will behave as open circuit to ground and so no current will flow through transistors and so the output will be **high**. Therefore, the working of NAND gate in these two cases will be similar to case 1. Figures 11.15(b) and 11.15(c) shows voltage and current condition of NAND gate when any one of the inputs is **high**.

11.3.4 RTL AND Gate

The AND logic can be obtained by inverting the NAND logic. The inverted output of NAND gate can be obtained by connecting the biasing resistance on emitter side and taking output from emitter terminal as shown in Fig. 11.16. The voltage and current conditions for various inputs are shown in Figures 11.17(a) to 11.17(d).

Truth Table of AND Gate

Inputs		Output
A	B	Y
0	0	0
0	1	0
1	0	0
1	1	1

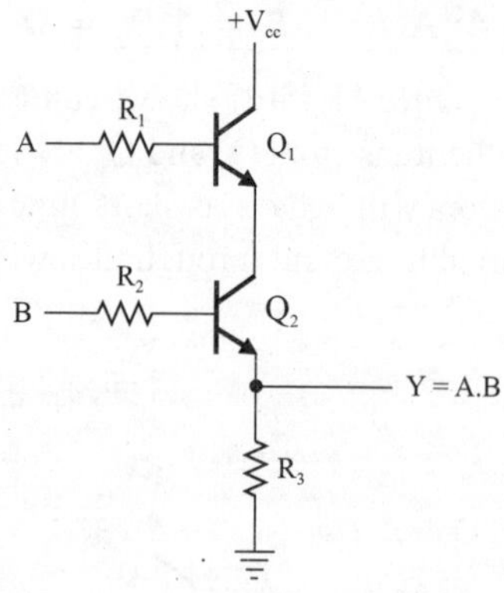

Fig. 11.16: Two input RTL AND gate.

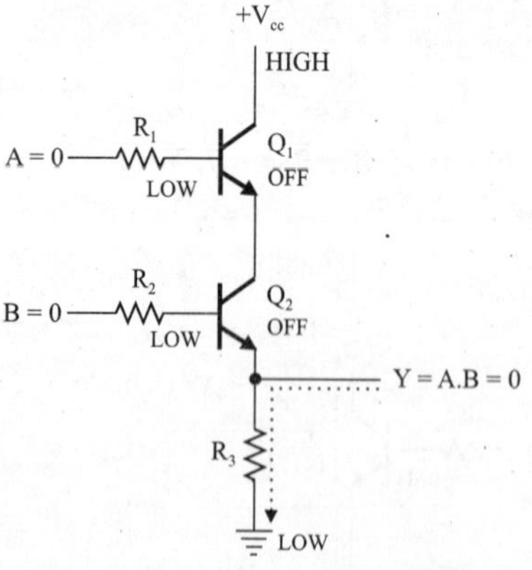

***Fig. a:** Case 1, A = B = 0, Y = 0.*

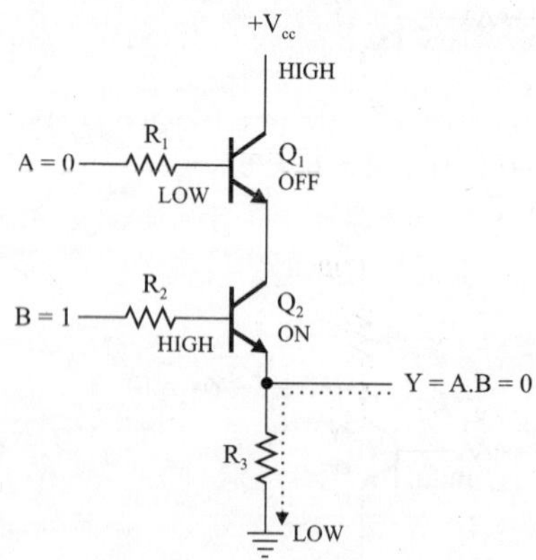

***Fig. b:** Case 2, A = 0, B = 1, Y = 0.*

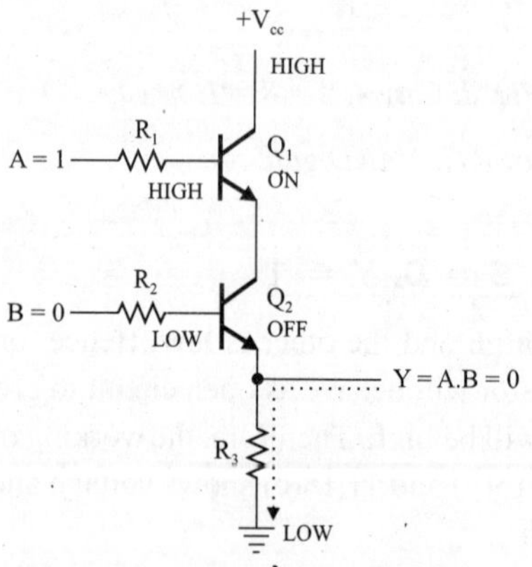

***Fig. c:** Case 3, A = 1, B = 0, Y = 0.*

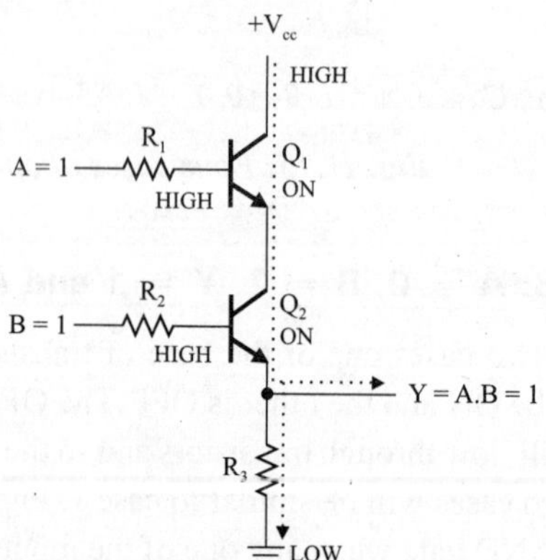

***Fig. d:** Case 4, A = B = 1, Y = 1.*

***Fig. 11.17:** Four cases of operation of RTL AND gate.*

11.3.5 RTL NOT Gate

(AU, Apr/May'19, 6 Marks)

Basically, a transistor in common emitter configuration will behave as inverter. Hence, NOT gate can be constructed using a single NPN transistor as shown in Fig. 11.18. The voltage and current conditions during two cases of input are shown in Figs. 11.19(a) and 11.19(b).

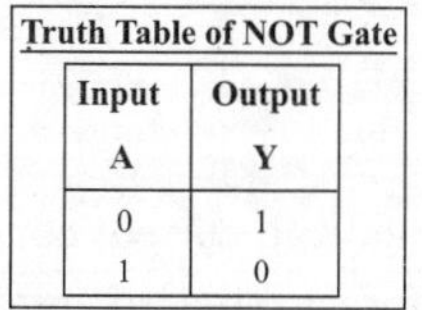

Truth Table of NOT Gate

Input A	Output Y
0	1
1	0

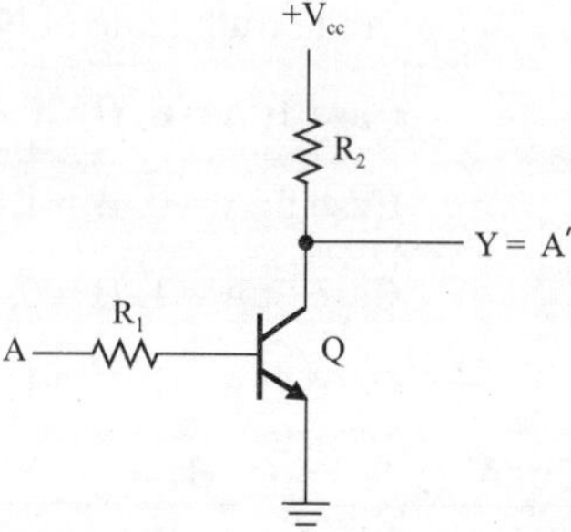

Fig. 11.18: *RTL NOT gate.*

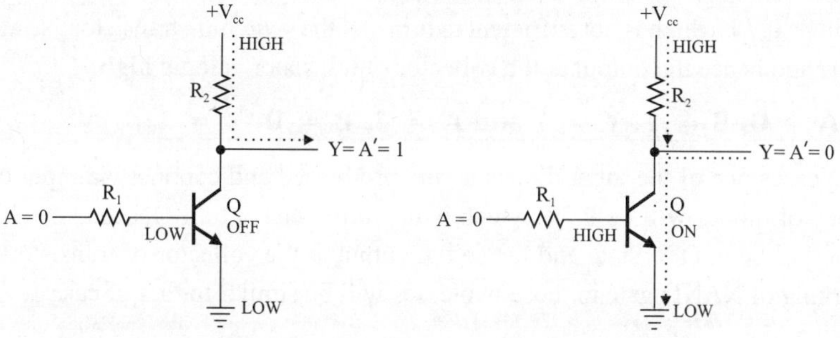

Fig. a: *Case 1, A = 0, Y = 1.* ***Fig. b:*** *Case 2, A = 1, Y = 0.*

Fig. 11.19: *Two cases of operation of RTL NOT gate.*

11.4 DTL (Diode-Transistor Logic)

(AU, Nov/Dec'18, 2 Marks)

The DTL devices are developed as an improvement over RTL devices by replacing the input resistors in RTL devices by diodes in DTL devices. The replacement of resistors by diodes reduces the power loss. Therefore, in DTL devices the diodes are input devices and a single switching transistor is the output device. This also eliminates the series connection of transistors in the RTL output which has a drawback of variable high state voltage with increase in series connected transistors.

The basic gate in DTL is NAND gate. The logic **low** is 0 to 0.2 *V* and logic **high** is 4 to 5 *V*.

11.4.1 DTL NAND Gate

A two inputs DTL NAND gate is shown in Fig. 11.20. The input circuit has two diodes with their anodes tied together and connected to V_{cc}. The inputs are applied to cathode of diodes. The output is obtained from the collector of switching transistor.

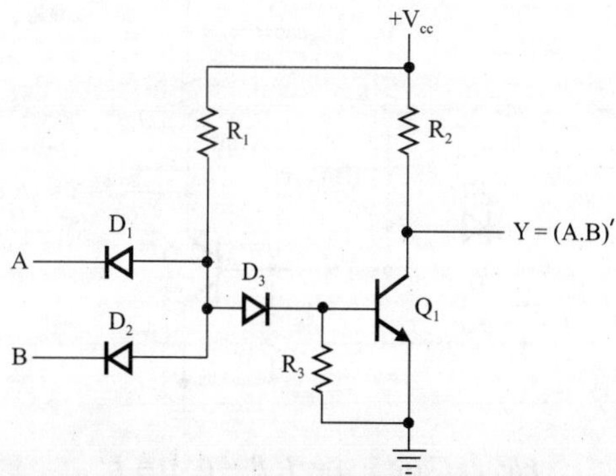

Fig. 11.20: *Two input DTL NAND gate.*

From the truth table of NAND gate we get the following four cases of inputs.

Case 1: A = 0, B = 0, Y = 1

Case 2: A = 0, B = 1, Y = 1

Case 3: A = 1, B = 0, Y = 1

Case 4: A = 1, B = 1, Y = 0

Truth Table of NAND Gate

Inputs		Output
A	B	Y
0	0	1
0	1	1
1	0	1
1	1	0

Case 1: A = 0, B = 0, Y = 1

Figure 11.21(a) shows the voltage and current condition of NAND gate when both the inputs are **low**. The input diodes D_1 and D_2 are forward biased and conduct via input circuit and the voltage at the anode of diodes will be 0.7 *V* which is not sufficient to turn ON the switching transistor. Now, the transistor will be in OFF state and hence the output at the collector of transistor is logic **high**.

Case 2 and 3: A = 0, B = 1, Y = 1 and A = 1, B = 0, Y = 1

In these two cases one of the input diodes is forward biased and conduct via input circuit and the voltage at the anode of diodes will be 0.7 *V* which is not sufficient to turn ON the switching transistor. Now, the transistor will be in OFF state and hence the output at the collector of transistor is logic **high**. Therefore, the working of NAND gate in these two cases will be similar to that of case 1.

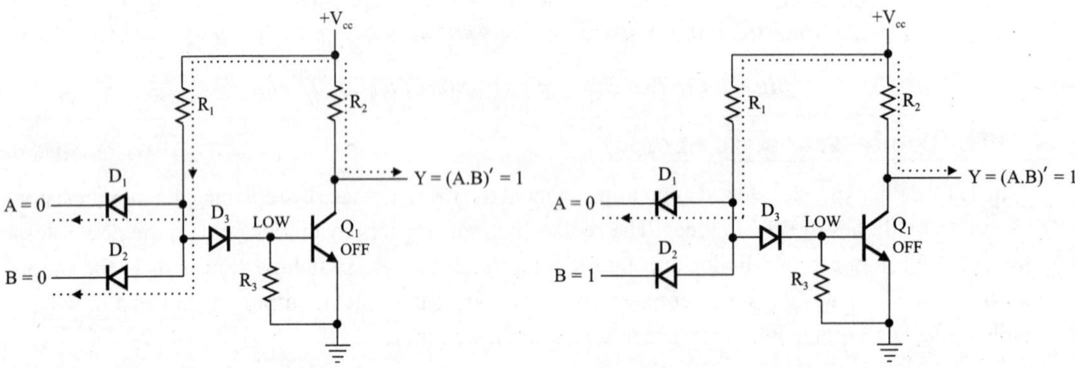

Fig. a: Case 1, A = B = 0, Y = 1.

Fig. b: Case 2, A = 0, B = 1, Y = 1.

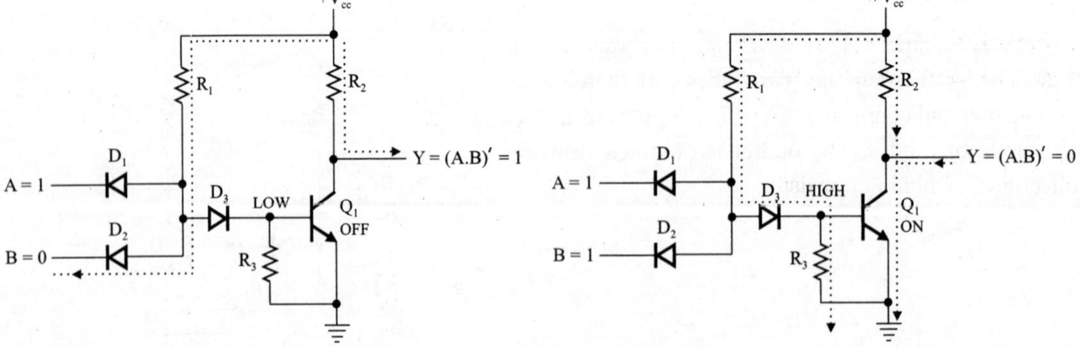

Fig. c: Case 3, A = 1, B = 0, Y = 1.

Fig. d: Case 4, A = B = 1, Y = 0.

***Fig. 11.21:** Four cases of operation of DTL NAND gate.*

Case 4: A = 1, B = 1, Y = 0

Figure 11.21(d) shows the voltage and current condition of NAND gate when both the inputs are **high**. The input diodes D_1 and D_2 are reverse biased and hence the diodes will not conduct. The voltage at anode of diode D_3 will be logic **high** and turn ON the switching transistor and hence the output at the collector of transistor is logic **low**.

11.4.2 DTL NOR Gate

A two inputs DTL NOR gate is shown in Fig. 11.22. The input circuit has two diodes with their cathodes tied together and connected to ground via a resistor. The inputs are applied to anode of diodes. The output is obtained from the collector of switching transistor.

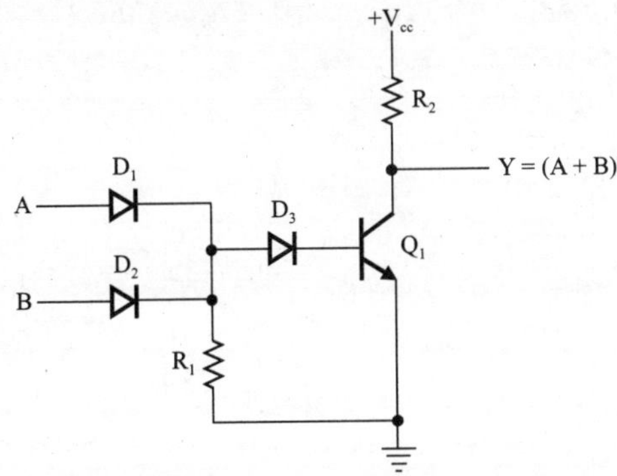

***Fig. 11.22:** Two input DTL NOR gate.*

From the truth table of NOR gate we get the following four cases of inputs.

Case 1: A = 0, B = 0, Y = 1

Case 2: A = 0, B = 1, Y = 0

Case 3: A = 1, B = 0, Y = 0

Case 4: A = 1, B = 1, Y = 0

Truth Table of NOR Gate

Inputs		Output
A	B	Y
0	0	1
0	1	0
1	0	0
1	1	0

Case 1: A = 0, B = 0, Y = 1

Figure 11.23(a) shows the voltage and current condition of NOR gate when both the inputs are **low**. There is biasing voltage for diodes D_1 and D_2 and hence they will not conduct and the voltage at base of switching transistor will be **low**. Now, the transistor will be in OFF state and hence the output at the collector of transistor is logic **high**.

Case 4: A = 1, B = 1, Y = 0

Figure 11.23(d) shows the voltage and current condition of NOR gate when both the inputs are **high**. The input diodes D_1 and D_2 are forward biased and hence the diodes will conduct via input circuit and the voltage across R_3 is sufficient to turn ON the switching transistor. Now, the transistor will be in ON state and hence the output at the collector of transistor is logic **low**.

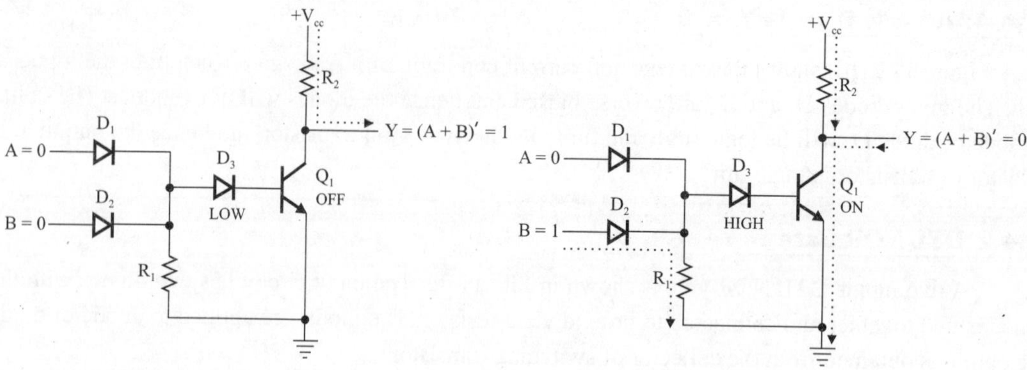

Fig. a: Case 1, A = B = 0, Y = 1.

Fig. b: Case 2, A = 0, B = 1, Y = 0.

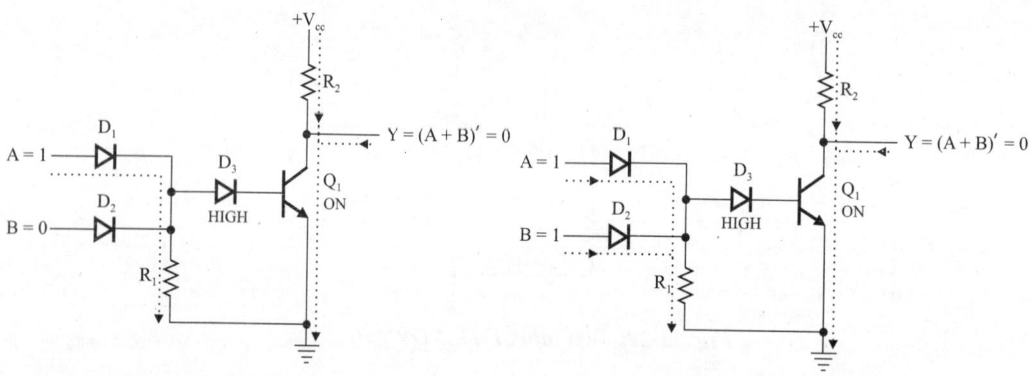

Fig. c: Case 3, A = 1, B = 0, Y = 0.

Fig. d: Case 4, A = B = 1, Y = 0.

***Fig. 11.23:** Four cases of operation of DTL NOR gate.*

Case 2 and 3: A = 0, B = 1, Y = 1 and A = 1, B = 0, Y = 1

In these two cases one of the input diodes is forward biased and conduct via input circuit and the voltage across R_3 is sufficient to turn ON the switching transistor. Now, the transistor will be in ON state and hence the output at the collector of transistor is logic **low**. Therefore, the working of NOR gate in these two cases will be similar to that of case 4.

11.4.3 DTL NOT Gate

A DTL NOT gate is shown in Fig. 11.24. The input circuit has two diodes with anode connected to V_{cc}. The input is applied to cathode of diode. The output is obtained from the collector of switching transistor.

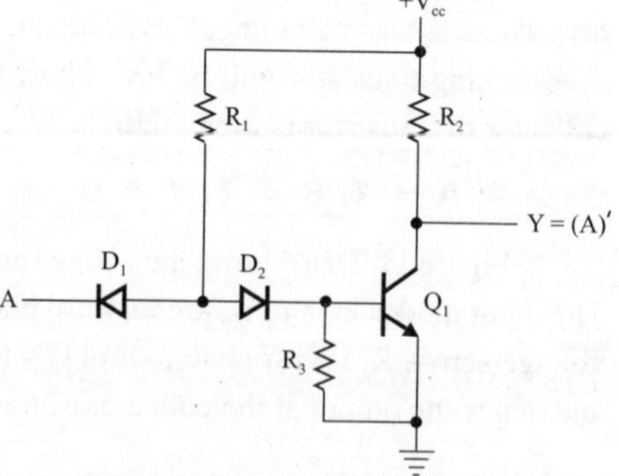

Fig. 11.24: DTL NOT gate.

From the truth table of NOT gate we get the following two cases of inputs.

Case 1: A = 0, Y = 1

Case 2: A = 1, Y = 0

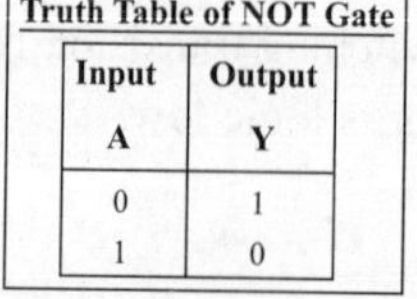

Truth Table of NOT Gate

Input A	Output Y
0	1
1	0

Case 1: A = 0, Y = 1

Figure 11.25(a) shows the voltage and current condition of NOT gate when the input is **low**. The input diode is forward biased and conduct via input circuit and the voltage at the anode of diode will be 0.7 *V* which is not sufficient to turn ON the switching transistor. Now, the transistor will be in OFF state and hence the output at the collector of transistor is logic **high**.

Case 2: A = 1, Y = 0

Figure 11.25(b) shows the voltage and current condition of NOT gate when the input is **high**. The input diode is reverse biased and hence the diode will not conduct. The voltage at anode of diode D_2 will be logic **high** and turn ON the switching transistor and hence the output at the collector of transistor is logic **low**.

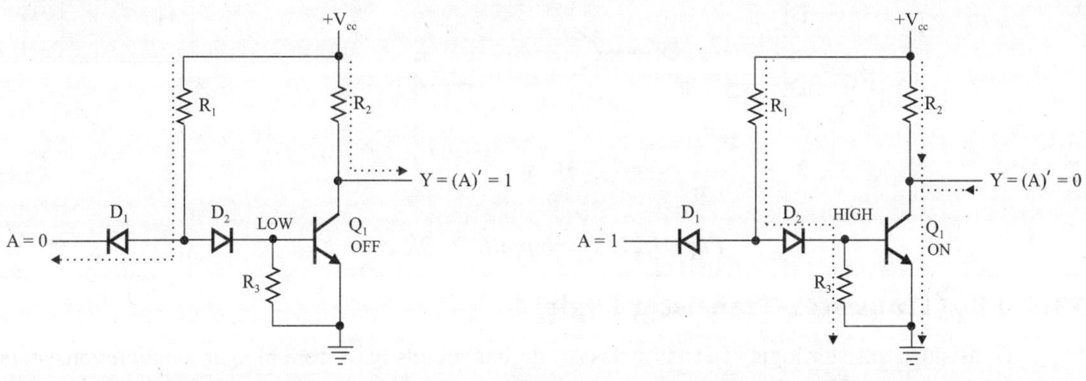

Fig. a: Case 1, A = 0, Y = 1. *Fig. b: Case 2, A = 1, Y = 0.*

***Fig. 11.25:** Two cases of operation of DTL NOT gate.*

11.4.4 DTL AND Gate

The AND logic can be obtained by inverting the output of NAND logic. Therefore, the DTL AND gate is obtained by connecting a NOT gate at the output of NAND gate as shown in Fig. 11.26.

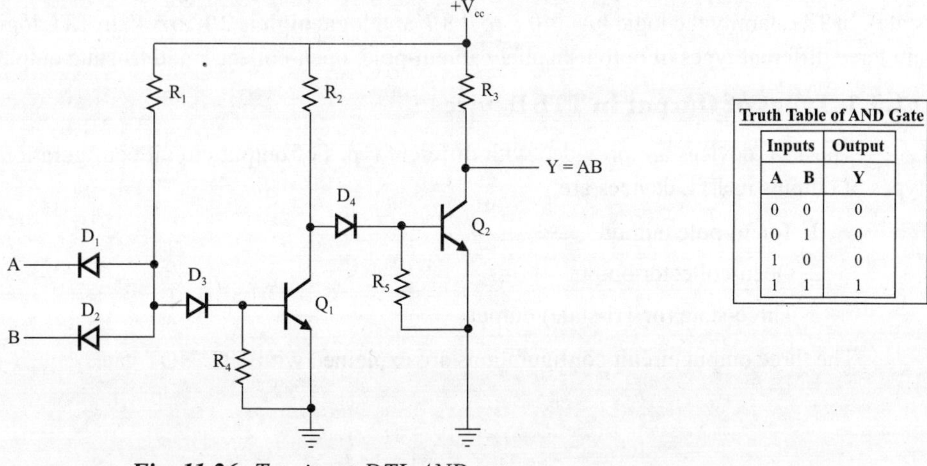

Truth Table of AND Gate

Inputs		Output
A	B	Y
0	0	0
0	1	0
1	0	0
1	1	1

***Fig. 11.26:** Two input DTL AND gate.*

11.4.5 DTL OR Gate

The OR logic can be obtained by inverting the output of NOR logic. Therefore, the DTL OR gate is obtained by connecting a NOT gate at the output of NOR gate as shown in Fig. 11.27.

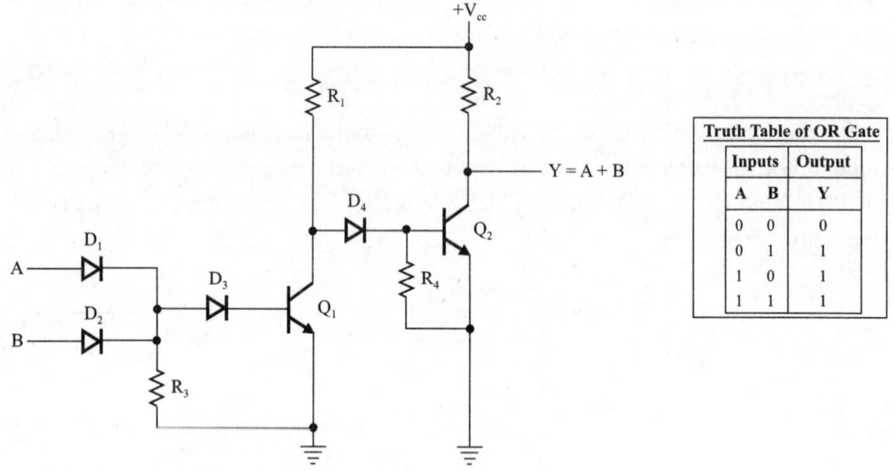

Truth Table of OR Gate

Inputs		Output
A	B	Y
0	0	0
0	1	1
1	0	1
1	1	1

***Fig. 11.27:** Two input DTL OR gate.*

11.5 TTL (Transistor-Transistor Logic)

Transistor-transistor logic (TTL) is a class of digital circuits built from bipolar junction transistors (BJT) and resistors. In TTL (Transistor-Transistor Logic) family, input network is made of transistors and switching devices are also transistors and so the name TTL.

In TTL family, the inputs are applied through emitters of a multi-emitter transistor. As every emitter is nothing but a diode, this logic eliminates the use of all diodes. This is the major advantage because transistors ON and OFF much rapidly than a diode and so switching time will be faster.

TTL family of devices are fastest of saturated logic family and so, widely used bipolar logic family. The basic circuit in TTL family is NAND gate. Transistors operate in cut-off (OFF state) or saturation (ON state). In TTL family the logic **low** is 0 *V* to 0.8 *V* and logic **high** is 2 *V* to 5 *V*. In TTL logic devices, there are three different types of outputs namely, totem-pole, open-collector and tri-state output.

11.5.1 Types of Output in TTL Devices

The TTL devices are provided with different types of output circuit configurations. The different types of outputs in TTL devices are,

1. Totem-pole output
2. Open-collector output
3. Three-state (or Tri-state) output

The three output circuit configurations are explained with TTL NOT gate.

Totem-Pole Output

(CSVTU, Nov/Dec'19, 7 Marks) (AU, Nov/Dec'18, 13 Marks) (IKGPTU, May'20, 2 Marks)

A TTL NOT gate with totem-pole output is shown in Fig. 11.28. In this configuration the output circuit has two transistors Q_3 and Q_4 such that one sit over the other and hence the name totem-pole. In this arrangement, the transistor Q_3 will be in common emitter configuration and hence during turn ON the output impedance will be **low**.

This will provide a fast turn ON and lesser propagation delay. During turn OFF the common emitter transistor configuration of Q_3 provides very high output impedance. The transistor Q_4 acts as an active pull-up to collector of transistor Q_3.

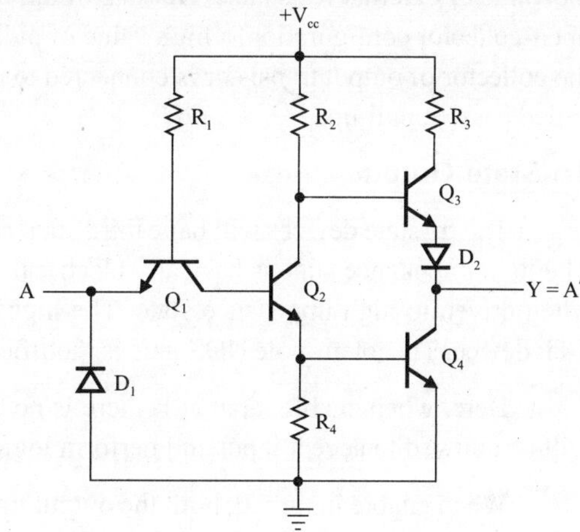

Fig. 11.28: TTL NOT gate with totem-pole output.

Totem-pole outputs consist of both an NPN and a PNP transistor, arranged in a configuration that allows for both sourcing and sinking current. This output structure will lead to short circuit between supply rails (or short circuit between positive and ground supply lines) when two totem-pole outputs are connected together and one output is driven low and the other is driven high. This situation can lead to excessive current flow, potentially damaging the transistors or other components in the circuit.

Moreover, when one output tries to drive the voltage high while the other tries to drive it low. This can result in unpredictable behavior and can cause the outputs to latch up or oscillate. In order to prevent these problems, TTL logic typically uses a "wired-AND" or "wired-OR" configuration when multiple outputs need to be connected together. In these configurations, output signals are combined using external resistors and diodes.

Open-Collector Output

A TTL NOT gate with open-collector output is shown in Fig. 11.29. In this configuration the collector of transistor Q_3 is left open and the output is obtained from this open collector. In order to close the output circuit, an external resistor called pull-up resistor should be connected between open collector and power supply V_{cc}.

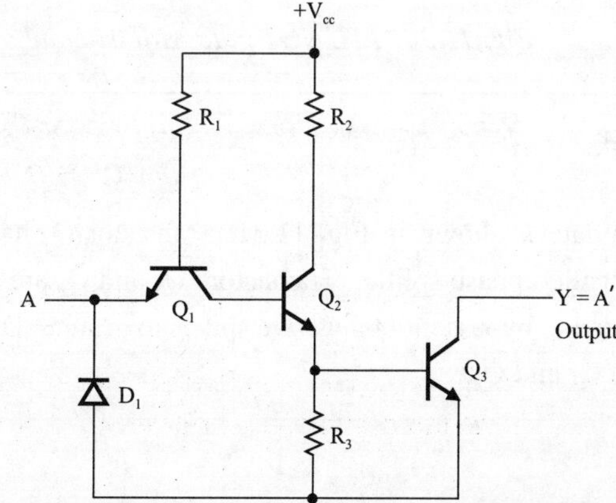

Fig. 11.29: TTL NOT gate with open-collector output.

This will provide flexibility in altering the value of current sourcing and sinking which depends on the value of external resistance. But maximum current is subject to current rating of output transistor. In open-collector configuration, a high value of pull-up resistance will increase the propagation delay. Here, the collector of output transistor is connected to power supply through a resistor and so this connection is called passive pull-up.

Tri-State Output

(AU, Apr/May'19, 5 Marks)

The tri-state devices will have third state called high impedance state apart from low and high state. The high impedance state is basically electrical open circuit condition so that the device will be isolated when driven to high impedance state. The high impedance is achieved by providing an enable input to TTL device. The totem-pole NOT gate is modified as shown in Fig. 11.30 to convert it to a tri-state device.

Here, when enable input = 1, there is no effect on input and output transistors. Hence, the device will be allowed to accept input and perform logical operation.

When enable input = 0, both the output transistors in totem-pole configuration are turned OFF and hence a floating output is created which is a physical connection with very high impedance. The output will behave as an open circuit. This arrangement helps in connecting the outputs in parallel, without interacting with one another.

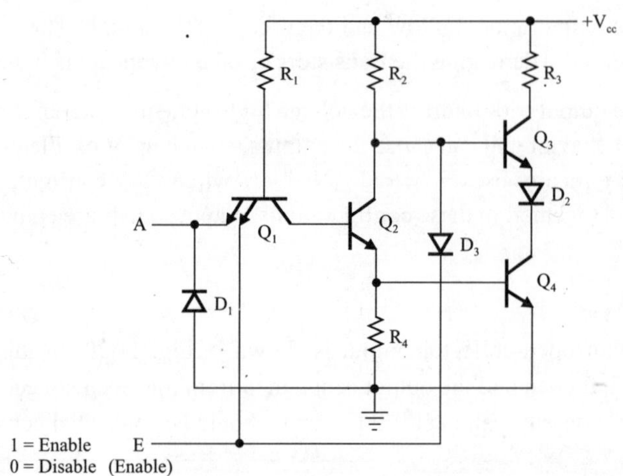

Fig. 11.30: TTL NOT gate with tri-state output.

11.5.2 TTL NAND Gate

(AKTU, Nov/Dec'22, 10 Marks) (IKGPTU, May'20, 5 Marks) (GTU, May'19, 4 Marks) (BATU, Jun'19, 6 Marks) (AU, Apr/May'19, 7 Marks)

A two input NAND gate is shown in Fig. 11.31. Transistor Q_1 has two emitters to accept two inputs. The transistor Q_2 is called phase splitter. Transistors Q_3 and Q_4 are totem-pole output transistors. Diodes D_1 and D_2 are provided to bypass input negative spikes to ground. Diode D_3 is provided to prevent simultaneous conduction of Q_3 and Q_4.

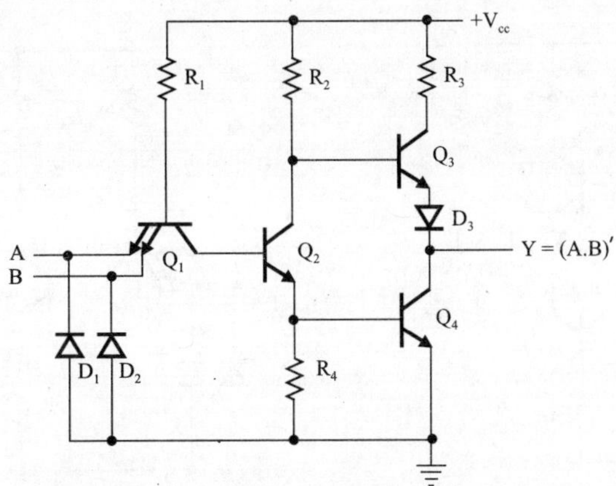

Fig. 11.31: *Two input TTL NAND gate.*

From the truth table of NAND gate, we get the following four cases of inputs.

Case 1: A = 0, B = 0, Y = 1

Case 2: A = 0, B = 1, Y = 1

Case 3: A = 1, B = 0, Y = 1

Case 4: A = 1, B = 1, Y = 0

Truth Table of NAND Gate

Inputs		Output
A	B	Y
0	0	1
0	1	1
1	0	1
1	1	0

Case 1: A = 0, B = 0, Y = 1 .

Figure 11.32(a) shows voltage and current condition of NAND gate when both inputs are **low**. Base-emitter junction of Q_1 is forward biased and collector-base junction is reverse biased and so Q_1 turns ON. Current flows from collector power supply to ground via base-emitter junction. Now, the voltage at base of Q_1 will be 0.7 *V* and this voltage cannot turn ON transistor Q_2. Hence, Q_2 is OFF and its collector voltage will be 5 *V* (V_{cc}) which will turn ON transistor Q_3.

Since Q_2 is OFF, no current flows through R_3 and so no biasing voltage at base of Q_4 and hence, Q_4 is OFF and act as open circuit to ground. Now Q_3 will be in saturation state and the voltage across the collector-emitter terminals will be negligible and so a high voltage appears in output. The collector power supply will source current to load at output through Q_3.

Case 4: A = 1, B = 1, Y = 0

Figure 11.32(b) shows voltage and current condition of NAND gate when both inputs are **high**. Base-emitter junction of Q_1 is reverse biased and collector-base junction is forward biased and so Q_1 turns OFF. Current cannot flow through base-emitter junction of Q_1. Now, the voltage at base of Q_1 will be 5 *V* (V_{cc}) and this voltage will turn ON transistor Q_2. Hence, Q_2 is ON and its collector voltage will be V_{sat} which is negligible which will turn OFF transistor Q_3. Since, Q_3 is OFF it will act as open circuit to collector supply.

Since Q_2 is ON, current flows through R_3 and the voltage drop across R_3 will turn ON transistor Q_4 and so Q_4 act as short circuit to ground and so a **low** will appear on the output. Also, the load at output is shorted to ground and so current will sink to output.

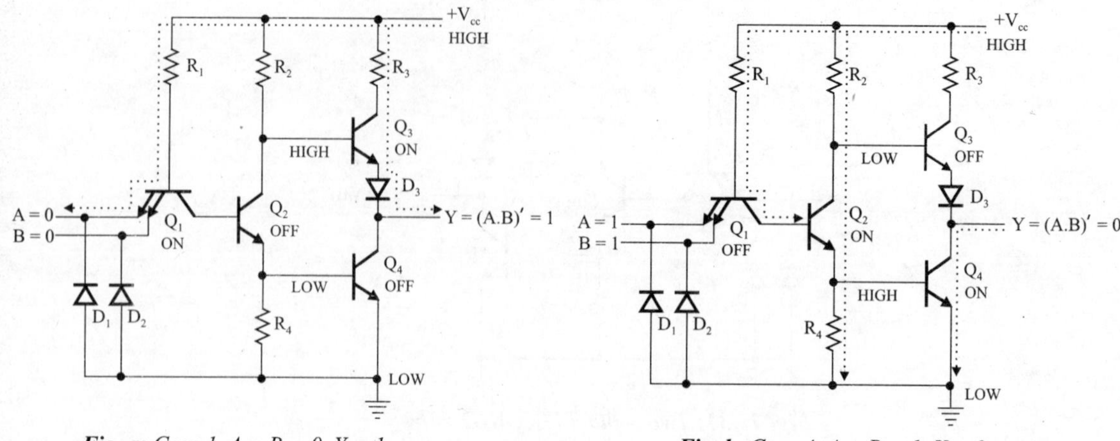

Fig. a: Case 1, A = B = 0, Y = 1. *Fig. b: Case 4, A = B = 1, Y = 0.*

Fig. 11.32: Two cases of operation of TTL NAND gate.

Case 2 and 3: A = 0, B = 1, Y = 1 and A = 1, B = 0, Y = 1

In these two cases one of the emitters of transistor Q_1 is **high**, which will turn ON Q_1 and turn OFF Q_2. Therefore, the working of NAND gate in these two cases will be similar to case 1. The voltage and current condition of NAND gate when any one of the input is **high** is similar to Fig. 11.32(a).

11.5.3 TTL NOT Gate

From the truth table of NAND gate, it can be observed that when the inputs are same then the output is complemented. Therefore, when the input emitters of NAND gate are shorted it will behave as NOT gate, which is equivalent to single input NAND gate as shown in Fig. 11.33. The voltage and current conditions of NOT gate during two cases of input are shown in Figs. 11.34(a) and 11.34(b).

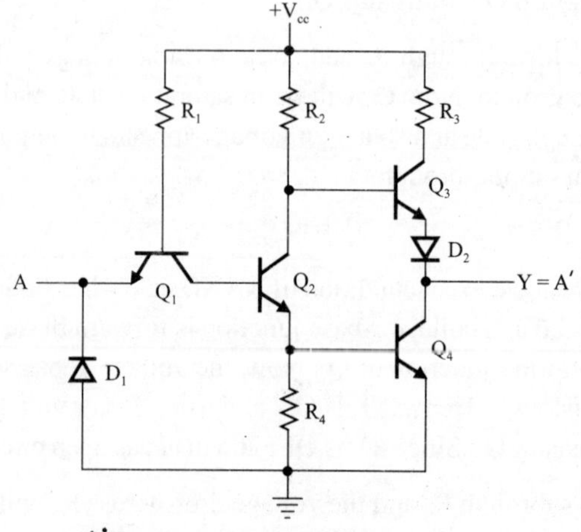

Fig. 11.33: TTL NOT gate.

Truth Table of NOT Gate

Input	Output
A	Y
0	1
1	0

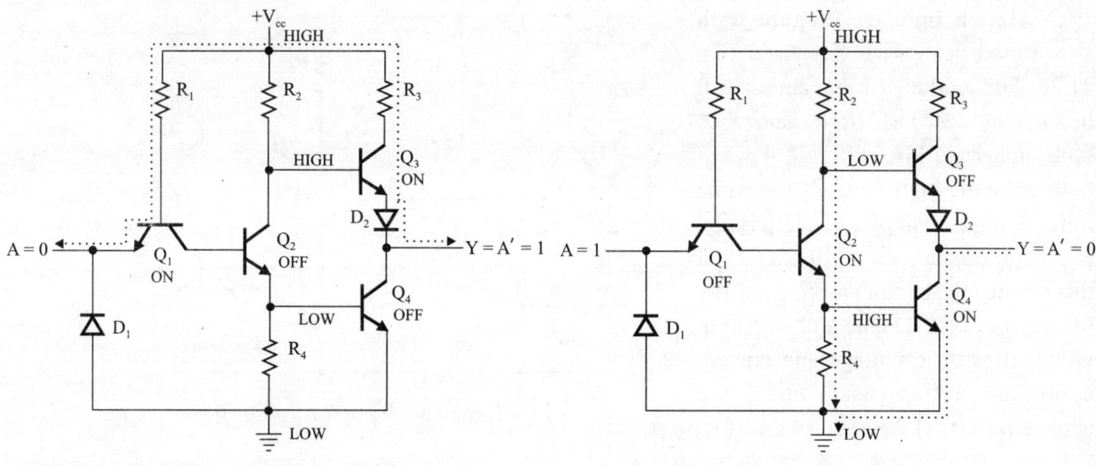

Fig. a: *Case 1, A = 0, Y = 1.* *Fig. b:* *Case 2, A = 1, Y = 0.*

Fig. 11.34: Two cases of operation of TTL NOT gate.

11.5.4 TTL AND Gate

The AND logic can be obtained by inverting the NAND logic. On observing the circuit of NAND gate, the NAND logic actual output is available at the collector of Q_2 and so an additional transistor Q_4 can be inserted between Q_2 and totem-pole transistors as shown Fig. 11.35. The transistor Q_4 will invert the output of Q_2 and send to totem pole output transistors. The diode D_3 is provided to prevent reverse current flow and the transistor Q_3 is provided to ensure logic **low** at the base of Q_4 during ON state of Q_2.

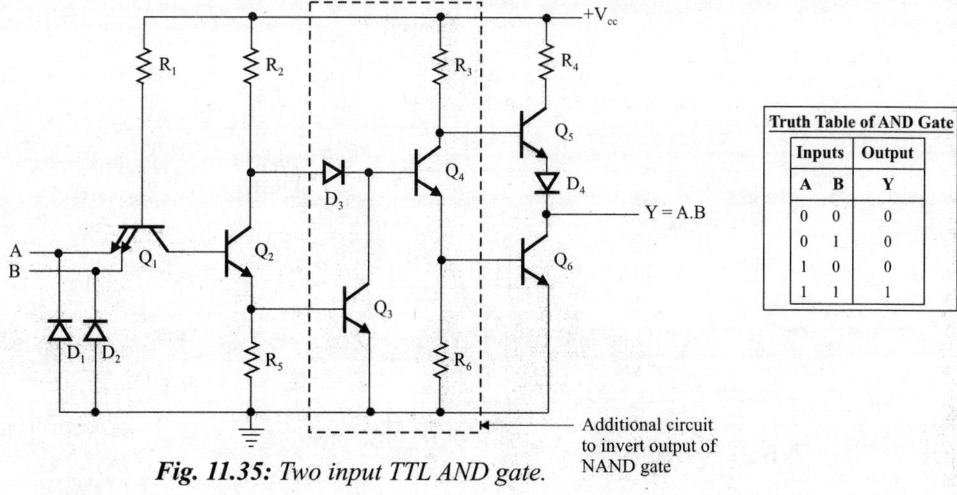

Truth Table of AND Gate

Inputs		Output
A	**B**	**Y**
0	0	0
0	1	0
1	0	0
1	1	1

Fig. 11.35: Two input TTL AND gate.

11.5.5 TTL NOR Gate

For NOR logical operation, two changes are made in the basic TTL NAND gate. The multi-emitter transistor at the input is replaced by individual transistors. The phase splitter Q_2 is replaced by parallel transistors.

A two input NOR gate with above modifications is shown in Fig. 11.36. The working of this circuit will be similar to case 1 of NAND gate when both inputs are **low** and case 4 when both inputs are **high**. But, when any one input is **high**, then one of the parallel phase splitter transistors will be ON and this ensures the circuit working similar to case 4 of NAND gate and so output will be **low**. The voltage and current conditions during case 1 and 2 are shown in Figs. 11.37(a) and 11.37(b).

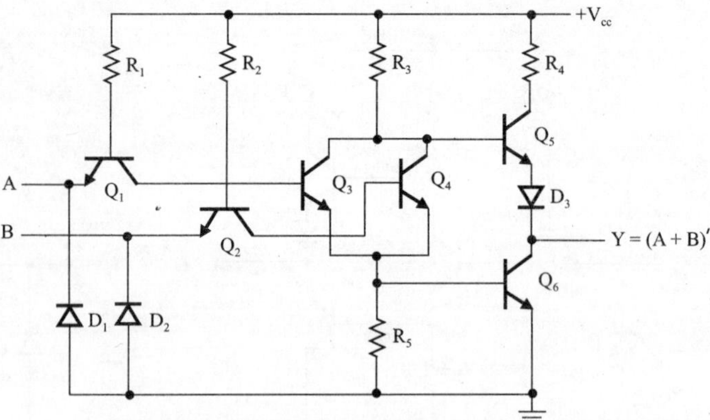

Fig. 11.36: Two input TTL NOR gate.

Truth Table of NOR Gate

Inputs		Output
A	B	Y
0	0	1
0	1	0
1	0	0
1	1	0

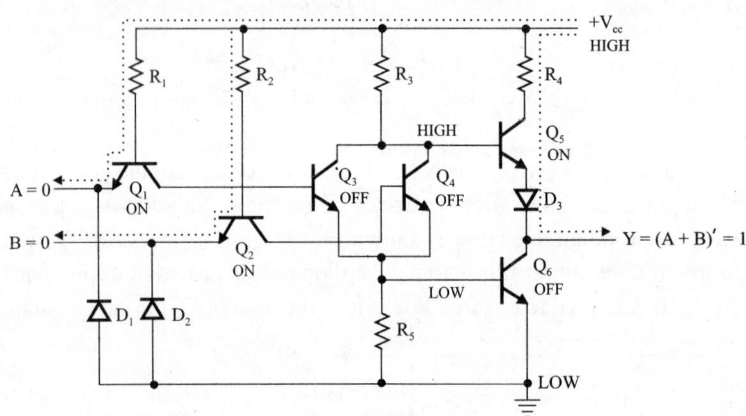

Fig. a: Case 1, A = B = 0, Y = 1.

Fig. b: Case 2, A = 0, B = 1, Y = 0.

Fig. 11.37: Two cases of operation of TTL NOR gate.

11.5.6 TTL OR Gate

The OR logic can be obtained by inverting the NOR logic. On observing the circuit of NOR gate, the NOR logic actual output is available at the collector of parallel combination of Q_3 and Q_4 and so an additional transistor Q_6 can be inserted between parallel and totem-pole transistors as shown Fig. 11.38.

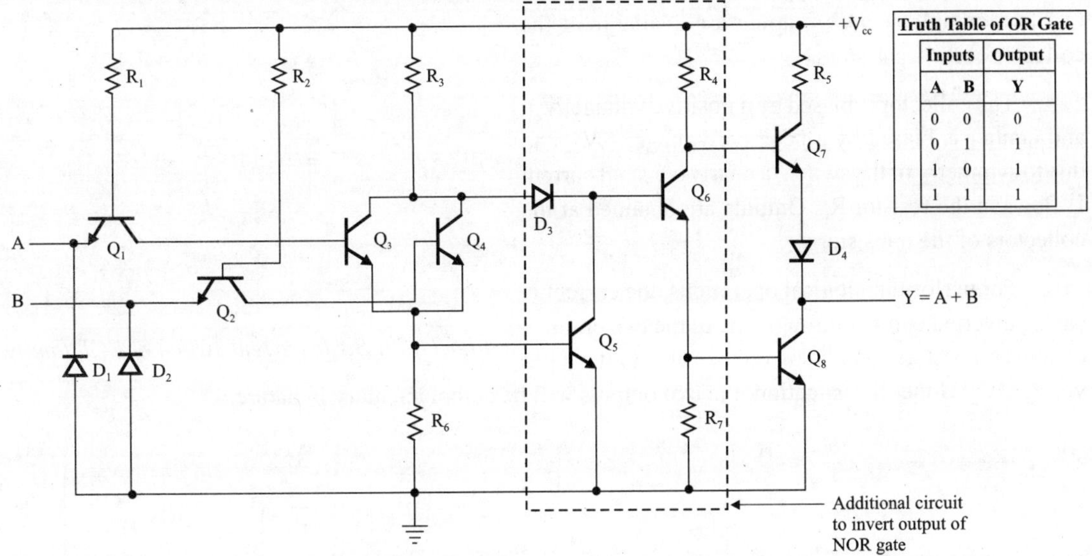

Truth Table of OR Gate		
Inputs		Output
A	B	Y
0	0	0
0	1	1
1	0	1
1	1	1

Fig. 11.38: Two input TTL OR gate.

The transistor Q_6 will invert the output at the collector of parallel transistors and send to totem-pole output transistors. The diode D_3 is provided to prevent reverse current flow and the transistor Q_5 is provided to ensure logic **low** at the base of Q_6 during ON state of Q_3 and Q_4.

11.6 ECL (Emitter Coupled Logic)

(CSVTU, Nov/Dec'19, 7 Marks)

Emitter-coupled logic (ECL) circuits are the fastest available bipolar logic gates. When compared to other logic families, ECL circuits are less susceptible to capacitive loading because of their extremely low output impedance.

Here the important distinction is that TTL uses bipolar transistors as voltage mode switches that swing from cut-off to saturation. In contrast, ECL utilizes bipolar transistors as current mode switches that never saturate. This approach avoids the capacitive effects and saturation delays associated with excess minority carrier storage in the devices. Hence ECL devices are faster than TTL devices.

A second advantage of ECL is that small logic swings are used (less than 1 *V*), resulting in short rise and fall times. Finally, the output drivers used in ECL are emitter followers that are always ON and achieve minimal output impedance, therefore providing superior immunity to capacitive loading at the output.

The primary disadvantage of ECL is the high quiescent power dissipation. The current mode logic concept requires a steady DC current flow in an emitter-coupled current switch. The emitter follower output drivers also require large and steady emitter currents.

11.6.1 Emitter Coupled Current Switch

The basic circuit in an ECL gate is the emitter-coupled current switch as shown in Fig. 11.39. Here input is applied to base of transistor Q_1 and a constant reference voltage, V_{ref} is applied to base of transistor, Q_2. Two complementary outputs are available at the collector of the transistors.

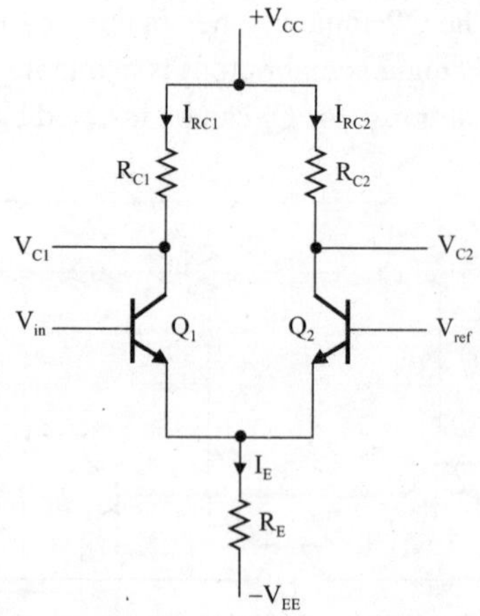

Fig. 11.39: Basic current switch of ECL family.

The collector is biased by a positive voltage, V_{CC} and emitter is biased by a negative voltage, $-V_{EE}$ and due to symmetry of the switch, a nearly constant current I_E flows in the resistor R_E. Outputs are obtained at the collectors of the transistors.

For performing logical operations, the current I_E can be diverted entirely through one of the two emitter-coupled transistors, depending on the value of the input voltage. V_{in}. Under this situation, the two outputs will be complementary in nature.

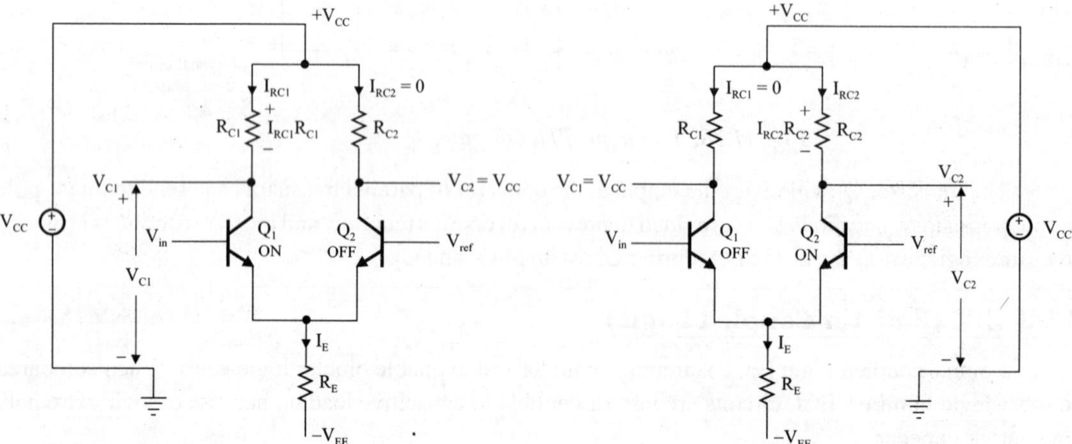

Fig. a: Voltage and current when $V_{in} > V_{ref}$. *Fig. b: Voltage and current when $V_{in} < V_{ref}$.*

Fig. 11.40: Voltage and current in basic current switch of ECL family.

If V_{in} is slightly greater than V_{ref}, then the current I_E flows through Q_1 as shown in Fig. 11.40(a). The following equations for output voltages are obtained using Kirchhoff's Voltage Law (KVL).

$$V_{C1} + I_{RC1} R_{C1} = V_{CC}$$

$$\therefore V_{C1} = V_{CC} - I_{RC1} R_{C1} \Rightarrow V_{OL}$$

$$V_{C2} = V_{CC} \Rightarrow V_{OH}$$

$$\therefore V_{OH} - V_{OL} = V_{CC} - (V_{CC} - I_{RC1} R_{C1})$$
$$= I_{RC1} R_{C1}$$
$$= I_E R_{C1} \qquad \boxed{I_E \approx I_{RC1}}$$

Now, the collector of Q_1 will be at logic **low** voltage and collector of Q_2 at logic **high** voltage.

If V_{in} is slightly lesser than V_{ref}, then the current I_E flows through Q_2 as shown in Fig. 11.40(b). The following equations for output voltages are obtained using Kirchhoff's Voltage Law (KVL).

$$V_{C2} + I_{RC2} R_{C2} = V_{CC}$$
$$\therefore V_{C2} = V_{CC} - I_{RC2} R_{C2} \Rightarrow V_{OL}$$
$$V_{C1} = V_{CC} \Rightarrow V_{OH}$$
$$\therefore V_{OH} - V_{OL} = V_{CC} - (V_{CC} - I_{RC2} R_{C2})$$
$$= I_{RC2} R_{C2}$$
$$= I_E R_{C2} \qquad \boxed{I_E \approx I_{RC2}}$$

Now, the collector of Q_1 will be at logic **high** voltage and collector of Q_2 at logic **low** voltage.

Also, from the above equations we can infer that the voltage difference between logic **high** and **low** levels depends on emitter current, I_E which can be decided by proper choice of resistances.

In the actual gate circuit, the outputs are provided with common emitter driver transistors and V_{CC} is replaced by ground so that the gate is operated using a single negative power supply. Also, from stability point of view, the best choice of V_{CC} is ground.

11.6.2 Emitter Coupled Basic Gate Circuit

An Emitter coupled basic gate circuit is shown in Fig. 11.41. It consists of an emitter coupled current switch with common emitter driver transistors for outputs. The collector end of all transistors is grounded and emitters are biased by a negative power supply of value – 5.2 *V*.

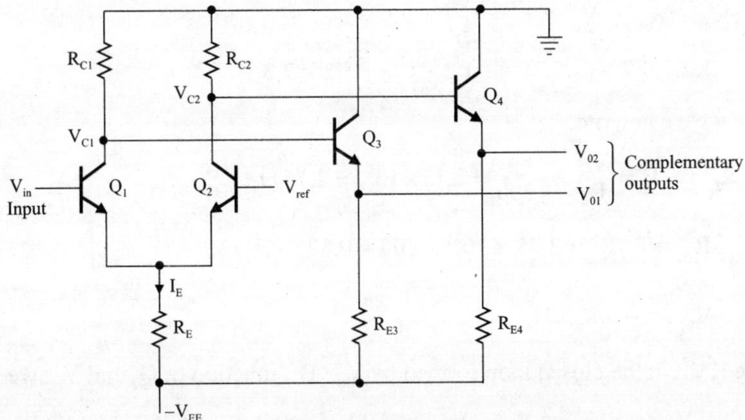

***Fig. 11.41:** Emitter coupled basic gate circuit.*

Emitter coupled transistors operate in active region for both OFF and ON state with a slight difference between logic voltage levels. In ECL family the logic 0 or **low** is –1.5 to –1.7 *V* and logic 1 or **high** is – 0.6 *V* to – 0.9 *V*. The voltages and currents in the ECL basic gate circuit when the inputs are logic **low** and **high** are shown respectively, in Figs. 11.42 and 11.43.

Analysis of ECL basic Gate Circuit when $V_{in} < V_{ref}$

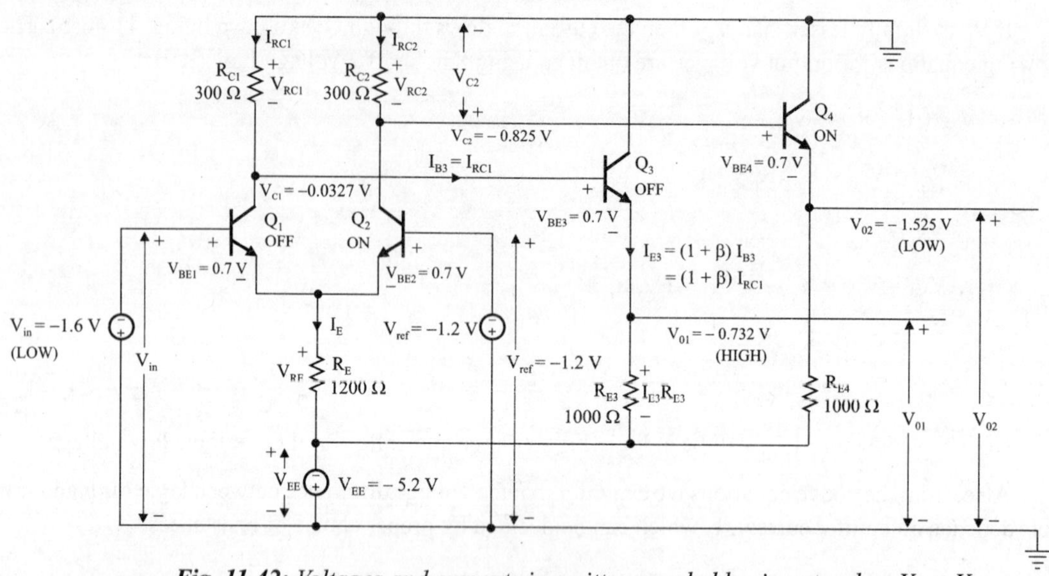

Fig. 11.42: Voltages and currents in emitter coupled basic gate when $V_{in} < V_{ref}$.

In the ECL gate circuit shown in Fig. 11.42, the logic **high** and **low** voltages can be computed using Kirchoff's voltage law and Ohm's law as shown below:

On applying Kirchoff's voltage law (KVL) in the closed loop formed by V_{ref}, BE junction of Q_2, R_E and V_{EE} we can write,

$$V_{ref} = V_{BE2} + V_{RE} + V_{EE}$$

$$\therefore\ V_{RE} = V_{ref} - V_{BE2} - V_{EE}$$

$$= -1.2 - 0.7 - (-5.2) = -1.2 - 0.7 + 5.2 = 3.3\ V$$

By Ohm's law we can write,

$$V_{RE} = I_E R_E \Rightarrow I_E = \frac{V_{RE}}{R_E} = \frac{3.3}{1200} = 2.75 \times 10^{-3} = 2.75\ mA$$

$$V_{RC2} = I_{RC2} R_{C2} = I_E R_{C2} = 2.75 \times 10^{-3} \times 300 = 0.825\ V \qquad \boxed{I_{RC2} \approx I_E}$$

Now, $V_{C2} = -V_{RC2} = -0.825\ V$

On applying KVL in the closed loop formed by V_{02}, BE junction of Q_4 and V_{C2} we get,

$$V_{02} + V_{BE4} = V_{C2} \Rightarrow V_{02} = V_{C2} - V_{BE4} = -0.825 - 0.7 = -1.525\ V \text{ (Logic } \textbf{low}\text{)}$$

On applying KVL in the closed loop formed by R_{C1}, BE junction of Q_3, R_{E3} and V_{EE} we get,

$$V_{RC1} + V_{BE3} + I_{E3} R_{E3} + V_{EE} = 0$$

$$I_{RC1} R_{C2} + V_{BE3} + I_{RC1}(1+\beta) R_{E3} + V_{EE} = 0$$

Let, $\beta = 40$

$$I_{RC1} R_{C1} + I_{RC1}(1+\beta) R_{E3} = -V_{BE3} - V_{EE}$$

$$\therefore I_{RC1} = \frac{-V_{BE3} - V_{EE}}{R_{C1} + (1+\beta) R_{E3}} = \frac{-0.7-(-5.2)}{300+(1+40)1000}$$

$$= 0.109 \times 10^{-3} = 0.109\ mA$$

By Ohm's law,

$$V_{RC1} = I_{RC1} R_{C1} = 0.109 \times 10^{-3} \times 300 = 0.0327\ V$$

Now, $V_{C1} = -V_{RC1} = -0.0327\ V$

On applying KVL, in the closed loop formed by V_{01}, BE junction of Q_3 and R_{C1} we get,

$$V_{01} + V_{BE3} + V_{RC1} = 0$$

$$\therefore V_{01} = -V_{RC1} - V_{BE3}$$

$$= -0.0327 - 0.7$$

$$= -0.7327\ V \text{ (Logic **high**)}$$

Analysis of ECL basic Gate Circuit when $V_{in} > V_{ref}$

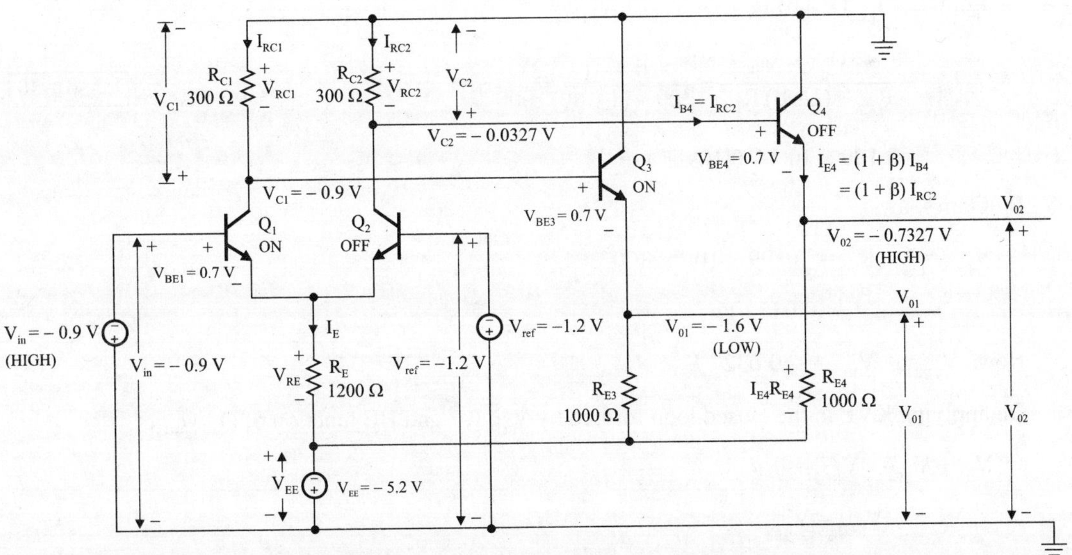

Fig. 11.43: Voltages and currents in emitter coupled basic gate when $V_{in} > V_{ref}$.

In the ECL gate circuit shown in Fig. 11.43, the logic **high** and **low** voltages can be computed using Kirchoff's voltage law and Ohm's law as shown below:

On applying Kirchoff"s voltage law (KVL) in the closed loop formed by V_{in}, BE junction of Q_1, R_E and V_{EE} we can write,

$$V_{in} = V_{BE1} + V_{RE} + V_{EE}$$

$$\therefore V_{RE} = V_{in} - V_{BE1} - V_{EE}$$

$$= -0.9 - 0.7 - (-5.2) = -0.9 - 0.7 + 5.2 = 3.6\ V$$

By Ohm's law we can write,

$$V_{RE} = I_E\, R_E \Rightarrow I_E = \frac{V_{RE}}{R_E} = \frac{3.6}{1200} = 3\times10^{-3} = 3\ mA$$

$$V_{RC1} = I_{RC1}\, R_{C1} = I_E\, R_{C1} = 3 \times 10^{-3} \times 300 = 0.9\ V$$ $\boxed{I_{RC1} \approx I_E}$

Now, $V_{C1} = -V_{RC1} = -0.9\ V$

On applying KVL in the closed loop formed by V_{01}, BE junction of Q_3 and V_{C1} we get,

$$V_{01} + V_{BE3} = V_{C1} \Rightarrow V_{01} = V_{C1} - V_{BE3} = -0.9 - 0.7 = -1.6\ V \text{ (Logic low)}$$

On applying KVL in the closed loop formed by R_{C2}, BE junction of Q_4, R_{E4} and V_{EE} we get,

$$V_{RC2} + V_{BE4} + I_{E4}\, R_{E4} + V_{EE} = 0$$

$$I_{RC2}\, R_{C2} + V_{BE4} + I_{RC2}(1 + \beta)\, R_{E4} + V_{EE} = 0$$

$$I_{RC2}\, R_{C2} + I_{RC2}(1 + \beta)\, R_{E4} = -V_{BE4} - V_{EE}$$

$$\therefore I_{RC2} = \frac{-V_{BE4} - V_{EE}}{R_{C2} + (1+\beta)\, R_{E4}} = \frac{-0.7 - (-5.2)}{300 + (1+40)\, 1000}$$ $\boxed{\text{Let, } \beta = 40}$

$$= 0.109 \times 10^{-3} = 0.109\ mA$$

By Ohm's law,

$$V_{RC2} = I_{RC2}\, R_{C2} = 0.109 \times 10^{-3} \times 300$$

$$= 0.0327\ V$$

Now, $V_{C2} = -V_{RC2} = -0.0327\ V$

On applying KVL in the closed loop formed by V_{02}, R_{C2} and BE junction of Q_4 we get,

$$V_{02} + V_{BE4} + V_{RC2} = 0$$

$$\therefore V_{02} = -V_{RC2} - V_{BE4}$$

$$= -0.0327 - 0.7$$

$$= -0.7327\ V \text{ (Logic high)}$$

11.6.3 ECL NOR Gate

A two input ECL NOR gate can be formed by connecting input transistors in parallel in the basic ECL gate circuit as shown in Fig. 11.44. Transistors Q_2 and Q_3 forms the emitter coupled current switch in which a reference voltage is applied to base of Q_3 and one of the inputs is applied to base of Q_2. An additional input transistor Q_1 is connected in parallel to Q_1. Here, Q_4 is emitter follower output driver transistor and output is obtained from the emitter of Q_4.

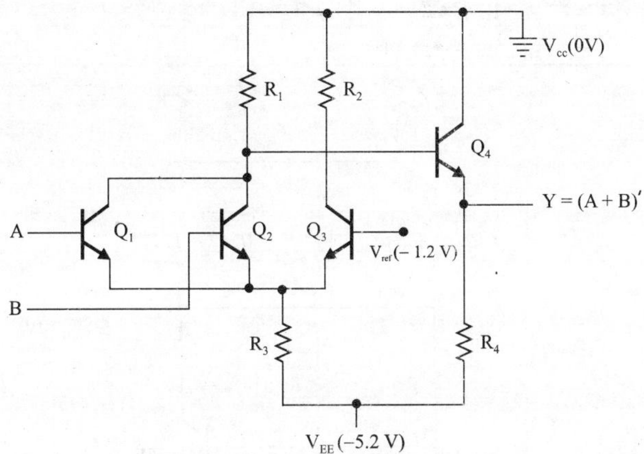

***Fig. 11.44:** Two input ECL NOR gate.*

From the truth table of NOR gate, we get the following four cases of inputs.

Case 1: A = 0, B = 0, Y = 1

Case 2: A = 0, B = 1, Y = 0

Case 3: A = 1, B = 0, Y = 0

Case 4: A = 1, B = 1, Y = 0

Truth Table of NOR Gate

Inputs		Output
A	B	Y
0	0	1
0	1	0
1	0	0
1	1	0

Case 1: A = 0, B = 0, Y = 1

Figure. 11.45(a) shows voltage and current condition of NOR gate when both inputs are **low**. Now, $V_{in} < V_{ref}$ and so forward bias voltage of BE junction of Q_3 is more than Q_2 and so Q_3 turns ON and Q_2 turns OFF. The collector of Q_2 will be at a logic **high** voltage which will turn ON Q_4. Hence a logic **high** voltage appear at the output emitter terminal of Q_4. The collector power supply will source current to load at output through Q_4.

Case 4: A = 1, B = 1, Y = 0

Figure. 11.45(d) shows voltage and current condition of NOR gate when both inputs are **high**. Now, $V_{in} > V_{ref}$ and so forward bias voltage of BE junction of Q_3 is less than Q_2 and so Q_3 turns OFF and Q_2 turns ON. The collector of Q_2 will be at a logic **low** voltage which will turn OFF Q_4. Hence a logic **low** voltage appears at the output emitter terminal of Q_4. The collector power supply will sink current from load at output through emitter resistance R_4.

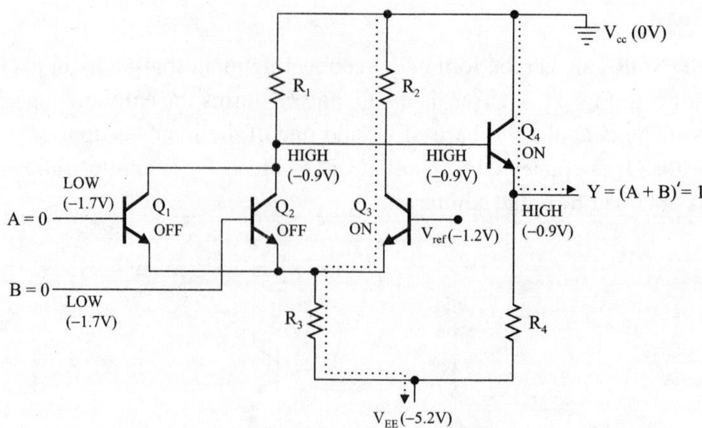

Fig. a: Case 1, A = B = 0, Y = 1.

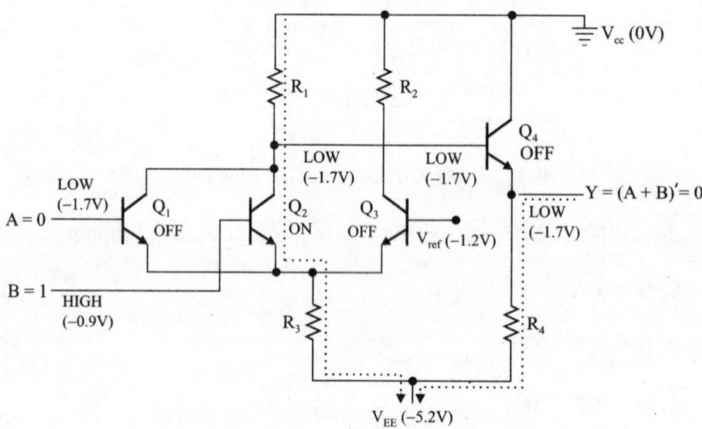

Fig. b: Case 2, A = 0, B = 1, Y = 0.

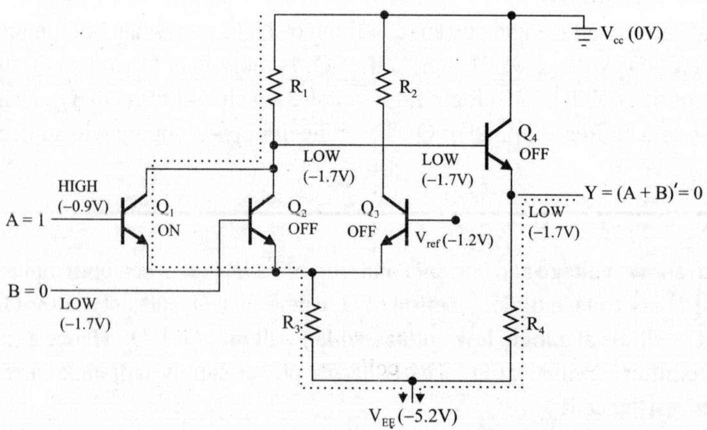

Fig. c: Case 3, A = 1, B = 0, Y = 0.

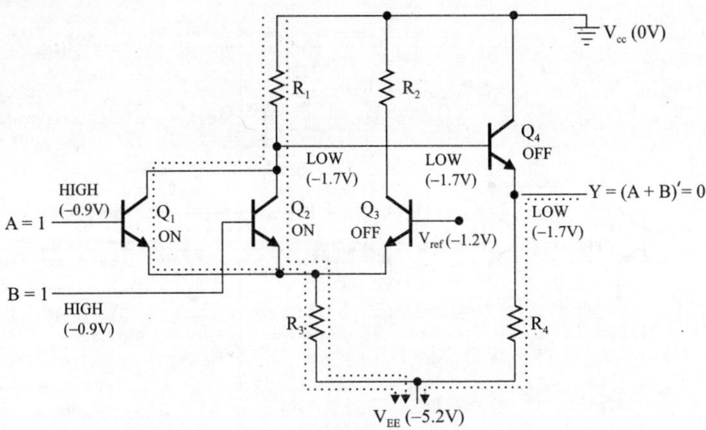

Fig. d: Case 4, A = B = 1, Y = 0.

Fig. 11.45: Four cases of operation of ECL NOR gate.

Case 2 and 3: A = 0, B = 1, Y = 0 and A = 1, B = 0, Y = 0

In these two cases one of the bases of input transistors is **high** and the transistor whose base is **high** will turn ON so that the voltage at the parallel connected collector terminal will at a logic **low** voltage. Therefore, the working of NOR gate in these two cases will be similar to case 4. Figures 11.45(b) and 11.45(c) shows voltage and current condition of NOR gate when any one of the inputs is **high**.

11.6.4 ECL OR Gate

A two input ECL OR gate is shown in Fig. 11.46. The OR logic can be obtained by inverting the NOR logic and so emitter coupled switch output is obtained from collector of Q_3 which will be invert logic of NOR operation. The voltage and current conditions of ECL OR gate during four cases of inputs is shown in Fig. 11.47.

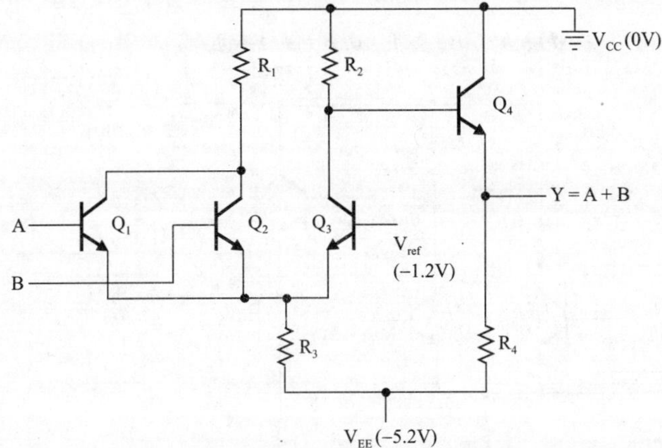

Truth Table of OR Gate

Inputs		Output
A	B	Y
0	0	0
0	1	1
1	0	1
1	1	1

Fig. 11.46: Two input ECL OR gate.

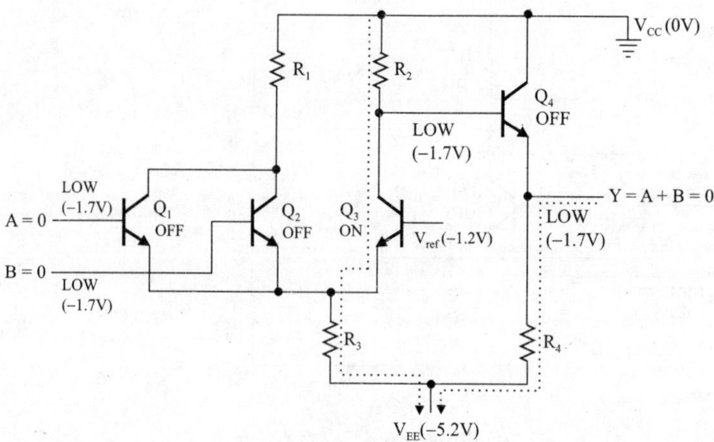

Fig. a: Case 1, A = B = 0, Y = 0.

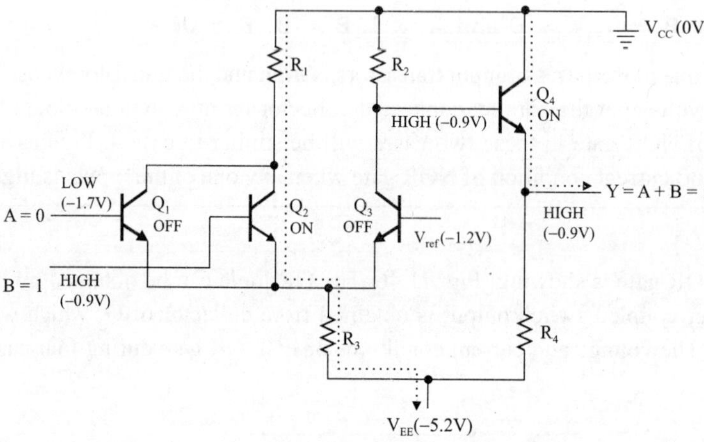

Fig. b: Case 2, A = 0, B = 1, Y = 1.

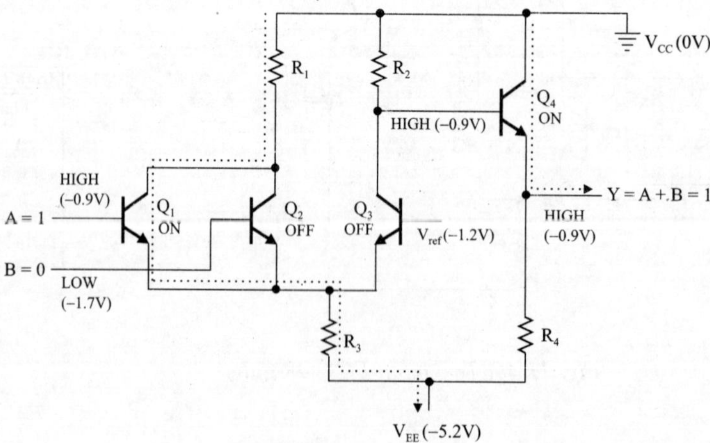

Fig. c: Case 3, A = 1, B = 0, Y = 1.

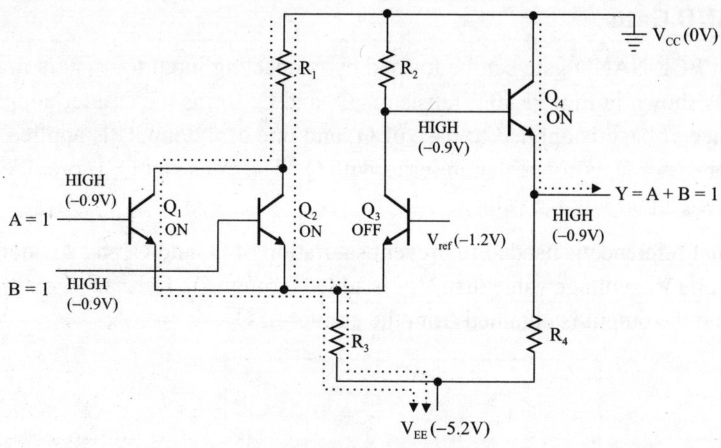

***Fig. d:** Case 4, A = B = 1, Y = 1.*

***Fig. 11.47:** Four cases of operation of ECL OR gate.*

11.6.5 ECL NOT Gate

Basically, the emitter coupled current switch has two outputs in which one of the output is inverted logic of input. Hence NOT gate is formed by connecting a emitter follower driver transistor to this inverted current switch output as shown in Fig. 11.48. The voltage and current conditions of ECL NOT gate during two cases of inputs is shown in Fig. 11.49.

Truth Table of NOT Gate

Input A	Output Y
0	1
1	0

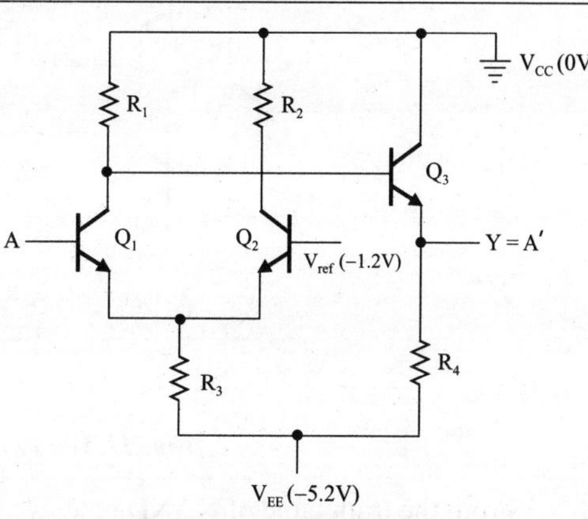

***Fig. 11.48:** ECL NOT gate.*

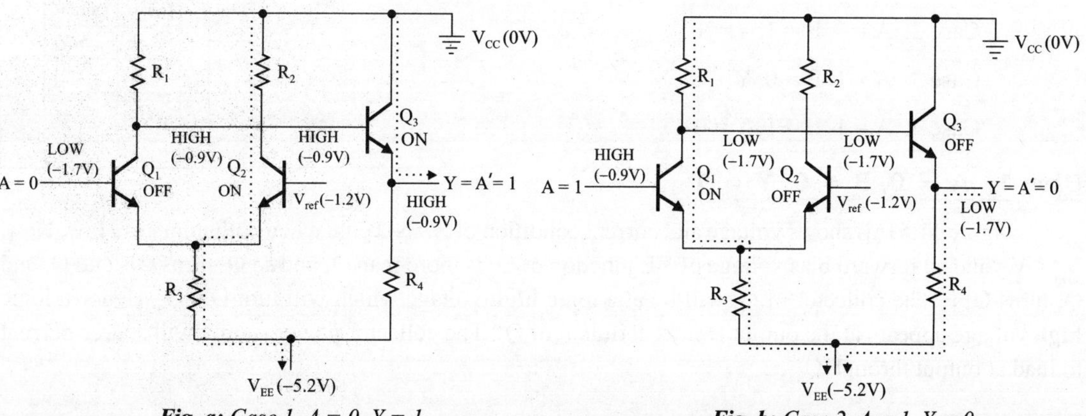

***Fig. a:** Case 1, A = 0, Y = 1.* ***Fig. b:** Case 2, A = 1, Y = 0.*

***Fig. 11.49:** Two cases of operation of ECL NOT gate.*

11.6.6 ECL NAND Gate

A two input ECL NAND gate can be formed by connecting input transistors in series in the basic ECL gate circuit as shown in Fig. 11.50. Transistors Q_1 and Q_2 forms the emitter coupled current switch in which a reference voltage is applied to base of Q_2 and one of the inputs is applied to base of Q_1. An additional input transistor Q_3 is connected in series with Q_1. The transistor Q_5 is provided to level shift the B input by one base-emitter voltage value.

An additional reference is needed to prevent saturation of Q_2 and Q_3 and so another reference V_{ref2} which is lower by one V_{BE} voltage value than V_{ref1} is added through Q_4. Here, Q_6 is emitter follower output driver transistor and the output is obtained from the emitter of Q_6.

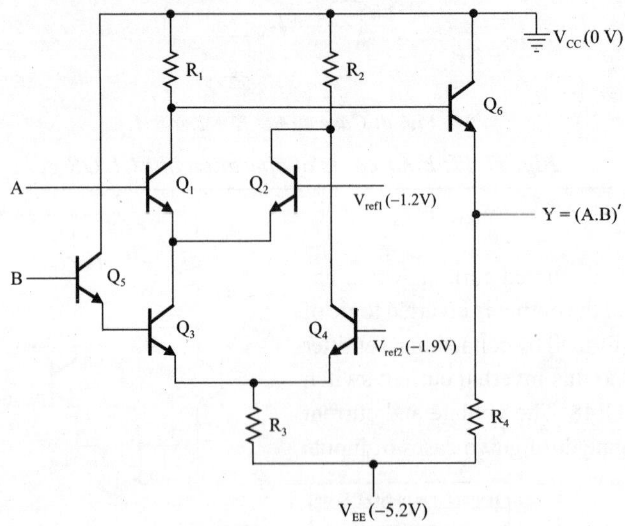

***Fig. 11.50:** Two input ECL NAND gate.*

From the truth table of NAND gate, we get the following four cases of inputs.

Case 1: A = 0, B = 0, Y = 1

Case 2: A = 0, B = 1, Y = 1

Case 3: A = 1, B = 0, Y = 1

Case 4: A = 1, B = 1, Y = 0

Truth Table of NAND Gate

Inputs		Output
A	B	Y
0	0	1
0	1	1
1	0	1
1	1	0

Case 1: A = 0, B = 0, Y = 1

Figure 11.51(a) shows voltage and current condition of NAND gate when both inputs are **low**. Now, $V_{in} < V_{ref}$ and so forward bias voltage of BE junction of Q_2 is more than Q_1 and so Q_2 turns ON and Q_1 and Q_3 turns OFF. The collector of Q_1 will be at a logic **high** voltage which will turn ON Q_6. Hence a logic **high** voltage appears at the output emitter terminal of Q_6. The collector power supply will source current to load at output through Q_6.

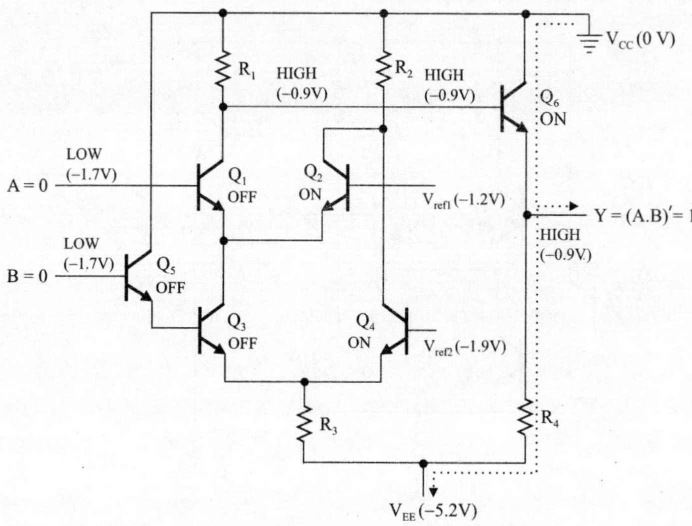

***Fig. a:** Case 1, A = B = 0, Y = 1.*

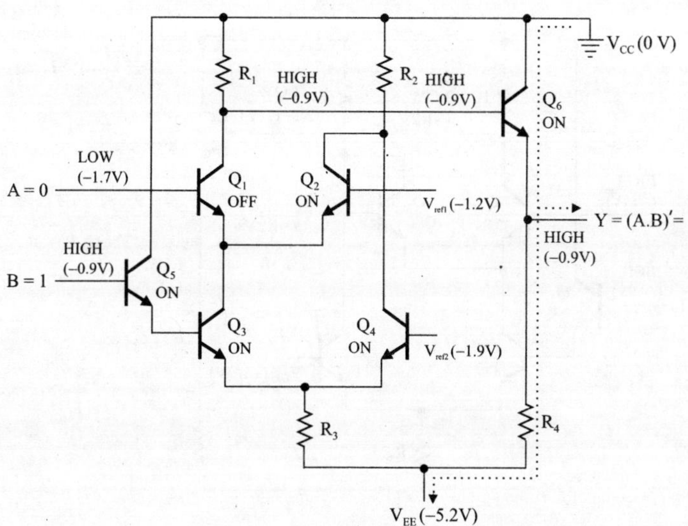

***Fig. b:** Case 2, A = 0, B = 1, Y = 1.*

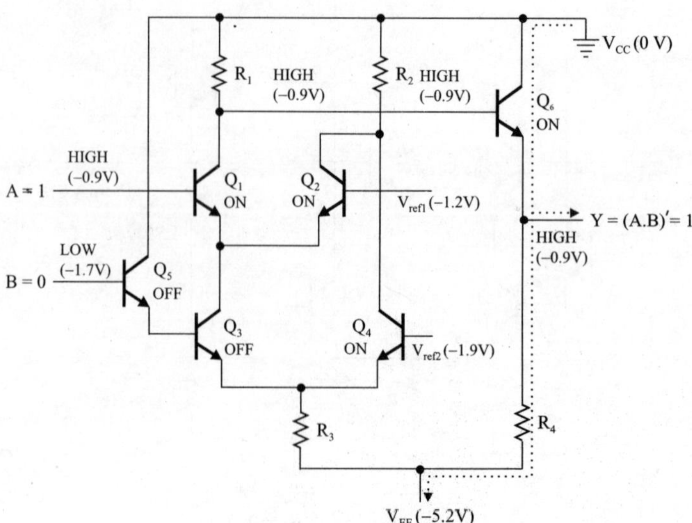

Fig. c: Case 3, A = 1, B = 0, Y = 1.

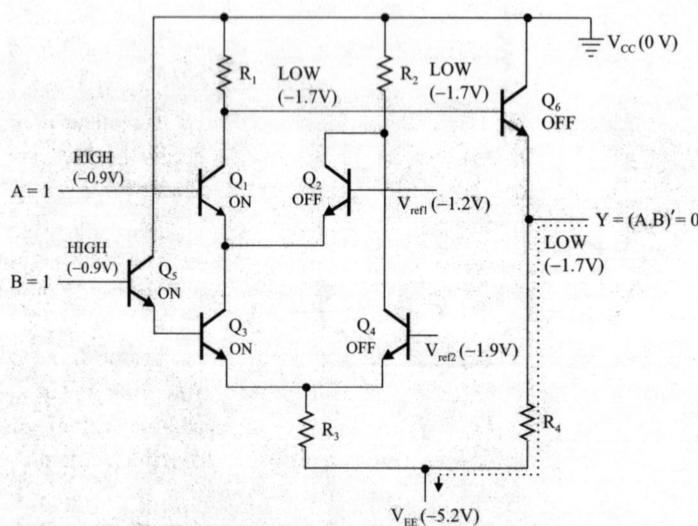

Fig. d: Case 4, A = B = 1, Y = 0.

Fig. 11.51: *Four cases of operation of ECL NAND gate.*

Case 4: A = 1, B = 1, Y = 0

Fig. 11.51(d) shows voltage and current condition of NAND gate when both inputs are **high**. Now, $V_{in} > V_{ref}$ and so forward bias voltage of BE junction of Q_2 is less than Q_1 and Q_3 and so Q_2 turns OFF and Q_1 and Q_3 turns ON. The collector of Q_1 will be at a logic **low** voltage which will turn OFF Q_6. Hence a logic **low** voltage appears at the output emitter terminal of Q_6. The collector power supply will sink current from load at output through emitter resistance R_4.

Case 2 and 3: A = 0, B = 1, Y = 1 and A = 1, B = 0, Y = 1

In these two cases one of the bases of input transistors is **high** and the corresponding transistor will turn ON and the other transistor will be OFF. Therefore, the voltage at the series connected collector terminal will at a logic **high** voltage. Therefore, the working of NAND gate in these two cases will be similar to case 1 Figs. 11.51(b) and 11.51(c) shows voltage and current condition of NAND gate when any one of the inputs is **high**.

11.6.7 ECL AND Gate

A two input ECL AND gate is shown in Fig. 11.52. The AND logic can be obtained by inverting the NAND logic and so emitter coupled switch output is obtained from collector of Q_2 which will be invert logic of NAND operation. The voltage and current conditions of ECL AND gate during four cases of inputs is shown in Fig. 11.53.

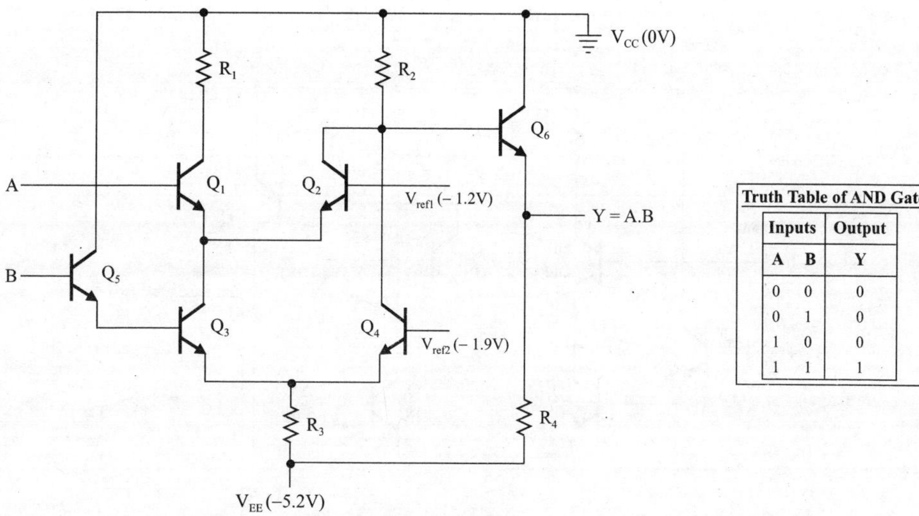

Inputs		Output
A	B	Y
0	0	0
0	1	0
1	0	0
1	1	1

Fig. 11.52: Two input ECL AND gate.

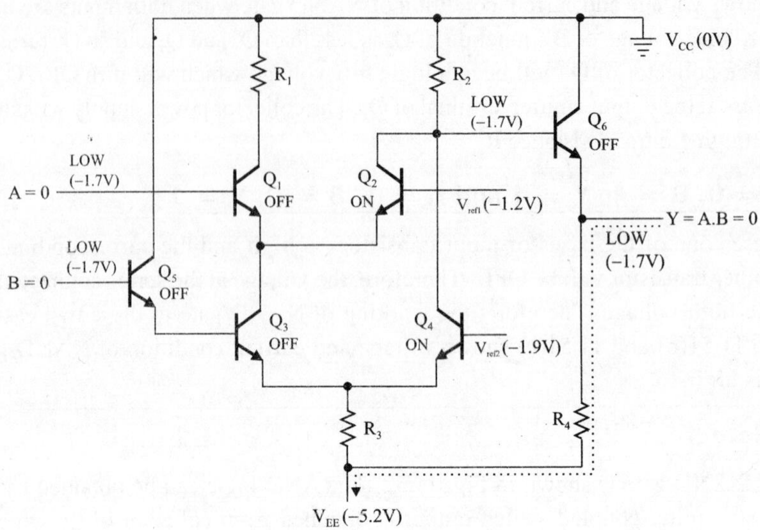

Fig. a: Case 1, A = B = 0, Y = 0.

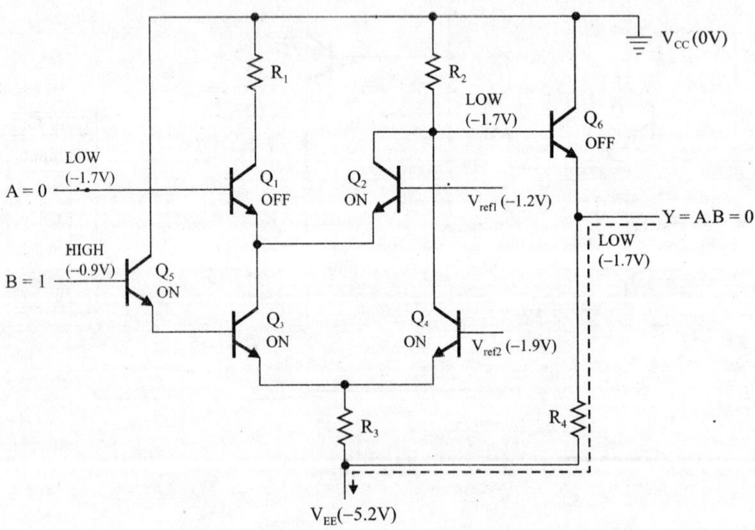

Fig. b: Case 2, A = 0, B = 1, Y = 0.

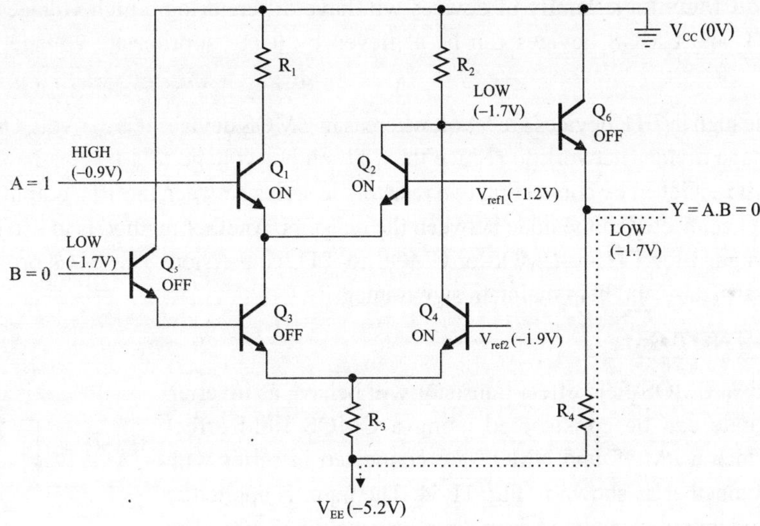

Fig. c: Case 3, A = 1, B = 0, Y = 0.

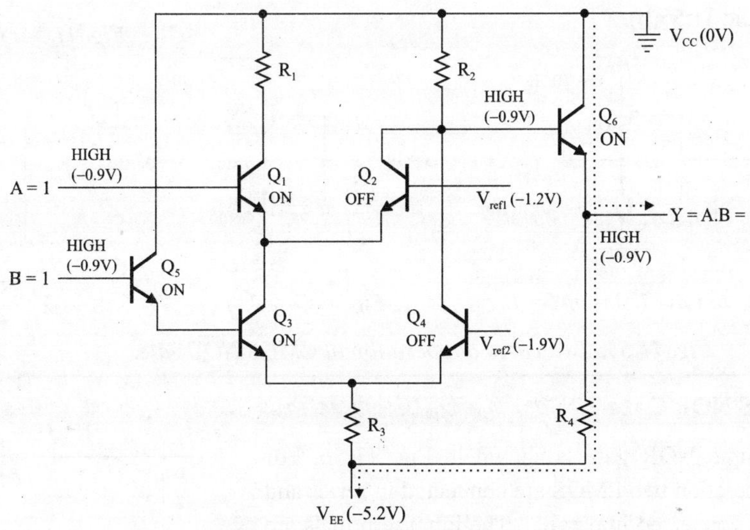

Fig. d: Case 4, A = B = 1, Y = 1.

Fig. 11.53: Four cases of operation of ECL AND gate.

11.7 CMOS (Complementary MOSFET)

(IKGPTU, May'20, 7 Marks)

In CMOS logic family of devices, the logic functions are realized by parallel or series connection of PMOS and NMOS field effect transistors.

In CMOS family, the inputs are applied to gate terminal of MOSFET and the output is obtained from the drain terminal of NMOS field effect transistor.

The MOSFETs operate in cut-off (OFF state) or saturation (ON state). In CMOS family, logic **low** is 0 *V* and logic **high** is 3.3 *V*.

TTL and CMOS logic family of devices will have different logic high voltage levels and hence interfacing TTL and CMOS devices can be achieved by using appropriate voltage level conversion techniques. *(CSVTU, Nov/Dec'19, 7 Marks) (AKTU, Dec'22, 10 Marks)*

The logic high in TTL devices are 5 Volt whereas in CMOS devices it is 3.3 Volt. One simple method is to use a voltage divider network to reduce the TTL output voltage to a level suitable for the CMOS input. This can be achieved by connecting two resistors in series between the TTL output and ground, with the CMOS input connected to the node between the resistors. Another method is to use level-shifting ICs specifically designed for TTL-to-CMOS or CMOS-to-TTL conversion. These ICs provide bidirectional voltage level translation and help maintain signal integrity.

11.7.1 CMOS NOT Gate

Basically, a CMOS field effect transistor will behave as inverter. Hence, NOT gate can be constructed using a CMOS field effect transistor in which a PMOS and NMOS are connected in series with their gates tied together as shown in Fig. 11.54. The input is applied to gate terminal and output is obtained from drain terminal of NMOS. The voltage and current conditions during two cases of inputs are shown in Figs. 11.55(a) and 11.55(b).

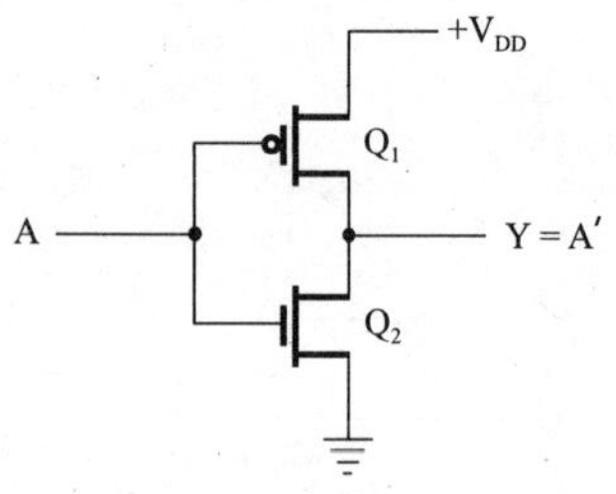

Fig. 11.54: CMOS NOT gate.

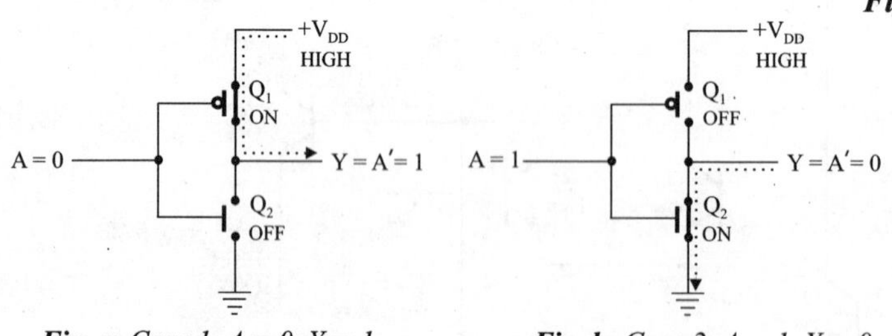

Fig. a: Case 1, A = 0, Y = 1. *Fig. b: Case 2, A = 1, Y = 0.*

Truth Table of NOT Gate

Input A	Output Y
0	1
1	0

Fig. 11.55: Two cases of operation of CMOS NOT gate.

11.7.2 CMOS NOR Gate *(AU, Apr/May'19, 7 Marks)*

A two input NOR gate is shown in Fig. 11.56. For NOR logical operation two PMOS are connected in series and two NMOS are connected in parallel. The input terminals are formed by combining the gates of one series connected PMOS and one parallel connected NMOS. The two NMOS field effect transistors Q_3 and Q_4 are connected in parallel such that drain terminals are tied together to form the output and source terminals are grounded.

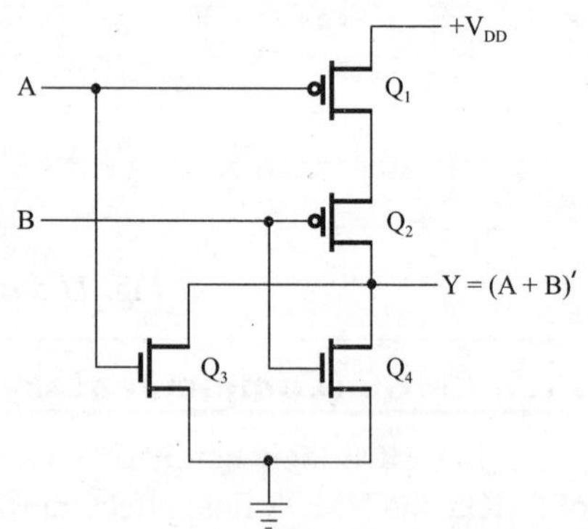

Fig. 11.56: Two input CMOS NOR gate.

From the truth table of NOR gate we get the following four cases of inputs.

Case 1: A = 0, B = 0, Y = 1

Case 2: A = 0, B = 1, Y = 0

Case 3: A = 1, B = 0, Y = 0

Case 4: A = 1, B = 1, Y = 0

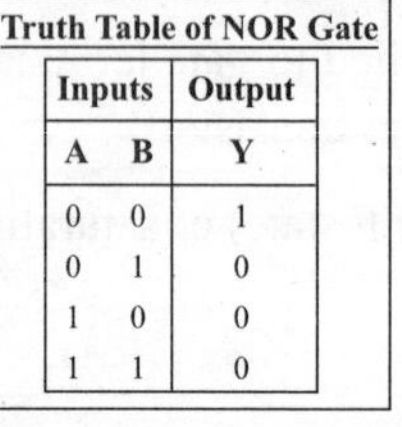

Truth Table of NOR Gate

Inputs A	Inputs B	Output Y
0	0	1
0	1	0
1	0	0
1	1	0

Case 1: A = 0, B = 0, Y = 1

Figure 11.57(a) shows voltage and current condition of NOR gate when both inputs are **low**. Now, the transistors Q_1 and Q_2 are in ON state and the transistors will behave as short circuit and Q_3 and Q_4 are in OFF state and behave as open circuit to ground. Hence, a logic **high** voltage appears in the output. The drain power supply will source current to load at output.

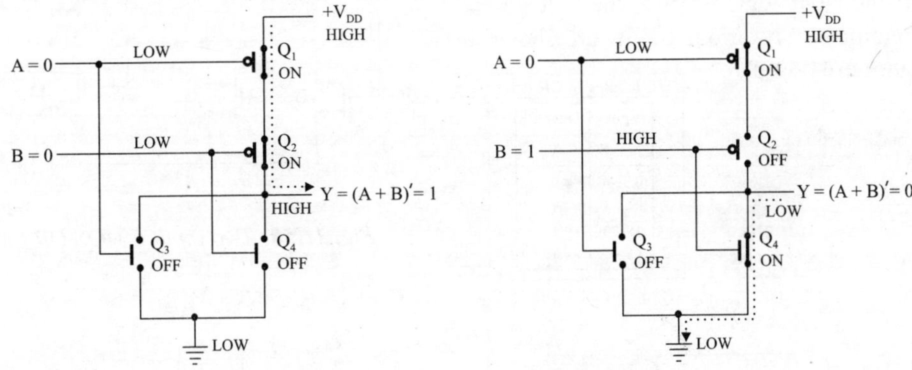

Fig. a: *Case 1, A = B = 0, Y = 1.*

Fig. b: *Case 2, A = 0, B = 1, Y = 0.*

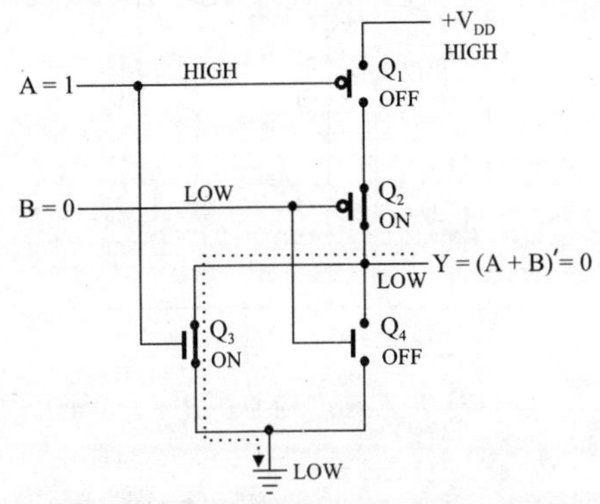

Fig. c: *Case 3, A = 1, B = 0, Y = 0.*

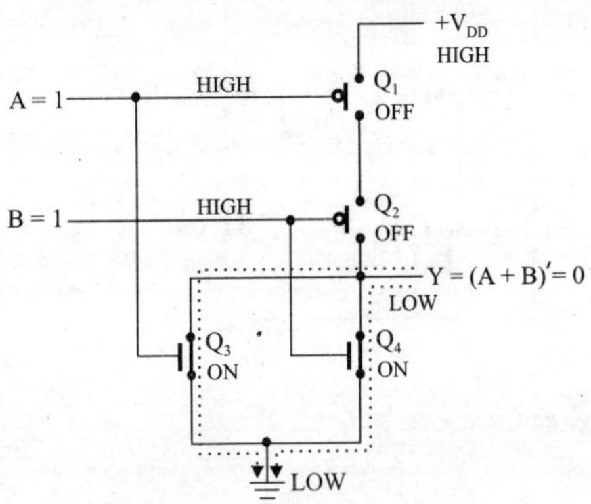

Fig. d: *Case 4, A = B = 1, Y = 0.*

Fig. 11.57: *Four cases of operation of CMOS NOR gate.*

Case 4: A = 1, B = 1, Y = 0

Fig. 11.57(d) shows voltage and current condition of NOR gate when both inputs are **high**. Now, the transistors Q_1 and Q_2 are in OFF state and so they behave as open circuit and the transistors Q_3 and Q_4 are in ON state and will behave as short circuit to ground. Hence a logic **low** voltage appears in the output. The switching transistors will sink current from output.

Case 2 and 3: A = 0, B = 1, Y = 0 and A = 1, B = 0, Y = 0

In these two cases one of the inputs is **high** and the other is **low**. Hence, one of the series transistors will be ON and the other is OFF. Similarly, one of the parallel connected transistors will be ON and the other will be OFF. Now, the ON transistor in parallel connection will behave as short circuit to ground and pull the output to **low**. Therefore, the working of NOR gate in these two cases will be similar to case 4. Figs. 11.57(b) and 11.57(c) shows voltage and current condition of NOR gate when any one of the inputs is **high**.

11.7.3 CMOS OR Gate

The OR logic can be obtained by inverting the NOR logic. The inverted output of NOR gate can be obtained by adding another CMOS inverter at the output of NOR gate as shown in Fig. 11.58 which will invert the NOR logic output. The voltage and current conditions for various inputs are shown in Figs. 11.59(a) to 11.59(d).

Truth Table of OR Gate

Inputs		Output
A	B	Y
0	0	0
0	1	1
1	0	1
1	1	1

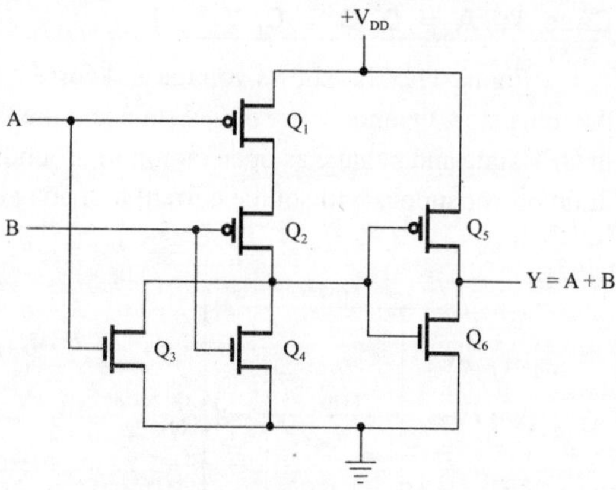

***Fig. 11.58:** Two input CMOS OR gate.*

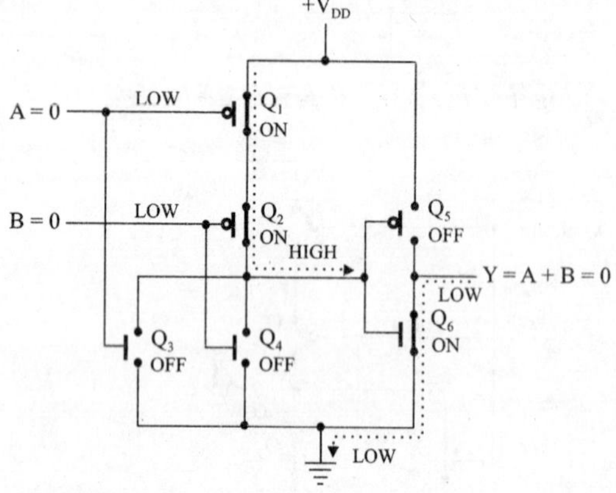

***Fig. a:** Case 1, A = B = 0, Y = 0.*

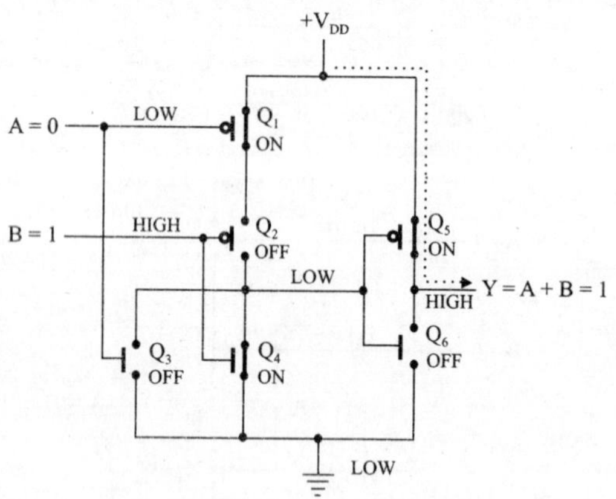

***Fig. b:** Case 2, A = 0, B = 1, Y = 1.*

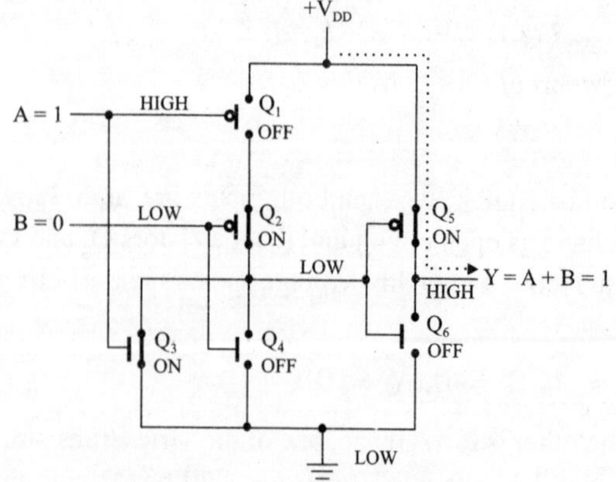

***Fig. c:** Case 3, A = 1, B = 0, Y = 1.*

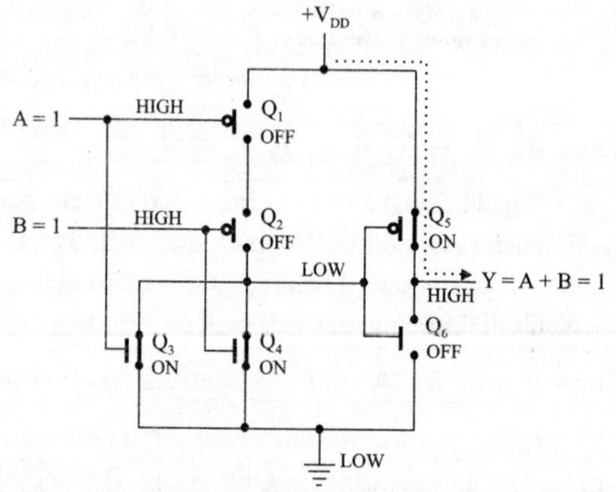

***Fig. d:** Case 4, A = B = 1, Y = 1.*

***Fig. 11.59:** Four cases of operation of CMOS OR gate.*

11.7.4 CMOS NAND Gate

(CSVTU, Nov/Dec'19, 7 Marks) (AKTU, Sep'20, 5 Marks)

A two input NAND gate is shown in Fig. 11.60. For NAND logical operation two PMOS are connected in parallel such that drain terminals are tied together and connected to the drain power supply and source terminals are connected together to form the output.

The two NMOS are connected in series such that the drain terminal of upper most series transistor is connected to output and the source terminal of lower most series connected transistor is connected to ground. The input terminals are formed by combining the gates of one series connected PMOS and one parallel connected NMOS.

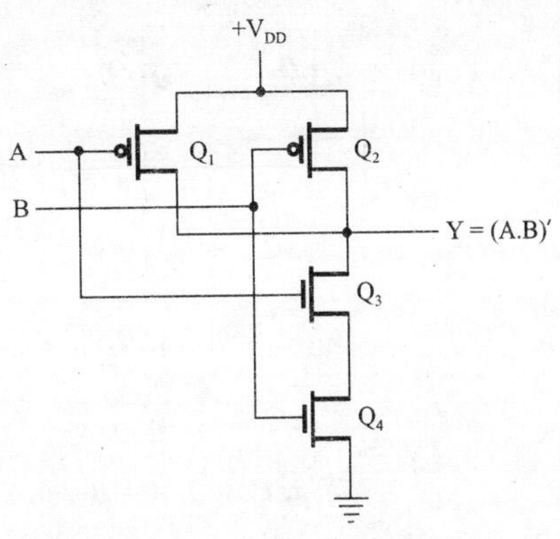

***Fig. 11.60:** Two input CMOS NAND gate.*

From the truth table of NAND gate, we get the following four cases of inputs.

Case 1: A = 0, B = 0, Y = 1

Case 2: A = 0, B = 1, Y = 1

Case 3: A = 1, B = 0, Y = 1

Case 4: A = 1, B = 1, Y = 0

Truth Table of NAND Gate

Inputs		Output
A	B	Y
0	0	1
0	1	1
1	0	1
1	1	0

Case 1: A = 0, B = 0, Y = 1

Figure 11.61(a) shows voltage and current condition of NAND gate when both inputs are **low**. Now, the transistors Q_1 and Q_2 are in ON state and behave as short circuit and Q_3 and Q_4 are in OFF state and behave as open circuit to ground. Hence, a logic **high** voltage appears in the output. The drain power supply will source current to load at output.

Case 4: A = 1, B = 1, Y = 0

Figure 11.61(d) shows voltage and current condition of NAND gate when both inputs are **high**. Now, the transistors Q_1 and Q_2 are in OFF state and behave as open circuit and the transistors Q_3 and Q_4 are in ON state and will behave as short circuit to ground. Hence a logic **low** voltage appears in the output. The switching transistors will sink current from output.

Case 2 and 3: A = 0, B = 1, Y = 1 and A = 1, B = 0, Y = 1

In these two cases one of the inputs is **high** and the other is **low**. Hence, one of the series connected transistors will be ON and the other is OFF. Similarly, one of the parallel connected transistors will be ON and the other will be OFF. The ON transistor in parallel connection will behave as short circuit and connect output to drain power supply and so output will be **high**.

Therefore, the working of NAND gate in these two cases will be similar to case 1. Figs. 11.61(b) and 11.61(c) shows voltage and current condition of NAND gate when any one of the inputs is **high**.

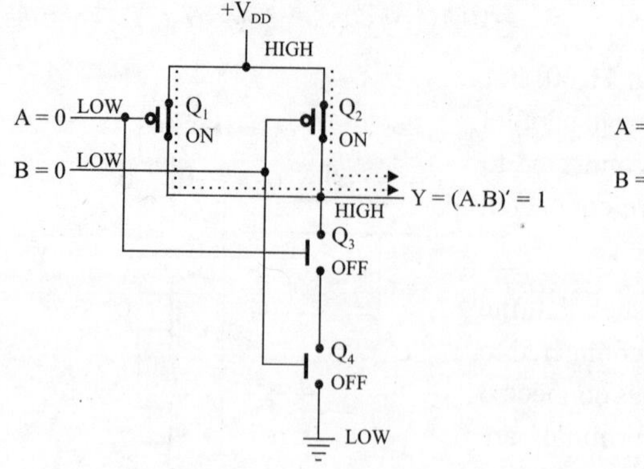

Fig. a: Case 1, A = B = 0, Y = 1.

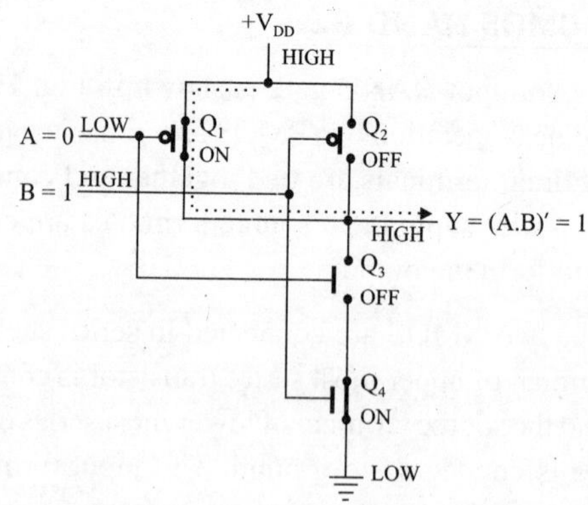

Fig. b: Case 2, A = 0, B = 1, Y = 1.

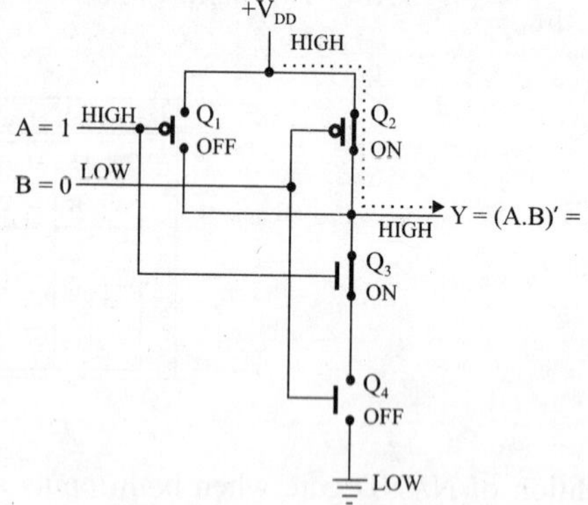

Fig. c: Case 3, A = 1, B = 0, Y = 1.

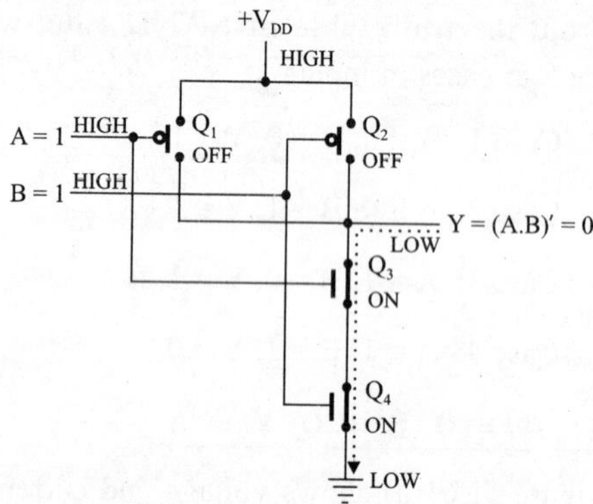

Fig. d: Case 4, A = B = 1, Y = 0.

***Fig. 11.61:** Four cases of operation of CMOS NAND gate.*

11.7.5 CMOS AND Gate

The AND logic can be obtained by inverting the NAND logic. The inverted output of NAND gate can be obtained by adding another CMOS inverter at the output of NAND gate as shown in Fig. 11.62 which will invert the NAND logic output. The voltage and current conditions for various inputs are shown in Figs. 11.63(a) to 11.63(d).

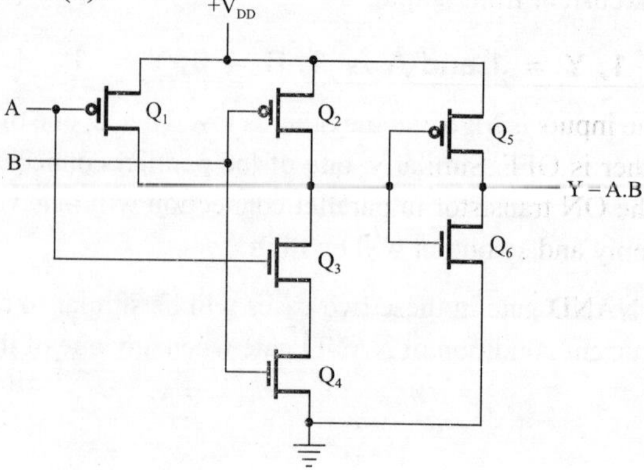

Truth Table of AND Gate

Inputs		Output
A	B	Y
0	0	0
0	1	0
1	0	0
1	1	1

***Fig. 11.62:** Two input CMOS AND gate.*

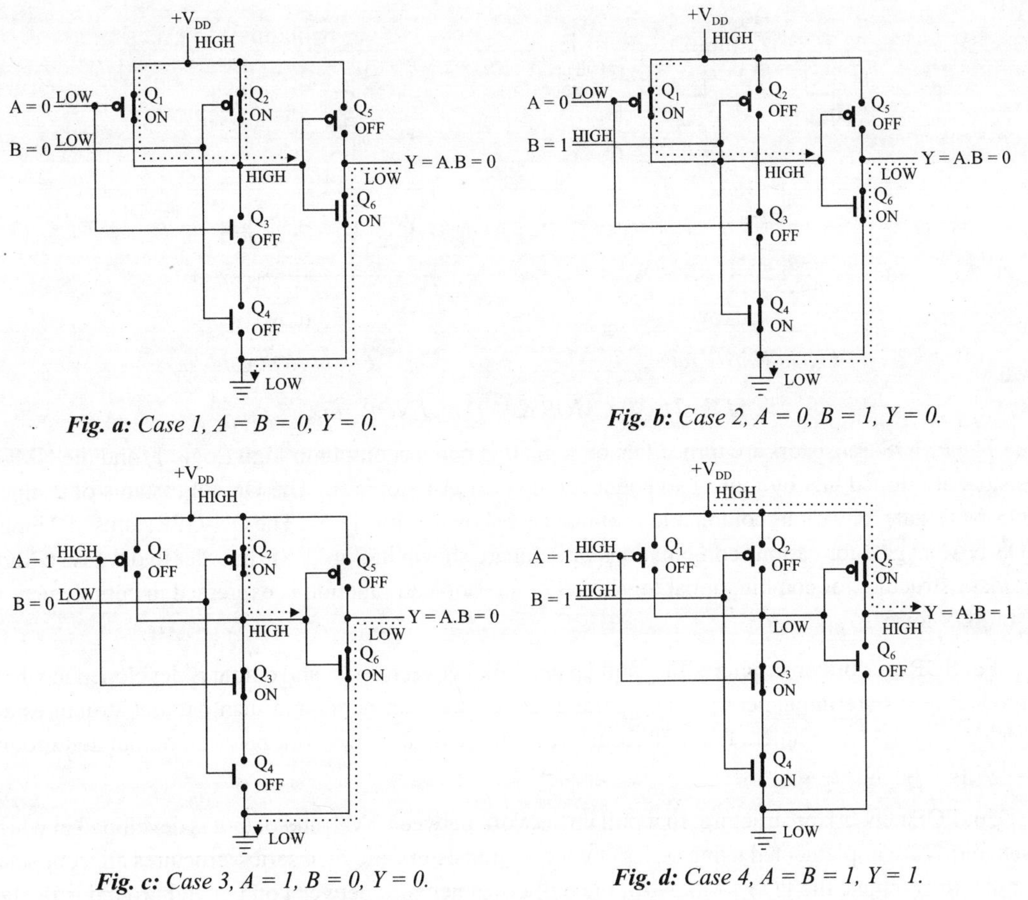

Fig. 11.63: Four cases of operation of CMOS AND gate.

11.7.6 Buffered and Unbuffered CMOS Devices

In a Buffered CMOS device, buffer stages are added at the output in order to improve the performance of the device for variations in load conditions. A buffer is an additional logic gate that isolates the output from the load, ensuring that the output voltage levels remain stable regardless of variations in load conditions.

The inclusion of buffer stages in CMOS devices helps to mitigate issues such as signal degradation, propagation delay, and loading effects. They offer improved signal integrity, faster operation, and enhanced robustness compared to simple CMOS designs.

11.7.7 Combinational Logic Implementation in CMOS

The fundamental CMOS gates are NAND and NOR gates. These NAND and NOR gates will have PMOS transistors which act as pull up network between $+V_{DD}$ (or logic **high**) and output and NMOS transistors which act as pull down network between ground (or logic **low**) and output as shown in Fig 11.64. In the pull up network, the AND operation is implemented by parallel connected transistors and OR operation is implemented by series connected transistors. The pull down network is dual of pull up network so that the AND operation is represented by series connected transistors and OR operation is represented by parallel connected transistors.

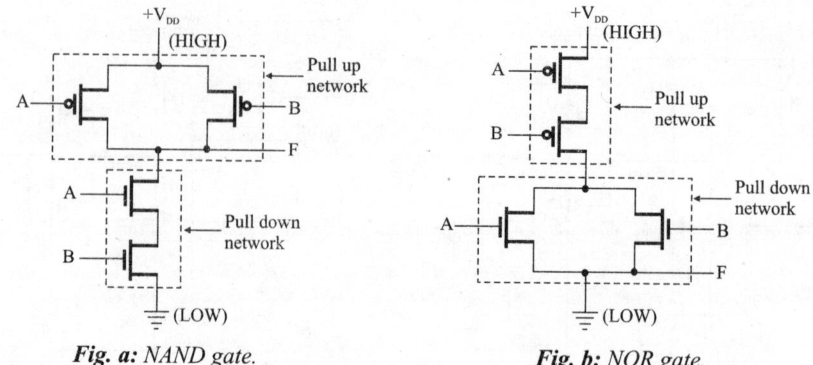

Fig. a: NAND gate.

Fig. b: NOR gate.

Fig. 11.64: CMOS NAND and NOR gate.

The PMOS transistors are turned ON by logic 0 to connect output to **high** (logic 1) and the NMOS transistors are turned ON by logic 1 to connect output to **low** (logic 0). The ON-OFF status of 2 inputs CMOS-NOR gate for various combinations of input are shown in Fig.11.57. The ON-OFF status of 2 inputs CMOS-NAND gate for various combinations of input are shown in Fig.11.61. In order to develop CMOS network or structure for combinational logic circuit, the Boolean function is expressed in complement of SOP or POS form.

For SOP network or structure, first pull up network between $+V_{DD}$ and output is developed in which the product terms are implemented using parallel connected transistors and then parallel structures are connected in series to complete the SOP form. Now, the pull down network between output and ground is dual of pull up network.

For POS network or structure, first pull up network between $+V_{DD}$ and output is developed in which the sum terms are implemented using series connected transistors and then series structures are connected in parallel to complete the POS form. Now, the pull down network between output and ground is dual of pull up network.

Example 11.1

Determine the Boolean equation represented by the CMOS network shown in Fig. 1 and hence draw the logic circuit.

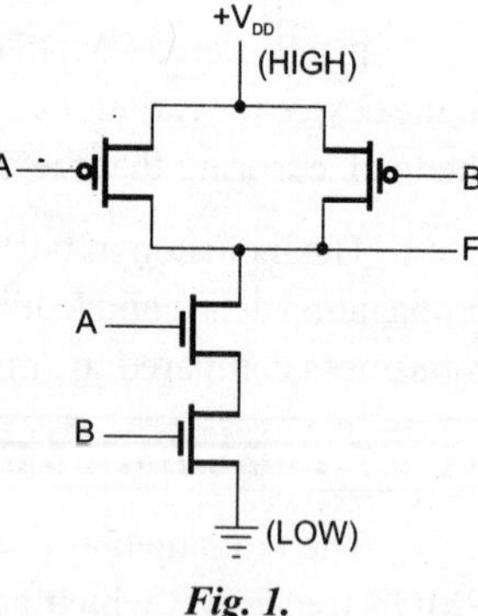

Fig. 1.

Solution

The transistors in the CMOS network are named Q_1 to Q_4 as shown in Fig. 2. The ON/OFF status of all transistors are determined for all possible combinations of inputs and listed in Table 1. The ON transistor is equivalent to short circuit and OFF transistor is equivalent to open circuit and using this concept, for every input, the connection of output to either V_{DD} or ground is determined and recorded in Table 1. For a particular input, if the output is connected to V_{DD} then the corresponding output is **high** and if the output is connected to ground then the corresponding output is **low**.

Table 1:

Inputs		Transistor ON/OFF State				Output	Minterm
A	B	Q_1	Q_2	Q_3	Q_4	F	
0	0	ON	ON	OFF	OFF	1	m_0
0	1	ON	OFF	OFF	ON	1	m_1
1	0	OFF	ON	ON	OFF	1	m_2
1	1	OFF	OFF	ON	ON	0	m_3

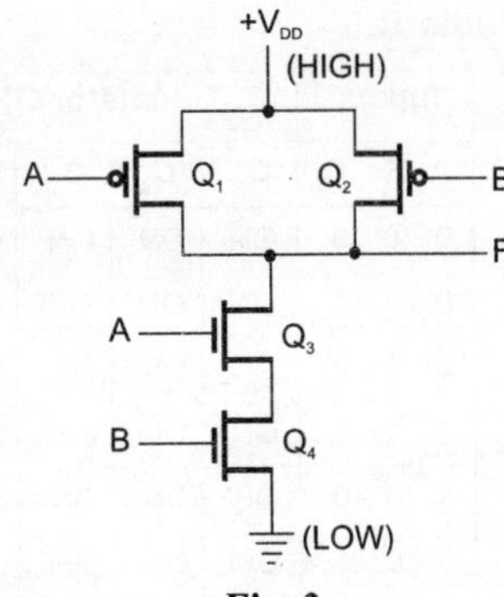

Fig. 2.

The K-map for output is constructed by taking inputs as minterms as shown in Fig. 3.

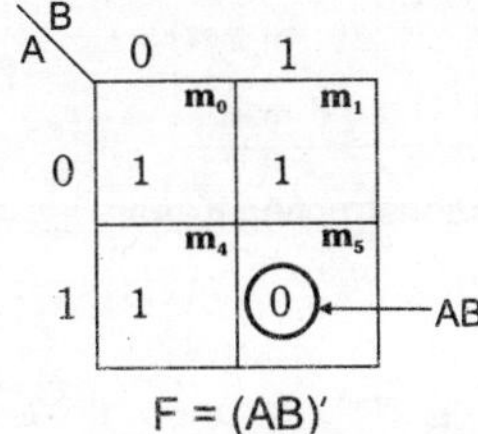

Fig. 3: K-map for F.

The Boolean equation for output is obtained by forming prime implicants using 0's the K-map.

$\therefore$ F = (AB)′

Using the above Boolean equation, the logic circuit represented by the given CMOS network is drawn as shown in Fig. 4.

A
B
F = (AB)′

Fig. 4: Logic circuit of CMOS network of Fig. 1.

Example 11.2

Determine the Boolean equation represented by the CMOS structure shown in Fig. 1 and hence draw the logic circuit.

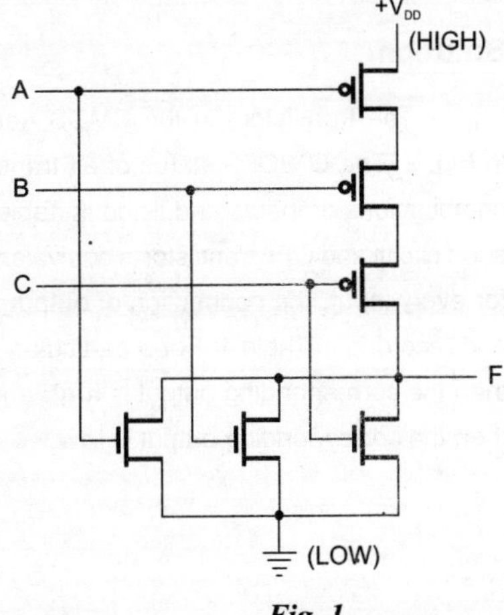

Fig. 1.

Solution

The transistors in the CMOS structure are named Q_1 to Q_6 as shown in Fig. 2. The ON/OFF status of all transistors are determined for all possible combinations of inputs and listed in Table 1. The ON transistor is equivalent to short circuit and OFF transistor is equivalent to open circuit and using this concept, for every input, the connection of output to either V_{DD} or ground is determined and recorded in Table 1. For a particular input, if the output is connected to V_{DD} then the corresponding output is **high** and if the output is connected to ground then the corresponding output is **low**.

Table 1:

Inputs			Transistor ON/OFF State						Output	Minterm
A	B	C	Q_1	Q_2	Q_3	Q_4	Q_5	Q_6	F	
0	0	0	ON	ON	ON	OFF	OFF	OFF	1	m_0
0	0	1	ON	ON	OFF	OFF	OFF	ON	0	m_1
0	1	0	ON	OFF	ON	OFF	ON	OFF	0	m_2
0	1	1	ON	OFF	OFF	OFF	ON	ON	0	m_3
1	0	0	OFF	ON	ON	ON	OFF	OFF	0	m_4
1	0	1	OFF	ON	OFF	ON	OFF	ON	0	m_5
1	1	0	OFF	OFF	ON	ON	ON	OFF	0	m_6
1	1	1	OFF	OFF	OFF	ON	ON	ON	0	m_7

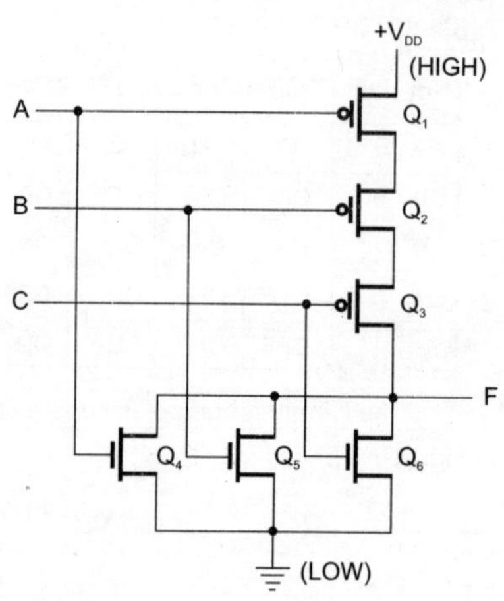

Fig. 2.

The Boolean equation for output is obtained by constructing K-map for output by taking inputs as minterms as shown in Fig. 3.

The Boolean equation obtained from K-map is,

$$F = A'B'C' = (A + B + C)'$$

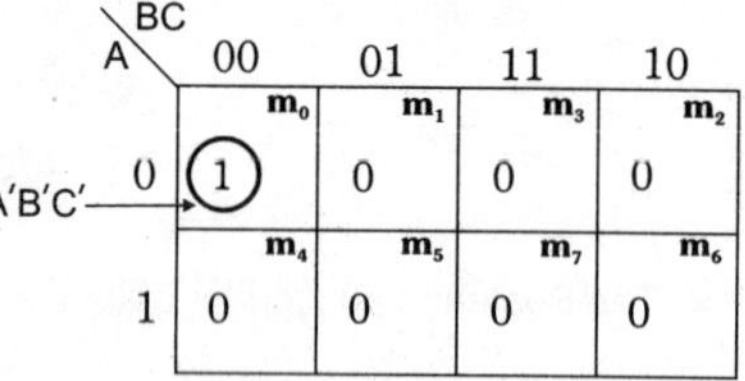

Fig. 3: K-map for F.

Using the above Boolean equation, the logic circuit represented by the given CMOS network is drawn as shown in Fig. 4.

A
B
C
F = (A+B+C)′

Fig. 4: Logic circuit of CMOS network of Fig. 1.

Example 11.3

Determine the Boolean equation represented by the CMOS network shown in Fig. 1 and hence draw the logic circuit.

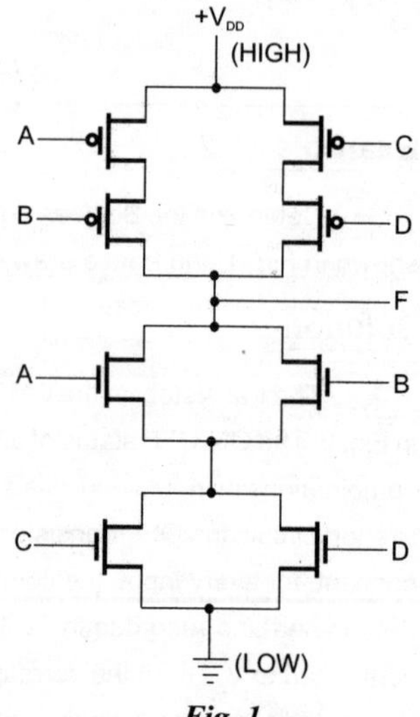

Fig. 1.

Solution

The transistors in the CMOS network are named Q_1 to Q_8 as shown in Fig. 2. The ON/OFF status of all transistors are determined for all possible combinations of inputs and listed in Table 1. The ON transistor is equivalent to short circuit and OFF transistor is equivalent to open circuit and using this concept, for every input, the connection of output to either V_{DD} or ground is determined and recorded in Table 1. For a particular input, if the output is connected to V_{DD} then the corresponding output is **high** and if the output is connected to ground then the corresponding output is **low**.

Table 1:

Inputs				Transistor ON/OFF State								Output	Minterm
A	B	C	D	Q_1	Q_2	Q_3	Q_4	Q_5	Q_6	Q_7	Q_8	F	
0	0	0	0	ON	ON	ON	ON	OFF	OFF	OFF	OFF	1	m_0
0	0	0	1	ON	ON	ON	OFF	OFF	OFF	OFF	ON	1	m_1
0	0	1	0	ON	ON	OFF	ON	OFF	OFF	ON	OFF	1	m_2
0	0	1	1	ON	ON	OFF	OFF	OFF	OFF	ON	ON	1	m_3
0	1	0	0	ON	OFF	ON	ON	OFF	ON	OFF	OFF	1	m_4
0	1	0	1	ON	OFF	ON	OFF	OFF	ON	OFF	ON	0	m_5
0	1	1	0	ON	OFF	OFF	ON	OFF	ON	ON	OFF	0	m_6
0	1	1	1	ON	OFF	OFF	OFF	OFF	ON	ON	ON	0	m_7
1	0	0	0	OFF	ON	ON	ON	ON	OFF	OFF	OFF	1	m_8
1	0	0	1	OFF	ON	ON	OFF	ON	OFF	OFF	ON	0	m_9
1	0	1	0	OFF	ON	OFF	ON	ON	OFF	ON	OFF	0	m_{10}
1	0	1	1	OFF	ON	OFF	OFF	ON	OFF	ON	ON	0	m_{11}
1	1	0	0	OFF	OFF	ON	ON	ON	ON	OFF	OFF	1	m_{12}
1	1	0	1	OFF	OFF	ON	OFF	ON	ON	OFF	ON	0	m_{13}
1	1	1	0	OFF	OFF	OFF	ON	ON	ON	ON	OFF	0	m_{14}
1	1	1	1	OFF	OFF	OFF	OFF	ON	ON	ON	ON	0	m_{15}

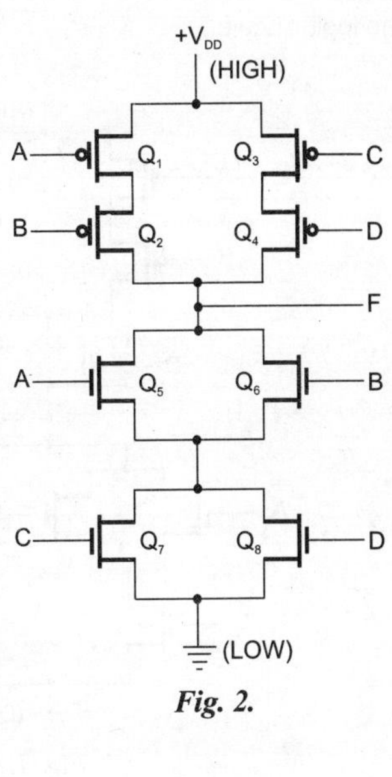

Fig. 2.

The Boolean equation for output is obtained by constructing K-map for output by taking inputs as minterms as shown in Fig. 3.

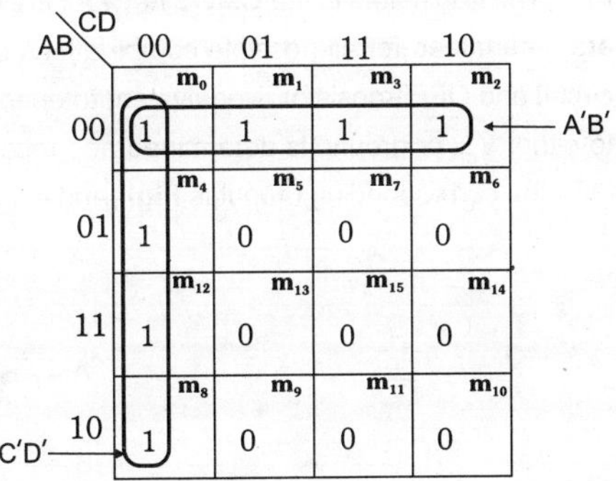

Fig. 3: K-map for F.

The Boolean equation obtained from K-map is,

$$F = A'B' + C'D'$$

$$= ((A'B' + C'D')')'$$

$$= ((A'B')'(C'D')')'$$

$$= ((A + B)(C + D))'$$

Using the above Boolean equation, the logic circuit represented by the given CMOS network is drawn as shown in Fig. 4.

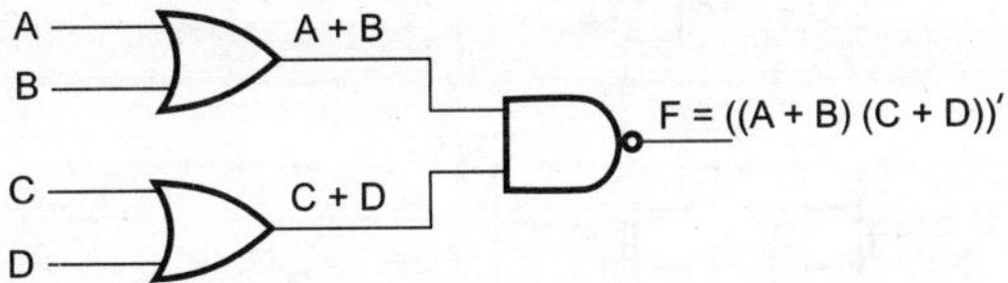

Fig. 4: Logic circuit of CMOS network of Fig. 1.

Example 11.4

(AKUB, Dec'20, 6 Marks)

Determine the Boolean equation represented by the CMOS network shown in Fig. 1 and 2 and hence draw the logic circuit.

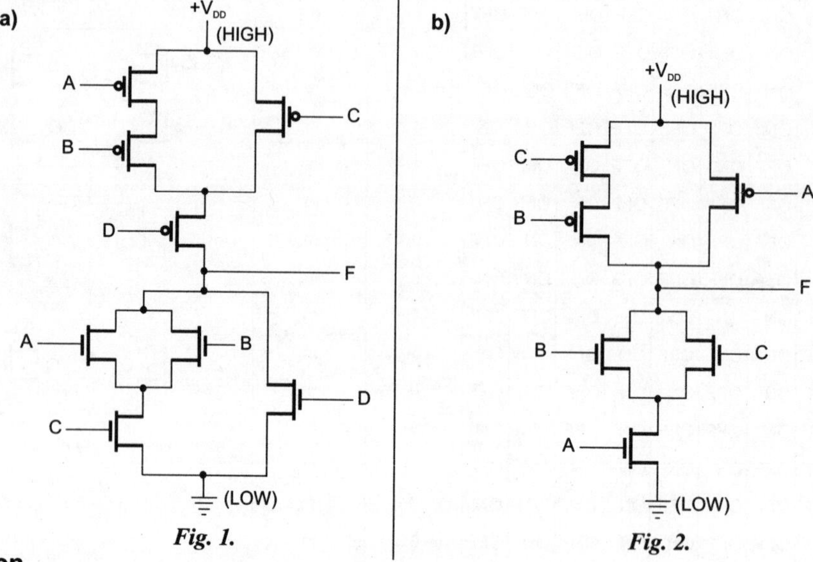

Fig. 1. *Fig. 2.*

Solution

a) The transistors in the CMOS network are named Q_1 to Q_8 as shown in Fig. 2. The ON/OFF status of all transistors are determined for all possible combinations of inputs and listed in Table 1. The ON transistor is equivalent to short circuit and OFF transistor is equivalent to open circuit and using this concept, for every input, the connection of output to either V_{DD} or ground is determined and recorded in Table 1. For a particular input, if the output is connected to V_{DD} then the corresponding output is **high** and if the output is connected to ground then the corresponding output is **low**.

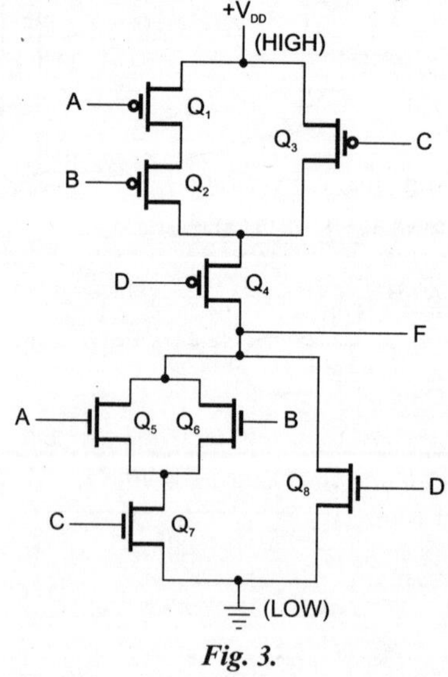

Fig. 3.

Table 1:

Inputs				Transistor ON/OFF State								Output	Minterm
A	B	C	D	Q_1	Q_2	Q_3	Q_4	Q_5	Q_6	Q_7	Q_8	F	
0	0	0	0	ON	ON	ON	ON	OFF	OFF	OFF	OFF	1	m_0
0	0	0	1	ON	ON	ON	OFF	OFF	OFF	OFF	ON	0	m_1
0	0	1	0	ON	ON	OFF	ON	OFF	OFF	ON	OFF	1	m_2
0	0	1	1	ON	ON	OFF	OFF	OFF	OFF	ON	ON	0	m_3
0	1	0	0	ON	OFF	ON	ON	OFF	ON	OFF	OFF	1	m_4
0	1	0	1	ON	OFF	ON	OFF	OFF	ON	OFF	ON	0	m_5
0	1	1	0	ON	OFF	OFF	ON	OFF	ON	ON	OFF	0	m_6
0	1	1	1	ON	OFF	OFF	OFF	OFF	ON	ON	ON	0	m_7
1	0	0	0	OFF	ON	ON	ON	ON	OFF	OFF	OFF	1	m_8
1	0	0	1	OFF	ON	ON	OFF	ON	OFF	OFF	ON	0	m_9
1	0	1	0	OFF	ON	OFF	ON	ON	OFF	ON	OFF	0	m_{10}
1	0	1	1	OFF	ON	OFF	OFF	ON	OFF	ON	ON	0	m_{11}
1	1	0	0	OFF	OFF	ON	ON	ON	ON	OFF	OFF	1	m_{12}
1	1	0	1	OFF	OFF	ON	OFF	ON	ON	OFF	ON	0	m_{13}
1	1	1	0	OFF	OFF	OFF	ON	ON	ON	ON	OFF	0	m_{14}
1	1	1	1	OFF	OFF	OFF	OFF	ON	ON	ON	ON	0	m_{15}

The Boolean equation for output is obtained by constructing K-map for output by taking inputs as minterms as shown in Fig. 4.

The Boolean equation obtained from K-map is,

$$F = A'B'D' + C'D'$$

$$= ((A'B'D' + C'D')')'$$

$$= ((A'B'D')(C'D')')'$$

$$= ((A + B + D)(C + D))'$$

$$= (((A + B)C) + D)'$$

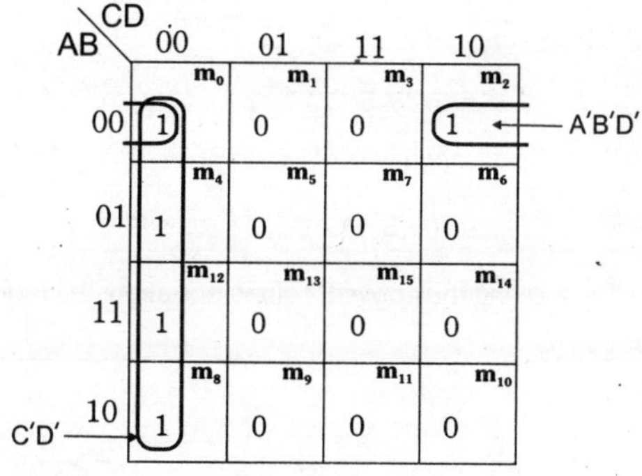

Fig. 4: K-map for F.

Using the above Boolean equation, the logic circuit represented by the given CMOS network is drawn as shown in Fig. 5.

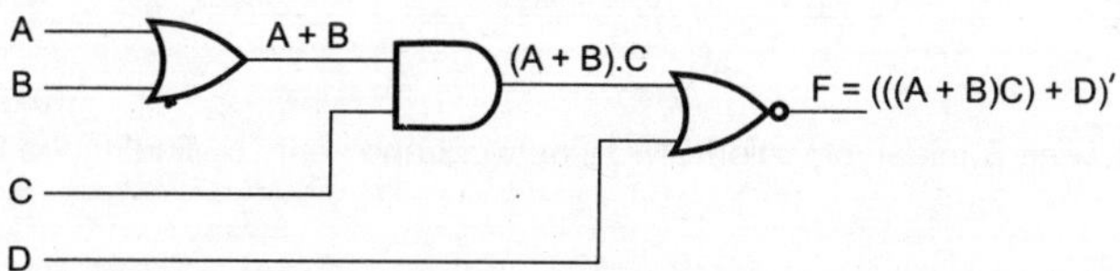

Fig. 5: Logic circuit of CMOS network of Fig. 1.

b) The transistors in the CMOS network are named Q_1 to Q_6 as shown in Fig. 2. The ON/OFF status of all transistors are determined for all possible combinations of inputs and listed in Table 1. The ON transistor is equivalent to short circuit and OFF transistor is equivalent to open circuit and using this concept, for every input, the connection of output to either V_{dd} or ground is determined and recorded in Table 1. For a particular input, if the output is connected to V_{dd} then the corresponding output is **high** and if the output is connected to ground then the corresponding output is **low**.

Table 2:

Inputs			Transistor ON/OFF State						Output	Minterm
A	B	C	Q_1	Q_2	Q_3	Q_4	Q_5	Q_6	F	
0	0	0	ON	ON	ON	OFF	OFF	OFF	1	m_0
0	0	1	OFF	ON	ON	OFF	ON	OFF	1	m_1
0	1	0	ON	OFF	ON	ON	OFF	OFF	1	m_2
0	1	1	OFF	OFF	ON	ON	ON	OFF	1	m_3
1	0	0	ON	ON	OFF	OFF	OFF	ON	1	m_4
1	0	1	OFF	ON	OFF	OFF	ON	ON	0	m_5
1	1	0	ON	OFF	OFF	ON	OFF	ON	0	m_6
1	1	1	OFF	OFF	OFF	ON	ON	ON	0	m_7

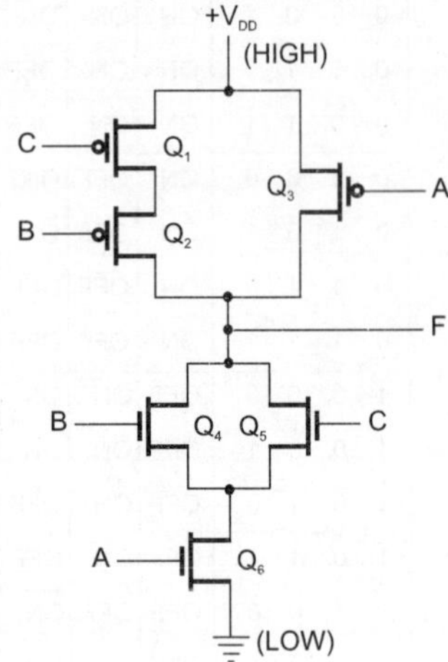

Fig. 6.

The Boolean equation for output is obtained by constructing K-map for output by taking inputs as minterms as shown in Fig. 7.

The Boolean equation obtained from K-map is,

$$F = A' + B'C'$$
$$= ((A' + B'C')')'$$
$$= (A(B'C')')'$$
$$= ((A(B + C))'$$

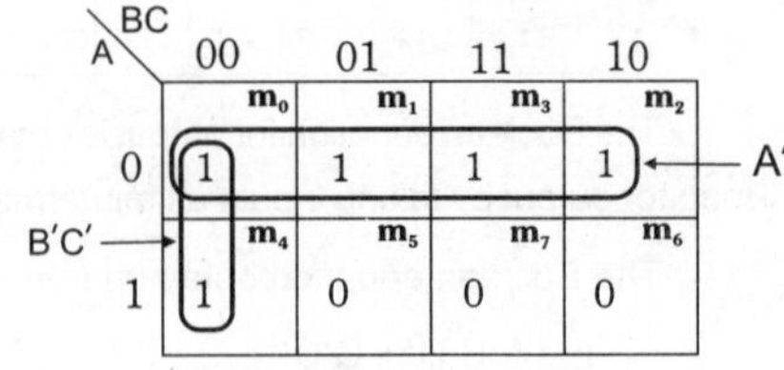

Fig. 7: K-map for F.

Using the above Boolean equation, the logic circuit represented by the given CMOS network is drawn as shown in Fig. 8.

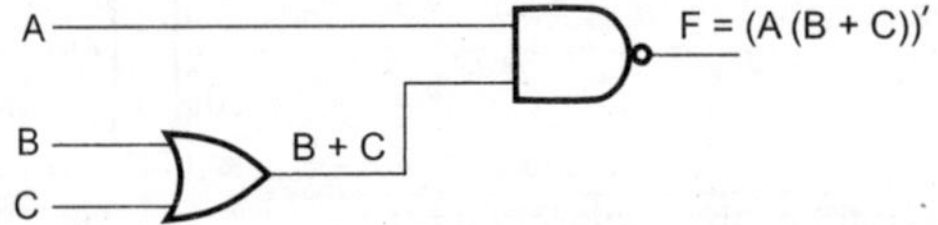

Fig. 8: Logic circuit of CMOS network of Fig. 2.

Example 11.5

(AKUB, Dec'20, 8 Marks)

Implement the following Boolean equation CMOS network and verify by finding the Boolean equation from the CMOS structure.

a) $(AB(A + B))'$

b) $(((CD) + B)A)'$

c) $((ABC) + D)'$

Solution

a) (AB (A + B))'

The CMOS network for the given Boolean equation will have pull up network constructed using PMOS transistors and pull down network constructed using NMOS transistors. First pull up network is developed and then the pull down the pull down network is dual of pull up network.

In pull up network the AND operation is represented by parallel connected PMOS transistors and OR operation is represented by series connected NMOS transistors as shown in Fig. 1.

The COMS network is verified by finding the Boolean equation represented by the CMOS network.

The transistors in the CMOS network are named Q_1 to Q_8 as shown in Fig. 1. The ON/OFF status of all transistors are determined for all possible combinations of inputs and listed in Table 1. The ON transistor is equivalent to short circuit and OFF transistor is equivalent to open circuit and using this concept, for every input, the connection of output to either V_{DD} or ground is determined and recorded in Table 1. For a particular input, if the output is connected to V_{DD} then the corresponding output is **high** and if the output is connected to ground then the corresponding output is **low**.

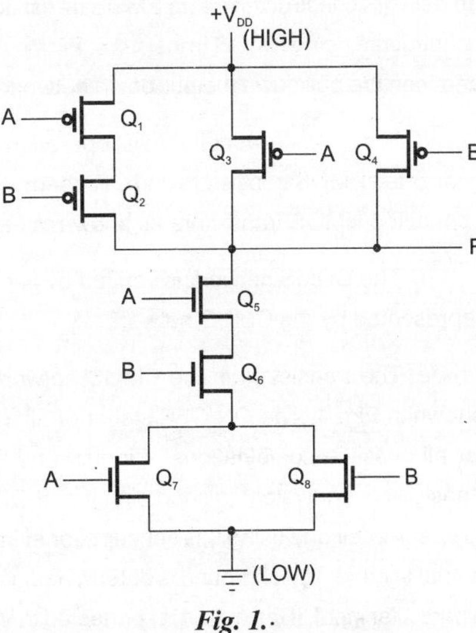

Fig. 1.

Table 1:

Inputs	Transistor ON/OFF State								Output	Minterm
A B	Q_1	Q_2	Q_3	Q_4	Q_5	Q_6	Q_7	Q_8	F	
0 0	ON	ON	ON	ON	OFF	OFF	OFF	OFF	1	m_0
0 1	ON	OFF	ON	OFF	OFF	ON	OFF	ON	1	m_1
1 0	OFF	ON	OFF	ON	ON	OFF	ON	OFF	1	m_2
1 1	OFF	OFF	OFF	OFF	ON	ON	ON	ON	0	m_3

The Boolean equation for output is obtained by constructing K-map for output by taking inputs as minterms as shown in Fig. 2.

The Boolean equation obtained from K-map is,

$$F = A' + B' = (AB)'$$

The given Boolean equation is, $F = (AB(A + B))'$

Let, $F = (AB(A + B))' = (ABA + ABB)' = (AB + AB)' = (AB)'$

A \ B	0	1
0	1 (m_0)	1 (m_1)
1	1 (m_4)	0 (m_5)

$F = A' + B' = (AB)'$

Fig. 2: K-map for F.

It is observed that the Boolean equation obtained from the CMOS network is same as given Boolean equation.

Using the above Boolean equation, the logic circuit represented by the given CMOS network is drawn as shown in Fig. 3.

A, B → $F = (AB)'$

Fig. 3: Logic circuit of CMOS network of Fig. 1.

b) (((CD) + B) A)′

The CMOS network for the given Boolean equation will have pull up network constructed using PMOS transistors and pull down network constructed using NMOS transistors. First pull up network is developed and then the pull down the pull down network is dual of pull up network.

In pull up network the AND operation is represented by parallel connected PMOS transistors and OR operation is represented by series connected NMOS transistors as shown in Fig. 4.

The COMS network is verified by finding the Boolean equation represented by the CMOS network.

The transistors in the CMOS network are named Q_1 to Q_8 as shown in Fig. 4. The ON/OFF status of all transistors are determined for all possible combinations of inputs and listed in Table 2. The ON transistor is equivalent to short circuit and OFF transistor is equivalent to open circuit and using this concept, for every input, the connection of output to either V_{DD} or ground is determined and recorded in Table 1. For a particular input, if the output is connected to V_{DD} then the corresponding output is **high** and if the output is connected to ground then the corresponding output is **low**.

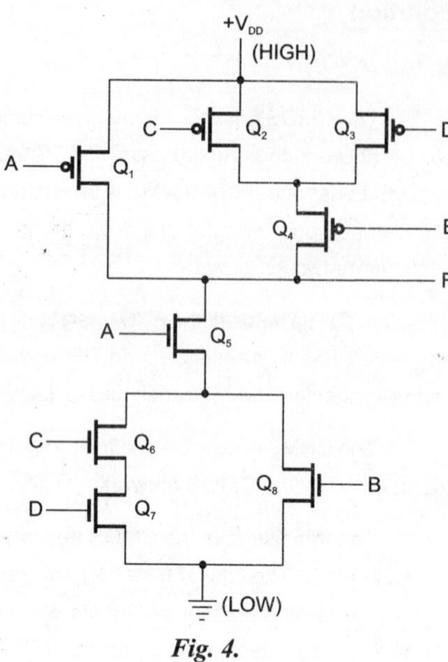

Fig. 4.

Table 2:

Inputs				Transistor ON/OFF State								Output	Minterm
A	B	C	D	Q_1	Q_2	Q_3	Q_4	Q_5	Q_6	Q_7	Q_8	F	
0	0	0	0	ON	ON	ON	ON	OFF	OFF	OFF	OFF	1	m_0
0	0	0	1	ON	ON	OFF	ON	OFF	OFF	ON	OFF	1	m_1
0	0	1	0	ON	OFF	ON	ON	OFF	ON	OFF	OFF	1	m_2
0	0	1	1	ON	OFF	OFF	ON	OFF	ON	ON	OFF	1	m_3
0	1	0	0	ON	ON	ON	OFF	OFF	OFF	OFF	ON	1	m_4
0	1	0	1	ON	ON	OFF	OFF	OFF	OFF	ON	ON	1	m_5
0	1	1	0	ON	OFF	ON	OFF	OFF	ON	OFF	ON	1	m_6
0	1	1	1	ON	OFF	OFF	OFF	OFF	ON	ON	ON	1	m_7
1	0	0	0	OFF	ON	ON	ON	ON	OFF	OFF	OFF	1	m_8
1	0	0	1	OFF	ON	OFF	ON	ON	OFF	ON	OFF	1	m_9
1	0	1	0	OFF	OFF	ON	ON	ON	ON	OFF	OFF	1	m_{10}
1	0	1	1	OFF	OFF	OFF	ON	ON	ON	ON	OFF	0	m_{11}
1	1	0	0	OFF	ON	ON	OFF	ON	OFF	OFF	ON	0	m_{12}
1	1	0	1	OFF	ON	OFF	OFF	ON	OFF	ON	ON	0	m_{13}
1	1	1	0	OFF	OFF	ON	OFF	ON	ON	OFF	ON	0	m_{14}
1	1	1	1	OFF	OFF	OFF	OFF	ON	ON	ON	ON	0	m_{15}

The K-map for output is constructed by taking inputs as minterms as shown in Fig. 5.

The Boolean equation for output is obtained by forming prime implicants using 0's in K-map.

$\therefore$ F = (AB + ACD)′

The given Boolean equation is, F = (((CD) + B) A)′

Let, F = (((CD) + B)A)′= ((CD)A + BA)′ = (AB + ACD)′

It is observed that the Boolean equation obtained from the CMOS network is same as given Boolean equation.

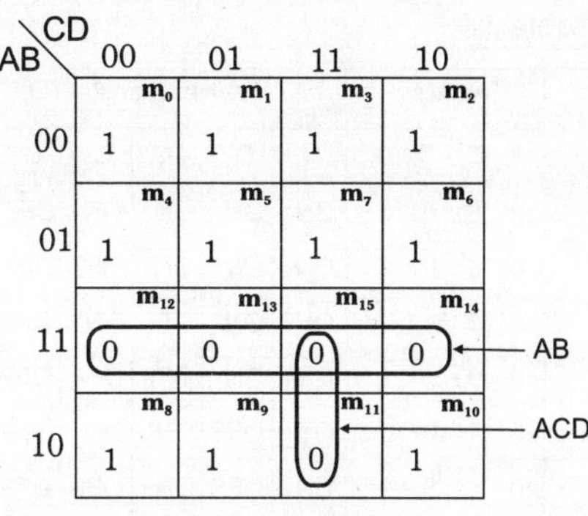

Fig. 5: K-map for F.

Using the above Boolean equation, the logic circuit represented by the given CMOS network is drawn as shown in Fig. 6.

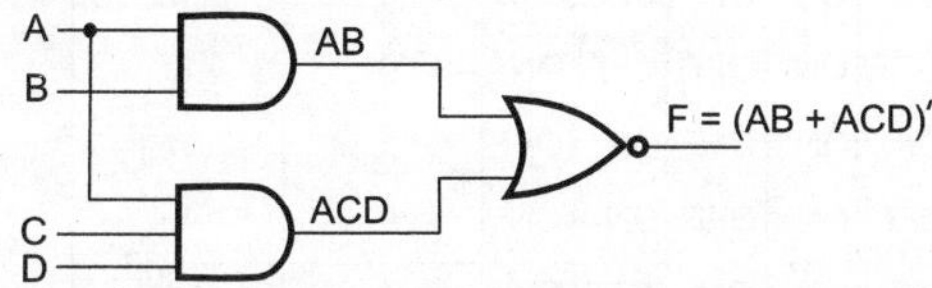

Fig. 6: Logic circuit of CMOS network of Fig. 4.

c) ((ABC) + D)′

The CMOS network for the given Boolean equation will have pull up network constructed using PMOS transistors and pull down network constructed using NMOS transistors. First pull up network is developed and then the pull down the pull down network is dual of pull up network.

In pull up network the AND operation is represented by parallel connected PMOS transistors and OR operation is represented by series connected NMOS transistors as shown in Fig. 7.

The COMS network is verified by finding the Boolean equation represented by the CMOS network.

The transistors in the CMOS network are named Q_1 to Q_8 as shown in Fig. 7. The ON/OFF status of all transistors are determined for all possible combinations of inputs and listed in Table 3. The ON transistor is equivalent to short circuit and OFF transistor is equivalent to open circuit and using this concept, for every input, the connection of output to either V_{DD} or ground is determined and recorded in Table 1. For a particular input, if the output is connected to V_{DD} then the corresponding output is **high** and if the output is connected to ground then the corresponding output is **low**.

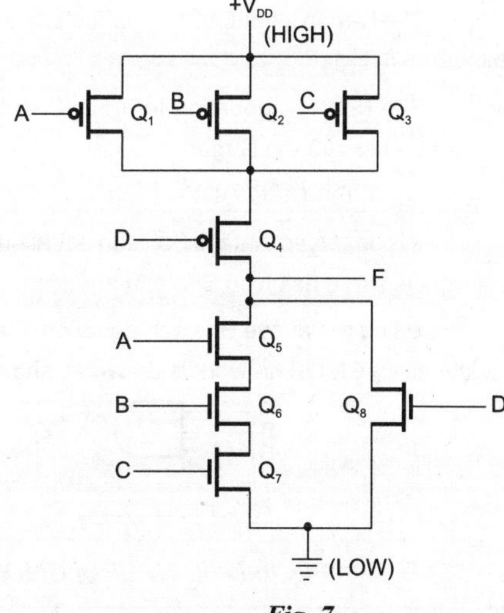

Fig. 7.

Table 3:

Inputs				Transistor ON/OFF State								Output	Minterm
A	B	C	D	Q_1	Q_2	Q_3	Q_4	Q_5	Q_6	Q_7	Q_8	F	
0	0	0	0	ON	ON	ON	ON	OFF	OFF	OFF	OFF	1	m_0
0	0	0	1	ON	ON	ON	OFF	OFF	OFF	OFF	ON	0	m_1
0	0	1	0	ON	ON	OFF	ON	OFF	OFF	ON	OFF	1	m_2
0	0	1	1	ON	ON	OFF	OFF	OFF	OFF	ON	ON	0	m_3
0	1	0	0	ON	OFF	ON	ON	OFF	ON	OFF	OFF	1	m_4
0	1	0	1	ON	OFF	ON	OFF	OFF	ON	OFF	ON	0	m_5
0	1	1	0	ON	OFF	OFF	ON	OFF	ON	ON	OFF	1	m_6
0	1	1	1	ON	OFF	OFF	OFF	OFF	ON	ON	ON	0	m_7
1	0	0	0	OFF	ON	ON	ON	ON	OFF	OFF	OFF	1	m_8
1	0	0	1	OFF	ON	ON	OFF	ON	OFF	OFF	ON	0	m_9
1	0	1	0	OFF	ON	OFF	ON	ON	OFF	ON	OFF	1	m_{10}
1	0	1	1	OFF	ON	OFF	OFF	ON	OFF	ON	ON	0	m_{11}
1	1	0	0	OFF	OFF	ON	ON	ON	ON	OFF	OFF	1	m_{12}
1	1	0	1	OFF	OFF	ON	OFF	ON	ON	OFF	ON	0	m_{13}
1	1	1	0	OFF	OFF	OFF	ON	ON	ON	ON	OFF	0	m_{14}
1	1	1	1	OFF	OFF	OFF	OFF	ON	ON	ON	ON	0	m_{15}

The K-map for output is constructed by taking inputs as minterms as shown in Fig. 8.

The Boolean equation for output is obtained by forming prime implicants using 0's in K-map.

$$\therefore F = (ABC + D)'$$

It is observed that the Boolean equation obtained from the CMOS network is same as given Boolean equation.

Using the above Boolean equation, the logic circuit represented by the given CMOS network is drawn as shown in Fig. 9.

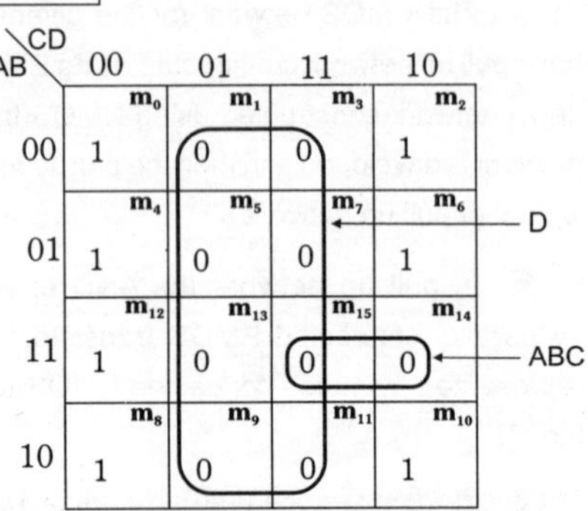

Fig. 8: K-map for F.

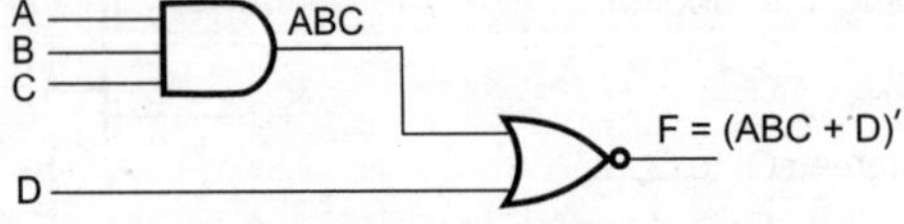

Fig. 9: Logic circuit of CMOS network of Fig. 7.

11.8 Comparison of Logic Families

(BATU, Jun'19, 6 Marks) (GTU, Jun'19, 7 Marks)

The popular logic families are RTL, DTL, TTL, ECL and CMOS. The RTL, DTL, TTL and ECL digital logic family of ICs are constructed using bipolar NPN transistors as fundamental switching device. The CMOS digital logic family of devices are constructed using unipolar MOSFET as fundamental switching device.

The important characteristics of digital logic family of integrated circuits (ICs) that are used to compare the performance of various families of devices are logic levels, propagation delay, power dissipation, noise margin, fan-out and fan-in.

11.8.1 Resistor-Transistor Logic (RTL) Family

RTL is the first logic family used to construct the logic gates. RTL is one of the saturated bipolar logic families. Resistors and transistors are the fundamental components in this family. Resistor is used as input component and transistor is used as output component. The switching transistors operate in cut-off (OFF state) or saturation (ON state). NOR gate is the basic logic device in RTL logic family and other logic gates are built using this basic device.

Characteristics of RTL Family

- The output of a RTL device can serve as an input to a maximum of 5 gates.
- Logic **low** voltage for a RTL is defined between 0 - 0.2 *V*.
- Logic **high** voltage for a RTL is at 1 to 3.6 *V*.
- Noise margin is at around 0.2 *V*.
- Propagation delay is about 12 *ns*.
- A typical RTL logic draws a power of about 30 *mW*.

Advantages of RTL Family

- Simple circuit when compared to other logic families.
- Minimum number of transistors is required to implement a logic device.
- Operation is very simple.
- Use silicon bipolar transistor as saturated switches.
- Bipolar transistors exhibit higher trans-conductance than MOSFETs.

Disadvantages of RTL Family

- Response time is poor.
- Propagation delay is high.
- Power dissipation is high due to resistor.
- Capacitive loading reduces fan-out capability.

11.8.2 Diode-Transistor Logic (DTL) Family

DTL is one of the saturated bipolar logic families and it developed to reduce the power dissipation of RTL family by replacing the input resistor by diodes. Diodes are used as input component and transistor used as output component. NAND gate is the basic logic device in RTL logic family and other logic gates are built using this basic device.

Characteristics of DTL Family

- The output of a DTL device can serve as an input to a maximum of 8 gates.
- Logic **low** voltage for a DTL is defined between 0 - 0.2 *V*.
- Logic **high** voltage for a DTL is at 4 to 5 *V*.
- Noise margin is at around 0.4 *V*.
- Propagation delay is about 30 *ns*.
- A typical TTL logic draws a power of about 12 *mW*.

Advantages of DTL Family

- Power loss is lesser than RTL family.
- Minimum number of components is required to implement a logic device.
- Operation is very simple.
- Use silicon bipolar transistor as saturated switches.
- Bipolar transistors exhibit higher trans-conductance than MOSFETs.

Disadvantages of DTL Family

- Propagation delay is more than RTL family due to diode capacitance.
- Usage of DTL logic is limited.
- Capacitive loading increases the OFF time.

11.8.3 Transistor-Transistor Logic (TTL) Family

In TTL family of devices the diodes of DTL family are replaced by transistors to minimize the input capacitance. In TTL, the base-emitter junction is used to receive inputs instead of diodes in DTL. Since transistors are used as input and switching device, this family of devices are called transistor-transistor logic. NAND gate is the basic logic device in TTL logic family and other logic gates are built using this basic device.

Characteristics of TTL Logic Families

- The output of a TTL device can serve as an input to a maximum of 10 gates.
- Logic **low** voltage for a TTL is defined between 0 - 0.2 V.
- Logic **high** voltage for a TTL is at 5 V.
- Noise margin is at around 0.4 V.
- Propagation delay is about 9 ns.
- A typical TTL logic draws a power of about 11 mW.

Advantages of TTL Logic Families

- TTL has better characteristics than RTL and DTL.
- It is least susceptible to electrical damage.
- Lesser immune to noise when compared to ECL, but more than CMOS.
- Fastest saturation, when compared to other logic families.
- Low output impedance for all states.

Disadvantages of TTL Logic Families

- TTL device dissipates a lot of power, when compared to CMOS devices.
- Not recommended in VLSI chips as it requires more space and isolation.
- Expensive compared to MOSFETs.

11.8.4 Emitter-Coupled Logic (ECL) Family

(GTU, Jun'19, 7 Marks)

The basic component in ECL family of devices is emitter coupled differential current switch operating in active region. ECL is the only non-saturated bipolar logic family. OR-NOR gate is the basic logic device in ECL logic family and other logic gates are built using this basic device.

Emitter coupled transistors operate in active region for both OFF and ON state with a slight difference between logic voltage levels.

Characteristics of ECL Logic Families

- Large fan-out among transistor family due to low output impedance.
- The propagation time is about 1ns, making it the fastest logic family.
- A logic **low** voltage for ECL is about – 1.7 to – 1.5 *V*.
- A logic **high** voltage for ECL is about – 0.9 to – 0.6 V.
- It is the fastest among all the logic families.
- It has an average propagation delay time of about 1 *ns*.

Advantages of ECL Logic Families

- Has a fan-out better than the TTL Logic family.
- Offers the highest speed in operation.
- ECL systems produce complementary outputs (OR-NOR, AND-NAND).
- Parameters do not vary much with temperature.

Disadvantages of ECL Logic Families

- Worst noise immunity compared to TTL and CMOS.
- Highest power consumption, when compared with TTL and CMOS.
- VLSI design difficult due to resistors, thus increasing system size.
- Capacitive loading reduces the fan-out capacity.
- Requires negative supply to emitter terminal.

11.8.5 Complementary MOSFET Logic (CMOS) Family

CMOS is a saturated MOSFET family of devices. Enhancement type MOSFETs are the basic components in CMOS. A CMOS field effect transistor is a pair of NMOS and PMOS field effect transistors with their channels connected in series and gate terminals tied together. The CMOS devices are **high** immunity to noise and have **low** static power consumption.

The MOSFETs operate in cut-off (OFF state) or saturation (ON state). NAND gate is the basic logic device in CMOS logic family other logic gates are built using this basic device.

Characteristics of CMOS Logic Families

- CMOS supports a very large fan-out, more than 50.
- It has excellent noise immunity among all families.
- A logic **low** voltage for CMOS is 0 *V*.
- A logic **high** voltage for CMOS is 3.3 *V*.
- Propagation delay is very large at about 20 to 50 *ns*.

Advantages of CMOS Logic Families

- Has the highest fan-out, when compared with TTL and ECL
- Works well over a wide range of temperature
- Noise immunity is better than TTL and ECL

Disadvantages of CMOS Logic Families

- Large propagation delay

Table 11.2: Comparison of Characteristics of Logic Families

Characteristics	Logic Family				
	RTL	**DTL**	**TTL**	**ECL**	**CMOS**
Basic Component	Resistor, NPN Transistor	Diode, NPN Transistor	NPN Transistor	NPN Transistor	PMOS, NMOS
Basic Gate	NOR	NAND	NAND	OR-NOR	NOR-NAND
Logic Low	0 to 0.2 *V*	0 to 0.2 *V*	0 to 0.8 *V*	– 1.7 to – 1.5 *V*	0 *V*
Logic High	1 to 3.6 *V*	4 to 5 *V*	2 to 5 *V*	– 0.9 to – 0.6 *V*	3.3 *V*
Propagation Delay	12 *ns*	30 *ns*	9 *ns*	1 *ns*	20 to 50 *ns*
Power Dissipation	30 *mW*	12 *mW*	10 *mW*	50 *mW*	0.01 *mW*
Fan-out	5	8	10	10	50
Fan-in	5	5	8	5	10
Noise Margin	0.2 *V*	0.4 *V*	0.4 *V*	0.25 *V*	5 *V*

11.9 Summary of Important Concepts

1. The collection of different logical devices that perform different logic functions but manufactured using same semiconductor technology is called a logic family.
2. The fundamental devices used in the logic families are either bipolar junction transistors or unipolar CMOS field effect transistors.
3. The RTL, DTL, TTL and ECL digital logic family of ICs are constructed using bipolar NPN transistors as fundamental switching device.

4. The CMOS digital logic family of devices are constructed using unipolar MOSFET as fundamental switching device.
5. The propagation delay is defined as the time required for a change in input to be available in the output.
6. The fan-out is defined as the maximum number of standard load that an output can drive reliably.
7. A standard load is usually defined as the amount of current needed by an input of another gate in the same logic family.
8. Fan-in is the number of inputs provided (or available) in a logic gate and it is decided during manufacturing itself which in turn depends on technology used for manufacturing the IC.
9. A quantitative measure of noise immunity in terms of voltage is called noise margin.
10. The characteristics of logic devices of a logic family depends on the fundamental devices used in that logic family.
11. In logic devices the transistors are operated either in saturation region/state or cut-off region/state.
12. In logic families, NPN transistors are widely used and the input is applied at base and output is taken either from collector or emitter of the transistors.
13. Enhancement type MOSFETs are the basic switching devices in CMOS field effect transistors.
14. A CMOS field effect transistor is a pair of NMOS and PMOS field effect transistors with their channels connected in series and gate terminals tied together.
15. In logic devices the MOSFETs are used as switches in which they are operated either in saturation region/state or cut-off region/state.
16. The RTL (Resistor-Transistor Logic) family of devices uses resistors to receive inputs and transistor as switching device and so the name resistor-transistor logic.
17. DTL (Diode-Transistor Logic) is a class of digital circuits built from diodes, resistors and bipolar junction transistors.
18. In TTL (Transistor-Transistor Logic) family, input network is made of transistors and switching devices are also transistors and so the name TTL.
19. The different types of outputs in TTL devices are totem-pole, open-collector and three-state (or Tri-state) output.
20. The basic circuit in an ECL gate is the emitter-coupled current switch.
21. ECL (Emitter-Coupled Logic) family of logic devices works in active region with a small difference in two logic voltage levels which is decided by the currents.
22. Emitter-coupled logic (ECL) circuits are the fastest available bipolar logic gates.
23. The primary disadvantage of ECL is the high quiescent power dissipation.
24. In CMOS logic family of devices, the logic functions are realized by parallel or series connection of PMOS and NMOS field effect transistors.

11.10 Short-Answer Questions

Q11.1 ***The input and output voltage specifications for the standard TTL is given as,***

$V_{OH} = 2.4\,V,\ V_{IH} = 2,\ V_{OL} = 0.4,\ V_{IL} = 0.8$

Find high state and low state noise margin.

Solution:

High state noise margin, $V_{NH} = V_{OH} - V_{IH}$

$= 2.4 - 2 = 0.4\ V$

Low state noise margin, $V_{NL} = V_{IL} - V_{OL}$

$= 0.8 - 0.4 = 0.4\ V$

Q11.2 ***The 7420 data sheet gives the following values.***

$I_{OH\_max} = 0.4\,mA,\ I_{OL\_max} = 8\,mA,\ I_{IH\_max} = 20\,\mu A,\ I_{IL\_max} = 0.1\,mA$

What is the fan-out?

Solution:

$$\text{Fan-out (High)} = \frac{I_{OH}}{I_{IH}} = \frac{0.4\ \text{mA}}{20\ \mu\text{A}} = \frac{400\ \mu\text{A}}{20\ \mu\text{A}} = 20$$

$$\text{Fan-out (Low)} = \frac{I_{OL}}{I_{IL}} = \frac{8\ \text{mA}}{0.1\ \text{mA}} = 80$$

The lower among the above values is fan-out. So fan out is 20.

Q11.3 ***How many 7400 NAND gate TTL inputs can driven by a 7400 NAND gate TTL output? The current specifications are,***

$I_{OL\_max} = 16\,mA,\ I_{IL\_max} = -1.6\,mA,\ I_{OH\_max} = -400\,\mu A,\ I_{IH\_max} = 40\,\mu A$

Solution:

$$\text{Fan-out (Low)} = \frac{I_{OL\_max}}{I_{IL\_max}} = \frac{16\ \text{mA}}{1.6\ \text{mA}} = 10$$

$$\text{Fan-out (High)} = \frac{I_{OH\_max}}{I_{IH\_max}} = \frac{400\ \mu\text{A}}{40\ \mu\text{A}} = 10$$

> ***Note:*** *Negative sign in I_{IL_max} and I_{OH_max} indicates that this current flows output of the input terminal. So this negative sign can be ignored.*

Q11.4 ***ECL gates are cascaded. The voltage levels are,***

$V_{IH} = -1.105\,V,\ V_{OH} = -0.96\,V,\ V_{IL} = -1.475\,V,\ V_{OL} = -1.65\,V$

What is high state and low state noise margin.

Solution:

Low state noise margin, $V_{NL} = V_{IL} - V_{OL}$

$= -1.475 - (-1.65) = 0.175\ V$

High state noise margin, $V_{NH} = V_{OH} - V_{IH}$

$= 0.96 - (-1.105) = 2.065\ V$

Q11.5 ***The specifications for the standard TTL NAND gate is given as,***

$V_{cc} = 5\ V, I_{CCH} = 1\ mA, I_{CCL} = 3\ mA$

Find average current and average power dissipation.

Solution:

The average current, $I_{CC\_av} = \dfrac{I_{CCH} + I_{CCL}}{2}$

$= \dfrac{1+3}{2} = 2\ mA$

Average power dissipation, $P_{av} = I_{CC\_av} \times V_{CC}$

$= 2 \times 5 = 10\ mW$

Q11.6 ***The specifications for the standard TTL NAND gate is given as,***

$t_{PLH} = 11\ ns, t_{PHL} = 7\ ns$

Find propagation delay.

Solution:

Propagation delay, $t_{pd} = \dfrac{t_{PHL} + t_{PLH}}{2}$

$= \dfrac{7+11}{2} = 9\ ns$

Q11.7 ***An IC family has an average propagation delay of 10 ns and an average power dissipation of 5 mW. Find the speed-power product.***

Solution:

Speed-power product = Average propagation delay × Average power dissipation

$= 10\ ns \times 5\ mW$

$= 10 \times 10^{-9} \times 5 \times 10^{-3}$

$= 50 \times 10^{-12}$ watt-second (or) Joules

= 50 pico-Joules (pJ)

Q11.8 ***Why totem-pole output cannot be connected together.*** ***(IKGPTU, May'20, 2 Marks)***

Totem-pole outputs consist of both an NPN and a PNP transistor, arranged in a configuration that allows for both sourcing and sinking current. This output structure will lead to short circuit between supply rails (or short circuit between positive and ground supply lines) when two totem-pole outputs are connected together and one output is driven low and the other is driven high. This situation can lead to excessive current flow, potentially damaging the transistors or other components in the circuit.

Q11.9 ***Why is ECL logic faster than TTL?*** ***(AKTU, Dec'22, 2 Marks)***

Here the important distinction is that TTL uses bipolar transistors as voltage mode switches that swing from cut-off to saturation. In contrast, ECL utilizes bipolar transistors as current mode switches that never saturate. This approach avoids the capacitive effects and saturation delays associated with excess minority carrier storage in the devices. Hence ECL devices are faster than TTL devices.

Q11.10 What is the basic difference between buffered and unbuffered CMOS devices? ***(IKGPTU, May'20, 2 Marks)***

Buffered and unbuffered CMOS devices differ in their internal circuitry and their ability to drive loads.

Buffered CMOS devices provide additional circuitry to isolate the output and load, offering advantages in signal integrity and load driving capability, while unbuffered CMOS devices are simpler and used in applications where buffering is not necessary.

11.11 Exercises

I. Fill in the blanks

1. _______ is defined as the time required for a change in input to be available in the output.
2. _______ is a class of digital circuits built from diodes, resistors and bipolar junction transistors.
3. The propagation delay of a TTL logic device is _______ .
4. The TTL output acts as a current sink in the _______ .
5. The basic TTL logic circuit is _______ .
6. The noise margin of a TTL gate is about _______ .
7. Digital ICs are manufactured using different _______ technologies.
8. The collection of different logic devices that perform different logic functions but manufactured using same semiconductor technology is called _______.
9. The CMOS digital logic family of device are constructed using _______ as fundamental switching device.
10. _______ is the first logic family used to construct the logic gates.
11. _______ and _______ are the fundamental components in this family.
12. _______ is the only non-saturated bipolar logic family.
13. The basic component in ECL family of devices is emitter coupled differential current switch operating in _______ region.
14. _______ gate is the basic logic device in ECL logic family.
15. A quantitative measure of noise immunity in terms of voltage is called _______.

Answers

1. propagation delay	6. 0.4 *V*	11. resistors and transistors
2. DTL	7. semiconductor	12. ECL
3. 9 *ns*	8. logic family	13. active
4. low state	9. unipolar MOSFET	14. OR-NOR
5. NAND	10. RTL	15. noise margin

II. State whether the following statements are True or False

1. Fan-in is defined as the maximum number of standard load that an output can drive reliably.
2. A quantitative measure of noise immunity in terms of voltage is called noise margin.
3. The maximum negative voltage that can be applied to a TTL input is $-0.2\ V$.
4. The output of a tristate logic state is high.
5. Emitter-coupled logic (ECL) circuits are the fastest available bipolar logic gates.
6. The primary advantage of ECL is the high quiescent power dissipation.

7. The RTL, DTL, TTL, and ECL digital logic family of ICs are constructed using bipolar NPN transistors as fundamental switching device
8. The disadvantages of RTL is its high power dissipation
9. ECL is saturated logic family of device and has low power dissipation and very low propagation delay
10. The tri-state devices will have third state called high impedance state apart from low and high state
11. RTL is one of the saturated bipolar logic families
12. NAND gate is the basic logic device in RTL logic family

Answers

1. False	5. True	9. False
2. True	6. False	10. True
3. False	7. True	11. True
4. False	8. True	12. False

III. Choose the right answer for the following questions

1. ***Propagation delay t_{PLH} and t_{PHL} of a standard TTL NAND gate is***
   a) 10 ns and 6 ns **b)** 11 ns and 7 ns
   c) 11 ns and 6 ns **d)** 10 ns and 7 ns

2. ***The worst case dc noise margin of a TTL gate is***
   a) 600 mV **b)** 800 mV **c)** 400 mV **d)** 100 mV

3. ***The (unconnected) floating input of a TTL circuit acts as a***
   a) Logic 0 **b)** Logic 1
   c) Either logic 0 (or) logic 1 **d)** None of the above

4. ***Totem pole output of TTL***
   a) Reduces propagation delay **b)** Increases the speed
   c) Reduces the power dissipation **d)** Increases the fan out

5. ***TTL gate which does not use multiple emitter is***
   a) NOR **b)** NAND **c)** AND **d)** NOR and OR

6. ***Multiple emitter is the characteristics of the***
   a) CMOS **b)** DTL **c)** TTL **d)** ECL

7. ***Which one of the following logic families is known as current mode logic (CML)***
   a) ECL **b)** TTL **c)** DTL **d)** CMOS

8. ***TTL circuits employ protective shunt diodes on each input to***
   a) Reduce power consumption **b)** Increase the noise margin
   c) To limit negative voltage excursions **d)** None of the above

9. ***Unused input of TTL is***
   a) Left disconnected (floating) **b)** Connected to the used input
   c) Connected to +5 V through a 1 kΩ resistor **d)** All the above

10. ***Basic circuit of CMOS is inverter, which consists of***

a) One p-channel and one n-channel transistor **b)** Both p-channel

c) Both n-channel **d)** None of the above

11. ***TTL open collector output is generally slower than totem pole output. Because,***

a) Open collector output does not use active pull up resistor

b) Open collector output use active pull up resistor

c) Open collector output uses pull down resistor

d) Totem pole output uses pull down resistor

12. ***A 5V battery has an internal resistance of 100 Ω and it is supplying N identical gates each having an input resistance of 3 KΩ. The gates are in parallel. The minimum voltage at each should not fall below 3.5 V. Then N is,***

a) 12 **b)** 10 **c)** 11 **d)** 11

13. ***The unused inputs of an AND gate can be connected to,***

a) used input **b)** to $+V_{CC}$ through 1 kΩ resistance

c) ground **d)** None of the above

14. ***In totem pole arrangement***

a) Only one transistor will be conducting **b)** Both the transistors will not conduct

c) Both the transistors will be conducting **d)** Above all

15. ***Which of the following TTL configuration outputs should not be tied together***

a) Open collector **b)** Totem pole **c)** Tristate **d)** None of the above

16. ***In ECL the switching speed is increased by***

a) Reducing the storage delay time **b)** By preventing the transitor to drive in to saturation

c) Using current switching **d)** All the above

17. ***In standard TTL, the totem pole stage refers to***

a) The multi-emitter input stage **b)** The phase splitter

c) The output buffer **d)** Open collector output stage

18. ***Commerically available ECL gate use two ground lines and one negative supply in order to***

a) reduce power dissipation

b) Increase fan-out

c) reduce loading effect

d) eliminate the effect of power line glitches or the biasing circuit

19. ***The full form of the abbreviations TTL and CMOS in reference to logic families are***

a) Triple Transistor logic and chip metal oxide semiconductor.

b) Tristate Transistor logic and chip metal oxide semiconductor.

c) Transistor Transistor logic and complementary metal oxide semiconductor.

d) Tristate Transistor logic and complementary metal ocide silicon.

Answers

1. b	6. c	11. a	16. d
2. c	7. a	12. a	17. c
3. b	8. c	13. b	18. d
4. a	9. b	14. a	19. c
5. d	10. a	15. b	

IV. Answer the following questions

1. Define logic family.
2. List out the important characteristics of digital logic family if ICs.
3. Define noise margin.
4. Advantages of RTL, DTL, TTL, ECL.
5. Disadvantages of RTL, DTL, TTL, ECL.
6. Advantages and disadvantages of CMOS logic familiy.
7. Define propagation delay.
8. Write a short note on noise margin.
9. Write the formula for speed-power product.
10. Define fan-out and fan-in.

V. Solve the following problems

E11.1 The input and output voltage specifications for the standard TTL is given as,

$V_{OH} = 2.6\ V,\ V_{IH} = 2.1,\ V_{OL} = 0.3,\ V_{IL} = 0.6$

Find high state and low state noise margin.

E11.2 The 7420 data sheet gives the following values.

$I_{OH\_max} = 0.2\ mA,\ I_{OL\_max} = 6\ mA,\ I_{IH\_max} = 16\ \mu A,\ I_{IL\_max} = 0.3\ mA$

What is the fan-out?

E11.3 How many 7400 NAND gate TTL inputs can driven by a 7400 NAND gate TTL output? The current specifications are,

$I_{OL\_max} = 18\ mA,\ I_{IL\_max} = -1.5\ mA,\ I_{OH\_max} = -300\ \mu A,\ I_{IH\_max} = 25\ \mu A$

E11.4 ECL gates are cascaded. The voltage levels are,

$V_{IH} = -1.95\ V,\ V_{OH} = -0.93\ V,\ V_{IL} = -1.432\ V,\ V_{OL} = -1.56\ V$

What is high state and low state noise margin.

E11.5 The specifications for the standard TTL NAND gate is given as,

$V_{cc} = 5\ V,\ I_{CCH} = 4\ mA,\ I_{CCL} = 12\ mA$

Find average current and average power dissipation.

E11.6 The specifications for the standard TTL NAND gate is given as,

$t_{PLH} = 9\ ns,\ t_{PHL} = 5\ ns$

Find propagation delay.

E11.7 An IC family has an average propagation delay of 15 ns and an average power dissipation of 4 mW. Find the speed-power product.

E11.8 **Determine the Boolean equation represented by the CMOS network shown in Fig. E11.8 and hence draw the logic circuit.**

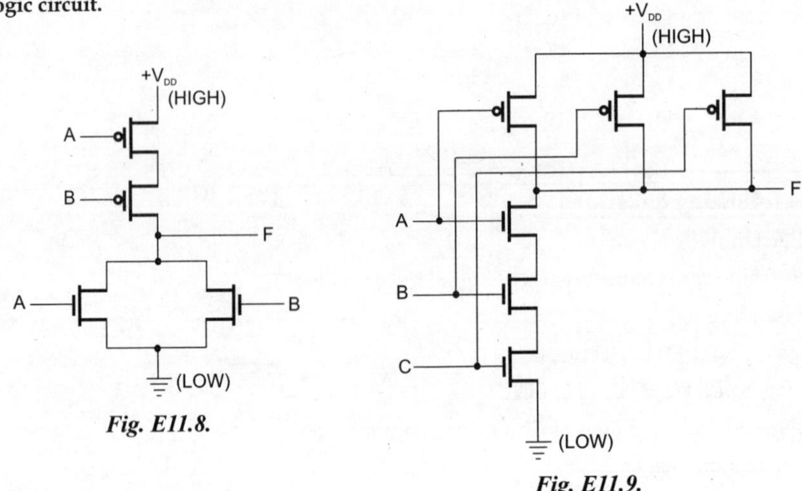

Fig. E11.8.

Fig. E11.9.

E11.9 **Determine the Boolean equation represented by the CMOS network shown in Fig. E11.9 and hence draw the logic circuit.**

E11.10 **Determine the Boolean equation represented by the CMOS network shown in Fig. E11.10 and hence draw the logic circuit.**

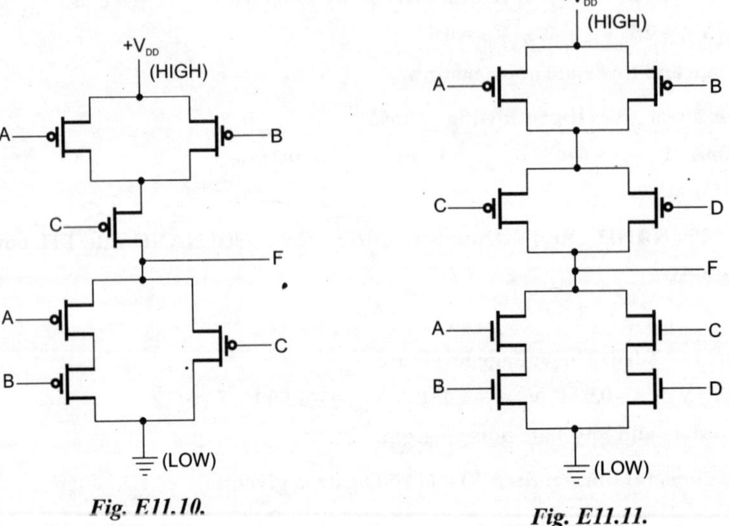

Fig. E11.10.

Fig. E11.11.

E11.11 **Determine the Boolean equation represented by the CMOS network shown in Fig. E11.11 and hence draw the logic circuit.**

E11.12 **Implement the following Boolean equation CMOS network and verify by finding the Boolean equation from the CMOS network.**

a) $(A(B + C) + D)'$ **b)** $((A + B + C) D)'$ **c)** $((A.B) + C(A + B))'$

Answers

E11.1 $V_{NH} = 0.5\ V,\ V_{NL} = 0.3\ V$	**E11.2** Fan-out = 12
E11.3 Fan-out (High) = 12, Fan-out (Low) = 12	**E11.4** $V_{NH} = 2.88\ V,\ V_{NL} = 0.128\ V$
E11.5 $I_{CC\_av} = 8\ mA,\ P_{av} = 40\ mW$	**E11.6** Propagation delay, $t_{pd} = 7\ ns$
E11.7 Speed-power product = 60 pico-Joules (pJ)	

E11.8

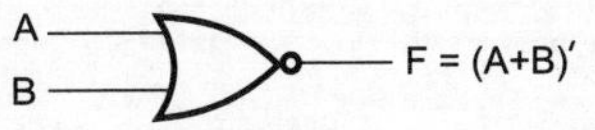

A, B → F = (A+B)′

Fig: Logic circuit of CMOS network.

E11.9

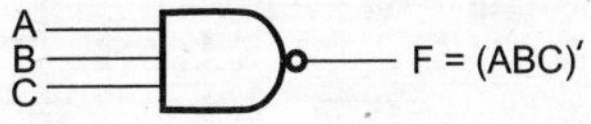

A, B, C → F = (ABC)′

Fig: Logic circuit of CMOS network.

E11.10

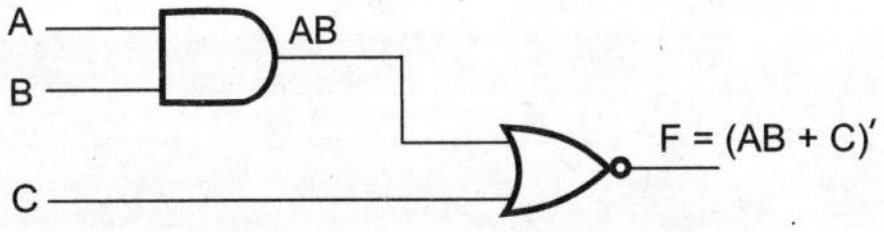

A, B → AB; AB, C → F = (AB + C)′

Fig: Logic circuit of CMOS network.

E11.11

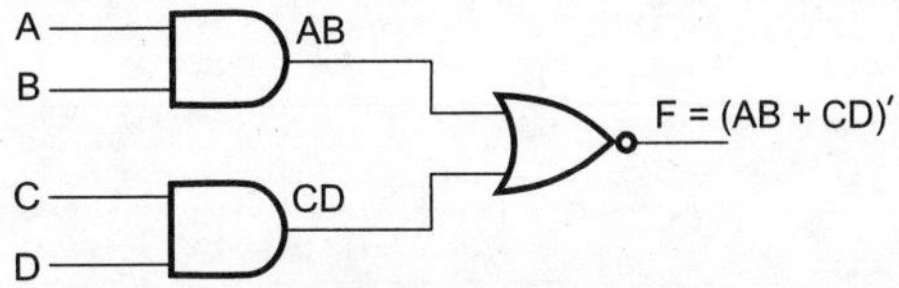

A, B → AB; C, D → CD; F = (AB + CD)′

Fig: Logic circuit of CMOS network.

E11.12 **a)**

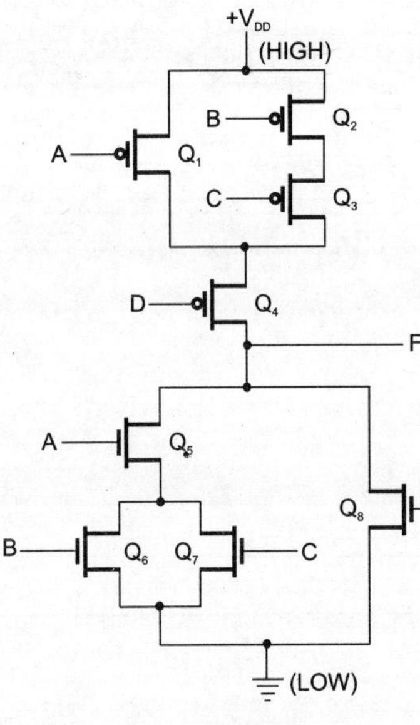

+V_{DD} (HIGH); A, Q_1; B, Q_2; C, Q_3; D, Q_4; F; A, Q_5; B, Q_6; Q_7, C; Q_8, D; (LOW)

Fig.

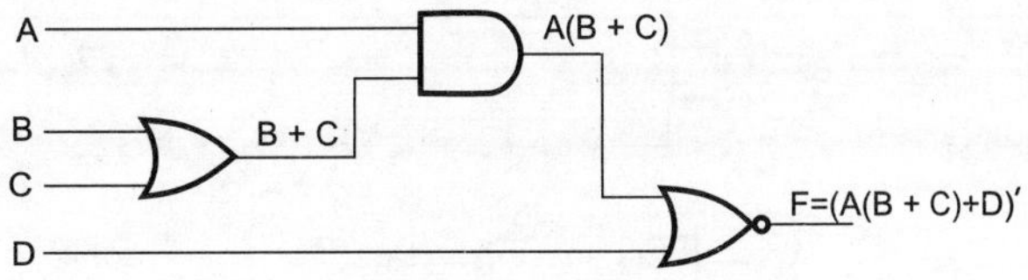

A; B, C → B + C; A(B + C); D; F=(A(B + C)+D)′

Fig: Logic circuit of CMOS network.

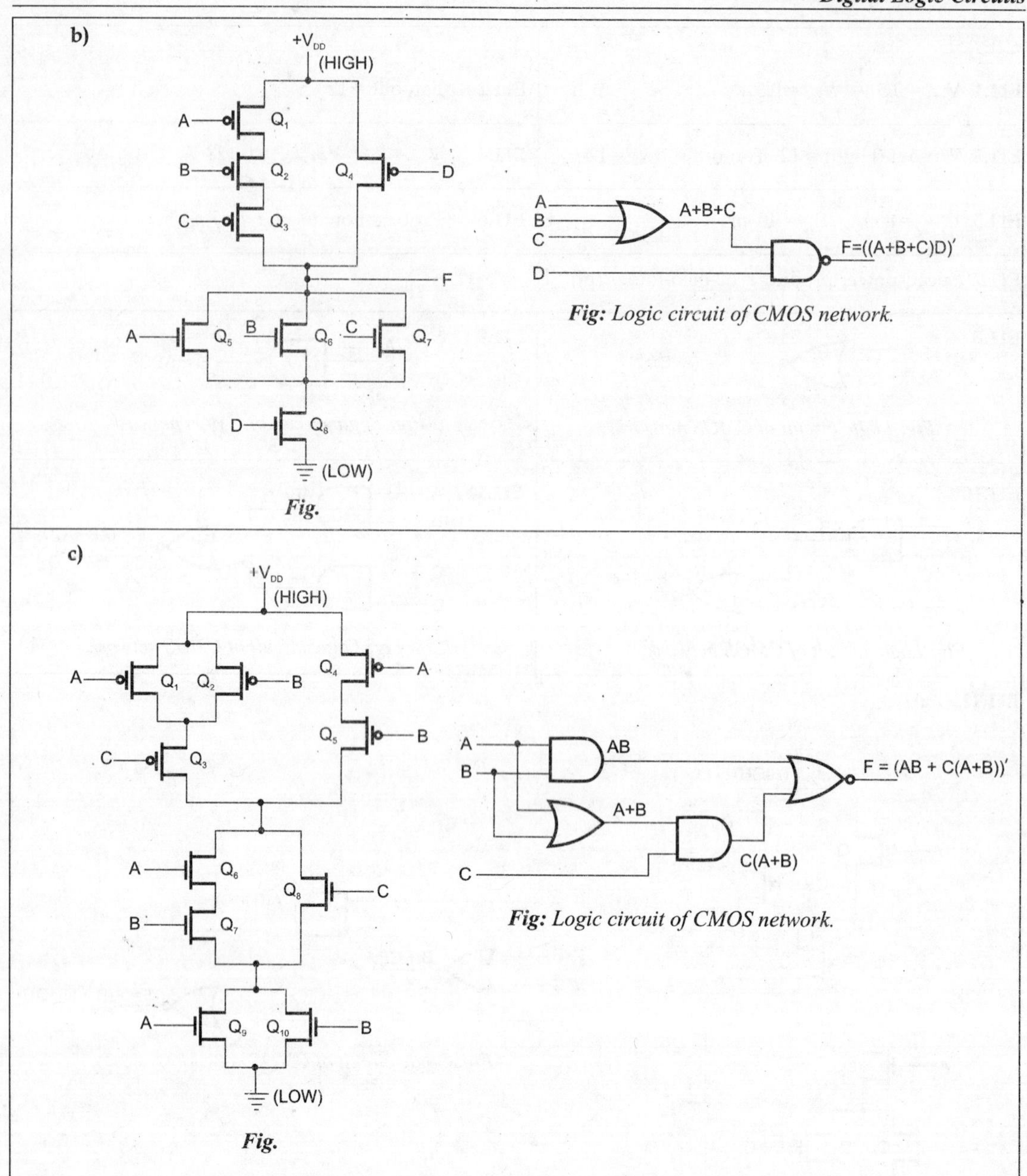

Fig.

Fig: Logic circuit of CMOS network.

Fig.

Fig: Logic circuit of CMOS network.

CHAPTER 12

MEMORY AND PROGRAMMABLE LOGIC DEVICES

12.1 Basic Concepts of Memory

(CSVTU, Nov/Dec'19, 7 Marks)

Memory is a storage device used to store binary information. The memory is manufactured using semiconductors in the form of ICs (Integrated Circuits). The basic storage unit in a memory device is called ***memory cell*** or ***binary cell***. A memory cell can store one bit of binary information.

Binary information is stored in group of bits called ***byte*** or ***word***. Byte is 8-bit binary information. Word is 16-bit binary information. Larger size binary information are multiples of a word.

Standard larger size binary information with reference to memory storage are 32 bits, 64 bits and 128-bits. The 32-bit binary is called ***double word*** and has 4 bytes. The 64-bit binary is called ***long word*** and has 8 bytes.

Byte becomes the standard size to denote memory capacity because the human-machine communication is performed using ASCII (**A**merican **S**tandard **C**ode for **I**nformation **I**nterchange) which is 8-bit binary code. A memory device will have large number of storage cells arranged in groups of 8 cells to store a byte and the group of 8 cells is called a ***location*** or ***memory location***. In general, a memory location can store m bits and the size of binary information that can be stored in a memory location is called ***memory word size*** and the stored binary information is called ***memory word***.

The process of storing binary information in a memory location is called ***write operation***. The process of reading back the stored information from the memory location is called ***read operation***. The read operation is non-destructive which means that after the read operation the memory content remains same until it is rewritten with new information.

Each memory location has a unique binary number called ***address***. The address is used to access the location for read and write operation. The number of addresses for memory locations depends on size of binary used to generate addresses. We know that with n-bit binary we can generate 2^n different binary codes. Therefore, with n-bit binary we can generate 2^n different binary addresses.

$n = 1,\ 2^1 = 2$	$n = 4,\ 2^4 = 16$	$n = 7,\ 2^7 = 128$
$n = 2,\ 2^2 = 4$	$n = 5,\ 2^5 = 32$	$n = 8,\ 2^8 = 256$
$n = 3,\ 2^3 = 8$	$n = 6\ \ 2^6 = 64$	$n = 9,\ 2^9 = 512$

$n = 10,\ 2^{10} = 1024 = 1\ kilo$

$n = 11,\ 2^{11} = 2048 = 2 \times 1024 = 2\ kilo$

$n = 12,\ 2^{12} = 4096 = 4 \times 1024 = 4\ kilo$

$n = 16,\ 2^{16} = 65536 = 64 \times 1024 = 64\ kilo$ and so on.

A memory with 1 *kilo* locations that can store one byte in each location is called 1 *kB (kilo-byte)* memory. This memory will have 8 data lines and 10 address lines. Apart from that it will have read control, write control and chip select/enable signal as shown in Fig. 12.1.

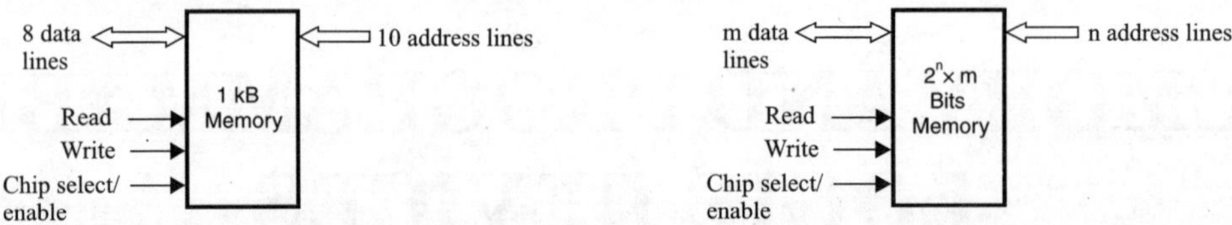

Fig. 12.1: *Block diagram of 1 kB memory.* ***Fig. 12.2:*** *Block diagram of $2^n \times$ m bits memory.*

In general, a memory will have m data lines and n address lines as shown in Fig. 12.2. The capacity of memory is $2^n \times$ m bits. Here, m is always multiple of 8, so that memory capacity can be expressed in bytes.

> ***Note:*** $2^n \times m$ *bits* $= \dfrac{2^n \times m}{8}$ *bytes* $= \dfrac{2^n \times m}{8 \times 1024}$ *kilo-bytes.*

12.1.1 Types of Semiconductor Memories

(GTU, Jun'19, 3 Marks)

The semiconductor memories can be broadly classified into two major types.

1. RAM (Random Access Memory)
2. ROM (Read Only Memory)

The RAM is also called RWM (Read/Write Memory). The RAM memory can be read or write in the system it is fitted or connected. But ROM memory can be only read in the system it is fitted or connected. For writing ROM memory a separate system called ROM programmer is needed.

The random access refer to accessing any location of memory with same time duration. Alternatively, the time to read or write any memory location is same in semiconductor memory. This feature is called ***random access***.

> ***Note:*** *In magnetic memories like hard disc, CD, etc., the time to read/write a memory location will not be same.*

All semiconductor memories has random access feature for read operation. For write operation only RAM memory has random access feature and so it is called ***RAM***. (ROM also has random access features for read operation but not for write operation)

The RAM memory can be again classified into DRAM and SRAM

DRAM: Dynamic RAM

SRAM: Static RAM

Both DRAM and SRAM are volatile memories. In ***volatile memories***, the information will be retained as long as power supply is ON. When power supply is OFF, the stored information will be lost in RAM memory.

In DRAM, the memory content has to be refreshed periodically to retain the stored information even when power is ON. Therefore, in all DRAM memory device an in-build refreshing logic will be available, which will read the memory content periodically and write into same location.

But in SRAM, refreshing is not required. Once information is stored in SRAM, it will be retained as long as power is ON.

The ROM memory can be classified into OTP-ROM, EPROM and EEPROM/EAPROM.

OTP-ROM is one-time programmable and once programmed the content cannot be altered and can only read by the system. For programming OTP-ROM separate programmer system is needed.

EPROM is Erasable Programmable ROM. For erasing the EPROM memory an UV (Ultra Violet) erasure is needed. It has to be exposed to UV light for 20 to 30 minutes. Then it can be reprogrammed using a separate programmer system.

Both ROM and EPROM can be read in the system it is fitted like a RAM memory, but differs only in write operation.

The EEPROM/EAPROM memory can be erased by passing an electrical current of specified value. They can be erased in the system they are fitted, and so no separate system is needed for erasing. The read and write operation is similar to RAM.

12.1.2 Functional Block Diagram of Semiconductor Memory

A typical semiconductor memory IC will have n address pins (lines) and m data pins (lines). The capacity of the memory will be $2^n \times$ m bits. Figure 12.3 shows a simplified functional block diagram of semiconductor memory. The functional blocks of semiconductor memory are Row address decoder, Column address decoder, Memory array, Input buffer and Output buffer.

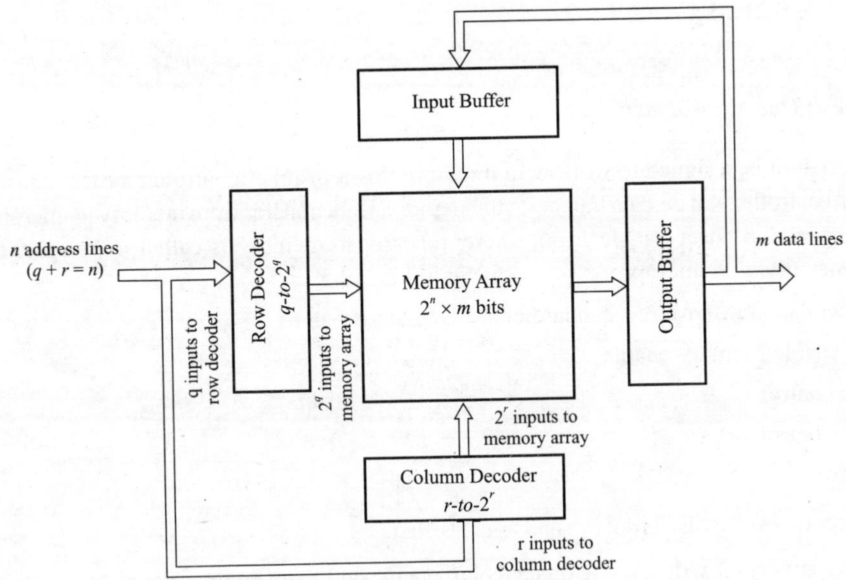

***Fig. 12.3:** A simplified functional block diagram of a typical semiconductor memory.*

The input and output buffers are used to hold the data until the valid time and also takes care of the signal current level matching (Impedance matching). The n address lines are split into q lines and r lines, such that $q + r = n$. The q address lines are applied as input to the row decoder and r address lines are applied as input to the column decoder.

The output lines of the row and column decoders are used to form a matrix array of size, $2^q \times 2^r$ consisting of 2^n crossing points as shown in Fig. 12.4. Each crossing point is called ***memory cell*** or ***binary cell*** and can store one bit of binary information. A typical memory array will consist of ***m*** layers of matrix array as that of Fig. 12.4 and all of them are wired in parallel.

When an address is send to memory IC, the row and column decoder will select one line each, which in turn select one memory cell in each layer. Thus, m memory cells are selected by an address. Then using read or write control signals, the data can be read or stored in the selected memory cells.

In the first version of semiconductor memory, the memory cells were made of passive elements like resistors and capacitors. Later, diodes were used instead of passive elements. With advancement in semiconductor technology, bipolar and MOS transistors were used to form memory cells. The latest technology used for fabricating memory cells are CMOS and HMOS which offer very low power and high-speed operation.

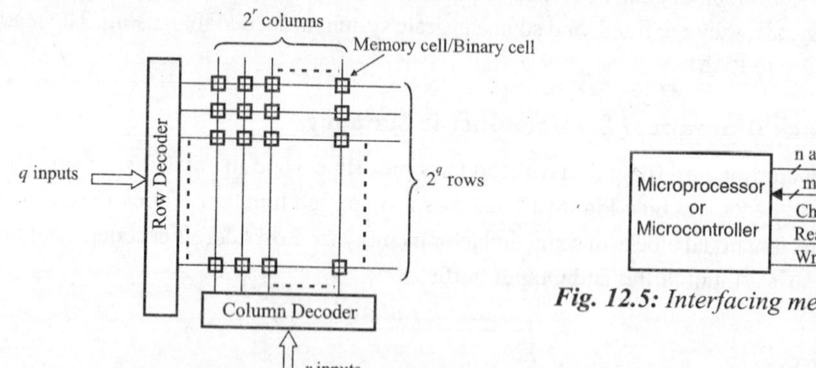

Fig. 12.5: *Interfacing memory to microprocessor.*

Fig. 12.4: *One layer of memory array.*

A memory unit is designed to work with microprocessor or microcontroller system. Basically the microprocessor/controller will do data transfer with memory. Data transfer from memory to microprocessor is called ***read operation***. Data transfer from microprocessor to memory is called ***write operation***. The connection of memory with microprocessor is shown in Fig. 12.5.

For proper data transfer three standard control signals are used,

1. Chip enable/Memory enable
2. Write control
3. Read Control

Read Operation

The microprocessor will output or send address first.

Then microprocessor will activate or send chip enable and Read control signal.

In response the memory will output data on data lines.

The microprocessor will read data from data lines.

All these operation are governed by a clock and completed in 3 to 4 clock pulses/periods.

Write Operation

The microprocessor will output or send address first.

Then microprocessor will activate or send chip enable and write control signal.

Then the microprocessor output data on data lines.

The memory will read data from data lines and store in memory location.

All these operations are governed by a clock and completed in 3 to 4 clock pulses/periods.

Memory Unit with Capacity 1024×16

A memory unit with 1024×16 capacity will have 1024 locations and each locations can store 16-bit binary.

$1024 = 2^{10} \Rightarrow$ Memory has 10 address lines

The range of 10-bit address is as follows.

Binary Address			Hexa Address
0 0	0 0 0 0	0 0 0 0	000_H
0 0	0 0 0 0	0 0 0 1	001_H
0 0	0 0 0 0	0 0 1 0	002_H
0 0	0 0 0 0	0 0 1 1	003_H
0 0	0 0 0 0	0 1 0 0	004_H
⋮	⋮	⋮	⋮
1 1	1 1 1 1	1 1 1 1	$3FF_H$

$3FF_H = 1023_{10}$

Here, memory address and data are both in binary. Hexa is short-hand form of binary.

A simplified schematic of memory is shown below:

Memory Address	Memory Content
$a_9\ a_8\ a_7\ a_6\ a_5\ a_4\ a_3\ a_2\ a_1\ a_0$	$d_{15}\ d_{12}\ d_{13}\ d_{12}\ d_{11}\ d_{10}\ d_9\ d_8\ d_7\ d_6\ d_5\ d_4\ d_3\ d_2\ d_1\ d_0$
0 0 0 0 0 0 0 0 0 0	× × × × × × × × × × × × × × × ×
0 0 0 0 0 0 0 0 0 1	× × × × × × × × × × × × × × × ×
0 0 0 0 0 0 0 0 1 0	× × × × × × × × × × × × × × × ×
0 0 0 0 0 0 0 0 1 1	× × × × × × × × × × × × × × × ×
⋮	⋮
1 1 1 1 1 1 1 1 1 1	× × × × × × × × × × × × × × × ×

Note: ×××× ×××× ×××× ×××× *is any 16-bit data or binary content available in memory location.*

12.1.3 Memory or Binary Cell (BC)

The basic storage unit to store a binary bit is called ***memory cell*** or ***binary cell*** (BC). To store 16-bit word we need 16-binary cells. A 1024×16-bit word memory will have 1024×16 binary cells. In the simplest form, a binary cell is constructed with 4 to 6 transistors. The binary cell can also be constructed using flip-flops. A binary cell using SR flip-flop is shown in Fig. 12.6.

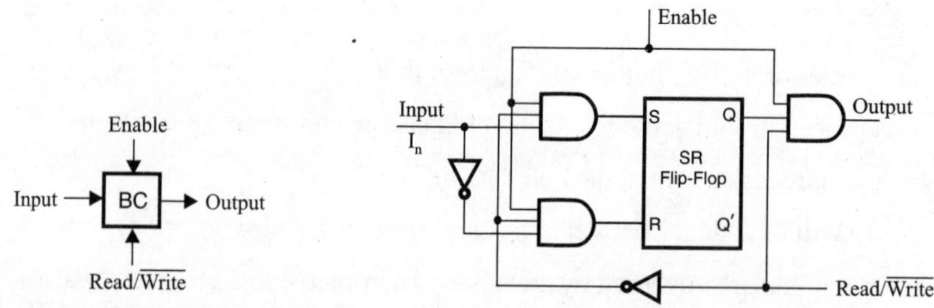

Fig. a: Symbolic representation.

Fig. b: SR flip-flop based BC.

Fig. 12.6: Read operation in SR based BC.

Here, If, Enable = 1; Read/Write operation is allowed or enabled

If, Enable = 0; Read/Write operation is not possible

If, Read/Write = 1; The binary value of Q is available on output and so read operation.

If, Read/Write = 0; The binary value of I_n is stored in Q and so write operation.

Read Operation in SR Flip-Flop based Binary Cell

In Fig. 12.7, when Enable = 1 and Read/Write = 1, one of the inputs to input AND gates is 0 and so input AND gates are disabled. Therefore, S = R = 0 and so no change in output Q. Output AND gate is enabled by both Enable and Read control signal. Therefore, the value of Q will be available on output lines.

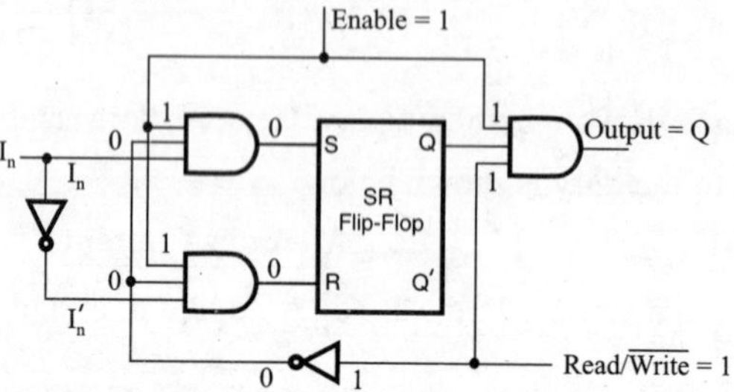

Fig. 12.7: Read operation in SR based BC.

Write Operation in SR Flip-Flop based Binary Cell

In Fig. 12.8, when Enable = 1 and Read/Write = 0, both the control inputs to input AND gates are 1.

Now, the inputs to flip-flops are I_n and complement of input, I_n'. Therefore, the value at I_n will be stored in Q. The output AND gate is disabled by Read/Write control signal.

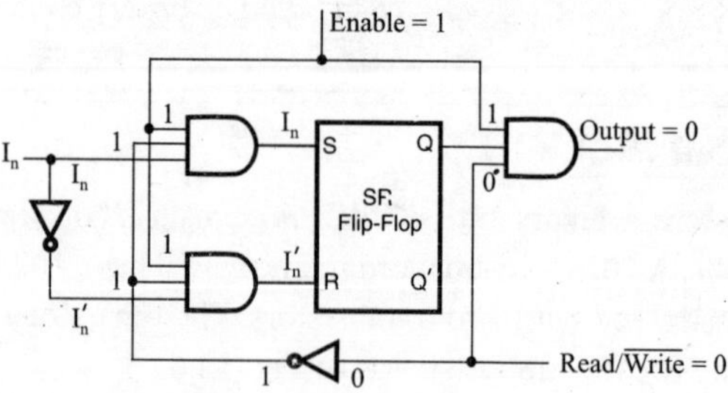

Fig. 12.8: Write operation in SR based BC.

Disable/(Enable = 0) Operation in SR Flip-Flop based Binary Cell

From Fig. 12.9 it can be observed that when enable = 0, both input AND gates and output AND gates are all disabled and there is no possibility of read/write operation.

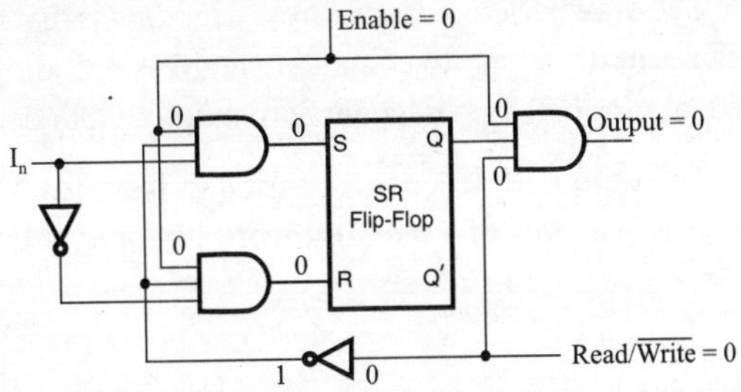

Fig. 12.9: Disable/(Enable = 0) operation in SR based BC.

12.1.4 Address Decoding in Memory

A $2^n \times$ m-bit memory will have 2^n storage locations that can be selected using n-bit address and each location will have m number of binary cells (BC). So that in each location m-bit memory word can be stored.

A schematic with linear address decoding scheme is shown in Fig. 12.10.

A n-to-2^n decoder is employed to generate 2^n control signals to select 2^n rows.

Each row has m number of BCs. When an n-bit address is send to decoder it will select only one of the 2^n rows and so data can be read/write only with the selected row.

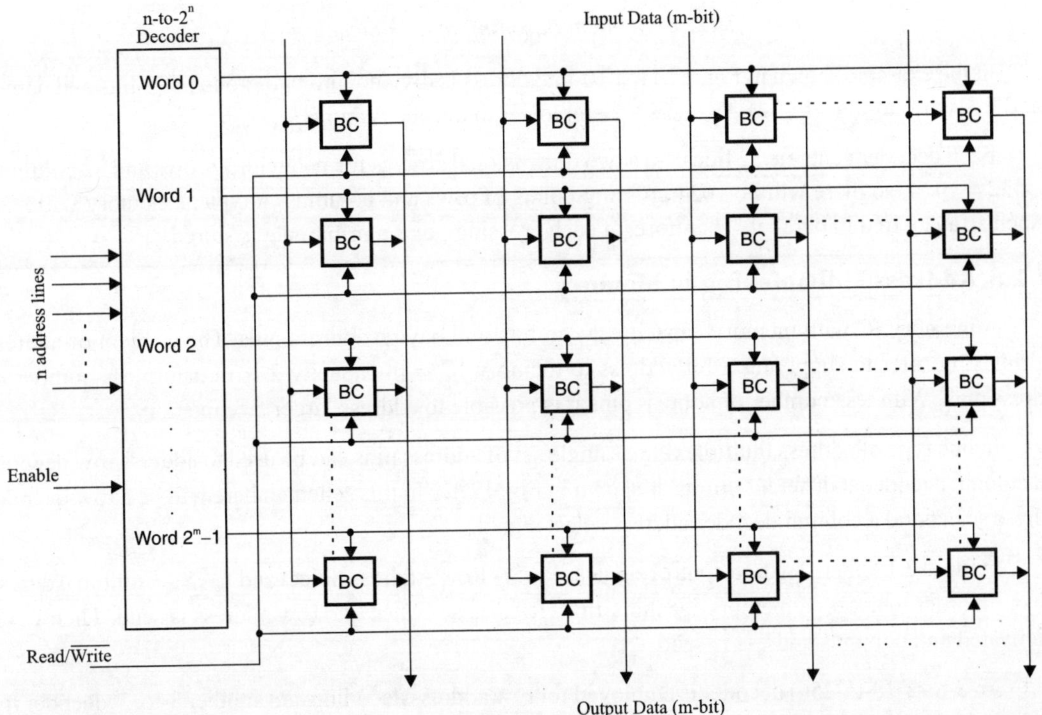

Fig. 12.10: Linear address decoding.

Coincident Decoding

The drawback in linear address decoding is that it require a single large decoder. This can be overcome in coincident decoding. In coincident address decoding the address bits are divided into two groups (nearly two equal groups). Two decoders are employed to decode the address bits.

The output of decoders are used to form rows and columns of a matrix array (or two dimensional array). The data bits of a memory word is stored at the intersection of a row and column.

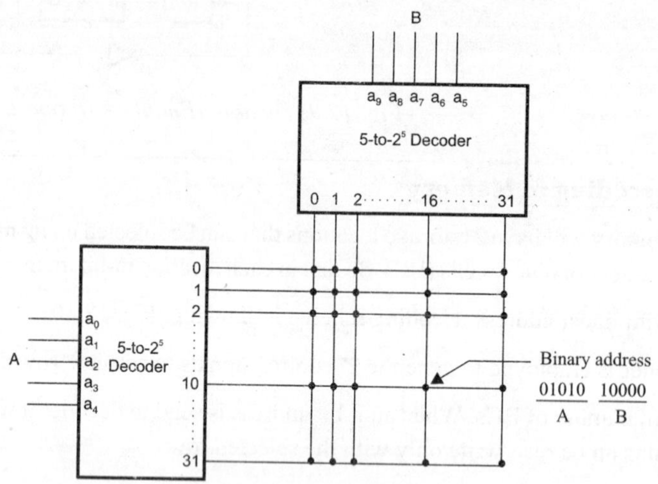

Fig. 12.11: Coincident decoding in a 1k memory.

In the example shown in Fig. 12.11, a 10-bit address is divided into two groups of 5-bits and named A and B. Two 5-to-2^5 (5-to-32) decoders are employed to decode A and B.

Each decoder generate 32 lines. So a two dimensional array is formed with 32 rows and 32 columns. $32 \times 32 = 1024$, so there will be 1024 crossing points of rows and columns. A typical memory will have m such arrays wired in parallel. Therefore, at each crossing point a m-bit data is stored.

12.1.5 Address Multiplexing in Memory

A memory IC with memory capacity $2^n \times m$ bits will have n address pins. The concept of address multiplexing refer to supplying n-bit address to memory IC with capacity $2^n \times m$ using less number of address pins. With less number of address pins it is possible to address larger size memory.

In one type of address multiplexing, a single set of address pins can be used to address row decoder and column decoder at different timing as shown in Fig. 12.12. In this scheme, there will be a row decoder address select and a column decoder address select signal.

In Fig. 12.12, there are two control signals: $\overline{RAS}$ - Row Address Select and $\overline{CAS}$ - Column Address Select. In order to access a memory location, first $\overline{RAS}$ is activated and row address is send. Then $\overline{CAS}$ is activated and columns address is send.

An 8-to-2^8 (8-to-256) decoder is employed for row address decoding and another 8-to-2^8 decoder for column address decoding. Using these two decoder outputs a 256×256 two dimentional array is formed. ($256 \times 256 = 2^8 \times 2^8 = 2^{8+8} = 2^{16} = 2^4 \times 2^{10} = 16 \times 1024 = 16k$). A typical memory will have m such arrays wired in parallel. At intersection of each row and column of the array a m-bit data is stored.

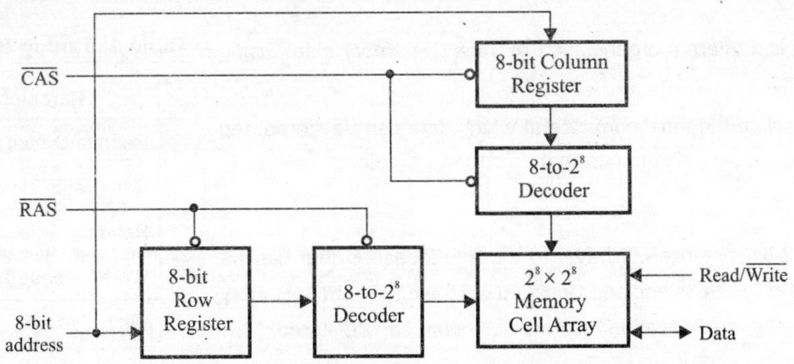

Fig. 12.12: Address multiplexing in a 64k memory.

Example 12.1

Design a 8 × 6 memory array using 4 × 3 memory array.

Solution

A 4 × 3 memory array will have 4 rows with 3 memory cells per row. For 8 × 6 memory array we require 8 rows with 6 memory cells per row. The 8 × 6 memory array can be formed with 4 numbers of 4 × 3 memory array as shown in Fig. 1 in which two 4 × 3 memory arrays are used for lower four locations of 8 × 6 memory array and another two 4 × 3 memory arrays are used for upper four locations of 8 × 6 memory array.

The 8 × 6 memory will have 3-bit address which can be decoded to 8 outputs using 3-to-8 (3-to-2^3) decoder. The lower four decoded lines are used to select 4 rows of one pair of 4 × 3 array and upper four decoded lines are used to select 4 rows another pair of 4 × 3 array.

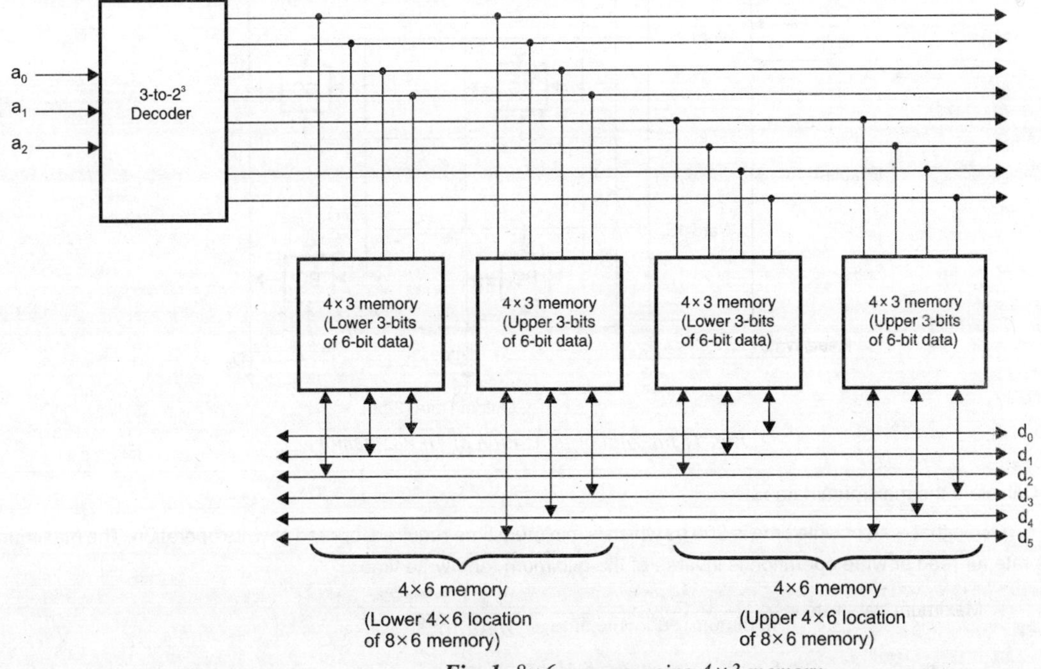

Fig. 1: 8×6 memory using 4×3 memory.

Example 12.2

(AKUB, Dec'20, 6 Marks)

a) Explain internal organization of 16 × 2 memory chips using suitable diagrams.

b) Calculate the maximum rate at which data can be stored and read for a memory having timing parameters shown in Table 1.

Table 1: Timing Parameters

Parameter	Time (ns)
Read to Output Time (t_{RD})	70
Data Setup Time (t_{DW})	120
Read to Cycle Time (t_{RC})	200
Write Release Time (t_{WR})	0
Write Cycle (t_{WC})	200

Solution

a) A 16 × 2-bit memory will have 16 storage locations that can be selected using 4-bit address and each location will have 2 number of binary cells (BC). So that in each location 2-bit memory word can be stored.

The organization of 16 × 2 memory with linear address decoding scheme is shown in Fig. 1.

A 4-to-16 decoder is employed to generate 16 control signals to select 16 rows.

Each row has 2 number of BCs. When an 4-bit address is send to decoder it will select only one of the 16 rows and so data can be read/write only with the selected row.

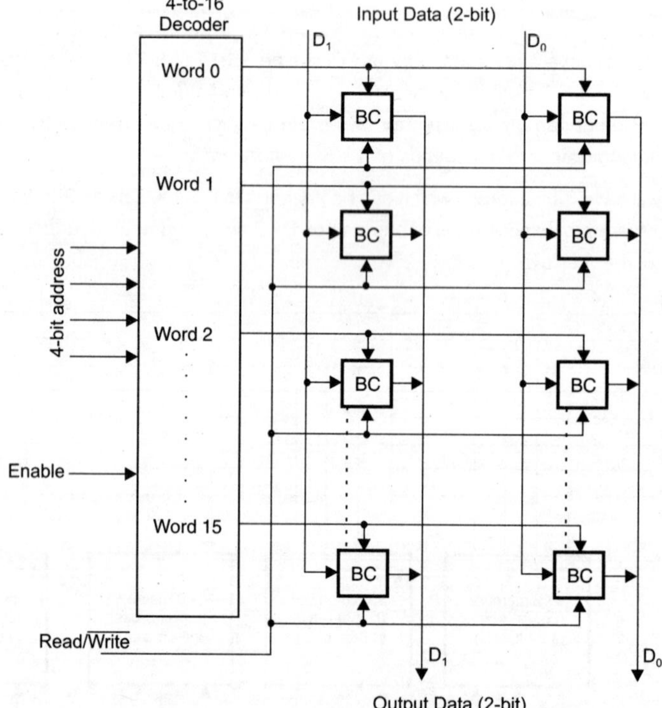

Fig. 1: Internal organization of 16 × 2 memory.

b) Calculate the maximum data rate

Given that read or write time is 200 *ns* which is minimum time required for read or write operation. The maximum data rate for read or write operation is inverse of the minimum read/write time.

$$\therefore \text{ Maximum data rate} = \frac{1}{\text{Minimum read/write time}} = \frac{1}{200 \times 10^{-9}}$$

$$= 5 \times 10^{6} \textit{ bit/second} = 5 \textit{ Mega-bit/second}$$

12.2 ROM and PROM

(RGPV, Jun'20, 4 Marks) (VTU, Aug/Sep'20, 5 Marks) (GTU, May'19, 4 Marks)

ROM is a semiconductor memory which permits only a read access. ROM functions as a memory array whose contents once programmed, are permanently fixed and cannot be altered by the microprocessor to which the memory is interfaced. Other names for this type of memory are dead memory, fixed memory, permanent memory and **R**ead-**O**nly **S**tore (ROS).

ROM is a nonvolatile memory, i.e., loss of power or system malfunction does not change the contents of the memory. Also, the ROM memories have the feature of random access for read operation, which means that the access time (or read time) for a given memory location is same as that for all other locations. The process of storing information in ROM is called ***programming***. The technique employed for storing information in the ROM provides a convenient method for classifying ROMs into one of the following three categories:

1. Custom programmed or Mask programmed ROM (ROM).
2. Programmable or Field programmable ROM (PROM).
3. Reprogrammable or Erasable-Programmable ROM (EPROM).

The custom programmed ROMs are programmed by the manufacturer as specified by the user during fabrication and the contents cannot be changed after packaging. The programmable ROMs are one-time programmable by the user.

The reprogrammable ROMs have facilities for programming as well as for erasing its content and reprogramming the memory. The reprogrammable ROMs are erased either by passing electrical current or Ultraviolet light.

The programming of ROMs can be carried out using ROM (EPROM) programmer. Usually the ROM programmer is a digital system interfaced to a **P**ersonal **C**omputer (PC). The information to be programmed are first stored as a file in the PC and converted to the required binary format using a conversion software. Then the information is transferred from the PC to ROM programmer.

12.2.1 Custom or Mask Programmed ROM

ROM devices are programmed during manufacturing as per customers requirements. After manufacturing the content of ROM connot be altered. The ROM device will have n-bit address input and m-bit data output as shown in Fig. 12.13.

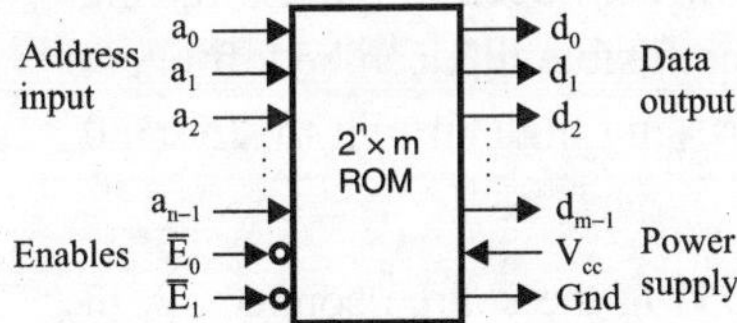

Fig. 12.13: Symbolic representation of ROM.

In mask programmed ROM memory, the memory cell (storage unit) will have a MOS transistor either with open gate or closed gate as shown in Fig. 12.14. The transistors with closed gate represent **1's** and with open gate represent **0's**. Since the configuration is fixed, they permanently store **1's** and **0's**.

A simple structure of a 4 × 4 bit ROM using a single MOS transistor for each shell is shown in Fig. 12.15.

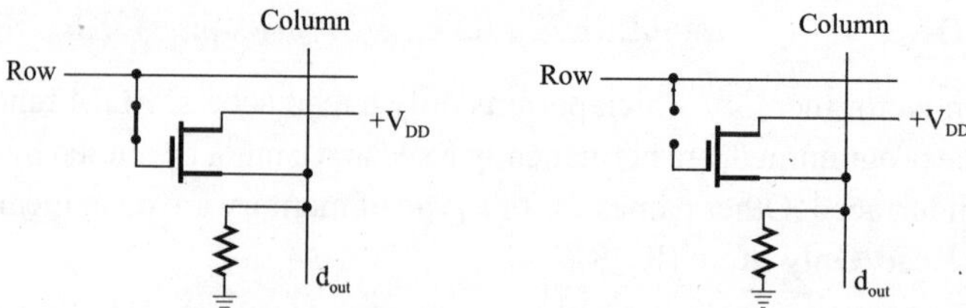

Fig. a: *A ROM cell with a 1 stored in it.*

Fig. b: *A ROM cell with a 0 stored in it.*

Fig. 12.14: *MOS transistor based ROM cell.*

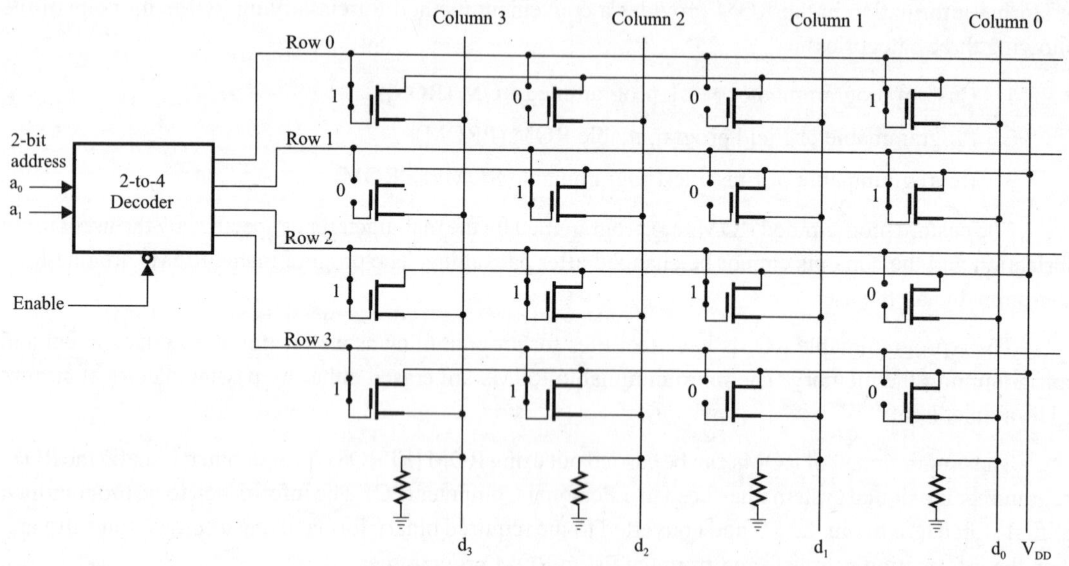

Fig. 12.15: *Structure of a 4 × 4 bit ROM.*

12.2.2 Programmable or Field Programmable ROM

(JNTUA, Jun/Jul'22, 7 Marks)

(CSVTU, Nov/Dec'19, 2 Marks)

The programmable ROMs are one-time programmable (OTP) by the user. Hence it is also called **OTP-ROM**. The structure of PROM cells will be same as ROM but instead of permanent open or close a fusible link is provided in each memory cell. The fusible link can be left intact to store a 1 and can be blown open by passing high current to store a 0 and this process is irreversible.

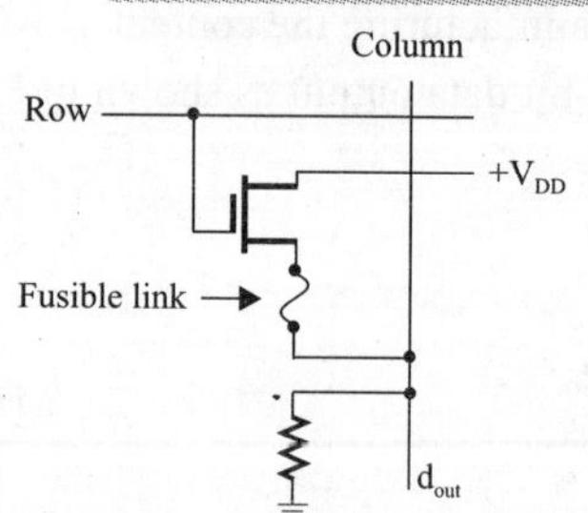

Fig. 12.16: *MOS transistor based PROM cell.*

Usually, the fusible link is provided in the source connection in order to pass higher current to blow the fuse. A simple MOS transistor based PROM cell is shown in Fig. 12.16. The PROM devices will have n-bit address input and m-bit data output similar to ROM device as shown in Fig. 12.13. The programming of PROM device is carried by fixing the PROM in a programming device and send an address to select a row and then passing current pulses through data lines where ever 0 has to be stored. This process is repeated for all rows one by one.

12.2.3 EPROM

The **R**ead **O**nly **M**emory (ROM) which has reprogrammable features is called EPROM (**E**rasable-**P**rogrammable **R**ead **O**nly **M**emory). The EPROM memory is nonvolatile and also has the feature of random access for read operation. In an EPROM, the binary informations are entered using electrical impulses and the stored information is erased using ultraviolet rays. Typical erase time varies between 10 to 30 minutes.

In EPROM, the memory cell (storage location of a bit) consists of a MOS transistor with isolated or floating gate. The isolated gate is located between the normal control gate and the source/drain region of a MOS transistor.

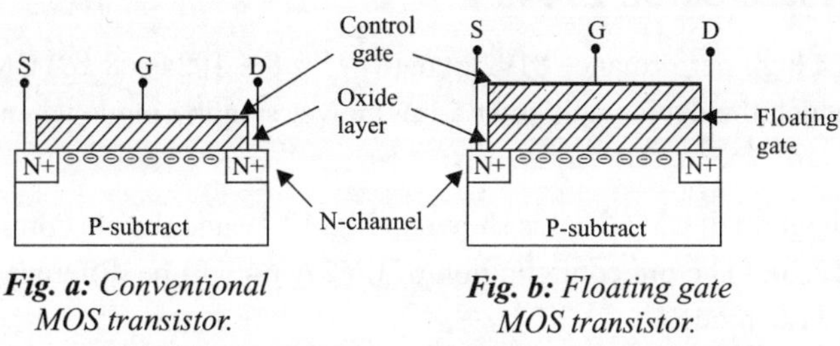

Fig. a: Conventional MOS transistor.

Fig. b: Floating gate MOS transistor.

Fig. 12.17: Cross-section of conventional and floating gate NMOS transistor.

The isolated gate has no electrical connection and so it can store electrical charge permanently. The 1 and 0 of data bits are represented by presence or absence of charge in the isolated gate.

The isolated gate may be charged with electrons during the programming operation and when charged with electrons, the transistor is permanently turned OFF. The state of the isolated gate, charged or uncharged, is permanent because the gate is isolated in an extremely pure oxide.

A charged isolated gate represents a 0 and uncharged isolated gate represents a 1. When 0 is stored the channel will be completely OFF but when 1 is stored the channel can be turned ON by applying a gate pulse. Hence for a read operation, select/address bit is applied to gate and the channel will turn ON if 1 is stored otherwise OFF. When channel is turned ON, the current flowing through the channel will create a voltage drop in an active load connected in series with the channel which is read as 1. When channel is OFF there is no voltage in active load which is read as 0. A single EPROM cell read operation is shown in Fig. 12.18.

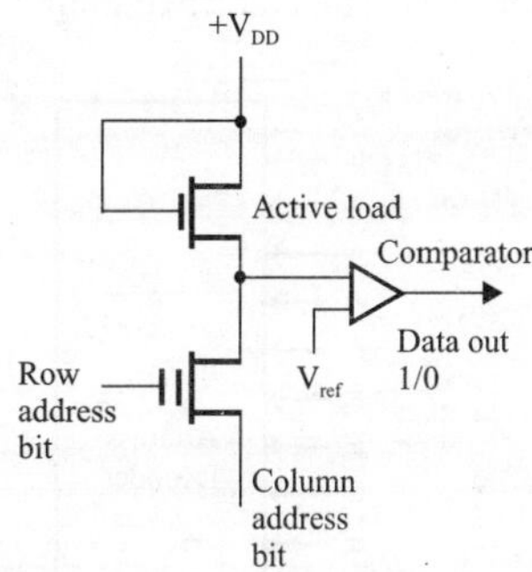

Fig. 12.18: Read operation in an isolated gate NMOS EPROM cell.

The cells are connected in rows and columns. The number of columns depends on the size of data stored in each row and hence each column will have an active load. The number of rows depends on the capacity of EPROM.

The charge on the isolated gate may be removed if the device is irradiated with ultraviolet light. The ultraviolet light allows the electrons to recombine and discharge through the control gate. The process of charging and discharging is repeatable.

The EPROM is programmed by inserting the EPROM chip into the socket of a PROM programmer and providing addresses and voltage pulses at the appropriate pins of the chip. Usually, the PROM programmer is interfaced to a Personal Computer (PC) and the information to be programmed is downloaded from PC.

EPROMs are manufactured by many semiconductor industries like INTEL, Hitachi, Toshiba, Cypress, etc. The manufacturers have a common industry standard, so that a product from different industry will be pin-to-pin compatible and slightly differ in electrical and switching characteristics.

The various features of 2764 (8 *kB* EPROM) manufactured by Cypress semiconductor corporation are discussed in this section.

CY27C64 (Cypress Make CMOS 2764)

The CY27C64 is a high performance 8192 byte ($8192 = 8 \times 1024 = 8\ kB$) CMOS EPROM. It has power down mode, in which the device will enter a low-power standby mode when it is not enabled (or deselected).

The logic block diagram of CY27C64 is shown in Fig. 12.19 and the pin configuration during read mode is shown in Fig. 12.20. [The pin configuration of CY27C64 will be different to that of Fig. 12.20 during programming or write mode.]

The chip has thirteen address inputs denoted as a_0-a_{12}. The address is used to access any one of the 8 *kilo* ($2^{13} = 2^3 \times 2^{10} = 8 \times 1024 = 8192_{10}$) locations within the chip.

The eight output lines, O_0 to O_7 are used to output data from the chip. The chip will be in standby mode when $\overline{CE}$ is inactive. The $\overline{CE}$ is activated for selecting the chip and $\overline{OE}$ is activated for enabling the output buffer during the read operation.

The CY27C64 EPROM is available with maximum access time of 70, 90, 120, 150 or 200 ns (nanosecond). The electrical characteristics or ratings of the EPROM are listed in Table 12.1.

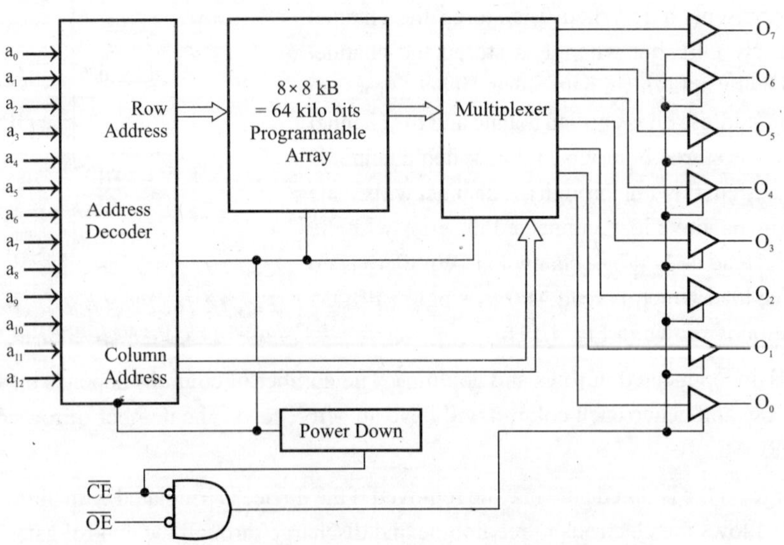

Fig. 12.19: Logic block diagram of a CY27C64.

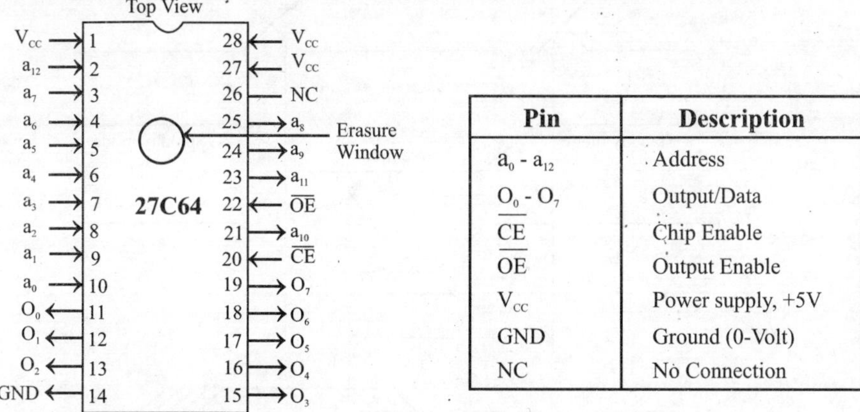

Pin	Description
a_0 - a_{12}	Address
O_0 - O_7	Output/Data
$\overline{CE}$	Chip Enable
$\overline{OE}$	Output Enable
V_{CC}	Power supply, +5V
GND	Ground (0-Volt)
NC	No Connection

Fig. 12.20: Pin configuration of a CY27C64 in read mode.

The timing diagram (or switching waveforms) of CY27C64 for read operation is shown in Fig. 12.20. Only four important timings are shown in this diagram. For detailed discussions on timing diagram refer to manufacturer's data sheet. The switching timings of various signals of CY27C64 are listed in Table 12.2.

Table 12.1: Electrical Characteristics of CY27C64

Description		Rating		Unit
		Min	Max	
Operating Current	Commercial		80	mA
	Military		100	mA
Standby Current			15	mA
Output High Voltage		2.4		V
Output Low Voltage			0.4	V
Input High Voltage		2.0		V
Input Low Voltage			0.8	V
Output Capacitance			10	pF
Input Capacitance			10	pF

The read operation is carried out in the following steps:

1. Place the address of the location to be read, on the address pins a_0 - a_{12}.
2. Enable the chip by asserting chip enable **low** ($\overline{CE}$ = 0).
3. Assert the output enable signal **low** ($\overline{OE}$ = 0).
4. The data can be read from the output lines (O_0 to O_7) after a delay time of t_{OE} (40 or 50 *ns*) after asserting $\overline{OE}$ signal **low**.

When the address is placed on the address lines, the memory will take a time of t_{AA} to place the data on the output lines, provided the $\overline{CE}$ and $\overline{OE}$ are both asserted **low**, at the appropriate time.

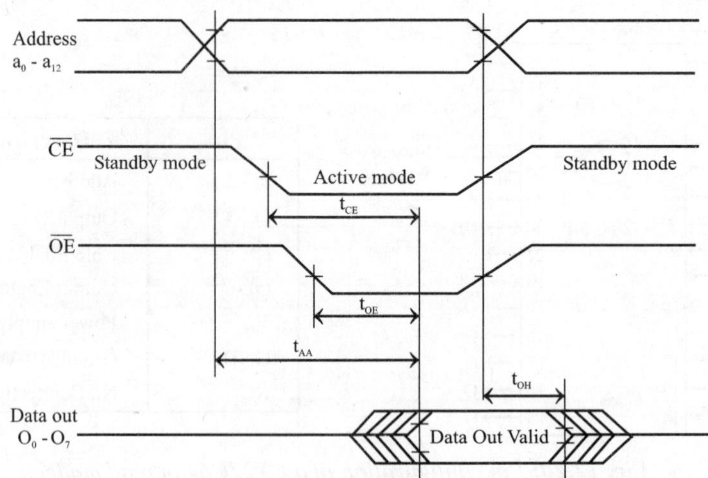

Fig. 12.21: Timing diagram of a CY27C64 for read operation.

Table 12.2: Switching Characteristics of CY27C64

Parameter	Description	Time		Unit
		Min	**Max**	
t_{AA}	Address to output valid	70	200	ns
t_{OE}	Output enable active to output valid	40	50	ns
t_{CE}	Chip enable active to output valid	70	200	ns
t_{OH}	Data hold from address change	3		ns

The CY27C64 EPROM is equipped with an erasure window. When the window is exposed to UV light, the contents of EPROM are erased and then it can be reprogrammed. Wavelengths of light less than 4000 A° (Angstrom unit) begin to erase the EPROM if the EPROM is exposed to sunlight or fluorescent lighting for a very long time. Hence, an opaque label should be placed over the window

The recommended dose of UV light for erasure is a wavelength of 2537 A° for a minimum dose (UV intensity multiplied by exposure time) of 25 *W-sec/cm*^{2}. For an UV lamp with a 12 *mW/cm*^{2} power rating, the exposure time would be approximately 35 minutes.

The EPROM has to be placed within a distance of 1 inch from the lamp during erasure. Permanent damage may result if EPROM is exposed to high intensity UV light for a very long time. (Maximum dosage is 7258 *W-sec/cm*^{2}).

12.2.4 EEPROM/EAPROM

(IKGPTU, Dec'19, 2 Marks)

EEPROM/EAPROM is a nonvolatile read/write memory. They are also called ***flash memory***. These memory devices are electrically erasable in the system, but require more time to erase than a static RAM. Therefore, they are called EEPROM (**E**lectrically **E**rasable **P**rogrammable **ROM**) or EAPROM (Electrically Alterable Programmable ROM).

The drawback in EEPROMs is that it takes longer time for the system to erase and write. The maximum number of write operations that can be performed in most of the EEPROMs is about 10,000 operations.

INTEL and XYCOR have released their versions of nonvolatile RAMs, which does not have the drawbacks of EEPROMs. (The drawbacks of EEPROM are high write time and limited number of write cycles.) This type of nonvolatile RAMs consists of a high speed static RAM and a corresponding EEPROM on a single chip. For this reason, it is also called ***shadow RAM***.

In these devices for each cell of static RAM, there is one EEPROM cell. A typical example of shadow RAM is INTEL 2004 and XYCOR's X2004. The 2004 has a special pin called ***nonvolatile enable*** (NE). Normally, this pin is **high** and read or write operation is performed with static RAM. When NE is asserted **low**, the data in the static RAM cells are written into the corresponding EEPROM cells.

12.3 RAM

(VTU, Aug/Sep'20, 5 Marks) (RGPV, Jun'22, 3 Marks)

12.3.1 Static RAM

The static RAM (**R**andom **A**ccess **M**emory) is a read/write memory. In static RAM, the memory cell (storage location for each bit of information) consists of a flip-flop or similar device constructed using four to eight MOS transistors. [Even though ROM memories are also technically random access memory, the read/write memories are called RAM.] Besides random access feature, the static RAMs are volatile in nature. The stored information is retained in the memory cell as long as power is supplied to the circuit.

A typical SRAM cell is made up of six NMOS transistors is shown in Fig. 12.22 and it is called a 6T SRAM cell. Four transistors M_1, M_2, M_3 and M_4 that form two cross-coupled inverters are used to store a data bit and its complement. This storage cell has two stable states which are used to denote 0 and 1.Two additional access transistors serve to control the access to a storage cell during read and write operations.

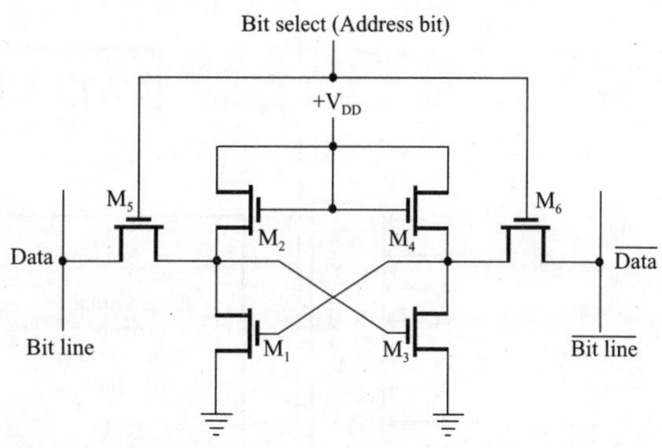

Fig. 12.22: A 6-transistor static RAM cell.

Access to the cell is enabled by the address bit which controls the two access transistors M_5 and M_6 which, decides whether the cell should be connected to the data bit lines and its complement. They are used to transfer data for both read and write operations. Although it is not strictly necessary to have two bit lines, both the signal and its inverse are typically provided in order to improve noise margin.

During read/write accesses, the bit lines are actively driven high and low by the inverters in the SRAM cell. This improves SRAM bandwidth.

The static RAMs are manufactured by many semiconductor companies like Motorola, Hitachi, Toshiba, Cypress, etc. The manufacturers have a common industry standard, so that a product from different companies will be pin-to-pin compatible and only slightly differ in electrical and switching characteristics. The various features of 6264 (8 *kB* RAM) manufactured by Cypress semiconductor corporation are discussed in this section.

CY6264 (Cypress Make CMOS 6264)

The CY6264 is a high performance CMOS static RAM organized as 8192 bytes ($8192 = 8 \times 1024 = 8\ kB$). The device has a power down mode. When CY6264 is not enabled (deselected), it will enter the power down mode and in this mode the power consumed is reduced to 30% of active mode power.

The logic block diagram and the pin configuration of CY6264 are shown in Figs. 12.23 and 12.24. The chip has 13 address inputs denoted as a_0 - a_{12}. The address is used to access any one of the 8 *kilo* (8192) locations within the chip. It has eight IO pins for reading/writing the data and they are denoted as IO_0-IO_7.

The chip has four control signals $\overline{CE}_1$, CE_2, $\overline{WE}$ and $\overline{OE}$. When $\overline{CE}_1$ and $\overline{WE}$ inputs are both **low** and CE_2 is **high**, data on the eight data pins (IO_0 through IO_7) is written into the memory location addressed by the address pins (a_0 through a_{12}).

When $\overline{CE}_1$ and $\overline{OE}$ are both **low** and CE_2 is **high**, the content of the memory location addressed by the address pins will be loaded on the eight data pins, IO_0 to IO_7.

The CY6264 RAM is available with maximum access time of 55 or 70 *ns* (nanosecond). The electrical characteristics or ratings of the RAM are listed in Table 12.3.

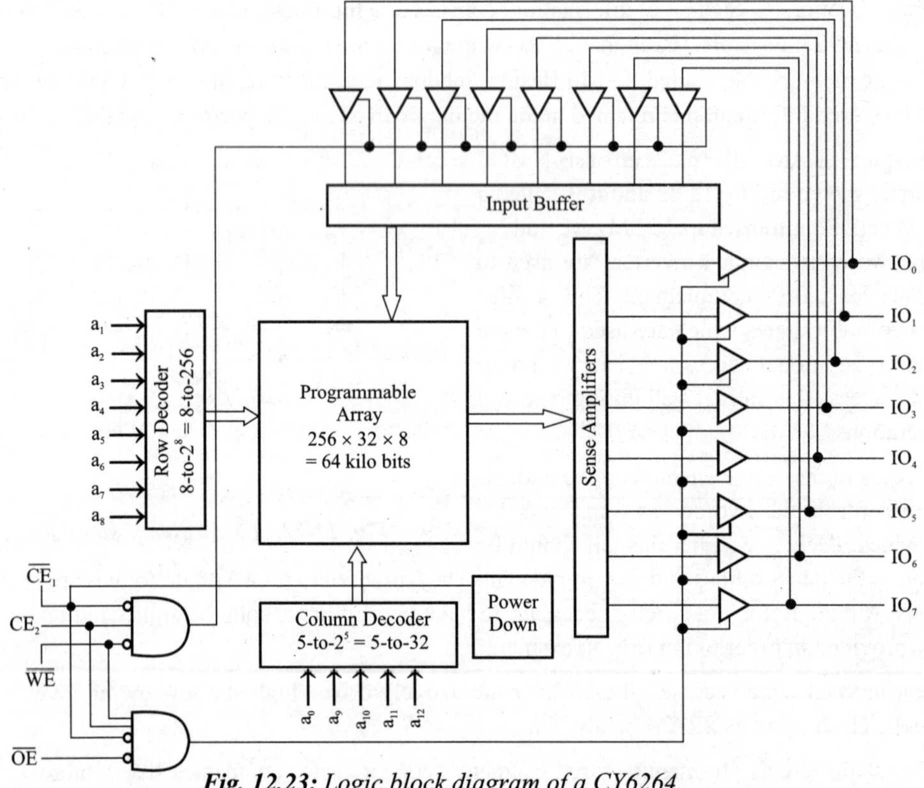

Fig. 12.23: Logic block diagram of a CY6264.

Top View

6264

Signal	Pin	Pin	Signal
NC	1	28	V_{CC}
a_4	2	27	$\overline{WE}$
a_5	3	26	CE_2
a_6	4	25	a_3
a_7	5	24	a_2
a_8	6	23	a_1
a_9	7	22	$\overline{OE}$
a_{10}	8	21	a_0
a_{11}	9	20	$\overline{CE_1}$
a_{12}	10	19	IO_7
IO_0	11	18	IO_6
IO_1	12	17	IO_5
IO_2	13	16	IO_4
GND	14	15	IO_3

Pin	Description
a_0 - a_{12}	Address
IO_0 - IO_7	Input/Output Data
$\overline{CE_1}$	Active **low** Chip Enable
CE_2	Active **high** Chip Enable
$\overline{WE}$	Active **low** Write Enable
$\overline{OE}$	Active **low** Output Enable
V_{CC}	Power supply, +5V
GND	Ground (0-Volt)
NC	No Connection

Fig. 12.24: Pin configuration of a CY6264.

Table 12.3: Electrical Characteristics of CY6264

Description	Value		Unit
	Min	Max	
Operating Current		100	mA
Standby Current	15	20	mA
Output High Voltage	2.4		V
Output Low voltage		0.4	V
Input High Voltage	2.2	V_{cc}	V
Input Low Voltage	−0.5	0.8	V
Output Capacitance		7	pF
Input Capacitance		7	pF

Table 12.4: Read Cycle Timings of CY6264

Parameter	Description	Time		Unit
		Min	Max	
t_{RC}	Read cycle time	70		ns
t_{AA}	Address to data valid		70	ns
t_{OHA}	Data hold from address change	5		ns
t_{ACE}	CE **low/high** to data valid		70	ns
t_{DOE}	$\overline{OE}$ **low** to data valid		35	ns

The timing diagram (or switching waveforms) of CY6264 for read operation is shown in Fig. 12.25. Only five important timings are shown in this diagram.

For detailed discussions on timing diagram, refer to manufacturer's data sheet. The timings of various signals of CY6264 are listed in Table 12.4.

The read operation is carried out in the following steps:

1. Place the address of the location to be read on the address pins a_0- a_{12}.
2. Enable the chip by asserting Chip Enable-1 ($\overline{CE}_1$) as **low** and Chip Enable-2 (CE_2) as **high**.
3. Assert the Output Enable ($\overline{OE}$) signal **low**.
4. When $\overline{OE}$ signal is asserted **low**, the data can be read from the input/output lines (IO_0 - IO_7) after a delay time of t_{DOE} (35 *ns*).

When the address is placed on the address line, the memory will take a time of t_{AA} to place the data on the output lines, provided the $\overline{CE}_1$ and $\overline{OE}$ are asserted **low** and CE_2 is asserted **high** at the appropriate time.

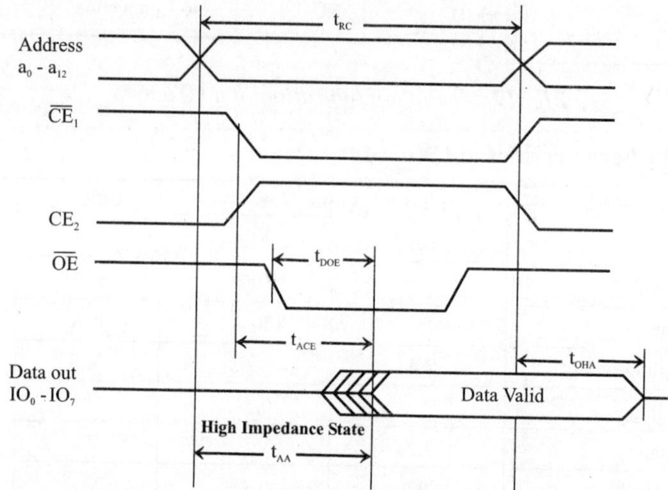

Fig. 12.25: Read cycle timings of a CY6264.

The timing diagram (or switching waveform) of CY6264 for write operation is shown in Fig. 12.26. The diagram shows some important timings of write cycle.

For detailed discussions on timing diagram, refer to the manufacturer's data sheet. The timings of various signals are listed in Table 12.5.

The write operation is carried out in the following steps:

1. Place the address of the location to be written on the address pins a_0 - a_{12}.
2. Enable the chip by asserting $\overline{CE}_1$ signal as **low** and after a small delay assert CE_2 signal as **high**.
3. Assert the write enable $\overline{WE}$ signal as **low**.
4. Place the data to be written on the IO_0-IO_7 lines immediately after $\overline{WE}$ is asserted **low**.

After the address is placed on the address lines and $\overline{CE}_1$, CE_2 and $\overline{WE}$ are asserted appropriately, the data has to be placed on the data lines within the time t_{SD} (data set-up to write end).

Table 12.5: Write Cycle Timings of CY6264

Parameter	Description	Minimum time	Unit
t_{WC}	Write cycle time	50	ns
t_{SCE1}	$\overline{CE}_1$ **low** to write end	40	ns
t_{SCE2}	CE_2 **high** to write end	30	ns
t_{AW}	Address setup to write end	40	ns
t_{PWE}	$\overline{WE}$ pulse width	25	ns
t_{SD}	Data set-up to write end	25	ns

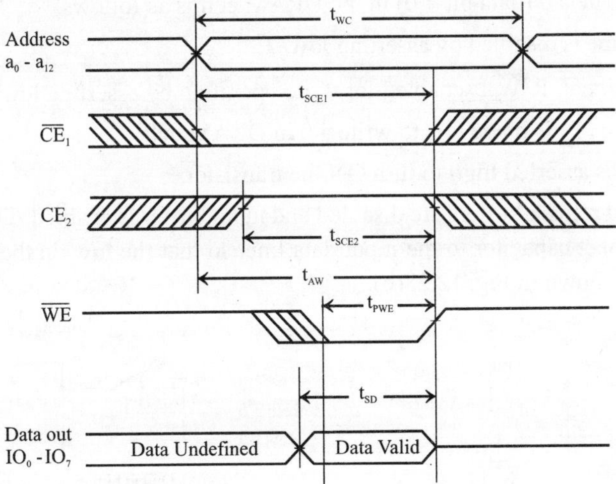

***Fig. 12.26:** Write cycle timings of CY6264.*

12.3.2 DRAM

DRAM (**D**ynamic **RAM**) is a read/write memory in which the information is stored in the form of electric charge on the gate-to-substrate capacitance of a MOS transistor. This charge dissipates in a few milliseconds and the element must be refreshed periodically. A simplified MOS transistor based DRAM cell is shown in Fig. 12.27. The write, read and refreshing operations on a DRAM cell are shown in Fig. 12.28.

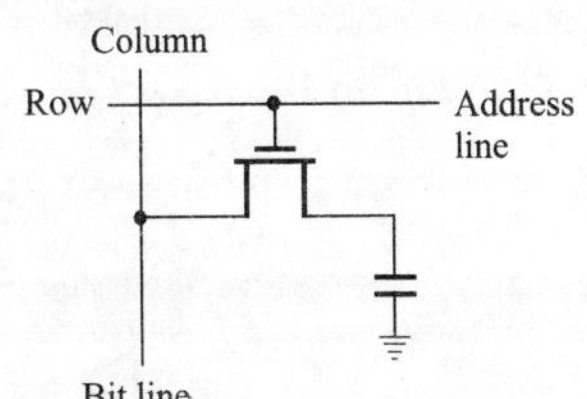

***Fig. 12.27:** Dynamic RAM cell.*

DRAMs are volatile and have random access feature. Dynamic RAMs are important because fewer elements are required to store a bit, so that more bits can be packed into an IC of a given physical area. They are also faster than the static RAM and consume less power in the quiescent state.

Refreshing of DRAMs needs extra circuitry and so the interfacing of DRAMs are more complex than the interfacing of static RAMs. The recent versions of DRAMs have internal refreshing circuit. The manufacturing of DRAMs will be cheaper only for very large capacity memories. Therefore, smaller memories are generally static elements (up to 256 *kB*) and large memories (> 1 *MB*) are typically dynamic.

Write Operation in DRAM

The procedure to write a 1 (data bit = 1) in the DRAM cell is as follows,

- The refresh line is disabled by asserting **low**.
- The control signal, $R/\overline{W}$ is enabled for write operation by asserting **low**.
- Input data line, d_{IN} is made **high** to write a 1 in DRAM cell.
- The row line is asserted **high** to turn ON the transistor.

Now, the output and refresh buffers are disabled and input buffer is enabled. The transistor will act as a short circuit and connect the input data line to capacitor, so that the **high** on the data input line will charge the capacitor as shown in Fig. 12.28(a).

The procedure to write a 0 (data bit = 0) in the DRAM cell is as follows,

- The refresh line is disabled by asserting **low**.
- The control signal, $R/\overline{W}$ is enabled for write operation by asserting **low**.
- Input data line, d_{IN} is made **low** to write a 0 in DRAM cell.
- The row line is asserted **high** to turn ON the transistor.

Now, the output and refresh buffers are disabled and input buffer is enabled. The transistor will act as a short circuit and connect capacitor to the input data line, so that the **low** on the data input line will discharge the capacitor as shown in Fig. 12.28(b).

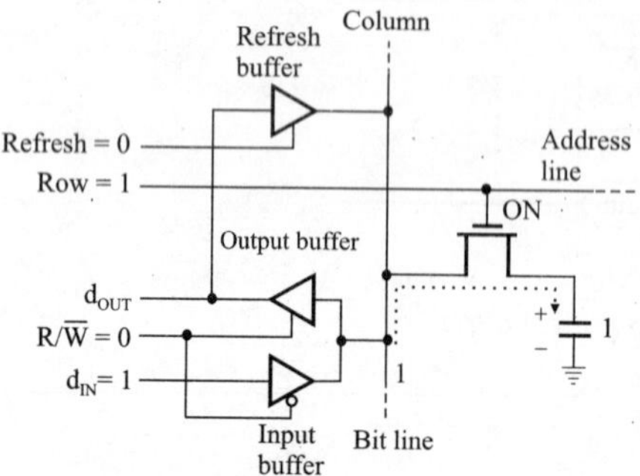

Fig. a: Writing a 1 into the memory cell.

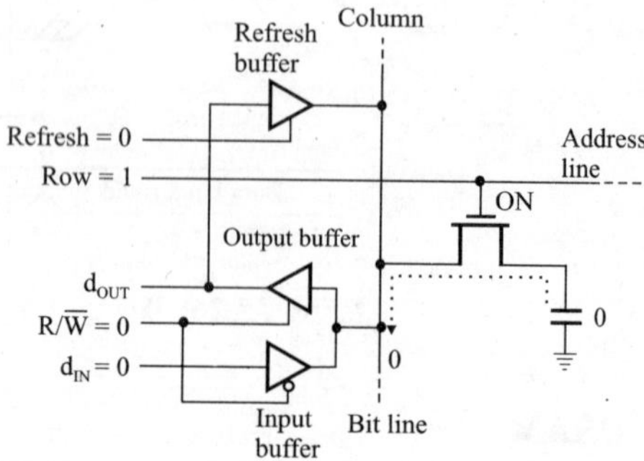

Fig. b: Writing a 0 into the memory cell.

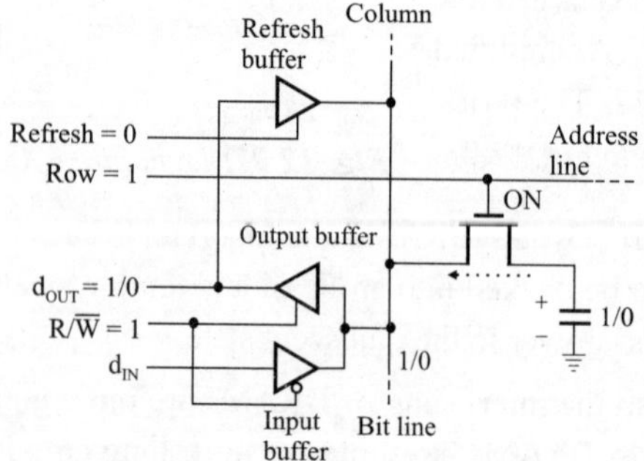

Fig. c: Reading a data bit from the memory cell.

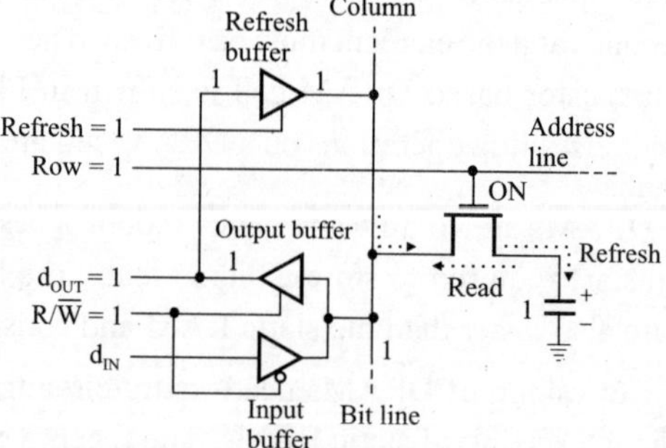

Fig. d: Refreshing a stored 1.

***Fig. 12.28:** Write, read and refreshing operations on a DRAM cell.*

Read Operation in DRAM

The procedure to read a data bit from the DRAM cell is as follows,

- The refresh line is disabled by asserting **low**.
- The control signal, $R/\overline{W}$ is enabled for read operation by asserting **high**.
- The row line is asserted **high** to turn ON the transistor.

Now, the input and refresh buffers are disabled and output buffer is enabled. The transistor will act as a short circuit and connect the output data line to capacitor, so that the **high/ low** voltage in the capacitor will appear on the output data as shown in Fig. 12.28(c). Here, the output buffer will only sense the voltage level on the capacitor and will not allow the capacitor to discharge and so the output buffer is also called sense amplifier.

Refreshing Operation in DRAM

The charge in the capacitor of DRAM cell is refreshed by reading the data bit and then writing into same cell. The procedure to refresh a stored 1 (data bit = 1) in the DRAM cell is as follows,

- The refresh line is enabled by asserting **high**.
- The control signal, $R/\overline{W}$ is enabled for read operation by asserting **high**.
- The row line is asserted **high** to turn ON the transistor.

Now, the input buffer is disabled and refresh and output buffers are enabled. The transistor will act as short circuit and connect the output data line to capacitor, so that the **high/low** voltage in the capacitor will appear on the output data line which is also input to refresh buffer. The refresh buffer output will charge the capacitor through the transistor as shown in Fig. 12.28(d).

12.3.3 Content-Addressable Memory (CAM)

CAM is a type of semiconductor memory that is designed for high-speed searching and retrieval of data. In a traditional memory, data is accessed by specifying an address, but in a CAM data retrieval is based on the actual content of the data itself.

In a standard semiconductor memory, the user supplies a memory address and the memory returns the data word stored at that address. But in a CAM the user supplies a data word and the CAM searches its entire memory and if the data word is found, then the CAM returns an address where the word is stored.

A CAM implements the lookup-table function in a single clock cycle using dedicated comparison circuitry. CAMs are especially popular in network routers for packet forwarding and packet classification, but they are also beneficial in a variety of other applications that require high-speed table lookup.

CAM is sometimes referred to as associative memory because it associates data content with its corresponding address, allowing for associative searches based on content rather than location.

CAM consists of a large array of memory cells, with each cell storing a bit of data (content). A CMOS binary CAM Cell consists of a 6T SRAM cell and in addition 4 comparison transistors as shown in Fig. 12.29. Every cell is connected to one data bit line (BL) and its complement (BL′) so that data can be stored in cells using write control signal. For search operation, every cell is connected to one data search line (SL) and its complement (SL′) so that data can be searched through read or search control signal.

When the data on the search lines (SL) differs from the data stored in the cell through the bit lines (BL), the match line (ML) will be pulled low to indicate a mismatch. If none of the cells on a match line indicate a mismatched bit, the match line will remain high at the precharge level to indicate a word match.

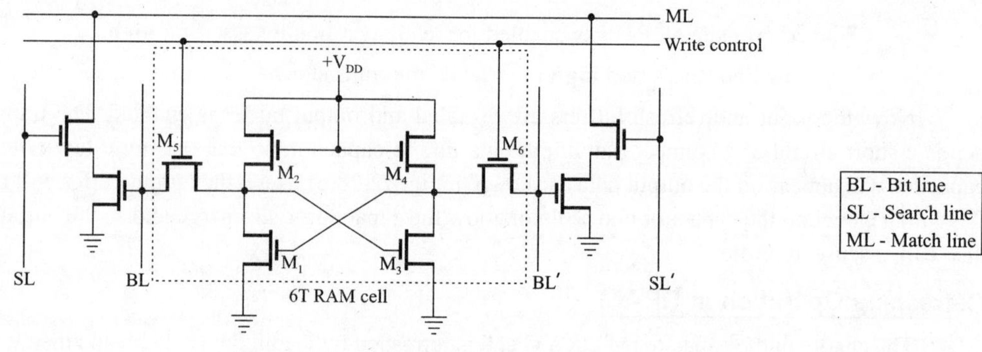

Fig. 12.29: CAM cell.

The architecture of CAM (Content-Addressable Memory) consists of conventional semiconductor memory (usually SRAM) with added comparison circuitry that enables a search operation to complete in a single clock cycle.

The basic data storage in a CAM consists of a large array of memory cells, with each cell storing a bit of data (content). The position of cell in the array is the address of the cell. Each CAM cell contains a data bit along with associated comparison circuitry. The cells are arranged in rows and columns with each row store a word and number columns in each row determine the word size. In general, m × n, CMA will store m words of size n-bits each. The simplified architecture of m × n CAM is shown in Fig. 12.30.

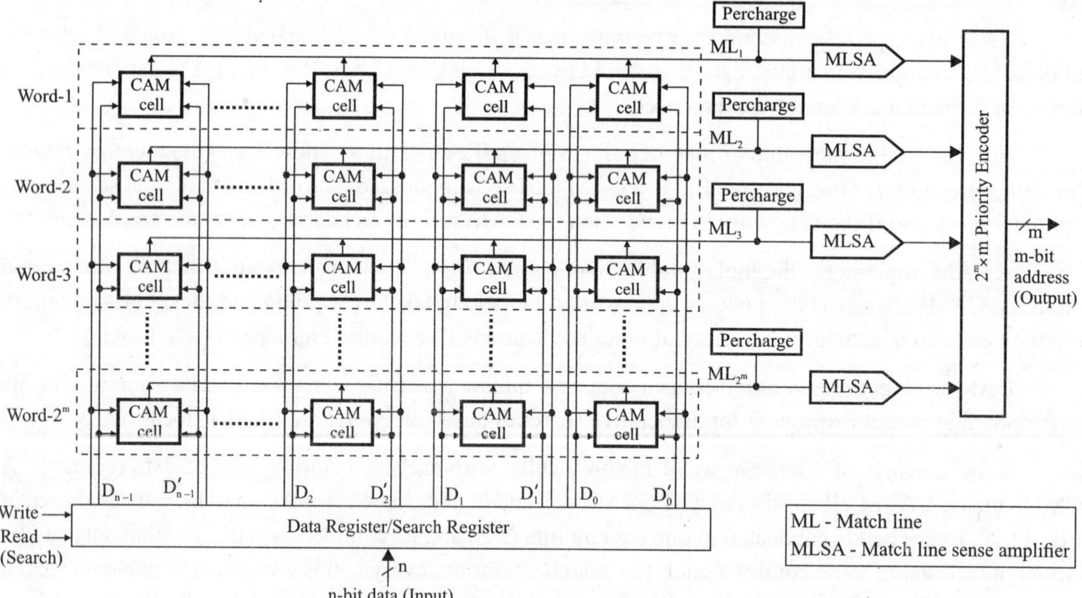

Fig. 12.30: Architecture of CAM

CAM uses match lines to indicate whether a match has been found during a search operation. These match lines are connected to the comparison circuitry and are used to identify memory cells where the stored data matches the input data.

When a search operation is initiated, the input data (search data) is loaded in search register of CAM. The comparison circuitry then compares this input data with the data stored in all memory cells in parallel. If a match is found, the corresponding match line is activated, and a priority encoder is employed to generate the address or location of the matching data.

Example 12.3

Explain the logical construction of a 256 × 8 RAM using 64 × 8 RAM.

Solution

The 256 × 8 memory will have 256 memory locations with 8-bit data per location. Since, 256 = 2^8, 8-bit address is required to address 256 memory locations.

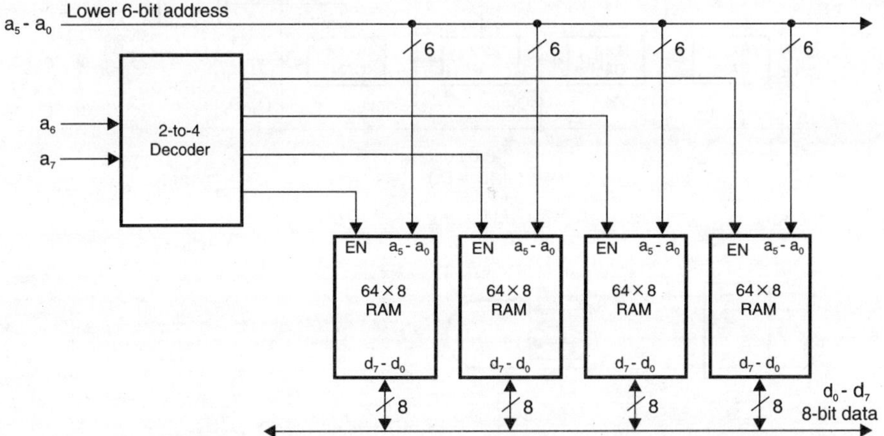

Fig. 1: Logical diagram of 256×8 memory using 64×8 memory.

The 64 × 8 memory will have 64 memory locations with 8-bit data per location. Here, 256/64 = 4, hence 4 numbers of 64 ×8 memory is required to implement 256 × 8 memory.

Since, 64 = 2^6, 6-bit address is required to address 64 memory locations. In the 8-bit address of 256 × 8 memory, the lower 6-bits are common to all the four 64 × 8 memory. The upper two bits are decoded using a 2-to-4 decoder to generate four chip enable signals which are used as enables of 64 × 8 memory so that only one 64 × 8 memory is selected at any time. The logical diagram of 256 × 8 RAM using 64 × 8 RAM is shown in Fig. 1.

Example 12.4

(AKUB, Dec'20, 6 Marks)

Design a 16 × 8 CAM using 8 × 2 CAM chips.

Solution

The 8 × 2 CAM has a capacity of 8 words each of size 2-bits. The 16 × 8 CAM has a capacity of 16 words each of size 8-bits. In order to increase the word bit size to 8-bits, 4 numbers of 8 × 2 CAM should be connected in parallel to achieve an 8 × 8 CAM. The word capacity can be increased to 16 words by using two numbers of 8 × 8 CAM. The organization of 16 × 8 CAM using 8 × 2 CAM is shown in Fig. 1 (d).

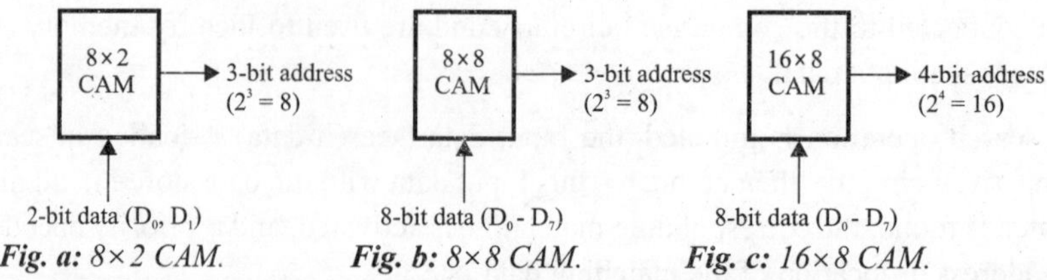

Fig. a: 8×2 CAM. *Fig. b: 8×8 CAM.* *Fig. c: 16×8 CAM.*

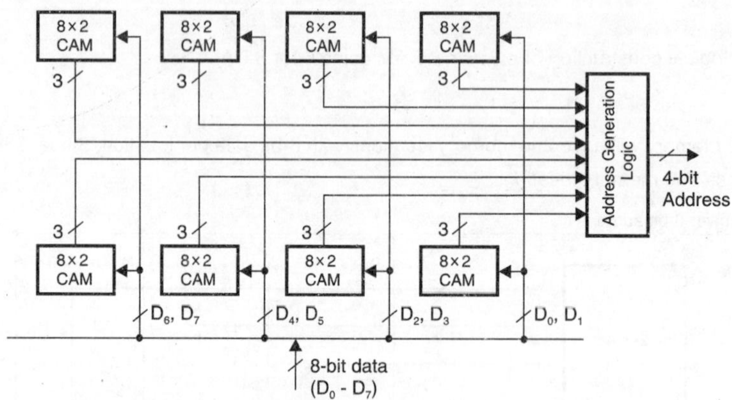

Fig. d: 16×8 CAM using 8 × 2 CAM.

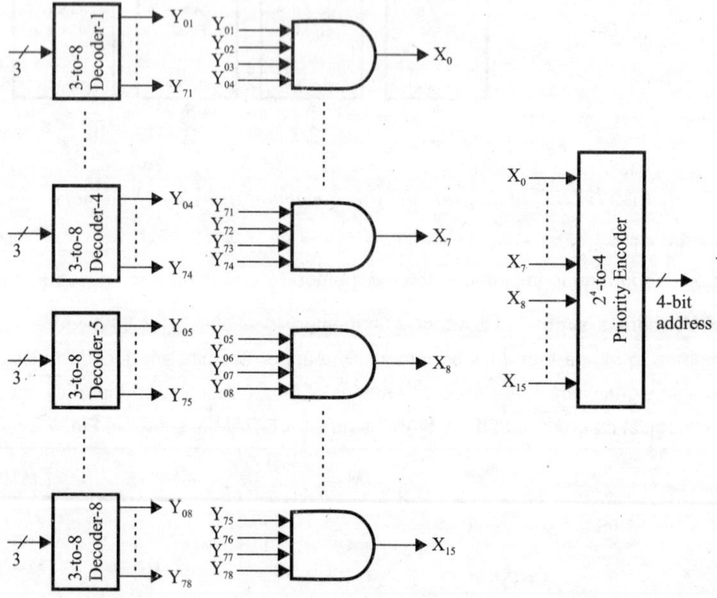

Fig. e: Address generation logic for 16×8 CAM using 8×2 CAM.

Fig. 1: Organization of 16×8 CAM using 8×2 CAM.

The 8-bit data is split into group of 2-bits and each group is connected to one of the 8 × 2 CAM. Each 8 × 2 CAM will provide a 3-bit address when a 2-bit match data is found in that CAM. The 3-bit addresses of all the 8 × 2 CAM has to be decoded to generate a 4-bit address to identify the row in which search data is available.

Address generation logic is developed as shown in Fig. 1 (e) which employs 8 numbers of 3-to-8 decoder to decode the 3-bit address of the 8 numbers of 8 × 2 CAM. The outputs corresponding to same row of CAMs are logically ANDed to combine the decoded output and then encoded using 16-to-4 encoder to generate a unique 4-bit address of the identified search data.

12.4 Programmable Logic Devices

The digital ICs (Integrated Circuits) can be broadly classified into fixed function ICs and programmable ICs. The fixed function ICs are manufactured to perform a fixed logical function. Examples of fixed function ICs are flip-flops, decoders, multiplexers, counters, etc. The programmable ICs are designed to be programmed by the user to implement a variety of logical functions.

Some of the programmable logical devices are,

- PROM (Programmable ROM)
- PLA (Programmable Logic Array)
- PAL (Programmable Array Logic)
- SPLD (Sequential Programmable Logic Device)
- CPLD (Complex Programmable Logic Device)
- FPGA (Field Programmable Gate Array)

The PROM, PAL and PLA are one-time programmable and consists of programmable logic gate arrays and can be used to implement only combinational logic functions. The SPLD, CPLD and FPGA devices are reprogrammable and consists of programmable logic gate arrays and flip-flops and so can be used to implement both combinational and sequential logic functions.

The advantages of using PLDs to implement a logical function are,

- PLDs require smaller chip and board area to implement a logic circuit.
- Logical functions can be easily altered or modified without rewiring and changing the board.
- Implementation is much faster when compared to fixed function ICs.

12.5 PROM as PLD

The inputs-outputs of a combinational circuit are fixed and can be listed in a truth table. Any fixed truth table can be implemented in a PROM device discussed in Section 12.2.2.

The inputs of a combinational circuit can be used as address of PROM. The output of combinational circuit can be permanently stored in fusible links of PROM in which a 0 is blown fuse and 1 is intact fuse.

An example of 16 × 8-bit PROM is shown in Fig. 12.31 in which 4 address lines ($16 = 2^4$) and 8 data lines are provided. ROM has a internal decoder (4-to-16 decoder) to decode the 4-bit address to 16 rows. Each row has 8 storage locations to store 8-bit data and 8 column lines are provided to access the 8-bit data. The columns are terminated on OR gates which act as output data buffers. Hence the memory will have 16 × 8 array. Each crossing point is a fusible link and 0's and 1's are permanently stored at these point.

A permanent 0 can be made by disconnecting the crossing point permanently which is called fuse opened or blown. A permanent 1 can be made by leaving the fussible link at the crossing point without open or kept intact. The 1 and 0 pattern stored in PROM will be specified by a truth table as shown in Table 12.6.

A typical truth table for programming the PROM is shown in Table 12.6 and the corresponding PROM is shown in Fig. 12.31. In Table 12.6, inputs are address bits (4-bit address), outputs are data bits (8-bit data) stored in PROM. A stored 1 is shown by × in Fig. 12.31. A stored 0 is shown by a simple crossing point of row and column.

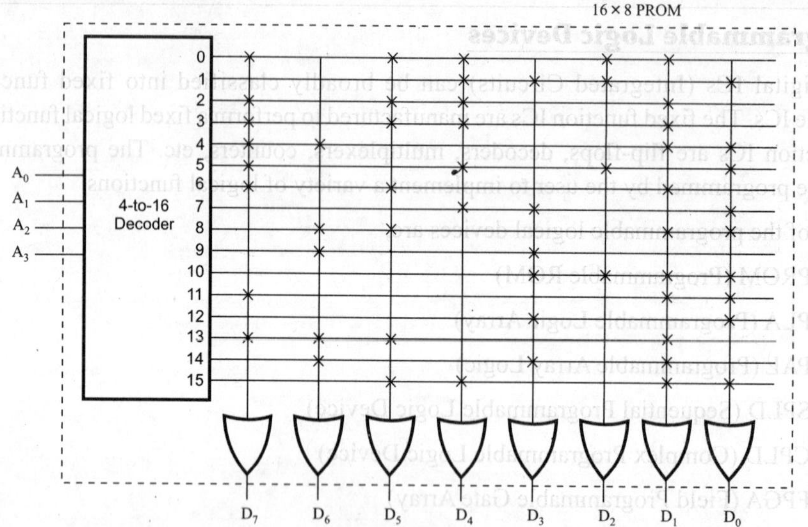

Fig. 12.31: Programming the PROM according to Table 12.6.

Table 12.6: Truth Table Implemented in PROM shown in Fig. 12.31

Inputs				**Outputs**							
A_3	A_2	A_1	A_0	D_7	D_6	D_5	D_4	D_3	D_2	D_1	D_0
0	0	0	0	1	0	1	1	0	1	1	0
0	0	0	1	0	0	0	1	1	1	0	1
0	0	1	0	1	0	1	1	0	0	1	0
0	0	1	1	1	0	1	1	0	0	1	0
0	1	0	0	1	1	0	0	0	0	1	1
0	1	0	1	1	0	0	1	0	1	0	1
0	1	1	0	1	0	1	0	0	0	1	0
0	1	1	1	0	0	0	1	1	0	0	1
1	0	0	0	0	1	1	0	0	0	1	1
1	0	0	1	0	1	0	0	1	0	0	0
1	0	1	0	0	0	0	0	1	1	1	1
1	0	1	1	1	0	0	0	0	0	1	1
1	1	0	0	0	0	0	0	1	0	0	1
1	1	0	1	1	1	1	0	0	0	1	0
1	1	1	0	0	1	0	0	1	0	1	0
1	1	1	1	0	0	1	1	0	0	1	1

Example 12.5

(AU, Apr/May'19, 6 Marks)

a) Derive the PROM programming table for the combinational circuit that squares a 4-bit number.

b) Minimize the number of product terms in the Boolean equations obtained as sum of minterms.

Solution

a)

Table 1: Truth Table

Inputs (4-Bit Binary Number)				Minterm	Decimal Value	Squared Decimal Value	Outputs (Binary Value of 4-bit Squared Number)							
A	B	C	D				B_7	B_6	B_5	B_4	B_3	B_2	B_1	B_0
0	0	0	0	m_0	0	0	0	0	0	0	0	0	0	0
0	0	0	1	m_1	1	1	0	0	0	0	0	0	0	1
0	0	1	0	m_2	2	4	0	0	0	0	0	1	0	0
0	0	1	1	m_3	3	9	0	0	0	0	1	0	0	1
0	1	0	0	m_4	4	16	0	0	0	1	0	0	0	0
0	1	0	1	m_5	5	25	0	0	0	1	1	0	0	1
0	1	1	0	m_6	6	36	0	0	1	0	0	1	0	0
0	1	1	1	m_7	7	49	0	0	1	1	0	0	0	1
1	0	0	0	m_8	8	64	0	1	0	0	0	0	0	0
1	0	0	1	m_9	9	81	0	1	0	1	0	0	0	1
1	0	1	0	m_{10}	10	100	0	1	1	0	0	1	0	0
1	0	1	1	m_{11}	11	121	0	1	1	1	1	0	0	1
1	1	0	0	m_{12}	12	144	1	0	0	1	0	0	0	0
1	1	0	1	m_{13}	13	169	1	0	1	0	1	0	0	1
1	1	1	0	m_{14}	14	196	1	1	0	0	0	1	0	0
1	1	1	1	m_{15}	15	225	1	1	1	0	0	0	0	1

The truth table of the combinational circuit that squares a 4-bit binary number is shown in Table 1. Using the truth table the PROM is programmed as shown in Fig. 1. The x indicates a 1 at the crossing points.

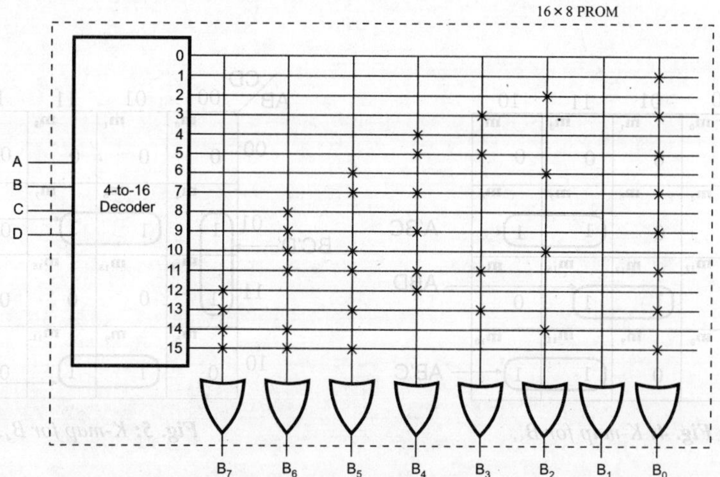

Fig. 1: *Programming the PROM according to Table 1.*

A16 × 8 PROM grid is used to implement the truth table. The PROM has an internal 4-to-16 decoder which is used to select one of the 16 lines of the PROM grid. The decoder will accept the 4-bit binary address and select the corresponding output line to generate the output.

b) To Minimize Number of Product Terms in Boolean Equations

The Boolean equation of an output of the combinational circuit can be written directly from the truth table by taking the sum of minterms for which the output is 1.

The Boolean equations are,

$B_0 = \Sigma(m_1, m_3, m_5, m_7, m_9, m_{11}, m_{13}, m_{15})$

$B_1 = 0$

$B_2 = \Sigma(m_2, m_6, m_{10}, m_{14})$

$B_3 = \Sigma(m_3, m_5, m_{11}, m_{13})$

$B_4 = \Sigma(m_4, m_5, m_7, m_9, m_{11}, m_{12})$

$B_5 = \Sigma(m_6, m_7, m_{10}, m_{11}, m_{13}, m_{15})$

$B_6 = \Sigma(m_8, m_9, m_{10}, m_{11}, m_{14}, m_{15})$

$B_7 = \Sigma(m_{12}, m_{13}, m_{14}, m_{15})$

The Boolean can be converted to minimum literal form by constructing K-map. The K-map for the outputs are drawn as shown in Figs. 2 to 7. Here $B_1 = 0$ and $B_0 = D$ and so K-maps are not drawn for there two outputs.

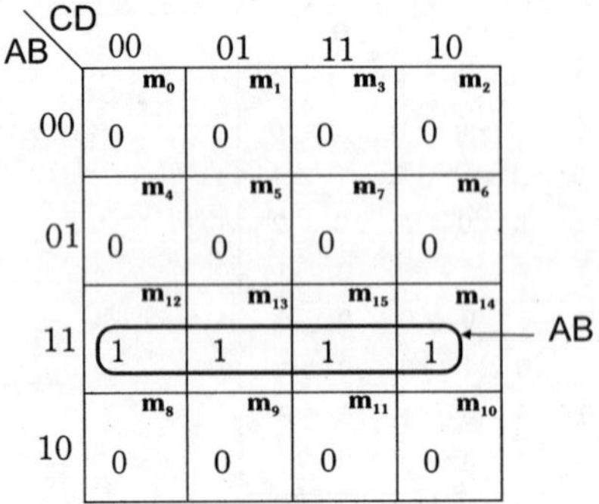

Fig. 2: K-map for B_7.

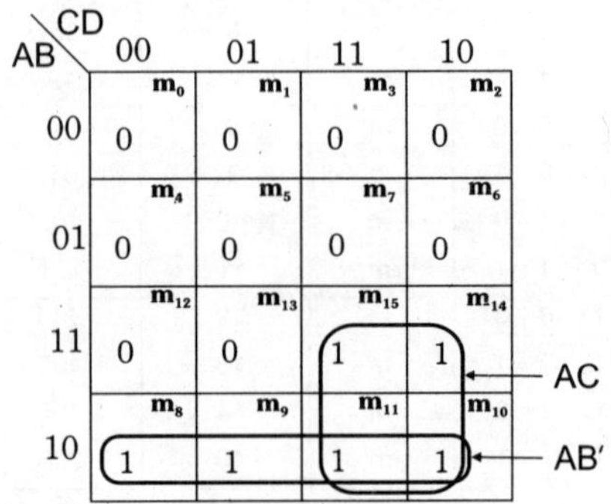

Fig. 3: K-map for B_6.

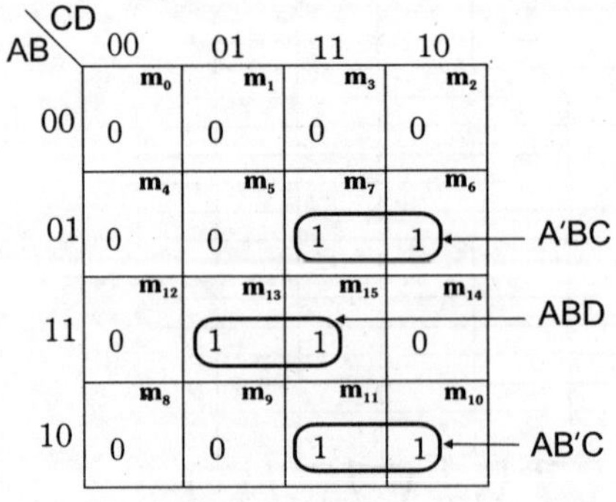

Fig. 4: K-map for B_5.

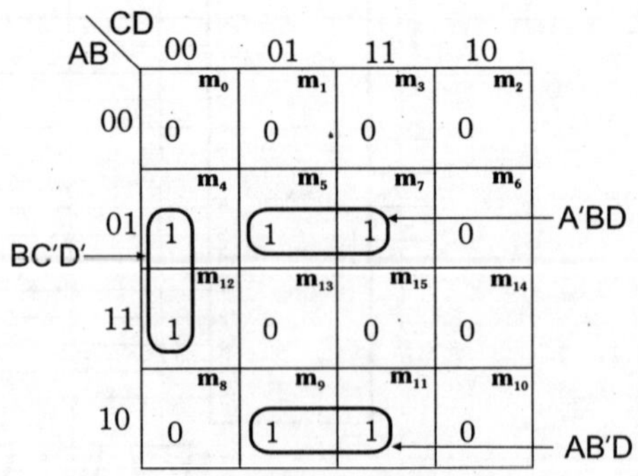

Fig. 5: K-map for B_4.

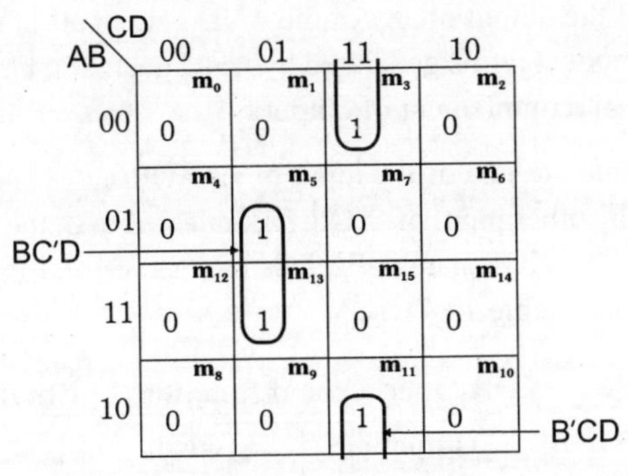

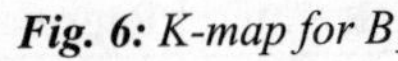

Fig. 6: K-map for B_3.

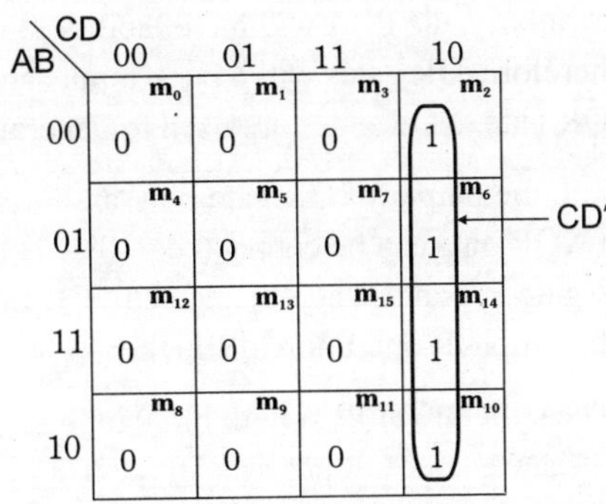

Fig. 7: K-map for B_2.

The following simplified Boolean functions of outputs are obtained from K-maps shown in Figs. 2 to 7.

$B_7 = AB$

$B_6 = AB' + AC$

$B_5 = A'BC + AB'C + ABD$

$B_4 = BC'D' + A'BD + AB'D$

$B_3 = BC'D + B'CD$

$B_2 = CD'$

The output column B_1 is all zero values.

$\therefore B_1 = 0$

The input column D and output column B_0 is same.

$\therefore B_0 = D$

Conclusion: It can be observed that the number of literals in the simplified Boolean equations are lesser than the Boolean equations obtained from truth table.

12.6 PLA (Programmable Logic Array)

(RGPV, June'20, 4 Marks) (RGPV, June'22, 4 Marks)

The PLA is a one-time programmable device to implement a Boolean function in sum of products form. The PLA will have an array of AND gates followed by an array of OR gates. In PLA, the AND gates are provided to implement product terms and OR gates are provided to implement sum of the product terms.

In order to minimize the gate requirements it is common practice to simplify the Boolean function to minimum literal form before implementation.

A typical PLA device will have,

- n number of inputs
- k number of AND gates
- m number of OR gates
- m number of XOR gates

Every input is provided with a buffer and an inverter. PLA will have facility to connect every input/inverted input to array of the AND gate. Therefore, AND gates will have 2n inputs.

Similarly, the PLA will have facility to connect the output of any of the AND gate to all of the OR gate. Therefore, OR gates will have k inputs and the product terms generated by AND gates can be routed to any OR gate. XOR gates are used to generate normal /complement of output.

Let, the output of OR gate be F and it is connected to one of the input of an XOR gate. The other input of XOR gate can be connected to 0 or 1. When the other input of XOR is connected to 0, the output of XOR gate is F and when connected to 1, the output of XOR gate is F′. This can be verified from the truth table of XOR operation of function, F with 0 and 1 (Table 12.7).

The structure of PLA with three inputs, five AND gates, three OR gates and three XOR gates is shown in Fig. 12.32. In PLA device the connections to AND gates and OR gates are programmable.

The structure of PLA in Fig. 12.32 can be simplified as shown in Fig. 12.33 by combining the input buffers to a single buffer with two outputs and multiple input lines to gates by a single input line.

Table 12.7: XOR Operation of function, F with 0 and 1

Input Fixed	F	Output of XOR Gate	Symbolic Representation
0	0	$0 \oplus 0 = 0 = F$	
0	1	$0 \oplus 1 = 1 = F$	
1	0	$1 \oplus 0 = 1 = F'$	
1	1	$1 \oplus 1 = 0 = F'$	

A bubble is placed in one of the output of input buffer to indicate that this output will be inverted value of input. A slash with a number in the input line to gate indicates the number of input lines in that gate.

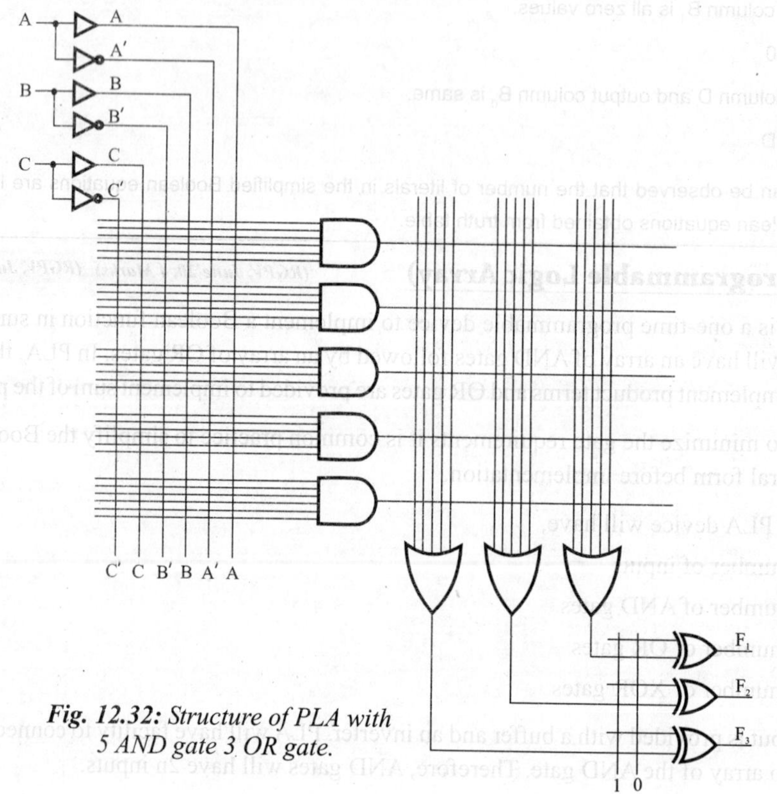

Fig. 12.32: Structure of PLA with 5 AND gate 3 OR gate.

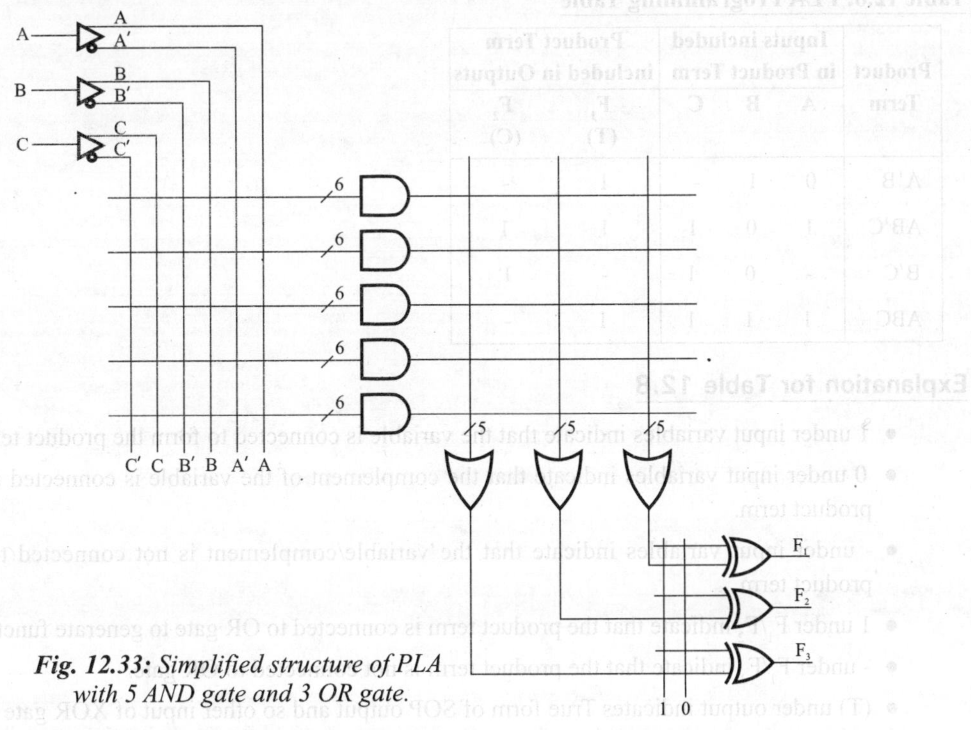

Fig. 12.33: Simplified structure of PLA with 5 AND gate and 3 OR gate.

12.6.1 Implementation of Combinational Circuit using PLA

Consider the following two Boolean equations of a combinational circuit.

$$F_1 = A'B + AB'C + ABC$$

$$F_2 = (AB'C + B'C)' \quad \Rightarrow \quad F_2' = AB'C + B'C$$

The Boolean equations has three variables and four product terms. When three input variables are complemented we get six inputs and so we require four numbers of six inputs AND gates to implement the four product terms. Here, there are two outputs, F_1 and F_2. Hence we require two OR gates to sum up the product terms to get two outputs.

The structure of PLA with four AND gates two OR gates and two XOR gates is shown in Fig. 12.34. The connected inputs to AND gate are marked by x. The connected output of AND gate to input of OR gate are marked by x. The connections to 0 or 1 of one of the inputs of XOR gate are also shown by x. Here, F_1 in expressed in SOP form and F_2 in complement of SOP form. Therefore, the output of OR gate that generate SOP form of F_1 is XORed with 0 to get F_1. The output of OR gate that generates SOP form of F_2 is XORed with 1 to get F_2.

The simplified structure of the combinational circuit implemented in PLA is shown in Fig. 12.35.

A PLA programming table can be formed as shown in Table 12.8 to provide informations regarding the implementation of combinational circuit. Here, T refers to True output of a function and C refers to complement output of a function.

Table 12.8: PLA Programming Table

Product Term	Inputs included in Product Term			Product Term included in Outputs	
	A	B	C	F_1 (T)	F_2 (C)
A'B	0	1	-	1	-
AB'C	1	0	1	1	1
B'C	-	0	1	-	1
ABC	1	1	1	1	-

Explanation for Table 12.8

- 1 under input variables indicate that the variable is connected to form the product term
- 0 under input variables indicate that the complement of the variable is connected to form the product term.
- - under input variables indicate that the variable/complement is not connected to form the product term.
- 1 under F_1/F_2 indicate that the product term is connected to OR gate to generate function output.
- - under F_1/F_2 indicate that the product term is not connected to OR gate.
- (T) under output indicates True form of SOP output and so other input of XOR gate is 0.
- (C) under output indicates complement form of SOP output and so other input of XOR gate is 1.

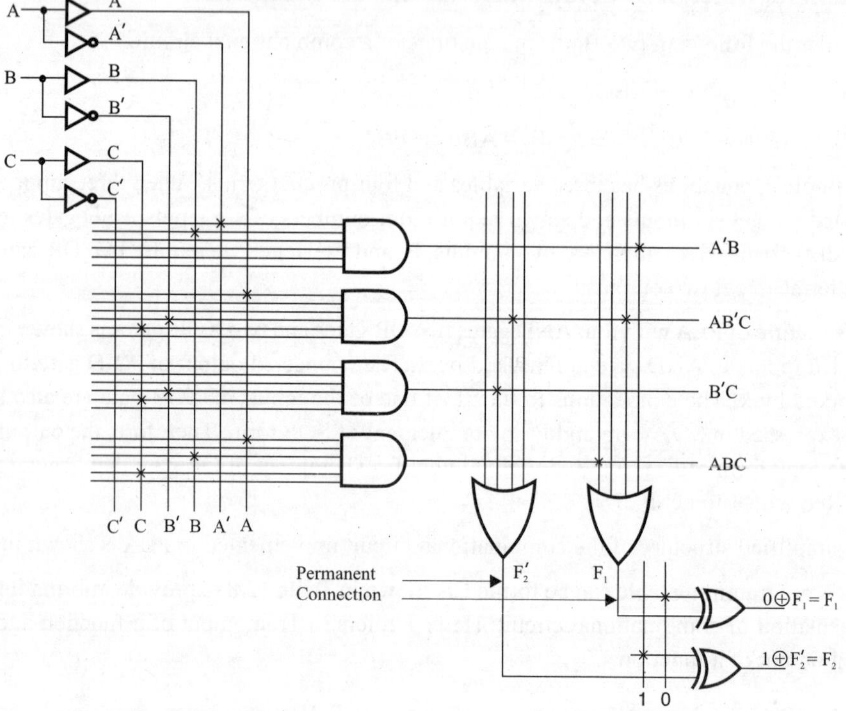

Fig. 12.34: Structure of a combinational circuit implemented in PLA.

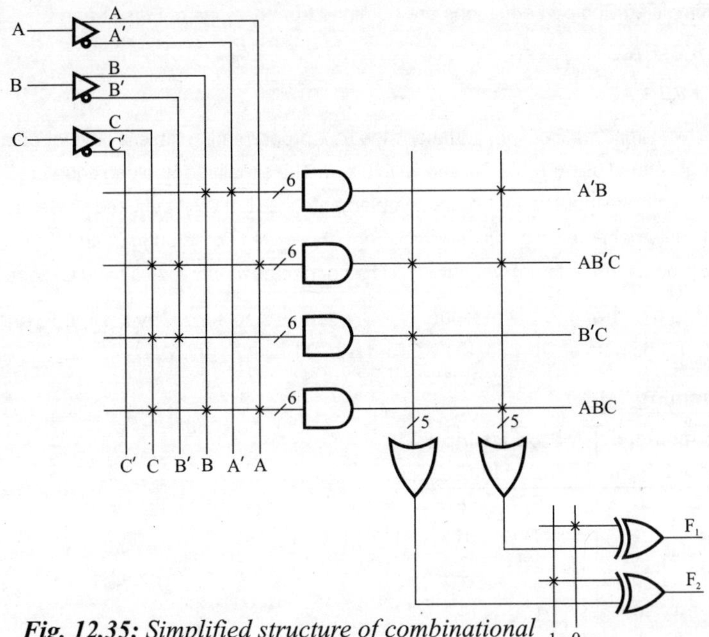

***Fig. 12.35:** Simplified structure of combinational circuit implemented in PLA.*

Example 12.6

(IKGPTU, May'20, 10 Marks)

Implement the given Boolean functions using PLA.

$F_1(A, B, C) = \Sigma m(0, 1, 2, 4)$ and $F_2(A, B, C) = \Sigma m(0, 5, 6, 7)$

Solution

The truth table of Boolean functions is formed as shown in Table 1. Here, the function output is 1 for the minterms included in that function. The simplified Boolean equations of the functions are obtained by constructing K-maps shown in Figs. 1 and 2.

Table 1: Truth Table

Inputs			Minterm	Outputs	
A	**B**	**C**		**F_1**	**F_2**
0	0	0	m_0	1	1
0	0	1	m_1	1	0
0	1	0	m_2	1	0
0	1	1	m_3	0	0
1	0	0	m_4	1	0
1	0	1	m_5	0	1
1	1	0	m_6	0	1
1	1	1	m_7	0	1

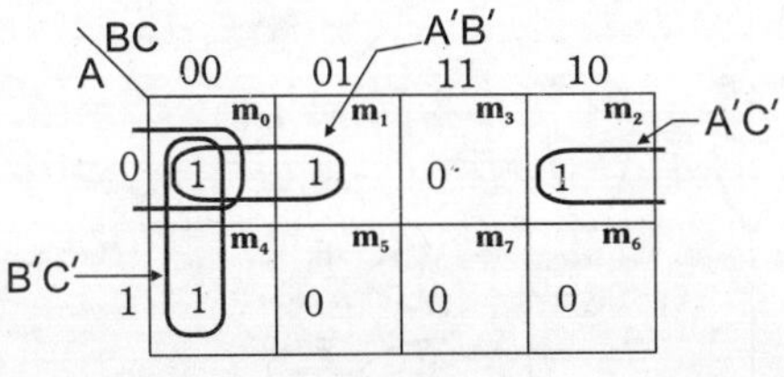

***Fig. 1:** K-map for F_1.*

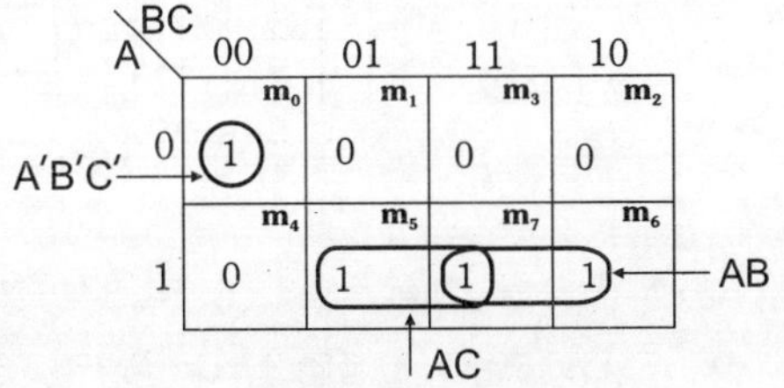

***Fig. 2:** K-map for F_2.*

The following simplified Boolean equations are obtained from K-maps of Figs. 1 and 2.

$F_1 = A'B' + B'C' + A'C'$

$F_2 = A'B'C' + AB + AC$

Using the above two simplified Boolean equations the PLA programming table is developed as shown in Table 2. The structure of PLA programmed using Table 2 is shown in Fig. 3. The simplified Boolean equations has three variables and six product terms. When three input variables are complemented we get six inputs and so we require six numbers of six inputs AND gates to implement the six product terms. Here, there are two outputs, F_1 and F_2. Hence we require two OR gates to sum up the product terms to get two outputs. The connected inputs to AND and OR gates are marked by x.

The connections to 0 or 1 of one of the inputs of XOR gate is also shown by x. Here F_1 and F_2 are expressed in SOP form. Therefore, the outputs of OR gate are XORed with 0 to get F_1 and F_2.

Table 2: PLA Programming Table

Product Term	Inputs included in Product Term			Product Term included in Outputs	
	A	B	C	F_1 (T)	F_2 (T)
A'B'	0	0	-	1	-
B'C'	-	0	0	1	-
A'C'	0	-	0	1	-
A'B'C'	0	0	0	-	1
AB	1	1	-	-	1
AC	1	-	1	-	1

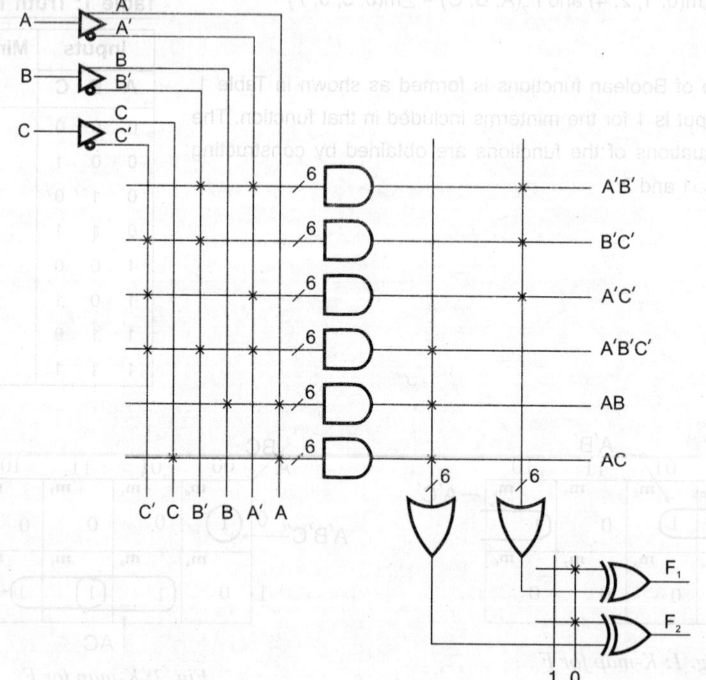

Fig. 3: Simplified structure of combinational circuit implemented in PLA.

Example 12.7

(JNTUH, Nov/Dec'20, 8 Marks)

Implement the given Boolean functions using PLA.

$F_1(A, B, C) = \Sigma m(0, 1, 3, 5)$ and $F_2(A, B, C) = \Sigma m(1, 2, 4, 7)$

Solution

The truth table of Boolean functions is formed as shown in Table 1. Here, the function output is 1 for the minterms included in that function. The simplified Boolean equations of the functions are obtained by constructing K-maps shown in Figs. 1 and 2.

Table 1: Truth Table

Inputs			Minterm	Outputs	
A	B	C		F_1	F_2
0	0	0	m_0	1	0
0	0	1	m_1	1	1
0	1	0	m_2	0	1
0	1	1	m_3	1	0
1	0	0	m_4	0	1
1	0	1	m_5	1	0
1	1	0	m_6	0	0
1	1	1	m_7	0	1

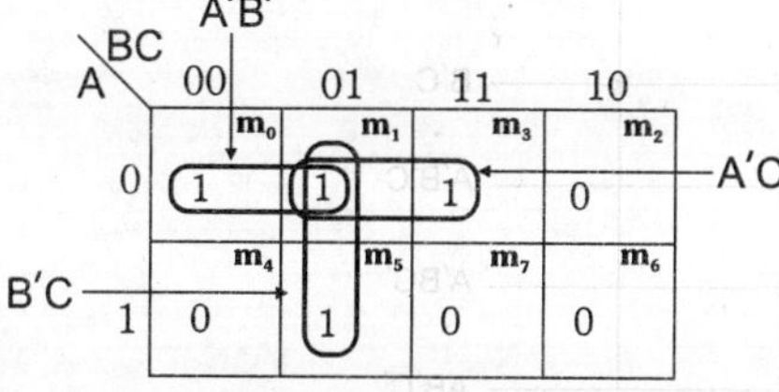

Fig. 1: K-map for F_1.

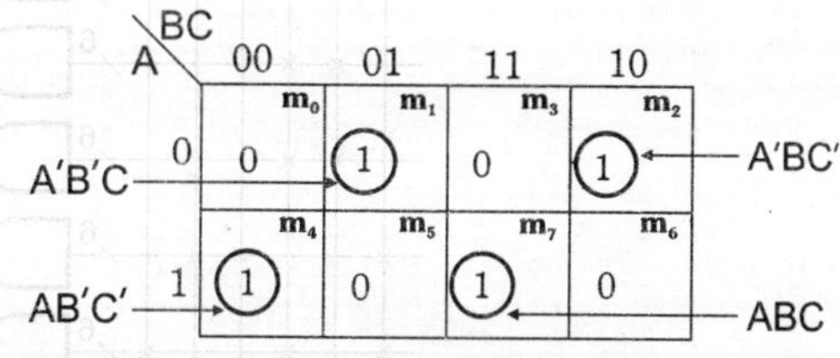

Fig. 2: K-map for F_2.

The following simplified Boolean equations are obtained from K-maps of Figs. 1 and 2.

$F_1 = A'B' + A'C + B'C$

$F_2 = A'B'C + A'BC' + AB'C' + ABC$

Using the above two simplified Boolean equations the PLA programming table is developed as shown in Table 2. The structure of PLA programmed using Table 2 is shown in Fig. 3. The simplified Boolean equations has three variables and seven product terms. When three input variables are complemented we get six inputs and so we require seven numbers of six inputs AND gates to implement the seven product terms. Here, there are two outputs, F_1 and F_2. Hence we require two OR gates to sum up the product terms to get two outputs. The connected inputs to AND and OR gates are marked by x.

The connections to 0 or 1 of one of the inputs of XOR gate is also shown by x. Here F_1 and F_2 are expressed in SOP form. Therefore, the outputs of OR gate are XORed with 0 to get F_1 and F_2.

Table 2: PLA Programming Table

Product Term	Inputs included in Product Term			Product Term included in Outputs	
	A	B	C	F_1 (T)	F_2 (T)
A'B'	0	0	-	1	-
A'C	0	-	1	1	-
B'C	-	0	1	1	-
A'B'C	0	0	1	-	1
A'BC'	0	1	0	-	1
AB'C'	1	0	0	-	1
ABC	1	1	1	-	1

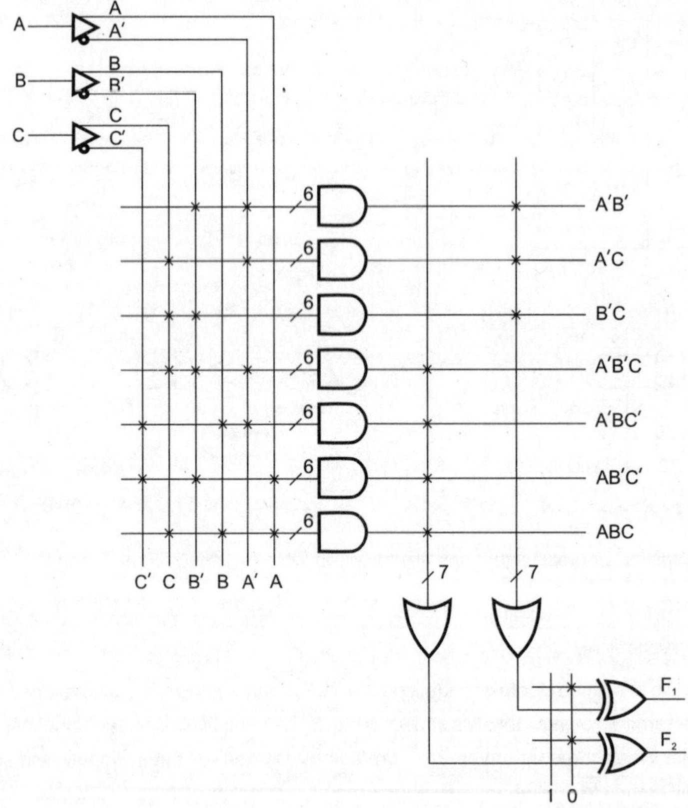

Fig. 3: Simplified structure of combinational circuit implemented in PLA.

Example 12.8

Implement the following two Boolean functions with a PLA.

Note: Alternatively, K-map can be used to simplify Boolean function.

$F_1 = AB'C' + ABC' + ABC$ and $F_2 = (A'BC + AB'C + ABC)'$

Solution

The given Boolean equations are simplified as shown below:

Adding an existing minterm will not alter the function value.

$F_1 = AB'C' + ABC' + ABC$

$= AB'C' + ABC' + ABC' + ABC = (ABC + ABC') + (ABC' + AB'C')$

$x + x' = 1$

$= AB(C + C') + AC'(B + B') = AB + AC'$

$F_2 = (A'BC + AB'C + ABC)'$

$= (A'BC + AB'C + ABC + ABC)' = ((ABC + A'BC) + (ABC + AB'C))'$

$= (BC(A + A') + AC(B + B'))' = (BC + AC)'$

Now the simplified Boolean equations are,

$F_1 = AB + AC'$

$F_2 = (BC + AC)' \Rightarrow F_2' = BC + AC$

Using the above two simplified Boolean equations the PLA programming table is developed as shown in Table 1. The structure of programmed PLA using Table 1 is shown in Fig. 1.

The simplified Boolean equations has three variables and four product terms. When three input variables are complemented we get six inputs and so we require four numbers of six inputs AND gates to implement the four product terms. Here, there are two outputs, F_1 and F_2.

Hence we require two OR gates to sum up the product terms to get two outputs. The connected inputs to AND and OR gates are marked by ×. The connections to 0 or 1 of one of the inputs of XOR gates are also shown by ×. Here, F_1 is expressed in SOP form and F_2 in complement of SOP form. Therefore, the output of OR gate that generates SOP form of F_1 is XORed with 0 to get F_1. The output of OR gate that generates SOP form of F_2 is XORed with 1 to get F_2.

Table 1: PLA Programming Table

Product Term	Inputs included in Product Term			Product Term included in Outputs	
	A	B	C	F_1 (T)	F_2 (C)
AB	1	1	-	1	-
AC'	1	-	0	1	-
BC	-	1	1	-	1
AC	1	-	1	-	1

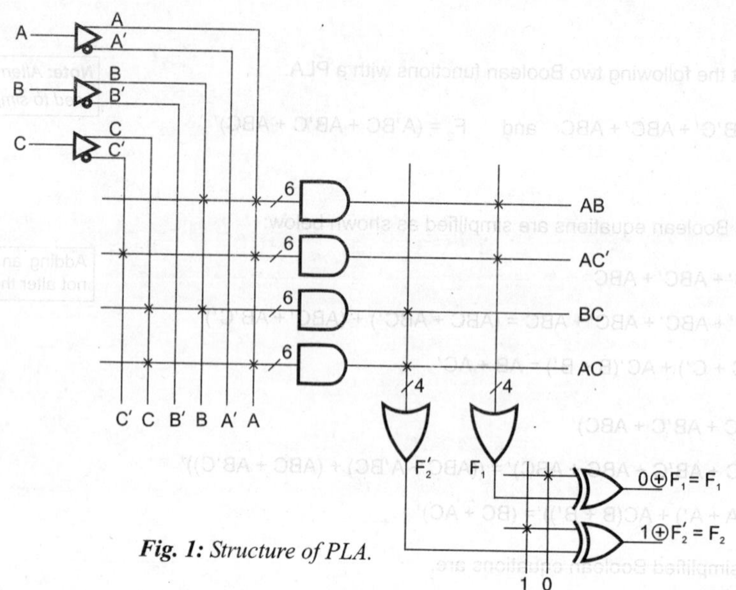

Fig. 1: Structure of PLA.

Example 12.9

(BATU, Jun'19, 6 Marks)

Implement the full adder circuit using PLA by deriving the PLA programming table.

Solution

The truth table of full adder is shown in Table 1. The simplified Boolean equations can be obtained by constructing K-maps for s and c_o as shown in Figs. 1 and 2.

Table 1: Truth Table of Full Adder

Inputs			Minterm	Outputs	
a	b	c_i		s	c_o
0	0	0	m_0	0	0
0	0	1	m_1	1	0
0	1	0	m_2	1	0
0	1	1	m_3	0	1
1	0	0	m_4	1	0
1	0	1	m_5	0	1
1	1	0	m_6	0	1
1	1	1	m_7	1	1

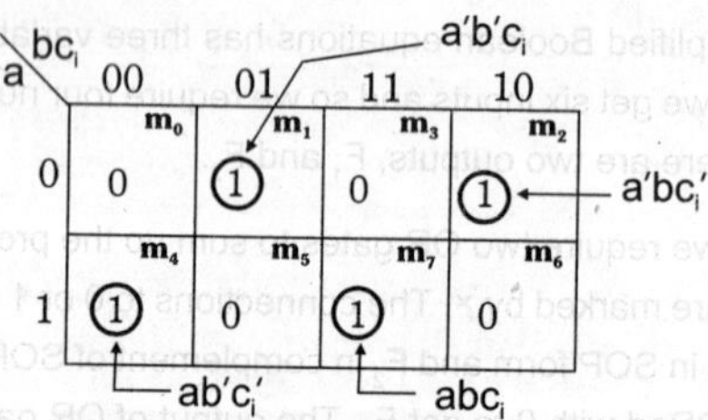

Fig. 1: K-map for s.

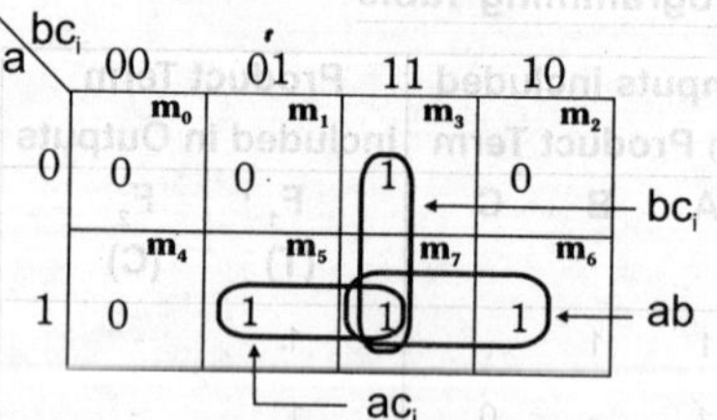

Fig. 2: K-map for c_o.

From the K-map the following Boolean equations are obtained for full adder.

$$s = a'b'c_i + a'bc_i' + ab'c_i' + abc_i$$

$$c_o = ab + ac_i + bc_i$$

Using the above Boolean equations the PLA programming table is developed as shown in Table 2. The structure of full adder implemented in PLA using Table 2 is shown in Fig. 3.

The simplified Boolean equations has three variables and seven product terms. When three input variables are complemented we get six inputs and so we require seven numbers of six input AND gates to implement the seven product terms. Here, there are two outputs, s and c_o. Hence we require two OR gates to sum up the product terms to get two outputs. The connected inputs to AND and OR gates are marked by ×.

The connections to 0 or 1 of one of the inputs of XOR gates are also shown by ×. Here, s and c_o are expressed in SOP form. Therefore, the outputs of OR gate are XORed with 0 to get s and c_o.

Table 2: PLA Programming Table

Product Term	Inputs included in Product Term			Product Term included in Outputs	
	a	b	c_i	s (T)	c_o (T)
$a'b'c_i$	0	0	1	1	-
$a'bc_i'$	0	1	0	1	-
$ab'c_i'$	1	0	0	1	-
abc_i	1	1	1	1	-
ab	1	1	-	-	1
ac_i	1	-	1	-	1
bc_i	-	1	1	-	1

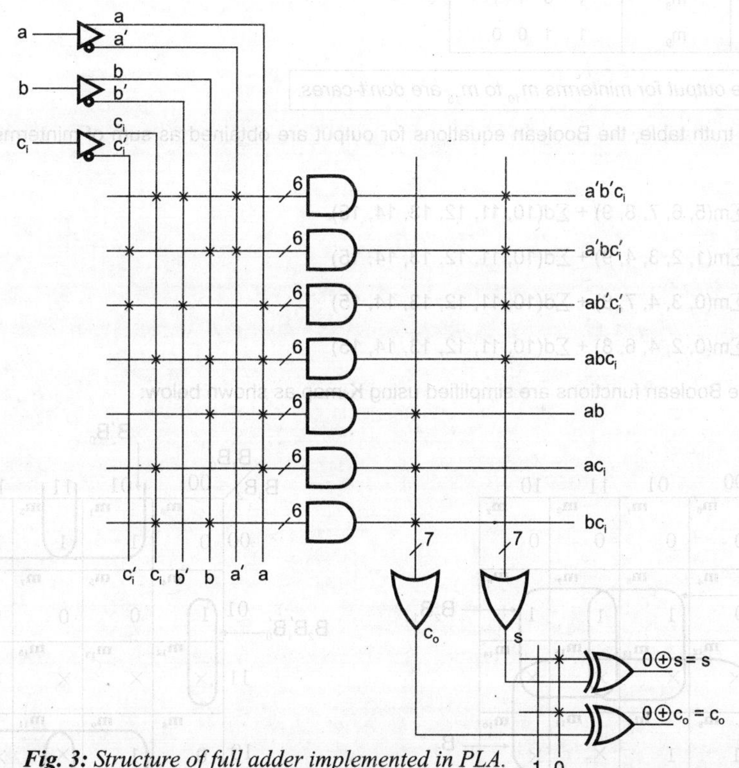

Fig. 3: Structure of full adder implemented in PLA.

Example 12.10

(AKTU, Dec'22, 10 Marks)

Implement a BCD to Excess-3 code conversion logic using PLA.

Solution

The truth table of 4-digit BCD to 4-digit excess-3 code converter is shown in Table 1.

Table 1: Truth Table of BCD to Excess-3 Code Converter

Inputs (BCD) $B_3\ B_2\ B_1\ B_0$	Minterm	Outputs (Excess-3 code) $F_3\ F_2\ F_1\ F_0$
0 0 0 0	m_0	0 0 1 1
0 0 0 1	m_1	0 1 0 0
0 0 1 0	m_2	0 1 0 1
0 0 1 1	m_3	0 1 1 0
0 1 0 0	m_4	0 1 1 1
0 1 0 1	m_5	1 0 0 0
0 1 1 0	m_6	1 0 0 1
0 1 1 1	m_7	1 0 1 0
1 0 0 0	m_8	1 0 1 1
1 0 0 1	m_9	1 1 0 0

Note: *The output for minterms m_{10} to m_{15} are don't-cares.*

From the truth table, the Boolean equations for output are obtained as sum of minterms and don't-cares as shown below:

$$F_3 = \Sigma m(5, 6, 7, 8, 9) + \Sigma d(10, 11, 12, 13, 14, 15)$$

$$F_2 = \Sigma m(1, 2, 3, 4, 9) + \Sigma d(10, 11, 12, 13, 14, 15)$$

$$F_1 = \Sigma m(0, 3, 4, 7, 8) + \Sigma d(10, 11, 12, 13, 14, 15)$$

$$F_0 = \Sigma m(0, 2, 4, 6, 8) + \Sigma d(10, 11, 12, 13, 14, 15)$$

The above Boolean functions are simplified using K-map as shown below:

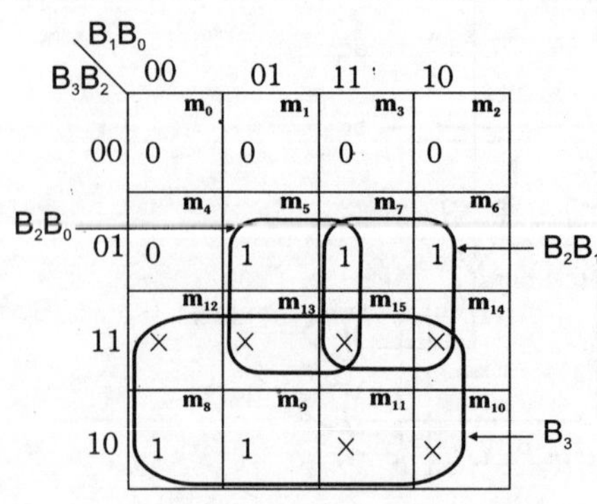

Fig. 1: K-map for F_3.

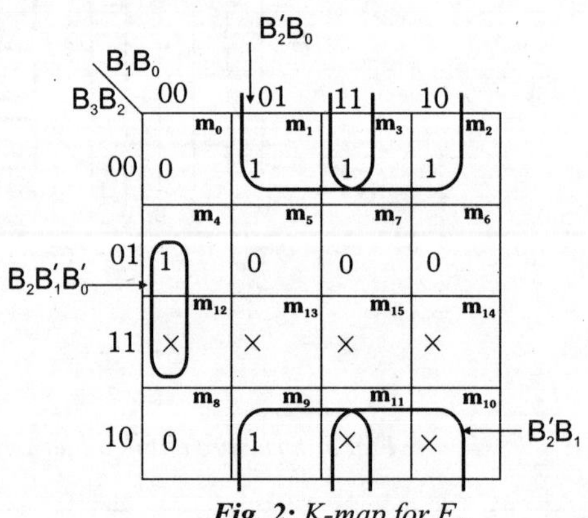

Fig. 2: K-map for F_2.

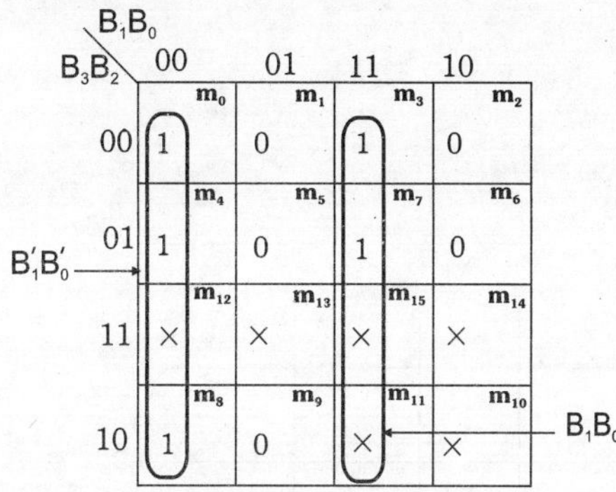

Fig. 3: K-map for F_1.

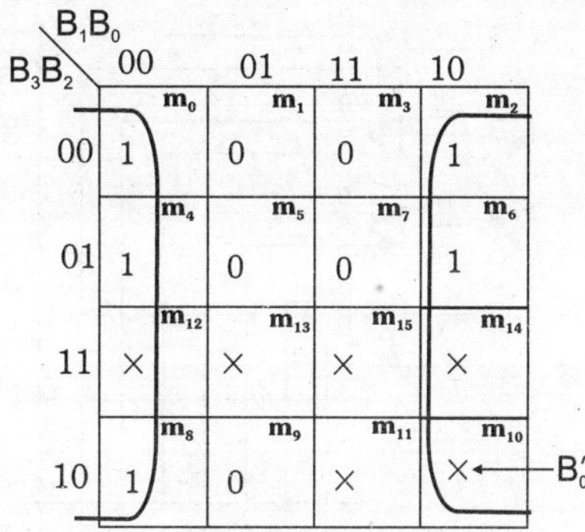

Fig. 4: K-map for F_0.

The following simplified Boolean equations are obtained from K-maps of Figs. 1 to 4.

$$F_3 = B_3 + B_2B_0 + B_2B_1$$

$$F_2 = B_2'B_0 + B_2B_1'B_0' + B_2'B_1$$

$$F_1 = B_1'B_0' + B_1B_0$$

$$F_0 = B_0'$$

Using the above two simplified Boolean equations the PLA programming table is developed as shown in Table 2. The structure of PLA programmed using Table 2 is shown in Fig. 5.

The simplified Boolean equations has four variables and nine product terms. When four input variables are complemented we get eight inputs and so we require nine numbers of eight inputs AND gates to implement the nine product terms.

Here, there are four outputs, F_3, F_2, F_1 and F_0. Hence we require four OR gates to sum up the product terms to get four outputs. The connected inputs to AND and OR gates are marked by x.

Table 2: PLA Programming Table

Product Term	Inputs included in Product Term				Product Term included in Outputs			
	B_3	B_2	B_1	B_0	F_3 (T)	F_2 (T)	F_1 (T)	F_0 (T)
B_2B_0	-	1	-	1	1	-	-	-
B_2B_1	-	1	1	-	1	-	-	-
B_3	1	-	-	-	1	-	-	-
$B_2'B_0$	-	0	-	1	-	1	-	-
$B_2B_1'B_0'$	-	1	0	0	-	1	-	-
$B_2'B_1$	-	0	1	-	-	1	-	-
$B_1'B_0'$	-	-	0	0	-	-	1	-
B_1B_0	-	-	1	1	-	-	1	-
B_0'	-	-	-	0	-	-	-	1

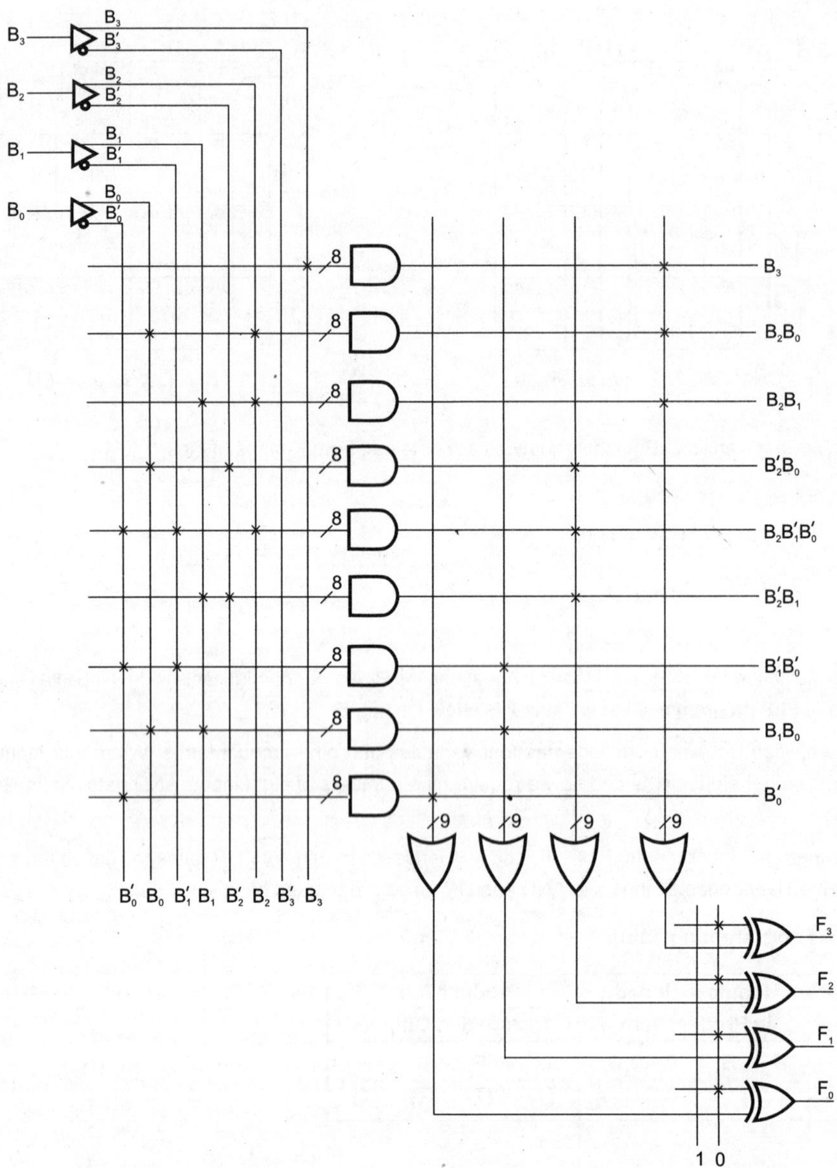

Fig. 5: Simplified structure of combinational circuit implemented in PLA.

12.7 PAL (Programmable Array Logic)

(RGPV, June'20, 4 Marks)

PAL is a programmable logic device to implement a Boolean function in sum of products form. PAL will have programmable AND gate array followed by a fixed OR gate. A typical PAL device will have m sets of AND-OR array with each array having k number of AND gates and one OR gate. In PAL device one set of AND-OR array combination is used to independently implement one function.

The only difference between PLA and PAL is that the OR gate array is programmable or not. In PLA device, the OR array is programmable but in PAL device the connections to OR array is fixed. In order to minimize the gate requirements, it is common practice to simplify the Boolean function to minimum literal form before implementation.

In PAL, the AND array is used to generate product terms. The product terms of AND array is routed to fixed OR gate in PAL, so that product terms of one function cannot be directly connected to another function. Whereas in PLA the product terms generated by AND gates can be routed to any OR gate. The PAL will have programmable facility to connect the output of an AND-OR array as input to AND gate of any other AND-OR array. This facility can be used to cascade the AND-OR array so that the cascade combination can be used to implement a larger number of product terms than permitted by a single AND-OR array.

A typical PAL device will have,

- n number of inputs
- k number of AND gates
- m number of OR gates
- m number of XOR gates

(AU, Apr/May'19, 2 Marks)

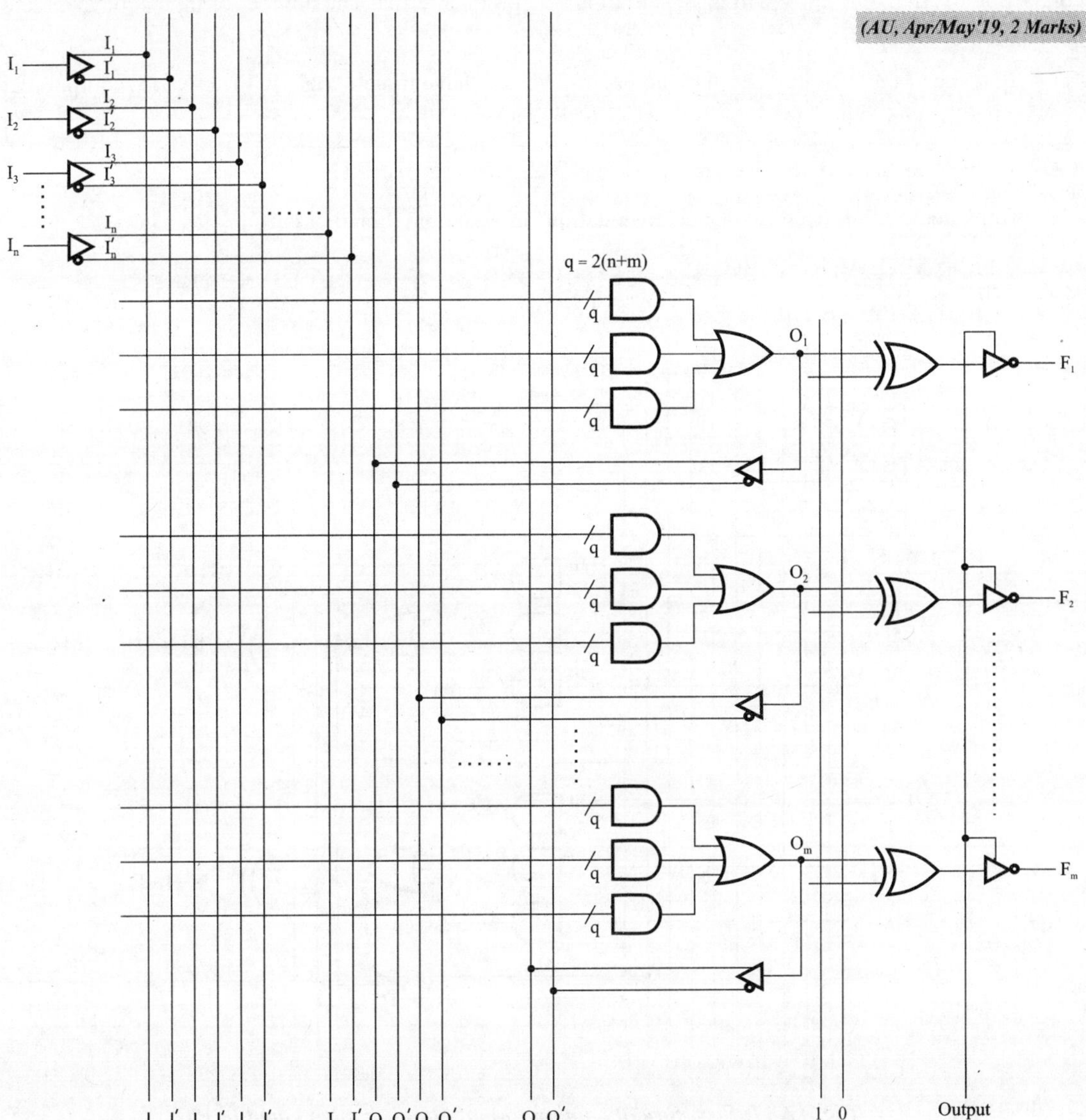

Fig. 12.36: *Architecture or structure of a PAL device.*

Every input is provided with a buffer and an inverter. PAL will have facility to connect every input/ inverted input to any of the AND gate. XOR gates are used to generate True/complement of output.

The output of each OR gate is connected to one of the inputs of an XOR gate. The other input of XOR gate can be connected to 0 or 1. When the other input of XOR is connected to 0 we get True output and when connected to 1 we get complemented output. This can be verified from Table 12.7.

The internal structure or architecture of PAL device with n inputs and m outputs is shown in Fig. 12.36. One set of AND-OR combination has 3 AND gates and so only three product terms can be generated. The three product terms are summed up using an OR gate. When we need a larger number of product terms, the output of an AND-OR array can be connected as input to any AND gate so that cascade connection of AND-OR array will have sum of all the product terms. The number of inputs to AND gate will be 2(n + m).

In some of the PAL devices, the outputs of PAL will have tristate logic with active low output and provided with complement facility using XOR gate similar to PLA.

12.7.1 Implementation of Combinational Circuit using PAL

Consider the following two Boolean equations of a combinational circuit.

$$F_1 = A'B + AB'C + ABC$$

$$F_2 = (AB'C + B'C)'$$

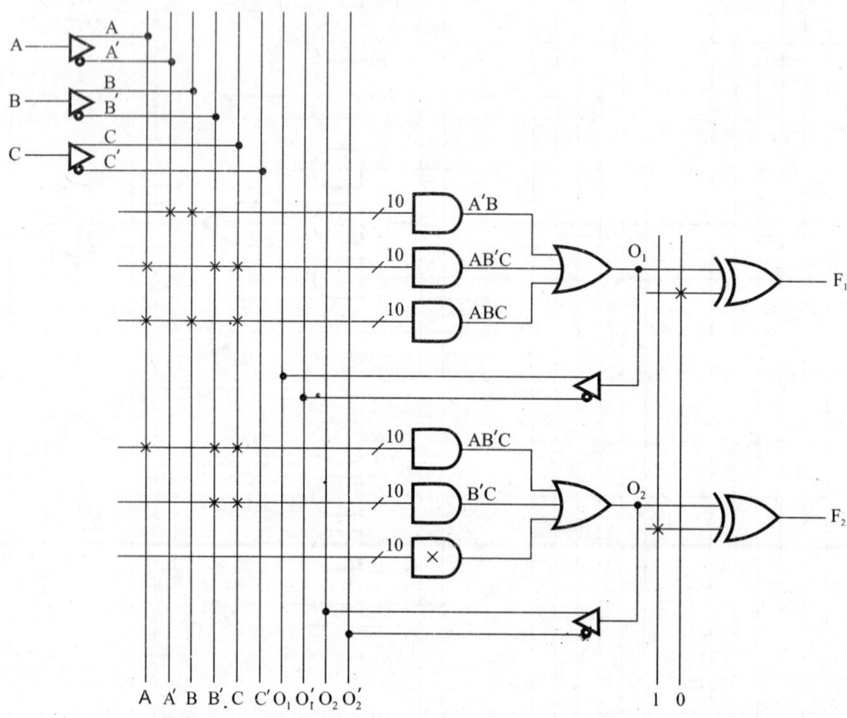

Fig. 12.37: Structure of combinational circuit implemented in PAL.

The Boolean equations has three variables and one equation has three product terms and the other has two product terms. In PAL implementation each equation has to be implemented individually in a separate AND-OR array. Hence we require two set of AND-OR array. Since the maximum number of product terms in Boolean equation is three, we require AND-OR array with three AND gates. When three input variables are complemented we get six inputs and when two outputs and their complements are feedback to input there will be additional four inputs and so we require ten input AND gates to implement the circuit.

The structure of PAL with two sets of AND-OR array and two XOR gates is shown in Fig. 12.37. The connected inputs to AND gate are marked by x. The connection to 0 or 1 of one of the inputs of XOR gates are shown by x.

A PAL programming table can be formed as shown in Table 12.9 to provide informations regarding the implementation of combinational circuit.

Table 12.9: PAL Programming Table

Product Term	Inputs included in Product Term			Product Term included in Outputs
	A	B	C	
A'B	0	1	-	$F_1 = A'B + AB'C + ABC$ (T)
AB'C	1	0	1	
ABC	1	1	1	
AB'C	1	0	1	$F_2 = (AB'C + B'C)'$ (C)
B'C	-	0	1	

Explanation for Table 12.9

- 1 under input variables indicate that the variable is connected to form the product term
- 0 under input variables indicate that the complement of the variable is connected to form the product term.
- - under input variables indicate that the variable/complement is not connected to form the product term.
- (T) under output indicates True form of SOP output and so other input of XOR gate is 0.
- (C) under output indicates complement form of SOP output and so other input of XOR gate is 1.

Example 12.11

Implement the following Boolean equations using PAL.

$F_1 = A'B + AC' + A'BC'$ and $F_2 = AB + AC + BC'$

Solution

The programming table to implement the given Boolean equations is shown in Table 1.

The Boolean equations has three variables and both equations has three product terms. In PAL implementation each equation has to be implemented individually in a separate AND-OR array. Hence we require two set of AND-OR array.

Since the maximum number of product terms in Boolean equation is three, we require AND-OR array with three AND gates. When three input variables are complemented we get six inputs and when two outputs and their complements are feedback to input there will be additional four inputs and so we require ten input AND gates to implement the circuit. The connected inputs to AND gate are marked by ×.

Here, outputs of OR gates are function outputs and so one of the inputs of XOR gates are connected to 0 and the connections are marked by × as shown in Fig. 1.

Table 1: PAL Programming Table

Product Term	Inputs included in Product Term A	B	C	Product Term included in Outputs
A'B	0	1	-	
AC'	1	-	0	F_1 = A'B + AC' + A'BC' (T)
A'BC'	0	1	0	
AB	1	1	0	
AC	1	-	1	F_2 = AB + AC + BC' (T)
BC'	-	1	0	

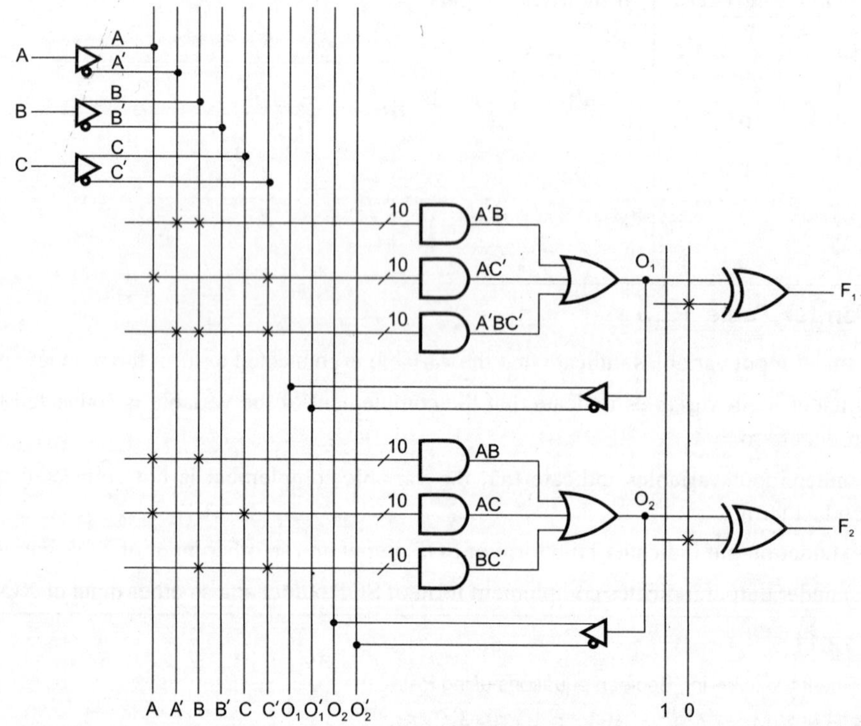

Fig. 1: Structure of PAL.

Example 12.12

(AU, Apr/May'19, 7 Marks)

Implement the following Boolean functions in a PAL device.

$F_1(X, Y, Z) = \Sigma m(1, 2, 4, 5)$

$F_2(X, Y, Z) = \Sigma m(0, 1, 3, 4)$

$F_3(X, Y, Z) = \Sigma m(2, 3, 6, 7)$.

Solution

The truth table of Boolean functions is formed as shown in Table 1. Here, the function output is 1 for the minterms included in that function. The simplified Boolean equations of the functions are obtained by constructing K-maps shown in Figs. 1 to 3.

Table 1: Truth Table

Inputs			Minterm	Outputs		
X	Y	Z		F_1	F_2	F_3
0	0	0	m_0	0	1	0
0	0	1	m_1	1	1	0
0	1	0	m_2	1	0	1
0	1	1	m_3	0	1	1
1	0	0	m_4	1	1	0
1	0	1	m_5	1	0	0
1	1	0	m_6	0	0	1
1	1	1	m_7	0	0	1

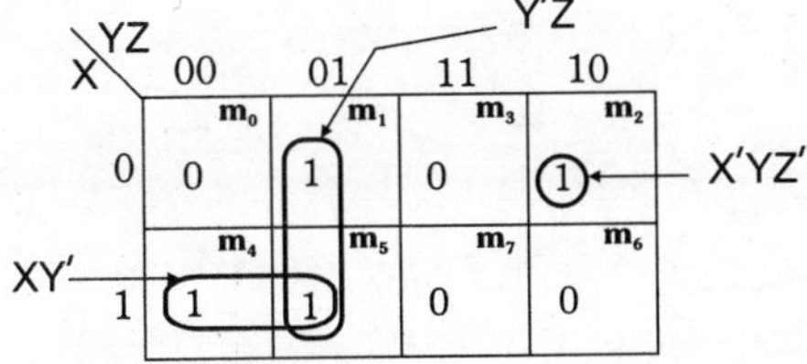

Fig. 1: K-map for F_1.

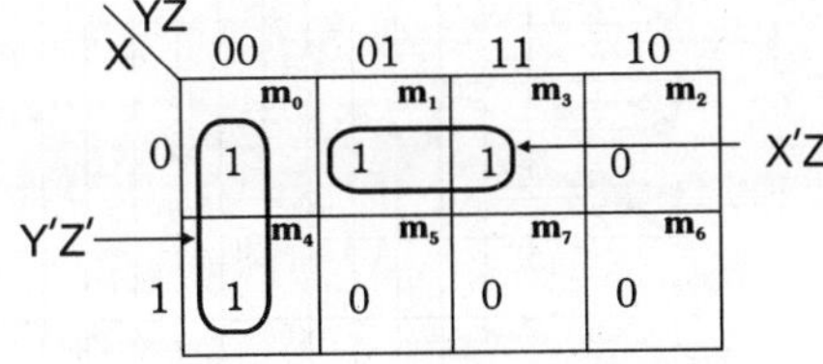

Fig. 2: K-map for F_2.

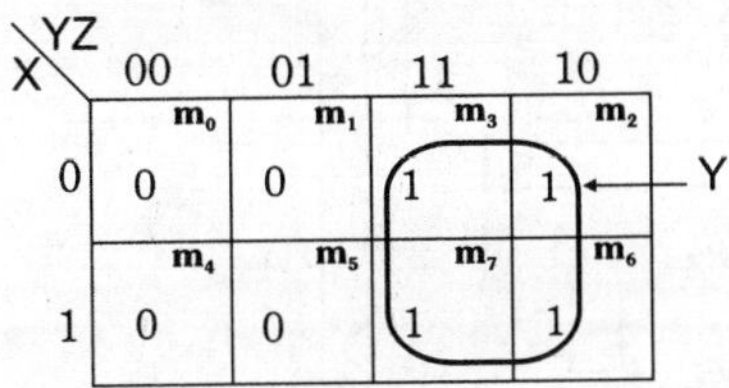

Fig. 3: K-map for F_3.

The following Boolean functions are obtained from K-map of Figs. 1 to 3.

$$F_1 = XY' + Y'Z + X'YZ'$$

$$F_2 = X'Z + Y'Z'$$

$$F_3 = Y$$

Using the above Boolean equations the PAL programming table is developed as shown in Table 2. The structure of programmed PAL using Table 2 is shown in Fig. 4.

The Boolean equations has three variables and first equation has three product terms, second equation has two product terms and the third equation has a single variable product term. In PAL implementation each equation has to be implemented individually in a separate AND-OR array. Hence we require three sets of AND-OR array. Since the maximum number of product terms in Boolean equation is three, we require AND-OR array with three AND gates. When three input variables are complemented we get six inputs and so when three outputs and their complements are feedback to input there will be additional six inputs and so we require twelve input AND gates to implement the circuit.

The connected inputs to AND gate are marked by ×. Here, ouptut of OR gates are function outputs and so one of the inputs of XOR gates are connected to 0 and the connections are marked by × as shown in Fig. 4.

Table 2: PAL Programming Table

Product Term	Inputs included in Product Term X	Y	Z	Product Term included in Outputs
XY'	1	0	-	F_1 = XY' + Y'Z + X'YZ' (T)
Y'Z	-	0	1	
X'YZ'	0	1	0	
X'Z	0	-	1	F_2 = X'Z + Y'Z' (T)
Y'Z'	-	0	0	
Y	-	1	-	F_3 = Y (T)

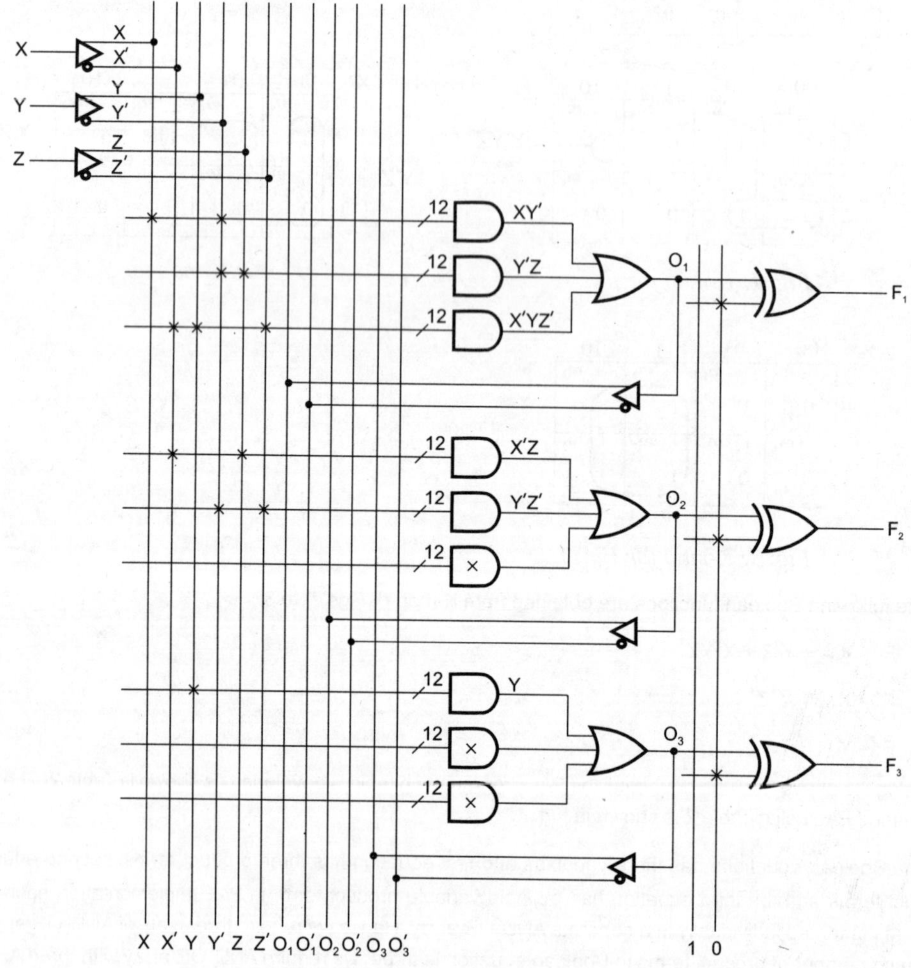

Fig. 4: Structure of PAL.

Example 12.13

(IKGPTU, Dec'19, 5 Marks)

Implement a 3-bit binary to gray conversion logic using PAL.

Solution

The truth table of 3-bit binary to gray code converter is shown in Table 1.

Table 1: Truth Table of 3-bit Binary to Gray Code Converter

Decimal Number	Inputs: 3-bit Binary B_2	B_1	B_0	Minterm	Outputs: Gray Code G_2	G_1	G_0
0	0	0	0	m_0	0	0	0
1	0	0	1	m_1	0	0	1
2	0	1	0	m_2	0	1	1
3	0	1	1	m_3	0	1	0
4	1	0	0	m_4	1	1	0
5	1	0	1	m_5	1	1	1
6	1	1	0	m_6	1	0	1
7	1	1	1	m_7	1	0	0

From the truth table, the Boolean equations for output are obtained as sum of minterms as shown below:

$G_2 = \Sigma m(4, 5, 6, 7)$; $G_1 = \Sigma m(2, 3, 4, 5)$; $G_0 = \Sigma m(1, 2, 5, 6)$

The above Boolean functions are simplified using K-map as shown below:

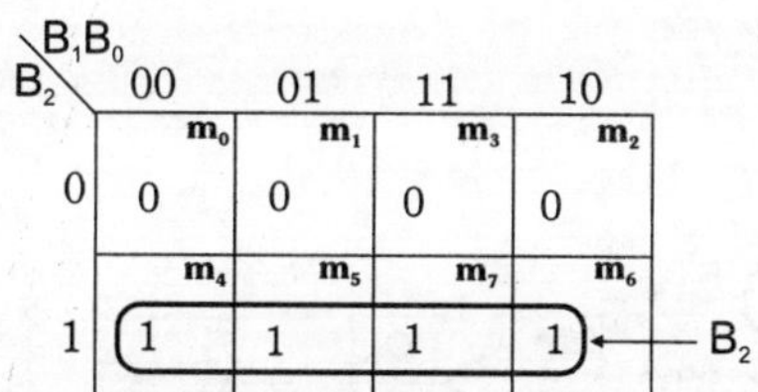

Fig. 1: K-map for G_2.

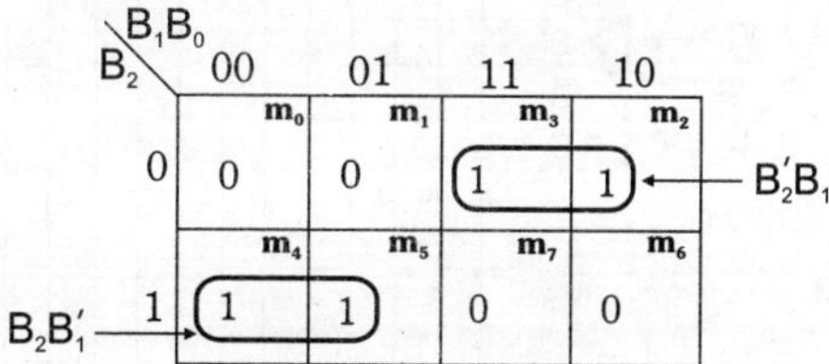

Fig. 2: K-map for G_1.

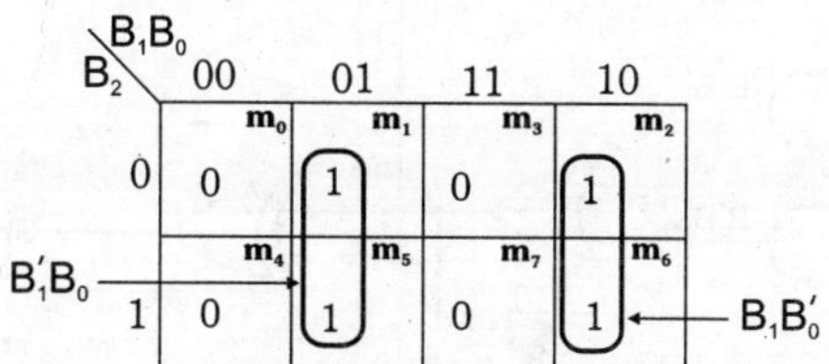

Fig. 3: K-map for G_0.

The following simplified Boolean equations are obtained from K-maps of Figs. 1 to 3.

$G_2 = B_2$

$G_1 = B_2B_1' + B_2'B_1$

$G_0 = B_1B_0' + B_1'B_0$

Using the above Boolean equations the PAL programming table is developed as shown in Table 2. The structure of programmed PAL using Table 2 is shown in Fig. 4.

The Boolean equations has three variables and first equation has single variable product term, second equation and third equation has a two product term. In PAL implementation each equation has to be implemented individually in a separate AND-OR array. Hence we require three sets of AND-OR array. Since the maximum number of product terms in Boolean equation is two, we require AND-OR array with two AND gates. When three input variables are complemented we get six inputs and so when three outputs and their complements are feedback to input there will be additional six inputs and so we require twelve input AND gates to implement the circuit.

The connected inputs to AND gate are marked by ×. Here, ouptut of OR gates are function outputs and so one of the inputs of XOR gates are connected to 0 and the connections are marked by × as shown in Fig. 4.

Table 2: PAL Programming Table

Product Term	Inputs included in Product Term			Product Term included in Outputs
	B_2	B_1	B_0	
B_2	1	-	-	$G_2 = B_2$ (T)
B_2B_1'	1	0	-	$G_1 = B_2B_1' + B_2'B_1$
$B_2'B_1$	0	1	-	(T)
B_1B_0'	-	1	0	$G_0 = B_1B_0' + B_1'B_0$
$B_1'B_0$	-	0	1	(T)

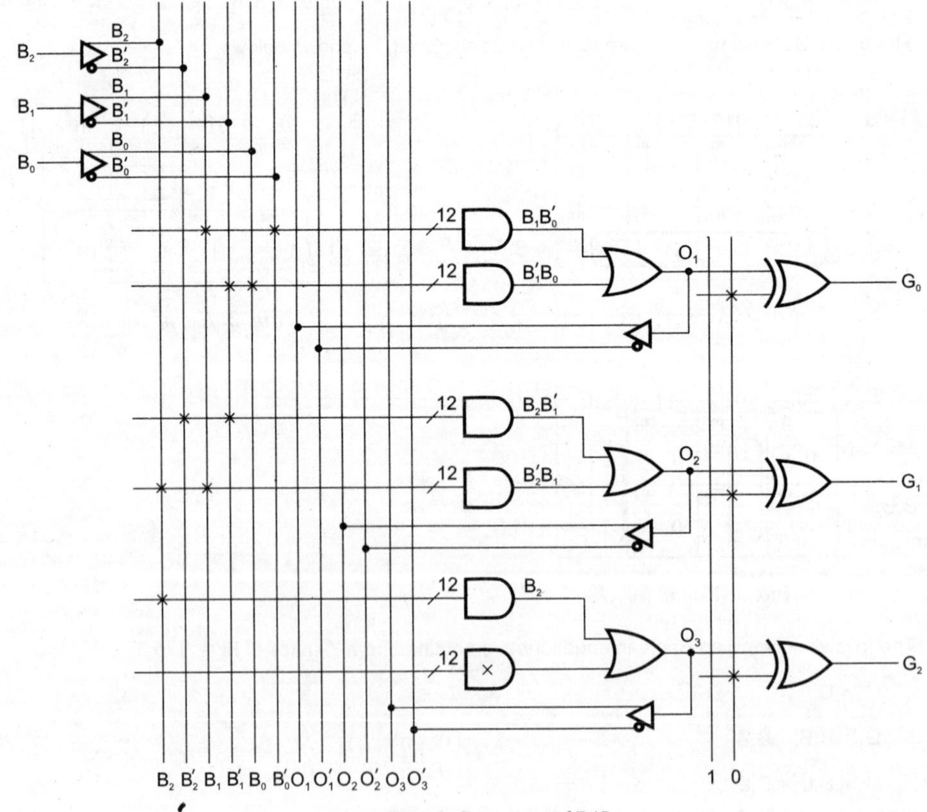

Fig. 4: Structure of PAL.

12.8 Sequential Programmable Logic Device (SPLD) *(JNTUH, Nov/Dec'20, 7 Marks)*

The SPLD includes flip-flops, in addition to the AND-OR array, within the integrated circuit chip. The block diagram of a SPLD is shown in Fig. 12.38. In a SPLD, a PLA or PAL is modified by including a number of flip-flops connected to form registers. The circuit outputs can be taken from the OR gates or from the outputs of the flip-flops. Additional programmable connections are available to include the flip-flop output in the product terms formed with the AND array. The flip-flops may be D or JK type.

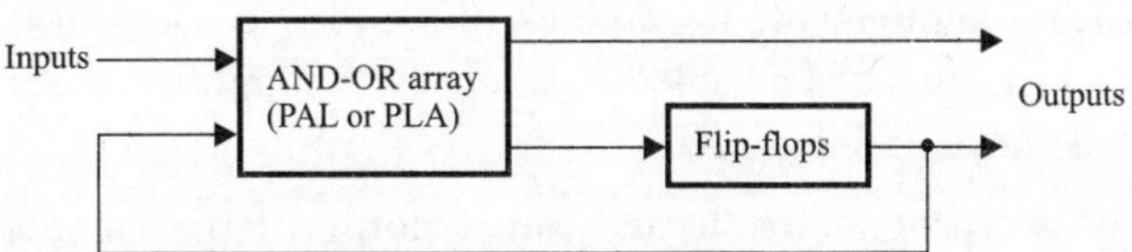

Fig. 12.38: *Sequential programmable logic device (SPLD).*

A PAL that includes flip-flops is also known as a registered PAL, to signify that the device contains flip-flops in addition to the AND-OR array. The SPLD will have a number of sections of registered PAL and these sections of a SPLD are called ***macrocells***.

Fig. 12.39 shows the logic of a basic macrocell. The AND-OR array is the same as in the combinational PAL. The output is driven by an edge-triggered D flip-flop connected to a common clock input. The ouput of the flip-flop is connected to a three-state buffer (or inverter) controlled by an output enable (OE) signal. The output of the flip-flop is fed back into one of the inputs of the programmable AND gates to provide the present-state condition for the sequential circuit. A typical SPLD has 8 to 10 macrocells.

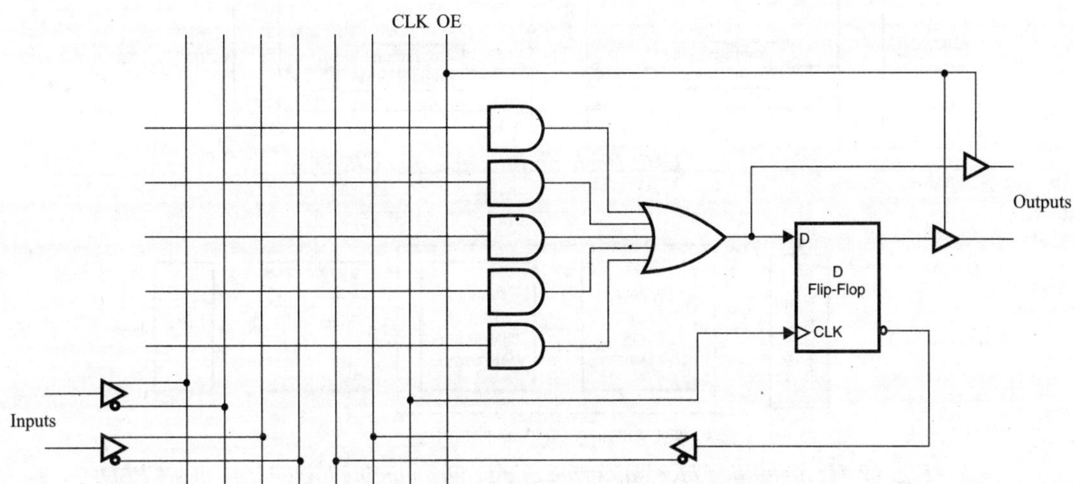

Fig. 12.39: *Basic macrocell logic of SPLD.*

In addition, a macrocell may have other programming features like ability to either use or bypass the flip-flop, the selection of clock edge polarity, the selection of preset and clear for the register, and the selection of the true value or complement of an output. An XOR gate is used to program a true/complement condition.

12.9 Complex Programmable Logic Device (CPLD)

(RGPV, Jun'22, 7 Marks)

A CPLD integrates a number of PLDs in a single device so that large complex combinational and sequential digital systems can be implemented in a CPLD. (AU, Apr/May'19, 6 Marks)

The CPLDs of different manufactures differ in architecture and functionality but have some common features. The common basic architecture of a CPLD is shown in Fig. 12.40.

A CPLD consists of three main components: programmable logic blocks, programmable interconnect channel and I/O block. Each programmable logic block of a CPLD consists of a fully programmable AND-OR array similar to a PAL/PLA and a bank of macrocells. A simplified block diagram of a typical programmable logic block is shown in Fig. 12.41.

The AND/OR array is reprogrammable and can perform a large number of logic functions. Macrocells are functional blocks with flip-flop, data selector and other control logic circuits that can perform combinatorial or sequential logic. Macrocells also have the added flexibility for true or complement output, along with feedback paths. Each macrocell has multiple configurations and each macrocell can be used in cascade, so that more complex combinational and sequential logic functions can be realized.

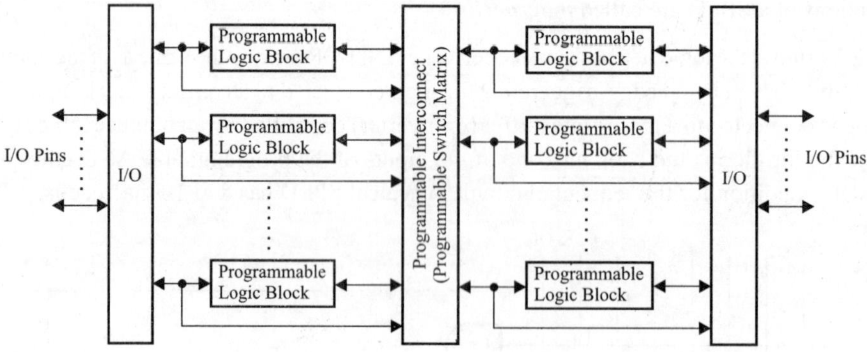

Fig. 12.40: Basic architecture of a CPLD.

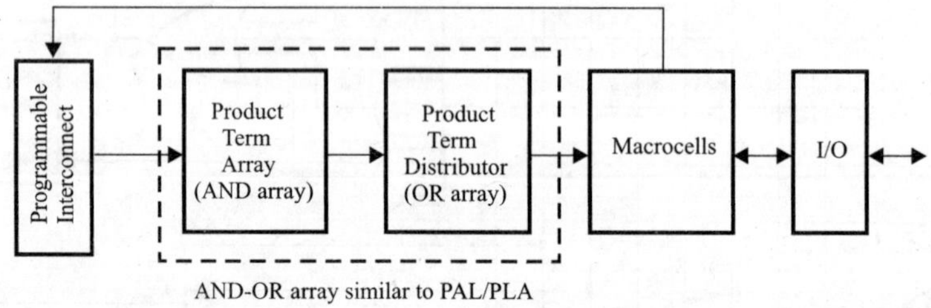

Fig. 12.41: Simplified block diagram of a programmable logic block in a CPLD.

Programmable interconnect channels provide interconnection networks between logic blocks, macrocells and input/output pins. The input/output blocks (I/O blocks) provide the interface between the internal logic to the I/O pins of the device. In advanced version of CPLDs, embedded array blocks with on-chip RAM/ROM are usually provided.

12.10 Field Programmable Gate Array (FPGA)

R.H.Freeman invented the first practical FPGA in 1985.

FPGA is a programmable semiconductor logic device consisting of large number of logic gates which can be configured by the user to implement any kind of digital circuit or system. FPGAs are manufactured with thousands to millions of logic gates. The programming of logic gates of FPGA is called configuration and FPGAs are reprogrammable (or reconfigurable).

The configuration of the FPGA is generally performed using a Hardware Description Language (HDL). The popular HDLs are Verilog and VHDL.

FPGAs store their customized configuration data in SRAM-type internal latches. Depending on the device size and user-design implementation options the number of configuration bits varies from 1 *Mb* to 33 *Mb* or even more in modern FPGAs.

The internal SRAM is volatile and so the configurations bits will be erased whenever power is turned OFF. Therefore, when an FPGA is powered up the configuration storage must be reloaded. Usually, the modern FPGA based systems will have serial EEPROM as boot loader, so that the configuration bits are permanently stored in boot loader and whenever power is turned ON the configuration bits are serially transmitted from EEPROM to configuration SRAM in FPGA.

Xilinx and Altera are two major manufacturers of FPGA holding nearly 90% of market share. The discussions in this section are with reference to Xilinx FPGAs. Altera FPGAs may also have resources similar to Xilinx FPGAs but there will be differences in architecture and functionality.

12.10.1 Basic Architecture of FPGA

FPGAs are evolved from gate arrays and early versions of FPGA basically have simple Configurable Logic Blocks (CLB) and small number of inputs and outputs as shown in Fig. 12.42. Each CLB will consist of a Look-Up Table (LUT) and a Flip-Flop (FF). The basic idea of FPGA's is to inter-connect small "truth tables" to emulate complex digital circuits. In an FPGA, these tables are called "Lookup Table" (LUT).

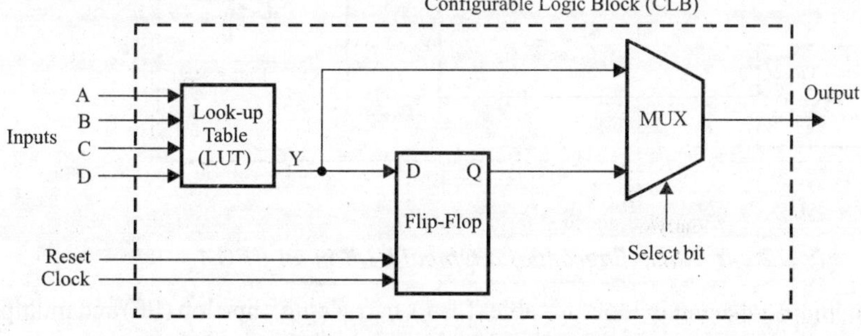

LUT	
Inputs A B C D	Output Y
0 0 0 0	0
0 0 0 1	1
0 0 1 0	1
⋮	⋮
1 1 1 1	0

Fig. 12.42: Simple CLB architecture.

The basic architecture of an FPGA is shown in Fig. 12.43.

The FPGA architecture consists of four major components

- Programmable/Configurable Logic Blocks
- Programmable switch blocks
- Programmable Routing (interconnect lines)
- I/O blocks

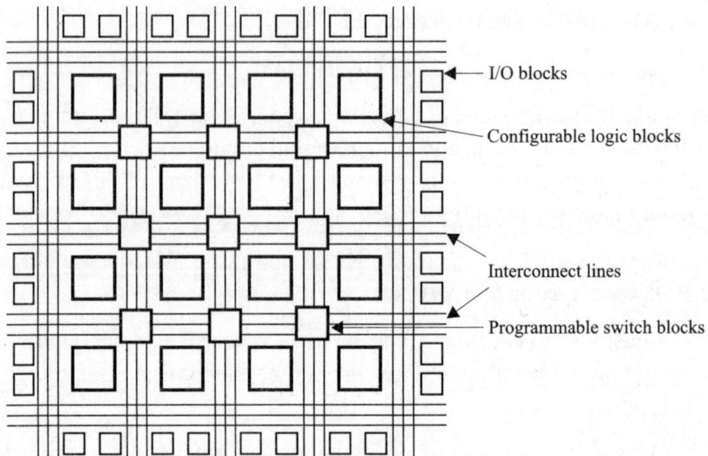

Fig. 12.43: Basic architecture of FPGA.

A configurable logic block (CLB) is the basic logic resource on an FPGA. The configurable logic block (CLB) includes logic and look-up tables (LUTs) that can be configured to implement combinational logic circuits. When CLBs are linked together by routing resources (carry chain), the components in CLBs execute complex logic functions, implement memory functions and synchronize code on the FPGA. The block diagram of a typical CLB in an FPGA is shown in Fig. 12.44.

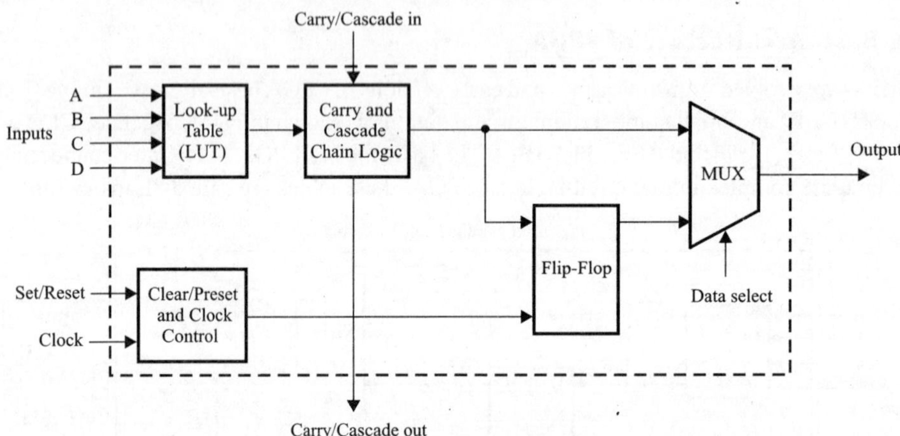

Fig. 12.44: Block diagram of a typical CLB in an FPGA.

The CLB shown in Fig. 12.44 contain look-up-table (LUT), carry logic, flip-flop (FF) and multiplexer. It also has a clear/preset and clock control unit. A LUT can speed up processing by providing the output for a given input without logical processing. A flip-flop is a primitive storage device that can store a single bit of information.

Carry look-ahead logic is implemented in each CLB with a combination of dedicated multiplexer and XOR gates that are used within the carry/cascade chain. The carry/cascade chain logic in an individual CLB is the circuit that directly connects a column of CLBs together. It is possible to cascade a carry chain across multiple CLBs in order to quickly implement addition, subtraction, and multiplication operations on operands that are too large to be processed by a single CLB.

The programmable routing are the wiring resources in FPGAs. The programmable routing establishes a connection between logic blocks and Input/Output blocks to complete a user-defined design unit. It consists of multiplexers pass transistors and tri-state buffers. Pass transistors and multiplexers are used in a logic cluster to connect the logic elements.

The ***programmable switches*** are user-programmable switch that can connect a logic element to an interconnect wire and/or one interconnect wire to another. The programmable I/O pads are used to interface the logic blocks and routing architecture to the external components. The I/O pad and the surrounding logic circuit form as an I/O cell. These cells consume a large portion of the FPGA's area. And the design of I/O programmable blocks is complex, as there are great differences in the supply voltage and reference voltage.

Modern FPGAs contain a heterogeneous mixture of different blocks like dedicated memory blocks, multiplexers. Configuration memory is used throughout the logic blocks to control the specific function of each element.

Advantages of FPGA

- FPGAs can implement faster and parallel processing.
- FPGAs are programmable IC, programmed by software and can be re-programmed or reused any number of times.
- Since FPGA ICs are programmable the solution is available faster to the market.
- FPGA development is cheaper due to non recurring expenses.
- In FPGA, software takes care of routing, placement and timing and so manual intervention is lesser.

Disadvantages of FPGA

- Programming the FPGA requires the knowledge of VHDL/Verilog programming languages and digital system fundamentals.
- Programmers do not have any control on power optimization.
- Development resources are more specific for specific FPGA.
- FPGAs are costly for large quantity production.

Comparison of CPLD and FPGA

- FPGA is more suitable for structures with large number of flip-flops, while CPLD is more suitable for structures with limited flip-flops and rich product terms.
- The continuous wiring structure of CPLD determines that its timing delays are uniform and predictable, while the segmented wiring structure of FPGA determines the unpredictability of its delays.
- CPLD is programmed by modifying the logic function with fixed interconnect circuit, while FPGA is programmed by changing the wiring of internal wires.
- FPGA can be programmed under the logic gate, while CPLD is programmed under the logic block. Hence, CPLDs are faster than FPGAs and have greater time predictability.
- FPGAs are more integrated than CPLDs and have more complex wiring structures and logic implementations.

• CPLD is programmed using internal EEPROM or FASTFLASH technology, no external memory chip is needed. While the FPGA programming information needs to be stored in external memory, and transferred to internal RAM during switch ON.

• CPLD is mainly based on EEPROM or FLASH memory programming, which can be programmed up to 10,000 times, and has the advantage that the programming information is not lost even when the system is powered off.

12.10.2 Xilinx FPGA Family of Devices

Xilinx has released a number of families of FPGA devices with a large number of internal logical devices in each family with enhanced and variable features. XC2000 series is the first family of FPGA devices released by Xilinx in the year 1985 manufactured using 2000 nm (nano-meter) process.

> ***Note:*** *The nano-meter manufacturing process refer to the size of fundamental transistor in the FPGA IC. Xilinx started with 2000 nm process and realized up to 7 nm process in their latest Versal family of FPGAs.*

This is followed by XC3000, XC4000, XC5200, XC6200 and XC8100 series of FPGA devices released over a period of 10 years, from 1985, with enhanced features in every family of devices. These FPGA devices are outdated and no longer manufactured by Xilinx.

The Family of Xilinx FPGA devices released after 1995 are named as Spartan, Artix, Kintex, Virtex and Versal. The evolution of various families of FPGA devices has made improvements in density, performance, power consumption, voltage levels, IO support and other functionality. The various Xilinx FPGAs are listed in Appendix 7. A brief discussion about Spartan-6 family of FPGAs is presented in the following section.

12.10.3 Spartan-6 Family of FPGA

The Spartan-6 family includes thirteen types of FPGA devices with logic cell densities ranging from 3,840 to 1,47,443 logic cells and manufactured using 45 nm low-power copper process technology. The specific features of Spartan-6 family members are summarized in Tables 12.10 and 12.11. The salient features of Spartan-6 family of FPGAs are,

- Dual-register 6-input lookup table (LUT) logic.
- 18 *kB* (2 x 9 *kB*) block RAMs.
- Second generation DSP48 slices.
- SDRAM memory controllers.
- Enhanced mixed-mode clock management blocks.
- Select IO technology.
- Power optimized high-speed serial transceiver blocks.
- PCI Express compatible Endpoint blocks.
- Advanced system-level power management modes.
- Auto-detect configuration options.
- Enhanced IP security with AES and Device DNA protection.

Table 12.9: Spartan-6 FPGA Feature Summary by Device

Device	Logic Cells	Configurable Logic Blocks (CLBs)			DSP48A1 Slices	Block RAM Blocks		Total I/O Banks	Max User I/O
		Slices	Flip-Flops	Max Distributed RAM (Kb)		18Kb	Max (Kb)		
XC6SLX4	3,840	600	4,800	75	8	12	216	4	132
XC6SLX9	9,152	1,430	11,440	90	16	32	576	4	200
XC6SLX16	14,579	2,278	18,224	136	32	32	576	4	232
XC6SLX25	24,051	3,758	30,064	229	38	52	936	4	266
XC6SLX45	43,661	6,822	50,576	401	58	116	2,088	4	358
XC6SLX75	74,637	11,662	93,296	692	132	172	3,096	6	408
XC6SLX100	101,261	15,822	126,576	976	180	268	4,824	6	480
XC6SLX150	127,443	23,038	184,304	1,355	180	268	4,824	6	576
XC6SLX25T	24,051	3,758	30,064	229	38	52	936	4	250
XC6SLX45T	43,661	6,822	54,576	401	58	116	2,088	4	296
XC6SLX75T	74,637	1,662	93,296	692	132	172	3,096	6	348
XC6SLX100T	101,261	15,822	126,576	976	180	268	4,824	6	498
XC6SLX150T	147,443	23,038	184,304	1,355	180	268	4,824	6	540

Table 12.10: Spartan-6 FPGA Logic Resources

Device	Logic Cells	Total Slices	SLICEMs	SLICELs	SLICEXs	Number of 6-Input LUTs	Maximum Distributed RAM (Kb)	Shift Register (Kb)	Number of Flip-Flops
XC6SLX4	3,840	600	300	0	300	2,400	75	38	4,800
XC6SLX9	9,152	1,430	360	355	715	5,720	90	45	11,440
XC6SLX16	14,579	2,278	554	595	1,139	9,112	136	68	18,224
XC6SLX25	24,051	3,758	916	963	1,879	15,032	229	115	30,064
XC6SLX45	43,661	6,822	1,602	1,809	3,411	27,288	401	200	54,576
XC6SLX75	74,637	11,662	2,768	3,063	5,831	46,648	692	346	93,296
XC6SLX100	101,261	15,822	3,904	4,007	7,911	63,288	976	488	126,576
XC6SLX150	127,443	23,038	5,420	6,099	11,519	92,152	1,355	678	184,304
XC6SLX25T	24,051	3,758	916	963	1,879	15,032	229	115	30,064
XC6SLX45T	43,661	6,822	1,602	1,809	3,411	27,288	401	200	54,576
XC6SLX75T	74,637	1,662	2,768	3,063	5,831	46,648	692	346	93,296
XC6SLX100T	101,261	15,822	3,904	4,007	7,911	63,288	976	488	126,576
XC6SLX150T	147,443	23,038	5,420	6,099	11,519	92,152	1,355	678	184,304

12.10.4 Spartan-6 FPGA Architecture

Spartan-6 FPGA architecture is shown in Fig. 12.45 and it consists of following elements.

- Configurable logic blocks (CLB)
- Embedded memories (Block RAM) for distributed data storage.
- Clock management tiles (CMT) with PLLs for driving the FPGA fabric at different clock rates.
- On-chip memory controller blocks (MCB).
- Multiply-accumulate blocks (DSP 48-bit blocks).
- High-speed serial transceivers.
- PCIE endpoint
- I/O banks

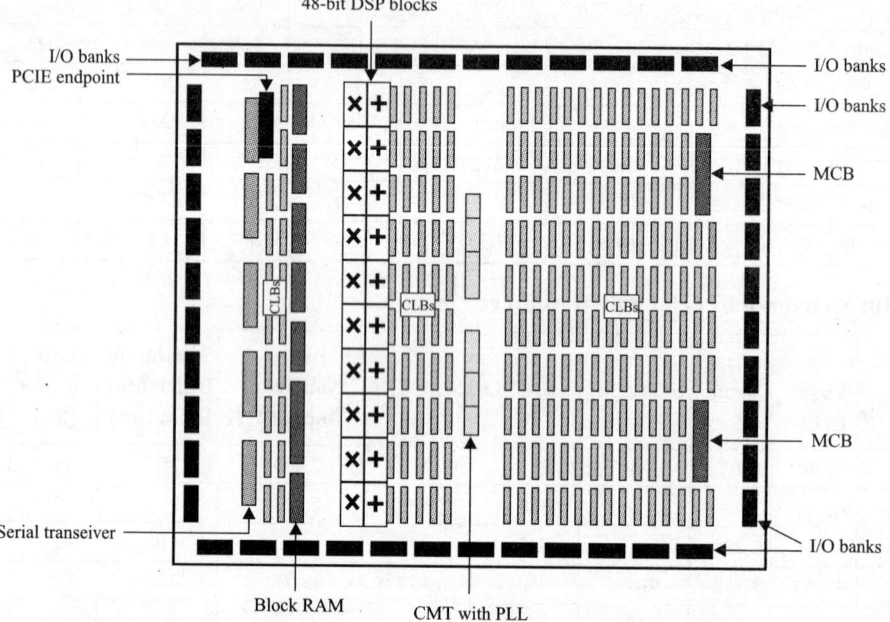

Fig. 12.45: *Architecture of Spartan-6 FPGA.*

Spartan-6E FPGA contains the following basic resources.

1. Logic resources
   - Slices (grouped into CLB)
   - Memory
   - Multipliers
2. Interconnect resources
   - Programmable interconnect
   - I/O banks

3. Other resources
   - Global clock buffers
   - Boundary scan logic

CLBs, Slices and LUTs

Each configurable logic block (CLB) in Spartan-6 FPGAs consists of two slices, arranged side-by-side as two vertical columns. There are three types of CLB slices in the Spartan-6 architecture: SLICEM, SLICEL and SLICEX. Each slice contains four LUTs, eight flip-flops and other control logic. The LUTs are used to implement combinational and sequential logic functions.

In the total slices in Spartan-6 FPGA, 25% of slices are SLICEMs. Each of the four SLICEM LUTs can be configured as either a 6-input LUT with one output, or as dual 5-input LUTs with identical 5-bit addresses and two independent outputs. These LUTs can also be used as distributed 64-bit or 32-bit RAM, or as a single 32-bit shift register, or as two 16-bit shift registers. Each LUT output can be registered in a flip-flop within the CLB. For arithmetic operations, a high-speed carry chain propagates carry signals in a column of slices.

In the total slices in Spartan-6 FPGA, 25% of slices are SLICELs, which contain all the features of the SLICEM except the memory/shift register function.

In the total slices in Spartan-6 FPGA, 25% of slices are SLICEXs. The SLICEXs have the same structure as SLICELs except the arithmetic carry option and the wide multiplexers.

Clock Management

Spartan-6 have up to six clock management tiles (CMTs), each consisting of two digital clock managers (DCMs) and one phase lock loop (PLL). The DCM provides four phase shift of the input frequency (CLKIN): 0°, 90°, 180°, and 270°. It also provides a doubled clock frequency output and its complement output.

The PLL can act as a frequency synthesizer and as a filter for input clock in combination with the DCMs. Voltage-Controlled Oscillator (VCO) is the heart of the PLL with a frequency range of 400 *MHz* to 1,080 *MHz*. PLL have of programmable frequency dividers to divide the clock generated clock to convert to clock at different clock rates.

Block RAM

Every Spartan-6 FPGA has between 12 and 268 dual-port block RAMs, each with a capacity of 18 *kB*. Each block RAM has two independent ports that share the stored data.

Each memory access is controlled by the clock. All the inputs, data, address, clock enables, and write enables are stored in registers. The data output retains data until the next operation.

The memory data width is programmable and hence, each port can be configured as 16K × 1, 8K × 2, 4K × 4, 2K × 9 (or 8), 1K × 18 (or 16), or 512 x 36 (or 32). Parity bits are included in the x9, x18, and x36 configurations.

Memory Controller Blocks

Spartan-6 devices have dedicated memory controller blocks (MCBs) for interfacing external DRAM and support access rates up to 800 *Mb/s*. The MCB has dedicated routing to predefined I/Os. If the MCB is not used, these I/Os are used as general purpose I/Os. The MCB can be connected to 4–, 8–, or 16-bit external DRAM. The MCB provides a faster DRAM interface.

The memory controller offers a complete multi-port arbitrated interface to the logic inside the Spartan-6 FPGA. Commands and data can be pushed to and poped from independent built-in FIFOs, using conventional FIFO control signals.

Digital Signal Processing-DSP48 Slice

Spartan-6 FPGAs have many dedicated DSP slices consisting of binary multipliers and accumulators to support DSP computations. Each DSP48 slice consists of a dedicated 18 × 18 bit two's complement multiplier and a 48-bit accumulator.

The DSP48 slice provides extensive pipelining and extension capabilities that enhance speed and efficiency of many applications, such as wide dynamic bus shifters, memory address generators, wide bus multiplexers and memory-mapped I/O register files. The accumulator can also be used as a synchronous up/down counter. The multiplier can perform barrel shifting.

Input/Output Banks

The number of I/O pins varies from 102 to 576, depending on device and package size. Each I/O pin is configurable and can comply with a large number of standards, using up to 3.3 *V*.

With the exception of supply pins and a few dedicated configuration pins, all other package pins have the same I/O capabilities. All user I/Os are bidirectional. All I/O pins are organized in banks, with four banks on the smaller devices and six banks on the larger devices.

Serial Transceiver

Spartan-6 LXT devices have 2 to 8 transceiver circuits. Each transceiver is a combined transmitter and receiver capable of operating at data rates up to 3.2 *Gb/s*. The transmitter and receiver are independent circuits that use separate PLLs to multiply the reference frequency input by certain programmable numbers between 2 and 25, to become the bit-serial data clock.

The transmitter is fundamentally a parallel-to-serial converter with a conversion ratio of 8, 10, 16, or 20. The transmitter output drives the PC board with a single-channel differential current-mode logic (CML) output signal. The receiver is fundamentally a serial-to-parallel converter, changing the incoming bit serial differential signal into a parallel stream of words, each 8, 10, 16, or 20 bits wide.

Integrated Endpoint Block for PCI Express Designs

The PCI Express standard is a packet-based, point-to-point serial interface standard. The PCI Express Base Specification 1.1 defines bit rate of 2.5 *Gb/s* per lane, per direction (transmit and receive). The Spartan-6 LXT devices include one integrated Endpoint block for PCI Express technology that is compliant with the PCI Express Base Specification Revision 1.1.

12.11 Summary of Important Concepts

1. The digital ICs (Integrated Circuits) can be broadly classified into fixed function ICs and programmable ICs.
2. PROM, PAL and PLA are one-time programmable and consists of programmable logic gate arrays and so can be used to implement only combinational logic functions.
3. CAM is a type of semiconductor memory that is designed for high-speed searching and retrieval of data.
4. CAM consists of a large array of memory cells, with each cell storing a bit of data (content).
5. FPGA devices are reprogrammable and consists of programmable logic gate arrays and flip-flops and so can be used to implement both combinational and sequential logic functions.
6. PLA is a one-time programmable device to implement a Boolean function in sum of products form.
7. PAL device will have m sets of AND-OR array with each array having k number of AND gates and one OR gate.
8. A CPLD integrates a number of PLDs in a single device.
9. A CPLD consists of three main components: programmable logic blocks, programmable inter connect channel and I/O block.
10. Each programmable logic block of a CPLD consists of a fully programmable AND-OR array similar to a PAL/PLA and a bank of macrocells.
11. FPGA is a programmable semiconductor logic device consisting of large number of logic gates.
12. The programming of logic gates of FPGA is called configuration and FPGAs are reprogrammable (or reconfigurable).
13. FPGAs store their customized configuration data in SRAM-type internal latches.
14. The basic idea of FPGA's is to inter-connect small "truth tables" to emulate complex digital circuits.
15. The FPGA architecture consists of four major components: programmable/configurable logic blocks, programmable switch blocks, programmable routing (interconnect lines), I/O blocks.
16. A configurable logic block (CLB) is the basic logic resource on an FPGA.
17. The configurable logic block (CLB) includes logic and look-up tables (LUTs) that can be configured to implement combinational logic circuits.
18. Carry look-ahead logic is implemented in each CLB with a combination of dedicated multiplexer and XOR gates.
19. The programmable routing are the wiring resources in FPGAs.
20. The programmable switches are user-programmable switch that can connect a logic element to an interconnect wire and/or one interconnect wire to another.
21. The digital ICs (Integrated Circuits) can be broadly classified into fixed function ICs and programmable ICs.
22. PROM, PAL and PLA are one-time programmable and consists of programmable logic gate arrays and so can be used to implement only combinational logic functions.
23. FPGA devices are reprogrammable and consists of programmable logic gate arrays and flip-flops and so can be used to implement both combinational and sequential logic functions.

24. PLA is a one-time programmable device to implement a Boolean function in sum of products form.
25. PAL device will have m sets of AND-OR array with each array having k number of AND gates and one OR gate.
26. The SPLD includes flip-flops, in addition to the AND-OR array, within the integrated circuit chip.
27. A CPLD integrates a number of PLDs in a single device.
28. A CPLD consists of three main components: programmable logic blocks, programmable inter connect channel and I/O block.
29. Each programmable logic block of a CPLD consists of a fully programmable AND-OR array similar to a PAL/PLA and a bank of macrocells.
30. FPGA is a programmable semiconductor logic device consisting of large number of logic gates.
31. The programming of logic gates of FPGA is called configuration and FPGAs are reprogrammable (or reconfigurable).
32. FPGAs store their customized configuration data in SRAM-type internal latches.
33. The basic idea of FPGA's is to inter-connect small "truth tables" to emulate complex digital circuits.
34. The FPGA architecture consists of four major components: programmable/configurable logic blocks, programmable switch blocks, programmable routing (interconnect lines), I/O blocks.
35. A configurable logic block (CLB) is the basic logic resource on an FPGA.
36. The configurable logic block (CLB) includes logic and look-up tables (LUTs) that can be configured to implement combinational logic circuits.
37. Carry look-ahead logic is implemented in each CLB with a combination of dedicated multiplexer and XOR gates.
38. The programmable routing are the wiring resources in FPGAs.
39. The programmable switches are user-programmable switch that can connect a logic element to an interconnect wire and/or one interconnect wire to another.

12.12 Short-Answer Questions

Q12.1 Distinguish between a memory address and memory's addressability?

The memory address is the identification number in binary for a memory location.

The memory's addressability is the size of binary that can be stored in a memory location.

For example in 6264 RAM memory, the memory address range is 0000_H to $1FFF_H$ with addressability of 8-bits i.e., each memory location can store 8-bit binary information.

Q12.2 A memory unit is 16-bit addressable and has 12 k locations. How many bits of storage does the memory consist of?

The storage capacity of the memory will be 12 k × 16-bit = 192 *kilo-Bits*.

Q12.3 Define memory address and memory word.

Each memory location has a unique binary number called address. The address is used to access the location for read and write operation.

In general, a memory location can store m bits and the size of binary information that can be stored in a memory location is called memory word size and the stored binary information is called memory word.

Q12.4 Give two advantages of machine language programming over high-level language programming?

i) Machine language program can be directly executed in a machine and does not require conversion using assembler/compiler. Hence execution of machine language program will be much faster than any other language program.

ii) The machine language program can be written using only two symbols '0' and '1'.

Q12.5 What are the types of semiconductor memories?

The semiconductor memories can be broadly classified into two major types,

1. RAM (Random Access Memory)
2. ROM (Read Only Memory)

Q12.6 What is a RAM?

RAM is a semiconductor memory. The RAM is also called RWM (Read/Write Memory). The RAM memory can be read or write in the system it is fitted or connected.

Q12.7 What is a ROM? ***(RGPV, Jun'22, 2 Marks)***

ROM is a semiconductor memory. ROM memory can be only read in the system it is fitted or connected. For writing ROM memory a separate system called ROM programmer is needed.

Q12.8 List the types of ROM. ***(RGPV, Jun'22, 5 Marks)***

1. Programmable Read Only Memory (PROM)
2. Erasable Programmable Read Only Memory (EPROM)
3. Electrically Erasable Programmable Read Only Memory (EEPROM).

Q12.9 What is Read and Write operation?

The binary information can be stored in a memory location which is called write operation.

The stored information can be taken back from the memory loaction which is called read operation.

Q12.10 Define memory location.

A memory device will have large number of storage cells arranged in groups of 8 cells to store a byte and the group of 8 cells is called location or memory location.

Q12.11 What are the characteristics of ROM memory ?

1. ROM is non-volatile memory.
2. The contents of ROM can be read by the processor but it cannot perform a write operation.
3. The ROM has the feature of random access for read operation.
4. The memory cell has a MOS transistor either with an open gate or a closed gate.

Q12.12 How are ROMs classified?

ROMs can be classified into the following three categories based on the method of programming.

1. Custom programmed or Mask programmed ROM.
2. Programmable or Field programmable ROM.
3. Reprogrammable or Erasable-programmable ROM.

Q12.13 List the characteristics of EPROM.

1. The EPROM is non-volatile.
2. It has random-access feature for read operation.
3. The contents of EPROM can be erased by passing UV light and then the device can be reprogrammed.
4. The EPROM is read only memory and for writing into EPROM, a separate hardware set up is required.

Q12.14 List the differences between RAM and ROM. ***(AKTU, Sep'20, 5 Marks)***

RAM	ROM
1. In RAM, the memory cell consists of a flip-flop or similar device constructed using four to eight MOS transistors.	1. In ROM, the memory cell consists of a MOS transistor with isolated or floating gate.
2. RAM is a volatile memory hence it can retain the stored information as long as power is ON.	2. ROM is a nonvolatile memory hence it can the stored information even after power is OFF.
3. RAM memory can be read or write in the system it is fitted or connected.	3. ROM memory can be only read in the system it is fitted or connected. For writing ROM memory a separate system called ROM programmer is needed.
4. RAM is used as a temporary storage device.	4. ROM is used as permanent storage device.

Q12.15 Write a short note on the memory cell of EPROM.

The memory cell of EPROM contains a MOS transistor with isolated or floating gate. The isolated gate is located between the normal control gate and the source/drain region of transistor. The information is stored as a charge or no charge in the isolated gate.

Q12.16 What is NVRAM?

The non-volatile read/write memories are called NVRAM. The various types of NVRAMs are flash memory, EEPROM and Shadow RAM.

Q12.17 List the features of static RAM.

1. Static RAMs are read/write memories.
2. They are volatile and have random access feature for both read and write operations.
3. The memory cell is a flip-flop constructed using 6 to 8 MOS transistors.

Q12.18 What is DRAM ?

DRAMs are read/write semiconductor memories in which the information is stored in the form of electric charge on the gate to substrate capacitance of a MOS transistor.

Q12.19 List the characteristics of DRAM.

1. The DRAMs are volatile and have random access feature.
2. They are read/write memories.
3. The contents of DRAM has to be refreshed periodically using refreshing circuits.
4. The memory cell of DRAM will have 3 to 4 MOS transistor.

Q12.20 A DRAM chip uses two dimensional address multiplexing. It has 13 common address pins with the row address having one bit more than the column address. What is the capacity of the memory?

The memory capacity is decided by the total address pins.

Given that the total address pins is 13.

Memory capacity = 2^{13} = 8192 bits = 8192 / 8 = 1024 bytes.

Q12.21 Compare the Static RAM with DRAM. *(AKTU, Dec'22, 2 Marks) (GTU, May'19, 3 Marks)*

Static RAM	DRAM
1. Information is stored as voltage level in a flip-flop. 2. Six to eight transistors are required to form one memory cell. 3. Packing density is low. 4. The contents of memory need not be refreshed.	1. Information is stored as a charge in the gate to substrate capacitance. 2. Three to four transistors are required to form one memory cell. 3. Packing density is high. 4. The contents of memory has to be refreshed periodically.

Q12.22 Compare ROM and PLA *(GTU, Jun'19, 4 Marks)*

ROM	PLA
1. ROM has memory cells to store 1's and 0's. 2. The output of truth table of Boolean function is stored in memory cells.	1. PLA has AND and OR array to form product and sum terms. 2. The Boolean functions in SOP form is used to program the PLA.

Q12.23 Give comparison between EPROM and FLASH memory. *(GTU, May'19, 4 Marks)*

EPROM memory is erasable using UV light whereas the flash memory is erasable using current.

EPROM requires separate systems for erasing and programming whereas flash can be erased and programmed in the system where the flash memory is fitted.

Q12.24 How many address bits are required for a 512 × 4 memory? *(IKGPTU, May'20, 2 Marks)*

A 512 × 4 memory has 512 locations with 4-bit data stored in each location. Here, 512 = 2^9, and hence with 9-bit address it is possible to generate 512 addresses in order to address all the 512 locations.

∴ Address bits = 9

Q12.25 A microprocessor uses RAM chips of 1024 × 1 capacity. *(IKGPTU, Dec'19, 5 Marks)*

a) How many chips will be required and how many address lines will be connected to provide a capacity of 1024 bytes?

b) How many chips will be required to obtain a memory of capacity of 16 K bytes?

Solution:

a) A 1024 × 1 memory will have 1024 storage locations with 1-bit stored in each location. A memory with 1024 bytes (1024 × 8) will have 1024 storage locations with 8-bit stored in each locations. Therefore, 8 memory ICs of capacity 1024 × 1 are connected to common address lines. Here, 1024 = 2^{10}, and hence 10 address lines are required to address all the 1024 locations.

b) Here, 1024 bytes = 1 kB, and it requires 8 numbers of 1024 × 1 memory IC. Therefore, 16 kB memory requires 16 × 8 = 128 numbers of 1024 × 1 memory IC.

Q12.26 What is programmable logic array? How it differs from ROM? ***(AU, Nov/Dec'18, 2 Marks)***

The PLA is a device to implement Boolean function in sum of products form. The PLA will have an array of AND gates to implement product terms followed by an array of OR gates to implement the sum of the products.

ROM is a storage device used to store the truth table of a Boolean function. But PLA is used to implement the SOP form of a Boolean function.

Q12.27 What is the difference between PLA and PAL?

The only difference between PLA and PAL is that the OR gate array is programmable or not. In PLA device, the OR array is programmable but in PAL device the connections to OR array is fixed.

Q12.28 Explain about combinational PLD's. ***(AKTU, Sep'20, 5 Marks)***

PLDs without flip-flops are also called combinational PLDs and they are used to implement only combinational logic circuits. The PROM, PAL and PLA are combinational PLDs.

Q12.29 What are the basic internal components of CPLD?

A CPLD consists of three main components.

- Programmable logic blocks.
- Programmable inter connect channel.
- I/O block.

Q12.30 What is FPGA?

FPGA is a programmable semiconductor logic device consisting of large number of logic gates which can be configured by the user to implement any kind of digital circuit or system. FPGAs are manufactured with thousands to millions of logic gates. The programming of logic gates of FPGA is called configuration and FPGAs are reprogrammable (or reconfigurable).

Q12.31 What are basic internal components of FPGA?

The FPGA architecture consists of four major components.

- Programmable/Configurable Logic Blocks
- Programmable switch blocks
- Programmable Routing (interconnect lines)
- I/O blocks

Q12.33 What is configuration in FPGA?

The programming of logic gates of FPGA is called configuration and FPGAs are reprogrammable (or reconfigurable). The configuration of the FPGA is generally performed using a Hardware Description Language (HDL). The popular HDLs are Verilog and VHDL.

Q12.34 What are the difference between CPLD and FPGA?

- FPGA is more suitable for structures with large number of flip-flops, while CPLD is more suitable for structures with limited flip-flops and rich product terms.
- The continuous wiring structure of CPLD determines that its timing delays are uniform and predictable, while the segmented wiring structure of FPGA determines the unpredictability of its delays.
- CPLD is programmed by modifying the logic function with fixed interconnect circuit, while FPGA is programmed by changing the wiring of internal wires.
- FPGA can be programmed under the logic gate, while CPLD is programmed under the logic block.

Q12.35 Differentiate between word capacity and word size. ***(AKUB, Dec'20, 2 Marks)***

The word capacity refers to number storage locations in a memory. The word size refers to the size of binary that can be stored in a memory location. For example, a memory of capacity 1024 × 16, has the word capacity of 1024 locations with word size of 16-bits, so that 1024 numbers of 16-bit binary words can be stored in 1024 × 16 memory.

12.13 Exercises

I. Fill in the blanks

1. _______ is a storage device used to store binary information.
2. The basic storage unit in a memory device is called _______ .
3. A memory cell is used to store _______ bit of binary information.
4. Each memory location has a unique binary number called _______ .
5. The memory which needs refreshing is _______ .
6. The RAM which uses capacitors to store 1 and 0 is _______ .
7. The number of address lines in a 64 $k \times 8$ ROM are _______ .
8. The static RAM (Random Access Memory) is a _______ memory.
9. Binary information is stored in group of bits called _______.
10. The process of storing binary information in a memory location is called _______.
11. The process of reading back the stored information from the memory location is called _______.
12. The address is used to access the location for _______ and _______ operation
13. A PLA can be used to realize a _______ .
14. A ______ integrates a number of PLDs in a single device.
15. FPGA is a programmable semiconductor logic device consisting of large number of ______.
16. FPGAs store their customized configuration data in ______ type internal latches.
17. ______ is the basic logic resource on an FPGA.
18. The ______ are the wiring resources in FPGAs.
19. The ______ are devices that connect a logic element to an interconnect wire and/or one interconnect wire to another.

20. ______ is a programmable logic device to implement a Boolean function in sum of products form

21. The PLA is a ______ programmable device to implement a Boolean function in sum of products form

Answers

1. memory	8. read/write	15. logic gates
2. binary cell	9. byte	16. SRAM
3. one	10. write operation	17. Configurable logic block (CLB)
4. address	11. read operation	18. programmable routing
5. DRAM	12. read, write	19. programmable swithces
6. dynamic RAM	13. combinational logic	20. PAL
7. 16	14. CPLD	21. one-time

II. State whether the following statements are True or False

1. Binary information is stored in group of bits called byte or word.
2. The process of storing binary information in a memory location is called read operation.
3. RAM is a semiconductor memory which permits only a read access.
4. A 16 × 8 ROM stores eight 16-bit data words.
5. EPROM memory is nonvolatile and also has the feature of random access for read operation.
6. EEPROM/EAPROM is a volatile read/write memory.
7. A dynamic RAM cell consists of 1 transistor and 1 capacitor.
8. OTP-ROM is one-time programmable and once programmed the content cannot be altered and can only read by the system
9. The drawback in linear address decoding is that it require a single large decoder
10. The process of storing information in ROM is called programming
11. PLA is a one-time programmable device to implement a Boolean function in sum of products form.
12. A CPLD integrates a number of PLDs in a single device.
13. FPGA is a programmable semiconductor logic device consisting of large number of logic gates.
14. FPGAs store their customized configuration data in DRAM-type internal latches.
15. A configurable logic block (CLB) is the basic logic resource on an CPLD.
16. Carry look-ahead logic is implemented in each CLB with a combination of dedicated multiplexer and XOR gates
17. The programmable switches are user-programmable switch that can connect a logic element to an interconnect wire and/or one interconnect wire to another.
18. CPLD is mainly based on SRAM memory programming, which can be programmed up to 10000 times.
19. The inputs of a combinational circuit can be used as address of ROM.

Answers

1. True	6. True	11. True	16. True
2. True	7. True	12. True	17. True
3. True	8. False	13. True	18. False
4. False	9. False	14. False	19. False
5. False	10. True	15. False	

III. Choose the right answer for the following questions

1. *Number of RAM needed to form a RAM with 4 bit word length using 256 × 1 RAM is,*

   a) 1 **b)** 2 **c)** 3 **d)** 4

2. *A memory has a capacity 16 k × 8. The total number of memory cells in the memory,*

   a) 16000 **b)** 16384 **c)** 131072 **d)** 8000

3. *Battery powered equipments use*

   a) PROM **b)** EPROM **c)** Static RAM **d)** Dynamic RAM

4. *The capacity of 2 k × 16 PROM is to be expanded to 16k × 16. The number of PROM chips required is*

   a) 2 **b)** 4 **c)** 8 **d)** 3

5. *A memory of 4 k means*

   a) 10^8 **b)** 8000 **c)** 4 × 1024 **d)** 2^4

6. *How many address lines are needed to address each memory location in a 2048 × 4 memory chip?*

   a) 10 **b)** 11 **c)** 12 **d)** 18

7. *A certain memory has a capacity of 8 k × 8. Then the number of address lines are*

   a) 8 **b)** 10 **c)** 13 **d)** 16

8. *A computer has a memory space of 2^{16} and word length of 24 bits. The memory chips available have 10 address and 8 data lines. The number of chips required for the computer memory space is,*

   a) 192 **b)** 256 **c)** 512 **d)** 1024

9. *How many 128 × 8 RAM chips are needed to access 2048 bytes?*

   a) 14 **b)** 15 **c)** 16 **d)** 11

10. *How many 128 × 8 RAM chips are needed to access 4096 bytes?*

    a) 14 **b)** 15 **c)** 16 **d)** 32

11. *The minimum number of MOS transistors required to make a dynamic RAM cell is,*

    a) 1 **b)** 2 **c)** 3 **d)** 4

12. *Each cell of a static random access memory contains.*

    a) 6 MOS transistors

    b) 4 MOS transistors and 2 capacitors

    c) 2 MOS transistors and 4 capacitors

    d) 1 MOS transistors and 1 capacitor

13. *Choose the correct statement(s) from the following*

a) PROM contains a programmable AND array and a fixed OR array.

b) PLA contains a fixed AND array and a programmable OR array.

c) PROM contains a fixed AND array and a programmable OR array.

d) PLA contains a programmable AND array and a programmable OR array.

14. *In FPGA store their customized configuration data in ______ type internal latches.*

a) SRAM

b) DRAM

c) RAM

d) ROM

15. *A configurable logic block (CLB) is the basic logic resource on an _______.*

a) CPLD

b) FPGA

c) PLD

d) LUT

16. *PAL refers to*

a) Programmable array loaded

b) Programmable logic array

c) Programmable array logic

d) Programmable AND logic

17. *PLA contains*

a) AND and OR arrays

b) NAND and OR arrays

c) NOT and AND arrays

d) NOR and OR arrays

18. *If a PAL has been programmed once*

a) Its logic capacity is lost

b) Its outputs are only active HIGH

c) Its outputs are only active LOW

d) It can be reprogrammed

Answers

1. d	6. b	11. a	16. c
2. c	7. c	12. b	17. a
3. c	8. a	13. d	18. d
4. c	9. c	14. a	
5. c	10. d	15. b	

IV. Answer the following questions

1. What is binary cell.
2. What is address multiplexing in memory.
3. Define bit, byte or word.
4. Differentiate volatile and non-volatile memory?
5. What is mask programmable ROM?
6. Why RAMs are called volatile?
7. Explain PROM
8. What is static RAM.
9. Compare PROM, PAL, PLA.
10. What is PAL.
11. Why was PAL developed?
12. Compare PROM, PAL, PLA.
13. Draw the basic architecture of CPLD.
14. What are the advantages of FPGA.
15. List the features of spartan-6 family of FPGA.
16. What are the elements of spartan-6 FPGA architecture.

V. Solve the following problems

E12.1 Design a 8 × 8 memory array using 4 × 4 memory array.

E12.2 Implement the given Boolean functions using PLA.

$F_1(A, B, C) = \sum m(0, 2, 6, 7)$ **and** $F_2(A, B, C) = \sum m(1, 3, 4, 5, 7)$

E12.3 Implement the given Boolean functions using PAL.

$F_1(A, B, C) = \sum m(0, 4, 6, 7)$ **and** $F_2(A, B, C) = \sum m(1, 2, 3, 5, 7)$

E12.4 Implement the following logical functions using PLA.

$Y_1 = AB + A'C + ABCD,\ Y_2 = AB'C + ABC' + AC',\ Y_3 = AB$

E12.5 Implement the following logical functions using PAL.

$Y_1 = AB + A'CB' + ABC',\ Y_2 = AB'C + AB + AC',\ Y_3 = AB + BC + CA$

E12.6 Implement the following functions using a single PROM grid.

$F_1(x, y, z) = \sum m(1, 2, 4, 5),\ F_2(x, y, z) = \sum m(0, 1, 3, 4),\ F_3(x, y, z) = \sum m(2, 3, 6, 7),\ F_4(x, y, z) = \sum m(1, 4, 5, 7)$

E12.7 Implement the following logical functions using PLA.

$F_1 = xz' + y',\quad F_2 = xyz' + x'yz' + x'y'z',\quad F_3 = xz' + x'y'z$

E12.8 Implement the full subtractor using PLA by deriving the PLA programming table.

E12.9 Implement the given Boolean functions using PLA.

$F_1(x, y, z) = \sum m(0, 3, 4, 7),\ F_2(x, y, z) = \sum m(0, 2, 4, 6)$ **and** $F_3(x, y, z) = \sum m(2, 5, 6, 7)$

Answers

E12.1

a_0
a_1
a_2
3-to-2^3 Decoder
4×4 memory (Lower 4-bits of 8-bit data)
4×4 memory (Upper 4-bits of 8-bit data)
4×4 memory (Lower 4-bits of 8-bit data)
4×4 memory (Upper 4-bits of 8-bit data)
d_0 d_1 d_2 d_3 d_4 d_5 d_6 d_7
4×8 memory (Lower 4×8 location of 8×8 memory)
4×8 memory (Upper 4×8 location of 8×8 memory)

***Fig:** 8 × 8 memory using 4 × 4 memory.*

E12.2

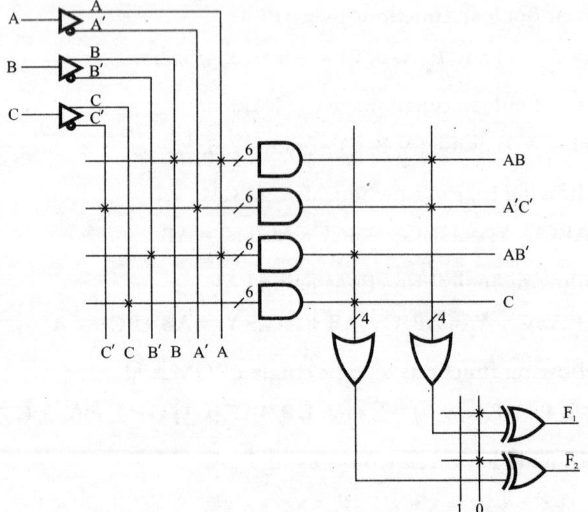

***Fig:** Simplified structure of combinational circuit implemented in PLA.*

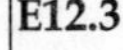

E12.3

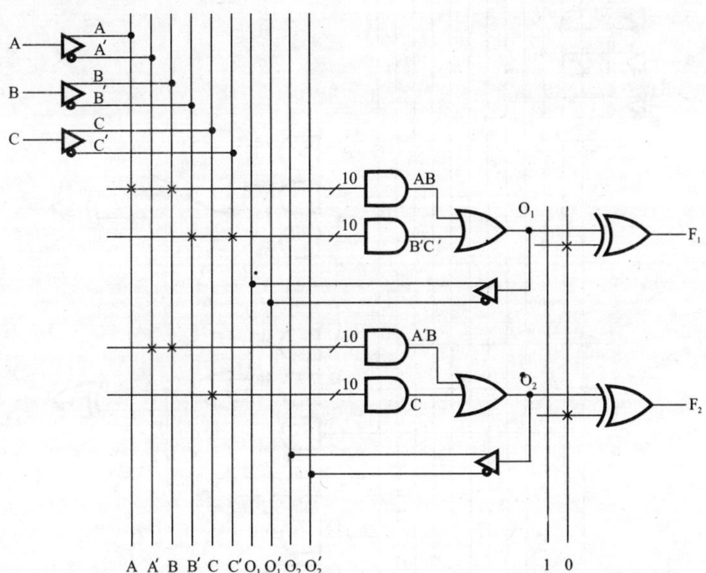

Fig: Structure of combinational circuit implemented in PAL.

E12.4

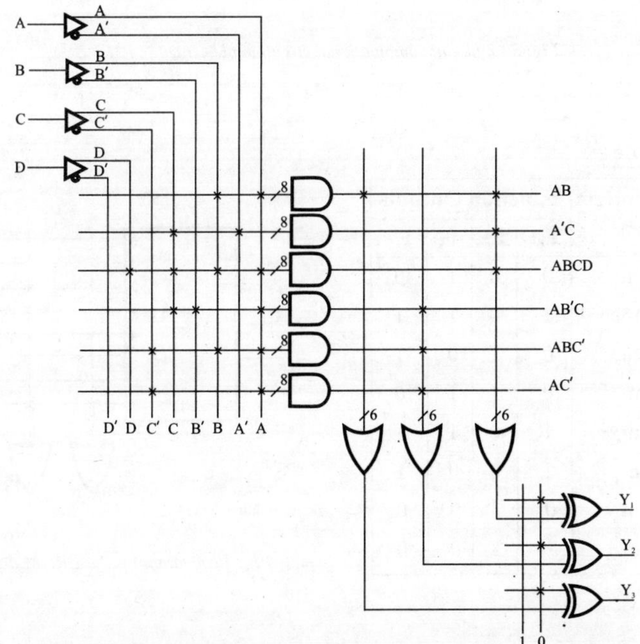

Fig: Simplified structure of combinational circuit implemented in PLA.

E12.5

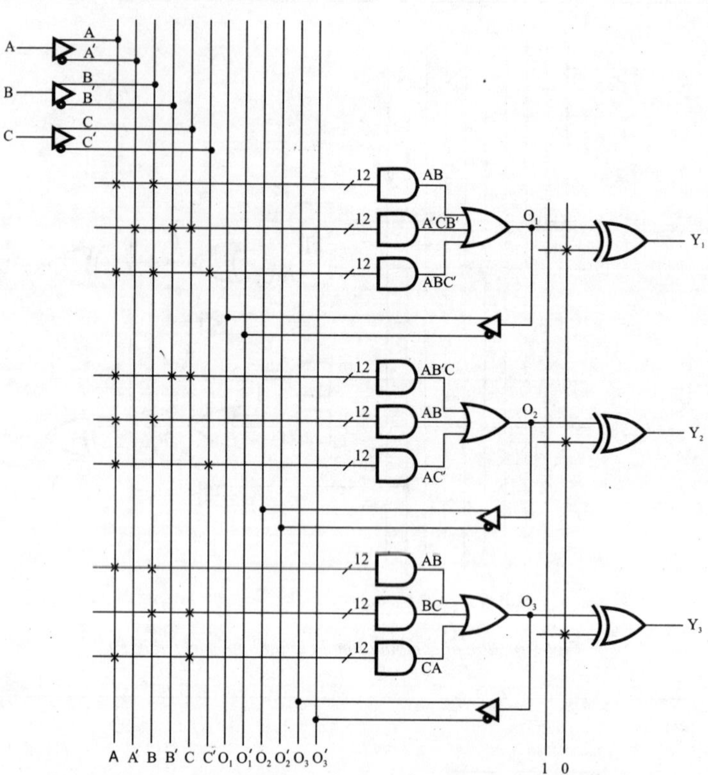

Fig: Structure of combinational circuit implemented in PAL.

E12.6

Table 1: Truth Table

Inputs			Minterm	Function Outputs			
x	y	z		F_4	F_3	F_2	F_1
0	0	0	m_0	0	0	1	0
0	0	1	m_1	1	0	1	1
0	1	0	m_2	0	1	0	1
0	1	1	m_3	0	1	1	0
1	0	0	m_4	1	0	1	1
1	0	1	m_5	1	0	0	1
1	1	0	m_6	0	1	0	0
1	1	1	m_7	1	1	0	0

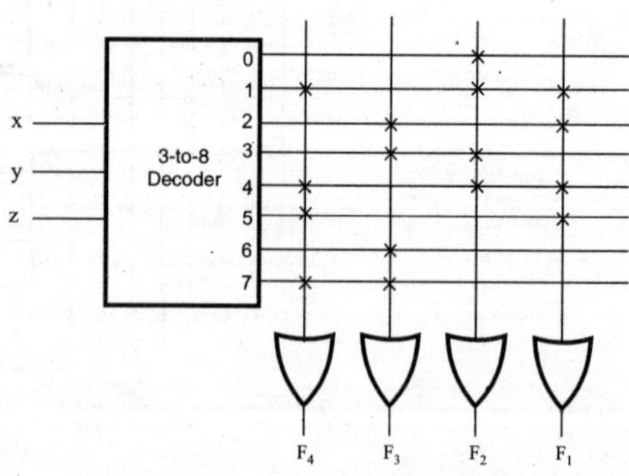

Fig: Programming the PROM according to Table 1.

E12.7

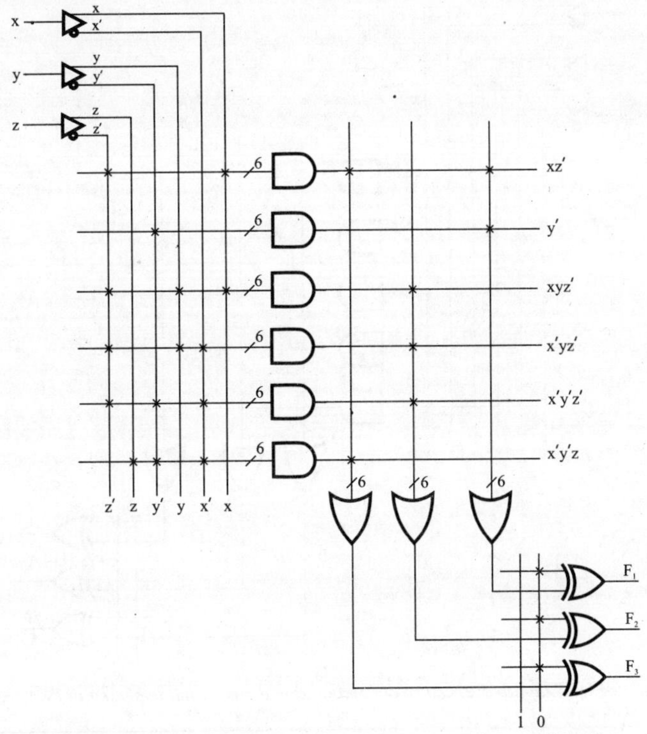

Fig: Simplified structure of combinational circuit implemented in PLA.

E12.8

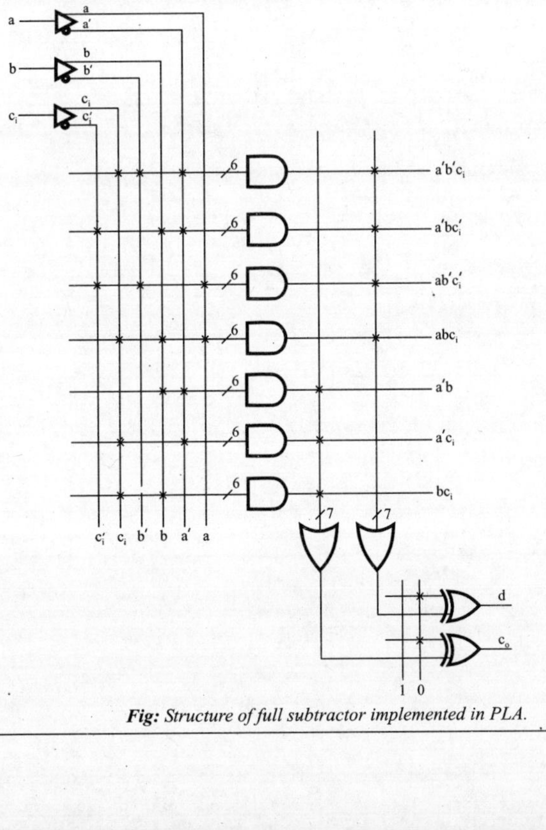

Fig: Structure of full subtractor implemented in PLA.

E12.9

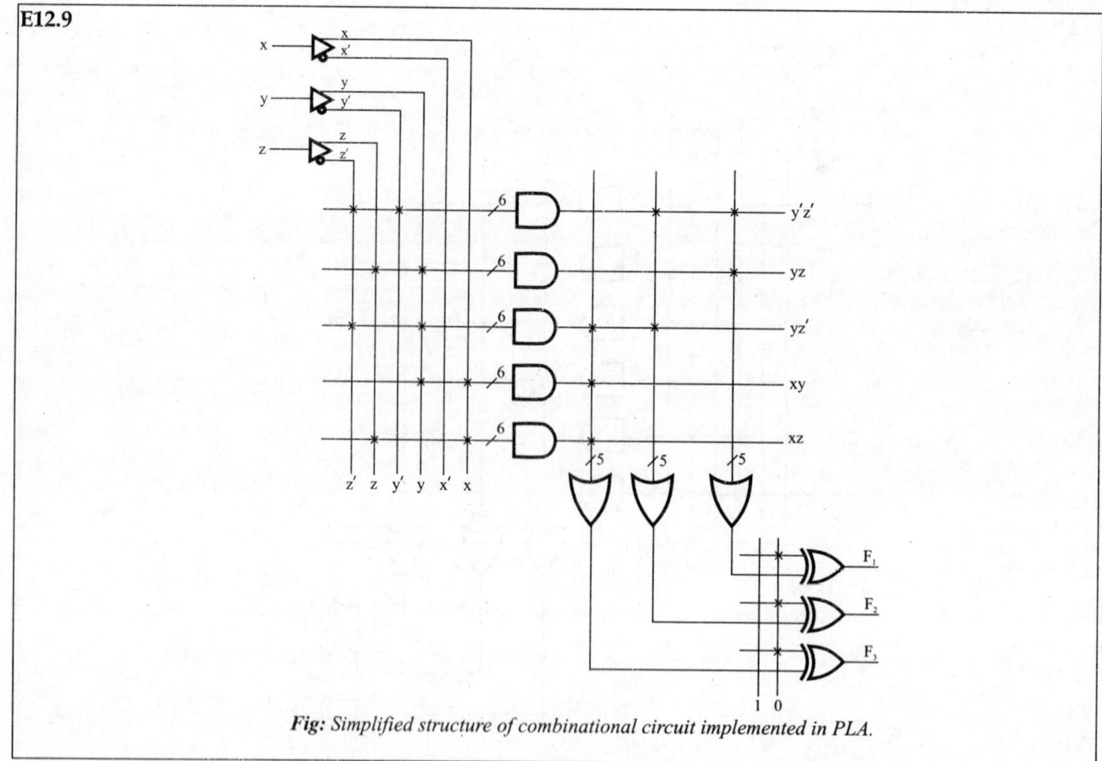

Fig: *Simplified structure of combinational circuit implemented in PLA.*

CHAPTER 13

DATA CONVERTERS

13.1 Introduction

Processing of various types of signals is an important requirement in engineering. Since the real world signals are analog in nature, before the invention of digital systems the signals were processed by analog systems. After the invention of digital systems, it becomes more convenient to process signals by digital systems.

In order to process signals by digital systems the analog signal has to be converted to digital signal and after processing, the processed digital signal has to be converted back to analog signal for real world application. The conversion of analog to digital signal is performed using a device called Analog to Digital Converter (ADC). The conversion of digital to analog signal is performed using a device called Digital to Analog Converter (DAC). In general, the ADC and DAC are referred to as data converters.

The ADC will generate an n-bit binary corresponding to analog input voltage signal. The n-bit binary can have 2^n different binary values. Every ADC will have a positive and negative voltage reference. The input analog voltage should be within this reference value. The ADC will divide the difference between positive and negative reference into $2^n - 1$ steps and it can generate a unique n-bit binary for each step of reference value. When the input analog value matches one of the steps of reference value the ADC will output the corresponding n-bit binary.

The DAC will generate an analog voltage or current signal corresponding to input n-bit binary. The n-bit binary can have 2^n different binary values. Every DAC will have a positive and negative voltage/ current reference. The generated output analog voltage/current will be within this reference value. The DAC will divide the difference between positive and negative reference into $2^n - 1$ steps which can be matched to $2^n - 1$ binary values of n-bit excluding zero. For every input n-bit binary, the DAC will generate the correspond step of analog value.

Note: *Since one of the 2^n binary value is 0, the reference value is divided into $2^n - 1$ steps and matched to $2^n - 1$ binary values.*

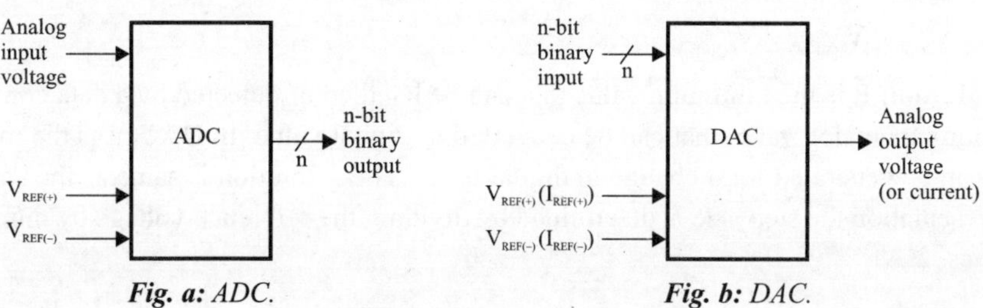

Fig. a: ADC. *Fig. b: DAC.*

Fig. 13.1: Data converters.

The ADC is also known as A/D converter and the DAC is known as D/A converter. The process of analog to digital conversion is also referred to as A/D process. Similarly the process of digital to analog conversion is referred to as D/A process.

Since, different methods are available for analog-to-digital conversion and vice-versa, there are large number of types of ADCs and DACs.

Some of the popular types of ADCs are,

- Flash type ADC
- Counter type ADC
- Tracking type ADC
- Dual slope ADC
- Successive Approximation ADC

Some of the different types of DACs are,

- R-2R ladder type DAC
- Weighted resistor type DAC
- Switched-current source type DAC
- Switched capacitor type DAC

13.1.1 Parameters of Data Converters

(AKUB, Dec'20, 4 Marks)

Quantization: The process of dividing the reference value is called quantization and each step of analog value is also called quantization step or simply step size. In n-bit binary representation the step size is determined by dividing the reference voltage by number of possible binary values, 2^n.

$$\therefore \text{ Step size} = V_{REF} \times \frac{1}{2^n}$$

Coding: Attaching a unique n-bit binary code to each quantization step is called coding.

Reference value: The data converter will have a positive and negative reference analog voltage/ current to indicate the maximum value of positive and negative analog value it can handle for conversion from analog to digital or vice-versa. The analog value handled by data converters should be within the range of positive and negative reference value. The difference between the positive and negative reference is the reference value. In unipolar data converters the negative reference is ground or 0 volt and hence reference value is equal to positive reference.

Let, $V_{REF(+)}$ = Positive reference value

$V_{REF(-)}$ = Negative reference value

$$\therefore V_{REF} = V_{REF(+)} - V_{REF(-)}$$

Resolution: It is the minimum value that can be handled or detected by a data converter. In ADC, it is the minimum analog value that can be converted to digital value. In DAC, it is the minimum analog value that can be generated for a change in digital input. The resolution is same as the step size. In n-bit binary representation the step size is determined by dividing the reference voltage by number of possible binary values, 2^n.

$$\therefore \text{ Resolution} = V_{REF} \times \frac{1}{2^n}$$

The resolution of a DAC is also defined as the product of reciprocal of the number of discrete steps and the full-scale output of the DAC.

$$\therefore \text{Resolution} = \frac{\text{Full-scale value}}{\text{Number of steps}} = \frac{\text{Full-scale value}}{2^n - 1}$$

Full scale value: It is the maximum value of analog voltage or current that can be represented in digital by a data converter. Full scale value will be one step lesser than reference value.

$$\therefore \text{Full scale value} = V_{REF} - \text{Resolution}$$

Since number of steps of analog values is $2^n - 1$ the full scale value is also given by product of number of steps and resolution.

$$\therefore \text{Full scale value} = \text{Number of steps} \times \text{Step size} = (2^n - 1) \times \text{Step size}$$

Conversion time: It is the time taken by data converter to convert an analogue signal to digital or vice-versa. The conversion time depends on the type of ADC/DAC. ***(GTU, May'19, 4 Marks)***

Settling Time: It is the minimum time required for converted signal to settle at a stable value for a change in input.

Accuracy: It refers to closeness of converted value to true value. With smaller resolution higher accuracy can be achieved. For smaller resolution the binary data size, n of the converter should be increased. For example, a 10-bit converter will have more accuracy than 8-bit converter.

Linearity: It refers to direct proportionality between input and output. For linear relationship, the relation between input and output should follow a straight line.

13.2 Digital to Analog Converter (DAC)

The DAC will accept a digital (binary) input and convert to analog voltage or current. Every DAC will have n input data lines and an analog output line. The DAC requires a reference analog voltage (V_{REF}) or current (I_{REF}) source. The reference value is the difference between positive and negative reference value. The smallest possible analog value that can be represented by the n-bit binary code is called resolution. The resolution of DAC with n-bit binary input is $\frac{1}{2^n}$ of the reference analog value.

$$\therefore \text{Resolution} = \frac{1}{2^n} \times V_{REF}$$

In general, for an n-bit DAC, the number of different binary values will be 2^n including zero and therefore, the number of analog steps will be $2^n - 1$. The greater the number of bits, the greater will be the number of steps and the smaller will be the step size, and therefore, the finer will be the resolution. Of course the cost of the DAC increases with the number of input bits.

Every analog output will be a multiple of the resolution. For example, consider an 4-bit DAC with reference analog voltage of 5 *Volts*. Now the resolution of the DAC is $(1/2^4) \times 5$ *Volts*. The 4-bit digital input can take, $2^4 = 16$ different values which is given by the product of decimal value of digital input and resolution. The analog values for all possible digital inputs are as shown in Table 13.1.

$$\text{Output voltage, } V_o = D_{IN} \times \frac{1}{2^4} \times 5\,Volts$$

Where, D_{IN} = Decimal equivalent of 4-bit binary input

Table 13.1: Analog Value for Digital Input

Digital Input	Decimal Value	Analog Output	Digital Input	Decimal Value	Analog Output
0000	0	$0 \times \frac{1}{2^4} \times 5 = 0\,V$	1000	8	$8 \times \frac{1}{2^4} \times 5 = 2.5\,V$
0001	1	$1 \times \frac{1}{2^4} \times 5 = 0.3125\,V$	1001	9	$9 \times \frac{1}{2^4} \times 5 = 2.8125\,V$
0010	2	$2 \times \frac{1}{2^4} \times 5 = 0.625\,V$	1010	10	$10 \times \frac{1}{2^4} \times 5 = 3.125\,V$
0011	3	$3 \times \frac{1}{2^4} \times 5 = 0.9375\,V$	1011	11	$11 \times \frac{1}{2^4} \times 5 = 3.4375\,V$
0100	4	$4 \times \frac{1}{2^4} \times 5 = 1.25\,V$	1100	12	$12 \times \frac{1}{2^4} \times 5 = 3.75\,V$
0101	5	$5 \times \frac{1}{2^4} \times 5 = 1.5625\,V$	1101	13	$13 \times \frac{1}{2^4} \times 5 = 4.0625\,V$
0110	6	$6 \times \frac{1}{2^4} \times 5 = 1.875\,V$	1110	14	$14 \times \frac{1}{2^4} \times 5 = 4.375\,V$
0111	7	$7 \times \frac{1}{2^4} \times 5 = 2.1875\,V$	1111	15	$15 \times \frac{1}{2^4} \times 5 = 4.6875\,V$

The maximum input digital signal will have an analog value which is equal to reference analog value minus resolution. The maximum analog value is also called full-scale value.

In the output of 4-bit DAC listed in Table 13.1, the resolution is given by,

$$\text{Resolution} = \frac{1}{2^n} \times V_{REF} = \frac{1}{2^4} \times 5 = 0.3125\ V$$

$$\therefore \text{ Maximum analog value} = V_{REF} - \text{Resolution} = 5 - 0.3125 = 4.6875\ V$$

(Full scale value)

13.2.1 R-2R Ladder Type DAC

(AKTU, Sep'20, 5 Marks) (JNTUH, Nov/Dec'20, 8 Marks)

The basic components of a R-2R ladder type DAC are resistive network with appropriate values, switches, a reference source, a current to voltage (I to V) converter and a sign changer as shown in Fig. 13.2.

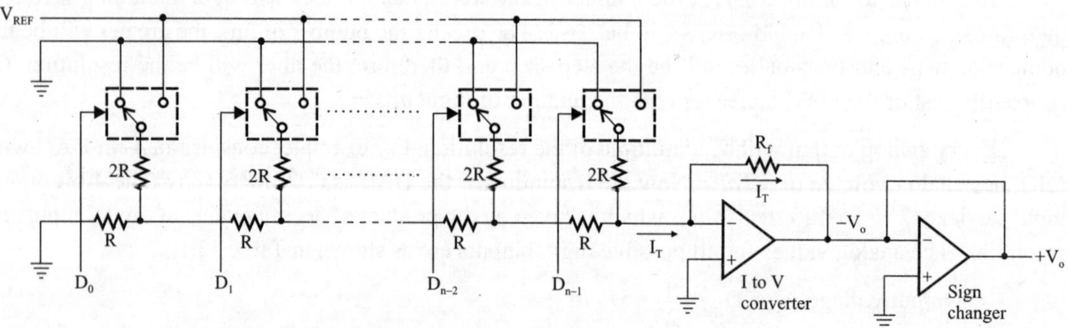

Fig. 13.2: A typical R-2R ladder resistive network as DAC.

The switches in the circuit of Fig. 13.2 can be transistors which connect the resistance either to ground or V_{REF}. The resistors are connected in such a way that for any possible binary input, the total current I_T is in binary proportion. The operational amplifier converts the current I_T to a voltage signal V_o which can be calculated from the following equation.

$$V_o = V_{REF} \frac{R_f}{R}\left(\frac{D_{n-1}}{2^1} + \frac{D_{n-2}}{2^2} + \dots\dots + \frac{D_1}{2^{n-1}} + \frac{D_0}{2^n}\right)$$

For a 3-bit DAC the output voltage is given by,

$$V_o = V_{REF} \frac{R_f}{R}\left(\frac{D_2}{2^1} + \frac{D_1}{2^2} + \frac{D_0}{2^3}\right)$$

Let, $V_{REF} = 5\text{ V}, \quad R_f = R$

$$\therefore\ V_o = 5 \times \left(\frac{D_2}{2} + \frac{D_1}{4} + \frac{D_0}{8}\right)$$

Here, $D_2D_1D_0$ will take 8 values 000, 001, 010, 011, 100, 101, 110 and 111. The 8 possible values of V_o are,

$$V_o = 5 \times \left(\frac{0}{2} + \frac{0}{4} + \frac{0}{8}\right) = 0\,V$$

$$V_o = 5 \times \left(\frac{0}{2} + \frac{0}{4} + \frac{1}{8}\right) = 0.625\,V$$

$$V_o = 5 \times \left(\frac{0}{2} + \frac{1}{4} + \frac{0}{8}\right) = 1.25\,V$$

$$V_o = 5 \times \left(\frac{0}{2} + \frac{1}{4} + \frac{1}{8}\right) = 1.875\,V$$

$$V_o = 5 \times \left(\frac{1}{2} + \frac{0}{4} + \frac{0}{8}\right) = 2.5\,V$$

$$V_o = 5 \times \left(\frac{1}{2} + \frac{0}{4} + \frac{1}{8}\right) = 3.125\,V$$

$$V_o = 5 \times \left(\frac{1}{2} + \frac{1}{4} + \frac{0}{8}\right) = 3.75\,V$$

$$V_o = 5 \times \left(\frac{1}{2} + \frac{1}{4} + \frac{1}{8}\right) = 4.375\,V$$

For an 8-bit DAC the output voltage is given by,

$$V_o = V_{REF} \frac{R_f}{R}\left(\frac{D_7}{2^1} + \frac{D_6}{2^2} + \frac{D_5}{2^3} + \frac{D_4}{2^4} + \frac{D_3}{2^5} + \frac{D_2}{2^6} + \frac{D_1}{2^7} + \frac{D_0}{2^8}\right)$$

The time required for converting the digital signal to analog signal is called **conversion time.** It depends on the response time of the switching transistors and the output amplifier. The digital data (signal) should remain at the input of DAC, until the conversion is complete. Hence, to hold the data a latch is provided at the input of DAC.

13.2.2 Weighted-Resistor DAC

(GTU, May'19, 7 Marks) (IKGPTU, May'20, 5 Marks)

The diagram of the weighted-resistor DAC is shown in Fig. 13.3. The operational amplifier act as an adder (or summer) and it generates a weighted sum of the digital inputs. The resistor values are weights which are proportional to the weights of the bit positions of inputs.

Since the operational amplifier is connected as an inverting amplifier, each input is amplified by a factor equal to the ratio of the feedback resistance divided by the input resistance to which it is connected. The output of operational amplifier will be a negative voltage and hence the output voltage is inverted by connecting an analog inverter or sign changer.

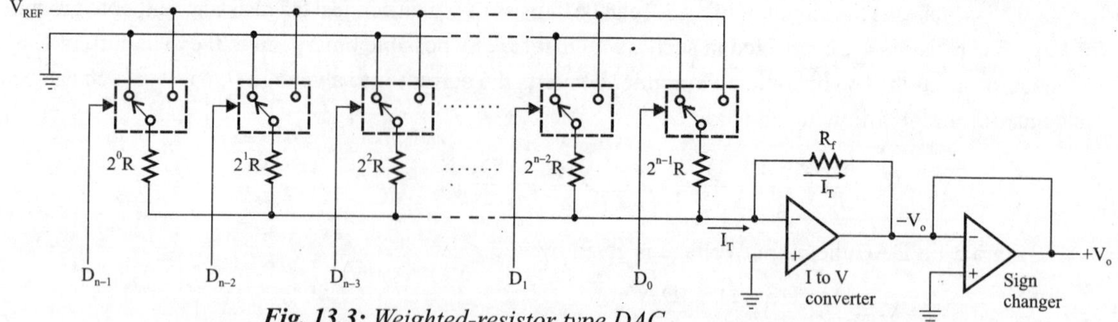

Fig. 13.3: Weighted-resistor type DAC.

The equation for output voltage, V_o of n-bit weighed resistor type DAC is given by,

$$V_o = V_{REF}\frac{R_f}{R}\left(\frac{D_{n-1}}{2^0}+\frac{D_{n-2}}{2^1}+\ldots\ldots+\frac{D_1}{2^{n-2}}+\frac{D_0}{2^{n-1}}\right)$$

For a 3-bit DAC the output voltage is given by,

$$V_o = V_{REF}\frac{R_f}{R}\left(\frac{D_2}{2^0}+\frac{D_1}{2^1}+\frac{D_0}{2^2}\right)$$

Let, $V_{REF} = 5$ V, $R_f = R$

$$\therefore\ V_o = 5\times\left(D_2+\frac{D_1}{2}+\frac{D_0}{4}\right)$$

Here, $D_2D_1D_0$ will take 8 values 000, 001, 010, 011, 100, 101, 110 and 111. The 8 possible values of V_o are,

$$V_o = 5\times\left(0+\frac{0}{2}+\frac{0}{4}\right)=0\,V$$

$$V_o = 5\times\left(0+\frac{0}{2}+\frac{1}{4}\right)=1.25\,V$$

$$V_o = 5\times\left(0+\frac{1}{2}+\frac{0}{4}\right)=2.5\,V$$

$$V_o = 5\times\left(0+\frac{1}{2}+\frac{1}{4}\right)=3.75\,V$$

$$V_o = 5\times\left(1+\frac{0}{2}+\frac{0}{4}\right)=5\,V$$

$$V_o = 5\times\left(1+\frac{0}{2}+\frac{1}{4}\right)=6.25\,V$$

$$V_o = 5\times\left(1+\frac{1}{2}+\frac{0}{4}\right)=7.5\,V$$

$$V_o = 5\times\left(1+\frac{1}{2}+\frac{1}{4}\right)=8.75\,V$$

The main disadvantage in this type of DAC is that the accuracy depends on the precision of the resistors and hence high precision resistor has to be employed for good accuracy.

The R-2R ladder network DAC and the weighted-resistor DAC can be regarded as switched voltage-source DACs, because when an input position goes **high**, the **high** voltage is effectively switched into the circuit, where it is summed up with other input voltages

13.2.3 Switched Current-Source Type DAC

The switched current source type DAC employs current switches which are faster than voltage switches. The currents are weighted according to the bit positions they represent and are summed up in an operational amplifier. Fig. 13.4 shows the simplified circuit of a current-switching type DAC.

The current that flows in the kth 2R resistor is given by, $I_k = (V_{REF}/R) \times 1/2^{n-k}$ where k = 0, 1,...., n – 1. Thus, each current is weighted by the bit position it represents.

The current switches are controlled by the digital input and connect the current either to ground or to the input of the operational amplifier. The amplifier also serves as a current to voltage converter. The output of operational amplifier will be a negative voltage and hence the output voltage is inverted by connecting an analog inverter or sign changer.

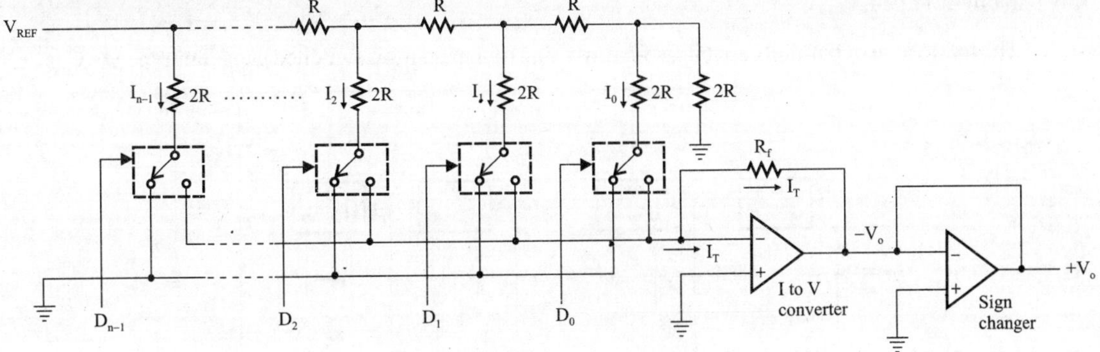

Fig. 13.4: *Switched current source DAC.*

The equation for output voltage, V_o of n-bit switched current source type DAC is given by,

$$V_O = V_{REF}\frac{R_f}{R}\left(\frac{D_{n-1}}{2^1} + \frac{D_{n-2}}{2^2} + ... + \frac{D_1}{2^{n-1}} + \frac{D_0}{2^n}\right)$$

For a 3-bit DAC the output voltage is given by,

$$V_O = V_{REF}\frac{R_f}{R}\left(\frac{D_2}{2^1} + \frac{D_1}{2^2} + \frac{D_0}{2^3}\right)$$

Let, $V_{REF} = 5\text{ V}$, $R_f = R$

$$\therefore\ V_O = 5 \times \left(\frac{D_2}{2} + \frac{D_1}{4} + \frac{D_0}{8}\right)$$

Here, $D_2D_1D_0$ will take 8 values 000, 001, 010, 011, 100, 101, 110 and 111. The 8 possible values of V_o are,

$$V_O = 5 \times \left(\frac{0}{2} + \frac{0}{4} + \frac{0}{8}\right) = 0\,V$$

$$V_O = 5 \times \left(\frac{0}{2} + \frac{0}{4} + \frac{1}{8}\right) = 0.625\,V$$

$$V_O = 5 \times \left(\frac{0}{2} + \frac{1}{4} + \frac{0}{8}\right) = 1.25\,V$$

$$V_O = 5 \times \left(\frac{0}{2} + \frac{1}{4} + \frac{1}{8}\right) = 1.875\,V$$

$$V_O = 5 \times \left(\frac{1}{2} + \frac{0}{4} + \frac{0}{8}\right) = 2.5\,V$$

$$V_O = 5 \times \left(\frac{1}{2} + \frac{0}{4} + \frac{1}{8}\right) = 3.125\,V$$

$$V_O = 5 \times \left(\frac{1}{2} + \frac{1}{4} + \frac{0}{8}\right) = 3.75\,V$$

$$V_O = 5 \times \left(\frac{1}{2} + \frac{1}{4} + \frac{1}{8}\right) = 4.375\,V$$

13.2.4 Switched Capacitor Type DAC

The switched-capacitor type DAC employs weighted capacitors. In this method, charged capacitors form a capacitive voltage divider whose output is proportional to the sum of the product of binary digits and its corresponding weights.

Fig. 13.5 shows an n-bit switched-capacitor DAC. Here, the capacitance values are proportional to binary weights. A two-phase clock is used to control the switching of the capacitors. When phase-1 of clock goes **high**, all capacitors are switched to ground and discharged. When phase-2 of clock goes **high**, those capacitors whose digital inputs are **high** are switched to V_{REF} and those capacitors whose inputs are **low** remain grounded.

The total value of parallel capacitors for any value of n in an n-bit switched capacitor type DAC is 2C.

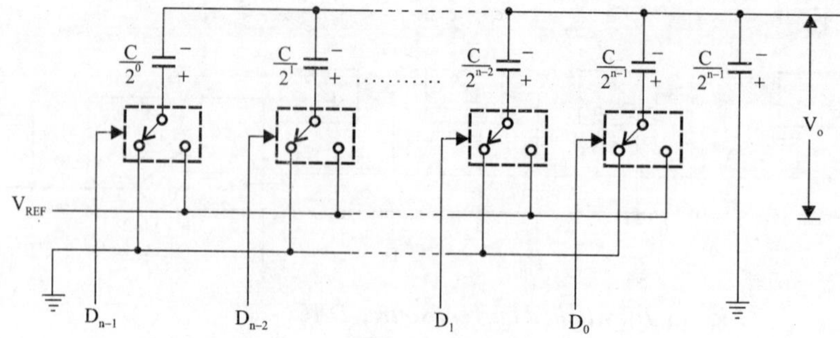

***Fig. 13.5:** Switched capacitor type DAC.*

The equation for output voltage, V_o of n-bit switched capacitor type DAC is given by,

$$V_O = V_{REF}\frac{1}{2C}\left(D_{n-1}\frac{C}{2^0} + D_{n-2}\frac{C}{2^1} + ... + D_1\frac{C}{2^{n-2}} + D_0\frac{C}{2^{n-1}}\right)$$

$$= V_{REF}\left(\frac{D_{n-1}}{2^1} + \frac{D_{n-2}}{2^2} + ... + \frac{D_1}{2^{n-1}} + \frac{D_0}{2^n}\right)$$

For a 3-bit DAC the output voltage is given by,

$$V_O = V_{REF}\left(\frac{D_2}{2^1} + \frac{D_1}{2^2} + \frac{D_0}{2^3}\right)$$

Let, $V_{REF} = 5$ V

$$\therefore\ V_O = 5\times\left(\frac{D_2}{2} + \frac{D_1}{4} + \frac{D_0}{8}\right)$$

Here, $D_2D_1D_0$ will take 8 values 000, 001, 010, 011, 100, 101, 110 and 111. The 8 possible values of V_o are,

$$V_O = 5\times\left(\frac{0}{2} + \frac{0}{4} + \frac{0}{8}\right) = 0\,V$$

$$V_O = 5\times\left(\frac{0}{2} + \frac{0}{4} + \frac{1}{8}\right) = 0.625\,V$$

$$V_O = 5\times\left(\frac{0}{2} + \frac{1}{4} + \frac{0}{8}\right) = 1.25\,V$$

$$V_O = 5\times\left(\frac{0}{2} + \frac{1}{4} + \frac{1}{8}\right) = 1.875\,V$$

$$V_O = 5\times\left(\frac{1}{2} + \frac{0}{4} + \frac{0}{8}\right) = 2.5\,V$$

$$V_O = 5\times\left(\frac{1}{2} + \frac{0}{4} + \frac{1}{8}\right) = 3.125\,V$$

$$V_O = 5\times\left(\frac{1}{2} + \frac{1}{4} + \frac{0}{8}\right) = 3.75\,V$$

$$V_O = 5\times\left(\frac{1}{2} + \frac{1}{4} + \frac{1}{8}\right) = 4.375\,V$$

13.2.5 DAC0800

The DAC0800 is an 8-bit, high speed, current output DAC with a typical settling time (conversion time) of 100 ns. It produces complementary current output which can be converted to voltage by using a simple resistor load. The DAC0800 employs binary weighted resistors to convert digital data to analog current. The DAC0800 is available as a 16-pin IC in DIP. The pin configuration of DAC0800 is shown in Fig. 13.6 and the internal block diagram of a DAC0800 is shown in Fig. 13.7. The reference voltage and the digital input will decide the analog output current, which can be converted to a voltage by simply connecting a resistor to output terminal or by using an op-amp I to V converter.

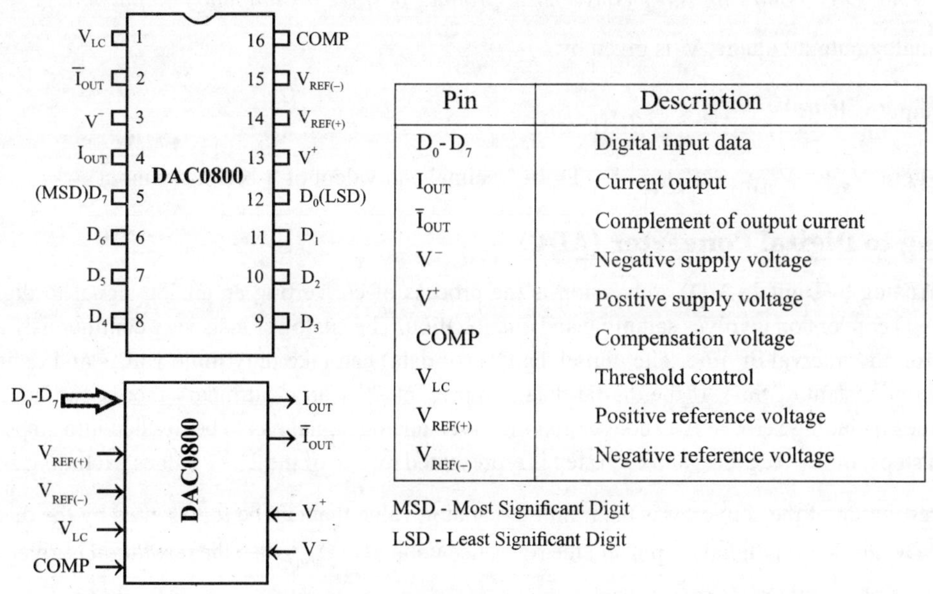

Pin	Description
D_0-D_7	Digital input data
I_{OUT}	Current output
$\bar{I}_{OUT}$	Complement of output current
V^-	Negative supply voltage
V^+	Positive supply voltage
COMP	Compensation voltage
V_{LC}	Threshold control
$V_{REF(+)}$	Positive reference voltage
$V_{REF(-)}$	Negative reference voltage

MSD - Most Significant Digit
LSD - Least Significant Digit

Fig. 13.6: Pin description of DAC0800.

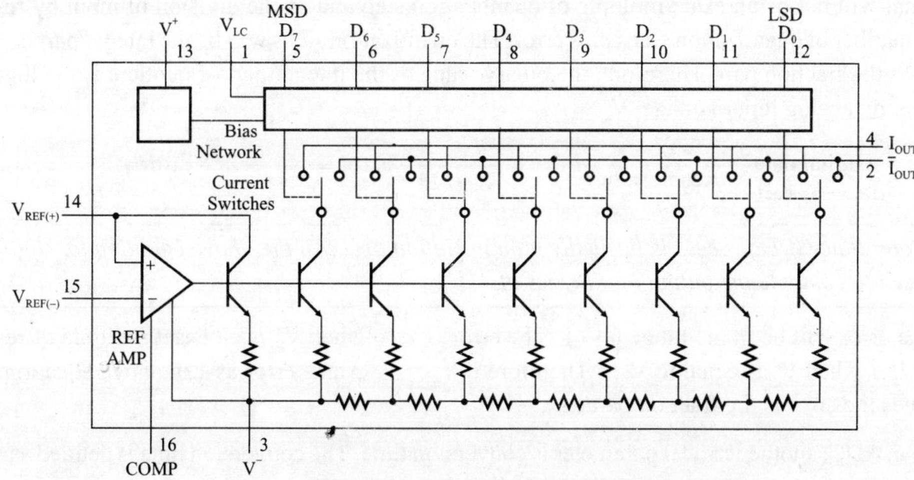

Fig. 13.7: Block diagram of DAC0800.

A typical example of generating a positive (or unipolar) voltage output using DAC0800 is shown in Fig. 13.8. Here, an 8-bit latch is provided at the input of DAC to hold the input till the conversion time.

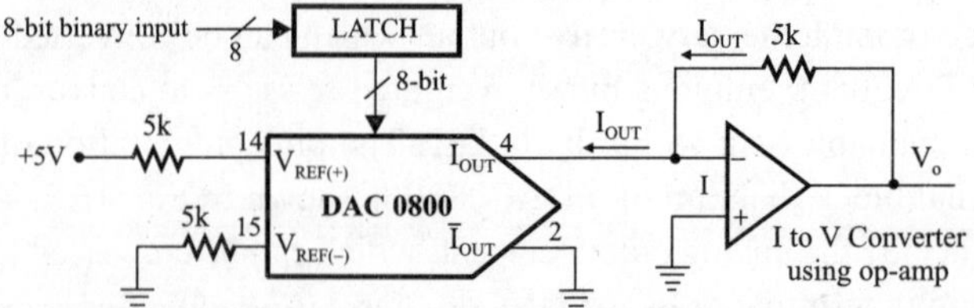

Fig. 13.8: DAC 0800 with I to V converter to produce positive (or unipolar) output voltage.

The analog output voltage, V_o is given by,

$$\text{Output voltage, } V_o = D_{IN} \times \frac{1}{2^8} \times V_{REF}$$

Where, $V_{REF} = V_{REF(+)} - V_{REF(-)}$; D_{IN} = Decimal equivalent of 8-bit binary input

5.3 Analog to Digital Converter (ADC)

The **A**nalog-to-**D**igital (A/D) conversion is the process of converting an analog signal to digital signal. The A/D conversion involves sampling and quantization. The analog signals vary continuously and are defined for any interval of time. The digital signals (or data) can take only finite values and defined only for discrete instant of time. If the digital data is represented by an n-bit binary then it can have 2^n different values including zero. In A/D conversion the given analog signal has to be divided into steps (or quantization steps) of $2^n - 1$ values, and each step is represented by one of the $2^n - 1$ values excluding zero.

The resolution of the converter is the minimum analog value that can be represented by the digital data. If the ADC gives n-bit digital output and the reference voltage is V_{REF}, then the resolution is given by,

$$\text{Resolution} = \frac{1}{2^n} V_{REF} \textit{ Volts}$$

The analog input signal can be divided by resolution to get number of quantization steps. The analog input signal will not be an exact multiple of quantization step and so the division of input by resolution will give number of quantization steps plus error. The quantization steps will be the integer part of quotient and error is the fraction part. Therefore, the binary value of the integer part of quotient is the digital data corresponding analog input voltage, V_{in}.

$$\therefore \underset{\text{(Decimal value)}}{\text{Digital data}} = \frac{V_{in}}{\text{Resolution}} - \text{Error} = \frac{V_{in}}{V_{REF}/2^n} - \text{Error} = \frac{V_{in}}{V_{REF}} \times 2^n - \text{Error}$$

> ***Note:*** *Since A/D process is basically a quantization process, the above equation for digital data can be used to compute the output of all types of ADC.*

The error will be in the range 0 to 1. The error is zero when V_{in} is an exact multiple of resolution and error is 1 when V_{in} is equal to V_{REF}. Therefore if $V_{in} = V_{REF}$ take error as 1 in above equation and for other values the error is the fraction part.

In an ADC, another critical parameter is conversion time. The conversion time is defined as the total time required to convert an analog signal into its digital equivalent. It depends on the conversion technique and the propagation delay in various circuits.

The analog-to-digital converters can be classified into two groups based on the technique involved for conversion. The first group includes successive approximation, counter and flash-type converters. The technique involved in these devices is that the given analog signal is compared with internally generated analog signal. The second group includes integrator converters. In the devices of the second group, the given analog signal is converted to time and compared with the known values to produce digital signal.

The trade-off between the two techniques is based on accuracy versus speed. The successive approximation and the flash type are faster but generally less accurate than the integrator type converters. Also, the flash type is costlier. The successive approximation type converters are used for high speed conversion and the integrating type converters are used for high accuracy.

13.3.1 Flash Type ADC

(JNTUH, Nov/Dec'20, 7 Marks)

The flash type A/D converter employs parallel differential comparators that compare various steps of reference voltages with the analog input voltage. The n-bit flash type converter will have 2^n resistors and $2^n - 1$ comparators. Fig. 13.9(a) shows n-bit flash type ADC. The 2^n resistors are connected in series with reference voltage as voltage divider network in order to divide the reference voltage into $2^n - 1$ steps. Every step voltage will be connected as one of the input to comparators and the input analog voltage is applied as another input to comparator. Now, the outputs of comparators will be **high** if the input is greater than the step value of reference otherwise **low**. All outputs of comparator are connected to a priority encoder so that the encoder will generate an n-bit binary corresponding to highest order output of comparator whose output is 1.

The flash type ADC is one of the fastest ADC converters. The main advantage of this type of converter is the conversion time is less, but the disadvantage is that, an n-bit converter of this type requires $2^n - 1$ comparators, 2^n resistors, and a 2^n to n priority encoder. The flash converter does not use a clock signal and the conversion takes place continuously. The conversion time depends on the delay in the comparators and the priority encoder.

3-bit Flash Type ADC

Fig. 13.9(b) shows a 3-bit flash type A/D converter which employs 7 comparators ($2^3 - 1 = 7$). A reference voltage V_{REF} is connected to a resistive voltage divider network with 8 equal value resistors that divides the reference voltage into seven equal increment levels. Each level is compared to the analog input by a voltage comparator. All comparator outputs are connected to an 8 to 3 priority encoder. The encoder produces a 3-bit digital output corresponding to the input having the highest priority. The higher priority input to encoder corresponds to output of comparator with largest value of input. The digital output for various values of analog input is listed in Table 13.2.

Table 13.2:

Analog Input V_{in}	Digital Output	Analog Input V_{in}	Digital Output
$\frac{7}{8}V_{REF} \le V_{in} < \frac{8}{8}V_{REF}$	111	$\frac{3}{8}V_{REF} \le V_{in} < \frac{4}{8}V_{REF}$	011
$\frac{6}{8}V_{REF} \le V_{in} < \frac{7}{8}V_{REF}$	110	$\frac{2}{8}V_{REF} \le V_{in} < \frac{3}{8}V_{REF}$	010
$\frac{5}{8}V_{REF} \le V_{in} < \frac{6}{8}V_{REF}$	101	$\frac{1}{8}V_{REF} \le V_{in} < \frac{2}{8}V_{REF}$	001
$\frac{4}{8}V_{REF} \le V_{in} < \frac{5}{8}V_{REF}$	100	$\frac{0}{8}V_{REF} \le V_{in} < \frac{1}{8}V_{REF}$	000

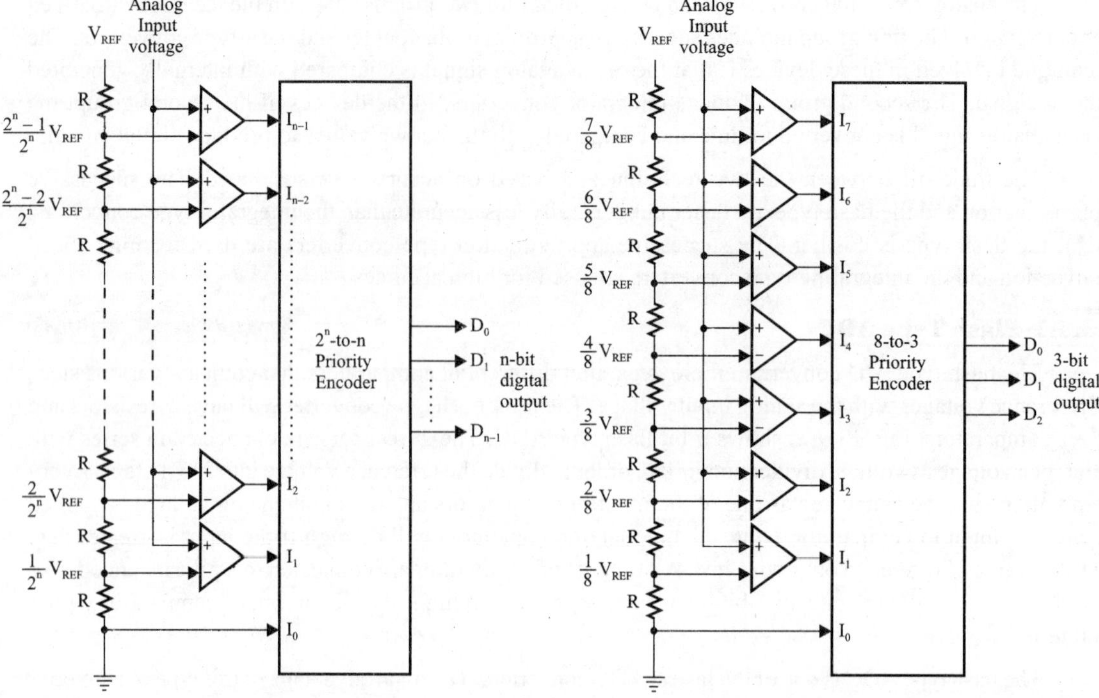

Fig. a: The n-bit flash-type ADC. *Fig. b: The 3-bit flash-type ADC.*

Fig. 13.9: The flash-type ADC.

8-bit Flash Type ADC

Fig. 13.10 shows the block diagram of a modified 8-bit flash type A/D converter in which 8-bit conversion is performed using two 4-bit flash type converters. An 8-bit flash type ADC require 255 comparators ($2^8 - 1 = 255$) but a 4-bit flash type ADC requires 13 comparators ($2^4 - 1 = 13$) so that two 4-bit flash type ADC require only 30 comparators to perform the task of 8-bit conversion.

In this method, one of the 4-bit flash type converters is used to generate the four most significant bits (MSBs) and output through an 8-bit latch as bits D_7 to D_4. Also, these four bits are converted back to an analog voltage by a DAC and this voltage is subtracted from the analog input. This difference in voltage is the analog voltage corresponding to the four least significant bits (LSBs). Therefore, this difference in voltage is converted to the four least significant bits by another 4-bit flash type converter and output through the 8-bit latch as bits D_3 to D_0.

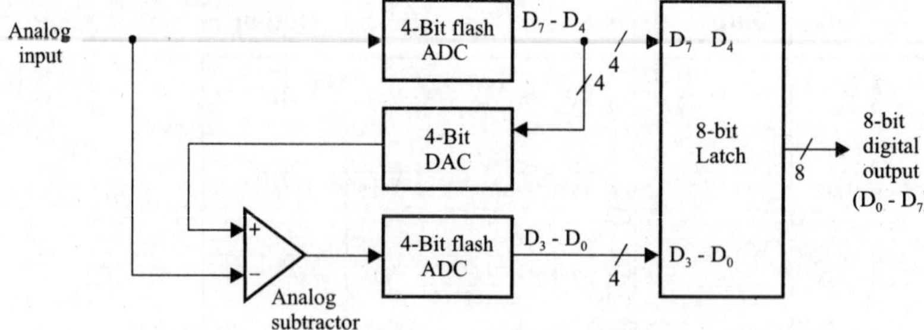

Fig. 13.10: The 8-bit modified flash type ADC

3-bit Bipolar ADC

(AKUB, Dec'20, 8 Marks)

The bipolar ADC can convert both positive and negative analog voltages within the range of positive and negative reference voltages into binary. In bipolar ADC the reference voltage will be divided into two equal half with opposite signs to set as positive reference, $V_{REF(+)}$ and negative reference $V_{REF(-)}$.

The 3-bit bipolar flash type converter is shown in Fig. 13.11 which is similar to unipolar 3-bit ADC except the voltage across each resistor in the resistive voltage divider network. Here, upper half of resistive network will have positive voltages and lower half will have negative voltages. The output of ADC will be similar to unipolar ADC but to get 2's complement output the most significant bit (MSB) alone is complemented. The digital output for various values of analog input in the 3-bit bipolar ADC is listed in Table 13.3.

Table 13.3:

Analog Input V_{in}	Digital Output	Two's Complement Output
$+\frac{3}{4}V_{REF} \le V_{in} < +\frac{4}{4}V_{REF}$	111	011
$+\frac{2}{4}V_{REF} \le V_{in} < +\frac{3}{4}V_{REF}$	110	010
$+\frac{1}{4}V_{REF} \le V_{in} < +\frac{2}{4}V_{REF}$	101	001
$0 \le V_{in} < +\frac{1}{4}V_{REF}$	100	000
$-\frac{1}{4}V_{REF} \le V_{in} < 0$	011	111
$-\frac{2}{4}V_{REF} \le V_{in} < -\frac{1}{4}V_{REF}$	010	110
$-\frac{3}{4}V_{REF} \le V_{in} < -\frac{2}{4}V_{REF}$	001	101
$-\frac{4}{4}V_{REF} \le V_{in} < -\frac{3}{4}V_{REF}$	000	100

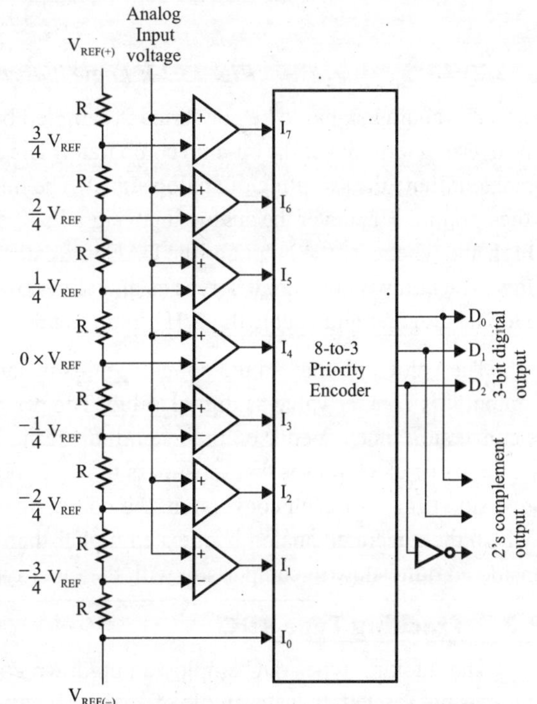

Fig. 13.11: The 3-bit bipolar ADC.

13.3.2 Counter Type ADC

The counter type ADC consists of a sample and hold (S/H) circuit, a comparator, a counter, a DAC, an AND gate, a latch and the control circuit as shown in Fig. 13.12. The binary counter generate all possible value of n-bit binary and convert every binary value to analog and compare with input analog signal. The binary count that generates matched analog value is output as digital value of the analog signal.

The counter is controlled by a clock so that the counter will keep counting as long as clock is active. Initially, the counter is reset and clock is allowed to start count. The output of counter is converted to analog by the DAC and compared with input analog.

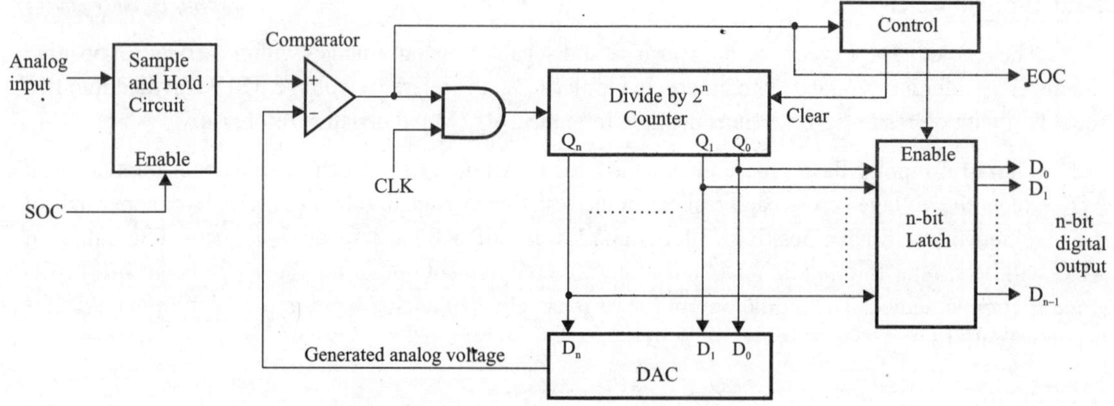

Fig. 13.12: Logic diagram of the counter-type ADC.

The analog signal to be converted is sampled by S/H (Sample and Hold) circuit and applied to the non-inverting terminal of the op-amp comparator. The S/H circuit is enabled by a start of conversion (SOC) signal to initiatc thc sampling of analog signal. The output of the DAC is applied to the inverting terminal of the op-amp. Whenever the analog input signal is greater than the DAC output, the output of the op-amp is **high** and whenever the output of the DAC is greater than the analog signal, the output of the comparator is **low**. The comparator output serves as an active-low end of the conversion signal (EOC) which is used to stop the counter and enable the S/H circuit to take next sample of analog input.

The value of binary in the counter at this instant is the value of digital signal and latch is enabled to output this counter value as digital value. The principal disadvantage of this type of converter is that, the conversion time depends on the magnitude of the analog input. The larger the input, the more will be the number of clock pulses that must pass to reach the proper count, and, therefore, the larger will be the conversion time. For each conversion, the counter has to start from reset only and count up to the point at which the generated analog is one step higher than the analog input voltage. This type of converter is considered quite slow in comparison with the other types.

13.3.3 Tracking Type ADC

The tracking-type ADC employ an up/down counter and is faster than the counter type ADC. The counter is not reset after each sample of analog signal, but counts up or down from its last count to its new count. Thus the total number of clock pulses required to perform a conversion is proportional to the change in the analog input between counts and hence it is called tracking type ADC.

Figure 13.13 shows the logic diagram of a tracking type ADC. As long as the DAC generated reference voltage is less than the analog input, the comparator output is **high** putting the counter in the up count mode. When the generated reference voltage is higher than the analog input, the comparator output switches **low** and puts the counter in the down count mode. Thus the counter effectively track the input.

The conversion time of this ADC is the time interval between the start of the conversion (SOC) and the time the generated analog voltage is equal to sampled analog input voltage. The end of conversion is indicated by either a high to low (for up count) or a low to high (for down count) comparator output. The control logic will detect this transition and generates an end of conversion (EOC) signal.

At the end of conversion, the output latch is enabled to latch the current count value as digital output.

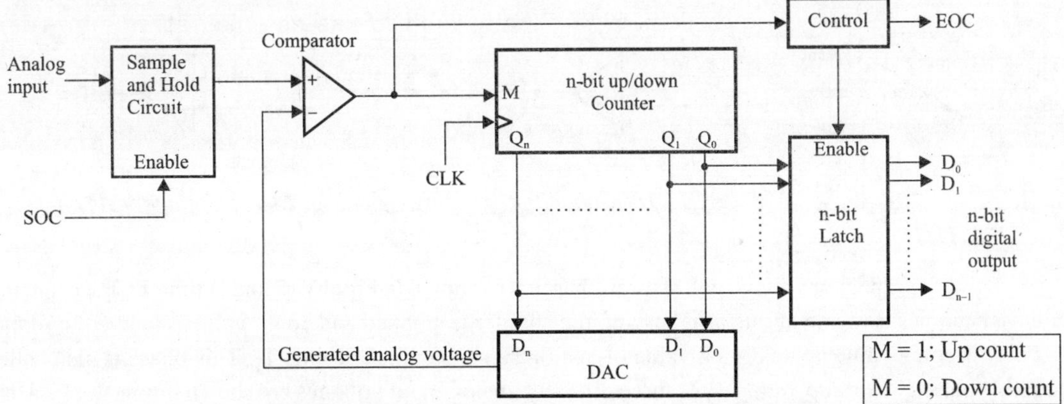

Fig. 13.13: Logic diagram of the tracking type ADC.

13.3.4 Dual-slope Type ADC

A dual-slope ADC uses an operational amplifier based integrator to generate a ramp signal and a binary counter to generate digital output. The integrator employs two components R and C to generate the ramp voltage. The ramp voltage controls the length of time that the binary counter is allowed to count and the binary value of the counter is proportional to input analog voltage.

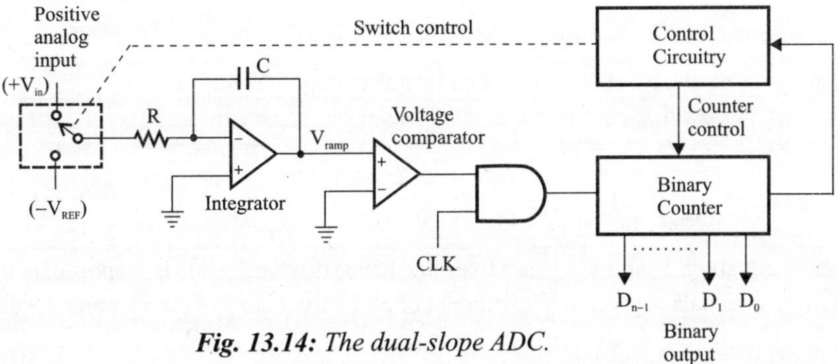

Fig. 13.14: The dual-slope ADC.

Figure 13.14 shows the functional block diagram of a dual-slope ADC. In dual slope type ADC, the integrator performs two integrations in order to generate two different ramps one with negative slope and the other with positive slope. Initially the counter is reset and the analog input V_{in} is connected to input of integrator. In unipolar dual slope ADC, the input is a positive analog voltage. In the first integration the integrator generates a ramp with negative slope proportional to analog input voltage, V_{in}. The capacitor is allowed to charge for fixed time T_1 as shown in Fig. 13.16. Let the capacitor charged to a voltage of V_{rx}. This phase is also called measuring phase.

At time, T_1 the integrator input is connected to a reference voltage V_{REF}, having a polarity opposite to that of the analog input which is in this case a negative reference voltage. Therefore, the output of the second integration is a positive going ramp (or ramp with positive slope) proportional to negative reference value. Also, the counter is reset at time T_1 and the AND gate is enabled to allow the clock to counter to start a new count.

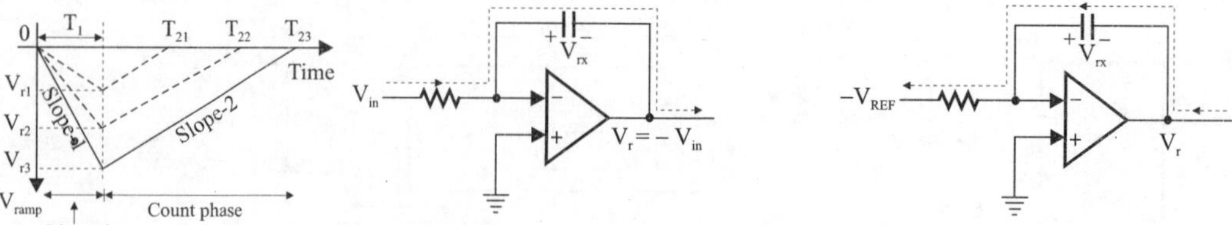

Fig. 13.15:

Fig. 13.16: Capacitor charging.

Fig. 13.17: Capacitor discharging.

Now, the capacitor voltage will start discharging as shown in Fig. 13.17 and at time T_{2x} the capacitor voltage becomes zero and at this time instant, the counter is stopped and count value is latched to output latch. This count value is the digital value of analog signal sampled at time T_1. This phase is also called counting phase. The ramp voltages for three different analog input voltages are shown in Fig. 13.13. Here it can be observed that the slope-1 is variable because input voltage is variable and slope-2 is fixed. The equations for slope are,

$$\text{Slope-1} = \frac{V_{in}}{RC} \quad ; \quad \text{Slope-2} = \frac{V_{REF(-)}}{RC}$$

The equation for capacitor voltage during charging is given by,

$$V_{rx} = V_{in} \frac{T_1}{RC}$$

The equation for capacitor voltage during discharging is given by,

$$V_{rx} = V_{REF(-)} \frac{T_{2x}}{RC}$$

On equating the above two equations of capacitor voltage we get,

$$V_{REF(-)} \frac{T_{2x}}{RC} = V_{in} \frac{T_1}{RC}$$

$$\therefore T_{2x} = \frac{V_{in}}{V_{REF(-)}} T_1$$

In the above equation, T_1 and $V_{REF(-)}$ are fixed and hence discharge time is proportional to input analog voltage. Now, the digital data is given by the product of clock frequency of counter and the discharge time.

$$\therefore \text{Digital data} = \text{Clock frequency} \times T_{2x}$$

The dual-slope converter is one of the slowest converters. But it is relatively inexpensive because it does not require a DAC. Another advantage of the dual-slope ADC is its low sensitivity to noise and to variations in its component values caused by temperature changes.

13.3.5 Successive Approximation ADC

(RGPV, Jun'20, 5 Marks) (GTU, Jun'19, 7 Marks) (AKTU, Sep'20, 5 Marks) (IKGPTU, May'20, 10 Marks)

A successive approximation ADC consists of D/A converter, successive approximation register and comparator. Figure 13.18 shows the functional blocks of a typical successive approximation A/D converter.

The analog signal to be converted is sampled by S/H (Sample and Hold) circuit and applied to the non-inverting terminal of the op-amp comparator. The S/H circuit is enabled by a start of conversion (SOC) signal to initiate the sampling of analog signal. On receiving the SOC, the control unit of the ADC will give a start command to the successive approximation register and it starts generating digital signal by the successive approximation method. The generated digital data is converted to analog signal by the D/A converter and then compared with the input analog signal.

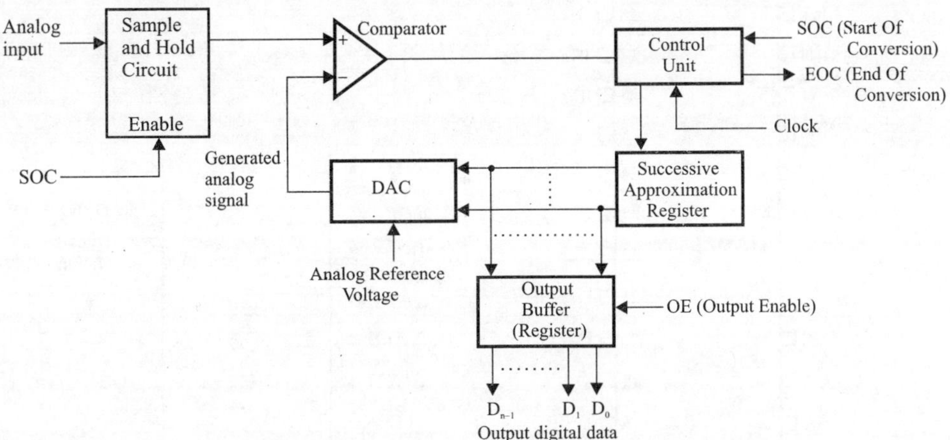

Fig. 13.18: Successive approximation A/D converter

When the analog signals are equal, the comparator output informs the control unit to stop generation of digital signal. The digital data available at this instant is given as output through output register. Also, the control unit generates a signal to indicate the **E**nd **O**f **C**onversion (EOC) process.

Successive Approximation Method of Conversion

In this method, the MSD (**M**ost **S**ignificant **D**igit) is first set to **"1"** and all other digits are reset to **"0"**. The analog signal generated for this digital data is compared with the given analog signal. (Initially, the comparator output will be **high**. After comparison the output of the comparator remains in **high** state if the input analog signal is higher than the generated analog signal. Otherwise, if the input signal is less than the generated signal then the output of the comparator changes from **high** to **low** state.) If the output state of the comparator changes then the MSD is reset to **"0"** otherwise it is retained as **"1"**.

Then the above process is repeated by setting the next higher order bit to **"1"**. The process is continued for each bit starting from MSD to LSD. (During a process, the higher order bits are the bits determined in earlier steps and the lower order bits are reset to "0".) After one complete cycle through MSD to LSD, the data available on the successive approximation register will be the digital equivalent of the input analog signal.

13.3.6 ADC0809

The ADC0809 is an 8-bit successive approximation type ADC with inbuilt 8-channel multiplexer. The ADC0809 is available as a 28-pin IC in DIP (**D**ual **I**n-line **P**ackage). The pin configuration of ADC0809 is shown in Fig. 13.19.

The internal block diagram of ADC0809 is shown in Fig. 13.20. The various functional blocks of ADC are 8-channel multiplexer, comparator, 256R resistor ladder, switch tree, successive approximation register, output buffer, address latch and decoder.

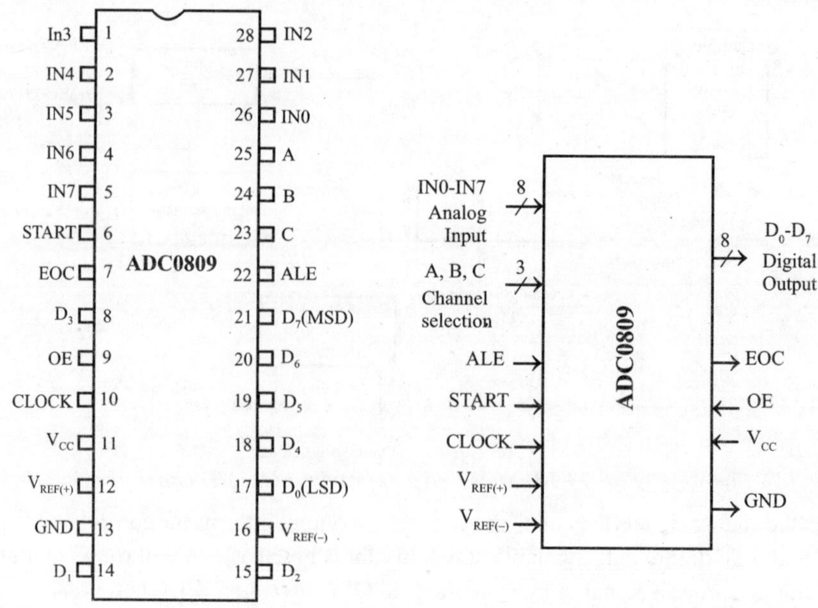

Fig. 13.19: Pin configuration of DAC0809/DAC0808.

Table 13.4: Signal Description of ADC0809

Signals	Description
IN0-IN7	Eight single-ended analog input to ADC.
A, B, C	3-bit binary input to select one of the eight analog signals for conversion at any one time.
ALE	Address latch enable. Used to latch the 3-bit address input to an internal latch.
START	Start of conversion pulse input. To start ADC process this signal should be asserted **high** and then **low**. This signal should remain **high** for atleast 100 *ns*.
CLOCK	Clock input and the frequency of clock can be in the range of 10 *kHz* to 1280 *kHz*. Typical clock input is 640 *kHz*.
$V_{REF(+)}$, $V_{REF(-)}$	Reference voltage input. The positive reference voltage can be less than or equal to V_{cc} and the negative reference voltage can be greater than or equal to ground.
D_0 - D_7	The 8-bit digital output. The reference voltages will decide the mapping of the analog input to the digital data.
EOC	End of conversion. This signal is asserted **high** by the ADC to indicate the end of conversion process.
OE	Output buffer enable. This signal is used to read the digital data from the output buffer after a valid EOC.
V_{cc}	Power supply, +5 *V*
GND	Power supply ground, 0 *V*

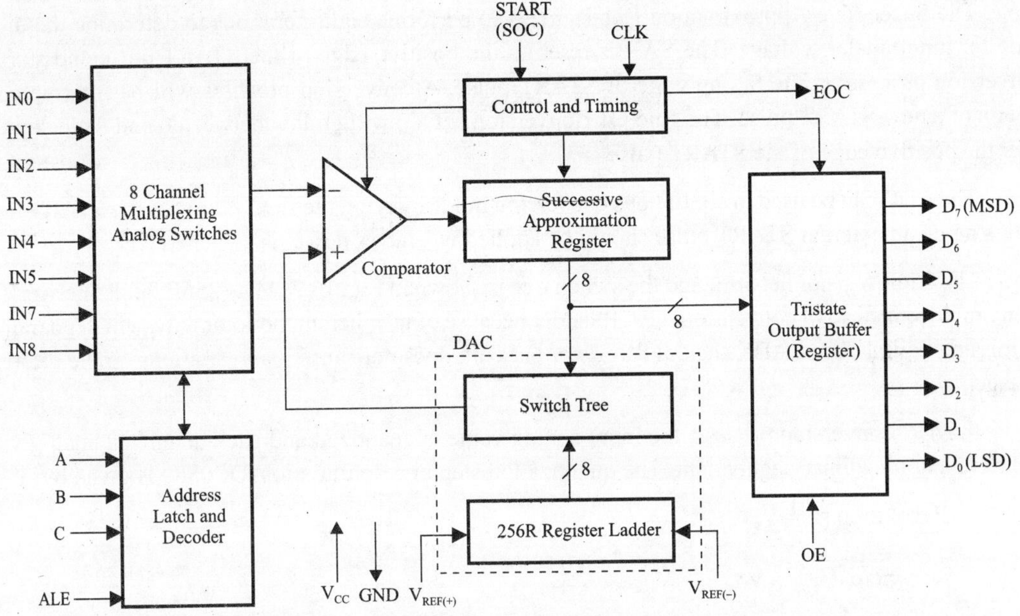

***Fig. 13.20:** Functional block diagram of ADC0809.*

The 8-channel multiplexer can accept eight analog inputs in the range of 0 to 5 V and allow one by one for conversion depending on the 3-bit address input. The channel selection logic is shown in Table 13.5.

Table 13.5

Address input			Selected channel
C	**B**	**A**	
0	0	0	IN0
0	0	1	IN1
0	1	0	IN2
0	1	1	IN3
1	0	0	IN4
1	0	1	IN5
1	1	0	IN6
1	1	1	IN7

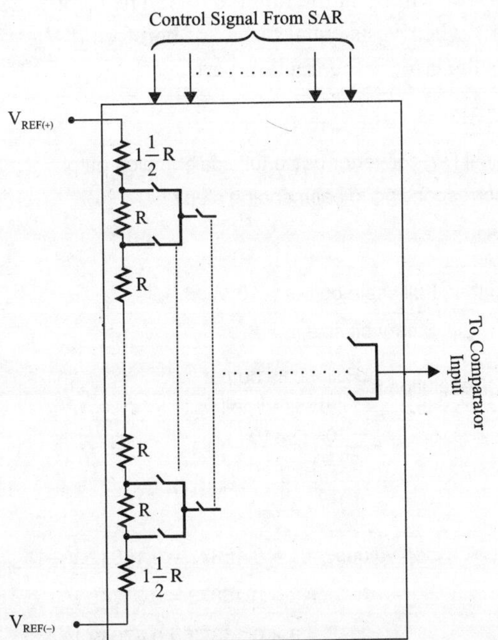

***Fig. 13.21:** 256R resistor network and switch tree.*

The Successive Approximation Register (SAR) performs eight iterations to determine the digital code for input analog voltage. The SAR is reset on the positive edge of the START pulse and start the conversion process on the falling edge of START pulse. A conversion processs will be interrupted on receipt of a new START pulse. The End Of Conversion (EOC) will go **low** between 0 and 8 clock pulses after the positive edge of the START pulse.

The ADC can be used in continuous conversion mode by tying the EOC output to the START input. In this mode an external START pulse should be applied whenever power is switched ON.

The 256R resistor network and the switch tree is shown in Fig. 13.21. The 256R ladder network has been provided instead of conventional R/2R ladder because of its inherent monotonicity, which guarantees no missing digital codes. Also, the 256R resistor network does not cause load variations on the reference voltage.

In ADC conversion process, the input analog value is quantized and each quantized analog value will have a unique binary equivalent. The quantization step or resolution in ADC0809 is given by,

$$\text{Resolution} = \frac{1}{2^8} \times V_{REF}$$

$$\text{where, } V_{REF} = V_{REF(+)} - V_{REF(-)}$$

The digital data is given by,

$$\underset{\text{(Decimal value)}}{\text{Digital data}} = \frac{V_{in}}{\text{Resolution}} - \text{Error}$$

The error will be in the range 0 to 1. The error is zero when V_{in} is an exact multiple of resolution and error is 1 when V_{in} is equal to V_{REF}. Therefore if $V_{in} = V_{REF}$ take error as 1 in above equation and for other values the error is the fraction part.

Example 13.1

(AKUB, Dec'20, 2 Marks)

A typical D/A converter has a full-scale analog output of 10 V and accepts 6 binary bits as input. What will be the voltage corresponding to each analog step?

Solution

Given that, Full-scale output = 10 V

Binary bit size, n = 6

$$\therefore \text{Resolution} = \frac{\text{Full-scale output}}{\text{Number of steps}}$$

$$= \frac{10}{2^n - 1} = \frac{10}{2^6 - 1}$$

$$= 0.15873\ V$$

$$\text{Analog step values} = 1 \times 0.15873 = 0.15872\ V$$

$$= 2 \times 0.15873 = 0.31746\ V$$

$$= 3 \times 0.15873 = 0.47619\ V$$

$$\vdots$$

$$= (2^6 - 1) \times 0.15873 = 10\ V$$

Example 13.2

What is the binary bit size of a DAC if the full-scale output is 12.6 *V* and resolution is 200 *mV*?

Solution

Full-scale output = Number of steps × Resolution

$$\therefore 12.6 = (2^n - 1) \times 200 \times 10^{-3}$$

$$\therefore 2^n - 1 = \frac{12.6}{200 \times 10^{-3}}$$

$$\therefore 2^n = \frac{12.6}{200 \times 10^{-3}} + 1 = 64$$

$$\therefore n = \log_2 64 = \log_2 (2^6)$$

$$= 6 \times \log_2 2$$

$$= 6 \times 1 = 6$$

Example 13.3

A 5-bit DAC produces an output of 0.1 V for a digital input of 00001. What is the full-scale output? Find the output for an input of 10101.

Solution

Since the output for minimum digital value (00001) is 0.1 *V*, the resolution or step size is 0.1.

Full-scale output = Step size × Number of steps

$$= 0.1 \times (2^5 - 1) = 3.1\ V$$

Given that, Digital data = $10101_2 = 21_{10}$

$$\therefore \text{Analog output} = 21_{10} \times 0.1 = 2.1\ V$$

Example 13.4

A 6-bit switched current source DAC has an output current of 20 mA for a digital input of 101100. What will be the output current for an input of 010110.

Solution

Given that, Digital data = $101100_2 = 44_{10}$

Analog output = Resolution × Digital input

$$\therefore \text{Resolution} = \frac{\text{Analog output}}{\text{Digital input}} = \frac{20}{44}\ mA$$

Decimal value of $010110_2 = 22_{10}$

$\therefore$ Analog output for 010110_2 = Resolution × Digital input

$$= \frac{20}{44} \times 010110_2$$

$$= \frac{20}{44} \times 22 = 10\ mA$$

Example 13.5

An 6-bit switched current source DAC has V_{REF} = 20 *V* and R_f = R. Find the output voltage when the input is 110100_2.

Solution

$$\text{Output voltage, } V_o = V_{REF} \times \left(\frac{D_5}{2} + \frac{D_4}{4} + \frac{D_3}{8} + \frac{D_2}{16} + \frac{D_1}{32} + \frac{D_0}{64}\right)$$

$$= 20 \times \left(\frac{1}{2} + \frac{1}{4} + \frac{0}{8} + \frac{1}{16} + \frac{0}{32} + \frac{0}{64}\right)$$

$$= 16.25\ V$$

Example 13.6

A 4-bit switched capacitor type DAC has V_{REF} = 10 *V*. Find the output voltage for a digital input of 1011.

Solution

$$\text{Output voltage, } V_o = V_{REF} \times \left(\frac{D_3}{2} + \frac{D_2}{4} + \frac{D_1}{8} + \frac{D_0}{16}\right)$$

$$= 10 \times \left(\frac{1}{2} + \frac{0}{4} + \frac{1}{8} + \frac{1}{16}\right) = 6.875\ V$$

Example 13.7

A flash-type 5-bit ADC has a reference voltage of 20 V. How many voltage comparators does it have? How many resistors does it have? What is the increment between the reference voltage applied to the comparators?

Solution

Comparators = $2^n - 1 = 2^5 - 1 = 31$ comparators

Resistors = $2^n = 2^5 = 32$ resistors

The increment reference voltage between two consecutive comparators is given by resolution.

$$\therefore \text{Resolution} = V_{REF} \times \frac{1}{2^n}$$

$$= 20 \times \frac{1}{2^5} = 0.625\ V$$

Example 13.8

The resolution of a 12-bit ADC is 10 mV. What is its full-scale range?

Solution

Number of steps = $2^n - 1 = 2^{12} - 1$

= 4095

Full-scale range = Number of steps × Resolution

= 4095 × 10 *mV* = 40.95 *V*

Example 13.9

The positive reference voltage in an 8-bit ADC is 3.84 *V*. What is the resolution and the digital data for an input of 2.56 *V*?

Solution

Given that, $V_{REF(+)} = 3.84\ V$

Let, $V_{REF(-)} = 0\ V$

$$\therefore \text{Resolution} = \frac{V_{REF(+)} - V_{REF(-)}}{2^8} = \frac{3.84}{256} = 0.013\ V = 13\ mV$$

Given that, $V_{in} = 2.56\ V$,

The digital data corresponding to 2.56 *V* is given by,

$$\text{Digital data} = \frac{V_{in}}{\text{Resolution}} - \text{Error} = \frac{2.56}{0.015} - \text{Error} = 170.66 - \text{Error}$$

Here, Error = 0.66

$$\therefore \text{Digital data} = 170.66 - 0.66 = 170_{10} = AA_H = 1010\ 1010_2$$

13.4 Summary of Important Concepts

1. Analog to Digital Converter (ADC) and Digital to Analog Converter (DAC) are referred to as data converters.
2. The ADC will convert analog signal to digital signal and the DAC will convert digital signal to analog signal.
3. The process of analog to digital conversion is also referred to as A/D process
4. The process of digital to analog conversion is referred to as D/A process.
5. The ADC will divide the reference voltage into $2^n - 1$ steps and it can generate a unique n-bit binary for each step of reference value.
6. The DAC will divide the reference value into $2^n - 1$ and for every input n-bit binary, the DAC will generate the corresponding step of analog value.
7. The process of dividing the reference value is called quantization and each step of analog value is also called quantization step or simply step size.
8. Attaching a unique n-bit binary code to each quantization step is called coding.
9. In data converters, the difference between the positive and negative reference is the reference value.
10. Resolution is the minimum value that can be handled or detected by a data converter.
11. In ADC, resolution is the minimum analog value that can be converted to digital value.
12. In DAC, resolution is the minimum analog value that can be generated for a change in digital input.
13. Full-scale value is the maximum value of analog voltage or current that can be represented in digital by a data converter.
14. Conversion time is the time taken by data converter to convert an analogue signal to digital or vice-versa.
15. Settling time is the minimum time required for converted signal to settle at a stable value for a change in input.

16. Accuracy in data converter, refers to closeness of converted value to true value.
17. Linearity in data converters refers to direct relationship between input and output signals.
18. The R-2R ladder type DAC employs only two values of resistors.
19. In weighed resistor type DAC, the resistor values are proportional to the weights of the bit positions of inputs.
20. In switched current source type DAC, the currents flows through resistors will be proportional to binary weights.
21. In switched capacitor type DAC, the capacitor values are proportional to binary weights of the bit positions of inputs.
22. The flash type A/D converter employs parallel differential comparators that compare reference voltages with the analog input voltage.
23. The n-bit flash type converter will have 2^n resistors, $2^n - 1$ comparators and 2^n-to-n priority encoder
24. The counter type ADC employs a counter and DAC to generate an analog signal internally and compare with input analog to determine the binary value of analog input.
25. The dual slope ADC employs integrator whose capacitor is charged by input voltage and the discharge time of the capacitor voltage is proportional to binary value of analog input.

13.5 Short-Answer Questions

Q13.1 What is resolution in DAC?

The resolution in DAC is the smallest possible analog value that can be generated by the n-bit binary input. If the reference voltage in n-bit DAC is V_{REF}, then the resolution is $(1/2^n) \times V_{REF}$ *Volts*.

Q13.2 What are the internal devices of a typical DAC?

The internal devices of a DAC are R-2R resistive network, an internal latch and current to voltage converting amplifier.

Q13.3 What is settling or conversion time in DAC?

The time taken by the DAC to convert a given digital data to a corresponding stable analog signal is called conversion or settling time.

Q13.4 What are the different types of ADC?

The different types of ADC are successive approximation ADC, counter-type ADC, flash-type ADC, tracking type ADC and dual slope ADC.

Q13.5 What is conversion time in ADC?

The conversion time in ADC is defined as the time required to convert a sample of analog signal into its digital equivalent.

Q13.6 What do meant by analog quantity?

An analog quantity or signal is a physical quantity like voltage, current, temperature, etc., that can take on continuous values with respect to independent variable time. The analog quantity is defined for any instant of time.

Q13.7 What do you meant by digital quantity?

A digital quantity or signal is a physical quantity like voltage, current, temperature, etc., that can take on discrete values with respect to discrete instants of independent variable time. The digital values are the quantized and coded value of discrete values represented by n-bit binary. The discrete instants of independent variable time is also represented by discrete instants of time called sampling time.

Q13.8 What is the need of ADC and DAC in controlling a process?

When a process is controlled by a digital system all the control variables should be in digital. But, the control variables of a process will be analog in nature. Hence, the analog control variables are converted into digital signals using ADC. These digital signals are processed by the digital system to generate processed digital control signals. But the process requires only processed analog signals and so the processed digital control signals are converted into processed analog control signals using DAC.

Q13.9 What is the function of ADC and DAC? ***(IKGPTU, Dec'19, 2 Marks)***

The function of ADC is to convert an analog signal to digital signal and the function of DAC is to convert digital signal to analog signal. Therefore, the ADC proforms the reverse function of DAC.

Q13.10 What is meant by D/A conversion?

The D/A conversion refer to the process of converting an digital signal represented in binary code into analog voltage or current proportional to the value of binary code.

Q13.11 Is the output of DAC a true analog quantity?

The output of a DAC will be a stepped analog quantity with each step represented by the quantized analog value. The output will be closer to true analog value if the quantized analog value is represented by large number of steps.

Q13.12 How many different output voltages can an 8-bit DAC produce?

An n-bit binary will take 2n different binary values which can be converted to $2^n - 1$ steps of analog value by a DAC. Therefore, an 8-bit DAC can produce $2^8 - 1 = 255$ different output voltages.

Q13.13 What is the advantage of smaller (finer) resolution?

The advantage of smaller (finer) resolution is that the analog value can be represented by large number of quantized steps and the quantized analog quantity will be closer to true analog value.

Q13.14 Distinguish between unipolar and bipolar DAC?

In unipolar DAC, the negative reference will be tied to ground or 0 Volt and hence it can generate an output analog voltage or current in the range 0 to positive reference ($V_{REF(+)}$ / $I_{REF(+)}$).

In bipolar DAC, the negative reference will be tied to a negative reference voltage/current ($V_{REF(-)}$/ $I_{REF(-)}$) and hence it can generate an output analog voltage or current in the range negative reference ($V_{REF(-)}$/ $I_{REF(-)}$) to positive reference ($V_{REF(+)}$ / $I_{REF(+)}$).

Q13.15 What is the advantage of the R-2R ladder DAC over the weighted resistor type DAC?

The advantage of R-2R ladder type DAC over the weighted resistor type DAC is that only two values of resistors are required for R-2Rladder DAC. Therefore standard resistors can be used. But in weighed resistor type DAC, different values of resistors proportional to binary weights are required. Therefore, high precision resistors are employed for better accuracy in weighed type DAC.

Q13.16 ***A certain 6-bit DAC uses binary weighted resistors. If the MSB resistor is 40 kΩ, what is the LSB resistor?***

The resistor in binary weighed type DAC should be proportional to binary weights. The binary weights for 6-bit DAC are 2^5, 2^4, 2^3, 2^2, 2^1 and 2^0. The MSB resistor is proportional to 2^0 and LSB resistor is proportional to 2^5.

$$\text{LSB resistor} = 2^5 \times 40 = 1280\ k\Omega$$

Q13.17 ***Why the current switch type DAC is faster than voltage switch type DAC?***

The ON and OFF characteristics of current operated switches are much faster than voltage operated switches and hence the current switch type DAC is generally faster than voltage switch type DAC

Q13.18 ***Which is the simplest type ADC? What is its other name?***

The counter-type ADC is the simplest type of ADC. It is also called a digital ramp ADC.

Q13.19 ***Write the advantage and disadvantage of a counter-type ADC.***

The advantage of counter-type ADC is that it is the simplest ADC.

The disadvantage of counter-type ADC is that, the conversion time depends on the magnitude of the analog input.

Q13.20 ***Why does the conversion time increase with the value of the analog input voltage in a counter type ADC?***

In a counter-type ADC, the conversion time increases with the value of the analog input voltage because the larger the input, the more will be the number of clock pulses that must pass to reach the proper count.

Q13.21 ***Why is the tracking-type ADC faster than the counter type ADC?***

The counter-type ADC is slow because the counter resets itself after each conversion. The tracking-type ADC uses an up-down counter and is faster than the counter type ADC because the counter is not reset after each sample, but tends to track the analog input, i.e. counts up or down from its last count to its new count.

Q13.22 ***Why is the name tracking-type ADC and what is it other name?***

Since the counter starts count from previous value for any change in input, the ADC is called the tracking-type ADC. The tracking-type ADC is also called the continuous conversion type ADC.

Q13.23 ***What is the main disadvantage of a flash-type ADC?***

The main disadvantage of a flash-type ADC is the complexity of the circuit. An n-bit converter requires $2^n - 1$ comparators, 2^n resistors and a 2^n-to-n priority encoder.

Q13.24 ***Give two advantages of a dual-slope type ADC?***

The advantages of dual-slope ADC are,

- It is relatively inexpensive
- It is less sensitive to variation in its component values.

Q13.25 ***Which is the fastest ADC and why?*** ***(IKGPTU, May'20, 2 Marks)***

The flash type ADC is the fastest ADC because it does not have a clock and counter-DAC circuit to generate an internal analog value and compare with input. The flash type ADC will continuously convert the analog input and the only delay is the delay in comparator and priority encoder.

13.6 Exercises

I. Fill in the blanks

1. The time taken by the DAC to convert a given digital data to a corresponding analog signal is called _______ time.
2. _______ signal of ADC is used to indicate the end of conversion of analog to digital conversion process.
3. Digital processing of analog signals requires _______ and _______ converters.
4. _______ and _______ function as interfaces between a digital system and the analog world.
5. The output of DAC is not a _______ quantity but it is a _______ quantity.
6. The _______ of a DAC is equal to the step size.
7. The _______ DAC is the most popular DAC.
8. The R-2R ladder network DAC and the weighted resister DAC are _______ type DAC
9. The process of approximation used in digitizing samples is called _______.
10. The _______ type ADC is the simplest and the ___ type the fastest.
11. The _______ type ADC is also called the continuous conversion type ADC.
12. The _______ ADC is the slowest type, but it is cheap.
13. The _______ type ADC is the most widely used type of ADC.
14. _______ switch type DAC is generally slower than _______ switch type DAC
15. The _______ type ADC has fixed conversion time.
16. The counter-type ADC is also called a _______ ADC.

Answers

1. conversion	7. R-2R ladder type	13. successive approximation
2. End of Conversion (EOC)	8. switched voltage	14. voltage, current
3. A/D, D/A	9. quantization	15. successive approximation
4. ADC, DAC	10. counter, flash	16. digital ramp
5. true analog, stepped analog	11. flash	
6. resolution	12. dual slope	

II. State whether the following statements are True or False

1. ADC0809 uses successive approximation method to convert analog to digital.
2. The ADC conversion is also called as quantization.
3. The process of analog to digital conversion is also referred to as A/D process.
4. ADC will divide the reference voltage into $2^n - 1$ steps and it can generate a unique n-bit binary for each step of reference value.
5. The DAC will divide the reference value into 2^n and for every input n-bit binary, the DAC will generate the corresponding step of analog value.
6. In data converters, the sum of the positive and negative reference is the reference value.

7. In data converters, resolution and step size are same.
8. Full-scale value is the maximum value of analog voltage or current that can be represented in digital by a data converter.
9. Conversion time is the time taken by data converter to convert an analogue signal to digital or vice-versa.
10. Settling time is the maximum time required for converted signal to settle at a stable value for a change in input.
11. In data converters, accuracy and resolution are same.
12. In linear data converters the output is directly proportional to input signal.

Answers

1. True	5. False	9. True
2. False	6. False	10. False
3. True	7. True	11. False
4. True	8. True	12. True

III. Choose the right answer for the following questions

1. *The minimum analog value that can be represented by the digital data is*
   a) average b) resolution c) percentage d) none of the above

2. *Most popular DAC is*
   a) R-2R ladder type b) Weighted-resister type
   c) Switched-current source type d) Switched-capacitor type

3. *The simplest type of ADC is*
   a) Counter-type b) Flash-type
   c) Successive-approximation type d) Dual-slope type

4. *The fastest ADC is*
   a) Counter-type b) Flash-type
   c) Successive-approximation type d) Dual-slope type

5. *Most expensive ADC is*
   a) Counter-type b) Flash-type
   c) successive-approximation type d) Dual-slope type

6. *The slowest ADC is*
   a) Counter-type b) Flash-type
   c) Successive-approximation type d) Dual-slope type

7. *Most widely used type of ADC is*
   a) Counter-type b) Flash-type
   c) Successive-approximation type d) Dual-slope type

8. ***The ADC which has fixed conversion time is***

a) Counter-type b) Flash-type

c) Successive-approximation type d) Dual-slope type

9. ***A certain 4-bit DAC uses binary weighted resistors. If the MSB resistor is 100 kΩ, the LSB resistor will be***

a) 400 *k*Ω b) 25 KΩ

c) 800 *k*Ω d) 12.5 *k*Ω

10. ***The ADC used in digital voltmeters and multimeters is***

a) Counter-type b) Flash-type

c) Successive-approximation type d) Dual-slope type

11. ***In which type of ADC, the conversion time depends on the magnitude of the analog input?***

a) Counter-type b) flash-type

c) Successive-approximation type d) dual-slope type

Answers

1. b	5. b	9. c
2. a	6. d	10. d
3. a	7. c	11. a
4. b	8. c	

IV. Answer the following questions

1. Explain the concept analog to digital conversion with an example.
2. Explain the concept digital to analog conversion with an example.
3. Draw a simple sketch to explain the process of A/D and D/A in a digital system.
4. Draw the circuit of R-2R resistive network and explain its operation.
5. What is weighed resistor type DAC? Explain its working.
6. Explain the working of switched current type DAC.
7. What is switched capacitor type DAC? Explain with simple sketch.
8. What is unipolar and bipolar ADC? Draw the circuit of 4-bit unipolar and bipolar flash type ADC.
9. Draw the block diagram of a counter type ADC and explain its working.
10. Explain the working of dual slope ADC with a suitable sketch.
11. What is meant by successive approximation ADC? Explain its operation.

V. Solve the following problems

E13.1 What is the binary bit size of a DAC if the full-scale output is 6.2 *V* and resolution is 200 *mV*?

E13.2 A 5-bit DAC produces an output of 0.3 V for a digital input of 00011. What is the full-scale output? Find the output of an input of 10101.

E13.3 A 6-bit switched current source DAC has an output current of 16 *mA* for a digital input of 101110. What will be the output current for an input of 010110.

E13.4 **An 6-bit switched current source DAC has V_{REF} = 18 *V* and R_f = R. Find the output voltage when the input is 110110_2.**

E13.5 **A 4-bit switched capacitor type DAC has V_{REF} = 12 *V*. Find the output voltage for a digital input of 1101.**

E13.6 **A flash-type 6-bit ADC has a reference voltage of 15 *V*. How many voltage comparators does it have? How many resistors does it have? What is the increment between the reference voltage applied to the comparators?**

E13.7 **The resolution of a 8-bit ADC is 20 *mV*. What is its full-scale range?**

E13.8 **The positive reference voltage in an 8-bit ADC is 5 *V*. What is the resolution and the digital data for an input of 1.3 *V*?**

E13.9 **A typical D/A converter has a full-scale analog output of 12 *V* and accepts 6 binary bits as input. What will be the voltage corresponding to each analog step?**

Answers
E13.1 n = 5
E13.2 Full-scale output = 9.3 *V*, Analog output = 6.3 *V*
E13.3 Analog output = 7.652 *mA*
E13.4 V_o = 15.1875 *V*
E13.5 V_o = 9.75 *V*
E13.6 Resolution = 0.234375 *V*
E13.7 Full-scale range = 5.1 *V*
E13.8 Resolution = 0.01953125 *V*, Digital data = $0100\ 0010_2$
E13.9 Analog step values = 12 *V*

CHAPTER 14

VHDL

14.1 Introduction to VHDL

VHDL is an acronym for VHSIC (Very High Speed Integrated Circuits) **H**ardware **D**escription **L**anguage.

In 1983, VHDL was originally developed by the direction of the U.S. Department of Defense to the ASICs (Application Specific ICs) manufacturers in order to standardize the hardware description. Since 1987, VHDL has been standardized by the Institute of Electrical and Electronics Engineers (IEEE) as **IEEE Standard 1076**. The latest version of VHDL is **IEEE standard 1076-2019**.

VHDL is used to model a digital system at multiple levels of abstraction. An abstraction is a description of a system with a set of minimum details or components required for modelling the system and hiding all other associated details or components of the system. VHDL supports the following four levels of abstraction for digital system design.

- Transistor level
- Gate level
- Register-transfer level (RTL)
- Processor level

The various levels of abstraction will differ in the following aspects of system design.

- Basic building block
- Signal representation
- Time representation
- Behavioral representation
- Physical representation

The basic building blocks are the parts or components used in various levels of abstraction. Signal and time representations refer to expressing signal values and changes in signal over time. Behavioral and physical representations are descriptions concerned with behavioral and physical views.

In transistor level abstraction, basic building blocks are basic electronic components like transistors, resistors, capacitors, etc. In gate level abstraction, basic building blocks are logic gates, flip-flops, multiplexers, etc. In register-transfer level (RTL) abstraction, the basic building blocks are modules developed using registers, adders, multiplexers, etc. Alternatively, the RTL abstraction can be called module-level abstraction. In processor level abstraction, the basic building blocks are processors, memory modules, bus interfaces, etc.

14.2 Elements of VHDL Language

The VHDL language has been developed by integrating the following languages:

- Sequential language
- Concurrent language
- Net-list language
- Timing specifications
- Waveform generation language

Therefore, VHDL language has constructs to express the concurrent or sequential behaviour of a digital system with or without timing, to model the system as an interconnection of components and to generate test waveforms in a single model of the system.

The basic elements of the VHDL language are literals, comments, identifiers, data objects and operators.

14.2.1 Literals

Explicit data values that are assigned to objects or used within expressions are called literals.

Literals can be classified into following six general categories.

- Character literals
- String literals
- Bit string literals
- Numeric literals
- Based numeric literals
- Physical literals

Character literals: ***Character literals*** are 1-character ASCII values that are enclosed in single-quotes.

Examples of character literals

'0', 'X', '$'

String literals: ***String literals*** are collections of one or more ASCII characters enclosed in double-quote characters.

Examples of string literals

"my_state", "count", "result"

Bit string literals: ***Bit string literals*** are special forms of string literals that are used to represent binary string in 1's and 0's in binary, octal, or hexadecimal numeric data values. Base specifier B, O and X are used for binary, octal and hexadecimal respectively.

Examples of bit string literals

```
B"1111_1101"    -- decimal value 253
O"6144"         -- decimal value 428
O"146_231"      -- decimal value 52,377
X"C300"         -- decimal value 49,920
```

Numeric literals: Numeric literals are decimal numbers. There are two basic forms of numeric literals in VHDL and they are ***integer literals*** and ***real literals***.

Integer literals are signed decimal numbers without fraction. The general range of integers is −2,147,483,647 to +2,147,483,647 (32 bits of precision, including the sign bit).

Real literals are decimal numbers with an integer part and a fraction part separated by a decimal point. For large numbers, scientific notation is also allowed using the character 'e', where the number to the left of the 'e' represents the mantissa of the real number, while the number to the right of the 'e' represents the exponent. The general range of real numbers is – 1.0e38 to +1.0e38.

Examples of numeric literals

```
5.0
– 12.9
1.6e10
1.2e – 20
```

Based numeric literals: ***Based numeric literals*** are another form of integer or real values, with a base specification such as 2, 8 or 16 and enclose the integer or mantissa with a pair of '# ' characters.

Examples of based numeric literals

```
2#10010001#    --integer value 145
16#FFCC#       --integer value 65,484
2#101.0#E10    --real value 5,120.0
```

Physical literals: ***Physical literals*** are special types of literals used to represent physical quantities such as time, voltage, current, distance, etc. Physical literals include both an integer numeric part and a unit specification.

Examples of physical literals

300 ns *--300 nanoseconds*

900 ps *--900 picoseconds*

40 ma *--40 mille-amperes*

14.2.2 Comments

Comments are texts that are added as explanations, wherever necessary. A comment starts with a minimum of two dashes (--) followed by the text to explain the concepts. The comment is provided for explaining the various concepts of VHDL program and hence, anything after two or more dashes will be ignored while compiling the code.

14.2.3 Identifiers

Identifiers are name of objects that are used in VHDL programs. Identifiers are formed using English letters/alphabets, decimal numbers and underscore. Special characters are not allowed in identifiers. Reserved words of VHDL language cannot be used as identifiers. The list of reserved words is listed in Table 14.1. Identifiers are case insensitive and hence uppercase and lower case letters are equivalent.

The identifiers have to satisfy the following rules.

- Identifiers can have only English letters, decimal numbers and underscores.
- First character of an identifier must be an alphabet.

- Last character of an identifier cannot be an underscore.
- Identifiers cannot have two underscore in succession.

Table 14.1: List of Reserved Words

abs	access	after	alias
all	and	architecture	array
asserts	attribute		
begin	block	body	buffer
bus			
case	component	configuration	constant
disconnect	downto		
else	elsif	end	entity
exit			
file	for	function	
generate	generic	guarded	
if	in	inout	is
label	library	linkage	loop
map	mod		
nand	new	next	nor
not	null		
of	on	open	or
others	out		
package	port	procedure	process
range	record	register	rem
report	return		
select	severity	signal	subtype
then	to	transport	type
units	until	use	
variable			
wait	when	while	with
xor			

14.2.4 Predefined Data Types

The VHDL IEEE standard package defines a number of standard data types which are known as predefined data types. The predefined data types in VHDL are,

- Integer
- Boolean
- Bit
- Character
- String
- Bit_vector
- Std_logic
- Std_logic_vector

The predefined data types of VHDL are listed in Table 14.2.

Integer data type is integer decimal number in the range $-(2^{31}-1)$ to $+(2^{31}-1)$.

Boolean data type is true, false.

Bit data type is ‘0’, ‘1’.

Bit_vector is one dimensional binary array.

The std_logic and std_logic_vector are defined in the std_logic_1164 package of IEEE standard 1164. The std_logic data type has nine possible values listed in Table 14.3.

The std_logic_vector is an array of elements with std_logic data type. A group of n-bit binary signals or an n-bit bus in a logic circuit is represented by std_logic_vector.

Example of std_logic_vector data type

```
signal A: std_logic_vector (7 downto 0);
```

In the above example, signal A is declared as an 8-bit binary array indexed from 7 to 0, with most significant bit as index 7 and least significant bit as index 0. Each bit of signal A can be accessed as A(0), A(1), etc. A part of signal A can be accessed as A(3 downto 0).

Table 14.2: Predefined VHDL Data Types

VHDL Data Type	Value
Bit	'0' or '1'
Boolean	false or true
Integer	$-(2^{31}-1) \leq$ Integer value $\leq (2^{31}-1)$
Positive	$1 \leq$ Integer value $\leq (2^{31}-1)$
Natural	$0 \leq$ Integer value $\leq (2^{31}-1)$
Real	$-$ 1.0e38 $\leq$ Floating point value $\leq$ 1.0e38
Character	Alphabetical characters (a .... z, A .... Z), digits (0 .... 9), special characters (e.g.,%) each enclosed in single quotes
Time	Integer with units fs, ps, ns, ms, sec, min, hr

Table 14.3: Std_logic Data Types of the IEEE std_logic_1164 Package

Logic Symbols	Description
'U'	Uninitialized
'X'	Strong drive, unknown logic value
'0'	Strong drive, logic 0
'1'	Strong drive, logic 1
'Z'	High impedence
'W'	Weak drive, unknown logic value
'L'	Weak drive, logic 0
'H'	Weak drive, logic 1
'–'	Don't-care

14.2.5 Data Types and Subtypes

The data types and subtypes other than predefined data types that can be declared in VHDL programs are discussed in this section. In VHDL, each data object has data type and associated operators. The data type is declared using a type statement.

Syntax for type declaration

```
type object_name is data_type;
```

Example of type declaration

```
type my_count is ('2', '4', '6', '8');
```

Subtype is a part or subset of a data type that is specified by a condition or constraint.

Syntax for type and subtype declaration

```
type object_name is data_type;
subtype subtype_object_name is object_name data_type;
```

Example of type and subtype

```
type pin_count is range 0 to 40;
subtype address_pin is pin_count range 25 to 40;
```

The data types that are used in VHDL programs can be classified into following four groups.

- Scalar types
- Composite types
- Access types
- File types

14.2.6 Scalar Data Types

The scalar data types can be again classified into following four groups.

- Enumeration
- Integer
- Physical
- Floating point

Enumeration type is associating user defined set of values called enumeration literals to an object. The values of enumeration literals are formed using alphabets and numeric character literals.

Example of enumeration data type

```
type my_state is (S0, S1, S2, S3, S4);
type rain_sevirity is (green, yellow, orange, red);
```

Integer type is associating an integer with values within a specified range to an object.

Example of integer data type

```
type count_int is range 0 to 64;
```

Physical type associates numeric value of a physical quantity such as time, length, mass, voltage, current, etc to an object. Each value of a physical type has a position number that is an integer value. Any value of a physical type is a straight multiple of the primary unit of measurement for that type.

Syntax for physical type

```
type type_name is range left_bound to right_bound
units
    primary_unit_name;
    secondary_unit_name = number primary_unit_name;
    secondary_unit_name = number primary_unit_name;
    ...
end units type_name;
```

Example of physical data type

```
type length is 0 to 1e3
    units
    mm;
    cm   = 10 mm;
    m    = 1000 mm;
    inch = 25 mm;
    ft   = 12 inch;
end units;
```

Floating point type is associating a real decimal number within a specified range to an object.

Example of floating point data type

```
type probability is range 0.0 to 1.0;
type signal_a_is range – 10.0 to +10.0;
```

14.2.7 Composite Data Types

A composite data type represents a collection of signals or objects. Composite types can be further classified into array types and record types. An array type represents a collection of signals or objects belong to a same data type. A record type represents a collection of signals or objects belong to same or different data types.

Examples of array data type

```
type word is array (15 downto 0) of bit:
type food is array (1 to 10) of character;
```

Example of record type

```
type uart_transmit is record
    rx  : std_logic:
    cts : std_logic;
    tx  : std_logic;
    rts : std_logic;
end record uart_transmit;
```

14.2.8 Access Data Types

Access type is a pointer to an object of some other data type.

Example of access data type
type *fifo_stack* **is array** *(0* **to** *63, 0* **to** *7)* **of bit;** **type** *fifo_ptr* **is access** *fifo_stack;*

File Data Types

Data stored as a file in the host environment (the system that simulate and synthesize the code) is called file data type.

Syntax of a file data type declaration
type *file_type_name* **is file of** *data_type;*
Example of file data type
type *input_data* **is file of bit_vector;** **type** *employee_names* **is file of string;**

14.2.9 Data Objects (or Objects)

Data object or simply object is an identifier that holds a specific data type, which is declared while creating the object. A data object is created by declaring the class, identifier/name, data type and optional initial value.

The objects in the VHDL language are classified into the following four categories.

- Constants
- Signals
- Variables
- Files

Constant: A constant object is assigned a fixed value and cannot be changed in other part of the code.

Syntax for declaring constant
constant *name_of_constant: data_type := value_of_constant;*
Examples of constants
constant *address_size :* **integer** *:= 16;* **constant** *sram :* **bit_vector** *(15* **downto** *0) := X"FOFO";* **constant** *message :* **string** *:= "this is a string";* **constant** *error_flag :* **boolean** *:= true;*

Signal: A signal object is created for intermediate signals in the design entity apart from input and output signals. Signals are declared in the architecture and outside the process. Initial value is meaning less in signal objects that are assigned a fixed value and cannot be changed in other part of the code.

Syntax for declaring signal

signal *name_of_signal : data_type := initial_value_of_signal;*

Examples of signals

signal *propagation_delay* : **time** := *12ns;*

signal *data_size* : **bit_vector** *(0* **to** *15);*

Variable: An object which takes different values during program execution is called variable. When an object is declared as a variable, the value of the variable can be changed in any part of the code using an assignment statement. Variables are declared inside a process and they are local to a process.

Syntax for declaring variable

variable *name_of_variable : data_type := initial_value_of_variable;*

Examples of variables

variable *propagation_delay:* **time** := *12ns;*

variable *data_size:* **bit_vector** *(0* **to** *15);*

File: A file with a list of input and/or output data can be declared as an object. The file can used to read the stored input data and write the output data while simulating a test bench.

Syntax for declaring a file

type *file_type_name* **is file of** *data_type;*

type *file_name: file_type_name* **is mode** *"file path";*

Note: *mode is either* **in** *or* **out.**

Examples of file

type *input_data* **is file of bit_vector;**

file *sel_values: input_data* **is in** *"/usr/home/vhdl_ex";*

type *employee_names* **is file of** *string;*

file *sel_employee: employee_names* **is out** *"/usr/home/vhdl_ex"*

14.2.10 Operators

The predefined operators in the VHDL language are classified into the following five categories.

- Logical operators
- Relational operators
- Adding operators
- Multiplying operators
- Miscellaneous operators

The operators of VHDL are listed in Table 14.4 and the precedence of the operators is listed in Table 14.4.

Table 14.4: VHDL Operators

Operator Group	Operator Symbol	Operation / Operator Name
Binary logical	**and**	Logical and operation
	or	Logical or operation
	nand	Logical nand operation
	nor	Logical nor operation
	xor	Logical xor operation
	xnor	Logical xnor operation
Relational	=	Equality operator
	/ =	Inequality operator
	<	Less than
	>	Greater than
	<=	Less than or equal to
	>=	Greater than or equal to
Shift operators	**sll**	Shift-left logical
	srl	Shift-right logical
	sla	Shift-left arithmetic
	sra	Shift-right arithmetic
	rol	Rotate left
	ror	Rotate right
Addition operators	**+**	Addition
	−	Subtraction
Concatenation operator	**&**	Concatenation
Unary sign operator	**+**	Positive sign
	−	Negative sign
Multiplication operators	*	Multiplication
	/	Division
	mod	Modulus
	rem	Remainder
Miscellaneous operators	**not**	Complement
	abs	Absolute
	**	Exponentiation
Assignment operators	<=	Assign data of one object to another object
	:=	Assign a value to object
	=>	Associate one object to another object

Table 14.14: Precedence of the VHDL Operators

Precedence	Operators
Highest	****, abs, not**
	***, /, mod, rem**
	+, – (identity and negation)
	&, +, – (addition and subtraction)
	sll, srl, sla, sra, rol, ror
↓	=, /=, <, <=, >, >=
Lowest	**and, or, nand, nor, xor, xnor**

14.3 Process and Process Statement

The executable functionality of an entity or a component is defined as a process in behavioral modelling. Hence the process is the key programming structure in behavioral modelling. All components in the model must be defined using one or more processes inside the architecture.

The executable statements of a process are enclosed in a process statement. In a single process model, the process statement includes all executable statements of the model. In a multi-process model each process is defined by a process statement that includes executable statement of the individual process.

Initially, the process will be executed once using the current value of objects in the sensitivity list and the process execution will be repeated only when a change occurs in one or more objects in the sensitivity list. The process label is optional and made to identify a process in a multi-process model. In process without sensitivity list, a wait statement should be provided inside the process statement.

Syntax of process statement

```
process_label : process (sensitivity list)
            -- declaration statements;
            begin
            -- sequential statements;
            end process (process_label);
```

Example of process statement

```
--program segment showing only a process statement
--a, b, c are declared as inputs and z as output in entity
   process (a, b, c)
      variable m1, m2 : std_logic;
   begin
      m1 := a nand b;
      m2 := m1 nand c;
      z  <= m1 nand m2;
   end process;
```

***Note:** All VHDL statement/program examples in this book are simulated and verified in license free Xilinx VHDL tool Vivado 2019.2 web edition which can be directly downloaded from the following link.*

https://www.xilinx.com/support/download/index.html/content/xilinx/en/downloadNav/vivado-design-tools/archive.html

***File name:** Vivado HLx 2019.2: All OS installer Single-File Download (TAR/GZIP-26.1414GB).*

14.4 Sequential Statements

Sequential statements allow the user to describe signal assignments in an algorithmic fashion. The sequential statements are written inside a process statement. All statements in a process are evaluated sequentially and therefore the order of the statement is important.

The various types of sequential statement available in VHDL are,

- Variable assignment statement
- Signal assignment statement
- Wait statement
- If statement
- Case statement
- Loop statement
- Null statement
- Exit statement
- Next statement

14.4.1 Variable Assignment Statement

The variable objects will take different values whenever the process is executed and the variable assignment statements are used to assign the next value to variable objects.

Syntax of variable assignment statement

```
variable_name := value_expression;
```

Example of variable assignment statement inside a process

```
--program segment showing only a process statement
--a, b, c are declared as inputs and y, z as outputs in entity
   process (a, b, c)
     variable m1, m2 : integer range 0 to 10 ;
   -- variables are declared inside the process
   begin
       m1 := (a + b) – c;  -- variable assignment statement
       y   <= – m1
       m2 := m1 * 2;       -- variable assignment statement
       z   <= m2;
   end process;
```

14.4.2 Signal Assignment Statement

Signals are assigned values using a signal assignment statement.

Syntax of signal assignment statement

```
signal_name <= value expression;
```

Example of signal assignment statement inside a process

```
--program segment showing only a process statement
--a, b, c are declared as inputs and y as output in entity
    signal m1, m2 : std_logic;   -- signals are declared outside the process
        begin
            process (a, b, c)
                begin
                    m1 <= a or c;        -- signal assignment statement
                    m2 <= a and b;       -- signal assignment statement
                    y <= m1 and m2;
            end process;
```

14.4.3 Wait Statement

The wait statement is used to suspend a process without sensitivity list. The wait statement will have a condition and the process will resume only when the condition is true. There are three basic forms of wait statement.

Syntax of a wait statement

```
    wait on sensitivity_list;
    wait until boolean_expression;
    wait for time_expression;
```

Examples of wait statements

```
    wait on a, b, c;       --wait until a change occurs in one or more object ( a, b, c)
    wait until (a = b);    --wait until object a is equal to object b
    wait for 30 ns;        --Resume the process after a delay of 30 ns
```

Example of a process with wait statement

```
--program segment showing only a process statement
--a, b, c, d are declared as inputs and y as output in entity
    process
        variable m1, m2 : std_logic;
    begin
        m1 := a and b;
        m2 := c and d;
        y  <= m1 or m2;
        wait on a, b, c, d;
    end process;
```

14.4.4 If Statement

If statement is used to execute one or more statements when a condition specified in the if statement is true. The general syntax of the if statement shown below can be used to execute a number of statements based on different conditions.

An if statement will have one **if-condition-then** statement, one or more optional **elsif-condition-then** statement and one optional **else** statement. At any one time only one condition will be true and hence the statements followed by the true condition are executed. If no condition is true then the statements followed by else will be executed and if there is no else statement then no statement will be executed

Syntax of if statement when multiple conditions have to be checked

```
if condition_1 then
    sequential_statements ;
elsif condition_2 then
    sequential_statements;
    .
    .
    .
elsif condition_n then
    sequential_statements;
else
    sequential_statements;
end if;
```

Syntax of if statement for only one condition

```
if condition then
    sequential_statements;
else
    sequential_statements;
end if;
```

Syntax of if statement without else statement for only one condition

```
if condition then
    sequential_statements;
end if;
```

Example of if statement inside a process

```
--program segment of 4:1 multiplexer based on if statement
--i0, i1, i2, i3 are declared as inputs and y as output in entity
--s is declared as 2-bit std_logic_vector input in entity
    process (i0, i1, i2, i3)
        begin
            if (s = "00") then
                y <= i0;
            elsif (s = "01") then
                y <= i1;
            elsif (s = "10") then
                y <= i2;
            else
                y <= i3;
            end if;
    end process;
```

14.4.5 Case Statement

A case statement will check the value of an expression or object that can take multiple values. At any onetime only one value will be true and one (or a set of) sequential statement after the true case is executed. The values of the expression or object are denoted as choice_1 to choice_n.

Syntax of case statement

```
case_expression is
    when choice_1 =>
        sequential_statements;
    when choice_2 =>
        sequential_statements;
        .
        .
        .
    when choice_n =>
        sequential_statements;
end case;
```

Example of case statement inside a process

```
--program segment of 4:1 multiplexer based on case statement
--i0, i1, i2, i3 are declared as inputs and y as output in entity
--sel_in is declared as 2-bit std_logic_vector input in entity
    process (i0, i1, i2, i3)
        begin
        case sel_in is
            when "00" =>
                y <= i0;
            when "01" =>
                y <= i1;
            when "10" =>
                y <= i2;
            when others =>
                y <= i3;
        end case;
    end process;
```

14.4.6 Loop Statement

The loop statement is used to execute a sequential statement (or a set of sequential statements) repeatedly with in a process. There are three types of loop statements in VHDL.

- For-loop
- While-loop
- Loop terminated using an exit or next statement

For-loop

Syntax of for-loop statement

```
(loop_label:) for index in loop_range loop
        sequential_statements;
        end loop (loop_label);      -- loop label is optional
```

Example of for-loop

```
sum_int := 0;
for int_num in 1 to 10 loop
    sum_int := sum_int + int_num;
end loop;
```

VHDL Program 14.1

```
--VHDL program using for-loop to find sum of integers
library ieee;
use ieee.std_logic_1164.all;
    entity test_for_loop is
        port (y : out integer range 0 to 60);
    end test_for_loop;
    architecture bm_test_for_loop of test_for_loop is
    begin
        process
            variable sum_int : integer range 0 to 60;
        begin
            sum_int := 0;
            for int_num in 1 to 10 loop
            sum_int := sum_int + int_num;
        end loop;
                y <= sum_int;
            wait for 100ns;
        end process;
    end bm_test_for_loop;
```

While-loop

Syntax of while-loop statement

```
while Boolean_expression loop
    --sequential_statements;
end loop;
```

Example of while-loop

```
sum_int := 0;
int_num := 1;
while int_num <= 10 loop
    sum_int := sum_int + int_num;
    int_num:=int_num+1;
end loop;
```

Note: *The example program given for for-loop can be run with while-loop, by replacing the statements corresponding to for-loop with statements for while-loop.*

Loop terminated using an exit or next statement

Syntax of loop statement

```
loop_label: loop
        --sequential_statements;
        exit when boolean_expression;
        end loop loop_label;            -- loop label is optional
```

Example of simple loop

```
sum_int := 0;
int_num := 1;
L1: loop
    sum_int := sum_int + int_num;
    int_num := int_num + 1;
    exit when int_num>10;
end loop L1;
```

***Note:** The example program given for for-loop can be run with simple loop, by replacing the statements corresponding to for-loop with statements for simple loop.*

14.4.7 Exit Statement

The exit statement is a sequential statement that can be used only inside a loop. The exit statement will check the specified condition/Boolean_expression and if the condition / Boolean_expression is true then It stops execution of loop statement immediately and jump out of the loop. Hence it is used to terminate the iterations in the loop statement.

Syntax of exit statement

```
exit when Boolean_expression:
```

***Note:** Refer example of simple loop for usage of exit statement.*

14.4.8 Next Statement

The next statement is also a sequential statement that can be used only inside a loop. It is used to skip a part of statements in the iteration. When a next statement is encountered and if the specified condition is true then the execution jump to start of next iteration by skipping the statements after the next statement.

Syntax of next statement

```
next when Boolean_expression:
```

Example of next statement

```
sum_int := 0;
int_num := 0;
while int_num < 10 loop
    int_num := int_num + 1;
    next when int_num = 14;      --integer 14 will not be added in sum
    sum_int := sum_int + int_num;
end loop;
```

14.4.9 Null Statement

The null statement is a sequential statement and used in places where no action is required for a particular condition.

Example of null statement

```
--program segment showing usage of null statement inside a case statement
--a, b, c are declared as inputs and y as output in entity
--opcode is declared as signal in architecture
    case opcode is
        when "000"   => y <= a and b;
        when "000"   => y <= a or b;
        when "000"   => y <= a xor b;
        when others => null;
    end case;
```

14.4.10 Assertion Statement

An assertion statement is used to verify a specified condition and to report if the condition is violated. If the condition is false then the string_expression in the report statement is displayed and the string_expression in the severity statement is used to display the severity clause.

Note: *During simulation in vivado, the assert statement is included in test bench. The string_expression in report statement is displayed in log file and the string_expression in severity statement is displayed in TCL console.*

Syntax of assertion statement

```
assert condition report "string_expression " severity "string_expression";
```

Example of assertion statement

```
assert (s = '0') report "se1" severity "error";
assert (co = '1') report "ce1" severity "error";
```

VHDL Program 14.2

```
--VHDL program example showing usage of assert statement
--code for half adder
library ieee;
use ieee.std_logic_1164.all;
    entity test_ha is
        port ( a, b  : in std_logic;
               s, co : out std_logic );
    end test_ha;

    architecture dfm_test_ha of test_ha is
        begin
            s  <= a xor b;
            co <= a and b;
    end dfm_test_ha;
```

```
--code for test bench of half adder
library ieee;
use ieee.std_logic_1164.all;
```

```
entity test_tb_ha is
    --port ()
end test_tb_ha;

architecture bm_test_tb_ha of test_tb_ha is
    component test_ha is
        port ( a, b  : in std_logic;
               s, co : out std_logic );
    end component;
    signal a, b, s, co : std_logic;

    begin
        mapping : test_ha port map (a, b, s, co);
    process
        variable e_c  : integer := 0;
    begin
```

```
--test1
    a <= '1';
    b <= '0';
wait for 10 ns;

    assert (s = '0') report "se1";          --without severity clause
    assert (co = '1') report "ce1";         --without severity clause

--assert (s = '0') report "se1" severity "error";
--assert (co = '1') report "ce1" severity "error";

if (s /= '1' or co /= '0') then
    e_c := e_c + 1;
end if;
```

```
--test2
    a <= '1';
    b <= '1';
wait for 10 ns;

    assert (s = '0') report "se2";          --without severity clause
    assert (co = '1') report "ce2";         --without severity clause

--assert (s = '0') report "se2" severity "error";
--assert (co = '1') report "ce2" severity "error";

    if (s /= '0' or co /= '1') then
        e_c := e_c + 1;
    end if;
end process;
end bm_test_tb_ha;
```

14.5 Concurrent Statements

The concurrent statements are used in dataflow architecture. VHDL supports the following concurrent statements.

- Concurrent signal assignment statement
- Conditional signal assignment statement
- Selected signal assignment statement
- Concurrent assertion statement

14.5.1 Concurrent Signal Assignment Statement

Signal assignment statements that appear outside of a process are called concurrent signal assignment statements and they are used to assign a value to an object. The syntax for concurrent signal assignment statement is same as that of signal assignment statement discussed under sequential statements. An architecture body can have any number of concurrent signal assignment statements. Concurrent signal assignment statements are event triggered and hence, they are executed whenever there is an event or change in a signal that appears in the expression in the statement.

14.5.2 Conditional Signal Assignment Statement

Conditional signal assignment statement assigns different values for the target signal based on the specified conditions similar to if statement.

Syntax of conditional signal assignment statement

```
signal_name <= value_exp_1 when condition_1 else
               value_exp_2 when condition_2 else
                    .
                    .
               value_exp_n
```

Example of conditional assignment statement in a dataflow architecture

```
--program segment of 4:1 multiplexer based on
--conditional assignment statement
--i0, i1, i2, i3 are declared as inputs and y as output in entity
--sel_in is declared as 2-bit std_logic_vector input in entity
   y <= i0 when (sel_in = "00") else
        i1 when (sel_in = "01") else
        i2 when (sel_in = "10") else
        i3;
```

14.5.3 Selected Signal Assignment Statement

Selected signal assignment statement assigns different values for a target signal based on the value of a select_expression similar to case statement.

Syntax of selected signal assignment statement

```
with select_expression select
signal_name <= value_exp_1 when choice_1,
               value_exp_2 when choice_2,
                    .
               value_exp_n when choice_n;
```

Example of selected signal assignment statement in a dataflow architecture

```
--program segment of 4:1 multiplexer based on
--selected signal assignment statement
--i0, i1, i2, i3 are declared as inputs and y as output in entity
--sel_in is declared as 2-bit std_logic_vector input in entity
    with sel_in select
        y <= i0 when "00",
             i1 when "01",
             i2 when "10",
             i3 when others;
```

14.5.4 Concurrent Assertion Statement

Assertion statements that appear outside of a process are called concurrent assertion statements and they are used to report the error conditions. The syntax for concurrent assertion statement is same as that of assertion statement discussed under sequential statements.

14.6 VHDL Design Entity (or Model)

In order to describe a design entity (or model), VHDL provides the following five different types of primary constructs called design units.

- Entity declaration
- Architecture body
- Configuration declaration
- Package declaration
- Package body

In the simplest form, a VHDL design entity (or model) includes an entity declaration and at least one architecture body.

The entity declaration describes the input and output signal names. The architecture body contains a set of interconnected components which represents the structure of the entity, or as a set of concurrent or sequential statements that represents the behaviour of the entity. Figure 14.1 shows a simple VHDL design entity or model.

Entity declaration: Describes the input output parameter

Architecture body: Describes the functionality of the design

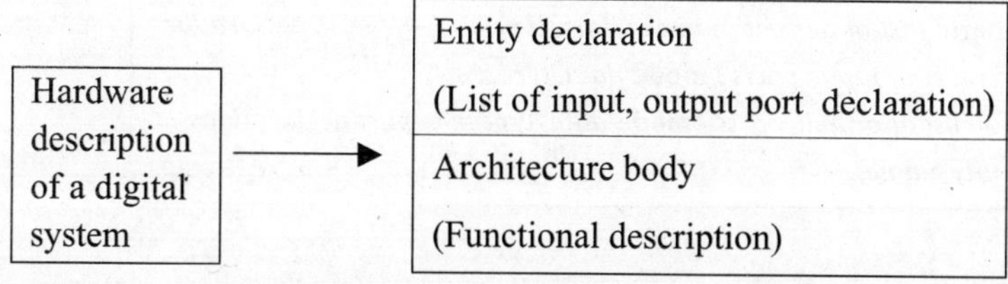

Fig. 14.1: VHDL design entity (or model)

VHDL separates entity declaration and architecture body into two independent design units. ***Configuration*** is the process of binding a component with an entity and architecture. In a simple VHDL program configuration is not needed but configuration will be useful in development and testing of large systems.

A ***package declaration*** is used to store a set of common declarations like components, types, procedures, and functions. These declarations can then be imported into other design units.

A package body is always associated with a package declaration. A ***package body*** is used to store the definitions of functions and procedures that were declared in the corresponding package declaration and also the constant declarations for any constants that appear in the package declaration.

14.6.1 Entity Declaration

Entity declaration specifies the name of the entity and a set of interface ports. Ports are signals through which the entity communicates with the other models in its external environment.

Syntax for entity declaration

```
entity entity_name is
    port ( list of input_port  : mode data_type;
           list of output_port : mode data_type );
end entity_name;
```

Note: **mode** *is in, out, inout, buffer.*

Port

Each input and output signal in an entity is called a port. Each port should have a name, data type and direction (or mode). The mode can be assigned one of the four possible choices: **in, out, inout, buffer**. The set of ports defined in an entity is called port declaration.

Generic

The generic construct of VHDL is provided to pass information into an entity and a component. Generics are like parameters. They are first declared in entity and component declaration and later assigned a value during component instantiation. In order to use generics, an additional declaration called generic declaration is added in entity.

Syntax for entity declaration with generics

```
entity entity_name is
    generic (list of generic_names : data_types);   --generic declarations
    port (list of input_ports : mode data_types;
          list of output_ports : mode data_types);  --port declarations
end entity_name;
```

14.6.2 Architecture Body

The internal details of a design/model are specified in architecture body. The different types of architecture models are,

- Dataflow model
- Behavioral model
- Structural model
- Combination of the above three.

Dataflow Model

A dataflow model specifies the functionality of a design entity or model without specifying the structure of the design. It shows the flow of information through the design entity

Syntax for architecture body of dataflow model

```
architecture architecture_name of entity_name is
                --architecture item declarations
begin
                --concurrent statements;
end architecture_name;
```

Behavioral Model

The behavior of the entity is defined using a set of sequentially executed statements .

A process statement is the main mechanism used to model the behavior of an entity.

Architecture Body

The architecture body consist of a set of interconnected components which represents the structure of the entity, or as a set of concurrent or sequential statements that represents the behavior of the entity

Syntax for architecture body of behavior model

```
architecture architecture_name of entity_name is
        --architecture item declarations
begin
        --concurrent statements ;
                process (sensitivity list)
                begin
                   --statements
                   .
                   .
                   .
                end process
end architecture_name;
```

Structural Model

In structural model, a design entity is designed with the set of components connected by signals.

Components are VHDL modules compiled as separate entity and can be used in any other structural model of digital systems.

Instead of coding a complex design in single VHDL Code, it can be divided into submodules as components and combine them using port map technique in a main module.

The component declaration and component instantiation statements are the primary structures used to describe the design entity.

14.6.3 Component Declaration

A component declaration declares the name of the component and a set of interface ports.

The port interface specifies the mode and the type of ports, similar to an entity declaration.

Syntax of component declaration

```
component component-name
    port (list_of_interface_ports) ;
end component;
```

14.6.4 Component Instantiation

A component instantiation statement is used to define the subcomponent of the entity in which it is appearing.

It connects the signals in the entity with the ports of the subcomponent.

The ***association-list*** associates signals in the entity called ***actuals***, with the ports of a component, called ***locals***.

Syntax for component instantiation

```
component_label: component_name port map (association_list)
```

Syntax for architecture body of Structural model

```
architecture architecture_name of entity_name is
    component component_name1
        port (list_of_interface_ports) ;
    end component;

    component component_name2
        port ( list_of_interface_ports ) ;
    end component;

        begin
            component_label1: component_name1 port map ( association_list1 );
            component_label2: component_name2 port map ( association_list2 );
            end architecture_name;
```

14.6.5 VHDL Programs for Half Adder with Three Different Architecture

The Boolean equations for half adder are, ***(AU, Nov/Dec'18, 13 Marks)***

$s = a \oplus b$

$c_o = a . b$

Where, a and b are 1-bit input data sum and carry are 1-bit output data. The VHDL codes for dataflow model of half adder are developed using the above equations. (Refer Chapter 5, Section 5.9.1 for design of half adder).

VHDL Program 14.3

VHDL Program for Dataflow Model of Half Adder

```
library ieee;
use ieee.std_logic_1164.all;
    entity half_adder is
        port ( a, b  : in std_logic;
               s, co : out std_logic );
    end half_adder;

    architecture dfm_ha of half_adder is
    begin
        s  <= a xor b;
        co <= a and b;
    end dfm_ha;
```

VHDL Program for Behavioral Model of Half Adder

```
library ieee;
use ieee.std_logic_1164.all;
    entity half_adder is
        port ( a, b  : in std_logic;
               s, co : out std_logic );
    end half_adder;

    architecture bm_ha of half_adder is
    begin
        process ( a, b )
            begin
                s  <= a xor b;
                co <= a and b;
        end process;
    end bm_ha;
```

Program Logic for Structural Model

The VHDL code for structural model of half adder is developed using the logic circuit shown in Fig. 14.1. In structural model, each component involved in the half adder are coded separately and then the components are added in the code for half adder.

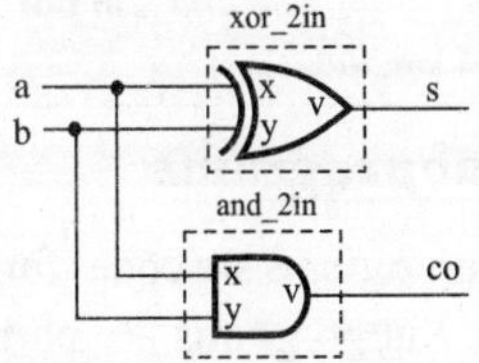

Fig. 14.1: Logic circuit of half adder.

VHDL Program for Structural Model of Half Adder

```
--code for 2-input XOR gate
library ieee;
use ieee.std_logic_1164.all;
    entity xor_2in is
        port ( x, y : in std_logic;
               v    : out std_logic );
    end xor_2in;

    architecture dfm_xor_2in of xor_2in is
    begin
        v <= x xor y;
    end dfm_xor_2in;
```

```
--code for 2-input AND gate
library ieee;
use ieee.std_logic_1164.all;
    entity and_2in is
        port ( x, y : in std_logic;
                v   : out std_logic );
    end and_2in;

    architecture dfm_and_2in of and_2in is
        begin
            v <= x and y;
    end dfm_and_2in;
```

```
--code for half adder
library ieee;
use ieee.std_logic_1164.all;
    entity half_adder is
        port ( a, b  : in std_logic;
                s, co : out std_logic );
    end half_adder;

    architecture sm_ha of half_adder is
        component xor_2in is
            port ( x, y : in std_logic;
                    v   : out std_logic );
        end component;

        component and_2in is
            port ( x, y : in std_logic;
                    v   : out std_logic );
        end component;
            begin
            a1 : xor_2in port map (a, b, s);
            a2 : and_2in port map (a, b, co);
end sm_ha;
```

14.7 Subprograms

Subprograms are codes for repeated operations in a VHDL program, similar to any high level language programs. Subprograms are part of a main program and cannot be compiled separately. Procedures and functions are two types of subprograms supported by VHDL. Procedures and functions are first declared in the declaration part of architecture and called later any number of times in the execution part of architecture.

14.7.1 Procedure

Procedures are subprograms written for repeated operations in a main program. Procedure requires a declaration which includes inputs, outputs and return values. Any number of values can be returned from a procedure.

VHDL Program 14.4

```
--example program for usage of procedure
library ieee;
use ieee.std_logic_1164.all;
use ieee.numeric_std.all;
```

```
entity pro_ex is
   generic (N : integer := 7);
   port ( x, y        : in signed (N – 1 downto 0);
          sum, dif : inout signed (N – 1 downto 0));
end pro_ex;

architecture arch_pro_ex of pro_ex is
signal p : signed (2 * N – 1 downto 0);

--procedure to add and subtract two numbers
procedure sum_dif ( signal a     : in signed (N – 1 downto 0);
                    signal b     : in signed (N – 1 downto 0);
                    signal s, d : out signed (N – 1 downto 0)) is
begin
   s <= a + b;
   d <= a – b;
end sum_dif;                  -- procedure ends, procedure returns two values s and d


begin
   sum_dif ( a => x, b => y, d => dif, s => sum);     -- procedure call
                 p <= sum * dif;                       -- product of sum and difference
end arch_pro_ex;
```

14.7.2 Functions

Functions are subprograms written for repeated operations in a main program. Function requires a declaration which includes inputs and a return value. In a function outputs need not be declared and it returns only one value.

The differences between procedure and function are,

- In a function, outputs need not be declared whereas in a procedure, the outputs have to be declared.
- A function will return only one value, whereas a procedure can return multiple values.

VHDL Program 14.5

```
--example program for usage of function
library ieee;
use ieee.std_logic_1164.all;
use ieee.numeric_std.all;
   entity func_ex is
      generic ( N : integer := 7);
      port ( x, y : in signed (N – 1 downto 0);
             sum : inout signed (N – 1 downto 0);
             dif   : inout signed (N – 1 downto 0) );
   end func_ex;

   architecture arch_func_ex of func_ex is
   signal p : signed (2 * N – 1 downto 0);

 --function to find sum of two integers
   function sum2int ( signal a : in signed (N – 1 downto 0);
                      signal b : in signed (N – 1 downto 0) );
   return signed is variable s : signed (N –1 downto 0);
```

```
begin
    s := a + b;
    return s;
end sum2int;     -- end of function, function returns the value of s.

--function to find difference of two integers
function dif2int ( signal a : in signed (N – 1 downto 0);
                   signal b : in signed (N – 1 downto 0));
return signed is variable d : signed (N – 1 downto 0);
begin
    d := a – b;
    return d;
end dif2int;                    --end of function, function returns the value of d;
begin
    sum <= sum2int (a => x, b => y);  --function call
    dif  <= dif2int (a => x, b => y);  --function call
    p    <= sum * dif;                  --product of sum and difference
end arch_func_ex;
```

14.8 Introduction to Packages

A package provides a convenient method to store and share declarations that are common to many design units. A Package is a VHDL file, contains data types, constants, functions, procedures, etc, that can be shared across many VHDL designs.

A package is created by a package declaration and package body. A package declaration contains a set of declarations that can be shared by many design units. A package body contains the functions and procedures that are declared in the corresponding package.

A package can be compiled or simulated separately and stored in a default library or in any user defined library and called in any other program using package name. The default library in Vivado is "xil_defaultlib". Also, the global include should be activated in Vivado before simulating the package.

Syntax of a package declaration

```
package package_name is
    -- subprogram declarations
    -- subtype declarations
    -- constant declarations
    -- signal declarations
    -- file declarations
    -- alias declarations
    -- component declarations
    -- attribute declarations
    -- attribute specifications
    -- disconnection specifications
    -- use clauses
end package [ package_name ]; -- package name is optional
```

14.8.1 Package Body

A package body primarily contains the behavior of the subprograms and the values of the deferred constants declared in a package declaration. It may contain other declarations as well, as shown by the following syntax of a package body.

Syntax of a package declaration

```
package body package_name is
    - subprogram bodies
    - subprogram declarations
    - type and subtype declarations
    - file and alias declarations
    - use clauses
end package body [ package_name ];  -- package name is optional
```

The package body name must be the same as the name of its corresponding package declaration. A package body is not necessary, if its associated package declaration does not have any subprogram or deferred constant declarations.

A package body is used to store local declarations that cannot be access by other design, while a package declaration is used to store public declarations which can be accessed by other design units.

VHDL Program 14.6

```
--example program for package declaration and package body
library ieee;
use ieee.std_logic_1164.all;
use ieee.numeric_std.all;
-- package declaration
    package mypackage3 is
-- declare a constant with value 1
    constant k : signed (3 downto 0) := "0001";
-- procedure to add two integers
    procedure sum_2int ( signal a     : in signed (3 downto 0);
                         signal b     : in signed (3 downto 0);
                         signal sum : out signed (3 downto 0));
-- declare signals for if loop
    signal m : unsigned (2 downto 0):=  (others => '0');
    type state_type is (start_state, continue_state, stop_state);
    end package mypackage3;
-- package body declaration
    package body mypackage3 is
-- procedure to add two integers
    procedure sum_2int ( signal a     : in signed (3 downto 0);
                         signal b     : in signed (3 downto 0);
                         signal sum : out signed (3 downto 0)) is
begin
    sum <= a + b;
end sum_2int;
end package body mypackage3;
```

```
--example program that calls the above package (mypackage3)
library ieee;
library xil_defaultlib;                       --mypackage3 is compiled in xil_defaultlib
use ieee.std_logic_1164.all;
use ieee.numeric_std.all;
use xil_defaultlib.mypackage3.all;            --use clause for mypackage3
   entity mypackage3_call is
      port ( clk                 : std_logic;
             x                   : in unsigned (4 downto 0);
             count               : out unsigned (2 downto 0);
             num1, num2          : in signed (3 downto 0);
             sum_num1_constant : out signed (3 downto 0);
             sum_num1_2          : inout signed (3 downto 0) );
   end mypackage3_call;

architecture arch_mypackage3_call of mypackage3_call is
signal current_state : state_type;            --state_type is defined in package
begin
-- add constant k to num1
   sum_num1_constant <= num1 + k;             --k is defined in package
-- add num1 and num2
   sum_2int(num1, num2, sum_num1_2);          --procedure sum_2int is defined in package

--process for start-stop control of a count sequence
   process (clk, current_state, m)
   begin
      if (m <= x) then                         --m is defined in package
         current_state <= continue_state;     --continue_state is defined in package
      else
         current_state <= stop_state;         --stop_state is defined in package
      end if;
   end process;

--process to generate a count sequence
   process (clk, current_state)
   begin
      if (current_state = continue_state) then
         m <= m + 1;
      else
         m <= (others => '0');
      end if;                                  --if loop end
         count <= m;
      end process;
      end arch_mypackage3_call;
```

14.9 Library

A library is a collection of packages that are compiled and stored in a directory in the host environment (or personal computer) in which the VHDL software tool is installed.

A library is a directory and each package is a file in that directory. The package file is a database containing information about the components in the package.

The two most commonly used libraries are IEEE and WORK. The IEEE library is a standard library standardized by IEEE and has a large number of packages. The WORK library is always available for user defined library packages. In vivado, the default library for user defined library and packages is "xil_defaultlib". The programmer or user can select either WORK library or "xil_defaultlib" library.

The keyword "library" identifies a library in VHDL followed by the logical library name.

Example

```
library IEEE;
library WORK;
library xil_defaultlib;
library mylib1;
```

A library may have large number of packages and all the packages may not be required for a design unit. Normally, the library statement shown above has to be combined with use statements to access the required packages in the library. In general programming practice, the library and use statements are placed in the beginning of the source file.

```
use ieee.std_logic_1164.all;
    --makes all components in std_logic_1164 library from ieee library visible
use ieee.std_logic-arith.all;
    --makes all components in std_logic_arith library from ieee library visible
use ieee.std_logic_unsigned.all;
    --makes all components in std_logic_unsigned library from ieee library visible
use ieee.std_logic_signed.all;
    --makes all components in std_logic_signed library from ieee library visible
use ieee.std_logic_textio.all;
    --makes all components in std_logic_textio library from ieee library visible
use ieee.numeric_bit.all;
    --makes all components in numeric_bit library from ieee library visible
use ieee.numeric_std.all;
    --makes all components in numeric_std library from ieee library visible
```

Design Library

(AU, Apr/May'19, 6 Marks)

Users can create small design units (with a design entity and architecture) and compile and store in a user defined library so that they can be accessed in any other larger design units. A design unit can be compiled or simulated separately and stored in a default library or WORK library or in any user defined library and called in any other program using the design entity name.

The default library in Vivado is "xil_defaultlib". Also, the global include should be activated in Vivado before simulating the design units. In order to use a design entity in another design, the library name in which the design unit is available should be included in design in which the library is called and use clause statement is not required.

VHDL Program 14.7

```
--example of VHDL file compiled in the library named "mylib1"
library ieee;
use ieee.std_logic_1164.all;
    entity not_1in is
        port ( a : in std_logic;
               y : out std_logic );
    end not_1in;
```

```
architecture dfm_not_1in of not_1in is
begin
    y <= not a;
end dfm_not_1in;
```

```
--example of VHDL file that use above library the compiled in the library named "mylib1"
library ieee;
use ieee.std_logic_1164.all;
library mylib1;              --not_1in is compiled in the user library "mylib1"
    entity top is
        port ( a : in std_logic;
               y : out std_Logic);
    end top;
architecture dfm of top is
begin
    u0 : entity mylib1.not_1in
    port map (a => a, y => y);
end dfm;
```

14.10 Test Bench

A test bench is also a VHDL program that is used to verify the correctness of the VHDL program of the hardware model. A test bench VHDL program has three main purposes:

- Generation of necessary inputs and clock for simulation.
- Apply this inputs and clock to the entity under test and to monitor the output
- Comparing the output of VHDL program model with expected output.

The application of inputs and clock to the entity under test is accomplished automatically by instantiating the entity in the test bench model and then specifying the appropriate interface signals.

Format of a test bench that drives an entity under test

```
entity test_bench is
    --port
end test_bench;

architecture tb_behavior of test_bench is
    component entity_under_test
        port (list_of_ports_their_types_and_modes);
    end component;
    local_signal_declarations;
begin
        generate_waveforms_using_behavioral_constructs;
    apply_to_entity_under_test;
        eut: entity_under_test port map (port_associations);
        monitor_values_and_compare_with_expected_values;
end tb_behavior;
```

VHDL program examples without test bench are presented in Section 14.12 and VHDL programs with test bench along with the output waveform are presented in Section 14.13. All the programs are simulated in Vivado 2019.2, which is a VHDL tool from XILINX.

14.11 RTL Design

In the VLSI design flow, the designer describes the behavior of the circuit using a hardware description language (HDL) such as VHDL or Verilog. These descriptions are then synthesized into a gate-level netlist, which is a representation of the circuit at the level of gates and interconnections. This netlist can be used to create a physical layout of the circuit on a FPGA IC using computer-aided design (CAD) tools.

(AU, Apr/May'19, 7 Marks)

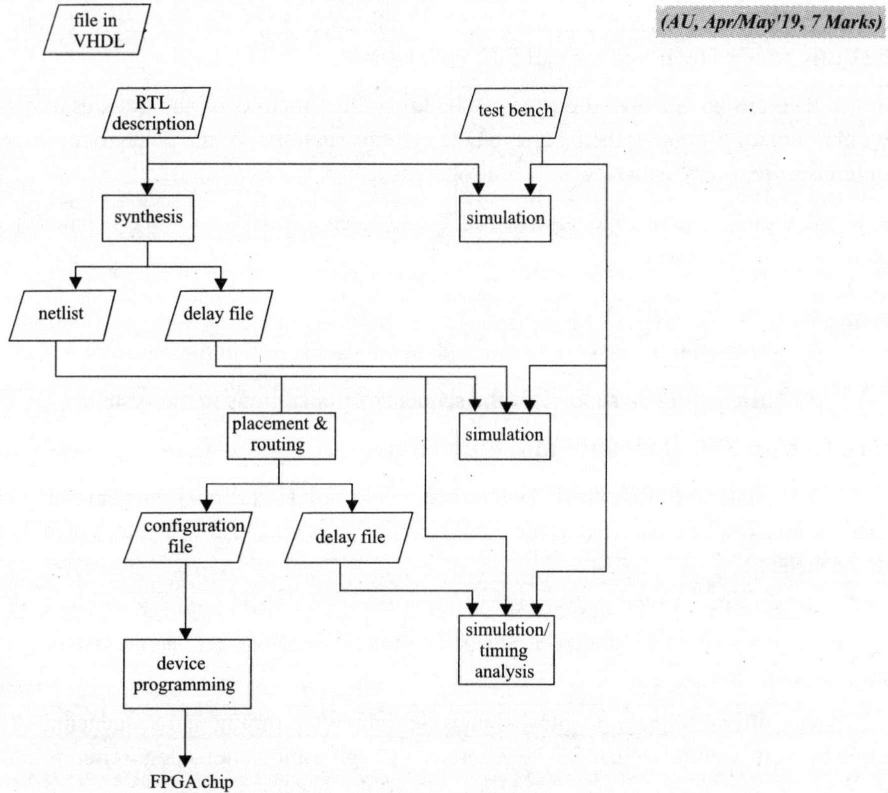

***Fig. 14.2:** Flowchart for IC design using VHDL.*

The various steps in the design of a an IC using VHDL and implementing in FPGA are,

- Develop the design file and testbench in VHDL.
- Use the design file as the circuit description, and perform a simulation to verify the functionality of the design.
- Perform a synthesis and generate a netlist.
- Use the output netlist file of the synthesizer as the circuit description, and perform a simulation of timing analysis.
- Perform placement and routing.
- Update the timing information to the netlist, and perform a simulation and timing analysis to verify the correctness of the placement and routing.

- Generate the configuration file and program the FPGA device.
- Verify the working of FPGA device.

A flowchart for IC design using VHDL and implementing in FPGA is shown in Fig. 14.1.

In IC design, it is important to consider various factors such as timing, power and area constraints. Timing constraints refer to the speed at which the circuit must operate, power constraints refer to the amount of power the circuit consumes, area constraints refer to the physical size of the IC. These requirements can be easily realized by the concept of RTL coding style.

RTL design is a digital design methodology that focuses on the transfer of data between registers within a digital system. At the RTL level, the system functionality and behavior of the system are described in terms of registers, data flow and control signals.

A digital system is represented at the register transfer level (RTL) when it is specified by the following three elements:

1. The set of registers in the system.
2. The operations that are performed on the data stored in the registers.
3. The control that supervises the sequence of operations in the system.

RTL Design Key Concepts and Elements

- **Registers:** Registers are the building blocks of RTL design. Registers can hold binary values, and their outputs can be connected to the inputs of other registers, enabling the flow of data between different components.

- **Data Flow:** RTL design focuses on the transfer of data between registers. Data flow is typically defined using a data path diagram, illustrating the connections between registers and the combinational logic blocks.

- **Control Signals:** Control signals determine the timing and sequencing of operations within a digital system. Control signals can be generated through combinational logic circuits or finite state machines, ensuring proper synchronization and operation of the system.

- **Combinational Logic:** Combinational logic circuits perform operations on input data to produce desired output values. Combinational logic circuits include various components such as arithmetic and Boolean operations, multiplexers, decoders, encoders, and logic gates.

The programming structure of RTL design can be broadly classified into data path and control path. The basic structure of RTL design is shown in Fig. 14.3. The data path has routing network, combinational logic and output registers. The control path has next state logic, registers and output logic.

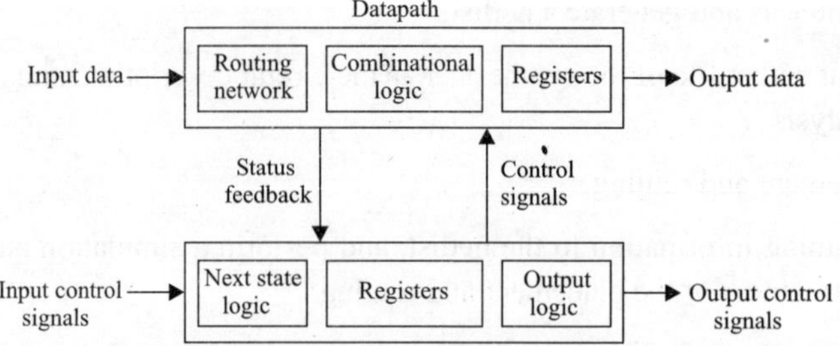

Fig. 14.3: Basic structure of RTL design.

Advantages of RTL Design

- **Performance Optimization:** RTL design allows for performance optimization by carefully managing the data flow, control signals and interconnections between registers.

- **Design Validation:** RTL design provides a structured approach to validate the functionality of a digital system before proceeding to physical implementation.

- **Reusability and Modularity:** Designers can create libraries of RTL blocks, such as adders, multipliers, memory controllers, or communication interfaces. These reusable blocks can be easily integrated into new designs.

RTL Design Methodologies

- **Manual RTL Design:** In manual RTL design, designers explicitly specify the behavior and functionality of the digital system using a hardware description language (HDL) such as Verilog or VHDL. The designer defines the registers, data flow, control signals, and interconnections. Manual RTL design allows for precise control over the system implementation and provides in-depth understanding of the design details. Some examples of manual RTL design are presented in Section 14.14.

- **RTL Synthesis:** RTL synthesis is an automated process that transforms a behavioral description written in VHDL into RTL description using a VLSI software tool. The synthesis tool translates the algorithmic representation into a register-level implementation, optimizing the design for area, power, and timing considerations. The Vivado tool used to synthesize the VHDL codes in this book will follow this automated process of RTL simulation and synthesis. All the example VHDL programs in Sections 14.12, 14.13 and 14.14 are simulated in Vivado 2019.2 as RTL project and outputs are verified.

Table 14.1: Differences between RTL Design and Sequential Logic Design

RTL Design	Sequential Logic Design
1. In RTL design the basic building blocks are registers, multiplexers, adders.	1. In sequential logic design the basic building blocks are the logic gates, flip-flops.
2. RTL design is much closer to the behavioural design of a logic circuit as it models the data flow among different registers, and hence more realistic.	2. Sequential logic design process is more mechanical in nature as compared to RTL design process.
3. RTL modelling allows us to synthesize complex circuits with a large number of states with ease as compared to sequential logic design.	3. The sequential logic design techniques are only applicable to circuits having a small number of states.

14.11.1 Combinational Logic

A combinational circuit, by definition, is a circuit whose output, is a function of current input. It has no internal state and therefore, it is system without memory about the past events (or past inputs).

Combinational RTL design involves the specification of circuits that perform logical operations on input signals to produce output signals, without any internal memory or state. Even though a digital system requires a clock for functioning, it is not mandatory to include clock signals in combinational logic circuit design. But the combinational digital system working will depend on FPGA IC default clock.

Combinational logic circuits can be designed using VHDL concurrent and sequential statements in dataflow or behavioral or structural architectures. Examples of combinational logic circuits are half adder, full adder, multiplexers, demultiplexers, etc. The RTL synthesisable VHDL codes for some combinational logic circuits are presented in Example programs in Section 14.12 and 14.13. All the example programs are simulated in Vivado 2019.2 as RTL project and outputs are verified.

14.11.2 Sequential Circuit

A sequential circuit has an ***internal state***, or ***memory***. Its output is a function of current input as well as the internal state. The internal state essentially "memorizes" the effect of the past input values. Sequential RTL design involves the specification of circuits that perform logical operations on input signals to produce output signals, with internal memory or states.

In addition, the operations of a sequential circuit depend on a clock signal. The "event" attribute of VHDL can be attached to a signal named "clk" to create a clock dependent process in a sequential circuit design. An example VHDL program of D flip-flop using positive edge triggered clock signal is given below:

> ***Note:*** *The attribute is an advanced feature in VHDL. An attribute is a value or function or type or range or signal or a constant that can be associated with certain names in a VHDL description. The name can be an entity name or an architecture name or a label or a signal.*

VHDL Program 14.8

```
--VHDL program for positive edge triggered D flip-flop
library ieee;
use ieee. std_logic_1164.all;
entity pos_dff is
    port ( clk : in std_logic;
           d   : in std_logic;
           q   : out std_logic );
end pos_dff ;

architecture arch of pos_dff is
    begin
        process (clk)
            begin
                if (clk'event and clk = '1') then
                        q <= d;
                end if;
        end process;
end arch;
```

In the above program, the key expression that creates clock sensitive process is,

```
if (clk'event and clk = '1') then
        q <= d;
end if;
```

The 'event term is a VHDL attribute returning true when there is a change in clk signal value (i.e., an event). When the clk = '1' expression is true, it means that the new value of clk is '1'. If statement states that at the rising of the clk signal, q gets the value of d. Since there is no else branch, it means that q keeps its previous value otherwise.

In the above program, the clock can be modified to have negative edge sensitivity by using the following if statement.

```
if (clk'event and clk = '0') then
        q <= d;
end if;
```

Sequential logic circuits can be designed using VHDL concurrent and sequential statements in dataflow or behavioral or structural architectures. Examples of sequential logic circuits are flip-flops, counters, shift registers, etc. The RTL synthesisable VHDL codes for some sequential logic circuits are presented in Example programs in Section 14.12 and 14.13. All the example programs are simulated in Vivado 2019.2 as RTL project and outputs are verified.

14.12 VHDL Programs Without Test Bench

VHDL Program 14.9

Simplify the Boolean function F by means of K-map and write a dataflow and structural HDL to implement the reduced Boolean function.

$f(a, b, c, d) = \sum m(0, 6, 7, 9, 11, 13, 15)$, $d(a, b, c, d) = \sum m(2, 5, 8, 10)$

To find Boolean Equation using K-map

The given function has four variables and so a 4-variable K-map with 16 squares ($2^4 = 16$) is drawn as shown in Fig. 1. The function is defined as sum of minterms m_0, m_6, m_7, m_9, m_{11}, m_{13} and m_{15} and so a "1" is filled in the corresponding squares in K-map. The function also has don't-care outputs for the minterms m_2, m_5, m_8 and m_{10} and so a "x" is filled in the corresponding squares in K-map and the remaining squares are filled with "0".

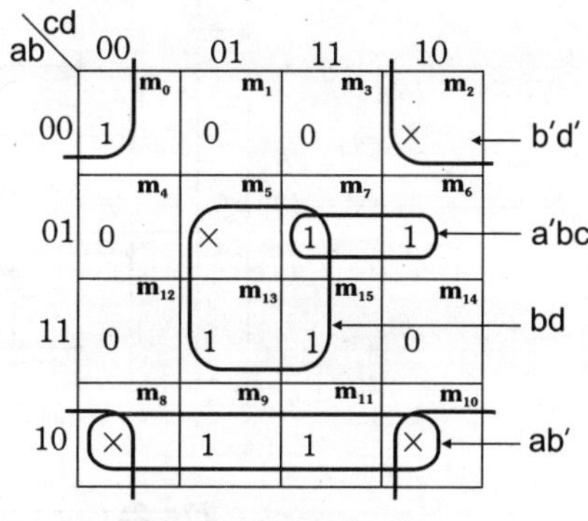

Fig. 1: K-map for f.

For NAND implementation, the Boolean function should be formed in SOP form. In the K-map of Fig. 1, by considering the don't-care condition in m_2 and m_5 as "1", three prime implicants can be formed with four adjacent 1's and the corresponding product terms are ab', bd, a'bc and b'd'. The simplified Boolean function is given by sum of these three product terms.

$\therefore$ f = ab' + bd + a'bc + b'd'

Using the above Boolean function the logic circuit using only NAND gates is drawn as shown in Fig. 2.

Program Logic for Dataflow Model

The Boolean function for K-map is,

f = ab' + bd + a'bc + b'd'

The VHDL code for dataflow model of Boolean function is developed using the above equations.(Refer Chapter 4, Section 4.6 for design of Boolean function).

VHDL Program for Dataflow Model of Boolean Function

```
library ieee;
use ieee.std_logic_1164.all;
    entity boolean_func is
        port ( a, b, c, d : in std_logic;
                f         : out std_logic );
    end boolean_func;
```

```
architecture dfm_boolean_func of boolean_func is
signal m1, m2, m3 : std_logic;
   begin
      m1 <= not a;
      m2 <= not b;
      m3 <= not d;
      f  <= ((a and m1) or (b and d) or (m1 and b and c) or (m2 and m3));
end dfm_boolean_func;
```

Program Logic for Structural Model

The VHDL code for structural model of Boolean function is developed using the logic circuit shown in Fig. 2. In structural model, each component involved in the logic circuit of Fig. 2 are coded separately and then the components are added in the code for function, f.

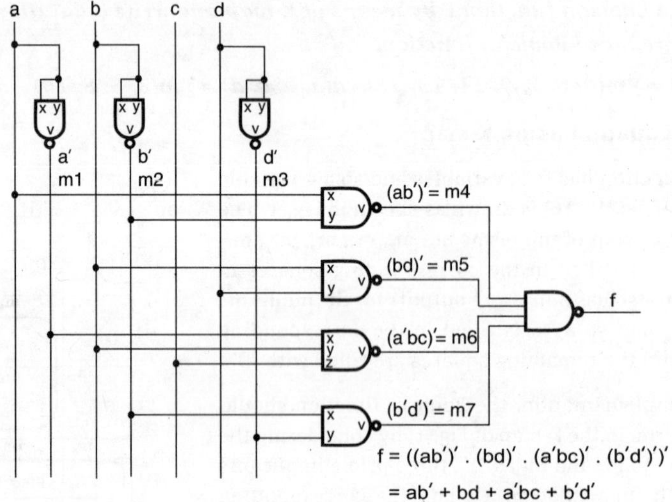

Fig. 2: Logic circuit of function, f using only NAND gates.

VHDL Program for Structural Model of Boolean Function

```
--code for 2-input NAND gate
library ieee;
use ieee.std_logic_1164.all;
   entity nand_2in is
      port ( x, y : in std_logic;
             v    : out std_logic );
   end nand_2in;

   architecture dfm_nand_2in of nand_2in is
      begin
         v <= x nand y;
   end dfm_nand_2in;
```

```
--code for 3-input NAND gate
library ieee;
use ieee.std_logic_1164.all;
```

```
    entity nand_3in is
        port ( x, y, z : in std_logic;
               v       : out std_logic );
    end nand_3in;

    architecture dfm_nand_3in of nand_3in is
        begin
            v <= not (x and (y and z));
    end dfm_nand_3in;
```

```
--code for 4-input NAND gate
library ieee;
use ieee.std_logic_1164.all;
    entity nand_4in is
        port ( w, x, y, z : in std_logic;
               v          : out std_logic );
    end nand_4in;

    architecture dfm_nand_4in of nand_4in is
        begin
            v <= not (w and (x and (y and z)));
    end dfm_nand_4in;
```

```
--code for Boolean function
library ieee;
use ieee.std_logic_1164.all;
    entity boolean_func is
        port ( a, b, c, d : in std_logic;
               f          : out std_logic;
    end boolean_func ;

    architecture sm_boolean_func of boolean_func is
    signal m1, m2, m3, m4, m5, m6, m7 : std_logic;
        component nand_2in is
            port ( x, y : in std_logic;
                   v    : out std_logic );
        end component;

        component nand_3in is
            port ( x, y, z : in std_logic;
                   v       : out std_logic );
        end component;

        component nand_4in is
            port ( w, x, y, z : in std_logic;
                   v          : out std_logic );
        end component;

            begin
                a1  : nand_2in port map (a, a, m1);
                a2  : nand_2in port map (b, b, m2);
                a3  : nand_2in port map (d, d, m3);
                a4  : nand_2in port map (a, m2, m4);
                a5  : nand_2in port map (b, d, m5);
                a6  : nand_3in port map (m1, b, c, m6);
```

```
        a7 : nand_2in port map (m2, m3, m7);
        a8 : nand_4in port map (m4, m5, m6, m7, f);
end sm_boolean_func ;
```

VHDL Program 14.10

Design a combinational circuit that accepts a 3-bit binary number as input and generates square of the input number as output. Write a high-level behavior VHDL description for the circuit.

To find Boolean Equation using K-map

The truth table of the given problem is developed as shown in Table 1. Using the truth table the K-maps are constructed as shown in Figs. 1 to 5.

Table 1: Truth Table

Inputs					Outputs						
Binary			Decimal	Minterm	Decimal	Binary					
a	b	c				b_5	b_4	b_3	b_2	b_1	b_0
0	0	0	0	m_0	0	0	0	0	0	0	0
0	0	1	1	m_1	1	0	0	0	0	0	1
0	1	0	2	m_2	4	0	0	0	1	0	0
0	1	1	3	m_3	9	0	0	1	0	0	1
1	0	0	4	m_4	16	0	1	0	0	0	0
1	0	1	5	m_5	25	0	1	1	0	0	1
1	1	0	6	m_6	36	1	0	0	1	0	0
1	1	1	7	m_7	49	1	1	0	0	0	1

***Note:** Since b_1 is always 0, K-map is not constructed for b_1.*

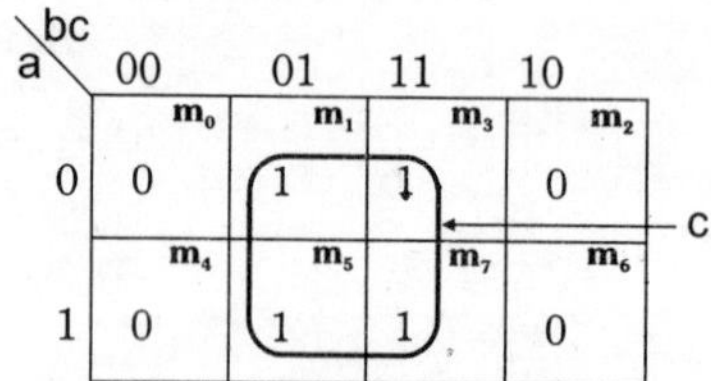

Fig. 1: K-map for b_0.

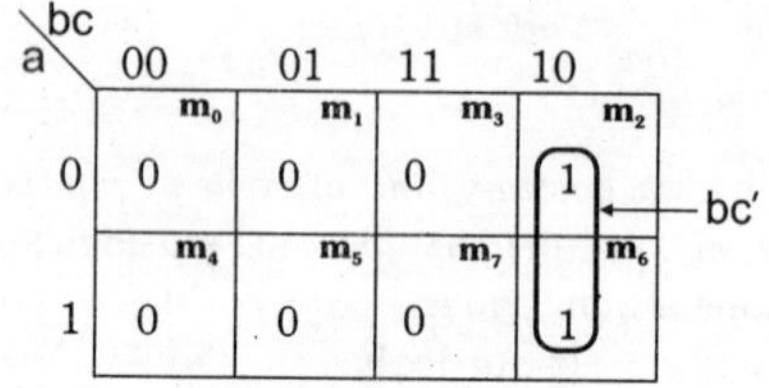

Fig. 2: K-map for b_2.

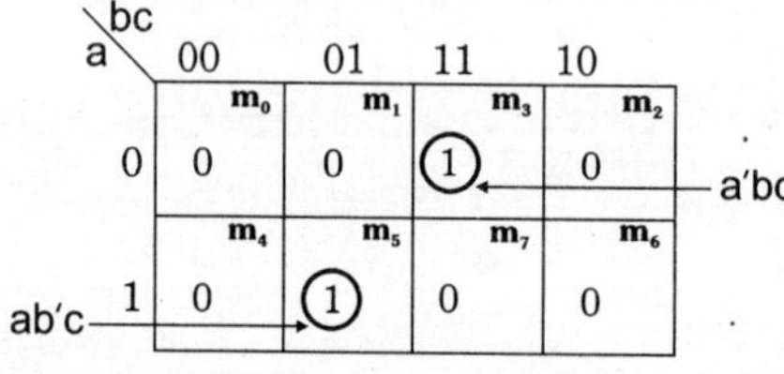

Fig. 3: K-map for b_3.

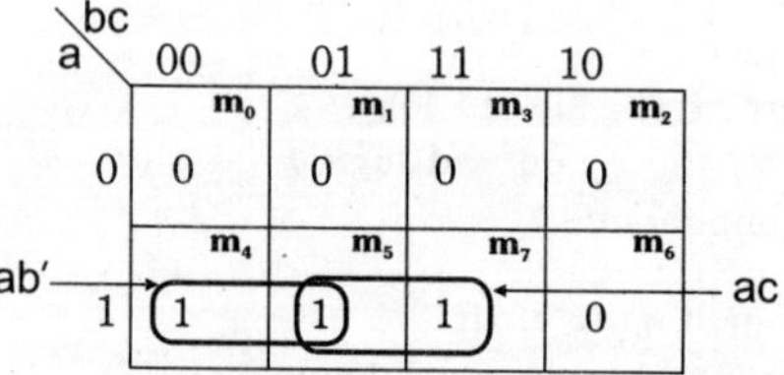

Fig. 4: K-map for b_4.

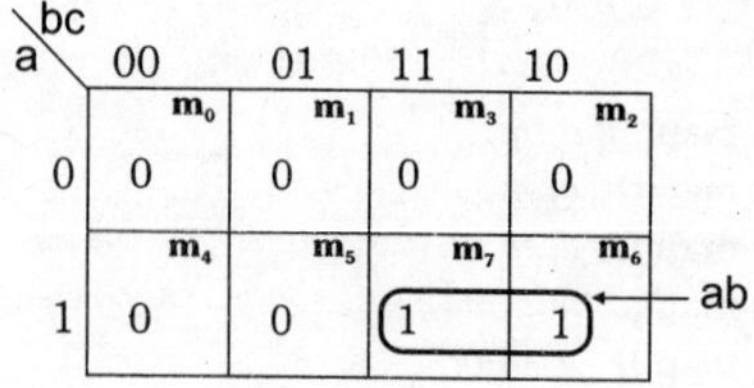

Fig. 5: K-map for b_5.

From the K-map the following Boolean equations are obtained and using these equations the logic circuit is drawn in Fig. 6.

$b_0 = c$	$b_3 = a'bc + ab'c$
$b_1 = 0$	$b_4 = ac + ab'$
$b_2 = bc'$	$b_5 = ab$

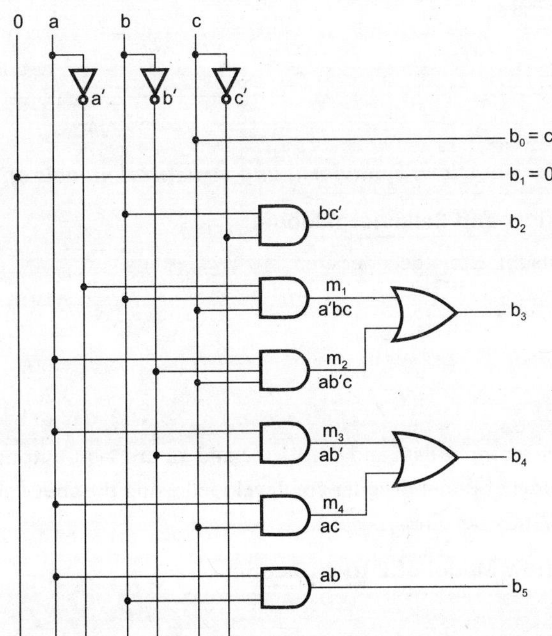

Fig. 6: Logic circuit to generate square of a 3-bit binary number.

Program Logic for Behavioral Model of Square of 3-bit Binary

Using Fig. 6, the VHDL codes for behavioral model of combinational circuit is developed.

VHDL Program for Behavioral Model of Square of 3-bit Binary

```
library ieee;
use ieee.std_logic_1164.all;
    entity square_3bit is
        port ( a, b, c                      : in std_logic;
               b0, b1, b2, b3, b4, b5 : out std_logic );
    end square_3bit ;

    architecture bm_square_3bit of square_3bit is
    signal abar, bbar, cbar, m1, m2, m3, m4 : std_logic;
        begin
            process (a ,b, c)
                begin
                    b0    <= c;
                    b1    <= '0';
                    abar <= not a;
                    bbar <= not b;
                    cbar <= not c;
```

```
            m1  <= abar and b and c;
            m1  <= a and bbar and c;
            m3  <= a and bbar;
            m4  <= a and c;
            b2  <= b and cbar;
            b3  <= m1 or m2;
            b4  <= m3 or m4;
            b5  <= a and b;
      end process;
  end bm_square_3bit;
```

VHDL Program 14.11

Write VHDL code for dataflow, behavioral and structural models of a 2-to-4 decoder.

Program Logic for Dataflow and Behavioral Model

The Boolean equations for 2-to-4 decoder are,

Y0 = I1'.I0'

Y1 = I1'.I0

Y2 = I1.I0'

Y3 = I1.I0

Where, I1 and I0 are 1-bit input data and Y0, Y1, Y2 and Y3 are 1-bit output data. The VHDL codes for dataflow and behavioral model of 2-to-4 decoder are developed using the above equations. (Refer Chapter 5, Section 5.4.1 for design of 2-to-4 decoder).

VHDL Program for Dataflow Model of 2-to-4 Decoder

```
library ieee;
use ieee.std_logic_1164.all;
   entity decoder_2_4 is
      port ( i1, i0          : in std_logic;
          y0, y1, y2, y3 : out std_logic );
   end decoder_2_4;

   architecture dfm_decoder_2_4 of decoder_2_4 is
   signal m1, m2 : std_logic;
      begin
         m1 <= not i1;
         m2 <= not i0;
         y0  <= m1 and m2;
         y1  <= m1 and i0;
         y2  <= i1 and m2;
         y3  <= i1 and i0;
   end dfm_ decoder_2_4;
```

VHDL Program for Behavioral Model of 2-to-4 Decoder using "case" Statement

```
library ieee;
use ieee.std_logic_1164.all;
   entity decoder_2_4 is
   port ( i             : in std_logic_vector (1 downto 0);
        y0, y1, y2, y3 : out std_logic );
   end decoder_2_4;
```

```
architecture bm_decoder_2_4 of decoder_2_4 is
signal m1, m2 : std_logic;
    begin
        process (i)
            begin
                m1 <= not i(0);
                m2 <= not i(1);
            case i is
                when "00"    => y0 <= m1 and m2;
                when "01"    => y1 <= m1 and i(1);
                when "10"    => y2 <= i(0) and m2;
                when others => y3 <= i(0) and i(1);
            end case;
        end process;
end bm_decoder_2_4;
```

VHDL Program for Behavioral Model of 2-to-4 Decoder using "if elsif" Statement

```
library ieee;
use ieee.std_logic_1164.all;
    entity decoder_2_4 is
        port ( i                  : in std_logic_vector (1 downto 0);
               y0, y1, y2, y3 : out std_logic );
    end decoder_2_4;

    architecture bm_decoder_2_4 of decoder_2_4 is
    signal m1, m2 : std_logic;
        begin
            process (i)
                begin
                    m1<= not i(0);
                    m2<= not i(1);
                if i = "00" then
                    y0 <= m1 and m2;

                elsif i = "01" then
                    y1 <= m1 and i(1);
                elsif i = "10" then
                    y2 <= i(0) and m2;
                else
                    y3 <= i(0) and i(1);
                end if;
            end process;
    end bm_decoder_2_4;
```

Program Logic for Structural Model

The VHDL code for structural model of 2-to-4 decoder is developed using the logic circuit shown in Fig. 1. In structural model, each component involved in the 2-to-4 decoder are coded separately and then the components are added in the code for 2-to-4 decoder.

Program for Structural Model of 2-to-4 Decoder

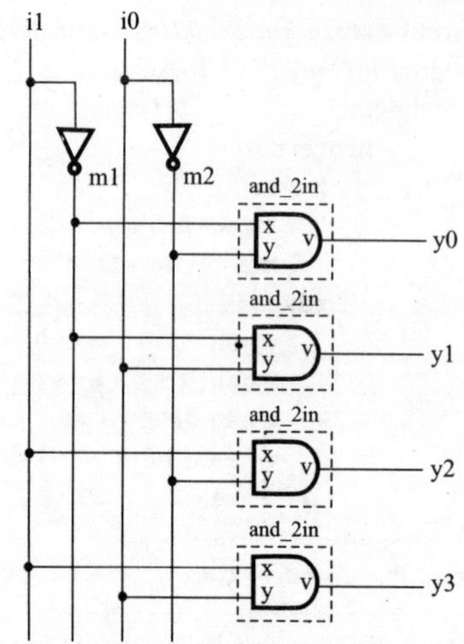

Fig. 1: Logic circuit for 2-to-4 decoder.

```
--code for 1-input NOT gate
library ieee;
use ieee.std_logic_1164.all;
    entity not_1in is
        port ( x : in std_logic;
               v : out std_logic );
    end not_1in;

    architecture dfm_not_1in of not_1in is
        begin
            v <= not x;
    end dfm_not_1in;
```

```
--code for 2-input AND gate
library ieee;
use ieee.std_logic_1164.all;
    entity and_2in is
        port ( x, y : in std_logic;
               v    : out std_logic );
    end and_2in;

    architecture dfm_and_2in of and_2in is
        begin
            v <= x and y;
    end dfm_and_2in;
```

```
--code for 2*4 decoder
library ieee;
use ieee.std_logic_1164.all;
    entity decoder_2_4 is
        port ( i1, i0           : in std_logic;
               y0, y1, y2, y3 : out std_logic );
    end decoder_2_4;

    architecture sm_decoder_2_4 of decoder_2_4 is
    signal m1, m2 : std_logic;

        component not_1in is
            port ( x : in std_logic;
                   v : out std_logic );
        end component;

        component and_2in is
            port ( x, y : in std_logic;
                   v    : out std_logic );
        end component;
    begin
        a1 : not_1in port map (i1, m1);
        a2 : not_1in port map (i0, m2);
        a3 : and_2in port map (m1, m2, y0);
        a4 : and_2in port map (m1, i0, y1);
        a5 : and_2in port map (m2, i, y2);
        a6 : and_2in port map (i1, i0, y3);
    end sm_decoder_2_4;
```

VHDL Program 14.12

Write a VHDL code for behavioral model of a 3-to-8 decoder.

Program Logic for Behavioral Model

The VHDL code for behavioral model of 3-to-8 decoder are developed using the truth table of 3-to-8 decoder. In the truth table $I_2I_1I_0$ is 3-bit input and Y_0 and Y_7 are 8 numbers of 1-bit outputs. (Refer Chapter 5, Section 5.4.7 for design of 3-to-8 decoder).

Truth Table of Logic High 3-to-8 Decoder

Inputs	Minterm	Outputs							
I_2 I_1 I_0		Y_0	Y_1	Y_2	Y_3	Y_4	Y_5	Y_6	Y_7
0 0 0	m_0	1	0	0	0	0	0	0	0
0 0 1	m_1	0	1	0	0	0	0	0	0
0 1 0	m_2	0	0	1	0	0	0	0	0
0 1 1	m_3	0	0	0	1	0	0	0	0
1 0 0	m_4	0	0	0	0	1	0	0	0
1 0 1	m_5	0	0	0	0	0	1	0	0
1 1 0	m_6	0	0	0	0	0	0	1	0
1 1 1	m_7	0	0	0	0	0	0	0	1

VHDL Program for Behavioral Model of 3-to-8 Decoder

```
library ieee;
use ieee.std_logic_1164.all;
use ieee.std_logic_arith.all;
use ieee.std_logic_unsigned.all;
   entity decoder3_8 is
      port ( I  : in bit_vector   (2 downto 0);
            y : out bit_vector (7 downto 0) );
   end decoder3_8;
   architecture bm_decoder3_8 of decoder3_8 is
      begin
         process (I)
            begin
               if I ="000" then
                  y <= "10000000";
               elsif I = "001" then
                  y <= "01000000";
               elsif I = "010" then
                  y <= "00100000";
               elsif I = "011" then
                  y<="00010000";
               elsif I = "100" then
                  y <= "00001000";
               elsif I = "101" then
                  y <= "00000100";
               elsif I = "110" then
                  y <= "00000010";
               else
                  y <= "00000001";
               end if;
         end process;
   end bm_decoder3_8;
```

VHDL Program 14.13

Write a VHDL code for dataflow and structural models of a 4-to-2 encoder.

Program Logic for Dataflow Model

The Boolean equations for 4-to-2 encoder are,

Y1 = I2 + I3 ; Y0 = I1 + I3

Where, I0, I1, I2 and I3 are 1-bit input data, Y0 and Y1 are 1-bit output data. The VHDL code for dataflow model of 4-to-2 encoder is developed using the above equations. (Refer Chapter 5, Section 5.5.1 for design of 4-to-2 encoder).

VHDL Program for Dataflow Model of 4-to-2 Encoder

```
library ieee;
use ieee.std_logic_1164.all;
   entity encoder_4_2 is
      port ( i0, i1, i2, i3 : in std_logic;
               y0, y1        : out std_logic);
   end encoder_4_2;

   architecture dfm_encoder_4_2 of encoder_4_2 is
      begin
         y1 <= i2 or i3;    y0 <= i1 or i3;
   end dfm_encoder_4_2;
```

Program Logic for Structural Model

The VHDL code for structural model of 4-to-2 encoder is developed using the logic circuit shown in Fig. 1. In structural model, each component involved in the 4-to-2 encoder are coded separately and then the components are added in the code for 4-to-2 encoder.

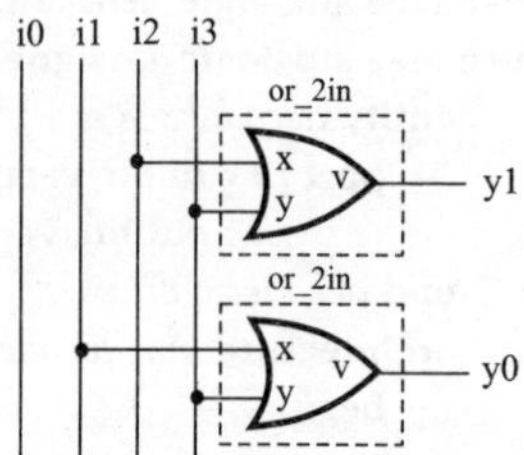

Fig. 1: 4-to-2 encoder.

Program for Structural Model of 4-to-2 Encoder

```
--code for 2-input OR gate
library ieee;
use ieee.std_logic_1164.all;
   entity or_2in is
      port ( x, y : in std_logic;
               v   : out std_logic );
   end or_2in;

   architecture dfm_or_2in of or_2in is
      begin
         v <= x or y;
   end dfm_or_2in;
```

```
--code for 4*2 encoder
library ieee;
use ieee.std_logic_1164.all;
   entity encoder_4_2 is
      port ( i0, i1, i2, i3 : in std_logic;
               y0, y1        : out std_logic );
   end encoder_4_2;
```

```
architecture sm_encoder_4_2 of encoder_4_2 is
    component or_2in is
        port ( x, y : in std_logic;
               v    : out std_logic );
    end component;

    begin
        a1 : or_2in port map (i2, i3, y1);
        a2 : or_2in port map (i1, i3, y0);
end sm_encoder_4_2;
```

VHDL Program 14.14

Write a VHDL code for dataflow model of priority encoder.

Program Logic for Dataflow Model

Boolean equations for priority encoder with I_3 having highest priority and I_0 having lowest priority are,

Y1 = I2 + I3 ; Y0 = I1.I2′ + I3 ; v = I0 + I1 + I2 + I3

Where, I0, I1, I2 and I3 are 1-bit input data and Y0, Y1 and v are 1-bit output data. The VHDL code for dataflow model of priority encoder is developed using the above equations. (Refer Chapter 5, Section 5.6.2 for design of priorioty encoder).

VHDL Program for Dataflow Model of Priority Encoder

```
library ieee;
use ieee.std_logic_1164.all;
    entity priority_encoder is
        port ( i0, i1, i2, i3 : in std_logic;
               y0, y1, v      : out std_logic );
    end priority_encoder;

    architecture dfm_priority_encoder of priority_encoder is
    signal m1 : std_logic;

        begin
            m1 <= not i2;
            y1 <= i2 or i3;
            y0 <= (i1 and m1) or i3;
            v  <= i0 or i1 or i2 or i3;
    end dfm_priority_encoder;
```

VHDL Program 14.15

Write a VHDL code for behavioral and structural model of a 4*1 multiplexer.

Program Logic for Behavioral Model

The VHDL code for behavioral model of 4*1 multiplexer is developed using the truth table of 4*1 multiplexer. In the truth table S0 and S1 are 1-bit selection input and Y is 1-bit output data. . (Refer Chapter 5, Section 5.7 for design of 4*1 multiplexer).

Truth Table of 4:1 Multiplexer

Selection Input		Output
S_1	S_0	Y
0	0	I_0
0	1	I_1
1	0	I_2
1	1	I_3

VHDL Program for Behavioral Model of 4*1 Multiplexer using 'if' Statement

```
library ieee;
use ieee.std_logic_1164.all;
  entity mux_4_1 is
      port ( s              : in std_logic_vector (1 downto 0);
             i0, i1, i2, i3 : in std_logic;
             y              : out std_logic);
  end mux_4_1;

  architecture bm_mux_4_1 of mux_4_1 is
      begin
         process (s, i0, i1, i2, i3)
            begin
               if s = "00" then
                  y <= i0;
               elsif s = "01" then
                  y <= i1;
               elsif s = "10" then
                  y <= i2;
               else
                  y <= i3;
               end if;
         end process;
  end bm_mux_4_1;
```

Program Logic for Structural Model

The VHDL code for structural model of 4*1 multiplexer is developed using the logic circuit shown in Fig. 1. In structural model, each component involved in the multiplexer are coded separately and then the components are added in the code for multiplexer.

VHDL Program for Structural Model of 4*1 Multiplexer

```
--code for 1-input NOT gate
library ieee;
use ieee.std_logic_1164.all;
  entity not_1in is
      port ( x : in std_logic;
             v : out std_logic );
  end not_1in;

  architecture dfm_not_1in of not_1in is
      begin
         v <= not x;
  end dfm_not_1in;
```

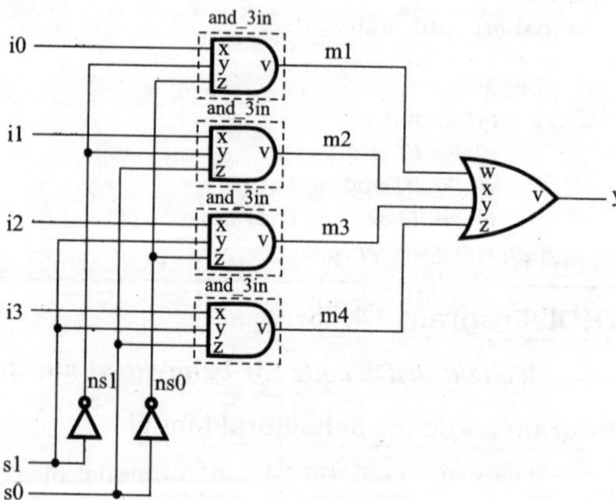

Fig. 1: Logic circuit of 4:1 multiplexer.

```
--code for 3-input AND gate
library ieee;
use ieee.std_logic_1164.all;
  entity and_3in is
      port ( x, y, z : in std_logic;
             v       : out std_logic );
  end and_3in;
```

```
architecture dfm_and_3in of and_3in is
    begin
        v <= x and y and z;
end dfm_and_3in;
```

```
--code for 4-input OR gate
library ieee;
use ieee.std_logic_1164.all;
    entity or_4in is
        port ( w, x, y, z : in std_logic;
               v          : out std_logic );
    end or_4in;

    architecture dfm_or_4in of or_4in is
        begin
            v <= w or x or y or z;
    end dfm_or_4in;
```

```
--code for 4*1 multiplexer
library ieee;
use ieee.std_logic_1164.all;
    entity mux_4_1 is
        port ( s0, s1         : in std_logic;
               i0, i1, i2, i3 : in std_logic;
               y              : out std_logic );
    end mux_4_1;

    architecture sm_mux_4_1 of mux_4_1 is
    signal ns0, ns1          : std_logic;
    signal m1, m2, m3, m4 : std_logic;
        component not_1in is
            port ( x : in std_logic;
                   v : out std_logic );
        end component;

        component and_3in is
            port ( x, y, z : in std_logic;
                   v       : out std_logic );
        end component;

        component or_4in is
            port ( w, x, y, z : in std_logic;
                   v          : out std_logic );
        end component;

            begin
                a1 : not_1in port map (s0, ns0);
                a2 : not_1in port map (s1, ns1);
                a3 : and_3in port map (i0, ns1, ns0, m1);
                a4 : and_3in port map (i1, ns1, s0, m2);
                a5 : and_3in port map (i2, s1, ns0, m3);
                a6 : and_3in port map (i3, s1, s0, m4);
                a7 : or_4in port map (m1, m2, m3, m4, y);
    end sm_mux_4_1;
```

VHDL Program 14.16

(AU, Apr/May'19, 6 Marks)

Write a VHDL code for behavioral model of a 8*1 multiplexer.

Program Logic for Behavioral Model

The VHDL code for behavioral model of 8*1 multiplexer is developed using the truth table of 8*1 multiplexer. In the truth table S0, S1 and S2 are 1-bit selection input and Y is 1-bit output data. (Refer Chapter 5, Section 5.7 for design of 8*1 multiplexer).

Truth Table of 8:1 Multiplexer

Selection Inputs			Output Y
S_2	S_1	S_0	
0	0	0	I_0
0	0	1	I_1
0	1	0	I_2
0	1	1	I_3
1	0	0	I_4
1	0	1	I_5
1	1	0	I_6
1	1	1	I_7

VHDL Program for Behavioral of 8*1 Multiplexer

```
library ieee;
use ieee.std_logic_1164.all;
   entity mux_8_1 is
      port ( s                          : in std_logic_vector (2 downto 1);
             i0, i1, i2, i3, i4, i5, i6, i7 : in std_logic;
             y                          : out std_logic );
   end mux_8_1;

   architecture bm_mux_8_1 of mux_8_1 is
      begin
         process (s, i0, i1, i2, i3, i4, i5, i6, i7)
            begin
               if s = "'000" then
                  y <= i0;
               elsif s = "001" then
                  y <= i1;
               elsif s = "010" then
                  y <= i2;
               elsif s = "011" then
                  y <= i3;
               elsif s = "100" then
                  y <= i4;
               elsif s = "101" then
                  y <= i5;
               elsif s = "110" then
                  y <= i6;
               else
                  y <= i7;
               end if;
         end process;
   end bm_mux_8_1;
```

VHDL Program 14.17

Write a VHDL code for dataflow and structural model of a 1*4 demultiplexer.

Program Logic for Dataflow Model

The Boolean equations for 1*4 demultiplexer are,

Y0 = Ii.S1'.S0' ; Y1 = Ii.S1'.S0 ; Y2 = Ii.S1.S0' ; Y3 = Ii.S1.S0

Where, S1 and S0 are 1-bit selection input and Y0, Y1, Y2 and Y3 are 1-bit output data. The VHDL code for dataflow model of 1*4 demultiplexer is developed using the above equations. (Refer Chapter 5, Section 5.8 for design of 1*4 demultiplexer).

VHDL Program for Dataflow Model of 1*4 Demultiplexer

```
library ieee;
use ieee.std_logic_1164.all;
   entity demux_1_4 is
      port ( ii, s0, s1       : in std_logic;
             y0, y1, y2, y3 : out std_logic );
   end demux_1_4;

   architecture dfm_demux_1_4 of demux_1_4 is
   signal m1, m2 : std_logic;
      begin
         m1 <= not s0;
         m2 <= not s1;
         y0  <= ii and (m1 and m2);
         y1  <= ii and (m1 and s1);
         y2  <= ii and (s0 and m2);
         y3  <= ii and (s0 and s1);
   end dfm_demux_1_4;
```

Program Logic for Structural Model

The VHDL code for structural model of 1*4 demultiplexer is developed using the logic circuit shown in Fig. 1. In structural model, each component involved in the demultiplexer are coded separately and then the components are added in the code for demultiplexer.

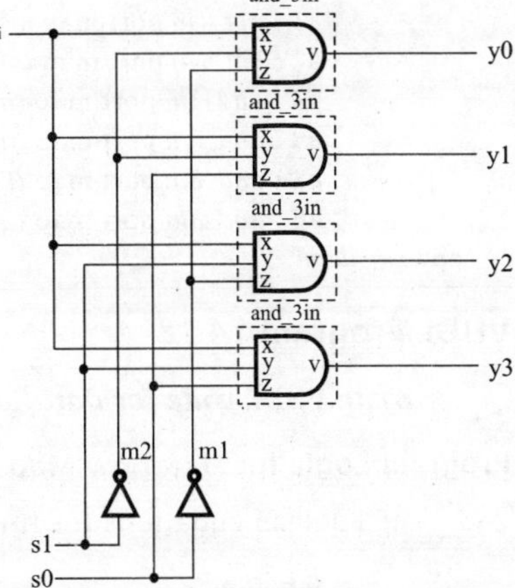

***Fig. 1:** Logic circuit of 1:4 demux.*

VHDL Program for Structural Model of 1*4 Demultiplexer

```
--code for 1-input NOT gate
library ieee;
use ieee.std_logic_1164.all;
   entity not_1in is
      port ( x : in std_logic;
             v : out std_logic );
   end not_1in;

   architecture dfm_not_1in of not_1in is
      begin
         v <= not x;
   end dfm_not_1in;
```

```
--code for 3-input AND gate
library ieee;
use ieee.std_logic_1164.all;
    entity and_3in is
        port ( x, y, z : in std_logic;
               v       : out std_logic );
    end and_3in;

    architecture dfm_and_3in of and_3in is
        begin
            v <= x and y and z;
    end dfm_and_3in;
```

```
--code for 1*4 demultiplexer
library ieee;
use ieee.std_logic_1164.all;
    entity demux_1_4 is
        port ( ii, s1, s0      : in std_logic;
               y0, y1, y2, y3 : out std_logic );
    end demux_1_4;

    architecture sm_demux_1_4 of demux_1_4 is
    signal m1, m2 : std_logic;
        component not_1in is
            port ( x : in std_logic;
                   v : out std_logic );
        end component;

        component and_3in is
            port ( x, y, z : in std_logic;
                   v       : out std_logic );
        end component;

            begin
                a1 : not_1in port map (s0, m1);
                a2 : not_1in port map (s1, m2);
                a3 : and_3in port map (ii, m1, m2, y0);
                a4 : and_3in port map (ii, m2, s0, y1);
                a5 : and_3in port map (ii, s1, m1, y2);
                a6 : and_3in port map (ii, s1, s0, y3);
    end sm_demux_1_4;
```

VHDL Program 14.18

(AKTU, Sep'20, 5 Marks) *(AU, Apr/May'19, 7 Marks)*

Write VHDL code for dataflow and structural models of a full adder.

Program Logic for Dataflow Model

The Boolean equations for full adder are,

$$s = a \oplus b \oplus c_i$$

$$c_o = (a\,b) + (b\,c_i) + (c_i\,a)$$

Where, a, b and ci are 1-bit input data, sum (s) and carry (c_o) are 1-bit output data. The VHDL code for dataflow model of full adder is developed using the above equations. (Refer Chapter 5, Section 5.9.2 for design of full adder).

VHDL Program for Dataflow Model of Full Adder

```
library ieee;
use ieee.std_logic_1164.all;
   entity full_adder is
      port ( a, b, ci  : in std_logic;
             s, co     : out std_logic );
   end full_adder;

   architecture dfm_fa of full_adder is
   signal m1, m2, m3 : std_logic;
      begin
         s   <= a xor b xor ci;
         m1 <= a and b;
         m2 <= b and ci;
         m3 <= ci and a;
         co  <= m1 or m2 or m3;
   end dfm_fa;
```

Program Logic for Structural Model

The VHDL code for structural model of full adder is developed using the logic circuit shown in Fig. 1. In structural model, each component involved in full adder are coded separately and then the components are added in the code for full adder.

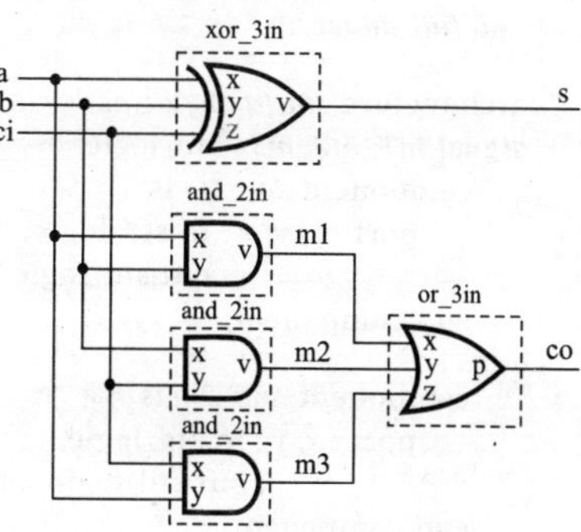

***Fig. 1:** Logic circuit of full adder.*

VHDL Program for Structural Model of Full Adder

```
--code for 3-input XOR gate
library ieee;
use ieee.std_logic_1164.all;
   entity xor_3in is
      port ( x, y, z : in std_logic;
             v       : out std_logic );
   end xor_3in;

   architecture dfm_xor_3in of xor_3in is
      begin
         v <= x xor y xor z;
   end dfm_xor_3in;
```

```
--code for 2-input AND gate
library ieee;
use ieee.std_logic_1164.all;
   entity and_2in is
      port ( x, y : in std_logic;
             v    : out std_logic );
   end and_2in;

   architecture dfm_and_2in of and_2in is
      begin
         v <= x and y;
   end dfm_and_2in;
```

```
--code for 3-input OR gate
library ieee;
use ieee.std_logic_1164.all;
    entity or_3in is
        port ( x, y, z : in std_logic;
               v       : out std_logic );
    end or_3in;

    architecture dfm_or_3in of or_3in is
        begin
            v <= x or y or z;
    end dfm_or_3in;
```

```
--code for full adder
library ieee;
use ieee.std_logic_1164.all;
    entity full_adder is
        port ( a, b, ci : in std_logic;
               s, co    : out std_logic );
    end full_adder;

    architecture sm_fa of full_adder is
    signal m1, m2, m3 : std_logic;
        component xor_3in is
            port ( x, y, z : in std_logic;
                   v       : out std_logic );
        end component;

        component and_2in is
            port ( x, y : in std_logic;
                   v    : out std_logic );
        end component;

        component or_3in is
            port ( x, y, z : in std_logic;
                   v       : out std_logic );
        end component;
            begin
                a1 : xor_3in port map (a, b, ci, s);
                a2 : and_2in port map (a, b, m1);
                a3 : and_2in port map (b, ci, m2);
                a4 : and_2in port map (ci, a, m3);
                a5 : or_3in port map (m1, m2, m3, co);
    end sm_fa;
```

VHDL Program 14.19

Write VHDL code for structural model of a 4-bit parallel adder.

Program Logic for Structural Model

The Boolean equations for 4-bit parallel adder performs arithmetic sum of two 4-bit binary numbers. A full adder can add two 1-bit binary along with previous carry. In order to add two 4-bit binary numbers, 4 full adders are required. Each ful adder will add one bit of binary numbers. Consider two 4-bit binary numbers a and b shown below:

Binary number 1 : a3, a2, a1, a0 (data-a) ; Binary number 2 : b3, b2, b1, b0 (data-b)

Now, the logic circuit of 4-bit parallel adder will be as shown in Fig. 1. (Refer Chapter 5, Section 5.9.3 for design of 4-bit parallel adder).

The VHDL code for structural model of 4-bit parallel adder is developed using the logic circuit shown in Fig. 1. In structural model, each component involved in 4-bit adder are coded separately and then the components are added in the code for 4-bit adder.

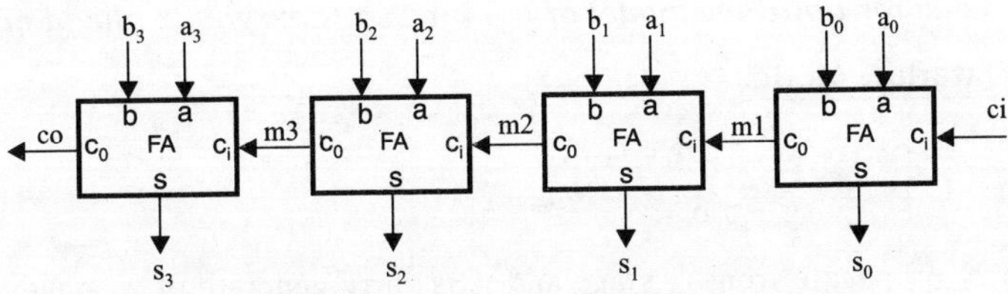

***Fig. 1:** 4-bit binary adder using full adders.*

VHDL Program for Structural Model of 4-bit Parallel Adder

```
--code for full adder
library ieee;
use ieee.std_logic_1164.all;
    entity full_adder is
        port ( a, b, ci  : in std_logic;
               s, co     : out std_logic );
    end full_adder;

    architecture dfm_fa of full_adder is
    signal m1, m2, m3 : std_logic;
        begin
            s    <= a xor b xor ci;
            m1 <= a and b;
            m2 <= b and ci;
            m3 <= ci and a;
            co   <= m1 or m2 or m3;
    end dfm_fa;
```

```
--code for 4-bit parallel adder
library ieee;
use ieee.std_logic_1164.all;
    entity parallel_adder is
        port ( a, b : in std_logic_vector (3 down to 0);
               ci   : in std_logic;
               s    : out std_logic_vector (3 down to 0);
               co   : out std_logic );
    end parallel_adder;

    architecture sm_parallel_adder of parallel_adder is
    signal m1, m2, m3 : std_logic;
        component full_adder is
            port ( a, b, ci : in std_logic;
                   s, co    : out std_logic );
        end component;

        begin
            fa0 : full_adder port map (a(0), b(0), ci, s(0), m1);
```

```
        fa1 : full_adder port map (a(1), b(1), s(1), m1, m2);
        fa2 : full_adder port map (a(2), b(2), s(2), m2, m3);
        fa3 : full_adder port map (a(3), b(3), s(3), m3, co);
    end sm_parallel_adder;
```

VHDL Program 14.20

Write a VHDL code for dataflow model of a 4-bit CLA (Carry look ahead adder).

Program Logic for Dataflow Model

$$s_i = p_i \oplus c_i \quad ; \quad c_{i+1} = g_i + p_i c_i$$

$$p_i = a_i \oplus b_i \quad ; \quad g_i = a_i b_i$$

Where, p_i is carry propagate from i_{th} stage and g_i is carry generate of i_{th} stage. The VHDL code for dataflow model of 4-bit CLA is developed using the above equations. (Refer Chapter 5, Section 5.9.4 for design of 4-bit CLA).

VHDL Program for Dataflow Model of 4-bit CLA

```
library ieee;
use ieee.std_logic_1164.all;
   entity cla_4bit is
      port ( a3, a2, a1, a0    : in std_logic;
             b3, b2, b1, b0    : in std_logic;
             c0                : in std_logic;
             c4, s3, s2, s1, s0 : out std_logic);
   end cla_4bit;

   architecture dfm_cla_4bit of cla_4bit is
   signal m1, m2, m3, m4, m5, m6, m7 : std_logic;
   signal p0, p1, p2, p3 : std_logic;
   signal g0, g1, g2, g3 : std_logic;

      begin
         p0 <= a0 xor b0;  p1 <= a1 xor b1;  p2 <= a2 xor b2;  p3 <= a3 xor b3;

         g0 <= a0 and b0;  g1 <= a1 and b1;  g2 <= a2 and b2;  g3 <= a3 and b3;

         m1 <= g0 or (p0 and c0);
         m2 <= g1 or (p1 and g0) or (p1 and p0 and c0);
         m3 <= g2 or (p2 and g1) or (p2 and p1 and g0) or (p2 and p1 and p0 and c0);
         s0 <= p0 xor c0;
         s1 <= p1 xor m1;
         s2 <= p2 xor m2;
         s3 <= p3 xor m3;
         m4 <= p3 and g2
         m5 <= p3 and p2 and g1
         m6 <= p3 and p2 and p1 and g0
         m7 <= p3 and p2 and p1 and p0 and c0
         c4 <= g3 or m4 or m5 or m6 or m7;
   end dfm_cla_4bit;
```

VHDL Program 14.21

Write VHDL code for dataflow and structural models of a half subtractor.

Program Logic for Dataflow Model

The Boolean equations for half subtractor are,

$$d = a \oplus b = ab' + a'b$$

$$c_o = a'b$$

Where, a and b are 1-bit input data, difference (d) and borrow (c_o) are 1-bit output data. The VHDL code for dataflow model of half subtractor is developed using the above equations. (Refer Chapter 5, Section 5.10.1 for design of half subtractor).

VHDL Program for Dataflow Model of Half Subtractor

```
library ieee;
use ieee.std_logic_1164.all;
   entity half_subtractor is
      port ( a, b  : in std_logic;
             d, co : out std_logic );
   end half_subtractor;

   architecture dfm_hs of half_subtractor is
   signal m1 : std_logic;
      begin
         d   <= a xor b;
         m1 <= not a;
         co  <= m1 and b;
   end dfm_hs;
```

Program Logic for Structural Model

The VHDL code for structural model of half subtractor is developed using the logic circuit shown in Fig. 1. In structural model, each component involved in half subtractor are coded separately and then the components are added in the code for half subtractor.

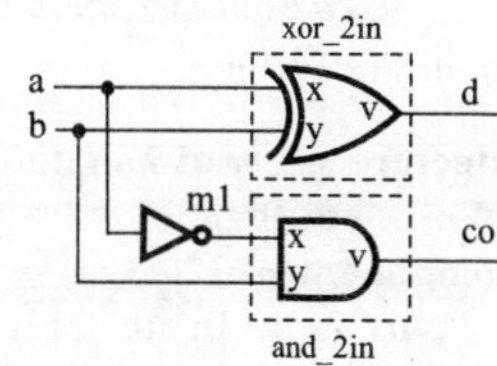

Fig. 1: Logic circuit of half subtractor.

VHDL Program for Structural Model of Half Subtractor

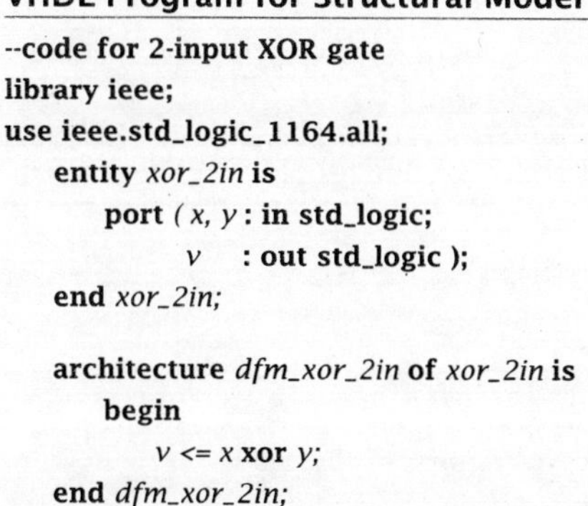

```
--code for 2-input XOR gate
library ieee;
use ieee.std_logic_1164.all;
   entity xor_2in is
      port ( x, y : in std_logic;
             v    : out std_logic );
   end xor_2in;

   architecture dfm_xor_2in of xor_2in is
      begin
         v <= x xor y;
   end dfm_xor_2in;
```

```
--code for 2-input AND gate
library ieee;
use ieee.std_logic_1164.all;
   entity and_2in is
      port ( x, y : in std_logic;
             v    : out std_logic );
   end and_2in;

   architecture dfm_and_2in of and_2in is
      begin
         v <= x and y;
   end dfm_and_2in;
```

```
--code for 1-input NOT gate
library ieee;
use ieee.std_logic_1164.all;
   entity not_1in is
      port ( x : in std_logic;
             v : out std_logic );
   end not_1in;

   architecture dfm_not_1in of not_1in is
      begin
         v <= not x;
   end dfm_not_1in;
```

```
--code for half subtractor
library ieee;
use ieee.std_logic_1164.all;
   entity half_subtractor is
      port ( a, b  : in std_logic;
             d, co : out std_logic );
   end half_subtractor;

   architecture sm_hs of half_subtractor is
   signal m1 : std_logic;
      component xor_2in is
         port ( x, y : in std_logic;
                v    : out std_logic );
      end component;

      component and_2in is
         port ( x, y : in std_logic;
                v    : out std_logic );
      end component;

      component not_1in is
         port ( x : in std_logic;
                v : out std_logic );
      end component;

         begin
            a1 : xor_2in port map (a, b, d);
            a2 : not_1in port map (a, m1);
            a3 : and_2in port map (m1, b, co);
   end sm_hs;
```

VHDL Program 14.22

Write VHDL code for dataflow and structural models of a full subtractor.

Program Logic Dataflow Model

The Boolean equations for full subtractor are,

$$d = a'b'c_i \oplus a'bc_i' \oplus a'bc_i' \oplus abc_i$$

$$c_o = (a'b) + (b\,c_i) + (c_i\,a')$$

Where, a, b and c_i are 1-bit input data, difference (d) and borrow (c_o) are 1-bit output data. The VHDL code for dataflow model of full subtractor is developed using the above equations. (Refer Chapter 5, Section 5.10.2 for design of full subtractor).

VHDL Program for Dataflow Model of Full Subtractor

```
library ieee;
use ieee.std_logic_1164.all;
    entity full_subtractor is
        port ( a, b, ci  : in std_logic;
               d, co     : out std_logic );
    end full_subtractor;

    architecture dfm_fs of full_subtractor is
    signal m1 : std_logic;

        begin
            d    <= a xor b xor ci;
            m1 <= not a;
            co   <= (m1 and b) or (b and ci) or (ci and m1);
    end dfm_fs;
```

Program Logic for Structural Model

The VHDL code for structural model of full subtractor is developed using the logic circuit shown in Fig. 1. In structural model, each component involved in full subtractor are coded separately and then the components are added in the code for full subtractor.

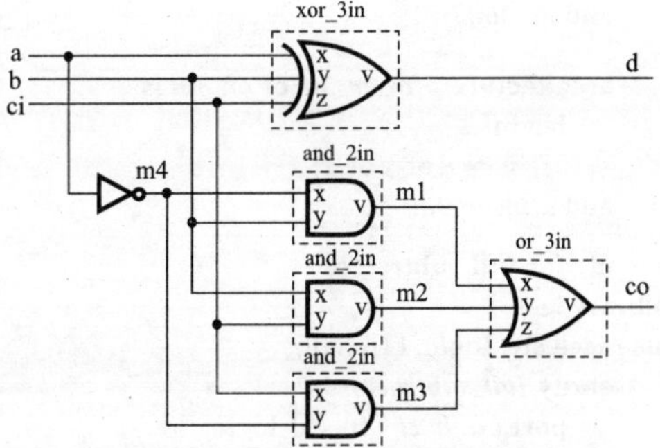

Fig. 1: Logic circuit of full subtractor.

VHDL Program for Structural Model of Full Subtractor

```
--code for 3-input XOR gate
library ieee;
use ieee.std_logic_1164.all;
    entity xor_3in is
        port ( x, y, z : in std_logic;
               v        : out std_logic );
    end xor_3in;

    architecture dfm_xor_3in of xor_3in is
        begin
            v <= x xor y xor z;
    end dfm_xor_3in;
```

```
--code for 1-input NOT gate
library ieee;
use ieee.std_logic_1164.all;
    entity not_1in is
        port ( x : in std_logic;
               v : out std_logic );
    end not_1in;

    architecture dfm_not_1in of not_1in is
        begin
            v <= not x;
    end dfm_not_1in;
```

```
--code for 2-input AND gate
library ieee;
use ieee.std_logic_1164.all;
    entity and_2in is
        port ( x, y : in std_logic;
               v    : out std_logic );
    end and_2in;

    architecture dfm_and_2in of and_2in is
        begin
            v <= x and y;
    end dfm_and_2in;
```

```
-- code for 3-input OR gate
library ieee;
use ieee.std_logic_1164.all;
    entity or_3in is
        port ( x, y, z : in std_logic;
               v       : out std_logic );
    end or_3in;

    architecture dfm_or_3in of or_3in is
        begin
            v <= x or y or z;
    end dfm_or_3in;
```

```
-- code for full subtractor
library ieee;
use ieee.std_logic_1164.all;
    entity full_subtractor is
        port ( a, b, ci : in std_logic;
               d, co    : out std_logic );
    end full_subtractor;

    architecture sm_fs of full_subtractor is
    signal m1, m2, m3, m4 : std_logic;
        component xor_3in is
            port ( x, y, z : in std_logic;
                   v       : out std_logic );
        end component;
```

```
        component not_1in is
            port ( x  : in std_logic;
                   v  : out std_logic );
        end component;

        component and_2in is
            port ( x, y : in std_logic;
                   v    : out std_logic );
        end component;

        component or_3in is
            port ( x, y, z : in std_logic;
                   v       : out std_logic );
        end component;
            begin
                a1 : xor_3in port map (a, b, ci, d);
                a2 : not_1in port map (a, m4);
                a3 : and_2in port map (b, m4, m1);
                a4 : and_2in port map (b, ci, m2);
                a5 : and_2in port map (ci, m4, m3);
                a6 : or_3in port map (m1, m2, m3, co);
    end sm_fs;
```

VHDL Program 14.23

Write a VHDL code for dataflow model of 1-bit magnitude comparator.

Program Logic for Dataflow Model

Let a and b are 1-bit binary numbers.

The Boolean equations for 1-bit magnitude comparator are,

if $a = b$, then $(a \oplus b)' = 1$

if $a > b$, then $a\,b' = 1$

if $a < b$, then $a'\,b = 1$

The VHDL codes for dataflow model of 1-bit magnitude comaparator are developed using the Boolean equations. (Refer Chapter 5, Section 5.12 for design of 1-bit magnitude comparator).

VHDL Program for Dataflow Model of 1-bit Magnitude Comparator

```
library ieee;
use ieee.std_logic_1164.all;
    entity comparator_1bit is
        port ( a, b                  : in std_logic;
               a_eq_b, a_lt_b, a_gt_b : out std_logic );
    end comparator_1bit;

    architecture dfm_comparator_1bit of comparator_1bit is
    signal a_bar, b_bar : std_logic;
        begin
            a_bar  <= not a;
            b_bar  <= not b;
            a_eq_b <= not (a xor b);
            a_gt_b <= a and b_bar;
            a_lt_b  <= a_bar and b;
    end dfm_comparator_1bit;
```

VHDL Program 14.24

Write a VHDL code for dataflow model of a 3-bit comparator.

Program Logic for Dataflow Model

Let a and b are 3-bit binary numbers.

The Boolean equations for 3-bit magnitude comparator are,

if a = b, then $(a_0 \oplus b_0)' . (a_1 \oplus b_1)' . (a_2 \oplus b_2)' = 1$

if a > b, then $(a_2 . b_2') + ((a_2 \oplus b_2)' . a_1 . b_1') + ((a_2 \oplus b_2)' . (a_1 \oplus b_1)' . a_0 . b_0') = 1$

if a < b, then $(a_2' . b_2) + ((a_2 \oplus b_2)' . a_1' . b_1) + ((a_2 \oplus b_2)' . (a_1 \oplus b_1)' . a_0' . b_0) = 1$

The VHDL codes for dataflow model of 3-bit magnitude comparator are developed using the Boolean equations. (Refer Chapter 5, Section 5.12.2 for design of 3-bit magnitude comparator).

VHDL Program for Dataflow Model of 3-bit Comparator

```
library ieee;
use ieee.std_logic_1164.all;
   entity comparator_3bit is
      port ( a0, a1, a2, b0, b1, b2  : in std_logic;
             a_gt_b, a_lt_b, a_eq_b  : out std_logic );
   end comparator_3bit;

   architecture dfm_comparator_3bit of comparator_3bit is
   signal a0bar, a1bar, a2bar, b0bar, b1bar, b2bar : std_logic;
   signal m0, m1, m2, m3, m4, m5, m6, m7, m8 : std_logic;

      begin
         m0 <= not (a0 xor b0);
         m1 <= not (a1 xor b1);
         m2 <= not (a2 xor b2);
         m3 <= a2 and b2bar;
         m4 <= m2 and a1 and b1bar;
         m5 <= m2 and m1 and a0 and b0bar;
         m6 <= a2bar and b2;
         m7 <= m2 and a1bar and b1;
         m8 <= m2 and m1 and a0bar and b0;

         a0bar <= not a0;  a1bar <= not a1;  a2bar <= not a2;
         b0bar <= not b0;  b1bar <= not b1;  b2bar <= not b2;

         a_eq_b <= m0 and m1 and m2;
         a_gt_b  <= m3 or m4 or m5;
         a_lt_b  <= m6 or m7 or m8;
   end dfm_comparator_3bit;
```

VHDL Program 14.25

Write a VHDL code for structural model of a 2-bit multiplier.

Program Logic for Structural Model

Let a_1a_0 and b_1b_0 be the two input data and $P_3P_2P_1P_0$ be the 4-bit product.

The VHDL code for structural model of 2-bit multiplier is developed using the logic circuit shown in Fig. 1.

The Boolean equations for half adder are,

Sum, $s = a \oplus b$

Carry out, $c_o = a.b$

where, a and b are 1-bit inputs. (Refer Chapter 5, Section 5.9.1)

In structural model, each component involved in digital system are coded separately and then the components are added in the code for 2-bit multiplier. (Refer Chapter 5, Section 5.13.1 for design of 2-bit multiplier).

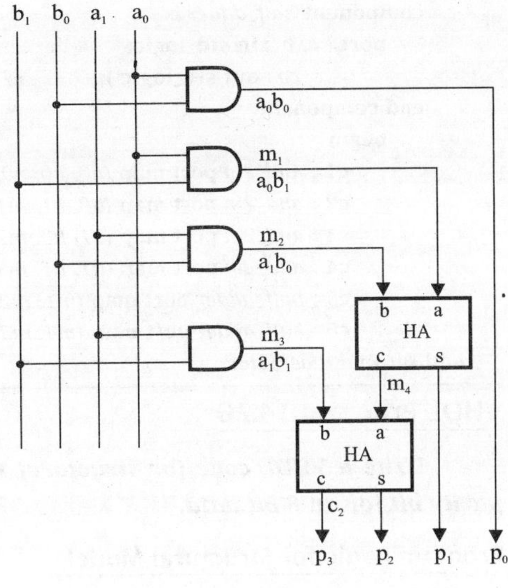

Fig. 1: *Logic circuit of 2-bit multiplier.*

VHDL Program for Structural Model of 2-bit Multiplier

```
--code for 2-input AND gate
library ieee;
use ieee.std_logic_1164.all;
   entity and_2in is
      port ( x, y : in std_logic;
             v    : out std_logic );
   end and_2in;

   architecture dfm_and_2in of and_2in is
      begin
         v <= x and y;
   end dfm_and_2in;
```

```
--code for half adder
library ieee;
use ieee.std_logic_1164.all;
   entity half_adder is
      port ( a, b  : in std_logic;
             s, co : out std_logic );
   end half_adder;

   architecture dfm_ha of half_adder is
      begin
         s  <= a xor b;          co <= a and b;
   end dfm_ha;
```

```
library ieee;
use ieee.std_logic_1164.all;
use ieee.std_logic_arith.all;
use ieee.std_logic_unsigned.all;
   entity multiplier_2bit is
      port ( a1, a0, b1, b0 : in std_logic;
             p0, p1, p2, p3 : out std_logic);
   end multiplier_2bit;

   architecture sm_multiplier_2bit of multiplier_2bit is
   signal m1, m2, m3, m4 : std_logic;
      component and_2in is
         port ( x, y : in std_logic;
                v    : out std_logic );
      end component;
```

```
component half_adder is
    port ( a, b  : in std_logic;
           s, co : out std_logic );
end component;
    begin
        c1 : and_2in port map (a0, b0, p0);
        c2 : and_2in port map (a0, b1, m1);
        c3 : and_2in port map (a1, b0, m2);
        c4 : and_2in port map (a1, b1, m3);
        c5 : half_adder port map (m1, m2, p1, m4);
        c6 : half_adder port map (m4, m3, p2, p3);
end sm_multiplier_2bit;
```

VHDL Program 14.26

Write a VHDL code for structural model of parity bit generator to generate even and odd parity bit for an 8-bit data.

Program Logic for Structural Model

The VHDL code for structural model of parity bit generator for an 8-bit data is developed using the logic circuit shown in Fig. 1. In structural model, each component involved in digital system are coded separately and then the components are added in the code for parity bit generator. (Refer Chapter 5, Section 5.15.4 for design of parity bit generator).

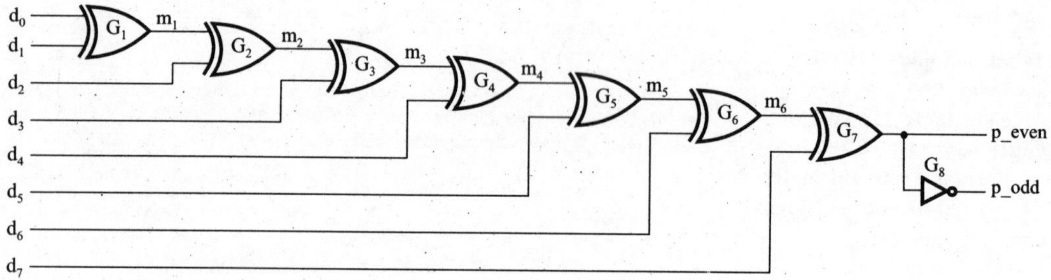

Fig. 1: Logic circuit of parity bit generator for an 8-bit data.

VHDL Program for Structural Model of Parity Bit Generator

```
--code for 2-input xor gate
library ieee;
use ieee.std_logic_1164.all;
    entity xor_2in is
        port ( x, y : in std_logic;
               v    : out std_logic );
    end xor_2in;

    architecture dfm_xor_2in of xor_2in is
        begin
            v <= x xor y;
    end dfm_xor_2in;
```

```
-- code for 1-input not gate
library ieee;
use ieee.std_logic_1164.all;
    entity not_1in is
        port ( x : in std_logic;
               v : out std_logic );
    end not_1in;

    architecture dfm_not_1in of not_1in is
        begin
            v <= not x;
    end dfm_not_1in;
```

```
-- code for parity generator
library ieee;
use ieee.std_logic_1164.all;
    entity parity_bit_generate is
        port ( d0, d1, d2, d3, d4, d5, d6, d7 : in std_logic;
               p_even, p_odd                  : inout std_logic );
    end parity_bit_generate;

    architecture sm_parity_bit_generate of parity_bit_generate is
    signal m1, m2, m3, m4, m5, m6 : std_logic;
        component xor_2in is
            port ( x, y : in std_logic;
                   v    : out std_logic );
        end component;

        component not_1in is
            port ( x : in std_logic;
                   v : out std_logic);
        end component;

    begin
        a1 : xor_2in port map (d0, d1, m1);
        a2 : xor_2in port map (m1, d2, m2);
        a3 : xor_2in port map (m2, d3, m3);
        a4 : xor_2in port map (m3, d4, m4);
        a5 : xor_2in port map (m4, d5, m5);
        a6 : xor_2in port map (m5, d6, m6);
        a7 : xor_2in port map (m6, d7, p_even);
        a8 : not_1in port map (p_even, p_odd);
    end sm_parity_bit_generate;
```

VHDL Program 14.27

Write a VHDL code for dataflow, behavioral and structural model of a SR flip-flop using NAND gates.

Program Logic

The following Boolean equations for SR flip-flop are developed from Fig. 1.

$m_1 = (s \cdot clk)'$

$m_2 = (r \cdot clk)'$

$q = (m_1 \cdot q')'$

$q' = (m_2 \cdot q)'$

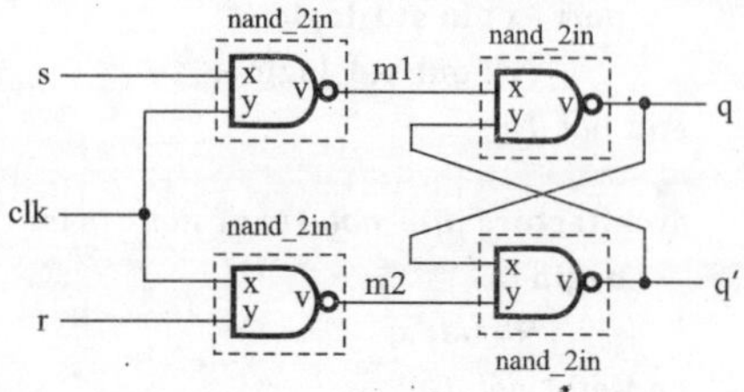

Fig. 1: NAND latch based SR flip-flop.

Where, s and r are 1-bit input data, q and q'(qbar) are 1-bit output data. The VHDL code for dataflow model of SR flip-flop is developed using the above equations. The behavioral model of SR flip-flop is developed using the truth table of SR flip-flop. The VHDL code for structural model of SR flip-flop is developed using the logic circuit shown in Fig. 1. In structural model, each component involved in the SR flip-flop are coded separately and then the components are added in the code for SR flip-flop. (Refer Chapter 6, Section 6.2.2 for design of SR flip-flop).

Truth Table of SR Flip-Flop

Inputs		Outputs	
S	R	Q	Q'
0	0	No change	
0	1	0	1
1	0	1	0
1	1	Indeterminate	

VHDL Program for Dataflow Model of SR Flip-Flop

```
library ieee;
use ieee.std_logic_1164.all;
    entity sr_ff is
        port ( s, r, clk   : in std_logic;
               q, qbar     : inout std_logic);
    end sr_ff;

    architecture dfm_sr_ff of sr_ff is
    signal m1, m2, m3, m4 : std_logic;
    signal q1      : std_logic := '0';
    signal q1bar : std_logic := '1';

        begin
            m1 <= not(s and clk);
            m3 <= not(m1 and q1bar);
            m2 <= not(r and clk);
            m4 <= not(m2 and q1);
            q <= m3;
            qbar <= not(q);
    end dfm_sr_ff;
```

VHDL Program for Behavioral Model of SR Flip-Flop using 'if' Statement

```
library ieee;
use ieee.std_logic_1164.all;
    entity sr_ff is
        port ( s, r, clk : in std_logic;
               q, qbar : inout std_logic);
    end sr_ff;
```

```
architecture bm_sr_ff of sr_ff is
signal q_temp, qbar_temp : std_logic;

    begin
        process (s, r, clk) begin
            if (clk' event and clk = '1') then
            if (s = '0' and r = '0' ) then
                q_temp       <= q_temp;
                qbar_temp <= qbar_temp;
            elsif (s = '0' and r = '1' ) then
                q_temp  <= '0';                 qbar_temp <= '1';
            elsif (s = '1' and r = '0' ) then
                q_temp       <= '1';            qbar_temp <= '0';
            elsif (s = '1' and r = '1') then
                q_temp       <='z';             qbar_temp <= 'z';
            end if;
            end if;
        end process;
                q       <= q_temp;
                qbar <= qbar_temp;
    end bm_sr_ff;
```

Program for Structural Model of SR Flip-Flop

```
--code for 2-input NAND gate
library ieee;
use ieee.std_logic_1164.all;
    entity nand_2in is
        port ( x, y : in std_logic;
               v    : out std_logic);
    end nand_2in;

    architecture dfm_nand_2in of nand_2in is
        begin
            v <= x nand y;
    end dfm_nand_2_in;
```

```
--code for SR flip-flop
library ieee;
use ieee.std_logic_1164.all;
    entity rs_ff is
        port ( r, s, clk : in std_logic;
                q, qbar : inout std_logic );
    end rs_ff;

    architecture sm_ rs_ff of rs_ff is
    signal m1, m2 : std_logic;

        component nand_2in is
            port ( x, y : in std_logic;
                    v    : out std_logic );
        end component;
```

```
        begin
            a1 : nand_2in port map (s, clk, m1);
            a2 : nand_2in port map (r, clk, m2);
            a3 : nand_2in port map (m1, qbar, q);
            a4 : nand_2in port map (m2, q, qbar);
    end sm_ rs_ff;
```

VHDL Program 14.28

Write a VHDL code for dataflow, behavioral and structural model of a JK flip-flop using NAND gates.

Program Logic for Dataflow and Behavioral Model

The following Boolean equations for JK flip-flop are developed drom Fig. 1.

$m_1 = (j \,.\, clk \,.\, q')'$

$m_2 = (k \,.\, clk \,.\, q)'$

$q = (m_1 \,.\, q')'$

$q' = (m_2 \,.\, q)'$

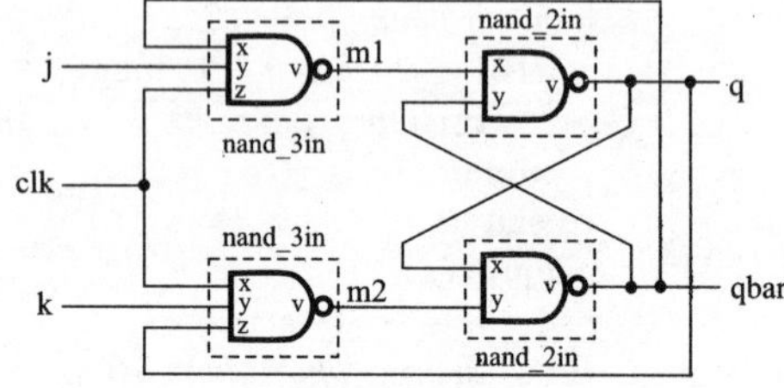

Fig. 1: JK flip-flop.

Where, j and k are 1-bit input data, q and q'(qbar) are 1-bit output data. The VHDL code for dataflow model of JK flip-flop is developed using the above equations. The behavioral model of JK flip-flop is developed using the truth table of JK flip-flop. The VHDL code for structural model of JK flip-flop is developed using the logic circuit shown in Fig. 1. In structural model, each component involved in the JK flip-flop are coded separately and then the components are added in the code for JK flip-flop. (Refer Chapter 6, Section 6.2.3 for design of JK flip-flop).

Truth Table of JK Flip-Flop

Inputs		Outputs	
J	**K**	**Q**	**Q'**
0	0	No change	
0	1	0	1
1	0	1	0
1	1	Toggle	

VHDL Program for Dataflow Model of JK Flip-Flop

```
library ieee;
use ieee.std_logic_1164.all;
    entity jk_ff is
        port ( j, k, clk : in std_logic;
                q, qbar  : out std_logic );
    end jk_ff;

    architecture dfm_jk_ff of jk_ff is
    signal m1, m2, m3, m4 : std_logic;
    signal q1 : std_logic := '0';
    signal q1bar : std_logic := '1';

        begin
            m1 <= not(j and clk and q1bar);
            m3 <= not(m1 and q1bar);
            m2 <= not(k and clk and q1);
            m4 <= not(m2 and q1);
            q <= m3;
            qbar <= not(q);
    end dfm_jk_ff;
```

VHDL Program for Behavioral Model of JK Flip-Flop

```
library ieee;
use ieee.std_logic_1164.all;
    entity jk_ff is
        port ( j, k         : in std_logic;
               clk, reset : in std_logic;
               q, qbar    : out std_logic);
    end jk_ff;
    architecture bm_jk_ff of jk_ff is
        begin
            qbar <= not q;
        process (clk, reset)
            begin
                if (reset = '1') then
                    q <= 0;          qbar <= 1;
                elsif (clk' event and clk = '1') then
                if (j = '0' and k = '0') then
                    q <= q;          qbar <= qbar;
                if (j = '0' and k = '1') then
                    q <= 0;          qbar <= 1;
                if (j = '1' and k = '0') then
                    q <= 1;          qbar <= 0;
                else
                    q      <= qbar;    qbar <= not q;
                end if;
                end if;
                end if;
                end if;
        end process;
    end bm_jk_ff;
```

VHDL Program for Structural Model of JK Flip-Flop

```
--code for 3-input NAND gate
library ieee;
use ieee.std_logic_1164.all;
    entity nand_3in is
        port ( x, y, z : in std_logic;
               v         : out std_logic );
    end nand_3in;

    architecture dfm_nand_3in of nand_3in is
        begin
            v <= not (x and (y and z);
    end dfm_nand_3in;
```

```
--code for 2-input NAND gate
library ieee;
use ieee.std_logic_1164.all;
    entity nand_2in is
        port ( x, y : in std_logic;
               v     : out std_logic );
    end nand_2in;
```

```
architecture dfm_nand_2in of nand_2in is
    begin
        v <= x nand y;
end dfm_nand_2in;
```

```
--code for JK flip-flop
library ieee;
use ieee.std_logic_1164.all;
    entity jk_ff is
        port (j, k, clk : in std_logic;
              q, qbar : inout std_logic);
    end jk_ff;

    architecture sm_jk_ff of jk_ff is
    signal m1, m2 : std logic;

        component nand_2in is
            port ( x, y : in std_logic;
                   v    : out std_logic);
        end component;

        component nand_3in is
            port ( x, y, z : in std_logic;
                   v       : out std_logic );
        end component;

            begin
                a1 : nand_3in port map (j, clk, qbar, m1);
                a2 : nand_3in port map (k, clk, q, m2);
                a3 : nand_2in port map (m1, qbar, q);
                a4 : nand_2in port map (m2, q, qbar);
    end sm_jk_ff;
```

VHDL Program 14.29

Write a VHDL code for behavioral model of a D flip-flop.

Program Logic for Behavioral Model

The Boolean equations for D flip-flop are,

q = d ; qbar = q'

Where, d is 1-bit input data q and q'(qbar) are 1-bit output data. The VHDL code for behavioral model of D flip-flop is developed using the above equations. (Refer Chapter 6, Section 6.2.5 for design of D flip-flop).

VHDL Program for Behavioral Model of D Flip-Flop

```
library ieee;
use ieee.std_logic_1164.all;
    entity d_ff is
        port ( d, clk, reset : in std_logic;
               q, qbar       : inout std_logic);
    end d_ff;

    architecture bm_d_ff of d_ff is
        begin
            process (d, clk, reset)
                begin
```

```
            qbar <= not q;
        if reset = '1' then
            q <= '0';   qbar <= not q;
        elsif (clk' event and clk= '1') then
            q <= d;   qbar <= not q;
        end if;
    end process;
end bm_d_ff;
```

VHDL Program 14.30

Write a VHDL code for structural model of a T flip-flop.

Program Logic for Structural Model

The VHDL code for structural model of T flip-flop is developed using the logic circuit shown in Fig. 1. In structural model, each component involved in the T flip-flop are coded separately and then the components are added in the code for T flip-flop.

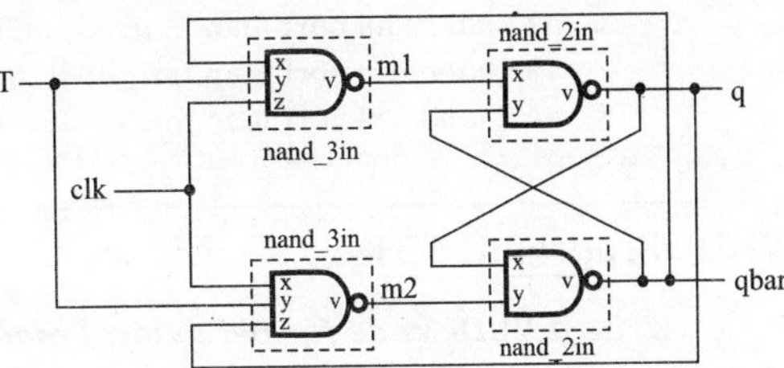

Fig. 1: *T flip-flop.*

VHDL Program for Structural Model of T Flip-Flop

```
--code for 3-input NAND gate
library ieee;
use ieee.std_logic_1164.all;
    entity nand_3in is
        port ( x, y, z : in std_logic;
               v       : out std_logic );
    end nand_3in;

    architecture dfm_nand_3in of nand_3in is
        begin
            v <= not (x and (y and z));
    end dfm_nand_3in;
```

```
--code for 2-input NAND gate
library ieee;
use ieee.std_logic_1164.all;
    entity nand_2in is
        port ( x, y : in std_logic;
               v    : out std_logic );
    end nand_2in;

    architecture dfm_nand_2in of nand_2in is
        begin
            v <= x nand y;
    end dfm_nand_2in;
```

```
--code for T flip-flop
library ieee;
use ieee.std_logic_1164.all;
    entity t_ff is
        port ( t, clk  : in std_logic;
               q, qbar : inout std_logic);
    end t_ff;
```

```
architecture sm_t_ff of t_ff is
signal m1, m2 : std logic;

    component nand_2in is
        port ( x, y : in std_logic;
               v    : out std_logic );
    end component;

    component nand_3in is
        port ( x, y, z : in std_logic;
               v       : out std_logic );
    end component;
        begin
            a1 : nand_3in port map (t, qbar, clk, m1);
            a2 : nand_3in port map (t, q, clk, m2);
            a3 : nand_2in port map (m1, qbar, q);
            a4 : nand_2in port map (m2, q, qbar);
end sm_t_ff;
```

VHDL Program 14.31

Write a VHDL code for behavioral model of a 2-bit up/down counter.

Program Logic for Behavioral Model

A selection bit, a is required to select up count when a = 1 and down count when a = 0. Initial count is kept 0 and increment by 1 for up count and decrement by 1 for down count for every clock pulse. (Refer Chapter 8, Section 8.2.3 for design of up/down counter).

VHDL Program for Behavioral Model of 2-bit up/down Counter

```
library ieee;
use ieee.std_logic_1164.all;
use ieee.std_logic_arith.all;
use ieee.std_logic_unsigned.all;
  entity updown_2bit is
      port ( clk, a : in std_logic;
             dout   : out std_logic_vector (1 downto 0) );
  end updown_2bit;

  architecture bm_updown_2bit of updown_2bit is
  signal m1 : std_logic_vector (1 downto 0) := "00";

      begin
          process (clk)
              begin
                  if (clk = '1') then
                  if a = '1' then
                      m1   <= m1 + "01";
                      dout <= m1;
                  else
                      m1   <= m1 - "01";
                      dout <= m1;
                  end if;
                  end if;
          end process;
  end bm_updown_2bit;
```

14.13 VHDL Programs With Test Bench

VHDL Program 14.32

Write VHDL code for dataflow model of a binary to gray code converter.

Program Logic for Dataflow Model

The following procedure will be useful for bit-by-bit conversion of n-bit binary to n-bit gray code.

Step 1: The most significant bit of binary and gray code are same. Hence retain the most significant bit of binary as most significant bit of gray code.

Step 2: Starting from most significant bit of binary, add two consecutive bits of binary and record the sum as bit of gray code and discard the carry in every addition.

The VHDL code for dataflow model of binary to gray code converter is developed using the above procedure.(Refer Chapter 5, Section 5.3 for design of binary gray code).

VHDL Program for Dataflow Model of Binary Gray Code

```
library ieee;
use ieee.std_logic_1164.all;
    entity binary_gray is
        port ( b3, b2, b1, b0 : in std_logic;
               g3, g2, g1, g0  : out std_logic );
    end binary_gray;

    architecture dfm_binary_gray of binary_gray is
        begin
            g3 <= b3;
            g2 <= b3 xor b2;
            g1 <= b2 xor b1;
            g0 <= b1 xor b0;
    end dfm_binary_gray;
```

Test Bench

```
library ieee;
use ieee.std_logic_1164.all;
    entity tb_binary_gray is
        --port ( );
    end tb_binary_gray;

    architecture dfm_tb_binary_gray of tb_binary_gray is
        component binary_gray
            port ( b3, b2, b1, b0 : in std_logic;
                   g3, g2, g1, g0 : out std_logic );
        end component;

    signal b3 : std_logic := '0';
    signal b2 : std_logic := '0';
    signal b1 : std_logic := '0';
    signal b0 : std_logic := '0';
    signal g3 : std_logic;
    signal g2 : std_logic;
    signal g1 : std_logic;
    signal g0 : std_logic;
```

```
begin
    uut : binary_gray port map
        ( b3 => b3,  b2 => b2,  b1 => b1,  b0 => b0,
          g3 => g3,  g2 => g2,  g1 => g1,  g0 => g0 );
    stimulus_process : process

        begin
            b3 <= '0';    b2 <= '1';    b1 <= '0';    b0 <= '1';
        wait for 10ns;

            b3 <= '0';    b2 <= '1';    b1 <= '1';    b0 <= '1';
        wait for 10ns;

            b3 <= '1';    b2 <= '0';    b1 <= '0';    b0 <= '1';
        wait for 10ns;

            b3 <= '1';    b2 <= '0';    b1 <= '1';    b0 <= '0';
        wait for 10ns;
    end process;
end dfm_tb_binary_gray;
```

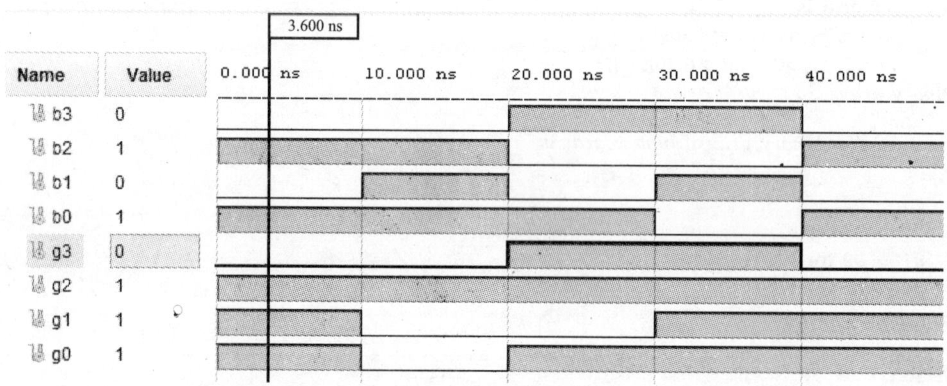

***Fig. a:** Inputs: b3 b2 b1 b0 = 0101; Outputs: g3 g2 g1 g0 = 0111.*

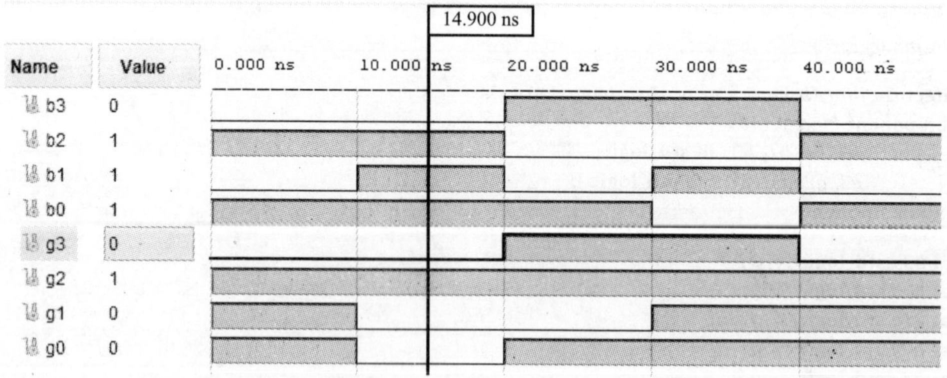

***Fig. b:** Inputs: b3 b2 b1 b0 = 0111; Outputs: g3 g2 g1 g0 = 0100.*

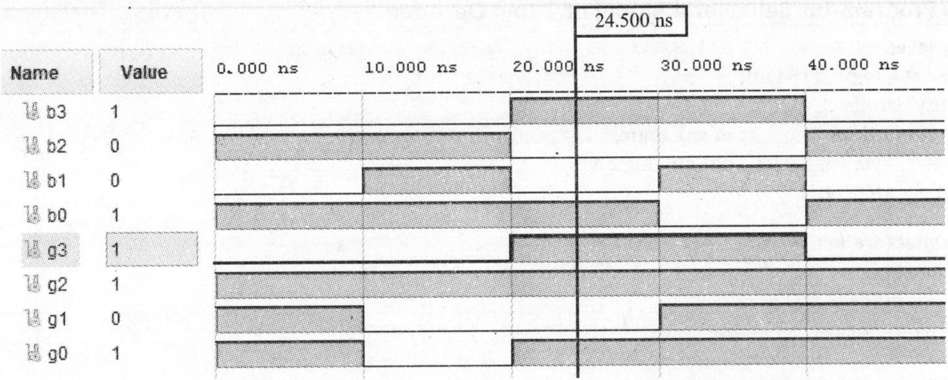

***Fig. c:** Inputs: b3 b2 b1 b0 = 1001; Outputs: g3 g2 g1 g0 = 1101.*

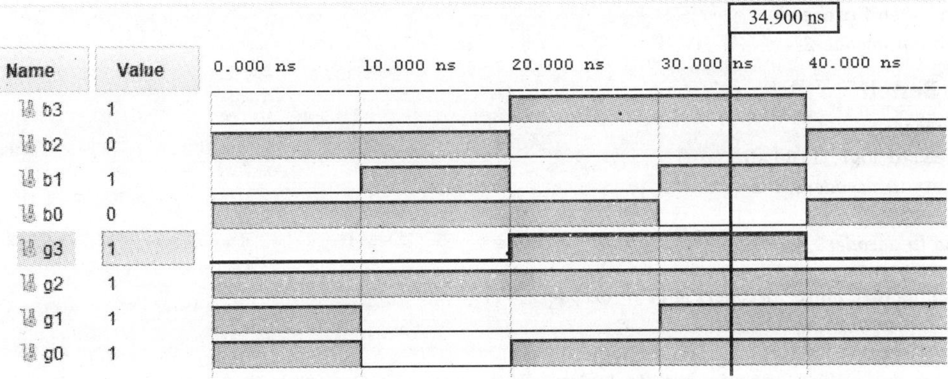

***Fig. d:** Inputs: b3 b2 b1 b0 = 1010; Outputs: g3 g2 g1 g0 = 1111.*

***Fig. 1:** Input-output waveform of binary to gray code converter.*

VHDL Program 14.33

Write VHDL code for behavioral model of a 2-to-4 decoder.

Program Logic for Behavioral Model

The Boolean equations for 2-to-4 decoder are,

Y0 = I1'.I0'

Y1 = I1'.I0

Y2 = I1.I0'

Y3 = I1.I0

Where, I1 and I0 are 1-bit input data, Y0, Y1, Y2 and Y3 are 1-bit output data. The VHDL code for behavioral model of 2-to-4 decoder is developed using the above equations. (Refer Chapter 5, Section 5.4.1 for design of 2-to-4 decoder).

VHLD Program for Behavioral Model of 2-to-4 Decoder

```
library ieee;
use ieee.std_logic_1164.all;
    entity decoder2_4 is
        port ( i1, i0              : in std_logic;
               y0, y1, y2, y3 : out std_logic );
    end decoder2_4;

    architecture bm_decoder2_4 of decoder2_4 is
        begin
            process ( i1, i0 )
                begin
                    y0 <= not i1 and not i0;
                    y1 <= not i1 and i0;
                    y2 <= i1 and not i0;
                    y3 <= i1 and i0;
            end process;
    end bm_decoder2_4;
```

Test Bench

```
library ieee;
use ieee.std_logic_1164.all;
    entity tb_decoder2_4 is
        --port ( );
    end tb_decoder2_4;

    architecture bm_tb_decoder2_4 of tb_decoder2_4 is
        component decoder2_4
            port ( i1, i0              : in std_logic;
                   y0, y1, y2, y3 : out std_logic);
        end component;

    signal i1 : std_logic := '0';
    signal i0 : std_logic := '0';
    signal y0 : std_logic;
    signal y1 : std_logic;
    signal y2 : std_logic;
    signal y3 : std_logic;

    begin
        uut : decoder2_4 port map
            ( i1 => i1,   i0 => i0,  y0 => y0,
              y1 => y1,  y2 => y2,  y3 => y3 );
        stimulus_process : process

                begin
                    i1 <= '0';          i0 <= '0';
                wait for 20ns;

                    i1 <= '0';          i0 <= '1';
                wait for 20ns;

                    i1 <= '1';          i0 <= '0';
                wait for 20ns;
```

```
            i1 <= '1';          i0 <= '1';
        wait for 20ns;
    end process;
end bm_tb_decoder2_4;
```

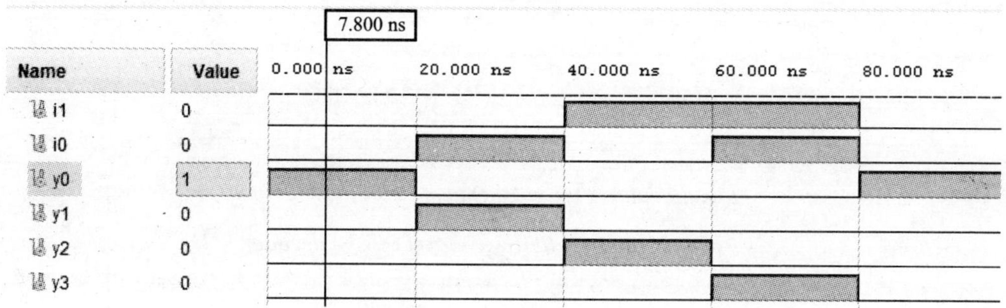

Fig. a: *Inputs: i0 i1 = 00; Outputs: y0 y1 y2 y3 = 1000.*

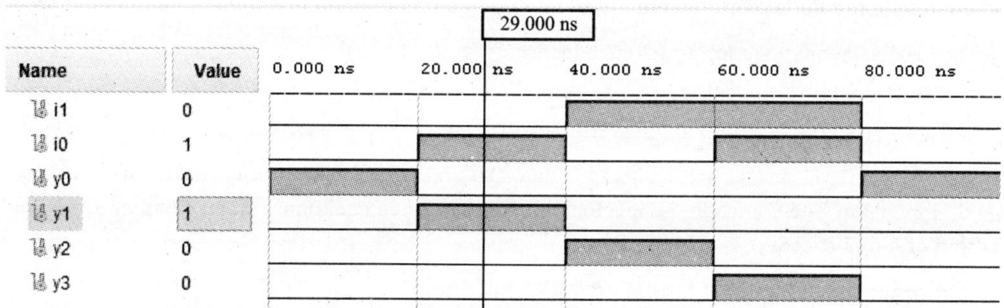

Fig. b: *Inputs: i0 i1 = 10; Outputs: y0 y1 y2 y3 = 0100.*

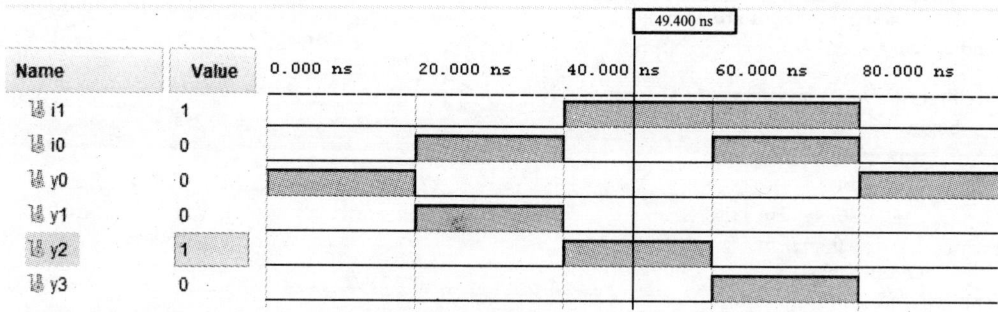

Fig. c: *Inputs: i0 i1 = 01; Outputs: y0 y1 y2 y3 = 0010.*

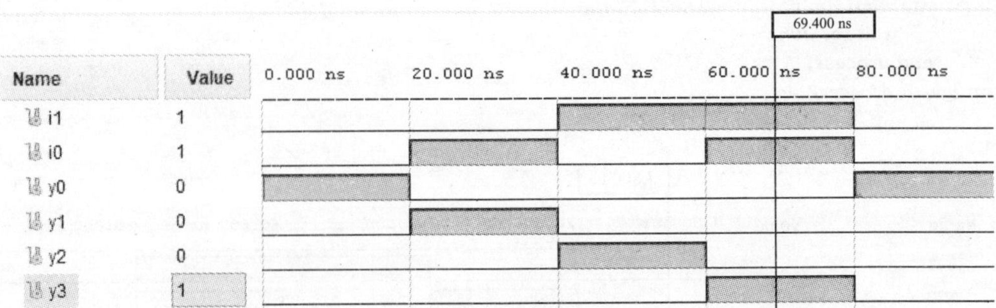

Fig. d: Inputs: i0 i1 = 11; Outputs: y0 y1 y2 y3 = 0001.

*Fig. 1: Input-output waveform of 2*4 decoder.*

VHDL Program 14.34

Write VHDL code for behavioral model of a 4-to-2 encoder.

Program Logic for Behavioral Model

The Boolean equations for 4-to-2 encoder are,

$$Y1 = I2 + I3$$

$$Y0 = I1 + I3$$

Where, I0, I1, I2 and I3 are 1-bit input data, Y0 and Y1 are 1-bit output data. The VHDL code for behavioral model of 4-to-2 encoder is developed using the above equations. (Refer Chapter 5, Section 5.5.1 for design of 4-to-2 encoder).

VHDL Program for Behavioral Model of 4-to-2 Encoder

```
library ieee;
use ieee.std_logic_1164.all;
    entity encoder_4_2 is
        port ( i0, i1, i2, i3 : in std_logic;
               y0, y1         : out std_logic);
    end encoder_4_2;

    architecture bm_ encoder_4_2 of encoder_4_2 is
        begin
            process (i0, i1, i2, i3)
                begin
                    y1 <= i2 or i3;
                    y0 <= i1 or i3;
            end process;
    end bm_encoder_4_2;
```

Test Bench

```
library ieee;
use ieee.std_logic_1164.all;
    entity tb_encoder_4_2 is
        --port ( );
    end tb_encoder_4_2;
```

```
architecture bm_tb_encoder_4_2 of tb_encoder_4_2 is
    component encoder_4_2
        port ( i0, i1, i2, i3 : in std_logic;
               y0, y1         : out std_logic );
    end component;

signal i0     : std_logic := '0';
signal i1     : std_logic := '0';
signal i2     : std_logic := '0';
signal i3     : std_logic := '0';
signal y0, y1 : std_logic;
begin
    uut : encoder_4_2 port map
        ( i0 => i0, i1 => i1, i2 => i2, i3 => i3, y0 => y0, y1 => y1 );
    stimulus_process : process
    begin
        i0 <= '1';    i1 <= '0';    i2 <= '0';    i3 <= '0';
    wait for 20ns;

        i0 <= '0';    i1 <= '1';    i2 <= '0';    i3 <= '0';
    wait for 20ns;

        i0 <= '0';    i1 <= '0';    i2 <= '1';    i3 <= '0';
    wait for 20ns;

        i0<= '0';     i1 <= '0';    i2 <= '0';    i3 <= '1';
    wait for 20ns;
end process;
end bm_tb_encoder_4_2;
```

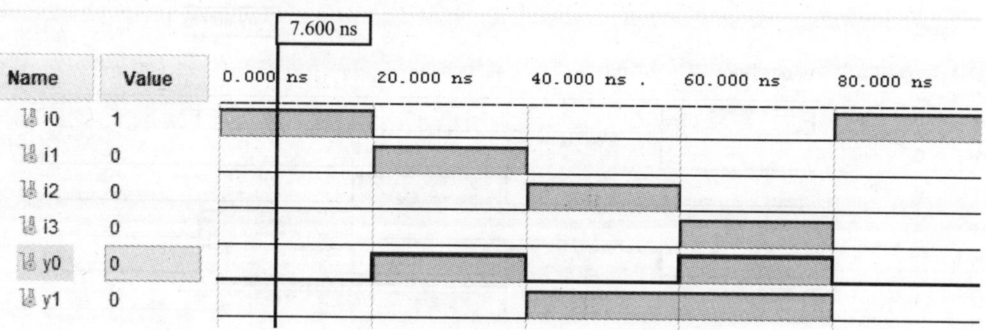

Fig. a: *Inputs: i0 i1 i2 i3 = 1000; Outputs: y0 y1 = 00.*

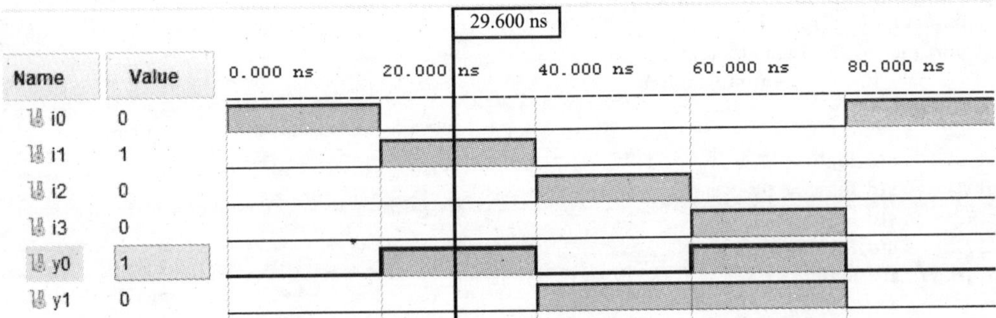

***Fig. b:** Inputs: i0 i1 i2 i3 = 0100; Outputs: y0 y1 = 10.*

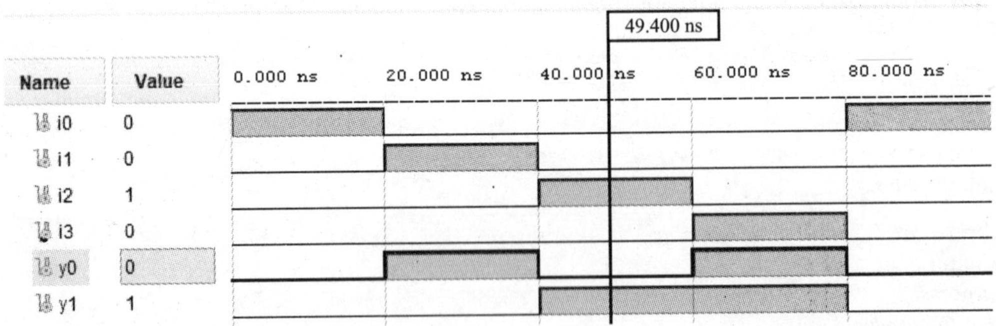

***Fig. c:** Inputs: i0 i1 i2 i3 = 0010; Outputs: y0 y1 = 01.*

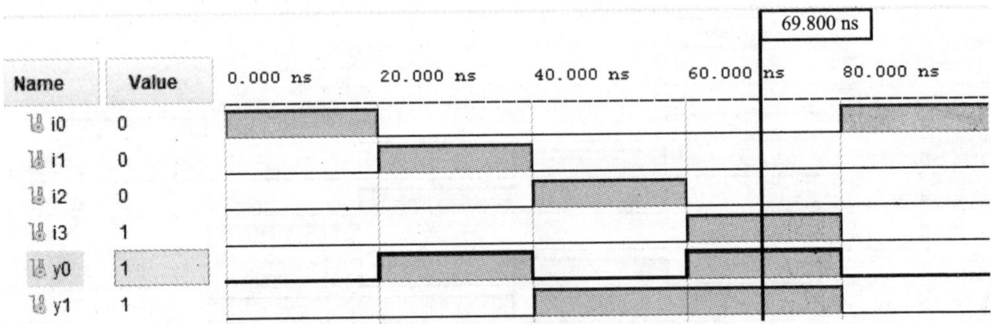

***Fig. D:** Inputs: i0 i1 i2 i3 = 0001; Outputs: y0 y1 = 11.*

***Fig. 1:** Input-output waveform of 4*2 encoder.*

VHDL Program 14.35

Write VHDL code for dataflow model of a 4*1 multiplexer.

Program Logic for Dataflow Model

The Boolean equations for 4*1 multiplexer are,

Y = S0'.S1'.I0 + S0'.S1.I1 + S0.S1'.I2 + S0.S1.I3

Where, S0 and S1 are 1-bit selection input, Y is 1-bit output data. The VHDL code for dataflow model of 4*1 multiplexer is developed using the above equations. (Refer Chapter 5, Section 5.7 for design of 4*1 multiplexer).

VHDL Program for Dataflow Model of 4*1 Multiplexer

```
library ieee;
use ieee.std_logic_1164.all;
    entity mux_4_1 is
        port ( s0, s1        : in std_logic;
               i0, i1, i2, i3 : in std_logic;
               y              : out std_logic );
    end mux_4_1;

    architecture dfm_mux_4_1 of mux_4_1 is
        signal ns0, ns1          : std_logic;
        signal m1, m2, m3, m4 : std_logic;
            begin
                ns0 <= not s0;
                ns1 <= not s1;
                m1  <= ns0 and ns1 and i0;
                m2  <= ns0 and s1 and i1;
                m3  <= s0 and ns1 and i2;
                m4  <= s0 and s1 and i3;
                y   <= m1 or m2 or m3 or m4;
    end dfm_mux_4_1;
```

Test Bench

```
library ieee;
use ieee.std_logic_1164.all;
    entity tb_mux_4_1 is
        --port ( );
    end tb_mux_4_1;

    architecture dfm_tb_mux_4_1 of tb_mux_4_1 is
        component mux_4_1 is
            port ( i0, i1, i2, i3, s0, s1 : in std_logic;
                   y                      : out std_logic);
        end component;

    signal i0 : std_logic := '0';
    signal i1 : std_logic := '0';
    signal i2 : std_logic := '0';
    signal i3 : std_logic := '0';
    signal s0 : std_logic := '0';
    signal s1 : std_logic := '0';
    signal y  : std_logic;
```

```
begin
    uut : mux_4_1 port map
        ( i0 => i0,  i1 => i1,  i2 => i2,  i3 => i3,  s0 => s0,  s1 => s1,  y  => y );
    stimulus_process : process

        begin
            i0 <= '1';    i1 <= '0';    i2 <= '0';    i3 <= '0';    s0 <= '0';    s1 <= '0';
        wait for 10ns;

            i0 <= '0';    i1 <= '0';    i2 <= '0';    i3 <= '0';    s0 <= '0';    s1 <= '1';
        wait for 10ns;

            i0 <= '0';    i1 <= '0';    i2 <= '1';    i3 <= '0';    s0 <= '1';    s1 <= '0';
        wait for 10ns;

            i0 <= '0';    i1 <= '0';    i2 <= '0';    i3 <= '0';    s0 <= '1';    s1 <= '1';
        wait for 10ns;
    end process;
end dfm_tb_mux_4_1;
```

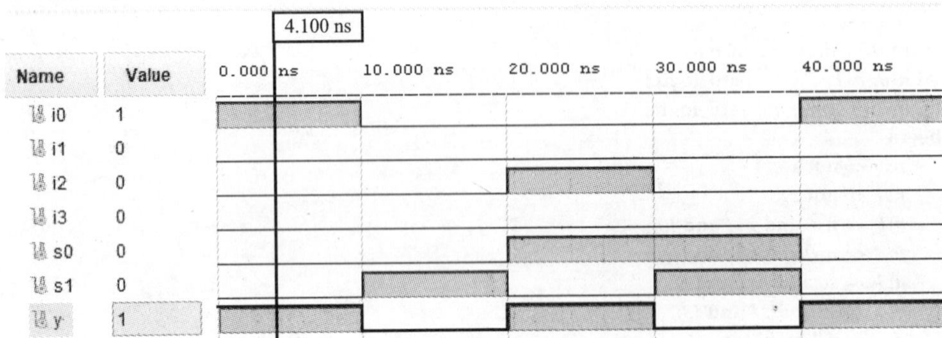

Fig. a: *Inputs: i0 i1 i2 i3 = 1000, s0 s1 = 00; Output: y = 1.*

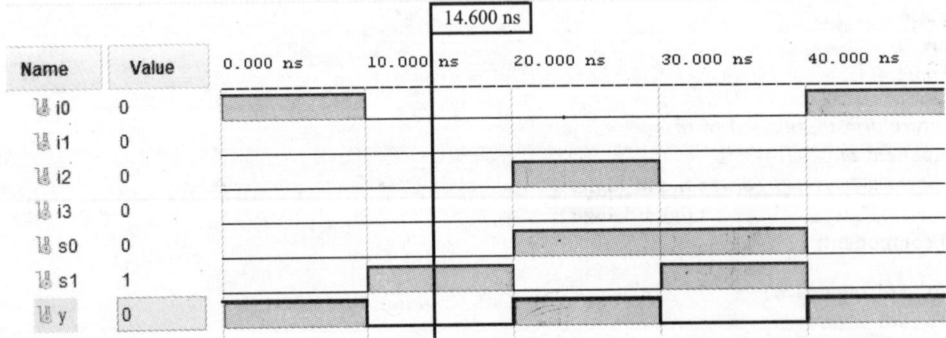

Fig. b: *Inputs: i0 i1 i2 i3 = 0000, s0 s1 = 01; Output: y = 0.*

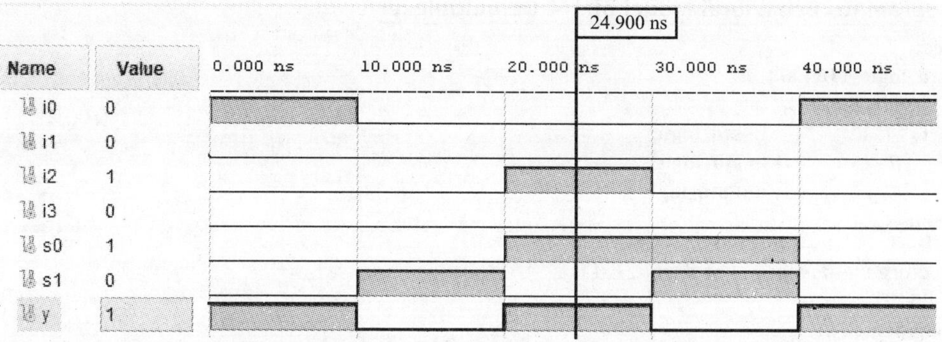

Fig. c: Inputs: i0 i1 i2 i3 = 0010, s0 s1 = 10; Output: y = 1.

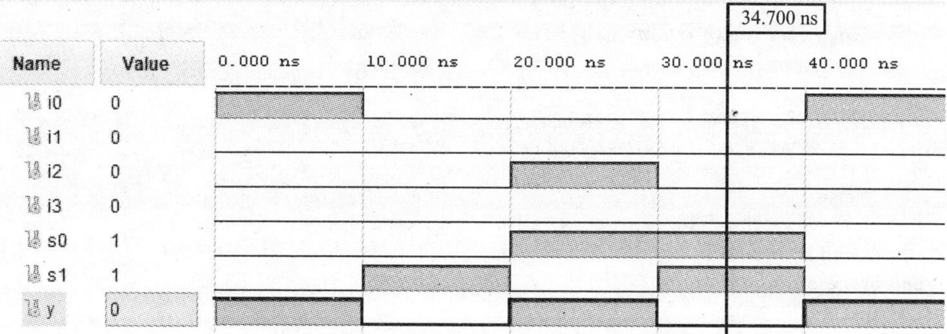

Fig. d: Inputs: i0 i1 i2 i3 = 0000, s0 s1 = 11; Output: y = 0.

*Fig. 1: Input-output waveform of 4*1 multiplexer.*

VHDL Program 14.36

Write VHDL code for behavioral model of a 1*4 demultiplexer.

Program Logic for Behavioral Model

The Boolean equations for 1*4 demultiplexer are,

Y0 = Ii.S1'.S0'

Y1 = Ii.S1'.S0

Y2 = Ii.S1.S0'

Y3 = Ii.S1.S0

Truth Table of 1:4 Demux

S_1	S_0	Y_0	Y_1	Y_2	Y_3
0	0	I_i	0	0	0
0	1	0	I_i	0	0
1	0	0	0	I_i	0
1	1	0	0	0	I_i

Where, Ii, S1 and S0 are 1-bit selection input, Y0, Y1, Y2 and Y3 are 1-bit output data. The VHDL code for behavioral model of 1*4 demultiplexer is developed using the above equations. The behavioral models of 1*4 demultiplexer is developed using the truth table of 1*4 demultiplexer. (Refer Chapter 5, Section 5.8 for design of 1*4 demultiplexer).

VHDL Program for Behavioral Model of 1*4 Demultiplexer

```
library ieee;
use ieee.std_logic_1164.all;
    entity demux_1_4 is
        port ( s1, s0          : in std_logic;
               ii              : in std_logic;
               y0, y1, y2, y3  : out std_logic );
    end demux_1_4;

    architecture bm_demux_1_4 of demux_1_4 is
        begin
            process ( s1, s0, ii )
                begin
                    if s0 = '0' and s1 = '0' then
                        y0 <= ii;     y1 <= '0';     y2 <= '0';     y3 <= '0';

                    elsif s0 = '0' and s1 = '1' then
                        y0 <= '0';    y1 <= ii;      y2 <= '0';     y3 <= '0';

                    elsif s0 = '1' and s1 = '0' then
                        y0 <= '0';    y1 <= '0';     y2 <= ii;      y3 <= '0';

                    else
                        y0 <= '0';    y1 <= '0';     y2 <= '0';     y3 <= ii;
                    end if;
            end process;
    end bm_demux_1_4;
```

Test Bench

```
library ieee;
use ieee.std_logic_1164.all;
    entity tb_demux is
        --port ( );
    end tb_demux;

    architecture bm_tb_demux of tb_demux is
        component demux_1_4
            port ( s1, s0          : in std_logic;
                   ii              : in std_logic;
                   y0, y1, y2, y3  : out std_logic );
        end component;

    signal s1, s0          : std_logic;
    signal ii              : std_logic := '0';
    signal y0, y1, y2, y3  : std_logic;
        begin
            uut : demux_1_4 port map
                ( s1  => s1,  s0 => s0,  ii  => ii,  y0 => y0,  y1 => y1,  y2 => y2,  y3 => y3 );
            stimulus_process : process
            begin
                s0 <= '0';      s1 <= '0';      ii  <= '1';
            wait for 10ns;
```

```
            s0 <= '0';      s1 <= '1';      ii <= '1';
        wait for 10ns;

            s0 <= '1';      s1 <= '0';      ii <= '1';
        wait for 10ns;

            s0 <= '1';      s1 <= '1';      ii <= '1';
        wait for 10ns;
    end process;
end bm_tb_demux;
```

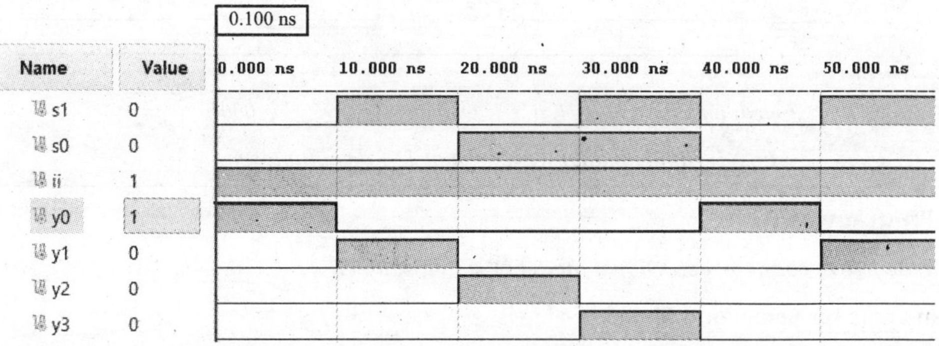

Fig. a: Inputs: s0 s1 = 00, ii = 1; Outputs: y0 y1 y2 y3 = 1000.

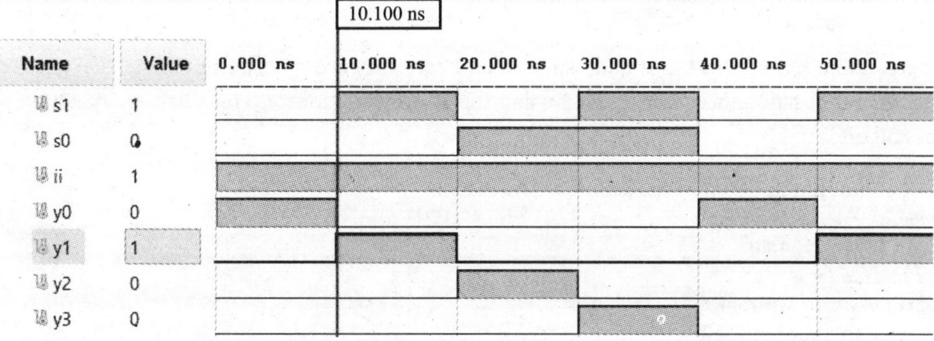

Fig. b: Inputs: s0 s1 = 01, ii = 1; Outputs: y0 y1 y2 y3 = 0100.

Fig. c: Inputs: s0 s1 = 10, ii = 1; Outputs: y0 y1 y2 y3 = 0010.

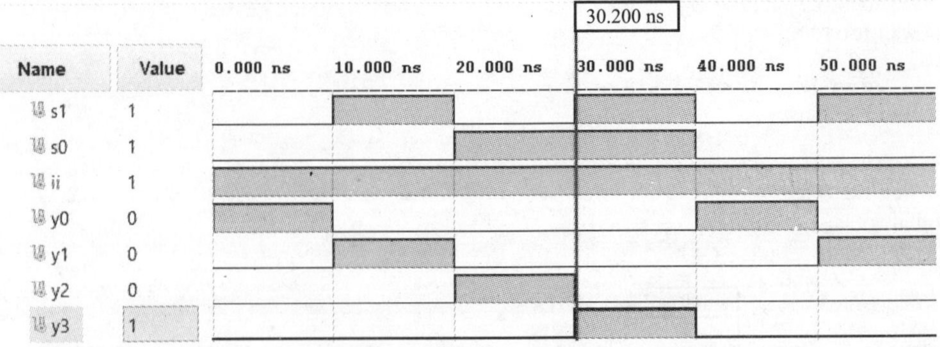

Fig. d: Inputs: s0 s1 = 11, ii = 1; Outputs: y0 y1 y2 y3 = 0001.

*Fig. 1: Inputs-output waveform of 1*4 demultiplexer.*

VHDL Program 14.37

Write VHDL code for behavioral model of a half adder.

Program Logic for Behavioral Model

The Boolean equations for half adder are,

$$s = a \oplus b$$

$$c_o = a\,b$$

Where, a and b are 1-bit input data, sum (s) and carry (c_o) are 1-bit output data. The VHDL code for behavioral model of half adder is developed using the above equations. (Refer Chapter 5, Section 5.9.1 for design of half adder).

VHDL Program for Behavioral Model of Half Adder

```
library ieee;
use ieee.std_logic_1164.all;
   entity half_adder is
      port ( a, b  : in std_logic;
             s, co : out std_logic );
   end half_adder;

   architecture bm_ha of half_adder is
      begin
         process ( a, b )
            begin
               s <= a xor b;     co <= a and b;
         end process;
   end bm_ha;
```

Test Bench

```
library ieee;
use ieee.std_logic_1164.all;
   entity tb_halfadder is
      --port ( );
   end tb_halfadder;
```

```
architecture bm_tb_halfadder of tb_halfadder is
    component half_adder
        port ( a, b  : in std_logic;
               s, co : out std_logic );
    end component;

signal a  : std_logic := '0';
signal b  : std_logic := '0';
signal s  : std_logic;
signal co : std_logic;

    begin
        uut : half_adder port map
            ( a  => a,  b  => b,  s  => s,  co => co );
        stimulus_process : process

            begin
                a <= '0';     b <= '0';
            wait for 10ns;

                a <= '0';     b <= '1';
            wait for 10ns;

                a <=' 1';     b <= '0';
            wait for 10ns;

                a <= '1';     b <= '1';
            wait for 10ns;
        end process;
end bm_tb_halfadder;
```

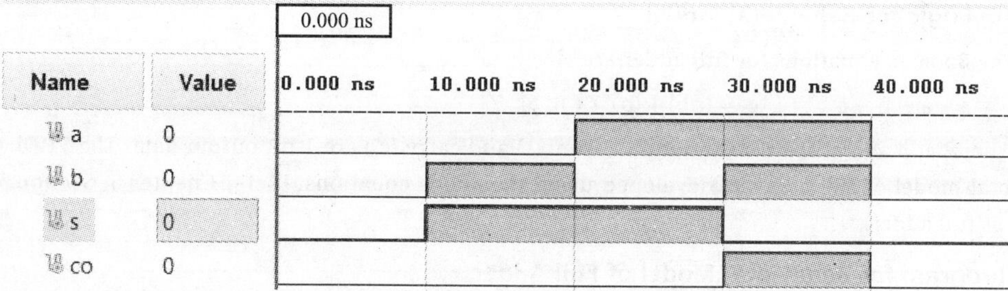

***Fig. a:** Inputs: a = 0, b = 0; Outputs: s = 0, co = 0.*

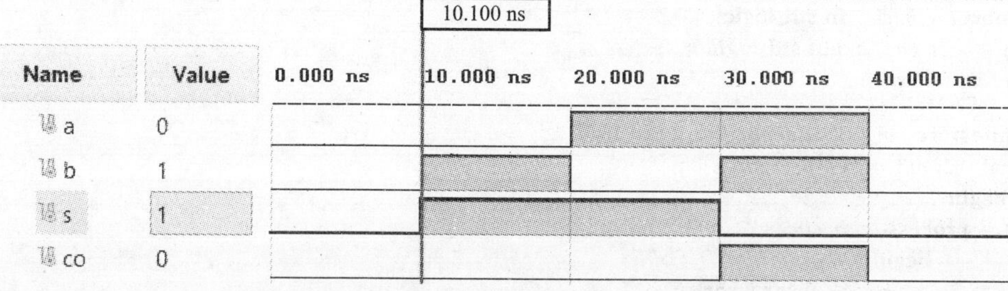

***Fig. b:** Inputs: a = 0, b = 1; Outputs: s = 1, co = 0.*

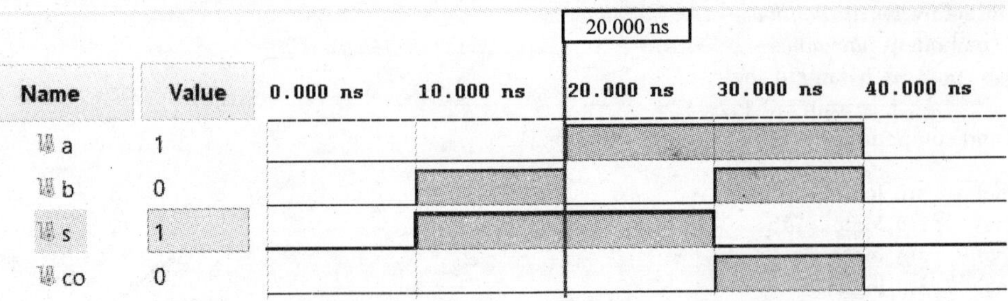

Fig. c: Inputs: a = 1, b = 0; Outputs: s = 1, co = 0.

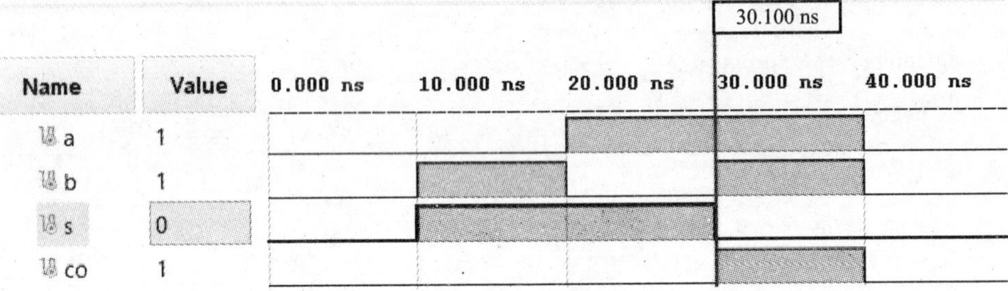

Fig. d: Inputs: a = 1, b = 1; Outputs: s = 0, co = 1.

***Fig. 1:** Input-output waveform of half adder.*

VHDL Program 14.38

Write VHDL code for behavioral model of a full adder.

Program Logic for Behavioral Model

The Boolean equations for full adder are,

$$s = a \oplus b \oplus c_i \quad ; \quad c_o = (a\,b) + (b\,c_i) + (c_i\,a)$$

Where, a, b and ci are 1-bit input data, sum (s) and carry (c_o) are 1-bit output data. The VHDL code for behavioral model of full adder is developed using the above equations. (Refer Chapter 5, Section 5.9.2 for design of full adder).

VHDL Program for Behavioral Model of Full Adder

```
library ieee;
use ieee.std_logic_1164.all;
   entity full_adder is
      port ( a, b, ci  :  in std_logic;
              s, co    :  out std_logic );
   end full_adder;

   architecture bm_full_adder of full_adder is
   signal m1, m2, m3 : std_logic := '0';
      begin
         process (a, b, ci)
            begin
               s   <= a xor b xor ci;
               m1 <= a and b;
               m2 <= b and ci;
```

```
            m3 <= ci and a;
            co  <= m1 or m2 or m3;
        end process;
    end bm_full_adder;
```

Test Bench

```
library ieee;
use ieee.std_logic_1164.all;
    entity tb_full_adder is
        --port ( );
    end tb_full_adder;

    architecture bm_tb_full_adder of tb_full_adder is
        component full_adder
            port ( a, b, ci  :  in std_logic;
                   s, co     :  out std_logic );
        end component;

    signal a  : std_logic := '0';
    signal b  : std_logic := '0';
    signal ci : std_logic := '0';
    signal s  : std_logic := '0';
    signal co : std_logic := '0';

    begin
        uut : full_adder port map
            ( a  => a,  b  => b,  ci  => ci,  s  => s,  co => co );
        stimulus_process : process

            begin
                a <= '0';     b <= '0';     ci <= '0';
            wait for 10ns;

                a <= '0';     b <= '0';     ci <= '1';
            wait for 10ns;

                a <= '0';     b <= '1';     ci <= '0';
            wait for 10ns;

                a <= '0';     b <= '1';     ci <= '1';
            wait for 10ns;

                a <= '1';     b <= '0';     ci <= '0';
            wait for 10ns;

                a <= '1';     b <= '0';     ci <= '1';
            wait for 10ns;

                a <= '1';     b <= '1';     ci <= '0';
            wait for 10ns;

                a <= '1';     b <= '1';     ci <= '1';
            wait for 10ns;
        end process;
    end bm_tb_full_adder;
```

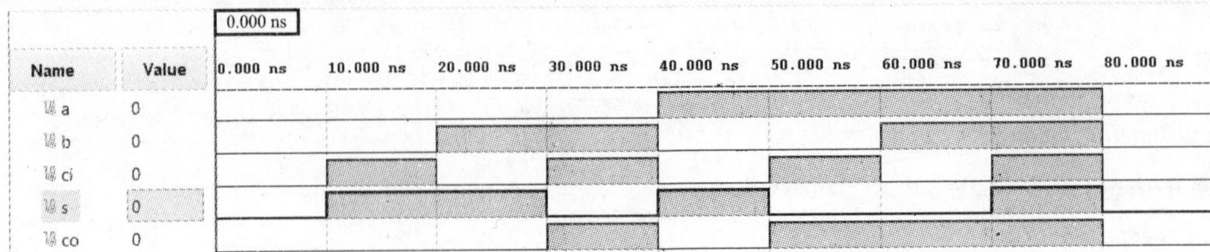

Fig. a: *Inputs: a = 0, b = 0, ci = 0; Outputs: s = 0, co = 0.*

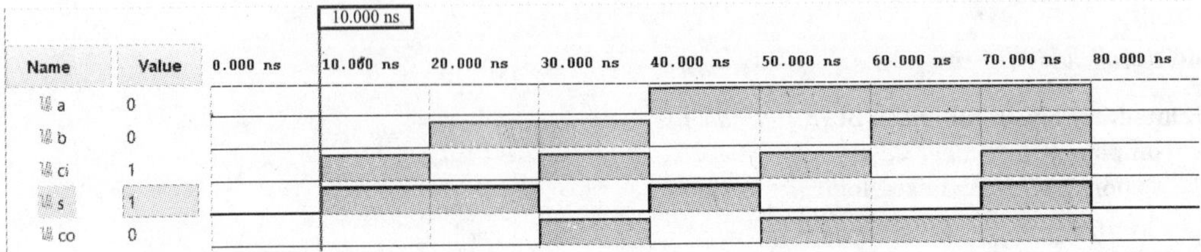

Fig. b: *Inputs: a = 0, b = 0, ci = 1; Outputs: s = 1, co = 0.*

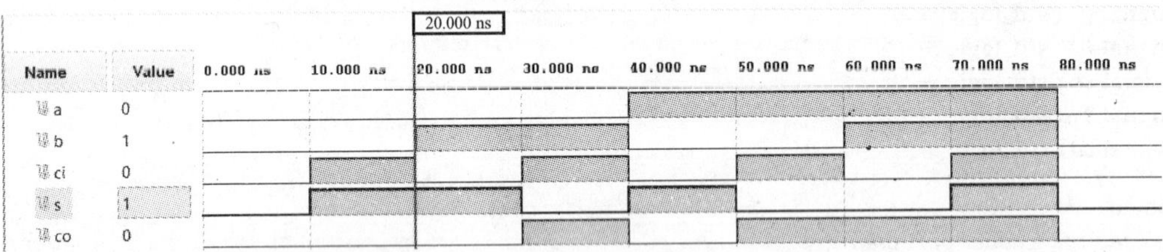

Fig. c: *Inputs: a = 0, b = 1, ci = 0; Outputs: s = 1, co = 0.*

Fig. d: *Inputs: a = 0, b = 1, ci = 1; Outputs: s = 0, co = 1.*

Fig. e: *Inputs: a = 1, b = 0, ci = 0; Outputs: s = 1, co = 0.*

Fig. f: *Inputs: a = 1, b = 0, ci = 1; Outputs: s = 0, co = 1.*

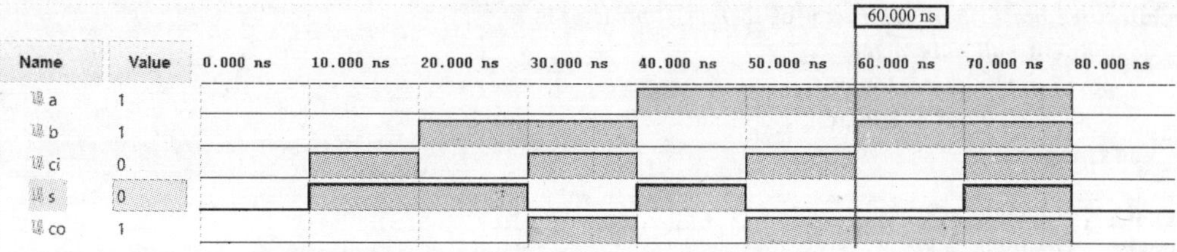

Fig. g: Inputs: a = 1, b = 1, ci = 0; Outputs: s = 0, co = 1.

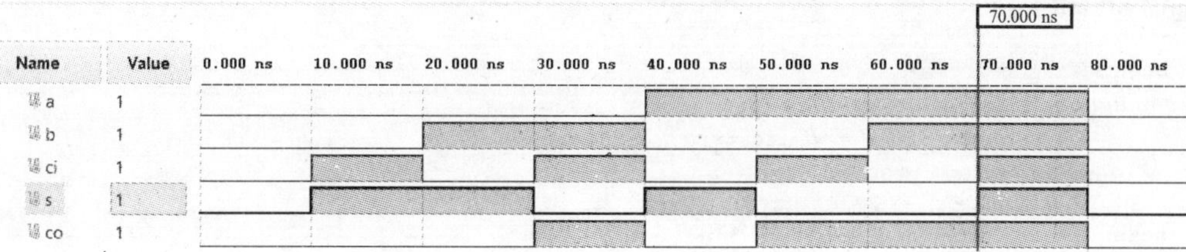

Fig. h: Inputs: a = 1, b = 1, ci = 1; Outputs: s = 1, co = 1.

Fig. 1: Input-output waveform of full adder.

VHDL Program 14.39

Write VHDL code for behavioral model of a half subtractor.

Program Logic for Behavioral Model

The Boolean equations for half subtractor are,

$$d = a \oplus b \quad ; \quad c_o = a'b$$

Where, a and b are 1-bit input data, difference (d) and borrow (c_o) are 1-bit output data. The VHDL code for behavioral model of half subtractor is developed using the above equations. (Refer Chapter 5, Section 5.10.1 for design of half subtractor).

VHDL Program for Behavioral Model of Half Subtractor

```
library ieee;
use ieee.std_logic_1164.all;
   entity half_subtractor is
      port ( a, b  : in std_logic;
             d, co : out std_logic );
   end half_subtractor;

   architecture bm_half_subtractor of half_subtractor is
   signal m1 : std_logic;
      begin
         process ( a, b )
            begin
               d <= a xor b;      m1 <= not a;      co <= m1 and b;
         end process;
   end bm_half_subtractor;
```

Test Bench

```
library ieee;
use ieee.std_logic_1164.all;
   entity tb_half_subtractor is
      --port ( );
   end tb_half_subtractor;
```

```
architecture bm_tb_half_subtractor of tb_half_subtractor is
    component half_subtractor
        port ( a, b  : in std_logic;
               d, co : out std_logic );
    end component;

signal a  : std_logic := '0';
signal b  : std_logic := '0';
signal d  : std_logic;
signal co : std_logic;

    begin
        uut : half_subtractor port map
            ( a => a,   b => b,   d => d,   co => co );
        stimulus_process : process

    begin
        a <= '0';     b <= '0';
    wait for 20ns;

        a <= '0';     b <= '1';
    wait for 20ns;

        a <= '1';     b <= '0';
    wait for 20ns;

        a <= '1';     b <= '1';
    wait for 20ns;
end process;
end bm_half_subtractor;
```

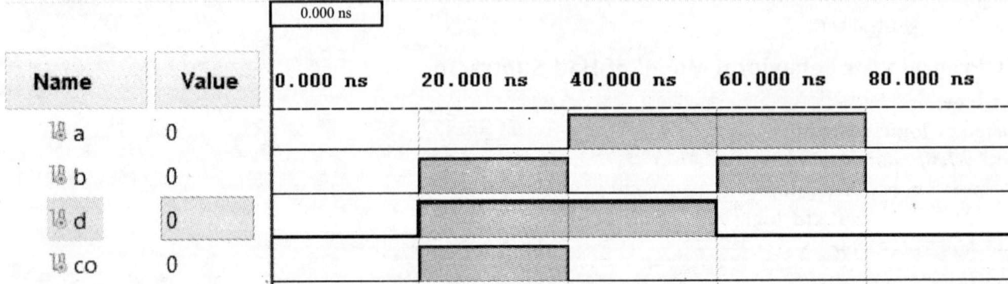

Fig. a: *Inputs: a = 0, b = 0; Outputs: d = 0, co = 0.*

Fig. b: *Inputs: a = 0, b = 1; Outputs: d = 1, co = 1.*

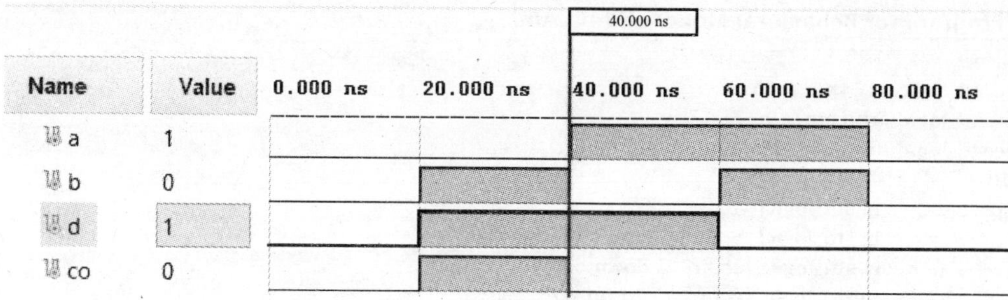

***Fig. c:** Inputs: a = 1, b = 0; Outputs: d = 1, co = 0.*

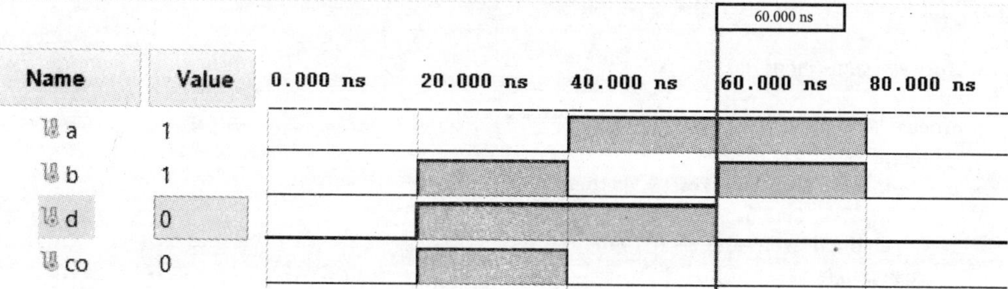

***Fig. d:** Inputs: a = 1, b = 1; Outputs: d = 0, co = 0.*

***Fig. 1:** Input-output waveform of half subtractor.*

VHDL Program 14.40

Write VHDL code for behavioral model of a 4-bit ALU.

Program Logic for Behavioral Model

The VHDL codes for behavioral model of 4-bit ALU are developed to perform 8 different operations listed in Table 1. (Refer Chapter 5, Section 5.14.1 for design of 4-bit ALU).

Table 1: Operation Select Codes for 4-bit ALU

Operation Select Inputs			Logical/Arithmetic Operations	Comments
s_2	s_1	s_0		
0	0	0	CLEAR	Clear output
0	0	1	B – A	Arithmetic subtraction
0	1	0	A – B	Arithmetic subtraction
0	1	1	A + B	Arithmetic addition
1	0	0	$A \oplus B$	Logical XOR operation
1	0	1	A + B	Logical addition
1	1	0	AB	Logical AND operation
1	1	1	PRESET	Set output

VHDL Program for Behavioral Model of 4-bit ALU

```
library ieee;
use ieee.std_logic_1164.all;
use ieee.std_logic_arith.all;
use ieee.std_logic_unsigned.all;
   entity alu_4bit is
      port ( clk    : in std_logic;
             reset : in std_logic;
             a, b  : in std_logic_vector   (3 downto 0);
             ci    : in std_logic_vector   (1 downto 0);
             sel   : in std_logic_vector (2 downto 0);
             f     : out std_logic_vector (3 downto 0) );
   end alu_4bit;

   architecture bm_alu_4bit of alu_4bit is
      begin
         process (clk, reset)
            begin
               if (reset'event and reset = '0') then
                  f <= "0000";
               elsif (clk'event and clk = '1') then
            case (sel) is
               when "000" => f <= "0000";
               when "001" => f <= b – a;
               when "010" => f <= a – b;
               when "011" => f <= a + b;
               when "100" => f <= a xor b;
               when "101" => f <= a or b;
               when "110" => f <= a and b;
               when others => f <= "1111";
            end case;
         end if;
         end process;
      end bm_alu_4bit;
```

Test Bench

```
library ieee;
use ieee.std_logic_1164.all;
   entity alu_4bit_tb is
      --port ( );
   end alu_4bit_tb;

   architecture bm_alu_4bit_tb of alu_4bit_tb is
      component alu_4bit is
         port ( clk    : in std_logic
                reset : in std_logic
                a, b  : in std_logic_vector (3 downto 0);
                sel   : in std_logic_vector (2 downto 0);
                f     : out std_logic_vector (3 downto 0));
      end component;

      signal clk    : std_logic;
      signal reset : std_logic;
```

```
signal a      : std_logic_vector (3 downto 0);
signal b      : std_logic_vector (3 downto 0);
signal sel    : std_logic_vector (2 downto 0);
signal f      : std_logic_vector (3 downto 0);

begin
uut: alu_4bit port map
    (clk => clk, reset => reset, sel => sel, a => a, b => b, f => f);

clk_process : process
    begin
        clk <= '0'; wait for 5 ns;
        clk <= '1'; wait for 5 ns;
end process;

stimulus_process : process
    begin
        reset <= '1';   wait for 5 ns;
        reset <= '0';   wait for 5 ns;
        sel <= "000";     a <= "1001";     b <= "0011"; wait for 10ns;
        sel <= "001"; wait for 10ns;
        sel <= "010"; wait for 10ns;
        sel <= "011"; wait for 10ns;
        sel <= "100"; wait for 10ns;
        sel <= "101"; wait for 10ns;
        sel <= "110"; wait for 10ns;
        sel <= "111"; wait for 10ns;
        sel <= "000";     a <= "0000";     b <= "0000"; wait for 10ns;
    wait;
end process;
end bm_alu_4bit_tb;
```

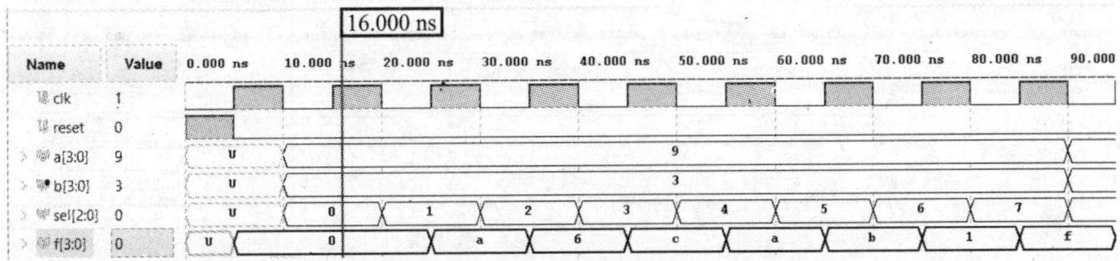

Fig. a: *Input: sel = 000; Output: f = 0000.*

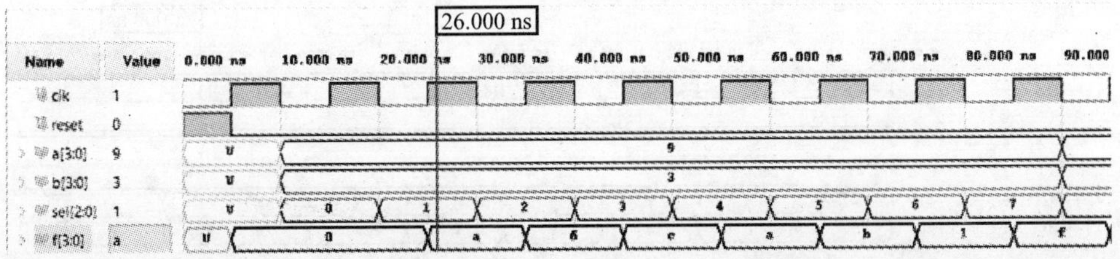

Fig. b: *Inputs: a = 9 = 1001; b = 3 = 0011, sel = 1 = 001; Output, f = b – a = A_H = 1010.*

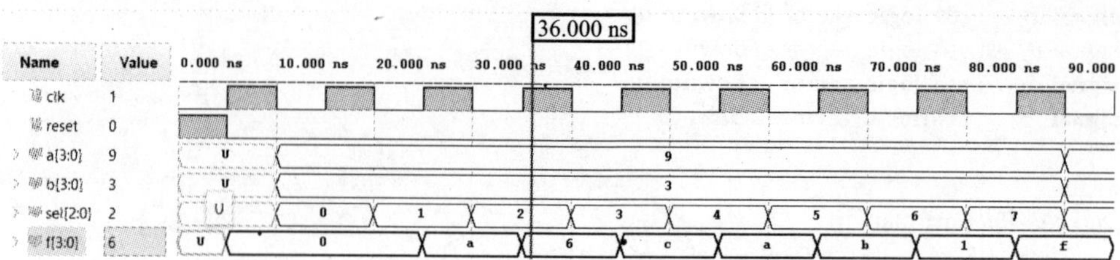

Fig. c: *Inputs: a = 9 = 1001; b = 3 = 0011, sel = 2 = 010; Output, f = a – b =* 6_H *= 0110.*

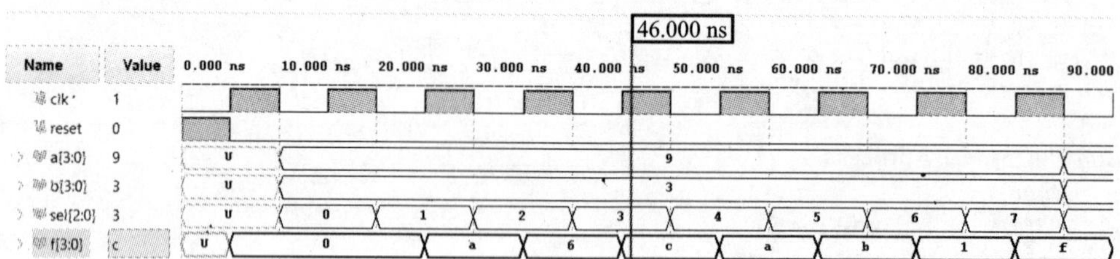

Fig. d: *Inputs: a = 9 = 1001; b = 3 = 0011, sel = 3 = 011; Output, f = a + b =* C_H *= 1100.*

Fig. e: *Inputs: a = 9 = 1001; b = 3 = 0011, sel = 4 = 100; Output, f = a xor b =* A_H *= 1010.*

Fig. f: *Inputs: a = 9 = 1001; b = 3 = 0011, sel = 5 = 101; Output, f = a or b =* B_H *= 1011.*

Fig. g: *Inputs: a = 9 = 1001; b = 3 = 0011, sel = 6 = 110; Output, f = a and b =* 1_H *= 0001.*

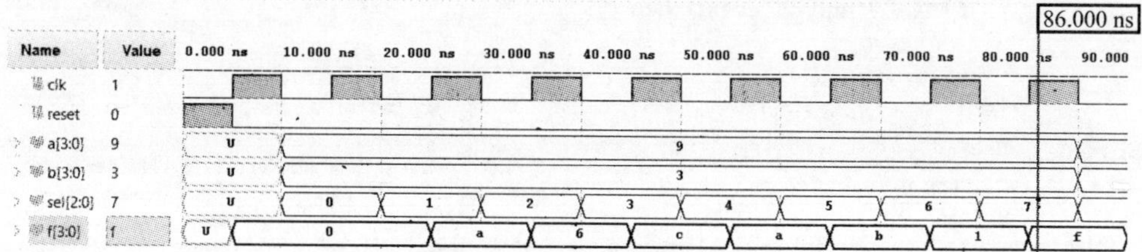

Fig. h: *Input: sel = 7 = 111; Output,* $f = F_H = 1111$.

Fig. 1: *Input-output waveform of 4-bit ALU.*

VHDL Program 14.41

Write VHDL code for behavioral model of the Moore FSM given below.

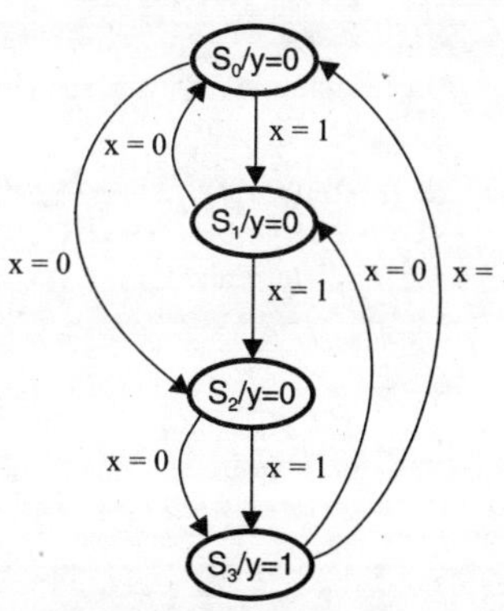

Fig. 1: *State diagram.*

State Table

Current_State	Next_State		Output	
	x = 1	x= 0	x = 1	x = 0
			y	y
S0	S1	S2	0	0
S1	S2	S0	0	0
S2	S3	S3	0	0
S3	S0	S1	1	1

Program Logic for Behavioral Model

The given Moore FSM has four states. Here, case statement is used to select a current state and the next state is decided based on output and input in each case. (Refer Chapter 7, Section 7.2.1 for design of Moore_fsm).

VHDL Program for Behavioral Model of Moore_FSM

```
library ieee;
use ieee.std_logic_1164.all;
   entity moore_fsm is
      port ( clk : in std_logic;   -- clock signal
             x   : in std_logic;   -- binary sequence input
             y   : out std_logic; -- output of the VHDL sequence detector);
   end moore_fsm;

   architecture bm_moore_fsm of moore_fsm is
   type moore_type is (s0, s1, s2, s3);
   signal current_state, next_state : moore_type;

     begin
        process (clk, current_state, x)
           begin
              case (current_state) is
              when s0 =>
```

```
            if (clk = '1' and clk' event) then
                y <= '0';
            if (x = '1') then
                next_state <= s1;
        else
            next_state <= s2;
        end if;
        end if;

        when s1 =>
            if (clk = '1' and clk' event) then
                y <= '0';
            if (x = '1') then
                next_state <= s2;
        else
            next_state <= s0;
        end if;
        end if;

        when s2 =>
            if (clk = '1' and clk' event) then
                y <= '0';
            if (x = '1') then
                next_state <= s3;
        else
            next_state <= s3;
        end if;
        end if;

        when s3 =>
            if (clk = '1' and clk' event) then
                y <= '1';
            if (x = '1') then
                next_state <= s0;
        else
            next_state <= s1;
        end if;
        end if;
    end case;
        current_state <= next_state;
  end process;
end bm_moore_fsm;
```

Test Bench

```
library ieee;
use ieee.std_logic_1164.all;
  entity tb_moore_fsm is
      --port ( );
  end tb_moore_fsm;

  architecture bm_tb_moore_fsm of tb_moore_fsm is
      component moore_fsm
          port ( clk : in std_logic;     x : in std_logic;     y : out std_logic );
      end component;
```

```
signal clk : std_logic;
signal x   : std_logic;
signal y   : std_logic;
type moore_type is (s0, s1, s2, s3);
signal current_state, next_state : moore_type;
constant clk_period : time := 10 ns;

begin
    uut : moore_fsm port map
        ( clk => clk,      x => x,      y => y );
    clock_process : process

    begin
            clk <= '1';
        wait for clk_period/2;

            clk <= '0';
        wait for clk_period/2;
 end process;

    stimulus_process : process
        begin
            x <= '1';     current_state <= s0;     next_state    <= s1;
            wait for 5 ns;

            x <= '0';     current_state <= s0;     next_state <= s2;
            wait for 5 ns;

            x <= '1';     current_state <= s1;     next_state <= s2;
            wait for 5 ns;

            x <= '0';     current_state <= s1;     next_state <= s0;
            wait for 5 ns;

            x <= '1';     current_state <= s2;     next_state <= s3;
            wait for 5 ns;

            x <= '0';     current_state <= s2;     next_state <= s3;
            wait for 5 ns;

            x <= '1';     current_state <= s3;     next_state <= s0;
            wait for 5 ns;

            x <= '0';     current_state <= s3;     next_state <= s1;
            wait for 5 ns;

  end process;
end bm_tb_moore_fsm;
```

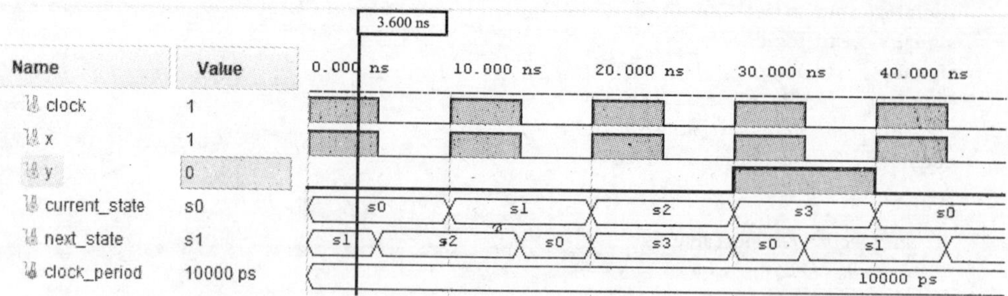

Fig. a: *Input: x = 1; Output: y = 0; Current_state = s0; Next_state = s1.*

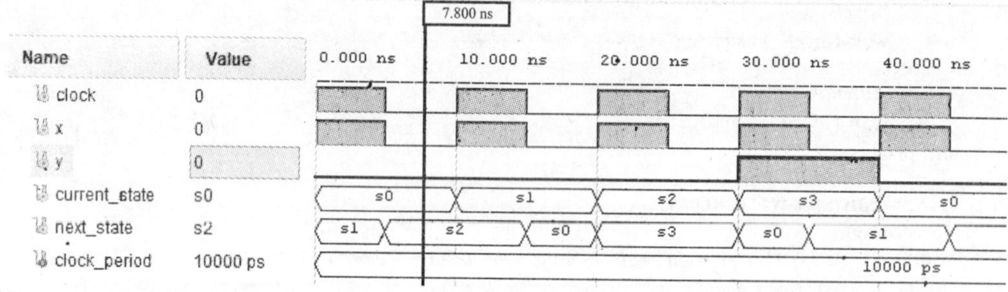

Fig. b: *Input: x = 0; Output: y = 0; Current_state = s0; Next_state = s2.*

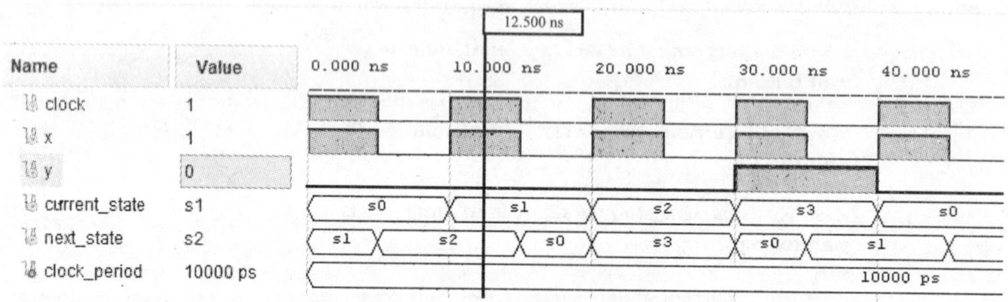

Fig. c: *Input: x = 1; Output: y = 0; Current_state = s1; Next_state = s2.*

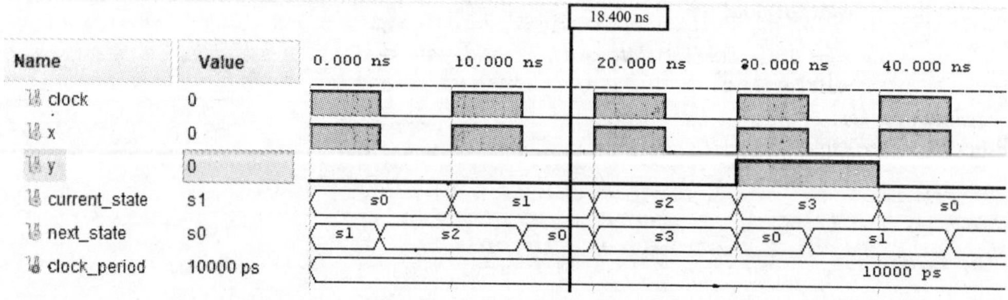

Fig. d: *Input: x = 0; Output: y = 0; Current_state = s1; Next_state = s0.*

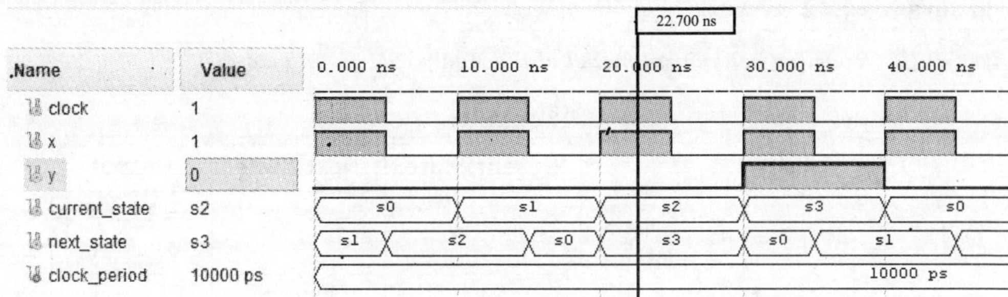

***Fig. e:** Input: x = 1; Output: y = 0; Current_state = s2; Next_state = s3.*

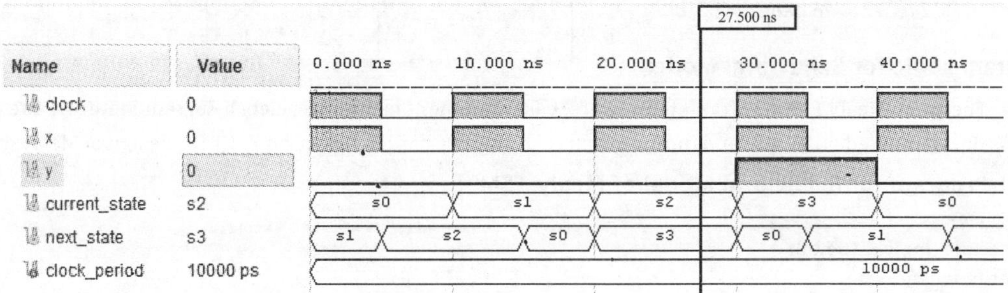

***Fig. f:** Input: x = 0; Output: y = 0; Current_state = s2; Next_state = s3.*

33.700 ns
Name | Value | 0.000 ns | 10.000 ns | 20.000 ns | 30.000 ns | 40.000 ns
clock 1
x 1
y 1
current_state s3
next_state s0
clock_period 10000 ps
s0 s1 s2 s3 s0
s1 s2 s0 s3 s0 s1
10000 ps

***Fig. g:** Input: x = 1; Output: y = 1; Current_state = s3; Next_state = s0.*

37.400 ns
Name | Value | 0.000 ns | 10.000 ns | 20.000 ns | 30.000 ns | 40.000 ns
clock 0
x 0
y 1
current_state s3
next_state s1
clock_period 10000 ps
s0 s1 s2 s3 s0
s1 s2 s0 s3 s0 s1
10000 ps

***Fig. e:** Input: x = 0; Output: y = 1; Current_state = s3; Next_state = s1.*

***Fig. 1:** Input-output waveform of Moore_FSM.*

VHDL Program 14.42

Write VHDL code for behavioral model of the Mealy FSM given below.

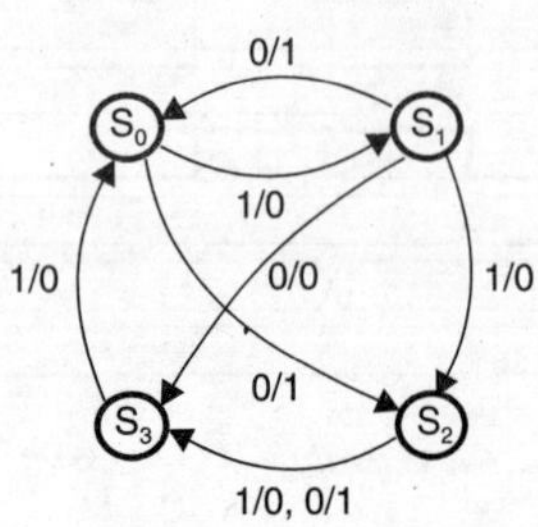

Fig. 1: State diagram.

State Table

Present_State	Next_State		Output	
	x = 1	x= 0	x = 1	x = 0
			y	y
S0	S1	S2	0	1
S1	S2	S0	0	1
S2	S3	S3	0	1
S3	S0	S1	0	1

Program Logic for Behavioral Model

The given Mealy FSM has four states. Here, case statement is used to select a current state and the next state is decided based on output and input in each case. (Refer Chapter 7, Section 7.2.1 for design of Melay_fsm).

VHDL Program for Behavioral Model of Mealy_FSM

```
library ieee;
use ieee.std_logic_1164.all;
    entity melay_fsm is
        port ( x, clk : in bit;
            y      : out bit );
    end melay_fsm;

    architecture bm_melay_fsm of melay_fsm is
    type s_type is (s0, s1, s2, s3);
    signal c_s, n_s : s_type;

        begin
            process (clk)
                begin
                    if (clk = '1' and clk' event) then
                        c_s <= n_s;
                    end if;
            end process;

            process (x, c_s, n_s)
                begin
                    case c_s is
                        when s0 => if x = '1' then
                            n_s <= s1;      y <= '0';
                        else
                            n_s <= s2;      y <= '1';
                    end if;

                        when s1 => if x = '1' then
                            n_s <= s2;      y <= '0';
                    else
                        n_s <= s0;          y <= '1';
                    end if;
```

```
            when s2=> if x='1' then
                n_s <= s3;      y <= '0';
            else
                n_s <= s3;      y <= '1';
            end if;

            when s3 => if x = '1' then
                n_s <= s0;      y <= '0';
            else
                n_s <= s1;      y <= '1';
            end if;
        end case;
    end process;
end bm_melay_fsm;
```

Test Bench

```
library ieee;
use ieee.std_logic_1164.all;
    entity tb_melay_fsm is
        --port ( );
    end tb_melay_fsm;

    architecture bm_tb_melay_fsm of tb_melay_fsm is
        component melay_fsm
            port ( x, clk : in bit;
                   y      : out bit);
        end component;

    type s_type is (s0, s1, s2, s3);
    signal c_s, n_s : s_type;
    signal x   : bit;
    signal clk : bit;
    signal y   : bit;
    constant clk_period : time := 20 ns;
        begin
            uut : melay_fsm port map
                ( x => x,   clk => clk,   y => y );
            clock_process : process
                begin
                    clk <= '1';
                wait for clk_period/2;
                    clk <= '0';
                wait for clk_period/2;
            end process;

            stimulus_process : process
                begin
                    x <= '1';    c_s <= s0;    n_s <= s1;
                wait for 10ns;

                    x <= '0';    c_s <= s0;    n_s <= s2;
                wait for 10ns;

                    x  <= '1';   c_s <= s1;    n_s <= s2;
                wait for 10ns;
```

```
        x <= '0';      c_s <= s1;      n_s <= s0;
    wait for 10ns;

        x <= '1';      c_s <= s2;      n_s <= s3;
    wait for 10ns;

        x <= '0';      c_s <= s2;      n_s <= s3;
    wait for 10ns;

        x <= '1';      c_s <= s3;      n_s <= s0;
    wait for 10ns;

        x <= '0';      c_s <= s3;      n_s <= s1;
    wait for 10ns;
  end process;
end bm_tb_melay_fsm;
```

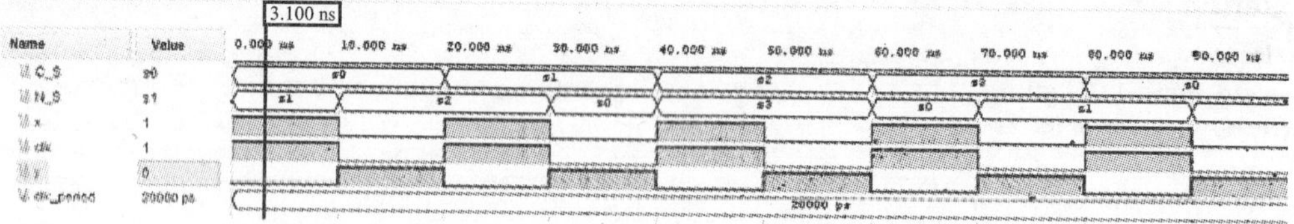

Fig. a: *Input: x = 1; Output: y = 0; Current_state = s0; Next_state = s1.*

Fig. b: *Input: x = 0; Output: y = 1; Current_state = s0; Next_state = s2.*

Fig. c: *Input: x = 1; Output: y = 0; Current_state = s1; Next_state = s2.*

Fig. d: *Input: x = 0; Output: y = 1; Current_state = s1; Next_state = s0.*

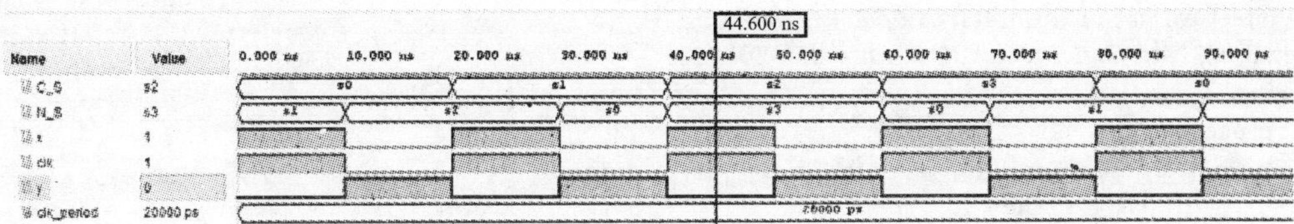

Fig. e: Input: x = 1; Output: y = 0; Current_state = s2; Next_state = s3.

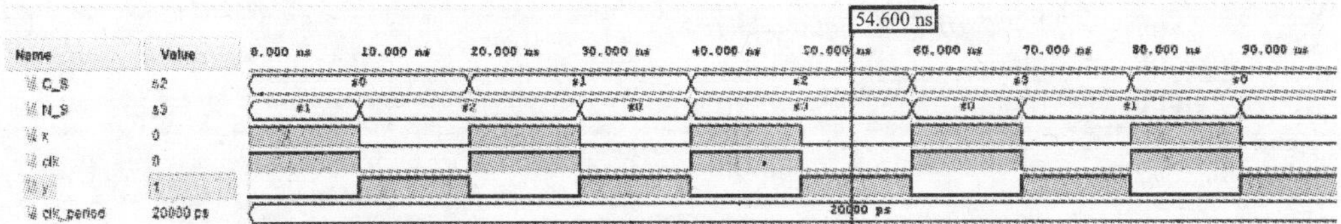

Fig. f: Input: x = 0; Output: y = 1; Current_state = s2; Next_state = s3.

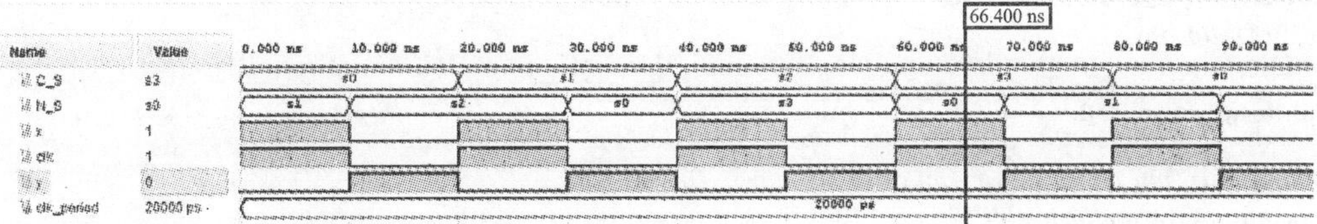

Fig. g: Input: x = 1; Output: y = 0; Current_state = s3; Next_state = s0.

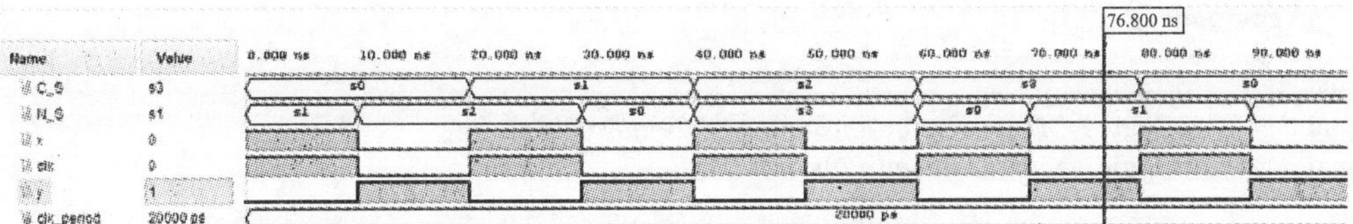

Fig. h: Input: x = 0; Output: y = 1; Current_state = s3; Next_state = s1.

Fig. 1: Input-output waveform of Mealy_FSM.

VHDL Program 14.43

Write VHDL code for behavioral model of a 4-bit up/down counter.

Program Logic for Behavioral Model

A selection bit, a is required to select up count when a = 1 and down count when a = 0. Initial count is kept 0 and increment by 1 for up count and decrement by 1 for down count for every clock pulse. (Refer Chapter 8, Section 8.2.3 for design of up/down counter).

VHDL Program for Behavioral Model of 4-bit up/down Counter

```
library ieee;
use ieee.std_logic_1164.all;
use ieee.std_logic_arith.all;
use ieee.std_logic_unsigned.all;
    entity updown_4bit is
        port ( clk, a : in std_logic;        dout : inout std_logic_vector (3 downto 0) );
    end updown_4bit;
```

```
architecture bm_updown_4bit of updown_4bit is
signal m1 : std_logic_vector (3 downto 0) := "0000";
begin
    process (clk)
        begin
            if (clk = '1') then
            if a ='1' then
                m1 <= m1 + "0001";        dout <= m1;
            else
                m1 <= m1 - "0001";        dout <= m1;
            end if;
            end if;
    end process;
end bm_updown_4bit;
```

Test Bench

```
library ieee;
use ieee.std_logic_1164.all;
    entity tb_up_down_4bit is
        --port ( );
    end tb_up_down_4bit;

    architecture bm_tb_up_down_4bit of tb_up_down_4bit is
        component updown_4bit
            port ( clk, a : in std_logic;
                   dout   : out std_logic_vector (3 downto 0) );
        end component;

    signal clk   : std_logic := '0';
    signal a     : std_logic := '1';  -- :='1' for upcount, :='0' for downcount,
    signal dout : std_logic_vector (3 downto 0);


        begin
            uut : updown_4bit port map
                ( clk   => clk,
                  a     => a,
                  dout => dout );
            stimulus_process : process
            begin
                clk <= '1';
            wait for 1ns;

                clk <= '0';
            wait for 1ns;
        end process;
    end bm_tb_up_down_4bit;
```

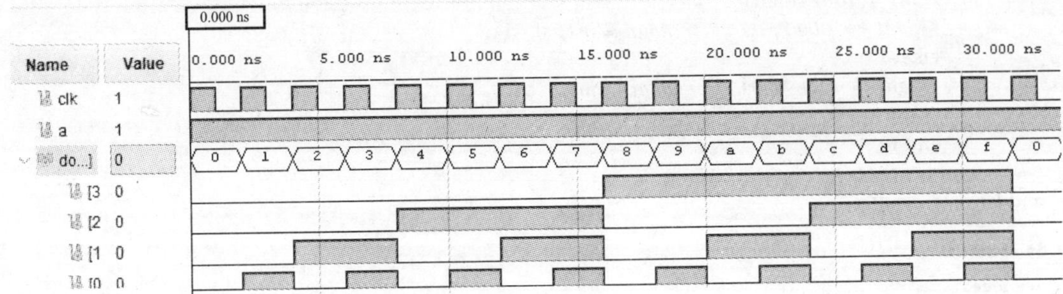

***Fig. 1:** Input-output waveform of up counter.*

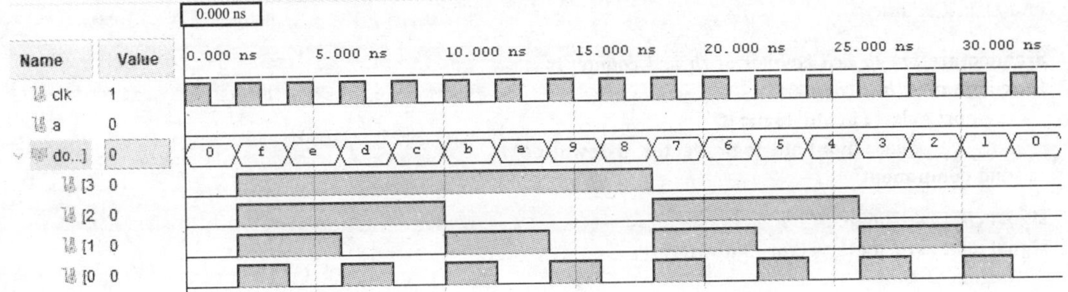

***Fig. 2:** Input-output waveform of down counter.*

VHDL Program 14.44

Write VHDL code for behavioral model of a BCD counter/MOD10 counter.

Program Logic for Behavioral Model

Initially the count value is kept 0. BCD counter is basically an up counter and hence the count value is incremented by 1 for every clock pulse. When the count reaches 1001 the count value is clear to 0.

(Refer Chapter 8, Section 8.2.5 for design of BCD counter/MOD10 counter).

VHDL Program for Behavioral Model of BCD Counter/MOD10 Counter

```
library ieee;
use ieee.std_logic_1164.all;
use ieee.std_logic_arith.all;
use ieee.std_logic_unsigned.all;
    entity bcd_counter is
            port ( clk    : in std_logic;
                   dout : inout std_logic_vector (3 downto 0) );
    end bcd_counter;

    architecture bm_bcd_counter of bcd_counter is
        signal m1 : std_logic_vector (3 downto 0) := "0000";
            begin
                process (clk)
            begin
                if (clk = '1') then
```

```
            if m1 = "1001" then
                m1 <= "0000";                dout <= m1;
            else
                m1<= m1 + "0001";            dout <= m1;
            end if;
            end if;
        end process;
    end bm_bcd_counter;
```

Test Bench

```
library ieee;
use ieee.std_logic_1164.all;
    entity tb_bcd_counter is
        --port ( );
    end tb_bcd_counter;

    architecture bm_tb_bcd_counter of tb_bcd_counter is
        component bcd_counter
            port ( clk   : in std_logic;
                dout : inout std_logic_vector (3 downto 0) );
        end component;

    signal clk   : std_logic := '0';
    signal dout : std_logic_vector (3 downto 0);
        begin
            uut : bcd_counter port map
                ( clk    => clk,      dout => dout);
            stimulus_process : process
        begin
            clk <= '1';
            wait for 10ns;

            clk <= '0';
            wait for 10ns;
        end process;
    end bm_tb_bcd_counter;
```

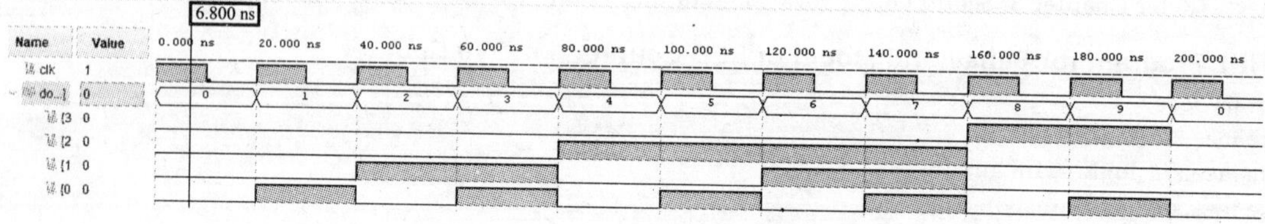

Fig. a: Counter output: dout = 0000.

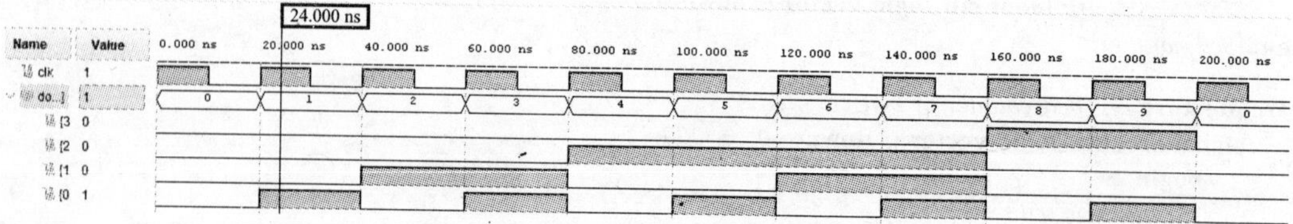

Fig. b: Counter output: dout = 0001.

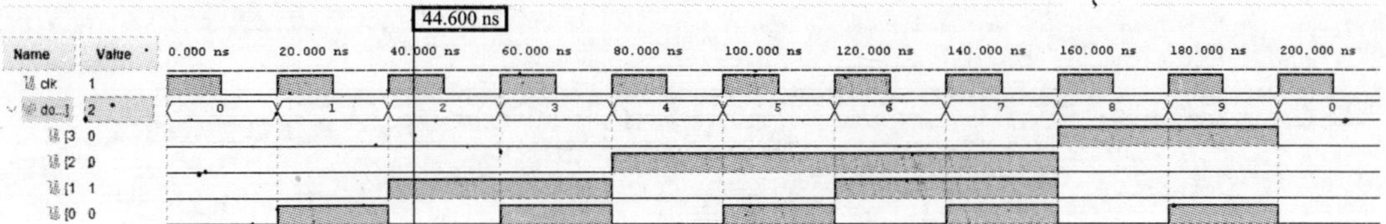

Fig. c: *Counter output: dout = 0010.*

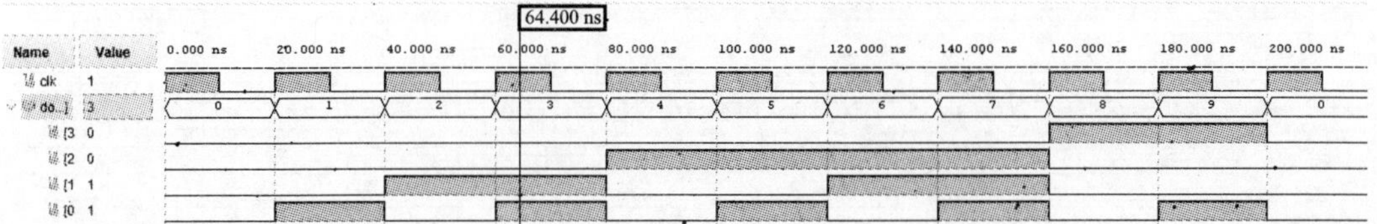

Fig. d: *Counter output: dout = 0011.*

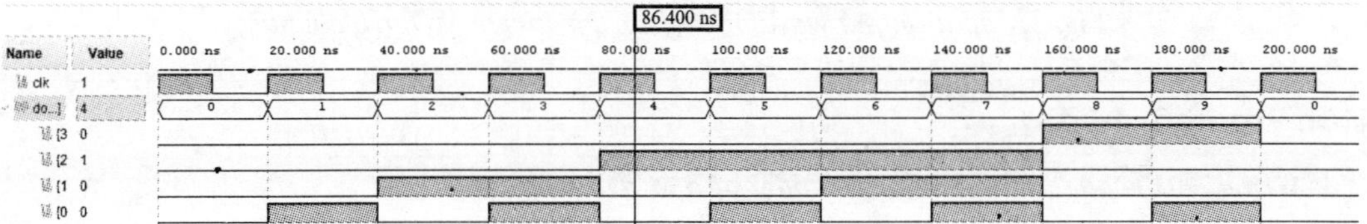

Fig. e: *Counter output: dout = 0100.*

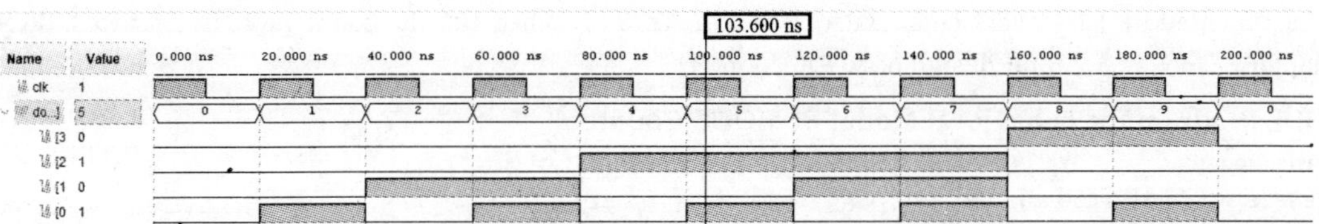

Fig. f: *Counter output: dout = 0101.*

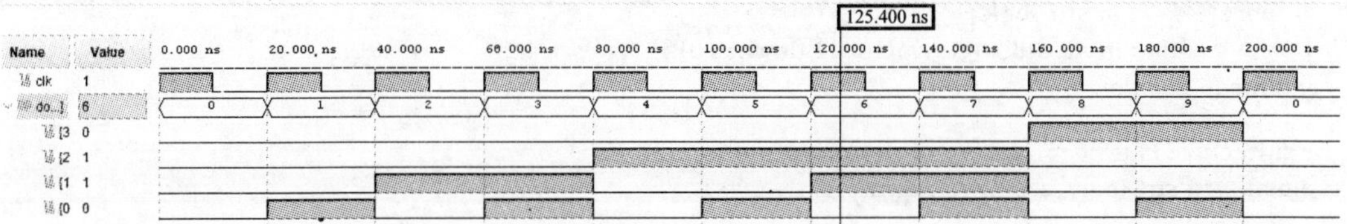

Fig. g: *Counter output: dout = 0110.*

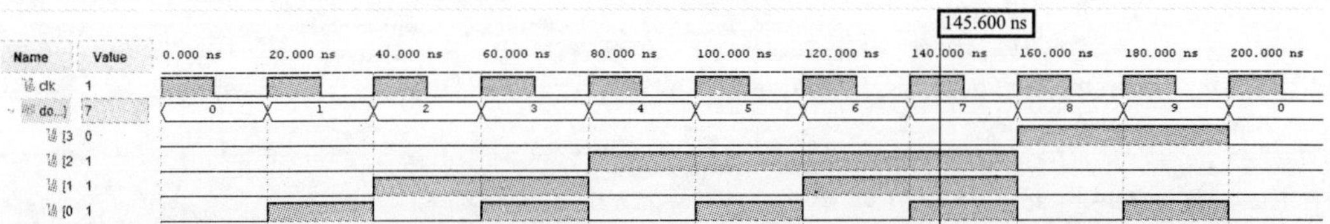

Fig. h: *Counter output: dout = 0111.*

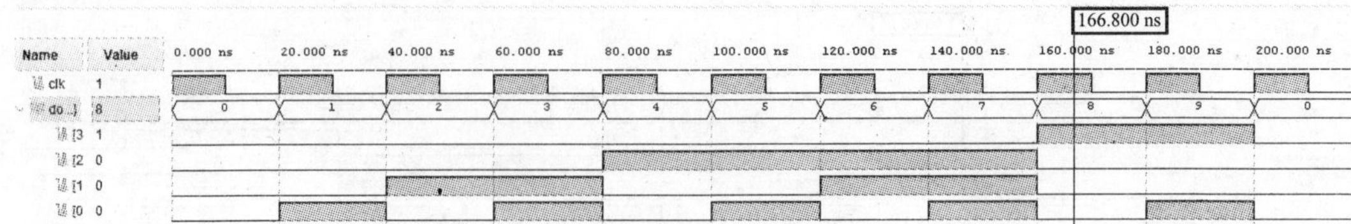

Fig. i: Counter output: dout = 1000.

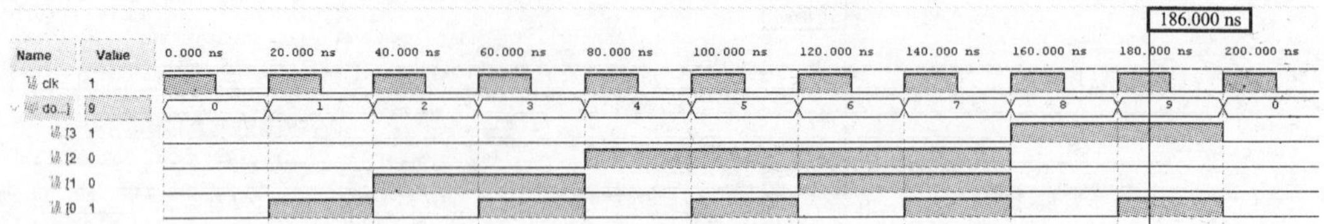

Fig. j: Counter output: dout = 1001.

***Fig. 1:** Input-output waveform of BCD counter/MOD10 counter.*

VHDL Program 14.45

Write VHDL code for behavioral model of a MOD6 counter.

Program Logic for Behavioral Model

Initially the count value is kept 0. MOD6 counter is basically an up counter and hence the count value is incremented by 1 for every clock pulse. When the count reaches 101 the count value is clear to 0. (Refer Chapter 8, Section 8.2.6 for design of MOD6 counter).

VHDL Program for Behavioral Model of MOD6 Counter

```
library ieee;
use ieee.std_logic_1164.all;
use ieee.std_logic_arith.all;
use ieee.std_logic_unsigned.all;
    entity mod6_counter is
        port ( clk   : in std_logic;
               dout : inout std_logic_vector (2 downto 0));
    end mod6_counter;

    architecture bm_mod6_counter of mod6_counter is
    signal m1 : std_logic_vector (2 downto 0) := "000";
        begin
            process (clk)
                begin
                    if (clk = '1') then
                    if m1 = "101" then
                        m1 <= "000";          dout <= m1;
                     else
                        m1 <= m1 + "001";   dout <= m1;
                    end if;
                    end if;
            end process;
    end bm_mod6_counter;
```

Test Bench

```
library ieee;
use ieee.std_logic_1164.all;
    entity tb_mod6_counter is
        --port ( );
    end tb_mod6_counter;

    architecture bm_tb_mod6_counter of tb_mod6_counter is
        component mod6_counter
            port ( clk   : in std_logic;
                   dout : inout std_logic_vector (2 downto 0));
        end component;

    signal clk    : std_logic := '0';
    signal dout : std_logic_vector (2 downto 0);

        begin
            uut : mod6_counter port map
                ( clk    => clk,       dout => dout );
            stimulus_process : process
                begin
                        clk <= '1';

                    wait for 10ns;
                        clk <= '0';
                    wait for 10ns;
            end process;
    end bm_tb_mod6_counter;
```

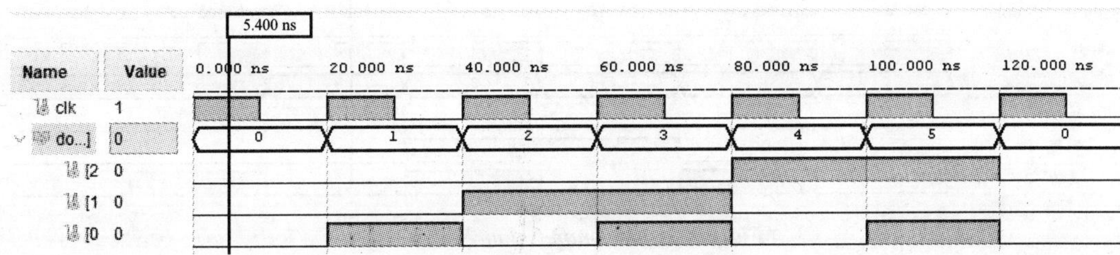

***Fig. a:** Counter output: dout = 000.*

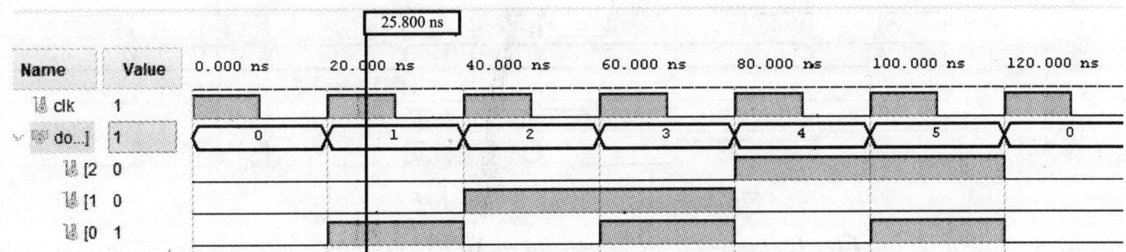

***Fig. b:** Counter output: dout = 001.*

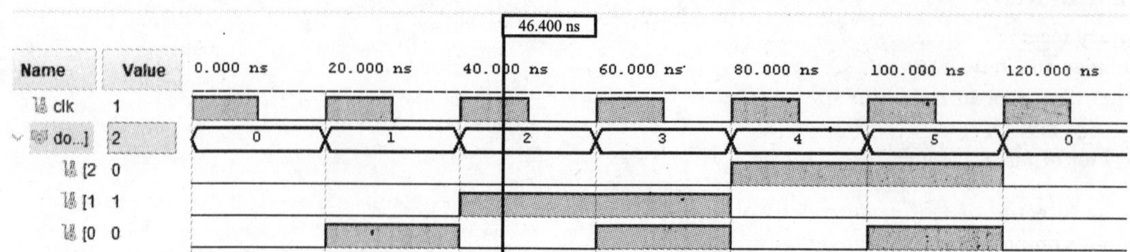

Fig. c: Counter output: dout = 010.

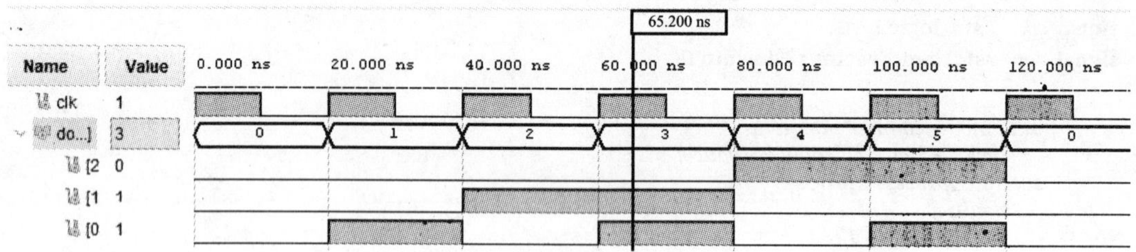

Fig. d: Counter output: dout = 011.

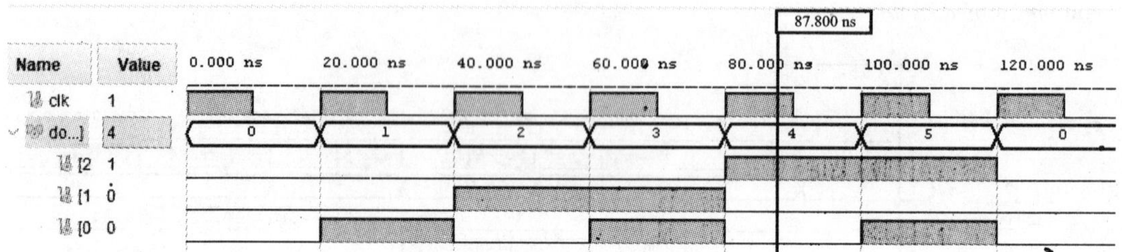

Fig. e: Counter output: dout = 100.

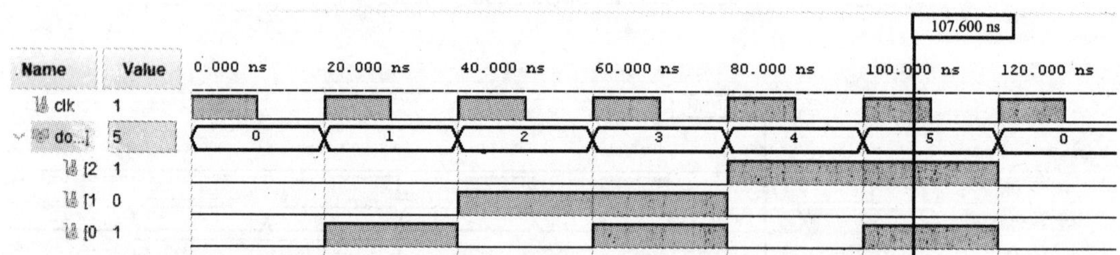

Fig. f: Counter output: dout = 101.

***Fig. 1:** Input-output waveform of MOD6 counter.*

VHDL Program 14.46

Write a VHDL code for behavioral model of 8-bit SISO right shift register.

Program Logic for Behavioral Model

In the VHDL code, an 8-bit buffer is declared to store the 8-bit data. The LSB (Least Significant Bit) of 8-bit buffer is output as serial output data. The shift operation is performed by concatenation of input data bit as MSB (Most Significant Bit) and upper 7-bit of buffer. (Refer Chapter 8, Section 8.6.1 for right shift SISO register.

VHDL Program for Behavioral Model of 8-bit SISO Right-Shift Register

```
library ieee;
use ieee.std_logic_1164.all;
  entity RS_reg is
    port ( clk            : in std_logic;
          reset           : in std_logic;
          ser_data_in     : in std_logic;
          ser_data_out    : out std_logic;
          rightshift_reg  : buffer std_logic_vector(7 downto 0));
  end RS_reg;

  architecture bm_RS_reg of RS_reg is
    begin
      process (clk, reset)
        begin
          if (reset'event and reset = '0') then
              rightshift_reg <= "11110000"
          elsif (clk'event and clk = '1') then
              ser_data_out <= rightshift_reg(0);
              rightshift_reg <= ser_data_in & rightshift_reg(7 downto 1);
          end if;
        end process;
      end bm_RS_reg;
```

Test Bench

```
library ieee;
use ieee.std_logic_1164.all;
use ieee.numeric_std.all;
  entity RS_reg_tb is
      --port( );
  end RS_reg_tb;

  architecture bm_RS_reg_tb of RS_reg_tb is
  component RS_reg
    port ( clk            : in std_logic;
          reset           : in std_logic;
          ser_data_in     : in std_logic;
          ser_data_out    : out std_logic;
          rightshift_reg  : buffer std_logic_vector (7 downto 0));
  end component;

  signal clk          : std_logic;
  signal reset        : std_logic;
  signal ser_data_in  : std_logic;
```

```
signal ser_data_out  : std_logic;
signal rightshift_reg : std_logic_vector (7 downto 0);
   begin
      uut: RS_reg port map
         ( clk             => clk,
           reset           => reset,
           ser_data_in     => ser_data_in,
           ser_data_out    => ser_data_out,
           rightshift_reg  => rightshift_reg );

      clk_process : process
         begin
            clk <= '0';   wait for 5 ns;
            clk <= '1';   wait for 5 ns;
      end process;

      stimulus_process : process
      begin
         reset <= '1';          wait for 5 ns;
         reset <= '0';          wait for 5 ns;
         ser_data_in <= '1';    wait for 10 ns;
         ser_data_in <= '0';    wait for 10 ns;
         ser_data_in <= '1';    wait for 10 ns;
         ser_data_in <= '0';    wait for 10 ns;
         ser_data_in <= '1';    wait for 10 ns;
         ser_data_in <= '0';    wait for 10 ns;
         ser_data_in <= '1';    wait for 10 ns;
         ser_data_in <= '0';    wait for 10 ns;
      wait;
   end process;
   end bm_RS_reg_tb;
```

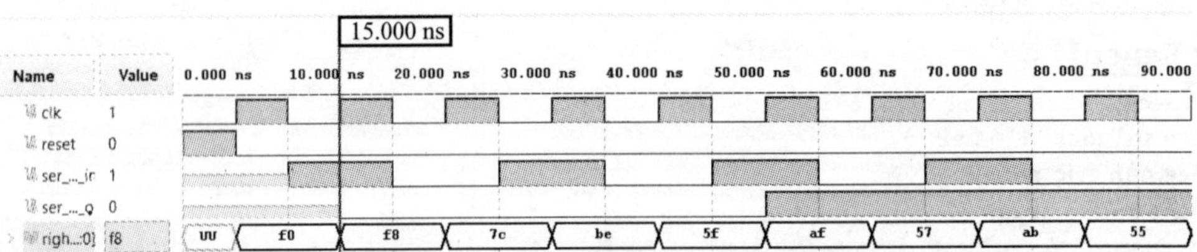

Fig. a: ser_data_in = 1; rightshift_reg: 1111 0000($F0_H$) $\xrightarrow{\text{Right shift}}$ 1111 1000($F8_H$); ser_data_out = 0.

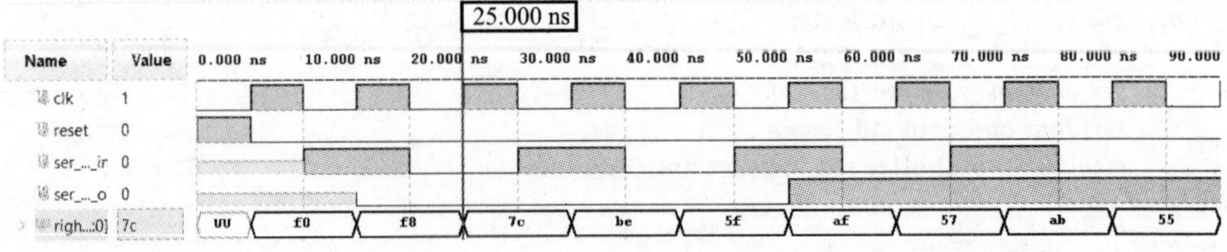

Fig. b: ser_data_in = 0; rightshift_reg: 1111 1000($F8_H$) $\xrightarrow{\text{Right shift}}$ 0111 1100($7C_H$); ser_data_out = 0.

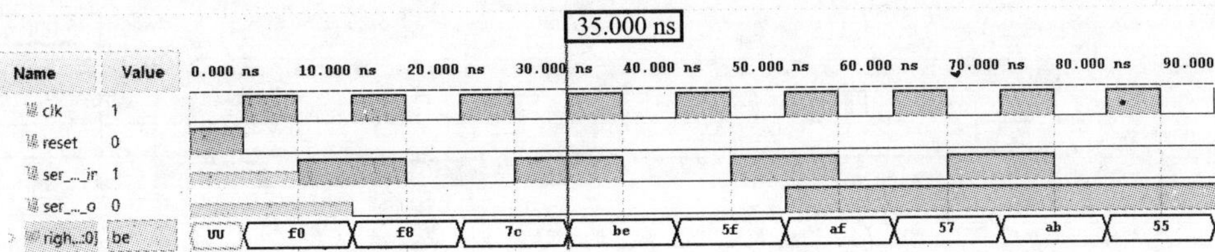

Fig. c: *ser_data_in = 1; rightshift_reg: 0111 1100($7C_H$) $\xrightarrow{\text{Right shift}}$ 1011 1110(BE_H); ser_data_out = 0.*

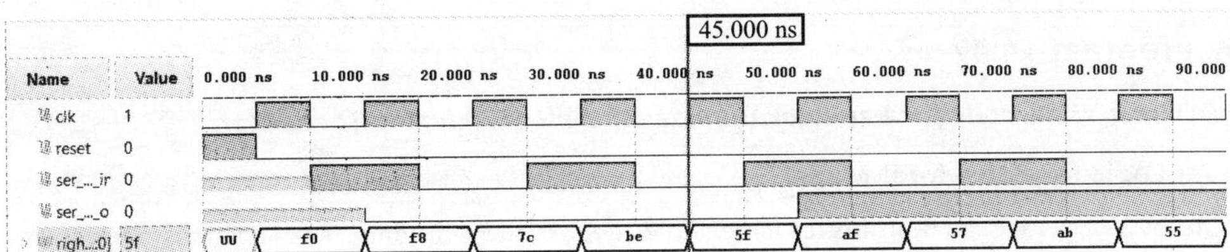

Fig. d: *ser_data_in = 0; rightshift_reg: 1011 1110(BE_H) $\xrightarrow{\text{Right shift}}$ 0101 1111($5F_H$); ser_data_out = 0.*

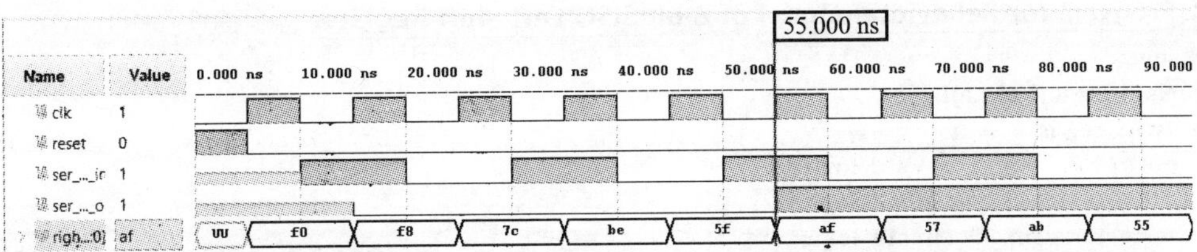

Fig. e: *ser_data_in = 1; rightshift_reg: 0101 1111($5F_H$) $\xrightarrow{\text{Right shift}}$ 1010 1111(AF_H); ser_data_out = 1.*

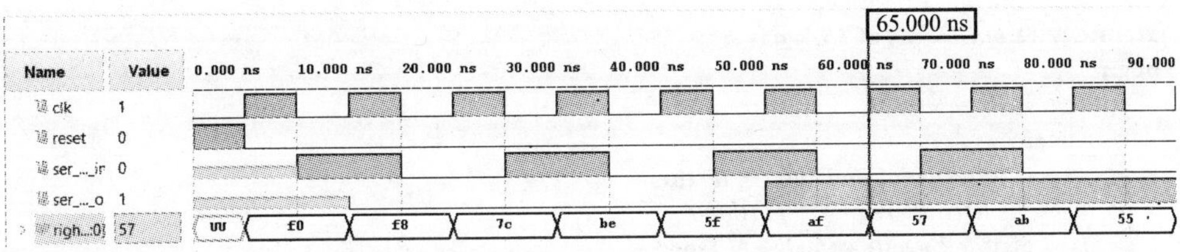

Fig. f: *ser_data_in = 0; rightshift_reg: 1010 1111(AF_H) $\xrightarrow{\text{Right shift}}$ 0101 0111(57_H); ser_data_out = 1.*

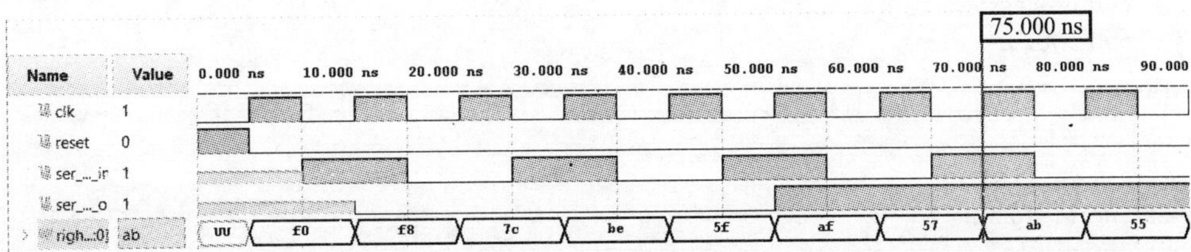

Fig. g: *ser_data_in = 1; rightshift_reg: 0101 0111(57_H) $\xrightarrow{\text{Right shift}}$ 1010 1011(AB_H); ser_data_out = 1.*

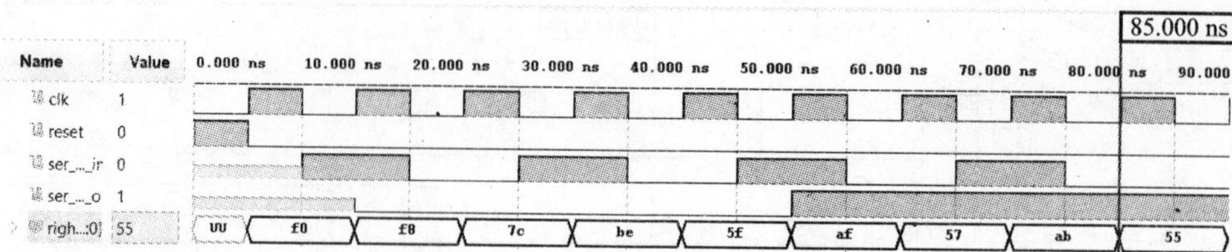

***Fig. h:** ser_data_in = 0; rightshift_reg: 1010 1011(AB_H) $\xrightarrow{\text{Right shift}}$ 0101 0101(55_H); ser_data_out = 1.*

***Fig. 1:** Input-output waveform of 8-bit SISO right shift register.*

VHDL Program 14.47

Write a VHDL code for behavioral model of 8-bit SISO left shift register.

Program Logic for Behavioral Model

In the VHDL code, an 8-bit buffer is declared to store the 8-bit data. The MSB (Most Significant Bit) of 8-bit buffer is output as serial output data The shift operation is performed by concatenation of input data bit as LSB (Least Significant Bit) and lower 7-bit of buffer. (Refer Chapter 8, Section 8.6.1 for left shift SISO register.

VHDL Program for Behavioral Model of 8-bit SISO Left Shift Register

```
library ieee;
use ieee.std_logic_1164.all;
    entity LS_reg is
        port ( clk          : in std_logic;
               reset        : in std_logic;
               ser_data_in  : in std_logic;
               ser_data_out : out std_logic;
               leftshift_reg : buffer std_logic_vector (7 downto 0));
        end LS_reg;

        architecture bm_LS_reg of LS_reg is
        begin
            process(clk, reset)
                begin
                    if (reset'event and reset = '0') then
                        leftshift_reg <= "11110000";
                    elsif (clk'event and clk = '1') then
                        ser_data_out <= leftshift_reg(7);
                    leftshift_reg <= leftshift_reg(6 downto 0) & ser_data_in;
                end if;
            end process;
            end bm_LS_reg;
```

Test Bench

```
library ieee;
use ieee.std_logic_1164.all;
use ieee.numeric_std.all;
    entity LS_reg_tb is
        --port( );
    end LS_reg_tb;
```

```
architecture bm_LS_reg_tb of LS_reg_tb is
component LS_reg
   port ( clk           : in std_logic;
          reset         : in std_logic;
          ser_data_in   : in std_logic;
          ser_data_out  : out std_logic;
          leftshift_reg : buffer std_logic_vector (7 downto 0));
end component;

signal clk            : std_logic;
signal reset          : std_logic;
signal ser_data_in    : std_logic;
signal ser_data_out   : std_logic;
signal leftshift_reg  : std_logic_vector (7 downto 0);
   begin
      uut: LS_reg port map
         ( clk           => clk,
           reset         => reset,
           ser_data_in   => ser_data_in,
           ser_data_out  => ser_data_out,
           leftshift_reg => rightshift_reg );

 clk_process : process
         begin
            clk <= '0';   wait for 5 ns;
            clk <= '1';   wait for 5 ns;
      end process;

      stimulus_process : process
      begin
         reset <= '1';          wait for 5 ns;
         reset <= '0';          wait for 5 ns;
         ser_data_in <= '1';    wait for 10 ns;
         ser_data_in <= '0';    wait for 10 ns;
         ser_data_in <= '1';    wait for 10 ns;
         ser_data_in <= '0';    wait for 10 ns;
         ser_data_in <= '1';    wait for 10 ns;
         ser_data_in <= '0';    wait for 10 ns;
         ser_data_in <= '1';    wait for 10 ns;
         ser_data_in <= '0';    wait for 10 ns;
      wait;
   end process;
   end bm_LS_reg_tb;
```

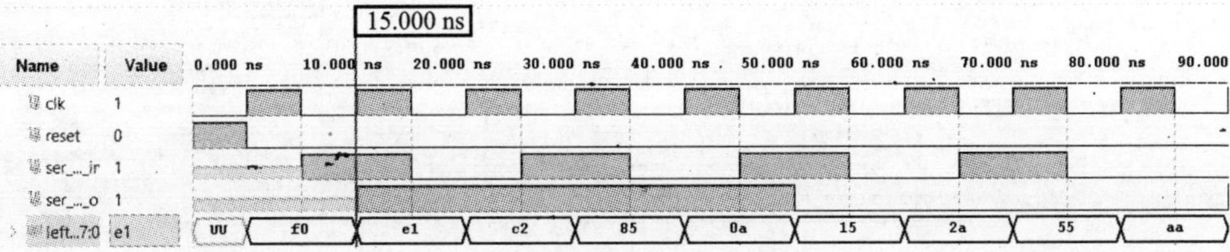

Fig. a: *ser_data_out = 1; 1110 0001($E1_H$)* ← *Left shift* — *leftshift_reg:1111 0000($F0_H$); ser_data_in = 1.*

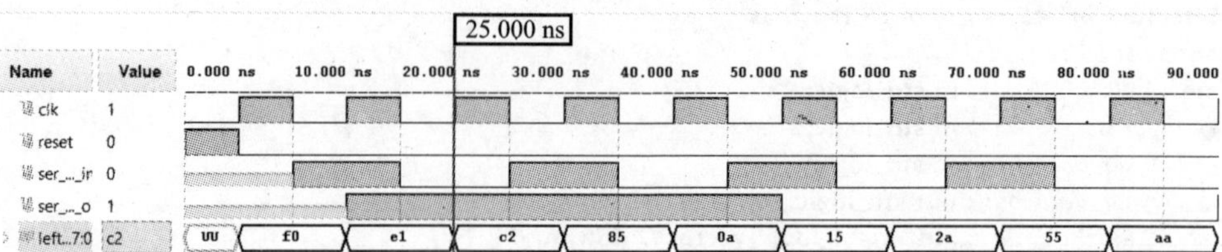

Fig. b: *ser_data_out = 1; 1100 0010($C2_H$)* $\xleftarrow{\text{Left shift}}$ *leftshift_reg:1110 0001($E1_H$); ser_data_in = 0.*

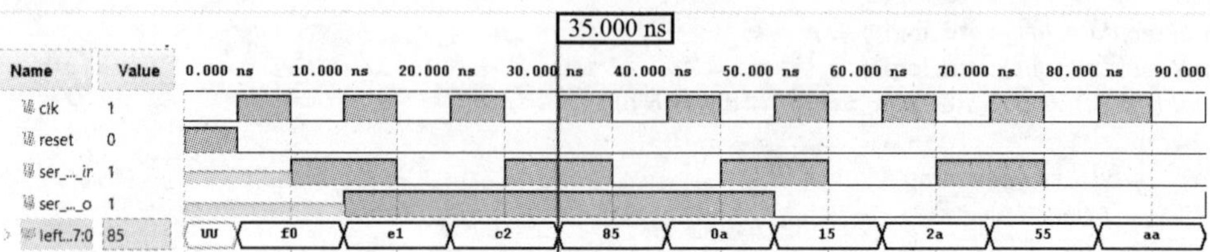

Fig. c: *ser_data_out = 1; 1000 0101(85_H)* $\xleftarrow{\text{Left shift}}$ *leftshift_reg:1100 0010($C2_H$); ser_data_in = 1.*

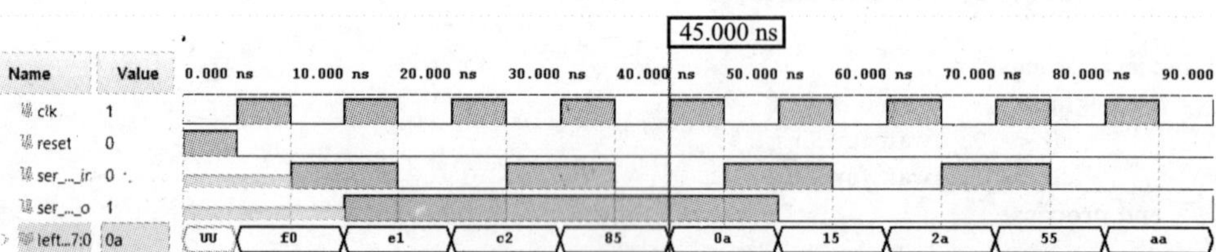

Fig. d: *ser_data_out = 1; 0000 1010($0A_H$)* $\xleftarrow{\text{Left shift}}$ *leftshift_reg:1000 0101(85_H); ser_data_in = 0.*

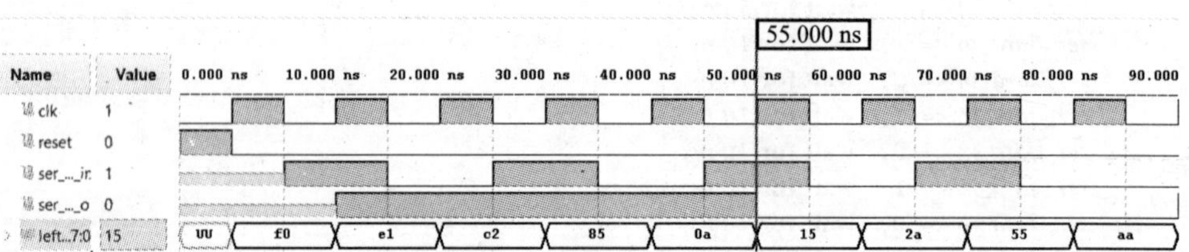

Fig. e: *ser_data_out = 0; 0001 0101(15_H)* $\xleftarrow{\text{Left shift}}$ *leftshift_reg:0000 1010($0A_H$); ser_data_in = 1.*

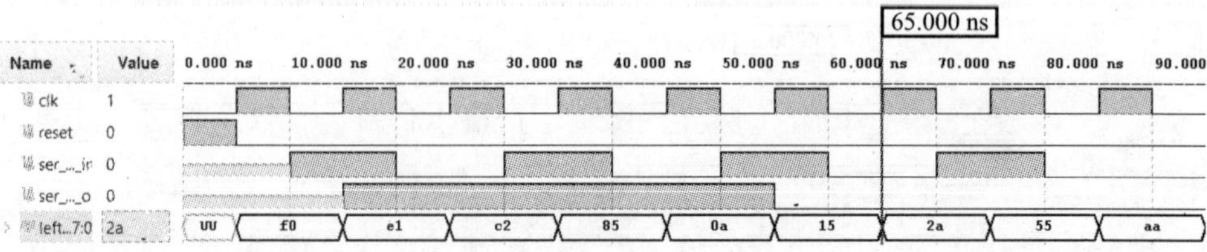

Fig. f: *ser_data_out = 0; 0010 1010($2A_H$)* $\xleftarrow{\text{Left shift}}$ *leftshift_reg: 0001 0101(15_H); ser_data_in = 0.*

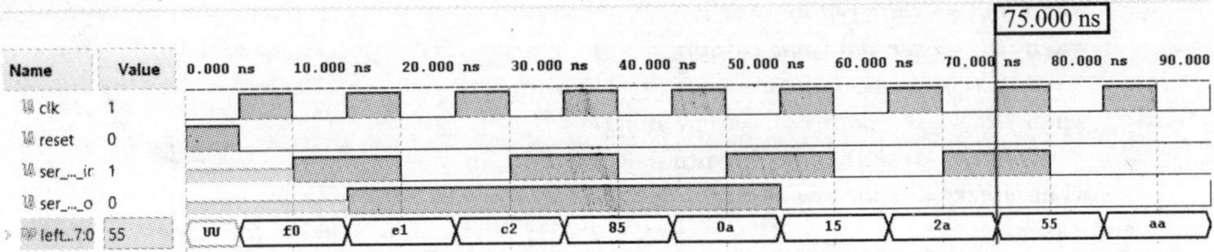

Fig. g: *ser_data_out = 0; 0101 0101(55$_H$)* ← *Left shift* — *leftshift_reg: 0010 1010(2A$_H$); ser_data_in = 1.*

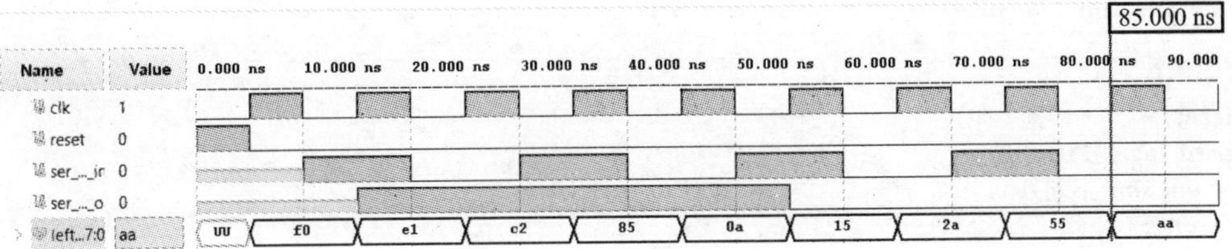

Fig. h: *ser_data_out = 0; 1010 1010(AA$_H$)* ← *Left shift* — *leftshift_reg: 0101 0101(55$_H$); ser_data_in = 0.*

Fig. 1: *Input-output waveform of 8-bit SISO left shift register.*

VHDL Program 14.48

Write a VHDL code for behavioral model of a universal shift register.

Program Logic for Behavioral Model

In the VHDL code an 8-bit buffer is used to store 8-bit data. The VHDL code for universal shift register is written to perform the following four operations. (Refer Chapter 8, Section 8.6.5 for the design of universal shift register).

- Clear buffer
- Left shift buffer
- Right shift buffer
- Load parallel data to buffer

Program for Behavioral Model of Universal Shift Register

```
library ieee;
use ieee.std_logic_1164.all;
   entity uni_shift_reg is
      port ( clk           : in std_logic;
             reset         : in std_logic;
             sel           : in std_logic_vector (1 downto 0);
             ser_data_in   : in std_logic;
             paral_data_in : in std_logic_vector(7 downto 0);
             ser_data_out  : out std_logic;
             shift_reg     : buffer std_logic_vector(7 downto 0));
   end uni_shift_reg;

   architecture bm_uni_shift_reg of univ_shift_reg is
   begin
      process (clk, reset)
         begin
            if (reset'event and reset = '0') then
            shift_reg <= "11110000";
            elsif (clk'event and clk = '1') then
            case (sel) is
```

```
                when "00" => shift_reg <= shift_reg;
                when "01" => ser_data_out <= shift_reg(0);
                    shift_reg <= ser_data_in & shift_reg(7 downto 1);
                when "10" => ser_data_out <= shift_reg(7);
                    shift_reg <= shift_reg(6 downto 0) & ser_data_in;
                when others => shift_reg <= paral_data_in(7 downto 0);
            end case;
            end if;
        end process;
    end bm_uni_shift_reg;
```

Test Bench

```
library ieee;
use ieee.std_logic_1164.all;
    entity uni_shift_reg_tb is
        --port( );
    end uni_shift_reg_tb;

    architecture bm_uni_shift_reg_tb of uni_shift_reg_tb is
        component uni_shift_reg
            port ( clk            : in std_logic;
                   reset          : in std_logic;
                   sel            : in std_logic_vector(1 downto 0);
                   ser_data_in    : in std_logic;
                   paral_data_in  : in std_logic_vector(7 downto 0);
                   ser_data_out   : out std_logic;
                   shift_reg      : buffer std_logic_vector(7 downto 0));
        end component;

        signal clk            : std_logic;
        signal reset          : std_logic;
        signal sel            : std_logic_vector (1 downto 0);
        signal ser_data_in    : std_logic;
        signal paral_data_in  : std_logic_vector(7 downto 0);
        signal ser_data_out   : std_logic;
        signal shift_reg      : std_logic_vector(7 downto 0);

        begin
        uut: uni_shift_reg port map
            ( clk    => clk,
              reset => reset,
              sel    => sel,
              ser_data_in    => ser_data_in,
              paral_data_in  => paral_data_in,
              ser_data_out   => ser_data_out,
              shift_reg      => shift_reg );
    clk_process : process
    begin
        clk <= '0';  wait for 5 ns;
        clk <= '1';  wait for 5 ns;
    end process;
```

```
stimulus_process : process
   begin
      reset <= '1';  wait for 5 ns;
      reset <= '0';  wait for 5 ns;
      sel <= "00";  wait for 10ns;
      sel <= "01";
      ser_data_in <= '1';  wait for 10 ns;
      ser_data_in <= '0';  wait for 10 ns;
      sel <= "10";
      ser_data_in <= '1';  wait for 10 ns;
      ser_data_in <= '0';  wait for 10 ns;
      sel <= "11";
      paral_data_in <= "10101010";  wait for 10ns;
      paral_data_in <= "00000000";  wait for 10ns;
      sel <= "01";
      ser_data_in <= '1';  wait for 10 ns;
      ser_data_in <= '0';  wait for 10 ns;
   wait;
end process;
end bm_uni_shift_reg_tb;
```

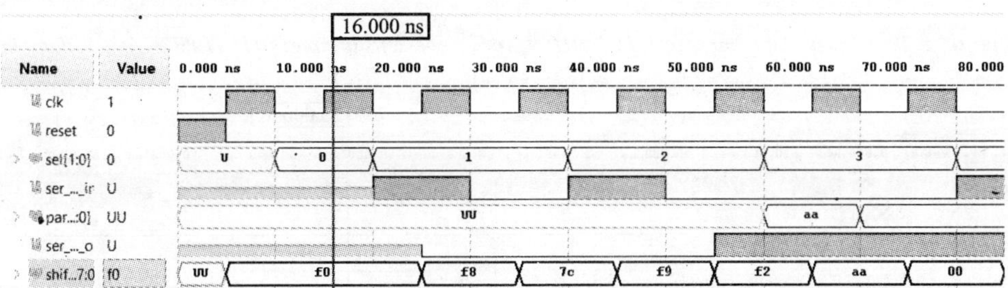

Fig. a: *Input: sel = 0 = 00; No change in shift register.*

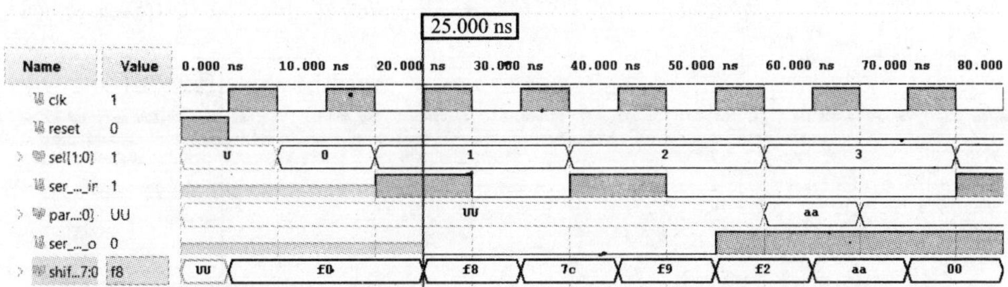

Fig. b: *Input: sel = 1 = 01; ser_data_in = 1; rightshift_reg: 1111 1111*$(F0_H) \xrightarrow{\text{Right shift}}$ *1111 1000*$(F8_H)$*; ser_data_out = 0.*

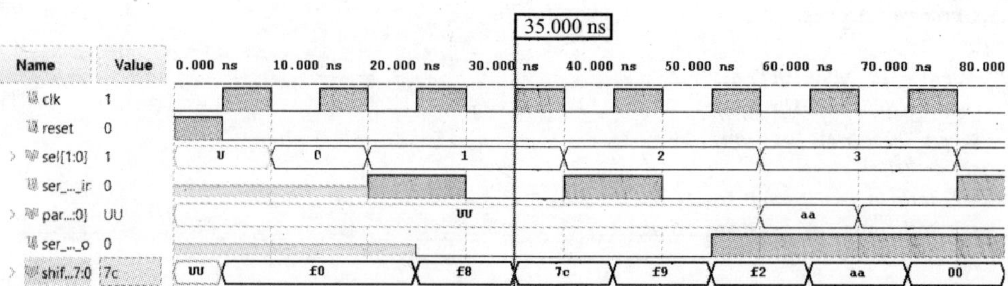

***Fig. c:** Input: sel = 1 = 01; ser_data_in = 0; rightshift_reg: 1111 1000(F8$_H$) $\xrightarrow{\text{Right shift}}$ 0111 1100(7C$_H$); ser_data_out = 0.*

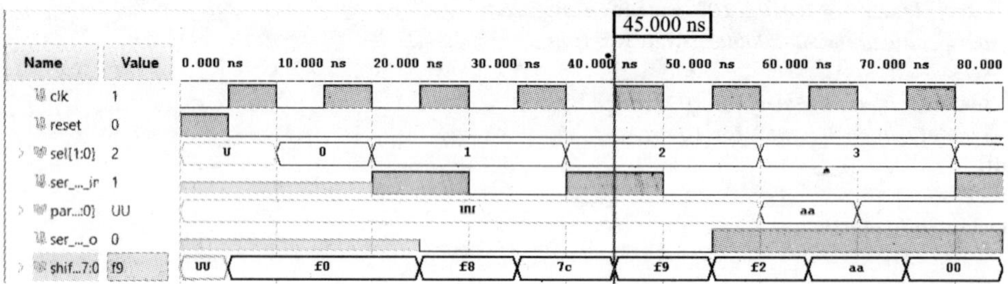

***Fig. d:** Input: sel = 2 = 01; ser_data_out = 0; 1111 1001(F9$_H$) $\xleftarrow{\text{Left shift}}$ leftshift_reg: 0111 1100(7C$_H$); ser_data_in = 1.*

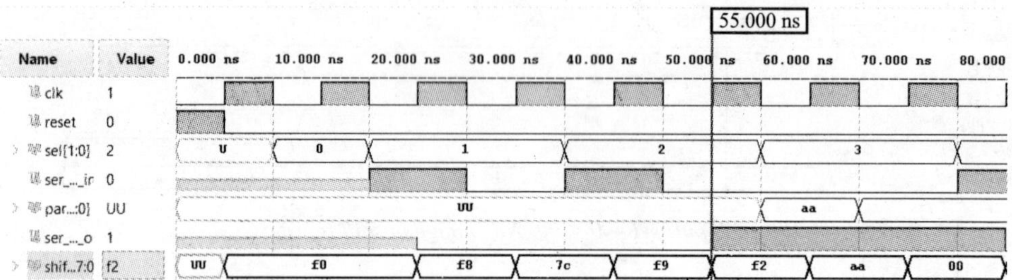

***Fig. e:** Input = sel = 2 = 10; ser_data_out = 1; 1111 0010(F2$_H$) $\xleftarrow{\text{Left shift}}$ leftshift_reg: 1111 1001(F9$_H$); ser_data_in = 0.*

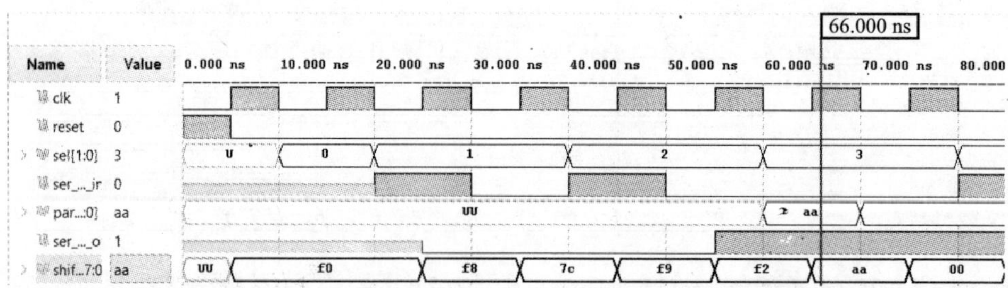

***Fig. f:** Inputs: Parallel in = AA$_H$, Sel = 3 = 11; Output: Shift register = AA$_H$.*

***Fig. 1:** Input-output waveform of universal shift register.*

14.14 VHDL Programs in RTL Methodology

VHDL Program 14.49

Write a VHDL program to perform addition of four 8-bit unsigned integer numbers by using RTL design methodology.

RTL Design Program Logic for Sum of Four Integers

Let a, d, c and d is the four integers.

Now, sum of four integers can be expressed as.

Sum = ((a + b) + c) + d

In RTL implementation, the computation task is breakdown into a sequence of steps, each one involving only a single addition and each operation is executed sequentially.

sum = 0	sum = sum + a
sum = sum + b	sum = sum + c
sum = sum + d	

Now the logic components required in data path module are one adder, a register to hold the value of sum in-between operations, a multiplexer to select the input to be added, and a circuit to clear sum at the start of computation. A simple block diagram showing data and control path of RTL design is shown in Fig. 1.

The RTL design is implemented in three modules: rtl_adder_data_path, rtl_adder_control_path , rtl_adder. The rtl_adder_data_path and rtl_adder_control_path are compiled separately and used as in rtl_adder which is the top module. The VHDL programs for the three modules are given below:

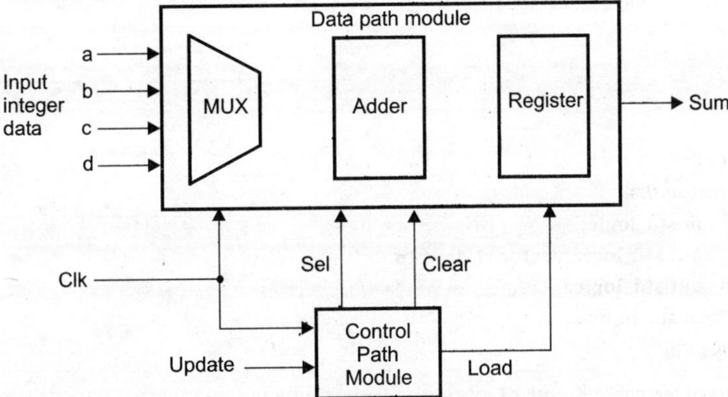

Fig. 1: Block diagram of RTL design for adding four 8-bit intergers.

VHDL Program for Adder

```
--rtl_adder_data_path.vhd
library ieee;
use ieee.std_logic_1164.all;
use ieee.std_logic_arith.all;
    entity rtl_adder_data_path is
        port ( a, b, c, d        : in unsigned (7 downto 0);
               sum               : out unsigned (7 downto 0);
               sel               : in std_logic_vector (1 downto 0);
               load, clear, clk  : in std_logic );
    end rtl_adder_data_path;
```

```
architecture arch_rtl_adder_data_path of rtl_adder_data_path is
signal mux_out, sum_reg, next_sum_reg : unsigned (7 downto 0);
constant sum_zero                     : unsigned (7 downto 0) := "00000000";
begin

-- mux to select input data
    with sel select mux_out <=
        a when "00",
        b when "01",
        c when "10",
        d when others;

-- mux to select register input
    next_sum_reg <=
        sum_reg + mux_out when load = '1' else
        sum_zero when clear         = '1' else
        sum_reg;

-- process to load sum_reg
process (clk)
    begin
        if clk'event and clk = '1' then
            sum_reg <= next_sum_reg ;
        end if;
end process;

--output reg_sum
    sum <= sum_reg ;
end arch_rtl_adder_data_path;
```

```
--rtl_adder_control_path.vhd
library ieee;
use ieee.std_logic_1164.all;
    entity rtl_adder_control_path is
        port ( update      : in std_logic;
               sel         : out std_logic_vector (1 downto 0);
               load, clear : out std_logic;
               clk         : in std_logic );
    end rtl_adder_control_path;

    architecture arch_rtl_adder_control_path of rtl_adder_control_path is
    type states is (clr, add_a, add_b, add_c, add_d, hold);
    signal cs, holdns, ns : states;
    signal tmp            : std_logic_vector (3 downto 0);
    begin
        -- select next state
        with cs select ns <=
            add_a  when clr,
            add_b  when add_a,
            add_c  when add_b,
            add_d  when add_c,
            hold   when add_d,
            holdns when others;
```

```
--process to decide current state (cs)
process (clk)
   begin
      if clk'event and clk = '1' then
         cs <= ns;
      end if;
end process;

--output control signal to select mux input
   with cs select sel<=
      "00" when add_a,       "01" when add_b,
      "10" when add_c,       "11" when others;

--output control signal to clear register
   holdns <=
      clr when update = '1' else
      hold;
   load  <= '0' when cs = clr or cs = hold else '1';
   clear <= '1' when cs = clr else '0';
end arch_rtl_adder_control_path;
```

```
--rtl_adder.vhd, top module of VHDL program
--rtl_adder_data_path.vhd and rtl_adder_control_path.vhd are added as components in this top module
library ieee;
use ieee.std_logic_1164.all;
use ieee.std_logic_arith.all;
   entity rtl_adder is
      port ( a, b, c, d   : in unsigned (7 downto 0);
            sum           : out unsigned (7 downto 0);
            update, clk : in std_logic );
   end rtl_adder;

   architecture arch_rtl_adder of rtl_adder is
   signal sel          : std_logic_vector (1 downto 0);
   signal load, clear : std_logic;

      component rtl_adder_data_path
         port ( a, b, c, d          : in unsigned (7 downto 0);
               sum                  : out unsigned (7 downto 0);
               sel                  : in std_logic_vector (1 downto 0);
               load, clear, clk : in std_logic );
      end component;
      component rtl_adder_control_path
         port ( update       : in std_logic;
               sel           : out std_logic_vector (1 downto 0);
               load, clear  : out std_logic;
               clk           : in std_logic );
      end component;

      begin
         d1 : rtl_adder_data_path port map (a => a, b => b, c => c, d => d, sum => sum,
                                          sel => sel, load => load, clear => clear, clk => clk);
         c1: rtl_adder_control_path port map (update => update, sel => sel, load => load,
                                          clear => clear, clk => clk );
   end arch_rtl_adder;
```

VHDL Program 14.50

Write a VHDL program to generate a pulse of width equal to width of four input clock pulse using RTL design methodology.

RTL Design Program Logic for Pulse Genereation

A simple block diagram showing data and control path of RTL design is shown in Fig. 1. The data path has a count register that counts the input clock pulses and compares the count with a specified pulse width count. Three input control signals start, stop and reset are used control the pulse generation process. The reset can be used as general system reset. The stop signal is used to stop the process at any desired time. The program is designed to generate an output pulse for a specified width whenever the start pulse goes high.

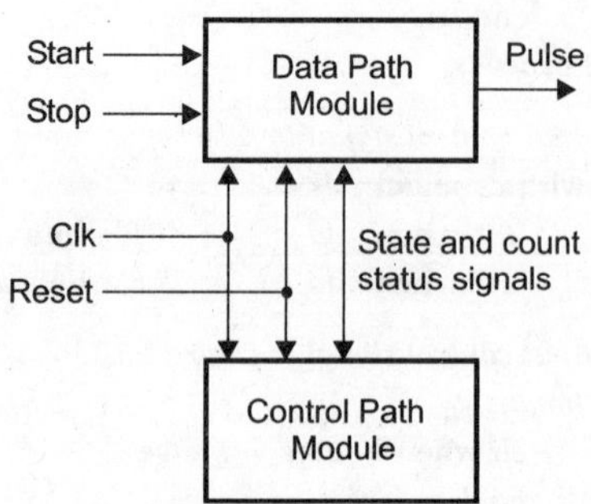

Fig. 1: *Block diagram of RTL design for FSMD implementation of one-shot pulse generator.*

The RTL design is implemented in three modules: rtl_pulse_gen_data_path, rtl_pulse_gen_control_path, rtl_pulse_gen. The rtl_pulse_gen_data_path and rtl_pulse_gen_control_path are compiled separately and used as in rtl_pulse_gen which is the top module. The VHDL programs for the three modules are given below:

VHDL Program for Pulse Genereation

```
--rtl_pulse_gen_data_path.vhd
library ieee;
use ieee.std_logic_1164.all;
use ieee.std_logic_arith.all;
use ieee.std_logic_unsigned.all;
   entity rtl_pulse_gen_data_path is
        port ( clk, reset  : in std_logic;
               start, stop : in std_logic;
               pulse       : out std_logic );
   end rtl_pulse_gen_data_path;

   architecture arch_rtl_pulse_gen_data_path of rtl_pulse_gen_data_path is
   constant pulse_width : integer := 4;
   type state_type is (idle, delay);
   signal state_reg, state_next   : state_type;
   signal count_reg, count_next : unsigned (3 downto 0);
   begin
   -- next_state logic & datapath functional units/routing
        process (state_reg, start, stop, count_reg)
            begin
                pulse <= '0';       count_next <= count_reg;
            case state_reg is
            when idle =>
                if (start = '1') then
                    state_next <= delay;
                else
                    state_next <= idle;
```

```
            end if;
                count_next <= (others => '0');

        when delay =>
            if (stop = '1') then
                state_next <= idle;
            elsif (count_reg = pulse_width - 1) then
                state_next <= idle;
            else
                state_next  <= delay;      count_next <= count_reg + 1;
            end if;
                pulse <= '1';
            end case;
        end process;
    end arch_rtl_pulse_gen_data_path;
```

```
--rtl_pulse_gen_control_path.vhd
library ieee;
use ieee.std_logic_1164.all;
use ieee.std_logic_arith.all;
use ieee.std_logic_unsigned.all;
    entity rtl_pulse_gen_control_path is
        port ( clk, reset : in std_logic );
    end rtl_pulse_gen_control_path;

    architecture arch_rtl_pulse_gen_control_path of rtl_pulse_gen_control_path is
    type state_type is (idle, delay);
    signal state_reg, state_next   : state_type;
    signal count_reg, count_next : unsigned (3 downto 0);
    begin
        process (clk, reset)
            begin
            if (reset = '1') then
                state_reg  <= idle;        count_reg <= (others => '0');
            elsif (clk'event and clk = '1') then
                state_reg  <= state_next;     count_reg <= count_next;
            end if;
        end process;
    end arch_rtl_pulse_gen_control_path;
```

```
--rtl_adder.vhd, top module of VHDL program
--rtl_pulse_gen_data_path.vhd and rtl_pulse_gen_control_path.vhd are added as components in this top module
library ieee;
use ieee.std_logic_1164.all;
use ieee.std_logic_arith.all;
use ieee.std_logic_unsigned.all;
    entity rtl_pulse_gen is
        port ( clk, reset  : in std_logic;
            start, stop : in std_logic;
            pulse       : out std_logic );
    end rtl_pulse_gen;
```

```
architecture arch_rtl_pulse_gen of rtl_pulse_gen is
constant pulse_width : integer := 4;
type state_type is (idle, delay);
signal state_reg, state_next   : state_type;
signal count_reg, count_next : unsigned (3 downto 0);
    component rtl_pulse_gen_data_path
        port ( clk, reset    : in std_logic;
               start, stop  : in std_logic;
               pulse          : out std_logic );
    end component;

    component rtl_pulse_gen_control_path
        port ( clk, reset : in std_logic );
    end component;
   begin
       d1 : rtl_pulse_gen_data_path port map (clk => clk, reset => reset, start => start, stop => stop,
                                                pulse => pulse);
       c1 : rtl_pulse_gen_control_path port map (clk => clk, reset => reset);
   end arch_rtl_pulse_gen;
```

VHDL Program 14.51

Write a VHDL program to design a UART receiver using RTL design methodology.

RTL Design Program Logic for UART receiver

Universal asynchronous receiver and transmitter (UART) is a scheme that sends bytes of data through a serial line. Normally, the serial line is in the '1' state when it is idle and the start of reception is indicated by asserting this line '0', which is considered as start bit. The start bit is followed by eight data bits and ended with a stop bit, which is ' 1'.

UART receiver uses an oversampling scheme to ensure that the data bits are retrieved at the right point. This scheme utilizes a high-frequency sampling signal to estimate the middle point of a data bit and then retrieve data bits at these points. Here, the sampling rate is 16 times the baud rate 1200 and hence, there are 16 sampling pulses for each received bit. The program uses a sampling pulse counter of size 16 bits.

The procedure to retrieve the incoming data bits is follows:

Step-1: When the serial line goes '0' indicating beginning of the start bit, initiate the sampling pulse counter.

Step-2: When the counter reaches 7, clear it to 0 and restart. At this point, the incoming signal reaches about a half of the start bit (or the middle point of the start bit).

Step-3: When the counter reaches 15, clear it to 0 and restart. At this point, the incoming signal progresses for one bit and reaches the middle of the first data bit. The data in the serial line should be retrieved and shifted into a register.

Step-4: Repeat Step 3 seven times to retrieve the remaining seven data bits.

Step-5: Repeat Step 3 one more time but without shifting. The incoming signal should reach the middle of the stop bit at this point, and its value should be '1'.

Here, the system clock is assumed as 1 MHz and the baud rate is 1200 baud. The frequency of the sampling enable pulse should be 16 * 1200, which can be obtained by a mod-52 counter ($1 \times 10^6 \div (16 \times 1200) = 52$). The syn_pulse signal is the enable pulse whose frequency is 16 times that of the baud rate.

The program uses four states idle, start, data and stop in order to keep track of various activities in the data bit receiving process. Note that the system, remains in the same state unless the syn-pulse signal is activated.

There are two counters namely, syn_pulsecount and syn_bitcount registers. The syn_pulsecount register keeps track of the number of sampling pulses.

The syn_bitcount register keeps track of the number of data bits received in the data state. The retrieved bits are shifted into and reassembled in the data bit register.

The RTL design is implemented in three modules: rtl_uart_rec_data_path, rtl_ uart_rec _control_path, rtl_ uart_rec. The rtl_ uart_rec _data_path and rtl_ uart_rec _control_path are compiled separately and used as in rtl_ uart_rec which is the top module. The VHDL programs for the three modules are given ahead:

VHDL Program for UART Receiver

```
--rtl_uart_rec_data_path.vhd
library ieee;
use ieee.std_logic_1164.all;
use ieee.std_logic_arith.all;
use ieee.std_logic_unsigned.all;
    entity rtl_uart_rec_data_path is
        port ( clk, reset, rec_bit : in std_logic;
                ready              : out std_logic;
                pout               : out std_logic_vector (7 downto 0));
    end rtl_uart_rec_data_path;

    architecture arch_rtl_uart_rec_data_path of rtl_uart_rec_data_path is
    type state_type is (idle, start, data, stop);
    signal syn_pulse                     : std_logic;
    signal state_reg, state_next         : state_type;
    signal syn_pulsecount_reg, pulse_next : unsigned (3 downto 0);
    signal syn_bitcount_reg, bit_next    : unsigned (2 downto 0);
    signal databit_reg, databit_next     : std_logic_vector (7 downto 0);
    begin
        process (state_reg, syn_pulsecount_reg, syn_bitcount_reg, databit_reg, syn_pulse, rec_bit)
            begin
                pulse_next   <= syn_pulsecount_reg;
                bit_next     <= syn_bitcount_reg;
                databit_next <= databit_reg;
                ready        <= '0';
            case state_reg is
                when idle =>
                    if (rec_bit ='0') then
                        state_next <= start;
                    else
                        state_next <= idle;
                    end if;
                        ready <= '1';

                when start =>
                    if (syn_pulse = '0') then
                        state_next <= start;
                    elsif (syn_pulsecount_reg = 7) then
```

```
                state_next <= data;       pulse_next <= (others => '0');
            else
                state_next <= start;      pulse_next <= syn_pulsecount_reg + 1;
            end if;
        when data =>
            if (syn_pulse = '0') then
                state_next <= data;
            elsif (syn_pulsecount_reg = 15) then
                pulse_next    <= (others => '0');
                databit_next <= rec_bit & databit_reg (7 downto 1);
            if (syn_bitcount_reg = 7) then
                state_next <= stop;
                bit_next    <= (others => '0');
            else
                state_next <= data;       bit_next <= syn_bitcount_reg + 1;
            end if;
            else
                state_next <= data;       pulse_next <= syn_pulsecount_reg + 1;
            end if;

        when stop =>
            if (syn_pulse = '0') then
                state_next <= stop;
            elsif (syn_pulsecount_reg = 15) then
                state_next <= idle;
                pulse_next <= (others => '0');
            else
                state_next <= stop;       pulse_next <= syn_pulsecount_reg + 1;
            end if;
      end case;
   end process;
      pout <= databit_reg;
end arch_rtl_uart_rec_data_path;
```

```
--rtl_uart_rec_control_path.vhd
library ieee;
use ieee.std_logic_1164.all;
use ieee.std_logic_arith.all;
use ieee.std_logic_unsigned.all;
   entity rtl_uart_rec_control_path is
      port (clk, reset : in std_logic);
   end rtl_uart_rec_control_path;

   architecture arch_rtl_uart_rec_control_path of rtl_uart_rec_control_path is
   type state_type is (idle, start, data, stop);
   signal syn_pulse                         : std_logic;
   signal clk16_reg, clk16_next             : unsigned (5 downto 0);
   constant dvsr                            : integer := 52;
   signal state_reg, state_next             : state_type;
   signal syn_pulsecount_reg, pulse_next : unsigned (3 downto 0);
   signal syn_bitcount_reg, bit_next        : unsigned (2 downto 0);
   signal databit_reg, databit_next         : std_logic_vector (7 downto 0);
```

```
begin
-- free running mod-52 counter, independent of fsmd.
    process (clk, reset)
        begin
            if (reset = '1') then
                clk16_reg <= (others => '0');

            elsif (clk'event and clk = '1') then
                clk16_reg <= clk16_next;
            end if;

            if (reset = '1') then
                state_reg           <= idle;
                syn_pulsecount_reg <= (others => '0');
                syn_bitcount_reg    <= (others => '0');
                databit_reg         <= (others => '0');
            elsif (clk'event and clk = '1') then
                state_reg           <= state_next;
                syn_pulsecount_reg <= pulse_next;
                syn_bitcount_reg    <= bit_next;
                databit_reg         <= databit_next;
            end if;
    end process;

--next state/output logic
        clk16_next <= (others => '0') when clk16_reg = (dvsr - 1)
    else clk16_reg + 1;
        syn_pulse <= '1' when clk16_reg = 0 else '0';
end arch_rtl_uart_rec_control_path;
```

```
--rtl_rtl_uart_rec.vhd top module of VHDL program
--rtl_ uart_rec._data_path.vhd and rtl_ uart_rec. control_path.vhd are added as components in this top module
library ieee;
use ieee.std_logic_1164.all;
use ieee.std_logic_arith.all;
use ieee.std_logic_unsigned.all;
  entity rtl_uart_rec is
      port ( clk, reset, rec_bit : in std_logic;
          ready            : out std_logic;
          pout             : out std_logic_vector (7 downto 0) );
  end rtl_uart_rec;
  architecture arch_rtl_uart_rec of rtl_uart_rec is
      component rtl_uart_rec_data_path
          port ( clk, reset, rec_bit : in std_logic;
              ready            : out std_logic;
              pout             : out std_logic_vector (7 downto 0) );
      end component;

      component rtl_uart_rec_control_path
          port (clk, reset: in std_logic);
      end component;
```

```
begin
    d1 : rtl_uart_rec_data_path port map (clk => clk, reset => reset, rec_bit => rec_bit,
                                          ready => ready, pout => pout);
    c1 : rtl_uart_rec_control_path port map (clk => clk, reset => reset);
end arch_rtl_uart_rec;
```

14.15 Summary of Important Concepts

1. VHDL is an acronym for VHSIC (Very High Speed Integrated Circuits) Hardware Description Language.
2. VHDL is used to model a digital system at multiple levels of abstraction.
3. VHDL supports four levels of abstraction for digital system design: Transistor level, gate level, register-transfer level (RTL), processor level.
4. In transistor level abstraction, basic building blocks are basic electronic components like transistors, resistors, capacitors, etc.
5. In gate level abstraction, basic building blocks are logic gates, flip-flops, multiplexers, etc.
6. In register-transfer level (RTL) abstraction, the basic building blocks are modules developed using registers, adders, multiplexers, etc.
7. The RTL abstraction can be called as module-level abstraction.
8. In processor level abstraction, basic building blocks are processors, memory modules, bus interfaces, etc.
9. The basic elements of the VHDL language are literals, comments, identifiers, data objects and operators.
10. Explicit data values that are assigned to objects or used within expressions are called literals.
11. Physical literals are special types of literals used to represent physical quantities such as time, voltage, current, distance, etc.
12. A comment starts with a minimum of two dashes (--) followed by the text to explain the concepts.
13. The VHDL IEEE standard package defines a number of standard data types which are known as predefined data types.
14. Integer data type is integer decimal number in the range $-(2^{31}-1)$ to $+(2^{31}-1)$.
15. Enumeration type is associating user defined set of values called enumeration literals to an object.
16. A composite data type represents a collection of signals or objects.
17. Composite types can be further classified into array types and record types.
18. An array type represents a collection of signals or objects belong to a same data type.
19. A record type represents a collection of signals or objects belong to same or different data types.
20. Data stored as a file in the host environment (the system that simulate and synthesize the code) is called file data type.
21. A constant object is assigned a fixed value and cannot be changed in other part of the code.
22. A file with a list of input and/or output data can be declared as an object.
23. The predefined operators in the VHDL language are classified into five categories: logical operators, relational operators, adding operators, multiplying operators, miscellaneous operators.
24. The concurrent statements are used in dataflow architecture.
25. Signal assignment statements that appear outside of a process are called concurrent signal assignment statements and they are used to assign a value to an object.

26. Assertion statements that appear outside of a process are called concurrent assertion statements and they are used to report the error conditions.
27. Configuration is the process of binding a component with an entity and architecture.
28. A package declaration is used to store a set of common declarations like components, types, procedures, and functions.
29. A package body is always associated with a package declaration.
30. A dataflow model specifies the functionality of a design entity or model without specifying the structure of the design.
31. The behavior of the entity is defined using a set of sequentially executed statements.
32. A process statement is the main mechanism used to model the behavior of an entity.
33. The association-list associates signals in the entity called actuals, with the ports of a component, called locals.
34. A package provides a convenient method to store and share declarations that are common to many design units.
35. A Package is a VHDL file, contains data types, constants, functions, procedures, etc, that can be shared across many VHDL designs.
36. A library is a collection of packages that are compiled and stored in a directory in the host environment (or personal computer) in which the VHDL software tool is installed.
37. The two most commonly used libraries are IEEE and WORK.
38. A test bench is also a VHDL program that is used to verify the correctness of the VHDL program of the hardware model.

14.16 Short-Answer Questions

Q14.1 What are components in VHDL?

Components are reusable VHDL modules compiled as separate entity and can be used in any other structural model of digital systems.

Instead of coding a complex design in single VHDL Code, it can be divided into sub-modules as components and combine them using port map technique in a main module.

Q14.2 List the operators in VHDL.

The predefined operators in the VHDL language are classified into the following five categories.

- Logical operators
- Relational operators
- Adding operators
- Multiplying operators
- Miscellaneous operators

Q14.3 Define data objects in VHDL.

Data object or simply object is an identifier that holds a specific data type, which is declared while creating the object. A data object is created by declaring the class, identifier/name, data type and optional initial value.

Q14.4 List the language elements of VHDL. ***(AU, Nov/Dec'18, 2 Marks)***

The VHDL language has been developed by integrating the following languages:

- Sequential language
- Concurrent language
- Net-list language
- Timing specifications
- Waveform generation language

Q14.5 Write a VHDL program for an 2-input AND gate.

```
library ieee;
use ieee.std_logic_1164.all;
entity and_2in is
    port ( x, y : in std_logic;
           v    : out std_logic );
end and_2in;

architecture dfm_and_2in of and_2in is
begin
    v <= x and y;
end dfm_and_2in;
```

Q14.6 Write a VHDL program for an 2-input NAND gate.

```
library ieee;
use ieee.std_logic_1164.all;
entity nand_2in is
    port ( x, y : in std_logic;
           v    : out std_logic );
end nand_2in;

architecture dfm_nand_2in of nand_2in is
begin
    v <= x nand y;
end dfm_nand_2in;
```

Q14.7 Write a VHDL program for an 2-input OR gate.

```
library ieee;
use ieee.std_logic_1164.all;
entity or_2in is
    port ( x, y : in std_logic;
           v    : out std_logic );
end or_2in;

architecture dfm_or_2in of or_2in is
begin
    v <= x or y;
end dfm_or_2in;
```

Q14.8 Write a VHDL program for an 2-input NOR gate.

```
library ieee;
use ieee.std_logic_1164.all;
entity nor_2in is
    port ( x, y : in std_logic;
           v    : out std_logic );
end nor_2in;
```

```
architecture dfm_nor_2in of nor_2in is
begin
    v <= x nor y;
end dfm_nor_2in;
```

Q14.9 Write a VHDL program for an 2-input XOR gate.

```
library ieee;
use ieee.std_logic_1164.all;
entity xor_2in is
    port ( x, y : in std_logic;
           v    : out std_logic );
end xor_2in;
architecture dfm_xor_2in of xor_2in is
begin
    v <= x xor y;
end dfm_xor_2in;
```

Q14.10 Write a VHDL program for an 2-input XNOR gate.

```
library ieee;
use ieee.std_logic_1164.all;
entity xnor_2in is
    port ( x, y : in std_logic;
           v    : out std_logic );
end xnor_2in;
architecture dfm_xnor_2in of xnor_2in is
begin
    v <= x xnor y;
end dfm_xnor_2in;
```

Q14.11 Write a VHDL program for an 2-input XNOR gate using behavioral model. ***(AU, Apr/May'19, 2 Marks)***

```
library ieee;
use ieee.std_logic_1164.all;
entity xnor_2in is
    port ( x, y : in std_logic;
           v    : out std_logic );
end xnor_2in;
architecture bm_xnor_2in of xnor_2in is
begin
  process (x, y)
    begin
        v <= x xnor y;
  end process;
end bm_xnor_2in;
```

Q14.12 Write a VHDL program for an 1-input NOT gate.

```
library ieee;
use ieee.std_logic_1164.all;
entity not_1in is
    port ( x : in std_logic;
           v : out std_logic );
end not_1in;
architecture dfm_not_1in of not_1in is
begin
    v <= not x;
end dfm_not_1in;
```

Q14.13 Formulate the syntax of process statement in VHDL.

Syntax of process statement

```
process_label : process (sensitivity list)
        -- declaration statements;
        begin
        -- sequential statements;
        end process (process_label);
```

Q14.14 Formulate the syntax of selected signal assignment statement in VHDL.

Syntax of selected signal assignment statement

```
with select_expression select
signal_name <= value_exp_1 when choice_1,
               value_exp_2 when choice_2,
               .
               .
               value_exp_n when choice_n;
```

Q14.15 Formulate the syntax of conditional signal assignment statement in VHDL.

Syntax of conditional signal assignment statement

```
signal_name <= value_exp_1 when condition_1 else
               value_exp_2 when condition_2 else
               .
               .
               value_exp_n
```

Q14.16 State the purpose of test bench. **(AU, Apr/May'19, 2 Marks)**

A test bench is also a VHDL program that is used to verify the correctness of the VHDL program of the hardware model. A test bench VHDL program has three main purposes:

1. Generation of necessary inputs and clock for simulation.
2. Apply this inputs and clock to the entity under test and to monitor the output
3. Comparing the output of VHDL program model with expected output.

Q14.17 Translate the following VHDL statement using if else statement.

A <= B1 when C1 = 1' else

B2 when C2 = 1' else

B3 when C3 = 1'

where all the signals are of type bit.

```
library ieee;
use ieee.std_logic_1164.all;
   entity signals_of_type_bit is
      port ( b1, b2, b3, c1, c2, c3 : in std_logic;
         a                          : out std_logic );
   end signals_of_type_bit;

   architecture bm_signals_of_type_bit of signals_of_type_bit is
      begin
         process (c1, c2, c3)
```

```
        begin
            if c1 = '1' then
                a <= b1;
            elsif c2 = '1' then
                a <= b2;
            elsif c3 = '1' then
                a <= b3;
            end if;
        end process;
    end bm_signals_of_type_bit;
```

14.17 Exercises

I. Fill in the blanks

1. The RTL abstraction can be called ______.
2. Explicit data values that are assigned to objects or used within expressions are called ______.
3. The VHDL IEEE standard package defines a number of standard data types which are known as _____.
4. Integer data type is integer decimal number in the range ______.
5. A composite data type represents a collection of ______ or ______.
6. Composite types can be further classified into ______ and ______.
7. Data stored as a file in the host environment is called ______.
8. The concurrent statements are used in ______.
9. Signal assignment statements that appear outside of a process are called ______ statements.
10. ______ is the process of binding a component with an entity and architecture.
11. A ______ is always associated with a package declaration.
12. A ______ is the main mechanism used to model the behavior of an entity.
13. A ______ is a collection of packages that are compiled and stored in a directory in the host environment (or personal computer) in which the VHDL software tool is installed.
14. The two most commonly used libraries are ______ and ______.
15. A ______ is also a VHDL program that is used to verify the correctness of the VHDL program of the hardware model.

Answers

1. module-level abstraction	6. array types, record types	11. package body
2. literals	7. file data type	12. process statement
3. predefined data types	8. data flow architecture	13. library
4. $-(2^{31}-1)$ to $+(2^{31}-1)$	9. concurrent signal assignment	14. IEEE, WORK
5. signals or objects	10. configuration	15. test bench

II. State whether the following statements are True or False

1. VHDL is used to model a digital system at multiple levels of abstraction.
2. A comment starts with a minimum of two dashes (--) followed by the text to explain the concepts.
3. Character literals are 1-character ASCII values that are enclosed in double-quotes.
4. The general range of integers is – 2,147,483,647 to + 2,147,483,647.
5. The general range of real number is – 1.0e39 to + 1.0e39.

6. An array type represents a collection of signals or objects of same data type.
7. A record type represents a collection of signals or objects belong to different data types.
8. Reserved words of VHDL language can be used as identifiers.
9. String literals are collections of one or more ASCII characters enclosed in double-quote characters.
10. Special characters are allowed in identifiers.
11. An entity can have more than one architecture.
12. A process is the basic unit of execution in VHDL.
13. Functions and procedures are both same.
14. Procedure can return any number of values.

Answers

1. True	6. True	11. True
2. True	7. True	12. True
3. False	8. False	13. False
4. True	9. True	14. True
5. False	10. False	

III. Choose the right answer for the following questions

1. *The general range of integers is ______*

   a) – 2,147,483,645 to + 2,147,483,645 b) – 2,147,483,646 to + 2,147,483,646
   c) – 2,147,483,647 to + 2,147,483,647 d) – 2,147,483,648 to + 2,147,483,648

2. *The general range of real number is ______*

   a) – 1.0e37 to + 1.0e37 b) – 1.0e38 to + 1.0e38
   c) – 1.0e39 to + 1.0e39 d) – 1.0e40 to + 1.0e40

3. *Which one of the following is a VHDL relational operator.*

   a) NOT b) NAND c) AND d) Equality

4. *What is the standard form of SLL?*

   a) Shift left logical b) Shift line logical c) Side left logical d) None of the above

5. *Loop is a ______ statement.*

   a) Concurrent b) Sequential c) Assignment d) Functional

6. *The addition operators in VHDL are categorized into _____ types?*

   a) Two b) Three c) Four d) Five

7. *What is the expansion of HDL?*

   a) Hardware Describe Language b) Hardware Definition Language
   c) Hardware Description Language d) None of the above

8. *Which one of the following is not a reserverd word in VHDL?*

   a) Constant b) Identifier c) Variable d) Signal

9. *signal x : std_logic; In this statement x is*

   a) Variable b) Identifier c) Name d) Literal

10. ***Which of the following keyword is not associated with IF statement?***

    a) else b) then c) elsif d) when

11. ***What is the correct syntax for using assertion statement?***

    a) **assert** *condition* **report** *"string_expression"* **severity** *"string_expression"*;

    b) *condition* **report** *"string_expression"* **severity** *"string_expression"* **assert**;

    c) *condition* **severity** *"string_expression"* **report** *"string_expression"* **assert** ;

    d) **assert** *condition* **severity** *"string_expression"* **report** *"string_expression "* ;

12. ***Which of the following is not a type of VHDL modeling?***

    a) Behavioral modeling b) Dataflow modeling c) Structural modeling d) Component modeling

13. ***Which of the following is basic unit of behavioral description?***

    a) Structure b) Sequence c) Process statement d) Dataflow

14. ______ ***modeling uses logic gates and basic blocks to describe the functionality of system.***

    a) Behavioral modeling b) Dataflow modeling c) Structural modeling d) Component modeling

Answers

1. c	6. a	11. a
2. b	7. c	12. d
3. d	8. b	13. c
4. a	9. b	14. b
5. b	10. d	

IV. Answer the following questions

1. List the modules of HDL.
2. Explain the various data objects supported by VHDL. Give the necessary examples.
3. Define the following terms relevant to VHDL.

   1. Parameters 2. Keywords 3. Constants 4. Identifiers
4. Explain the difference in program structure of VHDL and any other procedural language. Give an example.
5. Explain various architectural bodies/modeling styles in VHDL with examples.
6. Explain the various data types supported by VHDL. Give the necessary examples.
7. Explain the program structure of VHDL with the help of block diagram.
8. List out the different operators available in VHDL. Explain with example.
9. Explain the different concurrent statements and sequential statements in VHDL?
10. Discuss about the difference between CASE and IF statements in VHDL with examples?
11. List the various advantages and features of VHDL.
12. Describe RTL in HDL with example.
13. How the package declaration is different from entity declaration? Give the syntax for each?
14. Give the syntax for package declaration and package body in VHDL.

V. Write Solution for Following VHDL Programs

E14.1 Design a 4-bit unsigned adder in VHDL.

E14.2 Translate the 4:1 mux if else statement to case statement.

```
if s = "00" then
    y <= i0;
elsif s = "01" then
    y <= i1;
elsif s = "10" then
    y <= i2;
else
    y <= i3;
end if;
```

E14.3 Translate the following code to a case-when statement.

```
with s select
    d <= "0000" when "00",
         "0001" when "01",
         "0010" when "10",
         "0011" when "11",
         "1111" when others;
```

E14.4 Translate the following code to, two with-select-when statements.

```
case state is
    when "00"     => a <= "1111"; b <= "0000";
    when "01"     => a <= "0001"; b <= "0010";
    when "10"     => a <= "1000"; b <= "1001";
    when others => a <= "1100"; b <= "1010";
end case;
```

E14.5 Find the errors in the following code.

```
architecture has_errors of design
begin
    p1 : process
        begin
            if clk'event and clk = '1' then
                q <= a or b and c;
            end if;
        end;
end;
```

E14.6 Write the VHDL code for T flip-flop in behavioral model.

E14.7 Write a VHDL code to demonstrate the use of variables in counter design.

E14.8 Write a VHDL code to perform bit-by-bit logical and of an 8-bit data using variables.

E14.9 Write a VHDL code for D flip-flop with rising-edge clock and asynchronous reset and preset.

E14.10 Write a VHDL code for traffic light sequencer.

E14.11 Write a VHDL code for the following Boolean expressions.

$$E = A + BC + B'D \quad ; \quad F = B'C + BC'D'$$

E14.12 Write an example VHDL code to demonstrate the use of for-loop statement.

E14.13 Write a VHDL code for logical not gate using variables.

E14.14 Write a VHDL program to generate sum of continuous integer input data.

E14.15 Write VHDL code for dataflow model of a 2-bit comparator.

E14.16 Write VHDL code for dataflow model of a 2*1 multiplexer.

CHAPTER 15
Verilog

15.1 Introduction to Verilog

Hardware **D**escription **L**anguages (HDLs) are modelling language specially developed to describe digital hardware structures and to study the behavior of logic circuits. It can be used to implement logic circuits, truth tables, Boolean functions and state tables/diagrams as computer programs for simulation and verification.

Verilog is a standardized **H**ardware **D**escription **L**anguage (HDL). The name Verilog is the combination of two words "verification" and "logic".

Verilog was first developed as a proprietary HDL by Gateway automation system in the period 1983 to 1990 and later acquired by Cadence design system in 1990. In order to popularize the Verilog, the Cadence system, transfer the Verilog to a public domain under the name Open Verilog International (OVI). Verilog was later submitted to IEEE for standardization and became IEEE Standard 1364-1995. Since 1995, IEEE keeps on updating the Verilog and latest update released in December-2023.

The basic structure of Verilog language is a module. Every Verilog program file will have a module which starts with keyword ***module*** and end with keyword ***endmodule***. The entire Verilog code or program is enclosed by the keywords ***module*** and ***endmodule*** and every line of the code is terminated by a semicolon (;).

Verilog language supports designing a module using several coding styles. Depending on the needs of a design, internals of each module can be defined at different levels of abstractions. Irrespective of the internal abstraction level, the module would behave exactly in the similar way to the external environment.

The various levels of abstraction are common to VHDL and Verilog and hence refer Chapter 14, Section 14.1 for various levels of abstraction.

15.2 Elements of Verilog Language

The coding style of Verilog mostly resembles C and C++ language. Verilog is case sensitive language and hence lower case and upper case characters are distinguishable. The blank spaces are ignored but blank spaces are not allowed in identifiers, keywords and numbers.

The basic elements of Verilog language are,

- Identifiers
- Keywords
- Comments
- Logic values
- Numbers
- Operators
- Data types

15.2.1 Identifiers

Identifiers are name of objects that are used in Verilog programs. Identifiers are formed using English letters/alphabets, decimal numbers and underscore. Special characters except dollar sign ($) are not allowed in identifiers. Reserved words of Verilog language cannot be used as identifiers. The list of reserved words is listed in Table 15.1.

Identifiers are case sensitive and hence uppercase and lower case letters are not equivalent. Identifiers that start with a dollar sign refer to built-in system functions.

The identifiers have to satisfy the following rules.

- Identifiers can have only English letters, decimal numbers, dollar and underscore.
- First character of an identifier must be an alphabet or an underscore.
- Last character of an identifier cannot be an underscore.
- Identifiers cannot have two underscore in succession.

15.2.2 Keywords

Keywords are predefined lowercase identifiers that define the language constructs. Verilog defines many special character strings, called reserved words or keywords and they are listed in Table 15.1.

Table 15.1 List of Keywords

always	inout	rtranif0	and	input	rtranif1
assign	integer	scalared	begin	join	small
buf	large	specify	buffif0	macromodule	specparam
buffif1	medium	strength	case	module	strong0
casex	nand	strong1	casez	negedge	supply0
cmos	nmos	supply1	deassign	nor	table
default	not	task	defparam	notif0	time
disable	notif1	tran	edge	or	tranif0
else	output	tranif1	and	parameter	tri
endcase	pmos	tri0	andmodule	posedge	tri1
endfunction	primitive	triand	endprimitive	pull0	trior
endspecify	pull1	trireg	endtable	pulldown	vectored
endtask	pullup	wait	event	rcmos	wand
for	real	weak0	force	realtime	weak1
fork	reg	while	function	release	wire
highz0	repeat	nor	highz	rnmos	xnor
if	rpmos	xor	initial	rtran	

15.2.3 Comments

Comments are texts that are added as explanations, wherever necessary. Comments can be single line comment or multiline comments. A single line comment starts with a minimum of two slashes (//) followed by the text to explain the concepts. Multiline begin with /* and terminate with */. The comments are provided for explaining the various concepts of Verilog program and hence, comments are ignored while compiling the code.

15.2.4 Logic Values in Verilog

For any logical variable or signal, Verilog supports four logic values: 0, 1, x, z. The meaning of logic values are listed in Table 15.2.

Table 15.2: Logic Values Supported by Verilog

Logic Value	Description
0	Logic zero or false condition
1	Logic one or true condition
x, X	Unknown or uninitialized
z , Z	High impedance, tri-stated or floating

15.2.5 Number Representation in Verilog

In Verilog, the binary, octal, decimal and hexadecimal numbers are represented by the following format.

Number format: n'Bv

where, n = Number of binary digits used to represent the number

B = Base of number

v = Value of the number

The letters b, o, d and h are used to represent the base of binary, octal, decimal and hexadecimal respectively. Some examples of numbers are listed in Table 15.3.

Table 15.3: Examples of Number Representation in Verilog

Number System	Example	Explanation
Binary	8'b00001111	8-bit binary number 00001111
Octal	12'o4532	4 digit octal number 4532 represented by 12 bit binary
Decimal	6'd48	2 digit decimal number 48 represented by 6 bit binary
Hexadecimal	16'h4A7C	4 digit hexa number 4A7C represented by 16 bit binary

15.3 Operators in Verilog Language

The predefined operators in the Verilog HDL are classified into the following seven categories. The operators of Verilog are listed in Table 15.4 and the precedence of Verilog operators are listed in Table 15.5.

1. Bitwise/Reduction logical operators
2. Logical operators
3. Arithmetic operators
4. Relational operators
5. Shift operators
6. Equality operators
7. Other operators

Table 15.4: Operators of Verilog

<table>
<tr><th>Operator Type</th><th>Operator Symbol</th><th>Operation Performed</th></tr>
<tr><td rowspan="8">Bitwise</td><td>~</td><td>Bitwise NOT (1's complement)</td></tr>
<tr><td>&</td><td>Bitwise AND (or reduction AND)</td></tr>
<tr><td>|</td><td>Bitwise OR (or reduction OR)</td></tr>
<tr><td>^</td><td>Bitwise XOR (or reduction XOR)</td></tr>
<tr><td>~&</td><td>Bitwise NAND (or reduction NAND)</td></tr>
<tr><td>~|</td><td>Bitwise NOR (or reduction NOR)</td></tr>
<tr><td>~^</td><td rowspan="2">Bitwise XNOR (or reduction XNOR)</td></tr>
<tr><td>^~</td></tr>
<tr><td rowspan="3">Logical</td><td>!</td><td>NOT</td></tr>
<tr><td>&&</td><td>AND</td></tr>
<tr><td>||</td><td>OR</td></tr>
<tr><td rowspan="6">Arithmetic</td><td>+</td><td>Addition</td></tr>
<tr><td>-</td><td>Subtraction</td></tr>
<tr><td>-</td><td>2's complement</td></tr>
<tr><td>*</td><td>Multiplication</td></tr>
<tr><td>/</td><td>Division</td></tr>
<tr><td>**</td><td>Exponentiation</td></tr>
<tr><td rowspan="4">Relational</td><td>></td><td>Greater than</td></tr>
<tr><td><</td><td>Less than</td></tr>
<tr><td>>=</td><td>Greater than or equal to</td></tr>
<tr><td><=</td><td>Less than or equal to</td></tr>
<tr><td rowspan="4">Shift</td><td>>></td><td>Logical right shift</td></tr>
<tr><td><<</td><td>Logical left shift</td></tr>
<tr><td>>>></td><td>Arithmetic right shift</td></tr>
<tr><td><<<</td><td>Arithmetic left shift</td></tr>
<tr><td rowspan="4">Equality</td><td>= =</td><td>Equality</td></tr>
<tr><td>!=</td><td>Inequality</td></tr>
<tr><td>= = =</td><td>Case equality</td></tr>
<tr><td>!= =</td><td>Case inequality</td></tr>
<tr><td>Concatenation</td><td>{, }</td><td>Concatenation</td></tr>
<tr><td>Replication</td><td>{n{m}}</td><td>Replicate value m for n times</td></tr>
<tr><td>Conditional</td><td>? :</td><td>Conditional</td></tr>
</table>

The bitwise logical operators will perform bit by bit logical operation of two n-bit data and generates an n-bit result. The reduction operators will perform logical operation on n-bit data and generates 1-bit result either true or false. The bitwise and reduction operations are used in assignment statements. The logical operators are used in conditional statements to operate on two variables and determine the true or false condition.

The arithmetic operators are used to perform binary arithmetic. The concatenation operator is used to combine or join two binary data. The conditional operator is used to check true or false condition of a variable and execute a statement for true condition or execute an alternative statement for false condition. The format of condition statement with condition operator is,

A ? statement_B : statement_C ;

If A is true then statemet_B is executed and if A is false then statement_C is executed. This condition statement is another form of if-else statement.

Table 15.5: Verilog Operator Precedence

<table>
<tr><th>Precedence</th><th>Operators</th></tr>
<tr><td>Highest</td><td>+ – ! ~ & ~& | ~| ^ ~^ ^~ (unary)</td></tr>
<tr><td></td><td>**</td></tr>
<tr><td></td><td>* / %</td></tr>
<tr><td></td><td>+ – (binary)</td></tr>
<tr><td></td><td><< >> <<< >>></td></tr>
<tr><td></td><td>== != === !==</td></tr>
<tr><td></td><td>& (binary) (Bitwise, Reduction AND)</td></tr>
<tr><td></td><td>^ ^~ ~^ (Binary) (Bitwise, Reduction OR)</td></tr>
<tr><td></td><td>| (binary)</td></tr>
<tr><td></td><td>&& (logical AND)</td></tr>
<tr><td></td><td>|| (logical OR)</td></tr>
<tr><td></td><td>?: (conditional operator)</td></tr>
<tr><td>Lowest</td><td>{}, {{, }} (concatenation)</td></tr>
</table>

15.4 Data Types in Verilog

In Verilog every signal, constant, variable and function must be assigned a data type. The data type defines the nature of information that a signal or variable can hold. Verilog provides a variety of predefined data types. The data types of Verilog can be classified into two groups: Net type and Variable type.

The following are some of the common data types in Verilog.

- wire
- wand
- wor
- tri
- trior
- tri0
- tril
- supply0
- vector
- real
- Integer
- reg

- trireg
- triand
- supply1
- bit
- time
- parameter

Scalars and Vectors

The data types can be a single bit or group of bits. A single bit data type is called scalar and group of bits are called vectors. A vector is n-bit wide binary. All data types net, variables and constants can be declared as vectors.

Syntax of vector

```
data_type [msb : lsb] vector_name;
```

Example of vector

```
reg [3 : 0] P_state;    // P_state is a 4-bit vector of data type reg.
wire [7 : 0] sum;       // sum is an 8-bit vector of data type wire.
```

Individual bit of a vector can be accessed by its bit index. For example, if a variable named count is declared as 4-bit vector then each bit of count can be accessed as count[3], count[2], count[1] and count[0].

Array

An array is a set of variables of the same type. In Verilog, arrays allow grouping of multiple variables of the same type under a single name and each variable accessible by an index. Arrays are useful for storing collections of related data and individual data is accessed by an array index. The elements of array can be 1-bit or n-bit wide.

Syntax of array

```
reg             identifier [start : end];
reg [msb : lsb] identifier [start : end];
integer         identifier [start : end];
```

Example of array

```
integer [3 : 0] count [0 : 7];    // array of 8 variables named count of size 4-bits.
reg [7 : 0] port_id [0 : 7];      // array of 8 variables named port_id of size 8-bits.
```

Individual elements of an array can be accessed by its array index. For example, if a variable named port_id is declared as an array of 8 variables then each element of the array port_id can be accessed as port_id [7], port_id [6], port_id [5]……… port_id [0].

15.4.1 Net Data Types

Net type signals represent physical connections between the hardware components in a digital circuit. The simplest of net data type is ***wire***. The signals in an input/output port list of a module in Verilog code is by default ***wire*** and it can be changed by declaring other net data type. The net data types can be either single bit data or n-bit vector data. The net data types in Verilog are listed in Table 15.6.

Table 15.6: Net Data Types in Verilog

Net Data Type	Description
wire	Represent a physical connection
wand	Multiple connections resolved by AND logic in case of contention
wor	Multiple connections resolved by OR logic in case of contention
tri	Represent a physical connection with tristate logic
trireg	Bidirectional connection with tristate logic
triand	Tristate multiple connections resolved by AND logic in case of contention
trior	Tristate multiple connections resolved by OR logic in case of contention
tri0	Tristate connection with pull up to logic 1.
tri1	Tristate connection with pull down to logic 0.
supply0	Connection to power supply ground.
supply1	Connection to power supply Vcc.

15.4.2 Variable Data Types

Variables are used for computation, manipulation and temporary storage of data during program execution. A variable's value can be changed only within procedural code of a module and it cannot be changed outside the module. Therefore, input and inout ports cannot be variables. But the output port can be either net data type or variable data type. The various variable data types are ***reg, integer, real, string, time, realtime*** and ***parameter***.

Data type reg, integer and real

The ***reg*** data type is the basic and most fundamental data type to hold or store 1-bit or n-bit binary in different formats. The variables declared as ***reg*** data type can be used to exchange data during program execution and output data to output port.

The ***integer*** data type is used to hold or store decimal values as signed or unsigned 32 bit integer.

The ***real*** data type is used to hold or store decimal values with integer and fraction part separated by decimal point. The ***real*** data type is also used to hold or store decimal values in floating point format.

Example of reg, integer and real data types

```
reg sum;              // sum is 1-bit reg variable
reg [3 : 0] sum;      // sum is 4-bit reg variable
integer count;        // count is 32-bit integer variable
real power;           // power is a real variable
```

String

The ***reg*** variables are used to store strings in Verilog. The reg variable should be declared with enough size to hold the string. Each character in a string represents an ASCII value and requires 1 byte or 8-bit.

Example of string

```
reg [11*7:0] test_string; // test_string is 11 character string with each character 8-bit size
reg [11*7:0] test_string = "Hello World";   // test_string is initialized with "Hello World"
```

Time and Realtime

The time and realtime are special data types used for representing time-related values in simulations. They are useful for modeling delays, timing checks, and other time-sensitive operations. A ***time*** variable is unsigned 64 bits and can be used to store simulation time quantities for debugging purposes. A ***realtime*** variable simply stores time as a floating point quantity.

Example of time, realtime

```
time delay_time;                    // declare a variable delay_time with integer time values.
time delay_time = 20ns;             // time variable initialized with a value 20 nano-seconds.
realtime start_time, end_time;      // two time variables are declared to record the start
                                    // and end time of simulation in real number format
```

Parameter

A parameter data object defines an integer constant. The parameter declaration includes one or more identifier with a constant value assigned to each identifier and the identifier can be used in place of the constant throughout the module.

Syntax of parameter

```
parameter identifier = value;   // declaring single parameter
parameter identifier = value; identifier = value;
          ..................................
          identifier = value;   // declaring multiple parameters
```

Example of parameter

```
parameter address_size = 16;   // declaring an 16-bit address size.
parameter bus_size = 32;       // declaring a 32-bit width bus size.
```

15.5 Verilog Primitives

Verilog primitives are predefined basic building blocks in Verilog HDL that can be used to model or design the structure of large digital circuits. Verilog includes 12 basic logic gates as predefined primitives and they are,

and, nand, or, nor, xor, xnor, not, buf, bufif1, bufif0, notif1, notif0

The logic gates employs four valued logic system and the truth table of basic gates and, or, xor and not are listed in Table 15.7.

Table 15.7: Truth Table of Predefined Primitive Gates

AND

Input 1	Input 2 0	1	x	z
0	0	0	0	0
1	0	1	x	x
x	0	x	x	x
z	0	x	x	x

OR

Input 1	Input 2 0	1	x	z
0	0	1	x	x
1	1	1	1	1
x	x	1	x	x
z	x	1	x	x

XOR

Input 1	Input 2 0	1	x	z
0	0	1	x	x
1	1	0	x	x
x	x	x	x	x
z	x	x	x	x

NOT

Input	Output
0	1
1	0
x	x
z	x

The primitive gates can be instantiated in Verilog program by the following statements.

```
and G (out, in1, in2);
nand G (out, in1, in2);
or G (out, in1, in2);
nor G (out, in1, in2);
xor G (out, in1, in2);
xnor G (out, in1, in2);
not G (out1, in);
buf G (out1, in);
bufif1 G (out, in, ctrl);
bufif0 G (out, in, ctrl);
notif1 G (out, in, ctrl);
notif0 G (out, in, ctrl);
```

15.5.1 User Defined Primitive (UDP)

Truth tables can be implemented as user defined primitives in Verilog program.

The user defined primitive starts with the keyword ***primitive*** and ends with the keyword ***endprimitive*** and do not have a module. Therefore, the primitive has to be inserted before the module in a program in which the truth table has to be instantiated.

The keyword primitive is followed by a name and port list and declaration of output and inputs. Then the truth table is created in the same format followed in this book and enclosed by the keyword ***table*** and ***endtable***. One UDP can have only one output and hence, when there are multiple outputs separate UDP has to be created for each output.

Verilog Program 15.1

```
//Example of user defined primitive
//Consider the Boolean function, F1= Σ m(1,3,5,7)
//User defined primitive for the truth table of the above Boolean function
primitive UDP_F1 (F1, A, B, C);
output F1;
input A, B, C;
table
    // A B C : F1  --Header for truth table
      0 0 0 :  0;
      0 0 1 :  1;
      0 0 0 :  0;
      0 0 1 :  1;
      0 0 0 :  0;
      0 0 1 :  1;
      0 0 0 :  0;
      0 0 1 :  1;
    endtable
    endprimitive
    //Example of Verilog program using user defined primitive
    //Program to implement Boolean function, F2 = F1 + B
    //The user defined primitive UDP_F1 is instantiated in the Verilog code
    module F2_with_UDF_F1 (F2, F1, A, B, C);
        output F2, F1;
        input A, B, C;
        UDP_F1 (F1, A, B, C);
        or G1(F2, F1, B);
    endmodule
```

15.6 Verilog Statements and Loops

Verilog statements can be classified as follows,

- Continuous assignment statement
- Procedural statements
  - Continuous assignment, Blocking and nonblocking statements.
- Conditional Statements
  - If-else and case statements
- Loop Statements
  - While, for, forever and repeat statements

15.6.1 Continuous Assignment Statement

Continuous assignment statements can be used to develop code for combinational circuit in terms of the flow of data and operations on the circuit. A continuous assignment statement evaluates the value of the expression on its right hand side and assigns it to the variable on left hand side. The continuous assignment statements is prefixed with keyword ***assign***.

Syntax of continuous assignment statement

```
assign net_name = expression;
assign net_name [bit_index] = expression;
assign net_name [msb : lsb] = expression;
```

Verilog Program 15.2

```
//Example of program with continuous assignment statement
//Refer Chapter 5, Section 5.9.2 for design of full adder
module full_adder (a, b, cin, sum, cout);
    input wire  a, b, cin;
    output wire  sum, cout;

    assign sum = a ^ b ^ cin;
    assign cout = (a & b) | (b & cin) | (cin & a);
end module
```

15.6.2 Procedural Assignment Statement

Procedural assignment occur within procedures such as always, initial, task and functions and are used to attach values to variables. They are used to evaluate an expression and assign the value obtained by evaluation of the expression to a variable.

There are two types of procedural assignment statements: Blocking and Nonblocking assignment statements. The blocking statements are evaluated sequentially and values assigned sequentially. The nonblocking statements are evaluated concurrently and assigned after evaluation of all the expressions.

Syntax of procedural statement

```
variable_name = expression;    // blocking assignment
variable_name <= expression; // nonblocking assignment
```

Example of blocking procedural statement

```
-----------------------------------------------
-----------------------------------------------
always @ (posedge clk) begin
   A = B;
   C = A;
end
-----------------------------------------------
-----------------------------------------------
```

***Note:** In this example, first value of B is assigned to A and then value of A is assigned to C. Hence, C will have the value of B.*

Example of nonblocking procedural statement

```
-----------------------------------------------
always @ (posedge clk) begin
   A <= B;
   C <= A;
end
-----------------------------------------------
```

***Note:** In this example, the value of B is assigned to A and the value of A is assigned to C concurrently. Hence, C will have the old value of A.*

15.6.3 If and If-else Statement

The ***if*** statement is used to execute one or more statements when a condition specified in the ***if*** statement is true. When one or more statements are to be executed for false condition then ***else*** statement can be used after ***if*** statement. When more than one statement are to be executed after if condition or else condition then the statements has to be enclosed by keywords ***begin*** and ***end***.

The ***if-else*** statement will have one ***if*** (condition) statement, one or more ***else if*** (condition) statement and one optional ***else*** statement. At any one time only one condition will be true and hence the statements followed by the true condition are executed. If no condition is true then the statements followed by else will be executed and if there is no else statement then no statement will be executed.

Syntax of if and if-else statement

Syntax-1

```
if (condition)
   statement;
```

Syntax-2

```
if (condition)
   statement1;
      else
      statement2;
```

Syntax-3

```
if (condition)
   begin
      statement1;
      statement2;
      .............................
   end
```

Syntax-4

```
if (condition)
   begin
      statement11;
      statement12;
      .............................
   end
else
   begin
      statement21;
      statement22;
      .............................
   end
```

Syntax of if and else-if statement

```
if   (condition_1)
    begin
        statement11;
        statement12;
        .............................
    end
else if (condition_2)
    begin
        statement21;
        statement22;
        .............................
    end
    --------------------------------------------------
    --------------------------------------------------
else if (condition_n)
    begin
        statementN1;
        statementN2;
        .............................
    end
else
    begin
        statement1;
        statement2;
        .............................
    end
```

Verilog Program 15.3

```
//Example of program with if and else-if statement
//Refer Chapter 5, Section 5.6.1, for design of priority encoder
module priority_encoder (sel, code);
    input [3:0] sel;   output [1:0] code;   reg [1:0] code;

always @ (sel)
    begin
        if (sel[0])
            code = 2'b00;
        else if (sel[1])
            code = 2'b01;
        else if (sel[2])
            code = 2'b101;
        else if (sel[3])
            code = 2'b11;
        else
            code = 2'bxx;
end
endmodule
```

15.6.4 Case Statement

A ***case*** statement will check the value of an expression or object that can take multiple values. At any one time only one value will be true and one (or a set of) sequential statement after the true case is executed. The values of the expression or object are denoted as choice_1 to choice_n and an optional default choice. If none of the case is true then the default choice is executed. This statement begins with the keyword is ***case*** and terminate with keyword ***end case***.

The case statement has two other variants ***casex*** and ***casez***. The ***casex*** treats x and z logic values as x and ***casez*** treats z as x while checking the value of expression or object specified in the statement.

Syntax of case statement

```
case (expression)
    choice_1: procedural_statements;
    choice_2: procedural_statements;
    ....................................
    ....................................
    choice_n: procedural_statements;
    default : procedural_statement
endcase
```

Note: *The syntax for casex and casez statements is same as case statement except that case is replaced by* ***casex*** *or* ***casez***.

Syntax of casex statement

```
casex (expression)
    choice_1: procedural_statements;
    choice_2: procedural_statements;
    ....................................
    ....................................
    choice_n: procedural_statements;
    default : procedural_statement
endcase
```

Syntax of casez statement

```
casez (expression)
    choice_1: procedural_statements;
    choice_2: procedural_statements;
    ....................................
    ....................................
    choice_n: procedural_statements;
    default : procedural_statement
endcase
```

Verilog Program 15.4

```
//Example of program with case statement
//Refer Chapter 5, Scetion 5.6.1, for design of priority encoder
module encoder_4_2 (d_in, d_out);
    input [4:0] d_in;
    output [1:0] d_out;
    reg [1:0] d_out;
always @ (d_in)
    case (d_in)
        4'b0001  : d_out = 2'b00;
        4'b0010  : d_out = 2'b01;
        4'b0100  : d_out = 2'b10;
        4'b1000  : d_out = 2'b11;
        default  : d_out = 2'b00;
    end case
endmodule
```

15.6.5 While Loop

The ***while*** loop will test a condition and executes the statements enclosed by while loop as long as the condition is true. When the condition is false the program execution will come out of the while loop.

Syntax of while loop

```
while (logical-expression)
begin
    procedural_statements;
end
```

Example of while loop

```
while (count < 10)          // increment count value as long as
    begin                   // count is less than 10
        count = count + 1;
    end
```

15.6.6 For Loop

The ***for*** loop will have a loop index to keep track of the number of times the statements enclosed by for loop has to be executed. The loop index will have an initial value, final value and increment value.

The loop index is initialized with initial value and incremented each time after execution of the statements enclosed by for loop. The execution of statements enclosed by for loop continues till the loop index reaches final value.

Syntax of for loop

```
for (index = initial_value; final_value_of_index; index = index + increment value;)
            procedural_statement;
end
```

Example of for loop

```
for (j = 0 ; j < 16; j = j + 2)   // loop index j is initialized with a value 0 and incremented
      a <= a + 1;                 // by 2 after each execution of the statement enclosed by
end                               // for loop. The execution continues until j is less than 16
```

15.6.7 Forever and Repeat Loop

The ***forever*** loop is used to execute the statements enclosed by forever loop continuously until the program execution is stopped. The forever loop is used to generate clock signals. The ***repeat*** loop is used to execute the statements enclosed by repeat loop a finite number of times specified in the statement. The forever and repeat loop will execute the statements without checking any condition.

Syntax of forever loop

```
forever
   begin
      procedural_statements;
   end
```

Example of forever loop

```
initial
   begin
      clock = 0;                  // clock signal is initialized to 0 (low)
      forever
         #5 clock = ~clock;  // clock signal is continuously complemented in a
   end                            // time of 5 nano-sec so that clock signal will be
                                  // high and low alternatively for 5 nano-sec and
                                  // time period of clock will be 10 nano-sec and
                                  // frequency of clock is 1/10 nano-sec = 10 MHz
```

Syntax of repeat loop

```
   repeat (integer_constant)
   begin
      procedural_statements;
   end
```

Example of repeat loop

```
   if (rotate = = 1) begin
   repeat(8)                      // if rotate = 1, execute the statements enclosed by
   begin                          // repeat loop 8 times
      tmp = data[15];
      data = data << 1 + tmp;
   end
end
```

15.7 Function in Verilog

The ***function*** in Verilog is like a subroutine in a programming language. The repeated computation of a variable value using a group of statements are coded separately as a function and used any number times in a main program.

Functions in Verilog are used to code a block of procedural statements that computes a value based on the inputs and return a single value (bit, integer, vector, etc.). In Verilog functions, timing control statements and continuous assignment statements are not allowed. Functions are typically used for combinatorial logic that involves repeated computations.

Syntax of function declaration

```
function result_type/size function_name;
    input declarations;
    local variable declarations;
    parameter declarations;
    procedural_statements;
end function
```

Verilog Program 15.5

```
//Example of program with function
//Program for full adder using half adder
//Refer Chapter 5, Section 5.9.2, for design of full adder using half adder
module FA_with_function_HA
    ( input wire a, b, ci,
     output reg s, co);

     reg s1, c1, c2;

     always @ (a, b, ci)
     begin
        s1 = ha_s (a, b);
        c1 = ha_co (a, b);
        s = ha_s (s1, ci);
        c2 = ha_co(s1, ci);
        co = c1 | c2;
     end

     function ha_s;
     input wire a, b;
        ha_s = a ^ b;
     endfunction

     function ha_co;
     input wire a, b;
        ha_co = a & b;
     endfunction
endmodule
```

15.8 Task in Verilog

The ***task*** is a subroutine equivalent to a module that is written inside the main module and can be called in a Verilog main module code any number of times. Task is a complete Verilog code similar to a module and enclosed with keyword ***task*** and ***endtask*** instead of ***module*** and ***endmodule***.

Whenever a component of a digital system are repeatedly used in the system, then the code of that component is developed separately as a task and used any number times in a main program. Tasks are used for modeling components of both combinational and sequential logic that are used in the system a number of times.

```
Syntax of task declaration
    task task_name;
        Input declarations;
        Inout declarations;
        Output declarations;
        Variable declarations;
        Parameter declarations;
        Procedural_statements;
    end task
```

Verilog Program 15.6

```
//Example of program with task
//Program for full adder using half adder
//Refer Chapter 5, Section 5.9.2, for design of full adder using half adder
module FA_with_task_HA
    ( input wire a, b, ci,
     output reg s, co);

     reg s1, c1, c2;

     always @ (a, b, ci)
     begin
        ha (a, b, s1, c1);
        ha (s1, ci, s, c2);
        co = c1 | c2;
     end

     task ha;
          input wire a, b;
          output reg ha_s, ha_so;
          begin
               ha_s = a^ b;
               ha_co = a & b;
          end
     endtask
endmodule
```

15.9 Timescale and Timing Control

The Verilog has a timescale compiler directive.

```
Syntax of timescale directive
    `timescale    time_unit/time_precision
```

The default time scale in Vivado HDL simulation tool is ***`timescale 1 ns/1 ps***.

The default time scale is followed in all Verilog programs in this book.

The timing control refers to various methods used to control the timing of the behavior of signals and events in a digital system. The Verilog language provides the following three methods of timing control during execution of procedural statements.

- **delay control**
- **event control**
- **wait** statement

15.9.1 Delay Control (#)

A ***delay control*** expression specifies the time duration to be delayed for execution of the statement.

> **Syntax for delay control:**
>
> *#<time> <statement>;*
>
> **Example of delay control:**
>
> *#10 a = a + 2; // a = a + 2 will be executed after a delay time of 10 time units*
> *// default time unit is nano-second*

15.9.2 Event Control

An **event** control is used to control the execution of a procedural statement at the occurrence of the named event.

> **Syntax for event control:**
>
> *@ (event_name);*
>
> **Examples of event control:**
>
> *@r begin a = b & c; end //a = b & c is executed only if there is a change in r*
> *@(posedge clock) a = b & c; //a = b & c is executed only at positive edge of clock*
> *@(negedge clock) a = b & c; //a = b & c is executed only at negative edge of clock*

15.9.3 Wait Statement

The **wait** statement allows a procedural statement or a block of statements to be delayed until a condition become true.

> **Syntax for wait statement:**
>
> wait *<statement> ;*
>
> **Example of wait statement:**
>
> wait *(r == 3) begin a = b & c; end //a = b & c is executed only when r = 3*

15.10 Verilog Program Structure

A Verilog program starts with a keyword ***module*** and ends with a keyword ***endmodule***. The module will include the following elements.

- Declarations
- Procedural blocks
  - □ Initial and always blocks
    - ■ Procedural statements

15.10.1 Module

The module is the main unit of the Verilog program and one program can have only one module. The entire Verilog code or program is enclosed by the keywords ***module*** and ***endmodule*** and every line of the code is terminated by a semicolon (;).

The module represents a component of a digital system or the entire digital system. For large digital systems, Verilog supports modular programming in which the system with large number of components are coded such that each component is coded as a separate module and compiled and then all the modules of components are instantiated in a main module.

A name is attached to every module in order to identity the digital component or system represented by the module. Every module requires a port list (or list of inputs and outputs) that act as the interface between program and the external environment. Each port has a mode or direction and they are ***input, output*** or ***inout***.

Syntax for module:

```
module <module name> <port list>;
<input and output declaration of port list>
<module blocks and statements>
endmodule
```

Example of module:

```
module full_adder (a, b, ci, s, co);
    input wire a, b, ci;
    output wire s, co;
    <module blocks and statements>
endmodule
```

The module includes one or more ***always*** and ***initial*** block. The statements of both these blocks are enclosed by the keywords ***begin*** and ***end***. Each initial block is executed once and always block is executed again and again until the program execution is stopped.

15.10.2 Declarations

The declarations refer to input and output declarations and their data type declarations. The declarations are also made for intermediate variables, constants, functions and task.

Syntax of input and output declarations

```
input wire   identifier, identifier, ..., identifier;          //declaring 1-bit inputs and outputs
output wire  identifier, identifier, ..., identifier;
output reg   identifier, identifier, ..., identifier;
inout wire   identifier, identifier, ..., identifier;

input wire   [msb:lsb]  identifier, identifier, ..., identifier; //declaring vector inputs and outputs
output wire  [msb:lsb]  identifier, identifier, ..., identifier;
output reg   [msb:lsb]  identifier, identifier, ..., identifier;
inout wire   [msb:lsb]  identifier, identifier, ..., identifier;
```

15.10.3 Initial Block

The initial blocks are primarily used to initialize variables and ports with specific values. An initial block executes only once, at the beginning of simulation. A module can have any number of initial blocks. When initial block has multiple statements, they are enclosed by the keywords ***begin*** and ***end***.

Syntax of initial block with single statement

```
initial
    procedural_assignment_statement;
```

Syntax of initial block with multiple statements

```
initial
    begin
        procedural_assignment_statement;
        ..................................................
        procedural_assignment_statement;
    end
```

Example of initial block

```
module stimulus ( );
reg a, b, m;
initial
    m = 1'b0;   // variable m is initialized with value 0.
initial
    begin
        a = 1'b1;    // variable a is initialized with value 1.
        b = 1'b0;    // variable b is initialized with value 0.
    end
end module
```

15.10.4 Always Block

The always blocks are used to enclose the executable statements of the Verilog code. The always block is executed repeatedly throughout the simulation. A module can have any number of always blocks. When always block has multiple statements, they are enclosed by the keywords ***begin*** and ***end***.

Event Control in Always Block

Since always block is executed continuously, an event control is specified to control the execution of always block. Mostly event controls are changes in clock, reset and inputs. Most popular events are positive edge (posedge) of clock signal or negative edge (negedge) of clock signal.

Syntax for always block:

```
always @ (event1, event2, .......) begin
    <procedural statements>
end
```

Example of always block:

```
always @ (posedge clock) begin
    < procedural statements>
end
```

15.11 Types of Program Modeling in Verilog

Verilog supports the following three basic models for design of a digital system.

- Dataflow model
- Behavior model
- Structural model

The input, output and signal declarations are common for all the above three basic types of Verilog program models of digital systems.

15.11.1 Dataflow Model

The dataflow model describes how data flows through the design. In data flow model, the relationship between input and output signals is expressed using Boolean equations.

In dataflow model, the continuous assignment (assign statement) is used to assign values to outputs and these assignment statements are executed concurrently.

Dataflow models are useful for design of combinational logic circuits, where the output depends only on the current values of the inputs and not on any previous values.

15.11.2 Behavior Model

The behavior model of digital system employs a set of sequentially or concurrently executed statements. The statements like ***if-else, case, for, while,*** etc., are used to describe the behavior of the system. The statements of behavior model also enclosed by an initial block which is executed once or enclosed by an always block which is executed repeatedly.

15.11.3 Structural Model

In structural model, a digital system is designed with the set of components connected by signals. Components are separate Verilog modules and compiled separately and can be used in any other structural model of digital systems.

Instead of coding a complex design in single Verilog module, it can be divided into a number of separate modules as components and combine them using port map technique in a main module. The component declaration and component instantiation statements are the primary structures used to describe the design in main module.

In structural model, the primitive gates can be directly used without a code as separate module and its instantiation.

Verilog Program 15.7

```
//Verilog program to demonstrate dataflow, behavioral and structural model

//Program for half adder

//Refer Chapter 5, Section 5.9.1, for design of half adder
```

Verilog Program for Dataflow Model of Half Adder

```
module dfm_half_adder (a, b, s, co);
        input wire a, b;
        output wire s, co;

    assign s = a ^ b;
    assign co = a & b;
endmodule
```

Verilog Program for Behavioral Model of Half Adder

```
module bm_half_adder (a, b, s, co);
        input wire a, b;
        output reg s, co;

    always @(a or b)
    begin
        s  = a ^ b;
        co = a & b;
    end
endmodule
```

Program Logic for Structural Model

The Verilog code for structural model of half adder is developed using the logic circuit shown in Fig. 15.1. In structural model, each component involved in the half adder are coded separately and then the components are added in the code for half adder.

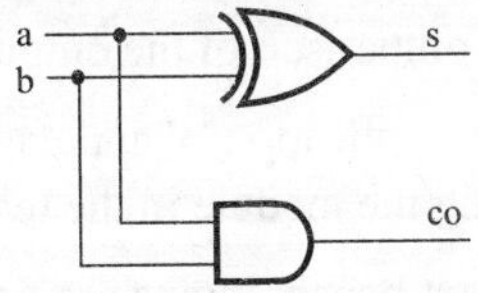

Fig. 15.1: Logic circuit of half adder.

Verilog Program for Structural Model of Half Adder

```
//code for 2-input xor gate
module dfm_xor_gate (x, y, v);
    input wire x, y;
    output wire v;
        assign v = x ^ y;
endmodule
```

```
//code for 2-input and gate
module dfm_and_gate (x, y, v);
    input wire x, y;
    output wire v;
  assign v = x & y;
endmodule
```

```
//code for half adder
module sm_half_adder (a, b, s, co);
    input wire a, b;
    output wire s, co;

// Instatiation of XOR and AND gate modules
        dfm_xor_gate dut1 (.x(a), .y(b), .v(s));
        dfm_and_gate dut2 (.x(a), .y(b), .v(co));
```

```
// port mapping of XOR and AND gates
        dfm_xor_gate G1(a, b, s);
        dfm_and_gate G2(a, b, co);
endmodule
```

```
//Alternate code for structural model of half adder using primitive gates
module sm_half_adder (a, b, s, co);
    input wire a, b;
    output wire s, co;

    xor G1 (s, a, b);   // Here XOR and AND gates are primitve gates in
    and G2 (co, a, b);  // Verilog and hence separate code is not needed
                        // for gates
endmodule
```

15.12 Test Bench

A test bench is also a Verilog program that is used to verify the correctness of the Verilog program of the hardware model. A test bench Verilog program has three main purposes:

- Generating inputs and clock signal for simulation of the hardware model.
- Apply this inputs and clock to the module under test and to monitor the output
- Comparison of the output of Verilog program model with expected output.

The application of inputs and clock to the module under test is accomplished automatically by instantiating the module in the test bench model and then specifying the appropriate interface signals.

In test bench, inputs are declared as ***reg*** and the signals connected to output are declared as ***wire***. The ***initial*** block is used specify or apply the inputs to generate the outputs.

In test bench the system command ***$finish*** is used to terminate the clock and test inputs.

15.13 Verilog Programs Without Test Bench

Verilog Program 15.8

Simplify the Boolean function F by means of K-map and write a dataflow and structural HDL to implement the reduced Boolean function.

$$f(a, b, c, d) = \sum m(0, 6, 7, 9, 11, 13, 15) \quad ; \quad d(a, b, c, d) = \sum m(2, 5, 8, 10)$$

To find Boolean Equation using K-map

The given function has four variables and so a 4-variable K-map with 16 squares ($2^4 = 16$) is drawn as shown in Fig. 1. The function is defined as sum of minterms m_0, m_6, m_7, m_9, m_{11}, m_{13} and m_{15} and so a "1" is filled in the corresponding squares in K-map. The function also has don't-care outputs for the minterms m_2, m_5, m_8 and m_{10} and so a "x" is filled in the corresponding squares in K-map and the remaining squares are filled with "0".

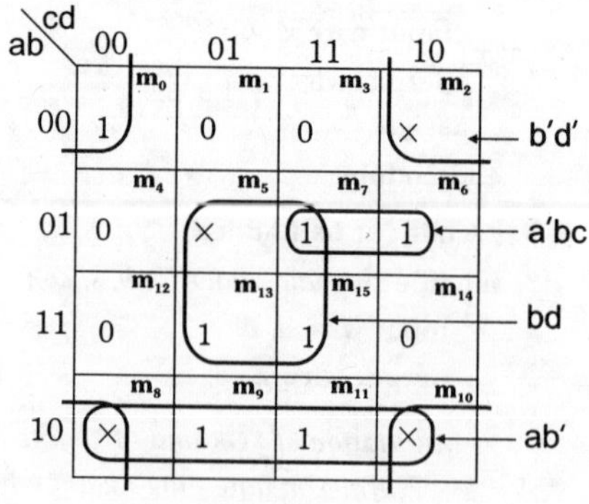

Fig. 1: K-map for f.

For NAND implementation, the Boolean function should be formed in SOP form. In the K-map of Fig. 1, by considering the don't-care condition in m_2 and m_5 as "1", three prime implicants can be formed with four adjacent 1's and the corresponding product terms are ab', bd, a'bc and b'd'. The simplified Boolean function is given by sum of these three product terms.

$$\therefore\ f = ab' + bd + a'bc + b'd'$$

Using the above Boolean function the logic circuit using only NAND gates is drawn as shown in Fig. 2.

Program Logic for Dataflow Model

The Boolean function for K-map is,

$$f = ab' + bd + a'bc + b'd'$$

The Verilog code for dataflow model of Boolean function is developed using the above equations. (Refer Chapter 4, Section 4.6 for design of Boolean function).

Verilog Program for Dataflow Model of Boolean Function

```
module dfm_boolean_func
   ( input a, b, c, d,
   output f);
   wire m1, m2, m3;
   wire f1, f2, f3, f4;

   assign m1 = ~a;         assign m2 = ~b;        assign m3 = ~d;
   assign f1 = a & m2;     assign f2 = b & d;

   assign f3 = m1 & b & c; assign f4 = m2 & m3;
   assign f = f1 | f2 | f3 | f4;
endmodule
```

Program Logic for Structural Model

The Verilog code for structural model of Boolean function is developed using the logic circuit shown in Fig. 2. The NAND gates in the logic circuit of Fig. 2 are modeled using primitive NAND gates in verilog.

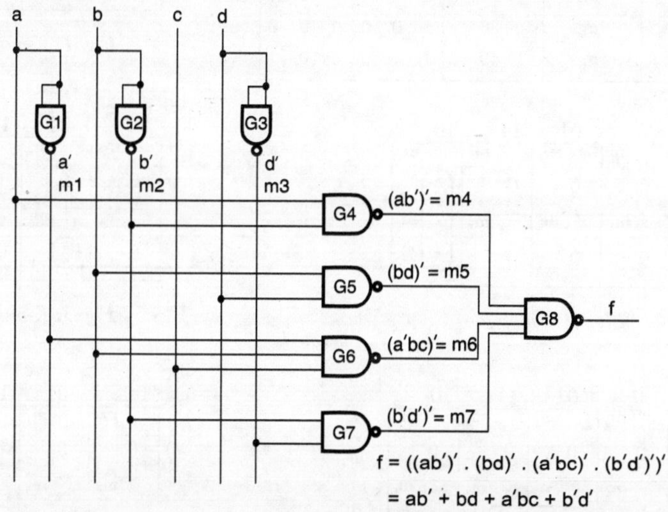

Fig. 2: Logic circuit of function, f using only NAND gates.

Verilog Program for Structural Model of Boolean Function

```
module sm_boolean_func
   ( input a, b, c, d,
     output f );
     wire m1, m2, m3, m4, m5, m6, m7;

     nand G1 (m1, a, a);       nand G2 (m2, b, b);     nand G3 (m3, d, d);
     nand G4 (m4, a, m2);      nand G5 (m5, b, d);
     nand G6 (m6, m1, b, c);   nand G7 (m7, m2, m3);
     nand G8 (f, m4, m5, m6, m7);
endmodule
```

Verilog Program 15.9

Design a combinational circuit that accepts a 3-bit binary number as input and generates square of the input number as output. Write a high-level behavior Verilog description for the circuit.

To find Boolean Equation using K-map

The truth table of the given problem is developed as shown in Table 1. Using the truth table the K-maps are constructed as shown in Figs. 1 to 5.

Table 1: Truth Table

Inputs					Outputs						
Binary			Decimal	Minterm	Decimal	Binary					
a	b	c				b_5	b_4	b_3	b_2	b_1	b_0
0	0	0	0	m_0	0	0	0	0	0	0	0
0	0	1	1	m_1	1	0	0	0	0	0	1
0	1	0	2	m_2	4	0	0	0	1	0	0
0	1	1	3	m_3	9	0	0	1	0	0	1
1	0	0	4	m_4	16	0	1	0	0	0	0
1	0	1	5	m_5	25	0	1	1	0	0	1
1	1	0	6	m_6	36	1	0	0	1	0	0
1	1	1	7	m_7	49	1	1	0	0	0	1

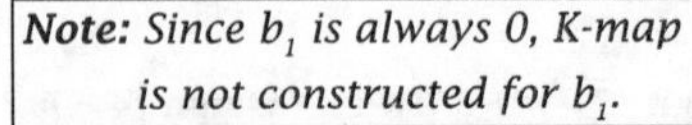
Note: *Since b_1 is always 0, K-map is not constructed for b_1.*

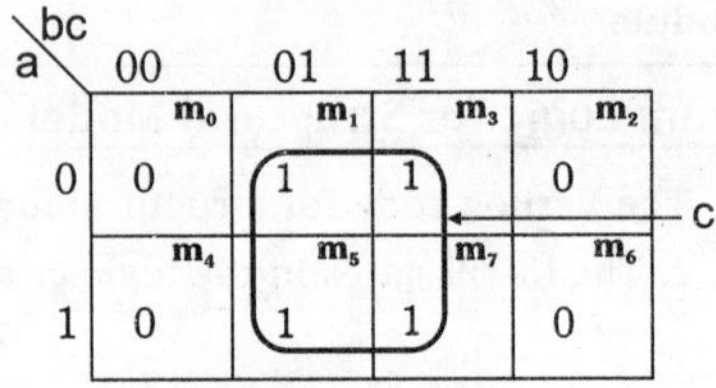

Fig. 1: *K-map for b_0.*

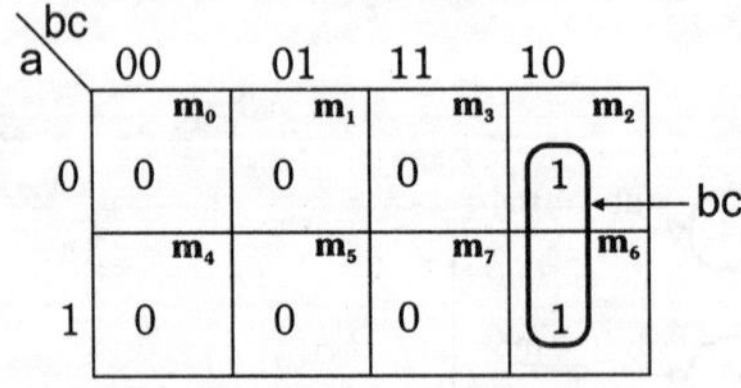

Fig. 2: *K-map for b_2.*

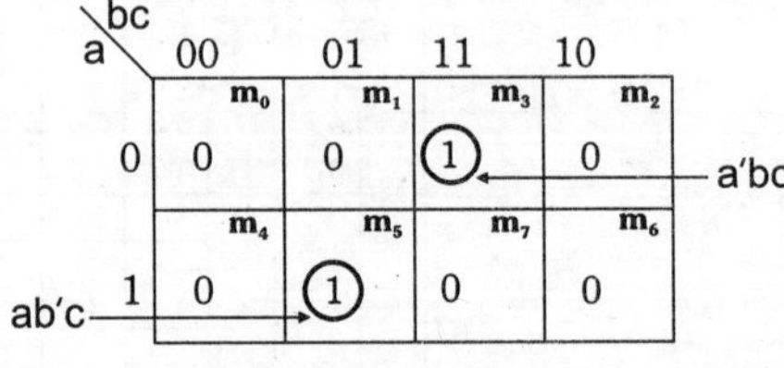

Fig. 3: *K-map for b_3.*

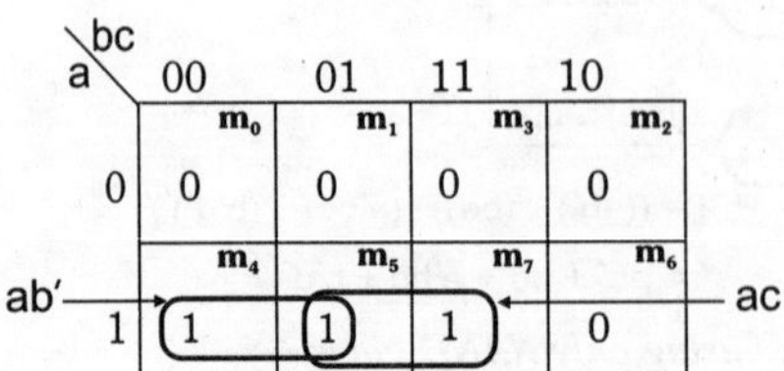

Fig. 4: *K-map for b_4.*

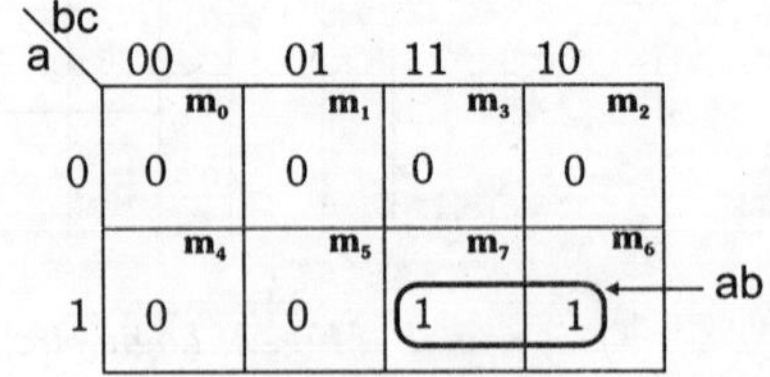

Fig. 5: *K-map for b_5.*

From the K-map the following Boolean equations are obtained and using these equations the logic circuit is drawn in Fig. 6.

$b_0 = c$	$b_3 = a'bc + ab'c$
$b_1 = 0$	$b_4 = ac + ab'$
$b_2 = bc'$	$b_5 = ab$

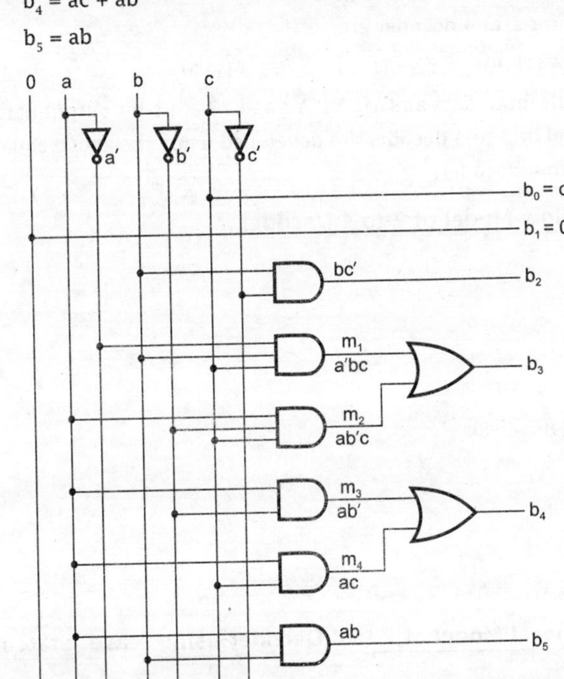

Fig. 6: Logic circuit to generate square of a 3-bit binary number.

Program Logic for Behavioral Model of Square of 3-bit Binary

Using Fig. 6, the Verilog codes for behavioral model of combinational circuit is developed.

Verilog Program for Behavioral Model of Square of 3-bit Binary

```
module bm_square_3bit
  ( input a, b, c,
    output b0, b1, b2, b3, b4, b5 );

    reg b0, b1, b2, b3, b4, b5;
    reg m1, m2;

  always @ (a or b or c) begin
      m1 = (~a) & b;
      m2 = a & (~b);

      b0 = c;
      b1 = 1'b0;
      b2 = b & (~c);
      b3 = (m1 & c) | (m2 & c);
      b4 = (a & c) | m2;
      b5 = a & b;
   end
endmodule
```

Verilog Program 15.10

Write Verilog code for dataflow, behavioral and structural models of a 2-to-4 decoder.

Program Logic for Dataflow and Behavioral Model

The Boolean equations for 2-to-4 decoder are,

$$Y0 = I1'.I0' \;;\; Y1 = I1'.I0 \;;\; Y2 = I1.I0' \;;\; Y3 = I1.I0$$

Where, I1 and I0 are 1-bit input data and Y0, Y1, Y2 and Y3 are 1-bit output data. The Verilog codes for dataflow and behavioral model of 2-to-4 decoder are developed using the above equations. (Refer Chapter 5, Section 5.4.1 for design of 2-to-4 decoder).

Verilog Program for Dataflow Model of 2-to-4 Decoder

```
module dfm_decoder_2_4
   ( input i1, i0,
     output y0, y1, y2, y3 );

    wire m1, m2;
    wire y0, y1, y2, y3;

    assign  m1 = ~i1;    assign m2 = ~i0;

    assign y0 = m1 & m2;
    assign y1 = m1 & i0;
    assign y2 = i1 & m2;
    assign y3 = i1 & i0;
endmodule
```

Verilog Program for Behavioral Model of 2-to-4 Decoder using "case" Statement

```
module bm_decoder_2_4
   ( input [1:0] i,
     output [3:0] y );

    reg[3:0] y;

    always @ (i) begin
    case (i)
       2'b00 : y = 4'b0001;
       2'b01 : y = 4'b0010;
       2'b10 : y = 4'b0100;
       2'b11 : y = 4'b1000;
       // default : y = 4'b0000;
   endcase
   end
endmodule
```

Verilog Program for Behavioral Model of 2-to-4 Decoder using "if elsif" Statement

```
module bm_decoder_2_4
   ( input [1:0] i,
     output [3:0] y);
     reg [3:0] y;

     always @ (i)
     begin
```

```
        if (i == 2'b00)       y = 4'b0001;
        else if (i == 2'b01)  y = 4'b0010;
        else if (i == 2'b10)  y = 4'b0100;
        else                  y = 4'b1000;
    end
endmodule
```

Program Logic for Structural Model

The Verilog code for structural model of 2-to-4 decoder is developed using the logic circuit shown in Fig. 1. The AND and NOT gates in the logic circuit of Fig. 1 are modeled using primitive AND and NOT gates in verilog.

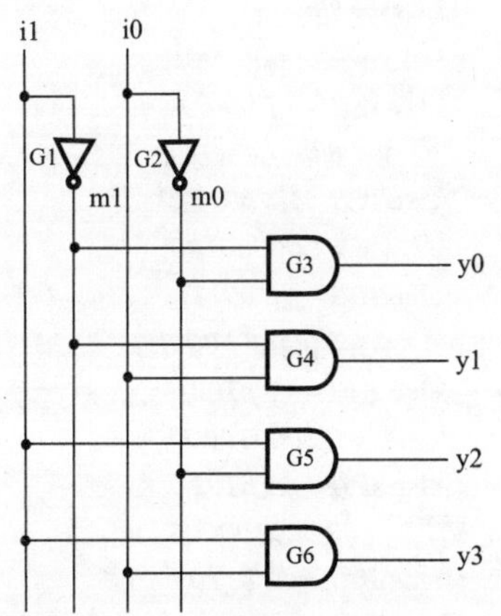

Fig. 1: Logic circuit for 2-to-4 decoder.

Program for Structural Model of 2-to-4 Decoder

```
module sm_decoder_2_4
  ( input [1:0] i,
    output [3:0] y );

    wire m0, m1;

    not G1 (m1, i[1]);
    not G2 (m0, i[0]);
    and G3 (y[0], m1, m0);
    and G4 (y[1], m1, i[0]);
    and G5 (y[2], i[1], m0);
    and G6 (y[3], i[1], i[0]);
endmodule
```

Verilog Program 15.11

Write a Verilog code for behavioral model of a 3-to-8 decoder.

Program Logic for Behavioral Model

The Verilog code for behavioral model of 3-to-8 decoder are developed using the truth table of 3-to-8 decoder In the truth table $I_2I_1I_0$ is 3-bit input and Y_0 and Y_7 are 8 numbers of 1-bit outputs. (Refer Chapter 5, Section 5.4.7 for design of 3-to-8 decoder).

Truth Table of Logic High 3-to-8 Decoder

Inputs	Minterm	Outputs							
I_2 I_1 I_0		Y_0	Y_1	Y_2	Y_3	Y_4	Y_5	Y_6	Y_7
0 0 0	m_0	1	0	0	0	0	0	0	0
0 0 1	m_1	0	1	0	0	0	0	0	0
0 1 0	m_2	0	0	1	0	0	0	0	0
0 1 1	m_3	0	0	0	1	0	0	0	0
1 0 0	m_4	0	0	0	0	1	0	0	0
1 0 1	m_5	0	0	0	0	0	1	0	0
1 1 0	m_6	0	0	0	0	0	0	1	0
1 1 1	m_7	0	0	0	0	0	0	0	1

Verilog Program for Behavioral Model of 3-to-8 Decoder

```
module bm_decoder_3_8
   ( input [2:0] I,
     output reg [7:0] y);

   always @ (I)
      begin
      if (I == 3'b000)
            y = 8'b00000001;
      else if (I == 3'b001)
            y = 8'b00000010;
      else if (I == 3'b010)
            y = 8'b00000100;
      else if (I == 3'b011)
            y = 8'b00001000;
      else if (I == 3'b100)
            y = 8'b00010000;
      else if (I == 3'b101)
            y = 8'b00100000;
      else if (I == 3'b110)
            y = 8'b01000000;
      else
            y = 8'b10000000;
      end
endmodule
```

Verilog Program 15.12

Write a Verilog code for dataflow and structural models of a 4-to-2 encoder.

Program Logic for Dataflow Model

The Boolean equations for 4-to-2 encoder are,

$$Y1 = I2 + I3 \quad ; \quad Y0 = I1 + I3$$

Where, I0, I1, I2 and I3 are 1-bit input data, Y0 and Y1 are 1-bit output data. The Verilog code for dataflow model of 4-to-2 encoder is developed using the above equations. (Refer Chapter 5, Section 5.5.1 for design of 4-to-2 encoder).

Verilog Program for Dataflow Model of 4-to-2 Encoder

```
module dfm_encoder_4_2
   ( input [3:0] i,
     output [1:0] y);

     assign y[1] = i[2] | i[3];
     assign y[0] = i[1] | i[3];
endmodule
```

Program Logic for Structural Model

The Verilog code for structural model of 4-to-2 encoder is developed using the logic circuit shown in Fig. 1. The OR gates in the logic circuit of Fig. 1 are modeled using primitive OR gates in verilog.

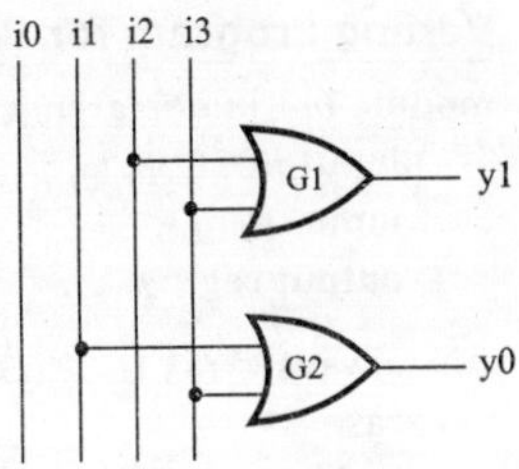

Fig. 1: 4-to-2 encoder.

Program for Structural Model of 4-to-2 Encoder

```
module sm_encoder_4_2
    ( input [3:0] i,
      output [1:0] y );

    or G1 (y[1], i[2], i[3]);
    or G2 (y[0], i[1], i[3]);
endmodule
```

Verilog Program 15.13

Write a Verilog code for dataflow model of priority encoder.

Program Logic for Dataflow Model

Boolean equations for priority encoder with I_3 having highest priority and I_0 having lowest priority are,

Y1 = I2 + I3

Y0 = I1.I2' + I3

v = I0 + I1 + I2 + I3

Where, I0, I1, I2 and I3 are 1-bit input data and Y0, Y1 and v are 1-bit output data. The Verilog code for dataflow model of priority encoder is developed using the above equations. (Refer Chapter 5, Section 5.6.2 for design of prioroty encoder).

Verilog Program for Dataflow Model of Priority Encoder

```
module dfm_priority_encoder
    ( input i0, i1, i2, i3,
      output y0, y1, v );

    wire m1;

    assign m1 = ~i2;
    assign y1 = i2 | i3;
    assign y0 = ( i1 & m1 ) | i3 ;
    assign v  = i0 | i1 | i2 | i3;
endmodule
```

Verilog Program 15.14

Write a Verilog code for behavioral and structural model of a 4*1 multiplexer.

Program Logic for Behavioral Model

The Verilog code for behavioral model of 4*1 multiplexer is developed using the truth table of 4*1 multiplexer. In the truth table S0 and S1 are 1-bit selection input and Y is 1-bit output data. (Refer Chapter 5, Section 5.7 for design of 4*1 multiplexer).

Truth Table of 4:1 Multiplexer

Selection Input		Output
S_1	S_0	Y
0	0	I_0
0	1	I_1
1	0	I_2
1	1	I_3

Verilog Program for Behavioral Model of 4*1 Multiplexer using 'case' Statement

```
module bm_multiplexer_4_1
   ( input i0, i1, i2, i3,
    input  [1:0] s,
    output reg y );

    always @ (i0, i1, i2, i3, s)
    case (s)
       2'b00 : y <= i0;
       2'b01 : y <= i1;
       2'b10 : y <= i2;
       2'b11 : y <= i3;
    endcase
endmodule
```

Program Logic for Structural Model

The Verilog code for structural model of 4*1 multiplexer is developed using the logic circuit shown in Fig. 1. The AND, OR and NOT gates in the logic circuit of Fig. 1 are modeled using primitive AND, OR and NOT gates in verilog.

Verilog Program for Structural Model of 4*1 Multiplexer

```
module sm_multiplexer_4_1
   ( input s0, s1, i0, i1, i2, i3,
    output y );

    wire ns0, ns1;
    wire m1, m2, m3, m4;

    not G1 (ns0, s0);
    not G2 (ns1, s1);

    and G3 (m1, ns1, ns0, i0);
    and G4 (m2, ns1, s0,  i1);
    and G5 (m3, s1,  ns0, i2);
    and G6 (m4, s1,  s0,  i3);
    or  G7 (y, m1, m2, m3, m4);
endmodule
```

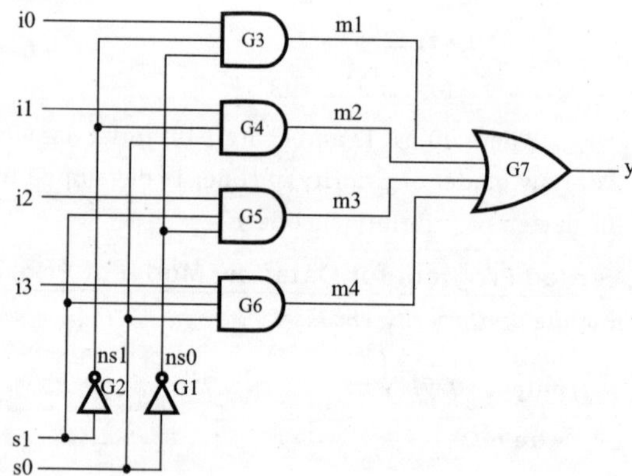

Fig. 1: Logic circuit of 4:1 multiplexer.

Verilog Program 15.15

Write a Verilog code for behavioral model of a 8*1 multiplexer.

Program Logic for Behavioral Model

The Verilog code for behavioral model of 8*1 multiplexer is developed using the truth table of 8*1 multiplexer. In the truth table S0, S1 and S2 are 1-bit selection input and Y is 1-bit output data. (Refer Chapter 5, Section 5.7 for design of 8*1 multiplexer).

Truth Table of 8:1 Multiplexer

Selection Inputs			Output Y
S_2	S_1	S_0	
0	0	0	I_0
0	0	1	I_1
0	1	0	I_2
0	1	1	I_3
1	0	0	I_4
1	0	1	I_5
1	1	0	I_6
1	1	1	I_7

Verilog Program for Behavioral of 8*1 Multiplexer

```
module bm_multiplexer_8_1
    ( input [2:0] s,
      input [7:0] i,
      output reg y );

    always @ (s, i)
    begin
            if(s == 3'b000)  y = i[0];
       else if(s == 3'b001)  y = i[1];
       else if(s == 3'b010)  y = i[2];
       else if(s == 3'b011)  y = i[3];
       else if(s == 3'b100)  y = i[4];
       else if(s == 3'b101)  y = i[5];
       else if(s == 3'b110)  y = i[6];
       else                  y = i[7];
    end
endmodule
```

Verilog Program 15.16

Write a Verilog code for dataflow and structural model of a 1*4 demultiplexer.

Program Logic for Dataflow Model

The Boolean equations for 1*4 demultiplexer are,

Y0 = Ii.S1'.S0' ; Y1 = Ii.S1'.S0

Y2 = Ii.S1.S0' ; Y3 = Ii.S1.S0

Where, S1 and S0 are 1-bit selection input and Y0, Y1, Y2 and Y3 are 1-bit output data. The Verilog code for dataflow model of 1*4 demultiplexer is developed using the above equations. (Refer Chapter 5, Section 5.8 for design of 1*4 demultiplexer).

Verilog Program for Dataflow Model of 1*4 Demultiplexer

```
module dfm_demux_1_4 (ii, s1, s0, y0, y1, y2, y3);
    input ii, s1, s0;
    output y0, y1, y2, y3;
    wire m0, m1;

    assign m0 = ~s0;
    assign m1 = ~s1;
    assign y0 = ii & m1 & m0;
    assign y1 = ii & m1 & s0;
    assign y2 = ii & s1 & m0;
    assign y3 = ii & s1 & s0;
endmodule
```

Program Logic for Structural Model

The Verilog code for structural model of 1*4 demultiplexer is developed using the logic circuit shown in Fig. 1. The AND and NOT gates in the logic circuit of Fig. 1 are modeled using primitive AND and NOT gates in verilog.

Verilog Program for Structural Model of 1*4 Demultiplexer

```
module sm_demux_1_4 (ii, s1, s0, y0, y1, y2, y3);

    input ii, s1, s0;
    output y0, y1, y2, y3;
    wire m0, m1;

    not G1 (m0, s0);
    not G2 (m1, s1);

    and G3 (y0, ii, m1, m0);
    and G4 (y1, ii, m1, s0);
    and G5 (y2, ii, s1, m0);
    and G6 (y3, ii, s1, s0);
endmodule
```

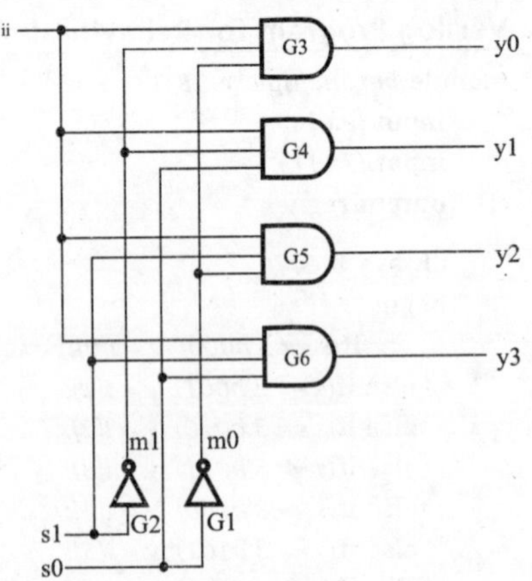

Fig. 1: Logic circuit of 1:4 demux.

Verilog Program 15.17

Write Verilog code for dataflow and structural models of a full adder.

Program Logic for Dataflow Model

The Boolean equations for full adder are,

$$s = a \oplus b \oplus c_i$$

$$c_o = (a\,b) + (b\,c_i) + (c_i\,a)$$

Where, a, b and c_i are 1-bit input data, sum (s) and carry (c_o) are 1-bit output data. The Verilog code for dataflow model of full adder is developed using the above equations. (Refer Chapter 5, Section 5.9.2 for design of full adder).

Verilog Program for Dataflow Model of Full Adder

```
module dfm_full_adder (a, b, s, ci, co);
    input a, b, ci;
    output s, co;
    wire s, co;

    assign s = a ^ b ^ ci;
    assign co = (a & b) | (b & ci) | (ci & a);
endmodule
```

Program Logic for Structural Model

The Verilog code for structural model of full adder is developed using the logic circuit shown in Fig. 1. The AND, OR and XOR gates in the logic circuit of Fig. 1 are modeled using primitive AND, OR and XOR gates in verilog.

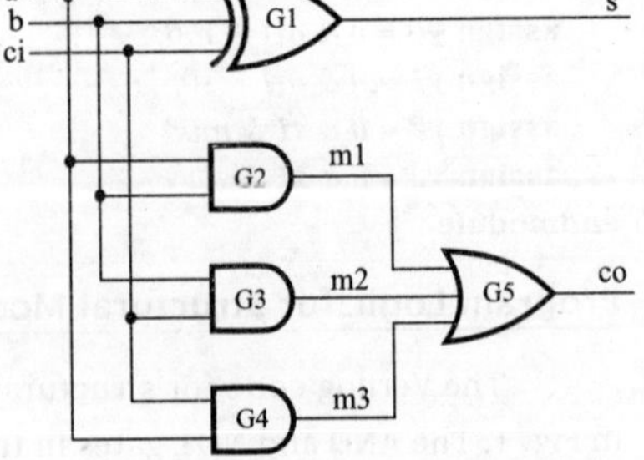

Fig. 1: Logic circuit of full adder.

Verilog Program for Structural Model of Full Adder

```
module sm_full_adder (a, b, s, ci, co);
    input a, b, ci;
    output s, co;
    wire m1, m2, m3;
```

```
    xor G1 (s, a, b, ci);
    and G2 (m1, a, b);
    and G3 (m2, b, ci);
    and G4 (m3, a, ci);
    or  G5 (co, m1, m2, m3);
endmodule
```

Verilog Program 15.18

Write Verilog codes for structural model of a 4-bit parallel adder designed using 4 full adders by considering full adder as,

a) a component b) a function c) a task

a) Program Logic for Structural Model

The Boolean equations for 4-bit parallel adder performs arithmetic sum of two 4-bit binary numbers. A full adder can add two 1-bit binary along with previous carry. In order to add two 4-bit binary numbers, 4 full adders are required. Each ful adder will add one bit of binary numbers. Consider two 4-bit binary numbers A and B shown below:

Binary number 1 : A3, A2, A1, A0 (data-A)

Binary number 2 : B3, B2, B1, B0 (data-B)

Now, the logic circuit of 4-bit parallel adder will be as shown in Fig. 1. (Refer Chapter 5, Section 5.9.3 for design of 4-bit parallel adder).

The Verilog code for structural model of 4-bit parallel adder is developed using the logic circuit shown in Fig. 1. In structural model, the code for full adder is developed separately and 4 full adders are added as components in the code for 4-bit parallel adder.

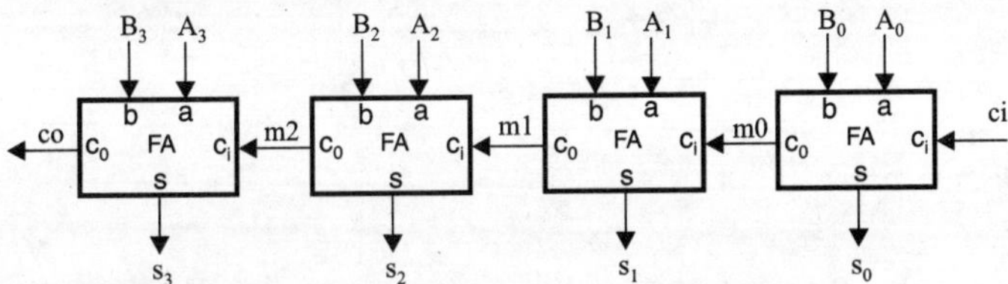

Fig. 1: *4-bit binary adder using full adders.*

Verilog Program for Dataflow Model of Full Adder

```
module dfm_full_adder (a, b, s, ci, co);
  input wire a, b, ci;
  output wire s, co;

  assign s = a ^ b ^ ci;
  assign co = (a & b) | (b & ci) | (ci & a);
endmodule
```

Verilog Program for Structural Model of 4-bit Parallel Adder

```
module sm_parllel_adder_4_bit
   ( input wire [3:0] A,
     input wire [3:0] B,
     input wire ci,
     output wire [3:0] s,
     output wire co );
```

```
    wire m0, m1, m2;

    dfm_full_adder fao(A[0], B[0], ci, s[0], m0);
    dfm_full_adder fa1(A[1], B[1], m0, s[1], m1);
    dfm_full_adder fa2(A[2], B[2], m1, s[2], m2);
    dfm_full_adder fa3(A[3], B[3], m2, s[3], co);
endmodule
```

b) Verilog program for structural model of 4-bit parallel adder using function.

```
module par_add_4bit_function
    ( input wire [3:0] A,
    input wire [3:0] B,
    input wire ci,
    output reg [3:0] s,
    output reg co );

    reg m0, m1, m2;
    always @ (A, B, ci)
    begin
        s[0] = fa_s (A[0], B[0], ci);
        m0  = fa_co (A[0], B[0], ci);

        s[1] = fa_s (A[1], B[1], m0);
        m1  = fa_co (A[1], B[1], m0);

        s[2] = fa_s (A[2], B[2], m1);
        m2  = fa_co (A[2], B[2], m1);
        s[3] = fa_s (A[3], B[3], m2);
        co   = fa_co (A[3], B[3], m2);
    end

    function fa_s;
        input wire a, b, ci;
        fa_s = a ^ b ^ ci;
    endfunction

    function fa_co;
        input wire a, b, ci;
        fa_co = (a & b) | (b & ci) | (ci & a);
    endfunction
endmodule
```

c) Verilog program for structural model of 4-bit parallel adder using task.

```
module par_add_4_bit_using_task_
    ( input wire [3:0] A,
    input wire [3:0] B,
    input wire ci,
    output reg [3:0] s,
    output reg co );

    reg m0, m1, m2;

    always @ (A, B, ci)
    begin
        fa (A[0], B[0], ci, s[0], m0);
        fa (A[1], B[1], m0, s[1], m1);
```

```
        fa (A[2], B[2], m1, s[2], m2);
        fa (A[3], B[3], m2, s[3], co);
    end

    task fa;
        input wire a, b, ci;
        output reg s, co;
            begin
                s = a ^ b ^ ci;
                co = (a & b) | (b & ci) | (ci & a);
            end
        endtask
endmodule
```

Verilog Program 15.19

Write a Verilog code for dataflow model of a 4-bit CLA (Carry look ahead adder).

Program Logic for Dataflow Model

$$s_i = p_i \oplus c_i$$

$$c_{i+1} = g_i + p_i c_i$$

$$p_i = a_i \oplus b_i$$

$$g_i = a_i b_i$$

Where, p_i is carry propagate from i_{th} stage and g_i is carry generate of i_{th} stage. The Verilog code for dataflow model of 4-bit CLA is developed using the above equations. (Refer Chapter 5, Section 5.9.4 for design of 4-bit CLA).

Verilog Program for Dataflow Model of 4-bit CLA

```
module dfm_CL_Adder_4bit(a, b, c0, s, c4);
    input [3:0] a;
    input [3:0] b;
    input c0;
    output c4;
    output [3:0] s;

    wire m1, m2, m3, m4, m5, m6, m7;
    wire p0, p1, p2, p3;
    wire g0, g1, g2, g3;

    assign p0 = a[0] ^ b[0],  p1 = a[1] ^ b[1],  p2 = a[2] ^ b[2],  p3 = a[3] ^ b[3];
    assign g0 = a[0] & b[0],  g1 = a[1] & b[1],  g2 = a[2] & b[2],  g3 = a[3] & b[3];
    assign m1 = g0 | (p0 & c0);
    assign m2 = g1 | (p1 & g0) | (p1 & p0 & c0);
    assign m3 = g2 | (p2 & g1) | (p2 & p1 & g0) | (p2 & p1 & p0 & c0);
    assign s[0] = p0 ^ c0,  s[1] = p1 ^ m1,  s[2] = p2 ^ m2,  s[3] = p3 ^ m3;
    assign m4 = (p3 & g2),  m5 = (p3 & p2 & g1),  m6 = (p3 & p2 & p1 & g0);
    assign m7 = (p3 & p2 & p1 & p0 & c0);
    assign c4 = g3 | m4 | m5 | m6 | m7 ;
endmodule
```

Verilog Program 15.20

Write Verilog code for dataflow and structural models of a half subtractor.

Program Logic for Dataflow Model

The Boolean equations for half subtractor are,

$$d = a \oplus b = ab' + a'b$$

$$c_o = a'b$$

Where, a and b are 1-bit input data, difference (d) and borrow (c_o) are 1-bit output data. The Verilog code for dataflow model of half subtractor is developed using the above equations. (Refer Chapter 5, Section 5.10.1 for design of half subtractor).

Verilog Program for Dataflow Model of Half Subtractor

```
module dfm_Half_Subtractor (a, b, d, co);
   input a, b;
   output wire d, co;

   assign d  = a ^ b;
   assign co = (~a) & b;
endmodule
```

Program Logic for Structural Model

The Verilog code for structural model of half subtractor is developed using the logic circuit shown in Fig. 1. The AND, NOT and XOR gates in the logic circuit of Fig. 1 are modeled using primitive AND, NOT and XOR gates in verilog.

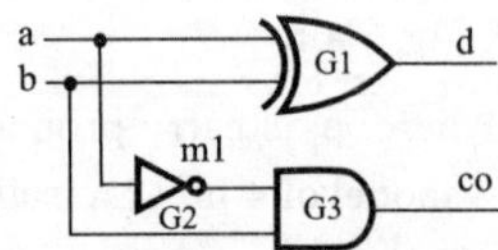

Fig. 1: Logic circuit of half subtractor.

Verilog Program for Structural Model of Half Subtractor

```
module sm_Half_Subtractor(a, b, d, co);
   input  a, b;
   output d, co;
   wire m1;

   xor G1 (d, a, b);
   not G2 (m1, a);
   and G3 (co, m1, b);
endmodule
```

Verilog Program 15.21

Write Verilog code for dataflow and structural models of a full subtractor.

Program Logic Dataflow Model

The Boolean equations for full subtractor are,

$$d = a'b'c_i \oplus a'bc_i' \oplus a'bc_i' \oplus abc_i$$

$$c_o = (a'b) + (b\,c_i) + (c_i\,a')$$

Where, a, b and c_i are 1-bit input data, difference (d) and borrow (c_o) are 1-bit output data. The Verilog code for dataflow model of full subtractor is developed using the above equations. (Refer Chapter 5, Section 5.10.2 for design of full subtractor).

Verilog Program for Dataflow Model of Full Subtractor

```
module dfm_Full_Subtractor(a, b, ci, d, co);
    input a, b, ci;
    output wire d, co;

    assign d  = a ^ b ^ ci;
    assign co = ((~a) & b) | (b & ci) | ((~a) & ci);
endmodule
```

Program Logic for Structural Model

The Verilog code for structural model of full subtractor is developed using the logic circuit shown in Fig. 1. The AND, OR, NOT and XOR gates in the logic circuit of Fig. 1 are modeled using primitive AND, OR, NOT and XOR gates in verilog.

Verilog Program for Structural Model of Full Subtractor

```
module sm_Full_Subtractor (a, b, ci, d, co);
    input a, b, ci;
    output d, co;
    wire m1, m2, m3, m4;

    xor G1 (d, a, b, ci);
    not G2 (m4, a);
    and G3 (m1, b, m4);
    and G4 (m2, b, ci);
    and G5 (m3, ci, m4);
    or  G6 (co, m1, m2, m3);
endmodule
```

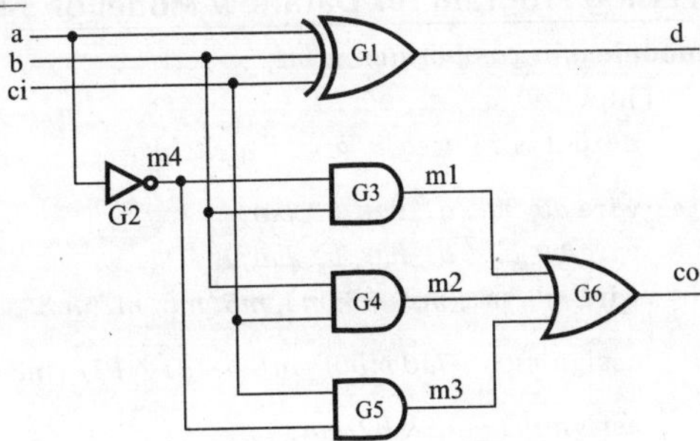

Fig. 1: Logic circuit of full subtractor.

Verilog Program 15.22

Write a Verilog code for dataflow model of 1-bit magnitude comparator.

Program Logic for Dataflow Model

Let a and b are 1-bit binary numbers.

The Boolean equations for 1-bit magnitude comparator are,

if $a = b$, then $(a \oplus b)' = 1$

if $a > b$, then $a b' = 1$

if $a < b$, then $a' b = 1$

The Verilog codes for dataflow model of 1-bit magnitude comaparator are developed using the Boolean equations. (Refer Chapter 5, Section 5.12 for design of 1-bit magnitude comparator).

Verilog Program for Dataflow Model of 1-bit Magnitude Comparator

```
module dfm_comparator_1_bit
    ( input a, b,
      output wire a_eq_b, a_gt_b, a_lt_b );
      wire a_bar, b_bar;

      assign a_bar = ~a, b_bar = ~b;
      assign a_eq_b = ~(a ^ b);
      assign a_gt_b = a & b_bar;
      assign a_lt_b = a_bar & b;
endmodule
```

Verilog Program 15.23

Write a Verilog code for dataflow model of a 3-bit comparator.

Program Logic for Dataflow Model

Let a and b are 3-bit binary numbers.

The Boolean equations for 3-bit magnitude comparator are,

$$\text{if } a = b, \text{ then } (a_0 \oplus b_0)' \,.\, (a_1 \oplus b_1)' \,.\, (a_2 \oplus b_2)' = 1$$

$$\text{if } a > b, \text{ then } (a_2 \,.\, b_2') + ((a_2 \oplus b_2)' \,.\, a_1 \,.\, b_1') + ((a_2 \oplus b_2)' \,.\, (a_1 \oplus b_1)' \,.\, a_0 \,.\, b_0') = 1$$

$$\text{if } a < b, \text{ then } (a_2' \,.\, b_2) + ((a_2 \oplus b_2)' \,.\, a_1' \,.\, b_1) + ((a_2 \oplus b_2)' \,.\, (a_1 \oplus b_1)' \,.\, a_0' \,.\, b_0) = 1$$

The Verilog codes for dataflow model of 3-bit magnitude comparator are developed using the Boolean equations. (Refer Chapter 5, Section 5.12.2 for design of 3-bit magnitude comparator).

Verilog Program for Dataflow Model of 3-bit Comparator

```
module dfm_comparator_3_bit
  ( input a0, a1, a2, b0, b1, b2,
   output wire a_eq_b, a_gt_b, a_lt_b);

   wire a0_bar, a1_bar, a2_bar;
   wire b0_bar, b1_bar, b2_bar;
   wire m0, m1, m2, m3, m4, m5, m6, m7, m8;

   assign m0 = ~(a0 ^ b0),  m1 = ~(a1 ^ b1),  m2 = ~(a2 ^ b2);

   assign m3 = a2 & b2_bar;
   assign m4 = m2 & a1 & b1_bar;
   assign m5 = m2 & m1 & a0 & b0_bar;

   assign m6 = a2_bar & b2;
   assign m7 = m2 & a1_bar & b1;
   assign m8 = m2 & m1 & a0_bar & b0;

   assign a0_bar = ~a0, a1_bar = ~a1, a2_bar = ~a2;
   assign b0_bar = ~b0, b1_bar = ~b1, b2_bar = ~b2;

   assign a_eq_b = m0 & m1 & m2;
   assign a_gt_b = m3 | m4 | m5;
   assign a_lt_b = m6 | m7 | m8;
endmodule
```

Verilog Program 15.24

Write a Verilog code for structural model of a 2-bit multiplier designed using AND gates and 2 half adders by considereing half adder as,

a) a component b) a function c) a task

a) Program Logic for Structural Model

Let a_1a_0 and b_1b_0 be the two input data and $P_3P_2P_1P_0$ be the 4-bit product.

The Verilog code for structural model of 2-bit multiplier is developed using the logic circuit shown in Fig. 1.

The Boolean equations for half adder are,

Sum, s = a ⊕ b

Carry out, c_o = a.b

where, a and b are 1-bit inputs. (Refer Chapter 5, Section 5.9.1)

In structural model, the code for half adder is developed separately and 2 half adders are added as components in 2-bit multiplier. The AND gates in the logic circuit of Fig. 1 are modeled using primitive AND gates in verilog. (Refer Chapter 5, Section 5.13.1 for design of 2-bit multiplier).

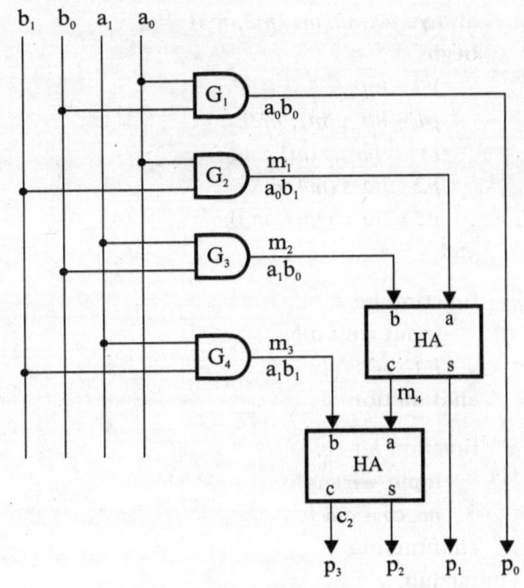

Fig. 1: Logic circuit of 2-bit multiplier.

Verilog Program for Dataflow Model of Half Adder

```
module dfm_half_adder (a, b, s, co);
    input wire a, b;
    output wire s, co;

    assign s = a ^ b;
    assign co = a & b;
endmodule
```

Verilog Program for Structural Model of 2-bit Multiplier

```
module sm_multiplier_2_bit
    ( input wire [1:0] a,
    input wire [1:0] b,
    output wire [3:0] p );

    wire m1, m2, m3, m4;

    and G1 ( p[0], a[0], b[0] );
    and G2 ( m1, a[0], b[1] );
    and G3 ( m2, a[1], b[0] );
    and G4 ( m3, a[1], b[1] );
    dfm_half_adder ha0 ( m1, m2, p[1], m4 );
    dfm_half_adder ha1 ( m4, m3, p[2], p[3]);
endmodule
```

b) Verilog program for structural model of 2-bit multiplier using function.

```
module multiplier_2_bit_with_function
    ( input wire a1, a0,
    input wire b1, b0,
    output reg p0, p1, p2, p3 );
    wire m0, m1, m2, m3;
    reg m4;

    and G1 (m0, a0, b0);
    and G2 (m1, a1, b1);
    and G3 (m2, a1, b0);
    and G4 (m3, a1, b1);
```

```
    always @ (m0, m1, m2, m3)
    begin
        p0 = m0;
        p1 = ha_s (m1, m2);
        m4 = ha_co (m1, m2);
        p2 = ha_s (m4, m3);
        p3 = ha_co (m4, m3);
    end

    function ha_s;
        input wire a,b;
        ha_s = a ^ b;
    endfunction

    function ha_co;
        input wire a,b;
        ha_co = a & b;
    endfunction
endmodule
```

c) Verilog program for structural model of 2-bit multiplier using task.

```
module multiplier_2_bit_with_task
    ( input wire a1, a0,
      input wire b1, b0,
      output reg p0, p1, p2, p3 );

      wire m0, m1, m2, m3;
      reg m4;

      and G1 (m0, a0, b0);
      and G2 (m1, a1, b1);
      and G3 (m2, a1, b0);
      and G4 (m3, a1, b1);

      always @ (m1, m2, p1, p2, p3)
      begin
         p0 = m0;
         ha (m1, m2, p1, m4);
         ha (m4, m3, p2, p3);
      end

      task ha;
         input wire a, b;
         output reg s, co;
         begin
            s  = a ^ b;
            co = a & b;
         end
      endtask
endmodule
```

Verilog Program 15.25

Write a Verilog code for structural model of parity bit generator to generate even and odd parity bit for an 8-bit data.

Program Logic for Structural Model

The Verilog code for structural model of parity bit generator for an 8-bit data is developed using the logic circuit shown in Fig. 1. The XOR and NOT gates in the logic circuit of Fig. 1 are modeled using primitive XOR and NOT gates in verilog. (Refer Chapter 5, Section 5.15.4 for design of parity bit generator).

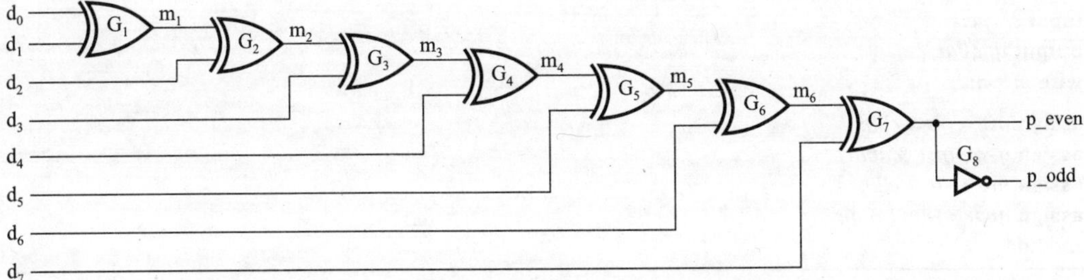

Fig. 1: Logic circuit of parity bit generator for an 8-bit data.

Verilog Program for Structural Model of Parity Bit Generator

```
module sm_parity_bit_generate
   ( input wire [7:0] d,
    output wire p_even,
    output wire p_odd );

    wire m1, m2, m3, m4, m5, m6;

    xor G1 (m1, d[0], d[1]);
    xor G2 (m2, m1, d[2]);
    xor G3 (m3, m2, d[3]);
    xor G4 (m4, m3, d[4]);
    xor G5 (m5, m4, d[5]);
    xor G6 (m6, m5, d[6]);
    xor G7 (p_even, m6, d[7]);
    not G8 (p_odd, p_even);
endmodule
```

Verilog Program 15.26

Write a Verilog code for dataflow, behavioral and structural model of a SR flip-flop using NAND gates.

Program Logic

The following Boolean equations for SR flip-flop are developed from Fig. 1.

$m_1 = (s \, . \, clk)'$

$m_2 = (r \, . \, clk)'$

$q = (m_1 \, . \, q')'$

$q' = (m_2 \, . \, q)'$

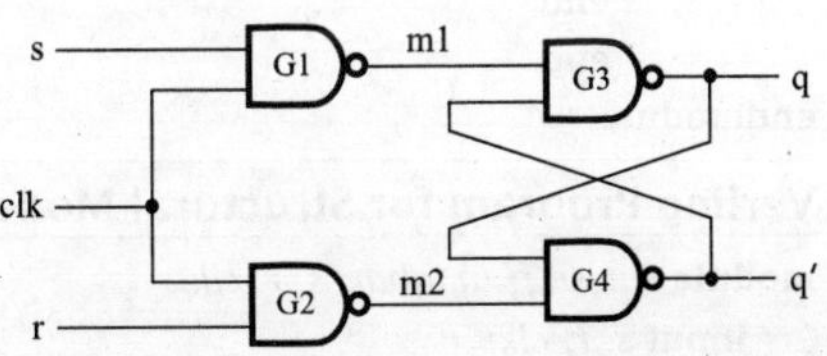

Fig. 1: NAND latch based SR flip-flop.

Where, s and r are 1-bit input data, q and q'(qbar) are 1-bit output data. The Verilog code for dataflow model of SR flip-flop is developed using the above equations. The behavioral model of SR flip-flop is developed using the truth table of SR flip-flop. The Verilog code for structural model of SR flip-flop is developed using the logic circuit shown in Fig. 1. The NAND gates in the logic circuit of Fig. 1 are modeled using primitive NAND gates in verilog. (Refer Chapter 6, Section 6.2.2 for design of SR flip-flop).

Truth Table of SR Flip-Flop

Inputs		Outputs	
S	R	Q	Q'
0	0	No change	
0	1	0	1
1	0	1	0
1	1	Indeterminate	

Verilog Program for Dataflow Model of SR Flip-Flop

```
module dfm_sr_ff (s, r, clk, q, qbar);
   input s, r, clk;
   output q, qbar;
   wire m1, m2;

   assign m1 = ~(s & clk);
   assign q  = ~(m1 & qbar);
   assign m2 = ~(r & clk);
   assign qbar = ~(m2 & q);
endmodule
```

Verilog Program for Behavioral Model of SR Flip-Flop using 'if' Statement

```
module bm_sr_ff (s, r, clk, q, qbar);
   input wire s, r, clk;
   output reg q, qbar;

   initial begin
      q = 0;  qbar = 1;
   end

   always @ (posedge (clk))
      begin
         if(s==0 & r==0)
            begin
               q = q;  qbar = qbar;
            end
         else if (s==0 & r==1)
            begin
               q = 0;  qbar = 1;
            end
         else if (s==1 & r==0)
            begin
               q = 1;  qbar = 0;
            end
         else begin
               q = q;    qbar = q;
         end
         end
endmodule
```

Verilog Program for Structural Model of SR Flip-Flop

```
module sm_sr_ff (q, qbar, s, r, clk);
   input s, r, clk;
   output q, qbar;
   wire m1, m2;
```

```
    nand G1(m1, s, clk);
    nand G2(m2, r, clk);
    nand G3(q, m1, qbar);
    nand G4(qbar, m2, q);
endmodule
```

Verilog Program 15.27

Write a Verilog code for dataflow, behavioral and structural model of a JK flip-flop using NAND gates.

Verilog Program Logic for Dataflow and Behavioral Model

The following Boolean equations for JK flip-flop are developed from Fig. 1.

$$m_1 = (j\,.\,clk\,.\,q')'$$

$$m_2 = (k\,.\,clk\,.\,q)'$$

$$q = (m_1\,.\,q')'$$

$$q' = (m_2\,.\,q)'$$

Where, j and k are 1-bit input data, q and q'(qbar) are 1-bit output data. The Verilog code for dataflow model of JK flip-flop is developed using the above equations. The behavioral model of JK flip-flop is developed using the truth table of JK flip-flop. The Verilog code for structural model of JK flip-flop is developed using the logic circuit shown in Fig. 1. The NAND gates in the logic circuit of Fig. 1 are modeled using primitive NAND gates in verilog. (Refer Chapter 6, Section 6.2.3 for design of JK flip-flop).

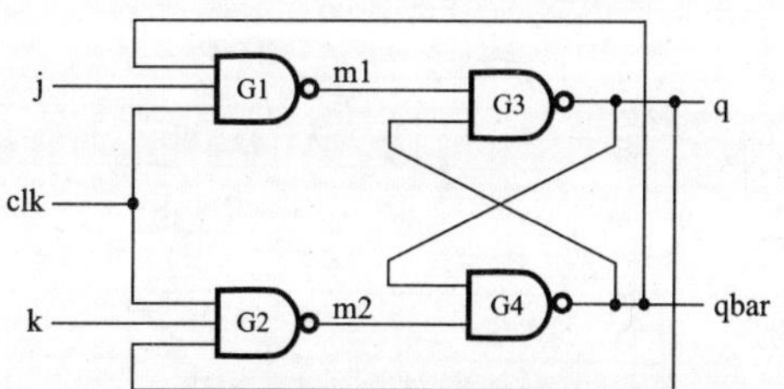

Fig. 1: JK flip-flop.

Truth Table of JK Flip-Flop

Inputs		Outputs	
J	K	Q	Q'
0	0	No change	
0	1	0	1
1	0	1	0
1	1	Toggle	

Verilog Program for Dataflow Model of JK Flip-Flop

```
module dfm_jk_ff (j, k, clk, q, qbar);
    input wire j, k, clk;
    output wire q, qbar;
    wire m1, m2, m3, m4;
    reg q1, q1bar;

    initial begin
        q1 = 0;  q1bar = ~q;
    end

    assign m1 = ~(j & clk & q1bar);
    assign m3  = ~(m1 & q1bar);
    assign m2 = ~(k & clk & q1);
    assign m4 = ~(m2 & q1);
    assign q = m3;
    assign qbar = ~q;
endmodule
```

Verilog Program for Behavioral Model of JK Flip-Flop

```
module bm_jk_ff (j, k, clk, q, qbar);
   input j, k, clk;
   output q, qbar;
   reg q, qbar;

   initial begin
      q = 0;   qbar = 1;
   end

   always @ (posedge (clk))
      begin
         if (j==0 & k==0)  begin
            q = q;   qbar = qbar; end
         else if (j==0 & k==1) begin
            q = 0;   qbar = 1;      end
         else if (j==1 & k==0) begin
            q = 1;   qbar = 0;      end

         else begin
            q = qbar; qbar = ~q; end
      end
endmodule
```

Verilog Program for Structural Model of JK Flip-Flop

```
module sm_jk_ff (j, k, clk, q, qbar);
   input wire j, k, clk;
   output wire q;
   output wire qbar;

   wire m1, m2, m3, m4;
   reg q1;   reg q1bar;

   initial begin
      q1 = 0;  q1bar = ~q1;
   end

   nand G1 (m1, j, clk, q1bar);
   nand G2 (m2, k, clk, q1);
   nand G3 (m3, m1, q1bar);
   nand G4 (m4, m2, q1);

   assign q = m3;
   assign qbar = ~q;
endmodule
```

Verilog Program 15.28

Write a Verilog code for behavioral model of a D flip-flop.

Program Logic for Behavioral Model

The Boolean equations for D flip-flop are,

$$q = d \quad ; \quad qbar = q'$$

Where, d is 1-bit input data q and q'(qbar) are 1-bit output data. The Verilog code for behavioral model of D flip-flop is developed using the above equations. (Refer Chapter 6, Section 6.2.5 for design of D flip-flop).

Verilog Program for Behavioral Model of D Flip-Flop

```
module bm_d_ff (d, clk, q, qbar);
   input  d, clk;
   output q, qbar;
   reg q, qbar;

   always @ (posedge clk)
      begin
         q = d; qbar = ~q;
   end
endmodule
```

Verilog Program 15.29

Write a Verilog code for structural model of a T flip-flop.

Program Logic for Structural Model

The Verilog code for structural model of T flip-flop is developed using the logic circuit shown in Fig. 1. The NAND gates in the logic circuit of Fig. 1 are modeled using primitive NAND gates in verilog.

The following Boolean equations for T flip-flop are developed from Fig. 1.

$m_1 = (T.clk.q')'$

$m_2 = (T.clk.q)'$

$q = (m_1.q')'$

$q' = (m_2.q)'$

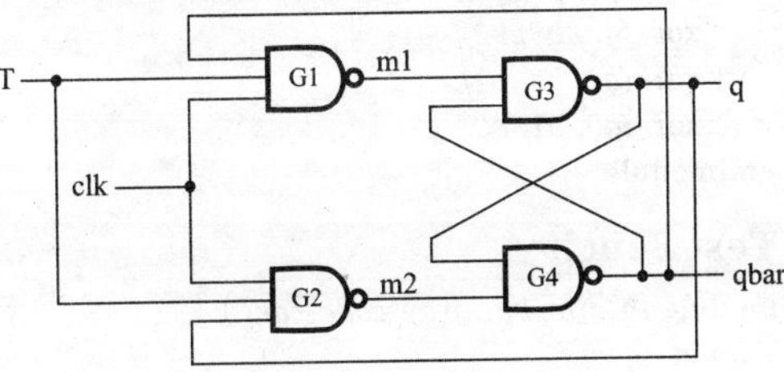

Fig. 1: T flip-flop.

Verilog Program for Structural Model of T Flip-Flop

```
module sm_t_ff (t, clk, q, qbar);
   input wire t, clk;
   output wire q;
   output wire qbar;

   wire m1, m2, m3, m4;
   reg q1;
   reg q1bar;

   initial begin
      q1 = 0;
      q1bar = ~q1;
   end

   nand G1 (m1, t, clk, q1bar);
   nand G2 (m2, t, clk, q1);
   nand G3 (m3, m1, q1bar);
   nand G4 (m4, m2, q1);

   assign q = m3;
   assign qbar = ~q;
endmodule
```

15.14 Verilog Programs With Test Bench

Verilog Program 15.30

Write Verilog code for dataflow model of a binary to gray code converter.

Program Logic for Dataflow Model

The following procedure will be useful for bit-by-bit conversion of n-bit binary to n-bit gray code.

Step 1: The most significant bit of binary and gray code are same. Hence retain the most significant bit of binary as most significant bit of gray code.

Step 2: Starting from most significant bit of binary, add two consecutive bits of binary and record the sum as bit of gray code and discard the carry in every addition.

The Verilog code for dataflow model of binary to gray code converter is developed using the above procedure. (Refer Chapter 5, Section 5.3 for design of binary to gray code).

Verilog Program for Dataflow Model of Binary to Gray Code

```
module dfm_bin_gray_code_con
   ( input  b3, b2, b1, b0,
    output g3, g2, g1, g0 );

    and (g3, b3, b3);
    xor  (g2, b3, b2);
    xor  (g1, b2, b1);
    xor  (g0, b1, b0);
endmodule
```

Test Bench

```
module tb_dfm_bin_gray_code_con( );
//Inputs
   reg b3;   reg b2;   reg b1;   reg b0;

//Outputs
   wire g3;  wire g2;  wire g1;  wire g0;

//Instantiate the unit under test
   dfm_bin_gray_code_con uut
       (.b3(b3), .b2(b2), .b1(b1), .b0(b0), .g3(g3), .g2(g2), .g1(g1), .g0(g0));

//Stimulus inputs
   initial begin
            b3 = 1'b0;   b2 = 1'b1;   b1 = 1'b0;   b0 = 1'b1;
      #10   b3 = 1'b0;   b2 = 1'b1;   b1 = 1'b1;   b0 = 1'b1;
      #10   b3 = 1'b1;   b2 = 1'b0;   b1 = 1'b0;   b0 = 1'b1;
      #10   b3 = 1'b1;   b2 = 1'b0;   b1 = 1'b1;   b0 = 1'b0;
      #10 $finish;
      end
endmodule
```

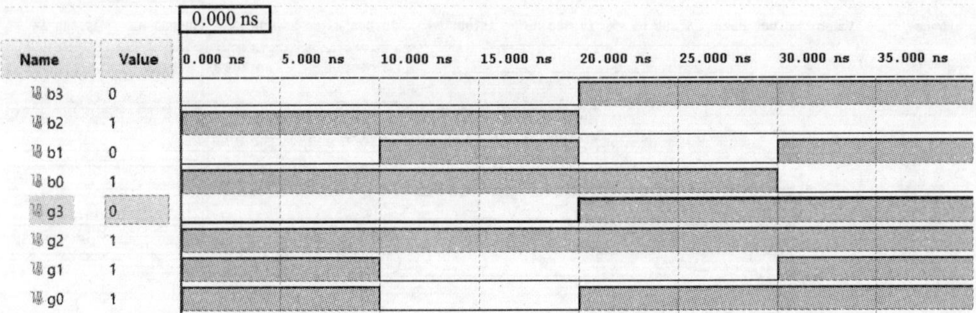

Fig. a: *Inputs: b3 b2 b1 b0 = 0101; Outputs: g3 g2 g1 g0 = 0111.*

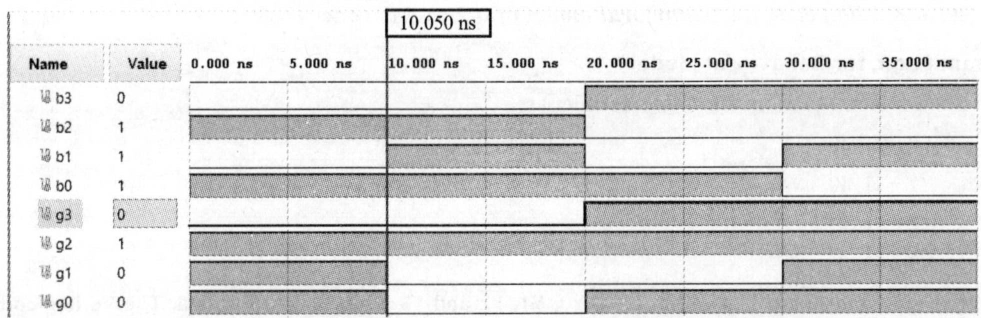

Fig. b: *Inputs: b3 b2 b1 b0 = 0111; Outputs: g3 g2 g1 g0 = 0100.*

Fig. c: *Inputs: b3 b2 b1 b0 = 1001; Outputs: g3 g2 g1 g0 = 1101.*

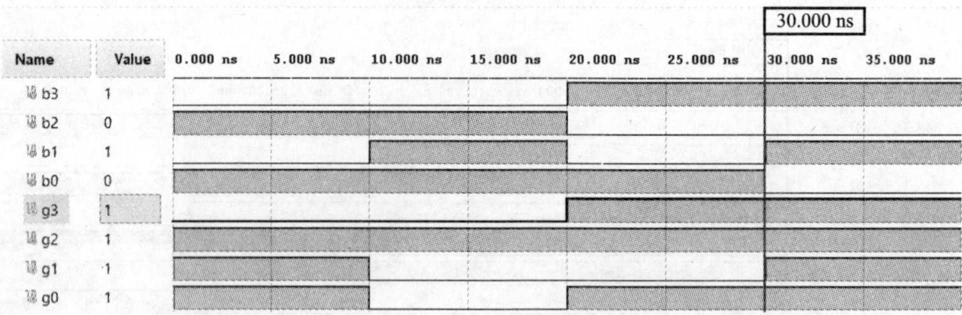

Fig. d: Inputs: b3 b2 b1 b0 = 1010; Outputs: g3 g2 g1 g0 = 1111.

Fig. 1: Inputs-output waveform of binary to gray code converter.

Verilog Program 15.31

Write Verilog code for behavioral model of a 2-to-4 decoder.

Program Logic for Behavioral Model

The Boolean equations for 2-to-4 decoder are,

Y0 = I1'.I0'

Y1 = I1'.I0

Y2 = I1.I0'

Y3 = I1.I0

Where, I1 and I0 are 1-bit input data, Y0, Y1, Y2 and Y3 are 1-bit output data. The Verilog code for behavioral model of 2-to-4 decoder is developed using the above equations. (Refer Chapter 5, Section 5.4.1 for design of 2-to-4 decoder).

Verilog Program for Behavioral Model of 2-to-4 Decoder

```
module bm_decoder2_4
  ( input  i1, i0,
    output y0, y1, y2, y3 );

   reg y0, y1, y2, y3;

always @(i1 or i0)
   begin
       y0 <= (!i1) && (!i0);
       y1 <= (!i1) && i0;
       y2 <= i1    && (!i0);
       y3 <= i1    && i0;
   end
endmodule
```

Test Bench

```
module tb_bm_decoder2_4( );
//Inputs
   reg i1;   reg i0;
```

```
//Outputs
    wire y0;  wire y1;  wire y2;  wire y3;

//Instantiate the unit under test
    bm_decoder2_4 uut
        (.i1(i1),  .i0(i0),  .y0(y0),  .y1(y1),  .y2(y2),  .y3(y3));

//Stimulus inputs
    initial begin
            i1 = 1'b0;  i0 = 1'b0;
        #10  i1 = 1'b0;  i0 = 1'b1;
        #10  i1 = 1'b1;  i0 = 1'b0;
        #10  i1 = 1'b1;  i0 = 1'b1;
        #10 $finish;
    end
endmodule
```

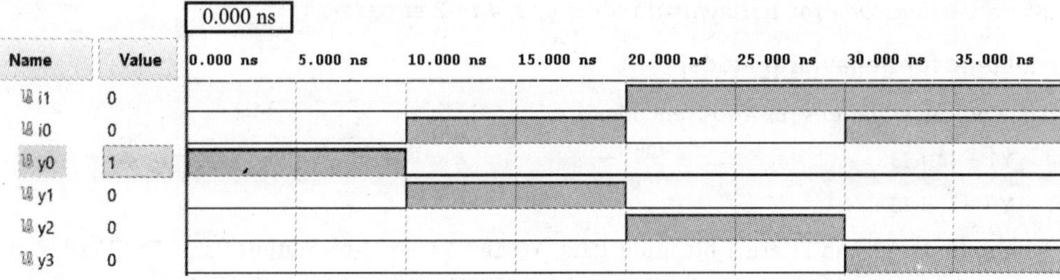

Fig. a: Inputs: i0 i1 = 00; Outputs: y0 y1 y2 y3 = 1000.

Fig. b: Inputs: i0 i1 = 10; Outputs: y0 y1 y2 y3 = 0100.

Fig. c: Inputs: i0 i1 = 01; Outputs: y0 y1 y2 y3 = 0010.

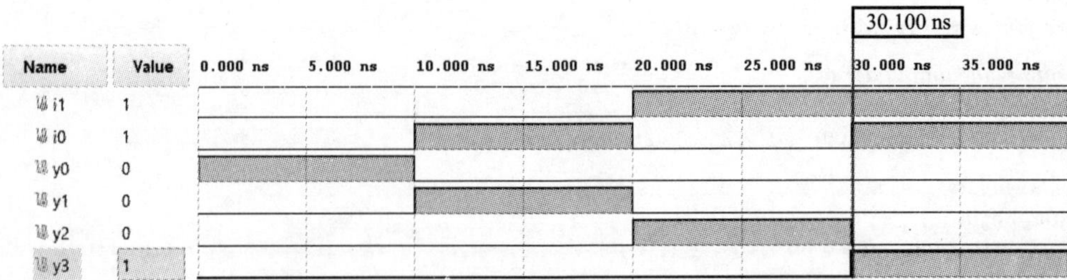

Fig. d: Inputs: i0 i1 = 11; Outputs: y0 y1 y2 y3 = 0001.

***Fig. 1:** Inputs-output waveform of 2*4 decoder.*

Verilog Program 15.32

Write Verilog code for behavioral model of a 4-to-2 encoder.

Program Logic for Behavioral Model

The Boolean equations for 4-to-2 encoder are,

Y1 = I2 + I3

Y0 = I1 + I3

Where, I0, I1, I2 and I3 are 1-bit input data, Y0 and Y1 are 1-bit output data. The Verilog code for behavioral model of 4-to-2 encoder is developed using the above equations. (Refer Chapter 5, Section 5.5.1 for design of 4-to-2 encoder).

Verilog Program for Behavioral Model of 4-to-2 Encoder

```
module bm_encoder_4_2
   ( input  i0, i1, i2, i3,
    output y0, y1 );

   reg y0,y1;

always @(i0 or i1 or i2 or i3)
   begin
      y0 <= i1 | i3;
      y1 <= i2 | i3;
   end
endmodule
```

Test Bench

```
module tb_bm_encoder_4_2( );
//Inputs
   reg i0;    reg i1;    reg i2;    reg i3;
//Outputs
   wire y0;  wire y1;
//Instantiate the unit under test
   bm_encoder_4_2 uut
      (.i0(i0),  .i1(i1),  .i2(i2),  .i3(i3),  .y0(y0),  .y1(y1));
```

```
//Stimulus inputs
    initial begin
             i0 = 1'b1;  i1 = 1'b0;  i2 = 1'b0;  i3 = 1'b0;
       #10   i0 = 1'b0;  i1 = 1'b1;  i2 = 1'b0;  i3 = 1'b0;
       #10   i0 = 1'b0;  i1 = 1'b0;  i2 = 1'b1;  i3 = 1'b0;
       #10   i0 = 1'b0;  i1 = 1'b0;  i2 = 1'b0;  i3 = 1'b1;
       #10 $finish;
      end
endmodule
```

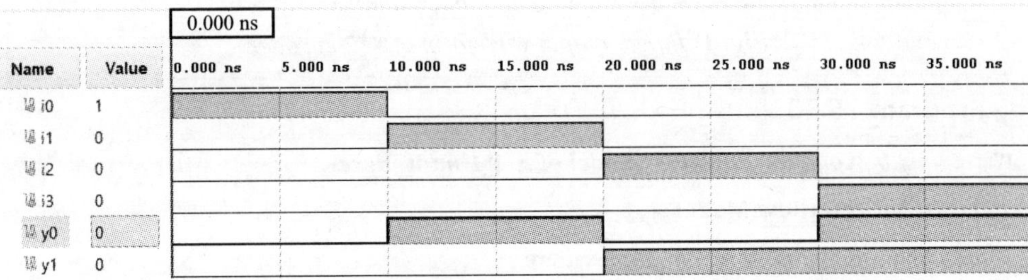

Fig. a: *Inputs: i0 i1 i2 i3 = 1000; Outputs: y0 y1 = 00.*

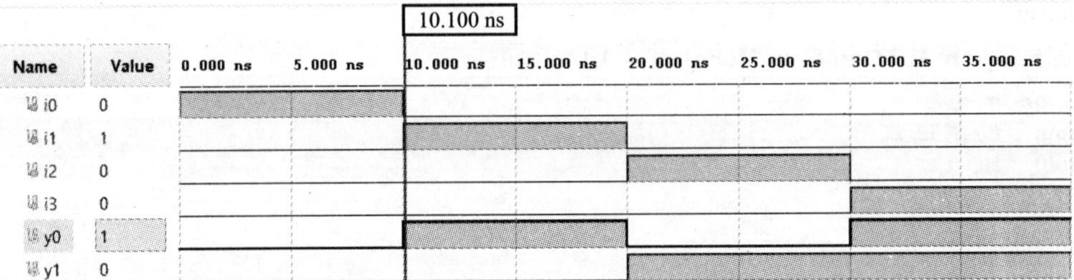

Fig. b: *Inputs: i0 i1 i2 i3 = 0100; Outputs: y0 y1 = 10.*

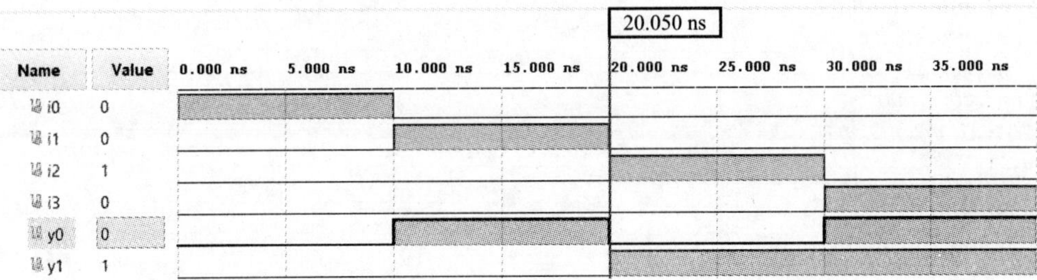

Fig. c: *Inputs: i0 i1 i2 i3 = 0010; Outputs: y0 y1 = 01.*

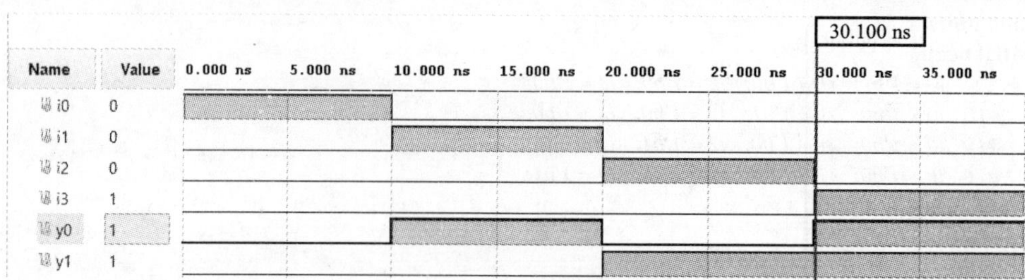

Fig. d: Inputs: i0 i1 i2 i3 = 0001; Outputs: y0 y1 = 11.

*Fig. 1: Inputs-output waveform of 4*2 encoder.*

Verilog Program 15.33

Write Verilog code for dataflow model of a 4*1 multiplexer.

Program Logic for Dataflow Model

The Boolean equations for 4*1 multiplexer are,

Y = S0'.S1'.I0 + S0'.S1.I1 + S0.S1'. I2 + S0.S1.I3

Where, S0 and S1 are 1-bit selection input, Y is 1-bit output data. The Verilog code for dataflow model of 4*1 multiplexer is developed using the above equations. (Refer Chapter 5, Section 5.7 for design of 4*1 multiplexer).

Verilog Program for Dataflow Model of 4*1 Multiplexer

```
module dfm_mux_4_1
   ( input   i0, i1, i2, i3,
     input   s0, s1,
     output  y );

   wire ns0;  wire ns1;
   wire m1;  wire m2;  wire m3;  wire m4;

   not (ns0, s0);
   not (ns1, s1);

   and (m1, ns0, ns1, i0);
   and (m2, ns0, s1, i1);
   and (m3, s0, ns1, i2);
   and (m4, s0, s1, i3);
   or   (y, m1, m2, m3, m4);
endmodule
```

Test Bench

```
module tb_dfm_mux_4_1( );
//Inputs
   reg i0;  reg i1;  reg i2;  reg i3;  reg s0;  reg s1;

//Outputs
   wire y;

//Instantiate the unit under test
   dfm_mux_4_1 uut
      (.i0(i0),  .i1(i1),  .i2(i2),  .i3(i3),  .s0(s0),  .s1(s1),   .y(y));
```

```
//Stimulus inputs
    initial begin
            i0 = 1'b1;  i1 = 1'b0;  i2 = 1'b0;  i3 = 1'b0;   s0 = 1'b0;  s1 = 1'b0;
        #10 i0 = 1'b0;  i1 = 1'b0;  i2 = 1'b0;  i3 = 1'b0;   s0 = 1'b0;  s1 = 1'b1;
        #10 i0 = 1'b0;  i1 = 1'b0;  i2 = 1'b1;  i3 = 1'b0;   s0 = 1'b1;  s1 = 1'b0;
        #10 i0 = 1'b0;  i1 = 1'b0;  i2 = 1'b0;  i3 = 1'b0;   s0 = 1'b1;  s1 = 1'b1;
        #10 $finish;
    end
endmodule
```

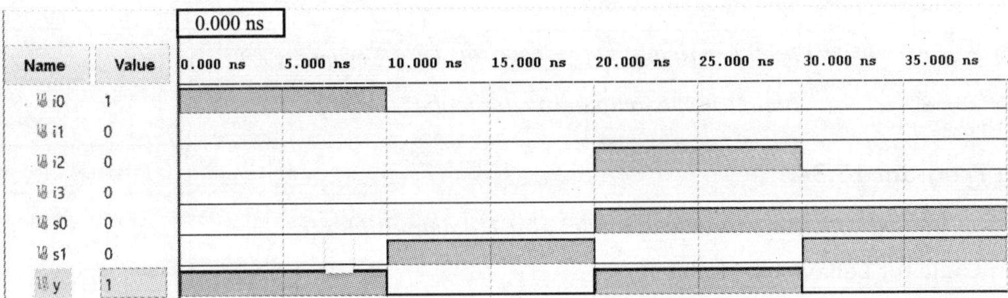

Fig. a: *Inputs: i0 i1 i2 i3 = 1000, s0 s1 = 00; Output: y = 1.*

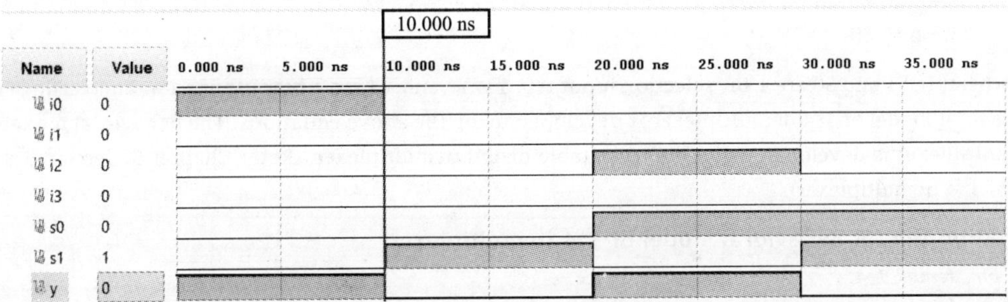

Fig. b: *Inputs: i0 i1 i2 i3 = 0000, s0 s1 = 01; Output: y = 0.*

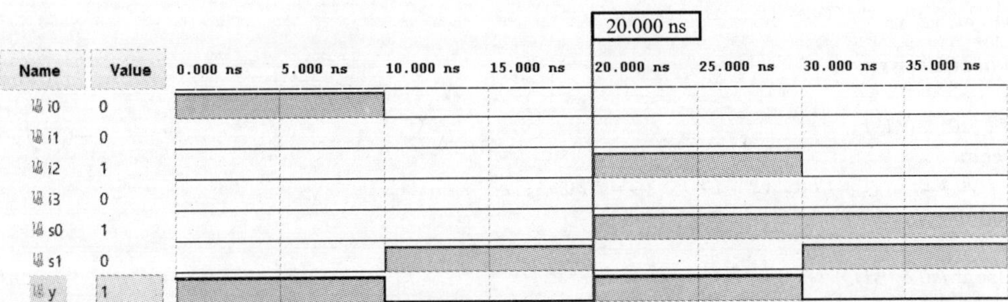

Fig. c: *Inputs: i0 i1 i2 i3 = 0010, s0 s1 = 10; Output: y = 1.*

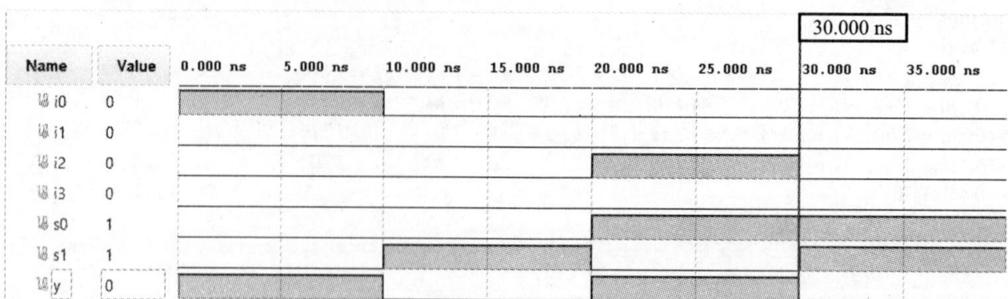

Fig. d: Inputs: i0 i1 i2 i3 = 0000, s0 s1 = 11; Output: y = 0.

*Fig. 1: Inputs-output waveform of 4*1 multiplexer.*

Verilog Program 15.34

Write Verilog code for behavioral model of a 1*4 demultiplexer.

Program Logic for Behavioral Model

The Boolean equations for 1*4 demultiplexer are,

Y0 = Ii.S1'.S0'

Y1 = Ii.S1'.S0

Y2 = Ii.S1.S0'

Y3 = Ii.S1.S0

Where, Ii, S1 and S0 are 1-bit selection input, Y0, Y1, Y2 and Y3 are 1-bit output data. The Verilog code for behavioral model of 1*4 demultiplexer is developed using the above equations. The behavioral models of 1*4 demultiplexer is developed using the truth table of 1*4 demultiplexer. (Refer Chapter 5, Section 5.8 for design of 1*4 demultiplexer).

Verilog Program for Behavioral Model of 1*4 Demultiplexer

```
module bm_demux_1_4
   ( input   ii,
     input   s1, s0,
     output  y0, y1, y2, y3 );

   reg y0, y1, y2, y3;

always @(ii or s0 or s1)
   begin
      if ( (!s0) & (!s1) )
      begin
         y0 = ii;   y1 = 1'b0;  y2 = 1'b0;   y3 = 1'b0;
      end

      else if (s0 & (!s1) )
      begin
         y0 = 1'b0;  y1 = ii;    y2 = 1'b0;   y3 = 1'b0;
      end
```

```
        else if ( (!s0) & s1 )
        begin
            y0 = 1'b0;  y1 = 1'b0;  y2 = ii;      y3 = 1'b0;
        end
        else if ( s0 & s1)
        begin
            y0 = 1'b0;  y1 = 1'b0;  y2 = 1'b0;  y3 = ii;
        end
        end
endmodule
```

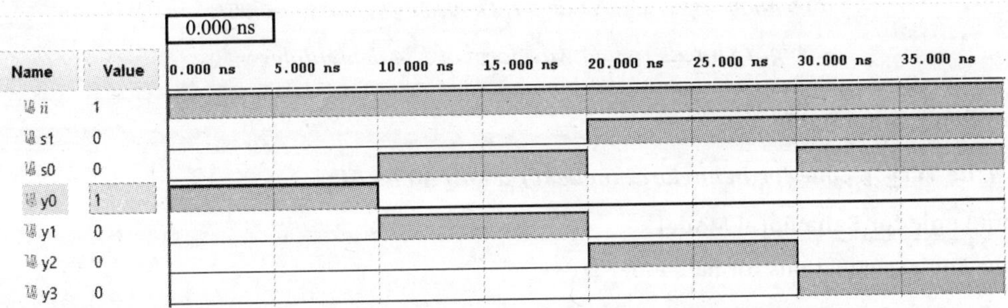

Fig. a: *Inputs: s0 s1 = 01, ii = 1; Outputs: y0 y1 y2 y3 = 1000.*

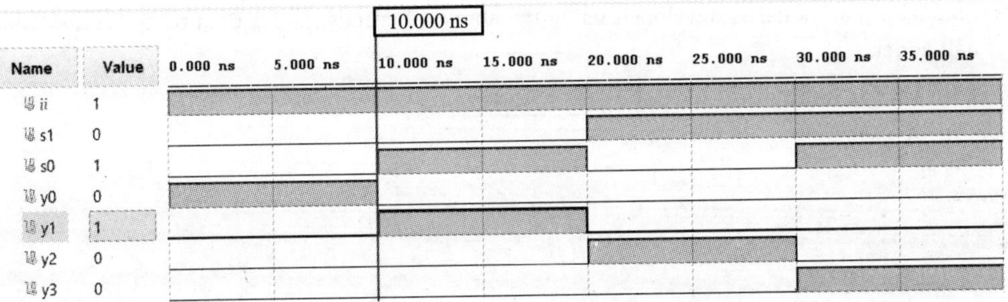

Fig. b: *Inputs: s0 s1 = 10, ii = 1; Outputs: y0 y1 y2 y3 = 0100.*

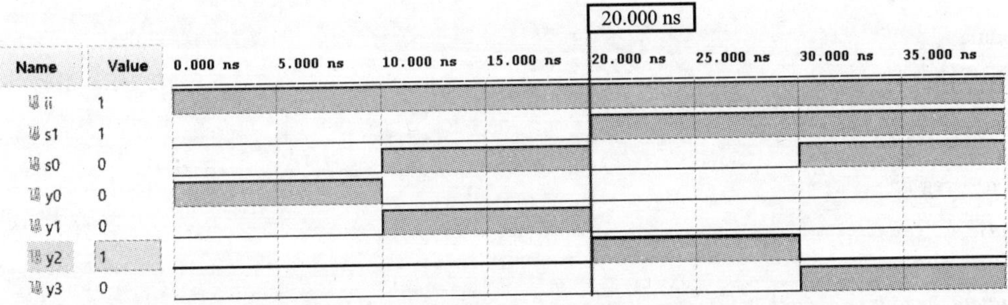

Fig. c: *Inputs: s0 s1 = 01, ii = 1; Outputs: y0 y1 y2 y3 = 0010.*

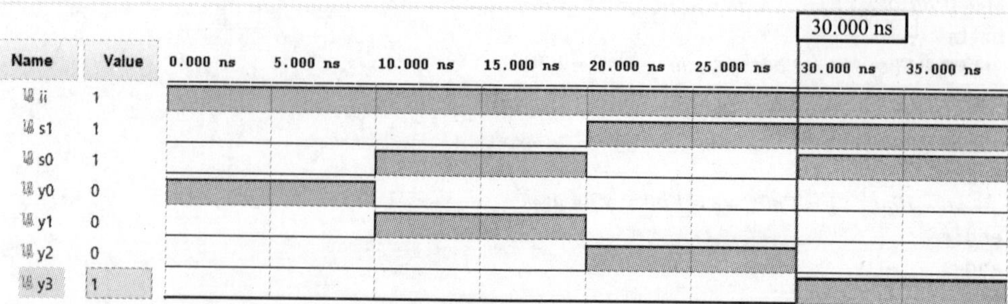

Fig. d: Inputs: s0 s1 = 11, ii = 1; Outputs: y0 y1 y2 y3 = 0001.

*Fig. 1: Inputs-output waveform of 1*4 demultiplexer.*

Verilog Program 15.35

Write Verilog code for behavioral model of a half adder.

Program Logic for Behavioral Model

The Boolean equations for half adder are,

$$s = a \oplus b$$

$$c_o = a\,b$$

Where, a and b are 1-bit input data, sum (s) and carry (c_o) are 1-bit output data. The Verilog code for behavioral model of half adder is developed using the above equations. (Refer Chapter 5, Section 5.9.1 for design of half adder).

Verilog Program for Behavioral Model of Half Adder

```
module bm_half_adder
   ( input  a, b,
     output s, co );

     reg s, co;

always @ ( a or b )
   begin
      s  = a ^ b;
      co = a & b;
   end
endmodule
```

Test Bench

```
module tb_bm_half_adder( );
//Inputs
   reg a;   reg b;

//Outputs
   wire s;  wire co;

//Instantiate the unit under test
   bm_half_adder uut
      (.a(a),  .b(b),   .s(s),  .co(co));
```

```
//Stimulus inputs
  initial begin
          a = 1'b0;  b = 1'b0;
      #10 a = 1'b0;  b = 1'b1;
      #10 a = 1'b1;  b = 1'b0;
      #10 a = 1'b1;  b = 1'b1;

      #10 $finish;
  end
endmodule
```

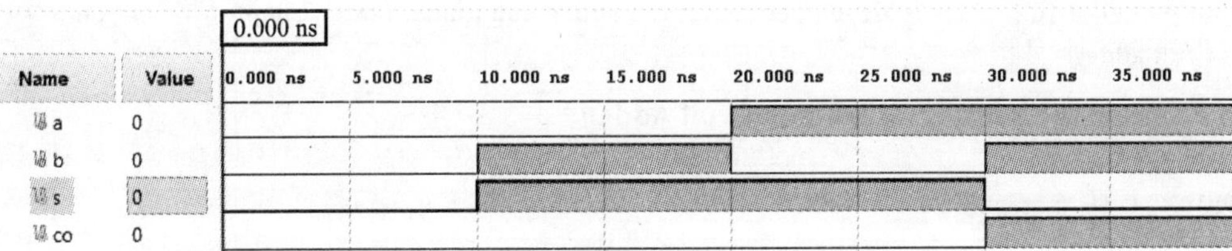

Fig. a: Inputs: a = 0, b = 0; Outputs: s = 0, co = 0.

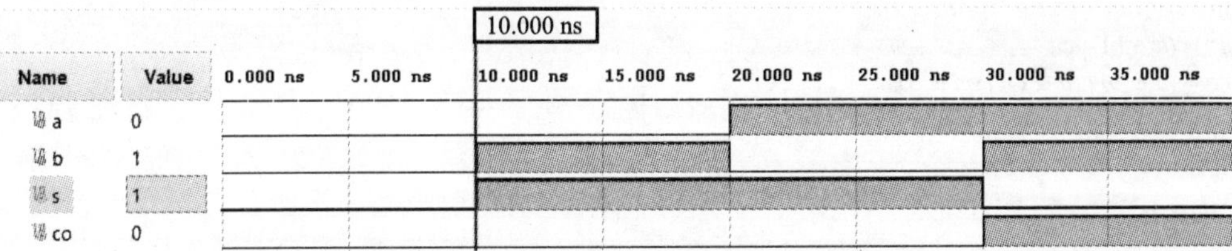

Fig. b: Inputs: a = 0, b = 1; Outputs: s = 1, co = 0.

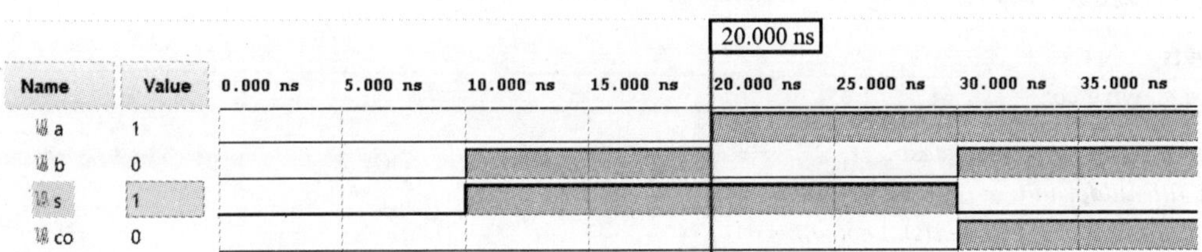

Fig. c: Inputs: a = 1, b = 0; Outputs: s = 1, co = 0.

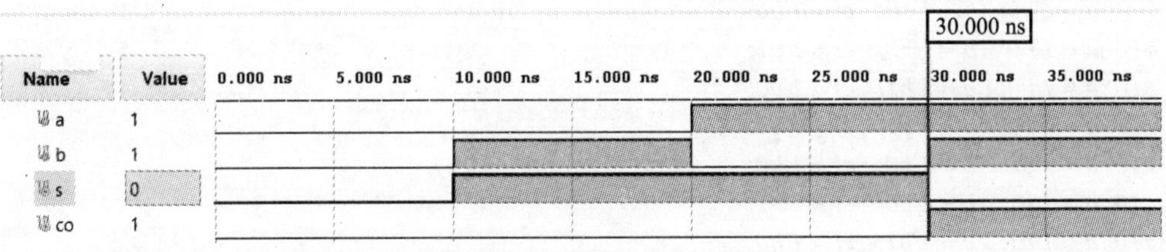

Fig. d: Inputs: a = 1, b = 1; Outputs: s = 0, co = 1.

Fig. 1: Inputs-output waveform of half adder.

Verilog Program 15.36

Write Verilog code for behavioral model of a full adder.

Program Logic for Behavioral Model

The Boolean equations for full adder are,

$$s = a \oplus b \oplus c_i$$

$$c_o = (a\,b) + (b\,c_i) + (c_i\,a)$$

Where, a, b and c_i are 1-bit input data, sum (s) and carry (c_o) are 1-bit output data. The Verilog code for behavioral model of full adder is developed using the above equations. (Refer Chapter 5, Section 5.9.2 for design of full adder).

Verilog Program for Behavioral Model of Full Adder

```
module bm_full_adder
   ( input   a, b, ci,
     output s, co );

   reg s, co;

always @ (a or b or ci)
   begin
      s   = (a ^ b) ^ ci;
      co = (a & b) | (b & ci) | (ci & a);
   end
endmodule
```

Test Bench

```
module tb_bm_full_adder ( );
//Inputs
   reg a;   reg b;   reg ci;

//Outputs
   wire s; wire co;

//Instantiate the unit under test
   bm_full_adder uut
      (.a(a), .b(b), .ci(ci), .s(s), .co(co));

//Stimulus inputs
   initial begin
              a = 1'b0;  b = 1'b0;  ci = 1'b0;
      #10  a = 1'b0;  b = 1'b0;  ci = 1'b1;
      #10  a = 1'b0;  b = 1'b1;  ci = 1'b0;
      #10  a = 1'b0;  b = 1'b1;  ci = 1'b1;
      #10  a = 1'b1;  b = 1'b0;  ci = 1'b0;
      #10  a = 1'b1;  b = 1'b0;  ci = 1'b1;
      #10  a = 1'b1;  b = 1'b1;  ci = 1'b0;
      #10  a = 1'b1;  b = 1'b1;  ci = 1'b1;
      #10 $finish;
   end
endmodule
```

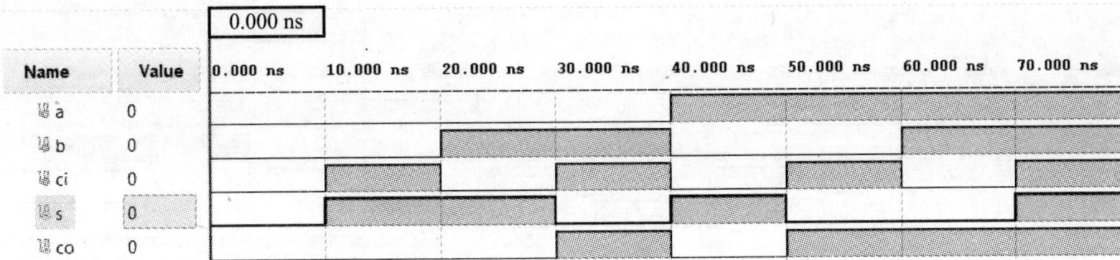

***Fig. a:** Inputs: a = 0, b = 0, ci = 0; Outputs: s = 0, co = 0.*

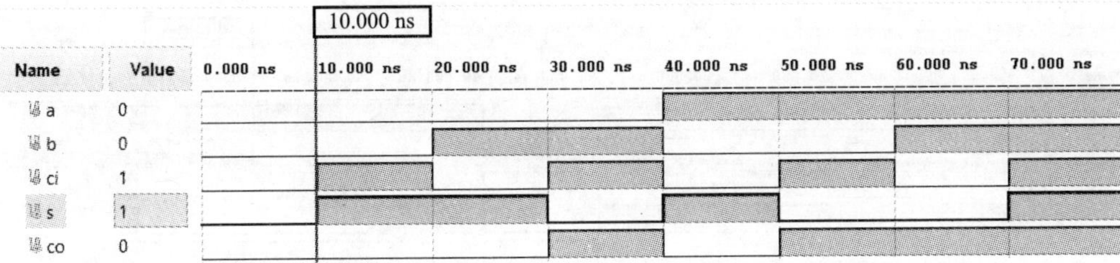

***Fig. b:** Inputs: a = 0, b = 0, ci = 1; Outputs: s = 1, co = 0.*

***Fig. c:** Inputs: a = 0, b = 1, ci = 0; Outputs: s = 1, co = 0.*

***Fig. d:** Inputs: a = 0, b = 1, ci = 1; Outputs: s = 0, co = 1.*

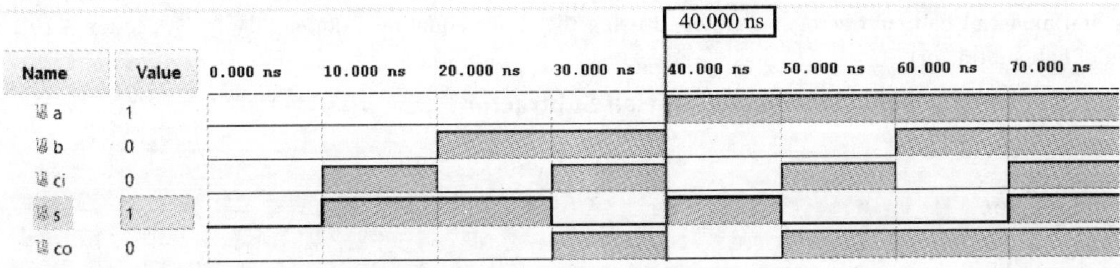

***Fig. e:** Inputs: a = 1, b = 0, ci = 0; Outputs: s = 1, co = 0.*

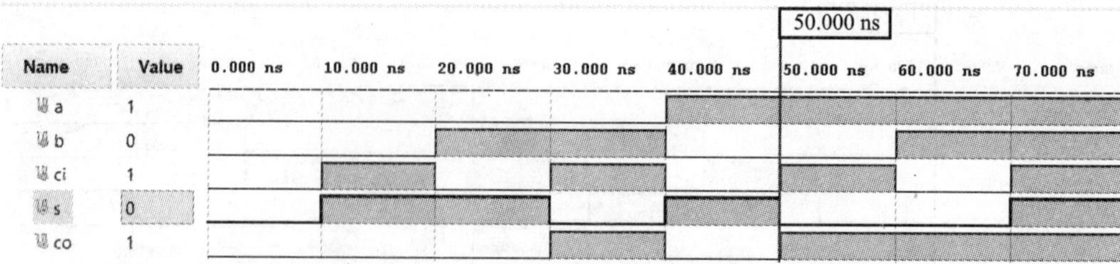

Fig. f: Inputs: a = 1, b = 0, ci = 1; Outputs: s = 0, co = 1.

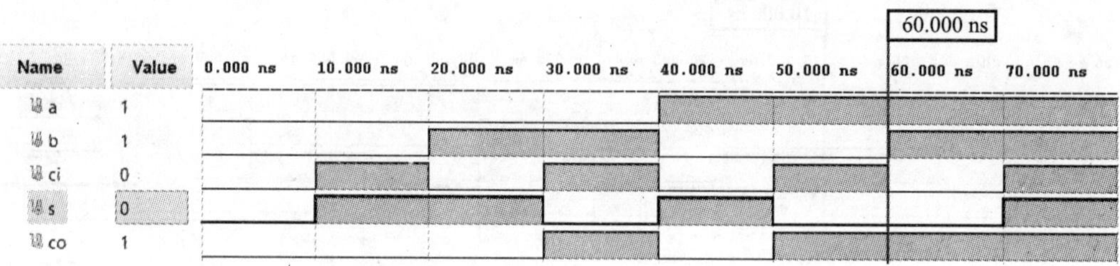

Fig. g: Inputs: a = 1, b = 1, ci = 0; Outputs: s = 0, co = 1.

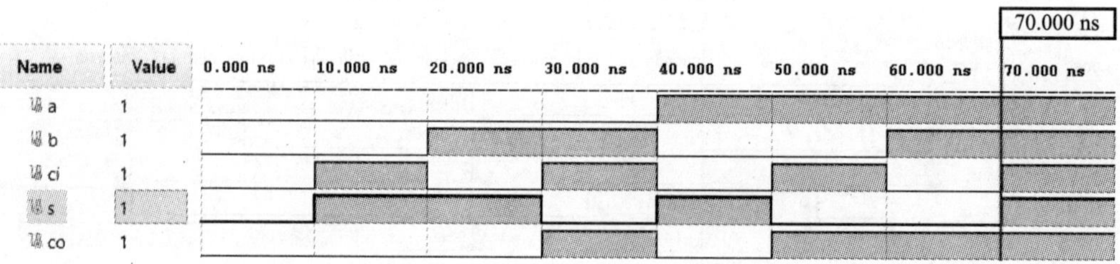

Fig. h: Inputs: a = 1, b = 1, ci = 1; Outputs: s = 1, co = 1.

Fig. 1: Inputs-output waveform of full adder.

Verilog Program 15.37

Write Verilog code for behavioral model of a half subtractor.

Program Logic for Behavioral Model

The Boolean equations for half subtractor are,

$$d = a \oplus b \quad ; \quad c_o = a'b$$

Where, a and b are 1-bit input data, difference (d) and borrow (c_o) are 1-bit output data. The Verilog code for behavioral model of half subtractor is developed using the above equations. (Refer Chapter 5, Section 5.10.1 for design of half subtractor).

Verilog Program for Behavioral Model of Half Subtractor

```
module bm_half_subtractor
   (input a, b,
    output d, co);
    reg d, co;
always @ (a or b)
```

```
    begin
        d  = a ^ b;
        co = (!a) & b;
    end
endmodule
```

Test Bench

```
module tb_half_subtractor( );
//Inputs
    reg a;   reg b;

//Outputs
    wire d;  wire co;

//Instantiate the unit under test
    bm_half_subtractor uut
        (.a(a),  .b(b),  .d(d),  .co(co));

//Stimulus inputs
    initial begin
            a = 1'b0;  b = 1'b0;
        #10 a = 1'b0;  b = 1'b1;
        #10 a = 1'b1;  b = 1'b0;
        #10 a = 1'b1;  b = 1'b1;
        #10 $finish;
    end
endmodule
```

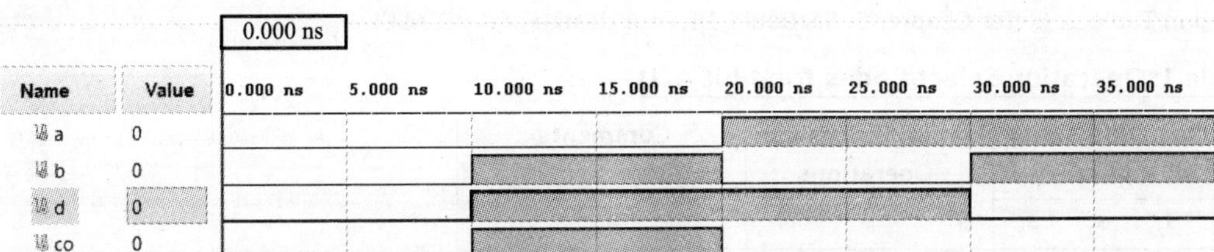

Fig. a: *Inputs: a = 0, b = 0; Outputs: d = 0, co = 0.*

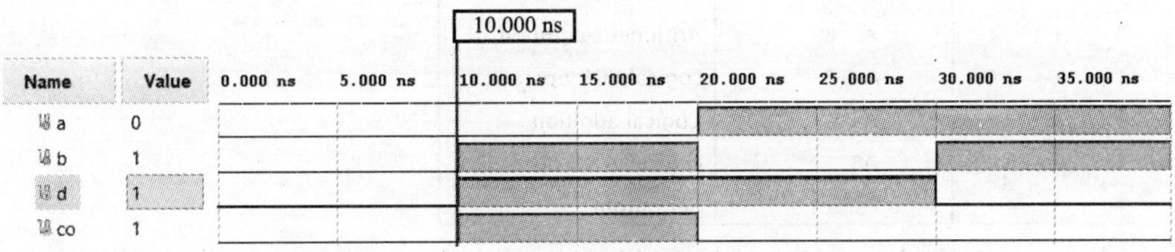

Fig. b: *Inputs: a = 0, b = 1; Outputs: d = 1, co = 1.*

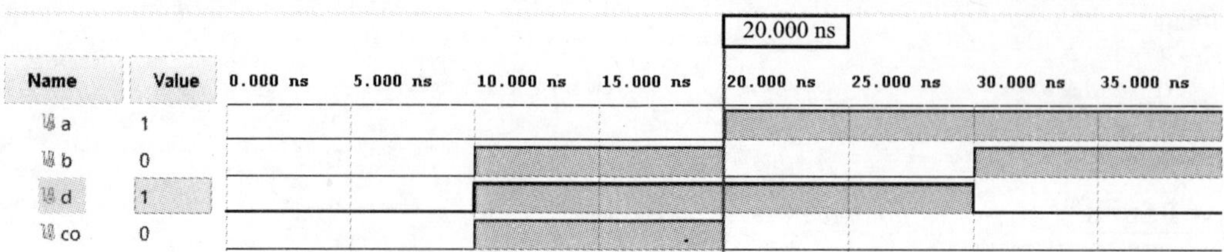

Fig. c: Inputs: a = 1, b = 0; Outputs: d = 1, co = 0.

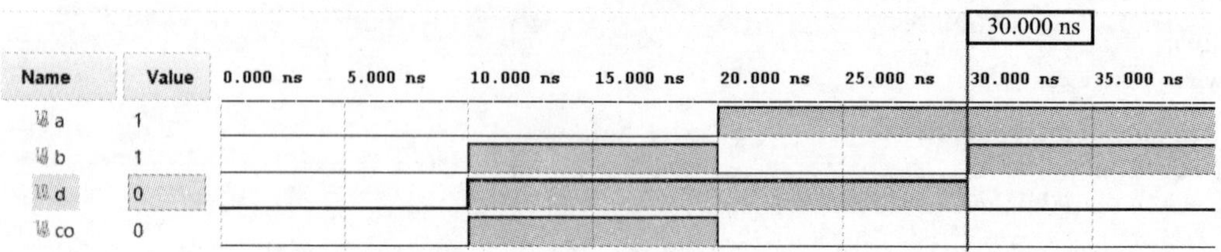

Fig. d: Inputs: a = 1, b = 1; Outputs: d = 0, co = 0.

Fig. 1: Inputs-output waveform of half subtractor.

Verilog Program 15.38

Write verilog code for behavioral model of a 4-bit ALU.

Program Logic for Behavioral Model

The verilog codes for behavioral model of 4-bit ALU are developed to perform 8 different operations listed in Table 1. (Refer Chapter 5, Section 5.14.1 for design of 4-bit ALU).

Table 1: Operation Select Codes for 4-bit ALU

Operation Select Inputs			Logical/Arithmetic Operations	Comments
S_2	S_1	S_0		
0	0	0	CLEAR	Clear output
0	0	1	B – A	Arithmetic subtraction
0	1	0	A – B	Arithmetic subtraction
0	1	1	A + B	Arithmetic addition
1	0	0	$A \oplus B$	Logical XOR operation
1	0	1	A + B	Logical addition
1	1	0	AB	Logical AND operation
1	1	1	PRESET	Set output

Verilog Program for Behavioral Model of 4-bit ALU

```
module bm_alu_4bit
  ( input clk,
    input reset,
    input [3:0] a,
    input [3:0] b,
    input [2:0] sel,
```

```
    output reg [3:0] f);
initial begin
   if (reset)
      f = 4'b0000;  //Clear output register
   end

always @(posedge clk) begin
   if (reset==0) begin
   case (sel)
      0 : f = 4'b0000;
      1 : f = b - a;
      2 : f = a - b;
      3 : f = b + a;
      4 : f = a ^ b;
      5 : f = a | b;
      6 : f = a & b;
      7 : f = 4'b1111;
   endcase
   end
end
endmodule
```

Test Bench

```
module tb_bm_alu_4bit( );
//Inputs
   reg clk;
   reg reset;
   reg [2:0] sel;
   reg [3:0] a;
   reg [3:0] b;

//Outputs
   wire [3:0] f;

//Instantiate the unit under test
   bm_alu_4bit uut
      (.clk(clk),
      .reset(reset),
      .sel(sel),
      .a(a),
      .b(b),
      .f(f));

// Create 100 MHz clock
   initial begin
      clk = 0; reset = 0;
      forever #5 clk = ~clk; // 5ns ON, 5ns OFF, Time period 10ns, 1/10ns = 100 HHz
   end
//Stimulus inputs
   initial begin
      reset = 1'b1;
      #10 reset = 1'b0;
      sel = 0;  a = 4'b1001;  b = 4'b0011;
```

```
        #10 sel = 1;
        #10 sel = 2;
        #10 sel = 3;
        #10 sel = 4;
        #10 sel = 5;
        #10 sel = 6;
        #10 sel = 7;
        #10 sel = 0;  a = 4'b0000;  b = 4'b0000;

        #10 $finish;
    end
endmodule
```

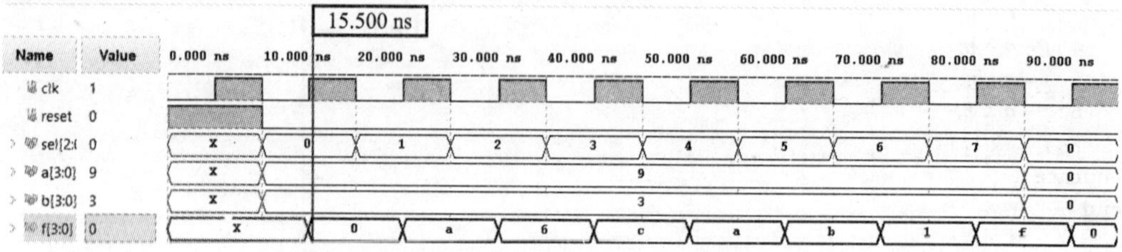

Fig. a: *Input: sel = 000; Output: f = 0000.*

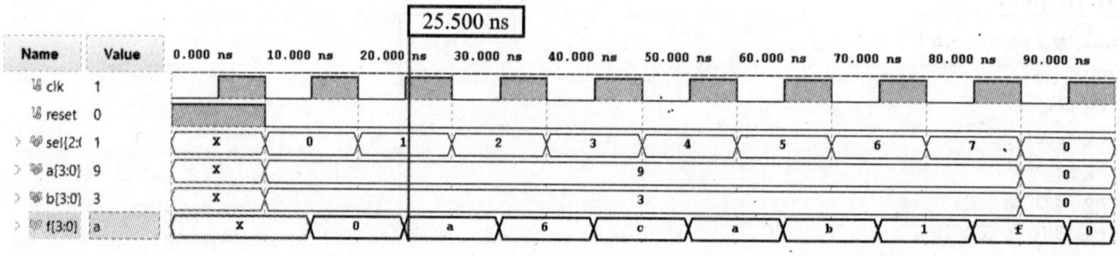

Fig. b: *Inputs: a = 9 = 1001, b = 3 = 0011, sel = 1 = 001; Output:* $f = b - a = A_H = 0000$.

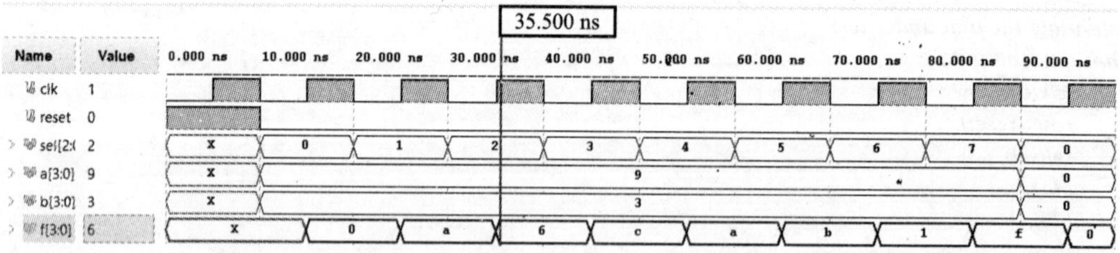

Fig. b: *Inputs: a = 9 = 1001, b = 3 = 0011, sel = 2 = 010; Output:* $f = a - b = 6_H = 0110$.

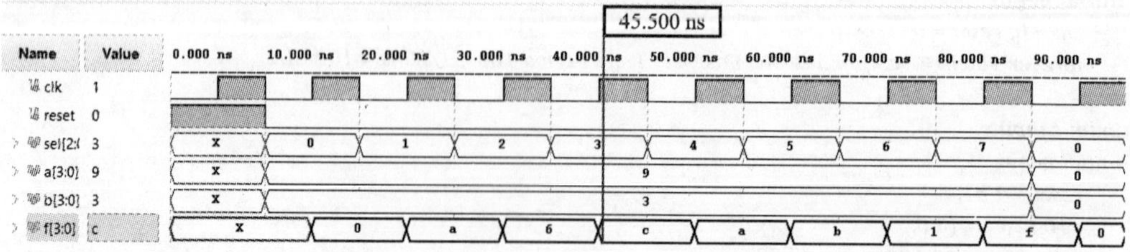

Fig. d: *Inputs: a = 9 = 1001, b = 3 = 0011, sel = 3 = 011; Output:* $f = b + a = C_H = 1100$.

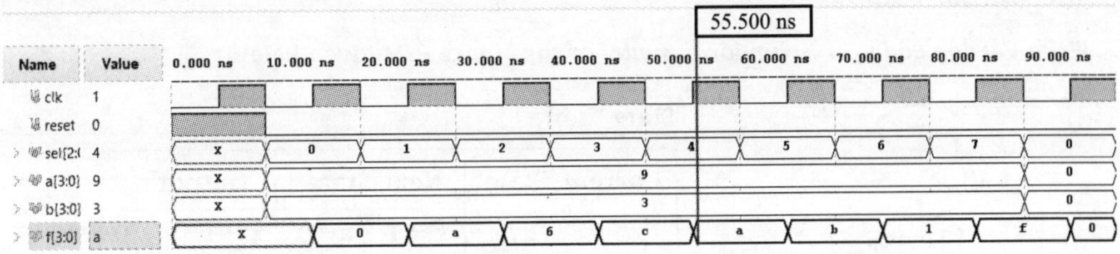

Fig. e: *Inputs:* $a = 9 = 1001$, $b = 3 = 0011$, $sel = 4 = 100$; *Output:* $f = a \wedge b = A_H = 1010$.

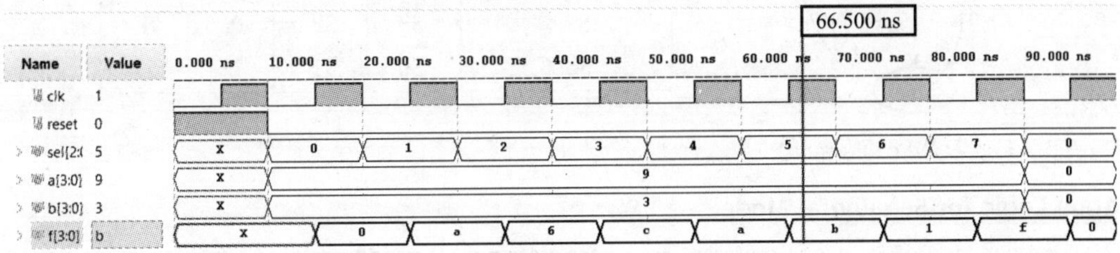

Fig. f: *Inputs:* $a = 9 = 1001$, $b = 3 = 0011$, $sel = 5 = 101$; *Output:* $f = a \mid b = B_H = 1011$.

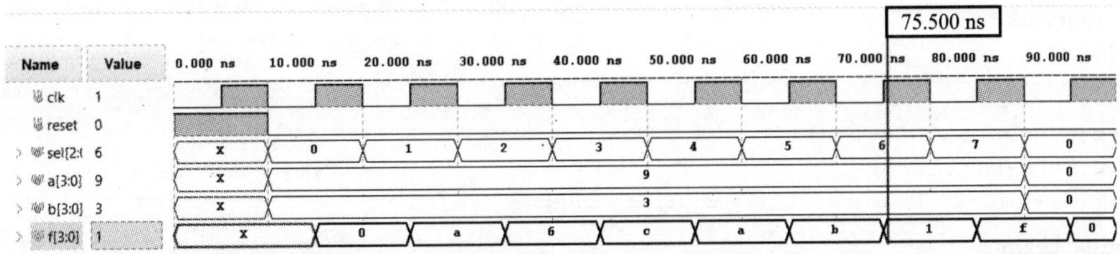

Fig. g: *Inputs:* $a = 9 = 1001$, $b = 3 = 0011$, $sel = 6 = 110$; *Output:* $f = a \,\&\, b = 1_H = 0001$.

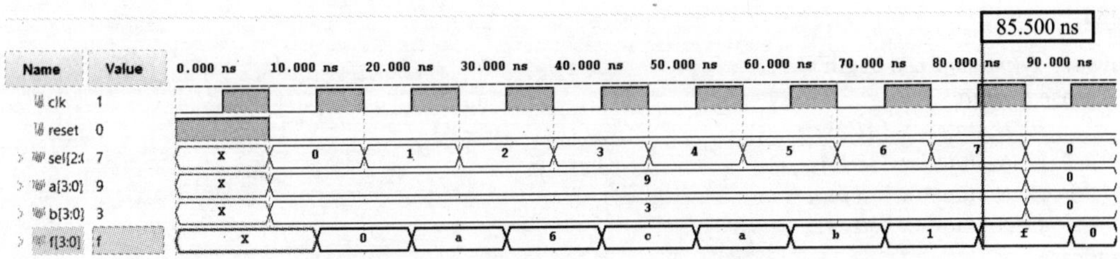

Fig. h: *Inputs:* $a = 9 = 1001$, $b = 3 = 0011$, $sel = 7 = 111$; *Output:* $f = F_H = 1111$.

Fig. 1: *Inputs-output waveform of 4-bit ALU.*

Verilog Program 15.39

Write Verilog code for behavioral model of the Moore FSM given below.

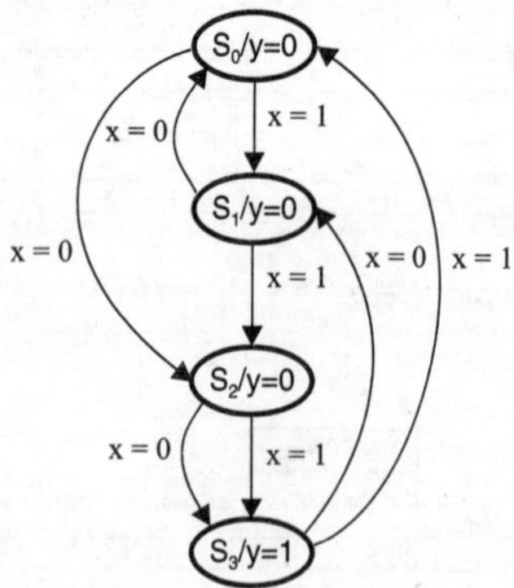

Fig. 1: State diagram.

State Table

Current_State	Next_State		Output	
	x = 1	x= 0	x = 1	x = 0
			y	y
S0	S1	S2	0	0
S1	S2	S0	0	0
S2	S3	S3	0	0
S3	S0	S1	1	1

Program Logic for Behavioral Model

The given Moore FSM has four states. Here, case statement is used to select a current state and the next state is decided based on output and input in each case. (Refer Chapter 7, Section 7.2.1 for design of Moore_fsm).

Verilog Program for Behavioral Model of Moore_FSM

```
module bm_moore_fsm
  ( input clk,
    input x,
    output y );

  reg y;
  reg [1:0] P_state, N_state;
  parameter s0 = 2'b00,  s1 = 2'b01,  s2 = 2'b10, s3 = 2'b11;

  initial begin
      P_state <= s0;  y <= 1'b0;
  end

  always @ (posedge clk) begin
      P_state <= N_state;
  end

  always @ (posedge clk) begin
     case (P_state)
         s0: N_state = x ? s1 : s2;
         s1: N_state = x ? s2 : s0;
         s2: N_state = x ? s3 : s3;
         s3: N_state = x ? s0 : s1;
  endcase

  case (P_state)
      s0: y = 1'b0;
      s1: y = 1'b0;
      s2: y = 1'b0;
      s3: y = 1'b1;
```

```
    endcase
    end
endmodule
```

Test Bench

```
module bm_tb_moore_fsm( );
//Inputs
    reg clk;  reg x;

//Output
    wire y;

    reg [1:0] P_state, N_state;
    parameter s0=2'b00,  s1=2'b01,  s2=2'b10, s3=2'b11;

//Instantiate the unit under test
    bm_moore_fsm uut
        (.clk(clk),
         .x(x),
         .y(y));

//Create 100 MHz clock
    initial begin
    clk = 0;
    forever #5 clk = ~clk; // 5ns ON, 5ns OFF, Time period 10ns, 1/10ns = 100 HHz
    end

//Stimulus inputs
    initial begin
        x = 1'b1; P_state <= s0; N_state <= s1;
    #10 x = 1'b0; P_state <= s0; N_state <= s2;

    #10 x = 1'b1; P_state = s1; N_state = s2;
    #10 x = 1'b0; P_state = s1; N_state = s0;
    #10 x = 1'b1; P_state = s2; N_state = s3;
    #10 x = 1'b0; P_state = s2; N_state = s3;

    #10 x = 1'b1; P_state = s3; N_state = s0;
    #10 x = 1'b0; P_state = s3; N_state = s1;

    #10 $finish;
    end
endmodule
```

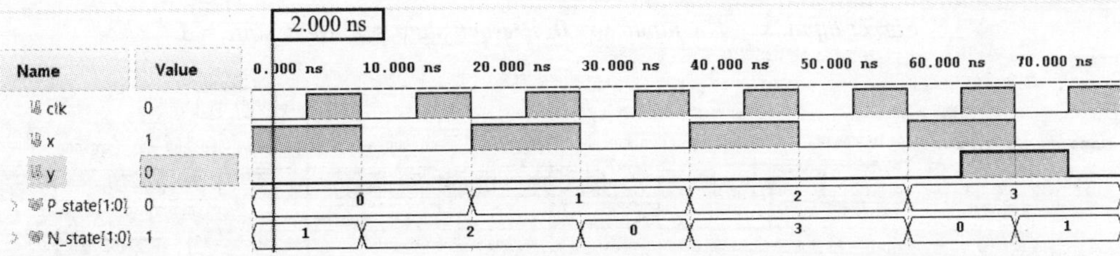

Fig. a: *Input: x = 1; Output: y = 0; Present_state = 0; Next_state = 1.*

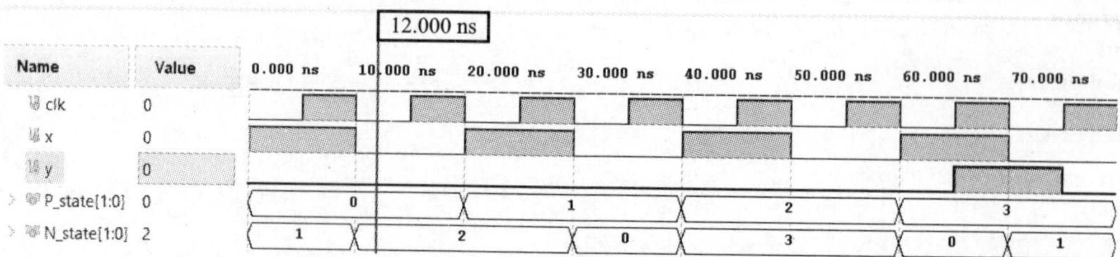

Fig. b: *Input: x = 0; Output: y = 0; Present_state = 0; Next_state = 2.*

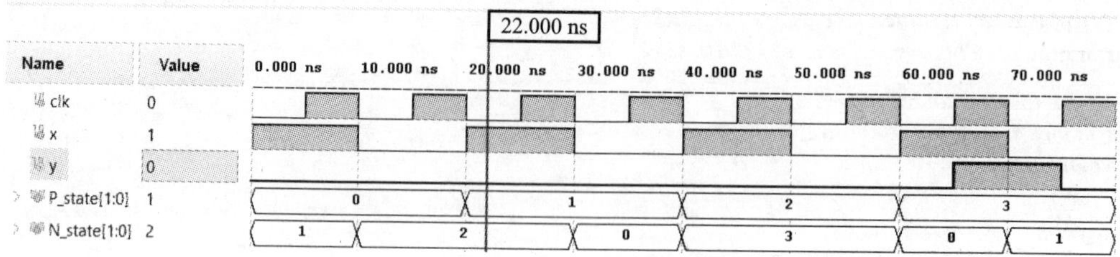

Fig. c: *Input: x = 1; Output: y = 0; Present_state = 1; Next_state = 2.*

Fig. d: *Input: x = 0; Output: y = 0; Present_state = 1; Next_state = 0.*

Fig. e: *Input: x = 1; Output: y = 0; Present_state = 2; Next_state = 3.*

Fig. f: *Input: x = 0; Output: y = 0; Present_state = 2; Next_state = 3.*

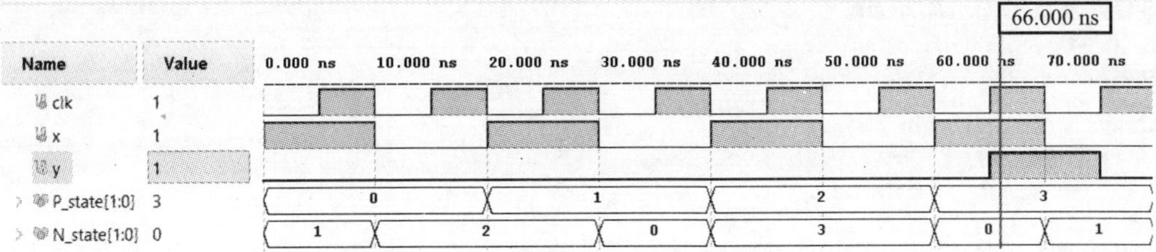

Fig. g: Input: x = 1; Output: y = 1; Present_state = 3; Next_state = 0.

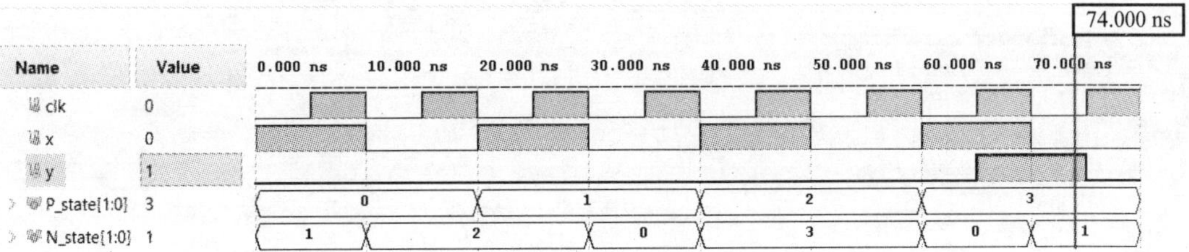

Fig. h: Input: x = 0; Output: y = 1; Present_state = 3; Next_state = 1.

Fig. 1: Inputs-output waveform of Moore_FSM.

Verilog Program 15.40

Write Verilog code for behavioral model of the Mealy FSM given below.

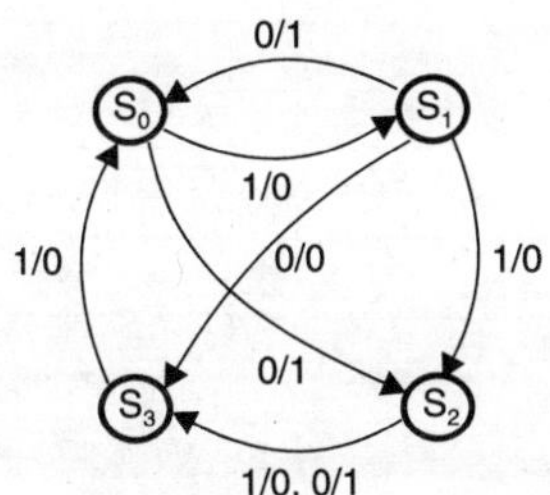

Fig. 1: State diagram.

State Table

Present_State	Next_State		Output	
	x = 1	x= 0	x = 1	x = 0
			y	y
S0	S1	S2	0	1
S1	S2	S0	0	1
S2	S3	S3	0	1
S3	S0	S1	0	1

Program Logic for Behavioral Model

The given Mealy FSM has four states. Here, case statement is used to select a current state and the next state is decided based on output and input in each case. (Refer Chapter 7, Section 7.2.1 for design of Melay_fsm).

Verilog Program for Behavioral Model of Mealy_FSM

```
module bm_mealy_fsm
   ( input clk,
     input x,
     output reg y );

   reg [1:0] P_state, N_state;
   parameter s0 = 2'b00,  s1 = 2'b01,  s2 = 2'b10,  s3 = 2'b11;

   initial begin
        P_state <= s0;
   end
```

```
always @ (posedge clk) begin
    P_state <= N_state;
end

always @ (P_state) begin
    case (P_state)
        s0: N_state = x ? s1 : s2;
        s1: N_state = x ? s2 : s0;
        s2: N_state = x ? s3 : s3;
        s3: N_state = x ? s0 : s1;
    endcase

    case (P_state)
        s0: y = x ? 1'b0 : 1'b1;
        s1: y = x ? 1'b0 : 1'b1;
        s2: y = x ? 1'b0 : 1'b1;
        s3: y = x ? 1'b0 : 1'b1;
    endcase
end
endmodule
```

Test Bench

```
module tb_bm_mealy_fsm( );
//Inputs
    reg clk;  reg x;

//Output
    wire y;

    reg [1:0] P_state, N_state;

    parameter s0 = 2'b00,  s1 = 2'b01,  s2 = 2'b10, s3 = 2'b11;

//Instantiate the unit under test
    bm_mealy_fsm uut
        (.clk(clk),
         .x(x),
         .y(y));

// Create 100 MHz clock
    initial begin
        clk = 0;
        forever #5 clk = ~clk; // 5ns ON, 5ns OFF, Time period 10ns, 1/10ns = 100 HHz
    end
//Stimulus inputs
    initial begin
            x = 1'b1; P_state <= s0; N_state <= s1;
        #10 x = 1'b0; P_state <= s0; N_state <= s2;

        #10 x = 1'b1; P_state = s1; N_state = s2;
        #10 x = 1'b0; P_state = s1; N_state = s0;

        #10 x = 1'b1; P_state = s2; N_state = s3;
        #10 x = 1'b0; P_state = s2; N_state = s3;

        #10 x = 1'b1; P_state = s3; N_state = s0;
        #10 x = 1'b0; P_state = s3; N_state = s1;
```

```
    #10 $finish;
  end
endmodule
```

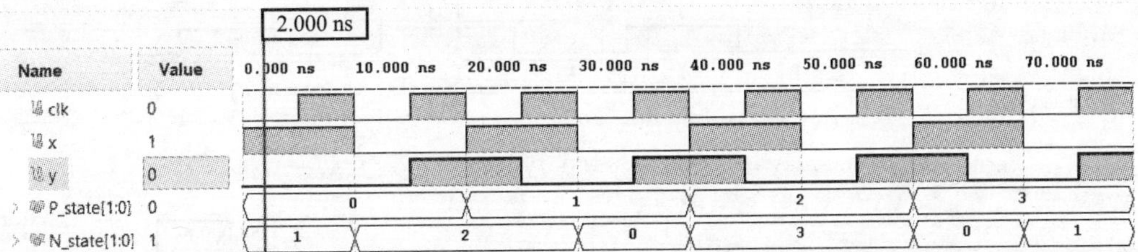

Fig. a: *Input: x = 1; Output: y = 0; Present_state = 0; Next_state = 1.*

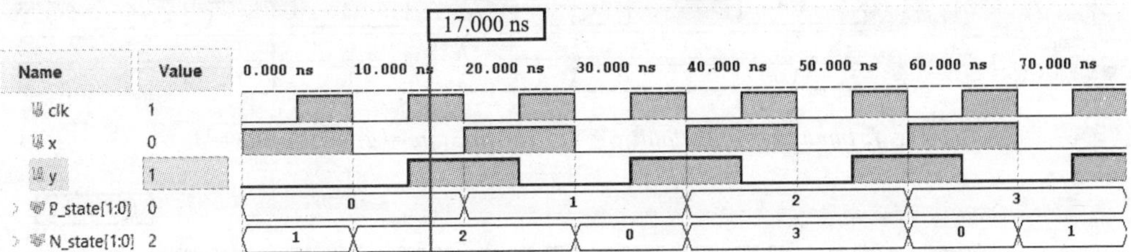

Fig. b: *Input: x = 0; Output: y = 1; Present_state = 0; Next_state = 2.*

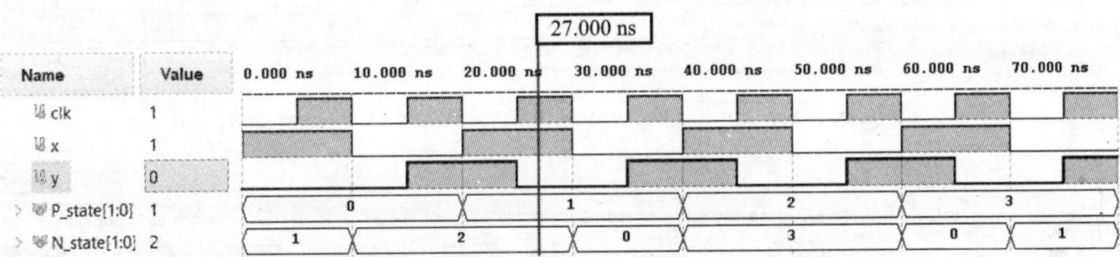

Fig. c: *Input: x = 1; Output: y = 0; Present_state = 1; Next_state = 2.*

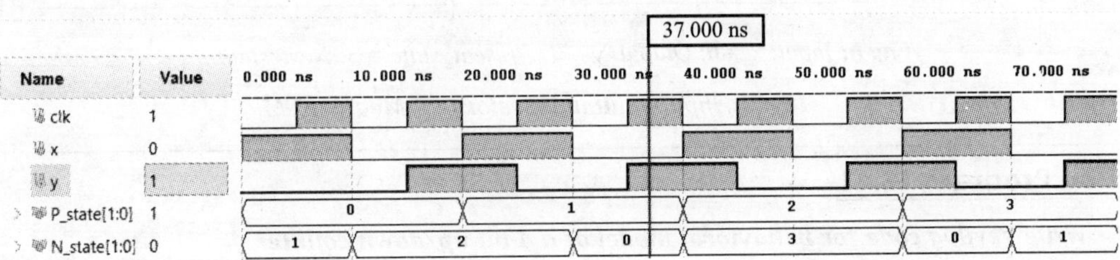

Fig. d: *Input: x = 0; Output: y = 1; Present_state = 1; Next_state = 0.*

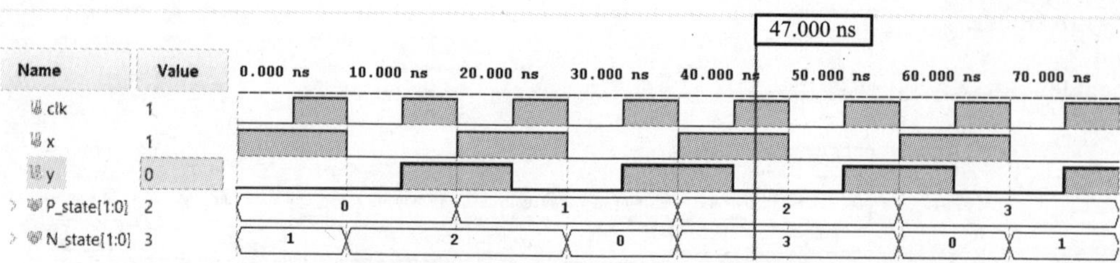

Fig. e: Input: x = 1; Output: y = 0; Present_state = 2; Next_state = 3.

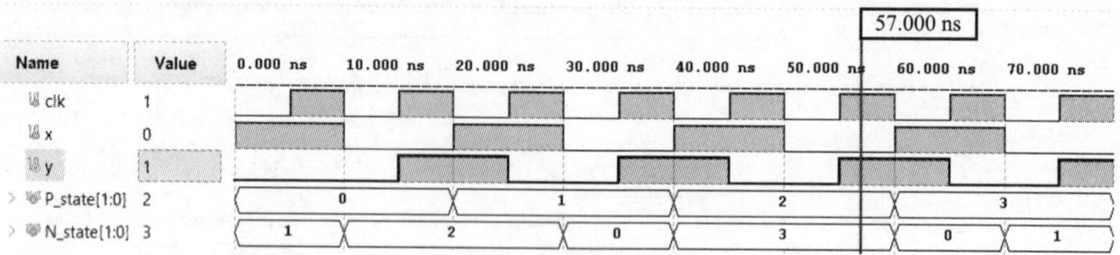

Fig. f: Input: x = 0; Output: y = 1; Present_state = 2; Next_state = 3.

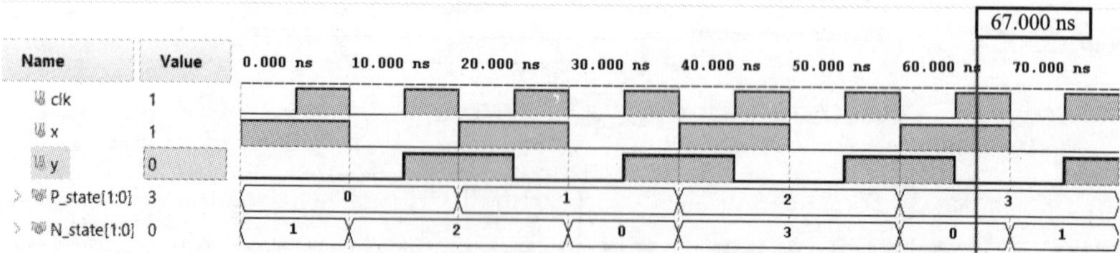

Fig. g: Input: x = 1; Output: y = 0; Present_state = 3; Next_state = 0.

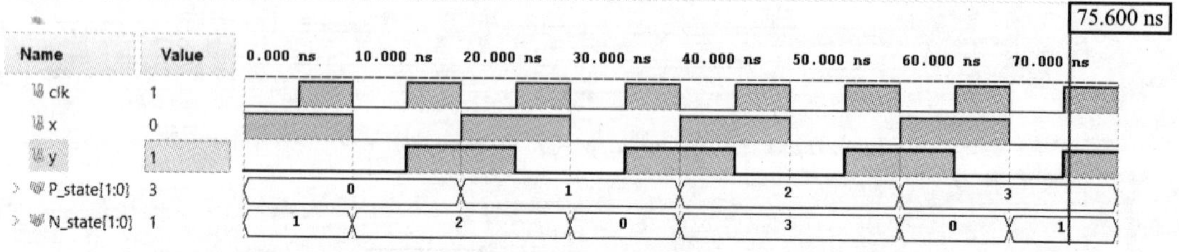

Fig. h: Input: x = 0; Output: y = 1; Present_state = 3; Next_state = 1.

Fig. 1: Inputs-output waveform of Mealy_FSM.

Verilog Program 15.41

Write Verilog code for behavioral model of a 4-bit up/down counter.

Program Logic for Behavioral Model

A selection bit, a is required to select up count when a = 1 and down count when a = 0. Initial count is kept 0 and increment by 1 for up count and decrement by 1 for down count for every clock pulse. (Refer Chapter 8, Section 8.2.3 for design of up/down counter).

Verilog Program for Behavioral Model of 4-bit up/down Counter

```
module bm_updown_4bit
    ( input clk, a,
     output reg [3:0] dout );

    initial begin
        dout = 4'b0000;
    end
always @(posedge clk) begin
    if ((a==1) & (clk ==1))
        begin
            dout = dout + 4'b0001;
        end

        else if ((a==0) & (clk ==1))
        begin
            dout = dout - 4'b0001;
        end
        else dout = dout;
    end
endmodule
```

Test Bench

```
module bm_tb_updown_4bit( );
//Inputs
    reg clk;   reg a;

//Output
    wire [3:0] dout;

//Instantiate the unit under test
    bm_updown_4bit uut
        (.clk(clk),  .a(a),  .dout(dout));

//Stimulus inputs
    initial begin
        a = 1'b1;          // a=1 for upcount, a = 0 for down count
        #5 clk = 1'b0;
        #5 clk = 1'b1;  #5 clk = 1'b0;  #5 clk = 1'b1;  #5 clk = 1'b0;
        #5 clk = 1'b1;  #5 clk = 1'b0;  #5 clk = 1'b1;  #5 clk = 1'b0;
        #5 clk = 1'b1;  #5 clk = 1'b0;  #5 clk = 1'b1;  #5 clk = 1'b0;
        #5 clk = 1'b1;  #5 clk = 1'b0;  #5 clk = 1'b1;  #5 clk = 1'b0;
        #5 clk = 1'b1;  #5 clk = 1'b0;  #5 clk = 1'b1;  #5 clk = 1'b0;
        #5 clk = 1'b1;  #5 clk = 1'b0;  #5 clk = 1'b1;  #5 clk = 1'b0;
        #5 clk = 1'b1;  #5 clk = 1'b0;  #5 clk = 1'b1;  #5 clk = 1'b0;
        #5 clk = 1'b1;  #5 clk = 1'b0;  #5 clk = 1'b1;  #5 clk = 1'b0;
        #5 $finish;
    end
endmodule
```

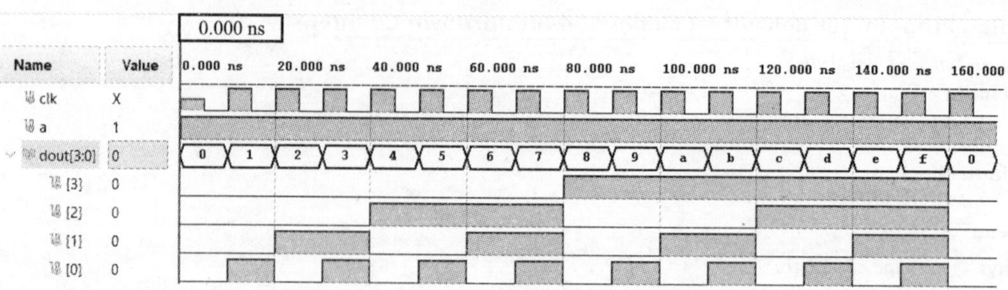

Fig. a: Input-output waveform of up counter.

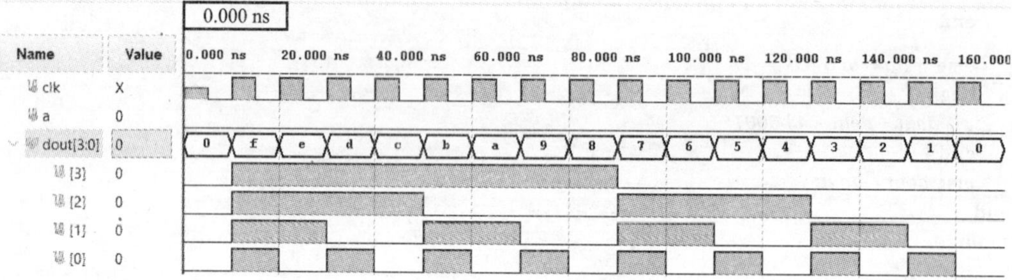

Fig. b: Input-output waveform of down counter.

Verilog Program 15.42

Write Verilog code for behavioral model of a BCD counter/MOD10 counter.

Program Logic for Behavioral Model

Initially the count value is kept 0. BCD counter is basically an up counter and hence the count value is incremented by 1 for every clock pulse. When the count reaches 1001 the count value is clear to 0.

(Refer Chapter 8, Section 8.2.5 for design of BCD counter/MOD10 counter).

Verilog Program for Behavioral Model of BCD Counter/MOD10 Counter

```
module bm_bcd_counter
   ( input clk,
   output reg [3:0] dout );

   initial begin
      dout = 4'b0000;
   end

   always @(posedge clk) begin
      if ((clk ==1) & (dout == 4'b1001))
         dout <= 4'b0000;

      else if (clk ==1)
         dout = dout + 4'b0001;

      else dout = dout;
   end
endmodule
```

Test Bench

```
module tb_bm_bcd_counter( );
//Inputs
    reg clk;

//Output
    wire [3:0] dout;

//Instantiate the unit under test
    bm_bcd_counter uut
        (.clk(clk),    .dout(dout));

//Stimulus inputs
    initial begin
        clk = 1'b0;
        #5 clk = 1'b1;  #5 clk = 1'b0;  #5 clk = 1'b1;  #5 clk = 1'b0;
        #5 clk = 1'b1;  #5 clk = 1'b0;  #5 clk = 1'b1;  #5 clk = 1'b0;
        #5 clk = 1'b1;  #5 clk = 1'b0;  #5 clk = 1'b1;  #5 clk = 1'b0;
        #5 clk = 1'b1;  #5 clk = 1'b0;  #5 clk = 1'b1;  #5 clk = 1'b0;
        #5 clk = 1'b1;  #5 clk = 1'b0;  #5 clk = 1'b1;  #5 clk = 1'b0;
        #5 $finish;
    end
endmodule
```

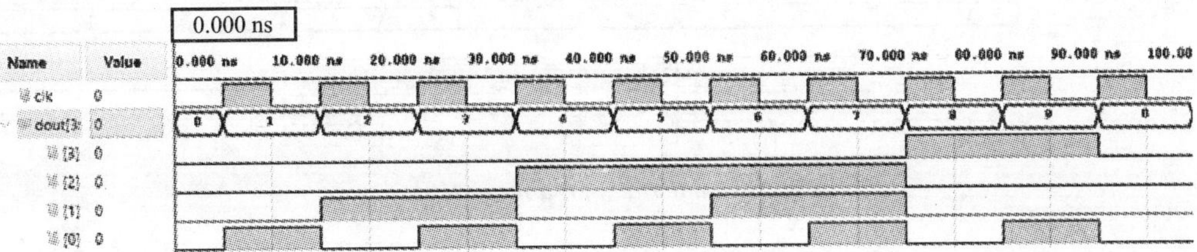

Fig. a: Counter output: dout = 0000.

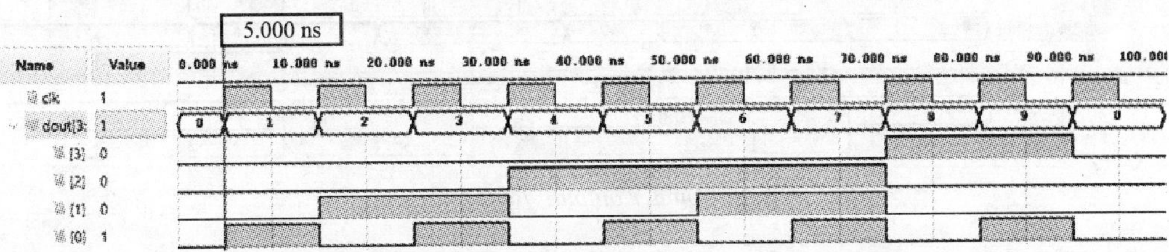

Fig. b: Counter output: dout = 0001.

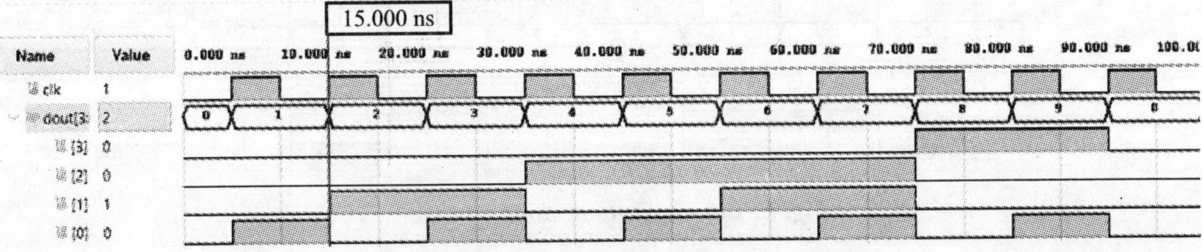

Fig. c: Counter output: dout = 0010.

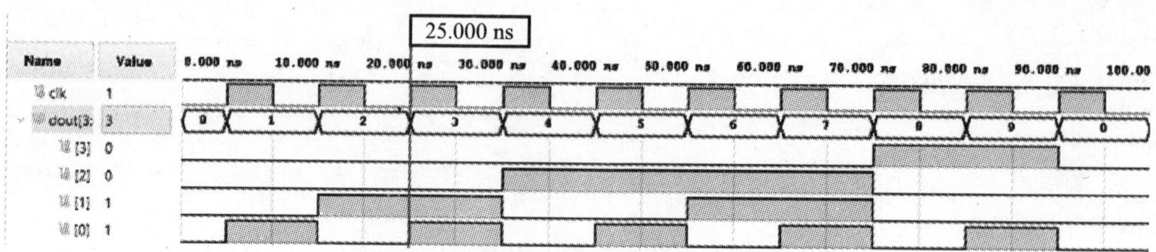

Fig. d: Counter output: dout = 0011.

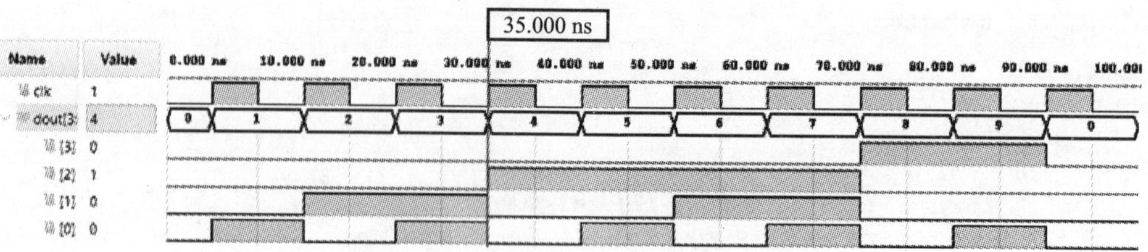

Fig. e: Counter output: dout = 0100.

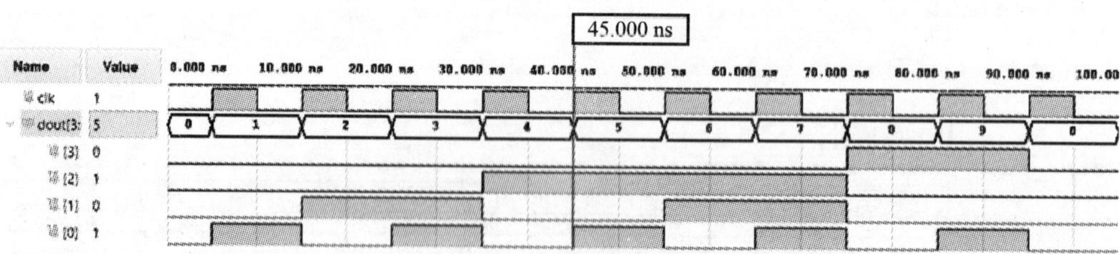

Fig. f: Counter output: dout = 0101.

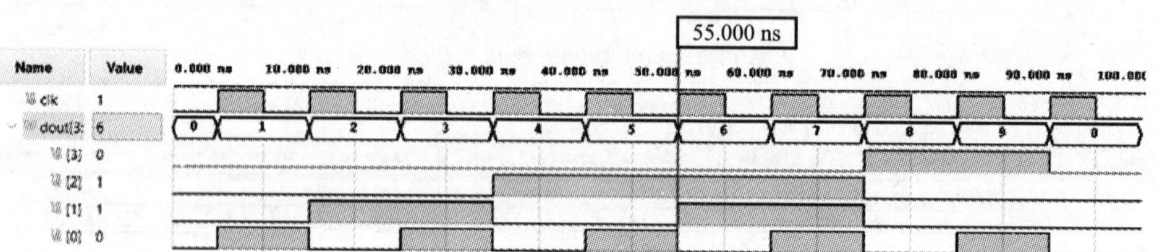

Fig. g: Counter output: dout = 0110.

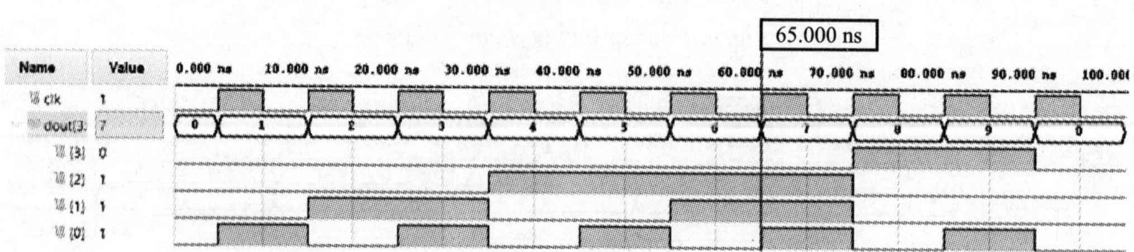

Fig. h: Counter output: dout = 0111.

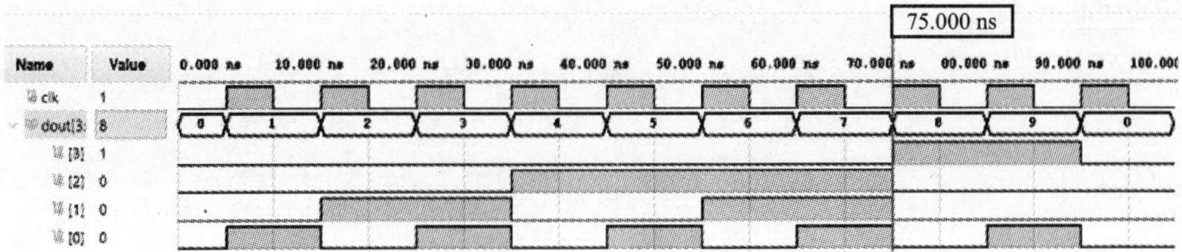

Fig. i: Counter output: dout = 1000.

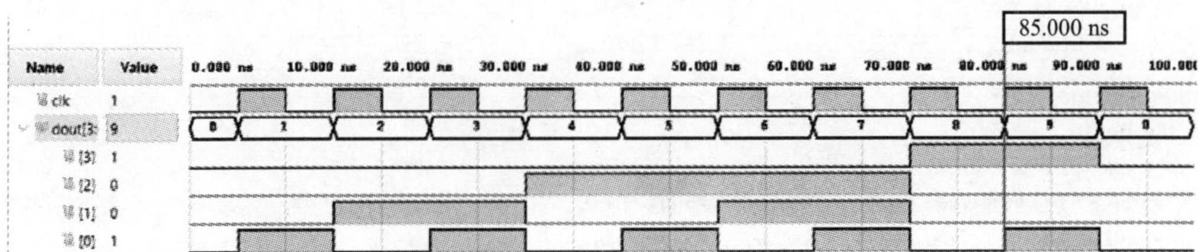

Fig. j: Counter output: dout = 1001.

Fig. 1: Inputs-output waveform of BCD counter/MOD10 counter.

Verilog Program 15.43

Write Verilog code for behavioral model of a MOD6 counter.

Program Logic for Behavioral Model

Initially the count value is kept 0. MOD6 counter is basically an up counter and hence the count value is incremented by 1 for every clock pulse. When the count reaches 101 the count value is clear to 0. (Refer Chapter 8, Section 8.2.6 for design of MOD6 counter).

Verilog Program for Behavioral Model of MOD6 Counter

```
module bm_mod6_counter
  ( input clk,
   output reg [2:0] dout);

   initial begin
     dout = 3'b000;
   end

   always @(posedge clk) begin
     if ((clk ==1) & (dout == 3'b101))
     dout <= 3'b000;

     else if (clk ==1)
     dout = dout + 3'b001;

     else dout = dout;
   end
endmodule
```

Test Bench

```
module tb_bm_mod6_counter( );
//Inputs
    reg clk;

//Output
    wire [2:0] dout;

//Instantiate the unit under test
    bm_mod6_counter uut
        (.clk(clk),
         .dout(dout));

//Stimulus inputs
    initial begin
        clk = 1'b0;
      #5 clk=1'b1;  #5 clk=1'b0;  #5 clk=1'b1;  #5 clk=1'b0;
      #5 clk=1'b1;  #5 clk=1'b0;  #5 clk=1'b1;  #5 clk=1'b0;
      #5 clk=1'b1;  #5 clk=1'b0;  #5 clk=1'b1;  #5 clk=1'b0;
      #5 $finish;
    end
endmodule
```

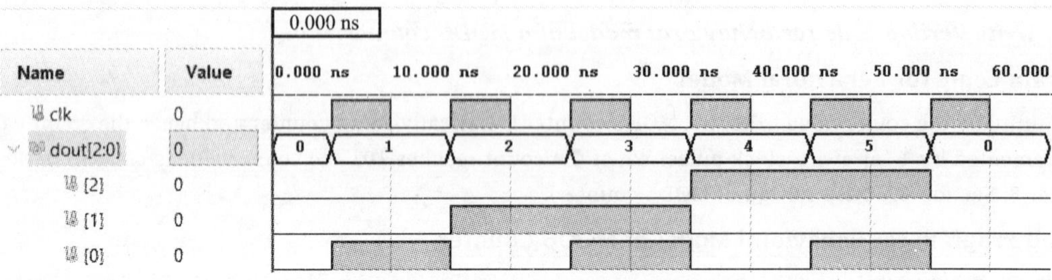

Fig. a: Counter output: dout = 000.

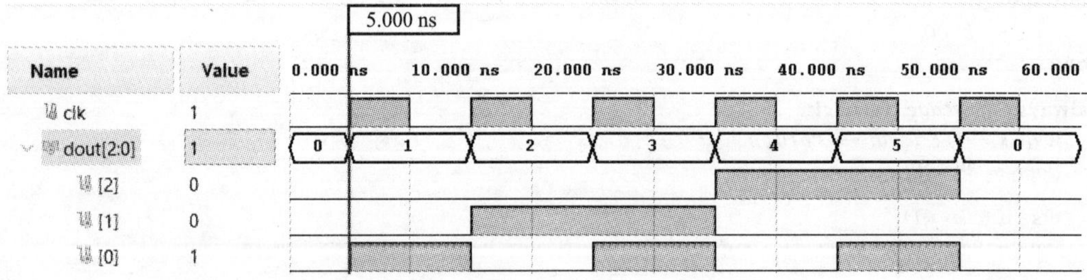

Fig. b: Counter output: dout = 001.

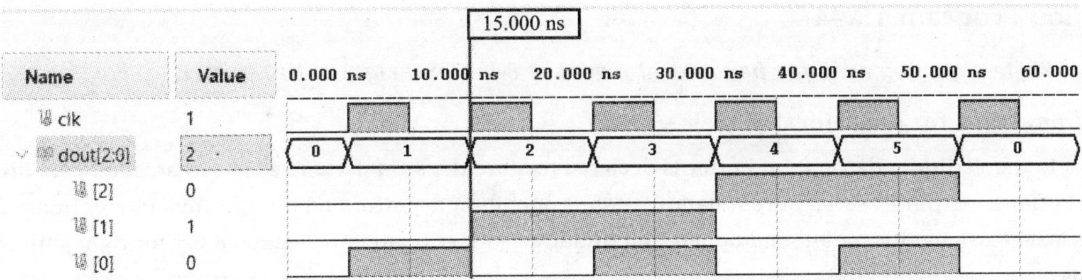

Fig. c: *Counter output: dout = 010.*

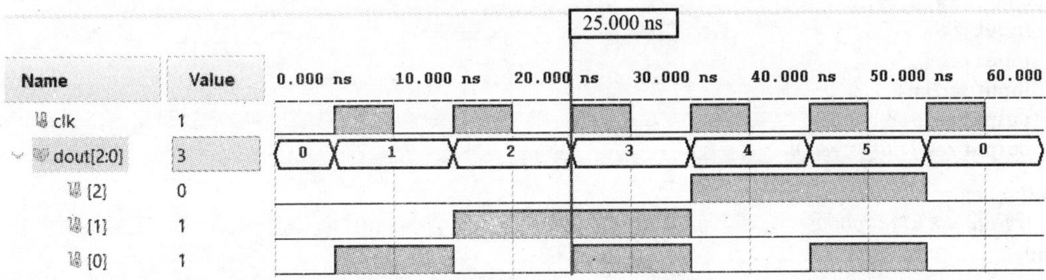

Fig. d: *Counter output: dout = 011.*

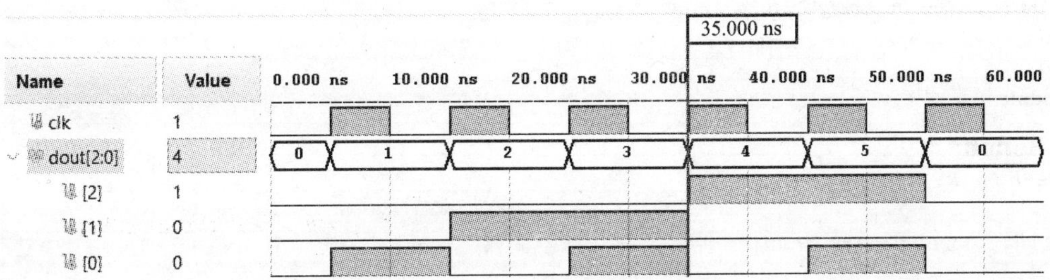

Fig. e: *Counter output: dout = 100.*

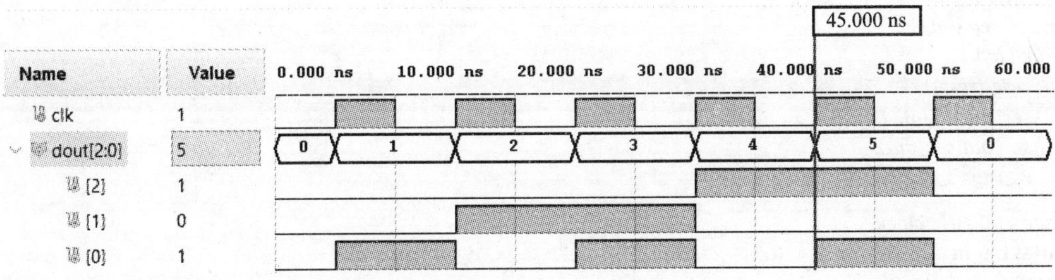

Fig. f: *Counter output: dout = 101.*

Fig. 1: *Inputs-output waveform of MOD6 counter.*

Verilog Program 15.44

Write a Verilog code for behavioral model of 8-bit SISO right shift register.

Program Logic for Behavioral Model

In the Verilog code, an 8-bit buffer is declared to store the 8-bit data. The LSB (Least Significant Bit) of 8-bit buffer is output as serial output data. The shift operation is performed by concatenation of input data bit as MSB (Most Significant Bit) and upper 7-bit of buffer. (Refer Chapter 8, Section 8.6.1 for right shift SISO register.

Verilog Program for Behavioral Model of 8-bit SISO Right Shift Register

```
module bm_RS_reg
      ( input clk,
        input reset,
        input ser_in,
        output reg ser_out,
        output reg [7:0] rs_reg );

    initial begin
        rs_reg = 8'b11110000;
    end

    always @(posedge clk) begin
        if (reset==0) begin
        ser_out = rs_reg[0];
        rs_reg = {ser_in, rs_reg[7:1]};
    end
    end
endmodule
```

Test Bench

```
module tb_bm_RS_reg( );
//Inputs
    reg clk;   reg reset;   reg ser_in;

//Outputs
    wire ser_out;  wire [7:0] rs_reg;

//Instantiate the unit under test
    bm_RS_reg uut
       (.clk(clk),
        .reset(reset),
        .ser_in(ser_in),
        .ser_out(ser_out),
        .rs_reg(rs_reg));

// Create 100 MHz clock
    initial begin
        clk = 0; reset=0;
        forever #5 clk = ~clk; // 5ns ON, 5ns OFF, Time period 10ns, 1/10ns = 100 HHz
        end
//Stimulus inputs
    initial begin
        reset = 1'b1;
        #10 reset = 1'b0; ser_in = 1'b1;
```

```
        #10 ser_in = 1'b0;
        #10 ser_in = 1'b1;
        #10 ser_in = 1'b0;
        #10 ser_in = 1'b1;
        #10 ser_in = 1'b0;
        #10 ser_in = 1'b1;
        #10 ser_in = 1'b0;
        #15 $finish;
    end
endmodule
```

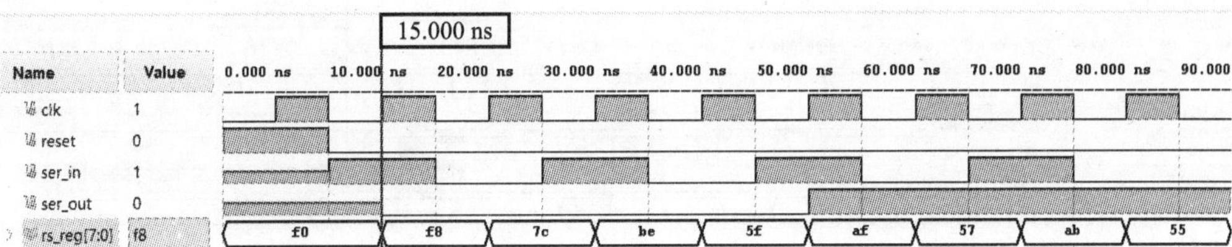

Fig. a: *ser_data_in = 1; rightshift_reg: 1111 0000(F0$_H$) $\xrightarrow{\text{Right shift}}$ 1111 1000(F8$_H$); ser_data_out = 0.*

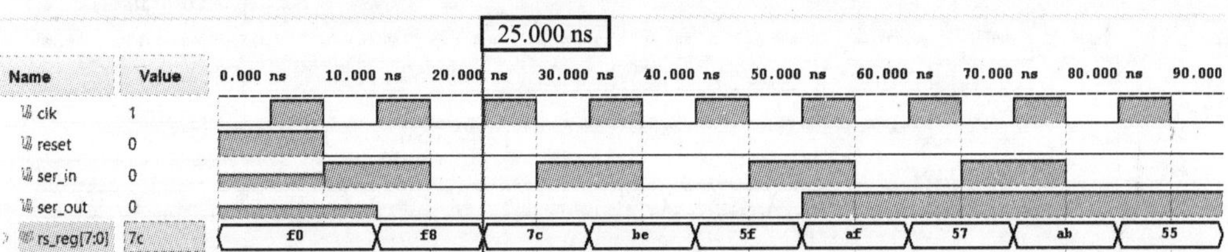

Fig. b: *ser_data_in = 0; rightshift_reg: 1111 1000(F8$_H$) $\xrightarrow{\text{Right shift}}$ 0111 1100(7C$_H$); ser_data_out = 0.*

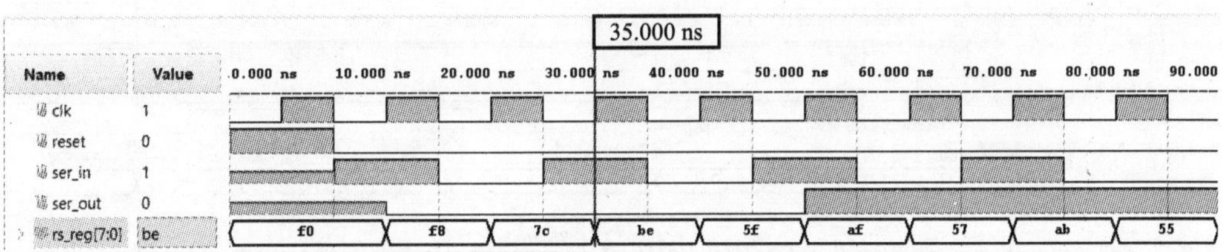

Fig. c: *ser_data_in = 1; rightshift_reg: 0111 1100(7C$_H$) $\xrightarrow{\text{Right shift}}$ 1011 1110(BE$_H$); ser_data_out = 0.*

Fig. d: *ser_data_in = 0; rightshift_reg: 1011 1110(BE$_H$) $\xrightarrow{\text{Right shift}}$ 0101 1111(5F$_H$); ser_data_out = 0.*

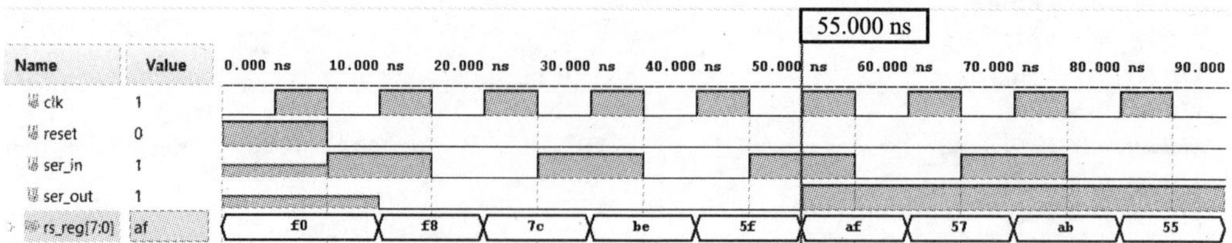

Fig. e: *ser_data_in = 1; rightshift_reg: 0101 1111(5F$_H$)* $\xrightarrow{\text{Right shift}}$ *1010 1111(AF$_H$); ser_data_out = 1.*

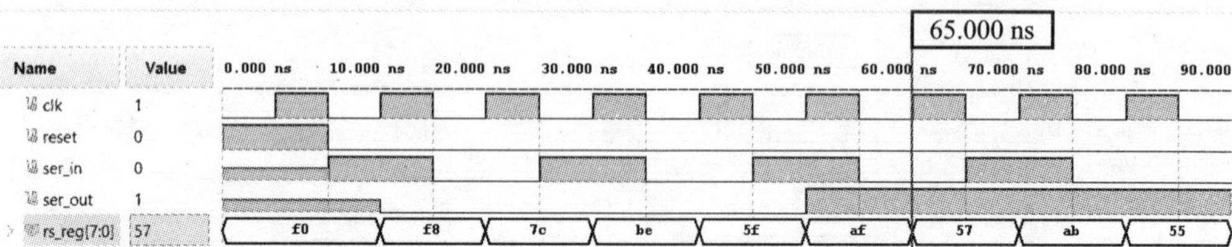

Fig. f: *ser_data_in = 0; rightshift_reg: 1010 1111(AF$_H$)* $\xrightarrow{\text{Right shift}}$ *0101 0111(57$_H$); ser_data_out = 1.*

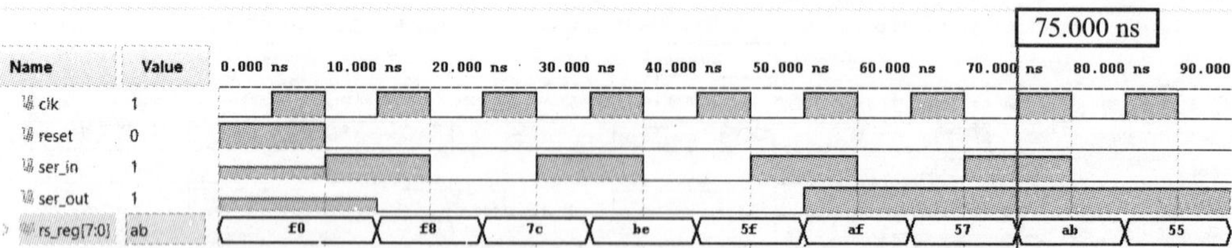

Fig. g: *ser_data_in = 1; rightshift_reg: 0101 0111(57$_H$)* $\xrightarrow{\text{Right shift}}$ *1010 1011(AB$_H$); ser_data_out = 1.*

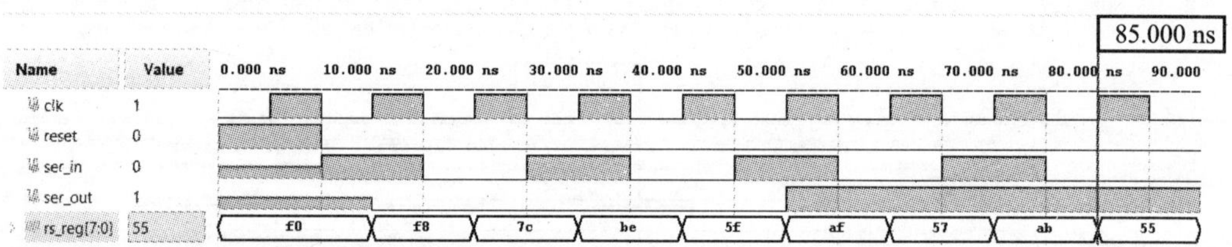

Fig. h: *ser_data_in = 0; rightshift_reg: 1010 1011(AB$_H$)* $\xrightarrow{\text{Right shift}}$ *0101 0101(55$_H$); ser_data_out = 1.*

Fig. 1: *Inputs-output waveform of 8-bit SISO right shift register.*

Verilog Program 15.45

Write a Verilog code for behavioral model of 8-bit SISO left shift register.

Program Logic for Behavioral Model

In the Verilog code, an 8-bit buffer is declared to store the 8-bit data. The MSB (Most Significant Bit) of 8-bit buffer is output as serial output data The shift operation is performed by concatenation of input data bit as LSB (Least Significant Bit) and lower 7-bit of buffer. (Refer Chapter 8, Section 8.6.1 for left shift SISO register.

Verilog Program for Behavioral Model of 8-bit SISO Left Shift Register

```
module bm_LS_reg
   ( input clk,
     input reset,
     input ser_in,
     output reg ser_out,
     output reg [7:0] LS_reg );

   initial begin
       LS_reg = 8'b11110000;
   end

   always @(posedge clk) begin
       if (reset==0) begin
       ser_out = LS_reg[7];
       LS_reg = {LS_reg[6:0], ser_in};
   end
   end
endmodule
```

Test Bench

```
module bm_tb_LS_reg( );
//Inputs
   reg clk;  reg reset;  reg ser_in;

//Outputs
   wire ser_out;  wire [7:0] LS_reg;

//Instantiate the unit under test
   bm_LS_reg uut
      (.clk(clk),
       .reset(reset),
       .ser_in(ser_in),
       .ser_out(ser_out),
       .LS_reg(LS_reg));

// Create 100 MHz clock
   initial begin
       clk = 0; reset = 0;
       forever #5 clk = ~clk; // 5ns ON, 5ns OFF, Time period 10ns, 1/10ns = 100 HHz
   end

//Stimulus inputs
   initial begin
           reset = 1'b1;
       #10 reset = 1'b0;  ser_in = 1'b1;
       #10 ser_in = 1'b0;
       #10 ser_in = 1'b1;
       #10 ser_in = 1'b0;
       #10 ser_in = 1'b1;
       #10 ser_in = 1'b0;
       #10 ser_in = 1'b1;
       #10 ser_in = 1'b0;
       #15 $finish;
   end
endmodule
```

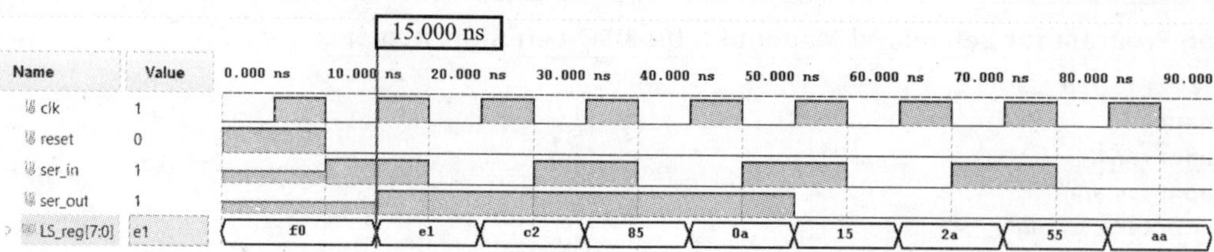

Fig. a: *ser_data_out = 1; 1110 0001($E1_H$)* $\xleftarrow{\text{Left shift}}$ *leftshift_reg: 1111 0000($F0_H$); ser_data_in = 1.*

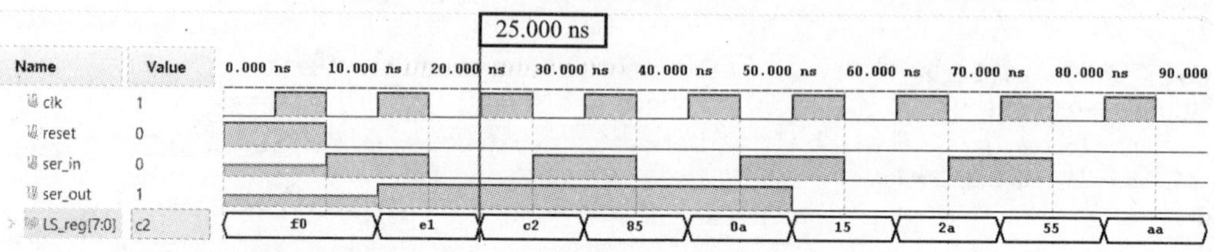

Fig. b: *ser_data_out = 1; 1100 0010($C2_H$)* $\xleftarrow{\text{Left shift}}$ *leftshift_reg: 1110 0001($E1_H$); ser_data_in = 0.*

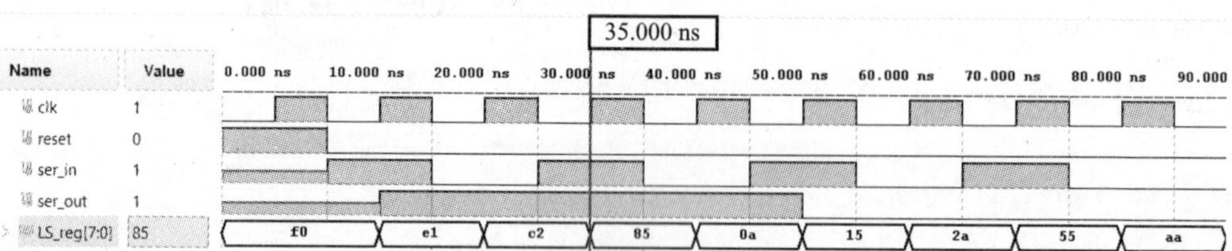

Fig. c: *ser_data_out = 1; 1000 0101(85_H)* $\xleftarrow{\text{Left shift}}$ *leftshift_reg: 1100 0010($C2_H$); ser_data_in = 1.*

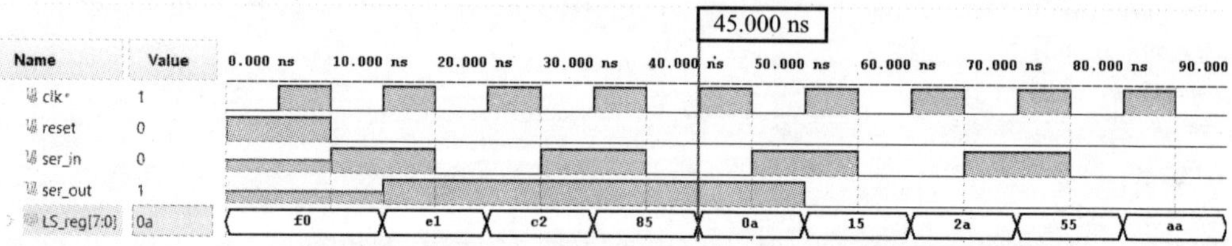

Fig. d: *ser_data_out = 1; 0000 1010($0A_H$)* $\xleftarrow{\text{Left shift}}$ *leftshift_reg: 1000 0101(85_H); ser_data_in = 0.*

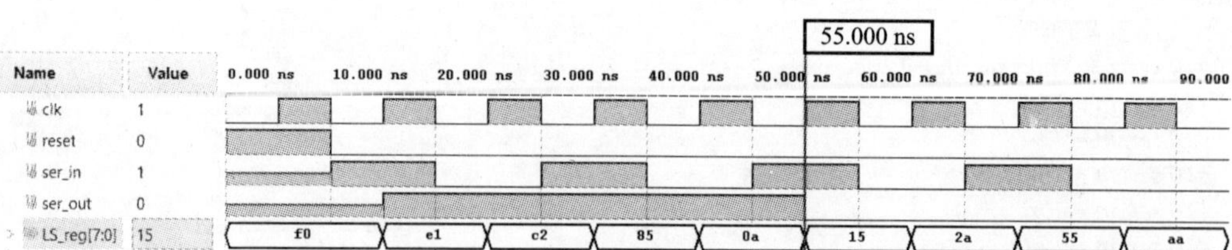

Fig. e: *ser_data_out = 0; 0001 0101(15_H)* $\xleftarrow{\text{Left shift}}$ *leftshift_reg: 0000 1010($0A_H$); ser_data_in = 1.*

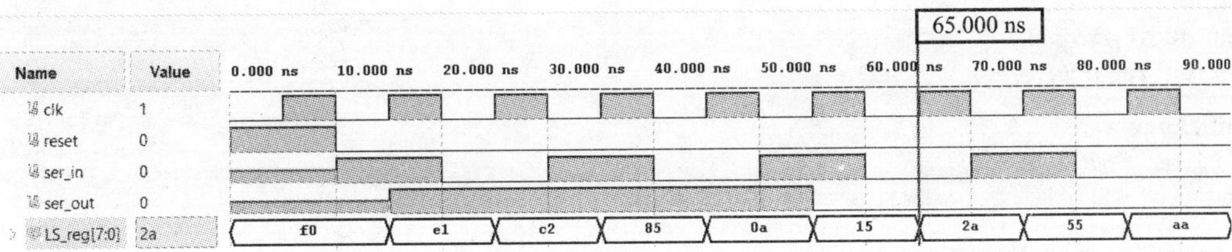

***Fig. f:** ser_data_out = 0; 0010 1010(2A$_H$) ←(Left shift)— leftshift_reg: 0001 0101(15$_H$); ser_data_in = 0.*

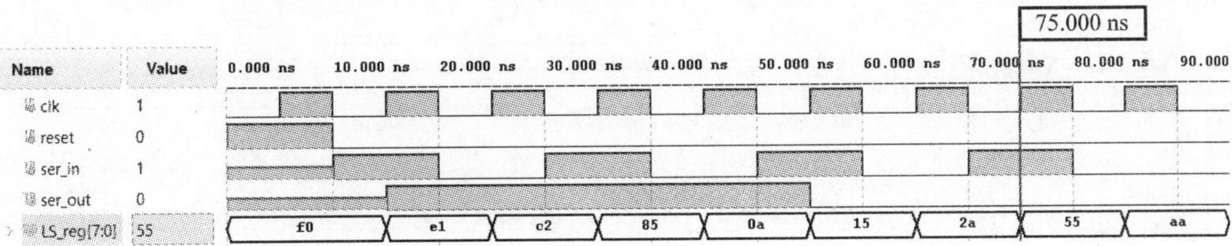

***Fig. g:** ser_data_out = 0; 0101 0101(55$_H$) ←(Left shift)— leftshift_reg: 0010 1010(2A$_H$); ser_data_in = 1.*

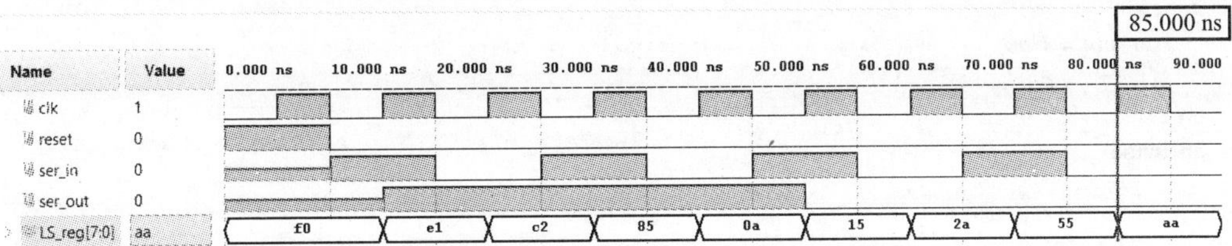

***Fig. h:** ser_data_out = 0; 1010 1010(AA$_H$) ←(Left shift)— leftshift_reg: 0101 0101(55$_H$); ser_data_in = 0.*

***Fig. 1:** Inputs-output waveform of 8-bit left SISO shift register.*

Verilog Program 15.46

Write a Verilog code for behavioral model of a universal shift register.

Program Logic for Behavioral Model

In the Verilog code an 8-bit buffer is used to store 8-bit data. The Verilog code for universal shift register is written to perform the following four operations. (Refer Chapter 8, Section 8.6.5 for the design of universal shift register).

- Clear buffer
- Left shift buffer
- Right shift buffer
- Load parallel data to buffer

Verilog Program for Behavioral Model of Universal Shift Register

```
module bm_uni_shift_reg
   ( input clk,
     input reset,
     input [2:0] sel,
     input ser_in,
```

```
    input [7:0] parl_in,
    output reg ser_out,
    output reg [7:0] S_reg);

  initial begin
      S_reg = 8'b11110000;
  end

  always @(posedge clk) begin
  case (sel)
      0: S_reg = S_reg;

      1: begin
          ser_out = S_reg[0];
          S_reg   = {ser_in, S_reg[7:1]};
      end

      2: begin
          ser_out = S_reg[7];
          S_reg   = {S_reg[6:0], ser_in};
      end

      3: begin
          ser_out = 1'b0;
          S_reg   = parl_in;
      end
      endcase
end
endmodule
```

Test Bench

```
module bm_tb_uni_shift_reg( );
//Inputs
  reg clk;  reg reset;  reg ser_in;  reg [2:0] sel;  reg [7:0] parl_in;

//Outputs
  wire ser_out;  wire [7:0] S_reg;
//Instantiate the unit under test
  bm_uni_shift_reg uut
     (.clk(clk),
      .reset(reset),
      .sel(sel),
      .ser_in(ser_in),
      .parl_in(parl_in),
      .ser_out(ser_out),
      .S_reg(S_reg));

// Create 100 MHz clock
  initial begin
      clk = 0; reset = 0;
      forever #5 clk = ~clk; // 5ns ON, 5ns OFF, Time period 10ns, 1/10ns = 100 HHz
  end

//Stimulus inputs
  initial begin
          reset = 1'b1;
```

```
        #10 reset = 1'b0;
        sel = 0;

        #10 sel = 1; ser_in = 1'b1;
        #10 ser_in = 1'b0;

        #10 sel = 2; ser_in = 1'b1;
        #10 ser_in = 1'b0;

        #10 sel = 3; parl_in = 8'b10101010;

        #10 parl_in = 8'b00000000;

        #10 sel = 1; ser_in = 1'b1;
        #10 ser_in = 1'b0;

        #5 $finish;
    end
endmodule
```

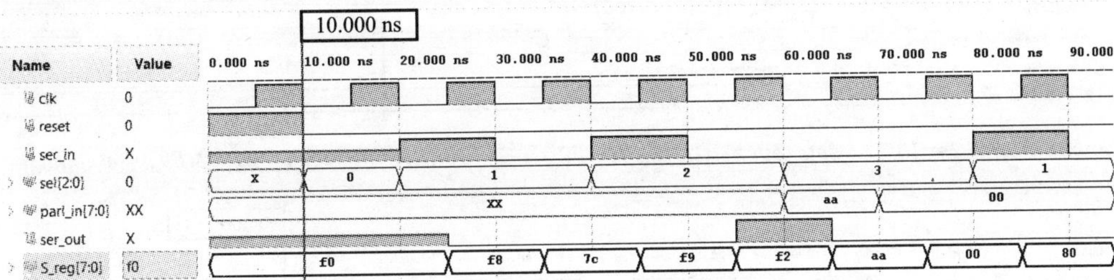

Fig. a: *Input: sel = 0 = 00; No change in shift register.*

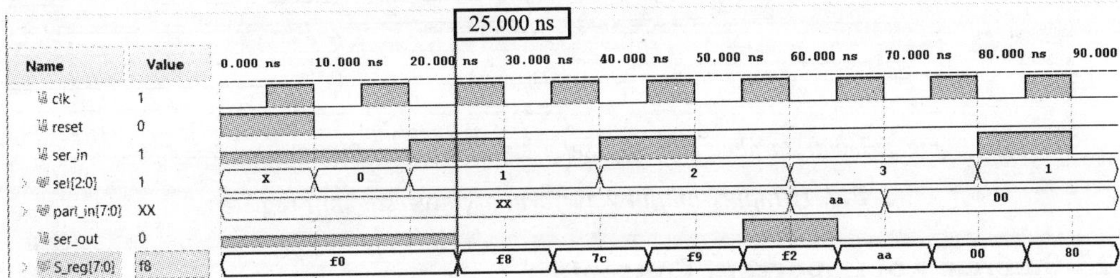

Fig. b: *Input: sel = 1 = 01; ser_data_in = 1; rightshift_reg: 1111 1111($F0_H$)* $\xrightarrow{\text{Right shift}}$ *1111 1000($F8_H$); ser_data_out = 0.*

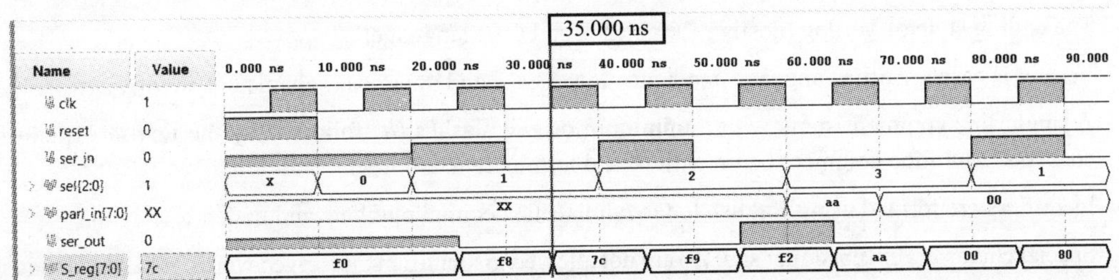

Fig. c: *Input: sel = 1 = 01; ser_data_in = 0; rightshift_reg: 1111 1000($F8_H$)* $\xrightarrow{\text{Right shift}}$ *0111 1100($7C_H$); ser_data_out = 0.*

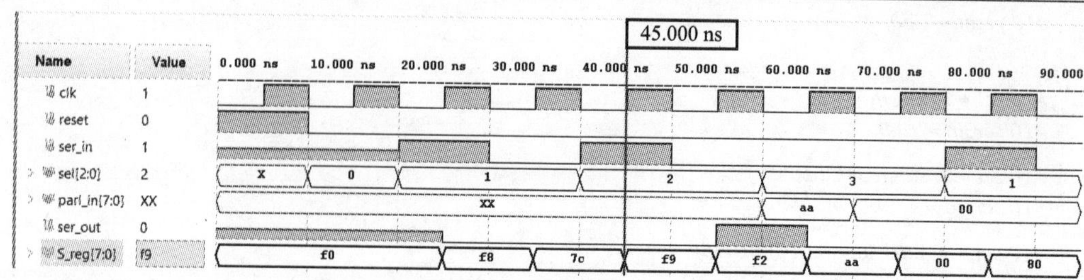

***Fig. d:** Input: sel = 2 = 01; ser_data_out = 0; 1111 1001(F9$_H$)* ←Left shift— *leftshift_reg: 0111 1100(7C$_H$); ser_data_in = 1.*

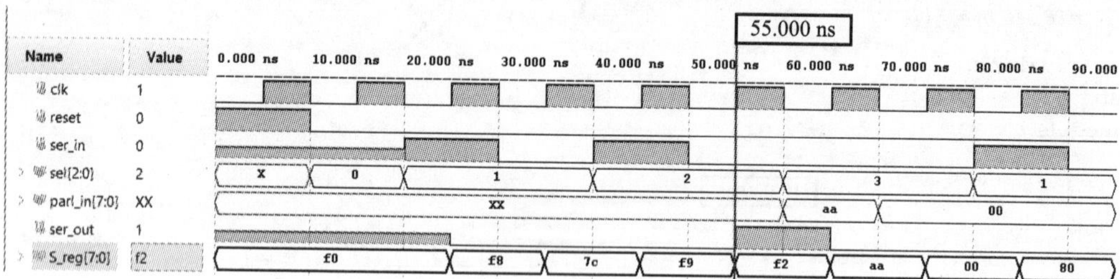

***Fig. e:** Input: sel = 2 = 10; ser_data_out = 1; 1111 0010(F2$_H$)* ←Left shift— *leftshift_reg: 1111 1001(F9$_H$); ser_data_in = 0.*

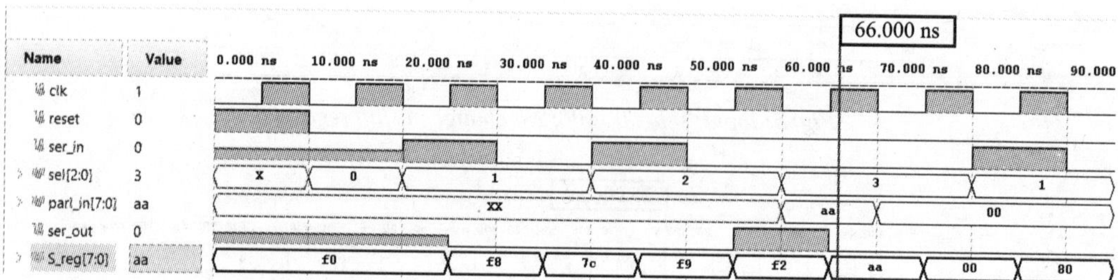

***Fig. f:** Input: Parallel_in = AA$_H$, Sel = 3 = 11; Output: Shift register = AA$_H$.*

***Fig. 1:** Inputs-output waveform of universal shift register.*

15.15 Summary of Important Concepts

1. The name Verilog is the combination of two words "verification" and "logic".
2. The basic structure of Verilog language is a module.
3. The coding style of Verilog mostly resembles C and C++ language.
4. Verilog is case sensitive language and hence lower case and upper case characters are distinguishable.
5. A single line comment starts with a minimum of two slashes (//) followed by the text to explain the concepts. Multiline comment begin with /* and terminate with */.
6. Identifiers are formed using English letters/alphabets, decimal numbers and underscore.
7. Special characters except dollar sign ($) are not allowed in identifiers. Reserved words of Verilog language cannot be used as identifiers.
8. Verilog supports four logic values: 0, 1, x, z.
9. The data types of Verilog can be classified into two groups: Net type and Variable type.

10. Net type signals represent physical connections between the hardware components in a digital circuit.
11. Variables are used for computation, manipulation and temporary storage of data during program execution.
12. A parameter data object defines a constant and only integer constants can be attached to a parameter.
13. Verilog primitives are predefined basic building blocks in Verilog HDL.
14. The 12 Verilog primitive gates are and, nand, or, nor, xor, xnor, not, buf, bufif1, bufif0, notif1, notif0.
15. Truth tables can be implemented as user defined primitives in Verilog program.
16. Verilog behavioral statements are declared inside an always or initial block.
17. The procedural assignment statements are used to evaluate an expression and assign the value obtained by evaluation of the expression to a variable.
18. The two types of procedural assignment statements in Verilog are blocking and nonblocking assignment statements.
19. The blocking statements are evaluated sequentially and values assigned sequentially.
20. The nonblocking statements are evaluated concurrently and assigned after evaluation of all the expressions.
21. The conditional statements in Verilog are if-else and case statements
22. The loop statements in Verilog are while, for, forever and repeat statements.
23. The ***function*** in Verilog is like a subroutine in a programming language.
24. Timing control statements and continuous assignment statements are not allowed in functions.
25. Task is a complete Verilog code similar to a module and enclosed with keyword ***task*** and ***endtask***.
26. The default time scale in Vivado HDL simulation tool is ***`timescale 1 ns/1 ps***.
27. A ***delay control*** expression specifies the time duration to be delayed for execution of the statement.
28. An ***event*** control is used to control the execution of a procedural statement at the occurrence of the named event.
29. The ***wait*** statement allows a procedural statement or a block of statements to be delayed until a condition become true.
30. The module is the main unit of the Verilog program and one program can have only one module.
31. The entire Verilog code or program is enclosed by the keywords ***module*** and ***endmodule***.
32. Three possible mode or direction a port in Verilog and ***input, output*** or ***inout***.
33. The initial blocks are primarily used to initialize variables and ports with specific values.
34. The always blocks are used to enclose the executable statements of the Verilog code.
35. The three types of programming model in Verilog are dataflow, behavioral and structural models.
36. The dataflow model is direct implementation of Boolean equations of a system.
37. The behavioral model is implementation of a digital system using procedural statements.
38. In structural model, a digital system is designed with the set of components connected by signals.
39. A test bench is also a Verilog program that is used to verify the correctness of the Verilog program of the hardware model.
40. In test bench the system command ***$finish*** is used to terminate the clock and test inputs.

15.16 Short-Answer Questions

Q15.1 List the language elements of Verilog.

The basic elements of Verilog language are,

- Identifiers
- Keywords
- Comments
- Logic values
- Numbers
- Operators
- Data types

Q15.2 What are the operators group in Verilog?

The predefined operators in the Verilog HDL language are classified into the following seven groups.

- Bitwise/Reduction operators
- Logical operators
- Arithmetic operators
- Relational operators
- Shift operators
- Equality operators
- Other operators

Q15.3 Write a Verilog program for AND gate.

```
module dfm_and_gate (m, n, p);
    input wire m, n;
    output wire p;
    assign p  = m & n;
endmodule
```

Q15.4 Write a Verilog program for OR gate.

```
module dfm_or_gate (m, n, p);
    input wire m, n;
    output wire p;
    assign p  = m | n;
endmodule
```

Q15.5 Write a Verilog program for NOT gate.

```
module dfm_not_gate (m, p);
    input wire m;
    output wire p;
    assign p  = ~m;
endmodule
```

Q15.6 Write a Verilog program for NAND gate.

```
module dfm_nand_gate (m, n, p);
    input wire m, n;
    output wire z;
    assign z  = ~(x & y);
endmodule
```

Q15.7 Write a Verilog program for NOR gate.

```
module dfm_nor_gate (m, n, p);
    input wire m, n;
    output wire p;
    assign p  = ~(m | n);
endmodule
```

Q15.8 Write a Verilog program for an 2-input XOR gate using behavioral model.

```
module dfm_xor_gate (m, n, p);
    input wire m, n;
    output wire p;
    assign p  = m ^ n;
endmodule
```

Q15.9 Write a Verilog program for an XNOR gate.

```
module dfm_xnor_gate (m, n, p);
    input wire m, n;
    output wire p;
    assign p  = ~(m ^ n);
endmodule
```

Q15.10 How numbers are represented in Verilog?

In Verilog, the binary, octal, decimal and hexadecimal numbers are represented by the following format.

Number format: n'Bv

where, n = Number of binary digits used to represent the number

B = Base of number

v = Value of the number

The letters b, o, d and h are used to represent the base of binary, octal, decimal and hexadecimal respectively.

Q15.11 Give one example of binary, octal, decimal and hexadecimal representation in Verilog.

Binary : 8'b'11101101 - 8-bit binary number 11101101

Octal : 12'o2467 - 4 digit octal number 2467 represented by 12 bit binary

Decimal : 6'd32 - 2 digit decimal number 32 represented by 6 bit binary

Hexadecimal : 16'h3B8C - 4 digit hexa number 3B8C represented by 16 bit binary

Q15.12 Explain the conditional operator in Verilog.

The conditional operator is used to check true or false condition of a variable and execute a statement for true condition or execute an alternative statement for false condition. The format of condition statement with condition operator is,

A ? statement_B : statement_C ;

If A is true then statemet_B is executed and if A is false then statement_C is executed.

Q15.13 What are Verilog primitives?. List the Verilog primitive logic gates.

Verilog primitives are predefined basic building blocks in Verilog HDL that can be used to model or design the structure of large digital circuits. Verilog includes 12 basic logic gates as predefined primitives and they are,

and, nand, or, nor, xor, xnor, not, buf, bufif1, bufif0, notif1, notif0

Q15.14 Write a short note on user defined primitives.

Truth tables can be implemented as user defined primitives in Verilog program.

The user defined primitive starts with the keyword ***primitive*** and ends with the keyword ***endprimitive*** and do not have a module. Therefore, the primitive has to be inserted before the module in a program in which the truth table has to be instantiated.

Q15.15 What are function and task in Verilog?

The ***function*** in Verilog is like a subroutine in a programming language. The repeated computation of a variable value using a group of statements are coded separately as a function and used any number times in a main program.

The ***task*** is a subroutine equivalent to a module that is written inside the main module and can be called in a Verilog main module code any number of times. Task is a complete Verilog code similar to a module and enclosed with keyword ***task*** and ***endtask***.

Q15.16 Write a short note on initial and always blocks in Verilog.

The initial blocks are primarily used to initialize variables and ports with specific values. An initial block executes only once, at the beginning of simulation.

The always blocks are used to enclose the executable statements of the Verilog code. The always block is executed repeatedly throughout the simulation.

A module can have any number of initial and always blocks. When initial and always block has multiple statements, they are enclosed by the keywords ***begin*** and ***end***.

15.17 Exercises

I. Fill in the blanks

1. The basic structure of Verilog language is a ______.
2. ______ are name of objects that are used in Verilog programs.
3. The ______ operators will perform logical operation on n-bit data and generates 1-bit result either true or false.
4. The ______ operator is used to combine or join two binary data.
5. The two groups of data types in Verilog are ______ and ______.
6. ______ are used for temporary storage of data during program execution.
7. Verilog ______are predefined basic building blocks in Verilog HDL.
8. The continuous assignment statements is prefixed with keyword ______.
9. The ______ statements are evaluated sequentially and values assigned sequentially.
10. The ______ loop is used to execute the statements enclosed by the loop continuously.
11. ______ in Verilog are used to code a block of procedural statements that computes a value based on the inputs and return a single value.
12. Task is a complete Verilog code similar to a module and enclosed with keyword ______ and ______.
13. A ______expression specifies the time duration to be delayed for execution of the statement.
14. The ______ statement allows a procedural statement or a block of statements to be delayed until a condition become true.

15. A Verilog program starts with a keyword ______ and ends with a keyword ______.
16. When initial and always block has multiple statements, they are enclosed by the keywords ______ and ______. (begin, end).

Answers

1. module	5. net , variables	9. blocking	13. delay control
2. Identifiers	6. variables	10. forever	14. wait
3. reduction	7. primitives	11. Functions	15. module, endmodule
4. concatenation	8. assign	12. task, endtask	16. begin, end

II. State whether the following statements are True or False

1. The entire Verilog code or program is enclosed by the keywords module and endmodule.
2. In Verilog, lower case and upper case characters are distinguishable.
3. In Verilog, special characters are allowed in identifiers.
4. Verilog supports three logic values: 0, 1, x.
5. Net type signals represent physical connections between the hardware components in a digital circuit.
6. The signals in an input/output port list of a module in Verilog code is by default reg.
7. Truth tables can be implemented as user defined primitives in Verilog program.
8. The condition statement " A ? statement_B : statement_C " is another form of if-else statement.
9. The nonblocking statements are evaluated sequentially.
10. The repeat loop is used to execute the statements enclosed by repeat loop continuously.
11. In Verilog functions, timing control statements and continuous assignment statements are not allowed.
12. An event control is used to control the execution of a procedural statement at the occurrence of the named event.
13. A Verilog program can have any number of modules.
14. Each initial and always block are executed once during program execution.
15. A module can have any number of initial and always block.
16. In structural model, the primitive gates can be used directly without instantiation.

Answers

1. true	5. true	9. false	13. false
2. true	6. false	10. false	14. false
3. false	7. true	11. true	15. true
4. false	8. true	12. true	16. true

III. Choose the right answer for the following questions

1. *Find the illegal identifier in Verilog?*

   a) sum b) case

   c) count d) product

2. *Which one of the following is not a keyword in Verilog?*

   a) casex b) strong1

   c) supply1 d) weakx

3. *The logic values supported by Verilog are*

   a) 0, 1, x b) 0, 1, z c) 0, 1, x, z d) 0, 1

4. *Which is the legal representation of binary number in Verilog?*

   a) 8'b00001111 b) '8b'00001111

   c) 8'b_00001111 d) 8'b'00001111

5. *Identify the operator with highest precedence in Verilog.*

   a) ?: b) % c) ! d) *

6. *Which one of the following is not a valid data type in Verilog?*

   a) wire b) wnand

   c) wand d) wor

7. *Which of the following is a not a valid port declaration in Verilog?*

   a) Input b) output

   c) bidirectional d) inout

8. *Which of the following is not allowed in Verilog function?*

   a) input declaration b) continuous assignment statement

   c) variable declaration d) parameter declaration

9. *Which of the following statement is true in Verilog program?*

   a) Only one Initial block is allowed in a Verilog program

   b) Initial block is executed only one time

   c) Always block is executed only one time

   d) A Verilog program can have any number of modules

10. *Which of the following statement is false in Verilog program?*

    a) Dataflow model will have continuous assignment statement

    b) Behavioral model will have procedural assignment statement

    c) Structural model will have primitive logic gates as components

    d) Dataflow model will have initial and always blocks

Answers

1. b	6. b
2. d	7. c
3. c	8. b
4. a	9. b
5. c	10. d

IV. Answer the following questions

1. What are the rules to form identifiers?
2. What are keywords? Give some examples of keywords.
3. List the logic values supported by Verilog.
4. Explain the bitwise and reduction logical operators with examples.
5. How data types are classified in Verilog?
6. Differentiate vectors and arrays in Verilog.
7. Write a short note on continuous assignment statement.
8. What are blocking and nonblocking statements in Verilog?
9. Give the formats of conditional statements in Verilog.
10. Explain the difference between forever and repeat loop.
11. Differentiate between function and task in Verilog.
12. Explain the event control techniques in Verilog.
13. Write a short note on dataflow model in Verilog, Give an example.
14. Explain the behavioral model in Verilog with an example.
15. What is structural model in Verilog? Explain with an example.

V. Write Solution for Following Verilog Programs

E15.1 Simplify the Boolean function F by means of K-map and write a dataflow and structural Verilog code to implement the reduced Boolean function.

$$F(A, B, C, D) = \Sigma m(1, 3, 4, 6, 8, 11, 15) + d(0, 5, 7)$$

E15.2 Simplify the Boolean function F by means of K-map and write a structural HDL to implement the reduced Boolean function.

a) Using only NAND gates

b) Using only NOR gates

E15.3 Write the Verilog code for 3-to-8 decoder using case statement.

E15.4 Write a structural model of Verilog code for 8-to-3 encoder.

E15.5 Develop Verilog code for 4:1 multiplexer using 2:1 multiplexer by considering 2:1 multiplexer as

a) a component

b) a function

c) a task

E15.6 **Write the behavioral model of 4:1 multiplexer using if-else statement.**

E15.7 **Develop Verilog code for dataflow model of 8:1 multiplexer.**

E15.8 **Write Verilog code for 2-bit comparator.**

E15.9 **Write Verilog code for parity detector to detect parity in 8-bit data with 1-bit parity.**

E15.10 **Write Verilog code for dataflow model of T flip-flop.**

APPENDIX

APPENDIX 1: List of Some Standard IC's

Table A1.1: List of Some Standard IC's

IC Number	IC Name	Number of Inputs in each Device	Number of Devices in each IC
7404	NOT Gate	1	6
7408	Quad 2-input AND Gate	2	4
7411	Triple 3-input AND Gate	3	3
7421	Dual 4-input AND Gate	4	2
7400	Quad 2-input NAND Gate	2	4
7410	Triple 3-input NAND Gate	3	3
7422	Dual 4-input NAND Gate	4	2
7430	Single 8-input NAND Gate	8	1
7432	Quad 2-input OR Gate	2	4
744075	Triple 3-input OR Gate	3	3
744072	Dual 4-input OR Gate	4	2
744078	Single 8-input OR Gate	8	1
7402	Quad 2-input NOR Gate	2	4
7427	Triple 3-input NOR Gate	3	3
7429	Dual 4-input NOR Gate	4	2
74260	Dual 5-input NOR Gate	5	2
744078	Single 8-input NOR Gate	8	1
74157	Quad 2:1 Multiplexer	2	4
74153	Dual 4:1 Multiplexer	4	2
74151	8:1 Multiplexer	8	1
74150	16:1 Multiplexer	16	1
74237	Logic High 3-to-8 Decoder	3	1
74138	Logic Low 3-to-8 Decoder	3	1
7474	Dual D Flip-Flop	2	2
74171	Quad D Flip-Flop	1	4
74112	Dual JK Flip-Flop	2	2
7471	RS Master Slave Flip-Flop	2	1

APPENDIX 2: Pin Configuration of Some Standard IC's

a_1	1	14	V_{cc}(+5V)
y_1	2	13	a_6
a_2	3	12	y_6
y_2	4	11	a_5
a_3	5	10	y_5
y_3	6	9	a_4
(0V) GND	7	8	y_4

7404

Pin	Description
a_1 - a_6	Inputs
y_1 - y_6	Outputs
V_{cc}	Supply voltage (+5V)
GND	Ground (0V)

Fig. A2.1: *Inverter gate IC 7404.*

a_1	1	14	V_{cc}(+5V)
b_1	2	13	b_4
y_1	3	12	a_4
a_2	4	11	y_4
b_2	5	10	b_3
y_2	6	9	a_3
(0V) GND	7	8	y_3

7408

Pin	Description
a_1 - a_4	Inputs
b_1 - b_4	Inputs
y_1 - y_4	Outputs
V_{cc}	Supply voltage (+5V)
GND	Ground (0V)

Fig. A2.2: *Quad 2-input AND gate IC 7408.*

a_1	1	14	V_{cc}(+5V)
b_1	2	13	c_1
a_2	3	12	y_1
b_2	4	11	c_3
c_2	5	10	b_3
y_2	6	9	a_3
(0V) GND	7	8	y_3

7411

Pin	Description
a_1 - a_3	Inputs
b_1 - b_3	Inputs
c_1 - c_3	Inputs
y_1 - y_3	Outputs
V_{cc}	Supply voltage (+5V)
GND	Ground (0V)

Fig. A2.3: *Triple 3-input AND gate IC 7411.*

Note: *The suffix 1, 2, 3, 4, 5 and 6 in pin details refer to device number in the IC's.*

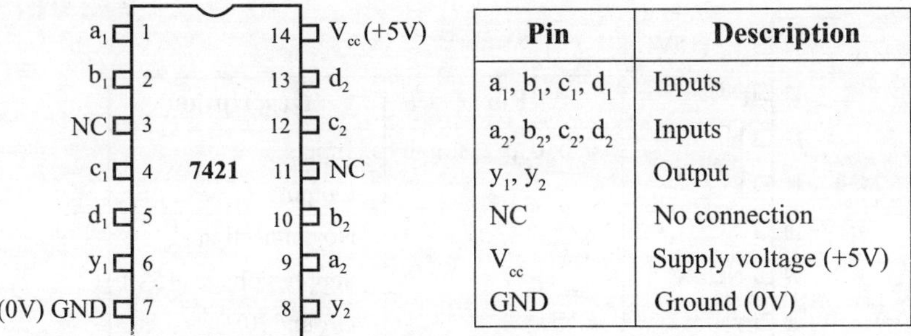

Pin	Description
a_1, b_1, c_1, d_1	Inputs
a_2, b_2, c_2, d_2	Inputs
y_1, y_2	Output
NC	No connection
V_{cc}	Supply voltage (+5V)
GND	Ground (0V)

Fig. A2.4: *Dual 4-input AND gate IC 7421.*

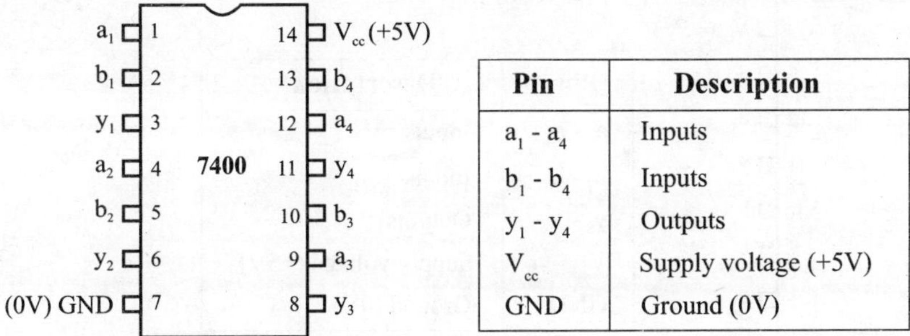

Pin	Description
a_1 - a_4	Inputs
b_1 - b_4	Inputs
y_1 - y_4	Outputs
V_{cc}	Supply voltage (+5V)
GND	Ground (0V)

Fig. A2.5: *Quad 2-input NAND gate IC 7400.*

Pin	Description
a_1 - a_3	Inputs
b_1 - b_3	Inputs
c_1 - c_3	Inputs
y_1 - y_3	Outputs
V_{cc}	Supply voltage (+5V)
GND	Ground (0V)

Fig. A2.6: *Triple 3-input NAND gate IC 7410.*

Pin	Description
a, b, c, d	Inputs
y_1, y_2	Outputs
NC	No connection
V_{cc}	Supply voltage (+5V)
GND	Ground (0V)

Fig. A2.7: *Dual 4-input NAND gate IC 7422.*

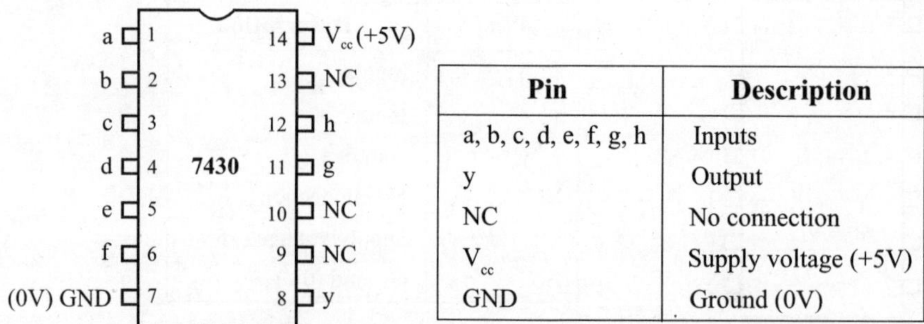

Pin	Description
a, b, c, d, e, f, g, h	Inputs
y	Output
NC	No connection
V_{cc}	Supply voltage (+5V)
GND	Ground (0V)

Fig. A2.8: *Single 8-input NAND gate IC 7430.*

Pin	Description
a_1 - a_4	Inputs
b_1 - b_4	Inputs
y_1 - y_4	Outputs
V_{cc}	Supply voltage (+5V)
GND	Ground (0V)

Fig. A2.9: *Quad 2-input OR gate IC 7432.*

Pin	Description
a_1 - a_3	Inputs
b_1 - b_3	Inputs
c_1 - c_3	Inputs
y_1 - y_3	Outputs
V_{cc}	Supply voltage (+5V)
GND	Ground (0V)

Fig. A2.10: *Triple 3-input OR gate IC 744075.*

Pin	Description
a_1, b_1, c_1, d_1	Inputs
a_2, b_2, c_2, d_2	Inputs
y_1, y_2	Output
NC	No connection
V_{cc}	Supply voltage (+5V)
GND	Ground (0V)

Fig. A2.11: *Dual 4-input OR gate IC 744072.*

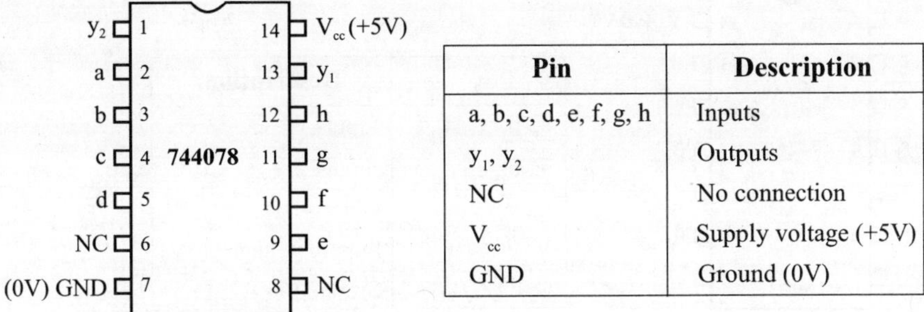

Pin	Description
a, b, c, d, e, f, g, h	Inputs
y_1, y_2	Outputs
NC	No connection
V_{cc}	Supply voltage (+5V)
GND	Ground (0V)

Fig. A2.12: *Single 8-input OR/NOR gate IC 744078.*

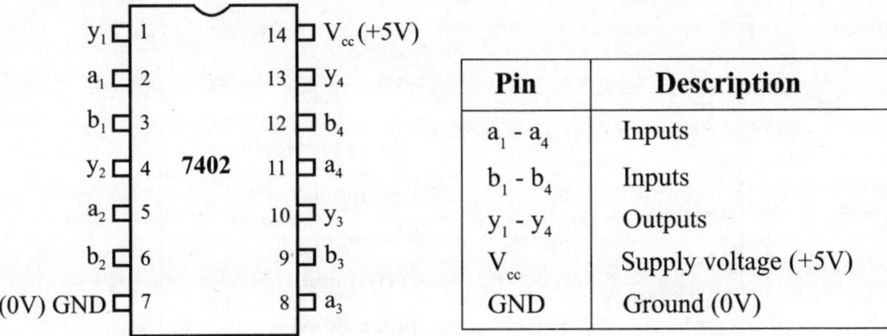

Pin	Description
a_1 - a_4	Inputs
b_1 - b_4	Inputs
y_1 - y_4	Outputs
V_{cc}	Supply voltage (+5V)
GND	Ground (0V)

Fig. A2.13: *Quad 2-input NOR gate IC 7402.*

Pin	Description
a_1 - a_3	Inputs
b_1 - b_3	Inputs
c_1 - c_3	Inputs
y_1 - y_3	Outputs
V_{cc}	Supply voltage (+5V)
GND	Ground (0V)

Fig. A2.14: *Triple 3-input NOR gate IC 7427.*

Pin	Description
a_1, b_1, c_1, d_1	Inputs
a_2, b_2, c_2, d_2	Inputs
y_1, y_2	Output
NC	No connection
V_{cc}	Supply voltage (+5V)
GND	Ground (0V)

Fig. A2.15: *Dual 4-input NOR gate IC 7429.*

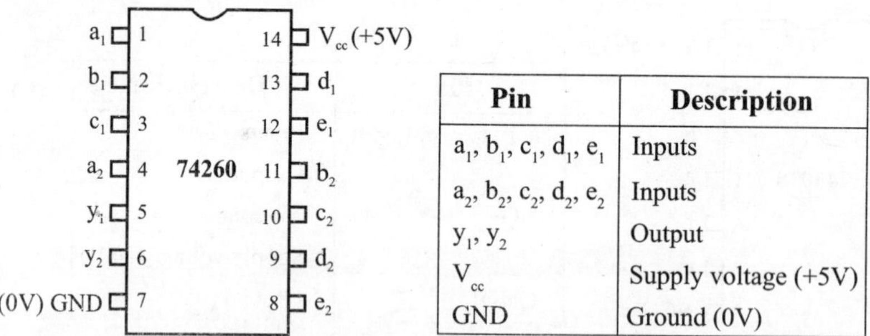

Pin	Description
a_1, b_1, c_1, d_1, e_1	Inputs
a_2, b_2, c_2, d_2, e_2	Inputs
y_1, y_2	Output
V_{cc}	Supply voltage (+5V)
GND	Ground (0V)

Fig. A2.16: *Dual 5-input NOR gate IC 74260.*

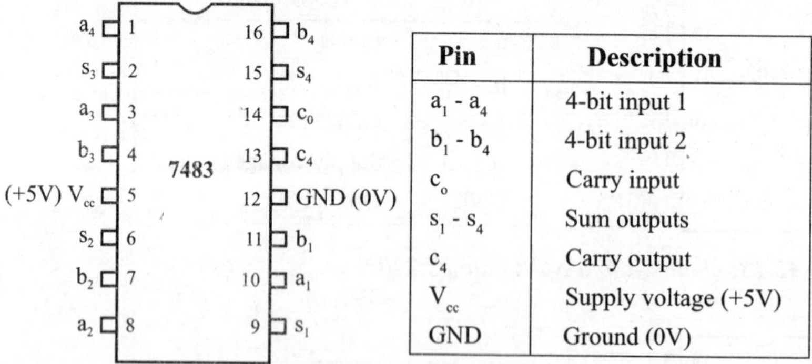

Pin	Description
a_1 - a_4	4-bit input 1
b_1 - b_4	4-bit input 2
c_o	Carry input
s_1 - s_4	Sum outputs
c_4	Carry output
V_{cc}	Supply voltage (+5V)
GND	Ground (0V)

Fig. A2.17: *4-bit binary full adder IC 7483.*

74182

Pin	Description
1	g_1
2	p_1
3	g_0
4	p_0
5	g_3
6	p_3
7	p
8	(0V) GND
16	V_{cc}(+5V)
15	p_2
14	g_2
13	c_0
12	c_1
11	c_2
10	c_4
9	c_3

Pin	Description
g_0 - g_3	Carry generate inputs
p_0 - p_3	Carry propagate inputs
c_o	Carry input
c_1, c_2, c_3	Carry outputs
c_4	Carry generate output
p	Carry propagate output
V_{cc}	Supply voltage (+5V)
GND	Ground (0V)

Fig. A2.18: *Look ahead carry generator IC 74182.*

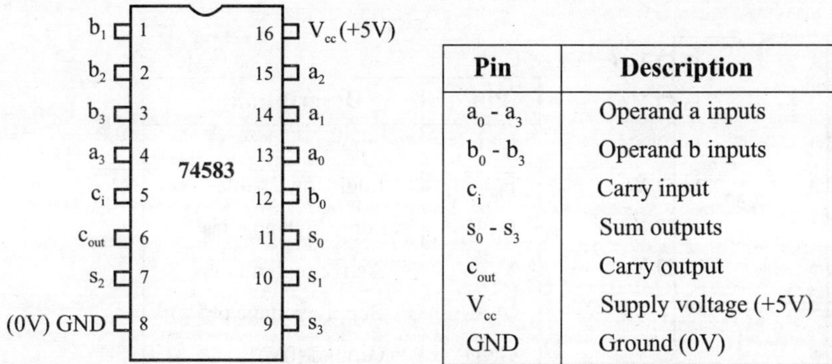

Pin	Description
a_0 - a_3	Operand a inputs
b_0 - b_3	Operand b inputs
c_i	Carry input
s_0 - s_3	Sum outputs
c_{out}	Carry output
V_{cc}	Supply voltage (+5V)
GND	Ground (0V)

Fig. A2.19: *4-bit BCD adder IC 74583.*

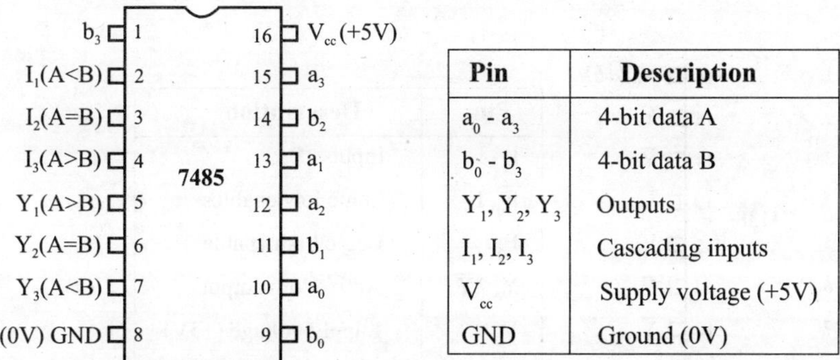

Pin	Description
a_0 - a_3	4-bit data A
b_0 - b_3	4-bit data B
Y_1, Y_2, Y_3	Outputs
I_1, I_2, I_3	Cascading inputs
V_{cc}	Supply voltage (+5V)
GND	Ground (0V)

Fig. A2.20: *4-bit magnitude comparator IC 7485.*

Pin	Description
I_{0a}, I_{1a}	Inputs
I_{0b}, I_{1b}	Inputs
$\overline{Y}_{0a}$ - $\overline{Y}_{3a}$	Active low outputs
$\overline{Y}_{0b}$ - $\overline{Y}_{3b}$	Active low outputs
$\overline{E}_a$, $\overline{E}_b$	Enables
V_{cc}	Supply voltage (+5V)
GND	Ground (0V)

Fig. A2.21: *Dual 2-to-4 decoder IC 74139.*

74237

	Pin	Pin	
I_0	1	16	V_{cc}(+5V)
I_1	2	15	Y_0
I_2	3	14	Y_1
$\overline{E}_0$	4	13	Y_2
E_1	5	12	Y_3
E_2	6	11	Y_4
Y_7	7	10	Y_5
(0V) GND	8	9	Y_6

Pin	Description
I_0 - I_2	Inputs
$\overline{E}_0$	Logic low enable
E_1, E_2	Logic high enables
Y_0 - Y_7	Active high outputs
V_{cc}	Supply voltage (+5V)
GND	Ground (0V)

Fig. A2.22: *Logic high 3-to-8 decoder IC 74237.*

74138

	Pin	Pin	
I_0	1	16	V_{cc}(+5V)
I_1	2	15	$\overline{Y}_0$
I_2	3	14	$\overline{Y}_1$
$\overline{E}_0$	4	13	$\overline{Y}_2$
$\overline{E}_1$	5	12	$\overline{Y}_3$
E_2	6	11	$\overline{Y}_4$
$\overline{Y}_7$	7	10	$\overline{Y}_5$
(0V) GND	8	9	$\overline{Y}_6$

Pin	Description
I_0 - I_2	Inputs
$\overline{E}_0$, $\overline{E}_1$	Logic low enables
E_2	Logic high enable
$\overline{Y}_0$ - $\overline{Y}_7$	Active low outputs
V_{cc}	Supply voltage (+5V)
GND	Ground (0V)

Fig. A2.23: *Logic low 3-to-8 decoder IC 74138.*

74148

	Pin	Pin	
I_4	1	16	V_{cc}(+5V)
I_5	2	15	EO
I_6	3	14	v
I_7	4	13	I_3
EI	5	12	I_2
Y_2	6	11	I_1
Y_1	7	10	I_0
(0V) GND	8	9	Y_0

Pin	Description
I_0 - I_7	Inputs
Y_0 - Y_2	Outputs
v	Outputs
V_{cc}	Supply voltage (+5V)
GND	Ground (0V)
EI	Input Enable
EO	Output Enable

Fig. A2.24: *8-to-3 priority encoder IC 74148.*

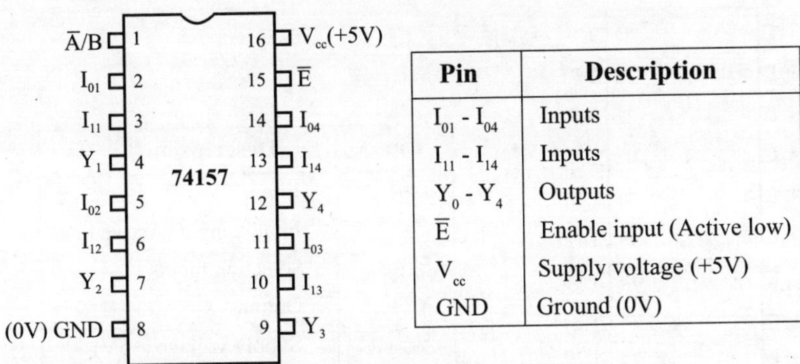

Pin	Description
I_{01} - I_{04}	Inputs
I_{11} - I_{14}	Inputs
Y_0 - Y_4	Outputs
$\overline{E}$	Enable input (Active low)
V_{cc}	Supply voltage (+5V)
GND	Ground (0V)

Fig. A2.25: *Quad 2:1 multiplexer IC 74157.*

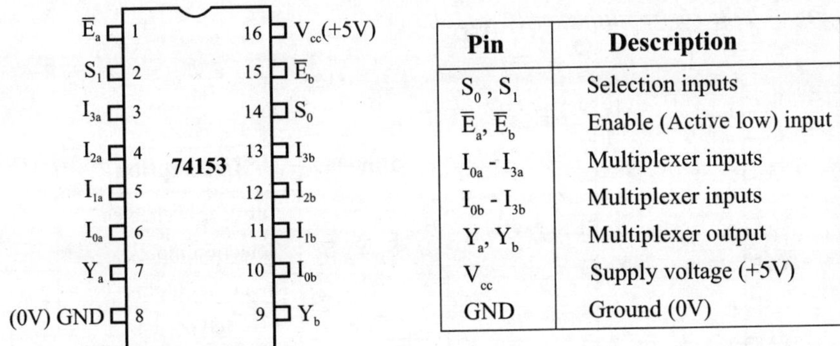

Pin	Description
S_0 , S_1	Selection inputs
$\overline{E}_a$, $\overline{E}_b$	Enable (Active low) input
I_{0a} - I_{3a}	Multiplexer inputs
I_{0b} - I_{3b}	Multiplexer inputs
Y_a, Y_b	Multiplexer output
V_{cc}	Supply voltage (+5V)
GND	Ground (0V)

Fig. A2.26: *Dual 4:1 multiplexer IC 74153.*

74151

Left pins		Right pins	
I_3	1	16	V_{cc} (+5V)
I_2	2	15	I_4
I_1	3	14	I_5
I_0	4	13	I_6
Y	5	12	I_7
$\overline{Y}$	6	11	S_0
$\overline{E}$	7	10	S_1
(0V) GND	8	9	S_2

Pin	Description
S_0 - S_2	Selection inputs
$\overline{E}$	Enable (Active low) input
I_0 - I_7	Multiplexer inputs
Y	Multiplexer output
$\overline{Y}$	Complement of output
V_{cc}	Supply voltage (+5V)
GND	Ground (0V)

Fig. A2.27: *8:1 multiplexer IC 74151.*

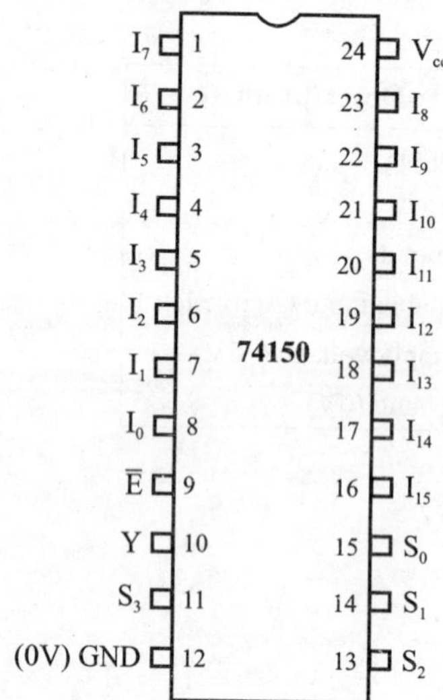

Pin	Description
I_0 - I_{15}	Inputs
$\overline{E}$	Enable input (Active low)
S_0 - S_3	Selection inputs
Y	Output
V_{cc}	Supply voltage (+5V)
GND	Ground (0V)

Fig. A2.28: *16:1 multiplexer IC 74150.*

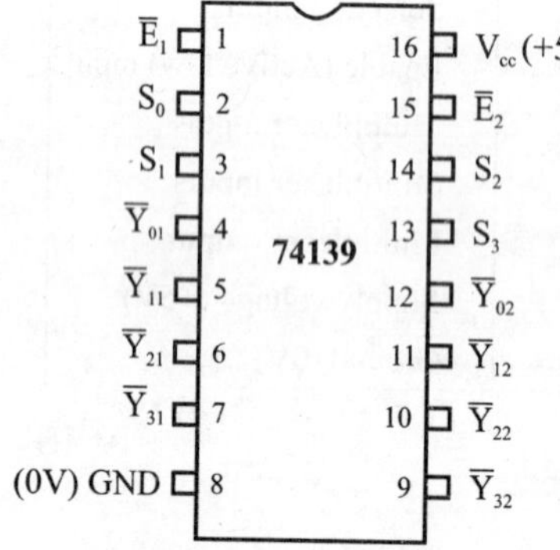

Pin	Description
$\overline{E}_1$, $\overline{E}_2$	Active low enables
S_0, S_1, S_2, S_3	Selection inputs
$\overline{Y}_{01}$ - $\overline{Y}_{31}$	Inverted outputs
$\overline{Y}_{02}$ - $\overline{Y}_{32}$	Inverted outputs
V_{cc}	Supply voltage (+5V)
GND	Ground (0V)

Fig. A2.29: *Dual 1:4 demultiplexer IC 74139.*

Pin	Description
$\overline{S}_{11}$ - $\overline{S}_{12}$	Active low set inputs
$\overline{S}_{21}$ - $\overline{S}_{23}$	Active low set inputs
$\overline{R}_1$ - $\overline{R}_4$	Active low reset inputs
Q_1 - Q_4	Outputs
V_{cc}	Supply voltage (+5V)
GND	Ground (0V)

Fig. A2.30: *Quad SR latch IC 74279.*

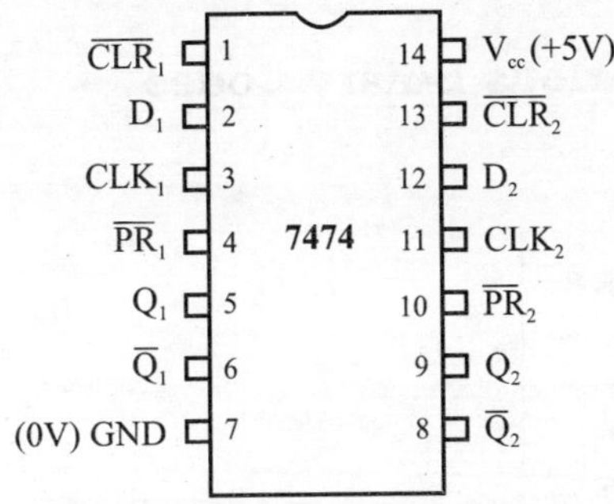

Pin	Description
D_1, D_2	Inputs
$\overline{PR}_1$, $\overline{PR}_2$	Preset inputs
$\overline{CLR}_1$, $\overline{CLR}_2$	Clear inputs
CLK_1, CLK_2	Clock inputs
Q_1, Q_2	Outputs
$\overline{Q}_1$, $\overline{Q}_2$	Complement of outputs
V_{cc}	Supply voltage (+5V)
GND	Ground (0V)

***Fig. A2.31:** Dual D flip-flop IC 7474.*

74171	Pin		Pin	
$\overline{Q}_1$	1		16	V_{cc} (+5V)
$\overline{Q}_2$	2		15	Q_1
Q_2	3		14	D_1
D_2	4		13	$\overline{CLR}$
D_3	5		12	CLK
Q_3	6		11	D_4
$\overline{Q}_3$	7		10	Q_4
(0V) GND	8		9	$\overline{Q}_4$

Pin	Description
D_1 - D_4	Inputs
$\overline{CLR}$	Clear input
CLK	Clock input
Q_1 - Q_4	Outputs
$\overline{Q}_1$ - $\overline{Q}_4$	Complement of outputs
V_{cc}	Supply voltage (+5V)
GND	Ground (0V)

***Fig. A2.32:** Quad D flip-flop IC 74171.*

74112	Pin		Pin	
CLK_1	1		16	V_{cc} (+5V)
K_1	2		15	$\overline{CLR}_1$
J_1	3		14	$\overline{CLR}_2$
$\overline{PR}_1$	4		13	CLK_2
Q_1	5		12	K_2
$\overline{Q}_1$	6		11	J_2
$\overline{Q}_2$	7		10	$\overline{PR}_2$
(0V) GND	8		9	Q_2

Pin	Description
J_1, K_1, J_2, K_2	Inputs
$\overline{PR}_1$, $\overline{PR}_2$	Preset inputs
$\overline{CLR}_1$, $\overline{CLR}_2$	Clear inputs
CLK_1, CLK_2	Clock inputs
Q_1, Q_2	Outputs
$\overline{Q}_1$, $\overline{Q}_2$	Complement of outputs
V_{cc}	Supply voltage (+5V)
GND	Ground (0V)

***Fig. A2.33:** Dual JK flip-flop IC 74112.*

7471	Pin		Pin	
NC	1		14	V_{cc} (+5V)
$\overline{CLR}$	2		13	$\overline{PRE}$
S_1	3		12	CLK
S_2	4		11	R_3
S_3	5		10	R_2
$\overline{Q}$	6		9	R_1
(0V) GND	7		8	Q

Pin	Description
S_1 - S_3	Active high set inputs
R_1 - R_3	Active high reset inputs
$\overline{CLR}$	Clear input
CLK	Clock input
Q, $\overline{Q}$	Outputs
V_{cc}	Supply voltage (+5V)
GND	Ground (0V)

***Fig. A2.34:** RS master slave flip-flop IC 7471.*

APPENDIX 3: Summary of Various Binary Codes

Table A3.1: BCD Codes

Decimal Digit	8421 BCD	2421 BCD	5421 BCD	8 4 –2 –1 BCD	Excess-3 BCD
0	0000	0000	0000	0000	0011
1	0001	0001	0001	0111	0100
2	0010	0010	0010	0110	0101
3	0011	0011	0011	0101	0110
4	0100	0100	0100	0100	0111
5	0101	1011	1000	1011	1000
6	0110	1100	1001	1010	1001
7	0111	1101	1010	1001	1010
8	1000	1110	1011	1000	1011
9	1001	1111	1100	1111	1100

Table A3.2: Gray Code

Decimal Number	1-bit Gray Code	2-bit Gray Code	3-bit Gray Code	4-bit Gray Code
0	0	00	000	0000
1	1	01	001	0001
2		11	011	0011
3		10	010	0010
4			110	0110
5			111	0111
6			101	0101
7			100	0100
8				1100
9				1101
10				1111
11				1110
12				1010
13				1011
14				1001
15				1000

Table A3.3: ASCII Codes

ASCII in Hex	Symbol	ASCII in Hex	Symbol	ASCII in Hex	Symbol	ASCII in Hex	Symbol
00	NUL	20	SP	40	@	60	`
01	SOH	21	!	41	A	61	a
02	STX	22	“	42	B	62	b
03	ETX	23	#	43	C	63	c
04	EOT	24	$	44	D	64	d
05	ENQ	25	%	45	E	65	e
06	ACK	26	&	46	F	66	f
07	BEL	27	‘	47	G	67	g
08	BS	28	(	48	H	68	h
09	HT	29	)	49	I	69	i
0A	LF	2A	*	4A	J	6A	j
0B	VT	2B	+	4B	K	6B	k
0C	FF	2C	,	4C	L	6C	l
0D	CR	2D	–	4D	M	6D	m
0E	SO	2E	.	4E	N	6E	n
0F	SI	2F	/	4F	O	6F	o
10	DLE	30	0	50	P	70	p
11	DC1	31	1	51	Q	71	q
12	DC2	32	2	52	R	72	r
13	DC3	33	3	53	S	73	s
14	DC4	34	4	54	T	74	t
15	NAK	35	5	55	U	75	u
16	SYN	36	6	56	V	76	v
17	ETB	37	7	57	W	77	w
18	CAN	38	8	58	X	78	x
19	EM	39	9	59	Y	79	y
1A	SUB	3A	:	5A	Z	7A	z
1B	ESC	3B	;	5B	[	7B	{
1C	FS	3C	<	5C	\	7C	|
1D	GS	3D	=	5D	]	7D	}
1E	RS	3E	>	5E	^	7E	~
1F	US	3F	?	5F	_	7F	DEL

APPENDIX 4: Unsigned and Signed Binary Number System

Table A4.1: Decimal Number and its Equivalent Unsigned Binary Number

Decimal Number	Unsigned Binary Number	8-bit Unsigned Binary Number
0	0	0000 0000
1	1	0000 0001
2	10	0000 0010
3	11	0000 0011
4	100	0000 0100
5	101	0000 0101
6	110	0000 0110
7	111	0000 0111
8	1000	0000 1000
9	1001	0000 1001
10	1010	0000 1010
11	1011	0000 1011
12	1100	0000 1100
13	1101	0000 1101
14	1110	0000 1110
15	1111	0000 1111

Table A4.2: Decimal and Sign-Magnitude form of Binary Number

Decimal Number	4-bit Sign-magnitude Binary Number	Standard 8-bit Sign-magnitude Binary Number
+0	0000	0000 0000
+1	0001	0000 0001
+2	0010	0000 0010
+3	0011	0000 0011
+4	0100	0000 0100
+5	0101	0000 0101
+6	0110	0000 0110
+7	0111	0000 0111
–0	1000	1000 0000
–1	1001	1000 0001
–2	1010	1000 0010
–3	1011	1000 0011
–4	1100	1000 0100
–5	1101	1000 0101
–6	1110	1000 0110
–7	1111	1000 0111

Table A4.3: Decimal and One's Complement form of Binary Number

Decimal Number	4-bit One's Complement Binary Number	Sign-extended One's Complement 8-bit Binary Number
+0	0000	0000 0000
+1	0001	0000 0001
+2	0010	0000 0010
+3	0011	0000 0011
+4	0100	0000 0100
+5	0101	0000 0101
+6	0110	0000 0110
+7	0111	0000 0111
–7	1000	1111 1000
–6	1001	1111 1001
–5	1010	1111 1010
–4	1011	1111 1011
–3	1100	1111 1100
–2	1101	1111 1101
–1	1110	1111 1110
–0	1111	1111 1111

Table A4.4: Decimal and Two's Complement form of Binary Number

Decimal Number	4-bit Two's Complement Binary Number	Sign-extended Two's Complement 8-bit Binary Number
+0	0000	0000 0000
+1	0001	0000 0001
+2	0010	0000 0010
+3	0011	0000 0011
+4	0100	0000 0100
+5	0101	0000 0101
+6	0110	0000 0110
+7	0111	0000 0111
–8	1000	1111 1000
–7	1001	1111 1001
–6	1010	1111 1010
–5	1011	1111 1011
–4	1100	1111 1100
–3	1101	1111 1101
–2	1110	1111 1110
–1	1111	1111 1111

APPENDIX 5: Boolean Theorems

Theorem 1: $x + x = x$

$x \cdot x = x$

Theorem 2: $x + 1 = 1$

$x \cdot 0 = 0$

Theorem 3: $(x')' = x$
(Involution)

Theorem 4: $x + (y + z) = (x + y) + z$

(Associative) $x \cdot (y \cdot z) = (x \cdot y) \cdot z$

Theorem 5: $(x + y)' = x' \cdot y'$

(DeMorgan's Law or DeMorgan's Theorem) $(x \cdot y)' = x' + y'$

Theorem 6: $x + (x \cdot y) = x$

(Absorption) $x \cdot (x + y) = x$

(y is absorbed)

APPENDIX 6: Two Variable Boolean Function

Table A6.1: Two-Variable Boolean Functions

Functions	x	y	F	Functions	x	y	F
Null operation	0	0	0	NOR operation	0	0	1
$F = F_0 = 0$	0	1	0	$F = F_8 = (x + y)'$	0	1	0
	1	0	0		1	0	0
	1	1	0		1	1	0
AND operation	0	0	0	Exclusive NOR	0	0	1
$F = F_1 = x . y$	0	1	0	(or equivalence)	0	1	0
	1	0	0	$F = F_9 = (x \oplus y)'$	1	0	0
	1	1	1	$= x . y + x' . y'$	1	1	1
Inhibition	0	0	0	Complement of y	0	0	1
$F = F_2 = x . y'$	0	1	0	$F = F_{10} = y'$	0	1	0
(x true, but y false)	1	0	1		1	0	1
	1	1	0		1	1	0
Transfer	0	0	0	Implication	0	0	1
$F = F_3 = x$	0	1	0	$F = F_{11} = x + y'$	0	1	0
(Transfer x)	1	0	1	(If $y = 1, F = x$	1	0	1
	1	1	1	If $y = 0, F = y'$)	1	1	1
Inhibition	0	0	0	Complement of x	0	0	1
$F = F_4 = x' . y$	0	1	1	$F = F_{12} = x'$	0	1	1
(y true, but x false)	1	0	0		1	0	0
	1	1	0		1	1	0
Transfer	0	0	0	Implication	0	0	1
$F = F_5 = y$	0	1	1	$F = F_{13} = x' + y$	0	1	1
(Transfer y)	1	0	0	(If $x = 1, F = y$	1	0	0
	1	1	1	If $x = 0, F = x'$)	1	1	1
XOR operation	0	0	0	NAND operation	0	0	1
$F = F_6 = x \oplus y$	0	1	1	$F = F_{14} = (x . y)'$	0	1	1
$- x . y' + x' . y$	1	0	1		1	0	1
	1	1	0		1	1	0
OR operation	0	0	0	Identity	0	0	1
$F = F_7 = x + y$	0	1	1	$F = F_{15} = 1$	0	1	1
	1	0	1		1	0	1
	1	1	1		1	1	1

APPENDIX 7: Summary of Flip-Flop Tables

SR Flip-Flop

Table A7.1: Truth Table

Inputs		Outputs	
S	R	Q	Q′
0	0	No change	
0	1	0	1
1	0	1	0
1	1	Indeterminate	

Table A7.2: Characteristics Table

Inputs S R	Present State Q	Next State Q(t+1)
0 0	0	0
0 0	1	1
0 1	0	0
0 1	1	0
1 0	0	1
1 0	1	1
1 1	0	×
1 1	1	×

Table A7.3: Excitation Table

Present State Q(t)	Next State Q(t+1)	Flip-Flop Inputs S	R
0	0	0	×
0	1	1	0
1	0	0	1
1	1	×	0
0	×	1	1
1	×	1	1

JK Flip-Flop

Table A7.4: Truth Table

Inputs		Outputs		Comment
J	K	Q	Q′	
0	0	No change		-
0	1	0	1	Reset state
1	0	1	0	Set state
1	1	Toggle		Toggle previous output

Table A7.5: Characteristics Table

Inputs J K	Present State Q	Next State Q(t+1)
0 0	0	0
0 0	1	1
0 1	0	0
0 1	1	0
1 0	0	1
1 0	1	1
1 1	0	1
1 1	1	0

Table A7.6: Excitation Table

Present State Q(t)	Next State Q(t+1)	Flip-Flop Inputs J	K
0	0	0	×
0	1	1	×
1	0	×	1
1	1	×	0

T Flip-Flop

Table A7.7: Truth Table

T	CLK	Q	Q′
0	↑	No change	
1	↑	Toggles previous output	

Table A7.8: Characteristics Table

Input T	Present State Q	Next State Q(t+1)
0	0	0
0	1	1
1	0	1
1	1	0

Table A7.9: Excitation Table

Present State Q(t)	Next State Q(t+1)	Flip-Flop Input T
0	0	0
0	1	1
1	0	1
1	1	0

D Flip-Flop

Table A7.10: Truth Table

Input D	Outputs Q	Q′
0	0	1
1	1	0

Table A7.11: Characteristics Table

Input D	Present State Q	Next State Q(t+1)
0	0	0
0	1	0
1	0	1
1	1	1

Table A7.12: Excitation Table

Present State Q(t)	Next State Q(t+1)	Flip-Flop Input D
0	0	0
0	1	1
1	0	0
1	1	1

APPENDIX 8: List of Xillinx FPGAs

Table A8.1: List of Xillinx FPGAs

Generation	Family Name	Year of Launch	Manufacturing Process	Internal Operating Voltage
XC2000	XC2000	1985	2000 nm	5 V
	XC2000L	1993		3.3 V
	XC3000	1988	1200 nm	5 V
	XC3000A	1993	800 nm	
XC3000	XC3000L	1993		3.3 V
	XC3100	1992	800 nm	5 V
	XC3100A	1994	500 nm	
	XC3000L	1995		3.3 V
	XC4000	1991		5 V
	XC4000A	1991		5 V
	XC4000D	1994		5 V
	XC4000H	1993		5 V
	XC4000E	1995	500 nm	5 V
XC4000	XC4000L	1995		3.3 V
	XC4000EX	1996	500 nm	5 V
	XC4000XL	1997	350 nm	3.3 V
	XC4000XLA	1998	350 nm, 250 nm	3.3 V
	XC4000XV	1998	250 nm	2.5 V
	Spartan	1998	500 nm, 350 nm	5 V
	Spartan XL	1998	350 nm, 250 nm	3.3 V
XC5200	XC5200	1994	600 nm	5 V
	XC5200L		500 nm	3.3 V
XC6200		1995	650 nm	5 V
XC8100		1995		5 V or 3.3 V
	Virtex	1998	220 nm	2.5 V
	Spartan-II	2000		
Virtex	Virtex E	1999		
	Virtex EM	2000	180 nm	1.8 V
	Spartan-IIE	2001		
	Virtex-II	2001	150 nm	1.5 V
Virtex-II	Virtex-II Pro	2002	130 nm	1.2 V
	Virtex-II Pro X	2003		
	Spartan-3	2003		
	Spartan-3E	2004		
Spartan-3	Spartan-3A	2006	90 nm	1.2 V
	Spartan-3 AN	2007		
	Spartan-3A DSP			

Continued Table A8.1...

Generation	Family Name	Year of Launch	Manufacturing Process	Internal Operating Voltage
	Virtex-4 LX			
Virtex-4	Virtex-4 SX	2004	90 nm	1.2 V
	Virtex-4 FX			
	Virtex-5 LX			
	Virtex-5 LXT			
Virtex-5	Virtex-5 SXT	2006	65 nm	1.0 V
	Virtex-5 FXT			
	Virtex-5 TXT			
	Virtex-6 LX			
	Virtex-6 LXT			
Virtex-6	Virtex-6 SXT	2009	40 nm	0.9 V or 1.0 V
	Virtex-6 HXT			
	Virtex-6 CXT			
Spartan-6	Spartan-6 LX	2009	45 nm	1.0 V or 1.2 V
	Spartan-6 LXT			
	Spartan 7	2017		
	Artix-7	2010		
	Kintex-7	2010		
7 Series	Virtex-7	2010	28 nm	0.9 V, 0.95 V or 1.0 V
	Virtex-7 3D	2011		
	Virtex-7 HT	2012		
	Zynq-7000	2011		
		2013		
	Kintex Ultrascale	2013		
Ultrascale	Kintex Ultrascale	2013	20 nm	0.9 V, 0.85 V or 1.0 V
	Virtex Ultrascale	2014		
	Artix Ultrascale+	2021		
	Kintex Ultrascale	2015		
	Virtex Ultrascale+	2016		
	Virtex Ultrascale+ 58G			
Ultrascale+	Virtex Ultrascale+ HBM		16 nm	0.72 V, 0.85 V or 0.9 V
	Zynq Ultrascale+ MPSoC 2015			
	Zynq Ultrascale+ RFSoC 2017			
	Alveo	2018		
	Versal Prime	2019		
Versal	Versal AI Core	2019	7 nm	0.7 V, 0.8 V or 0.88 V
	Versal Premium			

APPENDIX 9: Spartan-6 FPGA Feature and Logic Resources

Table A9.1: Spartan-6 FPGA Feature Summary by Device

Device	Logic Cells	Configurable Logic Blocks (CLBs)			DSP48A1 Slices	Block RAM Blocks		Total I/O Banks	Max User I/O
		Slices	Flip-Flops	Max Distributed RAM (Kb)		18Kb	Max (Kb)		
XC6SLX4	3,840	600	4,800	75	8	12	216	4	132
XC6SLX9	9,152	1,430	11,440	90	16	32	576	4	200
XC6SLX16	14,579	2,278	18,224	136	32	32	576	4	232
XC6SLX25	24,051	3,758	30,064	229	38	52	936	4	266
XC6SLX45	43,661	6,822	50,576	401	58	116	2,088	4	358
XC6SLX75	74,637	11,662	93,296	692	132	172	3,096	6	408
XC6SLX100	101,261	15,822	126,576	976	180	268	4,824	6	480
XC6SLX150	147,443	23,038	184,304	1,355	180	268	4,824	6	576
XC6SLX25T	24,051	3,758	30,064	229	38	52	936	4	250
XC6SLX45T	43,661	6,822	54,576	401	58	116	2,088	4	296
XC6SLX75T	74,637	1,662	93,296	692	132	172	3,096	6	348
XC6SLX100T	101,261	15,822	126,576	976	180	268	4,824	6	498
XC6SLX150T	147,443	23,038	184,304	1,355	180	268	4,824	6	540

Table A9.2: Spartan-6 FPGA Logic Resources

Device	Logic Cells	Total Slices	SLICEMs	SLICELs	SLICEXs	Number of 6-Input LUTs	Maximum Distributed RAM (Kb)	Shift Register (Kb)	Number of Flip-Flops
XC6SLX4	3,840	600	300	0	300	2,400	75	38	4,800
XC6SLX9	9,152	1,430	360	355	715	5,720	90	45	11,440
XC6SLX16	14,579	2,278	554	595	1,139	9,112	136	68	18,224
XC6SLX25	24,051	3,758	916	963	1,879	15,032	229	115	30,064
XC6SLX45	43,661	6,822	1,602	1,809	3,411	27,288	401	200	54,576
XC6SLX75	74,637	11,662	2,768	3,063	5,831	46,648	692	346	93,296
XC6SLX100	101,261	15,822	3,904	4,007	7,911	63,288	976	488	126,576
XC6SLX150	147,443	23,038	5,420	6,099	11,519	92,152	1,355	678	184,304
XC6SLX25T	24,051	3,758	916	963	1,879	15,032	229	115	30,064
XC6SLX45T	43,661	6,822	1,602	1,809	3,411	27,288	401	200	54,576
XC6SLX75T	74,637	1,662	2,768	3,063	5,831	46,648	692	346	93,296
XC6SLX100T	101,261	15,822	3,904	4,007	7,911	63,288	976	488	126,576
XC6SLX150T	147,443	23,038	5,420	6,099	11,519	92,152	1,355	678	184,304

APPENDIX 10: Reserved Words in VHDL

Table A10.1: Reserved Words in VHDL

abs all asserts	access and attribute	after architecture	alias array
begin bus	block	body	buffer
case	component	configuration	constant
disconnect	downto		
else exit	elseif	end	entity
file	for	function	
generate	generic	guarded	
if	in	inout	is
label	library	linkage	loop
map	mod		
nand not	new null	next	nor
of others	on out	open	or
package	port	procedure	process
range report	record return	register	rem
select	severity	signal	subtype
then	to	transport	type
units	until	use	
variable			
wait	when	while	with
xor			

APPENDIX 11: Data Types and Operators in VHDL

Table A11.1: Std_logic Data Types of the IEEE std_logic_1164 Package

Logic Symbols	Description
'U'	Uninitialized
'X'	Strong drive, unknown logic value
'0'	Strong drive, logic 0
'1'	Strong drive, logic 1
'Z'	High impedence
'W'	Weak drive, unknown logic value
'L'	Weak drive, logic 0
'H'	Weak drive, logic 1
'–'	Don't-care

Table A11.2: Predefined VHDL Data Types

VHDL Data Type	Value
bit	'0' or '1'
boolean	FALSE or TRUE
integer	$-(2^{31}-1) \leq$ INTEGER VALUE $\leq (2^{31}-1)$
positive	$1 \leq$ INTEGER VALUE $\leq (2^{31}-1)$
natural	$0 \leq$ INTEGER VALUE $\leq (2^{31}-1)$
real	– 1.0e38 $\leq$ FLOATING POINT VALUE $\leq$ 1.0E38
character	Alphabetical characters (a .... z, A .... Z), digits (0, ... 9), special characters (e.g.,%) each enclosed in single quotes
time	integer with units fs, ps, ns, ms, sec, min or hr

Table A11.3: VHDL Operators

Operator Group	Symbol	Operation/Operator
Binary logical	**and**	Logical and operation
	or	Logical or operation
	nand	Logical nand operation
	nor	Logical nor operation
	xor	Logical xor operation
	xnor	Logical xnor operation
Relational	=	Equality operator
	/=	Inequality operator
	<	Less than
	>	Greater than
	<=	Less than or equal to
	>=	Greater than or equal to
Shift operators	**sll**	Shift-left logical
	srl	Shift-right logical
	sla	Shift-left arithmetic
	sra	Shift-right arithmetic
	rol	Rotate left
	ror	Rotate right
Addition operators	**+**	Addition
	–	Subtraction
Concatenation operator	&	Concatenation
Unary sign operator	**+**	Positive sign
	–	Negative sign
Multiplication operators	*	Multiplication
	/	Division
	mod	Modulus
	rem	Remainder
Miscellaneous operators	**not**	Complement
	abs	Absolute
	**	Exponentiation
Assignment operators	<=	Assign data of one object to another object
	:=	Assign a value to object
	=>	Associate one object to another object

Table A11.4: Precedence of the VHDL Operators

Precedence	Operators
Highest	** abs not
	* / mod rem
	+ – (identity of negation)
	& + – (addition and subtraction)
	sll srl sla sra rol ror
	= /= < <= > >=
Lowest	and or nand nor xor xnor

APPENDIX 12: Keywords in Verilog

Table A12.1: List of Keywords

always	inout	rtranif0	and	input	rtranif1
assign	integer	scalared	begin	join	small
buf	large	specify	buffif0	macromodule	specparam
buffif1	medium	strength	case	module	strong0
casex	nand	strong1	casez	negedge	supply0
cmos	nmos	supply1	deassign	nor	table
default	not	task	defparam	notif0	time
disable	notif1	tran	edge	or	tranif0
else	output	tranif1	and	parameter	tri
endcase	pmos	tri0	andmodule	posedge	tri1
endfunction	primitive	triand	endprimitive	pull0	trior
endspecify	pull1	trireg	endtable	pulldown	vectored
endtask	pullup	wait	event	rcmos	wand
for	real	weak0	force	realtime	weak1
fork	reg	while	function	release	wire
highz0	repeat	nor	highz	rnmos	xnor
if	rpmos	xor	initial	rtran	

APPENDIX 13: Operators and Net Data Types in Verilog

Table A13.1: Operators of Verilog

<table>
<tr><th>Operator Type</th><th>Operator Symbol</th><th>Operation Performed</th></tr>
<tr><td rowspan="8">Bitwise</td><td>~</td><td>Bitwise NOT (1's complement)</td></tr>
<tr><td>&</td><td>Bitwise AND (or reduction AND)</td></tr>
<tr><td>|</td><td>Bitwise OR (or reduction OR)</td></tr>
<tr><td>^</td><td>Bitwise XOR (or reduction XOR)</td></tr>
<tr><td>~&</td><td>Bitwise NAND (or reduction NAND)</td></tr>
<tr><td>~|</td><td>Bitwise NOR (or reduction NOR)</td></tr>
<tr><td>~^</td><td rowspan="2">Bitwise XNOR (or reduction XNOR)</td></tr>
<tr><td>^~</td></tr>
<tr><td rowspan="3">Logical</td><td>!</td><td>NOT</td></tr>
<tr><td>&&</td><td>AND</td></tr>
<tr><td>||</td><td>OR</td></tr>
<tr><td rowspan="6">Arithmetic</td><td>+</td><td>Addition</td></tr>
<tr><td>-</td><td>Subtraction</td></tr>
<tr><td>-</td><td>2's complement</td></tr>
<tr><td>*</td><td>Multiplication</td></tr>
<tr><td>/</td><td>Division</td></tr>
<tr><td>**</td><td>Exponentiation</td></tr>
<tr><td rowspan="4">Relational</td><td>></td><td>Greater than</td></tr>
<tr><td><</td><td>Less than</td></tr>
<tr><td>>=</td><td>Greater than or equal to</td></tr>
<tr><td><=</td><td>Less than or equal to</td></tr>
<tr><td rowspan="4">Shift</td><td>>></td><td>Logical right shift</td></tr>
<tr><td><<</td><td>Logical left shift</td></tr>
<tr><td>>>></td><td>Arithmetic right shift</td></tr>
<tr><td><<<</td><td>Arithmetic left shift</td></tr>
<tr><td rowspan="4">Equality</td><td>= =</td><td>Equality</td></tr>
<tr><td>!=</td><td>Inequality</td></tr>
<tr><td>= = =</td><td>Case equality</td></tr>
<tr><td>!= =</td><td>Case inequality</td></tr>
<tr><td>Concatenation</td><td>{, }</td><td>Concatenation</td></tr>
<tr><td>Replication</td><td>{n{m}}</td><td>Replicate value m for n times</td></tr>
<tr><td>Conditional</td><td>? :</td><td>Conditional</td></tr>
</table>

Table A13.2: Verilog Operator Precedence

<table>
<tr><th>Precedence</th><th>Operators</th></tr>
<tr><td>Highest</td><td>+ – ! ~ & ~& | ~| ^ ~^ ^~ (unary)</td></tr>
<tr><td></td><td>**</td></tr>
<tr><td></td><td>* / %</td></tr>
<tr><td></td><td>+ – (binary)</td></tr>
<tr><td></td><td><< >> <<< >>></td></tr>
<tr><td></td><td>== != === !==</td></tr>
<tr><td></td><td>& (binary) (Bitwise, Reduction AND)</td></tr>
<tr><td></td><td>^ ^~ ~^ (Binary) (Bitwise, Reduction OR)</td></tr>
<tr><td></td><td>| (binary)</td></tr>
<tr><td></td><td>&& (logical AND)</td></tr>
<tr><td></td><td>|| (logical OR)</td></tr>
<tr><td></td><td>?: (conditional operator)</td></tr>
<tr><td>Lowest</td><td>{}, {{, }} (concatenation)</td></tr>
</table>

Table A13.3: Net Data Types in Verilog

Net Data Type	Description
wire	Represent a physical connection
wand	Multiple connections resolved by AND logic in case of contention
wor	Multiple connections resolved by OR logic in case of contention
tri	Represent a physical connection with tristate logic
trireg	Bidirectional connection with tristate logic
triand	Tristate multiple connections resolved by AND logic in case of contention
trior	Tristate multiple connections resolved by OR logic in case of contention
tri0	Tristate connection with pull up to logic 1.
tri1	Tristate connection with pull down to logic 0.
supply0	Connection to power supply ground.
supply1	Connection to power supply Vcc.

University Question Papers

CHHATTISGARH SWAMI VIVEKANAND TECHNICAL UNIVERSITY

Fourth Semester, B.E. Degree Examination, November/December-2019

DIGITAL ELECTRONICS AND LOGIC DESIGN - EEE

Time: Three hours **Maximum:** 80 marks

Note: Attempt all questions. Part (a) of each unit is compulsory and carries 2 marks. Attempt any two parts from part (b), (c) and (d) of each question carrying 7 marks each.

UNIT - I

1. a) What is binary weighted code? (2)

   Chapter 3, Sec 3.2 ***[Page No - 3.4]***

   b) Justify the following statements with example: (7)

   (i) Excess-3 code is self complementary code. Chapter 3, Sec 3.2 ***[Page No - 3.4]***

   (ii) Gray code is a reflected code. Chapter 3, Sec 3.3 ***[Page No - 3.25]***

   c) What is self complementing code? Explain with suitable example. (7)

   Chapter 3, Sec 3.2 ***[Page No - 3.4]***

   d) The message 0010001 is coded in the 7-bit Hamming code and transmitted through noisy channel. Find correct code if error occured assume that even parity has been used. (7)

   Chapter 3, Example 3.17 ***[Page No - 3.48]***

UNIT - II

2. a) Why K-map is arranged in gray code? (2)

   Chapter 4, SA - Q4.30 ***[Page No - 4.98]***

   b) Minimize the function using K-map and implement the simplified function using only NAND gates. (7)

   $F(A, B, C, D) = \sum m(0, 1, 2, 9, 11, 15) + \sum d(8, 10, 14)$

   Chapter 4, Example 4.35 ***[Page No - 4.72]***

   c) Minimize the following boolean expression using Karnaugh map method: (7)

   $Y(A, B, C, D) = \prod M(1, 3, 5, 9, 11, 13)$

   Chapter 4, Example 4.30 ***[Page No - 4.67]***

   d) Minimize the given Boolean function using Quine-McCluskey method. (7)

   $F(A, B, C, D) = \sum m(0, 2, 3, 6, 7, 8, 9, 10)$

   Chapter 4, Example 4.38 ***[Page No - 4.80]***

UNIT - III

3. a) What is parity bit? (2)

Chapter 3, Sec 3.5.1 ***[Page No - 3.36]***

b) Draw the logic diagram of look-ahead carry generator and explain its working. (7)

Chapter 5, Sec 5.9.4 ***[Page No - 5.115]***

c) Explain full adder and design a full adder circuit using 3-to-8 decoder and two OR gates. (7)

Chapter 5, Example 5.38 ***[Page No - 5.109]***

d) Realize $F(A, B, C, D) = \Sigma m(0, 1, 3, 4, 8, 9, 15)$ using 8-to-1 Mux, select A as an input. (7)

Chapter 5, Example 5.26 ***[Page No - 5.82]***

UNIT - IV

4. a) Define race around condition. (2)

Chapter 9, SA - Q9.13 ***[Page No - 9.93]***

b) Explain the operation of master slave flip-flop and show how the race around condition is eliminated? (7)

Chapter 6, Sec 6.3 ***[Page No - 6.34]***

c) Design a synchronous BCD counter with JK flip-flops. (7)

Chapter 8, Example 8.7 ***[Page No - 8.60]***

d) Explain the operation of 4-bit PISO (parallel-in-serial-out) register. Draw the timing diagram and give its truth table. (7)

Chapter 8, Sec 8.6.3 ***[Page No - 8.116]***

UNIT - V

5. a) Define fan-in and fan-out. (2)

Chapter 11, Sec 11.2.4 ***[Page No - 11.5]***

b) Give the classification of logic families. Also list the characteristics of digital IC. (7)

Chapter 11, Sec 11.1 ***[Page No - 11.1]***

c) Explain the emitter-coupled logic (ECL) in detail. (7)

Chapter 11, Sec 11.6 ***[Page No - 11.27]***

d) Explain two inputs CMOS NAND gate with neat diagram. (7)

Chapter 11, Sec 11.7.4 ***[Page No - 11.47]***

CHHATTISGARH SWAMI VIVEKANAND TECHNICAL UNIVERSITY

Third Semester, B.E. Degree Examination, November/December-2019

DIGITAL ELECTRONICS AND LOGIC DESIGN - CSE

Time: Three hours **Maximum:** 80 marks

Note: Attempt all questions. Part (a) of each unit is compulsory and carries 2 marks. Attempt any two parts from part (b), (c) and (d) of each question carrying 7 marks each.

UNIT - I

UNIT - I

1. a) Discuss weighted code with examples. (2)

Chapter 3, Sec 3.2 ***[Page No - 3.4]***

b) Simplify the product of sums boolean expression using K-map. (7)

$Y = (A + B + C + D')(A + B + C' + D)(A + B' + C + D')(A + B' + C' + D)(A' + B + C' + D)$

Chapter 4, Example 4.28 ***[Page No - 4.60]***

c) What do you understand by error correcting code? Construct the even parity seven hamming code for a word 1011. (7)

Chapter 3, Sec 3.5 ***[Page No - 3.36]***

d) Simplify the function (7)

$F = \Sigma(0, 1, 2, 3, 5, 9, 11) + d(4, 7, 15)$ using Quine McCluskey method.

Chapter 4, Example 4.40 ***[Page No - 4.84]***

UNIT - II

2. a) Define propagation delay, noise margin and fan-out. (7)

Chapter 11, Sec 11.2.2 ***[Page No - 11.3]***

Chapter 11, Sec 11.2.5 ***[Page No - 11.5]***

Chapter 11, Sec 11.2.4 ***[Page No - 11.5]***

b) Explain CMOS-TTL and TTL-CMOS interfacing. Why is it required. (7)

Chapter 11, Sec 11.7 ***[Page No - 11.44]***

c) Explain transistor-transistor-logic (TTL) in totem-pole output arrangement. (7)

Chapter 11, Sec 11.5.1 ***[Page No - 11.21]***

d) Explain the operation CMOS NAND gate with circuit diagram. (7)

Chapter 11, Sec 11.7.4 ***[Page No - 11.47]***

UNIT - III

3. a) Design a half adder using NOR gates only. (2)

Chapter 5, Example 5.34 ***[Page No - 5.106]***

b) Draw a circuit for adding BCD numbers and discuss its working. (7)

Chapter 5, Sec 5.9.5 ***[Page No - 5.118]***

c) Explain how a comparator circuit works giving the design aspect. (7)

Chapter 5, Sec 5.12 ***[Page No - 5.131]***

d) Implement the boolean function: (7)

F (A, B, C, D) = Σm (1, 3, 4, 11, 12, 13, 14, 15) using 8:1 multiplexer.

Chapter 5, Example 5.27 ***[Page No - 5.84]***

UNIT - IV

4. a) Write the function of preset and clear inputs of a flip-flop. (2)

Chapter 6, Sec 6.2.6 ***[Page No - 6.33]***

b) Explain the principle of operation of clocked T flip-flop with truth table and output waveform. Why it is called a toggle switch?. (7)

Chapter 6, Sec 6.2.4 ***[Page No - 6.29]***

c) What is edge triggering and level triggering? Explain edge triggered SR flip-flop with waveform. (7)

Chapter 6, Sec 6.2.1 ***[Page No - 6.15]***

d) Design a synchronous mod-6 counter using JK flip-flops and draw the timing diagram of this counter. (7)

Chapter 6, Example 8.10 ***[Page No - 8.71]***

UNIT - V

5. a) Fusible link is associated with which kind of memory. (2)

Chapter 12, Sec 12.2.2 ***[Page No - 12.12]***

b) What do you understand by memory? How are they classified? What is the difference between serial and random accessing of memory? (7)

Chapter 12, Sec 12.1 ***[Page No - 12.1]***

c) What are programmable logic arrays? Compare the working of a FIFO register and a conventional shift register. (7)

Chapter 12, Sec 12.6 ***[Page No - 12.31]***

d) Explain R-2R ladder type DAC. (7)

Chapter 13, Sec 13.2.1 ***[Page No - 13.4]***

GUJARAT TECHNOLOGICAL UNIVERSITY

Third semester, B.E. Degree Examination, June 2019

DIGITAL LOGIC CIRCUITS - CSE

Time: Three hours **Maximum:** 70 marks

Note : (i) Attempt all questions.

(ii) Make suitable assumptions wherever necessary.

(iii) Figures to the right indicate full marks.

Q.1 a) Perform subtraction using 2's complement of 10010 – 10011. (3)

Chapter 2, Example 2.19 ***[Page No - 2.33]***

b) Explain Associative Laws and DeMorgan's theorems with necessary diagram, truth table. (4)

Chapter 4, Sec 4.1.3 ***[Page No - 4.5]***

c) Design a combinational circuit for full adder and full subtractor. (7)

Chapter 5, Sec 5.9.2 ***[Page No - 5.104]***

Chapter 5, Sec 5.10.2 ***[Page No - 5.123]***

Q.2 a) Show that $AB'C + B + BD' + ABD' + A'C = B + C$ (3)

Chapter 4, SA - Q4.19 ***[Page No - 4.94]***

b) Convert SR flip-flop into T flip-flop. (4)

Chapter 6, Sec 6.5.3 ***[Page No - 6.45]***

c) Simplify the following equation using K-map and implement using logic gates:

$F(A, B, C, D) = \Sigma(0, 1, 2, 3, 5, 7, 8, 9, 11, 14)$ (7)

Chapter 4, Example 4.24 ***[Page No - 4.55]***

(OR)

c) Simplify the following Boolean function by using tabulation method. (7)

$F = \Sigma(0, 1, 2, 8, 10, 11, 14, 15)$

Chapter 4, Example 4.38 ***[Page No - 4.80]***

Q.3 a) Convert the following number $(110011.011)_2$ to decimal and octal. (3)

Chapter 2, Example 2.2 ***[Page No - 2.6]***

b) Design 4-bit gray code to binary code converter. (4)

Chapter 5, Example 5.8 ***[Page No - 5.9]***

c) Compare various logic families. (7)

Chapter 11, Sec 11.8 ***[Page No - 11.60]***

(OR)

Q.3 a) List out different types of memories used in digital logic circuits and define them. (3)

Chapter 12, Sec 12.1.1 ***[Page No - 12.2]***

b) What is canonical form and standard form of equation? Give examples. (4)

Chapter 4, Sec 4.4.4 ***[Page No - 4.26]***

Chapter 4, Sec 4.4.6 ***[Page No - 4.28]***

c) Design a 3-bit binary counter using T flip-flop. (7)

Chapter 8, Example 8.1 ***[Page No - 8.32]***

Q.4 a) What is state diagram? Explain with example. (3)

Chapter 7, Sec 7.3.1 ***[Page No - 7.4]***

b) Design two inputs EX-OR gate using 4:1 Multiplexer. (4)

Chapter 5, Example 5.23 ***[Page No - 5.76]***

c) State the charcteristics and disadvantages of emitter coupled logic family. (7)

Chapter 11, Sec 11.8.4 ***[Page No - 11.63]***

(OR)

Q.4 a) Explain D flip-flop. (3)

Chapter 6, Sec 6.2.5 ***[Page No - 6.31]***

b) Explain 4-bit magnitude comparator in detail with necessary Boolean expression. (4)

Chapter 5, Sec 5.12.1 ***[Page No - 5.131]***

c) What is meant by demultiplexer? Give any example of demultiplexer. (7)

Chapter 5, Sec 5.8 ***[Page No - 5.95]***

Q.5 a) Draw the diagram of 2-to-4 line decoder. (3)

Chapter 5, Sec 5.4.1 ***[Page No - 5.17]***

b) Compare ROM and PLA. (4)

Chapter 12, SA - Q12.22 ***[Page No - 12.67]***

c) Draw the block diagram of successive approximation type ADC and explain its operation. (7)

Chapter 13, Sec 13.3.5 ***[Page No - 13.16]***

(OR)

Q.5 a) Draw the diagram of 3-to-8 line decoder. (3)

Chapter 5, Sec 5.4.7 ***[Page No - 5.31]***

b) Explain arithmetic, logic and shift micro operations. (4)

Chapter 5, Sec 5.14.1 ***[Page No - 5.142]***

c) Explain R-2R ladder type DAC. (7)

Chapter 13, Sec 13.2.1 ***[Page No - 13.4]***

GUJARAT TECHNOLOGICAL UNIVERSITY

Fourth semester, B.E. Degree Examination, May 2019

DIGITAL ELECTRONICS - EEE

Time: Three hours **Maximum:** 70 marks

Note : (i) Attempt all questions.

(ii) Make suitable assumptions wherever necessary.

(iii) Figures to the right indicate full marks.

Q.1 a) Find the XS-3 code of following decimal numbers (3)

(i) 26 (ii) 42 (iii) 63

Chapter 3, Example 3.8 ***[Page No - 3.18]***

b) Differentiate between combinational logic circuit and sequential logic circuit. (4)

Chapter 7, SA - Q7.9 ***[Page No - 7.75]***

c) Explain why NAND and NOR are known as universal gates a nd construct AND, OR and NOT using the universal gates. (7)

Chapter 4, Sec 4.3.1 ***[Page No - 4.13]***

Q.2 a) Convert the following Binary to gray code (3)

(i) 1001 (ii) 1010 (iii) 1011

Chapter 3, Example 3.14 ***[Page No - 3.28]***

b) Convert the following (4)

(i) $(4CD)_{16} = (\quad)_2$ (ii) $(26.24)_8 = (\quad)_{10}$

Chapter 2, Example 2.7 ***[Page No - 2.13]***

Chapter 2, Example 2.6 ***[Page No - 2.13]***

c) Simplify Y = A'BCD' + BCD' + BC'D' + BC'D and implement using NAND gate only. (7)

Chapter 4, Example 4.18 ***[Page No - 4.36]***

(OR)

c) State and explain DeMorgan's theorems with truth tables. (7)

Chapter 4, Sec 4.1.3 ***[Page No - 4.6]***

Q.3 a) Explain minterm and maxterm. (3)

Chapter 4, Sec 4.4.2 ***[Page No - 4.24]***

Chapter 4, Sec 4.4.3 ***[Page No - 4.25]***

b) List down the various types of ROMs and discuss and two of them. (4)

Chapter 12, Sec 12.2 ***[Page No - 12.11]***

c) Minimize using K-Map f(A, B, C, D) = Σ(1, 3, 4, 6, 8, 11, 15) + d(0, 5, 7) also draw MSI circuit for the output. (7)

Chapter 4, Example 4.29 ***[Page No - 4.62]***

(OR)

Q.4 a) Explain parity checking method of error detection. (3)

Chapter 3, Sec 3.5 ***[Page No - 3.36]***

b) Perform the decimal subtraction 206.7 – 147.8 in 8421 BCD code. (4)

Chapter 3, Example 3.2 ***[Page No - 3.9]***

c) Discuss 4-bit magnitude comparator in detail. (7)

Chapter 5, Sec 5.12.1 ***[Page No - 5.131]***

Q.5 a) Explain full adder. (3)

Chapter 5, Sec 5.9.2 ***[Page No - 5.104]***

b) Expalin R-2R ladder DAC with necessary diagram. (4)

Chapter 13, Sec 13.2.1 ***[Page No - 13.4]***

c) Draw 4 bit down counter, explain its working with timing diagram and truth table. (7)

Chapter 8, Sec 8.2.2 ***[Page No - 8.4]***

(OR)

Q.6 a) Discuss multiplexer with suitable diagram. (3)

Chapter 5, Sec 5.7 ***[Page No - 5.62]***

b) Explain terms accuracy and settling time for DAC. (4)

Chapter 13, Sec 13.1.1 ***[Page No - 13.3]***

c) Describe 3-to-8 line decoder with logic diagram and truth table. (7)

Chapter 5, Sec 5.4.7 ***[Page No - 5.31]***

Q.7 a) Compare SRAM with DRAM. (3)

Chapter 12, SA - Q12.21 ***[Page No - 12.67]***

b) Draw the two input TTL NAND gate circuit with totem-pole output. (4)

Chapter 11, Sec 11.5.2 ***[Page No - 11.22]***

c) Describe operation of D/A converter with binary-weighted resistors. (7)

Chapter 13, Sec 13.2.2 ***[Page No - 13.5]***

(OR)

Q.8 a) Compare SOP and POS. Chapter 4, SA - Q4.18 ***[Page No - 4.49]*** (3)

b) Give comparision between EPROM and FLASH memory. (4)

Chapter 12, SA - Q12.23 ***[Page No - 12.67]***

c) Describe the working of look-ahead-carry adder. (7)

Chapter 5, Sec 5.9.4 ***[Page No - 5.115]***

I K GUJRAL PUNJAB TECHNICAL UNIVERSITY

Third Semester, B.E. Degree Examination, May 2020

DIGITAL CIRCUITS AND LOGIC DESIGN - ECE

Time: Three hours **Maximum:** 60 marks

Instructions to Candidates:

1. SECTION - A is compulsory consisting of TEN questions carrying TWO marks each.
2. SECTION - B contains FIVE questions carrying FIVE marks each and students have to attempt any four questions.
3. SECTION - C contains THREE questions carrying TEN marks each and students have to attempt any two questions.

SECTION - A

1. **Answer briefly:**

a) Perform the subtraction $10111_2 - 11111_2$ using 1's complement method of subtraction.

Chapter 2, Example 2.19 ***[Page No - 2.33]***

b) Convert the decimal number 359_{10} to BCD.

Chapter 3, Example 3.1 ***[Page No - 3.6]***

c) Give the difference between SOP and POS forms.

Chapter 4, SA - Q4.18 ***[Page No - 4.94]***

d) Draw symbol and construct the truth table for three input Ex-OR gate.

Chapter 4, Example 4.7 ***[Page No - 4.22]***

e) Why totem-pole outputs cannot be connected together.

Chapter 11, SA - Q11.8 ***[Page No - 11.67]***

f) How multiplexer differs from decoder?

Chapter 5, SA - Q5.6 ***[Page No - 5.168]***

g) What is the basic difference between buffered and unbuffered CMOS devices?

Chapter 11, SA - Q11.10 ***[Page No - 11.68]***

h) What is totem-pole output stage? What are its advantages?

Chapter 11, Sec 11.5.1 ***[Page No - 11.21]***

i) Which is the fastest ADC and why? Chapter 13, SA - Q13.25 ***[Page No - 13.26]***

j) How many address bits are required for a 512 × 4 memory?

Chapter 12, SA - Q12.24 ***[Page No - 12.67]***

SECTION - B

2. Using the boolean algebra, simplify the expression:

   A'B'C + (A + B + C')' + A'B'C'D Chapter 4, SA - Q4.20 ***[Page No - 4.94]***

3. What is meant by Decoder? Explain 3-to-8 line decoder with diagram and truth table.

   Chapter 5, Sec 5.4 ***[Page No - 5.16]***

   Chapter 5, Sec 5.4.7 ***[Page No - 5.31]***

4. Design a counter with the following binary sequence : 0, 4, 2, 1, 6 and repeat. Use JK flip-flops.

   Chapter 8, Example 8.6 ***[Page No - 8.56]***

5. Explain the working of a basic totem-pole TTL 2 input NAND gate.

   Chapter 11, Sec 11.5.2 ***[Page No - 11.22]***

6. Realize F(w, x, y, z) = $\sum$ m (1, 4, 6, 7, 8, 9, 10, 11, 15) using 8-to-1 Mux.

   Chapter 5, Example 5.27 ***[Page No - 5.84]***

SECTION - C

7. Find a minimal SOP representation for f(A, B, C, D, E) = $\sum$ m (1, 4, 6, 10, 20, 22, 24, 26) + d(0, 11, 16, 27) using K-map method. Draw the circuit of the minimal expression using only NAND gates.

   Chapter 4, Example 4.36 ***[Page No - 4.74]***

8. a) Implement the following function using PLA F1 = $\sum$ m (0, 1, 2, 4) and F2 = $\sum$ m (0, 5, 6, 7).

   Chapter 12, Example 12.6 ***[Page No - 12.35]***

   b) Describe with the help of a schematic diagram the principle of operation of a successive type A/D converter.

   Chapter 13, Sec 13.3.5 ***[Page No - 13.16]***

9. Write short notes on any two :

   a) Weighted register D/A converter. Chapter 13, Sec 13.2.2 ***[Page No - 13.5]***

   b) Edge triggered JK flip-flop. Chapter 6, Sec 6.2.1 ***[Page No - 6.15]***

   c) CMOS logic family. Chapter 11, Sec 11.7 ***[Page No - 11.43]***

I K GUJRAL PUNJAB TECHNICAL UNIVERSITY

Third Semester, B.E. Degree Examination, December 2019

DIGITAL CIRCUITS AND LOGIC DESIGN - CSE

Time: Three hours **Maximum:** 60 marks

Instructions to Candidates:

1. SECTION - A is compulsory consisting of TEN questions carrying TWO marks each.
2. SECTION - B contains FIVE questions carrying FIVE marks each and students have to attempt any four questions.
3. SECTION - C contains THREE questions carrying TEN marks each and students have to attempt any two questions.

SECTION - A

Answer briefly:

1. Perform 2's complement subtraction of $(7)_{10} - (11)_{10}$.

   Chapter 2, SA - Q2.8 ***[Page No - 2.44]***

2. What is race around condition? How it can be avoided?

   Chapter 9, SA - Q9.13 ***[Page No - 9.93]***

   Chapter 9, Sec 9.1.5 ***[Page No - 9.8]***

3. How many states can an n-bit ring counter and an n-bit Johnson's counter have?

   Chapter 8, Sec 8.3.6 ***[Page No - 8.29]***

4. What is meant by the term edge triggered? Chapter 6, Sec 6.2.1 ***[Page No - 6.15]***
5. How many flip-flops are required to design a mod-7 up down counter?

   Chapter 8, Sec 8.2.6 ***[Page No - 8.11]***

6. What is difference between static RAM and dynamic RAM?.

   Chapter 12, SA - Q12.21 ***[Page No - 12.67]***

7. What is EEPROM? Chapter 12, Sec 12.2.4 ***[Page No - 12.16]***
8. What are the advantages of ring counter? Chapter 8, Sec 8.3.5 ***[Page No - 8.27]***
9. Differentiate between A/D and D/A conversion techniques. Chapter 13, SA-Q13.9 ***[Page No - 13.25]***
10. Write short note on: SOP and POS. Chapter 4, Sec 4.4.5 ***[Page No - 4.27]***

SECTION - B

11. Prove the following identities using boolean algebra:

    $(A + B)(A + (AB)')C + A'(B + C') + A'B + ABC = C(A + B) + A'(B + C')$

    Chapter 4, Example 4.3 ***[Page No - 4.13]***

12. A microprocessor uses RAM chips of 1024 × 1 capacity.

   a) How many chips will be required and how many address lines will be connected provide capacity of 1024 bytes?

   b) How many chips will be required to obtain a memory of capacity of 16 K bytes?

   Chapter 12, SA - Q12.25 ***[Page No - 12.67]***

13. What are the characteristics of digital ICs used to compute their performance?

   Chapter 11, Sec 11.2 ***[Page No - 11.2]***

14. Design an FPGA circuit, programmed to implement a 3-bit binary to gray conversion.

   Chapter 12, Example 12.13 ***[Page No - 12.51]***

15. Design a sequence detector to detect the sequence 1010 (overlapping of the sequence is permitted). Use D flip-flop to design the circuit. Show the intermediate design steps

   (a) State Diagram

   (b) Truth table/Excitation table

   (c) Logic Diagram of the circuit.

   Chapter 7, Example 7.10 ***[Page No - 7.36]***

SECTION - C

16. A combinational circuit has 3 inputs A, B, C and output F. F is true for following input combinations

   A is False, B is True

   A is False, C is True

   A, B, C are False

   A, B, C are True

   a) Write the truth table for F. Use the convention True = 1 and False = 0.

   b) Write the simplified expression for F in SOP form.

   c) Write the simplified expression for F in POS form.

   Chapter 4, Example 4.34 ***[Page No - 4.71]***

17. Design a 8 to 1 multiplexer by using four variable function given by

   $F(A, B, C, D) = \Sigma m(0, 1, 3, 4, 8, 9, 15)$

   Chapter 5, Example 5.26 ***[Page No - 5.82]***

18. What are synchronous counters? Design a 3-bit gray code counter using T flip-flop.

   Chapter 8, Example 8.4 ***[Page No - 8.45]***

VISVESVARAYA TECHNOLOGICAL UNIVERSITY

Third Semester, B.E. Degree Examination, August/Sepetember 2020

DIGITAL SYSTEM DESIGN - ECE

Time: Three hours **Maximum:** 100 marks

Note: Answer any FIVE full questions. Choosing ONE full question from each module.

Module - I

1\. a) Place the following equations into proper canonical forms:

i) $f(a, b, c) = ab' + ac' + bc$

ii) $f(a, b, c, d) = (a + b')(a + b' + d)$ (6)

Chapter 4, Example 4.17 ***[Page No - 4.32]***

b) Identify all prime implicants and essential prime implicants of Boolean function using K-map.

$f(a, b, c, d) = \sum m(0, 1, 2, 5, 6, 7, 8, 9, 10, 13, 14, 15)$ (6)

Chapter 4, Example 4.38 ***[Page No - 4.80]***

c) Find the minimal sum and minimal product for the function using K-map.

$f(a, b, c, d) = \sum m(6, 7, 9, 10, 13) + \sum d(1, 4, 5, 11, 15)$ (8)

Chapter 4, Example 4.33 ***[Page No - 4.70]***

(OR)

2\. a) Represent the number of days in a month for a non-leap year by a truth table, indicating the output of invalid input if any by '0' (5)

Chapter 4, Example 4.41 ***[Page No - 4.86]***

b) Find all the prime implicants of the function using Quine-Mccluskey method.

$f(a, b, c, d) = \sum (7, 9, 12, 13, 14, 15) + \sum d(4, 11)$ (10)

Chapter 4, Example 4.40 ***[Page No - 4.84]***

c) Simpilfy the given Boolean equation using K-map:

$f(a, b, c, d) = \prod (1, 2, 3, 4, 9, 10) + \prod d(0, 14, 15)$ (5)

Chapter 4, Example 4.32 ***[Page No - 4.69]***

Module - II

3\. a) Design 2-bit magnitude comparator. (8)

Chapter 5, Sec 5.12.3 ***[Page No - 5.138]***

b) Implement the boolean function using 8 : 1 MUX treat a, b, s as select lines:

$f(a, b, c, d) = \sum (0, 1, 5, 6, 7, 9, 10, 15)$ (6)

Chapter 5, Example 5.26 ***[Page No - 5.82]***

(OR)

4. a) Implement the boolean function f(a, b, c, d) = $\Sigma m(0, 2, 4, 5, 7, 9, 10, 14)$ using multiplexers with two 4:1 MUX with variable a, d connected to their select lines in the first level and one 2:1 MUX with variable 'C' connected to its select lines in the second level. (10)

Chapter 5, Example 5.24 ***[Page No - 5.78]***

b) Implement boolean function f(a, b, c, d) = Σm (4, 5, 7, 8, 10, 12, 15) using 4:1 MUX and external gates:

i) a, b are connected to select line also respectively.

ii) c, d are connected to select lines also respectively. (10)

Chapter 5, Example 5.25 ***[Page No - 5.80]***

Module - III

5. a) Explain the operation of switch debouncer using SR latch with the help of circuit and waveforms. (7)

Chapter 6, Sec 6.1.5 ***[Page No - 6.11]***

b) Explain Master slave JK flip-flop with the help of circuit diagram and waveforms. (7)

Chapter 6, Sec 6.3 ***[Page No - 6.34]***

c) Design a 4-bit binary ripple up counter using negative edge triggered JK flip-flop. (6)

Chapter 8, Sec 8.2.1 ***[Page No - 8.2]***

(OR)

6. a) Explain positive edge triggered D flip-flop with the help of circuit diagram and waveforms. (8)

Chapter 6, Sec 6.2.5 ***[Page No - 6.32]***

b) Design a 4-bit universal shift register using positive edge triggered D flip-flop and multiplexers to operate as indicated below :

Mode select	**Operation**
00	Hold
01	Right shift
10	Left shift
11	Parallel load

(8)

Chapter 8, Sec 8.6.5 ***[Page No - 8.117]***

c) Write the difference between ripple counter and synchronous counter. (4)

Chapter 8, SA - Q8.10 ***[Page No - 8.127]***

Module - IV

7. a) Design 3-bit synchronous up-counter using JK flip-flop. (10)

Chapter 8, Example 8.9 ***[Page No - 8.68]***

b) Design a MOD-6 synchronous counter using D flip-flop for the sequence 0-2-3-6-5-1. (10)

Chapter 8, Example 8.11 ***[Page No - 8.77]***

(OR)

8. a) Draw and explain block diagram of Moore model and Mealy model. (6)

Chapter 7, Sec 7.2.1 ***[Page No - 7.2]***

b) Design a synchronous circuit using positive edge triggered JK flip-flop with minimal combinational gating to generate the sequence: (8)

0 - 1 - 2 - 0 if input x = 0 ; 0 - 2 - 1 - 0 if input x = 1

Provide an output which goes high to indicate the non-zero state in the sequence 0 - 1 - 2 - 0.

Chapter 8, Example 8.8 ***[Page No - 8.65]***

c) Design mod-5 synchronous counter using T flip-flop. (6)

Chapter 8, Sec 8.2.6 ***[Page No - 8.14]***

Module - V

9. a) A sequential circuit has one input (x) and output (z) the circuit examines groups of four consecutive inputs and produces an output z = 1 if the input sequence 0101 or 1001 occurs. The circuit resets after every four inputs. Find the mealy state graph typical sequence is 0101 0010 1001 0100. (10)

Chapter 7, Example 7.11 ***[Page No - 7.40]***

b) Explain with block diagram design and serial adder with accumulator. (10)

Chapter 8, Sec 8.6.6 ***[Page No - 8.121]***

(OR)

10. a) Write a short note on 4 × 4 bit binary parallel multiplication. (10)

Chapter 5, Sec 5.13.2 ***[Page No - 5.140]***

b) List the guidelines for construction of state graphs. (10)

Chapter 7, Sec 7.3.1 ***[Page No - 7.4]***

VISVESVARAYA TECHNOLOGICAL UNIVERSITY

Third semester, B.E. Degree Examination, August/September 2020

DIGITAL SYSTEM DESIGN - EEE

Time: Three hours **Maximum:** 100 marks

Note: Answer any FIVE full questions. Choosing ONE full question from each module.

Module - I

1. a) With a basic block diagram, explain the combinational logic circuit. (4)

Chapter 5, Sec 5.2 ***[Page No - 5.1]***

b) Convert the following in the proper canonical formula and write the decimal notation.

i) R = f(x, y, z) = (x + y) (x' + z) into maxterm canonical formula.

ii) Z = f(a, b, c) = ab + b'c + ac into minterm canonical formula. (8)

Chapter 4, Example 4.16 ***[Page No - 4.32]***

c) Reduce the following expression using K-map and implement using basic gates.

i) $f(a, b, c, d) = \sum m(0, 1, 2, 3, 8, 9)$

ii) $f(A, B, C, D) = \prod m(0, 1, 4, 5, 14, 15) + d(12, 13)$ (8)

Chapter 4, Example 4.24 ***[Page No - 4.55]***

(OR)

2. a) Find the minimal sum and minimal product expression for the function:

$f(a, b, c, d) = \sum m(4, 5, 12, 13, 14, 15) + d(10, 11)$ (6)

Chapter 4, Example 4.33 ***[Page No - 4.70]***

b) Simplify using Quine-Mccluskey method

$f(a, b, c, d) = \sum m(0, 2, 8, 10)$ (7)

Chapter 4, Example 4.39 ***[Page No - 4.82]***

c) Simplify using K-map

$f(a, b, c, d) = \sum m(0, 1, 2, 3, 4, 5, 16, 17, 18, 19, 24, 25) + d(26, 27)$ (7)

Chapter 4, Example 4.32 ***[Page No - 4.69]***

Module - II

3. a) Design 3 inputs, a, b and c and output y combinational circuit which has an output equal to 1 when majority of its inputs equal to 1 and ouptut is 0 otherwise. (6)

Chapter 5, Example 5.3 ***[Page No - 5.4]***

b) Design a full adder by constructing the truth table and simplify the output equations. (6)

Chapter 5, Sec 5.9.2 ***[Page No - 5.104]***

c) Implement the function:

$f(a, b, c, d) = \sum m(4, 5, 7, 9, 11, 12, 13,15)$ using:

i) 8 : 1 MUX with a, b, c as select lines.

ii) 4 : 1 MUX with a, b as select lines. (8)

Chapter 5, Example 5.26 ***[Page No - 5.82]***

(OR)

4. a) What is a comparator? Design a 2-bit magnitude comparator using logic gates. (10)

Chapter 5, Sec 5.12.3 ***[Page No - 5.138]***

b) Explain Master slave JK flip-flop with the help of circuit diagram and waveforms. (7)

Chapter 6, Sec 6.3 ***[Page No - 6.34]***

c) Configure a 16:1 MUX using 4:1 MUX. (4)

Chapter 5, Example 5.21 ***[Page No - 5.72]***

Module - III

5. a) Explain the operation of Master-Slave JK flip-flop with logic diagram, truth table, symbol and timing diagram. (10)

Chapter 6, Sec 6.3 ***[Page No - 6.34]***

b) Differentiate latches and flip-flops. Derive the characteristics equation of SR flip-flop, JK flip-flop, T flip-flop and D flip-flop. (10)

Chapter 6, Sec 6.4 ***[Page No - 6.36]***

(OR)

6. a) Explain the operation of a gated SR latch using NAND logic. (8)

Chapter 6, Sec 6.1.3 ***[Page No - 6.5]***

b) Explain the working of a switch debouncer using SR latch with waveforms. (8)

Chapter 6, Sec 6.1.5 ***[Page No - 6.11]***

c) Convert a JK flip-flop to T flip-flop. (4)

Chapter 6, Sec 6.5.4 ***[Page No - 6.46]***

Module - IV

7. a) Differentiate synchronous and asynchronous counter. (4)

Chapter 8, SA - Q8.10 ***[Page No - 8.127]***

b) Design a MOD-10 ripple counter using JK flip-flop. (6)

Chapter 8, Sec 8.2.5 ***[Page No - 8.10]***

c) Draw the logic diagram of a 4 bit shift register with four D flip-flop and four 4:1 MUX with mode select inputs S_1 and S_0. The resistor operates as follows: (10)

S_1	S_0	Register Operation
0	0	No change
0	1	Compliment
1	0	Clear to zero
1	1	Load parallel data

Chapter 8, Sec 8.6.5 ***[Page No - 8.117]***

8. a) Mention the four different modes of operation shift register. With a neat block diagram, explain the operation of a 4-bit ring counter and Johnson counter. (10)

Chapter 8, Sec 8.6 ***[Page No - 8.113]***

b) Design a MOD 6 synchronous upcounter using T flip-flop. (10)

Chapter 8, Example 8.15 ***[Page No - 8.94]***

Module - V

9. a) With a neat block diagram, explain and distinguish between Moore and Mealy model in a sequential circuit analysis. (10)

Chapter 7, Sec 7.2.1 ***[Page No - 7.2]***

b) Design a synchronous counter using JK flip-flop to count the following sequence $7 \rightarrow 4 \rightarrow 3 \rightarrow 1 \rightarrow 6 \rightarrow 0 \rightarrow 7$. (10)

Chapter 8, Example 8.3 ***[Page No - 8.41]***

(OR)

10. a) What are the different types of RAM and ROM? Explain. (10)

Chapter 12, Sec 12.3 ***[Page No - 12.17]***

Chapter 12, Sec 12.2 ***[Page No - 12.11]***

b) Construct a sequential circuit by obtaining the state and excitation table for the given diagram using JK flip-flop. (10)

Chapter 7, Example 7.17 ***[Page No - 7.62]***

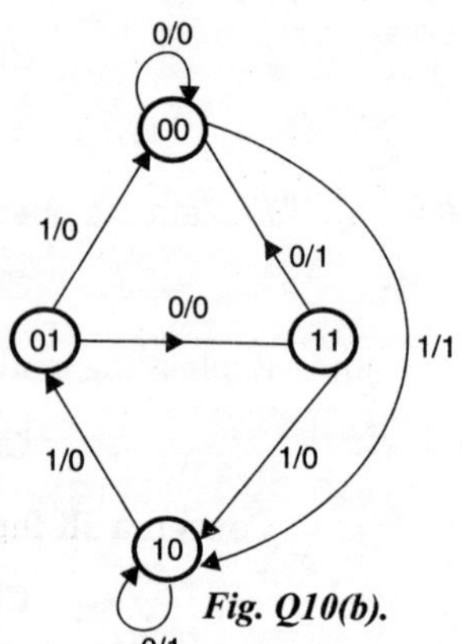

Fig. Q10(b).

JAWAHARLAL NEHRU TECHNOLOGICAL UNIVERSITY KAKINADA

II B. Tech II Semester Supplementary Examinations, February - 2022

DIGITAL ELECTRONICS - EEE

Time: 3 hours **Max.marks:** 75

Answer any FIVE full questions each question from each unit
All questions carry equak marks.

1. a) Express a Boolean function F(p, q, r) = (pq + r) (q + pr) in a sum of minterms and a product of maxterms. (7)

Chapter 4, Example 4.15 ***[Page No - 4.31]***

b) Obtain the 9's and 10's complement of $(389.61)_{10}$. (8)

Chapter 2, Example 2.22 ***[Page No - 2.36]***

(OR)

2. a) What do you understand by minterm and maxterm. What is the difference between canonical form and standard form? (7)

Chapter 4, Sec 4.4.4 ***[Page No - 4.26]***

b) Realize 2 input XOR gate and 2 input XNOR gates using NOR gates only. (8)

Chapter 4, Example 4.8 ***[Page No - 4.22]***

Chapter 4, Table 4.20 ***[Page No - 4.15]***

3. a) Prove the following identity

(i) AB + A(B + C) + B(B + C) = B + AC (ii) AB + A′B + A′B′ = A′ + B

(iii) AB′C + A′BC + ABC = AC + BC (7)

Chapter 4, Example 4.2 ***[Page No - 4.12]***

b) Simplify the following function using K-map,

F = ABCD + AB′C′D′ + AB′C + AB (8)

Chapter 4, Example 4.28 ***[Page No - 4.60]***

(OR)

4. a) Explain Boolean algebra with Boolean therorems. (7)

Chapter 4, Sec 4.1 ***[Page No - 4.1]***

Chapter 4, Sec 4.1.3 ***[Page No - 4.5]***

b) Simplify the function F(A, B, C, D, E) = $\sum$m(0, 2, 4, 6, 9, 11, 13, 15, 17, 21, 25, 27, 29, 31) using Karnaugh map. (8)

Chapter 4, Sec 4.6.5 ***[Page No - 4.53]***

(OR)

5. a) Draw logic diagram of a look-ahead carry generator and describe 4-bit full adder with look-ahead carry in detail. (7)

Chapter 5, Sec 5.9.4 ***[Page No - 5.115]***

b) Implement the following function with a multiplexer:

$F(A, B, C, D) = \Sigma(0, 1, 3, 4, 8, 9, 15)$ (8)

Chapter 5, Example 5.27 ***[Page No - 5.84]***

6. a) Implement the Boolean function $F(A, B, C, D) = \Sigma m(0, 1, 3, 4, 8, 9, 15)$ using 8:1 multiplexer. (7)

Chapter 5, Example 5.26 ***[Page No - 5.82]***

b) With neat sketch explain 4-bit binary subtractor. (8)

Chapter 5, Sec 5.11.2 ***[Page No - 5.130]***

7. a) Convert SR flip-flop into JK flip-flop and also draw excitation table. (7)

Chapter 6, Sec 6.5.1 ***[Page No - 6.43]***

b) Explain BCD ripple counter with logic diagram and timing diagram. (8)

Chapter 8, Sec 8.2 ***[Page No - 8.2]***

(OR)

8. a) Explain Master-Slave flip-flop and also draw excitation table. (7)

Chapter 6, Sec 6.3 ***[Page No - 6.34]***

b) Design a counter that has repeated sequence of six states 0, 1, 2, 4, 5, 6 using JK flip-flop. (8)

Chapter 8, Example 8.10 ***[Page No - 8.73]***

9. a) With neat sketch explain about Moore FSM. (7)

Chapter 7, Sec 7.2.1 ***[Page No - 7.2]***

b) For the given Mealy machine transition table convert into Moore machine. (8)

Present State	**Next State**			
	a = 0		**a = 1**	
	State	**Output**	**State**	**Output**
q^0	q^3	0	q^{11}	1
q^{10}	q^0	1	q^3	0
q^{11}	q^0	1	q^3	0
q^{20}	q^{21}	1	q^{20}	0
q^{21}	q^{21}	1	q^{20}	0
q^3	q^{10}	0	q^0	1

Chapter 7, Example 7.18 ***[Page No - 7.66]***

10. a) Compare Mealy and Moore machine (7)

Chapter 7, SA - Q7.2 ***[Page No - 7.74]***

b) With neat sktch explain state diagram, state table and state assignment. (8)

Chapter 7, Sec 7.3.1 ***[Page No - 7.4]***

Chapter 7, Sec 7.3.3 ***[Page No - 7.6]***

JAWAHARLAL NEHRU TECHNOLOGICAL UNIVERSITY ANANTAPUR

II B. Tech II Semester Regular Examinations, June/July - 2022

DIGITAL ELECTRONICS - EEE

Time: 3 hours **Max.Marks:** 70

Answer any FIVE full questions each question from each unit

All questions carry equal marks.

UNIT-I

1. a) Given the two binary numbers X = 10101111 and Y = 10000101, perform the subtraction.
   i) X – Y and ii) Y – X by using 2's complement technique? (7)

   Chapter 2, SA - Q2.15 ***[Page No - 2.47]***

   b) i) Express the Boolean function F = xy + x'z as a product of maxterms. (7)

   Chapter 4, Example 4.14 ***[Page No - 4.30]***

   ii) Find the complement of F = wx + yz, then show that FF = 0 and F + F = 1. (7)

   Chapter 4, SA - Q4.3 ***[Page No - 4.89]***

(OR)

2. a) Convert the following expressions into sum of products and product of sums: (7)

   i) (AB + C'D) (B + C'D) ii) x' + x(x + y') (y + z')

   Chapter 4, Example 4.17 ***[Page No - 4.32]***

   b) Draw the logic diagram to the following Boolean expressions without simplifying them: (7)

   i) BC' + AB + ACD ii) (A + B) (C + D) (A' + B + D)

   Chapter 4, SA - Q4.23, Q4.25 ***[Page No - 4.95, 4.96]***

UNIT-II

3. a) Simplify the following Boolean function, using three-variable maps: (7)

   $F(x, y, z) = \sum (0, 2, 3, 4, 6)$

   Chapter 4, Example 4.25 ***[Page No - 4.56]***

   b) Draw the circuit diagram of a 2-bit adder-subtractor and explain the function? (7)

   Chapter 5, Sec 5.11.1 ***[Page No - 5.128]***

(OR)

4. a) Draw the block diagram of a full adder using two half adders and one OR gate. (7)

   Chapter 5, Sec 5.9.2 ***[Page No - 5.104]***

b) Simplify the following Boolean function, using three-variable maps: (7)

$F(x, y, z) = xy + xy'z' + x'yz'$

Chapter 4, Example 4.26 ***[Page No - 4.57]***

UNIT-III

5. a) Desing a combinational circuit that will accomplish the multiplication of the 2-bit binary number X1 X0 by the 2-bit binary number Y1 Y0. Is a two-level circuit the most economical? Justify? (7)

Chapter 5, Sec 5.13.1 ***[Page No - 5.139]***

b) What is advantage of a PROM compared to the PLA and PALs? Explain. (7)

Chapter 12, Sec 12.2.2 ***[Page No - 12.12]***

(OR)

6. a) Construct a 4-to-16 decoder using five 2-to-4 decoder modulus. Explain with a neat schematic diagram. (7)

Chapter 5, Example 5.11 ***[Page No - 5.36]***

b) What is the difference between a digital multiplexer and a digital demultiplexer? Explain with an example? (7)

Chapter 5, SA - Q5.16 ***[Page No - 5.171]***

UNIT-IV

7. a) Define D and T flip-flop with the help of truth table? Also design the D and T flip-flop using JK flip-flop? (7)

Chapter 6, Sec 6.2.5, 6.2.4 ***[Page No - 6.31, 6.29]***

b) Design a 4-bit asynchronous decade counter and draw the timing diagram. (7)

Chapter 8, Sec 8.2.5 ***[Page No - 8.8]***

(OR)

8. a) Draw the timing diagram of a 4-bit asynchronous counter and explain? (7)

Chapter 8, Sec 8.2 ***[Page No - 8.3]***

b) Draw and explain briefly an asynchronous MOD-12 counter? (7)

Chapter 8, Example 8.8 ***[Page No - 8.89]***

UNIT-V

9. a) Design a 5 state sequential machine whose sequential states are: 000, 001, 010, 110, 111, 000 ..... Assume initial state is 000. (7)

Chapter 7, Example 7.12 ***[Page No - 7.41]***

(OR)

b) Explain the design procedure of synchronous sequential circuits. (7)

Chapter 7, Sec 7.2 ***[Page No - 7.2]***

10. a) Why state reduction is necessary in sequential circuit design? What are the different methods of state reduction? Example implication table method of state reduction with an example. (7)

Chapter 7, Sec 7.3.2 ***[Page No - 7.5]***

Chapter 9, Sec 9.2 ***[Page No - 9.32]***

b) A synchronous counter is controlled by two input signals A and B. The counter does not operate, if A = 0 and B = 0. When A = 0 and B = 1, the counter operates as a mod 4 counter. If A = 1 and B = 0 the counter operates as a mod 8 counter. Draw an FSM chart and design a circuit? (7)

Chapter 8, Example 8.16 ***[Page No - 8.97]***

JAWAHARLAL NEHRU TECHNOLOGICAL UNIVERSITY HYDERABAD

B. Tech II Year II Semester Examinations, November/December - 2020

DIGITAL ELECTRONICS - EEE

Time: 3 hours **Max.Marks:** 75

Answer any five full questions

All questions carry equak marks.

1. a) Explain various number systems and codes and their conversion with examples for each.

   Chapter 2, Sec 2.1 ***[Page No - 2.1]*** (9)

   b) Simplify the following Boolean expressions to minimum number of literals: (6)

   i) ABC + A'B + ABC' ii) xy + x(wz + w'z)

   Chapter 4, Example 4.23 ***[Page No - 4.43]***

2. a) Express the following numbers in decimal: $(10110.0101)_2$, $(16.5)_{16}$, $(26.24)_8$. (8)

   Chapter 2, Example 2.2 ***[Page No - 2.6]***

   Chapter 2, SA - Q2.4 ***[Page No - 2.43]***

   Chapter 2, Example 2.4 ***[Page No - 2.10]***

   b) Demonstrate by means of truth tables the Boolean Associative law and distributive law. Simplify the Boolean expression to minimum number of literals: (A + B)'(A' + B'). (9)

   Chapter 4, Sec 4.1.3 ***[Page No - 4.6]***

   Chapter 4, Sec 4.1.2 ***[Page No - 4.4]***

3. a) Simplify the Boolean functions, using a four variable Karnaugh map method and implement the simplified function using NAND gates. (9)

   $F(A, B, C, D) = \Sigma(0, 2, 4, 5, 6, 7, 8, 10, 13, 15)$

   Chapter 4, Example 4.35 ***[Page No - 4.72]***

   b) Show that the dual of the exclusive OR is also its complement. (6)

   Chapter 4, SA - Q4.21 ***[Page No - 4.94]***

4. a) Draw the multiple level NAND circuit for the following expression:

   (AB' + CD')E + BC(A + B) (7)

   Chapter 4, SA - Q4.24 ***[Page No - 4.95]***

   b) Construct a 4-bit BCD adder-subtractor circuit using BCD adder and 9's complemented. (8)

   Chapter 5, Sec 5.11.3 ***[Page No - 5.131]***

5. a) Design a 4-bit shift register using D flip-flops.

   Chapter 8, Sec 8.6.5 ***[Page No - 8.117]***

b) Design a counter with the following repeated binary sequence: 0, 1, 2, 3, 4, 5, 6, use JK flip-flop. (6+9)

Chapter 8, Example 8.9 ***[Page No - 8.68]***

6. a) Draw the circuit diagram of a 4-bit binary counter with parallel load and explain its working with its function table.

Chapter 8, Sec 8.2 ***[Page No - 8.2]***

b) Design a 4-bit synchronous counter with D flip-flops and explain its working. (8+7)

Chapter 8, Example 8.6 ***[Page No - 8.56]***

7. a) Expalin the working of R-2R ladder DAC with neat circuit diagram and mention its limitations.

Chapter 13, Sec 13.2.1 ***[Page No - 13.4]***

b) Draw and explain the working principle of flash type ADC. (8+7)

Chapter 13, Sec 13.3.1 ***[Page No - 13.11]***

8. a) Explain the functions and applications PLAs in memory addressing and implement the following two Boolean function with a PLA:

$F_1(A, B, C) = \Sigma(0, 1, 3, 5)$ and $F_2(A, B, C) = \Sigma(1, 2, 4, 7)$

Chapter 12, Example 12.7 ***[Page No - 12.37]***

b) What are sequential programmable devices? Draw the sequential programmable logic for a basic microcell logic. (8+7)

Chapter 12, Sec 12.8 ***[Page No - 12.53]***

ARYABHATTA KNOWLEDGE UNIVERSITY - BIHAR

B. Tech 3rd Semester Exam, 2020

DIGITAL ELECTRONICS

Time: 3 hours **Max.Marks:** 70

Instructions:

i) The marks are indicated in the right-hand margin.

ii) There are NINE questions in this paper.

ii) Attempt FIVE questions in all.

iv) Question No.1 is compulsory.

1. Choose the correct answer of the following (any seven): (2 × 7 = 14)

a) Dynamic RAM employs

(i) Capacitor or MOSFET (ii) FET and JFET

(iii) Capacitor or BJT (iv) BJT or MOS **[Ans : (i)]**

b) The resolution of a 10-bit AD converter for an input range of 10 V is approximately

(i) 1 V (ii) 1 mV

(iii) 10 mV (iv) 100 mV **[Ans : (iii)]**

c) The evolution of PLD begins with

(i) EROM (ii) RAM

(iii) PROM (iv) EEPROM **[Ans : (iii)]**

d) The parameter through which 16 distinct values can be represented is known as

(i) bit (ii) byte

(iii) word (iv) nibble **[Ans : (iii)]**

e) The number of full and half adders required to add 16-bit number is

(i) 8 HA, 8 FA (ii) 1 HA, 15 FA

(iii) 16 HA, 0 FA (iv) 4 HA, 12 FA

(f) If we record any music in any recorder, such type of process is called

(i) multiplexing (ii) encoding

(iii) decoding (iv) demultiplexing **[Ans : (ii)]**

(g) The number of D flip-flop required to form a 5-bit ring counter is

(i) 3 (ii) 4

(iii) 5 (iv) None of the above **[Ans : (iii)]**

(h) An overflow is a/an

(i) hardware problem (ii) software problem

(iii) user-input problem (iv) input-output problem **[Ans : (iii)]**

(i) The systematic reduction of logic circuits is accomplished by

(i) symbolic reduction (ii) TTL logic

(iii) Boolean algebra (iv) truth table **[Ans : (iii)]**

(j) A latch is an example of a/an

(i) monostable multivibrator (ii) astable multivibrator

(iii) bistable multivibrator (iv) 555 timer **[Ans : (iii)]**

2. a) Design an excess-3 to BCD code converted using minimum number of NAND gates. (8)

Chapter 5, Example 5.7 ***[Page No - 5.7]***

b) Prove the following: (3 × 2 = 6)

i) $A \oplus B = A' \oplus B'$ ii) $A' \oplus B' = A \oplus B' = A' \oplus B$

Chapter 4, Example 4.4 ***[Page No - 4.13]***

3. a) How can we implement preset and clear inputs in a flip-flop? Explain with the help of a diagram and list their uses. (3 + 3 = 6)

Chapter 6, Sec 6.2.6 ***[Page No - 6.33]***

b) Design a mod 9 counter using T flip-flops. Chapter 8, Example 8.13 ***[Page No - 8.83]***

4. a) Explain internal organization of 16 × 2 memory chips using suitable diagrams. Calculate the maximum rate at which data can be stored and read for a memory having following timing parameters: (4 + 2 = 6)

Parameter	Time (ns)
Read to Output Time (t_{RD})	70
Data Setup Time (t_{DW})	120
Read to Cycle Time (t_{RC})	200
Write Release Time (t_{WR})	0
Write Cycle (t_{WC})	200

Chapter 12, Example 12.2 ***[Page No - 12.10]***

b) Differentiate between word capacity and word size. Design a 16 × 8 CAM, using 8 × 2 CAM chips. (2 + 6 = 8)

Chapter 12, SA - Q12.35 ***[Page No - 12.69]***

Chapter 12, Example 12.4 ***[Page No - 12.25]***

5. a) Define resolution, linearity, accuracy and settling time of D/A converters. A typical D/A converter has a full-scale analog output of 10 *V* and accepts 6 binary bits as input. What will be the voltage corresponding to each analog step? (4 + 2 = 6)

Chapter 13, Sec 13.1.1 ***[Page No - 13.2]***

Chapter 13, Example 13.1 ***[Page No - 13.20]***

b) Design a 3-bit parallel comparator A/D converter that provides output in 2's complement format.

Chapter 13, Sec 13.3.1 ***[Page No - 13.16]*** (8)

6. a) Design a BCD to 7-segment display decoder circuit using logic gates. (8)

Chapter 5, Sec 5.15.5 ***[Page No - 5.160]***

b) Design a full adder using the following: (3 × 2 = 6)

(i) 8:1 MUX (ii) 4:1 MUX

Chapter 5, Example 5.39 ***[Page No - 5.110]***

7. a) On the following graph, inputs CLK and D are shown:

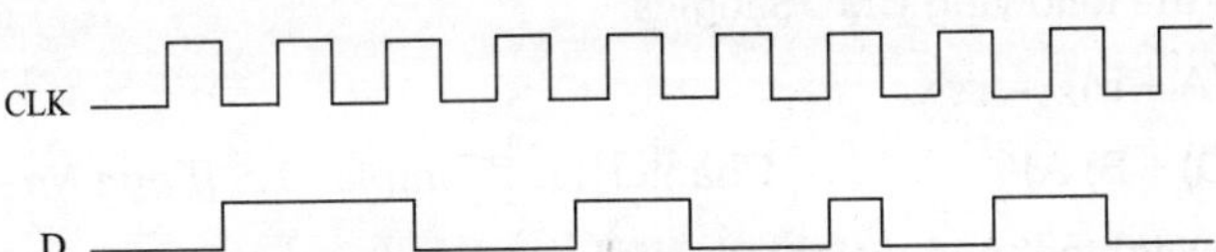

They are inputs to a D latch and a positive edge triggered D flip-flop. Assuming initial output 0, draw the output waveform for flip-flop and latch. Do the two outputs differ? If so, why?

Chapter 6, Example 6.7 ***[Page No - 6.58]*** (2 + 2 + 2 = 6)

b) Explain SIPO and SISO operations of shift register with relevant logic diagrams and truth tables. (8)

Chapter 8, Sec 8.6.2 ***[Page No - 8.116]***

Chapter 8, Sec 8.6.1 ***[Page No - 8.113]***

8. a) Identify the following logic functions implemented at F: (6)

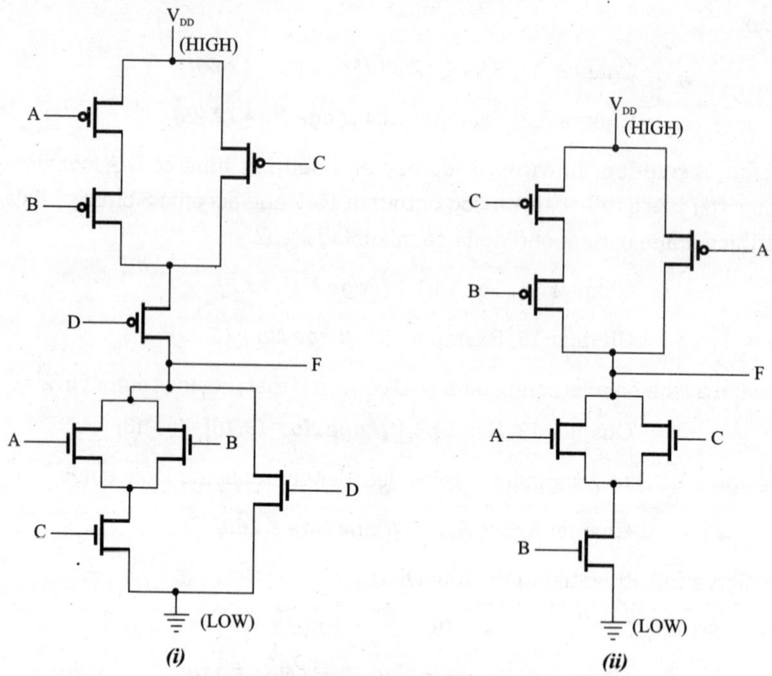

Chapter 11, Example 11.4 ***[Page No - 11.54]***

b) Implement the following CMOS logics: (4 × 2 = 8)

i) (AB(A + B))'

ii) (((CD) + B) A)' Chapter 11, Example 11.5 ***[Page No - 11.56]***

9. a) What are weighted, non-weighted, cyclic and self-complemntary codes? Explain each with examples. Chapter 3, Sec 3.2 ***[Page No - 3.4]*** (8)

b) Find the values of X in the following conversions:

i) $(95.10)_{10}$ to $(X)_2$ Chapter 2, Example 2.1 ***[Page No - 2.5]***

ii) $(45.70)_8$ to $(X)_2$ Chapter 2, Example 2.9 ***[Page No - 2.15]***

iii) $(168.16)_8$ to $(X)_{16}$ Chapter 2, Example 2.9 ***[Page No - 2.15]***

Dr.APJ.ABDUL KALAM TECHNICAL UNIVERSITY

B. Tech, 3rd Semester Examination, 2021-2022

DIGITAL ELECTRONICS

Time: 3 hours **Max.Marks:** 70

Notes:

1. Attempt all Section and Assume any missing data.
2. Appropriate marks are allotted to each question, answer accordingly.

SECTION A - Attempt **All** of the following Questions in brief (10 × 2 = 20)

Q1. a) How are binary digits used to express the integer and fractional parts of a number?

Chapter 2, Sec 2.1.3 ***[Page No - 2.4]***

b) Explain how BCD addition is carried out.

Chapter 3, Sec 3.2.2 ***[Page No - 3.6]***

c) Implement a 4:1 multiplexer using 2:1 multiplexer.

Chapter 5, Sec 5.7.1 ***[Page No - 5.67]***

d) Demultiplexer is decoder circuit with an additional enabling input. Do you agree with the above statement?

Chapter 5, Sec 5.8.1 ***[Page No - 5.101]***

e) Give the difference between positive and negative edge triggering.

Chapter 6, SA - Q6.9 ***[Page No - 6.70]***

f) A flip-flop has 5 ns delay from the time the clock edge occurs to the time the output is complemented. What is the maximum delay in a 10-bit binary ripple counter that uses these flip-flops? What is maximum frequency the counter can operate reliably?

Chapter 8, SA - Q8.14 ***[Page No - 8.128]***

g) Define critical race and non-critical race.

Chapter 9, Sec - Q9.14 ***[Page No - 9.93]***

h) What is the significance of state assignment?

Chapter 7, Sec 7.3.3 ***[Page No - 7.6]***

i) Why is ECL logic faster than TTL?

Chapter 11, SA - Q11.9 ***[Page No - 11.67]***

j) Compare static RAM and dynamic RAM.

Chapter 12, SA - Q12.21 ***[Page No - 12.67]***

SECTION B - Attempt **ANY THREE** of the following Questions (3 × 10 = 30)

Q2. a) Realize a 3-input gate using 2-input gates for the following gates:

(i) AND (ii) OR (iii) NAND (iv) NOR

Chapter 4, Example 4.9 ***[Page No - 4.22]***

b) i) Implement a full subtractor circuit using only NAND gates.

Chapter 5, Example 5.43 ***[Page No - 5.127]***

ii) Using 4:1 multiplexers, implement the following function

$F(A, B, C) = \Sigma m(0, 2, 3, 5, 7)$ Chapter 5, Example 5.22 ***[Page No - 5.74]***

c) Define bi-directional shift register. Draw and explain 3 bit bi-directional shift register using D flip-flop.

Chapter 8, Sec 8.6.5 ***[Page No - 8.117]***

d) Design a 4-bit synchronous counter with D flip-flops and explain its working.

Chapter 8, Example 8.6 ***[Page No - 8.56]***

e) i) Write a note on interfacing TTL with CMOS.

Chapter 11, Sec 11.7 ***[Page No - 11.44]***

ii) Explain the parameters used to characterize logic families.

Chapter 11, Sec 11.2 ***[Page No - 11.2]***

SECTION C - Attempt **ANY ONE** following Questions (1 × 10 = 10)

Q3. a) Minimize the following using Tabular method

$F(A, B, C, D, E) = \Sigma m(0, 1, 2, 3, 6, 7, 14, 15, 16, 19, 31)$

Chapter 4, Example 4.39 ***[Page No - 4.82]***

b) i) Reduce the expression $f = \Sigma m(0, 1, 2, 3, 5, 7, 8, 9, 10, 12, 13)$ using K-maps and implement the real minimal expression using NAND logic.

Chapter 4, Example 4.35 ***[Page No - 4.72]***

ii) Design the logic circuit for a BCD to decimal decoder.

Chapter 5, Example 5.10 ***[Page No - 5.13]***

SECTION C - Attempt **ANY ONE** following Questions (1 × 10 = 10)

Q4. a) Construct BCD adder using two 4-bit binary parallel adder and logic gates.

Chapter 5, Sec 5.9.5 ***[Page No - 5.118]***

b) Explain 4-bit magnitude comparator.

Chapter 5, Sec 5.12.1 ***[Page No - 5.131]***

SECTION C - Attempt **ANY ONE** following Questions (1 × 10 = 10)

Q5. a) A sequential circuit has one flip-flop Q, two inputs x and y, and one output S. It consists of a full adder circuit connected to a D flip-flop, as shown in given figure. Derive the state table and state diagram of the sequential circuit. What does this circuit represent?

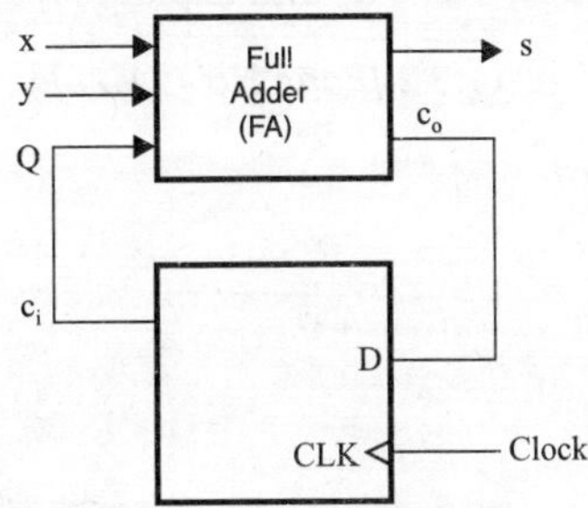

Chapter 7, Example 7.3 ***[Page No - 7.11]***

b) Design a 3-bit up/down counter with a direction control M, using JK flip-flops.

Chapter 8, Example 8.9 ***[Page No - 8.68]***

SECTION C - Attempt **ANY ONE** following Questions (1 × 10 = 10)

Q6. a) Derive the transition table for the asynchronous sequential circuit shown in given figure. Determine the sequence of internal states Y_1 Y_2 for the following sequence of inputs x_1 x_2: 00, 10, 11, 01, 11, 10, 00.

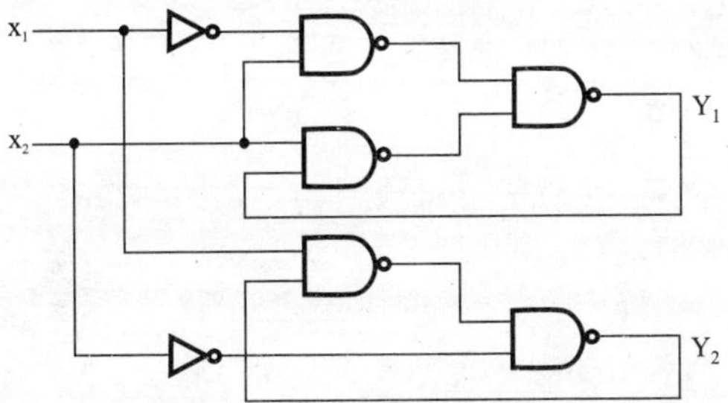

Chapter 9, Example 9.10 ***[Page No - 9.30]***

b) i) Distinguish between static and dynamic hazard. How will you determine hazard in combinational circuits?

Chapter 10, Sec 10.5 ***[Page No - 10.12]***

ii) Draw the logic diagram of the product-of-sums expression $Y = (x_1 + x_2')(x_2 + x_3)$. Show that there is a static-0 hazard when x_1 and x_3 are equal to 0 and x_2 goes from 0 to 1. Find a way to remove the hazard by adding one more OR gate.

Chapter 10, Sec 10.2.1 ***[Page No - 10.3]***

Chapter 10, Sec 10.5.1 ***[Page No - 10.13]***

SECTION C - Attempt **ANY ONE** following Questions (1 × 10 = 10)

Q7. a) Design a BCD to Excess-3 code converter and implement it using a suitable PLA.

Chapter 12, Example 12.10 ***[Page No - 12.42]***

b) Draw a neat diagram of TTL NAND gate and explain its operation.

Chapter 11, Sec 11.5.2 ***[Page No - 11.22]***

APJ ABDULKALAM TECHNOLOGICAL UNIVERSITY

Third Semester B.E. Degree Examination, September 2020

DIGITAL SYSTEM DESIGN - ECE

Time: 3 hours **Max.Marks:** 70

Answer all questions, each carries 5 marks

PART A

1. Convert

a) $(2469)_{10}$ into BCD. (1)

Chapter 2, Example 2.1 ***[Page No - 2.5]***

b) $(735)_8$ to decimal. (1)

Chapter 2, Example 2.4 ***[Page No - 2.10]***

c) $(650)_{10}$ to hexa decimal, gray and binary. (3)

Chapter 2, Example 2.5 ***[Page No - 2.12]***

2. Using boolean algebra prove that (A + B) (A'+ C) = AC + A'B. (5)

Chapter 4, SA - Q4.19 ***[Page No - 4.94]***

3. Design a full subtractor logic circuit. (5)

Chapter 5, Sec 5.10.2 ***[Page No - 5.123]***

4. Explain SISO and SIPO shift registers. (5)

Chapter 8, Sec 8.6.1 ***[Page No - 8.113]***

Chapter 8, Sec 8.6.2 ***[Page No - 8.116]***

5. Draw the logic diagram and timing sequence of a 4-bit ring counter. (5)

Chapter 8, Sec 8.3.5 ***[Page No - 8.26]***

6. Prepare the state table and derive the logic expression for each flip-flop input for a 3-bit binary synchronous down counter using T flip flop? (5)

Chapter 8, Sec 8.3.2 ***[Page No - 8.19]***

7. Explain the working of R-2R ladder type DAC. (5)

Chapter 13, Sec 13.2.1 ***[Page No - 13.4]***

8. Compare PAL and PLA. (5)

Chapter 12, SA - Q12.27 ***[Page No - 12.68]***

PART B

9. (a) Given X = 38_{10} and Y = 105_{10}. Using 2's complement method calculate

(i) X – Y (ii) Y – X (5)

Chapter 2, SA - Q2.15 *[Page No - 2.47]*

(b) How is the error detection and correction carried out using parity method in digital data transmission? (5)

Chapter 3, Sec 3.5 *[Page No - 3.36]*

10. (a) Using K map, minimize the expression

$F(A, B, C, D) = \Sigma m (1, 2, 3, 8, 14, 15) + d (0, 4, 6, 10)$.

Chapter 4, Example 4.32 *[Page No - 4.69]*

(b) Realize the boolean expression Z = ABC + AD + CD' using NAND gates only. (5)

Chapter 4, SA - Q4.12 *[Page No - 4.92]*

11. (a) Explain a CMOS NAND gate. (5)

Chapter 11, Sec 11.7.4 *[Page No - 11.47]*

(b) Find the standard Product of sum (POS) for the logic expression

F = (A + B'C) C (5)

Chapter 4, Example 4.20 *[Page No - 4.38]*

PART C

12. Develop a 3-stage carry look ahead adder and implement using basic gates. (10)

Chapter 5, Sec 5.9.4 *[Page No - 5.115]*

13. Realize thye following function $F(A, B, C, D) = \Sigma m (1, 3, 4, 10, 11, 12, 13)$ using

(i) 4 × 1 MUX (ii) 8 × 1 MUX (10)

Chapter 5, Example 5.2.2 *[Page No - 5.74]*

14. (a) Explain a 3-bit asynchronous up counter. Draw the timing diagram and truth table. (5)

Chapter 8, Sec 8.2.1 *[Page No - 8.2]*

(b) Draw the logic diagram of JK flip-flop and explain it. What is the advantage of JK flip-flop over SR flip-flop. (5)

Chapter 6, Sec 6.2.3 *[Page No - 6.22]*

PART D

15. Design a 3-bit gray code synchronous counter using JK flip-flop and explain the steps in detail. (10)

Chapter 8, Example 8.4 ***[Page No - 8.45]***

16. (a) Compare Mealy and Moore state machine models with example. (5)

Chapter 7, Sec 7.2.1 ***[Page No - 7.2]***

(b) Differentiate between ROM and RAM. (5)

Chapter 12, SA - Q12.14 ***[Page No - 12.66]***

17. (a) Implement a full adder circuit using VHDL. (5)

Chapter 14, VHDL Program 14.14 ***[Page No - 14.45]***

(b) Explain the working of successive approximation ADC. Mention the advantages and disadvantages. (5)

Chapter 13, Sec 13.3.5 ***[Page No - 13.16]***

RAJIV GANDHI PROUDYOGIKI VISHWAVIDYALAYA

EE/EX-403 (GS)

B. Tech IV Semester Examination, June 2022

Grading System (GS)

DIGITAL ELECTRONICS AND LOGIC DESIGN (DELD)

Time: 3 hours **Max.Marks:** 70

1. a) Explain in detail about parity generator and checkers. (7)

   Chapter 5, Sec 5.15.4 ***[Page No - 5.155]***

   b) Why NAND and NOR gates are called universal gates? Realize AND, OR with the help of NAND and NOR gates. (7)

   Chapter 4, Sec 4.3.1 ***[Page No - 4.13]***

2. a) What is demultiplexer? Design the given expression using 4:1 multiplexer $F(A, B) = A' + B'$

   Chapter 5, SA - Q5.10 ***[Page No - 5.168]*** (7)

   b) What is half adder? Write its truth table and design the logic circuit. (7)

   Chapter 5, Sec 5.9.1 ***[Page No - 5.103]***

3. a) What do you mean by flip-flop? Explain with the help of diagrams and tables? (7)

   Chapter 6, Sec 6.2 ***[Page No - 6.14]***

   b) What are latches? Write applications of flip-flop. (7)

   Chapter 6, Sec 6.1 ***[Page No - 6.1]***

4. a) What is asynchronous and synchronous counter? Give some applications of counter. (7)

   Chapter 8, Sec 8.1, 8.3.7 ***[Page No - 8.1, 8.104]***

   b) Write about ring counter and universal shift registers? (7)

   Chapter 8, Sec 8.3.5, 8.6.5 ***[Page No - 8.26, 8.117]***

5. a) What is ROM and its types? (7)

   Chapter 12, SA - Q12.7, Q12.8 ***[Page No - 12.65]***

   b) Explain about combinational PLD's. (7)

   Chapter 12, Sec 12.9 ***[Page No - 12.54]***

6. a) Define encoder and design 2-bit magnitude comparator. (7)

   Chapter 5, SA - Q5.8 ***[Page No - 5.168]***

   Chapter 5, Sec 5.12.3 ***[Page No - 5.138]***

b) Convert the following (7)

i) $(1001.0101)_2$ to decimal Chapter 2, Example 2.2 ***[Page No - 2.6]***

ii) $(137.21)_8$ to decimal Chapter 2, Example 2.4 ***[Page No - 2.10]***

7. a) Convert SR flip-flop to JK flip-flop. (7)

Chapter 6, Sec 6.5.1 ***[Page No - 6.43]***

b) Explain BCD counter with the help of diagram. (7)

Chapter 8, Sec 8.2.5 ***[Page No - 8.8]***

8. Write short notes on the following: (14)

a) DeMorgan's theorem Chapter 4, Sec 4.1.3 ***[Page No - 4.6]***

b) RAM Chapter 12, Sec 12.3 ***[Page No - 12.17]***

c) PLA and PAL Chapter 12, Sec 12.6 ***[Page No - 12.31]***

Chapter 12, Sec 12.7 ***[Page No - 12.44]***

RAJIV GANDHI PROUDYOGIKI VISHWAVIDYALAYA

EE/EX-226-CBCS

B.E. IV Semester Examination, June 2020

Choice Based Credit System (CBCS)

DIGITAL ELECTRONICS LOGIC DESIGN

Time: 3 hours **Max.Marks:** 60

Note : (i) Attempt any five questions.

(ii) All questions carry equal marks.

1. a) State and prove DeMorgan's theorems.

   Chapter 4, Sec 4.1.3 ***[Page No - 4.6]***

   b) Convert the following:

   i) $(204)10 = (\ )_8$ Chapter 2, Example 2.3 ***[Page No - 2.9]***

   ii) $(3A7)H = (\ )_2$ Chapter 2, Example 2.7 ***[Page No - 2.13]***

   iii) $(105)10 = (\ )_2$ Chapter 2, Example 2.1 ***[Page No - 2.5]***

   iv) $(11001)2 = (\ )_{10}$ Chapter 2, Example 2.2 ***[Page No - 2.6]***

2. a) Describe the operation of logic gate with the help of truth table and standard logic diagram.

   Chapter 4, Sec 4.3 ***[Page No - 4.9]***

   b) Define and explain minterm, maxterm, sum-of-product (SOP), product-of-sum (POS).

   Chapter 4, Sec 4.4.2 ***[Page No - 4.24]***
   Chapter 4, Sec 4.4.3 ***[Page No - 4.25]***
   Chapter 4, Sec 4.4.5 ***[Page No - 4.27]***

3. a) Design a full adder circuit with the help of truth table.

   Chapter 5, Sec 5.9.2 ***[Page No - 5.104]***

   b) Implement a NAND gate using NOR gates. Chapter 4, Table 4.20 ***[Page No - 4.15]***

4. a) Design a BCD adder giving rules of BCD addition.

   Chapter 3, Sec 3.2.2 ***[Page No - 3.6]***

   b) Design and explain the working of look ahead carry generator.

   Chapter 5, Sec 5.9.4 ***[Page No - 5.115]***

5. a) What are flip-flops? Construct a SR flip-flop and explain its operation.

   Chapter 6, Sec 6.2.2 ***[Page No - 6.16]***

b) Define and explain:

i) State table Chapter 7, Sec 7.3.1 ***[Page No - 7.4]***

ii) State diagram Chapter 7, Sec 7.3.1 ***[Page No - 7.4]***

iii) Excitation table Chapter 6, Sec 6.4 ***[Page No - 6.36]***

iv) Characteristic Equation Chapter 6, Sec 6.4 ***[Page No - 6.36]***

6. a) Design a synchronous 4-bit up/down counter.

Chapter 8, Sec 8.3.3 ***[Page No - 8.21]***

b) What is a shift register? Explain serial-in-parallel-out shift register.

Chapter 8, Sec 8.6, 8.6.2 ***[Page No - 8.113, 8.116]***

7. a) What do you mean by Random Access Memory (RAM)? Draw the logic diagram of a 4 × 4 RAM and describe the operations.

Chapter 12, Sec 12.3 ***[Page No - 12.17]***

b) With the help diagram explain the working of successive approximation type analog to digital converter.

Chapter 13, Sec 13.3.5 ***[Page No - 13.16]***

8. Write short notes on any two of the following:

i) Multiplexer Chapter 5, Sec 5.7 ***[Page No - 5.62]***

ii) Magnitude comparator Chapter 5, Sec 5.12 ***[Page No - 5.131]***

iii) BCD counter Chapter 8, Sec 8.2.5 ***[Page No - 8.8]***

iv) Types of ROM and their comparison Chapter 12, Sec 12.2 ***[Page No - 12.11]***

v) Programmable Logic Array (PLA) Chapter 12, Sec 12.6 ***[Page No - 12.31]***

vi) Digital to analog converter Chapter 14, Sec 14.1 ***[Page No - 14.1]***

DR. BABASAHEB AMBEDKAR TECHNOLOGICAL UNIVERSITY

Third semester B.E. Degree Examination, June 2019

DIGITAL ELECTRONICS - EEE

Time: Three hours **Maximum:** 60 marks

Note : (i) Attempt all questions.

(ii) Use of non-programmable scientific calculators is allowed.

(iii) Assume suitable data wherever necessary and mention it clearly.

Q.1 a) Design a 4-bit binary to Gray code converter. (6)

Chapter 5, Example 5.9 ***[Page No - 5.12]***

b) Minimize the given Boolean expression using Quine-McCluskey method. (6)

$F(A, B, C, D) = \sum m\ (0, 2, 3, 4, 5, 6, 7, 8, 9, 10, 11, 12, 13)$

Chapter 4, Example 4.39 ***[Page No - 4.82]***

Q.2 a) Give comparison between combinational and sequential circuit. (6)

Chapter 7, SA - Q7.9 ***[Page No - 7.75]***

b) Write down next state equation and excitation table for SR, JK, T and D flip-flop. (6)

Chapter 6, Sec 6.4 ***[Page No - 6.36]***

Q.3 a) Draw and explain the serial-in parallel-out shift register. (12)

Chapter 8, Sec 8.6.2 ***[Page No - 8.116]***

b) Design the circuit to generate the sequence: (6)

0....2....5....4...7...3 Chapter 8, Example 8.6 ***[Page No - 8.54]***

Q.4 a) Explain two input TTL NAND gate. (6)

Chapter 11, Sec 11.5.2 ***[Page No - 11.22]***

b) Give comparison between TTL and CMOS logic family. (6)

Chapter 11, Sec 11.8 ***[Page No - 11.60]***

Q.5 a) Implement full adder circuit using PLA. (6)

Chapter 12, Example 12.9 ***[Page No - 12.40]***

b) Explain VHDL three modelling styles with example. (6)

Chapter 14, Sec 14.6.2 ***[Page No - 14.23]***

Q.6 a) Design 4 : 16 decoder using 2 : 4 decoder. (6)

Chapter 5, Example 5.11 ***[Page No - 5.36]***

b) Implement 3-bit asynchronous up counter. Draw clock waveform. (6)

Chapter 8, Sec 8.2 ***[Page No - 8.2]***

ANNA UNIVERSITY

B.E DEGREE EXAMINATION APRIL/MAY 2019

Third semester

Electrical and Electronics Engineering

EE 8351 - DIGITAL LOGIC CIRCUITS

(Common to Electronics and Instrumentation Engineering/ Instrumentation and Control Engineering)

(Regulation 2017)

Time: Three hours Maximum: 100 Marks

Answer All questions

PART A - (10 × 2 = 20 Marks)

1. Convert $(101.01)_2$ to decimal number.

   Chapter 2, SA - Q2.3 ***[Page No - 2.42]***

2. Give each one example for error detecting code and error correcting code.

   Chapter 3, Example 3.15 ***[Page No - 3.43]***

3. Determine the exact number of half adders and full adders required for performing the addition of two binary numbers of 5-bits length each.

   Chapter 5, SA - Q5.23 ***[Page No - 5.172]***

4. Find the result of A + A' D + A C'.

   Chapter 4, SA - Q4.11 ***[Page No - 4.92]***

5. Write down the characteristic table of JK flip-flop.

   Chapter 6, Sec 6.4.2, Table 6.23 ***[Page No - 6.39]***

6. What is FSM? List its two basic types.

   Chapter 7, SA - Q7.3 ***[Page No - 7.74]***

7. Define metastable state.

   Chapter 9, SA - Q9.8 ***[Page No - 9.93]***

8. Draw the structure of PAL.

   Chapter 12, Sec 12.7 ***[Page No - 12.45]***

9. State the purpose of test bench.

   Chapter 14, SA - Q14.16 ***[Page No - 14.125]***

10. Write a VHDL program for an EX-NOR gate using behavioural coding.

    Chapter 14, SA - Q14.11 ***[Page No - 14.124]***

PART - B (5 × 13 = 65 Marks)

11. a) i) Design a 3-input NAND gate circuit using TTL logic. (7)

Chapter 11, Sec 11.5.2 ***[Page No - 11.22]***

ii) Explain in detail the generation of Hamming code for 4-bit data. (6)

Chapter 3, Sec 3.2.5 ***[Page No - 3.37]***

(OR)

b) i) Design a 2 input NOR gate using CMOS logic. (7)

Chapter 11, Sec 11.7.2 ***[Page No - 11.44]***

ii) Explain the operation of RTL inverter circuit with relevant diagrams. (6)

Chapter 11, Sec 11.3.5 ***[Page No - 11.15]***

2. a) i) Design a 3 × 8 decoder using 2 × 4 decoders. Draw the truth table. (7)

Chapter 5, Example 5.13 ***[Page No - 5.39]***

ii) Design Full adder circuit using logic gates. (6)

Chapter 5, Sec 5.9.2 ***[Page No - 5.104]***

(OR)

b) i) Simplify and implement the logic function $F(A, B, C) = \sum m(0, 1, 4, 5, 7)$ using logic gates.

Chapter 5, Example 4.24 ***[Page No - 4.55]*** (7)

ii) Design a 4 × 2 priority encoder using logic gates. (6)

Chapter 5, Sec 5.5.1 ***[Page No - 5.55]***

13. a) i) Design a 2-bit synchronous sequential down counter. (7)

Chapter 8, Sec 8.3.2 ***[Page No - 8.19]***

ii) Explain the operation of a 3-bit universal shift register. (6)

Chapter 8, Sec 8.6.5 ***[Page No - 8.117]***

(OR)

b) i) Explain Moore and Mealy models with the help of block diagrams. (7)

Chapter 7, Sec 7.2.1 ***[Page No - 7.2]***

ii) Draw the state table for the following state diagram. (6)

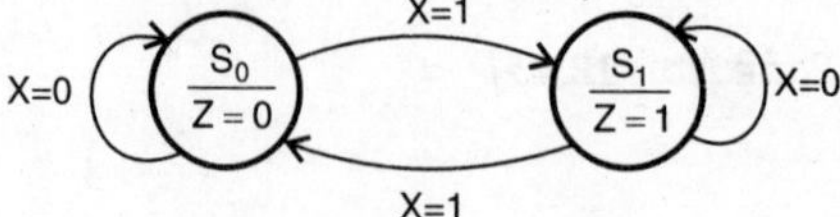

Chapter 7, Example 7.1 ***[Page No - 7.8]***

14. a) i) Design a Modulo-6 asynchronous binary up counter. (7)

Chapter 8, Sec 8.2.6 ***[Page No - 8.14]***

ii) Implement the functions $F_1(X, Y, Z) = \Sigma m(1, 2, 4, 5)$, $F_2(X, Y, Z) = \Sigma m(0, 1, 3, 4)$ and $F_3(X, Y, Z) = \Sigma m(2, 3, 6, 7)$ using a single PROM grid. (6)

Chapter 12, Example 12.5 ***[Page No - 12.29]***

(OR)

b) i) Diffentiate PAL and PLA implementations with the help of the same example $F_2(a, b, c) = \Sigma m(0, 1, 3, 4, 6, 7)$. (7)

Chapter 12, Example 12.12 ***[Page No - 12.48]***

ii) Explain the structure of CPLD with the help of a block diagram. (6)

Chapter 12, Sec 12.9 ***[Page No - 12.54]***

15. a) i) Draw the VLSI design flow chart used for IC design and fabrication. (7)

Chapter 14, Sec 14.11, Fig 14.2 ***[Page No - 14.33]***

ii) Write down a VHDL code for 8:1 multiplexer. (6)

Chapter 14, VHDL Program 14.11 ***[Page No - 14.42]***

(OR)

b) i) Illustrate the two approaches used in VHDL coding with full adder design as your examples. (7)

Chapter 14, VHDL Program 14.14 ***[Page No - 14.45]***

ii) What are components in VHDL? Show step-by-step how a NOR gate component can be created added in the library. (6)

Chapter 14, Sec 14.9 ***[Page No - 14.31]***

PART - C (1 × 15 = 15 Marks)

16. a) Design a synchronous sequential logic circuit that goes through the sequence 0, 2, 4, 6, 8, 10, 12, 14 repeatedly. Use D flip-flops for your design. (15)

Chapter 8, Example 8.6 ***[Page No - 8.56]***

(OR)

b) Simplify the following function and implement it using NAND gates only:

$F(w, x, y, z) = \Sigma m(1, 3, 5, 7, 9, 11, 13, 15)$, with don't-care states

$d(w, x, y, z) = \Sigma m(0, 2, 4, 6, 8)$. (15)

Chapter 4, Example 4.35 ***[Page No - 4.72]***

ANNA UNIVERSITY

B.E DEGREE EXAMINATION NOVEMBER/DECEMBER 2018

Third semester

Electrical and Electronics Engineering

EE 8351 - DIGITAL LOGIC CIRCUITS

(Common to Electronics and Instrumentation Engineering/ Instrumentation and Control Engineering)

(Regulation 2017)

Time: Three hours Maximum: 100 Marks

Answer All questions

PART A - (10 × 2 = 20 Marks)

1. Draw the DTL based NAND gate.

   Chapter 11, Sec 11.4.1 ***[Page No - 11.15]***

2. Perform subtraction on the following unsigned binary numbers using the 2's complement of the subtrahend **(a)** 11011 – 11001 **(b)** 110100 – 10101

   Chapter 2, SA - Q2.6 ***[Page No - 2.43]***

3. Mention the dependency of output in combinational circuits.

   Chapter 5, SA - Q5.1 ***[Page No - 5.167]***

4. Draw the NAND gate circuit using NOT, AND and OR gates.

   Chapter 4, SA - Q4.9 ***[Page No - 4.91]***

5. Write the role of master clock generator in synchronous circuits.

   Chapter 5, SA - Q5.15 ***[Page No - 5.170]***

6. Comment about a presettable counter and ripple counter.

   Chapter 8, SA - Q8.9 ***[Page No - 8.127]***

7. Draw the block diagram of asynchronous sequential circuit.

   Chapter 9, SA - Q9.5 ***[Page No - 9.92]***

8. Outline about PLA.

   Chapter 12, SA - Q12.26 ***[Page No - 12.68]***

9. Draw the basic structure of MOS transistor.

   Chapter 11, Sec 11.2.1 ***[Page No - 11.8]***

10. List the languages that are combined together to get VHDL language.

    Chapter 14, SA - Q14.4 ***[Page No - 14.122]***

PART - B (5 × 13 = 65 Marks)

11. a) Assume a 3-input AND gate with output F and a 3-input OR gate with G output. Show the signals of the outputs F and G as functions of the three inputs ABC. Use all 8 possible combinations of inputs ABC. (13)

Chapter 5, Example 5.4 ***[Page No - 5.5]***

(OR)

b) Show that a positive logic NAND gate is a negative logic NOR gate and vice versa. (13)

Chapter 4, Sec 4.3.2 ***[Page No - 4.17]***

12. a) Given the following Boolean function F = A'C + A'B + AB'C + BC (13)

i) Express it in sum of minterms.

ii) Find the minimal sum of products expression.

Chapter 4, SA - Q4.29 ***[Page No - 4.97]***

(OR)

b) Draw the logic diagram of 2-to-4 decoder using NOR gates only. Include an enable input. (13)

Chapter 5, Sec 5.4.1 ***[Page No - 5.17]***

13. a) Explain the operation, state diagram and characteristics of a T flip-flop and master slave JK flip-flop. (13)

Chapter 6, Sec 6.3 ***[Page No - 6.34]***

(OR)

b) Describe the design procedure with neat diagram about 4-bit bidirectional shift register with parallel load. (13)

Chapter 8, Sec 8.6 ***[Page No - 8.113]***

14. a) Discuss the operation of SR Latch with NOR and NAND gates analysis. (13)

Chapter 6, Sec 6.1.2 ***[Page No - 6.2]***

Chapter 6, Sec 6.1.3 ***[Page No - 6.5]***

(OR)

b) Illustrate about hazards in sequential circuits and the steps to avoid hazards in it. (13)

Chapter 10, Sec 10.5.2 ***[Page No - 10.18]***

15. a) Explain the structure and working principles of TTL based Totem-pole output configuration.

Chapter 11, Sec 11.5 ***[Page No - 11.21]*** (13)

(OR)

b) Write a VHDL code to realize a half adder using behavioral modeling and structural modeling.

Chapter 14, VHDL Program 14.2 ***[Page No - 14.24]*** (13)

PART - C (1 × 15 = 15 Marks)

16. a) Design a sequential circuit with two D flip-flops A and B and one input x. When x = 0, the state of the circuit remains the same. When x = 1, the circuit goes through the state transitions from 00 to 01 to 11 to 10 back to 00, and repeats. (15)

Chapter 7, Example 7.13 ***[Page No - 7.44]***

(OR)

b) Design a combinational circuit with three inputs, x,y and z, and the three outputs, A, B, and C, when the binary input is 0, 1, 2, or 3, the binary output is one greater than the input. When the binary input is 1, 5, 6 or 7 the binary output is one less than the input. (15)

Chapter 5, Example 5.2 ***[Page No - 5.3]***

INDEX

D

E

F

G

H

I

J

K

L

M

N

O

P

Q

R

S

T

U

V

W